A Textbook of Practical Organic Chemistry
including Qualitative Organic Analysis

VOGEL'S | TEXTBOOK OF PRACTICAL ORGANIC CHEMISTRY

including

QUALITATIVE ORGANIC ANALYSIS

Fourth edition
Revised by the following members of
The School of Chemistry
Thames (formerly Woolwich) Polytechnic

B. S. Furniss, B.Sc., Ph.D., Dip.F.E., C.Chem., F.R.I.C.
Senior Lecturer in Organic Chemistry

A. J. Hannaford, B.Sc., Ph.D., C.Chem., M.R.I.C.
Senior Lecturer in Organic Chemistry

V. Rogers, B.Sc., Ph.D.
Senior Lecturer in Organic Chemistry

P. W. G. Smith, B.Sc., Ph.D., A.R.C.S., D.I.C., C.Chem., M.R.I.C
Senior Lecturer in Organic Chemistry

A. R. Tatchell, M.Sc., Ph.D., C.Chem., F.R.I.C.
Principal Lecturer in Organic Chemistry

Longman   London and New York

Longman Group Limited London

*Associated companies, branches and representatives
throughout the world*

*Published in the United States of America
by Longman Inc., New York*

First published 1948
*New impression with minor
corrections, October 1948*
Second Edition 1951
*New Impression with addition of
Chapter XII on Semimicro Technique 1954*
Third Edition, 1956
New impression with corrections and additions 1957
New Impressions 1959, 1961, 1962, 1964, 1965, 1967
Fourth Edition, 1978

Library of Congress Cataloging in Publication Data

Vogel, Arthur
    Vogel's Practical organic chemistry, including
qualitative organic analysis.

    Third ed. published in 1956 under title: A text-book
of practical organic chemistry, including qualitative
organic analysis.
    Includes bibliographies and index.
    1. Chemistry, Organic—Laboratory manuals.
2. Chemistry, Analytic—Qualitative. I. Tatchell,
Austin Robert. II. Title. III. Title: Practical
organic chemistry, including qualitative organic
analysis.
QD261.V63 1978   547   77–23559
ISBN 0–582–44250–8

Filmset in 'Monophoto' Times 10 on 11 pt and
printed in Great Britain by
Richard Clay (The Chaucer Press), Ltd,
Bungay, Suffolk

CONTENTS

CHAPTER I EXPERIMENTAL TECHNIQUES

CONTENTS

CONTENTS

CHAPTER IV AROMATIC HYDROCARBONS

CHAPTER V   SOME ALICYCLIC COMPOUNDS

CHAPTER VI   SOME HETEROCYCLIC COMPOUNDS

CHAPTER VII QUALITATIVE ORGANIC ANALYSIS

CHAPTER VIII PHYSICAL CONSTANTS OF ORGANIC COMPOUNDS

Appendices

Table of Atomic Weights

Indexes

ACKNOWLEDGEMENTS

WE are grateful to the following for permission to reproduce copyright material:

American Chemical Society for two tables giving 'Schematic Representation of Summation Bands', by C. W. Young, R. B. Duval and N. Wright in *Analytical Chemistry* (1951), Vol. 23, and 'Graphical Display of Chemical Shifts for Classes of Paraffins', by L. P. Lindemann and J. Q. Adams in *Analytical Chemistry* (1971), Vol. 32 (10); Butterworths for figures giving 'Decoupled Spectrum of Crotonaldehyde', by W. MacFarlane and R. F. M. White, from *Techniques of High Resolution Nuclear Magnetic Resonance Spectroscopy* (1972); Heyden & Son Ltd, for Figs. 1 and 2 in *Organic Magnetic Resonance* (1972), Vol. 4; Pergamon Press Ltd, for figures from *Organic Chemistry, Applications of NMR* (1969), by L. M. Jackman and S. Sternhell (2nd edn); the authors for tables taken from *Principles of Organic Mass Spectroscopy*, by D. H. Williams and I. Howe; Verlag Chemie GMBH for figures by E. Breitmaier, G. Jung and W. Voelter in *Angewandte Chemie* (1971), Vol. 10; John Wiley & Son Inc for a table by R. M. Silverstein, C. G. Bassler and T. C. Merrill from *Spectroscopic Identification of Organic Compounds* (1974), 3rd edn; the authors for tables taken from *Carbon-13 Nuclear Magnetic Resonance for Organic Chemists* (1972) by G. C. Levy and G. L. Nelson.

PREFACE TO FOURTH EDITION

IT is now twenty-eight years since the first appearance of Vogel's *Textbook of Organic Chemistry* and twenty years since the publication of the previous edition, and it is a tribute to the thoroughness and farsightedness of the late Dr A. I. Vogel that the book continues to be used as a standard text in the organic chemistry laboratory.

The period since the appearance of the third edition has seen considerable changes in the practice and theory of organic chemistry. Among these may be included the ready availability of a much wider range of substrates and reagents; the development of a whole host of new synthetic methods; a greater awareness of the hazards associated with handling of organic chemicals; the now routine use of chromatographic and spectroscopic methods; and the use of mechanistic concepts to rationalise and predict the outcome of organic reactions. In preparing this new edition it has been our intention to reflect these changes in the subject whilst at the same time maintaining the essential character of the work, which has become an invaluable one-volume reference text both for undergraduates and postgraduates, and for the practising organic chemist.

The text has been completely revised in the light of current theory and practice. Major sections have been completely rewritten or reorganised and a substantial amount of new material has been added. In order to keep the size of the volume to realistic proportions it has been necessary to make room for much of the new material by discarding some of the previous contents which have either been superseded, or are not now directly relevant to the practising chemist. The major casualty in this respect was the chapter dealing with the Theory of General Technique, but this is now available to the reader in the second edition of Vogel's *Elementary Practical Organic Chemistry*.

The book now commences with the chapter on Experimental Techniques which has been completely revised and is arranged under the following subheadings: Apparatus and Reaction Procedures (incorporating most of the special techniques previously in the chapter entitled Miscellaneous Reactions); Isolation and Purification Procedures; and the Determination of Physical Constants. Major introductions in this chapter are the Sections on Safe Working in Organic Chemical Laboratories (**I,3**), Chromatography (**I,31**) and Spectroscopic Methods (**I,39**). We have included examples where appropriate of the use of chromatographic techniques and of spectroscopic methods in some of the newly incorporated experimental sections. Aspects of the interpretation of infrared, ultraviolet–visible, nuclear magnetic resonance and mass spectra are developed in Chapter VII, Qualitative Organic Analysis, so that the reader can gain experience in the use and appreciate the value of these techniques. Chapter I also includes an extended bibliography to allow the interested reader to obtain further information if required. Substantially expanded and up-dated Sections on the Purification of Common Organic Solvents, and the Preparation and Purification of Reagents now constitutes Chapter II.

The preparative sections are now organised into four chapters: III, Aliphatic Compounds; IV, Aromatic Compounds; V, Some Alicyclic Compounds; and VI, Some Heterocyclic Compounds. We have retained a substantial proportion of the reactions from the previous edition which continue to be of value either as examples for student exercises, or as standard preparative procedures. The examples included in the Alicyclic and Heterocyclic chapters are mainly restricted to methods for the formation of the cyclic system; any functionally group modifications in these systems are included in the appropriate sections of the other chapters. The preparations previously included in the chapters entitled Miscellaneous Reactions; Organic Reagents in Inorganic and Organic Chemistry; Dyestuffs, Indicators and Related Compounds; Some Physiologically Active Compounds, have now been incorporated as appropriate into the four new Chapters.

The past two decades have seen the publication in the literature of a vast number of new synthetic methods. We have sought in our selection of new preparative material to exemplify some of these new techniques and methods of synthesis where the value and generality has been well proven. It is not possible here to give a comprehensive list of all the additions; the following are some of the important reactions or reagents, examples of the use of which have been introduced into this edition: lithium dialkyl cuprates (Section **III,5**); phosphorus ylides (Section **III,14**); sulphur ylides (Section **VI,35**); formation of allenes (Sections **III,17** and **III,18**); selective oxidation (Sections **III,87** and **IV,25**); hydroboronation followed by oxidation (Sections **III,43** and **III,89**); selective reductions (Sections **IV,57**, **IV,65** and **VI,1**); acetylenic coupling reactions (Sections **III,22** and **III,23**); hydration of alkynes (Section **III,91**); photochemical reactions (Sections **III,30**, **IV,17**, **V,16** and **V,17**); oxymercuration-demercuration (Section **III,44**); use of aprotic solvents (Section **III,161**); enzymic resolution (Section **III,198**); benzyne intermediates (Section **IV,9**); use of triphenylphosphine dibromide (Section **IV,34**); formylation reactions (Section **IV,129**); esterification procedures (Sections **III,148** and **IV,179**); carbene and phase transfer reactions (Section **V,15**). In addition we have broadened the range of the preparative reactions included, particularly by the expansion of the sections on alicyclic compounds, heterocyclic compounds, amino acids and carbohydrates. A total of about 120 new experiments has been introduced, all of which have been checked in these laboratories. A literature reference is given for modifications or extensions to existing methods or techniques which we have not ourselves checked experimentally.

All the experiments which have been retained have been carefully scrutinised and any errors or omissions which were apparent, or known to us, have been corrected. Throughout, quantities of reagents and reactants have been expressed in molar amounts and yields calculated as a percentage of theory. The theoretical discussions and mechanistic interpretations which formed the introductions to major sections in the preparative chapters of the previous edition have been completely rewritten. In revising and up-dating the nomenclature we have been guided by the IUPAC rules for the Nomenclature of Organic Compounds, particularly with regard to the retention of established trivial names.

Every laboratory worker, and particularly those in supervisory roles, should be fully aware of the hazards associated with particular compounds and procedures, and should ensure that safe working habits are adopted. We have endeavoured to point out in the text any hazards associated with specific experiments; a comprehensive general introduction to safety is to be found in Section I,3. This, together with the cross-referencing provided by the indexes, should enable the dangers associated with the use of any particular material or operation to be ascertained.

The chapter on Qualitative Organic Analysis (Chapter VII) has been rewritten and now includes all the reactions for the characterisation of functional groups and the preparation of derivatives previously scattered throughout the book. This chapter also includes substantial new Sections on the interpretation of infrared, ultraviolet–visible nuclear magnetic resonance, and mass spectra of organic compounds together with a range of illustrative spectra. The Tables of Physical Constants previously dispersed throughout the book have been gathered together in Chapter VIII and Spectroscopic Correlation Tables are included in Appendices 2–4. Appendix 1, The Literature of Organic Chemistry, surveys the range of journals, monographs, texts, compilations of data, etc., which are available to the organic chemist.

We wish to thank Dr G. H. Jeffery, C.Chem., F.R.I.C., formerly Acting Head of Department who was instrumental in initiating our involvement in this project, and to Dr T. C. Downie, C.Chem., F.R.I.C. and Dr B. R. Currell, C.Chem., F.R.I.C., successive Heads of School of Chemistry in this Polytechnic for allowing the use of laboratory facilities during the work of revision. We also wish to express our thanks to Mr V. Kyte who was responsible for many of the diagrams of the previous edition, and willingly undertook the complete redrawing of retained and new material to the total of some 180 illustrations. The considerable work involved in typing from the manuscript and retyping was shared by Mrs V. Rogers and the late Mrs G. E. Tatchell; our thanks are due to both for their patience and perseverance which was such as to lighten the work involved in this extensive revision.

B.S.F   A.J.H.   V.R.   P.W.G.S.   A.R.T.

Thames Polytechnic London, S.E.16 6 PF

September 1976

PREFACE TO FIRST EDITION

THE present volume is an attempt to give to students of practical organic chemistry the benefit of some twenty years' experience in research and teaching of the subject. The real foundations of the author's knowledge of the subject were laid in 1925–1929 when, as a research student at the Imperial College under the late Professor J. F. Thorpe, F.R.S., he was introduced to the methods and experimental technique employed in a large and flourishing school of research in organic chemistry. Since that period the author and his students have been engaged *inter alia* in researches on *Physical Properties and Chemical Constitution* (published in the Journal of the Chemical Society) and this has involved the preparation of over a thousand pure compounds of very varied type. Many of the new procedures and much of the specialised technique developed and employed in these researches are incorporated in this book. Furthermore, new experiments for the elementary student have emanated from these researches; these have been tried out with large classes of undergraduate students over several sessions with gratifying success and have now been included in the present textbook.

In compiling this book, the author has drawn freely from all sources of information available to him—research notes, original memoirs in scientific journals, reference works on organic chemistry, the numerous textbooks on practical organic chemistry, and pamphlets of manufacturers of specialised apparatus. Whilst individual acknowledgement cannot obviously be made—in many cases the original source has been lost track of—it is a duty and a pleasure to place on record the debt the writer owes to all these sources. Mention must, however, be made of *Organic Syntheses*, to which the reader is referred for further details of many of the preparations described in the text.

The book opens with a chapter on the theory underlying the technique of the chief operations of practical organic chemistry: it is considered that a proper understanding of these operations cannot be achieved without a knowledge of the appropriate theoretical principles. Chapter II is devoted to a detailed discussion of experimental technique; the inclusion of this subject in one chapter leads to economy of space, particularly in the description of advanced preparations. It is not expected that the student will employ even the major proportion of the operations described, but a knowledge of their existence is thought desirable for the advanced student so that he may apply them when occasion demands.

Chapters III and IV are confined to the preparation and properties of

Aliphatic Compounds and Aromatic Compounds respectively. This division, although perhaps artificial, falls into line with the treatment in many of the existing theoretical textbooks and also with the author's own lecture courses. A short theoretical introduction precedes the detailed preparations of the various classes of organic compounds: it is recommended that these be read concurrently with the student's lecture course and, it is hoped, that with such reading the subject will become alive and possess real meaning. The partition of the chapters in this manner provides the opportunity of introducing the reactions and the methods of characterisation of the various classes of organic compounds; the foundations of qualitative organic analysis are thus laid gradually, but many teachers may prefer to postpone the study of this subject until a representative number of elementary preparations has been carried out by the student. The division into sections will facilitate the introduction of any scheme of instruction which the teacher considers desirable.

Chapters V–X deal respectively with Heterocyclic and Alicyclic Compounds; Miscellaneous Reactions; Organic Reagents in Inorganic and Organic Chemistry; Dyestuffs, Indicators and Related Compounds; Some Physiologically-Active Compounds; and Synthetic Polymers. Many of these preparations are of course intended for advanced students, but a mere perusal of the experimental details of selected preparations by those whose time for experimental work is limited may assist to impress them on the memory. Attention is particularly directed to the chapter on Organic Reagents in Inorganic and Organic Chemistry. It is always a good plan to set advanced students or adequately-trained laboratory assistants on the preparation of those compounds which are required in the laboratory for organic and inorganic analysis; the resulting cost is comparatively low (for o-phenanthroline, for example, it is less than one-tenth of the commercial price) and will serve to promote the use of these, otherwise relatively expensive, organic reagents in the laboratory.

Chapter XI is devoted to Qualitative Organic Analysis. The subject is discussed in moderate detail and this, coupled with the various Sections and Tables of Physical Constants of Organic Compounds and their Derivatives in Chapters III and IV, will provide a satisfactory course of study in this important branch of chemistry. No attempt has been made to deal with Quantitative Organic Analysis in this volume.

The textbook is intended to meet the requirements of the student of chemistry throughout the whole of his training. Considerable detail is given in those sections of particular interest to the elementary student; in the author's opinion it is the duty of a writer of a practical textbook to lay a secure foundation of sound experimental technique for the beginner. The subject matter of the book is sufficiently comprehensive to permit the teacher to cover any reasonable course of instruction. It will be observed that the scale of the preparations varies considerably; the instructor can easily adapt the preparation to a smaller scale when such a step is necessary from considerations of cost and time or for other reasons. Quantities of liquid reagents are generally expressed as weights and volumes: the latter refer to a temperature of 20°. The book will be suitable for students preparing for the Pass and Honours (General and Special) B.Sc. of the Universities, the A.R.I.C. and the F.R.I.C. (Organic Chemistry). It will also provide an introduction to research methods in organic chemistry and, it is hoped, may serve as an intermediate reference book for practising organic chemists.

Attention is directed to the numerous references, particularly in Chapter II on Experimental Technique, to firms supplying specialised apparatus. The author has usually had first-hand experience with this apparatus and he feels that some readers may wish to know the present source of supply and also from whom to obtain additional information. It must be mentioned that most of the specialised apparatus has been introduced to the market for the first time by the respective firms after much development research and exhaustive tests in their laboratories. A reference to such a firm is, in the writer's opinion, equivalent to an original literature reference or to a book. During the last decade or two much development work has been carried out in the laboratories of the manufacturers of chemical apparatus (and also of industrial chemicals) and some acknowledgement of the great help rendered to practical organic chemists by these industrial organisations is long overdue; it is certainly no exaggeration to state that they have materially assisted the advancement of the science. A short list of the various firms is given on the next page.

ARTHUR I. VOGEL.

Woolwich Polytechnic, London, S.E.18.   *December* 1946.

MANUFACTURERS AND SUPPLIERS OF LABORATORY APPARATUS EQUIPMENT AND CHEMICALS

Aldrich Chemical Co Ltd, The Old Brickyard, New Road, Gillingham, Dorset, SP84JL, UK.

Anderman & Co Ltd, Central Avenue, East Molesey, Surrey, KY8 0QZ, UK.

Applied Photophysics Ltd, 20 Albemarle Street, London, W1X 3HA, UK.

Applied Research Laboratories, Wingate Road, Luton, Bedfordshire, UK.

Applied Science Laboratories, Inc, UK. Suppliers Field Instruments Co Ltd, Tetrapak House, Orchard Road, Richmond, Surrey, UK.

Baird & Tatlock (London) Ltd, PO Box 1, Romford, RM1 1HA, UK.

W. & R. Balston Ltd, Maidstone, ME14 2LE, Kent, UK. (Whatman— registered trademark); Sales Agents, H. Reeve Angel & Co Ltd.

Baskerville & Lindsay Ltd, 324c Barlow Moor Road, Chorlton-cum-Hardy, Manchester, M21 2AX, UK.

Bausch & Lomb, Rochester 2, New York, USA; British Agents, **V. A. Howe & Co Ltd,** 88 Peterborough Road, London, SW6, UK.

B.D.H. Chemicals Ltd, Poole, Dorset, BH12 4NN, UK.

Beckmann-R11C Ltd, Eastfield Industrial Estate, Glenrothes, Fife, KY7 4NG, Scotland.

British American Optical Co Ltd, Instrument Group, 820, Yeovil Road, Slough, Bucks, UK.

Buchi Laboratoriums-Technik AG, CH 9230 Flawil, Switzerland.

Donald Brown (Brownall) Ltd, Lower Moss Lane, Chester Road, Manchester, M15 4JH, UK.

Cambrian Chemicals Ltd, Beddington Farm Road, Croydon, CR0 4XB, UK.

Camlab (Glass) Ltd, Nuffield Road, Cambridge, CB4 1TH, UK.

W. Coles & Co Ltd, PO Box 42, Plastic Works, 47/49 Tanner Street, London SE1, UK.

C. W. Cook & Sons Ltd, 97 Walsall Road, Perry Barr, Birmingham, B42 1TT, UK.

Corning Glass International SA, Corning Glass Works, Corning, New York, USA; UK Branch Office, 1 Cumberland House, Kensington Court, London, W8 5NP.

Decon Laboratories Ltd, Ellen Street, Portslade, Brighton, BN4 1EQ, UK.

Delmar Scientific Laboratories, 317 Madison Street, Maywood, Illinois, USA.

Draeger Normalair Ltd, Kitty Brewster, Blyth, Northumberland, UK.

Eastman Kodak Co, Rochester, New York, 14650, USA; UK Distributors, **Kodak Ltd,** Kirkby, Liverpool, L33 7UF, UK.

Edwards High Vacuum International, Manor Royal, Crawley, Sussex, UK.

Electrothermal Engineering Ltd, 270 Neville Road, London E7, UK.

Fluka-A.G., Chemische Fabrik, CH-9470, Buchs, Switzerland.

Fisher Scientific Co, 711 Forbes Ave, Pittsburgh, Pa, 15219, USA.

Fisons Scientific Apparatus, Bishop Meadow Road, Loughborough, Leicester-shire, LE11 0RG, UK.

A. Gallenkamp & Co Ltd, PO Box 290, Technico House, Christopher Street, London, EC2 2ER, UK.

Glass-Col Apparatus Co Inc, 711 Hulman Street, Terre Haute, Indiana, USA.

Griffin and George Ltd, Ealing Road, Alperton, Wembley, Middlesex, HA0 1HJ, UK.

Hanovia Lamps Ltd, Bath Road, Slough, Buckinghamshire SL1 6BL, UK.

Hewlett-Packard Ltd, King Street Lane, Winnersh, Berkshire, RG1 15AR, UK.

Hopkin and Williams Ltd, Freshwater Road, Chadwell Heath, Essex, UK.

Imperial Chemical Industries, Millbank, London WC1, UK.

Isopad Ltd, Electric Surface Heaters, Barnet-by-Pass, Boreham Wood, Hertfordshire, UK.

Japan Spectroscopic Co Ltd; British Agents: **Laser Associates Ltd,** Paynes Lane, Rugby, UK.

Jencons Scientific Ltd, Mark Road, Hemel Hempstead, Hertfordshire, HP2 7DE, UK.

JJ's (Chromatography) Ltd, Hardwick Trading Estate, King's Lynn, Norfolk, UK.

James A. Jobling & Co Ltd, Laboratory Division Headquarters, Stone, Staffordshire, ST15 0BG, UK.

Koch-Light Laboratories Ltd, Colnbrook, Buckinghamshire, SL3 0BZ, UK.

Laport Industries Ltd, Moorfield Road, Widnes, Lancashire, UK.

The Matheson Co, East Rutherford, New Jersey, USA.

Manostat Division of Greiner Scientific Corporation, 22 North Moore Street, New York, NY 10013, USA.

May & Baker Ltd, Liverpool Road, Barton Moss, Eccles, Manchester, UK.

E. Merck, 61 Darmstadt, Germany.

Midland Silicones Ltd, (Subsidiary of Dow Corning Corp.), Reading Bridge House, Reading, Berkshire, RG1 8PW, UK.

Perkin-Elmer Corp, Norwalk, Connecticut, 06856, USA; UK Division **Perkin-Elmer Ltd,** Beaconsfield, Buckinghamshire, UK.

Phase Separations; R. and D. Chemical Co, Deeside Industrial Estate, Queensferry, Flintshire, CH5 20LR, UK.

D. A. Pitman Ltd, Jessamy Road, Weybridge, Surrey, KT13 8LE, UK.

Pye-Unicam Ltd, York Street, Cambridge, CB1 2PX, UK.

P. B. Radley & Co Ltd, 53 London Road, Sawbridgeworth, Hertfordshire, UK.

Rank Precision Industries Ltd, Westwood, Margate, CT9 4JL, UK.

H. Reeve Angel & Co Ltd, 70 Newington Causeway, London, SE1 6BD, UK.

Satchell-Sunvic Ltd, Watling Street, Motherwell, Lanarkshire ML1 3SA, Scotland.

Scientific Furnishings Ltd, Industrial Estate, Chichester, Sussex, UK.

Scientific Glass Co Inc, 735 Broad Street, Bloomfield, New Jersey, 07003, USA.

Searle Diagnostic (Gurr Products), PO Box No. 53, Lane End Road, High Wycombe, Buckinghamshire, HP12 4HL, UK.

Shandon Southern Instruments Ltd, Lysons Avenue, Aldershot, Hampshire, GU12 5QP, UK.

The Southern New England Ultraviolet Co, 954 Newfield Street, Middletown, Connecticut, 06457, USA.

Stuart and Turner Ltd, Engineers, Henley-on-Thames, Oxfordshire, RG9 2AD, UK.

Techmation Ltd, 58 Edgware Way Edgware, Middlesex, HA8 8JP, UK.

Townson & Mercer Ltd, Beddington Lane, Croydon, Surrey, CR9 4EG, UK.

Union Carbide Chemicals Co, 270 Park Ave, New York 17, USA.

Union Carbide Ltd, Cryogenics Dept., Redworth Way, Darlington, County Durham, DL5 6HE, UK.

Van Waters & Rogers Scientific, PO Box 428, Newton Upper Falls, Massachusetts 02164, USA.

Varian Associates Ltd, Russell House, Molesey Road, Walton-on-Thames, UK.

Ventron Corp Metals Hydrides Division, Congress Street, Beverley, Massachusetts, USA.

Waters Associates Inc, Maple Street, Milford, Massachusetts, 01757, USA; **Waters Associates (Instruments) Ltd,** 324 Cheshire Road, Hartford, Northwich, Cheshire, CW8 2AH, UK.

M. Woelm, D-344 Eschwege, West Germany; UK distributors, **Koch-Light Laboratories Ltd**; USA distributors, **Alupharm Chemicals**, PO Box 30628, New Orleans, La, 70130.

Wright Scientific Co Ltd, 3 Lower Road, Kenley, Surrey, CR2 5NH, UK.

J. Young (Scientific Glassware) Ltd, 11 Colville Road, Acton, London, W3 8BS, UK.

The Zenith Electric Co Ltd, Cranfield Road, Wavedon, Milton Keynes, Buckinghamshire, UK.

CHAPTER I
EXPERIMENTAL TECHNIQUES

I,1. INTRODUCTION The synthesis of organic compounds is traditionally an important part of the training of the organic chemist. By undertaking the preparation of a varied range of compounds, and using a representative selection of reaction processes and practical techniques, the prospective organic chemist becomes familiar with the chemical and physical properties of organic substances and begins to understand more clearly the factors which govern their reactivity. The synthesis of quite simple compounds is of considerable educational value particularly if the reactions involved are of a general nature which may be applied, with suitable modifications if necessary, to more complex systems. Such elementary illustrative syntheses are described in full detail in the text, enabling the beginner to become familiar with standard laboratory operations.

The catalogues of most suppliers of organic chemicals are today very extensive and comprehensive, and it is not usually economic to synthesise in the laboratory compounds which are commercially available at a reasonable cost, and which may be needed as starting materials for more elaborate synthetic projects. Their preparation however often provides much of intrinsic interest and value. Furthermore the successful completion of a somewhat more difficult synthesis resulting in a high quality product in good yield, can afford much satisfaction and provide the developing worker with an increased confidence in his practical ability.

The last decades have occasioned many significant and exciting developments in organic synthesis. During this time the structures of an increasing range of biologically important complex natural products have been elucidated, largely as a result of the almost universal adoption of a powerful armoury of spectroscopic techniques. The challenge to the synthetic organic chemist to attempt the preparation in the laboratory of these compounds and their analogues has become correspondingly more demanding. Many of the successes gained have resulted from, and been the cause of, the development of a large number of newer, selective reagents; a knowledge of the applications and limitations of these reagents and the ability to fully utilise their selectivity is an essential part of the expertise of the advanced worker.

The isolation of the products from a selection of relatively simple organic reactions provides the beginner with the necessary experience in the standard procedures for solvent extraction, distillation, crystallisation and so on (Section **I,18**, *et seq.*). These long-established methods are frequently adequate for the isolation of a product in a desired degree of purity; it is usually essential for research purposes to confirm the purity and identity of the compound by spectroscopic investigation. For the separation of intractable mixtures chromatographic methods are invaluable, and experience in their use should be gained with the aid of the exercises included in the text. In a research situation, the best combination of isolation and purification techniques requires the application of the mature judgement acquired by the competent practical worker.

I,2. GENERAL INSTRUCTIONS FOR WORK IN THE LABORATORY

Before commencing any experiment in the laboratory it is essential that the worker should carefully study the complete details of the experiment as well as the underlying theory, and so have a clear idea of what is to be done and how he proposes to do it.

An able laboratory chemist will always utilise his time as efficiently as possible. Quite a number of reactions require rather prolonged periods of heating, stirring or standing, during which the whole of his attention is not engaged. This provides an opportunity for writing up the experimental details, cleaning up the work area, cleaning and drying apparatus and planning other experiments.

Writing up. This is an important part of any chemical experiment, since careful observation, allied to accurate reporting, is the very essence of any scientific exercise.

The guiding principle in writing up an experiment is to record all details which would enable another person to understand what was done and to repeat the entire experiment exactly without prior knowledge. Thus, in addition to a written account of the work done, including notes on any special apparatus used, details of all volumes, weights, temperatures, times, thin-layer and gas–liquid chromatography (TLC and GLC) results, etc., must all be recorded. The writing up of all laboratory work must be done at the time of the work, in a stiff-covered notebook of adequate size; a loose-leaved notebook is *not* suitable. It is important that numerical results such as yields, titration volumes, melting points and boiling points, etc., are entered directly into the notebook and not on scraps of paper. The latter are liable to be lost and their use encourages untidy practical habits.

The recommended format is to use a fresh double-sided page for each separate experiment or part of an experiment. The right-hand page should be used for a descriptive account of what was done and what was observed at the time; this account being continued, if necessary, overleaf on the next right-hand page. The left-hand page should be reserved for equations, calculation of yields, melting points, reaction mechanisms, etc., and possibly a later commentary. A properly written experimental account should be generously spaced out so that the different sections are readily discernible at a glance. It is not a good idea to stick spectra and GLC traces into the notebook, since it will rapidly become very bulky and the binding will be damaged. Rather they should be kept in a separate folder and cross-referenced with the (numbered) pages of the notebook.

During a reaction, all unexpected happenings and anything not understood should be carefully recorded at the time. Experiments sometimes go wrong, even with well-known procedures. (Note that there are often mistakes in the instructions published in the chemical literature, e.g., decimal points in the wrong places, leading to the use of incorrect weights and volumes.) In such a case, the worker should always try to unravel the reason for the failure of the reaction and to make suitable changes in the procedure, rather than just hurrying to repeat the experiment without modification.

Cleanliness. Some indication of a chemist's practical ability is apparent from the appearance of his working bench. The bench should always be kept clean and dry; this can easily be done if suitable wet and dry rags are kept at hand. Apparatus not immediately required should be kept as far as possible in a cupboard beneath the bench; if it must be placed on the bench, it should be arranged in a neat and orderly manner. Dirty apparatus can be placed in a plastic bowl away from the working area until it can be cleaned and put away. Solid waste and filter papers must not be thrown into the sink and all operations requiring the handling of unpleasant and noxious material should be carried out in a fume cupboard ('hood').

Cleaning glassware. All glassware should be scrupulously clean and, for most purposes, dry before being employed in preparative work in the laboratory. It is well to develop the habit of cleaning all glass apparatus immediately after use; the nature of the contamination will, in general, be known at the time, and, furthermore, the cleaning process becomes more difficult if the dirty apparatus is allowed to stand for any considerable period, particularly if volatile solvents have evaporated in the meantime.

It must be emphasised that there is no universal cleaning mixture. The chemist must take into account the nature of the substance to be removed and act accordingly. Thus if the residue in the flask is known to be basic in character, dilute hydrochloric or sulphuric acid may dissolve it completely; similarly, dilute sodium hydroxide solution may be employed for acidic residues. If the residue is known to dissolve in a particular inexpensive organic solvent, this should be employed.

The simplest method, when access by a test-tube brush is possible, is to employ a commercial household washing powder containing an abrasive which does not scratch glass (e.g., 'Vim', 'Ajax', etc.). The washing powder is either introduced directly into the apparatus and moistened with a little water or else it may be applied to the dirty surface with a wet test-tube brush which has been dipped into the powder; the glass surface is then scrubbed until the dirt has been removed. The operation should be repeated if necessary. Finally, the apparatus is thoroughly rinsed with distilled water. If scrubbing with the water–washing powder mixture is not entirely satisfactory, the powder may be moistened with an organic solvent, such as acetone.*

Excellent results are often obtained with warm 15 per cent trisodium phosphate solution to which a little abrasive powder, such as pumice, has been added; this reagent is not suitable for the removal of tar.

The most widely used cleansing agent is the **'chromic acid' cleaning mixture**. It is essentially a mixture [for precautions in its use see Section **I,3,B,***4*(v)] of chromium trioxide (CrO_3) and concentrated sulphuric acid, and possesses powerful oxidising and solvent properties. Two methods of preparation are available:

(1) Five grams of sodium dichromate are dissolved in 5 ml of water in a 250 ml beaker; 100 ml of concentrated sulphuric acid are then added slowly with

---

* Scrubbing or soaking with detergent solutions may be beneficial; the most recently introduced surface acting agent is Decon 90 concentrate, manufactured by Decon Laboratories Ltd, and is claimed to be more effective than chromic acid when used as a 5 per cent aqueous solution.

constant stirring. The temperature will rise to 70–80 °C. The mixture is allowed to cool to about 40 °C and then transferred to a dry, glass-stoppered bottle.

(2) One hundred ml of concentrated sulphuric acid, contained in a 250 ml Pyrex beaker, are cautiously heated to about 100 °C, and 3 g of sodium or potassium dichromate gradually added with stirring. Stirring should be continued for several minutes in order to prevent the resulting chromic acid from caking together. The mixture is allowed to cool to about 40 °C, and transferred to a dry, glass-stoppered bottle. The chromic acid mixture, prepared by shaking an excess of sodium dichromate or of finely-powdered potassium dichromate with concentrated sulphuric acid at the laboratory temperature, is not as efficient as that prepared by method (1) or (2); it is, however, useful for cleaning glassware for volumetric analysis. Exhaustion of a chromic acid cleaning solution is readily recognised by the change in colour from reddish-brown to green.

Before using the chromic acid mixture for cleaning, the vessel should be rinsed with water to remove organic matter and particularly reducing agents as far as possible. After draining away as much of the water as practicable, a quantity of the cleaning mixture is introduced into the vessel, the soiled surface thoroughly wetted with the mixture, and the main quantity of the cleaning mixture returned to the stock bottle. After standing for a short time with occasional rotation of the vessel to spread the liquid over the surface, the vessel is thoroughly rinsed successively with tap and distilled water [see Section I,3,B,4(v)].

The procedure which exploits the reaction which ensues when a little ethanol is rapidly added to the vessel, supported in the fume cupboard and containing some concentrated nitric acid, *should never be used*. The reaction is generally preceded by a short induction period and is extremely dangerous. *Under no circumstances* should concentrated nitric acid be added to ethanol—a violent explosion may result.

Small glass apparatus may be dried by leaving it in an electrically-heated oven maintained at 100–120 °C for about 1 hour. However, much organic apparatus is too bulky for oven drying and, moreover, is generally required soon after washing; other methods of drying are therefore used. All these methods depend upon the use of a current of air, which should preferably be warm. If the apparatus is wet with water, the latter is drained as completely as possible, then rinsed with a little industrial spirit or acetone. For reasons of economy, the wet industrial spirit or acetone should be collected in suitably labelled Winchester bottles for future recovery by distillation and re-use. After rinsing with the organic solvent, the subsequent drying is more conveniently done with the aid of the warm air blower * shown in Fig. I,1. It consists of a power-driven blower which draws air through a filter, passes it through a heater and forces it through the apparatus support tubes which are specially constructed to accommodate flasks and cylinders having narrow necks which make other means of drying difficult; each apparatus support tube has a number of holes at the end to ensure good distribution of heated air. Cold air may also be circulated if required.

---

* Glassware dryers manufactured and supplied by A. Gallenkamp & Co. Ltd, who kindly provided the photograph.

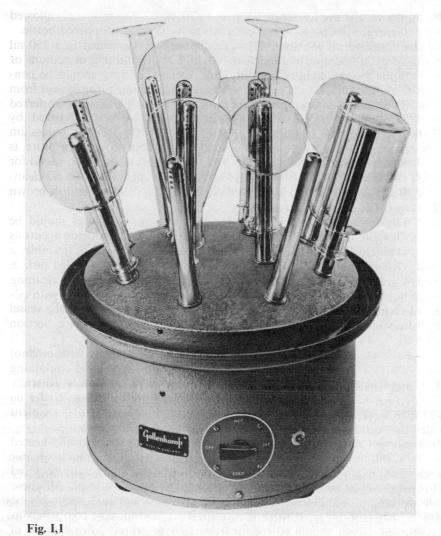

Fig. I,1

I,3. SAFE WORKING IN ORGANIC CHEMICAL LABORATORIES
A. General safety considerations
1. Introduction.
2. Conduct in the laboratory.
3. Tidiness.
4. Personal protection.
5. Accident procedure.
6. After-hours working.
7. Storage of chemicals.
8. Glassware.
9. Waste disposal.

B. Explosion and fire hazards
1. General aspects.
2. Explosive compounds.
3. Potentially dangerous mixtures.
4. Some specific dangers of explosives.
5. Fire hazards.
6. The dangerous operations laboratory.
7. The conduct of explosive or violent reactions.

C. Reactive inorganic reagents
1. Strong acids.
2. Strong bases.
3. Halogens.
4. Reactive halides.

D. Hazards due to toxic chemicals
1. Highly toxic solids.
2. Dangerously toxic gases.
3. Dangerously toxic liquids and severe irritants.
4. Other harmful substances.
5. Carcinogenic substances.
6. Substances with very harmful cumulative effects.

E. Electrical safety

F. Ultraviolet radiation

A. General safety considerations (see also Refs. 1–5)

1. Introduction. Chemistry laboratories need not be dangerous places in which to work, despite the many potential hazards associated with them, provided that certain elementary precautions are taken and that each worker conducts himself with commonsense and alertness.

There will almost invariably be a senior person assigned to be in charge of a chemical laboratory, irrespective of the nature of the work done there. However, it must be emphasised that the exercise of care and the adoption of safe working procedures is the responsibility of each and every individual person in that laboratory. If there is any doubt as to the safety of a proposed experiment, advice should be sought from an experienced person rather than just hoping for the best.

Every worker must adopt a responsible attitude to his work and avoid any thoughtless, ignorant or hurried behaviour which may lead to an accident and possible harm to himself or to others. He should always pay attention to what is going on around him and be aware of any possible dangers arising from the work of others as well as from his own experiments.

Laboratory accidents are very often caused by attempts to obtain results in too great a hurry. The laboratory chemist must therefore always adopt a thoughtful, careful and methodical approach to what he is doing. In particular, he must concentrate on the job in hand and not allow his attention to be distracted. Similarly he should not unnecessarily distract others. In routine experiments and operations it is important to remember the truth of the adage 'familiarity breeds contempt' and to be on one's guard against the feeling that 'it cannot happen to me'.

2. Conduct in the laboratory.   Except in an emergency, running, or any over-hurried activity, should be forbidden in and around laboratories, as should be practical jokes or other irresponsible behaviour. It is strongly recommended that no worker should ever eat, drink or smoke in a laboratory; these constitute a further, avoidable, risk of the ingestion of toxic substances, and in the case of smoking an additional fire hazard.

3. Tidiness.   Coupled with a general consideration for the safety of others, tidiness is a major factor in laboratory safety; the laboratory must be kept clean and tidy at all times. Passageways between and around benches and near exits must not be blocked with equipment or furniture. Floors must be kept clean and in good condition to prevent slipping or tripping, i.e., they must be kept free from oil or water, or from any protrusion. Any spillage on a floor or bench should be thoroughly cleaned up immediately.

4. Personal protection.   In the laboratory, each worker should wear a full-length protective coat, preferably white, since spillages and stains are then more easily detected. *It is strongly recommended that in all organic chemistry laboratories, all personnel, including visitors, should wear safety spectacles or goggles at all times.* There are several good lightweight goggles and spectacles available for routine use which provide good coverage of the eyes and upper face and which may be worn over prescription glasses. However, it should be noted that in the event of an accident, conventional safety glasses provide varying degrees of protection against flying fragments,* but very little protection against splashing or spraying of hot, corrosive or toxic liquids and gases. Close-fitting safety goggles or, preferably, a visor covering the whole face afford a much greater measure of protection in these circumstances.

5. Accident procedure.   Every person working in a laboratory should ensure that he knows where the exits and fire escapes are situated and that there is free access to them. The worker should also note the positions of fire extinguishers, fire blankets and drench showers, and make sure he knows how to use them. He should also know the location of the first aid equipment provided for emergency use, the position of the nearest telephone and the numbers of the appropriate medical teams, hospitals and fire brigades.

6. After-hours working.   No one should ever work alone in a laboratory. Experiments which must be left running overnight are best sited in a special overnight room (see Section **I,16**), but if this is not possible, the precautions recommended in Section **I,16** should still be adopted and the apparatus labelled clearly as to the nature of the reaction and likely hazards. Clear instructions must be left so that an unqualified person can terminate the experiment in an emergency. 'Please leave on' notices should be left alongside any service which is to be left running (water, electricity).

7. Storage of chemicals.   Chemicals should never be allowed to accumulate on benches or in fume cupboards, but should always be returned to their proper places on robust storage shelves; incompatible chemicals should be separated from one another. Heavy containers and bottles of dangerous chemicals should be stored on the floor or on the lowest shelf. Winchester bottles of

---

* Goggles and spectacles are available in laminated glass and in a range of polymeric materials. Of these, polycarbonate is the best; it has a high impact resistance and does not shatter into fragments as methacrylate (e.g., Perspex) may do. Cellulose acetate is also generally satisfactory.

dangerous chemicals (e.g., strong acids) must never be carried by the neck, but in the special carriers available for the purpose.

Fume cupboards must be kept free from surplus chemicals and discarded apparatus. If stocks of noxious chemicals have to be stored in a fume cupboard they should all be assigned to one which is set aside for this purpose and is properly fitted with shelving.

All chemical reagent bottles, sample bottles and vials must carry a label clearly indicating the nature of the contents and possible hazards. In the case of vessels which have lost their labels, the contents should be positively identified and relabelled, or disposed of safely. Since gummed labels readily dry out and drop off, it is a good idea to seal them to the vessel with transparent adhesive tape. Since many chemicals deteriorate with age, it is also a good idea to write the date on the label when the bottle arrives from the distributor.

8. *Glassware.* Glass apparatus should be carefully examined before use and any which is cracked, chipped, flawed or dirty should be rejected. Minute cracks in glassware for use in evacuated systems are particularly dangerous.

Many apparently simple manipulations such as the cutting of glass tubing or rod, the insertion of glass tubing or thermometers into bungs or corks, or the removal of tight stoppers from bottles, can lead to serious cuts. Care should be taken to adopt the correct procedures (Sections **I,9** and **I,10**). All apparatus and glassware not in use should be stored away and not allowed to accumulate on benches. Dirty glassware should always be cleaned as soon as possible while the nature of the residues is still known and suitable washing procedure can be selected before any hardening process occurs (Section **I,2**).

9. *Waste disposal.* Waste material must never be allowed to accumulate in the laboratory; it should be removed regularly from the working area for storage in suitable containers so that it can be disposed of appropriately. There should be separate bins with properly fitting lids for broken glassware and for inflammable materials such as paper or cloths which may have been used to mop up inflammable liquids. Waste solids should be placed in bins provided (toxic solids can be sealed in plastic bags first) and waste solvents should be placed in suitable containers appropriately labelled, but indiscriminate mixing of contaminated solvents should be avoided. Halogenated solvents in particular should be kept apart from other solvents and may with advantage be purified for re-use.

Most large laboratory complexes will have arrangements by which the accumulated waste material is disposed of appropriately; they may for example have facilities for the combustion of quantities of inflammable organic material. Smaller establishments however may have to rely on the services of specialised contractors. The problems associated with the disposal of relatively small quantities of toxic or hazardous unwanted chemicals can be lessened by the individual laboratory worker taking intelligent action. A useful summary of suitable disposure procedures is available.* Frequently it is possible to render the material for disposal relatively innocuous by appropriate treatment so that it can be washed down the drain with an excess of water. Local regulations may however prevent the disposal of some materials in this way, and in any case untreated wastes and water-insoluble organic solvents should never be thrown down the sink.

---

* The Aldrich Catalogue and Handbook of Organic and Biochemicals 1977/78.

B. Explosion and fire hazards

1. General aspects. Explosive and highly inflammable substances or mixtures of substances quite commonly have to be used in organic chemistry laboratories. Ignorance of the hazards which are likely to be encountered all too frequently leads to explosions and fires, but these may usually be avoided and the experiment conducted with a reasonable measure of safety if, in addition to the general rules for laboratory practice mentioned under **A**, the following guidelines are followed.

(a) The use of a substance known to be explosive should be avoided if a safer alternative can be used.

(b) If an explosive or dangerously reactive substance has to be used, then it should be used in the smallest possible quantity and with all the appropriate precautions which are indicated in sub-section **B,7** below.

(c) Workers should try to foresee and avoid the situation where a dangerously reactive chemical is likely to come into contact with combustible material, or where an explosive substance is likely to be subjected to the stimulus of shock or excess heat.

(d) Reactions known or likely to involve explosion or fire hazards should always be tried out on a small scale first, and only then carefully scaled up in stages if no warning signs of danger are apparent (e.g., no undue rise in temperature or evolution of gas, etc.). Since for a reaction vessel the surface area per unit volume decreases with increasing volume, scaled-up reactions may exhibit unexpectedly large and possibly dangerous temperature rises. If a small scale reaction procedure is known to be safe, it is better to repeat it several times to acquire the required stock of product, rather than to attempt to scale up the process to achieve this in one step.

(e) For notably exothermic reactions or reactions involving dangerously active reagents, the safest procedure is to add the reagent dropwise, with rapid stirring, at the same rate as it is used up. *Overcooling* must be avoided since this may inhibit the reaction sufficiently to allow a dangerous accumulation of the reagents; if the temperature is then allowed to rise, a violent reaction may occur. It may actually be safer to heat such a reaction to ensure complete consumption of each drop of reagent as it is being added.

2. Explosive compounds. The following compounds or groups of compounds are likely to be dangerously explosive in their own right. They may explode under the stimulus of heat, impact or friction, or apparently spontaneously.

(a) Acetylene gas and the acetylide salts of heavy metals; silver and copper acetylides are extremely shock-sensitive. Polyacetylenes and some halogenated acetylenes.

(b) Hydrazoic acid and all azides, both organic and inorganic (only sodium azide is safe); aryl azides and silver azide may be inadvertently formed during some reactions [see sub-section **B,4**(vi) below].

(c) Diazonium salts (when solid) and diazo compounds.

(d) Inorganic nitrates, especially ammonium nitrate. The nitrate esters of polyhydric alcohols.

(e) Polynitro compounds, e.g., picric acid (and heavy metal picrates), tri-nitrobenzene (TNB), trinitrotoluene (TNT); all these substances are safe when damp with water.

(f) Metal salts of nitrophenols.

(g) Peroxides; these are a common cause of explosions due to their formation in ether solvents [see sub-section **B**,*4*(i)]. Concentrated aqueous hydrogen peroxide solution, see Section **II**,2,*34*.

(h) Nitrogen tribromide, trichloride and triiodide; these are all highly sensitive and violently explosive, and should never be prepared or used unless absolutely necessary.

3. Potentially dangerous mixtures. Powerful oxidants are particularly dangerous when mixed with easily oxidised organic substances such as simple alcohols, polyhydric alcohols, carbohydrates and cellulose containing materials such as paper, cloth or wood. They are also dangerous when mixed with elements such as sulphur and phosphorus, and with finely divided metals such as magnesium powder. The following are common examples:

(a) Perchloric acid, chlorates and perchlorates.

(b) Chromium trioxide ('chromic anhydride'), chromates and dichromates. Concentrated nitric acid and nitrates.

(c) Permanganates.

(d) Concentrated hydrogen peroxide.

(e) Liquid oxygen and liquid air.

4. Some specific dangers of explosion. (i) *Peroxides in ether solvents.* This is one of the commonest causes of explosions in organic chemistry laboratories. Simple dialkyl ethers such as diethyl ether and di-isopropyl ether, and cyclic ethers such as 1,4-dioxan and tetrahydrofuran, form less volatile peroxides on exposure to air and light. If therefore one of these solvents is purified by distillation, the peroxide content in the residue is progressively increased and eventually a violent explosion may occur. In view of this, (a) such solvents should not be stored for long periods or in half empty bottles; containers should be of dark glass, (b) before the solvents are distilled a peroxide test should be carried out, and, if positive, the peroxide must be removed (Sections **II**,1,*11* and *15*), and (c) since purified ethers in contact with air rapidly peroxidise again (10 minutes in the case of tetrahydrofuran) they should be re-tested for peroxides and purified if necessary immediately before use.

(ii) *Solid sodamide and potassium metal.* Both of these substances undergo surface oxidation to give oxide films which may initiate explosions when the samples are handled. In the case of potassium, surface oxidation occurs even when the metal is stored under oil, and the act of paring off the oxide film with a knife may initiate an explosion. Samples of potassium which are heavily encrusted with oxide should not be used but should be carefully destroyed by adding the lumps to a large excess of propan-2-ol. Similarly, old or obviously encrusted (yellow) lumps of sodamide (Section **II**,2,*54*) should not be ground in a pestle and mortar, but should be destroyed by mixing with solid ammonium chloride.

(iii) *Alkali metals with chlorinated solvents.* The alkali metals sodium, potassium and lithium (and also other metals, e.g., aluminium and magnesium, especially when finely divided), are all violently reactive towards halogenated organic compounds, notably the common chlorinated solvents such as carbon tetrachloride. Lumps or chips of these metals should *never* be washed with halogenated solvents—a violent explosion can result.

(iv) *Perchloric acid.* This can react violently with organic material such as cork, cloth, rubber or wood. In addition the fumes which are readily evolved from the liquid acid are easily absorbed by these substances which are thus rendered violently inflammable or explosive. For this reason, perchloric acid should not be stored in a wood-framed fume cupboard or near to any organic material.

(v) *Chromic acid and nitric acid as cleaning agents.* Violent explosions have ensued when attempts have been made to remove tarry residues from reaction flasks by adding chromic acid mixtures, or concentrated nitric acid, and heating. If such residues are not removed by chromic acid mixtures in the cold (even after prolonged treatment with several changes of acid) followed by scrubbing with scouring powder, then the only safe course is to throw the flask away. The old fashioned method of cleaning glassware by mixing concentrated nitric acid and ethanol is *extremely dangerous* and must not be used under any circumstances.

(vi) *Azides.* Explosive aryl azides may be formed inadvertently during the Sandmeyer and other diazonium reactions.

Explosive silver azide forms when solutions of ammoniacal silver nitrate (Tollen's reagent) are allowed to stand before use. This is extremely dangerous; Tollen's reagent should always be freshly prepared following the procedure given on p. 1072, taking care not to exceed the recommended concentrations. Unused reagent should be destroyed by the addition of aqueous sodium chloride.

(vii) *Liquid nitrogen.* Liquid nitrogen (b.p. $-196\,^\circ$C) contains some liquid oxygen (b.p. $-183\,^\circ$C) as an impurity and therefore evaporation leads to an increasing proportion of liquid oxygen, so that before complete evaporation occurs the residual liquid may contain *up to 80 per cent of liquid oxygen.* Contact of this residue with organic or combustible material of almost any sort is likely to cause an explosion. If Dewar flasks containing liquid nitrogen have been used as cooling baths, great care must be taken to ensure that all liquid nitrogen and oxygen has evaporated *completely* before the Dewar flask is used for another purpose (e.g., as an acetone–solid carbon dioxide cooling bath).

(viii) *Glass vacuum assemblies.* Before using any glass apparatus for vacuum distillation or sublimation it should be examined to ensure that, (*a*) it is of the correct thickness and type (thin-walled glassware and conical shaped flasks are not suitable), and (*b*) that it is free from cracks and flaws. Vacuum desiccators should always be used in the smallest suitable size and should be encased in wire safety cages. Dewar flasks can cause considerable damage since they may collapse violently ('implode') if they are maltreated. All Dewar flasks should therefore be bound, over their entire length, with adhesive plastic tape to contain flying fragments of glass in the event of an implosion.

(ix) *Opening glass ampoules.* Ampoules of volatile chemicals must be thoroughly cooled before opening. Cooling must be effected with care, particularly if the contents are highly reactive (e.g., boron trichloride). If cooling is mishandled the glass may crack and the release of the contents into the cooling bath may lead to a violent explosion. Ampoules should not be cooled to a low temperature *too quickly.* Cooling in ice-water initially, followed by ice-salt, will usually be satisfactory; cooling to solid carbon dioxide temperatures is not necessary.

If the contents have a tendency to decompose, considerable pressure may develop in the sealed ampoule on storage, and great care should be taken during opening. The cooled ampoule should be removed from the cooling bath and wrapped in strong cloth behind a safety screen. A clean scratch should be made in the neck of the ampoule with a sharp file or glass knife and the neck cracked off by touching the scratch with the molten end of a thin red-hot glass rod.

Ampoules should be well cooled before resealing. Resealing should be avoided if possible however; it is best to obtain a smaller size of ampoule and use the whole of the contents for one experiment.

(x) *Compressed gas cylinders.* Under certain circumstances, cylinders of compressed gas may constitute major explosion and fire hazards and, despite their apparent robust construction, they should always be handled with care (Ref. 6).

In view of the high pressures involved, any possibility of slow leakage from a cylinder of an inflammable or toxic gas should be carefully guarded against. Thus, gas should never be drawn from a cylinder unless the appropriate reduction valve has been correctly fitted. The main cylinder valve should never be opened more than is necessary to provide the required gas flow (two full turns of the spindle at the most *); when the cylinder is not in use the gas should be shut off at the main valve and not at the regulator, which should then be bled of surplus pressure and closed. A suspected leak may be tested by brushing with a 5 per cent aqueous detergent solution.

The valves and screw threads of cylinders and regulators should *never* be greased since this may lead to an explosion. If a cylinder has a very stiff spindle valve or if the screw threads are damaged, it should be returned to the suppliers for replacement. Similarly, defective regulators and pressure gauges should never be used.

The possibility of the sudden release of the entire contents of a cylinder must be guarded against. Apart from obvious dangers in the release of inflammable or toxic gases, the sudden release of any gas can transform a cylinder into a lethal jet-propelled missile. Thus any weakening of a cylinder by damage, particularly to the valve, must be prevented. Cylinders must never be allowed to stand free in an upright position where they might be knocked over. They should be supported either by strapping to a bench or wall, or kept in one of the special mobile trolleys available for the purpose. Cylinders should only be moved by the use of these trolleys.

Gas cylinders should always be properly stored in areas of moderate temperature which are adequately ventilated but protected from the elements. These areas should be kept free from water and any corrosive fluids or vapours. Large notices labelled FULL and EMPTY should be prominently displayed on cylinders to prevent confusion and mistakes.

5. *Fire hazards.* Fire hazards in organic chemistry laboratories are often considerable due to the quantities of volatile and inflammable chemicals, particularly solvents, which are commonly used. Many of the general precautions

---

* Spindles should never be fully unscrewed since some are not captive and will be blown out by the full gas pressure if they are unscrewed completely.

recommended to minimise these fire hazards have been dealt with under **A** but specific methods for dealing with the more notable hazards are given below.

(i) *Inflammable solvents.* Particular care should be taken when handling inflammable solvents (and other chemicals) which are also *highly volatile.* The vapour may drift to a distant ignition source and burn back to ignite the main bulk of the liquid. An important rule is never to allow any vapour of a volatile chemical to escape into the open laboratory (in addition to fire hazards many vapours are toxic; see **D** below). All containers of volatile chemicals should be kept tightly closed and the quantity stored limited to foreseeable short-term requirements; large quantities of solvent should not be allowed to accumulate in the laboratory. If spillage of solvent or accidental release of inflammable vapour occurs, the whole laboratory should be ventilated as soon as possible.

A measure of the inflammability of a compound is given by the *'flash point'* (the temperature at which the liquid gives rise to ignitable vapour). Any liquid with a flash point of less than 15 °C should be regarded as dangerously inflammable and treated accordingly. If a solvent also has a low *'autoignition temperature'* (the temperature at which the vapour will spontaneously ignite in air), it should be treated with particular care.

Some highly inflammable common solvents are given in Table I,1 in order of increasing flash point. This does not include other dangerously inflammable substances not commonly used as solvents.

Table I,1   Flash points of common solvents

	(°C)		(°C)
Pentane and light petroleum	49	Butan-2-one	−7°
(b.p. 40–60 °C)		Ethyl acetate	−4
Diethyl ether	−45	Heptane	−4
Cyclopentane	−37	Methylcyclohexane	−4
Carbon disulphide	−30*	Toluene	4
Di-isopropyl ether	−28	1,2-Dimethoxyethane	4.5
Hexane and light petroleum	−23	Acetonitrile	6
(b.p. 60–80 °C)		Pentan-2-one	7
Cyclohexane	−20	Methanol	10
Acetone	−18	1,4-Dioxan	12
Tetrahydrofuran	−17	Propan-2-ol	12
Benzene	−11	Ethanol	12
Methyl acetate	9	Ethylbenzene	15

* Carbon disulphide has the very low autoignition temperature of 100 °C. The vapour may therefore ignite on contact with steam pipes or with boiling water baths.

(ii) *Ignition sources.* Naked flames should rarely be used in organic chemistry laboratories. The heating of reaction mixtures is much more safely accomplished by means of a steam bath, an electric heating mantle or an oil bath heated by means of a small electric immersion heater (or, less safely, by a hot plate). If Bunsen flames have to be used, they should only be lit after a careful survey of neighbouring apparatus and chemicals has revealed no fire hazard. The flame should be turned out whenever it is not actually in use; a gas–air Bunsen flame may be invisible in bright sunlight and thus the cause of a fire or burning accident.

If inflammable vapour is allowed to accumulate in the vicinity of electrical devices such as thermostats, stirrer motors,* vacuum pumps, drying ovens, etc., it may be ignited by sparking from electrical contacts; this may be minimised by good laboratory ventilation and the prevention of the local build-up of solvent vapours. Sparking of contacts has caused serious explosions and fires when domestic-type refrigerators have been used to store volatile substances, even in small quantities.† Since a volatile compound can have an appreciable vapour pressure at 0 °C a dangerous concentration of vapour can accumulate within the cabinet of a refrigerator. Volatile compounds should therefore be stored in the refrigerator in glass containers, with well-fitting stoppers.

(iii) *Leaking oxygen cylinders.* If a cylinder containing oxygen is allowed to leak over a period when normal laboratory ventilation is turned off, the concentration of oxygen in the air may become great enough to cause a very fierce fire in the event of an ignition source being present and of there being inflammable materials in the vicinity. All compressed oxygen cylinders should be tested for leaks by brushing the valve joints with a dilute solution of detergent. Leaking cylinders should be suitably labelled and removed from the laboratory.

(iv) *Sodium residues.* Bottles containing sodium wire previously used for solvent drying constitute a fire and explosion hazard. The sodium, sometimes heavily coated with hydroxide or oxide film, should be covered with propan-2-ol and set aside with occasional swirling until all the sodium particles are destroyed (at least 2 hours). The contents of the bottle should then be poured into a large excess of water (water should *not* be added to the bottle) and the bottle washed out several times with industrial spirit. Only then can the bottle be safely rinsed with water.

Fires involving sodium metal are very hot and localised and are best dealt with by smothering with sand or by using a dry powder extinguisher, *NOT* a carbon tetrachloride or carbon dioxide extinguisher.

(v) *Metal hydrides.* Lithium hydride, sodium hydride and lithium aluminium hydride all react violently with water liberating hydrogen; the heat of reaction may cause explosive ignition. Excess metal hydride from a reaction must be destroyed by careful addition of ethyl acetate or acetone.

6. *The dangerous operations laboratory.* It is strongly recommended that all reactions involving any possible hazard from explosive, inflammable, dangerously reactive or highly toxic substances should be carried out in a special laboratory solely designed for the purpose. Such a laboratory should not be used for any routine teaching or research purposes, or for the storage of chemicals or apparatus apart from those required for specific hazardous reactions (e.g., autoclaves, furnaces for Carius tubes, etc.). Ideally the laboratory should be purpose-built and should incorporate the following safety features:

(*a*) Water-proof and vapour-proof electric lamps, switches and power points.

(*b*) Fume cupboards fitted with powerful extractor fans capable of rapidly changing all the air in the laboratory.

(*c*) Fire-resistant doors and walls.

(*d*) An adequate supply of protective clothing including safety visors and goggles, protective gloves, rubber aprons and boots.

---

* An air-driven stirrer as supplied by the Fisher Scientific Co is safer in this context.

† This danger is avoided if a specially designed laboratory refrigerator (e.g., of the type supplied by the Fisher Scientific Co) is used.

(e) Good quality safety shields and screens for guarding potentially violent reactions.

(f) An automatic carbon dioxide fire-extinguishing system.

(g) The following should also be provided immediately outside the laboratory: storage facilities for gas masks and self-contained breathing apparatus; fire blankets, and buckets of sand for spilled liquids and for smothering fires; large carbon dioxide and dry powder extinguishers if no automatic fire extinguishing system is installed; a telephone with a clear notice beside it listing procedures and numbers to be dialled in an emergency.

If no laboratory or other room is available for conversion to a dangerous operations laboratory, then a semi-permanent structure of adequate design sited at a safe distance from regularly occupied laboratories and offices may be constructed from lightweight and fireproof building materials (cf. unattended operations, Section **I,16**).

7. *The conduct of explosive or violent reactions.* There is a common tendency to regard fume cupboards as the proper sites for potentially explosive or violently reactive processes. This is not to be recommended since the glass windows of fume cupboards are very rarely of sufficient quality and thickness to withstand an explosion, and the confinement of gaseous reaction products by the sides and top of the fume cupboard increases the severity of the blast. All potentially violent reactions should therefore be conducted on an open bench, with the apparatus surrounded by safety shielding on all sides but open at the top. It has been shown that even flimsy protection at the top dramatically reduces the efficiency of such side screens to contain an explosion. The best design of safety shield is a flat plate of polycarbonate (minimum thickness 3 mm) suspended in a vertical plane from above, and heavily weighted along the bottom edge. The performance of conventional curved free-standing shields may be considerably improved by heavily weighting the bottom edge to prevent the whole shield being blown over in an explosion.

C. Reactive inorganic reagents Many inorganic reagents used in organic chemistry laboratories are highly reactive (and hence have 'corrosive' properties), causing immediate and severe damage if they are splashed or spilled on to the skin, or when they are inhaled as vapours, dusts or mists. In addition, their high reactivity may cause a rapid evolution of heat when they are mixed with other chemicals, including water, resulting in a corrosive and possibly toxic mixture being sprayed and splashed about; sometimes fires or explosions follow. When using such chemicals suitable protective clothing including gloves should be worn. Adequate protection of the eyes is absolutely essential and safety spectacles, or preferably goggles or a visor, must always be worn. When there is any possibility of inhalation of reactive vapours or dusts, all operations should normally be conducted in a fume cupboard. Additional protection may be provided by a gas mask or well-fitting dust mask.

If any corrosive liquid or solid is spilled on to the skin it should be immediately washed off with copious quantities of water; in cases of splashes in the eyes, every second counts. Any spillages should be cleaned up without delay, preferably with the aid of sand. Flooding a spillage on a floor or bench with water is not always advisable if this is likely to spread the corrosive material and cause it to lodge in crevices and between floorboards. In cleaning up extensive spillage where noxious fumes are involved, full protective clothing including respirators should be used.

Some highly reactive chemicals and their dangerous properties are listed below. Those which give off highly corrosive irritant and/or toxic vapours or, if solids, are similarly hazardous in the form of dusts, are marked with an asterisk and should only be used in fume cupboards. More details concerning the properties of many of these are given in Section **II,2**. For specific information on the hazardous properties of individual chemicals, the comprehensive works cited in Refs. 1–5 are recommended.

1. Strong acids. All of the following react violently with bases and most give off very harmful vapours.

* Hydrobromic acid and hydrogen bromide.

* Hydrochloric acid and hydrogen chloride.

* Hydrofluoric acid and hydrogen fluoride—both react readily with glass and quickly destroy organic tissue. New thick rubber or plastic gloves should be worn after carefully checking that no holes are present. Skin burns must receive immediate and specialised medical attention.

* Nitric acid (concentrated and fuming).

* Perchloric acid (explosion danger, see sub-section **B,4**).

Sulphuric acid (concentrated and 'oleum')—should always be mixed with water very carefully, by pouring into cold water as a thin stream to prevent acid splashes or spray. 'Chromic acid' cleaning mixtures have the corrosive properties of concentrated sulphuric acid as well as the dangerous oxidising properties of the chromic acid.

* Chlorosulphonic acid—this is a highly corrosive liquid which reacts violently with water.

2. Strong bases. Calcium oxide, potassium hydroxide and sodium hydroxide—these react violently with acids, generate heat on contact with water, and have a powerful corrosive action on the skin, particularly the corneal tissue of the eye.

* Ammonia (gas and concentrated aqueous solution, *d* 0.880). Concentrated hydrazine solutions (and hydrazine salts); * hydrazine vapour is harmful.

* Sodamide—usually obtained in a granular form which reacts violently with water; it is irritant and corrosive in a finely divided form. Old and highly coated samples should not be crushed for use but should be destroyed (see subsection **B,4(ii)**).

3. Halogens. All are toxic and corrosive. Great care should be exercised when working with fluorine, which is violently reactive towards a wide range of substances. The interhalogen compounds are also powerfully reactive.

4. Reactive halides. All of the following are highly reactive, particularly towards water; ampoules of liquids should be opened in a fume cupboard after cooling, observing the precautions detailed in sub-section **A,4(ix)**.

* Boron trichloride; * phosphorus tribromide, * trichloride and * pentachloride; * silicon tetrachloride. * Aluminium chloride and * titanium tetrachloride are rather less reactive.

5. Chromium trioxide, chromates and dichromates. All these form corrosive dusts; those from water-soluble chromates are particularly dangerous

---

* Highly corrosive, irritant and/or toxic vapours or dusts.

since they dissolve in nasal fluid and in perspiration. Long-term exposure can lead to ulceration and cancer.

D. Hazards due to toxic chemicals A very large number of compounds encountered in organic chemistry laboratories are poisonous (i.e., 'toxic'). Indeed, nearly all substances are toxic to some extent and the adoption of safe and careful working procedures which prevent the entry of foreign substances into the body is therefore of paramount importance, and should become second nature to all laboratory workers. Toxic substances can enter the body by the following routes:

(a) *Ingestion* (*through the mouth*). This is fortunately not common in laboratories, but can occur through the accidental contamination of food, drink or tobacco, and by misuse of mouth pipettes. It is strongly recommended that no one should ever eat, drink or smoke in a laboratory. The practice of storing bottles of milk or beer in laboratory refrigerators is to be strongly condemned.

Workers should always wash their hands thoroughly on leaving a laboratory and before eating. All pipetting by mouth should be avoided since there are excellent rubber bulb and piston-type pipette fillers available commercially.

In addition to the ingestion hazard associated with smoking, the vapours of many volatile compounds yield toxic products on pyrolysis when drawn through a lighted cigarette or pipe (e.g., carbon tetrachloride yields phosgene).

(b) *Inhalation* (*into the lungs*). This is a more common pathway for the absorption of toxic chemicals; these may be in the form of gases, vapours, dusts or mists. All toxic powders, volatile liquids and gases should only be handled in efficient fume cupboards. The practice of sniffing the vapours of unknown compounds for identification purposes should be conducted with caution.

(c) *Direct absorption* (*through the skin into the bloodstream*). This is also a common route for the absorption of a toxic substance whether liquid, solid or gaseous. The danger may be reduced by wearing rubber or plastic gloves, in addition to the usual laboratory white coat. However, clean and careful working procedures are still necessary despite these precautions. Protective gloves are often permeable to organic solvents and are easily punctured; they should therefore be frequently inspected and replaced when necessary. If a toxic substance is accidentally spilled on the skin, it should be washed off with copious quantities of cold water with the aid of a little soap where necessary. The use of solvents for washing spilled chemicals off the skin is best avoided since this may hasten the process of absorption through the skin.

Repeated contact of solvents and many other chemicals with the skin may lead to dermatitis, an unsightly and irritating skin disease which is often very hard to cure. In addition, sensitisation to further contact or exposure may occur.

The toxic effects of chemical compounds can be classified as either 'acute' (short-term) or 'chronic' (long-term). Acute effects, as exemplified by powerful and well-known poisons such as hydrogen cyanide and chlorine, are immediately obvious, well appreciated by most laboratory workers, and are therefore fairly easily avoided. However, many chemicals exhibit chronic toxic effects which may only come to light after long-term exposure to *small quantities*. This type of insidious poisoning is harder to detect (and therefore prevent) since the results may only manifest themselves after months or even years of exposure (or even long after exposure has ceased). Chronic poisoning may also cause symptoms which are not easily recognisable as such, e.g., sleeplessness, irritability, memory lapses and minor personality changes. It must be stressed, however, that the final results of chronic poisoning may be very serious and can lead to premature death. Every effort should be made by the laboratory worker to guard against these possibilities by adopting a rigorous approach to the avoidance of breathing all vapours and dusts, and of any contact between the skin and liquids or powders.* The guiding principle for all workers should be to treat all chemicals as potentially harmful unless one has positive knowledge to the contrary. Compounds with acute toxic properties† which are likely to be encountered in organic laboratories are listed in the following sub-sections *1* and *2*, and those which give rise to particularly severe chronic effects in *4* and *5*. Substances marked **C** are also known to be carcinogenic, see **D,5** below.

For detailed information on the toxicological properties of individual substances reference should be made to the specialist monographs on the subject (Refs. 7–10).

1. Highly toxic solids. Even small quantities of these substances are likely to rapidly cause serious illness or even death. Particular care should be taken to avoid inhalation of dusts and absorption through the skin as well as the more obvious hazards of direct ingestion.

	TLV mg m^{-3}
Arsenic compounds	0.5 (as As)
Inorganic cyanides	5 (as CN)
Mercury compounds, particularly alkyl mercurials	0.01
Osmium tetroxide (hazardous vapour)	0.002
Oxalic acid (and its salts)	1
Selenium and its compounds	0.2 (as Se)
Thallium salts	0.1 (as Tl)
Vanadium pentoxide	0.5

2. Dangerously toxic gases. All operations involving the use or liberation of these substances must be carried out in an efficient fume cupboard. In most cases contact with the skin must be prevented.

	TLV p.p.m.
Boron trifluoride	1
Carbon monoxide	50
Chlorine	1
Cyanogen	10
Diazomethane, **C** †	0.2
Fluorine	0.1
Hydrogen cyanide	10
Hydrogen fluoride	3
Hydrogen sulphide	10

---

* For monitoring the concentrations of a wide range of toxic vapours in the atmosphere, inexpensive portable detectors are available from Draeger Normalair Ltd, or Pitman Instruments.

† An indication of the hazard associated with the use of a toxic material in the form of vapour or dispersed dust is given by the threshold limit value (TLV), expressed as parts per million or as milligrams per cubic metre. These represent maximum concentrations to which it is believed most workers can be repeatedly exposed without harmful effects. Currently recommended TLV data is available in Ref. 11.

Nitrogen dioxide (nitrous fumes) and nitrosyl chloride	5
Ozone	0.1
Phosgene	0.1
Phosphine	0.3

3. Dangerously toxic liquids and severe irritants. These substances have dangerously toxic vapours and are also harmful through skin absorption. Prolonged exposure to small amounts is likely to give rise to chronic effects. The vapours of many are powerful irritants particularly to the respiratory system and to the eyes.

	TLV p.p.m.
Acetyl chloride	—
Acrylaldehyde (acrolein)	0.1
Alkyl (and aryl) nitriles	—
Allyl alcohol.	2
Allyl chloride	1
Benzene,* C	10
Benzyl bromide (and chloride)	1
Boron tribromide (and trichloride)	1
Bromine	0.1
Bromomethane (methyl bromide).	15
Carbon disulphide	20
2-Chloroethanol (ethylene chlorohydrin)	1
3-Chloropropionyl chloride	—
Crotonaldehyde	2
Diketen	—
Dimethyl sulphate, C (and diethyl sulphate). . . .	1
Fluoroboric acid	—
Hydrofluoric acid.	3
Isocyanatomethane (methyl isocyanate)	0.02
Nickel carbonyl, C	0.001
Oxalyl chloride	—
Pentachloroethane	—
Tetrabromoethane	1
Tetrachloroethane	5
Trimethylchlorosilane	—

* Dangerous chronic toxic effects (see p. 22).

4. Other harmful substances. The following compounds and groups of compounds have generally harmful effects when inhaled as vapours or dusts, or absorbed through the skin, or both; some are also corrosive. All should be regarded as potentially harmful by long-term exposure to small quantities.

(i) Many *simple alkyl bromides and chlorides*, and poly-halogenated methanes and ethanes fall into this category, including some common solvents. All should be treated as potentially harmful, but in addition to those listed in sub-section **D,3** above the following are some of the more dangerous.

	TLV p.p.m.
Bromoethane (ethyl bromide)	200
Bromoform	0.5
3-Bromopropyne (propargyl bromide)	—
Carbon tetrachloride	10
Chloroform	10
Dichloromethane	200

1,2-Dibromoethane (ethylene dibromide) 20
1,2-Dichloroethane (ethylene dichloride) 50
Iodomethane, C (methyl iodide) 5

(ii) *Aromatic and aliphatic amines.* Simple aliphatic primary, secondary and tertiary amines have toxic vapours, e.g., diisopropylamine (TLV 5 p.p.m.), dimethylamine (TLV 10 p.p.m.), ethylamine (TLV 10 p.p.m.) and triethylamine (TLV 25 p.p.m.).

Likewise many aromatic amines are extremely harmful as vapours and by skin absorption. The following list includes some representative examples, but *all* aromatic amines, including alkoxy-, halogeno- and nitro-amines should be treated as potentially harmful. In addition many aromatic amines are known to be powerful cancer-causing agents (carcinogens, see sub-section D,5 below) and the use of some of these is legally controlled.

	TLV
Aniline	5 p.p.m.
Anisidines (aminoanisoles)	0.5 mg m^{-3}
Chloroanilines	—
Chloronitroanilines	—
N,N-Diethylaniline.	—
N,N-Dimethylaniline	5 p.p.m.
N-Ethylaniline	—
N-Methylaniline	2 p.p.m.
p-Nitroaniline (and isomers)	1 p.p.m.
p-Phenylenediamine (and isomers).	0.1 mg m^{-3}
o-Toluidine (and isomers)	5 p.p.m.
Xylidines	5 p.p.m.

(iii) *Phenols and aromatic nitro compounds.* As with aromatic amines, very many phenolic compounds and aromatic nitro compounds exhibit toxic properties. They give off harmful vapours, are readily absorbed through the skin and, particularly the phenols, have corrosive properties. All phenols and aromatic nitro compounds should therefore be handled with care and assumed to have the properties listed above.

	TLV
Phenol	5 p.p.m.
Cresols	5 p.p.m.
Catechol and resorcinol.	—
Chlorophenols and dichlorophenols	—
Nitrobenzene	1 p.p.m.
p-Nitrotoluene (and isomers)	5 p.p.m.
m-Dinitrobenzene (and isomers)	1 mg m^{-3}
2,4-Dinitrotoluene (and isomers)	1.5 mg m^{-3}
p-Chloronitrobenzene (and isomers)	1 mg m^{-3}
Dichloronitrobenzenes	—
Nitrophenols	—
Dinitrophenols and dinitrocresols	0.2 mg m^{-3}
Picric acid	0.1 mg m^{-3}

5. Carcinogenic substances. Many organic compounds have been shown to cause cancerous tumours in man, although the disease may not be detected for several years. The manufacture and use of some of these substances is forbidden in factories in Great Britain (according to *The Carcinogenic Substances Regulations* 1967, Ref. 12) and in the USA. However, at the time of writing, the use of these compounds is allowed in academic and research laboratories.

In addition to the compounds known to be carcinogenic to man, many substances have been shown to cause cancer in experimental animals and must therefore be presumed to be active in man also.

When handling known or suspected carcinogens, every effort should be made to avoid inhalation of their vapours and contamination of the skin. They must only be handled in fume cupboards using protective gloves. It is essential that bottles or vials containing the compounds should be properly labelled with suitable warnings. Supplies of carcinogenic compounds should be kept in a locked container, preferably in a fume cupboard.

In the following sub-sections the most dangerous known carcinogens are noted (see Ref. 13).

(i) *Aromatic amines and their derivatives.* These should *all* be treated as potentially carcinogenic, and as a group probably constitute the greatest danger to the organic chemist since even slight exposure (possibly even on a single occasion) may initiate the formation of tumours (Ref. 14). The following compounds and their salts are proven powerful carcinogens and are legally controlled in Great Britain or the USA or both.

2-Acetylaminofluorene
4-Aminobiphenyl (and 4-nitrobiphenyl)
Auramine and Magenta
4,4'-Diaminobiphenyl (benzidine)
4,4'-Diamino-3,3'-dichlorobiphenyl
4,4'-Diamino-3,3'-dimethylbiphenyl (*o*-tolidine)
4,4'-Diamino-3,3'-dimethoxybiphenyl (*o*-dianisidine)
Dimethylaminoazobenzene ('Butter Yellow')
2-Naphthylamine
1-Naphthylamine

4-Aminostilbene and o-aminoazotoluene are notably carcinogenic but are not at the moment legally controlled.

(ii) N-*Nitroso compounds.* *All* nitrosamines [R'·N(NO)·R] and nitrosamides [R'·N(NO)CO·R] should be regarded as potentially powerful carcinogens, since most compounds of these types have been shown to possess high activity in experimental animals. The following are some of the more likely to be encountered in the laboratory.

N-Methyl-N-nitrosoaniline
N-Methyl-N-nitrosourea *
N-Methyl-N-nitrosourethane *
N-Nitrosodimethylamine (legally controlled)
N-Nitrosopiperidine

(iii) *Alkylating agents.*
Aziridine (and some derivatives)
Bis(chloromethyl) ether

---

* Diazomethane precursors. It is recommended that N-methyl-N-nitrosotoluene-*p*-sulphonamide be used in their place (Section **II,2,**_19_).

Chloromethyl methyl ether
Diazomethane
Dimethyl sulphate
Methyl iodide
Nitrogen mustards [i.e., $R \cdot N \cdot (CH_2 \cdot CH_2Cl)_2$]
β-Propiolactone (legally controlled)

(iv) *Polycyclic aromatic hydrocarbons.*
Benz[a]pyrene
Dibenz[a,h]anthracene
Dibenz[c,g]carbazole
7,12-Dimethylbenz[a]anthracene

(v) *Sulphur containing compounds.*
Thioacetamide (which has been recommended as a source of hydrogen sulphide).
Thiourea

(vi) *Asbestos dust.* Inhalation of asbestos dust and fibres can cause '*asbestosis*', a crippling and eventually fatal lung disease which often becomes lung cancer in its later stages. The industrial use of asbestos is strictly controlled in Great Britain by the *Asbestos Regulations* 1969 (Ref. 15, see also Ref. 16), but there is little danger associated with the routine laboratory use of asbestos insulating materials, provided that the following precautions are observed:

(*a*) All apparatus and fittings incorporating asbestos should utilise the best quality long fibred white asbestos ('chrysotile') and should be maintained in good condition, avoiding any damage to their surfaces which could be a source of asbestos dust. All asbestos surfaces should preferably be *sealed* with a heat-resistant paint or varnish. Where soft asbestos *string* or *tape* is used for insulation purposes (e.g., around distillation columns), it should preferably be made a permanent feature and the surface sealed as above. Other soft asbestos equipment such as fire blankets and protective gloves should preferably be replaced with the glassfibre equipment now available.

(*b*) Any operation which involves the cutting, grinding or machining of any type of asbestos should be carried out in a well ventilated area with the operator wearing a dust mask. The asbestos should preferably be in a damp condition to minimise airborne dust. Any accumulated dust should be collected by vacuum suction and disposed of in plastic bags.

6. *Substances with very harmful cumulative effects.* (i) *Benzene* (TLV 10 p.p.m.). Inhalation of benzene vapour has a chronic cumulative effect leading to acute anaemia and may lead to leukaemia. Very few people can smell benzene in vapour concentrations of less than 75 p.p.m. (i.e., three times the TLV). If therefore one can smell benzene, it is being inhaled in harmful quantities. For general solvent use, benzene can in nearly all cases be replaced by the less volatile and less toxic toluene (TLV 100 p.p.m.).

(ii) *Lead compounds.* These are powerful cumulative poisons and ingestion of even small amounts must be guarded against. Organic lead compounds (e.g., lead tetra-ethyl) are volatile and inhalation of their vapours must be avoided; they are also dangerous by skin absorption.

(iii) *Mercury and mercury compounds.* These vary greatly in toxicity. Generally, mercury(II) salts are more toxic than mercury(I) salts. Liquid organic mercury compounds are often highly poisonous and dangerous by inhalation and absorption through the skin, whereas solid organomercurials are less toxic. However, all mercury compounds should be treated with caution and any long-term exposure avoided.

Elemental mercury readily evolves the vapour which constitutes a severe cumulative and chronic hazard.* No mercury surface should ever be exposed to the atmosphere but should be covered with water. All manipulations involving mercury should be carried out in a fume cupboard and over a tray to collect possible spillage. Spilt mercury is best collected using a glass nozzle attached to a water suction pump via a bottle trap; the contaminated areas should be spread with a paste of sulphur and lime.

A severe mercury vapour hazard may occur through misuse of mercury-containing vacuum gauges (e.g., the 'vacustat', Section **I,30**) attached to oil vacuum pumps. If the gauge is turned about its axis too quickly, mercury may be sucked into the pump and circulated with the hot oil to release large quantities of mercury vapour into the atmosphere. If there is any possibility of this having happened, the pump must not be used and should be stripped down and cleaned as soon as possible.

E. Electrical safety Concern with the hazards associated with the use of inflammable and toxic chemicals in the laboratory often causes the dangers from electrical equipment to be overlooked. However, many accidents are caused by the malfunctioning of electric appliances and by thoughtless handling.

New equipment should be carefully inspected to check that the plug has been correctly fitted, otherwise a 'live' chassis will result. International standards for Great Britain and Europe stipulate the following colours for electric cables:

Live, Brown; *Neutral*, Blue; *Earth*, Green/yellow.

In the USA (and for equipment imported from the USA) the colours are:

Live, White; *Neutral*, Black; *Earth*, Green.

Before any electric appliance is used, it should be inspected to ensure that (*a*) it is in good condition with no loose wires or connections, (*b*) it is properly earthed, (*c*) it is connected to the correct type of plug by good quality cable with sound insulation and (*d*) that it is protected by a fuse of the correct rating. Loose or trailing electric cables should be avoided and if the appliance has to be sited some way from the power source, the cable should run neatly along the side of a bench and preferably be secured with adhesive tape. Cable hanging across the aisle between two benches should never be permitted. Any items of equipment (e.g., stirrer motors or heating mantles) which have had any chemicals spilled on them should not be used until they have been thoroughly cleaned and dried.

In the handling and setting up of electrical equipment, the operator must ensure that the apparatus is set up on a dry bench. It is best to assemble the apparatus first, and only then to plug into the mains and switch on. The

---

* The normal vapour pressure of mercury at room temperature is many times the TLV value of 0.05 mg m^{-3} for the vapour.

apparatus should be switched off before any attempts are made to move or adjust it.

High voltage equipment (e.g., for use in electrophoresis, or in the generation of ozone) requires special precautions. Ideally, such apparatus should be isolated within an enclosure equipped with an interlocking device so that access is possible only when the current is switched off.

F. Ultraviolet radiation Ultraviolet lamps, arcs and other high intensity light sources which emit ultraviolet radiation should never be viewed directly or eye damage will result. Special close-fitting goggles which are opaque to u.v. radiation should be worn, and protective screens placed around the apparatus assembly (e.g., in a photochemical reaction) which incorporates the ultraviolet source; the need to avoid the inadvertent viewing of reflected ultraviolet light should also be borne in mind and the viewing of chromatographic columns or plates may be hazardous. Exposure of the skin to intense u.v. radiation gives rise to burns (cf., sunburn) and prolonged exposure may give rise to more extensive tissue damage. Protective gloves should therefore be worn during work involving such exposure risks. Adequate ventilation must also be provided to prevent possible build-up of the highly irritant and toxic ozone which is produced when oxygen is irradiated with u.v. light in the 185 nm region.

APPARATUS AND REACTION PROCEDURES

I,4. INTERCHANGEABLE GROUND GLASS JOINTS The commercial development of glass manipulation, coupled with the use of glasses of low expansion coefficient, has made available truly interchangeable ground glass joints at moderate cost. These, fitted to apparatus of standard and special types marketed by various firms,* have made possible a new outlook on the assembly of apparatus required for practical organic chemistry and, indeed, of many other branches of practical chemistry. The advantages of the use of ground glass joints include:

1. No corks or rubber stoppers are, in general, required and the selection, boring and fitting of corks is largely eliminated, thus resulting in a considerable saving of time. Furthermore contamination of chemicals as the result of contact with corks or bungs is therefore avoided.
2. Corrosive liquids and solids (concentrated acids, acid chlorides, bromine, phosphorous pentachloride, etc.) are easily manipulated, and no impurities are introduced into the product from the apparatus.
3. As all joints are made to accurate standards they should all fit well; this is particularly valuable for systems operating under reduced pressure.
4. By employing a few comparatively simple units, most of the common operations of organic chemistry may be carried out.

---

* For example: Jobling Laboratory Division; A. Gallenkamp & Co. Ltd; Jencons (Scientific) Ltd; Scientific Glass Apparatus Co. Inc.; Corning Glass Works. For precise details concerning joint size and specialised apparatus, the reader should consult the latest catalogues of the various manufacturers; for example those of the Jobling Laboratory Division ('Quickfit'), of Scientific Glass Apparatus Co. Inc., and of Corning Glass Works.

5. Wider passages are provided for vapours, thus diminishing the danger in violent reactions and reducing the dangers of flooding from condensing vapours in distillations especially under reduced pressure.

As illustrated later (Fig. I,2 and Fig. I,8), the types of ground glass joints which are manufactured to precise specifications are either **conical joints** or **spherical joints** respectively.

The interchangeability of conical joints (cone and socket joints) is ensured by the use of a standard taper of 1 in 10 on the diameter in accordance with the recommendations of the International Organisation for Standardisation and of the various national standardising authorities, e.g., the British Standards Institution (BS572 and 2761) or the National Bureau of Standards (Commercial Standard CS-21).* In the USA interchangeable ground glass joints conforming to these specifications are designated by the symbol **Ŧ**. It should be

Table I,2   Dimensions of British Standard interchangeable ground glass conical joints (supplied by Jobling Laboratory Division)

Size designation	Nominal diameter of wide end (mm)	Nominal diameter of narrow end (mm)	Nominal length of engagement (mm)	Former size designations
5/13	5.0	3.7	13	B5 and B5/13
7/11	7.5	6.4	11	C7 and C7/11
7/16	7.5	5.9	16	B7 and B7/16
10/13	10.0	8.7	13	C10 and C10/13
10/19	10.0	8.1	19	B10 and B10/19
12/21	12.5	10.4	21	B12 and B12/21
14/15	14.5	13.0	15	C14 and C14/15
14/23	14.5	12.2	23	B14 and B14/23
19/17	18.8	17.1	17	C19 and C19/17
19/26	18.8	16.2	26	B19 and B19/26
24/10	24.0	23.0	10	D24 and D24/10
24/20	24.0	22.0	20	C24 and C24/20
24/29	24.0	21.1	29	B24 and B24/29
29/32	29.2	26.0	32	B29 and B29/32
34/35	34.5	31.0	35	B34 and B34/35
40/13	40.0	38.7	13	D40 and D40/13
40/38	40.0	36.2	38	B40 and B40/38
45/40	45.0	41.0	40	B45 and B45/40
50/14	50.0	48.6	14	D50 and D50/14
50/42	50.0	45.8	42	B50 and B50/42
55/29	55.0	52.1	29	C55 and C55/29
55/44	55.0	50.6	44	B55 and B55/44
60/46	60.0	55.4	46	B60 and B60/46

noted that the term 'cone' is used for the part which is inserted and the term 'socket' for that part into which the cone is inserted.

The dimensions of conical joints are indicated by a numerical code which incorporates the nominal diameter of the wide end and the length of the

---

* The specifications are obtainable from the British Standards Institution, British Standards House, 2 Park St, London, W1, and from the Superintendent of Documents, Washington, DC, respectively.

Table I,3   Dimensions of USA standard interchangeable ground glass joints

Size designation	Approximate diameter at small end (mm)	Computed diameter at large end of ground zone (mm)	Approximate length of ground zone (mm)
Full-length joints			
5/20	3	5.0	20
7/25	5	7.5	25
10/30	7	10.0	30
12/30	9	12.0	30
14/35	11	14.5	35
19/38	15	18.8	38
24/40	20	24.0	40
29/42	25	29.2	42
34/45	30	34.5	45
40/50	35	40·0	50
45/50	40	45·0	50
50/50	45	50·0	50
55/50	50	55.0	50
60/50	55	60.0	50
71/60	65	70.0	60
Medium-length joints			
5/12	3.8	5.0	12
7/15	6.0	7.5	15
10/18	8.2	10.0	18
12/18	10.2	12.0	18
14/20	12.5	14.5	20
19/22	16.6	18.8	22
24/25	21.5	24.0	25
29/26	26.6	29.2	26
34/28	31.7	34.5	28
40/35	36.5	40.0	35

Table I,4   Dimensions of spherical joints (supplied by Jobling Laboratory Division)

Size designation	Nominal diameter (mm)	Minimum diameter of wide end (mm)	Maximum diameter of narrow end (mm)
S 13	12.700	12.5	7.0
S 19	19.050	18.7	12.5
S 29	28.575	28.0	19.0
S 35	34.925	34.3	27.5
S 41	41.275	40.5	30.0
S 51	50.800	50.0	36.0

ground zone, e.g., in British usage 14/23 indicates a wide end diameter of nominally 14.5 mm, and a length of engagement of 23 mm. Although the length of engagement in the joints differs somewhat in the British and USA specifications the range of nominal diameters which are manufactured are the same. The dimensions of British and USA interchangeable conical joints are listed in Table I,2 and Table I,3.

The sizes of standard spherical joints (semi-ball or ball and socket joints) are designated by a code which indicates approximately the nominal diameter of the ground hemisphere, e.g., in British usage S35 designates a spherical joint of nominal diameter 34.925 mm. Sizes of spherical joints are collected in Tables I,4 and I,5.

Table I,5   Dimensions of semi-ball connections
(these dimensions taken from USA sources)

Semi-ball size	Ball diameter (mm)	Tube bore (mm)	Semi-ball size	Ball diameter (mm)	Tube bore (mm)
12/2	12	2	28/12	28	12
12/3	12	3	28/15	28	15
12/5	12	5	35/20	35	20
18/7	18	7	35/25	35	25
18/9	18	9	40/25	40	25
28/11	28	11	50/30	50	30

I,5. TYPES OF GROUND GLASS JOINTS   All ground glass joints are usually constructed of high-resistance glass such as Pyrex. The most common form of conical joint is shown in Fig. I,2 and is the type largely encountered in practice. That shown in Fig. I,3 is similar but has reinforcing glass bands about the female joint which greatly add to the mechanical strength of the walls. Figure I,4 depicts a ground joint with glass hooks, to which light springs may be attached. Figures I,5 and I,6 illustrate drip cones for condensers and the like; the latter is generally employed for joints larger than 29 mm in diameter, the orifice being reduced to about 18 mm. Figure I,7 illustrates a cone joint' with stem for use, for example, as a gas or steam inlet.

The spherical joint or semi-ball joint is illustrated in Fig. I,8, which includes one type of special clamp for holding the two halves of the joint together. This connection cannot freeze or stick (as conical ground joints sometimes do) and it introduces a degree of flexibility into the apparatus in which it is used. The area of contact between the ground surfaces is relatively small so that the joints are not intended to provide for considerable angular deflection. The main application is in conjunction with conical joints rather than as a substitute for them. The conical-spherical adapters shown in Fig. I,9 provide a means of inserting a spherical joint whilst retaining the standard conical joint principle.

I,6. CARE AND MAINTENANCE OF GROUND GLASS JOINTS   Great care must be taken to keep all ground surfaces free from grit and dust. For work at atmospheric pressure, no lubricant should be required; it is advisable, however, in order to reduce the danger of sticking to apply a slight smear

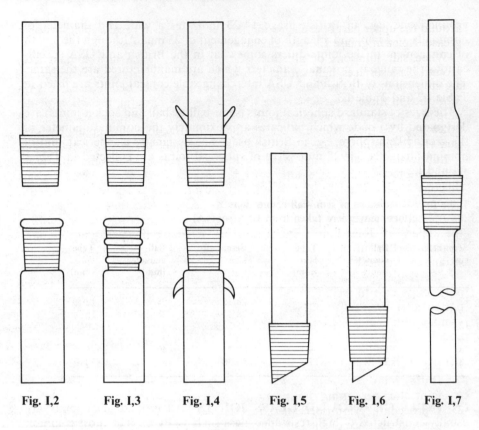

Fig. I,2   Fig. I,3   Fig. I,4   Fig. I,5   Fig. I,6   Fig. I,7

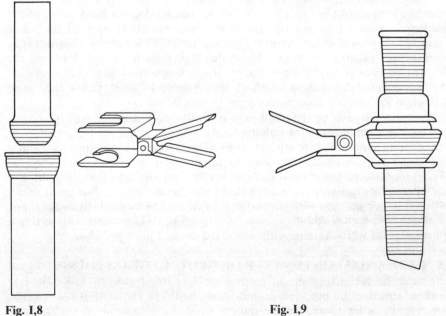

Fig. I,8   Fig. I,9

around the upper part of each ground joint of Vaseline (if permissible), a rubber grease, Apiezon grease L or M, or Silicone stop-cock grease.

When salt solutions or alkaline substances may come into contact with ground glass surfaces, light lubrication of the surfaces is essential. When greasing stop-cocks, only the outer parts of the plug should be lightly smeared with lubricant; in those cases where the lubricant is not harmful, the whole of the plug may be given a very thin smear of the grease but particular care must be taken to avoid the entrance of the lubricant into the bore of the plug. If necessary the bore of a stop-cock can be cleaned conveniently with a pipe-cleaner.

Lubrication of all ground glass surfaces is essential for distillations under reduced pressure. Suitable lubricants are Apiezon grease L, M or N and Silicone stop-cock grease. The use of synthetic joint linings particularly in distillation assemblies is a very useful alternative to lubrication with grease and completely removes any possibility of lubricant contaminating the contents of the apparatus. Joint linings (or sleeves) are available (from Fisons Scientific Apparatus Ltd) in polytetrafluoroethylene (PTFE), a polymer extremely resistant to chemical attack and to heat.

Seizing of ground glass joints.   Provided adequate care is exercised to use only joints that fit well, and the ground surfaces are suitably lubricated and parted after use whilst still warm, sticking will rarely occur. If, however, a ground joint should seize up or freeze the following suggestions may be found useful:

1.   Set the joint in a vertical position and apply a layer of glycerine or penetrating oil to the upper surface. The glycerine will slowly penetrate into the joint, thus permitting the separation of the ground surfaces.
2.   If procedure (1) is unsuccessful, direct a stream of hot air from a blower on to the outer surface of the joint for a few seconds and gently draw the members apart with a twisting action; gentle tapping on the edge of a wooden bench is sometimes helpful.
3.   Introduce the joint into a small luminous Bunsen flame for a few seconds, and then gently draw the ground surfaces apart. If the glass is of Pyrex (or of any other heat-resisting variety), there is very little danger of a crack resulting from this process. The object of the heating is to cause the glass of the socket to expand before any appreciable change has occurred in the inner cone. More even, gentle and local heating of the joint, which is particularly suitable for stoppered flasks containing volatile, inflammable or corrosive liquids, may be achieved by the following procedure. Wrap the flask in several layers of cloth with the neck protruding and hold securely within a fume cupboard; wrap, with one turn, a piece of fibrous string around the outside of the ground glass joint, and with a to-and-fro pulling action on the two string ends (gently at first until the rhythm of motion is acquired), allow the circle of string around the joint to move smoothly along its length. Two operators are required, the one holding the flask steady should ensure that the stopper is directed away from both in case of accidents; both should be wearing safety spectacles. After a few minutes the joint will have heated sufficiently by friction and to a temperature leading to a smaller risk of decomposition of substances held between the joint surfaces than by the flame method; frequently the stopper may be removed by pulling with a twisting motion.

I,7. APPARATUS WITH INTERCHANGEABLE GROUND GLASS JOINTS SUITABLE FOR GENERAL USE IN PREPARATIVE ORGANIC CHEMISTRY   In considering the following typical standard units of equipment fitted with ground glass joints, it must be borne in mind that whilst a particular piece of glass equipment of certain capacity or dimensions may be fitted with alternative joint sizes, the range is usually restricted in relation to their relative proportions. When equipping a laboratory, it is usually convenient to limit the range of socket sizes thus permitting interchangeability with the minimum number of adapters. For example 14/23, 19/26, 24/29 and 34/35 joints are suitable for macro scale experiments, and 10/19 and 14/23 for semimicro scale experiments.

In Fig. I,10(a–c), the various designs of **flasks** are collected. Type (a) is a pear-shaped flask, the capacity range being usually 5 ml to 100 ml, the joint sizes are in the range 10/19 to 24/29. Type (b) is a round-bottomed flask (short-necked), the capacity range being 5 ml to 10 l, joint sizes being in proportion; medium- and long-necked designs are also available. Type (c) illustrates a range of wide-necked reaction flasks which are useful in semimicro and in pilot scale experiments and which are fitted with large diameter flat-flange joints, the capacities range from 50 ml to 20 l, the flange bore being 50 mm to 100 mm respectively; the multi-socket lids are illustrated in Fig. I,20(a and b). The advantages of this type of reaction vessel are that, (i) the lids are easily detachable, (ii) large stirrers are readily accommodated, (iii) the vessels are cleaned readily and (iv) the removal or addition of solids and viscous fluids is facilitated; the ground flange joints are fully interchangeable. Special clamps are available for the support of such flasks.

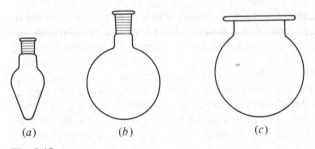

(a)   (b)   (c)

Fig. I,10

Various types of **multi-necked round-bottomed flasks** are illustrated in Fig. I,11(a–d); designs with pear-shaped flasks are available. The centre socket is usually the larger and the side sockets are generally smaller; type (d) shows the side socket being employed for the insertion of a capillary tube necessary in a vacuum distillation assembly (see Section I,27).

Ground glass **stoppers** of all standard sizes are available and may be of the design shown in Fig. I,12; the flat head is preferred since the stopper may be stood on end when not in use, thus avoiding contamination of the ground surface; an additional refinement is the provision of a finger grip.

Often in the assembly of apparatus, joint adapters are required if the joint sizes of the various parts are not compatible. A **reduction adapter** is illustrated in Fig. I,13 and an **expansion adapter** in Fig. I,14; numerous combinations are of course possible, but it must be emphasised, however, that in a well designed

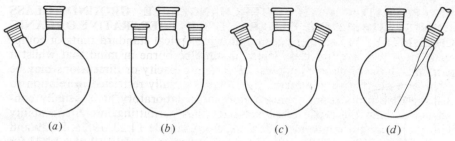

(a)   (b)   (c)   (d)

Fig. I,11

assembly of apparatus the number of adapters should be reduced to a minimum and, best of all, completely eliminated.

Distillation heads (or **still-heads**) are shown in Fig. I,15(a–c). Type (a) is a bend ('knee-tube') which is frequently employed for those distillations which merely require the removal of solvent. Type (b) is a simple distillation head; when fitted into a flask with a ground glass socket, the assembly is virtually a distillation flask. For some purposes, a thermometer may be fitted into a one-hole rubber stopper of correct taper and then inserted into the socket; the area

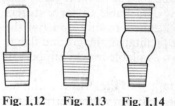

of rubber which is exposed to the organic vapour is relatively so small that the amount of contamination thus introduced is negligible. If, however, all rubber stoppers must be absent because of the highly corrosive nature of the vapour, a thermometer fitted with an appropriate size cone is employed. Alternatively the socket of a distillation head may be fitted with a **screw-capped adapter**

Fig. I,12   **Fig. I,13**   **Fig. I,14**

(see Fig. I,56(b)) through which a thermometer may be inserted. Type (c) is a Claisen distillation head; the left-hand socket accommodates the capillary tube for use in distillations under vacuum (see Section **I,27**) and the right-hand socket a suitable thermometer.

Frequently for semimicro and micro work it is more convenient to use the pear-shaped flask designs which incorporate the distillation heads (e.g., Fig. I,16(a and b)).

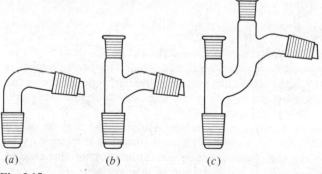

(a)   (b)   (c)

Fig. I,15

Multiple adapters provide for additional entries into a single-necked flask when a multi-necked flask is not available. Either double-necked or triple-necked adapters (Fig. I,17 and Fig. I,18(a and b)) are commonly used having a

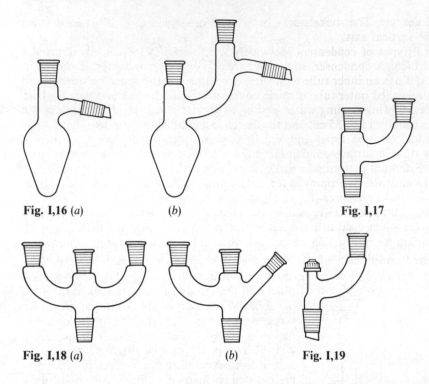

Fig. I,16 (a)   (b)   Fig. I,17

Fig. I,18 (a)   (b)   Fig. I,19

range of socket and cone sizes. The 'swan-neck adapter' of Fig. I,19 is useful for
vacuum distillations as it permits the insertion of a capillary tube through the
screw thread joint. This joint may also be used for insertion of a thermometer
or a gas inlet in the narrow neck and a reflux condenser into the ground joint;
this device virtually converts a three-necked flask into a four-necked flask.

Multiple socket lids for fitment to the flange flasks (illustrated in Fig. I,10(c))
are shown in Fig. I,20(a and b). These allow for the introduction of a
great variety of standard equipment for stirring, temperature measurement, the

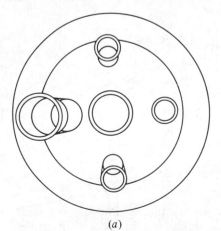

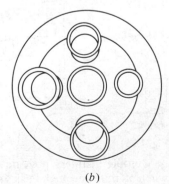

(a)   (b)

Fig. I,20

inlet of gas, etc. The sockets may be vertical or angled at 5, 10 or 15 degrees from the vertical axis.

Several types of **condensers** are widely used (Figs. I,21–I,25). An improved form of Liebig's condenser, sometimes termed a West condenser, is shown in Fig. I,21; it has an inner tube with very thin walls and the space between it and the heavy-walled outer tube is small, consequently there is a rapid heat transfer to the fast-flowing cooling water leading to greater efficiency. The length of the jacket is usually 15 to 35 cm and the design is available in a range of joint sizes. Figure I,22 (Davies type) and Fig. I,23 (double coil type) are examples of efficient double surface condensers. Figure I,24 depicts a 'screw' type of condenser (Friedrich pattern); the jacket is usually 10, 15 or 25 cm long: this highly efficient condenser is employed for both reflux and for downward distillation. The ice-cooled condenser (Fig. I,25) is useful for volatile liquids.

Various forms of **receiver adapters** or **connecters** for attachment to the end of condensers when used in a distillation assembly are shown in Figs. I,26–I,28. The simplest form (Fig. I,26) carries glass hooks for securing it to the condenser by means of a rubber band from the side tube to the hook; an improved form,

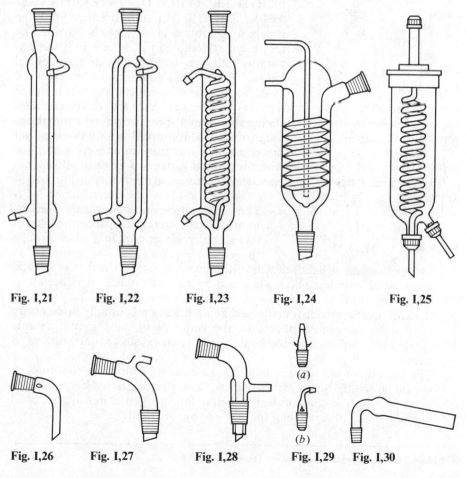

Fig. I,21   Fig. I,22   Fig. I,23   Fig. I,24   Fig. I,25

(a)

(b)

Fig. I,26   Fig. I,27   Fig. I,28   Fig. I,29   Fig. I,30 incorporating two glass joints, is shown in Fig. I,27. A useful adapter is illustrated in Fig. I,28; when employed at atmospheric pressure, a drying tube may be attached to the side tube, if desired; in a distillation under reduced pressure, the side tube is connected to a vacuum pump.

Cone/rubber tubing adapters ('*take-off*' adapters), shown in Fig. I,29(*a* and *b*), fulfil a number of useful purposes in preparative organic operations, for example where small volumes of solvents need to be rapidly removed without the necessity of solvent recovery.

A **calcium chloride guard-tube** is illustrated in Fig. I,30, which is widely used for protecting apparatus assemblies from the ingress of moisture.

For many operations the globular form of **separatory funnel** having a suitable cone joint fitted to the stem is convenient, but when required on either a multiple-necked flask or with a multiple adapter, the cylindrical design (Fig. I,31) is preferred; this is similarly provided with a cone on the stem and a ground socket. Figure I,32 illustrates a cylindrical funnel with pressure-equalising tube; this is invaluable for reactions which are conducted in an atmosphere of inert gas.

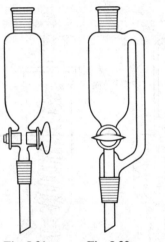

Fig. I,31 **Fig. I,32**

I,8. OTHER TYPES OF INTERCHANGE-ABLE JOINTS An improvement on the widely used ground glass joints is provided by the recent introduction of a new range of apparatus with interchangeable clear glass conical joints.* These cones and sockets are precision made to the same specifications which define ground glass joints, and are therefore interchangeable with them. They retain the versatility of the conventional ground joints but overcome many of the limitations associated with the ground surfaces. The main advantages may be enumerated as follows:

1. The apparatus incorporating 'Clearfit' joints is completely transparent and hence thermometer scales remain visible through the joint.

2. The mechanical strength of 'Clearfit' joints is greater than that of ground joints which may have been weakened by micro-scratches introduced during manufacture.

3. 'Clearfit' joints provide a better seal and lubrication is usually unnecessary even under vacuum; furthermore, the better fit of these joints prevents seepage of solutions into the joints which often causes ground surfaces to stick.

4. The smooth surfaces of the joints make cleaning simple yet thorough. It should be emphasised however that the joint surfaces should be wiped with a lint-free cloth or chamois leather before fitting together in order to avoid any possibility of scratching the surfaces by dirt particles.

---

* Manufactured by Jobling Laboratory Division, 'Clearfit'.

The **O-ring joint*** illustrated in Fig. I,33 is particularly suitable for incorporation into vacuum line assemblies (e.g., Fig. I,115). The joint consists of a slightly tapered cone, with a terminal annular indentation carrying a replaceable PTFE ring seal, which is inserted into a suitably-sized socket. The PTFE ring provides the seal under vacuum and also allows for a degree of flexibility in the vacuum line which facilitates assembly.

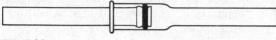

Fig. I,33

I,9. THE USE OF CORKS AND RUBBER STOPPERS Although these have been largely replaced by ground glass joints, corks and rubber stoppers still find occasional use in the laboratory.

Two points must be borne in mind when selecting a cork stopper. In the first place, the cork should be examined for flaws; unless corks of the highest quality are employed, they are liable to have deep holes, which render them useless. In the second place, the cork should originally fit as shown in Fig. I,34(*a*) and not as in Fig. I,34(*b*). It should then be softened by rolling in a cork press or by wrapping it in paper and rolling under the foot.

To bore a cork, a borer should be selected which gives a hole only very slightly smaller than that desired. The cork borer is moistened with water or

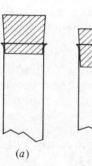

(*a*)   (*b*)

Fig. I,34   **Fig. I,35**

alcohol or best with glycerine; it is convenient to keep a small bottle (*ca.* 25 ml capacity) containing glycerine, Fig. I,35, for this purpose. The borer is held in the right hand and the cork in the left hand. The hole is started at the narrow end with a continuous rotary motion. Beginners should bear in mind that the borer is a cutting instrument and not a punch, and on no account should it be

---

* Manufactured and supplied by J. Young & Co Ltd.

allowed to burst its way through the cork because the borer, upon emerging, will almost invariably tear the surface of the cork. It is a good plan to examine the borer from time to time as it advances through the cork to see that it is cutting a straight hole. Boring should be stopped when it is half through the cork* and the tool removed from the hole. The cork plug is pushed out with the aid of the solid metal rod supplied with the set of borers, and the remainder of the hole is bored from the other end. If the holes are carefully aligned, a clean cut hole is obtained. Experienced laboratory workers frequently complete the whole boring operation from one side, but beginners usually tear the edges of the cork by this method, which is therefore not recommended. A well-fitting cork should slide over the tube (side arm of distilling flask, thermometer, lower end of condenser, etc.) which is to pass through it with only very moderate pressure. The bored cork should be tested for size; if it is too small, the hole should be enlarged to the desired diameter with a small round file. When the correct size is obtained, the tube is held near the end and inserted into the cork. The tube is then grasped near the cork and cautiously worked in by gentle twisting. Under no circumstances should the tube be held too far from the cork nor should one attempt to force a tube through too small an opening in a cork; neglect of these apparently obvious precautions may result in a severe cut in the hand from the breaking of the glass tube. The sharp edges of freshly cut glass tubing must be smoothed by fire polishing (Section I,10).

For consistently successful results in cork boring, a sharp cork borer must be used. The sharpening operation will be obvious from Fig. I,36. The borer is pressed gently against the metal cone, whilst slight pressure is applied with the cutter A at B; upon slowly rotating the borer a good cutting edge will be obtained. If too great pressure is applied either to the borer or to the 'cutter', the result will be unsatisfactory and the cutting circle of the borer may be damaged. To maintain a cork borer in good condition, it should be sharpened every second or third time it is used.

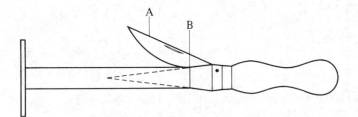

Fig. I,36

To bore a rubber stopper, it is essential to employ a very sharp cork borer of the same size as the tube to be inserted into the hole. The borer is lubricated with a little glycerol or alcohol (Fig. I,35) and steadily rotated under only very slight pressure. The operation requires a good deal of patience and time and frequent lubrication may be necessary; if too much pressure be exerted on the borer, a hole of irregular shape and diminishing size will result.

---

* With a little experience this can usually be accomplished in one operation without the necessity of stopping to see whether a straight hole is being cut.

The insertion of a glass tube into a rubber stopper or into rubber tubing is greatly facilitated by moistening the rubber with a little alcohol. Some grades of synthetic polymer tubing are only semi-flexible and are best fitted on to a glass tube after softening by immersion of the ends in a boiling water bath.

After some use rubber may stick to glass and great care must be taken not to break the glass tube when removing it. Frequently the exertion of gentle pressure on the rubber stopper by means of the two thumbs whilst the end of the tube (or thermometer) rests vertically on the bench will loosen the stopper; this operation must, however, be conducted with great care. Another method is to slip the smallest possible cork borer, lubricated with a little glycerol, over the tube, and to gradually rotate the borer so that it passes between the stopper and the glass tube without starting a new cut.

I,10. CUTTING AND BENDING OF GLASS TUBING Many students tend to forget the practical details learnt in elementary courses of chemistry; they are therefore repeated here. **To cut a piece of glass tubing,** a clean scratch is first made with a triangular file, sharp glass knife or diamond pencil. The tubing is held in both hands with the thumbs on either side of the scratch, but on the side opposite to it. The tubing is then pulled gently as though one wanted to stretch the tube and also open the scratch. A break with a clean edge will result. The cut edge must then be rounded or smoothed by **fire polishing**. The end of the tube is heated in the Bunsen flame * until the edges melt and become quite smooth; the tube is steadily rotated all the time so as to ensure even heating. Overheating should be avoided as the tube will then partially collapse.

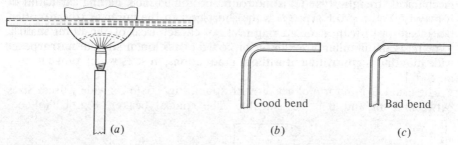

Good bend   Bad bend

(a)   (b)   (c)

Fig. I,37

A 'batswing' or 'fish-tail' burner is generally used for the **bending of soda glass tubing**. The tube is held with both hands in the length of the flame (Fig. I,37(a)) so that 5–8 cm are heated: the tube must be slowly rotated about its axis so as to heat all sides equally. As soon as the glass is felt to be soft, it is bent to the required shape. This is best done by removing it from the flame and allowing one end to fall gradually under its own weight, whilst being guided so that it is in the same plane as the rest of the tube. The glass must never be forced, otherwise a bad bend with a kink will be obtained as in Fig. I,37(c).

I,11. GENERAL LABORATORY APPARATUS Apart from the ground glass apparatus discussed in the previous sections, the student should also be

---

* Manipulations with Pyrex glass tubing, either fire polishing, bending, or drawing into a capillary leak for vacuum distillations, have to be carried out in the flame of an oxygen-gas blowpipe.

aware of the range of other pieces of general laboratory equipment which are used in the course of preparative and analytical work.* It should be remembered however that often the practical worker has to design, from the available equipment, pieces of apparatus to carry out a specific operation. This is particularly necessary when handling semimicro and micro quantities of material. Instances are cited in the later sections on isolation and purification procedures where designs are suggested.

Various types of **flasks** are shown in Fig. I,38(a–e). Types (a) and (b) are flat-

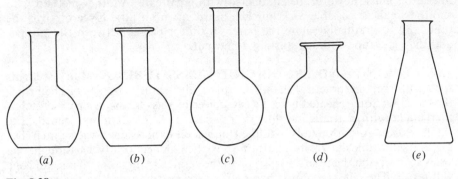

| (a) | (b) | (c) | (d) | (e) |

Fig. I,38

bottomed flasks (the Florence flask) with or without a wide neck, of capacities between 50 ml and 20 l, the larger sizes having a tooled ring neck (b) to increase mechanical strength; type (c) is the round-bottomed flask having capacities of between 50 ml to 20 l; type (d) is the short-necked boiling flask (the so-called bolt-head flask) with a tooled ring neck, of capacities between 50 ml to 10 l. Type (e) is the familiar Erlenmeyer or conical flask obtainable in narrow and wide mouth designs, with and without graduations, in sizes which range from 5 ml to 6 l.

The usual Griffin form of **beaker** with spout, Fig. I,39(a), is widely used. Sizes between 5 ml and 6 l are available. The conical beaker, Fig. I,39(b), oc-

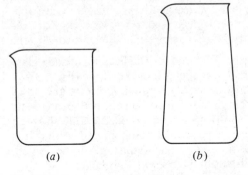

| (a) | (b) |

Fig. I,39

---

* Available, for example, from Baird and Tatlock (London) Ltd; Corning Glass International SA; Fisher Scientific Co; Fisons Scientific Apparatus Ltd; A. Gallenkamp & Co Ltd; Griffin and George Ltd; Jencons (Scientific) Ltd; Jobling Laboratory Division; Scientific Glass Apparatus Co Inc; Townson & Mercer Ltd.

casionally finds use in preparative work. Some designs have a Teflon (PTFE) rim to enable aqueous solutions to be poured safely, drop by drop, if necessary. Polypropylene and polyethylene beakers are available in various sizes but have more limited use being unable to withstand temperatures above 120 °C and being unsuitable for use with many organic solvents. PTFE beakers are very much more expensive but are able to withstand temperatures up to 300 °C and are inert to most chemicals.

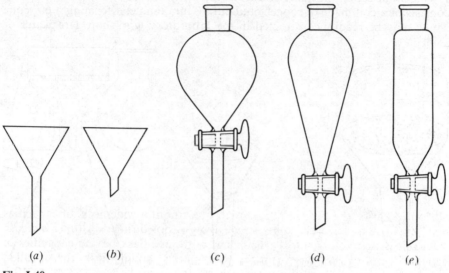

(a) (b) (c) (d) (e)

Fig. I,40

Various kinds of **funnels** are depicted in Fig. I,40(a–e). Type (a) is an ordinary filtration funnel having a 60° angle. It is convenient in many operations to employ funnels with a short stem (0.5–1 cm long) as in (b); a wide-stemmed design of (b) is useful when transferring powders. Types (c), (d) and (e) are known as **separatory funnels**; type (c) is the globular form (obtainable in capacities between 50 ml to 6 l) and is the most widely used; types (d) and (e) are the pear-shaped and cylindrical separatory funnels respectively (60 ml to 2 l); the

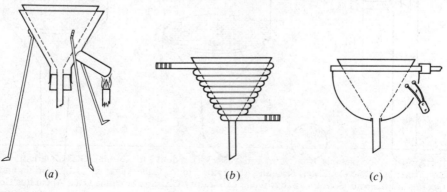

(a)   (b)   (c)

Fig. I,41

latter is available in a graduated form. The use of separatory funnels fitted with Teflon stop-cocks reduces the problems associated with tap-seizure and also avoids the use of lubricants which may contaminate the products.

Two varieties of **hot water jackets for funnels** are illustrated in Fig. I,41(a and b). Type (a) consists of a double-walled copper jacket to house the funnel; it is mounted on a tripod. The space between the walls is almost completely filled with water and the water may be heated to any desired temperature below 100 °C by directing the flame of a Bunsen burner on to the side tube. Type (b) consists of a coil made of copper of about 10 mm diameter forming a 60° cone which may be readily constructed in the laboratory workshop. Hot water or

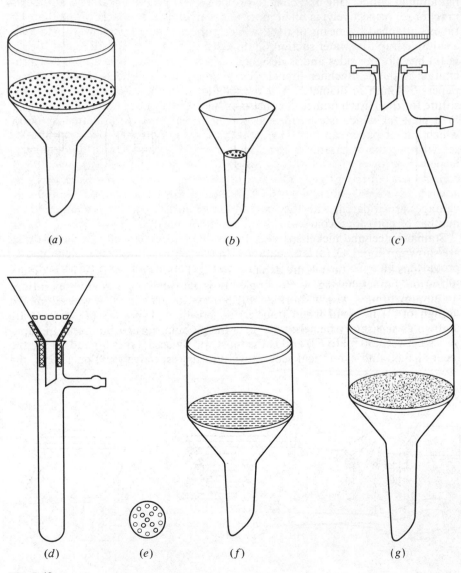

(a) (b) (c)

(d) (e) (f) (g)

Fig. I,42

steam is passed through the coil, hence it is very suitable for the filtration of inflammable liquids. The coil may also be used in 'cold filtration' and in some forms of sublimation apparatus (Section **I,21**) by circulating cold water through the coil. An electrically heated mantle designed to take a funnel is illustrated in Fig. I,41(c).

Funnels which are suitable for **filtration by suction** are illustrated in Fig. I,42 (a–f). The **Buchner funnel** shown in (a) is made of porcelain and has a perforated porcelain plate to support a filter-paper. A Buchner funnel (and other funnels described below) is used in conjunction with a **filter (suction) flask** or **tube** into which it is fitted by means of a rubber stopper; alternatively the use of either a flat annular rubber ring or a cone to provide a seal between flask and funnel (as in (c) or (d) respectively) is often more convenient. The side-arm of the flask or tube is attached by means of thick-walled rubber tubing ('pressure tubing') via a suitable trap to a water suction pump (Section **I,19**). The **Hirsch funnel** shown in (b) has sloping sides and is designed to deal with a smaller amount of precipitate than the Buchner funnel. The smallest size will accommodate filter-papers 3–4 mm in diameter. The funnel shown in the assembly (d) is a substitute for the Hirsch funnel. It consists of an ordinary glass funnel fitted with a **Witt plate** (e), which is a perforated porcelain plate 1–4 cm in diameter, upon which a filter-paper can rest. The great advantage of (d) is that it is possible to see whether the apparatus is clean; with porcelain funnels it is impossible to inspect the lower side of the perforated plate. The **'slit sieve' funnel** (f) is constructed entirely of glass (Jena or Pyrex) and therefore possesses obvious advantages over the opaque (porcelain) Buchner or Hirsch funnel. Similar advantages are apparent with the sintered glass funnel (g), which is available in a number of porosities (coarse, medium and fine).

Stainless steel and nickel spatulas are available commercially. Some of these are shown in Fig. I,43; (a) is a spatula with a flexible stainless steel blade and is provided with a wooden handle; (b) is a nickel spatula and has a turned-up end to facilitate the handling of small quantities of material—it is available in a number of sizes; (c) is a nickel double spoon-ended spatula. Flexible horn and moulded polythene spatulas are also available.

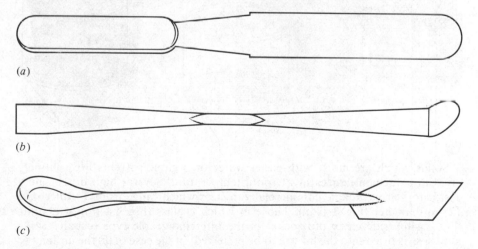

(a)

(b)

(c)

Fig. I,43

Fig. I,44

(a).

(b)   (c)

Solids which are moist with either water or organic solvents are routinely dried in a **vacuum desiccator** at room temperature. Several high dome safety desiccators are available commercially in sizes which range from 3″ (micro) to 12″ in diameter (Fig. I,44(a)). Those in which a glass (Fig. I,44(b)) or ebonite (Fig. I,44(c)) side entry tap socket of the interchangeable type (usually 34/35 joint size) is fitted into the lid, are to be preferred. In the case of (b) the air inlet to the desiccator terminates in a hooked extension which serves to ensure that the air flow when the vacuum is released is directed in an even upward spread to prevent dispersal of the sample. The joint between the lid and the base may be an interchangeable ground flange and this joint needs lubrication (e.g., with Apiezon grease) before the desiccator is evacuated. In this type the lid is removed by side-pressure after the vacuum has been released. Recently a 'Dry-Seal' joint * has been introduced (Fig. I,45(a)) in which a groove in the top flange of the desiccator base (A, Fig. I,45(b)) accommodates a removable elastomer sealing ring B. This ring becomes flattened by the lid C, when vacuum is applied, the design being such that over-compression is avoided.

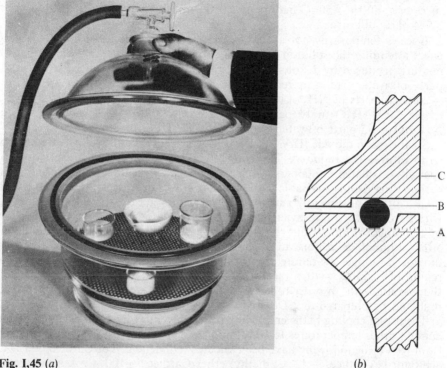

Fig. I,45 (a)   (b)

No lubricants are required, and when the vacuum is released the lid is simply lifted off—the removal of desiccator lids in the ground flange type can cause considerable difficulty. Ordinary (i.e., atmospheric pressure) desiccators are available in the Dry-Seal or ground flange range and have limited use for storage of samples in a dry atmosphere.

The nature of the charge in a desiccator, which is placed in the lower compartment below the metal gauze plate, is dependent on whether water or organic solvents are to be removed and whether acidic or basic vapours are likely to be evolved during the drying process. Suitable charges are discussed in Section **I,20**.

---

* Developed and marketed by Jencons (Scientific) Ltd. These manufacturers also supply a twinring sealing unit for fitment to standard desiccators of the ground flange type. All the photographs of desiccators have been kindly supplied by Jencons (Scientific) Ltd.

I,12. COOLING OF REACTION MIXTURES   It is often necessary to obtain temperatures below that of the laboratory. Finely crushed ice is used for maintaining the temperature at 0–5 °C; it is usually best to use a slush of crushed ice with sufficient water to provide contact with the vessel to be cooled and to stir frequently. It is of course essential to insert a thermometer into the reaction mixture to ensure that the desired temperature is attained. For temperatures below 0 °C, the commonest freezing mixture is an intimate mixture of common salt and crushed ice: a mixture of one part of common salt and three parts of ice will produce a temperature of about −5 to −18 °C. Greater cooling may be obtained by the use of crystalline calcium chloride; temperatures of −40 to −50 °C may be reached with five parts of $CaCl_2,6H_2O$ and 3.5–4 parts of crushed ice.

If ice is temporarily not available, advantage may be taken of the cooling effect attending the solution of certain salts or salt mixtures in water. Thus a mixture produced by dissolving 1 part of NH_4Cl and 1 part of $NaNO_3$ in 1–2 parts of water causes a reduction in temperature from 10 to −15 °C to −20 °C; 3 parts of NH_4Cl in 10 parts of water from 13 to −15 °C; 11 parts of $Na_2S_2O_3,5H_2O$ in 10 parts of water from 11 to −8 °C; and 3 parts of NH_4NO_3 in 5 parts of water from 13 to −13 °C.

Solid carbon dioxide (Dry Ice, Drikold, Cardice) is employed when very low temperatures are required. The commercially available blocks are stored in specially insulated containers. Since frostbite may result from handling solid carbon dioxide it is advisable to either wear gloves or to cover the hands with a thick cloth. Conveniently small-sized lumps may be obtained by hammering with a wooden or polyethylene mallet a suitable large piece wrapped in a cloth or contained within a stout canvas bag. The small pieces are carefully added to either ethanol or acetone in a plastic bowl until the lumps of solid carbon dioxide no longer evaporate vigorously. The temperatures attained are in the region of −50 to −70 °C according to the efficiency of the lagging around the freezing bath. In order to keep the freezing mixture for hours or overnight, it should be prepared in a Dewar flask.

The use of cooling baths employing other solvents with solid carbon dioxide enables other temperatures to be attained. An extensive list is given in Ref. 17, from which the following have been selected: ethylene glycol/Cardice, −15 °C; acetonitrile/Cardice, −42 °C; diethyl ether/Cardice, −100 °C. A steady state temperature cooling bath may also be obtained by adding solid carbon dioxide to o-xylene : m-xylene mixtures (Ref. 18); the volume fraction of o-xylene determines the temperature of the bath. For example, m-xylene/Cardice, −72 °C; o-xylene (0.4) : m-xylene (0.6), −58 °C; o-xylene (0.8) : m-xylene (0.2), −32 °C.

The attainment of temperatures lower than −100 °C requires the use of baths employing liquid nitrogen (Ref. 19).

I,13. HEATING OF REACTION MIXTURES   For temperatures up to 100 °C, a **water bath** or **steam bath** is generally employed. The simplest form is a beaker or an enamelled iron vessel mounted on a suitable stand; water is placed in the vessel, which is heated by means of a flame. This arrangement may be used for non-inflammable liquids or for refluxing liquids of low boiling point. Since numerous liquids of low boiling point are highly inflammable, the presence of a naked flame will introduce considerable risk of fire. For such liquids an electrically heated water bath provided with a constant-level device must be used. The circular type* (Fig. I,46) is provided with a series of concentric rings in order to accommodate flasks and beakers of various sizes. The rectangular type (Fig. I,47) has several holes each fitted with a series of concentric rings. In both cases the water bath is fitted with an immersion heating element controlled by a suitable regulator.

Fig. I,46

For temperatures above 100 °C, **oil baths** are generally used. Medicinal paraffin may be employed for temperatures up to about 220 °C. Glycerol and dibutyl phthalate are satisfactory up to 140–150 °C; above these temperatures fuming is usually excessive and the odour of the vapours is unpleasant. For temperatures up to about 250 °C, 'hard hydrogenated' cotton seed oil, m.p. 40–60 °C, is recommended: it is clear, not sticky and solidifies on cooling; its advantages are therefore obvious. Slight discoloration of the 'hard' oil at high temperature does not affect its value for use as a bath liquid. The Silicone fluids, e.g., MS550,† are probably the best liquids for oil baths but are somewhat expensive for general use. The MS550 fluid may be heated to 250 °C without appreciable loss or discoloration. Oil baths should be set up in the

---

* Electrically heated stainless steel or copper water baths are supplied by the principal laboratory equipment manufacturers.

† Supplied by Midland Silicones Ltd, a subsidiary of the Dow Corning Corporation; also by Imperial Chemical Industries, Ltd, Millbank, SW1.

Fig. I,47

fume cupboard wherever possible. A thermometer should always be placed in the bath to avoid excessive heating. Flasks, when removed from an oil bath, should be allowed to drain for several minutes and then wiped with a rag. Oil baths may be heated by a gas burner but the use of an electric immersion heater is safer and is to be preferred. A satisfactory bath suitable for temperatures up to about 250 °C may be prepared by mixing four parts by weight of 85 per cent ortho-phosphoric acid and one part by weight of meta-phosphoric acid; the mixed components should first be heated slowly to 260 °C and held at this temperature until evolution of steam and vapours has ceased. This bath is liquid at room temperatures. For temperatures up to 340 °C, a mixture of two parts of 85 per cent ortho-phosphoric acid and one part of meta-phosphoric acid may be used: this is solid (or very viscous) at about 20 °C.

High temperatures may be obtained also with the aid of **baths of fusible metal alloys**, e.g., **Woods metal**—4 parts of Bi, 2 parts of Pb, 1 part of Sn and 1 part of Cu—melts at 71 °C; **Rose's metal**—2 of Bi, 1 of Pb and 1 of Sn—has a melting point of 94 °C; a eutectic mixture of lead and tin, composed of 37 parts of Pb and 63 parts of Sn, melts at 183 °C. Metal baths should not be used at temperatures much in excess of 350 °C owing to the rapid oxidation of the alloy. They have the advantage that they do not smoke or catch fire; they are, however, solid at ordinary temperature and are usually too expensive for general use. It must be remembered that flasks or thermometers immersed in the molten metal must be removed before the metal is allowed to solidify.

One of the disadvantages of oil and metal baths is that the reaction mixture cannot be observed easily; also for really constant temperatures, frequent adjustment of the source of heat is necessary. These difficulties are overcome when comparatively small quantities of reactants are involved, in the apparatus shown in Fig. I,48 (not drawn to scale).

A liquid of the desired boiling point is placed in the flask A which is heated with an electric mantle (see below). The liquid in A is boiled gently so that its vapour jackets the reaction tube BC; it is condensed by the reflux condenser at D and returns to the flask through the siphon E. Regular ebullition in the flask

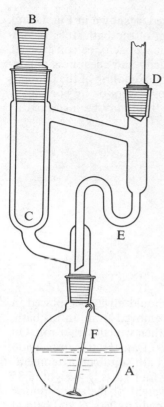

Fig. I,48

is ensured by the bubbler F. The reaction mixture in C may be stirred mechanically. It is convenient to have a number of flasks, each charged with a different liquid; changing the temperature inside C is then a simple operation. A useful assembly consists of a 50 ml flask A with 19/26 joint, a vapour jacket about 15 cm long, a 34/35 joint at B and a 19/26 or 24/29 joint at D.

The following liquids may be used (boiling points are given in parentheses): pentane (35 °C); acetone (56 °C); methanol (65 °C); carbon tetrachloride (77 °C); trichloroethylene (86 °C); toluene (110 °C); chlorobenzene (132–3 °C); bromobenzene (155 °C); p-cymene (176 °C); o-dichlorobenzene (180 °C); methyl benzoate (200 °C); tetralin (207 °C); ethyl benzoate (212 °C); 1,2,4-trichlorobenzene (213 °C); isopropyl benzoate (218 °C); methyl salicylate (223 °C); propyl benzoate (231 °C); diethyleneglycol (244 °C); butyl benzoate (250 °C); diphenyl ether (259 °C); dimethyl phthalate (282 °C); diethyl phthalate (296 °C); benzophenone (305 °C); benzyl benzoate (316 °C).

A shallow metal vessel containing sand, the so-called **sand bath**, heated by means of a flame, was formerly employed for heating flasks and other glass apparatus. Owing to the low heat conductivity of sand, the temperature control is poor; the use of sand baths is therefore not recommended for routine work in the laboratory. They may occasionally be employed where high temperatures are required, e.g., in thermal decomposition; as a rule, graphite or nickel shot is preferable for this purpose.

An **air bath** may be readily constructed from a commercial circular tin can (that from tinned fruit or food is quite suitable), and is very satisfactory for most work involving the heating of liquids of boiling point above 80 °C (or below this temperature if the liquid is non-inflammable). The top edge of the can is first smoothed and any ragged pieces of metal removed. A series of holes is then punched through the bottom, and a circular piece of asbestos board * (about 2–3 mm thickness) of the same diameter as the can inserted over the holes. The body of the can is then wrapped with asbestos cloth which is bound securely in position by two wires near the top and bottom of the can respectively. A piece of asbestos board (2–3 mm thickness) of diameter slightly larger than the top of the can is then obtained and a hole of suitable diameter made in its centre; the asbestos is then cut diametrically. The two halves, which constitute the cover of the air bath, will have the shape shown in Fig. I,49(b). The diameter of the hole in the asbestos lid should be approximately equal to the diameter of the neck of the largest flask that the air bath will accommodate. The air bath, supported on a tripod, is heated by means of a Bunsen burner; the

---

* Refer to Section I,3, p. 22, for precautions to be adopted in the handling of asbestos articles.

position of a distilling flask, which should be clamped, is shown in Fig. I,49(*a*). The flask should not, as a rule, rest on the bottom of the bath. It is recommended that three air baths of say 50, 100 and 250 ml capacity be constructed. The advantages of the above air bath are: (1) simplicity and cheapness of construction; (2) ease of temperature control; (3) rapidity of cooling of the contents of the flask effected either by removing the asbestos covers or by completely removing the air bath; and (4) the contents of the flask may be inspected by removing the asbestos covers.

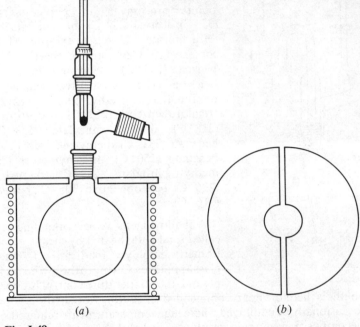

(*a*)   (*b*)

Fig. I,49

Heating mantles provide one of the most convenient means of controlled heating of reaction vessels when used in conjunction with a variable transformer;* they consist essentially of a knitted glass-fibre fabric containing an insulated nickel-chrome resistance wire which serves as the heating element; this is usually supplied fitted within an aluminium metal housing for bench operation (Fig. I,50).† Sizes suitable for use with round-bottomed flasks from 50 ml to 20 l capacity are available. Alternatively, the use of the fully enclosed flexible heating mantle with an elastic neck-entry (Fig. I,51)† may be more convenient when the apparatus assembly does not allow the satisfactory support of the metal incased type. Heating mantles in general are unsatisfactory for distillations when a suitable heating bath is much preferred (Section **I,24**).

---

* Variable transformers are manufactured by The Zenith Electrical Co Ltd ('Variac'), and also by those companies who manufacture the heating mantles, i.e., Glas-Col Apparatus Co, Electrothermal Engineering Ltd, Isopad Ltd.

† Photographs kindly supplied by Electrothermal Engineering Ltd.

Fig. I,50

Fig. I,51

Electric hot plates may also be employed and are provided with either a three-heat switch or with a thermostatic control. The diameter of the top may vary between 5″ to 10″. It is often advisable to interpose a sheet of asbestos board between the metal top and the vessel to be heated, particularly if the contents of the latter are liable to 'bump'. Electric hot plates should not be used with low boiling, inflammable liquids (e.g., ether, light petroleum) contained in open beakers since ignition can frequently occur when the heavier vapour spills over on to the heated metal.

I,14. MECHANICAL AGITATION Mechanical stirring is not necessary in work with homogeneous solutions except when it is desired to add a substance portion wise or dropwise and to bring it as rapidly as possible into intimate contact with the main bulk of the solution. This applies particularly in those cases where a precipitate is formed and adsorption may occur, or where heat is generated locally which may decompose a sensitive preparation. In such cases the solution must be continuously agitated by manual shaking or, preferably, by mechanical stirring. When large quantities of material are to be dealt with, it is much easier and much more efficient to employ mechanical stirring. The importance of mechanical agitation cannot be over-estimated where hetero-geneous mixtures are involved. In many preparations the time required for completion of the reaction is shortened, temperatures are more readily controlled, and the yields are improved when mechanical agitation is em-

Fig. I,52

ployed. No apology is therefore needed for discussing this subject in some detail.

Stirring in open vessels, such as beakers or flasks, can be effected with the aid of a stirrer attached directly to a small electric motor by means of a chuck or a short length of 'pressure' tubing. Excellent **stirring units** are available commercially, but only two of them will be described.

The photograph (Fig. I,52) illustrates a general purpose, variable speed, direct-drive stirrer * which may be fitted to any retort stand. The built-in rotary rheostat controls the speed in ten fixed steps, up to 2500 rev/min; the chuck will accept stirrers having shafts up to 6.35 mm diameter. With long stirrers, and particularly those fitted to the drive shaft with 'pressure' tubing, it is frequently necessary to employ a short length of glass tubing of the appropriate bore fitted into a cork, or a commercially available stirrer-guide, to reduce 'whip'.

A geared-drive model † (Fig. I,53) frequently gives better control of stirring

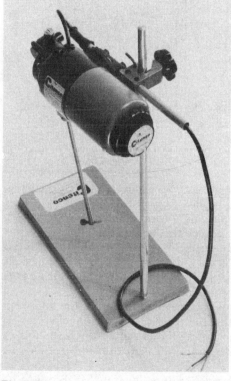

Fig. I,53

performance. These stirrers may often be fitted with a flexible drive shaft so that the motor may be placed at a distance from the stirrer head and reaction vessel, thus enabling the assembly to be used for inflammable, corrosive or fuming liquids without damage to the motor.

---

* Available from A. Gallenkamp & Co Ltd, who kindly supplied the photograph.
† A Citenco stirrer available from Jencons (Scientific) Ltd, who kindly supplied the photograph.

Stirrers are usually made of glass, but those of monel metal, stainless steel or Teflon also find application in the laboratory. An important advantage of a stirrer with a Teflon blade is that it is comparatively soft and merely bends if it hits the glass even at high speed; furthermore, it can be shaped to fit the bottom of the vessel, thus rendering the stirring of small volumes of liquid in a large flask possible. A few typical stirrers are shown in Fig. I,54(a–e); types (a) and (b) may be easily constructed from a glass rod. Types (c) and (d) are the Teflon and the glass-link stirrer respectively; because of the flexible end they possess the advantage that they may be inserted through a narrow neck; in (c) the half-moon shape allows it to be employed for stirring liquids in round-bottomed or flat-bottomed vessels (the latter by turning the blade over). Type (e) is a stainless steel propeller blade stirrer.

A useful stirrer—sometimes termed a **Hershberg stirrer**—for efficient agi-

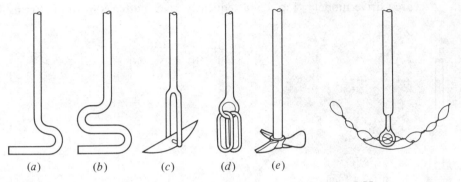

(a)	(b)	(c)	(d)	(e)

Fig. I,54   **Fig. I,55**

tation in round-bottomed vessels, even of pasty mixtures, is presented in Fig. I,55. It consists of a glass rod to which a glass ring is sealed; the glass ring is threaded with chromel or nichrome or tantalum wire (about 1 mm diameter). By sealing another glass ring at right angles to the first and threading this with wire, even better results will be obtained. The stirrer is easily introduced through a narrow opening, and in operation follows the contour of the flask; it is therefore particularly valuable when it is desired to stir a solid which clings obstinately to the bottom of a round-bottomed flask.

Some form of suitable **stirrer seal** must be provided in any of the following operations: (1) simultaneous stirring and refluxing of a reaction mixture; (2) stirring the contents of a closed vessel; (3) agitation with the prevention of the escape of gas or vapour; and (4) stirring in an inert atmosphere, such as nitrogen.

A simple rubber sleeve gland (the **Kyrides seal**) is illustrated in Fig. I,56(a); the short length of rubber tubing attached to the ground glass cone projects to form a tight seal around the stirrer shaft (5–6 mm). Glycerol (or Silicone grease) is applied at the point of contact of the glass and rubber to act as a lubricant and sealing medium. This seal may be used under reduced pressures down to about 10–12 mm Hg; it is not dependable for stirring operations lasting several hours since the rubber tubing may stick to the shaft and may also be attacked by organic vapours, causing it to swell and allow the escape of vapours. In these circumstances the screwcap gland, Fig. I,56(b), may be used when the

Silicone rubber ring should be similarly lubricated; this type of seal is not suitable for use at high speeds or under vacuum.

Precision ground stirrer/stirrer guide units * eliminate the need for any form of additional seal and can be used very satisfactorily under vacuum, in which case a trace of Silicone grease as lubricant is recommended. The long length of the ground bearing substantially reduces the problem of 'whipping' and vibration. Figure I,57(*a*) shows the general purpose type with a cone joint; a design in which a water-cooling jacket surrounds the precision ground tubes is also available. A useful variant is illustrated in Fig. I,57(*b*) in which a vapour 'take off' outlet is provided; it is also of use for reactions performed under an inert atmosphere.

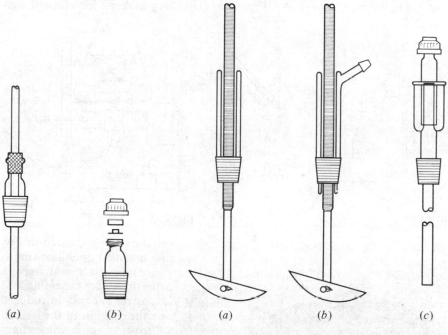

(*a*)   (*b*)   (*a*)   (*b*)   (*c*)

Fig. I,56   **Fig. I,57**

Ground sleeve glands have largely replaced the conventional **mercury-sealed stirrer**; one design of the latter which is suitable for general laboratory use is illustrated in Fig. I,57(*c*). It is not as versatile as the ground sleeve type; high speeds introduce the hazard arising from spattering mercury† and the seal is not suitable for use under vacuum.

A type of stirrer, known as a **Vibro-mixer**,‡ and of particular value for closed systems, is illustrated in Fig. I.58, fitted into the central neck of a flask. The

---

* A selection of stirrer/stirrer guide units is available from Jencons (Scientific) Ltd.
    † An alternative safer liquid seal is glycerol.
    ‡ Supplied by Shandon Southern Instruments Ltd. Model E1 is recommended for general laboratory purposes.

enclosed motor, operating on alternating current, vibrates the stirrer shaft at the same frequency as the a.c. mains, moving up and down in short, powerful strokes. A control knob at the top of the stirrer housing is provided for adjust-

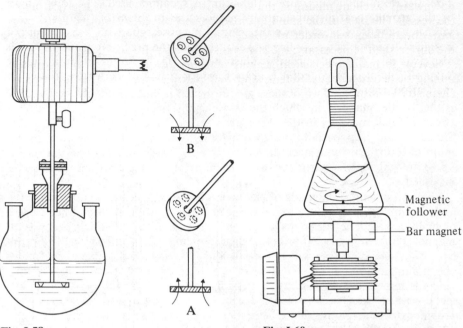

Fig. I,58

Fig. I,60

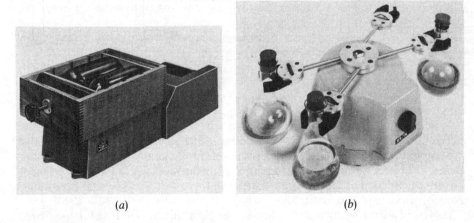

(a)   (b)

Fig. I,59

ing the stroke length from gentle strokes (0.2 mm) to powerful strokes (*ca.* 2 mm in thrust). As it is a non-rotating stirrer, a hermetic seal with the reaction flask can be made easily. Several types of stirrer blades are available; two 'plate' stirrers are shown in Fig. I,58(A and B). In A the holes taper upward; the liquid will (on the downstroke) flow up through the wide lower orifices to be violently expelled through the narrower orifices at the top. The principle involved is similar to what happens when water flowing through a pipe suddenly enters a narrower pipe; the speed of flow is greatly increased. In B the holes taper downward; excellent mixing is thus obtained for solids, etc., at the bottom of the vessel. The base of the stand upon which the stirrer is mounted rests upon anti-vibration mats in order to reduce vibration to a minimum. No guide for the stirrer shaft is necessary; the stirring is very efficient.

Several forms of **mechanical shaking machines** are employed for automatic mixing of heterogeneous systems and find many applications in the organic chemistry laboratory. Two of these are illustrated in Fig. I,59(a and b). Type (a) is the 'Bottle-Shaker'* in which the bottle carrier will accommodate sizes between 125 ml and 2.5 l in the Winchester series. Type (b) is the 'B.T.L. Flask Shaker' † and has specially designed clamps to accommodate up to a maximum of four flasks with neck diameters of between $\frac{3}{4}''$ and $1\frac{1}{2}''$ and capacities up to 500 ml each; a particular merit is that not all the positions need to be occupied.

Mention must also be made of **magnetic stirring**. A rotating field of magnetic force is employed to induce variable speed stirring action within either closed or open vessels. The stirring is accomplished with the aid of small permanent magnets ('followers') sealed in Pyrex glass, polypropylene or Teflon.‡ The principle of magnetic stirring will be evident from Fig. I,60. A permanent bar magnet, mounted horizontally, is attached to the shaft of an electric motor; the whole is mounted in a cylindrical housing with flat metal top and heavy cast metal base. A resistance (which is frequently incorporated in the housing) is provided to control the rate of stirring. To use the apparatus, the rheostat, which is initially in the off-position, is slowly rotated (this increases the motor speed) until the required rate of stirring is attained. When the experiment is complete, the rheostat is returned to the zero position, the 'stirrer' allowed to come to rest and removed with the aid of a pair of forceps.§

Magnetic stirring has many obvious applications, but the most important are probably to stirring in closed systems, e.g., (a) where gas volume changes must be observed as in catalytic hydrogenations, (b) where exclusion of air is desirable to prevent oxidation, (c) where reactions are to be carried out in an anhydrous environment and (d) where small containers are used and the introduction of a propeller shaft is inconvenient.

Many forms and sizes of magnetic stirring apparatus are available commercially. These include those fitted with an electric hot plate attached to the flat top; the hot plate is controlled by an energy regulator or variable transformer (Variac). Electric heating mantles are also available in which magnetic stirrers are incorporated so that both heating and stirring in an apparatus assembly using round-bottomed flasks may be controlled within one unit

---

* Available from A. Gallenkamp & Co Ltd, who kindly supplied the photograph.

† Manufactured by Baird & Tatlock Ltd, who kindly supplied the photograph.

‡ The 'followers' are available in different lengths and of varying designs to fulfil a range of stirring requirements, such as would be necessary with differently shaped vessels or with fluids of different viscosities; e.g., those available from Van Waters and Rogers Scientific.

§ A useful recovery device for followers is made by forcing a small bar magnet into the end of stiff polymer tubing.

(Rotamantle; manufactured by Electrothermal Engineering, Ltd). Magnetic stirring is not effective if the medium is excessively viscous, or if substantial amounts of solid are present. In such cases mechanical stirrers must be used.

I,15. TYPICAL GROUND GLASS JOINT ASSEMBLIES FOR STANDARD REACTION PROCEDURES It is hoped that the account of the interchangeable ground glass joint apparatus already given will serve as an introduction to the subject. For the numerous applications of such apparatus, the reader is referred to the catalogues of the manufacturers listed in Section **I,4**. Most of the simpler operations in practical organic chemistry may be carried out with a set of apparatus which can be purchased for a comparatively modest sum. Indeed, such a set is the first in a progressive series of 'Quickfit' sets which are specifically designed to cover the experimental requirements of practical chemistry in many different fields and in scales ranging from gramme to kilogramme quantities. For student work, in either introductory or advanced multi-stage experiments, the glassware is supplied in boxes fitted with integral plastic trays so that each item fits into a moulded well marked with the appropriate catalogue number. Checking equipment and its storage is therefore greatly simplified.

Some typical assemblies are collated in the following diagrams, which represent most of the basic reaction procedures which are adopted in the later experimental sections.

An assembly for **heating a reaction mixture under reflux** is illustrated in Fig. I,61. The precise design of condenser depends upon the volatility of the reaction liquid, low boiling liquids ($<60\,°C$) require the use of a double surface condenser. Additionally a calcium chloride tube may be inserted at the end of the condenser if the reaction mixture contains moisture-sensitive components. It is important that 'boiling chips' (carborundum, or broken porcelain) are added *before* heating (whether by gas or by electrical means, or whether using an oil or a water bath) is commenced.

For the purpose of **heating a reaction mixture under reflux with addition of liquid** the assembly shown in Fig. I,62 is suitable. The two-way adapter and flask shown here may, of course, be replaced by a two-necked flask. Manual agitation may be required at intervals to ensure good mixing, alternatively the vigour of boiling may automatically aid the mixing process. The selection of condenser design, the presence of calcium chloride protecting tubes (on condenser and separatory funnel) and the mode of heating depends upon the nature of the reactants.

It should be emphasised that the apparatus assemblies described above are for those cases where the reaction mixture is a homogeneous liquid. With reaction mixtures having suspended solid components or with reaction mixtures involving immiscible liquids stirring is essential. An assembly for **heating under reflux with the addition of liquid and with stirring** is shown in Fig. I,63. The stirrer shown may of course be omitted and agitation effected with a magnetic stirrer incorporated in an electric hot plate or heating mantle (Section **I,14**).

In an experiment where the **addition of a moisture sensitive solid** (e.g., anhydrous aluminium chloride) in small portions to a reaction mixture is required, but

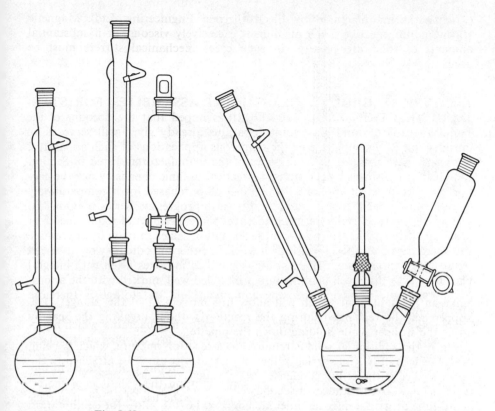

Fig. I,61   **Fig. I,62**   **Fig. I,63**

where no requirement of heating under reflux is necessary, the simplest assembly is that shown in Fig. I,64. Here the side socket is fitted with a length of wide, thin-walled rubber tubing and a 100 or 250 ml conical flask containing the reagent is inserted into the other end of the tubing. The solid is readily added in portions by raising the flask; the latter can be cut off from the reaction mixture by 'kinking' the rubber tube. The diagram also shows that the central cone has been fitted with a suitable stirrer (see Section **I,14**) and the third socket may be fitted either with a calcium chloride tube or with an appropriate nitrogen inlet system (see Fig. I,65) if an inert atmosphere is to be preserved. The stirring is of course essential and is far more efficient and convenient than manual agitation.

An alternative design for the addition of solid is shown in Fig. I,65 which also illustrates a nitrogen inlet system. It may be constructed, if desired, from a small pear-shaped flask, and a broken pipette; the connections are bored rubber of a size appropriate to fit into the sockets. The solid is charged into the conical reservoir; by raising the plunger to the appropriate height, any desired amount of solid may be made to flow into the reaction vessel, and the flow can be completely stopped by merely twisting down the plunger until the rubber ring seals the opening. The rubber ring should be, say, 3 mm thick and 5 mm wide; if solvents which attack rubber are present a neoprene gasket may be used. The rubber tubing at the top is lubricated with glycerine (Kyrides seal) to

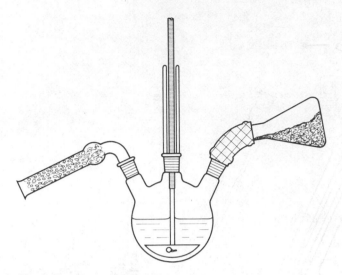

Fig. I,64

make an air-tight joint which will allow free movement of the plunger. The hopper may be recharged during the reaction without breaking the seal by lifting the stopper while holding down the pipette.

Where the **addition of solid** is required to a reaction mixture, which is being **heated under reflux with stirring**, either of these devices may be attached to the upper socket of the reflux condenser, which must, of course, be of the Liebig type. In these circumstances a second reflux condenser must be fitted with a guard-tube or with a nitrogen inlet system (Fig. I,67) to allow for equilibration of pressure within the apparatus.

The **addition of a gas to a reaction mixture** (commonly the hydrogen halides, fluorine, chlorine, phosgene, boron trifluoride, carbon dioxide, ammonia, gaseous unsaturated hydrocarbons, ethylene oxide) requires the provision of safety precautions which may not be immediately apparent. Some of these gases may be generated *in situ* (e.g., diborane in hydroboration reactions), some may be commercially available in cylinders, and some may be generated by chemical or other means (e.g., carbon dioxide, ozone). An individual description of the convenient sources of these gases will be found under Section **II,2**.

For those gases which are generated *in situ*, no further points need be noted here as the apparatus is described in the appropriate section.

For those gases which are available in cylinders, the manufacturer's notes about precautions against any hazards in their use should be carefully followed and the appropriate antidotes should be available for immediate use. It is important to realise that cylinders of all gases are potentially very dangerous owing to the possibility of the valve being broken off should the cylinder be knocked over; cylinders must therefore always be securely strapped, or supported, in frames whether in storage or in use. Cylinders of toxic gases (e.g., chlorine, sulphur dioxide) are often of such dimensions that they may be easily accommodated and suitably supported in an efficient fume cupboard containing the reaction apparatus. With the larger cylinders of toxic or hazardous gases (e.g., ammonia, acetylene) it is recommended that these should be located on the laboratory wall outside the building and piped as appropriate through

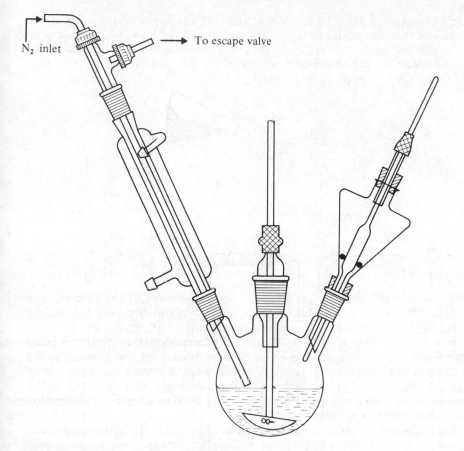

N₂ inlet

To escape valve

Fig. I,65

holes in the wall directly into the fume cupboard.* Cylinders so positioned should have suitable protection from the weather and should be easily accessible. Large cylinders of inert gases (e.g., nitrogen) should be supported in suitable stands, and although best located outside the laboratory, this may not always be feasible.

For gases which are generated by chemical or other means, it is essential that the flow of gas be maintained at as steady a rate as possible, by suitable control of the generating apparatus. Such apparatus should not be left unattended, even for a comparatively short space of time.

The gases provided by the above sources should be led to the reaction vessel via a train of Drechsel bottles, suitably charged to effect prior drying or purification, and to provide a trap if 'sucking back' of the reaction mixture does occur (e.g., in the rapid uptake of hydrogen chloride). Should 'suck-back' occur the experiment is not then ruined, nor does the reaction mixture either cause a potential hazard with the chemicals in the generating apparatus, or cause irreparable damage to the cylinder valve. A suitable apparatus set-up is shown in

---

* See Section **I,3** for recommendations on the provision of a Dangerous Operations Laboratory.

Fig. I,66, where it should be observed that the gas is released just under the surface of the reaction liquid by means of a glass tube fitted with a wide pore size glass frit; this improves gas–liquid contact and aids the absorption of the gaseous reactant, but should not be used if a solid reaction product that may block the pores is likely to be formed.

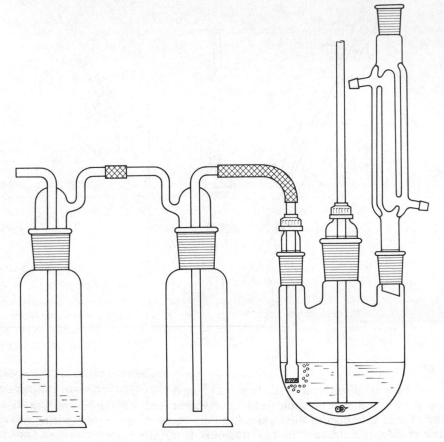

Fig. I,66

It is sometimes necessary (e.g., in reactions involving organolithium compounds or in certain Grignard preparations) to carry out a **reaction in an atmosphere of an inert gas**, such as nitrogen. A suitable set-up is shown in Fig. I,67. Dry nitrogen is introduced at the top of the condenser and initially can be allowed to sweep through the apparatus and escape at the mouth of the dropping funnel; it will be noted that the latter has a pressure-equalising side tube. After a few minutes the flow of inert gas may be reduced. The level of mercury (alternatively mineral oil or a high boiling ester, e.g., dibutyl phthalate) in the escape valve should be such that a slight pressure of gas within the apparatus is maintained when the funnel is closed. This arrangement* is economical in

---

* This nitrogen inlet system may also be used with other apparatus assemblies where an inert atmosphere is to be preserved in the reaction vessel, for example as illustrated in Fig. I,65.

N₂ →

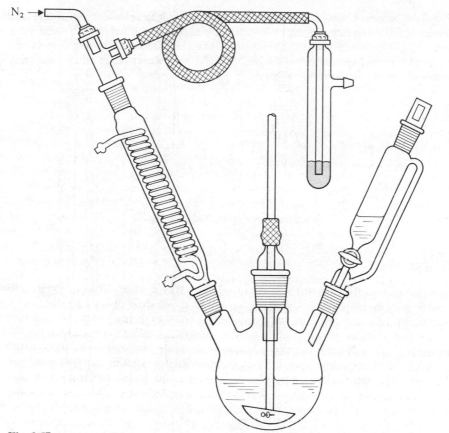

Fig. I,67

nitrogen, obviates the evaporation of solvent and is to be preferred to the use of a continuous stream of inert gas.

In some reactions a gas is evolved which may be of an irritant or corrosive nature (e.g., hydrogen chloride in Friedel–Crafts reactions, sulphur dioxide–hydrogen chloride in acid chloride preparations) and it is advisable to employ a suitable **gas absorption trap**. Either of the gas traps depicted in Fig. I,68(*a* and *b*) are used when limited quantities of water-soluble gases are to be absorbed. For larger volumes of gas, or where the gas is rapidly evolved, the gas traps shown in Fig. I,68(*c* and *d*) are very satisfactory. In (*c*) the gas is passed into a wide tube through which a stream of water (usually from a reflux condenser) flows into a large filter flask and overflows at constant level, which is above the lower end of the wide tube; a water seal is thus provided which prevents the escape of gas into the atmosphere, and the heat of solution of the gas is dissipated. A convenient size for (*d*) is a tube 80–100 cm long and 25 mm diameter.

A highly efficient gas-absorption apparatus* is depicted in Fig. I,69. The over-all length is about 40 cm; two inlets for obnoxious gases are provided, but

---

* Designed in the Research Laboratories of May and Baker Ltd, Dagenham.

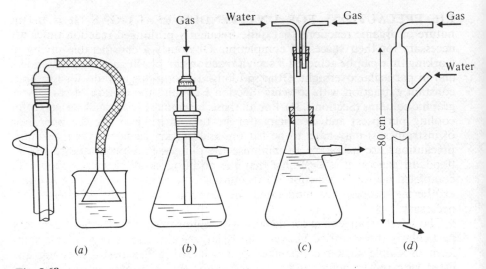

Fig. I,68

one can be readily closed if not required. The waste water from a water condenser may be employed. The water enters in the middle of the apparatus and
passes up the outer annulus, spraying out at the top of the tower on to 9-mm
Raschig or similar rings. It then passes down the column and through the
water trap at the bottom of the apparatus to waste through a side tube fitted
with a siphon-breaking device. The contaminated gas enters at either side of
the two inlet connections and is absorbed by the water passing down the
column.

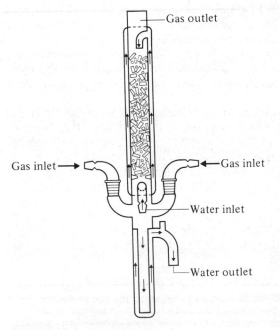

Fig. I,69

I,16. PRECAUTIONS FOR UNATTENDED REACTIONS It is in the nature of organic reactions that quite frequently prolonged reaction times are necessary for their successful completion. Obviously a considerable saving in working time can be achieved if such a reaction can be left running unattended, and in particular overnight. Other prolonged operations include, for example, constant extraction with solvents (Section I,22) and the elution of chromatographic columns (Section I,31). For all these, essential services such as water (for cooling purposes) and electricity (for heating, stirring or for the operation of instruments) may have to be left on, and certain elementary but essential precautions are necessary to minimise the danger of damage due to fire or flood. In general it is desirable that the addition of all reactants should be completed before the reaction is left unattended overnight; once any obviously exothermic process has moderated the reaction can usually be safely left if necessary.

The safest arrangement for dealing with the problem of overnight reactions, particularly those carried out on a fairly large scale, is to have available some form of simple shelter constructed of angle iron and asbestos sheeting and fitted with power points and a water supply; this may be sited on any convenient flat area on the roof of the laboratory building. Alternatively a small room fitted with efficient extractor fans and with an automatic fire extinguishing system, or even a large tiled fume cupboard kept free of extraneous apparatus, could be set aside for such reactions. Adequate provision must be made for drainage should flooding occur; in the event, for example, of condenser leads splitting and becoming unattached.*

For reactions which merely require stirring without heating it is necessary to ensure that the stirrer shaft is rotating freely in a rigidly clamped stirrer guide to avoid the possibility of the stirrer breaking the reaction vessel. If a stirrer seal of the Kyrides type (Section I,14) has to be used it must be well lubricated to avoid seizure. An electromagnetic stirrer can of course be used if the stirring it provides is adequate. Although modern shakers are designed so as to prevent the possibility of their 'travelling' over a bench or floor surface, nevertheless it is advisable to take steps to prevent movement and to ensure that the moving parts cannot accidentally touch other apparatus in the vicinity, or move against a wall surface.

The commonest operation which needs to be carried out during an overnight period is probably a reaction which requires heating under reflux, with or without stirring. Any form of gas heating is hazardous and electric heating mantles should be used; it is preferable that the controls should be sited well away from the reaction assembly. Rubber tubing used for connecting reflux condensers, etc., to the water supply should be inspected for latent defects and should be securely wired on to the taps and glass apparatus inlets and outlets. Waste water should be led away through a rubber tube which has fitted at its end a glass tube, which projects well into the main drain. Before the reaction is finally left, one should check that the cooling water is flowing freely but not violently through all condensers.

---

* A clear notice should be displayed near the apparatus indicating the type of reaction being carried out, the nature of the solvent (if any) being used, and whether the water and/or electricity supply is required to be left on. In certain cases it may be desirable to indicate what steps should be taken if any of the essential services fail.

I,17. APPARATUS FOR SPECIAL REACTION TECHNIQUES

I,17,1. Catalytic hydrogenation   Many classes of organic compounds may be efficiently reduced by molecular hydrogen in the presence of a suitable catalyst (*catalytic hydrogenation*). Depending upon the nature of the functional group which it is required to reduce, the experimental conditions necessary for hydrogenation may vary widely; for example, hydrogenation may be carried out either at room temperature or at temperatures up to about 300 °C, with the use of hydrogen at atmospheric pressure or at pressures up to about 350 atmospheres. The successful hydrogenation of a particular functional group depends also upon the correct choice of a suitable catalyst, and a wide range of formulations is available for the preparation of catalysts in a suitably active form. The main commonly used catalyst preparations together with the functional groups which they most effectively reduce are discussed briefly below; their preparation is dealt with under Section **II,2**. The nature of the solvent used may also influence the course of the hydrogenation. Acceptable solvents must not of course themselves be reduced under the hydrogenation conditions; cyclohexane, ethanol, acetic acid or ethyl acetate are widely used. Small quantities of acids or bases added to neutral solvents may have a significant effect on the course of some hydrogenations (for an excellent general review see Ref. 20).

Hydrogenation catalysts   **Platinum metal group.**   These are powerfully active catalysts which are used at normal or slightly elevated temperatures and pressures.

Platinum in a finely divided form is obtained by the *in situ* reduction of hydrated platinum dioxide (*Adams catalyst*); finely divided platinum may also be used supported on an inert carrier such as decolourising carbon. Finely divided *palladium* prepared by reduction of the chloride is usually referred to as *palladium black*. More active catalysts are obtained however when the palladium is deposited on decolourising carbon, barium or calcium carbonate, or barium sulphate. Finely divided *ruthenium* and *rhodium*, usually supported on decolourising carbon or alumina, may with advantage be used in place of platinum or palladium for some hydrogenation reactions.

The platinum metal group of catalysts readily reduce most olefinic and acetylenic multiple bonds at normal temperatures and pressures. The selective semi-hydrogenation of an alkyne to an alkene is efficiently carried out using deactivated palladium catalysts (e.g., *Lindlar's catalyst*). The reduction of aliphatic aldehydes and ketones to alcohols is a little more difficult to achieve; palladium is virtually ineffective but platinum which has been promoted by the addition of a little iron(II) sulphate works well. On the other hand, aryl aldehydes and ketones are readily reduced over palladium, but the products are the corresponding hydrocarbons rather than alcohols (i.e., *hydrogenolysis* of the intermediate alcohols, which are benzylic in nature, occurs). Such alcohols are best prepared using ruthenium, which does not promote hydrogenolysis. The platinum metal group are not the catalysts of choice for the reduction of aromatic ring systems, although ruthenium and rhodium are very effective, and Adams catalyst in the presence of acetic acid may be used. This latter catalyst system is also useful for the reduction of pyridine rings; somewhat elevated pressures of hydrogen are used. An important use of deactivated palladium catalysts is in the selective hydrogenolysis of acid chlorides to aldehydes (the *Rosenmund* procedure).

Nickel. Nickel catalysts for hydrogenation can be prepared in a range of activities (*Raney nickel catalysts*; see Section II,2,*41*). The most active grades are comparable with platinum and palladium in many of the reductions mentioned above. The less active grades are frequently used however and usually require moderate or high temperatures and pressures, depending upon the nature of the group which is to be reduced. These catalysts are particularly useful for the reduction of nitrogen-containing functional groups such as $-NO_2$, $-C\equiv N$, $>C=NOH$ to primary amines; nickel catalysts are not deactivated (poisoned) by amino compounds as are the platinum metal group. Nickel also effects the hydrogenation of benzene rings; for catalysts of a moderate degree of activity, temperatures about 100 °C and pressures of about 100 atmospheres are usually adequate. An important application of Raney nickel, which is related to its insensitivity to the action of catalyst poisons, is in the reductive fission of $C-S$ bonds, a process which is exploited in the desulphurisation of organic sulphur-containing compounds.

Copper–chromium oxide. This is a catalyst of uncertain composition prepared by the ignition of basic copper(II) ammonium chromate. It is an approximately equimolar combination of copper(II) chromite and copper(II) oxide ($CuCr_2O_4 \cdot CuO$) but is evidently not a simple mixture of these two components. The catalytic activity is enhanced by the incorporation of some barium chromite; hydrogenations require however the use of relatively high temperatures and pressures. This catalyst, which may be regarded as complementary to Raney nickel, is generally useful for the reduction of oxygen-containing functions, and is the catalyst of choice for converting esters into primary alcohols. It may be used also for the reduction of amides to amines; it does not usually reduce an aromatic ring unless the conditions are exceptionally severe.

Apparatus for catalytic hydrogenation **Hydrogenation at atmospheric pressure.** A suitable apparatus is shown semi-diagrammatically in Fig. I,70(*a*); it is supported on a suitable metal rod framework. The essential features are a long-necked hydrogenation flask A fitted to the apparatus with sufficient flexible tube to allow shaking, a series of water-filled burettes and reservoirs (two are shown, B and C), a manometer D and mercury safety trap E. The sizes of the burettes will be appropriate to the scale of operation for which the apparatus is designed; a suitable combination of 2 litre, 1 litre, 250 ml or 100 ml sizes may be used. The various parts of the apparatus are connected as far as is possible by ground glass joints lubricated with Silicone grease. Flexible tubing in contact with hydrogen is of polyvinyl chloride (PVC); the hydrogenation flask should be screened by a laminated safety glass shield. *No flames should be allowed in the laboratory while the hydrogenation apparatus is in use.*

The procedure for conducting the hydrogenation is as follows:

1. Disconnect the hydrogenation flask A, open taps H, K and L, and fill the burettes with water by raising the reservoirs. Close taps K and L and lower the reservoirs.
2. Charge the hydrogenation flask with the catalyst, and with the solution to be hydrogenated, taking care that the solution washes down traces of catalyst which might be adhering to the sides of the flask so that finally all of the catalyst is covered by solution. Attach the flask to the apparatus.

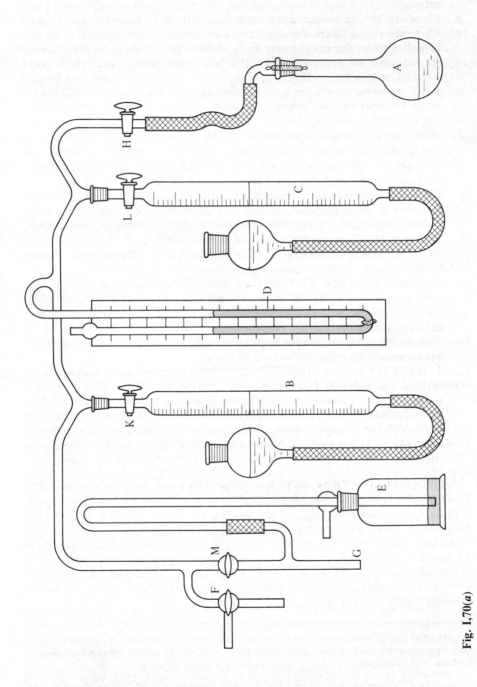

Fig. I,70(a)

Connect the apparatus to a water pump via a trap (not shown) and the three-way tap F and attach a hydrogen cylinder to G via a reducing valve.

3. Close tap M and evacuate the apparatus via tap F.

4. Close tap F and fill the apparatus with hydrogen to atmospheric pressure, as indicated by the manometer D by slowly opening tap M; close tap M.

5. Re-evacuate the apparatus via tap F and then close tap F; repeat steps 4 and 5 once more.

6. Refill the apparatus with hydrogen via tap M, open taps K and L and allow the burettes to fill with hydrogen, if necessary lowering the reservoirs further. Close tap M.

7. With taps K, L and H open, adjust the levels of water in the reservoirs to just above those in the burettes and momentarily open the three-way tap to the atmosphere so that the pressure of hydrogen in the system reaches atmospheric. Record the water levels in the burettes. Close tap L.

8. Shake the flask A to initiate the hydrogenation, and adjust the reservoir of the burette B periodically so that the pressure of hydrogen is slightly above atmospheric. When the hydrogen in B is all used up, close tap K and open tap L to use the hydrogen in burette C.

9. When hydrogen uptake ceases adjust the level of the reservoir for burette C and read the burette. Close tap L.

10. Stop the shaker and swirl the flask manually to wash down below the surface of the solution all traces of catalyst which may be adhering to the sides of the flask and evacuate the apparatus via tap F. Admit air* through F and detach the hydrogenation flask.

11. Correct the total volume of hydrogen used to standard temperature and pressure to determine the uptake in moles.

12. Filter off the spent catalyst on a small Hirsch funnel and wash it with a little of the solvent. The damp used catalyst should be transferred immediately to a residues bottle for subsequent recovery (Sections **II,2,**44 and 51); used hydrogenation catalysts should not be allowed to become dry on the filter-paper as they are liable to inflame. The filtrate should then be worked up in a manner appropriate to the nature of the product.

A convenient alternative procedure for carrying out atmospheric pressure hydrogenations involves the use of the **Brown**[2] **Hydrogenator,**† which is available as a standard assembly for the hydrogenation of about 1 to 100 g of material, or in a larger version for the hydrogenation of 100 to 1000 g. The procedure uses the reaction of acetic acid with sodium borohydride to provide a convenient source of pure hydrogen; the apparatus is designed to allow the automatic generation of the gas. The catalyst (usually a highly active form of platinum) is prepared *in situ* immediately before use by the reduction of the metal salt with sodium borohydride.

---

* Provided that the catalyst is covered with solution there is little danger of an explosion occurring when air is admitted to the apparatus; however, it is wise to ensure that appropriate precautions have been taken.

† Introduced by C. A. Brown and H. C. Brown and available from Delmar Scientific Laboratories; UK supplier, Perkin-Elmer Ltd. Reference 21 should also be consulted for a definitive account of this hydrogenation method.

The apparatus can be used in two ways:

(i) where hydrogen is generated in one flask and hydrogenation of the substrate in the presence of the catalyst is effected in another ('*external hydrogenation*');

(ii) where hydrogen generation and hydrogenation are effected in the same flask ('*internal hydrogenation*').

External hydrogenation. The apparatus for this operation is shown in Fig. I,70(*b*). It consists basically of three glass vessels, a hydrogen generator A, a hydrogenation flask B and a pressure control bubbler C, which are connected in series by means of air-tight O-ring joints.

The mercury bubbler C acts as a safety vent and controls the pressure in the apparatus. A ball valve near the top of the inlet tube prevents the mercury being sucked into flask B in the event of the automatic control valve D becoming blocked.

The hydrogenation vessel B is an Erlenmeyer flask with a slightly convex base, which is attached by means of wire springs to the inlet adapter E which

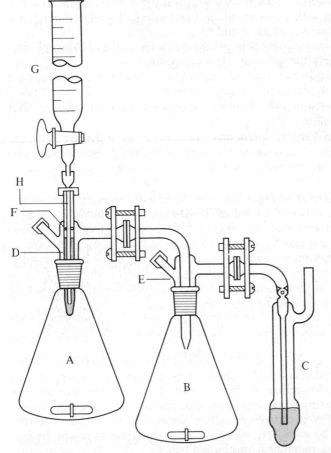

Fig. I,70(*b*)

incorporates a port closed with a serum cap to allow the introduction of appropriate solutions from a syringe.

The hydrogen generator flask A is a similarly shaped Erlenmeyer flask. The inlet adapter to which this flask is attached incorporates a mercury valve D which controls the rate at which the sodium borohydride solution contained in the 250 ml burette G is allowed to flow into the flask A via the syringe needle H (gauge 17 or 19) and the vent holes F in the control valve D.

Efficient stirring in the flasks A and B is provided by means of suitably sized Teflon-covered magnetic followers fitted with half-inch Teflon collars; the magnetic stirrer units serve also to further support the apparatus which should be securely clamped to a rack.

The following stock solutions are required:

Stabilised sodium borohydride solutions. (a) *1.00* M-*Aqueous solution:* dissolve 0.8 g of sodium hydroxide in 150 ml of water, add 7.71 g of sodium borohydride (assuming 98% purity) and stir to dissolve. Dilute the solution to 200 ml and filter. (b) *2.50* M-*Aqueous solution:* repeat the procedure under (a) exactly but increasing the amount of sodium borohydride to 19.25 g. (c) *1.00* M-*Ethanolic solution:* dissolve 0.8 g of sodium hydroxide in 10 ml of water, dilute to 200 ml with absolute ethanol and add 7.71 g of sodium borohydride. Stir until solution is effected and filter.

0.2 M-*Ethanolic chloroplatinic acid solution:* dissolve 1.00 g of chloroplatinic acid (40% platinum metal) in 10 ml of absolute ethanol.

The procedure for conducting the hydrogenation of 0.5 mol of a compound is described below. (Flask sizes and reagent quantities suitable for the hydrogenation of other molar amounts of substrate are listed in Table I,6.)

CAUTION: *All the following operations involving catalyst preparation and hydrogenation should be conducted in the fume cupboard; in particular, large quantities of hydrogen are evolved in step 2.*

1. Remove the flask B (500 ml capacity) from the apparatus assembly and add 100 ml of absolute ethanol, 5.0 ml of 0.2 M-ethanolic chloroplatinic acid and 5 g of decolourising charcoal; insert a $1\frac{1}{2}''$ Teflon-covered follower bar.
2. Place the flask on a magnetic stirrer unit and stir vigorously while adding 25 ml of 1.0 M-ethanolic sodium borohydride as rapidly as possible without allowing the contents of the flask to foam over. After about one minute, add 20 ml of glacial acetic acid or concentrated hydrochloric acid to destroy excess of sodium borohydride.
3. Add 0.5 mol of the compound to be hydrogenated either neat or in ethanolic solution and reconnect flask B to the apparatus but do not commence stirring.
4. Charge the burette G with 1.00 M-aqueous sodium borohydride solution.
5. Place 20 ml of glacial acetic acid in flask A (250 ml capacity) equipped with a Teflon-covered follower bar. Reconnect flask A to the apparatus and stir magnetically whilst injecting from a syringe through the side port of the inlet adapter 30 ml of the aqueous sodium borohydride solution. The rate of addition should be such as to effectively flush the apparatus with hydrogen without ejecting mercury from the bubbler C.
6. Open the stop-cock of burette G; the depth of mercury in the control valve

Table I,6 Specifications for representative hydrogenations (external)

Hydrogenation flask							
Compound (mmol)	Flask (ml)	Absolute ethanol (ml)	H_2PtCl_6 ml, 0.2 M	Charcoal (g)	1.00 M Ethanolic $NaBH_4$ (ml)	Conc. acid (ml)	Length of stirring bar (inch)
2000	2000	400	20	20	100	80	2
1000	1000	200	10	10	50	40	2
500	500	100	5	5	25	20	2 or $1\frac{1}{2}$
250	250	50	2.5	2.5	12.5	10	$1\frac{1}{2}$
100	125	25	1.0	1.0	5.0	4	$1\frac{1}{2}$
50	125	25	1.0	1.0	5.0	4	1

D is sufficient to support the column of borohydride solution. Now begin vigorous magnetic stirring of the contents of flask B when hydrogenation will begin. As the pressure drops in the system the valve allows sodium borohydride solution to be drawn into flask A via the vent holes F and the syringe needle H (gauge 17). The hydrogenation will then continue automatically until it has been completed. Finally note the volume of sodium borohydride solution which has run in from the burette.

7. Disconnect flask B, remove the catalyst by filtration and isolate the reduction product by suitable work-up procedures.

8. Calculate the uptake of hydrogen from the recorded volume of sodium borohydride solution used.

$$250 \text{ ml } 1.00 \ M\text{-NaBH}_4 \equiv 1.00 \text{ mol } H_2$$

If the substrate is insoluble in ethanol or if it is sensitive to protic media another solvent must of course be used. Ethyl acetate, tetrahydrofuran or diglyme are suitable alternatives but not dimethylformamide or acetonitrile which poison the catalyst. If an alternative solvent is needed the catalyst is prepared in ethanol as in step 1 and 2, but the procedure thereafter is modified as follows.

(a) Pour the contents of flask B into a sintered glass Buchner funnel, and remove most of the ethanol with gentle suction until the catalyst is left covered by an approximately 3 mm layer of the solvent.

(b) Add 50 ml of ethanol to the catalyst in the funnel, stir with a spatula and remove most of the ethanol by suction as in (a).

(c) Similarly wash the catalyst three times with 50 ml portions of the new solvent. The catalyst must not be allowed to become dry at any time during the above filtration procedure.

(d) Wash the catalyst into the hydrogenation flask with 100 ml of new solvent with the aid of a wash bottle. Continue with step 3 of the standard procedure described above.

Internal hydrogenation. For this mode of operation the Erlenmeyer flask B and the inlet adapter E are omitted from the assembly shown in Fig. I,70(*b*) and the pressure control bubbler C is connected directly to the inlet adapter fitted to flask A. Catalyst preparation, hydrogen generation and hydrogenation

Hydrogen generator				
Flask (ml)	Acetic acid (ml)	Aqueous NaBH₄ to flush (ml)	Molarity of aqueous NaBH₄ for hydrogenation	Needle gauge
500	80	40 (2.5 M)	2.5	17
500	40	25 (2.5 M)	2.5	17
250	20	30 (1.0 M)	1.0	17
125	10	15 (1.0 M)	1.0	19
125	10	10 (1.0 M)	1.0	19
125	10	10 (1.0 M)	1.0	19

are all carried out in flask A. The procedure for the hydrogenation of 0.5 mol of a compound is described below; this may be modified for other molar quantities of substrate as indicated in Table I,7. **CAUTION:** the entire operation should be conducted in the fume cupboard.

1. Place 100 ml of absolute ethanol, 5.0 ml of 0.2 M-ethanolic chloroplatinic acid and 5 g of decolourising charcoal in flask A (1000 ml capacity); insert a $1\frac{1}{2}''$ Teflon-covered follower bar.
2. Connect the flask to the inlet adapter and support it on a magnetic stirrer unit.
3. Charge the burette G with 1.00 M-ethanolic sodium borohydride and stir the contents of the flask vigorously. To prepare the catalyst rapidly inject 40 ml of 1.00 M-ethanolic sodium borohydride solution through the inlet port by means of a syringe and after one minute inject 25 ml of glacial acetic acid to destroy excess sodium borohydride. (**CAUTION:** a large volume of hydrogen is evolved.)
4. Open the stop-cock of the burette, inject from a syringe through the inlet port 0.5 mol of the compound to be hydrogenated as a liquid or as an ethanolic solution when hydrogenation will proceed automatically as in the external hydrogenation technique.
5. At the conclusion of the hydrogenation, record the volume of sodium borohydride solution used, remove the catalyst by filtration and isolate the product by suitable work-up procedures.

Hydrogenation under pressure. The following account refers primarily to commercial apparatus suitable for conducting hydrogenations under pressure; the apparatus can of course be employed for other reactions under pressure (Section **I,17,2**), but some modifications of experimental procedure will then be necessary.

The apparatus shown in the photograph (Fig. I,70(*c*)*) is designed for use at temperatures up to 70 °C and at working pressures up to 60 p.s.i. when using a glass reaction bottle (available with a capacity of either 500 ml or 1 litre). Stainless steel reaction bottles can be used at pressures up to 300 p.s.i. and if

---

* Designed and manufactured by Chas. W. Cook & Sons Ltd, who kindly supplied the photograph.

Table I,7 Specifications for representative hydrogenations (internal)

Compound (mmol)	Flask (ml)	Absolute ethanol (ml)	H_2PtCl_6 ml, 0.2 M	Charcoal (g)
1000	2000	200	10.0	10.0
500	1000	100	5.0	5.0
250	500	50	2.5	5.0
100	250	50	1.0	1.0
50	125	25	1.0	1.0

necessary at temperatures up to 200 °C. The bottle fits into an aluminium carrier fitted with an aluminium alloy cover carrying a sulphur-free rubber sealing ring. A metallic heating unit which surrounds the bottle is provided; the lid is fitted with a thermocouple well which dips into the reaction bottle. The carrier is pivoted in a support frame to allow controlled rocking by a geared motor with an eccentric drive. The reaction bottle is connected by a flexible

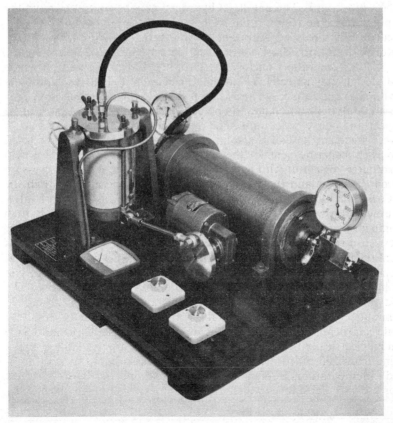

Fig. I,70(c)

1.00 M Ethanolic NaBH$_4$ for catalyst prep. (ml)	Conc. acid (ml)	Length of stirring bar (inch)	Needle gauge
80	50	2	17
40	25	2 or $1\frac{1}{2}$	17
20	10	$1\frac{1}{2}$	19
10	10	$1\frac{1}{2}$	19
5	5	1	19

PTFE tube to a mild steel hydrogen reservoir with a capacity of 4.3 litres and which as normally supplied has a maximum operating pressure of 200 p.s.i. (Reservoirs operating at pressures up to 500 p.s.i. for hydrogenations in stainless steel reaction bottles are also available.) At the forward end a double valve provides for (a) the evacuation of the bottle using a water pump, (b) the controlled charging of the bottle with hydrogen with the aid of a pressure gauge marked in pounds and (c) the release of pressure in the bottle without loss of gas from the storage tank. The uptake of hydrogen may be computed from the change in pressure as the result of the hydrogenation.

An excellent high-pressure autoclave is illustrated in Fig. I,70(d).* The special feature of this apparatus, constructed almost entirely of stainless steel, is the incorporation of a totally-enclosed agitator in the form of a plunger which is operated electro-magnetically; agitation efficiency is at least as high as is achieved with shaking autoclaves and is very effective for hydrogenation purposes. The apparatus is stationary, has no external moving parts, and can be made compact and convenient to use. The reaction vessel B is made of F.M.B. stainless steel machined out of the solid and is provided with a cover fitted respectively with a thermometer or thermocouple pocket T, a central vertical tube, and an outer vessel nut with compression screws for making the pressure joint between the cover and the vessel. Sd is a solenoid operated through the contactor C, Bd is a bursting disc, G is a pressure gauge, V$_1$ is a control valve, V$_2$ is an evacuation valve (the last-named is connected through VP to a vacuum pump for complete evacuation of the apparatus). The agitator A consists of a stainless steel rod at the lower end of which is secured a circular stainless steel plate; at the upper end of the rod passing through the centre of the vertical tube is a stainless steel sheathed armature which, in its lowest position, just enters the lower end of the solenoid coil surrounding the central tube. The solenoid Sd through the contactor C operates at a rate between 20 and 90 cycles per minute controlled by an adjustable screw on the contactor, resulting in a vertical reciprocating movement of the agitator rod. The whole autoclave is placed in an electrically-heated air bath H. Autoclaves are available in capacities ranging from 200 ml to 2 litres for use with pressures up to 350 atmospheres and to temperatures as high as 300 °C; special liners of Pyrex glass are supplied for use with substances which attack stainless steel or are affected by it.

---

* Manufactured by Baskerville and Lindsay Ltd.

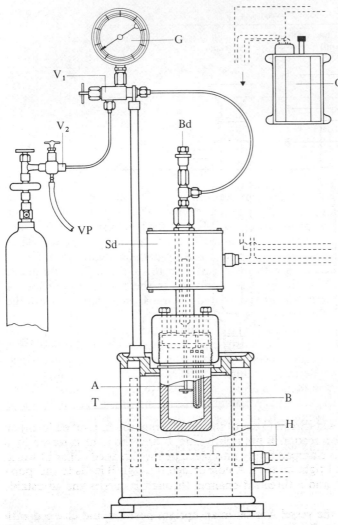

Fig. I,70(d)

I,17,2. Reactions under pressure Reactions which require the use of substantially increased pressures are usually carried out in a high-pressure metal autoclave. The apparatus described under *Catalytic hydrogenation* (Section **I,17,1**) is suitable for many reactions on a moderate scale, i.e., when the total volume of reactants is compatible with the size of the vessels available. Similar specially designed assemblies for small-scale reactions (volumes from 5 to 20 ml) are also available. Reactions involving corrosive materials require vessels provided with a resistant lining such as an acid-resisting enamel or Pyrex glass.

A cheap and effective small scale pressure vessel designed and constructed in these laboratories is shown in Fig. I,71(*a*). The main sections of the apparatus are constructed from a high grade of stainless steel (EN 58J). The main cylindri-

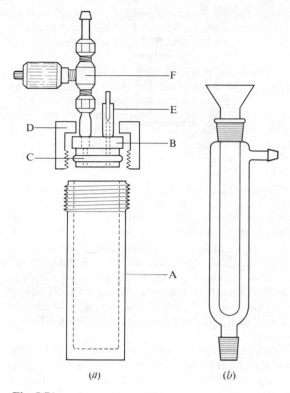

(a)   *(b)*

Fig. I,71

cal vessel A has a raised thread (3 cm of thread of 3 mm pitch, Unified form) on the open end. The top section B fits tightly into A and the joint is sealed by a 'Viton' O ring, C; B is clamped firmly into place by the threaded collar D which is screwed home hand-tight on to A. A hole drilled through B leads to the 'pop'-type safety valve E* and a threaded channel through B carries the adjustable 'Hoke' valve F.

To use the bomb, the vessel A is set in an upright position and charged with reactants and solvent; B is then pushed home into A with the valve F open and the securing collar D is screwed down hand-tight.† The vessel is evacuated with a vacuum pump attached to the nozzle of F, the valve is closed and the vacuum line removed. The bomb is then positioned behind adequate safety shielding, preferably in an isolated position, and heated in an oil bath or preferably by an electrical heating tape wound around the barrel A.

On completion of the reaction, the bomb should be allowed to cool to room

---

* This valve may be set to vent at various pressures by changing the strength of the internal spring; in this apparatus, the valve is set at about 300 p.s.i. For insurance purposes, the vessel must be subjected to a hydrostatic pressure test under the supervision of a Chartered Engineer. The vessel described above was subjected to a pressure of 60 bar (70 kgf cm^{-2}) at ambient temperature and at the maximum operating temperature (200 °C using a 'Viton' O-ring seal).

† The securing collar D must not be overtightened since this will produce an unnecessary additional load on the thread.

temperature and the barrel cooled further to about $-15\,°C$ in an ice-salt or acetone-Cardice bath. The valve F may then be opened to release any pressure (fume cupboard), and the bomb dismantled and the contents removed for work-up.

When one or more of the reactants is highly volatile the barrel A must be cooled thoroughly before the reactants are added. This may be achieved by standing A upright in an ice-salt or, if necessary, an acetone-Cardice bath. During cooling it is desirable to prevent the condensation of atmospheric moisture on the inside surface of A by closing the opening of A with a rubber bung carrying a calcium chloride guard-tube. After cooling the bung is removed, the cold reactants and solvent are added, and the apparatus is quickly assembled and evacuated.

When a reactant is a gas at room temperature (e.g., 1,3-butadiene, Section V,14) the following procedure, which should be conducted in the fume cupboard, may be adopted to liquefy and transfer it to the pre-cooled vessel A. The apparatus consists of a purpose built acetone-Cardice condenser Fig. I,71(b) with a 34/35 upper socket and a 24/29 cone at the lower end on to which is fitted a two-necked round-bottomed flask. The side arm of the flask carries a screw cap adapter through which is passed a length of glass tubing so that it just protrudes into the flask. The length of the tube should be such that it can be repositioned with the end reaching to the bottom of the flask. The outlet of this tube is connected to a calcium chloride guard-tube via polyethylene tubing.

Add a small quantity of an appropriate drying agent in granular form to the to the flask and grease the joints lightly and protect from the ingress of moisture by wrapping with absorbent cotton wool and sealing with adhesive tape.

Charge the inner vessel of the condenser with acetone-Cardice, surround the flask with a cooling bath of acetone-Cardice, and allow the gaseous reagent to flow slowly through the condenser inlet from a preparative assembly or from a compressed gas cylinder.

When sufficient reagent has been condensed in the flask shut off the supply of gas and connect the condenser inlet to a supply of nitrogen. Loosen the screw cap slightly, push the glass tube to the bottom of the flask, and re-tighten the screw cap. Remove the drying tube and apply a slight pressure of nitrogen to the condenser inlet to drive the condensed reagent directly into the precooled pressure vessel A via the plastic tubing. The pressure vessel is then quickly sealed and evacuated.

For small-scale work, when pressures not in excess of about 30 atmospheres are developed, the reaction may be carried out in a sealed thick-walled glass (Carius) tube (1), which is enclosed in a strong metal protecting jacket and heated in a suitable tube furnace. Since the sealed tube under pressure may explode with great violence, suitable precautions must be taken while conducting the whole operation.

The tube should be of 2–2.5 mm Pyrex glass, about 60 cm long and 2.5 cm outside diameter, evenly sealed at one end. The total volume of reactants introduced into the tube should not exceed one third of the volume of the tube. The tube and reactants are cooled in ice while the open end, which must be clean and free of contaminating reactants, is carefully sealed in an oxygen-gas blowpipe by drawing out the tube about 6–8 cm from the open end into a uniformly thick-walled capillary which is then sealed off.* The sealed end is then carefully annealed by rotating it in a small neat-gas flame until it is evenly coated with soot, and finally allowed to cool to room temperature. The tube is then wrapped in asbestos paper, enclosed in a screw-capped heavy metal protecting jacket, and heated to the required temperature in a furnace (2). After reaction the furnace should be allowed to cool until the following morning when the tube and the enclosing case may be removed. The sealed tube should then be opened cautiously behind a suitable safety screen in the following way. The uncapped protecting tube is first tilted carefully to allow the sealed Carius tube to slide a little way out of the case until the capillary seal is accessible. The latter is then heated, gently at first and then more strongly in a fine blow-pipe flame, until the glass softens sufficiently to allow the compressed gases to blow out and escape. When all of the pressure has been released the end of the tube may be cracked off by making a deep file cut at an appropriate point and touching this with the tip of a red-hot glass rod.

Notes. (1) The Carius method is a classical procedure, now largely superseded, for the determination of halogen and sulphur in organic compounds, and involves the decomposition of the compound by heating with fuming nitric acid in a sealed tube. Halogen and sulphur are then determined as halide and sulphate ions respectively.

(2) A Carius Low Temperature Furnace is available commercially;† the steel protecting tubes are open-ended. A perforated metal box at the rear and also the door are filled with glass wool to reduce the effect of blast should the reaction tube explode.

I,17,3. Uncatalysed and catalysed vapour phase reactions Thermal decompositions (pyrolyses) and catalysed reactions in the vapour phase are widely used large-scale industrial techniques. These vapour phase reactions often lead to more economic conversions than the smaller batchwise laboratory methods, because relatively inexpensive catalyst preparations (compared to the often expensive reagents required in laboratory procedures) may be used, and because the technique lends itself to automated continuous production. In undergraduate and research laboratories, the technique has not achieved widespread use; however, the various apparatus designs to meet a range of experimental conditions are not complex and the use of this technique could be more readily exploited.

Examples which are cited in the following preparative sections are the depolymerisation of dicyclopentadiene to cyclopentadiene (Section **V,16**), the thermal decomposition (pyrolysis) of acetone to keten (Section **III,140**), the pyrolysis of 1,5-diacetoxypentane to 1,4-pentadiene (Section **III,11**), the dehydrogenation of primary alcohols with a copper-chromium catalyst to aldehydes (Section **III,80**) and the formation of symmetrical and unsymmetrical ketones by reaction of carboxylic acid vapours with a manganese(II) oxide catalyst (Section **III,92**). In each case the apparatus incorporates a reservoir containing the reactants, a heated reaction chamber which may or may not contain a

---

* It may be advantageous prior to introducing the reactants to partially draw out the open end of the Carius tube (cf. Section **I,33**, the filling and sealing of ampoules).

† A. Gallenkamp & Co. Ltd.

catalyst and into which reactant vapours are led, and a collection flask in which the product is trapped. In some cases provision is made for separating product from unchanged reactant and returning the latter to the reservoir for recycling.

The simplest apparatus is that shown in Fig. I,72(a) where decomposition of the reactant occurs at reflux temperature without the aid of a catalyst, the

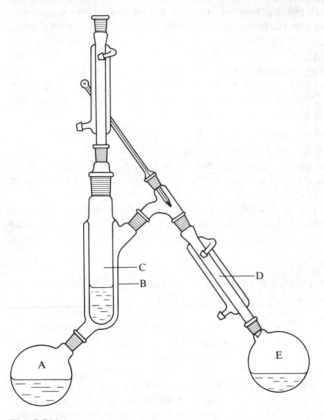

Fig. I,72(a)

products being more volatile than the reactant. This apparatus assembly uses the component parts of the vacuum drying pistol illustrated in Fig. I,84. The reactant in the reservoir A is heated to gentle reflux by means of a heating mantle or oil bath. The vapour, which consists of product and reactant, passes into the chamber B, when the undecomposed reactant condenses on the finger C and returns to A; the more volatile product passes to the condenser D to collect in the receiver flask E. The liquid in the finger C must have a boiling point above that of the product but well below that of the reactant; it will boil and reflux during the progress of the experiment and hence boiling chips in C will be required.

When higher temperatures for pyrolysis are required in an uncatalysed vapour phase reaction, the apparatus illustrated in Fig. I,72(b) could be used. This was originally designed for the pyrolysis of acetone vapour which when passed over a nichrome filament heated at 700–750 °C gives keten in yields exceeding 90 per cent.

The construction of the filament will be apparent from the enlarged inset. About 350 cm of 24 gauge nichrome wire* is formed into a tight spiral by winding the wire round a glass rod 3 mm in diameter and stretching the coil so formed to a length of 70 cm. The filament is held in position on 1.5-cm-long platinum hooks A sealed into the Pyrex glass rod B which supports them. The three platinum hooks at the bottom of the rod are placed 120° apart; two platinum hooks support the filament at a distance of 11 cm above the lower

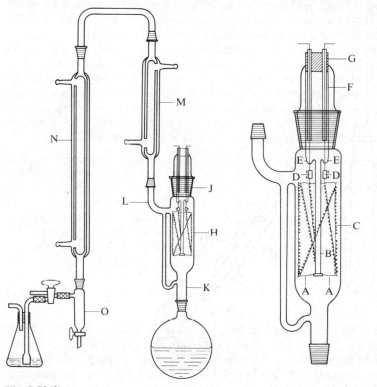

Fig. I,72(b)

end. The ends of the filament C are connected to tungsten leads by means of nickel or brass connectors D, 10 mm in length and 3.5 mm in internal diameter. The tungsten leads (24 gauge) are sealed into the glass at the points E; the leads are insulated by means of 6 mm glass tubing F, which are held by a cork stopper G. If desired, the tungsten leads may be soldered immediately above the glass seal to copper wires (24 gauge) which are passed through the glass tubing F. The tungsten or copper wire leads are connected to the 240–250 volt a.c. mains through a variable transformer (Variac).

All the glass in the apparatus is in Pyrex and connections are made by means of standard glass joints of appropriate size. Chamber H is constructed from a

---

* US: B. and S. gauge 24 Chromal A wire, an alloy of 80 per cent Ni and 20 per cent Cr.

25-cm length of glass tubing of 70 mm internal diameter; the joint J is 55/44. The connecting tube K is in 12–15 mm tubing, the side arm L is of 15 mm tubing; the condensers M and N are efficient double surface condensers, 50 cm and 90 cm long respectively (the sizes are not critical); O is a liquid trap, constructed of 35 mm tubing and is 120 mm long, with side tube of 8 mm diameter; the stop-cock is for the removal of liquid from the trap.

The operation of the apparatus for the preparation of keten (1) is as follows. Acetone is placed in the flask which is heated in an electric mantle until the liquid gently refluxes from the condenser M. After a few minutes the U-tube attached to K will fill with acetone and this provides a liquid trap which ensures that all the acetone vapour passes through H. After heating under reflux for a further five minutes to drive air from the chamber H, the filament current is switched on so that the filament C attains a dull red glow (700–750 °C). Keten is formed almost immediately and is allowed to pass directly via the three-way tap shown into the reaction flask (2). The apparatus requires little attention apart from occasionally removing the condensed acetone from the trap O. At the end of the run, the following operations must be carried out rapidly in this order: (i) remove the source of heat from the flask, (ii) turn off the filament current and (iii) open the stop-cock on O.

Notes.   (1) *Keten is a poisonous gas having a toxicity comparable with phosgene*; leaks from the apparatus, which must be contained in a fume cupboard, are recognised by a pungent odour resembling acetic anhydride. For this reason it is customary to attach a second receiver flask containing a compound which readily reacts with keten, e.g., aniline, to the other arm of the three-way stop-cock. In this way when the reaction in the main flask is complete the keten gas may be diverted to the second receiver flask whilst the apparatus is switched off and allowed to cool. Escape to the atmosphere of keten is thereby avoided.

(2) The yield of keten may be determined by weighing the acetanilide formed by passing keten through excess for a measured period of time.

For pyrolyses which proceed best in a heated tube, which may with advantage be packed with glass beads or with porcelain chips to increase the heated surface area, and for reactions which occur on the surface of a heated catalyst, the basic apparatus shown in Fig. I,72(c) is often suitable. A pressure-equalising funnel A allows the reactant or reactants to be dropped at a constant rate into the combustion tube B, which is about 100 cm long and made of Pyrex tube (23 mm outer diameter and 16 mm inner diameter) and contains either glass beads or a suitable catalyst. Frequently it is necessary to conduct the reaction in the absence of air, and the adapter with T-connection C provides a means of displacing the air in the apparatus with nitrogen and of sweeping the products of the reaction through the combustion tube into the condenser and thence to the collection flask D. C also provides the means of introducing a gaseous reactant into the combustion tube should this be required. The Drechsel bottles E may serve a number of different purposes according to the nature of the experiment; for example, (a) they may monitor the flow of nitrogen gas from the inlet C, (b) they may be used to check the complete displacement of air from the apparatus by using Fieser's solution (Section II,2,43) and (c) they may be used to absorb unwanted gaseous products to prevent contamination of the atmosphere. The furnace F may be a commercial

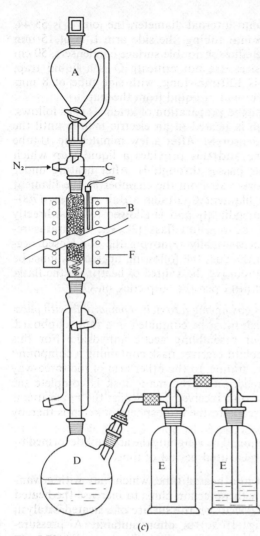

N₂ →

Fig. I,72(c)

cylindrical furnace about 70 cm in length in which B may be supported with
the aid of glass wool or an asbestos plug (**CARE:** see p.22); it is excellent practice however and certainly much cheaper to construct it from simple materials.

    Construction of the electric tube furnace. Obtain a thin-walled iron tube,
78 cm long and 2.8 cm in internal diameter, and securely wrap it with asbestos
cloth (*ca.* 2 mm thickness). Wind the central 70 cm evenly with 10 metres of
nichrome wire of No. 30 S.W.G., and cover it with two thicknesses of asbestos
cloth held in position by copper wire ligatures. Complete the insulation by
wrapping a further two thicknesses of the asbestos cloth round the tube. Attach
the two ends of the nichrome wire to two insulated connectors and then to a
power point (e.g., 230–240 volt mains). The temperature inside a Pyrex glass
tube placed close to the walls of the furnace is about 350 °C: some adjust-

ment of temperature may be achieved by removing one of the outer coverings of asbestos cloth. It is, however, preferable to connect the two ends of the nichrome wire to the mains through a small variable resistance (e.g., 25 ohms carrying 2.5 amperes),* or to a variable ratio transformer (e.g., Variac†), or to an energy regulator‡. The combustion tube may then be placed in the furnace and the temperature determined in various positions in the tube either with a long nitrogen-filled thermometer or preferably with a thermocouple; the temperature will be found to be constant over the central 40–50 cm of the tube. A graph may be constructed with temperatures as ordinates and instrument (ammeter, Variac or energy regulator unit) readings as abscissae; such a calibration is well worth while as it considerably extends the utility of the furnace.

General procedure for operation. The 100-cm combustion tube B is packed with glass beads or catalyst held in position with plugs of purified glass wool (1), and inserted centrally into the furnace. After fitting the remaining apparatus components, the air in the apparatus is displaced with nitrogen, the furnace is allowed to heat to the required temperature, and the combustion tube is allowed to reach temperature equilibrium. In those cases where the catalyst requires heat treatment (as in the case of manganese(II) carbonate on pumice) adequate time must be allowed for the activation process to reach completion. The reactant is then allowed to drop into the combustion tube (the flow of nitrogen must be stopped if the rate of formation of gaseous products is to be observed) at a rate of about one drop every 3–4 seconds. The apparatus subsequently requires little attention and the passage of say 750 ml of reactant requires a period of addition of between 48 and 72 hours.

The isolation and purification of the reaction products which collect in D will of course be determined by their chemical nature, and details are given in the appropriate Sections.

Note. (1) Purified glass wool is prepared by boiling a little glass wool with concentrated nitric acid for 30 minutes, washing thoroughly with distilled water to remove adhering acid, and drying in an oven.

In some cases a single pass of reactants through the combustion tube gives only a low conversion into products and hence it is necessary to provide a means of recycling unreacted material whilst continuously removing product to avoid its decomposition. Some adaption to the above apparatus is then required and one such assembly is shown in Fig. 1,72(*d*). The reactant is heated in the flask A and the vapour passes upwards through the combustion tube B; the reactant and products are swept by a slow nitrogen flow into the Vigreux fractionating column. Unreacted material is collected in the Dean and Stark side tube, and returned as appropriate to the flask A. The Vigreux column is surmounted by a still head fitted with a condenser and collector flask.

I,17,4. Ozonolysis The cleavage of a carbon–carbon double bond by oxidation with ozone (as ozonised oxygen) followed by hydrolysis to yield carbonyl compounds is a reaction sequence of considerable importance. This reaction,

---

* An ammeter, reading to 1.5–2.0 amperes, should be placed in the circuit.

† Manufactured by the Zenith Electric Co. Ltd, use Type V5H-M.

‡ For example those manufactured by Satchell-Sunvic Ltd, Sunvic energy control unit, type ERX/L.

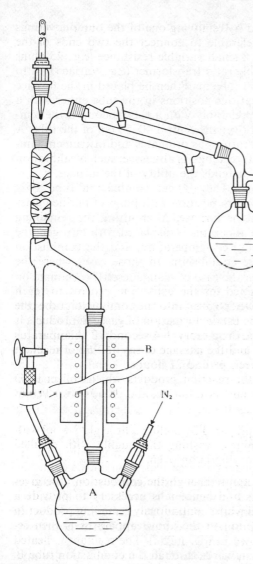

Fig. I,72(d)

for example, can be used for the determination of the structure of an unsaturated compound by identification of the carbonyl fragments, or it may be used in suitable cases for the preparation of aldehydes or ketones which are not readily available by other means. Ozonisation of carbon–carbon multiple bonds, unlike oxidation with excess potassium permanganate or chromic acid which for example will also oxidise primary and secondary alcohols, is a relatively specific process.

When ozonised oxygen is passed through a solution of an ethylenic compound in an inert solvent (e.g., methanol, ethyl acetate, glacial acetic acid, chloroform or hexane) preferably at a low temperature (-20 to $-30\,°C$), ozone adds on readily and quantitatively to the double bond to give an ozonide (I):

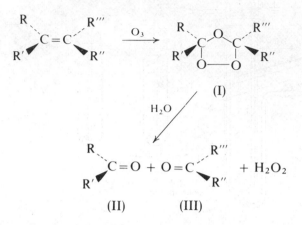

Excess ozone should be avoided since further oxidation may occur. Conveniently a wash bottle charged with potassium iodide solution and acetic acid is attached to the outlet of the reaction vessel; the completion of ozonolysis is indicated by the sudden extensive separation of iodine. Alternatively the flow of ozonised oxygen may be interrupted, and a drop of the solution removed and placed upon a white porcelain tile and allowed to mix with a drop of tetranitromethane when the production of a yellow colouration is indicative of the presence of unreacted alkene.

The ozonides are usually not isolated since they are generally viscid oils or glasses, *frequently with violently explosive properties, particularly upon warming.* They can however be smoothly converted into carbonyl compounds (II) and (III) by hydrolysis, preferably under reducing conditions (e.g., zinc dust and aqueous acetic acid), or by hydrogenation over platinum on calcium carbonate. These conditions prevent the further oxidation (by the hydrogen peroxide formed during hydrolysis) of any aldehydic products to the corresponding carboxylic acids; if the acids are in fact the desired products, the decomposition of the ozonides may be carried out oxidatively, e.g., in the presence of hydrogen peroxide or potassium permanganate.

A simple semimicro *laboratory ozoniser* is illustrated in Fig. I,73; this gives reasonably satisfactory results for small quantities (2–4 g) of organic compounds. It consists of a wash bottle or small bubbler A to indicate the rate of flow of the oxygen from a cylinder fitted with a reducing valve, a Berthelot tube B for the generation of ozone, a vessel C to hold the solution of the compound to be ozonised, and a flask D containing 5 per cent potassium iodide in aqueous acetic acid. Since ozone is markedly *toxic* and is also a *lung irritant* the outlet from D should be led by means of PVC tubing to the extraction vent of the fume cupboard. The Berthelot tube is charged with dilute copper(II) sulphate solution and is connected by a copper or stainless steel wire (2–4 mm in diameter) to the high voltage terminal of a transformer (7500–10 000 volts). The second electrode is the earthed aluminium foil covering most of the exterior of the Berthelot tube and is bound with insulating tape. As a precaution all high-voltage connections are heavily insulated with rubber tape and the lead to

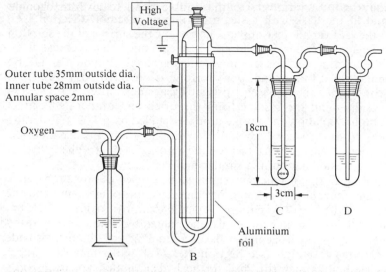

Outer tube 35mm outside dia.
Inner tube 28mm outside dia.
Annular space 2mm

Oxygen

18cm

3cm

C   D

Aluminium
foil

A   B

Fig. I,73

the top of the electrode is covered with PVC or equivalent tubing. The main dimensions are shown in the figure. The ozoniser should be constructed of soft soda-glass tubing (Pyrex glass is unsatisfactory): the glass should be thoroughly cleaned and the annular space through which the oxygen passes should be as uniform as possible. The complete apparatus should be placed in a fume cupboard (hood) behind a shatter-proof screen of laminated safety glass.

Commercially available ozonisers use either air or oxygen and are capable of yielding about 0.005 mol O_3/hour (with air) or 0.01 mol O_3/hour (with oxygen). A commercial ozoniser may be incorporated, by means of PVC tubing, in place of the Berthelot tube on an apparatus assembly similar to that above.

Should it be necessary to estimate the amount of ozone produced by the ozone generator, the vessel C is charged with 50 ml of a 5 per cent solution of potassium iodide in aqueous acetic acid (1:1 v/v) and the ozonised oxygen allowed to pass for a set period, say 1 hour, at a steady and measured flow rate (e.g., bubbles/second counted by means of the bubbler A). The iodine which is liberated is determined by washing the contents of C into a conical flask and titrating the liberated iodine with 0.1 M-sodium thiosulphate solution; the yield of ozone may be calculated in mol/hour for the particular flow rate selected.

$$O_3 + 2I^{\ominus} + 2H_3O^{\oplus} \longrightarrow O_2 + I_2 + 3H_2O$$
$$I_2 + 2S_2O_3{}^{2\ominus} \longrightarrow 2I^{\ominus} + S_4O_6{}^{2\ominus}$$

On completion of the ozonisation of the olefinic material the method of decomposition of the resulting ozonide and the subsequent work-up procedure will be determined by whether the object of the reaction is preparative in nature, or whether it is required to identify the carbonyl compounds produced as an aid to the determination of the structure of the alkene.

In the latter case the following procedure is recommended. Wash the contents of the reaction vessel into a round-bottomed flask, add zinc dust and acetic acid and fit the flask with a steam distillation assembly (Section I,25) ensuring that the receiver adapter outlet is just below the surface of the solution of 2,4-dinitrophenylhydrazine reagent (p. 210) contained in a conical flask. Steam distil the solution and collect the volatile carbonyl compounds until no further precipitate of 2,4-dinitrophenylhydrazone is observed with fresh portions of reagent. Extract the combined distillate–reagent solutions with methylene chloride and submit the residue from the dried and evaporated extract to chromatographic isolation and purification as detailed on p. 209. Extract the residual liquors from the steam distillation with ether or methylene chloride, wash, dry, evaporate and convert the residue into a 2,4-dinitrophenylhydrazone derivative for examination in a similar manner.

When ozonolysis of the olefinic material is to be carried out for preparative purposes, the initial ozonisation should be conducted in dry methanol and the ozonide decomposed by hydrogenation over palladium hydroxide on calcium carbonate in the following manner. Rinse the contents of the reaction vessel with methanol into the hydrogenation flask containing palladium hydroxide on calcium carbonate catalyst (see Section II,2,44,e) and a magnetic stirrer follower (Section I,14), attach the flask to the hydrogenation apparatus (Fig. I,70(a)) and immerse the hydrogenation flask in an ice bath placed upon a magnetic stirrer plate. This cooling is essential to avoid an undue rise in temperature of the solution during hydrogenation, which is exothermic, since this may lead to the alternative formation of a carboxylic acid at the expense of aldehyde. Charge the apparatus with hydrogen and hydrogenate the solution as detailed in Section I,17,1. Emphasis should be placed upon the importance of placing the hydrogenation vessel behind appropriate shatter proof screens. When hydrogenation is complete, filter off the catalyst, remove the solvent on a rotary evaporator, and purify the product by crystallisation or distillation as appropriate.

I,17,5. Organic photochemistry   Although it has long been recognised that chemical change can be effected by means of ultraviolet (200–400 nm) and visible light (400–750 nm), studies in this area of chemistry have until quite recently been largely the province of the physical chemist. However, a rapidly increasing number of investigations since 1960 have shown that many novel and synthetically useful reactions including dimerisation, cycloaddition, rearrangement, oxidation, reduction, substitution and elimination may be consequent upon the absorption of light by organic molecules. Many chemical transformations can be effected which would otherwise require a large number of steps by standard chemical procedures. This progress in synthetic organic photochemistry has been aided by the commercial development of suitable light sources, by advances in procedures available for the separation and identification of the components of mixtures and, not least, by the realisation that many photochemical reactions occur quite cleanly to give good yields of the desired product. In addition the photochemical experiments can often be carried out much more simply than many standard chemical reactions.

An understanding of organic photochemistry requires a knowledge of the energy transitions which a molecule may undergo following irradiation with electromagnetic radiation. Some consideration of these energy transitions are given in Section I,39 in relation to the use of u.v. and i.r. spectroscopy in qualitative and quantitative analysis. The following account is intended to provide sufficient theoretical background to allow some appreciation of photochemical reactions, of which illustrative practical examples are given in Sections **III,30, V,16**. A detailed treatment of photochemical processes may be found in a number of recent books on photochemistry; some examples are given in Ref. 22, in particular the comprehensive monograph of Calvert and Pitts (Ref. 23) deals extensively with both the theoretical and practical aspects of the subject.

The total energy of a molecule is the sum of its electronic, vibrational, rotational and translational energies. Whereas the translation energy increases continuously with the temperature of the system, the first three energy states are quantised and excitation to higher energy levels requires the absorption of discrete amounts of energy (quanta) which can be supplied by electromagnetic radiation. The amount of energy associated with such radiation depends on its wavelength, the longer the wavelength the smaller the energy (Section **I,39**, p. 245). Excitation of a molecule to higher rotational and vibrational energy levels can thus occur on absorption of radiation in the far infrared, and in the infrared regions of the spectrum respectively (i.e., the low energy portion of the spectrum), and is associated with relatively small increases in the energy of the molecule ($\sim 0.5 - 42 \, \text{kJ mol}^{-1}$). Absorption of ultraviolet (200–400 nm) and visible (400–750 nm) radiation by a molecule is associated with an increase in energy in the range 600–160 kJ mol^{-1} and results in the excitation of its valence electrons to higher energy levels. The energy associated with a photon of radiation in the ultraviolet region is of the same order as the bond energies of many of the bonds present in organic molecules (e.g., C–H, 410 kJ mol^{-1}). It is thus not surprising that absorption of light in this region can result in chemical reactions and that the reactions of molecules in such electronically excited states are often quite novel.

Excitation of a molecule to a higher energy level involves promotion of an electron from a bonding (σ or π) or a non-bonding (n) orbital to an antibonding (σ^* or π^*) orbital. Four types of transitions are possible and the energy associated with each (which can be represented diagrammatically in Fig. I,74(a)) decreases in the order $\sigma \rightarrow \sigma^* > n \rightarrow \sigma^* > \pi \rightarrow \pi^* \approx n \rightarrow \pi^*$.

The $\sigma \rightarrow \sigma^*$ and $n \rightarrow \sigma^*$ transitions are of little significance in organic photochemical synthesis as they occur in the far ultraviolet (< 200 nm), a region which is not readily accessible practically owing to the absorption of radiation in this region by oxygen. The $\pi \rightarrow \pi^*$ and $n \rightarrow \pi^*$ transitions occur in the ultraviolet region and are responsible for the vast majority of useful photochemical reactions; in simple ketones the $n \rightarrow \pi^*$ transition occurs at ~ 270 nm with an associated energy of 443.1 kJmol^{-1}, and the $\pi \rightarrow \pi^*$ transition of butadiene occurs at 217 nm with an associated energy of 551.5 kJmol. Whilst the overall magnitude of the energy required to effect these electron transitions explains why bonds may be broken during irradiation, a more detailed consideration of these electronically excited states is necessary to understand the various possible ways in which energy absorbed may be dissipated. In particular the importance of the concept of singlet and triplet states must be considered.

Most organic molecules have an even number of electrons and these are paired (spins in opposite direction); energy states with paired electrons are called singlet (*S*) states (no net electronic magnetic moment and hence only *one* possible energy state in a magnetic field). The ground state of a molecule is referred to as S_0 and the higher excited singlet states as S_1, S_2, S_3, etc. Inversion of the spin of one electron results in the formation of a different electronic state having two unpaired electrons (same spin); this is referred to as a triplet (T) state (a net electronic magnetic moment and hence *three* possible energy states in a magnetic field). For each possible excited singlet state (Sx) there is a corresponding lower energy triplet state (Tx). According to quantum mechanical theory, transitions between states of the same multiplicity are *allowed* whereas transitions between states of different multiplicity are formally *forbidden*. The ground state and lowest singlet and triplet states are represented schematically in Fig. I,74(b).

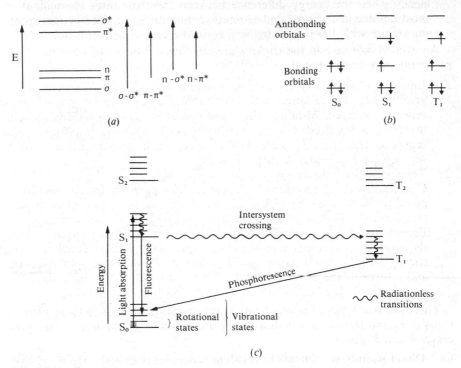

Fig. I,74(a–c)

Following the absorption of radiation and the promotion of an electron (π or n) to the vibrationally excited singlet state S_1† (which occurs very rapidly in $\sim 10^{-15}$ sec), the dissipation of this energy may take place in a variety of ways itemised below, some of which may be represented diagrammatically by means of a Jablonski diagram (Fig. I,74(c)).

1. Initially excess vibrational energy is rapidly lost by radiationless processes, such as collision with solvent molecules, to give the thermally equilibrated

---

† Absorption of sufficient energy can also cause excitation to higher singlet states such as S_2; however, these generally decay very rapidly by radiationless transitions to the highest vibrationally excited state S_1. This is possible since the S_0 and S_1 states are separated by the greatest energy gap, whereas the higher energy states have progressively smaller energy differences and overlapping potential energy surfaces which allow radiationless loss of energy. In certain cases the energy input may of course be sufficient to cause immediate bond dissociation.

excited singlet molecule S_1. This has a short lifetime ($\sim 10^{-8}$ sec) and may then lose its energy by any of the processes 2–5 below.

2. Emission of light from the excited molecule may occur which then returns to the ground state, i.e., fluorescence is observed.
3. Thermal dissipation of the energy to surrounding molecules may occur; this is a radiationless process (*internal conversion*).
4. Chemical reaction can occur.
5. Conversion to the lower energy triplet (T_1) by spin inversion (*intersystem crossing*); although formally forbidden this can occur with very high efficiency when the energy difference between the two states is small. It is most notable in carbonyl and aromatic compounds (e.g., intersystem crossing occurs with 100 per cent efficiency in the case of benzophenone).

An excited molecule in the triplet state also has a number of ways in which its energy may be dissipated.

1′. Emission of light from the excited molecule may occur with return to the ground state, but at longer wavelength than fluorescence, i.e., phosphorescence is observed. Although this transition is formally forbidden, as spin inversion is involved, it does eventually occur with the important consequence that the T_1 state has a very much longer lifetime (10^{-6} sec → several seconds) than the S_1 state.
2′. The species may decay by internal conversion.
3′. Chemical reaction may occur—the longer lifetime of the triplet state compared to the S_1 state means that chemical reaction is a much more important feature, and is of prime importance in synthetic photochemistry.
4′. Energy transfer to a neighbouring (different) molecule may occur so that the acceptor molecule is promoted to a triplet state of either equal or lower energy than the donor triplet species, which itself undergoes spin inversion and returns to the ground state S_0. Such a transfer will occur only if the acceptor molecule has an available lower energy excited level.

There are two types of photochemical processes which lead to these various transitions and thence to a realisation of the synthetic possibilities of the processes 4 and 3′ above.

(a) **Direct photolysis**—where the incident radiation is directly absorbed by a substrate X, which is thus promoted to the excited singlet state X* which then loses its energy by the processes outlined above.

$$X \xrightarrow{h\nu} X^* \longrightarrow \text{product(s) etc.}$$

(b) **Indirect or sensitised photolysis**—where a photo-excited donor molecule (D^*) in the singlet or triplet state, referred to as a *sensitiser* and produced by absorption of the incident radiation, transfers its energy to the substrate X which is thereby promoted to an excited state (e.g., see 4′ above). In this process the sensitiser returns to the ground state, is chemically unchanged, and may be further excited by incident radiation.

$$D \xrightarrow{h\nu} D^*$$
$$D^* + X \longrightarrow D + X^*$$
$$X^* \longrightarrow \text{product(s) etc.}$$

Many compounds (e.g., alkenes) do not undergo intersystem crossing from the singlet state to the synthetically more useful triplet state as the energy difference between the two states is large. However, provided that the energy of the triplet state of the sensitiser molecule is about 20.9 kJmol^{-1} greater than that of the triplet state of the substrate, energy may be transferred to provide excited molecules in the triplet state which may then undergo chemical reaction. The procedure is also useful for populating triplet states of a compound whose singlet state is in an inaccessible part of the ultraviolet spectrum (i.e., <200 nm).* This sensitising process is represented schematically in Fig. I,74(d).

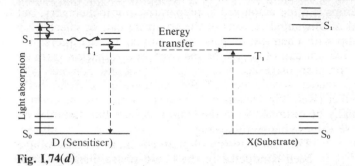

Fig. I,74(d)

Photochemical apparatus and experimental procedures.   Prior to a study of the possible photochemical reaction of a compound, its spectrum in the ultraviolet or visible region must be determined in order that a light source emitting the appropriate wavelength of radiant energy may be selected. In the case of a sensitised photochemical reaction the spectrum of the sensitiser should be determined.

Light sources.   In early photochemical work sunlight was the source of radiant energy, and it still remains a useful and cheap source in favourable climates for reactions requiring irradiation at wavelengths down to 320 nm. Ordinary high wattage tungsten lamps may also be used for reactions proceeding under the influence of visible light. However, photochemical reactions on a preparative scale are most often effected with radiant energy of wavelength 220–380 nm, and for these purposes mercury arc lamps are used almost exclusively. There are essentially three types available.

1. *Low-pressure mercury arc lamps*, which operate at a mercury vapour pressure of about 10^{-3} mmHg and emit mainly at 254 nm and 184 nm (about 80–95 per cent of this radiation is produced at 254 nm). Low-pressure mercury arc lamps with phosphor coatings on the interior walls are available which give maximum light emission at longer wavelengths over broad selected regions, e.g., centering at 300 nm or 350 nm.

2. *Medium-pressure mercury arc lamps*, which operate at internal pressures of from 1 to 10 atmospheres and emit radiation over the region 200–1400 nm, with particularly intense emission at 313 nm, 366 nm, 435.8 nm and 546.1 nm.

3. *High-pressure mercury arc lamps*, which operate at internal pressures of from 100 to several hundred atmospheres and give almost continuous

---

* Triplet and singlet excited state energy values for a large number of substances have recently been collected, see Ref. 24.

emission over the whole spectrum from about 220–1400 nm. The radiant energy is particularly rich in visible light.

Low-pressure mercury arc lamps operate at near room temperature. Much of the energy input of medium and especially high pressure lamps however is converted into heat so that these lamps must be cooled. The medium-pressure lamps have been used most extensively for synthetic work on account of their high light output, ease of handling and broad spectrum emission. The full arc spectrum of these lamps is often employed in preparative photochemistry, but if necessary removal of unwanted regions of the spectrum can be effected by surrounding the lamp with chemical or glass (e.g., Corex, Vycor) filters; unwanted light below 300 nm can of course be removed by irradiation through Pyrex glass. For a comprehensive account of light filters and commercially available lamps, the reader should consult the monograph of Schönberg, Schenck and Neumüller (Ref. 25). Details of the energy output at the emitted wavelengths can usually be obtained from the lamp suppliers but it should be appreciated that these values change on ageing.

Apparatus assembly.

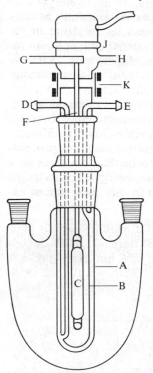

Fig. I,74(e)

The vast majority of photochemical syntheses have been conducted in the liquid phase, hence the apparatus assembly for a photochemical reaction must take into account the light transmission characteristics of the material from which the reaction vessels are made. Pyrex glass transmits most of the incident light above about 300 nm and may be used in the construction of apparatus for reactions which require light above this wavelength. Quartz vessels, transparent down to 200 nm, must be used for reactions which require light below 300 nm. Certain types of quartz allow transmission below 200 nm.

There are basically two assemblies for carrying out preparative photochemical syntheses. The light source may either be placed outside the vessel containing the solution of substrate (*external irradiation*), or it may be placed inside the vessel containing the solution (*internal irradiation*); in this latter case the solution is subjected to the full output of the lamp and therefore this is to be preferred. Both assemblies are available commercially;* a convenient apparatus for internal irradiation which has been used in these laboratories will be described and is shown diagrammatically in Fig. I,74(e). The apparatus consists of a three-necked Pyrex reaction flask of approximately 1 litre capacity with a central 45/50 socket to which is fitted the light source unit. The two side-necks of the flask allow the flushing of the reaction mixture with nitrogen and the attachment of a reflux con-

---

* Available from Hanovia Lamps Ltd for internal irradiation in reactor sizes of 1 and 10 litres; this manufacturer also supplies the 'Reading' photochemical reactor which is designed for external irradiation. Apparatus designed for preparative photochemistry is also available from Applied Photophysics Ltd, and the Southern New England Ultraviolet Co.

denser and/or a protective drying tube; the reactants are stirred magnetically. The light source unit has an outer quartz jacket A, an inner quartz jacket B and a mercury arc lamp C. Both jackets are made of fused quartz and each has a 75-mm-long zone made from pure synthetic quartz sited opposite the light-emitting region of the arc lamp; these zones allow about 80 per cent light transmission for every 1 mm thickness of quartz. Cooling water or air may be passed through the annular space between the two jackets via the inlet tube D and outlet tube E which are built into the inner jacket. Similar jackets are available in Pyrex for reactions which proceed on irradiation above 300 nm.

The arc lamp C is attached by metal clips to the hollow support tube F which also allows flushing of the lamp area by nitrogen via inlet G and outlet H. The support tube F, the electrical leads to the arc lamp (not shown) and terminals for the cable from the power control unit (also not shown) are all attached to the terminal block J which is fitted with an insulating cover. The terminal block is attached to the inner quartz jacket by means of a rubber sleeve K and two sleeve clamps, so that all the live parts are protected. The lamp and inner quartz jacket B can be withdrawn from the 40/38 socket attached to the outer quartz jacket A shown, and the entire unit can be withdrawn from the central socket of the reaction flask.

Two mercury arc lamps with associated power units are available: (a) a 2 watt low-pressure U-shaped lamp which emits mainly at 186 nm and 254 nm, and (b) a 100 watt medium-pressure straight-tube lamp emitting predominantly at 254 nm, 265 nm, 297 nm, 313 nm and 366 nm with intense emission also in the visible region; both lamps have synthetic quartz envelopes.

Water cooling via D and E is essential with the medium-pressure lamp whereas with the low-pressure lamp gas cooling is usually sufficient. The latter may be conveniently effected by drawing filtered air through the annular space between the jackets by connecting H to a water pump, which should be situated in the fume cupboard in order to vent any ozone formed. Light-filter solutions may replace the cooling water if it is required to remove any particular regions of light emission, and these must of course be circulated and cooled in an arrangement external to the apparatus.

Both lamps generate ozone and oxides of nitrogen in air, hence the inner lamp area should be flushed slowly with nitrogen via G and H as described above. It should be noted however that low-pressure lamps give maximum light output at a wall temperature of 40 °C so excessive cooling in the lamp region is to be avoided; this is particularly important when flushing with gas from a cylinder as the gas is likely to be cold due to expansion. In addition, it is of course essential that only dry gas be used because of the live connections in this region.

Reactor vessels may readily be constructed to cater for smaller volumes of reactants. Figure I,74(f) shows such a vessel having a capacity of about 110 ml. The outer quartz jacket A of the light source unit (Fig. I,74(e)) fits into the 45/40 centre socket, a reflux condenser and/or drying tube may be attached to one of the side 14/23 sockets while the other allows nitrogen flushing via the Teflon tube M attached to the drawn-out cone. Stirring is by means of the magnetic follower bar N. Even smaller volumes may be accommodated by taping suitably sized tubes to the side of the outer jacket A of the light source unit and surrounding the whole with aluminium foil.

A number of important aspects should be borne in mind when planning and executing a photochemical synthesis and these are outlined below.

1. *Safety.* Ultraviolet light is extremely dangerous to the eyes and also harmful to the skin so that proper precautions must be taken when conducting a photochemical experiment (see Section **I,3,F**). Ideally the apparatus assembly should be situated in the fume cupboard and aluminium foil wrapped around the reaction vessel (which also additionally serves as a light reflector) when using an internal irradiation arrangement; the whole should be surrounded by a light shield made from board. Rubber tubes for cooling water should be wired on and a suitable cut-out device for the lamp incorporated into the circuit. This latter precaution is essential in case the water supply should fail since the heat generated by a medium-pressure lamp could lead to fracture of the apparatus, loss of material and possibly fire when inflammable solvents are used. Suitable precautions should also be observed with regard to the electrical equipment.

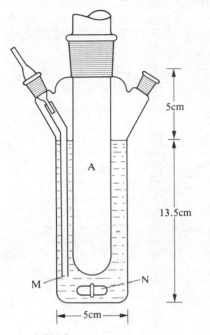

Fig. I,74(*f*)

2. *Degassing.* Dissolved oxygen should normally be removed by passing nitrogen or other inert gas through the reaction solution for about 0.5 hour prior to irradiation and a nitrogen atmosphere should be maintained throughout the experiment.

3. *Stirring.* Relatively concentrated solutions are often used in preparative photochemistry and in consequence most of the light is absorbed by a very thin layer of solution adjacent to the lamp. Some appreciation of this fact may be obtained from the following. Consider a 0.1 cm layer of a 0.01 M-solution of a compound having a molar absorptivity (ε) of 1000 l mol^{-1} cm^{-1}. From the Beer–Lambert Law (Section **I,39**), the light energy transmitted (I) through this layer is given by the expression:

$$ I = \frac{I_0}{10^{\varepsilon cl}} = \frac{I_0}{10^{1000 \times 0.01 \times 0.1}} = \frac{I_0}{10}, $$

where I_0 is the intensity of the incident radiation, c is the molar concentration of the compound and l the length of the absorbing solution in centimetres. Thus it can be seen that 90 per cent of the light is absorbed in this 0.1 cm layer of solution; vigorous stirring is therefore essential to change this layer continually. For volumes up to 1 litre this can usually be effected adequately by magnetic stirring. Stirring may also be effected by means of nitrogen introduced through a medium porosity fritted glass plate sealed to the bottom of the gas inlet tube; this is particularly useful with small volumes of reactants. It is important, using internal irradiation with equipment such as described above, that the solution level should be above the light-emitting region of the lamp.

4. *Time of reaction.* A measure of the efficiency of a photochemical

synthesis is given by the quantum yield (φ) for the product, which is defined as:

$$\varphi = \frac{\text{number of molecules of product formed}}{\text{number of quanta absorbed}}$$

Quantum yields for reactions which proceed by way of a free radical chain mechanism may be as high as many thousands, whereas reactions which do not occur through such a mechanism have quantum yields in the region 0–1. Thus a knowledge of φ, and the number of useful quanta emitted per second by the lamp, could allow calculation of the approximate time of photolysis. However, since these values are often not known it is usual to monitor the progress of the reaction, e.g., by using a suitable chromatographic procedure (Section **I,31**), or by noting the disappearance of a characteristic absorption band in the u.v. spectrum of the starting material.

Depending on the concentration of the substrate(s) the reaction time required may be of the order of days rather than hours. This can often of course be shortened by irradiating with a higher-intensity lamp; 500-watt medium-pressure lamps are commonly used.

To ensure as rapid a reaction as possible it is essential that all light-transmitting surfaces are kept absolutely clean, and handling of the quartz envelope of the mercury lamp should be avoided as finger marks will gradually 'burn' into the surface thereby reducing transmission. A thin film of polymer may occasionally be deposited on the outer surface of A adjacent to the lamp and this should be removed, otherwise light input to the reaction medium will be severely reduced. An ingenious apparatus has been described (Ref. 26) using fluorescent tubes as light source, which allows the radiation to fall directly on to a moving thin film of the reaction solution thus avoiding this difficulty and also obviating the need for expensive quartz apparatus.

Irradiation of a solid substrate often gives a single product, whereas in solution a number of isomers may be formed (e.g., in dimerisation reactions); this is presumably due to the ordered arrangement of the substrate molecules in the solid phase. Such reactions may be carried out in a variety of ways. When irradiation by a mercury arc lamp is necessary, the material may be deposited as a thin film on the inside wall of a container (such as a large glass gas jar) by evaporating its concentrated solution in a volatile solvent; the lamp unit may then be inserted into the jar. Alternatively the finely powdered material may be placed in petri dishes under an arc lamp and stirred occasionally to provide a fresh surface for irradiation. When sunlight is a suitable source of radiant energy, exposure of the powdered material contained in large petri dishes may be employed, or a round-bottomed flask may be coated on the inside with a thin layer of material by the evaporation procedure and exposed to the sun, occasionally rotating the flask to ensure even exposure.

Details of a very large number of photochemical reactions are given in Ref. 25 and the first part of a series of books dealing exclusively with preparative photochemistry has been published (Ref. 27).

I,17,6. Electrolytic (anodic) syntheses (*the Kolbé reaction*).   Electro-organic chemistry is the study of the oxidation and reduction of organic molecules and ions, dissolved in a suitable solvent, at an anode and cathode respectively in an electrolysis cell, and the subsequent reactions of the species so formed. The first experiment of this type was reported in 1849 by Kolbé, who described the electrolysis of an aqueous solution of a carboxylate salt and the isolation of a hydrocarbon. The initial step involves an anodic oxidation of the carboxylate anion to a radical which then dimerises to the alkane.

$$R \cdot CO_2 \overset{\ominus}{\longrightarrow} \overset{-e}{\longrightarrow} R \cdot + CO_2$$
$$2R \cdot \longrightarrow R - R$$

Following the study of the simple coupling of radicals derived from the salt of a single carboxylic acid, it was found that the electrolysis of a mixture of carboxylate anions or of the salts of half esters of dicarboxylic acids increased the synthetic value of the method. This arises from the possibility of the formation of symmetrical and unsymmetrical coupled products of the derived radicals. These anodic syntheses are illustrated in the synthesis of hexacosane (Section **III,6**), sebacic acid (decanedioic acid), octadecanedioic acid and myristic acid (tetradecanoic acid), in Section **III,131**).

The electrolysis cell used for these conversions may be readily constructed in the laboratory (Fig. I,75) and provides a simple introduction to the technique

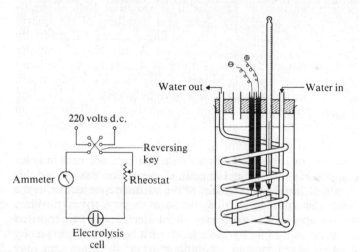

Fig. I,75

of electro-organic chemistry which is of increasing importance in organic synthesis. The cell consists of a cylindrical Pyrex glass vessel (16 cm × 7 cm) fitted with an internal cooling coil so that the temperature of the electrolyte may be controlled; a working temperature range of 30 to 35 °C is usually satisfactory, but excessive cooling may cause some of the product to crystallise. With a smaller electrolytic cell constructed from a large boiling tube the internal coil may be omitted and cooling effected by external means only. In either case no stirring is required since the evolution of carbon dioxide during electrolysis provides adequate agitation of the electrolyte. The electrodes consist of two platinum plates (4 cm × 2.5 cm × 0.3 mm) set about 2 mm apart to each of which is attached a platinum wire sealed into a glass tube containing mercury by which contact with the d.c. electrical circuit is effected. The carboxylic acid is dissolved in methanol (containing sufficient sodium methoxide to convert about 2 per cent of the added acids into the carboxylate salts), and the solution electrolysed with a current of between 1 and 2 amperes until the electrolyte becomes slightly alkaline. The length of time required to complete the electrolysis may be roughly estimated for carboxylic acids and half acid methyl esters to be between 20 and 50 per cent in excess of the theoretical value calculated from the number of Faradays required in relation to the amount of acid employed, i.e., from the equation $It/96\,500 = \text{mol } R \cdot CO_2^{\ominus}$, where I is the current in amperes and t the time in seconds. It is often advantageous to occasionally reverse the direction of the current to dislodge insoluble deposits on one or other of the electrodes; if this is not done the current will be observed to drop prematurely, leading to a slowing down of the electrolysis.

Some general considerations which require variations in the simple electrolysis cell construction described above to meet the requirements for electrolytic oxidations and reductions of a wide range of organic compounds may be briefly summarised, but attention is drawn to the very extensive surveys given in Refs. 28–31.

The first general comment relates to the solvent system. In those cases where the electrolysis substrate does not exist in an aqueous-ethanolic or methanolic solution in a suitable ionic form, it is necessary to provide a solvent system of low electrical resistance which will dissolve the substrate, and also a supporting electrolyte whose function is to carry the current between the electrodes. Examples of such solvents are dioxan, glyme, acetonitrile, dimethylformamide and dimethyl sulphoxide; supporting electrolytes include the alkali metal halides and perchlorates, and the alkylammonium salts (e.g., perchlorates, tetrafluoroborates, toluene-p-sulphonates). With these electrolysis substrates, mass transfer to the electrode surface is effected by efficient stirring.

Although not relevant to the Kolbé reaction, a second comment relates to the necessity of ensuring that products formed at the working electrode (either anode or cathode) do not migrate to, and react at, the counter electrode or indeed react with compounds formed at the counter electrode. Recent cell designs therefore incorporate anode and cathode compartments separated by a rigid porous membrane. The difficulties of so selecting a glass frit of porosity sufficient to ensure transport of current, but not of electrolysis substrate or products, has in part been overcome by the use of a frit which supports a gel membrane (Ref. 32). This gel is formed by impregnating the frit with a hot solution of methylcellulose in dimethylformamide and allowing to cool. These membranes appear to be stable for the length of time required for an electrolysis when acetonitrile and 1,2-dimethoxyethane are used as solvent systems.

The electrode material frequently has crucial consequences on the course of electrolytic oxidation and reduction processes. Although platinum is the commonest electrode material, carbon, mercury and copper have all been used in numerous specific conversions. Selection of electrode material should therefore be based upon previously established characteristics when new conversions are to be studied.

I,17,7. Liquid ammonia techniques Many important synthetic organic reactions are carried out in liquid ammonia (b.p. $-33\,°C$); this is a good solvent for many organic compounds having a range of polarities, and also for the metals lithium, potassium, sodium and calcium.

Solutions of these metals in liquid ammonia effect (i) the reduction of a range of functional groups such as carbonyl and acetylenic and also conjugated and aromatic systems, and (ii) cleavage of benzyl and allyl ethers and thioethers. These reactions are usually carried out by the general procedure of adding the metal to a solution of the substrate in liquid ammonia to which dry methanol or ethanol or t-butanol has been added to provide a ready proton source (alcohols are more acidic than ammonia) (see Ref. 33).

A second principal use of liquid ammonia involves forming a suspension of an alkali metal amide ($LiNH_2$, KNH_2 or $NaNH_2$) by adding the appropriate metal to liquid ammonia containing a trace of iron(III) ions (added as iron(III) nitrate) as a catalyst.

$$2Na + 2NH_3 \longrightarrow 2NaNH_2 + H_2$$

The amide ions are powerful bases and may be used (i) to dehydrohalogenate halo-compounds to alkenes and alkynes, and (ii) to generate reactive anions from terminal acetylenes, and compounds having reactive α-hydrogens (e.g., carbonyl compounds, nitriles, 2-alkylpyridines, etc.); these anions may then be used in a variety of synthetic procedures, e.g., alkylations, reactions with carbonyl components, etc. A further use of the metal amides in liquid ammonia is the formation of other important bases such as sodium triphenylmethide (from sodamide and triphenylmethane).

Although these amides are frequently used as a suspension in liquid ammonia, an inert co-solvent (such as ether or tetrahydrofuran) may be added should the organic substrate not be readily soluble in liquid ammonia. Alternatively after amide formation the liquid ammonia may be allowed to evaporate completely during simultaneous addition of the inert solvent; subsequently the organic substrate may be added in the same solvent to the alkali amide.

Liquid ammonia is supplied in cylinders (Section II,2,4) which incorporate a simple tap valve with a screw-thread wide-bore outlet. Although a special gas-reducing valve may be obtained from the suppliers, for most purposes it is adequate to screw on to the outlet a wide-bore metal-tube adaptor (Fig. I,76(a)) to which may be fitted wide-bore rubber or stout polyethylene tubing, secured by copper wire. When the cylinder is upright only ammonia gas will be released when the valve is opened. To obtain liquid ammonia, the cylinder needs to be supported valve downwards at an angle of about 60° from the vertical position, with the outlet valve above the level of the vessel into which the liquid ammonia is to be discharged. The cylinder should be securely supported in a purpose-designed scaffolding (an example is shown in Fig. I,76(b)) preferably located within a fume cupboard.*

The valve tap on the cylinder is very tightly closed; it is best released by attaching the valve lever and gently tapping the lever end with a hammer in short sharp blows with gradually increasing force until the ammonia starts to escape. This method is easier and is to be preferred to continuously applied hand pressure.

The rubber or plastic outlet tube should be depression-free and lead via an

---

* It has been found convenient in these laboratories to locate the cylinder outside the laboratory building along the wall adjoining the fume cupboard. The inlet tube is led through the hole in the wall directly into the fume cupboard.

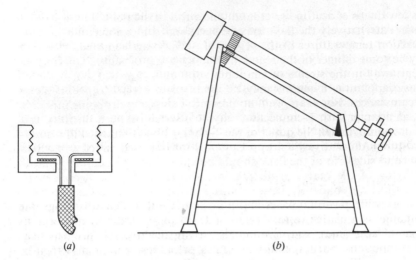

Fig. I,76(a–b)

adapter into the receiver flask (see below) which must be sited in the fume
cupboard, and it is advised that this delivery tube be additionally supported by
means of a retort stand and clamp. This latter precaution is necessary since the
initial force of ammonia release may cause it to flex with considerable thrust
which may lead to apparatus damage.

Until the cylinder valve, valve outlet and adapter, and rubber or plastic
tubing have cooled to $-33\ °C$, only ammonia gas emerges. Eventually however
liquid ammonia will flow into the flask and when this has cooled to $-33\ °C$ it
may be filled with the appropriate quantity; this may be gauged by pre-
marking the vessel to the volume of liquid required. The flask selected should
be of such size that it is only half full. This cooling effect leads to the final liquid
ammonia sample containing traces of moisture (between 0.1 and 0.5 g/l), but
this impurity is not harmful in the subsequent reactions. Such traces of moisture may be detected and removed by the addition of a little sodium metal with
stirring; a rapid disappearance of the initial blue colour indicates some water
present whilst a persistent blue colour indicates its absence. The laborious and
often wasteful procedure of redistilling liquid ammonia using an acetone-
Cardice charged condenser is not usually worthwhile.

The simplest apparatus assembly is shown in Fig. I,76(c). It consists of a
three-necked round-bottomed flask of appropriate size equipped with a
mechanical stirrer unit fitted with a Hershberg wire stirrer or a Teflon or a
glass stirrer sited in the central neck. A pressure-equalising dropping funnel
fitted with a soda-lime-filled guard-tube (not calcium chloride) is placed in one
side-neck. The second side-neck accommodates the inlet adapter for the liquid
ammonia supply, and this may subsequently be closed with a glass stopper; the
addition of metals, other solid reagents, or gaseous reagents such as acetylene
is also made through this side-neck. The flask is surrounded by a box containing cork chips, vermiculite or other insulating material; the outside of the
flask rapidly acquires a coating of ice when being charged with liquid ammonia
and this provides additional insulation. The flask contents may be viewed by
the removal from time to time of a section of the external ice coating by

pouring a few drops of acetone or ethanol on to the outside of the flask from a wash bottle. Alternatively the flask may be surrounded by an acetone-Cardice bath if reaction temperatures in the region of $-78\,^{\circ}C$ are required, or where there may be some danger of the reaction product being of such a volatility as to be swept away in the stream of ammonia vapour.

This basic apparatus assembly may be modified to meet a range of alternative circumstances. For example, if no pressure-equalising dropping funnel is available, an adapter with T-connection may be inserted between the flask and a normal dropping funnel; the outlet of the T-connection is attached to a soda-lime guard-tube. Alternatively a dropping funnel may be used directly, in which case the second side arm of the flask should be closed with a soda-lime guard-tube; if stirring of the reaction mixture is vigorous it may be necessary to interpose a short air condenser between this guard-tube and the flask.

A further modification may be required if the reagent to be added from the dropping funnel is sensitive to the action of ammonia vapour. In such a case not only should the atmosphere around the reagent be of some inert gas (e.g. nitrogen) but also the reagent should be added below the surface of the liquid ammonia. The modification shown in Fig. I,76(d) may be employed in this case. The adapter illustrated is a stirrer guide with gas inlet connection; the dropping funnel is of the pressure-equalising type; a normal dropping funnel requires

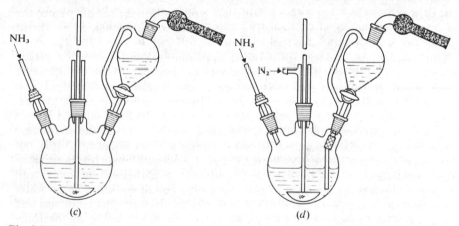

(c) (d)

Fig. I,76(c–d)

that the second side-neck of the flask permits the escape of gas to the atmosphere via a guard-tube.

The individual sequence of operations subsequent to half-filling the flask with liquid ammonia depends on the nature of the reactions involved. Specific details are given in Sections **III,19; III,21; III,42**. A general comment however is worth noting that during the course of the reaction ammonia gas is continuously escaping, and in all but prolonged reaction sequences it is advisable to ensure an initial adequate supply of liquid ammonia in the reaction flask to avoid the necessity of topping up the level. In some reactions it has been noted that considerable foaming occurs during the addition of reagents. This may be controlled to some extent by the addition of a little ether, or by raising the stirrer blade so that it agitates the surface of the liquid and hence assists in the rapid breakdown of foam. If these measures are unsuccessful, reagent addition should be stopped until foaming abates, and continued addition should then take place more slowly.

The simplest work-up procedure for non-volatile reaction products and products which are obtained in the form of alkali metal salts is to allow the ammonia to evaporate overnight through the guard-tube into a vented fume cupboard. Usually in these cases an inert solvent such as ether or tetrahydrofuran is added to the flask before evaporation commences. If time is important the flask may be placed on a warm water bath (*ca.* 45–50 °C) and the flask contents stirred; this should also be carried out in a fume cupboard. In either case a stream of nitrogen may be introduced into the flask if the product is sensitive to atmospheric oxidation. The subsequent treatment of the residue will depend upon the particular reaction and individual experiments should be consulted for typical isolation procedures.

With volatile products which may arise in reactions which do not lead to the initial formation of metal salts (i.e., dehydrohalogenations, alkylations leading to terminal acetylenes) the procedure recommended in Ref. 34 is clearly useful. Here the three-necked reaction flask is fitted with a stop-cock in one side-neck, a stopper in the second side-neck and a wide-bore glass tube (4–5 mm internal diameter) fitted by means of a rubber bung or screw-capped adapter to the central neck. This glass tube, which reaches to within a few millimetres from the flask bottom, is fitted with a plastic tube which terminates below the surface of an ample quantity of crushed ice contained in a large conical flask. Some extraction solvent which is to be used in the work-up is added to the crushed ice. When the stop-cock is turned off the pressure build-up forces the liquid ammonia solution into the ice-solvent mixture, the flow being controlled by periodically opening the stop-cock. Additional ice is added to the conical flask as required until the transfer of liquid ammonia solution is complete. The reaction flask is rinsed with extraction solvent and this is added to the main bulk. If the reaction mixture contains large amounts of salts (e.g., sodium halides) it is necessary to swirl the flask in order to keep the salts in suspension and so prevent the tube becoming blocked—if this should happen the stop-cock must be opened immediately. An alternative procedure is to allow the salts to settle to the bottom of the reaction flask and gradually lower the glass tube through the liquid as transfer is effected; when all the supernatant liquid has been transferred crushed ice and extraction solvent is added to the residue which is then combined with the main bulk.

ISOLATION AND PURIFICATION TECHNIQUES

I,18. GENERAL CONSIDERATIONS   At the conclusion of a reaction the pure product must be isolated from the reaction mixture by a sequence of operations collectively termed the 'work-up'. As well as the required product the reaction mixture may contain, for example, solvent which has been used as the reaction medium, excess reactants or reagents, unwanted reaction products (by-products) arising from alternative reaction pathways and so on. The planning of the isolation operations and application to such complex mixtures is therefore an exacting test of the expertise of the chemist. Frequently a student fails to bring a successful reaction to a fruitful conclusion by using an ill-considered work-up procedure, which results in loss of the required product either by decomposition during attempted isolation, or from a premature discard of product because of lack of appreciation of its physical or chemical properties. It should be emphasised that even when a detailed published procedure is being followed it is unwise to discard any liquid or solid fractions separated during work-up until the final product has been isolated and adequately characterised.

Because of the length of time that a complete isolation process often takes, it is wise practice, particularly with new syntheses carried out for the first time, to monitor the progress of the reaction. Thus the disappearance from a reaction mixture of one of the reactants or the build-up of the reaction product, measured on small aliquot portions removed at convenient time intervals from the bulk reaction mixture, can yield valuable information on the progress of a reaction. Usually the former is to be preferred since the physical properties (e.g., spectroscopic information, Section I,39), chemical reactivity (e.g., characteristic tests of functional groups, Section VII,5) and chromatographic behaviour (Section I,31) of the reactant, and the influence of solvents or other reactants on the reliability of the chosen monitoring processes may be readily checked before the reaction is commenced.

The adoption of a particular isolation procedure will depend to a large extent upon the physical and chemical properties of the product. Some guidelines for useful general approaches may however be given with regard to the physical state at ambient temperature of the crude mixture resulting from the reaction, i.e., whether it is a one-phase (either solid or liquid) or a two-phase (solid/liquid or liquid/liquid) system.

In the case of the one-phase solid system, if the organic product is neutral and insoluble in water, washing with water may be used to remove soluble impurities such as inorganic salts. Alternatively the crude solid may be extracted with a suitable organic solvent (Section I,22), filtered, and the extract washed with water. Further washing successively with dilute aqueous acid and dilute aqueous alkali removes basic and acidic impurities. Removal of solvent after drying (Section I,23 and Section I,24) leads to the recovery of the purified solid for recrystallisation from a suitable solvent (Section I,20). Continuous extraction of the solid (e.g., in a Soxhlet apparatus) may be necessary if the required product is only sparingly soluble in convenient organic solvents.

If the crude solid product contains the required product in the form of a salt (e.g., the alkali metal salt of a phenol) and is therefore water soluble, acidification of the aqueous solution (or basification in the case, for example, of amine salts) liberates the free acidic compound (or base) which may be recovered by filtration or solvent extraction as appropriate.

The one-phase liquid system is more frequently encountered since many organic reactions are carried out in solution. Direct fractional distillation may separate the product, if it is a liquid, from the solvent and other liquid reagents, or concentration or cooling may lead to direct crystallisation of the product if this is a solid. However, it is often more appropriate, whether the required product is a liquid or solid, to subject the solution to the acid/base extraction procedure outlined above and considered in detail on p. 134. This acid/base extraction procedure can be done directly if the product is in solution in a water-immiscible solvent. A knowledge of the acid/base nature of the product and of its water solubility is necessary to ensure that the appropriate fraction is retained for product recovery. In those cases where the reaction solvent is water miscible (e.g., methanol, ethanol, dimethyl sulphoxide, etc.) it is necessary to remove all or most of the solvent by distillation and to dissolve the residue in an excess of a water-immiscible solvent before commencing the extraction procedure. The removal of solvent from fractions obtained by these extraction procedures is these days readily effected by the use of a rotary evaporator (p. 161) and this obviates the tedium of removal of large volumes of solvent by conventional distillation.

A crude reaction mixture consisting of **two phases** is very common. In the case of a **solid/liquid** system, it will of course be necessary to make certain in which phase the required product resides. A simple example is where the product may have crystallised out from the reaction solvent; the mixture therefore only requires to be cooled and filtered for the bulk of the product to be isolated. The filtrate should then routinely be subjected to suitable concentration or extraction procedures to obtain the maximum yield of product.

Direct filtration would also be employed when the solid consists of unwanted reaction products, in which case the filtrate would be treated as the single-phase liquid system above. Where it is evident that the product has crystallised out admixed with contaminating solid material a separation might be effected if the mixture is reheated and filtered hot (p. 108).

Liquid/liquid two-phase systems are often encountered; for example, they result from the frequent practice of quenching a reaction carried out in an organic solvent by pouring it on to ice or into dilute acid. A further instance of a liquid/liquid system arises from the use of steam distillation (Section **I,25**) as a preliminary isolation procedure. This is particularly suitable for the separation of relatively high boiling liquids and steam volatile solids from inorganic contaminants, involatile tars, etc. The subsequent work-up procedure normally presents no additional problems since the phases are usually readily separable and can be treated in a manner appropriate to the chemical or physical properties of the required product by procedures already outlined.

All these preliminary procedures give solid or liquid products which are rarely of high purity; the degree of purity may be checked by chromatographic and spectroscopic methods. Purification may often be successfully accomplished by recrystallisation or sublimation for solids (Section **I,20** and Section **I,21**); fractional distillation under atmospheric or reduced pressure for liquids or low melting solids (Section **I,26** and Section **I,27**); molecular distillation for high boiling liquids (Section **I,28**). In those cases where the use of these traditional methods does not yield product of adequate purity, resort must be made to preparative chromatographic procedures. Here a knowledge of the chromatographic behaviour obtained from small-scale trial experiments will be particularly valuable.

The final assessment of the purity of a known product is made on the basis of its physical constants (Sections **I,34** to **I,39**) in comparison with those cited in the literature. In the case of a new compound the purity should be assessed and the structural identity established by appropriate chromatographic and spectroscopic methods.

I,19. FILTRATION TECHNIQUES   Filtration of a mixture after completion of a reaction will often be necessary either to isolate a solid product which has separated out or to remove insoluble impurities or reactants, in which case the desired product remains in solution. In this section the filtration of cold solutions is described; the filtration of hot solutions is considered in Section **I,20**.

When substantial quantities of a solid are to be filtered from suspension in a liquid, a Buchner funnel of convenient size is employed. The ordinary Buchner funnel (Fig. I,42(a)) consists of a cylindrical porcelain funnel carrying a fixed, flat, perforated porcelain plate. It is fitted by means of a rubber stopper, rubber cone or flat rubber ring into the neck of a thick-walled filtering flask (also termed a filter flask, Buchner flask or suction flask) (Fig. I,42(c)), which is connected by means of thick-walled rubber tubing (rubber 'pressure' tubing) to a similar flask or safety bottle, and the latter is attached by rubber 'pressure' tubing to a filter pump; the safety bottle or trap is essential since a sudden fall in water pressure may result in the water being sucked back and contaminating the filtrate. The use of suction renders rapid filtration possible and also results in a more complete removal of the mother-liquor than filtration under atmospheric pressure. A filter-paper is selected (and trimmed, if necessary) of such size that it covers the entire perforated plate, but its diameter should be slightly less than the inside diameter of the funnel; the filter-paper should never be folded up against the sides of the funnel. The filter-paper is moistened with a few drops of clear supernatant liquor and the suction of the pump applied, when the filter-paper should adhere firmly to, and completely cover, the perforated plate of the funnel and thus prevent any solid matter from passing under the edge of the paper into the flask below. As much of the supernatant liquor as possible is now poured into the funnel and filtered before the bulk of the residual slurry is transferred to the filter funnel, this procedure is often quicker than initially bringing all the solid into suspension and pouring it directly on to the filter. Furthermore initial gentle suction often leads to more effective filtration than powerful suction since in the latter case the finer particles of solid may reduce the rate of filtration by being drawn into the pores of the filter-paper. Any solid remaining in the reaction flask is easily transferred by rinsing with a little of the filtrate * and well stirring to remove solids which may be adhering to the sides of the reaction flask. This operation may be repeated until all solid material has been transferred to the filter. The suction is continued until most of the liquid has passed through and this is facilitated by pressing the solid down with a wide glass stopper to leave a uniformly flat, pressed surface. The filter cake is then washed with an appropriate solvent and again sucked dry. If the filter cake is the required product, then this must be subjected to purification using suitable recrystallisation procedures (Section **I,20**). If the filtrate contains the reaction product, further suitable isolation procedures would then of course be adopted (e.g., Section **I,22**; Section **I,25**; Section **I,31**).

Some modification to the above general technique of isolation by filtration may be necessary in the light of the chemical nature of the reaction mixture, of the particle size of the solid, or of the ratio of the amount of solid to liquid material to be filtered.

For example, strongly alkaline or strongly acidic reaction mixtures weaken cellulose filter-papers. Acid-hardened grades which are more chemically

---

* The filter flask must be disconnected from the pump before the latter is turned off otherwise suckback will occur which may ruin the reaction products should these be present in the filtrate.

resistant are commercially available,* but for maximum resistance to chemical attack, glass fibre paper † or a glass funnel fitted with a fixed sintered glass plate (Section **I,11**) may be used.

The filtration of very finely divided suspended material is often very tedious as a result of the filter-paper pores becoming clogged. In such a case the addition of a suitable filter-aid (e.g., a high grade diatomaceous earth such as Celite 545 ‡) to the suspension overcomes the problem; alternatively the suspension may be filtered through a bed of filter-aid prepared by pouring a slurry of it in a suitable solvent into the filter funnel fitted with the required size of filter-paper. The initial application of gentle suction in the filtration is in this case vital. A glass fibre filter-paper, supported on a conventional filter-paper in a Buchner or Hirsch funnel, is useful for the rapid removal of finely divided solid impurities from a solution.

The selection of a funnel appropriate to the amount of solid rather than the total volume of liquor to be filtered is important. When the volume of liquid is large relative to the amount of solid, the apparatus shown in Fig. I,77 may be

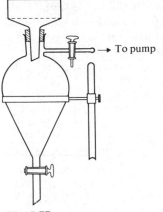

→ To pump

Fig. I,77

used; § here the funnel should be of a size appropriate to the amount of solid to be finally collected. When the receiver is about two-thirds full, atmospheric pressure is restored by suitably rotating the 'three-way' stop-cock; the filtrate may then be removed by opening the tap at the lower end. The apparatus is again exhausted and the filtration continued.

For the suction filtration of small quantities of solid (less than 5 g) contained in a small volume of liquid, a small conical Buchner funnel, known as a Hirsch funnel, is employed (see Fig. I,42(b)); the filtrate is collected either in a small filter flask or in a test-tube with side arm, the arrangement being illustrated in Fig. I,42(d). A small sintered funnel, a slit-sieve funnel or a glass funnel fitted with a Witt plate (see Section **I,11**) may also be employed. The procedure for filtration is similar to that already given for the Buchner funnel.

---

* Whatman Acid Hardened Grades 50–54, 540–542, manufactured by W. & R. Balston Ltd.
† Manufactured by W. & R. Balston Ltd; Sales Agents, H. Reeve Angel & Co Ltd.
‡ Manufactured by Johns-Manville Products Corp.; available from suppliers of chemicals.
§ Supplied in capacities of 1, 2 and 4 litres by Scientific Glass Apparatus Co.

Small volumes of solution (up to 2 ml in one operation) may conveniently be filtered through a dropping (Pasteur) pipette into the constriction of which has been rammed a small piece of paper tissue (about 3 cm square). The pipette is supported vertically and the solution is added from a second Pasteur pipette. Pressure to accelerate the filtration process may then be applied from a rubber bulb attached to the top of the pipette.

This method is particularly useful for the preparation of solutions of samples for spectroscopic examination where it is important to remove all insoluble impurities. The method may be adapted for the decolourisation of small samples by incorporating a short column of decolourising carbon above the paper plug (cf. Section I,20).

I,20. RECRYSTALLISATION TECHNIQUES Solid organic compounds when isolated from organic reactions are seldom pure; they are usually contaminated with small amounts of other compounds ('impurities') which are produced along with the desired product. The purification of impure crystalline compounds is usually effected by crystallisation from a suitable solvent or mixture of solvents.

The purification of solids by crystallisation is based upon differences in their solubility in a given solvent or mixture of solvents. In its simplest form, the crystallisation process consists of: (i) dissolving the impure substance in some suitable solvent at or near the boiling point, (ii) filtering the hot solution from particles of insoluble material and dust, (iii) allowing the hot solution to cool thus causing the dissolved substance to crystallise out and (iv) separating the crystals from the supernatant solution (or mother-liquor). The resulting solid, after drying, is tested for purity (usually by a melting point determination, Section I,34, but also by spectroscopic methods, Section I,39, or by thin-layer chromatography, Section I,31), and if found impure is again recrystallised from fresh solvent. The process is repeated until the pure compound is obtained; this often means until the melting point is unchanged, but confirmation by the other methods specified above is desirable.

The theory underlying the removal of impurities by crystallisation may be understood from the following considerations. It is assumed that the impurities are present in comparatively small proportion—usually less than 5 per cent of the whole. Let the pure substance be denoted by A and the impurities by B, and let the proportion of the latter be assumed to be 5 per cent. In most instances the solubilities of A (S_A) and of B (S_B) are different in a particular solvent; the influence of each compound upon the solubility of the other will be neglected. Two cases will arise for any particular solvent: (i) the impurity is more soluble than the compound which is being purified ($S_B > S_A$) and (ii) the impurity is less soluble than the compound ($S_B < S_A$). It is evident that in case (i) several recrystallisations will give a pure sample of A, and B will remain in the mother-liquors. Case (ii) can be more clearly illustrated by a specific example. Let us assume that the solubility of A and B in a given solvent at the temperature of the laboratory (15 °C) are 10 g and 3 g per 100 ml of solvent respectively. If 50 g of the crude material (containing 47.5 g of A and 2.5 g of B) are dissolved in 100 ml of the hot solvent and the solution allowed to cool to 15 °C, the mother-liquor will contain 10 g of A and 2.5 g (i.e., the whole) of B; 37.5 g of pure crystals of A will be obtained.

The most desirable characteristics of a solvent for recrystallisation are:

(a) a high solvent power for the substance to be purified at elevated temperatures and a comparatively low solvent power at the laboratory temperature or below;

(b) it should dissolve the impurities readily or to only a very small extent;

(c) it should yield well-formed crystals of the purified compound;

(d) it must be capable of easy removal from the crystals of the purified compound, i.e., possess a relatively low boiling point.

It is assumed, of course, that the solvent does not react chemically with the substance to be purified. If two or more solvents appear to be equally suitable for recrystallisation, the final selection will depend upon such factors as ease of manipulation, toxicity, inflammability and cost.

Some common solvents available for the recrystallisation are collected in Table I,8, broadly in the order of decreasing polarity. Their purification is included in Section **II,1**.

Table I,8   Common solvents for recrystallisation

Solvent	b.p. (°C)	
Water (distilled)	100	To be used whenever suitable
Methanol†	64.5	Inflammable; toxic
Ethanol	78	Inflammable
Industrial spirit	77–82	Inflammable
Rectified spirit	78	Inflammable
Acetone	56	Inflammable
Ethyl acetate	78	Inflammable
Acetic acid (glacial)	118	Not very inflammable, pungent vapours
Dichloromethane (methylene chloride)†	41	Non-inflammable; toxic
Chloroform†	61	Non-inflammable; vapour toxic
Diethyl ether	35	Inflammable, avoid whenever possible
Benzene†‡	80	Inflammable, vapour highly toxic
Dioxan†	101	Inflammable, vapour toxic
Carbon tetrachloride†	77	Non-inflammable, vapour toxic
Light petroleum	40–60	Inflammable*
Cyclohexane	81	Inflammable

* Other fractions available have b.p. 60–80, 80–100 and 100–120 °C; when the boiling point exceeds 120 °C the fraction is usually called 'ligroin'. Pentane, b.p. 36 °C, and heptane, b.p. 98 °C, are also frequently used recrystallisation solvents.

† The vapours of these solvents are toxic and therefore recrystallisations involving their use must be conducted in an efficient fume cupboard; excessive inhalation of any vapour should be avoided. For notes on cumulative toxic effects refer to Section **I,3**.

‡ Toluene is much less toxic than benzene and should be used in place of the latter whenever possible.

The use of ether as a solvent for recrystallisation should be avoided wherever possible, partly owing to its great inflammability and partly owing to its tendency to creep up the walls of the containing vessel, thus depositing solid matter by complete evaporation instead of preferential crystallisation. Carbon disulphide, b.p. 46 °C, should never be used if an alternative solvent can be found; it has a dangerously low flash point and forms very explosive mixtures with air.

Other recrystallisation solvents include tetrahydrofuran (THF), b.p. 65–66 °C; butan-2-one (ethyl methyl ketone), b.p. 80 °C; 1,2-dichloroethane * (ethylene chloride), b.p. 84 °C; acetonitrile* (methyl cyanide), b.p. 80 °C; toluene* b.p. 110 °C; pyridine,* b.p. 115.5 °C; chlorobenzene,* b.p. 132 °C; cellosolve* (2-ethoxyethanol), b.p. 134.5 °C; dibutyl ether, b.p. 141 °C; s-tetrachloroethane,* b.p. 147 °C; dimethylformamide* (DMF; formdimethylamide), b.p. 153 °C; dimethyl sulphoxide, b.p. 189 °C (d); nitrobenzene,* b.p. 209.5 °C; and ethyl benzoate, b.p. 212 °C.

The following rough generalisations may assist the student in the selection of a solvent for recrystallisation, but it must be clearly understood that numerous exceptions are known (for a further discussion see Section **VII,3**):

1. A substance is likely to be most soluble in a solvent to which it is most closely related in chemical and physical characteristics.
2. In ascending a homologous series, the solubilities of the members tend to become more and more like that of the hydrocarbon from which they may be regarded as being derived.
3. A polar substance is more soluble in polar solvents and less soluble in non-polar solvents. The solvents in Table I,8 have been listed broadly in order of decreasing polar character.

In practice the choice of a solvent for recrystallisation must be determined experimentally if no information is already available. About 0.1 g of the powdered substance ‡ is placed in a small test-tube (75 × 11 mm or 110 × 12 mm) and the solvent is added a drop at a time with continuous shaking of the test-tube. After about 1 ml of the solvent has been added, the mixture is heated to boiling, due precautions being taken if the solvent is inflammable. If the sample dissolves easily in 1 ml of cold solvent or upon gentle warming, the solvent is unsuitable. If all the solid does not dissolve, more solvent is added in 0.5 ml portions, and again heated to boiling after each addition. If 3 ml of solvent is added and the substance does not dissolve on heating, the substance is regarded as sparingly soluble in that solvent, and another solvent should be sought. If the compound dissolves (or almost completely dissolves§) in the hot solvent, the tube is cooled to determine whether crystallisation occurs. If crystallisation does not take place rapidly, this may be due to the absence of suitable nuclei for crystal growth. The tube should be scratched below the surface of the solution with a glass rod; the fine scratches on the walls (and the minute fragments of glass produced) may serve as excellent nuclei for crystal growth. If crystals do not separate, even after scratching for several minutes and cooling in an ice-salt mixture, the solvent is rejected. If crystals separate,

---

* The vapours of these solvents are toxic and therefore recrystallisations involving their use must be conducted in an efficient fume cupboard; excessive inhalation of any vapour should be avoided. For notes on cumulative toxic effects refer to Section **I,3**.

† Toluene is much less toxic than benzene and should be used in place of the latter whenever possible.

‡ With practice the student should be able readily to perform trial recrystallisations with much smaller quantities of material (e.g., 5 mg) using a small ignition or centrifuge tube and correspondingly smaller quantities of solvents.

§ If the crude substance contains an insoluble impurity, difficulty may be experienced at a later stage in estimating how much solute has crystallised from the cold solution. The hot solution should therefore be filtered into another tube through a very small fluted filter-paper contained in a small short-stemmed funnel. The solution must always be clear before cooling is attempted.

the amount of these should be noted. The process may be repeated with other possible solvents, using a fresh test-tube for each experiment, until the best solvent is found; the approximate proportions of the solute and solvent giving the most satisfactory results should be recorded.

If the substance is found to be far too soluble in one solvent and much too insoluble in another solvent to allow of satisfactory recrystallisation, **mixed solvents** or **'solvent pairs'** may frequently be used with excellent results. The two solvents must, of course, be completely miscible.* Recrystallisation from mixed solvents is carried out near the boiling point of the mixture. The compound is dissolved in the solvent in which it is very soluble, and the hot solvent, in which the substance is only sparingly soluble, is added cautiously until a slight turbidity is produced. The turbidity is then just cleared by the addition of a small quantity of the first solvent and the mixture is allowed to cool to room temperature; crystals will separate. Pairs of liquids which may be used include: alcohols and water; alcohols and benzene; benzene and light petroleum; acetone and light petroleum; diethyl ether and pentane; glacial acetic acid and water; dimethylformamide with either water or benzene.

When the best solvent or solvent mixture and the appropriate proportions of solute and solvent have been determined by these preliminary tests or have been obtained from reference books containing solubility data,† the solid substance is placed in a round-bottomed flask of suitable size fitted with a reflux condenser (Fig. I,61) and slightly less than the required quantity of solvent is added together with a few pieces of porous porcelain to prevent 'bumping' (see Section **I,24**). The mixture is heated to boiling on a water bath (if the solvent boils below 80 °C) or with an electric heating mantle, and more solvent is added down the condenser until a clear solution, apart from insoluble impurities,‡ is produced. If the solvent is not inflammable, toxic or expensive, recrystallisation may be carried out in a conical flask, into the neck of which a funnel with a short stem is inserted, which is heated on an electric plate.

Filtration of the hot solution. The boiling or hot solution must be rapidly filtered before undue cooling has occurred. (If an inflammable solvent has been used, all flames in the vicinity must be extinguished.) This is usually done through a fluted filter-paper (see below) supported in a relatively large funnel with a short wide stem; separation of crystals in and clogging of the stem is thus reduced to a minimum. The funnel should be warmed in an electric or steam oven before filtration is started, when it should be supported in a conical flask of sufficient size to hold all the solution; the conical flask is stood on an electric hot plate or steam bath and the filtrate is kept boiling gently so that the warm solvent vapours maintain the temperature of the solution undergoing filtration, and thus prevent premature deposition of crystals on the filter or in the neck of the funnel. If solid does separate out on the filter it must be scraped back into the first flask, redissolved and refiltered. The filtered solution is covered with a watch- or clock-glass, and then set aside to cool undisturbed. If large crystals are desired, any solid which may have separated from the filtered

---

* Solvent pairs selected from the extremes of the list in Table I,8 are not usually sufficiently miscible to be satisfactory, e.g., methanol and light petroleum.

† For example, see Refs. 35–37.

‡ The undissolved material will be readily recognised if preliminary solubility tests have been correctly interpreted.

solution should be redissolved by warming (a reflux condenser must be used for an inflammable solvent), the flask wrapped in a towel or cloth, and allowed to cool slowly. If small crystals are required, the hot saturated solution should be stirred vigorously and cooled rapidly in a bath of cold water or of ice. It should be noted that large crystals are not necessarily purer than small ones; generally very impure substances are best purified by slow recrystallisation to give large crystals, followed by several rapid recrystallisations to give small crystals.

If large quantities of hot solution are to be filtered, the funnel (and fluted filter-paper) should be warmed externally during the filtration (hot water funnel). Three types of hot water funnel are illustrated in Fig. I,41(a–c); no flames should be present whilst inflammable solvents are being filtered through the funnel of type (a). When dealing with considerable volumes of aqueous or other solutions which do not deposit crystals rapidly on cooling, a Buchner funnel preheated in an oven may be used for filtration (see Section I,19). The filter-paper should be of close-grained texture and should be wetted with solvent before suction is applied; the solution may then be poured on to the filter.

Preparation of a fluted filter-paper. The filter-paper is first folded in half and again in quarters, and opened up as shown in Fig. I,78(a). The edge 2,1 is then folded on to 2,4 and edge 2,3 on to 2,4, producing, when the paper is opened, new folds at 2,5 and 2,6. The folding is continued, 2,1 to 2,6 and 2,3 to 2,5, thus producing folds at 2,7 and 2,8 respectively (Fig. I,78(b)); further 2,3 to 2,6 giving 2,9, and 2,1 to 2,5 giving 2,10 (Fig. I,78(c)). The final operation consists in making a fold in each of the eight segments—between 2,3 and 2,9, between 2,9 and 2,6, etc.—in a direction *opposite* to the first series of folds, i.e., the folds are made outwards instead of inwards as at first. The result is a fan arrangement (Fig. I,78(d)), and upon opening, the fluted filter-paper (Fig. I,78(e)) is obtained.

Use of decolourising carbon. The crude product of an organic reaction may contain a coloured impurity. Upon recrystallisation, this impurity may dissolve in the boiling solvent and be partly adsorbed by the crystals as they

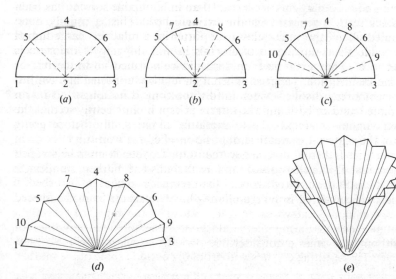

(a)   *(b)*   *(c)*

(d)   *(e)*

Fig. I,78

separate upon cooling, yielding a coloured product. Sometimes the solution is slightly turbid owing to the presence of a little resinous matter or a very fine suspension of an insoluble impurity, which cannot always be removed by simple filtration. These impurities can be removed by boiling the substance in solution with a little decolourising charcoal for 5–10 minutes, and then filtering the solution while hot as described above. The decolourising charcoal adsorbs the coloured impurity and holds back resinous, finely-divided matter, and the filtrate is usually free from extraneous colour, and therefore deposits pure crystals. The decolourisation takes place most readily in aqueous solution but can be performed in almost any organic solvent; the process is least effective in hydrocarbon solvents. It must be pointed out that boiling in a solvent with decolourising carbon is not always the most effective method of removing the colour; if this is only partially effective, it is often worth while to pass the cold solution of the substance (preferably in an organic solvent such as ethanol) through a small amount of decolourising carbon supported on a wad of cotton wool in the stem of a funnel—this is effectively a chromatographic procedure.

An excessive quantity of decolourising agent must be avoided, since it may also adsorb some of the compound which is being purified. The exact quantity to be added will depend upon the amount of impurities present; for most purposes 1–2 per cent by weight of the crude solid will be found satisfactory. If this quantity is insufficient, the operation should be repeated with a further 1–2 per cent of fresh decolourising charcoal. Sometimes a little charcoal passes through the close-grained filter-paper: the addition, before filtration of a filter-aid (filter-paper pulp or Celite), will give a clear filtrate. Attention is directed to the fact that the decolourising charcoal should not be added to a superheated solution as the latter may foam excessively and boil over.

The most widely known form of decolourising carbon is animal charcoal (also known as bone black or bone charcoal); it is the least expensive, but by no means the best. It has limited adsorptive power, and contains a large proportion of calcium phosphate and other calcium salts; it should not be used with acidic solutions, particularly if the desired compound is to be subsequently obtained by a process of neutralisation. This difficulty may be overcome, and all risk of the introduction of impurities into the hot solution avoided, by boiling the commercial animal charcoal with dilute hydrochloric acid (1:1) for 2–3 hours. The mixture is then diluted with hot distilled water, filtered through a fine-grained filter-paper supported on a Buchner funnel and washed repeatedly with boiling distilled water until the filtrate is no longer acid. It is then well drained, and dried in an evaporating basin in an electric oven. Acid-washed decolourising charcoal is also available commercially; before use, a suspension of the material in water should be checked for neutrality.

On the whole it is better to employ the activated decolourising charcoal prepared from wood. Excellent decolourising carbons are marketed under the trade names 'Norit'* (from birch wood), 'Darco' and 'Nuchar'.

Difficulties encountered in recrystallisation. The separation of a second liquid phase, commonly known as an 'oil', instead of the expected crystalline solid, sometimes occurs during recrystallisation. The oil often solidifies on standing, although at times a considerable period may elapse before crystallisation occurs. The resulting crystals will probably occlude some of the mother-

---

* Obtainable, for example, from Eastman Kodak Company.

liquor, and the purity will therefore not be high. The separation of the oil may be avoided by diluting the solution considerably, but this will lead to large losses. It is probably best to re-heat the mixture until a clear solution is obtained, and allow it to cool spontaneously; immediately the oil commences to separate, the mixture is vigorously stirred so that the oil is well dispersed in the solution. Eventually, crystals will separate and these will grow in the bulk of the solution and not in a pool of oil, so that occlusion of the mother-liquor is considerably reduced. When all the oil has disappeared, stirring may be stopped and the crystals allowed to accumulate. Sometimes the addition of a minute quantity of the crude compound in order to 'seed' the solution may facilitate the initial crystallisation.

Occasionally substances form supersaturated solutions from which the first crystals separate with difficulty; this is sometimes caused by the presence of a little tar or viscous substance acting as a protective colloid. The following methods should be tried in order to induce crystallisation:

1. By scratching the inside of the vessel with a glass rod. The effect is attributed to the breaking off of small particles of glass which may act as crystal nuclei, or to the roughening of the surface, which facilitates more rapid orientation of the crystals on the surface.

2. By inoculating ('seeding') the solution with some of the solid material or with isomorphous crystals, crystallisation frequently commences and continues until equilibrium is reached. The 'seed crystals' may be obtained by cooling a very thin film of liquid to a low temperature. Several drops of the solution are placed in a test-tube or beaker and spread into a thin film by rotating the container; the latter is then cooled in a mixture of ice and salt or in some other suitable freezing mixture. A better procedure, which avoids the necessity of subsequently scraping the surface to remove the 'seed crystals' and the attendant melting if the compound is impure or of low melting point, is to moisten a small glass bead with the supersaturated solution, place it in a test-tube, cool the latter in a freezing mixture and thus form crystals on the surface of the bead. The glass bead can then be rolled out of the tube into the vessel containing the main bulk of the solution. Seed crystals may sometimes be formed when a few drops of the solution are placed on a watch glass and the solvent is gradually allowed to evaporate whilst at the same time the film is rubbed with a glass rod.

3. By cooling the solution in a freezing mixture (ice and salt, ice and calcium chloride, or solid carbon dioxide and acetone). It must be borne in mind that the rate of crystal formation is inversely proportional to the temperature; cooling to very low temperatures may render the mass very viscous and thus considerably hinder crystallisation. In such a case, the mixture should be allowed to warm slowly so that it may be given the opportunity to form crystals if it passes through an optimum temperature region for crystal formation. Once minute crystals have been formed, it is very probable that their size will be increased by keeping the mixture at a somewhat higher temperature.

4. By adding a few lumps of solid carbon dioxide; this produces a number of cold spots here and there, and assists the formation of crystals.

5. If all the above methods fail, the solution should be left in an ice chest (or a refrigerator) for a prolonged period. The exercise of considerable patience

is sometimes necessary so as to give the solute every opportunity to crystallise.

The product of a chemical reaction which has been isolated by solvent extraction and subsequent removal of solvent (see Sections **I,22** and **I,24**), and which normally should be crystalline, is sometimes an oil due to the presence of impurities. It is usually advisable to attempt to induce the oil to crystallise before purifying it by recrystallisation. Methods 1 and 2 (previous paragraph) may be applied; method 2 cannot always be used because of the difficulty of securing the necessary seed crystals, but should these be available, successful results will usually be obtained. Another procedure is to add a small quantity of an organic solvent in which the compound is sparingly soluble or insoluble, and then to rub with a stirring rod or grind in a mortar until crystals appear; it may be necessary to continue the rubbing for an hour before signs of solidification are apparent. Another useful expedient is to leave the oil in a vacuum desiccator over silica gel or some other drying agent. If all the above methods fail to induce crystallisation, direct recrystallisation may be attempted: the solution should be boiled with decolourising carbon as this may remove some of the impurities responsible for the difficulty of crystal formation. Occasionally, conversion into a simple crystalline derivative is applicable; subsequent regeneration of the original compound will usually yield a pure, crystalline solid. Instances will occur however when assessment by thin-layer chromatography (Section **I,31**) of the number of probable impurities in the isolated reaction mixture and of their relative amounts is advisable. It may then be judged whether some prior purification by suitable preparative chromatography (Section **I,31**) or by solvent extraction (Section **I,22**), should be performed before crystallisation is attempted.

The technique for the removal of solids by filtration with suction has already been described (Section **I,19**). The same technique will of course be applied to the collection of recrystallised compounds. Additionally, however, it should be noted that the mother-liquor from a recrystallisation is often of value for the recovery of further quantities of product, and should be transferred to another vessel after the crystals have been drained and washed with solvent. The mother-liquor may be then subsequently concentrated (Section **I,24**; suitable precautions being taken, of course, if the solvent is inflammable), and a further crop of crystals obtained. Occasionally yet another crop may be produced. The crops thus isolated are generally less pure than the first crystals which separate, and they should be combined and recrystallised from fresh solvent; the purity is checked by a melting point determination.

After the main filtrate has been removed for such treatment, the crystals on the filter pad should be washed to remove remaining traces of mother-liquor which, on drying, would contaminate the crystals. The wash liquid will normally be the same solvent or solvent mixture used for recrystallisation and must be used in the smallest amount compatible with efficient washing, in order to prevent appreciable loss of the solid. With the suction discontinued the crystals are treated with a small volume of the chilled solvent and cautiously stirred with a spatula or with a flattened glass rod (without loosening the filter-paper) so that the solvent wets all the crystals. The suction is then applied again, and the crystals are pressed down with a wide glass stopper as before. The washing is repeated, if necessary, after connection to the filter pump has again been broken.

If the solvent constituting the crystallisation medium has a comparatively high boiling point, it is advisable to wash the solid with a solvent of low boiling point in order that the ultimate crystalline product may be easily dried; it need hardly be added that the crystals should be insoluble or only very sparingly soluble in the volatile solvent. The new solvent must be completely miscible with the first, and should not be applied until the crystals have been washed at least once with the original solvent.

Recrystallisation at very low temperatures. This technique is necessary either when the solubility of the compound in the requisite solvent is too high at ordinarily obtained temperatures (refrigerator to room temperatures) for recovery to be economic, or when handling compounds which are liquid at room temperature but which may be recrystallised from a solvent maintained at much lower temperatures (say -10 to $-40\,°C$). In this latter case, after several successive low temperature recrystallisations the compound will revert to a liquid on storage at room temperature, but the purification process by recrystallisation will have been achieved.

The following crude, but nonetheless relatively effective, procedure may be adopted in those cases where the compound is not moisture sensitive, and where the amount of product is relatively large (say 5 to 50 g). A round-bottomed or conical flask protected with a calcium chloride tube and containing the solvent in which the compound has been previously dissolved is placed in a suitable cooling mixture (see Section **I,12**) until crystallisation is complete. A second flask, also protected by a calcium chloride tube and containing the washing solvent, is also chilled in the same cooling bath. A Buchner funnel is fitted with a suitable filter-paper, attached to the filter flask and filled with powdered solid carbon dioxide. Immediately prior to filtration the solid carbon dioxide is tipped out (for very low temperature filtrations some proportion of the solid carbon dioxide may be retained in the funnel since this helps to maintain the low temperature of the filtering mixture) and the solution is filtered as rapidly as possible using the previously chilled solvent for rinsing and washing. One must work as rapidly as possible, returning the flasks to the cooling bath at every opportunity and ensuring that before pouring from either flask the outside is wiped with a cloth, otherwise some drops of the cooling-bath mixture may drain on to the filter cake.

It will be clear from the above account that recrystallisation and filtration at low temperatures is attended by two inherent difficulties, (1) moisture is rapidly deposited on the chilled compound, the solvent and the Buchner funnel, and (2) it is difficult to maintain the apparatus, product and solvent at the required temperature throughout the filtration process.

To overcome these difficulties some ingenuity in apparatus design is necessary, and Fig. I,79 illustrates one possible assembly.

Here a three-necked, pear-shaped flask A is fitted with a condenser and calcium chloride tube B, a filter stick * C (this being a glass tube having a sintered glass frit at its end), a stirrer guide D with flexible rubber or polythene tubing to allow vertical and circular movements of the filter stick, a two-necked

---

* Filter sticks of a range of dimensions and porosities are available from Jobling Laboratory Division. The choice of size of flasks and filter sticks will be governed by the scale of the recrystallisation. Broadly speaking pear-shaped flasks are more suitable for the range 100 mg to 5 g, but may be replaced by round-bottomed flasks for operations on a larger scale.

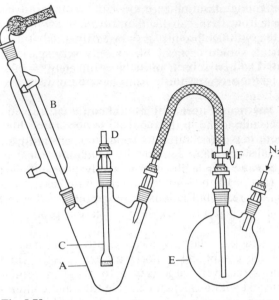

Fig. I,79

flask E fitted with a nitrogen gas inlet system, and a delivery tube (with stop-cock) F. The thoroughly dried assembled apparatus, containing the compound to be recrystallised in flask A and with the filter stick drawn into the upper part of the flask, is flushed through with nitrogen. The nitrogen inlet tube is re-moved, solvent is introduced into flask E, the nitrogen inlet tube is replaced, and by suitable control of the stop-cock F (1) the required amount of solvent is allowed to flow into flask A (2). Flask A is now heated by suitable means until solution of the compound is complete, and the liquors are allowed to cool to near room temperature before both flasks A and E are immersed in the ap-propriate cooling bath. When recrystallisation is complete, the filter stick C is lowered to the level of the mother-liquors and suction applied slowly (3). The horizontal and circular movement allowed by the flexible connection D enables the crystalline solid to be pressed down into a filter cake and the mother-liquor to be efficiently removed. *Without disconnecting the suction*, washing solvent is allowed to flow into the flask A; by adjustment of the position of the delivery tube it is possible to direct solvent flow on to the outside of the filter-stick tube in order to remove contaminating mother-liquor. These initial washings also serve the purpose of rinsing the inside of the filter stick so that subsequent drainage which may occur when suction is discontinued will not cause con-tamination of the crystalline material. If further washing is required the suction is discontinued and the solvent allowed to flow on to the crystals which are stirred with the filter stick; finally suction is reapplied. Further recrystalli-sations may of course be carried out without removal of material from flask A. The entire assembly should be allowed finally to reach room temperature with the dry atmosphere maintained within, so that when the apparatus is discon-nected the purified product will not be contaminated with condensing water vapour from the laboratory atmosphere.

   Notes. (1) The nitrogen inlet system should be of the type suggested in

Fig. I,67 with a sufficiently great enough head of mercury or mineral oil in the escape valve to force the solvent from flask E to A. Control of solvent flow should be by stop-cock F, and the outlet of the solvent delivery tube should be above the final level of solution.

(2) The amount of solvent used will have been estimated from the trial recrystallisations carried out in ignition or test-tubes using a cooling bath to effect crystallisation.

(3) The filter stick should be attached to a suction pump via a filter trap so that the mother-liquor may be collected, and, if need be, concentrated for further crystal crops; furthermore it is essential that the suction be carefully controlled (by the use of additional stop-cocks which are not shown) so that the filter stick may be lowered at the same rate as the level of mother-liquor falls. Undue immersion of the filter stick leads to some troublesome difficulties in removing contaminating mother-liquor from the outside of the filter stick.

Recrystallisation in an inert atmosphere. Substances which decompose (or otherwise undergo structural modification, on contact with air must be recrystallised in an indifferent atmosphere, which is usually nitrogen but may on occasions be carbon dioxide, or rarely, hydrogen. The apparatus assembly shown in Fig. I,79 is suitable with the modification that the calcium chloride tube is replaced by a second nitrogen inlet system. The apparatus is flushed with nitrogen, and the solid material is quickly transferred to flask A; the apparatus is flushed with nitrogen before solvent transfer from flask E to A is carried out. Subsequent operations are as described above using such cooling methods as are appropriate to the recrystallisation process. Finally the material is thoroughly dried under nitrogen in flask A by allowing the gas to pass through the system before the apparatus is disconnected. Short periods of exposure of the crystalline material to the atmosphere is not always harmful and hence it may usually be transferred rapidly to a suitable container for storage under nitrogen (e.g., a nitrogen-filled desiccator).

In those cases where even a short exposure to the atmosphere is harmful, the recrystallisation and filtration processes may be carried out in a nitrogen-filled manipulator glove box,* which has been adapted to accommodate the services required for a normal recrystallised procedure. The size of the glove box itself and the dimensions of the outlet panels will naturally limit the scale on which recrystallisation can be carried out in this manner.

Technique of semimicro and micro recrystallisations. The student in the later stages of his training will certainly be required to recrystallise quantities of solid material within the range of 1 g to fractions of a milligram. These small quantities could arise from (1) small-scale preparations involving very expensive materials, (2) preparations of derivatives of small amounts of natural products, (3) byproducts isolated from a reaction process, (4) chromatographic separation procedures (column and thin-layer techniques), etc. For convenience the experimental procedure to be adopted for recrystallisations of small quantities may be described under three groups:

(i) a scale in the range of 1 g to 20 mg;
(ii) a scale in the range of 20 mg to 1 mg;
(iii) a scale below 1 mg.

---

* Available from, for example, A. Gallenkamp Ltd.

The scale of the recrystallisation envisaged in **group (i)** means that the operations are carried out in the conventional manner but in apparatus of reduced size. Thus small conical flasks (5 to 20 ml), pear-shaped flasks (5 to 20 ml), semimicro test-tubes (75 × 10 mm or 100 × 12 mm) or centrifuge tubes (1 to 5 ml) are employed; it is best that the solutions in these receptacles be heated on a water or oil bath rather than directly with a semimicro burner since the heating process can thus be better controlled. Operations involving hot inflammable solvents should be performed under reflux using semimicro interchangeable ground glass joint apparatus (7/11 and 10/19). The crystals which separate on cooling are removed by filtration using a small Hirsch funnel, or a small conical glass funnel fitted with a perforated or sintered glass filtration plate; it is often advantageous to place a small filter-paper upon the sintered glass plate since complete removal of the crystalline material is thus facilitated. Typical filtration assemblies are shown in Fig. I,80(*a* and *b*). In Fig. I,80(*a*) a rubber cone is used with the Hirsch funnel and boiling tube (150 × 25 mm) having a side arm for attachment to the suction pump; the filtrate is collected in a centrifuge tube or in a semimicro test-tube resting upon a wad of cotton wool. In Fig. I,80(*b*) the assembly incorporates ground glass joints and is suitable for the 20–100 mg scale.

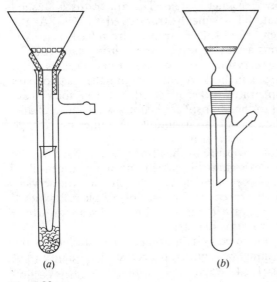

(*a*)   (*b*)

Fig. I,80

An apparatus for the filtration of quantities of crystals within the scale of both **groups (i) and (ii)** incorporates the so-called **Willstätter 'filtration nail'**. The latter consists of a thin glass rod flattened at one end. It is readily constructed by heating the end of a short glass rod in the blowpipe flame and pressing vertically upon an asbestos board. The nail is fitted into a small glass funnel (Fig. I,81(*a*)) which is attached to the filter tube by a rubber ring or cone, or the funnel may incorporate a ground glass joint. The nail head may be covered with a circle of filter-paper cut with the aid of a cork borer of appropriate size; alternatively, with a good fit between the nail-head edge and the funnel surface and with well-formed crystals, an initial layer of crystals is held at the join and provides a filter medium. The latter technique is valuable for the final stages in the preparation of dust- and fibre-free samples for subsequent elemental analysis. The dimensions given in Fig. I,81(*a*) are a guide to the construction of a 'filtration nail' and funnel capable of handling up to 1 g of solid; a smaller size will handle correspondingly smaller quantities (20 mg) just as efficiently. It is often convenient to make several 'nails' with different sized heads which all fit a single funnel so that the 'nail' and filter-paper diameter appropriate to the amount of solid to be collected may be employed. For the larger sizes of 'nail' which accommodate filter-papers of 15–25 mm diameter, it is advisable to corrugate the head of the 'nail' as shown (somewhat exaggerated) in Fig. I,81(*b*) in order to permit drainage of filtrate over the entire paper; these corrugations are easily produced by pressing the hot glass on the surface of an old single-cut file of coarse grade.

Should it be necessary, before crystallisation takes place, to filter the hot or boiling solutions to remove dust, fibre, etc., the Pyrex micro filter (8 ml capacity) shown in Fig. I,82 is suitable since the long cylindrical tube reduces evaporation. The filter would be used in an assembly similar to that shown in Fig. I,80.

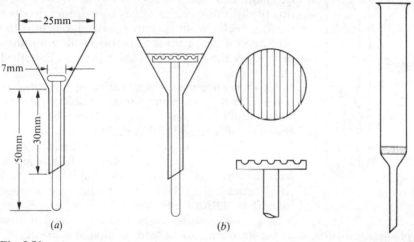

Fig. I,81   Fig. I,82

It is clear that when the scale of the operation falls within **group (ii)** (20 mg to 1 mg), losses arising from transference of material from flask to filter become more serious and contamination by dirt and filter-paper fibre more likely. For these reasons one of the following variations in technique should be employed.

The solution and crystallisation processes may be conducted in a centrifuge tube; when crystallisation is complete the tube and its contents should be centrifuged* to pack the crystalline mass. Filtration is now performed by introducing the end of a dropper pipette (previously drawn out to a capillary of approximately 0.1 mm diameter), down the inside surface of the centrifuge tube until it reaches the bottom; in this way there is less chance of breakage of the

* Suitable hand-operated or electric centrifuges may be obtained from most laboratory equipment suppliers.

capillary because of the support provided by the side of the tube. The mother-liquor is now drawn into the pipette, the capillary end serving as the filter. Drops of washing solvent from another dropper pipette are now directed, first to the inside of the tube contaminated with mother-liquor, and then on to the surface of the crystalline mass. The solvent percolates through the crystalline mass and is drawn up into the capillary—the filled capillary pipette is withdrawn and returned empty as frequently as is necessary. Further recrystallisation may be performed without removing the solid from the centrifuge tube, and finally the centrifuge tube containing the purified compound is placed in a vacuum desiccator or drying pistol. When completely solvent-free it will be found that the product may be cleanly removed from the centrifuge tube.

When the volume of solvent required for the recrystallisation is large, compared to the quantity of solid material, initial removal of mother-liquor from the centrifuge tube by this capillary pipette method may be tedious. In such a case, centrifugation should be more prolonged to get more effective packing of the solid, and the bulk of mother-liquor may be drained without disturbance of the solid by inverting the tube over another receptacle in a smooth jerk-free motion. Whilst the tube is in the inverted position the lip is now rinsed with solvent from a dropper pipette; finally wash solvent is introduced into the tube and on to the solid. Subsequent filtration may now be by the capillary-ended pipette as described above.

An alternative means of filtration of quantities of material in the 1 to 5 mg region is to effect solution and crystallisation in a narrow glass tube (6 mm) sealed at one end and with a slight constriction sited at such a position to give the required capacity. After crystallisation is complete, a glass rod flattened at one end to give a type of Willstätter 'nail' (Fig. I,83(a)) is introduced, the tube is inverted into a centrifuge tube (Fig. I,83(b)) and the whole is centrifuged. Washing may be effected by removing the nail from the crystallisation tube, introducing wash solvent with a capillary pipette, reintroducing the 'nail' and filtering by centrifuging in the inverted position as before.

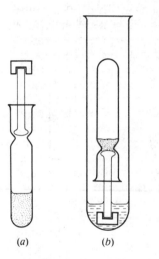

(a)   (b)

Fig. I,83

Where a recrystallisation has been conducted in a centrifuge or small diameter tube and too much solvent has been added initially, concentration of the solution may be effected by the following technique. A small carborundum chip is introduced and the tube is heated very carefully over a water or oil bath of suitable temperature to give gentle boiling. A flow of nitrogen is directed to the surface of the boiling liquid by means of a capillary-end pipette; the process is continued until the required concentration is effected. The technique of concentration may also be used in those cases where further crystalline crops are required from the mother-liquor.

Where it is clear that the solid to be purified contains dust, or insoluble impurities, then it is necessary to use initially a greater volume of solvent (to prevent prior crystallisation) and to filter the solution in the micro-filter shown in Fig. I,82. The filtered solution is concentrated using the technique described above until the required smaller volume is reached using first, if necessary, larger test-tubes and transferring in stages the concentrated solutions by dropper pipettes to finally the centrifuge or small diameter tube of an appropriate size.

It should be emphasised that in all these operations the laboratory worker should have readily to hand racks, retort stands, clamps and bosses to hold and support the centrifuge tubes and dropping pipettes, and watch glasses to cover tube ends and prevent the entry of dust.

Recrystallisation of quantities of materials less than 1 mg (**group (iii)**) is carried out in melting point or capillary tubes (e.g., 1.5 mm diameter), and is a much simpler operation than might be imagined. All manipulations should be carried out on a bench top covered with a piece of white glazed card or opaque glass. This ensures that all dust and dirt may be immediately observed and removed, and that the glass apparatus does not accidentally pick up dirt, bench polish, etc., small quantities of which could seriously contaminate the product.

The solid is introduced into one of the two open ends of the capillary tube (*ca.* 1 mm i.d.) by pushing the end into the solid contained in a specimen tube or on a watch glass. The *solvent* is introduced by holding the same open end of the capillary tube against a drop of solvent suspended from the end of a capillary dropping pipette. The amount of solvent taken up depends on the balance between surface tension and capillary forces, and the force of gravity, which in turn is determined by the angle at which the capillary tube is held and by the portion of the suspended drop which is touched by the capillary end. In general a horizontally held tube rapidly takes up a column of several centimetres of liquid, whilst a tube held vertically over the top of a suspended drop takes up only a few millimetres of liquid. Practice with empty tubes results in the technique of solvent introduction being easily acquired. The solute/solvent-free end is now sealed in the flame of a micro-burner, allowed to cool, and the tube contents transferred to the newly sealed end by placing the capillary tube in a centrifuge tube and centrifuging. It is advisable to cover the centrifuge tube with a soft rubber cap through which some pin holes have been pierced; the capillary tubes inserted through these holes are then supported adequately to prevent breakage. The open end is now washed clean by the introduction of more solvent. This is accomplished by passing the midsection of the tube through a microburner flame and then introducing the end into a drop of liquid; as the tube cools liquid is drawn into the tube. This additional liquid may be centrifuged down, or if not required to augment the volume of solvent already present it may be removed from the tube by wrapping the open end in a filter or paper tissue and passing the midsection of the tube again through the microburner flame. The cleaned end is now sealed, and solution of the solute in the solvent effected by immersing the end of the tube in an appropriate heating bath. The tube is now cooled and when crystallisation is complete [a cooling bath (1) may be necessary to induce crystallisation], the tube is centrifuged to compact the crystalline mass. Removal of mother-liquor may be achieved by cutting off the top of the capillary tube (2) and using a very fine capillary pipette; washing solvent is introduced by another capillary pipette. Finally the end of the tube containing the purified solid may be severed, dried in a vacuum desiccator, and the contents removed by holding the glass section in a pair of tweezers and using a short piece of platinum wire sealed into a glass rod as a spatula.

Notes. (1) For the support of these tubes in heating and cooling baths it is recommended that the top of the bath be covered with a wire mesh large enough to allow for insertion of the tube; the upper end of the tube should be pushed into a small rubber collar which then allows the tube to be suspended satisfactorily.

(2) Scratching of tube of this size is accomplished by a carborundum glass cutter, which is easily constructed by fusing a carborundum chip in the end of a glass rod. Alternatively a diamond pencil may be used.

Drying of recrystallised material. The conditions for drying recrystallised material depend upon the quantity of product, the nature of the solvent to be removed and the sensitivity of the product to heat and to the atmosphere.

With large-scale preparations of stable compounds, moist with non-toxic solvents which are volatile at room temperature (e.g., water, ethanol, ethyl acetate, acetone), the Buchner funnel is inverted over two or three thicknesses of drying paper (i.e., coarse-grained, smooth-surfaced filter-paper) resting upon a pad of newspaper, and the crystalline cake removed with the aid of a clean spatula; several sheets of drying paper are placed on top and the crystals are pressed firmly. If the sheets become too damp with solvent, the crystals should be transferred to fresh paper. The crystals are then covered by a piece of filter-paper perforated with a number of holes or with a large clock glass or sheet of glass supported upon corks. The air drying is continued until only traces of solvent remain (usually detected by smell or appearance) and final drying is accomplished by placing the solid in an electric oven controlled at a suitable temperature.* The disadvantage of this method of drying is that the crystallised product is liable to become contaminated with filter-paper fibre.

With smaller amounts (e.g., 1–20 g) of more valuable recrystallised material the filter cake is transferred to a tared watch glass, broken down into small fragments without damaging the crystalline form, and air dried under another suitably supported watch glass before being placed into a temperature-controlled oven.

With low melting solids, the best method of drying is to place the crystals on a watch glass in a desiccator (Section **I,11**) charged with an appropriate substance to absorb the solvent. For general purposes, water vapour is absorbed by a charge of granular calcium chloride, concentrated sulphuric acid † or silica gel.‡ Methanol and ethanol vapours are absorbed by granular calcium chloride or silica gel. Vapours from diethyl ether, chloroform, carbon tetrachloride, benzene, toluene, light petroleum and similar solvents are absorbed by a charge

---

* Many students place carefully recrystallised samples into a heated oven maintained at a temperature higher than the melting point of their solid with inevitable results; this leads to undue waste of effort and chemicals. Even if the melting point is known, it is always advisable to make a trial with a small quantity on a watch glass. In fact, a temperature of about 50° over a period of 1–2 hours is usually adequate for the removal of the common organic solvents mentioned in Table I,8. If the material can be left overnight at this temperature complete removal of water will occur.

† If a solution of 18 g of barium sulphate in 1 litre of concentrated sulphuric acid is employed, a precipitate of barium sulphate will form when sufficient water has been absorbed to render it unfit for drying; recharging will then, of course, be necessary. Alternatively Merck Laboratory Chemicals manufacture a granular drying agent (supplied by Anderman & Co Ltd), which consists of an inert mineral carrier coated with sulphuric acid; the inclusion of an indicator, which turns from reddish-purple to colourless as water is absorbed, shows when recharging is necessary.

‡ It is usual to employ blue self-indicating silica gel crystals which turn a pale pink colour when regeneration by heating in an electric oven is required.

of freshly cut shavings of paraffin wax; since the sample may contain traces of moisture, it is advisable to insert also a dish containing a suitable desiccant. If the compound is moist with glacial acetic acid (e.g., from a recrystallisation of some 2,4-dinitrophenylhydrazones), or with concentrated hydrochloric acid (e.g., from a recrystallisation of an amine hydrochloride), a dual charge of silica gel or concentrated sulphuric acid together with a separate receptacle containing flake sodium hydroxide is necessary to absorb the water and acid vapours respectively. Samples which are to be used subsequently in reactions requiring anhydrous conditions are best dried in a desiccator charged with phosphoric oxide.*

Drying is more rapid in a *vacuum desiccator* of either of the designs shown in Section **I,11**. When exhausting a desiccator a filter flask trap should always be inserted between the desiccator and the pump. The vacuum should be applied gradually and should not exceed about 20 mm of mercury unless the precaution be taken of surrounding the desiccator by a cage of fine-mesh steel wire; the collapse of the desiccator will then do no harm.†

When using a vacuum desiccator, the vessel containing the substance (clock glass, etc.) should be covered with an inverted clock glass. This will protect the finer crystals from being swept away should the air, accidentally, be rapidly admitted to the desiccator. In actual practice the tube inside the desiccator leading from the stop-cock is bent so that the open end points in the direction of the lid, hence if the tap is only slightly opened and air allowed to enter slowly, there is little danger of the solid being blown from the clock glass or other receptacle.

Frequently the water or other solvent is so firmly held that it cannot be completely removed in a vacuum desiccator at ordinary temperatures. Large quantities of material (100 g upwards) must therefore be dried in a vacuum oven at higher temperatures, using one of the commercial designs which are available. For smaller amounts of recrystallised material a convenient laboratory form of vacuum oven is the so-called '*drying pistol*'. An interchangeable glass joint assembly ‡ is shown in Fig. I,84 where vapour from a boiling liquid in the flask A rises through the jacket surrounding the drying chamber B (holding the substance) and returns to the flask from the condenser; the drying chamber B is connected to the vessel C containing the drying agent; C is attached to a suction pump. The liquid in A is selected according to the temperature required, e.g., chloroform (61 °C), trichloroethylene (86 °C), water (100 °C), perchloroethylene (120 °C), s-tetrachloroethane (147 °C), etc. The charge in C consists of phosphoric oxide distributed on glass wool§ (to prevent 'caking') when water is to be removed, potassium hydroxide for removal of acid vapours, paraffin wax for removal of organic solvents such as chloroform, carbon tetrachloride, benzene, etc.

---

* Phosphoric oxide coated on a mineral carrier (Merck Laboratory Chemicals) is available from Anderman & Co Ltd, and is an efficient agent for desiccators. An indicator is incorporated which turns from colourless to blue as water absorption increases.

† The Desiguard, supplied by Fisher Scientific Co; also desiccator cages manufactured by Jencons (Scientific) Ltd.

‡ Available from Jobling Laboratory Division; a similar apparatus is also manufactured by Corning Glass Works. This apparatus is also referred to as the *Abderhalden vacuum drying apparatus*.

§ Alternatively the granular desiccant of phosphoric oxide coated on a mineral carrier may be used, supplied by Anderman & Co Ltd.

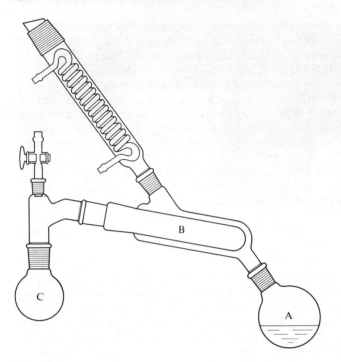

Fig. I,84

An electric vacuum drying pistol (Fig. I,85) is commercially available* and when the temperature control has been calibrated provides a convenient alternative.

It is essential that samples for elemental analysis are thoroughly dried in the above manner before submission.

Exercises in recrystallisation   To gain experience in recrystallisation technique the student should carry out the following experiments.

   Choice of solvent for recrystallisation.   Obtain small samples (about 0.5 g) of the following compounds from the storeroom: (i) salicylic acid, (ii) acetanilide, (iii) *m*-dinitrobenzene, (iv) naphthalene and (v) toluene-*p*-sulphonamide. Use the following solvents: distilled water, industrial spirit, rectified spirit, acetone, toluene, glacial acetic acid and hexane.

   Place 0.1 g of the substance in a semimicro test-tube (75 × 10 mm or 100 × 12 mm) and proceed systematically with the various solvents as detailed on p. 107. Finally, summarise your results, and indicate the most suitable solvent or solvents for the recrystallisation of each of the above compounds.

   1. **Acetanilide from water.**   Weigh out 4.0 g of commercial acetanilide into a 250 ml conical flask. Add 80 ml of water and heat nearly to the boiling point on an electric hot plate. The acetanilide will appear to melt and form an 'oil' in the solution. Add small portions of hot water, whilst stirring the mixture and boiling gently, until all the solid has dissolved (or almost completely dissolved). If the solution is not colourless, allow to cool slightly, add about 0.1 g of

---

* Available from A. Gallenkamp & Co Ltd, who kindly supplied the photograph.

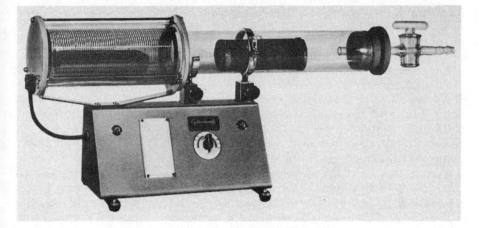

Fig. I,85

decolourising carbon and continue the boiling for a few minutes in order to remove the coloured impurities. Filter the boiling solution through a fluted filter-paper (for preparation, see p. 109) supported in a short-necked funnel; if the solution cannot be filtered in a single operation, keep the unfiltered portion hot by returning the conical flask to the hot plate. Alternatively, the solution may be filtered through a hot water funnel (Fig. I,41(a)). Collect the filtrate in a 250 ml conical flask. When all the solution has been filtered, cover the flask containing the hot filtrate with a clock glass and cool rapidly with swirling. Allow to stand for about 30 minutes to complete the separation of the solid. Filter with suction through a small Buchner funnel (Section I,19), wash the crystals twice with 5 ml portions of cold water (to remove the adhering mother-liquor) and press them in the funnel with a spatula or the back of a flat glass stopper. Remove the funnel from the filter flask, invert it on two thicknesses of filter or absorbent paper resting upon a pad of newspaper and if necessary dislodge the pad of crystals by tapping the funnel; allow the crystals to dry in the air. It is advisable in air drying to cover the crystals with a large clock glass resting upon corks, or the crystals may be covered with a large filter-paper perforated with a number of holes in order to allow the solvent to evaporate. For more rapid drying, the crystals may be placed on a clock glass or in an evaporating basin in an oven held at a temperature of about 80 °C. Weigh the yield of recrystallised material and determine the melting point. If the recrystallised product is not sufficiently pure (melting point low or melting over a range of several degrees), repeat the recrystallisation. Pure acetanilide has m.p. 114 °C.

If an m.p. determination is required soon after recrystallisation, a small quantity may be rapidly dried by pressing it several times upon a pad of several thicknesses of filter or absorbent paper and placing it upon a watch glass in a warm place. A piece of unglazed porous plate may also be used.

Optional or alternative experiments are the recrystallisation of 3.0 g of crude benzoic or salicylic acid from water.

2. **Naphthalene from alcohol (crystallisation from an inflammable solvent).** Weigh out 5.0 g of commercial naphthalene into a 100 ml round-

bottomed flask. Add 25 ml of rectified spirit (or of industrial spirit), 2–3 fragments of porous porcelain, and fit a reflux condenser (compare Fig. I,61; a guard-tube is not required here). Heat the mixture on a water or steam bath or in an electric heating mantle until the solvent boils. Add successive small volumes (each 2–3 ml) of the solvent, and boil gently after each addition, until the naphthalene has dissolved (apart from insoluble impurities). If the solution is coloured, remove it from the heat source, and when it has cooled somewhat add 0.2–0.3 g of decolourising charcoal and mix thoroughly. Boil the mixture for several minutes. Filter the hot solution through a fluted filter-paper or through a hot water funnel (**CAUTION**: all flames in the vicinity must be extinguished), and collect the filtrate in a conical flask. Cover the receiver with a watch glass and cool it in cold water. Stir or shake the solution as cooling proceeds. After 30 minutes, filter off the crystals through a small Buchner funnel at the water pump; wash all the crystals into the funnel by rinsing the flask with some of the filtrate. Discontinue the suction and wash the crystals with two 5-ml portions of chilled rectified or industrial spirit. Continue the suction and press the crystals down firmly with a spatula or a flat glass stopper. Dry the crystals on filter paper as in *1*. When dry, determine the weight and also the m.p. of the purified naphthalene. Pure naphthalene has m.p. 80 °C.

Alternative experiments: (*a*) Recrystallisation of crude benzoic acid (5.0 g) from methanol (30 ml); the wash liquid should be 50 per cent aqueous methanol. (*b*) Recrystallisation of acetanilide (5 g) from toluene (100 ml); filter through a preheated funnel.

3. **Sulphanilic acid from water.** Use 5.0 g of crude (grey) sulphanilic acid and proceed as in *1*. Add 0.2 g of decolourising carbon to the solution at 70–80 °C, and continue the boiling for several minutes. If the filtered solution is not colourless it must be boiled with a further 0.2 g of decolourising carbon. Filter the cold solution at the water pump, wash with a little cold water, dry and weigh the yield of recrystallised product.

I,21. SUBLIMATION TECHNIQUES—FREEZE DRYING Purification of some organic compounds may frequently be achieved by the technique of sublimation as an alternative, or, in addition, to recrystallisation. The success of the method depends upon the compound having a high enough vapour pressure at a temperature below the melting point, so that the rate of vaporisation from the solid will be rapid and the vapour may be condensed back to the solid upon a cooled surface. Impurities should have materially different vapour pressures to the compound undergoing purification so that they may be either removed with the initial sublimate or allowed to remain in the residue. The yield of sublimate will be greatly improved if the sublimation is carried out under reduced pressure, and further under these conditions the lower temperature employed reduces the possibility of thermal degradation. Substances having low vapour pressures at their melting points can only be sublimed under greatly diminished pressures (10^{-3} to 10^{-6} mmHg).

The theory of the sublimation process has been discussed in detail elsewhere (Ref. 38); the following describes the practical aspects of this technique which is applicable down to a few milligrams of material.

The simplest form of apparatus for the sublimation at atmospheric pressure of quantities of material in the region of 10 to 25 g consists of a porcelain dish covered with a filter-paper which has been perforated with a number of small holes; a watch glass of the same size, convex side uppermost, is placed upon the filter-paper. The substance is placed inside the dish, and the latter heated with a minute flame on a wire gauze or sand bath in a fume cupboard. The sublimate collects on the watch glass, and the filter-paper below prevents the sublimate from falling into the residue. The watch glass may be kept cool by covering it with several pieces of damp filter-paper and moistening these from time to time. A modification, for use with larger quantities of material, employs an inverted glass funnel with a plug of glass wool in the stem in place of the watch glass and supported on the porcelain dish by a narrow ring of asbestos board fitted near the rim. Upon heating the dish gently the vapour of the pure compound passes through the holes in the filter-paper and condenses on the inside walls of the funnel; care must be taken that the heat supply is adjusted so that the funnel does not become more than luke warm. An inverted water jacket (cf. Fig. I,41(*a*)), filled with cold water, gives excellent results.

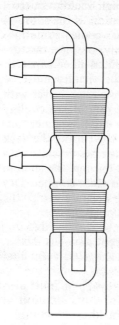

Fig. I,86

For sublimation under reduced pressure (**vacuum sublimation**), several designs of apparatus are available to suit a range of differently scaled operations. The sublimation of quantities of materials in the region of 5 g is usually conducted in an apparatus having the basic design shown in Fig. I,86.* The impure substance is placed in the bottom of the wider tube, the cold finger is inserted and connected to the water supply; the T-connection is attached to an oil vacuum pump. Evacuation of the system should be carried out slowly since the sudden removal of traces of moisture, solvent or air from the crystalline mass could cause spattering of the solid on to the cold finger, leading to contamination of the final sublimate. The outer tube is then heated by immersion in a bath of liquid paraffin or silicone oil, not by a direct flame, which would be difficult to control and which may lead to decomposition due to too high a temperature being attained. The temperature should be allowed to rise slowly and held at that level at which sublimation is seen to occur. Usually a 'misting' of the cold finger—provided that the sample was completely dry—is the first indication of sublimation. The process should not be hurried, either by raising the temperature too rapidly in the initial stages or by using finally too high a temperature. As a general guide the temperature should be in the region of 30 °C below the melting point of the solid or lower if sublimation takes place reasonably smoothly.

With care a mixture may be fractionated by sublimation; when the amount of sublimate formed at a particular temperature no longer seems to increase the sublimation process should be stopped and the sublimate removed. The cleaned cold finger is then reintroduced and the sublimation is continued at a higher temperature when further less volatile fractions may be obtained. The temperature and pressure of sublimation should be recorded, together with the melting point of the impure substance and of the sublimate, for reference purposes.

---

* The apparatus has been constructed to the editor's specification by R. B. Radley & Co Ltd.

Other designs are illustrated in Figs. I,87 and I,88 and may be purchased complete from some manufacturers* or alternatively made to specification; the sizes of the pot and of the cold finger are appropriate to the quantity of material to be sublimed, which may be as low as 20 mg. Frequently these assemblies, sometimes with slight modification, may be used for high vacuum micro distillation of viscous liquids (Section **I,28**).

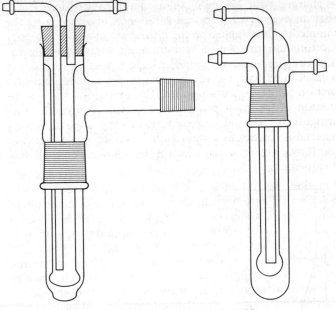

Fig. I,87   **Fig. I,88**

For the sublimation of quantities of materials in the region of 2 to 20 mg the commercially available **'micro-sublimation apparatus'**† is particularly suitable. Here the sample is loaded into a small glass capsule which is then located at the bottom of the closed end of a clean dry glass tube of 9 mm outside diameter being of approximately 15 cm length. A slight constriction is made near the open end of the glass tube by rotation in a suitable flame, and after cooling a small plug of cotton wool is inserted. The closed end of the tube is then introduced into the drilled cavities of the aluminium heater block, the open end of the tube is connected to an oil pump and the tube slowly evacuated to about 0.01 mmHg. The sample may then be heated slowly; an energy regulator allows any temperature between 50 to 350 °C (± 2 °C) to be selected. The sublimate will be seen to collect in the cooler part of the tube which lies outside the heater block. At the conclusion of the experiment the vacuum may be slowly released and the sublimate retrieved by cutting the glass tube appropriately.

Freeze drying. This process, frequently called lyophilisation, is necessary when water is to be removed from solutions containing heat-labile materials so that conventional distillation, even under reduced pressure, would cause exten-

---

* For example, Scientific Glass Apparatus Co.
† Manufactured by A. Gallenkamp & Co.

sive losses by decomposition. Examples are to be found in the removal of water from aqueous solutions of enzymes, polysaccharides, peptides, etc. In principle the aqueous solution is frozen in a suitable solid carbon dioxide freezing mixture (see Section I,12), and the ice is sublimed off to leave a dry residue (see Ref. 39).

Figure I,89 illustrates a commercially available* Quickfit lyophiliser accommodating four round-bottomed flasks (1 litre) for processing 1 litre of aqueous solution. Each flask is charged with 250 ml of solution and rotated in a Card-ice-acetone bath so that an even layer of frozen solution is obtained over the inside. The flasks are immediately attached to the refrigerant chamber which is filled with a Cardice-acetone mixture. An oil vacuum pump is connected to the refrigerant chamber via the supplementary trap, which if possible should be immersed in a Dewar flask filled with liquid nitrogen; such a cooled trap provides maximum protection for the vacuum pump. Vacuum is applied to the apparatus and sublimation of ice takes place over a period of several hours (best carried out overnight). Air should be readmitted to the apparatus very slowly as the dried material is frequently a light 'fluffy' powder and is liable to become dispersed. The flasks are removed and the ice allowed to melt and drain off through the stop-cock.

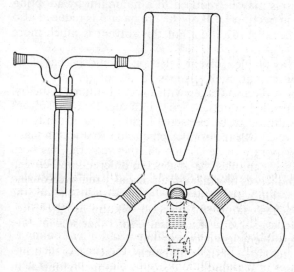

Fig. I,89

Other designs of apparatus, which are variations on this pattern, are available for freeze-drying smaller quantities of solution using test-tubes having ground glass joints or round-bottomed flasks of up to 50 ml capacity.

I,22. SOLVENT EXTRACTION As was pointed out in Section I,18, the crude products of most organic reactions are multi-component mixtures, and a convenient initial isolation procedure, for the first stages of both the separation of such mixtures and of the purification of the components, may involve sol-

---

* Manufactured by Jobling Laboratory Division.

vent extraction processes. The general cases which are discussed below to illustrate the technique of solvent extraction are selected to cover many of the commonly met systems. The student is however recommended to refer to the comments in Section **I,18** on the necessity of assessing the chemical and physical nature of the components of a particular reaction mixture with regard to their solubilities in solvents, and to their acidic, basic or neutral characteristics.

Extraction of liquids **Batch-extraction processes.** Perhaps one of the most frequent cases that is encountered is the separation of a neutral organic compound (or compounds) from a solution or suspension (as either a solid or liquid) in an aqueous medium, by shaking with an organic solvent in which the compound is soluble and which is immiscible (or nearly immiscible) with water.

The solvents generally employed for extraction are diethyl ether or di-isopropyl ether, benzene or toluene, chloroform, methylene chloride and light petroleum. The solvent selected will depend upon the solubility of the substance to be extracted in that solvent and upon the ease with which the solvent can be separated from the solute. Diethyl ether, owing to its powerful solvent properties and its low boiling point (35 °C), thus rendering its removal extremely facile, is very widely used; its chief disadvantage lies in the great fire hazard attending its use, but this may be reduced to a minimum by adopting the general precautions given in Section **I,3,B,5**. The fire hazard is reduced also by employing di-isopropyl ether (b.p. 67.5 °C), but this solvent is much more expensive than diethyl ether.

If prior information is not available, solvent selection should be based on some small-scale trials. A few millilitres of suspension or solution to be extracted are placed in a small test-tube and shaken with an equal volume of diethyl ether, when dissolution of suspended material clearly indicates that the solvent would be satisfactory. If the solution to be extracted is homogeneous initially, then the ether solution is removed with a dropper pipette on to a watch glass, and the ether is allowed to evaporate to determine whether material has been extracted. A little experience soon enables the student to differentiate between organic liquids so extracted and traces of water simultaneously removed during the extraction process. If extraction with diethyl ether proves unsatisfactory the experiment is repeated with a fresh sample of reaction mixture using chloroform (b.p. 61 °C) as the extraction solvent. If necessary, the other solvents are tried similarly until a suitable solvent has been selected.

By way of illustration, the technique of the bulk batch-extraction of an aqueous solution with diethyl ether* is as follows. A separatory funnel (globular or pear-shaped with short stem and fitted with a ground glass interchangeable stopper) is selected of about twice the volume as that to be extracted and mounted in a ring on a stand with a firm base. The barrel and plug of the stop-cock are dried with a linen cloth and in the case of glass stop-cocks lightly treated with a suitable lubricant (Vaseline, Cello-grease,† 'Silicone stop-cock grease‡). The solution and the extraction solvent (usually about one-third of

---

* The diethyl ether (frequently abbreviated to ether) should be reasonably free from 'peroxides', see Section **II,1,15**.

† Cello-grease marketed by Fisher Scientific Co.

‡ 'Silicone' stop-cock grease—Midland Silicones Ltd, a subsidiary of Dow Corning; ICI Silicones Ltd.

the volume of the solution, but see theory of extraction, below) are introduced into the funnel, and the latter stoppered. All naked flames in the immediate vicinity should be extinguished. The funnel with the stopper firmly held in place is then shaken gently (so that the excess vapour pressure* will be developed slowly), inverted and the stop-cock opened in order to relieve the excess pressure. The stop-cock is again closed, the funnel again shaken and the internal pressure released. When the atmosphere inside the funnel is saturated with ether vapour, further shaking develops little or no additional pressure. At this stage, the funnel is vigorously shaken for 2–3 minutes to ensure the maximum possible transfer of the organic substance to the ether layer, and then returned to the stand in order to allow the mixture to settle. When two sharply defined layers have formed, the lower aqueous layer is run off and separated as completely as possible. The residual ethereal layer is then poured out through the upper neck of the funnel; contamination with any drops of the aqueous solution still remaining in the stem of the funnel is thus avoided. The aqueous solution may now be returned to the funnel and the extraction repeated, using fresh ether on each occasion until the extraction is complete. Not more than three extractions are usually required, but the exact number of extractions will naturally depend upon the partition coefficient of the substance between water and ether. The completeness of the extraction can always be determined by evaporating a portion of the last extract on the water bath and noting the amount of residue. The combined ethereal solutions are dried with an appropriate reagent (Section **I,23**), and the ether removed on a water bath (Sections **I,24** and **I,27**). The residual organic material is now further purified, depending upon its properties and the organic impurities removed in the extraction, by chromatography, by recrystallisation or by distillation. It is also important to retain the aqueous solution until the final purified product is isolated so that incorrect observations on the solubility characteristics of the required product do not lead to premature discarding of the product.

Occasionally **emulsions** are formed in the extraction of aqueous solution by organic solvents, thus rendering a clean separation impossible. Emulsion formation is particularly liable to occur when the aqueous solution is alkaline, and when benzene or chloroform is the extracting solvent. The emulsion may be broken by any of the following devices, but in general its occurrence may be minimised by using a very careful swirling action in the shaking of the separatory funnel during the initial extraction; only in the final extraction is a more vigorous action adopted.

1. Mechanical means, such as agitating the end of a glass rod at the interface of the emulsion with the growing liquid phase, or alternatively gentle rocking of the funnel or gentle swirling, may be successful. Slow filtration through a compacted pad of glass wool in a Hirsch or Buchner funnel is often satisfactory.

2. An increase in concentration of ionic species may be helpful as the result of

---

* When ether is poured into a funnel containing an aqueous solution, a two liquid-phase system is formed. If the funnel is stoppered and the mixture shaken the vapour pressure of the ether (300–500 mm according to the temperature) is ultimately added to the pressure of the air (about 760 mm) plus water vapour, thus producing excess of pressure inside the funnel. Hence the necessity for shaking gently and releasing the pressure from time to time until the air has been expelled.

the addition of sodium chloride, sodium sulphate or potassium carbonate, for example. With extractions involving alkaline solutions the addition of dilute sulphuric acid may be helpful, providing that complete neutralisation or acidification does not take place since this may result in a change in the chemical nature of some of the components (see below).

3. Emulsions may sometimes be broken by the addition of a few drops of alcohol or other suitable solvent from a dropper pipette, the outlet of which is sited at the emulsion–liquid interface.

4. A satisfactory separation is frequently obtained if the mixture is simply allowed to stand for some time.

In the isolation of organic compounds from aqueous solutions, use is frequently made of the fact that the solubility of many organic substances in water is considerably decreased by the presence of dissolved inorganic salts (sodium chloride, calcium chloride, ammonium sulphate, etc.). This is the so-called **salting-out effect**. A further advantage is that the solubility of partially miscible organic solvents, such as ether, is considerably less in the salt solution, thus reducing the loss of solvent in extractions.

The process of extraction is concerned with the **distribution law** or **partition law** which states that if to a system of two liquid layers, made up of two immiscible or slightly miscible components, is added a quantity of a third substance soluble in both layers, then the substance distributes itself between the two layers so that the ratio of the concentration in one solvent to the concentration in the second solvent remains constant at constant temperature. It is assumed that the molecular state of the substance is the same in both solvents.* If c_A and c_B are concentrations in the layers A and B, then, at constant temperature:

$$c_A/c_B = \text{constant} = K$$

The constant K is termed the **distribution or partition coefficient**. As a very rough approximation the distribution coefficient may be assumed equal to the ratio of the solubilities in the two solvents. Organic compounds are usually relatively more soluble in organic solvents than in water, hence they may be extracted from aqueous solutions. If electrolytes, e.g., sodium chloride, are added to the aqueous solution, the solubility of the organic substance is lowered, i.e., it will be **salted out**: this will assist the extraction of the organic compound.

The problem that arises in extraction is the following. Given a limited quantity of the solvent, should this be used in one operation or divided into several portions for repeated extractions in order to secure the best result? A general solution may be derived as follows. Let the volume v ml of the aqueous solution containing w_0 grams of the dissolved substance be repeatedly extracted with fresh portions of s ml of the organic solvent, which is immiscible with water. If w_1 grams is the weight of the solute remaining in the aqueous phase after the first extraction, then the concentrations are w_1/v g per ml in the aqueous

---

* For a theoretical treatment involving association or dissociation in one solvent, see Refs. 40 and 41.

phase and $(w_0 - w_1)/s$ g per ml in the organic solvent layer. The partition coefficient K is given by:

$$\frac{w_1/v}{(w_0 - w_1)/s} = K$$

or $\qquad w_1 = w_0 \dfrac{Kv}{Kv + s}$

Let w_2 grams remain in the aqueous layer after the second extraction, then:

$$\frac{w_2/v}{(w_1 - w_2)/s} = K$$

or $\qquad w_2 = w_1 \dfrac{Kv}{Kv + s}$

$$= w_0 \left(\frac{Kv}{Kv + s}\right)^2$$

Similarly if w_n grams remain in the aqueous layer after the nth extraction:

$$w_n = w_0 \left(\frac{Kv}{Kv + s}\right)^n \tag{1}$$

We desire to make w_n as small as possible for a given weight of solvent, i.e., the product of n and s is constant, hence n should be large and s small; in other words, the best results are obtained by dividing the extraction solvent into several portions rather than by making a single extraction with the whole quantity. It must be emphasised that the expression deduced above applies strictly to a solvent which may be regarded as completely immiscible with water, such as benzene, chloroform or carbon tetrachloride; if the solvent is slightly miscible, e.g., ether, the equation (1) is only approximate, but is nevertheless useful for indicating the qualitative nature of the results to be expected.

Let us consider a specific example, viz., the extraction of a solution of 4.0 g of butyric acid in 100 ml of water at 15 °C with 100 ml of benzene at 15 °C. The partition coefficient of the acid between benzene and water may be taken as 3 (or $\frac{1}{3}$ between water and benzene) at 15 °C. For a single extraction with benzene, we have:

$$w_n = 4 \left(\frac{\frac{1}{3} \times 100}{\frac{100}{3} + 100}\right) = 1.0 \text{ g}$$

For three extractions with 33.3 ml portions of fresh benzene:

$$w_n = 4 \left(\frac{\frac{1}{3} \times 100}{\frac{100}{3} + 33.3}\right)^3 = 0.5 \text{ g}$$

Hence one extraction with 100 ml of benzene removes 3.0 g (or 75%) of the butyric acid, whilst three extractions remove 3.5 g (or 87.5%) of the total acid. This clearly shows the greater efficiency of extraction obtainable with several extractions when the total volume of solvent is the same. Moreover, the smaller the distribution coefficient between the organic solvent and the water, the larger the number of extractions that will be necessary.

The above considerations apply also to the removal of a soluble impurity by extraction (or washing) with an immiscible solvent. Several washings with portions of the solvent give better results than a single washing with the same total volume of the solvent.

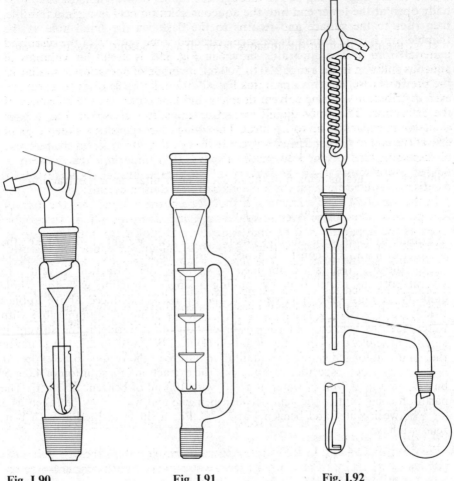

Fig. I,90   Fig. I,91   Fig. I,92

Continuous extractions of liquids. When the organic compound is more soluble in water than in the organic solvent (i.e., the distribution coefficient between the organic solvent and water is small), very large quantities of organic solvent must be employed to obtain even a moderately efficient extraction. This may be avoided by the use of an apparatus for continuous extraction where only relatively small volumes of solvent are required. Various designs of apparatus* are available according to whether the aqueous solution is to be extracted with an organic solvent which is heavier or lighter than water.

Figures I,90, I,91 and I,92 illustrate apparatus employed for the extraction of

* Available from Jobling Laboratory Division.

aqueous solution by solvents lighter than water, such as ether or benzene (*liquid/liquid extraction by upward displacement*). That shown in Fig. I,90 is suitable for small amounts of aqueous solution (6–8 ml); the solvent distils from a flask (attached to the lower end) and condenses in a reflux condenser (attached to the upper end), passes through the funnel down a narrow tube partially open at the lower end into the aqueous solution held in a glass thimble, then rises to the surface and returns to the flask via the small hole at the thimble top, having during its passage extracted some portion of the dissolved material from it. The apparatus shown in Fig. I,91 is useful for volumes of aqueous solution in the range 100 to 500 ml; its mode of operation is similar to the previous case, but this apparatus has additionally baffle discs to assist the even distribution of rising solvent droplets and hence increase the efficiency of the extraction. The liquid–liquid extraction apparatus shown in Fig. I,92 is available in capacities up to 2.5 litres. This design incorporates a sintered glass disc at the end of the condensed solvent delivery tube, which serves the purpose of dispersing the solvent into small droplets. It is important that solvent is poured continuously down the delivery tube to escape through the sinter whilst the disc is lowered through the aqueous solution during assembly.

In the use of all these extractors, it should be borne in mind that the extraction process takes place over several hours and due precautions should be taken as the apparatus will be unattended (see Section **I,16**). Furthermore as the extraction solvent is more than likely to be inflammable, due attention to fire hazards should be taken. The flasks of solvent should be heated with a heating mantle and the level of the solvent in the flask should be above the ring of contact between the mantle element and the outer glass surface, otherwise there is danger of prolonged and possibly harmful overheating of material left as a ring above the solvent level. A further point to note with the use of all these types of apparatus is that the level of aqueous solution should be substantially below the thimble holes, or the extractor side arms, even when the baffle discs, etc., are in position. This is necessary since the volumes of aqueous solution may increase, either owing to the small solubility in water of the extracting solvent (and this is particularly noticeable with diethyl ether) or to a small rise in the temperature of the aqueous solution during extraction leading to expansion; this could lead to some of the aqueous solution being carried over into the solvent flask.

Figures I,93, I,94 and I,95 illustrate apparatus employed for the extraction of aqueous solutions by solvent heavier than water such as methylene chloride or carbon tetrachloride (*liquid–liquid extraction by downward displacement*).

That shown in Fig. I,93, when fitted with flask and condenser, is suitable for extractions of about 10 ml of solution. The condensed solvent drops through the funnel and thence down through the solution and escapes via the side arm sealed into the bottom of the extractor thimble. When assembling the apparatus it is advisable to pour a few millimetres of the extracting solvent into the thimble before pouring in the liquid to be extracted; in this way contamination of the solvent in the flask by a carry over of solution to be extracted is minimised. The apparatus shown in Fig. I,94, when fitted with a flask and condenser, is suitable for the extraction of about 50 ml of aqueous solution; the baffle discs improve the dispersion of solvent into droplets. Some of the solvent should be placed in the extraction vessel first, then the baffle plates and finally the aqueous solution; further addition of solvent to prevent the passage of

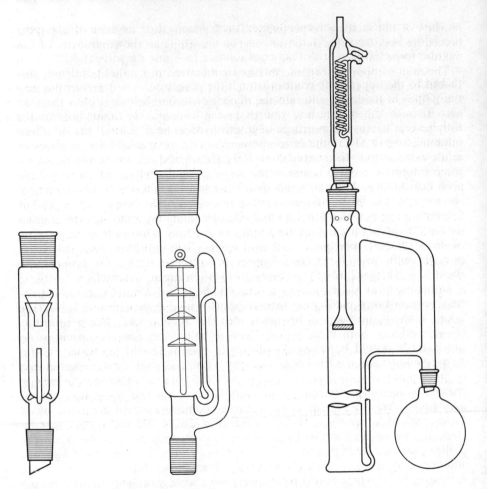

Fig. I,93   **Fig. I,94**   **Fig. I,95**

aqueous solution down the solvent return tube may be necessary. The apparatus shown in Fig. I,95 is suitable for the extraction of up to 1 litre of aqueous solution. Here the solvent vapours pass through the holes at the top of the solvent delivery tube to condense in the reflux condenser, and pass through the delivery tube and thence through the sintered disc which should be located just under the surface of the solution to be extracted. The fine droplets collect at the base of the extractor which should have been previously loaded with sufficient solvent so that with the weight of aqueous solution, the solvent level in the side arm coincides with the horizontal portion of the solvent return tube.

Extraction by chemically active solvents. Not infrequently the crude organic product from a reaction may contain a mixture of acidic (phenolic and carboxylic acids), basic and neutral components in various combinations. Some of these components may of course be impurities, but none the less, whether as a preliminary purification stage or as a means of separating the mixture, a carefully planned solvent extraction procedure may be adopted using acidic and basic reagents which react chemically with the basic and acidic components of the mixture respectively. The following full account of a typical procedure may be abbreviated in practice according to the complexity of the mixture to be handled.

The multicomponent organic mixture is dissolved in a suitable solvent; this should be diethyl ether if at all possible for the reasons stated above, but any low boiling, water-immiscible solvent (light petroleum, methylene chloride, etc.) may be used. This solution is now shaken in a separatory funnel (see above) with several successive portions of dilute hydrochloric acid (1 M) or dilute sulphuric acid (1 M; (1)). Basic components are thus extracted into the aqueous acidic extract, and the combined extracts are washed once with the clean organic solvent to remove traces of the original organic phase which may have been carried over into the aqueous extract (this is called a 'back-extraction' process, (2)). The basic components are recovered (unless they only represent known impurities which are not required), by cooling the aqueous extract in an ice bath, basifying it carefully by adding an aqueous solution of sodium hydroxide (5 M; see (3)) dropwise and with stirring, extracting the precipitated oil or solid with organic solvent, and drying (Section **I,23**) and evaporating (Section **I,27**) the extract. The original organic solution from which the basic components have been removed is now extracted with several successive portions of dilute aqueous sodium hydroxide or sodium carbonate solution (1 M). Acidic components will be extracted into the aqueous alkaline layer. After 'back-extraction' with fresh organic solvent, the acidic components may be recovered (if necessary) by cooling the alkaline extract, acidifying by the careful dropwise addition of hydrochloric acid (5 M; (4)), extracting the precipitated solid or liquid with an organic solvent, and drying and evaporating the extract (5). The original organic solution now only contains neutral components; these may be recovered by washing the solution first with a little dilute aqueous hydrochloric acid to remove traces of alkali, then with distilled water until the washings are neutral. The organic solution is finally dried and the solvent is evaporated.

Notes. (1) The volumes of organic and aqueous solvents to be used depend, of course, on the quantities of material to be handled. As a guide, a 5 g mixture may be dissolved in 30 ml of organic solvent and extracted with three successive portions of 10 ml of aqueous acid. The student should always check the completeness of the extraction by removing a little of the final extract into a test-tube and adding a little concentrated sodium hydroxide solution to make the solution alkaline; a cloudiness suggests that further extractions of the original solution with aqueous acid are necessary. Because of the conversion of the base component into its water-soluble salt, almost complete removal of the base from the mixture is achieved in relatively few extractions.

(2) The solvent washings are best returned to the original solvent solution which in any case may require 'topping-up' in a prolonged extraction process.

(3) The reason for using a concentrated solution of alkali is to keep the final total volume of aqueous solution to a minimum to facilitate the subsequent recovery of the basic components. If no precipitate is visible but the aroma of an amine is noticeable, this implies some degree of water solubility; recovery is then best attempted using one of the continuous extraction techniques.

(4) If aqueous sodium carbonate has been used then considerable effervescence will accompany this acidification process. It is advisable that a flask

which is large compared to the volume of solution to be treated should be employed and that the solution be shaken vigorously during the addition of acid.

(5) Extraction of the original solution with sodium hydroxide will have removed phenols, enols and carboxylic acids. Separation of these may be readily accomplished by redissolving the acidic components in diethyl ether (or other suitable solvent). Extraction with saturated aqueous sodium hydrogen carbonate will remove the carboxylic acids, enabling the phenolic (or enolic) components to be recovered by evaporating the dried organic phase. Acidification of the aqueous extract will liberate the carboxylic acid components which may then be isolated by extraction in the usual way.

Whilst the above details provide a general procedure for handling mixtures of acidic, basic and neutral components, other selective extraction reagents may be utilised in certain special instances. For example, cold concentrated sulphuric acid will remove unsaturated hydrocarbons (alkenes and alkynes) present in saturated hydrocarbons, or alcohols and ethers present in alkyl halides. In the former case soluble sulphonated products are formed whilst in the latter case alkyl hydrogen sulphates or addition complexes that are soluble in the concentrated acid are produced. Another example is provided by the removal of contaminating benzaldehyde from the benzyl alcohol obtained by the Cannizzaro reaction (Section **IV,145**).

Extraction of solids The process is generally applied to the removal of natural products from dried tissue originating from plants, fungi, seaweed, mammals, etc. The steam volatile natural products (e.g., those occurring in the essential oils) such as the alcohols, esters and carbonyl compounds of the aliphatic (both acyclic and alicyclic) and the simpler aromatic systems, are removed by steam distillation (Section **I,25**). The non-steam-volatile compounds may be removed by solvent extraction using a batch or continuous process. Not infrequently a comprehensive study of the range of organic substances in a particular tissue requires extraction with a succession of solvents starting with light petroleum (b.p. $\approx 40\,°C$) for the removal of the least polar components (e.g., higher homologues of terpenes, steroids, etc.), progressing through to more polar solvents such as diethyl ether, acetone, ethanol and finally water for the sequential removal of the more polar compounds (e.g., amino acids, carbohydrates, etc.).

The **batch process**, which tends to be less efficient than the continuous extraction process, involves macerating the tissue with the appropriate solvent in a Waring Blender, soaking for a short time (1), filtering in a suitable size of Buchner funnel and then returning the residue to fresh solvent for further extraction. The combined solvent extracts are then evaporated, usually under reduced pressure and the residue submitted to appropriate fractionation procedures (2).

Notes. (1) Warming the suspended solid in the solvent may be necessary by removing the 'porridge' to a suitable flask and heating under reflux. Care must be taken if such is the case to carefully supervise this operation as there may be considerable tendency towards 'bumping'. It should also be borne in mind that this batch-extraction process uses open type vessels and usually large volumes of solvents; precautions must therefore be taken in relation to the possible fire and toxic hazards involved in the use of a particular solvent.

(2) As a first step, this procedure would involve solvent extraction procedures to divide the multi-component mixture into acidic, basic and neutral fractions (see above). Subsequently chromatography, fractional crystallisation, etc., would be employed as appropriate.

For the **continuous extraction of a solid** by a hot solvent, it is better to use a Soxhlet extraction apparatus such as that shown in Fig. I,96. The solid substance is placed in the porous thimble A (made of tough filter-paper) and the

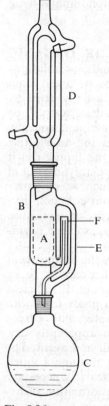

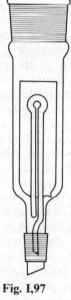

Fig. I,96   **Fig. I,97**

latter is placed in the inner tube of the Soxhlet apparatus. The apparatus is then fitted to a round-bottomed flask C of appropriate size containing the solvent and boiling chips, and to a reflux condenser D (preferably of the double surface type). The solvent is boiled gently; the vapour passes up through the tube E, is condensed by the condenser D, and the condensed solvent falls into the thimble A and slowly fills the body of the Soxhlet.* When the solvent reaches the top of the tube F, it siphons over into the flask C, and thus removes that portion of the substance which it has extracted in A. The process is repeated automatically until complete extraction is effected. The extracted com-

---

* For solids of low density, the top of the porous thimble A should.be above the siphon tube F, otherwise the solid may tend to float out of the thimble and pass down the siphon tube; a plug of glass wool may also be placed within the top of the thimble.

pound may be isolated from its solution in C by any of the usual methods. One minor disadvantage of this apparatus is that the temperature of the liquid in A differs considerably from the boiling point of the solvent; extraction is thus effected by the lukewarm liquid and is therefore relatively slow, particularly if the solubility of the substance increases markedly with temperature. This disadvantage is absent in the modifications illustrated in Fig. I,97 in which the part of the apparatus housing the extraction thimble is surrounded by the vapour of the solvent: extraction is accordingly effected by the hot solvent. The capacity of the Soxhlet extractor is quoted in terms of the siphoning volume; sizes ranging from 6 ml to 5 litre are available.

1,23. DRYING OF LIQUIDS OR OF SOLUTIONS OF ORGANIC COMPOUNDS IN ORGANIC SOLVENTS

Organic liquids, or solutions of organic substances in organic solvents such as would be obtained from solvent extraction procedures described in Section I,22, are usually dried by direct contact with a solid drying agent. The selection of the desiccant will be governed by the following considerations: (i) it must not combine chemically with the organic compound; (ii) it should have a rapid action and an effective drying capacity; (iii) it should not dissolve appreciably in the liquid; (iv) it should be as economical as possible; and (v) it should have no catalytic effect in promoting chemical reactions of the organic compound, such as polymerisation, condensation reactions and auto-oxidation. The various common drying agents are discussed in detail below; the drying of solvents, including the use of molecular sieves, together with other special techniques of solvent purification is discussed in Section II,1.

It is generally best to shake the liquid with small amounts of the drying agent until no further action appears to take place; too large an excess is to be avoided in order to keep adsorption losses down to a minimum. If sufficient water is present to cause the separation of a small aqueous phase (e.g., with calcium chloride), this must be removed* and the liquid treated with a fresh portion of the desiccant. If time permits, the liquid, when apparently dry, should be filtered and left overnight in contact with fresh drying agent. The desiccant should, in general, be separated by filtration (best through a fluted filter-paper) before the distillation of the liquid. This is particularly necessary with many reagents whose drying action depends upon the formation of hydrates (e.g., sodium sulphate, magnesium sulphate and calcium chloride): at higher temperatures the vapour pressures above the salts become appreciable and unless the salts are removed, much, if not all, of the water may be returned to the distillate. However, with some desiccating agents (e.g., calcium oxide, phosphoric oxide), the reaction products with water are quite stable and filtration is not essential.

A **list of the common drying agents** with their practical limitations and their important applications follows.

Anhydrous calcium chloride. This reagent is widely employed because of its high drying capacity and its cheapness. It has a high water-absorption capacity (since it forms $CaCl_2,6H_2O$ below 30 °C) but is not very rapid in its action;

---

* The aqueous phase may be removed rapidly and conveniently by filtration through a Phase Separating Paper (manufactured by W. & R. Balston Ltd) and washing with a small quantity of the dry solvent. The water-repellent paper retains the aqueous layer.

ample time must therefore be given for desiccation. The slowness of the action is attributed to the blanketing of the particles of calcium chloride with a thin layer of the solution formed by the extraction of the water present; on standing, the water combines forming a solid lower hydrate, which is also a desiccating agent.

The industrial process for preparing the reagent usually permits a little hydrolysis to occur, and the product may contain some free calcium hydroxide or basic chloride. It cannot therefore be employed for drying acids or acidic liquids. Calcium chloride combines with alcohols, phenols, amines, amino acids, amides, ketones and some aldehydes and esters, and thus cannot be used with these classes of compounds.

Magnesium sulphate. The most effective commercially available form of this desiccant is the monohydrate; a cheaper grade contains from 30 to 40 per cent of water but this retains useful desiccating action (the fully hydrated form is the heptahydrate). It is an excellent neutral desiccant, rapid in its action, chemically inert and fairly efficient, and can be employed for most compounds including those (e.g., esters, aldehydes, ketones, nitriles, amides) to which calcium chloride is not applicable.

Anhydrous sodium sulphate. This is a neutral drying agent, is inexpensive, and has a high water-absorption capacity (forming $Na_2SO_4,10H_2O$, below 32.4 °C). It can be used on almost all occasions, but the drying action is slow and not thorough. The desiccant is valuable for the preliminary removal of large quantities of water. Sodium sulphate is an inefficient drying agent for solvents such as benzene and toluene and is useless as a desiccant above 32.4 °C, at which temperature the decahydrate begins to lose water of crystallisation.

Anhydrous calcium sulphate. When the dihydrate $CaSO_4,2H_2O$ or the hemihydrate $2CaSO_4,H_2O$ is heated in an oven at 230–240 °C for about three hours, anhydrous calcium sulphate is obtained. It is sold commercially under the name of '*Drierite*' (not to be confused with '*Dehydrite*', which is anhydrous magnesium perchlorate). The reagent is extremely rapid and efficient in its action, is chemically inert, and is insoluble in organic solvents; it may therefore be used with most organic compounds. The only disadvantage is its limited capacity for absorption of water since it passes into the hemihydrate $2CaSO_4,H_2O$, and should theoretically absorb only 6.6 per cent of its weight of water to retain its maximum efficiency; where extreme desiccation is not essential, the porous commercial product may absorb up to about 10 per cent of its weight of water. It is recommended that the solution or liquid be subjected to a preliminary drying with magnesium or sodium sulphate, before using anhydrous calcium sulphate.

Anhydrous potassium carbonate. This drying agent possesses a moderate efficiency and drying capacity (the dihydrate is formed). It is applied to the drying of nitriles, ketones, esters and some alcohols, but cannot be employed for acids, phenols and other acidic substances. It also sometimes replaces sodium hydroxide or potassium hydroxide for amines, when a strongly alkaline reagent is to be avoided. Potassium carbonate frequently finds application in the salting-out of water-soluble alcohols, amines and ketones, and as a preliminary drying agent. In many cases it may be replaced by the desiccant magnesium sulphate.

Sodium and potassium hydroxides. The use of these efficient reagents should usually be confined to the drying of amines (soda lime, barium oxide or calcium oxide may also be employed), potassium hydroxide is somewhat superior to the sodium compound. These bases react with many organic compounds (e.g., acids, phenols, esters and amides) in the presence of water, and with some common solvents (e.g., chloroform) so that their use as desiccants is very limited.

Calcium oxide. This reagent is commonly used for the drying of alcohols of low molecular weight; its action is improved by preheating to 700–900 °C in an electric furnace. Both calcium oxide and calcium hydroxide are insoluble in the solvents, stable to heat, and practically non-volatile, hence the reagent need not be removed before distillation.* Owing to its high alkalinity, it cannot be used for acidic compounds or for esters; the latter would undergo hydrolysis.

Phosphoric oxide. This is an extremely efficient reagent and is rapid in its action. Phosphoric oxide is difficult to handle, channels badly, is expensive and tends to form a protective syrupy coating on its surface. A preliminary drying with anhydrous magnesium sulphate, etc., should precede its use. Phosphoric oxide is only employed when extreme desiccation is required. It may be used for hydrocarbons, ethers, alkyl and aryl halides and nitriles, but not for alcohols, acids, amines and ketones.

Drying by distillation. In most cases the distillation of organic preparations before drying is regarded as bad technique, but in a number of instances of solvents or liquids, which are practically insoluble in water, the process of distillation itself effects the drying. In short, advantage is taken of the formation of binary and ternary mixtures of minimum boiling point. Thus if moist benzene is distilled, the first fraction consists of a mixture of benzene and water (the constant boiling point mixture, b.p. 69 °C, contains 9 per cent of water); after the water has been removed, dry benzene distils. Other solvents which may be dried in this manner include carbon tetrachloride, toluene, xylene, hexane, heptane, light petroleum, 1,4-dioxan and ethylene dichloride. The dry solvent should not be collected until after about 10 per cent of the main bulk has passed over, since it is necessary to eliminate also the moisture adsorbed by the walls of the flask and the condenser. If moist aniline (b.p. 184 °C) or moist nitrobenzene (b.p. 210 °C) is distilled, the moisture is rapidly removed in the first portion of the distillate and the remainder of the liquid passes over dry. Sometimes a moist liquid preparation, which is sparingly soluble in water, is dried by admixture with a solvent (frequently benzene †) immiscible with water, and the resulting mixture is distilled. Thus when a mixture of pentanoic acid, water and benzene is distilled, the mixture of benzene and water passes over first (b.p. 69.3 °C), this is followed by dry benzene (b.p. 80 °C), and finally by dry pentanoic acid (b.p. 186 °C). This method has been used for the drying of commercial preparations of 3-methylbutanoic acid and higher aliphatic carboxylic acids by distillation with about 40 per cent of the weight of benzene until the temperature of the vapours reaches 100 °C. The dehydration of crystallised oxalic acid by distillation with carbon tetrachloride is sometimes regarded as another example of the use of a binary mixture for the removal of water (see also pinacol from pinacol hydrate, Section III,29).

---

* Some finely divided particles of solid may be carried over during the distillation from calcium oxide. It is recommended that the head of the ground glass distillation assembly leading to the condenser be filled with purified glass wool in order to retain the finely-divided solid. The purified glass wool is prepared by boiling commercial glass wool with concentrated nitric acid for about 15 minutes, washing thoroughly with distilled water, and drying at 120 °C. Alternatively it may be more convenient to use a splash head (Fig. I,101).

† Benzene may with advantage be replaced by the much less toxic toluene; this forms a binary azeotrope with water, b.p. 85 °C, containing 20 per cent of water.

The following is an example of the use of a ternary mixture in the drying of a solid. D-Fructose (laevulose) is dissolved in warm absolute ethyl alcohol, benzene is added, and the mixture is fractionated. A ternary mixture, alcohol–benzene–water, b.p. 64 °C, distils first, and then the binary mixture, benzene–alcohol, b.p. 68.3 °C. The residual, dry alcoholic solution is partially distilled and the concentrated solution is allowed to crystallise: the anhydrous sugar separates.

Table I,9   Common drying agents for organic compounds

Alcohols	Anhydrous potassium carbonate; anhydrous calcium sulphate or magnesium sulphate; calcium oxide.
Alkyl halides **Aryl halides**	Anhydrous calcium chloride; anhydrous calcium sulphate or magnesium sulphate; phosphoric oxide.
Saturated and aromatic hydrocarbons **Ethers**	Anhydrous calcium chloride; anhydrous calcium sulphate; phosphoric oxide.
Aldehydes	Anhydrous calcium sulphate; magnesium sulphate or anhydrous sodium sulphate.
Ketones	Anhydrous calcium sulphate; magnesium sulphate or anhydrous sodium sulphate; anhydrous potassium carbonate.
Organic bases (amines)	Solid potassium or sodium hydroxides; calcium oxide or barium oxide.
Organic acids	Anhydrous calcium sulphate, magnesium sulphate or anhydrous sodium sulphate.

I,24. DISTILLATION AT ATMOSPHERIC PRESSURE A typical assembly for the purification of liquids by simple distillation at atmosphere pressure is shown in Fig. I,98. The flask may be of any appropriate size, although small quantities (between 3 and 25 ml) are best distilled in pear-shaped flasks; the flask when charged with liquid should be one-half to two-thirds full. The screw-cap adapter on the still-head allows the bulb of the thermometer to be located slightly below the level of the side tube. If the boiling point of the liquid is likely to be above 150 °C the water-cooled condenser shown is replaced by an unjacketed tube fitted with ground glass joints at each end to act as an air-cooled condenser. A drying tube attached to the side-arm adapter may be filled with anhydrous calcium chloride held in position by loose plugs of cotton wool if it is desired to protect the distillate from moisture in the atmosphere. When the distillation flask has been charged with liquid, a few fragments of unglazed porous porcelain (porous pot) to promote regular ebullition in the subsequent heating are added; they should never be added to the hot liquid.* The flask may be heated on an asbestos-centred wire gauze or preferably in a

---

* Other aids to regular boiling include the addition of the following: fragments of pumice stone or of carborundum; small strips of Teflon tape, $\frac{3}{4}$ in. wide or of shredded Teflon (these may be washed with an organic solvent, dried and reused); small pieces of platinum wire (use is made of the well-known property of platinum in absorbing large quantities of gases).

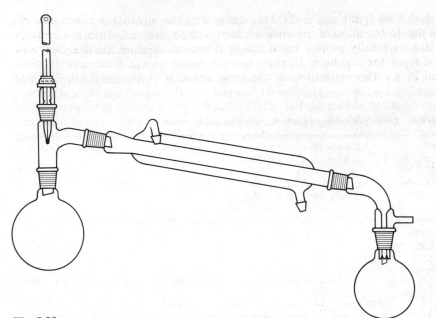

Fig. I,98

bath (**Section I,13**) appropriate to the boiling point of the liquid undergoing distillation. For small quantities of liquid contained in pear-shaped flasks careful heating with a semi-luminous bunsen flame may be used. Heating may be rather rapid until boiling commences; the rate of heating must then be reduced and the source of heat adjusted so that the distillate is collected at the rate of one or two drops per second. It must be borne in mind that at the commencement of the distillation it takes an appreciable time for the vapour to heat the upper part of the flask and the thermometer. The distillation should not be conducted too slowly, for the thermometer may momentarily cool from lack of a constant supply of fresh vapour on the bulb, and an irregular thermometer reading will result.

It will be found that the temperature will first rise rapidly until it is near the boiling point of the liquid, then slowly, and finally will remain practically constant. At this point a clean, weighed receiver should be connected to the apparatus and the distillate collected until only a small volume of liquid remains in the flask; the temperature should be noted at regular intervals. If the liquid being distilled is not grossly impure most of it will pass over within a narrow temperature range (within 2–3 degrees).

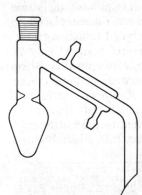

Should the temperature rise steadily, instead of remaining virtually constant, it is then clear that this simple distillation procedure is unsuitable for the purification of the sample and some form of fractional distillation (Section **I,26**) will have to be used.

For the distillation of quantities of liquid in the range 0.5 to 3.0 ml the pear-shaped flask with fixed side-arm condenser (Fig. I,99) is available in capacities of

Fig. I,99

2 ml and 5 ml (joint size 7/11). This design has the advantage of minimising losses due to retention of the distillate as a film on the glass surface. The side-arm has a specially ground tip at the drip end to facilitate the drainage and collection of the distillate. Heating is by a micro-burner and carefully controlled by a rotary movement of the semi-luminous flame around the base of the flask.

The assembly shown in Fig. I,100 is useful for distilling off solvent from solutions, as would be obtained for example from solvent extraction procedures. The solution is placed in the separatory funnel and is allowed to drop

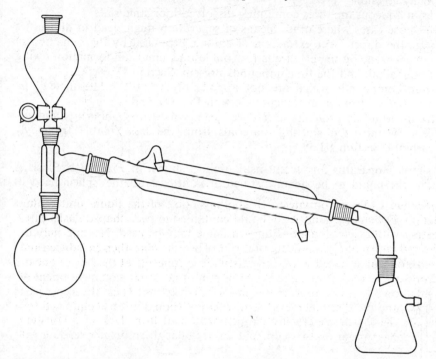

Fig. I,100

into the flask, initially about half-full with the solution, at approximately the same rate as that at which the solvent distils into the receiver. The use of a large flask for distilling the solution is thus rendered unnecessary. The distilling flask (alternatively, a Claisen flask with fractionating side-arm—see Fig. I,110—may be used, particularly if the residue is to be ultimately distilled under diminished pressure) should have a capacity of about twice the estimated volume of the residue after the removal of the solvent. The removal of the solvent in this manner is sometimes termed **flash distillation**.

Relatively large volumes of solvents are conveniently removed by 'stripping' under reduced pressure using a rotary evaporator (Section **I,27**, Fig. I,114).

I,25. STEAM DISTILLATION   Steam distillation is a means of separating and purifying organic compounds. Essentially the operation consists in volatilising a substance by passing steam into a mixture of the compound and water. Provided the organic compound has an appreciable vapour pressure (at least 5–10 mm at 100 °C), it will distil with the steam. Steam distillation takes place at a temperature below the boiling point of water * and hence, in numerous cases, well below the boiling point of the organic substance. This renders possible the purification of many substances of high boiling point by low-temperature distillation, and is particularly valuable when the substances undergo decomposition when distilled alone at atmospheric pressure. It is also of importance in the separation of the desired organic compound:

(a) from non-volatile tarry substances which are formed as by-products in many reactions;

(b) from aqueous mixtures containing dissolved inorganic salts;

(c) in those cases where other means of separation might lead to difficulties (e.g., the direct ether extraction of aniline, produced by the reduction of nitrobenzene by tin, etc., leads to troublesome emulsion formation owing to the alkali and the tin compounds present: Section **IV,54**);

(d) from compounds which are not appreciably volatile in steam (e.g., o-nitrophenol from p-nitrophenol: Section **IV,114**); and

(e) from certain by-products which are steam volatile (e.g., biphenyl and excess of unreacted starting materials from the less volatile triphenyl-carbinol: Section **III,38**).

A simple apparatus for steam distillation is shown in Fig. I,101. Flask A contains the liquid to be steam distilled; it is fitted with the 'splash-head' B

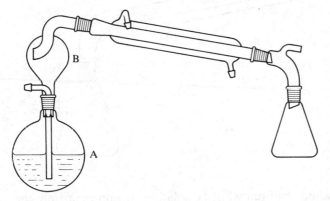

Fig. I,101

which prevents the carry-over of the contents of the flask A into the receiver. To carry out a steam distillation, the solution (or mixture of the solid with a little water) is placed in the flask A, and the apparatus is completely assembled. Steam is passed into flask A, which is itself heated by means of a flame to prevent too rapid an accumulation of water. If the substance crystallises in the condenser and tends to choke it, the water should be run out of the condenser for a few minutes until the solid material has been melted and carried by the steam into the receiver; the water should then be cautiously readmitted to the hot condenser. It is best to use a condenser of the double surface type if the rate at which the steam distillation is carried out is rapid; if necessary two such

---

* See Refs. 43 and 44 for theory of steam distillation.

condensers connected in series may be used since in most steam distillations best results are obtained when the process of distillation is carried out rapidly. The passage of steam is continued until no appreciable amount of water-insoluble material is detectable in the distillate (1). To discontinue the distillation the supply of steam is disconnected from the splash-head and the source of heat removed from flask A. The method of isolation of the organic compound from the distillate will depend upon the physical state and upon its water solubility. For example, a solid compound which is virtually insoluble in water would be removed by filtration; liquids and water-soluble solids would be isolated by batch or continuous solvent extraction procedures as described in Section I,22.

A compact apparatus for the steam distillation of small quantities of material is depicted in Fig. I,102. It is designed to be fitted into a standard 100 ml Pyrex Kjeldahl flask: if desired, the dimensions may be reduced proportionally for use with a 50 ml flask. A screw clip is attached by rubber tubing to the side-arm E; the latter may be replaced by a thin glass tube (3–4 mm external diameter)

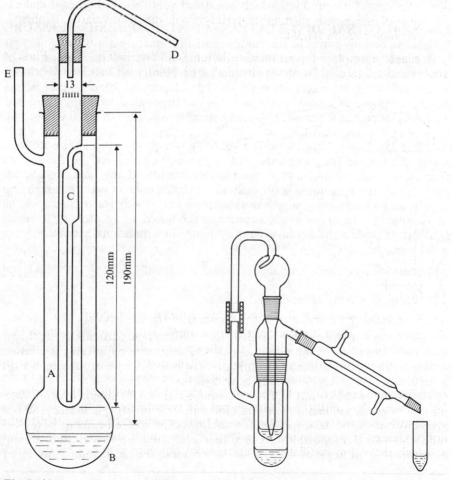

Fig. I,102   Fig. I,103

145

inserted through the upper cork, thus permitting the use of a Kjeldahl flask without modification. The liquid to be steam distilled is placed in the tube A and water in the outer vessel B. Upon heating the latter, steam passes through the inlet tube C and steam-volatile compounds pass out through D; a condenser is attached at D and the steam distillate is collected in a suitable receiver. The functions of the bulb in the inlet tube C are to reduce the danger of spurting and also the 'sucking back' of the liquid in A; the latter can be prevented by the timely opening of the screw clip on E.

Note. (1) With water-insoluble materials distillation will usually be continued until the distillate is quite clear. For water-soluble materials a suitable chemical procedure for detection must be used, e.g., for aldehydic or ketonic compounds, portions of the distillate would be tested with an *aqueous* acidic solution of 2,4-dinitrophenylhydrazine.

An alternative apparatus for the steam distillation of small quantities of material which incorporates ground glass joints is available commercially,* and is illustrated in Fig. I,103.

I,26. FRACTIONAL DISTILLATION AT ATMOSPHERIC PRESSURE

Unless the boiling points of the components of a mixture are widely different it is usual to employ a fractionating column to attempt the separation of liquid mixtures by distillation. Apparatus for precision fractionation, which can successfully separate mixtures in which the components have boiling points which differ by only a few degrees, is available, although careful operation and an appreciation of the factors which influence the efficiency of the fractionating column chosen are needed.

A fractionating column consists essentially of a long vertical tube through which the vapour passes upward and is partially condensed; the condensate flows down the column and is returned eventually to the flask. Inside the column the returning liquid is brought into intimate contact with the ascending vapour and a heat interchange occurs whereby the vapour is enriched with the more volatile component at the expense of the liquid, in an attempt to reach equilibrium within the liquid-vapour system. The conditions necessary for a good separation are:

(*a*)   comparatively large amounts of liquid continually returning through the column;

(*b*)   thorough mixing of liquid and vapour;

(*c*)   a large active surface of contact between liquid and vapour.

Excessive cooling should be avoided; this difficulty is particularly apparent with liquids of high boiling point and may be overcome by suitably insulating or lagging the outer surface of the column or, if possible, by surrounding it with a vacuum jacket or an electrically heated jacket.

The assembly shown in Fig. I,104 illustrates a set-up for simple fractionation using a **Vigreux column** which has moderate fractionating efficiency and is probably one of the most widely used columns. The column consists of a glass tube with a series of indentations such that alternate sets of indentations point downwards at an angle of 45° in order to promote the redistribution of liquid

---

* Manufactured by Scientific Glass Apparatus Co.

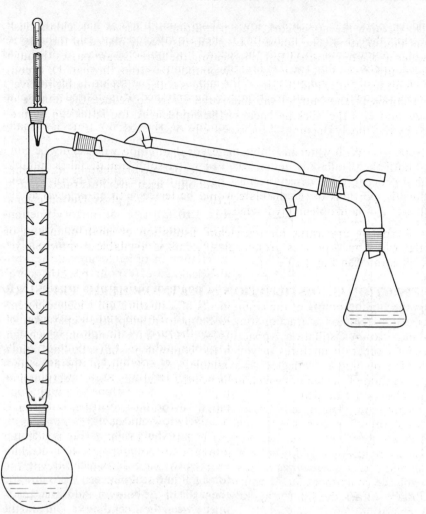

Fig. I,104

from the walls to the centre of the column. The mixture to be fractionated is placed in a flask of convenient size (it should be one-third to one-half full), a few fragments of porous porcelain added and a water condenser attached to the side arm. The distillate is collected in small flasks or in test-tubes. The bulb of the thermometer should be just below the level of the side arm. It is advisable to lag the column by wrapping asbestos cloth round it; this will minimise the effect of draughts in producing excessive cooling. Lagging of the column is essential if the boiling point of any of the components exceeds 100°. The flask is then heated in an air bath or in an oil bath (Section **I,13**), to ensure a uniform heating. The initial heating must not be hurried as, owing to the considerable extra condensation which occurs whilst the column is warming up, the latter may easily choke with liquid. Once distillation has commenced, the rate of heating is adjusted so that the liquid passes over at the rate of one drop every two or three seconds. Under these conditions fairly efficient fractionation should be obtained. When the low boiling point fraction has passed over, distillation should cease. The heating is then slowly increased, and a sharp rise in boiling point should occur as the second fraction commences to distil; it is assumed, of course, that the fractionating system is capable of effecting a sharp fractionation of the components of the mixture. If the set-up is inefficient, a relatively large intermediate fraction may be obtained. It is desired to emphasise the fact that the distillation must be conducted slowly; no time is usually saved by distilling rapidly since a second fractionation will then be necessary.

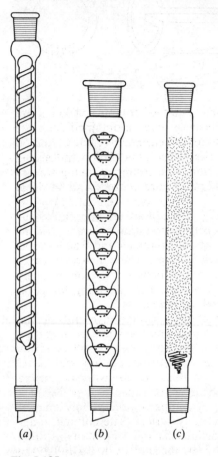

Other designs of fractionating columns commonly used are illustrated in Fig. I,105(a–c). The **all-glass Dufton column** (Fig. I,105(a)) is a satisfactory fractionating column for general use. The glass spiral must be carefully ground to fit the outer tube in order to prevent appreciable leakage of vapour past the spiral. The length of the spiral is usually 15, 20 or 30 cm, the internal diameter of the tube is 15–20 mm and the distance between the turns of the spiral is 9–13 mm; the cone and socket are 19/26 or 24/29. This type of column has the advantage of a small hold-up (i.e., a low volume of liquid is retained within the column compared to the flask charge), but it is of relatively low efficiency.

The **pear-bulb column** (Fig. I,105(b)) is a precision-bore tube with accurately fitting removable bulbs which fit closely to form a liquid seal between the bulbs and the walls of the column. Additional mixing between the ascending vapour and the descending liquid is provided by inserting small glass 'bubbles' between the separate pear-bulbs.

(a)   *(b)*   *(c)*

Fig. I,105

The **Hempel column** (Fig. I,105(c)) is a single glass tube, 25 to 75 cm long and 15 to 25 mm diameter, fitted with either 24/29, 29/32 or 34/35 ground glass joints, which may be filled to within 5 cm of the top with a suitable packing. This packing is supported by a small glass spiral of appropriate size. A number of excellent column packings are available commercially. The simplest and cheapest, yet very efficient, packing consists of hollow glass rings (**Raschig rings**, Fig. I,106(a)) of 6 or 9 mm length and 6 or 9 mm diameter; similar hollow porcelain rings are almost equally effective. **Lessing rings** (Fig. I,106(b)) consist of hollow cylinders of approximately equal height and diameter with a central partition, made of porcelain. When constructed of metal (Fig. I,106(c)) (aluminium, copper or nickel give very good results where no chemical action occurs between the metal and the vapour of the liquid), there is a gap in the circumference and a more or less diametrical partition connected on one side with the cylinder but out of touch on the opposite side.

The provision of the central partition in the Lessing rings increases the efficiency of the packing material by adding to the contact surface available and

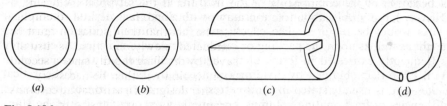

    (a)   (b)   (c)   (d)

Fig. I,106

enhancing the drainage capacity. Both these types of column packings are comparatively cheap. Some idea of their efficiency may be obtained from the fact that one fractionation of a mixture of 50 ml of benzene (b.p. 80 °C) and 50 ml of toluene (b.p. 110 °C) through a simple Hempel column filled with Raschig porcelain rings or with Lessing aluminium rings (the column packing was 35 cm long and 18 mm in diameter) gave 43–47 ml of pure benzene and 44–46 of pure toluene.

Single-turn glass helices (**Fenske rings**) (Fig. I,106(d)) are an alternative but somewhat more expensive packing for Hempel fractionating columns. A convenient size for single-turn helices is 4.0 mm external diameter and 0.50 mm rod thickness: one kilo of these occupies almost 2 litres. These helices form a closely-spaced packing providing maximum contact between the continuously moving liquid and vapour streams and at the same time allowing for good 'through-put' of descending liquid and ascending vapour.

A detailed discussion of the theory of fractional distillation is outside the scope of this volume,* but a brief description of the terms used in discussing fractionating columns and the chief desiderata of efficient columns will be given. The **capacity** of a column is a measure of the quantity of vapour and liquid which can be passed counter-current to each other in a column without causing it to choke or flood. The **efficiency** of a column is the separating power of a definite length of the column; it is measured by comparing the performance of the column with that calculated for a theoretically perfect plate column under similar conditions. A **theoretical plate** is defined as the length of distilling column such that the vapour leaving the plate has the same composition as the vapour which would be in stationary equilibrium with liquid at that temperature, as obtained from the vapour–liquid phase diagram. Since the efficiency of the column depends on the establishment of equilibrium conditions between ascending vapour and descending liquid by thorough and intimate mixing, it is clear that the removal of the more volatile component from the top of the column should be as slow as possible. The number of theoretical plates cannot be determined from the dimensions of the fractionating column; it is computed from the separation effected by distillation of a liquid mixture (e.g., benzene and toluene; benzene and carbon tetrachloride; benzene and dichloroethane; heptane and methylcyclohexane), the vapour and liquid compositions of which are

---

* See, for example, Ref. 46.

accurately known.* An ordinary 1 cm tube 1 metre long might be equivalent to only one theoretical plate, whilst the same tube filled with a suitable packing can give the equivalent of twenty or more theoretical plates. A column with twelve theoretical plates is satisfactory for the practical separation of a mixture of benzene and toluene (Δb.p. 30 °C); where the two components of a mixture differ in b.p. by only about 3°, a column with approximately 100 theoretical plates would be required. The effectiveness of a column depends upon the height as well as upon the packing or internal construction, hence the efficiency is frequently expressed in terms of the **height equivalent per theoretical plate (H.E.T.P.)**. It is obtained by dividing the height by the number of theoretical plates, and is usually stated in centimetres. For the comparison of the relative efficiencies of fractionating columns, the operating procedure should be standardised.

The ideal fractionation yields a series of sharply defined fractions, each distilling at a definite temperature. After each fraction has distilled, the temperature rises rapidly, no liquid being distilled as an intermediate fraction. If the temperature is plotted against the volume of the distillate in such an ideal fractionation, the graph obtained is a series of alternate horizontal and vertical lines resembling a staircase. A more or less sloping break reveals the presence of an intermediate fraction and the amount of such fraction can be used as a qualitative criterion of the performance of different columns. The ultimate aim in the design of efficient fractionating columns is to reduce the proportion of the intermediate fractions to a minimum. The most important factors which influence the separation of mixtures into sharp fractions are the following:

1. **Time of distillation.** For any column there is always an optimum time of distillation below which accuracy is sacrificed and above which the slightly improved separation does not justify the extra time taken. For most laboratory columns this will vary between 1 hour and 8–10 hours.

2. **Hold-up of column.** The hold-up of liquid should be reduced to a minimum compatible with scrubbing effectiveness and an adequate column capacity. The ratio of charge of the still to the hold-up of the column should be as large as possible; in general, the still charge should be at least twenty times the hold-up.

3. **Thermal insulation.** Even slight heat losses considerably disturb the delicate equilibrium of an efficient column, and almost perfect thermal insulation is required for the separation of compounds with boiling points only a few degrees apart. Theoretically, the greatest efficiency is obtained under adiabatic conditions. If the components boil below 100 °C, a silvered vacuum jacket is satisfactory; the efficiency of such a jacket will depend upon the care with which it is cleaned, silvered and exhausted. In general, the most satisfactory insulation is provided by the application of heat to balance the heat loss. An electrically-heated jacket is fitted round the column; the temperature of the jacket, which should be controlled by means of an external resistance or a variable voltage transformer (Variac), should be adjusted within 5° of the temperature of the vapour condensing at the upper end of the column.

An electrically-heated jacket is easily constructed from two pieces of Pyrex glass tubing of such a length as to extend from the bottom of the head to just

---

* For experimental details, see, for example, Refs. 47 and 48.

above the lower end of the column—the latter may carry a ground joint. The inner tube may be of 35 mm bore and the outer tube of 55 mm bore: this allows room for the column with attached thermometer inside the inner tube (compare Fig. I,108). The narrow tube is wound with 20 mm electric heating tape; heat input is controlled by a resistance, energy regulator or a variable transformer. The ends of the jacket are closed with asbestos or other insulating board of convenient size and shape.

4. **Reflux ratio.** This is defined as the ratio between the number of moles of vapour returned as refluxed liquid to the fractionating column and the number of moles of final product (collected as distillate), both per unit time. The reflux ratio should be varied according to the difficulty of fractionation, rather than be maintained constant; a high efficiency of separation requires a high reflux ratio.*

Otherwise expressed, the number of theoretical plates required for a given separation increases when the reflux ratio is decreased, i.e., when the amount of condensed vapour returned to the column is decreased and the amount distilled off becomes greater. The variation in reflux ratio is achieved by the use of a suitable take-off head (or still-head), usually of the **total condensation variable take-off** type. In use, all the vapour is condensed and the bulk of the condensate is returned to the fractionating column, small fractions of the condensate being allowed to collect in a suitable receiver. One commercially available design is shown in Fig. I,107 but others are illustrated in the distillation units described below.

Fig. I,107

Figure I,108 is a photograph which shows a generally useful fractional distillation unit † employing a packing of glass helices. The column is provided with an electrically-heated jacket the temperature of which may be adjusted with a variable transformer. The still-head is of the total-condensation variable take-off type; all the vapour at the top of the column is condensed, a portion of the condensate is returned to the column by means of the special stop-cock (which permits fine adjustment of the reflux ratio) and the remainder is collected in the receiver. The advantages of the still-head are that true equilibrium conditions can be established before any distillate is collected; this is particularly important when the jacket temperature must be controlled. Furthermore changing from a lower to a higher boil-

---

* The more difficult the fractionation, the greater the reflux ratio to be employed. Thus for compounds differing only slightly in boiling point, this may be as high as 50 to 1; for liquids of wider boiling point range, thus permitting of fairly easy separation, a reflux ratio of 5 or 10 to 1 may be used.

Beyond certain limits increase of the reflux ratio does not appreciably increase the separating power or efficiency of the column. As a rough guide, if the column has an efficiency of n plates at total reflux, the reflux ratio should be between $2n/3$ and $3n/2$.

† Supplied by A. Gallenkamp & Co. Ltd, who kindly supplied the photograph.

ing point fraction is comparatively easy. The stop-cock is closed and the liquid is allowed to reflux until the thermometer records the lowest temperature possible; at this point the column is effecting its maximum degree of separation and an equilibrium condition is reached. The tap is then partially opened and the distillate is collected in the receiver until the temperature begins to rise. The stop-cock is then closed and equilibrium conditions again established, and a further fraction is removed. In this way sharper separations may be obtained. Further improvement results from the use of a capillary tube to drain the condensate into the receiver. The reflux ratio may be measured approximately by counting the number of drops of liquid which fall back into the column as compared with the number of drops which fall into the receiver flask (the liquid drops falling off the slanting ends of the drip tubes are readily observable).

The vacuum distillation adapter also shown in Fig. I,108 allows the collection of fractions when distilling under reduced pressure. Its operation is similar to the 'Perkin triangle' or equivalent device ('intermediate receiver adapter') described in Section **I,27**. The general technique of conducting a fractional distillation is as follows:

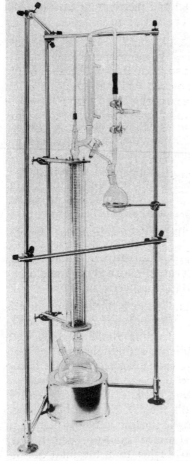

1. Charge the flask with the mixture, and attach to the column. Set the still-head for total reflux and heat the flask until the material begins to reflux into the column. Then heat the column very slowly until the refluxing liquid reaches the top of the column and the boiling point registers on the thermometer. Adjust the temperature in (or near the top of) the jacket, as recorded on the thermometer (not visible in photograph), adjacent to the column until it is just below (i.e., within 5 °C) the boiling point recorded in the vapour. With random packings, such as glass helices or gauze rings, the column should first be flooded in order to coat the packing completely with liquid; it is then operated under total reflux until the equilibrium is attained (about 1 hour per ten theoretical plates).

2. When the column has reached equilibrium, adjust the head to give the desired reflux ratio, change the receiver and collect the lowest boiling point component over an appropriate distillation range, say 1–2 °C. During the distillation, maintain as high a rate of reflux as possible consistent with prevention of flooding the column; under these conditions the reflux ratio is controlled by the rate of take-off. As the lowest boiling component is removed, the proportion of it in the distilla-

Fig. I,108

tion flask gradually decreases and eventually a mixture of two components reaches the top of the column, and this will be indicated by a slight rise in boiling point. When this occurs, gradually increase the reflux ratio, i.e., decrease the rate of take-off: this will make it possible to collect the lowest boiling point fraction over a narrow range; eventually a point will be reached when even with a high reflux ratio the boiling point rises. At this stage, change the receiver and commence the collection of the intermediate fraction.

3. During the distillation of the intermediate fraction, keep the rate of take-off very slow. The boiling point will rise and eventually either remain constant or increase very slowly. At this point, change the receiver, adjust the temperature of the heating jacket again and collect the second fraction over a narrow distillation range—rapidly so long as the temperature remains essentially constant, then more slowly until finally the second intermediate fraction is reached again whilst distilling very slowly. Change the receiver, collect the intermediate fraction and proceed as before for the third component, etc.

The following general comments upon situations which may arise during fractionation may be helpful:

(a) The sharper the fractionation, the smaller, of course, is the intermediate fraction. If the difference in boiling points of the components being separated is considerable, the separation will be so facile that practically all the lower boiling point component will be removed whilst the boiling point remains essentially constant. Eventually the upper part of the column will begin to run dry, distillation will slow up and finally stop, whilst the reflux at the bottom of the column will be heavy. The vapour temperature may begin to fall until it is below the temperature at the top of the heating jacket. Mere increase of the bath temperature may result in the flooding of the column: the power input to the heating jacket must be gradually increased until reflux again reaches the top of the column, the boiling point begins to rise and eventually becomes constant; the temperature in the jacket is maintained just below the boiling point of the vapour.

(b) As the rate of take-off is reduced near the end of a fraction, a slight lowering of the bath temperature may be necessary to avoid flooding of the column. Also as the boiling point rises during the collection of the intermediate fraction, the power input to the jacket must be increased in order to hold its temperature just below the boiling point.

(c) If the column is flooding near the top and there is little reflux at the bottom, the jacket temperature is too high. If there is normal heavy reflux at the bottom of the column and there is flooding at the top, the bath temperature is probably too high. If the column is flooding near the bottom and there is little reflux near the top, the jacket temperature is too low.

(d) If it is desired to collect the liquid remaining in the column at the end of the fractionation (constituting the 'hold-up'), the column may be stripped by the addition of a 'chaser' at the beginning of the fractional distillation in a quantity somewhat greater than the estimated 'hold-up'. The boiling point of the 'chaser' should be at least 20° higher than the final boiling point of the material being fractionated. For this operation the bath temperature is kept sufficiently high to distil the end component, and the jacket temperature is carefully and slowly raised above the boiling point of the component. 'Chasers' should be chemically inert, inexpensive, and should not form azeotropic mixtures with the components of the mixture undergoing fractionation; examples are: toluene, b.p. 110 °C; *p*-cymene, b.p. 175 °C; tetralin, b.p. 207 °C; diphenyl ether, b.p. 259 °C.

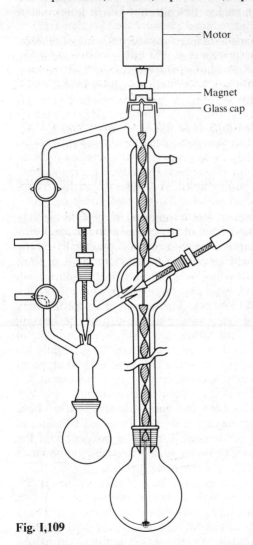

Motor

Magnet

Glass cap

Fig. I,109

When it is required to separate by fractional distillation components of a mixture which differ in their boiling points by only a few degrees, the **spinning-band fractionating column** * offers the best chance of success. The fractionating column consists of a vertical glass tube into which is accurately fitted throughout its length a spiral of Teflon or metal gauze (platinum, stainless steel or Monel) which is fixed to a central Teflon or metal rod and which has a diameter very slightly less than the internal bore diameter of the tube. The spiral extends through the reflux condenser of the specially designed, integral, total-condensation variable take-off still-head, and may be spun by means of either direct or magnetic couplings to an electric motor (Fig. I,109). The central rod of the spinning band extends into the distillation flask and terminates in a Teflon stirrer to ensure smooth boiling. The column may be jacketed with a nichrome heating element or alternatively the column and still-head may be vacuum jacketed and silvered. The advantage of the spinning band (which rotates at selected speeds between 600 and 3000 r.p.m.) is that it increases vapour-liquid contact upon which the column efficiency depends, by causing the vapour to be thrown on to the column walls, and into contact with the liquid descending as a thin film. Furthermore these columns have little tendency to flood and have a low hold-up which therefore allows their high efficiency to be realised to the full. When deliberately flooded they rapidly clear and liquid–vapour equilibrium is re-attained. The spinning band has the further advantage of assisting the passage of vapours through the column. This reduces the pressure difference which always

---

* For example, the Nester-Faust Spinning Band Column manufactured by Perkin-Elmer Ltd.

exists between the top (lower pressure) and bottom (higher pressure) regions of a fractionating column. This pressure difference depends upon the dimensions of the column, the nature of the column packing and the rate of distillation. A large pressure drop is undesirable since it leads to a higher heat input at the distillation flask being needed to sweep the vapours to the still-head. With a spinning-band column this difference may be as low as 0.23 mmHg. This small pressure drop is a feature which makes this column design particularly suitable for fractional distillations under reduced pressure (see Section **I,27**).

The spinning-band fractional distillation unit is available in a range of sizes suitable for use on a micro or semi-micro scale (e.g., 1–5 ml), a laboratory scale (e.g., 250 ml) or a pilot scale (12 l). The mode of operation is essentially similar to that described above for the conventional fractionating column packed with Fenske rings.

I,27. DISTILLATION UNDER DIMINISHED PRESSURE ('VACUUM' DISTILLATION)

Many organic substances cannot be distilled satisfactorily under atmospheric pressure because they undergo partial or complete decomposition before the normal boiling point is reached. By reducing the external pressure to 0.1–30 mm of mercury, the boiling point is considerably reduced and the distillation may usually be conducted without danger of decomposition.

In a **vacuum distillation** apparatus certain features should be present to facilitate the ease of operation and these have been incorporated into the assembly illustrated in Fig. I,110. A is a pear-shaped Claisen-Vigreux flask, the left-hand neck of which carries a screw-cap adapter through which is inserted a glass

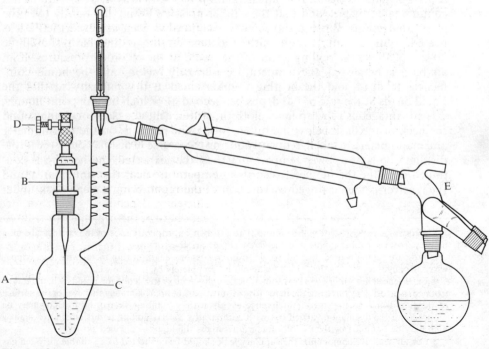

Fig. I,110

tube B of appropriate diameter drawn out to a capillary C, at its lower end (1). The tube B carries at its upper end a short piece of pressure tubing and a screw clip D.* The condenser carries a three-limbed multiple receiver and adapter E, frequently called a '*pig*', the outlet being connected via a suitable trap and manometer (Section I,30) to either a water or an oil pump (Section I,29). The pig adapter permits the collection of three individual fractions without breaking the vacuum and interrupting the progress of the distillation. The flask is heated either by means of an air bath or by means of a water or oil bath as appropriate† (Section I,13); in the latter case the bulb of the flask is immersed at least two-thirds into the bath, which should contain a thermometer.

To carry out a distillation (under the reduced pressures obtainable with a water pump), the liquid is poured into the Claisen flask so that it is about one-half full, the apparatus is completely assembled as in Fig. I,110, and the water supply to the condenser is turned on. The water pump is then allowed to reach its maximum capacity with the screw clip D almost fully closed (2). The latter is then adjusted so that a fine stream of air bubbles passes through the liquid in order to minimise 'bumping' in the subsequent distillation.‡ (The introduction of a gas (air) tends to prevent a delay in the appearance of vapour and thus to prevent superheating; the volume of air introduced in the form of minute bubbles is small so that the effect of the partial pressure upon the boiling point will usually be negligible.) When the mercury level in the manometer (Section I,30) is steady, the pressure in the system is noted. If the pressure is unsatisfactory, the apparatus must be carefully tested for leaks and these eliminated before the distillation can be commenced: special attention should be paid to ensure that all the glass joints are firmly in position and not contaminated by grit, and that the rubber pressure tubing fits tightly over the glass connections. When a satisfactory vacuum has been achieved the flask is heated. With a water or oil bath, the temperature of the bath should be 20–25 °C above the boiling point of the liquid at the recorded pressures. If an air bath is employed, the temperature is slowly raised until the liquid commences to distil, and the heating is maintained at this intensity so that the liquid distils at the rate of 1–2 drops per second. (For high boiling point liquids, it is advantageous to wrap linen cloth or, better, asbestos cloth or string round the neck of the flask below the outlet tube.) The readings on the thermometer and manometer are taken frequently during the course of distillation. If the initial distillate boils at a lower temperature than that expected, the heating is continued until the thermometer records a temperature near that anticipated, and the receiver is then changed by rotation of the pig to bring a clean flask under

---

* After some experience it will be found that a drawn-out capillary tube of the correct size may be prepared; the rubber tubing and the screw clip D are then omitted.

If pressure tubing is used, it is advisable to insert a short length of thin metal wire (e.g., copper wire, 22 gauge) to prevent the tubing being closed completely by the screw clip.

† Experienced laboratory workers sometimes employ a large free flame for liquids which tend to froth considerably; by directing the flame for the most part at approximately the level of the surface of the liquid and heating the circumference evenly with a 'rotating' flame, the frothing may be reduced and the distillation carried out with comparative safety. Boiling points which are slightly high may be obtained by the use of a free frame unless the liquid is distilled slowly.

‡ For air-sensitive compounds the capillary leak should be connected to a suitable nitrogen gas supply.

the condenser outlet. For a pure compound the boiling point will not rise more than a degree or two during the whole of the distillation, even when the bath temperature has to be raised considerably towards the end to drive off the last of the liquid. At the conclusion of the distillation the heating bath is removed, the 'vacuum' is gradually released and the screw clip on D is fully opened (this will prevent any liquid entering the capillary).

If the pressure during distillation is not exactly that given in the recorded boiling point, it may be estimated *very approximately* for the working pressures of a water pump (10–25 mm) by assuming that a difference of 1 mm in pressure corresponds to one degree difference in the boiling point. Table I,10 may be

Table I,10   Approximate boiling points (°C) at reduced pressures

Pressure (mmHg)	Water	Chlorobenzene	Benzaldehyde	Ethyl salicylate	Glycerol	Anthracene
760	100	132	179	234	290	354
50	38	54	95	139	204	225
30	30	43	84	127	192	207
25	26	39	79	124	188	201
20	22	34.5	75	119	182	194
15	17.5	29	69	113	175	186
10	11	22	62	105	167	175
5	1	10	50	95	156	159

found useful as a guide to the *approximate* boiling point under diminished pressure when the boiling point under atmospheric pressure is known: it will enable the student to select the thermometer employed in the distillation.

Notes. (1) The capillary when drawn out should be sufficiently robust so that it is not broken during the vigorous boiling but should have a degree of flexibility to permit some movement of the capillary during distillation; this is particularly advantageous when round-bottomed flasks are used. Furthermore the bore should be such that only a fine stream of bubbles is admitted to the flask when the vacuum is initially applied. The successful construction of a suitable capillary requires some practice and it is usually helpful to perform the operation in two stages. Initially a length of Pyrex tubing (*ca.* 15 cm × 5 mm) is rotated in the flame of an oxygen-gas burner so that about 2 cm in the middle of the tube is heated to dull redness. The softened glass is allowed to gently thicken before it is removed from the flame and extended by a few centimetres (Fig. I,111(*a*)). The second stage is to reheat with a needle flame a narrow section of the thickened portion which when pliable is extended by a steady pulling action (Fig. I,111(*b*)). Experience will determine how the speed and length of extension affects the dimensions of the final capillary. The

(*a*)   (*b*)

Fig. I,111

capillary is then cut to the length required so that when it is inserted into the flask the end comes within 1–2 mm of the bottom.

(2) If the material in the flask contains traces of volatile solvents, it is advisable to allow the passage of a comparatively large volume of air through the liquid while warming the flask slightly; this drives off the last traces of volatile solvents, which are carried down the water pump. If this is not done, the pressure obtained when testing out the apparatus will be above the real capacity of the pump, and the student will erroneously assume either that the pump is not functioning efficiently or that leaks are present in the apparatus. When all traces of volatile solvents have been removed, the screw clip D is almost completely closed or otherwise adjusted.

When it is necessary to use an oil vacuum pump to attain lower pressures, it is essential to prevent large volumes of solvent vapour from passing into the pumping system. The oil pump should therefore be guarded with a suitable trap; furthermore distillation at water pump pressures should first be used to remove the bulk of low boiling solvent before the oil pump is brought into operation.

For the **fractional** distillation of mixtures under diminished pressure, when a more efficient fractionating column is necessary, the pig type of adapter should be replaced by the more versatile *Perkin-type receiver adapter* to enable the more numerous fractions to be collected conveniently. The complete apparatus for vacuum distillation is depicted in Fig. I,112. The two-necked round-

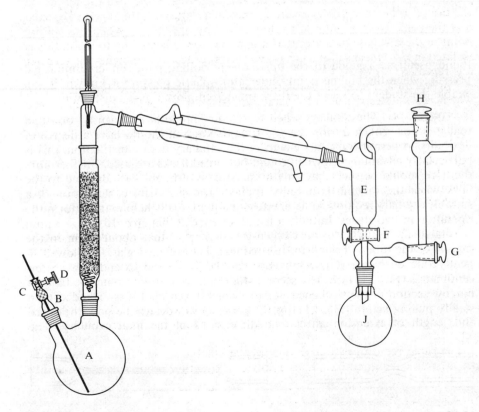

Fig. I,112

bottomed flask A is fitted with a Hempel column (packed with Fenske rings) connected via a still-head with thermometer to a water-cooled condenser which terminates in the Perkin receiver adapter. The capillary leak shown illustrates a simple alternative arrangement to that described above. Here the take-off adapter B (see also Fig. I,29(a)) carries a short length of pressure tubing C, fitted with a screw clip D. A convenient length of fine capillary tube, which must be flexible yet reasonably robust, is threaded through the tubing and located in position by tightening the clip D.

To carry out the distillation the flask is charged and the apparatus assembled. Before evacuating the system the tap F is closed to isolate the receiver E from the flask J, but the taps H and G are turned so as to connect pump to the receiver E (via H) and to the flask J (via G). The apparatus is now evacuated until a steady pressure reading is obtained on the manometer (checking for leaks as previously) and the flask is heated by an air or oil bath until distillation commences. The initial distillate is allowed to run directly into the flask J by turning the tap F; as soon as observation of the thermometer shows that a steady boiling fraction is distilling, the tap F is closed and the required fraction is allowed to collect in E, noting the thermometer and manometer readings. While distillation is taking place, the three-way tap G is turned to admit air to the flask J, which is then removed and replaced by a clean receiver. The vacuum in J is now restored, by first isolating the distillation unit from the pumping system by closing the tap H, and then connecting J to the pump via the tap G. When a steady pressure is once more attained, the tap H is opened and the contents of E allowed to flow into J by opening F. When the boiling point of the distillate indicates that a new fraction is beginning to distil, this is isolated in E by means of the tap F and the procedure for changing the receiver J is repeated.

For vacuum fractional distillation of liquids having close boiling points, which necessitates the use of a total condensation variable take-off head, the receiver adapter modification noted in Fig. I,108 is employed. Its operation is similar to that of the Perkin triangle; the column is operated under total reflux while the receivers are being changed.

The high efficiency and small pressure drop of the spinning-band columns (Fig. I,109) makes them very suitable for precision vacuum fractional distillation. The still-head is provided with a type of Perkin triangle assembly which allows the receivers to be changed without disturbing the column equilibrium.

Vacuum distillation on the semimicro scale (1–8 ml) is conveniently carried out using apparatus of the types shown in Fig. I,113(a* and b†). The former (a) consists essentially of a two-necked distillation flask with a fixed side-arm condenser and a specially designed vacuum receiver which can be rotated to allow the collection of several individual fractions from a flask charge of between 1 and 4 ml. The latter (b) has a short vacuum jacketed fractionating head and is useful for the distillation of somewhat larger volumes (5–8 ml).

The rapid removal of a large quantity of volatile solvent from a solution of an organic compound (i.e., from a solvent extraction process) is conveniently

---

* Supplied by Jobling Laboratory Division.
† Supplied to specification by Scientific Furnishings Ltd.

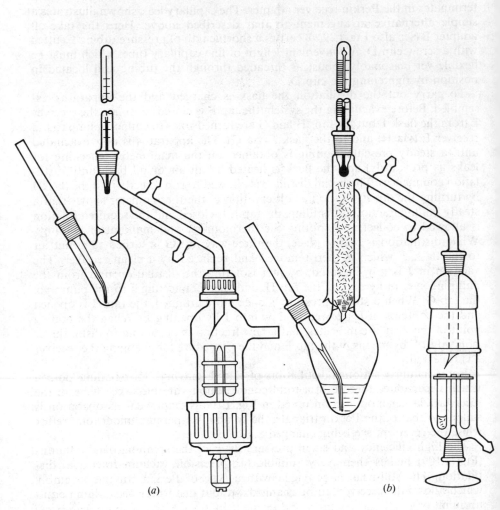

(a)   (b)

Fig. I,113

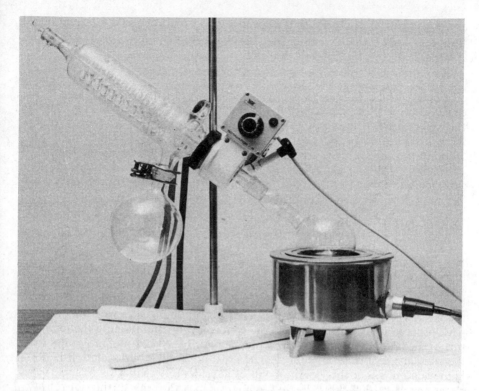

Fig. I,114

effected by using the **Rotary Film Evaporator** (Fig. I,114).* Evaporation is conducted under reduced pressure (a water pump is the most convenient) and therefore at relatively low temperatures. The evaporator flask is heated to the appropriate temperature on a water bath, and is rotated during evaporation; this ensures thorough mixing, prevents bumping and also exposes a relatively large liquid film. The solvent distils from the evaporating surface, is condensed by the spiral condenser and runs off into the receiving flask. Further quantities of the solution may be added through the feed-tube controlled by the stop-cock. A quick-action jack with built-in tension spring is sometimes fitted; the operator may lift or lower the assembly and secure it in any desired position within preset upper and lower limits. Various sizes of specially shaped flasks are available or alternatively round-bottomed flasks with appropriate ground glass joints may be used satisfactorily. The special sleeve gland is accurately ground.

I,28. HIGH VACUUM DISTILLATION—MOLECULAR DISTILLATION The apparatus designs which have been described in the previous section for distillation under reduced pressure are virtually useless for the distillation of compounds having very high boiling points, which need to be distilled at pressures in the region of 10^{-5} mmHg if decomposition is to be avoided. Successful distillation is however achieved in a greatly simplified distillation

---

* Permission to use this photograph has been kindly provided by Buchi Laboratories Techniques Ltd.

unit in which the chief feature is the short direct path between a heated liquid surface and the cooled condensing area (**molecular distillation, short path distillation**) (see also Refs. 50 and 51).

In molecular distillation, the permanent gas pressure is so low (less than 0.001 mm of mercury) that it has very little influence upon the speed of the distillation. The distillation velocity at such low pressures is determined by the speed at which the vapour from the liquid being distilled can flow through the enclosed space connecting the still and condenser under the driving force of its own saturation pressure. If the distance from the surface of the evaporating liquid to the condenser is less than (or of the order of) the mean free path of a molecule of distillate vapour in the residual gas at the same density and pressure, most of the molecules which leave the surface will not return. The mean free path of air at various pressures is as follows:

Pressure (mmHg)	1.0	0.1	0.01	0.001
Mean free path (cm)	0.0056	0.0562	0.562	5.62

The mean free path of large organic molecules is shorter; it is evident, therefore, that the condenser must be quite close to the evaporating surface. Strictly speaking, a molecular still may be defined as a still in which the distance between the evaporating surface and the cold condensing surface is less than the mean free path of the molecules. The escaping molecules will, for the most part, proceed in a straight path to the condenser; by maintaining the temperature of the latter comparatively low, the amount of reflection of molecules from the condensing surface is reduced. The great advantage of distillation under a high vacuum is that the 'boiling point' is considerably reduced—in some cases by as much as 200–300 °C—thus rendering possible the distillation of substances which decompose at higher temperatures, of substances which are very sensitive to heat, and also of compounds of very high boiling point and large molecular weight.

When the evaporating liquid is a single substance, the rate of evaporation will be $\rho c/s$ grams per square cm per second, where ρ is the density of the saturated vapour at the given temperature, c is the mean molecular velocity and s the mean free path of a distillate molecule. If the liquid is a mixture, the rate of evaporation of the rth component will be $\rho_r c_r/s$ grams per square cm per second. The separation obtained in a molecular distillation thus depends upon the quantity $\rho_r c_r$, unlike the separation obtained in ordinary distillation, where the vapour is in equilibrium with the liquid, which depends upon ρ_r. Since c_r is inversely proportional to the square root of the molecular weight, and the magnitude of ρ_r is in general greatest for the components of least molecular weight, $\rho_r c_r$ is greatest for constituents of least molecular weight. Molecular distillation (sometimes termed *evaporative distillation*) is the only method by which substances of high molecular weight can be distilled without decomposition. According to Langmuir (1917) the theoretical rate of distillation can be written in the form:

$$w = \rho \sqrt{\frac{1}{2\pi MRT}}$$

where w is the weight of substance evaporating per square cm of liquid surface per second, M is the molecular weight of the liquid, R the gas constant and T the absolute temperature. In practice, lower values are obtained because of the reflection of molecules from the condensing surface.

The vacuum sublimation apparatus (Fig. I,88) is particularly suitable when only small quantities (10–50 mg) of fairly viscous high boiling liquids need to be distilled. The design offers the least hindrance to the flow of vapour from the evaporating to the condensing surface. The rate of distillation is determined by the rate at which the liquid surface is able to produce vapour. Since a liquid sample may almost certainly contain dissolved gases, or solvents which have been used to aid its transference to the distillation chamber, even greater care must be taken in applying the vacuum than is the case with the sublimation of solids. Initially a stopper should be used in place of the cold finger. To avoid excessive frothing and splashing the vacuum must be reduced *very gradually* and the temperature increased in careful stages. Initially the vacuum attainable with a water pump is employed and the temperature increased slowly by immersion of the distillation unit in a water bath at a suitably controlled temperature; gentle agitation of the unit during the heating will aid the removal of solvent and keep frothing to a minimum. When ebullition has ceased the water pump is replaced by an oil pump and the vacuum slowly reapplied and the gentle heating continued. Only when it is clear that no further volatile material is being removed (and often this may take up to an hour or so) is the stopper replaced by the cold finger, the apparatus connected to the source of high vacuum and the molecular distillation commenced. The vacuum required is that provided by a suitable vapour diffusion pump, and the complete assembly required for the distillation is illustrated schematically in Fig. I,115. The in-

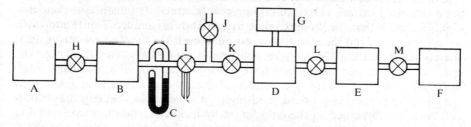

Fig. I,115

dividual components are the distillation unit A, the vapour traps B, a simple uncalibrated mercury manometer C, the vapour diffusion pump D (Section **I,29**), a phosphoric oxide moisture trap E, an oil immersion rotatory 'backing' pump F (Section **I,29**) and the McLeod gauge G (Section **I,30**).* These components are connected with wide-bore glass tubing having the minimum number of bends and fitted with ground glass joints sealed with Apiezon wax W, or better with joints of the O-ring seal type (Section **I,8**) as appropriate. The wide-bore vacuum taps H–M are sited at the points indicated to enable sections of the assembly to be isolated and to facilitate vacuum control and measurement. The entire apparatus, without of course the unit A, is usually permanently assembled on laboratory scaffolding with the section between taps K and M, together with the McLeod gauge, kept permanently under vacuum.

In use, with all the taps in the closed position, the distillation unit A is attached to the ground glass joint which is fitted to tap H; the Dewar flasks

---

* It is highly desirable that a portable safety screen suitably located should be provided between the operator and the distillation unit and vacuum system illustrated.

surrounding the vapour traps are filled with suitable coolant (Cardice-acetone, or liquid nitrogen), and the condenser water to the vapour diffusion pump and to the cold finger (or condenser) of A turned on. The backing pump is switched on and the taps M, L, K and H are opened in sequence so that the system is evacuated to the pressure attainable with this pump. The pressure in the system is indicated by the auxiliary manometer C, which can also be used to check for leaks in the apparatus by closing M and noting any fluctuations in its mercury level. When the backing pump has been reconnected to the system by turning tap M, the heat supply to the vapour diffusion pump is brought into operation and the system allowed to reach the minimum pressure as indicated by the McLeod gauge (see Section I,30 for the operation of this gauge). The distillation unit A is now heated slowly in an oil bath until misting of the cold finger is observed when the temperature of the oil bath should be noted and maintained at this level. The reading on the McLeod gauge should be checked periodically during the progress of the distillation. At the conclusion of the distillation the heat supply to the vapour diffusion pump is disconnected and the unit A isolated by closing tap H. After allowing several minutes to elapse to allow the temperature of the diffusion pump fluid* and the distillation unit A to drop substantially, tap K is closed and the backing pump switched off after connecting it to the atmosphere via the three-way tap M. The pin hole three-way tap I is opened carefully to admit air to the traps B which are then dismantled for cleaning, drying and reassembly. Tap H is opened and the unit A dismantled when it has reached room temperature. The water supply to the condensers is finally turned off.

Fig. I,116

In cases where the quantity of material to be distilled is such that there is a danger of drainage of droplets of condensed liquid from the cold finger, a small collection cup may be attached to the cold finger by means of platinum wire suitably fused to the two glass surfaces (Fig. I,116).

An apparatus for the high vacuum distillation of larger quantities is the Hickman vacuum still† shown in Fig. I,117; it is about 600 mm in diameter, 45 mm high and will hold about 40 ml of liquid. The roof of the still is filled with ice-water or any appropriate freezing mixture. A modification which permits continuous flow of cooling liquid over the roof of the still is shown in Fig. I,118.

Small quantities (0.1–2 g) of material may be distilled using the distillation unit shown in Fig. I,119 which is readily constructed from 9-mm-diameter Pyrex tubing, the bulbs being made to a size appropriate to the size of sample. The material to be distilled is diluted with a little solvent so that it can be introduced in stages directly into the distillation bulb, without contaminating the sides of the tube, by means of a capillary pipette. After each addition the

---

* This is particularly important in those cases where the fluid in the vapour diffusion pump is Apiezon oil and is heated by electrical means. The capacity of the heating element is such that overheating of the oil occurs leading to 'cracking' with the formation of lower boiling components which diminish the efficiency of the vapour diffusion pump in any subsequent operation.

† Supplied, for example, by Scientific Glass Apparatus Co.

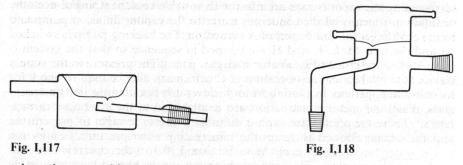

Fig. I,117   Fig. I,118

solvent is removed in the usual careful manner and the residue prepared for the final high vacuum distillation as in the former cases. The unit is supported by a retort stand and clamp so that the receiver tube slopes slightly downward and the distillation bulb is encased in an air bath or immersed in an oil bath. The vacuum is allowed to reach a steady value before heating of the air or the oil bath by means of a controlled flame is commenced. Liquid distils into the first indentation and the temperature and pressure is noted. The indentations further along the tube prevent further distilled material flowing to waste.

A design of apparatus* which has found particular use in these laboratories for the distillation of viscous high boiling monosaccharide derivatives under high vacuum (10^{-5} mmHg) in relatively large amounts (up to 100 g) is shown in Fig. I,120. Preliminary removal of volatile solvents from the material is carried out in a round-bottomed flask on a rotary evaporator under water-pump pressure and then by means of an oil vacuum pump which is fitted with a series of suitable cooled solvent traps. The hot solvent-free material, whilst still in an adequately fluid state, is then poured into the distillation retort using a pre-warmed, long-necked, wide-tube funnel, the end of which reaches into the distillation bulb. In this way contamination of the inside of the unit is avoided, and removal of last traces of solvent, which would be tiresome on this scale,

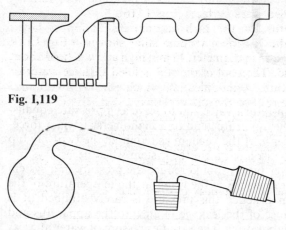

Fig. I,119

Fig. I,120

* This apparatus has been made to specification by R. B. Radley & Co. Ltd; the size of the distillation bulb is either 200 ml, 50 ml or 10 ml.

obviated. A pine splint is inserted into the flask to prevent 'bumping' during the distillation, which is effected by evacuating the flask with the aid of a suitable high vacuum source and heating it in an oil bath.

I,29. VACUUM PUMPS Water Pumps. The high-pressure water supply service is employed for the operation of the ordinary 'filter pump', which finds so many applications in the laboratory. Several types of water-jet pumps of glass, plastic or metal construction are available from most laboratory suppliers.* These are often fitted with a suitable non-return valve to prevent the apparatus being flooded as the result of fluctuating water pressure. Connection to the water tap in the case of the metal pump is by a direct screw-threaded joint; with the plastic or glass models high-pressure rubber tubing of suitable bore is wired to the tap and to the pump.

It is routinely desirable to interpose a large pressure bottle A (Fig. I,121) fitted with a rubber bung between the pump and the apparatus to act as a trap

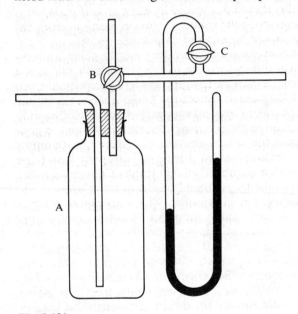

Fig. I,121

in the event of failure of the non-return valve and to serve as a pressure equalising reservoir. Connection to the apparatus and to a manometer is via a three-way tap B which allows for release of the vacuum as required; the two-way tap C permits the manometer to be isolated from the system when necessary.

Theoretically, an efficient filter pump should reduce the pressure in a system to a value equal to the vapour pressure of the water at the temperature of the water of the supply mains. In practice this pressure is rarely attained (it is usually 4–10 mm higher) because of the leakage of air into the apparatus and the higher temperature of the laboratory. The vapour pressure of water at 5, 10, 15, 20 and 25 °C is 6.5, 9.2, 12.8, 17.5 and 23.8 mm respectively. It is evident

---

* For example, those manufactured by Donald Brown (Brownhall) Ltd.

that the 'vacuum' obtained with a water pump will vary considerably with the temperature of the water and therefore with the season of the year; in any case a really good 'vacuum' cannot be produced by a filter pump.

Oil immersion rotatory pumps. Water pumps are not always satisfactory, particularly in the summer or if the pressure on the water mains is not too high; they are sometimes erratic in action and cannot be used if low pressures are required. Oil immersion rotary pumps now find extensive use either as individual units or as a large-capacity unit connected to numerous points situated at convenient positions in the laboratory.* Commercially available single-stage pumps may evacuate down to about 0.1 mmHg; somewhat higher pressures are quite satisfactory for many purposes. To take advantage of the low pressure produced by a good oil pump, narrow-bore connections in the apparatus assembly should be avoided by using ground glass joints wherever possible; rubber tubing connections from the pump should be as short as possible.

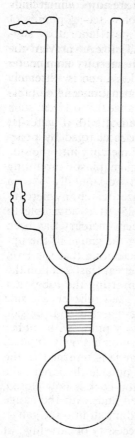

For convenience in laboratory use, the pump is mounted on a suitable mobile trolley,† which also houses a manometer (Section **I,30**) and a pair of glass vapour traps (Fig. I,122); these protect the pump against the intake of moisture or chemical vapours‡ either of which would be harmful to the pumping efficiency if allowed to contaminate the oil of the pump. Before the pump is brought into operation, therefore, the trapping vessels are filled with a suitable coolant (usually Cardice-acetone). After use the traps should be cleaned and dried.

Fig. I,122

Vapour diffusion pumps. To attain pressures lower than that produced by the oil immersion rotatory pump, a vapour diffusion pump is employed which gives pressures down to 5×10^{-6} mmHg. The principles of the operation of a vapour diffusion pump are illustrated by reference to the schematic diagram shown in Fig. I,123. Vapour molecules ascending from the boiler emerge in a downward direction through the various orifices sited under baffle plates attached to the central tube. Gas molecules diffuse into this descending stream and are thereby propelled downwards and removed by a subsidiary backing pump of the oil immersion rotatory type, after the vaporising fluid has been condensed on the cooled jacket which surrounds the unit. The condensed fluid drains into the boiler to be revaporised.

Mercury-charged vapour diffusion pumps in which the boiler unit is constructed of quartz and designed to be heated with a gas flame or by an electric element, were formerly used rather extensively; the remainder of the unit is

---

* Satisfactory vacuum installation units are supplied by Edwards High Vacuum Ltd.
† One such type of trolley is available from A. Gallenkamp & Co. Ltd.
‡ If there is the possibility that the vapours are corrosive then it is essential that a more elaborate trapping system should be employed.

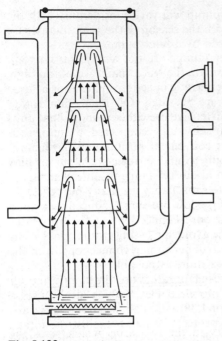

Fig. I,123

constructed in Pyrex glass. More robust and highly efficient vapour diffusion pumps of all-metal construction designed for use with either mercury or suitable grades of Apiezon or Silicone oil* and electrically heated are now commercially available.

I,30. MANOMETERS AND MANO-STATS A simple mercury manometer is shown in Fig. I,124 and is extremely suitable for the measurement of pressures within the range 0.5 to 17 cm such as would be obtained with a water-jet pump. The gauge is charged by pouring about 8 ml of mercury into the side arm whilst the gauge (plastic mounting removed) is held horizontally over a tray of adequate size to catch any mercury spillage (Fig. I,125). It is advisable to check that sufficient mercury has been added by holding the gauge in the upright position and noting that the level of mercury registers at least 2 cm on the scale. Great care should be taken in the operation of inverting the tube, that the mercury is allowed to flow down the inside of the glass tube on the side remote from the side arm. The vacuum gauge is now re-inverted and the side arm connected to a good vacuum source, e.g., oil rotatory pump with rubber tubing into which has been inserted a stop-cock. The vacuum is slowly applied, and when evacuation is thought to be complete, the gauge is returned to the upright position when the levels of mercury inside and outside the inner tube should be equal and should read at least 1.5 cm. The stop-cock is now closed, the rubber tubing from the vacuum source disconnected and, with the gauge in the upright position, the stop-cock opened slowly to admit air into the gauge; the mercury will be seen to rise in the inner tube. The necessity of carrying out this latter operation slowly cannot be too strongly advised, since sudden admission of air will cause the mercury to rise with such force that the glass top of the gauge may be broken. At the conclusion of this operation the stop-cock is disconnected and the gauge mounted in the plastic base; it is then ready for use.

Another type of manometer which is also frequently employed for the measurement of pressures in the range 0.5–17 cm is the U-tube design illustrated in Fig. I,126. It consists of a U-tube with mercury and mounted in a wooden stand. The scale B, graduated in mm,† between the two arms of the U-tube is movable; this enables adjustment of the scale so that one of the mercury levels coincides with a convenient point on the scale, and facilitates the reading

---

* When first charged with fresh oil the vapour diffusion pump does not usually reach its lowest pressures until it has been under continuous operation for about 24 hours, during which time dissolved gases, etc., are removed from the fluid.

† This is sometimes made of mirror glass in order to eliminate the error due to parallax.

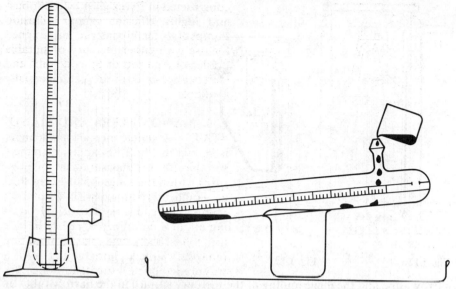

Fig. I,124   Fig. I,125

of the difference in height of the mercury in the two arms which gives the pressure directly. A tap A is usually provided so that the manometer may be isolated from the distillation apparatus. Great care should be exercised when using this manometer; if air is allowed to enter the exhausted apparatus rapidly when tap A is open, the mercury may rise to the top of the closed end with sufficient velocity to break it. It is advisable, therefore, to open tap A only when the pressure needs to be measured during an experiment, and at the conclusion of the distillation to open the tap *very slowly* after the pressure in the apparatus has been restored to atmospheric.

The **'vacustat'** * (Fig. I,127) is another useful gauge usually employed in conjunction with an oil pump; two models which cover the ranges 10 to 0.01 mmHg and 1 to 0.001 mmHg are available. It is direct reading, compact, and is charged with only about 8 ml of mercury. The gauge must be rotated carefully to the vertical position when a reading of the pressure is required; it is then returned to the horizontal position equally carefully otherwise there is a tendency for some of the mercury to be spilled over into the tubing, which connects the vacustat to the apparatus, and hence into the pump (p. 23). The gauge does not automatically record a variable pressure.

The **Zimmerli vacuum gauge** † covers a wide range of pressure (0–100 mmHg) and is depicted in Fig. I,128. It is an improvement on the U-tube gauge (Fig. I,126). The chief disadvantages of the latter are: (i) the necessity for boiling the mercury to remove the air from the closed reference tube when filling the gauge, (ii) the tendency for air to enter the closed limb after a period of time and (iii) the difficulty of precision reading due to the capillary action in the

---

* Manufactured by Edwards High Vacuum Ltd. This is essentially a form of McLeod gauge, described later.

† Supplied by the Scientific Glass Apparatus Company; a precision model, reading to 0.1 mm, is also marketed.

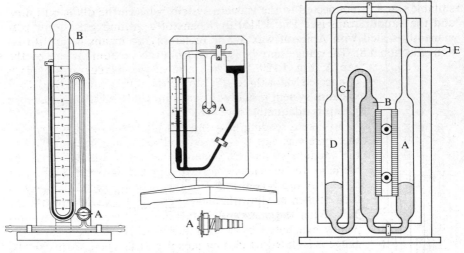

Fig. I,126   Fig. I,127   Fig. I,128

narrow tube and the facile fouling of the mercury surface in the narrow tube. In the Zimmerli gauge A and B are the limbs of a ∪-tube each having a diameter of 16 mm. Tube A is the indicating limb; tube B, the reference limb, is not sealed at the top but is connected to a capillary tube C, which is in turn joined to a wide tube D: both A and D are connected to the vacuum line at E. Thus the indicating limb and the reference limb are both connected to the same vacuum line; this facilitates the filling of the gauge and its maintenance in perfect working condition. The gauge is filled with mercury as shown in the figure; the levels in A and D should be about 2 cm above the bottom so as to form an effective seal.

When the manometer is connected to the apparatus in which the pressure is to be measured, it will be found that as soon as the pressure is reduced to a value corresponding to the difference in height of the mercury levels in A and B (or C and D) the mercury will separate at the top of the bend (between B and C), and as the pressure diminishes each part will recede in B and C until the levels become constant. The difference in height of the mercury levels in A and B, read on the scale between A and B, indicates the absolute pressure.

The reading of the pressure gauge is improved by providing blackened metal sleeves which can be moved up and down over the limbs of A and B; when viewed against diffused light, the lower edges of the sleeves and the meniscus of the mercury show up against a white background as sharply defined straight and curved lines. When the sleeves are adjusted so that they seem to touch the top of the mercury columns, the absolute pressure is represented by the difference in height between the edges of the sleeves.

The McLeod gauge, * illustrated in Fig. I,129, is used for the measurement of pressures down to about 5×10^{-6} mmHg that would be obtained by the use of a mercury or oil diffusion pump (Section **I,29**). The gauge, mounted on a

---

* Gauges of this type may be obtained, *inter alia*, from Manostat Division of Greiner Scientific Corp., and from Edwards High Vacuum Ltd.

suitable stand, is connected to the vacuum system between the diffusion pump and the vapour traps (see Fig. I,115) by means of a ground glass joint permanently sealed with Apiezon wax W, or preferably by means of an O-ring joint (Section **I,8**). The gauge may be isolated from the vacuum system by the tap A, Fig. I,129. The side arm of the mercury reservoir is connected via a three-way stop-cock B to a suitable auxiliary vacuum system (usually the vacuum achieved with a water-jet pump is adequate).

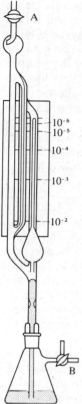

To take a reading after the gauge has been newly installed, the three-way tap B is closed and the gauge is connected to the vacuum system by opening the tap A and allowing the mercury to be partially drawn upward into the bulb. The three-way tap B connection to the auxiliary vacuum supply is then opened to allow the mercury to be drawn down from the bulb. When the gauge and the mercury reservoir have been completely evacuated by several such successive operations, the gauge is isolated by closing taps A and B. The pressure in the system may now be recorded by carrying out the following sequence. The tap A is opened carefully, and after a pause of a minute or so to allow the gauge to be finally evacuated to the pressure in the remainder of the system, tap B is cautiously opened to admit air into the reservoir which allows mercury to rise into the bulb. As the mercury approaches the bottom end of the closed capillary the three-way tap should be adjusted so that mercury rises at a very slow rate until the level in the reference capillary is coincident with the zero on the scale, at which point tap B is closed. The pressure reading is now recorded by the level of the mercury in the closed capillary. The gauge is then isolated once more by closing tap A and the mercury reservoir is evacuated via the three-way tap B to withdraw mercury well below the bulb in the gauge. Whilst the bulb-trap at the top of the gauge prevents overflow of mercury into the vacuum system it is advisable to carry out all the operations needed to record the pressure in the system carefully and methodically, since violent 'bumping' in the gauge due to the incautious inlet of air could lead to breakage and the hazard associated with spilling mercury. Finally it is customary to keep the gauge and reservoir permanently under vacuum.

Fig. I,129

The maintenance of a constant pressure in a system during distillation under diminished pressure is of great practical importance if trustworthy boiling points are desired. Devices which maintain a constant pressure in a system that is higher than the minimum pressure that the pump will give are termed **manostats**. A simple manostat, due to M. S. Newman, is illustrated in Fig. I,130. The underlying principle is that the gas in the system whose pressure is to be controlled must overcome the pressure of a column of liquid before it can be pumped out. When the pump is first started the bulk of the air in the system is removed through the open stop-cock. When the pressure, as measured on a mercury manometer, has almost reached the desired value, the stop-cock is closed, thus forcing the remaining gas to be pumped through the head of liquid in the manostat. The end of the gas inlet tube is constricted (say, to 1–1.5 mm), so that when the system has reached equilibrium the constant leaks therein (including the distillation capillary) cause a steady stream of bubbles instead of the more intermittent larger bubbles that result if a capillary is absent. Before admitting air when the distillation has been completed, the stop-cock is opened in order to avoid violent splashing of the liquid in the pressure regulator. The device is essentially one for maintaining a constant pressure between the pump system and the distillation system; the difference in pressure is equal to the head of liquid in the regulator. The liquid in the regulator may be dibutyl phthalate; this permits a pressure range of 1 to 15 mm of mercury to be covered, provided of course that the minimum pressure delivered by the pump is less than 1 mm. If higher pressures are desired, it is more convenient to use two or more of these regulators in series than to employ a larger one; alternatively, the manostat may be charged with a suitable volume of mercury, in which case the base should be of the Dreschsel bottle type.

A mechanical device embodying a bellows-sealed needle valve with a lever reduction movement for fine control* is shown in Fig. I,131. The needle is of stainless steel. This valve assembly is useful for providing a controlled leak to regulate the pressure in a vacuum system.

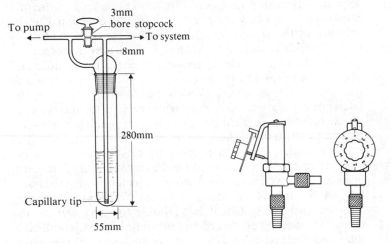

Fig. I,130   Fig. I,131

An excellent manostat, based on the Cartesian diver principle, is marketed under the name of **Cartesian manostat**. The action of the manostat may be explained with the aid of Fig. I,132. Mercury is introduced into the container until the disc of the float just makes contact with the orifice, when the pressure is equalised inside and outside the float. The device is connected to the pump and to the system by way of a large reservoir and a manometer. With the stop-cock open, the pressure in the system is reduced by way of a by-pass between the pump and the system until the desired value as read on the manometer is reached, then both the stop-cock and by-pass are closed; the device will automatically maintain the desired pressure. If the system is vacuum tight, the pressure will maintain itself; a slight leak, which may be introduced in-

---

* Supplied by Edwards High Vacuum Ltd.

tentionally, will cause the pressure to rise slightly. This will produce a displacement of the mercury level downward outside the float and a corresponding displacement upward inside the float; the buoyant force on the float is consequently diminished and when this reduction in buoyancy becomes sufficient to overcome the suction force at the orifice due to the pressure differential, the disc will break away from the orifice and permit the pump to evacuate sufficient gas from the system to restore the original pressure. When the original pressure is restored, the disc will return to its former position and seal off the orifice. The cycle is repeated indefinitely, if the size of the leak in the system does not exceed the capacity of the gas flow that is possible through the orifice and the pump is of sufficient rating to carry the load.

The commercial form of **Cartesian manostat,* model 7**, is depicted in Fig. I,133; it is normally charged with mercury except for very low pressures when dibutyl phthalate is employed. The manostat is highly sensitive in its action; furthermore, once the pressure has been set in the instrument, the system may be shut down without disturbing the setting.

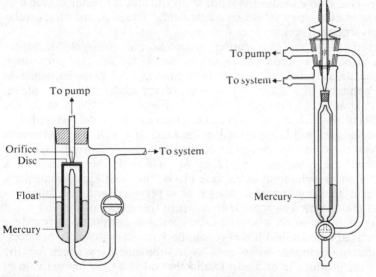

Fig. I,132   **Fig. I,133**

I,31. CHROMATOGRAPHY Chromatography is a separation process which depends on the differential distributions of the components of a mixture between a mobile bulk phase and an essentially thin film stationary phase. The stationary phase may be either in the form of a packed column (*column chromatography*) through which a mobile phase is allowed to flow, or in the form of a thin-layer adhering to a suitable form of backing material (*thin-layer chromatography*) over which the mobile phase is allowed to ascend by capillary action.

The thin film stationary phase may be either a liquid or a solid, and the

---

* Supplied by the Manostat Division of Greiner Scientific Corp. of New York; an all-metal Cartesian manostat is also marketed.

mobile phase a liquid or a gas. Possible combinations of these phases then give rise to the principal chromatographic techniques in general use.

In **partition chromatography** the stationary phase is a thin liquid film adsorbed on the surface of an essentially inert support. The mobile phase may be either a liquid (*liquid–liquid partition chromatography*) or a gas (*gas–liquid partition chromatography* or *gas chromatography*, frequently abbreviated to GLC). In either system the separation depends largely upon partition between the two phases (compare Section **I,32**), although the separation process may be complicated by the incursion of adsorption effects involving the inert support and the compounds undergoing chromatographic separation. *Paper chromatography* is an important example of partition chromatography in which filter-paper serves as a support for the immobile liquid phase.

In **adsorption chromatography** the mobile phase is usually a liquid and the stationary phase is a finely-divided solid adsorbent (*liquid–solid chromatography*). Separation here depends on the selective adsorption of the components of a mixture on the surface of the solid. Separations based on gas–solid chromatographic processes are of limited application to organic mixtures. The use of *ion-exchange resins* as the solid phase constitutes a special example of liquid–solid chromatography in which electrostatic forces augment the relatively weak adsorption forces.

Apart from partition and adsorption processes, chromatographic separations may also be based upon differences in molecular size (*gel permeation chromatography*, or *gel filtration*). In this technique gel-like material, which is commercially available in a range of porosities, serves as the stationary phase, and separation is achieved through differential diffusion into the pores of the matrix, of molecules which are not large enough to be completely excluded.

The basic techniques will be described in the following sections; the reader's attention is also directed to a selection of monographs on chromatographic procedures which are listed in Refs. 52 to 67. However, the selection of a particular technique which might be expected to be the most appropriate for a given situation is to some extent a matter of experience, and the following considerations may be used as a guide to assist in the choice of method.

In general preparative work, when the extraction and purification procedure follows a prescribed and detailed literature method, it will not usually be necessary to use chromatographic techniques as a supplementary check on the purity of the final product, in addition to that provided by the comparison of physical constants (e.g., m.p., b.p., η_D, $[\alpha]_D^t$, etc.) and spectral data with those quoted in the literature. Nevertheless a chromatographic examination of the purity of such a product may sometimes be desirable (e.g., for the detection of isomers, as in the preparation of *p*-nitroacetanilide, Section **IV,76**); thin-layer chromatography provides the most convenient and rapid procedure in the case of solid products. Gas chromatography would usually be used to evaluate the purity of a liquid product or volatile solid; this may be particularly desirable in those cases where some measure of fractional separation of the distilled product is necessary, since a single simple fractional distillation rarely yields a product of high purity. It should also be borne in mind that commercially produced starting materials may not always be entirely free from impurities of a chemically similar nature (i.e., isomers, homologues), the presence of which may lead to impurities in the final product whose presence might not otherwise be expected; the purity of such starting materials may be routinely checked by chromatography.

Chromatographic techniques may also be used on occasions to monitor the progress of a reaction for which the optimum conditions are uncertain, as may be the case when an established published procedure is used as the basis for carrying out other preparations of a similar nature. In general, reactions are monitored by the periodic removal from the reaction mixture of test portions for suitable chromatographic study. In the case of solid products and starting materials it is usually convenient and sufficient to load a sequence of appropriately sized samples of the homogeneous reaction mixture, taken at various times during the progress of the reaction, directly on to a thin-layer chromatographic plate.* The chromatographic behaviour of the components on development with a few selected solvent systems (see below) is investigated; the results will allow the progress of the reaction to be assessed. With liquid reactants and products, some form of simple purification and extraction into a suitable solvent for gas chromatographic investigation must be developed. By way of illustration the following simple example of a suitable procedure may be cited. In the preparation of an alkyl halide from the alcohol (Section **III,54**) using constant boiling hydrobromic acid and sulphuric acid, an aliquot portion of the reaction mixture is removed by means of a capillary pipette after a suitable time interval from the start of the reaction (say 30 minutes) and transferred to a micro test-tube. A few drops of chloroform are added and the contents of the tube shaken. The lower chloroform layer is removed to another micro test-tube with a capillary pipette and washed successively by shaking it with dilute aqueous sodium hydrogen carbonate and water, the aqueous layers being removed with the aid of a pipette. The chloroform layer is dried by the addition of a little magnesium sulphate desiccant, the tube centrifuged if necessary, and a sample of the clear upper layer submitted to gas chromatographic investigation.

When the success or outcome of a reaction is uncertain or unknown, chromatographic methods are invaluable both for monitoring the progress of the reaction and also for assessing the success of the purification process. They may well reveal that the procedures adopted are unsuited to the required isolation and purification of the reaction product. In these cases the chromatographic behaviour of the components in the system which has been revealed by these preliminary small-scale studies provides a basis upon which purification by preparative chromatographic methods may be achieved. Thin-layer chromatographic behaviour may be reasonably closely duplicated by employing a similar stationary phase and a similar mobile phase in either a 'wet' or a 'dry' column technique, which can then be readily scaled up to accommodate the bulk of the reaction product. Preparative gas chromatography has the advantage that usually little further investigational work is required to accomplish the separation of larger quantities, but as will be seen later a preparative scale separation may require the use of an automated apparatus using a multiple cycling procedure.

Chromatographic techniques *1*. **Thin-layer chromatography (TLC).** In this technique it is usual to employ glass plates coated with layers of the solid stationary phase, which adhere to the plates, generally by virtue of a binding

---

* Some pretreatment of the sample may of course be necessary before it is loaded on to the chromatographic plate.

agent, such as calcium sulphate, which is incorporated. The prepared thin-layer on glass is often called a *chromaplate*.

The most commonly used stationary phases, which are available in grades specially prepared for TLC use,* include silica gel, alumina, kieselguhr and cellulose powder; many of these are available with a fluorescent compound (e.g., zinc sulphide) incorporated in order to facilitate the detection of the re-solved components of the mixture which is then achieved by viewing the plates under ultraviolet light. Other materials suitable for special applications are polyamides, modified celluloses with ion exchange properties and the various forms of organic gel having molecular sieving properties (e.g., Sephadex, Bio-Gel P).

Preparation of plates.   Before glass plates are coated with adsorbent they must be carefully cleaned with laboratory detergent, using a test-tube brush to remove adhering particles, rinsed thoroughly with distilled water, placed in a suitable metal rack and dried in an oven. Subsequent to the treatment with detergent solution the plates should only be handled by the edges or by the under-surface which is not to be coated with adsorbent. Failure to observe this precaution may result in the formation of a mechanically unstable layer which is liable to flaking due to grease spots on the glass surface. In severe cases of grease contamination it may be necessary to use a chromic acid cleaning mix-ture (Section I,2).

Small plates suitable for preliminary exploration of the chromatographic process with regard to the selection of a suitable stationary phase or the selec-tion of a solvent system are conveniently prepared from microscope slides using a dipping technique; this operation is conducted in a fume cupboard.

A slurry is prepared by the slow addition with shaking of 30 g of adsorbent (most usually silica gel or alumina) to 100 ml of carbon tetrachloride contained in a wide-necked capped bottle. A pair of microscope slides is held together and dipped into the slurry, slowly withdrawn and allowed to drain momentarily whilst held over the bottle. The slides are parted carefully and placed horizon-tally in a rack sited in a fume cupboard to dry for approximately 10 minutes. The surplus adsorbent is then removed by means of a razor blade drawn down the glass edges. It may be desirable to activate the adsorbent further by heating it at 110 °C; since the activity of the adsorbent varies with the heat treatment and the subsequent storage conditions of the prepared plate this further treat-ment should be carefully standardised. For the attainment of a high degree of reproducibility it is usually best to activate the adsorbent and allow the plate to cool in a desiccator cabinet immediately before use.

Larger single-glass plates (i.e., 20 × 5 cm) may be coated conveniently using the easily assembled apparatus shown in Fig. I,134. It consists of a sheet of plate glass (20 × 30 cm) at the upper and lower ends of which two glass plates (20 × 5 cm) are secured by means of a cement for glass: the plate to be coated is placed in the central depression and is held in position by two uncoated plates one on either side. The thickness of the layer can be adjusted to the

---

* Although most laboratory suppliers market a range of adsorbents for use in thin-layer chro-matography, the following is a selection of the specialist firms: Camlab (Glass) Ltd, Cambridge; Eastman Kodak Co., New York; Fluka AG., Switzerland; Johns-Manville, New York; E. Merck Products through Anderman & Co. Ltd; H. Reeve Angel and Co. Ltd, London; M. Woelm, through Koch-Light Laboratories, Ltd, Buckinghamshire.

thickness required (say 0.25–0.3 mm) by wrapping both ends of a glass rod 14 cm long and 7.5 mm diameter with equal lengths (12.5 cm) of 2.5 cm Sellotape: this is the 'spreader'.

The exact composition of a suitable slurry for spreading depends on the nature of the adsorbent; this should therefore generally be prepared according to the procedure recommended by the supplier. The composition of the slurry

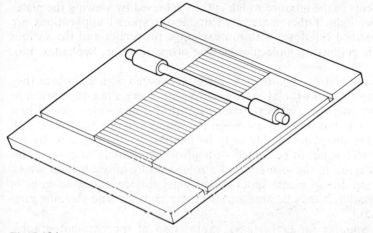

Fig. I,134

may have to be varied, as for example in the preparation of buffered silica gel or kieselguhr plates, or of plates for reversed phase thin-layer chromatography which require coatings of hydrophobic layers. A slurry from about 2 g of dry adsorbent is sufficient to coat one plate of the size stated here. The slurry is poured on to the upper end of the central glass plate (Fig. I,134) and spread evenly over the plate with the special applicator or spreader: this should be completed in 15 seconds. The layer is allowed to stand for 5 minutes in order to set; cellulose and polyamide plates are allowed to dry at room temperature and are then stored in a dust-free cabinet—they are not normally heated. Inorganic adsorbents are activated and stored under standard conditions (see above).

Figure I,135 illustrates diagrammatically one form of commercially available spreader * which consists of a flat frame capable of holding rigidly five 20 × 20 cm glass plates or the equivalent number of 20 × 10 cm or 20 × 5 cm plates. The construction of the frame is such that when the plates have been positioned edge to edge the upper surfaces are all aligned in the same plane. This flat surface, free from ridges at plate joints, ensures a uniform thickness of adsorbent layer over all the plates and provides a smooth path for the metal hopper (Fig. I,136) when drawn across the surface. It is usually best to place at either end of the line of plates to be coated a 20 × 5 cm glass plate to provide an area upon which the coating process may be started and finished, since frequently the layer thickness at the immediate start and finishing points is not uniform. The prepared slurry is poured evenly into the rectangular well of the hopper located on the end plate and this is then drawn steadily over the glass plate surface so that the slurry flows evenly through the gap provided on the

---

* Shandon Southern Instruments Ltd; other models are available, e.g., Camlab (Glass) Ltd.

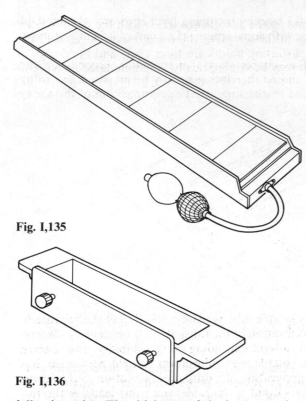

Fig. I,135

Fig. I,136

following edge. The thickness of the layer may be preselected by loosening the nuts and adjusting the accurately machined metal gate by means of a 'feeler' gauge, the hopper resting on a flat glass surface. This enables layers of thickness 200–2000 μm to be selected; in fact a thickness of 250 μm is usually the most suitable for routine use. When the full capacity of the spreader is to be utilised and when thick layers are to be spread, necessitating a fairly large volume of slurry, it is sometimes helpful to place a glass rod of suitable length immediately along the gap between the glass surface and the metal gate to prevent excessive flow of slurry whilst the hopper is being charged prior to the spreading operation.

Loading of plates. In order to load the prepared chromaplate (say 20 × 5 cm size) with the sample to be investigated the following procedure should be followed; this may be modified appropriately when the larger plates and the micro plates are used. Wipe any excess adsorbent from the back and edges of the plate. Form a sharp boundary by scoring a line with a metal scriber paralleled with and 5 mm from the shorter edge (in the case of 5- and 10-cm-wide plates) of the plate and carefully remove surplus adsorbent with the aid of the flat side of a spatula and blowing carefully. Align the lower edge of the commercially available template (Fig. I,137(A)) along this edge of adsorbent layer; draw the metal scriber through the adsorbent layer using the upper edge (B) of the template as a guide. Carefully blow surplus adsorbent from this cut which then provides a finishing line for the subsequent solvent front; if not clear of adsorbent the solvent will flow unevenly across the line and lead to an unsatisfactory evaluation of the chromatogram. With the template still in position, lightly mark a series of starting points on the adsorbent surface with the metal scriber through the application holes in the template; these are usually located 15 cm from the finishing line. (These starting points are 1 cm apart and the design of the template is such that four may be symmetrically accommodated on a 5-cm-width plate.)

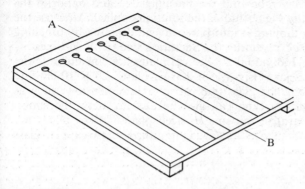

Fig. I,137

In a plastic-stoppered glass sample tube prepare a solution of the mixture to be investigated having a concentration in the range 0.5–3.0 per cent by dissolving 3 mg in from 0.1–0.6 ml of solvent depending on its solubility. The selected solvent should be reasonably volatile (e.g., chloroform or light petroleum, b.p. 40–60 °C); in the subsequent application of a drop of this solution to the adsorbent rapid evaporation of the solvent is desirable, since this leads to the formation of a small-diameter spot which results in a better separation of the components in the subsequent chromatographic development process. Similar volumes of aqueous solutions give larger spots which result in a more diffuse chromatogram—if aqueous solutions need to be used a technique of multiple application (see below) may be necessary. Solutions of pure compounds thought to be present in the mixture (e.g., starting materials, possible reaction products) are similarly prepared for application to the adsorbent on the same plate and alongside the mixture spot—these act as reference compounds to permit more ready interpretation of the chromatogram.

The solutions are applied individually to the marked points on the adsorbent layer by means of a sample applicator. This is prepared by drawing out a melting point capillary tube in a micro-bunsen flame and snapping the drawn-out portion in two after scratching with the edge of a fragment of unglazed porcelain to ensure a clean break. The applicator is charged by dipping the capillary end into the solution and after withdrawing, touching the end on a piece of filter-paper until the volume is reduced to about 0.5 μl. Using the template as a hand rest, the solution is transferred to the plate by touching the tip of the capillary on to the adsorbent layer, taking care not to disturb the surface unduly. If the 0.5-μl volume has been estimated with reasonable accuracy the diameter of the spot should be about 3 mm diameter. Alternatively the use of commercially available disposable micro-pipettes which contain standard volumes of from 1 to 5 μl and which can easily be partly filled to contain the required amount of solution, or the use of calibrated syringes delivering from 0.1 to 5 μl, allows the volume of solution to be applied to the plate to be judged more accurately. With more dilute solutions or where a heavier loading of material is required, larger volumes may be applied by allowing the solvent to evaporate during the intermittent addition of the solution to the plate (the use of a commercial electric warm-air blower is recommended) so that the diameter of the spot never exceeds 3 mm. A suitable cipher is inscribed on the adsorbent layer beyond the finishing line and opposite the point of application to identify the nature of the solution applied. After use the capillary may be cleaned by dipping it into pure solvent, draining by touching the tip on to a filter-paper and repeating the operation two or three times.

Development of plates. Individual 20 × 5 cm plates are conveniently developed in a cylindrical glass jar * (Fig. I,138). Larger plates, 20 × 10 cm and 20 × 20 cm, require a rectangular glass tank of suitable dimensions such as that shown in Fig. I,139; such a tank can also be used to allow the simultaneous development of several of the smaller-sized plates. Micro plates are easily accommodated individually in 4-oz wide-mouthed screw-capped glass bottles.

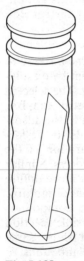

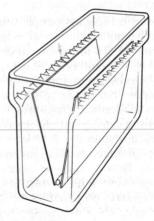

| Fig. I,138 | Fig. I,139 |

Line the inside of the jar with filter-paper, leaving a gap for viewing the chromaplate. Saturate the filter-paper with the developing solvent (see below), close the jar and allow to stand for about 10 minutes so that the atmosphere in the jar becomes saturated with solvent vapour. Insert the plate with the origin spots towards the bottom of the jar, tilted as shown, so that the uncoated face is uppermost. Carefully pour down the side of the jar more of the developing solvent so that the bottom of the adsorbent layer is well-immersed; the solvent level should not however reach as far as the spots. Recap the jar and allow the solvent to ascend by capillary action to the finishing line which has been scored across the plate. The time required to complete this development varies greatly with the composition of the solvent and the nature of the adsorbent. If the

---

* For example a 'Chromajar', manufactured and supplied by Shandon Southern Instruments Ltd.

system is inconveniently slow running the development process may be terminated before the solvent reaches the finishing line, provided that the position of the solvent front is marked on the adsorbent layer *immediately* the plate is removed from the development tank.

After removal, the plate is dried suitably depending upon the volatility and toxicity of the solvent system; for example, dry the plate in the fume cupboard (if necessary) with a warm-air blower or dry in a temperature-controlled oven, etc.

Location of spots. The positions of coloured components can of course usually be seen without any difficulty providing that the concentration in the initial spot is sufficiently high and that excessive spreading of the component during development has not occurred. Viewing the plate under an ultraviolet lamp* will reveal u.v. fluorescent compounds the positions of which must of course be marked with the scriber on the surface of the adsorbent. Non-fluorescent compounds can be detected by virtue of their fluorescent quenching effect when they are chromatographed on adsorbents into which a fluorescent indicator has been incorporated (e.g., Silica Gel G.F. 254). Routine inspection of plates under ultraviolet light is to be recommended before any further detection processes are applied.

A useful general, but unspecific, detecting agent for most organic compounds is iodine vapour. The dried plate is allowed to stand in a closed tank containing a good supply of iodine crystals scattered over the tank bottom; usually the spots are revealed as brown stains. Their positions should be marked as soon as the plate has been removed from the iodine tank since standing in air for a short while causes the iodine to evaporate and the stains to disappear.

Another general locating procedure applicable in the main only to plates coated with inorganic adsorbents and for the detection of organic material is to spray† the plate with concentrated sulphuric acid or with a solution of concentrated sulphuric acid (4 ml) in methanol (100 ml), and then to heat the plate in an oven to about 200 °C until the organic materials are revealed as dark charred spots.

This spraying operation must of course be carried out with considerable care and it is advisable to place the plate at the bottom of a large rectangular glass tank, placed on its side and located in a fume cupboard before spraying is attempted. It is also good practice to wear a suitable protective face mask. It is essential to cover the plate evenly with spray but without so saturating the adsorbent layer that the liquid visibly flows over the surface, since this will cause distortion of the zones.

Chemical methods for the detection of colourless compounds by the use of a suitable chromogenic spray reagent are widely used. Many of these are selective for a particular functional group or groups and may be extremely sensitive; e.g., the ninhydrin reagent for the detection of amino acids. Other such spray reagents are more general in their application; e.g., indicators may be used in sprays for the detection of acids and bases. Such chemical locating agents may

---

* Purpose-designed units are available from most suppliers of chromatographic equipment, e.g., the Hanovia Chromatolite supplied by A. Gallenkamp & Co. Ltd.

† Spray guns operated by an aerosol propellent are available commercially; alternatively the cheaper all-glass spray units (such as that supplied by Jobling Laboratory Division, Fig. I,140), operated by compressed air or a rubber bulb-type hand blower, may be used.

be usually applied with advantage after successively viewing the plate under ultraviolet light, exposing the plate to iodine vapour and allowing the iodine to evaporate. This extended treatment gives a much more comprehensive picture of the composition of the mixture of components on the chromatogram than does a single non-selective method.

The selectivity and sensitivity of a wide range of spray formulations may be found in the many specialist monographs on thin-layer chromatography, a selection of which is listed in Refs. 57–60.

Provided that the experimental conditions are reproducible the movement of any substance relative to the solvent front in a given chromatographic system is constant and characteristic of the substance. The constant is the R_F value and is defined as:

$$R_F = \frac{\text{distance moved by substance}}{\text{distance moved by the solvent front}}$$

Figure I,141 indicates the method of measurement of R_F values of each of the components of a typical chromatogram:

$$R_F = \frac{a}{b}; \ R_F{}' = \frac{a'}{b}; \ R_F{}'' = \frac{a''}{b}$$

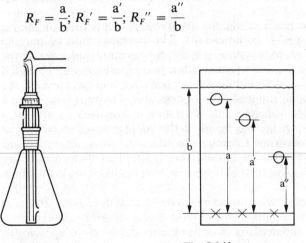

Fig. I,140   **Fig. I,141**

True reproducibility in R_F values is however rarely achieved in practice due to minor changes in a number of variables such as:

(i)   the particle size of different batches of adsorbent;
(ii) the solvent composition and the degree of saturation of the tank atmosphere with solvent vapour;
(iii) prior activation and storage conditions of the plates;
(iv) the thickness of adsorbent layer, etc.

It is therefore *not* desirable to use an R_F value in isolation as a criterion of identity; marker spots, when reference compounds are available, should always be run on the same plate as the mixture to substantiate the identification of the components.

If the chromatographic behaviour of the substances under investigation is unknown, the most satisfactory developing solvent must be ascertained by preliminary trial runs using the micro plates in screw-capped bottles. It is convenient to set up a series of such bottles containing solvent systems of decreasing polarity selected from Table I,8 (cf. *Selection of Solvents for Adsorption Chromatography*, p. 202). Identically loaded micro plates are developed separately using the chosen solvents, dried and sprayed with the appropriate reagent and the chromatographic mobility of the individual compounds noted. Solvents which cause all the components to remain near to the spot origin or to move near to the solvent front are clearly unsatisfactory. If it is seen that no single pure solvent gives a satisfactory chromatogram with well-spaced compact spots, it is necessary to examine the effect of using mixtures of solvents to provide systems having a range of intermediate polarities.

Precoated plates. Glass plates precoated with most of the commonly used adsorbents, with or without fluorescent indicators, can be purchased from the specialist suppliers. Having been prepared under carefully standardised conditions these have a much higher degree of reproducibility in the subsequent analysis than hand-coated plates, since some of the factors noted earlier which cause variation in R_F values, e.g., particle size, layer thickness, etc., have been eliminated.

Reference must also be made to the use of layers precoated on to flexible sheets which are also available commercially. The backing material may be either of aluminium foil or more usually a solvent-resistant polyester sheet. These sheets can be cut to the desired size with a pair of scissors and activated if necessary. Several millimetres'-width of adsorbent are then scraped off the sheet, on the sides which will be parallel to the direction of solvent flow, using a spatula guided by the edge of a steel rule. This step is necessary to prevent a solvent film travelling along the edge and in the space created between the adsorbent layer and the backing sheet by the scissor cutting. The prepared sheet is then used as one would a normal thin-layer glass plate, but it has the added advantage that the adsorbent layer is less likely to be accidentally damaged. A further advantage with these flexible plates is that the developed chromatogram may be stored in a note book, etc., after use.

Preservation of chromatograms in the case of layers on glass plates may if required be conveniently achieved by treating the developed plate with a suitable lacquer* which after drying allows the layer to be peeled off.

Exercises in thin-layer chromatography **Separation of amino acids.** Prepare solutions of DL-alanine, L-leucine and L-lysine hydrochloride by dissolving 5 mg of each separately in 0.33 ml of distilled water, measured with a graduated 1 ml pipette (leucine may require warming to effect solution). Mix one drop of each solution to provide a mixture of the three amino acids and dilute the remainder of each solution to 1 ml to give solutions of the respective amino acids. The latter will contain about 5 μg of amino acid per μl: the mixture will contain about 5 μg of each amino acid per μl. Apply approximately 0.5 μl of each of the solutions to a Silica Gel G plate and allow to dry in the air (i.e., until the spots are no longer visible).

Prepare the developing solvent by mixing 70 ml of propan-1-ol with 30 ml of concentrated aqueous ammonia (d 0.88). Line the inside of the jar with filter-

---

* Ducofilm lacquer, supplied by Camlab (Glass) Ltd.

paper reaching to within 3 cm of the bottom and moisten with the developing solvent. Insert the prepared plate into the jar and carefully introduce by means of a pipette sufficient of the developing solvent so that the lower edge of the adsorbent layer is immersed in the solvent; put the cover in position in the mouth of the jar, and allow the chromatogram to develop.

Remove the chromatogram, dry it at 100 °C for 10 minutes and spray with ninhydrin reagent [0.2% solution in butan-1-ol, (1)]; heat at 110 °C for 5–10 minutes in order to develop the colour. Mark the centre of each spot with the metal scriber and evaluate and record the R_F values.

Note. (1) Ninhydrin (I) is the 2-hydrate of indane-1,2,3-trione. It reacts with α-amino acids to yield highly coloured products. Contact with the skin should be avoided since it produces a rather long-lasting purple discolouration.

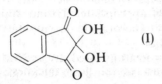

Separation of 2,4-dinitrophenylhydrazones. The solutions are prepared by dissolving 10 mg of each of the 2,4-dinitrophenylhydrazones of acetone, butan-1-one and hexan-3-one (or hexan-2-one) in 0.5 ml of ethyl acetate. Prepare a flexible silica gel* sheet of dimensions 20 × 5 cm in the manner already described and apply *ca.* 0.5 μl of each of the three solutions to give the marker spots of a diameter of between 2 and 3 mm. A mixed spot is conveniently obtained by loading sequentially to the same area further 0.5 μl aliquot portions of each of the solutions and allowing the solvent to evaporate completely between each addition.

Charge the paper-lined jar with the developing solvent (toluene : light petroleum b.p. 40–60 °C, 3 : 1), insert the loaded flexible sheet and allow the development to proceed. Air dry the developed chromatogram and record directly the R_F values of the components.

2. **Paper chromatography.** Paper chromatography is a form of partition chromatography in which the stationary phase is the absorbed water always present in filter-paper (*ca.* 22%), supported by the cellulose molecules of the paper.

Although ordinary filter-paper can be used for paper chromatographic work, specially selected grades are available commercially (e.g., Whatman papers †) having a higher degree of uniformity, and these are to be preferred. Whatman No. 1 paper is the most suitable for general use, but other grades for specific purposes include No. 4 when speed of development is a consideration, No. 3 when a thick grade for heavier loading is required and No. 20 when particularly good resolution is required. Modified cellulose papers, such as silicone-treated papers to provide a support for non-aqueous phases when employing the technique of reversed phase chromatography, or ion-exchange cellulose

---

* The Editors used Eastman Chromatogram silica gel sheets with added fluorescent indicator in this experiment and also for many routine laboratory studies; these are available from Eastman Kodak Co.

† Manufactured by W. & R. Balston Ltd, available through H. Reeve Angel & Co. Ltd.

papers for the separation of ionic organic and inorganic substances are also available.

The mobile phase is a mixture of one or more organic solvents and water. Water-miscible organic solvents such as phenol, acetic acid or pyridine are frequently used as one of the organic components to increase the proportion of water in the mobile phase. Many of the development solvent mixtures recommended form two-phase systems in which case the prepared mixture is allowed to settle and the organic phase, which is therefore saturated with water, is removed and used as the mobile phase, the aqueous layer being used to saturate the atmosphere of the vessel in which the chromatogram is subsequently developed.

For the chromatography of amino acids and other amphoteric substances it is often advantageous, in order to obtain well-resolved, compact spots, to dip the paper into an aqueous buffer solution (of appropriate pH and approximately 0.066 M), and then allow it to air dry. The organic developing phase must be pre-equilibrated with the buffer solution.

Drops of solutions of substances to be chromatographed are individually applied to the paper by means of a capillary tube (see thin-layer technique) and the paper is dried. The paper is then placed in a suitable container so that it can be irrigated with the mobile phase (downward by gravity *descending technique*, or upwards by capillarity—*ascending technique*) without losses by evaporation. When the solvent has travelled the required distance, the paper or papers are removed from the container, the position of the solvent front marked, and the paper or papers dried. If the spots are not coloured their location is determined by spraying with a suitable chromogenic reagent; sometimes the solutes exhibit fluorescence in ultraviolet light and their presence can be detected this way. As for thin-layer chromatography the R_F is defined as:

$$R_F = \frac{\text{distance moved by substance}}{\text{distance moved by the solvent front}}$$

Chromatogram development.   In the **ascending technique** the paper is supported vertically in a closed jar or tank (Fig. I,142(a–c)) so that the paper edge nearest to the spot origin just dips into the developing mobile phase. The paper support arrangements in Fig. I,142(a) are self-explanatory but in (b) it should be noted that the loaded paper is suitably curled into a cylinder and secured by means of three staples* so placed that the two paper ends do not touch. A boiling tube may be conveniently used as a small-scale version of (c).

In the **descending technique** a specially designed tank or jar such as that illustrated in Fig. I,143 is employed. The essential requirement is a trough adequately supported near the top of the tank and which contains the mobile solvent phase. The edge of the paper is held securely within the trough by means of a glass rod, one end of which is angled for convenient handling. The paper passes over another glass rod sited close to the trough edge so that (i) it is not torn by the glass edge of the trough, and (ii) the solvent flows regularly and evenly down the paper. The descending technique may be used with advantage when a two-phase solvent system is to be employed. In such a case the aqueous phase is placed in the bottom of the tank and allowed to saturate the atmo-

---

* Special plastic staples are available, e.g., from Shandon Scientific Co. Ltd.

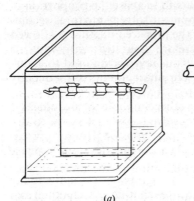
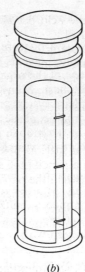
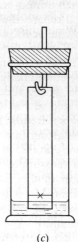

(a)   (b)   (c)

Fig. I,142

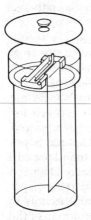

Fig. I,143

sphere; the loaded paper is then appropriately supported in the empty trough, the tank reclosed and the paper surface allowed to equilibrate with tank atmosphere. Finally the mobile phase is poured into the trough from a separatory funnel via the capped hole in the tank lid.

Exercises in paper chromatography   **Separation of amino acids by the descending technique.**   Solutions of DL-alanine, glycine, L-leucine and L-lysine hydrochloride are prepared as follows. Dissolve 5 mg of each amino acid separately in 0.25 ml of water; use a graduated 1 ml pipette to measure out the volume of water (leucine may require warming to effect solution). Mix 1 drop of each solution to provide a test mixture: dilute the remainder of each solution separately to 1 ml to give solutions of the respective amino acids. The latter will contain 5 μg of amino acid per μl: the test mixture will contain about 5 μg of each amino acid per μl.

Prepare the developing solvent by mixing 40 ml of butan-1-ol, 10 ml of glacial acetic acid and 50 ml of water in a separatory funnel. Shake the mixture and allow the two phases to settle and separate the upper and lower layers. This developing solvent should only be prepared as required since the composition changes on standing due to esterification.

Use the special jar (or tank) designed for chromatography such as that illustrated in Fig. I,143. Cut 50 × 10 cm filter-paper strips from Whatman No. 1 paper; handle them as little as possible, and always handle by the edges. Mark a starting line using a pencil (not a ball-point or similar type pen) 7.5 cm from one end (i.e., so that when the paper is supported in the trough, this line is clear of the supporting rod); with other designs of chromatography tanks the position of this line will have to be appropriately sited. Then mark the central point M on the starting line together with four others in pairs each 1.5 cm apart on either side of M to act as points of application of the drops of the solution.

To apply spots of each amino acid solution and of the mixture to the paper, use a 2 μl micro-pipette completely filled and touch the chromatogram paper at the marked position so that the sample drains into the paper: the spots will be about 6 mm diameter. Rinse the tube with distilled water in the manner previously described (under thin-layer chromatography) and similarly apply the remainder of the samples.

Place some of the lower (aqueous) layer of the developing solvent in the bottom of the jar, anchor the paper in the trough, align it vertically (Fig. I,143) cover with the lid and allow to stand for 1–2 hours so that the system becomes equilibrated with solvent vapour. Introduce through the hole in the cover sufficient of the upper layer of the development solvent to half-fill the trough and insert the stopper in the lid. Allow development to proceed for 12–13 hours. Remove the lid of the jar, cut the paper along the trough edge and remove the paper from the jar by lifting the supporting rod. Fold several thicknesses of filter-paper over the centre of the supporting rod, attach a bulldog clip and suspend the paper from a horizontal rod fixed to a retort stand sited in a fume cupboard. Mark the solvent front with a lead pencil and dry the paper with a warm-air blower; alternatively the paper may be suspended vertically in an oven and dried at 100 °C for about 5 minutes. To locate the positions of the amino acids, spray the paper in a fume cupboard lightly and evenly with ninhydrin (0.2% solution in butan-2-ol), and develop the colour by placing the paper in an oven at 100 °C for 5–10 minutes. The positions of the four individual amino acids are clearly shown by the colour of their zones or spots; the mixture should have separated into four distinct spots. Mark the position of the spots by encircling them with a thin pencil line. Calculate the R_F values of the individual amino acids.

Separation of monosaccharides by the descending technique. Prepare solutions of D-xylose, D-arabinose, D-glucose, D-galactose and D-fructose by dissolving 30 mg of each monosaccharide in 0.25 ml of water. Apply spots of each solution to a Whatman No. 1 filter-paper strip, cut and prepared as in the previous experiment, using two 0.5 μl portions in each case and allowing the solvent from the first application to completely evaporate (use an electric warm-air blower) before loading the second portion; the spots will be about 3 mm diameter. Rinse the micro-pipette with distilled water between the application of each monosaccharide solution. Locate the paper in the glass trough, replace the lid and introduce through the hole in the lid the homogeneous solvent mixture prepared from butan-1-ol : acetone : water, 4 : 5 : 1. Allow the development to proceed for about 12–14 hours. The solvent will in this case drip from the paper end during the latter stages of chromatogram development. Remove the paper, suspend in a fume cupboard and dry with a warm-air blower. Spray the paper with an ethanolic solution of naphthoresorcinol (1) and place the paper in an oven at 100 °C for about 3–5 minutes, or until the spots become visible. Calculate the R_g values of each monosaccharide which is defined as:

$$R_g = \frac{\text{distance moved by monosaccharide}}{\text{distance moved by D-glucose}}$$

The R_F values of monosaccharides are rather low, and in order to reveal differences in chromatographic behaviour a protracted solvent development process is required, which therefore necessitates allowing the solvent to drip from the paper end. With no solvent front position the chromatographic mobility of each monosaccharide is therefore compared to the behaviour of a standard, D-glucose; hence the R_g value.

Note. (1) The spray reagent is 0.2 per cent of naphthoresorcinol in ethanol containing 10 per cent v/v of orthophosphoric acid.

Two-dimensional chromatography. When complex mixtures are to be studied the R_F values of the individual components may be so close that a clear-cut separation of the components is not achieved. In such a case a two-dimensional thin-layer or paper chromatographic separation can be used with advantage. In the former a 20 × 20 cm plate is employed; in the latter case a conveniently sized square of paper (12 × 12 in; 18 × 18 in; 24 × 24 in) appropriate to the tank available is used.

A single spot of the mixture is applied near to one of the corners of the plate or paper and the chromatogram developed in one direction as usual. The plate or paper is then removed and dried and the chromatogram re-developed in a second solvent system so that the direction of solvent flow is at right angles with respect to the first (Fig. I,144(a and b)). These illustrations point the need for correctly placing the origin spot with respect to the edge of the chromatogram so that the solvent levels in both development processes do not cover the

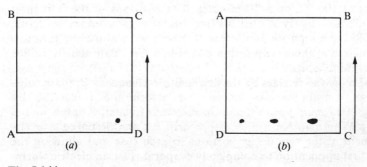

Fig. I,144

applied spot or the individual components which separate in the first development.

It is usual to employ two solvents systems in both of which the individual components have an adequately wide range of R_F values since in this way a good separation of components over the whole plate is observed. The spot location is achieved by the means previously described; each component will be characterised by two R_F values. The incorporation of an individually placed marker reference spot cannot of course be applied in the two-dimensional chromatogram. If the chromatographic behaviour of reference compounds under the operating conditions needs to be determined, these must be run on a second two-dimensional chromatogram which is developed along with the chromatogram of the unknown mixture.

3. **Gas–liquid chromatography.** Providing that the components are adequately volatile gas–liquid chromatography (GLC) is perhaps the most powerful technique for the rapid and convenient analysis of the composition of mixtures of organic compounds. It is based upon the partition of the components between a mobile gas phase and a stationary liquid phase retained as a surface layer on a suitable solid supporting medium.

For routine analytical work the supporting medium, impregnated with stationary phase, is packed into a metal or glass column, approximately 2–3 m length and 2–4 mm internal diameter, which is in the form of a circular spiral of three or four turns, and is located in a temperature-controlled oven. The mobile gas phase (usually, but not necessarily nitrogen) enters the column at one end, which also incorporates an injection port for the introduction of the sample. The components of the sample are carried down the column, separation being dependent upon their individual partition coefficients between the stationary and the mobile phase, and emerge in turn into a detector system which is attached to the column end. The signal from the detector is suitably amplified and fed to a pen recorder. The qualitative (and if necessary quantitative) composition of the mixture is assessed from an examination of the graphical traces so produced; most recent models may be also provided with a digital print-out system which greatly facilitates the accurate quantitative evaluation of the mixture composition.

A typical chromatogram (Fig. I,145) for a homologous mixture of ketones

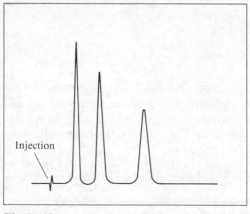

Injection

Fig. I,145

illustrates the powerful nature of this analytical tool. The chromatographic behaviour of each component of a mixture under any given experimental conditions is usually recorded in terms of its retention time (t_R). This represents the time required for the component to emerge from the column after injection and is evaluated by measuring the distance on the chart paper from the point of injection to the centre of the peak and dividing this by the chart speed. A point to note from this trace is that an increase in retention times leads in general to a broadening of the peaks. From the practical point of view this fact is of value in the interpretation of chromatograms of mixtures having an unknown number of components. If insufficient time has been allowed for the slowest running component to emerge and a second injection is made, the appearance of a peak which is anomalously broad in relation to its retention time strongly suggests that it represents a slow running component from a previous injection.

Provided that the detector is equally sensitive to each of the components of the mixture, and this is a reasonable assumption when the same type of functional compounds are being studied, and provided that all the components are volatile and have been eluted, the percentage of each component can be determined by measuring the area under each peak (see below for methods) and expressing this as a proportion of the total area.

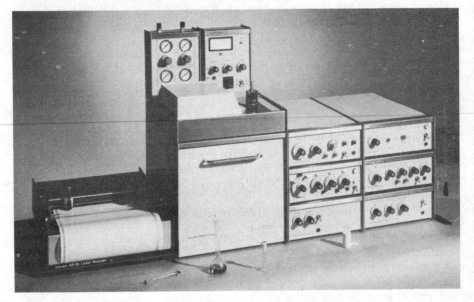

Fig. I,146

When the detector is not equally sensitive to all of the components, additional studies must determine the correction factors which need to be applied; for details a specialist monograph should be consulted (for example, Refs. 63–65).

A widely used instrument for gas–liquid chromatography is the Pye-Series 104 Chromatograph * and is shown in Fig. I,146; many other excellent instru-

---

* Available from Pye Unicam Ltd, who kindly supplied the photograph.

ments are available;* a schematic diagram of the principal features of the apparatus is shown in Fig. I,147.

The most usual **solid support material** is either kieselguhr or firebrick,† which are available from commercial suppliers ‡ in various standard particle sizes. Pre-treated grades are available which have either been acid washed to remove acid-soluble minerals from the support surface, or which have been both acid and base washed to remove additionally any acidic organic contaminants. The polar character of the support material, which is due to the presence of surface hydroxyl groups, may be modified by treatment with dimethyldichlorosilane.

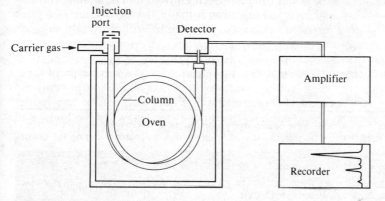

Fig. I,147

The choice of **stationary phase** will be influenced by the polar character of the components of the mixture. In general mixtures with components of high polarity separate better on chromatography when the more polar stationary phases are used. The chromatographic separation of a mixture is judged to be successful if the peaks are well separated (i.e., good resolution) and the peak shape is symmetrical and lacking in extensive trailing. The oven temperature has to be related to the overall volatility of the components and this in turn also influences the choice of the stationary phase; for example, it would be useless to endeavour to separate a mixture of high boiling compounds on a column with a stationary phase of low thermal stability.

The range of stationary phases which is now available commercially is very extensive and offers an adequate choice of polar character and thermal stability. Table I,11 lists some of the more generally useful stationary phases, their maximum operating temperatures and the classes of compounds for which satisfactory separations have most usually been found.

It is frequently desirable when assessing the purity of new compounds or evaluating the complexity of liquid mixtures to observe the chromatographic behaviour on more than one type of column packing material, since rarely will

---

* For example those manufactured by Varian Associates Ltd; Hewlett-Packard Ltd; Perkin-Elmer Ltd.

† Porapak manufactured by Waters Associates Inc., a polymeric ethylvinylbenzene-divinylbenzene is available in a range of modified structures and is used as column packing directly without the need for a stationary phase.

‡ For example, Applied Science Laboratories Inc., UK suppliers Field Instruments Co. Ltd; Phase Separations Ltd; J.J.'s (Chromatography) Ltd.

Table I,11   A selection of commonly used stationary phases for gas–liquid chromatography columns

Stationary phase	Temperature limit (°C)	Solvent* code	Applications
Squalane	150	A	General; hydrocarbons, halogenated compounds
Benzylbiphenyl	100	A	
Apiezons	150–300	A	
Silicone GE SE-30	350	A	High temperature general use; when on
Silicone GE SE-52	300		silanised support material used for trimethylsilyl ethers of polyhydroxy compounds, polyamines, etc.
Carbowax, grades 550–20 M	100–200	B	Alcohols, aldehydes, ketones, ethers
Diglycerol	150	C	Alcohols, carbonyl compounds, etc.
Dinonyl phthalate	150	D	Alcohols, carbonyl compounds, etc.
Diethylene glycol succinate	200	B	Esters; general use with polar compounds
Dexsil 300 GC†	500	B	High temperature work

* Solvent code: these solvents have been found to be the most suitable for the preparation of solutions of the stationary phase prior to the addition of support material during the coating process.

A = toluene; B = methylene chloride; C = methanol; D = acetone.

† The polycarboranesiloxane developed by the Orlin Corp., available from suppliers of chromatography accessories, e.g., Phase Sep. Ltd, etc.

two compounds have identical retention times on two substantially different types of stationary phases.

Whilst the **resolution** obtainable for a given mixture is largely determined by the nature of the stationary phase, the length of the column and the efficiency of the column packing process, some modifications to the appearance of the final chromatogram to enable a better evaluation of retention times and peak areas may be achieved by certain simple operations. The best resolution that a particular column is capable of achieving may only be realised with the lowest sample loading, e.g., of the order of 0.1 μl, with appropriate adjustment of the sensitivity controls to lead to peak responses which utilise the available chart space. Lowering the temperature of the column increases the retention times of all components and may make marginal improvements to the resolution of the peaks; this delayed emergence of components leads to broader peaks and this in turn may simplify the calculation of peak areas. Reduction in the flow rate of gas also causes an increase in retention time. Where adjustment of the chart speed is possible, a fast speed for components of short retention times is to be preferred. Most recent instruments now incorporate a temperature programming device which enables the more volatile components of a mixture to be eluted with good resolution at an appropriate low initial column temperature, the column temperature is then raised at a pre-determined rate to a higher level to allow the less volatile components to be eluted more quickly. The chromatogram profile obtained in this way is superior to that obtained at constant temperature.

A provisional **identification** of the components of the mixture may be made from a comparison of the retention times with those obtained for the pure components, if available (e.g., a solvent or reactant used in the original preparation, etc.). Identification must however be confirmed by careful co-chromatography of the mixture with each of the suspected components, added in turn. An enhancement of the appropriate peak will confirm the presence in

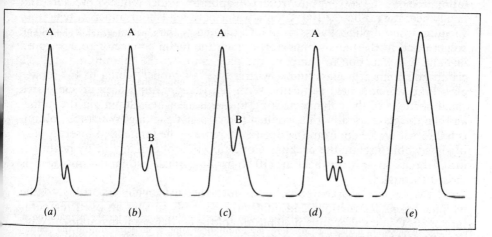

Fig. I,148

the mixture of the added component. It must be emphasised however that the amount of reference compound added should be related to the amount suspected to be present in the mixture, and that several separate additions of say 20, 50 and 100 per cent should be planned. If care is not taken in this way the peak corresponding to a trace component whose identity is required may be swamped by that of an added reference compound and its identification made unreliable. Figure I,148(a–e) illustrates a sequence of results which might be expected to arise from adding increasing amounts of component B to the mixture, consisting of a major component A with a trace impurity, the original chromatogram of which is shown in (a); (b) and (c) are chromatograms which could be expected on the addition of compound B, the suspected trace impurity, in 50 and 100 per cent of the amount of trace component, and the fact that the peak enhancement is proportional to the added component confirms that the trace component is in fact B; had B not been identical with the trace component, addition may have given a chromatogram such as (d) revealing the separation of the trace component from the added component B. The effect of adding too great a proportion of B is illustrated in (e), where it is clear that it cannot be established that the coincidence of the peaks of the trace impurity and the added reference compound B is exact.

Exercises in gas chromatography **Analysis of a mixture of ketones.** (a) **Qualitative analysis.** In most laboratories which routinely use the gas chromatographic technique a range of columns packed with a selection of stationary phases on appropriate solid support material will of course be available. The following instructions for the packing of a standard 5 ft length, 4 mm diameter glass column with a 10 per cent dinonyl phthalate on Chromosorb W support for subsequent fitting to a Pye-Series '104' Chromatograph can be taken to be indicative of the general technique appropriate to the packing of a column.

Dissolve 1.5 g of dinonyl phthalate in 40 ml of acetone in a 250 ml round-bottomed flask and add 13.5 g of Chromosorb W, mesh 60/80, slowly and with swirling. Allow the mixture to stand for two hours with occasional gentle agitation, remove the solvent on a rotary evaporator under reduced pressure and finally heat the residue at 100 °C on a water bath for 1 hour under water-pump vacuum. Insert a plug of glass wool into the end of the shorter glass limb of the column and fit the metal connector. Attach the metal connector to a vacuum line and apply suction to compress the plug. Support the column in a suitable clamp and connect a glass funnel with a piece of rubber tubing to the longer glass limb which is held vertically. With the vacuum applied, pour successive small portions of the column packing material into the column via the funnel and tap the glass spiral with a wooden rod to assist packing. When the column is full to within 5.5 cm from the open end, remove the funnel and insert a plug of glass wool to retain the packing. Condition the column for use by heating it for 24 hours in an oven held at 110 °C while a steady stream of nitrogen is passed through (1).

Fit two similarly prepared columns into the instrument according to the manufacturer's instructions and adjust the controls so that an oven temperature of 80 °C and a flow rate of 40 ml/min of the carrier gas is maintained. With the attenuator control of the instrument set at a relatively high value (say, 20×10^4, which represents a fairly low sensitivity) and with the recorder switched on, prepare to load the front column with a 1 μl sample of acetone in the following way. Insert the needle of a 1 μl micro-syringe into a sample of acetone, and withdraw the plunger to beyond the 1 μl calibrated mark. Remove the syringe and adjust the plunger lead to the mark, wipe the outside of the needle with paper tissue and insert the full needle length into the column via the injection port—some resistance will be felt during passage through the rubber septum. Simultaneously with the injection, activate the pen recorder by operating the jet polarity switch on the amplifier to indicate the injection time. After a few seconds pause remove the syringe needle and evaporate residual acetone by operating the plunger a few times. Both during insertion and withdrawal of the needle some support should be given to it with the fingers to prevent damage by bending.

Observe the peak obtained and estimate any necessary adjustments to the sensitivity controls to modify its size so that its maximum is between one-half and three-quarters of the chart paper width. Check this adjustment by injecting a further 1 μl sample and then repeat until a reproducible peak is obtained. Finally increase the sensitivity setting by a factor of 10 and reduce the injection volume to 0.1 μl—similarly check the reproducibility of this injection by repeating several times (2).

Inject 0.1 μl samples of each of the following straight chain aliphatic ketones; butan-2-one, pentan-3-one, pentan-2-one and heptan-3-one. After each injection clean the syringe by filling it with acetone and expelling the latter several times and finally evaporating the residual acetone as before.

Plot a graph of $\log_{10} t_R$ against the molecular weight or alternatively the carbon number (i.e., the number of carbon atoms in the molecule) of the ketone. Estimate the retention time of a ketone containing six carbon atoms and check your result by a suitable injection. Finally examine the chromatographic behaviour of 3-methylbutan-2-one, 4-methylpentan-2-one and 5-methylhexan-2-one.

(b) **Quantitative analysis.** Prepare a standard series of mixtures of pentan-2-one (A) and heptan-3-one (B) having the approximate composition 20 : 80, 50 : 50 and 75 : 25, w/w, by weighing accurately appropriate quantities into a semi-micro test-tube.

Inject successively and in duplicate 0.1 μl samples of each of the mixtures into the column and record the chromatographic traces. Determine the areas under the peaks by both of the following procedures:

(i)   measure the heights of each peak from the extrapolated base line, and the width of the peak at half peak height; the approximate area is the product of these two values;

(ii) trace each peak carefully on to another sheet of paper, cut out and weigh the area enclosed between the trace and the base line.

Establish that there is an equal response by the detector to each component (3) by showing that for each mixture:

$$\frac{\text{area of peak A}}{\text{area of peak B}} = \frac{\text{weight of A}}{\text{weight of B}}$$

Inject a 0.1 μl sample of pentan-2-one and heptan-3-one in unknown proportions; measure the area of each peak.

The percentage composition of each mixture is given by the equation:

$$\%A = \frac{\text{area of peak A}}{\text{area of [peak A + peak D]}} \times 100$$

$$\text{or} \quad \frac{\text{weight of peak A}}{\text{weight of [peak A + peak B]}} \times 100$$

Notes.   (1) With columns having other stationary phases the temperature of the column conditioning process is usually 10–15° above that of the temperature at which the column will be subsequently operated; it should not exceed the temperature limit value quoted in Table I,11.

(2) If difficulty is experienced in achieving the required reproducibility it is possible that the injection septum is leaking, thus allowing some of the injected sample to escape—a variation in retention times may also be noticed as a result of leakage of carrier gas. The septum should be replaced. Unreproducible results may also be due to leakage around the syringe plunger. If no response is obtained within the expected time this could be due to a variety of causes: initially check that the syringe is not blocked; that in the case of a flame ionisation detector the flame is still alight (note the condensation on a piece of glass held over the detector head after removing the ignitor cap); that the sensitivity controls are set at an adequate level.

(3) With other mixtures, particularly of compounds of different functional types, the detector response to each component may well not be equivalent. In such cases the methods of internal normalisation or internal standardisation should be used; these methods are described in detail in Ref. 64.

Preparative gas chromatography.   The purified fractions from a gas chromatographic column can in principle be collected by interposing between the column end and the detector a splitting device which diverts most of the effluent through suitably cooled traps (Fig. I,149) in which the components are individually condensed. The microlitre sample injection employed in a conventional analytical chromatogram does not of course yield useful amounts of purified components. However, with the aid of longer columns (7, 15 or 30 ft) of somewhat larger diameter (*ca.* 1 cm) and larger sample loads (20–100 μl) sufficient of the isolated components for direct spectroscopic examination can usually be obtained from a single injection. Larger quantities can be obtained by a repeating cycle of manual injection and collection using a suitably adapted analytical instrument. Separation on a truly preparative scale is achieved most conveniently using one of the commercially available automatic preparative gas chromatographic instruments (e.g., Pye-Series '105' Gas Chromatograph).

Fig. I,149

4. **Liquid–solid chromatography.** Separations on a preparative scale employing this technique are generally accomplished by loading the substance on to a cylindrical column of the solid stationary phase, and developing the chromatogram by allowing the liquid mobile phase contained in a suitable reservoir to flow through the column under gravity, or under slight pressure applied to the top of the solvent reservoir.

In the conventional technique, continuous passage of a single eluting solvent through the column may eventually result in the emergence from the bottom of the column of the individual components of the mixture so that they can be individually collected and recovered. A refinement is to progressively increase the polarity of the mobile phase to assist in the displacement of the individual fractions from the stationary phase and hence to speed up the overall time of operation. If, however, actual elution from the column of the components of the mixture by such means is impracticable (e.g., because the time required for elution is too prolonged), it may be necessary to drain off surplus solvent so that the column packing material may be extruded in one piece on to a glass sheet and portions of the column cut off and separately extracted and examined. In the so-called 'dry-column' technique described below (p. 204—which differs from the conventional technique in details of chromatographic development) this column cutting and extraction is the method employed for recovery of separated components.

If the desired compound or compounds are coloured (or strongly fluorescent under ultraviolet light), their location on the column or in selected eluent fractions presents no problems. Hence suitable fractions are combined and concentrated to recover the purified material.

Colourless compounds in eluate fractions are usually detected by one of the following generally applicable procedures:

1. Provided that the mobile phase is a relatively volatile organic solvent, the simplest procedure for assessing the progress of the chromatographic separation is to collect the eluate as a series of fractions of equal volume* and to evaporate to dryness each fraction in a rotary evaporator and to weigh the

---

* Large numbers of fractions having pre-selected volumes of between less than 1 ml to 50 or 100 ml are most easily collected with the aid of one of the many designs of automatic collectors.

residues obtained. A graphical plot of weight versus fraction number then gives a profile of the chromatographic separation which has been achieved; the total weight eluted from the column at any one time should always be compared to the amount of mixture loaded on to the column to provide a guide to recovery of loaded material. The homogeneity of the residues should be further examined by TLC to decide whether any further fractionation is required or whether any of the residues contain essentially one and the same component and may therefore be combined.

2. Each of the individual fractions collected could of course be examined directly by TLC (using one of the non-selective detecting agents, e.g., iodine vapour), but it may be somewhat difficult to estimate the possible concentration of material in the eluate, and hence to determine the loading required on the chromaplate. Multiple application of some of the individual fractions to the same area of the plate may be necessary to detect compounds in relatively high dilution. These TLC results will then determine in what manner the fractions may be combined and if necessary further treated.

Either of these methods could be expected to cover the majority of cases encountered in qualitative or rough quantitative studies applicable to the development of preparatively useful laboratory synthetic procedures. Nevertheless in this context the monitoring of a column chromatographic separation by ultraviolet spectroscopic methods may frequently be the less tedious method provided that certain conditions are met. Firstly the components of the mixture should adsorb in the ultraviolet, and the wavelengths used for screening the fractions should be selected so that all of them are detected. Where measurement of the absorption of fractions at a single wavelength is inadequate determination of the absorption at two or more wavelengths should be made. Secondly the method is only conveniently used with a single solvent development, and is only applicable when solvents which are transparent at the wavelength of ultraviolet light selected have been employed for chromatogram development. The reference cell will contain the pure solvent and in preparative work aliquot portions of the fractions containing components will need to be diluted with pure solvent in order to obtain an on-scale absorbance reading. The optical activity of suitable organic compounds may also be exploited for their detection in chromatographic fractions, particularly in the case of the separation of mixtures of natural products.

In the detection of colourless and non-fluorescent compounds on extruded column packing material, the technique of applying a thin streak of a suitable test reagent, if appropriate, lengthwise down the extruded column may be used. The colours which appear on the surface of the column at the place touched by the reagent indicate the positions of the zones; that part of the column packing containing colour test reagent can be readily shaved off so that it may be discarded before the bulk of material is separated and the component recovered by extraction. Alternatively a strip of Sellotape may be placed momentarily lengthwise down the extruded column surface; the thin layer so removed is then sprayed with the chromogenic reagent and the revealed bands located on the main column by suitable alignment. When no such colour test is available the extruded column must simply be divided into arbitrary segments, each of which must be extracted, the solvent removed by evaporation and the residue examined further (i.e., weight, TLC).

The technique of conventional column chromatography. The essential part of the apparatus consists of a long narrow glass tube (10–90 cm long and 1–4.8 cm diameter); these dimensions give columns which hold between from 25 to 400 g of column packing material. Figure I,150 depicts an assembly with one of the smaller columns (10–40 cm long and 1–1.8 cm diameter) fitted with ground glass joints at its ends to allow for the attachment of a separatory funnel (to act as a solvent reservoir) and a Buchner flask via an adapter with tap for the collection of the eluate fractions. This design incorporates a sintered glass disc (porosity 0) to retain the column packing. Figure I,151 illustrates the larger chromatography column which has ground glass joints fitted to both ends; in use the bottom of the column is closed with a ground glass joint having a sealed sintered glass disc incorporated (see inset). This design allows for the column packing to be easily extruded.

When substances which are oxidised by air need to be handled, the separatory funnel is additionally fitted with a screw-capped adapter with delivery tube attached to a suitably controlled supply of an inert gas (e.g., nitrogen—see

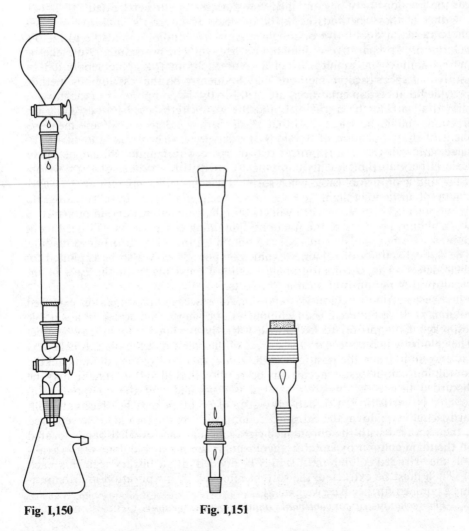

Fig. I,150   **Fig. I,151**

Fig. I,67); additionally the receiver should be a three-necked flask of suitable size, the two side joints enabling a flow of inert gas to be passed through the receiver.

It is not usually desirable to attach a vacuum supply to the receiver since this results in tight packing of the column material which in turn slows the rate of solvent flow. Application of slight pressure to the top of the separatory funnel by means of a cone adapter attached to a suitably controlled supply of compressed air is the more satisfactory means of increasing the rate of solvent flow.

Column packing materials. The selection of suitable column packing materials * is made on the basis of the chromatographic process which needs to be employed for a particular separation, e.g., adsorption or ion-exchange processes or gel filtration. The most widely used and generally applied process in preparative organic chemistry, where the specific need is frequently the purification of starting materials (when only impure technical products are available) and the complete or partial resolution of components of reaction mixtures, is adsorption chromatography. Chromatographic separation on ion-exchange columns is a useful analytical and preparative technique for the resolution of mixtures of acids or bases (e.g., amino acids, amino-phenols, etc.) and for the isolation of neutral organic material from aqueous solutions containing cationic and anionic substances (the technique of 'de-salting'; see Section III,182 for the removal of chloride ion by a resin in the isolation of amino acids). The procedures for packing columns in both these groups differ little in principle. Gel filtration is an invaluable chromatographic process for the quantitative analysis of mixtures of naturally occurring compounds of high molecular weight (e.g., proteins, peptides, enzymes, hormones, nucleic acids, etc.) available in small-sized samples; its use with these groups of compounds on a preparative scale is becoming increasingly important but a discussion of the specialised apparatus used and examples of specific applications are outside the terms of reference of this volume.

The most widely used column packing for adsorption chromatography is aluminium oxide (alumina).† The particle size of the commercially available grade is in the range 50–200 μm (70–290 mesh) which allows for relatively even packing of adsorbent during column packing, for reasonable solvent flow under the force of gravity, and for the rapid attainment of equilibrium distribution of the adsorbate between the surface of the adsorbent and the mobile liquid phase. Alumina may be obtained in basic (pH 10), neutral (pH 7) and acidic (pH 4) forms (1), and it is important to ensure that the correct type is employed because of catalytically induced reactions which each may cause with particular functional compounds. For example, basic alumina may lead to hydrolysis of esters, acidic alumina may lead to dehydration of alcohols (particularly tertiary alcohols) or may cause isomerisation of carbon–carbon double bonds; in these circumstances neutral alumina is to be recommended. The activity of all three forms of alumina, which is broadly regarded as relating

---

* Column packing materials are readily available from most suppliers of laboratory chemicals; firms specialising in materials for chromatography include Applied Science Laboratories Ltd; J. J. Chromatography Ltd; Phase Separations Ltd; H. Reeve Angel and Co. Ltd; M. Woelm.

* For example, 'CAMAG' manufactured by Camag, Chemie-Erzeugnisse und Adsorptionstechnik, available through Hopkin and Williams Ltd. The Spence Chromatographic Alumina, Type 'H' and 'O', is manufactured and available from Laporte Industries Ltd.

both to the magnitude of the attractive forces between the surface groups on the adsorbent and the molecules being adsorbed, and to the number of sites at which such attraction takes place, is classified into five grades (*the Brockmann scale*, Ref. 68). Grade I is the most active (i.e., it retains polar compounds most strongly), and is obtained by heating the alumina at about 300–400 °C for several hours. Successively less active grades, II–V, are then obtained by the addition of appropriate amounts of water (II, 3–4%; III, 5–7%; IV, 9–11%; V, 15–19%) (2). The activity grade is assessed by determining the chromatographic behaviour of specified dyes, loaded in pairs on to an alumina column (5 cm long and 1–5 cm diameter) under carefully standardised conditions and developing the chromatogram with benzene–light petroleum (b.p. 60–80 °C) (3). By comparing the results with those given in Table I,12 the grade of alumina may be assigned.

Table I,12   Grading of activated alumina

	Grade of activity		
Dye position	I	II	III
Column, top 1 cm	p-methoxy-azobenzene		Sudan Red*
Column, bottom 1 cm	azobenzene	p-methoxy-azobenzene	Sudan Yellow†	Sudan Yellow
Eluate	—	azobenzene	—	p-methoxy-azobenzene

Dye position	IV	V
Column, top 1 cm		p-amino azobenzene	p-hydroxy-azobenzene
Column, bottom 1 cm	Sudan Red	Sudan Red	p-amino-azobenzene
Eluate	Sudan Yellow	—	

* Sudan Red (Sudan III) has the structure

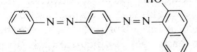

† Sudan Yellow is 1-(phenylazo)-2-naphthol and is available from Searle Diagnostic (Gurr Products); see also Section **IV,91**.

Silica gel (pH 7) may also be graded according to the amount of water added to the most active grade, obtained by heating for several hours at temperatures not exceeding 300 °C; these are II (5%), III (15%), IV (25%), V (38%). These activity gradings are assigned using the same dye pairs as for the grading of alumina (Ref. 69). Addition of larger amounts of water leads to the formation of a substantial film of surface water so that a column prepared from such material may be used to effect separations by partition rather than adsorption (see paper chromatography).

Less frequently employed adsorbents include magnesium silicate, magnesium oxide, magnesium carbonate, calcium carbonate, barium carbonate, calcium hydroxide, calcium sulphate, lactose, starch, cellulose and Fuller's earth.

Activation by drying obviously requires careful control in many of these cases and clearly some would be quite unsuitable for separations involving compounds possessing certain functional groups. The inorganic adsorbents of this group are usually obtained as very fine powders and the solvent flow through columns prepared from such material is extremely slow; the column performance in this respect may be improved by mixing the adsorbent before column preparation with diatomaceous earth filter-aids (trade names: Filter-Cel, Super-Cel, Clara-Cel, etc.), which have only low adsorbent activity. These diatomaceous earths are frequently used as support material for chromatography on liquid partition columns.

Table I,13, due largely to Strain (Ref. 70), gives a list of adsorbents in increasing order of adsorption.

Table I,13 Graded series of adsorbents

1. Sucrose, starch
2. Inulin
3. Talc
4. Sodium carbonate
5. Calcium carbonate
6. Calcium phosphate
7. Magnesium carbonate
8. Magnesium hydroxide
9. Calcium hydroxide
10. Silica gel
11. Magnesium silicate (Florisil)
12. Alumina
13. Fuller's earth

Notes. (1) Neutral alumina may be obtained from the basic form by stirring it with excess water and heating to 80 °C; dilute hydrochloric acid is added dropwise with stirring until slightly acid (pH 6.5) and the heating maintained for 1 hour. The supernatant liquor is decanted and the alumina stirred with aqueous ammonia (2%) at 70–80 °C for 30 minutes. The alumina is recovered by filtration, washed with distilled water until the filtrate is free of chloride ion and dried at 120 °C.

(2) The deactivation of an adsorbent by the addition of water is achieved by simply mixing the appropriate quantities in a stoppered flask and shaking in a mechanical shaker for about 1 hour to ensure equilibration.

(3) Prepare two glass columns of approximately 10 cm length from 1.5-cm-diameter glass tubing as follows: select a 22 cm length of tubing and heat the mid-section in a broad blow-pipe flame (see Section I,10) until it has softened and thickened. Draw the ends apart by about 5 cm to give a constricted portion of about 0.5 cm diameter with walls sufficiently thick so that when cut into two portions these constricted ends are not fragile; anneal both ends of each glass column. Plug the constricted end of one tube with a little glass wool, clamp vertically in a retort and add alumina, whilst at the same time tapping the glass tube gently with a wooden rod, to give finally a 5 cm length of adsorbent. Place a small conical beaker under the column outlet. Dissolve 5 mg of each of the pair of dyes selected in 5 ml of benzene (warming on a water bath may be necessary) and when solution is complete add 20 ml of light petroleum

(b.p. 60–80 °C); fill a 10 ml pipette, fitted with a rubber suction bulb, with the dye solution, and with the pipette top touching the inner part of the glass column about 1 cm from the adsorbent surface, allow the liquid to discharge from the pipette so as not to disturb the adsorbent surface. (It is frequently desirable to place a piece of filter-paper on to the top of the adsorbent surface to prevent it being disturbed by the flow of solution—in this case the pipette may be allowed to discharge directly on to the filter-paper cover.) While the solution soaks into the column fill a clean 10 ml pipette with eluting solvent (benzene : light petroleum, 1 : 4) and as the column liquid level just falls to the level of the adsorbent, carefully rinse the inside of the glass tube with a few ml of the solvent. When these washings have drained into the adsorbent surface add the remainder of the 10 ml portion of developing solvent. The chromatogram is assessed when the solvent portion has drained. The operation is repeated in precisely the same manner with a fresh alumina column and with another dye pair if necessary until the activity grading of the alumina has been established.

Selection of solvents for adsorption chromatography.   The choice of solvent for transferring the mixture to be chromatographed to the column will naturally depend upon the solubility characteristics of the mixture. If it is already in solution, for example as an extract, this is usually evaporated to dryness under reduced pressure and the residue dissolved in the minimum volume of the most non-polar solvent suitable. As concentrated a solution as possible is desirable to achieve a compact band at the top of the column of adsorbent, so that during subsequent development the separation will hopefully proceed with formation of discrete bands.

Generally adsorption on to the adsorbent takes place most readily from non-polar solvents, such as light petroleum or benzene, and least from more highly polar solvents such as esters and alcohols. Frequently the most non-polar solvent for introducing the mixture on to the column and the initial solvent for chromatogram development are the same. Initial adsorption therefore takes place rapidly and development may if necessary be accelerated by progressively increasing the polarity of the eluting solvent using the 'eluotropic' series given below as a guide to sequential solvent selection; all these solvents have sufficiently low boiling points to permit ready recovery of eluted material.

Hexane, cyclohexane, carbon tetrachloride, trichloroethylene, toluene, benzene, methylene chloride, chloroform, diethyl ether, ethyl acetate, acetone, propanol, ethanol, methanol.

Rather than effecting a sharp change in solvent composition it is usual to introduce on to the column gradually increasing concentrations of the more polar solvent until a complete change has been effected. This may be carried out in practice by using successively mixtures of a non-polar and a polar solvent in which the proportions of the components are in a ratio of, say, 90 : 10, 70 : 30, 50 : 50, 30 : 70, 10 : 90, 0 : 100, or by continuously dripping the more polar solvent into the reservoir containing the non-polar solvent, which should be fitted with a stirring device so that the composition changes gradually.

The order in which components of a mixture are eluted from a column is related to their relative polarity. Thus with a mixture of two components of differing polarity, e.g., a hydrocarbon and a ketone, separation is achieved because the more polar ketone is adsorbed more strongly on the adsorbent and hence the hydrocarbon may be eluted with a relatively non-polar solvent; the ketone is then eluted by changing to a more polar solvent. The ease of elution of the adsorbate may be broadly in the following order:

Saturated hydrocarbons > alkenes, alkynes, aromatic hydrocarbons > esters, aldehydes and ketones > amines, alcohols, thiols > phenols, carboxylic acids.

In a comprehensive study of a mixture having unknown chromatographic characteristics it is frequently desirable to be initially guided in the selection of adsorbents and solvents from information obtained by TLC analysis using alumina or silica gel on microscope slides. Only if these prove unsatisfactory would recourse be made to the other adsorbents.

It should be noted that the resolution obtained on a TLC plate is rather better than would be obtained on a conventional adsorption column (see however dry column technique below) and hence further trials should be made with the various activity grades and with controlled solvent composition changes before the bulk of material is submitted to this type of separation. It is in these trials that careful attention to the chromatographic profile obtained from suitable analysis of the eluate fractions, and to the total recovery of material from the column, is so important.

Adsorption column preparation and loading. In order to obtain satisfactory results, the tube must be uniformly packed with the adsorbent; uneven distribution may lead to the formation of cracks and channels and to considerable distortion of adsorption band shapes. If there is any doubt concerning the uniformity of particle size of the adsorbent powder it should be sifted before use to remove the larger particles; fines are removed from the adsorbent using a sedimentation procedure immediately prior to column packing. In this the alumina or silica gel adsorbent is stirred into between five to ten times its volume of the selected solvent or solvent system, allowed to settle for five minutes and the supernatant liquor decanted off; the procedure is repeated until the supernatant liquid is clear.

As a rough guide the amount of adsorbent used should normally be 25–50 times the weight of the material to be separated. A slurry of the adsorbent in the solvent (approximately 1 : 10) is poured through a funnel into a clean dry column clamped vertically, in a position away from draughts or warm air currents from a radiator or electric oven, etc. The adsorbent will settle evenly and free of air bubbles if assisted by gentle tapping of the tube with a wooden rod. For packing large columns the slurry is best contained in a separatory funnel and stirred with a link-type stirrer while it is allowed to flow into the column. Solvent is removed via the tap fitted to the adapter at the column end and more slurry is added until the required length of column is obtained. Some workers recommend that a second column is fitted to the top of the first so that all the slurry can be added in one portion thereby yielding a more perfectly uniform column of adsorbent on settling. Fresh solvent is allowed to flow through the column under the hydrostatic pressure that is envisaged for subsequent chromatographic development, until no further settling is apparent. At no time during the column preparation nor in subsequent use should the level of liquid fall below the level of adsorbent.

The top of the column is frequently covered with a circle of filter-paper or a layer of clean sand to prevent disturbance of the surface during subsequent loading. A suitably concentrated solution of the mixture is added from a pipette, the liquid is allowed to drain just to the surface of the adsorbent and the inside of the tube is rinsed with a small quantity of the solvent which is again allowed to drain just on to the column. Finally the column space above the adsorbent is filled with solvent and a dropping funnel filled with solvent is attached.

The subsequent chromatographic development, analysis of fractions and recovery of separated components is as described above.

Dry-column chromatography. This technique has been developed by Loev *et al.* (Ref. 71) following the observation that the resolution of mixtures on TLC plates is far superior to the resolution obtained in conventional column chromatography using the same adsorbents and solvent systems. In this method alumina (100–200 μm; activity II or III, with incorporated fluorescent indicator) or silica gel* (100–250 μm; activity III, with incorporated fluorescent indicator) (1) is packed dry (using an approximate ratio of 1 g adsorbate : 300 g of adsorbent) into a glass column (2 to 5 cm diameter) fitted with an adapter incorporating a sintered disc (Fig. I,151) or into nylon tubing† (2 to 5 cm in diameter) which has been welded or sealed at the lower end and into which a piece of glass wool is then placed. The packing process in the case of a glass column is assisted by the use of an ultra-vibrator moved alongside the tubing; to fill the nylon tube the bag is filled to about one-third its length and then allowed to fall vertically from a distance of a few inches on to the bench surface to assist good packing; an ultra-vibrator may be used but is not essential and the column obtained is sufficiently rigid to be handled as a filled glass tube. During packing and until the chromatogram development is complete the tap in the glass assembly is kept open; with the nylon tubing the bottom end is perforated by a needle since the packing is retained by the glass wool plug.

The column may be loaded by the technique of dissolving the liquid or solid in the minimum volume of solvent, and by means of a capillary pipette distributing this solution evenly on to the top of the column and allowing the solvent to completely drain. An alternative, and indeed preferable, method is to dissolve the material to be chromatographed in a suitable volatile solvent (ether, light petroleum, methylene chloride) adding column adsorbent equal to about five times its weight, and evaporating the solvent. The loaded adsorbent is then added to the column top and packed into an even layer. With both techniques the column surface is covered with clean, acid-washed sand to a depth of between 0.5 and 1 cm.

The column is developed by gravity flow with a solvent head of between 3 to 5 cm. This is conveniently achieved by placing the solvent (2) in a dropping funnel, closing with a stopper, and with the outlet tube touching the surface layer of sand, carefully opening the stop-cock. Solvent will slowly escape until the drop in pressure in the funnel prevents further flow. The funnel outlet is allowed to remain under the level of liquid in the column, since as the level falls air bubbles will rise into the funnel and allow further portions of solvent to escape. Development is complete when the solvent front reaches the bottom of the column (between 15 and 30 minutes).

---

* These special grades are available from M. Woelm.
† Nylon film tubing available from Chemische Fabrik Budenhelm and from W. Coles & Co. Ltd.

Progress of the development is readily observable by viewing the nylon columns under ultraviolet light which penetrates the nylon covering; the results are less satisfactory with glass columns. The column is extruded from the glass assembly; with the nylon columns the tube is laid horizontally on a sheet of glass and the nylon covering slit lengthways with a razor blade. Detection of the zones and extraction of the separated components is as described previously (p. 197).

Notes. (1) The success of the dry-column technique for the resolution of mixtures is completely dependent on the use of the correct activity grade of adsorbent. It should not be assumed that the grade purchased is activity I, to which the appropriate amount of water could be added (see earlier). It is therefore essential to determine the activity by the use of the method previously described, or by using the elegant micro-method described by Loev *et al.* in their definitive paper (Ref. 71).

(2) The solvent selected will be that which has proved successful with the trial examination on TLC analysis using microscope slides. Preferably this should be a single solvent system. If a mixed solvent system is necessary Loev *et al.* suggest that the deactivated adsorbent should be mixed with about 10 per cent by weight of the solvent system equilibrated by shaking before being used to prepare the dry column.

Partition column chromatography. As was discussed in the technique of paper chromatography, separation of the components of a mixture by liquid partition chromatography is based upon the different partition coefficients which each component has between two immiscible solvents. The more polar solvent system is retained as a stationary film on a suitable support medium and the less polar system acts as the mobile liquid phase. In reversed phase partition chromatography the less polar solvent acts as the stationary phase and the more polar solvent as the mobile phase; in this case the support material is treated with dichlorodimethylsilane which reacts with the surface hydroxyl groups and thus makes the support surface non-wettable.

The most usual support materials are silica gel, usually of the same particle size as is used in adsorption chromatography, kieselguhr (Celite) or cellulose powder. The stationary phase is frequently water or aqueous buffer solutions, dilute sulphuric or hydrochloric acids or methanol; the mobile phase may be butanol–chloroform mixtures, butanol–benzene mixtures, carbon tetrachloride, ethyl acetate, hexane or 2,2,4-trimethylpentane. Examples of solvent systems in reversed phase procedures are octane (stationary)—60 per cent aqueous methanol (mobile) and toluene-formamide.

The solvent system is selected from preliminary trials in which a range of solvent pairs are shaken with the components of the mixture to be separated (or the mixture itself) and each solvent portion analysed appropriately to determine roughly the partition coefficients of individual components. That solvent system in which the major component has a partition coefficient in the region of 1 is the system of choice (cf. counter-current distribution, Section **I,32**).

The two liquid phases are shaken together in a separatory funnel, an appropriate amount of solvent which is to act as the stationary phase is removed and stirred with the support material (usually about 0.5–1 ml/g), and the prepared support material is then made into a slurry with the equilibrated solvent which is to act as the mobile phase. The slurry is poured into a suitable glass column in a similar way to the formation of an adsorption column. However, the action of gravity will rarely be sufficient to give effective packing, and some assistance will be needed by the use of a perforated metal disc (of a slightly smaller diameter than the inside diameter of the glass tube) attached to a long metal rod. The slurry may be stirred in the column by using long smooth vertical strokes; packing is achieved by moving the disc slowly downward through the column of slurry so as to entrain the support material into a packed portion at the bottom. By successively removing solvent from the column end, adding more slurry and alternately stirring and packing, a column largely devoid of striations and air bubbles may be obtained.

The subsequent loading of the partition column, development and analysis of fractions is as described formerly for adsorption columns.

The applications of partition column chromatography are largely to be found in the separation of homologues of a particular functional group which would not be separable by adsorption chromatography. Thus the separation of mixtures of the simpler alcohols, of the fatty acids, of dibasic acids, of phenols has been achieved by these means as well as separation of proteins, purines, steroids and lipids.

High performance liquid chromatography. The conventional techniques of liquid–solid and liquid–liquid column chromatography described in the foregoing paragraphs have recently been modified so that analysis of mixtures of solid compounds (using sample sizes in the microgramme region) may be achieved with a resolution and rapidity comparable with that obtained with gas–liquid chromatographic systems. These improved techniques are particularly suitable for the analysis of relatively involatile or thermally unstable compounds which, of course, are not amenable to GLC analysis.

This new technique (HPLC) uses stainless steel columns, 50 to 100 cm in length, having internal diameters similar to the columns used in GLC systems. Specially prepared stationary phases for adsorption, partition, ion exchange or gel permeation chromatographic processes are used for packing the columns. The uniformity of grading of the very fine particles of the column packing is one particularly important feature upon which the resolving power of the system depends.

A consequence of the column dimensions, and the physical nature of the packing material, is the need for a pressurised liquid flow through the column. The pressures are maintained by suitable pumps and may be pre-set as required at values between 1000 to 6000 p.s.i. The detectors employed measure the differential absorption at selected wavelengths, or differential refraction, between a reference signal from the solvent system employed and that from the eluent from the sample column.

A range of instruments is commercially available,* and it may be reasonably expected that the next decade will see the enormous potential of this chromatographic technique for analytical and preparative purposes being greatly exploited.

Exercises in adsorption chromatography   **Purification of anthracene.**   Dissolve, with warming if necessary, 50 mg of crude anthracene (usually yellowish in

---

* For example, those manufactured by Waters Associates Ltd.

colour) in 50 ml of hexane. Prepare a 20 cm column of activated alumina using a slurry containing 50 g of activated alumina (Grade II) in hexane in a glass chromatographic column of 1.8 mm diameter and 40 cm length fitted with a sintered disc, adapter with tap and separatory funnel. Add the anthracene solution to the top of the column from a separatory funnel, rinse the funnel with a little hexane and then develop the chromatogram with 200 ml of hexane. Examine the column from time to time in the light of an ultraviolet lamp (Section **I,3,F**). A narrow, deep blue fluorescent zone (due to carbazole) will be observed near the top of the column; the next zone down the column is a yellow non-fluorescent zone due to naphthacene, the intensity of which will depend upon the purity of the sample of anthracene used. The anthracene forms a broad, blue-violet fluorescent zone in the lower part of the column. Continue to develop the chromatogram until the anthracene begins to emerge from the column, reject the first runnings since these contain the less strongly adsorbed paraffin-like impurities and change to a clean receiver. Continue to elute the column with hexane until the removal of the anthracene is complete (1); the yellow zone should not reach to the bottom of the column. Concentrate the eluate fraction containing the anthracene under reduced pressure on a rotary evaporator to about 2 ml, cool the flask in an ice–salt bath and by means of ice-cold hexane quantitatively transfer the crystals and solution to a filter funnel. Wash the pure anthracene crystals with chilled hexane; the product (30 mg), which is fluorescent in daylight, has m.p. 215–216 °C.

Note. (1) When the anthracene band begins to emerge from the bottom of the column its elution may be accelerated by changing to a mixture of hexane : benzene (1 : 1). In this case the fractions containing the anthracene are evaporated to dryness under reduced pressure and then redissolved in hexane and concentrated to low bulk as described in the main text. This additional operation is necessary owing to the solubility of anthracene in benzene.

Separation of cholestenone from cholesterol.*

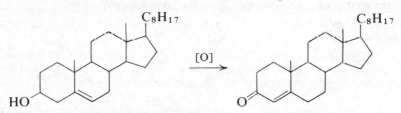

Place a mixture of 1.0 g of purified cholesterol and 0.2 g of cupric oxide in a test-tube clamped securely at the top, add a fragment of Cardice in order to displace the air by carbon dioxide and insert a plug of cotton wool in the mouth of the tube. Heat in a metal bath at 300–315 °C for 15 minutes and allow to cool; rotate the test-tube occasionally in order to spread the melt on the sides. Warm the cold residue with a few ml of benzene and pour the black suspension directly into the top of a previously prepared chromatographic

---

* The experimental details were kindly supplied by Professor D. H. R. Barton, F.R.S. and Dr W. Rigby; because of the toxicity of benzene this chromatographic column should be set up in the fume cupboard.

column (1); rinse the test-tube with a little more benzene and pour the rinsings into the column. With the aid of slight pressure (*ca.* 3–4 cm of mercury), allow the solution to drain into the alumina column; stir the top 0.5 cm or so with a stout copper wire at frequent intervals to prevent blockage by the finely divided copper compounds. When all the black liquid has run in, there should be free flow without the necessity of further stirring. Continue the development with benzene until a distinctly yellowish diffuse zone approaches the bottom of the column; some 150 ml of liquid will have been collected. Now collect 5 ml fractions until the yellow band is completely removed. Evaporate each of these fractions separately; the earlier ones yield oils (giving a yellow 2 : 4-dinitrophenylhydrazone) and the later ones will crystallise upon rubbing (cholestenone). Continue the elution with a further 400 ml of benzene; the latter upon evaporation yields most of the cholestenone. Isolate the remaining cholestenone by continuing the elution with benzene containing 0.5 per cent of absolute ethanol until a dark brown band approaches the bottom of the column. Collect all the crystalline residues with the aid of a little light petroleum, b.p. 40–60 °C, into a small flask and remove the solvent. Dissolve the residue in 40–50 ml of hot methanol, add 0.2 g of decolourising carbon, filter through a small bed of alumina (6 mm × 6 mm), concentrate to about 20 ml and leave to crystallise overnight. The yield of cholestenone, m.p. 82 °C, is 0.5 g.

Note. (1) Prepare the column for chromatography by mixing 90 g of chromatographic alumina (Spence) with sufficient benzene to form a thin slurry when stirred. Pour this, stirring briskly, into a tube (40 cm long and 18 mm internal diameter) having a sintered glass disc, and rinse with a little more benzene. An evenly packed column, about 35 cm long, should result. Allow to drain until the supernatant benzene is within 1 cm of the alumina before adding the solution to be chromatographed. Under no circumstances should the column be allowed to drain so that the liquid level falls below that of the alumina.

Preparation of *cis*-azobenzene from the *trans*-isomer.

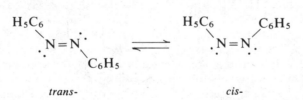

trans-   cis-

Dissolve 1.0 g of azobenzene (Section **IV,102**; this is the *trans*-form) in 50 ml of light petroleum, b.p. 40–60 °C, in a 200-ml beaker. Irradiate the solution for 30 minutes with ultraviolet light; this is conveniently carried out by supporting a Hanovia fluorescent lamp, model 16, about 13 cm above the surface of the liquid in the beaker. Meanwhile prepare a chromatographic column from 50 g of activated acid alumina (Grade I) as a slurry in light petroleum, b.p. 40–60 °C, to give a column of approximate dimensions 20 cm × 1.8 cm. After the column has been formed in this way, place a well-fitting filter-paper at the top of the column and pour the solution, immediately after it has been irradiated, slowly down a glass rod on to the filter-paper until the column is filled with liquid; take great care not to disturb the upper portion of the column. Develop the chromatogram with 100 ml of light petroleum, b.p. 40–60 °C. A sharp coloured band (cis-form), ca. 2 cm in length, makes its appearance at the top of the column whilst a diffuse coloured region (containing the trans-form) moves down the column. The upper portion of the column should be screened from light by wrapping it with black paper, held in position by a rubber band, during the development process; this will largely prevent the reconversion of the cis- into the trans-form. Extrude the column and remove the coloured 2 cm band from the column and shake it with 150 ml of light petroleum, b.p. 40–60 °C, containing 1.5 ml of absolute methanol (1); filter off the alumina, with suction, and wash the filtrate with two 15 ml portions of water to remove the methanol present. Dry the light petroleum extract by shaking it with about 1 g of anhydrous sodium sulphate for 10 minutes, filter and evaporate the solvent under reduced pressure. The residual coloured solid, m.p. 71.5 °C, is practically pure cis-azobenzene. Its individuality and its purity may be confirmed by recording the ultraviolet absorption spectrum in ethanol solution as soon as possible after its isolation; cis-azobenzene has λ_{max} 281 nm, ε 5260, trans-azobenzene has λ_{max} 320 nm, ε 21 300 in ethanol solution.

Note. (1) Alternatively the cis-isomer may be removed from the column by changing the eluting solvent to one of light petroleum, b.p. 40–60 °C, containing 1 per cent of methanol. The eluant fraction is then washed with water to remove the methanol in the manner described in the main text.

Exercise in partition column chromatography **The separation of 2 : 4-dinitrophenylhydrazones by reversed phase chromatography.** The advantage of the following method is that it may be applied to the quantitative analysis of a mixture of aldehyde and ketone derivatives. These may only be available in milligram amounts, such as would be obtained from the ozonolysis of olefinic material. The following experiment uses a mixture of homologous ketones.

Using a 1 ml graduated pipette add 0.02 ml portions each of acetone, butan-2-one, pentan-2-one and 4-methylpentan-2-one to 250 ml of an aqueous acidic solution of 2 : 4-dinitrophenylhydrazine (1) contained in a conical flask, stopper the flask and shake vigorously for 2–3 minutes to mix the contents thoroughly. Warm the solution on a water bath to 35–40 °C and maintain at this temperature for half an hour; cool and extract the solution with four 25 ml portions of methylene chloride. Combine the extracts, wash with water until free of acid, dry the organic layer over sodium sulphate, filter and evaporate the solution to dryness. Dissolve the residue in 10 ml of toluene and transfer this solution to an alumina column (10 cm × 1.8 cm) prepared from a slurry of 25 g of alumina (Grade II) in toluene. Elute the column with toluene until a fast running broad orange band has been eluted, leaving a dark red band at the top of the column which is unreacted 2,4-dinitrophenylhydrazine. Combine the eluate fractions and evaporate to dryness using a rotary evaporator. Dissolve the residue in 5 ml of toluene and quantitatively transfer the solution (using a dropping pipette and small additional quantities of toluene) to a 10-ml graduated flask; make up to the mark with toluene.

Equilibrate 500 ml of formamide (2) with 50 ml of toluene by shaking together in a separatory funnel; transfer each layer to a stoppered conical flask. Weigh out in a beaker 28 g of silanised Celite 545 (3) and add 14 ml of equilibrated toluene; stir the mixture to disperse the toluene on the support material give a slurry. If air bubbles are present these must be removed by centrifuging the slurry, removing the scum and resuspending the precipitate—usually two such treatments give a slurry free of air bubbles. The slurry is poured in portions into a suitable chromatographic column, stirred by vertical strokes with a metal perforated disc attached to a metal rod, and the packing consolidated by slowly moving the disc downward through the slurry and entraining support material; additional slurry is added to give a column of between 26–30 cm length. Finally the column is allowed to settle under a pressure of 6 cm of mercury, which is maintained during subsequent development and elution: the prepared column should be protected from currents of cold or warm air and should remain stable for at least three days.

With a calibrated 1 ml pipette drain a 0.5 ml aliquot portion of the prepared toluene solution of the 2,4-dinitrophenylhydrazones into a small beaker, add 1 g of silanised Celite and stir to disperse the solution on the support. Allow the level of formamide to fall in the column to within 0.5 cm of the support material; transfer the entire solid material from the beaker to the top of the column; rinse the beaker with 1 ml portions of equilibrated formamide, adding each to the column with appropriate draining so that the quantitative transfer of derivatives to the column is achieved. Finally develop the column with equilibrated formamide and collect equal fractions within the range 1.5–2.0 ml in an automatic fraction collector.

Determine the chromatographic profile by plotting graphically the fraction number against the light absorption measured for each fraction at a suitable wavelength. If a Spekker photoelectric absorptiometer* is available this is used with an Ilford Filter No. 602, and with a reference cell filled with equilibrated formamide. If a manual or recording ultraviolet and visible spectrophotometer is available it is excellent practice to determine for each of the 2,4-dinitrophenylhydrazones the absorption curve in equilibrated formamide and hence select the wavelength most suitable for scanning all the fractions.

Combine the fractions corresponding to each component in separate (usually 100 ml) graduated flasks and dilute to the mark with equilibrated formamide; measure the light absorption of these solutions. Dilute each solution separately with 20 ml of water and extract with methylene chloride; wash the methylene chloride extracts with water to remove entrained formamide, dry over sodium sulphate, filter and evaporate to dryness. Recrystallise each residue (*ca.* 4 mg) from a few drops of methanol using the techniques described in Section **I,20** and identify the components by melting point determination on a microscope hot stage apparatus (Section **I,34**). When the relevant peaks in the chromatographic profile have thus been identified, the quantitative analysis of the mixture should be completed by comparing the absorbance reading of the combined and diluted fractions with standard graphs obtained from the absorbance measurements of solutions of known concentrations of each derivative in equilibrated formamide, determined upon the same instrument and at the same wavelength.

Notes. (1) The 2 : 4-dinitrophenylhydrazine reagent for this experiment is prepared as follows: add 600 mg of 2,4-dinitrophenylhydrazine to a mixture of 42 ml of concentrated hydrochloric acid and 50 ml of water, warm in a water bath and then dilute the cooled solution to 200 ml with distilled water.

---

* Available from Rank Hilger.

(2) Formamide is very hygroscopic; distillation at atmospheric pressure (b.p. 193 °C) leads to decomposition. It is purified sufficiently for this experiment by distillation under reduced pressure. (See Section II,1 for further purification methods.)

(3) Celite may be silanised by mixing with a 5 per cent solution of dichloro-dimethylsilane in carbon tetrachloride solution and allowing the slurry to stand for 24 hours; the Celite is recovered by filtration and washed with carbon tetrachloride and then with acetone until acid free (moist universal indicator paper). After drying in the air and then in an oven at 100 °C a small sample when shaken vigorously with water in a test-tube and allowed to settle should form a non-wettable layer at the liquid surface. If any solid settles at the bottom of the tube (and this is very unlikely) the treatment should be repeated.

I,32. COUNTERCURRENT DISTRIBUTION In Section I,22 the partition coefficient (K) was defined as the ratio of the concentrations of a solute when distributed between two immiscible liquid phases and was used in the discussion of the theory of batchwise solvent extraction. It is used here to explain the theoretical basis of the technique of countercurrent distribution (or *multiple fractional solvent extraction*) which has developed into a practical method for the separation of components of mixtures and for establishing the purity of compounds. The method is of particular value for mixtures which contain thermally labile components where most gas chromatographic procedures would be unsuitable, or for mixtures having components which are liable to decompose during conventional liquid–solid chromatography as a result of contact with the surface of the adsorbent. An account of the numerous applications of this technique is to be found in the monograph by Craig and Craig (Ref. 72).

Since an understanding of the application of the theoretical principles to the actual experimental results is essential in this separative method, in order to obtain the maximum information about the efficiency of the separation and the purity of the components of the mixture, an introductory treatment is included here. For a more detailed treatment of the mathematical background to the theory, the reviews cited in Refs. 73–76 should be consulted.

The practical essentials of countercurrent distribution may be illustrated and the basic theoretical principles broadly explained by reference to the following simple case. Consider an assembly of eleven separatory funnels, numbered *0, 1, 2, 3, ... 10*, containing equal volumes of two immiscible solvents. Introduction into funnel *0* of a pure substance *A* (say 160 mg for mathematical convenience), having $K = 1$ in this particular solvent system, followed by shaking (*equilibration*) to mix the contents will result in 80 mg ($160 \times \frac{1}{2}$) dissolving in the upper phase (U_0) and 80 mg dissolving in the lower phase (L_0) (1). Removal of the upper phases of each funnel and transference to neighbouring funnels such that $U_0 \to$ funnel *1*, $U_1 \to$ funnel *2*, ... $U_{10} \to$ funnel *0*, means that *A* will be distributed between funnel *0* and funnel *1* (80 mg in each). This *transfer* leads to the definition of the upper phase as the *mobile phase* and the lower phase as the *stationary phase*. Equilibration of the contents of funnel *0* and funnel *1*, followed by a similar transfer (*transfer 2*), gives a distribution pattern of the total amounts of solute in funnels *0, 1* and *2* of 40, 80 and 40 mg respectively. Continuation of successive equilibration/transfer stages results after ten transfers in the upper phase which originally started in funnel *0* being in funnel *10*. The distribution and then gradually add with continuous stirring the equilibrated formamide to patterns of A at each transfer stage is shown in Table I,14; Fig. I,152 gives a graphical plot of the distribution pattern after ten transfers.

An essential point to note is that this theoretical distribution pattern would

Table I,14 Distribution of compound A ($K = 1$) at individual transfer stages

Transfer	Funnel number										
	0	1	2	3	4	5	6	7	8	9	10
0	160										
1	80	80									
2	40	80	40								
3	20	60	60	20							
4	10	40	60	40	10						
5	5	25	50	50	25	5					
6	2.5	15	37.5	50	37.5	15	2.5				
7	1.25	8.75	26.25	43.75	43.75	26.25	8.75	1.25			
8	0.62	4.99	17.57	34.99	43.75	34.99	17.57	4.99	0.62		
9	0.31	2.80	11.27	26.27	39.36	39.36	26.27	11.27	2.80	0.31	
10	0.15	1.55	7.03	18.76	32.81	39.36	32.81	18.76	7.03	1.55	0.15

exactly fit an experimental distribution for a compound having $K = 1$, using equal volumes of two immiscible solvents, over ten transfers provided that the concentration of the compound in any funnel (and this is particularly important in the early stages of the distribution) is low (i.e., normally between 1 to 2 per cent) and provided that there is complete absence of differential physical interactions (e.g., solvation effects, dissociations, etc.) of the solute in each solvent phase (2).

Notes. (1) The fraction of solute in the upper phase and in the lower phase may be denoted by p and q respectively. In a solvent system where the volumes of the upper and lower phases are equal, K may be re-expressed as:

$$K = \frac{p}{q}$$

Since $p + q = 1$, the fraction p may be expressed only in terms of K since:

$$\frac{p}{1 - p} = K$$

therefore: $p = K - Kp$

and: $$p = \frac{K}{K + 1}$$

Similarly $q = \frac{1}{K + 1}$

Therefore, in the example of A ($K = 1$) quoted above p and q are each equal to $\frac{1}{2}$.

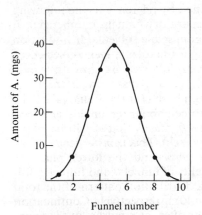

Fig. I,152

Funnel number

Amount of A. (mgs)

(2) With a mixture of solutes S_1, S_2, S_3, etc., it is an experimentally verified fact that for dilute solutions, and providing that there is no chemical or physical interaction between solutes and solvents, each solute partitions between the liquid phases according to individual partition coefficients (K_1, K_2, K_3, etc.). Such partitioning is independent of the presence of other solutes. Hence

$$\frac{p_1}{q_1} = K_1, \frac{p_2}{q_2} = K_2, \text{ etc.}$$

If the volumes of the upper phases and lower phases are not equal but are V_U and V_L respectively, then

$$K = \frac{p/V_U}{q/V_L}$$

Clearly a calculation of a theoretical distribution pattern in this manner over a large number of funnels and with solutes having partition coefficients other than unity would be excessively tedious. However the fractions of solute in each funnel after n transfers is given by the terms of the binomial expansion $(p + q)^n$ (see Table I,15). The fraction of solute ($T_{n,r}$) in funnel r may therefore be calculated from the appropriate individual terms, i.e.,

$$T_{n,r} = \frac{n!}{(n - r)!\, r!} \cdot p^r \cdot q^{n-r} \tag{1}$$

To illustrate the use of equation 1 with the case of compound A above, the fraction of A in funnel 4 after ten transfers is:

$$T_{10,4} = \frac{10!}{6!\, 4!} \cdot p^4 \cdot q^6$$
$$= 210 p^4 q^6$$

Since $p = \frac{1}{2}$ and $q = \frac{1}{2}$, then:

$$T_{10,4} = 210(\tfrac{1}{2})^4(\tfrac{1}{2})^6 = 0.205$$

Since the 160 mg of A is being distributed the total amount of A in funnel 4 is 32.81 mg (0.205×160).

Figure I,153 gives a graphical plot of the theoretical distribution patterns of each of two compounds B (100 mg) and C (200 mg) ($K_B = 0.45$ and $K_C = 2.2$ respectively) after sixteen transfers. Had these quantities of B and C been

Table I,15   Terms of the binomial expansion $(p + q)^n$

Transfer	Funnel number						
	0	1	2	3	4	5	6
0	1						
1	q	p					
2	q^2	$2pq$	p^2				
3	q^3	$3pq^2$	$3p^2q$	p^3			
4	q^4	$4pq^3$	$6p^2q^2$	$4p^3q$	p^4		
5	← $(p + q)^5$ → etc.						

introduced together into the first funnel, and the mixture submitted to sixteen transfers, the concentration of material in each funnel would have been that shown by the dotted line. It is clear that removal of the contents of funnels *0*, *1*,

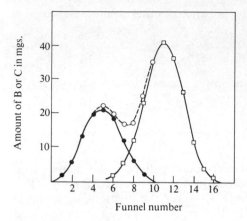

Fig. I,153

2, *3*, *4*, and of funnels *11*, *12*, *13*, *14*, *15* and *16* followed by appropriate work-up procedures would have resulted in the isolation of the separated components *B* and *C* respectively. The graph also shows that funnels *5* to *10* inclusive contain varying proportions of *B* and *C* in the overlap region of the two distribution patterns. Complete separation of *B* and *C* could be achieved by increasing the number of transfers.

These simple cases illustrate the value of this method for the separation of the components of mixtures based upon differences in their partition coefficients. The ratio of the partition coefficients (the separation factor, β) is important for evaluating the effectiveness of the separation of two components having partition coefficients K_1 and K_2,

$$\beta = \frac{K_2}{K_1} \quad \text{where } K_2 > K_1$$

If $\beta = 1$, separation would be impossible; the greater the numerical value of β the fewer will be the number of transfers required to effect separation. Broadly speaking, for a relatively small number of transfers (20–50), β should be about 3 to 4, depending upon the relative amounts of the two components; for a large number of transfers (> 100), $\beta \approx 2$ is adequate.

Separation by countercurrent distribution on a practical scale, even with only a few transfers, needs at least an apparatus in which the equilibration and transfer stages can be carried out semi-automatically. Figure I,154 * is a photograph of a hand-operated countercurrent apparatus and Fig. I,155 * is a photograph of the fully automatic model.

In the hand-operated model, the apparatus consists of twenty glass units or cells each comprising an extraction or partition tube (which represents the

---

* These models are manufactured and supplied by Wright Scientific Ltd, who kindly provided the photographs.

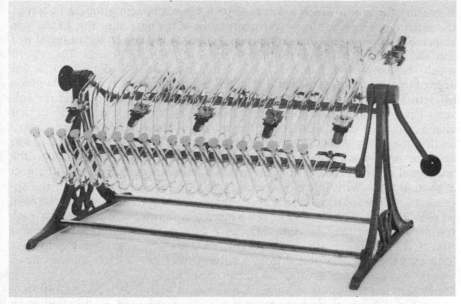

Fig. I,154

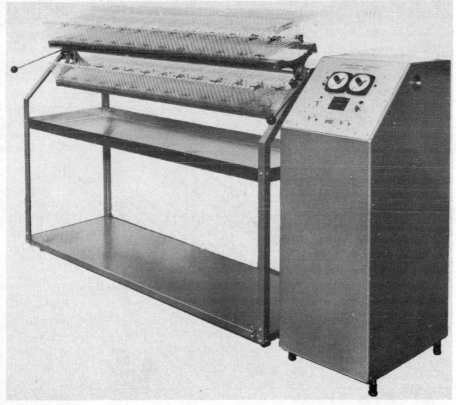

Fig. I,155

separatory funnel) to which is attached by a glass connecting tube a reservoir, the outlet of which leads to the extraction tube of the next unit (Fig. I,156). The units are in continuous banks of five which may be connected by means of ball and socket glass joints to give a multiple series of 10, 15 or 20 units. The rack to which the bank of extraction tubes is fitted may be rocked by hand a few degrees on either side of the horizontal position; this allows the contents of the extraction tubes (during equilibration) to be smoothly agitated. After the solvents have separated into two layers again, the rack is tilted manually so that all the extraction tubes are vertical. The solvent volumes are previously selected so that the phase boundary now coincides with the side-arm leading to the reservoir into which the upper phase flows (1). On returning the rack to the horizontal position the upper phase in each reservoir discharges into the next extraction tube. With this model each extraction tube is charged with the appropriate volume of equilibrated lower phase (2) and the appropriate volume of equilibrated upper phase may be introduced portionwise into tube *0* after each transfer (3). This may be done manually with a small number of transfers (10–20), but the apparatus shown has the means of automatically delivering fresh upper phase into tube *0* from a solvent reservoir at each transfer stage; tube *19* has a drain tube attached to the reservoir.

In operation the solute mixture, in an amount such that a concentration of about 2 per cent in the total tube contents is not exceeded, is dissolved in an appropriate volume of upper solvent phase and added to tube *0*. The contents are then equilibrated and transferred in the manner previously described (4). At

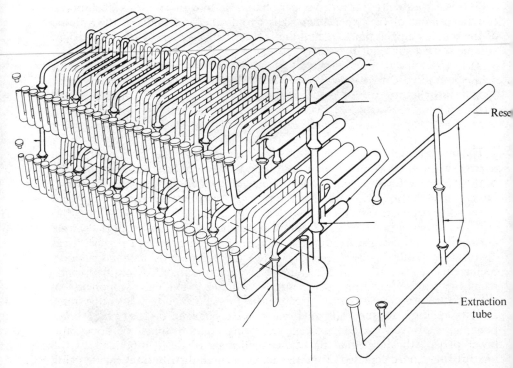

Rese

Extraction
tube

Fig. I,156

the conclusion of the selected number of transfers the amount of solute in each tube is determined by suitable quantitative means. The actual method selected will of course be governed by the nature of the solutes and by the solvent system employed; some methods are considered in general terms below. It is usually only necessary, and indeed more convenient for the purpose of fitting of practical with theoretical curves, to determine the amount of solute in either the upper or the lower phase in each tube after the final equilibration stage. A graphical plot of the amounts of solute in each tube so determined against the tube number (first tube is numbered *0*) gives a profile from which (*a*) the purity of a particular component may be evaluated, (*b*) the effectiveness of further transfer stages on incomplete separations may be assessed and (*c*) tubes containing essentially separated components may be selected and the contents combined and submitted to suitable isolation procedures.

Notes. (1) In the apparatus illustrated here the volume of lower phase in each extraction tube is 20 ml; in most cases it is usual to employ an equal volume of mobile (upper) phase, though separations can be carried out using a smaller or larger volume of upper phase.

(2) It is essential in all countercurrent distribution experiments to shake the two selected solvents together in bulk to ensure their complete equilibration before the apparatus is filled with the appropriate phases. Failure to do this will result in considerable variations in the relative volumes of the two solvents during the subsequent experiment which makes the fitting of the practical and the theoretical distribution curves impossible.

(3) It is advisable to introduce upper phase also into tubes *1* and *2* before the commencement of the experiment. This avoids the possibility of the volume of the leading upper phase containing solute being progressively reduced in volume during subsequent transfers.

(4) For handling larger quantities of solute one can, on a machine with a greater number of tubes, usefully load the first five or even ten tubes with an appropriate quantity of the solute mixture. This makes the evaluation of the theoretical distribution curve much more complex, but it does allow the separation of larger quantities of material to be effected more readily.

The fully automatic instrument allows for a greater number of transfers to be effected without effort and enables optimum times to be selected for the equilibration process, for the separation of phases after equilibration, and for allowing complete drainage into the reservoirs, etc. Partition tubes having different lower solvent phase capacities are available.

The fitting of practical and theoretical distribution patterns. In the simple illustrative examples discussed at the beginning of this Section, the theoretical distribution patterns have been calculated for compounds having partition coefficients of 1.00, 0.45 and 2.2. Quite clearly in an actual distribution experiment the precise partition coefficients for all the individual components of the mixture may be unknown (see below for selection of solvent systems and its relation to the estimated partition coefficient of the major components). However, for distributions involving equal volumes of upper and lower phases, the individual partition coefficients of the components of the mixture may be calculated from the experimental distribution curve, using one of the following two relationships depending upon the total number of transfers (n) which have been effected and upon the position of the maximum (r_{max}).

(i) $\dfrac{nK}{K+1} = r_{max}$

This formula is only exact if K is unity or if there are an infinite number of transfers. In actual practice it may be applied satisfactorily for distributions of greater than 20 transfers and for components having a K value close to unity. The value of r_{max} need not necessarily be an integer and its fractional value is estimated from the position of the maximum on the smooth curve drawn through the appropriate points of the distribution curve.

(ii) $K = \dfrac{r_{max} + 1}{n - r_{max}}$

This formula is suitable for experiments in which n lies between 8 to 20 transfers. The value of r_{max} is again estimated from the position of the maximum in the smooth distribution curve.

Using these values of K for each component, the subsequent calculations to deduce the theoretical distribution patterns may be approached in two ways: the first, (a), is appropriate to a small number of transfers (15–25); the second, (b), to those instances where a larger number of transfers (>25) has been employed.

(a) In this case, for each of the components (S_1, S_2, S_3, etc.) theoretical distribution patterns are determined, using the values of K_1, K_2, K_3, etc., as appropriate, to calculate the fraction of a solute in the upper phase of each of the relevant tubes from the equation:

Fraction of solute in upper phase of tube $r = T_{n,\,r} \cdot p = \dfrac{n!}{(n-r)!\,r!}\, p^r \cdot q^{n-r} \cdot p$

Each of these values is then multiplied by the ratio w/pT_{max}, where w is the analytical value of the amount of solute in the upper phase of tube r_{max} * (e.g., a weight, an absorbance reading, etc.) and pT_{max} is the calculated fraction of the solute in the upper phase of the tube r_{max}. A plot of these calculated values is superimposed upon the experimentally determined upper phase distribution curve; a good fit of profiles for each of the components establishes their purity. Deviations in the leading or following edges of the curves (making allowances for known overlapping of adjacent curves) reveals the presence of unresolved components. In such cases a greater number of transfers are required to effect separation.

(b) In those cases where the number of transfers is large, ($n > 20$), the calculation of the distribution pattern for each component of a mixture by method (a) would be somewhat prolonged. The theoretical pattern may be evaluated from the distribution of each solute in the upper phase from the expression:

$$y_x = y_0 \div \frac{\text{antilog } 0.434x^2}{2nK/(K+1)^2}$$

---

* In cases where adjacent peaks of two components overlap to a small extent it may be possible to estimate by graphical extrapolation the extent to which w in the tube r_{max} must be corrected to compensate for the contribution from the underlying portion of the adjacent peak. After calculation of both theoretical curves on this preliminary estimate of mutual contributions at the maximum, further refinements may be made to these estimates, followed by recalculations to determine the exactitude of fitting in each case.

where y_0 = amount of relevant solute (e.g., by weight, by absorbance reading, etc.) in the upper phase of the tube having the maximum concentration, y_x = amount of solute to be calculated for tube x numbering the peak tube as 0 (see Fig. I,157) and n = number of transfers. A plot of these theoretical distribution curves should again match exactly the experimental curve if the components have been resolved and are pure. Minor deviations could suggest the

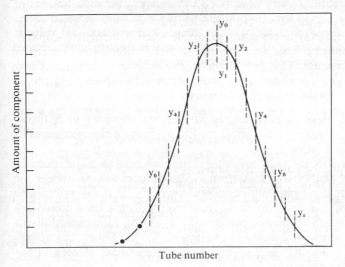

Fig. I,157

presence of components having partition coefficients which are only marginally different from the major component.

This method of calculation is well adapted to the estimation of the percentage of the components present in a mixture since:

amount (P_l) of solute (S_l) present in all of the upper phases
$$= y_0 + 2(y_1 + y_2 + \ldots + y_x)$$

$\therefore$ total amount (Q_l) of S_l in all upper and lower phases $= \dfrac{P_l(K_l + 1)}{K_l}$

$\therefore$ Percentage of $S_l = \dfrac{100Q_l}{Q_1 + Q_2 +_{\ 3}, \text{ etc.}}$

Selection of solvent systems. The successful application of the technique of countercurrent distribution to the separation of a particular mixture depends vitally upon the selection of an appropriate solvent system. Whilst it is rarely possible to select a system which satisfies all of the following criteria, it is essential to spend time on small-scale trial experiments to evaluate a range of possible solvent combinations (see selection of a solvent system in *Partition Column Chromatography*, Section **I,31**, p. 205). The final choice of a suitable system will be governed by the need to strike a balance between separation efficiency and factors such as solvent toxicity and inflammability, freedom from emulsion formation upon solvent mixing, etc.

1. The major component of the mixture should have a partition coefficient of approximately 1 in the selected two-phase solvent system (cf. partition

chromatography). The value of β should be greater than 2 for distribution experiments of less than 100 transfers.

2. The individual partition coefficients for the components should be constant over the concentration range to be expected in the relevant tubes. If this is not so the distribution patterns will be distorted.

3. The solvents should not react with each other or with the components of the mixture. The presence of solute should not lead to an alteration in the mutual miscibility of the two solvent phases, otherwise the volumes of the phases in the tubes may become progressively altered during the counter-current distribution process, with the result that the fitting of theoretical curves becomes invalid. This phenomenon is particularly associated with those solvent systems (e.g., butanol–water–acetic acid) which yield phases which have considerable mutual miscibility.

4. The solvents should be of low toxicity and economic to use in large quantities. The solvent pair should readily separate after equilibration either alone or in the presence of solute otherwise a considerable time lag will be experienced between each distribution stage. A tendency for the formation of emulsions should be particularly noted.

5. The solvent system should be such as to allow the quantitative analysis for solute in each phase of each tube a relatively simple process. The method of analysis selected (as in the case of eluate fractions from liquid–solid chromatographic columns, p. 196) will be determined by the chemical and physical nature of the solute. Methods based upon (a) weighing the residue after solvent removal, (b) absorbance measurements at a suitable wavelength in the visible or ultraviolet region using equilibrated solvent as a reference, and (c) titration with standard acid or base as appropriate, are all methods suitable for quantitative estimations. For a qualitative assessment of the effectiveness of the separation, loading of appropriately sized samples on to a thin-layer plate and subsequent chromatographic development has many advantages.

6. Following the completion of the distribution sequence and analysis of the contents of each tube, the solvent system selected must allow for facile isolation of the separated components from suitably bulked tube contents.

In the case of neutral compounds, immiscible solvent pairs selected from the eluotropic series (p. 202) would provide a suitable starting point for preliminary experiments. Such experiments also enable a suitable analytical method to be developed and this in turn would lead to rough estimates of the K value for the major component. (Note: this value would not be used in subsequent calculations of the theoretical distribution curve; the K value must be calculated from the distribution pattern.)

With mixtures of acidic or basic components, an aqueous buffer solution with an immiscible organic phase is frequently highly satisfactory. Adjustment of the pH of the buffer and/or the addition to the aqueous phase of methanol or ethanol can lead to fine control of the K value which can be extremely useful.

Exercises in countercurrent distribution The following two experiments have been found useful for the demonstration of many aspects of countercurrent distribution discussed above. A twenty-tube apparatus is employed; equal volumes (20 ml) of upper and lower phases are used.

Distribution of Light Green dye. In a 2 litre separating funnel shake thoroughly a mixture of 440 ml of butanol, 560 ml of water and 62 ml of glacial acetic acid. Allow the two phases, which should be of approximately equal volume, to separate, and run off and retain each layer.

Remove the polyethylene stoppers from the first ten extraction tubes and into each pour just over 40 ml of the lower aqueous phase. Recap the tubes and tip the extraction tubes into the vertical position to decant the surplus lower phase of each tube into its respective reservoir tube; after complete decantation return the extraction tubes to the horizontal position to allow the reservoirs to drain. Repeat this operation a sufficient number of times to fill all of the extraction tubes with the correct volume of lower phase; ensure that the drain tube is attached to the last extraction tube to collect surplus lower phase.

Into each of the second and third extraction tubes introduce 20 ml of equilibrated upper phase; reserve a further 20 ml portion of upper phase in a conical flask. Pour the remaining upper phase into the upper phase reservoir sited at the back of the apparatus. Attach the upper phase dispenser to connect this reservoir to the first tube.

Dissolve about 3 mg of Light Green dye in the 20 ml of upper phase previously placed upon one side and add the solution to the first tube. Carry out successive equilibrations and transfer operations using a shaking period of about 3 minutes (involving about 50 rocking motions). Little difficulty with the formation of emulsions during the equilibration stages should be experienced using this particular solvent system and the two phases should separate quite easily.

After twenty transfer operations and a final equilibration of the extraction tube contents, remove individually the contents of each tube to a separating funnel and segregate the layers into suitably labelled stoppered conical flasks. Rinse the apparatus thoroughly with water, using the transfer operations to wash the reservoirs, then with industrial spirit and allow to drain and air dry.

Remove a sample of one upper and one lower phase which visibly contains the dye and determine the visible spectrum (400–700 nm) using equilibrated upper and lower phases respectively in the reference beam. From an inspection of each spectrum select a suitable wavelength at which to determine the absorbance of all the upper and lower phases against the appropriate reference. Plot the distribution curve of absorbance value vs. tube number.

Using the mathematical procedures discussed above, calculate K for Light Green dye and also the theoretical distribution pattern.

Separation of 3,4,5-trihydroxybenzoic acid (gallic acid) and p-hydroxybenzoic acid. Use a solvent system of thoroughly equilibrated ether–water; fill the extraction tubes with lower phase and the main reservoir and the second and third extraction tubes with upper phase as described in the previous example. Dissolve about 60 mg of each of gallic acid and p-hydroxybenzoic acid in the 20 ml of equilibrated ether set aside for this purpose, and transfer the solution to the first tube.

After twenty equilibrations and transfers remove the contents of each tube and separate the upper and lower phases into labelled stoppered flasks. Determine the absorbance of each phase against the appropriate reference at λ 275 nm, at which wavelength both acids have a similar ε value (1). Plot a graph of absorbance against tube number for both upper and lower phases. Calculate (a) the value of K for each acid, and (b) the theoretical distribution pattern of each acid using either the upper or lower phase values.

Finally select those tubes where overlap of the distribution patterns of the acid does not occur, bulk the contents and evaporate the solvent (ether and water) on a rotary evaporator. Identify each acid by m.p. determination.

Note. (1) Solutions of each acid in ether or water having a concentration of approximately 10 p.p.m. are suitable for recording absorption curves from which ε values may be calculated.

I,33. STORAGE OF SAMPLES After preparation, all pure products should be stored in suitable clearly and permanently labelled containers. The bulk samples of stable solids and liquids may be stored in screw-capped and ground glass stoppered bottles respectively. Hygroscopic samples, or those liable to decomposition by contact with atmospheric moisture, should either be stored in a desiccator or in a bottle which is sealed by painting over the closure with molten paraffin wax. Where there is a possibility of photochemical decomposition of chemicals it is generally good practice to keep them out of direct sunlight and to store them in brown bottles.

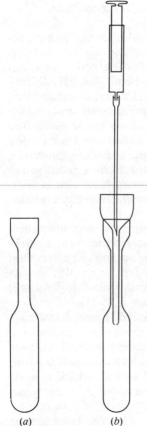

Small specimens of all products, including reaction intermediates isolated from reaction sequences, and particularly samples of fractions isolated as the result of lengthy chromatographic or other purification procedures, should invariably be retained for reference purposes. The commercially available straight-sided specimen tubes with polyethylene plug seals, which are available in a range of sizes, are suitable in the case of solid samples. It is usually advantageous to label them with the name and a code reference to enable physical data (elemental analysis, spectroscopic information, etc.) to be located in the laboratory notebooks.

Liquid samples may be sealed in specially prepared glass ampoules for prolonged storage or when specimens need to be sent away, for example, for elemental analysis. Ready made ampoules may be obtained commercially in a range of sizes or can be made by the following method. A short length of moderately thick-walled tubing of internal diameter suited to the sample size is cleaned by immersion in a narrow cylinder containing chromic acid cleaning mixture (Section **I,2**), thoroughly washed with distilled water, followed by a little acetone and is then dried by passing a current of warm air through it. One end is then sealed off in the blowpipe flame as in Fig. I,158(*a*). The constriction is then made by carefully rotating the tube in a small blowpipe flame; it is important that the wall of the tubing remains uniformly thick at this point. The sample is most conveniently introduced into the ampoule from a capillary pipette or micro-syringe needle passing through a protective tube previously inserted into the ampoule via the constriction (Fig. I,158(*b*)). It is advantageous for the end of this protective tube, which should not reach below the

(*a*)   (*b*)

Fig. I,158

final level of the liquid to be introduced, to be slightly rounded off. This prevents any of the liquid sample which adheres to the syringe or capillary pipette from contaminating the outer side of the protecting tube. This may be safely withdrawn without danger of introducing sample on to the inside surface of the constriction. The ampoule may then be sealed off in the usual way using a small flame.

If the sample needs to be sealed in an atmosphere of nitrogen, the inlet of the protecting tube is connected to a low-pressure supply of the gas when the syringe or pipette has been withdrawn. When all the air has been displaced the tube may be slowly withdrawn and the ampoule sealed. Filled ampoules containing volatile samples should be thoroughly chilled in a suitable cooling bath before such air displacement, and before sealing. Care must be exercised when sealed ampoules, particularly those which have been stored for some time, are reopened, and the precautions outlined in Section I,3,B,4(ix) should be noted.

DETERMINATION OF PHYSICAL CONSTANTS

I,34. DETERMINATION OF MELTING POINT—MIXED MELTING POINTS A pure crystalline organic compound has, in general, a definite and sharp melting point; that is, the melting point range (the difference between the temperature at which the collapse of the crystals is first observed and the temperature at which the sample becomes completely liquid) does not exceed about 0.5 °C. The presence of small quantities of miscible, or partially miscible, impurities will usually produce a marked increase in the melting point range and cause the commencement of melting to occur at a temperature lower than the melting point of the pure substance. The melting point is therefore a valuable criterion of purity for an organic compound.

A sharp melting point is usually indicative of the high purity of a substance. There are, however, some exceptions. Thus a eutectic mixture of two or more compounds may have a sharp melting point, but this melting point may be changed by fractional crystallisation from a suitable solvent or mixture of solvents. The number of exceptions encountered in practice is surprisingly small, hence it is reasonable to regard a compound as pure when it melts over a range of about 0.5 °C (or less) and the melting point is unaffected by repeated fractional crystallisation.

The experimental method in most common use is to heat a small amount (about 1 mg) of the substance in a capillary tube inserted into a suitable melting point apparatus and to determine the temperature at which melting occurs. **The capillary melting point tubes are prepared** either from soft glass test-tubes or from wide glass tubing (ca. 12 mm diameter).* A short length of glass tubing or glass rod is firmly fused to the closed end of the test-tube. The test-tube (or wide glass tubing) must first be thoroughly washed with distilled water to remove dust, alkali and products of devitrification which remain on the surface

---

* Pyrex glass is preferable, but this requires an oxygen-gas blowpipe for manipulation. Suitable melting point tubes may be purchased from dealers in scientific apparatus or chemicals. It is, however, excellent practice for the student to learn to prepare his own capillary tubes.

of the glass, and then dried. The closed end of the test-tube is first heated whilst being slowly rotated in a small blowpipe flame; the glass rod or tube is simultaneously heated in the same manner (Fig. I,159(a)). When the extremities of both pieces of glass are red hot, they are firmly fused together, twisting of the joint being avoided, and then removed momentarily from the flame until the seal is just rigid enough that no bending occurs. The test-tube is then immediately introduced into a large 'brush' flame (Fig. I,159(b)) so that a length of about 5 cm is heated, and the tube is rotated uniformly in the flame. When the heated portion has become soft and slightly thickened as the result of the heating, the tube is removed from the flame and, after a second or two, drawn, slowly at first and then more rapidly, as far apart as the arms will permit (or until the external diameter of the tube has been reduced to 1–2 mm). If the operation has been successfully performed, a long capillary of regular bore throughout most of its length will be obtained. The long thin tube is then cut into lengths of about 8 cm by touching it lightly with a file and then tapping gently with the flat portion of the file; after a little practice, no difficulty should be experienced in dividing the long capillary into suitable lengths without crushing the fragile tubing. It will be found that a short length of tubing ('glass spindle'), sufficiently rigid to act as a holder, will remain attached to the test-tube after the long capillary has been cut off. The operation may then be repeated. When the test-tube becomes too short to be handled at the open end, a piece of glass tubing or rod may be fused on, in the manner previously described, to act as a convenient handle. In this way a large number of capillary tubes may be prepared from one test-tube. One end of each of the capillary tubes should be sealed by inserting it horizontally into the extreme edge of a small Bunsen flame for a few seconds, and the capillary tube rotated meanwhile; the formation of a glass bead at the end of the tube should be avoided. The prepared capillary tubes should be stored either in a large specimen tube or in a test-tube closed with a cork.

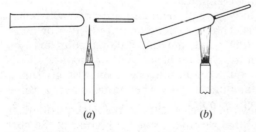

(a)   (b)

Fig. I,159

The capillary tube is then filled as follows. About 25 mg of the dry substance is placed on a glass slide or upon a fragment of clean porous porcelain plate and finely powdered with a clean metal or glass spatula, and then formed into a small mound. The open end of the capillary tube is pushed into the powder, 'backing' the latter, if necessary, with a spatula. The solid is then shaken down the tube by tapping the closed end on the bench or by gently drawing the flat side of a triangular file (a pocket nail file is quite effective) along the upper end of the tube. The procedure is repeated until the length of lightly-packed material is 3–5 mm, and the outside of the tube is finally wiped clean.

The two principal types of **melting point apparatus** in common use are those in which heating of the capillary tube is by means of a heated liquid bath and those in which heating is carried out in, or on, an electrically heated metal block.

Three convenient forms of bath are shown in Fig. I,160(*a*, *b* and *c*). The first consists of a long-necked, round-bottomed flask (a long-necked Kjeldahl flask of 100 ml capacity is quite satisfactory) supported securely but not tightly by

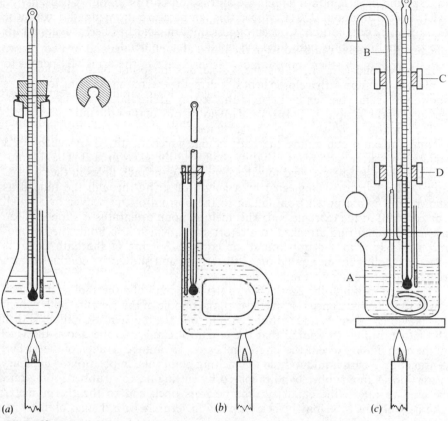

(*a*)   (*b*)   (*c*)

Fig. I,160

means of a clamp immediately below the bevelled edge of the flask as shown in the diagram. A thermometer (1) is fitted through a cork, a section of the cork having been cut away (see inset) so that the thermometer scale is visible and also to allow for expansion of the air in the apparatus. The bulb is about three-quarters filled with a suitable heating fluid (2).

Notes. (1) For melting point (and also for boiling point) determinations, it is convenient to use thermometers which have been calibrated by partial immersion to a distance marked on the stem (30 mm is suitable for melting points, 80 mm for boiling points). If a thermometer calibrated by total immersion is used, an error is introduced resulting from the cooling of the mercury thread which is not heated in the apparatus but is exposed to the cooler laboratory atmosphere. The necessary *stem correction* to be added to the observed melting point to give the 'corrected' value is given by the expression

225

0.00016 N ($t_1 - t_2$), where N is the length of the exposed mercury thread in degrees, t_2 is mean temperature of the exposed mercury thread determined on an auxiliary thermometer placed alongside with its bulb at the middle of the exposed thread, and t_1 is the observed temperature on the thermometer scale.

(2) The safest and most satisfactory bath liquids are the highly stable and heat-resistant Silicone oils. A cheaper alternative is medicinal paraffin; it has a low specific heat, is non-inflammable and is non-corrosive, but it can only be safely heated to about 220 °C; above this temperature it begins to decompose and becomes discoloured. Concentrated sulphuric acid has been suggested for use as a melting point bath fluid but is NOT recommended.

In the second (Fig. I,160(b)), the Thiele melting point bath, the liquid is contained in a tube with a closed bent side arm, the thermometer being located as shown. On heating the bent side arm, the heated liquid circulates and raises the temperature of the sample. In both (a) and (b) no stirring of the bulk fluid is required.

The apparatus shown in Fig. I,160(c) consists of a small Pyrex beaker (e.g., of 100 ml capacity) containing the bath liquid, which is stirred by means of a small glass stirrer A so placed that its shaft is in the glass tube B. The thermometer and glass tube are held together by passing through holes in the corks C and D. The stirrer is connected by a length of string through the tube B as shown, and is prevented from falling to the bottom of the beaker by a small cork or knot at the extreme end. This melting point apparatus is supported on a gauze-covered ring attached to a retort stand, which also holds the thermometer and tube in a clamp round the cork C. Stirring of the bath liquid is effected by suitable manipulation of the string, and should be conducted at a regular rate throughout the heating.

The filled melting tube is attached to the lower end of a thermometer in such a way that the substance is at the level of the middle of the mercury bulb (which has previously been wetted with the bath liquid); the moistened capillary is then slid into position. Providing that the length of capillary tube above the level of the bath liquid exceeds the length immersed, advantage is taken of the surface tension of the bath liquid to hold the melting point tube in position by capillary attraction. A thin rubber band prepared by cutting narrow rubber tubing may be used to attach the capillary tube near its open end to the thermometer; alternatively the tube may be held in position securely with the aid of fine wire. The thermometer, with the tube attached, is inserted into the centre of the bath.

The melting point apparatus is heated comparatively rapidly with a small flame until the temperature of the bath is within 15 °C of the melting point of the substance, and then slowly and regularly at the rate of about 2° per minute until the compound melts completely. The temperature at which the substance commences to liquefy and the temperature at which the solid has disappeared, i.e., the melting point range, is observed. For a pure compound, the melting point range should not exceed 0.5–1 °C; it is usually less. Any sintering or softening below the melting point should be noted as well as any evolution of gas or any other signs of decomposition.* If the approximate melting point is not known, it is advisable to fill two capillaries with the substance. The tem-

---

* A substance which commences to soften and pull away from the sides of the capillary tube at (say) 120 °C, with the first appearance of liquid at 121 °C, and complete liquefaction at 122 °C with bubbling, would be recorded as m.p. 121–122 °C (decomp.), softens at 120 °C.

perature of the bath may then be raised fairly rapidly using one capillary tube in order to determine the melting point approximately; the bath is then allowed to cool about 30 °C, the second capillary substituted for the first and an accurate determination made.

It should be noted that a second determination of the melting point should not be made as the bath liquid cools by observing the temperature at which the molten material in the capillary tube solidifies, or by reheating the bath after the solidification has occurred. This is because, in many cases, the substance may partially decompose, and in some instances it may undergo a change into another crystalline form possessing a different melting point. A freshly-filled capillary tube should always be employed for each subsequent determination. Substances which sublime readily are sometimes heated in melting point capillaries sealed at both ends. For compounds which melt with decomposition, difficulties sometimes arise in the melting point determination; it is best to insert the capillary tube into the bath when the temperature is only a few degrees below the melting and decomposition point of the material. This avoids decomposition, with consequent lowering of the melting point, during the time that the temperature of the bath liquid is being raised.

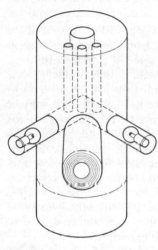

Fig. I,161

A liquid heating bath may be dispensed with by the use of an **apparatus employing electrical heating**. An electrically heated aluminium or copper block is very convenient for this purpose. The essential features of a commercial apparatus* are shown in Fig. I,161. The large hole at the centre is for the thermometer and the three smaller holes are for the melting point capillaries which can be observed simultaneously. The block is heated electrically and the rate of heating can be controlled from the front of the apparatus. The melting point tubes are illuminated from two sides, making observation of the melting point by means of the lens housed in the eye-piece very easy. With this apparatus, it is important to raise the temperature slowly near the melting point since heat transfer is less efficient than with liquid baths.

The **Kofler hot bench**,† illustrated in Fig. I,162, consists of a metal alloy band with a corrosion-free steel surface, 36 cm long and 4 cm wide, heated electrically at one end, the other remaining unheated to give a moderate almost constant temperature gradient. Fluctuations in the mains voltage are compensated for by a built-in stabiliser. The graduations cover the range 50 to 260 °C in 2°. Provision is made for variations in room temperature by adjustment on the reading device moving over the scale. The current must be switched on at least 1 hour before the apparatus is required and the latter should be surrounded by a screen to protect it from draughts.

---

* The apparatus marketed by A. Gallenkamp & Co. Ltd is both relatively inexpensive and very satisfactory.

† Manufactured by C. Reichert Co., Vienna, and obtainable in the UK from British American Optical Co. Ltd.

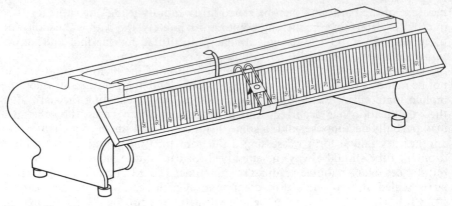

Fig. I,162

The hot bench should be calibrated before use with the aid of several of the substances supplied by the manufacturers. The test substances include: azobenzene, m.p. 68 °C; benzil, m.p. 95 °C; acetanilide, m.p. 114 °C; phenacetin, m.p. 135 °C; benzanilide, m.p. 163 °C; *p*-acetamidophenyl salicylate (salophene), m.p. 190 °C; and saccharin, m.p. 228 °C. The melting point is readily determined by sprinkling a few small crystals of the substance on the hot bench; these may be moved along the bench by the brass lancet attached to the reading device. Usually a sharp division occurs between the solid and liquid, and the temperature corresponding to the line of demarcation is read off on the scale. For maximum accuracy, the apparatus should be recalibrated with two test substances with melting points close to that of the unknown. The procedure is clearly rapid and is very useful for substances which tend to decompose upon gradual heating. It is important not to allow any of the molten substance to remain on the corrosion-resistant steel surface for long periods; it should be wiped away with paper tissues immediately after the experiment.

The **microscope hot stage type** of melting point apparatus * (essentially an electrically-heated block on a microscope stage) is of particular value when the melting point of a very small amount (e.g., of a single crystal) has to be determined. Further advantages include the possibility of observation of any change in crystalline form of the crystals before melting. The main features of a commercial form of apparatus are shown in Fig. I,163. The apparatus also incorporates a polariser which facilitates the observation of the process of melting. The rate of heating is controlled by means of a rheostat.

A considerably simplified version is also available commercially† (**micro melting point apparatus**). The hot stage may be fitted to any standard microscope and the rate of heating is controlled by a transformer unit. The sample is placed on a cover glass in the cavity in the heating block which may be taken to a temperature of 325 °C.

Mixed melting points.   In the majority of cases the presence of a 'foreign substance' will lower the melting point of a pure organic compound. This fact is utilised in the so-called mixed melting point test for the identification of organic compounds. Let us suppose that an organic compound *X* having a melt-

---

* Manufactured by C. Reichert Co.
† A. Gallenkamp & Co. Ltd.

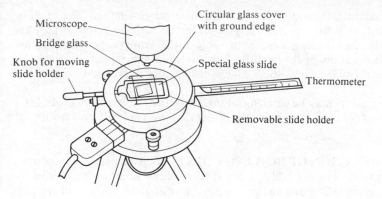

Microscope

Bridge glass

Knob for moving
slide holder

Circular glass cover
with ground edge

Special glass slide

Thermometer

Removable slide holder

Fig. I,163

ing point of 140 °C is suspected to be *o*-chlorobenzenoic acid. Its identity may be established by performing a melting point determination on a mixture containing approximately equal weights of *X* and of an authentic specimen of *o*-chlorobenzoic acid (*A*). If the melting point of the mixture is 140 °C, then *X* is *o*-chlorobenzenoic acid, but if the melting point is depressed by several degrees *A* and *X* cannot be identical. It is recommended that at least three mixtures containing, say, 20 per cent *X* + 80 per cent *A*; 50 per cent *X* + 50 per cent *A*; and 80 per cent *X* + 20 per cent *A* be prepared, and the melting points be determined.

Cases may arise in which the melting point of certain mixtures are higher than the individual components, e.g., if an addition compound of higher melting point is formed or if the two compounds are completely soluble in the solid state forming solid solutions. The mixed melting point test therefore, although of great practical value, is not infallible and should accordingly be used with reasonable regard to these possibilities.

Exercises in the determination of melting points and mixed melting points   The melting points of pure samples of the following compounds are determined using a melting point bath or an electrically heated apparatus, in the manner detailed above. The correct melting points of the pure substances are given in parentheses.

(*a*) *p*-Nitrotoluene (54 °C) or azobenzene (68 °C).

(*b*) 1-Naphthol (96 °C) or catechol (104 °C) or benzil (95 °C).

(*c*) Benzoic acid (122 °C) or 2-naphthol (123 °C) or urea (133 °C).

(*d*) Salicylic acid (159 °C) or phenylurea (mono) (148 °C).

(*e*) Succinic acid (185 °C) or *p*-tolylurea (mono) (180 °C).

(*f*) *p*-Nitrobenzoic acid (239 °C) or *s*-diphenylurea (242 °C).

By working in the above order, it will not be necessary to wait for the apparatus to cool between consecutive determinations.

In order to gain experience in the determination of mixed melting points the following simple experiment should be carried out.

Determine the melting point of pure cinnamic acid (133 °C) and pure urea (133 °C). Approximately equal weights (*ca.* 50 mg) of the two compounds are placed on a clean porous porcelain tile. These are now ground together and intimately mixed with the aid of the flat side of a micro-spatula. The melting point tube filled with this mixture is placed in the melting point apparatus alongside melting point tubes filled with each of the two components. In this way careful observation of the melting behaviour of the mixture and of the pure components will clearly show the considerable depression of melting point.

Similar experiments may be carried out on a mixture of benzoic acid (122 °C) and 2-naphthol (123 °C), or a mixture of acetanilide (113 °C) and antipyrin (113 °C).

I,35. DETERMINATION OF BOILING POINT When reasonable amounts of liquid compounds are available (> 5 ml) the boiling point is readily determined by slowly distilling the material from a pear-shaped flask in an apparatus assembly shown in Fig. I,98, and recording the temperature at which the bulk of the compound distils. Due attention should be paid to the experimental procedure which was discussed in detail in Section **I,24**.

For smaller quantities of liquid compounds (0.5–3.0 ml) the material should be distilled in the apparatus assembly shown in Fig. I,99.

When only minute quantities of liquid are available, either of the two **micro methods for the determination of the boiling point** may be used.

Method 1 (**Siwoloboff's method, 1886**). Two tubes, closed at one end, are required; one, an ordinary melting point capillary, 90–110 mm long and 1 mm in diameter, and the other, 80–100 mm long and 4–5 mm in diameter. The latter may be prepared from 4–5 mm glass tubing and, if desired, a small thin bulb, not exceeding 6 mm in diameter, may be blown at one end. A small quantity of the liquid, 0.25–0.5 ml (depending upon the boiling point), is placed in the wider tube, and the capillary tube, with sealed end uppermost, is introduced into the liquid. The tube is then attached to the thermometer by a rubber band (Fig. I,164) and the thermometer is immersed in the bath of a melting point apparatus. As the bath is gradually heated there will be a slow escape of bubbles from the end of the capillary tube, but when the boiling point of the liquid is attained a rapid and continuous escape of bubbles will be observed. The reading of the thermometer when a rapid and continuous stream of bubbles first emerges from the capillary tube is the boiling point of the liquid. Unless the temperature is raised very slowly in the vicinity of the boiling point of the liquid, the first determination may be slightly in error. A more accurate result is obtained by removing the source of heat when the rapid stream of bubbles rises from the end of the capillary tube; the speed at which bubbles are given off will slacken and finally, when the last bubble makes its appearance and exhibits a tendency to suck back, the thermometer is read immediately. This is the boiling point of the liquid because it is the point at which the vapour pressure of the liquid is equal to that of the atmosphere. As an additional check on the latter value, the bath is allowed to cool a few degrees and the temperature slowly raised; the thermometer is read when the first continuous series of bubbles is observed. The two thermometer readings should not differ by more than 1°. It should, however, be remembered that the Siwoloboff method gives trustworthy results only for comparatively pure liquids; small amounts of volatile impurities such as ether or water may lead to boiling points being recorded which approximate to those of the volatile component.

Method 2 (**Emrich's method**).   A capillary tube about 10 cm long and of about 1 mm bore is used. One end is drawn out by means of a micro flame into a capillary with a very fine point and about 2 cm long as in Fig. I,165(*a*). Such a capillary pipette may also be constructed by suitably drawing out soft glass tubing of 6–7 mm diameter. The tube (*a*) is then dipped into the liquid of which

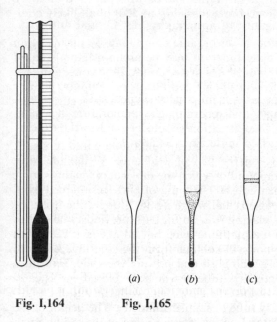

(*a*)   (*b*)   (*c*)

Fig. I,164   Fig. I,165

the boiling point is to be determined; the liquid will rise slowly by capillary attraction, and the tube is removed when the liquid has filled the narrow conical portion. The capillary end is then sealed by merely touching with a minute flame. A small air bubble is formed in the point of the capillary; it should be examined with a lens to make sure that it is not too large. A convenient size is 1–3 mm long (Fig. I,165(*b*)). The prepared capillary tube is then attached to a thermometer as in a melting point determination (Section **I,34**) and slowly heated in an open bath, which is kept well stirred. The capillary is best observed with a lens. When the bubble enlarges (as in Fig. I,165(*c*)) and begins to exhibit signs of upward motion, the flame is removed or considerably lowered. The temperature at which the bubble reaches the surface of the bath liquid is the boiling point of the liquid. The bath is allowed to cool about 10° below the first observed boiling point, then slowly heated again, and a second determination of the boiling point is made.

Exercises in boiling point determination   The following pure liquids offer a convenient selection of compounds having a range of boiling points: (*a*) carbon tetrachloride (77 °C); (*b*) ethylene dibromide (132 °C) or chlorobenzene (132 °C); (*c*) aniline (184.5 °C); and (*d*) nitrobenzene (211 °C). The boiling points may be determined using conventional distillation apparatus assemblies. The compounds could also be employed to give practice in the determination of boiling point by the Siwoloboff or Emrich methods.

I,36. DETERMINATION OF MOLECULAR WEIGHT Mass spectrometry offers the most refined method for the evaluation of the molecular weight of those compounds having vapour pressures higher than 0.1 mmHg at 350 °C. With instruments of high resolving power, the molecular weight is obtained to an accuracy of ±5 p.p.m. These accurate molecular weights may be used to deduce possible molecular formulae with the aid of Mass and Abundance Tables (Refs. 77–78) which list the accurate mass (up to 500) of all likely combinations of C, H, O and N (see also Section **VII,4,3**).

Occasionally the characterisation of an organic compound by means of an appproximate molecular weight determination may be useful. Methods based upon ebullioscopic or cryoscopic procedures are often too time consuming for routine use. However, the high freezing point depression of camphor permits molecular weights to be determined rapidly and with reasonable accuracy (between 1 and 5%) using an ordinary melting point apparatus (**Rast's camphor method**).

Support a small clean test-tube (e.g., 75 × 10 mm) in a hole bored in a cork so that it will stand conveniently on the pan of a balance. Weigh the tube. Introduce about 50 mg of the compound of which the molecular weight is to be determined and weigh again. Then add 500–600 mg of pure, resublimed camphor (e.g., the micro-analytical reagent) and weigh again. Stopper the test-tube loosely and melt the contents by placing it in an oil bath previously heated to about 180 °C;* stir the liquid with a platinum wire, but do not heat the liquid for more than 1 minute or camphor will sublime from the solution. Allow to cool, transfer the solid to a clean watch glass and powder the solid. Introduce some of the powder into a thin capillary tube of which the closed end is carefully rounded; press the solid down into the closed end with the aid of a platinum wire or with a closed capillary tube of smaller diameter. The height of the solid should not exceed 2 mm. Determine the melting point of the mixture in a liquid melting point bath or in an electrically heated apparatus (Section **I,34**) using, preferably, a 100–200° thermometer graduated in 0.1° or 0.2°; good illumination and very careful control of the rate of heating is essential. The melting point is taken as that temperature at which the last fragment of solid disappears. To make sure that the mixture is homogeneous repeat the melting point determination with a second sample; if the two differ appreciably, prepare a new mixture. Then determine the melting point of the original camphor. The difference in melting points gives the depression of the melting point of camphor caused by the addition of the compound. The molecular weight M can then be calculated from the formula:

$$M = \frac{K \times w \times 1000}{\Delta T \times W}$$

where K is the molecular depression constant of camphor (39.7), w is the weight of the compound, W is the weight of the camphor and ΔT is the depression of the melting point.

Note. The solute concentration should be above $0.2M$; in dilute solution K increases from 39.7 to about 50.

---

* If very great care is taken, the mixture may be melted by heating over a very small flame for about 30 seconds; the technique described in the text is to be preferred.

The Rast camphor method, although very simple, is nevertheless liable to some limitations. One serious difficulty is that the melting point of camphor is itself rather high and this may lead to decomposition of the compound whose molecular weight is to be determined. Another difficulty is the limited solubility of many classes of compound in liquid camphor, and this severely restricts its general applicability. Some useful alternative solvents having high molar freezing point depression constants are given in Table I,16.

Table I,16   Solvents for molecular weight determination by depression of freezing point

Compound	Melting point (°C)	Molar depression constant
Cyclohexanol	24.7	42.5
Camphene	49	31
Cyclopentadecanone	65.6	21.3
Bornylamine	164	40.6
Borneol	202	35.8
cis-4-Aminocyclohexane-1-carboxylic acid lactam	196	40*

* This solvent has been found in these laboratories to be of value for the molecular weight determination of peptides.

I,37. DETERMINATION OF OPTICAL ROTATORY POWER Compounds which rotate the plane of polarised light around its axis, whether they are in the gaseous, liquid or molten state, or in solution, are said to be **optically active**. This property arises from the lack of certain elements of symmetry in the molecule (i.e., a centre, a plane or an n-fold alternating axis of symmetry) with the result that the molecule and its mirror image are non-superimposable. Although first observed with compounds having one or more chiral carbon atoms (i.e., a carbon substituted with four different groups), optically active compounds having chiral centres including atoms of silicon, germanium, nitrogen, phosphorus, arsenic, sulphur, etc., have also been prepared. Molecular dissymmetry, and hence optical activity, also arises in molecules, such as certain substituted biphenyls, allenes, etc., which have chiral axes or chiral planes rather than chiral atoms as such. The study of optically active coordination complexes is a more recent, important and expanding field of study. The reader is referred to the many excellent monographs and articles which are currently available (Ref. 79) and which fully explore the current aspects of the stereochemistry of molecules and the importance of stereochemical considerations of appropriate reaction processes. This Section is devoted to the experimental determination of optical rotatory power.

When a beam of monochromatic light is passed through a crystal of Iceland spar, two beams are transmitted, each vibrating in one plane which is perpendicular to the other. A **Nicol prism** is composed of two sections of Iceland spar so cut, and again sealed with Canada balsam, that one of these rays is refracted to the side (this is absorbed by the black surroundings of the prism) so that the light which finally passes through the prism is vibrating in one plane only. This light is said to be **plane polarised**. This polarised light is allowed to pass through another Nicol prism similarly orientated and the light viewed from a point remote from, but in line with, the light source. It will be now found that on rotating the second prism the field of view appears alternately light and dark and the minimum of brightness follows the maximum as the prism is rotated through an angle of 90°; the field of view will appear dark when the axes of the two prisms are at right angles to one another. The prism by which the light is polarised is termed the **polariser**, and the second prism, by which the light is examined, is called the **analyser**.

If, when the field of view appears dark, a tube containing a solution of an optically active compound is placed between the two prisms, the field lights up; one of the prisms must then be turned through a certain angle α before the original dark field is restored. Since the plane of vibration of polarised light may have to be rotated either clockwise or anti-clockwise, it is necessary to observe a convention to designate the direction of rotation. When, in order to obtain darkness, the analyser has to be turned clockwise (i.e., to the right), the optically active substance is said to be **dextrorotatory**, or (+); it is **laevo-rotatory**, or (−), when the analyser must be rotated anti-clockwise (i.e., to the left).

The obvious disadvantage of the above simple instrument (**polarimeter**) is the difficulty of determining the precise 'end-point' or the point of maximum darkness. The human eye is a poor judge of absolute intensities, but is capable of matching the intensities of two simultaneously viewed fields with great accuracy. For this reason all precision polarimeters are equipped with an optical device that divides the field into two or three adjacent parts (half shadow or triple-shadow polarimeter; Ref. 80) such that when the 'end-point' is reached the sections of the field become of the same intensity. A very slight rotation of the analyser will cause one part to become lighter and the other darker. The increase in sensitivity so attained is illustrated by the fact that an accuracy of at least 0.01° is easily obtained with the use of an 'end-point' device, whereas with the unaided eye the settings are no more accurate than 4–5°.

A half-shadow polarimeter (**Lippich type**) is illustrated diagrammatically in Fig. I,166. Here two polarised rays are produced by means of the main Nicol prism P and a small Nicol prism P'; the latter covers half the field of the larger

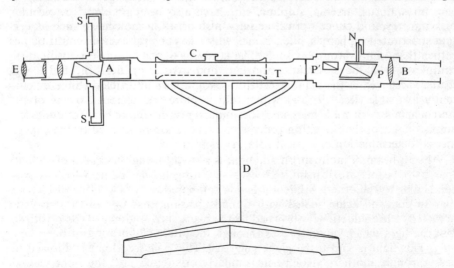

Fig. I,166

polariser P and its plane of polarisation is slightly inclined to that of P. The angles between the planes of polarisation may be altered by a slight rotation of the polariser P. Upon rotating the analyser A, a position will be found at which one beam will be completely, the other only partially, extinguished; the one half of the field of view will therefore appear dark, while the other will still remain light when viewed with the eyepiece E as in Fig. I,167(a). Upon rotating the analyser A still further, a second position will be found at which only the second beam will be extinguished and the field will have the appearance shown in (c). When, however, the analyser occupies an intermediate position, the field of view will appear of uniform brightness (as in b), and this is the position to which the analyser must be set, and the reading from the circular scale S which also incorporates a vernier carefully noted.

In Fig. I,166, B is a collimator, T the trough (shown without cover) which houses the polarimeter tube C, E the eyepiece and D the heavy support stand

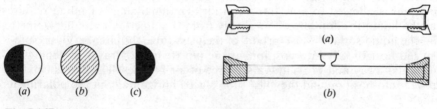

(a)

(b)

(a)   (b)   (c)

Fig. I,167   Fig. I,168

for the apparatus. N is a device for moving P and thus altering the 'half-shadow angle'; this has the advantage of increasing the intensity of light which is transmitted by the polarising prism and this may be essential when the optical activity of coloured solutions is to be determined. On the other hand the reproducibility with which the accurate position of uniform brightness of field may be ascertained diminishes.

Two forms of polarimeter tube are shown in Fig. I,168. The common type (a) consists of a tube of thick glass with accurately ground ends: the tube is closed by means of circular plates of glass with parallel sides, which are pressed against the ends of the tube by means of screw caps. The caps must not be screwed so tightly as to cause strain in the glass end plates as this would cause a rotation; the glass plates at the end must be clear and free from finger marks and the exposed surface must be dry. In a modification, the tube is surrounded by a jacket to permit the circulation of water at constant temperature by means of a pump. Tube (b) has a cup opening in the centre of the glass tube; the glass plate ends are in this case sealed on to the tube with a cement which is inert to most solvents.

The unit of length in polarimetry is 1 dm, hence the tubes are generally made in lengths which are fractions or multiples of this quantity, e.g., 0.5, 1, 2 or 4 dm. The tube bore is usually 8 mm and hence the capacity of the 1 dm tube is 5.02 ml. Frequently 'neat' liquids or solutions having volumes in the range 1–2 ml need to be examined for rotatory power and in these cases polarimeter tubes (usually 0.5 dm) having a tube bore of 4 or 5 mm are employed.*

* Polarimeter tubes having smaller bores (1, 2 or 3 mm) are available, together with the other sizes of tubes, from Bellingham & Stanley Ltd.

The filling of these polarimeter tubes requires some attention to detail. Tubes which have the wider bore can conveniently be of the centre-filling type (Fig. I,168(b)) and it is merely necessary to carefully pour the solution or liquid * into the central opening, carefully rocking the tube to disperse air-locks; both the tube and the liquid should be at the temperature of the laboratory before filling and the final level of liquid should be within the central cup. The end-filling tubes (Fig. I,168(a)), no matter which size of bore, require a little practise to avoid the presence of air-bubbles in the finally sealed tube which would hamper the field of view. Having ensured that the tube and the circular glass plates (which should be handled by the edges) are clean and dry, one of the ends of the tube is securely capped. The tube is then placed vertically on a bench surface, sealed end downwards, and the tube is filled nearly to the top of the open end with liquid. In the case of the narrow-bore tube a capillary pipette is used. In both cases, to avoid air-locks, the end of the pipette is carefully lowered to the bottom of the tube (without touching the glass end plate surface) and raised as the liquid is allowed to flow into the tube. The final stage of filling requires that the liquid is allowed to flow slowly from the pipette or capillary pipette, until the liquid surface is just 'proud' of the glass tube; the circular glass plate is then slid horizontally into position, as the pipette is withdrawn, sweeping surplus liquid away and providing a seal free of air bubbles. The cap is carefully screwed into position and the filled tube placed horizontally in the polarimeter trough T.

For accurate work it is essential to determine the 'zero' position of the instrument with the empty polarimeter tube; its position in the polarimeter and the exact position of the end plates should be registered by suitable markings. The readings on the circular scale (using the vernier) should be noted on about ten successive determinations in which the analyser prism is returned to the position of uniform brightness both from a clockwise and an anti-clockwise direction and the results averaged. The solution in the filled tube, exactly orientated as previously, is then examined and the average of ten successive readings of the position of uniform brightness from clockwise and anti-clockwise direction determined. Subtraction gives the optical rotation of the liquid or solution.

The magnitude of the optical rotation depends upon (i) the nature of the substance, (ii) the length of the column of liquid through which the light passes, (iii) the wavelength of the light employed, (iv) the temperature and (v) the concentration of the optically active substance, if a solute. In order to obtain a measure of the rotatory power of a substance, these factors must be taken into account. As a rule the wavelength employed is either that for the sodium D line, 5893 Å (obtained with a sodium vapour lamp) or the mercury green line, 5461 Å (produced with a mercury vapour lamp provided with a suitable filter). The temperature selected is 20° or that of the laboratory t °C. The **specific rotation for a neat active liquid** at a temperature t for the sodium line is given by:

$$[\alpha]_D^t = \frac{\alpha}{ld}$$

---

* The neat liquids or solutions must be completely free of suspended particles or droplets of immiscible solvents.

where α is the angular rotation, l is the length of the column of liquid in decimetres and d is the density at a temperature t. The **specific rotation for a solution of an optically active substance** is likewise given by:

$$[\alpha]_D^t = \frac{100\alpha}{lc} = \frac{100\alpha}{lpd}$$

where l is the length of the column of liquid in decimetres, c is the number of grams of the substance dissolved in 100 ml of the solution, p is the number of grams of the substance dissolved in 100 g of the solution and d is the density of the solution at the temperature t. In expressing the specific rotation of a substance in solution, the concentration and the solvent used (the nature of which has an influence on the rotation) must be clearly stated.

The **molecular rotation** is given by the expression:

$$[M]_D^t = [\varphi]_D^t = \frac{[\alpha]_D^t \times M}{100} = \frac{\alpha}{lc \ (\text{mol}/100 \ \text{ml})}$$

where M is the molecular weight. For example, natural camphor has $[\alpha]_D^{20} - 44.3$ (c 3.6 in EtOH) and $[M]_D^t - 67.3$.

Measurement of optical rotation is much more convenient and accurate using one of the modern photoelectric instruments. These detect the balance point electronically and the optical rotation at a single wavelength may be read on a micrometer scale or may be displayed on a digital readout. Such instruments (e.g., the Perkin-Elmer Model 141) can measure the optical rotation of a few milligrammes of sample to an accuracy of 0.001°. The Model 141 M can be used at the sodium D wavelength or at any one of the twenty-one available mercury emission wavelengths. Using alternative light sources, such an instrument may be used to measure optical rotation at any point in the range 250–650 nm enabling optical rotatory dispersion (ORD) curves to be plotted. An instrument which is used to record ORD curves is described as a 'spectropolarimeter'. Examples of automatic recording spectropolarimeters for the measurement of ORD and circular dichroism (CD) curves in the wavelength range 185–650 nm are the Cary Model 60* and the Jasco Model J-20.† Reference 81 cites a selection of authoritative accounts of the value of ORD and CD data in studies on the structure of organic molecules.

I,38. DETERMINATION OF DENSITY AND OF REFRACTIVE INDEX

The density and the refractive index of a liquid are frequently quoted physical constants, which together with the boiling point, chromatographic and spectral characteristics, provide the means which aid the characterisation of organic liquids.

Density.   The density of a liquid is conveniently determined with the aid of a pycnometer (Fig. I,169). The bulb has a capacity of 1–2 ml and the capillary arms have a bore of about 0.5 mm; a mark A is made with a fine file or diamond (in the position indicated) for the adjustment of the level of the liquid in the pycnometer. A thin silver loop is provided for supporting the pycnometer on the hook over the balance pan. This pycnometer is readily filled by

---

* Varian Associates Ltd.
   † Japan Spectroscopic Co. Ltd, British Agents, Laser Associates Ltd.

means of the device shown in Fig. I,170 (colloquially known as a 'snake'); it consists of a short length of narrow (about 3 mm bore) rubber tubing into one end of which is fitted a piece of glass tubing as shown in the figure. The pycnometer * is cleaned by successively drawing into and expelling from the bulb a solution of detergent, distilled water, alcohol and finally sodium-dried ether. It is dried by attaching it to a filter pump and sucking a current of air filtered by passage through a tube containing a short plug of cotton wool through it for a period of 10–15 minutes; the outside of the pycnometer is then

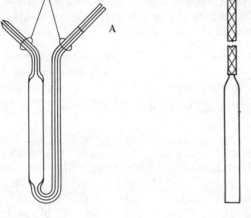

Fig. I,169   Fig. I,170

carefully polished with a paper tissue and the pycnometer is weighed. The pycnometer is filled slightly beyond the file mark A with the liquid whose density it is required to determine, by inserting the tip under the surface of the liquid which is contained in a small glazed crucible or beaker (3–5 ml capacity) and by applying gentle suction at the glass tube of the 'snake' by appropriate means; † no air bubbles should be present if the pycnometer has been carefully filled. The pycnometer is now placed in a boiling tube full of water so that the pycnometer arms rest on the rim, and the tube is suspended in a thermostat maintained at 20 °C or other suitable temperature, so that the pycnometer arms are just clear of the water in the thermostat. As the temperature of the liquid rises expansion will occur and droplets are carefully removed with filter-paper. After about 10–15 minutes the level of liquid in the pycnometer is adjusted to the mark A by touching the short arm with a filter-paper. This will cause the liquid to move along the capillary; immediately it reaches the mark A, the filter-paper is sharply removed. With a little practice, no difficulty will be experienced in filling the pycnometer accurately to the mark A. The pycnometer is removed from the thermostat, dried and polished with a paper tissue, and weighed. The liquid is then emptied into the crucible or other vessel by attaching the snake to the longer arm and blowing gently. After cleaning and

---

* It will be necessary to clean the pycnometer with chromic acid cleaning mixture at frequent intervals.
† An empty wash-bottle, equipped with a two-holed cork, may be interposed between the suction device and the pycnometer for corrosive or poisonous liquids.

drying, the operation is repeated using distilled water recently boiled to remove air. The density of the liquid can now be calculated in the following way.

The relative density (d_t^t) of a liquid may be defined as the ratio of the weight of the liquid to that of an equal volume of water at the same temperature. Thus:

$$\text{Relative density, } d_{20^\circ}^{20^\circ} = \frac{\text{Weight of liquid at 20 }^\circ\text{C}}{\text{Weight of an equal volume of water at 20 }^\circ\text{C}} = \frac{W_l^{20^\circ}}{W_w^{20^\circ}}$$

The density $(d_{4^\circ}^{20^\circ})$ may be regarded as the relative density referred to an equal volume of water at 4°, i.e.,

$$d_{4^\circ}^{20^\circ} = \frac{W_l^{20^\circ}}{W_w^{20^\circ}} \times D^{20^\circ} = W_l^{20^\circ} \times \left[\frac{0.9982}{W_w^{20^\circ}} \right]$$

where D^{20° is the density of water at 20 °C, i.e., 0.9982 g cm^{-3}. The quantity enclosed in the bracket is the constant for the pycnometer and should be recorded permanently. In all subsequent determinations of the density, only the weight of the liquid filling the pycnometer will be required. It is advisable, however, to redetermine the constant periodically.

Refractive index. The refractive index of a liquid is conveniently determined with an **Abbé refractometer**. This refractometer possesses the following advantages:

(a) The refractive index (1.3000 to 1.7000) may be read directly on a scale with an accuracy of about 0.0002.

(b) It requires only a drop of the sample.

(c) A source of monochromatic light is not essential; by means of a compensator the observed refractive index corresponds to that obtained with the D line of sodium even though white light is used as a source of illumination.

The principle of the instrument is the observation of the 'critical angle' for total reflection between glass of high refractive index (e.g., flint glass, n_D 1.75) and the substance to be examined. The glass is in the form of a right-angled prism upon the hypotenuse face AB of which the compound to be investigated is placed as a thin film (about 0.15 mm thick) and then covered by a second similar prism (Fig. I,172). The face AC of the prism plays a part in the refraction of the light, and it is the angle of emergence (α) from this face which is measured, the scale of the instrument being, however, divided to read the refractive index directly. The ray shown in Fig. I,171 and in Fig. I,172 is that which enters the face AB at grazing incidence, and corresponds to the edge of

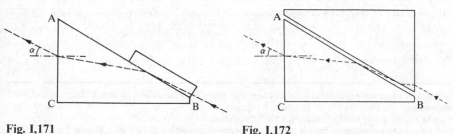

Fig. I,171   **Fig. I,172**

the dark part of the field of view of the instrument. The direction of the ray after entering the face AB depends upon its wavelength, and thus the scale of refractive index will vary with the light employed. That selected is for sodium light, but in order to permit the use of white light, the resultant dispersion of the light emerging from the face AC is neutralised by means of a dispersion compensator situated at the base of the telescope. It consists of two direct

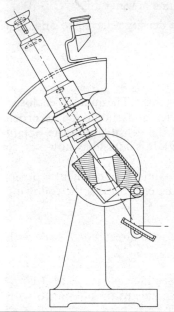

vision prisms, made accurately direct for the D sodium line, which are capable of rotation at equal rates and in opposite directions about the axis of the telescope of the refractometer. They form a system of variable dispersion which can be made equal in amount and of opposite direction to the resultant dispersion of the refractometer prism and the substance investigated.

Figure I,173* is a line diagram of the Hilger Abbé refractometer, whilst Fig. I,174* is a photograph of the instrument showing the prism box open.

To determine the refractive index of a liquid at 20 °C, circulate water at 20 °C from a thermostat through the jacket† surrounding the two prisms until the temperature on the thermometer has remained steady for at least 10 minutes. Separate the prism jackets by opening the clamp, and move the index arm, if necessary, until the face of the prism is horizontal. Wipe the latter with a paper tissue, place a drop or two of the liquid on

Fig. I,173

the ground glass prism face and then clamp it to the upper polished prism. Focus the cross-wires of the telescope by rotating the eyepiece, and then adjust the mirror so as to give good illumination from a suitably placed frosted electric lamp. By means of the rack and pinion controlling the arm at the side of the apparatus, turn the prism box until the field of view becomes partly light and partly dark. When white light is used, the edge of the light band will show a coloured fringe. By means of the milled screw at the base of the telescope, rotate the dispersion compensator until the coloured fringe disappears and the light (or dark) band is bounded by a sharp edge. Now rotate the prism box slowly until the sharp edge coincides with the intersection of the cross-wires in the telescope, and read off directly the refractive index for the D sodium line on the divided arc by means of the magnifying lens. Immediately the determination has been completed, wipe off the organic liquid with a paper tissue and clean the prism surfaces with a tissue soaked in acetone. The accuracy of the instrument may be checked by measuring the refractive index of distilled water ($n_D^{10°}$ 1.3337, $n_D^{20°}$ 1.3330, $n_D^{30°}$ 1.3320, $n_D^{40°}$ 1.3307.

---

* The author is indebted to Rank Hilger for these two figures. A detailed description of the instrument will be found in their booklet *Instructions for the Use of the Abbé Refractometer*, and to which the reader is referred for further particulars.

† The author employs a Stuart centrifugal pump supplied by Stuart and Turner Ltd.

Fig. I,174

The refractive index of a liquid is recorded as $n_D^{t°}$ where t is the temperature usually 20° or 28° at which the measurement is made, and D refers to the wavelength of the D line of sodium.

I,39. SPECTROSCOPIC METHODS The recognition of the presence of functional groups and of the nature of structural and stereochemical features in chemical compounds has been revolutionised by the availability of instruments able to measure rapidly and conveniently the absorption by molecules of electromagnetic radiation in the ultraviolet, visible, infrared and radiofrequency regions, the relevant instrumental techniques being **ultraviolet-visible spectroscopy** (or *electronic spectroscopy*), **infrared spectroscopy** (or *vibrational spectroscopy*) and **nuclear magnetic resonance spectroscopy**. Absorption of ultraviolet and visible radiation is associated with changes in the energy states of the valence electrons of the molecules, infrared absorption is associated with the energy differences between the possible vibrational states of the molecules, and energy changes involved in nuclear magnetic resonance absorption are associated with changes in the orientation of atomic nuclei in an applied magnetic field.

The availability of low-cost spectrophotometers* for the measurement of absorption of electromagnetic radiation in the ultraviolet-visible and in the infrared regions have resulted in the wide use in recent years of these instruments in most laboratories. Intermediates in synthetic sequences and purified reaction products are routinely examined to adequately characterise them by comparison of their spectra with those of authentic material, or by careful interpretation of the significance of the main absorption frequencies with reference to correlation tables, which list the characteristic frequencies associated with the most commonly encountered bonding systems. In qualitative analysis the recognition of structural features by these spectroscopic methods supplements the information derived from chemical tests and from derivative preparations; the use of these techniques extends further to include monitoring the progress of chemical reactions, quantitative analysis of mixtures of chemical compounds and studies on the effects of structural modifications on the characteristic group absorption frequencies (e.g., solvent interactions, etc.).

Until recently instruments for the determination of nuclear magnetic resonance spectra were costly and required considerable skill to yield satisfactory spectra; the laboratory worker usually therefore submitted his samples to some central service for routine examination. However, a new series of instruments† offers promise that in a few years nuclear magnetic resonance spectrometers will, with the ultraviolet-visible and infrared spectrophotometers, be readily available for routine measurements. Most qualified laboratory workers nevertheless need to be able to interpret appropriate n.m.r. spectra in relation to the structural features of the compounds under study.

In a similar way the routine interpretation of mass spectrometric data by the laboratory worker is usually to be expected, even though the measurement of such spectra requires the facilities of a central expert service.

For these reasons, this section will be devoted to the techniques of measure-

---

* Those available from, for example, Beckmann-RIIC Ltd; Perkin-Elmer Ltd; Pye-Unicam Ltd; Rank Hilger.
† Those now developed by, for example, Varian Associates Ltd.

ment of ultraviolet-visible and infrared spectra; the interpretive aspects of spectroscopic techniques will be deferred to Section **VII,4,***1–4*.

The electromagnetic spectrum. Units. The wavelengths of electromagnetic radiation of interest in qualitative and quantitative analysis vary from metres for the radiofrequency range to about 10^{-8} cm for X-rays. A wave has associated with it both wavelength, λ, and frequency, ν, which are related by the equation:

$$\nu\lambda = c$$

where c is the velocity of the electromagnetic radiation (3×10^{10} cm sec^{-1}). Hence the wavelength is the distance between adjacent crests while the frequency is the number of crests which pass a fixed point in a given time.

The specific regions and the phenomena they produce are correlated with the wavelength and the frequency in Table I,17.

Table I,17   Regions of the electromagnetic spectrum

Spectral region	Wavelength	Frequency in wave numbers (cm^{-1})	Special phenomena
Gamma rays	0.0001–0.01 nm		Nuclear reactions
X-rays	0.01–2 nm		Inner electron transitions
Vacuum ultraviolet	2–200 nm	5 000 000–50 000	Ionisation of atoms and molecules
Ultraviolet	200–400 nm	50 000–25 000 ⎱	Outer electron transitions
Visible	400–750 nm	25 000–13 333 ⎰	
Infrared	0.75–25 μm	13 333–400	⎧ (Stretching) Molecular vibrations
Far infrared	25 μm–1 mm	400–10	⎩ (Bending)
Microwave	1 mm–30 cm	10–0.033	⎰ Molecular rotation ⎱ Electron spin resonance
Radio ⎧ Short wave	10–50 m		Nuclear magnetic resonance
Medium wave	190–555 m		Nuclear quadrupole
Long wave	1000–2000 m		resonance

The limitations on the extent of the various regions given above are, of course, arbitrary.

Because of the great difference in wavelength of the various regions, it is inconvenient to use the same units throughout to specify a particular position in the spectrum. In the ultraviolet-visible regions the wavelengths are expressed in nanometers (nm, 10^{-9} m; formerly this wavelength unit was called a millimicron, mμ). In the infrared region the wavelengths are expressed in micrometers (μm, 10^{-6} m; formerly this wavelength unit was called a micron, μ), or as the reciprocal wavelength in centimetres, $1/\lambda$, termed the wave number, $\bar{\nu}$. In the radiofrequency region absolute frequencies are used rather than the wave numbers. For example, a wavelength of 5 metres corresponds to a frequency of c/λ or 6×10^7 Hz (hertz, defined as cycles per second) which may be written as 60 MHz.

Instrumental features of ultraviolet-visible and infrared spectrometers. The essential components of any spectrometer are a *source* of *radiant energy* covering the entire region to be measured; a *monochromator* and *slit system* to isolate monochromatic or narrow wavelength bands of radiant energy derived from the source; a compartment to hold both a *sample cell* and a *reference cell* (frequently the sample is dissolved in a suitable solvent and the reference cell therefore contains the neat solvent); a *detector* to differentiate between the intensity of the signals emerging from the reference cell and the sample cell; an *amplifier* for increasing the magnitude of the resultant signal to such a level that it may be registered on a *meter* (in the case of manual instruments), or converted by means of a *pen recorder* to a graphical trace (in the case of automatic recording instruments). In the latter case a revolving drum or movable table upon which the *chart paper* is mounted is mechanically geared to the monochromator or its associated mirror system, so that the positions of the pen at specified wavelengths on a properly located chart paper correspond to the wavelength of radiant energy passing through the sample and reference cells.

Commercial instruments for ultraviolet spectrophotometry usually also cover the visible region and therefore have two light sources—a deuterium or hydrogen discharge tube for the region 200–370 nm, and a tungsten filament lamp for the region 325–750 nm; with recording instruments there is an automatic interchange at 370 nm. The monochromator incorporates a quartz prism or diffraction grating. In normal circumstances the lower limit of measurement is 190 nm owing to the fact that oxygen absorbs radiation below 190 nm, and quartz becomes less transparent in this region. Below 190 nm, measurements require the use of diffraction gratings and special vacuum techniques. The cell compartment is a light-proof box which contains, in the manual instruments, a platform carrying a cell holder by which the cells containing the sample solution and reference solvent may be successively brought in turn into the beam of light. The cells are made entirely of quartz and are available in sizes which provide a path length (i.e., the length of sample in the beam of radiation) varying from 0.5 to 10 cm. In automatic recording instruments (the so-called *double beam recording spectrophotometers*), the light beam is either split into two parallel beams which pass through the sample and reference cells and thence to the detector system, or is allowed to pass alternately and automatically through first one cell and then the other and then to the detector system. The detector is a photoelectric cell (manual instruments) or photomultiplier (recording instruments). Typical commercial recording instruments which have found satisfactory use in these laboratories are the **Unicam SP.800*** and the **Spectronic 505.**†

The source of radiant energy in infrared spectrophotometers is a glowing ceramic rod maintained at approximately 1700 °C (Nernst filament). Although many spectrophotometers which are used for the routine examination of samples employ a sodium chloride prism,‡ the best results for high resolution spectra are achieved by the use of two diffraction gratings to cover the ranges 4000–1300 cm^{-1} (2.5–7.7 μm) and 2000–650 cm^{-1} (5.0–15.4 μm) which may be selected either at the control panel of the instrument or by an automatic interchange at 2000 cm^{-1}. All modern instruments are double beam recording spectrophotometers and the energy difference of the beams emerging from the

---

* Supplied by Pye Unicam Ltd.

† Supplied by Bausch & Lomb.

‡ Lithium fluoride, calcium fluoride and potassium bromide prisms are used to study with high resolution the absorption characteristics of compounds in specified regions (usually in conjunction with diffraction gratings), e.g., 4000–1700, 4200–1300, 1100–385 cm^{-1} respectively.

reference cell and the sample cell is measured by the optical null method. In this method the resultant signal from the detector is amplified and used to drive mechanically a comb device (known as an attenuator) into the reference beam to reduce its intensity to that of the sample beam; at this point the detector emits no signal and movement of the attenuator stops. The detector is a device capable of measuring small differences in temperature of the two beams and may be either a thermistor, a thermocouple or a Golay cell. The degree of compensation to the reference beam to balance it with the sample beam is of course a measure of the absorption by the sample. This method is in contrast to the ratio-recording method in which the detector compares directly the intensity of the two emergent beams. The cells are of a different design to those used in ultraviolet spectroscopy and are described later; the cell windows are made of compressed sodium chloride (most usually), or of potassium bromide, silver chloride or caesium bromide. Typical commercial recording instruments which have found satisfactory use in these laboratories are the **Unicam SP.200,*** the **Perkin-Elmer 257** and the **Perkin-Elmer 337.†**

Determination of ultraviolet-visible spectra. When a molecule absorbs ultraviolet or visible light of frequency v or wavelength λ, an electron undergoes a transition from a lower to a higher energy level in the molecule. The energy difference ΔE is given by the expression:

$$\Delta E = hv = \frac{hc}{\lambda}$$

where h is Planck's constant and c the velocity of the radiation. Multiplication by Avogadro's number, N_A $(6.02 \times 10^{23} \text{ mol}^{-1})$, will express the energy absorbed per mole. By inserting the numerical values for h $(6.63 \times 10^{-27} \text{ erg sec})$, for c $(3 \times 10^{10} \text{ cm sec}^{-1})$ and using the conversion factor (4.184×10^{10}) to convert ergs into kcal, the expression becomes:

$$\Delta E_{\text{kcal mol}^{-1}} = \frac{N_A hc}{\lambda} = \frac{6.02 \times 10^{23} \times 6.63 \times 10^{-27} \times 3 \times 10^{10}}{4.184 \times 10^{10} \times \lambda_{\text{cm}}}$$

$$= \frac{28.6 \times 10^{-4}}{\lambda_{\text{cm}}}$$

$$= \frac{28.6 \times 10^{3}}{\lambda_{\text{nm}}}$$

$$\text{or } \Delta E_{\text{kJ mol}^{-1}} = \frac{28.6 \times 4.184 \times 10^{3}}{\lambda_{\text{nm}}}$$

For the region 200–750 nm, therefore, the energy required for electron transitions is in the range 600–160 kJ mol^{-1}; for the ultraviolet region 200–400 nm, the energy is of the same order of magnitude as the bond energies of common covalent bonds (e.g., $C-H$ bond energy is ≈ 410 kJ mol^{-1}). For this reason prolonged exposure of the sample to ultraviolet radiation during measurement should be avoided to minimise possible decomposition of a proportion of the sample.

---

* Supplied by Pye Unicam Instruments Ltd.
† Supplied by Perkin-Elmer Ltd.

Energies of these magnitudes are associated with the promotion of an electron from a non-bonding (n) orbital or a π-orbital, to an antibonding π-orbital (π^*) or to an antibonding σ-orbital (σ^*). The most important transitions in organic compounds are:

(i) $\pi \to \pi^*$ transitions; these are usually associated with the multiple bonds of carbon with carbon, nitrogen, oxygen, sulphur, etc., and they generally give rise to high intensity absorption;

(ii) $n \to \pi^*$ transitions; these are usually associated with groups such as carbonyl, thiocarbonyl, nitroso, etc., and generally the intensity of absorption is very much lower than that arising from (i).

In the spectra of simple molecules the absorptions due to $n \to \pi^*$ transitions lie at longer wavelengths than those arising from the $\pi \to \pi^*$ excitations which is of course a measure of the lower energy required for electron promotion in the former case. The absorption occurs over a range of wavelengths about discernible maxima leading most frequently in solution to a broad absorption curve. This is because the spacing between the rotational and vibrational transitions of a polyatomic molecule are relatively small (about 0.5–4.2 kJ mol^{-1}) and electron transitions occur (with corresponding slight differences of energy) from a range of vibrational-rotational levels in the ground state to a range of such levels in the excited state.

Laws of light absorption. The Beer–Lambert law states that the proportion of light absorbed by a solute in a transparent solvent is independent of the intensity of the incident light and is proportional to the number of absorbing molecules in the light path:

$$\log_{10}\left(\frac{I_0}{I}\right) = A = \varepsilon c l$$

where

I_0 = intensity of incident light
I = intensity of transmitted light
ε = molar absorptivity or molar extinction coefficient
c = concentration of solute in moles/litre
l = cell (path) length (cm)
A = absorbance

It will therefore be seen that ε is a measure of the absorbance of the solution at a concentration of 1 mole per litre in a 1 cm cell. Beer's law is a limiting law and is strictly valid only at low concentrations.

When the molecular weight (M) of the absorbing substance is unknown, the extinction coefficient of a 1 per cent solution in a 1 cm cell ($A_{1\,cm}^{1\%}$) is generally used for comparison of absorption intensities:

$$A_{1\,cm}^{1\%} = \frac{A}{cl}$$

where c is now in grams per 100 ml and l is in cm. It is related to the molar absorptivity by the expression

$$\varepsilon = A_{1\,cm}^{1\%} \times \frac{M}{10}$$

Both ε and $A_{1\,cm}^{1\%}$ are independent of concentration or cell length provided the Beer–Lambert law is obeyed; the latter constant does not involve the molecular weight and is therefore used for compounds of unknown or uncertain constitutions.

The ultraviolet spectrum is a plot of the wavelength or frequency of absorption against the absorbance ($\log_{10} I_0/I$) or the transmittance (I/I_0). Spectral data are also presented in which the absorbance is expressed as the molar absorptivity, ε or $\log \varepsilon$, i.e., as a graphical plot of λ versus ε or $\log \varepsilon$. The intensity of an absorption band in the ultraviolet spectrum is usually expressed as the molar absorptivity (ε) at maximum absorption (λ_{max}). The smaller the difference between ground and excited states the longer will be the wavelength of absorption. The latter follows from the expression $\lambda = hc/\Delta E$. Thus absorption of light in the visible region, which is responsible for the colour of certain compounds, involves a lower energy transition as compared with the absorption of light in the ultraviolet region.

Solvents for ultraviolet spectroscopy. Organic compounds generally absorb too strongly for their ultraviolet spectra to be directly determined, and dilute solutions must be prepared with solvents that are transparent to ultraviolet light over the wavelengths of interest. Fortunately a suitable solution of a compound whose ultraviolet spectrum is required to be recorded can usually be prepared, since a reasonably wide selection of solvents which are transparent down to about 205 nm is available. These include hexane,* heptane,* cyclohexane,* iso-octane (2,4,4-trimethylpentane),* chloroform,* tetrahydrofuran, 1,4-dioxan,* propan-2-ol,* ethanol,* methanol* and water.* In the far ultraviolet, suitable solvents are hexane and heptane. Those solvents which are commercially available as 'spectroscopically pure' are marked with an asterisk, although for many purposes analytical grade reagents are satisfactory if the cell length is small. It should be noted that the value of λ_{max} may be dependent upon solvent polarity, and hence the solvent used for the spectral determination should always be specified.

Solution preparation. To obtain accurate absorbance values in the region of maximum sensitivity of the spectrophotometer it is usual to prepare a solution having a concentration which would give an A value at λ_{max} in the region of about 0.5 for the manual instrument, or about 0.9 for the automatic recording instrument. For a compound having an ε value of the order of 15 000, as is found for example in the case of crotonaldehyde (λ_{max} 220 nm, ε 15 000, M 70) and using a path length cell of 1 cm, substitution in the Beer–Lambert equation:

$$\log_{10} I_0/I = \varepsilon c l$$

gives $\quad 0.5 = 15\,000 \times c \times 1;$

when $\quad c = \dfrac{0.5}{15\,000} = 3.33 \times 10^{-5} \text{ mol dm}^{-3}$

or $\quad = \dfrac{70 \times 0.5}{15\,000} = 2.33 \times 10^{-3} \text{ g dm}^{-3}$

A solution of this concentration is most usually prepared by weighing (say) 23 mg (or 2.3 mg) of the substance and dissolving it in 100 ml (or 10 ml) of solvent using a graduated flask; 1 ml of the solution is then diluted exactly to 100 ml with the same solvent.

Optical cells and their care. Although cells constructed of glass are suitable for the determination of spectra in the visible range, this material is not sufficiently transparent in the ultraviolet region and quartz cells must be used. These commonly have path lengths of from 0.5–10 cm; a 1-cm-square cell requires about 3 ml of solution.

Cells may become contaminated as a result of evaporation of solvent from solutions, and also by acquiring a film of grease as a result of careless handling. Such films and dust particles decrease transmission and can also contaminate liquids placed subsequently in the cell. Immediately after use therefore cells must be emptied and rinsed with clean solvent and then cleaned with a suitable detergent solution and stored in distilled water. A brush which might scratch the optical surface should *never* be used. Solid contaminants must always be removed by the following wet cleaning procedure:

(a)   stand the cell in cold detergent for 15 minutes;
(b)   rinse several times with distilled water;
(c)   rinse with ethanol and store in covered containers containing distilled water or dry under a radiant lamp.* Do not allow the cell to dry until the cleaning procedure is complete.

The following precautions in handling cells should also be observed:

(a)   cells should only be handled by means of the etched surfaces;
(b)   when wiping the outside surfaces of the cell, prior to placing in the instrument, paper tissues only should be used;
(c)   cells should be removed from storage under water with the aid of suitably protected tongs.

The cells are filled bearing in mind the following points. The clean dry cell is rinsed with the appropriate solvent and then with the prepared solution before being finally filled. For solutions made up with volatile solvents the filled cell should be closed with the fitted lid provided; a reference cell is similarly filled with the neat solvent. If a cell has been stored under water it is first rinsed well with the appropriate solvent and then with a little of the prepared solution before being finally filled. If the solvent used is immiscible with water a preliminary washing with ethanol is necessary.

For accurate work a pair of matched cells should be used and each should be placed in the instrument so that the incident radiation enters via the same optical face every time.

It is not considered appropriate here to give the details of the operation of the spectrometer; the manual provided by the manufacturers for a particular instrument available should be consulted.

Determination of infrared spectra. Infrared radiation refers broadly to the wavelength region 0.5–1000 μm. The limited portion of infrared radiation between 2 and 15 μm (5000–660 cm^{-1}) is of greatest practical use to the organic chemist.

Infrared radiation is absorbed and converted by an organic molecule into energy of molecular vibration. There are two types of molecular vibration;

---

* Supplied by A. Gallenkamp & Co. Ltd.

stretching and bending. A stretching vibration is a motion along a bond axis such that the distance between the two atoms is continuously decreasing and increasing. A bending vibration involves a change in bond angles.

For interaction to be possible between the electromagnetic radiation and the bonding system of a molecule, leading to uptake of energy and therefore to an increase in the amplitude of the appropriate stretching or bending vibration, two conditions have to be met:

(a) there must be a change in the charge distribution within the bond under-going stretching or bending, i.e., the dipole moment of the bond must vary during vibration so that interaction with the alternating electric field of the radiation is possible;
(b) the frequency of the incident radiation must exactly correspond to the frequency of the particular vibrational mode.

As the frequency range is scanned the various infrared active vibrations (i.e., those involving a dipole moment change) will sequentially absorb radiation as the energy equivalence of the radiation and the particular vibrational mode is met, giving rise to a series of absorptions.

A single vibrational energy change is accompanied by a number of rotational energy changes and consequently vibrational spectra occur as bands rather than as lines. Absorption band maxima are presented either in wavelengths (micrometers, μm) or in wave numbers ($\bar{v}$, expressed as reciprocal centimetres, cm^{-1}). Band intensities are expressed either as the transmittance T (the ratio of the radiant power transmitted by a sample to the radiant power incident on the sample I/I_0) or the absorbance A ($\log_{10} I_0/I$). The intensities of the absorption bands in an infrared spectrum are usually indicated qualitatively, e.g., as very strong (vs), strong (s), medium (m) or weak (w), etc.

Sample preparation. The infrared spectrum of a *liquid* may conveniently be recorded as a thin film of the substance held in the infrared beam between two infrared-transparent discs without the need for a diluting solvent. It is customary to use polished plates of sodium chloride as the support material; this material is adequately transparent in the region 2–15 μm. Spectra in the longer wavelength region (12–25 μm) can be recorded using potassium bromide plates. Sealed cells (p. 254) should be used for volatile liquids.

Great care must be exercised in the handling and use of these plates since in particular traces of moisture will cause polished surfaces to become 'fogged', thus causing undesirable scattering of the transmitted radiation. Plates are stored in air-tight containers containing small bags of a suitable desiccant, e.g., silica gel; they should only be handled by the edges and as far as possible under a radiant heater. After use the plates should be first wiped with a paper tissue, rinsed with a jet of methylene chloride, wiped again with a tissue and finally allowed to dry under a radiant heater. Washings should be collected in a suitable container for subsequent recovery. Plates which have become fogged through misuse need to be carefully repolished using one of the commercially available polishing kits.

In order to determine the spectrum of a neat liquid sample, a capillary film of the pure dry material is formed between a pair of plates by carefully placing three small drops on the polished surface of one plate, covering them with the second plate and exerting gentle pressure with a slight rotatory motion to ensure that the film contains no air bubbles. The prepared plates are then placed in the demountable cell holder* (Fig. I,175), ensuring that the gaskets are properly located, and the quick release nuts are firmly screwed down (but not too tightly, otherwise the liquid will exude from between the plates). The whole assembly is then located in the sample beam path of the infrared spectrophotometer.

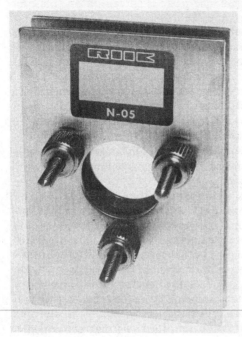

Fig. I,175

With most instruments, before the spectrum is actually recorded, it can readily be ascertained whether the thickness of the film is adequate to provide a satisfactory spectrum by scanning the spectrum rapidly, noting the movement of the pen recorder without allowing it to make a permanent trace. This may sometimes be done by moving the chart paper manually. If it is judged that the band intensities are too high, the cell must be dismantled and a thinner film prepared. When the spectrum has been recorded, calibration of the wavelength scale may be checked by superimposing on the recorded spectrum characteristic peaks from the spectrum of a polystyrene film which is inserted into the instrument in place of the sample cell. Suitable intense bands of the spectrum occur at 3027, 2851, 1602, 1028 and 907 cm^{-1}.

Solids are generally examined as a **mull** or in a pressed **alkali halide disc** (usually potassium bromide). For mulls, Nujol (a high boiling fraction of petroleum) is most commonly used, although when it is desired to study frequency ranges in which Nujol absorption bands appear, Fluorolube (perfluorokerosene, a mixture of fluorinated hydrocarbons), or hexachlorobutadiene is

---

* Supplied by Beckmann-RIIC Ltd who kindly provided the photograph.

employed. The infrared spectra of the mulling agents should be recorded and kept available for reference purposes.

Mulls are prepared by grinding about 2.5 mg of the solid sample with one or two drops of Nujol in a small agate pestle and mortar. The mixture must be thoroughly ground for at least five minutes to ensure that a fine particle size is obtained, so that the light scattering is reduced to a minimum and damage to the halide plates by scratching is avoided. The paste is spread on one plate of the demountable cell, covered with the other and the sample thickness varied by rotating and squeezing the plates to force out excess material. The plates are inserted into the cell holder and placed in the sample beam path of the spectrophotometer as described above. Rapid scanning of the spectrum in conjunction with a knowledge of the spectrum of the relevant mulling agent will give an indication as to whether the concentration of the sample in the mull is sufficient to provide a satisfactory spectrum. If necessary the concentration should be adjusted either by adding more of the sample followed by regrinding or by diluting with more of the mulling agent.

In the pressed disc technique a known weight of sample is intimately ground with pure, dry potassium bromide and the mixture inserted into a special die and subjected to pressure under vacuum. The concentration of sample in the disc is usually in the region of 1.0 per cent. The disc so produced may be mounted directly in the sample beam path of the spectrophotometer and the spectrum recorded. This method has the advantage that the spectrum so produced is entirely due to the sample since pure dry potassium bromide is infrared transparent in the 2–25 μm region. To eliminate the possibility of impurities in the potassium bromide, however, a blank disc (no sample) can be made and mounted in the reference beam path of the spectrophotometer. Care should be taken to ensure that both discs are of equal thickness otherwise inverse peaks may occur if the potassium bromide is damp or impure, and this will be particularly noticeable if the reference disc is thicker than the sample disc.

The halide used must be of AnalaR grade and should be pre-powdered to a particle size which will pass through a 70-mesh sieve; sieving is not absolutely necessary provided that each batch of powder is tested to show that it does subsequently produce good discs. Pre-powdering may be carried out in a mechanical grinding machine or by hand in an agate pestle and mortar. Drying of the powder is best done by leaving it in a shallow dish in an oven at 120 °C for at least 24 hours. It may then be transferred to a loosely stoppered bottle which should be kept in a desiccator.

As a general procedure, 500 mg of pre-ground potassium bromide is weighed and mixed with the appropriate quantity of the sample (i.e., 5 mg for a 1% disc) whose spectrum is to be recorded. The mixture is further intimately ground in a vibration mill;* the time required for the grinding depends on the degree of pre-powdering but is usually in the region of 1–2 minutes.

Pressing of the disc is usually carried out in a commercially available stainless steel disc die assembly† under pressure of the order of 8–9 tons/square inch. The die is normally kept in a vacuum desiccator and placed under a radiant heater for at least 30 minutes before use. Dies must be scrupulously

---

* The Vibromill supplied by Beckman-RIIC Ltd.
† The Evacuable KBr Die supplied by Beckman-RIIC Ltd.

cleaned after use to remove all traces of alkali halide which may corrode the stainless steel. To prepare the disc proceed as follows:

(a)   push the body of the die (A) on to the base (B) (Fig. I,176);

(b)   insert one of the stainless steel pellets (C) into the barrel of the die, polished face *upwards*;

(c)   introduce a measured quantity of the ground sample (150–250 mg) into the bore of (A);

(d)   distribute the powder evenly over the polished surface of the pellet (C) by slowly introducing the plunger (D) into the barrel with careful rotation, and then rotating it a few times whilst simultaneously exerting gentle pressure with the fingers;

(e)   slowly withdraw the plunger taking care not to disturb the powder which should show a surface which is perfectly smooth and free of pits or cracks;

(f)   insert the second pellet into the barrel of the die, polished face *downwards*, and bring into contact with the powder by pressing down lightly with the plunger; complete the assembly by pushing the O-ring seal on the plunger into contact with the die surface (Fig. I,177);

(g)   connect the die to the vacuum pump and evacuate for a period of at least 3 minutes; place the die (whilst still attached to the vacuum pump) under the hydraulic press* and apply a pressure not exceeding 8–9 tons for 1 minute;

(h)   release the pressure and vacuum after this time, invert the die, and remove the base (B) (the plunger is kept in position with the fingers);

(i)   place the perspex cylinder on the top of the assembly, which is then returned to the hydraulic press, and apply sufficient pressure to lift the lower pellet and the disc clear of the barrel (Fig. I,178);

(j)   finally release the press and remove the die body A and the upper steel pellet from the face of the disc and then remove the disc itself with tweezers (the disc should never be handled with the fingers), and mount it in the specially designed holder. These latter operations are best conducted under a radiant heater. The disc holder is located in position in the sample beam path of the spectrophotometer; if required a blank potassium bromide disc, similarly prepared, is introduced into the reference beam path.

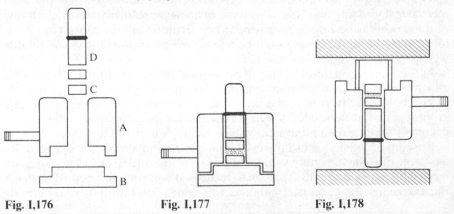

Fig. I,176   **Fig. I,177**   **Fig. I,178**

* Beckman-RIIC Ltd.

Providing that care has been taken in the disc preparation the final disc should be slightly opaque due to the presence of the sample (the blank disc should be transparent). Should the disc show a number of white spots, it is probable that the mixture has been unevenly ground. If the disc shows a tendency to flake, then excessive grinding of the powder is indicated. If after being removed from the die the disc becomes cloudy this is indicative of the uptake of water; to avoid this difficulty it is necessary to ensure that the die is evacuated for a sufficiently long period and that the removal of the disc from the die is carried out under a radiant heater.

Several commercially available small-scale, manually operated presses are available for use when the number of halide discs which require to be prepared does not warrant the purchase of the more elaborate hydraulic press. They operate on the principle of achieving the necessary pressure on the powdered sample (about 50–100 mg) by spreading it evenly between the optically polished faces of two bolts which are then screwed into opposite ends of a cylinder and tightened relative to each other. One commercially available design* in which the cylinder is evacuable is illustrated in the photograph (Fig. I,179). After removal of the bolts, the barrel containing the disc is mounted in the specially designed cell slide in the sample beam path of the spectrophotometer. A suitably adjusted metal comb attenuator must be introduced into the reference beam path to compensate for the fact that the barrel structure of the press which is held in the sample beam restricts the amount of radiation which can pass through the halide disc, otherwise even at a wavelength in which no absorption is taking place the reference and sample beams will be out of balance. The barrel and bolts are cleaned by a stream of tap water, rinsed with ethanol, then with methylene chloride and finally dried with paper tissue and

Fig. I,179

---

* The Mini-press (Wilks Scientific Corp.), available in the UK from Techmation Ltd, who kindly supplied the photograph.

placed under a radiant heater. Care should be taken not to damage the polished surfaces of the bolts by scraping off adhering particles of halide with a spatula or by allowing the faces of the bolts to come into contact within the barrel.

Solutions of either solids or liquids are normally handled in cells of 0.025 mm to 1 mm thickness using concentrations of 20 per cent to 0.5 per cent respectively in cases of compounds having a molecular weight of about 150. Compounds of higher molecular weight are examined at correspondingly higher concentrations.

Pure dry chloroform, carbon tetrachloride or carbon disulphide (Section **II,1,**6,7,32) are the solvents most commonly employed. Their selection is based upon the fact that they exhibit relatively few intense absorption bands in the region 5000–650 cm^{-1} (Fig. I,180(*a*, *b* and *c*)).

In the region of these intense absorption bands it is not possible to record the absorption due to the solute, even with the compensating effect of an identical path length of the pure solvent in the reference beam, since virtually no radiation reaches the detector system and the pen recorder is not activated.

In the region of the less intense solvent absorption bands, the use of pure solvent and solution in matched fixed path length cells enables the absorption due to the solute to be recorded satisfactorily, providing a sufficiently concentrated solution requiring a short path length cell is used. (Clearly a more dilute solution necessitating longer path length cells causes the less intense solvent absorption bands to become more prominent, thus reducing the overall transmittance in this region.)

For routine use liquid cells are of two types: (*a*) the demountable cells in which the path length may be varied by utilising spacers of lead or Teflon of appropriate thickness; these cells have the advantage that they may be easily dismantled after use for cleaning and if necessary for repolishing of the cell windows; (*b*) the fully assembled sealed cells of fixed path length (available as matched pairs) which use an amalgamated lead spacer forming a permanent leak-proof seal; these cells are particularly useful for volatile samples but they require much greater care in use since deterioration of the inside plate surface necessitates an expensive overhaul. The overall design of these two types is similar and one such example is illustrated in Fig. I,181.* The cell design incorporates inlet and outlet ports by means of which the liquid may be introduced. For this operation the cell is placed horizontally on the bench with the ports uppermost, a syringe needle is inserted into one of the ports, and the solution is injected. The passage of fluid across the cavity is easily observed and care must be taken to avoid the presence of trapped air bubbles. Teflon stoppers are inserted into the ports to keep the liquid within the cell which is then mounted in the sample beam path of the spectrophotometer. The neat solvent is similarly introduced into the matched cell and placed in the reference beam path.

The cells are emptied by attachment of one of the ports to a vacuum line incorporating a suitable trapping system, cleaned by several rinses with neat solvent and then with methylene chloride. Solvent is removed by a short period of suction and the cell is finally dried under a radiant heater and stored in a desiccator. Prolonged passage of air through the cell must be avoided otherwise fogging of the cell windows from atmospheric moisture may occur.

The most accurate way for compensating for solvent absorption is to use the

---

* Supplied by Beckman-RIIC Ltd, who kindly provided the photographs.

Fig. I,180(*a–c*) Spectra of solvents for infrared spectroscopy.

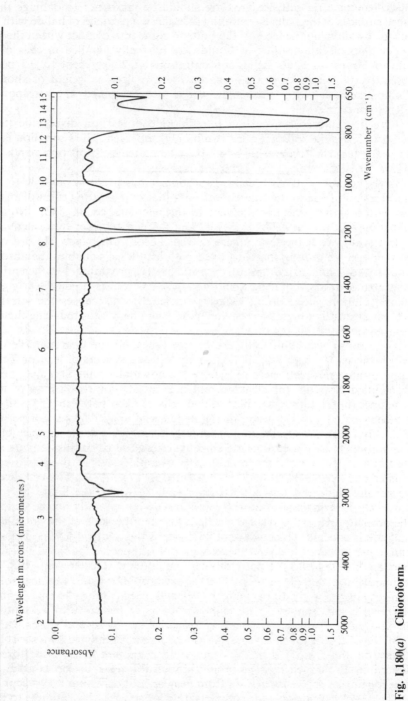

Fig. I,180(*a*) Chloroform.

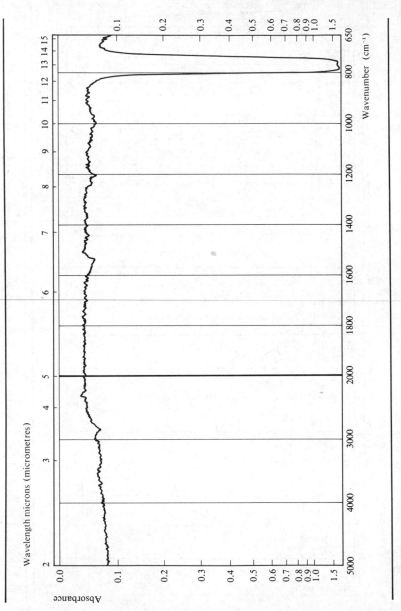

Fig. I,180(b) Carbon tetrachloride.

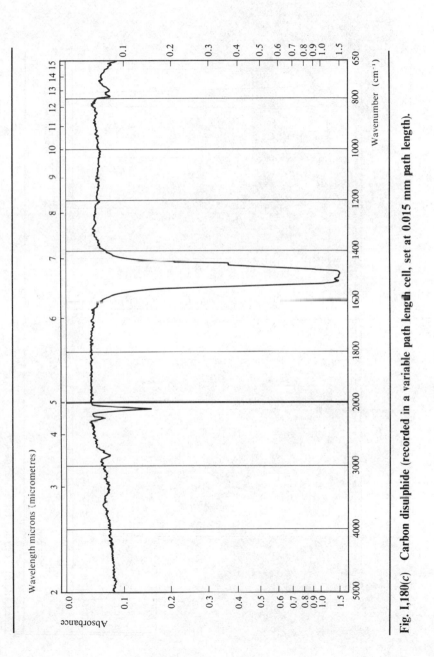

Fig. I,180(*c*) Carbon disulphide (recorded in a variable path length cell, set at 0.015 mm path length).

Fig. I,181

Fig. I,182

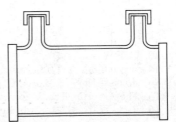

Fig. I,183

more expensive variable path length cells (Fig. I,182*) in which one of the plates which constitute the liquid cell can be moved with the aid of a micrometer device to allow adjustment to any required path length. This allows the accurate matching in the spectrophotometer of two such cells filled with the appropriate solvent. One cell may then be emptied, cleaned and refilled with solution so that the spectrum of the solute may be recorded.

On occasions it is desirable to investigate the spectroscopic properties of solutions at high dilution (i.e., 0.005 M), as for example in the study of inter- and intra-molecular hydrogen bonding of compounds containing hydroxyl groups. Such concentrations require cell path lengths in the region of 1.5 cm. The construction of a cell of these dimensions is relatively simple and a suitable design which has found use in these laboratories is shown in Fig. I,183.

BIBLIOGRAPHY AND REFERENCES

I,3. Safe working in organic chemical laboratories

1. *Hazards in the Chemical Laboratory* (1972). 2nd edn. Ed. G. D. Muir, London; The Chemical Society.
2. *Fisher Manual of Laboratory Safety* (1972). Pittsburgh; The Fisher Scientific Co.
3. *Safety in the Laboratory*. Proceedings of the Interlab Symposium at Shell Centre, Department of Trade and industry, London, June 1973.
4. *Safety in Chemical Laboratories and in the Use of Chemicals* (1971). 2nd edn. London; Imperial College of Science and Technology.
5. *Safety Manual of the University of Manchester Institute of Science and Technology*, July 1973.
6. *Safety in the Use of Compressed Gas Cylinders* (1972). London; The British Oxygen Co.
7. Ethel Browning (1965). *Toxicity and Metabolism of Industrial Solvents*. Amsterdam; Elsevier.
8. R. H. Driesbach (1963). *Handbook of Poisons*. 4th edn. Los Altos, California; Large Medical Publications.
9. N. I. Sax (1968). *Dangerous Properties of Industrial Materials*. 3rd edn. New York; Reinhold.
10. *Handbook of Toxicology* (1956-9). Vols. 1-5. Philadelphia; Saunders.
11. *Threshold Limit Values for 1973: Technical Data Note No. 2*. The Department of Trade and Industry, London; HMSO. Available from The Health and Safety Inspectorate.
12. *The Carcinogenic Substances Regulations*, 1967, No. 879, London; HMSO.
13. C. F. Searle (1970). *Chem. Brit.*, **6**, 5.
14. *Precautions for Laboratory Workers who handle Carcinogenic Aromatic Amines* (1966). London; Chester Beatty Research Institute.
15. The Asbestos Regulations, 1969, No. 690, London; HMSO.
16. P. J. Warren, 'Hazards involved in the use of Asbestos', University of London Safety Sub-Committee Memorandum, 1969. Available from The Department of Biochemistry, The London Hospital Medical College.

I,12. Cooling of reaction mixtures.

17. A. J. Gordon and R. A. Ford (1972). *The Chemists Companion*. New York; Wiley-Interscience, p. 451.
18. A. M. Phipps and D. N. Hume (1968). *J. Chem. Ed.*, **45**, 664.
19. R. E. Randeau (1966). *J. Chem. Eng. Data*, **11**, 124.

I,17,1. Catalytic hydrogenation

20. R. L. Augustine (1965). *Catalytic Hydrogenation: Techniques and Applications in Organic Synthesis*. London; Arnold. See also H. O. House (1972). *Modern Synthetic Reactions*. 2nd edn. California; Benjamin, p. 1.
21. C. A. Brown and H. C. Brown (1966). *J. Org. Chem.*, **31**, 3989.

I,17,5. Organic photochemistry

22. For example: A. Cox and T. J. Kemp (1971). *Introductory Photochemistry*. London, McGraw-Hill; R. B. Cundall and A. Gilbert (1970). *Photochemistry*. London, Nelson; N. J. Turro (1965). *Molecular Photochemistry*. New York, Benjamin.
23. J. G. Calvert and J. N. Pitts (1966). *Photochemistry*. New York; Wiley.
24. Reference 17, p. 351.
25. A. Schönberg, G. O. Schenck and O. A. Neumüller (1968). *Preparative Organic Photochemistry*. Berlin; Springer-Verlag.
26. S. D. Cohen, M. V. Mijovic, G. A. Newmann and E. Pitts (1967). *Chem. and Ind.*, 1079.
27. *Organic Photochemical Syntheses* (1971). Ed. R. Srinivasan. New York; Wiley-Interscience, Vol. 1.

I,17,6. Electrolytic (anodic) syntheses

28. S. Swann (1956). 'Electrolytic Reactions', in *Technique of Organic Chemistry*. 2nd edn. Ed. A. Weissberger, New York; Interscience, Vol. II, p. 385.
29. B. C. L. Weedon (1960). 'The Kolbé Electrolytic Synthesis', in *Advances in Organic Chemistry*. Ed. R. A. Raphael, E. C. Taylor and H. Wynberg. New York; Interscience, Vol. I, Chap. 1.
30. A. J. Fry (1972). *Synthetic Organic Electrochemistry*. New York; Harper and Row.
31. *Organic Electrochemistry* (1973). Ed. M. M. Baizer, New York.
32. G. Dryhurst and P. J. Elving (1967). *Anal. Chem.*, **39**, 607.

I,17,7. Liquid ammonia techniques

33. H. O. House (1972). *Modern Synthetic Reactions*. 2nd edn. California; Benjamin, p. 145.
34. L. Brandsma (1971). *Preparative Acetylenic Chemistry*. Amsterdam; Elsevier.

I,20. Recrystallisation techniques

35. *Dictionary of Organic Compounds* (1965). 4th edn. Ed. Sir Ian Heilbron, A. H. Cook, H. M. Bunbury and D. H. Hey. London; Eyre and Spottiswoode and E. and F. N. Spon.
36. *Handbook of Chemistry* (1961). 10th edn. Ed. N. A. Lange, New York; McGraw-Hill.
37. *Handbook of Chemistry and Physics* (1971–2). 52nd edn. Ed. R. C. Weast, Ohio; The Chemical Rubber Co.

I,21. Sublimation techniques

38. (a) A. I. Vogel (1966). *Elementary Practical Organic Chemistry*, Pt. I, Small Scale Preparations, 2nd edn. Longmans, p. 33; see also (b) R. S. Tipson (1965). 'Sublimation', in *Technique of Organic Chemistry*. Ed. A. Weissberger, Interscience, Vol. IV, p. 661, and (c) N. D. Cheronis (1954). 'Distillation, Sublimation and Extraction', in *Technique of Organic Chemistry*. Ed. A. Weissberger, Interscience, Vol. VI, p. 84.

39. G. Broughton (1956). 'Freeze Drying', in *Technique of Organic Chemistry*. 2nd edn. Ed. A. Weissberger, New York; Interscience, Vol. III, p. 831.

I,22. Solvent extraction

40. S. Glasstone and D. Lewis (1963). *Elements of Physical Chemistry*, revised 2nd edn. London; Macmillan & Co., pp. 379–82.

41. L. C. Craig and D. Craig, 'Laboratory Extraction and Countercurrent Distribution'. Reference 39, p. 149.

I,24. Distillation at atmospheric pressure

42. E. Knell (1963). *Handbook of Laboratory Distillation*. 2nd edn. Ed. E. C. Lumb, Amsterdam; Elsevier.

I,25. Steam distillation

43. C. S. Carlson and J. Stewart (1965). 'Extractive and Azeotropic Distillation', in *Technique of Organic Chemistry*. 2nd edn. Ed. A. Weissberger, New York; Wiley-Interscience, Vol. IV, p. 497.

44. Reference 38 (a), p. 16.

45. L. H. Horsley (1962). 'Azeotropic Data-II', in *Advances in Chemistry Series*. American Chemical Society Applied Publication.

I,26. Fractional distillation at atmospheric pressure

46. F. E. Williams (1965). 'Ordinary Fractional Distillation', in *Technique of Organic Chemistry*. 2nd edn. Ed. A. Weissberger, New York; Interscience, Vol. IV, p. 299.

47. F. Daniels, J. W. Williams, P. Bender, R. A. Alberty and C. D. Cromwell (1962). *Experimental Physical Chemistry*. 6th edn. New York; McGraw-Hill, p. 60.

48. F. L. J. Sixma and H. Wynberg (1964). *A Manual of Physical Methods in Organic Chemistry*. New York; Wiley, p. 107.

I,27. Distillation under diminished pressure

49. R. S. Tipson, *Distillation under Moderate Vacuum*. Reference 46, p. 511.

I,28. High vacuum distillation—molecular distillation

50. E. S. Perry. *Distillation under High Vacuum*. Reference 46, p. 535.

51. P. Ridgeway Watt (1963). *Molecular Stills*. London; Chapman & Hall.

I,31. Chromatography

General monographs

52. H. G. Cassidy (1963). 'Fundamentals of Chromatography', in *Technique of Organic Chemistry*. Ed. A. Weissberger, New York; Interscience, Vol. X.
53. *Chromatography* (1974). 3rd edn. Ed. E. Heftmann, New York; Reinhold.
54. R. Stock and C. Rice (1967). *Chromatographic Methods*. 2nd edn. London; Chapman and Hall.
55. O. Mikes (1970). *Laboratory Manual of Chromatographic Methods*. London; Van Nostrand.
56. B. L. Karger, L. R. Snyder, C. Horvath (1973). *An Introduction to Separation Science*. New York; Wiley-Interscience.

Specialised monographs

Thin-layer chromatography
57. J. G. Kirchner (1967). 'Thin-Layer Chromatography', in *Technique of Organic Chemistry*. Ed. A. Weissberger, New York; Interscience, Vol. XII.
58. K. Randerath (1966). *Thin-layer Chromatography*. 2nd edn. New York; Academic Press.
59. E. Stahl (1969). *Thin-layer Chromatography: A Laboratory Handbook*. 2nd edn. Berlin; Springer-Verlag, London; Allen and Unwin.
60. *T.L.C. Visualisation Reagents and Chromatographic Solvents*. Eastman Organic Chemicals, Rochester, New York (Kodak Publication No. JJ-5).

Paper chromatography
61. I. M. Hain and K. Macek (1963). *Paper Chromatography*. New York; Academic Press.
62. I. Smith and J. G. Feinberg (1972). *Paper and Thin-Layer Chromatography*. London; Longman Group Ltd.

Gas–liquid chromatography
63. O. E. Schupp, III (1968). 'Gas Chromatography', in *Technique of Organic Chemistry*. Ed. E. S. Perry and A. Weissberger. New York; Interscience, Vol. XIII.
64. *Practical Manual of Gas Chromatography* (1969). Ed. J. Tranchant. Amsterdam; Elsevier.
65. D. A. Leathard and B. C. Shurlock (1970). *Identification Technique in Gas Chromatography*. London; Wiley-Interscience.

Gel permeation chromatography
66. H. Determann (1968). *Gel Chromatography*. New York; Springer-Verlag.

Ion-exchange chromatography
67. F. Hefferich (1962). *Ion Exchange*. New York; McGraw-Hill.
68. H. Brockmann and H. Schodder (1941). *Ber.*, **74B**, 73.
69. R. Hernandez, R. Hernandez and L. R. Axelrod (1961). *Anal. Chem.*, **33**, 370.
70. H. H. Strain (1945). *Chromatographic Adsorption Analysis*. 2nd edn. New York; Interscience.
71. B. Loev and Marjorie M. Goodman (1967). *Chem. and Ind.*, 2026.

I,32. Countercurrent distribution

72. L. C. Craig and D. Craig, in *Technique of Organic Chemistry* (1956). 2nd edn. Ed. A. Weissberger. New York; Interscience, Vol. III, Part 1, p. 149.
73. L. C. Craig (1944). *J. Biol. Chem.*, **155**, 519.
74. L. C. Craig and B. Williamson (1947). *J. Biol. Chem.*, **168**, 687.
75. L. Alders (1959). *Liquid-Liquid Extraction*. 2nd edn. Amsterdam; Elsevier.
76. R. E. Treybal (1963). *Liquid Extraction*. 2nd edn. New York; McGraw-Hill.

I.36. Molecular weight

77. J. H. Benyon and A. E. Williams (1963). *Mass and Abundance Tables for Use in Mass Spectrometry*. New York; Elsevier.
78. J. Lederberg (1964). *Computation of Molecular Formulae for Mass Spectrometry*. San Francisco; Holden-Day.

I,37. Optical rotatory power

79. For example: J. F. Stoddart (1973). 'Stereochemistry', in *MTP International Review of Science*. Ed. W. D. Ollis, London; Butterworths, Vol. 1; E. L. Eliel (1962). *Stereochemistry of Carbon Compounds*. New York, McGraw-Hill; *Progress in Stereochemistry*. Ed. B. J. Aylett, M. M. Harris, P. B. D. de la Mare, W. Klyne, London, Butterworths, 1954–68, Vols. 1–4; *Topics in Stereochemistry*. Ed. E. L. Eliel and N. L. Allinger, Interscience, 1967–74, Vols. 1–8.
80. For example: Reference 47, p. 239; *Findlay's Practical Physical Chemistry* (1972). 9th edn. Ed. B. P. Levitt, London; Longman, p. 206.
81. For example: W. Klyne (1960). 'Optical Rotatory Dispersion and the Study of Organic Structures', in *Advances in Organic Chemistry*. Ed. R. A. Raphael, E. C. Taylor, H. Wynberg. New York, Interscience, Vol. 1, p. 239; *Optical Rotatory Dispersion and Circular Dichroism in Organic Chemistry* (1967). Ed. G. Snatzke. Heyden & Son; L. Velluz, M. Legrand, M. Grosjean (1965). *Optical Circular Dichroism*. Verlag-Chemie, GmbH, Weinheim/Bergstr, and New York, Academic Press; P. Crabbé and A. C. Parker (1972). 'Optical Rotatory Dispersion and Circular Dichroism', in *Techniques of Chemistry*. Ed. A. Weissberger and B. W. Rossiter. Wiley, Vol. 1, Part IIIc, p. 183.

I,39. Spectroscopic methods

82. The following, together with those cited on p. 1152, are examples of some of the important monographs on theoretical, practical and interpretative aspects. C. N. Banwell (1965). *Fundamentals of Molecular Spectroscopy*. London, McGraw-Hill; D. H. Whiffen (1972). *Spectroscopy*. London, Longman Group Ltd; A. D. Cross and R. A. Jones (1969). *An Introduction to Practical Infrared Spectroscopy*. 3rd edn. London, Butterworths; M. J. de Faubert Maunder (1971). *Practical Hints on Infrared Spectrometry*. London, A. Hilger; J. R. Edisbury (1967). *Practical Hints on Absorption Spectrometry (u.v. and visible)*. London, A. Hilger.

CHAPTER II SOLVENTS AND REAGENTS

II,1. THE PURIFICATION OF COMMON ORGANIC SOLVENTS
Commercially available grades of organic solvents are of adequate purity for use in many reactions provided that the presence of small quantities of water (the most widespread impurity in all organic solvents) is not harmful to the course of the reaction, and also that the presence of other impurities (e.g. ethanol in diethyl ether) are unlikely to participate in undesirable side reactions. The commercially available grades for general use are often accompanied by specifications indicating the amount and nature of any impurities present.

When however the levels of impurities, including moisture, are unacceptable for particular reactions, and when large volumes of such solvents are likely to be required, it is frequently more economic to purify the commercial grades than to purchase the more expensive AnalaR grades. Solvents of the appropriate grade of purity should also be used in isolation (extraction) and purification (recrystallisation) processes, particularly in the latter stages immediately prior to spectroscopic and/or elemental analysis.

An account of the more important common drying agents and their use in drying solutions of organic compounds has already been given (Section **I,23**). Although the drying efficiency of the individual desiccants cited differs considerably, many of them are of value for the preliminary drying procedures for the vast majority of organic solvents. This preliminary treatment is in fact essential, unless it is certain that the water content is very low, before using the more powerful drying agents (such as a reactive metal, e.g., sodium, or a metal hydride, e.g., calcium hydride, lithium aluminium hydride). These latter drying agents remove the remaining traces of water from solvents which are required for reactions necessitating strictly anhydrous conditions. Attention should be drawn to the considerable fire or explosion hazards of these highly reactive drying agents, particularly at the end of a solvent distillation when residual material has to be disposed of. Recommended methods of safely destroying sodium metal and metal hydrides are given in Section **I,3**.

While the drying agent selected for preliminary and final drying must of course have no chemical action on the organic solvent, with some solvents a specific chemical treatment is necessary to remove impurities other than water before drying is attempted. Apart from impurities arising during the manufacturing processes, many organic solvents undergo autoxidation on standing with the formation of dangerously explosive peroxides. Such solvents should always be tested for the presence of peroxides, and if present these should be removed according to the methods detailed below under individual examples, before other purification processes are attempted.

The purified and dried solvent is then distilled; separation from contaminants of similar boiling point may require the use of an efficient fractionating column, and high boiling solvents may need to be distilled under reduced pressure. It must be remembered that many of the common organic solvents are markedly toxic—benzene and the halogenated hydrocarbons are notorious examples, but many others are potentially hazardous to varying degrees (see Section I,3), and the inhalation of vapour should always be avoided. Almost all organic solvents are also inflammable; the more volatile compounds, notably ether and carbon disulphide, are particularly hazardous in consequence. Apart from taking the obvious precautions of avoiding all flames in the vicinity of a solvent distillation, it must be remembered that faulty electrical connections or even contact with hot metal surfaces may ignite the vapour of volatile solvents. The capacity of the condensing system used must be fully adequate to cope with the volume of solvent distilled, and double surface condensers are essential in many cases.

Rigorously dried organic solvents are frequently markedly hygroscopic. The distillation assembly should therefore be protected by a suitable drying tube. The redistilled material should be stored in a clean dry bottle with a well fitting stopper which can, if necessary, be sealed with paraffix wax. If exceptionally hygroscopic compounds need to be stored, this is best done in sealed glass ampoules, but it is usually best in such cases to use the material immediately after drying and purification.

It is often convenient to remove final traces of water with the aid of a molecular sieve and to store the dried solvent in the presence of the sieve. The descriptive term, **molecular sieve**, applies to a group of dehydrated synthetic sodium and calcium aluminosilicate adsorbents (*zeolites*)* which have a crystal lattice structure incorporating uniformly sized holes or pores which are able to accept molecules smaller than a limiting dimension; the larger molecules do not diffuse into the lattice structure. This selectivity, based upon molecular shape and dimension, accounts for the sieving action and is particularly valuable for the removal from gases and liquids of water, which readily diffuses into the pores and is retained by a strong adsorptive attraction.† The pore size is determined by the nature of the manufacturing process and currently five principal types are available in bead, pellet or powder forms, namely Types 3A, 4A, 5A, 10X and 13X, representing an effective pore diameter of approximately 0.3, 0.4, 0.5, 0.8 and 1.0 nm respectively. All these are stable over a pH range 5–11 but interaction with strong acids is to be avoided; grades are available which are more resistant to the action of acid (e.g., AW-300, AW-500).

Types 3A, 4A and 5A are those which are most usually employed for drying purposes; 5A is also capable of adsorbing the higher homologues of straight chain alkanes, alkenes and alcohols. Their adsorptive capacity for water is higher than that of silica gel, alumina or activated charcoal. After use, molecular sieves may

---

* The first synthetic zeolites were known as Linde Molecular Sieves but are now marketed as 'Union Carbide' Molecular Sieves; they are available from Union Carbide International Company, USA, or Union Carbide (UK) Ltd directly, or through the usual chemical suppliers.

† Booklets giving detailed information on the structure, action and applications of molecular sieves are available from most suppliers of laboratory chemicals, for example BDH, Fison's Scientific Apparatus Ltd, etc.

be readily regenerated by heating between 150 and 300 °C in a suitable oven or in a stream of dry air and then cooling in a desiccator.

Solvents used for the preparation of solutions for spectroscopic examination (particularly infrared measurements) need to be rigorously purified or spectroscopic grades must be purchased (see Section I,39). Even in this latter case, and particularly with the more hygroscopic solvents, the solvent may become contaminated with moisture during usage of the solvent from a previously opened bottle. It is therefore advisable to dry the solvent immediately before use by means of a molecular sieve.

Saturated aliphatic hydrocarbons *1.* **Light petroleum.*** The fractions of refined petroleum which are commonly used have b.p. 40–60, 60–80, 80–100 and 100–120 °C. It is not advisable to employ a fraction with a wider b.p. range than 20 °C, because of possible loss of the more volatile portion during its use in recrystallisation, etc., and consequent different solubility relationships with the higher boiling residue. For some purposes the presence of unsaturated (chiefly aromatic) hydrocarbons in light petroleum is undesirable. Most of the unsaturated hydrocarbons may be removed by shaking two or three times with 10 per cent of the volume of concentrated sulphuric acid (for details, see under *Benzene*); vigorous shaking is then continued with successive portions of a concentrated solution of potassium permanganate in 10 per cent sulphuric acid until the colour of the permanganate remains unchanged. The solvent is then thoroughly washed with water, dried over anhydrous calcium chloride and distilled. If required perfectly dry, it should be allowed to stand over sodium wire (see *15. Diethyl ether* below).

More recently a convenient method of purification has been recommended † which is to decant the solvent, previously treated with sulphuric acid, directly on to a basic alumina (Grade I) column using about 50 g of adsorbent for each 100 ml of solvent; the first 5 per cent of eluate is discarded. The column receiver should be suitably protected from the ingress of moisture by the attachment of a calcium chloride tube.

Light petroleum fractions free from aromatic hydrocarbons are marketed, as are the pure homologues, **pentane**, **heptane**, **hexane**, **octane**, *etc.* While some of these latter are available in spectroscopically pure grades, their purification for spectroscopic use may be readily achieved by passing through a chromatographic column having silica gel (Grade I) in the lower section and basic alumina (Grade I) in the upper section.

Similar purification procedures apply to **cyclohexane**, **methylcyclohexane** and the **decalins**.

The purity of all these hydrocarbon solvents may be checked by gas–liquid chromatography (Section I,31) using an Apiezon, a Silicone oil or an SE-52 silicone rubber gum chromatographic column.

Aromatic hydrocarbons *2.* **Benzene.** The analytical reagent grade benzene is satisfactory for most purposes; if required dry, it is first treated with anhydrous

---

* Sometimes termed ligroin when referring to fractions of b.p. above 100 °C.

† Details abstracted from M. Woelm information leaflets. It should be particularly noted that peroxides remain unchanged on alumina, which must therefore not be heated; the used alumina should be thoroughly wetted before throwing away.

calcium chloride, filtered and then placed over sodium wire (for experimental details, see under *15. Diethyl ether*) or a Type 5A molecular sieve.

Commercial benzene may contain thiophen C_4H_4S, b.p. 84 °C, which cannot be separated by distillation or by fractional crystallisation. The presence of thiophen may be detected by shaking 3 ml of benzene with a solution of 10 mg of isatin in 10 ml of concentrated sulphuric acid and allowing the mixture to stand for a short time: a bluish-green colouration is produced if thiophen is present. The thiophen may be removed from benzene by shaking with concentrated sulphuric acid, advantage being taken of the fact that thiophen is more readily sulphonated than benzene. The technical benzene is shaken repeatedly with about 15 per cent of its volume of concentrated sulphuric acid in a stoppered separatory funnel* until the acid layer is colourless or very pale yellow on standing, or until the thiophen test is negative. After each shaking lasting a few minutes, the mixture is allowed to settle and the lower layer is drawn off. The benzene is then shaken twice with water in order to remove most of the acid, once with 10 per cent sodium carbonate solution, again with water and finally dried with anhydrous calcium chloride. After filtration, the benzene is distilled through an efficient column and the fraction, b.p. 80–81 °C, collected. If required perfectly dry the distilled benzene may either be stored over sodium wire or left in the presence of a Type 5A molecular sieve. Pure benzene has b.p. 80 °/760 mmHg and m.p. 5.5 °C.

3. **Toluene.** Toluene free from sulphur compounds may be purchased. Commercial toluene may contain methyl thiophens (thiotolenes), b.p. 112–113 °C, which cannot be removed by distillation. It may be purified with concentrated sulphuric acid in a similar manner to the purification of benzene, but care must be taken that the temperature is not allowed to rise unduly (<30 °C) as toluene is sulphonated more easily than benzene. If required perfectly dry the distilled toluene may be stored over sodium wire or left in the presence of a Type 5A grade of molecular sieve. Pure toluene has b.p. 110.5 °C/760 mmHg.

4. **Xylenes.** For solvent purposes various grades of xylenes (the mixture of isomers and ethylbenzene) are available; purification and drying procedures are similar to those used for benzene and toluene. For chemical purposes the commercially available pure isomeric xylenes are usually available in at least 99 per cent purity.

Halogenated hydrocarbons *5.* **Dichloromethane (methylene chloride).** The commercial grade is purified by washing with 5 per cent sodium carbonate solution, followed by water, dried over anhydrous calcium chloride and then fractionated. The fraction, b.p. 40–41 °C, is collected.

Methylene chloride is a useful substitute for diethyl ether in extraction processes when it is desired to employ a solvent which is heavier than water.

6. **Chloroform.** The commercial product contains up to 1 per cent of ethanol which is added as a stabiliser. The ethanol may be removed by any of the following procedures:

---

* Alternatively, the mixture may be stirred mechanically for 20–30 minutes. After three such treatments, the acid usually has only a pale colour.

(a) The chloroform is shaken five or six times with about half its volume of water, then dried over anhydrous calcium chloride for at least 24 hours, and distilled.

(b) The chloroform is shaken two or three times with a small volume (say 5 per cent) of concentrated sulphuric acid, thoroughly washed with water, dried over anhydrous calcium chloride or anhydrous potassium carbonate and distilled.

(c) The chloroform is passed through a column of basic alumina (Grade I; 10 g per 14 ml of solvent), a procedure which also removes traces of water and acid; the eluate may be used directly.

Pure chloroform has b.p. 61 °C/760 mmHg. *It must not be dried by standing with sodium or an explosion may occur. The solvent, when free of alcohol, should be kept in the dark in order to avoid the photochemical formation of dangerous quantities of phosgene.*

7. **Carbon tetrachloride.** The analytical reagent product is sufficiently pure for most purposes; the carbon disulphide content does not usually exceed 0.005 per cent. The technical product may contain up to 4 per cent of carbon disulphide; this may be removed by the following method. One litre of commercial carbon tetrachloride is treated with potassium hydroxide (1.5 times the quantity required to combine with the carbon disulphide) dissolved in an equal weight of water and 100 ml of rectified spirit, and the mixture is shaken vigorously for 30 minutes at 50–60 °C. After washing with water, the process is repeated with half the quantity of potassium hydroxide. Ethanol is then removed by shaking several times with 500 ml of water, followed by shaking with small portions of concentrated sulphuric acid until there is no further colouration. The carbon tetrachloride is then washed with water, dried over anhydrous calcium chloride and distilled. Further purification may be effected, if necessary, by passing the distilled solvent through a column of alumina and then allowing it to stand in the presence of a Type 5A molecular sieve and finally distilling before use. The pure compound has b.p. 76.5 °C/760 mmHg. *Carbon tetrachloride must not be dried over sodium, as an explosion may result. Fire extinguishers containing this solvent cannot be applied to a fire originating from sodium or similarly reactive metals.*

Carbon tetrachloride is one of the solvents which may be dried relatively efficiently by simple distillation, rejecting the first 10 per cent of distillate, until the distillate is clear (compare Section **I,23**).

Aliphatic alcohols 8. **Methanol.** The synthetic methanol now available is suitable for most purposes without purification: indeed some manufacturers claim a purity of 99.85 per cent with not more than 0.1 per cent by weight of water and not more than 0.02 per cent by weight of acetone.

Most of the water may be removed from commercial methanol by distillation through an efficient fractionating column (Fig. I,108); no constant boiling point mixture is formed as is the case with ethanol. Anhydrous methanol can be obtained from the fractionally distilled solvent by standing over a Type 4A molecular sieve or by treatment with magnesium metal using the procedure given for super-dry ethanol described below. Pure methanol has b.p. 65 °C/760 mmHg.

If the small proportion of acetone present in synthetic methanol is objectionable it may be removed when present in quantities up to 1 per cent by the following procedure (Morton and Mark, 1934). A mixture of 500 ml of methanol, 25 ml of furfural and 60 ml of 10 per cent sodium hydroxide solution is refluxed in a 2-litre round-bottomed flask, fitted with a double surface condenser, for 6–12 hours. A resin is formed which carries down all the acetone present. The alcohol is then fractionated through an efficient column, the first 5 ml which may contain a trace of formaldehyde being rejected. The recovery of methanol is about 95 per cent.

9. **Ethanol.** Ethanol of a high degree of purity is frequently required in preparative organic chemistry. For some purposes ethanol of *ca.* 99.5 per cent purity is satisfactory; this grade may be purchased (the 'absolute alcohol' of commerce), or it may be conveniently prepared by the dehydration of rectified spirit with calcium oxide. *Rectified spirit* is the constant boiling point mixture which ethanol forms with water, and usually contains 95.6 per cent of ethanol by weight. Whenever the term rectified spirit is used in this book, approximately 95 per cent ethanol is to be understood. Ethanol which has been denatured by the incorporation of certain toxic additives, notably methanol, to render it unfit for consumption, constitutes the *industrial spirit* (*industrial methylated spirit, IMS*) of commerce; it is frequently a suitable solvent for recrystallisations.

Dehydration of rectified spirit by calcium oxide. Pour the contents of a Winchester bottle of rectified spirit (2–2.25 litres) into a 3-litre round-bottomed flask and add 500 g of calcium oxide which has been freshly ignited in a muffle furnace and allowed to cool in a desiccator. Fit the flask with a double surface condenser carrying a calcium chloride guard-tube, reflux the mixture gently for 6 hours (preferably using a heating mantle) and allow to stand overnight. Reassemble the condenser for downward distillation via a splash head adapter to prevent carry-over of the calcium oxide in the vapour stream. Attach a receiver flask with a side arm receiver adapter which is protected by means of a calcium chloride guard-tube. Distil the ethanol gently discarding the first 20 ml of distillate. Preserve the absolute ethanol (99.5%) in a bottle with a well fitting stopper.

'*Super-dry*' *ethanol.* The yields in several organic preparations (e.g., malonic ester syntheses, reductions involving sodium and ethanol, etc.) are considerably improved by the use of ethanol of 99.8 per cent purity or higher. This very high grade ethanol may be prepared in several ways from commercial absolute alcohol or from the product of dehydration of rectified spirit with calcium oxide.

The method of Lund and Bjerrum depends upon the reactions:

$$Mg + 2C_2H_5OH \longrightarrow H_2 + Mg(OC_2H_5)_2 \tag{1}$$

$$Mg(OC_2H_5)_2 + 2H_2O \longrightarrow Mg(OH)_2 + 2C_2H_5OH \tag{2}$$

Reaction (1) usually proceeds readily provided the magnesium is activated with iodine and the water content does not exceed 1 per cent. Subsequent interaction between the magnesium ethanolate and water gives the highly insoluble magnesium hydroxide; only a slight excess of magnesium is therefore necessary.

Fit a dry 1.5- or 2-litre round-bottomed flask with a double surface condenser and a calcium chloride guard-tube. Place 5 g of clean dry magnesium turnings and 0.5 g of iodine in the flask, followed by 50–75 ml of commercial absolute ethanol. Warm the mixture until the iodine has disappeared: if a lively evolution of hydrogen does not set in, add a further 0.5 g portion of iodine. Continue heating until all the magnesium is converted into ethanolate, then add 900 ml of commercial absolute ethanol and reflux the mixture for 30 minutes. Distil off the ethanol directly into the vessel in which it is to be stored, using an apparatus similar to that described for the dehydration of rectified spirit. The purity of the ethanol exceeds 99.95 per cent provided adequate precautions are taken to protect the distillate from atmospheric moisture. The super-dry ethanol is exceedingly hygroscopic; it may with advantage be stored over a Type 4A molecular sieve.

If the alcohol is required for conductivity or other physico-chemical work and traces of bases are objectionable, these may be removed by redistillation from a little 2,4,6-trinitrobenzoic acid. This acid is selected because it is not esterified by alcohols, consequently no water is introduced into the alcohol.

10. **Propan-1-ol.** The purest available commercial propan-1-ol (*propyl alcohol*) should be dried with anhydrous potassium carbonate or with anhydrous calcium sulphate, and distilled through an efficient fractionating column. The fraction, b.p. 96.5–97.5 °C/760 mmHg, is collected. If the propan-1-ol is required perfectly dry, it may be treated with magnesium activated with iodine by the method described above for ethanol.

11. **Propan-2-ol.** Two technical grades of propan-2-ol (*isopropyl alcohol*) are usually marketed having purities of 91 per cent and 99 per cent respectively. The former has a b.p. of about 80.3 °C and is a constant boiling point mixture with water. Propan-2-ol may contain peroxide, which if present must be removed before dehydration is attempted. Therefore *test for peroxide* by adding 0.5 ml of propan-2-ol to 1 ml of 10 per cent potassium iodide solution acidified with 0.5 ml of dilute (1 : 5) hydrochloric acid and mixed with a few drops of starch solution just prior to the test: if a blue (or blue-black) colouration appears in one minute, the test is positive. *To remove peroxide* heat under reflux 1 litre of propan-2-ol with 10–15 g of solid tin(II) chloride for half an hour. Test a portion of the cooled solution for peroxide: if iodine is liberated, add further 5 g portions of tin(II) chloride and heat under reflux for half-hour periods until the test is negative. Add about 200 g of calcium oxide and heat under reflux for 4 hours, and then distil, discarding the first portion of distillate. The water content may be further reduced by allowing the distillate to stand over calcium metal or a Type 5A molecular sieve for several days, followed by further fractionation. Anhydrous propan-2-ol has b.p. 82–83 °C/760 mmHg; however, peroxide generally redevelops during several days.

12. **Higher alcohols.** These may be purified by drying with anhydrous potassium carbonate or with anhydrous calcium sulphate, and fractionated after filtration from the desiccant in apparatus with ground glass joints. The boiling points of the fractions to be collected are as follows:

Butan-1-ol (*butyl alcohol*), b.p. 116.5–118 °C/760 mmHg.
2-Methylpropan-1-ol (*isobutyl alcohol*), b.p. 106.5–107.5 °C/760 mmHg.
Butan-2-ol (*s-butyl alcohol*), b.p. 99–100 °C/760 mmHg.
2-Methylpropan-2-ol (*t-butyl alcohol*), b.p. 81.5–82.5 °C/760 mmHg, m.p. 25.5 °C.
Pentan-1-ol (*amyl alcohol*), b.p. 136–137.5 °C/760 mmHg.
3-Methylbutan-1-ol (*isoamyl alcohol*), b.p. 130–131 °C/760 mmHg.
Hexan-1-ol (*hexyl alcohol*), b.p. 156.5–157.5 °C/760 mmHg.

If perfectly anhydrous alcohols are required these may in general be obtained by treatment with sodium followed by addition of the corresponding alkyl succinate or phthalate. Sodium alone cannot be used for the complete removal of water in an alcohol owing to the equilibrium between the resulting sodium hydroxide and the alcohol:

$$NaOH + ROH \rightleftharpoons RONa + H_2O$$

The purpose of adding the ester is to remove the sodium hydroxide by the saponification reaction:

$$
\begin{array}{ll}
CH_2 \cdot CO_2R & CH_2 \cdot CO_2Na \\
| \qquad\quad + 2NaOH \longrightarrow | \qquad\quad + 2ROH \\
CH_2 \cdot CO_2R & CH_2 \cdot CO_2Na
\end{array}
$$

Typically 7 g of sodium metal are added to 1 litre of butan-2-ol (having no more than 0.5% of water) contained in a two-necked flask fitted with a double surface condenser. When all the metal has reacted (some warming may be necessary to increase the speed of reaction), 33 g of pure 2-butyl succinate or 41 g of pure 2-butyl phthalate are added and the mixture is heated under gentle reflux for two hours. Distillation through a Vigreux column affords a distillate containing not more than 0.05 per cent of water.

13. **Mono-alkyl ethers of ethylene glycol, $R \cdot O \cdot CH_2 \cdot CH_2OH$.** The mono-methyl, ethyl and butyl ethers are inexpensive and are known as *methyl cellosolve, cellosolve* and *butyl cellosolve* respectively. They are completely miscible with water, and are excellent solvents. The commercial products are purified by drying over anhydrous potassium carbonate or anhydrous calcium sulphate, followed by fractionation after the removal of the desiccant. The boiling points of the pure products are:

Ethylene glycol monomethyl ether (or *2-methoxyethanol*), b.p. 124.5 °C/760 mmHg

Ethylene glycol monoethyl ether (or *2-ethoxyethanol*), b.p. 135 °C/760 mmHg

Ethylene glycol monobutyl ether (or *2-butoxyethanol*), b.p. 171 °C/760 mmHg

14. **Mono-alkyl ethers of diethylene glycol, $R \cdot O \cdot CH_2 \cdot CH_2 \cdot O \cdot CH_2 \cdot CH_2OH$.** The monomethyl, ethyl and butyl ethers are inexpensive commercial products and are known as *methyl carbitol, carbitol* and *butyl carbitol* respectively. They are all completely miscible with water and are purified as already described for the 'cellosolves' above. The boiling points of the pure compounds are:

Diethylene glycol monomethyl ether, b.p. 194 °C/760mmHg

Diethylene glycol monoethyl ether, b.p. 198.5 °C/760 mmHg

Diethylene glycol monobutyl ether, b.p. 230.5 °C/760 mmHg

Note. The cellosolve and carbitol solvents may contain traces of peroxide. These can be removed either by heating under reflux over anhydrous tin(II) chloride (see *11. Propan-2-ol*) or by filtration under slight pressure through a column of activated basic alumina (Grade I); the used alumina should be saturated with water before being discarded.

Ethers *15.* **Diethyl ether** (*Ether*).* The chief impurities in commercial ether (d 0.720) are water and ethanol. Furthermore, when ether is allowed to stand for some time in contact with air and exposed to light, slight oxidation occurs with the formation of the highly explosive diethyl peroxide, $(C_2H_5)_2O_3$. The danger from this unstable compound becomes apparent at the conclusion of the distillation of impure ether, when the comparatively non-volatile peroxide becomes concentrated in the residue in the distillation flask, and a serious explosion may then result if an attempt is made to evaporate the ether to dryness. It is perhaps worthy of comment in this connection that in the extraction of an organic compound with ether and the subsequent removal of the solvent, the presence of the residual compound seems largely to eliminate the danger due to traces of peroxide, due presumably to its catalytic effect which leads to a more controlled decomposition. Nevertheless ether which has been standing for several months in a partially filled bottle exposed to light and air should be tested for peroxide by the procedure described under *11. Propan-2-ol.* If present, the peroxide may be removed by shaking 1 litre of ether with 10–20 ml of a concentrated solution of an iron(II) salt prepared either by dissolving 60 g of iron(II) sulphate in a mixture of 6 ml of concentrated sulphuric acid and 110 ml of water, or by dissolving 100 g of iron(II) chloride in a mixture of 42 ml of concentrated hydrochloric acid and 85 ml of water.†

Peroxide may also be removed by shaking with an aqueous solution of sodium sulphite or with solid tin(II) chloride (see *11. Propan-2-ol*) or by passage through a column of alumina. *It is worthy of note that all dialkyl ethers have a tendency to form explosive peroxides and they should be routinely tested before further purification leading to a final distillation process is attempted*

Apart from the dangers inherent in the use of diethyl ether due to the presence of peroxide, attention must be directed to the fact that ether is *highly inflammable* and also extremely volatile (b.p. 35 °C), and great care should be taken that there is no naked flame in the vicinity of the liquid (see Section I,3). *Under no circumstances should ether be distilled over a bare flame*, but always from a steam bath or an electrically heated water bath and with a highly efficient double surface condenser. Ether vapour has been known to ignite on contact with a hot plate or even a hot tripod upon which a water bath has previously been heated.

Purification of commercial ether. Divide the contents of a Winchester bottle of ether into approximately two equal volumes and shake each in a large separatory funnel with 10–20 ml of the above iron(II) sulphate solution diluted with 100 ml of water. Remove the aqueous solution and combine the two ether portions in a clean dry Winchester bottle and add 100–200 g of anhydrous calcium chloride. Allow this mixture to stand for 24 hours with occasional shaking; the water and ethanol are largely removed during this period. Filter the ether through a large fluted filter-paper into another clean dry Winchester bottle (**CAUTION**: *all flames in the vicinity must be extinguished*). Fine sodium wire (about 7 g) is then introduced into the ether with the aid of a sodium press ‡

---

* In this book, and in most others, the term *ether* implies reference to diethyl ether; other homologues are referred to by their systematic or trivial names.

† Traces of aldehydes are produced. If ether of high purity is required, it should be further shaken with 0.5 per cent potassium permanganate solution (to convert the aldehyde into acid), then with 5 per cent sodium hydroxide solution, and finally with water.

‡ Supplied by Griffin & George Ltd.

(Fig. II,1). The latter consists of a rigid metal framework, which can be attached to the bench by means of a single bolt (as in the figure). An adjustable bottle stand is provided so that bottles up to a capacity of one Winchester quart can be used and their necks brought up to the underside of the mould. The plunger is of stainless steel as is also the one-piece mould and die. (A number of dies of various sizes, thus giving sodium wire of different diameters, are usually available for alternative use.) The die is nearly filled with lumps of clean sodium (Section **II,2,***55*), then placed in position in the press, and the plunger slowly screwed down. As soon as the sodium wire emerges from the die, the Winchester bottle containing the ether is held immediately beneath the die, and the plunger is gradually lowered until all the sodium has been forced as a fine wire into the ether. The Winchester bottle is then closed by a rubber stopper carrying a calcium chloride guard-tube to exclude moisture and to permit the escape of hydrogen; the ether is allowed to stand for about 24 hours. The steel die must be removed from the press after use, any residual sodium destroyed by immersion in industrial spirit, and then thoroughly washed with water and finally dried, preferably in a heated oven; the plunger should also be swabbed with a rag or filter-paper soaked in industrial spirit. If, on the following day, no bubbles of hydrogen rise from the sodium in the ether and the latter still possesses a bright surface, the Winchester bottle is closed by its own screw-capped stopper or by a rubber stopper, and preserved in the dark (to check the formation of peroxide as far as possible) in a cool place remote from flames. If, however, the surface of the sodium wire is badly attacked, due to insufficient drying with the calcium chloride, the ether must be filtered through a fluted filter-paper into another clean, dry Winchester bottle and the treatment with sodium repeated.*

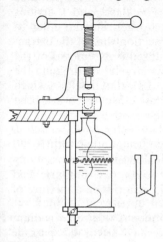

Fig. II,1

The absolute diethyl ether thus prepared is suitable for use, for example, for Grignard reactions. If a fresh supply of high-grade ether of analytical reagent quality is available, the treatment with ferrous salt solution prior to the final sodium drying procedure may be omitted.

16. **Di-isopropyl ether.** The commercial product usually contains appreciable quantities of peroxide; this should be removed by treatment with an acidified solution of an iron(II) salt or with a solution of sodium sulphite (see under *15. Diethyl ether*). The di-isopropyl ether is then dried over anhydrous calcium chloride and distilled, the fraction b.p. 68.5 °C/760 mmHg being collected.

17. **Dibutyl ether.** Technical dibutyl ether does not usually contain appreciable quantities of peroxide, unless it has been stored for a prolonged period. It should, however, be tested for peroxide, and, if the test is positive, the solvent should be shaken with an acidified solution of an iron(II) salt or with a solution of sodium sulphite (see under *15. Diethyl ether*). The dibutyl ether is dried with anhydrous calcium chloride and distilled through a fractionating

---

* The sodium residues in the bottle should be destroyed as recommended on p. 14.

column: the portion b.p. 140–141 °C is collected. If a fraction of low boiling point is obtained, the presence of butan-1-ol is indicated and may be removed by shaking twice with an equal volume of concentrated hydrochloric acid (see, however, Section **III,76**), followed by washing with water and drying. Pure dibutyl ether has b.p. 142 °C/760 mmHg.

18. **Di-alkyl ethers of monoethylene and diethylene glycol.** The dimethyl ether of ethylene glycol (dimethoxyethane $CH_3 \cdot O \cdot CH_2 \cdot CH_2 \cdot O \cdot CH_3$, frequently referred to as DME or *dimethyl cellosolve* or *glyme*) has b.p. 85 °C/760 mmHg, is miscible with water, is a good solvent and an excellent inert reaction medium. The diethyl ether of ethylene glycol (*diethyl cellosolve*) is partially miscible with water (21% at 20 °C) and has b.p. 121.5 °C/760 mmHg.

The dimethyl ether of diethylene glycol ($CH_3 \cdot O \cdot (CH_2 \cdot CH_2 \cdot O)_2 CH_3$, *diglyme*) has b.p. 62 °C/17 mmHg; the corresponding diethyl ether (*diethyl carbitol*) has b.p. 186 °C/760 mmHg.

All these are excellent solvents for organic compounds; they are purified by initial storage over sodium hydroxide pellets and then heated under reflux with calcium hydride, lithium aluminium hydride, sodium hydride, or sodium, before being fractionally distilled (under reduced pressure if necessary) in an atmosphere of nitrogen.

19. **Tetrahydrofuran.** The commercial grade of this solvent is obtainable in greater than 99.5 per cent purity, in which water and peroxides are the major impurities; an inhibitor for peroxide formation may have been added by the manufacturers. Peroxide, if present, *must* be removed by passage through a column of alumina (see *1. Light petroleum* for footnote on the disposal of used alumina), or by shaking with iron(II) sulphate solution as described under diethyl ether before drying and further purification is attempted. If the latter method is employed the solvent should then be dried initially over calcium sulphate or solid potassium hydroxide,* before being heated under reflux over calcium hydride or lithium aluminium hydride. The solvent is finally fractionally distilled. Pure tetrahydrofuran has b.p. 65–66 °C/760 mmHg; it should be stored over calcium hydride.

20. **Dioxan** (*1,4-dioxan; diethylene dioxide*). The commercial grade usually contains small quantities of acetaldehyde and appreciable amounts of glycol acetal (ethylene acetal), $CH_3 \cdot CH \overset{\displaystyle O - CH_2}{\underset{\displaystyle O - CH_2}{|}}$, together with some water. Upon keeping, the acetal tends to undergo hydrolysis and the liberated acetaldehyde leads to some peroxide formation. Purification may be effected by decomposing the acetal with dilute acid, followed by drying and fractionation. One litre of technical dioxan, 14 ml of concentrated hydrochloric acid and 100 ml of water are heated in a fume cupboard (dioxan vapour is highly toxic) under reflux for 6–12 hours whilst a slow stream of nitrogen is bubbled through the solution to remove the acetaldehyde formed. The cold solution is treated with excess potassium hydroxide pellets with shaking until some remain undissolved, and the strongly alkaline aqueous layer is run off; most of the re-

---

* It has been reported that a serious explosion may occur when impure THF containing peroxide is treated with solid potassium hydroxide or with a concentrated aqueous solution of potassium hydroxide (see Ref. 2).

sidual water is removed by keeping the dioxan over fresh potassium hydroxide pellets for 24 hours. This treatment is followed by heating the decanted solvent under reflux over excess of sodium for 6–12 hours, i.e., until reaction ceases and the sodium remains bright. Finally, the dioxan is distilled from sodium preferably into a receiver encased in black paper; it should be stored out of contact with air and in the dark. The pure compound has b.p. 101.5 °C/760 mmHg and m.p. 12 °C.

Distillation of the purified dioxan from lithium aluminium hydride (1), or preferably passage through a column of basic activated alumina before use, ensures the removal of any peroxide which may develop on storage.

Dioxan is a very useful solvent for a variety of organic compounds; its solvating properties are similar to, and often better than, those of diethyl ether. It is miscible with water in all proportions and is extremely hygroscopic.

Note. (1) The use of lithium aluminium hydride for the final purification has recently been criticised; an alternative to passage through alumina consists of heating the dioxan under reflux over sodium followed by the addition of benzophenone which gives a deep-blue purple solution of a disodium benzophenone complex when the dioxan becomes anhydrous. The solvent should be distilled from this mixture under nitrogen, the fractionating column and receiving flask being covered with black polythene film.

Ketones *21. Acetone.* Although major impurities in the commercial grades of acetone are methanol, acetic acid and water, the analytical reagent generally contains less than 0.1 per cent of the organic impurities although the water content may be as high as 1 per cent.

Commercial acetone may be purified in several ways:

(a) The acetone is heated under reflux with successive quantities of potassium permanganate until the violet colour persists. It is then dried with anhydrous potassium carbonate or anhydrous calcium sulphate,* filtered from the desiccant and fractionated; precautions are taken to exclude moisture.

(b) To 700 ml of acetone, b.p. 56–57 °C, contained in a litre bottle, a solution of 3 g of silver nitrate in 20 ml of water is added, followed by 20 ml of 1 *M*-sodium hydroxide solution, and the mixture is shaken for about 10 minutes. The mixture is then filtered, dried with anhydrous calcium sulphate and distilled.

(c) When only a relatively small quantity of pure, dry acetone is required, it may be purified through the bisulphite complex: the latter is decomposed with sodium carbonate solution (for details, see under *22. Ethyl methyl ketone*), dried over anhydrous calcium sulphate and distilled. A more convenient procedure is to make use of the addition compound with sodium iodide ($NaI,3C_3H_6O$), which decomposes on gentle heating and is particularly well adapted for the preparation of pure acetone. One hundred grams of finely powdered sodium iodide are dissolved under reflux in 440 g of boiling commercial acetone, and the solution is cooled in a mixture of ice and salt (-8 °C). The crystals are filtered off and quickly transferred to a dry distilling flask, connected to an efficient condenser and to a receiver cooled in ice. Upon gentle warming, the acetone distils rapidly.

---

* Anhydrous calcium chloride should not be used as some chemical combination occurs.

Acetone purified by these means, or the analytical reagent grade, may have the water content reduced by storage over a Type 4A molecular sieve. Silica gel or alumina should not be used as an aldol type reaction is initiated with the formation of water as a by-product. Pure acetone has b.p. 56.2 °C/760 mmHg and is highly inflammable.

22. **Ethyl methyl ketone** (*butan-2-one*). This excellent solvent has properties similar to those of acetone but it has a somewhat higher boiling point and is therefore less inflammable. A preliminary purification is effected by drying the commercial product with anhydrous potassium carbonate or anhydrous calcium sulphate, filtering from the desiccant, and fractionating through an efficient column; the fraction, b.p. 79–80 °C, is collected separately, and is quite satisfactory for recrystallisations. This may be further purified either through the bisulphite addition compound or through the sodium iodide addition compound. The ethyl methyl ketone, b.p. 79–80 °C, is shaken with excess of saturated sodium bisulphite solution until reaction ceases, cooled to 0 °C, the bisulphite complex filtered off, the filter cake well drained, washed with a little ether, and then dried in the air. The dry bisulphite complex is decomposed with a slight excess of sodium carbonate solution, and distilled in steam. The ketone is salted out from the distillate with potassium carbonate, separated, dried with anhydrous potassium carbonate (this will also remove traces of sulphur dioxide and carbon dioxide present) and, after filtration, allowed to stand for several hours over anhydrous calcium sulphate. It is then distilled. In the sodium iodide method, the ketone is saturated with sodium iodide by boiling under reflux, the solution is filtered through a hot water funnel, cooled in a freezing mixture and white crystals (which have a m.p. of 73–74 °C) filtered off. Gentle heating of the crystals in a fractional distillation assembly gives pure ethyl methyl ketone of b.p. 79.5 °C/760 mmHg.

Esters *23.* **Methyl acetate.** An anhydrous product of 99 per cent purity (b.p. 56.5–57 °C) is available commercially; this is comparatively cheap so that purification of inferior products is not worthwhile. It is appreciably soluble in water (*ca.* 24% at 20 °C). The pure compound has b.p. 57 °C/760 mmHg.

If it is desired to purify an inferior product, 1 litre is heated under reflux for 6 hours with 85 ml of acetic anhydride and then distilled through a fractionating column: the liquid passing over at 56–57 °C is collected. The distillate is shaken with 20 g of anhydrous potassium carbonate for 10 minutes, filtered and redistilled. The resulting methyl acetate has a purity of 99.9 per cent.

24. **Ethyl acetate.** Various grades of ethyl acetate are marketed. The anhydrous compound, b.p. 76–77 °C, is of 99 per cent purity, is inexpensive and is suitable for most purposes. The 95–98 per cent grade usually contains some water, ethanol and acetic acid, and may be purified in the following manner. A mixture of 1 litre of the commercial ethyl acetate, 100 ml of acetic anhydride and 10 drops of concentrated sulphuric acid is heated under reflux for 4 hours and then fractionated. The distillate is shaken with 20–30 g of anhydrous potassium carbonate, filtered and redistilled. The final product has a purity of about 99.7 per cent and boils at 77 °C/760 mmHg.

Nitrogen-containing solvents *25.* **Formamide.** Formamide is an excellent solvent for many polar organic compounds and for a selection of inorganic salts. It is very hygroscopic and readily hydrolysed by acids or bases. The commercial product frequently contains formic acid, water and ammonium formate. Purification may be effected by passing ammonia gas into the solvent until a slight alkaline reaction is obtained; addition of dry acetone then precipitates the ammonium formate. The filtered solution is dried over magnesium sulphate and fractionally distilled under reduced pressure; distillation at atmospheric pressure causes decomposition. Pure formamide has b.p. 105 °C/11 mmHg.

26. **N,N-Dimethylformamide** (*DMF*). *N*,*N*-Dimethylformamide is a widely used solvent for many recently developed synthetic procedures because of its powerful solvating properties and its chemical stability in the absence of acidic or basic catalysts. However, distillation at atmospheric pressure (b.p. 149–156 °C) or contact with desiccants such as solid sodium or potassium hydroxide or calcium hydride causes varying degrees of decomposition.

Commercial grades of DMF may be initially purified by azeotropic distillation with benzene. Distil a mixture of 1 litre of DMF and 100 ml of benzene at atmospheric pressure and collect the water : benzene azeotrope which distils between 70 and 75 °C. Shake the residual solvent with powdered barium oxide or with activated alumina (Grade I), filter and distil under nitrogen at reduced pressure; collect the fraction having b.p. 76 °C/39 mmHg or 40 °C/10 mmHg. The distillate is best stored over a Type 4A molecular sieve.

27. **Acetonitrile.** This is another highly versatile solvent for many synthetic procedures. Although some commercial grades may be available having purities equal to or greater than 99.5 per cent, the usual contaminants in lower purity grades are water, acetamide, ammonium acetate and ammonia. Water may be removed with activated silica gel or Type 4A molecular sieves (not solid potassium hydroxide since this causes decomposition; calcium sulphate and calcium chloride are inefficient in this instance). This partially dried solvent is stirred with calcium hydride which is added portionwise until hydrogen evolution ceases. The solvent is decanted from the solid and fractionally distilled at atmospheric pressure using a high efficiency column (Fig. I,108); the pure solvent has b.p. 81–82 °C/760 mmHg.

28. **N-Methylpyrrolidone.** This versatile solvent has good chemical stability in the absence of acids and bases which catalyse the cleavage of the lactam ring. It is most conveniently dried by initial azeotropic distillation with previously dried benzene as described for DMF, and the residual liquid is shaken with barium oxide, the desiccant is removed and the solvent is fractionally distilled under reduced pressure (*ca.* 20 mm). The pure solvent has b.p. 94–96 °C/20 mmHg, or 202 °C/760 mmHg.

29. **Pyridine.** The analytical reagent grade (>99.5% purity) will satisfy most requirements. If required perfectly dry, it should be heated under reflux over potassium hydroxide or sodium hydroxide pellets or over barium oxide, and then distilled with careful exclusion of moisture. As it is hygroscopic and forms a hydrate of b.p. 94.5 °C it should be stored over solid potassium hydroxide pellets. Pure pyridine has b.p. 115.3 °C/760 mmHg.

Pure pyridine may be prepared from technical coal-tar pyridine in the following manner. The technical pyridine is first dried over solid sodium hydroxide, distilled through an efficient fractionating column, and the fraction, b.p. 114–116 °C, collected. Four hundred ml of the redistilled pyridine are added to a reagent prepared by dissolving 340 g of anhydrous zinc chloride in a mixture of 210 ml of concentrated hydrochloric acid and 1 litre of absolute ethanol. A crystalline precipitate of an addition compound (probable composition $2C_5H_5N,ZnCl_2,HCl*$) separates and some heat is evolved. When cold, this is collected by suction filtration, and washed with a little absolute ethanol. The yield is about 680 g. It is recrystallised from absolute ethanol to constant m.p. (151.8 °C). The base is liberated by the addition of excess of concentrated sodium hydroxide solution (ca. 40%) to the complex followed by steam distillation until the distillate is no longer alkaline to litmus (ca. 1000 ml). The steam distillate is treated with 250 g of solid sodium hydroxide, the upper layer separated and the aqueous layer extracted with two 250 ml portions of ether. The combined upper layer and ether extracts are dried with anhydrous potassium carbonate, the ether removed on a water bath and the pyridine distilled through a fractionating column. Further drying is effected as described for the analytical grade reagent. The pure pyridine has b.p. 115.3 °C/760 mmHg.

30. **Quinoline.** Quinoline is little used as a solvent. It may be dried if required with potassium hydroxide pellets and fractionally distilled under reduced pressure; b.p. 114 °C/17 mmHg.

31. **Nitrobenzene.** Nitrobenzene of analytical reagent quality is satisfactory for most purposes. The technical product may contain dinitrobenzene, the nitrotoluenes and aniline. Most of the impurities are retained in the residue after addition of dilute sulphuric acid and steam distillation: the nitrobenzene in the distillate is separated, dried with calcium chloride and distilled under reduced pressure. The pure solvent has b.p. 210 °C/760 mmHg and m.p. 5.7 °C.

Nitrobenzene is an extremely versatile solvent, and may sometimes be employed for the crystallisation of compounds which do not dissolve appreciably in the common organic solvents. The vapour is very toxic, so that recrystallisations must be carried out in the fume cupboard. After the crystals have been collected, they should be washed with a volatile solvent, such as ethanol or ether, to remove the excess of nitrobenzene (compare Section I,20). A disadvantage of nitrobenzene as a solvent is its pronounced oxidising action at the boiling point.

Sulphur-containing solvents *32.* **Carbon disulphide.** *When working with this solvent, its toxicity (it is a blood and nerve poison) and particularly its high inflammability should be borne in mind.* Distillation of appreciable quantities of carbon disulphide should be carried out in a water bath at 55–65 °C; it has been known to ignite as the result of being overheated on a steam bath.

The analytical reagent grade is suitable for most purposes. The commercial substance may be purified by shaking for 3 hours with three portions of potassium permanganate solution (5 g per litre), twice for 6 hours with mercury and finally with a solution of mercury(II) sulphate (2.5 g per litre). It is then dried over anhydrous calcium chloride, and fractionated from a water bath at 55–65 °C. The pure compound boils at 46.5 °C/760 mmHg.

33. **Dimethyl sulphoxide** (*DMSO*).† Dimethyl sulphoxide has recently

---

* There appear to be at least two zinc chloride complexes of pyridine, one of m.p. 207 °C and composition $2C_5H_5N,ZnCl_2$, and the other of m.p. 152 °C and probable composition $2C_5H_5N,ZnCl_2,HCl$. The former is slightly soluble in water and in hot ethanol: the latter passes into the former in aqueous solution, is readily soluble in hot ethanol and can therefore be readily recrystallised from this solvent.

† There have been reports of several serious explosions resulting from dimethyl sulphoxide being allowed to come into contact with periodic acid, magnesium perchlorate or perchloric acid. A mixture of sodium hydride (4.5 mol) and dimethyl sulphoxide (18.4 mol) may also explode after about 1 hour.

come to be recognised as a highly useful water-miscible solvent for many preparative procedures and for spectroscopic work; a study of its chemistry has revealed its potential as an important synthetic reagent. It is hygroscopic and distillation at atmospheric pressures causes some decomposition. The commercial grade may be dried by standing overnight over freshly activated alumina, barium oxide or calcium sulphate. The filtered solvent is then fractionally distilled over calcium hydride under reduced pressure (ca. 12 mmHg) and stored over a Type 4A molecular sieve. An additional stage involving cooling the solvent to about 5 °C, filtering the partially crystallised mass and fractionally distilling the melted crystals has been recommended. The pure solvent has b.p. 75–76 °C/12 mmHg, m.p. 18–19 °C.

34. **Sulfolane** (*tetrahydrothiophen-1,1-dioxide*). This is a further useful water-miscible aprotic solvent of moderately high dielectric constant but with weak acidic and basic characteristics. Although not soluble in saturated hydrocarbon solvents it is a good solvent for most other classes of organic compounds. Purification is effected by passage through a column of activated alumina and distillation under reduced pressure or by repeated vacuum distillation from sodium hydroxide pellets. The pure compound has b.p. 113–117 °C/6 mmHg and m.p. 27 °C; some decomposition occurs at the boiling point (287 °C) at atmospheric pressure. The solvent may be stored over a Type 4A molecular sieve.

Phosphorus-containing solvents *35*. **Hexamethylphosphoric triamide** (*HMPT*). This solvent is miscible both with water and with many polar and non-polar organic solvents with the exception of saturated aliphatic hydrocarbons. It forms a complex with chlorinated solvents by which means it may be removed from aqueous solutions. The solvent may be dried by shaking with calcium or barium oxide followed by distillation under reduced pressure and storage over a Type 4A molecular sieve. The pure solvent has b.p. 127 °C/20 mmHg.

II,2. PREPARATION AND PURIFICATION OF REAGENTS An account will be given in this Section of the inorganic and organic reagents which are required for the preparations described later. The preparation of those reagents which can be purchased at reasonable cost will not normally be described, but in those cases where the purified reagents are somewhat expensive, the methods of purification of the technical products will be outlined. Some comments on precautions for handling hazardous reagents are also included. Attention is drawn to those gaseous reagents which are commercially available in cylinders;* the comments in Section **I,3,B,4**(x) regarding the hazards of storage and handling of gas cylinders and of the precautions to be taken in their use should be noted.

1. **Aluminium alkoxides.** (*a*) **Aluminium t-butoxide.** In a 500-ml round-bottomed flask fitted with a reflux condenser protected by a calcium chloride guard-tube, place 16 g (0.59 mol) of aluminium foil, 50 g (63.5 ml, 0.67 mol) of anhydrous t-butyl alcohol and 2 g of aluminium isopropoxide (see (*b*) below, to remove traces of water). Heat the mixture (electric mantle) to boiling, add

---

* A wide range of gaseous reagents in conveniently sized cylinders is available from, for example, BDH Laboratory Gas Service, Poole, Dorset; Cambrian Chemicals Ltd, Croydon, London; The Matheson Company, East Rutherford, NJ, USA.

about 0.1 g of mercury(II) chloride and shake vigorously: the object of the shaking is to distribute the mercury(II) chloride and thus assist an even amalgamation of the aluminium. Continue heating; the colour of the reaction mixture gradually changes from clear to milky to black and hydrogen is evolved. When the mixture is black, allow the reaction to proceed for an hour without further heating, and then add 61 g (77 ml, 0.82 mol) of anhydrous t-butyl alcohol and 50 ml of anhydrous benzene. Heat gently to restart the reaction; it will continue vigorously without further heating for about 2 hours: when the reaction subsides heat the mixture under reflux for 12 hours. Remove the benzene and unreacted t-butyl alcohol by distillation under reduced pressure (water pump) using a rotary evaporator, taking care to remove the final traces of solvent as far as possible. Add 250 ml of anhydrous ether; dissolve the solid aluminium t-butoxide by heating under reflux for a short time. After cooling, add 9 ml of undried ether and immediately shake vigorously; the small amount of water thus introduced forms aluminium hydroxide, which assists the precipitation of the black suspended material. Allow to stand for 2 hours, centrifuge the mixture for 30 minutes to remove aluminium hydroxide, unused aluminium and mercury. After centrifugation the solution should be colourless or almost so; if it is still dark in colour, add a further 6 ml of undried ether and centrifuge again. Now remove the solvent under reduced pressure (water pump) using a rotary evaporator. Allow the flask to cool with drying tube attached, crush the product with a spatula and transfer it to a small bottle: seal the latter against moisture. The yield of white or pale grey aluminium t-butoxide is 105 g (72%).

(b) **Aluminium isopropoxide.** Place 27 g (1 mol) of clean aluminium foil in a 1-litre round-bottomed flask containing 235 g (300 ml, 3.91 mol) of dried isopropyl alcohol (*Propan-2-ol*, Section **II,1,*11***) and 0.5 g of mercury(II) chloride. Attach an efficient double surface reflux condenser carrying a calcium chloride guard-tube. Heat the mixture on a water bath or heating mantle. When the liquid is boiling, add 2 ml of carbon tetrachloride (a catalyst for the reaction between aluminium and dry alcohols) through the condenser, and continue the heating. The mixture turns grey and, within a few minutes, a vigorous evolution of hydrogen commences. Discontinue the heating: it may be necessary to moderate the reaction by cooling the flask in ice-water or in running tap water. After the reaction has moderated heat the mixture under reflux until all the metal has reacted (6–12 hours). The mixture becomes dark because of the presence of suspended particles.* Pour the hot solution into a 500-ml flask fitted with a still-head, a condenser and a 250-ml receiving flask. Add a few fragments of porous porcelain and heat the flask in an oil bath at 90 °C under slightly diminished pressure (water pump). When nearly all the isopropyl alcohol has distilled, raise the temperature of the bath to 170 °C and lower the pressure gradually to the full vacuum of the water pump. Immediately the temperature of the distillate rises above 90 °C, stop the distillation and remove the condenser. Attach a 500-ml receiving flask with adapter directly to the still-head, add a few fresh boiling chips and distil; use either an oil bath, at 180–190 °C, or an air bath. The aluminium isopropoxide passes over, as a colourless viscid liquid, at 140–150 °C/12 mmHg; the yield is 190 g (90%). Pour the molten aluminium isopropoxide into a wide-mouthed, glass-stoppered bottle and seal the bottle

---

* This crude solution may be used directly for some preparations (see later). In general the presence of these particles has no influence on reactions using aluminium isopropoxide.

with paraffin wax (or with sealing tape) to exclude moisture. Generally the alkoxide (m.p. 118 °C) crystallises out, but the substance exhibits a great tendency to supercool and it may be necessary to cool to 0 °C for 1–2 days before solidification occurs.

The reagent is conveniently stored as a solution in isopropyl alcohol. The molten (or solid) alkoxide is weighed out after distillation into a glass-stoppered bottle or flask and is dissolved in sufficient dry isopropyl alcohol to give a one-molar solution. This solution may be kept without appreciable deterioration provided the glass stopper is sealed with paraffin wax or sealing tape. Crystals of aluminium isopropoxide separate on standing, but these may be redissolved by warming the mixture to 65–70 °C.

For many reductions it is not necessary to distil the reagent. Dilute the dark solution, prepared as above to the point marked with an asterisk, to 1 litre with dry isopropyl alcohol; this gives an approximately one-molar solution. Alternatively, prepare the quantity necessary for the reduction, using the appropriate proportions of the reagents.

2. **Aluminium amalgam.** Place 100 g of thin aluminium foil (*ca.* 0.05 mm thickness), cut into strips about 15 cm long and 2.5 cm wide and loosely folded, in a 3-litre flask and cover with a 10 per cent solution of sodium hydroxide; warm the flask on a water bath until a vigorous evolution of hydrogen has taken place for several minutes (**CAUTION!**). Wash the foil thoroughly with water and with rectified spirit; this operation produces an exceptionally clean surface for amalgamation. Add sufficient of a 2 per cent solution of mercury(II) chloride to cover the aluminium completely and allow to react for about 2 minutes; pour off the supernatant solution and wash the amalgam with water, rectified spirit and finally with moist ether. Cover the amalgam with about 1.5 litres of moist ether, when it is then ready for immediate use. If another solvent, e.g., methyl or ethyl acetate, is to be employed in the reduction with moist aluminium amalgam, the ether may of course be replaced by this solvent.

3. **Aluminium chloride (anhydrous).** This reagent may be purchased in the form of small pellets which are easily and quickly crushed in a mortar immediately prior to addition to the reaction flask (as for example in a Friedel–Crafts reaction, Section **IV,6**), and covered with solvent. In these cases the small amount of hydrolysis which occurs during the grinding operation is not unduly harmful to the reaction yield. This pelleted material does unfortunately deteriorate fairly rapidly on reaction with atmospheric moisture on continued opening and closing of the reagent bottle. The material should therefore be carefully inspected before use and if a large amount of powdery white material is present a fresh bottle should be used.

However, in some reactions a very high grade of anhydrous aluminium chloride may be required. This is conveniently prepared by placing the crushed pellets in a suitably sized round-bottomed flask fitted with a simple distillation bend to which is attached a two-necked round-bottomed receiver flask; the second outlet is connected to a water pump via a drying tower similar to that shown in Fig. **II,2** and filled with granular calcium chloride. The distillation flask is heated cautiously with a brush flame and the aluminium chloride sublimes under reduced pressure. It is inadvisable to use an oil immersion rotary pump because of possible corrosion damage even with suitably placed protection traps.

4. **Ammonia.** Gaseous ammonia is conveniently obtained from a cylin-

der of the liquefied gas; the cylinder must be equipped with a reducing valve. The rate of flow of the gas may be determined by passage through a bubble counter containing a small quantity of concentrated potassium hydroxide solution (12 g of KOH in 12 ml of water). A safety bottle should be inserted between the cylinder and the bubble counter and the reaction vessel; the gas may be dried by passage through a column loosely packed with soda lime or calcium oxide lumps (cf. Fig II,2). For reactions which require the use of liquid ammonia see Section **I,17,7**.

Small quantities of ammonia may if necessary be prepared with the aid of the apparatus depicted in Fig. II,2. Concentrated ammonia solution (d 0.88) is

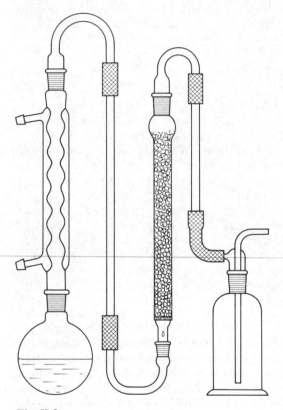

Fig. II,2

gently heated in the flask surmounted by an efficient reflux condenser. The gas is further dried by passage through a column which is loosely packed with soda lime or calcium oxide lumps,* and is then passed through a Drechsel bottle to act as a safety trap.

5. **Benzoyl peroxide.** The commercial product is cheap and is usually supplied moistened with approximately 25 per cent of water. Small quantities

---

* The column packing is retained by a plug of glass wool resting on the sintered glass disc fitted at the base of the glass column; a plug of glass wool is also inserted on the top of the column packing.

may be prepared in the laboratory from benzoyl chloride and hydrogen peroxide in the presence of alkali.

$$2C_6H_5 \cdot COCl + H_2O_2 \xrightarrow{\ominus OH} C_6H_5 \cdot CO \cdot O \cdot O \cdot CO \cdot C_6H_5 + 2HCl$$

Immerse a 600-ml beaker containing 50 ml (0.175 mol) of 12 per cent (40 volume) hydrogen peroxide and equipped with a mechanical stirrer, in an ice bath sited in a fume cupboard. Support two dropping funnels, containing respectively 30 ml of 4 M-sodium hydroxide solution and 30 g (25 ml, 0.214 mol) of redistilled benzoyl chloride (*lachrymatory*), with their stems inside the beaker. Add the two reagents alternately a few drops at a time, taking care that the temperature does not rise above 5–8 °C and that the solution is maintained faintly alkaline throughout. When all the reagents have been added, stir the solution for a further half an hour; by this time the odour of the benzoyl chloride should have disappeared. Filter off the flocculent precipitate at the pump, wash it with a little cold water and air dry upon filter-paper. The yield of benzoyl peroxide is 12 g (46%). It may be purified by dissolving in chloroform at *room temperature* and adding twice the volume of methanol. Benzoyl peroxide should *not* be recrystallised from hot chloroform, as a serious explosion may result. The compound melts at 106 °C with decomposition. Like all organic peroxides, benzoyl peroxide should be handled with care behind shatterproof screens, and horn or moulded polyethylene (*not* nickel) spatulas should be used. It is very shock sensitive.

To determine the exact peroxide content of benzoyl peroxide (and of other organic peroxides), the following procedure may be employed. Dissolve about 0.5 g, accurately weighed, of benzoyl peroxide in 15 ml of chloroform in a 350-ml conical flask. Cool to −5 °C, and add 25 ml of a 0.1 M solution of sodium methoxide in methanol in one portion with cooling and shaking. After 5 minutes at −5 °C, add 100 ml of iced water, 5 ml of 10 per cent sulphuric acid and 2 g of potassium iodide in 20 ml of 10 per cent sulphuric acid, in the order mentioned, with vigorous stirring. Titrate the liberated iodine with standard 0.10 M-sodium thiosulphate solution:

1 ml of 0.10 M-$Na_2S_2O_3 \equiv 0.0121$ g of benzoyl peroxide

6. **Boron trifluoride.** This gas, b.p. −101 °C, is available in cylinders and can be bubbled directly into the reaction mixture through a suitable safety trap. It is highly irritant and toxic and should only be handled in a suitable fume cupboard. For many preparative purposes it is convenient to use the commercially available boron trifluoride-etherate [$BF_3 \cdot (C_2H_5)_2O$] which contains 48 per cent w/w of boron trifluoride. This reagent is a colourless liquid which frequently becomes very dark on storage, but is readily purified before use by adding 2 per cent by weight of dry ether, to ensure an excess, and distilling from calcium hydride under reduced pressure (b.p. 46 °C/10 mmHg). Further convenient sources of boron trifluoride are the liquid boron trifluoride–acetic acid complex * which contains about 40 per cent w/w of boron trifluoride, and boron trifluoride–methanol † containing *ca.* 51 per cent w/w BF_3.

7. **Bromine.** CAUTION: Bromine is extremely corrosive and must be handled with great care (preferably using protective gloves) and always in the

---

* Available from Koch-Light Laboratories Ltd.
† Available from Aldrich Chemical Co. Ltd.

fume cupboard. The liquid produces painful burns and the vapour is an extremely powerful irritant. Bromine burns should be treated immediately with a liberal quantity of glycerol. If the vapour is inhaled, relief may be obtained by soaking a handkerchief in ethanol and holding it near the nose. The commercial product may be dried (and partially purified) by shaking with an equal volume of concentrated sulphuric acid, and then separating the acid. Chlorine, if present, may be removed by fractionation in an all-glass apparatus from pure potassium bromide: the b.p. of pure bromine is 59 °C/760 mmHg.

8. **N-Bromosuccinimide** $CH_2 \cdot CO \cdot NBr \cdot CO \cdot CH_2$. This reagent is available commercially and may be further purified by recrystallisation as rapidly as possible from ten times its weight of hot water or from glacial acetic acid.

It can be prepared from succinimide by dissolving the latter in a slight molar excess of chilled sodium hydroxide solution (of approximately 3 M strength) and adding rapidly with vigorous stirring one molar proportion of bromine dissolved in an equal volume of carbon tetrachloride. A finely crystalline white product is obtained which may be collected, washed with ice-cold water, dried and used directly or recrystallised as detailed above.

9. **Carbon dioxide.** This gas is conveniently generated from calcium carbonate chips (marble) and dilute hydrochloric acid (1 : 1) in a Kipp's apparatus; it should be passed through a Drechsel bottle containing water or sodium hydrogen carbonate solution to remove acid spray and, if required dry, through two further Drechsel bottles charged with concentrated sulphuric acid.

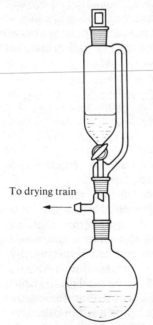

Large quantities of carbon dioxide may be obtained from a cylinder of the liquefied gas; the gas should be dried by passing it through two Drechsel bottles containing concentrated sulphuric acid. A little air is present in the gas.

For some purposes (e.g., in the Grignard reaction) solid carbon dioxide marketed as 'Cardice', 'Dry Ice' or 'Drikold' may be employed. Lumps of solid carbon dioxide should not be picked up with unprotected hands otherwise frostbite may result. If powdered material is required, the larger lumps should be wrapped in cloth and hammered. Solid carbon dioxide affords a convenient regulated supply of the gas when suitably sized lumps are allowed to evaporate in a Buchner flask attached to a sulphuric acid drying train.

To drying train

10. **Carbon monoxide.** Carbon monoxide is available in cylinders. *It is very poisonous and all operations involving its generation and use must be carried out in an efficient fume cupboard.* The gas may be prepared in the laboratory by the action of concentrated sulphuric acid upon concentrated formic acid (d 1.2; about 90 per cent w/w $H \cdot CO_2H$) at 70–80 °C. The apparatus shown in Fig. II,3 employs a 500-ml round-

Fig. II,3

bottomed flask fitted with a socket/cone adapter with 'T' connection and a stoppered pressure-equalising funnel. The gas is dried by passage through two Drechsel bottles containing concentrated sulphuric acid. The round-bottomed flask is charged with 125 g of concentrated sulphuric acid and 85 g of formic acid is slowly added from the dropping funnel; the pressure-equalising side arm allows the addition process to be carefully controlled and this minimises excessive frothing. Frothing may be further controlled by the addition of a small quantity of liquid paraffin. The steady stream of carbon monoxide which is evolved may contain traces of carbon dioxide and sulphur dioxide; these impurities may be removed, if desired, by passage upwards through a glass column filled with potassium hydroxide pellets (cf. Fig. II,2).

11. Chlorine. *Chlorine is a very toxic and irritant gas and all operations involving its generation and use must be carried out in an efficient fume cupboard.* For relatively large quantities of chlorine, cylinders of the gas fitted with an appropriate reducing valve should be used. The gas should be dried by passage through two Drechsel bottles containing concentrated sulphuric acid and then through a Drechsel bottle filled with glass wool to remove acid spray. Small quantities of chlorine are readily prepared by the action of concentrated hydrochloric acid upon potassium permanganate in the apparatus assembly shown in Fig. II,3. The calculated quantity of potassium permanganate (0.367 g $KMnO_4 \equiv 0.412$ g Cl_2) is placed in the round-bottomed flask and a slight excess of concentrated hydrochloric acid is placed in the pressure-equalising funnel which should then be stoppered and the stopper secured with an elastic band (1.000 g $KMnO_4$ requires 6.2 ml of concentrated hydrochloric acid). The chlorine is passed through a Drechsel bottle containing water to remove hydrogen chloride and is then dried by means of another Drechsel bottle charged with concentrated sulphuric acid; it is advisable to insert an empty Drechsel bottle to act as a safety trap between the reaction vessel and the dry chlorine generating source. The acid is allowed to drop slowly upon the permanganate crystals; the flask should be shaken from time to time. When about half the acid has been added, the evolution of gas tends to slow down; the flask should then be warmed slightly; after all the acid has been added the mixture is boiled gently to complete the evolution of chlorine.

12. Chlorosulphonic acid ($Cl \cdot SO_3H$). *This reagent must be handled with great care*; it is very corrosive to the skin and clothing, and reacts with water with great violence. If the specimen is impure, or discoloured, it should be distilled in an all-glass apparatus and the fraction, b.p. 148–150 °C (760 mmHg), collected; due precautions should be taken to protect the distillate from moisture.

13. Copper. (*a*) **Copper powder.** One hundred grams of recrystallised copper(II) sulphate are dissolved in 350 ml of hot water in a 1-litre beaker; a magnetic stirrer is provided. After cooling to the laboratory temperature, the stirrer is set in motion and 35 g (or more, if necessary) of purified zinc powder (see *67. Zinc*) are gradually added until the solution is decolourised. The precipitated copper is washed by decantation with water. Dilute hydrochloric acid (5%) is added to the precipitate in order to remove the excess of zinc, and stirring is continued until the evolution of hydrogen ceases. The copper powder is filtered, washed with water and kept in a moist condition (as a paste) in a stoppered bottle.

(*b*) **Activated copper bronze.** Commercial copper bronze does not always give satisfactory yields in the Ullmann reaction, but uniform results can be obtained by the following activation process. One hundred grams of copper bronze are treated with 1 litre of a 2 per cent solution of iodine in acetone for 5–10 minutes. This results in the production of a rather greyish colour due to the formation of copper iodide. The product is filtered off on a Buchner funnel, removed and washed by stirring with 500 ml of 1 : 1 solution of concentrated hydrochloric acid in acetone. The copper iodide dissolves, and the residual copper bronze is filtered and washed with acetone. It is then dried in a vacuum desiccator. The activitated copper bronze should be used immediately after preparation.

14. (*a*) **Copper–chromium oxide catalyst ('copper chromite' catalyst).** A reactive form of copper–chromium oxide suitable for hydrogenations (see Section **I,17,1**) may be obtained by the decomposition of basic copper ammonium chromate; the main reactions may be written as:

$$2Cu(NO_3)_2 + Na_2Cr_2O_7 + 4NH_3 + 3H_2O$$

$$\xrightarrow{55\,°C} 2CuNH_4(OH)CrO_4 + 2NaNO_3 + 2NH_4NO_3$$

$$\xrightarrow{300\,°C} CuO·CuCr_2O_4 + N_2 + 5H_2O$$

The most active forms contain barium chromite, which is incorporated by adding barium nitrate to the reaction mixture. The barium in the catalyst gives protection against sulphate poisoning and is said to stabilise the catalyst against reduction.

Dissolve 15.5 g of barium nitrate [AnalaR] and 130 g of copper(II) nitrate trihydrate [AnalaR] in 450 ml of water at 80 °C. Prepare a solution of sodium chromate by dissolving 89 g of recrystallised sodium dichromate dihydrate in 200 ml of water and adding 112.5 ml of concentrated ammonia solution (*d* 0.880). Add the warm solution (80 °C) of nitrates in a thin stream, with stirring, to the sodium chromate solution (at 25 °C). Collect the orange precipitate by suction filtration, wash it with two 50 ml portions of water, drain well and dry at 75–80 °C for 12 hours; powder finely.

Equip a 500-ml three-necked flask with a funnel for introducing a solid, a wide air condenser and a stainless steel stirrer with crescent blade, 1 cm long and 8 cm wide, so shaped that it conforms to the bottom of the flask. Immerse the flask in a metal bath at 350 °C. Add the powder through the funnel, with rapid stirring, during a period of 15 minutes. Heat with stirring at a bath temperature of 350 °C for 20 minutes after all the solid has been added. Leach the product by stirring for 30 minutes with 300 ml of 10 per cent acetic acid at room temperature (1). Allow to settle, decant the solution and wash the residue with six 50–60 ml portions of water. Filter with suction on a Buchner funnel, dry at 125 °C for 12 hours and grind finely in a mortar. The yield of catalyst (a brownish-black powder) is 85 g. No special precautions are necessary in handling or storing the catalyst since it is unaffected by exposure to air or moisture.

Note. (1) This treatment removes a substantial proportion of the excess of cupric oxide and renders the catalyst more active for hydrogenation purposes.

(*b*) **Copper–chromium oxide on pumice.** This form of catalyst has particular value for the dehydrogenation of primary and secondary alcohols to the corresponding carbonyl compounds (see Section **III,80**). Dissolve 10.4 g of barium nitrate [AnalaR] in 280 ml of water at about 80 °C and add to this hot solution 87 g of copper(II) nitrate trihydrate [AnalaR]; stir the mixture and heat until a homogeneous solution results. Prepare a solution of 50.4 g of recrystallised ammonium dichromate in a mixture of 200 ml of water and 75 ml of concentrated ammonia solution (d 0.880). To the ammonium chromate solution at 25–30 °C add the hot (80 °C) nitrate solution in a thin stream with stirring. Allow the mixture to cool and filter off the yellowish-brown precipitate with suction; press with a glass stopper and suck as dry as possible. Transfer the precipitate of copper barium ammonium chromate to a large evaporating dish, add sufficient water to form a moderately thick paste and introduce pumice (4–8 mesh) with stirring until most of the paste has been transferred to the pumice; about 300 g of pumice are required. Heat on an electric hot plate until the particles of pumice no longer adhere one to another. Remove the impregnated pumice (yellowish-brown) to a small evaporating dish and heat, by means of a Bunsen flame, with stirring until the colour changes through brown to a uniform black.

15. **Copper(I) bromide.** Dissolve 45 g (0.18 mol) of copper(II) sulphate pentahydrate and 19 g (0.19 mol) of sodium bromide in 150 ml of water. Add, with stirring, a solution of 11.8 g of sodium metabisulphite in 120 ml of water to the hot solution during 5 minutes (1). If the blue colour is not completely discharged, add a little more sodium metabisulphite. Cool the mixture and decant the supernatant liquor. Wash the precipitate by decantation with water containing a little dissolved sulphur dioxide to prevent oxidation. A solution of copper(I) bromide may be prepared by dissolving the moist solid in 30 ml of constant boiling point hydrobromic acid (48% w/w HBr). If however the solid salt is required collect the precipitate in a Buchner funnel, wash with water containing a little dissolved sulphur dioxide, followed by ethanol and then ether each containing sulphur dioxide and press with a glass stopper. Remove residual solvent and finally dry the solid in a vacuum desiccator over sulphuric acid and potassium hydroxide.

$$4CuSO_4 + 4NaCl + Na_2S_2O_5 + 3H_2O \longrightarrow 4CuCl + 6NaHSO_4$$

Note. (1) Alternatively pass a stream of sulphur dioxide through the heated solution (60 °C) for about 2 hours.

An alternative means of preparation of a solution of copper(I) bromide involves heating under reflux a mixture of 63 g (0.25 mol) of copper(II) sulphate pentahydrate, 20 g (0.314 mol) of copper turnings, 114 g (1.109 mol) of sodium bromide, 30 g (16.3 ml) of concentrated sulphuric acid and 1 litre of water for 3–4 hours. If the colour of the solution has not become yellowish after this period of heating, a few grams of sodium sulphite should be added to complete the reduction.

$$CuSO_4 + Cu + 2NaBr \longrightarrow 2CuBr + Na_2SO_4$$

16. **Copper(I) chloride.** Dissolve 35 g (0.14 mol) of copper(II) sulphate pentahydrate and 9.2 g (0.157 mol) of pure sodium chloride in 125 ml of water; warming may be necessary. Add a solution of 8.4 g (0.044 mol) of sodium metabisulphite (1) in 90 ml of water to the hot solution during about 5 minutes with constant shaking. Cool to room temperature (use an ice bath if necessary) and decant the supernatant liquor from the colourless copper(I) chloride. Wash the precipitate twice by decantation with water containing a little dissolved sulphur dioxide, the latter to prevent oxidation. For most purposes a solution of copper(I) chloride is required: the moist copper(I) chloride is dissolved in 60 ml of concentrated hydrochloric acid. The solution should be used within 24 hours of its preparation as it tends to oxidise (and therefore darken) on keeping. If it is not to be used immediately, the solution is kept in a tightly stoppered bottle (2). If the dry solid copper(I) chloride is required, the moist solid should be washed several times by decantation with water containing sulphur dioxide, collected on a Buchner funnel, washed several times with small portions of glacial acetic acid and dried in an air oven at 100–120 °C until the odour of acetic acid has disappeared. The copper(I) chloride thus obtained has a pure white colour and should be kept in a tightly stoppered bottle. The yield is almost quantitative.

$$4CuSO_4 + 4NaCl + Na_2S_2O_5 + 3H_2O \longrightarrow 4CuCl + 6NaHSO_4$$

Notes. (1) The reduction may also be effected with an alkaline solution of sodium sulphite prepared from 8.4 g of sodium metabisulphite and 7.0 g of sodium hydroxide pellets in 90 ml of water. The reaction is probably:

$$2CuSO_4 + 4NaCl + Na_2SO_3 + H_2O \longrightarrow 2CuCl + 3Na_2SO_4 + 2HCl$$

(2) An alternative means of preparation of a solution of copper(I) chloride is given under Section **IV,80**.

17. **Copper(I) cyanide.** The preparation is based essentially upon the reaction formulated below. It should be carried out in the fume cupboard and great caution exercised in the handling of the cyanides.

$$4CuSO_4 + 4NaCN + Na_2S_2O_5 + 3H_2O \longrightarrow 4CuCN + 6NaHSO_4$$

Place 500 g (2 mol) of powdered copper(II) sulphate pentahydrate in a 3-litre beaker or round-bottomed flask equipped with a stirrer and dissolve in 1600 ml of water at 40–50 °C. Prepare solutions of 140 g (0.74 mol) of commercial sodium metabisulphite in 400 ml of water (*A*) and of 140 g (2.15 mol) of potassium cyanide (96–99% purity) in 400 ml of water (*B*), and filter if necessary from small amounts of insoluble matter. Warm solutions *A* and *B* separately to 60 °C. Make the copper(II) sulphate solution faintly acid to Congo red with dilute sulphuric acid and add solution *A* with mechanical stirring during 1–2 minutes, followed immediately by solution *B*. There is a slight frothing, a little sulphur dioxide is evolved, but no appreciable amount of cyanogen or hydrogen cyanide. After about 10 minutes, filter the hot solution and wash the product thoroughly with boiling water, and finally with rectified spirit. Dry at 100–110 °C to a fine soft powder (24–36 hours). The yield is 167 g (93%).

Copper(I) cyanide solution. The most satisfactory method of preparation is to dissolve the copper(I) cyanide (90 g, 1 mol) in a solution of sodium cyanide (125 g, 2.5 mol) in 600 ml of water. If it is desired to avoid the preparation of solid copper(I) cyanide, the following procedure may be adopted. Copper(I) chloride, prepared from 35 g of copper(II) sulphate pentahydrate as described under *16* above, is suspended in 60 ml of water contained in a 500-ml round-bottomed flask, which is fitted with a mechanical stirrer. A solution of 18.5 g of sodium cyanide (96–98%) in 30 ml of water is added and the mixture is stirred. The copper(I) chloride passes into solution with considerable evolution of heat. As the copper(I) cyanide is usually employed in reactions with solutions of aryl diazonium salts it is usual to cool the resulting copper(I) cyanide solution in ice.

18. **Dialkyl sulphates.** (*a*) **Dimethyl sulphate** $[(CH_3)_2SO_4]$ is a liquid of b.p. 188.5 °C and is practically without odour. *Both vapour and liquid dimethyl sulphate are highly poisonous and the substance should only be used in a fume cupboard with a good draught, and rubber gloves should be worn.* Inhalation of vapour may lead to giddiness or even more serious results. The liquid itself is readily absorbed through the skin, with toxic results. If the liquid is accidentally splashed upon the hands, wash immediately with much concentrated ammonia solution in order to hydrolyse the compound before it can be absorbed through the skin; then rub gently with a wad of cotton wool soaked in ammonia solution.

Commercial dimethyl sulphate may be purified by, (*a*) allowing it to stand over anhydrous potassium carbonate until it is neutral to Congo red paper, or (*b*) by washing, just before use, with an equal volume of ice-water, followed by one-third of its volume of cold, saturated sodium hydrogen carbonate solution, and finally drying over calcium oxide. In both cases the purified dimethyl sulphate is fractionally distilled under reduced pressure from calcium oxide and the fraction having b.p. 72–73 °C/13 mmHg collected.

(*b*) **Diethyl sulphate** $[(C_2H_5)_2SO_4]$. Although diethyl sulphate is somewhat less poisonous than dimethyl sulphate, similar precautions in the use and handling of this reagent should be observed; all operations should be conducted in the fume cupboard and rubber gloves should be worn. If the diethyl sulphate is dark in colour, it should be placed in a separatory funnel of suitable size and washed first with ice-water, then with aqueous sodium hydrogen carbonate solution until free of acid and finally dried over successive portions of calcium oxide. The dried diethyl sulphate is fractionally distilled from calcium oxide and has b.p. 93 °C/13 mmHg.

19. **Diazomethane.** Liquid diazomethane, CH_2N_2, b.p. -24 °C, is an explosive compound and explosions may also occur in the gaseous state if the substance is dry and undiluted. The gas may be handled with safety by diluting it with nitrogen. For synthetic work, a dry dilute ethereal solution of the gas is employed and this can be handled with reasonable safety, although such solutions have also been known to explode. *However, due regard must be paid to the highly toxic character of the gas by carrying out all operations in an efficient fume cupboard.*

An ethereal solution of diazomethane is usually prepared immediately before it is required for reaction. A convenient precursor for the generation of diazomethane is *N*-methyl-*N*-nitrosotoluene-*p*-sulphonamide which is prepared by the action of nitrous acid on *N*-methyltoluene-*p*-sulphonamide. This latter is formed from toluene-*p*-sulphonyl chloride and methylamine in alkaline solution.

$$p\text{-}CH_3 \cdot C_6H_4 \cdot SO_2Cl + CH_3NH_2 + NaOH \longrightarrow$$
$$p\text{-}CH_3 \cdot C_6H_4 \cdot SO_2NH \cdot CH_3 + NaCl + H_2O$$

$$p\text{-}CH_3 \cdot C_6H_4 \cdot SO_2NH \cdot CH_3 + HNO_2 \longrightarrow$$
$$p\text{-}CH_3 \cdot C_6H_4 \cdot SO_2N(NO) \cdot CH_3 + H_2O$$

N-methyl-*N*-nitrosotoluene-*p*-sulphonamide

The methylnitrosamide affords diazomethane on reaction with potassium hydroxide solution.

$$p\text{-}CH_3 \cdot C_6H_4 \cdot SO_2N(NO)CH_3 + KOH \longrightarrow$$
$$p\text{-}CH_3 \cdot C_6H_4 \cdot SO_3K + CH_2N_2 + H_2O$$

N-*Methyl*-N-*nitrosotoluene*-p-*sulphonamide*.* Divide 320 g (1.68 mol) of purified toluene-p-sulphonyl chloride, m.p. 68–69 °C (Section II,2,65) into three portions of 190, 90 and 40 g. Prepare a solution of 70 g (1.75 mol) of sodium hydroxide in 70 ml of water and cool to room temperature. Place 210 ml (2.25 mol) of 33 per cent aqueous methylamine solution (or 174 ml of the 40% aqueous solution) in a 1-litre round-bottomed flask and add the 190 g of toluene-p-sulphonyl chloride in portions with swirling during about 5 minutes. The mixture becomes warm. Allow the temperature to rise to 80–90 °C in order to maintain the N-methyltoluene-p-sulphonamide (m.p. 78 °C) in a molten condition, otherwise the latter may form a hard cake and reaction may be incomplete; also do not permit the temperature to rise above 90 °C as appreciable loss of methylamine may result. The mixture should be acid to litmus within 5 minutes after the completion of the first addition of the sulphonyl chloride (1). Then add 50 ml of the 50 per cent sodium hydroxide solution carefully with swirling, followed immediately by 90 g of the sulphonyl chloride in portions as before. When the mixture has again become acidic (1), introduce 25 ml of the sodium hydroxide solution, followed by 40 g of toluene-p-sulphonyl chloride with vigorous swirling. After the mixture has again become acidic, add the remainder of the sodium hydroxide solution. The liquid phase of the final mixture should be alkaline; if it is acidic, indicating excessive loss of methylamine, add sufficient methylamine solution to render the mixture basic.

Rinse the walls of the flask with a little water and complete the reaction by heating the mixture (which consists of two layers and a precipitate of sodium chloride) on a boiling water bath for 15 minutes with vigorous mechanical stirring. Pour the hot reaction mixture into 1500 ml of glacial acetic acid contained in a 4-litre round-bottomed flask; rinse the flask with 250 ml of acetic acid. Cool the solution in an ice bath to 5 °C (2), stir mechanically and add a solution of 125 g (1.8 mol) of sodium nitrite in 250 ml of water from a dropping funnel during about 45 minutes: maintain the temperature below 10 °C and continue the stirring for 15 minutes after the addition is complete. The nitroso derivative separates as a yellow crystalline solid during the reaction. Add 1 litre of water to the reaction mixture and collect the precipitate by suction filtration; press it on the funnel and wash with about 500 ml of water. Transfer the product to a beaker, stir it well with about 400 ml of water, then filter and wash again on the funnel until the odour of acetic acid is no longer apparent. Dry to constant weight in a vacuum desiccator over concentrated sulphuric acid. The yield of N-methyl-N-nitrosotoluene-p-sulphonamide, m.p. 58–60 °C, is 325 g (90%). This is sufficiently pure for the preparation of diazomethane. It should be kept in a dark bottle. It may be recrystallised by dissolution in boiling ether (1 ml/g), addition of an equal volume of light petroleum, b.p. 40–60 °C, and cooling in a refrigerator. This reagent may be stored for long periods at room temperature in a dark bottle without appreciable decomposition.

Notes. (1) Occasionally the liquid may not become acidic after the first or second addition, even though the sulphonyl chloride has reacted completely (this is due to a smaller loss of methylamine than is expected). If such is the case, no more than 5 minutes should be allowed between successive additions

---

* These experimental conditions are reproduced by kind permission of Professor H. J. Backer.

of sulphonyl chloride and alkali. The whole procedure occupies about 30 minutes.

(2) A reaction temperature below 0 °C should be avoided because the total volume of acetic acid is just sufficient to keep the N-methyltoluene-p-sulphonamide in solution above 0 °C.

Preparation of diazomethane. CAUTION: This preparation should be carried out only in a fume cupboard provided with a powerful exhaust system. The use of a screen of safety glass in addition to wearing safety spectacles is strongly recommended. Since the explosive decomposition of diazomethane is frequently initiated by rough or sharp surfaces the use of apparatus with ground glass joints and of boiling chips in the subsequent distillation is *NOT* recommended. New glass apparatus with rubber stopper connections carrying fire polished glass connecting tubes, or glass apparatus with Clearjoints should be used. A further precaution against explosion is to ensure that the apparatus assembly, located in a fume cupboard, is not exposed to direct sunlight or placed near a strong artificial light.

The following procedures may be used for the preparation of ethereal solutions of diazomethane containing ethanol; they differ slightly according to whether large or small quantities are required. The presence of ethanol is not harmful in many uses of diazomethane.

Method 1. Add 50 ml of 96 per cent ethanol to a solution of 10 g (0.18 mol) of potassium hydroxide in 15 ml of water. Place this solution in a 200-ml distillation flask equipped with a dropping funnel and an efficient double surface condenser. Fit by means of rubber bungs and glass tube connectors two receiving flasks (of 500 ml and 100 ml capacity) in series to the end of the condenser; cool both the flasks in an ice–salt mixture and arrange that the inlet tube leading into the second smaller receiver flask dips below the surface of a charge of 40 ml of ether. This will serve to trap any uncondensed diazomethane–ether vapour escaping from the first receiver. Heat the distilling flask in a water bath at 60–65 °C; place a solution of 43 g (0.2 mol) of N-methyl-N-nitrosotoluene-p-sulphonamide in about 250 ml of ether in the dropping funnel and introduce it into the flask over a period of 45 minutes. Adjust the rate of addition so that it is about equal to the rate of distillation. When the dropping funnel is empty, add more ether (*ca.* 30 ml) gradually until the ether distilling over is colourless. The combined ethereal solutions in the receivers contain 5.9–6.1 g (70%) of diazomethane (1).

Method 2. For smaller quantities of diazomethane, the use of a dropping funnel is unnecessary. Dissolve 2.14 g of N-methyl-N-nitrosotoluene-p-sulphonamide in 30 ml of ether, cool in ice and add a solution of 0.4 g of potassium hydroxide in 10 ml of 96 per cent ethanol. If a precipitate forms, add more ethanol until it just dissolves. After 5 minutes, distil the ethereal diazomethane solution from a water bath. The ethereal solution contains 0.32–0.35 g of diazomethane (1).

Method 3. An ethereal solution of diazomethane free from ethanol may be prepared by this method: such a solution is required, for example, in the Arndt–Eistert reaction with acid chlorides (compare Section **III,130**). In a 100-ml distilling flask provided with a dropping funnel and an efficient downward condenser, place a solution of 6 g of potassium hydroxide in 10 ml of water, 35 ml of carbitol (diethyleneglycol monoethyl ether) and 10 ml of water: connect the condenser to two conical flasks in series containing 10 and 35 ml of ether respectively and cooled in an ice–salt bath (see *Method 1*). Heat the mixture on a water bath at 70–75 °C in a beaker of water placed upon a hot plate incorporating a magnetic stirring unit with a Teflon-coated follower situated in the flask. As soon as the ether commences to distil, add a solution of 21.5 g of *N*-methyl-*N*-nitrosotoluene-*p*-sulphonamide in 125 ml of ether through the dropping funnel during a period of about 15 minutes, stirring electromagnetically. After the addition of the nitrosoamide, add 30–40 ml of ether through the dropping funnel and distil until the distillate is colourless. The ethereal solution in the conical flasks contains about 3.4 g of diazomethane (1).

The diazomethane–ether solutions prepared by all these methods should be dry. If in doubt, it may be dried with potassium hydroxide pellets. The anhydrous ethereal solution may be stored in the fume cupboard in a smooth glass flask or bottle for a day or so at − 70 °C; since slow decomposition occurs with liberation of gas, the containing vessel should be protected by a calcium chloride guard-tube.

Note. (1) **To determine the exact diazomethane content,** allow an aliquot portion of the ethereal diazomethane solution to react with an accurately weighed amount (say, about 1 g) of benzoic acid [AnalaR] in 50 ml of anhydrous ether. The solution should be completely decolourised, thus showing that the benzoic acid is present in excess. Dilute the solution with water and titrate the excess of benzoic acid with standard 0.1 *M* alkali using phenolphthalein as indicator.

General methylation procedure with diazomethane. *The reaction must be carried out in the fume cupboard.* Dissolve 2–3 g of the compound (say, a phenol or a carboxylic acid) in a little anhydrous ether or absolute methanol, cool in ice and add the ethereal solution of diazomethane in small portions until gas evolution ceases and the solution acquires a pale yellow colour. Test the coloured solution for the presence of excess of diazomethane by removing a few drops into a test-tube and introducing a glass rod moistened with glacial acetic acid: immediate evolution of gas should occur. Evaporate the solvent, and purify the product by distillation or crystallisation.

20. **Diborane** (solution in tetrahydrofuran). Although a 1 *M*-BH$_3$ (borine) solution in tetrahydrofuran is available commercially it is often more economic to prepare the reagent as required by adding a solution of sodium borohydride to boron trifluoride-etherate and sweeping the resulting diborane (B$_2$H$_6$) into tetrahydrofuran with the aid of a nitrogen gas stream.

$$3NaBH_4 + 4BF_3 \longrightarrow 4BH_3 + 3NaBF_4$$

Fit the central neck of a 500-ml round-bottomed two-necked flask (the generating flask), supported on a magnetic stirrer unit, with a pressure-equalising dropping funnel connected to a nitrogen supply via a suitable mercury safety valve (cf. Fig. I,67). Connect with Tygon or polyethylene tubing the side-neck of the generating flask to a Drechsel wash bottle, the outlet of which leads into a 500-ml two-necked round-bottomed flask via a gas inlet tube terminating in a glass frit (cf. Fig. I,66). Charge this flask with 400 ml of dry tetrahydrofuran (Section II,1,*19*) and close the second neck with a calcium chloride guard-tube. Dissolve 13.3 g (0.35 mol) of sodium borohydride (Section II,2,*58*) in 350 ml of dry diglyme (Section II,1,*18*) and place 300 ml (0.3 mol) of this solution in the dropping funnel. Use the remainder of the solution to adequately charge the

Drechsel wash bottle so that the diborane subsequently generated is scrubbed free of boron trifluoride. Place 85 g (0.6 mol) of redistilled boron trifluoride etherate (Section **II,2,6**) in the generating flask, insert a follower bar and allow nitrogen to pass through the assembly for a period of about 15 minutes to displace all the air. Generate the diborane by adding the sodium borohydride solution steadily to the stirred boron trifluoride-etherate. When all the solution has been added continue to pass the nitrogen gas for a further period of 15 minutes to ensure that all the diborane has been swept into the tetrahydrofuran solution. This will also ensure that little diborane gas will escape to the atmosphere when the apparatus is disconnected; diborane is highly toxic and may contain impurities which can lead to spontaneous ignition.

The final tetrahydrofuran solution prepared in this way is approximately 1 M with respect to borine; the solution is reasonably stable if stored between 0 and 5 °C.

An alternative technique (which uses essentially the same apparatus assembly as described above) involves passage of diborane directly into a solution of the organic substrate in tetrahydrofuran, diglyme or ether to effect immediate reaction. This requires the additional facility of magnetic stirring and of cooling the reaction mixture.

21. **N,N-Dicyclohexylcarbodiimide**(DCC; $C_6H_{11}\cdot N = C = N\cdot C_6H_{11}$). The reagent is a potent skin irritant and must be handled with care. Material of good quality (*ca.* 99% pure) is available commercially as a waxy low-melting solid, m.p. 34–35 °C. It is most convenient to liquefy the reagent by the use of a hot-air blower or by standing the reagent bottle in a little warm water to assist weighing out.

In most of its applications the reagent acts as a dehydrating condensing agent and is often recovered as dicyclohexylurea. Although the reagent is not unduly expensive, it may be desirable to make use of the recovered urea; the latter may be converted into the diimide by treatment in pyridine solution with toluene-*p*-sulphonyl chloride, phosphorus oxychloride or phosphoric oxide.

Preparation from dicyclohexylurea. Recrystallise the recovered urea from ethanol; m.p. 234 °C. Add dropwise with stirring 17.1 g (47 ml, 0.11 mol) of phosphorus oxychloride to 22.5 g (0.1 mol) of dicyclohexylurea in 50 ml of pyridine at 50 °C, and heat at 60–90 °C for 1.5 hours. Pour the reaction product on to crushed ice, extract with light petroleum (b.p. 60–80 °C) and dry the extract over anhydrous sodium sulphate. Remove the solvent using a rotary evaporator and distil the residual oil under reduced pressure. The yield of diimide, b.p. 157–159 °C/15 mmHg (131 °C/3–4 mmHg), is about 14 g (68%).

22. **Ethylene oxide.** The reagent has b.p. 11 °C and is supplied either in 100-ml sealed tubes or in 100-ml cylinders equipped with an appropriate valve. The gas, which is highly inflammable, has no very distinctive smell and must be regarded as a hazardous toxic reagent which must not be inhaled or allowed to come into contact with the skin and eyes. Precautions in the use of ethylene oxide are described in Section **III,34**, which may be regarded as typical.

23. **Formaldehyde.** Commercial formalin is an aqueous solution containing 37–40 per cent w/v of formaldehyde (0.37–0.40 g HCHO per ml), stabilised by the addition of 12 per cent of methanol.

When dry gaseous formaldehyde is required it may be obtained by the depolymerisation of paraformaldehyde at 180–200 °C. For reactions at elevated temperatures dried paraformaldehyde may be used as the *in situ* source of formaldehyde (for details of these techniques see Section **III,33**, Note (2)).

24. Girard's reagents 'T' and 'P'. Girard's reagent 'T' is carbohydrazido-methyltrimethylammonium chloride (I) (trimethylaminoacetohydrazide chloride) and is prepared by the reaction of the quaternary ammonium salt formed from ethyl chloroacetate and trimethylamine with hydrazine hydrate in alcoholic solution:

$$(CH_3)_3N + ClCH_2 \cdot CO_2C_2H_5 \longrightarrow \overset{\ominus}{Cl}\{(CH_3)_3\overset{\oplus}{N} \cdot CH_2 \cdot CO_2C_2H_5$$

$$\overset{\ominus}{Cl}\{(CH_3)_3\overset{\oplus}{N} \cdot CH_2 \cdot CO_2C_2H_5 + H_2N \cdot NH_2 \longrightarrow$$

$$\overset{\ominus}{Cl}\{(CH_3)_3\overset{\oplus}{N} \cdot CH_2 \cdot CO \cdot NH \cdot NH_2 + C_2H_5OH$$
$$(I)$$

Girard's reagent 'P' is the corresponding pyridinium compound, prepared by replacing the trimethylamine by pyridine. The reagent 'T', unlike the reagent 'P', is very deliquescent, but is nevertheless widely used for laboratory work because of its greater solubility. The quaternary ammonium grouping imparts water solubility.

The main use of the Girard reagents 'T' and 'P' is for the isolation of small amounts of ketones from admixture with other organic matter contained in, for example, various natural products; the carbonyl derivatives are water soluble. The ketonic material, dissolved in ethanol containing 10 per cent acetic acid, is heated for 30–60 minutes with the reagent in slight excess, the volume being adjusted to give a 5 to 10 per cent solution of the reagent. The cooled solution is diluted with water containing enough alkali to neutralise 90 per cent of the acid and to give an alcohol content of 10–20 per cent. It is then exhaustively extracted with ether to remove non-ketonic compounds; the water-soluble hydrazone derivatives are decomposed by the addition of mineral acid up to a concentration of 0.5 M and, after about 1 hour at room temperature, the liberated ketonic compound is isolated by extraction with ether.

$$\text{Girard's reagent 'T'} + RR'C = O \underset{HCl}{\overset{C_2H_5OH,HOAc}{\rightleftarrows}}$$

$$(CH_3)_3\overset{\oplus}{N} \cdot CH_2 \cdot CO \cdot NH \cdot N = CRR'\}\overset{\ominus}{Cl} + H_2O$$

Girard's reagent 'T'. Place a solution of 98.5 g (84.5 ml, 0.8 mol) of ethyl chloroacetate and 200 ml of absolute ethanol in a 1-litre, three-necked flask, fitted with a thermometer, stirrer and an ice-cooled condenser (Fig. **I,25**). Cool the solution to 0 °C by stirring in an ice–salt bath, stop the stirrer and add 49 g (74 ml, 0.83 mol), measured after pre-cooling to − 5 °C, of trimethylamine all at once. Control the exothermic reaction sufficiently by external cooling so that the temperature of the mixture rises to 60 °C during about 1 hour. When there is no further evolution of heat, allow the reaction mixture to stand at room temperature for 20–24 hours. Remove the condenser, replace the thermometer by a dropping funnel and add 40 g (0.8 mol) of 100 per cent hydrazine hydrate (Section **II,2,**25) with stirring during 10–15 minutes. Stir for a further 45 minutes, cool the solution slightly and, unless crystallisation commences spontaneously, scratch the walls of the vessel with a glass rod to induce crystallisation. The product separates in fine colourless needles. Cool in an ice bath, collect the highly hygroscopic salt rapidly on a Buchner funnel, wash with 150 ml of cold absolute ethanol and press dry. Dry the product in a vacuum desiccator over concentrated sulphuric acid; the yield is 105 g (85%), m.p. 175–180 °C (decomp.). (This material (1) contains a small amount of the symmetrical dihydrazide, but is quite satisfactory as a reagent for the separation of ketones.) A further crop of 12 g may be obtained after distilling off 200–300 ml of solvent from the mother-liquor and washings at the pressure of the water pump.

Note. (1) The deliquescent solid must be stored in a dry, tightly-stoppered container. If exposed to the air it deteriorates rapidly, developing an unpleasant odour. Samples that have been kept for some time are best recrystallised from absolute ethanol before use.

Girard's reagent 'P' ($C_5H_5\overset{\oplus}{N}\cdot CH_2\cdot CO\cdot NH\cdot NH_2\}\overset{\ominus}{Cl}$).   In a 1-litre, three-necked flask, equipped with a sealed stirrer unit, a dropping funnel and a reflux condenser, place 200 ml of absolute ethanol, 63 g (64.5 ml, 0.8 mol) of pure anhydrous pyridine and 98.5 g (84.5 ml, 0.8 mol) of ethyl chloroacetate. Heat the mixture under reflux for 2–3 hours until the formation of the quaternary salt is complete; acidify a small test-portion with dilute sulphuric acid; it should dissolve completely and no odour of ethyl chloroacetate should be apparent. Cool the mixture in ice and salt. Run in rapidly from a dropping funnel a solution of 40 g (0.8 mol) of 100 per cent hydrazine hydrate in 50 ml of absolute ethanol all at once. A vigorous exothermic reaction soon develops and is accompanied by vigorous effervescence. The product separates almost immediately. When cold, filter with suction, wash with ice-cold ethanol and dry in the air. The yield of Girard's reagent 'P' is 135 g (90%); this is satisfactory for the isolation of ketones. A pure product may be obtained by recrystallisation from methanol.

25. **Hydrazine hydrate.**   Hydrazine hydrate containing 60 per cent w/w and 98–100 per cent w/w $NH_2\cdot NH_2\cdot H_2O$ is available commercially (1). A hydrazine solution, 60 per cent w/w of hydrazine hydrate, may be concentrated as follows. A mixture of 150 g (144 ml) of the solution and 230 ml of xylene is distilled from a 500-ml round-bottomed flask through a well-lagged Hempel (or other efficient fractionating) column in an atmosphere of nitrogen. All the xylene passes over with about 85 ml of water. Upon distillation of the residue, about 50 g of 90–95 per cent hydrazine hydrate (2) are obtained.

Notes. (1) Solutions of hydrazine hydrate are extremely corrosive; protective gloves and safety spectacles should always be worn when handling this reagent.

(2) Hydrazine hydrate may be titrated with standard acid using methyl orange as indicator or, alternatively, against standard iodine solution with starch as indicator. In the latter case about 0.1 g, accurately weighed, of the hydrazine hydrate solution is diluted with about 100 ml of water, 2–3 drops of starch indicator added, and immediately before titration 5 g of sodium bicarbonate are introduced. Rapid titration with iodine gives a satisfactory end-point.

$$5NH_2\cdot NH_2\cdot H_2O + 2I_2 \longrightarrow 4NH_2\cdot NH_2\cdot HI + 5H_2O + N_2$$

Anhydrous hydrazine may be obtained by refluxing 100 per cent hydrazine hydrate (or the 95% material which has previously been allowed to stand overnight over 20% w/w of potassium hydroxide and filtered) with an equal weight of sodium hydroxide pellets for 2 hours and then distilling in a slow stream of nitrogen; b.p. 114–116 °C. Distillation in air may lead to an explosion.

26. **Hydriodic acid.** This is supplied as an azeotrope with water (constant boiling hydriodic acid), b.p. 125.5–126.5 °C/760 mmHg, d 1.70, which contains 55–57 per cent w/w HI (0.936 to 0.99 g HI per ml). Additional grades available contain 45 per cent w/w HI and 67 per cent w/w HI; the latter is stabilised by the addition of 0.03 per cent w/w of hypophosphorous acid.

The constant boiling azeotrope may be conveniently prepared in the laboratory, if required, by the following procedure. In a fume cupboard a 1.5-litre, three-necked flask is charged with a mixture of 480 g of iodine and 600 ml of water. The central socket is fitted with an efficient mechanical stirrer which leads almost to the bottom of the flask, and the smaller sockets respectively with a lead-in tube for hydrogen sulphide extending to well below the surface of the liquid and with an exit tube attached to an inverted funnel just dipping into 5 per cent sodium hydroxide solution. The mixture is vigorously stirred and a stream of hydrogen sulphide (either from a freshly-charged Kipp's apparatus or from a cylinder of the gas) passed in as rapidly as it can be absorbed. After several hours the liquid assumes a yellow colour (sometimes it is almost colourless) and most of the sulphur sticks together in the form of a hard lump (1). The sulphur is removed by filtration through a funnel plugged with glass wool (or through a sintered glass funnel), and the filtrate is boiled until the lead acetate paper test for hydrogen sulphide is negative. The solution is filtered again, if necessary. The hydriodic acid is then distilled from a 500-ml Claisen flask, and the fraction, b.p. 125.5–126.5 °C/760 mmHg, is collected. The yield of the constant boiling acid containing 57 per cent w/w HI is 785 g (90%).

$$H_2S + I_2 \longrightarrow 2HI + S$$

Note. (1) The hard lump of sulphur remaining in the flask is best removed by boiling with concentrated nitric acid in the fume cupboard.

27. **Hydrobromic acid.** This is supplied as an azeotrope with water (constant boiling hydrobromic acid), b.p. 126 °C, d 1.46–1.49, which contains 47–48 per cent w/w HBr (0.695 to 0.715 g HBr per ml). A grade of hydrobromic acid is available containing 60 per cent w/w HBr (1.007 g HBr per ml).

28. **Hydrochloric acid.** This is supplied in grades containing either 32 per cent w/w HCl or 36 per cent w/w HCl (0.371 g HCl per ml and 0.424 g HCl per ml respectively).

29. **Hydrogen.** Cylinders of compressed hydrogen (99.9998% purity) may be purchased or hired; the impurities comprise traces of nitrogen, oxygen, moisture and hydrocarbons. Less pure grades containing slightly larger amounts of contaminating oxygen are however quite suitable for most purposes. Since hydrogen is chiefly employed for catalytic reductions and oxygen has, in general, no harmful effect upon the reduction, no purification is usually necessary. If, however, oxygen-free hydrogen is required, the oxygen may be removed by passage through Fieser's solution (see under 43. Nitrogen), and then through a wash bottle containing concentrated sulphuric acid to which some silver sulphate has been added; the latter will detect and remove

any hydrogen sulphide that may have formed from the decomposition of Fieser's solution.

30. **Hydrogen bromide.** Hydrogen bromide is most conveniently prepared by the action of bromine upon tetrahydronaphthalene (tetralin) if the quantity required does not justify the purchase of a cylinder of the gas.

$$C_{10}H_{12} + 4Br_2 \longrightarrow C_{10}H_8Br_4 + 4HBr$$

Only half of the added bromine is recovered as hydrogen bromide; a yield of about 45 per cent of hydrogen bromide is evolved computed on the weight of bromine taken. It is essential that the tetralin is pure and perfectly dry; it should therefore be allowed to stand over magnesium sulphate or anhydrous calcium sulphate for several hours, filtered and distilled under reduced pressure. The tetralin is placed in a round-bottomed flask fitted with a socket/cone adapter with 'T' connection and a dropping funnel (this may be of the pressure-equalising type (cf. Fig. II,3)). Bromine is allowed to drop in from the funnel at a regular rate and the contents of the flask are gently swirled from time to time to ensure a steady evolution of hydrogen bromide. The traces of bromine carried over with the gas may be removed by allowing it to bubble through a Drechsel bottle charged with dry tetralin. A safety trap should always be interposed between the generating apparatus and the reaction vessel.

A solution of **hydrogen bromide in glacial acetic acid** (45% w/v) is a convenient commercially available source, which is suitable for many preparations.

31. **Hydrogen chloride.** *Method 1 (from concentrated sulphuric acid and fused ammonium chloride).* The most convenient procedure is to allow concentrated sulphuric acid to react with lumps of fused ammonium chloride in a Kipp's apparatus.[*] The gas may be dried by passage through a Drechsel bottle containing concentrated sulphuric acid; the latter should be followed by an empty Drechsel bottle as a precaution against 'sucking back' of the contents of the reaction vessel.

Method 2 (from concentrated sulphuric acid and concentrated hydrochloric acid). The apparatus shown in Fig. II,4 is employed. The upper funnel has a capacity of 100 ml and has an appropriate length of capillary tubing fused to the outlet tube; the lower funnel has a capacity of 500 ml. When the capillary tube is filled with concentrated hydrochloric acid, there is sufficient hydrostatic pressure to force the hydrochloric acid into the sulphuric acid. The Drechsel bottle contains concentrated sulphuric acid and the other acts as a safety trap; the whole apparatus must be mounted on a heavy stand.

About 150 ml of concentrated sulphuric acid is placed in the larger funnel and 100 ml of concentrated hydrochloric acid in the smaller separatory funnel. The latter is raised from the adapter until the capillary tube is above the sulphuric acid, the capillary tube is carefully filled with concentrated hydrochloric acid, and the funnel is then lowered into the adapter socket. The rate of evolution of hydrogen chloride is controlled by regulation of the supply of hydrochloric acid: this will continue until a volume of hydrochloric acid equal

---

[*] An alternative and quickly assembled apparatus employs a Buchner flask fitted with a ground glass joint to which is attached a dropping funnel. Ammonium chloride is placed in the flask, moistened with concentrated hydrochloric acid, and concentrated sulphuric acid added dropwise from the funnel at such a rate that the evolution of gas may be controlled.

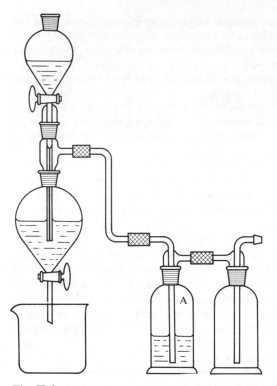

Fig. II,4

to that of the concentrated sulphuric acid has been used. The diluted sulphuric acid should then be removed and the apparatus recharged. The yield is 31–33 g of hydrogen chloride per 100 ml of concentrated hydrochloric acid.

32. **Hydrogen cyanide.**   Great care must be exercised in the preparation of this gas for it is a dangerous poison; all operations must be conducted in a fume cupboard provided with an efficient draught. The apparatus used is similar to, but generally on a smaller scale than, that described for *Hydrogen chloride, Method 2*, in which the larger separating funnel is replaced by a round-bottomed flask of suitable size. A saturated solution of sodium cyanide (prepared by dissolving 50 g of commercial sodium cyanide, about 96 per cent purity, in sufficient water to make 125 ml of solution) is added 1 cm below the surface of 125 ml of aqueous sulphuric acid (50% by volume) contained in the flask. Any residual hydrogen cyanide may be expelled by warming the flask on a water bath. The gas may be used directly, or may be collected in the liquid form (b.p. 26 °C) by passing through a glass coil, 4–5 cm bore and 50 cm long, surrounded by ice (Fig. I,25); a freezing mixture must not be used as this may result in solidification of the hydrogen cyanide (m.p. −15 to −14.5 °C) and consequent clogging of the apparatus. If the hydrogen cyanide is required anhydrous, it should be passed through a train of three large U-tubes filled with anhydrous calcium chloride, the tubes being immersed in a water bath maintained at 30–40 °C.

33. **Hydrogen iodide.**   This gas may be conveniently prepared by allowing a solution of two parts of iodine in one part of hydriodic acid (*d* 1.7) to drop on to excess of red phosphorus; the apparatus assembly employed could be of the general type illustrated in Fig. II,3. The evolution of hydrogen iodide takes place in the cold; when the evolution of gas slackens considerably, the mixture should be gently warmed.

34. **Hydrogen peroxide.** Aqueous solutions of hydrogen peroxide having concentrations of 6, 12, 30, 50 and 70 per cent w/v are available from most chemical suppliers. Higher concentrations of hydrogen peroxide, i.e., 86 per cent w/v, are also available* ('High Test Peroxide'). Care should be taken with all concentrations of hydrogen peroxide since explosions can occur on contact with organic material or with transition metals.

Even the highly concentrated solutions of hydrogen peroxide ($>50\%$) may be handled safely providing certain basic precautions are observed. Firstly it is advisable to wear protective spectacles and rubber or plastic gloves. Since these high-strength solutions may ignite textiles, it is also strongly advised that a rubber or plastic apron be worn and that all operations involving transfer of such solutions be conducted in the fume cupboard, with the additional precaution that all the apparatus is sited in a plastic or aluminium tray containing water to catch accidental spillage. Inhalation of vapour arising from use at elevated temperatures from these higher concentrations may cause inflammation of the nose and throat, and exposure of the eyes leads to ulceration of the cornea. Solutions of hydrogen peroxide spilt on the skin should be washed off immediately with water, when only temporary discomfort will then be experienced. Indeed, a ready supply of water should be to hand to wash away splashes and leakages.†

Frequently the strengths of solutions of hydrogen peroxide are quoted according to available oxygen content; thus 30 per cent w/v $\equiv$ 100 volumes, i.e., 1 ml of the H_2O_2 solution when fully decomposed by heat gives 100 ml of oxygen at s.t.p. This solution is approximately 8.82 M and hence 1 mol is contained in 113 ml. The strength of an aqueous solution of hydrogen peroxide is conveniently measured by the volumetric procedure of titration with standard sodium thiosulphate of the iodine released when aqueous hydrogen peroxide is treated with acidified potassium iodide (see Ref. 4).

35. **Hydrogen sulphide.** This very poisonous gas should only be prepared and used in an efficient fume cupboard. It is usually generated from iron(II) sulphide and dilute hydrochloric acid (1 : 3) in a Kipp's apparatus; it should be washed with water to remove acid spray. The resulting hydrogen sulphide contains hydrogen because of the presence of free iron in commercial iron(II) sulphide.

Pure hydrogen sulphide (99.6%) can be obtained commercially in cylinders.

36. **Hydrogen tetrafluoborate** (*fluoroboric acid*). A purified grade of the reagent is available commercially containing 42 per cent w/w HBF_4. It is usually more convenient to prepare the reagent before use; the details are given in Section **IV,85**. Great caution should be exercised in handling this reagent.

37. **Iodine monochloride.** In a fume cupboard, pass dry chlorine gas into 127 g (0.5 mol) of iodine in a distilling flask until the weight has increased by

---

* From Laporte Industries Ltd, who supply notes on the safe handling of these concentrated solutions, and advise on the use of hydrogen peroxide and peroxyacids (Section **II,2,46**).

† A freshly prepared solution (5%) of sodium ascorbate should be readily available for use as an eye wash.

34.5 g (0.49 mol). The chlorine should be fed in via a glass frit located at or below the surface of the iodine whilst the flask is gently shaken. Distil the iodine chloride in an ordinary distillation apparatus; protect the receiver flask from the atmospheric moisture by a calcium chloride guard-tube. Collect the fraction, b.p. 97–105 °C; the yield is 140 g (88%). Preserve the iodine mono-chloride in a dry, glass-stoppered bottle. Care should be taken when handling this compound since the liquid is corrosive and gives off a harmful vapour. If it should come into contact with the skin, an effective antidote is dilute hydrochloric acid (1 : 1).

38. Lead tetra-acetate. The reagent as supplied by manufacturers is moistened with glacial acetic acid to prevent hydrolytic decomposition. It may be prepared in the laboratory by warming lead oxide (red lead, Pb_3O_4) with acetic acid in the presence of sufficient acetic anhydride to combine with the water formed.

$$Pb_3O_4 + 8(CH_3 \cdot CO)_2O \longrightarrow (CH_3 \cdot CO_2)_4Pb + 2(CH_3 \cdot CO_2)_2Pb + 4H_2O$$

The filtrate which contains lead acetate may be treated with chlorine:

$$2(CH_3 \cdot CO_2)_2Pb + Cl_2 \longrightarrow (CH_3 \cdot CO_2)_4Pb + PbCl_2$$

and the resulting lead tetra-acetate separated from the accompanying lead chloride by recrystallisation from glacial acetic acid.

A mixture of 550 g of glacial acetic acid and 185 g of acetic anhydride is placed in a 1-litre, three-necked flask provided with a thermometer and a sealed stirrer unit. The liquid is vigorously stirred, heated to 55–60 °C and 300 g of dry red lead powder are added in portions of 15–20 g. A fresh addition is made only after the colour due to the preceding portion has largely disappeared. The temperature should not be allowed to rise above 65 °C. Towards the end it may be necessary to warm the flask cautiously to about 80 °C in order to complete the reaction. At the end of the reaction, the thick and somewhat dark solution is cooled, and the precipitated lead tetra-acetate is filtered off (the mother-liquor, *A*, is put aside) and washed with glacial acetic acid. The crude product, without being dried, is dissolved in hot glacial acetic acid containing a little acetic anhydride, the solution treated with a little decolourising carbon, filtered through a hot water funnel and cooled. The colourless crystalline product is filtered off and dried in a vacuum desiccator over potassium hydroxide pellets. The yield is about 150 g (78%).

A further 100 g of lead tetra-acetate may be obtained from the mother-liquor (*A*) by returning it to the original flask, heating to about 75 °C with stirring, and passing through it a stream of dry chlorine. When the reaction is complete a few grams of decolourising carbon are added, the mixture is maintained at 75 °C for a few minutes, and the hot suspension is filtered with suction through a preheated Buchner funnel. The residue, consisting largely of lead chloride, is washed with hot glacial acetic acid. The filtrate, on cooling, deposits lead tetra-acetate in colourless needles; it is collected and dried as described above. Although contaminated with a little (< 5%) lead chloride the resulting lead tetra-acetate is satisfactory for most purposes; if required perfectly pure, it should be recrystallised as detailed above.

39. Lithium aluminium hydride. The reagent is usually supplied in powdered form contained in individual plastic bags sealed within a metal can. Lithium aluminium hydride reacts violently with water, liberating hydrogen, and any contact with even traces of moisture must be avoided since the heat of reaction which ensues may cause the material to ignite.* Great caution should therefore be employed in handling the reagent, taking particular care not to inhale the powder during weighing or transfer operations, etc. Whenever possible these operations should be conducted in a fume cupboard with the front partly pulled down leaving an opening just sufficient for the hands to enter; it is advisable to wear plastic gloves. The availability of the powdered form makes unnecessary the process frequently mentioned in the earlier literature of grinding the lumps of hydride before use.

Lithium aluminium hydride reductions are usually carried out in a suitable ethereal solvent, such as rigorously dried diethyl ether (Section II,1,*15*) or tetrahydrofuran (Section II,1,*19*); the solubility of the reagent is 25–30 g and 13 g per 1000 g of solvent respectively. These solutions invariably contain a voluminous insoluble residue arising from impurities which may be formed by the action of moisture on the hydride during handling operations; this however constitutes not more than 1 per cent of the added reagent, and does not normally interfere with the subsequent reduction processes. In those cases which require the use of accurately known quantities of hydride it is necessary to prepare a standardised ethereal solution of the reagent. An ethereal solution (about 0.5 M) may be prepared by heating under reflux for 1 hour in an atmosphere of nitrogen about 20 g of lithium aluminium hydride with 1 litre of rigorously dried ether in a 2-litre, round-bottomed flask fitted with a double surface reflux condenser protected with a calcium chloride tube. The solution is cooled and filtered through a sintered glass Buchner funnel previously rinsed with dried ether and the residue is washed with a small portion of dried ether. The filtrate is transferred to the reservoir of a suitable automatic burette previously flushed with nitrogen from which aliquot portions may be dispensed for standardisation, and for use in subsequent reductions. Care should be exercised in the treatment of the residue in the Buchner funnel; it is safely destroyed by adding it to ethyl acetate contained in a beaker; the funnel is then rinsed with further quantities of ethyl acetate. The other vessels (Buchner flask, round-bottomed flask, etc.) are best rinsed with ethanol.

To standardise the ethereal solution of lithium aluminium hydride, an aliquot portion is added to an excess of a standardised solution of iodine in benzene, when the following reaction ensues:

$$2I_2 + LiAlH_4 \longrightarrow LiAlI_4 + 2H_2$$

Unreacted iodine is then titrated with standard sodium thiosulphate solution.

Add 5 ml of an approximately 0.5 M ethereal solution of lithium aluminium hydride to 50 ml of a standardised approximately 0.2 M iodine solution in benzene. Shake the solution gently and allow to stand for 5 minutes, add 50 ml of distilled water followed by about 2 ml of glacial acetic acid. Titrate the excess of iodine with a standardised 0.2 M solution of sodium thiosulphate using starch solution in the final stages to determine the end-point.

$$1 \text{ ml } M \text{ Na}_2\text{S}_2\text{O}_3 \equiv 0.12692 \text{ g iodine}$$
$$0.03795 \text{ g LiAlH}_4 \equiv 0.50768 \text{ g iodine}$$

---

* Solid lithium aluminium hydride fires should only be extinguished using dry sand or powdered limestone. Water or carbon dioxide or chemically filled extinguishers must never be used.

The experimental conditions for reductions using lithium aluminium hydride are similar to those for the Grignard reaction. For compounds which are readily soluble in ether, a solution of the compound in dry ether is added to an ethereal solution of lithium aluminium hydride (slight excess) at such a rate that the reaction mixture boils gently. For compounds which are slightly or sparingly soluble in ether, a Soxhlet apparatus (Fig. I,96) is inserted between the reaction flask and the reflux condenser and the compound is placed in the Soxhlet thimble. When the reduction is complete, the excess of reagent is decomposed by the cautious addition of moist ether, or an ethanol–ether mixture or by the dropwise addition of cold water with vigorous stirring; when water is used, it is desirable to employ a large flask because of the foaming which takes place. On the whole it is best to employ ethyl acetate, as its reduction product (ethanol) does not interfere in the subsequent isolation and no hydrogen is evolved. The reaction mixture is then poured gradually into excess of ice-cold dilute sulphuric acid to decompose the complex aluminium compounds and to dissolve the precipitated aluminium hydroxide; the product is usually in the ethereal layer but, if it is water-soluble, it must be isolated from the aqueous solution. For bases, after extraction of any neutral or acidic products, the solution is rendered alkaline with 10 M-sodium hydroxide and the whole (including the precipitated aluminium hydroxide) is extracted with ether. As an alternative decomposition procedure in this latter case, and also for use with products which are acid sensitive, the excess of lithium aluminium hydride is best decomposed by the careful dropwise addition of just sufficient water to produce a granular precipitate of lithium aluminate ($LiAlO_2$) which is readily filterable. An excess of water produces a gelatinous precipitate of aluminium hydroxide and should be avoided.

40. **Manganese dioxide.** The principal organic use for manganese dioxide is for the selective oxidation of allylic and benzylic primary and secondary alcohols to the corresponding carbonyl compounds. The activity of the reagent varies with its method of preparation; material of good activity may be obtained by oxidising manganese(II) ions with an excess of permanganate under alkaline conditions: *

$$3Mn^{2\oplus} + 2MnO_4^{\ominus} + 4\overset{\ominus}{O}H \longrightarrow 5MnO_2 + 2H_2O$$

Add simultaneously during 1 hour (*a*) a solution of 223 g (1 mol) of manganese(II) sulphate tetrahydrate in 300 ml of water, and (*b*) 240 ml (2.5 mol) of 40 per cent aqueous sodium hydroxide solution to a hot stirred solution of 190 g (1.2 mol) of potassium permanganate in 1200 ml of water. Continue stirring for a further 1 hour, isolate the fine brown precipitate of manganese dioxide (preferably by centrifugation) and wash it thoroughly with water until the washings are colourless. Dry the product at 100–120 °C and grind finely. Alternatively remove as much of the wash-water as possible by prolonged suction and activate the damp cake (in 25 g portions) by removing most of the remaining water by azeotropic distillation with 150 ml of benzene (see Ref. 5).

Some commercial samples of precipitated manganese dioxide may be active enough for use directly in an oxidation process. **To assess the activity of a sample of manganese dioxide,** dissolve 0.25 g of pure cinnamyl alcohol

---

* For the preparation of an active manganese dioxide/charcoal catalyst see Ref. 6.

in 50 ml of dry light petroleum (b.p. 40–60 °C) and shake the solution at room temperature for 2 hours with 2 g of the sample of manganese dioxide (previously dried over phosphoric oxide). Filter, remove the solvent by evaporation and treat the residue with an excess of 2,4-dinitrophenylhydrazine sulphate in methanol * (Section **VII,6,***13***). Collect the cinnamaldehyde 2,4-dinitrophenylhydrazone and crystallise it from ethyl acetate. An active dioxide should give a yield of the derivative, m.p. 255 °C (decomp.), in excess of 0.35 g (60%).

41. **Nickel.** The introduction by Raney (1927) of a new form of catalyst (the **Raney nickel catalyst**) with enhanced activity for hydrogenation at low pressures and temperatures in comparison with the usual form of nickel catalyst as employed by Sabatier and Senderens opened up a new field of controlled catalytic hydrogenation. A special alloy, prepared essentially by the fusion of approximately equal parts of aluminium and nickel at 1200–1500 °C, is treated with alkali which dissolves the aluminium and leaves the nickel as a finely-divided black suspension. The catalyst is thoroughly washed to free it from alkali, is stored under absolute ethanol in an air-free container and is measured in the form of the suspension; it must be handled under a solvent at all times as it is highly pyrophoric.

The advantages of this catalyst are that it is cheaper and less delicate than platinum, fairly large quantities of organic substrate may be hydrogenated and the process is reasonably rapid. The following method gives a catalyst of moderate activity (*W2 Raney nickel*).

Place a solution of 190 g of sodium hydroxide in 750 ml of water in a 2-litre beaker equipped with an efficient stirrer (1), cool in an ice bath to 10 °C and add 150 g of nickel–aluminium alloy in small portions, with stirring, at such a rate that the temperature does not rise above 25 °C. If excessive foaming is encountered, add 1 ml of octan-1-ol. When all the alloy has been introduced (about 2 hours), stop the stirrer, remove the beaker from the ice bath and allow the contents to attain room temperature. When the evolution of hydrogen becomes slow, heat the reaction mixture gradually (2) on a water bath, until the evolution again becomes slow (about 8–12 hours); add distilled water to restore the original volume, stir the mixture, allow to settle and decant the supernatant liquid. Transfer the nickel to a stoppered graduated cylinder with the aid of distilled water, and decant the water again. Add a solution of 25 g of sodium hydroxide in 250 ml of water, shake to disperse the catalyst thoroughly, allow to settle and decant the alkali solution. Wash the nickel by suspension in distilled water and decantation until the washings are neutral to litmus, then ten times more to remove the alkali completely (25–40 washings are required) (3). Repeat the washing process three times with 100 ml of rectified spirit (95% ethanol) and three times with absolute ethanol. Store the catalyst in bottles which are completely filled with absolute ethanol and tightly stoppered; the product is highly pyrophoric and must be kept under liquid at all times. The Raney nickel contained in this suspension weighs about 75 g.

In the practical applications of Raney nickel it is more convenient to measure the catalyst than to weigh it. The product, prepared as above, contains about 0.6 g of the catalyst per millilitre of settled material: a level teaspoonful is about 3 g of nickel.

---

* For the determination of the extent of oxidation by measurement of i.r. absorption at 3 μm see Ref. 7.

Notes. (1) The stirrer should be provided with a motor which will not ignite the hydrogen—an induction motor or an air stirrer is suitable. The stirrer itself may be of glass, Monel metal or stainless steel (cf. Fig. I,54).

(2) The heating should not be too rapid initially or the solution may froth over.

(3) The number of washings may be reduced to about twenty, if time is allowed for diffusion of the alkali from the surface of the catalyst into the surrounding wash water. Use 750 ml of water in each washing, allow diffusion to proceed for 3–10 minutes, stir again and decant the supernatant liquid as soon as the catalyst settles to the bottom.

42. **Nitric acid.**   The commercial concentrated acid, d 1.42, is a constant boiling azeotrope with water, b.p. 120.5 °C/760 mmHg, containing about 70 per cent w/w HNO_3 (0.989 g HNO_3 per ml). Colourless concentrated acid may be obtained free of coloured impurities (due to oxides of nitrogen or to nitrous acid) by warming to about 60 °C and passing in a stream of dust-free nitrogen; the addition of a little urea considerably accelerates the process.

$$2HNO_2 + (NH_2)_2CO \longrightarrow CO_2 + 2N_2 + 3H_2O$$

The so-called fuming nitric acid, d 1.5, contains about 95 per cent w/w HNO_3 (1.419 g HNO_3 per ml) and is available commercially; it has a yellow colour due to the presence of oxides of nitrogen which may be removed as detailed above. The acid may be prepared by distilling a mixture of equal volumes of concentrated nitric acid and concentrated sulphuric acid in a distillation assembly which incorporates a splash-head fitment to the distillation flask to act as a trap for acid spray. The volume of distillate collected should be slightly less than one-half of concentrated nitric acid originally used.

43. **Nitrogen.**   Cylinders of compressed nitrogen may be purchased or hired. The gas may contain traces of oxygen which may be removed, if necessary, by passage either through an alkaline solution of pyrogallol (15 g of pyrogallol dissolved in 100 ml of 50% sodium hydroxide solution) or through Fieser's solution, which consists of an alkaline solution of sodium dithionite to which sodium anthraquinone-2-sulphonate is added. *Fieser's solution* is prepared by dissolving 20 g of potassium hydroxide in 100 ml of water, and adding 2 g of sodium anthraquinone-2-sulphonate and 15 g of commercial sodium dithionite (*ca.* 85%) to the warm solution and stirring until dissolved: the blood-red solution is ready for use when it has cooled to room temperature, and will absorb about 750 ml of oxygen. The exhaustion of this solution is indicated by the change in colour to dull-red or brown, or when a precipitate appears. Oxygen-free nitrogen in cylinders is available commercially, but is, of course, more expensive than the normal commercial compressed gas.

44. **Palladium catalysts.**   (*a*) **Palladium on charcoal (5 per cent Pd).** Prepare a solution of 1.7 g of palladium chloride (1) in 1.7 ml of concentrated hydrochloric acid and 20 ml of water by heating on a water bath for 2 hours or until solution is complete, and add this to a solution of 30 g of sodium acetate trihydrate in 200 ml of water contained in a 500-ml hydrogenation flask. Add 20 g of acid-washed activated charcoal (2) and hydrogenate in an atmospheric hydrogenation apparatus (Fig. I,70(*a*)) until absorption ceases. Collect the catalyst on a Buchner funnel and wash it with five 100 ml portions of water and suck as dry as possible. Dry the catalyst at room temperature (3) over potassium hydroxide pellets or anhydrous calcium chloride in a vacuum desiccator. Powder the catalyst (about 20 g) and store in a tightly stoppered bottle.

(b) **Palladium on charcoal (30 per cent Pd).** Prepare a solution of 8.25 g of palladium chloride (1) in 5 ml of concentrated hydrochloric acid and dilute with 50 ml of distilled water. Cool the solution in an ice–salt bath and add 50 ml of 40 per cent formaldehyde solution and 11 g of acid-washed activated charcoal (2). Stir the mixture mechanically and add a solution of 50 g of potassium hydroxide in 50 ml of water, keeping the temperature below 5 °C. When the addition is complete, raise the temperature to 60 °C for 15 minutes. Wash the catalyst thoroughly by decantation with water and finally with dilute acetic acid, collect on a suction filter and wash with water until free from chloride or alkali. Dry at 100 °C and store in a desiccator.

(c) **Palladium black.** Dissolve 5 g of palladium chloride in 30 ml of concentrated hydrochloric acid and dilute with 80 ml of water; cool in an ice–salt bath and add 35 ml of 40 per cent formaldehyde solution. Add a cold solution of 35 g of potassium hydroxide in 35 ml of water dropwise during 30 minutes to the vigorously stirred palladium solution. Warm to 60 °C for 30 minutes and then wash the palladium precipitate six times by decantation with water. Filter on a sintered crucible, wash with 1 l of water and suck dry and transfer to a desiccator charged with silica gel. The yield is 3.1 g.

(d) **Palladium on barium sulphate catalyst (5 per cent Pd).** Dissolve 4.1 g of palladium chloride (1) in 10 ml of concentrated hydrochloric acid and dilute with 25 ml of water. Add all at once 60 ml of 3 M-sulphuric acid to a rapidly stirred, hot (80 °C) solution of 63.1 g of barium hydroxide octahydrate in 600 ml of water contained in a 2-litre beaker. Add more 3 M-sulphuric acid to render the suspension just acid to litmus. Introduce the palladium chloride solution and 4 ml of 40 per cent formaldehyde solution into the hot mechanically-stirred suspension of barium sulphate. Render the suspension slightly alkaline with 30 per cent sodium hydroxide solution, continue the stirring for 5 minutes longer and allow the catalyst to settle. Decant the clear supernatant liquid, replace it by water and resuspend the catalyst. Wash the catalyst by decantation 8–10 times and then collect it on a medium-porosity sintered glass funnel, wash it with five 25 ml portions of water and suck as dry as possible. Dry the funnel and contents at 80 °C, powder the catalyst (48 g) and store it in a tightly stoppered bottle.

For use in the Rosenmund reduction (Section IV,133) the catalyst is moderated by the addition of the appropriate quantity of a quinoline–sulphur poison prepared in the following manner. Heat under reflux 1 g of sulphur with 6 g of quinoline for 5 hours and dilute the resulting brown liquid to 70 ml with xylene which has been purified by distillation over anhydrous aluminium chloride. Thiourea (about 20% by weight of the palladium–barium sulphate catalyst) may also be used as a catalyst poison.

The preparation of **Lindlar's catalyst** (palladium on calcium carbonate moderated by treating with lead acetate and quinoline) is described in detail in Ref. 8; it is used for affecting the partial reduction of an acetylenic bond to an olefin.

Where it is advantageous to maintain the neutrality of the hydrogenation mixture, **palladium on barium carbonate** catalyst is recommended. For the preparation of this catalyst the experimental details noted above for the barium sulphate based catalyst are used, but the barium hydroxide and sulphuric acid are replaced by 46.5 g of precipitated barium carbonate and the volume of hydrochloric acid is reduced to 4.1 ml.

(e) **Palladium hydroxide on calcium carbonate.** Mix hot solutions of 55 g of anhydrous calcium chloride and 53 g of anhydrous sodium carbonate each dissolved in 150 ml of distilled water. Filter the resulting precipitate of calcium carbonate, wash it well with water and suspend it in 200 ml of distilled water. Dissolve 1 g of palladium chloride in 2.4 ml of concentrated hydrochloric acid and dilute with 30 ml of distilled water. Adjust the pH of the solution (4) to 4.0–4.5 by the cautious addition with stirring of 3 M-sodium hydroxide solution. No permanent precipitate should be obtained at this stage. Add this solution to the calcium carbonate suspension, warm to 80 °C with stirring until conversion to the insoluble palladium hydroxide is complete, i.e., until the supernatant liquors are colourless. Wash several times with distilled water by decantation, filter with suction and wash sparingly with distilled water until the washings are chloride free. Dry over silica gel in a vacuum desiccator and preserve in a tightly stoppered bottle.

Notes. (1) Alternatively, the equivalent quantity of palladium chloride dihydrate may be used.

(2) Any of the commercial forms of activated carbon ('Norit', 'Darco', etc.) may be employed; the carbon should be heated on a steam bath with 10 per cent nitric acid for 2–3 hours, washed free from acid with water and dried at 100–110 °C before use.

(3) Heating may cause ignition of the carbon.

(4) The pH adjustment is most conveniently followed with the aid of a pH meter.

45. **Periodic acid.** Periodic acid has a selective oxidising action upon compounds having two hydroxyl groups or a hydroxyl and an amino group attached to adjacent carbon atoms, which is characterised by the cleavage of the carbon–carbon bond (**Malaprade reaction**):

$$R \cdot CH(OH) \cdot CH(OH) \cdot R' + HIO_4 \longrightarrow R \cdot CHO + R' \cdot CHO + HIO_3 + H_2O$$
$$R \cdot CH(OH) \cdot CH(NH_2) \cdot R' + HIO_4 \longrightarrow R \cdot CHO + R' \cdot CHO + HIO_3 + NH_3$$

No oxidation occurs unless the hydroxyl groups or a hydroxyl and an amino group are attached to adjacent carbon atoms, hence the reaction may be employed for testing for the presence of contiguous hydroxyl groups (e.g., 1,2-diols) and hydroxyl and amino groups. Carbonyl compounds in which the carbonyl group is contiguous to a hydroxyl group or a second carbonyl group are also oxidised, e.g., α-hydroxy-aldehydes or -ketones, 1,2-diketones and α-hydroxy acids:

$$R \cdot CH(OH) \cdot CO \cdot R' + HIO_4 \longrightarrow R \cdot CHO + R' \cdot CO_2H + HIO_3$$
$$R \cdot CO \cdot CO \cdot R' + HIO_4 + H_2O \longrightarrow R \cdot CO_2H + R' \cdot CO_2H + HIO_3$$

The oxidation may proceed through the hydrated form of the carbonyl group $> CH(OH)_2$. The rate of oxidation is 1,2-glycols $>$ α-hydroxy aldehydes $>$ α-hydroxy ketones $>$ α-hydroxy acids.

Periodic acid is available in two grades, one contains a minimum of 95 per cent w/w $HIO_4 \cdot 2H_2O$, the other is an aqueous solution of periodic acid containing 50 per cent w/w $HIO_4 \cdot 2H_2O$.

Sodium periodate, and less frequently potassium periodate, are employed in oxidations which are to be carried out within the pH range 3–5. **Sodium metaperiodate**, ($NaIO_4$), has a solubility in water of *ca.* 0.07 g per ml; the addition of alkali leads to the precipitation of the far less soluble **sodium paraperiodate** ($Na_2H_3IO_6$, disodium trihydrogen orthoperiodate, water solubility *ca.* 0.20%). Sodium paraperiodate is available commercially and may be converted into the metaperiodate salt by dissolving 100 parts in a mixture of 150 parts of water and 45 parts of concentrated nitric acid, warming to effect solution, filtering through a sintered glass funnel if necessary, and allowing the sodium metaperiodate to crystallise overnight at room temperature. Potassium metaperiodate has a lower solubility in water than sodium metaperiodate.

In cases where the organic substrate is insoluble in water it may be necessary to carry out the oxidation in ethanol, methanol, dioxan or acetic acid which have been diluted with water.

It is essential that only a slight excess of the calculated amount of reagent required to effect the oxidation is used otherwise oxidation of the reaction products may become a significant undesirable side reaction. The course of the oxidation may be followed by removing aliquot portions of the reaction mixture and determining the amount of unused oxidant. Details of an iodimetric procedure are given in Section III,117, Note (1).

46. Peroxyacids. CAUTION—All reactions involving hydrogen peroxide solutions should be carried out behind a shatter-proof safety screen (see also Section II,2,*34*).

(*a*) **Performic acid.** A solution of the reagent is prepared as required by treating 30 per cent hydrogen peroxide solution ('100-volume') with excess 88–90 per cent formic acid.

(*b*) **Peracetic acid.** For many purposes a reagent containing varying amounts of peracetic acid can be prepared *in situ* if convenient by adding 30 per cent hydrogen peroxide (1 part) to glacial acetic acid (3 parts) in the presence of a catalytic amount of sulphuric acid. Alternatively an approximately 40 per cent solution of peracetic acid in acetic acid containing a little sulphuric acid is available commercially.* The peracetic acid content of the solution may be determined by the iodometric procedure described under perbenzoic acid. In use it is sometimes desirable to neutralise the sulphuric acid with a stoichiometric amount of sodium acetate; the neutralised reagent should not be allowed to stand but should be used immediately.

(*c*) **Pertrifluoroacetic acid.** The reagent may be prepared by the reaction of hydrogen peroxide with trifluoroacetic acid. The following procedure gives an anhydrous solution of the reagent. Add trifluoroacetic anhydride (25 ml, 0.18 mol) dropwise to a stirred suspension of 86 per cent hydrogen peroxide (4.1 ml, 0.15 mol) in ice-cold methylene chloride (70 ml). On completion of the addition stir at 0 °C for a further 10 minutes, dry with anhydrous sodium sulphate and use the solution without delay.

$$(F_3C \cdot CO \cdot)_2O + H_2O_2 \longrightarrow F_3C \cdot CO_3H + F_3C \cdot CO_2H$$

(*d*) **Perbenzoic acid.** Place 5.2 g (0.225 mol) of sodium in a 500 ml dry conical flask provided with a reflux condenser and add 100 ml of absolute meth-

---

anol; slight cooling may be necessary to moderate the vigour of the reaction. Cool the resulting solution of sodium methoxide to $-5\,°C$ in a freezing mixture of ice and salt: remove the condenser. Add a solution of 50 g (0.206 mol) of freshly recrystallised benzoyl peroxide (Section II,2,5) (1) in 200 ml of chloroform, with shaking and cooling, at such a rate that the temperature does not rise above $0\,°C$. Keep the mixture in the ice–salt bath for 5 minutes with continuous shaking; it turns milky but no precipitate appears. Transfer the reaction mixture to a 1-litre separatory funnel and extract the sodium perbenzoate with 500 ml of water containing much crushed ice. It is essential that the separation be carried out as rapidly as possible and the temperature kept as near $0\,°C$ as feasible, especially before the free acid is liberated from the sodium salt. Separate the chloroform layer, and extract the aqueous layer twice with 100 ml portions of cold chloroform to remove the methyl benzoate. Liberate the perbenzoic acid from the aqueous solution by the addition of 225 ml of ice-cold 0.5 M-sulphuric acid and extract it from solution with three 100 ml portions of cold chloroform. Dry the moist chloroform solution (about 308 ml) with a little anhydrous sodium sulphate, transfer to a polyethylene container (*not* glass), and keep it in an ice box or a refrigerator until required (2); it contains about 24 g (84%) of perbenzoic acid.

To determine the exact perbenzoic acid content of the solution, proceed as follows. Dissolve 1.5 g of sodium iodide in 50 ml of water in a 250-ml reagent bottle, and add about 5 ml of glacial acetic acid and 5 ml of chloroform. Introduce a known weight or volume of the chloroform solution of perbenzoic acid and shake vigorously. Titrate the liberated iodine with standard 0.1 M-sodium thiosulphate solution in the usual manner.

$$1 \text{ ml of } 0.1 \ M \ Na_2S_2O_3 \equiv 0.0069 \text{ g of perbenzoic acid}$$

To obtain **crystalline perbenzoic acid**, the following procedure may be adopted, *the operation being conducted behind a shatter-proof screen.* Dry the moist chloroform solution with a little anhydrous sodium or magnesium sulphate for an hour, filter and wash the desiccant with a little dry chloroform. Remove the chloroform under reduced pressure at the ordinary temperature whilst carbon dioxide is introduced through a capillary tube. Dry the white or pale yellow residue for several hours at 30–$35\,°C$ under 10 mm pressure. The yield of crystalline perbenzoic acid, m.p. about $42\,°C$, which is contaminated with a little benzoic acid, is 22 g.

Perbenzoic acid may be recrystallised by dissolving it in a mixture of 3 parts of light petroleum (b.p. 40–60 °C; freed from alkenes, Section II,1,1) and 1 part ether using about 4–5 ml per gramme, seeding and cooling to $-20\,°C$. [*Use a shatter-proof screen.*] Long white needles, m.p. 41–42 °C, are obtained. It is moderately stable when kept in the dark at low temperatures ($-20\,°C$); it is very soluble in chloroform, ethyl acetate and ether, but only slightly soluble in cold water and in cold light petroleum.

Notes. (1) It is essential to use freshly recrystallised benzoyl peroxide since the commercial material usually gives poor results. The material may be assayed as described in Section II,2,5.

(2) Perbenzoic acid is used for the conversion of olefinic compounds into epoxides.

$$\begin{array}{c} \diagup C = C \diagdown \end{array} + C_6H_5 \cdot C \diagup^{\displaystyle O}_{\displaystyle O \cdot OH} \longrightarrow \begin{array}{c} \diagdown \\ C \end{array}\!\!-\!\!\begin{array}{c} \diagup \\ C \end{array} + C_6H_5 \cdot CO_2H$$

The number of olefinic linkages in a given compound can be established with accuracy by quantitative titration with perbenzoic acid. A solution of the substance and excess perbenzoic acid in chloroform is allowed to stand for several hours at a low temperature and the amount of unreacted perbenzoic acid in solution is determined: a blank experiment is run simultaneously.

A convenient general method for the conversion of aliphatic and aromatic carboxylic acids into the corresponding peroxyacids involves reaction with 70 per cent hydrogen peroxide in the presence of methanesulphonic acid (Ref. 9).

47. **Phosgene (carbonyl chloride).** Phosgene may be purchased in cylinders or in the form of a solution (*ca.* 12.5% by weight) in toluene in glass ampoules. *Owing to the very poisonous character of the gas (b.p. 8 °C), all operations with it must be conducted in a fume cupboard provided with a powerful draught,* and all excess of phosgene must be absorbed in 20 per cent sodium hydroxide solution.* The preparation of the gas is rarely undertaken in the laboratory, but small quantities may be prepared by the following procedure. The apparatus (assembled in a fume chamber) consists of a flask with a short reflux condenser carrying in the top cone a socket/cone adapter with 'T' connection fitted with a dropping funnel. The 'T' connection is attached to a train of Drechsel bottles as shown in Fig. II,5; A and C act as safety traps, the phosgene is absorbed in the toluene contained in B, and the hydrogen chloride and traces of phosgene are absorbed in the 20 per cent aqueous sodium hydroxide solution contained in D. Concentrated sulphuric acid, to which 2 per cent by weight of ignited kieselguhr has been added, is placed in the flask; carbon tetrachloride is introduced into the dropping funnel. The sulphuric acid is heated to 120–130 °C in an oil bath and the carbon tetrachloride is allowed to drop in slowly; the resulting phosgene is absorbed in the toluene (B), whilst the hydrogen chloride is retained in D.

$$3CCl_4 + 2H_2SO_4 \longrightarrow 3COCl_2 + 4HCl + S_2O_5Cl_2$$

48. **Phosphoric acid.** Commercial syrupy orthophosphoric acid has an approximate composition of 88–90 per cent w/w H_3PO_4 (*d* 1.75; 1.57 g H_3PO_4 per ml; 65% P_2O_5 equivalent). An approximately 100 per cent w/w H_3PO_4 (anhydrous orthophosphoric acid) is also marketed (72% P_2O_5 equivalent) but may be prepared from the 90 per cent H_3PO_4 by mixing with cooling four parts by weight of 90 per cent H_3PO_4 with one part of phosphoric oxide.

Polyphosphoric acid (tetraphosphoric acid), having an approximate composition $2P_2O_5 \cdot 3H_2O$, has a phosphoric oxide equivalent of 82–84 per cent. As

---

* It is usually advisable to suspend in the fume cupboard in which operations involving phosgene are being carried out, several filter-papers dipped in an ethanolic solution containing 5 per cent of *p*-dimethylaminobenzaldehyde and 5 per cent of colourless diphenylamine. A dangerous quantity of phosgene in the atmosphere is indicated by a colour change from yellow to deep orange. An additional precaution which is recommended is that the worker should wear a suitable gas mask; this is essential when the apparatus is disconnected for cleaning.

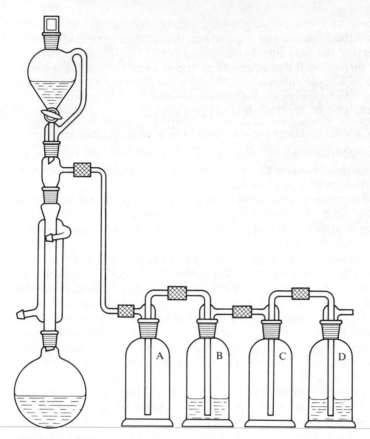

Fig. II,5

supplied commercially it is a very viscous liquid and difficult to handle at laboratory temperature. Warming on a steam bath produces a mobile liquid which can readily be poured. Alternatively a satisfactory reagent can be prepared in the laboratory by dissolving 1.8 parts by weight of phosphoric oxide in 1 part by weight of 88–90 per cent H_3PO_4; this reagent has a phosphoric oxide equivalent of 87 per cent.

49. **Phosphorus (red).** Commercial red phosphorus is usually contaminated with small quantities of acidic products. It should be boiled for 15 minutes with distilled water, allowed to settle, decanted through a Buchner funnel and then washed two or three times with boiling water by decantation. Finally the phosphorus is completely transferred to the Buchner funnel and washed with hot water until the washings are neutral. It is dried at 100 °C, and kept in a desiccator or in a tightly stoppered bottle.

50. **Phosphorus tribromide.** Into a 500-ml three-necked flask, provided with a sealed mechanical stirrer (1), a dropping funnel and a reflux condenser, are placed 28 g of purified red phosphorus (*49.* above), and 200 ml of carbon tetrachloride (dried over anhydrous calcium chloride). Dry bromine (198 g; 63.5 ml) is placed in the dropping funnel and added to the vigorously stirred contents of the flask at the rate of about 3 drops per second. A little hydrogen bromide is evolved and the preparation should be carried out in a fume cupboard. After all the bromine has been added, the mixture is refluxed for 15 minutes by immersing the flask in a water bath at 80–90 °C. The clear solution is then decanted through a fluted filter-paper, and the carbon tetrachloride is distilled off through a short column (e.g., the all-glass Dufton column, Section I,26); the residue, upon distillation through the well-lagged column, boils at 166–169 °C (mainly at 168 °C). The yield is 190 g (78%).

Note. (1) A precision-ground glass stirrer unit is most satisfactory.

51. **Platinum dioxide** (*Adams' Catalyst*). Platinum dioxide for use in hydrogenations is available commercially. It may alternatively be prepared by either of the following methods.

Method 1. **From ammonium chloroplatinate.** Place 3.0 g of ammonium chloroplatinate and 30 g of sodium nitrate (AnalaR) (1) in a Pyrex beaker or porcelain dish and heat gently at first until the rapid evolution of gas slackens and then more strongly until a temperature of about 300 °C is reached. This operation occupies about 15 minutes, and there is no spattering. Maintain the fluid mass at 500–530 °C for 30 minutes, and allow the mixture to cool. Treat the solid mass with 50 ml of water. The brown precipitate of platinum oxide ($PtO_2.H_2O$) settles to the bottom. Wash it once or twice by decantation, filter through a hardened filter-paper and wash on the filter until practically free from nitrates. Stop the washing process immediately the precipitate tends to become colloidal (2): traces of sodium nitrate do not affect the efficiency of the catalyst. Dry the oxide in a desiccator, and weigh out portions of the dried material as required.

Method 2. **From chloroplatinic acid** Dissolve 3.5 g of the purest commercial chloroplatinic acid in 10 ml of water contained in a 250-ml Pyrex beaker or porcelain basin, and add 35 g of sodium nitrate (AnalaR) (1). Evaporate the mixture to dryness by heating gently over a Bunsen flame whilst stirring with a glass rod. Then raise the temperature to 350–370 °C within about 10 minutes: fusion will occur accompanied by the evolution of brown oxides of nitrogen and the gradual separation of a precipitate of brown platinum oxide. If foaming occurs, stir the mixture more vigorously and direct an additional flame at the top of the reaction mixture, if necessary. If the burner beneath the beaker is removed when frothing commences, the top of the fused mass solidifies and material may be carried over the sides of the vessel. After 15 minutes, when the temperature has reached about 400 °C, the evolution of gas decreases considerably. Continue the heating until at the end of 20 minutes the temperature is 500–550 °C; at this stage the evolution of oxides of nitrogen has practically ceased and there is a gentle evolution of gas. Maintain the temperature at this point (best with the full force of a Bunsen burner) for about 30 minutes, by which time fusion is complete. Allow the mass to cool (the Pyrex beaker may crack), add 50 ml of water and proceed as in *Method 1*.

Notes. (1) The use of an equivalent quantity of potassium nitrate (AnalaR) is said to produce a more active catalyst.

(2) It is advisable to test a small portion of the filtrate for platinum by acidifying with hydrochloric acid and adding a few drops of tin(II) chloride solution: a yellow or brown colour develops according to the quantity of platinum present. The yellow colour is soluble in ether, thus rendering the test more

sensitive. If platinum is found, treat the filtrate with excess of formaldehyde and sodium hydroxide solution and heat; platinum black separates on standing and may be filtered and worked up with other platinum residues.

Platinum residues from hydrogenation reactions should be carefully preserved and subsequently recovered by conversion into ammonium chloroplatinate by the following method. Dissolve the platinum or platinum residues in *aqua regia*, evaporate just to dryness several times with concentrated hydrochloric acid, dissolve the final residue in a little water and filter. Precipitate ammonium chloroplatinate from the filtrate by addition of excess of a saturated solution of ammonium chloride. Filter and dry the precipitate at 100 °C.

52. **Selenium dioxide.** The preparation of any quantity of selenium dioxide from selenium is hardly worth while, and it is better to purify the commercially available dioxide. *Extreme care should be taken in this operation, which should be conducted in an efficient fume cupboard, because of the very poisonous properties of selenium compounds.* The crude dioxide is sublimed in one of the assemblies described in Section I,21, appropriate to the quantity of material needed. The sublimate of pure selenium dioxide is obtained in colourless, long, needle-like crystals which should be stored in a tightly stoppered bottle.

53. **Silver nitrite.** Warm concentrated solutions of silver nitrate (containing 48 g of $AgNO_3$) and potassium nitrite (containing 30 g of KNO_2) are mixed, and the mixture is allowed to cool. The silver nitrite which separates is filtered off and washed with water. It may be recrystallised from water at 70 °C, and is dried either in a vacuum desiccator or in an air oven at about 40 °C; the yield is about 90 per cent. Silver nitrite should be stored in a tightly-stoppered amber bottle.

54. **Sodamide.** Sodamide is available in a granular form having a purity in the region of 80–90 per cent, but it is rather difficult to pulverise which in any case must be done by grinding in a glass mortar under an inert hydrocarbon solvent (toluene, xylene, etc.). Powdered sodamide of high activity, free from sodium and oxygenated components, is also commercially available* in 25-g bottles or 500-g tins. Sodamide should never be stored in a stoppered bottle from which samples are to be removed intermittently, since dangerous mixtures may result when the substance is exposed for 2–3 days to even limited amounts of air at ordinary temperature. As a safe practice, sodamide should be used immediately the container is opened and should not be kept longer than 12–24 hours unless under an inert solvent. In all cases where the sodamide has been seen to become yellowish or brown in colour, due to the formation of oxidation products, *the resulting mixture should not be used,* as it may be highly explosive; it should be destroyed by covering with toluene and slowly adding, with stirring, ethanol diluted with toluene. Small amounts of unused sodamide should be destroyed by the addition of methanol or ethanol, see also p. 10.

In many reactions involving the use of sodamide it is frequently more convenient to prepare the reagent *in situ* by the iron(III) ion catalysed reaction of sodium (see *55.* below) with liquid ammonia. In a 500-ml three-necked flask, equipped with a removable glass stopper, sealed stirrer unit of the Hershberg type and a reflux condenser with soda-lime guard-tube, place 300 ml of an-

---

* For example from May & Baker, Ltd, Dagenham.

hydrous liquid ammonia (see Section I,17,7). Add just sufficient sodium (0.5 g) to the stirred liquid to produce a permanent blue colour, then 0.5 g of powdered iron(III) nitrate (to catalyse the conversion of sodium into sodamide), followed by 13.3 g of clean sodium metal (cut into small pieces) over a period of 30 minutes. When the sodium has been converted into sodamide (as indicated by the change from a blue solution to a grey suspension) allow the ammonia to evaporate (Section I,17,7) adding sufficient anhydrous ether through a dropping funnel to keep the volume of the liquid at about 300 ml. After practically all the ammonia has evaporated, stir the suspension of sodamide and heat under reflux for 5 minutes, and then cool to room temperature. A suspension of 23.4 g of sodamide in dry ether is thus obtained; the conversion is practically quantitative.

55. Sodium. CAUTION: *Sodium must be handled with great care and under no circumstances should the metal be allowed to come into contact with water as a dangerous explosion may result.* Sodium is stored under solvent naphtha (petroleum distillates, b.p. 152–204 °C) or xylene; it should *not* be handled with the fingers but with tongs or tweezers. Waste or scrap pieces of sodium should be placed in a bottle provided for the purpose and containing solvent naphtha or xylene; they should never be thrown into the sink or into the waste box. If it is desired to destroy scrap sodium, it should be added in small portions to a rather large quantity of industrial spirit. The preparation of **sodium wire** is described under the technique of drying of ether, Section II,1,*15*. **Granulated sodium** (also termed **molecular sodium, powdered sodium** or **sodium sand**) may be conveniently prepared by either of the following methods which require clean sodium. Commercial sodium is invariably covered with a non-metallic crust which is usually shaved off with a knife under a dry inert solvent (e.g., ether, xylene) before use. As this procedure is rather wasteful the method recommended in Ref. 10 is to be preferred. Here, lumps of sodium metal are immersed in dry xylene contained in a wide-mouthed conical flask and heated carefully on an electric hot plate with gentle swirling until the sodium just melts and flows away from the contaminating surface oxide. The flask is then removed from the hot plate and upon cooling the sodium melt solidifies in globules which may then be removed with a pointed spatula to be immediately reimmersed under fresh inert solvent. The residual crust, after decantation of the xylene, is covered with industrial spirit to ensure the safe destruction of the remaining traces of sodium metal, see also p. 14.

Method 1. Twenty-three grams of clean sodium* are introduced into a 750- or 1000-ml, round-bottomed Pyrex flask containing about 200 ml of sodium-dried xylene, or at least sufficient xylene to cover the sodium completely, and the flask is placed on a sand bath. Two or three thicknesses of dry cloth (or a thick towel), sufficient to envelop the whole flask, are placed in a convenient position on the bench. The sand bath is heated cautiously and the ring of condensed vapour of the xylene is carefully watched. When the ring of condensed vapour has risen to within one inch from the neck of the flask the flame beneath the burner is extinguished. A well-fitting rubber stopper or good quality bark cork is rapidly fitted into the flask neck and the flask is completely wrapped in the previously prepared cloth. The stopper is held firmly in place

---

* Clean sodium should be weighed under a dry inert solvent; sodium-dried ether or light petroleum (b.p. 60–80 °C) is usually used.

through the enveloping cloth and the flask is shaken vigorously for 30–60 seconds or until the molten sodium is converted into a fine dispersion. The flask is then placed on a cork ring and the stopper immediately removed. The sodium is thus obtained in the form of small spheres the size of which is controlled by the time and rapidity of the shaking. Any particles of sodium adhering to the sides of the flask are washed under the xylene. When the contents of the flask have cooled to room temperature, the xylene may be decanted and the sodium washed twice with 100 ml of sodium-dried ether to remove traces of xylene; finally the finely-divided sodium is covered with absolute ether. A bucket, half full of dry sand, should be kept at hand in case of breakage of the flask. Not more than 25–30 g of sodium should be powdered at one time by this procedure.

Method 2. It is often convenient to prepare the powdered sodium in the flask in which the subsequent reaction is to be carried out; this is usually a three-necked flask. Into a 1-litre three-necked flask fitted with a reflux condenser (protected by a drying tube containing soda lime), a sealed stirrer and a dropping funnel are placed 23 g of clean sodium and 150–200 ml of sodium-dried xylene. The flask is surrounded by a mantle and heated until the sodium has melted. The stirrer is started and, after the sodium is suitably granulated, the mantle is removed. When the contents of the flask have cooled to the laboratory temperature, the stirrer is stopped. The xylene may then be decanted, and the sodium washed with two 100 ml portions of sodium-dried ether to remove traces of xylene as in *Method 1*. Large quantities of molecular sodium may be prepared by this method.

56. **Sodium acetate.** The anhydrous salt is prepared from the crystallised sodium acetate, $CH_3 \cdot CO_2Na.3H_2O$, by heating in a large porcelain evaporating basin over a small free flame. The salt first liquefies, steam is evolved and the mass solidifies as soon as most of the water of crystallisation has been driven off. To remove the residual water, the solid is carefully heated with a larger flame, the burner being constantly moved until the solid just melts. Care must be taken that the solid is not overheated; too strong heating will be recognised by the evolution of combustible gases and charring of the substance. The fused salt is allowed to solidify and is removed from the vessel whilst still warm with a knife or spatula. It is immediately powdered and stored in a tightly stoppered bottle.

Commercial fused sodium acetate is usually satisfactory, but if necessary it can be melted and maintained in the fused state for several minutes in order to remove the water absorbed during storage.

57. **Sodium amalgam.** The amalgam which is generally employed for reductions contains from 1 to 3 per cent of sodium. Amalgams with a sodium content greater than 1.2 per cent are solid at ordinary temperature and can be powdered in a mortar; the 1.2 per cent amalgam is semi-solid at room temperature, but is completely fluid at 50 °C. Clean sodium (22.8 g) is placed in a 500-ml round-bottomed flange flask, provided with a dropping funnel (containing 750 g of mercury) in the central socket, and inlet and outlet tubes for dry nitrogen in two side sockets. The air is displaced by nitrogen. About 10 ml of mercury are added and the flask is warmed gently with a free flame until reaction commences. The flame is then removed and the reaction is maintained by the slow addition of the mercury. When about half the mercury has been introduced, the amalgam will commence to solidify; it should be kept molten by heating with occasional shaking. After the addition of all the mercury, the hot molten amalgam is poured on to a uralite board, powdered in a mortar* and stored in a tightly stoppered bottle. Amalgams of 1 and 2 per cent strength may be prepared similarly.

58. Sodium borohydride. Commercial material is now available in at least 98 per cent purity and it is not therefore necessary to purify it by recrystallisation from diglyme as was formerly the case. Unlike lithium aluminium hydride, sodium borohydride is insoluble in ether (but soluble in dioxan) and is normally used as a reducing agent in aqueous or aqueous alcoholic solution. The stability of the aqueous solution is increased by the addition of alkali; indeed, a stabilised 12 per cent solution of sodium borohydride in 43 per cent aqueous sodium hydroxide is available commercially as *Sodium Borohydride—SWS.*†

The hydride is rather readily decomposed by methanol or ethanol; isopropyl alcohol and t-butyl alcohol are the preferred alcoholic solvents.

An alkaline borohydride solution may be assayed by addition of excess standard potassium iodate solution leading to its decomposition in accordance with the equation:

$$3\overset{\ominus}{B}H_4 + 4IO_3^{\ominus} \longrightarrow 4I^{\ominus} + 3H_2BO_3^{\ominus} + 3H_2O$$

Unreacted potassium iodate is then determined by addition of potassium iodide followed by acidification when the iodine liberated is estimated by titration with standard sodium thiosulphate solution.

Reaction of sodium borohydride with lithium bromide in diglyme (Section **II,1,***18*) gives a solution containing lithium borohydride. This reagent, which is also available commercially, is more soluble in ethereal solvents than sodium borohydride although less stable towards hydroxylic solvents. It is intermediate in reducing power between sodium borohydride and lithium aluminium hydride and is, for example, a useful reducing agent for the conversion of esters into the corresponding alcohols.

59. Sodium hydride. Sodium hydride is a white, crystalline, free-flowing powder; it must be kept in air-tight containers for protection against moisture and oxygen. If exposed to the air unduly, traces of sodium hydroxide formed on the surface render the material hygroscopic; rapid absorption of moisture may then take place, and the heat generated by the reaction with water may suffice to ignite the solid. Great care must be taken therefore in handling sodium hydride and all operations involving the manipulation of the dry solid material should be conducted in a manipulator glove box in an atmosphere of dry nitrogen. Sodium hydride residues are best destroyed by careful treatment with ethyl acetate [Section **I,3,B,***5*(vi)].

For most purposes it is much more convenient to use the commercially available 55–60 per cent dispersion of the hydride‡ in white mineral oil, which being less susceptible to atmospheric moisture is much less hazardous. For

---

* The mortar should be provided with a tightly fitting rubber cover to protect the powdered material from oxidation in air. Alternatively a manipulator glove box previously flushed with nitrogen could be used.

† Available from Ventron Corp., Metals Chemical Division or through Aldrich Chemical Co.

‡ A. A. Hinckley, Sodium Hydride Dispersions, Metal Hydrides Inc. (now Ventron Corp.), July 1964.

reactions which are performed in those organic solvents in which the oil is soluble (e.g., light petroleum, benzene, toluene, etc.) the presence of the mineral oil does not impair the reactivity of the hydride and the dispersion may be used directly. If removal of the oil prior to reaction is necessary, as for example in those cases where its separation from the reaction product may prove difficult, this can be accomplished by washing the dispersion by decantation with a rigorously dried solvent (e.g., light petroleum), the operation being conducted in a manipulator glove box.

60. **Tin(II) chloride.** The anhydrous reagent is prepared from the hydrate as follows. Crystalline tin(II) chloride, $SnCl_2 \cdot 2H_2O$, is heated for one hour in an oil bath at 195–200 °C, the cooled melt is powdered and kept in a desiccator or a tightly stoppered bottle. The resulting product, although satisfactory in many instances, is not entirely dependable. The following procedure (Stephen, 1930) invariably gives an excellent product. In a 400-ml beaker are placed 102 g (89.5 ml; 1 mol) of redistilled acetic anhydride, and 123 g of analytical reagent grade tin(II) chloride dihydrate (0.5 mol) are added whilst the liquid is stirred either manually or mechanically; dehydration is almost instantaneous. The operation must be conducted in the fume cupboard as much heat is evolved and the acetic anhydride may boil. After about one hour, the anhydrous tin(II) chloride is filtered off on a Buchner or sintered glass funnel, washed free from acetic acid with two 30 ml portions of anhydrous ether and dried overnight in a vacuum desiccator. Anhydrous tin(II) chloride may be kept for an indefinite period in a desiccator; it may also be stored in a tightly stoppered bottle.

The anhydrous compound is not appreciably hygroscopic, is readily soluble in acetone and pentan-1-ol, and insoluble in benzene, toluene, xylene and chloroform; it is also readily soluble in absolute methanol or ethanol, but a trace of water causes immediate hydrolysis with the formation of an opalescent precipitate.

61. **Sulphur dioxide.** The liquefied gas is commercially available* in aluminium canisters (net weight 500 g) which are provided with screw operated valves.

62. **Sulphuric acid.** Ordinary concentrated acid, d 1.84, is a constant boiling mixture, b.p. 338 °C/760 mmHg, and it contains 98 per cent w/w H_2SO_4 (1.799 g H_2SO_4 per ml). The 100 per cent acid may be obtained by addition of the calculated quantity of oleum (Ref. 11); it is also available commercially.

Oleum is marketed in a range of strengths up to *ca.* 65 per cent SO_3. From 0 to 30 per cent free SO_3, it is a liquid; from 30 to 55 per cent free SO_3 it is a solid (maximum m.p. 35 °C at 45% free SO_3); from 60 to 70 per cent free SO_3, it is a liquid. The acid must be kept in ground glass stoppered thick-walled bottles. If it is required to melt the acid, the stopper is removed, a watch glass placed on the mouth of the bottle, and the bottle is placed on an asbestos mat in a warm temperature-controlled oven at 40 °C. The liquid should be removed from the bottle with the aid of an automatic dispenser fitted into the neck; this procedure is more satisfactory than that of pouring the liquid acid from the bottle.

63. **Sulphuryl chloride.** The technical product should be fractionated in an all-glass apparatus; the fraction, b.p. 69–70 °C, is collected. The pure substance has b.p. 69 °C/760 mmHg.

---

* For example, from BDH Ltd.

64. **Thionyl chloride.** The technical product frequently contains traces of acids, sulphur chlorides and sulphuryl chloride; it is essential to remove these before using the reagent for the preparation of acid chlorides, etc. Commercial purified thionyl chloride is satisfactory for most purposes. A colourless product of high purity may be obtained by either of the following methods.

Method 1. Commercial thionyl chloride is first fractionated in an all-glass apparatus from quinoline in order to remove acid impurities (50 g of thionyl chloride from 10 ml of quinoline); the receiver is protected from the entrance of moisture by a guard-tube, filled with anhydrous calcium chloride. The distillate is then refractionated as before from boiled linseed oil (50 g of thionyl chloride from 20 ml of linseed oil), the fraction, b.p. 76–78 °C, being collected.

Method 2. This recent method of purification is the more economic, the original details being described in Ref. 12. In this method technical thionyl chloride is placed in a distilling flask and not more than 5 per cent w/w of dipentene (*p*-mentha-1,8-diene) added with swirling. The mixture is immediately distilled at atmospheric pressure (not reduced pressure), using a gas burner (not a heating mantle), without a fractionating column, and with a thermometer dipping into the liquid mixture. Distillation is stopped when the temperature of the liquid reaches 84–86 °C; between 80 and 90 per cent of the original quantity of thionyl chloride having b.p. 76–78 °C will be collected. Redistillation of the product, to which 1–2 per cent of linseed oil has been added, through a short fractionating column gives pure colourless thionyl chloride having b.p. 77 °C/760 mmHg. This reagent must be stored in a well-fitting glass stoppered bottle.

65. **Toluene-*p*-sulphonyl chloride.** Unless the reagent has been recently purchased it may contain substantial amounts of toluene-*p*-sulphonic acid. The most satisfactory procedure for the purification of the chloride involves dissolving it in the minimum amount of chloroform (about 2.5 ml per g) and diluting with 5 volumes of light petroleum (b.p. 40–60 °C), which precipitates impurities. The filtered solution is treated with decolourising charcoal, filtered and concentrated to small volume when colourless crystals of the pure reagent, m.p. 68 °C, are obtained; these should be washed with chilled light petroleum (b.p. 40–60 °C).

66. **Triphenylmethyl halides (trityl halides).** **Triphenylmethyl chloride** is commercially available but may contain a proportion of triphenylmethanol formed by hydrolysis during storage. It may be purified by dissolving in about one half its weight of hot benzene containing 10–20 per cent of acetyl chloride, diluting with two volumes of light petroleum, b.p. 60–80 °C, and cooling. The product is filtered rapidly, washed with light petroleum and dried in a vacuum desiccator over paraffin wax shavings and silica gel to remove solvent. The pure compound has m.p. 112–113 °C; it should be stored in a well-stoppered bottle sealed against the ingress of moisture.

Alternatively it may be prepared (1) from triphenylmethanol (10 g) by heating under reflux in dry benzene (5 ml) with redistilled acetyl chloride (6.0 ml) for 30 minutes. The mixture is cooled, diluted with light petroleum (10 ml, b.p. 40–60 °C), chilled on an ice bath and the crystals collected and recrystallised and stored as described above.

Triphenylmethyl bromide, m.p. 153–154 °C, may be prepared in a similar manner from triphenylmethanol and acetyl bromide.

Note. (1) For the preparation of triphenylmethyl chloride from carbon tetrachloride and benzene see Section **IV,8**.

67. **Zinc.** Commercial **zinc powder** is usually about 90 per cent pure and requires acid treatment to remove surface oxide which reduces its activity. The zinc may be activated by stirring, say, 400 g of powder with 150 ml of 10 per cent hydrochloric acid for 2 minutes, filtering and washing with 300 ml of water followed by 100 ml of acetone. The zinc powder should be analysed, if the amount of zinc required in a reaction procedure is critical, by one of the methods described in Ref. 13. Zinc is also available in the form of sheet, wire and wool.

One method of preparation of a **zinc–copper couple** is described in Section **IV,169**. It may also be prepared by stirring activated zinc dust with 2 per cent aqueous copper sulphate solution and washing successively and thoroughly with water, absolute ethanol and dry ether, Section **III,17**.

Zinc amalgam (*for Clemmensen reduction*) may be prepared by either of the following two methods.

Method 1. Two hundred grams of zinc wool are placed in a 2-litre, three-necked flask and covered with a 10–15 per cent solution of sodium hydroxide. The flask is gently warmed on a water bath until hydrogen is vigorously evolved (**CAUTION!**); the sodium hydroxide solution is then immediately poured off (it may be necessary to dilute with water first in order to moderate the vigour of the reaction), and the zinc is washed repeatedly with distilled water until most, if not all, of the sodium hydroxide has been removed. The zinc is then covered with a 1 per cent solution of mercury(II) chloride and allowed to stand for 30–60 minutes with occasional shaking. The mercury(II) chloride solution is then poured off, and the amalgamated zinc is washed twice with distilled water. The amalgamated zinc is then covered with 500 ml of concentrated hydrochloric acid and 100 ml of water. The compound (about 0.3–0.4 mol) to be reduced is then added, and the reaction is allowed to proceed whilst a current of hydrogen chloride gas is passed through the liquid.

Method 2 (Ref. 14). A mixture of 200 g of zinc wool, 15 g of mercury(II) chloride, 10 ml of concentrated hydrochloric acid and 250 ml of water is stirred or shaken for 5 minutes. The aqueous solution is decanted, and the amalgamated zinc is covered with 150 ml of water and 200 ml of concentrated hydrochloric acid. The material (about 0.3–0.4 mol) to be reduced is then added immediately and the reaction is commenced.

68. **Zinc cyanide.** The preparation of zinc cyanide should be carried out in the fume cupboard and great caution exercised in the handling of the cyanides. Prepare solutions containing 100 g of technical sodium cyanide (97–98% NaCN) in 125 ml of water and 150 g of anhydrous zinc chloride in the minimum volume of 50 per cent ethanol (1). Add the sodium cyanide solution rapidly, with agitation, to the zinc chloride solution. Filter off the precipitated zinc cyanide at the pump, drain well, wash with ethanol and then with ether. Dry the product in a desiccator or in an air bath at 50 °C, and preserve in a tightly stoppered bottle. The yield is almost quantitative and the zinc cyanide has a purity of 95–98 per cent. It has been stated that highly purified zinc cyanide does not react in the Adams' modification of the Gattermann reaction (compare Section **IV,126**). The product, prepared by the above method, is, however, highly satisfactory. Commercial zinc cyanide may also be used.

Note. (1) It is important in this preparation to ensure an excess of zinc chloride over sodium cyanide. If the latter is in excess, the zinc cyanide generally precipitates as a sticky mass, which is difficult to filter and unsatisfactory for the preparation of phenolic aldehydes.

BIBLIOGRAPHY AND REFERENCES

II,1. The purification of common organic solvents

1. Methods for the purification of all solvents of value are collected in J. A. Riddick and W. B. Burger (1970). 'Organic Solvents', in *Techniques of Chemistry*. 3rd edn. Ed. A. Weissberger. New York; Wiley-Interscience, Vol. II.
2. *Organic Syntheses* (1963). Wiley, Coll. Vol. 4, pp. 474 and 792.

II,2. Preparation and purification of reagents

3. (a) L. F. Fieser and M. Feiser (1967–75). *Reagents for Organic Synthesis*. New York; Wiley-Interscience, Vols. 1–5.
   (b) D. D. Perrin, W. L. F. Armarego and D. R. Perrin (1966). *Purification of Laboratory Chemicals*. Oxford; Pergamon Press.
4. A. I. Vogel (1962). *Text-Book of Quantitative Inorganic Analysis: Theory and Practice*. 3rd edn. London; Longmans, p. 363.
5. L. A. Carpino (1970). *J. Org. Chem.*, **35**, 3971.
6. I. M. Goldman (1969). *J. Org. Chem.*, **34**, 1979.
7. R. J. Gritter and T. J. Wallace (1959). *J. Org. Chem.*, **24**, 1051.
8. R. L. Augustine (1965). *Catalytic Hydrogenation; Techniques and Application in Organic Synthesis*. London; Arnold, p. 69.
9. L. S. Silbert, E. Siegel and D. Swern (1962). *J. Org. Chem.*, **27**, 1336.
10. Reference 3 (a), p. 1022.
11. Reference 4, p. 248.
12. W. Rigby (1969). *Chem. and Ind.*, **42**, 1508.
13. Reference 4, p. 1091.
14. E. L. Martin (1936). *J. Am. Chem. Soc.*, **58**, 1438.

CHAPTER III ALIPHATIC COMPOUNDS

A ALKANES

The synthesis of alkanes is exemplified by the following typical procedures.

1. The hydrogenation of alkenes (Section **III,1**).
2. The reduction of aldehydes and ketones (Section **III,2**).
3. The hydrolysis of alkylmagnesium halides (Section **III,3**).
4. Coupling reactions (*a*) using organometallic compounds (Sections **III,4** and **III,5**) and (*b*) at an anode (Section **III,6**).

   A,1. The hydrogenation of alkenes. Conversion of an alkene (or alkyne) into an alkane is readily achieved by shaking it under hydrogen at room temperature and at atmospheric pressure in the presence of a platinum or palladium catalyst. With a nickel catalyst somewhat higher temperatures and pressures are employed (see Section **I,17,1**).

$$R^1 \cdot CH = CH \cdot R^2 \xrightarrow[\text{Pt}]{H_2} R^1 \cdot CH_2 \cdot CH_2 \cdot R^2$$

$$R^1 \cdot C \equiv C \cdot R^2 \xrightarrow[\text{Pt}]{H_2} R^1 \cdot CH_2 \cdot CH_2 \cdot R^2$$

The method is illustrated by the conversion of 2-methylbut-2-ene (Section **III,7**) into 2-methylbutane (Section **III,1**). The experiment has been incorporated to illustrate the handling of low boiling point liquids, and also to illustrate some aspects of interpretative infrared spectroscopy.

III,1. 2-METHYLBUTANE

$$CH_3 \cdot CH = C(CH_3)_2 \xrightarrow[\text{Pt}]{H_2} CH_3 \cdot CH_2 \cdot CH(CH_3)_2$$

Place 100 mg of Adams' platinum dioxide catalyst (Section **III,2,**51) and 9.8 g (0.14 mol) of 2-methylbut-2-ene in a 100-ml hydrogenation flask (Section **I,17,1**). Attach the flask to the adapter of the atmospheric hydrogenation apparatus (Fig. I,70(*a*)) and cool the lower part in an ice-water bath. Fill the flask and gas burettes with hydrogen by the procedure discussed in Section **I,17,1**;

320

note the volumes in the gas burettes, remove the cooling bath and gently agitate the flask. When uptake of hydrogen ceases (the catalyst often coagulates and collects at the bottom of the flask at this stage), note the total volume of hydrogen absorbed; this should be in the region of 3 litres. Cool the flask contents and follow the procedure for replacing the hydrogen in the apparatus with air. Disconnect the hydrogenation flask and with a suitably sized dropper pipette transfer the liquid to a small distillation flask leaving the catalyst in the hydrogenation flask (1). Distil the 2-methylbutane (b.p. 30 °C) using a small ice-cooled water condenser with the receiver flask immersed in an ice–salt cooling bath (2). The yield is 7 g (70%). Record the i.r. spectrum and compare it with the spectrum of the starting material (3). Note (a) the disappearance upon hydrogenation of the absorption bands at 810 and 1675 cm^{-1} (due to the out-of-plane deformation of the $=C-H$ bond and the stretching of the carbon–carbon double bond respectively) in the alkene, and (b) the replacement of the peak at 1380 cm^{-1} in the alkene ($-CH_3$ bending mode) by a doublet at 1385 and 1375 cm^{-1} (($CH_3)_2C-$ plus terminal $-CH_3$).

Notes. (1) This technique, rather than conventional filtration, is to be preferred in this case owing to the high volatility of the hydrocarbon. For details on the disposal and recovery of the catalyst see Section **I,17,1**, *Hydrogenation at atmospheric pressure* (*12*).

(2) An alternative and convenient arrangement is to supply the condenser with cooled water from an ice/water reservoir by means of a peristaltic pump.

(3) In both cases a fixed path length cell (0.025 mm) should be employed.

A,2. The reduction of aldehydes and ketones. One procedure for the conversion of a carbonyl group to a methylene group is the **Clemmensen reduction**, and involves the use of zinc amalgam in the presence of concentrated hydrochloric acid.

$$R^1 \cdot CO \cdot R^2 \xrightarrow[\text{HCl}]{\text{Zn/Hg}} R^1 \cdot CH_2 \cdot R^2$$

This represents a somewhat unusual reduction of a carbonyl group, which might be expected to give an alcohol with the metal–acid reducing system (cf. p. 353, **III,D,1**). Alcohols are stable to the Clemmensen conditions and are not therefore intermediates in the reduction process, which is thought to proceed by a mechanism involving the formation of organo-zinc intermediates. A typical example is provided by the conversion of heptan 2-one to heptane (Section **III,2**).

The method is often used for the reduction of aromatic carbonyl compounds (see p. 600, **IV,A,1**) which are reduced in good yield, especially if the procedure is modified by the addition of a solvent immiscible with hydrochloric acid, e.g., toluene.

When the Clemmensen method fails, or when strongly acidic conditions are precluded owing to the presence of acid-sensitive functional groups, the **Wolff–Kishner reduction** or the Huang–Minlon modification of it may succeed. The latter method is also discussed on p. 600, **IV,A,1**, and illustrated in Section **IV,5**, *Method B*.

III,2. HEPTANE

$$(CH_3 \cdot CH_2 \cdot CH_2)_2 CO \longrightarrow CH_3 \cdot (CH_2)_5 \cdot CH_3$$

Place 100 g (1.53 mol) of zinc wool in a 1-litre three-necked flask and amalgamate it in accordance with *Method 1* in Section **II,2,67**. Fit the flask with a sealed stirrer unit, an efficient double surface condenser and a lead-in tube dipping almost to the bottom of the flask for the introduction of hydrogen chloride gas (compare Figs. II,4 and I,66); insert an empty wash bottle between the hydrogen chloride generator and the flask. Introduce through the condenser 250 ml of concentrated hydrochloric acid and 50 ml of water, set the stirrer in motion and then add 40 g (0.35 mol) of heptan-4-one (Section **III,92,A**). Pass a slow current of hydrogen chloride through the mixture; if the reaction becomes too vigorous, the passage of hydrogen chloride is temporarily stopped. After 2–3 hours most of the amalgamated zinc will have reacted. Leave the reaction mixture overnight, but disconnect the hydrogen chloride gas supply first. Remove the stirrer and the condenser from the flask. Arrange for direct steam distillation from the flask by fitting a stopper into one neck, a knee tube connected to a downward condenser in the central aperture and connect the lead-in tube to a source of steam. Stop the steam distillation when the distillate passes over as a clear liquid. Separate the upper layer, wash it twice with distilled water, dry with magnesium sulphate or anhydrous calcium sulphate and distil through a short fractionating column. Collect the fraction, b.p. 97–99 °C (1). The yield of heptane is 26 g (74%).

Note. (1) The products of most Clemmensen reductions contain small amounts of unsaturated hydrocarbons. These can be removed by repeated shaking with 10 per cent of the volume of concentrated sulphuric acid until the acid is colourless or nearly so; each shaking should be of about 5 minutes duration. The hydrocarbon is washed with water, 10 per cent sodium carbonate solution, water (twice), dried with magnesium sulphate or anhydrous calcium sulphate and finally fractionally distilled over sodium.

Cognate preparation   Octane. Use 45 g (0.35 mol) of octan-2-one (Section **III,87**) and 100 g of amalgamated zinc. Collect the fraction, b.p. 124–126 °C; the yield is 23 g (58%).

A,3. The hydrolysis of alkylmagnesium halides. Although there are several methods available for the reduction of alkyl halides to alkanes, a very satisfactory procedure to effect this conversion is the hydrolysis of the corresponding alkylmagnesium halide (Grignard reagents, p. 363, **III,D,2**).

$$RX \xrightarrow[\text{ether}]{Mg} RMgX \xrightarrow{H_2O} RH + Mg(OH)X$$

The conversion is quantitative and this is made use of in the **Zerewitinoff method** for the determination of 'active hydrogen' (e.g., $\equiv C - H$, $- OH$, $- SH$, $- NH_2$, $- CO_2H$, etc.). The compound is allowed to react with an excess of methylmagnesium iodide and the methane evolved is measured in a gas burette. The preparative example is provided by the formation of hexane (Section **III,3**).

III,3. HEXANE

$$C_6H_{13}Br \xrightarrow[\text{ether}]{Mg} C_6H_{13}MgBr$$

$$C_6H_{13}MgBr \xrightarrow[\text{(ii) aq. HCl}]{\text{(i) NH}_4\text{Cl}} C_6H_{14}$$

Fit a 500- or 750-ml three-necked flask with a sealed stirrer unit, a 100-ml dropping funnel and an efficient double surface condenser (Figs. I,22 or I,23); place a calcium chloride guard-tube on the funnel and on the condenser respectively. All parts of the apparatus must be thoroughly dry. Arrange the flask so that it can be heated in a bath of hot water. Place 12.0 g (0.5 mol) of magnesium turnings (1), 100 ml of sodium-dried ether and a crystal of iodine in the flask. Weigh out 82.5 g (70.5 ml, 0.5 mol) of dry 1-bromohexane (Section III,55) and introduce it into the dropping funnel. Run about 10 g of the 1-bromohexane into the magnesium and ether. Warm the flask by surrounding it with hot water; remove the hot water immediately reaction sets in. Start the stirrer and add the remainder of the bromide slowly and at such a rate that the reaction is under control. Continue the stirring, heating if necessary to maintain gentle reflux of the ether, until most of the magnesium has passed into solution (about 4 hours). Add 27 g of AnalaR ammonium chloride, and leave the reaction mixture overnight. Cool the flask in ice and add slowly a large excess of dilute hydrochloric acid; the precipitate will dissolve completely. Separate the upper ethereal layer, and wash it successively with dilute hydrochloric acid and water; dry with magnesium sulphate or anhydrous calcium sulphate. Distil the ethereal solution through an efficient fractionating column (e.g., a Hempel column filled with 6-mm glass or porcelain rings, or a 30-cm all-glass Dufton column; see Figs. I,105(c) and I,105(a)). After the ether has passed over, hexane will distil at 67–70 °C (13–14 g, 30–33%).

Note. (1) Commercial magnesium turnings for the Grignard reaction should be washed with sodium-dried ether to remove any surface grease which may be present, dried at 100 °C and allowed to cool in a desiccator.

A,4(a). Coupling reactions using organometallic compounds. A long established method for the preparation of alkanes involves heating an alkyl halide with sodium metal (the **Wurtz synthesis**).

$$2RX + 2Na \longrightarrow R-R + 2NaX$$

This type of symmetrical coupling of two molecules of an alkyl halide will obviously give better yields than the alternative unsymmetrical coupling process which must be used, for example, to make a straight chain alkane having an odd number of carbon atoms.

$$R^1X + R^2X + 2Na \longrightarrow R^1-R^2 + R^1-R^1 + R^2-R^2 + 2NaX$$

The symmetrical Wurtz coupling has been found to give particularly good yields in the case of the higher alkane homologues. Selected illustrative examples (hexane, octane, decane and dodecane) are to be found in Section III,4.

Coupling of alkyl groups by the formation of carbon–carbon σ-bonds can also be achieved by the use of copper 'ate' complexes. The organocopper reagents are prepared most conveniently by the reaction of two mols of an organolithium reagent with one mol of the appropriate copper(I) halide. Reaction of the organocuprate with an alkyl halide results in the displacement of halogen by an S_N2 mechanism.

$$2R^1Li + CuX \longrightarrow R^1{}_2CuLi + LiX$$
$$\text{Lithium dialkyl}$$
$$\text{cuprate}$$
$$R^1{}_2CuLi + R^2X \longrightarrow R^1 - R^2 + R^1Cu + LiX$$

An example of this coupling reaction is provided by the preparation of undecane (Section III,5).

III,4. OCTANE

$$2C_4H_9Br + 2Na \longrightarrow C_8H_{18} + 2NaBr$$

Weigh out 23 g (1 mol) of clean sodium under sodium-dried ether (Section II,1,*15*), cut it up rapidly into small pieces and introduce the sodium quickly into a dry 750- or 1000-ml round-bottomed flask. Fit a dry 30-cm double surface condenser (e.g., of the Davies type) into the flask and clamp the apparatus so that the flask can be heated on a wire gauze. Weigh out 68.5 g (53 ml, 0.5 mol) of butyl bromide (Section III,54) previously dried over anhydrous sodium sulphate. Introduce about 5 ml of the bromide through the condenser into the flask. If no reaction sets in, warm the flask gently with a small luminous flame; remove the flame immediately reaction commences (the sodium will acquire a blue colour). When the reaction subsides, shake the contents of the flask well; this will generally produce further reaction and some of the sodium may melt. Add a further 5 ml of butyl bromide, and shake the flask. When the reaction has slowed down, repeat the above process until all the alkyl bromide has been transferred to the flask (about 1.5 hours). Allow the mixture to stand for 1–2 hours. Then add down the condenser by means of a dropping funnel 50 ml of rectified spirit dropwise over 1.5 hours, followed by 50 ml of 50 per cent aqueous ethanol during 30 minutes, and 50 ml of distilled water over 15 minutes; shake the flask from time to time. Add 2–3 small pieces of porous porcelain and reflux the mixture for 3 hours; any unchanged butyl bromide will be hydrolysed. Add a large excess (500–750 ml) of water, and separate the upper layer of crude octane (17–18 g). Wash it once with an equal volume of water, and dry it with magnesium sulphate. Distil through a short fractioning side-arm (cf. Fig. I,104) and collect the fraction, b.p. 123–126 °C (15 g, 52%) (1).

Note. (1) All hydrocarbons prepared by the Wurtz reaction contain small quantities of unsaturated hydrocarbons. These may be removed by shaking repeatedly with 10 per cent of the volume of concentrated sulphuric acid until the acid is no longer coloured (or is at most extremely pale yellow); each shaking should be of about 5 minutes' duration. The hydrocarbon is washed with water, 10 per cent sodium carbonate solution, water (twice), and dried with magnesium sulphate or anhydrous calcium sulphate. It is then distilled from sodium; two distillations are usually necessary.

Cognate preparations   Hexane. Use 23 g (1 mol) of sodium and 61.5 g (45.5 ml, 0.5 mol) of propyl bromide (Section III,54). It is advisable to employ two efficient double surface condensers in series. Collect the fraction, b.p. 68–70 °C (10 g, 47%).

Decane. Use 23 g (1 mol) of sodium and 75.5 g (62 ml, 0.5 mol) of 1-bromopentane (Section **III,54**) or 99 g (65.5 ml, 0.5 mol) of 1-iodopentane (Section **III,58**). Collect the fraction, b.p. 171–174 °C (28 g, 79%).

Dodecane. Use 23 g (1 mol) of sodium and 82.5 g (70.5 ml, 0.5 mol) of 1-bromohexane (Section **III,55**). Collect the fraction, b.p. 94 °C/13 mmHg (37 g, 87%).

III,5. UNDECANE

$$C_4H_9Li + CuI \longrightarrow (C_4H_9)_2CuLi + LiI$$

$$C_7H_{15}I \xrightarrow[-78\ °C]{(C_4H_9)_2CuLi} C_{11}H_{24}$$

Place 9.53 g (0.05 mol) of copper(I) iodide in a two-necked, 500-ml round-bottomed flask containing a glass-covered magnetic follower bar. Fit a rubber septum to one of the necks of the flask and connect the other neck to one arm of a three-way stop-cock. Connect the second arm of the stop-cock to a supply of dry, oxygen-free nitrogen (cf. Fig. I,67) and the remaining arm to a vacuum pump. Evacuate the flask by opening the stop-cock to the vacuum pump (1) and then, with a rapid stream of nitrogen flowing, *carefully* open the stop-cock to the nitrogen supply to fill the flask with nitrogen. Repeat this process twice; flame the flask gently on the final occasion it is evacuated. Maintain a static atmosphere of nitrogen in the flask throughout the reaction by passing a slow stream of nitrogen through the nitrogen line. Cool the flask in an acetone/Cardice bath and transfer 100 ml of dry tetrahydrofuran (2) to the flask via the septum using a hypodermic syringe (3). Stir the cooled (− 78 °C) suspension and add 52.0 ml of a 1.92 molar solution of butyl lithium in hexane (4) from a hypodermic syringe (5). Stir the flask solution at − 78 °C for 1 hour, and then add a solution of 3.39 g (0.015 mol) of 1-iodoheptane (Section **III,59**) in 10 ml of dry tetrahydrofuran dropwise from the syringe. Stir the solution at − 78 °C for 1 hour and then at 0 °C (ice-bath) for a further 2 hours (6, 7). Hydrolyse the reaction mixture by pouring it carefully into 100 ml of 1 M-hydrochloric acid. Separate the upper organic layer and extract the aqueous layer with two 50 ml portions of pentane. Wash the combined organic layers with water (50 ml), dry over magnesium sulphate and evaporate the solvents on the rotary evaporator. Distil the residue at atmospheric pressure using a semi-micro scale distillation unit fitted with a short fractionating side arm packed with glass helices (cf. Fig. I,113(a)). Undecane has b.p. 194–197 °C; the yield is 1.24 g (53%).

Notes. (1) This operation should be carried out behind a safety screen.

(2) Tetrahydrofuran should be freshly distilled from lithium aluminium hydride. It is convenient to store a supply of peroxide-free tetrahydrofuran (Section **II,2,19**) under nitrogen over lithium aluminium hydride and distil appropriate quantities as required.

(3) The syringes and long flexible needles can be obtained from Aldrich Chemical Co.

(4) Solutions of butyl lithium, and other lithium reagents, may be purchased from Alfa Inorganics.

(5) See below for notes on the storage and handling of alkyl lithium reagents.

(6) The progress of the reaction may be conveniently followed by GLC. Remove samples (*ca*. 2–3 ml) from the reaction mixture by means of a hypodermic syringe and submit them to a small-scale hydrolysis and extraction procedure similar to that described for the main reaction mixture. Analyse the organic layer on a 10 per cent squalane on Chromosorb W column held at 140 °C. Under these conditions 1-iodoheptane has a slightly longer retention time than undecane.

(7) Complete reaction of the alkyl lithium can be tested by means of the Gilman test described in Section III,15, Note (3).

Notes on the procedures for handling solutions of alkyl lithiums Alkyl lithium reagents must be handled with care and protective clothing should be worn. Solutions of alkyl lithiums are extremely inflammable and may ignite on contact with moist air; they should therefore always be stored and handled in an inert atmosphere (argon or nitrogen). They are measured and transferred most conveniently by means of hypodermic syringes. A number of alkyl lithiums are available commercially and the manufacturers' instructions for their handling should be followed carefully. The lithium reagent may be stored in the original container by placing a rubber septum on the neck and removing portions of the reagent as required by means of a hypodermic syringe; the volume of reagent removed should then be replaced by an equal volume of nitrogen similarly introduced by means of the syringe. An alternative means of transferring the reagent to a two- or three-necked flask is as follows. Equip the necks with stop-cocks for nitrogen inlet and outlet. The outlet should be of such a design that when it is open the needle of a hypodermic syringe can be passed through its centre into the flask. The transfer of the lithium reagent to the flask is then achieved by inserting into the neck of the reagent bottle a tight fitting rubber bung previously fitted with two glass tubes: a short inlet tube for connection to a supply of inert gas, and an exit tube which reaches to the bottom of the reagent bottle. The exit tube is connected by polythene tubing to the flask which has been previously dried and flushed with nitrogen. The reagent is then transferred by means of a slight positive pressure of inert gas. Whenever the reagent is removed a stream of inert gas should be passed through the flask. Butyl lithium can be stored for an extended period at room temperature under nitrogen. The concentration of the reagent should be checked periodically by the method described below.

Notes on the estimation of solutions of alkyl lithiums and Grignard reagents Alkyl lithium reagents and Grignard reagents form coloured charge transfer complexes with certain aromatic nitrogen heterocycles such as 2,2'-bipyridyl, 1,10-phenanthroline and 2,2'-biquinolyl. When the organometallic reagent is decomposed by compounds having an acidic hydrogen, in this case butan-2-ol, the colour disappears. The complexes thus form suitable indicators for the titration of the organometallic reagents. The titration is performed under an atmosphere of dry, oxygen-free nitrogen in a 50-ml, three-necked flask fitted with a nitrogen inlet and outlet and containing a glass-covered magnetic stirrer follower bar. The third neck is fitted with a bung (or rubber septum) through the centre of which passes the delivery tip of the burette; the tip should be well clear of the bottom of the bung.

Dry the flask in the oven and allow to cool whilst passing a stream of nitrogen (fit a stopper in the third neck during this operation). Place about 2 mg of 2,2'-bipyridyl (1) and 10 ml of anhydrous benzene in the flask and flush the flask with nitrogen for about 20 minutes. Whilst maintaining a rapid flow of nitrogen, remove the nitrogen exit briefly and add a 5.00 ml aliquot portion (2) of the organometallic reagent to the solution. Titrate the resulting solution with a 1 M standard solution of butan-2-ol in p-xylene (3) until the colour of the charge transfer complex is discharged (4).

1 mol butan-2-ol ≡ 1 mol of organometallic reagent.

1 1,10-Phenanthroline or 2,2'-biquinolyl may be used as alternative indicators. The colour of the complex is dependent upon the complexing agent. Thus butyl lithium gives a greenish-yellow colour with 2,2'-biquinolyl, reddish-brown with 1,10-phenanthroline and reddish-purple with 2,2'-bipyridyl.

2 Measure the aliquot portion of the organometallic reagent with a pipette which has been previously warmed in the oven and allowed to cool with a stream of nitrogen flowing through it. The use of a suction bulb is essential.

3 Depending on the anticipated concentration of the organometallic reagent, a more or a less concentrated solution may be required. Both the p-xylene and butan-2-ol should be freshly dried and distilled (Sections **II,1,**4 and **II,1,**12 respectively).

4 The solution remains clear throughout the titration since the lithium but-2-oxide which is formed in the reaction is soluble in the solvent mixture.

A,4(b). Anodic coupling reactions. The synthesis of alkanes which involves the electrolysis of salts of carboxylic acids was first reported by Kolbé in 1849. The technique and the apparatus has been described in Section **I,17,6**, and the illustrative example relevant to this section is that of the preparation of hexacosane (the experimental details were originally supplied by Dr R. P. Linstead, C.B.E., F.R.S.). (See also p. 485, **III,M,5.**)

III,6. HEXACOSANE

$$2CH_3 \cdot (CH_2)_{12} \cdot CO_2^{\ominus} \xrightarrow{-2e} C_{26}H_{54} + 2CO_2$$

Dissolve 5.0 g of pure myristic acid (tetradecanoic acid) in 25 ml of absolute methanol to which 0.1 g of sodium has been added. Place the solution in a cylindrical cell (25 cm long, 3 cm diameter) provided with two platinum plate electrodes (2.5 × 2.5 cm) set 1–2 mm apart (cf. Section **I,17,6**). Electrolyse at about 1 amp until the electrolyte is just alkaline (pH 7.5–8). Cool the cell in an ice bath during the electrolysis. Reverse the current from time to time; this will help to dislodge the coating of insoluble by-products on the electrodes. Neutralise the cell contents by adding a few drops of glacial acetic acid, and evaporate most of the solvent under reduced pressure using a rotary evaporator. Pour the residue into water and extract the crude product with ether. Wash the ethereal solution with dilute sodium hydroxide solution, dry (magnesium sulphate) and evaporate the solvent. Recrystallise the residue from light petroleum (b.p. 40–60 °C). The yield of hexacosane, m.p. 57–58 °C, is 2.4 g (65%).

B ALKENES

The introduction of a carbon–carbon double bond into a molecule may be effected by the following typical procedures.

1. 1,2-Elimination processes (β-eliminations) applied to alcohols, alkyl halides, quaternary ammonium salts and acetate or xanthate esters (Sections **III,7** to **III,12**).
2. The partial catalytic hydrogenation of alkynes (Section **III,13**).
3. The reaction between an alkylidenephosphorane and a carbonyl compound (Section **III,14**).
4. Wurtz-type coupling reactions involving allylic halides (Sections **III,15** and **III,16**).
5. Rearrangement of alkynes (Sections **III,17** and **III,18**).

B,1. 1,2-Elimination (β-elimination) processes. Many frequently used methods for the introduction of carbon–carbon double bonds into a saturated carbon chain involve the removal of two atoms or groups from adjacent carbon atoms. Usually, but not invariably, one of these groups is hydrogen and the process may take place by either a step-wise (E1) or a synchronous (E2) pathway.

(i) The dehydration of alcohols under acidic conditions is widely used. In the laboratory phosphoric acid is the reagent of choice; sulphuric acid, which is also often used, can lead to extensive charring and oxidation and hence to lower yields of alkene.

Tertiary alcohols undergo elimination the more readily, and almost certainly by an E1 process.

$$
\underset{\underset{OH}{|}}{R_2C}-CR_2 \quad \xrightarrow{H^\oplus} \quad \underset{\underset{\oplus OH_2}{|}}{R_2C}-CR_2 \quad \longrightarrow \quad \underset{\underset{\oplus}{|}}{R_2C}-CR_2
$$

$$
H_2O: \quad R_2C-CR_2 \quad \longrightarrow \quad R_2C=CR_2
$$

In the preparative example, the dehydration of 2-methylbutan-2-ol (Section **III,7**), loss of a proton from the intermediate carbonium ion may take place from either of the two adjacent positions giving a mixture of isomeric alkenes, 2-methylbut-2-ene (I) and 2-methylbut-1-ene (II).

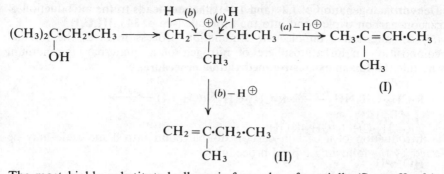

$$(CH_3)_2C \cdot CH_2 \cdot CH_3$$

The most highly substituted alkene is formed preferentially (**Saytzeff rule**); in this case the ratio of products, I : II, is 4 : 1 (GLC analysis). The dehydration of the secondary alcohol 2-methylcyclohexanol similarly yields a mixture of isomeric alkenes in which the more highly substituted alkene again predominates.

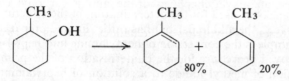

The dehydration of pinacol (Section **III,29**) affords a convenient synthesis of the diene, 2,3-dimethylbutadiene (Section **III,8**), although some concomitant rearrangement to pinacolone (Section **III,97**) occurs under the acidic conditions employed.

(*ii*) The dehydrohalogenation of alkyl halides to alkenes is effected under strongly basic conditions, e.g., a concentrated alcoholic solution of sodium or potassium hydroxide or alkoxide. These conditions favour an E2 process in which the participating sites adopt an antiperiplanar conformation leading to an *anti*-elimination process.

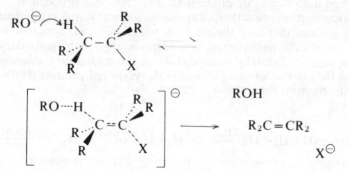

Bimolecular substitution leading, for example, to the formation of ethers is a competing reaction and in fact is the main result when simple primary alkyl halides are employed. For example, ethyl bromide with ethanolic potassium hydroxide gives very little ethylene, the major product being diethyl ether.

An example illustrating the dehydrohalogenating process is provided by the ready conversion of 3-bromocyclohexene (Section **III,65**) into cyclohexa-1,3-diene (Section **III,9**) where the base employed is quinoline.

Dehydrohalogenation of 1,2- (and 1,1-) dihalides leads to the introduction of a carbon–carbon triple bond into the molecule (see p. 345, **III,C,1**).

(*iii*) An alternative bimolecular elimination process involves the thermal decomposition in an atmosphere of nitrogen of a quaternary ammonium hydroxide (**Hofmann exhaustive methylation procedure**).

$$R \cdot CH_2 \cdot CH_2 NH_2 \xrightarrow{3CH_3I} R \cdot CH_2 \cdot CH_2 \overset{\oplus}{N}(CH_3)_3 \} \overset{\ominus}{I} \xrightarrow{\text{'AgOH'}}$$

$$R \cdot CH_2 \cdot CH_2 \overset{\oplus}{N}(CH_3)_3 \} \overset{\ominus}{O}H$$

$$HO \overset{\ominus}{\underset{}{:}} \curvearrowright H$$
$$R \cdot CH \overset{\downarrow}{\curvearrowleft} CH_2 \longrightarrow R \cdot CH = CH_2 + N(CH_3)_3 + H_2O$$
$$\overset{\curvearrowleft}{\underset{\oplus}{N}}(CH_3)_3$$

The reaction has been extensively used for the determination of the structure of naturally occurring bases (e.g., the alkaloids), but has rather limited preparative value. The illustrative example is the conversion of heptylamine into hept-1-ene (Section **III,10**). An improved procedure for the quaternisation of the primary amine involves the addition of methyl iodide to a solution of heptylamine in dimethylformamide, in the presence of an adequately basic but weakly nucleophilic acceptor base (e.g., tributylamine) for the release of the free amines from their hydrohalide salts to promote complete alkylation (see p. 570, **III,R,4**). The conversion of the quaternary ammonium halide into the hydroxide is still most generally effected by the use of moist silver oxide despite its expense, and, in some cases, the incidence of oxidative side reactions. Alternative procedures include the use of thallium(I) hydroxide, which is expensive and toxic but does not lead to an oxidation effect, and the use of a basic ion exchange resin. The aqueous solutions of quaternary ammonium hydroxide obtained by this latter method are often more dilute than with the other procedures.

The elimination step occurs without any rearrangement of the skeletal structure. A further general feature of the reaction, which becomes apparent when the Hofmann exhaustive methylation procedure is applied to secondary or tertiary amines, is that thermal decomposition of the quaternary ammonium hydroxide gives the least substituted alkene as the preferred product (**Hofmann rule**; contrast the Saytzeff rule).

$$R \cdot CH_2 \cdot CH_2 \cdot NH \cdot CH_2 \cdot CH_3 \xrightarrow{CH_3I} R \cdot CH_2 \cdot CH_2 \cdot \overset{\overset{\displaystyle CH_3}{|}}{\underset{\underset{\displaystyle CH_3}{|}}{\overset{\oplus}{N}}} \cdot CH_2 \cdot CH_3 \} X^{\ominus} \xrightarrow{\text{'AgOH'}}$$

$$R \cdot CH_2 \cdot CH_2 \cdot \overset{\overset{\displaystyle CH_3}{|}}{\underset{\underset{\displaystyle CH_3}{|}}{\overset{\oplus}{N}}} \cdot CH_2 \cdot CH_3 \} \overset{\ominus}{O}H \longrightarrow$$

$$R \cdot CH_2 \cdot CH_2 \cdot N(CH_3)_2 + CH_2 = CH_2 + H_2O$$

(*iv*) Two further illustrative examples of alkene syntheses are provided by the pyrolysis of acetate esters and by the thermal decomposition of the related xanthate esters which occurs at a lower temperature. Both these are mechanistically classified as *syn*-elimination processes.

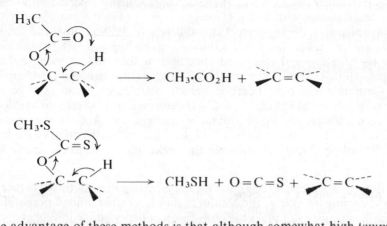

The advantage of these methods is that although somewhat high temperatures are employed the product is not usually contaminated by compounds arising from double bond migration or from skeletal rearrangement.

Acetate pyrolysis is illustrated by the synthesis of penta-1,4-diene (Section **III,11**) from 1,5-diacetoxypentane. This is prepared from 5-chloropentyl acetate which is conveniently obtained from tetrahydropyran by treatment with acetyl chloride, this ring opening reaction offers interesting comparison with the methods for the preparation of α,ω-dihalides discussed on p. 386, **III,E,1**(*b*).

The thermal decomposition of *O*-[1,2,2-trimethylpropyl] *S*-methyl dithiocarbonate (Section **III,207**) gives a satisfactory yield of 3,3-dimethylbut-1-ene (Section **III,12**). In contrast, the acid-catalysed dehydration of the corresponding alcohol, proceeding by way of an E1 mechanism and a carbonium ion intermediate, would inevitably result in rearrangement and formation of 2,3-dimethylbut-2-ene as the main product.

$$(CH_3)_3C \cdot CH(OH) \cdot CH_3 \xrightarrow{H_3O^\oplus} (CH_3)_3C \cdot \overset{\oplus}{C}H \cdot CH_3 \xrightarrow[\substack{\text{shift of methyl} \\ \text{group}}]{\text{1,2-nucleophilic}}$$

$$(CH_3)_2\overset{\oplus}{C} \cdot CH(CH_3)_2 \longrightarrow (CH_3)_2C = C(CH_3)_2$$

The product from the xanthate reaction is homogeneous on GLC analysis and the i.r. spectrum clearly confirms the unrearranged skeletal structure.

III,7. 2-METHYLBUT-2-ENE (*in admixture with* 2-*METHYLBUT-1-ENE*)

$$(CH_3)_2 \cdot C(OH) \cdot CH_2 \cdot CH_3 \xrightarrow{H_3PO_4}$$

$$(CH_3)_2 \cdot C = CH \cdot CH_3 + CH_2 = C(CH_3) \cdot CH_2 \cdot CH_3$$

Place 25.0 g (31 ml, 0.28 mol) of 2-methylbutan-2-ol and 10 ml of 85 per cent orthophosphoric acid in a 100-ml, round-bottomed flask and swirl to mix thoroughly. Fit the flask with a 20-cm fractionating column filled with glass helices, a Claisen still-head and a condenser leading to a 50-ml receiving flask cooled in a beaker of iced water (Fig. I,104). Add a few pieces of porous porcelain and heat the reaction mixture gently with a Bunsen burner. Collect the alkene fraction which distils in the range 35–38 °C during a period of 30 minutes. Dry the distillate with 1–2 g of magnesium sulphate. Wash and dry the distillation apparatus, decant the dried distillate into a 50-ml flask and redistil in the reassembled apparatus. Collect the fraction boiling at 37–38 °C; the yield is 12.5 g (64%). Record the infrared spectrum of the product using a fixed path-length cell (0.025 mm). The stretching bands of the terminal (1645 cm^{-1}) and non-terminal (1670 cm^{-1}) carbon–carbon double bonds can both be observed; bands at 890 and 805 cm^{-1} ($=C-H$ deformation) also establish the presence of both terminal and non-terminal olefinic systems. Analyse the product by GLC on a Silicone oil column at 30 °C; 2-methylbut-1-ene appears first, closely followed by 2-methylbut-2-ene; the areas under the peaks are in the ratio of 1 : 4.

Cognate preparations   Cyclohexene.   Fit a 500-ml three-necked flask with a fractionating column (e.g., a Hempel column filled with 6-mm glass or porcelain rings) carrying a thermometer at its upper end, and a separatory funnel; close the third neck with a stopper. Attach an efficient double surface condenser to the column; use a filter flask, cooled in ice, as receiver. Place 50 g of 85 per cent orthophosphoric acid in the flask and heat it in an oil bath at 160–170 °C. Add, through the funnel, 250 g (2.5 mol) of cyclohexanol over a period of 1.5–2 hours. When all the cyclohexanol has been introduced, raise the temperature of the bath to about 200 °C and maintain it at this temperature for 20–30 minutes. The temperature at the top of the column should not rise above 90 °C. Saturate the distillate with salt, separate the upper layer and dry it with magnesium sulphate. Distil the crude cyclohexene through an efficient column and collect the fraction boiling at 81–83 °C; the residue is largely cyclohexanol. The yield of cyclohexene is 165 g (80%).

1-Methylcyclohexene.   Use 20 g (0.18 mol) of 2-methylcyclohexanol and 5 ml of phosphoric acid. Proceed as for 2-methylbut-2-ene but use a shorter fractionating column (12 cm); it is not necessary to cool the receiving flask. Collect the fraction boiling between 100 and 112 °C and dry and redistil using a simple distillation apparatus. Collect the purified alkene fraction boiling at 103–110 °C; the yield is 14.2 g (76%). Analyse the product by GLC on a Silicone oil column at 60 °C (Section I,31); 3-methylcyclohexene (20%) appears first followed by 1-methylcyclohexene (80%). The major product, 1-methylcyclohexene, b.p. 110 °C, can be separated from the 3-methyl isomer, b.p. 103 °C, by careful fractionation through a spinning band column (Fig. I,109).

III,8. 2,3-DIMETHYLBUTA-1,3-DIENE

$$(CH_3)_2C(OH)\cdot C(OH)(CH_3)_2 \xrightarrow{HBr}$$

$$CH_2 = C(CH_3)\cdot C(CH_3) = CH_2 + (CH_3)_3C\cdot CO\cdot CH_3$$

In a 1-litre round-bottomed flask, surmounted by an efficient fractionating column (e.g., one of the Hempel type, see Section I,26) place 177 g (1.5 mol) of anhydrous pinacol (Section **III,29**), 5 ml of constant boiling point hydrobromic acid (*ca.* 47–48% w/w HBr) and a few fragments of porous porcelain. Attach a condenser and a receiver to the column. Heat the flask gently in an oil bath; the rate of distillation should be 20–30 drops per minute. Collect the distillate until the temperature at the top of the column is 95 °C (60–70 minutes). Separate the upper non-aqueous layer, wash it twice with 50 ml portions of water, add 0.25 g of hydroquinone as an inhibitor and dry it overnight with 7–8 g of anhydrous calcium chloride. Transfer to a 500-ml flask and distil through the same column. Collect the following fractions: (*a*) 69–70.5 °C (70 g), (*b*) 70.5–105 °C (7 g) and (*c*) 105–106 °C (35 g). Fraction (*a*) is pure dimethylbutadiene (yield 57%), (*b*) is an intermediate fraction and (*c*) is pinacolone.

Dimethylbutadiene may be kept for a limited period in an ice box or in a refrigerator; it is advisable to add about 0.2 g of hydroquinone as an inhibitor.

III,9. CYCLOHEXA-1,3-DIENE

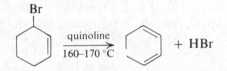

In a 100-ml round-bottomed flask place 16.1 g (0.10 mol) of 3-bromocyclohexene (Section **III,65**), 38.7 g of dried, redistilled quinoline and a magnetic stirrer follower. Attach to the flask a Claisen still-head fitted with a thermometer and a condenser set for downward distillation. Protect the apparatus from atmospheric moisture by a calcium chloride guard-tube attached to the side arm of the receiver. Support the reaction flask in an oil bath which rests on a magnetic stirrer/hotplate, commence stirring rapidly and heat the oil bath to 160–170 °C. The cyclohexadiene steadily distils as colourless liquid, b.p. 80–82 °C, over a period of about 30 minutes, yield 5.4 g (68%). The product is 99 per cent pure by GLC analysis; use either a 9-ft column of 10 per cent polyethyleneglycol adipate on Chromosorb W held at 60 °C with a flow rate of carrier gas of 40 ml/minute, t_R 2.4 minutes, or, a 5-ft column of 10 per cent Silicone oil on Chromosorb W held at 60 °C with a flow rate of carrier gas of 40 ml/minute, t_R 1.5 minutes.

III,10. HEPT-1-ENE

$$C_5H_{11}\cdot CH_2\cdot CH_2NH_2 \xrightarrow[\text{(C}_4\text{H}_9)_3\text{N; DMF}]{\text{excess CH}_3\text{I}} C_5H_{11}\cdot CH_2\cdot CH_2\overset{\oplus}{N}(CH_3)_3\}\overset{\ominus}{I} \xrightarrow{\text{'AgOH'}}$$

$$C_5H_{11}\cdot CH_2\cdot CH_2\overset{\oplus}{N}(CH_3)_3\}\overset{\ominus}{O}H \xrightarrow{\text{pyrolysis}} C_5H_{11}\cdot CH=CH_2$$

CAUTION: *methyl iodide is carcinogenic; handle appropriately in a fume cupboard.*

Heptyltrimethylammonium iodide. To a mixture of 11.4 g (0.1 mol) of redistilled heptylamine (Section **III,191**), 37 g (0.2 mol) of redistilled tributylamine and 30 ml of dry redistilled dimethylformamide (Section **II,1,**_26_), add dropwise 40 g (0.28 mol) of methyl iodide, with stirring and sufficient cooling to moderate the exothermic reaction. During the addition the two-phase mixture becomes homogeneous. Allow the reaction mixture to stand overnight at room temperature, and then add 100 ml of sodium-dried ether, shake and place in a refrigerator. Filter off the crystals, wash them with ether and recrystallise from absolute ethanol. The recrystallised quaternary iodide has m.p. 145–6 °C, the yield is 18.5 g (84%).

Heptyltrimethylammonium hydroxide.   Dissolve 17 g (0.1 mol) of silver nitrate in 170 ml of distilled water, warm to 85 °C and add a solution of 3.90 g (0.097 mol) of sodium hydroxide (AnalaR) in 170 ml of water similarly warmed to 85 °C. Shake vigorously, and allow to stand until coagulation of the precipitated silver oxide is complete. Wash the silver oxide by decantation with five 100 ml portions of hot distilled water. Drain the final residual wash liquid as completely as possible and then add to the damp silver oxide a solution of 14.2 g (0.05 mol) of heptyltrimethylammonium iodide in 80 ml of distilled water containing 10 ml of methanol. Stir the reaction mixture magnetically under nitrogen for 2 hours, filter off the silver salts and wash the precipitate with water. Transfer the filtrate to a rotary evaporator and remove the water under reduced pressure (bath temperature _ca._ 40 °C). The quaternary ammonium hydroxide is obtained as a clear viscous syrup and should be stored under nitrogen.

Hept-1-ene.   Assemble the following apparatus in a fume cupboard; a 250-ml round-bottomed flask fitted with a nitrogen inlet, a knee-bend adapter and a condenser, and a receiver adapter with the outlet suitably extended to reach near to the bottom of a two-necked, 100-ml round-bottomed receiver flask. Connect a second two-necked receiving flask to the first by means of a bent glass tube and screw-capped adapters. The outlet of the second receiver flask should lead directly into the fume cupboard vent. Cool the first receiver flask in an ice-water bath and the second in acetone-Cardice. Flush the apparatus with nitrogen, dissolve the quaternary hydroxide prepared above in a little water and transfer to the 250-ml flask. With the nitrogen stream flowing slowly but continuously heat the flask in an oil bath. Water distils at a bath temperature of 160 °C; pyrolysis sets in (frothing is observed) when the bath temperature is at about 190 °C. When there is little or no observable residue, transfer the contents of the receiver flasks to a separating funnel, using ether to effect quantitative transfer of the alkene which forms the upper oily layer. Remove the lower aqueous layer and wash the ether layer first with hydrochloric acid (to remove remaining trimethylamine) and then with water, and finally dry over magnesium sulphate. Carefully remove the ether by flash distillation and distil the residue in a semi-micro distillation apparatus fitted with a short fractionating column. Collect the hept-1-ene at 93–94 °C; the yield is 3.0 g (60%). The purity of the sample may be verified by GLC using a column of 10 per cent squalane on Chromosorb W held at 50 °C; t_R is 5 min 24 sec. Infrared absorptions are clearly visible at 1645 (C=C str.), 995 and 915 (=C−H bend, 1820 overtone) and 720 cm^{-1} [(CH$_2$)$_n$ rocking].

III,11. PENTA-1,4-DIENE (*Divinylmethane*)

$$\text{[cyclic } O] + CH_3 \cdot COCl \longrightarrow CH_3 \cdot CO \cdot O \cdot CH_2 \cdot (CH_2)_3 \cdot CH_2Cl \xrightarrow{CH_3 \cdot \overset{\ominus}{CO_2} \overset{\oplus}{K}}$$

$$CH_3 \cdot CO \cdot O \cdot CH_2 \cdot (CH_2)_3 \cdot CH_2 \cdot O \cdot CO \cdot CH_3 \longrightarrow CH_2 = CH \cdot CH_2 \cdot CH = CH_2$$

1,5-Diacetoxypentane. Equip a 5-litre three-necked flask with an efficient sealed stirrer unit of the Hirschberg type, a double-surface reflux condenser and screw-capped adapter carrying a thermometer. Add 516 g (582 ml, 6.0 mol) of tetrahydropyran, 480 g (434 ml, 6.0 mol) of acetyl chloride and 4 g of powdered fused zinc chloride. Heat the mixture with a heating mantle with vigorous stirring under reflux, and continue to heat until the temperature of the reaction mixture reaches 150 °C (about 3–5 hours). Allow the reaction mixture to cool to near room temperature and add 980 g (10 mol) of solid potassium acetate and 20 g of sodium iodide. Heat the mixture again with stirring to 160 °C and maintain the temperature for 12 hours. Cool to room temperature, add 1 litre of light petroleum (b.p. 40–60 °C), filter under suction and wash the residue thoroughly with more light petroleum. Combine the filtrate and washings and

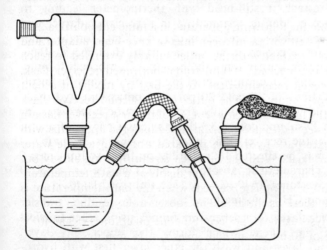

Fig. III,1

remove the solvent on a rotary evaporator. Fractionally distil the residue under reduced pressure and collect the 1,5-diacetoxypentane at 108–114 °C/3.5–4 mmHg. The yield is about 790 g (70%); the product is sufficiently pure for use in the next stage, but a sample may be purified by redistillation and then has b.p. 102–104 °C/3 mmHg (or 92–94 °C/1.0 mmHg).

Penta-1,4-diene. Assemble the pyrolysis apparatus shown in Fig. I,72(*c*) with the pyrolysis tube packed with glass beads but with the acetone-Cardice cooling trap arrangement shown in Fig. III,1 attached to the water condenser. Pass a gentle stream of nitrogen gas through the apparatus and heat the combustion tube to 580 °C when 900 g of 1,5-diacetoxypentane is added dropwise during about 12 hours (1). Fractionally distil the total pyrolysate carefully and slowly at atmospheric pressure and collect the fraction boiling below 50 °C which should be refractionated to give penta-1,4-diene of b.p. 24–27 °C. The yield is 170 g (52%).

Note. (1) Solidified acetic acid often blocks the flow of nitrogen in the cooling trap. It can be dislodged by careful warming with a warm-air blower.

III,12. 3,3-DIMETHYLBUT-1-ENE

$$(CH_3)_3C \cdot CH \cdot CH_3 \xrightarrow{\text{heat}} (CH_3)_3C \cdot CH = CH_2 + S = C = O + CH_3SH$$
$$|$$
$$O \cdot CS \cdot SCH_3$$

This preparation should be conducted in an efficient fume cupboard.

From 51 g (63 ml, 0.5 mol) of 3,3-dimethylbutan-2-ol prepare O-[1,2,2-trimethylpropyl]S-methyl dithiocarbonate (the xanthate ester) using the procedure described in Section **III,207**, up to and including the removal of toluene and residual alcohols with a rotary evaporator. Transfer the red residual crude xanthate ester to a 250-ml round-bottomed flask. Fit a vertical air condenser (about 12 inches), and attach a still-head with thermometer leading to an efficient double surface condenser terminating in a receiver adapter and flask. Immerse the receiver flask in an ice–salt bath and arrange for the water supply to the condenser to be cooled in an ice–salt bath. Heat the xanthate ester to boiling using a Bunsen burner; smooth decomposition occurs forming the alkene which slowly distils. On completion of the pyrolysis wash the distillate with three 10 ml portions of ice-cold 20 per cent potassium hydroxide solution, then with ice-cold water and finally dry the organic layer with anhydrous calcium chloride. Distil and collect 3,3-dimethylbut-1-ene as a fraction having b.p. 40–42 °C (cool the receiver flask in ice). The yield is 23.5 g (53%) (1). Further purification may be effected by redistillation from sodium metal. The product shows the characteristic absorption bands in the i.r. at 3050 ($= C - H$), 1820 ($= CH_2$ overtone), 1640 ($C = C$), 1385 (m), 1365 (s) (($CH_3)_3C$), 995 and 915 cm^{-1} (CH and CH_2 deformation).

Note. (1) GLC analysis using 10 per cent Silicone oil column (5′) at 40 °C with a nitrogen flow rate of 40 ml per minute gives t_R 1.18 minutes.

B,2. The partial catalytic hydrogenation of alkynes. Alkynes readily undergo catalytic hydrogenation to form first the corresponding alkene and then the alkane.

$$R^1 \cdot C \equiv C \cdot R^2 \longrightarrow R^1 \cdot CH = CH \cdot R^2 \longrightarrow R^1 \cdot CH_2 \cdot CH_2 \cdot R^2$$

With most hydrogenation catalysts, a mixture of products is obtained even if an attempt is made to stop the reaction at the half-way stage, but the alkene is obtained in good yield if deactivated palladium catalysts are used. A highly effective formulation is Lindlar's catalyst (Section **II,2,44**); palladium-on-barium sulphate in the presence of quinoline is also recommended. With these catalyst preparations hydrogenation ceases abruptly when 1 mol of hydrogen has been taken up.

The partial hydrogenation process is stereospecific; *cis*-alkenes are obtained in a virtually stereochemically pure form. The reaction is illustrated by the conversion of but-2-ynoic acid into *cis*-but-2-enoic acid (Section **III,13**), wherein GLC analysis shows the crude product to be 95 per cent stereochemically pure, the remainder being the *trans*-isomer.

III,13. *CIS*-BUT-2-ENOIC ACID

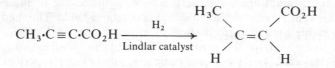

$$CH_3 \cdot C \equiv C \cdot CO_2H \xrightarrow[\text{Lindlar catalyst}]{H_2}$$

In a 250-ml hydrogenation flask (Section **I,17,1**) place 100 mg of Lindlar catalyst (1) and a solution of 2 g (0.024 mol) of but-2-ynoic acid (Section **III,211**) in 100 ml of ethanol. Attach the flask to the adapter of the atmospheric hydrogenation apparatus (Fig. I,70(*a*)), and fill the flask and gas burettes with hydrogen by the procedure discussed in Section **I,17,1**; note the volumes in the gas burettes. Agitate the flask and note the burette readings when uptake of hydrogen practically ceases (the volume of hydrogen absorbed should be in the region of 540 ml). Follow the procedure given in Section **I,17,1** for replacing the hydrogen in the apparatus with air. Disconnect the hydrogenation flask, and filter the solution to remove the catalyst which should then be transferred to a residue bottle. Evaporate the ethanol on a rotary evaporator; a white solid is obtained if the temperature of the flask is below 15 °C. GLC analysis (using a 2′ column of 10 per cent diisodecyl phthalate containing 1 per cent phosphoric acid on Chromosorb W held at 90 °C with a nitrogen flow rate of 20 ml per minute) of a solution of the acid in ether shows the crude product to contain 95 per cent of *cis*-but-2-enoic acid (2). The pure acid, m.p. 15 °C, may be obtained by low temperature recrystallisation (Section **I,20**) using light petroleum (b.p. 40–60 °C).

Notes. (1) See Section **II,2,44(*d*)**; a suitable preparation is available from Fluka-A.G.

(2) Under these conditions isomerisation to the more stable *trans* isomer is not apparent. The retention times for *cis*-but-2-enoic acid, *trans*-but-2-enoic acid and but-2-ynoic acid are 6.8, 10.0 and 19.8 minutes respectively.

B,3. The reaction between an alkylidenephosphorane and a carbonyl compound (the Wittig reaction). Quaternisation of triphenylphosphine with an alkyl halide gives a quaternary phosphonium halide (I) which under the influence of a strong base eliminates hydrogen halide to give an alkylidenephosphorane ((II), an **ylid**). The latter reacts readily with an aldehyde or a ketone to give an intermediate betaine (III), which under the reaction conditions eliminates triphenylphosphine oxide to form an alkene.

$$(C_6H_5)_3P \colon \curvearrowright CH_2 \overset{\frown}{\underset{\underset{R^1}{|}}{-X}} \longrightarrow (C_6H_5)_3\overset{\oplus}{P} - CH_2 \cdot R^1\}\overset{\ominus}{X} \xrightarrow{\text{base}}$$

$$(I)$$

$$(C_6H_5)_3\overset{\oplus}{P} - \overset{\ominus}{\ddot{C}}H \cdot R^1 \longleftrightarrow (C_6H_5)_3P = CH \cdot R^1$$

$$(II)$$

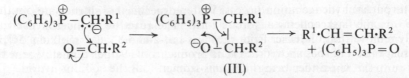

(III)

This is one example of the Wittig olefin synthesis. The reaction proceeds under mild conditions which do not normally promote structural isomerisation, thus enabling an alkene of unambiguous structure to be synthesised; it has been widely exploited for syntheses in the steroid and carotenoid fields. The method is particularly valuable for the introduction of an exocyclic double bond; a process which is illustrated by the synthesis of methylenecyclohexane from cyclohexanone (Section **III,14**). A variety of basic reagents have been used in the past to effect the conversion of the phosphonium salt into the ylid; the preparative details given involve the use of sodium hydride in dry dimethyl sulphoxide which generates the powerfully basic methylsulphinyl carbanion,

$CH_3 \cdot SO_2 \cdot \overset{\ominus}{C}H_2$. Alternatively phenyl lithium in ether solution may be used as the base, as in the synthesis of 3,3,5-trimethyl-1-methylenecyclohexane. The formation and subsequent reaction of the ylid is carried out under nitrogen since these compounds are sensitive to the presence of oxygen.

III,14. METHYLENECYCLOHEXANE

$$(C_6H_5)_3P + CH_3Br \longrightarrow (C_6H_5)_3P\cdot CH_3\}Br \xrightarrow{CH_3\cdot SO_2\cdot\overset{\ominus}{C}H_2} (C_6H_5)_3P = CH_2$$

$$(C_6H_5)_3P = CH_2 + \langle\,\rangle = O \longrightarrow \langle\,\rangle = CH_2 + (C_6H_5)_3PO$$

Triphenylmethylphosphonium bromide. Dissolve 68.1 g (0.26 mol) of triphenylphosphine in 60 ml of dry benzene. Place the solution in a stout-walled wide-necked glass bottle of about 1-litre capacity and cool in an ice–salt bath. Add 33.4 g (20 ml, 0.35 mol) of methyl bromide (1) and seal the bottle with a rubber bung secured with wire. Remove the bottle from the cooling bath and leave for one day at room temperature. Cool the bottle in an ice–salt bath before opening, filter off the white crystalline solid under suction and wash with hot benzene (fume cupboard). Dry the product in the oven at 100 °C and store in a desiccator over phosphorus pentoxide. The yield is 91.5 g (99%), m.p. 232–233 °C.

Note. (1) Methyl bromide (b.p. 4.5 °C) is supplied in sealed vials. The vial must be thoroughly cooled in an ice–salt bath before any attempt is made to open it. It is advisable to wear a face mask and thick gloves during the operation. Wrap a cloth around the vial and score around the neck with a glass knife. Place several layers of cloth around the scored neck and break off the end of the neck.

Methylenecyclohexane. Fit a 250-ml three-necked flask containing a magnetic stirring bar with a condenser, nitrogen inlet and a thermometer; surround the flask with a water bath supported on a hot-plate/magnetic stirrer unit. Pass a steady stream of dry nitrogen through the flask and maintain the flow throughout the reaction. Place 125 ml of dry dimethyl sulphoxide (Section **II,1,**33) in the flask and stir vigorously for 30 minutes to remove oxygen gas. Briefly remove the condenser from the flask and add 4.8 g (0.1 mol) of a 50 per cent suspension of sodium hydride in oil by means of a solids funnel. Warm the mixture to 65 °C on the water bath and stir until all the sodium hydride dissolves (about 1.5 hours). Cool the mixture to room temperature in a cold-water bath. Remove the condenser, place a solids funnel in the central neck of the flask and add 35.7 g (0.1 mol) of triphenylmethylphosphonium bromide in portions over 15 minutes; the reaction is slightly exothermic. Replace the condenser and stir at room temperature for 45 minutes to complete the formation of the methylenephosphorane. Add 9.8 g (0.1 mol) of cyclohexanone in one portion; an exothermic reaction ensues and the temperature rises to about 70 °C. Stir the mixture at 50 °C for 1 hour and then transfer to a single-neck 250-ml flask, add a boiling stick and attach a still-head and condenser arranged for vacuum distillation. A calcium chloride drying tube should be placed in the line between the water pump and the receiving flask to prevent water vapour condensing in the flask. Distil and collect the volatile products boiling below 75 °C at 50 mmHg in a receiver cooled in an acetone–Cardice bath; the distillate consists of methylenecyclohexane contaminated with a small amount of benzene. Redistil the crude product at atmospheric pressure through a 7-cm fractionating column packed with glass helices. Reject the material (mainly benzene) which boils below 99 °C, and collect the pure methylenecyclohexane as a fraction of b.p. 99–100 °C; the yield is 6.2 g (64%). Confirm its homogeneity by GLC analysis on a Silicone oil column at 60 °C.

Cognate preparation 3,3,5-Trimethyl-1-methylenecyclohexane. Prepare an ethereal solution of phenyl lithium by the procedure described in Section **VI,19**, using 2.7 g (0.39 mol) of lithium shavings, 26 g (17.5 ml, 0.16 mol) of dry redistilled bromobenzene and 85 ml of anhydrous ether. After the conversion to phenyl lithium is complete dilute the solution with a 15 ml portion of anhydrous ether and decant the solution from any unreacted lithium [Section **VI,19**, Note (2)] into a clean three-necked 250-ml flask equipped with a nitrogen inlet, a solids addition funnel and a reflux condenser. Add at room temperature 53.5 g (0.15 mol) of triphenylmethylphosphonium bromide in portions over 15 minutes; temporarily replace the solids addition funnel with a stopper if the reaction mixture shows a tendency to boil. Stir the suspension at room temperature for 4 hours and then add 21 g (0.15 mol) of 3,3,5-trimethylcyclohexanone; stir vigorously to effect solution. Heat the reaction mixture with stirring overnight and then add 25 ml of water. Remove the precipitate by filtration, wash it with ether and dry the combined ether extracts. Remove the solvent on a rotary evaporator and distil the residue through a fractionating column; collect the fraction having b.p. 155–160 °C. Refractionate and collect the pure product having b.p. 156–158 °C; the yield is 11.3 g (53%). The purity may be checked by GLC analysis using a Carbowax 20 M (10%) column held at 80 °C.

B,4. Wurtz type coupling reactions involving allylic halides. A useful method for the preparation of terminal alkenes (e.g., hept-1-ene, Section **III,15**) is based upon the reaction of allyl bromide with a Grignard reagent.

$$RMgBr + BrCH_2 \cdot CH = CH_2 \longrightarrow R \cdot CH_2 \cdot CH = CH_2 + MgBr_2$$

The replacement of the bromine in the allyl bromide probably involves an S_N2 process and the reaction is formally similar to the second stage of a Wurtz coupling reaction (see p. 323, III,A,4(a)).

Application of the Wurtz process to allyl halides (for preference the iodide) readily yields the unconjugated diene, thus hexa-1,5-diene (Section III,16) is obtained from allyl bromide (Section III,54).

III,15. HEPT-1-ENE

$$CH_3 \cdot (CH_2)_2 \cdot CH_2 MgBr + Br \cdot CH_2 \cdot CH = CH_2 \longrightarrow$$
$$CH_3 \cdot (CH_2)_3 \cdot CH_2 \cdot CH = CH_2 + MgBr_2$$

In a 1-litre three-necked flask prepare the Grignard reagent, butyl magnesium bromide, from 12.2 g (0.5 mol) of dry magnesium turnings, a small crystal of iodine, 68.5 g (53 ml, 0.5 mol) of butyl bromide and 260 ml of anhydrous ether, following the experimental details given in Section III,34. Equip a 500-ml three-necked flask with a sealed stirrer unit, a 100-ml separatory funnel and a double surface condenser. Force the solution of the Grignard reagent with the aid of pure, dry nitrogen and a tube containing a plug of purified glass wool (1) into the 500-ml flask through the top of the double surface condenser. Charge the separatory funnel with a solution of 50 g (35 ml, 0.42 mol) of allyl bromide (Section III,54) in 25 ml of anhydrous ether; place calcium chloride drying tubes into the top of the double surface condenser and of the dropping funnel. Immerse the flask containing the Grignard reagent in cold water, stir vigorously, and add the allyl bromide at such a rate that the ether boils gently; cool momentarily in ice if the reaction becomes too vigorous. It is important that the allyl bromide reacts when added, as indicated by gentle boiling of the solution (2). When all the allyl bromide has been introduced, continue stirring for 45 minutes whilst refluxing gently by immersing the flask in a bath of warm water. Allow to cool (3). Pour the reaction mixture cautiously on to excess of crushed ice contained in a large beaker. Break up the solid magnesium complex and decompose it with ice and dilute sulphuric acid or concentrated ammonium sulphate solution. Separate the ether layer, wash it with ammoniacal ammonium sulphate solution to remove any dissolved magnesium salts and dry over magnesium sulphate. Distil the dry ethereal solution through a fractionating column: after the ether has passed over, collect the hept-1-ene at 93–95 °C. The yield is 29 g (71%).

Notes. (1) Solid magnesium must be absent to avoid the formation of biallyl via allyl magnesium bromide; the insertion of a short plug of glass wool effectively removes any finely divided magnesium or alternatively use a tube terminating in a glass frit.

(2) If reaction does not occur when a little allyl bromide is first introduced, further addition must be discontinued until the reaction has commenced. Remove 2–3 ml of the Grignard solution with a dropper pipette, add about 0.5 ml of allyl bromide and warm gently to start the reaction; after this has reacted well, add the solution to the main portion of the Grignard reagent.

(3) A slight excess of Grignard reagent should be present at this stage. The **test for** the presence of **a Grignard reagent** is as follows. Remove 0.5 ml of the clear liquid with a dropper pipette and add 0.5 ml of a 1 per cent solution of Michler's ketone (4,4'-bis(dimethylamino)benzophenone) in benzene, followed by 1 ml of water and 3–4 drops of 0.01 M iodine in glacial acetic acid; shake. A

greenish-blue colour results if a Grignard reagent is present. In the absence of iodine, the colour fades.

A dye of the diphenylmethane type is produced:

$$p\text{-}(CH_3)_2N\cdot C_6H_4\cdot CO\cdot C_6H_4\cdot N(CH_3)_2\text{-}p \xrightarrow{RMgX}$$

$$p\text{-}(CH_3)_2N\cdot C_6H_4\cdot CR(OMgX)\cdot C_6H_4\cdot N(CH_3)_2\text{-} \xrightarrow{H^{\oplus}}$$

$$p\text{-}(CH_3)_2\overset{\oplus}{N}=C_6H_4=CR\cdot C_6H_4\cdot N(CH_3)_2\text{-}p$$
(coloured cation)

III,16. HEXA-1,5-DIENE (*Biallyl*)

$$2CH_2=CH\cdot CH_2I + 2Na \longrightarrow CH_2=CH\cdot CH_2\cdot CH_2\cdot CH=CH_2 + 2NaI$$

Place 55 g (2.4 mol) of clean sodium cut into small pieces (Section **II,2,55**) in a 500-ml round-bottomed flask fitted with two 25- or 30-cm double surface condensers in series. Weigh out 136 g (72 ml, 0.8 mol) of freshly distilled allyl iodide, b.p. 99–101 °C (Section **III,56**). Introduce about one-quarter of the allyl iodide through the condensers. Warm the flask gently until the sodium *commences* to melt and immediately remove the flame. A vigorous reaction sets in and a liquid refluxes in the condensers. Add the remainder of the allyl iodide in small portions over a period of 2 hours. Allow the mixture to cool during 3 hours and arrange the flask for distillation (compare Fig. I,98). Distil from an oil bath maintained at 90–100 °C when most of the hydrocarbon will pass over; finally raise the temperature of the bath to 150 °C in order to recover the product as completely as possible. Redistil from a 50-ml flask with fractionating side arm and containing a little sodium and collect the fraction of b.p. 59–60 °C; the yield of hexa-1,5-diene is 26 g (79%).

B,5. Rearrangement of alkynes. Unsaturated compounds containing two cumulated double bonds, i.e., $>C=C=C<$, are known as allenes. In addition to methods involving elimination reactions and other reactions analogous to those used in the synthesis of alkenes, compounds of this type may be prepared by the rearrangement of suitably substituted alkynes. A particularly useful method of preparing simple allenic hydrocarbons involves the treatment of secondary or tertiary propargyl halides (e.g., (III)) with a zinc–copper couple in boiling ethanol. The mechanism of the reaction, which involves replacement of a halogen by hydrogen with propargylic bond migration, is not fully understood but a cyclic mechanism involving an organozinc intermediate (V) and the solvent has been postulated.

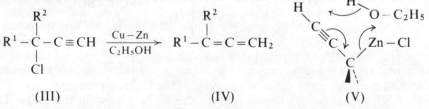

(III)　　　　　　(IV)　　　　　　(V)

The process is illustrated by the conversion of 3-chloro-3-methylpent-1-yne into 3-methylpenta-1,2-diene (Section **III,17**). Isolation of the allene is facilitated by the fact that it forms an azeotrope with ethanol, which may be removed by washing with water. The required tertiary chloride is obtained by shaking 3-methylpent-1-yn-3-ol (VI) with cold concentrated hydrochloric acid. Some of the 1-chloroallene (VIII) is additionally formed via the resonance stabilised carbonium ion (VII). However, 1-chloroallenes are reduced by a zinc–copper couple to the corresponding allenes and hence the crude unsaturated chloride may be used directly for the allene synthesis.

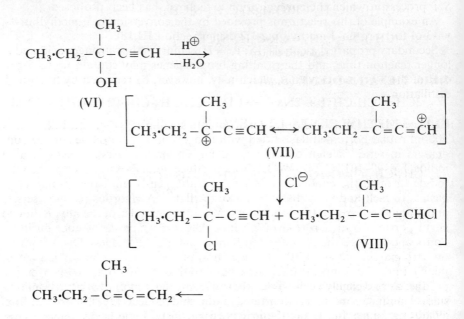

Many secondary and tertiary propargyl alcohols give 1-haloallenes in good yield when treated with concentrated aqueous halo-acid in the presence of the corresponding copper(I) halide. Thus tertiary propargyl alcohols (IX) react

$$HBr + Cu^IBr \rightleftharpoons H^{\oplus} + \overset{\ominus}{Cu^I}Br_2$$

with 48 per cent w/w hydrobromic acid in the presence of a catalyst consisting of copper(I) bromide, copper powder and ammonium bromide over a temperature range of 0–35 °C to form the corresponding 1-bromoallene (X) in yields up to 80 per cent. A mechanism involving the intermediacy of an acetylene–copper(I) π-complex from which the 1-bromoallene is formed by a stereospecific $S_N i'$ process in which the configuration is retained, has been proposed.

An example of this reaction is provided by the conversion of 3-methylbut-1-yn-3-ol to 1-bromo-3-methylbuta-1,2-diene (Section **III,18**).

Secondary propargyl alcohols (IX; $R^2 = H$) require rather stronger acid and a longer reaction time, and the resulting bromoallenes may contain up to 5 per cent of the 3-bromoacetylenes, which may, however, be removed by fractional distillation.

III,17. 3-METHYLPENTA-1,2-DIENE

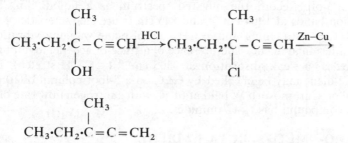

3-Chloro-3-methylpent-1-yne. Place 73.6 g (0.75 mol) of 3-methylpent-1-yn-3-ol and 750 ml of concentrated hydrochloric acid (d 1 18) in a 2 litre separating funnel and shake the mixture vigorously. Periodically remove portions of the upper organic layer and record the infrared spectrum as a liquid film; continue the shaking until the broad band centred at 3390 cm^{-1}, arising from the O–H stretching vibration of the alcohol, has completely disappeared (about 15 minutes). Discard the dark lower aqueous acid layer and dry the pale brown organic layer over anhydrous potassium carbonate for about 2 hours; a fresh portion of anhydrous potassium carbonate should be added after 30 minutes. Remove the drying agent by filtration. The yield of deep straw-coloured liquid, which consists of 3-chloro-3-methylpent-1-yne (*ca.* 73%) and 1-chloro-3-methylpenta-1,2-diene together with traces of other rearrangement products, is about 70 g; it can be used without purification for the next stage. The chloroacetylene has $t_R \sim 2.2$ minutes on a 5-foot 10 per cent Silicone oil on Chromosorb W column held at 70 °C with a nitrogen flow rate of 40 ml/minute.

3-Methylpenta-1,2-diene. Prepare a zinc–copper couple as follows. Place 65.4 g (1 mol) of zinc dust in a 250-ml conical flask and wash successively with four 50 ml portions of 3 per cent hydrochloric acid, two 50 ml portions of distilled water, two 100 ml portions of 2 per cent aqueous copper(II) sulphate, two 50 ml portions of distilled water, 100 ml of 95 per cent ethanol and 100 ml of absolute ethanol. Each washing operation should last for about 1 minute and the liquid should then be drained as thoroughly as possible by decantation before continuing with the next liquid wash. Finally transfer the zinc–copper couple by washing with 100 ml of absolute ethanol into a 250-ml three-necked, round-bottomed flask fitted with a reflux condenser, separatory funnel and an efficient (e.g., paddle-type) sealed stirrer. Heat the stirred zinc–ethanol suspension to boiling, remove the source of heat and then add the crude chloroacetylene (69.5 g) dropwise from the separatory funnel at such a rate so as to maintain gentle reflux; this should take about 50 minutes. Then heat the stirred reaction mixture under gentle reflux for 45 minutes, allow to cool slightly and rearrange the condenser for downward distillation. Distil the reaction mixture with stirring and collect the allene–ethanol azeotrope which passes over between 64 and 78 °C. Wash the colourless distillate with three portions of water (1 × 150 ml; 2 × 50 ml) and then separate the upper organic layer and dry over magnesium sulphate. Distil the crude product through a fractionating column (15 cm × 2 cm) filled with glass helices. Reject any material distilling below 70 °C, which contains appreciable quantities of acetylenic material, and collect 3-methylpenta-1,2-diene (⩾95% purity) as a colourless liquid, b.p. 70–72 °C; the yield is about 18.9 g (approximately 31% overall yield from 3-methylpent-1-yn-3-ol). Record the infrared spectrum as a liquid film. The strong absorption bands at 1964 cm^{-1} and 840 cm^{-1} are characteristic of the asymmetric $C=C=C$ stretching and $=CH_2$ out-of-plane wagging vibrations respectively of a terminal allene. The presence of a trace of acetylenic impurity may be observed as a weak absorption at 3305 cm^{-1} ($\equiv C - H$ stretch). The purity of the 1,2-diene may be checked by GLC on a 5-foot column of 10 per cent Silicone oil on Chromosorb W held at 61 °C with a nitrogen flow rate of 40 ml/minute; the compound has t_R 1.7 minutes.

III,18. 1-BROMO-3-METHYLBUTA-1,2-DIENE

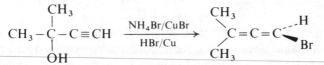

In a 250-ml two-necked, round-bottomed flask fitted with a stirrer and a dropping funnel, place a mixture of 14.3 g (0.1 mol) of powdered copper(I) bromide, 9.8 g (0.1 mol) of powdered ammonium bromide, 0.64 g (0.01 mol) of copper powder and 80 ml (*ca.* 0.66 mol) of 46–48 per cent w/w concentrated hydrobromic acid (*d* 1.46–1.49). Stir the dark green mixture vigorously at room temperature and add 25.2 g (0.3 mol) of 3-methylbut-1-yn-3-ol dropwise during 5 minutes; maintain the vigorous stirring for 1.5 hours (1). Filter the reaction mixture through a sintered glass Buchner funnel and rinse the reaction vessel and funnel with 10 ml of light petroleum (b.p. 40–60 °C). Transfer the filtrate to a separatory funnel, discard the lower aqueous layer and remove any unchanged alcohol and copper salts from the almost colourless organic layer by shaking with three 10 ml portions of concentrated hydrobromic acid. In the final washing the lower acid layer should show no trace of violet colouration. Dry the combined organic layers over a mixture of solid sodium hydrogen carbonate and magnesium sulphate, filter through a sintered glass Buchner funnel; rinse the flask and the inorganic solids with 5 ml of light petroleum (b.p. 40–60 °C). Transfer the filtrate to a 50-ml distillation flask fitted with a 9-cm fractionating column filled with glass helices and distil under reduced pressure, using a nitrogen bleed, from a bath held at 80–90 °C. Collect the 1-bromo-3-methylbuta-1,2-diene as a fraction, b.p. 52–54 °C/50 mmHg, in a flask cooled in ice-water (2). The yield is 28.1 g (64%). The purity of the product may be checked by GLC analysis using a 5-ft column of 10 per cent Silicone oil on Chromosorb W held at 80 °C with a nitrogen flow rate of 40 ml/minute; the material has t_R 3.5 minutes. The infrared spectrum (liquid film) shows the characteristic strong allene (C=C=C) absorption band at 1950 cm^{-1}. On standing 1-bromo-3-methylbuta-1,2-diene tends to isomerise to 1-bromo-3-methylbuta-1,3-diene and polymerise.

Notes. (1) The progress of the reaction may be monitored by occasionally halting the stirring and removing a portion of the clear upper organic layer for infrared examination. The reaction is complete when a sample does not show hydroxyl and alkynyl hydrogen absorption at 3400 and 3300 cm^{-1} respectively.

(2) In common with the distillation of all unsaturated compounds in general, the residue in the distillation flask should be cooled to room temperature before admitting air into the system.

C ALKYNES

The synthesis of acetylenic compounds is exemplified by the following typical procedures.

1. The dehydrohalogenation of 1,2-dihalides (Sections **III,19** and **III,20**).
2. Alkylation of a terminal alkyne (Section **III,21**).
3. Coupling of terminal alkynes (Sections **III,22** and **III,23**).

C,1. The dehydrohalogenation of 1,2-dihalides. A simple method of introducing a triple bond into an organic compound is to treat a vicinal or geminal dihalide with a strong base, e.g., ethanolic potassium hydroxide or sodamide in liquid ammonia.

$$R^1 \cdot CHBr \cdot CHBr \cdot R^2 \xrightarrow{-2HBr} R^1 \cdot C \equiv C \cdot R^2$$

$$R^1 \cdot CBr_2 \cdot CH_2 \cdot R^2 \xrightarrow{-2HBr} R^1 \cdot C \equiv C \cdot R^2$$

The products from this reaction are not invariably homogeneous since the first formed alkyne may isomerise, probably by way of an intermediate allene. With alcoholic potassium hydroxide at 170–180 °C, for example, terminal alkynes tend to rearrange to internal alkynes.

$$R \cdot CH_2 C \equiv CH \rightleftharpoons R \cdot CH = C = CH_2 \rightleftharpoons R \cdot C \equiv C \cdot CH_3$$

The rearrangement process may be minimised by using sodamide in liquid ammonia as the reagent.

The most readily available starting materials are the 1,2-dibromo compounds which are obtainable by the addition of bromine to a carbon–carbon double bond. An example is given in which undecylenic acid, commercially obtainable from castor oil, is brominated and then dehydrobrominated with hot concentrated aqueous potassium hydroxide to undec-10-ynoic acid (Section **III,19**). In a similar manner ethyl cinnamate is converted into phenylpropynoic acid.

$$C_6H_5 \cdot CH = CH \cdot CO_2C_2H_5 \xrightarrow{Br_2} C_6H_5 \cdot CHBr \cdot CHBr \cdot CO_2C_2H_5 \xrightarrow[\text{(ii) } H_3O^{\oplus}]{\text{(i) KOH}}$$

$$C_6H_5 \cdot C \equiv C \cdot CO_2H$$

The use of the sodamide–liquid ammonia reagent is illustrated in the conversion of styrene into phenylacetylene via its dibromide.

$$C_6H_5 \cdot CH = CH_2 \xrightarrow{Br_2} C_6H_5 \cdot CHBr \cdot CH_2Br \xrightarrow[NH_3]{NaNH_2} C_6H_5 \cdot C \equiv CH$$

The loss of a bromide ion when situated in a β-position to a carboxylate group occurs very readily under mildly basic conditions. Thus in an alternative synthesis of phenylacetylene (Section III,20), dibromocinnamic acid is converted into β-bromostyrene under the influence of hot aqueous sodium carbonate solution. Dehydrobromination to yield the acetylenic compound is then achieved in the usual manner with potassium hydroxide.

$$C_6H_5 \cdot CH - CHBr \xrightarrow{-CO_2} C_6H_5 \cdot CH = CHBr \xrightarrow{KOH} C_6H_5 \cdot C \equiv CH$$

III,19. UNDEC-10-YNOIC ACID

$$CH_2 = CH \cdot (CH_2)_8 \cdot CO_2H \xrightarrow{Br_2} CH_2Br \cdot CHBr \cdot (CH_2)_8 \cdot CO_2H \xrightarrow{KOH}$$

$$CH \equiv C \cdot (CH_2)_8 \cdot CO_2H$$

Purify commercial undecylenic acid by distillation of, say, 250 g under diminished pressure and collect the fraction, b.p. 152–154 °C/6 mmHg; this has a freezing point of 23 °C. Dissolve 108 g (0.58 mol) of the purified undecylenic acid in 285 ml of dry carbon tetrachloride (1) in a 1-litre three-necked flask provided with a sealed stirrer unit, a dropping funnel and a reflux condenser. Cool the flask in a freezing mixture of ice and salt, stir the solution and add 96 g (31 ml, 0.6 mol) of dry bromine (Section II,2,7) during a period of 1 hour: allow the mixture to gradually warm up to the temperature of the laboratory. Remove the carbon tetrachloride with a rotary evaporator and pour the residue into a large evaporating dish (fume cupboard). Upon standing 1–2 days (more rapidly when left in a vacuum desiccator over silica gel), the dibromo acid crystallises completely. The yield is quantitative.

Transfer the solid dibromo acid to a 2-litre round-bottomed flask attached to a reflux condenser, add a solution of 263 g of potassium hydroxide in 158 ml of water, and heat in an oil bath at 150–160 °C for 8 hours. Considerable frothing occurs, but this is reduced by the addition of small quantities (about 0.1 g) of a suitable detergent, e.g., sodium dodecyl benzenesulphonate, from time to time. Allow the mixture to stand overnight, add 1500 ml of water, shake until all the solid dissolves and acidify with dilute sulphuric acid to Congo red. A solid cake of acid separates on the surface of the liquid after standing for several hours. Extract with four 250 ml portions of ether, dry with anhydrous sodium or magnesium sulphate and remove the ether on a water bath. Transfer the residue to a 250-ml flask fitted with a Claisen still-head and distil cautiously under diminished pressure using a free flame. A little ether and water pass over first and the temperature rises rapidly to 175 °C/15 mmHg. Collect separately the fractions (a) b.p. 177–182 °C/15 mmHg (52 g) and (b) 182–200 °C/15 mmHg (15 g). The flask contains a large residue, which is discarded. Fraction (a) solidifies completely on cooling and has m.p. 37–41 °C; upon recrystallisation from light petroleum, b.p. 60–80 °C, 34 g (32%) of pure undec-10-ynoic acid, m.p. 41–42 °C, are obtained. A further quantity of product is obtained from fraction (b), which solidifies to a slightly sticky solid: upon recrystallisation from light petroleum, b.p. 60–80 °C, a sticky solid separates, which, after spreading upon a porous tile, becomes colourless and has m.p. 41–42 °C (3 g).

Note. (1) Dry carbon tetrachloride may be prepared by distillation of the commercial product and rejection of the first 20 per cent of the distillation (see Section I,23). This distillation must be conducted in the fume cupboard.

Cognate preparations   Phenylpropynoic acid.   Place a solution of 88 g (84 ml, 0.5 mol) of ethyl cinnamate (Section IV,149) in 50 ml of carbon tetrachloride in a 500-ml round-bottomed flask. Immerse the flask in ice and add 80 g (25.5 ml, 0.5 mol) of bromine from a separatory funnel slowly with frequent shaking. The halogen will disappear rapidly at first, but more slowly towards the end of the reaction; no hydrogen bromide is evolved and the time of the addition is about 20–25 minutes. Allow the mixture to stand for 1 hour, pour the solution into a large evaporating dish and permit the excess of bromine and the carbon tetrachloride to evaporate spontaneously in the fume cupboard. The crude ethyl 2,3-dibromo-3-phenylpropanoate will remain as a solid cake; this can be dried by pressing between large filter papers. The yield of crude ester, m.p. 66–71 °C, is 140 g (83%) (1).

Dissolve 85 g of potassium hydroxide in 400 ml of rectified spirit by heating in a 1500-ml round-bottomed flask, provided with a reflux condenser, on a water bath. Cool to 40–50 °C, and add 112 g (0.33 mol) of the crude dibromo ester; when the initial exothermic reaction has subsided, heat the mixture on a water bath for 5–6 hours. Pour the contents of the flask into a large beaker and, when cold, add concentrated hydrochloric acid with stirring until neutral to litmus. Cool, filter the precipitated solids at the pump and wash with a little alcohol. Set the solids (A) aside. Transfer the filtrate to the original flask and distil the liquid until the temperature of the vapour reaches 95 °C. Combine the residue in the flask with the precipitated solids (A), dissolve in 270 ml of water, add about 300 g of crushed ice and cool the flask in an ice bath. Stir the mixture mechanically, and add 20 per cent sulphuric acid slowly until the solution is strongly acid to Congo red. Allow to stand for 20 minutes, filter off the dark-coloured crude phenylpropynoic acid at the pump and wash it with three 15 ml portions of 2 per cent sulphuric acid. Dissolve the solid in about 300 ml of 5 per cent sodium carbonate solution, add 6 g of decolourising charcoal and heat on a water bath for 30 minutes with occasional shaking. Filter through a fluted filter-paper, cool the filtrate in ice and then add 70 g of crushed ice. Stir the solution mechanically and add 20 per cent sulphuric acid slowly until acid to Congo red. After 20 minutes, filter the precipitated acid by suction, wash with 15 ml of 2 per cent sulphuric acid, then with a little water, and dry in the air. The yield of pure phenylpropynoic acid, m.p. 134–135 °C, is 23 g (47%).

Note.   (1) To obtain the pure dibromo ester, recrystallise from light petroleum, b.p. 60–80 °C; the recovery of the pure ester, m.p. 75 °C, is 85 per cent.

Phenylacetylene.   Place a solution of 208 g (228 ml, 2 mol) of freshly distilled styrene (1) in 200 ml of dry chloroform in a litre beaker, cooled in an ice bath, and provided with a mechanical stirrer. Support a dropping funnel over the beaker and charge the former with a solution of 320 g (103 ml, 2 mol) of dry bromine in 200 ml of chloroform. Add the bromine solution with stirring at a rate to conform with the discharge of colour from red to pale yellow. This preparation is advantageously carried out in bright sunlight. When all the bromine has been added, continue the stirring until the reaction is complete. Evaporate the chloroform on a water bath; the residual crude styrene dibromide weighs 510 g (97%). This may be used directly for the preparation of phenylacetylene (2).

Support a 5-litre glass Dewar flask in a wooden case. Equip the flask with a lid of clear Perspex, provided with suitable apertures for a mechanical stirrer, introducing solids (e.g., sodium) or liquids, a calibrated dip stick for measuring the volume of liquid in the Dewar vessel, a gas inlet tube and an ammonia inlet: arrange for an electric light to shine downwards into the flask.

Charge the Dewar flask with 3 litres of liquid ammonia (Section I,17,7), set the stirrer into operation and introduce 1.5 g of powdered iron(III) nitrate followed by 5 g of clean sodium. After 2 minutes, introduce 160 g of clean sodium in 3 g lumps during 30 minutes (3). Allow to stand until the initially deep blue reaction mixture assumes a light grey colour (about 20 minutes). Add a solution of 510 g of styrene dibromide in 1500 ml of dry ether slowly during 2 hours: a vigorous reaction ensues, accompanied by the loss of some ammonia by evaporation. Allow to stand for 4 hours, add 180 g of finely-powdered ammonium chloride to the pasty mass (to decompose the sodio derivative), followed by 500 ml of ether and continue the stirring for several minutes. Pour the contents of the Dewar flask with the aid of a purpose-made plastic spout into a cold beaker. Allow the ammonia to evaporate overnight. Add ether, filter off the inorganic salts and wash well with ether; keep the filtrate (A). Dissolve the inorganic salts in water, extract the solution with ether and combine the ethereal extracts with the filtrate (A). Wash with dilute sulphuric acid until acid to Congo red paper, then with water, dry with magnesium sulphate, distil off the ether on a water bath with the aid of a short but efficient column and fractionate the residue through a well-lagged efficient fractionating column. Collect the phenylacetylene at 142–143 °C; the yield is 156 g (79%). Alternatively, distil the residue under reduced pressure and collect the phenylacetylene at 82 °C/80 mmHg (4). CAUTION, see Section III,33, Note (1).

Notes.   (1) Styrene has b.p. 42–43 °C/18 mmHg. Distillation at atmospheric pressure, b.p. 145–146 °C, causes considerable loss by polymerisation.

(2) Styrene dibromide (1,2-dibromophenylethane) may be purified by recrystallisation from aqueous ethanol; m.p. 73–74 °C. CAUTION: the compound is a skin irritant and any contact with it should be avoided. Plastic gloves should be worn.

(3) See also Section III,21 concerning the preparation of sodamide in liquid ammonia.

(4) This pressure is readily obtained by placing an air leak between the water pump and the apparatus or, better, with the aid of a manostat (Section I,30).

II,20. PHENYLACETYLENE

$$C_6H_5 \cdot CH = CH \cdot CO_2H \xrightarrow{Br_2} C_6H_5 \cdot CHBr \cdot CHBr \cdot CO_2H \xrightarrow{Na_2CO_3}$$

$$C_6H_5 \cdot CH = CHBr \xrightarrow{KOH} C_6H_5 \cdot C \equiv CH$$

Dissolve 74 g (0.5 mol) of cinnamic acid in 300 ml of hot chloroform in a 500-ml flask and cool the solution in ice-water with shaking. As soon as the solid begins to crystallise out, add a solution of 80 g (26 ml, 0.5 mol) of bromine (care in handling) in 50 ml of chloroform rapidly in three portions with vigorous shaking and cooling. Allow the flask and contents to stand in an ice bath for 30 minutes to allow complete crystallisation of the product; collect the latter by filtration. Obtain a pure specimen of 2,3-dibromo-3-phenylpropionic acid (m.p. 204 °C decomp.) by recrystallising a small sample of the crude product from aqueous ethanol. Boil the bulk of the crude bromo-acid under reflux with 750 ml of 10 per cent aqueous sodium carbonate solution, cool and separate the layer of crude β-bromostyrene. Extract the aqueous phase with two 75 ml portions of ether, combine this extract with the organic phase, dry over anhydrous calcium chloride and remove the ether on a rotary evaporator. About 65–70 g of crude β-bromostyrene is obtained.

Place 100 g of potassium hydroxide pellets in a 500-ml flask, moisten the pellets with about 2 ml of water and fit the flask with a still-head carrying a dropping funnel and a condenser set for downward distillation. Heat the flask in an oil bath maintained at 200 °C and add the crude β-bromostyrene dropwise on to the molten alkali at a rate of about 1 drop per second. Phenylacetylene begins to distil over; slowly raise the bath temperature to about 220 °C and keep it at this point until the addition is complete. Then continue to heat at about 230 °C until no more product distils over. Separate the upper layer of the distillate, dry it over potassium hydroxide pellets and redistil. Collect the phenylacetylene at 142–144 °C; the yield is 25 g (49%).

C,2. Alkylation of terminal alkynes. Terminal alkynes by virtue of the presence of an acidic hydrogen atom can be converted into a corresponding alkynyl sodium or alkynylmagnesium halide (Section III,40) which may be alkylated to give a homologous alkyne.

$$R^1 \cdot C \equiv CH + \overset{\oplus}{Na} \overset{\ominus}{N}H_2 \longrightarrow R^1 \cdot C \equiv \overset{\ominus}{C} \overset{\oplus}{Na} \xrightarrow{R^2Br} R^1 \cdot C \equiv C \cdot R^2$$

Acetylene itself can form both the mono or disodium salt; the former is the main product when a large excess of acetylene is used in reaction with sodamide in liquid ammonia.

Primary alkyl halides (the bromide for preference) should be used as the alkylating agents, since secondary and tertiary halides undergo extensive olefin-forming elimination reactions in presence of the strongly basic acetylide ion. A typical synthesis is that of hex-1-yne (Section III,21).

III,21. HEX-1-YNE (Butylacetylene)

$$CH \equiv CH + \overset{\oplus}{Na} \overset{\ominus}{:}NH_2 \xrightarrow{liq. NH_3} CH \equiv \overset{\ominus}{C} \overset{\oplus}{Na} + NH_3$$

$$C_4H_9Br + \overset{\ominus}{:}C \equiv CH \xrightarrow{liq. NH_3} C_4H_9 \cdot C \equiv CH + \overset{\ominus}{:}Br$$

Equip a 5-litre three-necked flask (or a 5-litre flange flask fitted with a multiple head adapter) with a Herschberg or other efficient stirrer, a soda-lime guard-tube and a liquid ammonia inlet tube as described in Section I,17,7; Fig. I,76(c). Observing the precautions noted in this Section, run in liquid ammonia until the flask is about two-thirds full (ca. 3.5 litres). Prepare a suspension of sodamide from 138 g (6 mol) of clean sodium (Section II,2,55) using 0.5 g of finely-powdered crystallised iron(III) nitrate as the catalyst as described in detail in Section II,2,54. It may be necessary to add further quantities of liquid ammonia to maintain the volume of 3.5 litres. When the conversion of sodium into sodamide is complete, replace the ammonia-addition tube by a wide tube reaching almost to the bottom of the flask to allow for the passage of acetylene through the suspension of sodamide. The acetylene gas from a cylinder should be freed from acetone by passing through two Drechsel bottles half-filled with concentrated sulphuric acid: when the acid in the second wash bottle becomes discoloured the wash bottles should be recharged with fresh acid. An empty Drechsel bottle (to act as a safety trap) and a mercury escape valve (cf. Fig. I,67) should be interposed between the reaction flask and the Drechsel bottles charged with sulphuric acid. Surround the reaction flask with an acetone–Cardice cooling bath and pass acetylene rapidly (2–3 litres per minute) into the sodamide suspension until a uniformly black liquid is formed (usually 4–5 hours) (1). Carefully watch the mercury escape valve in case the wide entry tube becomes blocked by deposition of solid. If this should happen, temporarily stop the acetylene flow and clear the tube by inserting a glass rod of appropriate diameter. It may also become necessary to introduce some more liquid ammonia. Replace the soda-lime guard-tube with a pressure-equalising separatory funnel charged with 685 g (538 ml, 5 mol) of butyl bromide which is introduced over a period of 1.5–2 hours with stirring and whilst a slow stream of acetylene (ca. 500 ml per minute) continues to pass through the reaction mixture. The reaction is exothermic so that it will be necessary to maintain the cooling bath at −50 °C by continued addition of solid carbon dioxide. When all the alkyl bromide has been added, discontinue the supply of acetylene and allow the ammonia to evaporate. Before evaporation is complete add cautiously 60 g of ammonium chloride with stirring to decompose the excess of sodium acetylide (or sodamide) if present. Now introduce 500 g of crushed ice followed by about 1.5 litres of distilled water. Steam distil the reaction mixture, separate the hydrocarbon layer in the distillate, dry over magnesium sulphate and fractionally distil through a Hempel column filled with Fenske rings. Collect the hex-1-yne, b.p. 71–72 °C. The yield is 280 g (68%).

Note. (1) Occasionally the reaction mixture does not become completely black nor free from suspended solid; here the acetylide is an insoluble (or sparingly soluble) form, but it gives satisfactory results in the preparation of hex-1-yne. The saturated solution of the soluble forms of sodium acetylide in liquid ammonia at −34 °C is about 4.1 M.

C,3. Coupling of terminal alkynes.   Oxidative coupling of alkynes is a particularly easily performed carbon–carbon σ-bond forming reaction, which results in a good yield of the symmetrical diacetylene. A widely used procedure involves oxidation of the alkyne with air or oxygen in aqueous ammonium chloride in the presence of a copper(I) chloride catalyst (**Glaser oxidative coupling**).

$$2R \cdot C \equiv CH + [O] \xrightarrow{\text{Cu}^{\oplus}} R \cdot C \equiv C \cdot C \equiv C \cdot R + H_2O$$

A modified procedure illustrated by the synthesis of a diynediol (Section **III,22**) from the acetylenic alcohol, 2-methylbut-3-yn-2-ol, uses methanol and pyridine as the solvent, the latter acting also as a complexing agent for the copper(I) ions. An alternative effective coupling system involves the use of a copper(II) acetate in pyridine which does not require the use of air or oxygen.

A coupling procedure particularly suited to the synthesis of unsymmetrical diacetylenes involves the reaction of a terminal acetylene with a 1-bromoacetylene in the presence of a catalyst consisting of a solution of copper(I) chloride in a primary amine to which small quantities of hydroxylamine hydrochloride is added (the **Cadiot–Chodkiewicz coupling**).

$$R^1 \cdot C \equiv CH + Cu^{\oplus} \longrightarrow R^1 \cdot C \equiv CCu + H^{\oplus}$$
$$R^1 \cdot C \equiv CCu + BrC \equiv C \cdot R^2 \longrightarrow R^1 \cdot C \equiv C \cdot C \equiv C \cdot R^2 + Cu^{\oplus} + Br^{\ominus}$$

The organic base serves to remove the liberated protons and to assist solution of the copper(I) catalyst by the formation of a complex. Hydroxylamine hydrochloride helps to maintain an adequate concentration of copper(I) ions, which however is best kept rather low otherwise unwanted self-coupling of the bromoalkyne occurs. The reaction is illustrated in Section **III,23** by the coupling of ω-bromophenylacetylene with 2-methylbut-3-yn-2-ol.

III,22. 2,7-DIMETHYLOCTA-3,5-DIYNE-2,7-DIOL

$$(CH_3)_2C(OH) \cdot C \equiv CH \xrightarrow[\text{pyridine}]{O_2; \ CuCl} (CH_3)_2C(OH) \cdot C \equiv C \cdot C \equiv C \cdot (OH)C(CH_3)_2$$

Place in a 250-ml conical flask 17.4 g of 2-methylbut-3-yn-2-ol (20 ml, 0.207 mol), 20 ml of methanol, 6 ml of dry pyridine (0.074 mol), 1 g of copper(I) chloride (0.010 mol) and a magnetic following bar. Then, with fairly vigorous stirring, bubble oxygen through the solution for 2 hours, at a rate of at least 10 l/hour. The solution becomes warm and rapidly turns dark green. Cool the reaction mixture using an ice-water cooling bath and add 10 ml of concentrated hydrochloric acid to neutralise the pyridine and to keep copper salts in solution. Add 50 ml of a saturated aqueous sodium chloride solution, cool and filter off the precipitated product. Wash the filtered solid with a little ice-cold water to remove any remaining colour and recrystallise it from dry toluene. This also serves to dry the product if the toluene/water azeotrope is removed by using a Dean and Stark water-separator. Cool the toluene solution and filter the colourless crystalline product; 12 g (70%) of alkynol, m.p. 132–133 °C, is obtained. Further recrystallisation from toluene raises the m.p. to 133–133.5 °C.

III,23. 2-METHYL-6-PHENYLHEXA-3,5-DIYN-2-OL

$$C_6H_5 \cdot C \equiv CH + HOBr \longrightarrow C_6H_5 \cdot C \equiv CBr + H_2O$$

$$(CH_3)_2C(OH) \cdot C \equiv CH + BrC \equiv C \cdot C_6H_5 \xrightarrow[\substack{(CH_3)_2NH \\ NH_2OH}]{H_2O, \ CuCl}$$

$$(CH_3)_2C(OH) \cdot C \equiv C \cdot C \equiv C \cdot C_6H_5 + HBr$$

ω-**Bromophenylacetylene.** Prepare a solution of sodium hypobromite as follows. To a 250-ml conical flask fitted with a ground glass joint add 60 g of ice, 30 ml of 10 *M*-aqueous sodium hydroxide (0.30 mol) and 21.8 g (7.0 ml, 0.136 mol) of bromine. Swirl for a few seconds to dissolve the bromine, then add 13.0 g (14.0 ml, 0.127 mol) of phenylacetylene (Section **III,20**) (1). Securely stopper the flask, and shake vigorously for 5 hours on a mechanical shaker. Extract the product from the reaction mixture with three 50 ml portions of ether, wash the combined extracts with water and dry the ethereal layer with anhydrous sodium sulphate. Filter the solution and remove the ether on a rotary evaporator. ω-Bromophenylacetylene (2) sufficiently pure for use in the following reaction is obtained as a light yellow oil (3); the yield is 20.4 g (89%).

2-Methyl-6-phenylhexa-3,5-diyn-2-ol. In a 100-ml conical flask place 40 ml of 25–30 per cent aqueous dimethylamine solution (0.22–0.27 mol), 200 mg (0.002 mol) of copper(I) chloride (Section **II,2,***16*) and 650 mg (0.009 mol) of hydroxylamine hydrochloride. Add 12 ml (10.42 g, 0.124 mol) of 2-methylbut-3-yn-2-ol, add a magnetic follower bar and chill in an ice-water bath. Then, whilst stirring magnetically, and with continued cooling, add from a dropping pipette 20.4 g (0.113 mol) of ω-bromophenylacetylene and ensure complete transfer of the bromo compound by rinsing several times the pipette and container with reaction mixture. When the addition has been completed, continue stirring and cooling for a further 10 minutes and then add 25 ml of 5 per cent aqueous potassium cyanide to complex the copper salts. Extract the solution with four 40 ml portions of ether and dry the combined extracts with anhydrous sodium sulphate. Filter and evaporate the ether solution using a rotary with a nitrogen flow rate of 45 ml/minute, gives retention times for phenylacetylene and ω-bromophenylacetylene of 2.16 and 0.76 minutes respectively. 40–60 °C). The yield of pure diynol (4), m.p. 55–58 °C, is 15.3 g (73%).

Notes. (1) Phenylacetylene has a penetrating and persistent odour and is mildly lachrymatory. It is recommended that it be handled in the fume cupboard and that protective gloves be worn.

(2) ω-Bromophenylacetylene has a very penetrating and persistent odour and is mildly lachrymatory. It should be stored in the fume cupboard and care taken during subsequent handling.

(3) GLC analysis using a 5-ft S.E. 52 chromatographic column, at 120 °C with a nitrogen flow rate of 45 ml/minute, gives retention times for phenylacetylene and ω-bromophenylacetylene of 2.16 and 0.76 minutes respectively.

(4) The purity of this alkynol may be checked by GLC analysis using an S.E. 52 chromatographic column. With a 5-ft column, at 170 °C and with a nitrogen gas flow rate of 45 ml/minute, the retention time is 3.84 minutes. Similar conditions using an S.E. 30 column give a retention time of 3.16 minutes. The two reactants have very much shorter retention times [see Note (3) above].

D ALIPHATIC ALCOHOLS

The synthesis of alcohols is exemplified by the following typical procedures.

1. The reduction of aldehydes, ketones or esters (Sections **III,24** to **III,32**).

2. The interaction of carbonyl-containing compounds with organometallic reagents (Sections **III,33** to **III,42**).
3. The hydroboronation–oxidation of alkenes (Section **III,43**).
4. The oxymercuration–demercuration of alkenes (Section **III,44**).
5. The hydroxylation of alkenes (Sections **III,45** and **III,46**).

D,1. The reduction of aldehydes, ketones or esters. Primary and secondary alcohols may be synthesised by the reduction of the corresponding carbonyl compounds by a variety of dissolving metal systems.

$$R \cdot CHO \xrightarrow{[H]} R \cdot CH_2OH$$

$$R^1 \cdot CO \cdot R^2 \xrightarrow{[H]} R^1 \cdot CH(OH) \cdot R^2$$

For example, heptan-1-ol is formed by the action of iron and glacial acetic acid on heptanal (Section **III,24**). The reduction of ketones may be effected with sodium and absolute ethanol or less commonly with sodium and moist ether (Section **III,25**). As an alternative the use of zinc and aqueous sodium hydroxide is illustrated by the reduction of benzophenone (Section **III,26**).

These dissolving metal reductions have the disadvantage of being relatively unselective. Potassium borohydride and sodium borohydride however each show a considerable degree of selectivity; thus aldehydes and ketones may be reduced to alcohols whilst halogeno, cyano, nitro, amido and alkoxycarbonyl groups remain unaffected. The reductions of chloral and *m*-nitrobenzaldehyde (Section **III,27**) to the corresponding substituted alcohols are illustrative of this selectivity. The reagents are used in aqueous or aqueous ethanolic solution. The essential step in the mechanism of this reaction involves a hydride ion transfer to the carbonyl carbon from the borohydride anion, which is capable of reducing 4 mols of carbonyl compound. Decomposition of the resulting anionic complex with water or dilute acid liberates the required alcohol.

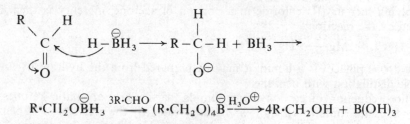

Aldehydes and ketones can be selectively reduced to the corresponding alcohols by aluminium alkoxides. The most satisfactory alkoxide for general use is aluminium isopropoxide.

$$3R^1 \cdot CO \cdot R^2 + [(CH_3)_2CHO]_3Al \rightleftharpoons [R^1R^2CHO]_3Al + 3CH_3 \cdot CO \cdot CH_3$$

The carbonyl compound to be reduced is heated with aluminium isopropoxide in excess isopropyl alcohol (propan-2-ol) under a simple fractionating column with provision for slow distillation until no more acetone is detected in the distillate; the alcoholic reduction product is recovered from the reaction mixture after acidification. The process is usually termed the **Meerwein–**

Ponndorf–Verley reduction. It is a mild method of reducing carbonyl compounds in good yield, and is particularly valuable since other groups, e.g., a conjugated double bond, a nitro group or a halogen atom, are unaffected. Experimental details for the reduction of crotonaldehyde (Section **III,28**) are given. The above reversible equation indicates that 1 mol of aluminium isopropoxide will reduce directly 3 mols of the carbonyl compound. It is generally desirable to use excess of the reductant except for aromatic aldehydes; for the latter, side reactions (e.g., $2R \cdot CHO \rightarrow R \cdot CO_2CH_2 \cdot R$; Tishchenko reaction) tend to occur with excess of the reagent.

The mechanism of the reaction involves the coordination of the carbonyl compound with the aluminium atom in aluminium isopropoxide followed by an intramolecular transfer of a hydride ion:

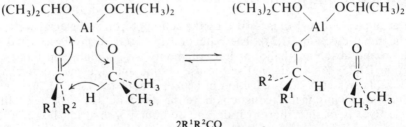

$$[(CH_3)_2CHO]_2Al \cdot OCHR^1R^2 \xrightarrow{2R^1R^2CO} [R^1R^2CHO]_3Al + 2(CH_3)_2CO$$

$$[R^1R^2CHO]_3Al \xrightarrow{H_3O^{\oplus}} 3R^1R^2CHOH + Al^{3\oplus}$$

Acetone is reduced by amalgamated magnesium largely to a bimolecular reduction product, tetramethylethylene glycol or pinacol $(CH_3)_2 \cdot C(OH) \cdot C(OH)(CH_3)_2$; some propan-2-ol is also formed (Section **III,29**). Pinacol possesses the unusual property of forming a crystalline hexahydrate, m.p. 45 °C, and the pinacol is separated in this form from the unreacted acetone and the propan-2-ol. The magnesium is conveniently amalgamated by dissolving mercury(II) chloride in a portion of acetone; mercury is then liberated by the reaction:

$$HgCl_2 + Mg \longrightarrow Hg + MgCl_2$$

Anhydrous pinacol is a liquid; it may be prepared from the hydrate by azeotropic distillation with benzene.

Photoreduction of carbonyl compounds, particularly aromatic ketones, is often a useful route to 1,2-diols and is illustrated by the conversion of benzophenone to benzopinacol (Section **III,30**) by irradiation in propan-2-ol. Sunlight or a medium pressure mercury arc lamp (Section **I,17,5**) may be used; the reaction is completed more rapidly using the arc lamp. The mechanism involved in this transformation is probably as follows, the propan-2-ol acting as a hydrogen donor and itself becoming oxidised to acetone.

$$(C_6H_5)_2C = O \xrightarrow{h\nu} (C_6H_5)_2C = O^* \text{ (singlet)} \longrightarrow (C_6H_5)_2C = O^* \text{ (triplet)}$$

$$(C_6H_5)_2C = O^* \text{ (triplet)} + (CH_3)_2CHOH \longrightarrow (C_6H_5)_2\dot{C}OH + (CH_3)_2\dot{C}OH$$

$$(CH_3)_2\dot{C}OH + (C_6H_5)_2C = O \longrightarrow (CH_3)_2C = O + (C_6H_5)_2\dot{C}OH$$

$$2(C_6H_5)_2\dot{C}OH \longrightarrow (C_6H_5)_2C(OH) \cdot C(OH)(C_6H_5)_2$$

Primary alcohols may be conveniently prepared by the reduction of esters with sodium and absolute ethanol (the **Bouveault–Blanc reduction**, Section **III,31**).

$$R \cdot CO_2C_2H_5 + 4[H] \xrightarrow{Na/C_2H_5OH} R \cdot CH_2OH + C_2H_5OH$$

The method may also be applied to the esters of dicarboxylic acids for the preparation of α,ω-diols.

Esters may alternatively be reduced to primary alcohols using hydrogen under pressure in the presence of a copper–chromium oxide catalyst (Section **III,32**, *Method A*) or lithium aluminium hydride (Section **III,32**, *Method B*), but not with sodium borohydride which is insufficiently reactive.

III,24. HEPTAN-1-OL

$$CH_3 \cdot (CH_2)_4 \cdot CH_2 \cdot CHO \xrightarrow[Fe/H^{\oplus}]{2[H]} CH_3 \cdot (CH_2)_4 \cdot CH_2 \cdot CH_2OH$$

Place into a 3-litre three-necked flask, fitted with a mechanical stirrer (1) and a short reflux condenser, 450 g of grease-free iron filings, 750 ml of glacial acetic acid, 750 ml of water and 114 g (140 ml, 1.0 mol) of freshly distilled heptanal, b.p. 154–156 °C. Stopper the third neck and heat the mixture on a water bath, with stirring (1), for 2–4 hours; if the frothing is considerable, remove the mixture momentarily from the water bath. Steam distil the reaction product directly from the flask until no more oily drops pass over (*ca.* 2 litres of distillate). Separate the oil; a further small quantity may be obtained by saturating the aqueous layer with salt. Heat the crude heptanol with 250 ml of 20 per cent sodium hydroxide solution with stirring or vigorous hand shaking on the water bath for 2 hours; this will hydrolyse the small proportion of heptyl acetate which is present. Allow to cool and separate the oil (2). Dry it with a little anhydrous potassium carbonate or anhydrous calcium sulphate and distil. Collect the fraction, b.p. 173–176 °C, as pure heptan-1-ol. The yield is 90 g (78%).

Notes. (1) Stirring is not essential, but is advantageous since it reduces considerably the danger of frothing or foaming over. If mechanical stirring is not employed, the mixture must be shaken by hand from time to time, and the period of heating on the water bath increased to 6–8 hours.

(2) A further small quantity of heptan-1-ol may be obtained from the alkaline solution by diluting it with 50 ml of water and distilling: the distillate is saturated with salt, and the oil is separated, dried and distilled from a small flask.

III,25. HEPTAN-2-OL

$$CH_3 \cdot (CH_2)_4 \cdot CO \cdot CH_3 \xrightarrow[Na/C_2H_5OH]{2[H]} CH_3 \cdot (CH_2)_4 \cdot CHOH \cdot CH_3$$

Place a mixture of 114 g (140 ml, 1.0 mol) of heptan-2-one (1), 300 ml of rectified spirit (95% ethanol) and 100 ml of water in a 1.5-litre three-necked flask fitted with an efficient double surface condenser and a thermometer dipping into the reaction mixture. Through the third neck add 65 g (2.8 mol) of clean sodium, preferably in the form of wire (Section **II,2,55**) although small pieces may be used with somewhat inferior results, gradually and at such a rate that the reaction is under control; cool the flask in running water or in ice during the addition. The temperature should not rise above 30 °C. When the sodium has *completely* reacted, add 1 litre of water and cool the mixture to about 15 °C. Separate the upper layer, wash it with 25 ml of dilute hydrochloric acid (1 : 1), then with 25 ml of water, and dry with anhydrous potassium carbonate or anhydrous calcium sulphate. Distil through an efficient fractionating column and collect the heptan-2-ol at 156–158 °C. The yield is 75 g (65%).

Note. (1) The ketone may be synthesised as in Section **III,94**; it is also available commercially. The latter should first be dried, redistilled, and the fraction, b.p. 150–152 °C, collected.

Cognate preparations   Hexan-2-ol. Dissolve 100 g (123 ml, 1 mol) of hexan-2-one (Section **III,94**) in 750 ml of ether, add 150 ml of water and stir the mixture vigorously. Introduce 69 g (3 mol) of clean sodium in the form of wire (or small pieces) as rapidly as possible; the reaction must be kept under control and, if necessary, the flask must be cooled in ice or in running water. When all the sodium has reacted, separate the ethereal layer, wash it with 25 ml of dilute hydrochloric acid (1 : 1), then with water, dry with anhydrous potassium carbonate or with anhydrous calcium sulphate and distil through a fractionating column. Collect the fraction of b.p. 136–138 °C. The yield of hexan-2-ol is 97 g (95%).

Cyclopentanol. Use cyclopentanone and proceed as for hexan-2-ol. Collect the cyclopentanol as a fraction of b.p. 139–142 °C.

III,26. BENZHYDROL (*Diphenylmethanol*)

$$C_6H_5 \cdot CO \cdot C_6H_5 \xrightarrow[\text{Zn}/\ominus OH]{2[H]} C_6H_5 \cdot CHOH \cdot C_6H_5$$

In a 1-litre three-necked flask, equipped with a reflux condenser, a mechanical stirrer and a thermometer dipping into the reaction mixture, place 50 g (0.275 mol) of benzophenone (Section **IV,134**), 500 ml of rectified spirit, 50 g of sodium hydroxide and 50 g (0.76 mol) of zinc powder. Stir the mixture; the temperature slowly rises to about 70 °C. After 3 hours, when the temperature has commenced to fall, filter the reaction mixture with suction and wash the residue twice with 25 ml portions of hot rectified spirit. Do not allow the residual zinc powder to become dry as it is inflammable. Pour the filtrate into 2 litres of ice water acidified with 100 ml of concentrated hydrochloric acid. The benzhydrol separates as a white crystalline mass. Filter at the pump and dry in the air. The yield of crude benzhydrol, m.p. 65 °C, is 49 g. Recrystallise from 50 ml of hot ethanol and cool in a freezing mixture of ice and salt. Collect the colourless crystals and dry in the air; 36 g of pure benzhydrol, m.p. 68 °C, are obtained. Dilute the mother-liquor with water to precipitate the residual benzhydrol, and recrystallise this from a small quantity of hot alcohol.

III,27. 2,2,2-TRICHLOROETHANOL

$$CCl_3 \cdot CH(OH)_2 \xrightarrow[\text{NaBH}_4]{2[H]} CCl_3 \cdot CH_2OH + H_2O$$

Dissolve 16.5 g (0.1 mol) of chloral hydrate in 20 ml of water in a 200-ml beaker. Place a solution of 1.3 g (0.03 mol) of sodium borohydride in 20 ml of cold water in a small dropping funnel. Cool the chloral hydrate in an ice-water bath; add the borohydride solution dropwise (whilst stirring with a thermometer) at such a rate that the temperature of the solution is maintained at 20–30 °C—the reaction is strongly exothermic. When the borohydride solution has been added, allow the reaction mixture to stand at room temperature for 10 minutes; stir occasionally. Then add 2 ml of 2.5 M-hydrochloric acid dropwise and with stirring to destroy any residual borohydride and finally add a further 5 ml of the acid. Add sufficient ether to form two distinct layers, separate the ether layer, wash it with a little water and dry over magnesium sulphate. Remove the ether on a rotary evaporator and distil the residue from an air bath. Collect the 2,2,2-trichloroethanol at 151–153 °C. The yield is 9.8 g (65%).

Cognate preparation   *m*-**Nitrobenzyl alcohol.*** Clamp a 500-ml three-necked flask, equipped with a mechanical stirrer, a thermometer and a burette, above the bench so that an ice bath can be placed beneath it. Place a solution of 15.1 g (0.1 mol) of *m*-nitrobenzaldehyde (Section **IV,122**) in 100 ml of methanol in the flask and, whilst stirring, add a solution of sodium borohydride (1.4 g, 0.037 mol $NaBH_4$ in 2 ml of 2 M-sodium hydroxide diluted with 18 ml of water) at the rate of 0.5 ml per minute, with occasional cooling to keep the reaction at 18–25 °C. When about three-quarters of the solution has been added, there is no further tendency for the temperature to rise, and the addition is stopped. Treat a small portion of the reaction mixture with dilute sulphuric acid: hydrogen should be evolved.

Remove most of the methanol by distillation on a steam bath, and dilute the residue with 100 ml of water. Extract the mixture with ether, wash the upper layer with water and dry it rapidly with a little anhydrous magnesium sulphate. Remove the ether by flash distillation and distil the residual pale yellow oil under diminished pressure. Collect the *m*-nitrobenzyl alcohol at 183–185 °C/17 mmHg; it solidifies to a pale yellow solid, m.p. 30 °C, when cooled in ice. The yield is 13 g (85%).

III,28. BUT-2-EN-1-OL *(Crotyl alcohol)*

$$CH_3CH{=}CH{\cdot}CHO + CH_3{\cdot}CHOH{\cdot}CH_3 \xrightarrow[\text{isopropoxide}]{\text{Aluminium}}$$

$$CH_3{\cdot}CH{=}CH{\cdot}CH_2OH + CH_3{\cdot}CO{\cdot}CH_3$$

Prepare a solution of aluminium isopropoxide (see Section **II,2,*1(b)***) from 23.5 g (0.83 mol) of aluminium, 0.5 g of mercury(II) chloride and 250 ml of dry iso propyl alcohol (Section **II,1,*11***); add 105 g (1.5 mol) of redistilled crotonaldehyde, b.p. 102–103 °C, and 500 ml of dry isopropyl alcohol. Attach an efficient fractionating column to the flask and arrange for distillation from an oil bath so that the acetone distils as it is formed. Maintain the temperature of the bath at about 110 °C and the temperature at the top of the column at 60–70 °C. When the distillate no longer gives a test for acetone (8–9 hours) (1), distil off most of the remaining isopropyl alcohol, preferably under reduced pressure. Cool the residue to 40 °C and add 450 ml of cold 3 M-sulphuric acid (from 72.5 ml of

---

* The experimental details were kindly provided by the Research Laboratories, May & Baker Ltd.

concentrated sulphuric acid and 395 ml of water); cooling is necessary. Separate the upper oily layer, wash it once with water and distil at 60–70 °C whilst lowering the pressure slowly from about 275 mm to 60 mm; then continue the distillation to 100 °C and 20 mm. In this way the crotyl alcohol (A) is separated from the higher boiling polymerisation products. Combine the aqueous layers and distil until the distillate no longer gives a test for unsaturation with a dilute solution of bromine in carbon tetrachloride. Saturate the aqueous distillate with potassium carbonate, separate the oily layer and add it to (A). Dry with 5 g of anhydrous potassium carbonate, decant the oil and distil through an efficient fractionating column. Collect the crotyl alcohol at 119–121 °C. The yield is 55 g (50%).

Note. (1) The **acetone test reagent** consists of a 0.1 per cent solution of 2,4-dinitrophenylhydrazine and is prepared as follows: Dissolve 0.25 g of 2,4-dinitrophenylhydrazine in 50 ml of water and 42 ml of concentrated hydrochloric acid by warming on a water bath; cool the clear yellow solution and dilute to 250 ml with water. The acetone test is considered negative when 5 ml of the reagent and 4–5 drops of the distillate give no cloudiness or precipitate of acetone 2,4-dinitrophenylhydrazone within 30 seconds. After a negative test is obtained, it is strongly recommended that the mixture in the flask be refluxed for 5–10 minutes with complete condensation and then to collect a few drops of distillate for another test. If no acetone is now detected, the reduction is complete.

The above test will detect 1 part of acetone in 500–1000 parts of isopropyl alcohol. The reagent should not be kept for more than 1–2 months since it deteriorates upon keeping.

III,29. 2,3-DIMETHYLBUTANE-2,3-DIOL (Pinacol)

$$2CH_3 \cdot CO \cdot CH_3 \xrightarrow{Mg/Hg} \underbrace{(CH_3)_2C - C(CH_3)_2}_{\substack{| \quad | \\ O^{\ominus} \ O^{\ominus} \\ Mg^{2\oplus}}} \xrightarrow{H_2O}$$

$$(CH_3)_2C(OH) \cdot C(OH)(CH_3)_2 \cdot 6H_2O$$

This preparation should be conducted in an efficient fume cupboard.

Pinacol hydrate. Place 20 g (0.83 mol) of dry magnesium turnings and 200 ml of anhydrous benzene in a dry, 1-litre two-necked flask, fitted with a dropping funnel and an efficient double surface condenser (Fig. I,62) and carrying calcium chloride guard-tubes. Place a solution of 22.5 g of mercury(II) chloride (POISONOUS) in 100 g (127 ml, 1.72 mol) of dry AnalaR acetone in the funnel and run in about one-quarter of this solution; if the reaction does not commence in a few minutes, as indicated by a vigorous ebullition, warm the flask on a water bath and be ready to cool the flask in running water to moderate the reaction. Once the reaction has started, no further heating is required. Add the remainder of the solution at such a rate that the reaction is as vigorous as possible and yet under control. When all the mercuric chloride solution has been run in and whilst the mixture is still refluxing, add a mixture of 50 g (63.5 ml, 0.86 mol) of dry AnalaR acetone and 50 ml of dry benzene. When the reaction slows down, warm the flask on a water bath for 1–2 hours.

During this period the magnesium pinacolate swells up and nearly fills the flask. Cool slightly, disconnect the flask from the condenser and shake until the solid mass is well broken up: it may be necessary to use a stirrer. Attach the condenser and reflux for about 1 hour, or until the magnesium has disappeared.

Now add 50 ml of water through the dropping funnel and heat again on the water bath for 1 hour with occasional shaking. This converts the magnesium pinacolate into pinacol (soluble in benzene) and a precipitate of magnesium hydroxide. Allow the reaction mixture to cool to 50 °C and filter at the pump. Return the solid to the flask and reflux with a fresh 125 ml portion of benzene for 10 minutes in order to extract any remaining pinacol; filter and combine with the first filtrate. Distil the combined extracts to one-half the original volume in order to remove the acetone: treat the residual benzene solution with 75 ml of water and cool in an ice bath, or to at least 10–15 °C, with good stirring. After 30–60 minutes, filter the pinacol hydrate which has separated at the pump and wash it with benzene to remove small quantities of mercury compound present as impurities. Dry the pinacol hydrate by exposure to air at the laboratory temperature. The yield is 90 g (48%), m.p. 45.5 °C. This product is sufficiently pure for most purposes. The crude pinacol hydrate may be purified by dissolving it in an equal weight of boiling water, treating with a little decolourising charcoal if necessary, filtering the hot solution and cooling in ice; the recovery is over 95 per cent.

Pinacol. Pinacol hydrate may be dehydrated in the following manner (compare Section I,23, *Drying by distillation*). Mix 100 g of pinacol hydrate with 200 ml of benzene and distil; a mixture of water and benzene passes over. Separate the lower layer and return the upper layer of benzene to the distilling flask. Repeat the process until the benzene distillate is clear. Finally distil the anhydrous pinacol and collect the fraction boiling at 169–173 °C (50 g). The pure pinacol has m.p. 43 °C, but on exposure to moist air the m.p. gradually falls to 29–30 °C and then rises to 45–46 °C when hydration to the hexahydrate is complete.

III,30. BENZOPINACOL

$$2(C_6H_5)_2CO + (CH_3)_2CHOH \xrightarrow{hv} \begin{matrix} C_6H_5 & C_6H_5 \\ | & | \\ HO-C & C-OH \\ | & | \\ C_6H_5 & C_6H_5 \end{matrix} + (CH_3)_2CO$$

Method A. **Irradiation with sunlight.** Dissolve 10 g (0.055 mol) of benzophenone in 50 ml of propan-2-ol in a 100-ml round-bottomed Pyrex flask by slight warming, and add one drop of glacial acetic acid. Add further quantities of propan-2-ol, cooling to room temperature, until the solution is about $\frac{1}{4}$ inch below the bottom of the flask joint. Stopper the flask, taking care that none of the solution contaminates the joint, cover the stopper and joint with aluminium foil and place the flask in direct sunlight. Colourless crystals begin to separate within 24 hours. Allow the flask to remain in the sunlight until no further solid appears to separate (about 8 days). Cool the solution in ice-water and collect the product by suction filtration, wash it with about 10 ml of ice-cold propan-2-ol and dry. About 9.28 g (92%) of almost pure benzopinacol, m.p. 180–182 °C, is obtained. It may be recrystallised from glacial acetic acid (about 80 ml) from which it separates as colourless needles, m.p. 185–186 °C; the yield is 8.1 g (81%).

Method B. **Irradiation with a mercury arc lamp.** Use the 100 watt medium-pressure mercury arc lamp with the Pyrex outer and inner jackets (Section **I,17,5**) and a reaction vessel of approximately 110 ml capacity (Fig. I,74(f)). Make up a solution of 10 g (0.055 mol) of benzophenone in about 110 ml of propan-2-ol containing one drop of glacial acetic acid as described above. Place this solution in the reactor vessel, together with a magnetic follower bar, and irradiate the vigorously stirred solution under an atmosphere of nitrogen. Benzopinacol begins to separate within half an hour. As the quantity of product increases it gradually collects on the surface of the Pyrex jacket which thus restricts the amount of light reaching the solution. Therefore, after about 2–3 hours, switch off the lamp, raise the lamp insert from the reaction mixture and carefully scrape off the solid into the reactor vessel. Collect the product (about 4 g) by vacuum filtration, wash it with a few ml of cold propan-2-ol and dry. Return the filtrate to the reaction vessel. Clean the lamp insert with paper tissue moistened with ethanol and dry, and replace it in the reaction vessel. Continue the irradiation as above until a further appreciable quantity of product has separated (about 2 hours) and collect as before. Repeat this procedure until no further solid separates from the reaction mixture. About four crops of material will be obtained during a total irradiation period of 8 hours. The yield of benzopinacol, m.p. 178–182 °C, is about 8.65 g (75%). It may be purified by recrystallisation from glacial acetic acid as described above.

III,31. PENTAN-1-OL (*Amyl alcohol*)

$$CH_3 \cdot (CH_2)_2 \cdot CH_2 \cdot CO_2C_2H_5 \xrightarrow[Na/C_2H_2OH]{4[H]}$$
$$CH_3 \cdot (CH_2)_2 \cdot CH_2 \cdot CH_2OH + C_2H_5OH$$

Fit the central neck of a 1-litre two-necked flask with an efficient double surface condenser and close the side-neck with a stopper. Place 52 g (59.5 ml, 0.4 mol) of ethyl valerate (Section **III,153**) and 800 ml of super-dry ethanol (Section **II,1,9**) in the flask. Add 95 g (4.1 mol) of clean sodium (Section **II,2,55**) in small pieces through the aperture at such a rate that the vigorous refluxing is continuous (20–30 minutes). Reflux the mixture in an oil bath for 1 hour in order to be certain that all the sodium has dissolved. Replace the reflux condenser by an efficient fractionating column (e.g., Hempel or all-glass Dufton column, etc.) and set the condenser for downward distillation. Fractionate the mixture from an oil bath; about 250 ml of absolute ethanol are thus recovered. Treat the residue, consisting of pentanol and sodium ethoxide, with 330 ml of water and continue the distillation (oil bath at 110–120 °C) until the temperature at the top of the column reaches 83 °C, indicating that practically all the ethanol has been removed; about 600 ml of approximately 90 per cent ethanol are recovered. Remove the fractionating column and steam distil the mixture (Fig. I,101); about 200 ml must be collected before all the pentan-1-ol is removed. Separate the crude pentanol, dry it over anhydrous potassium carbonate or anhydrous calcium sulphate and distil through a short column.

Collect the fraction boiling at 137–139 °C. The yield of pentan-1-ol is 25 g (71%).

Note. (1) The ethanol must be absolute; a lower grade gives a poor yield.

Cognate preparations 2-Phenylethanol. Prepare a suspension of 42 g (1.83 mol) of sodium in 120 ml of sodium-dried toluene in a 3-litre three-necked flask following the procedure described in *Method 2* under *Sodium*, Section **II,2,55**. Do not decant the toluene; when the mixture has cooled to about 60 °C, add a solution of 50 g (0.30 mol) of ethyl phenylacetate (Section **III,153**) in 150 g (190 ml) of super-dry ethanol (Section **II,1,9**) as rapidly as possible without allowing the reaction to get out of control. Then add a further 200 g (253 ml) of super-dry ethanol. When the reaction has subsided, heat the flask in a water bath until the sodium is completely dissolved. Distil off the ethanol and toluene under reduced pressure using a rotary evaporator. Dilute the residue with water and extract the phenylethanol with ether, dry the extract with magnesium sulphate, remove the solvent and distil the residual oil under reduced pressure. Collect the 2-phenylethanol at 116–118 °C/25 mmHg. The yield is 25 g (67%).

The alcohol may be purified by conversion into the calcium chloride addition compound. Treat it with anhydrous calcium chloride; much heat is evolved and the addition compound is formed. After several hours, remove any oil which has not reacted by washing with petroleum ether (b.p. 60–80 °C). Decompose the solid with ice-water, separate the alcohol, dry and distil.

Butane-1,4-diol (*Tetramethylene glycol*). Place 60 g (2.6 mol) of clean sodium in a 3-litre three-necked flask fitted with two efficient double surface condensers and a dropping funnel protected by a calcium chloride tube. Add from the dropping funnel a solution of 35 g (0.2 mol) of diethyl succinate (Section **III,145**) in 700 ml of super-dry ethanol (Section **II,1,9**) as rapidly as possible consistent with the reaction being under control; it may be necessary to immerse the flask momentarily in a freezing mixture. When the vigorous action has subsided, warm the mixture on a water bath or in an oil bath at 130 °C until all the sodium has reacted (30–60 minutes). Allow to cool and cautiously add 25 ml of water (1); reflux for a further 30 minutes to bring all the solid into solution and to hydrolyse any remaining ester. Add 270 ml of concentrated hydrochloric acid to the cold reaction mixture, cool in ice, filter off the precipitated sodium chloride and treat the filtrate with 300 g of anhydrous potassium carbonate to free it from water and acid. Filter the alcoholic solution through a large sintered glass funnel, and extract the solid twice with boiling ethanol. Distil off the ethanol from the combined solutions; towards the end of the distillation solid salts will separate. Add dry acetone, filter and distil off the acetone. Distil the residue under diminished pressure, and collect the butane-1,4-diol at 133–135 °C/18 mmHg. The yield is 13 g (72%).

Note. (1) Alternatively, the following procedure for isolating the diol may be used. Dilute the partly cooled mixture with 250 ml of water, transfer to a distilling flask, and distil from an oil bath until the temperature reaches 95 °C. Transfer the hot residue to an apparatus for continuous extraction with ether (e.g., Fig. I,92). The extraction is a slow process (36–48 hours) as the diol is not very soluble in ether. (Benzene may also be employed as the extraction solvent.) Distil off the ether and, after removal of the water and ethanol, distil the diol under reduced pressure.

III,32. HEXANE-1,6-DIOL

$$\begin{array}{ccc}
CO_2C_2H_5 & & CH_2OH \\
| & & | \\
(CH_2)_4 & \xrightarrow[\text{or } B \text{ LiAlH}_4]{A \text{ H}_2/\text{Cu–CrO}} & (CH_2)_4 \\
| & & | \\
CO_2C_2H_5 & & CH_2OH
\end{array}$$

Method A (*hydrogenation using a copper–chromium oxide catalyst*). In a stainless steel high pressure autoclave having an adequate safety factor (Section I,17,1) and possessing a capacity of at least 250 ml, place 101 g (0.52 mol) of diethyl adipate (Sections III,145 and III,146) and 10 g of copper–chromium oxide catalyst (Section II,2,14(a)). Close the reaction vessel, make it gas tight, remove the air and introduce hydrogen until the pressure is about 2000 p.s.i. (1). Start the agitation, and heat as rapidly as possible to 255 °C and maintain this temperature. Continue the hydrogenation until hydrogen absorption is complete. As the hydrogenation proceeds, the pressure drops as indicated on the gauge; the progress of the reaction may be followed by the change in pressure readings, and the reaction is complete (after 6–12 hours) when the pressure is constant (2). Stop the agitation, allow to cool and release the pressure. Transfer the reaction mixture to a 400-ml beaker with the aid of four 12 ml portions of rectified spirit. Add 25 ml of 40 per cent sodium hydroxide solution to the combined alcoholic solutions, and reflux the mixture for 2 hours in order to hydrolyse any unchanged ester. Transfer the reaction mixture to a 500-ml flask and distil until the temperature of the vapour reaches 95 °C: this will remove ethanol. Transfer the hot residue with the aid of 25 ml of water to a continous extraction apparatus (Fig. I,91) and exhaustively extract the solution with ether (36–48 hours). Remove the ether and any remaining ethanol and distil the residue under reduced pressure. Collect the hexane-1,6-diol at 146–149 °C/17 mmHg; it solidifies on cooling (m.p. 41–42 °C). The yield is 52 g (88%).

Notes.   (1) The original pressure should not be more than 2000 p.s.i. if the maximum working pressure of the autoclave is 5000 p.s.i. The full operating pressure is not applied at the beginning because the pressure will rise as the bomb is heated: thus at 255 °C the pressure will be 1.8 times that at 20 °C.

(2) Unless a high pressure is used initially or the reaction vessel is large (about 1 litre), it will be necessary to introduce more hydrogen into the reaction vessel; the pressure should not be allowed to fall below 1400–1500 p.s.i. if the reaction is to run smoothly to completion.

Method B (*reduction with lithium aluminium hydride*). All the apparatus and reagents must be thoroughly dry. Set up a dry bowl to serve later as a cooling bath in a fume cupboard, a 1500-ml three-necked flask with a sealed stirrer unit, a 250-ml dropping funnel and a double surface condenser (see Fig. I,63); attach guard-tubes containing calcium chloride to the open ends of the condenser and the dropping funnel. The mechanical stirrer should be a powerful one. It must be emphasised that all operations, including weighing, with solid lithium aluminium hydride must be conducted in the fume cupboard; during weighing, etc., the front of the fume chamber is pulled down so that there is a narrow opening to allow the hands to enter (see also Section II,2,39, for additional precautions and methods for removal of traces of reagent).

Remove the dropping funnel from the flask neck and replace it by a funnel with a very short wide stem and introduce 10.5 g (0.263 mol) of powdered lithium aluminium hydride into the flask through this funnel, and use about 300 ml of sodium-dried ether to transfer the last traces into the flask. Replace the dropping funnel and guard-tube. Set the stirrer in motion, and place a solution of 50.5 g (0.25 mol) of freshly distilled diethyl adipate, b.p. 133–135 °C/14 mmHg (Sections **III,145** and **III,146**) in 150 ml of anhydrous ether in the dropping funnel. After stirring for 10 minutes (some of the lithium aluminium hydride may remain undissolved), add the diethyl adipate solution so that the ether refluxes gently; the reaction mixture rapidly becomes viscous and four 50 ml portions of anhydrous ether must be added during the reduction to facilitate stirring. Continue the stirring for 10 minutes after the diethyl adipate has been added. Decompose the excess of lithium aluminium hydride by the dropwise addition, with stirring, either of 75 ml of water, or, preferably, by the more rapid addition of 22 g (24.5 ml) of ethyl acetate (1). Filter the reaction product from the sludge through a sintered glass funnel; dry the ethereal solution with magnesium sulphate and distil off the ether with a rotary evaporator. The colourless viscous residue (18.5 g) solidifies completely on cooling and has m.p. 41–42 °C, i.e., is pure hexane-1,6-diol. Dissolve the sludge remaining in the filter funnel in 20 per cent sulphuric acid, extract the resulting solution with six 100 ml portions of ether, or use a continuous ether extractor (Fig. I,91). Remove the ether by means of a rotary evaporator; the residue (6 g) crystallises completely on cooling, m.p. 41–42 °C. The total yield of hexane-1,6-diol is 24.5 g (91%).

Note. (1) Before adding water, remove the calcium chloride tubes and fit the reflux condenser with a long tube extending to the duct at the top of the fume cupboard; this will carry the escaping hydrogen above the motor of the stirrer. A spark-proof stirring motor is recommended and should be used, if available. The dropwise addition of water must be conducted whilst the mixture is stirred vigorously; foaming may occur and the reaction may be moderated by filling the bath surrounding the reaction vessel with cold water.

D,2. The interaction of carbonyl-containing compounds with organometallic reagents. Primary, secondary or tertiary alcohols may be prepared by the reaction of a suitable carbonyl-containing compound with an alkylmagnesium halide (a **Grignard reagent**). The reagent is usually obtained quite readily by adding an alkyl halide (the bromide is frequently preferred) to a suspension of magnesium turnings in anhydrous ether. Initiation of the reaction may require the addition of a few crystals of iodine, the purpose of which may be to form a catalytic amount of magnesium iodide or possibly simply to etch the metal surface. As it is important that the iodine should be concentrated at the metal surface the mixture should not be stirred at this stage.

$$RX + Mg \xrightarrow{\text{ether}} RMgX$$

The resulting alkylmagnesium halide is soluble in the ether solvent as a result of coordination of two ether molecules on to the magnesium, and may be represented as follows:

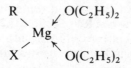

However, the true nature of the reactive species in solution is uncertain and for convenience the reagent may be represented as a polarised species, $\overset{\delta-}{R} - \overset{\delta+}{MgX}$. The reaction with a carbonyl group may then be represented as a nucleophilic addition process in the following way:

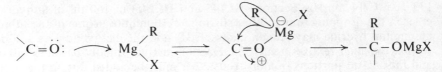

Primary alcohols may be prepared by the reaction of the Grignard reagent with anhydrous formaldehyde, obtained by depolymerising paraformaldehyde, and decomposing the resulting complex with mineral acid (Section **III,33**).

$$RMgX + H \cdot CHO \longrightarrow R \cdot CH_2OMgX \xrightarrow{H_3O^{\oplus}} R \cdot CH_2OH$$

The use of ethylene oxide in place of formaldehyde results in the formation of a primary alcohol having two additional carbon atoms (Section **III,34**).

$$RMgX + \underset{\underset{O}{\diagdown \diagup}}{CH_2 - CH_2} \longrightarrow R \cdot CH_2 \cdot CH_2OMgX \xrightarrow{H_3O^{\oplus}} R \cdot CH_2 \cdot CH_2OH$$

Secondary alcohols are similarly prepared by reaction of a Grignard reagent with an aldehyde (Section **III,35**).

$$R^1MgX + R^2 \cdot CHO \longrightarrow R^1 \cdot CH(OMgX) \cdot R^2 \xrightarrow{H_3O^{\oplus}} R^1 \cdot CH(OH) \cdot R_2$$

The reaction of two mols of the Grignard reagent with ethyl formate yields a symmetrical secondary alcohol (Section **III,36**).

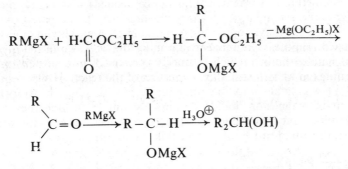

Tertiary alcohols are formed by the reaction of either one, two or three mols of a Grignard reagent with (*a*) a ketone, (*b*) an ester (other than a formate) or (*c*) diethyl carbonate, respectively (Sections **III,37** to **III,39**).

$$(a) \quad R^1MgX + R^2 \cdot CO \cdot R^3 \longrightarrow \underset{OMgX}{R^1 - \overset{R^2}{\underset{|}{\overset{|}{C}}} - R^3} \xrightarrow{H_3O^{\oplus}} \underset{OH}{R^1 - \overset{R^2}{\underset{|}{\overset{|}{C}}} - R^3}$$

(b) $R^1MgX + R^2 \cdot C \cdot OC_2H_5 \longrightarrow R^1 - \overset{\overset{\displaystyle R^2}{|}}{\underset{\underset{\displaystyle OMgX}{|}}{C}} - OC_2H_5 \xrightarrow{-Mg(OC_2H_5)X}$

$\quad\qquad\qquad \overset{\displaystyle \|}{O}$

$R^1 - \overset{\overset{\displaystyle R^2}{|}}{C} = O \xrightarrow{R^1MgX} R^1 - \overset{\overset{\displaystyle R^2}{|}}{\underset{\underset{\displaystyle OMgX}{|}}{C}} - R^1 \xrightarrow{H_3O^{\oplus}} R^1 - \overset{\overset{\displaystyle R^2}{|}}{\underset{\underset{\displaystyle OH}{|}}{C}} - R^1$

(c) $RMgX + C_2H_5O - \overset{\overset{\displaystyle }{}}{\underset{\underset{\displaystyle O}{\|}}{C}} - OC_2H_5 \longrightarrow C_2H_5O - \overset{\overset{\displaystyle R}{|}}{\underset{\underset{\displaystyle OMgX}{|}}{C}} - OC_2H_5 \xrightarrow{-Mg(OC_2H_5)X}$

$R - \overset{\overset{\displaystyle O}{\|}}{C} - OC_2H_5 \xrightarrow{RMgX} R - \overset{\overset{\displaystyle R}{|}}{\underset{\underset{\displaystyle OMgX}{|}}{C}} - OC_2H_5 \xrightarrow{-Mg(OC_2H_5)X} R - \overset{\overset{\displaystyle R}{|}}{C} = O \xrightarrow{RMgX}$

$R - \overset{\overset{\displaystyle R}{|}}{\underset{\underset{\displaystyle OMgX}{|}}{C}} - R \xrightarrow{H_3O^{\oplus}} R - \overset{\overset{\displaystyle R}{|}}{\underset{\underset{\displaystyle OH}{|}}{C}} - R$

Alkynols may be similarly prepared by using an acetylenic Grignard reagent (an alkynylmagnesium halide) which is readily prepared from a simple alkyl-magnesium halide and a terminal alkyne (Sections **III,40** and **III,41**).

$R^1 \cdot C \equiv CH + C_2H_5MgBr \longrightarrow R^1 \cdot C \equiv C \cdot MgBr + C_2H_6 \!\uparrow$

$R^2 - \overset{\overset{\displaystyle }{}}{\underset{\underset{\displaystyle O}{\|}}{C}} - R^3 + R^1 \cdot C \equiv CMgBr \longrightarrow R^3 - \overset{\overset{\displaystyle R^2}{|}}{\underset{\underset{\displaystyle OMgX}{|}}{C}} - C \equiv C \cdot R^1 \xrightarrow{H_3O^{\oplus}}$

$R^2R^3C(OH) \cdot C \equiv C \cdot R^1$

An alternative route to alkynols involves the use of an alkynyl sodium which is formed by the action of sodamide on a terminal alkyne (Section **III, 42**).

$R \cdot C \equiv CH + \overset{\oplus}{Na} \overset{\ominus}{:} NH_2 \longrightarrow R \cdot C \equiv \overset{\ominus}{C} \overset{\oplus}{:} Na + NH_3$

III,33. CYCLOHEXYLMETHANOL (*Cyclohexylcarbinol*)

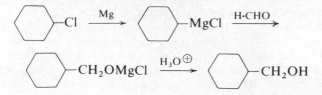

Fit a 2-litre three-necked flask with a sealed stirrer unit, a 500-ml separatory funnel and an efficient double surface condenser; place calcium chloride guard-tubes on the top of the funnel and on the condenser. All parts of the apparatus must be thoroughly dry. Arrange the apparatus so that the flask can be heated in a bath of hot water. Place 26.7 g (1.1 mol) of magnesium turnings (1) and a crystal of iodine in the flask. Measure out in separate dry vessels 118.5 g (121 ml, 1.0 mol) of cyclohexyl chloride (Section III,49) and 450 ml of sodium-dried ether. Introduce about 100 ml of the ether and 15 ml of the chloride into the flask. Heat the water bath so that the ether refluxes gently in order to start the reaction. When the reaction has commenced remove the water bath (cool, if necessary), set the stirrer in motion and add sufficient ether to cover the magnesium; then introduce the remainder of the cyclohexyl chloride dissolved in the residual ether during 30–45 minutes. If the reaction becomes too vigorous, cool the flask in ice-water. Continue the refluxing and stirring for 15–20 minutes to complete the formation of the Grignard reagent.

Replace the separatory funnel by a wide rubber tube fitted over the neck of the flask, and attach to this a small conical flask (cf. Fig I,64) charged with 50 g (*ca.* 1.67 mol HCHO) of paraformaldehyde, which has been previously dried in a vacuum desiccator over phosphorus pentoxide (2). Stir the mixture vigorously and gradually add the paraformaldehyde by suitably inclining the conical flask. After 2 hours cool the mixture, add 300 g of finely crushed ice all at once and vigorously agitate the mixture until the decomposition is complete. Add twice the theoretical quantity of 30 per cent sulphuric acid to dissolve the magnesium hydroxide, and then steam distil the mixture until no more oil passes over (2000–2500 ml). CARE: ether is distilled. Saturate the distillate with sodium chloride and separate the upper ether–alcohol layer. Dry with anhydrous potassium carbonate and distil off the ether on a water bath. Add 5 g of freshly dehydrated calcium oxide and heat on a water bath for 30 minutes; this will remove the last traces of water and give a halogen-free product. Filter, remove the ether and distil the residual alcohol under diminished pressure. Collect the fraction of b.p. 88–93 °C/18 mmHg; most distils at 91 °C/18 mmHg. The yield is 50 g (44%). The boiling point of cyclohexylmethanol at atmospheric pressure is 182 °C.

Notes. (1) See Section III,3, Note (1).

(2) An improved yield (*ca.* 75 g) may be obtained by substituting gaseous formaldehyde for paraformaldehyde. The former is obtained by placing 50 g of paraformaldehyde, previously dried for 2 days over phosphorus pentoxide, in a 500-ml round-bottomed two-necked flask provided with an inlet tube for admitting dry nitrogen. The flask is heated in an oil bath at 180–200 °C, and the formaldehyde vapour (produced by depolymerisation) is carried into the Grignard reagent by a slow stream of nitrogen through a wide glass tube (12

mm in diameter) fitted into the neck of the flask. The entry tube should terminate about 1 cm above the surface of the solution; clogging, due to repolymerised formaldehyde, is thus largely avoided.

By using dibutyl ether (Section III,76) as solvent, paraformaldehyde may be employed instead of gaseous formaldehyde without appreciable influence upon the yield. The high boiling point (141 °C) of butyl ether obviates the necessity of depolymerising the paraformaldehyde as a separate operation. The Grignard reagent is prepared [Section III,34, cognate preparation nonan-1-ol, Note (1)] with butyl ether as solvent using the proportions of reagents given above. The solution is heated to 100–110 °C in an oil bath, and 100 g of dry paraformaldehyde is added in small portions (compare Figs. I,64 or I,65) to the well-stirred solution over 2 hours. The product is isolated as above; 70 g of cyclohexylmethanol, b.p. 88–93 °C/18 mmHg, are obtained.

III,34. HEXAN-1-OL

$$CH_3 \cdot (CH_2)_2 \cdot CH_2 MgBr + CH_2 \underset{O}{-} CH_2 \longrightarrow$$

$$CH_3 \cdot (CH_2)_4 \cdot CH_2 OMgBr \xrightarrow{H_3O^{\oplus}} CH_3 \cdot (CH_2)_4 \cdot CH_2 OH$$

Owing to the toxicity of ethylene oxide this preparation must be carried out in an efficient fume cupboard.

The apparatus required, which must be perfectly dry, is identical with that described in Section III,33. Place 37.5 g (1.54 mol) of magnesium turnings (1) and 300 ml of sodium-dried ether in the flask, and add a small crystal of iodine. Prepare a solution of 205.5 g (161 ml, 1.50 mol) of pure dry butyl bromide (Section III,54) in 300 ml of dry ether in the separatory funnel and introduce about 25 ml of the solution into the flask. As soon as the reaction commences (2), set the stirrer in motion and add the remainder of the butyl bromide solution at such a rate that steady refluxing of the reaction mixture is maintained (if the reaction becomes too vigorous it should be moderated by momentarily cooling the flask in an ice bath). When the addition is complete, maintain gentle reflux, with the aid of a hot-water bath if necessary, until most of the magnesium has reacted (about 15–30 minutes). Cool the flask in a freezing mixture of ice and salt. Remove the separatory funnel and replace it by a tube, 4 mm in diameter, the end of which is about 2 cm above the surface of the liquid. Attach this delivery tube to a flask fitted with 'wash bottle' tubes, the long tube being nearer the three-necked flask and the other end being connected to a supply of dry nitrogen. Cool this flask in a mixture of ice and salt and introduce rapidly 90 g (2.02 mol) of ethylene oxide (3) from a 100 g sealed bulb of the reagent (Section II,2,22); the latter must, of course, be cooled in an ice and salt mixture before opening (4). Gradually introduce the ethylene oxide into the reaction flask over a period of 1.5–2 hours; the temperature should not rise above 10 °C. When all has been added, remove the freezing mixture surrounding the three-necked flask. The temperature of the mixture will gradually rise and the reaction mixture will boil gently. When boiling ceases, reflux on a water bath for 30 minutes. Allow to cool, insert a thermometer into a neck of the flask, arrange the condenser for downward distillation and collect 250 ml of ether in a measuring cylinder; do not collect a larger volume of ether as a violent reaction may set in, apparently due to a rearrangement of the initial reaction product, and considerable loss may ensue. Change the receiver, and introduce 250 ml of sodium-dried benzene into the reaction mixture. Continue the distillation with stirring until the temperature of the distilling vapour reaches 65 °C. Then boil the mixture under reflux for 30 minutes; generally by this time the mixture has become so viscous that stirring is no longer very effective. Allow to cool. Decompose the reaction mixture with 500 ml of an ice-water mixture, and dissolve the precipitated magnesium hydroxide with 30 per cent sulphuric acid; add sufficient finely-crushed ice to keep the mixture cold. Steam distil and collect about 2 litres of distillate. Separate the oily layer (*A*), and distil the aqueous layer until free of hexan-1-ol; add the oil so obtained to (*A*). Stir the crude hexan-1-ol on a water bath with 250 ml of 20 per cent sodium hydroxide solution, and steam distil again as before. Dry the oil with a little anhydrous calcium sulphate, distil through an efficient fractionating column and collect the fraction, b.p. 154–157 °C. The yield of hexan-1-ol is 90 g (49%).

Notes. (1) See Section **III,33**, Note (1).

(2) The onset of the reaction is accompanied by the disappearance of the iodine colour and the development of cloudiness or opalescence, and the spontaneous boiling of the solvent with bubbles originating on the surface of the magnesium. If the reaction does not start after a few minutes, warm the flask gently on a water bath and if necessary add a further crystal of iodine.

(3) It is advisable to cool and open the ampoule of ethylene oxide behind a safety screen in a fume cupboard, and to wear plastic gloves and goggles.

(4) Instead of adding the liquid ethylene oxide (b.p. 10.5 °C), the latter may be dissolved in 100 ml of ice-cold anhydrous ether; this solution is added directly to the reaction mixture during 15–30 minutes. The yield however is somewhat lower.

Cognate preparation   Nonan-1-ol.   This preparation is an example of the use of dibutyl ether as a solvent in the Grignard reaction. Prepare a Grignard reagent from 24.5 g (1 mol) of magnesium turnings, 179 g (157 ml, 1 mol) of 1-bromoheptane (Section **III,55**) and 300 ml of dibutyl ether (1). Cool the solution to 0 °C and, with vigorous stirring, add an excess of ethylene oxide. Maintain the temperature to 0 °C for 1 hour after the ethylene oxide has been introduced, then allow the temperature to rise to 40 °C and maintain the mixture at this temperature for 1 hour. Finally heat the mixture on a water bath for 2 hours. Decompose the addition product by pouring the cooled reaction mixture into ice-water, acidify with sulphuric acid to dissolve the precipitated magnesium hydroxide and isolate the reaction products as described in the procedure for hexan-1-ol. Collect the nonan-1-ol at b.p. 95–100 °C/12 mmHg; the yield is 95 g (69%).

Note.   (1) Commercial dibutyl ether is purified by washing with sodium hydroxide solution, water, drying with anhydrous calcium chloride and then fractionating. The fraction, b.p. 140–142 °C, is collected.

The general procedure for the preparation of Grignard reagents in dibutyl ether solution may be adapted from the following description of a small-scale experiment. A 200-ml three-necked flask is fitted with a mechanical stirrer, separatory funnel, reflux condenser and thermometer. A mixture of 40 ml of dibutyl ether, 1.5 g of magnesium turnings and a small crystal of iodine is placed in the flask. The theoretical amount of the halogen compound, dissolved in sufficient dibutyl ether to make a total volume of 30 ml is placed in the funnel. A small amount of the solution of the halogen compound is added and the flask is heated until the reaction commences—the exact temperature varies according to the nature of the halogen compound. Once the reaction has started, stirring is begun, and the remainder of the solution is added at a rate which permits the reaction to proceed smoothly. After the addition of the halogen compound, stirring is continued until the mixture cools to the temperature of the laboratory.

III,35. 3-METHYLBUTAN-2-OL (*Methylisopropylmethanol*)

$$(CH_3)_2CHMgBr \xrightarrow{CH_3 \cdot CHO} (CH_3)_2CH \cdot \overset{|}{C}H \cdot CH_3 \xrightarrow{H_3\overset{\oplus}{O}}$$
$$\qquad\qquad\qquad OMgBr$$
$$(CH_3)_2CH \cdot CH(OH) \cdot CH_3$$

Prepare a solution of isopropylmagnesium bromide from 37.5 g (1.54 mol) of magnesium turnings and 185 g (142 ml, 1.50 mol) of isopropyl bromide (Section **III,53**) in 200 ml of anhydrous ether; use a 1-litre flask equipped as in Section **III,33**, and follow the broad experimental details given in Section **III,34**. Cool the resulting Grignard reagent to -10 to $-5\,^\circ C$ in a freezing mixture of crushed ice and anhydrous calcium chloride. Remove the separatory funnel and re-attach to the flask via a two-necked adapter to allow the introduction of a thermometer for monitoring the reaction temperature. Add a solution of 67 g (83.5 ml, 1.53 mol) of acetaldehyde (1) in 90 ml of anhydrous ether over a period of 30 minutes. Do not allow the temperature to rise above $-5\,^\circ C$. When all the acetaldehyde has been added, pour the reaction product upon 700 g of crushed ice; any excess of magnesium should remain in the flask. Dissolve the basic magnesium bromide by the addition of 350 ml of 15 per cent sulphuric acid. Separate the ethereal solution and extract the aqueous layer with four 50 ml portions of ether. Dry the combined ethereal solutions over 8 g of anhydrous potassium carbonate (or the equivalent quantity of anhydrous calcium sulphate), and fractionally distil through an all-glass Dufton (or other efficient fractionating) column. Collect the 3-methylbutan-2-ol at 110–111.5 °C. The yield is 70 g (52%).

Note. (1) The acetaldehyde should be freshly distilled (b.p. 20.5–21 °C). It can be conveniently prepared by depolymerising pure dry paraldehyde (see Section **III,81**).

III,36. NONAN-5-OL (*Dibutylmethanol*)

$$2CH_3 \cdot (CH_2)_2 \cdot CH_2MgBr + H \cdot CO_2C_2H_5 \longrightarrow$$
$$(CH_3 \cdot CH_2 \cdot CH_2 \cdot CH_2)_2CHOMgBr \xrightarrow{H_3\overset{\oplus}{O}} (CH_3 \cdot CH_2 \cdot CH_2 \cdot CH_2)_2CHOH$$

Prepare a solution of butylmagnesium bromide from 12.2 g (0.50 mol) of magnesium turnings, 69 g (54 ml, 0.50 mol) of butyl bromide (Section **III,54**) and 250 ml of dry ether; use a 1-litre flask equipped as in Section **III,33**, and follow the broad experimental details given in Section **III,34**. Cool the flask containing the resulting Grignard reagent in an ice bath. Place a solution of 18.5 g (20 ml, 0.25 mol) of pure ethyl formate (1) in 40 ml of anhydrous ether in the separatory funnel. Stir the solution of the Grignard reagent and run in the ethyl formate solution at such a rate that the ether refluxes gently (10–15 minutes). Remove the ice bath and continue the stirring for 10 minutes.

Place 35 ml of water in the separatory funnel and run it into the *vigorously stirred* reaction mixture at such a rate that rapid refluxing occurs. Follow this by a cold solution of 15.5 ml of concentrated sulphuric acid in 135 ml of water. Two practically clear layers will now be present in the flask. Decant as much as possible of the ethereal layer (*A*) into a 500-ml round-bottomed flask. Transfer the remainder, including the aqueous layer, into a separatory funnel: wash the residual solid with two 10 ml portions of ether and combine these washings with the liquid in the separatory funnel. Separate the ethereal portion and combine it with (*A*). Distil off the ether through an efficient fractionating column until the temperature of the vapour rises to about 50 °C. The residual crude nonanol contains a little of the corresponding formic ester. Remove the latter by refluxing for 3 hours with 25 ml of 15 per cent aqueous potassium hydroxide, and then isolate the purified nonanol by steam distillation (volume of distillate about 500 ml). Separate the upper layer of the secondary alcohol, dry it over anhydrous potassium carbonate or anhydrous calcium sulphate, and distil from a flask carrying a Claisen still-head under reduced pressure. Collect the pure nonan-5-ol at 97–98 °C/20 mmHg; the yield is 30 g (83%). The boiling point under atmospheric pressure is 195 °C.

Note. (1) Freshly distilled ethyl formate must be used. **Commercial ethyl formate may be purified** as follows. Allow the ethyl formate to stand for 1 hour with 15 per cent of its weight of anhydrous potassium carbonate with occasional shaking. Decant the ester into a dry flask containing a little fresh anhydrous potassium carbonate and allow to stand for a further hour. Filter into a dry flask and distil through an efficient fractionating column, and collect the fraction, b.p. 53–54 °C; protect the receiver from atmospheric moisture.

III,37. 2-METHYLHEXAN-2-OL (*Butyldimethylmethanol*)

$$CH_3 \cdot (CH_2)_2 \cdot CH_2 MgBr + CH_3 \cdot CO \cdot CH_3 \longrightarrow$$

$$CH_3 \cdot (CH_2)_2 \cdot CH_2 \cdot C(OMgBr)(CH_3)_2 \xrightarrow{H_3O^{\oplus}}$$

$$CH_3 \cdot (CH_2)_2 \cdot CH_2 \cdot C(OH)(CH_3)_2$$

The apparatus and experimental details are similar to those given in the previous Sections. Prepare a Grignard reagent from 24.5 g (1 mol) of magnesium turnings, 137 g (107 ml, 1 mol) of butyl bromide and 450 ml of sodium-dried ether. Add slowly with rapid stirring, and cooling with ice if necessary, a solution of 58 g (73.5 ml, 1 mol) of dry acetone in 75 ml of anhydrous ether. Allow the reaction mixture to stand overnight. Decompose the product by pouring it on to 500 g of crushed ice; dissolve the precipitated magnesium compounds by the addition of 10 per cent hydrochloric acid or of 15 per cent sulphuric acid. Transfer to a separatory funnel, remove the ether layer and extract the aqueous solution with three 50 ml portions of ether. Dry the combined ethereal solutions over anhydrous potassium carbonate or anhydrous calcium sulphate, filter, distil off the ether, and fractionally distil the residue. Collect the 2-methylhexan-2-ol at 137–141 °C. The yield is 105 g (90%).

Cognate preparation 2-Methylpentan-2-ol (*Dimethylpropylmethanol*). From propylmagnesium bromide and acetone. Collect the tertiary alcohol at 121–124 °C.

III,38. 2-METHYLPENTAN-2-OL (*Dimethylpropylmethanol*)

$$CH_3 \cdot CH_2 \cdot CH_2 \cdot CO_2C_2H_5 \xrightarrow{CH_3MgI} [CH_3 \cdot CH_2 \cdot CH_2 \cdot CO \cdot CH_3] \xrightarrow{CH_3MgI}$$

$$CH_3 \cdot CH_2 \cdot CH_2 \cdot C(OMgI)(CH_3)_2 \xrightarrow{H_3O^{\oplus}} CH_3 \cdot CH_2 \cdot CH_2 \cdot C(OH)(CH_3)_2$$

The broad experimental details will be evident from those described in the previous experiments, particularly Section **III,34**. Place 49 g (2 mol) of dry magnesium turnings and 100 ml of sodium-dried ether in a 1-litre three-necked flask and a solution of 284 g (124.5 ml, 2 mol) of dry methyl iodide in 300 ml of anhydrous ether in the separatory funnel protected by a calcium chloride guard-tube. Run in about 15 ml of the iodide solution. The reaction should start within a few minutes: if it does not, warm gently on a water bath and add a crystal of iodine, if necessary. Once the reaction has commenced, remove the water bath, add the iodide solution, with stirring, at such a rate that the mixture refluxes gently; if the reaction becomes too vigorous, cool the flask in ice-water. Finally reflux the reaction mixture until all, or most, of the magnesium has reacted. Allow to cool, and slowly add a solution of 116 g (132 ml, 1 mol) of ethyl butyrate (1) in 100 ml of anhydrous ether into the vigorously stirred solution of the Grignard reagent. Reflux the mixture on a water bath for 1 hour to complete the reaction. Pour the ethereal solution into a mixture of 200 ml of approximately 2 *M*-sulphuric acid and 750 g of crushed ice. Separate the upper ethereal layer and extract the aqueous solution with two 150 ml portions of ether. Wash the combined ethereal extracts with dilute sodium hydrogen carbonate solution, followed by a little water, then dry with anhydrous potassium carbonate or anhydrous calcium sulphate, distil off the ether on a water bath and distil the residue through a short fractionating column. Collect the 2-methylpentan-2-ol at 117–120 °C. A further small quantity of the tertiary alcohol may be obtained by redrying the low-boiling distillate, filtering and redistilling. The yield is 90 g (88%).

Note. (1) Ethyl butyrate may be prepared as described in Section **III,142**.

Cognate preparations   2-Methylbutan-2-ol (*t-pentyl alcohol*). From ethyl propionate and methylmagnesium iodide. Collect the tertiary alcohol at 100–102 °C.

Triphenylmethanol. Prepare a solution of phenylmagnesium bromide from 14 g (0.57 mol) of magnesium turnings, 90.5 g (60.5 ml, 0.57 mol) of dried redistilled bromobenzene and 250 ml of anhydrous ether. Treat this Grignard reagent with a solution of 37.5 g (36 ml, 0.25 mol) of dry ethyl benzoate (Section **IV,177**) in 100 ml of sodium-dried benzene following the procedure described above. Decompose the reaction product by pouring it slowly, with constant stirring, into a mixture of 750 g of crushed ice and 25 ml of concentrated sulphuric acid. Continue the stirring until all the solid dissolves; it may be necessary to add 25 g of solid ammonium chloride to facilitate the decomposition of the magnesium complex, and also a little more benzene to dissolve all the product. When all the solids have passed into solution, separate the benzene layer and wash it successively with 100 ml of water, 100 ml of 5 per cent sodium bicarbonate solution and 100 ml of water. Remove the benzene as completely as possible from a 1-litre round-bottomed flask using a rotary evaporator: steam distil the residue (Fig. I,101) in order to separate unchanged bromobenzene and biphenyl (by-product). Filter the cold residue in the flask at the pump, wash it with water and dry. The resulting crude triphenylmethanol weighs 62 g. Recrystallise it from carbon tetrachloride (4 ml per gram of solid): the first crop of crystals, after drying in air to remove the solvent of crystallisation, weighs 56 g (86%) and melts at 162 °C. Treat the mother-liquid with 1 g of decolourising charcoal, concentrate to one-quarter of the original volume and cool in ice: a further 3 g of pure triphenylmethanol is obtained.

In **an alternative method of preparation**, benzophenone is used. Prepare the Grignard reagent as above, cool in cold water and add a solution of 91 g (0.5 mol) of benzophenone (Section **IV,134**) in 200 ml of dry benzene at such a rate that the mixture refluxes gently. Reflux the mixture for 60 minutes, and isolate the triphenylmethanol in the manner described above. The yield is of the same order.

III,39. 3-ETHYLPENTAN-3-OL (*Triethylmethanol*)

$$3C_2H_5MgBr + (C_2H_5O)_2CO \longrightarrow (C_2H_5)_3COMgBr + 2C_2H_5OMgBr$$

$$\xrightarrow[NH_4Cl]{H_2O} (C_2H_5)_3COH + 2C_2H_5OH$$

In a 1-litre three-necked flask, equipped as in Section **III,33**, prepare a solution of ethylmagnesium bromide from 36.5 g (1.50 mol) of magnesium turnings, 163 g (112 ml, 1.50 mol) of ethyl bromide (Section **III,54**) and 600 ml of anhydrous ether, following the general procedure outlined in Section **III,34**. Run into the resulting ethereal Grignard reagent a solution of 52 g (53.5 ml, 0.44 mol) of pure diethyl carbonate (1) in 70 ml of anhydrous ether, with rapid stirring, over a period of about 1 hour. A vigorous reaction sets in and the ether refluxes continually. When the diethyl carbonate has been added, heat the flask on a water bath with stirring for another hour. Pour the reaction mixture, with frequent shaking, into a 2-litre round-bottomed flask containing 500 g of crushed ice and a solution of 100 g of ammonium chloride in 200 ml of water. Transfer to a separatory funnel, remove the ether layer and extract the aqueous solution with two 175 ml portions of ether. Dry the combined ethereal extracts with anhydrous potassium carbonate or with anhydrous calcium sulphate, and remove the ether on a water bath. Distil the alcohol through a short fractionating column and collect the fraction boiling at 139–142 °C as pure 3-ethylpentan-3-ol. A further small quantity may be obtained by drying the low-boiling fraction with 2 g of anhydrous potassium carbonate or anhydrous calcium sulphate, filtering and redistilling. The total yield is 44 g (86%).

Note. (1) **Commercial diethyl carbonate may be purified** by the following process. Wash 100 ml of diethyl carbonate successively with 20 ml of 10 per cent sodium carbonate solution, 20 ml of saturated calcium chloride solution and 25 ml of water. Allow to stand for 1 hour over anhydrous calcium chloride with occasional shaking, filter into a dry flask containing 5 g of the same desiccant, and allow to stand for a further hour. Distil and collect the fraction boiling at 125–126 °C. Diethyl carbonate combines with anhydrous calcium chloride slowly and prolonged contact should therefore be avoided. Anhydrous calcium sulphate may also be used.

Cognate preparations The following tertiary alcohols may be prepared from the appropriate Grignard reagent and diethyl carbonate in yields of 75–80 per cent.

4-Propylheptan-4-ol. B.p. 89–92 °C/20 mmHg.
5-Butylnonan-5-ol. B.p. 129–131 °C/20 mmHg.
6-Pentylundecan-6-ol. B.p. 160–163 °C/19 mmHg.

III,40. *TRANS*-1-ETHYNYL-3,3,5-TRIMETHYLCYCLOHEXAN-1-OL

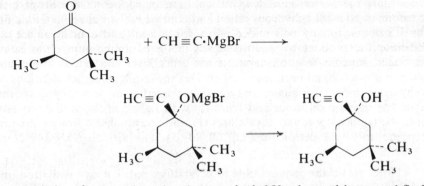

Assemble in a fume cupboard a three-necked, 250-ml round-bottomed flask mounted on a magnetic stirrer unit and equipped with a gas inlet tube terminating in a glass frit for the passage of acetone-free acetylene (Section **III,21**), and a dropping funnel protected with a calcium chloride guard-tube; insert a calcium chloride tube in the third neck. Place in the flask 60 ml of pure tetrahydrofuran (Section **II,1,19**) and a magnetic follower. Saturate the tetrahydrofuran with acetylene by passing a rapid stream of the gas through the solvent. Then with the continued passage of acetylene, add dropwise from the funnel over 2 hours a previously prepared solution of ethylmagnesium bromide [from 16.4 g (0.15 mol) of ethyl bromide, 3.7 g (0.154 mol) of magnesium in 100 ml of tetrahydrofuran]. Cool the reaction mixture to 0 °C and add with stirring a solution of 7 g (0.05 mol) of 3,3,5-trimethylcyclohexanone in 20 ml of tetrahydrofuran; the slow passage of acetylene gas should be continued. Stir the mixture for 1 hour and then pour it into a mixture of 80 g of ammonium chloride and 200 g of crushed ice. Separate the organic layer and extract the aqueous layer with three 25 ml portions of tetrahydrofuran. Evaporate the tetrahydrofuran from the combined organic layers, dissolve the residue in 100 ml of ether, dry the solution over magnesium sulphate and evaporate the solvent on a rotary evaporator. Distil (Fig. 1,110) the residue and collect the alkynol at b.p. 78–79 °C/8 mmHg; this crystallises on standing and has m.p. 26–27 °C. The yield is 5.6 g (70%). The i.r. spectrum clearly shows characteristic group frequencies at v_{max} 3450 (OH), 3300 ($\equiv$C–H) and 2110 cm^{-1} (C$\equiv$C).

III,41. 1,1,3-TRIPHENYLPROP-2-YN-1-OL

$$C_6H_5 \cdot C \equiv CH \xrightarrow{C_2H_5MgBr} C_6H_5 \cdot C \equiv CMgBr \xrightarrow{(C_6H_5)_2CO}$$

$$C_6H_5 \cdot C \equiv C \cdot C(C_6H_5)_2OMgBr \xrightarrow{H_3O^{\oplus}} C_6H_5 \cdot C \equiv C \cdot C(C_6H_5)_2OH$$

Equip a 500-ml three-necked flask with a reflux condenser, a sealed stirrer unit and a dropping funnel protected by a calcium chloride guard-tube. Prepare in the flask a solution of ethylmagnesium bromide in 50 ml of anhydrous ether from 27.3 g (19 ml, 0.25 mol) of ethyl bromide, 6.0 g (0.25 mol) of magnesium and a trace of iodine, following the general procedure described in Section **III,34**. Cool the solution and add dropwise a solution of 25.5 g (27 ml, 0.25 mol) of phenylacetylene (Section **III,20**) in 30 ml of anhydrous ether. Boil the reaction mixture gently under reflux for 2 hours and cool to room temperature. Start the stirrer, add slowly a solution of 45.5 g (0.25 mol) of benzophenone in 50 ml of anhydrous ether, and continue to stir at room temperature for 1.5 hours. Finally boil under reflux for 1 hour and cool in an ice bath. Liberate the product by adding slowly 55 g of ammonium chloride as a saturated aqueous solution, separate the ether layer and extract the aqueous phase with two 20 ml portions of ether. Dry the combined ether solutions over anhydrous sodium sulphate, and remove the ether on a rotary evaporator. Cool the residual oil in ice and triturate with light petroleum (b.p. 60–80 °C) until the triphenylpropynol crystallises (1), and recrystallise it from a mixture of benzene and light petroleum (b.p. 60–80 °C). The yield is 35 g (49%), m.p. 78–80 °C.

Note. (1) If the product fails to crystallise, purify it by distillation under reduced pressure, b.p. 190 °C/0.05 mmHg.

III,42. 1-ETHYNYLCYCLOHEXANOL

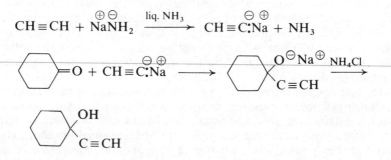

$$CH \equiv CH + \overset{\oplus}{Na}\overset{\ominus}{NH_2} \xrightarrow{\text{liq. NH}_3} CH \equiv \overset{\ominus}{C} \overset{\oplus}{:Na} + NH_3$$

Use the same technique as detailed for Section **III,21** but with a 3-litre three-necked flask. Charge the flask with 1.5 litres of liquid ammonia. Prepare the sodamide using 0.7 g of iron(III) nitrate and 2 g of sodium, followed by 46 g (total 2.1 mol) of sodium, and convert it into a solution of sodium acetylide as before. Add, with stirring, a solution of 196 g (206 ml, 2 mol) of dry, redistilled cyclohexanone (1) in 256 ml of dry ether during 1 hour and continue the stirring for a further 2 hours. Decompose the sodium derivative of the product by the gradual addition of a slight excess (118 g) of powdered ammonium chloride. Allow to stand overnight, preferably with stirring, by which time all the ammonia will have evaporated. Extract the residue repeatedly with ether, i.e., until all the alkynol has been separated from the inorganic material (2). Wash the ethereal extract successively with water, dilute sulphuric acid and sodium hydrogen carbonate solution, dry with magnesium sulphate and distil. Collect the 1-ethynylcyclohexanol at 83 °C/20 mmHg (3); the yield is 210 g (85%).

Notes. (1) Dry the cyclohexanone over excess of anhydrous calcium chloride before distillation.

(2) A continuous ether extractor (Fig. I,92) is recommended.

(3) The product has m.p. *ca.* 25 °C, but the m.p. depends upon the purity of the cyclohexanone and the efficiency of the distillation. Pure 1-ethynylcyclohexanol has m.p. 32 °C.

D,3. The hydroboronation–oxidation of alkenes. Diborane, as a solution in tetrahydrofuran (Section **II,2,**20) or generated *in situ* by the reaction of a metal hydride with boron trifluoride etherate, adds readily to alkenes to yield trialkylboranes. In the case of a terminal alkene a primary trialkylborane is formed following an anti-Markownikoff mode of addition as the result of the electrophilic character of the borine reagent.

$$R \cdot CH = CH_2 \longrightarrow R \cdot CH_2 \cdot CH_2 \cdot BH_2 \xrightarrow{R \cdot CH = CH_2}$$

$$H - B - H$$
$$\overset{|}{\underset{|}{B}}$$
$$H$$

$$(R \cdot CH_2 \cdot CH_2)_2 BH \xrightarrow{R \cdot CH = CH_2} (R \cdot CH_2 \cdot CH_2)_3 B$$

The trialkylborane is not usually isolated but is converted *in situ* to an alcohol by oxidation with alkaline hydrogen peroxide.

$$(R \cdot CH_2 \cdot CH_2)_3 B \xrightarrow[\underset{:OH}{\ominus}]{H_2O_2} 3R \cdot CH_2 \cdot CH_2 OH + B(OII)_3$$

This reaction is illustrated by the conversion of hex-1-ene into hexan-1-ol (Section **III,43**).

An important consequence of the mechanism of the oxidation step is that the stereochemistry resulting from the *cis*-addition step and formulated above is retained in the final product.

$$\begin{array}{ccc}
R & R & R \\
| & | & | \\
R-B-R & (R-)B\overset{\ominus}{-}R & B-R \\
:O \quad OH & O-OH & R \cdot O
\end{array} + \ominus OH \xrightarrow{H_2O} ROH + R_2BOH$$

III,43. HEXAN-1-OL

$$3NaBH_4 + 4BF_3 \cdot (C_2H_5)_2O \longrightarrow 4BH_3 + 3NaBF_4 + 4(C_2H_5)_2O$$

$$12CH_3 \cdot (CH_2)_3 \cdot CH = CH_2 + 4BH_3 \longrightarrow 4[CH_3 \cdot (CH_2)_5]_3B$$

$$4[CH_3 \cdot (CH_2)_5]_3B + 12H_2O_2 + 4NaOH \longrightarrow$$
$$12CH_3 \cdot (CH_2)_4 \cdot CH_2OH + 4NaB(OH)_4$$

This experiment should be carried out in an efficient fume cupboard.

Equip a 500-ml three-necked round-bottomed flask (1) with a sealed stirrer unit, a 100-ml pressure-equalising dropping funnel fitted with an inlet adapter to allow flushing with dry nitrogen, and a two-necked adapter carrying a condenser fitted with a calcium chloride guard-tube, and a thermometer reaching to the bottom of the flask. Arrange the apparatus so that it may be cooled occasionally in an ice-water bath supported on a laboratory jack.

Maintain a slow stream of dry nitrogen through the apparatus and place in the flask 90 ml of dry diglyme (Section **II,1,**_18_), and 3.41 g (0.090 mol, 20% excess over the theoretical requirement of 0.075 mol) of powdered sodium borohydride. Stir until most of the borohydride has dissolved then add a solution of 25.2 g (0.30 mol) of hex-1-ene in 50 ml of dry diglyme. Place 17.0 g (15.1 ml, 0.12 mol; 20% excess over the theoretical requirement of 0.10 mol) of purified boron trifluoride–etherate (48%, w/w, Section **II,2,**_6_) in the dropping funnel followed by 25 ml of dry diglyme. Adjust the flow of dry nitrogen so that a slow stream is maintained throughout the experiment. Add the boron trifluoride–etherate in portions to the rapidly stirred mixture during 30 minutes whilst maintaining the temperature at 20–25 °C by occasional cooling with the ice-water bath. Continue stirring at room temperature for 1 hour to ensure completion of the hydroboronation reaction. Add 20 ml of water dropwise from the dropping funnel to the reaction mixture during about 25 minutes to decompose excess sodium borohydride; vigorous hydrogen evolution may cause foaming during this addition. When hydrogen evolution has stopped, place 40 ml (0.35 mol) of 30 per cent hydrogen peroxide in the dropping funnel. Add 40 ml (0.12 mol) of 3 *M*-aqueous sodium hydroxide in one portion down the condenser to the reaction mixture and then add the hydrogen peroxide dropwise keeping the temperature at 30–50 °C by cooling in a cold-water bath; this addition should take about 25 minutes. Stir the reaction mixture at room temperature for a further 1 hour to ensure oxidation is complete and then pour it on to 250 ml of ice-water in a separatory funnel. Rinse the reaction vessel with 50 ml of water and add to the contents of the funnel. Extract the aqueous mixture with two 200 ml portions of ether (some insoluble inorganic material may separate in the aqueous layer at this stage) and then wash the combined ethereal extracts with eight 50 ml portions of water to remove diglyme (2). Dry the ether solution over magnesium sulphate, filter and remove the solvent by flash distillation or on the rotary evaporator. Transfer the residual colourless liquid to a 100-ml round-bottomed flask and fractionally distil through a well-lagged 14 cm column filled with glass helices, collecting the fraction, b.p. 154–157 °C. The product is hexan-1-ol containing approximately 6 per cent of hexan-2-ol (3); the yield is 24.7 g (81%). A complete separation of the isomers may be effected by using a more efficient fractionating column such as a spinning band column; hexan-1-ol and hexan-2-ol boil at 155–156 °C and 137–138 °C respectively at 760 mmHg.

Notes. (1) The apparatus should be thoroughly dried in an oven and assembled under a stream of dry nitrogen and allowed to cool.

(2) Diglyme, which has b.p. 162 °C at 760 mmHg, must be completely removed from the ether extract—otherwise it will contaminate the product. Its presence in the extract may be conveniently checked by GLC, using a 5-ft column of 10 per cent Silicone oil on Chromosorb W held at 100 °C, with a nitrogen flow rate of 40 ml/minute, t_R 3.2 minutes.

(3) GLC analysis under the conditions specified in Note (2) gives hexan-2-ol and hexan-1-ol with t_R 1.4 minutes and t_R 2 minutes respectively.

Cognate preparation Octan-1-ol. Use 1.70 g (0.045 mol, 20% excess over the theoretical requirement of 0.0375 mol) of sodium borohydride, 45 ml dry diglyme and a solution of 16.8 g (0.15 mol) of oct-1-ene in 25 ml dry diglyme in a 250-ml three-necked round-bottomed flask (1). In the pressure-equalising dropping funnel place 8.5 g (7.55 ml, 0.06 mol, 20% excess over the theoretical requirement of 0.05 mol) of purified boron trifluoride–etherate. Add the boron trifluoride in portions over 30 minutes, stir at room temperature for 30 minutes and finally destroy excess borohydride by dropwise addition of 10 ml of water during 15 minutes. To oxidise the organoborane use 20 ml of 3 M-aqueous sodium hydroxide and 20 ml of 30 per cent hydrogen peroxide added dropwise over 30 minutes, with a temperature of 30–50 °C. After stirring at room temperature for 1 hour, pour the reaction mixture on to 70 ml of ice-water in a separating funnel and rinse the reaction vessel with 30 ml water and add to the funnel. Extract the mixture with two 100 ml portions of ether and wash the ether solution with six 50 ml portions of water. Dry over magnesium sulphate, remove the ether on the rotary evaporator and fractionally distil the residue from a 25-ml pear-shaped flask with a fractionating side arm (*ca.* 8 cm) filled with glass helices. Collect the fraction distilling at 191–193 °C. A product, 15.8 g (81%), is obtained which consists of octan-1-ol contaminated with about 7 per cent of octan-2-ol (2). These may be separated, if required, by using a more efficient'fractionating column. At 760 mmHg octan-1-ol has b.p. 194–195 °C and octan-2-ol has b.p. 179 °C.

Notes. (1) Alternatively the apparatus may be assembled with the funnel in the centre neck of the flask and the thermometer and condenser in the side-necks and stirred using a magnetic stirrer. However, the follower bar must be substantial as the reaction mixture becomes viscous towards the end of the experiment.

(2) The purity of the product may be determined by GLC using a 5-ft column of 10 per cent Silicone oil on Chromosorb W held at 128 °C with a nitrogen flow rate of 40 ml/minute. Octan-1-ol has t_R 3.3 minutes and octan-2-ol has t_R 2.4 minutes.

D,4. The oxymercuration–demercuration of alkenes. A mild and highly convenient procedure for the hydration of a carbon–carbon double bond involves the initial reaction of an alkene with mercury(II) acetate in aqueous tetrahydrofuran, the resulting mercurial intermediate is reduced *in situ* by alkaline sodium borohydride solution. The yields of alcohols which are obtained with a wide variety of alkenes are usually excellent and the reaction proceeds exclusively in the Markownikoff sense, e.g., in the preparation of hexan-2-ol from hex-1-ene none of the isomeric hexan-1-ol is detectable by GLC analysis of the product.

$$R\cdot CH = CH_2 \quad Hg - O\cdot CO\cdot CH_3 \qquad R\cdot \overset{\oplus}{C}H\cdot CH_2\cdot Hg\cdot O\cdot CO\cdot CH_3 \longrightarrow$$

$$O\cdot CO\cdot CH_3 \qquad \qquad \underset{\ominus}{:}O\cdot CO\cdot CH_3$$

$$R\cdot CH\cdot CH_2\cdot Hg\cdot O\cdot CO\cdot CH_3 \xrightarrow[\text{NaBH}_4]{\ominus OH} R\cdot CH(OH)\cdot CH_3 + Hg + 2CH_3\cdot CO_2^{\ominus}$$

$$O\cdot CO\cdot CH_3$$

Complicating side reactions may occasionally occur—as in the oxymercuration–demercuration of styrene to 1-phenylethanol for which experimental details are also given. In this case evidently some organomercurial compounds survive the reductive stage, and their subsequent decomposition during final distillation complicates the isolation of the pure product.

III,44. HEXAN-2-OL

$$CH_3 \cdot (CH_2)_3 \cdot CH = CH_2 + Hg(O \cdot CO \cdot CH_3)_2 \longrightarrow$$

$$CH_3 \cdot (CH_2)_3 \cdot CH(OH) \cdot CH_2 \cdot Hg \cdot O \cdot CO \cdot CH_3 \xrightarrow[\text{NaBH}_4]{\text{[H]}}$$

$$CH_3 \cdot (CH_2)_3 \cdot CH(OH) \cdot CH_3 + Hg + CH_3 \cdot CO_2H$$

Place 31.9 g (0.1 mol) of mercury(II) acetate and 100 ml of water in a 1-litre three-necked flask fitted with an efficient mechanical stirrer, a dropping funnel and a thermometer. Stir until the acetate has dissolved and then run in rapidly 100 ml of tetrahydrofuran; an orange-yellow suspension forms almost immediately. After stirring for a further 15 minutes, add 8.4 g (12.5 ml, 0.1 mol) of hex-1-ene, whereupon the colour is rapidly discharged. Stir the mixture at room temperature for 1 hour to ensure completion of the oxymercuration step. Next add with vigorous stirring 100 ml of 3 *M*-sodium hydroxide solution, followed by a solution of 1.9 g (0.05 mol) of sodium borohydride in 100 ml of 3 *M*-sodium hydroxide. Control the rate of addition of both solutions so that the temperature of the reaction mixture remains at about 25 °C, cooling the flask in cold water from time to time if necessary. Reduction occurs readily with the separation of elemental mercury. Finally stir vigorously at ambient temperature for 3 hours and then allow the reaction mixture to remain overnight in a separating funnel supported over a large empty conical flask. Separate the mercury layer (19.25 g, 96%) and then the aqueous alkaline phase, retaining the organic layer. Saturate the aqueous phase with sodium chloride, remove the additional organic layer which separates and extract the aqueous phase with two 30 ml portions of ether. Combine both of the organic layers with the ether extracts and remove most of the organic solvent carefully under reduced pressure using a rotary evaporator; stop the evaporation when two phases begin to separate. Add 50 ml of ether and 20 ml of water, separate the ether layer and wash it with four 25 ml portions of water, and dry it over anhydrous calcium sulphate. Remove the ether by flash distillation and distil the residue, collecting the hexan-2-ol at 136–140 °C; the yield is 6.9 g (68%).

Check the purity by GLC on a 10 per cent Silicone oil column at 100 °C, nitrogen flow rate 40 ml per minute. The retention time is 1.42 minutes (cf. hexan-1-ol, 1.96 minutes).

Cognate preparation 1-Phenylethanol. Use 10.4 g (11.5 ml, 0.1 mol) of styrene, and carry out the oxymercuration and reduction as described above. The yield of recovered mercury is 17.5 g (87%), and traces continue to separate during the work-up procedure. Distil the final crude product under reduced pressure and collect the 1-phenylethanol at 110–115 °C/25 mmHg. Towards the end of the distillation the decomposition of residual organomercurial compounds ensues, and co-distillation of mercury contaminates the product; collect the contaminated fraction separately. The first fraction, yield 6.2 g (51%), is 92 per cent pure by GLC (retention time 5.33 minutes); the impurity is mainly styrene (t_R 2.16 minutes). The mercury-contaminated fraction (3.0 g, 25%) is 85 per cent pure by GLC.

D,5. The hydroxylation of alkenes. Hydroxylation of an alkene may be carried out using osmium tetroxide in an inert solvent (e.g., ether or dioxan), whereupon a cyclic osmate ester is formed. This undergoes hydrolytic cleavage under reducing conditions (e.g., aqueous sodium sulphite) to give the 1,2-diol resulting from a *cis*-hydroxylation process.

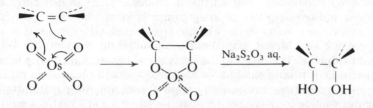

Alternatively and more conveniently this *cis*-hydroxylation process can be effected using only catalytic amounts of osmium tetroxide together with hydrogen peroxide, which cleaves the first formed osmate ester to the diol and regenerates the osmium tetroxide. The reaction is carried out in t-butyl alcohol and is illustrated by the conversion of cyclohexene into *cis*-cyclohexane-1,2-diol (Section **III,45**).

A 1,2-diol arising from a *trans*-hydroxylation process is formed from an alkene by way of an intermediate epoxide which is subjected to a ring-opening reaction and hydrolysis. The epoxides may be isolated when the alkene is reacted with perbenzoic acid (Section **II,2,**46(*d*)) and in a solvent such as chloroform. With performic acid (Section **II,2,**46(*a*)) isolation of the intermediate epoxide is not possible since it is converted by the formic acid solvent into the hydroxyformate ester. This product, which is formed by the nucleophilic ring opening of the epoxide, is hydrolysed by treatment with dilute alkali to give the 1,2-diol. The formation of *trans*-cyclohexane-1,2-diol (Section **III,46**) is illustrative of this reaction.

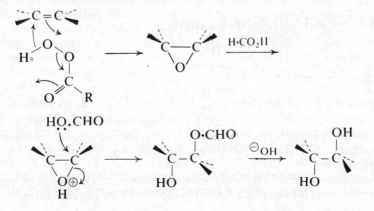

III,45. *CIS*-CYCLOHEXANE-1,2-DIOL

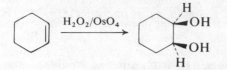

This preparation involving the toxic osmium tetroxide must be carried out in a fume cupboard. For the precautions to be followed when using 30 per cent hydrogen peroxide see Section II,2,34.

Prepare the *reagent* as follows. To a mixture of 100 ml of pure t-butyl alcohol (2-methylpropan-2-ol) and 25 ml of 30 per cent hydrogen peroxide, add anhydrous sodium sulphate or magnesium sulphate in small portions; two layers separate out. Remove the alcohol layer which contains most of the hydrogen peroxide, and dry it with magnesium sulphate, followed by anhydrous calcium sulphate. The resulting liquid is a stable solution of 6.3 per cent hydrogen peroxide in t-butyl alcohol.

Free cyclohexene from peroxides by treating it with a saturated solution of sodium metabisbisulphite, separate, dry and distil; collect the fraction having b.p. 81–83 °C. Mix 8.2 g (0.1 mol) of cyclohexene with 55 ml of the reagent (0.1 mol), add 3 ml of a 0.5 per cent solution of osmium tetroxide [**CAUTION:** (1)] in anhydrous t-butyl alcohol and cool the mixture to 0 °C. Allow to stand overnight, by which time the initial orange colouration will have disappeared. Remove the solvent and unreacted cyclohexene by distillation at atmospheric pressure and fractionate the residue under reduced pressure using an air condenser. Collect the fraction of b.p. 120–140 °C/15 mmHg; this solidifies almost immediately. Recrystallise from ethyl acetate. The yield of pure *cis*-cyclohexane-1,2-diol, m.p. 96 °C, is 5.0 g (45%).

Note. (1) Osmium tetroxide is extremely irritating and toxic and constitutes a severe eye injury hazard. It may be purchased in sealed ampoules, e.g., 100 mg; the solution in t-butyl alcohol must be prepared and dispensed in an efficient fume cupboard, with the added protection of gloves and goggles. This solution is reasonably stable (e.g., the decomposition after one month is about 20%), provided that no 2-methylprop-1-ene arising from the t-butyl alcohol is present as impurity. In the latter case formation of black colloidal osmium, which can catalyse the decomposition of hydrogen peroxide, is rapid.

III,46. *TRANS*-CYCLOHEXANE-1,2-DIOL

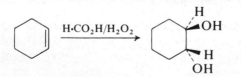

All preparations and reactions with hydrogen peroxide and organic peroxyacids must be conducted behind a safety screen, because these reactions sometimes proceed with violence.

In a 500-ml three-necked flask, equipped with a mechanical stirrer, a thermometer and a dropping funnel, place 300 ml (6 mol) of 88–90 per cent formic acid and add 70 ml (0.62 mol) of 30 per cent hydrogen peroxide (see Section II,2,*34*). Then introduce slowly 41 g (51 ml, 0.5 mol) of freshly distilled cyclohexene (Section III,7) over a period of 20–30 minutes; maintain the temperature of the reaction mixture between 40 and 45 °C by cooling with an ice bath and controlling the rate of addition. Keep the reaction mixture at 40 °C for 1 hour after all the cyclohexene has been added and then allow to stand overnight at room temperature. Remove most of the formic acid and water by distillation from a water bath under reduced pressure (rotary evaporator). Add an ice-cold solution of 40 g (1 mol) of sodium hydroxide in 75 ml of water in small portions to the residual mixture of the diol and its formate: take care that the temperature does not rise above 45 °C. Warm the alkaline solution to 45 °C and add an equal volume (*ca.* 200 ml) of ethyl acetate. Extract thoroughly, separate the lower layer and extract at 45 °C six times with equal volumes of ethyl acetate. Combine the ethyl acetate extracts (total volume about 1 l), distil off the solvent using a rotary evaporator until the resuidual volume is about 150 ml and solid commences to crystallise. Cool to 0 °C and separate the crude product (*ca.* 4.5 g) by suction filtration. Concentrate the mother-liquor to 30–40 ml, when more solid crystallises (*ca.* 8 g). Cool and filter the mixture as before. Distil the combined crude products under reduced pressure from an oil bath and using an air condenser (see Section I,27); the pure *trans*-cyclohexane-1,2-diol passes over at 128–132 °C/15 mmHg (or at 120–124 °C/4 mmHg) and solidifies immediately, m.p. 102–103 °C. The yield is 40 g (69%). It may be recrystallised from acetone or from ethyl acetate.

E ALKYL HALIDES

The synthesis of alkyl halides is exemplified by the following typical procedures.

1. The displacement of the hydroxyl group in an alcohol by halogen [(a) chlorides, Sections III,47 to III,52, (b) bromides, Sections III,53 to III,55, (c) iodides, Sections III,56 to III,59].
2. Halogen exchange reactions (Sections III,60 and III,61).
3. The addition of hydrogen halides or halogens to alkenes (Sections III,62 to III,64).
4. The replacement of reactive allylic hydrogen atoms by bromine (Sections III,65 and III,66).

E,1(a). **Preparation of alkyl chlorides from alcohols.** The hydroxyl group in tertiary alcohols is most readily replaced, and this is effected by simply allowing the alcohol to react with concentrated hydrochloric acid at room temperature. The reaction is a nucleophilic displacement of the S_N1 type involving the formation of a relatively stable carbonium ion intermediate.

$$R_3COH + H^{\oplus} \rightleftharpoons R_3C\overset{\oplus}{O}H_2 \rightleftharpoons R_3C^{\oplus} + H_2O$$

$$R_3\overset{\oplus}{C} + \overset{\ominus}{Cl} \longrightarrow R_3CCl$$

The reaction is illustrated by the preparation of t-butyl chloride (Section **III,47**).

Secondary, and to a greater extent primary, alcohols require more vigorous conditions to effect the substitution reaction, which is usually achieved by heating the alcohol–acid mixture with anhydrous zinc chloride. Illustrative examples are given in Section **III,48**.

In the case of alicyclic secondary alcohols, anhydrous calcium chloride is recommended as an alternative (e.g., the preparation of chlorocyclohexane, Section **III,49**).

The hydrochloric acid–zinc chloride reaction may be an S_N2 type displacement, particularly in the case of primary alcohols.

$$ROH + HCl + ZnCl_2 \longrightarrow Cl_2\overset{\ominus}{Zn}\!-\!Cl \;\; \nwarrow R\!-\!\overset{\oplus}{O}H_2 \longrightarrow$$

$$ZnCl_2 + RCl + H_2O$$

An S_N1 mechanism is also possible, however, i.e.,

$$ROH + ZnCl_2 \longrightarrow R\!-\!\overset{\oplus}{\underset{\underset{H}{|}}{O}}\!-\!\overset{\ominus}{Zn}Cl_2 \longrightarrow \overset{\oplus}{R} + HO\!-\!\overset{\ominus}{Zn}Cl_2 \longrightarrow$$

$$RCl + HOZnCl$$

This latter reaction pathway introduces a tendency for rearrangement of the alkyl group to occur, particularly when branching of the carbon chain occurs in the β-position.

Rearrangement may be avoided by preparing the chloride by reaction of the alcohol with thionyl chloride alone or in the presence of pyridine, which may be present either in catalytic amounts or in an equimolar proportion. A chlorosulphite ester is first formed which in the absence of pyridine decomposes to the alkyl chloride by a cyclic (S_Ni) mechanistic pathway.

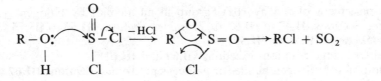

In the presence of pyridine, however, the chloride ion liberated in the first reaction step (chlorosulphite formation) effects an S_N2 displacement.

$$HCl + C_5H_5N \longrightarrow C_5H_5\overset{\oplus}{N}H + \overset{\ominus}{Cl}$$

$$\overset{\ominus}{Cl} \quad R\!-\!\overset{}{O}\!-\!S\!-\!Cl \longrightarrow RCl + SO_2 + \overset{\ominus}{Cl}$$
$$\underset{O}{\overset{\|}{}}$$

Examples of both reaction processes are to be found in Section **III,50** and **III,51**.

The unsaturated alcohol, allyl alcohol, gives a poor yield by the HCl–ZnCl₂ method. However, an alternative procedure using copper(I) chloride as catalyst gives an excellent yield (Section **III,52**).

III,47. t-BUTYL CHLORIDE (2-*Chloro-2-methylpropane*)

$$(CH_3)_3COH + HCl \longrightarrow (CH_3)_3CCl + H_2O$$

In a 250-ml separatory funnel place 25 g (0.34 mol) of 2-methylpropan-2-ol (t-butyl alcohol, b.p. 82–83 °C, m.p. 25 °C) and 85 ml of concentrated hydrochloric acid (1) and shake the mixture from time to time during 20 minutes. After each shaking, loosen the stopper to relieve any internal pressure. Allow the mixture to stand for a few minutes until the layers have separated sharply; draw off and discard the lower acid layer. Wash the halide with 20 ml of 5 per cent sodium hydrogen carbonate solution and then with 20 ml of water. Dry the preparation with 5 g of anhydrous calcium chloride or anhydrous calcium sulphate. Decant the dried liquid through a funnel supporting a fluted filter-paper into a 100-ml distilling flask, add 2–3 chips of porous porcelain and distil. Collect the fraction boiling at 49–51 °C. The yield of t-butyl chloride is 28 g (90%).

Note. (1) The addition of 10 g of anhydrous calcium chloride tends to concentrate the acid and assists the separation of the chloride; the yield is slightly improved.

Cognate preparation   2-Chloro-2-methylbutane. Use 22 g (27 ml, 0.25 mol) of 2-methylbutan-2-ol (t-pentyl alcohol) and 65 ml of concentrated hydrochloric acid. Distil the chloride twice from a Claisen flask with fractionating side arm or through a short column. Collect the 2-chloro-2-methylbutane at 83–85 °C; the yield is 18 g (68%).

III,48. BUTYL CHLORIDE (1-*Chlorobutane*)

$$CH_3 \cdot (CH_2)_2 \cdot CH_2OH + HCl \xrightarrow{ZnCl_2} CH_3 \cdot (CH_2)_2 \cdot CH_2Cl + H_2O$$

Fit a 250-ml round-bottomed flask with a reflux condenser, the top of which is connected to a device for absorbing hydrogen chloride (Fig. I,68(*a*)). Place 68 g (0.5 mol) of anhydrous zinc chloride and 40 ml (47.5 g) of concentrated hydrochloric acid in the flask, add 18.5 g (23 ml, 0.25 mol) of butan-1-ol and reflux the mixture gently for 2 hours. Arrange the condenser for downward distillation, and distil the reaction product, collecting the material which boils below 115 °C. Separate the upper layer of the distillate, mix it with an equal volume of concentrated sulphuric acid (1) and transfer the mixture to a 250-ml flask fitted with a reflux condenser. Reflux gently for 15–30 minutes, and then distil the chloride from the acid; it will pass over at 76–79 °C. Wash the distillate successively with 25 ml of water, 10 ml of 5 per cent sodium hydroxide solution and 25 ml of water; dry over 1–2 g of anhydrous calcium chloride, filter and distil from a small distilling flask. Collect the butyl chloride at 75–78 °C. The yield is 15–16 g (65–69%).

Note. (1) The sulphuric acid treatment removes high-boiling impurities which are not easily separated by distillation.

Cognate preparations The following alkyl chlorides may be prepared in similar yield by replacing the butan-1-ol in the above preparation by the appropriate quantity of the requisite alcohol:

1-Chloropentane, b.p. 104–107 °C, from 22 g of pentan-1-ol;
1-Chloro-3-methylbutane, b.p. 98–100 °C, from 22 g of 3-methylbutan-1-ol (isopentyl alcohol);
2-Chlorobutane, b.p. 67–69 °C, from 18.5 g of butan-2-ol;
2-Chloropentane, b.p. 96–98 °C, from 22 g of pentan-2-ol;
3-Chloropentane, b.p. 95–97 °C, from 22 g of pentan-3-ol.

III,49. CHLOROCYCLOHEXANE (*Cyclohexyl chloride*)

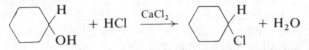

In a 1500-ml round-bottomed flask, carrying a reflux condenser, place 100 g (1 mol) of pure cyclohexanol, 250 ml of concentrated hydrochloric acid and 80 g of anhydrous calcium chloride: heat the mixture on a boiling water bath for 10 hours with occasional shaking (1). Some hydrogen chloride is evolved, consequently the preparation should be conducted in the fume cupboard. Separate the upper layer from the cold reaction product, wash it successively with saturated salt solution, saturated sodium hydrogen carbonate solution, saturated salt solution, and leave the crude chlorocyclohexane over an excess of anhydrous calcium chloride for at least 24 hours. Distil from a 150-ml flask through a fractionating side arm, and collect the pure product at 141.5–142.5 °C. The yield is 90 g (76%).

Note. (1) The refluxing period may be reduced to 6 hours and the yield improved slightly by mechanical stirring; a three-necked flask should be used.
An alternative method of conducting the preparation consists in treating 100 g of cyclohexanol with 250 ml of concentrated hydrochloric acid, refluxing slowly whilst a stream of hydrogen chloride gas is passed into the mechanically stirred mixture for 3 hours. (The apparatus required is similar to that described for a Clemmensen reduction in Section **III,2**.) Chlorocyclohexane, b.p. 141–143 °C, is isolated as above; the yield is 80 g.

Cognate preparation **Chlorocyclopentane** (*cyclopentyl chloride*). Use 43 g (0.5 mol) of cyclopentanol (Section **III,25**), 125 ml of concentrated hydrochloric acid and 50 g of anhydrous calcium chloride. The yield of chlorocyclopentane, b.p. 113–115 °C, is 30 g (57%).

III,50. 1-CHLOROHEXANE

$$CH_3 \cdot (CH_2)_4 \cdot CH_2OH + SOCl_2 \longrightarrow CH_3 \cdot (CH_2)_4 \cdot CH_2Cl + HCl + SO_2$$

Carry out this preparation in a fume cupboard using the apparatus described in the following section. Place 179 g (109.5 ml, 1.5 mol) of redistilled thionyl chloride in the flask and 51 g (62.5 ml, 0.5 mol) of hexan-1-ol, b.p. 156–158 °C, in the separatory funnel. Add the alcohol with stirring during 2 hours; there is a slight evolution of heat, sulphur dioxide is evolved and the liquid darkens considerably. When all the alcohol has been added, reflux the mixture for 2 hours. Rearrange the apparatus for distillation, and distil slowly; the excess of thionyl chloride passes over below 80 °C, followed by a small fraction up to 120 °C; and finally the crude 1-chlorohexane at 132–134 °C. Wash the last-named successively with water, 10 per cent sodium carbonate solution, and twice with water. Dry with anhydrous calcium chloride and distil through a short fractionating column. Pure 1-chlorohexane passes over at 133–134 °C. The yield is 36 g (60%).

Cognate preparations **1-Chloroheptane.** From 58 g (70.5 ml, 0.5 mol) of heptan-1-ol (b.p. 175–177 °C) (Section **III,24**) and 179 g (109.5 ml, 1.5 mol) of redistilled thionyl chloride; refluxing period, 4 hours. The yield of 1-chlorohep-tane, b.p. 159–160 °C, is 52 g (77%).

1-Chlorododecane. From 46.5 g (0.25 mol) of dodecan-1-ol (lauryl alcohol), m.p. 24 °C, and 119 g (73 ml, 1 mol) of redistilled thionyl chloride; refluxing period, 6 hours. The crude chloride passes over at 252–257 °C, mainly at 255–257 °C. Upon redistillation under reduced pressure 35 g (68%) of 1-chlorododecane, b.p. 116.5 °C/5 mmHg, are obtained.

1,4-Dichlorobutane. Place 22.5 g (0.25 mol) of redistilled butane-1,4-diol and 3 ml of dry pyridine in the flask in an ice bath. Add 119 g (73 ml, 1 mol) of redistilled thionyl chloride dropwise to the vigorously stirred mixture at such a rate that the temperature remains at 5–10 °C. When the addition is complete, remove the ice bath, keep the mixture overnight and then reflux for 3 hours. Cool, add ice-water cautiously and extract with ether. Wash the ethereal extract successively with 10 per cent sodium hydrogen carbonate solution and water, and dry with magnesium sulphate. Remove the ether by flash distillation and distil the residue under reduced pressure. Collect the 1,4-dichloro-butane at 55.5–56.5 °C/14 mmHg; the yield is 18 g (58%). The b.p. under atmospheric pressure is 154–155 °C.

III,51. ISOBUTYL CHLORIDE (*1-Chloro-2-methylpropane*)

$$(CH_3)_2CH\cdot CH_2OH + SOCl_2 \xrightarrow{C_5H_5N} (CH_3)_2CH\cdot CH_2Cl + SO_2$$

Assemble in a fume cupboard a 500-ml three-necked flask equipped with a sealed stirrer unit, a double surface reflux condenser and a separatory funnel; fit the condenser and the funnel with calcium chloride guard-tubes. Place 37 g (46 ml, 0.5 mol) of 2-methylpropan-1-ol (b.p. 106–108 °C) and 40 g (41 ml, 0.5 mol) of pure pyridine in the flask, and 119 g (73 ml, 1.0 mol) of redistilled thionyl chloride in the funnel. Introduce the thionyl chloride with stirring during 3–4 hours; a white solid separates, which partially dissolves as the reaction proceeds. Reflux for 45 minutes: the solid will dissolve completely. Allow to cool and remove the upper layer (1). Wash the latter cautiously with water, 5 per cent sodium hydroxide solution, and twice with water; dry with anhydrous calcium chloride. Distil the product through a short fractionating column and collect the isobutyl chloride at 68–69 °C. The yield is 26 g (56%).

Note. (1) The lower pyridine layer contains most of the excess of thionyl chloride; it may be recovered by distillation through an efficient fractionating column.

III,52. ALLYL CHLORIDE (*1-Chloroprop-2-ene*)

$$CH_2 = CH \cdot CH_2OH + HCl \xrightarrow[\text{CuCl}]{H_2SO_4} CH_2 = CH \cdot CH_2Cl + H_2O$$

Place 87 g (100 ml, 1.5 mol) of allyl alcohol, 150 ml of concentrated hydrochloric acid and 2 g of freshly prepared copper(I) chloride (Section **II,2,***16*) in a 750-ml round-bottomed flask equipped with a reflux condenser. Cool the flask in ice and add 50 ml of concentrated sulphuric acid dropwise through the condenser with frequent shaking of the flask. A little hydrogen chloride may be evolved towards the end of the reaction. Allow the turbid liquid to stand for 30 minutes in order to complete the separation of the allyl chloride. Remove the upper layer, wash it with twice its volume of water, and dry over anhydrous calcium chloride. Distil and collect the allyl chloride which passes over at 46–47 °C; the yield is about 100 g (87%).

E,1(b). Preparation of alkyl bromides from alcohols. The formation of alkyl bromides is more ready than that of alkyl chlorides, and hence secondary as well as tertiary bromides can be obtained directly from the corresponding alcohols by heating with constant boiling point hydrobromic acid (e.g., Section **III,53**).

$$ROH + HBr \longrightarrow RBr + H_2O$$

Especially in the case of primary alcohols the presence of sulphuric acid results, as a rule, in a more rapid reaction and in improved yields. The method is readily adapted for the preparation of dibromides from diols. Typical examples are provided in Section **III,54**. The cyclic ethers tetrahydrofuran and tetrahydropyran are readily cleaved by the hydrobromic acid–sulphuric acid medium, and this provides an alternative and convenient preparation of the corresponding α,ω-dihalides.

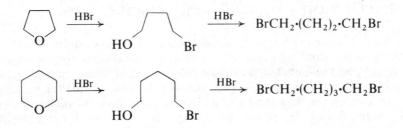

Alkyl bromides may also be readily obtained by the addition of liquid bromine to a warm suspension of purified red phosphorus in the appropriate alcohol.

$$10R \cdot CH_2OH + 2P + 5Br_2 \longrightarrow 10R \cdot CH_2Br + 2H_3PO_4 + 2H_2O$$

The reaction is of general application: with primary alcohols (isobutyl alcohol to hexadecan-1-ol) the yields are over 90 per cent of the theoretical, but with secondary alcohols the yields are in the range 50–80 per cent.

III,53. ISOPROPYL BROMIDE (*2-Bromopropane*)

$$CH_3 \cdot CH(OH) \cdot CH_3 + HBr \longrightarrow CH_3 \cdot CHBr \cdot CH_3 + H_2O$$

Mix 40 g (51 ml, 0.67 mol) of propan-2-ol (isopropyl alcohol) with 460 g (310 ml) of constant boiling point hydrobromic acid in a 500-ml flask fitted with a double surface condenser, add a few boiling chips and distil slowly (1–2 drops per second) until about half of the liquid has passed over. Separate the lower alkyl bromide layer (70 g), and redistil the aqueous layer when a further 7 g of the crude bromide will be obtained (1). Shake the crude bromide in a separatory funnel successively with an equal volume of concentrated hydrochloric acid (2), water, 5 per cent sodium hydrogen carbonate solution and water, and dry with anhydrous calcium chloride. Distil from a 100-ml flask; the isopropyl bromide passes over at 59 °C. The yield is 66 g (81%).

Notes. (1) The residue in the flask may be mixed with the aqueous layer of the first distillate, 40 g of propan-2-ol added, and the slow distillation repeated. The yield of crude isopropyl bromide in the second distillation is only slightly less than that obtained in the original preparation. Subsequently most of the residual hydrobromic acid may be recovered by distillation as the constant boiling point acid (126 °C).

(2) The hydrochloric acid washing removes any unchanged alcohol which may be present.

Cognate preparations   Bromocyclohexane.   Use 50 g (0.5 mol) of cyclohexanol and 260 g (176 ml) of 48 per cent hydrobromic acid and distil *all* the mixture slowly (6 hours). Add a little water to the distillate, separate the lower layer of crude bromide, and purify as above. Collect the bromocyclohexane at 163–165 °C (60 g, 74%).

Bromocyclopentane.   Use 43 g (0.5 mol) of cyclopentanol (Section **III,25**) and 260 g (176 ml) of 48 per cent hydrobromic acid. Collect the bromocyclopentane at 135–137 °C (55 g, 74%).

III,54. BUTYL BROMIDE (*1-Bromobutane*)

$$CH_3 \cdot (CH_2)_2 \cdot CH_2OH + HBr \xrightarrow{H_2SO_4} CH_3 \cdot (CH_2)_2 \cdot CH_2Br + H_2O$$

To 250 g of 48 per cent hydrobromic acid contained in a 500-ml round-bottomed flask add 75 g (41 ml) of concentrated sulphuric acid in portions with shaking; some hydrogen bromide may be evolved. Add 88 g (110 ml, 1.2 mol) of butan-1-ol, followed by 60 g (32.5 ml) of concentrated sulphuric acid in several portions with shaking, and finally a few chips of porous porcelain. Attach a reflux condenser to the flask and reflux the mixture gently on a wire gauze for 2–3 hours; during this period the formation of butyl bromide is almost complete and a layer separates above the acid (1). If the preparation is carried out in the open laboratory, fit an absorption device (Fig. I,68(*a* or *b*)) to the top of the condenser in order to absorb any hydrogen bromide and sulphur dioxide which may be evolved. Allow the contents of the flask to cool, remove the condenser and set it for downward distillation. Distil the mixture until no more oily drops of butyl bromide pass over (30–40 minutes). Transfer the distillate to a separatory funnel and remove the halide which forms the lower layer. Wash it successively with water, an equal volume of concentrated hydrochloric acid (2), water, 5 per cent sodium hydrogen carbonate or sodium carbonate solution, and water. Separate the water as completely as possible and dry with 2–3 g of anhydrous calcium chloride or magnesium sulphate; the desiccant should be left in contact with the bromide for at least 30 minutes and shaken occasionally. Filter the dried product through a small funnel supporting a fluted filter-paper into a 200-ml flask, add a few chips of porous porcelain and distil either from an air bath (Fig. I,49(a)) or on an asbestos-centred wire gauze. Collect the portion boiling at 100–103 °C. The yield is 155 g (95%).

Notes. (1) A suitable hydrogen bromide medium can be prepared by dissolving 240 g of potassium bromide in 400 ml of warm water, cooling and adding 200 ml of concentrated sulphuric acid slowly and with constant stirring, so that the temperature does not rise above 40 °C. After further cooling to 15 °C, the mixture is filtered and the butan-1-ol is added to the filtrate. A further 120 ml of concentrated sulphuric acid is then added carefully and the mixture is heated under reflux for 3–4 hours.

(2) The crude bromide contains a little unchanged alcohol and is said to contain some dibutyl ether (b.p. 141 °C). The former is removed by washing with concentrated hydrochloric acid and this purification process is satisfactory for most purposes. Both the alcohol and the ether are removed by washing with 11–12 ml of concentrated sulphuric acid; the butyl bromide is not affected by this reagent.

Cognate preparations   s-Butyl bromide (2-bromobutane). The quantities required are as for butyl bromide but with butan-2-ol (b.p. 99–100 °C) replacing the butan-1-ol. Two to three washings with concentrated hydrochloric acid are necessary, i.e., until the volume of the acid layer remains unchanged on shaking with halide. The yield of s-butyl bromide, b.p. 90.5–92.5 °C, is 150 g (92%).

Ethyl bromide (bromoethane). Place 415 g (281 ml) of 48 per cent hydrobromic acid in the flask and add 120 g (65 ml) of concentrated sulphuric acid in portions, with shaking. When cold, add 100 g (145 ml, 2.06 mol) of rectified spirit (95% ethanol) and attach the reflux condenser. Introduce 200 g (109 ml) of concentrated sulphuric acid slowly from the tap funnel down the condenser. Reassemble the apparatus for downward distillation using a modified receiver adapter which will extend below the surface of some water contained in a 500-ml receiver flask; cool the latter with ice. Distil the reaction mixture slowly, and separate the resulting layer of ethyl bromide. Wash the crude ethyl bromide with an equal volume of concentrated hydrochloric acid, then with water, a little 5 per cent sodium hydrogen carbonate solution, and finally with water. Dry with anhydrous calcium chloride. Distil the dry bromide, to which a few chips of porous porcelain have been added, from a water bath and collect the ethyl bromide, b.p. 38–39 °C, in a receiver cooled in ice. The yield is 205 g (91%).

Propyl bromide (1-bromopropane). Use the procedure described for *ethyl bromide*, substituting the following quantities of reagents: 500 g (338 ml) of 48 per cent hydrobromic acid and 150 g (82 ml) of concentrated sulphuric acid; 144 g (179 ml, 2.4 mol) propan-1-ol (b.p. 96.5–97.5 °C). Introduce 120 g (65 ml) of concentrated sulphuric acid slowly down the condenser and then distil slowly until no more oily drops pass over. The yield of propyl bromide, b.p. 70–72 °C, is 255 g (86%).

1-Bromopentane (*pentyl bromide*). Use 210 g (142 ml) of 48 per cent hydrobromic acid, 60 g (33 ml) of concentrated sulphuric acid, followed by 88 g (108 ml, 1 mol) pentan-1-ol (b.p. 135–136 °C) and 10 g (5.5 ml) of concentrated sulphuric acid. Distil the product through a short fractionating column, and collect the 1-bromopentane at 127–130 °C (135 g, 89%).

1-Bromo-3-methylbutane. Proceed as for *1-bromopentane*, but use 88 g (109 ml, 1 mol) of 3-methylbutan-1-ol, b.p. 129.5–131 °C. Distil the purified product through a fractionating column and collect the 1-bromo-3-methyl-butane at 117–120 °C (125 g, 83%).

2-Bromopentane. Proceed as for *1-bromopentane*, but use 88 g (108 ml, 1 mol) of pentan-2-ol, b.p. 118.5 °C. During the washing with concentrated hydrochloric acid, difficulty may be experienced in separating the acid layer; this is overcome by adding a little water to decrease the density of the acid. Distil the purified product through a fractionating column; some alkene passes over first, followed by the 2-bromopentane at 115–118 °C (120 g, 79%).

3-Bromopentane. Proceed as for *1-bromopentane*, but use 88 g (108 ml, 1 mol) of pentan-3-ol, b.p. 115.5–116 °C. The experimental observations are similar to those given for *2-bromopentane*. Collect the 3-bromopentane at 116–119 °C (120 g, 79%).

1,3-Dibromopropane (*trimethylene dibromide*). In a 1-litre round-bottomed flask place 500 g (338 ml) of 48 per cent hydrobromic acid and add 150 g (82 ml) of concentrated sulphuric acid in portions, with shaking. Then add 91 g of propane-1,3-diol (b.p. 210–215 °C), followed by 240 g (130.5 ml) of concentrated sulphuric acid slowly and with shaking. Attach a reflux condenser to the flask and reflux the mixture for 3–4 hours. Arrange for downward distillation and distil, using a wire gauze, until no more oily drops pass over (30–40 minutes). Purify the 1,3-dibromopropane as detailed for *butyl bromide* above. About 220 g (91%) of the pure dibromide, b.p. 162–165 °C, are obtained.

Allyl bromide. Introduce into a 1-litre three-necked flask 250 g (169 ml) of 48 per cent hydrobromic acid and then 75 g (40.5 ml) of concentrated sulphuric acid in portions, with shaking; finally add 58 g (68 ml, 1 mol) of pure allyl alcohol. Fit the flask with a separatory funnel, a mechanical stirrer and an efficient condenser (preferably of the double surface type) set for downward distillation. Place 75 g (40.5 ml) of concentrated sulphuric acid in the separatory funnel, set the stirrer in motion and allow the acid to flow slowly into the warm solution. The allyl bromide will distil over (< 30 minutes). Wash the distillate with 5 per cent sodium carbonate solution, followed by water, dry over anhydrous calcium chloride, and distil from a flask through a short fractionating column. The yield of allyl bromide, b.p. 69–72 °C, is 112 g (93%). There is a small high-boiling fraction containing 1,2-dibromopropane.

1,4-Dibromobutane (*from butane-1,4-diol*). In a 500-ml three-necked flask fitted with a stirrer, reflux condenser and dropping funnel, place 154 g (105 ml) of 48 per cent hydrobromic acid. Cool the flask in an ice bath. Add slowly, with stirring, 130 g (71 ml) of concentrated sulphuric acid. To the resulting ice-cold solution add 30 g (0.33 mol) of redistilled butane-1,4-diol dropwise. Leave the reaction mixture to stand for 24 hours; heat for 3 hours on a steam bath. The reaction mixture separates into two layers. Separate the lower layer, wash it successively with water, 10 per cent sodium carbonate solution and water, and then dry with magnesium sulphate. Distil and collect the 1,4-dibromobutane at 83–84 °C/12 mmHg. The yield is 55 g (76%).

1,4-Dibromobutane (*from tetrahydrofuran*). Place a mixture of 250 g (170 ml) of 48 per cent hydrobromic acid and 75 g (41 ml) of concentrated sulphuric acid in a 500-ml round-bottomed flask, add 18.1 g (20.5 ml, 0.25 mol) of redistilled tetrahydrofuran (Section II,1,*19*) (b.p. 65–66 °C), attach a reflux condenser and reflux gently for 3 hours. Separate the lower layer of dibromide and purify as in the previous preparation. The yield of 1,4-dibromobutane, b.p. 83–84 °C/12 mmHg is 40 g (74%).

1,5-Dibromopentane (*from pentane-1,5-diol*). Proceed as for 1,4-dibromobutane but use 35 g (0.33 mol) of redistilled commercial pentane-1,5-diol. The yield of 1,5-dibromopentane, b.p. 99 °C/13 mmHg, is 39 g (51%).

1,5-Dibromopentane (*from tetrahydropyran*). Proceed as for 1,4-dibromobutane (from tetrahydrofuran) but use 21.5 g (24.4 ml, 0.25 mol) of redistilled tetrahydropyran (b.p. 86.5–87.5 °C). The yield of 1,5-dibromopentane, b.p. 99 °C/13 mmHg, is 46 g (80%).

III,55. ISOBUTYL BROMIDE (*1-Bromo-2-methylpropane*)

$$10(CH_3)_2CH\cdot CH_2OH + 2P + 5Br_2 \longrightarrow$$
$$10(CH_3)_2CH\cdot CH_2Br + 2H_3PO_4 + 2H_2O$$

Place 92.5 g (115 ml, 1.25 mol) of isobutyl alcohol (2-methylpropan-1-ol) and 8.55 g (0.275 mol) of purified red phosphorus (Section II,2,*49*) in a 500-ml three-necked flask fitted with a sealed mechanical stirrer, a reflux condenser and a dropping funnel containing 100 g (32 ml, 0.62 mol) of bromine (*for precautions in the use of bromine*, see Section II,2,*7*). Start the stirrer, heat the flask (e.g., in an oil bath) so that the contents reflux gently, and introduce the bromine at such a rate that it appears to react completely so that there is little bromine vapour above the surface of the reaction mixture, and the reaction is under control. When all the bromine has been added, reflux the mixture gently for 15–30 minutes more. Remove the stirrer, arrange the condenser for downward distillation and distil off most of the isobutyl bromide (1). Then add about 50 ml of water through the dropping funnel and continue the distillation to remove the remainder of the product. Separate the crude bromide and wash it successively with water, an approximately equal volume of concentrated hydrochloric acid, water, 10 per cent sodium carbonate solution, and finally water. Dry the product over anhydrous calcium chloride and distil, collecting the isobutyl bromide which passes over at 91–94 °C. The yield is 150 g (91%).

Note. (1) It is not advisable to distil the mixture almost to dryness since the formation of inflammable alkene may then occur. This is avoided by conducting the distillation in two stages as described.

Cognate preparations* **1-Bromohexane.** Use 152.5 g (186.5 ml, 1.49 mol) of hexan-1-ol, 9.3 g (0.3 mol) of purified red phosphorus and 120 g (38.5 ml, 0.95 mol) of bromine.† B.p. 154–156 °C.

---

* Unless otherwise stated, the yields exceed 90 per cent of the theoretical.

† The slight excess of bromine over the theoretical equivalent to the alcohol in the preparation of high boiling point bromides ensures the absence of unchanged alcohol in the product; any excess of bromine may be removed by the addition of a little sodium metabisulphite.

1-Bromoheptane. Use 173 g (209 ml, 1.49 mol) of heptan-1-ol, 9.3 g (0.30 mol) of purified red phosphorus and 120 g (38.5 ml, 0.95 mol) of bromine. B.p. 180 °C.

1-Bromooctane. Use 81 g (98.5 ml, 0.623 mol) of octan-1-ol (b.p. 193–194 °C), 5.18 g (0.167 mol) of purified red phosphorus and 55 g (18 ml, 0.343 mol) of bromine. B.p. 198–201 °C.

1-Bromododecane. Use 116 g (0.623 mol) of dodecan-1-ol (lauryl alcohol), m.p. 24 °C, 5.18 g (0.167 mol) of purified red phosphorus and 55 g (18 ml, 0.343 mol) of bromine. Heat the alcohol–phosphorus mixture to about 250 °C with vigorous stirring and add the bromine slowly. Allow the mixture to cool after all the bromine has been introduced. Add ether, filter off the excess of phosphorus and wash the ethereal solution of the bromide with water and dry over anhydrous potassium carbonate. Remove the ether on a water bath, and distil the residue under reduced pressure. B.p. 149–151 °C/18 mmHg.

1-Bromotetradecane. Use 107 g (0.5 mol) of tetradecan-1-ol (m.p. 38 °C), 3.41 g (0.11 mol) of purified red phosphorus and 44 g (14.5 ml, 0.275 mol) of bromine and proceed as under *1-bromododecane*. B.p. 178.5–179.5 °C/20 mmHg, m.p. 5 °C.

1-Bromohexadecane. Use 121 g (0.5 mol) of hexadecan-1-ol (cetyl alcohol), m.p. 48 °C, 3.41 g (0.11 mol) of purified red phosphorus and 44 g (0.275 mol) of bromine, and proceed as for *1-bromododecane*; filter off the excess of phosphorus at 16–20 °C. B.p. 202–203 °C/21 mmHg; m.p. 14 °C.

1-Bromo-2-phenylethane. Use 152.5 g (148 ml, 1.25 mol) of 2-phenylethanol (Section **III,31**), b.p. 216.5–217 °C, 10.35 g (0.33 mol) of purified red phosphorus and 110 g (35.5 ml, 0.68 mol) of bromine. Isolate the 1-bromo-2-phenylethane as detailed for *1 bromododecane*. B.p. 98 °C/12 mmHg.

1,4-Dibromobutane (*from butane-1,4-diol*). Use 45 g (0.5 mol) of redistilled butane-1,4-diol, 6.84 g (0.22 mol) of purified red phosphorus and 80 g (26 ml, 0.5 mol) of bromine. Heat the glycol–phosphorus mixture to 100–150 °C and add the bromine slowly; continue heating at 100–150 °C for 1 hour after all the bromine has been introduced. Allow to cool, dilute with water, add 100 ml of ether and remove the excess of red phosphorus by filtration. Separate the ethereal solution of the dibromide, wash it successively with 10 per cent sodium thiosulphate solution and water, then dry over the anhydrous potassium carbonate. Remove the ether on a water bath and distil the residue under diminished pressure. Collect the 1,4-dibromobutane at 83–84 °C/12 mmHg; the yield is 73 g (67%).

1,6-Dibromohexane. Proceed as for 1,4-dibromobutane but use 58 g (0.49 mol) of hexane-1,6-diol. The yield of 1,6-dibromohexane, b.p. 114–115 °C/12 mmHg, is 85 g (71%).

1,4-Dibromobutane (*from tetrahydrofuran*). Place 18.1 g (20.5 ml, 0.25 mol) of redistilled tetrahydrofuran (b.p. 65–66 °C), 3.41 g (0.11 mol) of purified red phosphorus and 4.5 g of water in the flask. Heat the mixture gently and add 40 g (13 ml, 0.25 mol) of bromine at such a rate that there is little bromine vapour above the surface of the reaction mixture. Heat at 100–150 °C for 45–60 minutes after all the bromine has been introduced. Work up as for the butane-1,4-diol preparation. The yield of 1,4-dibromobutane, b.p. 83–84 °C/12 mmHg, is 42 g (72%).

1,5-Dibromopentane (*from tetrahydropyran*). Proceed as in the previous preparation but replace the tetrahydrofuran by 21.5 g (24.4 ml, 0.25 mol) of redistilled tetrahydropyran (b.p. 86.5–87.5 °C). The yield of 1,5-dibromopentane, b.p. 99 °C/13 mmHg, is 43 g (75%).

E,1(c). Preparation of alkyl iodides from alcohols. Alkyl iodides are the most easily formed of the alkyl halides and the slow distillation of the alcohol with constant boiling point hydriodic acid is a general method of preparation (e.g., Section **III,56**).

$$ROH + HI \longrightarrow RI + H_2O$$

An alternative reagent, which is particularly effective for the conversion of diols into diiodo compounds, is a mixture of potassium iodide and 95 per cent orthophosphoric acid (Section **III,57**). The reagent also cleaves tetrahydrofuran and tetrahydropyran to yield the corresponding α,ω-diiodo compounds [cf. the hydrobromic acid–sulphuric acid reagent, p. 386, **E,1(b)**].

$$HOCH_2 \cdot (CH_2)_n \cdot CH_2OH + 2KI + 2H_3PO_4 \longrightarrow$$
$$ICH_2 \cdot (CH_2)_n \cdot CH_2I + 2KH_2PO_4 + 2H_2O$$

A further generally applicable method for the preparation of alkyl iodides involves the addition of iodine to a gently boiling suspension of purified red phosphorus in the corresponding alcohol. The yields from primary and secondary alcohols are excellent (e.g., Section **III,58**).

$$10ROH + 2P + 5I_2 \longrightarrow 10RI + 2H_3PO_4 + 2H_2O$$

Reaction of an alcohol with the reagent *o*-phenylene phosphorochloridite followed by treatment of the alkyl *o*-phenylene phosphite so obtained with iodine in methylene chloride at room temperature results in a good yield of alkyl iodide. This method, exemplified by the preparation of 1-iodoheptane (Section **III,59**), is the preferred procedure when acid-sensitive functional groups are present.

III,56. ISOPROPYL IODIDE (*2-Iodopropane*)

$$CH_3 \cdot CH(OH) \cdot CH_3 + HI \longrightarrow CH_3 \cdot CHI \cdot CH_3 + H_2O$$

Mix 30 g (38 ml, 0.5 mol) of propan-2-ol with 450 g (265 ml) of constant boiling point hydriodic acid (57%) (Section **II,2,26**) in a 500-ml distilling flask, attach a condenser for downward distillation, and distil slowly (1–2 drops per second) from an oil or air bath. When about half the liquid has passed over, stop the distillation. Separate the lower layer of crude iodide (70 g, 82%). Redistil the aqueous layer and thus recover a further 5 g of iodide from the first quarter of the distillate (1). Wash the combined iodides with an equal volume of concentrated hydrochloric acid, then, successively, with water, 5 per cent sodium carbonate solution and water. Dry with anhydrous calcium chloride and distil. The isopropyl iodide distils constantly at 89 °C.

Note. (1) A further quantity of isopropyl iodide, only slightly less than that obtained in the first distillation, may be prepared by combining the residues in the distilling flask, adding 30 g (38 ml) of propan-2-ol, and repeating the distillation. Finally, the residues should be distilled and the 57 per cent constant boiling point acid recovered.

Cognate preparations Isobutyl iodide (*1-iodo-2-methylpropane*). Use 30 g (37.5 ml, 0.37 mol) of 2-methylpropan-1-ol and 273 g (161 ml) of 57 per cent hydriodic acid; 65 g (96%) of the crude iodide are obtained. If the crude iodide is dark in colour, add a little sodium metabisulphite. B.p. 119–120 °C.

s-Butyl iodide (*2-iodobutane*). Use 30 g (37.5 ml, 0.37 mol) of butan-2-ol and 273 g (161 ml) of 57 per cent hydriodic acid; 63 g of crude iodide are obtained. B.p. 117.5–119 °C.

Iodocyclopentane. Use 43 g (45.5 ml, 0.5 mol) of cyclopentanol and 340 g (200 ml) of 57 per cent hydriodic acid; 89 g (91%) of crude iodide are obtained. B.p. 58 °C/22 mmHg.

Allyl iodide. Use 29 g (34 ml, 0.5 mol) of allyl alcohol and 340 g (200 ml) of 57 per cent hydriodic acid; 74 g (88%) of crude iodide are obtained. Upon adding 29 g (34 ml) of allyl alcohol to the combined residue in the flask and the aqueous layer and distilling as before, a further 72 g of crude allyl iodide may be isolated. B.p. 99–101 °C (mainly 100 °C). The compound is very sensitive to light; the distillation should therefore be conducted in a darkened room and preferably in the presence of a little silver powder.

III,57. 1,4-DIIODOBUTANE

$$HOCH_2 \cdot CH_2 \cdot CH_2 \cdot CH_2OH + 2KI + 2H_3PO_4 \longrightarrow$$
$$ICH_2 \cdot CH_2 \cdot CH_2 \cdot CH_2I + 2KH_2PO_4 + 2H_2O$$

In a 500-ml three-necked flask, equipped with a thermometer, a sealed stirrer unit and a reflux condenser, place 32.5 g of phosphorus pentoxide and add 115.5 g (67.5 ml) of 85 per cent orthophosphoric acid (1). When the stirred mixture has cooled to room temperature, introduce 166 g (1 mol) of potassium iodide and 22.5 g (0.25 mol) of redistilled butane-1,4-diol (b.p. 228–230 °C or 133–135 °C/18 mmHg). Heat the mixture with stirring at 100–120 °C for 4 hours. Cool the stirred mixture to room temperature and add 75 ml of water and 125 ml of ether. Separate the ethereal layer, decolourise it by shaking with 25 ml of 10 per cent sodium thiosulphate solution, wash with 100 ml of cold saturated sodium chloride solution, and dry with magnesium sulphate. Remove the ether by flash distillation (Fig. I,100) on a steam bath and distil the residue from a flask with fractionating side arm under diminished pressure. Collect the 1,4-diiodobutane at 110 °C/6 mmHg, the yield is 65 g (84%).

Alternatively, add 18 g (20 ml, 0.25 mol) of redistilled tetrahydrofuran (b.p. 65–66 °C) to a mixture of 32.5 g of phosphorus pentoxide, 115.5 g (67.5 ml) of 85 per cent orthophosphoric acid and 166 g of potassium iodide, heat for 3–4 hours, cool and isolate the 1,4-diiodobutane as above. The yield of product, b.p. 110 °C/6 mmHg, is 70 g (90%).

Note. (1) The orthophosphoric acid must be adjusted to a concentration of 95 per cent H_3PO_4. Alternatively, the commercial 100 per cent orthophosphoric acid may be diluted with water to this concentration. The 95 per cent acid is claimed to be the most efficient for the preparation of iodides from alcohols and glycols, and for effecting cleavage of tetrahydrofuran and tetrahydropyran. Anhydrous orthophosphoric acid does not give such good results because of the limited solubility of hydrogen iodide in the reagent.

Cognate preparations **1,5-Diiodopentane** (*from pentane-1,5-diol*). Proceed as for 1,4-diiodobutane but use 26 g (26.5 ml, 0.25 mol) of redistilled pentane-1,5-diol (b.p. 238–239 °C) in place of the butane-1,4-diol. The yield of 1,5-diiodopentane, b.p. 142–143 °C/16 mmHg, is 65 g (80%).

1,5-Diiodopentane (*from tetrahydropyran*).   Use 21.5 g (24.4 ml, 0.25 mol) of redistilled tetrahydropyran (b.p. 86.5–87.5 °C) in place of the tetrahydrofuran, otherwise proceed as for 1,4-diiodobutane. The yield of 1,5-diiodopentane, b.p. 142–143 °C/16 mmHg, is 71 g (88%).

1,6-Diiodohexane.   Proceed exactly as detailed for 1,4-diiodobutane but replace the butane-1,4-diol by 29.5 g (0.25 mol) hexane-1,6-diol, m.p. 41–42 °C. The yield of 1,6-diiodohexane, b.p. 150 °C/10 mmHg, m.p. 10 °C, is 70 g (83%).

Butyl iodide.   Use 37 g (46 ml, 0.5 mol) of butan-1-ol together with the quantities of the other reactants used above; a 2-hour reaction time is sufficient. The yield of butyl iodide, b.p. 129–130 °C, is 64 g (70%).

Iodocyclohexane.   Proceed as for butyl iodide using 50 g (0.5 mol) of redistilled cyclohexanol (b.p. 160–161 °C). Distil the iodocyclohexane under reduced pressure, b.p. 67–69 °C/9 mmHg; the yield is 90 g (86%).

III,58. BUTYL IODIDE (*1-Iodobutane*)

$$10CH_3 \cdot (CH_2)_2 \cdot CH_2OH + 2P + 5I_2 \longrightarrow$$
$$10CH_3(CH_2)_2 \cdot CH_2I + 2H_3PO_4 + 2H_2O$$

Place 12.5 g (0.4 mol) of purified red phosphorus (Section **II,2,**49) and 78 g (96 ml, 1.05 mol) of butan-1-ol in a 250-ml round-bottomed flask fitted with a Liebig-type reflux condenser. Heat the mixture to gentle refluxing with a Bunsen flame over a wire gauze. Remove or lower the flame and add 127 g (0.5 mol) of coarsely powdered iodine in approximately 2 g portions down the centre of the condenser. Complete the entire addition of the iodine fairly rapidly (i.e., in about 20–30 minutes), but allow the mildly exothermic reaction to subside after the addition of each portion. Finally continue heating under reflux for 30–60 minutes; little or no iodine should then be visible. Arrange the condenser for downward distillation and distil off most of the crude product. When the volume of liquid in the flask has been reduced to about 15–20 ml, add about 40 ml of water and continue the distillation until no more oily drops pass over into the receiver (1). Separate the crude alkyl iodide and wash it successively with approximately equal volumes of water, concentrated hydrochloric acid (2), water, 10 per cent sodium carbonate solution and water (3). Dry the product over anhydrous calcium chloride and distil, collecting the butyl iodide which passes over at 129–131 °C; the yield is 165 g (90%) (4).

Notes.   (1) See Section **III,55**, Note (1).

(2) The washing with concentrated hydrochloric acid removes any unchanged alcohol which may be present.

(3) If the washed organic phase is darker than a pale brown-pink, add a few crystals of sodium thiosulphate to the final wash-water to remove traces of iodine.

(4) The purified iodide may be preserved in a bottle containing a short coil of clean copper wire.

Cognate preparations   The following may be prepared by the above procedure in similar yield using the same quantities of red phosphorus and of iodine:

Propyl iodide (*1-iodopropane*)—from 63 g (78 ml, 1.05 mol) of propan-1-ol. B.p. 102–103 °C.

Isopropyl iodide (*2-iodopropane*)—from 63 g (80 ml, 1.05 mol) of propan-2-ol. A little hydrogen iodide is evolved. B.p. 89–90 °C.

s-Butyl iodide (*2-iodobutane*)—from 78 g (97 ml, 1.05 mol) of butan-2-ol. A little hydrogen iodide is evolved. B.p. 118–120 °C.

1-Iodopentane—from 92 g (113 ml, 1.05 mol) of pentan-1-ol. B.p. 153–156 °C.

2-Iodopentane—from 92 g (113 ml, 1.05 mol) of pentan-2-ol. B.p. 142–144 °C.

1-Iodohexane—from 107 g (130 ml, 1.05 mol) of hexan-1-ol. B.p. 178–180 °C.

1-Iodoheptane—from 122 g (148 ml, 1.05 mol) of heptan-1-ol. B.p. 198–201 °C, 62.5 °C/3.5 mmHg.

1-Iodooctane—from 137 g (166 ml, 1.05 mol) of octan-1-ol. B.p. 219–222 °C, 86.5 °C/5 mmHg.

Iodocyclohexane—from 105 g (110 ml, 1.05 mol) of cyclohexanol. Carry out the reaction as above, dilute the cooled reaction product with ether and filter. Wash and dry the organic phase, remove the ether and distil under reduced pressure. B.p. 81–83 °C/20 mmHg.

1-Iodo-2-phenylethane (*2-phenylethyl iodide*)—from 149 g (149 ml, 1.05 mol) of 2-phenylethanol. Proceed as for iodocyclohexane; a little hydrogen iodide is evolved towards the end of the reaction. B.p. 114–116 °C/12 mmHg.

1,3-Diodopropane—from 40 g (38 ml, 0.525 mol) of propane-1,3-diol (trimethylene glycol). Proceed as for iodocyclohexane; stop heating as soon as all the iodine has been added. B.p. 88–89 °C/6 mmHg.

III,59. 1-IODOHEPTANE

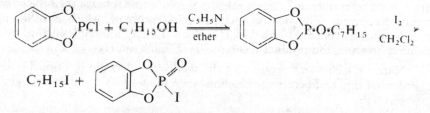

Heptyl o-phenylene phosphite. Place 87.0 g (0.5 mol) of o-phenylene phosphorochloridite (Section **III,70**), 39.5 g (0.5 mol) of dry redistilled pyridine and 500 ml of dry ether, in a 2-litre conical flask and cool to 0 °C. Add 58.1 g (0.5 mol) of heptan-1-ol dissolved in 400 ml of dry ether during about 5 minutes with occasional shaking. Stopper the flask and allow the reaction to proceed at room temperature overnight. Filter off the precipitated pyridinium chloride under suction, wash well with dry ether and remove the ether from the combined filtrate and washings on a rotary evaporator to obtain 127.0 g (100%) of heptyl o-phenylene phosphite as a colourless oil of sufficient purity for use in the next stage.

1-Iodoheptane. In a conical flask, stir magnetically at room temperature a solution of 50.9 g (0.20 mol) of heptyl o-phenylene phosphite in 500 ml of dry methylene chloride and add 50.8 g (0.20 mol) of iodine. After 1 hour (1) transfer the solution to a separatory funnel and wash the organic layer with 400 ml of 10 per cent aqueous sodium thiosulphate solution to remove any residual unreacted iodine. Wash the organic layer twice with 400 ml portions of 5 per cent aqueous sodium hydroxide solution to remove the o-phenylene phosphoroiodate by-product, followed by two 400 ml portions of 5 per cent aqueous sodium metabisulphite, and finally with 400 ml of a saturated solution of sodium chloride. Dry the organic layer with anhydrous calcium sulphate, filter and remove the solvent on a rotary evaporator. The yield of crude 1-iodoheptane is 41.8 g (92%). Fractionally redistil the product under reduced pressure using a nitrogen capillary leak when 39 g (86%) of the pure material having b.p. 96–104 °C/46–48 mmHg is obtained.

Note.   (1) The reaction may be monitored by GLC by the direct loading of a sample (0.1 μl) on to an S.E. 30-column, 5 ft, held at 170 °C (nitrogen flow 40 ml/min.). The retention time of 1-iodoheptane is 52 seconds; that of heptyl o-phenylene phosphite under the same conditions is 4 minutes.

Cognate preparation 1-Chloro-3-iodopropane.   Prepare 3-chloropropyl o-phenylene phosphite from 34.9 g (0.2 mol) of o-phenylene phosphorochloridite 15.8 g (0.2 mol) of pyridine, and 19 g (0.2 mol) of trimethylene chlorohydrin in 200 ml of dry ether. Allow the mixture to stand for 12 hours at room temperature, decant the ether layer from the sticky solid and evaporate the ether (rotary evaporator) to obtain 38.4 g (83%) of product as a mobile colourless liquid.

To 33.9 g (0.15 mol) of the crude 3-chloropropyl o-phenylene phosphite in 300 ml of methylene chloride, add 38 g (0.15 mol) of iodine and stir at room temperature for 12 hours. Wash the resulting brown solution successively with 200 ml of 10 per cent aqueous sodium thiosulphate solution, two 300 ml portions of 5 per cent aqueous sodium hydroxide solution, 300 ml of 5 per cent aqueous sodium metabisulphite solution and then twice with 350 ml water. Dry the organic layer over anhydrous calcium sulphate and remove the solvent on a rotary evaporator. The yield of 1-chloro-3-iodopropane, which is sufficiently pure for most purposes, is 26 g (84%) (1); distillation if required is effected under reduced pressure, the product has b.p. 70–72 °C/28 mmHg.

Note.   (1) The neat product or the methylene chloride solution becomes discoloured due to liberation of iodine when allowed to stand in sunlight or artificial light.

E,2. Halogen exchange reactions.   Alkyl fluorides may be prepared in moderate yield by interaction of an alkyl bromide with anhydrous potassium fluoride in the presence of dry ethylene glycol as a solvent for the inorganic fluoride (e.g., Section **III,60**). A little alkene accompanies the alkyl fluoride produced and is readily removed by treatment with KBr/Br_2 solution.

$$RBr + KF \xrightarrow[\text{glycol}]{\text{ethylene}} RF + KBr$$

The most widely used example of halogen exchange is provided by the preparation of alkyl iodides from chlorides or bromides using sodium iodide in a solvent, such as acetone, in which sodium iodide is soluble but sodium chloride or bromide is relatively less so (e.g., Section **III,61**).

$$RCl + NaI \rightleftharpoons RI + NaCl\downarrow$$

III,60. 1-FLUOROHEXANE

$$CH_3 \cdot (CH_2)_4 \cdot CH_2Br + KF \xrightarrow[\text{glycol}]{\text{ethylene}} CH_3 \cdot (CH_2)_4 \cdot CH_2F + KBr$$

Alkyl fluorides are said to be highly toxic. Great care should be taken not to inhale the vapours; conduct the entire operation in an efficient fume cupboard.

In a dry 500-ml three-necked flask, equipped with a sealed stirrer unit, a 100-ml dropping funnel and a short fractionating column (1), place a mixture of 116 g (2 mol) of anhydrous, finely powdered potassium fluoride (2) and 200 g of dry ethylene glycol (3). Connect the fractionating column (which carries a thermometer) to a downward double surface condenser fitted with a receiving flask with the aid of a side-arm adapter. Heat the flask in an oil bath at 160–170 °C and introduce 165 g (141 ml, 1 mol) of 1-bromohexane (Section III,55) dropwise, with stirring, during 5 hours. A liquid passes over intermittently at 60–90 °C. When the addition is complete, allow the bath temperature to fall to 110–120 °C; replace the dropping funnel by a tube of narrow bore dipping just below the surface of the liquid, attach the side arm of the receiver adapter to a water pump, and draw a slow stream of air through the apparatus whilst maintaining the stirring. It is advisable to interpose a trap (e.g., a Drechsel bottle) cooled in ice between the water pump and receiver in order to recover any uncondensed liquid. Distil the combined distillates through an efficient fractionating column (4); after a small forerun (0.5 g) of hex-1-ene collect the crude 1-fluorohexane at 92–97 °C. Purify the crude product by cooling in ice and adding 1 ml portions of a solution containing 9.0 g of bromine and 6.0 g of potassium bromide in 50 ml of water until the organic layer acquires an orange colour: shake the mixture vigorously for a minute or so after each addition. The volume of KBr/Br_2 solution required is usually less than 5 ml. Separate the aqueous layer, wash the organic layer with saturated aqueous potassium bromide solution until colourless, and finally with water. Dry the liquid with magnesium sulphate and fractionate. Collect the fraction 92–94 °C: the yield is 44 g (42%). The colourless liquid keeps unchanged for long periods.

Notes. (1) Any fractionating column of moderate efficiency is satisfactory, e.g., a Dufton column (20 cm long containing a spiral 10 cm in length, 2 cm in diameter with 8 turns of the helix) or a Vigreux column (20–25 cm long).

(2) Grind finely pure laboratory grade, anhydrous potassium fluoride, and heat it in an electrically heated oven at 180–210 °C; store in a desiccator. Before use, dry the powdered salt at 180 °C for 3 hours and grind again in a warm (*ca.* 50 °C) glass mortar.

(3) Redistil laboratory grade ethylene glycol under reduced pressure and collect the fraction of b.p. 85–90 °C/7 mmHg for use as a solvent for the potassium fluoride.

(4) A Widmer column (spiral 18 cm in length, 1.5 cm in diameter with 20 turns of the helix) is satisfactory.

Cognate preparation   1-Fluoropentane.   Use 116 g (2 mol) of dry potassium fluoride in 200 g of dry ethylene glycol: heat in an oil bath at 140–150 °C and add 151 g (124 ml, 1 mol) of 1-bromopentane during 5 hours with stirring. The reaction product distils intermittently at 50–85 °C. The yield of 1-fluoropentane, b.p. 63.5–65 °C, is 25 g (28%).

III,61. 1-IODO-3-METHYLBUTANE

$(CH_3)_2CH\cdot CH_2\cdot CH_2Br + NaI \longrightarrow (CH_3)_2CH\cdot CH_2\cdot CH_2I + NaBr$

Dissolve 37.5 g (0.25 mol) of dry sodium iodide (1) in 250 ml of dry acetone in a 500-ml flask fitted with a reflux condenser protected by a calcium chloride guard-tube, and add 30.2 g (25 ml, 0.2 mol) of 1-bromo-3-methylbutane. A precipitate of sodium bromide soon begins to form; leave the reaction mixture at room temperature for 30 minutes, and then boil under reflux for 45 minutes to complete the reaction. Allow to cool and filter off the sodium bromide, washing the residue with a little acetone. Remove the acetone from the filtrate on a rotary evaporator, and shake the residual organic halide with 100 ml of water. Separate the lower dark-coloured layer and wash it twice more with 50 ml portions of water; incorporate sufficient crystals of sodium thiosulphate into the first portion of wash-water to decolourise the organic phase. Dry the product over anhydrous calcium sulphate, filter and distil, collecting the 1-iodo-3-methylbutane at 145–147 °C. The yield is 26 g (66%).

Note. (1) Dry the sodium iodide for 4 hours at 100 °C under reduced pressure (oil pump).

E,3. Addition of hydrogen halides or halogens to alkenes. Direct addition of a hydrogen halide to an alkene gives rise to an alkyl halide, the order of reactivity being HI > HBr > HCl. In the case of an unsymmetrical alkene, addition proceeds in the Markownikoff manner via that positively charged intermediate which is stabilised to the greatest extent by charge dispersal.

$$R\cdot CH_2 \cdot CH = CH_2 \quad H - X \longrightarrow R\cdot CH_2 \cdot \overset{\oplus}{C}H \cdot CH_3 + X^{\ominus}$$
$$\longrightarrow R\cdot CH_2 \cdot CHX\cdot CH_3$$

Addition of hydrogen halide (1 mol) to a diene (1 mol) is a method of greater preparative value. This reaction is illustrated by the addition of hydrogen bromide to isoprene (Section **III,62**); the overall 1,4-addition process, as opposed to the 1,2-addition, predominates under the conditions specified.

$$CH_2 = CH - C \overset{CH_3}{=} CH_2 \quad H - Br \longrightarrow \left[\begin{array}{c} CH_3 \\ | \\ CH_2 = CH - \overset{\oplus}{C} - CH_3 \\ CH_3 \\ | \\ \overset{\oplus}{C}H_2 - CH = C - CH_3 \end{array} \right] + Br^{\ominus}$$

$$\underset{BrCH_2\cdot CH = C - CH_3}{\overset{CH_3}{\underset{|}{}}} \longleftarrow$$

The addition of hydrogen bromide (but not the iodide or chloride) in the presence of an added peroxide catalyst proceeds by a radical mechanism giving rise to the so-called anti-Markownikoff mode of addition. Hence a terminal alkene gives a primary bromide (e.g., the preparation of 11-bromoundecanoic acid, Section **III,63**).

$$C_6H_5\cdot C\overset{O}{\diagdown}\underset{O-O}{\overset{O}{\diagup}}C\cdot C_6H_5 \longrightarrow 2C_6H_5\cdot C\overset{O}{\underset{\underset{..}{\overset{..}{O}}\cdot}{\diagdown}} \longrightarrow 2\dot{C}_6H_5 + CO_2$$

$$\dot{C}_6H_5 + HBr \longrightarrow C_6H_6 + \dot{B}r$$

$$R\cdot CH = CH_2 \quad \dot{B}r \longrightarrow R\cdot \dot{C}H\cdot CH_2Br \overset{HBr}{\longrightarrow} R\cdot CH_2\cdot CH_2Br + \dot{B}r$$

Halogens add to alkenes to give vicinal dihalides.

$$R \cdot CH = CH_2 + X_2 \longrightarrow R \cdot CHX \cdot CH_2 X$$

The addition of bromine usually proceeds the most smoothly, and is conveniently carried out in a solvent such as carbon tetrachloride. The examples of the addition of bromine to allyl bromide and the addition of bromine to undec-10-enoic acid are illustrative (Sections **III,64** and **III,19**).

III,62. 1-BROMO-3-METHYLBUT-2-ENE

$$\overset{\overset{\displaystyle CH_3}{|}}{CH_2 = C - CH = CH_2} + HBr \longrightarrow \overset{\overset{\displaystyle CH_3}{|}}{CH_3 - C = CH - CH_2 Br}$$

The entire preparation must be carried out in an efficient fume cupboard since the product is highly lachrymatory.

Weigh a 100-ml three-necked flask fitted with stoppers. Remove the stoppers and attach to the flask a mechanical stirrer, a calcium chloride guard-tube to the outlet of which is connected a tube leading to the fume cupboard drain, and a gas inlet tube terminating in a glass frit and attached to a dry hydrogen bromide gas generator (Section **II,2,*30***). Charge the flask with 34 g (0.5 mol) of redistilled 2-methylbuta-1,3-diene (*isoprene*) (1), cool in an ice–salt bath and pass dry hydrogen bromide gas slowly through the reaction mixture until an increase in weight of 40 g is obtained (2); this may require about 6 hours. Now fit the flask with a Vigreux column in the central joint and with a stopper and a suitable capillary air leak in the side joints. Fractionally distil the crude 1-bromo-3-methylbut-2-ene under reduced pressure (ensure that suitable potassium hydroxide traps are sited between the apparatus and the oil immersion rotatory pump) and collect the pure product of b.p. 56–57 °C/25 mmHg; the yield is 58 g (78%).

Notes. (1) Isoprene is purchased in sealed capsules. These are usually stored in a refrigerator and only removed just prior to opening and redistillation of the isoprene. A simple distillation unit may be employed using a double surface condenser, the rubber tubing water leads being immersed in a large container of ice to effect more efficient cooling. The distillation receiver should be cooled and the outlet protected with a calcium chloride guard-tube. Pure isoprene has a b.p. 33–34 °C.

(2) The reaction flask should be removed and replaced by a similar flask to ensure that the stirrer and gas inlet tube do not become unnecessarily contaminated with moisture. The reaction flask should be stoppered with the same stoppers used in the original weighing.

III,63. 11-BROMOUNDECANOIC ACID

$$CH_2 = CH \cdot (CH_2)_8 \cdot CO_2 H + HBr \xrightarrow{(C_6H_5CO)_2O_2} BrCH_2 \cdot (CH_2)_9 \cdot CO_2 H$$

Equip a 500-ml three-necked round-bottomed flask with a sealed stirrer unit, a wide-bore gas inlet tube reaching to the bottom of the flask and a two-necked multiple adapter fitted with a thermometer and a condenser protected with a calcium chloride guard-tube; arrange the apparatus so that occasional cooling can be effected with an ice-water bath. In the flask place a solution of 27.6 g (0.15 mol) of undec-10-enoic acid in 220 ml of dry light petroleum (b.p. 40–60 °C) together with 1.5 g (0.006 mol) of benzoyl peroxide (air dried; CAUTION: see Section II,2,*5*). Pass a rapid stream of dry hydrogen bromide [from a cylinder or from 20 ml of tetralin and 17 ml of bromine (Section II,2,*30*)] through the stirred mixture until it is saturated (about 0.75 hour) whilst maintaining the temperature between 10 and 20 °C by occasional cooling. Should a white solid separate and tend to block the inlet tube towards the end of the reaction, either add a further small portion of dry light petroleum or maintain the temperature nearer to 20 °C when the material should dissolve.

Decant the pale straw-coloured solution into a conical flask and rinse the reaction vessel with 40 ml of light petroleum (b.p. 40–60 °C) and combine with the main solution. Cool the solution to −10 °C and collect the solid which separates by filtration under suction and wash with about 40 ml of similarly cooled light petroleum. A further quantity of the product may be obtained by concentrating the filtrate to about 40 ml and cooling. Dissolve the crude product in approximately 200 ml of boiling light petroleum (b.p. 40–60 °C), add decolourising charcoal, filter, concentrate to about 150 ml and cool. Collect the 11-bromoundecanoic acid which separates as microcrystalline platelike needles, m.p. 49–50 °C. A further quantity of slightly less pure acid may be obtained by concentrating the mother-liquors to about 30 ml and cooling. The yield is 27.9 g (70%).

III,64. 1,2,3-TRIBROMOPROPANE

$$CH_2 = CH \cdot CH_2 Br + Br_2 \longrightarrow CH_2 Br \cdot CHBr \cdot CH_2 Br$$

Provide a 1-litre three-necked flask with a dropping funnel carrying a calcium chloride guard-tube, a mechanical stirrer and a thermometer reaching almost to the bottom of the flask, and cool the flask in a mixture of ice and salt. Place in the flask 182 g (132 ml, 1.5 mol) of allyl bromide (1) and 250 ml of dry carbon tetrachloride (Section II,1,*7*), and introduce 255 g (80 ml, 1.6 mol) of dry bromine (Section II,2,*7*) into the dropping funnel. Set the stirrer in motion and when the temperature has fallen to −5 °C, drop the bromine in slowly at such a rate that the temperature does not rise above 0 °C (about 90 minutes). Allow the orange-coloured solution (the colour is due to a slight excess of bromine) to warm to room temperature with constant stirring (about 30 minutes) and then remove the solvent under reduced pressure on a rotary evaporator. Distil the residue under reduced pressure; the residual carbon tetrachloride passes over first, followed by 1,2,3-tribromopropane at 92–93 °C/10 mmHg or 100–103 °C/18 mmHg as an almost colourless liquid. The yield is 400 g (95%).

Note. (1) The allyl bromide (Section III,*54*) should be dried over anhydrous calcium chloride and redistilled; the fraction, b.p. 69–72 °C, is collected for use in this preparation.

E,4. Allylic halogenation. The direct introduction of bromine into the allylic position of an alkene using *N*-bromosuccinimide is known as the **Wohl–Ziegler reaction**. Bromination is carried out in anhydrous reagents (to avoid hydrolysis of the bromoimide), usually boiling carbon tetrachloride or chloroform solution. The progress of the reaction can be followed by the fact that at first the dense *N*-bromosuccinimide is at the bottom of the flask and is gradually replaced by succinimide, which rises to the surface: the reaction is complete when all the crystals are floating at the surface (detected by stopping the boiling momentarily). This can be confirmed (when equimolar amounts are used) by transferring a drop of the solution to acidified potassium iodide–starch solution: iodine should not be liberated. After cooling, the insoluble succinimide is filtered off, washed with solvent and the product isolated, after removal of the solvent, by distillation or crystallisation.

The specific substitution into the allylic position is the result of a radical process which requires the generation of a low concentration of molecular bromine, probably by way of the action of traces of hydrogen bromide on the bromoimide.

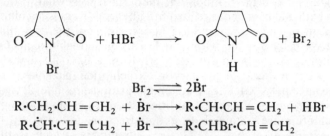

$$Br_2 \rightleftharpoons 2\dot{Br}$$
$$R \cdot CH_2 \cdot CH = CH_2 + \dot{Br} \longrightarrow R \cdot \dot{CH} \cdot CH = CH_2 + HBr$$
$$R \cdot \dot{CH} \cdot CH = CH_2 + \dot{Br} \longrightarrow R \cdot CHBr \cdot CH = CH_2$$

Two simple applications may be mentioned. With cyclohexene, 3-bromo-cyclohexene (I) is obtained in a satisfactory yield (Section III,65), the latter upon dehydrobromination with quinoline affords an 80–90 per cent yield of cyclohexa-1,3-diene (II) (Section III,9).

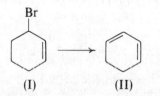

(I)   (II)

Methyl crotonate yields the valuable synthetic reagent methyl γ-bromo-crotonate (III) (Section III,66); this latter compound permits the introduction (in moderate yield) of a four-carbon atom chain at the site of the carbonyl group by the use of the Reformatsky reaction (compare Section III,173):

$$R^1 \cdot CO \cdot R^2 + BrCH_2 \cdot CH = CH \cdot CO_2CH_3 \xrightarrow[\text{benzene}]{Zn}$$
(III)
$$R^1R^2C(OH) \cdot CH_2 \cdot CH = CH \cdot CO_2CH_3$$

III,65. 3-BROMOCYCLOHEXENE

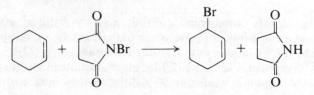

Since many simple unsubstituted allylic compounds are powerful irritants by inhalation and by skin contact, this preparation should be carried out in an efficient fume cupboard and the product treated with appropriate caution.

In a 500-ml round-bottomed flask place 39.4 g (0.20 mol) of N-bromo-succinimide (Section II,2,8), 49.2 g (0.60 mol) of redistilled cyclohexene, 150 ml of carbon tetrachloride and about 500 mg of benzoyl peroxide (**CAUTION**: Section II,2,5), previously dried by pressing between filter-papers. Attach a double surface reflux condenser and allow the mixture to stand at room temperature when after a short induction period the reaction begins. The reaction mixture becomes warm and the heavy yellowish N-bromosuccinimide begins to be transformed into the light colourless succinimide which becomes suspended in the reaction mixture. When the reaction has moderated somewhat, transfer the flask to a steam bath and heat under reflux until all the N-bromosuccinimide has been converted to succinimide (about 1.5 hours) (1). Cool the flask, filter under suction and wash the residue with a little carbon tetrachloride. Distil the filtrate and washings on a boiling water bath from a flask fitted with a Claisen still-head to remove most of the carbon tetrachloride (b.p. 77 °C) and unreacted cyclohexene (b.p. 83 °C). Then fractionally distil the residue (2) under reduced pressure (water pump). Further quantities (about 50 g) of cyclohexene and carbon tetrachloride are obtained as a first fraction, followed by the main fraction (about 24 g) of b.p. 72–77 °C/32–35 mmHg, which is only about 50 per cent pure by GLC analysis and which rapidly goes brown on standing. Subsequent redistillation under reduced pressure and using a short Vigreux column (12 cm) gives 14.4 g (45%) of colourless mobile 3-bromo-cyclohexene of b.p. 66–67 °C/20 mmHg, which is 99 per cent pure by GLC analysis (10% Silicone oil on Chromosorb W, 5-ft column held at 85 °C, nitrogen flow rate 40 ml/minute, t_R 10.4 minutes).

Notes. (1) The completion of the reaction may be confirmed by treating starch-iodide paper with a drop of the reaction solution. At the end of the reaction no colouration is observed.

(2) The flask and still-head should preferably have at least 24/29 sized joints to prevent losses due to foaming.

III,66. METHYL γ-BROMOCROTONATE (*Methyl 4-bromobut-2-enoate*)

$$CH_3 \cdot CH = CH \cdot CO_2H + CH_3OH \longrightarrow CH_3 \cdot CH = CH \cdot CO_2CH_3$$

$$CH_3 \cdot CH = CH \cdot CO_2CH_3 + \begin{matrix} CH_2-CO \\ | \quad\quad NBr \\ CH_2-CO \end{matrix} \longrightarrow$$

$$BrCH_2 \cdot CH = CH \cdot CO_2CH_3 + \begin{matrix} CH_2-CO \\ | \quad\quad NH \\ CH_2-CO \end{matrix}$$

Methyl crotonate. Purify commercial crotonic acid by distilling 100 g from a flask attached to an air condenser; use an air bath (Fig. I,49(*a*)). The pure acid passes over at 180–182 °C and crystallises out on cooling, m.p. 72–73 °C; the recovery is about 90 per cent. Place 75 g (2.34 mol) of absolute methanol, 5 g (2.7 ml) of concentrated sulphuric acid and 50 g (0.58 mol) of pure crotonic acid in a 500-ml round-bottomed flask and heat under reflux for 12 hours. Add water, separate the precipitated ester and dissolve it in ether; wash with dilute sodium carbonate solution until effervescence ceases, dry with magnesium sulphate and remove the ether on a water bath. Distil and collect the methyl crotonate at 118–120 °C; the yield is 40 g (69%).

Methyl γ-bromocrotonate. Mix 36 g (0.2 mol) of N-bromosuccinimide, 20 g (0.2 mol) of methyl crotonate and 60 ml of dry, redistilled carbon tetrachloride in a 500-ml round-bottomed flask. Reflux on a water bath for 12 hours; by this time all the solid should have risen to the surface of the liquid. Filter off the succinimide at the pump and wash it with a little dry carbon tetrachloride. Remove the solvent on a water bath and distil the residue under reduced pressure through a short fractionating column. Collect the methyl γ-bromocrotonate at 77–78 °C/8 mmHg; the yield is 31 g (86%).

F ESTERS OF INORGANIC ACIDS

Alkyl hypochlorites. These are obtained by the interaction of the alcohol, sodium hydroxide and chlorine at low temperatures.

$$ROH + NaOH + Cl_2 \longrightarrow ROCl + H_2O + NaCl$$

Primary and secondary hypochlorites, however, readily explode when exposed to light, and even in the absence of light rapid decomposition occurs at room temperature. t-Butyl hypochlorite, the preparation of which is illustrated in Section III,67, is a relatively stable compound although purification by distillation is not normally attempted nor is it necessary. Its preparation is not hazardous provided that the temperature during reaction is carefully controlled. An alternative procedure involving the reaction of t-butyl alcohol with sodium hypochlorite in the presence of acetic acid is described in Ref. 27. The compound is a useful chlorinating agent.

Alkyl sulphites. Symmetrical esters of sulphurous acid (e.g., dibutyl sulphite, Section III,68) are best prepared by the action of thionyl chloride on the alcohol.

$$2ROH + SOCl_2 \longrightarrow O = S(OR)_2 + 2HCl$$

Alkyl sulphates. The preparation of these compounds in the laboratory is not recommended and is rarely worthwhile because of the cheapness of the commercial products. The dimethyl and diethyl esters may be prepared *inter alia*, however, by the interaction of chlorosulphonic acid with the anhydrous alcohol, followed by distillation of the resulting alkyl hydrogen sulphate under diminished pressure, for example:

$$CH_3OH + ClSO_3H \longrightarrow CH_3 \cdot O \cdot SO_3H + HCl$$
$$2CH_3 \cdot O \cdot SO_3H \longrightarrow CH_3 \cdot O \cdot SO_2 \cdot O \cdot CH_3 + H_2SO_4$$

These widely used alkylating agents are dangerously toxic. Reference should be made to Section II,2,*18* for details of purification procedures and for precautions relating to their use.

Alkyl phosphites. Trialkyl phosphites [P(OR)$_3$] are readily prepared by the interaction of an alcohol with phosphorus trichloride in the presence of pyridine (e.g., tributyl phosphite, Section **III,69**).

$$3ROH + PCl_3 + 3C_5H_5N \longrightarrow (RO)_3P + 3C_5H_5\overset{\oplus}{N}H\}\overset{\ominus}{Cl}$$

The presence of pyridine is necessary otherwise action of the liberated hydrochloric acid on the trialkyl phosphite gives rise to the alkyl chloride.

$$(RO)_3P + HCl \longrightarrow (RO)_2HPO + RCl$$

Phenols give rise to triaryl phosphites on reaction with the appropriate amount of phosphorus trichloride and in the absence of added base; formation of the aryl halide is not a significant competing reaction. Reaction of catechol with phosphorus trichloride in molar proportions gives o-phenylene phosphorochloridite (Section **III,70**), the use of which in the synthesis of alkyl iodides has already been illustrated (Section **III,59**). Hydrolysis of o-phenylene phosphorochloridite with a limited amount of water gives rise to the anhydride, bis-o-phenylene pyrophosphite (Section **III,71**).

Alkyl phosphates. These may be readily prepared from phosphorus oxychloride and the alcohol in the presence of pyridine (e.g., tributyl phosphate, Section **III,72**).

$$3ROH + POCl_3 + 3C_5H_5N \longrightarrow (RO)_3PO + 3C_5H_5\overset{\oplus}{N}H\}\overset{\ominus}{Cl}$$

One or two molar proportions of the alcohol with phosphorus oxychloride gives rise to the mono-alkyl phosphorochloridate or the dialkyl phosphorochloridate respectively, which can then be hydrolysed to the corresponding mono or dialkyl phosphates.

$$ROH + POCl_3 \longrightarrow RO{\cdot}POCl_2 \xrightarrow{H_2O} RO{\cdot}PO(OH)_2$$

$$2ROH + POCl_3 \longrightarrow (RO)_2POCl \xrightarrow{H_2O} (RO)_2PO(OH)$$

Alkyl borates. These are formed by the reaction of boric acid with an excess of the boiling alcohol. In the case of esters of the higher alcohols the water formed in the reaction is conveniently removed by fractional distillation as an azeotropic mixture with the alcohol (e.g., tributyl borate, Section **III,73**).

$$3ROH + B(OH)_3 \longrightarrow B(OR)_3 + 3H_2O$$

Alkyl thiocyanates. The preparation of propyl thiocyanate (Section **III,74**) illustrates a general route to this group of compounds.

$$RBr + KSCN \xrightarrow{heat} R{\cdot}SCN + KBr$$

Alkyl nitrites. These are readily prepared by the interaction at 0 °C of the alcohol with sodium nitrite in the presence of excess of concentrated sulphuric acid.

$$ROH + HONO \xrightarrow{H_2SO_4} RONO + H_2O$$

Alkyl nitrites are valuable nitrosating agents (e.g., Section **III,101**); isopentyl nitrite is the common commercially available example, but several illustrative preparations are given in Section **III,75**.

III,67. t-BUTYL HYPOCHLORITE

$$(CH_3)_3COH + Cl_2 + NaOH \longrightarrow (CH_3)_3C \cdot OCl + NaCl + H_2O$$

Assemble in a fume cupboard a 1-litre three-necked flask fitted with a gas inlet tube extending almost to the bottom of the flask, a sealed stirrer unit and a gas outlet adapter (cf. Fig. IV,2) incorporating a thermometer extending well into the flask. Place in the flask a solution of 40 g (1 mol) of sodium hydroxide in 250 ml of water, and add 37 g (47 ml, 0.5 mol) of t-butyl alcohol followed by sufficient water (about 250 ml) to give a homogeneous solution. Surround the flask with a bath of iced water, and pass chlorine (from a cylinder of the gas) steadily into the solution, with stirring, as long as the gas is readily absorbed, and then slowly for about 30 minutes more. Ensure that the temperature of the reaction mixture does not exceed 20 °C throughout the addition. Transfer the product to a separating funnel, remove and discard the aqueous phase, and wash the upper yellow layer with 25 ml portions of 10 per cent sodium carbonate solution until the washings are no longer acidic to Congo red indicator. Finally wash the product three times with cold water and dry over anhydrous calcium chloride. The yield of t-butyl hypochlorite, which is usually about 98 per cent pure, is 40–50 g (74–92%). It should be stored in a securely stoppered dark bottle in the cold; the vapour is powerfully lachrymatory.

III,68. DIBUTYL SULPHITE

$$2CH_3 \cdot (CH_2)_2 \cdot CH_2OH + SOCl_2 \longrightarrow$$
$$CH_3 \cdot (CH_2)_3 \cdot O \cdot SO \cdot O \cdot (CH_2)_3 \cdot CH_3 + 2HCl$$

In a fume cupboard, set up the apparatus described in Section III,51, using a flask of 250 ml capacity. Place 40 g (24.5 ml, 0.33 mol) of redistilled thionyl chloride in the flask and 50 g (62 ml, 0.67 mol) of dry butan-1-ol (b.p. 116–117 °C) in the dropping funnel. Cool the flask in ice and add the butan-1-ol, with stirring over 1 hour. Reflux the mixture gently for 1 hour to complete the reaction and to remove the residual hydrogen chloride. Distil the reaction mixture under normal pressure until the temperature rises to 120 °C and then continue to distil under diminished pressure. Collect the dibutyl sulphite at 116–118 °C/20 mmHg; the yield is 55 g (85%).

Cognate preparations **Dimethyl sulphite.** From 21 g (27 ml, 0.67 mol) of anhydrous methanol and 40 g (24.5 ml, 0.33 mol) of thionyl chloride. B.p. 126 °C. Yield: 30 g (84%).

Diethyl sulphite. From 31 g (40 ml, 0.67 mol) of absolute ethanol and 40 g (24.5 ml, 0.33 mol) of thionyl chloride. B.p. 156–157 °C. Yield: 31 g (68%).

Dipentyl sulphite. From 59 g (73 ml, 0.67 mol) of pentan-1-ol and 40 g (24.5 ml, 0.33 mol) of thionyl chloride. B.p. 111.5 °C/5 mmHg. Yield: 57 g (78%).

III,69. TRIBUTYL PHOSPHITE

$$3CH_3 \cdot CH_2 \cdot CH_2 \cdot CH_2OH + PCl_3 + 3C_5H_5N \longrightarrow$$
$$(CH_3 \cdot CH_2 \cdot CH_2 \cdot CH_2O)_3P + 3C_5H_6\overset{\oplus}{N}\}\overset{\ominus}{Cl}$$

Place 74 g (91 ml, 1.0 mol) of butan-1-ol, 79 g (81 ml, 1.0 mol) of pyridine (1) and 250 ml of dry ether in a 1-litre three-necked flask fitted with a stirrer.

Attach a dropping funnel containing a mixture of 46 g (29 ml, 0.33 mol) of phosphorus trichloride and 30 ml of dry ether protected by a calcium chloride guard-tube. Cool the flask in an ice bath and run in the phosphorus trichloride solution slowly with stirring, maintaining the temperature of the reaction mixture at about 0 °C. When all has been added, stir for 15 minutes more, and then filter off the pyridinium chloride with suction and wash it with ether. Dry the combined filtrate and washings over anhydrous sodium sulphate, remove the ether on a rotary evaporator and distil the residue under reduced pressure. Collect the tributyl phosphite as a fraction of b.p. 105–110 °C/0.5 mmHg (122 °C/12 mmHg); the yield is 75 g (90%).

Note. (1) An equivalent amount of *N,N*-dimethylaniline may alternatively be used as the base.

III,70. *o*-PHENYLENE PHOSPHOROCHLORIDITE

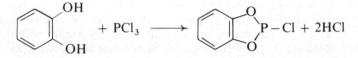

Place 110 g (1.0 mol) of catechol (pyrocatechol) in a 500-ml three-necked flask supported on a water bath and moisten it with about 2 ml of water. Equip the flask with an efficient mechanical stirrer, a large double surface reflux condenser and a 250-ml dropping funnel. Connect the mouth of the condenser to an efficient gas absorption trap (Fig. I,68(*d*)). Add 206 g (131 ml, 1.5 mol) of phosphorus trichloride from the dropping funnel during 15 minutes with vigorous stirring. There is a brisk evolution of hydrogen chloride and the contents of the flask soon solidify, stopping the stirrer. Leave the reaction mixture standing at room temperature for 1 hour, and then heat the water bath to boiling. The solid mass soon melts and stirring again becomes possible. Continue to heat on the boiling water bath with stirring for 2 hours. Arrange the flask for distillation under reduced pressure (water pump); insert a trap cooled in an acetone–Cardice mixture between the receiver and the pump. Distil the product and collect the *o*-phenylene phosphorochloridite as a fraction of b.p. 98 °C/25 mmHg (91 °C/18 mmHg). The yield is 165 g (95%); the product crystallises in the refrigerator, m.p. 30 °C. About 50 g of unreacted phosphorus trichloride is collected in the trap.

III,71. BIS-*o*-PHENYLENE PYROPHOSPHITE

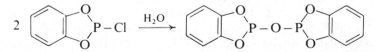

Fit a 250-ml three-necked flask with a sealed stirrer unit. Insert into one of the side-necks a gas inlet tube constricted at the end by drawing it out to about 1 mm internal diameter and extending almost to the bottom of the flask. Connect the open end of the inlet tube by means of a polyethylene tubing sleeve to a short glass connecting tube bent at a right angle, and connect this via a tap and size 19/26 ground glass joint to a 30-ml test-tube containing 4 g (0.22 mol) of water. Place 87 g (0.5 mol) of *o*-phenylene phosphorochloridite

(Section III,70) in the flask, surround the latter with a water bath at 40 °C and stir vigorously. Evacuate the flask by means of a water pump connected to the remaining side-neck, and open the tap leading to the tube containing water to allow the introduction of the water as vapour into the reaction flask during about 5 hours; maintain the temperature of the water bath at 40 °C throughout.

Distil the product under reduced pressure, initially using a water pump, to remove dissolved hydrogen chloride and unreacted starting material (b.p. 98 °C/25 mmHg, 12 g). Continue to distil using an oil pump and collect the pyrophosphite as a fraction of b.p. 146–150 °C/0.6 mmHg. The product solidifies, m.p. 68–70 °C; the yield is 58 g (86% allowing for recovered starting material).

III,72. TRIBUTYL PHOSPHATE

$$3CH_3 \cdot (CH_2)_2 \cdot CH_2OH + POCl_3 \xrightarrow{C_5H_5N}$$
$$(CH_3 \cdot CH_2 \cdot CH_2 \cdot CH_2 \cdot O)_3PO + 3HCl$$

The apparatus required is a 1-litre multi-necked flange flask fitted with a double surface condenser, a sealed stirrer unit, a short-stemmed dropping funnel and a thermometer. Calcium chloride guard-tubes should be provided for the dropping funnel and the reflux condenser. Place 111 g (137 ml, 1.5 mol) of dry butan-1-ol, 130 g (132.5 ml, 1.65 mol) of dry pyridine and 140 ml of dry benzene in the flask, set the stirrer in motion and cool the flask in an ice–salt mixture until the temperature falls to − 5 °C. Introduce 76.5 g (40.5 ml, 0.5 mol) of redistilled phosphorus oxychloride (b.p. 106–107 °C) dropwise from the funnel at such a rate that the temperature does not rise above 10 °C. Reflux gently for 2 hours, and then allow to cool to room temperature. Add 250 ml of water to dissolve the pyridine hydrochloride, separate the benzene layer, wash it several times with water until the washings are neutral and dry over 10 g of anhydrous sodium sulphate. Remove most of the benzene by distillation under normal pressure and finally distil the product under diminished pressure. Collect the tributyl phosphate at 160–162 °C/15 mmHg (or 138–140 °C/6 mmHg). The yield is 95 g (72%).

The above is a general procedure for preparing trialkyl orthophosphates. Similar yields are obtained for **trimethyl phosphate**, b.p. 62 °C/5 mmHg; **triethyl phosphate**, b.p. 75.5 °C/5 mmHg; **tripropyl phosphate**, b.p. 107.5 °C/5 mmHg; **tri-isopropyl phosphate**, b.p. 83.5 °C/5 mmHg; **tri-isobutyl phosphate**, b.p. 117 °C/5.5 mmHg; and **tripentyl phosphate**, b.p. 167.5 °C/5 mmHg.

III,73. TRIBUTYL BORATE

$$3CH_3 \cdot CH_2 \cdot CH_2 \cdot CH_2OH + H_3BO_3 \longrightarrow$$
$$(CH_3 \cdot CH_2 \cdot CH_2 \cdot CH_2 \cdot O \cdot)_3B + 3H_2O$$

The apparatus required consists of a 1-litre two-necked flask carrying a dropping funnel and a 30-cm Hempel column filled with 6-mm glass or porcelain rings (Fig. I,105(c); cf. Fig. III,3) or a 30-cm all-glass Dufton column (Fig. I,105(a)) connected to an efficient double surface condenser. The fractionating column is fitted with a thermometer. Place 62 g (1 mol) of 'AnalaR' boric acid, 333 g (412 ml, 4.5 mol) of butan-1-ol and a few chips of porous porcelain in the flask. Heat the reaction mixture (e.g., in an air bath, or heating mantle) so that it boils gently, and adjust the rate of heating so that 45–50 ml of distillate are collected in 30 minutes. The temperature of the vapour at the top of the column remains at 91 °C over a period of about 2 hours whilst an azeotropic mixture of water and butan-1-ol distils; the latter separates into two layers and contains about 72 per cent of wet alcohol. After 1 hour separate the upper layer of butan-1-ol in the distillate, dry it with anhydrous potassium carbonate or magnesium sulphate and return it to the flask through the dropping funnel; repeat this process after 90 minutes of heating. Subsequently the temperature at the top of the column rises slowly as most of the water is removed, and when the temperature has risen to 110–112 °C (after 2 hours or so) stop the heating. Replace the dropping funnel by a capillary air leak connected to a nitrogen supply, fit a side-arm adapter between the condenser and the receiving flask, and distil under reduced pressure (the fractionating column should be well lagged). The unreacted butanol passes over first, and the temperature then rises sharply. The receiver is changed, and the tributyl borate is collected at 114–115 °C/15 mmHg, or 103–105 °C/8 mmHg. The yield is 210 g (91%) (1).

Cognate preparation Tripentyl borate. Use 62 g (1 mol) of 'AnalaR' boric acid and 396 g (490 ml, 4.5 mol) pentan-1-ol. During the first hour the azeotropic mixture, containing approximately 44 per cent of pentan-1-ol and 56 per cent of water, passes over at 95 °C: subsequently the temperature rises slowly to 136–137 °C. It is unnecessary to return the recovered alcohol to the reaction mixture. The yield of tripentyl borate, b.p. 146–148 °C/16 mmHg, is 260 g (95%).

Note. (1) All alkyl borates are rapidly hydrolysed by moisture and they should therefore be stored in vessels completely sealed from the atmosphere by rubber septum caps. They are usually dispensed subsequently by means of a syringe.

III,74. PROPYL THIOCYANATE

$$CH_3 \cdot CH_2 \cdot CH_2Br + KSCN \longrightarrow CH_3 \cdot CH_2 \cdot CH_2SCN + KBr$$

Fit a 1-litre three-necked flask with a sealed stirrer unit, a reflux condenser and a 250-ml separatory funnel. Place 133 g (1.37 mol) of 'AnalaR' potassium thiocyanate and 310 ml of rectified spirit in the flask, stir the mixture vigorously and heat to boiling. Run in from the separatory funnel 154 g (113.5 ml, 1.25 mol) of propyl bromide (Section **III,54**) during the course of 15–20 minutes; potassium bromide separates. Reflux the mixture, with vigorous stirring, for 5 hours; the stirring must be vigorous otherwise bumping occurs. Filter off the precipitated potassium bromide from the cold reaction mixture (1) and wash it with 75 ml of rectified spirit. Distil off as much of the alcohol as possible on a water bath through a short column. Treat the residue in the flask with 125 ml of water, and separate the upper layer of propyl thiocyanate. Extract the aqueous layer with two 50 ml portions of ether. Combine the ether extracts with the crude thiocyanate, dry with anhydrous sodium or magnesium sulphate and remove most of the ether on a water bath. Distil the residue through an efficient fractionating column (e.g., a long all-glass Dufton column, Fig. I,105(*a*); or a Hempel column filled with 6-mm porcelain or glass rings, Fig. I,105(*c*)); the column must be well lagged with asbestos cloth and preferably heated electrically. A little ether, ethanol and water pass over first (to 110 °C);

the temperature then rises rapidly to 164–165 °C and the remainder distils steadily at 165 °C. The yield of propyl thiocyanate is 93 g (74%).

Note. (1) The evil-smelling residue in the reaction flask is best removed by the *cautious* addition of concentrated nitric acid.

Cognate preparation   Butyl thiocyanate.   Use 133 g (1.37 mol) of 'AnalaR' potassium thiocyanate, 310 ml of rectified spirit and 172 g (135 ml, 1.25 mol) of butyl bromide. The yield of butyl thiocyanate, b.p. 183–184 °C, is 126 g (87%).

III,75. PENTYL NITRITE

$$CH_3 \cdot (CH_2)_3 \cdot CH_2OH + HNO_2 \xrightarrow{H_2SO_4} CH_3 \cdot (CH_2)_3 \cdot CH_2O \cdot NO + H_2O$$

Equip a 1-litre three-necked flask with a powerful mechanical stirrer, a separatory funnel with stem extending to the bottom of the flask and a thermometer. Cool the flask in a mixture of ice and salt. Place a solution of 95 g (1.38 mol) of 'AnalaR' sodium nitrite in 375 ml of water in the flask and stir. When the temperature has fallen to 0 °C (or slightly below) introduce slowly from the separatory funnel a mixture of 25 ml of water, 62.5 g (34 ml) of concentrated sulphuric acid and 110 g (135 ml, 1.25 mol) of pentan-1-ol, which has previously been cooled to 0 °C. The rate of addition must be controlled so that the temperature is maintained at ±1 °C; the addition takes 45–60 minutes. Allow the mixture to stand for 1.5 hours and then filter from the precipitated sodium sulphate (1). Separate the upper yellow pentyl nitrite layer, wash it with a solution containing 1 g of sodium hydrogen carbonate and 12.5 g of sodium chloride in 50 ml of water, and dry it with 5–7 g of magnesium sulphate. The resulting crude pentyl nitrite (107 g, 73%) is satisfactory for many purposes (2). Upon distillation, it passes over largely at 104 °C with negligible decomposition. The b.p. under reduced pressure is 29 °C/40 mmHg.

Notes. (1) Care must be exercised in handling pentyl and the other alkyl nitrites; inhalation of the vapour may cause severe headache and heart excitation. The preparation must therefore be conducted in an efficient fume cupboard.

(2) Alkyl nitrites decompose slowly upon standing and should be kept in a cool place. They should preferably be used within a few days or, at most, within two weeks of their preparation. The decomposition products include water, oxides of nitrogen, the alcohol and polymerisation products of the aldehyde.

Cognate preparations   Hexyl nitrite.   Use 95 g (1.38 mol) of 'AnalaR' sodium nitrite in 375 ml of water; a mixture of 25 ml of water, 62.5 g (34 ml) of concentrated sulphuric acid and 127.5 g (156 ml, 1.25 mol) of hexan-1-ol. The yield of crude product is 124 g (76%). B.p. 129–130.5 °C or 52 °C/44 mmHg.

Butyl nitrite.   Use quantities as for *hexyl nitrite*, but with 92.5 g (114 ml, 1.25 mol) of butan-1-ol replacing the hexan-1-ol. The yield of crude product is 110 g (85%). Butyl nitrite boils at 76.5–77.5 °C at atmospheric pressure with slight decomposition, but distils unchanged at 27 °C/43 mmHg.

Ethyl nitrite.   Dissolve 38 g (0.55 mol) of sodium nitrite in 120 ml of water in a 500-ml flask equipped as above. Dilute 23 g (29 ml, 0.5 mol) of ethanol with an equal volume of water, carefully add 25 g (13.5 ml) of concentrated sulphuric acid and dilute to 120 ml with water. Cool both solutions to $-10\,°C$ in an ice–salt bath and add the acid–alcohol mixture to the nitrite solution slowly with constant stirring during about 30 minutes. Transfer the reaction mixture to a cooled separating funnel, run off the lower aqueous phase, wash the ethyl nitrite layer rapidly with ice-cold 2 per cent sodium hydrogen carbonate solution and dry over anhydrous sodium sulphate. The product may be kept at $0\,°C$ as a 50 per cent solution in absolute ethanol if required but should be used as soon as possible. The b.p. of pure ethyl nitrite is $17\,°C$.

G ALIPHATIC ETHERS

The synthesis of ethers is exemplified by two main general procedures.

1. The elimination of water from two molecules of the alcohol under acidic conditions (Section **III,76**).
2. The interaction of an alkyl halide with a sodium alkoxide (**Williamson's synthesis**, Section **III,78**).

G,1. The dehydration of alcohols to ethers.   Diethyl ether $[(C_2H_5)_2O]$ can be prepared by heating ethanol with sulphuric acid at about $140\,°C$, and adding more alcohol as the ether distils out of the reaction medium. A similar 'continuous etherification process' is used industrially. A more general procedure for the preparation of ethers from primary alcohols (e.g., dibutyl ether, Section **III,76**) is to arrange for the water formed in the reaction to be removed azeotropically.

$$2R \cdot CH_2OH \xrightarrow{H_2SO_4} R \cdot CH_2 \cdot O \cdot CH_2 \cdot R + H_2O$$

Excessive heating of the reaction mixture must be avoided otherwise an alkene-forming elimination reaction is induced; this is particularly the case with secondary and especially tertiary alcohols.

The synthesis of dichloromethyl methyl ether has been included because of its usefulness as a reagent for the preparation of aromatic aldehydes (Section **IV,129**). It is readily obtained by the reaction of phosphorus pentachloride in admixture with phosphorus oxychloride with methyl formate (Section **III,77**).

Chloromethyl methyl ether and bis-chloromethyl ether ($ClCH_2 \cdot O \cdot CH_2Cl$) in common with several alkylating agents possess carcinogenic properties in experimental animals and have been listed as assumed human carcinogens (see also Section **IV,33**); similar hazardous properties associated with dichloromethyl methyl ether have not been reported but it would be prudent to handle this compound with due care.

III,76. DIBUTYL ETHER

$$2CH_3 \cdot CH_2 \cdot CH_2 \cdot CH_2OH \xrightarrow{H_2SO_4} (CH_3 \cdot CH_2 \cdot CH_2 \cdot CH_2)_2O + H_2O$$

The success of this preparation depends upon the use of the Dean and Stark apparatus which permits the automatic separation of the water produced in the reaction. The quantity of water which should be eliminated, assuming a quantitative conversion of the alcohol (0.67 mol) into the ether, is 6.0 g (0.33 mol).

Assemble the apparatus illustrated in Fig. III,2; B is a 250-ml flask. Fill the water-separator tube A of the Dean and Stark apparatus with water and remove 6.0 ml (the quantity to be formed in the reaction). Place 49 g (61 ml, 0.67 mol) of butan-1-ol together with 16 g (9 ml) of concentrated sulphuric acid in the flask B. Heat the flask gently on a wire gauze so that the liquid refluxes from the condenser. Water and butan-1-ol will collect in the tube A and when this is full automatic separation of the two liquids will commence; the water will fall to the bottom of the tube A and the lighter butan-1-ol will pass back into the flask. Continue the heating until the temperature inside the flask rises to 134–135 °C (after about 30–40 minutes). At this stage 5–6 ml of additional water will have collected in A and the reaction may be regarded as complete. Further heating will merely result in considerable darkening of the mixture in the flask and the formation of the highly inflammable but-1-ene. Allow the reaction mixture to cool or cool the flask under running water from the tap. Pour the contents of the flask and water-separator tube into a separatory funnel containing 100 ml of water, shake well and remove the upper layer containing the crude ether mixed with a little unchanged butanol. Shake the crude ether with 25 ml of cold 50 per cent by weight sulphuric acid (from 20 ml of concentrated acid and 35 ml of water) (1) for 2–3 minutes, separate the upper layer and repeat the extraction with another 25 ml of the acid. Finally wash twice with 25 ml portions of water; dry with 2 g of anhydrous calcium chloride (2). Filter through a fluted filter-paper into a 50-ml flask and distil. Collect the dibutyl ether at 139–142 °C. The yield is 15 g (34%).

Notes. (1) This separation utilises the fact that butan-1-ol is soluble in 50 per cent sulphuric acid by weight, whereas dibutyl ether is only slightly soluble.

(2) An alternative method for isolating dibutyl ether utilises the fact that butan-1-ol is soluble in saturated calcium chloride solution whilst dibutyl ether is slightly

Fig. III,2

soluble. Cool the reaction mixture in ice and transfer to a separatory funnel. Wash cautiously with 100 ml of 2.5–3 M-sodium hydroxide solution; the washings should be alkaline to litmus. Then wash with 30 ml of water, followed by 30 ml of saturated calcium chloride solution. Dry with 2–3 g of anhydrous calcium chloride.

Cognate preparations Dipentyl ether. Use 58.7 g (72 ml, 0.67 mol) of pentan-1-ol (b.p. 136–137 °C) and 8 g (4.5 ml) of concentrated sulphuric acid. The calculated volume of water (6 ml) is collected when the temperature inside the flask rises to 157 °C (after 90 minutes). Steam distil the reaction mixture, separate the upper layer of the distillate and dry it with anhydrous potassium carbonate. Distil from a 50-ml flask and collect the fractions of boiling point (i) 145–175 °C (15 g), (ii) 175–185 °C (9 g) and (iii) 185–190 °C (largely 185–185.5 °C) (15 g). Combine fractions (i) and (ii), reflux for 1 hour over 3 g of sodium and distil from the sodium alkoxide and excess of sodium; this yields 11 g of fairly pure dipentyl ether which combined with fraction (iii) gives a total yield (26 g) of 49 per cent. A perfectly pure product, b.p. 184–185 °C, is obtained by further distillation from a little sodium. (See Section I,3,B,5(iv) for the destruction of sodium residues.)

Di-isopentyl ether.   Proceed as for dipentyl ether, using 58.7 g (73 ml, 0.67 mol) of 3-methylbutan-1-ol; the temperature rises to 148–150 °C. Collect the fractions of b.p. 135–150 °C (16 g), 150–168 °C (12 g) and 168–174 °C (12 g). After distillation from sodium the yield of di-isopentyl ether, b.p. 170–171.5 °C, is 28 g (53%).

Dihexyl ether.   Use 68 g (83 ml, 0.67 mol) of hexan-1-ol (b.p. 156–157 °C); heat until the temperature rises to 180 °C. Pour the reaction mixture into water, separate the upper layer, wash it twice with 5 per cent sodium hydroxide solution, then with water, and dry over anhydrous potassium carbonate. Distil and collect the fractions of b.p. (i) 160–221 °C (23 g) and (ii) 221–223 °C (23 g). Reflux fraction (i) with 5 g of sodium and distil from the excess of sodium when a further quantity of fairly pure dihexyl ether (13 g, fraction (iii)) is thus obtained. Combine fractions (ii) and (iii) and distil from a little sodium; collect the pure dihexyl ether (26 g, 42%) at 221.5–223 °C.

III,77. DICHLOROMETHYL METHYL ETHER (1,1-dichlorodimethyl ether)

$$H \cdot CO_2 CH_3 \xrightarrow{PCl_5/POCl_3} Cl_2 CH \cdot O \cdot CH_3$$

In a fume cupboard equip a 500-ml three-necked flask with a dropping funnel, a mechanical stirrer, a thermometer and a reflux condenser, using a double neck adapter (Fig. I,17) to accommodate the last two items. Protect the condenser and dropping funnel with calcium chloride guard-tubes. Place 50 ml of phosphorus oxychloride and 156 g (0.75 mol) of phosphorus pentachloride in the flask and 48 g (49 ml, 0.8 mol) of methyl formate in the dropping funnel. Cool the mixture to 10 °C and add the methyl formate dropwise at such a rate that the reaction temperature does not rise above 20 °C (about 1 hour). When the addition is complete, remove the ice bath, stir the mixture until all the phosphorus pentachloride has dissolved, keeping the temperature below 30 °C by occasional cooling. Remove the stirrer, condenser, thermometer and dropping funnel, stopper the two side-necks and insert a simple distillation head into the central neck. Introduce a pine splint to serve as an anti-bumping device (1) and attach a condenser leading to a receiver flask via a receiver adapter for vacuum distillation. Distil the reaction mixture under reduced pressure (water pump) on a water bath at about 60 °C and collect the distillate in a flask cooled to −25 to −30 °C in a cooling bath (acetone–Cardice or ice–salt). The distillate weighs 209 g and consists of a mixture of dichloromethyl methyl ether (b.p. 85 °C) and phosphorus oxychloride (b.p. 105 °C). Fractionally distil the mixture at atmospheric pressure through a 50-cm column filled with glass helices and surrounded by a heating jacket at 60 °C, using a reflux ratio of about 1 : 8 (Section I,26). Collect the fraction which boils between 82 and 95 °C and refractionate to give 70 g (76%) pure dichloromethyl methyl ether, b.p. 85 °C. Protect the product from moisture.

Note. (1) A conventional capillary air leak, inserted into a side-neck, should only be used in this case if the air supply is pre-dried since the product is susceptible to decomposition by moisture.

G,2. The Williamson synthesis. This involves the direct nucleophilic displacement of halogen in an alkyl halide by an alkoxide ion.

$$RO\overset{\ominus}{:} \longrightarrow R\overset{\frown}{-I} \longrightarrow R\cdot O\cdot R + I^{\ominus}$$

The method is particularly useful for the preparation of mixed ethers. However, the action of the strongly basic alkoxide ions on secondary or tertiary alkyl halides gives rise to extensive alkene formation. Secondary or tertiary alkyl groups can therefore only be incorporated into ethers by the Williamson synthesis by way of the corresponding alkoxide ions in reaction with a primary halide.

$$R_3C\cdot O\overset{\ominus}{:} + R'\cdot CH_2Br \longrightarrow R_3C\cdot O\cdot R' + Br^{\ominus}$$

III,78. ETHYL HEXYL ETHER

$$C_5H_{11}\cdot CH_2OH \xrightarrow{Na} C_5H_{11}\cdot CH_2O\overset{\ominus}{:}$$
$$\overset{\oplus}{Na} \xrightarrow{CH_3\cdot CH_2I} C_5H_{11}\cdot CH_2\cdot O\cdot CH_2\cdot CH_3$$

Place 204 g (250 ml, 2 mol) of dry hexan-1-ol in a 500-ml round-bottomed flask fitted with a Liebig-type reflux condenser and introduce 5.75 g (0.25 mol) of clean sodium (Section II,2,55) in small pieces and warm under reflux until all the sodium has reacted (*ca.* 2 hours). Introduce 39 g (20 ml, 0.25 mol) of ethyl iodide down the condenser from a dropping funnel and reflux gently for 2 hours; sodium iodide gradually separates. Arrange the apparatus for downward distillation and collect the crude ether at 143–148 °C (27 g). When cold, refit the reflux condenser, add a further 5.75 g (0.25 mol) of clean sodium and warm until all has reacted: alternatively, allow the reaction to proceed overnight, by which time all the sodium will have reacted. Introduce a further 39 g (20 ml, 0.25 mol) of ethyl iodide and reflux for 2 hours; distil off the crude ether and collect the fraction passing over at 143–148 °C. Combine the two distillates. Remove most of the hexan-1-ol still present in the crude ether by heating under reflux for 2 hours with a large excess of sodium and then distil until no more liquid passes over. Distil the resulting liquid from a few grammes of sodium using a short fractionating column, and collect the ethyl hexyl ether at 140–143 °C. The yield is 30 g (46%). If the sodium is appreciably attacked, indicating that all the alcohol has not been completely removed, repeat the distillation from a little fresh sodium. [See Section I,3,B,5(iv) for instructions in the destruction of sodium residues.]

Cognate preparations Hexyl methyl ether. Use 204 g (250 ml, 2 mol) of hexan-1-ol, 2 × 5.75 g (total 0.5 mol) of clean sodium and 2 × 35.5 g (2 × 15.6 ml, total 0.5 mol) of methyl iodide. The yield of hexyl methyl ether, b.p. 125–126 °C, is 42 g (72%).

Butyl methyl ether. Use 148 g (183 ml, 2 mol) of butan-1-ol, 2 × 5.75 g (0.5 mol) of clean sodium and 2 × 35 g (2 × 15.6 ml, total 0.5 mol) of methyl iodide. The yield of butyl methyl ether, b.p. 70–71 °C, is 31 g (70%).

Butyl phenyl ether. Weigh out 11.5 g (0.5 mol) of clean sodium into a dry, 1-litre round-bottomed flask, provided with a double surface condenser, and add 250 ml of absolute ethanol. If the reaction becomes so vigorous that the alcohol cannot be held back by the condenser, direct a stream of cold water or place a wet towel on the outside of the flask until it is again under control: do not cool the alcohol unduly otherwise the last traces of sodium will take a considerable time to dissolve. Add a solution of 47 g (0.5 mol) of pure phenol in 50 ml of absolute ethanol and shake. Into a small separatory funnel supported in the top of the condenser, place 133 g (82.5 ml, 0.72 mol) of butyl iodide (Section **III,58**) or an equivalent quantity of butyl bromide (Section **III,54**) and add it, with shaking, during 15 minutes. Boil the solution gently for 3 hours, arrange the apparatus for downward distillation and distil off as much as possible of the alcohol on a water bath; this process is facilitated by wrapping the exposed part of the flask in a cloth. Add water to the residue in the flask, separate the organic layer and wash it twice with 25 ml portions of 10 per cent sodium hydroxide solution, then successively with water, dilute sulphuric acid and water: dry with magnesium sulphate. Distil and collect the butyl phenyl ether at 207–208 °C. The yield is 60 g (80%).

H ALIPHATIC ALDEHYDES

The synthesis of aliphatic aldehydes is exemplified by the following typical procedures.

1. The controlled oxidation or dehydrogenation of primary alcohols (Sections **III,79** and **III,80**).
2. The oxidative cleavage of 1,2-diols (Section **III,82**).
3. The ozonolysis of suitably substituted alkenes (Section **III,83**).
4. The reduction of nitriles by way of the aldimine hydrochlorides (Section **III,84**).
5. The reaction of Grignard reagents with triethyl orthoformate (Section **III,85**).
6. The hydrolysis and decarboxylation of α,β-epoxy esters (glycidic esters) (Section **III,86**).

H,1. The controlled oxidation or dehydrogenation of primary alcohols. Simple aldehydes may be obtained in reasonably good yield by oxidation of the corresponding primary alcohol with sodium dichromate in dilute sulphuric acid solution (e.g., butyraldehyde, Section **III,79**). To avoid further oxidation to the corresponding acid, the aldehyde is removed as rapidly as possible by distillation through a fractionating column.

$$\text{R·CH}_2\text{OH} \xrightarrow{\text{[O]}} \text{R·CHO}$$

The main by-product is an ester which arises as the result of the oxidation of an intermediately formed hemi-acetal:

$$R \cdot CH_2OH + R \cdot CHO \overset{H^{\oplus}}{\rightleftharpoons} R \cdot CH_2O \cdot CH(OH) \cdot R \longrightarrow R \cdot CH_2O \cdot CO \cdot R$$

Satisfactory yields of aldehydes are usually obtained when the vapour of the primary alcohol is dehydrogenated by passage over a heated catalyst of copper–chromium oxide deposited on pumice (Section **III,80**).

Acetaldehyde may be obtained by the depolymerisation of paraldehyde (Section **III,81**). Similarly, depolymerisation of paraformaldehyde offers a convenient source of anhydrous formaldehyde as illustrated in Section **III,33**.

III,79. BUTYRALDEHYDE

$$CH_3 \cdot (CH_2)_2 \cdot CH_2OH \xrightarrow[Na_2Cr_2O_7/H_2SO_4]{[O]} CH_3 \cdot (CH_2)_2 \cdot CHO$$

Fit up the apparatus shown in Fig. III,3. The flask is of 500 ml capacity and the Hempel column is filled with 6-mm glass or porcelain rings (1); the receiver

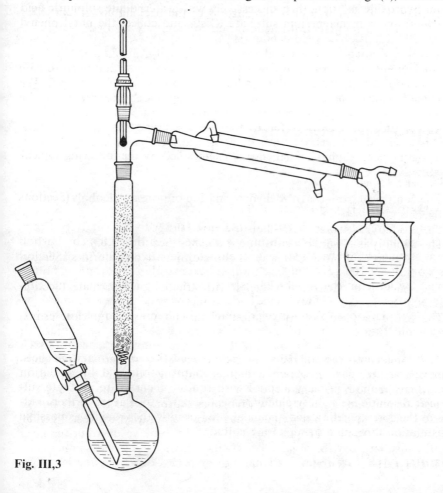

Fig. III,3

is cooled in crushed ice or in cold water. Dissolve 56 g (0.188 mol) of sodium dichromate dihydrate in 300 ml of water and add cautiously, with stirring, 40 ml of concentrated sulphuric acid. Place 41 g (51 ml, 0.55 mol) of butan-1-ol together with a few small chips of porous porcelain in the flask, heat the butan-1-ol to boiling and run in the dichromate solution via the dropping funnel during about 20 minutes. The oxidation to butyraldehyde proceeds with the evolution of heat, but it is necessary to continue to heat the flask so that the mixture boils vigorously to maintain steady distillation. The temperature at the top of the column, however, should not exceed 80–85 °C. When all the oxidising agent has been added, continue heating the mixture for 15 minutes and collect all that passes over below 90 °C. Separate the water from the distillate and dry the residue (29 g) for 30–60 minutes with 3–4 g of anhydrous sodium sulphate. Meanwhile detach the fractionating column from the apparatus and dry the glass or porcelain rings by washing with acetone and blowing hot air through them. Fit the column into a 100-ml flask and arrange for distillation as before. Distil the dried distillate slowly (1–2 drops per second) through the column and collect as fairly pure butyraldehyde all that distils below 76 °C. The yield is 13 g (32%). Pure butyraldehyde boils at 74.5 °C.

Note. (1) The approximate dimensions of the packing are 15 cm × 1.5 cm. Any other form of efficient fractionating column may be used.

Cognate preparation   Propionaldehyde.   Use 34 g (42.5 ml, 0.567 mol) of propan-1-ol and a solution containing 56 g (0.188 mol) of sodium dichromate dihydrate, 300 ml of water and 40 ml of concentrated sulphuric acid. The experimental details are identical with those for butyraldehyde, except that the temperature at the top of the column is not allowed to rise above 70–75 °C, and during the subsequent heating for 15 minutes the liquid passing over below 80 °C is collected; the receiver must be cooled in ice. The yield of propionaldehyde, b.p. 47–50 °C, is 12 g (36%).

III,80.   HEXANAL

$$CH_3 \cdot (CH_2)_4 \cdot CH_2OH \xrightarrow{-H_2} CH_3 \cdot (CH_2)_4 \cdot CHO$$

Assemble the apparatus illustrated in Fig. I,72(c) with the catalyst of copper–chromium oxide deposited on pumice packed into the Pyrex combustion tube in sections of about 25 cm long, each being separated by a small plug of glass wool. Place 100 g (122 ml, 0.98 mol) of hexan-1-ol in the dropping funnel. The gas outlet from the Drechsel bottle, E, should be led into a fume cupboard or to an outside window, since hydrogen is evolved in the reaction. Place 0.1 g of hydroquinone in the receiver to act as a 'stabiliser' for the aldehyde. Pass a gentle flow of nitrogen gas through the combustion tube and adjust the temperature of the furnace to 330 °C. After 2 hours, allow the alcohol to pass into the combustion tube at the rate of 1 drop every 3–4 seconds. The commencement of the dehydrogenation will be indicated by the production of white fumes at the point where the combustion tube enters the condenser; it will also be indicated by a gas flow (hydrogen) through the Drechsel bottles which continues after the nitrogen flow has been temporarily stopped. When all the hexanol has passed through the catalyst tube, remove the aqueous layer from the distillate, dry the organic layer with a little magnesium sulphate and distil from a flask carrying a fractionating side arm. Collect the fraction (30 g)

(1) having b.p. 125–135 °C, and redistil to obtain 21 g (21%) of hexanal (2) having b.p. 127–129 °C.

Notes. (1) If the high boiling residue is transferred to a smaller flask and fractionally distilled, some hexanol passes over first, followed by hexyl hexanoate ($CH_3 \cdot (CH_2)_4 \cdot CO_2(CH_2)_5 \cdot CH_3$) (2 g) at 240–250 °C (mainly at 245 °C).

(2) About 0.1 per cent of hydroquinone should be added as a stabiliser since hexanal exhibits a marked tendency to polymerise. To obtain pure hexanal, treat the 21 g of the product with a solution of 42 g of sodium metabisulphite in 125 ml of water and shake; much bisulphite derivative will separate. Steam distil the suspension of the bisulphite complex until about 50 ml of distillate have been collected; this will remove any non-aldehydic impurities together with a little aldehyde. Cool the residual aldehyde–bisulphite solution to 40–50 °C, and add slowly a solution of 32 g of sodium hydrogen carbonate in 80 ml of water, and remove the free aldehyde by steam distillation. Separate the upper layer of hexanal, wash it with a little water, dry with magnesium sulphate and distil: the pure aldehyde passes over at 128–128.5 °C.

Cognate preparations   Valeraldehyde. Use 100 g (123 ml, 1.14 mol) of pentan-1-ol, and fractionate the dried distillate. Collect the fraction of b.p. 98–110 °C (23 g); upon redistillation 20 g (20%) of valeraldehyde, b.p. 101–105 °C, are obtained. From the high boiling fractions 25 g of pentan-1-ol (b.p. 135–139 °C) may be recovered, together with 1.5 g of pentyl valerate (b.p. 205–210 °C).

Butyraldehyde. Use 100 g (123.5 ml, 1.35 mol) of butan-1-ol. The yield of butyraldehyde, b.p. 70–75 °C, is 38 g (39%), and of butyl butyrate (b.p. 165–170 °C) is 2 g; 40 g of butan-1-ol are recovered.

Propionaldehyde. Use 100 g (125 ml, 1.67 mol) of propan-1-ol and surround the receiver by a freezing mixture. The yield of propionaldehyde, b.p. 48–49.5 °C (mainly 49 °C), is 35 g (36%), and of propyl propionate, b.p. 120–125 °C, is 1 g; 30 g of propan-1-ol are recovered.

III,81. ACETALDEHYDE

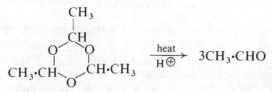

Assemble a fractional distillation apparatus consisting of a 250-ml round-bottomed flask, a Hempel (or Dufton) fractionating column and a condenser leading to a 100-ml receiving flask. Place 50 ml of paraldehyde in the flask together with 0.5 ml of concentrated sulphuric acid (which acts as the depolymerising agent) (1) and a few small fragments of porous porcelain. Cool the receiver in crushed ice to diminish losses due to evaporation. Warm the flask very gently on a wire gauze (or, better, in a water bath at 50–60 °C); do not allow the temperature at the head of the column to rise above 30–32 °C. The distillation must be conducted very slowly in order that the fractionation may be efficient, since acetaldehyde and paraldehyde form a constant boiling point mixture, b.p. 42 °C (53.4 and 46.6 mol per cent respectively). In practice it is found that most of the acetaldehyde distils at 21–25 °C. Stop the distillation when 10 ml of liquid remain in the flask: distillation to dryness may result in an explosion. The resulting acetaldehyde, produced in excellent yield, is sufficiently pure for most purposes. If it is not required immediately, stopper the flask and keep it in the ice chest or in a refrigerator until required.

To obtain pure acetaldehyde, the product must be redistilled. Clean and dry the 250-ml flask first used, immerse it in cold or ice water, pour in the crude acetaldehyde rapidly, attach the fractionating column, etc. Immerse the receiver in crushed ice. Heat the flask gently in a water bath and adjust the temperature so that the aldehyde distils slowly and at a uniform temperature. The temperature recorded at the top of the column may depend partly upon the temperature of the laboratory, if this is above 21 °C. Pure acetaldehyde boils at 21 °C.

Note. (1) The sulphuric acid may be replaced by 1–2 g of toluene-p-sulphonic acid.

H,2. The oxidative cleavage of 1,2-diols. Cleavage of 1,2-diols using either lead tetra-acetate or sodium metaperiodate is a general reaction which has some preparative applications.

$$R^1 \cdot CH(OH) \cdot CH(OH) \cdot R^2 \xrightarrow[\text{or } NaIO_4/H^{\oplus}]{Pb(O \cdot CO \cdot CH_3)_4} R^1 \cdot CHO + R^2 \cdot CHO$$

Cleavage of open-chain symmetrical diols, which must yield a single aldehydic product, is clearly of most value. The preparation of butyl glyoxylate from dibutyl (+)-tartrate (Section **III,82**) using lead tetra-acetate in benzene solution is an interesting example.

Periodic acid (or sodium metaperiodate) is the reagent of choice when aqueous media are preferred, as in the carbohydrate field; an example of its use is to be found in Section **III,117**.

III,82. BUTYL GLYOXYLATE

$$C_4H_9O_2C \cdot CH(OH) \cdot CH(OH) \cdot CO_2C_4H_9 + Pb(O \cdot CO \cdot CH_3)_4 \longrightarrow$$
$$2C_4H_9O_2C \cdot CHO + Pb(O \cdot CO \cdot CH_3)_2 + 2CH_3 \cdot CO_2H$$

In view of the quantities of benzene required, the entire preparation must be carried out in the fume cupboard. Place a mixture of 125 ml of pure benzene and 32.5 g (0.123 mol) of dibutyl (+)-tartrate (1) in a 500-ml three-necked flask, equipped with a Hershberg stirrer (Fig. I,55) and a thermometer. Stir the mixture rapidly and add 58 g (0.13 mol) of lead tetra-acetate (Section **II,2,**38) in small portions over a period of 20 minutes whilst maintaining the temperature below 30 °C by occasional cooling with cold water. Continue the stirring for a further 60 minutes. Separate the salts by suction filtration and wash with two 25 ml portions of benzene. Remove the benzene and acetic acid from the filtrate by flash distillation and distil the residue under diminished pressure, preferably in a slow stream of nitrogen. Collect the butyl glyoxylate (2) at 66–69 °C/5 mmHg. The yield is 26 g (81%).

Notes. (1) The purified commercial dibutyl (+)-tartrate, m.p. 22 °C, may be used. It may be prepared by using the procedure described under isopropyl lactate (Section **III,147**). Place a mixture of 75 g of (+)-tartaric acid, 10 g of Zerolit 225/H$^{\oplus}$, 110 g (135 ml) of redistilled butan-1-ol and 150 ml of sodium-dried benzene in a 1-litre three-necked flask equipped with a sealed stirrer, a double surface condenser and an automatic water separator (cf. Fig. III,2).

Reflux the mixture with stirring for 10 hours: about 21 ml of water collects in the water separator. Filter off the ion-exchange resin at the pump and wash it with two 30–40 ml portions of hot benzene. Wash the combined filtrate and washings with two 75 ml portions of saturated sodium hydrogen carbonate solution, followed by 100 ml of water, and dry over magnesium sulphate. Remove the benzene by distillation under reduced pressure (water pump) and finally distil the residue. Collect the dibutyl (+)-tartrate at 150 °C/1.5 mmHg. The yield is 90 g (69%).

(2) Store the butyl glyoxylate under nitrogen; it undergoes autoxidation in air. The product decomposes on boiling (159–161 °C) at atmospheric pressure.

H,3. The ozonolysis of suitably substituted alkenes. Oxidation of alkenes with ozone followed by cleavage of the resulting ozonides to carbonyl compounds is widely used for the determination of structure of unsaturated compounds. The ozonolysis technique is described in detail in Section **I,17,4**.

For preparative purposes the cleavage of the ozonide is best carried out by catalytic hydrogenation over palladium hydroxide-on-calcium carbonate, a catalyst system which does not hydrogenate the aldehydic products; the yield of the latter are usually fairly good. Two examples of preparative ozonolysis are given (Section **III,83**). In the first cyclohexene is subjected to ozonolysis in ethyl acetate to provide a convenient synthesis of the α,ω-dicarbonyl compound, adipaldehyde. In the second, the substrate is oleic acid, which yields two aldehydic fragments, nonanal and azelaic hemialdehyde.

$$CH_3 \cdot (CH_2)_7 \cdot CH = CH \cdot (CH_2)_7 \cdot CO_2H \longrightarrow$$
$$CH_3 \cdot (CH_2)_7 \cdot CHO + OCH \cdot (CH_2)_7 \cdot CO_2H$$

The solvent used in the latter case is dry ethyl chloride, which must be replaced by methanol before hydrogenation. Methanol when cooled to about − 20 °C is not attacked by ozone and can in many cases be used as the ozonisation solvent. In such cases the main product of addition of ozone to the alkene is not the usual ozonide, but is the methoxyhydroperoxide which, however, is also reductively cleaved to the aldehyde on hydrogenation. This ozonisation process may be mechanistically summarised as follows:

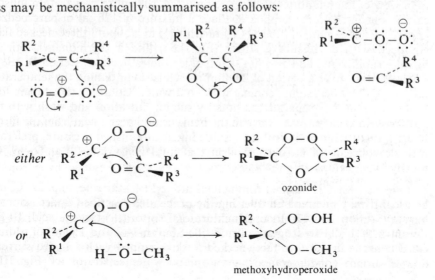

419

III,83. ADIPALDEHYDE (*Hexanedial*)

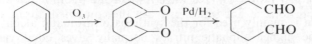

Dissolve 8.2 g (0.1 mol) of cyclohexene (Section **III,7**) in 200 ml of pure dry ethyl acetate (Section **II,1,24**) contained in a 500-ml glass-stoppered wash bottle, cool the solution to -20 to $-30\,°C$ or below (e.g., with solid carbon dioxide–acetone) and attach the wash bottle through a calcium chloride drying tube to another containing acidified potassium iodide solution. Pass ozonised oxygen (**CAUTION:** see Section **I,17,4**) until the reaction is complete, i.e., until iodine is abundantly liberated. Then add 0.5 g of palladium hydroxide–calcium carbonate catalyst (Section **II,2,44**(*e*)) (1) and hydrogenate the cold solution of the ozonide in the usual manner (see Section **I,17,4**; cool the hydrogenation vessel in ice. Filter off the catalyst and remove the solvent by distillation at normal pressure. Distil the residue under reduced pressure and collect the adipaldehyde at 92–94 $°C/12$ mmHg. The yield is 7 g (61%). This aldehyde oxidises readily and should be kept in a sealed tube in an atmosphere of nitrogen or carbon dioxide. It may be converted into the dioxime by warming with aqueous hydroxylamine acetate solution: after recrystallisation from water, the dioxime has m.p. 172 $°C$.

Cognate preparation   Azelaic hemialdehyde semicarbazone. Dissolve 7 g (0.025 mol) of pure oleic acid in 30 ml of dry ethyl chloride (chloroform may be used but is less satisfactory), and ozonise at about $-30\,°C$. Remove the bulk of the solvent under reduced pressure, dissolve the residue in 50 ml of dry methanol and hydrogenate as for adipaldehyde in the presence of 0.5 g of palladium–calcium carbonate (2). Warm the resulting solution for 30 minutes with a sight excess of semicarbazide acetate and pour into water. Collect the precipitated semicarbazones and dry: the yield is 8.5 g. Separate the mixture of semicarbazones by either of the following methods: (*a*) Treat with dilute sodium hydrogen carbonate solution to extract the semicarbazone of azelaic hemialdehyde; upon acidifying the extract with dilute sulphuric acid, the semicarbazone of azelaic hemialdehyde is precipitated (4.4 g, 77%, m.p. 162 $°C$, after recrystallisation from methanol). The residue from the sodium hydrogen carbonate extraction consists of the semicarbazone of nonanal and melts at 101 $°C$ after recrystallisation from methanol: yield 3.8 g (77%).

(*b*) Extract the dry mixture of semicarbazones with ether: only the semicarbazone of nonanal dissolves easily.

Notes. (1) The Adams platinum oxide catalyst gives satisfactory results in the reduction of ozonides.

(2) An alternative work up procedure after hydrogenation is to isolate the nonanal by steam distillation and to recover the azelaic hemialdehyde from the residual aqueous layer (Ref. 1).

H,4. The reduction of nitriles by way of the aldimine hydrochlorides. This is achieved with anhydrous tin(II) chloride dissolved in ether or ethyl acetate saturated with dry hydrogen chloride (the **Stephen reaction**). The resulting aldimine hydrochloride (probably in the form of a complex with tin(IV) chloride) is then hydrolysed with warm water.

$$R \cdot C \equiv N + HCl \longrightarrow [R \cdot C \equiv \overset{\oplus}{N}H]\overset{\ominus}{Cl} \xrightarrow{[H]} [R \cdot CH = \overset{\oplus}{N}H_2]\overset{\ominus}{Cl} \xrightarrow{H_2O} R \cdot CHO$$

The Stephen reaction is mainly used for the synthesis of aromatic aldehydes but reduction of the higher aliphatic nitriles normally gives good yields; the method is illustrated by the synthesis of hexanal and octanal (Section **III,84**).

The catalytic reduction of carboxylic acid chlorides by the Rosenmund procedure may be used for the preparation of aliphatic aldehydes but its application is mainly to the synthesis of aromatic aldehydes (e.g., Section **IV,133**).

III,84. HEXANAL

$$CH_3 \cdot (CH_2)_3 \cdot CH_2 \cdot CN \xrightarrow[SnCl_2]{HCl} CH_3 \cdot (CH_2)_3 \cdot CH_2 \cdot CH = \overset{\oplus}{N}H_2 \} \overset{\ominus}{Cl} \xrightarrow{H_2O}$$

$$CH_3 \cdot (CH_2)_3 \cdot CH_2 \cdot CHO$$

Into a 500-ml three-necked flask, provided with a mechanical stirrer, a gas inlet tube and a reflux condenser, place 57 g of anhydrous tin(II) chloride (Section **II,2,60**) and 200 ml of anhydrous ether. Pass in dry hydrogen chloride gas (Section **II,2,31**) until the mixture is saturated and separates into two layers; the lower viscous layer consists of tin(II) chloride dissolved in ethereal hydrogen chloride. Set the stirrer in motion and add 19.5 g (0.2 mol) of hexanenitrile (Section **III,163**) through the separatory funnel. Separation of the crystalline aldimine hydrochloride commences after a few minutes; continue the stirring for 15 minutes. Filter off the crystalline solid, suspend it in about 50 ml of water and heat under reflux until it is completely hydrolysed. Allow to cool and extract with ether; dry the ethereal extract with anhydrous calcium sulphate or magnesium sulphate and remove the ether slowly by distillation through a short fractionating column. Finally, distil the residue and collect the hexanal at 127–129 °C. The yield is 19 g (95%).

Cognate preparation   Octanal. Use 25 g of octanenitrile (Section **III,161**), b.p. 87 °C/10 mmHg, 57 g of anhydrous tin(II) chloride and 200 ml of anhydrous ether. Isolate the aldehyde by steam distillation and ether extraction. An almost quantitative yield of octanal, b.p. 65 °C/11 mmHg, is obtained.

H,5. The reaction of Grignard reagents with triethyl orthoformate.
Although the addition of a Grignard reagent to a formate ester yields an aldehyde, further addition of the reagent results in the formation of the corresponding secondary alcohol (cf. synthesis of secondary alcohols, p. 363, **III,D,2**). However, the reaction may be stopped at the aldehyde stage by an inverse addition process. More conveniently, the formation of a secondary alcohol may be prevented by using triethyl orthoformate in reaction with a Grignard reagent, when the product is an acetal which is unreactive towards the organometallic compound.

$$R \cdot MgX + HC(OC_2H_5)_3 \longrightarrow R \cdot CH(OC_2H_5)_2 + Mg(OC_2H_5)X$$

The acetal is isolated in the crude state and hydrolysed by heating with aqueous sulphuric acid. The reaction is illustrated by the preparation of hexanal (Section **III,85**).

Recently the procedure has been improved by the introduction of the use of phenyl diethyl orthoformate, which may be prepared by an ester exchange reaction between equimolar amounts of triethyl orthoformate and phenol (Ref. 2).

III,85. HEXANAL

$$C_4H_9 \cdot CH_2Br \xrightarrow{Mg} C_4H_9 \cdot CH_2MgBr \xrightarrow{CH(OC_2H_5)_3}$$

$$C_4H_9 \cdot CH_2 \cdot CH(OC_2H_5)_2 + C_2H_5OMgBr$$

$$C_4H_9 \cdot CH_2 \cdot CH(OC_2H_5)_2 \xrightarrow[H_2O]{H_2SO_4} C_4H_9 \cdot CH_2 \cdot CHO + 2C_2H_5OH$$

The apparatus required is a 1-litre three-necked flask, provided with a dropping funnel, a sealed stirrer and a double surface condenser (carrying a calcium chloride guard-tube). Place 15 g (0.675 mol) of dry magnesium turnings, 25 ml of sodium-dried ether and a small crystal of iodine in the flask. Add 3 g (2.5 ml) of dry 1-bromopentane (Section III,54) and set the stirrer in motion. As soon as the reaction commences, add 100 ml of sodium-dried ether, followed by a solution of 91.5 g (76 ml, total 0.625 mol) of dry 1-bromopentane in 100 ml of anhydrous ether at such a rate that the ether refluxes steadily (about 20 minutes). If the reaction becomes too vigorous, cooling in ice-water may be necessary. Reflux the solution for 30 minutes in order to complete the reaction. Remove the source of heat, cool the flask to about 5 °C and add 74 g (83 ml, 0.5 mol) of triethyl orthoformate (Section III,155) during about 10 minutes. Reflux the mixture for 6 hours; then arrange the condenser for distillation and remove the ether on a water bath. Allow the reaction mixture to cool. Add 375 ml of ice-cold 6 per cent hydrochloric acid with stirring; keep the contents of the flask cool by the occasional addition of a little crushed ice. When all the white solid has passed into solution, transfer to a separatory funnel and remove the upper layer of hexanal diacetal. Hydrolyse the acetal by distilling it with a solution of 50 g (27.5 ml) of concentrated sulphuric acid in 350 ml of water; collect the aldehyde, which distils over as an oil, in a solution of 50 g of sodium metabisulphite in 150 ml of water. Remove the oily layer (largely pentan-1-ol) insoluble in the bisulphite solution and discard it. Steam distil the bisulphite solution until 100 ml of the distillate have been collected: this will separate the remainder of the pentan-1-ol and other impurities. Cool the residual bisulphite solution to about 45 °C, cautiously add a suspension of 40 g of sodium hydrogen carbonate in 100 ml of water and separate the resulting free aldehyde by steam distillation. Remove the upper layer (crude aldehyde) of the distillate, wash it with three 25 ml portions of water and dry it with 10 g of anhydrous sodium or magnesium sulphate. Distil through a short fractionating column, and collect the hexanal at 127–129 °C. The yield is 25 g (50%).

H,6. The hydrolysis and decarboxylation of α,β-epoxy esters (glycidic esters). Glycidic esters are prepared by the reaction of an aldehyde or ketone with an α-haloester in the presence of base (sodium ethoxide, sodamide, finely divided sodium or potassium t-butoxide) (the **Darzens glycidic ester condensation**).

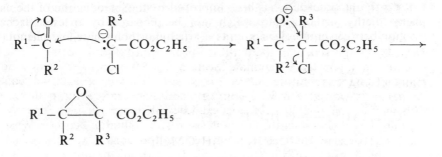

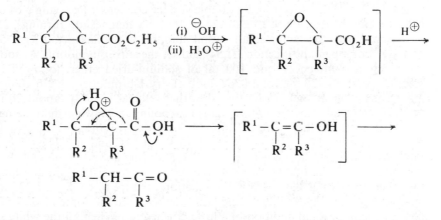

With aqueous alkali the glycidic ester is hydrolysed to the corresponding epoxy acid which decarboxylates under acidic conditions to give a carbonyl compound.

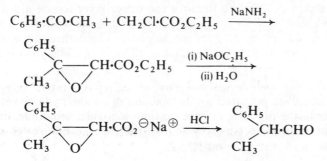

The use of ethyl chloroacetate as the starting ester results in the formation of aldehydes, as in the preparation of 2-phenylpropionaldehyde (Section **III,86**), whereas ketones are obtained using homologous ethyl α-chloroesters. A special procedure for the hydrolysis of a glycidic ester is used in Section **III,86**. This involves treatment of the ester with one equivalent of sodium ethoxide in absolute ethanol followed by the addition of one equivalent of water when precipitation of the sodium salt of the glycidic acid occurs; the precipitation of the sodium salts of glycidic acids is frequently facilitated by the addition of ether.

III,86. 2-PHENYLPROPIONALDEHYDE (*Hydratropaldehyde*)

$$C_6H_5 \cdot CO \cdot CH_3 + CH_2Cl \cdot CO_2C_2H_5 \xrightarrow{NaNH_2}$$

Ethyl 3-phenyl-2,3-epoxybutanoate. In a 500-ml three-necked flask, fitted with a mechanical stirrer and a thermometer, place a mixture of 60 g (58.5 ml, 0.5 mol) of acetophenone, 61.5 g (53 ml, 0.5 mol) of ethyl chloroacetate (b.p. 142–143 °C) and 100 ml of sodium-dried benzene. Add, with stirring, 23.6 g (0.6 mol) of finely powdered sodamide (Section **II,2,**54) over a period of 2 hours; maintain the temperature at 15–20 °C with the aid of external cooling. Ammonia is evolved. Stir for 2 hours at room temperature and pour the reddish mixture on to 350 g of crushed ice with hand stirring. Separate the organic layer and extract the aqueous layer with 100 ml of benzene. Wash the combined benzene solutions with three 150 ml portions of water, the last one containing 5 ml of acetic acid, and then dry with magnesium sulphate. After removal of the benzene on a rotary evaporator, distil the residue under reduced pressure through a short fractionating column. Collect the fraction of b.p. 111–114 °C/3 mmHg as the glycidic ester; the yield is 67 g (65%).

2-Phenylpropionaldehyde (*hydratropaldehyde*). Prepare a solution of sodium ethoxide in a 500-ml round-bottomed flask from 7.6 g (0.33 mol) of clean sodium and 150 ml of absolute ethanol (for experimental details see Section **III,94**). Add 66.5 g of the above glycidic ester slowly and with shaking. Cool the flask externally to 15 °C and add 6 ml of water slowly; much heat is evolved and the sodium salt soon separates. Keep the reaction mixture overnight. Collect the salt by suction filtration, wash it with 25 ml of ethanol followed by 25 ml of ether.

Add the salt to dilute hydrochloric acid (prepared from 28 ml of the concentrated acid and 150 ml of water) contained in a 500-ml flask fitted with a reflux condenser. Warm the mixture gently; carbon dioxide is evolved and an oil separates. Heat on a steam bath for 90 minutes, cool and extract the oil with 75 ml of benzene. Wash the extract with 100 ml of water, and distil the benzene solution under reduced pressure. Collect the 2-phenylpropionaldehyde at 90–93 °C/10 mmHg; the yield is 30 g (70%).

Cognate preparation **Ethyl 2,1′-epoxycyclohexylacetate.** Add a mixture of 55 g (48 ml, 0.45 mol) of ethyl chloroacetate and 43 g (0.44 mol) of cyclohexanone dropwise to a suspension of finely divided sodium (11 g, 0.48 mol) in anhydrous xylene (165 ml) (Section **II,1,**4) with stirring and cooling in an ice–salt bath. Regulate the rate of addition so that the temperature of the reaction mixture does not exceed 8 °C. Pour the resulting dark-red clear solution into water, wash the organic layer repeatedly with water, dry with magnesium sulphate and distil. Collect the glycidic ester at 81–83 °C/0.04 mmHg or at 115–117 °C/10 mmHg. The yield is 37 g (46%).

J ALIPHATIC KETONES

The synthesis of aliphatic ketones is exemplified by the following typical procedures.

1. The oxidation of secondary alcohols (Sections **III,87** to **III,90**).
2. The hydration of alkynes (Section **III,91**).

3. The thermal decarboxylation of acids over a metal oxide catalyst (Section **III,92**).
4. The interaction of cadmium dialkyls with carboxylic acid chlorides (Section **III,93**).
5. The hydrolysis and decarboxylation of β-keto esters (Sections **III,94** and **III,95**), and the hydrolysis of β-diketones (Section **III,96**).
6. The acid-catalysed rearrangement of 1,2-diols (Section **III,97**).

J,1. The oxidation of secondary alcohols. The oxidation of secondary alcohols with sodium dichromate in dilute sulphuric acid gives acceptable yields of ketones since these do not normally undergo extensive further oxidation under the reaction conditions (cf. p. 414, **III,H,1**, the oxidation of primary alcohols to aldehydes).

$$3R_2CHOH + Na_2Cr_2O_7 + 4H_2SO_4 \longrightarrow$$
$$3R_2CO + Na_2SO_4 + Cr_2(SO_4)_3 + 7H_2O$$

An excellent method for the conversion of ether-soluble secondary alcohols to the corresponding ketones is by chromic acid oxidation in a two-phase ether–water system. The reaction is carried out at 25–30 °C with the stoichiometric quantity of chromic acid calculated on the basis of the above equation, and is exemplified by the preparation of octan-2-one and cyclohexanone (Section **III,87**). The success of this procedure is evidently due to the rapid formation of the chromate ester of the alcohol, which is then extracted into the aqueous phase, followed by formation of the ketone which is then extracted back into the ether phase and is thus protected from undesirable side reactions.

A slightly modified procedure—oxidation with 100 per cent excess of chromic acid at 0 °C for a short period—is adopted for strained bicyclic alcohols (e.g., the oxidation of $(-)$-borneol to $(-)$-camphor, Section **III,88**) and gives excellent yields of the corresponding ketones. Cyclic ketones which are susceptible to acid-catalysed epimerisation are moreover obtained by this procedure in a high degree of epimeric purity.

The conversion of an alkene into the corresponding ketone may be effected by means of a convenient sequence which involves hydroboronation followed by oxidation with chromic acid of the resulting organoborane (cf. p. 375, **III,D,3**); isomeric purity is dependent upon the regiospecificity of the hydroboronation step. The sequence is illustrated by the conversion of 1-methylcyclohexene into 2-methylcyclohexanone (Section **III,89**).

Secondary alcohols may be oxidised to the corresponding ketones by the use of an aluminium alkoxide, frequently the t-butoxide, in the presence of a large excess of acetone (the **Oppenauer oxidation**). The reaction involves an initial alkoxy-exchange process followed by a hydride ion transfer from the so-formed aluminium alkoxide of the secondary alcohol by a mechanism analogous to that of the Meerwein–Ponndorf–Verley reduction (see p. 353, **III,D,1**).

$$3R_2CHOH + [(CH_3)_3CO]_3Al \rightleftharpoons [R_2CHO]_3Al + 3(CH_3)_3COH$$
$$[R_2CHO]_3Al + 3(CH_3)_2CO \rightleftharpoons 3R_2CO + [(CH_3)_2CHO]_3Al$$

Acetone in conjunction with benzene as a solvent is widely employed. Alternatively cyclohexanone as the hydrogen acceptor, coupled with toluene or xylene as solvent, permits the use of higher reaction temperatures and consequently the reaction time is considerably reduced; the excess of cyclohexanone can be easily separated from the reaction product by steam distillation. Usually at least 0.25 mol of aluminium alkoxide per mol of secondary alcohol is employed. However, since an excess of alkoxide has no detrimental effect, the use of 1 to 3 mol of alkoxide is desirable, particularly as water, either present in the reagents or formed during secondary reactions, will remove an equivalent quantity of the reagent. It is recommended that 50 to 200 mol of acetone or 10 to 20 mol of cyclohexanone be employed. Other oxidisable groups are usually unaffected in the Oppenauer oxidation and the reaction has found wide application in the steroid field.

The reaction is illustrated by the oxidation of cholesterol to cholest-4-en-3-one (Section **III,90**): the migration of the double bond from the β,γ to the α,β position is a commonly occurring side reaction associated with unsaturated steroids of this structural type.

III,87. OCTAN-2-ONE (*Hexyl methyl ketone*)

$$CH_3 \cdot (CH_2)_5 \cdot CH(OH) \cdot CH_3 \xrightarrow[\text{(C}_2\text{H}_5)_2\text{O}]{\text{Na}_2\text{Cr}_2\text{O}_7/\text{H}_2\text{O}/\text{H}_2\text{SO}_4} CH_3 \cdot (CH_2)_5 \cdot CO \cdot CH_3$$

Equip a 500-ml three-necked flask with a mechanical stirrer, a thermometer and a two-way adapter carrying a dropping funnel and condenser; arrange for the occasional cooling of the flask in an ice-water bath. Place a solution of 32.5 g (0.25 mol) of octan-2-ol in 100 ml of ether in the flask and 125 ml (0.083 mol) of chromic acid solution (1) in the dropping funnel. Add the chromic acid solution dropwise during 15 minutes to the vigorously stirred ether solution of the ketone, keeping the temperature between 25 and 30 °C by cooling as necessary. Continue stirring at room temperature for a further 2 hours and then transfer the reaction mixture to a separating funnel. Separate the ether layer and extract the dark green aqueous layer with four 60 ml portions of ether. Combine the ether extracts and wash with 40 ml of saturated sodium hydrogen carbonate solution and then with 40 ml of saturated sodium chloride solution; finally dry the ether extract over anhydrous sodium sulphate. Filter, remove the ether on a rotary evaporator and distil the residue at atmospheric pressure. Collect the octan-2-one (2) at 170–172 °C. The yield is 26 g (81%). The purity may be checked by GLC using a 5-foot, 10 per cent Silicone oil column at 110 °C and with a nitrogen flow rate of 40 ml/minute. The ketone has a retention time of 3.1 minutes.

Cognate preparation   Cyclohexanone. Proceed as for octan-2-one using 25 g (0.25 mol) of cyclohexanol in 100 ml of ether and adding 125 ml (0.083 mol) of the chromic acid solution (1) during 15–20 minutes. Distil the crude product at atmospheric pressure and collect the cyclohexanone at 154–156 °C. The yield is 10.25 g (79%); the product is homogeneous on GLC, t_R 1.8 minutes.

Notes. (1) The chromic acid solution may be prepared as follows. Dissolve 100 g (0.33 mol) of sodium dichromate dihydrate in 300 ml of water and slowly add 134 g (73 ml, 1.34 mol) of concentrated sulphuric acid (98%, *d* 1.84). Cool the solution and dilute to 500 ml with water in a graduated flask.

(2) The ketone has a cheese-like, rather pervading, odour.

III,88. (−)-CAMPHOR

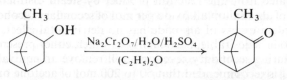

Assemble the apparatus described in Section **III,87**. Place a solution of 7.7 g (0.05 mol) of (−)-borneol in 25 ml of ether in the flask and cool to 0 °C. Precool 50 ml (0.033 mol) of chromic acid solution (1) to 0 °C before placing 25 ml in the dropping funnel. Add the cold chromic acid solution to the vigorously stirred borneol solution during 5 minutes, keeping the remainder cold in an ice-water bath. Add the remaining chromic acid over a further 5 minutes. Stir the mixture with cooling in the ice-water bath for a further 5 minutes and then transfer to a separating funnel. Carefully remove the ether layer (2) and extract the aqueous solution with two 25 ml portions of ether. Combine the ether extracts, and wash with 30 ml of 5 per cent sodium carbonate solution followed by four 25 ml portions of water. Dry the ether solution over anhydrous sodium sulphate, filter and remove the ether on the rotary evaporator. The yield of crude (−)-camphor, m.p. 159–164 °C (sealed tube), is 6.3 g (83%). It may be purified by sublimation at 80–90 °C/12 mmHg (Section **I,21**); after two sublimations 5.76 g (76%) of material is obtained, which melts at 173–175 °C (sealed tube).

Notes. (1) Prepared as described in Section **III,87**, Note (1).

(2) Both the ether and aqueous solutions are almost black and it is difficult to see the interface; it is best judged by noting the change in flow of the solutions through the tap.

III,89. 2-METHYLCYCLOHEXANONE

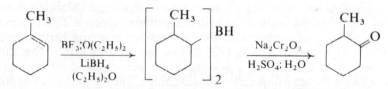

This experiment should be carried out in a fume cupboard.

Equip a 250-ml three-necked round-bottomed flask with a thermometer, a reflux condenser protected by a calcium chloride guard-tube and (in the central neck) a 100-ml pressure-equalising dropping funnel fitted with an inlet adapter to allow flushing with a stream of dry nitrogen. The apparatus should be thoroughly dried in an oven and assembled under a stream of dry nitrogen. Adjust the nitrogen flow to a slight trickle, which should be maintained throughout the reaction, and in the flask place a magnetic follower bar, 30 ml of dry ether, 4.8 g (0.050 mol) of 1-methylcyclohexene (Section **III,7**) and 0.5 g (0.0225 mol, 20% excess) of lithium borohydride (1). In the dropping funnel place 5 ml of dry ether and 0.95 ml (0.075 mol, 20% excess) of boron trifluoride etherate (2). Add the boron trifluoride etherate solution to the stirred mixture dropwise during 15 minutes, keeping the temperature at 25–30 °C by cooling in a water bath. Stir for 2 hours at room temperature and then destroy excess hydride by adding carefully 5 ml of water. Place in the dropping funnel a solution of chromic acid, prepared from 11.0 g (0.0369 mol) of sodium dichromate dihydrate and 8.1 ml (0.1474 mol) of 98 per cent sulphuric acid made up to 45 ml with water. Add the chromic acid solution portionwise to the reaction mixture during 15 minutes whilst maintaining the temperature at 25–30 °C by cooling in an ice-water bath. Heat the dark mixture under reflux for 2 hours, cool and separate the upper ether layer. Extract the aqueous acid layer with two 15 ml portions of ether, combine the organic extracts, wash once with 5 ml of saturated sodium chloride solution and dry over magnesium sulphate. Filter, remove the ether from the filtrate by flash distillation and carefully distil the residual straw-coloured liquid. Collect the 2-methylcyclohexanone having b.p. 160–164 °C. The yield is about 3.5 g (63%). The purity may be checked by GLC on a 5-foot column of 10 per cent Silicone oil on Chromosorb W held at 128 °C, with a nitrogen flow rate of 40 ml/minute; t_R is about 1.7 minutes.

Notes. (1) Reaction of lithium borohydride with water may be rapid and violent; do not expose to high humidity and avoid contact with eyes, skin and clothing (contact with cellulosic material may cause combustion). It should be handled with the same caution as is afforded to lithium aluminium hydride (Section **II,2,**_39_).

(2) Boron trifluoride etherate should be purified as described in Section **II,2,**_6_.

III,90. CHOLEST-4-EN-3-ONE

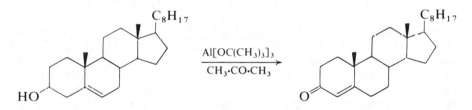

Place a mixture of 20 g (0.052 mol) of pure cholesterol (m.p. 147–150 °C; dried to constant weight at 80–100 °C), 150 ml of dry acetone and 200 ml of sodium-dried benzene in a dry, 1-litre, two-necked round-bottomed flask fitted with a reflux condenser and dropping funnel both protected with calcium chloride tubes. Add a few carborundum chips and heat to boiling in an oil bath at 75–85 °C. Add from the dropping funnel, in one portion, a solution of 16 g (0.065 ml) of aluminium t-butoxide (Section **II,2,**_1(a)_) in 100 ml of anhydrous benzene. The reaction mixture becomes cloudy and develops a yellow colour in about 10 to 15 minutes. Continue gentle boiling at a bath temperature of 75–85 °C for 8 hours. Treat the cold mixture with 40 ml of water and then with 100 ml of 10 per cent sulphuric acid. Shake vigorously and transfer to a 1-litre separating funnel. Dilute the mixture with 300 ml of water, shake for 5 minutes (filter, if necessary), then run off the yellow aqueous layer into a second separating funnel and extract the latter with 25 ml of benzene. Wash the combined benzene extracts thoroughly with water, dry with magnesium sulphate and remove the solvent on a rotary evaporator under reduced pressure. The yellow oily residue solidifies when it is cooled in an ice–salt bath and scratched with a glass rod; keep a small portion for seeding in the subsequent crystallisation. Dissolve the solid in a warm mixture of 14 ml of acetone and 20 ml of methanol, allow the solution to cool very slowly and seed, if necessary. When the bulk of the solid has crystallised, keep the mixture at 0 °C for 24 hours, filter with suction, wash with 20 ml of ice-cold methanol and dry in a vacuum desiccator. The yield of almost colourless cholest-4-en-3-one, m.p. 79–80 °C, is 17 g (85%).

J,2. The hydration of alkynes. The direct hydration of a terminal alkyne, with dilute sulphuric acid in the presence of a mercury salt as a catalyst, results in a Markownikoff mode of addition, yielding initially an enol which re-arranges to the more stable ketone. The conversion of undec-10-ynoic acid to 10-oxoundecanoic acid (Section **III,91**) is illustrative of this reaction.

$$R \cdot C \equiv CH \xrightarrow[Hg^{2\oplus}]{H_2SO_4} [R \cdot C(OH) = CH_2] \longrightarrow R \cdot CO \cdot CH_3$$

In the case of acetylene itself the product is acetaldehyde and this reaction was formerly used extensively for its manufacture.

With non-terminal alkynes the preparative applications are more limited since a mixture of ketones is usually obtained, the proportions being dependent upon the nature of R^1 and R^2.

$$R^1 \cdot C \equiv C \cdot R^2 \longrightarrow R^1 \cdot CO \cdot CH_2 \cdot R^2 + R^1 \cdot CH_2 \cdot CO \cdot R^2$$

III,91. 10-OXOUNDECANOIC ACID

$$HC \equiv C \cdot (CH_2)_8 \cdot CO_2H \xrightarrow[H_3O^\oplus]{Hg(O \cdot CO \cdot CH_3)_2} CH_3 \cdot CO \cdot (CH_2)_9 \cdot CO_2H$$

Heat under reflux for 4 hours a solution of 3 g (0.0165 mol) of undec-10-ynoic acid (Section **III,19**) in 240 ml of glacial acetic acid containing 13 ml of concentrated sulphuric acid and 1.4 g of mercury(II) acetate. Dilute the dark coloured solution with 300 ml of water, filter and extract the solution with methylene chloride using a continuous extraction apparatus (Fig. I,95) (1). Wash the methylene chloride extract carefully with distilled water until the washings are neutral, dry the extract and evaporate on a rotary evaporator. Recrystallise the solid residue from light petroleum (b.p. 60–80 °C) using a little decolourising charcoal. The yield of keto-acid, m.p. 56–57 °C, is 1.65 g (50%).

Note. (1) Batchwise extraction leads to the formation of stable emulsions which frequently take several days to break. When filling the continuous extraction apparatus, care should be taken to avoid forming an emulsion.

J,3. The thermal decarboxylation of acids over a metal oxide catalyst. A long established method of formation of symmetrical ketones (albeit in rather low yield) involves the pyrolysis of certain salts (usually calcium or barium) of carboxylic acids.

$$(R \cdot CO_2)_2 Ca \xrightarrow{heat} R \cdot CO \cdot R + CaCO_3$$

The method has been extended to include the synthesis of aldehydes or of unsymmetrical ketones by using appropriate mixtures of carboxylate salts.

$$(R^1 \cdot CO_2)_2 Ca + (R^2 \cdot CO_2)_2 Ca \xrightarrow{heat} 2R^1 \cdot CO \cdot R^2 + 2CaCO_3$$

The unsymmetrical carbonyl compound will clearly be accompanied by both possible symmetrical products, but the yield can be improved by the use of an excess of one of the carboxylate salts.

A more satisfactory process for the preparation of either symmetrical or unsymmetrical ketones involves the passage of the vapour of the acid or mixture of acids over heated manganese(II) oxide or thorium oxide deposited upon pumice. The yields of unsymmetrical ketones are satisfactory (*ca.* 50%) and the technique is particularly well suited for laboratory preparations on a reasonably large scale. Several examples are included in Section III,92,*A* and *B*.

III,92,*A*. DIETHYL KETONE (*Pentan-3-one*)

$$2CH_3 \cdot CH_2 \cdot CO_2 H \xrightarrow{\text{catalyst}} CH_3 \cdot CH_2 \cdot CO \cdot CH_2 \cdot CH_3$$

Preparation of manganese(II) carbonate–pumice catalyst. Dissolve 70 g (0.35 mol) of manganese(II) chloride tetrahydrate in 100 ml of water and add a solution of 38 g (0.35 mol) of anhydrous sodium carbonate in 120 ml of water with mechanical stirring. Filter the precipitated manganese(II) carbonate and wash well with distilled water. Transfer the solid to a large evaporating basin and add sufficient water to form a thick paste. Add sufficient pumice (4–8 mesh) with stirring so that most of the paste has been transferred to the pumice and heat cautiously (avoid local overheating) on an electric hotplate until the pumice lumps no longer cling together. If too much water is added to form the paste of manganese(II) carbonate initially, it will not adhere to the pumice satisfactorily.

Diethyl ketone. Pack the catalyst into the Pyrex glass combustion tube of the pyrolysis apparatus illustrated in Fig. I,72(c) and assemble the remaining components. Displace the air in the apparatus with nitrogen, and while maintaining a continued gentle gas flow, heat the pumice for 8 hours at 360–400 °C in order to convert the manganese(II) carbonate into manganous oxide. If necessary the catalyst may be allowed to cool in a stream of nitrogen if the preparation needs to be interrupted at this point. Place 740 g (746 ml, 10 mol) of redistilled propionic acid, b.p. 139–141 °C, in the dropping funnel, and with the furnace at about 350 °C, add the acid to the catalyst dropwise (about 30 drops per minute); the stream of nitrogen is stopped when addition of acid is commenced. The apparatus requires little attention and the addition of acid occupies 48–72 hours. The distillate consists of two layers. Separate the lower aqueous layer, salt out the ketone with solid potassium carbonate and add it to the main ketonic layer. Treat the combined ketone fractions with small quantities of anhydrous potassium carbonate (1) until effervescence ceases (this both removes the excess of acid and dries the ketone), filter and distil through a short fractionating column. Collect the diethyl ketone at 101–103 °C. The yield is 252 g (29%). An improved yield may be obtained by recirculating the distillate over the catalyst, but in practice this is rarely worth while. It must be remembered that on each occasion that the catalyst is allowed to cool a slow stream of nitrogen must be passed through the apparatus to prevent the oxidation of the manganese(II) oxide catalyst.

Note. (1) An alternative method of working up the distillate, which has its advantages when dealing with volatile ketones or when it is suspected that conversion into the ketone is incomplete, is to treat the combined fractions of ketones with sodium hydroxide pellets until the mixture is alkaline. Should solids separate, these may be dissolved by the addition of a little water. The ketone is then separated, dried over anhydrous potassium carbonate and fractionated.

Cognate preparations Heptan-4-one (*dipropyl ketone*). Use 880 g (920 ml, 10 mol) of butyric acid, b.p. 162–164 °C. The yield of ketone, b.p. 142–143 °C, is 285 g * (46%).

Pentan-2-one (*methyl propyl ketone*). Use 360 g (6 mol) of glacial acetic acid and 176 g (184 ml, 2 mol) of butyric acid. The yield of methyl propyl ketone, b.p. 102–104 °C, is 75 g (43%); 75 g of acetone, b.p. 56–57 °C,† are also obtained.

Hexan-3-one (*ethyl propyl ketone*). Use 296 g (298 ml, 4 mol) of propionic acid and 352 g (368 ml, 4 mol) of butyric acid. The yield is 214 g (53%) of ethyl propyl ketone, b.p. 122–124 °C; the by-products are 98 g of diethyl ketone, b.p. 100–102 °C, and 66 g of dipropyl ketone, b.p. 144–146 °C.

Undecan-6-one (*dipentyl ketone*). Use 400 g (428 ml, 3.45 mol) of hexanoic acid, b.p. 204–206 °C. The yield of ketone, b.p. 222–226 °C, is 225 g (76%).

III,92,B. BENZYL METHYL KETONE

$$C_6H_5 \cdot CH_2 \cdot CO_2H + CH_3 \cdot CO_2H \xrightarrow{\text{catalyst}} C_6H_5 \cdot CH_2 \cdot CO \cdot CH_3$$

Preparation of thorium carbonate–pumice catalyst. Dissolve 294 g (0.5 mol) of thorium nitrate hexahydrate in the minimum of water (*ca.* 450 ml) and add slowly a solution of 106 g (1 mol) of anhydrous sodium carbonate in 400 ml of water with stirring. Allow the thorium carbonate to settle, decant as much as possible of the mother-liquor and wash the sediment once by decantation with 500 ml of water. Make the resulting moist solid into a thick paste with distilled water and stir in pumice (4–8 mesh) until most of the suspension appears to be absorbed. Dry the impregnated pumice in quantities of 200 g by heating in a large evaporating dish upon an electric hotplate and stirring constantly with a glass rod. Stop the heating when the pumice particles no longer cling together. Sieve the resulting pumice, 250 g of a white powder (consisting largely of thorium carbonate but containing some oxide) are recovered and can be used for impregnating more pumice. The total weight of pumice catalyst thus prepared is about 1400 g; the exact weight will depend upon the grade of pumice used.

Benzyl methyl ketone. Fill the Pyrex combustion tube with catalyst and proceed as in Section **III,92,A** but use a temperature of 400–450 °C for the conversion of thorium carbonate into the corresponding oxide; 6–12 hours are usually required for complete conversion and a slow stream of nitrogen should

---

* All the yields given refer to one circulation of the acid (or acids) over the catalyst, but can be improved by recirculating the product, from which the water layer has been removed, over the catalyst. With the higher ketones, the second circulation may result in carbonisation of the catalyst, thus rendering it inefficient.

† The symmetrical ketones, produced as by-products in the preparation of mixed ketones, are separated by distillation through an efficient fractionating column. If acetone is a by-product (as in the preparation of pentan-2-one; methyl propyl ketone), some is lost in the washing process.

be maintained through the combustion tube. Place a solution of 170 g (1.25 mol) of pure phenylacetic acid (m.p. 77 °C) in 225 g (3.75 mol) of glacial acetic acid in the funnel, and adjust its rate of flow into the catalyst tube to 1 drop every 2 or 3 seconds. Also pass a slow stream of nitrogen (1 bubble per second) through the apparatus in order to keep the gases in motion; the rate of flow may be estimated by passing the inert gas through a concentrated sulphuric acid wash-bottle or 'bubbler' before it enters the furnace. When all the acid mixture has passed through the catalyst tube, separate the lower aqueous layer of the product and treat the organic layer with 10–20 per cent sodium hydroxide solution until the washings are alkaline to litmus and then twice with water. Extract the aqueous layer twice with 50 ml portions of ether, wash the extracts successively with sodium hydroxide solution (until alkaline) and water, and add the resulting ether solution to the main product. Dry with magnesium sulphate, remove the ether on a rotary evaporator and distil the residue under reduced pressure preferably through a fractionating column. Collect the benzyl methyl ketone at 102–102.5 °C/20 mmHg; the yield is 85 g (51%). The residue in the flask is dibenzyl ketone; it may be purified by transferring to a smaller flask and redistilling (b.p. 200 °C/21 mmHg; m.p. 34–35 °C).

Cognate preparations   Benzyl ethyl ketone.   Use 204 g (1.5 mol) of phenylacetic acid (m.p. 77 °C) and 333 g (335.5 ml, 4.5 mol) of propionic acid (b.p. 139–141 °C), but omit the extraction with ether when working up the distillate. Distil the dried product from a 500-ml round-bottomed flask through an efficient fractionating column. Collect the diethyl ketone at 99.5–102.5 °C (160 g), and when the temperature rises to 130 °C (b.p. 103–130 °C: 7 g) transfer the residue to a 250-ml flask and distil fractionally under reduced pressure. The benzyl ethyl ketone passes over mainly at 118–123 °C/22 mmHg (105 g, 47%); the residue of high boiling point (34 g) consists largely of dibenzyl ketone. Pure benzyl ethyl ketone may be obtained by redistilling the fraction, b.p. 118–123 °C/22 mmHg, and collecting the fraction of b.p. 113–115 °C/17 mmHg.

Benzyl propyl ketone.   Use 204 g (1.5 mol) of pure phenylacetic acid and 396 g (414 ml, 4.5 mol) of butyric acid (b.p. 161–164 °C). Upon working up as for benzyl ethyl ketone, 180 g of dipropyl ketone, b.p. 140–145 °C (mainly 143–145 °C), 108 g (45%) of crude benzyl propyl ketone, b.p. 240–260 °C, and 49 g of crude dibenzyl ketone (residue in flask) are obtained. Redistil the fraction of b.p. 240–260 °C and collect the benzyl propyl ketone at 243–247 °C (the pure ketone boils at 244 °C).

4-Phenylbutan-2-one (*methyl 2-phenylethyl ketone*).   Use 100 g (0.66 mol) of hydrocinnamic acid (3-phenylpropionic acid) (m.p. 49–50 °C) (Section III,132 and IV,150) and 160 g (2.66 mol) of glacial acetic acid. The yield of methyl 2-phenylethyl ketone, b.p. 230–235 °C, is 70 g (71%) (the pure ketone boils at 234 °C).

1-Phenylpentan-3-one (*ethyl 2-phenylethyl ketone*).   Use 100 g (0.66 mol) of pure hydrocinnamic acid and 200 g (201.5 ml, 2.7 mol) of pure propionic acid. Fractionation of the distillate yields 70 g of diethyl ketone (b.p. 100–102 °C), 72 g (67%) of ethyl 2-phenylethyl ketone (b.p. 245–249 °C; the pure ketone boils at 248 °C) and 18 g of crude 1,5-diphenylpentan-3-one (high b.p. residue).

1-Phenylhexan-3-one (*propyl 2-phenylethyl ketone*).   Use 100 g (0.66 mol) of pure hydrocinnamic acid and 235 g (245.5 ml, 2.66 mol) of pure butyric acid. Upon working up as for benzyl ethyl ketone the following yields are obtained: 98 g of dipropyl ketone, b.p. 140–144 °C; 65 g (55%) of propyl 2-phenylethyl ketone, b.p. 139–143 °C/17 mmHg; and 22 g of crude 1,5-diphenylpentan-3-one (high b.p. residue). The required ketone, upon redistillation, boils almost completely at 138–139 °C/16 mmHg.

J,4. The interaction of cadmium dialkyls with carboxylic acid chlorides. Cadmium dialkyls may be prepared by the action of anhydrous cadmium chloride upon the corresponding Grignard reagents.

$$2R \cdot MgBr + CdCl_2 \longrightarrow R_2Cd + MgBr_2 + MgCl_2$$

The cadium chloride is added to a boiling ethereal solution of the Grignard reagent and the resulting mixture is stirred and heated under reflux until a negative test (Section **III,15**, Note (3)) is obtained, thus indicating the complete conversion of the Grignard reagent; the cadmium dialkyl does not give a positive test.

The main use of cadmium dialkyls is for the preparation of ketones and keto esters (Ref. 3), and their special merit lies in the fact that they react vigorously with acid chlorides of all types but, in contrast to the behaviour of Grignard reagents, add sluggishly or not at all to ketones.

$$R_2{}^1Cd + 2ClCO \cdot CH_2 \cdot R^2 \longrightarrow 2R^1 \cdot CO \cdot CH_2 \cdot R^2 + CdCl_2$$
$$R_2Cd + 2ClCO \cdot (CH_2)_n \cdot CO_2CH_3 \longrightarrow 2R \cdot CO \cdot (CH_2)_n \cdot CO_2CH_3 + CdCl_2$$

The success of the last reaction depends upon the inertness of the ester carbonyl group towards the cadmium dialkyl: with its aid and the use of various ester acid chlorides, a carbon chain can be built up to any reasonable length whilst retaining a reactive functional group (the ester group) at one end of the chain.

For most purposes the use of 1.0 mol of an alkyl or aryl bromide (for the preparation of the cadmium dialkyl (or diaryl) through the Grignard reagent) to 0.8 mol of the acid halide is recommended. This results in nearly equivalent molar ratios of the cadmium dialkyl and acid halide, since the overall yield of the former is usually about 80 per cent. It is generally advantageous to replace the ether solvent by benzene before the addition of the acid halide: a higher reflux temperature is possible, thus reducing the time required for the reaction. The entire preparation can be carried out in one flask without isolation of intermediates. Experimental details are given for the preparation of 1-chlorohexan-2-one (Section **III,93**).

III,93. 1-CHLOROHEXAN-2-ONE

$$2CH_3 \cdot (CH_2)_2 \cdot CH_2Br \xrightarrow[\text{ether}]{Mg} 2CH_3 \cdot (CH_2)_2 \cdot CH_2MgBr \xrightarrow{CdCl_2}$$

$$(C_4H_9)_2Cd \xrightarrow{2CH_2Cl \cdot COCl} C_4H_9 \cdot CO \cdot CH_2Cl$$

Equip a 1-litre three-necked flask with a sealed Hershberg stirrer (preferably of tantalum wire) (see Fig. I,55), a reflux condenser and a 250-ml dropping funnel (1). All apparatus must be thoroughly dry. Place 8.1 g (0.33 mol) of dry magnesium turnings in the flask, add 60 ml of anhydrous ether through the dropping funnel and charge the latter with a solution of 46 g (35.5 ml, 0.33

mol) of butyl bromide in 110 ml of dry ether. Start the stirrer and prepare the Grignard reagent in the usual manner (compare Sections **III,33** and **III,34**). When the formation of the Grignard reagent is complete, cool the flask in an ice bath with stirring, remove the dropping funnel and, when cold, add 32.7 g (0.178 mol) of anhydrous cadmium chloride (2) in portions from a small conical flask during 5–10 minutes. Replace the dropping funnel, remove the ice bath, stir for 5 minutes and then heat the mixture under reflux with stirring for 45 minutes; at this point a test for the presence of Grignard reagent is made (3): continue stirring and refluxing until the test is negative. Replace the reflux condenser by a bend adapter connected to a condenser set for distillation, distil off the ether as stirring is continued; continue the distillation, with stirring, on a water bath until it becomes very slow and dark viscous residue remains. At this point add 120 ml of anhydrous benzene from the dropping funnel, and continue the distillation until a further 35 ml of liquid has passed over. Then add 120 ml of dry benzene and replace the reflux condenser: reflux the mixture with vigorous stirring in order to break up the cake inside the flask. Remove the heating bath, cool the mixture to about 5 °C in an ice bath and add a solution of 38 g (25.5 ml, 0.33 mol) of chloroacetyl chloride (b.p. 105 °C) in 70 ml of anhydrous pure benzene from the dropping funnel during 2–3 minutes. After completion of the addition, stir the reaction mixture and hold the temperature at 15–20 °C for 3 hours and then at 20–25 °C for a further 1.5 hours. Add excess of crushed ice (*ca.* 200 g) and dilute sulphuric acid. Separate the benzene and aqueous layers; extract the aqueous phase with two 30 ml portions of benzene. Wash the combined benzene layers successively with 70 ml of water, 70 ml of saturated sodium hydrogen carbonate solution, 70 ml of water and 35 ml of saturated sodium chloride solution. Filter the benzene solution through a little anhydrous sodium sulphate (this separates most of the suspended water), remove the benzene by flash distillation at atmospheric pressure and distil the residue under reduced pressure through a short fractionating column. Collect the 1-chlorohexan-2-one at 71–72 °C/15 mmHg; the yield is 24 g (54%).

Notes. (1) It is best to conduct the preparation in a nitrogen atmosphere; the apparatus shown in Fig. I,67 may be used.

(2) Dry hydrated cadmium chloride (AnalaR) to constant weight at 110 °C; grind finely, dry again for 2–3 hours at 110 °C and then place in a screw-capped bottle and keep in a desiccator over calcium chloride. It is important to realise that cadmium compounds are *very toxic* and a suitable face mask should be worn to prevent inhalation of dust.

(3) See Section **III,15**, Note (3).

J,5. The hydrolysis and decarboxylation of β-keto esters and the hydrolysis of β-diketones. A general route to methyl ketones involves the alkylation of ethyl acetoacetate followed by hydrolysis and decarboxylation. Alkylation is effected by the interaction of an alkyl halide with the sodio-derivative of ethyl acetoacetate under anhydrous conditions. Hydrolysis and decarboxylation is best brought about by the action of dilute alkali in the cold followed by acidification and boiling. The free alkylated β-keto acid is produced which readily undergoes decarboxylation. The reaction is illustrated by the preparation of hexan-2-one (Section **III,94**), and also in Section **III,176** (the forced Claisen ester condensation).

$$CH_3 \cdot CO \cdot CH_2 \cdot CO_2C_2H_5 + NaOC_2H_5 \xrightarrow{-C_2H_5OH}$$

$$[CH_3 \cdot CO \cdot \overset{\ominus}{C}H \cdot CO_2C_2H_5]\overset{\oplus}{Na} \xrightarrow{RBr} CH_3 \cdot CO \cdot CH(R) \cdot CO_2C_2H_5 \xrightarrow[\text{(ii) } H_3O^{\oplus}]{\text{(i) } {}^{\ominus}OH}$$

$$CH_3 \cdot CO \cdot CH(R) \cdot CO_2H \xrightarrow{heat} CH_3 \cdot CO \cdot CH_2 \cdot R + CO_2$$

The preparation of many β-keto esters, including ethyl acetoacetate, is conveniently carried out by the Claisen ester condensation (p. 537, **III,P,4**(*b*)) and their alkylation and subsequent hydrolysis and decarboxylation enables a range of ketones to be synthesised. A more general ketone synthesis, which utilises the β-keto ester system as an intermediate, involves the acylation of a malonate ester by way of the ethoxymagnesium derivative. Hydrolysis and decarboxylation to the ketone is accomplished by heating in acid solution; the synthesis of cyclohexyl methyl ketone is the illustrative example (Section **III,95**).

$$R \cdot COCl + [CH(CO_2C_2H_5)_2]MgOC_2H_5 \longrightarrow$$

$$R \cdot CO \cdot CH(CO_2C_2H_5)_2 \xrightarrow{H_3O^{\oplus}} R \cdot CO \cdot CH_3$$

A related synthesis of methyl ketones involves the preparation and alkaline cleavage of a 3-alkylpentane-2,4-dione which can be readily achieved in one step by refluxing a mixture of pentane-2,4-dione and the alkyl halide in ethanolic potassium carbonate.

$$RBr + CH_3 \cdot CO \cdot CH_2 \cdot CO \cdot CH_3 \xrightarrow[K_2CO_3]{C_2H_5OH}$$

$$[CH_3 \cdot CO \cdot CH(R) \cdot CO \cdot CH_3] \longrightarrow CH_3 \cdot CO \cdot CH_2 \cdot R + CH_3 \cdot CO_2C_2H_5$$

The method is illustrated by the preparation of 5-methylhex-5-en-2-one (Section **III,96**).

III,94. HEXAN-2-ONE (*Butyl methyl ketone*)

$$CH_3 \cdot CO \cdot CH_2 \cdot CO_2C_2H_5 \xrightarrow[\text{(ii) } CH_3 \cdot CH_2 \cdot CH_2Br]{\text{(i) } NaOC_2H_5}$$

$$CH_3 \cdot CO \cdot CH \cdot CO_2C_2H_5 \xrightarrow{NaOH} CH_3 \cdot CO \cdot CH \cdot CO_2^{\ominus}Na^{\oplus}$$
$$\quad | \qquad\qquad\qquad\qquad\qquad\qquad\qquad | $$
$$CH_2 \cdot CH_2 \cdot CH_3 \qquad\qquad\qquad\qquad CH_2 \cdot CH_2 \cdot CH_3$$

$$\xrightarrow[H_3O^{\oplus}]{heat} CH_3 \cdot CO \cdot CH_2 \cdot CH_2 \cdot CH_2 \cdot CH_3 + CO_2$$

Fit a 2-litre three-necked flask with an efficient double surface condenser and a separatory funnel; close the central neck with a stopper. The apparatus must be perfectly dry. Place 34.5 g (1.5 mol) of clean sodium (Section **II,2**,*55*) cut into small pieces in the flask and clamp the flask by the wide central neck. Measure out 1 litre of super-dry ethanol (Section **II,1**,*9*), and place about 500 ml in the separatory funnel; insert calcium chloride guard-tubes at the top of the condenser and the separatory funnel respectively. Place a large bowl beneath the flask and have a large wet towel in readiness to control the vigour of the subsequent reaction. Run in about 200 ml of the absolute ethanol on to the sodium (1); a vigorous reaction takes place. If the ethanol refluxes violently in the condenser, cool the flask by wrapping it in the wet towel and also, if necessary, run a stream of cold water over it. As soon as the reaction moderates somewhat, introduce more alcohol to maintain rapid, but controllable, refluxing. In this manner most of the sodium reacts rapidly and the time required to produce the solution of sodium ethoxide is considerably reduced. Finally add the remainder of the ethanol and reflux the mixture on a water bath until the sodium has reacted completely. Remove the stopper in the central neck and introduce a sealed mechanical stirrer. Add 195 g (190 ml, 1.5 mol) of pure ethyl acetoacetate, stir the solution and heat to gentle boiling, then run in 205 g (151 ml, 1.66 mol) of propyl bromide (Section III,54) over a period of about 60 minutes. Continue the refluxing and stirring until a sample of the solution is neutral to moist litmus paper (6–10 hours); the reaction is then complete.

Cool the mixture and decant the solution from the sodium bromide: wash the salt with two 20 ml portions of absolute ethanol and add the washings to the main solution. Distil off the ethanol, which contains a slight excess of propyl bromide, through a short fractionating column from a water bath. The residue (A) of crude ethyl propylacetoacetate may be used directly in the preparation of hexan-2-one. If the fairly pure ester is required, distil the crude product under diminished pressure and collect the fraction boiling at 109–113 °C/27 mmHg (183 g, 71%) (B).

To prepare hexan-2-one add the crude ester (A) or the redistilled ethyl propylacetoacetate (B) to 1500 ml of a 5 per cent solution of sodium hydroxide contained in a 4-litre flask equipped with a mechanical stirrer. Continue the stirring at room temperature for 4 hours; by this time the mono-substituted acetoacetic ester is completely hydrolysed and passes into solution. Transfer the mixture to a large separatory funnel, allow to stand and remove the small quantity of unsaponified material which separates as an upper oily layer. Place the aqueous solution of sodium propylacetoacetate in a 3-litre two-necked flask fitted with a small separatory funnel and a wide bent delivery tube connected to a condenser set for downward distillation. Add 150 ml of 50 per cent by weight sulphuric acid (d 1.40) slowly through the separatory funnel with shaking; a vigorous evolution of carbon dioxide occurs. When the latter has subsided, heat the reaction mixture slowly to the boiling point and distil slowly until the total volume is reduced by about one-half; by this time all the hexan-2-one should have passed over. The distillate contains the ketone, ethanol and small quantities of acetic and valeric acids. Add small portions of solid sodium hydroxide to the distillate until it is alkaline and redistil the solution until 80–90 per cent has been collected; discard the residue.

Separate the ketone layer from the water, and redistil the latter until about one-third of the material has passed over. Remove the ketone after salting out any dissolved ketone with potassium carbonate (2). Wash the combined ketone fractions four times with one-third the volume of 35–40 per cent calcium chloride solution in order to remove the alcohol. Dry over 15 g of anhydrous calcium chloride; it is best to shake in a separatory funnel with 1–2 g of the anhydrous calcium chloride, remove the saturated solution of calcium chloride as formed, and then allow to stand over 10 g of calcium chloride in a dry flask. Filter and distil. Collect the hexan-2-one at 126–128 °C. The yield is 71 g (67%).

Notes. (1) The addition of the ethanol to the sodium, although attended by a very vigorous reaction which must be carefully controlled, is preferable to the reverse procedure of adding the sodium in small pieces to the ethanol. The latter method is longer and has the further disadvantage that it necessitates frequent handling and exposure to the air of small pieces of sodium.

(2) A more complete recovery of the ketone from the aqueous solution may be obtained by repeated distillation of the aqueous layer until no appreciable amount of ketone is found in the distillate. The procedure outlined is, however, quite satisfactory.

Cognate preparation   Heptan-2-one.   Use 34.5 g (1.5 mol) of sodium, 1 litre of super-dry absolute ethanol, 195 g (1.5 mol) of redistilled ethyl acetoacetate and 225 g (177 ml, 1.63 mol) of dry butyl bromide (Section **III,54**). This yields 280 g of crude or 200 g (72%) of pure ethyl butylacetoacetate, b.p. 112–116 °C/16 mmHg. Upon hydrolysis 105 g (80%) of heptan-2-one, b.p. 149–151 °C, are isolated.

III,95. CYCLOHEXYL METHYL KETONE

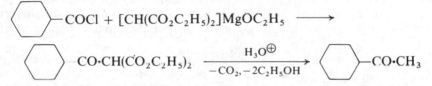

Place 10.7 g (0.44 mol) of magnesium turnings (1) in a 1-litre three-necked round-bottomed flask, equipped with a sealed stirrer unit, a dropping funnel and a double surface reflux condenser each protected with a calcium chloride guard-tube. Add in one portion a mixture of 10 ml of absolute ethanol and 1 ml of carbon tetrachloride. Allow the reaction, which commences almost immediately, to proceed for about 5 minutes and then add carefully 150 ml of sodium-dried ether (Section **II,1,*15***). Site the flask in a warm-water bath and allow the reaction mixture to reflux gently while a solution of 70 g (0.44 mol) of diethyl malonate in 50 ml of dry ether is added with stirring. On completion of the addition, heat the mixture under reflux for about 3 hours or until all the magnesium has reacted. Then add with vigorous stirring a solution of 58 g (0.4 mol) of cyclohexanecarbonyl chloride (Section **III,138**) in 50 ml of dry ether. Heat the reaction mixture under reflux for 2 hours and then cool and acidify with 50 ml of dilute sulphuric acid. Separate the ether layer and extract the residual aqueous solution with two 50 ml portions of ether. Wash the combined ether extracts with water and evaporate the solvent on a rotary evaporator. To the residue add a solution of 120 ml of glacial acetic acid, 15 ml of concentrated sulphuric acid and 80 ml of water and heat under reflux for 5 hours. Cool the reaction mixture, basify by the careful addition of 100 ml of 20 per cent sodium hydroxide solution and extract the solution with four 50 ml portions of ether. Dry the combined ether extracts over sodium sulphate and remove the ether on a rotary evaporator. Distil the crude product at atmospheric pressure through a short fractionating column. The yield of cyclohexyl methyl ketone of b.p. 178–180 °C is 35 g (70%).

Note.   (1) See Section **III,3**, Note (1).

III,96. 5-METHYLHEX-5-EN-2-ONE

$$CH_3 \cdot CO \cdot CH_2 \cdot CO \cdot CH_3 + CH_2 = \overset{\overset{\displaystyle CH_3}{|}}{C} - CH_2Cl \xrightarrow{K_2CO_3}$$

$$CH_2 = \overset{\overset{\displaystyle CH_3}{|}}{C} \cdot CH_2 \cdot CH_2 \cdot CO \cdot CH_3 + CH_3 \cdot CO_2C_2H_5$$

Equip a 1-litre two-necked round-bottomed flask with a sealed stirrer unit and a reflux condenser protected with a guard-tube containing anhydrous calcium sulphate. Place in the flask 500 ml of anhydrous ethanol, 75 g (0.75 mol) of freshly distilled pentane-2,4-dione (b.p. 136–137 °C) (Section **III,102**), 63.4 g (0.70 mol) of 3-chloro-2-methylpropene (methallyl chloride) and 96.8 g (0.70 mol) of anhydrous potassium carbonate. Heat the stirred mixture under gentle reflux for 16 hours. Allow the mixture to cool a little and replace the condenser by a still-head and condenser arranged for downward distillation. Distil the stirred mixture until about 370 ml of ethanol and the ethyl acetate formed during the reaction has collected, then cool the residue and add sufficient ice-water to dissolve the suspended salts (about 550 ml is required). Transfer to a separatory funnel and extract with three 200 ml portions of ether. Wash the combined extracts with two 100 ml portions of saturated aqueous sodium chloride and then dry the ethereal solution over anhydrous sodium sulphate. Filter, and remove the ether by flash distillation. Fractionally distil the residue using a well-lagged fractionating column of about 12 cm length filled with glass helices. Collect the unsaturated ketone as a fraction of b.p. 148–153 °C; (mainly 148–150 °C) (1). The yield of 5-methylhex-5-en-2-one is 33.1 g (39%); its purity may be checked by GLC on a 10 per cent Silicone oil on Chromosorb W 5-ft column, held at 82 °C, nitrogen flow rate 40 ml/minute, t_R 1 minute.

Note. (1) The forerun consists of residual ethanol and ethyl acetate together with some of the unsaturated ketone. A substantial high boiling residue remains.

J,6. The acid-catalysed rearrangement of 1,2-diols. The conversion of pinacol (Section **III,29**) to t-butyl methyl ketone (pinacolone, Section **III,97**) under acid conditions exemplifies a general reaction of 1,2-diols (the **pinacol–pinacolone rearrangement**). The mechanism, formulated below, involves loss of water from the protonated 1,2-diol accompanied by a 1,2-nucleophilic shift of a methyl group.

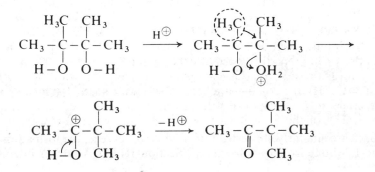

III,97. t-BUTYL METHYL KETONE (*Pinacolone*)

$$(CH_3)_2C(OH)\cdot C(OH)(CH_3)_2 \xrightarrow{H^{\oplus}} CH_3\cdot CO\cdot C(CH_3)_3$$

In a 500-ml round-bottomed flask carrying a dropping funnel and a connection to a condenser set for distillation, place 50 g of pinacol hydrate (Section **III,29**) and 130 ml of 3 *M*-sulphuric acid. Distil the mixture until the upper layer of the distillate no longer increases in volume (15–20 minutes). Separate the pinacolone layer from the water and return the latter to the reaction flask. Then add 12 ml of concentrated sulphuric acid to the water, followed by a second 50 g portion of pinacol hydrate. Repeat the distillation. Repeat the process twice more until 200 g of pinacol hydrate have been used.

Dry the combined pinacolone fractions over magnesium sulphate and distil. Collect the pinacolone at 103–107 °C. The yield is 62 g (70%).

K DICARBONYL COMPOUNDS

1. Synthesis of 1,2-dicarbonyl compounds ($R\cdot CO\cdot CHO$ and $R^1\cdot CO\cdot CO\cdot R^2$, Sections **III,98** to **III,101**).
2. Synthesis of 1,3-dicarbonyl compounds ($R\cdot CO\cdot CH_2\cdot CO\cdot R$, Sections **III,102** and **III,103**).
3. Synthesis of 1,4-dicarbonyl compounds ($R\cdot CO\cdot CH_2\cdot CH_2\cdot CO\cdot R$, Section **III,104**)

K,1. 1,2-Dicarbonyl compounds. The synthesis of 1,2-ketoaldehydes can be achieved from methyl ketones by oxidation with selenium dioxide in a suitable solvent (e.g., ethanol or dioxan); an illustrative example is the preparation of phenylglyoxal from acetophenone (Section **III,98**).

$$C_6H_5\cdot CO\cdot CH_3 \xrightarrow{SeO_2} C_6H_5\cdot CO\cdot CHO$$

The mechanism probably involves the formation of a selenite ester of the enol form of the carbonyl compound.

Cyclic ketones are similarly converted into 1,2-diketones (e.g., the formation of cyclohexane-1,2-dione from cyclohexanone, Section **III,99**). A further interesting application is provided by the synthesis of ninhydrin (Section **III,100**) from indane-1,3-dione (Section **V,8**) in which the methylene group is activated by two adjacent carbonyl groups. Ninhydrin is the stable monohydrate of the triketone, indane-1,2,3-trione, and it is a well-known colorimetric reagent for amino acids (Section **I,31**).

Selenium dioxide oxidation of unsymmetrically substituted ketones (e.g., $R \cdot CH_2 \cdot CO \cdot CH_3$), is complicated by the presence of two alternative sites of oxidation; in practice methyl groups appear to be oxidised in preference to methylene groups for reasons which have not yet been adequately clarified, but in any case the yields are usually poor.

Carbonyl compounds undergo nitrosation in the reactive α-position when treated with nitrous acid or an alkyl nitrite.

$$R^1 \cdot CO \cdot CH_2 \cdot R^2 + R^3 ONO \xrightarrow{H^{\oplus}} R^1 \cdot CO \cdot \underset{\underset{N=O}{|}}{CH} \cdot R^2 + R^3 OH$$

The presence of hydrogen on the α-carbon atom permits tautomeric rearrangement to the oxime of a 1,2-dicarbonyl compound. Acidic hydrolysis of the oxime, which is best carried out in the presence of a hydroxylamine acceptor such as laevulinic acid (Ref. 4), affords a further useful route to the 1,2-dicarbonyl system.

$$R^1 \cdot CO \cdot \underset{\underset{N=O}{|}}{CH} \cdot R^2 \rightleftharpoons R^1 \cdot CO \cdot \underset{\underset{NOH}{\|}}{C} \cdot R^2 \xrightarrow{H_3O^{\oplus}} R^1 \cdot CO \cdot CO \cdot R^2 + NH_2OH$$

In the example included here (Section **III,101**) ethyl methyl ketone is nitrosated to butane-2,3-dione monoxime, which is then reacted with hydroxylamine to give the dioxime, dimethylglyoxime, the well-known reagent for nickel.

A further general route to the 1,2-dicarbonyl system involves the oxidation of α-ketols (acyloins), which are usually prepared from carboxylate esters by treatment with finely divided sodium metal in anhydrous ether or benzene (Ref. 5). An example of this oxidation is provided by the preparation of benzil from benzoin (Section **IV,155**).

III,98. PHENYLGLYOXAL

$$C_6H_5 \cdot CO \cdot CH_3 + SeO_2 \longrightarrow C_6H_5 \cdot CO \cdot CHO + Se + H_2O$$

*Selenium and its compounds are toxic (Section **I,3**); carry out this and the following preparation in an efficient fume cupboard.*

Fit a 500-ml three-necked flask with a sealed stirrer, a reflux condenser and a thermometer. Place 300 ml of dioxan (1), 55.5 g (0.5 mol) of pure selenium dioxide (Section **II,2,52**) and 10 ml of water in the flask, heat the mixture to 50–55 °C and stir until the solid has dissolved. Remove the thermometer momentarily and add 60 g (0.5 mol) of acetophenone (Section **IV,134**) in one lot; replace the thermometer. Reflux the mixture, with stirring, for 4 hours; after about 2 hours the solution becomes clear and little further precipitation of selenium is observable. Decant the hot solution from the precipitated selenium through a fluted filter-paper, and remove the dioxan and water by distillation through a short column. Distil the residual phenylglyoxal under reduced pressure and collect the fraction boiling at 95–97 °C/25 mmHg. The yield of pure phenylglyoxal (a yellow liquid) is 48 g (72%); this sets to a stiff gel on standing, probably as a result of polymerisation, but may be recovered without appreciable loss by distillation. The aldehyde is best preserved in the form of the hydrate, which is prepared by dissolving the yellow liquid in 3.5–4 volumes of hot water and allowing to crystallise. Phenylglyoxal hydrate (m.p. 91 °C) also crystallises from chloroform, ethanol or ether–light petroleum (b.p. 60–80 °C); upon distillation under diminished pressure, the free aldehyde is obtained.

Note. (1) Rectified spirit can also be used as solvent. The dioxan can, however, be recovered and used in a subsequent run (cf. Section **II,1,**20).

III,99. CYCLOHEXANE-1,2-DIONE AND ITS DIOXIME (*Nioxime*)

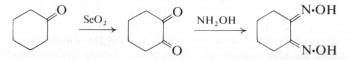

Cyclohexane-1,2-dione. Equip a 1-litre, three-necked flask with a reflux condenser, thermometer and dropping funnel. Place 250 g (2.55 mol) of pure cyclohexanone in the flask, heat to 70–80 °C and add a solution of 280 g (2.52 mol) of pure selenium dioxide (Section **II,2,**52) in 1500 ml of rectified spirit from the dropping funnel over a period of 2 hours, maintaining the temperature at 70–80 °C. Reflux the reaction mixture for a further 2 hours. Distil off as much of the alcohol as possible and decant the liquid residue from the elemental selenium. Wash the latter several times with ether, and combine the ether extracts with the decanted liquid. Remove the ether by distillation and distil the residue under reduced pressure (*ca.* 25 mmHg): about 200 g of an oil, consisting of cyclohexane-1,2-dione, cyclohexanone and water, is obtained. Dissolve the oil in 1 litre of ether, and extract thrice with ice-cold 10 per cent potassium hydroxide solution; the total amount of potassium hydroxide solution should be equivalent to 1.5 times that necessary to react with the oil assumed to be the pure dione in the monoenol form (about 1.5 l). Shake the alkaline extract once with ether to remove cyclohexanone, acidify with ice-cold hydrochloric acid and then saturate with salt. Extract the hydrochloric acid solution with ether, dry the ethereal extract with magnesium sulphate, remove the ether by distillation at normal pressure and distil the residue under reduced pressure. Collect the cyclohexane-1,2-dione (a pale green liquid) at 96–97 °C/25 mmHg; the compound decomposes slightly on keeping. The yield is 55–56 g (19%).

It is important that the synthesis should be carried out as quickly as possible, particularly the washing with alkali at 0 °C, since the latter tends to convert the product into 1-hydroxycyclopentanecarboxylic acid.

Cyclohexane-1,2-dione dioxime. Dissolve 55 g (0.5 mol) of the freshly distilled dione in 500 ml of water, cool the solution to 0 °C and dissolve 170 g of pure hydroxylamine hydrochloride in it. Add a solution of 225 g of potassium hydroxide in 1 litre of water at 0 °C dropwise over a period of 15 minutes with constant mechanical stirring. Heat the mixture on a steam bath for 2 hours, cool to 0 °C, neutralise with powdered solid carbon dioxide, saturate with salt, filter off the precipitated dioxime and wash with a little ice-cold water.

Recrystallise the crude dioxime from water. The yield of pure cyclohexane-1,2-dione dioxime (white needles), m.p. 187–188 °C (decomp.), is 39 g (56%).

III,100. INDANE-1,2,3-TRIONE HYDRATE (*Ninhydrin*)

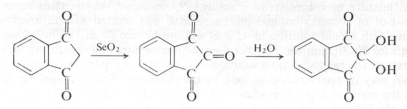

CAUTION: *conduct this preparation in a fume cupboard.* In a 500-ml three-necked flask, fitted with a reflux condenser and mechanical stirrer, place 11 g (0.1 mol) of pure selenium dioxide (Section **II,2,**52) dissolved in 240 ml of dioxan and 5 ml of water. Heat the stirred solution to 60–70 °C, remove the source of heat, add 15 g (0.1 mol) of crude indane-1,3-dione (Section **V,8**) and reflux the resulting mixture for 6 hours. A solid separates during this period. Filter the mixture, transfer the filtrate to a distilling flask and distil off about 180 ml of dioxan; then add 100 ml of water, boil the solution to co-agulate the red tarry precipitate and remove it by filtration. Concentrate the filtrate to about 50 ml and filter. Boil the filtrate with 0.2–0.3 g of decolourising carbon, filter again, concentrate to 20–25 ml and keep at room temperature. Collect the crystals of crude ninhydrin by suction filtration, and recrystallise from hot water with the addition of a little decolourising carbon, if necessary (1). The yield of colourless ninhydrin is 6 g (34%); the crystals turn red between 125 and 130 °C and melt at 242–243 °C.

Note. (1) Recrystallisation from water with the use of decolourising carbon should yield almost colourless crystals; selenium-containing contaminants may, however, give rise to discolouration. In this case a further recrystallisation with the addition of decolourising carbon and a little tin(II) chloride should be carried out.

III,101. BUTANE-2,3-DIONE DIOXIME (*Dimethylglyoxime*)

$$\text{CH}_3\text{·CO·CH}_2\text{·CH}_3 + \text{R·O·NO} \xrightarrow{\text{H}^\oplus} \text{CH}_3\text{·CO·C·CH}_3$$
$$\underset{\text{N·OH}}{\overset{\|}{}}$$

$$\underset{\text{N·OH}}{\overset{\|}{\text{CH}_3\text{·CO·C·CH}_3}} + \text{NH}_2\text{OH} \longrightarrow \underset{\text{HO·N·N·OH}}{\overset{\|\ \|}{\text{CH}_3\text{·C·C·CH}_3}}$$

In a 500-ml three-necked flask, supported on a water bath in a fume cupboard and provided with a dropping funnel, a reflux condenser and a thermometer, place 72 g (90 ml, 1 mol) of dried and redistilled ethyl methyl ketone (Section **II,1,**22). Introduce a follower bar and stir magnetically. Add 3 ml of concentrated hydrochloric acid and warm the liquid to 40 °C. Then add 103 g

(115 ml, 1 mol) of butyl nitrite (b.p. 76–79 °C) or 117 g (134 ml, 1 mol) of isopentyl nitrite (b.p. 96–99 °C) (Section III,75) slowly, maintaining the temperature at 40–50 °C; the mixture must be stirred vigorously. Heat is generated in the reaction so that cooling may now be required. Continue the stirring, without cooling, for 30 minutes after all the nitrite has been added. The reaction mixture now consists of a solution of butane-2,3-dione monoxime in butan-1-ol or 3-methylbutan-1-ol. To remove any unused ketone, treat the mixture with a cold solution of 45 g of sodium hydroxide in 100 ml of water and stir for 20–30 minutes. Transfer the reaction mixture to a separatory funnel and extract the reddish-brown solution twice with 50 ml portions of ether: the alcohol may be recovered, if desired, by fractionation of the ethereal extracts. Keep the aqueous layer; it contains the sodium salt of butane-2,3-dione monoxime (1). Prepare a solution of 70 g (1 mol) of hydroxylamine hydrochloride or of 82 g (1 mol) of hydroxylamine sulphate in about three times its weight of water, and add sodium hydroxide solution until the solution is neutral to litmus. Place the aqueous solution of the sodium salt of the monoxime in a 1-litre round-bottomed flask and add the hydroxylamine solution with stirring. Heat the mixture on a water bath for about 45 minutes. Filter off the precipitated dioxime (2) whilst the solution is still hot, wash it with hot water and drain well. Recrystallise the crude product from about 10 times its weight of rectified spirit. The yield of pure dimethylglyoxime (a white, crystalline solid, m.p. 240 °C) is 55 g (47%).

Notes. (1) If it is desired to isolate the monoxime, cool a portion (say one-fifth part) of the aqueous solution of the sodium salt in an ice–salt bath and carefully neutralise by adding concentrated hydrochloric acid (about 20 ml) with vigorous stirring, keeping the temperature below 15 °C. During the addition, the resulting slurry becomes difficult to stir efficiently; filter off the solid product at this stage and continue to neutralise the filtrate to obtain the remainder of the crude monoxime. The pale brown solid thus obtained contains some sodium chloride; crystallisation from water gives almost colourless crystals, m.p. 73–74 °C.

(2) If the product is coloured, dissolve it in 2 M-sodium hydroxide solution on a water bath. Filter the hot almost saturated solution, and to the hot filtrate add a concentrated solution of ammonium chloride in excess of the amount required to precipitate all the dimethylglyoxime, i.e., employ an amount greater than the equivalent of the sodium hydroxide used. Filter at once with suction, and wash with boiling water. Recrystallise the white product from rectified spirit.

K,2. 1,3-Dicarbonyl compounds. The preparation of the simplest 1,3-diketone (pentane-2,4-dione, Section III,102), is best achieved by the acylation of acetone with acetic anhydride in the presence of boron trifluoride. This yields the diketone in the form of a boron difluoride coordination complex from which it is liberated by steam distillation after the addition of aqueous sodium acetate. The diketone, which is appreciably water-soluble, is conveniently isolated by way of its characteristic copper complex (I). In the cognate preparation, which illustrates the generality of the procedure, the less water-soluble 2-acetylcyclohexanone may be isolated directly from the reaction mixture by solvent extraction after decomposition of the boron difluoride complex with sodium acetate.

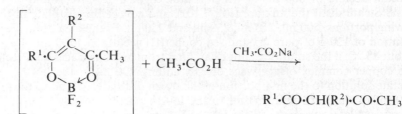

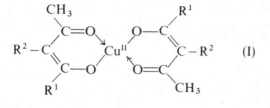

An alternative acylation procedure involves the base catalysed condensation of a ketone with an ester (**Claisen condensation**). This reaction is illustrated by the formation of benzoylacetone from acetophenone and ethyl acetate (Section **III,103**) and may be mechanistically represented in the following way (cf. p. 537, **III,P,4**(*b*)).

$$C_6H_5 \cdot CO \cdot CH_2 \overset{|}{\underset{H \leftarrow :OC_2H_5}{}} \rightleftharpoons [C_6H_5 \cdot CO \cdot \overset{\ominus}{\ddot{C}}H_2] + HOC_2H_5$$
mesomeric anion

$$CH_3 \cdot \overset{O}{\overset{\|}{C}} \overset{\frown}{\underset{\underset{OC_2H_5}{\curvearrowright}}{}} \overset{\ominus}{:}CH_2 \cdot CO \cdot C_6H_5 \rightleftharpoons CH_3 \cdot CO \cdot CH_2 \cdot CO \cdot C_6H_5 + \overset{\ominus}{:}OC_2H_5$$

$$\longrightarrow [CH_3 \cdot CO \cdot \overset{\ominus}{\ddot{C}}H \cdot CO \cdot C_6H_5] \xrightarrow{H_3O^{\oplus}} CH_3 \cdot CO \cdot CH_2 \cdot CO \cdot C_6H_5$$

III,102. PENTANE-2,4-DIONE (*Acetylacetone*)

$$CH_3 \cdot CO \cdot CH_3 + (CH_3 \cdot CO)_2O \xrightarrow{BF_3} CH_3 \cdot CO \cdot CH_2 \cdot CO \cdot CH_3 + CH_3 \cdot CO_2H$$

Fit a 1-litre three-necked flask with a gas inlet tube and a gas outlet leading to a gas absorption device (Fig. I,68) charged with aqueous alkali to trap excess boron trifluoride, and stopper the third neck. Place 58 g (73 ml, 1 mol) of pure, anhydrous acetone (Section **II,1,21**) and 255 g (236 ml, 2.5 mol) of acetic anhydride in the flask and cool in a freezing mixture of ice and salt. Connect the gas inlet tube through an empty wash bottle to a cylinder of commercial boron trifluoride (*TOXIC*: see Section **II,2,6**), and bubble the gas through the reaction mixture at such a rate that 250 g is absorbed in about 5 hours (2 bubbles per second). Pour the reaction mixture into a solution of 400 g of crystallised sodium acetate in 800 ml of water contained in a 2.5-litre round-bottomed flask. Steam distil the mixture (Fig. I,101), and collect the distillate in the following portions: 500 ml, 250 ml, 250 ml and 250 ml. In the meantime prepare a solution of 120 g of pure crystallised copper(II) acetate in 1500 ml of water at about 85 °C; if the solution is not clear, filter from any basic acetate. Precipitate the copper complex of acetylacetone by adding 700 ml of the hot copper(II) acetate solution to the first portion of the steam distillate, 350 ml to the second, 250 ml to the third and 200 ml to the fourth portion. Allow to stand for 3 hours, or better overnight, in the ice chest. Filter off the salt at the pump, wash once with water and suck as dry as possible. Transfer the copper complex to a separatory funnel, add 400 ml of 20 per cent by weight sulphuric acid and 400 ml of ether, and shake. Remove the ether layer. Extract the aqueous layer with two 150 ml portions of ether. Dry the combined extracts with 125 g of anhydrous sodium sulphate (or the equivalent quantity of magnesium sulphate), and distil off the ether. Distil the residue through a short fractionating column and collect the acetylacetone at 134–136 °C. The yield is 80 g (80%).

Cognate preparation **2-Acetylcyclohexanone.** Place a mixture of 24.5 g (0.25 mol) of cyclohexanone (regenerated from the bisulphite compound) and 51 g (47.5 ml, 0.5 mol) of acetic anhydride in a 500-ml three-necked flask, fitted with an efficient sealed stirrer, a gas inlet tube reaching to within 1–2 cm of the surface of the liquid, and in the third neck a thermometer immersed in the liquid, combined with a gas outlet tube leading to a trap (1). Immerse the flask in a bath of Cardice–acetone, stir the mixture vigorously and pass in the boron trifluoride as fast as possible (10–20 minutes) until the mixture, kept at 0–10 °C, is saturated (copious evolution of white fumes when the outlet tube is disconnected from the trap). Replace the Cardice–acetone bath by an ice bath and pass the gas in at a slower rate to ensure maximum absorption. Stir for 3.5 hours whilst allowing the ice bath to attain room temperature slowly. Pour the reaction mixture into a solution of 136 g of hydrated sodium acetate in 250 ml of water, reflux for 60 minutes (or until the boron fluoride complexes are hydrolysed), cool in ice and extract with three 50 ml portions of light petroleum, b.p. 40–60 °C (2), wash the combined extracts free of acid with sodium hydrogen carbonate solution, dry over anhydrous calcium sulphate, remove the solvent by flash distillation and distil the residue under reduced pressure. Collect the 2-acetylcyclohexanone at 95–97 °C/10 mmHg. The yield is 27 g (77%).

Notes. (1) Alternatively the reaction may be effected by adding the ketone and the acetic anhydride to 100 g (0.75 mol) of a 1 : 1 acetic acid–boron trifluoride complex (Section **II,2,6**).

(2) Light petroleum is preferable to ether because it removes smaller amounts of acetic acid from the aqueous phase.

III,103. BENZOYLACETONE

$$C_6H_5 \cdot CO \cdot CH_3 + CH_3 \cdot CO_2C_2H_5 \xrightarrow{\overset{\oplus}{Na}\ \overset{\ominus}{OC_2H_5}} C_6H_5 \cdot CO \cdot CH_2 \cdot CO \cdot CH_3$$

Sodium ethoxide. Prepare a suspension of 11.5 g (0.5 mol) of granulated sodium (Section **II,2,55**) in 75 ml of dry xylene, transfer it to a 1-litre three-necked flask, and decant the xylene. Wash the sodium by decantation with two 20 ml portions of dry ether and cover with 200 ml of dry ether. Set the flask on a water bath and fit it with a sealed stirrer unit, and with a reflux condenser and a dropping funnel, each protected by a calcium chloride guard-tube. Start the stirrer and run in 23 g (29 ml, 0.5 mol) of absolute ethanol from the dropping funnel during 1–2 hours with gentle refluxing, and continue to reflux the mixture with stirring until nearly all of the sodium has reacted (up to 6 hours; a little residual sodium does no harm). Stop the stirrer, set the condenser for downward distillation and distil off the ether as completely as possible. The residual sodium ethoxide should be white and finely divided. All moisture must be excluded during the preparation.

Benzoylacetone. Return the condenser (protected by the calcium chloride guard-tube) to the reflux position, surround the flask with ice and introduce 200 ml (2 mol) of pure, dry ethyl acetate (Section **II,1,**24). Start the stirrer and add 60 g (58 ml, 0.5 mol) of acetophenone (Section **IV,134**) from the dropping funnel; the reaction commences with the separation of the sodium salt of benzoylacetone. Continue stirring for 2 hours and then allow to stand in an ice box overnight. Filter the solid at the pump with the aid of the addition of a little dry ether. Dissolve the air-dried solid in cold water, and acidify the solution with acetic acid. Filter off the crude benzoylacetone, and dry in the air. Purify by distillation under reduced pressure; collect the benzoylacetone at 128–130 °C/10 mmHg. It solidifies on cooling to a colourless crystalline solid, m.p. 61 °C. The yield is 50 g (62%).

K,3. 1,4-Dicarbonyl compounds. Symmetrical 1,4-diketones are readily prepared from the sodio-derivatives of β-keto esters, or their mono-alkyl derivatives, by treatment with iodine.

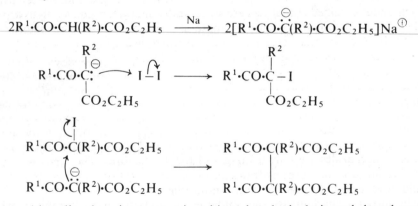

The resulting diacylsuccinate ester is subjected to hydrolysis and decarboxylation by heating with aqueous potassium carbonate (cf. p. 434, **III,J,5**). The reaction is illustrated by the preparation of hexane-2,5-dione (Section **III,104**).

III,104. HEXANE-2,5-DIONE (*Acetonylacetone*)

$$2CH_3 \cdot CO \cdot CH_2 \cdot CO_2C_2H_5 \xrightarrow{2Na} 2[CH_3 \cdot CO \cdot CH \cdot CO_2C_2H_5]^{\ominus} \overset{\oplus}{Na} \xrightarrow{I_2}$$

$$\begin{array}{c} CH_3 \cdot CO \cdot CH \cdot CO_2C_2H_5 \\ | \\ CH_3 \cdot CO \cdot CH \cdot CO_2C_2H_5 \end{array} \xrightarrow[\text{decarboxylation}]{\text{hydrolysis;}} CH_3 \cdot CO \cdot CH_2 \cdot CH_2 \cdot CO \cdot CH_3$$

Place 11.5 g (0.5 mol) of granular sodium (Section **II,2,**55) covered with 250 ml of dry ether in a 1-litre three-necked flask fitted with a sealed stirrer unit, a reflux condenser and a dropping funnel. Start the stirrer and add gradually a solution of 65 g (63.5 ml, 0.5 mol) of redistilled ethyl acetoacetate (Section **III,**176) in 250 ml of dry ether, cooling the flask if the reaction becomes too vigorous. Continue stirring until all of the sodium has reacted, and then add steadily a solution of 63.5 g (0.25 mol) of powdered iodine in 350 ml of dry ether until the iodine colour persists. Filter off the sodium iodide, wash it with ether and evaporate the combined filtrate and washings (rotary evaporator). A somewhat sticky residue of diethyl 2,3-diacetosuccinate remains; crystallise a small portion from 50 per cent aqueous acetic acid to obtain a specimen, m.p. 88 °C. Boil the bulk of the crude product under reflux for 1 hour with 250 ml of 20 per cent aqueous potassium carbonate solution and treat the cooled clear yellow solution with 50 g of anhydrous potassium carbonate. Separate the organic phase, and extract the aqueous layer with four 25 ml portions of ether. Combine the ether extracts with the original organic layer and dry over anhydrous sodium sulphate. Remove the ether by flash distillation and distil the residue, collecting the hexanedione as a fraction of b.p. 185–192 °C. The yield is 7.5 g (26%).

L CARBOHYDRATES

INTRODUCTION The simplest carbohydrates are the monosaccharides which under specified conditions are structurally characterised as polyhydroxy-aldehydes or polyhydroxy-ketones; these are termed aldoses and ketoses respectively. Aldoses and ketoses are sub-classified, according to the number of carbon atoms present in each molecule, into aldotetroses, aldopentoses, aldohexoses, etc., or ketotetroses, ketopentoses, etc. The structural relationships of the aldoses and of the ketoses are exemplified in Table III,1 and Table III,2 respectively (Fischer projection formulations). By convention the configuration at the chiral centre farthest from the carbonyl function is designated D when the hydroxyl group lies on the right side of the Fischer projection formulae, and relates the D-series of monosaccharides to D-glyceraldehyde. The L-series of monosaccharides are mirror images of these structures. The D,L-convention is still retained in the carbohydrates although the chiral sites in these molecules could be designated by the R- and S-notation.

The monosaccharides are the fundamental units for more complex carbohydrates. Thus disaccharides are compounds which yield two monosaccharide molecules upon dilute acid hydrolysis; trisaccharides give three monosaccharide molecules upon hydrolysis; tetrasaccharides give four monosaccharide molecules, etc. The upper limit in this group (the oligosaccharides) is reached in the case of a polymer having ten monosaccharide units. Mono- and oligosaccharides are characterised by having an invariable molecular weight, high water solubility, and are usually sweet to the taste; they are frequently known as the sugars. The non-sugar group of carbohydrates are polymers (polysaccharides) having more numerous monosaccharide units and the molecular weights of individual molecules in a given sample will not necessarily be of the same

Table III,1

Aldoses

Aldotriose

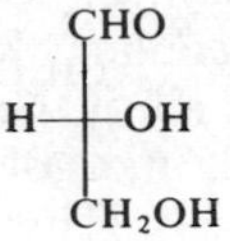

```
        CHO
      H—|—OH
        CH₂OH
```
D-glyceraldehyde

Aldotetroses

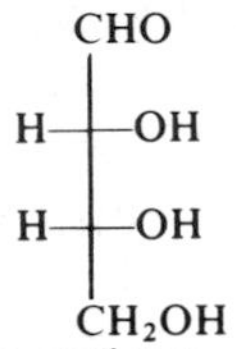

```
   CHO              CHO
 H—|—OH          HO—|—H
 H—|—OH           H—|—OH
   CH₂OH            CH₂OH
 D-erythrose       D-threose
```

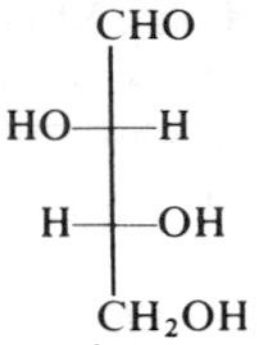

Aldopentoses

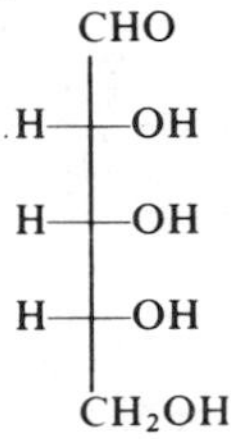

```
   CHO         CHO          CHO          CHO
 H—|—OH      HO—|—H       H—|—OH       HO—|—H
 H—|—OH       H—|—OH      HO—|—H       HO—|—H
 H—|—OH       H—|—OH       H—|—OH       H—|—OH
   CH₂OH        CH₂OH        CH₂OH        CH₂OH
 D-ribose     D-arabinose   D-xylose     D-lyxose
```

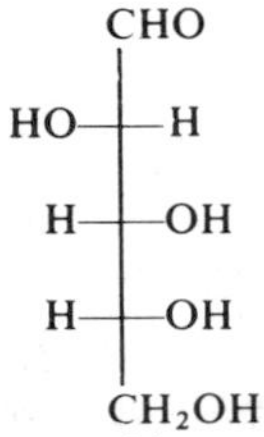

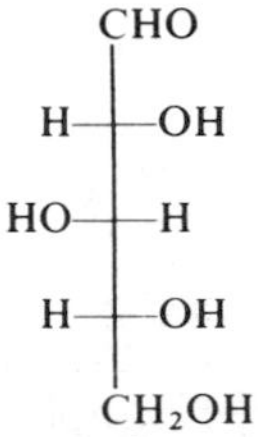

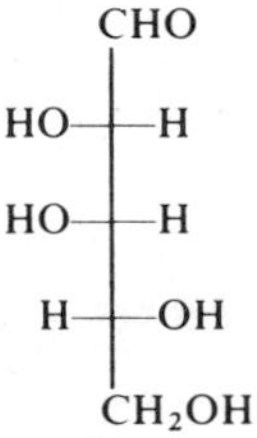

Aldohexoses

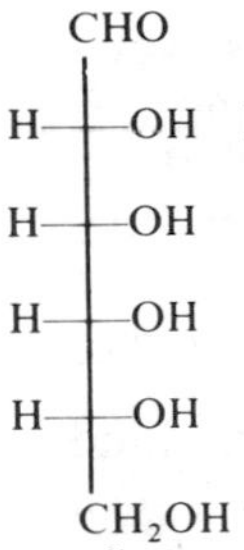

```
   CHO       CHO       CHO       CHO       CHO       CHO       CHO       CH
 H—|—OH    HO—|—H    H—|—OH    HO—|—H    H—|—OH    HO—|—H    H—|—OH    HO—|
 H—|—OH    H—|—OH   HO—|—H    HO—|—H    H—|—OH    H—|—OH   HO—|—H    HO—|
 H—|—OH    H—|—OH    H—|—OH    H—|—OH   HO—|—H    HO—|—H   HO—|—H    HO—|
 H—|—OH    H—|—OH    H—|—OH    H—|—OH    H—|—OH    H—|—OH    H—|—OH    H—|
   CH₂OH     CH₂OH     CH₂OH     CH₂OH     CH₂OH     CH₂OH     CH₂OH     C
 D-allose  D-altrose D-glucose D-mannose D-gulose  D-idose  D-galactose D-ta
```

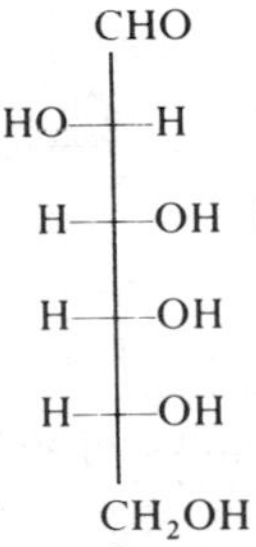

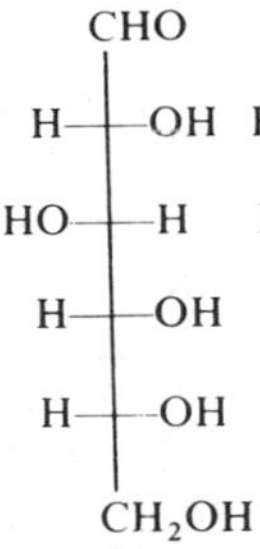

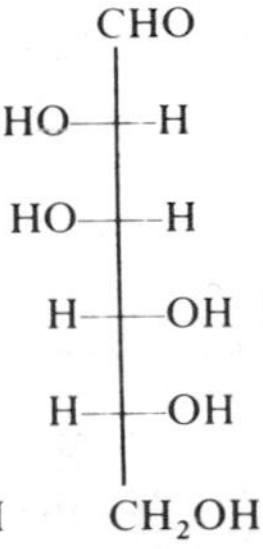

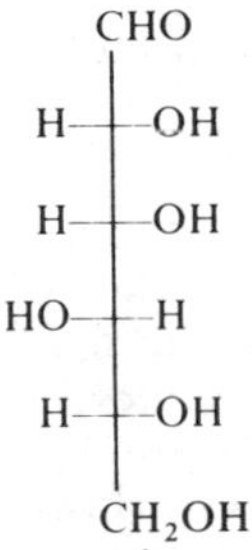

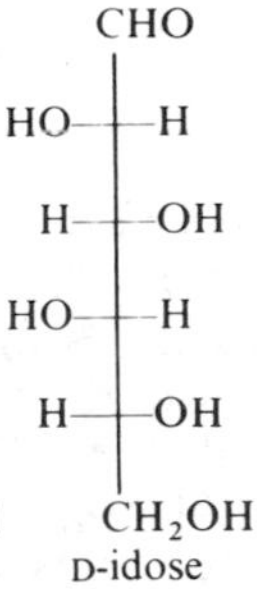

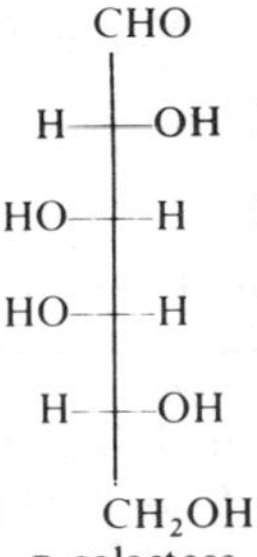

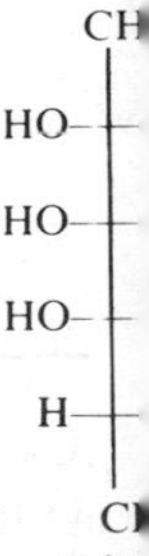

magnitude; acid hydrolysis yields the appropriate number of monosaccharide units.

Detailed studies on the structure of D-glucose based upon physical data and chemical reactivity have revealed that the open-chain formulation is an over-simplification. In *solution* D-glucose exists as an equilibrium mixture of five forms. The cyclic structures are those which arise from hemi-acetal formation involving the carbonyl function of C1 of the open-chain form IIIa/b with either (i) the hydroxyl group on C5 (in IIIa) to give two pyranose ring structures (I), α-D-glucopyranose, and (II), β-D-glucopyranose, or (ii) the hydroxyl group on

Table III,2

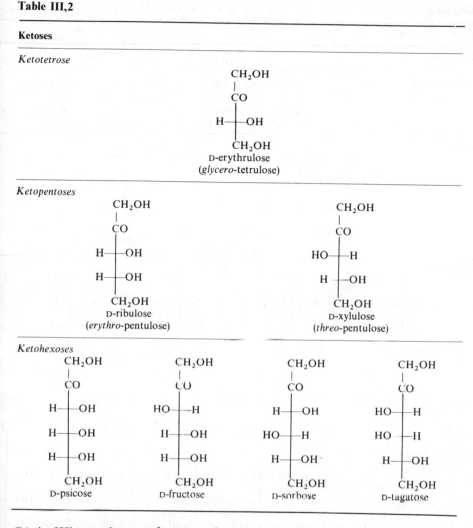

Ketoses

Ketotetrose

CH₂OH
|
CO
|
H——OH
|
CH₂OH
D-erythrulose
(*glycero*-tetrulose)

Ketopentoses

D-ribulose (*erythro*-pentulose) and D-xylulose (*threo*-pentulose)

Ketohexoses

D-psicose, D-fructose, D-sorbose, D-tagatose

C4 (in IIIb) to give two furanose ring structures (IV), α-D-glucofuranose, and (V), β-D-glucofuranose. These Haworth representations (I, II, IV and V) must be viewed as if the rings are perpendicular to the plane of the paper with the ring oxygens on the side remote from the viewer, and the groups attached to the ring carbons lying either above or below the plane of the ring.

Structures I and II, and structures IV and V, differ from one another only in the configuration at the anomeric carbon atom (C1), the hydroxyl group being either below (α) or above (β) the plane of ring. In solution the equilibrium ratio of the four cyclic and the acyclic structures is dependent upon the nature of the solvent; in aqueous solution at room temperature the percentage composition is (I), 36 per cent; (II), 64 per cent; (III), 0.0026 per cent; (IV) + (V), <1 per cent.

D-Glucose as commonly isolated in the crystalline state exists in the α-D-glucopyranose form. X-ray crystallographic studies have further shown that the ring structure adopts the more stable chair conformation (VI) in which the

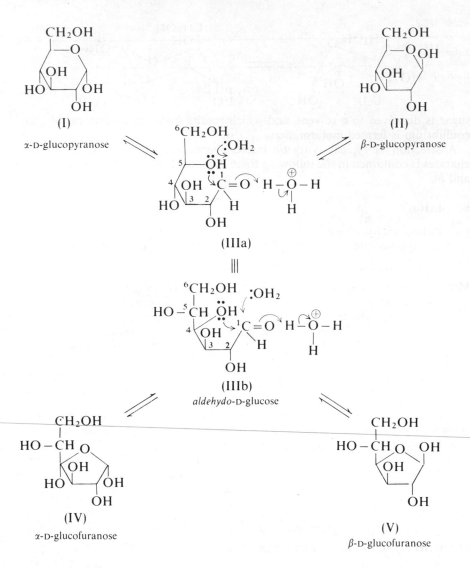

(I)

α-D-glucopyranose

(II)

β-D-glucopyranose

(IIIa)

|||

(IIIb)

aldehydo-D-glucose

(IV)

α-D-glucofuranose

(V)

β-D-glucofuranose

hydroxyl groups on C2, C3 and C4, and the hydroxymethyl group on C5 occupy equatorial positions, rather than the alternative chair conformation (VII) in which these groups occupy axial positions.*

Most, if not all, of the stable forms of crystalline aldose and ketose monosaccharides exist in the pyranose structure. Each in solution, as with D-glucose, exists as an equilibrium mixture of open-chain and of α- and β-anomers of the cyclic forms. The change in optical rotation, which may be observed when a

---

* There has been a considerable amount of discussion over recent years on the nomenclature to be used to describe the various conformations which these ring systems may adopt. An authoritative discussion is given in Ref. 6.

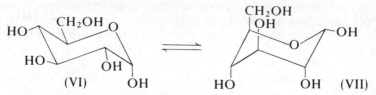

(VI)   (VII)

sugar is dissolved in a solvent, and which results from the attainment of this equilibrium is termed **mutarotation**.

A selection of typical synthetic interconversions undergone by monosaccharides is contained in the following three sections, **L**,1(*a* and *b*), **L**,2 and **L**,3(*a* and *b*).

L,1(*a*).

Potassium hydrogen
D-glucarate   ⟵ D-Glucose ⟶   1,2,3,4,6-Penta-*O*-acetyl-
α (or β)-D-glucopyranose

Methyl 2,3,4,6-tetra-*O*-acetyl   ⟵   2,3,4,6-Penta-*O*-acetyl-
β-D-glucopyranoside   α-D-glucopyranosyl bromide

Methyl β-D-glucopyranoside

L,1(*b*).

D-Galactose ⟶ Methyl α-D-galactopyranoside

Vigorous oxidation of D-glucose with dilute nitric acid converts both the terminal hydroxymethyl group and the potential aldehydic group into carboxyl groups, to give a polyhydroxydicarboxylic acid, D-glucaric acid (formerly termed saccharic acid). The acid is isolated as the sparingly water-soluble potassium hydrogen salt (Section **III,105**); attempted liberation of the free acid by neutralisation of an aqueous solution of this salt with sulphuric acid followed by evaporation of the solution leads however to a mixture of two monolactones the 1,4- (VIII) and the 3,6- (IX) (Ref. 7). These lactones may be separated by virtue of their differential solubility in acetone; they may each be converted into dilactones by heating in a vacuum (Ref. 8).

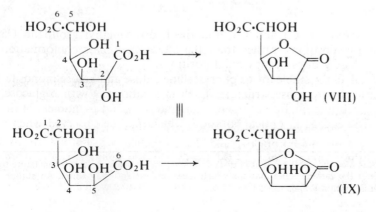

D-Galactose may be similarly converted into D-galactaric acid (formerly termed mucic acid) (Section **III,105**). In practice it is more economic to use lactose, a disaccharide containing D-glucose and D-galactose moieties, which on oxidation gives a mixture of D-glucaric acid and D-galactaric acid. The latter is sparingly water soluble and may be preferentially removed whilst the filtrate may be appropriately treated to yield the potassium hydrogen D-glucarate. D-Galactaric acid shows little tendency towards lactone formation.

The hydroxyl groups at both the anomeric and non-anomeric carbon atoms may be readily acylated. For example, treatment of D-glucose with acetic anhydride in the presence of zinc chloride gives a reasonable yield of the penta-acetate, 1,2,3,4,6-penta-O-acetyl-α-D-glucopyranose (X) (Section **III,106**). On the other hand the β-anomer predominates when D-glucose is treated with acetic anhydride in the presence of sodium acetate (Section **III,107**). Interestingly the β-anomer may be converted into the α-anomer by heating with an acetic anhydride/zinc chloride mixture. These reactions are reasonably interpreted on the basis that the β-anomer is the kinetically controlled product, being initially formed from the more rapid acetylation of the equatorial hydroxyl group at C1 of β-D-glucopyranose. In the presence of the Lewis acid catalyst, zinc chloride, the β-penta-acetate (XI) is thought to be readily converted into the mesomerically stabilised carbonium ion (XII) by loss of the C1-acetoxy group, which would then recombine to yield the thermodynamically more stable α-anomer.

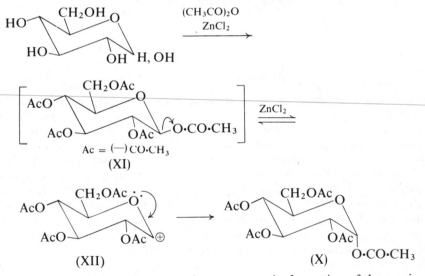

(XI)

Ac = (—)CO·CH₃

(XII)   (X)   O·CO·CH₃

Sodium acetate would not be expected to promote the formation of the species (XII) and hence the β-anomer would accumulate if this catalyst were used.

D-Glucose may be benzoylated with benzoyl chloride in pyridine/chloroform solution (Section **III,108**). Here temperature control is essential to prevent the formation of an anomeric mixture.

The acetate (or benzoate) groups attached to the anomeric site may be readily replaced by a bromine atom to yield the synthetically useful acylglycosyl halides. Thus the formation of 2,3,4,6-tetra-O-acetyl-α-D-glucopyranosyl bromide (XIII) from D-glucose involves the formation of the mixed penta-acetates (X and XI) followed by treatment of the reaction mixture with hydrogen bromide generated *in situ* from red phosphorus, bromine and water (Section **III,109**). Alternatively a solution of hydrogen bromide in glacial acetic acid is added to a solution of the acetylated or benzoylated monosaccharide (e.g., Section **III,110**, the preparation of 2,3,4,6-tetra-*O*-benzoyl-α-D-glucopyranosyl bromide). In either case the thermodynamically more stable anomer is formed. In the case of glucose (and of the other aldohexoses together with xylose and lyxose) this is the α-anomer, whilst in the case of arabinose and ribose the β-anomer predominates.

The halogen in the acylglycosyl halide is reactive and may be readily displaced, for example, by an alkoxy group on reaction with an alcohol under anhydrous conditions in the presence of a silver or mercury(II) salt. In this case the products are **glycosides** which are the mixed cyclic acetals related to the cyclic hemiacetal forms of the monosaccharides. In the case of the D-glucose derivative shown below (and of other 1,2-*cis* acylglycosyl halides) the replacement involves inversion of configuration at the anomeric site and the α-glucosyl halide (XIII) yields a β-glucoside (XIV) (e.g., the formation of methyl 2,3,4,6-tetra-*O*-acetyl-β-D-glucopyranoside, Section **III,111**).

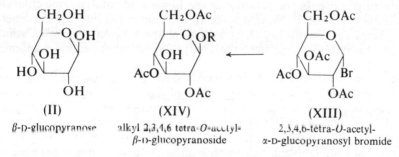

(II)	(XIV)	(XIII)
β-D-glucopyranose	alkyl 2,3,4,6 tetra-*O*-acetyl-β-D-glucopyranoside	2,3,4,6-tetra-*O*-acetyl-α-D-glucopyranosyl bromide

The presence of the protecting acyl groups in the acylglycosyl halide ensures that the pyranose ring structure is retained. Removal of these protecting groups (either acetyl or benzoyl) is readily effected by base (e.g., **Zemplen's method** which uses methanol containing a small amount of sodium methoxide) to give the alkali-stable glycoside (e.g., methyl β-D-glucopyranoside, Section **III,112**).

The above, indirect, procedure for glycoside synthesis is the **Koenigs–Knorr method.** Glycosidation may be effected from monosaccharides directly by treatment with an alcohol in the presence of a mineral acid catalyst. For example, when D-galactose is heated in methanolic solution containing 2 per cent of hydrogen chloride the thermodynamically more stable methyl α-D-galactopyranoside (Section **III,113**) is formed preferentially, and may be isolated from the reaction product by crystallisation as the monohydrate. The less abundant β-anomer may be recovered from the mother-liquors.

The commercially available methyl α-D-glucopyranoside may be obtained similarly. If however the glycosidation reaction of D-glucose is conducted at room temperature with a lower proportion of mineral acid, then the predominant products are the methyl α(and β)-D-glucofuranosides. It is considered that the furanosides are the products arising from kinetic control and the pyranosides from thermodynamic control. Clearly the composition of reaction mixtures involving glycosidation of other monosaccharides will be dependent upon the nature of the monosaccharide and the proportions may be rationalised by the application of conformational analysis.

III,105. POTASSIUM HYDROGEN D-GLUCARATE

$$\text{D-Glucose} \longrightarrow [^{\ominus}O_2C\cdot[CH(OH)]_4\cdot CO_2^{\ominus}]H^{\oplus}, K^{\oplus}$$

Carefully mix together in a large evaporating basin 10 g (0.056 mol) of D-glucose and 100 ml of aqueous nitric acid (d 1.15) and heat on a water bath in an efficient fume cupboard (1). A vigorous evolution of brown fumes may occur during the early stages of the oxidation reaction and it is advisable at such a time to remove the basin from the water bath temporarily. Allow the solution to concentrate on the steam bath until a syrupy residue remains, add 30 ml of water, heat to boiling and add solid potassium carbonate until no further evolution of carbon dioxide occurs. Adjust the pH of the solution to 4 with acetic acid, concentrate the solution to about 15 ml and refrigerate for several days at 5 °C to complete crystallisation of the product. Recrystallise the potassium hydrogen D-glucarate from a small quantity of hot water (employ decolourising charcoal if necessary); the yield is 4 g (40%).

Note. (1) Alternatively use the filtrate (A) from the oxidation of lactose (cognate preparation below) and evaporate this to a syrupy residue prior to dissolution in water and neutralisation with potassium carbonate.

Cognate preparation D-Galactaric acid. Use 10 g of lactose with 100 ml of aqueous nitric acid (d 1.15). On evaporation of the reaction mixture, when the volume has been reduced to about 20 ml, the mixture becomes thick and pasty owing to the separation of D-galactaric acid. Cool, add 30 ml of water, filter the product and use the filtrate (A) as described in Note (1), above. Purify the D-galactaric acid by dissolving it in the minimum volume of dilute aqueous sodium hydroxide and reprecipitating with dilute hydrochloric acid; do not allow the temperature to rise above 25 °C. Dry the purified acid (about 5 g) which melts with decomposition at 212–213 °C.

III,106. 1,2,3,4,6-PENTA-*O*-ACETYL-α-D-GLUCOPYRANOSE (α-D-*Glucopyranose penta-acetate*)

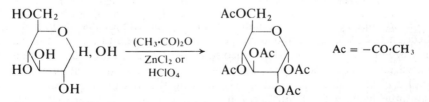

Method A. Into a 100-ml round-bottomed flask place 0.5 g of anhydrous zinc chloride (1) and 13.5 g (12.5 ml, 0.13 mol) of acetic anhydride; attach a Liebig reflux condenser and heat the mixture on a boiling water bath for 5–10 minutes with occasional shaking until the zinc chloride has largely dissolved. Add slowly 2.5 g (0.014 mol) of powdered α-D-glucose, shaking the mixture gently during the addition to control the vigorous reaction which ensues. Finally heat the flask for 1 hour on a boiling water bath (2). Pour the contents of the flask into 125 ml of ice-water and stir vigorously to assist the hydrolysis of unreacted acetic anhydride. After about 30 minutes the oil which first separates will gradually solidify. Filter, wash well with cold water and recrystallise several times from industrial spirit until the m.p. is constant. The pure product melts at 110–111 °C; the yield is 3.5 g (63%).

Method B. To a mixture of 16.25 g (15 ml, 0.16 mol) of acetic anhydride and 25 ml of glacial acetic acid in a conical flask add 5 g (0.028 mol) of powdered glucose. Add dropwise (Pasteur pipette) and with shaking 1 ml of perchloric acid–acetic anhydride catalyst (3), at such a rate that the temperature of the mixture does not exceed 35 °C. Leave at room temperature for 30 minutes and then pour the liquid into a mixture of ice and water. Filter off the crystalline solid which separates on vigorous stirring and wash it thoroughly with cold water. Recrystallise from industrial spirit until the m.p. is constant; the pure product has m.p. 110–111 °C, $[\alpha]_D^{18}$ +101.6° (c 0.28 in $CHCl_3$); the yield is 8 g (72%).

Notes. (1) Zinc chloride is extremely deliquescent and it must therefore be introduced into the flask as rapidly as possible. Place a small stick of zinc chloride in a glass mortar, powder rapidly, and weigh out the required amount.

(2) Although the time of heating may be reduced to 30 minutes by heating in an air bath the product is somewhat discoloured and requires at least one recrystallisation involving the use of decolourising charcoal.

(3) The perchloric acid–acetic anhydride catalyst may be prepared by adding 1.0 g of 60 per cent perchloric acid to 2.3 g of acetic anhydride maintained at 0 °C.

III,107. 1,2,3,4,6-PENTA-*O*-ACETYL-*β*-D-GLUCOPYRANOSE (*β*-D-*Glucopyranose penta-acetate*)

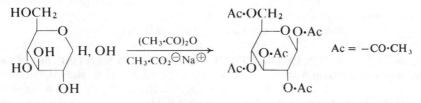

Grind together in a porcelain mortar 4 g of anhydrous sodium acetate (Section **II,2,***56*) and 5 g (0.028 mol) of dry α-D-glucose and place the powdered mixture in a 200-ml round-bottomed flask. Add 27 g (25 ml, 0.26 mol) of acetic anhydride, attach a double surface condenser and heat on a boiling water bath until a clear solution is obtained (1), shaking the mixture from time to time. Continue heating for a further 2 hours after a clear solution has been obtained and then pour the reaction mixture on to 250 ml of crushed ice. Allow to stand for 1 hour, stirring occasionally to break up the solid lumps which separate. Filter off the crystals, wash well with cold water and recrystallise from industrial spirit (or from methanol or ethanol) until the purified material has m.p. 131–132 °C, $[\alpha]_D^{18}$ +4.0° (c 4.5 in $CHCl_3$). The yield is 6.2 g (56%).

Conversion of *β*- into α-D-glucose penta-acetate. Add 0.5 g of anhydrous zinc chloride rapidly to 25 ml of acetic anhydride in a 100-ml round-bottomed flask, fitted with a Liebig condenser, and heat on a boiling water bath to dissolve the solid. Add 5 g of pure β-D-glucose penta-acetate, continue heating for 30 minutes, pour the mixture on to ice and purify the solid which separates as described above. The effectiveness of the conversion may be monitored by TLC on silica gel plates using cyclohexane/acetone (7 : 3) and locating the two closely running spots by immersing the developed and dried plate in a tank of iodine vapour.

Note. (1) It is dangerous to scale up this experiment without modifying the preparative procedure. If 50 g of glucose is to be acetylated, a 2-litre round-bottomed flask should be fitted with two wide-bore Liebig condensers in series, and a large vessel filled with ice-water should be readily available to plunge the reaction flask into, should the vigorous reaction which ensues on heating need controlling. With a scale using 100 g of glucose the details cited in Ref. 9 should be consulted; this describes a procedure involving the addition of α-D-glucose to a preheated sodium acetate–acetic anhydride mixture at such a rate as to keep the mixture under reflux but without the reaction getting out of control.

III,108. 1,2,3,4,6-PENTA-*O*-BENZOYL-α-D-GLUCOPYRANOSE
(*α-D-Glucopyranose pentabenzoate*)

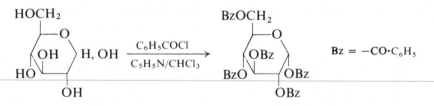

In a 1-litre flange flask fitted with a multiple socket head carrying a mechanical stirrer, a calcium chloride guard-tube, a 250-ml dropping funnel and a thermometer, place 126 ml of dry pyridine and 105 ml of dry chloroform. Cool the flask well in an ice–salt bath and add from the dropping funnel, with stirring, a previously prepared and cooled solution of 127 g (105 ml, 0.9 mol) of benzoyl chloride in 105 ml of dry chloroform. Remove the dropping funnel and add 50 g (0.28 mol) of dry powdered α-D-glucose portion-wise to the vigorously stirred benzoylating reagent at a rate which maintains the temperature of the reaction below 10 °C (1). Allow the pink-coloured solution to stand at 0 °C for 24 hours, dilute with 400 ml of chloroform and transfer the solution to a 2-litre separatory funnel. Wash the solution successively with several 300 ml portions of dilute aqueous sulphuric acid (2 *M*), water, saturated aqueous sodium hydrogen carbonate and water. Dry over anhydrous sodium sulphate and remove the chloroform on a rotary evaporator to give a yellow solid which is ground up with industrial spirit, filtered and washed well with spirit. Recrystallise the solid from acetone–water to give the pure product, m.p. 184–186 °C, $[\alpha]_D^{20}$ +184.4° (*c* 1.75 in CHCl$_3$). The yield is 149 g (77%).

Note. (1) When the experiment is performed without effective cooling, i.e., if the temperature rises to 40–50 °C, a mixture of anomers is obtained which cannot be separated by simple recrystallisation. Frequently in large-scale preparations it is advisable to replace the ice–salt bath with one of acetone–Cardice to ensure good temperature control.

III,109. 2,3,4,6-TETRA-*O*-ACETYL-α-D-GLUCOPYRANOSYL BROMIDE (α-*Acetobromoglucose*)

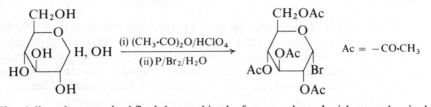

Fit a 1-litre three-necked flask located in the fume cupboard with a mechanical stirrer unit using a Kyrides seal (Fig. I,56(*a*)), a dropping funnel and a thermometer to read the temperature of the reaction mixture. Immerse the flask in an ice–salt bath supported on a laboratory jack so that it may be easily removed if the reaction conditions so demand. Place 432 g (400 ml, 4.24 mol) of acetic anhydride in the flask, cool to 4 °C and add dropwise and with stirring 2.4 ml of 60 per cent perchloric acid. Remove the cooling bath and allow the reaction mixture to warm to room temperature; then add 100 g (0.56 mol) of dry powdered α-D-glucose in portions with stirring so that the temperature of the reaction mixture is maintained at between 30 and 40 °C. Cool to about 20 °C and add 31 g (1 mol) of red phosphorus followed by 181 g (58 ml, 2.26 mol) of bromine dropwise at a rate that the temperature does not exceed 20 °C. Then add 36 ml of water over a period of about half an hour, the stirring and cooling being continued and the temperature being maintained below 20 °C. Allow the reaction mixture to stand for 2 hours at room temperature, transfer to a fume cupboard and dilute with 300 ml of methylene chloride, and filter through a large 60° glass funnel having a glass wool plug inserted not too tightly into the outlet (1). Finally rinse the reaction flask and funnel with small portions of methylene chloride, transfer the filtrate and washings to a 3-litre separatory funnel and wash rapidly by shaking vigorously with two 800 ml portions of iced water (2). Run the lower methylene chloride layer from the second washing into 500 ml of a stirred saturated solution of aqueous sodium hydrogen carbonate to which has also been added some crushed ice. When the vigorous evolution of carbon dioxide has subsided transfer the mixture to a separatory funnel, run the methylene chloride layer into a large conical flask containing 10 g of powdered activated silica gel and filter after about 10 minutes (the bulk of the solution may be decanted from the silica gel and the remainder filtered under reduced pressure using a sintered glass funnel). Remove the solvent under reduced pressure using a rotary evaporator on a water bath maintained at 60 °C. Towards the conclusion of this operation the syrupy mass crystallises as a thick layer around the inside of the flask. At this stage remove the flask from the evaporator, break the crystalline cake away from the sides of the flask and remove the remaining solvent under reduced pressure without heating further. Transfer portions of the solid to a mortar and grind with a 2 : 1 mixture of light petroleum (b.p. 40–60 °C) and dry ether. Filter the combined slurry and wash the filter cake first with a light petroleum–ether solvent mixture and then with

50 ml of previously chilled (0 °C) dry ether. The crude product is obtained in a yield of 210 g (92%), and when recrystallised from ether–light petroleum (b.p. 40–60 °C) has m.p. 88–89 °C, $[\alpha]_D^{20}$ +197.5° (c 2 in CHCl$_3$). The glucosyl halide should be stored in a desiccator over sodium hydroxide pellets; whenever possible it should be used without delay.

Notes. (1) If care is used most of the solution may be decanted from the solid deposit so that the glass wool does not become blocked with material and hence slow down the filtration process. This filtration is best conducted in a fume cupboard.

(2) All the isolation operations must be conducted with the minimum of delay and under conditions which reduce the contact of the solutions of unstable glucosyl halide with moisture. Solutions to be used for washing the organic layer should have been previously prepared and contain sufficient ice to ensure that the temperature of the liquid is approximately 4 °C. To obtain good yields and to ensure that vessels do not become unduly 'sticky' as the result of residual carbohydrate deposits, the separatory funnels, receiver vessels and aqueous extracts before being discarded should be rinsed with methylene chloride at each stage and these washings combined with the main organic solution.

Cognate preparations 2,3,4,6-Tetra-O-acetyl-α-D-galactopyranosyl bromide. Use 100 g (0.56 mol) of dry D-galactose under precisely the same conditions; the product is obtained in a yield of 202 g (88%). When recrystallised from ether–light petroleum (b.p. 40–60 °C) it has m.p. 84–85 °C, $[\alpha]_D^{20}$ +214° (c 1.2 in CHCl$_3$).

2,3,4-Tri-O-acetyl-β-L-arabinopyranosyl bromide. For this preparation use 10 g (0.067 mol) of L-(+)-arabinose, 40 ml (0.424 mol) of acetic anhydride, 0.24 ml of 60 per cent perchloric acid, 3.0 g (0.1 mol) of red phosphorus, 18.1 g (5.8 ml, 0.226 mol) of bromine and 3.6 ml of water. The yellow syrup which is obtained after the appropriate isolation procedure gives 21 g of crude crystalline product. Recrystallisation is effected by dissolving it in a mixture of benzene/ether (5 : 95), warming and adding light petroleum (b.p. 40–60 °C) until a slight cloudiness is apparent, and then allowing the solution to cool. The pure product is obtained in a yield of 11 g (48%), m.p. 136–138 °C, $[\alpha]_D^{22}$ +280° (c 3.13 in CHCl$_3$).

III,110. 2,3,4,6-TETRA-O-BENZOYL-α-D-GLUCOPYRANOSYL BROMIDE

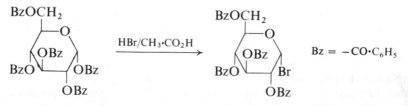

In a 250-ml conical flask fitted with a ground glass stopper place 40 ml of 1,2-dichloroethane and 20 g (0.029 mol) of α-D-glucopyranose pentabenzoate (Section **III,108**) (1). When all the solid has dissolved add 40 ml (0.225 mol)

of a solution of hydrogen bromide in glacial acetic acid (45% w/v HBr), stopper the flask and allow the reaction mixture to stand in the refrigerator overnight or at room temperature for about 2 hours. Pour the mixture into ice-water, rinse the flask with 1,2-dichloroethane, separate the organic layer and shake it with several portions of a saturated aqueous solution of sodium hydrogen carbonate until no further effervescence occurs. Wash the organic layer with water, dry over magnesium sulphate, filter and remove the 1,2-dichloro-ethane under reduced pressure on a rotary evaporator. Dissolve the crystalline solid which remains in dry ether, heating to 35 °C, and slowly add with further heating light petroleum (b.p. 40–60 °C) until a slight persistent cloudiness develops; then add a little more ether to give a clear solution, which is left to cool slowly to room temperature and finally refrigerated. Filter off the purified product and allow it to dry in the air; the yield is 16.5 g (88%), m.p. 129–130 °C, $[\alpha]_D^{20}$ +125° (c 2.0 in CHCl₃).

Note. (1) A mixture of anomeric glucose pentabenzoates such as might be obtained from a benzoylation reaction on glucose without careful temperature control gives equally good results.

III,111. METHYL 2,3,4,6-TETRA-O-ACETYL-β-D-GLUCOPYRANOSIDE

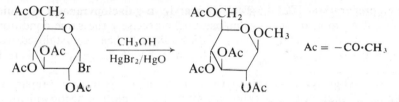

In a 500-ml two-necked flask fitted with a mechanical stirrer and calcium chloride guard-tube place 110 ml of dry methanol, 110 ml of pure chloroform, 22 g of anhydrous calcium sulphate, 7.2 g of yellow mercury(II) oxide and 0.55 g of mercury(II) bromide. Stir the suspension for 30 minutes and add 16.5 g (0.04 mol) of 2,3,4,6-tetra-O-acetyl-α-D-glucopyranosyl bromide (Section **III,109**) in one portion. The temperature of the mixture will rise to about 25–30 °C, the pH of the solution will fall from 7 to 2 and the yellow colouration of the mercury(II) oxide will disappear (1). Stir the suspension for a further 90 minutes, filter through a pad of Celite filter-aid and evaporate the filtrate on a rotary evaporator under reduced pressure. Dissolve the viscous oil which remains in 10 ml of chloroform, remove the inorganic salts which are precipitated by filtration and wash the residue well with further portions of chloroform. Evaporate the chloroform and triturate the resulting viscous oil with methanol until it solidifies. Recrystallise from methanol to give pure methyl 2,3,4,6-tetra-O-acetyl-α-D-glucopyranoside, m.p. 104–105 °C, $[\alpha]_D^{20}$ −18.2° (c 1 in CHCl₃). The yield is 13.7 g (95%).

Note. (1) The yellow colouration in the solution disappears within a few minutes of addition of the glucosyl halide and TLC analysis (solvent system benzene–methanol, 98 : 2) reveals virtual completion of the reaction.

Cognate preparation Methyl 2,3,4,6-tetra-O-acetyl-β-D-galactopyranoside. Use 13.5 g (0.033 mol) of 2,3,4,6-tetra-O-acetyl-α-D-galactopyranosyl bromide (Section **III,109**), 19 g of anhydrous calcium sulphate, 5.6 g of yellow mercury(II) oxide, 0.5 g of mercury(II) bromide, 90 ml of dry chloroform and 90 ml of dry methanol under the reaction conditions and subsequent isolation procedure described above; 7.5 g (63%) of methyl 2,3,4,6-tetra-*O*-acetyl-β-D-galactopyranoside, m.p. 96–97 °C, $[\alpha]_D^{20}$ −28.0° (*c* 2.5 in $CHCl_3$), is obtained after several recrystallisations from ethanol.

III,112. METHYL β-D-GLUCOPYRANOSIDE

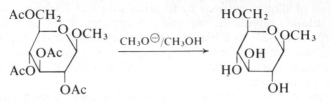

In a 100-ml conical flask place 5.5 g (0.015 mol) of methyl 2,3,4,6-tetra-*O*-acetyl-β-D-glucopyranoside (Section **III,111**), 50 ml of dry methanol and 10 ml of a solution of sodium methoxide in methanol previously prepared by the cautious addition of 0.1 g of sodium to 20 ml of methanol. Stopper the flask and allow the solution to stand for 1 hour, then add sufficient ion exchange resin [Zeolite 225 ($H^\oplus$)] to render the solution neutral to moist universal indicator paper. Remove the resin by filtration, wash with methanol and evaporate the combined filtrate and washings under reduced pressure (rotary evaporator). Triturate the colourless syrup with absolute ethanol to cause it to solidify and recrystallise from absolute ethanol. The pure methyl β-D-glucopyranoside has m.p. 108–109 °C, $[\alpha]_D^{20}$ −30.2° (*c* 2.8 in H_2O); the yield is 2.4 g (83%).

Cognate preparation **Methyl β-D-galactopyranoside.**   Use 3.6 g (0.01 mol) of methyl 2,3,4,6-tetra-*O*-acetyl-β-D-galactopyranoside (Section **III,111**) and proceed as above. After recrystallisation from absolute ethanol, 1.4 g (73%) of the methyl galactopyranoside, m.p. 174–175 °C, $[\alpha]_D^{20}$ +1.3° (*c* 1 in H_2O), is obtained.

III,113. METHYL α-D-GALACTOPYRANOSIDE

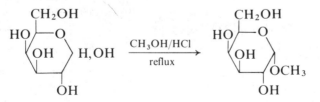

In a 2-litre flask fitted with a reflux condenser place 100 g (0.56 mol) of dry D-galactose and 700 ml of an anhydrous methanolic solution of hydrogen chloride (about 0.6 *M*) (1). Heat the mixture under reflux for 14 hours, cool, add 150 ml of distilled water and treat the light brown solution with solid lead carbonate until all the acid has been neutralised (2). Filter off the inorganic salts and remove the solvent on a rotary evaporator under reduced pressure. Triturate the resulting brown syrup with absolute ethanol with cooling in ice to cause the product to crystallise and recrystallise it from the minimum quantity of absolute ethanol to obtain 62 g of crude material, m.p. 85–90 °C. Isolate the methyl α-D-galactopyranoside as the hydrate by dissolving the crude product in 30 ml of water and allowing the solution to stand for one day at room temperature and two days at 4 °C. Repeat the recrystallisations several times using proportionate amounts of water until pure hydrated product, m.p. 109–110 °C, $[\alpha]_D^{20}$ +173.4° (c 1 in H$_2$O), is obtained; the yield is 38 g (35%) (3).

Notes. (1) Dry hydrogen chloride gas is passed into dry methanol (Section II,1,8) (contained in a flask protected by a calcium chloride guard-tube) until analysis of aliquot portions by titration with standard aqueous sodium hydroxide solution reveals the required concentration has been reached. It is usually more convenient to initially prepare a smaller volume of a more concentrated solution and dilute it to the appropriate concentration with dry methanol. The aliquot portions (say 5 ml) should be diluted with distilled water (20 ml) before titration.

(2) Universal indicator paper moistened with distilled water gives a satisfactory indication of neutralisation.

(3) Pure methyl β-D-galactopyranoside (m.p. 177–180 °C) may be isolated from the combined aqueous filtrates of these several crystallisations by removal of water and recrystallisation of the residue from absolute ethanol.

L,2.

D-Glucose ⟶ 1,2 : 5,6-Di-O-cyclohexylidene ⟶ 1,2-O-Cyclohexylidene
α-D-glucofuranose   α-D-glucofuranose

3-O-Benzyl-1,2 : 5,6-di-O-   1,2-O-Cyclohexylidene
cyclohexylidene-α-D-   α-D-xylofuranose
glucofuranose

Reaction of an aldehyde or ketone with two molar proportions of a monohydric alcohol or with one molar proportion of a 1,2- or 1,3-diol in the presence of an acid catalyst yields an acetal (XIV) or cyclic acetal (XV or XVI) respectively.

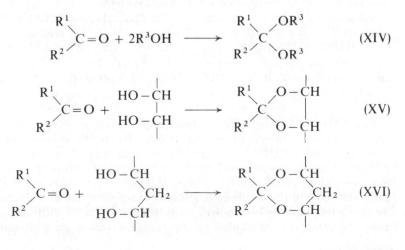

461

The formation of acetals of the type XV and XVI is an important reaction in the monosaccharide series. In general, ketones react with vicinal diol groups to give the five-membered cyclic products, and aldehydes react with 1,3-diols to give the six-membered cyclic products (see **L,3**), although exceptions can be found with suitable, partially protected, monosaccharide derivatives. An example of the reaction of a monosaccharide with a ketone is the formation of 1,2 : 5,6-di-*O*-cyclohexylidene-α-D-glucofuranose from D-glucose and cyclohexanone in the presence of sulphuric acid (Section **III,114**). The formation of a furanose derivative in this case is promoted by the two favourable vicinal diol groups which are present in α-D-glucofuranose (IV) (note the *cis*-orientation of the hydroxyl groups at C1 and C2), whereas there is only one vicinal *cis*-diol grouping in α-D-glucopyranose. Furthermore fusion of the five-membered acetal ring to the six-membered monosaccharide ring in (VI) would be expected to introduce considerable ring strain.

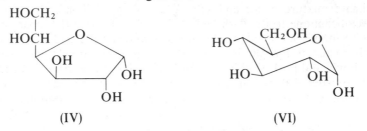

(IV)   (VI)

This reaction offers a convenient route to glucofuranose derivatives, particularly because the 5,6-acetal group may be selectively cleaved by hydrolysis to give the 1,2-mono acetal derivative (Section **III,115**). The reaction conditions need to be carefully controlled to avoid extensive hydrolysis to the parent sugar. This is particularly important with the partial hydrolysis of 3-*O*-benzyl-1,2 : 5,6-di-*O*-cyclohexylidene-α-D-glucofuranose (cognate preparation in Section **III,115**), since the product is then contaminated with 3-*O*-benzyl-D-glucose. Purification of the hydrolysis product, 3-*O*-benzyl-1,2-*O*-cyclohexylidene-α-D-glucofuranose, may be effected by column chromatography or by conversion of it into the more readily purified crystalline 5,6-dibenzoate derivative; the acyl groups may be then removed by the Zemplen method. The conversion of 1,2 : 5,6-di-*O*-cyclohexylidene-α-D-glucofuranose into its 3-*O*-benzyl derivative has been included (Section **III,116**) to illustrate the protection of hydroxyl groups in monosaccharides by conversion into ethers. The benzylation is effected by heating the diacetal with benzyl chloride in the presence of potassium hydroxide. The formation of a benzyl ether is useful in synthetic interconversions with monosaccharides since it is easily and selectively cleaved by catalytic hydrogenolysis. Other means for the protection of hydroxyl groups include their conversion into methyl ethers (Section **III,121**), trimethylsilyl ethers, triphenylmethyl ethers and allyl ethers.

The preparation of 1,2-*O*-cyclohexylidene-α-D-xylofuranose (Section **III,117**) from 1,2-*O*-cyclohexylidene-α-D-glucofuranose illustrates the use of sodium metaperiodate for the cleavage of carbon–carbon bonds in α-diols (see also Section **II,2,45**). In this case C-6 is lost as formaldehyde and C-5 is converted into an aldehyde group. This aldehydic product is isolated as a dimer, which is then reduced in methanol solution with sodium borohydride to the xylofuranose derivative.

III,114. 1,2 : 5,6-DI-O-CYCLOHEXYLIDENE-α-D-GLUCOFURANOSE

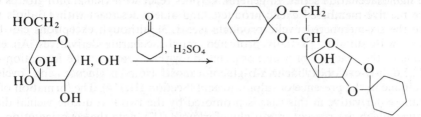

Fit a 3-litre flange flask with a multiple socket head carrying a mechanical stirrer capable of effecting vigorous agitation, a calcium chloride guard-tube, a 100-ml dropping funnel and a stoppered opening wide enough to allow for the addition of solid. Immerse the flask in a large plastic or metal container filled with an intimate mixture of ice and salt. Add 1000 g (1050 ml, 10 mol) of redistilled cyclohexanone to the flask and cool to 0 °C. Charge the separatory funnel with 62.5 ml of concentrated sulphuric acid and run the acid slowly into the vigorously stirred cyclohexanone; the final solution should be a light straw colour. Add slowly and portionwise with continued vigorous stirring 450 g (2.5 mol) of finely powdered dried α-D-glucose (1). Remove the cooling bath and allow the reaction mixture to reach ambient temperature with continual stirring; over a period of 8 hours the reaction mixture becomes progressively more viscous and finally sets into a solid off-white crystalline mass. Some caution should be exercised to prevent the stirrer motor from being overstrained. Allow the reaction mixture to stand at room temperature overnight, break up the crystalline mass, add 750 ml of heptane and heat the mixture under reflux for a few minutes on a boiling water bath. Decant as much of the upper heptane layer as possible from undissolved solid and an oily dark red layer which appears at the bottom of the flask. Add a further 750 ml portion of heptane to the residue and heat under reflux until the remainder of the solid dissolves; decant the clear heptane layer from the oily lower layer and cool the combined heptane extracts in the refrigerator. Filter off the crystalline material, m.p. 121–124 °C, and recrystallise from heptane (2) using decolourising charcoal to clear the hot solution of traces of dark oily droplets. The purified 1,2 : 5,6-di-O-cyclohexylidene-α-D-glucofuranose has m.p. 131–132 °C, $[\alpha]_D^{20}$ −2.2° (c 1.8 in C_2H_5OH); the yield is 380 g (47%).

Notes. (1) If adequate cooling and stirring is not employed the final solution is dark yellow; furthermore the addition of glucose leads to an unacceptable local rise in temperature, and the final appearance of the reaction mixture is a dark red intractable oily mass. The powdered glucose should be dried in a vacuum desiccator over phosphorus pentoxide, not in an oven which apparently causes changes on the surface of the glucose particles which render them unreactive.

(2) Another solvent for recrystallisation is methylcyclohexane (0.17 g/ml).

Cognate preparation   1,2 : 4,5-Di-O-cyclohexylidene-D-fructopyranose. Add 200 g (1.11 mol) of finely powdered dry D-fructose with vigorous stirring to 419 g (440 ml, 4.49 mol) of ice-cooled cyclohexanone containing 30 ml of concentrated sulphuric acid; the reaction mixture becomes solid within 30 minutes.

Leave the mixture overnight at room temperature, dissolve the product in 500 ml of chloroform and wash the solution with dilute aqueous sodium hydroxide, dilute hydrochloric acid and water and finally dry and evaporate. Solidify the residue by trituration with heptane and recrystallise from heptane to give the pure product, m.p. 145–156 °C, $[\alpha]_D^{20}$ −133.5° (c 1 in $CHCl_3$). The yield is 142 g (37%).

III,115. 1,2-O-CYCLOHEXYLIDENE-α-D-GLUCOFURANOSE

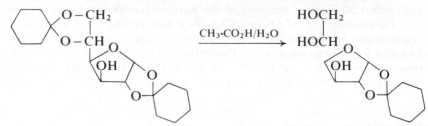

Heat a solution of 20 g (0.06 mol) of 1,2 : 5,6-di-O-cyclohexylidene-α-D-glucofuranose (Section **III,114**) in 100 ml of aqueous acetic acid (75% v/v) in a round-bottomed flask immersed in a hot-water bath held at 70–80 °C for 90 minutes with intermittent shaking; then remove the solvent under reduced pressure on a rotary evaporator. To the residual syrup add 20 ml of hot water, then sufficient solid sodium hydrogen carbonate to neutralise the remaining acetic acid, and finally 90 ml of heptane. Heat the heterogeneous liquid mixture until two clear layers are obtained and then remove the upper heptane layer by careful decantation (1). Cool the aqueous layer to 0 °C, filter off the crystals of 1,2-O-cyclohexylidene-α-D-glucofuranose which separate and recrystallise from water to give the pure product, m.p. 149–150 °C, $[\alpha]_D^{20}$ +5.9° (c 1 in $(CH_3)_2CO$). The yield is 11.5 g (75%).

Note. (1) The solid which separates from the cooled heptane layer may be shown to be unchanged starting material by TLC analysis on silica gel plates using methanol–benzene (4 : 96) as the developing solvent.

Cognate preparation   3-O-Benzyl-1,2-O-cyclohexylidene-α-D-glucofuranose. Dissolve 100 g (0.23 mol) of 3-O-benzyl-1,2 : 5,6-di-O-cyclohexylidene-α-D-glucofuranose (Section **III,116**) in 400 ml of aqueous acetic acid (75% v/v) maintained at 70–80 °C for 3 hours, remove the solvent under reduced pressure and dissolve the residual oil in 500 ml of chloroform. Wash this solution with aqueous sodium hydrogen carbonate and with water, dry over calcium sulphate and remove the chloroform by evaporation under reduced pressure with a rotary evaporator. Remove the last traces of solvent using an oil rotary immersion pump, transfer the warm fluid yellow syrup to the retort of a molecular still (Section **I,28**) and distil using a vapour diffusion pump to give a pale yellow glass, b.p. 195–200 °C/2 × 10⁻³ mmHg, $[\alpha]_D^{20}$ −36.4° (c 4 in $CHCl_3$). The yield is 76 g (94%), and the product is pure enough for most purposes. However TLC analysis on silica gel plates (solvent system; benzene–methanol 9 : 1) reveals one major and two minor components. Purification may be effected by either of the two methods described below.

  Chromatographic purification of 3-O-benzyl-1,2-O-cyclohexylidene-α-D-glucofuranose.   Prepare a silica gel column using benzene as a solvent; use 10 g of adsorbent for each 1 g of monosaccharide derivative to be chromatographed. Dissolve the latter in the smallest volume of benzene and transfer the solution to the chromatographic column with a pipette. Elute the column with benzene and collect suitable-sized fractions; evaporate the solvent from each fraction and weigh the residues (Section I,31) which consist of 3-O-benzyl-1,2 : 5,6-di-O-cyclohexylidene-α-D-glucofuranose. When all of this has been eluted continue the development with methanol which elutes the required product. Evaporate the methanol and distil the residual syrup using a molecular still; about 50 per cent recovery of the purified product may be expected.

Purification of 3-O-benzyl-1,2-O-cyclohexylidene-α-D-glucofuranose by benzoylation. Dissolve 5 g of the crude product in 10 ml of pure dry pyridine and add 5 g of benzoyl chloride. Leave the reaction mixture overnight at room temperature, pour it on to ice and stir thoroughly. Extract the oil which separates into 50 ml of chloroform and wash the chloroform solution successively with 30 ml of ice-cold dilute aqueous hydrochloric acid (2 M), 30 ml of saturated aqueous sodium hydrogen carbonate, and 2 × 30 ml of water. Dry over sodium sulphate and evaporate the chloroform. The viscous oil crystallises on trituration with methanol and is recrystallised from methanol to give the pure derivative 5,6-di-O-benzoyl-3-O-benzyl-1,2-O-cyclohexylidene-α-D-glucofuranose, m.p. 104–106 °C, $[\alpha]_D^{20}$ −26.4° (c 1 in $CHCl_3$), in a yield of 3.6 g (57%). Remove the benzoyl groups by dissolving 3 g of the foregoing product in 20 ml of methanol and adding 20 ml of a solution of sodium methoxide in methanol (0.5%). After 2 hours neutralise the solution by adding ion exchange resin Zeolite 225 (H), filter and evaporate. Distil the resulting colourless oil in a molecular still to obtain chromatographically pure 3-O-benzyl-1,2-O-cyclohexylidene-α-D-glucofuranose; the yield is 1.6 g (70% from the dibenzoate).

III,116. 3-O-BENZYL-1,2 : 5,6-DI-O-CYCLOHEXYLIDENE-α-D-GLUCO-FURANOSE

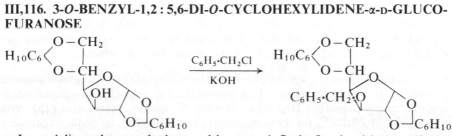

In a 1-litre three-necked round-bottomed flask fitted with an efficient mechanical stirrer, a reflux condenser and a thermometer the bulb of which dips into the reaction mixture, place 385 g (350 ml, 3.05 mol) of redistilled benzyl chloride, 100 g (0.3 mol) of 1,2 : 5,6-di-O-cyclohexylidene-α-D-glucofuranose (Section III,114) and 250 g (4.5 mol) of potassium hydroxide pellets. Raise the temperature of the vigorously stirred mixture to 150 °C (use a carefully controlled heating mantle) and maintain at this level for 4 hours. Cool the reaction mixture, dilute with 700 ml of iced water and separate the organic layer. Extract the aqueous solution with five 100 ml portions of chloroform, wash the combined organic phases with three 100 ml portions of water and dry the organic layer over calcium chloride. Remove the chloroform and excess benzyl chloride from the dried filtered solution by distillation under reduced pressure (4 mmHg), and then distil the viscous yellow residue in a molecular still (Section I,28), using a vapour diffusion pump and oil bath to obtain two fractions: (a) that which distils up to 100 °C at 10^{-3} mmHg and is benzyl alcohol, and (b) the product, 3-O-benzyl-1,2 : 5,6-di-O-cyclohexylidene-α-D-glucofuranose which distils at a bath temperature of 210–225 °C (10^{-3} mmHg), and has $[\alpha]_D^{20}$ −13.0° (c 5 in CHCl$_3$). The yield is 103 g (82%).

III,117. 1,2-O-CYCLOHEXYLIDENE-α-D-XYLOFURANOSE

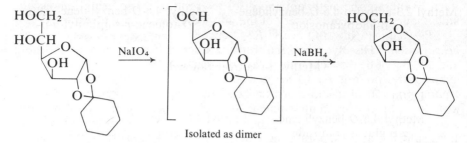

Isolated as dimer

Di-(1,2-O-cyclohexylidene-α-D-xylo-pentodialdofuranose-5-hydrate)-5,5′ : 3′,5-dianhydride. Add a solution of 14.3 g (0.061 mol) of sodium meta-periodate in 220 ml of water dropwise to a well-stirred solution of 17.4 g (0.067 mol) of 1,2-O-cyclohexylidene-α-D-glucofuranose (Section **III,115**) in 50 ml of water (1). Stir for a further 30 minutes and remove the water at a temperature below 50 °C by evaporation under reduced pressure. Extract the solid residue with three 75 ml portions of chloroform and dry the combined extracts over magnesium sulphate. Filter and evaporate to give a residue which crystallises spontaneously. After recrystallisation from acetone the dimer has m.p. 182–183 °C, the yield is 7.9 g (51%).

1,2-O-Cyclohexylidene-α-D-xylofuranose. In a 250-ml two-necked round-bottomed flask fitted with a mechanical stirrer and dropping funnel, place 2 g (0.009 mol) of the dimer dissolved in 50 ml of aqueous ethanol (75% v/v). Add a solution of 1 g (0.053 mol) of sodium borohydride in 50 ml of ethanol from the dropping funnel with stirring. After a further 30 minutes remove the ethanol under reduced pressure (rotary evaporator) and extract the aqueous slurry which remains with three 25 ml portions of chloroform. Dry the combined extracts over magnesium sulphate and remove the solvent by evaporation under reduced pressure to give a viscous oil which crystallises on trituration with light petroleum (b.p. 40–60 °C). Recrystallise from benzene–light petroleum (b.p. 40–60 °C) to give the purified product m.p. 84–85 °C, $[\alpha]_D^{20}$ −12° (c 1 in CH$_3$OH). The yield is 1.7 g (82%).

Note. (1) The progress of the oxidation may be followed by iodimetry. In this method unreacted periodate is reduced by arsenite solution in the presence of iodide at about pH 8. Excess arsenite is then determined by back titration with standard iodine solution.

$$IO_4^{\ominus} + AsO_2^{\ominus} \longrightarrow IO_3^{\ominus} + AsO_3^{\ominus}$$

For the oxidation described, remove 1 ml of solution, add 10 ml of saturated sodium hydrogen carbonate solution followed immediately by 10 ml of 0.05 M-sodium arsenite solution and 1 ml of 20 per cent potassium iodide solution. Stand the solution in the dark for 15 minutes and titrate excess arsenite with 0.05 M-iodine solution using a starch indicator.

A spectrophotometric method has been developed which utilises the stronger absorption of periodate ions at 222.5 nm compared to the iodate ion (Ref. 10).

L,3(a).

Methyl α-D-glucopyranoside ⟶ Methyl 4,6-*O*-benzylidene-α-D-glucopyranoside

↓

Methyl 2,3-anhydro-4,6-*O*-benzylidene-α-D-allopyranoside ⟵ Methyl 4,6-*O*-benzylidene-2,3-di-*O*-toluene-*p*-sulphonyl-α-D-glucopyranoside

↘ Methyl α-D-altropyranoside

L,3(b).

Methyl 4,6-*O*-benzylidene-α-D-glucopyranoside ⟶ Methyl 4,6-*O*-benzylidene-2,3-di-*O*-methyl-α-D-glucopyranoside

⟶ Methyl 2,3-di-*O*-methyl-α-D-glucopyranoside

The sequence L,3(a) serves to exemplify the formation and aspects of reactivity of toluene-*p*-sulphonate esters in monosaccharide systems, and further to illustrate the selective protection afforded to hydroxyl groups by the formation of cyclic acetals by reaction with carbonyl compounds. Thus reaction of methyl-α-D-glucopyranoside (XVII) with benzaldehyde in the presence of zinc chloride gives the 4,6-acetal (XVIII) (Section **III,118**), wherein two fused six-membered rings of the *trans*-decalin type are present.

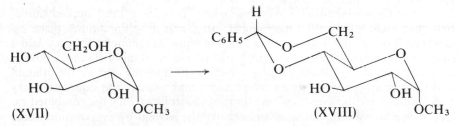

As a cognate preparation the reaction of benzaldehyde with methyl-α-D-galactopyranoside results in a similar conversion to a 4,6-acetal, but in this case the product is the conformationally flexible system of the *cis*-decalin type, the most likely conformation being that shown below.

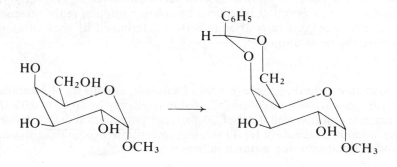

Since the 4,6-acetal grouping and the glycosidic grouping (XVIII) are stable under basic conditions, though unstable in the presence of acid, the remaining two hydroxyl groups may be suitably protected by reactions that are base catalysed. Thus (XVIII) may be converted into the 2,3-di-*O*-toluene-*p*-sulphonyl derivative by reaction with toluene-*p*-sulphonyl chloride in the presence of pyridine (Section **III,119**); this reaction is analogous to the formation of sulphonate esters of alcohols or phenols.

$$R \cdot CH_2OH + ClSO_2 \cdot C_6H_4 \cdot CH_3 \longrightarrow R \cdot CH_2 \cdot O \cdot SO_2 \cdot C_6H_4 \cdot CH_3$$

The value of the sulphonate esters in carbohydrate chemistry lies in their ability to undergo cleavage by reaction with nucleophiles in one of two possible ways, (i) cleavage of the S — O bond with regeneration of the hydroxyl group as a result of S_N2 attack at the sulphur atom, and (ii) cleavage of the C — O bond as a result of S_N2 attack at the carbon atom.

$$R \cdot CH_2 \overset{(ii)}{\underset{|}{}}O \overset{(i)}{\underset{|}{}}SO_2 \cdot C_6H_4 \cdot CH_3$$

(i) $\nearrow R \cdot CH_2O\overset{\ominus}{:} + CH_3 \cdot C_6H_4 \cdot SO_2X$
$X^{\ominus}$

(ii) $\searrow R \cdot CH_2X + CH_3 \cdot C_6H_4 \cdot SO_2 \cdot O^{\ominus}$
$X^{\ominus}$

Both these processes are illustrated by the reaction of the 2,3-di-*O*-toluene-*p*-sulphonyl derivative with sodium methoxide in methanol (Section **III,120**). Here the sulphonyloxy group at position 2 undergoes cleavage (i) to yield the anion (XIX); this anionic site then participates in reaction at position 3 whereby cleavage of the sulphonyloxy group takes place by pathway (ii) with the formation of an oxirane ring system. This neighbouring group effect must proceed via the unfavourable boat conformation (XX) in order that the stereoelectronic re-

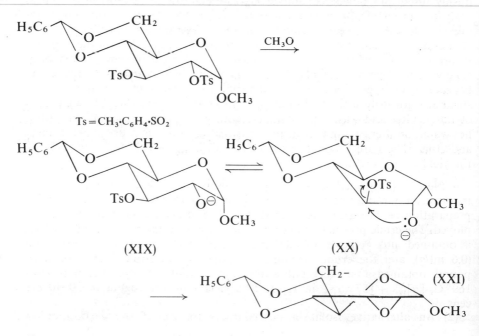

Ts = CH₃·C₆H₄·SO₂

(XIX) (XX)

(XXI)

quirements for reaction of the participating groups be met (i.e., *anti*-periplanar). This reaction has led to an inversion of configuration at C3 to yield the allose derivative (XXI). Ring opening of the oxirane system by treatment with alkali in aqueous media results in regeneration of a 2,3-vicinal diol but having the hydroxyl groups in the diaxial orientation, as a result of inversion of configuration at C2 (Section **III,120**). Finally the isolation of methyl-α-D-altropyranoside is described following the selective removal of the 4,6-acetal grouping by treatment with very dilute sulphuric acid, conditions which are insufficiently severe to result in hydrolysis of the glycosidic group.

The synthesis of the partially methylated glucoside, methyl 2,3-di-*O*-methyl-α-D-glucopyranoside (sequence **L,3**(*b*), Section **III,121**) also utilises the 4,6-*O*-benzylidene derivative (XVIII), which is first converted into the 2,3-di-*O*-methyl derivative by reaction with dimethyl sulphate in the presence of sodium hydroxide. Selective removal of the acetal grouping is achieved by mild acid hydrolysis; the methyl ether groups are stable under both acidic and basic conditions.

III,118. METHYL 4,6-*O*-BENZYLIDENE-α-D-GLUCOPYRANOSIDE

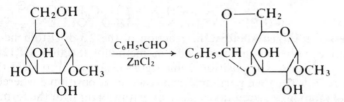

Shake vigorously a mixture of 105 g (1.0 mol) of purified benzaldehyde (Section **IV,22**), 38.8 g (0.2 mol) of methyl α-D-glucopyranoside (1) and 29.5 g of freshly fused and powdered anhydrous zinc chloride (0.22 mol) in a conical flask in a mechanical shaker for about 10 hours, until a clear solution is obtained. Allow the solution to stand at room temperature for a further period of 18 hours and then pour it into 700 ml of iced water (2). Stir the mixture vigorously, filter off the solid which separates and wash the compressed filter cake with light petroleum (b.p. 40–60 °C) to remove as much of the unreacted benzaldehyde as possible. Remove the solid from the Buchner funnel, stir or shake it vigorously with a solution of 12 g of sodium metabisulphite in 120 ml of water, filter and wash the filter cake with water. Crystallise the solid from hot water, or after drying in a vacuum desiccator, from a mixture of chloroform and ether. The pure product has m.p. 165 °C, $[\alpha]_D^{20}$ +112° (*c* 0.5 in $CHCl_3$), the yield is 29 g (52%).

Notes. (1) Methyl α-D-glucopyranoside is available commercially and it is not usually economic to prepare it in the laboratory; should, however, its preparation be necessary the method described in Section **III,113** may be employed. The crude product (containing a mixture of anomeric glucosides) which is obtained after removal of methanol is stirred with cold (10 °C) methanol (0.6 ml/g), and the crude α-anomer removed by filtration. The pure compound, obtained after recrystallisation from ethanol (10 ml/g), has m.p. 167–169 °C, $[\alpha]_D^{25}$ +157° (*c* 2 in H_2O). The yield is in the region of 30–40 per cent.

(2) An alternative' isolation procedure is to extract the clear solution

by shaking it with three successive 100 ml portions of light petroleum (b.p. 40–60 °C), which removes the unreacted benzaldehyde more effectively, and then to stir the viscous residue with ice-water until solidification occurs.

Cognate preparations   Methyl 4,6-*O*-benzylidene-α-D-galactopyranoside. Use 38.8 g (0.2 mol) of methyl α-D-galactopyranoside (Section **III,113**) under precisely the same conditions as described above. Recrystallise the crude product from ethanol/light petroleum (b.p. 60–80 °C) to give 38 g (68%) of the pure compound, m.p. 169–170 °C, $[\alpha]_D^{20}$ +168.2° (*c* 1.4 in CHCl$_3$).

Methyl 4,6-*O*-benzylidene-β-D-glucopyranoside. Use 38.8 g (0.2 mol) of methyl β-D-glucopyranoside (Section **III,112**) under the conditions specified above. Recrystallise the crude product from methanol to give 39 g (69%) of the pure compound, m.p. 194–196 °C, $[\alpha]_D^{20}$ −74° (*c* 1 in C$_2$H$_5$OH).

Methyl 4,6-*O*-benzylidene-β-D-galactopyranoside. Prepared under the conditions specified above for the α-anomer. The pure compound is obtained by recrystallisation from methanol and has m.p. 198–200 °C, $[\alpha]_D^{20}$ −35.5° (*c* 2 in CHCl$_3$); the yield is 39 g (69%).

III,119. METHYL 4,6-*O*-BENZYLIDENE-2,3-DI-*O*-TOLUENE-*p*-SULPHONYL-α-D-GLUCOPYRANOSIDE

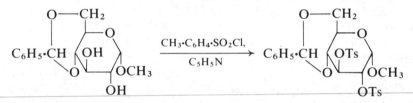

Place 40 ml of pure dry redistilled pyridine in a conical flask and add 14.9 g (0.066 mol) of toluene-*p*-sulphonyl chloride (1) with cooling. Allow the yellow solution to stand for about half an hour at room temperature (2) before adding 8.5 g (0.03 mol) of methyl 4,6-*O*-benzylidene-α-D-glucopyranoside (Section **III,118**) with shaking and cooling. Leave the reaction mixture in the stoppered flask for five days at room temperature and then pour on to 75 g of crushed ice. Stir vigorously, extract the mixture of syrup and water with three 25 ml portions of methylene chloride and wash the combined extracts successively with cold dilute hydrochloric acid (2 *M*), water, saturated aqueous sodium hydrogen carbonate, and water, and then dry over magnesium sulphate. Remove the methylene chloride by evaporation under reduced pressure and triturate the gummy residue with ether until solid. Recrystallise from chloroform–ether to obtain the pure product, m.p. 152–154 °C, $[\alpha]_D^{20}$ +11.8° (*c* 1 in CHCl$_3$). The yield is 13.5 g (75%).

Notes. (1) The toluene-*p*-sulphonyl chloride may be purified by following the procedure given in Section **II,2,65**.

(2) In all sulphonylations performed in these laboratories this procedure of allowing the mixture of toluene-*p*-sulphonyl chloride in pyridine to stand for about half an hour before the addition of the material to be sulphonylated has been found to be beneficial.

III,120. METHYL α-D-ALTROPYRANOSIDE

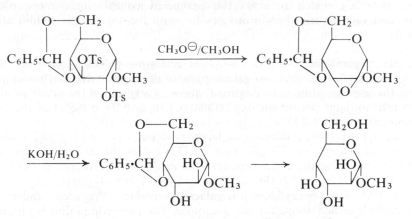

Methyl 2,3-anhydro-4,6-O-benzylidene-α-D-allopyranoside. In a two-necked, 250-ml round-bottomed flask fitted with a mechanical stirrer unit and a pressure-equalising funnel protected with a calcium chloride guard-tube place a solution of 11.7 g (0.0195 mol) of methyl 4,6-O-benzylidene-2,3-di-O-toluene-p-sulphonyl-α-D-glucopyranoside (Section III,119) in 150 ml of chloroform. Cool the solution to 0 °C by means of an ice–salt bath and add a solution of sodium methoxide in methanol [prepared from 2.3 g (0.1 mol) of sodium and 40 ml of methanol] dropwise with stirring. When the addition is complete remove the stirrer and funnel, stopper the flask and leave it in a refrigerator for 48 hours and then at room temperature for a further 24 hours. Extract the chloroform solution with water until the aqueous washings are neutral, dry over magnesium sulphate and remove the solvent on a rotary evaporator under reduced pressure. Crystallise the solid residue from chloroform to give the pure product, m.p. 195–199 °C, $[\alpha]_D^{20}$ +140° (c 2 in CHCl$_3$). The yield is 4.2 g (82%).

Methyl 4,6-O-benzylidene-α-D-altropyranoside. Triturate 4.0 g (0.015 mol) of the foregoing anhydro derivative in a mortar with a solution of 5 g of potassium hydroxide dissolved in 140 ml of water. Transfer the suspension to a round-bottomed flask and heat the mixture under reflux until all the solid has dissolved (about 28 hours). During this period solid material tends to creep up the inside of the flask surface; shake periodically to re-suspend material. Remove the trace of insoluble matter which remains and neutralise the cooled filtrate with carbon dioxide (use phenophthalein as an indicator). Extract the solution with five 25 ml portions of chloroform, wash the combined extracts with a little cold water, dry over anhydrous sodium sulphate and remove the solvent under reduced pressure (rotary evaporator). Crystallise the syrup by scratching a small portion on a watch glass with ether; stir the bulk syrup with ether and the seed crystals. Filter off and recrystallise the product from a small quantity of methanol to obtain 3.5 g (83%) of methyl 4,6-O-benzylidene-α-D-altropyranoside, m.p. 174 °C, $[\alpha]_D^{20}$ +115° (c 2 in CHCl$_3$).

Methyl α-D-altropyranoside. Hydrolyse the benzylidene protecting group by heating, at 60 °C for 1 hour, 3.5 g (0.025 mol) of the foregoing compound in a mixture of 140 ml of warm water and 7 ml of 0.05 M-sulphuric acid. Concentrate the residual solution to 50 ml using a rotary evaporator under reduced pressure (benzaldehyde is removed during the process) and make just alkaline (phenolphthalein) with 0.1 M-barium hydroxide solution (prepared by diluting a cold saturated solution with an equal volume of water). Remove the barium sulphate by filtration (Whatman No. 42 paper), wash the residue with water and concentrate the filtrate and washings to a syrup (rotary evaporator). Dissolve the residue in a little methanol, add ether until a slight turbidity is observed and set the solution on one side to crystallise. The resulting methyl α-D-altroside may be recrystallised from methanol/ether and has m.p. 107–108 °C, $[\alpha]_D^{20}$ +126° (c 3 in H_2O); the yield is 2.1 g (88%).

III,121. METHYL 2,3-DI-O-METHYL-α-D-GLUCOPYRANOSIDE

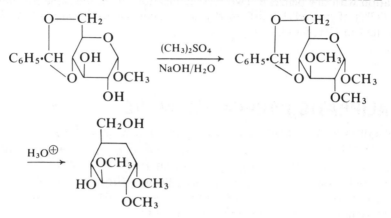

Methyl 4,6-O-benzylidene-2,3-di-O-methyl-α-D-glucopyranoside. Owing to the hazards in the use of dimethyl sulphate this experiment must be carried out in an efficient fume cupboard (see Section **II,2,**18 for precautions in the use of dimethyl sulphate). Mount a 500-ml, three-necked round-bottomed flask in an electrically heated water bath and fit an efficient stirrer unit in the central neck. Attach two 100-ml dropping funnels to the side-necks; one containing 38 g (29 ml, 0.30 mol) of purified dimethyl sulphate and the other containing 60 ml (0.6 mol) of 40 per cent aqueous sodium hydroxide solution. Place 14.1 g (0.05 mol) of methyl 4,6-O-benzylidene-α-D-glucopyranoside and 150 ml of acetone in the flask, run in 15 ml of sodium hydroxide solution, commence fairly rapid stirring and raise the temperature of the water bath to about 50 °C. Add dropwise and simultaneously the remainder of the sodium hydroxide solution and the dimethyl sulphate over a period of about 1.5 hours. Continue stirring at 50 °C for a further half an hour. Remove the dropping funnels, fit a condenser set for downward distillation into one of the side-necks and a nitrogen inlet tube into the other. Remove the acetone by distillation during 1 hour whilst maintaining a steady nitrogen flow. Pour the contents of the flask into 1500 ml of ice-cold water and collect the solid product by filtration. Wash the product with cold water until the washings are neutral to litmus, dry overnight at 50 °C and recrystallise twice from light petroleum (b.p. 60–80 °C), to give 12.5 g (81%) of product having m.p. 122–123 °C, $[\alpha]_D^{20}$ +94° (c 2 in $CHCl_3$), or $[\alpha]_D^{20}$ +97° (c 4 in $CH_3 \cdot CO \cdot CH_3$).

Methyl 2,3-O-di-O-methyl-α-D-glucopyranoside. Dissolve 6.2 g (0.02 mol) of the foregoing derivative in 100 ml of acetone containing 0.3 per cent of concentrated sulphuric acid in a 250-ml round-bottomed flask fitted with a reflux condenser. Boil the solution under gentle reflux on a steam bath. Periodically remove the flask, cool to room temperature in a stream of cold water, remove an aliquot of suitable size and follow the progress of the hydrolysis by measuring the optical rotation, which changes from about 6.12 to about 5.74° for a 1-dm cell. When the reaction is complete (under 1 hour), add 100 ml of water and neutralise the solution with solid barium carbonate. Filter the reaction mixture and evaporate the filtrate on a rotary evaporator. If an odour of benzaldehyde remains, add 100 ml of water and repeat the evaporation. Distil the crude syrupy product under reduced pressure to give a colourless glass of b.p. 130–135 °C/0.1 mmHg which may be crystallised by trituration with dry benzene. Two recrystallisations from the same solvent give 3.2 g (72%) of methyl 2,3-di-O-methyl-α-D-glucopyranoside, m.p. 85 °C, $[\alpha]_D^{20}$ +146° (c 4 in $CH_3 \cdot CO \cdot CH_3$).

M ALIPHATIC CARBOXYLIC ACIDS

The synthesis of aliphatic carboxylic acids is exemplified by the following typical procedures.

1. Oxidative methods (Sections **III,122** to **III,127**).
2. The hydrolysis of nitriles (Section **III,128,**A and B).
3. The carboxylation of Grignard reagents (Section **III,129**).
4. The Arndt–Eistert method (Section **III,130**).
5. Electrolytic (anodic) coupling (Section **III,131**).
6. Methods utilising diethyl malonate (Sections **III,132** to **III,137**).

M,1. Oxidative methods. Saturated primary alcohols are readily oxidised to aldehydes, which in turn are further oxidised to monocarboxylic acids having the same number of carbon atoms.

$$R \cdot CH_2OH \xrightarrow{[O]} R \cdot CHO \xrightarrow{[O]} R \cdot CO_2H$$

Reagents which are frequently used are alkaline potassium permanganate solution or aqueous sodium dichromate/sulphuric acid mixtures (e.g., Sections **III,122** and **III,123**). With the latter reagent, yields from higher alcohols (e.g., propan-1-ol and higher homologues) are not always satisfactory because of the attendant production of appreciable amounts of esters (cf. the preparation of aldehydes, **III,H,1**). In such cases it may be advantageous to isolate the ester by-product and subject it to hydrolysis with methanolic sodium hydroxide to increase the overall yield of the required carboxylic acid.

Secondary alcohols on oxidation give ketones which may be cleaved under vigorous oxidative conditions to a mixture of carboxylic acids.

$$R^1 \cdot CH_2 \cdot CH(OH) \cdot CH_2 \cdot R^2 \longrightarrow R^1 \cdot CH_2 \overset{a}{\underset{|}{-}} CO \overset{b}{\underset{|}{-}} CH_2 \cdot R^2$$

$$\xrightarrow{(a)} R^1 \cdot CO_2H + HO_2C \cdot CH_2 \cdot R^2$$

$$\xrightarrow{(b)} R^1 \cdot CH_2 \cdot CO_2H + HO_2C \cdot R^2$$

Clearly such a method is of limited preparative value, but an important exception is the oxidation of cyclic secondary alcohols which on oxidation with nitric acid give good yields of dicarboxylic acids by way of the intermediate cyclic ketone, e.g., adipic acid from cyclohexanone, Section **III,124**.

The oxidation of a methyl ketone into a carboxylic acid can be effected by the use of the **haloform reaction**. This involves treatment of the methyl ketone with an alkaline hypohalite reagent. A trihalomethyl ketone is initially formed which then undergoes hydrolysis under the basic conditions used.

$$2HO^{\ominus} + Br_2 \rightleftharpoons {}^{\ominus}OBr + {}^{\ominus}Br + H_2O$$

$$R \cdot CO \cdot CH_3 + {}^{\ominus}OBr \rightleftharpoons [R \cdot CO \cdot \overset{..}{C}H_2]^{\ominus} + HOBr$$

mesomeric anion

$$R \cdot CO \cdot \overset{\ominus}{\overset{..}{C}H_2} \overset{\frown}{} Br \overset{\frown}{} OH \longrightarrow R \cdot CO \cdot CH_2Br + {}^{\ominus}OH$$

$$R \cdot CO \cdot CH_2Br \longrightarrow R \cdot CO \cdot CBr_3 \xrightarrow{\ominus OH} R \cdot CO_2^{\ominus} + CHBr_3$$

In the examples given the preparative value depends upon the ready availability of the required methyl ketone; thus pinacolone (Section **III,97**), cyclopropyl methyl ketone (Section **V,10**) and mesityl oxide (Section **III,209**) are converted into 2,2-dimethylpropanoic acid, cyclopropane carboxylic acid and 3,3-dimethylacrylic acid respectively (Sections **III,125** and **III,126**).

A carboxylic acid group is formed when a carbon–carbon multiple bond is oxidatively cleaved. The main value of this is as a degradative process in structural elucidation, as for example in the ozonolysis of alkenes followed by oxidative decomposition of the ozonide (Section **I,17,4**). Oxidation of a symmetrical alkene or alkyne gives rise to a single carboxylic acid product which could, however, probably be more conveniently synthesised by other routes. Unsymmetrical non-terminal alkenes or alkynes are rarely used as substrates since a mixture of acidic products would be produced. The method has, however, found application using terminal alkynes since the terminal carbon is lost as carbon dioxide when, for example, the oxidation is carried out with potassium permanganate. In the illustrative example the acetylenic carbinol, 1-ethynyl-3,3,5-trimethylcyclohexanol (Section **III,40**), on oxidation gives a good yield of the corresponding hydroxy carboxylic acid (Section **III,127**).

III,122. ISOBUTYRIC ACID (2-Methylpropanoic acid)

$$(CH_3)_2CH \cdot CH_2OH \xrightarrow[KMnO_4]{[O]} (CH_3)_2CH \cdot CO_2H + H_2O$$

Place a mixture of 52 g (0.7 mol) of 2-methylpropan-1-ol and a solution of 15 g of sodium carbonate in 150 ml of water in a 5-litre round-bottomed flask. Add a solution of 142 g (0.9 mol) of potassium permanganate in 2750 ml of water, with vigorous stirring, during 3–4 hours, cooling the mixture to 4–5 °C by immersion in a bath of ice-water. Then allow the reaction mixture to attain room temperature gradually. After 12 hours, filter off (or preferably, centrifuge) the precipitated manganese dioxide, concentrate the filtrate to about 150 ml under reduced pressure and then cool. Cover the solution with a layer of ether and acidify with dilute sulphuric acid. Separate the ether layer and extract the aqueous layer two or three times with 50 ml portions of ether. Dry the combined ethereal extracts over anhydrous sodium sulphate, remove the ether on a water bath and fractionate the residual liquid. Collect the isobutyric acid at 153–155 °C. The yield is 45 g (76%).

III,123. ISOVALERIC ACID (*3-Methylbutanoic acid*)

$$(CH_3)_2CH \cdot CH_2 \cdot CH_2OH \xrightarrow[Na_2Cr_2O_7/H_2SO_4]{[O]} (CH_3)_2CH \cdot CH_2 \cdot CO_2H$$

Prepare dilute sulphuric acid by adding 140 ml of concentrated sulphuric acid cautiously and with stirring to 85 ml of water; cool and add 79 g (98 ml, 0.9 mol) of redistilled 3-methylbutan-1-ol (isoamyl alcohol). Place a solution of 200 g of crystallised sodium dichromate in 400 ml of water in a 1-litre (or 1.5-litre) round-bottomed flask and attach an efficient reflux condenser. Add the sulphuric acid solution of the isoamyl alcohol in *small* portions through the top of the condenser (1); shake the apparatus vigorously after each addition. No heating is required as the heat of the reaction will suffice to keep the mixture hot. It is important to shake the flask well immediately after each addition and not to add a further portion of alcohol until the previous one has reacted; if the reaction should become violent, immerse the flask momentarily in ice-water. The addition occupies 2–2.5 hours. When all the isoamyl alcohol has been introduced, reflux the mixture gently for 30 minutes, and then allow to cool. Arrange the flask for distillation and collect about 350 ml of distillate. The latter consists of a mixture of water, isovaleric acid and isoamyl isovalerate. Add 30 g of potassium (*not* sodium) hydroxide pellets to the distillate and shake until dissolved. Transfer to a separatory funnel and remove the upper layer of ester (16 g). Treat the aqueous layer contained in a beaker with 30 ml of dilute sulphuric acid (1 : 1 by volume) and extract the liberated isovaleric acid with two 50 ml portions of carbon tetrachloride. Keep the carbon tetrachloride extract (*A*).

To obtain a maximum yield of the acid it is necessary to hydrolyse the ester by-product: this is most economically effected with methanolic sodium hydroxide. Place a mixture of 20 g of sodium hydroxide pellets, 25 ml of water and 225 ml of methanol in a 500-ml round-bottomed flask fitted with a reflux (double surface) condenser, warm until the sodium hydroxide dissolves, add the ester layer and reflux the mixture for a period of 15 minutes. Rearrange the flask for distillation and distil off the methanol until the residue becomes pasty. Then add about 200 ml of water and continue the distillation until the temperature reaches 98–100 °C. Pour the residue in the flask, consisting of an aqueous solution of sodium isovalerate, into a 600-ml beaker and add sufficient water to dissolve any solid which separates. Add slowly, with stirring, a solution of 15 ml of concentrated sulphuric acid in 50 ml of water, and extract the liberated acid with 25 ml of carbon tetrachloride. Combine this extract with extract (*A*), dry with a little anhydrous calcium sulphate and distil off the carbon tetrachloride and then distil the residue. Collect the isovaleric acid, b.p. 172–176 °C. The yield is 55 g (60%).

Note. (1) If preferred, a 1.5-litre three-necked flask, equipped with a dropping funnel, mechanical stirrer and reflux condenser, may be used and the obvious modifications of technique introduced. This procedure is recommended.

III,124. ADIPIC ACID (*Hexanedioic acid*)

$$\text{(cyclohexanol)} \xrightarrow[\text{HNO}_3]{\text{[O]}} HO_2C \cdot (CH_2)_4 \cdot CO_2H$$

Into a 3-litre three-necked flask, fitted with a dropping funnel, a mechanical stirrer and an efficient reflux condenser, place 1900 ml (2700 g) of concentrated nitric acid, *d* 1.42. Since oxides of nitrogen are evolved in the subsequent oxidation, the reaction should be carried out in a fume cupboard, or the oxides of nitrogen are led by a tube from the top of the condenser to a water trap (Fig. I,68). Heat the nitric acid to boiling, set the stirrer in motion, add a few drops of cyclohexanol and *make certain that these are acted upon by the acid before adding more; an explosion may result if cyclohexanol is allowed to accumulate in the acid.* Once the reaction has started, add 500 g (5 mol) of cyclohexanol through the dropping funnel at such a rate that all is introduced in 4–5 hours. Keep the reaction mixture at the boiling point during the addition of the cyclohexanol and for a further period of about 15 minutes in order to complete the oxidation. Pour the warm reaction mixture into a beaker; upon cooling, the adipic acid crystallises. Filter on a large sintered glass funnel, and wash with 200 ml of cold water. Recrystallise the crude acid from 700 ml of concentrated nitric acid; filter and wash as above. The yield of recrystallised adipic acid, m.p. 152 °C, is 400 g (55%).

III,125. 2,2-DIMETHYLPROPIONIC ACID (*Trimethylacetic acid; pivalic acid*)

$$(CH_3)_3C \cdot CO \cdot CH_3 \xrightarrow[\text{(ii) } H_3O^{\oplus}]{\text{(i) 3NaOBr}} (CH_3)_3C \cdot CO_2H + CHBr_3$$

In a 3-litre three-necked flask, fitted with a thermometer, a mechanical stirrer and dropping funnel, place a solution of 160 g (4 mol) of sodium hydroxide in 1400 ml of water. Cool to 0 °C in an ice–salt bath. Add 240 g (77 ml, 1.5 mol) of bromine with vigorous stirring at such a rate as to keep the temperature below 10 °C (15–20 minutes). Cool again to 0 °C, introduce 50 g (0.5 mol) of pinacolone (Section III,97) keeping the temperature below 10 °C. After the solution is decolourised (*ca.* 1 hour), continue the stirring for 3 hours at room temperature. Replace the thermometer by a knee tube connected to a condenser for distillation and replace the stirrer by a steam inlet to allow the bromoform and carbon tetrabromide (if present) to be separated by steam distillation; heat the flask with a powerful Bunsen burner. Remove the burner, cool the reaction mixture to 50 °C and add 200 ml of concentrated sulphuric acid cautiously through the dropping funnel. Heat the flask again; the trimethylacetic acid passes over with about 200 ml of water. When all the trimethylacetic acid (the upper layer; 35–40 ml) has distilled, a liquid heavier than water (possibly brominated pinacolone) begins to pass over. Stop the distillation at this point, separate the trimethylacetic acid from the aqueous layer, and dry it by distillation with 25 ml of benzene (the latter carries over all the water) or with anhydrous calcium sulphate. Distil under reduced pressure and collect the trimethylacetic acid 75–80 °C/20 mmHg. The yield is 33 g (55%), m.p. 34–35 °C.

Cognate preparation   Cyclopropanecarboxylic acid.   Use 42 g (47 mol, 0.5 mol) of cyclopropyl methyl ketone (Section **V,10**) and react with alkaline hypo-bromite solution exactly as in the above preparation except that the final period of stirring at room temperature need be only 1.5 hours. After removal of bromoform by steam distillation, cool and cautiously acidify the solution to Congo red with 250 ml of concentrated hydrochloric acid. Discharge the pale yellow colour by adding a little sodium metabisulphite solution. Saturate the solution with salt and extract with four 300 ml portions of ether; dry the combined extracts with anhydrous sodium sulphate, and distil off the ether on a water bath through a short fractionating column. Distil the residue under reduced pressure and collect the pure cyclopropanecarboxylic acid (a colourless liquid) at 92 °C/22 mmHg. The yield is 33 g (76%).

III,126. 3,3-DIMETHYLACRYLIC ACID

$$(CH_3)_2C=CH\cdot CO\cdot CH_3 \xrightarrow[\text{(ii) } H_3O^{\oplus}]{\text{(i) } 3NaOCl} (CH_3)_2C=CH\cdot CO_2H + CHCl_3$$

Fit a 1-litre three-necked flask with two double surface condensers and a sealed stirrer unit. Place 25 g (29 ml, 0.25 mol) of mesityl oxide (Section **III,209**), 50 ml of dioxan and a cold (10 °C) solution of sodium hypochlorite in 750 ml of water (1) in the flask, and stir the mixture. Heat is evolved in the reaction and after about 5 minutes chloroform commences to reflux. As soon as the reaction becomes very vigorous, stop the stirrer and cool the flask with water so that the chloroform refluxes gently; after 20–30 minutes, when the reaction has subsided, resume the stirring and continue it until the temperature of the mixture has fallen to that of the laboratory (2–3 hours). Decompose the slight excess of hypochlorite by the addition of sodium metabisulphite (about 1 g), i.e., until a test-portion no longer liberates iodine from potassium iodide solution.

Replace one of the reflux condensers by a dropping funnel and add 50 per cent sulphuric acid (about 50 ml) with stirring and cooling until the solution is acid to Congo red paper. Extract the cold solution with eight 50 ml portions of ether (2) and shake the mixture well during each extraction. Dry the combined ethereal extracts with anhydrous calcium sulphate, and remove the ether and chloroform slowly on a water bath. Distil the residue from a flask fitted with a Claisen still-head and a short fractionating column under diminished pressure and collect the acid at 100–106 °C/20 mmHg; this fraction solidifies on cooling and melts at 60–65 °C. The yield is 13 g (51%). Recrystallise from hot water (1 g of acid in 10 ml of water) (3), cool the solution in ice for 2–3 hours, filter and dry overnight in a vacuum desiccator. Alternatively, recrystallise from light petroleum, b.p. 60–80 °C. Pure 3,3-dimethylacrylic acid has m.p. 68 °C.

Notes.   (1) This solution is prepared by diluting 300 ml of commercial sodium hypochlorite (containing 10–14% available chlorine) to 750 ml with water.

(2) A continuous ether extractor (Fig. I,92) gives more satisfactory results.

(3) Do not boil the aqueous solution for a long time as the acid is markedly steam-volatile.

III,127. 1-HYDROXY-3,3,5-TRIMETHYLCYCLOHEXANECARB-OXYLIC ACID

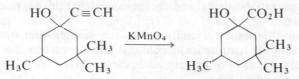

Equip a 100-ml round-bottomed two-necked flask with a magnetic follower bar, a dropping funnel and a thermometer positioned to dip into the reaction mixture. Site the flask in a plastic bowl on a magnetic stirrer unit. Place in the flask a solution of 1 g (0.0055 mol) of *trans*-1-ethynyl-3,3,5-trimethylcyclohexan-1-ol (Section **III,40**) in 10 ml of acetone and add a solution of 1.25 g (0.0079 mol) potassium permanganate in 25 ml of water from the dropping funnel with stirring over 1 hour. Ensure that the temperature of the reaction mixture is maintained below 20 °C by surrounding the reaction flask with ice-water.

When the addition is complete, continue stirring at room temperature for 1 hour, and then heat the solution under reflux for a further hour. Cool the reaction mixture, filter, decolourise the filtrate by the passage of sulphur dioxide and acidify the solution with hydrochloric acid. The product obtained by filtration is dried in a vacuum desiccator before being recrystallised from light petroleum (b.p. 60–80 °C). The yield of pure product is 0.7 g (58%); m.p. 134–135 °C.

M,2. Hydrolysis of nitriles. Since alkyl nitriles are readily available from the interaction of alkyl halides with sodium or potassium cyanide in aqueous alcoholic solution (Sections **III,160** and **III,161**), their hydrolysis to carboxylic acids is a valuable synthetic method. Aqueous alkaline or acidic conditions may be used. The reaction proceeds via the intermediate formation of an amide. Experimental conditions may be selected to interrupt the hydrolysis at the amide stage (Section **IV,182**).

$$R \cdot C \equiv N \quad \begin{array}{c} \overset{\ominus}{O}H/H_2O \nearrow \\ \\ H_3O^{\oplus} \searrow \end{array} \quad \begin{array}{c} R \cdot CONH_2 \longrightarrow R \cdot CO_2^{\ominus} + NH_3 \\ \\ R \cdot CONH_2 \longrightarrow R \cdot CO_2H + \overset{\oplus}{N}H_4 \end{array}$$

The probable mechanism for the complete hydrolysis under basic conditions may be formulated as:

$$R - C \equiv N \quad H - OH \xrightarrow{-\overset{\ominus}{O}H} R - C = NH \longrightarrow R \cdot CONH_2$$
$$\underset{\ominus}{:}OH \qquad\qquad\qquad\qquad \overset{|}{O}H$$

$$R - \overset{O}{\underset{\underset{\ominus}{:OH}}{\overset{\|}{C}}} - NH_2 \quad H - OH \xrightarrow[-NH_3\uparrow]{-OH^{\ominus}} R - C \overset{O}{\underset{OH}{\diagdown}} \xrightarrow{+\overset{\ominus}{O}H} R \cdot CO_2^{\ominus} + H_2O$$

As well as the illustrative synthesis of a simple carboxylic acid (valeric acid, Section **III,128,***A*), examples are given of the synthesis of some dicarboxylic acids, aryl substituted carboxylic acids, and the unsaturated acid, vinylacetic acid.

The nitrile group in the readily available α-hydroxynitriles (the cyano-hydrins) may also be similarly hydrolytically converted into a carboxyl group to afford a convenient synthesis of α-hydroxy acids (Section **III,172**).

III,128,*A*. **VALERIC ACID** (*Pentanoic acid*)

$$C_4H_9 \cdot CN + 2H_2O \xrightarrow{\ominus OH} C_4H_9 \cdot CO_2{}^{\ominus} + NH_3$$

$$C_4H_9 \cdot CO_2{}^{\ominus} \xrightarrow{H_3O^{\oplus}} C_4H_9 \cdot CO_2H$$

Place 100 g (125 ml, 1.2 mol) of valeronitrile (Section **III,161**) and a solution of 92 g of sodium hydroxide in 260 ml of water in a 1500-ml round-bottomed flask, attach a double surface condenser and boil under reflux until the nitrile layer disappears (5–10 hours). Add through the condenser 100 ml of water, then slowly, and with external cooling, 125 ml of 50 per cent (by volume) sulphuric acid. Separate the upper layer of valeric acid (it may be necessary to filter first from any solid present), and dry it with anhydrous calcium sulphate. Distil and collect the valeric acid at 183–185 °C (mainly 184 °C). The yield is 82 g (67%). A further 5 g of acid may be obtained by extracting the strongly acidified aqueous layer with ether, combining the ethereal extracts with the low and high boiling point fractions of the previous distillation, removing the ether on a water bath and distilling the residue.

III,128,*B*. **GLUTARIC ACID** (*Pentanedioic acid*)

$$NC \cdot CH_2 \cdot CH_2 \cdot CH_2 \cdot CN \xrightarrow[H_2O]{H_2SO_4}$$

$$HO_2C \cdot CH_2 \cdot CH_2 \cdot CH_2 \cdot CO_2H + (NH_4)_2SO_4$$

In a 2-litre round-bottomed flask, equipped with a double surface condenser, place 60 g (0.64 mol) of pentanedinitrile (Section **III,160**) and 900 g of 50 per cent sulphuric acid (by weight). Reflux the mixture for 10 hours and allow to cool. Saturate the solution with ammonium sulphate and extract with four 150 ml portions of ether; dry the ethereal extracts with anhydrous sodium sulphate. Distil off the ether on a water bath; the residual glutaric acid (69 g, 82%) crystallises on cooling and has m.p. 97–97.5 °C. Upon recrystallisation from chloroform, or benzene, the m.p. is 97.5–98 °C.

Cognate preparations **Suberic acid** (*octanedioic acid*). Heat a mixture of oc-tanedinitrile (Section **III,160**) with 15 times its weight of 50 per cent sulphuric acid by weight under reflux for 10 hours. The acid crystallises out on cooling. Filter off the suberic acid upon a sintered glass funnel, and recrystallise it from acetone: m.p. 141–142 °C. The yield is 90 per cent of the theoretical.

Pimelic acid (*heptanedioic acid*). Heat a mixture of 18 g (0.148 mol) of heptanedinitrile (Section **III,160**) and 250 g of 50 per cent sulphuric acid by weight in a 750-ml round-bottomed flask under reflux for 9 hours. Most of the pimelic acid separates from the cold reaction mixture. Filter off the crystalline acid upon a sintered glass funnel. Saturate the filtrate with ammonium sulphate and extract it with three 50 ml portions of ether. Dissolve the residue on the filter (which is slightly discoloured, but is fairly pure pimelic acid) in the combined ethereal extracts, dry with anhydrous sodium sulphate and remove the ether by distillation. Recrystallise the residual solid acid from benzene containing 5 per cent of ether. The yield of pure pimelic acid, m.p. 105–106 °C, is 22 g (93%).

Phenylacetic acid. Into a 500-ml round-bottomed flask, provided with a reflux condenser, place 100 ml of water, 100 ml of concentrated sulphuric acid and 100 ml of glacial acetic acid: add 100 g (98 ml, 0.85 mol) of benzyl cyanide. Heat under reflux for 45–60 minutes; hydrolysis is then complete. Pour the mixture into 2–3 volumes of water with stirring. Filter the crude acid at the pump. Melt the crude material under water, and wash it two or three times with small volumes of hot water; the acid solidifies on cooling. Test a small portion for the presence of phenylacetamide (m.p. 155 °C) by dissolving in sodium carbonate solution. If a clear solution results, phenylacetamide is absent: if the solution is not clear, shake the whole of the crude product with excess of sodium carbonate solution, filter and precipitate the phenylacetic acid from the clear filtrate by the addition of dilute sulphuric acid. Filter off the phenylacetic acid and recrystallise it from hot water or, better, light petroleum (b.p. 40–60 °C). The yield of pure acid, m.p. 77 °C, is 50 g (43%). Small quantities of acid may be recovered from the mother-liquors by extraction with ether, but this is rarely worth while. Alternatively the acid may be purified by distillation under reduced pressure, b.p. 140–150 °C/20 mmHg.

p-Nitrophenylacetic acid. Prepare a dilute solution of sulphuric acid by adding 150 ml of concentrated sulphuric acid cautiously to 140 ml of water. Place 50 g (0.31 mol) of p-nitrobenzyl cyanide (Section **IV,24**) in a 500-ml round-bottomed flask, pour in about two-thirds of the sulphuric acid and shake well until all the solid is moistened with the acid. Wash down any nitrile adhering to the walls of the flask into the liquid with the remainder of the acid. Attach a reflux condenser to the flask and boil under reflux for 15 minutes. Dilute the rather dark reaction mixture with an equal volume of cold water and cool to 0 °C. Filter with suction, and wash several times with ice-water. Dissolve the solid in 800 ml of boiling water (add decolourising carbon, if necessary) and filter rapidly through a hot-water funnel supporting a fluted filter-paper. If any solid remains on the filter, dissolve it in the minimum volume of boiling water and filter into the main filtrate. Collect the pale yellow needles of p-nitrophenylacetic acid which separate on cooling, and dry at 100 °C. The yield of acid, m.p. 151–152 °C, is 53 g (95%).

Vinylacetic acid. Place 134 g (161 ml, 2 mol) of allyl cyanide (1) and 200 ml of concentrated hydrochloric acid in a 1-litre round-bottomed flask attached to a reflux condenser. Warm the mixture cautiously with a small flame and shake from time to time. After 7–10 minutes, a vigorous reaction sets in and the mixture refluxes; remove the flame and cool the flask, if necessary, in cold water. Ammonium chloride crystallises out. When the reaction subsides, reflux the mixture for 15 minutes. Then add 200 ml of water, cool and separate the upper layer of acid. Extract the aqueous layer with three 100 ml portions of ether. Combine the acid and the ether extracts, and remove the ether under atmospheric pressure in a 250-ml flask fitted with a Claisen still-head and a short fractionating column: continue the heating on a water bath until the temperature of the vapour reaches 70 °C. Allow the apparatus to cool and distil under diminished pressure; collect the fraction (*a*) distilling up to 71 °C/14 mmHg and (*b*) at 72–74 °C/14 mmHg (chiefly at 72.5 °C/14 mmHg). A dark residue (about 10 ml) and some white solid (? crotonic acid) remains in the flask. Fraction (*b*) weighs 100 g (58%) and is analytically pure vinylacetic acid. Fraction (*a*) weighs about 50 g and separates into two layers: remove the water layer, dry the organic phase with anhydrous sodium sulphate and distil under reduced pressure; a further 15 g (8.7%) of reasonably pure acid, b.p. 69–70 °C/12 mmHg, is obtained.

Note. (1) Allyl cyanide may be prepared by the following procedure. Into a 2-litre three-necked flask, provided with a sealed stirrer and two long double surface condensers, place 293 g (210 ml, 2.42 mol) of freshly distilled allyl bromide, b.p. 70–71 °C (Section **III,54**) and 226 g (2.52 mol) of dry copper(I) cyanide (Section **II,2,17**). Warm the flask on a water bath so that the allyl bromide refluxes but do not stir at this stage. Immediately the vigorous reaction commences (after 15–30 minutes), remove the water bath and cool the flask in a bath of ice and water; the two double surface condensers will prevent any loss of product. When the reaction subsides, start the stirrer and heat the mixture on the water bath for 1 hour. Remove the condensers and arrange the apparatus for distillation: close one neck with a stopper. Heat the flask in an oil bath, and distil the allyl cyanide with stirring; it is advisable to reduce the pressure (water pump) towards the end of the distillation to assist the removal of the final portion of the allyl cyanide from the solid residue. Redistil and collect the pure allyl cyanide at 116–121 °C. The yield is 140 g (86%).

M,3. Carboxylation of Grignard reagents. The addition of a Grignard reagent to carbon dioxide gives the salt of the corresponding carboxylic acid, which on acidification yields the free carboxylic acid.

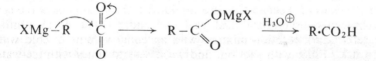

The reaction is best carried out by pouring the ethereal solution of the Grignard reagent directly on to an excess of coarsely powdered solid carbon dioxide. The alternative procedure of passing dry carbon dioxide gas into the Grignard reagent solution may give rise to the formation of ketonic by-products by further reaction of the Grignard reagent with the carboxylate salt.

$$R - \overset{\overset{O}{\parallel}}{C} - OMgX \longrightarrow R_2C(OMgX)_2 \xrightarrow{H_3O^{\oplus}} R_2CO$$
$$R - MgX$$

The synthesis of 2-methylbutanoic acid and valeric acid (Section **III,129**) are illustrative of the method. Other organometallic reagents undergo a similar carboxylation reaction, and examples of the use of organolithium and organosodium reagents are included in the section on the synthesis of aromatic carboxylic acids (**IV,N,3**).

III,129. 2-METHYLBUTANOIC ACID

$$CH_3 \cdot CH_2 \cdot CHCl \cdot CH_3 \xrightarrow[\text{ether}]{Mg} CH_3 \cdot CH_2 \cdot CH(CH_3)MgCl \xrightarrow{CO_2}$$

$$CH_3 \cdot CH_2 \cdot CH(CH_3) \cdot CO_2MgCl \xrightarrow{H_3O^{\oplus}} CH_3 \cdot CH_2 \cdot CH(CH_3) \cdot CO_2H$$

Fit a 1-litre three-necked flask with a mechanical stirrer, a double surface condenser and a separatory funnel and provide both the condenser and funnel with calcium chloride guard-tubes. Place 12.2 g (0.51 mol) of dry magnesium turnings, 50 ml of sodium-dried ether and a crystal of iodine in the flask: introduce 3 g (3.5 ml) of dry s-butyl chloride (Section III,48) (1). Warm the flask on a water bath or electric hot plate to start the reaction, and then allow it to proceed by its own heat for 20 minutes. Add a further 75 ml of anhydrous ether, followed by a solution of 43 g (49 ml, total 0.5 mol) of dry s-butyl chloride in 275 ml of anhydrous ether over a period of 20–25 minutes. If the reaction becomes too vigorous, cool the flask momentarily with cold water. The refluxing will continue for about 20 minutes after the addition of the halide solution owing to the heat of the reaction. When this subsides, reflux the mixture for 1 hour. Cool the flask in a mixture of ice and salt to − 12 °C and add a further 100 ml of anhydrous ether. Weigh out (rough balance) 125 g of Cardice (2) on a piece of stiff paper: wrap the Cardice in a stout cloth, and, by means of a pestle, break it into small lumps. Empty the Cardice into a dry 1500-ml beaker and at once pour in the Grignard reagent in a slow steady stream; any unreacted magnesium will adhere to the sides of the flask. A vigorous reaction occurs. Stir the mass well, and allow it to stand until all the Cardice has evaporated. Then add slowly a mixture of 300 g of crushed ice and 75 ml of concentrated hydrochloric acid. Stir until the gelatinous compound is decomposed and there is a clean separation into two layers. Pour the mixture into a separatory funnel; rinse the beaker with 50 ml of ether and transfer this to the funnel. Separate the upper layer and extract the aqueous layer with three 40 ml portions of ether. Cool the combined ether extracts by the addition of ice, and add cautiously 100 ml of 25 per cent sodium hydroxide solution; run off and keep the aqueous layer and repeat the extraction with a further 50 ml of alkali solution of the same strength. The organic acid is thus converted into the sodium salt and passes into the aqueous layer: test the extracts with phenolphthalein to make certain that all the acid has been removed. Distil the alkaline extract until its volume is reduced by about 10 per cent; this removes ether and other volatile impurities. Allow to cool, and cautiously acidify with concentrated hydrochloric acid; it is advisable to stir the mixture during the acidification process. Separate the upper layer of acid. Distil the water layer from a 1-litre flask until no more oily drops pass over; saturate the distillate with salt, remove the acid layer and combine it with the main product. Dry the combined acid fractions with anhydrous calcium sulphate, and distil. Collect the 2-methylbutanoic acid at 173–174 °C. The yield is 40 g (79%).

Notes. (1) s-Butyl chloride is employed in preference to the bromide because it is cheaper and the yield of acid is slightly higher.

(2) Cardice should be handled with gloves or with a dry towel; if Cardice is held for a long time in the hand, it may cause frost bite. The crushed Cardice

should be used immediately otherwise it may absorb water which would react with some of the Grignard reagent.

Cognate preparation Valeric acid. Prepare a Grignard reagent from 12.2 g (0.51 mol) of magnesium, a crystal of iodine, 69 g (54 ml, 0.5 mol) of butyl bromide and 250 ml of anhydrous ether (compare Section **III,34**). React the Grignard reagent with 125 g of Cardice and work up the product as described above for 2-methylbutanoic acid (1). Collect the valeric acid at 182–185 °C. The yield is 25 g (49%).

Note. (1) Alternatively the acid may be dried before distillation by adding benzene or toluene and distilling out the benzene (or toluene) water azeotrope (cf. Section **I,23**).

M,4. The Arndt–Eistert method. The Arndt–Eistert reaction is a comparatively simple method for converting an acid into its next higher homologue, or to a derivative of the homologous acid, such as an amide or an ester. The overall yield is generally good. The reaction is applicable to aliphatic, aromatic, alicyclic and heterocyclic acids. It involves three operations:

(*a*) Formation of the carboxylic acid chloride, e.g., with thionyl chloride or with phosphorus pentachloride.

$$R\cdot CO_2H \xrightarrow[\text{or PCl}_5]{\text{SOCl}_2} R\cdot COCl$$

(*b*) Formation of a diazoketone (I) by the gradual addition of the acid chloride to an excess of an ethereal solution of diazomethane. This reaction may be represented mechanistically as follows:

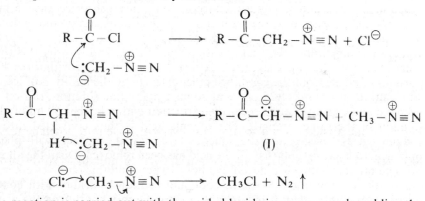

If the reaction is carried out with the acid chloride in excess, e.g., by adding the diazomethane solution slowly to the acid chloride, some halomethyl ketone is produced:

$$R\cdot CO\cdot \overset{\ominus}{C}H\cdot \overset{\oplus}{N}_2 + HCl \longrightarrow R\cdot CO\cdot CH_2Cl + N_2\uparrow$$

(*c*) Rearrangement of the diazoketone, with loss of nitrogen, in the presence of suitable reagents and a catalyst (colloidal silver–silver oxide, or silver nitrate in the presence of ammonia solution). An acid is formed in the presence of water, an amide results when ammonia or an amine is used, and an ester is produced in the presence of an alcohol.

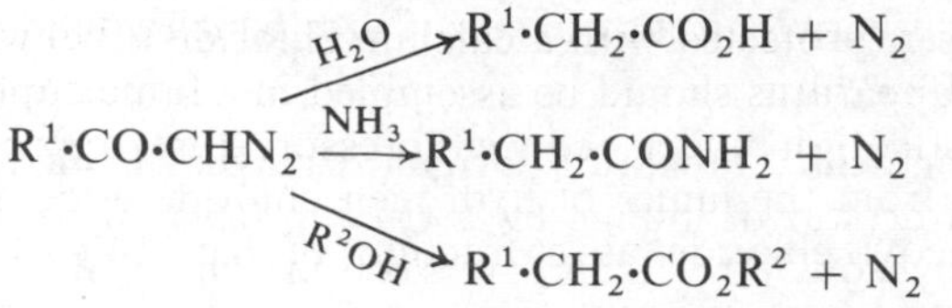

$$R^1 \cdot CO \cdot CHN_2 \begin{cases} \xrightarrow{H_2O} R^1 \cdot CH_2 \cdot CO_2H + N_2 \\ \xrightarrow{NH_3} R^1 \cdot CH_2 \cdot CONH_2 + N_2 \\ \xrightarrow{R^2OH} R^1 \cdot CH_2 \cdot CO_2R^2 + N_2 \end{cases}$$

This third operation, involving the conversion of the diazoketone into an acid or a simple derivative thereof, is known as the **Wolff rearrangement**. Loss of nitrogen from the diazoketone (I) is accompanied by a nucleophilic 1,2-shift of the alkyl group to yield a keten (II), which reacts with the solvent to give the carboxylic acid or the appropriate derivative.

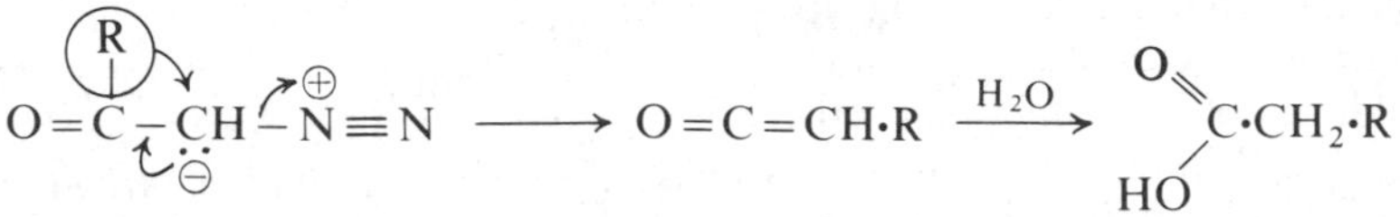

$$O = C - \overset{R}{\underset{|}{C}}H - \overset{\oplus}{N} \equiv N \longrightarrow O = C = CH \cdot R \xrightarrow{H_2O} \underset{HO}{\overset{O}{\diagdown}} C \cdot CH_2 \cdot R$$

(I)   (II)

In order to prepare an acid, a dioxan solution of the diazoketone is added slowly to a suspension of silver oxide in a dilute solution of sodium thiosulphate. If the conversion to the acid yields unsatisfactory results, it is usually advisable to prepare the amide or ester, which are generally obtained in good yields; hydrolysis of the derivative gives the free acid.

The conversion of a diazoketone to an acid amide may be accomplished by treating a warm solution in dioxan with 10–28 per cent aqueous ammonia solution containing a small amount of silver nitrate solution, after which the mixture is heated to 60–70 °C for some time. Precautions should be taken (by use of a safety glass shield) when heating mixtures containing ammoniacal silver nitrate.

Esters of the homologous acids are prepared by adding silver oxide in portions rather than in one lot to a hot solution or suspension of the diazoketone in an anhydrous alcohol (methanol, ethanol or propan-1-ol): methanol is generally used and the silver oxide is reduced to metallic silver, which usually deposits as a mirror on the sides of the flask. The production of the ester may frequently be carried out in a homogeneous medium by treating a solution of the diazoketone in the alcohol with a solution of silver benzoate in triethylamine.

The reaction is illustrated here by the overall conversion of the dicarboxylic acid, sebacic acid, into dodecanedioic acid by a bis-homologisation process (Section III,130).

III,130. DODECANEDIOIC ACID

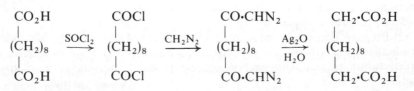

Sebacoyl chloride. Convert 20 g (0.1 mol) of sebacic acid (Section III,131) into the corresponding acid chloride by heating it on a water bath in a flask fitted with a reflux condenser (protected with a calcium chloride tube) with 20 ml of thionyl chloride; the apparatus should be assembled in a fume cupboard. Purify the product by distillation under reduced pressure (use appropriate traps to protect the pump from the fumes of hydrogen chloride and sulphur dioxide). Collect the sebacoyl chloride as a fraction of b.p. 140–143 °C/2 mmHg; the yield is 18 g (77%).

1,8-Bis-diazoacetyloctane. Dissolve 7.4 g (0.033 mol) of the resulting sebacoyl chloride in anhydrous ether and add the solution slowly to an ethereal solution containing about 6.8 g of diazomethane, i.e., two portions of the ethereal solution prepared as described in Section **II,2,**19, *Method 3*, CAUTION (1). Allow the mixture to stand overnight and remove any excess reagent together with some of the ether by distillation from a warm-water bath. To ensure that no undue hazard results from the possible presence of un-decomposed excess diazomethane use a distillation assembly as described for the distillation of diazomethane–ethereal solutions (Section **II,2,**19, *Method 1*). When the distillate is colourless, change the receivers, and complete the re-moval of solvent by distillation under reduced pressure (water pump). After recrystallisation from benzene the resulting 1,8 bis diazoacetyloctane has m.p. 91 °C; the yield is 6.4 g (83%).

Dodecanedioic acid. Add, with stirring, a solution of 5 g (0.02 mol) of the bis-diazoketone in 100 ml of warm dioxan to a suspension of 6.0 g of freshly precipitated silver oxide (2) in 250 ml of water containing 8 g of sodium thio-sulphate maintained at 75 °C. A brisk evolution of nitrogen occurs; after 1.5 hours at 75 °C, filter the liquid from the black silver residue. Acidify the almost colourless filtrate with nitric acid and extract the gelatinous precipitate with ether. Evaporate the dried ethereal extract: the residue of crude dodecanedioic acid weighs 3.3 g (72%), and has m.p. 116–117 °C. Recrystallisation from 20 per cent aqueous acetic acid raises the m.p. to 127–128 °C.

Alternatively, treat 3.9 g (0.0156 mol) of the bis-diazoketone in 50 ml of warm dioxan with 15 ml of 20 per cent aqueous ammonia and 3 ml of a 10 per cent aqueous silver nitrate solution under reflux in a 250- or 500-ml flask on a water bath. Nitrogen is evolved for a few minutes, followed by a violent re-action and the production of a dark brown opaque mixture. Continue heating for 30 minutes on the water bath and filter hot: the diamide of dodecanedioic acid is deposited on cooling. Filter the product and air dry: the yield is 3.1 g (87%), m.p. 182–184 °C, raised to 184–185 °C after recrystallisation from 20 per cent aqueous acetic acid. Hydrolyse the diamide (1 mol) by refluxing for 2–5 hours with a four molar excess of 3 *M*-potassium hydroxide solution. Acidify and recrystallise the precipitated acid from 20 per cent acetic acid. The yield of dodecanedioic acid, m.p. 127–128 °C, is almost quantitative.

Notes. (1) Precautions in the use of diazomethane are fully described in Section **II,2,**19, and should be carefully noted; the operations should be car-ried out in a fume cupboard.

(2) Prepare the silver oxide by adding dilute sodium hydroxide solution gradually to a stirred 10 per cent aqueous silver nitrate solution—until precipitation is just complete. Wash the product thoroughly with distilled water.

M,5. Electrolytic (anodic) coupling. Esters of long chain carboxylic acids may be conveniently prepared by electrolytic (anodic) syntheses (Section

I,17,6), a technique which has already been illustrated by the synthesis of hexacosane (Section III,6).

Simple anodic coupling by electrolysis in anhydrous methanolic solution (containing a little sodium methoxide) of methyl hydrogen adipate (Section III,149) gives dimethyl sebacate; methyl hydrogen sebacate (Section III,149) in turn yields dimethyl octadecanedioate (Section III,131, cognate preparations).

$$2CH_3O_2C\cdot(CH_2)_4\cdot CO_2^{\ominus} \xrightarrow{-2e} CH_3O_2C\cdot(CH_2)_8\cdot CO_2CH_3 + 2CO_2$$

$$2CH_3O_2C\cdot(CH_2)_8\cdot CO_2^{\ominus} \xrightarrow{-2e} CH_3O_2C\cdot(CH_2)_{16}\cdot CO_2CH_3 + 2CO_2$$

Electrolysis of a mixture of two carboxylic acids, $R^1\cdot CO_2H$ and $R^2\cdot CO_2H$, leads in addition to the products of normal coupling ($R^1 - R^1$ and $R^2 - R^2$) to the cross coupled product ($R^1 - R^2$). Similarly if a mixture of a saturated carboxylic acid and a half-ester of an α,ω-dicarboxylic acid is electrolysed, there are three main products, viz., a hydrocarbon (I), a mono-ester (II) and a di-ester (III). Normally the three products are readily separable by distillation. Furthermore, by increasing the molar proportion of the monocarboxylic acid, the yield of (II) is improved at the expense of (III).

$$R\cdot CO_2H + HO_2C\cdot(CH_2)_n\cdot CO_2CH_3 \longrightarrow R - R + R\cdot(CH_2)_n\cdot CO_2CH_3 +$$
$$\text{(I)} \qquad\qquad \text{(II)}$$

$$+ CH_3O_2C\cdot(CH_2)_{2n}\cdot CO_2CH_3 + CH_2 = CH\cdot(CH_2)_{n-2}\cdot CO_2CH_3$$
$$\text{(III)} \qquad\qquad\qquad \text{(IV)}$$

The unsaturated ester (IV) is also often present in small quantity and arises from the loss of a proton from the intermediate carbonium ion (VI), which is produced when the radical species (V) (which is involved in the coupling reaction) undergoes further anodic oxidation.

$$CH_3O_2C\cdot(CH_2)_{n-2}\cdot CH_2\cdot CH_2\cdot C \begin{smallmatrix} \nearrow O \\ \searrow \ddot{O}: \end{smallmatrix} \xrightarrow{-CO_2}$$

$$CH_3O_2C\cdot(CH_2)_{n-2}\cdot CH_2\cdot \dot{C}H_2 \xrightarrow{-e} CH_3O_2C\cdot(CH_2)_{n-2}\cdot CH_2\cdot \overset{\oplus}{C}H_2$$
$$\text{(V)} \qquad\qquad\qquad \text{(VI)}$$

$$\xrightarrow{-H^{\oplus}} CH_3O_2C\cdot(CH_2)_{n-2}\cdot CH = CH_2$$

Two alternative syntheses of methyl myristate, and thence myristic acid, are described (Section III,131). In *Method A* hexanoic acid (2 mol) is coupled with methyl hydrogen sebacate (1 mol), the products being methyl myristate, decane and dimethyl octadecanedioate.

In *Method B* decanoic acid (2 mol) is coupled with methyl hydrogen adipate (1 mol) and the products are methyl myristate, octadecane and dimethyl sebacate.

III,131. MYRISTIC ACID (*Tetradecanoic acid*)

Method A:

$$CH_3 \cdot (CH_2)_4 \cdot CO_2^{\ominus} + {}^{\ominus}O_2C \cdot (CH_2)_8 \cdot CO_2CH_3 \xrightarrow[-2CO_2]{-2e} CH_3 \cdot (CH_2)_8 \cdot CH_3 \; +$$

$$+ \; CH_3 \cdot (CH_2)_{12} \cdot CO_2CH_3 \; + \; CH_3O_2C \cdot (CH_2)_{16} \cdot CO_2CH_3$$

Method B:

$$CH_3 \cdot (CH_2)_8 \cdot CO_2^{\ominus} + {}^{\ominus}O_2C \cdot (CH_2)_4 \cdot CO_2CH_3 \xrightarrow[-2CO_2]{-2e} CH_3 \cdot (CH_2)_{16} \cdot CH_3 \; +$$

$$+ \; CH_3 \cdot (CH_2)_{12} \cdot CO_2CH_3 \; + \; CH_3O_2C \cdot (CH_2)_8 \cdot CO_2CH_3$$

Method A. Dissolve 23.2 g (0.184 mol) of redistilled hexanoic acid, b.p. 204.5–205.5 °C/760 mmHg, and 21.6 g (0.1 mol) of methyl hydrogen sebacate in 200 ml of absolute methanol to which 0.13 g of sodium has been added. Electrolyse at 2.0 amps (Section **I,17,6**), whilst maintaining the temperature between 30 and 40 °C, until the pH is about 8.0 (*ca.* 6 hours). Neutralise the contents of the electrolysis cell with a little acetic acid and distil off the methanol on a water bath. Dissolve the residue in 200 ml of ether, wash with three 50 ml portions of saturated sodium hydrogen carbonate solution, once with water, dry with magnesium sulphate, and distil through an efficient fractionating column (Sections **I,26** and **I,27**). Collect the decane at 60 °C/10 mmHg (3.0 g), the methyl myristate at 158–160 °C/10 mmHg (12.5 g, 52%) and dimethyl octadecanedioate at 215–230 °C/7 mmHg (1.5 g).

Reflux a mixture of 7.3 g of methyl myristate with a solution of 4.8 g of sodium hydroxide in 200 ml of 90 per cent methanol for 2 hours, distil off the methanol on a water bath, dissolve the residue in 400 ml of hot water, add 15 ml of concentrated hydrochloric acid to the solution at 50 °C in order to precipitate the organic acid, and cool. Collect the acid by suction filtration, wash it with a little water and dry in a vacuum desiccator. The yield of myristic acid (tetradecanoic acid), m.p. 57–58 °C, is 5.9 g (87%).

Method B. Dissolve 55.2 g (0.32 mol) of pure decanoic acid, m.p. 31–32 °C, and 25.6 g (0.16 mol) of methyl hydrogen adipate in 200 ml of absolute methanol to which 0.25 g of sodium has been added. Electrolyse at 2.0 amps at 25–35 °C until the pH of the electrolyte is 8.2 (*ca.* 9 hours). Neutralise the contents of the electrolytic cell with acetic acid, distil off the methanol on a water bath, dissolve the residue in about 200 ml of ether, wash with three 50 ml portions of saturated hydrogen carbonate solution and remove the ether on a water bath. Treat the residue with a solution of 8.0 g of sodium hydroxide in 200 ml of 80 per cent methanol, reflux for 2 hours and distil off the methanol on a water bath. Add about 600 ml of water to the residue to dissolve the mixture of sodium salts: extract the hydrocarbon with four 50 ml portions of ether, and dry the combined ethereal extracts with magnesium sulphate. After removal of the ether, 23.1 g of almost pure octadecane, m.p. 23–24 °C, remains. Acidify the aqueous solution with concentrated hydrochloric acid (*ca.* 25 ml), cool to 0 °C, filter off the mixture of acids, wash well with cold water and dry in a vacuum desiccator. The yield of the mixture of sebacic and myristic acids, m.p. 52–67 °C, is 26 g. Separate the mixture by extraction with six 50 ml portions of almost boiling light petroleum, b.p. 40–60 °C. The residue (5.2 g), m.p.

132 °C, is sebacic acid. Evaporation of the solvent gives 20 g (55%) of myristic acid, m.p. 52–53 °C; the m.p. is raised slightly upon recrystallisation from methanol.

Cognate preparations **Sebacic acid** (*decanedioic acid*). Dissolve 40 g (0.25 mol) of methyl hydrogen adipate in 100 ml of absolute methanol to which 0.1 g of sodium has been added. Pass a current of about 2.0 amps until the pH of the solution is about 8 (*ca.* 5 hours); test with narrow-range indicator paper. Transfer the contents of the electrolysis cell to a 500-ml round-bottomed flask, render neutral with a little acetic acid and distil off the methanol on a water bath. Dissolve the residue in 150 ml of ether, wash with three 50 ml portions of saturated sodium hydrogen carbonate solution, then with water, dry over magnesium sulphate and distil under reduced pressure. Collect the dimethyl sebacate at 155 °C/8 mmHg; it melts at 26 °C and the yield is 14.6–16.0 g (51–56%).

Reflux 14.6 g (0.064 mol) of the ester with a solution of 10 g of sodium hydroxide in 125 ml of 80 per cent methanol for 2 hours on a water bath. Add 200 ml of water to dissolve the solid which separates, extract·with two 30 ml portions of ether and warm the aqueous solution on a water bath to remove dissolved ether. Acidify the ice-cold aqueous solution to litmus by the addition of concentrated hydrochloric acid. Collect the precipitated acid by suction filtration, wash it with a little cold water and dry at 100 °C. The yield of sebacic acid, m.p. 133 °C, is 11.5 g (89%).

Octadecanedioic acid. Dissolve 31.5 g (0.145 mol) of methyl hydrogen sebacate in 140 ml of absolute methanol to which 0.4 g of sodium has been added. Electrolyse at 2.0 amps until the pH of the electrolyte is 7.8–8.0 (3.5–4 hours). Work up as described for sebacic acid. Upon distillation, an unsaturated ester passes over at 111–113 °C/20 mmHg (4.6 g), followed by dimethyl octadecanedioate at 212–219 °C/4 mmHg (mainly at 214–215 °C/4 mmHg), m.p. 56 °C (16.5 g, 66%).

Reflux 6.8 g of the dimethyl ester with a solution of 3.2 g of sodium hydroxide in 150 ml of 80 per cent methanol for 2 hours on a water bath. When cold, filter off the solid and wash it with a little cold methanol. Dissolve the solid in 350 ml of warm water, add concentrated hydrochloric acid to the solution at 60 °C until acidic to litmus, filter off the precipitated acid, wash with a little water and dry at 100 °C. The resulting octadecanedioic acid, m.p. 122 °C, weighs 5.3 g (84%). Recrystallisation from absolute methanol raises the m.p. to 124.5 °C.

M,6. Methods utilising diethyl malonate. (*a*) *Hydrolysis of alkylmalonic esters.* When treated with one equivalent of sodium ethoxide, diethyl malonate (Section **III,154**) is converted into the mono-sodio derivative, as the result of removal by base of one of the α-methylene protons to yield a mesomeric anion (I). This nucleophilic anion undergoes an S_N2 reaction with an alkyl halide to give a *C*-substituted malonic ester. A second, different, alkyl group can be similarly introduced on to the α-carbon atom, or alternatively two identical alkyl groups may be introduced in a one-step operation by using appropriate proportions of reactants.

Alkaline hydrolysis of an alkylmalonic ester followed by careful acidification at 0 °C gives the alkylmalonic acid (e.g., Section **III,133**). The alkylmalonic

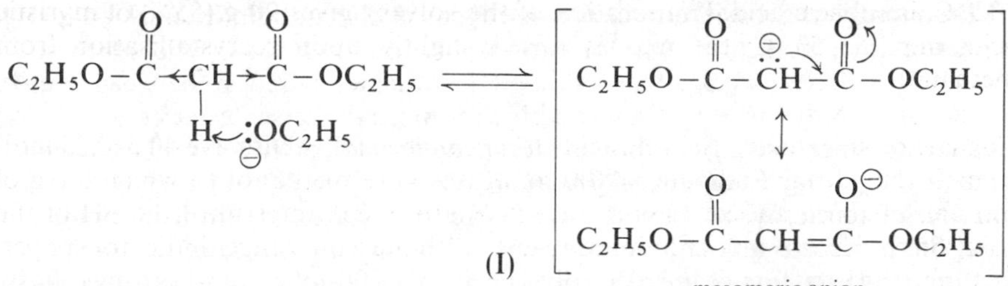

$$C_2H_5O-\overset{O}{\overset{\|}{C}}\text{-}CH\text{-}\overset{O}{\overset{\|}{C}}-OC_2H_5 \;\rightleftharpoons\; \left[\begin{array}{c} C_2H_5O-\overset{O}{\overset{\|}{C}}-\overset{\ominus}{\ddot{C}H}\text{-}\overset{O}{\overset{\|}{C}}-OC_2H_5 \\ \updownarrow \\ C_2H_5O-\overset{O}{\overset{\|}{C}}-CH=\overset{O^\ominus}{C}-OC_2H_5 \end{array} \right]$$

$$\overset{H\ \ :OC_2H_5}{\underset{\ominus}{}}$$

(I)   mesomeric anion

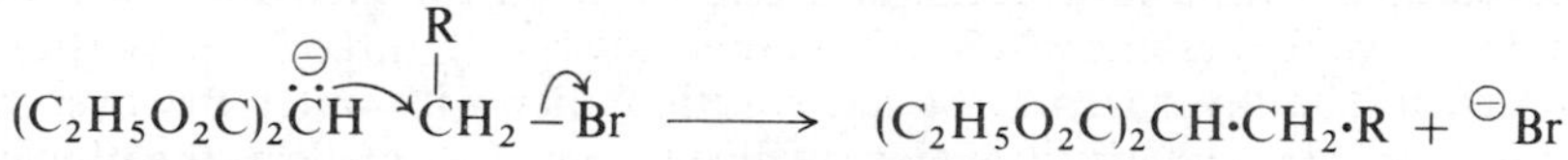

$$(C_2H_5O_2C)_2\overset{\ominus}{\ddot{C}H}\ \ \overset{R}{\underset{|}{CH_2}}-Br \longrightarrow (C_2H_5O_2C)_2CH\cdot CH_2\cdot R + {}^\ominus Br$$

$$CH_2(CO_2C_2H_5)_2 + 2RBr \xrightarrow{2\overset{\ominus}{O}C_2H_5} R_2C(CO_2C_2H_5)_2 + 2HOC_2H_5 + 2\overset{\ominus}{Br}$$

acids undergo smooth decarboxylation on heating under acidic conditions, thus providing a convenient synthesis of mono or disubstituted acetic acids (Section **III,132**).

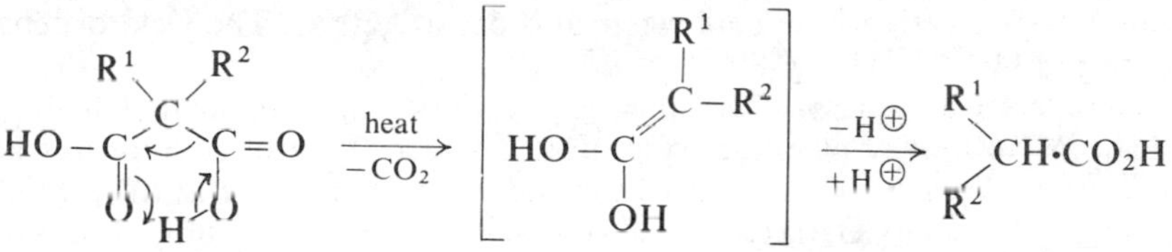

$$HO-\underset{\underset{H}{\overset{\|}{O}}}{C}\overset{R^1\ \ R^2}{\underset{}{\diagdown\overset{C}{\diagup}}}C=O \xrightarrow[-CO_2]{heat} \left[HO-\underset{OH}{C}\overset{R^1}{\diagdown}\overset{}{C-R^2} \right] \xrightarrow[+H^\oplus]{-H^\oplus} \overset{R^1}{\underset{R^2}{\diagdown\diagup}}CH\cdot CO_2H$$

(b) *The use of Michael additions of malonate ions.* Typically the addition of the mesomeric anion (I) to the α,β-unsaturated ester, diethyl fumarate, proceeds in a 1,4-manner (the **Michael reaction** or **Michael addition**).

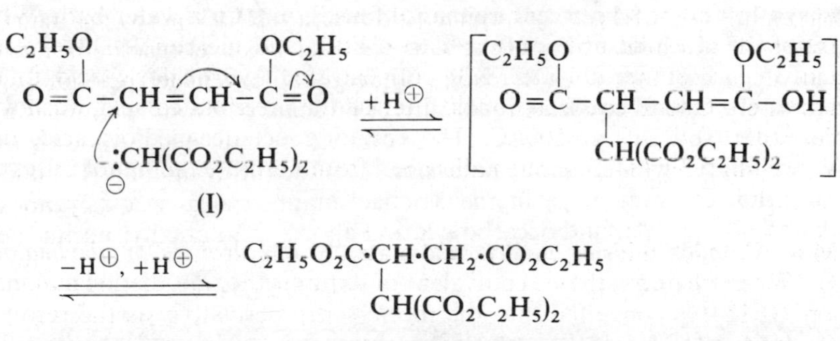

$$O=\underset{}{\overset{C_2H_5O}{\underset{|}{C}}}-CH=CH-\overset{OC_2H_5}{\underset{|}{C}}=O \xrightarrow{+H^\oplus} \left[O=\overset{C_2H_5O}{\underset{|}{C}}-CH-CH=\overset{OC_2H_5}{\underset{|}{C}}-OH \right]$$

$$\underset{\ominus}{:CH(CO_2C_2H_5)_2} \qquad\qquad CH(CO_2C_2H_5)_2$$

(I)

$$\xrightleftharpoons{-H^\oplus,\ +H^\oplus} \quad \overset{C_2H_5O_2C\cdot CH\cdot CH_2\cdot CO_2C_2H_5}{\underset{CH(CO_2C_2H_5)_2}{|}}$$

The resulting tetraethyl ester on hydrolysis and decarboxylation yields propane-1,2,3-tricarboxylic acid (Section **III,134**). In this example the malonate anion is generated by using one molar proportion of sodium ethoxide; this is Michael's original method. However, these conditions sometimes lead to competing side reactions and the formation of abnormal reaction products. Better yields of the required product are often obtained with small amounts of sodium ethoxide (the so-called catalytic method) or in the presence of a secondary amine (e.g., diethylamine, see below).

Frequently the basic conditions used cause the initial Michael adduct to undergo intramolecular transformations, as for example in the synthesis of dimedone (Section **V,3**). This involves a Michael reaction between mesityl oxide and diethyl malonate followed by an internal Claisen ester condensation.

The α,β-unsaturated component for a Michael reaction may be formed *in situ* by an initial **Knoevenagel reaction**. An example is provided by the formation of tetraethyl propane-1,1,3,3-tetracarboxylate (III) from formaldehyde and diethyl malonate in the presence of diethylamine. Diethyl methylene-malonate (II) is first formed by the simple Knoevenagel reaction and this is followed by the Michael addition process.

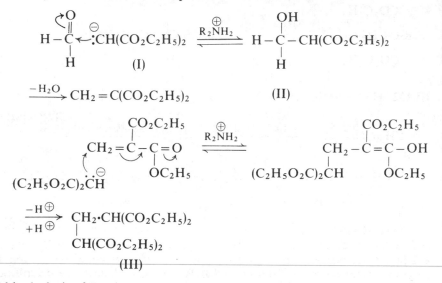

Acid hydrolysis of (III) is accompanied by decarboxylation to give glutaric acid (**Section III,135**).

The synthesis of 2,2-dimethylsuccinic acid (Section **III,136**) provides a further variant of the synthetic utility of the Knoevenagel–Michael reaction sequence. Ketones (e.g., acetone) do not readily undergo Knoevenagel reactions with malonic esters, but will condense readily in the presence of secondary amines with the more reactive ethyl cyanoacetate to give an α,β-unsaturated cyanoester (e.g., IV). When treated with ethanolic potassium cyanide the cyanoester (IV) undergoes addition of cyanide ion in the Michael manner to give a dicyanoester (V) which on hydrolysis and decarboxylation affords 2,2-dimethylsuccinic acid.

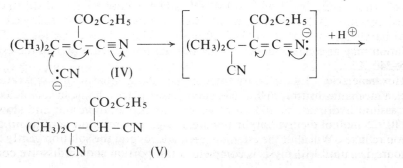

The synthesis of α-alkylglutaric acids (e.g., 2-propylglutaric acid, Section **III,137**), is conveniently achieved by allowing an alkylmalonic ester to react with the α,β-unsaturated nitrile, acrylonitrile, in the Michael manner and then subjecting the product to vigorous acidic hydrolysis.

$$
R-\underset{\underset{CO_2C_2H_5}{|}}{\overset{\overset{CO_2C_2H_5}{|}}{C\!:}}\curvearrowright CH_2=CH\curvearrowleft C\equiv N \longrightarrow \left[R-\underset{\underset{CO_2C_2H_5}{|}}{\overset{\overset{CO_2C_2H_5}{|}}{C}}-CH_2-CH=C=N\right]^{\ominus} \xrightarrow[20\,°C]{H_3O^{\oplus}}
$$

$$
R\cdot\underset{\underset{CO_2C_2H_5}{|}}{\overset{\overset{CO_2C_2H_5}{|}}{C}}\cdot CH_2\cdot CH_2\cdot CN \xrightarrow[-CO_2]{H_3O^{\oplus},\ heat} R\cdot\underset{\underset{CO_2H}{|}}{CH}\cdot CH_2\cdot CH_2\cdot CO_2H
$$

III,132. HEXANOIC ACID

$$
CH_2(CO_2C_2H_5)_2 \xrightarrow{C_2H_5O^{\ominus}} \overset{\ominus}{C}H(CO_2C_2H_5)_2 \xrightarrow{CH_3\cdot(CH_2)_2\cdot CH_2Br}
$$

$$
CH_3\cdot(CH_2)_2\cdot CH_2\cdot CH(CO_2C_2H_5)_2 \xrightarrow{\ominus OH} CH_3\cdot(CH_2)_3\cdot CH(CO_2H)_2
$$

$$
\xrightarrow{-CO_2} CH_3\cdot(CH_2)_4\cdot CO_2H
$$

Diethyl butylmalonate. Prepare a solution of sodium ethoxide from 34.5 g (1.5 mol) of clean sodium and 1 litre of super-dry ethanol (Section **II,1,9**) (1) in a 2-litre three-necked flask following the experimental conditions given for ethyl propylacetoacetate (Section **III,94**). When the sodium ethoxide solution, which is vigorously stirred, has cooled to about 50 °C, add 247.5 g (234.5 ml, 1.55 mol) of redistilled diethyl malonate slowly through the separatory funnel; to the resulting clear solution introduce gradually (60–90 minutes) 205.5 g (161.5 ml, 1.5 mol) of redistilled butyl bromide (Sections **III,54**). Reaction occurs almost immediately and much heat is evolved; if the reaction becomes violent, cool the flask by directing a stream of cold water over it. Reflux the reaction mixture on a water bath until it is neutral to moist litmus (about 2 hours). Remove as much of the ethanol as possible by distillation under reduced pressure (rotary evaporator) on a water bath. Cool the contents of the flask to about 20 °C, add 600 ml of water and shake well. Separate the upper layer of crude ester, dry it with anhydrous sodium sulphate and distil under reduced pressure. A low boiling point fraction passes over first, followed by diethyl butylmalonate at 130–135 °C/20 mmHg. The yield is 285 g (88%). The distillation may also be conducted under normal pressure; the b.p. of the ester is 235–240 °C.

Hexanoic acid. Into a 2-litre three-necked flask, fitted with a separatory funnel, a mechanical stirrer and a reflux condenser, place a hot solution of 200 g of potassium hydroxide in 200 ml of water. Stir the solution and add slowly 200 g (0.925 mol) of diethyl butylmalonate. A vigorous reaction occurs and the solution refluxes. When all the ester has been added, boil the solution gently for 2–3 hours, i.e., until hydrolysis is complete: a test portion should dissolve completely in water. Dilute with 200 ml of water and distil off 200 ml of liquid in order to ensure the complete removal of the alcohol formed in the hydrolysis (2). To the cold residue in the flask add a cold solution of 320 g (174 ml) of concentrated sulphuric acid in 450 ml of water: add the acid slowly with stirring in order to prevent excessive foaming. The solution becomes hot. Reflux the mixture for 3–4 hours and allow to cool. Separate the upper layer of the organic acid and extract the aqueous portion with four 150 ml portions of ether (3). Combine the acid layer with the ether extracts, wash it with 25 ml of water and dry with anhydrous sodium sulphate. Distil off the ether (rotary evaporator), transfer the residue to a flask fitted with a short fractionating column (the latter should be well lagged and, preferably, electrically heated) and distil the product from an air bath. Collect the hexanoic acid at 200–206 °C. The yield is 80 g (75%).

If desired, the distillation may be conducted under reduced pressure. The boiling points under various pressures are 99 °C/10 mmHg and 111 °C/20 mmHg; a 3° fraction should be collected.

Notes. (1) With commercial absolute ethanol, the yield is reduced to about 225 g.

(2) It is essential to remove the alcohol completely, otherwise some ethyl hexanoate, b.p. 168 °C, is formed which will contaminate the final product.

(3) Better results are obtained if a continuous extraction apparatus (e.g., Fig. I,92) is employed.

Cognate preparations **Diethyl propylmalonate.** Use 34.5 g (1.5 mol) of sodium and 345 g (440 ml) of super-dry ethanol, 240 g (227.5 ml, 1.5 mol) of diethyl malonate and 185 g (136.5 ml, 1.5 mol) of propyl bromide (Section **III,54**). The yield of diethyl propylmalonate, b.p. 218–225 °C, mainly 219.5–221.5 °C, is 220 g (72.5%).

Valeric acid. Convert the diethyl propylmalonate into valeric acid, b.p. 183–185 °C, following the procedure described for hexanoic acid. The yield is 75 per cent of theory.

3-Phenylpropionic acid (*hydrocinnamic acid*). Use 11.5 g (0.5 mol) of sodium and 250 ml of dry ethanol, 80 g (75 ml, 0.49 mol) of diethyl malonate and 64 g (58 ml, 0.51 mol) of redistilled benzyl chloride (Section **IV,31**). Follow the alkylation procedure described above and isolate the crude diethyl benzylmalonate. Hydrolyse the latter with a solution of 75 g of potassium hydroxide in 75 ml of water and isolate the resulting crude 3-phenylpropionic acid as described previously, using 180 ml of 5 M-sulphuric acid in the acidification stage. Purify the product by distillation under reduced pressure, collecting the fraction of b.p. 164–172 °C/25 mmHg which solidifies at room temperature. Recrystallise from light petroleum, b.p. 40–60 °C (or from water containing a little hydrochloric acid), to obtain 20 g (27%) of 3-phenylpropionic acid of m.p. 47–48 °C.

Nonanoic acid (*pelargonic acid*). Equip a 1-litre three-necked flask with a reflux condenser, a sealed stirrer unit and a thermometer. Place 23 g (1 mol) of sodium, cut in small pieces, in the flask, and add 500 ml of anhydrous butan-1-ol (1) in two or three portions: follow the experimental details given in Section **III,94** for the preparation of a solution of sodium ethoxide. When the sodium has reacted completely, allow the solution to cool to 70–80 °C and add 160 g (152 ml, 1 mol) of redistilled diethyl malonate rapidly and with stirring. Heat the solution to 80–90 °C, replace the thermometer with a dropping funnel and add 182.5 g (160 ml, 1.02 mol) of 1-bromoheptane (Section III,55) slowly at first until precipitation of sodium bromide commences, and subsequently at such a rate that the butanol refluxes gently. Reflux the mixture until it is neutral to moist litmus (about 1 hour).

Transfer the entire reaction mixture, including the precipitated sodium bromide and the small volume of water used to rinse the reaction flask, to a 3-litre flask. Add a solution of 140 g of potassium hydroxide in an equal quantity of water slowly and with shaking. Attach a reflux condenser to the flask, introduce a few fragments of porous porcelain and heat the mixture cautiously, with occasional shaking, until refluxing commences. Heat to gentle refluxing until hydrolysis is complete (about 5 hours, i.e., until a test portion is completely miscible with excess of water). Immediately equip the flask for steam distillation and steam distil the mixture until no more butanol passes over. Treat the residue cautiously with 270 ml of concentrated hydrochloric acid whilst shaking gently, and reflux the mixture for 1 hour; if sodium chloride separates as a solid cake, take care during the heating that the flask does not crack. When cold, transfer the mixture to a separatory funnel and remove the oil to a 750-ml round-bottomed flask. Heat it under an air-cooled reflux condenser in an oil bath at 180 °C until the evolution of carbon dioxide ceases (about 2 hours). Decant the oil into a Claisen flask with fractionating side arm (the latter should be well lagged) and distil under reduced pressure. Collect the pelargonic acid at 140–142 °C/12 mmHg. The yield is 115 g (73%).

Note. (1) This is conveniently prepared by drying commercial butan-1-ol with anhydrous potassium carbonate or anhydrous calcium sulphate, distilling through a column and collecting the fraction, b.p. 117–118 °C.

III,133. PROPYLMALONIC ACID

$$CH_3{\cdot}CH_2{\cdot}CH_2{\cdot}CH(CO_2C_2H_5)_2 \xrightarrow{\ominus OH} CH_3{\cdot}CH_2{\cdot}CH_2{\cdot}CH(CO_2H)_2$$

Dissolve 156 g (2.78 mol) of potassium hydroxide in 156 ml of water in a 1.5-litre round-bottomed flask and add 500 ml of rectified spirit to produce a homogeneous solution. Introduce 220 g (1.09 mol) of diethyl propylmalonate (Section III,132) slowly and with shaking. Attach a double surface reflux condenser and reflux the mixture for 3 hours; hydrolysis is then complete, i.e., a test portion dissolves completely in excess of water. Distil off as much ethanol as possible on a water bath, and dissolve the residue in a comparatively small volume of water. Cool the solution in a large beaker surrounded by ice; add dilute sulphuric acid slowly from a suitably supported dropping funnel, whilst stirring vigorously with a mechanical stirrer, until the solution is acid to Congo red paper. Extract the solution with three 150 ml portions of ether, dry the ethereal extract with anhydrous sodium sulphate and distil off the ether on a water bath. Spread the syrupy residue in thin layers upon large clock glasses (1); after 2–3 days, filter off the crystals at the pump, using light petroleum, b.p. 40–60 °C, to facilitate the transfer from the clock glasses to the sintered glass filter funnel. Spread the crystals on a porous tile to remove traces of oily impurities; the crude propylmalonic acid has m.p. 95–96 °C. Spread the filtrate and washings on large clock glasses as before and filter off the solid which crystallises after 1 day. Repeat the process until no further crystals are obtained. Recrystallise all the crystals from hot benzene. The yield of pure propylmalonic acid, m.p. 96 °C, is 110 g (69%).

Note. (1) An alternative procedure is to leave the syrupy residue in a vacuum desiccator over anhydrous calcium chloride and silica gel, and to filter off the successive crops of crystals as they separate. These are washed with light petroleum, b.p. 40–60 °C, spread on a porous tile and recrystallised.

Cognate preparations Butylmalonic acid. This acid may be similarly prepared from diethyl butylmalonate (Section **III,132**) and melts at 102 °C after recrystallisation from benzene.

s-Butylmalonic acid. From diethyl s-butylmalonate (Section **III,169**); the acid melts at 76 °C after recrystallisation from benzene.

III,134. PROPANE-1,2,3-TRICARBOXYLIC ACID (*Tricarballylic acid*)

$$C_2H_5O_2C{\cdot}CH = CH{\cdot}CO_2C_2H_5 + CH_2(CO_2C_2H_5)_2 \xrightarrow{C_2H_5O^{\ominus}}$$

$$\begin{array}{c} C_2H_5O_2C{\cdot}CH{\cdot}CH_2{\cdot}CO_2C_2H_5 \\ | \\ CH(CO_2C_2H_5)_2 \end{array} \xrightarrow[-CO_2]{H_3O^{\oplus}} \begin{array}{c} CH_2{\cdot}CO_2H \\ | \\ CH{\cdot}CO_2H \\ | \\ CH_2{\cdot}CO_2H \end{array}$$

Tetraethyl propane-1,1,2,3-tetracarboxylate. In a 1-litre three-necked flask, fitted with an efficient reflux condenser, mechanical stirrer and a dropping funnel, prepare a solution of sodium ethoxide from 18.4 g (0.8 mol) of clean sodium and 200 g (253 ml) of super-dry ethanol (Section **II,1,9**). Cool the flask and add, with stirring, 160 g (151 ml, 1 mol) of redistilled diethyl malonate through the condenser. Warm the mixture gently on a water bath and introduce 140 g (131.5 ml, 0.81 mol) of redistilled diethyl fumarate (Section **III,145**) from the dropping funnel at such a rate that the mixture boils gently. When the addition is complete, boil for 1 hour, cool and add 50 g (47.5 ml) of glacial acetic acid. Distil off most of the alcohol on the water bath under slightly reduced pressure and pour the residue into sufficient distilled water to dissolve all the solid. Separate the ester layer: extract the aqueous layer with four 25-ml portions of carbon tetrachloride. Wash the combined ester and carbon tetrachloride extracts twice with 25 ml of water. Distil off the carbon tetrachloride under atmospheric pressure through a short column (the moisture is carried over with the solvent) and distil the residue under reduced pressure. Collect the tetraethyl propane-1,1,2,3-tetracarboxylate at 182–184 °C/8 mmHg. The yield is 250 g (93%).

Propane-1,2,3-tricarboxylic acid. Place 228 g (204 ml, 0.685 mol) of tetraethyl propane-1,1,2,3-tetracarboxylate and 240 ml of 1 : 1-hydrochloric acid in a 1-litre three-necked flask, fitted with a mechanical stirrer and a fractionating column with condenser set for downward distillation; attach a receiver with side tube to the condenser and connect the side tube to a wash bottle containing water. Boil the mixture, with continual stirring, at such a rate that the alcohol is removed as fast as it is formed, but without undue removal of water from the flask. The progress of the reaction may be followed from the rate at which carbon dioxide passes through the wash bottle. When the temperature at the head of the column approaches 100 °C, adjust the heating of the flask so that very little liquid distils over: continue the heating until the evolution of carbon dioxide ceases (*ca.* 12 hours). Evaporate the solution to dryness under reduced pressure (rotary evaporator); redissolve the residue in water and again evaporate to dryness to remove the excess of hydrochloric acid. Dissolve the residue in distilled water, filter the solution through a short column of decolourising carbon and again evaporate to dryness under reduced pressure. Grind the dry residue to a fine powder, mix it to a paste with dry ether, filter by suction, wash with a little anhydrous ether and dry in an oven. The resulting tricarballylic acid, m.p. 160–161 °C, is practically pure and weighs 118 g (91%).

III,135. GLUTARIC ACID

$$H \cdot CHO + 2CH_2(CO_2C_2H_5)_2 \xrightarrow{(C_2H_5)_2NH}$$

$$(C_2H_5O_2C)_2CH \cdot CH_2 \cdot CH(CO_2C_2H_5)_2 \xrightarrow{H_3O^\oplus}$$

$$[(HO_2C)_2CH \cdot CH_2 \cdot CH(CO_2H)_2] \xrightarrow{-CO_2} HO_2C \cdot CH_2 \cdot CH_2 \cdot CH_2 \cdot CO_2H$$

Tetraethyl propane-1,1,3,3-tetracarboxylate. Cool a mixture of 320 g (302 ml, 2 mol) of redistilled diethyl malonate and 80 g (1 mol) of 40 per cent formaldehyde solution ('formalin') contained in a 1-litre round-bottomed flask to 5 °C by immersion in ice, and add 5 g (7 ml) of diethylamine. Keep the mixture at room temperature for 15 hours and then heat under a reflux condenser on a boiling water bath for 6 hours. Separate the aqueous layer, dry the organic layer with anhydrous sodium sulphate and distil under reduced pressure. Collect the tetracarboxylate ester at 200–215 °C/20 mmHg. The yield is 250 g (75%).

Glutaric acid. Heat a mixture of 125 g (0.376 mol) of the preceding ester and 250 ml of 1 : 1-hydrochloric acid under reflux with stirring in a 1-litre two-necked flask equipped with a mechanical stirrer and reflux condenser. Continue the heating until the mixture becomes homogeneous (6–8 hours). Evaporate the contents of the flask to dryness on a steam bath (rotary evaporator) and distil the residual glutaric acid under reduced pressure. Collect the fraction boiling at 185–195 °C/10 mmHg: it crystallises on cooling. Moisten with a little water (to convert any glutaric anhydride present into the acid), heat gently and dry at 30 °C. Recrystallise from chloroform (or benzene); the resulting practically pure glutaric acid, m.p. 96–97 °C, weighs 40 g (81%).

III,136. 2,2-DIMETHYLSUCCINIC ACID

$$CH_3 \cdot CO \cdot CH_3 + CH_2(CN) \cdot CO_2C_2H_5 \xrightarrow{base} (CH_3)_2C = C \cdot CO_2C_2H_5 \xrightarrow{\ominus CN}$$
$$\overset{|}{CN}$$

$$(CH_3)_2C \cdot \overset{\ominus}{C}(CN) \cdot CO_2C_2H_5 \xrightarrow{H_3O^\oplus} (CH_3)_2C \cdot CH(CO_2H)_2 \xrightarrow{-CO_2}$$
$$\overset{|}{CN} \qquad\qquad\qquad\qquad \overset{|}{CO_2H}$$

$$(CH_3)_2C \cdot CH_2 \cdot CO_2H$$
$$\overset{|}{CO_2H}$$

This preparation must be carried out in an efficient fume cupboard.

Into a 500-ml round-bottomed flask, provided with a double surface condenser, place 50 g (63 ml, 0.86 mol) of pure, dry acetone, 50 g (47 ml, 0.44 mol) of ethyl cyanoacetate (Section III,171) and 0.5 g of piperidine. Allow to stand for 60 hours and heat on a water bath for 2 hours. Treat the cold reaction mixture with 100 ml of ether, wash with dilute hydrochloric acid, then with water, and dry over anhydrous sodium sulphate. Distil under diminished pressure and collect the ethyl isopropylidenecyanoacetate (ethyl 2-cyano-3,3-dimethylacrylate) at 114–116 °C/14 mmHg (1). The yield is 39 g (58%).

Dissolve 20 g (0.13 mol) of the cyano ester in 100 ml of rectified spirit and add a solution of 19.2 g (0.295 mol) of pure potassium cyanide in 40 ml of water. Allow to stand for 48 hours, then distil off the alcohol on a water bath. Add a large excess of concentrated hydrochloric acid and heat under reflux for 3 hours. [CARE: hydrogen cyanide evolved.] Dilute with water, saturate the solution with ammonium sulphate and extract with four 75 ml portions of ether. Dry the combined ethereal extracts with anhydrous sodium sulphate, and distil off the ether. Recrystallise the residual acid from excess concentrated hydrochloric acid, and dry in the air. The yield of pure 2,2-dimethylsuccinic acid, m.p. 141–142 °C, is 12 g (63%).

Note. (1) Higher (including cycloaliphatic) ketones may be condensed with ethyl cyanoacetate under the following conditions. Mix 0.50 mol of ethyl cyanoacetate, 0.55–0.70 mol of the ketone, 0.02 mol of piperidine and 50 ml of dry benzene, and heat under reflux for 12–24 hours in an apparatus incorporating an automatic water separator (Fig. III,2). Piperidine may be replaced by a catalyst composed of 7.7 g (0.1 mol) of ammonium acetate and 24 g (0.4 mol) of glacial acetic acid. Wash the cold reaction mixture with three 25-ml portions of 10 per cent sodium chloride solution, and remove the benzene on a water bath under reduced pressure. Transfer the residue to a 1-litre bottle containing a solution of 65 g of sodium metabisulphite in 250 ml of water and shake mechanically for 2–6 hours. Dilute the turbid solution, which contains the sodium metabisulphite addition compound, with 400 ml of water, and extract the ethyl cyanoacetate with three 50 ml portions of benzene. Cool the bisulphite solution in ice, and add dropwise, with mechanical stirring, an ice-cold solution of 28 g of sodium hydroxide in 110 ml of water. Extract the regenerated unsaturated ester at once with four 25 ml portions of benzene, wash the extracts with 50 ml of 1 per cent hydrochloric acid and dry with anhydrous sodium sulphate. Filter and distil through a fractionating column under reduced pressure; the benzene may be conveniently removed by distilling at atmospheric pressure until the temperature rises to 90 °C. Diethyl ketone yields ethyl 2-cyano-3,3-diethylacrylate, b.p. 123–125 °C/12 mmHg or 96–97 °C/3 mmHg; dipropyl ketone gives ethyl 2-cyano-3,3-dipropylacrylate, b.p. 136–137 °C/11 mmHg or 116–117 °C/4 mmHg. The yield is 60–70 per cent.

The appropriate succinic acid can be prepared by condensation of the unsaturated cyano ester with alcoholic potassium cyanide and subsequent treatment with hydrochloric acid.

III,137. 2-PROPYLGLUTARIC ACID

$$CH_3 \cdot CH_2 \cdot CH_2 \cdot CH(CO_2C_2H_5)_2 + CH_2 = CH \cdot CN \longrightarrow$$

$$CH_3 \cdot CH_2 \cdot CH_2 \cdot \underset{\underset{CH_2 \cdot CH_2 \cdot CN}{|}}{C}(CO_2C_2H_5)_2 \xrightarrow{H_3O^\oplus} CH_3 \cdot CH_2 \cdot CH_2 \cdot \underset{\underset{CH_2 \cdot CH_2 \cdot CO_2H}{|}}{C}H \cdot CO_2H$$

Add 8.0 g (10.0 ml, 0.15 mol) of redistilled acrylonitrile (Section III,165, Note (1)) to a stirred solution of diethyl propylmalonate (30.2 g, 0.15 mol) (Section III,132) and of 30 per cent methanolic potassium hydroxide (4.0 g) in t-butyl alcohol (100 g). Keep the reaction mixture at 30–35 °C during the addition and stir for a further 3 hours. Neutralise the solution with 2 M-hydrochloric acid, dilute with water and extract with ether. Dry the ethereal extract with anhydrous sodium sulphate and distil off the ether: the residue [diethyl (2-cyanoethyl)-propylmalonate; 11 g] solidifies on cooling in ice, and melts at 31–32 °C after recrystallisation from ice-cold ethanol. Boil the cyanoethyl ester (10 g) under reflux with 40 ml of 48 per cent hydrobromic acid solution for 8 hours, and evaporate the solution almost to dryness under reduced pressure. Add sufficient water to dissolve the ammonium bromide, extract several times with ether, dry the ethereal extract and distil off the solvent. The residual oil (4.5 g, 66%) soon solidifies: upon recrystallisation from water, pure 2-propylglutaric acid, m.p. 70 °C, is obtained.

N CARBOXYLIC ACID DERIVATIVES

1. Acyl halides (Section III,138).
2. Acid anhydrides (Sections III,139 to III,141).
3. Esters (Sections III,142 to III,155).
4. Acid amides (Sections III,156 to III,159).

N,1. Acyl halides. The conversion of a carboxylic acid into the corresponding acyl chloride is usually achieved by heating the acid with phosphorus trichloride, or phosphorus pentachloride, or thionyl chloride.

$$3R \cdot CO_2H + PCl_3 \longrightarrow 3R \cdot COCl + H_3PO_3$$

$$R \cdot CO_2H + PCl_5 \longrightarrow R \cdot COCl + HCl + POCl_3$$

$$R \cdot CO_2H + SOCl_2 \longrightarrow R \cdot COCl + HCl + SO_2$$

Alternatively the anhydrous sodium salt of the acid may be heated with phosphorus oxychloride, a method which gives very pure products and is used commercially.

$$2R \cdot CO_2Na + POCl_3 \longrightarrow 2R \cdot COCl + NaCl + NaPO_3$$

The use of phosphorus pentachloride is illustrated in the preparation of aromatic acid chlorides (IV,O,1); generally speaking, however, for aliphatic acid chlorides, thionyl chloride is the most convenient reagent provided that the product has a boiling point which permits separation from an excess of reagent by fractional distillation. If the boiling point of the acyl chloride is too near that of thionyl chloride to render separation by distillation practicable, the excess of the reagent can be destroyed by the addition of pure formic acid.

$$H\cdot CO_2H + SOCl_2 \longrightarrow CO + SO_2 + 2HCl$$

III,138. BUTYRYL CHLORIDE (*Butanoyl chloride*)

$$CH_3\cdot(CH_2)_2\cdot CO_2H + SOCl_2 \longrightarrow CH_3\cdot(CH_2)_2\cdot COCl + SO_2 + HCl$$

Fit a 100-ml two-necked flask with a dropping funnel and a reflux condenser connected at the top to a gas absorption trap (Fig. I,68). Place 36 g (21.5 ml, 0.3 mol) of redistilled thionyl chloride in the flask and 22 g (23 ml, 0.25 mol) of butyric acid in the separatory funnel. Heat the flask gently on a water bath, and add the butyric acid during the course of 30–40 minutes (1). When all the acid has been introduced, heat on a water bath for 30 minutes. Rearrange the apparatus and distil: collect the crude acid chloride boiling between 70 and 110 °C. Finally, redistil from a flask provided with a short fractionating column and collect the butyryl chloride at 100–101 °C. The yield is 23 g (86%).

Note. (1) Wrap a piece of absorbent cotton wool around the stem of the reflux condenser above the joint of the reaction flask to prevent condensed moisture seeping into the flask.

Cognate preparations   Hexanoyl chloride. Place 58 g (62 ml, 0.5 mol) of hexanoic acid in the flask, heat on a water bath and add 72 g (43 ml, 0.6 mol) of redistilled thionyl chloride during 45 minutes; shake the flask from time to time to ensure mixing. Reflux for 30 minutes and isolate the hexanoyl chloride by distillation, b.p. 150–155 °C. The yield is 56 g (83%).

Valeryl chloride (*pentanoyl chloride*). Use 51 g (0.5 mol) of valeric acid and 72 g (0.6 mol) of redistilled thionyl chloride. Proceed as for hexanoyl chloride; the yield of valeryl chloride is 42 g (70%), b.p. 124–127 °C.

Isobutyryl chloride (*2-methylpropanoyl chloride*). Use 140 g (1.6 mol) of isobutyric acid and 236 g (2 mol) of redistilled thionyl chloride. Proceed as for hexanoyl chloride; the yield is 121 g (71%), b.p. 90–93 °C, after distillation through a Vigreux column (36 cm).

Isovaleryl chloride (*3-methylbutanoyl chloride*). Use 34 g (0.4 mol) of isovaleric acid and 47 g (0.5 mol) of thionyl chloride. Proceed as for hexanoyl chloride; the yield of isovaleryl chloride is 36 g (76%), b.p. 114–115 °C, after distillation through a Vigreux column.

Cyclohexanecarbonyl chloride. Use 91 g (0.7 mol) of cyclohexanecarboxylic acid and 166 g (1.4 mol) of thionyl chloride. Proceed as for hexanoyl chloride but heat under reflux for 2 hours. The yield of cyclohexanecarbonyl chloride is 100 g (78%), b.p. 76–78 °C/12 mmHg, after distillation through a Vigreux column.

N,2. Acid anhydrides. Carboxylic acid anhydrides may be prepared by any of the following procedures.

(a) By the reaction of the acyl chloride with the corresponding sodium salt.

$$R\cdot COCl + R\cdot CO_2Na \longrightarrow R\cdot CO\cdot O\cdot CO\cdot R + NaCl$$

An equivalent result may be obtained by treating excess of sodium salt with phosphorus oxychloride; the acyl chloride is an intermediate product.

$$4R{\cdot}CO_2Na + POCl_3 \longrightarrow R{\cdot}CO{\cdot}O{\cdot}CO{\cdot}R + 3NaCl + NaPO_3$$

An excellent alternative involves the interaction in benzene solution of the acyl chloride with the carboxylic acid in the presence of pyridine.

$$R{\cdot}COCl + R{\cdot}CO_2H + C_5H_5N \longrightarrow R{\cdot}CO{\cdot}O{\cdot}CO{\cdot}R + C_5H_5\overset{\oplus}{N}H\}\overset{\ominus}{C}l$$

The presence of the pyridine removes, as the insoluble pyridinium hydrochloride, the hydrogen chloride liberated, otherwise anhydride formation may be incomplete as a result of the following equilibrium:

$$R{\cdot}CO{\cdot}O{\cdot}CO{\cdot}R + HCl \rightleftharpoons R{\cdot}CO_2H + R{\cdot}COCl$$

Mixed anhydrides (i.e., $R^1{\cdot}CO{\cdot}O{\cdot}CO{\cdot}R^2$) can also be readily prepared by this general route by choosing appropriate reactants.

(b) By the action of keten (Section I,17,3) on the carboxylic acid. In this reaction a mixed anhydride is the first formed product.

$$R{\cdot}CO_2H + CH_2{=}C{=}O \longrightarrow R{\cdot}CO{\cdot}O{\cdot}CO{\cdot}CH_3$$

Slow distillation at atmospheric pressure of the mixed anhydride with a second molar proportion of the carboxylic acid yields the symmetrical anhydride and acetic acid.

$$R{\cdot}CO{\cdot}O{\cdot}CO{\cdot}CH_3 + R{\cdot}CO_2H \longrightarrow R{\cdot}CO{\cdot}O{\cdot}CO{\cdot}R + CH_3{\cdot}CO_2H$$

(c) Cyclic anhydrides of dibasic acids such as succinic or glutaric acid are readily prepared by dehydrating the acid with an excess of acetic anhydride, e.g., as in the preparation of succinic anhydride, Section III,141; cf. 3-nitrophthalic anhydride, Section IV,176.

III,139. ACETIC ANHYDRIDE

$$CH_3{\cdot}COCl + CH_3{\cdot}CO_2^{\ominus}Na^{\oplus} \longrightarrow (CH_3{\cdot}CO)_2O + NaCl$$

Place 20 g (18 ml, 0.25 mol) of acetyl chloride in a dropping funnel supported over the neck of a 100-ml round-bottomed flask containing 25 g (0.3 mol) of finely powdered anhydrous sodium acetate (Section II,2,56). Cool the flask in ice-water and add about half of the acetyl chloride drop by drop. Mix the contents thoroughly by cautious shaking and tapping of the flask against the palm of the hand, and run in the rest of the acetyl chloride drop by drop; do not allow the mixture to get so hot that it boils. Fit a condenser set for downward distillation, and heat the flask with a luminous Bunsen flame which is kept in constant motion round the base of the flask to ensure uniform heating and minimise the danger of cracking the flask. Continue the heating until no more liquid passes over. Add 2–3 g of finely powdered anhydrous sodium acetate to the distillate in order to convert any unchanged acetyl chloride into acetic anhydride, attach a condenser, and distil slowly. Collect the fraction which passes over at 135–140 °C as acetic anhydride. The yield is 20 g (77%).

Cognate preparation   Heptanoic anhydride (1). In a 250-ml, round-bottomed three-necked flask, provided with a dropping funnel, stirrer and thermometer, place 15.8 g (16.1 ml, 0.2 mol) of dry pyridine (Section II,1,29) and 25 ml of dry benzene. Stir and add rapidly 14.8 g (15.5 ml, 0.1 mol) of heptanoyl chloride (2): the temperature rises slightly and a pyridinium complex separates. Introduce 13.0 g (14.1 ml, 0.1 mol) of heptanoic acid with stirring, over a period of 5 minutes; the temperature rises to 60–65 °C and pyridine hydrochloride is formed. Continue the stirring for 10 minutes and collect the hygroscopic pyridine hydrochloride as rapidly as possible on a chilled Buchner or sintered glass funnel, and wash it with two 25-ml portions of dry benzene. Remove the benzene from the filtrate under reduced pressure on a water bath, and distil the residue through a short fractionating column. Collect the heptanoic anhydride at 170–173 °C/15 mmHg; the yield is 20 g (83%).

Notes. (1) This is an example of the acid chloride–pyridine–acid method referred to in the theoretical section.

(2) Prepare heptanoyl chloride from the acid by treatment with thionyl chloride as detailed for butyryl chloride (Section **III,138**); b.p. 173–175 °C.

III,140. HEXANOIC ANYDRIDE

$$2CH_3 \cdot (CH_2)_4 \cdot CO_2H + CH_2 = C = O \longrightarrow [CH_3 \cdot (CH_2)_4 \cdot CO]_2O + CH_3 \cdot CO_2H$$

Place 116 g (126 ml, 1 mol) of dry hexanoic acid in a 250-ml Drechsel bottle and cool in ice. Pass in 21–23 g of keten (Section **I,17,3**) (1). Carefully distil the reaction mixture through a highly efficient fractionating column (e.g., a well-lagged Widmer column; see Section **I,26**) (2), using an oil bath for heating. A fraction of low boiling point, containing acetone, keten, acetic acid and a little acetic anhydride, is thus removed at atmospheric pressure. Raise the temperature of the bath to 220 °C over a period of 1 hour and maintain it at this temperature for 3 hours from the time distillation commences: this time is necessary to ensure that the conversion of the mixed anhydride to hexanoic anhydride and acetic acid is complete and that the acetic acid is completely removed. Discontinue the distillation, allow to cool somewhat and distil the residue in the flask under reduced pressure (3–10 mmHg). Discard the small fraction (20 g) of low boiling point and collect the hexanoic anhydride at 118–121 °C/6 mmHg (or 109–112 °C/3 mmHg). The yield is 90 g (84%).

Notes. (1) Excess of keten over the calculated quantity does not increase the yield; it leads to more acetic anhydride being collected in the low boiling point fraction.

(2) The best results are obtained with a fractionating column surrounded by an electrically heated jacket but this is not essential for hexanoic anhydride. For the preparation of propionic or butyric anhydride, a highly efficient fractionating column must be used in order to obtain satisfactory results.

III,141. SUCCINIC ANHYDRIDE

$$\begin{array}{c} CH_2 \cdot CO_2H \\ | \\ CH_2 \cdot CO_2H \end{array} + (CH_3 \cdot CO)_2O \longrightarrow \begin{array}{c} CH_2 \cdot C{\nwarrow}^{\nearrow O} \\ | \qquad \quad O \\ CH_2 \cdot C{\diagdown}_{O}^{\diagup} \end{array} + CH_3 \cdot CO_2H$$

In a 500-ml round-bottomed flask, provided with a reflux condenser protected by a calcium chloride drying tube, place 59 g (0.5 mol) of succinic acid and 102 g (94·5 ml, 1 mol) of redistilled acetic anhydride. Reflux the mixture gently on a water bath with occasional shaking until a clear solution is obtained (ca. 1 hour), and then for a further hour to ensure the completeness of the reaction. Remove the complete assembly from the water bath, allow it to cool (observe the formation of crystals) and finally cool in ice. Collect the succinic anhydride on a Buchner funnel or a sintered glass funnel, wash it with two 40 ml portions of anhydrous ether and dry in a vacuum desiccator. The yield is 45 g (90%), m.p. 119–120 °C.

N,3. Esters. (a) *Direct esterification procedures.* The interaction between a carboxylic acid and an alcohol is a reversible process and proceeds very slowly:

$$R^1 \cdot CH_2 \cdot CO_2H + R^2OH \rightleftharpoons R^1 \cdot CH_2 \cdot CO_2R^2 + H_2O$$

Equilibrium is only attained after refluxing for several days. If, however, about 3 per cent (of the weight of the alcohol) of either concentrated sulphuric acid or of dry hydrogen chloride is added to the mixture, the same point of equilibrium can be reached after a few hours. When equimolecular quantities of the acid and alcohol are employed, only about two-thirds of the theoretically possible yield of ester is obtained. According to the law of mass action, the equilibrium may be displaced in favour of the ester by the use of an excess of one of the components. It is frequently convenient to use an excess of the acid, but if the acid is expensive a large excess of the alcohol is generally employed. This method of esterification, in general, gives good yields with primary alcohols and fairly good yields with secondary alcohols. The method is unsatisfactory for use with tertiary alcohols.

Esterification with alicyclic alcohols proceeds best when the alcohol is saturated with hydrogen chloride and treated with an excess of the carboxylic acid (the **Fischer–Speier method**); a very impure ester results if sulphuric acid is used as the catalyst.

Examples of the use of these methods for the preparation of simple esters are collected in Sections **III,142** and **III,143**.

Esters of formic acid (Section **III,144**) are most simply prepared from the alcohol and an excess of formic acid, which, being a comparatively strong acid, does not require the use of added mineral acid to catalyse the esterification reaction. Sulphuric acid in any case should not be added since it causes the decomposition of formic acid to carbon monoxide.

The acid-catalysed esterification reaction usually proceeds via an acyl–oxygen fission process. This involves the cleavage of the bond between the original carbonyl carbon atom and an oxygen of an hydroxyl group in the intermediate (II) arising from nucleophilic attack by an alcohol molecule on the protonated carboxylic acid group (I).

Several modifications of the simple direct esterification procedure described above have been developed. For example, it is sometimes convenient to prepare an ester by heating the organic acid, the alcohol and sulphuric acid in a solvent such as benzene. Upon the addition of water, followed by separation and distillation of the benzene layer (after washing and drying), benzene and alcohol pass over first, followed by the ester. This method is illustrated by the preparation of propyl valerate and also by a range of ethyl esters of dicarboxylic acids for which the procedure is particularly well suited (Section **III,145**).

$$R^1-\overset{O}{\overset{\|}{C}}-OH + H^{\oplus} \rightleftharpoons R^1-\overset{\overset{\oplus}{O}H}{\underset{\smile}{\overset{\|}{C}}-\ddot{O}H} \longleftrightarrow R^1-\overset{OH}{\underset{}{\overset{|}{C}}=\overset{\oplus}{O}H}$$

$$(I)$$

$$R^1-\overset{\overset{\oplus}{O}H}{\underset{H-\ddot{O}-R^2}{\overset{\|}{C}}-OH} \rightleftharpoons R^1-\overset{OH}{\underset{H-\underset{\oplus}{O}-R^2}{\overset{|}{C}}-OH} \underset{-H^{\oplus},\ +H^{\oplus}}{\rightleftharpoons} R^1-\overset{H-\overset{\oplus}{O}-H}{\underset{O-R^2}{\overset{|}{C}}-\ddot{O}H} \rightleftharpoons R^1-\overset{H-O-H}{\underset{O-R^2}{\overset{|}{C}}=\overset{\oplus}{O}H}$$

$$(II)$$

$$R^1-\overset{\overset{\oplus}{O}H}{\underset{O-R^2}{\overset{|}{C}}}= \rightleftharpoons R^1-\overset{}{\underset{O-R^2}{\overset{|}{C}}}=O + H^{\oplus}$$

The process of acid-catalysed esterification in the presence of benzene, or, better, of toluene, is greatly facilitated if the water produced in the reaction is removed by distillation as an azeotrope. When the reaction mixture is slowly fractionated a ternary azeotrope of the alcohol, toluene and water will pass over first, followed by a binary toluene–alcohol azeotrope. Continued distillation affords the required ester in good yield. Alternatively the reaction mixture is subjected to reflux under a Dean and Stark water separation unit. This allows the separation and removal of water from the azeotrope, the organic phase being returned continuously to the reaction flask. Examples are provided in Sections **III,146** and **III,147** respectively. The latter includes examples of the synthesis of esters where either the alcohol or the carboxylic acid component is 'acid-sensitive'. The mineral acid catalyst is then replaced by a cation exchange resin (e.g., Zerolit 225/H$^{\oplus}$), enabling good yields of the required esters to be obtained.

A recent procedure for the preparation of methyl esters involves refluxing the carboxylic acid with methanol and 2,2-dimethoxypropane in the presence of toluene-p-sulphonic acid as the catalyst (Section **III,148**). The water produced in the esterification process is effectively removed by acid-catalysed reaction with the ketal to give acetone and methanol.

$$R{\cdot}CO_2H + CH_3OH \overset{H^{\oplus}}{\rightleftharpoons} R{\cdot}CO_2CH_3 + H_2O$$
$$CH_3{\cdot}C(OCH_3)_2{\cdot}CH_3 + H_2O \longrightarrow CH_3{\cdot}CO{\cdot}CH_3 + 2CH_3OH$$

Methyl esters are conveniently prepared on the small scale using diazomethane; a general procedure is given in Section **II,2,**19.

$$R{\cdot}CO_2H + CH_2N_2 \longrightarrow R{\cdot}CO_2CH_3 + N_2$$

(b) *Preparation of acid esters of dicarboxylic acids.* The acid-catalysed reaction of a dicarboxylic acid with an excess of alcohol yields the diester. However, the process may be adapted to prepare acid esters of dicarboxylic acids by using molar proportions of the diacid and alcohol (e.g., methyl hydrogen adipate, Section **III,149**).

$$HO_2C \cdot (CH_2)_n \cdot CO_2H + ROH \xrightarrow{H^{\oplus}} HO_2C \cdot (CH_2)_n \cdot CO_2R + H_2O$$

Alternatively the acid ester may be prepared by subjecting the diester to controlled partial hydrolysis with one molar proportion of potassium hydroxide.

$$RO_2C \cdot (CH_2)_n \cdot CO_2R + KOH \longrightarrow K^{\oplus} \overset{\ominus}{O}_2C \cdot (CH_2)_n \cdot CO_2R + HOR$$

The acid esters of 1,2-dicarboxylic acids are conveniently prepared by heating the corresponding cyclic anhydride with one molar proportion of the alcohol.

$$
\begin{array}{c}
H_2C - C \\
\quad\quad\quad\backslash \\
\quad\quad\quad O + ROH \\
H_2C - C \\
\end{array}
\longrightarrow
\begin{array}{c}
CH_2 \cdot CO_2H \\
| \\
CH_2 \cdot CO_2R
\end{array}
$$

The preparation of alkyl hydrogen phthalates is described in Section **III,197** where the use of these derivatives in the resolution of racemic alcohols is elaborated.

Acid esters are useful synthetic intermediates. For example, their use in the synthesis of long-chain dicarboxylic esters by electrolytic (anodic) synthesis has already been noted (Section **III,131**). Furthermore the reaction of the acid ester with thionyl chloride in the usual way will convert the carboxylic acid grouping to an acyl chloride group thus yielding the synthetically useful ester–acyl chloride; the products are usually purified by distillation under reduced pressure.

$$HO_2C \cdot (CH_2)_n \cdot CO_2R + SOCl_2 \longrightarrow ClCO \cdot (CH_2)_n \cdot CO_2R + SO_2 + HCl$$

One of the uses of these ester–acyl chlorides is for the synthesis of ω-hydroxy esters which involves the selective reduction of the acyl chloride grouping with sodium borohydride (Ref. 11); the alkoxycarbonyl group is unaffected by this metal hydride reducing agent (cf. p. 353, **III,D,1**).

(c) *The use of acyl chlorides and acid anhydrides.* Acyl chlorides react readily with primary and secondary alcohols to give esters in very good yields. With tertiary alcohols the presence of a base (e.g., dimethylaniline) is essential to prevent acid-catalysed side reactions, such as dehydration or formation of the alkyl chloride.

$$R^1 \cdot COCl + R^2 \cdot CH_2OH \longrightarrow R^1 \cdot CO_2CH_2 \cdot R^2 + R^1 \cdot CO_2H$$

Acylation may also be carried out with acid anhydrides in the presence of a suitable catalyst; either an acidic catalyst, such as sulphuric acid or zinc chloride, or a basic catalyst such as pyridine, may be used.

$$(R^1 \cdot CO)_2O + R^2 \cdot CH_2OH \longrightarrow R^1 \cdot CO_2CH_2 \cdot R^2 + R^1 \cdot CO_2H$$

Examples of the use of these methods are given in Sections **III,150** and **III,151**. The use of an acyl chloride or acid anhydride is the method of choice for the synthesis of phenyl esters (e.g., phenyl cinnamate; see also Sections **IV,180** and **VII,6,6**), which cannot be prepared by the direct esterification methods described above.

The synthesis of ethyl 2-bromopropionate (Section **III,152**) illustrates the preparation of an acyl chloride and its ready bromination in the α-position in the presence of red phosphorus. The resulting bromoacyl chloride is converted into the α-bromoester on reaction with an alcohol.

$$R^1 \cdot CH_2 \cdot CO_2 H \xrightarrow{SOCl_2} R^1 \cdot CH_2 \cdot COCl \xrightarrow{P/Br_2} R^1 \cdot CHBr \cdot COCl \xrightarrow{R^2OH}$$

$$R^1 \cdot CHBr \cdot CO_2 R^2$$

(d) *The alcoholysis of nitriles.*   An ester is formed when a nitrile is heated with an alcohol in the presence of concentrated sulphuric acid, thus providing a two-step synthesis of an ester from an alkyl halide. Examples are to be found in Section **III,153**. The reaction proceeds by way of an intermediate imino-ester (I) which is not usually isolated.

$$R^1 - C \equiv N + R^2OH \xrightarrow{H^\oplus} \left[\begin{array}{c} R^1 - C = \overset{\oplus}{N}H_2 \\ | \\ OR^2 \end{array} \right] \xrightarrow{H_2O} R^1 - \underset{|\ OR^2}{C} = O + \overset{\oplus}{N}H_4$$

(I)

The method has been widely used for the preparation of polyfunctional compounds, a simple but important example being the preparation of diethyl malonate (Section **III,154**). The first step in the synthesis involves the preparation of the salt of cyanoacetic acid by heating sodium chloroacetate with potassium cyanide and then warming the crude product with ethanol in the presence of concentrated sulphuric acid. The synthetic uses of diethyl malonate are exemplified on p. 488, **III,M,6**.

(e) *Ortho-esters.*   These have the general formula $R^1 \cdot C(OR^2)_3$ and are stable derivatives of the unstable ortho acids $R \cdot C(OH)_3$. Important examples are the esters of orthoformic acid (the orthoformates), which may be readily prepared by the interaction of the appropriate sodium alkoxide with chloroform (Section **III,155**).

$$CHCl_3 + 3RO^\ominus Na^\oplus \longrightarrow CH(OR)_3 + 3NaCl$$

III,142. BUTYL ACETATE

$$CH_3 \cdot CO_2 H + CH_3 \cdot (CH_2)_2 \cdot CH_2 OH \xrightarrow{H^\oplus} CH_3 \cdot CO_2 \cdot (CH_2)_3 \cdot CH_3$$

Mix together 37 g (46 ml, 0.5 mol) of butan-1-ol and 60 g (60 ml, 1 mol) of glacial acetic acid in a 250- or 500-ml round-bottomed flask, and add cautiously 1 ml of concentrated sulphuric acid (use a small measuring cylinder or a calibrated dropper pipette). Attach a reflux condenser and reflux the mixture for 3–6 hours (1). Pour the mixture into about 250 ml of water in a separatory funnel, remove the upper layer of crude ester and wash it again with about 100 ml of water, followed by about 25 ml of saturated sodium hydrogen carbonate solution and 50 ml of water. Dry the crude ester with 5–6 g of anhydrous sodium sulphate. Filter through a small funnel containing a fluted filter-paper and distil on a wire gauze or from an air bath. Collect the pure butyl acetate at 124–125 °C. The yield is 40 g (69%).

Note. (1) A slightly better yield of ester can be obtained by increasing the quantity of acetic acid to 90–120 g and refluxing for 12–18 hours.

Cognate preparations   Propyl acetate.   Use 40 g (50 ml, 0.67 mol) of propan-1-ol, 160 g of glacial acetic acid and 2 g of concentrated sulphuric acid. Reflux for 12 hours. Add an equal volume of water, saturate with sodium chloride and isolate the crude ester. Treat the crude ester with saturated sodium hydrogen carbonate solution until effervescence ceases, saturate with salt, remove the ester and dry it with anhydrous sodium sulphate. B.p. 101–102 °C. Yield: 36 g (53%).

Isopropyl acetate.   Use 40 g (51 ml, 0.67 mol) of propan-2-ol, 160 g of glacial acetic acid and 2 g of concentrated sulphuric acid. Reflux for 18 hours. Proceed as for propyl acetate. B.p. 87–88 °C. Yield: 31 g (46%).

Pentyl acetate.   Use 44 g (54 ml, 0.5 mol) of pentan-1-ol, 120 g of glacial acetic acid and 2.5 g of concentrated sulphuric acid, and reflux for 20 hours. Isolate the ester as for butyl acetate. B.p. 146–148 °C. Yield: 50 g (77%).

Methyl acetate.   Use 48 g (61 ml, 1.5 mol) of absolute methanol, 270 g of glacial acetic acid and 3 g of concentrated sulphuric acid. Reflux for 5 hours. Distil the reaction mixture through a simple fractionating column (e.g., a Hempel column filled with $\frac{1}{4}$-in. glass or porcelain rings, or an all-glass Dufton column); the crude ester passes over at 55–56 °C (112 g) and the excess of acid, etc., remaining in the flask weighs 209 g. Wash once with a *little* water, saturate with salt, wash with saturated sodium hydrogen carbonate solution, saturate with salt, remove the ester layer and dry with anhydrous sodium sulphate, and distil. The methyl acetate passes over at 55–56 °C. The yield is 92 g (83%).

Ethyl acetate.   Use 58 g (73.5 ml, 1.25 mol) of absolute ethanol, 225 g of glacial acetic acid and 3 g of concentrated sulphuric acid. Reflux for 6–12 hours. Work up as for propyl acetate. B.p. 76–77 °C. Yield: 32 g (29%). Much ethyl acetate is lost in the washing process. A better yield may be obtained, and most of the excess of acetic acid may be recovered, by distilling the reaction mixture through an efficient fractionating column and proceeding as for methyl acetate.

Ethyl butyrate.   Use a mixture of 88 g (92 ml, 1 mol) of butyric acid, 23 g (29 ml, 0.5 mol) of ethanol and 9 g (5 ml) of concentrated sulphuric acid. Reflux for 14 hours. Pour into excess of water, wash several times with water, followed by saturated sodium hydrogen carbonate solution until all the acid is removed, and finally with water. Dry with anhydrous sodium sulphate, and distil. The ethyl butyrate passes over at 119.5–120.5 °C. Yield: 40 g (69%). An improved yield can be obtained by distilling the reaction mixture through an efficient fractionating column until the temperature rises to 125 °C, and purifying the crude ester as detailed above under methyl acetate.

Diethyl sebacate.   Reflux a mixture of 101 g (0.5 mol) of sebacic acid, 196 g (248 ml, 4.25 mol) of absolute ethanol and 20 ml of concentrated sulphuric acid for 12 hours. Distil off about half of the alcohol on a water bath, dilute the residue with 500–750 ml of water, remove the upper layer of crude ester and extract the aqueous layer with ether. Wash the combined ethereal extract and crude ester with water, then with saturated sodium hydrogen carbonate solution until effervescence ceases, and finally with water. Dry with magnesium sulphate or anhydrous sodium sulphate, remove the ether on a water bath and distil the residue under reduced pressure. B.p. 155–157 °C/6 mmHg. Yield: 110 g (85%).

Methyl crotonate. Use 43 g (0.5 mol) of redistilled crotonic acid (b.p. 180–182 °C, m.p. 72–73 °C), 75 g (95 ml, 2.33 mol) of absolute methanol, 3 ml of concentrated sulphuric acid and reflux for 12 hours. Isolate as for butyl acetate; the yield is 34 g (68%), b.p. 118–120 °C.

III,143. CYCLOHEXYL ACETATE

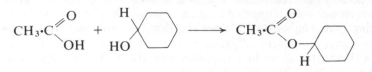

Pass dry hydrogen chloride (Section **II,2,31**) into 75 g (0.75 mol) of pure cyclohexanol until 1.5 g are absorbed, mix with 135 g (2.25 mol) of glacial acetic acid in a 500-ml round-bottomed flask, attach a reflux condenser and reflux for 14 hours. Pour into excess of water, wash the upper layer successively with water, saturated sodium hydrogen carbonate solution until effervescence ceases, and water. Dry with anhydrous calcium chloride. Distil through a well-lagged fractionating column (e.g., an all-glass Dufton column). A small fraction of low boiling point (containing cyclohexene) passes over first, followed by cyclohexyl acetate (57 g, 54%) at 168–170 °C. Upon redistillation, the boiling point is 170–172 °C, mainly 171–172 °C.

Cognate preparations **Cyclohexyl formate.** Use 103 g (84.5 ml, 2.24 mol) of formic acid (98/100%) and 75 g (0.75 mol) of cyclohexanol in which 1.5 g of dry hydrogen chloride gas are dissolved. Reflux for 14 hours. Work up as above and distil through a well-lagged column; 5.5 g of cyclohexene and 57 g (59%) of cyclohexyl formate, b.p. 156–158.5 °C (mainly 157–158.5 °C), are obtained. On redistillation the sample boils at 158–160 °C (mainly 159–160 °C).

s-Butyl acetate. Pass dry hydrogen chloride gas into 37 g (46 ml, 0.5 mol) of butan-2-ol until 1.5 g is absorbed. Mix the solution with 60 g (1 mol) of glacial acetic acid, and reflux for 10 hours. Isolate the ester as for butyl acetate (Section **III,142**). B.p. 110–112 °C. Yield: 35 g (60%).

Ethyl p-aminobenzoate. Saturate 80 ml (63.2 g, 1.37 mol) of absolute ethanol with dry hydrogen chloride, add 12 g (0.088 mol) of p-aminobenzoic acid and heat the mixture under reflux for 2 hours. Upon cooling, the reaction mixture sets to a solid mass of the hydrochloride of ethyl p-aminobenzoate. It is better, however, to pour the hot solution into excess of water (no hydrochloride separates) and add sodium carbonate to the clear solution until it is neutral to litmus. Filter off the precipitated ester at the pump and dry in the air. The yield of ethyl p-aminobenzoate, m.p. 91 °C, is 10 g (69%). Recrystallisation from rectified spirit does not affect the m.p.

III,144. BUTYL FORMATE

$$H\cdot CO_2H + CH_3\cdot(CH_2)_2\cdot CH_2OH \longrightarrow H\cdot CO_2(CH_2)_3\cdot CH_3 + H_2O$$

Into a 250- or 500-ml round-bottomed flask provided with a reflux condenser place 46 g (38 ml, 1 mol) of formic acid (98/100%) and 37 g (46 ml,

0.5 mol) of butan-1-ol. Reflux for 24 hours. Wash the cold mixture with small volumes of saturated sodium chloride solution, then with saturated sodium hydrogen carbonate solution in the presence of a little solid sodium hydrogen carbonate until effervescence ceases, and finally with saturated sodium chloride solution. Dry with anhydrous sodium sulphate, and distil through a short fractionating column. Collect the butyl formate at 106–107 °C. Yield: 38 g (74%).

Cognate preparations **Ethyl formate.** Reflux a mixture of 61 g (50 ml, 1.33 mol) of formic acid (98/100%) and 31 g (39.5 ml, 0.67 mol) of absolute ethanol for 24 hours. Attach a fractionating column to the flask, distil and collect the liquid passing over below 62 °C. Wash the distillate with saturated sodium hydrogen carbonate solution and saturate with salt before removing the ester layer. Dry with anhydrous sodium sulphate, filter and distil. The ethyl formate passes over at 53–54 °C. The yield is 36 g (72%).

Propyl formate. Use 46 g (38 ml, 1 mol) of formic acid (98/100%) and 30 g (37.5 ml, 0.5 mol) of propan-1-ol and reflux for 24 hours. Proceed as for ethyl formate, but collect the crude propyl formate up to 86 °C. B.p. 80.5–82 °C. Yield: 28 g (65%).

III,145. PROPYL VALERATE

$$CH_3 \cdot (CH_2)_3 \cdot CO_2H + CH_3 \cdot CH_2 \cdot CH_2OH \xrightarrow[\text{benzene}]{H^\oplus}$$

$$CH_3 \cdot (CH_2)_3 \cdot CO_2(CH_2)_2 \cdot CH_3 + H_2O$$

Place a mixture of 25.5 g (0.25 mol) of valeric acid (Sections **III,128** and **III,129**), 30 g (37.5 ml, 0.5 mol) of dry propan-1-ol, 50 ml of sodium-dried benzene and 10 g (5.5 ml) of concentrated sulphuric acid in a 250-ml round-bottomed flask equipped with a reflux condenser, and boil under reflux for 36 hours. Pour into 250 ml of water and separate the upper layer. Extract the aqueous layer with ether, and add the extract to the benzene solution. Wash the combined extracts with saturated sodium hydrogen carbonate solution until effervescence ceases, then with water, and dry with anhydrous sodium sulphate. Remove the low boiling point solvents by flash distillation and distil the residue; the temperature will rise abruptly and the propyl valerate will pass over at 163–164 °C. The yield is 28 g (78%).

Cognate preparations **Diethyl adipate.** Place 100 g (0.685 mol) of adipic acid in a 750-ml round-bottomed flask and add successively 100 g (127 ml, 2.18 mol) of absolute ethanol, 250 ml of sodium-dried benzene and 40 g (22 ml) of concentrated sulphuric acid (the last-named cautiously and with gentle swirling of the contents of the flask). Attach a reflux condenser and reflux the mixture gently for 5–6 hours. Pour the reaction mixture into excess of water (2–3 volumes), and isolate the product as above; distil under reduced pressure and collect the diethyl adipate at 134–135 °C/17 mmHg. The yield is 130 g (94%).

Diethyl oxalate. Reflux a mixture of 45 g (0.5 mol) of anhydrous oxalic acid (1), 81 g (102.5 ml, 1.76 mol) of absolute ethanol, 190 ml of sodium-dried benzene and 30 g (16.5 ml) of concentrated sulphuric acid for 24 hours. Work up as above and extract the aqueous layer with ether; distil under atmospheric pressure. The yield of diethyl oxalate, b.p. 182–183 °C, is 57 g (78%).

Note. (1) Anhydrous oxalic acid may be prepared by heating the finely powdered 'AnalaR' crystallised acid in carbon tetrachloride (fume cupboard) in a flask equipped with a Dean and Stark tube and reflux condenser. When no more water is collected, evaporate the carbon tetrachloride, powder the crystalline residue of anhydrous oxalic acid (*toxic*) and store in a tightly stoppered bottle.

Diethyl succinate.   Reflux a mixture of 58 g (0.5 mol) of succinic acid, 81 g (102.5 ml, 1.76 mol) of absolute ethanol, 190 ml of sodium-dried benzene and 20 g (11 ml) of concentrated sulphuric acid for 8 hours. Pour the reaction mixture into excess of water, separate the benzene–ester layer and extract the aqueous layer with ether. Work up the combined ether and benzene extracts as described above. B.p. 81 °C/3 mmHg. Yield: 75 g (86%). The boiling point under atmospheric pressure is 217–218 °C.

Diethyl fumarate.   Reflux a mixture of 145 g (1.25 mol) of fumaric acid (Section **III,213**), 185 g (236 ml, 4 mol) of absolute ethanol, 450 ml of benzene and 20 g of concentrated sulphuric acid for 12 hours. Pour into a large volume of water, separate the benzene layer, wash successively with water, saturated sodium hydrogen carbonate solution and water, dry with anhydrous sodium sulphate and remove the solvent on a steam bath. Distil the residue and collect the diethyl fumarate at 213–215 °C: the yield is 150 g (70%).

III,146. DIETHYL ADIPATE

$$HO_2C \cdot (CH_2)_4 \cdot CO_2H + 2C_2H_5OH \xrightarrow[\text{toluene}]{H^{\oplus}}$$

$$C_2H_5O_2C \cdot (CH_2)_4 \cdot CO_2C_2H_5 + 2H_2O$$

Place 146 g (1 mol) of adipic acid, 360 ml (285 g, 6.2 mol) of absolute ethanol, 180 ml of toluene and 1.5 g of concentrated sulphuric acid in a 1-litre round-bottomed flask, attach a *short* fractionating column connected to a downward condenser and heat in an oil bath at 115 °C. When the acid has dissolved, an azeotropic mixture of alcohol, toluene and water commences to distil at 75 °C; the temperature of the oil bath may then be lowered to 100–110 °C. Collect the distillate in a flask containing 150 g of anhydrous potassium carbonate. Continue the distillation until the temperature at the top of the column rises to 78 °C. Shake the distillate thoroughly with the potassium carbonate, filter through a Buchner funnel or fluted filter-paper and return the filtrate to the flask. Heat the flask again until the temperature rises to 78–80 °C (1). Transfer the warm residue to a flask of suitable size and distil under reduced pressure. Alcohol and toluene pass over first, the temperature rises abruptly and the diethyl adipate distils at 138 °C/20 mmHg (2). The yield is 195 g (96%).

Notes. (1) The distillate contains ethanol, toluene and water, and may be dried with anhydrous potassium carbonate and used again for esterification after the addition of the necessary quantity of alcohol.

(2) The b.p. may rise several degrees towards the end of the distillation owing to superheating.

Cognate preparation   Ethyl bromoacetate (1).   Fit a large modified Dean and Stark apparatus provided with a stop-cock at the lower end (cf. Fig. III,2) to a

1-litre flask containing 220 g (1.58 mol) of crude bromoacetic acid (Section III,168) and attach a double surface condenser to the upper end. Mix the acid with 155 ml (2.66 mol) of absolute ethanol, 240 ml of toluene and 1 ml of concentrated sulphuric acid. Heat the flask under gentle reflux; water, toluene and ethanol will collect in the special apparatus and separate into two layers. Run off the lower layer (ca. 75 ml), which includes all the water formed in the reaction together with excess of ethanol. When no more water separates, the reaction may be regarded as complete; add 20 ml of absolute ethanol to the reaction mixture and continue refluxing for a further 30 minutes. Run off the toluene which has collected in the trap. Transfer the reaction mixture to a separatory funnel, and wash it successively with 400 ml of water, 400 ml of 1 per cent sodium hydrogen carbonate solution and 400 ml of water. Dry over anhydrous sodium sulphate and distil through a short, well-lagged fractionating column (e.g., an all-glass Dufton column). Collect the ethyl bromoacetate at 154–155 °C. The yield is 205 g (78%).

Note. (1) Ethyl bromoacetate vapour is extremely irritating to the eyes. The preparation must therefore be conducted in a fume cupboard provided with a good draught: the material should be kept in closed vessels as far as possible.

III,147. ISOPROPYL LACTATE

$$CH_3 \cdot CH(OH) \cdot CO_2 H + (CH_3)_2 CHOH \xrightarrow[\text{benzene}]{H^{\oplus} \text{(resin)}}$$

$$CH_3 \cdot CH(OH) \cdot CO_2 CH(CH_3)_2 + H_2O$$

Place a mixture of 53 g (0.5 mol) of 'AnalaR' lactic acid (85–88% acid), 75 g (95.5 ml, 1.25 mol) of commercial anhydrous propan-2-ol (isopropyl alcohol), 300 ml of benzene and 20 g of Zerolit 225 (acid form) (1) in a 1-litre flask, equipped with an automatic water separator (e.g., a large modified Dean and Stark apparatus with a stop-cock at the lower end, see Fig. III,2) carrying an efficient reflux condenser at its upper end. Reflux the mixture using a magnetic stirrer/hotplate unit for 5 hours or until water no longer collects in appreciable amount in the water separator; run off the water from time to time. Filter off the resin at the pump and wash it with two 25 ml portions of benzene. Shake the combined filtrate and washings with about 5 g of precipitated calcium carbonate, filter, and wash with a little benzene. Distil the benzene solution under reduced pressure (water pump) through a short fractionating column; the isopropyl alcohol–benzene azeotrope (2) passes over first, followed by benzene. Collect the isopropyl lactate at 76 °C/24 mmHg; it is a colourless liquid and weighs 40 g (61%). The ester boils, with slight decomposition, at 157 °C/771 mmHg.

Notes. (1) This resin is available as the sodium form. It may be converted into the hydrogen form by treating it with about twice its volume of 1 M-sulphuric acid and stirring frequently: the resin is thoroughly washed by decantation with distilled water until the washings have a pH of 6–7, filtered and dried in the air.

(2) The b.p. of the propanol–benzene azeotrope at atmospheric pressure is 71–72 °C.

Cognate preparations   Butyl oleate.   Proceed as for isopropyl lactate using 28 g (0.1 mol) of redistilled oleic acid, 37.0 g (46 ml, 0.5 mol) of butan-1-ol (the excess of the latter acts as the water carrier) and 8.0 g of Zerolit 225/H$^\oplus$in a 250-ml flask. Reflux the mixture with magnetic stirring for 4 hours, allow to cool, separate the resin by suction filtration and wash it with three 5 ml portions of butan-1-ol. Remove the butanol from the combined filtrate and washings by distillation under reduced pressure (water pump); the residue consists of crude ester. Distil the residue under diminished pressure (oil pump) and collect the butyl oleate at 232 °C/9 mmHg. The yield is 27 g (85%).

Furfuryl acetate.   Reflux a mixture of 39.2 g (34.8 ml, 0.4 mol) of redistilled furfuryl alcohol, 48 g (0.67 mol) of glacial acetic acid, 150 ml of benzene and 20 g of Zerolit 225/H$^\oplus$in a 500-ml flask, using the apparatus described for isopropyl lactate. After 3 hours, when the rate of collection of water in the water separator is extremely slow, allow to cool, separate the resin by suction filtration and wash it with three 15 ml portions of benzene. Remove the benzene, etc., from the combined filtrate and washings under reduced pressure (water pump) and then collect the crude ester at 74–90 °C/10 mmHg; a small solid residue remains in the flask. Redistil the crude ester through a short fractionating column; pure furfuryl acetate passes over at 79–80 °C/17 mmHg. The yield is 14.5 g (26%).

III,148. DIMETHYL ADIPATE

$$HO_2C\cdot(CH_2)_4\cdot CO_2H + 2CH_3OH \xrightarrow{H^\oplus} CH_3O_2C\cdot(CH_2)_4\cdot CO_2CH_3 + 2H_2O$$

$$CH_3\cdot C(OCH_3)_2\cdot CH_3 + H_2O \longrightarrow CH_3\cdot CO\cdot CH_3 + 2CH_3OH$$

In a 500-ml single-necked flask containing a magnetic stirrer bar, place 58.5 g (0.4 mol) of adipic acid, 16 g (20 ml, 0.5 mol) of methanol, 83.2 g (0.8 mol) of 2,2-dimethoxypropane and 0.5 g of toluene-p-sulphonic acid. Fit a reflux condenser to the flask and stir the mixture magnetically for 4 hours in a water bath kept at 45 °C. Rearrange the condenser for distillation and distil off acetone (b.p. 56 °C) and methanol (b.p. 64 °C) on the water bath. Distil the residue under reduced pressure (water pump) and collect the dimethyl adipate, b.p. 130 °C/25 mmHg. The yield is 54.9 g (79%).

III,149. METHYL HYDROGEN ADIPATE

$$HO_2C\cdot(CH_2)_4\cdot CO_2H + CH_3OH \xrightarrow{H^\oplus} CH_3O_2C\cdot(CH_2)_4\cdot CO_2H + H_2O$$

Place 175 g (1.2 mol) of adipic acid, 50 ml (1.25 mol) of absolute methanol, 15 ml of concentrated hydrochloric acid and a few fragments of porous pot ('boiling chips') in a 500-ml round-bottomed flask provided with a reflux condenser (1). Heat cautiously at first until the mixture becomes homogeneous and then reflux for 8 hours. Transfer the mixture to a flask fitted with a fractionating column filled with glass helices and arrange to heat the column with a heating tape, the current to which is controlled by a Variac transformer. Careful fractionation under reduced pressure yields dimethyl adipate, b.p. 113–114 °C/6 mmHg (21 g), and methyl hydrogen adipate, b.p. 154–156 °C/6 mmHg (66 g, 34%). Unchanged adipic acid remains in the flask.

Cognate preparation   Methyl hydrogen sebacate.   Place 115 g (0.56 mol) of sebacic acid, 20 ml (0.5 mol) of absolute methanol, 6 ml of concentrated hydrochloric acid and a few fragments of porous pot in a 500-ml round-bottomed flask fitted with a reflux condenser. Warm the mixture on a water bath until it becomes homogeneous and then reflux gently for 8 hours. Transfer the mixture to a flask fitted with a fractionating column as for methyl hydrogen adipate and fractionate under reduced pressure; due precautions must be taken so that the distillate does not solidify in the condenser or receiver. Collect the dimethyl sebacate at 153–154 °C/6 mmHg (20 g, m.p. 26 °C) and the methyl hydrogen sebacate at 185–186 °C/6 mmHg (46 g, 43%, m.p. 37 °C). The residue in the flask consists of unchanged sebacic acid.

Note.   (1) The acid ester may also be prepared by either of two alternative procedures: (*a*) 1 mol of the diester is heated with 1 mol of the diacid for several hours (Ref. 12); or, (*b*) 1 mol of the diester is dissolved in 3 to 4 volumes of ethanol to which is added a solution of 1 mol of potassium hydroxide dissolved in the minimum amount of ethanol, and the solution allowed to stand at room temperature overnight. The ethanol is removed on a rotary evaporator, water is added to the residue and the solution extracted with ether to remove unreacted diester. The residual aqueous solution is then cautiously acidified at 0 °C, the acid ester extracted with ether and the ether extract washed, dried and evaporated. The residue is fractionally distilled under reduced pressure.

III,150. t-BUTYL ACETATE

$$(CH_3)_3COH \xrightarrow[\text{or } (CH_3 \cdot CO)_2O]{CH_3 \cdot COCl} CH_3 . CO_2 C(CH_3)_3$$

Method A.   Fit a 1-litre three-necked flask with a sealed stirrer, a reflux condenser and a dropping funnel. Place 57 g (73.5 ml, 0.77 mol) of dry 2-methylpropan-2-ol (t-butyl alcohol) (1), 101 g (106 ml, 0.84 mol) of pure dimethylaniline and 100 ml of anhydrous ether in the flask, set the stirrer in motion and heat the mixture to gentle refluxing on a water bath. Run in 62 g (56.5 ml, 0.79 mol) of redistilled acetyl chloride at such a rate that moderate refluxing continues after the source of heat is removed. When about two-thirds of the acetyl chloride has been introduced, the dimethylaniline hydrochloride commences to crystallise and the mixture refluxes very vigorously. Cool immediately in an ice bath, and, after refluxing ceases, add the remainder of the acetyl chloride; then heat the mixture on a water bath for 1 hour. Cool to room temperature, add about 100 ml of water and continue the stirring until all the precipitated solid has dissolved. Separate the ether layer and extract with 25-ml portions of cold 10 per cent sulphuric acid until the acid extract does not become cloudy when rendered alkaline with sodium hydroxide solution. Finally, wash with 15 ml of saturated sodium hydrogen carbonate solution and dry the ethereal solution with 5 g of anhydrous sodium sulphate overnight. Remove the ether by distillation through an efficient fractionating column and distil the residue through the same column. Collect the t-butyl acetate at 96–98 °C (mainly 97–98 °C). The yield is 55 g (62%).

Note.   (1) The t-butyl alcohol should be dried over calcium oxide or anhydrous calcium sulphate and distilled.

Cognate Preparation   t-Butyl propionate.   Use 85.5 g (110.5 ml, 1.15 mol) of t-butyl alcohol, 151.5 g (159 ml, 1.26 mol) of pure dimethylaniline and 110 g (103 ml, 1.19 mol) of propionyl chloride (compare Section **III,138**) and reflux for 3 hours. B.p. 117.5–118.5 °C. Yield: 92 g (62%).

Method B.   Fit a 500-ml round-bottomed flask with a reflux condenser carrying a calcium chloride guard-tube. Place 100 ml (108 g, 1.06 mol) of re-distilled acetic anhydride, 100 ml (1.07 mol) of dry t-butyl alcohol (see *Note* in *Method A*) and 0.3 g of anhydrous zinc chloride in the flask and shake. Heat the mixture gradually to the reflux temperature, maintain at gentle refluxing for 2 hours and then cool. Replace the reflux condenser by an efficient fractionating column and distil until the temperature reaches 110 °C. Wash the crude distillate, weighing 100–125 g, with two 25 ml portions of water, then with 25 ml portions of 10 per cent potassium carbonate solution until the ester layer is neutral to litmus, and finally dry with 10 g of anhydrous potassium carbonate. Filter off the desiccant, and distil through an efficient fractionating column (e.g., Widmer column) and collect the pure t-butyl acetate at 96–98 °C. The yield is 70 g (57%).

III,151. ETHYL VINYLACETATE

$$CH_2 = CH \cdot CH_2 \cdot CO_2H \xrightarrow{SOCl_2} CH_2 = CH \cdot CH_2 \cdot COCl \xrightarrow{C_2H_5OH}$$
$$CH_2 = CH \cdot CH_2 \cdot CO_2C_2H_5$$

Prepare vinylacetyl chloride from 50 g (31 ml, 0.42 mol) of thionyl chloride and 30 g (0.35 mol) of vinylacetic acid (Section **III,128,B**) following the procedure described for butyryl chloride (Section **III,138**); 27 g (0.26 mol) of the acid chloride, b.p. 98–99 °C, are obtained. Place 12.6 g (16.0 ml, 0.27 mol) of absolute ethanol in a 250-ml two-necked flask provided with a reflux condenser and dropping funnel. Cool the flask in ice and introduce the vinylacetyl chloride into the dropping funnel; insert a calcium chloride guard-tube into the mouth of the funnel. Add the acid chloride dropwise (45 minutes) to the alcohol with frequent shaking. Remove the ice and allow to stand for 1 hour. Pour the reaction mixture into water, wash with a little sodium hydrogen carbonate solution, then with water, and dry with anhydrous calcium sulphate. Distil from a 50-ml flask through a short fractionating column, and collect the ethyl vinylacetate at 125–127 °C. The yield is 22 g (75%).

Cognate preparation   Phenyl cinnamate.   Place 72 g (0.48 mol) of cinnamic acid (Section **IV,150**) and 60 g (37 ml, 0.5 mol) of thionyl chloride in a 250-ml flask, fitted with a reflux condenser which is connected to a gas absorption trap. Heat the mixture on a water bath, cautiously at first, until hydrogen chloride ceases to be evolved (about 1 hour), allow to cool and add 47 g (0.5 mol) of pure phenol. Heat the mixture on a water bath until no further evolution of hydrogen chloride is observed (about 1 hour). Then place the apparatus on an asbestos-centred wire gauze and heat the flask until the contents are brought just to the reflux temperature in order to complete the reaction: do not heat unduly long as prolonged heating leads to loss of product due to decomposition and polymerisation. Allow the reaction mixture to cool and distil under diminished pressure; collect the fraction of b.p. 190–210 °C/15 mmHg. This solidifies to a pale yellow solid, m.p. 66–69 °C, weighing 98 g. Grind it to a powder in a glass mortar and wash the powder with 250 ml of cold 2 per cent sodium hydrogen carbonate solution. Recrystallise from rectified spirit (150 ml): 81 g (72%) of pure phenyl cinnamate (white crystals) of m.p. 75–76 °C are obtained.

III,152. ETHYL 2-BROMOPROPIONATE

$$CH_3 \cdot CH_2 \cdot CO_2H + SOCl_2 \longrightarrow CH_3 \cdot CH_2 \cdot COCl + HCl + SO_2$$

$$CH_3 \cdot CH_2 \cdot COCl + Br_2 \xrightarrow{P} CH_3 \cdot CHBr \cdot COCl + HBr$$

$$CH_3 \cdot CHBr \cdot COCl + C_2H_5OH \longrightarrow CH_3 \cdot CHBr \cdot CO_2C_2H_5 + HCl$$

In a 1-litre, two-necked round-bottomed flask, equipped with a dropping funnel and a double surface reflux condenser to which is attached a gas absorption trap (Fig. I,68(c)), place 220 g (135 ml, 1.86 mol) of redistilled thionyl chloride, and heat to boiling. Add 125 g (126 ml, 1.69 mol) of pure propionic acid at such a rate that the mixture refluxes gently (ca. 1 hour). Reflux the mixture for a further 30 minutes to expel the dissolved sulphur dioxide, allow to cool and add 0.5 g of purified red phosphorus. Introduce 310 g (100 ml, 1.93 mol) of dry bromine during 5–7 hours to the gently boiling propionyl chloride, and then reflux the mixture for 7 hours, by which time the evolution of hydrogen bromide almost ceases. Add the crude 2-bromopropionyl chloride during 2 hours to 250 ml of absolute ethanol contained in a three-necked round-bottomed flask, equipped with a mechanical stirrer and a reflux condenser. Complete the reaction by heating on a water bath for 4 hours, when hydrogen chloride is slowly evolved. Filter the reaction liquid into 500 ml of distilled water, separate the oil and wash it successively with water, sodium hydrogen carbonate solution and water. Dry over calcium sulphate and distil at normal pressure to remove the low b.p. fraction (largely ethyl bromide: 75 g) and then under diminished pressure. Collect the ethyl 2-bromopropionate as a colourless liquid at 69–70 °C/25 mmHg; the yield is 221 g (72%).

III,153. ETHYL VALERATE

$$CH_3 \cdot (CH_2)_3 \cdot CN + C_2H_5OH + H_2O + H_2SO_4 \longrightarrow$$
$$CH_3 \cdot (CH_2)_3 \cdot CO_2C_2H_5 + NH_4 \cdot HSO_4$$

Place 200 g (250 ml) of rectified spirit in a 1-litre round-bottomed flask fitted with a reflux condenser. Cool in ice and run in, slowly and with frequent shaking, 200 g (109 ml) of concentrated sulphuric acid. Add 83 g (104 ml, 1 mol) of butyl cyanide (Section III,161) to the mixture and reflux the whole for 10 hours. Allow to cool, pour the reaction mixture into ice water, separate the upper layer of ester and alcohol, and dry over anhydrous calcium sulphate. Distil through a fractionating column and collect the ethyl valerate at 143–146 °C. A further amount of the pure ester may be obtained by redrying the fraction of low boiling point and redistilling. The yield is 100 g (85%).

Cognate preparation Ethyl phenylacetate. Place 75 g (74 ml, 0.64 mol) of benzyl cyanide (Section III,160), 125 g (153 ml) of rectified spirit and 150 g

513

(68 ml) of concentrated sulphuric acid in a round-bottomed flask, fitted with an efficient reflux condenser. Reflux the mixture, which soon separates into two layers, gently for 8 hours, cool and pour into 350 ml of water. Separate the upper layer. Dissolve it in about 75 ml of ether (1) in order to facilitate the separation of the layers in the subsequent washing process. Wash the ethereal solution carefully with concentrated sodium hydrogen carbonate solution until effervescence ceases and then with water. Dry over 10 g of anhydrous calcium sulphate for at least 30 minutes. Remove the solvent by flash distillation and distil the residue from an air bath. The ethyl phenylacetate passes over at 225–229 °C (mainly 228 °C) as a colourless liquid; the yield is 90 g (86%). Alternatively, the residue after removal of ether may be distilled under diminished pressure; collect the ester at 116–118 °C/20 mmHg.

Note. (1) Alternatively use 20 ml of carbon tetrachloride. The carbon tetrachloride solution then forms the lower layer in all washing operations.

III,154. DIETHYL MALONATE

$$CH_2Cl \cdot CO_2^{\ominus} \overset{\oplus}{Na} + KCN \longrightarrow CH_2(CN) \cdot CO_2^{\ominus} \overset{\oplus}{Na} + KCl$$

$$CH_2(CN) \cdot CO_2^{\ominus} \overset{\oplus}{Na} + 2C_2H_5OH + 2H^{\oplus} \longrightarrow$$

$$CH_2(CO_2C_2H_5)_2 + \overset{\oplus}{Na} + \overset{\oplus}{NH_4}$$

This preparation must be carried out in an efficient fume cupboard.

Dissolve 100 g (1.06 mol) of chloroacetic acid (Section **III,167**), contained in a large porcelain basin in 200 ml of water. Warm the solution to about 50 °C, using a 200 °C thermometer as a stirring rod. Introduce 90 g (1.07 ml) of pure, powdered sodium hydrogen carbonate in small quantities at a time with stirring: maintain the temperature at 50–60 °C until effervescence ceases. Now add 80 g (1.23 mol) of potassium cyanide [**CAUTION:** (1)] and stir the mixture without further warming until the somewhat vigorous reaction is complete. Evaporate the solution, preferably on an electrically heated hotplate, with vigorous and constant stirring, until the temperature rises to 130–135 °C. Protect the hand by a glove during this operation; arrange that the glass window of the fume cupboard is between the dish and the face during the period of heating. Stir the mass occasionally whilst the mixture cools and, immediately it solidifies, break up the solid mass coarsely in a mortar and transfer it to a 1-litre round-bottomed flask. Add 40 ml of absolute ethanol and attach a reflux condenser to the flask. Introduce through the condenser during 10 minutes in small portions and with frequent shaking, a cold mixture of 160 ml of absolute ethanol and 160 ml of concentrated sulphuric acid; some hydrogen chloride may be evolved during the final stages of the addition. Heat the flask on a water bath for 1 hour. Cool rapidly under the tap with shaking to prevent the formation of a solid mass of crystals. Add 200 ml of water, filter at the pump, wash the undissolved salts with about 75 ml of ether, shake up with the filtrate and transfer to a separatory funnel. Separate the upper layer, and extract the aqueous solution twice with 50 ml portions of ether. Place the combined ethereal extracts in a separatory funnel and shake *cautiously* with concentrated sodium carbonate solution until the latter remains alkaline and no more carbon dioxide is evolved. Dry the ethereal solution over magnesium sulphate or anhydrous calcium sulphate.

Remove the ether by flash distillation (Fig. I,100) and distil the residual ester under diminished pressure (Fig. I,110) collecting diethyl malonate at 92–94 °C/16 mmHg. The yield is 105 g (62%).

The b.p. under atmospheric pressure is 198–199 °C, but is attended by slight decomposition.

Note. (1) The technical grade of powdered potassium cyanide is adequate for this purpose.

III,155. TRIETHYL ORTHOFORMATE (*Triethoxymethane*)

$$CHCl_3 + 3C_2H_5ONa \longrightarrow CH(OC_2H_5)_3 + 3NaCl$$

Fit a 1500-ml round-bottomed flask with a large double surface reflux condenser. Make sure that the apparatus is thoroughly dry. Place 750 ml of super-dry ethanol (Section II,1,9) and 123 g (82 ml, 1.03 mol) of dry chloroform in the flask. Add 52 g (2.25 mol) of clean sodium, cut into small pieces, through the condenser in the course of 30 minutes; when the reaction becomes vigorous, cool the outside of the flask by running water from the condenser outlet. When all the sodium has reacted and the mixture has attained room temperature, filter off the sodium chloride through a sintered glass funnel. The filtration apparatus must be thoroughly dry, and a drying tube, filled with granular calcium chloride, should be placed between the filter flask and the pump. Wash the solid on the filter with 50 ml of absolute ethanol and allow the washings to run into the main filtrate. Distil the solution from a water bath through an efficient fractionating column in order to recover the excess of chloroform and most of the alcohol; collect the distillate (about 500 g) (*A*) in a filter flask protected by a drying tube. Decant the liquid remaining in the flask from a little salt which has separated, and distil it through an all-glass Dufton (or Widmer) column. A fraction (*B*) of low boiling point passes over first, followed by the triethyl orthoformate at 144–146 °C. The yield is 35 g (23%) but depends somewhat upon the efficiency of the fractionation.

Carry out a second run with the recovered chloroform–alcohol mixture (*A*): add 100 g of dry chloroform and sufficient super-dry ethanol (200–250 ml) to give a total volume of 750 ml. Add 52 g of sodium as before. Remove the excess of chloroform and alcohol as before on a water bath through a fractionating column, add the intermediate fraction (*B*) from the first run, and fractionate again. The yield of product, b.p. 144–146 °C, is 45 g (29%).

N,4. Amides. (*a*) *From carboxylic acids.* Primary aliphatic amides are obtained by heating the ammonium salt of the corresponding acid.

$$R \cdot CO_2NH_4 \longrightarrow R \cdot CONH_2 + H_2O$$

For preparative purposes it is best to heat the acid or its ammonium salt with urea. The reaction commences at about 120 °C: the carbamic acid ($NH_2 \cdot CO_2H$) formed decomposes immediately into carbon dioxide and ammonia. The latter may then interact with unreacted acid to yield the ammonium salt which then yields the amide as formulated above.

$$R \cdot CO_2H + NH_2 \cdot CO \cdot NH_2 \longrightarrow R \cdot CONH_2 + [NH_2 \cdot CO_2H]$$

The method is illustrated in the simple case of the preparation of acetamide (Section **III,156**) which is most conveniently isolated by distillation. Less water-soluble higher homologues are readily isolated by filtration following treatment of the final reaction mixture with aqueous sodium carbonate solution.

(b) *From acyl halides.* The reaction of an acyl chloride with an excess of ammonia represents one of the best procedures for the preparation of primary amides (Section **III,157**).

$$R \cdot COCl + 2NH_3 \longrightarrow R \cdot CONH_2 + NH_4Cl$$

The acyl chloride (the crude material prepared by the thionyl chloride method is quite satisfactory) is added dropwise to well-stirred concentrated aqueous ammonia cooled in a freezing mixture. The amides of the higher carboxylic acids crystallise out on standing and need only to be filtered and recrystallised. Water-soluble amides are isolated by extraction with hot ethyl acetate following removal of water on a rotary evaporator.

A milder procedure (Ref. 13) involves stirring a solution of the acyl chloride in acetone at room temperature with ammonium acetate. The filtered solution is evaporated to recover the required amide.

The use of primary or secondary amines in place of ammonia yields the corresponding secondary or tertiary amides in reaction with an acyl chloride. These compounds often serve as crystalline derivatives suitable for the characterisation of either the acyl chloride (and hence of the carboxylic acid itself) or the amine (Sections **VII,6,*16*** and **VII,6,*21***).

$$R^1 \cdot COCl + R^2NH_2 \longrightarrow R^1 \cdot CONH \cdot R^2$$
$$R^1 \cdot COCl + (R^2)_2NH \longrightarrow R^1 \cdot CON(R^2)_2$$

(c) *From esters.* Amides are very easily prepared by the interaction of carboxylic esters with concentrated aqueous ammonia (**ammonolysis**).

$$R^1 \cdot CO_2R^2 + NH_3 \longrightarrow R^1 \cdot CONH_2 + R_2OH$$

The reaction usually proceeds readily in the cold, particularly when the methyl esters of the lower molecular weight carboxylic acids are involved. Sparingly soluble amides crystallise out from the reaction mixture upon standing, as in the case of succinamide (Section **III,158**).

(d) *From nitriles.* The interruption of the hydrolysis of a nitrile at the amide stage can often be achieved in a preparative manner, as for example in the preparation of phenylacetamide (Section **III,159**), where the nitrile is dissolved in concentrated hydrochloric acid at 40 °C and subsequently poured into water. The use of hot polyphosphoric acid has also been recommended (Ref. 14).

$$R \cdot CN + H_2O \xrightarrow{H^\oplus} R \cdot CONH_2$$

Reaction conditions which are particularly applicable to aromatic nitriles involve the use of an aqueous solution of sodium hydroxide containing hydrogen peroxide, but alkyl cyanides do not always give good results; the method is illustrated by the preparation of toluamide (Section **IV,182**).

III,156. ACETAMIDE

$$CH_3 \cdot CO_2H + NH_2 \cdot CO \cdot NH_2 \longrightarrow CH_3 \cdot CONH_2 + CO_2 + NH_3$$

Place 25 g (0.42 mol) of glacial acetic acid and 25 g (0.42 mol) of urea in a 100-ml two-necked round-bottomed flask, fitted with a vertical air condenser and a 350 °C thermometer sited with the bulb dipping in the mixture and 1 cm from the bottom of the flask. Heat the mixture gently, either on a wire gauze or in an air bath (Section I,13), shaking the flask gently in order to mix the acid and urea layers. Gradually raise the temperature so that the liquid just refluxes in the condenser. The temperature is about 150 °C after 30 minutes and a white solid (presumably ammonium carbamate) commences to form in the condenser: push the solid back into the flask by means of a stout glass rod if complete blocking of the condenser appears likely. Continue the heating until the temperature of the liquid is 195–200 °C; this temperature is attained after a heating period of 3–3.5 hours. Both carbon dioxide and ammonia are evolved. Allow the apparatus to cool and fit the flask with a still-head and air condenser set for downward distillation. Heat the flask slowly at first; some ammonium carbamate first sublimes into the air condenser. When acetamide just reaches the air condenser, stop the distillation momentarily, replace the condenser by another of similar size and continue the distillation. Collect the acetamide at 200–216 °C (most of it passes over at 214–216 °C); if it crystallises in the condenser, it may be melted by the cautious application of a flame. The yield of almost pure, colourless acetamide, m.p. 80 °C, is 22 g (90%). Beautiful large crystals may be obtained by dissolving the acetamide (5 g) in warm methanol (4 ml), adding ether (180 ml) and allowing to stand.

III,157. HEXANAMIDE

$$CH_3 \cdot (CH_2)_4 \cdot COCl + 2NH_3 \longrightarrow CH_3 \cdot (CH_2)_4 \cdot CONH_2 + NH_4Cl$$

Place 125 ml of concentrated ammonia solution (d 0.88) in a 600-ml beaker and surround the latter with crushed ice. Stir the ammonia solution mechanically, and introduce 56 g (0.42 mol) of hexanoyl chloride (Section III,138) slowly by means of a suitably supported separatory funnel. The rate of addition must be adjusted so that no white fumes are lost. The amide separates immediately. Allow to stand in the ice-water for 15 minutes after all the acid chloride has been introduced. Filter off the amide at the pump; use the filtrate to assist the transfer of any amide remaining in the beaker to the filter (1). Spread the amide on sheets of filter or drying paper to dry in the air. The crude hexanamide (30 g, 63%) has m.p. 98–99 °C and is sufficiently pure for conversion into the nitrile (Section III,163) (2). Recrystallise a small quantity of the amide by dissolving it in the minimum volume of hot water and allowing the solution to cool; dry on filter-paper in the air. Pure hexanamide has m.p. 100 °C.

Notes. (1) The filtrate will deposit small amounts of hexanamide upon concentration to half its original volume.

(2) The process is of general application for higher (i.e., $>C_5$) fatty acids.

Cognate preparation **Isobutyramide** (*2-methylpropanamide*). Add 106 g (1 mol) of isobutyryl chloride (Section III,138) to 400 ml of concentrated ammonia solution (d 0.88) contained in a 2-litre two-necked flask fitted with a stirrer and dropping funnel, at such a rate that the temperature does not rise above 15 °C. Stir for 1 hour after the addition is complete and attach the flask to a rotary evaporator, connect to a water-jet pump and heat the flask on a boiling water bath until the crystalline deposit is quite dry. Boil the residue with 1200 ml of ethyl acetate, filter the hot solution and extract the residue with further quantities of hot ethyl acetate (2 × 750 ml). Combine the extracts, cool to 0 °C and collect the crystals of isobutyramide of m.p. 128–129 °C. The yield after recovery of further crops of crystalline material from the concentrated mother-liquors is 58 g (66%).

III,158. SUCCINAMIDE

$$\begin{array}{c} CH_2 \cdot CO_2CH_3 \\ | \\ CH_2 \cdot CO_2CH_3 \end{array} + 2NH_3 \longrightarrow \begin{array}{c} CH_2 \cdot CONH_2 \\ | \\ CH_2 \cdot CONH_2 \end{array} + 2CH_3OH$$

Add 5 g (4.8 ml, 0.034 mol) of dimethyl succinate to 25 ml of concentrated ammonia solution (d 0.88) in a 100-ml conical flask. Stopper the flask and shake the contents for a few minutes: allow to stand for 24 hours with occasional shaking. Filter off the crystals of succinamide, and wash with a little cold water. Recrystallise from a little hot water and dry in an oven. The yield is 3.5 g (88%). Pure succinamide melts at 254 °C with decomposition.

III,159. PHENYLACETAMIDE

$$C_6H_5 \cdot CH_2 \cdot CN \xrightarrow[40\,°C]{H_2O,\ HCl} C_6H_5 \cdot CH_2 \cdot CONH_2$$

In a 2-litre three-necked flask, provided with a thermometer, reflux condenser and *efficient* mechanical stirrer, place 100 g (98 ml, 0.85 mol) of benzyl cyanide (Section **III,160**) and 400 ml of concentrated hydrochloric acid. Immerse the flask in a water bath at 40 °C and stir the mixture vigorously: the benzyl cyanide passes into solution within 20–40 minutes and the temperature of the reaction mixture rises to about 50 °C. Continue the stirring for an additional 20–30 minutes after the mixture is homogeneous. Replace the warm water in the bath by tap water at 15 °C, replace the thermometer by a dropping funnel charged with 400 ml of cold distilled water and add the latter with stirring: crystals commence to separate after about 50–75 ml have been introduced. When all the water has been run in, cool the mixture externally with ice-water for 30 minutes (1), and collect the crude phenylacetamide by filtration at the pump. Remove traces of phenylacetic acid by stirring the wet solid for about 30 minutes with two 50 ml portions of cold water; dry the crystals at 50–80 °C. The yield of phenylacetamide, m.p. 154–155 °C, is 95 g (82%). Recrystallisation from rectified spirit raises the m.p. to 156 °C.

Note. (1) The suspension of phenylacetamide may be further hydrolysed to phenylacetic acid by refluxing with stirring until the solid dissolves. The mixture becomes turbid after 30 minutes and the product begins to separate as an oil: refluxing is continued for 6 hours, the mixture is cooled first with tap water and then by an ice-water bath for about 4 hours. The crude phenylacetic acid is filtered at the pump, washed with two 50 ml portions of cold water and dried in a desiccator. The resulting crude acid melts at 69–70 °C; it may be purified by recrystallisation from light petroleum (b.p. 40–60 °C) or, better, by vacuum distillation.

O ALIPHATIC NITRILES (ALKYL CYANIDES)

The preparation of aliphatic nitriles is illustrated by the following three procedures of which 1 and 2 are the most generally applicable.

1. The displacement of halogen by cyanide in an alkyl halide (Sections **III,160** and **III,161**).
2. The dehydration of amides (Sections **III,162** to **III,164**) and of aldoximes (Section **IV,185**).
3. Cyanoethylation procedures (Sections **III,165** and **III,166**).

O,1. Displacement of halogen by cyanide in an alkyl halide. The classical procedure for the reaction involves heating the alkyl halide (usually the chloride or bromide) with sodium cyanide in methanolic or ethanolic solution.

$$Br—R \overset{\frown}{\curvearrowleft} {:}\overset{\ominus}{CN} \longrightarrow R{\cdot}CN + {:}\overset{\ominus}{Br}$$

This method has been widely used for primary and secondary halides; tertiary halides undergo ready elimination under these conditions to yield the alkene and give little or no nitrile. A trace of the corresponding isonitriles may be formed during these displacement reactions but may be removed by virtue of their ready hydrolysis with aqueous mineral acid.

This general procedure has been significantly improved by the use of aprotic solvents such as dimethyl sulphoxide, in which primary alkyl chlorides are rapidly converted into nitriles in good yield. The use of dimethylformamide as solvent is equally effective.

The classical procedure is illustrated in Section **III,160** with reference to the preparation of a selection of dinitriles from the corresponding α,ω-dibromo-compounds. The use of dimethyl sulphoxide as solvent is illustrated by a range of examples in Section **III,161**.

III,160. PENTANEDINITRILE (*Glutaronitrile*)

$$BrCH_2{\cdot}CH_2{\cdot}CH_2Br + 2NaCN \longrightarrow NC{\cdot}CH_2{\cdot}CH_2{\cdot}CH_2{\cdot}CN + 2NaBr$$

Fit a 2-litre two-necked flask with a separatory funnel and a reflux condenser. Place 147 g (3 mol) of finely powdered sodium cyanide (1) and 150 ml of water in the flask and heat on a water bath until most of the solid passes into solution. Add a solution of 250 g (126 ml, 1.25 mol) of 1,3-dibromopropane (Section **III,54**) in 500 ml of rectified spirit through the separatory funnel over a period of 30 minutes. Reflux the mixture on a water bath for 35 hours; then remove the solvent (rotary evaporator) using a boiling-water bath. The residue in the flask consists of sodium bromide, unreacted sodium cyanide and the dinitrile: the last-named alone is soluble in ethyl acetate. Extract the residue with 200 ml of ethyl acetate. Filter the solution through a sintered glass funnel and wash the solid with about 50 ml of ethyl acetate. Dry the filtrate, after removing the aqueous layer, with anhydrous calcium sulphate, distil off the ethyl acetate at atmospheric pressure (about 245 ml are recovered), allow to cool somewhat and distil the liquid under reduced pressure. Collect the pentanedinitrile at 139–140 °C/8 mmHg. The yield is 95 g (82%).

Note. (1) See Section **III,161**, Note (1), for precautions in the use and disposal of alkali cyanides.

Cognate preparations   Heptanenitrile.   Use 30 g (0.61 mol) of sodium cyanide dissolved in 40 ml of water; 82 g (70 ml, 0.5 mol) of 1-bromohexane (Section **III,55**) in 150 ml of methanol. Remove the methanol through an efficient fractionating column, add 500 ml of water and separate the upper layer of crude nitrile. Purify the crude nitrile by shaking it twice with about half its volume of concentrated hydrochloric acid, and then successively with water, saturated sodium hydrogen carbonate solution and water. Dry with anhydrous calcium chloride and distil. Collect the heptanenitrile at b.p. 182–184 °C. The yield is 40 g (73%).

Heptanedinitrile.   In a 500-ml round-bottomed flask, equipped with a reflux condenser, place a solution of 29 g (0.446 mol) of potassium cyanide (or the equivalent quantity of powdered sodium cyanide) in 30 ml of warm water and add a solution of 45 g (0.195 mol) of 1,5-dibromopentane (Section **III,54**) in 75 ml of rectified spirit. Reflux the mixture on a water bath for 8 hours. Remove the solvent under diminished pressure, using a rotary evaporator. Extract the residue 4–5 times with 100 ml portions of ether, dry the combined ethereal extracts with anhydrous calcium chloride or anhydrous calcium sulphate, and distil off the ether under atmospheric pressure. Distil the residue under diminished pressure and collect the heptanedinitrile at 168–170 °C/15 mmHg. The yield is 18 g (75%).

Octanedinitrile.   Convert the 1,6-dibromohexane (Section **III,55**), b.p. 114–115 °C/12 mmHg, into octanedinitrile, b.p. 178–180 °C/15 mmHg, by refluxing it with a 20–25 per cent excess of aqueous–alcoholic sodium cyanide solution, distilling off the liquid under diminished pressure whilst heating on a water bath, adding water to the residue and exhaustively extracting with ether: upon evaporating the ether, and distilling the residue under diminished pressure, the dinitrile is obtained.

Benzyl cyanide.   Place 100 g (2 mol) of powdered sodium cyanide and 90 ml of water in a 1-litre round-bottomed flask provided with a reflux condenser. Warm on a water bath until the sodium cyanide dissolves. Add down the condenser a solution of 200 g (181.5 ml, 1.58 mol) of benzyl chloride (Section **IV,31**) in 200 g of rectified spirit during 30–45 minutes. Heat the mixture in a water bath for 4 hours, cool and filter off the precipitated sodium chloride with suction; wash with a little alcohol. Distil off as much as possible of the alcohol using a rotary evaporator. Cool the residual liquid, filter if necessary and separate the layer of crude benzyl cyanide. (Sometimes it is advantageous to extract the nitrile with ether.) Dry over a little magnesium sulphate, and distil under diminished pressure. Collect the benzyl cyanide at 102–103 °C/10 mmHg. The yield is 160 g (86%).

This product is sufficiently pure for most purposes but it contains some benzyl isocyanide and usually develops an appreciable colour on standing. The following procedure removes the isocyanide and gives a stable water-white compound. Shake the once-distilled benzyl cyanide vigorously for 5 minutes with an equal volume of warm (60 °C) 50 per cent sulphuric acid (prepared by adding 55 ml of concentrated sulphuric acid to 100 ml of water). Separate the benzyl cyanide, wash it with an equal volume of saturated sodium hydrogen carbonate solution and then with an equal volume of half-saturated sodium chloride solution. Dry with magnesium sulphate and distil under reduced pressure.

1-Naphthylacetonitrile. Place a mixture of 56 g (0.32 mol) of 1-(chloromethyl)naphthalene, 29 g (0.45 mol) of potassium cyanide, 125 ml of ethanol and 50 ml of water in a 500-ml round-bottomed flask fitted with a double surface reflux condenser, and reflux for 1 hour. Distil off the alcohol, transfer the residue to a separatory funnel, wash it with water, filter from a small amount of solid, transfer to a dish and dry under reduced pressure (vacuum desiccator charged with anhydrous calcium chloride). Distil under diminished pressure and collect the 1-naphthylacetonitrile at 155–160 °C/9 mmHg (1); the yield is 38 g (72%).

Note. (1) A little naphthalene may pass over first owing to impurities in the original chloromethylnaphthalene.

III,161. VALERONITRILE (Butyl cyanide)

$$CH_3 \cdot (CH_2)_2 \cdot CH_2Cl + NaCN \xrightarrow{DMSO} CH_3 \cdot (CH_2)_2 \cdot CH_2 \cdot CN$$

Set up on a water bath a 500-ml three-necked flask fitted with a mechanical stirrer, a thermometer and a two-way adapter fitted with a dropping funnel and a reflux condenser protected by a calcium chloride guard-tube. Place 150 ml of dry dimethyl sulphoxide in the flask and add 30 g (0.61 mol) of dry powdered sodium cyanide (1). Heat the mixture with stirring to 90 °C and then remove the water bath. Add 46.3 g (0.5 mol) of butyl chloride (Section **III,48**) slowly from the dropping funnel so that the temperature of the exothermic reaction does not rise above 150 °C (about 10 minutes); continue to stir until the temperature falls to 50 °C (about 30 minutes). Pour the mixture into water and extract with three 300 ml portions of ether. Wash the combined extracts with two 100 ml portions of saturated sodium chloride solution, dry over magnesium sulphate and distil off the ether on a water bath. Distil the residue at atmospheric pressure and collect the butyl cyanide at 138–140 °C; the yield is 35.2 g (85%). The product has a nauseating odour, and should be handled in an efficient fume cupboard.

Note. (1) Dry the powdered commercial material (ca. 98% pure) in a vacuum desiccator over potassium hydroxide pellets. *Sodium cyanide is very poisonous and must be handled with great care.* Residual solutions containing alkali cyanides should be rendered innocuous by the addition of an excess of sodium hypochlorite before being washed down the main drain of the laboratory with a liberal supply of water; they should never be treated with acid.

Cognate preparations   Octanenitrile. Use 250 ml of dry dimethyl sulphoxide, 89.6 g (0.5 mol) of 1-bromoheptane (Section **III,55**) and 30 g (0.61 mol) of dried sodium cyanide. The yield of octanenitrile is 50.8 g (81%), b.p. 199–203 °C.

Hexanedinitrile (adiponitrile). Use 150 ml of dry dimethyl sulphoxide, 21.2 g (0.17 mol) of 1,4-dichlorobutane (Section **III,50**) and 20 g (0.41 mol) of dried sodium cyanide. Maintain the reaction temperature at 90 °C for a further 15 minutes after the initial exothermic reaction has subsided. Add 150 ml of chloroform to the cooled reaction mixture and pour into an excess of saturated sodium chloride solution in a separatory funnel. Add just sufficient water to dissolve precipitated salts and separate the chloroform layer. Extract the aqueous layer once with chloroform, wash the combined extracts twice with salt solution, dry over magnesium sulphate and remove the solvent using a rotary evaporator. Fractionally distil the residue under reduced pressure and collect the adiponitrile as a fraction of b.p. 140–141 °C/1.5 mmHg. The yield is 14.6 g (81%). Some dimethyl sulphoxide (b.p. 40–42 °C/1.5 mmHg) may be obtained as a forerun.

O,2. Dehydration of amides. Conversion of amides into nitriles may be effected by heating them with a variety of dehydrating agents. Of these, phosphorus pentoxide and thionyl chloride are perhaps the most widely used, the latter having the advantage that the by-products are sulphur dioxide and hydrogen chloride and are hence easily removed. These procedures are illustrated by the preparation of isobutyronitrile (Section III,162) and hexanenitrile (Section III,163).

$$R \cdot CONH_2 \xrightarrow{-H_2O} R \cdot CN$$

An alternative procedure using phosphorus pentachloride is exemplified by the conversion of cyanoacetamide (itself formed from ethyl cyanoacetate and ammonia) into the unstable, but useful, synthetic intermediate malononitrile (Section III,164).

Dehydration of aldoximes proceeds under somewhat milder conditions, acetic anhydride being frequently used. The method is more particularly applicable to the synthesis of aromatic nitriles; an example is provided in Section IV,185.

III,162. ISOBUTYRONITRILE (*Isopropyl cyanide*)

$$(CH_3)_2CH \cdot CONH_2 \xrightarrow[-H_2O]{P_2O_5} (CH_3)_2CH \cdot CN$$

Equip a 2-litre round-bottomed flask with a simple distillation head carrying a capillary air leak, and attach a double surface condenser fitted with a receiver adapter with vacuum connection and a 100-ml receiver flask. Remove the still-head, wrap some glazed paper around a glass tube and insert it into the flask neck until the lower end enters the bulb of the flask; upon removing the glass tube, the paper roll expands and thus lines the neck of the flask. Weigh out on pieces of glazed paper first 65 g (0.75 mol) of 2-methylpropanamide (isobutyramide, Section III,157) and then, *as rapidly as possible* because of its extremely hygroscopic character, 114 g (0.8 mol) of phosphorus pentoxide (1). Immediately transfer, with the aid of a spatula, the phosphorus pentoxide down the glazed paper cylinder into the flask, then introduce the acid amide similarly, remove the paper, stopper the flask and mix the contents well by gentle shaking (2). Attach the flask to the distillation set-up and heat the flask cautiously with a luminous flame kept in constant motion and applied uniformly over the bottom of the flask. A reaction accompanied by much frothing takes place. After the reaction has subsided, apply the vacuum provided by a water-jet pump carefully and continue heating at such a rate that the isopropyl cyanide distils smoothly. Redistil the product from 10 g of phosphorus pentoxide in a flask fitted with a short Vigreux column and heated in an oil bath maintained at 145–155 °C; collect the pure product, b.p. 101–102 °C. The yield is 37 g (73%).

Notes. (1) Phosphorus pentoxide must be treated with great care since it produces painful burns if allowed to come in contact with the skin.

(2) Wet the papers thoroughly with water before throwing them away, as the residual phosphorus pentoxide may cause them to smoulder.

III,163. HEXANENITRILE (*Pentyl cyanide*)

$$C_4H_9 \cdot CH_2 \cdot CONH_2 + SOCl_2 \longrightarrow C_4H_9 \cdot CH_2 \cdot CN + SO_2 + 2HCl$$

Place 29 g (0.25 mol) of hexanamide (Section **III,157**) in a 100-ml flask fitted with a reflux condenser, add 45 g (27.5 ml, 0.38 mol) of redistilled thionyl chloride and connect the top of the condenser to a gas absorption device (Fig. I,68(*a* or *b*)). Boil the mixture gently under reflux for 1 hour. Remove the reflux condenser and arrange it for downward distillation. Distil from an oil bath or an air bath (Section **I,13**); the excess of thionyl chloride passes over below 90 °C (1) and the hexanenitrile as a fraction of b.p. 161–163 °C (2). The yield is 21 g (86%).

Notes. (1) If the residue at this stage is dark and contains some solid matter, it is advisable to add a little anhydrous ether and to filter the ethereal solution; the ether is first removed by distillation from a water bath and the nitrile is distilled from the residue.

(2) The nitrile is sometimes slightly turbid; the turbidity is readily removed by shaking with a little anhydrous calcium sulphate, filtering and redistilling.

Cognate preparation   Octanenitrile. Use octanamide and redistilled thionyl chloride in the proportion of 1 mol to 1.5 mols. Warm the mixture on a water bath for 1 hour, distil off the excess of thionyl chloride at atmospheric pressure and distil the residual octanenitrile under diminished pressure. B.p. 87 °C/10 mmHg. The yield is almost quantitative.

III,164. MALONONITRILE

$$CH_2 \cdot (CN) \cdot CO_2C_2H_5 + NH_3 \longrightarrow CH_2(CN) \cdot CONH_2 + C_2H_5OH$$
$$3CH_2(CN) \cdot CONH_2 + PCl_5 \longrightarrow 3CH_2(CN)_2 + HPO_3 + 5HCl$$

Cyanoacetamide. Place 150 ml of concentrated aqueous ammonia solution (*d* 0.88) in a 500-ml wide-mouthed conical flask and add 200 g (188 ml, 1.77 mol) of ethyl cyanoacetate (Section **III,171**). Shake the cloudy mixture: some heat is evolved and it becomes clear in about 3 minutes. Stand the loosely stoppered flask in an ice–salt mixture for 1 hour, filter rapidly with suction and wash the solid with two 25 ml portions of ice-cold ethanol. Dry in the air: the yield of pale yellow cyanoacetamide is 110 g (74%) (1). Recrystallise from 190 ml of 95 per cent ethanol; a colourless product, m.p. 119–120 °C, is deposited with practically no loss.

Note. (1) A further 25 g (17%) of cyanoacetamide may be obtained by evaporating the original mother-liquor to dryness under reduced pressure (rotary evaporator) whilst heating the flask on a steam bath. The residue is dissolved in 50 ml of hot ethanol, the solution shaken for a few minutes with decolourising carbon, filtered with suction whilst hot, and then cooled in ice. The resulting yellowish amide is recrystallised with the addition of decolourising carbon, if necessary.

Malononitrile.   Mix 75 g (0.89 mol) of cyanoacetamide intimately with 75 g (0.36 mol) of dry phosphorus pentachloride in a glass mortar (*FUME CUPBOARD!*). Transfer the mixture as rapidly as possible (with the aid of a large glass funnel with cut-off stem) to a 500-ml round-bottomed flask. Fit the latter with a Claisen still-head leading to a long air condenser terminating in a vacuum take-off adapter and a 200-ml receiving flask; insert a thermometer and a capillary air leak (of reasonably wide bore to reduce the danger of blocking) into the Claisen head. Connect the unit to a powerful water pump (or two glass water pumps in parallel) and a manometer. Evacuate the system to about 30 mmHg of mercury and immerse the flask in a boiling water bath. The mixture gradually melts, boiling commences about 15 minutes before the solid has melted completely and the pressure rises to about 150 mmHg owing to the liberation of hydrogen chloride and phosphorus oxychloride. The evolution of gas slackens in about 30–35 minutes, the boiling is then less vigorous and the pressure falls. At this point, connect a clean receiving flask and immerse it in ice-water. Remove the reaction flask immediately from the water bath, wipe it dry and immerse it in an oil bath at 140 °C to within 10 cm of the top of the flask. The malononitrile commences to pass over at 113 °C/30 mmHg (or 125 °C/50 mmHg): raise the temperature of the oil bath over a period of 25 minutes to 180 °C. Collect the dinitrile at 113–125 °C/30 mmHg; if it solidifies in the air condenser melt it with a hot air blower. Remove the oil bath when distillation has almost ceased; discolouration of the product is thus prevented. The yield of crude dinitrile is 45 g. Redistil and collect the pure malononitrile at 113–120 °C/30 mmHg as a colourless liquid (40 g, 68%); this quickly solidifies on cooling, m.p. 29–30 °C. Store in a brown bottle and protect it from the light.

O,3. Cyanoethylation procedures.   Cyanoethylation (i.e., the introduction of the $-CH_2 \cdot CH_2 \cdot CN$ group) ensues when acrylonitrile undergoes 1,4-addition of any of a range of nucleophilic addenda. Suitable nucleophiles are carbanions derived under basic conditions from compounds containing active hydrogens, and their use is illustrated by the reaction of acrylonitrile with diethyl malonate in the presence of sodium ethoxide to form diethyl (2-cyanoethyl) malonate (Section **III,165**).

$$C_2H_5O^{\ominus} H-CH(CO_2C_2H_5)_2 \rightleftharpoons C_2H_5OH + {}^{\ominus}CH(CO_2C_2H_5)_2$$

$$(H_5C_2O_2C)_2\ddot{C}H \quad CH_2 = CH - C \equiv N \rightleftharpoons$$

$$[(H_5C_2O_2C)_2CH \cdot CH_2 \cdot CH = C = \ddot{N}]^{\ominus} \xrightarrow{C_2H_5OH}$$

mesomeric

$$(H_5C_2O_2C)_2CH \cdot CH_2 \cdot CH_2 \cdot CN + C_2H_5O^{\ominus}$$

Other suitable carbanion sources include nitroalkanes, aldehydes and ketones, all of which must contain an α-hydrogen. Cyanoethylation with such addenda conveniently leads to a range of polyfunctional compounds; the cyanoethylation of alkylated malonic esters as a route to α-alkylglutaric acids is illustrated by an example in Section **III,137**.

Experimental conditions are given for the cyanoethylation of primary alcohols (Section **III,166,A**) and of secondary aliphatic amines (Section **III,166,B**); the reaction is equally applicable to phenols, thiols, ammonia, primary amines, etc.

$$R - \overset{\cdot\cdot}{\underset{\underset{H}{|}}{O}} : \curvearrowright CH_2 = CH - C \equiv N \longrightarrow RO \cdot CH_2 \cdot CH_2 \cdot CN$$

$$R_2 \overset{\cdot\cdot}{\underset{\underset{H}{|}}{N}} : \curvearrowright CH_2 = CH - C \equiv N \longrightarrow R_2 N \cdot CH_2 \cdot CH_2 \cdot CN$$

Anion exchange resins of the quaternary ammonium hydroxide type [e.g., Zerolit FF or Amberlite (IRA-400)] are strong bases and are useful catalysts for the cyanoethylation of alcohols. No additional basic catalyst is normally required in the case of the cyanoethylation of aliphatic amines.

III,165. DIETHYL (2-CYANOETHYL)MALONATE

$$CH_2(CO_2C_2H_5)_2 + CH_2 = CH \cdot CN \longrightarrow (C_2H_5O_2C)_2CH \cdot CH_2 \cdot CH_2 \cdot CN$$

Assemble in a fume cupboard a 1-litre three necked flask containing 480 g (455 ml, 3 mol) of diethyl malonate and fitted with a stirrer, a dropping funnel and a thermometer. Start the stirrer, add to the flask a solution of sodium ethoxide prepared from 3.5 g of sodium in 100 ml of absolute ethanol and then run in slowly 80 g (100 ml, 1.5 mol) of acrylonitrile (1) (**CAUTION**: toxic and lachrymatory vapour) at a rate such that the temperature does not exceed 35 °C. When the addition is complete, continue to stir for 1 hour more and then remove the stirrer. Equip the flask for distillation under reduced pressure through a short fractionating column (cf., Fig. I,112) and remove the ethanol by distillation using a water pump. Continue the distillation of the residue using an oil pump at a recorded pressure of about 0.2 mmHg, and collect (i) recovered diethyl malonate at 52 °C (240 g), and (ii) somewhat crude diethyl (2-cyanoethyl)malonate at 130–140 °C (200 g). Redistil the latter to obtain 175 g (55%) of the purified product of b.p. 102–106 °C/0.2 mmHg (127 130 °C/3 mmHg).

Note. (1) *Acrylonitrile vapour is highly toxic*; it should therefore be handled with due caution and all operations with it should be conducted in a fume cupboard provided with an efficient draught. Acrylonitrile forms an azeotropic mixture with water, b.p. 70.5 °C (12.5% water). The commercial product may contain the polymer; it should be redistilled before use and the fraction, b.p. 76.5–78 °C, collected separately as a colourless liquid.

III,166,A. 3-ETHOXYPROPIONITRILE

$$C_2H_5OH + CH_2=CH \cdot CN \xrightarrow{\ominus OH} C_2H_5O \cdot CH_2 \cdot CH_2 \cdot CN$$

Place 25 ml of 2 per cent aqueous sodium hydroxide and 26 g (33 ml, 0.565 mol) of ethanol in a 250-ml reagent bottle, add 26.5 g (33 ml, 0.5 mol) of acrylonitrile [CAUTION: see Section III,165, Note (1)] and close the mouth of the bottle with a well-fitting plastic stopper. Shake the resulting clear homogeneous liquid in a shaking machine for 2 hours. During the first 15 minutes the temperature of the mixture rises 15 to 20 °C and thereafter falls gradually to room temperature; two liquid layers separate after about 10 minutes. Remove the *upper* layer and add small quantities of 5 per cent acetic acid to it until neutral to litmus; discard the lower aqueous layer. Dry with anhydrous sodium sulphate, distil and collect the 3-ethoxypropionitrile at 172–174 °C. The yield is 32 g (65%).

The technique for using an *anion exchange resin as catalyst* is as follows. Regenerate the resin [Zerolit FF or Amberlite (IRA-400)] by washing it on a Buchner funnel with 5 per cent sodium hydroxide solution (5–6 times the volume of the resin); rinse the resin with distilled water until the washings are neutral and dry in the air. In a 500-ml three-necked flask equipped with a reflux condenser, stirrer and a dropping funnel, place 25 g of the regenerated resin and 46 g (58.5 ml, 1 mol) of ethanol. Immerse the flask in an ice bath to control the subsequent initial exothermic reaction and to hold the temperature below 15–20 °C throughout the experiment. Add 67 g (85 ml, 1.26 mol) of redistilled acrylonitrile slowly to the well-stirred mixture in the flask over a period of 1–2 hours; continue the stirring for a further 1.5 hours. Separate the resin by filtration. Distil the filtrate at atmospheric pressure to 100 °C in order to remove unreacted acrylonitrile and ethanol, and the residue under reduced pressure. Collect the 3-ethoxypropionitrile at 77–78 °C/25 mmHg. The yield is about 110 g (90%).

Cognate preparation   3-Propoxypropionitrile.   Introduce 0.15 g of potassium hydroxide and 33 g (41 ml, 0.55 mol) of dry propan-1-ol into a 250-ml, round-bottomed three-necked flask, warm gently until the solid dissolves and then cool to room temperature. Equip the flask with a dropping funnel, a mechanical stirrer and a thermometer. Introduce from the dropping funnel, with stirring, 26.5 g (33 ml, 0.5 mol) of pure acrylonitrile over a period of 25–30 minutes. Do not allow the temperature of the mixture to rise above 35–45 °C; immerse the reaction flask in a cold water bath when necessary. When all the acrylonitrile has been added, heat under reflux in a boiling water bath for 1 hour; the mixture darkens. Cool, filter and distil. Collect the 3-propoxypropionitrile at 187–189 °C. The yield is 38 g (67%).

III,166,B. 3-(DIETHYLAMINO)PROPIONITRILE

$$(C_2H_5)_2NH + CH_2=CH \cdot CN \longrightarrow (C_2H_5)_2N \cdot CH_2 \cdot CH_2 \cdot CN$$

Mix 42.5 g (60 ml, 0.58 mol) of freshly distilled diethylamine and 26.5 g (33 ml, 0.5 mol) of pure acrylonitrile [CAUTION: see Section III,165, Note (1)] in a 250-ml round-bottomed flask fitted with a reflux condenser. Heat at 50 °C in a water bath for 10 hours and then allow to stand at room temperature for 2 days. Distil off the excess of diethylamine on a water bath, and distil the residue under reduced pressure. Collect the 3-(diethylamino)propionitrile at 75–77 °C/11 mmHg; the yield is 54 g (86%).

Cognate preparation 3-Dibutylaminopropionitrile. Proceed as for the diethyl compound using 64.5 g (85 ml, 0.5 mol) of redistilled dibutylamine and 26.5 g (33 ml, 0.5 mol) of pure acrylonitrile. After heating at 50 °C and standing for 2 days, distil the entire product under diminished pressure (air bath); discard the low boiling point fraction containing unchanged dibutylamine and collect the 3-(dibutylamino)propionitrile at 120–122 °C/10 mmHg. The yield is 55 g (61%).

P SUBSTITUTED ALIPHATIC CARBOXYLIC ACIDS AND THEIR DERIVATIVES

1. Halogeno acids (Sections **III,167** to **III,170**).
2. Cyano acids (Section **III,171**).
3. Hydroxy acids (Sections **III,172** and **III,173**).
4. Keto acids (oxo acids) (Sections **III,174** to **III,178**).
5. Amino acids (Sections **III,179** to **III,186**).

P,1. Halogeno acids. (a) *Halogenation of the carboxylic acid.* Acetic acid can be chlorinated by gaseous chlorine in the presence of red phosphorus to yield successively mono- (I), di- (II) and trichloroacetic acid (III); the reaction proceeds better in bright sunlight. If the chlorination is stopped when approximately one molar proportion of chlorine is absorbed the main product is monochloroacetic acid (Section **III,167**).

$$CH_3 \cdot CO_2H \xrightarrow{Cl_2} CH_2Cl \cdot CO_2H \xrightarrow{Cl_2} CHCl_2 \cdot CO_2H \xrightarrow{Cl_2} CCl_3 \cdot CO_2H$$

$$\text{(I)} \qquad\qquad \text{(II)} \qquad\qquad \text{(III)}$$

Chlorination of other aliphatic carboxylic acids is not usually of preparative value since mixtures of several monochlorinated products are obtained. Bromination, on the other hand, is highly selective and only the α-bromoacid is obtained when the reaction is carried out in the presence of a reagent which yields an acyl bromide, such as red phosphorus (the **Hell–Volhard–Zelinsky reaction**) or phosphorus trichloride or tribromide. The acyl bromide undergoes halogenation (in the α-position) much more readily than the parent acid.

$$6R \cdot CH_2 \cdot CO_2H + 3Br_2 + 2P \longrightarrow 6R \cdot CH_2 \cdot COBr + 2H_3PO_3$$
$$R \cdot CH_2 \cdot COBr + Br_2 \longrightarrow R \cdot CHBr \cdot COBr + HBr$$
$$R \cdot CHBr \cdot COBr + R \cdot CH_2 \cdot CO_2H \longrightarrow R \cdot CHBr \cdot CO_2H + R \cdot CH_2COBr$$

Examples of the formation of a range of α-bromoacids are given in the cognate preparations in Section **III,168**; in the case of acetic acid, bromination proceeds readily in the presence of acetic anhydride and a trace of pyridine.

When it is required to prepare an α-bromo acid from a carboxylic acid which is not particularly readily available commercially, but which can be synthesised by the malonic ester route (**III,M,6**), advantage may be taken of the ease of bromination in the α-position of the intermediate alkylmalonic acid. The substituted bromomalonic acid undergoes ready decarboxylation on heating to yield the α-bromoacid (e.g., 2-bromo-3-methylvaleric acid, Section **III,169**).

$$RBr + CH_2(CO_2C_2H_5)_2 \xrightarrow{\;\overset{\ominus}{:}OC_2H_5\;} R \cdot CH(CO_2C_2H_5)_2 \xrightarrow{\;\overset{\ominus}{\;}OH\;}$$

$$R \cdot CH(CO_2H)_2 \xrightarrow{\;Br_2\;} R \cdot CBr(CO_2H)_2 \xrightarrow{\;heat\;} R \cdot CHBr \cdot CO_2H$$

(b) *Hydrogen halide addition to an unsaturated carboxylic acid.* A halogeno acid will be formed when an unsaturated carboxylic acid undergoes addition of a halogen acid. In the case of α,β-unsaturated acids, the addition process yields the β-halogeno acid; when the double bond is more remote from the carboxyl group a mixture of isomeric halogeno acids is likely to be obtained as the result of competing orientation effects and the process may have little preparative value. In the particular case of peroxide-catalysed addition of hydrogen bromide to a long-chain unsaturated acid having a terminal double bond, the reaction is essentially regiospecific and the product is an ω-bromo acid. This procedure is illustrated in Section **III,63**.

(c) *The ring-opening of lactones.* The ω-bromo acid, 6-bromohexanoic acid may be conveniently prepared by the action of concentrated hydrobromic acid in the presence of concentrated sulphuric acid on 6-hexanolide (ε-caprolactone) which is readily available commercially. In the procedure described in Section **III,170** the bromo acid is converted directly into the ethyl ester.

III,167. CHLOROACETIC ACID

$$CH_3 \cdot CO_2H + Cl_2 \xrightarrow{\;red\ P\;} CH_2Cl \cdot CO_2H + HCl$$

Assemble an apparatus consisting of a 1-litre three-necked flask carrying a thermometer, a gas distribution tube (a glass tube with a wide fritted disc sealed on at the bottom) and a reflux condenser. Connect the top of the reflux condenser to a gas absorption device (Fig. I,68(c)). Place 6 g (0.2 mol) of purified red phosphorus (Section **II,2,49**) and 150 g (2.5 mol) of glacial acetic acid in the flask and weigh the apparatus on a rough balance; heat the mixture to 100 °C. Pass chlorine from a cylinder, through two empty wash bottles, into the mixture and adjust the stream of chlorine so that a stream of fine bubbles issues through the gas distributor. Gradually increase the flow of chlorine and maintain the temperature inside the flask at 105–110 °C. Continue the passage of chlorine until the flask increases in weight by about 85 g; this roughly corresponds to the formation of monochloroacetic acid. The time required is 4–6 hours. The action of the chlorine is greatly facilitated by exposure of the apparatus to sunlight. Distil the reaction product from a 500-ml flask. Some acetyl chloride and acetic acid passes over first, the temperature then rises, and the fraction, b.p. 150–200 °C, is collected separately (1); run out the water from the condenser when the temperature reaches 150 °C. The fraction, b.p. 150–200 °C, solidifies on cooling. Drain off any liquid from the crystals as rapidly as possible, and redistil the solid using an air condenser. Collect the fraction, b.p.

182–192 °C: this sets to a solid mass on cooling and melts at 63 °C. The yield of chloroacetic acid is 150–175 g (64–74%).

Note. (1) Chloroacetic acid must be handled with great care as it causes blisters on the skin.

III,168. BROMOACETIC ACID

$$CH_3 \cdot CO_2H + Br_2 \xrightarrow[\text{pyridine}]{(CH_3 \cdot CO)_2O} CH_2Br \cdot CO_2H + HBr$$

Place a mixture of 262 g (250 ml, 4.4 mol) of glacial acetic acid, 54 g (50 ml, 0.53 mol) of acetic anhydride and 0.5 ml of pyridine in a 1-litre round-bottomed flask equipped with a reflux condenser (carrying a calcium chloride tube) and a dropping funnel, the stem of which reaches below the level of the liquid. Introduce a few glass beads into the flask and heat the mixture to boiling. Remove the flame, add about 1 ml of bromine and allow the reaction to proceed until the liquid becomes colourless; this takes about 10 minutes as there appears to be a time lag in the reaction. Add the remainder of the 281 g (90 ml, 1.75 mol) of dry bromine (Section **II,2,7**) as rapidly as it will react and avoiding loss through the condenser; during this period (about 2 hours) keep the acid gently boiling by means of a small flame beneath the flask. When about half of the bromine has been added, the liquid acquires a cherry red colour which it retains throughout the remainder of the bromination. Finally, heat the mixture until it becomes colourless.

Allow to cool and run in 20 ml of water slowly to destroy the acetic anhydride. Remove the excess of acetic acid and water with the aid of a rotary evaporator. The residue (220 g, 90%) crystallises on cooling and consists of almost pure bromoacetic acid (1). If it is required perfectly pure, distil the crude acid and collect the fraction of b.p. 202–204 °C. When distilled under diminished pressure, the acid boils at 117–118 °C/15 mmHg. Pure bromoacetic acid has m.p. 50 °C.

Note. (1) Bromoacetic acid must not be allowed to come into contact with the hands as it causes serious burns.

Cognate preparations 2-Bromohexanoic acid. Place 100 g (107 ml, 0.86 mol) of freshly distilled, dry hexanoic acid (b.p. 202–205 °C) (1) and 150 g (48 ml, 0.94 mol) of dry bromine (Section **II,2,7**) in a 500-ml flask equipped with a reflux condenser, the top of which is connected to a gas absorption device (compare Fig. I,68(c)). Momentarily remove the condenser and add cautiously 1.5 ml of phosphorus trichloride. Heat the mixture on a water bath to 65–70 °C, when reaction will commence and hydrogen bromide is smoothly evolved. Towards the end of the reaction allow the temperature of the bath to rise to 100 °C. The reaction is complete when all the bromine has reacted (about 4 hours). Distil the reaction mixture under reduced pressure using a water pump; much hydrogen bromide is evolved and a fraction of low boiling point passes over. When all the low boiling point fraction has distilled, connect the flask to an oil pump via a trap containing sodium hydroxide pellets and collect the 2-bromohexanoic acid at 116–125 °C/8 mmHg (or at 132–140 °C/15 mmHg). The yield is 145 g (86%). Upon redistillation the 2-bromohexanoic acid passes over almost entirely at 128–131 °C/10 mmHg.

Note. (1) See Section **III,129**, *Cognate preparation*, Note (1).

2-Bromopropionic acid. Proceed as detailed for 2-bromohexanoic acid using 64 g (64·5 ml, 0.86 mol) of freshly distilled, dry propionic acid (b.p. 139–142 °C), 150 g (48 ml, 0.94 mol) of dry bromine and 1.5 ml of phosphorus trichloride. The reaction commences on warming to about 50 °C. Collect the 2-bromopropionic acid at 95–97 °C/10 mmHg or at 100–102 °C/15 mmHg. The yield is 110 g (83%).

2-Bromovaleric acid. Use 88 g (0.86 mol) of valeric acid; allow reaction to proceed at 80 °C and finally heat for 2 hours at 100 °C. Collect the 2-bromovaleric acid at 145 °C/15 mmHg; the yield is 125 g (80%).

2-Bromoisovaleric acid (*2-bromo-3-methylbutyric acid*). Use 88 g of isovaleric acid; proceed as for 2-bromohexanoic acid, allowing the reaction to proceed at 70–80 °C for about 10–20 hours (until bromine vapour is no longer evident), then add a further 2.5 ml of bromine and heat at 100 °C for about 2 hours. Distil and collect 2-bromoisovaleric acid at 100–125 °C/15 mmHg. The yield is in the region of 80 per cent.

III,169. 2-BROMO-3-METHYLVALERIC ACID

$$CH_3 \cdot CH_2 \cdot CH(CH_3) \cdot CH(CO_2C_2H_5)_2 \xrightarrow{\ominus OH}$$

$$CH_3 \cdot CH_2 \cdot CH(CH_3) \cdot CH(CO_2H)_2 \xrightarrow{Br_2}$$

$$CH_3 \cdot CH_2 \cdot CH(CH_3) \cdot CBr(CO_2H)_2 \xrightarrow{heat}$$

$$CH_3 \cdot CH_2 \cdot CH(CH_3) \cdot CHBr \cdot CO_2H$$

Hydrolyse 108 g (0.5 mol) of diethyl s-butylmalonate (1) with aqueous ethanolic potassium hydroxide solution following the procedure described for the hydrolysis of diethyl propylmalonate (Section **III,133**). Transfer the dried ether extract of the product to a 1-litre three-necked flask fitted with a sealed stirrer unit, an efficient reflux condenser and a dropping funnel charged with 80 g (26 ml, 0.5 mol) of bromine. Stir the solution, add about 2–3 ml of bromine, and when this has reacted completely drop in the remainder of the bromine at such a rate that the ether refluxes gently. When all the bromine has reacted, add 100 ml of water cautiously from the dropping funnel to decompose any acyl bromide. Separate the ether layer, dry it over anhydrous sodium sulphate and remove the ether (rotary evaporator). Decarboxylate the residue by heating it for 5 hours in an oil bath maintained at 130 °C and then distil the resulting bromo acid under reduced pressure, collecting the fraction of b.p. 125–140 °C/20 mmHg. The yield is 65 g (67%).

Note. (1) Diethyl s-butylmalonate, b.p. 110–120 °C/20 mmHg, is prepared by reacting 2-bromobutane with diethyl malonate. Use the general procedure described for the butyl isomer (Section **III,132**) but extend the period of heating under reflux to 48 hours to complete the alkylation.

III,170. ETHYL 6-BROMOHEXANOATE

$$\text{[7-membered lactone ring, } O-C\overset{O}{=}] \xrightarrow[\text{H}_2\text{SO}_4]{\text{HBr}} Br(CH_2)_5 \cdot CO_2H \xrightarrow[\text{H}_2\text{SO}_4]{\text{C}_2\text{H}_5\text{OH}} Br(CH_2)_5 \cdot CO_2C_2H_5$$

Place a mixture of 140 ml of concentrated hydrobromic acid and 34 ml of concentrated sulphuric acid in a 500-ml, three-necked round-bottomed flask fitted with a sealed stirrer unit, a reflux condenser and a dropping funnel. Cool the flask contents in an ice–salt bath and add with stirring 28.5 g (0.25 mol) of 6-hexanolide. Allow the reaction mixture to reach room temperature and then to stand for 2 hours; heat on a water bath for a further 4-hour period. Cool the mixture and pour onto 300 g of crushed ice and separate the organic layer; saturate the aqueous layer with ammonium sulphate and extract the solution with four 50 ml portions of ether. Wash the combined organic phases with three 25 ml portions of a saturated aqueous solution of ammonium sulphate, dry (anhydrous sodium sulphate) and evaporate the ether on a rotary evaporator. Boil the residual bromo acid under reflux for 8 hours with 85 ml of absolute ethanol containing 2 ml of concentrated sulphuric acid and then remove the ethanol on a rotary evaporator. Dissolve the residue in 100 ml of ether and wash the ethereal solution first with water and then with 5 per cent aqueous sodium carbonate to remove mineral acid. Dry the ethereal solution over anhydrous sodium sulphate, remove the ether (rotary evaporator) and distil the residue under reduced pressure. Collect the ethyl 6-bromohexanoate having b.p. 120–124 °C/14 mmHg; the yield is 28 g (50%).

P,2. Cyano acids. The simplest route to the α-cyano acids is by the replacement of the halogen in a halogeno acid with a cyano group.

$$R \cdot CHX \cdot CO_2H \longrightarrow R \cdot CH(CN) \cdot CO_2H$$

For example, ethyl cyanoacetate (Section III,171), a substance of importance in synthetic work, is prepared by esterifying cyanoacetic acid, which is obtained by heating a solution of sodium chloroacetate with sodium cyanide and acidifying the reaction mixture. The esterification step is effected by reaction with ethanol in the presence of a small amount of concentrated sulphuric acid; these conditions should be contrasted with those employed in the related synthesis of diethyl malonate (Section III,154) where a much larger quantity of sulphuric acid is used to convert the nitrile group into an ethoxycarbonyl group as well as to esterify the carboxyl group.

An alternative route to α-cyano esters which is fairly generally applicable, although the yield in the case of ethyl cyanoacetate is poor, involves the α-ethoxycarbonylation of a nitrile. In this method the nitrile is heated with diethyl carbonate in toluene in the presence of anhydrous sodium ethoxide (Ref. 15).

$$R \cdot CH_2 \cdot CN + CO(OC_2H_5)_2 \xrightarrow{\ominus OC_2H_5} \left[R \cdot \overset{\ominus}{C} \cdot CN \atop CO_2C_2H_5 \right] \xrightarrow{CH_3CO_2H}$$

$$R \cdot CH(CN) \cdot CO_2C_2H_5$$

III,171. ETHYL CYANOACETATE

$$CH_2Cl \cdot CO_2^{\ominus} Na^{\oplus} \xrightarrow{NaCN} CH_2CN \cdot CO_2^{\ominus} Na^{\oplus} \xrightarrow{H^{\oplus}}$$

$$CH_2CN \cdot CO_2H \xrightarrow[H^{\oplus}]{C_2H_5OH} CH_2CN \cdot CO_2C_2H_5$$

This preparation must be carried out in an efficient fume cupboard.

Place 208 g (2.2 mol) of chloroacetic acid (see Section **III,167**) [CAUTION: do not allow the acid to come into contact with the hands] and 315 g of crushed ice in a large beaker and neutralise it accurately to litmus with a cold solution of sodium hydroxide (100 g in 300 ml of water; about 275 ml are required): do not allow the temperature to rise above 30 °C during the neutralisation. Prepare, in the fume cupboard, a solution of 125 g (2.5 mol) of sodium cyanide (97–98% pure) in 250 ml of water in a 3-litre flask: heat to about 55 °C for rapid solution and finally to boiling. Add to the resulting hot solution 100 ml of the solution of sodium chloroacetate and remove the flame immediately the reaction commences. When the vigorous reaction has subsided somewhat, add another 100 ml portion, followed by the remainder when the temperature commences to fall again. Boil the mixture for 5 minutes but no longer (otherwise some hydrogen cyanide may be lost and some sodium glycollate may form) and then cool with running water for 30 minutes, and filter the solution if it is not clear.

Liberate the cyanoacetic acid by cautiously adding with vigorous stirring 250 ml (290 g; a slight excess) of concentrated hydrochloric acid. [*CARE:* hydrogen cyanide is evolved.] Evaporate the solution under reduced pressure (water pump) using a rotary evaporator; do not heat above 75 °C as considerable loss may result owing to the decomposition of the cyanoacetic acid. Add 250 ml of rectified spirit to the residue, filter at the pump (1) from the sodium chloride and wash the residue with another 200 ml of rectified spirit. Evaporate the alcoholic solution under reduced pressure from a water bath (rotary evaporator), maintained at 50–60 °C (2) until no more liquid distils over: the residue weighs about 225 g. Add a mixture of 250 ml of absolute ethanol and 4.5 ml of concentrated sulphuric acid, and reflux on a water bath for 3 hours. Remove the excess of alcohol and some of the water formed by distillation under reduced pressure (rotary evaporator). Heat the residue again with 125 ml of absolute ethanol and 2 ml of concentrated sulphuric acid for 2 hours, and remove the excess of alcohol under diminished pressure as before. Allow the ester to cool to room temperature and neutralise the sulphuric acid with a concentrated solution of sodium carbonate. Separate the upper layer of ester, and extract the aqueous solution with ether (about 10% of the yield is in the extract). Dry the combined products with anhydrous sodium sulphate, remove the solvent by distillation under normal pressure and then distil the ester under reduced pressure. Collect the ethyl cyanoacetate at 97–98 °C/16 mmHg (or at 101–102 °C/19 mmHg or 107–108 °C/27 mmHg). The yield is 180 g (72%).

Notes. (1) The sodium chloride may also be removed by centrifugation. If this method is adopted, wash the salt first with 200 ml and then with 100 ml of ethanol.

(2) The solution containing mineral acid should not be heated above 50–60 °C or diethyl malonate will be formed.

P,3. Hydroxy acids. (*a*) *Hydrolysis of halogeno acids.* α-Halogeno acids are converted into α-hydroxy acids by hydrolysis; the standard conditions usually are to boil with aqueous sodium carbonate solution.

$$R \cdot CHX \cdot CO_2H \xrightarrow{\overset{\ominus}{:OH}} R \cdot CH(OH) \cdot CO_2H$$

The preparation of some α-hydroxy acids by this procedure, however, is not entirely satisfactory because the highly water-soluble and sometimes hygroscopic products are difficult to isolate efficiently. One method consists of treating an aqueous solution of the hydroxy acid with copper(II) acetate to precipitate the insoluble copper salt and decomposing an aqueous suspension of the latter with hydrogen sulphide (Ref. 16). The recovery of the α-hydroxy acid, however, is poor.

β-Halogeno acids on treatment with aqueous alkali give predominantly the α,β-unsaturated acid; the γ- and δ-analogues yield the corresponding lactones.

(*b*) *Hydrolysis of cyanohydrins.* A general method for the synthesis of cyanohydrins (I) is the gradual acidification of a mixture of the carbonyl compound and aqueous sodium cyanide (e.g., acetone cyanohydrin, Section **VI,8**). The hydrolysis (under acidic conditions) of the cyanohydrin constitutes a convenient general route to the α-hydroxy acids (II).

$$R^1R^2CO + HCN \longrightarrow R^1R^2C(OH) \cdot CN \xrightarrow[H_2O]{H^\oplus} R^1R^2C(OH) \cdot CO_2H$$
$$(I) \qquad\qquad\qquad (II)$$

An alternative procedure, suitable for the preparation of cyanohydrins of carbonyl compounds which readily form bisulphite complexes, is illustrated by the preparation of mandelic acid described in Section **III,172**. Here the procedure involves the addition of a saturated solution of sodium metabisulphite to a stirred solution of the carbonyl compound and aqueous sodium cyanide, and when applicable is usually to be preferred to the *in situ* generation of hydrogen cyanide or the use of the highly poisonous liquid hydrogen cyanide as a reagent.

(*c*) *The Reformatsky reaction.* This reaction leads to the ready formation of β-hydroxycarboxylic esters by the interaction of carbonyl compounds and an α-halogeno ester in the presence of zinc. The reaction sequence is similar to that of a Grignard reaction and involves the formation of an intermediate organozinc halide, its addition to the carbonyl group, and the final decomposition of the resulting complex with mineral acid (e.g., ethyl 3-phenyl-3-hydroxypropionate, Section **III,173**).

$$BrCH(R^1) \cdot CO_2C_2H_5 + Zn \longrightarrow BrZnCH(R^1) \cdot CO_2C_2H_5$$

$$R^2 \cdot CO \cdot R^3 + BrZnCH(R^1)CO_2C_2H_5 \longrightarrow$$
$$\underset{\underset{OZnBr}{|}}{R^2R^3C \cdot CH(R^1) \cdot CO_2C_2H_5} \longrightarrow R^2R^3C(OH) \cdot CH(R^1) \cdot CO_2C_2H_5$$

The product is normally the hydroxy ester when the reaction is carried out in a solvent such as ether or benzene, but occasionally dehydration in the course of the reaction or during distillation of the product results in the formation of an

α,β-unsaturated ester together with some of the β,γ-isomer. Dehydration can be deliberately effected with the aid of reagents such as fused potassium hydrogen sulphate or acetic anhydride. Catalytic hydrogenation of the mixture of unsaturated esters followed by ester hydrolysis yields a saturated carboxylic acid, and the whole procedure constitutes a useful two-carbon atom chain extension process.

Methyl γ-bromocrotonate ($BrCH_2 \cdot CH = CH \cdot CO_2CH_3$, a vinylogue of ethyl bromoacetate, Section III,66) can also be used in the Reformatsky reaction and its use permits an analogous four-carbon atom chain extension process.

III,172. MANDELIC ACID

$$C_6H_5 \cdot CHO \xrightarrow{NaHSO_3} C_6H_5 \cdot CH(OH) \cdot SO_3^{\ominus} Na^{\oplus} \xrightarrow{NaCN}$$

$$C_6H_5 \cdot CH(OH) \cdot CN \xrightarrow[\text{conc.}]{HCl} C_6H_5 \cdot CH(OH) \cdot CO_2H + NH_4Cl$$

This preparation must be carried out in an efficient fume cupboard.

Prepare a saturated solution of sodium metabisulphite by stirring 250 g of finely powdered sodium metabisulphite with 335 ml of water for half an hour and then filtering to remove excess of the salt. In a 1-litre three-necked flask equipped with a mechanical stirrer and a dropping funnel, place a solution of 25 g (0.5 mol) of sodium cyanide in 100 ml of water and 53 g (51 ml, 0.5 mol) of purified benzaldehyde (Section IV,22). Add the sodium metabisulphite solution from the dropping funnel, slowly at first and then more rapidly (the addition occupies 10–15 minutes). During the initial stages of the addition, add 150 g of crushed ice to the reaction mixture in several portions through the third neck. Transfer the two-layer liquid mixture to a separatory funnel and remove the crude mandelonitrile (1). Place the crude product at once (2) in a large evaporating dish, add 75 ml of concentrated hydrochloric acid, cover with a clock glass and allow the hydrolysis to proceed at room temperature for 12 hours. Evaporate the solution to dryness on a steam bath, stirring from time to time to break up the deposit of ammonium chloride and mandelic acid which separates. Grind the residue of slightly discoloured mandelic acid and inorganic salts to a fine powder and wash it with two portions of 125 ml of cold benzene (*fume cupboard*); this process will remove most of the colouring matter but a negligible quantity of mandelic acid. To separate the inorganic salts from the mandelic acid, extract the residue in a Soxhlet apparatus (Fig. I,97) with about 200 ml of benzene. Allow the hot benzene extract to crystallise, collect the crystals on a Buchner funnel and dry in air. The yield of pure ($\pm$)-mandelic acid, m.p. 118 °C, is 35 g (46%).

Notes. (1) For the safe disposal of the aqueous solution see Section **III,161**. A further small quantity of mandelonitrile may be obtained by extracting the aqueous solution with ether, evaporating the ether and adding the residue to the main portion of mandelonitrile. This extraction is hardly worth while except for large-scale preparations.

(2) It is important to mix the mandelonitrile with hydrochloric acid immediately it has been separated from the water. Standing results in rapid conversion to the acetal of benzaldehyde and mandelonitrile

$C_6H_5\cdot CH[OCH(CN)\cdot C_6H_5]_2$, the yield of mandelic acid will, in consequence, be reduced.

III,173. ETHYL 3-PHENYL-3-HYDROXYPROPIONATE

$$BrCH_2\cdot CO_2C_2H_5 + Zn \longrightarrow BrZnCH_2\cdot CO_2C_2H_5 \xrightarrow{C_6H_5\cdot CHO}$$

$$C_6H_5\cdot CH(OZnBr)\cdot CH_2\cdot CO_2C_2H_5 \xrightarrow{H_3O^\oplus} C_6H_5\cdot CH(OH)\cdot CH_2\cdot CO_2C_2H_5$$

It is essential that all the apparatus and the reagents be scrupulously dry for successful results (compare Grignard reaction). Equip a 500-ml three-necked flask with a 250-ml separatory funnel, a mechanical stirrer and a double surface condenser; insert calcium chloride guard-tubes in the funnel and condenser. Place 40 g (0.61 mol) of zinc dust (previously dried at 100 °C) (Section II,2,67) in the flask, and a solution of 83.5 g (55.5 ml, 0.5 mol) of ethyl bromoacetate (Section III,146, CAUTION: lachrymatory) (1) and 65 g (62 ml, 0.615 mol) of purified benzaldehyde (Section IV,22) in 80 ml of sodium-dried benzene and 20 ml of sodium-dried ether in the separatory funnel. Add about 10 ml of the solution to the zinc and warm the flask gently until the reaction starts. When the reaction has commenced, but not before, stir the mixture and add the remainder of the solution at such a rate that moderate refluxing occurs (about 1 hour). Reflux the reaction mixture on a water bath for a further 30 minutes. Cool the flask in an ice bath, and add 200 ml of cold 10 per cent sulphuric acid with vigorous stirring. Transfer to a separatory funnel, remove the aqueous layer, wash the benzene layer twice with 50 ml portions of 5 per cent sulphuric acid, once with 25 ml of 10 per cent sodium carbonate solution and finally with two 25 ml portions of water. Extract the combined acid solutions with 100 ml of ether, and dry the combined benzene and ether solution with 5 g of anhydrous calcium sulphate. Filter from the desiccant, remove the solvent by distillation under atmospheric pressure and distil the residue under reduced pressure. Collect the ethyl 3-phenyl-3-hydroxypropionate at 152–154 °C/12 mmHg. The yield is 60 g (62%).

Note. (1) Great care must be exercised in handling ethyl bromoacetate. Keep a 10 per cent aqueous ammonia solution available to react with any bromoester which may be spilled.

Cognate preparation Ethyl 1′-hydroxycyclohexylacetate. Place 65 g (1 mol) of clean dry zinc dust and a few crystals of iodine in a 2.5-litre three-necked flask, equipped with an efficient reflux condenser with drying tube, a mechanical stirrer and a dropping funnel. Prepare a mixture of 400 ml of sodium-dried benzene and 350 ml of sodium-dried toluene with 167 g (111 ml, 1 mol) of ethyl bromoacetate and 98 g (103.5 ml, 1 mol) of pure dried and redistilled cyclohexanone. Transfer 150 ml of this mixture to the flask, start the stirrer and heat the flask in a boiling water bath. A vigorous reaction soon sets in. Add the remainder of the mixture through the dropping funnel at such a rate that gentle refluxing is maintained. Continue the stirring for an additional 2 hours: practically all the zinc dissolves. Cool the mixture, add sufficient 10 per cent sulphuric acid with stirring to dissolve all the zinc hydroxide. Separate the benzene–toluene layer, dry it with anhydrous sodium sulphate, remove the solvent using a rotary evaporator and distil the residue under reduced pressure. Collect the ethyl 1'-hydroxycyclohexylacetate at 86–89 °C/2 mmHg. The yield is 125 g (67%).

P,4. Keto acids. (*a*) *α-Keto acids.* The simplest member of the series of aliphatic α-keto acids is pyruvic acid. It is conveniently prepared by the distillation of tartaric acid with a dehydrating agent such as potassium hydrogen sulphate (Section **III,174**). The reaction probably involves dehydration to the tautomeric oxaloacetic acid (I) intermediate, which then decarboxylates by virtue of its constitution as a *β*-keto acid.

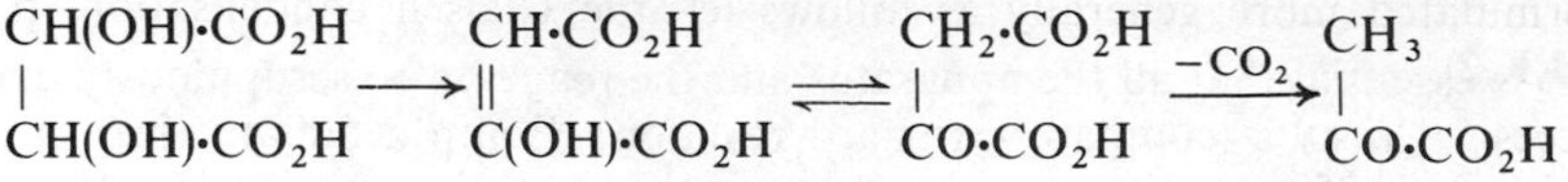

$$\begin{array}{c} CH(OH)\cdot CO_2H \\ | \\ CH(OH)\cdot CO_2H \end{array} \longrightarrow \begin{array}{c} CH\cdot CO_2H \\ || \\ C(OH)\cdot CO_2H \end{array} \rightleftharpoons \begin{array}{c} CH_2\cdot CO_2H \\ | \\ CO\cdot CO_2H \end{array} \xrightarrow{-CO_2} \begin{array}{c} CH_3 \\ | \\ CO\cdot CO_2H \end{array}$$

(I)

Substituted oxaloacetic esters (II), which may be hydrolysed and decarboxylated to homologues of pyruvic acid, may be synthesised by a mixed Claisen ester condensation (see (*b*) below) between a carboxylate ester and diethyl oxalate (cf. Section **VI,25**, ethyl phenyloxaloacetate).

$$R\cdot CH_2\cdot CO_2C_2H_5 + {}^{\ominus}OC_2H_5 \longrightarrow [R\cdot \overset{\ominus}{C}H\cdot CO_2C_2H_5] + C_2H_5OH$$

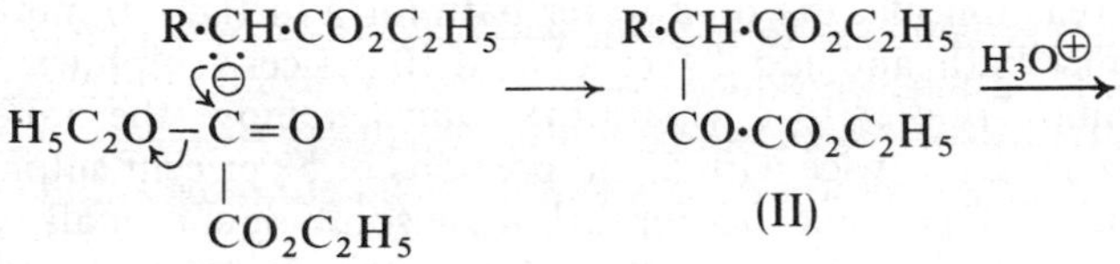

$$\begin{array}{c} R\cdot CH\cdot CO_2C_2H_5 \\ \overset{\overset{\ominus}{\text{C}}}{} \\ H_5C_2O\overset{\frown}{-}C=O \\ | \\ CO_2C_2H_5 \end{array} \longrightarrow \begin{array}{c} R\cdot CH\cdot CO_2C_2H_5 \\ | \\ CO\cdot CO_2C_2H_5 \end{array} \xrightarrow{H_3O^{\oplus}}$$

(II)

$$\begin{bmatrix} R\cdot CH\cdot CO_2H \\ | \\ CO\cdot CO_2H \end{bmatrix} \xrightarrow{-CO_2} R\cdot CH_2\cdot CO\cdot CO_2H$$

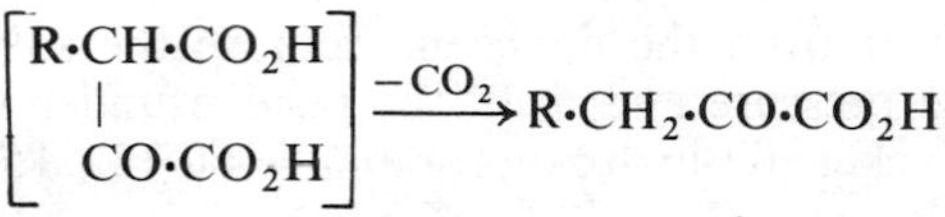

A general route to aryl-substituted pyruvic acids (e.g., phenylpyruvic acid, Section **III,175**) is the acid hydrolysis of 2-acetamido-3-arylacrylic acids (IV), which are themselves formed by hydrolysis of the corresponding azlactones (III) (cf. Section **VI,10**) with water.

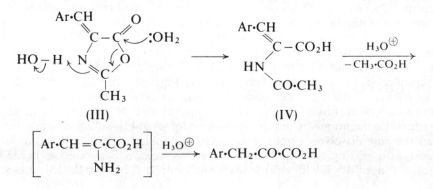

(III)   (IV)

$$\begin{bmatrix} Ar\cdot CH=C\cdot CO_2H \\ | \\ NH_2 \end{bmatrix} \xrightarrow{H_3O^{\oplus}} Ar\cdot CH_2\cdot CO\cdot CO_2H$$

(b) *β-Keto acids.*   Ethyl acetate, containing a trace of ethanol, reacts in the presence of sodium to give a self-condensation product, the sodio-derivative of ethyl acetoacetate, which after careful acidification yields the free β-keto ester (Section **III,176**).

$$2CH_3 \cdot CO_2C_2H_5 \xrightarrow[\text{(ii) } CH_3CO_2H]{\text{(i) Na, } C_2H_5OH \text{ (trace)}} CH_3 \cdot CO \cdot CH_2 \cdot CO_2C_2H_5$$

The overall reaction is an example of the **Claisen ester condensation** and may be formulated more generally as follows (cf. the Claisen condensation, p. 443, **III,K,2**).

$$R \cdot CH_2 \cdot \overset{\overset{\displaystyle O}{\|}}{C} - OC_2H_5 + H - \overset{\overset{\displaystyle R}{|}}{C}H \cdot CO_2C_2H_5 \xrightarrow{-C_2H_5OH}$$

$$R \cdot CH_2 \cdot CO \cdot CH(R) \cdot CO_2C_2H_5$$

The condensation proceeds under the influence of strong base catalysts of which sodium ethoxide is the most common example. This is usually formed from the ethanol present in ordinary samples of the ester by the action of the sodium used in the condensation. The first step in the mechanism, which is formulated below for the case of ethyl acetate, is the removal of the α-hydrogen in an ester molecule by the basic catalyst to produce the mesomerically stabilised α-carbanion (V). The nucleophilic carbanion so formed then attacks the carbonyl carbon of a second molecule of the ester to produce the anion (VI) which is converted into the β-keto ester (VII) by loss of an ethoxide ion. Finally (VII) reacts with the ethoxide ion to produce the mesomerically stabilised β-keto ester anion (VIII).

$$2Na + C_2H_5OH \longrightarrow 2\overset{\oplus}{Na}\overset{\ominus}{O}C_2H_5 + H_2$$

(1) $C_2H_5\overset{\ominus}{O:} \quad H - CH_2 \cdot CO_2C_2H_5 \rightleftharpoons$

$$C_2H_5OH + \left[\overset{\ominus}{C}H_2 - \underset{\underset{\displaystyle OC_2H_5}{|}}{C} = O \longleftrightarrow CH_2 = \underset{\underset{\displaystyle OC_2H_5}{|}}{C} - \overset{\ominus}{O} \right] \quad (V)$$

(2) $CH_3 \cdot \overset{\overset{\displaystyle O}{\|}}{\underset{\underset{\displaystyle OC_2H_5}{|}}{C}} \quad \overset{\ominus}{:}CH_2 \cdot CO_2C_2H_5 \rightleftharpoons CH_3 \cdot \underset{\underset{\displaystyle OC_2H_5}{|}}{\overset{\overset{\displaystyle :O^{\ominus}}{|}}{C}} CH_2 \cdot CO_2C_2H_5$

$$(VI)$$

$$\xrightarrow[]{-C_2H_5O^{\ominus}} CH_3 \cdot CO \cdot CH_2 \cdot CO_2C_2H_5 \quad (VII)$$

(3) $CH_3 \cdot CO \cdot \underset{\underset{\displaystyle H \quad :OC_2H_5}{|}}{C}H \cdot CO_2C_2H_5 \rightleftharpoons C_2H_5OH +$

$$\left[CH_3 \cdot \overset{\overset{\displaystyle O}{\|}}{C} - \overset{\ominus}{C}H \cdot CO_2C_2H_5 \longleftrightarrow CH_3 \cdot \overset{\overset{\displaystyle O}{|}}{C} = CH \cdot CO_2C_2H_5 \right] \quad (VIII)$$

The equilibrium in the last step (3) is far to the right because of the greater basic strength of the ethoxide ion compared to the anion (VIII), and this largely assists the forward reactions in (1) and (2). The reaction product is the sodio-derivative of the β-keto ester from which the free ester is obtained upon acidification. β-Keto esters may be hydrolysed in the cold with dilute aqueous sodium hydroxide whereupon careful acidification affords the free β-keto acid. These keto acids are, however, thermally unstable and aqueous solutions of the acids or their salts readily undergo decarboxylation on heating ('ketonic fission') to form the corresponding ketones (cf. p. 434, III,J,5).

Only esters containing at least two α-hydrogen atoms (ethyl acetate, propionate, butyrate, etc.) can be condensed with the aid of sodium alkoxides. For esters with only one α-hydrogen atom, such as ethyl isobutyrate, a more powerful base (e.g., sodium triphenylmethide, $(C_6H_5)_3C^{\ominus}: Na^{\oplus}$) is required to effect the condensation reaction (the *forced* Claisen ester condensation, e.g., the synthesis of ethyl isobutyrylisobutyrate, cognate preparation in Section III,176).

$$2(CH_3)_2CH\cdot CO_2C_2H_5 \xrightarrow{(C_6H_5)_3C^{\ominus}Na^{\oplus}} (CH_3)_2CH\cdot CO\cdot C(CH_3)_2\cdot CO_2C_2H_5$$
$$(IX)$$

In this case the reaction sequence is completed in the step corresponding to (2) above since the β-keto ester (IX) has no α-hydrogen for step (3), and the powerful base is required to force the equilibrium (1) to the right.

When a mixture of two dissimilar esters ($R^1\cdot CH_2\cdot CO_2C_2H_5$ and $R^2\cdot CH_2\cdot CO_2C_2H_5$) is treated with sodium as the ethoxide, a 'mixed' β-keto ester ($R^1\cdot CH_2\cdot CO\cdot CH(R^2)\cdot CO_2C_2H_5$ or $R^2\cdot CH_2\cdot CO\cdot CH(R^1)\cdot CO_2C_2H_5$) may be formed. However, the disadvantage which is common to all mixed reactions, i.e., the formation of both symmetrical and crossed products, reduces its preparative value. Exceptions are provided in those cases where one of the esters contains no α-hydrogen atoms. An important example is provided by the use of esters of oxalic acid in such mixed Claisen ester condensations (e.g., p. 536, III,P,4(a) and Section V,4).

Although a mixed Claisen ester condensation between ethyl acetate, for example, and another carboxylic ester does provide in principle a method for the synthesis of β-keto esters of the type $R\cdot CO\cdot CH_2\cdot CO_2C_2H_5$, these are more conveniently prepared starting from diethyl malonate. The magnesium enolate of diethyl malonate is acylated with an acyl chloride in benzene solution (cf. Section III,95), and the resulting acylmalonic ester undergoes loss of an ethoxycarbonyl group when it is heated to 200 °C with an arylsulphonic acid such as naphthalenesulphonic acid.

$$R\cdot COCl + [CH(CO_2C_2H_5)_2]Mg(OC_2H_5) \longrightarrow$$

$$R\cdot CO\cdot CH(CO_2C_2H_5)_2 \xrightarrow{\quad\text{SO}_3\text{H}\quad} R\cdot CO\cdot CH_2\cdot CO_2C_2H_5$$

An illustrative example is the preparation of ethyl propionylacetate (Section III,177).

(c) *γ-Keto acids*.   The reaction of aldehydes and ketones with succinic esters in the presence of potassium t-butoxide (or with sodium hydride) to give an alkylidenesuccinic acid ester (X) is known as the Stobbe condensation. The less basic sodium ethoxide is not satisfactory in this case since the reaction is then comparatively slow.

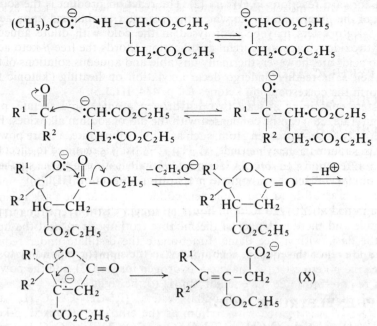

$$(CH_3)_3CO: \quad H-CH \cdot CO_2C_2H_5 \qquad :CH \cdot CO_2C_2H_5$$

One of the applications of the Stobbe condensation provides a convenient synthesis of γ-keto acids. In this case the alkylidenesuccinic acid ester (XI),

$$R \cdot CHO + CH_2 \cdot CO_2C_2H_5$$

$$R \cdot CH = C \cdot CO_2H$$

(XII) $CH_2 \cdot CO_2H$

$$\xrightarrow[+H^{\oplus}]{-CO_2} R \cdot CO \cdot CH_2 \cdot CH_2 \cdot CO_2H$$

formed from an aldehyde and diethyl succinate, is carefully hydrolysed to give the alkylidenesuccinic acid (XII). Photocatalysed addition of bromine yields the corresponding dibromoacid (XIII) which is converted into the γ-keto acid by treatment with alkali. This sequence is illustrated by the preparation of 4-oxo-decanoic acid (Section **III,178**).

III,174. PYRUVIC ACID

$$\begin{array}{l} CHOH \cdot CO_2H \\ | \\ CHOH \cdot CO_2H \end{array} \xrightarrow{KHSO_4} CH_3 \cdot CO \cdot CO_2H + CO_2 + H_2O$$

Grind together in a glass mortar 200 g (1.33 mol) of powdered tartaric acid and 300 g (2.2 mol) of freshly *fused* potassium hydrogen sulphate to form an intimate mixture. Place the mixture in a 1.5-litre round-bottomed flask, and fit the latter with a still-head and a long air condenser. Heat the flask in an oil bath maintained at 210–220 °C until liquid no longer distils over. If foaming is considerable and there is danger of the mixture frothing over, heat the upper part of the flask with a free flame. Fractionate the distillate under reduced pressure and collect the pyruvic acid at 75–80 °C/25 mmHg. The yield is 60 g (51%).

III,175. PHENYLPYRUVIC ACID

$$\begin{array}{l} C_6H_5 \cdot CH = C \cdot NH \cdot CO \cdot CH_3 + 2H_2O \longrightarrow \\ | \\ CO_2H \end{array}$$

$$C_6H_5 \cdot CH_2 \cdot CO \cdot CO_2H + CH_3 \cdot CO_2NH_4$$

Place 10.3 g (0.05 mol) of α-acetamidocinnamic acid (Section **VI,10**) and 200 ml of 1 *M*-hydrochloric acid in a 500-ml round-bottomed flask and boil the mixture steadily under reflux for 3 hours. Remove a small quantity of green oil by rapidly filtering the hot reaction mixture through a small plug of cotton wool loosely inserted into the stem of a pre-heated glass filter funnel, cool the filtrate to room temperature and leave it at 0 °C for 48 hours. Collect the crystalline product by filtration, wash it with a small quantity of ice-cold water and dry it in a vacuum desiccator over anhydrous calcium chloride and potassium hydroxide pellets. The yield of phenylpyruvic acid, which is sufficiently pure for most purposes, is 4.4 g, m.p. 157 °C (decomp.). A further 1.7 g of product of comparable purity (total yield 74%) separates from the aqueous acidic filtrate when this is set aside at 0 °C for about one week.

III,176. ETHYL ACETOACETATE

$$2CH_3 \cdot CO_2C_2H_5 \xrightarrow{Na} (CH_3 \cdot CO \cdot CH \cdot CO_2C_2H_5)^{\ominus} Na^{\oplus} \xrightarrow{H^{\oplus}}$$

$$CH_3 \cdot CO \cdot CH_2 \cdot CO_2C_2H_5$$

Into a 1-litre round-bottomed flask, fitted with a double surface condenser, place 250 g (277 ml, 2.84 mol) of dry ethyl acetate (1) and 23 g (1 mol) of clean sodium wire (2). Warm the flask on a water bath in order to start the reaction.

Once the reaction commences, it proceeds vigorously and cooling of the flask may be necessary in order to avoid loss of ethyl acetate through the condenser. When the vigorous reaction is over, warm the reaction mixture in a water bath until the sodium is completely dissolved (about 1.5 hours). Cool the resulting clear red solution and make it slightly acid to litmus paper by the addition of about 125 ml of 50 per cent acetic acid. Saturate the liquid with salt, separate the upper layer of ester and dry it with anhydrous calcium chloride. Distil under reduced pressure through a short fractionating column (3). After a fore-run of ethyl acetate, collect the ethyl acetoacetate at 76–80 °C/18 mmHg (4). The yield is 50 g (38% based on sodium).

Notes. (1) It is important to use dry ethyl acetate, but it should contain 2–3 per cent of ethanol. Commercial ethyl acetate is about 99.5 per cent pure and is usually satisfactory; it may be advantageous to dry it over anhydrous calcium sulphate. After filtering, 5 ml of absolute ethanol should be added to the 250 g portion of ester.

(2) Sodium wire, produced with a sodium press (Fig. II,1), is first collected in sodium-dried ether, the necessary quantity removed, rapidly dried between filter-paper, and transferred to the flask. Thin shavings of sodium although less satisfactory may also be employed, but it is important to avoid undue exposure of the sodium to the atmosphere which produces a surface film of sodium hydroxide.

(3) Ethyl acetoacetate decomposes slightly when distilled at atmospheric pressure. The extent of decomposition is reduced if the distillation is conducted rapidly. The b.p. is 180 °C/760 mmHg and a 6 °C fraction should be collected. Normal pressure distillation is not recommended if a pure product is desired.

(4) The boiling points of ethyl acetoacetate under various pressures are: 71 °C/12 mmHg; 73 °C/15 mmHg; 78 °C/18 mmHg; 82 °C/20 mmHg; 88 °C/30 mmHg; 92 °C/40 mmHg; 97 °C/60 mmHg and 100 °C/80 mmHg.

Cognate preparation Ethyl isobutyrylisobutyrate. *Triphenylmethyl sodium.* Prepare a 1.5 per cent sodium amalgam from 15 g (0.65 mol) of sodium and 985 g of mercury (Section II,2,57). Place a mixture of 1000 g of the amalgam and 74 g (0.265 mol) of triphenylchloromethane (Section II,2,66) in a 2-litre Pyrex glass-stoppered bottle and add 1500 ml of sodium-dried ether. Grease the glass stopper with a little Silicone grease, insert it firmly, clamp the bottle in a mechanical shaker and shake. The reaction is strongly exothermic; cool the bottle with wet rags and stop the shaking from time to time, if necessary. A characteristic red colour appears after about 10 minutes shaking. After shaking for 4 to 6 hours, *cool the bottle to room temperature*, remove it from the shaker, wire the stopper down and allow the mixture to stand undisturbed; sodium chloride and particles of mercury settle to the bottom.

Separate the ether solution of triphenylmethyl sodium as follows. Remove the glass stopper and replace it immediately by a tightly fitting two-holed bung carrying a short glass tube that protrudes about 1 cm into the bottle, and a long glass tube bent into an inverted U-shape. Connect the bottle through a drying train to a cylinder of nitrogen. Lead the other arm of the U-tube into a 2-litre, two-necked round-bottomed flask (which has been previously filled with nitrogen) via a suitable screw-capped adapter and fit a dropping funnel into the other neck. Open the stop-cock of the dropping funnel slightly and force the ether solution of triphenylmethyl sodium slowly and steadily into the nitrogen-filled flask by means of a small pressure of nitrogen from the cylinder. By carefully adjusting the depth of the siphon tube in the bottle, all but 50–75 ml of the clear ether solution may be removed.

If pure triphenylchloromethane and freshly prepared sodium amalgam are used, the yield of triphenylmethyl sodium should be almost quantitative and the concentration is usually 0.15 mol per litre (1). The reagent should be used as soon as possible after its preparation.

Ethyl isobutyrylisobutyrate. Add 24 g (28 ml, 0.21 mol) of ethyl isobutyrate, b.p. 110–111 °C, to the solution of *ca.* 0.21 mol of triphenylmethyl sodium in approximately 1400 ml of ether contained in the 2-litre two-necked flask. Stopper the flask, shake well to effect complete mixing and keep at room temperature for 60 hours. Acidify the reaction mixture by adding, with shaking, 15 ml of glacial acetic acid, and then extract with 100 ml of water. Wash the ethereal solution with 50 ml portions of 10 per cent sodium carbonate solution until free from excess acid, dry over anhydrous sodium sulphate; remove the ether under reduced pressure with a rotary evaporator. Distil the residue under reduced pressure through a short fractionating column. Collect the ethyl iso-butyrylisobutyrate at 95–96 °C/18 mmHg; the yield is 14.5 g (74%). The b.p. at atmospheric pressure is 201–202 °C.

Note. (1) The solution may be analysed approximately as follows. Remove 25 ml of the ether solution, run it into 25 ml of water contained in a small separatory funnel and shake. Run off the aqueous layer into a 250-ml conical flask and extract the ether layer with two 25 ml portions of water. Titrate the combined aqueous extracts with 0.05 M-sulphuric acid, using methyl red as indicator.

Ketonic hydrolysis to di-isopropyl ketone. Mix 14 g (0.075 mol) of the ester with 30 ml of glacial acetic acid, 10 ml of water and 10 ml of concentrated sulphuric acid, and boil under reflux until evolution of carbon dioxide ceases. Dilute the cooled solution with 180 ml of water, add 100 ml of ether and render alkaline to phenolphthalein with 20 per cent sodium hydroxide solution. Separate the ether layer, extract the aqueous layer with two 50 ml portions of ether, dry the combined ether layer and extracts with anhydrous sodium sulphate, distil off the ether and fractionate the residue. The yield of di-isopropyl ketone (2,4-dimethylpentan-3-one), b.p. 123–124 °C, is 6.5 g (76%).

III,177. ETHYL PROPIONYLACETATE

$$CH_3 \cdot CH_2 \cdot COCl + [CH(CO_2C_2H_5)_2]MgOC_2H_5 \longrightarrow$$

$$CH_3 \cdot CH_2 \cdot CO \cdot CH(CO_2C_2H_5)_2 \xrightarrow{\text{(naphthalene-}SO_3H)}$$

$$CH_3 \cdot CH_2 \cdot CO \cdot CH_2 \cdot CO_2C_2H_5$$

Prepare the ethoxymagnesium diethyl malonate derivative from 13 g (0.53 mol) of magnesium and 80 g (0.5 mol) of diethyl malonate following the procedure described in Section **III,95**. Then add with vigorous stirring a solution of 49 g (46 ml, 0.53 mol) of propionyl chloride in 50 ml of anhydrous ether. Reflux the reaction mixture for 30 minutes and then cool and acidify with 60 ml of dilute sulphuric acid. Separate the ether layer, extract the residual aqueous solution with two 50 ml portions of ether. Wash the combined organic phases with water, dry over anhydrous sodium sulphate and remove the ether on a rotary evaporator. Add to the residue 8 g of naphthalene-2-sulphonic acid monohydrate and heat the mixture slowly to 200 °C in an oil bath. A vigorous evolution of gas sets in at about 120 °C; when gas evolution has subsided, cool the reaction mixture and dissolve it in about 150 ml of ether. Wash the ethereal extract with four 25 ml portions of 10 per cent sodium carbonate solution and then with water; back extract the combined aqueous solutions with three 25 ml portions of ether. Dry the combined ether extracts with anhydrous sodium sulphate, remove the ether on a rotary evaporator and distil the residue under reduced pressure. Collect the ethyl propionylacetate as a fraction of b.p. 100–105 °C/22 mmHg; the yield is 34 g (47%).

III,178. 4-OXODECANOIC ACID

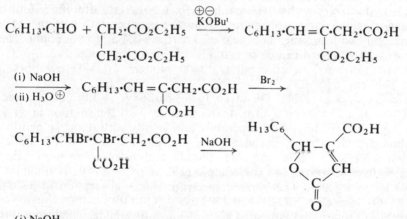

3-Carboxydec-3-enoic acid (*Heptylidenesuccinic acid*). *Potassium t-butoxide.* Prepare a solution of potassium t-butoxide in t-butyl alcohol using the following procedure. Equip a two-necked, round-bottomed 1-litre flask with a reflux condenser and a pressure-equalising dropping funnel. Attach to the top of the condenser a nitrogen inlet system with nitrogen escape valve (cf. Fig. I,67) and place the flask in a magnetic stirrer-heating mantle unit. Flush the flask with a stream of dry nitrogen, charge the flask with 375 ml of dry t-butanol (Section II,1,*12*) and add 19.6 g (0.5 mol) of potassium [CAUTION: (1)]. Heat the mixture under reflux until the potassium completely dissolves (*ca.* 4 hours). To the cooled solution add a mixture of 122 g (0.7 mol) of diethyl succinate and 63 g (0.5 mol) of heptanal over half an hour; the reaction is exothermic. Finally heat the mixture under reflux for 1 hour. Rearrange the condenser for distillation and remove the t-butyl alcohol by vacuum distillation using a water pump. Acidify the residue with dilute hydrochloric acid and extract with three 200 ml portions of ether. Extract the acidic compounds from the ethereal solution by shaking with 50 ml portions of saturated sodium hydrogen carbonate solution until no more carbon dioxide is evolved. Acidify the combined aqueous solutions by the careful addition of concentrated hydrochloric acid. Extract the 3-ethoxycarbonyldec-3-enoic acid which separates as an oil with ether, dry the ethereal extract over magnesium sulphate and evaporate the ether solution on a rotary evaporator. Saponify the crude half ester by heating it under reflux for 1 hour with 400 ml of 10 per cent sodium hydroxide solution. Cool the solution, acidify with concentrated hydrochloric acid and filter the precipitated acid with suction. Dissolve the crude acid in the minimum volume of ether (about 200 ml are required) and add the solution to an equal volume of light petroleum (b.p. 60–80 °C). Filter the precipitate with suction and recrystallise from benzene. The yield of 3-carboxydec-3-enoic acid, m.p. 128–130 °C, is 37.3 g (31.6%). The infrared spectrum shows absorptions at 3500–2300 cm^{-1} (O – H stretch of carboxylic acid), 1700 cm^{-1} (C=O) and 1640 cm^{-1} (C=C).

3-Carboxy-3,4-dibromodecanoic acid. *This reaction should be carried out in a fume cupboard.* Place 5.3 g (0.025 mol) of 3-carboxydec-3-enoic acid, 6.0 g (0.038 mol) of bromine and 60 ml of carbon tetrachloride in a 100-ml round-bottomed flask equipped with a magnetic stirrer and reflux condenser. Irradiate the stirred mixture with a 100-watt lamp for 6 hours; the dibromide forms and precipitates out during this period. Filter the product with suction and wash thoroughly with hexane. The yield of 3-carboxy-3,4-dibromodecanoic acid is 7.9 g (85%). The acid can be recrystallised from toluene, m.p. 142–143 °C. The infrared spectrum shows absorptions at 3400–2400 cm^{-1} (O – H stretch of CO$_2$H) and 1730 cm^{-1} (C=O).

4-Oxodecanoic acid. Dissolve 4.0 g of 3-carboxy-3,4-dibromodecanoic acid in 60 ml of 2 *M*-sodium hydroxide solution and heat the solution at 80–90 °C for 2 hours. Cool to room temperature and acidify with dilute sulphuric acid; carbon dioxide is evolved and a white precipitate is formed. Filter the precipitated keto acid from the cold solution and recrystallise from light petroleum (b.p. 40–60 °C). 4-Oxodecanoic acid, m.p. 68–69 °C, is obtained; the yield is 1.6 g (80%). The infrared spectrum shows absorption at 3400–2400 cm^{-1} (OH stretch of CO$_2$H) and 1700 cm^{-1} (C=O).

If the dibromo acid is treated with alkali under milder conditions, for example with 1 *M*-sodium hydroxide solution at 20–25 °C for 0.5 hour, the intermediate γ-hexylaconic acid (m.p. 123–125 °C) can be isolated after acidification. The infrared spectrum shows absorptions at 3110 cm^{-1} (C – H stretch, alkene), 1715 and 1745 cm^{-1} (C=O stretch of carboxylic acid and lactone) and 1630 cm^{-1} (C=C stretch).

Note. (1) *Great care must be taken in the handling of potassium* and the following precautions must be rigidly observed. Cut the metal under light petroleum (which has been dried over sodium wire) contained in a mortar: do not use a beaker or a crystallising dish because it is too fragile. Cut off the outer oxide-coated surface and immediately transfer the scraps with tweezers to a second mortar containing dry light petroleum. Weigh the freshly cut potassium by removing it with tweezers to a filter-paper, blot it rapidly and introduce it into a tared beaker containing dry light petroleum. Introduce the weighed potassium into the reaction mixture. The scraps of potassium should not be stored; they must be decomposed immediately by transferring the mortar to the rear of an empty fume cupboard and adding t-butyl alcohol (not methanol or ethanol) in small portions from a dropper pipette at such a rate that the reaction does not become vigorous. Keep a square sheet of asbestos, large enough to cover the mortar, at hand; if the liquid should catch fire, it may be extinguished easily by covering the mortar with the asbestos sheet. Add sufficient t-butyl alcohol to react completely with all the potassium. Any specks of potassium remaining in the first mortar used for the cutting operation or small scraps that adhere to the knife must be disposed of in the fume cupboard by cautious treatment with t-butyl alcohol as described above.

P,5. Amino acids. The α-aminocarboxylic acids are of particular importance as the result of their involvement in the fundamental structures of protein molecules, and their synthesis is illustrated here by a variety of procedures. The main general synthetic routes to the α-amino acids are included in the following methods; some of the natural amino acids can be isolated from the hydrolysates of suitable proteins.

(a) *The amination of α-halogenocarboxylic acids.* This method is generally suitable for those α-amino acids ($H_2N \cdot CH(R) \cdot CO_2H$) where R is an unsubstituted alkyl group. The examples included in Section **III,179** are alanine (R = CH_3), glycine (R = H), valine (R = $(CH_3)_2CH$), norvaline (R = $CH_3 \cdot CH_2 \cdot CH_2$), norleucine (R = $CH_3 \cdot CH_2 \cdot CH_2 \cdot CH_2$) and isoleucine (R = $CH_3 \cdot CH_2 \cdot CH(CH_3)$).

Chloroacetic acid is readily converted into glycine by treatment with concentrated aqueous ammonia solution, but in general an α-bromocarboxylic acid is preferred. This can usually be prepared in good yield by a Hell–Volhard–Zelinsky bromination (p. 257, **III,P,1**(a)) of the corresponding carboxylic acid; if the carboxylic acid is not readily available it can usually be obtained by the synthesis and bromination of the appropriate alkylmalonic acid (see Section **III,169**).

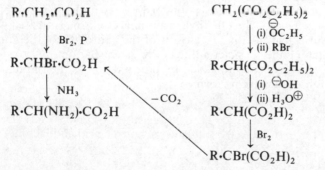

A large excess of concentrated ammonia has to be used to accomplish the amination step, otherwise the amino acid which is formed reacts further with the bromo acid to give substantial amounts of the secondary and tertiary amine derivatives. The use of an excess of ammonium carbonate in aqueous ammonia is also recommended. The reaction is usually carried out at room temperature, but temperatures of about 50 °C are advantageous when the substitution process is retarded as the result, for example, of steric effects. The higher homologous amino acids are not particularly water-soluble, and they may be isolated in the crude state by concentrating and filtering the reaction product. Care must be taken during the isolation of the readily water-soluble lower members of the series to free them from accompanying inorganic salts; in such cases advantage is taken of the solubility of ammonium bromide in methanol to achieve the required purification.

Although the formula $NH_2 \cdot CH(R) \cdot CO_2H$ is frequently used to represent an amino acid the molecule actually has a dipolar (zwitterionic) structure, i.e., $\overset{\oplus}{N}H_3 \cdot CH(R) \cdot CO_2^{\ominus}$, which more clearly indicates its salt-like properties.

(b) *The formation and hydrolysis of α-amino nitriles (the* **Strecker synthesis**). The original Strecker procedure is the reaction of an aldehyde with ammonia and then with hydrogen cyanide to form the α-amino nitrile (I). This intermediate may also be obtained by reacting the aldehyde cyanohydrin with ammonia, but a more convenient method is to treat the aldehyde in one step with ammonium chloride and sodium cyanide. The α-amino acid is obtained when the amino nitrile is hydrolysed under either acidic or basic conditions; the former are usually preferred. The preparation of α-phenylglycine ($R = C_6H_5$) from benzaldehyde is typical of the general procedure (cognate preparation in Section **III,180**).

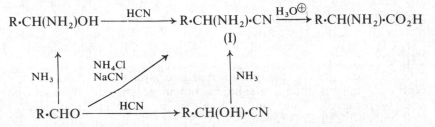

When applied to formaldehyde, however, the reaction is somewhat anomalous in that methyleneaminoacetonitrile ($CH_2 = N \cdot CH_2 \cdot CN$), the condensation product derived from the aldehyde and the amino nitrile, is formed (Section **III,180**). The free amino nitrile is obtained by careful basification of its sulphate salt, which is formed when methyleneaminoacetonitrile is treated with concentrated sulphuric acid in ethanol. Details of the hydrolysis of the amino nitrile (as the sulphate) under basic conditions are given. Barium hydroxide is used, the excess of which is finally removed by precipitation as the sulphate to facilitate the isolation of the glycine formed.

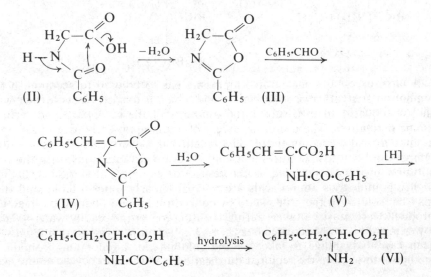

(c) *The reductive hydrolysis of arylideneoxazolones (azlactones—the* **Erlenmeyer synthesis**). This method is mainly restricted to the synthesis of amino acids with aromatic side-chains since the required unsaturated azlactones [e.g., (IV)] are most readily prepared using aromatic aldehydes. Typically, benzaldehyde condenses under the influence of base with the reactive methylene group in the azlactone (III) which is formed by the dehydration of benzoylglycine (II) when the latter is heated with acetic anhydride in the presence of sodium acetate (cf. Section **VI,10**). The azlactone ring is readily cleaved hydrolytically and compounds of the type (IV) yield substituted acylaminoacrylic acids [e.g., (V)] on boiling with water. Reduction and further hydrolysis yields the amino acid [e.g., phenylalanine, (VI)], but this result can be achieved in one step (Section **III,181**) by heating the azlactone (IV) in acetic anhydride solution with red phosphorus and hydriodic acid.

(d) *The formation and hydrolysis of C-substituted acylaminomalonic esters.* Acylaminomalonic esters and related reagents are widely used for the synthesis of α-amino acids. The method differs from those syntheses already discussed in that the amino group is incorporated into the system from the outset. A popular reagent is diethyl acetamidomalonate (IX). The acetamido group can readily be introduced into the reactive methylene position in diethyl malonate by first converting the latter into the hydroxy-imino derivative (VII) by reaction with nitrous acid or an alkyl nitrite (cf. Section **III,101**). This derivative is then reduced catalytically to diethyl aminomalonate (VIII) which is acetylated using acetic anhydride.

$$CH_2(CO_2C_2H_5)_2 \xrightarrow[\text{RONO}]{\text{HNO}_2 \text{ or}} \underset{\overset{|}{N=O}}{CH(CO_2C_2H_5)_2} \rightleftharpoons \underset{\overset{\|}{NOH}}{C(CO_2C_2H_5)_2} \quad \text{(VII)}$$

$$\xrightarrow{H_2;\ \text{catalyst}} \underset{\overset{|}{NH_2}}{CH(CO_2C_2H_5)_2} \xrightarrow{(CH_3\cdot CO)_2O} \underset{\overset{|}{NH\cdot CO\cdot CH_3}}{CH(CO_2C_2H_5)_2}$$

$$\text{(VIII)} \qquad\qquad\qquad\qquad\qquad \text{(IX)}$$

In common with other malonate derivatives, the α-hydrogen atom in the acetamidomalonate is reactive; on treatment with a base the reagent forms a mesomeric stabilised carbanion (X) from which a variety of substituted acetamidomalonic esters can be made. For example, C-alkylation ensues when the anion is allowed to react with an alkyl halide; the resulting product (XI) is then subjected to the hydrolytic and decarboxylative sequence shown to yield a simple α-amino acid.

The two examples illustrative of the section, however, involve alternative procedures for introducing the required substituent into the α-position of the acetamidomalonate reagent. In the first (Section **III,182**) the hydroxymethyl group is introduced by a simple base-catalysed condensation with formaldehyde; subsequent hydrolysis and decarboxylation yields serine (R = CH_2OH). In this case, acidic conditions are preferred for the final hydrolytic stage, and the use of a weakly basic ion exchange resin to obtain the halide-free amino acid from a solution of its hydrochloride is described.

The second example is an interesting synthesis of the heterocyclic amino acid

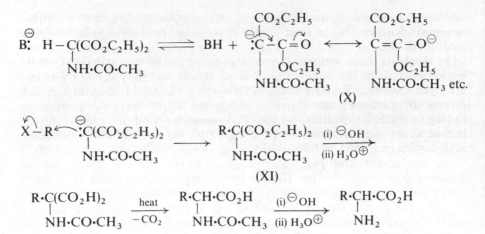

tryptophan (R = 3-indolylmethyl) which involves the initial base catalysed 1,4-addition (the Michael reaction, p. 488, **III,M,6**(*b*)) of diethyl acetamidomalonate to the α,β-unsaturated aldehyde, acrylaldehyde, yielding the aldehydic derivative (XII). The derived phenylhydrazone (XIII) is then cyclised under acidic conditions (see Section **VI,12**) to form the indolylacetamidomalonate derivative (XIV) which is then converted into the corresponding α-amino acid (i.e., tryptophan) in the usual way (Section **III,183**).

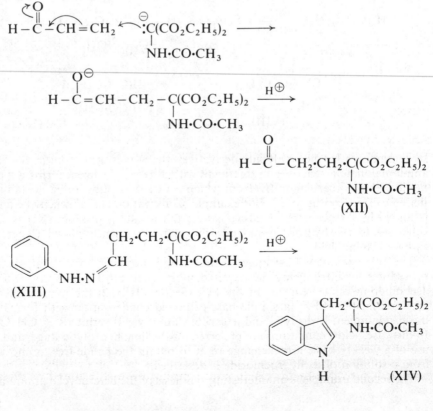

Although it does not involve the use of an acylaminomalonic ester a related convenient synthesis of the heterocyclic amino acid, proline, is included here (Section **III,184**).

The synthesis starts with the preparation of diethyl (2-cyanoethyl)malonate (Section **III,165**) by the Michael addition of diethyl malonate to acrylonitrile. Hydrogenation over Raney nickel converts the cyanoethyl compound to the corresponding primary amine (XV) which is converted into proline (isolated initially as the hydrochloride) by the reaction sequence shown. Liberation of the free amino acid from its salt is achieved in this case by treatment with triethylamine (cf. serine, Section **III,182**).

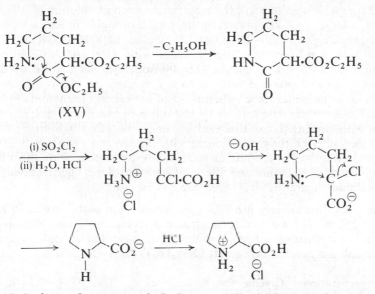

(XV)

(e) Isolation from protein hydrolysates. The hydrolysis of proteins yields a complex mixture of amino acids having closely related physical properties. Although a quantitative separation of the individual amino acids can be achieved by chromatographic methods, only in a few cases is the preparative isolation of an individual member worthwhile. Provided that racemisation is avoided during hydrolysis, however, amino acids are obtained in this way in the optically active form. The normal synthetic products are of course racemic, but they may be resolved by appropriate procedures (e.g., the enzymic method described in Section **III,198**).

The amino acid composition of keratin, the protein of hair and wool, includes a greater-than-average proportion of the sulphur-containing amino acid, cystine. Since this is the least soluble of the protein amino acids it can readily be isolated after carefully neutralising an acid hydrolysate of hair (Section **III,186**). Protein hydrolysis is usually effected by boiling for about 10–20 hours with 20 per cent hydrochloric acid. The hydrolysis of hair for the isolation of cystine is, however, best achieved using a mixture of hydrochloric and formic acids.

Many relatively specific reagents, particularly a variety of metal complexes, have been developed as an aid to the isolation of individual amino acids. An example is provided by the isolation of L-proline from a gelatine hydrolysate using the chromium complex ammonium rhodanilate {ammonium dianiline-

tetrathiocyanatochromate (III), $[Cr(CNS)_4(C_6H_5NH_2)_2]^{\ominus}\overset{\oplus}{NH_4}$}. Proline is isolated as the rhodanilate salt, which is purified and then treated with pyridine to
form the less soluble pyridine rhodanilate, thus liberating the proline into solution.

III,179.   DL-ALANINE (2-Aminopropionic acid)

$$CH_3 \cdot CHBr \cdot CO_2H \xrightarrow{NH_3} CH_3 \cdot CH(NH_2) \cdot CO_2H$$

Place 2 litres (1760 g, 36 mol) of concentrated ammonia solution (d 0.88, 35%
w/w) (1) in a large (e.g., Winchester) bottle, cool thoroughly in an ice bath and
pour in slowly 77 g (0.5 mol) of 2-bromopropionic acid (Section III,168). Close
the bottle with a rubber bung held in place with wire, and leave at room
temperature for 4 days. Concentrate the solution to about 250 ml by distillation under reduced pressure using a rotary evaporator; apply the vacuum
with caution in the initial stages when most of the excess of ammonia is being
removed. Filter, concentrate further to 150 ml, cool in ice and add 750 ml of
methanol with swirling. Leave the resulting suspension overnight in a refrigerator, and then filter off the crude alanine with suction and wash it with 200 ml
of methanol. Dissolve the product in 150 ml of water, reprecipitate the alanine
by adding 750 ml of methanol and filter and wash as before. The yield of almost
pure DL-alanine, m.p. 295–296 °C (decomp.), is 30 g (67%).

   Note.   (1) Alternatively use 225 g of 'ammonium carbonate', 175 ml of
water and 250 ml (4.5 mol) of concentrated ammonia solution. 'Ammonium
carbonate' is a mixture of roughly equimolar amounts of ammonium hydrogen
carbonate and ammonium carbamate ($NH_4HCO_3 \cdot NH_2CO_2NH_4$).

***

Cognate preparations   Glycine (*aminoacetic acid*).   Use 47 g (0.5 mol) of
chloroacetic acid (**CAUTION**: the compound causes blistering if it is allowed
to come into contact with the skin) and 2 litres of concentrated ammonia
solution. Allow the reaction to proceed for 2 days, concentrate to 60 ml and precipitate the crude glycine by adding 360 ml of methanol. This material contains
ammonium chloride as the chief impurity; remove most of this by stirring the
crystals with 150 ml of methanol and refiltering. Finally purify the glycine by
dissolving it in 50 ml of hot water and adding 250 ml of methanol; the yield is
25 g (67%), m.p. *ca.* 252–254 °C (decomp.).

   DL-**Valine** (*2-amino-3-methylbutyric acid*).   Use 60 g (0.33 mol) of 2-
bromo-3-methylbutyric acid (Section III,168) and 400 ml (7.25 mol) of concentrated ammonia solution. Allow reaction to proceed at room temperature
for 7 days. Concentrate the solution to 50 ml and filter the resulting thin paste.
Dissolve the solid in 150 ml of hot water, decolourise with 1 g of charcoal, filter
hot and dilute the filtrate with 150 ml of ethanol. Cool at 0 °C overnight, filter
off the purified DL-valine and wash with 10 ml of cold ethanol. The yield is 12.5 g
(32%); m.p. 280–282 °C (decomp.). A further 2 g may be isolated by concentrating the mother-liquor to about 25 ml and adding an equal volume of
ethanol.

   DL-**Norvaline** (*2-aminovaleric acid*).   Prepare as for valine, using 2-bromo-
valeric acid (Section III,168); m.p. *ca.* 300 °C (decomp.).

DL-**Norleucine** (*2-aminohexanoic acid*).   Use 65 g (0.33 mol) of 2-bromo-hexanoic acid (Section **III,168**) and 400 ml of concentrated ammonia. Ensure that the bung is securely wired to the reaction bottle and allow the latter to stand in a warm place (50–55 °C) for 30 hours. Filter the amino acid at the pump and keep the filtrate (A) separately. Wash the amino acid well with methanol to remove the ammonium bromide present. Concentrate the filtrate (A) almost to dryness and add 150 ml of methanol. A second crop of amino acid contaminated with ammonium bromide is thus obtained; wash it with methanol and recrystallise from hot water, thus affording a further 6 g of pure DL-norleucine. The total yield is 28 g (65%); the decomposition point is about 325 °C.

DL-**Isoleucine** (*2-amino-3-methylvaleric acid*).   Allow 65 g (0.33 mol) of 2-bromo-3-methylvaleric acid (Section **III,169**) to react with 400 ml of concentrated ammonia solution as for valine. Concentrate the resulting solution to about 130 ml, filter off a first crop of crude product and wash with 20 ml of ethanol. Further concentrate the aqueous filtrate to about 60 ml to obtain a second crop of crude product, and wash it with 10 ml of water followed by 10 ml of ethanol. Dissolve the combined product (28 g) in 400 ml of hot water, decolourise with charcoal and add 200 ml of rectified spirit. Cool well in ice, and filter off the pure DL-isoleucine; yield 16.5 g (38%), m.p. 278–280 °C (decomp.). A further 5 g may be recovered by concentrating the recrystallisation mother-liquor to 40 ml and diluting with an equal volume of ethanol.

III,180. AMINOACETONITRILE AND GLYCINE (*Aminoacetic acid*)

$$2H\cdot CHO + NaCN + NH_4Cl \longrightarrow$$
$$CH_2 = N\cdot CH_2\cdot CN + NaCl + 2H_2O$$

$$2CH_2 = N\cdot CH_2CN + 4C_2H_5OH + H_2SO_4 \longrightarrow$$
$$(H_3\overset{\oplus}{N}\cdot CH_2\cdot CN)_2SO_4{}^{2\ominus} + 2CH_2(OC_2H_5)_2$$

$$(H_3\overset{\oplus}{N}\cdot CH_2\cdot CN)_2SO_4{}^{2\ominus} + 2Na^{\oplus}\overset{\ominus}{O}CH_3 \longrightarrow$$
$$2NH_2\cdot CH_2CN + 2CH_3OH + Na_2SO_4$$

$$(H_3\overset{\oplus}{N}\cdot CH_2\cdot CN)_2SO_4{}^{2\ominus} \xrightarrow[H_2O]{Ba(OH)_2} (H_2N\cdot CH_2\cdot CO_2)_2Ba \xrightarrow{H_2SO_4}$$
$$2NH_2\cdot CH_2\cdot CO_2H$$

N-**Methyleneaminoacetonitrile.**   Place 160 g (3 mol) of ammonium chloride in a 2-litre flange or three-necked flask surrounded by a large bath containing a cooling mixture of ice and salt (*FUME CUPBOARD*). Add 450 ml (6 mol) of filtered 40 per cent w/v formaldehyde solution and stir the mixture with an efficient mechanical stirrer. When the temperature has reached 0 °C, begin the dropwise addition of a solution of 150 g (3 mol) of sodium cyanide (98% pure) in 250 ml of water; the addition should take about 5 hours and the temperature throughout should be kept between 0 and 5 °C. When half of the cyanide solution has been added and all of the ammonium chloride has dissolved begin the simultaneous gradual addition of 160 ml of glacial acetic acid, and adjust the rate so that addition is complete by the end of the remaining 2.5 hours. Stir the mixture for a further period of 1 hour, and then filter off the product and wash it with a little cold water. Transfer the filter-cake to a beaker, stir thoroughly with 500 ml of water to remove soluble salts, filter, wash with a little more water and dry in a vacuum desiccator. The yield of N-methyleneaminoacetonitrile (1) is 120 g (59%), m.p. 127–128 °C.

Aminoacetonitrile sulphate. Cautiously add 85 ml of concentrated sulphuric acid to 400 ml of rectified spirit in a 1-litre conical flask and adjust the temperature to 50 °C. Add rapidly 102 g (1.5 mol) of dried powdered N-methyleneaminoacetonitrile, shake vigorously until the solid has dissolved and continue shaking while the product crystallises. Cool in an ice bath for 4 hours, filter and wash the product with a little cold rectified spirit. Dissolve the crude product in a minimum of water, filter off traces of insoluble matter and run the solution with stirring into 400 ml of cold rectified spirit. Cool well, filter off the purified aminonitrile salt and wash it with a little cold rectified spirit. The yield is 105 g (67%), m.p. 164 °C (decomp.).

Hydrolysis to glycine. Boil a suspension of 79 g (0.25 mol) of barium hydroxide octahydrate in 175 ml of water in a 500-ml round-bottomed flask and add in portions 21 g (0.1 mol) of aminoacetonitrile sulphate. Fit a reflux condenser and continue boiling until no more ammonia is evolved (about 3 hours). Transfer the suspension to a beaker, add 50 per cent v/v aqueous sulphuric acid (about 20 ml) until precipitation of the barium is complete and the solution is slightly acidic, and digest on a steam bath. Filter the suspension through a medium speed filter-paper (e.g., Whatman No. 30) using gentle suction, and adjust the filtrate to neutrality by carefully adding saturated barium hydroxide solution. Digest further and decant the supernatant solution through a similar filter-paper (this is best done under gravity), suspend the precipitate in a little hot water and add this to the filter. Concentrate the filtrate under reduced pressure using a rotary evaporator until a thick suspension is obtained, and complete the precipitation of the glycine by adding 100 ml of methanol. Cool well and filter off the crude glycine (about 12 g). Dissolve the product in 25 ml of hot water, add gradually 125 ml of methanol, cool, filter and dry the crystals in an oven at 50 °C. The yield of glycine, m.p. *ca.* 250 °C (decomp.), is 11.2 g (75%).

Aminoacetonitrile. Stir a suspension of 88 g (0.4 mol) of the nitrile sulphate in 100 ml of dry methanol in a 500-ml three-necked flask cooled in crushed ice. Add a few crystals of phenolphthalein as an indicator, pass a slow stream of nitrogen through the flask and run in during 1 hour a solution of sodium methoxide in methanol prepared from 17 g (0.75 mol) of sodium and 350 ml of dry methanol; the suspension should at no time be allowed to become permanently alkaline to phenolphthalein. Filter, and remove the methanol and distil the residue under reduced pressure under nitrogen. The yield of aminoacetonitrile, b.p. 73.5 °C/15 mmHg, is 35 g (83% based on sodium). Store the product at 0 °C under nitrogen.

Note. (1) The product in fact has a trimeric structure.

Cognate preparation DL-2-Aminophenylacetic acid (α-*phenylglycine*). Dissolve 10 g (0.2 mol) of sodium cyanide in 40 ml of water, add 11 g (0.2 mol) of ammonium chloride and shake until dissolved. Add a solution of 21 g (0.2

mol) of redistilled benzaldehyde in 40 ml of methanol and shake vigorously. The mixture soon becomes warm; allow it to stand at ambient temperature for 2 hours, shaking occasionally. Then add 100 ml of water and shake well, and extract out the oily aminonitrile which separates using two portions (60 ml and 40 ml) of toluene. Combine the toluene layers, wash twice with water and extract with two 60 ml portions of 5 M-hydrochloric acid. Boil the acid extract under reflux for 22 hours (*CARE:* hydrogen cyanide is evolved). Cool and filter through a small plug of cotton wool to remove a little tarry matter. Basify the solution by adding about 40 ml of concentrated ammonia solution (d 0.88) with stirring and cooling, collect the resulting precipitate by filtration and wash it with 100 ml of cold water and then with 15 ml of warm ethanol. The crude material is almost colourless; the yield is 11.5 g after drying in an oven at 50 °C. To purify the phenylglycine, dissolve it in 80 ml of 1 M-sodium hydroxide, add 50 ml of ethanol and clarify the solution by adding a little decolourising charcoal, warming and filtering. Heat the filtrate almost to boiling and neutralise by slowly adding with stirring 16 ml of 5 M-hydrochloric acid. Filter off the purified DL-phenylglycine, wash it with 10 ml of ethanol followed by 20 ml of water and dry at 50 °C; the yield of colourless glistening plates is 9 g (30%). The compound has no m.p.; when placed in a rapidly heated melting point apparatus when the temperature reaches 275 °C, it sublimes between 300 and 310 °C, depending upon the rate of heating.

III,181. DL-PHENYLALANINE (*2-Amino-3-phenylpropionic acid*)

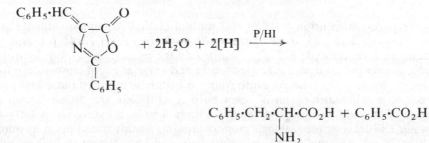

$$C_6H_5 \cdot CH_2 \cdot \underset{NH_2}{CH} \cdot CO_2H + C_6H_5 \cdot CO_2H$$

In a 1-litre three-necked flask, fitted with a reflux condenser, a sealed stirrer unit and dropping funnel, place 25 g (0.1 mol) of 4-benzylidene-2-phenyloxazol-5-one (Section **VI,10**) 20 g (0.65 mol) of purified red phosphorus (Section **II,2,49**) and 135 g (125 ml, 1.32 mol) of acetic anhydride. Add with stirring over a period of 1 hour 125 ml of hydriodic acid (d 1.56; 50%). Reflux the mixture for 3 hours, cool and filter with suction: wash the unreacted phosphorus on the filter with two 5 ml portions of glacial acetic acid. Evaporate the filtrate and washings on a water bath (rotary evaporator) and collect the distillate (which may be used for another reduction) in a flask cooled in ice. Add 100 ml of water to the dry residue and repeat the evaporation to dryness. Shake the residue in the flask with 150 ml of water and 150 ml of ether until solution is complete; separate the aqueous layer and extract it with three 75 ml portions of ether. Discard the ether extracts. Introduce 2–3 g of decolourising carbon and a trace of sodium sulphite into the aqueous phase, heat on a water bath until the dissolved ether has been removed, filter, heat the filtrate to boiling and neutralise to Congo red with ammonia solution (d 0.88; about 25 ml are required).

When cold, filter the colourless DL-phenylalanine at the pump and wash with two 30-ml portions of cold water and finally with a little cold ethanol. The yield is 11 g (67%), m.p. 284–288 °C (decomp., rapid heating).

III,182. DL-SERINE (*2-Amino-3-hydroxypropionic acid*)

$$CH_3 \cdot CO \cdot NH \cdot CH(CO_2C_2H_5)_2 + H \cdot CHO \xrightarrow{\ominus OH}$$

$$CH_3 \cdot CO \cdot NH \cdot \underset{\overset{|}{CH_2OH}}{C}(CO_2C_2H_5)_2 \xrightarrow{NaOH}$$

$$CH_3 \cdot CO \cdot NH \cdot \underset{\overset{|}{CH_2OH}}{C}(\overset{\ominus}{CO_2}\overset{\oplus}{Na})_2 \xrightarrow[\text{heat}]{CH_3 \cdot CO_2H} CH_3 \cdot CO \cdot NH \cdot \underset{\overset{|}{CH_2OH}}{CH} \cdot CO_2H$$

$$\xrightarrow[\text{heat}]{HCl} \overset{\ominus}{Cl}\overset{\oplus}{H_3}N \cdot \underset{\overset{|}{CH_2OH}}{CH} \cdot CO_2H \xrightarrow[\text{Resin}]{-HCl} NH_2 \cdot \underset{\overset{|}{CH_2OH}}{CH} \cdot CO_2H$$

Prepare a suspension of 43.5 g (0.2 mol) of diethyl acetamidomalonate in 25 ml of water and add in one portion 17 g of neutral (1) 37–41 per cent w/v aqueous formaldehyde (0.21–0.23 mol). Add 0.5 ml of 1 *M*-sodium hydroxide solution as the catalyst, shake vigorously and leave at room temperature for 2 hours. (Most of the solid goes into solution within 30 minutes; the mixture may require gentle warming on a steam bath to complete the dissolution of the solid.) Then add a solution of 18 g (0.45 mol) of sodium hydroxide in 350 ml of water and leave at room temperature overnight. Acidify the solution by adding 40 g (38 ml, 0.67 mol) of glacial acetic acid and heat almost to boiling, when brisk decarboxylation sets in. Continue to heat on a boiling water bath under reflux for 1 hour to complete the decarboxylation and then evaporate the solution to a syrup under reduced pressure on a rotary evaporator. Dissolve the syrup in 120 ml of concentrated hydrochloric acid, boil under reflux (*fume cupboard*) for 1 hour and evaporate to dryness under reduced pressure. Extract the dry residue with 200 ml of boiling absolute ethanol, filter off the sodium chloride and extract the latter with a further 100-ml portion of hot ethanol. Evaporate the ethanol extracts on the rotary evaporator and boil the residue under reflux with 100 ml of concentrated hydrochloric acid for 1 hour (*fume cupboard*). Evaporate again to a syrup, and dissolve the latter in water and re-evaporate twice more to remove most of the excess hydrochloric acid. Finally dissolve the residual gum in about 200 ml of distilled water and pass the solution through a 50 × 2.75 cm column of a weakly basic anion exchange resin (e.g., Amberlite IR 45, $\ominus$OH form, about 300 ml of moist granules). Continue to elute the column with distilled water (about 1.5 l) until the eluate gives no purple colouration when a portion is tested by boiling with a few mg of ninhydrin (2). Combine all the ninhydrin-positive eluates, which should be free from chloride ion, and evaporate to dryness under reduced pressure (rotary evaporator). Dissolve the straw-coloured residue in 150 ml of hot water, boil with a little decolourising charcoal, filter and add to the filtrate 750 ml of hot ethanol. Cool in ice, filter off the purified DL-serine which crystallises, wash it with a little cold ethanol and dry in an oven at 50 °C. The yield is 12 g (57%), m.p. *ca.* 235 °C (decomp.). The product thus obtained is homogeneous (R_F 0.57) on TLC (silica gel, 15 cm run; butan-1-ol-formic acid–water, 6 : 3 : 1) (3).

Notes. (1) If necessary, the formaldehyde solution should be neutralised (narrow range pH paper) by the careful dropwise addition of 1 *M*-sodium hydroxide solution.

(2) After use, regenerate the resin in the following way. Firstly exhaust the column by passing 0.25 *M*-hydrochloric acid through it until the pH of the eluate is about 2. Wash the column with 2 or 3 bed-volumes of distilled water, and then regenerate the resin by passing through it 0.25 *M*-sodium hydroxide solution until the eluate is strongly alkaline. Finally wash the column thoroughly with much distilled water until the pH of the eluate is within the range 5.5–6.5 (narrow range indicator paper).

(3) Using this solvent system, serine is not completely separable from glycine (R_F 0.58) which is a possible contaminant. The latter may be distinguished, however, by the characteristic brownish-pink spot which it gives on spraying with ninhydrin; that of serine is purple.

III,183. DL-TRYPTOPHAN [*2-Amino-3-(3-indolyl)propionic acid*]

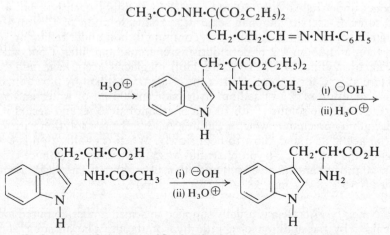

$$CH_3 \cdot CO \cdot NH \cdot CH(CO_2C_2H_5)_2 \; + \; CH_2 - CH \cdot CHO \quad \xrightarrow{\; C_2H_5O^{\ominus}\;}$$

$$CH_3 \cdot CO \cdot NH \cdot \underset{\overset{|}{CH_2 \cdot CH_2 \cdot CHO}}{C}(CO_2C_2H_5)_2 \quad \xrightarrow{\; C_6H_5 \cdot NH \cdot NH_2\;}$$

Phenylhydrazone of 4-acetamido-4,4-diethoxycarbonylbutanal. *Carry out this stage of the preparation in an efficient fume cupboard.* Place 43.5 g (0.2 mol) of diethyl acetamidomalonate and 70 ml of benzene in a 250-ml three-necked flask fitted with a stirrer and dropping funnel and surrounded by a bath of water at room temperature. Stir mechanically, add about 0.5 ml of a concentrated solution of sodium ethoxide in ethanol and then add slowly from the dropping funnel a solution of 12 g (14 ml, 0.215 mol) of acrylaldehyde [acrolein—CAUTION: highly toxic and irritant vapour, (1)] in 14 ml of benzene; adjust the rate of addition so that the temperature of the reaction mixture does not exceed 35 °C. When the addition is complete, stir for 2 hours more and filter off any traces of insoluble material. Add 5 ml of glacial acetic acid and 24 g (22 ml, 0.22 mol) of redistilled phenylhydrazine (Section IV,104; see cautionary note), warm to 50 °C and leave the resulting orange solution at room temperature for 2 days. Collect the crystalline phenylhydrazone by filtration and wash it thoroughly by trituration with two 40 ml portions of benzene. The yield of off-white crystals, m.p. 141 °C, is 50 g (69%). If the yield is low, warm the filtrate to 50 °C and set it aside for a further 2 days, when a further crop of the product may be obtained.

Diethyl (3-indolylmethyl)acetamidomalonate. Add 47 g (0.13 mol) of the phenylhydrazone to 300 ml of water containing 14 ml of concentrated sulphuric acid in a 500-ml two-necked flask fitted with a sealed stirrer unit and a reflux condenser. Boil the mixture under reflux with vigorous stirring for 4.5 hours; the suspended solid liquefies and then solidifies during this time. Cool, filter off the resulting product (in the form of hard nodules) and wash it thoroughly by grinding it with water and re-filtering. Recrystallise the product from 1 : 1 aqueous ethanol to obtain the purified malonate derivative; yield 32 g (71%). The product melts at 143 °C, re-solidifies and then melts at 159 °C.

DL-Tryptophan. Boil 31 g (0.09 mol) of the above product under reflux for 4 hours with a solution of 18 g (0.45 mol) of sodium hydroxide in 180 ml of water. Add a little decolourising charcoal, filter and cool the filtrate in an ice–salt bath. Acidify the filtrate by adding about 55 ml of concentrated hydrochloric acid slowly and with shaking, keeping the temperature below 20 °C. Cool the resulting suspension at 0 °C for 4 hours and then collect the crude (indolylmethyl)malonic acid, a pale buff solid, by filtration. Boil the crude product under reflux with 130 ml of water for 3 hours; decarboxylation ensues and some N-acetyltryptophan separates. Add a solution of 16 g (0.4 mol) of sodium hydroxide in 30 ml of water, continue to boil under reflux for 20 hours and then add about 1 g of decolourising charcoal and filter. Cool, acidify the filtrate by adding 24 g (23 ml, 0.4 mol) of glacial acetic acid and cool the mixture at 0 °C for 5 hours. Collect the crude tryptophan by filtration, dissolve it in a solution of 5 g of sodium hydroxide in 200 ml of water and warm to 70 °C. Dilute the solution with 100 ml of ethanol at 70 °C and decant it from a little gummy precipitate which separates. Acidify the hot solution with 7.5 ml of glacial acetic acid and allow to cool slowly. When crystallisation is complete, filter off the purified DL-tryptophan and wash it successively with ice-cold water (2 × 40 ml), ethanol (2 × 40 ml) and ether (2 × 40 ml). The yield of colourless plates is 15 g (82%), m.p. 283–284 °C (decomp.).

Note. (1) Acrolein is usually supplied in sealed amber-coloured ampoules stabilised by the addition of a little hydroquinone. The ampoule should be cooled thoroughly before being opened with great care. It has been recorded that opened samples of acrolein stored in screw-capped bottles may explode violently, presumably as the result of rapid exothermic polymerisation.

III,184. **DL-PROLINE** (DL-*Pyrrolidine-2-carboxylic acid*)

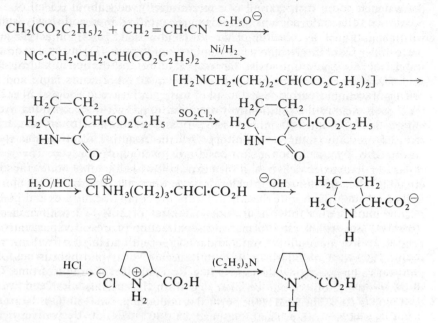

3-Ethoxycarbonyl-2-piperidone. Dissolve 160 g (0.75 mol) of diethyl (2-cyanoethyl)malonate (Section **III,165**) in 600 ml of ethanol and hydrogenate in the presence of about 4–5 g of Raney nickel catalyst (Section **II,2,41**) at 80 °C and 75 atmospheres pressure of hydrogen in an autoclave (Section **I,17,1**); uptake of hydrogen is complete in about 2 hours. Remove the catalyst by filtration and the solvent by distillation under reduced pressure (rotary evaporator) and pour the residue with stirring into 500 ml of light petroleum (b.p. 60–80 °C). Filter off the precipitated piperidone and allow it to dry in the air; yield 115 g (90%), m.p. 74 °C. A specimen crystallised from ethanol/light petroleum has m.p. 80 °C.

3-Chloro-3-ethoxycarbonyl-2-piperidone. Assemble in a fume cupboard a 500-ml three-necked flask fitted with a sealed stirrer unit, a dropping funnel and a reflux condenser protected by a calcium chloride guard-tube. Charge the flask with a solution of 111 g (0.65 mol) of 3-ethoxycarbonyl-2-piperidone in 175 ml of dry chloroform, and the dropping funnel with 90 g (54 ml, 0.67 mol) of redistilled sulphuryl chloride dissolved in 125 ml of dry chloroform; close the neck of the funnel with a calcium chloride guard-tube. Start the stirrer, and slowly run in the solution of sulphuryl chloride so that the reaction mixture refluxes gently. When the addition is complete, warm on a steam bath until hydrogen chloride evolution ceases. Remove the solvent under reduced pressure using a rotary evaporator, and cool the residue, scratching the sides of the flask to induce crystallisation. Dissolve the solid in 70 ml of hot ethyl acetate, add hot light petroleum (b.p. 80–100 °C) until the solution is slightly turbid, and cool while scratching the side of the vessel vigorously until the product crystallises (it helps to add a few crystals of the crude solid as a seed). Add 30 ml more of light petroleum, cool further and filter. The chloropiperidone is obtained as a somewhat sticky white solid, m.p. 64–68 °C; yield is 110 g (82%).

DL-**Proline hydrochloride.**   Boil 103 g (0.5 mol) of the above chloropiperidone under reflux with 200 ml of concentrated hydrochloric acid (*fume cupboard*) for 5 hours during which time decarboxylation ensues. Boil the solution with about 2 g of decolourising charcoal (acid-washed grade), filter and evaporate under reduced pressure using a rotary evaporator; dissolve the residue in water and re-evaporate to assist the removal of most of the excess hydrochloric acid. Dissolve the resulting pale golden syrup in 80 ml of water, add a solution of 60 g of sodium hydroxide in 120 ml of water and leave at room temperature for 2 days. Acidify the solution to Congo red paper with concentrated hydrochloric acid (about 130 ml are needed) and evaporate to dryness under reduced pressure (rotary evaporator). Dry the resulting solid completely by leaving it overnight in a vacuum desiccator over phosphorus pentoxide and potassium hydroxide pellets, and then extract it with 200 ml of boiling absolute ethanol. Filter off the sodium chloride and re-extract it with 150 ml more of boiling ethanol. Evaporate the ethanol from the combined filtrates and boil the residue under reflux for 1 hour with 1.25 litres of 2 M-hydrochloric acid to hydrolyse any proline ester formed in the extraction process. Evaporate to dryness, redissolve the residue in water and re-evaporate, and dry the resulting crude DL-proline hydrochloride in a vacuum desiccator over phosphorus pentoxide and potassium hydroxide; the yield is 70 g. Recrystallise the product from 175 ml of hot propan-2-ol, cool in ice, filter and wash the crystals with 20 ml of ice-cold propan-2-ol. The yield of the purified product, m.p. 148–150 °C, is 50 g (66%).

Conversion to DL-**proline.**   Suspend 15 g (0.1 mol) of dry proline hydrochloride in 70 ml of dry chloroform, stir vigorously and run in dropwise 15 g (0.15 mol) of dried, redistilled triethylamine. Continue stirring for 1 hour, filter off the product and wash it with a little cold chloroform. The yield of DL-proline is 10 g (87%); a sample crystallised from absolute ethanol has m.p. 206–207 °C (with preliminary sintering).

III,185. L-PROLINE (L-*Pyrrolidine-2-carboxylic acid*)

$$Cr^{3\oplus} + 6CN\overset{\ominus}{S} \longrightarrow [Cr(CNS)_6]^{3\ominus} \xrightarrow[\text{(ii) } CH_3 \cdot CO_2H]{\text{(i) } C_6H_5 \cdot NH_2}$$

$$[Cr(CNS)_4(C_6H_5 \cdot NH_2)_2]^{\ominus}\}C_6H_5 \cdot \overset{\oplus}{N}H_3 \xrightarrow{NH_3}$$

$$[Cr(CNS)_4(C_6H_5 \cdot NH_2)_2]^{\ominus}\overset{\oplus}{N}H_4$$

Ammonium rhodanilate

$$\text{GELATIN} \xrightarrow[H_2O]{HCl} \text{AMINO ACID MIXTURE} \longrightarrow \text{L-PROLINE RHODANILATE}$$

$$\xrightarrow{\text{pyridine}} \text{L-PROLINE} + \text{PYRIDINE RHODANILATE}$$

Ammonium rhodanilate.   Heat a mixture of 250 g (0.5 mol $Cr^{3\oplus}$) of hydrated chromium(III) potassium sulphate (chrome alum), 290 g (3.0 mol) of potassium thiocyanate and 250 ml of water in a 5-litre flask on a steam bath for 4 hours. Cool, add 235 g (230 ml, 2.53 mol) of redistilled aniline and heat in a water bath at 60 °C with stirring for 3 hours. Cool again, add while still stirring 3 litres of water containing 300 ml of glacial acetic acid and leave at 0 °C overnight. Filter off the purple precipitate with suction, wash it with water and suck as dry as possible. Extract the filter cake with 750 ml of methanol, filter and run into the filtrate, with stirring, 3 litres of water. Cool at 0 °C for 1 hour and filter off the crystalline purple mass of aniline rhodanilate. Dissolve the damp product in 400 ml of methanol, add 200 ml of concentrated ammonia solution (d 0.88), cool to 0 °C and then run in slowly, with stirring, 2 litres of water. Collect the precipitate by filtration, and again treat it with 250 ml of methanol, 125 ml of ammonia solution and 1250 ml of water in the above manner. Dry the resulting ammonium rhodanilate sesquihydrate in an oven at 50 °C; the yield is 130 g (50.5%).

L-**Proline.**   Place 150 g of good quality sheet gelatin (cut into conveniently sized pieces) in a 1-litre flask and add 450 ml of concentrated hydrochloric acid. Boil gently under reflux for 8 hours (*fume cupboard*), or boil under reflux for about 3 hours and leave on a steam bath overnight; complete hydrolysis is indicated by a negative biuret reaction (1). Concentrate the hydrolysate to a syrup by distillation under reduced pressure (rotary evaporator), and remove excess hydrochloric acid by dissolving the syrup in water and evaporating twice more. Dissolve the residual syrup finally in 500 ml of water, boil briefly with 3 g of decolourising charcoal, filter, cool and dilute with water to 1200 ml. Add this solution slowly with stirring to a filtered solution of 125 g of ammonium rhodanilate in 750 ml of methanol and keep at 0 °C for 2 hours to allow the complete separation of proline rhodanilate. Filter off the latter, wash it with water and suck as dry as possible. Dissolve the damp crude product in 400 ml of methanol, filter and add 800 ml of 0.5 M-hydrochloric acid slowly with stirring. Cool at 0 °C for 2 hours, filter off the purified proline rhodanilate and wash it with 250 ml of cold water. Dry the product in an oven at 50 °C; the yield is about 100 g, m.p. *ca.* 130 °C, with preliminary softening and blackening.

Suspend the purified salt in 850 ml of water in a stoppered bottle, add 25 ml of pure pyridine and shake the mixture for 4–5 hours. Remove the insoluble pyridine rhodanilate by filtration and wash it with 100 ml of cold water (2). Combine the pale pink filtrate and washings, and add glacial acetic acid dropwise until the formation of a small pink precipitate is complete. Filter, evaporate the almost colourless filtrate to dryness (rotary evaporator) and suspend the residue in absolute ethanol and re-evaporate twice. Dry the resulting faintly pink crude proline in a vacuum desiccator over silica gel; the yield is about 18 g. Recrystallise from the minimum volume of absolute ethanol to obtain 11 g (7.3% based on gelatin) of L-proline, m.p. 218–219 °C (decomp.), $[\alpha]_D^{18}$ $-85.6°$ (c 3.0 in H_2O). Check the purity of the product by TLC on silica gel using the solvent system butan-1-ol–acetic acid–water, 4 : 1 : 1; R_F 0.26 (yellow spot with ninhydrin).

Notes.   (1) Remove about 0.25 ml of the hydrolysate, cool it and basify it with 5 M-sodium hydroxide solution. To a portion add a few drops of very dilute copper(II) sulphate solution, and note the absence of any colour change. As a control, prepare a specimen of biuret ($NH_3\cdot CO\cdot NH\cdot CO\cdot NH_2$) by heating about 10 mg of urea just above its melting point for about 2 minutes. Add a little basified hydrolysate warm to dissolve, cool and add a trace of copper(II) sulphate. A deep pink colour superimposed upon the pale brown colour of the hydrolysate should be observed. If the hydrolysate gives a similar colour originally, it contains peptide material and hydrolysis should be continued until the biuret test is negative.

(2) About 100 g of pyridine rhodanilate is obtained. To convert this into the ammonium salt for re-use, suspend it in 175 ml of methanol and add 90 ml of concentrated ammonia solution. Stir at 0 °C for 30 minutes and then dilute gradually with 900 ml of water. Filter off the precipitated ammonium salt, wash it with water and repeat the methanol–ammonia treatment once more. The final yield of dried ammonium rhodanilate is about 80 g.

III,186. L-CYSTINE (*3,3′-Dithiodialanine*)

$$\text{Keratin (hair)} \xrightarrow[\text{H}_3\text{O}^{\oplus}]{\text{hydrolysis}} [\text{S·CH}_2\text{·CH(NH}_2\text{)·CO}_2\text{H}]_2$$

Place 500 g of washed, dried human hair (1) in a 3-litre round-bottomed flask and add 1 litre of a 1 : 1 mixture of concentrated hydrochloric acid and formic acid. Fit a reflux condenser and boil the mixture gently under reflux (fume cupboard) until the biuret test [Section **III,185**, Note (1)] is negative (about 20 hours). Treat the hydrolysate with about 12 g of acid-washed decolourising charcoal, filter hot and concentrate the filtrate to a thick syrup under reduced pressure using a rotary evaporator. Dissolve the residue in 250 ml of water and add with shaking warm 50 per cent aqueous sodium acetate until the solution is no longer acid to Congo red. Leave the mixture at room temperature for 3 days, and then filter off the precipitate of crude cystine and wash it with 50 ml of warm water. Dissolve the product in 750 ml of 1 *M*-hydrochloric acid, treat the hot solution with 5 g of acid-washed decolourising charcoal and filter. If the filtrate is more than faintly yellow, repeat the treatment with a further 5 g of charcoal. Neutralise the filtrate to Congo red with sodium acetate solution as previously and leave at room temperature for 5–6 hours (2). Filter off the colourless plates of purified L-cystine, wash with two 50 ml portions of hot water, then with ethanol and finally with ether. The yield is about 25 g (5%); m.p. 260–262 °C (decomp.); $[\alpha]_D^{20}$ −216° (*c* 0.69 in 1 *M*-HCl).

Notes. (1) Raw material can usually be obtained from barbers' shops. The washing procedure must remove effectively grease and natural oils. A warm aqueous solution of a good quality detergent is satisfactory; it is important that the washing medium should not be alkaline as this may markedly reduce the yield of cystine.

(2) The crude product contains some of the aromatic amino acid tyrosine which is present in the original hydrolysate. A portion of this is removed by the charcoal treatment and by the hot water washing, but the final recrystallised cystine may be contaminated with tyrosine if the suggested 5–6 hour period for crystallisation is greatly exceeded.

Q NITROALKANES

Laboratory routes to the nitroalkanes include the following procedures.

1. The displacement of a halogen by a nitro group in an alkyl halide (Section **III,187**).

2. The oxidation of oximes (Section **III,188**).
3. The preparation and decarboxylation of α-nitrocarboxylic acids (Section **III,189**).

Q,1. The displacement of a halogen by a nitro group in an alkyl halide. Primary nitroalkanes can be prepared in the laboratory by heating the alkyl bromide (or iodide, but not the chloride) with silver nitrite, frequently in anhydrous ether. The method is not satisfactory with secondary or tertiary halides.

$$\ddot{\underset{..}{O}} = N - \underset{\ominus}{\overset{..}{\underset{..}{O}}} \quad R \overset{\frown}{-} Br \quad \longrightarrow \quad R - N \overset{\oplus}{\underset{\underset{\ominus}{\overset{..}{\underset{..}{O}}}}{\overset{O}{\diagup}}} + \overset{\ominus}{:}Br$$

Under these conditions alkali metal nitrites give products containing substantial amounts of the isomeric alkyl nitrites (R·O·NO). It is worthy of comment, however, that the cheaper sodium nitrite has been employed to effect the conversion of primary alkyl halides into nitroalkanes, albeit in slightly lower yield, by using a solvent such as dimethylformamide or dimethylsulphoxide (Ref. 17), which enables a significant amount of sodium nitrite to be brought into solution. A further feature is that secondary halides may be converted in useful yields into secondary nitroalkanes, although even this modification fails with tertiary halides.

III,187. 1-NITROBUTANE

$$CH_3 \cdot (CH_2)_2 \cdot CH_2 Br + AgNO_2 \longrightarrow CH_3 \cdot (CH_2)_2 \cdot CH_2 \cdot NO_2 + AgBr$$

In a 200-ml round-bottomed flask place 64 g (50 ml, 0.47 mol) of dry butyl bromide and 80 g (0.52 mol) of dry silver nitrite (1), and attach a reflux condenser carrying a calcium chloride guard-tube. Allow the mixture to stand for 2 hours; heat on a steam bath for 4 hours (some brown fumes are evolved), followed by 8 hours in an oil bath at 110 °C. Then distil the mixture collecting the fraction of b.p. 149–151 °C as pure 1-nitrobutane (18 g, 39%).

Note. (1) The silver nitrite may be prepared as described in Section **II,2,**53. The commercial product should be washed with absolute methanol or ethanol followed by sodium-dried ether, and dried in an electrically heated oven at 100 °C for 30 minutes (longer heating results in darkening on the surface): the substance should be kept in a vacuum desiccator until required.

Cognate preparation   1-Nitrohexane. Use 41 g (0.26 mol) of dry silver nitrite, 51 g (0.24 mol) of 1-iodohexane (Section **III,58**) and 100 ml of sodium-dried ether. Reflux on a water bath for 8 hours; decant the ethereal solution and wash the solid well with sodium-dried ether. Distil the residue, after the removal of the ether from the combined extracts, from 5 g of dry silver nitrite, and collect the fraction of b.p. 190–192 °C (13 g, 41%) as 1-nitrohexane. The pure compound is obtained by distilling under diminished pressure: b.p. 81.5 °C/15 mmHg.

Q,2. The oxidation of oximes. This method offers an attractively simple route to nitroparaffins. The most effective reagent is peroxytrifluoroacetic acid in acetonitrile in the presence of sodium hydrogen carbonate as a buffer. Yields are improved by the addition of small quantities of urea to remove oxides of nitrogen. The reaction is illustrated by the conversion of dipropyl ketoxime into 4-nitroheptane (Section **III,188**).

$$R_2C=N\diagdown_{OH} \xrightarrow{[O]} R_2C=N\diagup^{O}\diagdown_{OH} \xrightarrow{-H^{\oplus},+H^{\oplus}} R_2CHN\diagup^{O}\diagdown_{O}$$

III,188.   4-NITROHEPTANE

$$(CF_3\cdot CO)_2O + H_2O_2 \longrightarrow CF_3\cdot CO_3H + CF_3\cdot CO_2H$$

$$(CH_3\cdot CH_2\cdot CH_2)_2CO \xrightarrow{NH_2OH} (CH_3\cdot CH_2\cdot CH_2)_2C=NOH \xrightarrow{[O]}$$

$$(CH_3\cdot CH_2\cdot CH_2)_2CHNO_2$$

The reactions involving hydrogen peroxide and peroxytrifluoroacetic acid should be carried out in a fume cupboard behind a safety screen. Adequate precautions should be observed in handling the hydrogen peroxide solution (1).

Dipropyl ketoxime (*heptan-4-one oxime*).   Heat a mixture of 20 g (0.18 mol) of heptan-4-one, 17.4 g (0.25 mol) of hydroxylamine hydrochloride, 19.6 g (20 ml, 0.25 mol) of pyridine and 150 ml of ethanol under reflux for 1 hour in a 500-ml round-bottomed flask. Rearrange the condenser for downward distillation and remove the ethanol on the water bath. Allow the residue in the flask to cool and add 150 ml of water. Extract the oxime with three 50 ml portions of ether. Wash the combined extracts with water and dry over magnesium sulphate. Remove the ether on the rotary evaporator and distil the residue at atmospheric pressure. Collect the fraction having b.p. 192–195 °C; the yield is 18.2 g (80%). The oxime shows infrared absorptions at 3300 cm⁻¹ (O−H stretch) and 1655 cm⁻¹ (C=N stretch).

4-Nitroheptane.   Prepare a solution of peroxytrifluoroacetic acid in acetonitrile as follows. Place 50 ml of acetonitrile in a two-necked, 250-ml round-bottomed flask fitted with a dropping funnel and a reflux condenser. Insert a plastic-covered magnetic stirrer follower bar and cool the flask in an ice bath sited on the stirrer unit. To the cooled and stirred solution add 5.8 ml (0.2 mol) of 85 per cent hydrogen peroxide (1) and then 39.0 ml (58.1 g, 0.24 mol) of trifluoroacetic anhydride. Stir the solution for 5 minutes and then allow to warm to room temperature. In a three-necked, 500-ml round-bottomed flask fitted with a sealed stirrer unit, dropping funnel and reflux condenser place 200 ml of acetonitrile, 47 g (0.56 mol) of sodium hydrogen carbonate, 2 g of urea and 12.9 g (0.1 mol) of dipropyl ketoxime. Heat the stirred suspension under reflux on the water bath and add dropwise over 90 minutes the prepared solution of peroxytrifluoroacetic acid. When the addition is complete heat the mixture under reflux for 1 hour. Pour the cooled reaction mixture into 600 ml of cold water and extract with four 100 ml portions of dichloromethane (note: the organic layer is the upper layer in the first extraction, but subsequently it is the lower layer). Wash the combined extracts with three 100 ml portions of saturated sodium hydrogen carbonate solution and dry over magnesium sulphate. Remove the solvent on the rotary evaporator and distil the residue under reduced pressure through a short fractionating column packed with glass helices. 4-Nitroheptane distils at 66–69 °C/2 mmHg; the yield is 8.4 g (58%).

A small amount of dipropyl ketone is obtained as a forerun. The infrared spectrum of the nitroalkane shows absorptions at 1555 and 1385 cm^{-1} attributable to the antisymmetric and symmetric stretching of the nitro group. The purity of the product may be investigated by GLC using a 10 per cent Silicone oil on Chromosorb W column held at 100 °C; the nitroalkane has a slightly shorter retention time than the oxime; the ketone is rapidly eluted from the column.

Note. (1) Considerable care should be exercised when carrying out reactions with 85 per cent w/v hydrogen peroxide (see also Section II,2,34). Rubber gloves and a face mask should be worn and reactions carried out behind a safety screen. Plenty of water should be at hand to wash away any spillages; fire may result if the peroxide is spilled on to combustible material. Any spillage on the skin similarly should be washed with plenty of water. Care should be taken to avoid the formation of potentially explosive emulsions with organic materials. Spirit thermometers and not mercury thermometers should be used; grease on taps and joints should be kept to the absolute minimum; safety pipettes must be used for pipetting. Disposal of high test peroxide solution may be effected by diluting with a large excess of water. The peroxide used in these experiments was supplied gratis by Laporte Industries Ltd, General Chemical Division.

Q,3. The preparation and decarboxylation of α-nitrocarboxylic acids. Although nitromethane is readily available commercially, its preparation is described (Section **III,189**) to illustrate a further general procedure for the synthesis of nitroalkanes. This involves heating together equimolar amounts of a sodium α-halogenocarboxylate and sodium nitrite in aqueous solution; the sodium α-nitrocarboxylate is intermediately formed and is decomposed into the nitroalkane and sodium hydrogen carbonate. The latter yields sodium carbonate and carbon dioxide at the temperature of the reaction.

$$R{\cdot}CHCl{\cdot}CO_2^{\ominus}Na^{\oplus} + NaNO_2 \longrightarrow \underset{\overset{|}{NO_2}}{R{\cdot}CH{\cdot}CO_2^{\ominus}Na^{\oplus}} + NaCl$$

$$\underset{\underset{O}{\overset{|}{\underset{\diagup}{N}}}\diagdown O}{R{\cdot}CH}\diagup\overset{O}{\underset{\ominus}{\overset{\diagup}{C}{\diagdown}\overset{..}{O}{:}}} \xrightarrow{-CO_2} R{\cdot}CH{=}N\overset{\overset{..}{\overset{\ominus}{O}}{:}}{\underset{O}{\diagdown}} \xrightarrow{H_2O} R{\cdot}CH_2NO_2 + {}^{\ominus}OH$$

An alternative and improved route (Ref. 18) to the required α-nitrocarboxylic acid involves treatment of the carboxylic acid with lithium di-isopropylamide in a hexamethylphosphoramide–tetrahydrofuran solvent system at low temperatures to yield the α-anion of the lithium carboxylate salt. This gives the lithium salt of the α-nitrocarboxylic acid on reaction with propyl nitrate.

$$R{\cdot}CH_2{\cdot}CO_2H \xrightarrow{Li[N(C_3H_7)_2]} \left[\underset{\overset{|}{Li}}{R{\cdot}CH{\cdot}CO_2Li}\right] \xrightarrow{C_3H_7ONO_2} \underset{\overset{|}{NO_2}}{R{\cdot}CH{\cdot}CO_2Li}$$

III,189. NITROMETHANE

$$CH_2Cl \cdot CO_2Na + NaNO_2 \longrightarrow CH_2(NO_2) \cdot CO_2Na + NaCl$$
$$CH_2(NO_2) \cdot CO_2Na + H_2O \longrightarrow CH_3NO_2 + NaHCO_3$$

To a mixture of 125 g (1.33 mol) of chloroacetic acid (Section III,167) and 125 g of crushed ice contained in a 1-litre, two-necked round-bottomed flask, add, with stirring or shaking, sufficient 40 per cent sodium hydroxide solution to render the solution faintly alkaline to phenolphthalein. About 90 ml are required; the temperature should not be allowed to rise above 20 °C, or else sodium glycolate will form. Introduce a solution of 91 g (1.33 mol) of pure sodium nitrite in 100 ml of water into the flask; insert a thermometer dipping well into the liquid by means of a screw-capped adapter fitted in the side-neck. Attach a still-head fitted with an efficient (e.g., double surface) condenser set for downward distillation; the receiver should preferably be cooled in ice-water. Heat the mixture slowly until the first appearance of bubbles of carbon dioxide; this occurs when the temperature has reached 80–85 °C. Immediately remove the flame. The reaction (decomposition of the sodium nitroacetate) sets in with liberation of heat and the temperature rises to almost 100 °C without further application of external heat. If heat is applied after the temperature of the reaction mixture reaches 85 °C, much frothing will occur and serious loss of nitromethane will result. If the reaction becomes too vigorous, it may be checked somewhat by applying a wet cloth to the flask. During the exothermic reaction about 30 ml of nitromethane, accompanied by about 40 ml of water, distil over. When the exothermic reaction apparently ceases (temperature below 90 °C), heat the mixture gently until the temperature rises to 110 °C. Transfer the distillate to a separatory funnel, allow to stand for at least 30 minutes to complete the separation of the two layers and remove the lower layer of nitromethane. Dry it with anhydrous calcium chloride or anhydrous calcium sulphate and distil: 30 g (37%) of nitromethane, b.p. 100–102 °C, are obtained. A further small quantity (3–4 g) may be isolated by mixing the aqueous layer with one-quarter of its weight of sodium chloride, distilling, and separating the nitromethane from the distillate.

R ALIPHATIC AMINES

Amines are classified as primary (RNH_2), secondary (R_2NH) or tertiary (R_3N). The alkyl groups in secondary and tertiary aliphatic amines can of course be the same or different; these latter classes of amines are usually prepared by appropriate alkylation at a nitrogen atom.

The more important general methods for the preparation of aliphatic amines are as follows.

1. The reduction of nitriles (Section III,190).
2. The reduction of nitrocompounds or oximes (Section III,191).
3. Reductive alkylation procedures (Section III,192).
4. The alkylation of ammonia or its derivative (Sections III,193 and III,194).
5. Molecular rearrangements of the Hofmann type (Section III,195).

R,1. The reduction of nitriles.   Nitriles in general are smoothly reduced to primary amines by the action of sodium and ethanol (e.g., the synthesis of pentylamine, Section **III,190**).

$$R{\cdot}CN \longrightarrow R{\cdot}CH_2NH_2$$

The use of metal-acid reducing media is unsatisfactory since extensive hydrolysis of the cyano group to the carboxyl group occurs. The preferred procedure for the reduction of a nitrile to an amine (e.g., 2-phenylethylamine, cognate preparation in Section **III,190**) is, however, catalytic hydrogenation in methanol solution over a Raney nickel catalyst (Section **II,2,***41*). It is necessary to carry out this hydrogenation in the presence of an excess of added ammonia, in order to suppress the formation of substantial amounts of secondary amines, which could arise from the interaction of the aldimine intermediate (I) with the primary amine product (II). The added ammonia reacts with the imine reversibly to yield the 1,1-diaminoalkane (III).

$$R{\cdot}CN \xrightarrow{\text{H}_2} [R{\cdot}CH = NH] \longrightarrow R{\cdot}CH_2NH_2$$
$$\qquad\qquad\quad \text{(I)} \qquad\qquad\qquad \text{(II)}$$

$$R{\cdot}CH = NH + R{\cdot}CH_2NH_2 \rightleftharpoons R{\cdot}CH - NH_2 \underset{}{\overset{-NH_3}{\rightleftharpoons}} R{\cdot}CH$$
$$\qquad\qquad\qquad\qquad\qquad\qquad\quad | \qquad\qquad\qquad\; \|$$
$$\qquad\qquad\qquad\qquad\qquad\quad NH{\cdot}CH_2{\cdot}R \qquad\quad N{\cdot}CH_2{\cdot}R$$

$$\xrightarrow{\text{H}_2} R{\cdot}CH_2{\cdot}NH{\cdot}CH_2{\cdot}R$$

$$RCH = NH + NH_3 \rightleftharpoons R{\cdot}CH(NH_2)_2$$
$$\qquad\qquad\qquad\qquad\qquad \text{(III)}$$

Lithium aluminium hydride has also been used for the reduction of nitriles to amines; a recommended procedure involves the slow addition of the nitrile to at least one molar proportion of the reducing agent in cooled ethereal solution (Ref. 19).

III,190. PENTYLAMINE

$$CH_3{\cdot}(CH_2)_2{\cdot}CH_2{\cdot}CN \xrightarrow{4[H]} CH_3{\cdot}(CH_2)_2{\cdot}CH_2{\cdot}CH_2NH_2$$

Equip a three-necked 1-litre flask with a dropping funnel, an efficient mechanical stirrer and a reflux condenser. Place 55 g (2.4 mol) of clean sodium and 200 ml of sodium-dried toluene in the flask, heat the mixture until the toluene commences to boil and then stir the molten sodium vigorously thus producing an emulsion. Run in through the dropping funnel a mixture of 33 g (41.5 ml) of butyl cyanide (Section **III,161**) and 60 g (76 ml) of absolute ethanol during 1 hour. During the addition and the subsequent introduction of ethanol and of water, the stirring should be vigorous and the temperature adjusted so that the refluxing is continuous; the heat of reaction will, in general, be sufficient to maintain the refluxing. After the butyl cyanide solution has been added, introduce gradually a further 60 g (76 ml) of absolute ethanol. In order to destroy any residual sodium, treat the reaction mixture slowly with 40 g (50 ml) of rectified spirit and then with 20 g of water. Steam distil the contents of the flask (compare Fig. I,101) (about 2 hours) and add 40 ml of concentrated hydrochloric acid to the distillate. Separate the toluene layer; evaporate the aqueous layer, which contains alcohol and amine hydrochloride, to dryness under reduced pressure (rotary evaporator). Treat the resulting amine hydrochloride with a solution of 40 g of sodium hydroxide in 200 ml of water. Separate the amine layer, dry it by shaking with sodium hydroxide pellets (prolonged contact is required for complete drying) and distil. Collect the fraction boiling at 102–105 °C as pure pentylamine. Dry the fraction of low boiling point again over sodium hydroxide and redistil; this gives an additional quantity of amine. The total yield is 30 g (86%).

Cognate preparation 2-Phenylethylamine. Saturate commercial absolute methanol with ammonia (derived from a cylinder) at 0 °C; the resulting solution is *ca.* 10 *M*. Dissolve 58 g (0.5 mol) of benzyl cyanide (Section **III,160**) (1) in 300 ml of the cold methanolic ammonia, and place the solution in a high-pressure hydrogenation bomb (Section **I,17,1**; add 10 ml of settled Raney nickel catalyst (Section **II,2,41**), securely fasten the cap and introduce hydrogen until the pressure is 500–1000 lb. Set the mechanical stirring device in motion, and heat at 100–125 °C until absorption of hydrogen ceases (about 2 hours). Allow the bomb to cool, open it and remove the contents. Rinse the bomb with two 100 ml portions of anhydrous methanol and pour the combined liquids through a fluted filter-paper to remove the catalyst; do not permit the catalyst to become dry since it is likely to ignite. Remove the solvent and ammonia by distillation (*fume cupboard!*), and fractionate the residue through a short column. Collect the 2-phenylethylamine at 92–93 °C/18 mmHg. The yield is 54 g (90%).

Note. (1) Minute amounts of halide have a powerful poisoning effect upon the catalyst; it is advisable to distil the benzyl cyanide from Raney nickel.

R,2. The reduction of nitro compounds or oximes. The reduction of nitro compounds to amines is mainly applied in the aromatic series using procedures discussed on p. 657, **IV,E,1**. These methods are also applicable for the reduction of nitroalkanes and may be used when the required nitroalkane is readily available.

$$R{\cdot}NO_2 \xrightarrow{\text{[H]}} RNH_2$$

The ready conversion of aldehydes and ketones into oximes and their subsequent reduction is more generally applicable to the synthesis of primary aliphatic amines. Reduction with sodium in ethanol is convenient and effective and two examples of its use are given in Section **III,191**.

$$R{\cdot}CH{=}NOH \xrightarrow{\text{[H]}} R{\cdot}CH_2NH_2$$

$$R^1{\cdot}C{\cdot}R^2 \xrightarrow{\text{[H]}} R^1{\cdot}CH(NH_2){\cdot}R^2$$
$$\underset{\text{NOH}}{\overset{\|}{}}$$

Other methods for effecting the reduction include the use of metal–acid systems, catalytic hydrogenation over Raney nickel or palladium-on-charcoal, or modified metal hydride reducing agents such as sodium dihydro bis-(2-methoxyethoxy) aluminate (Section **VI,1**, Note (1); Ref. 20).

III,191. HEPTYLAMINE

$$CH_3 \cdot (CH_2)_5 \cdot CH = NOH \xrightarrow{4[H]} CH_3 \cdot (CH_2)_5 \cdot CH_2NH_2$$

Heptaldoxime. Fit a 1-litre three-necked flask with an efficient mechanical stirrer, a double surface condenser and a thermometer. Place 115 g (141 ml, 1.25 mol) of heptanal (1) and a solution of 87 g of hydroxylamine hydrochloride in 150 ml of water in the flask, and stir the mixture vigorously (2). Introduce, from a separatory funnel down the reflux condenser, a solution of 67 g (0.63 mol) of anhydrous sodium carbonate in 250 ml of water at such a rate that the temperature of the reaction mixture does not rise above 45 °C. Continue the stirring for 1 hour at room temperature. Separate the upper layer and wash the oil with two 25 ml portions of water; dry with magnesium sulphate. Distil from a flask fitted with a short fractionating column. A small fraction of low boiling point (containing heptanenitrile and heptaldoxime) passes over first, and as soon as the temperature is constant the heptaldoxime is collected (e.g., at 103–107 °C/6 mmHg); the temperature of the oil bath is maintained at about 30 °C above the boiling point of the liquid. The yield is about 110 g, and the liquid slowly solidifies on cooling and melts at 44–46 °C; it is sufficiently pure for conversion into heptylamine. If required pure, the heptaldoxime may be recrystallised from 60 per cent ethanol (25 g of solid to 70 ml of solvent) and then melts at 53–55 °C (the m.p. depends somewhat upon the rate of heating).

Heptylamine. In a 3-litre round-bottomed flask, equipped with two large Liebig condensers (34/35 joints) joined in series, place a solution of 64.5 g (0.5 mol) of heptaldoxime in 1 litre of super-dry ethanol (Section **II,1,9**) and heat on a water bath. Immediately the alcohol boils, remove the flask from the water bath and introduce 125 g (5.4 mol) of sodium, cut in small pieces, as rapidly as possible through the condenser consistent with keeping the vigorous reaction under control. The last 30 g of sodium melts in the hot mixture and may be added very rapidly without appreciable loss of alcohol or of amine. As soon as the sodium has completely dissolved (some warming may be necessary), cool the contents of the flask and dilute with 1250 ml of water. At once equip the flask with a condenser set for downward distillation and arrange for the distillate to be collected in a solution of 75 ml of concentrated hydrochloric acid in 75 ml of water contained in a 3-litre flask. Continue the distillation as long as amine passes over. Towards the end of the reaction considerable frothing sets in; then add a further 750 ml of water to the distillation flask. The total distillate is 2–2.2 litres and contains alcohol, water and some unreacted oxime as well as the amine hydrochloride. Evaporate the solution under reduced pressure using a rotary evaporator; the amine hydrochloride will crystallise out in the flask. Cool the flask, attach a reflux condenser and introduce 250 ml of 40 per cent potassium hydroxide solution. Rotate the flask to wash down the hydrochloride from the sides of the flask, cool the mixture to room temperature and transfer it to a separatory funnel. Run off the lower alkaline layer and add solid potassium hydroxide to the amine in the funnel. Again remove the lower aqueous layer, add more solid potassium hydroxide and repeat the process until no further separation of an aqueous layer occurs. Finally, transfer the amine to a small flask and leave it in contact with potassium hydroxide pellets for 24 hours. Decant the amine into a flask and distil through a well-lagged fractionating column. Collect the heptylamine at 153–157 °C. The yield is 40 g (70%).

Notes. (1) The heptanal should be dried and redistilled: b.p. 150–156 °C or 54–59 °C/16 mmHg.

(2) The solution may be rendered homogeneous by the addition of ethanol but the yield appears to be slightly diminished and more high boiling point material is produced.

Cognate preparation Hexyl methyl ketoxime. From hexyl methyl ketone (Section **III,87**) in 90 per cent yield. B.p. 106–108 °C/12 mmHg.

1-Methylheptylamine. Reflux a solution of 50 g (0.35 mol) of the oxime in 200 ml of super-dry ethanol on a water bath whilst adding 75 g (3.25 mol) of sodium; introduce more alcohol (about 300 ml) to maintain a vigorous reaction. When all the sodium has passed into solution, cool, dilute with 250 ml of water and distil gently until the b.p. reaches 96 °C; add a further 200 ml of water and repeat the distillation to ensure the complete removal of the alcohol. The amine remains as a layer on the strongly alkaline solution: extract it with ether, dry the ethereal solution with sodium hydroxide or anhydrous calcium sulphate, remove the ether on a water bath, and distil the residue under diminished pressure. Collect the 1-methylheptylamine at 58–59 °C/13 mmHg; the b.p. under atmospheric pressure is 163–164 °C. The yield is 31 g (69%).

R,3. Reductive alkylation procedures. The process of reductive alkylation involves the treatment of ammonia with an aldehyde or ketone under reducing conditions. The conversion probably includes the following stages:

$$R^1 \cdot CO \cdot R^2 + NH_3 \longrightarrow \left[R^1 \cdot \underset{\underset{NH_2}{|}}{\overset{\overset{OH}{|}}{C}} \cdot R^2 \right] \qquad \left[R^1 \cdot \underset{\|}{\overset{\|}{C}} \cdot R^2 \atop NH \right] \xrightarrow{[H]} R^1 \cdot CH(NH_2) \cdot R^2$$

The reduction is usually effected catalytically in ethanol solution using hydrogen under pressure in the presence of Raney nickel. As in the reduction of nitriles (p. 565, **III,R,1** above), which also involves the formation of the intermediate imines, ammonia should be present in considerable excess to minimise the occurrence of undesirable side reactions leading to the formation of secondary amines. Reductive alkylation of a primary amine can in fact be used to prepare a secondary amine. Selected experimental conditions for these reductive alkylation procedures are to be found in Ref. 21.

$$R^1 \cdot CO \cdot R^2 + R^3 NH_2 \xrightarrow{[H]} R^1 \cdot CH(NH \cdot R^3) \cdot R^2$$

A major variation is the use of formic acid or one of its derivatives as the reductant (the **Leuckart reaction**). In the synthesis of 1-phenylethylamine (Section **III,192**), ammonium formate is heated with acetophenone while the water formed in the reaction is carefully removed by fractional distillation to give the required amine as its N-formyl derivative, (1-phenylethyl)formamide. This is then hydrolysed with acid to yield the primary amine. The procedure has been satisfactorily applied to many aliphatic–aromatic, alicyclic and aliphatic–heterocyclic ketones, some aromatic ketones and aldehydes, and to some aliphatic aldehydes and ketones boiling at about 100 °C or higher.

$$R^1 \cdot CO \cdot R^2 + H \cdot CO_2NH_4 \rightleftharpoons R^1 \cdot \underset{\underset{NH_2}{|}}{\overset{\overset{OH}{|}}{C}} \cdot R^2 + H \cdot CO_2H \longrightarrow$$

$$R^1 \cdot \underset{\underset{NH}{\|}}{C} \cdot R^2 + H_2O + H \cdot CO_2H \longrightarrow R^1 \cdot \underset{\underset{NH_2}{|}}{CH} \cdot R^2 + H_2O + CO_2$$

$$R^1 \cdot \underset{\underset{NH_2}{|}}{CH} \cdot R^2 + H \cdot CO_2NH_4 \longrightarrow R^1 \cdot \underset{\underset{NH \cdot CHO}{|}}{CH} \cdot R^2 + H_2O + NH_3$$

III,192. 1-PHENYLETHYLAMINE (α-Methylbenzylamine)

$$C_6H_5 \cdot CO \cdot CH_3 + 2H \cdot CO_2^{\ominus} \overset{\oplus}{N}H_4 \longrightarrow$$
$$C_6H_5 \cdot CH(CH_3) \cdot NH \cdot CHO + NH_3 + CO_2 + 2H_2O$$

$$C_6H_5 \cdot CH(CH_3) \cdot NH \cdot CHO \xrightarrow[H_2O]{HCl} C_6H_5 \cdot CH(CH_3) \cdot \overset{\oplus}{N}H_3\overset{\ominus}{Cl}$$

$$\xrightarrow{\overset{\ominus}{O}H} C_6H_5 \cdot CH(CH_3)NH_2$$

Place 126 g (2.0 mol) of ammonium formate, 72 g (0.6 mol) of acetophenone (Section **IV,134**) and a few chips of porous porcelain in a 250-ml flask fitted with a Claisen still-head carrying a short fractionating column; insert a thermometer extending nearly to the bottom of the flask, and attach a short condenser set for downward distillation to the side arm. Heat the flask with a heating mantle or in an air bath; the mixture first melts to two layers and distillation occurs. The mixture becomes homogeneous at 150–155 °C and reaction takes place with slight frothing. Continue the heating, more slowly if necessary, until the temperature rises to 185 °C (about 2 hours); acetophenone, water and ammonium carbonate distil. Stop the heating at 185 °C, separate the upper layer of acetophenone from the distillate and return it without drying to the flask. Heat the mixture for 3 hours at 180–185 °C and then allow to cool; the acetophenone may be recovered from the distillate by extraction with 20 ml portions of toluene (1). Transfer the reaction mixture to a 250-ml separatory funnel and shake it with two 75 ml portions of water to remove formamide and ammonium formate. Transfer the crude (1-phenylethyl)formamide into the original reaction flask; extract the aqueous layer with two 20 ml portions of toluene, transfer the toluene extracts to the flask, add 75 ml of concentrated hydrochloric acid and a few chips of porous porcelain. Heat the mixture cautiously until about 40 ml of toluene are collected, and boil gently under reflux for a further 40 minutes; hydrolysis proceeds rapidly to 1-phenylethylamine hydrochloride except for a small layer of unchanged acetophenone. Allow the reaction mixture to cool, remove the acetophenone by extraction with four

20 ml portions of toluene (1). Transfer the aqueous acid solution to a 500-ml round-bottomed flask equipped for steam distillation, cautiously add a solution of 62.5 g of sodium hydroxide in 125 ml of water, and steam distil: heat the distillation flask so that the volume remains nearly constant. Most of the amine is contained in the first 500 ml of distillate; stop the operation when the distillate is only faintly alkaline. Extract the distillate with five 25 ml portions of toluene, dry the extract with sodium hydroxide pellets and fractionally distil (2). Toluene distils over at 111 °C, followed by the phenylethylamine. Collect the latter as a fraction of b.p. 180–190 °C (the bulk of the product distils at 184–186 °C) (3); the yield is 43 g (59%).

Notes. (1) The acetophenone may be recovered by washing the toluene solution with dilute alkali, drying with anhydrous calcium sulphate and distilling; the fraction, b.p. 198–205 °C, is collected.

(2) Ground glass apparatus must be used as the amine attacks rubber (and cork). Since the product absorbs carbon dioxide from the air, attach the receiving flask to the condenser with a take-off adapter carrying a soda-lime guard-tube.

(3) The b.p. under diminished pressure is 80–81 °C/18 mmHg. To obtain a very pure sample of the amine, dissolve 1 part (by weight) of the above product in a solution of 1.04 parts of crystallised oxalic acid in 8 parts of hot water, add a little decolourising carbon and filter. The filtered solution deposits crystals of the oxalate salt; about 5 g of this salt remains in each 100 ml of mother-liquor, but most can be recovered by evaporation and further crystallisation. The amine may be liberated from its oxalate salt with sodium or potassium hydroxide, steam distillation and purification as described above. The salt provides a convenient method of obtaining a known weight of the amine in water, since it can be weighed out and decomposed with alkali hydroxide.

R,4. The alkylation of ammonia and its derivatives. A mixture of all three classes of amines is obtained, together with some of the quaternary ammonium salts, when an alkyl halide is heated under pressure with an alcoholic solution of ammonia. Although to some extent the use of excess ammonia minimises the occurrence of polyalkylation the use of this procedure for laboratory preparations is limited.

$$RX + NH_3 \rightleftharpoons R\overset{\oplus}{N}H_3\}\overset{\ominus}{X} \overset{NH_3}{\rightleftharpoons} RNH_2 + NH_4X$$

$$RX + RNH_2 \rightleftharpoons R_2\overset{\oplus}{N}H_2\}\overset{\ominus}{X} \overset{NH_3}{\rightleftharpoons} R_2NH + NH_4X$$

$$RX + R_2NH \rightleftharpoons R_3\overset{\oplus}{N}H\}\overset{\ominus}{X} \overset{NH_3}{\rightleftharpoons} R_3N + NH_4X$$

$$RX + R_3N \rightleftharpoons R_4\overset{\oplus}{N}\}\overset{\ominus}{X}$$

A similar ammonolysis of alcohols in the presence of certain metallic oxide catalysts is, however, extensively used on the large scale for the manufacture of all classes of amines.

In the laboratory pure primary amines are best prepared by the reaction between potassium phthalimide and an alkyl halide to give an N-alkylphthalimide, which is then cleaved to give the corresponding primary amine (the **Gabriel synthesis**). The preliminary preparation of potassium phthalimide (from a solution of phthalimide in absolute ethanol and potassium hydroxide in 75%

ethanol) may be avoided in some cases by boiling phthalimide with the halide in the presence of anhydrous potassium carbonate. The cleavage of the *N*-substituted phthalimide is best effected by reaction with hydrazine hydrate and then heating the reaction mixture with hydrochloric acid. The insoluble phthalyl hydrazide (I) is filtered off, leaving the amine hydrochloride in solution from which the amine may be liberated and isolated in the appropriate manner.

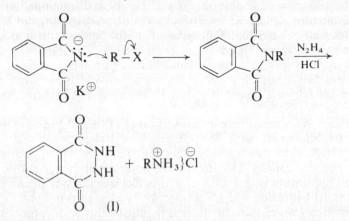

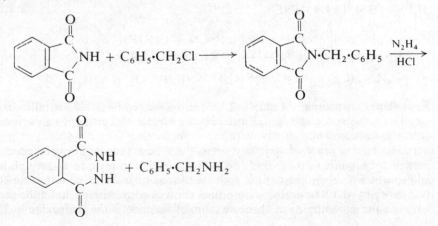

The Gabriel synthesis is illustrated by the preparation of benzylamine and 2-phenylethylamine (Section **III,193**).

The preparation of pure secondary amines (e.g., dibutylamine, Section **III,194**) is conveniently achieved by the hydrolysis of dialkyl cyanamides with dilute sulphuric acid. The appropriate dialkyl cyanamide is prepared by treating sodium cyanamide (itself obtained in solution from calcium cyanamide and aqueous sodium hydroxide solution) with an alkyl halide.

$$CaNCN + 2NaOH \longrightarrow Na_2NCN + Ca(OH)_2$$
$$Na_2N{\cdot}CN + 2RBr \longrightarrow R_2NCN + 2NaBr$$
$$R_2NCN + 2H_2O \xrightarrow{H^{\oplus}} R_2NH + CO_2 + \overset{\oplus}{N}H_4$$

III,193. BENZYLAMINE

N-Benzylphthalimide. Grind together 76 g (0.55 mol) of finely powdered, anhydrous potassium carbonate and 147 g (1 mol) of phthalimide (Section **IV,169**) in a glass mortar, transfer the mixture to a round-bottomed flask and treat it with 151 g (1.2 mol) of redistilled benzyl chloride. Heat in an oil bath at 190 °C under a reflux condenser for 3 hours. Whilst the mixture is still hot, remove the excess of benzyl chloride by steam distillation. The benzylphthalimide commences to crystallise near the end of the steam distillation. At this point, cool the mixture rapidly with vigorous swirling so that the solid is obtained in a fine state of division. Filter the solid with suction on a Bucher funnel, wash well with water and drain as completely as possible; then wash once with 200 ml of 60 per cent ethanol and drain again. The yield of crude product, m.p. 100–110 °C, is 180 g (76%). Recrystallise from glacial acetic acid to obtain pure benzylphthalimide, m.p. 116 °C: the recovery is about 80 per cent.

Benzylamine. Warm an alcoholic suspension of 118.5 g (0.5 mol) of finely powdered benzylphthalimide with 25 g (0.5 mol) of 100 per cent hydrazine hydrate (**CAUTION**: corrosive liquid, see Section **II,2,25**): a white, gelatinous precipitate is produced rapidly. Decompose the latter (when its formation appears complete) by heating with excess of hydrochloric acid on a steam bath. Collect the phthalyl hydrazide which separates by suction filtration, and wash it with a little water. Concentrate the filtrate by distillation on a rotary evaporator to remove alcohol, cool, filter from the small amount of precipitated phthalyl hydrazide, render alkaline with excess of sodium hydroxide solution and extract the liberated benzylamine with ether. Dry the ethereal solution with potassium hydroxide pellets, remove the solvent (rotary evaporator) and finally distil the residue. Collect the benzylamine at 185–187 °C: the yield is 50 g (94%).

Cognate preparation — 2-Phenylethylamine. Prepare 2-phenylethylphthalimide as above by substituting 2-phenylethyl bromide (Section **III,55**) for benzyl chloride: recrystallise the crude product from glacial acetic acid; m.p. 131–132 °C. Convert it into 2-phenylethylamine by treatment with hydrazine hydrate and hydrochloric acid as described for benzylamine. The yield of 2-phenylethylamine, b.p. 200–205 °C, is about 95 per cent.

III,194. DIBUTYLAMINE

$$CaNCN + 2NaOH \longrightarrow Na_2NCN + Ca(OH)_2$$
$$Na_2NCN + 2C_4H_9Br \longrightarrow (C_4H_9)_2NCN + 2NaBr$$
$$(C_4H_9)_2NCN + 2H_2O \longrightarrow (C_4H_9)_2NH + CO_2 + \overset{\oplus}{N}H_4$$

Dibutyl cyanamide. Equip a 2-litre three-necked flask with a reflux condenser and a sealed stirrer unit. Place 220 ml of water and 50 g of finely crushed ice in the flask and add slowly, with vigorous stirring, 70 g (0.46 mol) of commercial calcium cyanamide (1). As soon as the solid is thoroughly suspended, fit a separatory funnel into the third neck of the flask and introduce through it a cold solution of 34 g (0.85 mol) of sodium hydroxide in 70 ml of water; replace the funnel by a thermometer. Continue the vigorous stirring for 1 hour to complete the decomposition of the calcium cyanamide; if the temperature rises above 25 °C, add a little more ice. Add to the resulting solution of sodium cyanamide a solution of 134 g (105 ml, 1 mol) of butyl bromide (Sections III,54) in 220 ml of rectified spirit. Heat the mixture, with stirring, on a water bath until it refluxes gently; continue the refluxing and stirring for 2.5 hours. Replace the reflux condenser by one set for downward distillation and distil the mixture until 165–170 ml of liquid are collected: stir during distillation. Cool the residue in the flask and filter it, with suction, through a Buchner or sintered glass funnel, and wash the residue with rectified spirit. Extract the filtrate, which separates into two layers, first with 90 ml and then with 45 ml of benzene. Dry the combined benzene extracts with anhydrous calcium sulphate, and remove the benzene on a rotary evaporator. Distil under reduced pressure and collect the dibutyl cyanamide at 147–151 °C/35 mm. The yield is 33 g (47%).

Dibutylamine. Into a 1-litre round-bottomed flask furnished with a reflux condenser place a solution of 34 g (18.5 ml) of concentrated sulphuric acid in 100 ml of water: add 33 g (0.2 mol) of dibutyl cyanamide and a few fragments of porous porcelain. Reflux gently for 6 hours. Cool the resulting homogeneous solution and pour in a cold solution of 52 g of sodium hydroxide in 95 ml of water down the side of the flask so that most of it settles at the bottom without mixing with the solution in the flask. Connect the flask with a condenser for downward distillation and shake it to mix the two layers; the free amine separates. Heat the flask, when the amine with some water distils: continue the distillation until no amine separates from a test portion of the distillate. Estimate the weight of water in the distillate and add about half this amount of potassium hydroxide in the form of pellets so that it dissolves slowly. Cool the solution in ice while the alkali hydroxide is dissolving; some ammonia gas is evolved. When the potassium hydroxide has dissolved, separate the amine, and dry it for 24 hours over sodium hydroxide pellets. Filter and distil from a flask fitted with a Claisen still-head. Collect the dibutylamine at 157–160 °C. The yield is 21 g (75%).

Note. (1) Also known as 'nitrolim'. The fresh product contains approximately 55 per cent of calcium cyanamide, 20 per cent of lime, 12 per cent of graphite and small amounts of other impurities. It should be protected from moisture when stored in order to prevent slow polymerisation to dicyanodiamide.

R,5. Molecular rearrangements of the Hofmann type. By treatment of an amide with sodium hypobromite or sodium hypochlorite solution (or with halogen admixed with aqueous alkali), an amine having one less carbon atom is produced.

$$R \cdot CONH_2 + Br_2 + 2NaOH \longrightarrow RNH_2 + CO_2 + 2NaBr + H_2O$$

The conversion of an amide into an amine in this way is termed the **Hofmann reaction** or the **Hofmann rearrangement**. Although good yields are obtained when the reaction is applied to most monocarboxylic acids, it is not an economically viable method for the synthesis of simple aliphatic amines. For illustration purposes, however, the conversion of acetamide to methylamine, which is isolated as the hydrochloride salt, is included here (Section III,195). Examples of greater preparative value are provided by the preparation of anthranilic acid and 3-aminopyridine (Section IV,60).

The probable course of the reaction with sodium hydroxide and bromine involves the following stages:

(a) The formation of an N-bromoamide (I).

$$R \cdot CONH_2 + \overset{\ominus}{:}OH \longrightarrow [R \cdot CO \cdot \overset{\ominus}{N}H] + H_2O$$

$$R \cdot CO\overset{\ominus}{N}H \quad Br - Br \longrightarrow R \cdot CONHBr + \overset{\ominus}{:}Br$$
$$(I)$$

(b) Elimination of the bromide ion from the conjugate base (II) of the N-bromoamide. This is accompanied by a 1,2-nucleophilic shift of the alkyl group and an intermediate alkyl isocyanate (III) is thereby obtained.

$$\left[\begin{array}{c} R \diagdown \overset{..}{\underset{..}{N}} - Br \\ C \\ \parallel \\ O \end{array} \right] \xrightarrow{-Br^\ominus} O = \overset{\oplus}{C} - \overset{..}{\underset{..}{N}} - R \longleftrightarrow O = C = N - R$$

$$(II) \qquad\qquad (III)$$

(c) Hydrolysis of the alkyl isocyanate to the primary amine.

$$R - N = C = O \xrightarrow[H_2O]{\ominus OH} RNH_2 + CO_3^{2\ominus}$$

When the reaction involves treatment of the amide with bromine and a solution of sodium methoxide in methanol, the intermediate alkyl isocyanate is converted into the corresponding N-alkylurethane (IV), which may be isolated prior to its hydrolysis to the primary amine. This modification is the preferred procedure for the conversion of long-chain amides (C > 8) to the corresponding amines. The formation and isolation of the N-alkylurethane provides supporting evidence for the mechanistic pathway of the reaction outlined above.

$$CH_3 - O - H$$
$$R - N = C = O \longrightarrow R \cdot NH \cdot CO_2CH_3 \xrightarrow[H_2O]{\ominus OH}$$
$$CH_3 - \overset{..}{O}^\ominus \qquad (IV)$$

$$RNH_2 + CO_3^{2\ominus} + CH_3OH$$

III,195. METHYLAMINE HYDROCHLORIDE

$$CH_3 \cdot CONH_2 + Br_2 + 2NaOH \longrightarrow CH_3NH_2 + 2NaBr + CO_2 + H_2O$$

Place 25 g (0.42 mol) of dry acetamide in a 500-ml conical flask, and add 69 g (23 ml, 0.43 mol) of bromine (**CAUTION**): a deep red liquid is produced. Cool the flask in ice-water and add 10 per cent sodium hydroxide solution (about 210 ml) in small portions and with vigorous shaking until the solution acquires a pale yellow colour. At this stage the bromoacetamide is present in the alkaline solution. If any solid should crystallise out, add a little water.

Place a solution of 60 g (1.5 mol) of sodium hydroxide in 150 ml of water in a 1-litre, three-necked round-bottomed flask fitted with a dropping funnel, a thermometer reaching to within 1 cm of the bottom of the flask and a reflux condenser. Connect the top of the latter to the inverted funnel type of gas absorption trap (Fig. I,68(a)); charge the beaker with 100 ml of dilute (1 : 1) hydrochloric acid. Introduce a few fragments of porous pot and warm the solution to 60–70 °C. Run the bromoacetamide solution slowly into the flask at such a rate that the temperature does not rise above about 70 °C; heat is evolved in the reaction and if the temperature rises above 75 °C the flask should be surrounded momentarily by a bath of cold water. When all the solution has been added, maintain the temperature of the mixture in the flask for about 15 minutes at 65–70 °C; by this time the solution should be clear and colourless. Gently boil the solution and thus drive off the methylamine vapour into the dilute hydrochloric acid. As soon as the distillate is no longer alkaline (40–60 minutes), evaporate the hydrochloric acid solution on a rotary evaporator and dry the solid residue in an oven held at 100–105 °C. The yield of crude dry product (which is contaminated with some ammonium chloride) is about 24 g. Transfer the finely powdered, dry solid to a 250-ml round-bottomed flask fitted with a reflux condenser and calcium chloride guard-tube. Add about 120 ml of absolute ethanol (which dissolves only the methylamine hydrochloride) and boil the mixture for 10 minutes. Filter through a hot water funnel. Extract the residue with a further 50 ml of boiling absolute ethanol and filter again. Cool the combined ethanolic extracts when colourless crystals of methylamine hydrochloride will separate out. Filter rapidly at the pump, and transfer the crystals (which are deliquescent) to a stoppered bottle. Evaporate the filtrate to about one-third of the original volume, when a further crop of crystals will be obtained. Dry all the crystals in a desiccator. The yield is about 18 g (64%).

S RESOLUTION OF RACEMIC MODIFICATIONS

The most important general procedure for the resolution of a racemic modification involves its conversion into a pair of diastereoisomeric derivatives by reaction with an optically pure, optically active, reagent, e.g., the formation of a pair of diastereoisomeric salts (I) and (II).

$$2(\pm)\text{-A} + 2(+)\text{-B} \longrightarrow (+)\text{-A}(+)\text{-B} + (-)\text{-A}(+)\text{-B}$$
$$\text{(I)} \qquad\qquad \text{(II)}$$

Frequently the diastereoisomers have physical properties, e.g., solubility, boiling points, chromatographic behaviour, etc., which are sufficiently different to allow them to be separated. Resolution of the original racemic modification can then be achieved provided that one of the diastereoisomeric derivatives may be obtained in an optically pure state, and that regeneration from it of the pure enantiomorphous form of the racemic starting material is not accompanied by any degree of racemisation.

Pasteur's original chemical method of resolution, which is probably the most widely used at the present time, involves the formation of diastereoisomeric salts from racemic acids or bases by neutralisation with available optically pure bases or acids respectively. The required optically pure reactants are often available from natural sources and include tartaric, malic and mandelic acids, and alkaloids such as brucine, strychnine, morphine and quinine. Ideally, by appropriate choice of the resolving reagent, the diastereoisomeric salts are crystalline and have solubilities sufficiently different to permit the separation and ready purification of the less soluble salt by fractional crystallisation from a suitable solvent. The regeneration of the optically pure enantiomorph, and incidentally the recovery of the resolving reagent, normally presents no problems.

The procedure is illustrated in Section **III,196** by the resolution of (±)-α-methylbenzylamine (1-phenylethylamine) with the aid of tartaric acid.

The Pasteur method can also be applied to the resolution of neutral racemic modifications, if these can be first converted into an acidic or basic derivative from which eventually a mixture of crystalline diastereoisomeric salts may be prepared by appropriate neutralisation. Thus, a racemic alcohol (e.g., (±)-octan-2-ol, Section **III,197**) may be converted into the corresponding racemic hydrogen phthalate ester by heating with phthalic anhydride, and the ester is then resolved by the Pasteur procedure using an optically active base. The resulting optically active hydrogen phthalate ester is then carefully hydrolysed with aqueous sodium hydroxide to regenerate one of the optically active forms of the alcohol.

An alternative procedure for the resolution of a racemic alcohol is to con-

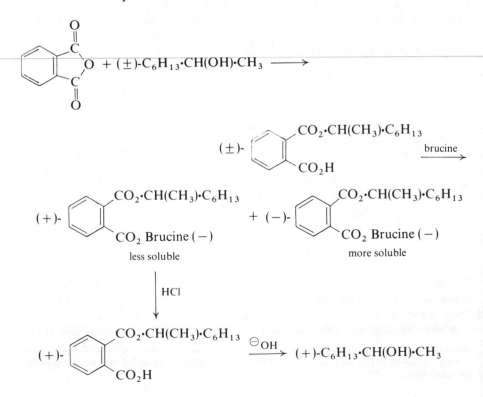

vert it into a mixture of diastereoisomeric esters by reaction with an optically active acyl halide (e.g., (−)-menthoxyacetyl chloride). The success of this method, however, depends upon the availability of a suitable procedure for separating the resulting mixture of diastereoisomeric esters. The methods employed may include fractional distillation, chromatographic procedures and so on.

A racemic aldehyde or ketone may similarly be resolved by conversion into a diastereoisomeric mixture of hydrazones (or semicarbazones) using, for example, the optically active 1-phenylethylhydrazine (Ref. 22).

Several useful methods of resolution, which are particularly applicable to the α-amino acids, involve the use of enzymes. Their success depends upon the fact that enzyme-catalysed reactions are stereospecific, only one of the enantiomorphous forms (actually the form having the 'natural', L, configuration) taking part in the reaction. For example, the conversion of an acyl-L-amino acid (best results are usually obtained with the benzoyl derivative) into the corresponding anilide by reaction with aniline in a buffered medium* is catalysed by the enzyme papain which occurs in papaya latex. Therefore, if the racemic acylamino acid is used, only the L-form is converted into the neutral anilide which can be readily separated from the unreacted acidic D-form.

$$2\text{DL-R}^2\text{·CONH·CH(R}^1\text{)·CO}_2\text{H} + \text{C}_6\text{H}_5\text{NH}_2 \xrightarrow{\text{papain}}$$

$$\text{L-R}^2\text{·CONH·CH(R}^1\text{)·CONHC}_6\text{H}_5 + \text{D-R}^2\text{·CONH·CH(R}^1\text{)·CO}_2\text{H}$$

This method represents a resolution by asymmetric enzymic synthesis (e.g., L-alanine, $H_2N\text{·CH(CH}_3\text{)·CO}_2\text{H}$, Section **III,198**). Related procedures involve other types of enzymic reactions (e.g., hydrolysis, oxidation, etc.). Asymmetric enzymic hydrolysis, for example, proceeds according to the following reaction sequence.

$$2\text{DL-H}_2\text{N·CH(R)·CO}_2\text{H} \xrightarrow{\text{(CH}_3\text{CO)}_2\text{O}}$$

$$2\text{DL-CH}_3\text{·CONH·CH(R)·CO}_2\text{H} \xrightarrow[\text{pigs kidney}]{\text{acylase } ex}$$

$$\text{L-H}_2\text{N·CH(R)·CO}_2\text{H} + \text{D-CH}_3\text{·CONH·CH(R)·CO}_2\text{H}$$

III,196. RESOLUTION OF (±)-α-METHYLBENZYLAMINE [(±)-*l*-Phenylethylamine]

(−)-α-**Methylbenzylamine.** Add 450 ml of methanol to 31.5 g (0.21 mol) of (+)-tartaric acid in a 1-litre conical flask and heat the mixture almost to boiling on a water bath. Then add cautiously with swirling 24.2 g (0.20 mol) of (±)-α-methylbenzylamine (Section **III,192**); too rapid an addition may cause the mixture to boil over. Allow the mixture to cool to room temperature and then to stand for 24 hours to allow slow separation of the (−)-amine-(+)-hydrogen tartrate as prismatic crystals (1). Filter off the product (17.9 g); concentrate the filtrate to 225 ml under reduced pressure on a rotary evaporator and allow it to stand at room temperature for 24 hours to obtain a second crop. The total yield of the (−)-amine-(+)-hydrogen tartrate is about 21 g (77%).

Shake the total product with 90 ml of water in a 250-ml separating funnel

---

* The optimum pH of the buffer varies within the approximate range 4.5–5.5 depending upon the substrate used.

and basify the mixture by cautiously adding 50 per cent aqueous sodium hydroxide. Extract out the liberated amine with three 40 ml portions of ether, dry the extract over anhydrous sodium sulphate, filter and concentrate to about 25 ml using a rotary evaporator. Remove the remainder of the ether in a distillation apparatus and fractionally distil the residue at atmospheric pressure (2), collecting the ($-$)-α-methylbenzylamine as a fraction of b.p. 184–186 °C; the yield is about 5 g (53%). Measure the optical rotation of the neat liquid and calculate the specific rotation (Section I,37). Pure ($-$)-α-methylbenzylamine has d_4^{22} 0.950, $[\alpha]_D^{22}$ $-40.3°$ (neat); $[\alpha]_D^{20}$ $-31.5°$ (c 3.2 in C_2H_5OH).

($+$)-α-**Methylbenzylamine.** This enantiomer may be recovered using the following procedure. Evaporate the methanolic filtrate from the isolation of the ($-$)-amine salt to dryness using a rotary evaporator. Convert the residual salt to the free amine by treatment with sodium hydroxide solution followed by ether extraction as described above. Do not distil the recovered amine but remove the last traces of ether completely by warming under reduced pressure. Weigh the resulting product (x g), measure its optical rotation and calculate the specific rotation. The ratio of this value to that of pure ($+$)-amine is the optical purity of the sample; the weight of ($+$)-amine in the sample in excess of that present in the racemic modification is given by:

$$\text{excess (+)-amine} = \left[\frac{x \times \text{observed } [\alpha]_D \text{ for neat liquid}}{40.3} \right] \text{grams}$$

For *each* gram of excess ($+$)-amine present add firstly 10.0 ml of rectified spirit, bring to the boil and then add, for each gram, a hot solution of 0.44 g of 98 per cent sulphuric acid (1.03 times the theoretical amount) in 21.5 ml of rectified spirit. Allow the solution to cool slowly to room temperature, filter off the crystalline ($+$)-amine sulphate and wash it with cold rectified spirit. The yield is about 1 g of sulphate per gram of ($+$)-amine (71%). Liberate the free ($+$)-amine from the sulphate as described for the ($-$)-amine from the tartrate, but use 4 ml of water and 0.5 ml of 50 per cent aqueous sodium hydroxide for each gram of sulphate. The yield of ($+$)-amine, b.p. 184–186 °C, is 60 per cent of theory; its optical purity is 95 per cent.

Notes. (1) If fine needles separate, the mixture should be warmed until they re-dissolve, and the solution allowed to cool. The solution should be seeded with the prismatic crystals if these are available.

(2) The free amine rapidly absorbs carbon dioxide. It is therefore essential to protect the distillation apparatus from the atmosphere with a guard-tube filled with soda-lime. As the product tends to foam excessively during distillation, the apparatus used should be larger than is customary for the volume of liquid to be distilled.

III,197. RESOLUTION OF ($\pm$)-OCTAN-2-OL*

Heat a mixture of 65 g (0.5 mol) of dry octan-2-ol (b.p. 178–180 °C), 74 g (0.5 mol) of pure phthalic anhydride (1) and 40 g of dry pyridine on a water bath for 1 hour, and allow to cool. Dissolve the resulting viscous mass in an

---

* The following experimental details were kindly supplied by the late Dr J. Kenyon F.R.S.

equal volume of acetone. Add slowly, preferably with stirring, 55 ml of concentrated hydrochloric acid diluted with an approximately equal volume of crushed ice: if an oil separates before all the hydrochloric acid has been added, introduce more acetone to render the mixture homogeneous. Add ice-water until the oil is completely precipitated; this usually sets to a hard mass within 1–2 hours. If the resulting mass is semi-solid or pasty (2), transfer it to a large flask and pass steam through it until the octan-2-one is removed, i.e., until the steam distillate is clear; pour the contents of the flask whilst still warm into a beaker. The (±)-2-octyl hydrogen phthalate solidifies completely on cooling. Filter the octyl hydrogen phthalate at the pump, wash it with water, grind it thoroughly in a mortar with water, filter again and dry in the air. The crude material is quite satisfactory for the subsequent resolution (3).

Introduce 197 g (0.5 mol) of anhydrous brucine (*CARE*: poisonous) or 215 g of the air-dried dihydrate (4) into a warm solution of 139 g of (±)-2-octyl hydrogen phthalate in 300 ml of acetone and warm the mixture under reflux on a water bath until the solution is clear. Upon cooling, the brucine salt [(+)-A, (−)-B] separates as a crystalline solid. Filter this off on a sintered glass funnel, press it well to remove mother-liquor and wash it in the funnel with 123 ml of acetone. Set the combined filtrate and washings (W) aside. Cover the crystals with acetone and add, slowly and with stirring, a slight excess (to Congo red) of dilute hydrochloric acid (1 : 1 by volume; about 60 ml); if the solution becomes turbid before the introduction of the acid is complete, add more acetone to produce a clear liquid. Add ice-water until the precipitation of the active 2-octyl hydrogen phthalate [crude (+)-A is complete; filter (5), wash with cold water and dry in the air. The yield is about half that of the (±)-ester originally taken (6).

Concentrate the combined filtrate and washings (W) to about half the original volume, and pour it into slightly more than the calculated amount of dilute hydrochloric acid (use a mixture of 30 ml of concentrated hydrochloric acid and 30 ml of ice-water); then add about 300 ml of water. Collect the active 2-octyl hydrogen phthalate (crude(−)-A) as above (5). The weight of the air-dried ester is about half that of the (±)-ester originally used (7).

Crystallise the two lots of crude active 2-octyl hydrogen phthalates separately twice from 90 per cent acetic acid; use 2 g of acetic acid to each gram of solid. The recrystallised esters, if optically pure (8), will melt sharply at 75 °C; if the melting points are below 75 °C, further recrystallisation is necessary. The yields of optically pure products, m.p. 75 °C, are 48 g and 49 g respectively.

To obtain optically pure (+)- and (−)-octan-2-ol, steam distil the respective esters with 30 per cent sodium hydroxide solution; use the proportions 1 mol of ester to 2 mols of sodium hydroxide. Separate the alcohols from the steam distillate, dry over anhydrous potassium carbonate and distil under diminished pressure. Both samples boil at 86 °C/20 mmHg (9) and have the following rotations:

$$[\alpha]_D^{17°} +9.9°, [\alpha]_{5461}^{17°} +11.8°; [\alpha]_D^{17°} -9.9°, [\alpha]_{5461}^{17°} -11.8°.$$

The yields from the 2-octyl hydrogen phthalates are almost quantitative.

Notes. (1) If the presence of phthalic acid is suspected, it may be readily removed by mixing with cold chloroform; phthalic anhydride dissolves readily, but the acid is insoluble.

(2) This is due to octan-2-one in the original octan-2-ol; it is most easily separated by steam distillation as described.

(3) The inactive 2-octyl hydrogen phthalate may be recrystallised from light petroleum, b.p. 60–80 °C, or from glacial acetic acid, and then melts at 55 °C. If the octan-2-ol is pure, the yield of pure material is almost quantitative.

(4) Commercial brucine is usually the tetrahydrate $C_{23}H_{26}O_4N_2,4H_2O$; upon air drying, this loses two molecules of water of crystallisation and passes into the dihydrate.

(5) The filtrates from the decomposition of the brucine salts with dilute hydrochloride acid should be carefully preserved. The brucine is recovered by the addition of an excess of dilute ammonia solution (1 : 4); if the solution becomes turbid before all the ammonia solution is added, introduce a little ethanol until the solution becomes clear. After several hours in an open beaker, filter off the brucine, wash it well with cold water and dry it in the air.

(6) The rotation in absolute ethanol is about $[\alpha]_D$ +44°, $[\alpha]_{5461}$ +47°.

(7) The rotation in absolute ethanol is about $[\alpha]_D$ −44°, $[\alpha]_{5461}$ −47°.

(8) The optically pure esters have rotations in ethanol of $[\alpha]_D$ −48.4°, $[\alpha]_{5461}$ −58.5°, and $[\alpha]_D$ +48.4°, $[\alpha]_{5461}$ +58.5° respectively. A preliminary check of the optical purity is, however, more simply made by a m.p. determination; the rotation is determined, if desired, when the m.p. is 75 °C.

(9) The boiling point under atmospheric pressure is 179 °C.

III,198. RESOLUTION OF DL-ALANINE

Benzoyl DL-alanine. Dissolve 100 g (1.1 mol) of DL-alanine (Section **III,179**) in 400 ml of water containing 44.5 g (1.1 mol) of sodium hydroxide and cool the solution in an ice bath. Add 175 g (1.2 mol) of benzoyl chloride and a solution of 49 g (1.2 mol) of sodium hydroxide in 200 ml of water to the stirred, cooled, amino acid solution, alternately and in portions during 2 hours; continue to stir for a further 2-hour period. Boil the reaction mixture with 10 g of decolourising charcoal, filter, cool the clear yellow filtrate to 0 °C and acidify carefully to Congo red with concentrated hydrochloric acid. Triturate a portion of the oil which separates with water to induce crystallisation and then seed the bulk of the acidified solution with crystals and leave in an ice bath to complete the crystallisation process. Filter off the product, wash the filter cake with 500 ml of ice-cold water and recrystallise from about 3.5 litres of boiling water. The yield of benzoyl-DL-alanine, m.p. 162–164 °C, is 194.5 g (90%).

Benzoyl-L-alanine anilide. Use freshly boiled, cooled, distilled water throughout this stage. Prepare an 0.1 molar citrate buffer solution by dissolving 48 g of anhydrous citric acid and 16.5 g of sodium hydroxide in 2.5 litres of water. Stir together 50 g of technical powdered papain (1) and 4 g of potassium cyanide in 500 ml of the buffer solution, adjust the pH to 5 (narrow range pH paper) with glacial acetic acid, and stir for a further 75 minutes. Filter this enzyme extract through a Celite filter bed. Dissolve 193 g (1 mol) of benzoyl-DL-alanine by warming it in 300 ml of the citrate buffer to which has been added 120 ml of 2.5 molar sodium hydroxide, 360 ml of 3 M-sodium acetate solution and 93 g (91 ml, 1 mol) of redistilled aniline; adjust the pH of this solution to 5 with sodium hydroxide solution. Cool the solution to 45 °C, add the filtered enzyme extract and transfer the mixture to several conical flasks of suitable size such that each is filled to the neck and tightly stoppered with a rubber bung to exclude air. Leave the flasks in an incubator held at 37 °C, shaking them occasionally during the early stages.

Product begins to separate within 5 minutes and the contents of the flasks becomes almost immobile within 2 hours. After 24 hours filter the mixture and return the filtrate to the incubator for a further 24 hours and remove the additional crop of solid which separates. Wash the combined solids with 250 ml of water and recrystallise from 1 litre of 50 per cent aqueous ethanol with the aid of decolourising charcoal. The yield of benzoyl-L-alanine anilide is 122 g (91%), m.p. 175–176 °C, $[\alpha]_D^{20}$ $-7.8°$ (c 5 in $CH_3 \cdot CO_2H$).

L-**Alanine.**   Heat a mixture of 50 g (0.187 mol) of benzoyl-L-alanine anilide and 250 ml of 6 M-hydrochloric acid under reflux for 5 hours. Leave the mixture at room temperature overnight, remove the precipitated benzoic acid by filtration and evaporate the filtrate to dryness under reduced pressure (rotary evaporator). Dissolve the brown oily residue in 100 ml of water and boil it with decolourising charcoal. Filter and pass the filtrate through 450 g of a weakly basic anion exchange resin, e.g., Amberlite IR4B (which has been washed free from soluble colour with dilute hydrochloric acid and regenerated with dilute aqueous ammonia) in the form of a column 60 cm long. Collect the effluent (in all about 2 litres) until it gives no colour when boiled with ninhydrin (Section **III,100**). Evaporate the effluent to dryness under reduced pressure (rotary evaporator) and boil the yellow solid residue with 20 ml of water and a little decolourising charcoal. Add ethanol to the hot filtered solution until crystallisation begins and cool in ice to complete the separation of the L-alanine. The yield is 13.9 g (76%), $[\alpha]_D^{20}$ $+12.0°$ (c 4 in 1 M-HCl).

Note.   (1) The inexpensive crude commercial product (dried papaya latex) was used; the activity was not determined. Recrystallised, highly active enzyme preparations may, however, be obtained

T ALIPHATIC SULPHUR COMPOUNDS

This section describes typical general methods for the preparation of miscellaneous sulphur-containing compounds.

T,1. Thiols via S-alkylisothiouronium salts.   The thiols are the sulphur analogues of the alcohols and were formerly called mercaptans. Although they may be prepared by the interaction of an alkyl bromide and sodium hydrosulphide in ethanolic solution, a better method involves the interaction of an alkyl bromide and thiourea to form an S-alkylisothiouronium salt (I), followed by hydrolysis of the latter with sodium hydroxide solution.

$$RBr + NaSH \longrightarrow RSH + NaBr$$

$$2\left[H_2N-C=\overset{\oplus}{N}H_2\right]Br^{\ominus} + NaOH \longrightarrow$$

$$2RSH + 2NaBr + 2H_2O + H_2N-C-NH\cdot CN$$

This preparative method is quite general as indicated by the range of examples included in Section **III,199**.

In many cases it is not necessary to prepare the alkyl bromide; the *S*-alkylisothiouronium salt may be prepared directly from the alcohol by heating it with thiourea and concentrated aqueous hydrobromic acid (Ref. 23).

The lower members of the thiol series have remarkably disagreeable odours, but the offensive odour diminishes with increasing carbon content; for example, dodecane-1-thiol is not noticeably unpleasant.

The preparation and isolation of *S*-benzylisothiouronium chloride by the interaction of benzyl chloride and thiourea is described in Section **III,200**. On recrystallisation the compound separates in either, or sometimes as both, of two dimorphic forms, m.p. 150 and 175 °C respectively. The former may be converted into the higher m.p. form by dissolving it in ethanol and seeding with crystals of the form, m.p. 175 °C: the low m.p. form when warmed to 175 °C gives, after solidification, a m.p. of 175 °C. The compound is of particular interest as a reagent for the characterisation of carboxylic acids or sulphonic acids with which it forms the *S*-benzylisothiouronium carboxylate (or sulphonate) salts (Sections **VII,6,***15*(*e*) and **VII,6,***26*(*b*)). Both dimorphic forms give identical derivatives.

III,199. HEXANE-1-THIOL

$$2CH_3 \cdot (CH_2)_4 \cdot CH_2Br + 2H_2N \cdot CS \cdot NH_2 \longrightarrow$$

$$2[CH_3 \cdot (CH_2)_4 \cdot CH_2 \cdot S \cdot C(NH_2) \overset{\oplus}{=} \overset{\ominus}{NH_2}]Br \xrightarrow{NaOH}$$

$$2CH_3 \cdot (CH_2)_4 \cdot CH_2 \cdot SH + H_2N \cdot C(=NH) \cdot NH \cdot CN$$

This preparation must be carried out in an efficient fume cupboard.

Into a 500-ml two-necked flask, equipped with a sealed stirrer unit and a reflux condenser, place 62.5 g (53.5 ml, 0.38 mol) of 1-bromohexane (Section **III,55**) and a solution of 38 g (0.5 mol) of thiourea in 25 ml of water. Connect a tube from the top of the condenser leading to an inverted funnel just immersed in potassium permanganate solution in order to prevent the escape of unpleasant odours. Stir the mixture vigorously and heat under reflux for 2 hours; the mixture becomes homogeneous after about 30 minutes and the additional heating ensures the completeness of the reaction. Add a solution of 30 g of sodium hydroxide in 300 ml of water and reflux, with stirring, for a further 2 hours; during this period the thiol separates since it is largely insoluble in the alkaline medium. Allow to cool and separate the upper layer of almost pure hexane-1-thiol (35 g). Acidify the aqueous layer with a cold solution of 7 ml of concentrated sulphuric acid in 50 ml of water, and extract it with 75 ml of ether. Combine the ethereal extract with the crude thiol, dry with anhydrous sodium sulphate and remove the ether on a water bath. Distil the residue using an air bath (Fig. I,49(*a* and *b*)) and collect the hexane-1-thiol at 150–152 °C. The yield is 37.5 g (84%).

Cognate preparation **Butane-1-thiol.** Use 51 g (40 ml, 0.372 mol) of butyl bromide (Section **III,54**), 38 g (0.5 mol) of thiourea and 25 ml of water. Reflux, with stirring, for 3 hours; the mixture becomes homogeneous after 1 hour. Allow to cool and separate the upper layer of the thiol (*A*). Acidify the aqueous layer with a cold solution of 7 ml of concentrated sulphuric acid in 50 ml of water, cool and saturate with salt; remove the upper layer of butane-1-thiol (*B*) and combine it with (*A*). Extract the aqueous liquid with 75 ml of ether, dry the ethereal extract with anhydrous sodium or calcium sulphate and distil off the ether from a water bath through a fractionating column. Combine the residue with (*A*) and (*B*), and distil. Collect the butane-1-thiol at 97–99 °C. The yield is 24 g (72%).

General remarks on the preparation of thiols. The above method is of quite general application. If the bromoalkane is inexpensive, the extraction with ether may be omitted, thus rendering the preparation of thiols a comparatively easy and 'not unduly unpleasant operation. The following thiols may be prepared in yields of the same order as hexane-1-thiol and butane-1-thiol; ethanethiol, b.p. 35–36 °C; propane-1-thiol, b.p. 66–67 °C; propane-2-thiol, b.p. 51–52 °C; 2-methylpropane-1-thiol, b.p. 87–88 °C; pentane-1-thiol, b.p. 124–125 °C; heptane-1-thiol, b.p. 175–176 °C; octane-1-thiol, b.p. 198–200 °C or 98–100 °C/22 mmHg; nonane-1-thiol, b.p. 220–222 °C or 98–100 °C/15 mmHg; decane-1-thiol, b.p. 96–97 °C/5 mmHg or 114 °C/13 mmHg; undecane-1-thiol, b.p. 103–104 °C/3 mmHg; dodecane-1-thiol, b.p. 111–112 °C/3 mmHg or 153–155 °C/24 mmHg; tetradecane-1-thiol, b.p. 176–180 °C/22 mmHg; α-toluenethiol, b.p. 195 °C.

III,200.   S-BENZYLISOTHIOURONIUM CHLORIDE

$$C_6H_5 \cdot CH_2Cl + H_2N - \underset{\underset{S}{\|}}{C} - NH_2 \longrightarrow \left[H_2N - C = \overset{\oplus}{N}H_2 \atop \underset{S \cdot CH_2 \cdot C_6H_5}{|} \right] Cl^{\ominus}$$

Method 1. Dissolve 76 g (1 mol) of thiourea in 200 ml of warm water in a 1-litre round-bottomed flask. Dilute the solution with 135 ml of rectified spirit and add 126.5 g (1 mol) of benzyl chloride. Heat the mixture under reflux on a water bath until the benzyl chloride dissolves (about 15 minutes) and for a further 30 minutes taking care that the mixture is well shaken from time to time. Cool the mixture in ice: there is a tendency to supersaturation so that it is advisable to stir (or shake) the cold solution vigorously, when the substance crystallises suddenly. Filter off the solid at the pump. Evaporate the filtrate to about half bulk in order to recover a further small quantity of product. Dry the compound upon filter-paper in the air. The yield of S-benzylisothiouronium chloride, m.p. 174 °C, is 200 g (99%). Recrystallise the salt from 400 ml of 0.2 *M*-hydrochloric acid; filter off the solid which separates on cooling. The yield of recrystallised salt, m.p. 175 °C, is 185 g (91%); some of the dimorphic form, m.p. 150 °C, may also separate.

Method 2. Place a mixture of 126.5 g (1 mol) of benzyl chloride, 76 g (1 mol) of thiourea and 150 ml of rectified spirit in a 500-ml round-bottomed flask fitted with a reflux condenser. Warm on a water bath. A sudden exothermic reaction soon occurs and all the thiourea passes into solution. Reflux the resulting yellow solution for 30 minutes and then cool in ice. Filter off the white crystals and dry in the air upon filter-paper. Concentrate the filtrate to half its original volume and thus obtain a further small crop of crystals. The yield of crude S-benzylisothiouronium chloride, m.p. 145 °C, is about the same as in *Method 1*. Recrystallise from 0.2 *M*-hydrochloric acid as in *Method 1*; the m.p. is raised to 150 °C, although on some occasions the form, m.p. 175 °C, separates.

T,2. Dialkyl sulphides (thioethers).   These are conveniently obtained by boiling alkyl halides with sodium sulphide in ethanolic solution.

$$2\,R\,Br + Na_2S \longrightarrow R\cdot S\cdot R + 2NaBr$$

This procedure can only yield a symmetrical sulphide (e.g., dipropyl sulphide, Section **III,201**). Mixed sulphides are prepared by alkylation of a thiolate salt (a *mercaptide*) with an alkyl halide (cf. Williamson's ether synthesis, p. 413, **III,G,2**).

$$R^1\cdot\overset{\ominus}{S}\!:\;\frown\;R^2\overset{\frown}{\text{—}}Br \longrightarrow R^1\cdot S\cdot R^2 + \overset{\ominus}{:}Br$$

This general procedure is illustrated by the preparation of isobutyl 2,4-dinitrophenyl sulphide (Section **III,202**), using the reactive aromatic halide, 1-chloro-2,4-dinitrobenzene. The alkyl 2,4-dinitrophenyl sulphides serve as useful solid derivatives for the characterisation of the thiols.

A variant of this procedure is provided by the preparation of S-benzyl-L-cysteine (Section **III,203**). The required thiolate salt is prepared by reductive cleavage with sodium in liquid ammonia of the disulphide linkage in the amino acid, L-cystine, and is alkylated *in situ* with benzyl chloride. The preparation of this S-benzyl derivative constitutes a method of protection of the thiol grouping in cysteine during procedures involved in peptide synthesis.

III,201. DIPROPYL SULPHIDE

$$2CH_3\cdot CH_2\cdot CH_2Br + Na_2S \longrightarrow (CH_3\cdot CH_2\cdot CH_2)_2S + 2NaBr$$

This preparation must be carried out in an efficient fume cupboard.

Place 56 g (0.5 mol) of finely powdered, fused sodium sulphide (1) and 100 ml of rectified spirit in a 500-ml round-bottomed flask equipped with a reflux condenser. To the boiling mixture add 46 g (34 ml, 0.374 mol) of propyl bromide (Section **III,54**) slowly and reflux for 6 hours. Distil off the ethanol on a water bath, and add a large excess of water to the distillate. Separate the upper layer of crude sulphide, wash it with three 40-ml portions of 5 per cent sodium hydroxide solution, then with water until the washings are neutral, and dry over anhydrous calcium chloride or anhydrous calcium sulphate. Distil, and collect the dipropyl sulphide at 141–143 °C. The yield is 20 g (91%). If the sulphide is required perfectly pure, it should be redistilled from a little sodium.

Cognate preparation   Dihexyl sulphide.   Use 83 g (71 ml, 0.5 mol) of 1-bromohexane (Section **III,55**), 56 g (0.5 mol) of finely powdered, fused sodium sulphide and 100 ml of rectified spirit. Reflux on a water bath for 20 hours. Distil off the alcohol from a water bath; very little sulphide is obtained upon adding excess of water to the distillate. Add excess of water to the residue in the flask and separate the upper layer of crude dihexyl sulphide. Purify as for dipropyl sulphide, but distil under reduced pressure. Collect the dihexyl sulphide at 113–114 °C/4 mmHg. The yield is 45 g (89%).

Dibenzyl sulphide.   Heat a solution of 63 g (0.5 mol) of benzyl chloride in 160 ml of rectified spirit on a steam bath and stir while adding a solution of 29 g (0.25 mol) of fused sodium sulphide in about 50–60 ml of water. Continue stirring and heating for 3 days, remove the ethanol on a rotary evaporator and pour the residue on to 350 g of crushed ice. Separate the oil and triturate with a little 70 per cent ethanol to crystallise the product. Recrystallise from the same solvent; the yield of dibenzyl sulphide is 26 g (83%), m.p. 49 °C.

Note. (1) The technical fused grade has the approximate composition of $Na_2S \cdot 2H_2O$.

III,202. ISOBUTYL 2,4-DINITROPHENYL SULPHIDE

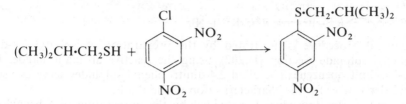

$$(CH_3)_2CH \cdot CH_2SH +$$

Dissolve 1 g (0.005 mol) of 1-chloro-2,4-dinitrobenzene in 5 ml of rectified spirit with warming and add a solution of 0.5 ml (0.005 mol) of 2-methyl-propane-1-thiol in 5 ml of rectified spirit containing 2 ml of 10 per cent aqueous sodium hydroxide. Heat under reflux for 10 minutes and decant the hot solution from any insoluble material into a clean conical flask. Allow the solution to cool, filter and recrystallise the sulphide twice from methanol. The product is obtained as yellow flakes, m.p. 75–76 °C; the yield is 440 mg (35%).

III,203. S-BENZYL-L-CYSTEINE [L-2-Amino-3-(benzylthio)propionic acid]

$$(NH_2 \cdot CH \cdot CH_2 - S)_2 \xrightarrow{Na/NH_3} 2NH_2 \cdot CH \cdot CH_2 \ S^{\ominus}Na^{\oplus} \xrightarrow[(ii)\ HCl]{(i)\ C_6H_5 \cdot CH_2Cl}$$
$$| \qquad\qquad\qquad\qquad\qquad |$$
$$CO_2H \qquad\qquad\qquad\qquad CO_2^{\ominus}Na^{\oplus}$$

$$2NH_2 \cdot CH \cdot CH_2 \cdot S \cdot CH_2 \cdot C_6H_5$$
$$|$$
$$CO_2H$$

Collect about 750 ml of liquid ammonia (Section **II,17,6**) in a 1-litre three-necked flask, surrounded by a lagging bath of cork chips and fitted with a sealed stirrer unit, a soda-lime guard-tube and a stopper. Weigh out 24 g (0.1 mol) of L-cystine (Section **III,186**), and about 10 g (0.48 mol) of sodium cut into small pieces under dry light petroleum. Start the stirrer, add about 2 g of the sodium followed by L-cystine in small portions until the blue colour has disappeared. Repeat this addition sequence until all of the cystine has been added and a permanent blue colour remains. Discharge the blue colour by gradually adding powdered ammonium chloride, and then add dropwise 25.3 g (23 ml, 0.2 mol) of benzyl chloride. Remove the stirrer and lagging bath and allow the ammonia to evaporate overnight. Dissolve the residue in 100 ml of cold water and add concentrated hydrochloric acid until the resulting mass is acid to Congo red. Heat the mixture gradually to boiling to dissolve the precipitated product and allow to cool. Filter off the long needles of S-benzyl-L-cysteine which separate, wash with a little cold water and allow to dry in the air. The yield is 38 g (90%), m.p. 214 °C (decomp.).

T,3. Dialkyl disulphides. The most important method for the preparation of this class of sulphur-containing compounds (named systematically as alkyldithioalkanes) is by the mild oxidation of thiols, usually with iodine in the presence of alkali (e.g., the preparation of diethyl disulphide, Section **III,204**).

$$2RSH + I_2 \xrightarrow{\overset{\ominus}{:}OH} R{\cdot}S{\cdot}S{\cdot}R + 2\overset{\ominus}{I}{:}$$

This preparative method is only applicable to the formation of symmetrical disulphides; unsymmetrical disulphides are formed when the symmetrical disulphide is heated with another thiol when the following equilibrium is established.

$$R^1{\cdot}S{\cdot}S{\cdot}R^1 + R^2SH \rightleftharpoons R^1{\cdot}S{\cdot}S{\cdot}R^2 + R^1SH$$

If the thiol R^1SH is the more volatile, its removal by fractional distillation allows the mixed disulphide to be obtained in good yield (Ref. 24).

III,204. DIETHYL DISULPHIDE

$$2C_2H_5\overset{\ominus}{S}\overset{\oplus}{Na} + I_2 \longrightarrow C_2H_5S{\cdot}SC_2H_5 + 2NaI$$

This preparation must be carried out in an efficient fume cupboard.

Fit a 500-ml three-necked flask with a mechanical stirrer and a double surface condenser. Cool the flask in ice, introduce 38.5 g (46 ml, 0.62 mol) of ethanethiol and 175 ml (0.66 mol) of 15 per cent sodium hydroxide solution, and stir the mixture. When all the thiol has reacted, add with constant stirring 67.5 g (0.266 mol) of iodine gradually (during about 2 hours) by momentarily removing the stopper from the third neck of the flask and replacing it immediately the iodine has been introduced. After each addition the iodine gradually disappears and an oily layer forms on the surface of the liquid. Stir the mixture (1) for a further 2.5 hours and allow to stand for 2 hours: transfer to a separatory funnel. Remove the colourless upper layer and extract the aqueous layer with ether. Combine the ethereal extract with the upper layer, wash it with one-third of its volume of 15 per cent sodium hydroxide solution, then twice with water, partially dry it with anhydrous calcium chloride and remove the ether on a water bath. The resulting colourless liquid usually has a slight odour of thiol. Wash it three times with one-third of its volume of 5 per cent sodium hydroxide solution, followed by water until free from alkali, and then dry with anhydrous calcium chloride or anhydrous calcium sulphate. Distil, and collect the diethyl disulphide at 151–152 °C. The yield is 27 g (83%).

Note. (1) The mixture should be colourless, otherwise difficulty will be experienced in the subsequent purification of the product. If the reaction mixture is coloured by iodine (due to volatilisation of some of the thiol), add just sufficient ethanethiol to decolourise it.

Cognate preparation Dibutyl disulphide. Use 45 g (53.5 ml, 0.5 mol) of butane-1-thiol (Section **III,199**), 135 ml (0.5 mol) of 15 per cent sodium hydroxide solution and 55 g (0.198 mol) of iodine. The iodine may be dissolved in 40 per cent potassium iodide solution, if desired. Wash the *colourless* upper layer (see Note (1) above) three times with one-third of its volume of 5 per cent sodium hydroxide solution, then with water until free from alkali, dry over anhydrous calcium chloride or anhydrous calcium sulphate and distil under reduced pressure. Collect the dibutyl disulphide at 84 °C/3 mmHg. The yield is 30 g (85%). The b.p. under atmospheric pressure is 230–231 °C.

T,4. Sulphoxides and sulphones.   Dialkyl sulphides may be oxidised to the sulphoxides (I) (alkylsulphinylalkanes) and thence to the sulphones (II) (alkylsulphonylalkanes).

$$R \cdot S \cdot R \xrightarrow{[O]} R \cdot SO \cdot R \xrightarrow{[O]} R \cdot SO_2 \cdot R$$
$$\text{(I)} \qquad\qquad \text{(II)}$$

Sulphoxides are obtained by using a variety of mild oxidising agents; hydrogen peroxide has been widely used but aqueous sodium metaperiodate solution is also recommended (Ref. 25) and its use is illustrated in Section **III,205**.

More vigorous oxidising agents yield the sulphones; an excess of potassium permanganate in aqueous acetic acid is usually employed. Oxidation of alkyl 2,4-dinitrophenyl sulphides (p. 584, **III,T,2**) to the corresponding sulphone (Section **III,206**) is often useful as a further aid to the characterisation of a thiol.

III,205. DIBENZYL SULPHOXIDE

$$(C_6H_5 \cdot CH_2)_2S \xrightarrow{NaIO_4} (C_6H_5 \cdot CH_2)_2SO$$

Stir 2.35 g (0.011 mol) of sodium metaperiodate (Section **II,2,45**) in 45 ml of a 1:1 mixture of water and methanol held at 0 °C. Add portionwise 2.14 g (0.01 mol) of dibenzyl sulphide (Section **III,200**) and continue to stir the mixture at 0 °C for several hours, preferably overnight. Extract the reaction mixture (which contains precipitated sodium iodate) with three 20 ml portions of chloroform. Dry the combined chloroform extracts over magnesium sulphate and remove the solvent on a rotary evaporator. Recrystallise the product from ethanol. The yield of dibenzyl sulphoxide is 2.2 g (96%), m.p. 135 °C.

III,206. ISOBUTYL 2,4-DINITROPHENYL SULPHONE

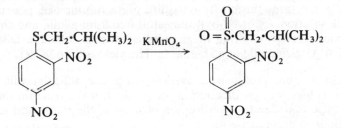

Dissolve 340 mg of isobutyl 2,4-dinitrophenyl sulphide (Section **III,202**) in 10 ml of acetic acid, warm to about 50 °C and add dropwise 8 ml of 3 per cent aqueous potassium permanganate solution. Maintain the solution at about 50 °C for 20 minutes, and then pass through the solution a stream of sulphur dioxide to decompose excess potassium permanganate. Add crushed ice to the yellow solution to precipitate the crude sulphone. Filter the precipitate, dry and recrystallise from rectified spirit to give the yellow crystalline product, m.p. 105–106 °C; the yield is 190 mg (50%).

T,5. *O*,*S*-Dialkyl dithiocarbonates (xanthate esters). Alkali metal salts of the *O*-alkyl dithiocarbonates (e.g., $RO \cdot CS \cdot S^{\ominus}K^{\oplus}$, the xanthates) are prepared by the reaction of carbon disulphide with an alcohol and an alkali metal hydroxide.

$$KOH + CS_2 + R^1OH \longrightarrow R^1O \cdot CS \cdot \overset{\ominus}{S} \overset{\oplus}{K} + H_2O$$

Alkylation of this xanthate salt with an alkyl halide in absolute ethanol leads to an *O*,*S*-dialkyl dithiocarbonate.

$$R^1O \cdot CS \cdot \overset{\ominus}{\underset{..}{S}} \overset{\frown}{} R^2 \underset{\downarrow}{-I} \longrightarrow R^1O \cdot CS \cdot SR^2 + \overset{\ominus}{\underset{..}{I}}$$

The xanthate esters are of interest in that if the *O*-alkyl group contains at least one α-hydrogen atom, pyrolysis produces an olefin, a thiol and carbon oxysulphide (the Chugaev reaction, p. 328, **III,B,1**).

Isolation of the intermediate xanthate salts in the preparation of the xanthate esters is not essential. The formation of the latter may be achieved, in one step, by converting the alcohol into the corresponding alkoxide by reaction with a potassium derivative of a tertiary alcohol (e.g., potassium t-pentoxide), followed by reaction successively with carbon disulphide and the alkyl halide. Both preparative procedures are illustrated in the following section.

III,207. *O*-ETHYL *S*-ETHYL DITHIOCARBONATE

$$C_2H_5O^{\ominus}K^{\oplus} + \underset{\underset{\|}{S}}{\overset{\overset{\|}{S}}{C}} \longrightarrow C_2H_5 \cdot O \cdot C\overset{\diagup S}{\diagdown S^{\ominus}K^{\oplus}} \xrightarrow{C_2H_5I} C_2H_5 \cdot O \cdot C\overset{\diagup S}{\diagdown SC_2H_5}$$

These preparations must be conducted in an efficient fume cupboard.

Potassium *O*-ethyl dithiocarbonate. Into a 500-ml round-bottomed flask, fitted with a reflux condenser, place 42 g (0.75 mol) of potassium hydroxide pellets and 120 g (152 ml) of absolute ethanol. Heat under reflux for 1 hour. Allow to cool and decant the liquid from the residual solid into another dry 500-ml flask; add 57 g (45 ml, 0.75 mol) of carbon disulphide (1) slowly and with constant shaking. Filter the resulting almost solid mass, after cooling in ice, on a sintered glass funnel at the pump, and wash it with three 25 ml portions of ether. Dry the potassium *O*-ethyl dithiocarbonate in a vacuum desiccator over silica gel. The yield is 74 g. If desired, it may be recrystallised from absolute ethanol but this is usually unnecessary.

O-Ethyl *S*-ethyl dithiocarbonate. Place 32 g (0.2 mol) of potassium *O*-ethyl dithiocarbonate and 50 ml of absolute ethanol in a 500-ml round-bottomed flask provided with a double surface condenser. Add 32 g (16.5 ml, 0.205 mol) of ethyl iodide. No reaction appears to take place in the cold. Heat on a water bath for 3 hours: a reaction sets in within 15 minutes and the yellow reaction mixture becomes white owing to the separation of potassium iodide. Add about 150 ml of water, separate the lower layer and wash it with water. Dry it with anhydrous calcium chloride or anhydrous calcium sulphate and distil collecting *O*-ethyl *S*-ethyl dithiocarbonate at 196–198 °C. The yield is 23 g (77%).

Note. (1) Carbon disulphide is toxic and has a dangerously low flash point (Section **I,3,B,**5(i)).

Cognate·preparation *O*-**Ethyl** *S*-**butyl dithiocarbonate.** Use 32 g (0.2 mol) of potassium *O*-ethyl dithiocarbonate, 37 g (23 ml, 0.2 mol) of butyl iodide (Section **III,58**) and 50 ml of absolute ethanol. Reflux on a water bath for 3 hours. Pour into 150 ml of water, saturate with salt (in order to facilitate the separation of the upper layer), remove the upper xanthate layer, wash it once with 25 ml of saturated salt solution and dry with anhydrous calcium chloride or anhydrous calcium sulphate. Distil under reduced pressure and collect the pale yellow *O*-ethyl *S*-butyl dithiocarbonate at 90–91 °C/4 mmHg. The yield is 34 g (95%).

O-[**1,2,2-Trimethylpropyl**] *S*-**methyl dithiocarbonate.** In a 2-litre three-necked flask equipped with a stirrer, reflux condenser and dropping funnel, prepare potassium t-pentoxide by dissolving 48.5 g (60 ml, 0.55 mol) of 2-methylbutan-2-ol (t-pentyl alcohol) in 750 ml of dry toluene and adding in portions 21.5 g (0.55 mol) of potassium metal (1) and refluxing gently until reaction is complete. Then add 51 g (63 ml, 0.5 mol) of 3,3-dimethylbutan-2-ol to the hot solution slowly and with stirring. Cool and add slowly 57 g (0.75 mol) of carbon disulphide. When the reaction has subsided, cool the resulting orange xanthate suspension to room temperature, add 78 g (34 ml, 0.55 mol) of methyl iodide and heat on a water bath for 4–5 hours. Filter, remove toluene and the residual alcohols on a rotary evaporator (*fume cupboard*) and distil the product under reduced pressure collecting the xanthate ester at 85–87 °C/6 mmHg (2). The yield is 63 g (70%).

Note. (1) Great care must be taken in the handling of potassium (see Section **III,178**, Note (1)) and of methyl iodide (p. 20, **I,3**).

(2) Use a 500-ml flask fitted with a short fractionating column; the distillation is accompanied by considerable frothing.

U SOME FURTHER UNSATURATED ALIPHATIC COMPOUNDS

1. Unsaturated aldehydes and ketones (Sections **III,208** and **III,209**).
2. Unsaturated carboxylic acids (Sections **III,210** to **III,213**).

U,1. Unsaturated aldehydes and ketones. Aldehydes possessing two α-hydrogen atoms undergo self-condensation under the influence of a catalyst (basic catalysts are usually employed) to give β-hydroxyaldehydes (the **aldol condensation**). All the steps in this reaction are reversible but the position of equilibrium is significantly in favour of the aldol, which generally may be obtained when the reaction is carried out at room temperature or below, followed by extraction and careful distillation under reduced pressure. The β-hydroxyaldehydes are not usually isolated, however, since they are readily converted into the conjugated unsaturated aldehydes by virtue of their ready dehydration (e.g., 2-ethylhex-2-enal, Section **III,208**). This may occur under the conditions employed in the initial condensation reaction (that is when the reaction mixture is

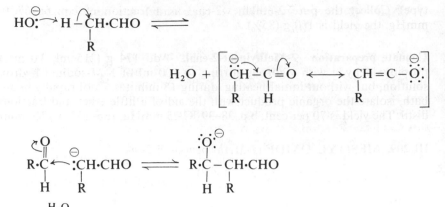

heated for some time) or more efficiently in the presence of a small amount of iodine.

$$HO-H \quad OH$$
$$R \cdot CH - C(R) \cdot CHO \longrightarrow R \cdot CH = C(R)CHO$$
$$H \qquad :OH_2$$

In the case of aldehydes with only one α-hydrogen atom, aldol formation occurs but the resulting α-hydroxyaldehyde cannot undergo the dehydration step.

Ketones form the corresponding ketols, but in this case the position of equilibrium is not in favour of the condensation product (which rapidly dissociates into the ketone in the presence of base), and a technique has to be employed which continuously removes the ketol product, as it is formed, from the presence of the base. A satisfactory procedure for converting acetone into diacetone alcohol is described in Section III,209. This ketol is readily dehydrated to mesityl oxide in the presence of a trace of iodine.

III,208. 2-ETHYLHEX-2-ENAL

$$2CH_3 \cdot CH_2 \cdot CH_2 \cdot CHO \xrightarrow{\ominus OH} CH_3 \cdot CH_2 \cdot CH_2 \cdot CH = C \cdot CHO$$
$$| \atop CH_2 \cdot CH_3$$

Place 100 ml of 1 *M*-sodium hydroxide solution in a 500-ml three-necked flask fitted with a sealed stirrer unit and an efficient reflux condenser. Heat the solution to 80 °C and attach to the flask a dropping funnel containing 216 g (264 ml, 3.0 mol) of redistilled butyraldehyde (compare Section III,79). With vigorous stirring, add the butyraldehyde as rapidly as the efficiency of the reflux condenser will allow and then boil the reaction mixture under reflux for 1 hour. Cool, separate the organic layer and distil it without further treatment under reduced pressure through a fractionating column (e.g., of the Vigreux type). Collect the pure 2-ethylhex-2-enal as a fraction of b.p. 66–67 °C/25 mmHg; the yield is 160 g (85%).

Cognate preparation 2-Methylpent-2-enal. Add 174 g (215 ml, 3.0 mol) of propionaldehyde with vigorous stirring to 100 ml of 1 M-sodium hydroxide solution, but without initial heating, during 15 minutes. Cool rapidly in an ice bath, isolate the organic product with the aid of a little ether and fractionally distil. The yield is 70 per cent; b.p. 38–39 °C/25 mmHg, 136–137 °C/760 mmHg.

III,209. MESITYL OXIDE (4-Methylpent-3-en-2-one)

$$(CH_3)_2CO + CH_3 \cdot CO \cdot CH_3 \xrightarrow{Ba(OH)_2} (CH_3)_2C(OH) \cdot CH_2 \cdot CO \cdot CH_3$$

$$(CH_3)_2C(OH) \cdot CH_2 \cdot CO \cdot CH_3 \xrightarrow[\text{heat}]{I_2} (CH_3)_2C = CH \cdot CO \cdot CH_3 + H_2O$$

4-Methyl-4-hydroxypentan-2-one (diacetone alcohol). Fit a 1-litre round-bottomed flask with a large Soxhlet extractor (Fig. I,96) and attach an efficient double surface condenser to the latter. Place 595 g (750 ml, 10.25 mol) of commercial acetone, preferably dried over anhydrous potassium carbonate, and a few fragments of porous porcelain in the flask. Select as large a Soxhlet thimble as the extractor will accommodate and three-quarters fill it with barium hydroxide (1). Fill the remaining space in the thimble with glass wool. Insert the charged thimble into the extractor. Heat the flask on a water bath or steam bath so that the acetone refluxes back into the extractor rather rapidly. Continue the heating until the acetone no longer refluxes when the flask is almost completely immersed in the boiling water bath (72–120 hours). The refluxing may be interrupted at any time for as long as desired without in-fluencing the preparation. Equip the flask with a fractionating column attached to an efficient double surface condenser set for downward distillation. Immerse the flask in an oil bath and raise the temperature gradually to 125 °C; maintain this temperature as long as acetone distils over. The recovery of acetone is complete when the temperature at the top of the column is about 70 °C. Distil the residue (2) under diminished pressure (3); a little acetone passes over first, followed by the diacetone alcohol at 71–74 °C/23 mmHg (or 62–64 °C/13 mmHg). The yield is 450 g (75%).

Mesityl oxide. Fit a 750-ml round-bottomed flask with a fractionating column attached to a condenser set for downward distillation. Place 400 g (3.44 mol) of diacetone alcohol (the crude product is quite satisfactory), 0.1 g of iodine and a few fragments of porous porcelain in the flask. Distil slowly with a small free flame (best in an air bath) and collect the following fractions: (a) 56–80 °C (acetone and a little mesityl oxide); (b) 80–126 °C (two layers, water and mesityl oxide); and (c) 126–131 °C, which is almost pure mesityl oxide. Separate the water from fraction (b), dry with anhydrous potassium carbonate or anhydrous sodium sulphate, fractionate from a small flask. A further quantity of mesityl oxide is thus obtained. The total yield is about 320 g (95%).

Notes. (1) If crystallised barium hydroxide $(Ba(OH)_2, 8H_2O)$ is employed, this becomes dehydrated after one run; the anhydrous compound is just as satisfactory and may be used repeatedly.

(2) The residual liquid contains about 95 per cent of diacetone alcohol and is satisfactory for the preparation of mesityl oxide.

(3) Diacetone alcohol partially decomposes when distilled under normal pressure.

U,2. Unsaturated acids. α,β-**Ethylenic acids** may be prepared by condensing an aldehyde with malonic acid in pyridine solution, often in the presence of a trace of piperidine (the **Doebner reaction**). The reaction mechanism is analogous to that of the aldol condensation and involves the addition of a malonate anion to the aldehydic carbonyl carbon atom followed by the elimination of water accompanied by decarboxylation.

$$R{\cdot}CHO + CH_2(CO_2H)_2 \longrightarrow [R{\cdot}CH(OH){\cdot}CH(CO_2H)_2] \xrightarrow[-CO_2]{-H_2O}$$
$$R{\cdot}CH{=}CH{\cdot}CO_2H$$

Examples given in Section **III,210** include the preparation of 3-hexylacrylic acid starting from hexanal, crotonic acid from acetaldehyde and also the dienoic acid, sorbic acid, starting from the conjugated aldehyde, crotonaldehyde.

A useful route to α,β-**acetylenic acids**, illustrated by the preparation of but-2-ynoic acid (Section **III,211**), involves the base-induced decomposition of the dibromopyrazolone which is obtained by brominating the pyrazol-5-one, which is itself prepared by the reaction of a β-keto ester with hydrazine.

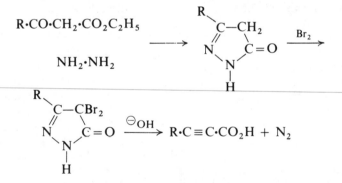

A likely mechanism for the last stage may be formulated as follows:

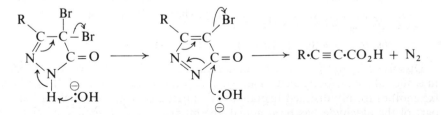

If the bromination of the pyrazol-5-one is interrupted when only one of the methylenic hydrogens has been substituted, and the monobromopyrazol-5-one is similarly treated with alkali, the α,β-ethylenic acid is obtained (Ref. 26).

The simplest unsaturated dicarboxylic acids are **maleic acid** and **fumaric**

acid. Commercially maleic anhydride (I) is prepared by the catalytic vapour phase oxidation of benzene with atmospheric oxygen in the presence of vanadium pentoxide at about 400 °C. Its laboratory preparation (Section **III,212**) may be achieved by warming malic acid (II) with acetyl chloride and distilling the mixture at atmospheric pressure. The anhydride is readily converted into maleic acid (III) by heating with water. Maleic acid has the *cis* (*Z*)-structure and may be converted into its geometric isomer, the *trans* (*E*)-form, fumaric acid (IV), by heating with hydrochloric acid. Fumaric acid, which does not form an anhydride, may be prepared from furfural (V) by oxidation with sodium chlorate in the presence of vanadium pentoxide as the catalyst (Section **III,213**). That the two acids are indeed geometrical isomers is confirmed by the fact that both acids yield succinic acid (VI) on catalytic hydrogenation. Maleic anhydride is widely used to form adducts with conjugated dienes (the **Diels–Alder reaction, V(d)**).

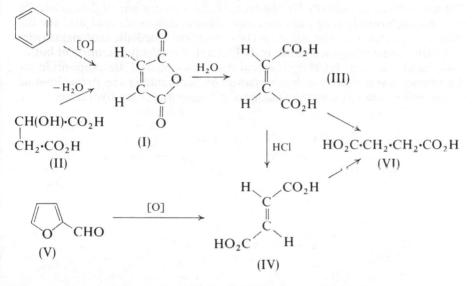

Most of these transformations are included in Section **III,212**.

III,210. 3-HEXYLACRYLIC ACID (*Non-2-enoic acid*)

$$CH_3 \cdot (CH_2)_5 \cdot CHO + CH_2(CO_2H)_2 \xrightarrow{\text{pyridine}}$$

$$CH_3 \cdot (CH_2)_5 \cdot CH = CH \cdot CO_2H + CO_2 + H_2O$$

Dissolve 57 g (0.55 mol) of malonic acid in 92.5 ml of dry pyridine contained in a 500-ml round-bottomed flask, cool the solution in ice and add 57 g (70 ml, 0.5 mol) of freshly distilled heptanal with stirring or vigorous shaking. After a part of the aldehyde has been added, the mixture yields a semi-solid slurry of crystals. Insert a calcium chloride tube into the mouth of the flask and allow the mixture to stand at room temperature for 60 hours with periodic shaking. Finally, warm the mixture on a water bath until the evolution of carbon dioxide ceases (about 8 hours) and then pour into an equal volume of water. Separate the oily layer and shake it with 150 ml of 25 per cent hydrochloric acid to remove pyridine. Dissolve the product in benzene, wash with water, dry with anhydrous sodium sulphate and distil under reduced pressure. Collect the nonenoic acid at 130–132 °C/2 mmHg. The yield is 62 g (79%).

Cognate preparations   Crotonic acid.   Mix together in a 250-ml flask carrying a reflux condenser and a calcium chloride drying tube 25 g (32 ml, 0.57 mol) of freshly distilled acetaldehyde (Section **III,81**) with a solution of 59.5 g (0.57 mol) of dry, powdered malonic acid in 67 g (68.5 ml, 0.85 mol) of dry pyridine to which 0.5 ml of piperidine has been added. Leave in an ice chest or refrigerator for 24 hours. Warm the mixture on a steam bath until the evolution of carbon dioxide ceases. Cool in ice, add 60 ml of 1 : 1 sulphuric acid (by volume) and leave in the ice bath for 3–4 hours. Collect the crude crotonic acid (*ca.* 27 g) which has separated by suction filtration. Extract the mother-liquor with three 25 ml portions of ether, dry the ethereal extract, and evaporate the ether; the residual crude acid weighs 6 g. Recrystallise from light petroleum, b.p. 60–80 °C; the yield of crude crotonic acid, m.p. 72 °C, is 20 g (41%).

Sorbic acid (*Hexa-2,4-dienoic acid*).   Place 40 g (46.5 ml, 0.57 mol) of crotonaldehyde (b.p. 101–103 °C), 60 g (0.575 mol) of malonic acid and 60 g (61 ml, 0.76 mol) of dry pyridine (b.p. 113–115 °C) in a 500-ml round-bottomed flask, attach a reflux condenser and heat on a water bath for 3 hours. At the end of this period the vigorous evolution of carbon dioxide will have ceased. Cool the mixture in ice and cautiously acidify it by the addition of a solution of 21.3 ml of concentrated sulphuric acid in 50 ml of water with shaking. Most of the sorbic acid separates out immediately; a more complete separation is obtained by cooling the solution in ice for 3–4 hours. Filter the acid at the pump and wash it with a little ice-cold water. Recrystallise from about 125 ml of boiling water; the maximum recovery of purified acid is achieved by leaving the solution in an ice chest or a refrigerator overnight and then filtering. The yield of sorbic acid, m.p. 134 °C, is 20 g (31%).

III,211. BUT-2-YNOIC ACID

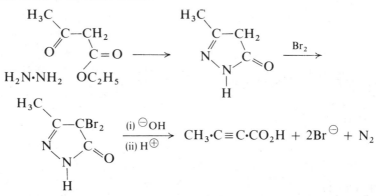

3-Methylpyrazol-5-one.   Place 65 g (0.5 mol) of ethyl acetoacetate in a conical flask and stir magnetically during the slow dropwise addition of a solution of 25 g (0.5 mol) of hydrazine hydrate (98–100%) in 40 ml of absolute ethanol. The temperature rises during this addition which should be regulated so that a temperature of about 60 °C is maintained; a crystalline deposit separates. After further stirring for 1 hour at room temperature, cool the reaction mixture in an ice bath to complete the crystallisation, and filter. Wash the product with ice-cold ethanol; it is then pure enough for use in the next stage. The yield is 43 g (90%), m.p. 222 °C (phase change at 195 °C; microscope m.p. apparatus).

4,4-Dibromo-3-methylpyrazol-5-one. Dissolve 20.0 g (0.2 mol) of 3-methylpyrazol-5-one in 80 ml of glacial acetic acid and stir magnetically during the slow dropwise addition of a solution of 32 g (0.2 mol) of bromine in 20 ml of glacial acetic acid (1). On completion of this addition, add 50 ml of water and continue the dropwise addition of a further 32 g (0.2 mol) of bromine dissolved in 20 ml of glacial acetic acid. On completion of this second addition of bromine solution allow the mixture to stand at room temperature overnight. Add water to precipitate the dibromopyrazolone, filter and wash the solid product under suction with distilled water until the washings are neutral. The air-dried product, sufficiently pure for use in the next stage, has m.p. 130–132 °C, the yield is 41 g (79%).

But-2-ynoic acid.   Prepare a solution of 20 g of sodium hydroxide in 500 ml of water and stir magnetically in an ice bath until the temperature reaches 0–5 °C. Add portionwise over 10 minutes 34 g (0.132 mol) of 4,4-dibromo-3-methylpyrazol-5-one. The bromoketone dissolves to give an orange-red solution which evolves nitrogen gas; the temperature of the solution during the addition shows only a slight tendency to rise. Stir the reaction mixture for 1 hour at 0–5 °C and then at room temperature for 1 hour. Cool the solution again and acidify it with concentrated hydrochloric acid. Continuously extract the acidified solution with ether overnight (Section I,22), dry the ethereal extract with magnesium sulphate and remove the solvent on a rotary evaporator. Place the flask containing the orange oil in a vacuum desiccator and allow to stand until it solidifies. Extract the orange crystalline deposit with successive portions of boiling light petroleum (b.p. 60–80 °C) and concentrate the combined extracts to about 50 ml. Filter the slightly off-white product, m.p. 74–75 °C; recrystallise by dissolving in the minimum volume of light petroleum (b.p. 80–100 °C), adding an equal volume of light petroleum (b.p. 40–60 °C) and allowing to cool. The pure but-2-ynoic acid has m.p. 75–76 °C, the yield is 5.9 g (54%). The i.r. spectrum shows absorption at 2950 (broad, $-OH$), 2240 (sharp, disubstituted $-C \equiv C-$), 1690 cm^{-1} (broad, $-C=O$ in carboxylic acid).

Note.   (1) Removal of a portion of the reaction mixture when 1 mol of bromine has been added and addition to it of water results in the precipitation of the monobromo compound, m.p. 180–182 °C.

III,212. MALEIC ACID

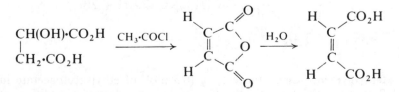

Place 45 g (0.33 mol) of dry malic acid in a 250-ml round-bottomed distilling flask and cautiously add 63 g (57 ml, 0.835 mol) of pure acetyl chloride. Attach a reflux condenser and connect the top of it to a gas absorption trap. Warm the flask gently on a water bath to start the reaction, which then proceeds exo-thermically. Hydrogen chloride is evolved and the malic acid passes into solu-tion. When the evolution of gas subsides, heat the flask on a water bath for 1–2 hours. Rearrange the apparatus and distil. A fraction of low boiling point passes over first and the temperature rises rapidly to 190 °C; at this point run out the water from the condenser. Continue the distillation and collect the maleic anhydride at 195–200 °C. Recrystallise the crude maleic anhydride from chloroform; 22 g (67%) of pure maleic anhydride, m.p. 54 °C, are obtained.

To obtain maleic acid, evaporate the maleic anhydride with one-half of its weight of water on a water bath; remove the last traces of water by leaving in a desiccator over concentrated sulphuric acid. The resulting maleic acid has m.p. 143 °C and is quite pure (1). It may be recrystallised, if desired, from acetone–light petroleum (b.p. 60–80 °C) and then melts at 144 °C (1).

Conversion of maleic acid into fumaric acid. Dissolve 10 g of maleic acid in 10 ml of warm water, add 20 ml of concentrated hydrochloric acid and boil gently under reflux for 30 minutes. Crystals of fumaric acid soon crystallise out from the hot solution. Allow to cool, filter off the fumaric acid and recrystallise it from hot 1 M-hydrochloric acid. The m.p. in a sealed capillary tube is 286–287 °C.

Note. (1) The melting point of pure maleic acid depends to a marked degree upon the rate of heating, and values between 133 °C and 143–144 °C may be observed. Slow heating (about 20 minutes) gives a value of 133–134 °C; with more rapid heating (about 10 minutes), the m.p. is 139–140 °C. If the acid is immersed in a bath at 140 °C, it melts sharply at 143 °C. The low melting points obtained by slow heating are evidently due to the formation of maleic anhydride and/or fumaric acid, which depress the melting point.

Hydrogenation of maleic acid. Place 20 mg of Adams' platinum dioxide catalyst (Section **II,2,***51*) in a hydrogenation flask, introduce a solution of 0.58 g (0.05 mol) of maleic acid in 15 ml of ethanol and attach the flask to the adapter of the atmospheric hydrogenation apparatus (Fig. I,70(*a*)). Fill the flask and gas burettes with hydrogen by the procedure discussed in Section **I,17,1**; note the volumes in the gas burettes and then gently agitate the flask contents by means of the shaker. When uptake of hydrogen ceases note the total volume of hydro-gen absorbed; this should be in the region of 1150 ml. Follow the procedure discussed in Section **I,17,1** for replacing the hydrogen in the apparatus with air; dis-connect the hydrogenation flask, filter off the catalyst and wash it with a little ethanol (do not allow the catalyst to become dry, but after the washing operation remove the filter-paper and rinse the catalyst into the residues bottle with water). Evaporate the ethanol to leave a residue of succinic acid, 0.58 g, m.p. 184 °C; the m.p. is unaffected after recrystallisation from 2.5 ml hot water.

III,213. FUMARIC ACID

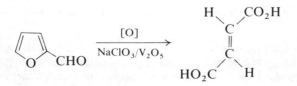

Place in a 1-litre three-necked flask, fitted with a reflux condenser, a mechanical stirrer and a thermometer, 112.5 g (1.06 mol) of sodium chlorate, 250 ml of water and 0.5 g of vanadium(V) oxide catalyst. Set the stirrer in motion, heat the flask on an asbestos-centred wire gauze to 70–75 °C and add 4 ml of 50 g (43 ml, 0.52 mol) of technical furfural (2-furaldehyde). As soon as the vigorous reaction commences (1) *but not before*, add the remainder of the furfural through a dropping funnel down the condenser at such a rate that the vigorous reaction is maintained (25–30 minutes). Then heat the reaction mixture at 70–75 °C for 5–6 hours (2) and allow to stand overnight at the laboratory temperature. Filter the crystalline fumaric acid with suction, and wash it with a little cold water (3). Recrystallise the crude fumaric acid from about 300 ml of 1 M-hydrochloric acid, and dry the crystals (26 g, 43%) at 100 °C. The m.p. in a sealed capillary tube is 282–284 °C. A further recrystallisation raises the m.p. to 286–287 °C.

Notes. (1) When the vigorous reaction commences, the temperature rises to about 105 °C and remains at this temperature for some time. The main quantity of furfural should not be added until the vigorous reaction has started: if this precaution is ignored, an explosion may result.

(2) A water bath may be used for this purpose.

(3) A small quantity (*ca.* 3 g) of fumaric acid may be recovered from the filtrate by heating it on a water bath with 15 ml of concentrated hydrochloric acid, evaporating to about 150 ml and then cooling with running water. The fumaric acid which separates is recrystallised from 1 M-hydrochloric acid.

REFERENCES

III. Aliphatic compounds

1. E. L. Jackson (1944). 'Periodic Acid Oxidation', in *Organic Reactions* (II). Ed. R. Adams. New York; Wiley, p. 363.
2. H. Stetter and E. Reske (1970). *Ber.*, **103**, 643.
3. J. Carson (1946). *J. Am. Chem. Soc.*, **68**, 2030.
4. C. H. Depuy and B. W. Ponder (1959). *J. Am. Chem. Soc.*, **81**, 4629.
5. S. M. McElvain (1948). 'The Acyloins', in *Organic Reactions* (IV). Ed. R. Adams. New York; Wiley, p. 256.
6. L. Hough and A. C. Richardson (1969). 'The Monosaccharides', in *Rodd's Chemistry of Carbon Compounds*. 2nd edn. Ed. S. Coffey. Amsterdam; Elsevier, p. 67.
7. R. J. Bose, T. L. Hullar, B. A. Lewis and F. Smith (1962). *J. Org. Chem.*, **26**, 1300.
8. F. Smith (1944). *J. Chem. Soc.*, 635.
9. M. L. Wolfrom and A. Thompson (1963). *Methods in Carbohydrate Chemistry*, Vol. II. Eds. R. L. Whistler and M. L. Wolfrom. New York; Academic Press, p. 21.
10. G. O. Aspinall and R. J. Ferrier (1957). *Chem. and Ind.*, 1216.
11. J. Dale (1965). *J. Chem. Soc.*, 72.
12. S. Swann, R. Oehler and R. J. Buswell. *Organic Syntheses*, Coll. Vol. II, p. 276.

13. P. A. Finnan and G. A. Fothergill (1962). *J. Chem. Soc.*, 2824.
14. H. R. Snyder and C. T. Elston (1954). *J. Am. Chem. Soc.*, **76**, 3039.
15. V. H. Wallingford, D. M. Jones and A. H. Homeyer (1942). *J. Am. Chem. Soc.*, **64**, 576.
16. C. S. Marvel, D. W. MacCorquodale, F. E. Kendall and W. A. Lazier (1924). *J. Am. Chem. Soc.*, **46**, 2838.
17. N. Kornblum (1962). 'The Synthesis of Aliphatic and Alicyclic Nitro Compounds', in *Organic Reactions* (12). Ed. A. C. Cope. New York; Wiley, p. 101.
18. P. E. Pfeffer and L. S. Silbert (1970). *Tetrahedron Letters*, 699.
19. L. H. Amunsden and L. S. Nelson (1951). *J. Am. Chem. Soc.*, **73**, 242.
20. M. Cerny, J. Malek and M. Capka (1969). *Coll. Czeck. Chem. Comm.*, **34**, 1033.
21. W. S. Emerson (1948). 'The Preparation of Amines by Reductive Alkylation', in *Organic Reactions* (IV). Ed. R. Adams. New York; Wiley, p. 174.
22. F. Nerdel and H. Henkel (1952). *Ber.*, **85**, 1138.
23. R. L. Frank and P. V. Smith (1946). *J. Am. Chem. Soc.*, **68**, 2103.
24. D. T. McAllan, T. V. Cullum, R. A. Dean and F. A. Fidler (1951). *J. Am. Chem. Soc.*, **73**, 3627.
25. N. J. Leonard and C. R. Johnson (1962). *J. Org. Chem.*, **27**, 282.
26. L. A. Carpino (1958). *J. Am. Chem. Soc.*, **80**, 601.
27. M. J. Mintz and C. Walling. *Organic Syntheses*, Coll. Vol. V, p. 184.

CHAPTER IV AROMATIC COMPOUNDS

A AROMATIC HYDROCARBONS

The synthesis of aromatic hydrocarbons (arenes) is for convenience organised under the following four headings.

1. Alkylbenzenes (Sections **IV,1** to **IV,6**).
2. Di- and tri-arylmethanes (Sections **IV,7** to **IV,9**).
3. Biphenyls (Sections **IV,10** to **IV,12**, see also Section **IV,87**).
4. Condensed polycyclic systems (Sections **IV,13** to **IV,17**).

A,1. Alkylbenzenes. (a) *The Wurtz–Fittig reaction* (cf. the Wurtz reaction, **III,A,4**(a)). The interaction of an aryl halide, an alkyl halide and sodium gives a reasonable yield of an alkylbenzene.

$$ArBr + RBr + 2Na \longrightarrow Ar\cdot R + 2NaBr$$

The by-products of the reaction, e.g., R–R and Ar–Ar, can usually readily be separated by distillation.

A likely mechanism for this reaction involves the formation of an aryl sodium and its subsequent reaction with the alkyl halide.

$$ArBr + 2Na \longrightarrow Ar\overset{\ominus}{:}Na^{\oplus} + NaBr$$

$$\overset{\ominus}{Ar:} \overset{\frown}{} R \overset{\frown}{-} Br \longrightarrow Ar\cdot R + \overset{\ominus}{:}Br$$

The reaction of the aryl sodium with the more reactive alkyl bromide occurs in preference to reaction with the less reactive aryl bromide, which would lead to the biaryl Ar–Ar. Furthermore the formation of the aryl anion, which is likely to be stabilised by electron delocalisation, probably occurs more readily than the formation of the alkyl anion by reaction of sodium with the alkyl bromide. For these reasons acceptable yields of the alkylbenzene (e.g., butylbenzene, Section **IV,1**) are usually obtained.

Somewhat better yields of alkylbenzene are, however, obtained if the aryl sodium is first prepared and then subjected to a suitable alkylation reaction. The aryl sodium is formed when the aryl halide is added to a suspension of sodium wire in dry light petroleum under an atmosphere of nitrogen. The preparation of *p*-xylene (Section **IV,2**) illustrates the alkylation of *p*-tolyl sodium using dimethyl sulphate; the use of an alkyl halide is illustrated by the ethylation of phenyl sodium in the preparation of ethylbenzene (cognate preparation in Section **IV,2**).

In an alternative and more efficient synthesis of butylbenzene (Section **IV,3**), benzyl sodium is prepared and alkylated with propyl bromide. The benzyl sodium is conveniently obtained by first forming phenyl sodium by reaction between sodium and chlorobenzene in a toluene medium, and then heating the toluene suspension of the phenyl sodium at 105 °C for about 35 minutes when a transmetalation process occurs (formulated at the beginning of Section **IV,3**).

The Grignard reagent derived from benzyl chloride undergoes ready alkylation with an alkyl toluene-*p*-sulphonate, a reaction which provides a further useful synthesis of an alkylbenzene (e.g., pentylbenzene, Section **IV,4**).

$$C_6H_5 \cdot CH_2Cl + Mg \longrightarrow C_6H_5 \cdot CH_2MgCl$$

$$C_6H_5 \cdot CH_2MgCl + 2RO \cdot SO_2 \cdot C_6H_4 \cdot CH_3(p) \longrightarrow$$

$$C_6H_5 \cdot CH_2 \cdot R + RCl + (p\text{-}CH_3 \cdot C_6H_4 \cdot SO_3)_2Mg$$

(*b*) *The reduction of aldehydes and ketones.*   Aromatic hydrocarbons are the main products when aromatic aldehydes or ketones are reduced with amalgamated zinc and concentrated hydrochloric acid (the Clemmensen reduction, e.g., ethylbenzene, Section **IV,5**, *Method A*).

$$Ar \cdot CO \cdot R \longrightarrow Ar \cdot CH_2 \cdot R$$

Some features of the Clemmensen reduction are discussed on p. 321, **III,A,2**. Purely aromatic ketones generally do not give satisfactory results: pinacols and resinous products often predominate. The reduction of ketonic compounds of high molecular weight and very slight solubility is facilitated by the addition of a solvent, such as ethanol, acetic acid or dioxan, which is miscible with aqueous hydrochloric acid. With some carbonyl compounds, notably keto acids, poor yields are obtained even in the presence of ethanol, etc., and the difficulty has been ascribed to the formation of insoluble polymolecular reduction products, which coat the surface of the zinc. The addition of a hydrocarbon solvent, such as toluene, is beneficial because it keeps most of the material out of contact with the zinc and the reduction occurs in the aqueous layer at such a high dilution that polymolecular reactions are largely inhibited (see Section **IV,135**).

Aromatic ketones are readily prepared by the Friedel–Crafts acylation process (see p. 770, **IV,L,1**) and their Clemmensen reduction constitutes a more efficient procedure for the preparation of monoalkylbenzenes than the alternative direct Friedel–Crafts alkylation reaction (see **IV,A,1**(*c*) below). Alternatively aldehydes and ketones may be reduced to the corresponding hydrocarbon by the Wolff–Kishner method which involves heating the corresponding hydrazone or semicarbazone with potassium hydroxide or with sodium ethoxide solution.

$$Ar \cdot CO \cdot R \xrightarrow{N_2H_4} Ar \cdot (R)C = N \cdot NH_2 \xrightarrow{KOH} Ar \cdot CH_2 \cdot R$$

The **Huang–Minlon modification** of the reaction has the following advantages: (i) the actual isolation of the hydrazone is unnecessary, (ii) the reaction time is considerably reduced, (iii) the reaction can be carried out at atmospheric pressure and on a large scale and (iv) the yields are usually excellent. The hydrazone is first formed *in situ* by refluxing a solution of the carbonyl compound in a moderate amount of diethylene glycol or triethylene glycol with the commercial 85 or 90 per cent hydrazine hydrate and about 3 equivalents of potassium hydroxide for 1 hour; the water and excess of hydrazine are removed by distillation until a favourable temperature for the decomposition of the hydrazone is attained (170–190 °C) and the solution is refluxed for 3–5 hours longer.

The reaction is illustrated by the preparation of ethylbenzene (Section **IV,5**, *Method B*) from acetophenone; the resulting hydrocarbon is quite pure and free from unsaturated compounds.

The disadvantages associated with the Clemmensen reduction of carbonyl compounds, which are absent in the modified Wolff–Kishner reduction, are: (i) the formation of small amounts of hydroxy compounds and unsaturated compounds as by-products, (ii) poor results with many compounds of high molecular weight, (iii) non-applicability to furan and pyrrole compounds owing to their sensitivity to acids and (iv) the sensitivity to steric hindrance.

The mechanism of the Wolff–Kishner reduction may involve the initial formation of the mesomeric anion (I) by removal of a proton by the base from the hydrazone. A prototropic shift gives the anion (II) which loses nitrogen to form the carbanion (III) which accepts a proton to yield the hydrocarbon.

$$R^1R^2C=N-N\diagup^{\!H}_{\diagdown H\frown :B} \quad \overset{-BH^\oplus}{\rightleftharpoons}$$

$$\left[R^1R^2\overset{\frown}{C}=N\overset{\ominus}{-}\overset{..}{N}H \longleftrightarrow R^1R^2\overset{\ominus}{C}-N=\overset{..}{N}H \right] \overset{-H^\oplus, +H^\oplus}{\rightleftharpoons}$$
$$(I)$$

$$R^1R^2\overset{\frown}{CH}-N\overset{\ominus}{=}N \xrightarrow{-N_2} R^1R^2\overset{\ominus}{CH} \xrightarrow{+BH^\oplus} R^1R^2CH_2 + B\overset{..}{:}$$
$$\quad (II) \qquad\qquad\qquad (III)$$

(*c*) *The Friedel–Crafts alkylation reaction.* An alkyl halide reacts with an aromatic hydrocarbon in the presence of aluminium chloride to yield in the first instance a hydrocarbon, thus:

$$ArH + RX \xrightarrow{AlCl_3} Ar\cdot R + HX$$

The reaction does not, however, stop at the stage of mono-substitution, since the alkylbenzene (Ar·R) initially produced undergoes alkylation more easily than the original hydrocarbon ArH, owing to the electron-releasing effect of the alkyl group. Mixtures of substances therefore often result and extensive purification may be required in order to isolate the monosubstituted compound. Some mono-alkylbenzenes may be prepared by using an excess of the hydrocarbon, which also acts as a diluent in moderating the violence of the reaction and prevents the undue formation of poly-alkylbenzenes.

The mechanism of the reaction is generally considered to proceed by way of carbonium ions which attack the aromatic nucleus:

$$R-X:\frown AlCl_3 \rightleftharpoons R^\oplus[XAlCl_3]^\ominus \rightleftharpoons R^\oplus + [XAlCl_3]^\ominus$$

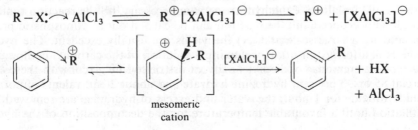

mesomeric
cation

In many cases the alkyl group may undergo extensive skeletal rearrangement under the reaction conditions employed. For example, when benzene is alkylated with propyl chloride in the presence of aluminium chloride over a wide temperature range, a mixture of propylbenzene and isopropylbenzene, in approximately 1 : 2 ratio, is obtained. On the other hand isobutyl bromide or chloride with benzene in the presence of aluminium chloride gives exclusively t-butylbenzene. This alkylbenzene is, however, more conveniently prepared by using t-butyl chloride as the alkylating reagent (Section **IV,6**).

The rearrangement processes mentioned above involve a 1,2-nucleophilic shift of a hydride ion to form a more stable secondary or tertiary carbonium ion.

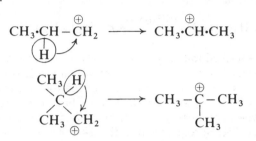

Other catalysts which may be used in the Friedel–Crafts alkylation reaction include iron(III) chloride, antimony pentachloride, zirconium tetrachloride, boron trifluoride, zinc chloride and hydrogen fluoride but these are generally not so effective in small-scale preparations. The alkylating agents include alkyl halides, alcohols and alkenes.

IV,1. BUTYLBENZENE (*1-Phenylbutane*)

$$C_6H_5Br + C_4H_9Br \xrightarrow{Na} C_6H_5 \cdot C_4H_9$$

Into a 1-litre round-bottomed flask, provided with a long (e.g., a 30-cm) double surface condenser, place 22.5 g (0.98 mol) of clean sodium cut into small pieces (see Section **II,2,55**) and mount the flask for heating on an asbestos-centred wire gauze. Prepare a mixture of 52 g (35 ml, 0.33 mol) of bromobenzene (Section **IV,27**) and 51 g (40 ml, 0.37 mol) of butyl bromide (Section **III,54**). Add 5–7 ml of the mixture through the condenser and warm the flask very gently with a small luminous flame. Immediately reaction commences (the sodium acquires a dark blue colour and much heat is evolved), remove the flame. Introduce the remainder of the mixture in small quantities during one hour; shake the mixture frequently and maintain a minute luminous flame beneath the flask. Reflux the reaction mixture for 1–1.5 hours using a *small luminous* flame; shake the fairly solid contents of the flask from time to time. Allow to cool and add 50 ml of rectified spirit during 30 minutes down the condenser from a small separatory funnel; then introduce a mixture of 25 ml of rectified spirit and 25 ml of water during 30 minutes, followed by 50 ml of water. This treatment will remove the excess of sodium. Reflux the resulting mixture for 2–3 hours. Add 500 ml of water and filter at the pump from some sludge which is generally present; it is advisable to wash the latter with a little ether. Transfer to a separatory funnel, remove the upper hydrocarbon layer and wash it successively with 25 ml of dilute sulphuric acid and 50 ml of water; dry over magnesium or anhydrous sodium sulphate and distil (Fig. I,98) using an air bath. Collect the butylbenzene at 178–188 °C (20 g, 46%); an appreciable dark residue containing biphenyl remains in the flask. Upon redistillation, the butylbenzene boils at 178–184 °C (1).

Note. (1) The butylbenzene contains some unsaturated hydrocarbons: these can be removed by repeated shaking with small quantities of concentrated sulphuric acid (see Section **III,4**, Note (1)).

IV,2. p-XYLENE

$$p\text{-}CH_3\cdot C_6H_4Cl \xrightarrow{2Na} p\text{-}CH_3\cdot C_6H_4Na \xrightarrow{(CH_3)_2SO_4} p\text{-}CH_3\cdot C_6H_4\cdot CH_3$$

Prepare p-tolyl sodium, as described in Section **IV,172**, *Method A*, using 76 g (0.6 mol) of p-chlorotoluene and 27.5 g (1.2 mol) of sodium wire in 250 ml of light petroleum, b.p. 40–60 °C, as the solvent. Allow the formation of the aryl sodium to go to completion at 30 °C by stirring for 2 hours but do not reflux the mixture. Introduce, with vigorous stirring, a mixture of 78.5 g (59 ml, 0.62 mol) of dimethyl sulphate (**CAUTION**: toxic, Section **II,2,18**) with 30 ml of dry benzene during 1 hour whilst maintaining the temperature at 30 °C. Add water to the colourless reaction mixture, separate the organic layer and fractionate it until the vapour temperature reaches 90 °C. Then separate the crude xylene by steam distillation in the presence of potassium hydroxide; dry the upper layer from the steam distillate (magnesium sulphate) and fractionate. Collect the p-xylene at 137–138 °C. The yield is 37 g (58%).

Cognate preparation   Ethylbenzene. Prepare a suspension of phenyl sodium from 23 g (1 mol) of sodium wire, 200 ml of light petroleum (b.p. 40–60 °C) and 56.3 g (50.9 ml, 0.5 mol) of chlorobenzene as described above for p-*xylene*. Add 43.5 g (30 ml, 0.4 mol) of ethyl bromide during 30–45 minutes at 30 °C and stir the mixture for a further hour. Add water slowly to decompose the excess of sodium and work up the product as detailed for *butylbenzene*, Section **IV,3**. The yield of ethylbenzene, b.p. 135–136 °C, is 23 g (54%).

IV,3. BUTYLBENZENE (*1-Phenylbutane*)

$$C_6H_5Cl + 2Na \longrightarrow C_6H_5Na + NaCl$$
$$C_6H_5Na + C_6H_5\cdot CH_3 \longrightarrow C_6H_6 + C_6H_5\cdot CH_2Na$$
$$C_6H_5\cdot CH_2Na + C_3H_7Br \longrightarrow C_6H_5\cdot CH_2\cdot C_3H_7 + NaBr$$

Equip a 500-ml three-necked flask as detailed for p-*toluic acid* (Section **IV,172**, *Method A*) and pass a slow stream of nitrogen through the apparatus. Charge the flask with 150 ml of sodium-dried, sulphur-free toluene (Section **II,1,3**) and 13.8 g (0.6 mol) of sodium wire. Place 34 g (31 ml, 0.3 mol) of chlorobenzene (Sections **IV,26** and **IV,80**) in the dropping funnel and add it dropwise through the condenser during 1 hour, with vigorous stirring, whilst maintaining the temperature inside the flask at 30–35 °C. The start of the reaction is indicated by the appearance of black specks on the sodium surface. (If the reaction is slow to start, it may be instantly initiated by a few drops of butanol.) Complete the formation of phenyl sodium by stirring for 2–3 hours at 30 °C. Attach a calcium chloride tube to the top of the reflux condenser and reflux the mixture for 40 minutes. The reflux temperature, initially 107 °C, gradually falls to 103 °C as benzene is formed by the exchange reaction. Remove the heating bath and add 27.6 g (20.5 ml, 0.224 mol) of redistilled propyl bromide during 20–25 minutes at 103–105 °C; the reaction is strongly exothermic. Allow the reaction mixture to cool to room temperature: maintain the stirring and the slow stream of nitrogen. Add water *slowly* to destroy the excess of sodium. Separate the toluene layer, dry it (magnesium sulphate) and distil it through a short, jacketed column filled with glass helices (19 cm packed length, 14 mm diameter; compare Fig. I,104). After removal of the toluene (up to 111 °C) and a small intermediate fraction (111–179 °C), pure butylbenzene passes over at 179.5–181 °C/752 mmHg (23 g, 77%). A brown residue (4 g) remains in the flask.

IV,4. PENTYLBENZENE (*1-Phenylpentane*)

$$C_6H_5 \cdot CH_2Cl \xrightarrow[\text{ether}]{Mg} C_6H_5 \cdot CH_2MgCl$$

$$C_6H_5 \cdot CH_2MgCl + 2p\text{-}CH_3 \cdot C_6H_4 \cdot SO_2 \cdot OC_4H_9 \longrightarrow$$
$$C_6H_5 \cdot CH_2 \cdot (CH_2)_3 \cdot CH_3 + C_4H_9Cl + (p\text{-}CH_3 \cdot C_6H_4 \cdot SO_2 \cdot O)_2Mg$$

Into a 1500-ml three-necked flask, equipped with a dropping funnel, a sealed stirrer unit and a double surface condenser (provided at the outlet with a guard-tube filled with a mixture of calcium chloride and soda-lime to prevent the ingress of moisture and carbon dioxide), place 24.3 g (1 mol) of clean, dry magnesium turnings, 100 ml of anhydrous ether and a small crystal of iodine (1). Charge the dropping funnel with a solution of 126.5 g (115 ml, 1 mol) of freshly distilled benzyl chloride (b.p. 177–179 °C) in 500 ml of sodium-dried ether. Allow about 12 ml of this solution to run into the flask; if the reaction does not commence within a minute or two, partially immerse the flask in a water bath at about 40 °C. Remove the flask from the bath immediately reaction sets in and commence stirring the mixture. Add the remainder of the benzyl chloride during 30 minutes; control the vigorous reaction by immersing most of the flask in ice-water. The reaction usually continues for about 15 minutes after all the benzyl chloride has been introduced, and is completed by refluxing for a further 15 minutes. Cool the flask by re-immersion in the bath of ice-water. Place a solution of 456 g (2 mol) of butyl toluene-*p*-sulphonate (Section **IV,51**) in about twice the volume of anhydrous ether in the dropping funnel, and add it slowly to the vigorously stirred benzylmagnesium chloride solution, at such a rate that the ether just boils; a white solid soon forms. The addition is complete after about 2 hours. Pour the reaction product slowly into a mechanically stirred mixture of 1 kg of finely crushed ice, 1 litre of water and 125 ml of concentrated hydrochloric acid contained in a 4- or 5-litre beaker; the precipitated magnesium toluene-*p*-sulphonate will ultimately pass into the solution. Separate the ether layer, extract the aqueous layer with 250 ml of ether and wash the combined ether solutions with about 100 ml of water. Dry the ether solution with about 10 g of anhydrous potassium carbonate. Distil off the ether on a rotary evaporator, add to the mixture 5–7 g of sodium cut into small pieces and heat under reflux for about 2 hours in order to remove any benzyl alcohol which may have formed by atmospheric oxidation of benzyl-magnesium chloride. Decant the solution and distil it from an air bath through a well-lagged and efficient fractionating column; collect the fraction, b.p. 190–210 °C. Redistil and collect the pentylbenzene at 198–203 °C. The yield is 90 g (61%).

Note. (1) For further general details on the preparation of Grignard reagents, see Section **III,34**.

IV,5. ETHYLBENZENE (*1-Phenylethane*)

$$C_6H_5 \cdot CO \cdot CH_3 \xrightarrow{\text{Zn/Hg, HCl}} C_6H_5 \cdot CH_2 \cdot CH_3$$

Method A (*Clemmensen reduction*). Prepare 200 g of amalgamated zinc in a 2-litre three-necked flask as detailed in Section **II,2,67**. Fit the flask with a reflux condenser, a sealed stirrer and a gas entry tube reaching to within 1 cm of the bottom; connect the last-named through an intermediate empty wash bottle to a Kipp's apparatus supplying hydrogen chloride gas (Section **II,2,31**). Place a mixture of 500 ml of concentrated hydrochloric acid and 100 ml of water in the flask and introduce 100 g (0.83 mol) of acetophenone (Section **IV,134**). Stir the mixture and pass in a slow stream of hydrogen chloride gas whilst warming the flask on an asbestos-centred wire gauze by means of a small flame. If the reaction becomes unduly vigorous, stop the supply of hydrogen chloride until it subsides somewhat. Most of the zinc dissolves after 6 hours, by which time the reaction is almost complete; allow to stand overnight. Arrange the apparatus for steam distillation (Fig. I,101) and pass steam into the flask, heated by means of a small flame, until the distillate is clear. Separate the upper hydrocarbon layer, wash it with 5 per cent sodium hydroxide solution, then with water, and dry over magnesium sulphate. Distil from a 100-ml Claisen flask and collect the ethylbenzene (1) at 134–135 °C. The yield is 50 g (57%).

Note. (1) The ethylbenzene contains some unsaturated compounds. These can be removed by repeated shaking with 5 per cent of the volume of concentrated sulphuric acid until the latter is colourless or, at most, very pale yellow. The hydrocarbon is then washed with 5 per cent sodium carbonate solution, then with water, and dried over magnesium sulphate. It is then distilled twice from sodium when pure ethylbenzene, b.p. 135 °C, is obtained.

Cognate preparations Butylbenzene. Use 225 g of amalgamated zinc and 100 g (0.675 mol) of ethyl benzyl ketone (Section **III,92,B**). The yield of butylbenzene, b.p. 180–183 °C, is 75 g (83%). With 200 g of amalgamated zinc and 75 g (0.5 mol) of butyrophenone (phenyl propyl ketone, Section **IV,134**), the yield of butylbenzene, b.p. 181–184 °C, is 40 g (60%).

Hexylbenzene. Use 200 g of amalgamated zinc and 100 g (0.57 mol) of 1-phenylhexan-3-one (Section **III,92,B**); the yield of crude hexylbenzene, b.p. 218–230 °C, is 55 g. This, when purified by treatment with concentrated sulphuric acid and distillation from sodium, yields 40 g (43%) of fairly pure hexylbenzene, b.p. 220–225 °C (mainly 222–224 °C).

Method B (*Huang–Minlon modification of Wolff–Kishner reduction*). Place 36.0 g (0.3 mol) of redistilled acetophenone, b.p. 201 °C (Section **IV,134**),

300 ml of diethylene glycol, 30 ml of 90 per cent hydrazine hydrate and 40 g of potassium hydroxide pellets in a 500-ml two-necked round-bottomed flask fitted with a reflux condenser; insert a thermometer supported in a screw-capped adapter in the side-neck so that the bulb dips into the reaction mixture. Warm the mixture on a boiling water bath until most of the potassium hydroxide has dissolved and then heat under reflux for 1 hour either by means of a free flame or by using a heating mantle. Remove the reflux condenser and fit a still-head and condenser for downward distillation. Distil until the temperature of the liquid rises to 175 °C (1). Separate the upper hydrocarbon layer from the distillate and extract the aqueous layer twice with 20 ml portions of ether. Dry the combined upper layer and ethereal extracts with magnesium sulphate, remove the ether on a water bath and distil the residue. Collect ethylbenzene at 135–136 °C; the yield is 20 g (62.5%).

Note.   (1) The reduction takes place at a comparatively low temperature and is fairly rapid for acetophenone. With higher ketones, the upper layer of the distillate should be returned to the contents of the flask and the heating under reflux continued for 3–5 hours. The reaction mixture and the aqueous distillate are then combined, extracted with ether and the ether extract treated as described above.

IV,6.   t-BUTYLBENZENE (2-Methyl-2-phenylpropane)

$$(CH_3)_3CCl + C_6H_6 \text{ (excess)} \xrightarrow{\text{AlCl}_3} C_6H_5 \cdot C(CH_3)_3$$

Into a 1-litre three-necked flask, equipped as in Section **IV,134**, place 50 g (0.33 mol) of anhydrous aluminium chloride (1) and 200 ml (2.25 mol) of dry benzene; cool in a bath of crushed ice. Stir the mixture and add 50 g (59 ml, 0.54 mol) of t-butyl chloride (Section **III,47**) from the dropping funnel during 4–5 hours; the first addition should be 3–4 ml in order to prevent the benzene from freezing. Maintain the mixture at a temperature of 0–5 °C by the addition of salt to the ice, if necessary. When all the t-butyl chloride has been run in, continue the stirring for 1 hour longer. Remove the separatory funnel and add 200 g of finely crushed ice in small portions with stirring; finally add 100 ml of cold water to complete the decomposition of the intermediate addition compound. Arrange the flask for steam distillation (Fig. I,101) and steam distil the resulting reaction mixture. Transfer the steam distillate to a separatory funnel, remove the upper hydrocarbon layer, extract the water layer with two 50 ml portions of ether and combine the extracts with the upper layer. Dry with magnesium sulphate, distil off the ether on a water bath and fractionally distil the residue twice, using a well-lagged column (Fig. I,104). Collect the t-butylbenzene at 165–170 °C. The yield is 45 g (62%). Pure t-butylbenzene boils at 168.5 °C.

Note.   (1) In an alternative procedure 25 g of anhydrous iron(III) chloride replace the aluminium chloride, the mixture is cooled to 10 °C and the 50 g of t-butyl chloride is added. The mixture is slowly warmed to 25 °C and maintained at this temperature until no more hydrogen chloride is evolved. The reaction mixture is then washed with dilute hydrochloric acid and with water, dried and fractionally distilled. The yield of t-butyl benzene, b.p. 167–170 °C, is 60 g.

A,2. Di- and triarylmethanes. Alkylation of benzene with benzyl chloride by the Friedel–Crafts procedure yields diphenylmethane (Section **IV,7**). The reactive nature of the halide obviates the need for heating the reaction mixture (see Ref. 1), and a smaller molar proportion of aluminium chloride is employed compared to the standard conditions as illustrated by Section **IV,6**.

A Friedel–Crafts reaction between chloroform and an excess of benzene affords a convenient route to triphenylmethane (Section **IV,8**). The use of carbon tetrachloride in place of chloroform leads to triphenylchloromethane (cognate preparation in Section **IV,8**), but this compound is more conveniently prepared from triphenylmethanol (Section **II,2,66**).

The synthesis of the triptycene system illustrates a synthetic use of the important reactive intermediate benzyne (III) which is generated by the thermal decomposition of benzenediazonium-2-carboxylate. The latter is produced when anthranilic acid (I) is treated with an alkyl nitrite in an aprotic solvent, and probably exists mainly as the zwitterionic form (II).

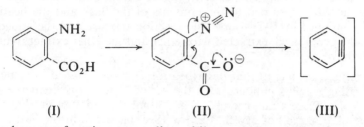

| (I) | (II) | (III) |

The benzyne functions as a dienophile towards reactive diene systems. The reactivity of the 9,10-positions in anthracene towards dienophiles is well known (**Diels–Alder reaction**, Section **V(d)**), and addition of benzyne to 9-bromoanthracene yields the interesting cage-ring alkyl halide: 9-bromotriptycene (9-bromo-9,10-*o*-benzenoanthracene). The reaction is incomplete and some unreacted 9-bromoanthracene remains in the crude reaction products, but may be removed by virtue of its ready conversion into a maleic anhydride adduct in a further Diels–Alder type reaction.

The environment of the halogen in the ring structure renders it almost unreactive in displacement reactions by nucleophiles.

IV,7. DIPHENYLMETHANE

$$C_6H_5{\cdot}CH_2Cl + C_6H_6 \text{ (excess)} \xrightarrow{AlCl_3} C_6H_5{\cdot}CH_2{\cdot}C_6H_5$$

Fit a 500-ml three-necked round-bottomed flask with a sealed mechanical stirrer, attach a gas absorption device to one of the side-necks and stopper the third neck. Place 38 g (35 ml, 0.3 mol) of redistilled benzyl chloride and 150 ml of dry benzene (Section **II,1,2**) in the flask. Weigh out 2 g (0.015 mol) of anhydrous aluminium chloride (Section **II,2,3**) into a dry capped specimen tube with the minimum exposure to the atmosphere. Cool the flask in a bath of crushed ice and add about one-fifth of the aluminium chloride. Stir the mixture; a vigorous reaction will set in within a few minutes and hydrogen chloride is evolved. When the reaction has subsided, add a further portion of the aluminium chloride and repeat the process until all has been introduced. The mixture should be kept well shaken and immersed in the ice bath during the addition. After 15 minutes cautiously add 100 g of crushed ice, followed by 100 ml of water in order to decompose the aluminium complex. Shake the mixture well, transfer to a separatory funnel and run off the lower aqueous layer. Wash the upper layer successively with dilute hydrochloric acid and water and dry it with anhydrous calcium chloride. Remove the benzene with the aid of the apparatus shown in Fig. I,100. Distil the remaining liquid through an air condenser either with a free flame or from an air bath. Collect the diphenylmethane at 250–275 °C (the pure substance boils at 262 °C) (1). The distillate should solidify on cooling in ice and scratching with a glass rod, or by seeding with a crystal of the pure material. If it does not crystallise, redistil from a small flask and collect the fraction, b.p. 255–267 °C; this generally crystallises on cooling and has m.p. 24–25 °C. The yield is 25 g (50%).

Note. (1) Alternatively the distillation may be conducted under diminished pressure; the fraction, b.p. 125–130 °C/10 mmHg, is collected.

IV,8. TRIPHENYLMETHANE

$$CHCl_3 + 3C_6H_6(\text{excess}) \xrightarrow{\text{AlCl}_3} (C_6H_5)_3CH$$

The apparatus is similar to that described for *Diphenylmethane* (see above) but incorporating a reflux condenser to the outlet of which is fitted the gas absorption device. Place a mixture of 200 g (230 ml, 2.57 mol) of dry benzene and 40 g (26 ml, 0.33 mol) of dry chloroform (Section II,1,6) in the flask, and add 35 g (0.26 mol) of anhydrous aluminium chloride in portions of about 6 g and at intervals of 5 minutes with constant stirring. The reaction sets in upon the addition of the aluminium chloride and the liquid boils with the evolution of hydrogen chloride. Complete the reaction by refluxing for 30 minutes on a water bath. When cold, pour the contents of the flask very cautiously on to 250 g of crushed ice and 10 ml of concentrated hydrochloric acid. Separate the upper benzene layer, dry it with anhydrous calcium chloride or with magnesium sulphate and remove the benzene by flash distillation (Fig. I,100). Attach a Claisen still-head connected to a short air condenser and distil the remaining oil under reduced pressure; collect the fraction, b.p. 190–215 °C/10 mmHg. This is crude triphenylmethane which solidifies on cooling. Recrystallise it from about four times its weight of ethanol; triphenylmethane separates in needles and melts at 92 °C. The yield is 30 g (37%).

Cognate preparation Triphenylchloromethane. Use the apparatus described above for the reaction of 125 g (145 ml, 1.62 mol) of sodium-dried benzene and 50 g (32 ml, 0.33 mol) of dry, pure carbon tetrachloride (Section II,1,7). Cool the flask in ice and add 35 g (0.26 mol) of finely powdered aluminium chloride gradually in small portions (see Fig. I,64), so that the reaction mixture does not reflux during the addition (*ca.* 1.5 hours). Remove the ice bath, allow the reaction to proceed without further cooling and finally heat under reflux on a steam bath until evolution of hydrogen chloride subsides. Cool to room temperature and pour the product in a thin stream on to a mixture of 200 g of crushed ice and 200 ml of concentrated hydrochloric acid with vigorous stirring. Separate the benzene layer and extract the aqueous layer with a little benzene, wash the combined extracts once with 125 ml of cold concentrated hydrochloric acid. Dry for at least 2 hours over about 15 g of anhydrous calcium chloride and remove the benzene by distillation. Add 5 ml of dry benzene and 3–4 ml of acetyl chloride (1) to the residue and heat the mixture nearly to the boiling point. Shake the solution vigorously whilst cooling rapidly to room temperature and then cool in ice for 2 hours. Crush the solid triphenylchloromethane and filter using a sintered glass funnel. Wash the product with three 20 ml portions of cold dry light petroleum (b.p. 60–80 °C) and dry in a vacuum desiccator charged with paraffin wax shavings or with silica gel to remove solvent. The resulting pale greenish-yellow crystals of triphenyl-chloromethane melt at 111–112 °C and weigh 55 g (60%). Store the product in a well-stoppered (or in a screw-capped) bottle sealed with paraffin wax; this is necessary since triphenylchloromethane is slowly hydrolysed to triphenyl-methanol by the moisture in the air (2).

Notes. (1) The acetyl chloride converts any triphenylmethanol which may be present into triphenylchloromethane.

(2) The partially hydrolysed product may be purified by recrystallisation from one-third its weight of pure benzene containing 10–20 per cent of acetyl chloride, and washing the crystals with light petroleum (b.p. 60–80 °C) to which a little acetyl chloride has been added.

IV,9. 9-BROMOTRIPTYCENE

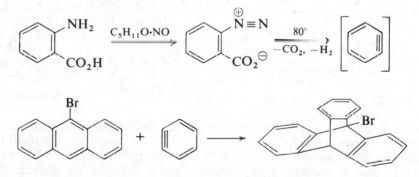

Place 2.3 g (0.02 mol) of isopentyl nitrite (isoamyl nitrite), 2.6 g (0.01 mol) of 9-bromoanthracene and 25 ml of 1,2-dimethoxyethane ('glyme', Section **II,1,**18) in a 250-ml round-bottomed flask fitted with a reflux condenser. Heat the mixture to gentle reflux on an electric mantle and add down the condenser, dropwise during 30 minutes, a solution of 3.4 g (0.025 mol) of anthranilic acid in 15 ml of glyme. Remove the mantle, cool the mixture to about 40 °C and add 2.3 g more of isopentyl nitrite dissolved in 5 ml of glyme. Resume the gentle refluxing and add during 15 minutes another solution of 3.4 g of anthranilic acid in 10 ml of glyme. Heat for an additional 15 minutes, add 10 ml of 95 per cent ethanol and pour the reaction mixture into a solution of 3.0 g of sodium hydroxide in 100 ml of water. Cool the resulting brown suspension thoroughly in ice-water and filter under suction. Wash the residue with a chilled methanol/water mixture (4 : 1 v/v) and transfer to a 100-ml round-bottomed flask which is then evacuated (rotary evaporator) on a steam bath until the weight is constant.

To the flask containing the crude product, add 2.0 g (0.02 mol) of maleic anhydride and 25 ml of triethyleneglycol dimethyl ether ('triglyme', b.p. 222 °C), fit a reflux condenser and boil under reflux for 10 minutes over a Bunsen flame. Cool the solution to about 100 °C, add 10 ml of 95 per cent ethanol and pour into a solution of 3.0 g of aqueous sodium hydroxide in 75 ml of water. Stir for a few minutes, then cool in ice-water and filter under suction. Wash the residue with chilled methanol/water (4 : 1 v/v) and recrystallise from methylcyclohexane or a chloroform/methanol mixture to give colourless plates of 9-bromotriptycene, m.p. 251–256 °C; yield is 1.5 g (45%). Further recrystallisation gives the pure product, m.p. 258–262 °C.

A,3. Biphenyl systems. A reaction which is reminiscent of the Wurtz coupling procedure, and which is particularly valuable in the synthesis of biphenyl and its derivatives, is that of **Ullmann**. It involves heating an aryl halide with copper powder, or better, with activated copper bronze.

$$2ArX + Cu \longrightarrow Ar-Ar + CuX_2$$

Aryl iodides and bromides are more reactive than the corresponding chlorides but the latter may be used when activating substituents (e.g., the nitro group) are present, as for example in the synthesis of 2,2'-dinitrobiphenyl (Section **IV,10**; see also 2,2'-bipyridyl, Section **VI,22**). The coupling reaction can be effectively carried out in the absence of a solvent, but the use of dimethylformamide as a solvent and diluent often results in an increase of yield, particularly in the case of reactive halides, when the vigour of the exothermic reaction is moderated.

The biphenyl system is also formed when aryl radicals generated from an appropriate precursor are allowed to undergo a self-coupling process. An interesting reaction involving the use of the diazonium salt derived from anthranilic acid affords an excellent method of preparing biphenyl-2,2'-dicarboxylic acid (diphenic acid, Section **IV,11**). The diazotised anthranilic acid is reduced with the aid of a cupro-ammonia reagent obtained by dissolving copper(II) sulphate in aqueous ammonia and treating the solution with hydroxylamine hydrochloride.

$$2Ar\cdot\overset{\oplus}{N} \equiv N \xrightarrow{+2e} 2Ar^{\cdot} + 2N_2$$

$$2Ar^{\cdot} \longrightarrow Ar - Ar$$

Reactive aromatic systems, in particular phenols, readily undergo oxidation by a single electron transfer process, leading to a mesomerically stabilised radical species, which then dimerises. The coupling of phenols in this way is frequently encountered in biogenetic pathways leading to naturally occurring molecules. A reaction illustrative of the process is the oxidation of 2-naphthol with iron(III) chloride (Section **IV,12**).

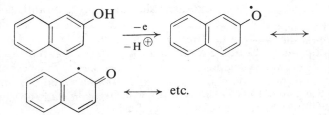

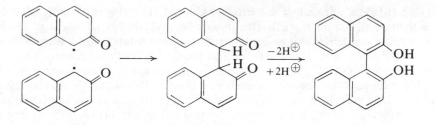

IV,10. 2,2′-DINITROBIPHENYL

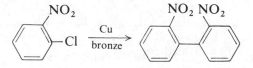

Place 50 g (0.32 mol) of o-chloronitrobenzene and 75 g of clean dry sand in a 250-ml three-necked flask equipped with a mechanical stirrer and an air reflux condenser. Heat the mixture in an oil or fusible metal bath to 215–225 °C and add, during 40 minutes, 50 g (0.78 mol) of copper bronze or, better, of activated copper bronze (Section II,2,*13*) (1). Maintain the temperature at 215–225 °C for a further 90 minutes and stir continuously. Pour the hot mixture into a Pyrex beaker containing 125 g of sand and stir until small lumps are formed; if the reaction mixture is allowed to cool in the flask, it will set to a hard mass, which can only be removed by breaking the flask. Break up the small lumps by powdering in a mortar, and boil them for 10 minutes with two 400 ml portions of ethanol; filter after each extraction. Cool the filtered extracts in ice, and collect the crude product on a Buchner funnel. Concentrate the filtrate to about half the original volume and thus obtain a second crop of crystals. The total yield of crude solid should be about 24 g; if it is less than this, a third extraction of the reaction product should be made. Dissolve the crude solid in about 400 ml of hot ethanol, add a little decolourising charcoal, boil for a few minutes, filter and cool in ice. Recrystallise again from hot ethanol. The yield of pure 2,2′-dinitrobiphenyl, m.p. 123–124 °C, is 20–22 g (54%).

The experimental conditions for conducting the above reaction in dimethylformamide as solvent are as follows. In a 250-ml three-necked flask, equipped with a reflux condenser and a tantalum wire Hershberg-type stirrer, place 20 g of o-chloronitrobenzene and 100 ml of dimethylformamide (dried over anhydrous calcium sulphate). Heat the solution to reflux and add 20 g of activated copper bronze in one portion. Heat under reflux for 4 hours, add another 20 g portion of copper powder and continue refluxing for a second 4-hour period. Allow to cool, pour the reaction mixture into 2 litres of water and filter with suction. Extract the solids with three 200 ml portions of boiling ethanol: alternatively, use 300 ml of ethanol in a Soxhlet apparatus. Isolate the 2,2′-dinitrobiphenyl from the alcoholic extracts as described above: the yield of product, m.p. 124–125 °C, is 11.5 g (75%).

Note. (1) If the temperature is allowed to rise above 240 °C, reduction of the nitro groups will occur and carbazole will be formed.

IV,11. BIPHENYL-2,2′-DICARBOXYLIC ACID (*Diphenic acid*)

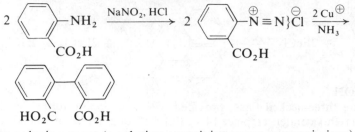

The reducing agent (a solution containing cupro-ammonia ions) is first prepared. Dissolve 63 g (0.25 mol) of crystallised copper(II) sulphate in 250 ml of water in a 1-litre beaker, add 100 ml of concentrated ammonium hydroxide solution (*d* 0.88) and cool the solution to 10 °C. Dissolve 17.8 g (0.256 mol) of hydroxylammonium chloride or 21 g (0.256 mol) of hydroxylammonium sulphate in 60 ml of water, cool to 10 °C and add 42.5 ml of 6 *M* sodium hydroxide solution; if the resulting solution of hydroxylamine is not clear, filter it at the pump. Without delay add the hydroxylamine solution, with stirring, to the ammoniacal copper(II) sulphate solution. Reduction occurs at once, a gas is evolved and the solution assumes a pale blue colour. Protect the reducing agent from the air if it is not used immediately.

Grind 25 g (0.18 mol) of anthranilic acid (Section **IV,60**) with 46 ml of concentrated hydrochloric acid and 75 ml of water in a glass mortar, and transfer the suspension to a 500-ml round-bottomed flask which is provided with a mechanical stirrer. Cool the contents of the flask in an ice bath to 0–5 °C, and add a solution of 13.0 g (0.19 mol) of sodium nitrite in 175 ml of water from a dropping funnel during about 20 minutes. Keep the diazonium solution below 5 °C and, if it is not clear, filter it by suction through a chilled Buchner funnel immediately before use.

Surround the reducing solution in the 1-litre beaker (which is equipped with a mechanical stirrer) with a bath of crushed ice so that the temperature of the solution is about 10 °C. Attach, by means of a short length of rubber tubing, to the stem of a dropping funnel a glass tube which dips well below the surface of the solution and is bent upwards at the end and constricted so that the opening is about 2 mm (this arrangement ensures that the diazonium solution reacts with the ammoniacal solution in the beaker and prevents the latter rising in the stem of the funnel). Place about 45 ml of the cold diazonium solution in the funnel and add it at the rate of about 10 ml per minute whilst the mixture is stirred. Add the remainder of the diazonium solution at the same rate; continue the stirring for 5 minutes after the addition is complete. Heat the solution rapidly to boiling and carefully acidify with 125 ml of concentrated hydrochloric acid; the diphenic acid precipitates as pale brown crystals. Allow to stand overnight and filter with suction; wash the crude diphenic acid with about 25 ml of cold water. Suspend the crude acid in 100 ml of water and add 20 g of solid sodium hydrogen carbonate. Filter the resulting solution by gravity, and then boil with about 0.5 g of decolourising carbon; filter and acidify the filtrate while still hot with excess of dilute hydrochloric acid (1 : 1). Collect the precipitated diphenic acid on a Buchner funnel, wash it with 20 ml of cold water and dry at 100 °C. The yield of diphenic acid is 18 g (82%); it melts at 227–228 °C and usually possesses a light cream colour.

IV,12. 2,2'-DIHYDROXY-1,1'-BINAPHTHYL

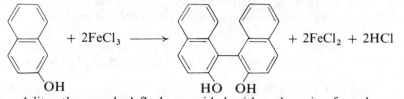

In a 1-litre three-necked flask, provided with a dropping funnel, a sealed stirrer and a reflux condenser, place 14.4 g (0.1 mol) of 2-naphthol and 600 ml of water, and heat to the boiling point. To the boiling liquid containing liquid 2-naphthol in suspension, add slowly through the dropping funnel and with vigorous stirring a solution of 28 g (0.1 mol) of crystallised iron(III) chloride in 60 ml of water. The oily drops of 2-naphthol will disappear and the bis-2-naphthol separates out in flakes. Boil for 5–10 minutes, filter the hot suspension at the pump through a previously warmed Buchner funnel, wash with boiling water and dry in the air upon filter-paper. The crude product weighs 9 g. Recrystallise from toluene (about 150 ml); almost colourless crystals (7.5 g, 52%), m.p. 218 °C, are obtained.

A,4. Condensed polycyclic systems. Dehydrogenation (the conversion of alicyclic or hydroaromatic compounds into their aromatic counterparts by removal of hydrogen—and also, in some cases, of other atoms or groups) has found wide application in the determination of structure of natural products containing complex hydroaromatic systems. Dehydrogenation is employed also for the synthesis of polycyclic hydrocarbons and their derivatives from readily accessible synthetic hydroaromatic compounds. The general procedure is illustrated by the conversion of tetralin into naphthalene (Section IV,13).

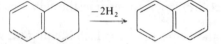

The principal dehydrogenation agents are (i) sulphur, (ii) selenium and (iii) catalytic metals. With *sulphur*, the general method is to heat the compound at 200–260 °C with the theoretical amount of sulphur required to convert it into an aromatic system. In the case of *selenium*, the substance is heated with a large excess of selenium at 280–350 °C for 36–48 hours. Better yields (and less side reactions) are usually obtained than with sulphur, but, owing to the higher temperature, rearrangements are more likely. Oxygen-containing groups are particularly prone to elimination. *Palladium* and *platinum catalysts* are generally employed with a charcoal or asbestos carrier. The dehydrogenation can be conducted in the vapour phase by distilling the compound through a tube containing the catalyst heated to 300–350 °C, but the liquid phase method is generally more convenient. Charcoal or asbestos is employed containing 10–30 per cent of the metal. It has been established that the best results are obtained by conducting the process in an actively boiling medium (e.g., mesitylene, b.p. 165 °C; *p*-cymene, b.p. 177 °C; naphthalene, b.p. 218 °C; and 1-methylnaphthalene, b.p. 242 °C) and to provide for the removal of the hydrogen as it is formed (e.g., by sweeping the system with a stream of carbon dioxide).

Further examples of dehydrogenation processes are to be found in the synthetic sequences described in Sections IV,14 and IV,15. In the former instance, 1-methylnaphthalene is synthesised from α-tetralone by first introducing a 1-

methyl substituent by reaction with methylmagnesium iodide. The resulting tertiary alcohol undergoes dehydration and dehydrogenation on heating with a palladium-on-charcoal catalyst. The preparation of the ketonic starting material from benzene is described in Section **IV,135** (see also p. 770, **IV,L,1**); the conversion of α-tetralone into 1-methylnaphthalene completes the reaction sequence known as the **Haworth procedure** for the synthesis of polycyclic aromatic hydrocarbons.

Intramolecular cyclisation of a γ-arylbutyric acid system is also an important step in a convenient synthesis of the polycyclic system, chrysene, which is formulated and described in Section **IV,15**. Here, methyl cinnamate is first subjected to reductive dimerisation to give methyl *meso-β,γ*-diphenyladipate, which is accompanied by some of the (±)-form. The *meso* isomer (I) is the most easily isolable and cyclisation occurs smoothly in sulphuric acid to yield the diketone 2,11-dioxo-1,2,9,10,11,18-hexahydrochrysene, which is obtained as the *trans* form II as shown in the following formulation. Clemmensen reduction of this ketone followed by dehydrogenation (in this case using selenium) completes the synthesis of chrysene.

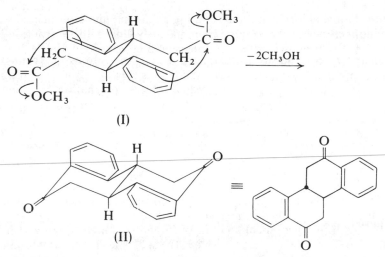

The synthesis of a tetraphenyl derivative (rubrene, Section **IV,16**) of the linearly fused tetracyclic aromatic hydrocarbon naphthacene involves an interesting intermolecular cyclisation process between two molecules of 1-chloro-1,3,3-triphenylpropa-1,2-diene. This substituted allene is formed *in situ* from 1,1,3-triphenylprop-2-yn-1-ol (Section **III,41**) when the latter is allowed to

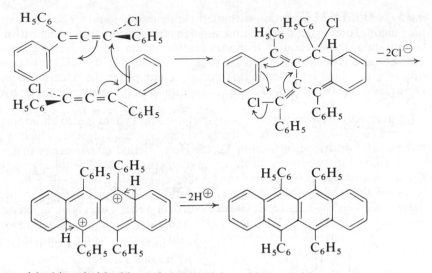

react with thionyl chloride and the resulting chlorosulphite ester heated with a little quinoline; cyclisation occurs spontaneously under these reaction conditions to give rubrene which has an intense red colour.

The synthesis of 9-phenylphenanthrene (Section **IV,17**) illustrates the formation of the phenanthrene system by the cyclisation of a 1,2-diphenylethylene (stilbene). The process involves an allowed photochemical cyclisation which gives initially a dihydrophenanthrene. This is readily dehydrogenated *in situ* by molecular oxygen in the presence of iodine.

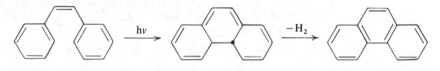

The required stilbene is usually readily prepared by dehydration of the appropriate tertiary alcohol obtained by a Grignard reaction, e.g.,

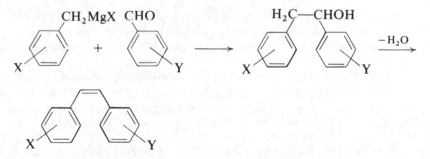

The method has been applied to the synthesis of a range of substituted phenanthrenes and has the merit that it involves fewer steps than, for example, the Haworth synthesis. Good results are not always obtained, however, when some electron-withdrawing substituents (e.g., NO_2, $CH_3 \cdot CO$) are present.

IV,13. NAPHTHALENE

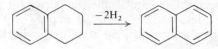

For small-scale dehydrogenations, the apparatus shown in Fig. IV,1 may be used. Place 2.5 g of purified tetralin (1) and 0.25 g of palladised charcoal (Section **II,2,**44) in the apparatus and heat to boiling for 4 hours in a slow current of dry carbon dioxide. Naphthalene, m.p. 81 °C, collects on the condenser in almost quantitative yield. If it is desired to follow the progress of the dehydrogenation, attach the side-tube through a U-tube packed with self-indicating soda lime to a nitrometer filled with potassium hydroxide solution: almost the theoretical quantity of hydrogen will be collected.

If the current of inert gas is omitted, the reaction is complete after about 22 hours.

Note. (1) Commercial tetralin may be purified as follows. Wash the technical product repeatedly with 10 per cent of its volume of concentrated sulphuric acid, then with 10 per cent sodium carbonate solution, followed by water, dry with anhydrous calcium sulphate, filter from the desiccant, reflux over sodium and finally distil from sodium. Collect the pure tetralin at 206–207 °C.

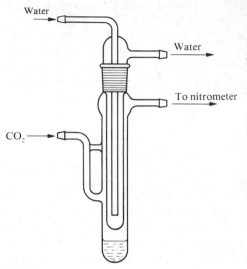

Fig. IV,1

IV,14. 1-METHYLNAPHTHALENE

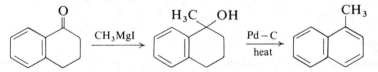

1-Hydroxy-1-methyl-1,2,3,4-tetrahydronaphthalene. Prepare an ethereal solution of methylmagnesium iodide (cf. Section **III,34**) from 1.8 g (0.075 mol) of magnesium, 10.6 g (4.7 ml, 0.075 mol) of methyl iodide and 30 ml of ether in a 100-ml two-necked flask fitted with a dropping funnel and a reflux condenser protected by a calcium chloride guard-tube. Slowly add a solution of 7.3 g (0.05 mol) of α-tetraione (Section **IV,135**) in 10 ml of ether, swirling the contents of the flask from time to time, and finally heat under reflux on a steam bath for 1 hour to complete the reaction. Cool in ice, and decompose the reaction mixture with a cold saturated aqueous solution of ammonium chloride. Separate the ether layer, and extract the aqueous phase with 10 ml of ether. Wash the combined ethereal solutions with aqueous ammonium chloride solution, dry over anhydrous sodium sulphate and evaporate off the ether. A solid residue of 1-hydroxy-1-methyl-1,2,3,4-tetrahydronaphthalene of sufficient purity for use in the next stage is obtained; the yield is 7.5 g (92%), m.p. 77–79 °C. Recrystallise a sample from light petroleum (b.p. 60–80 °C); the purified material has m.p. 86–87 °C.

1-Methylnaphthalene. Use the apparatus and follow the general procedure described under *naphthalene* (Section **IV,13**). Heat a mixture of 3.2 g (0.02 mol) of the above hydroaromatic compound with 0.3 g of palladised charcoal at 250–270 °C in a Silicone oil or fusible metal bath for 3 hours. Cool, dissolve the residue in ether and filter off the catalyst. Wash the extract with dilute aqueous sodium hydroxide and dry it over anhydrous sodium sulphate. Remove the ether and distil the residual oil under reduced pressure; use a small-scale distillation apparatus (cf. Fig. I,113). Collect the 1-methylnaphthalene, b.p. 121–123 °C/20 mmHg. The yield is 2.5 g (89%).

IV,15. CHRYSENE

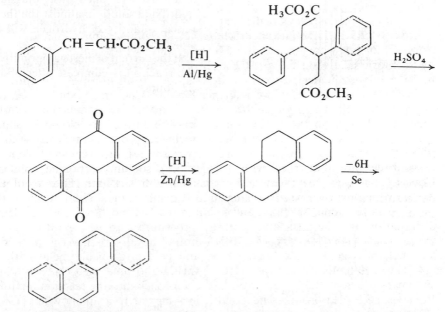

meso-Dimethyl 3,4-diphenyladipate. Prepare aluminium amalgam from 75 g of the foil (Section **II,2,2**) and cover with 1 litre of moist ether (1) in 2-litre flask fitted with a large double-surface reflux condenser. Immediately add a solution of 48.5 g (0.3 mol) of recrystallised methyl cinnamate (Section **IV,177**) in 100 ml of ether and shake the flask to mix the contents thoroughly. Evolution of hydrogen ceases and after an induction period of up to 1 hour the ether boils; moderate the reaction if necessary by cooling the flask. Towards the end of the reaction, which should be complete in 6 hours, warm the flask to maintain gentle reflux of the ether. Cool, filter off the aluminium hydroxide sludge (do not discard this residue) and dry the ethereal solution over anhydrous sodium sulphate. Remove ether on a rotary evaporator and cool the residual oil thoroughly in an ice bath (or overnight in a refrigerator). Filter the resulting crystalline mass and wash the crystals with a little cold ether; about 4 g of the crude *meso*-adipate, m.p. 170 °C, are obtained (2). Much of the product adheres to the aluminium hydroxide residue, however. Cover this residue with 1 litre of water and add 1 litre of concentrated hydrochloric acid slowly with stirring to dissolve most of the aluminium hydroxide. Extract with three 150 ml portions of chloroform, dry the extract over anhydrous sodium sulphate and remove the chloroform on a rotary evaporator; a further 7 g of the crude *meso*-adipate is obtained. Recrystallise the combined crops of crude material from rectified spirit (about 250 ml). Pure *meso*-dimethyl 3,4-diphenyladipate, m.p. 175 °C, is obtained; the yield is about 9 g (18%).

trans-**2,11-Dioxo-1,2,9,10,11,18-hexahydrochrysene.** Add 8.2 g (0.025 mol) of *meso*-dimethyl 3,4-diphenyladipate to a stirred mixture of 100 ml of concentrated sulphuric acid previously added to 33 ml of water and heat on a water bath for 3 hours. Cool the mixture and pour it carefully with stirring into 1 litre of water. Filter off the resulting precipitate and digest it on the steam bath with hot sodium carbonate solution. Filter again, wash the product with water and dry in an oven. The dioxohexahydrochrysene is obtained as a pale buff powder, m.p. 293–300 °C, yield 5.5 g (85%). Crystallise a small specimen from butan-1-ol; colourless plates, m.p. 303 °C (sealed tube), are obtained.

trans-**1,2,9,10,11,18-Hexahydrochrysene.** Prepare amalgamated zinc from 35 g of zinc wool (Section II,2,67) and cover it with 25 ml of concentrated hydrochloric acid. Add 5.2 g (0.02 mol) of the dioxohexahydrochrysene and boil the mixture under reflux for 8 hours. Add 20 ml more of concentrated hydrochloric acid and reflux for a further 8 hours. Cool the mixture, when the oily product solidifies. Extract the product, together with any solid material which has formed in the reflux condenser, with two 50 ml portions of hot benzene. Cool, separate the benzene layer and dry over anhydrous sodium sulphate. Remove the benzene with a rotary evaporator, transfer the residue to a small-scale distillation unit (Fig. I,113) and distil under reduced pressure. Collect the hexahydrochrysene which distils at 230 °C/14 mmHg as a colourless oil which readily crystallises, m.p. 113 °C, yield 3.5 g (75%).

Chrysene. Since selenium and selenium compounds are *toxic* this dehydrogenation and the associated work-up procedure *must be carried out in an efficient fume cupboard.* Mix 3.5 g (0.015 mol) of hexahydrochrysene with 16 g (0.2 mol) of selenium in a boiling tube and heat in a fusible metal bath at 300 °C for 20 hours (fume cupboard). From time to time, melt the crystalline sublimate which gradually forms so that it runs back into the reaction mixture. Remove the cooled product and grind it in a mortar to a fine powder. [*Care:* (3).] Extract by boiling under reflux for 30 minutes with 200 ml of benzene, filter and reflux the filtered extract over a little clean sodium wire (or thin narrow slices of sodium metal); this treatment removes traces of selenium. Evaporate the benzene solution using a rotary evaporator and crystallise the residue from toluene (about 20 ml per 1 g) (4). Colourless plates with a bluish fluorescence, m.p. 254 °C, are obtained. The yield of chrysene is about 2 g (59%).

Notes. (1) The ether should be saturated by shaking it with a little water in a separatory funnel.

(2) If the filtrate is distilled under reduced pressure, 30 g of methyl 3-phenylpropionate, b.p. 111–130 °C/15 mmHg, are obtained. When the distillation residue is dissolved in the minimum of hot ether and cooled, a further 1 g of the *meso*-adipate is obtained. The ether filtrate from this contains racemic dimethyl

3,4-diphenyladipate; the latter may be recovered by evaporating the ether and crystallising the residue from the minimum of methanol; m.p. 70–71 °C.

(3) Do not inhale any of the finely divided material; it is advisable to wear a face mask or alternatively to grind the material in an enclosed (i.e., glove) box.

(4) The crude product may also be purified by sublimation under reduced pressure (*ca.* 0.1 mmHg) from an oil bath maintained at 200 °C.

IV,16. 5,6,11,12-TETRAPHENYLNAPHTHACENE (*Rubrene*)

$$2C_6H_5 \cdot C \equiv C \cdot C(C_6H_5)_2OH \xrightarrow{SOCl_2} 2C_6H_5 \cdot CCl = C = C(C_6H_5)_2 \xrightarrow{-2HCl}$$

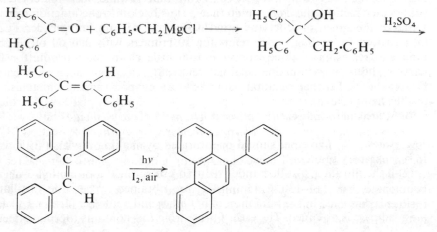

Add 28.5 g (0.1 mol) of 1,1,3-triphenyl-prop-2-yn-1-ol (Section **III,41**) in small portions with shaking to 24 g (14.5 ml, 0.2 mol) of redistilled thionyl chloride cooled to −10 °C in an ice–salt bath. Set the resulting solution aside at room temperature for 1 hour and then remove excess chloride by warming under reduced pressure (water pump); the residue solidifies (1). Add 0.5 ml of redistilled quinoline and heat the mixture in an oil bath at 120 °C under reduced pressure (water pump) for 2 hours. Wash the dark red residue with ether and then with boiling acetone and dissolve it in benzene. Filter off any insoluble impurity, dilute with light petroleum (b.p. 100–120 °C) and remove most of the benzene carefully by evaporation under reduced pressure. Cool and filter off the bright red crystals of rubrene; the yield is 7 g (26%), m.p. 320 °C. Store the product in a specimen tube protected from the light by a covering of black paper.

Note. (1) If the product does not solidify it must be purified by distillation under reduced pressure using an efficient oil-immersion pump; b.p. 190 °C/0.05 mmHg.

IV,17. 9-PHENYLPHENANTHRENE

$$\underset{H_5C_6}{\overset{H_5C_6}{>}}C=O + C_6H_5 \cdot CH_2MgCl \longrightarrow \underset{H_5C_6}{\overset{H_5C_6}{>}}C\underset{CH_2 \cdot C_6H_5}{\overset{OH}{<}} \xrightarrow{H_2SO_4}$$

$$\underset{H_5C_6}{\overset{H_5C_6}{>}}C=C\underset{C_6H_5}{\overset{H}{<}}$$

Triphenylethylene. Equip a 2-litre, three-necked round-bottomed flask with a sealed stirrer unit, a 250-ml dropping funnel and a double surface condenser. All the apparatus should be rigorously dried and the dropping funnel and reflux condenser protected by calcium chloride guard-tubes. Place in the flask 12.2 g (0.50 mol) of magnesium turnings and about 250 ml of ether previously dried over anhydrous calcium chloride. In the funnel place a solution of 57.5 ml (63.3 g, 0.50 mol) of freshly distilled benzyl chloride in about 100 ml of dry ether. Without stirring the contents of the flask run in about 20 ml of the benzyl chloride solution and then add a large crystal of iodine and allow the mixture to stand undisturbed. Reaction commences after about 5 minutes and the magnesium begins to dissolve. Add the benzyl chloride solution dropwise at such a rate as to maintain a steady reflux of the ether solvent. When the reaction is well established, start the stirrer motor and stir the reaction mixture at a moderate rate. When all the benzyl chloride solution has been added (about half an hour), heat the flask in an electric heating mantle and with continued stirring maintain a vigorous rate of reflux for a further half an hour. Turn off the heating mantle but continue stirring fairly rapidly, and add from the dropping funnel a solution of 91.1 g (0.50 mol) of benzophenone in about 250 ml of dry ether. The rate of addition should be sufficient to maintain rapid reflux. On completion of the addition (about 15 minutes) heat under reflux with stirring for a further half an hour before allowing the reaction mixture to stand at room temperature for at least 2 hours or preferably overnight. Cool the reaction mixture in an ice-water bath, add 400 g of crushed ice to the flask contents, followed by 250 ml of cold 2.5 M-sulphuric acid. Transfer the reaction mixture to a 3-litre separating funnel, run off the aqueous layer and extract with two 150 ml portions of ether. Combine the ether layer and extracts, dry over calcium sulphate and evaporate the ether on a rotary evaporator. To the syrupy residue add 100 ml of 2.5 M-sulphuric acid and boil vigorously under reflux for 2 hours. Cool the mixture to room temperature, add 50 ml of ether and shake vigorously. Transfer the resulting emulsion to a 1-litre separatory funnel, wash the flask thoroughly with a mixture of 50 ml of water and 50 ml of ether and add this to the separatory funnel. After a few minutes the emulsion separates into two layers and the aqueous fraction can be run off and extracted with two 50 ml portions of ether. Combine the ether layer and extracts (1), and evaporate the ether using a rotary evaporator. Distil the residue under reduced pressure collecting the main fraction (93 g, 73%) at 196–200 °C/2.5 mmHg. Dissolve the greenish-yellow syrupy distillate in 450 ml of hot 95 per cent aqueous ethanol and allow to cool slowly. Trituration is usually required to induce crystallisation. Complete the crystallisation process by cooling in ice-water and filter under suction to obtain 85 g (66%) of colourless crystals having m.p. 68–69 °C. Further recrystallisations from ethanol or acetic acid gives pure triphenylethylene, m.p. 72–73 °C.

9-Phenylphenanthrene. *The photochemical reactor should be completely screened with metal foil to avoid hazards from stray radiation* (see Section **I,17,5**). Dissolve 2.56 g (0.01 mol) of triphenylethylene and 0.127 g (0.0005 mol) of iodine in 1 litre of redistilled cyclohexane contained in a photochemical reactor vessel of 1-litre capacity and fitted with a 100-W medium-pressure mercury arc lamp in a water-cooled quartz immersion well (Section **I,17,5**; Fig. I,74). Place the apparatus on a magnetic stirrer unit, insert a magnetic follower and with both the side-necks of the vessel open to the atmosphere irradiate the stirred solution until all the triphenylethylene has reacted (2); this should take approximately 21–22 hours. There is no observable change in the colour of the cyclohexane solution during the irradiation. Remove the reaction mixture and evaporate to dryness under reduced pressure on a rotary evaporator to obtain the crude phenylphenanthrene as a pale fawn solid. Purify by column chromatography as follows. Dissolve the solid in 60 ml of warm cyclohexane, allow to cool and pour on to a short (*ca.* 2 cm × 7 cm) column of neutral alumina (*ca.* 16–17 g, Activity I, prepared as a slurry in cyclohexane). Collect the eluate, and when most of the solution has passed down the column rinse the flask which contained the solid with two 10 ml portions of cyclohexane and add to the column. When this has also been adsorbed, elute with a further 100–120 ml of cyclohexane or until no more phenylphenanthrene is obtained in the eluate. It is best to monitor the elution of the product by collecting and evaporating 40 ml fractions; a band of yellow material should remain near the top of the column and elution of this must be avoided. Crystallise the solid so obtained from 60 ml of ethanol and collect the crystals which separate by suction filtration; a second crop may be obtained by concentrating the filtrate to *ca.* 20 ml and cooling in ice. The yield of 9-phenylphenanthrene (white needles), m.p. 104.5–105.5 °C, is 2.16 g (85%) (3).

Notes. (1) Do not wash the ethereal fraction with water or with aqueous sodium hydrogen carbonate solution, since the presence of a trace of sulphuric acid during the final distillation is necessary to complete the dehydration.

(2) The irradiation time for complete reaction will depend on the power of the mercury arc used; in particular the light output of a lamp will vary with its age and use of a higher wattage lamp will, of course, shorten the reaction time required. The extent of the reaction may be most conveniently followed by GLC of 0.5 μl samples of the reaction solution on a 5-ft column of Methyl Silicone Gum S.E.30 on Chromosorb W held at 240 °C and with a nitrogen flow rate of 40 ml/minute. Triphenylethylene and 9-phenylphenanthrene have t_R 2.3 minutes and 5.2 minutes respectively, cyclohexane has t_R 0.5 minutes. Alternatively the reaction may be followed by evaporating 5–10 ml portions of the solution and recording the infrared spectrum of the residue as a mull in Nujol. Bands at 887(m), 750(m) and 738(s) cm^{-1} characteristic of 9-phenylphenanthrene gradually appear and increase in intensity relative to bands due to the stilbene at 872(w) and 757(s) cm^{-1} which gradually disappear. Other regions of the spectrum of the individual compounds are very similar and it is advisable to use the scale expansion adjustment of the instrument to enable the bands in the low wavenumber region to be readily identified.

(3) For the preparation of larger amounts of the phenanthrene the reaction should be repeated; photolysis of more concentrated solutions is to be avoided in order to minimise the possibility of photodimerisation to give a cyclobutane derivative.

B AROMATIC NITRO COMPOUNDS

The following procedures, of which the first is by far the most important, are available for the synthesis of aromatic nitro compounds.

1. Direct nitration (Sections **IV,18** to **IV,24**).
2. The oxidation of amines (Section **IV,25**).
3. The replacement of a diazo group by a nitro group (see Section **IV,86**).

B,1. Direct nitration. Aromatic hydrocarbons may be nitrated, i.e., the hydrogen atoms replaced by nitro (NO_2) groups, with concentrated nitric acid in the presence of concentrated sulphuric acid ('mixed acid reagent').

$$ArH + HNO_3/H_2SO_4 \longrightarrow Ar{\cdot}NO_2$$

The function of the sulphuric acid is to convert the nitric acid into the highly reactive, electrophilic, nitronium ion, $\overset{\oplus}{N}O_2$, which is the effective nitrating agent.

$$HONO_2 + H_2SO_4 \longrightarrow \overset{\oplus}{N}O_2 + H_3\overset{\oplus}{O} + 2HSO_4^{\ominus}$$

The mechanism of aromatic nitration, which is illustrated below in the case of benzene, is a two-step process involving electrophilic attack of the nitronium ion on the benzene molecule to form the intermediate mesomeric ion (I), followed by removal of a proton by the hydrogen sulphate ion, which is the most basic species in the reaction mixture.

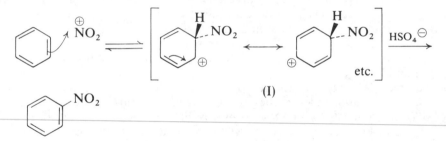

(I)

Nitration of aromatic hydrocarbons is usually carried out with the above mixed acid reagent at comparatively low temperatures (e.g., about 50 °C, as used in the preparation of nitrobenzene and 1-nitronaphthalene, Sections **IV,18** and **IV,19** respectively). Unnecessarily high temperatures should be avoided since polynitration is then more likely and oxidative breakdown of the aromatic ring system may occur.

The nitration of aromatic compounds containing an electron-withdrawing group (e.g., $-NO_2$, $-SO_3H$, $-CHO$, $-CO{\cdot}R$, $-CO_2H$, $-CO_2R$) does not occur readily under the above conditions, in which case forcing conditions which require the use of fuming nitric acid and concentrated sulphuric acid need to be employed. Nitrobenzene, for example, is converted by a mixture of fuming nitric acid and concentrated sulphuric acid into about 90 per cent of *m*-dinitrobenzene (Section **IV,20**) and small amounts of the *o*- and *p*-isomers; the latter are eliminated in the process of recrystallisation. *p*-Nitrotoluene is similarly converted largely into 2,4-dinitrotoluene (Section **IV,21**). A further example is the nitration of benzaldehyde to give *m*-nitrobenzaldehyde (Section **IV,22**) in reasonable yield in spite of the fact that the reaction conditions are strongly oxidising. The nitro group is thus seen to be *meta*-directing in common with the other deactivating groups specified above.

The deactivating effect of the nitro group is largely the result of its mesomeric interaction (-*M* effect) with the π-electron system of the benzene ring which is supplemented by the inductive (-*I*) effect (II).

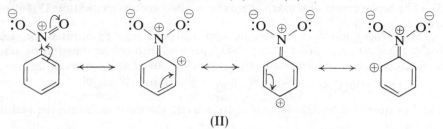

(II)

The overall electron-withdrawal from the ring system results in the rate of attack of the nitronium ion being substantially retarded compared to benzene. Moreover, the representations of the canonical forms of nitrobenzene formulated above show that the *ortho* and *para* positions are subject to the greatest reduction in electron density. In addition the reaction intermediate resulting from attack at a position *para* to the nitro group would be represented by the following hydrid species (similar formulations being possible as a result of *ortho* attack).

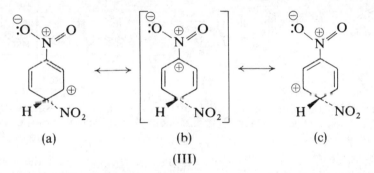

(III)

It follows that since (III(b)) is energetically unfavourable the mesomeric stabilisation of this intermediate is less than that of the corresponding intermediate resulting from attack in the *meta* position (IV).

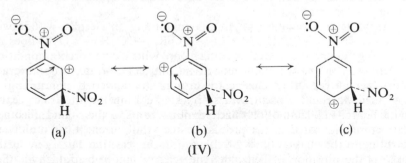

(IV)

The nitration of aromatic compounds containing electron-releasing groups (e.g., $-R$, $-OH$, $-NH_2$, $-NHCO\cdot CH_3$) needs to be conducted under conditions which are milder than those specified for benzene, since in these cases the aromatic nucleus is activated towards attack by the electrophilic species, and furthermore the ring system is more likely to be oxidatively cleaved. An example is provided by the nitration of acetanilide (Section **IV,76**) to yield a mixture of *o*- and *p*-nitroacetanilides (see also p. 682, **IV,F,2**); the overall nitration of phenol to a mixture of *o*- and *p*-nitrophenols (Section **IV,114**) is anomalous in that the reagent is dilute nitric acid alone; the mechanism is outlined on p. 742, **IV,J,2**. These activating groups are therefore seen to be *ortho* and *para* directing. By contrast to the effect of electron-withdrawing groups, activating and *ortho/para* directing groups will stabilise by electron release the reaction intermediate formed as a result of attack at any of the three possible positions, but particularly so when attack is at the *ortho* and *para* positions.

An interesting case is provided by the nitration of aryl halides where the effect of the halogen is to deactivate the aromatic nucleus (by the -*I* effect) but to direct the incoming nitronium ion to the *ortho* and *para* positions as a result of the mesomeric interaction of the halogen lone electron pair with the charge developed in the corresponding intermediates [e.g., (V), in the formation of *p*-bromonitrobenzene from bromobenzene, Section **IV,23**].

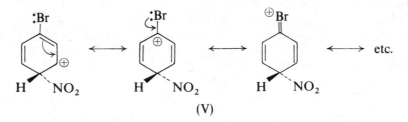

(V)

Experimental details for the nitration of benzyl cyanide are also included (Section **IV,24**). The product is largely *p*-nitrobenzyl cyanide (some of the *ortho* isomer is also formed): the cyanomethyl substituent (as a substituted alkyl group) is thus an *ortho/para* directing and weakly activating group.

The electronic effect of groups in aromatic electrophilic substitution processes is fully discussed in all standard organic chemistry textbooks.

IV,18. NITROBENZENE

$$C_6H_5 + HNO_3 \xrightarrow{H_2SO_4} C_6H_5 \cdot NO_2$$

Place 50 g (35 ml, *ca.* 0.5 mol) of concentrated nitric acid in a 500-ml round-bottomed flask, and add, in portions with shaking, 74 g (40 ml) of concentrated sulphuric acid. Keep the mixture cool during the addition by immersing the flask in cold water. Place a thermometer (110 °C range) in the acid mixture. Introduce 26 g (30 ml, 0.33 mol) of benzene in portions of 2–3 ml; shake the flask well, to ensure thorough mixing, after each addition of the benzene. Do not allow the temperature of the mixture to rise above 55 °C; immerse the flask, if necessary, in cold water or in ice-water. When all the benzene has been added, fit a reflux condenser to the flask and heat it in a water bath maintained at 60 °C (but not appreciably higher) for 40–45 minutes; remove the flask from time to time from the bath and shake it vigorously to ensure good mixing of the immiscible layers. Pour the contents of the flask into about 500 ml of cold water in a beaker, stir the mixture well in order to wash out as much acid as possible from the nitrobenzene and allow to stand. When the nitrobenzene has settled to the bottom, pour off the acid liquor as completely as possible, and transfer the residual liquid to a separatory funnel. Run off the lower layer of nitrobenzene and reject the upper aqueous layer; return the nitrobenzene to the separatory funnel and shake it vigorously with about 50 ml of water. Separate the nitrobenzene as completely as possible and run it into a small conical flask containing about 5 g of anhydrous calcium chloride. If the nitrobenzene does not become clear on shaking because of the presence of emulsified water, warm the mixture, with shaking, for a short period on a water bath; the cloudiness will soon disappear. Filter the cold product through a small fluted filter-paper into a small (50- or 100-ml) distilling flask and attach a still-head and air condenser. Heat the flask on an asbestos-centred wire gauze or preferably in an air bath, and collect the fraction which boils at 206–211 °C. Do not distil quite to dryness nor allow the temperature to rise above 214 °C, for there may be a residue of *m*-dinitrobenzene and higher nitro compounds and an explosion may result. The yield of nitrobenzene is 35 g (85%) (1). Pure nitrobenzene is a clear, pale yellow liquid, b.p. 210 °C.

Note.   (1) Nitrobenzene (and many other liquid organic compounds containing nitrogen) is appreciably toxic and its vapour should not be allowed to escape into the atmosphere of the laboratory; the delivery tube of the condenser should pass well into the mouth of the receiver flask. The liquid is also a skin poison; if it is accidentally spilled on the skin, it should be removed by washing with a little methylated spirit, followed by soap and warm water.

IV,19.   1-NITRONAPHTHALENE

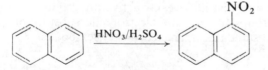

Prepare a mixture of 40 ml of concentrated nitric acid and 40 ml of concentrated sulphuric acid as detailed in the previous Section. Introduce 50 g (0.39 mol) of *finely powdered* naphthalene in small quantities at a time and with vigorous shaking: maintain the temperature at 45–50 °C and cool in ice-water if necessary. When all the naphthalene has been added, warm the mixture on a water bath at 55–60 °C for 30–40 minutes or until the smell of naphthalene has disappeared. Pour the mixture into 500 ml of cold water; the nitronaphthalene will sink to the bottom. Decant the liquid. Boil the solid cake with 200 ml of water for 20 minutes and pour the water away. Transfer the oil to a round-bottomed (250- or 500-ml) flask and subject it to steam distillation (Fig. I,101); any unattacked naphthalene will thus be removed. Pour the warm contents of the flask into a beaker containing a large volume of water which is vigorously stirred. Filter off the granulated 1-nitronaphthalene at the pump, press it well and recrystallise it from dilute alcohol. The yield of 1-nitronaphthalene, m.p. 61 °C, is 60 g (89%).

IV,20. *m*-DINITROBENZENE

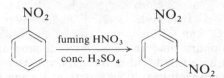

Place 37.5 g (21 ml) of concentrated sulphuric acid and 22.5 g (15 ml) of fuming nitric acid, *d* 1.5, in a 250- or 500-ml round-bottomed flask; add a few fragments of unglazed porcelain. Attach a reflux condenser and place the apparatus in a fume cupboard. Add slowly, in portions of about 3 ml, 15 g (12.5 ml, 0.122 mol) of nitrobenzene; after each addition, shake the flask to ensure thorough mixing. Heat the mixture, with frequent shaking, on a boiling water bath for 30 minutes. Allow the mixture to cool somewhat and pour it cautiously with vigorous stirring into about 500 ml of cold water; the dinitrobenzene soon solidifies. Filter with suction, wash thoroughly with cold water and allow to drain as completely as possible.

Transfer the crude dinitrobenzene to a 250-ml flask fitted with a reflux condenser, add 80–100 ml of industrial (or rectified) spirit and heat on a water bath until all the crystalline solid dissolves. If the resulting solution is not quite clear, filter it through a fluted filter-paper on a large funnel which has previously been warmed or through a warm Buchner funnel. Colourless crystals of *m*-dinitrobenzene (15 g, 73%) are deposited on cooling. If the m.p. is below 89–90 °C, recrystallisation is necessary.

IV,21. 2,4-DINITROTOLUENE

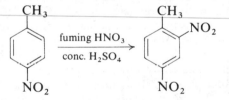

Place 18 g (12 ml, *ca.* 0.36 mol) of fuming nitric acid, *d* 1.5, and 30 g (16.5 ml) of concentrated sulphuric acid and a few fragments of unglazed porcelain in a 250- or 500-ml round-bottomed flask contained in a fume cupboard. Add gradually, in small portions, 14 g (0.1 mol) of *p*-nitrotoluene; do not allow the temperature to rise above 50 °C and cool the flask, if necessary, by immersion in cold water. Place a small funnel in the mouth of the flask and heat on a water bath at 90–95 °C for 30 minutes. Allow to cool almost to the laboratory temperature and pour the reaction mixture slowly into about 500 ml of ice-water containing a few small pieces of ice. Filter the crude dinitrotoluene through a Buchner funnel at the pump, wash it thoroughly with cold water and drain as completely as possible. Recrystallise from the minimum volume of hot methanol (flask, reflux condenser and water bath). The yield of pure 2,4-dinitrotoluene, m.p. 71 °C, is 12.5 g (69%).

IV,22. *m*-NITROBENZALDEHYDE

$$C_6H_5 \cdot CHO \xrightarrow[\text{conc. } H_2SO_4]{\text{fuming } HNO_3} m\text{-}O_2N \cdot C_6H_4 \cdot CHO$$

Place 250 ml of concentrated sulphuric acid and 21.5 ml of fuming nitric acid, *d* 1.5, in a 500-ml two-necked flask fitted with a mechanical stirrer (unsealed) and a dropping funnel. Stir and cool to 0 °C in a bath of ice and salt. Add 62.5 g (60 ml, 0.59 mol) of benzaldehyde (1) dropwise from the dropping funnel; do not allow the temperature to rise above 5 °C. Then warm the mixture gradually to 40 °C, cool to room temperature and pour in a thin stream with vigorous stirring on to finely crushed ice. Filter through a sintered glass funnel, wash with a little water, press out the oil with a wide glass stopper and dry the solid in the air upon absorbent paper. The resulting crude *m*-nitrobenzaldehyde weighs 55 g and melts at 48–50 °C. Melt the crude solid under excess of 10 per cent sodium carbonate solution, stir, cool, filter and dry in the air; the product has m.p. 51–52 °C. Dissolve the solid in 120 ml of hot benzene under reflux (*fume cupboard*), decant from any solid present and add light petroleum, b.p. 40–60 °C, until a slight turbidity results and cool. Collect the pure *m*-nitrobenzaldehyde and dry in the air; the yield is 45 g (50%), m.p. 58 °C.

Note. (1) The following details for the purification of impure technical grades of benzaldehyde may be found to be useful. Wash 50 g (48 ml) of technical benzaldehyde in a separatory funnel with 20 ml portions of 10 per cent sodium carbonate solution until no further carbon dioxide is evolved, then with water, and dry over 5 g of anhydrous magnesium sulphate or calcium chloride. Add 0.5 g of hydroquinone or catechol during the drying operation. Decant through a small fluted filter-paper (or through a *small* plug of glass wool) into a 100-ml flask, and distil under reduced pressure (Fig. I,110). A very fine capillary tube should be used. It is better to conduct the distillation in a stream of nitrogen gas. Collect the benzaldehyde over a 2 °C range, i.e., 1 °C on either side of the true b.p. The correct b.p. under the diminished pressure obtained in the apparatus may be interpolated from the following boiling point data: 79 °C/25 mmHg; 69 °C/15 mmHg; 62 °C/10 mmHg. Place about 0.05 g of hydroquinone or catechol in the product. Benzaldehyde is easily oxidised by atmospheric oxygen giving, ultimately, benzoic acid. This **auto-oxidation** is considerably influenced by catalysts; these are considered to react with the unstable 'peroxide' complexes which are the initial products of the oxidation. Catalysts which inhibit or retard auto-oxidation are termed **anti-oxidants**, and those that accelerate auto-oxidation are called **pro-oxidants**. Anti-oxidants find important applications in preserving many organic compounds, e.g., acrolein. For benzaldehyde, hydroquinone or catechol (considerably less than 0.1% is sufficient) are excellent anti-oxidants.

IV,23. *p*-BROMONITROBENZENE

$$C_6H_5Br + HNO_3 \xrightarrow{H_2SO_4} p\text{-}Br \cdot C_6H_4 \cdot NO_2$$

Prepare a mixture of 28.5 g (20 ml) of concentrated nitric acid and 37 g (20 ml) of concentrated sulphuric acid in a 250-ml round-bottomed flask (see

Section IV,18) and cool it to the laboratory temperature. Attach a reflux condenser to the flask. Add 16 g (10.5 ml, 0.1 mol) of bromobenzene (Section IV,27)
in portions of 2–3 ml during about 15 minutes; shake the flask vigorously
during the whole process and do not allow the temperature to rise above
50–60 °C by cooling in running water, if necessary. When the temperature no
longer tends to rise owing to the heat of reaction, heat the flask on a boiling
water bath for 30 minutes. Allow to cool to room temperature and pour the
reaction mixture with stirring into 200 ml of cold water. Filter the bromonitrobenzene at the pump, wash well with cold water and finally drain as far as
possible. Recrystallise from 100 to 125 ml of industrial spirit (flask, reflux condenser and water bath; see Section IV,20). When cold, filter the almost pure p-
bromonitrobenzene, m.p. 125 °C. The yield is 14 g (70%). The mother-liquor
contains the o-bromonitrobenzene, contaminated with some of the p-isomeride.

IV,24. p-NITROBENZYL CYANIDE

$$C_6H_5 \cdot CH_2 \cdot CN + HNO_3 \xrightarrow{H_2SO_4} p\text{-}O_2N \cdot C_6H_4 \cdot CH_2 \cdot CN$$

Place a mixture of 275 ml of concentrated nitric acid with an equal volume of
concentrated sulphuric acid in a 2-litre three-necked flask, fitted with a thermometer, a mechanical stirrer and a dropping funnel and assembled in the fume
cupboard. Cool the mixture to 10 °C in an ice bath, and run in 100 g (98 ml,
0.85 mol) of benzyl cyanide (Section III,160) at such a rate (about 1 hour) that
the temperature remains at about 10 °C and does not rise above 20 °C. Remove
the ice bath, stir the mixture for 1 hour and pour it on to 1200 g of crushed ice.
A pasty mass slowly separates; more than half of this is p-nitrobenzyl cyanide,
the other components being the *ortho* isomeride and a variable amount of an
oil. Filter the mass on a sintered glass funnel, press well to remove as much oil
as possible and then dissolve in 500 ml of boiling rectified spirit. The p-
nitrobenzyl cyanide crystallises on cooling. Filter this off at the pump and
recrystallise from 80 per cent ethanol. The yield of p-nitrobenzyl cyanide, m.p.
115–116 °C, is 75 g (54%). Another recrystallisation raises the m.p. to 116–
117 °C.

   B,2. The oxidation of amines.   Various reagents are available for the oxidation of an aromatic amine to the corresponding nitro compound. For example,
peroxymonosulphuric acid (Caro's acid) and other peroxyacids have been quite
widely used in the past, although the yields of nitro compounds are rather
variable owing to the concomitant formation of azoxycompounds. Peroxytrifluoroacetic acid is the reagent of choice since it generally gives improved
yields of purer products; its use is illustrated by the conversion of p-toluidine
into p-nitrotoluene (Section IV,25). The method is not, however, generally suitable for amines having ring systems which are highly activated as the result, for
example, of the presence of an alkoxy substituent.

$$ArNH_2 \xrightarrow{CF_3 \cdot CO_3H} ArNO_2$$

   B,3. The replacement of a diazo group by a nitro group.   This replacement
is achieved by the decomposition of the aryldiazonium fluoroborate with aqueous sodium nitrite in the presence of copper powder and is described in
Section IV,86. This procedure gives better yields and thus replaces the former

method of reacting an acidic aryldiazonium salt solution with nitrous acid in the presence of copper(I) oxide.

IV,25. p-NITROTOLUENE

$$(CF_3 \cdot CO)_2O + H_2O_2 \longrightarrow CF_3 \cdot CO_3H + CF_3 \cdot CO_2H$$

This reaction should be carried out in a fume cupboard behind a safety screen. Adequate precautions should be observed in handling the hydrogen peroxide solution (Section III,188, Note (1)).

Prepare a solution of peroxytrifluoroacetic acid in dichloromethane as follows. Place 50 ml of dichloromethane in a two-necked, 250-ml round-bottomed flask fitted with a reflux condenser and a dropping funnel. Insert a plastic-covered magnetic stirrer follower bar, and cool the flask in an ice bath sited on the stirrer unit. To the cooled and stirred solution add 3.5 ml (0.12 mol) of 85 per cent hydrogen peroxide and then 22.7 ml (33.9 g, 0.14 mol) of trifluoroacetic anhydride. Stir for 5 minutes after the addition is completed and then allow the solution to warm to room temperature. To this stirred solution add over 15 minutes a solution of 3.2 g (0.03 mol) of p-toluidine in 10 ml of dichloromethane. The reaction is exothermic but no cooling is necessary in this case. When the addition is complete heat the solution under reflux on the water bath for 1 hour. Cool, wash the dichloromethane solution with two 100 ml portions of water, dry (MgSO$_4$) and remove the solvent on a rotary evaporator to obtain p-nitrotoluene, 2.9 g (71%), having m.p. *ca.* 49 °C. Recrystallisation from a small volume of ethanol gives a purer product, m.p. 51–52 °C.

C AROMATIC HALOGEN COMPOUNDS

The following procedures are available for the preparation of aromatic halogen compounds.

1. Direct halogenation (Sections **IV,26** to **IV,32**).
2. Chloromethylation (Section **IV,33**).
3. The replacement of a diazo group by a halogen (see Sections **IV,79** to **IV,82**).
4. The replacement of a hydroxyl group by a halogen (Section **IV,34**).
5. Methods leading to polyvalent iodine compounds (Sections **IV,35** to **IV,38**).

C,1. Direct halogenation. (*a*) *Nuclear substitution.* Benzene does not react appreciably with chlorine and bromine in the cold, but in the presence of catalysts, such as aluminium amalgam, pyridine or iron, reaction takes place readily, affording in the first instance the mono-halogenated derivative as the main product. Di-substituted products (largely the *para* isomer) are obtained if the proportion of the halogen is increased. Typical procedures are given in the preparation of chlorobenzene (Section **IV,26**) and bromobenzene (Section **IV,27**). The function of the catalyst is to increase the electrophilic activity of the halogen and the mechanism of the chlorination of benzene can be represented by the following scheme (cf. the nitration of benzene, p. 622, **IV,B,1**).

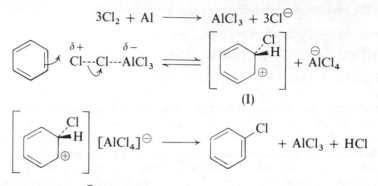

The base $[AlCl_4]^{\ominus}$ facilitates the removal of a proton from the mesomeric species (I). The reaction should preferably be carried out in the absence of direct sunlight, since this may promote direct addition of the halogen to the aromatic ring, particularly if the benzene is warm, to yield the hexachlorocyclohexane.

$$C_6H_6 + 3Cl_2 \longrightarrow C_6H_6Cl_6$$

Since iodine is the least active of the halogens, iodination does not occur unless the reaction is carried out in the presence of an oxidising agent such as fuming nitric acid, as in the preparation of iodobenzene (Section **IV,28**). The nature of the electrophile which functions as the $I^{\oplus}$ donor is uncertain and will depend on the nature of the oxidising agent employed; with nitric acid it is thought to be $[O = N(I)OH]^{\oplus}$ (Ref. 2).

Condensed aromatic hydrocarbons are more reactive towards electrophilic reagents, and naphthalene, for example, may be brominated quite readily in solution in carbon tetrachloride without the need for a catalyst; electrophilic attack takes place at the more reactive α-position to yield 1-bromonaphthalene (Section **IV,29**).

The effect of substituents in electrophilic aromatic substitution processes, which are outlined on p. 622, **IV,B,1**, are further illustrated here with reference to the bromination of nitrobenzene, where a substantially higher reaction temperature is required (the nitro group is deactivating) and the product is predominantly *m*-bromonitrobenzene (Section **IV,30**).

(*b*) *Side-chain halogenation.* In the absence of catalysts, treatment of toluene with chlorine (or bromine) at the boiling point, preferably with exposure to sunlight or other bright light source, results in halogenation in the side chain. The introduction of the first chlorine atom, for example, proceeds at a much faster rate than the introduction of the second chlorine atom so that in practice the major portion of the toluene is converted into benzyl chloride before appreciable chlorination of benzyl chloride occurs.

$$C_6H_5 \cdot CH_3 + Cl_2 \xrightarrow[\text{heat}]{h\nu} \underset{\text{Benzyl chloride}}{C_6H_5 \cdot CH_2Cl} \longrightarrow$$

$$\underset{\substack{\text{Benzylidene} \\ \text{chloride}}}{C_6H_5 \cdot CHCl_2} \longrightarrow \underset{\text{Benzotrichloride}}{C_6H_5 \cdot CCl_3}$$

The reaction proceeds by the radical mechanism shown; the first step is facilitated by the mesomeric stabilisation of the benzylic radical.

$$Cl_2 \rightleftharpoons 2Cl\cdot$$
$$C_6H_5 \cdot CH_3 + Cl\cdot \longrightarrow C_6H_5 \cdot \overset{\cdot}{C}H_2 + HCl$$
$$C_6H_5 \cdot \overset{\cdot}{C}H_2 + Cl_2 \longrightarrow C_6H_5 \cdot CH_2Cl + Cl\cdot$$

Rapid side-chain chlorination of toluene proceeds in the dark with sulphuryl chloride in the presence of benzoyl peroxide (0.001–0.005 mol per ml of SO_2Cl_2) as catalyst. With an excess of sulphuryl chloride, benzylidene chloride is formed, but in this case chlorination does not proceed beyond this stage.

These reactions are illustrated by the chlorination of toluene (Section **IV,31**, *Methods 1* and *2*) and by the bromination of *p*-nitrotoluene (Section **IV,32**).

IV,26. CHLOROBENZENE

$$C_6H_6 + Cl_2 \xrightarrow{Al/Hg} C_6H_5Cl$$

Into a 500-ml two-necked flask provided with an inlet tube (1) extending to within 1 cm of the bottom and a reflux (double surface) condenser connected with a device (Fig. I,68) for absorbing the hydrogen chloride gas subsequently evolved, place 220 g (250 ml, 2.83 mol) of sodium-dried benzene and 0.5 g of aluminium amalgam (2). Weigh the flask and contents. Immediately pass in dry chlorine from a cylinder through an intermediate empty wash bottle. An exothermic reaction occurs and much hydrogen chloride is evolved. Cool the flask by immersion in a bath of cold water and allow the chlorination to proceed until the liquid has increased in weight by 85 g. Pour the liquid into about 250 ml of cold water, separate the lower layer of chlorobenzene and wash it successively with dilute sodium hydroxide solution and water; dry with anhydrous calcium chloride or magnesium sulphate. Distil, using a well-lagged fractionating column (Section **I,26**), and collect the fraction, b.p. 127–135 °C (3). Redistil and collect the pure chlorobenzene at 131–132 °C. The yield is about 155 g (49%).

Notes. (1) A gas distribution tube, provided with a sintered glass plate at its lower end, is to be preferred.

(2) Prepared as described in Section **II,2,**2. After washing with water, the amalgam should first be washed with methanol and finally with dry benzene.

(3) The high boiling point residue contains *p*- (b.p. 173 °C, m.p. 53 °C) and *o*-dichlorobenzene (b.p. 179 °C). Upon cooling in ice, the moderately pure solid *para* isomer separates out.

IV,27. BROMOBENZENE

$$C_6H_6 + Br_2 \xrightarrow{\text{pyridine}} C_6H_5Br$$

Place 50 g (57 ml, 0.64 mol) of dry benzene and 0.5 ml of dry pyridine (1) (dried over potassium hydroxide pellets) in a 500-ml round-bottomed flask. Attach a reflux condenser to the flask and fit a device for absorbing the hydrogen bromide gas subsequently evolved (Fig. I,68(a)). Partially immerse the flask in a bath of cold water, supported upon a tripod and gauze. Carefully pour 125 g (40 ml) of bromine (for precautions to be taken with bromine, see Section II,2,7) through a condenser and immediately insert the absorption device into the upper end of the condenser. A vigorous reaction soon occurs and hydrogen bromide is evolved which is absorbed by the water in the beaker; when the reaction slackens, warm the bath to 25–30 °C for 1 hour. Finally raise the temperature of the bath to 65–70 °C for a further 45 minutes or until all the bromine has disappeared (no red vapours visible) and the evolution of hydrogen bromide has almost ceased. Transfer the dark-coloured reaction product to a separatory funnel and shake successively with water, with sufficient 5–10 per cent sodium hydroxide solution to ensure that the washings are alkaline to litmus, and finally with water. Dry with magnesium sulphate or anhydrous calcium chloride. Filter through a fluted filter-paper into a small distilling flask and distil slowly. Collect the crude bromobenzene at 150–170 °C; pour the residue whilst still hot into a small porcelain basin. Redistil the liquid of b.p. 150–170 °C (2) and collect the bromobenzene at 154–157 °C; the yield is about 60 g (60%).

Isolate the **pure p-dibromobenzene** from the residue in the basin by recrystallisation from hot ethanol with the addition of 1–2 g of decolourising charcoal; use about 4 ml of ethanol (industrial spirit) for each gram of material. Filter the hot solution through a fluted filter-paper, cool in ice and filter the crystals at the pump. The yield of p-dibromobenzene, m.p. 89 °C, is about 12 g.

Notes. (1) Other halogen carriers may be used, e.g., 1–2 g of iron filings, or 1 g of aluminium amalgam. The bromine must then be added slowly from a dropping funnel to the benzene warmed on a water bath; a suitable two- or three-necked flask should be used. After all the bromine has been introduced, the mixture is heated on a water bath until no red vapours are visible above the liquid. The subsequent procedure is as above.

(2) The best results are obtained by distillation from a small flask through a short fractionating column: a Hempel column filled with glass rings (Fig. I,105(c)) and lagged with asbestos cloth or several thicknesses of linen cloth is quite satisfactory.

IV,28. IODOBENZENE

$$2C_6H_6 + I_2 \xrightarrow[\text{HNO}_3]{\text{[O]}} 2C_6H_5I + H_2O$$

Equip a 500-ml three-necked flask with a reflux condenser, a sealed mechanical stirrer and separatory funnel, and support it on a water bath. Attach an absorption device (Fig. I,68) to the top of the condenser. Place 134 g (152 ml, 1.72 mol) of benzene and 127 g (0.5 mol) of iodine in the flask, and heat the water bath to about 50 °C; add 92 ml of fuming nitric acid, d 1.50, slowly from the separatory funnel during 30 minutes. Oxides of nitrogen are evolved in quantity. The temperature rises slowly without the application of heat until the mixture boils gently. When all the nitric acid has been introduced, reflux the mixture gently for 15 minutes. If iodine is still present, add more nitric acid to the warm solution until the purple colour (due to iodine) changes to brownish-red.

Separate the lower oily layer, mix with it an equal volume of 10 per cent sodium hydroxide solution and steam distil from a 1-litre flask until no more oil passes over. A yellow solid, consisting of nitro compounds, may collect towards the end of the distillation; remove this by mechanical stirring of the oil for about 3 hours with 7 ml of concentrated hydrochloric acid, 100 ml of water and 70 g of iron filings in a 1-litre three-necked flask connected with a reflux condenser. Allow the mixture to cool and filter. Render the filtrate distinctly acid to Congo red with hydrochloric acid and again steam distil. Separate the oil, dry it with anhydrous calcium chloride or magnesium sulphate, distil through a suitably lagged fractionating column and collect the fraction of b.p. 180–190 °C. Upon redistillation, pure iodobenzene, b.p. 184–186 °C, is obtained. The yield is 180 g (87%).

IV,29. 1-BROMONAPHTHALENE

$$C_{10}H_8 + Br_2 \xrightarrow{CCl_4} 1\text{-}C_{10}H_7Br + HBr$$

Use a 500-ml three-necked flask equipped as in Section **IV,30** but mounted on a water bath. Place 128 g (1 mol) of naphthalene and 45 ml of dry carbon tetrachloride in the flask, and 177 g (55 ml, 1.11 mol) of bromine in the separatory funnel. Heat the mixture to gentle boiling and run in the bromine at such a rate that little, if any, of it is carried over with the hydrogen bromide into the trap; this requires about 3 hours. Warm gently, with stirring, for a further 2 hours or until the evolution of hydrogen bromide ceases. Replace the reflux condenser by a condenser set for downward distillation, stir and distil off the carbon tetrachloride as completely as possible. Mix the residue with 8 g of sodium hydroxide pellets and stir at 90–100 °C for 3 hours; this treatment will remove impurities which gradually evolve hydrogen bromide. Distil the product under diminished pressure and collect the following fractions: (i) up to 131 °C/12 mmHg (or 144 °C/20 mmHg); (ii) 132–135 °C/12 mmHg (or 145–148 °C/20 mmHg); and (iii) above 135 °C/12 mmHg (or 148 °C/20 mmHg). Fraction (ii) is almost pure 1-bromonaphthalene. Fraction (i) contains unchanged naphthalene, whilst (iii) contains dibromonaphthalene. Cool fraction (i) in ice when most of the naphthalene will crystallise out; filter this off on a sintered glass funnel, combine the filtrate with fraction (iii), redistil and collect the 1-bromonaphthalene fraction separately. The total yield of colourless product is 150 g (72.5%).

IV,30. m-BROMONITROBENZENE

$$C_6H_5 \cdot NO_2 + Br_2 \xrightarrow{Fe} m\text{-}Br \cdot C_6H_4 \cdot NO_2 + HBr$$

Equip a 1-litre three-necked flask with a separatory funnel, a sealed mechanical stirrer (1) and a double surface reflux condenser carrying an outlet tube connected to a gas trap (Fig. I,68). Support the flask in an oil bath. Place 90 g (75 ml, 0.73 mol) of dry, freshly distilled nitrobenzene in the flask. Weigh out 10 g of pure iron powder (95%) 'reduced by hydrogen'. Heat the oil bath to 135–145 °C and introduce 3 g of the iron powder by temporarily removing the separatory funnel. Into the latter place 62.5 g (20 ml, 1.17 mol) of dry bromine (Section II,2,7) and run it into the flask at such a rate that bromine vapours do not rise appreciably in the condenser (ca. 20 minutes). Continue stirring and heating for 1 hour before adding a further 3 g of iron powder and 20 ml of dry bromine in a similar manner. Stir for a further hour, add another 3 g of iron powder and 20 ml of bromine. When there is no more bromine vapour in the condenser, make a final addition of 1 g of iron powder and heat for 1 hour longer.

Pour the resulting dark reddish-brown liquid into 500 ml of water to which 17 ml of saturated sodium metabisulphite solution has been added (the latter to remove the excess of bromine). Steam distil the resulting mixture (Fig. I,101); collect the first portion of the distillate, which contains a little unchanged nitrobenzene, separately. Collect about 4 litres of distillate. Filter the yellow crystalline solid at the pump, and press well to remove the adhering liquid. The resulting crude m-bromonitrobenzene, m.p. 51–52 °C, weighs 110 g (74%). If required pure, distil under reduced pressure (Section I,27) and collect the fraction of b.p. 117–118 °C/9 mmHg; it then melts at 56 °C and the recovery is about 85 per cent.

Note. (1) Mechanical stirring, although not essential and replaceable by occasional shaking by hand, is advantageous.

IV,31. BENZYL CHLORIDE

Method 1 $C_6H_5 \cdot CH_3 + Cl_2 \xrightarrow{h\nu} C_6H_5 \cdot CH_2Cl$

Method 2 $C_6H_5 \cdot CH_3 + SO_2Cl_2 \xrightarrow{peroxide} C_6H_5 \cdot CH_2Cl + SO_2 + Cl_2$

Method 1.   Fit a 500-ml three-necked flask with a thermometer (the bulb of which is within 2 cm of the bottom), an inlet tube extending to the bottom of the flask and a double surface condenser (1). Connect the top of the condenser through a calcium chloride guard-tube to two wash bottles containing 10 per cent sodium hydroxide solution: the long lead-in tubes in the wash bottles should be just above the surface of the alkali solution in order to avoid 'sucking back'. Place 100 g (115.5 ml, 1.09 mol) of dry toluene and a few chips of porous porcelain in the flask. Boil the toluene gently and pass in a stream of chlorine from a cylinder—interpose an empty wash bottle between the flask and the cylinder—until the thermometer registers 157–158 °C (2). The reaction time may be considerably shortened by exposing the mixture to bright sunlight or to a small mercury-vapour lamp; if neither of these is practicable, support a 200-watt lamp a few inches from the flask. Distil the reaction mixture first under atmospheric pressure until the temperature reaches 135–140 °C (3), and continue distillation under diminished pressure collecting the benzyl chloride at 64–69 °C/12 mmHg. The latter upon redistillation boils at 63–65 °C/12 mmHg. The yield of benzyl chloride is about 100 g (92.5%).

Notes. (1) Owing to the poisonous character of chlorine, the apparatus should be fitted up in the fume cupboard.

(2) An alternative method of determining the completion of the reaction is to weigh the flask and toluene, and to stop the passage of chlorine when the increase in weight is 37 g.

(3) The benzyl chloride may also be isolated by distillation under atmospheric pressure. The material boiling between 165 and 185 °C is collected and redistilled; the final product is collected at 178–182 °C (pure benzyl chloride has b.p. 179 °C). The resulting benzyl chloride is, however, of lower purity unless an efficient fractionating column is used.

Method 2. In a 500-ml round-bottomed flask, fitted with an efficient reflux condenser, place 92 g (106 ml, 1 mol) of toluene, 68 g (41 ml, 0.5 mol) of redistilled sulphuryl chloride and 1 g of benzoyl peroxide (Section **II,2,5**). Reflux gently, when a vigorous reaction takes place: the reaction is complete in 30 minutes. Isolate the benzyl chloride as described in *Method 1.* The yield is 50 g (79%).

Cognate preparation Benzylidene chloride (benzal chloride). Use 100 g of toluene and continue the passage of chlorine until the increase in weight of the flask and contents is 74 g or, alternatively, until the temperature rises to 187 °C. Collect the benzal chloride at 204–208 °C or at 104–105 °C/30 mmHg. Pure benzylidene chloride has b.p. 206 °C.

IV,32. *p*-NITROBENZYL BROMIDE

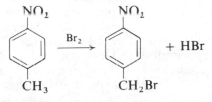

Place 150 g (1.1 mol) of *p*-nitrotoluene, m.p. 51–52 °C, in a 500-ml three-necked flask, fitted with a reflux condenser, a sealed mechanical stirrer and a separatory funnel with stem reaching nearly to the bottom of the flask. Attach a gas absorption trap (Fig. I,68) to the top of the condenser; the whole assembly should be sited in a fume cupboard. Heat the flask in an oil bath at 145–150 °C and add 184 g (59 ml, 1.15 mol) of bromine during 2 hours (1). Continue the stirring for an additional 10 minutes after all the bromine has been added. Pour the contents of the flask whilst still liquid (**CAUTION**) (2) into a 2.5 litre round-bottomed flask containing 2 litres of hot light petroleum, b.p. 80–100 °C, and 8 g of decolourising carbon. Attach a reflux condenser to the flask, heat it with the aid of a heating mantle until the material dissolves, boil for 10 minutes and filter rapidly through a pre-heated Buchner funnel. Cool the filtrate to 20 °C, filter the crystals with suction, press well and wash with two 25 ml portions of cold light petroleum. The crude *p*-nitrobenzyl bromide, m.p. 95–97 °C (150 g), is sufficiently pure for many purposes. Purify by dissolving in 1500–1700 ml of light petroleum, b.p. 80–100 °C, boil with 8 g of decolourising carbon, and filter through a pre-heated Buchner or sintered glass funnel. Cool the filtrate in ice, filter at the pump, drain well and wash with two 15 ml portions of cold light petroleum. The yield of pure *p*-nitrobenzyl bromide (pale yellow crystals, m.p. 98–99 °C) is 135 g (57%).

Notes.   (1) Improved yields may be obtained by exposing the flask to the light of two 300-watt tungsten lamps during the bromination.

(2) Care must be taken in manipulating the lachrymatory solutions of *p*-nitrobenzyl bromide. If the substance should come into contact with the skin, bathe the affected part with alcohol.

C,2. Chloromethylation.   This is the replacement of a hydrogen atom in an aromatic compound by a chloromethyl (CH_2Cl) group in a single operation. The reaction consists essentially of the interaction of formaldehyde and hydrogen chloride in the presence of a catalyst such as zinc chloride or aluminium chloride with an aromatic system (**Blanc chloromethylation reaction**). The reaction is similar in some respects to that of Friedel and Crafts (see **IV,A,1**(*c*) and **IV,L,1**) and involves the hydroxymethyl cation as the electrophilic species. This reacts with the aromatic ring to give the benzylic alcohol which is then converted into the chloromethyl derivative by hydrogen chloride.

$$H_2C=O + HCl \rightleftharpoons Cl^\ominus + [H_2C=\overset{\oplus}{O}H \longleftrightarrow H_2\overset{\oplus}{C}-OH]$$

$$ArH + \overset{\oplus}{C}H_2OH \longrightarrow Ar{\cdot}CH_2OH + H^\oplus$$

$$Ar{\cdot}CH_2OH + HCl \longrightarrow Ar{\cdot}CH_2Cl + H_2O$$

However, a by-product arising from the interaction of formaldehyde and hydrogen chloride is bis(chloromethyl)ether, a potent carcinogen. For this reason chloromethylation as a route to side-chain halogenated products should not be regarded as a desirable synthetic procedure, and in general should only be used if the required compound cannot be readily prepared by other methods. When used, as in Section **IV,33**, for the preparation of 1,2-bis(chloromethyl)-4,5-dimethylbenzene, effective precautions should be taken during the reaction and in the disposal of the reaction residues, as specified in the experimental description.

IV,33. 1,2-BIS(CHLOROMETHYL)-4,5-DIMETHYLBENZENE

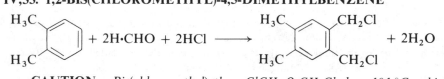

CAUTION.   *Bis(chloromethyl)ether, $ClCH_2{\cdot}O{\cdot}CH_2Cl$, b.p. 101 °C, which is known to be formed by the interaction of formaldehyde and hydrochloric acid under suitable conditions* (Ref. 3), *is a likely by-product of chloromethylation. Low concentrations cause severe irritation of the eyes and respiratory tract and the compound has been shown to be a potent cause of lung cancer in experimental animals* (Ref. 4).

Chloromethylation reactions should therefore only be performed with great caution and the entire procedure carried out well inside an efficient fume cupboard; plastic disposable gloves should be worn. Reaction residues should be carefully disposed of by alkaline hydrolysis as indicated in the note to the text.

Place 35 g (0.33 mol) of *o*-xylene, 30 g (1 mol HCHO) of paraformaldehyde and 200 ml of concentrated hydrochloric acid in a 500-ml two-necked flask fitted with a reflux condenser and a sealed stirrer unit. Fit a heating mantle and boil the mixture under reflux for 22 hours. Cool, decant the supernatant liquor, and wash the waxy residue twice by decantation with water (1). Add 100 ml of hexane, and heat under reflux until the residue has dissolved. Decant the hexane extract from the small quantity of water into a preheated conical flask containing about 2 g of magnesium sulphate desiccant, and filter while still hot through a fluted filter-paper in a preheated funnel. Cool the filtrate in ice and collect the product which separates (1) and wash it with a little cold hexane; the yield is 20 g (30%). After recrystallisation from hexane the m.p. is 100 °C.

Note. (1) The combined aqueous phase should be made strongly alkaline by the gradual addition, with stirring, or 40 per cent w/v aqueous sodium hydroxide, and then set aside for 24 hours before being flushed down the drain. The hexane mother-liquor should be shaken well with the strong sodium hydroxide solution, and allowed to stand before disposal.

C,3. The replacement of a diazo group by a halogen. The most generally applicable route to nuclear substituted aromatic halogen compounds involves decomposition of a diazonium salt under suitable conditions. These reactions are discussed on p. 687, **IV,G,1**.

C,4. The replacement of a hydroxyl group by a halogen. Replacement of the hydroxyl group in a phenol by halogen cannot be accomplished by reaction with the hydrogen halides as in the case of alcohols, and reaction with phosphorus halides gives only low yields of halogenobenzenes (except in the case of nitrophenols), the main product being a phosphite or phosphate ester.

$$ArOH + PCl_3 \longrightarrow (ArO)_3P + 3HCl$$

However, if the phenol is first treated with the complex formed from triphenylphosphine and a halogen in acetonitrile solution, an aryloxytriphenylphosphonium halide is formed which on thermal decomposition yields the aryl halide in good yield (e.g., the preparation of p-bromochlorobenzene, Section **IV,34**).

$$(C_6H_5)_3P + X_2 \longrightarrow (C_6H_5)_3PX_2 \xrightarrow{ArOH}$$

$$(C_6H_5)_3\overset{\oplus}{P}(OAr)\}\overset{\ominus}{X} + HX \longrightarrow ArX + (C_6H_5)_3P{=}O + HX$$

IV,34. p-BROMOCHLOROBENZENE

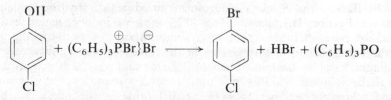

This preparation should be carried out in the fume cupboard.

Equip a 250-ml three-necked round-bottomed flask with a reflux condenser and a dropping funnel (both protected by calcium chloride guard-tubes), and a sealed stirrer unit. In the flask place 29 g (0.11 mol) of finely powdered triphenylphosphine and 100 ml of dry acetonitrile (Section **II,1,**27). Cool the flask in an ice-water bath and add with stirring over a period of 20 minutes 17.3 g (5.5 ml, 0.108 mol) of bromine. After the addition is complete remove the cooling bath, set the condenser for downward distillation and replace the dropping funnel with a stopper. Remove the acetonitrile under reduced pressure using a water pump; warm the flask to 40 °C on a water bath and use an oil pump to remove last traces of solvent. Add 10 g (0.078 mol) of redistilled, powdered *p*-chlorophenol to the solid residue, replace the condenser (together with its guard-tube) in the reflux position and site the flask on a sand bath. Raise the temperature of the sand bath to between 250 and 280 °C and slowly stir the contents of the flask which soon melt; hydrogen bromide is evolved over a period of 3 hours. When evolution of gas ceases, cool the residue, add water and steam distil. Collect the solid *p*-bromochlorobenzene by filtration of the distillate; it has m.p. 63–65 °C. Recrystallise from ethanol to obtain the pure product, m.p. 65–66 °C; the yield is 12.5 g (83%).

C,5. Methods leading to polyvalent iodine compounds. Aryl iodides are exceptional amongst the aromatic halogen compounds in that they form a series of derivatives in which the iodine exhibits a covalency greater than one. Some typical interconversions for which experimental conditions are given (Sections **IV,35** to **IV,38**) are summarised below.

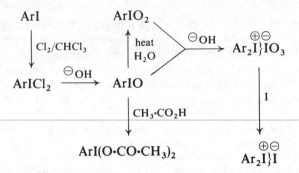

IV,35. (DICHLOROIODO)BENZENE (*Iodobenzene dichloride*)

$$C_6H_5I + Cl_2 \longrightarrow C_6H_5ICl_2$$

Equip a 500-ml three-necked flask with a mechanical stirrer, an adjustable inlet tube at least 10 mm in diameter for the introduction of chlorine (see Section **II,2,**11) and an outlet tube carrying a calcium chloride guard-tube. Charge the flask with 75 ml of chloroform (dried with anhydrous calcium chloride: see Section **II,1,**6) and 51 g (0.25 mol) of iodobenzene (Section **IV,28**); adjust the inlet tube so that it terminates about 5 mm above the surface of the liquid. Set up the apparatus in the fume cupboard and protect it from the light. Cool the flask in an ice–salt mixture and pass in dry chlorine as rapidly as the solution will absorb it until an excess is present (1.5–2 hours). Filter the yellow, crystalline iodobenzene dichloride at the pump, wash it sparingly with chloroform and dry it in the air upon filter-paper. The yield is 65 g (93.5%). The substance decomposes slowly upon standing; it may be kept unchanged for a short period in a well-fitting, ground glass stoppered bottle.

IV,36. IODOSYBENZENE   (*Iodosolbenzene*)   AND   IODOBENZENE DIACETATE   [(*Diacetoxyiodo*)*benzene*]

$$C_6H_5ICl_2 + 2NaOH \longrightarrow C_6H_5IO + 2NaCl + H_2O$$
$$C_6H_5IO + 2CH_3 \cdot CO_2H \longrightarrow C_6H_5I(O \cdot CO \cdot CH_3)_2 + H_2O$$

Cool a large glass mortar in ice and then place in it 50 g of anhydrous sodium carbonate, 55 g (0.2 mol) of iodobenzene dichloride (Section **IV,35**) and 100 g of finely crushed ice. Grind the mixture thoroughly until all the ice has melted and a thick paste results. Add 140 ml of 5 *M*-sodium hydroxide in 20 ml portions and triturate vigorously after each addition; finally add 120 ml of water and allow to stand overnight. Filter with suction, press well with a large flat glass stopper on the filter, transfer to a beaker and stir with 300 ml of water (1). Filter again at the pump, transfer again to a beaker containing 300 ml of water, filter and wash with about 200 ml of water on the filter. Dry in the air upon filter-papers, stir with a little chloroform (to dissolve a little iodobenzene which is present), filter with suction and dry on filter-paper in the air. The yield is 27 g (61%).

Iodosobenzene explodes violently at about 220 °C, so that determinations of the melting point should not be attempted. It may, however, be converted into **iodobenzene diacetate** in the following manner. Dissolve 2 g of iodosobenzene in 6 ml of glacial acetic acid; boiling is usually necessary. Cool. The resulting diacetate is readily soluble in acetic acid but is insoluble in ether. Add about 50 ml of ether in order to precipitate the iodobenzene diacetate. Filter and wash with ether. The yield is 2 g, m.p. 157 °C. It may be recrystallised from benzene, and will keep indefinitely (unlike the iodobenzene dichloride).

Note.   (1) The filtrate contains some diphenyliodonium salts; these may be recovered as the sparingly soluble diphenyliodonium iodide (about 8 g) (Section **IV,38**) by the addition of potassium iodide.

IV,37. IODYLBENZENE (*Iodoxylbenzene*)

$$2C_6H_5IO \xrightarrow{\text{heat}} C_6H_5IO_2 + C_6H_5I$$

Fit up a 1-litre round-bottomed flask for steam distillation (Fig. I,101) and place in it 22 g (0.1 mol) of iodosybenzene (Section **IV,36**) made into a thin paste with water (1). Steam distil until almost all the iodobenzene has been removed (about 9 g); cool the residue in the flask at once, filter the white solid with suction and dry in the air. Wash it with a little chloroform, filter with suction and dry in the air upon filter-paper. The yield is 10.5 g (89%). It may be recrystallised from 800–900 ml of water. Iodylbenzene melts with explosive decomposition at 237 °C.

Note.   (1) Iodosybenzene when heated directly may decompose with explosive violence, particularly when dry.

IV,38. DIPHENYLIODONIUM IODIDE

$$C_6H_5IO + C_6H_5IO_2 \longrightarrow [(C_6H_5)_2\overset{\oplus\ominus}{I}]IO_3$$
$$[(C_6H_5)_2\overset{\oplus\ominus}{I}]IO_3 \xrightarrow{\text{KI}} [(C_6H_5)_2\overset{\oplus\ominus}{I}]I$$

Grind together 12 g (0.05 mol) of iodylbenzene (Section **IV,37**), 11 g (0.05 mol) of iodosybenzene (Section **IV,36**) with 25 ml of water, add 100 ml of 1 *M*-

sodium hydroxide solution and stir for 24 hours in a 1-litre vessel. Dilute with 500 ml of cold water, stir thoroughly, allow to settle and decant the supernatant solution of diphenyliodonium iodate, through a fluted filter-paper. Extract the solid residue with two 250 ml portions of water, and decant the extract through a fluted filter-paper: a small tarry residue remains. To the combined filtrates add an aqueous solution containing 10 g of potassium iodide. Allow the bulky white precipitate of diphenyliodonium iodide to stand for 1.5 hours with occasional shaking, and then filter it with suction. Dry on a porous tile. The yield is 15 g (74%). The product melts at 173–175 °C with vigorous decomposition.

D AROMATIC SULPHONIC ACIDS AND THEIR DERIVATIVES

The reaction processes described in this section are as follows.

1. The preparation of arylsulphonic acids by direct sulphonation (Sections **IV,39** to **IV,44**).
2. The preparation of arylsulphonyl chlorides (Sections **IV,45** and **IV,46**).
3. The preparation of arylsulphonamides from arylsulphonyl chlorides (Sections **IV,47** to **IV,50**).
4. The preparation of arylsulphonate esters from arylsulphonyl chlorides (Section **IV,51**).
5. Reduction products from arylsulphonyl chlorides (Sections **IV,52** and **IV,53**).

D,1. Direct sulphonation. Aromatic hydrocarbons may be mono-sulphonated by heating with a slight excess of concentrated sulphuric acid; for benzene, oleum (7–8% SO_3) gives somewhat better results. The reaction is usually complete when all the hydrocarbon has dissolved.

$$ArH + H_2SO_4 \rightleftharpoons Ar \cdot SO_3H + H_2O$$

The mechanism of aromatic sulphonation is broadly analogous to that previously described for aromatic nitration and halogenation and may be represented in the following way, the neutral sulphur trioxide molecule functioning as the electrophilic species. Sulphonation differs from nitration and halogenation, however, in that the overall reaction is reversible.

$$2H_2SO_4 \rightleftharpoons SO_3 + H_3O^{\oplus} + HSO_4^{\ominus}$$

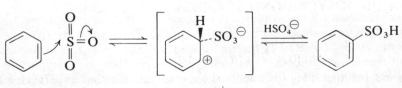

mesomeric

Because of their high solubility in water the sulphonic acids are not usually isolated from aqueous solution in the free state, but are converted into and isolated as their sodium salts. The simplest procedure is to partially neutralise the reaction mixture (say, with sodium hydrogen carbonate) and then to pour it into water and add an excess of sodium chloride, when the following equilibrium is established.

$$Ar \cdot SO_3H + NaCl \rightleftharpoons Ar \cdot SO_3Na + HCl$$

The high sodium ion concentration results in crystallisation of the sodium salt. This process of salting out with common salt may be used for recrystallisation, but sodium benzenesulphonate (Section **IV,39**) (and salts of other acids of comparable molecular weight) is so very soluble in water that the solution must be almost saturated with sodium chloride, and consequently the product is likely to be contaminated with it. In such a case a pure product may be obtained by crystallisation from, or Soxhlet extraction with, absolute ethanol; the sulphonate salt is slightly soluble but the inorganic salts are almost insoluble. Very small amounts of sulphones are formed as by-products, but since these are insoluble in water, they separate when the reaction mixture is poured into water.

$$Ar \cdot SO_3H + ArH \longrightarrow Ar \cdot SO_2 \cdot Ar + H_2O$$

The sulphonation of toluene (Section **IV,40**) with concentrated sulphuric acid at 100–120 °C results in the formation of toluene-*p*-sulphonic acid as the chief product, accompanied by small amounts of the *ortho* and *meta* isomers; these are easily removed by crystallisation of the sodium salt of the *para* isomer in the presence of sodium chloride. Sulphonation of naphthalene at about 160 °C yields largely the 2-sulphonic acid (the product of thermodynamic control) (Section **IV,41**), at lower temperatures (0–60 °C) the 1-sulphonic acid (the product of kinetic control) is produced almost exclusively. In both cases the product is isolated as its sodium salt. In anthraquinone the carbonyl groups deactivate the aromatic nucleus towards electrophilic attack and vigorous conditions of sulphonation are required, i.e., oleum at about 160 °C. The product is largely sodium anthraquinone-2-sulphonate (Section **IV,42**).

The free sulphonic acids (e.g., toluene-*p*-sulphonic acid, Section **IV,43**), as opposed to their sodium salts, may sometimes be obtained directly if the sulphonation reaction is carried out with continuous removal of the water formed in the reaction, conveniently by using a Dean and Stark water separator. *p*-Xylene-2-sulphonic acid (Section **IV,44**) is an example of a sulphonic acid whose solubility in water is such that it crystallises directly from the reaction medium and hence it may readily be isolated.

IV,39. SODIUM BENZENESULPHONATE

$$C_6H_6 + H_2SO_4 \longrightarrow C_6H_5 \cdot SO_3H$$

$$C_6H_5 \cdot SO_3H + NaCl \longrightarrow C_6H_5 \cdot \overset{\ominus}{S}O_3\overset{\oplus}{Na} + HCl$$

Into a 200-ml round-bottomed flask place 75 g (40 ml) of fuming sulphuric acid, *d* 1.88, containing 7–8 per cent of sulphur trioxide. Add, with frequent shaking, 20 g (22.5 ml, 0.25 mol) pure benzene (Section **II,1,2**) in portions of about 3 ml during about 15 minutes. Make sure that the first portion has dissolved before adding the next portion, etc.; maintain the temperature of the reaction mixture between 30 and 50 °C, and cool in a vessel of cold water if necessary. When all the benzene has completely reacted, cool and pour the reaction mixture slowly and with constant stirring into about 200 ml of water. Cool to the laboratory temperature and, if necessary, filter from any diphenyl-sulphone (a by-product) which may separate. Partially neutralise the acid solution by adding carefully and in small portions 24 g of sodium hydrogen carbonate: then add 40 g of sodium chloride and heat until it dissolves. Filter the hot solution with suction through a Buchner funnel (previously warmed in an oven or by pouring boiling water through it), transfer the warm filtrate to a beaker and cool rapidly (ice and cold water) with stirring. Filter the sodium benzenesulphonate which separates on a Buchner funnel and press well with a wide glass stopper; wash with about 30 ml of a filtered saturated sodium chloride solution and press the crystals as dry as possible. Finally wash with a little alcohol. Dry in the air upon filter-paper, powder and dry in the oven at 100–110 °C. The yield of the dry sodium benzenesulphate is about 20 g (44%). The product contains traces of sodium chloride and other salts, but is pure enough for most purposes. The impurities may be completely removed by recrystallisation from rectified spirit; about 18 ml are required for each gram of solid. The volume of alcohol for recrystallisation may be considerably reduced by the use of a Soxhlet extractor (Figs. I,96 and I,97).

Cognate preparation **Sodium p-bromobenzenesulphonate.** Equip a 500-ml three-necked flask with a separatory funnel, an unsealed stirrer guide fitted with a mechanical stirrer and a thermometer. Place 75 g (40 ml) of fuming sulphuric acid, d 1.88 (7–8% SO_3), in the flask and 40 g (27 ml, 0.25 mol) of bromobenzene (Section **IV,27**) in the separatory funnel. Add the bromobenzene in small portions so that the temperature does not rise above 100 °C. If any bromobenzene remains unattacked, warm the mixture on a water bath until all of it has passed into solution. Allow to cool, and pour the reaction mixture in a thin stream with stirring into 140 ml of cold water. If a precipitate separates (dibromodiphenylsulphone, a by-product), filter the warm solution at the pump. Add 55 g of sodium chloride to the filtrate and heat (with stirring) until the salt dissolves. Cool the solution rapidly with stirring, filter the separated crude sulphonate at the pump and press the crystals as dry as possible. Upon drying in the air, the yield is 47 g. To purify the crude sodium p-bromobenzenesulphonate, powder the crystals in a mortar, transfer to a beaker, add 75 ml of a filtered, saturated solution of sodium chloride, stir, heat on a water bath for 30 minutes, allow to cool, filter and press the crystals as dry as possible; finally wash with a little alcohol. Dry in the air by spreading upon filter-papers. The yield of purified sodium p-bromobenzenesulphonate is 45 g (68%). The product, although pure enough for most practical purposes, contains traces of sodium chloride and other salts: these can be removed either by recrystallisation from hot rectified spirit (1 g of salt requires *ca.* 25 ml of alcohol) or, more economically, by extraction with alcohol in a Soxhlet apparatus (Figs. I,96 and I,97).

IV,40. SODIUM TOLUENE-p-SULPHONATE

$$C_6H_5\cdot CH_3 + H_2SO_4 \longrightarrow p\text{-}CH_3\cdot C_6H_4\cdot SO_3H$$

$$p\text{-}CH_3\cdot C_6H_4\cdot SO_3H + NaCl \longrightarrow p\text{-}CH_3\cdot C_6H_4\cdot \overset{\ominus}{S}\overset{\oplus}{O_3}Na + HCl$$

Into a 500-ml three-necked flask, provided with a sealed mechanical stirrer and a reflux condenser, place 60 g (69 ml, 0.65 mol) pure toluene (Section **II,1,3**) and 60 g (33 ml) of concentrated sulphuric acid. Heat the mixture, with stirring, in an oil bath maintained at 110–120 °C. When the toluene layer has disappeared (ca. 1 hour), allow the reaction mixture to cool to room temperature. Pour it with stirring into 250 ml of cold water; filter from any solid substance which may separate. Partly neutralise the acid solution by adding cautiously and in small portions 30 g of sodium hydrogen carbonate. Heat the solution to boiling and saturate it with sodium chloride (about 100 g of salt are required), filter hot through a hot water funnel (Fig. I,41) or through a Buchner funnel previously warmed to about 100 °C. Transfer the hot filtrate to a beaker and cool the solution, with stirring, in ice. Filter the crystals at the pump (rinse any residual crystals out of the beaker with a little of the filtered mother-liquor), press well with a large glass stopper and wash with 30 ml of saturated salt solution. To recrystallise the crude sodium toluene-p-sulphonate, dissolve it in 200–250 ml of water, heat to boiling, saturate with salt, allow to cool somewhat, stir with 2–3 g of decolourising charcoal (if the solution is coloured) and filter the hot solution with suction through a previously warmed Buchner funnel. Transfer the warm filtrate to a beaker and cool in ice: collect the sulphonate with suction on a Buchner funnel, wash it with 20 ml of saturated sodium chloride solution, press well and finally wash with a little alcohol. Dry the hydrated crystals in air upon filter-papers, powder in a mortar and then dry in an oven or in an air oven at 100–110 °C. The yield of anhydrous sodium toluene-p-sulphonate is 50 g (40%). It still contains traces of sodium chloride and other salts; these can be removed by recrystallisation from rectified spirit (1 g of solid to about 40 ml of alcohol) or by extraction with boiling alcohol in a Soxhlet apparatus (Figs. I,96 and (I,97).

IV,41. SODIUM NAPHTHALENE-2-SULPHONATE

$$C_{10}H_8 + H_2SO_4 \longrightarrow 2\text{-}C_{10}H_7\cdot SO_3H$$

$$2\text{-}C_{10}H_7\cdot SO_3H + NaCl \longrightarrow 2\text{-}C_{10}H_7\cdot \overset{\ominus}{SO_3}\overset{\oplus}{Na} + HCl$$

Equip a 500-ml three-necked flask with a separatory funnel, a thermometer with its bulb about 2 cm from the bottom and a mechanical stirrer; the bearing for the stirrer consists of a glass tube lubricated with a little glycerine. Place 100 g (0.78 mol) of naphthalene in the flask and heat it either in an air bath or by means of a free flame. When the naphthalene melts, start the stirrer and adjust the heating so that the temperature is 160 ± 5 °C. Run in 166 g (90 ml) of concentrated sulphuric acid from the funnel during 5–6 minutes: take care to maintain the temperature at 160 °C and remove the flame if necessary. Stir for 5 minutes and pour the solution into 750 ml of cold water. If the sulphonation has been properly conducted, there will be no precipitate of naphthalene but about 4 g of insoluble di-2-naphthyl sulphone may separate. Boil with 3–4 g of decolourising carbon and filter with suction through a Buchner funnel. Partly neutralise the clear solution by carefully adding 40 g of sodium hydrogen carbonate in small portions. Heat the solution to the boiling point, saturate with sodium chloride (about 70 g are required) and then set aside to crystallise. Filter the crude sodium naphthalene-2-sulphonate at the pump and recrystallise from hot 10 per cent sodium chloride solution; dry by heating on a water bath or in an oven. The yield is 140 g (78%).

IV,42. SODIUM ANTHRAQUINONE-2-SULPHONATE

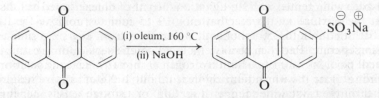

Place 50 g of fuming sulphuric acid (40–50% SO₃, Section **II,2,62**) in a 250- or 500-ml two-necked flask equipped with a thermometer with the bulb within 2 cm of the bottom and add 50 g (0.24 mol) of dry, finely powdered anthraquinone (Section **IV,136**). Fit an air condenser to the flask and heat the mixture slowly in an oil bath, with occasional shaking, so that at the end of 1 hour the temperature has reached 160 °C. Allow to cool and pour the warm mixture carefully into a 2-litre beaker containing 500 g of crushed ice. Boil for about 15 minutes and filter off the unchanged anthraquinone at the pump. Neutralise the filtrate by carefully adding concentrated (50% w/w) aqueous sodium hydroxide and allow to cool, when the greater part of the sodium anthraquinone-2-sulphonate separates as silvery glistening plates ('silver salt'). Filter these with suction and dry upon filter-paper or upon a porous plate. The yield is 40–45 g (54–60%).

IV,43. TOLUENE-*p*-SULPHONIC ACID

$$C_6H_5 \cdot CH_3 \xrightarrow{H_2SO_4} p\text{-}CH_3 \cdot C_6H_4 \cdot SO_3H$$

Use the apparatus employed for *Dibutyl ether* (Fig. III,2); it is advantageous to have the water separator tube calibrated (as in the Dean and Stark apparatus), otherwise place sufficient water in *A* so that with a further 9 ml the water level is at *B*. Place 87 g (100 ml, 0.95 mol) of pure toluene (Section **II,1,3**) and 37 g (20 ml) of concentrated sulphuric acid (92% H₂SO₄ by weight) in the 250- or 300-ml bolt-head flask and heat to gentle boiling. When 9 ml of water have been collected in the water separator tube (4–5 hours), extinguish the flame. The water is derived partly from the reaction (6.25 ml) and partly from the sulphuric acid. Add 6.3 ml of water to the cold contents of the flask: crystallisation then occurs. Transfer the product to a sintered glass funnel and remove the toluene and toluene-*o*-sulphonic acid by applying suction and pressing well with a glass stopper. Dissolve the residual solid (47 g) in about 22 ml of water and saturate the solution with hydrogen chloride gas; use any convenient device (e.g., a small funnel) to prevent 'sucking back'. After several hours the acid crystallises out as colourless prisms. Filter rapidly through a sintered glass funnel, wash with a little concentrated hydrochloric acid and dry in a vacuum desiccator charged with potassium hydroxide and anhydrous calcium chloride. The yield is 35 g (22%), m.p. 105–106 °C (sealed tube).

IV,44. 2,4-DIMETHYLBENZENESULPHONIC ACID (p-*Xylene-2-sulphonic acid*)

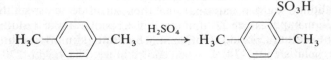

Place 5.2 g (6.0 ml, 0.05 mol) of *p*-xylene in a 25- or 50-ml round-bottomed flask and add, with gentle swirling agitation, 10 ml of concentrated sulphuric acid. Heat the mixture on a water bath for 10–15 minutes; remove the flask from the bath and mix the contents with a circular motion every two minutes. The reaction is complete when the xylene layer on the surface of the acid has disappeared. Cool to room temperature, add 5.0 ml of water cautiously with gentle swirling. Pour the warm reaction mixture into a 100-ml beaker and cool in ice. Filter off the crystalline solid with suction on a sintered glass funnel and press the crystals down with a glass stopper. Recrystallise the crude product from 5 ml of water and dry on filter-paper. The yield of pure 2,4-dimethyl-benzenesulphonic acid, m.p. 82 °C, is 8.2 g (90%).

D,2. The preparation of arylsulphonyl chlorides. Arylsulphonic acids, either free or in the form of their sodium salts, are converted into the acid chloride by reaction with phosphorus pentachloride (or phosphorus oxy-chloride).

$$3Ar\cdot SO_3Na + PCl_5 \xrightarrow{170-180\ °C} 3Ar\cdot SO_2Cl + NaPO_3 + 2NaCl$$

The arylsulphonyl chlorides may also be obtained from the aromatic hydro-carbon by reaction with an excess of chlorosulphonic acid **(chlorosulphona-tion)**.

$$ArH + 2Cl\cdot SO_3H \xrightarrow{20-25\ °C} Ar\cdot SO_2Cl + H_2SO_4 + HCl$$

These general procedures are illustrated by the preparation of benzenesul-phonyl chloride (Section **IV,45**). In the case of toluene, the mixture of toluene-*o*- and toluene-*p*-sulphonyl chlorides produced may be separated by cooling to −10 to −20 °C when most of the *para* isomer, which is a solid, m.p. 69 °C, separates out (Section **IV,46**).

IV,45. BENZENESULPHONYL CHLORIDE

Method 1 $C_6H_6 + 2Cl\cdot SO_3H \longrightarrow C_6H_5\cdot SO_2Cl + H_2SO_4 + HCl$

Method 2 $C_6H_5\cdot S\overset{\ominus}{O}_3\overset{\oplus}{Na} \xrightarrow[\text{or POCl}_3]{PCl_5} C_6H_5\cdot SO_2Cl$

Method 1. Equip a 1-litre three-necked flask with a separatory funnel, a sealed stirrer and a thermometer (with bulb within 2 cm from the bottom) attached by means of an adapter with T-connection (Fig. IV,2) leading to a gas absorption device (Fig. I,68). Place 700 g (400 ml, 6 mol) of chlorosulphonic acid (**CAUTION:** see Section **II,2,***12*) in the flask and add slowly, with stirring, 156 g (176 ml, 2 mol) of pure benzene (1); maintain the temperature between 20 and 25 °C by immersing the flask in cold water, if necessary. After the addition is complete (about 2.5 hours), stir the mixture for 1 hour, and then pour it on to 1500 g of crushed ice. Add 200 ml of carbon tetrachloride, stir and separate the lower layer as soon as possible (otherwise appreciable hydrolysis occurs); ex-tract the aqueous layer with 100 ml of carbon tetrachloride. Wash the com-bined extracts with dilute sodium carbonate solution, distil off most of the solvent under atmospheric pressure (2) and distil the residue under reduced pressure. Collect the benzenesulphonyl chloride at 118–120 °C/15 mmHg; it soli-difies to a colourless solid, m.p. 13–14 °C, when cooled in ice. The yield is 270 g

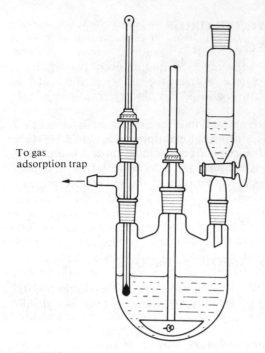

To gas
adsorption trap

Fig. IV,2

(76%). A small amount (10–20 g) of diphenylsulphone, b.p. 225 °C/10 mmHg, m.p. 128 °C, remains in the flask.

Notes. (1) A 50 per cent excess of chlorosulphonic acid is used; a smaller excess leads to increased formation of diphenylsulphone at the expense of the sulphonyl chloride.

(2) Any water present will distil with the carbon tetrachloride; the hydrolysis of the chloride is thus avoided.

Method 2. Place 90 g (0.5 mol) of sodium benzenesulphonate (Section **IV,39**) (previously dried at 130–140 °C for 3 hours) and 50 g (0.24 mol) of powdered phosphorus pentachloride (1) in a 500-ml round-bottomed flask furnished with a reflux condenser; heat the mixture in an oil bath at 170–180 °C for 12–15 hours. Every 3 hours remove the flask from the oil bath, allow to cool for 15–20 minutes, stopper and shake thoroughly until the mass becomes pasty. At the end of the heating period, allow the reaction mixture to cool. Pour on to 1 kg of crushed ice. Extract the crude benzenesulphonyl chloride with 150 ml of carbon tetrachloride and the aqueous layer with 75 ml of the same solvent. Remove the solvent under atmospheric pressure and proceed as in *Method 1.* The yield is about 70 g (80%) but depends upon the purity of the original sodium benzenesulphonate.

Note. (1) Alternatively a mixture of 90 g (0.5 mol) of sodium benzenesulphonate and 60 g (36 ml, 0.39 mol) of phosphorus oxychloride may be used. The experimental procedure is identical with that for phosphorus pentachloride, but the yield is slightly better.

IV,46. TOLUENE-p-SULPHONYL CHLORIDE

Method 1 $CH_3 \cdot C_6H_5 + 2Cl \cdot SO_3H \longrightarrow$
o- and *p*-$CH_3 \cdot C_6H_4 \cdot SO_2Cl + H_2SO_4 + HCl$

Method 2 p-$CH_3 \cdot C_6H_4 \cdot \overset{\ominus}{S}O_3\overset{\oplus}{N}a \xrightarrow[\text{or POCl}_3]{PCl_5} p$-$CH_3 \cdot C_6H_4 \cdot SO_2Cl$

Method 1. In a 750-ml three-necked flask equipped as in Fig. IV,2 (*Method 1*) place 400 g (228 ml, 3.44 mol) of chlorosulphonic acid (**CAUTION:** see Section **II,2,***12*) and cool to 0 °C in a freezing mixture of ice and salt. Introduce 100 g (115 ml, 1.09 mol) of pure dry toluene from the dropping funnel dropwise at such a rate that the temperature of the well-stirred mixture does not rise above 5 °C. When all the toluene has been added (about 3 hours), stir the reaction mixture for 4 hours, and then allow to stand overnight in a refrigerator. Pour the liquid on to 1 kg of crushed ice, separate the aqueous solution from the oily layer (mixture of toluene-*o*- and -*p*-sulphonyl chlorides) and wash the latter several times by decantation with cold water. To separate the *ortho* and *para* isomers, cool the oil at −10 to −20 °C (e.g., with ice and calcium chloride) for several hours; the almost pure toluene-*p*-sulphonyl chloride will crystallise out. Filter at the pump upon a sintered glass funnel. The crude toluene-*p*-sulphonyl chloride (30 g, 14%) may be purified by recrystallisation from light petroleum (b.p. 40–60 °C) and then melts at 69 °C. The filtrate consists largely of **toluene-*o*-sulphonyl chloride**: it may be obtained pure by dissolving it in carbon tetrachloride, removing the solvent and fractionating under reduced pressure; it is an oil, b.p. 126 °C/10 mmHg. The yield is about 120 g (58%).

Method 2. The procedure described under *benzenesulphonyl chloride*, *Method 2* (Section **IV,45**), may be used with suitable adjustment for the difference in molecular weights between sodium toluene-*p*-sulphonate (Section **IV,40**) and sodium benzenesulphonate. When the reaction product is poured on to ice, the toluene-*p*-sulphonyl chloride separates as a solid. This is filtered with suction; it may be recrystallised from light petroleum (b.p. 40–60 °C) and then melts at 69 °C.

D,3. The preparation of arylsulphonamides. Arylsulphonyl chlorides may be readily converted into the corresponding arylsulphonamides (e.g., by treatment with solid ammonium carbonate or with concentrated ammonia solution) or into a substituted arylsulphonamide (by reaction with the appropriate amino compound).

$Ar \cdot SO_2Cl + NH_3 \longrightarrow Ar \cdot SO_2NH_2 + HCl$
$Ar \cdot SO_2Cl + RNH_2 \longrightarrow Ar \cdot SO_2NHR + HCl$

These sulphonamides are highly crystalline, and in particular the derivatives from benzenesulphonyl chloride and toluene-*p*-sulphonyl chloride are used in the separation and identification of aromatic amines (Sections **VII,6,***21* and **VII,7,***1*).

Toluene-*o*-sulphonamide is an intermediate in the synthesis of saccharin (Section **IV,47**). Upon oxidising toluene-*o*-sulphonamide with potassium permanganate in alkaline solution, the sodium salt of *o*-sulphonamidobenzoic acid is formed, which on acidification with concentrated hydrochloric acid passes spontaneously into the cyclic imide of *o*-sulphobenzoic acid or saccharin.

Saccharin itself is sparingly soluble in cold water, but the imino hydrogen is acidic and the compound forms a water-soluble sodium salt. The latter is about 500 times as sweet as cane sugar.

When toluene-*p*-sulphonamide is dissolved in excess calcium hypochlorite solution and then acidified with acetic acid, the *N,N*-dichloro derivative [(I) dichloramine-T] separates rapidly. When this is heated with sodium hydroxide solution the sodium salt of the *N*-monochloro derivative [(II) chloramine-T)] is formed and crystallises out on cooling at a suitable concentration (Section **IV,48**).

$$Ar \cdot SO_2NH_2 + {}^{\ominus}OCl \longrightarrow Ar \cdot SO_2\overset{\ominus}{N}Cl + H_2O$$

$$Ar \cdot SO_2\overset{\ominus}{N}Cl + \overset{\ominus}{O}Cl + 2H^{\oplus} \longrightarrow Ar \cdot SO_2NCl_2 + H_2O$$

(I)

$$Ar \cdot SO_2NCl_2 + 2NaOH \longrightarrow Ar \cdot SO_2\overset{\ominus}{N}Cl\}\overset{\oplus}{N}a + NaOCl + H_2O$$

(II)

Both chloramine-T and dichloramine-T slowly liberate hypochlorous acid in contact with water and are therefore employed as antiseptics; the former is employed in the form of a dilute (e.g., 0.2%) aqueous solution, and the latter (which is insoluble in water) as a solution in an organic solvent, such as a chloroalkane.

The chlorosulphonation of acetanilide with excess chlorosulphonic acid affords mainly *p*-acetamidobenzenesulphonyl chloride (Section **IV,49**). This is the essential intermediate in the synthesis of a range of sulphanilamide drugs. The simplest example, *p*-aminobenzenesulphonamide (sulphanilamide), is obtained by converting *p*-acetamidobenzenesulphonyl chloride into the amide with aqueous ammonia, and then selectively removing the protecting acetyl grouping with boiling aqueous hydrochloric acid. 2-(*p*-Aminobenzenesulphonamido-pyridine (sulphapyridine, M and B 693, Section **IV,50**) is prepared by reacting *p*-acetamidobenzenesulphonyl chloride with 2-aminopyridine and then removing the acetyl group as before. The synthesis of the required 2-aminopyridine is of interest in that it represents an example of a nucleophilic aromatic substitution by the amide ion at the 2-position in a pyridine ring system, under the influence of the powerful electron-withdrawing influence of the heteroatom.

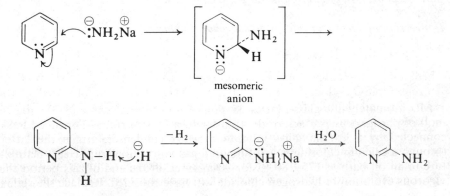

mesomeric
anion

IV,47. TOLUENE-o-SULPHONAMIDE AND SACCHARIN

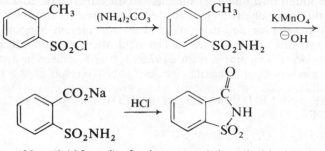

Place 20 g (0.105 mol) of toluene-o-sulphonyl chloride (Section **IV,46**) in a large evaporating dish mounted on a water bath. Add powdered ammonium carbonate cautiously with stirring until the mass is quite hard and solid and the unpleasant odour of the sulphonyl chloride has disappeared. Allow to cool, and extract with cold water to remove the excess of ammonium carbonate. Recrystallise the crude toluene-o-sulphonamide first from hot water (add a little decolourising carbon if it is dark in colour) and then from ethanol. The yield of pure product, m.p. 154 °C, is 16 g (89%).

Oxidation of toluene-o-sulphonamide to saccharin. In a 600-ml beaker, mounted on a wire gauze and provided with a mechanical stirrer, place 12 g (0.07 mol) of toluene-o-sulphonamide, 200 ml of water and 3 g of pure sodium hydroxide. Stir the mixture and warm to 34–40 °C until nearly all has passed into solution (about 30 minutes). Introduce 19 g (0.12 mol) of finely powdered potassium permanganate in small portions at intervals of 10–15 minutes into the well-stirred liquid. At first the permanganate is rapidly reduced, but towards the end of the reaction complete reduction of the permanganate is not attained. The addition occupies 4 hours. Continue the stirring for a further 2–3 hours, and then allow the mixture to stand overnight. Filter off the precipitated manganese dioxide at the pump and decolourise the filtrate by the addition of a little sodium metabisulphite solution. Exactly neutralise the solution with dilute hydrochloric acid (use methyl orange or methyl red as external indicator). Filter off any o-sulphonamidobenzoic acid (and/or toluene-o-sulphonamide) which separates at this point. Treat the filtrate with concentrated hydrochloric acid until the precipitation of the saccharin is complete. Cool, filter at the pump and wash with a little cold water. Recrystallise from hot water. The yield of pure saccharin, m.p. 228 °C, is 7.5 g (58%).

Conversion of saccharin into pseudosaccharin chloride.

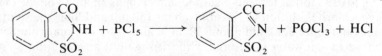

Mix intimately in a glass mortar 35 g (0.19 mol) of saccharin and 70 g (0.336 mol) of phosphorus pentachloride, transfer to a 250-ml round-bottomed flask connected by a ground glass joint to a reflux condenser; attach the latter through a calcium chloride guard-tube to a gas absorption trap. Heat the mixture in an oil bath at 175–180 °C for 90 minutes; at the end of this period the vigorous evolution of hydrogen chloride will have subsided. Replace the reflux condenser by a fractionating column, distil off the phosphorus oxychloride and pour the warm residue upon finely crushed ice. Extract the crude solid pseudosaccharin chloride with chloroform, dry the chloroform solution with anhydrous magnesium sulphate and distil off the solvent. Recrystallise the residue from chloroform or from dry benzene. The yield of pure pseudosaccharin chloride, m.p. 143–145 °C (decomp.), is 26 g (67.5%). It is best kept in a sealed glass tube or in a glass-stoppered bottle.

IV,48. TOLUENE-p-SULPHONAMIDE, DICHLORAMINE-T AND CHLORAMINE-T

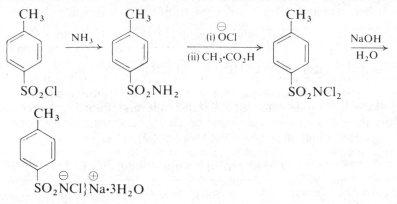

Toluene-p-sulphonamide. Grind together 10 g (0.0525 mol) of toluene-p-sulphonyl chloride (Section **IV,46**) and 20 g of ammonium carbonate in a mortar until a fine uniform powder is obtained. Heat the mixture in an evaporating dish on a water bath for 1–2 hours and stir the mixture frequently with a glass rod. Allow to cool and extract with a little cold water to remove the excess of ammonium salts. Recrystallise the crude toluene-p-sulphonamide from boiling water (200–250 ml), and dry the colourless crystals at 100 °C. The yield of pure product, m.p. 138 °C, is 7.9 g (88%).

Alternatively, grind 10 g of toluene-p-sulphonyl chloride to a fine powder and add it to 30 ml of concentrated ammonia solution (d 0.88). Heat the mixture to boiling (*FUME CUPBOARD*) and cool. Filter and recrystallise the toluene-p-sulphonamide from boiling water (add 1 g of decolourising carbon, if necessary). The yield of pure product, m.p. 138 °C, is almost theoretical.

Dichloramine-T (*N,N-dichlorotoluene-p-sulphonamide*). Prepare about 200 ml of a saturated solution of calcium hypochlorite by grinding a fresh sample of bleaching powder with water and filtering with slight suction. Dissolve 5 g (0.029 mol) of toluene-p-sulphonamide in as small a volume of the calcium hypochlorite solution as possible (about 150 ml) and filter the solution if necessary. Cool in ice, and add about 50 ml of a mixture of equal volumes of glacial acetic acid and water *slowly* and with stirring until precipitation is complete. The dichloramine-T separates out first as a fine emulsion, which rapidly forms colourless crystals. Filter the latter at the pump, wash with a little cold water, drain and dry immediately either between pads of filter-paper or upon a porous tile. The yield is 5.3 g (76%), m.p. 81 °C. Upon recrystallisation from light petroleum (b.p. 60–80 °C) or from chloroform–light petroleum, pure dichloramine-T, m.p. 83 °C, is obtained with negligible loss.

Chloramine-T (sodium *N*-chlorotoluene-*p*-sulphonamide).   For this preparation use dichloramine-T which has been prepared as above and thoroughly drained but not necessarily dried. Heat 45 ml of 10 per cent aqueous sodium hydroxide solution in a beaker to a temperature of about 80 °C, add 3.5 g (0.015 mol) of dichloramine-T in small quantities, stirring the mixture gently after each addition until a clear solution is obtained. When the addition is complete, filter the hot solution if turbid, and then allow it to cool spontaneously. Filter the crystals with suction, wash with a little saturated sodium chloride solution and dry upon filter-paper or in a desiccator over anhydrous calcium chloride. The resulting chloramine-T weighs 3 g (75%) and is almost pure. It may be recrystallised, if desired, from twice its weight of hot water.

Chloramine-T is a salt and has no definite m.p.: upon heating it loses water of crystallisation and decomposes violently at 175–180 °C.

IV,49. *p*-AMINOBENZENESULPHONAMIDE (*Sulphanilamide*)

$$CH_3 \cdot CO \cdot NH \cdot C_6H_5 \xrightarrow{Cl \cdot SO_3H} p\text{-}CH_3 \cdot CO \cdot NH \cdot C_6H_4 \cdot SO_2Cl \xrightarrow{NH_3}$$

$$p\text{-}CH_3 \cdot CO \cdot NH \cdot C_6H_4 \cdot SO_2NH_2 \xrightarrow{H_3O^{\oplus}} p\text{-}NH_2 \cdot C_6H_4 \cdot SO_2NH_2$$

p*-Acetamidobenzenesulphonyl chloride.**   Equip a 500-ml two-necked flask with a dropping funnel and a reflux condenser: attach the top of the latter to a device for the absorption of hydrogen chloride (e.g., Fig. I,68). Place 20 g (0.148 mol) of dry acetanilide in the flask and 50 ml (90 g, 0.77 mol) of a good grade of chlorosulphonic acid (**CAUTION:** see Section **II,2,*12) in the dropping funnel and insert a calcium chloride guard-tube into the latter. Add the chlorosulphonic acid in small portions and shake the flask from time to time to ensure thorough mixing (1). When the addition has been made, heat the reaction mixture on a water bath for 1 hour in order to complete the reaction. Allow to cool and pour the oily mixture in a thin stream with stirring into 300 g of crushed ice (or ice-water) contained in a 1-litre beaker. Carry out this operation carefully in the fume cupboard since the excess of chlorosulphonic acid reacts vigorously with the water. Rinse the flask with a little ice-water and add the rinsings to the contents of the beaker. Break up any lumps of solid material and stir the mixture for several minutes in order to obtain an even suspension of the granular white solid. Filter off the *p*-acetamidobenzenesulphonyl chloride at the pump and wash it with a little cold water; press and drain well. *Use the crude product (2) immediately in the next stage.*

***p*-Acetamidobenzenesulphonamide.**   Transfer the crude *p*-acetamidobenzenesulphonyl chloride to the rinsed reaction flask, and add a mixture of 70 ml of concentrated ammonia solution (*d* 0.88) and 70 ml of water. Mix the contents of the flask thoroughly, and heat the mixture with occasional swirling (*FUME CUPBOARD*) to just below the boiling point for about 15 minutes. The sulphonyl chloride will be converted into a pasty suspension of the corresponding sulphonamide. Cool the suspension in ice, and then add dilute sulphuric acid until the mixture is just acid to Congo red paper. Collect the product on a Buchner funnel, wash with a little cold water and drain as completely as possible. It is desirable, but not essential, to dry the crude *p*-acetamidobenzenesulphonamide at 100 °C: the yield is about 18 g. The material is sufficiently pure (3) for the next stage.

p-Aminobenzenesulphonamide. Transfer the crude p-acetamidobenzene-sulphonamide to a 500-ml flask, add 10 ml of concentrated hydrochloric acid and 30 ml of water. Boil the mixture gently under reflux for 30–45 minutes. The solution, when cooled to room temperature, should deposit no solid amide; if a solid separates, heat for a further short period. Treat the cooled solution with 2 g of decolourising carbon, heat the mixture to boiling and filter with suction through a hardened filter-paper. Place the filtrate (a solution of sulphanilamide hydrochloride) in a litre beaker and cautiously add 16 g of solid sodium hydrogen carbonate in portions with stirring. After the evolution of gas has subsided, test the suspension with litmus paper and if it is still acid, add more sodium hydrogen carbonate until neutral. Cool in ice, filter off the sulphanilamide with suction and dry. The yield is 15 g (59% overall yield), m.p. 161–163 °C. A pure product, m.p. 163–164 °C, may be obtained by recrystallisation from water or from alcohol.

Notes. (1) The reaction may be more easily controlled and the chlorosulphonic acid added all at once if the acetanilide is employed in the form of a hard cake. The latter is prepared by melting the acetanilide in the flask over a free flame and causing the compound to solidify over the lower part of the flask by swirling the liquid. If the reaction becomes too vigorous under these conditions, cool the flask momentarily by immersion in an ice bath.

(2) The crude sulphonyl chloride, even if dry, cannot be kept without considerable decomposition. It may be purified by triturating with dried acetone, filtering any undissolved p-acetamidosulphonic acid and evaporating the filtrate. The pure chloride has m.p. 149 °C.

(3) A small portion may be recrystallised from water, with the addition of a little decolourising carbon if necessary. The pure compound has m.p. 218 °C.

IV,50. 2-(p-AMINOBENZENESULPHONAMIDO)PYRIDINE(*Sulphapyridine*)

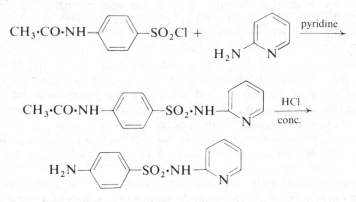

2-Aminopyridine. In a 1-litre four-necked flange flask, equipped with a sealed mechanical stirrer, reflux condenser, thermometer and inlet tube for nitrogen, place 300 ml of dry toluene (1) and 75 g (1.92 mol) of fine granular sodamide (2); bubble a steady stream of nitrogen through the toluene. Stir the mixture vigorously and heat the flask in an oil bath until the internal temperature is 110 °C (the bath temperature required is approximately 130 °C). Add 100 g (1.26 mol) of pure dry pyridine (Section II,1,29) dropwise through the condenser fitted with a pressure-equalising funnel over a period of 4 hours: maintain the very efficient stirring and the stream of nitrogen. After 1 hour the reaction mixture becomes black in colour, and after 3 hours becomes viscous, and bubbling and slight frothing occurs, due to liberation of hydrogen. When all the pyridine has been introduced, continue the heating for a further 5 hours whilst maintaining the internal temperature at 110 °C. Towards the end of the reaction, stirring may become difficult owing to the separation of a solid or viscous cake. Allow the reaction mixture to cool (without the stream of nitrogen and without stirring); then introduce 175 ml of water very slowly through the condenser over a period of 2 hours whilst continuing the passage of the stream of nitrogen. During the addition the temperature rises to about 50 °C; resume the stirring as soon as possible. Transfer the contents of the flask to a separatory funnel, separate the lower aqueous solution and extract it with two 150 ml portions of toluene. Dry the combined main toluene layer and toluene extracts over anhydrous potassium carbonate for 2 hours; filter and remove the toluene by distillation. Distil the syrupy residue from an oil bath under diminished pressure through an air condenser: adjust the bath temperature to 120–130 °C. Collect the 2-aminopyridine at 95 °C/10 mmHg; this solidifies on cooling to a colourless solid, m.p. 55 °C (3). The yield is about 80 g (67%).

Sulphapyridine. Dissolve 4.7 g (0.05 mol) of 2-aminopyridine in a mixture of 40 ml of anhydrous acetone and 6 ml of dry pyridine in a 250-ml flask, and add 11.7 g (0.05 mol) of pure *p*-acetamidobenzenesulphonyl chloride (4). The reaction mixture is set aside overnight and 5.5 g of the almost pure 2-(*p*-acetamidobenzenesulphonamide)pyridine (≡ acetylsulphapyridine) is filtered off; by diluting the filtrate with water a further crop (4 g) is obtained. The total product is recrystallised from acetone to give white needles of pure product, m.p. 224 °C. The yield is 8 g (55%).

The acetylsulphapyridine (7.3 g) is hydrolysed by heating it under reflux with 75 ml of ethanol containing 15 ml of concentrated hydrochloric acid for 20 min. The cooled solution is diluted with water and made just alkaline with concentrated ammonia solution (*d* 0.880). The sulphapyridine is isolated by filtration and recrystallised from ethanol; yield 4.9 g (75%), m.p. 190–191 °C.

Notes. (1) Technically pure toluene can be conveniently dried by distilling 350 ml from a litre flask and rejecting the first 50 ml.

(2) It is important to use recently prepared pure sodamide, which must be of fine granular form. Old material of irregular lumpy form, even if ground gives poor results, and should not be employed. The sodamide may be prepared as detailed in Section II,2,54. A satisfactory grade is marketed by May and Baker Ltd.

(3) The residue in the flask is said to contain 4-amino- and 2,6-diamino-pyridine, 4,4'-bipyridyl and di(1-pyridyl) amine in varying amounts.

(4) The *p*-acetamidobenzenesulphonyl chloride (Section IV,49) must be pure: under no circumstances should it contain more than 1–2 per cent of the corresponding sulphonic acid. This may be ensured by lixiviating the sulphonyl chloride with pure anhydrous acetone and filtering the solution from the acid.

D,4. The preparation of arylsulphonate esters.   These are prepared by the interaction of the arylsulphonyl chloride and an alcohol or phenol in the presence of sodium hydroxide solution or of pyridine (Section **IV,51**).

$$ROH + Ar \cdot SO_2Cl \xrightarrow{\text{NaOH}} RO \cdot SO_2Ar + ACl$$

These sulphonate esters cannot be made by direct esterification of the sulphonic acids since further reaction of the ester with the alcohol gives rise to an ether by means of alkyl–oxygen fission process.

$$R - O\!:\!\overset{\curvearrowright}{} R \overset{\curvearrowleft}{-} O \cdot SO_2Ar \longrightarrow R \cdot O \cdot R + HOSO_2 \cdot Ar$$
$$\underset{H}{|}$$

For this reason the arylsulphonate esters find use as alkylating agents (cf. dimethyl sulphate and dialkyl sulphates, Section **II,2,**_18_).

IV,51. BUTYL TOLUENE-_p_-SULPHONATE

$$p\text{-}CH_3 \cdot C_6H_4 \cdot SO_2Cl + C_4H_9OH \xrightarrow[\text{pyridine}]{\text{NaOH or}} p\text{-}CH_3 \cdot C_6H_4 \cdot SO_2 \cdot OC_4H_9$$

Equip a 1-litre three-necked flask with a separatory funnel, a mechanical stirrer and a thermometer, the bulb of which reaches within 2 cm from the bottom. Place 74 g (89 ml, 1 mol) of butan-1-ol and 105 g (0.55 mol) of toluene-_p_-sulphonyl chloride (Section **IV,46**) in the flask and 160 ml of 20 per cent sodium hydroxide solution in the separatory funnel; immerse the flask in a bath of cold water. Run in the sodium hydroxide solution, with stirring, at such a rate that the temperature does not rise above 15 °C (3–4 hours). Now add another portion of 105 g (0.55 mol) of toluene-_p_-sulphonyl chloride, and introduce 160 ml of 20 per cent sodium hydroxide solution slowly, keeping the temperature below 15 °C. Continue the stirring for 4 hours longer. Separate the oily layer and treat it with enough light petroleum (b.p. 60–80 °C) to cause it to float on water; then wash it well with 25 ml of 10 per cent sodium hydroxide solution, and dry by allowing it to stand over 10 g of anhydrous potassium carbonate. Filter and distil off the solvent using a 250-ml flask (rotary evaporator). Distil the residual ester under reduced pressure (1) (oil pump) and collect the butyl toluene-_p_-sulphonate at 132–133 °C/3 mmHg. The yield is 130 g (57%).

Note. (1) It is best to distil under greatly reduced pressure; slight decomposition occurs even at 10 mm pressure (b.p. 170–171 °C/10 mmHg).

Cognate preparation   Dodecyl toluene-_p_-sulphonate (_pyridine method_).   In a 500-ml three-necked flask, equipped with a stirrer and thermometer, place 46.5 g (0.25 mol) of dodecyl alcohol (lauryl alcohol), m.p. 22–23 °C, and 79 g (81 ml, 1 mol) of dry pyridine. Surround the flask by a bath sufficiently cold to lower the temperature of the mixture to 10 °C. Add 52.5 g of toluene-_p_-sulphonyl chloride in portions during 20 minutes, or at such a rate that the temperature does not rise above 20 °C. Stir the mixture for 3 hours at a temperature below 20 °C, then dilute with 150 ml of concentrated hydrochloric acid in 500 ml of ice-water. Collect the ester on a chilled Buchner funnel and suck as dry as possible. Transfer the solid to a 400-ml beaker, add 150 of methyl alcohol and warm the mixture on a steam bath until the ester melts. Cool in a freezing mixture whilst stirring vigorously; the ester separates in a finely divided state. Collect it on a chilled funnel and allow to dry in the air, preferably below 20 °C. The yield of ester, m.p. 24–25 °C, is 78 g (92%). Recrystallise by dissolving in 100 ml of light petroleum, b.p. 40–60 °C, drying the solution over magnesium sulphate to remove traces of water, and cool to 0 °C. Collect the pure dodecyl toluene-*p*-sulphonate, m.p. 29–30 °C, in a chilled funnel (Section **I,20**).

The pyridine procedure may be applied to the preparation of other esters; they are isolated by ether extraction. The yields are generally better than by the sodium hydroxide method.

D,5. Reduction products from arylsulphonyl chlorides. Reduction of an arylsulphonyl chloride with zinc dust and water affords the zinc salt of the sulphinic acid, converted by sodium carbonate to the sodium salt, in which form it is conveniently isolated (e.g., sodium toluene-*p*-sulphinate, Section **IV,52**) and thence by hydrochloric acid into the somewhat unstable sulphinic acid.

$$\text{Ar·SO}_2\text{Cl} \xrightarrow[\substack{\text{(ii) Na}_2\text{CO}_3 \\ \text{(iii) HCl}}]{\text{(i) Zn/H}_2\text{O}} \text{Ar·SO}_2\text{H}$$

Excessive drying of the free sulphinic acid must be avoided since it leads to partial conversion into the sulphonic acid and the thiolsulphonic ester.

$$3\text{Ar·SO}_2\text{H} \xrightarrow{\text{H}_2\text{O}} \text{Ar·SO}_3\text{H} + \text{Ar·SO}_2\text{·SAr}$$

An alternative synthesis of benzenesulphinic acid by way of the diazonium salt is given in Section **IV,83**.

More vigorous reduction of arylsulphonyl chlorides (or of sulphinic acids), for example with zinc and dilute sulphuric acid, gives rise to thiophenols, which are conveniently isolated by steam distillation. The example provided is that of the synthesis of thiophenol itself (Section **IV,53**).

$$\text{Ar·SO}_2\text{Cl} \xrightarrow{\text{Zn, HCl}} \text{ArSH}$$

IV,52. SODIUM TOLUENE-*p*-SULPHINATE

$$p\text{-CH}_3\text{·C}_6\text{H}_4\text{·SO}_2\text{Cl} \xrightarrow[\text{(ii) Na}_2\text{CO}_3]{\text{(i) Zn/H}_2\text{O}} p\text{-CH}_3\text{·C}_6\text{H}_4\text{·SO}_2\text{Na} \xrightarrow{\text{HCl}}$$

$$p\text{-CH}_3\text{·C}_6\text{H}_4\text{·SO}_2\text{H}$$

In a 2-litre bolt-necked flask place 300 ml of water and insert a mechanical stirrer. Warm on a boiling water bath until the temperature reaches 70 °C and add 40 g (0.61 mol) of zinc powder (90–100% pure). Stir the mixture and add 50 g (0.26 mol) of finely powdered toluene-*p*-sulphonyl chloride in portions during about 10 minutes; the temperature rises to about 80 °C. Stir for a further 10 minutes and then heat the mixture further until the temperature is 90 °C. Add

25 ml of 12 *M*-sodium hydroxide solution, followed by finely powdered sodium carbonate in 5 g portions until the mixture is strongly alkaline. Considerable frothing occurs. Filter at the pump and extract the residue by heating on a steam bath with 75 ml of water until excessive frothing occurs; continue stirring for a further 10 minutes. Filter with suction and add the filtrate to the main solution, concentrate to a volume of about 100 ml with a rotary evaporator and cool in ice-water. Filter at the pump and dry the crystals upon filter or drying paper until efflorescence just commences, then place in a tightly stoppered bottle. The yield of sodium toluene-*p*-sulphinate dihydrate is 35 g (63%).

The free sulphinic acid may be precipitated from a solution of the sodium salt in cold water by cautious acidification with hydrochloric acid. The toluene-*p*-sulphinic acid is filtered and sucked as dry as possible at the pump and dried rapidly between sheets of filter-paper, m.p. 85 °C.

IV,53. THIOPHENOL

$$C_6H_5{\cdot}SO_2Cl \xrightarrow[\text{Zn/H}_2\text{SO}_4]{\text{[H]},} C_6H_5SH + HCl + 2H_2O$$

This preparation must be carried out in the fume cupboard since thiophenol has an extremely unpleasant and repulsive odour; the substance should not be allowed to come into contact with the hands or clothing since the odour clings for days.

Place 720 g of crushed ice and 240 g (130 ml) of concentrated sulphuric acid in a 1500-ml three-necked flask equipped with a mechanical stirrer, a double surface condenser and a thermometer dipping into the reaction mixture. Immerse the flask in a freezing mixture of ice and salt and maintain the temperature at -5 to 0 °C throughout the preparation. Start the stirrer and add 60 g (0.34 mol) of benzenesulphonyl chloride (Section **IV,45**) in small portions over a period of half an hour. (Benzenesulphonyl chloride melts at 14 °C and hence it must be added slowly and with vigorous stirring in order that it may be as finely divided as possible for maximum reactivity in the subsequent reduction.) Then add 120 g (1.84 mol) of zinc powder (90–100% pure) as rapidly as possible without the temperature rising above 0 °C (about 30 minutes). Stir the mixture for a further 1.5 hours, then remove the ice–salt bath and allow the reaction mixture to warm up spontaneously, whilst continuing the stirring. Within 5 minutes or so, a rather violent reaction with the evolution of much hydrogen sets in; it may be necessary to cool the flask momentarily. When the energetic reaction has subsided, warm the mixture, with vigorous stirring, until the solution becomes clear (4–6 hours). Steam distil the thiophenol until organic material ceases to pass over. Separate the organic layer from the distillate, dry it with anhydrous calcium chloride or magnesium sulphate and distil. Collect the thiophenol at 166–169 °C; the yield is 34 g (91%).

Cognate preparation Thio-*p*-cresol (*p*-tolyl mercaptan), *p*-CH$_3$·C$_6$H$_4$SH. This compound may be similarly prepared from toluene-*p*-sulphonyl chloride (Section **IV,46**). The thio-*p*-cresol crystallises in the steam distillate and is collected and dried; m.p. 43 °C. The b.p. under normal pressure is 194–195 °C.

E AROMATIC AMINES

Aromatic amines may be divided into three classes:

(*a*) **Primary amines:** where the amino group is directly attached to the aromatic ring, e.g., aniline $C_6H_5NH_2$. Amines with the amino group in the side chain, e.g., benzylamine $C_6H_5 \cdot CH_2NH_2$, possess properties similar to those of simple aliphatic amines (p. 564, **III,R**).

(*b*) **Secondary amines:** (i) purely aromatic amines, e.g., diphenylamine $(C_6H_5)_2NH$, and (ii) aromatic–aliphatic amines, e.g., *N*-methylaniline, $C_6H_5NHCH_3$.

(*c*) **Tertiary amines:** (i) purely aromatic amines, e.g., triphenylamine $(C_6H_5)_3N$, and (ii) aromatic–aliphatic amines, e.g., *N,N*-dimethylaniline, $C_6H_5N(CH_3)_2$.

Primary aromatic amines are prepared by the following procedures:

1. The reduction of nitro compounds under appropriate conditions (Sections **IV,54** to **IV,59**).
2. Molecular rearrangements of the Hofmann type (Section **IV,60** and **IV,61**).
3. The replacement of an aromatic hydroxyl group by an amino group (Section **IV,62**).

Secondary and tertiary amines are prepared by:

4. Alkylation and reductive alkylation procedures (Section **IV,63** to **IV,65**).

E,1. The reduction of nitro compounds. Primary arylamines are generally prepared by the reduction of nitro compounds. When only small quantities are to be reduced and cost is a secondary consideration, tin and hydrochloric acid may be employed.

$$Ar \cdot NO_2 + 6[H] \longrightarrow ArNH_2 + H_2O$$

Theoretically 1.5 mol of tin are needed for the reduction of the nitro group, the metal being oxidised to the tin(IV) state.

$$2Ar \cdot NO_2 + 3Sn + 12H^\oplus \longrightarrow 2ArNH_2 + 3Sn^{4\oplus} + 4H_2O$$

When reduction is complete, a complex amine chlorostannate may separate from which the amine is liberated by basification, using enough alkali to dissolve the tin hydroxides formed (e.g., the preparation of aniline, Section **IV,54**).

$$[ArNH_3]_2^\oplus[SnCl_6]^{2\ominus} + 8^\ominus OH \longrightarrow ArNH_2 + SnO_3^{2\ominus} + 6Cl^\ominus$$

Alternatively decomposition of the complex is achieved by precipitation of the tin as its sulphide by passing hydrogen sulphide into the mixture at the end of the reaction (e.g., *m*-phenylenediamine, cognate preparation in Section **IV,54**).

Other metal–acid reducing systems may be used; reduction with iron and hydrochloric acid is illustrated in Section **IV,55** and is employed on the technical scale for the manufacture of aniline.

$$C_6H_5 \cdot NO_2 + 2Fe + 6H^\oplus \longrightarrow C_6H_5NH_2 + 2Fe^{3\oplus} + 2H_2O$$
$$\text{or } C_6H_5 \cdot NO_2 + 3Fe + 6H^\oplus \longrightarrow C_6H_5NH_2 + 3Fe^{2\oplus} + 2H_2O$$

In practice, however, the amount of hydrochloric acid used is less than 5 per cent of the amounts indicated by either of the above equations. Various explanations have been advanced to account for this; one is that the following reaction is catalysed by acid or by hydroxonium ions.

$$C_6H_5 \cdot NO_2 + 2Fe + 4H_2O \xrightarrow{\ H^\oplus\ } C_6H_5NH_2 + 2Fe(OH)_3$$

The use of iron and a little hydrochloric acid (or alternatively of iron and acetic acid) is to be preferred when the use of the more strongly acidic medium leads to the formation of undesirable by-products; nuclear chlorination, for example, often results when tin or zinc is used in association with concentrated hydrochloric acid.

Compounds which are sensitive to acidic conditions may sometimes be successfully reduced to amines under alkaline conditions; in these cases iron(II) sulphate is often used as the reducing agent. Reduction with metal under alkaline (or alternatively essentially neutral) conditions leads to the formation of intermediate products of the reduction of the nitro group (see p. 721, **IV,H,1**). Vigorous reduction under alkaline conditions using an excess of the reducing metal (e.g., zinc) may, however, lead to the complete reduction of the nitro group with the formation of the corresponding amine; thus o-phenylenediamine (Section **IV,56**) is conveniently obtained by reducing o-nitroaniline in this way.

Aromatic and heterocyclic nitro compounds are readily reduced in good yield to the corresponding amines (e.g., o-aminophenol, Section **IV,57**) by sodium borohydride in aqueous methanol solution in the presence of a palladium-on-carbon catalyst. In this reduction there is no evidence for the formation of intermediates of the azoxybenzene or azobenzene type, although if the reaction is carried out in a polar aprotic solvent, such as dimethyl sulphoxide, azoxy compounds may sometimes be isolated as the initial products.

Ammonium or alkali metal sulphides or polysulphides exhibit a useful selective reducing action in that they smoothly reduce one nitro group in a polynitro compound to yield the corresponding nitroamine (e.g., m-nitroaniline, Section **IV,58**).

The two methods described for the synthesis of ethyl p-aminobenzoate (Section **IV,59**) from p-nitrobenzoic acid offer interesting comparison. Prior esterification of the carboxylic acid group (*Method 2*) yields a nitroester which cannot be reduced by a metal–acid reducing system owing to the possibility of hydrolytic cleavage of the ester grouping; conversion of the nitro into the amino group in this case is therefore by catalytic hydrogenation with Adams' platinum dioxide catalyst. In *Method 1*, metal–acid reduction of the nitro group yields the amino acid, p-aminobenzoic acid, a reaction in which the final mixture requires careful basification to effect precipitation of the product, which is then esterified to give ethyl p-aminobenzoate. This compound (Benzocaine) is used as a local anaesthetic.

IV,54. ANILINE

$$C_6H_5 \cdot NO_2 + 6[H] \xrightarrow{Sn/HCl} C_6H_5 \cdot NH_2 + 2H_2O$$

Into a 500-ml round-bottomed flask equipped with a reflux condenser, place 25 g (21 ml, 0.25 mol) of nitrobenzene and 45 g (0.38 mol) of granulated tin. Measure out 100 ml of concentrated hydrochloric acid. Pour about 15 ml of this acid down the condenser and shake the contents of the flask steadily. The mixture becomes warm and before long the reaction should be quite vigorous; if it boils very vigorously, moderate the reduction somewhat by temporarily immersing the flask in cold water. When the initial reaction slackens of its own accord, pour another 15 ml of hydrochloric acid down the condenser, shake the flask steadily to ensure thorough mixing and cool again if the reduction becomes too violent. Do not cool more than is necessary to keep the reaction under control; keep the mixture well shaken. Proceed in this way until all the 100 ml of acid has been added. Finally heat the mixture on a boiling water bath for 30–60 minutes, i.e., until the odour of nitrobenzene is no longer perceptible and a few drops of the reaction mixture when diluted with water yield a perfectly clear solution. During the course of the reduction, particularly during the cooling, aniline chlorostannate may separate as a white or yellow crystalline complex.

Cool the reaction mixture to room temperature and add gradually a solution of 75 g of sodium hydroxide in 125 ml of water; if the mixture boils during the addition of the alkali, cool again. The hydroxide of tin which is first precipitated should all dissolve and the solution should be strongly alkaline: the aniline separates as an oil. Equip the flask for steam distillation as in Fig. I,101, and pass steam into the warm mixture until, after the distillate has ceased to pass over as a turbid liquid, a *further* 120 ml of clear liquid are collected. Since aniline is appreciably soluble (*ca.* 3%) in water, it must be 'salted out' by saturating the distillate with salt. Use about 20 g of commercial salt for each 100 ml of liquid. Transfer the distillate, saturated with salt, to a separatory funnel, add about 40 ml of ether and shake to ensure intimate mixing of the solution and the ether; relieve the pressure within the funnel by momentarily lifting the stopper. [All flames in the vicinity must be extinguished during the extraction.] Allow the two layers to separate; run off the lower aqueous layer into a beaker, and pour the remaining ethereal layer through the mouth of the funnel into a 200-ml flask. Return the aqueous solution to the funnel and extract with a further 40 ml of ether. Proceed as before, and pour the ethereal extract into the flask. Dry the combined ethereal solutions with a few grams of anhydrous potassium carbonate (1): shake the well-stoppered flask for several minutes.

Filter the ethereal solution through a fluted filter-paper and remove the ether by flash distillation (p. 143), using a 50-ml round-bottomed flask to which has been added a few boiling chips. Since ether is extremely volatile and also highly inflammable, the flask must be heated by means of an electrically heated water bath. When all the ethereal solution has been introduced into the flask, and no more ether distils on a boiling water bath, run out the water from the condenser, and distil the aniline either by direct heating over a wire gauze or, preferably, using an air bath (Fig. I,49). A small quantity of ether may pass over during the early part of the distillation; it is therefore advisable to interpose an asbestos or uralite board between the receiver and the flame. Collect the fraction, b.p. 180–184 °C, in a weighed conical flask. The yield of aniline is 18 g (97%).

Pure aniline has a b.p. of 184 °C. When freshly distilled it is a colourless liquid, but becomes discoloured on standing, particularly when exposed to light, owing to atmospheric oxidation. The colour may usually be removed by distillation from a little zinc dust.

Note. (1) Calcium chloride cannot be used to dry the ethereal solution because it combines with aniline (and other amines) to form molecular compounds. The best drying agent is sodium or potassium hydroxide (pellet form).

Cognate preparations *p*-**Toluidine.** Reduce *p*-nitrotoluene with tin and hydrochloric acid and isolate the amine by ether extraction. Since *p*-toluidine is a solid (m.p. 45 °C; b.p. 200 °C), it may crystallise in the condenser used for steam distillation: it is easily melted by stopping the current of cooling water in the condenser for a moment or two.

m-**Phenylenediamine.** In a 2-litre round-bottomed flask, provided with a reflux condenser, place 25 g (0.15 mol) of *m*-dinitrobenzene (Section **IV,20**) and 100 g (0.84 mol) of granulated tin; add 200 ml of concentrated hydrochloric acid in 15 ml portions. When all the acid has been introduced, complete the reduction by heating on a water bath for 1 hour. Dilute with 750 ml of water, heat nearly to boiling and pass hydrogen sulphide into the liquid until all the tin is precipitated as the sulphide. Filter a small quantity from time to time and test for completeness of precipitation with hydrogen sulphide. Allow the precipitate to settle overnight, decant the clear liquid and filter the residue through a large fluted filter-paper (1). Add sodium hydroxide solution to the filtrate until the latter is strongly alkaline, and extract several times with ether. Dry over anhydrous potassium carbonate or sodium hydroxide pellets, remove the ether and then distil the residue: use an air condenser after all the ether has passed over. Collect the portion boiling between 280 and 284 °C: this solidifies on standing to crystalline *m*-phenylenediamine, m.p. 63 °C. The yield is 13 g (74%).

Note. (1) The dihydrochloride may be obtained by evaporating the filtrate on a water bath until crystals appear, and then cooling in ice. The crystals are filtered at the pump, washed with a little concentrated hydrochloric acid and dried in a vacuum desiccator over sodium hydroxide.

IV,55. *o*-CHLOROANILINE

$$o\text{-Cl·}C_6H_4\text{·}NO_2 \xrightarrow[\text{Fe/HCl}]{\text{[H]}} o\text{-Cl·}C_6H_4\text{·}NH_2$$

The vapours of o-*chloroaniline are toxic and produce serious after-effects: the preparation must therefore be conducted in a fume cupboard.*

In a 2-litre three-necked flask, equipped with a mechanical stirrer, a reflux condenser and a glass funnel, the hole of which is plugged by means of a glass rod covered with a rubber tube, place 480 g (8.6 mol) of iron filings and 360 ml of water. Heat the mixture on a boiling water bath and, when hot, remove the water bath. While stirring vigorously, add 40 g (0.25 mol) of *o*-chloronitrobenzene through the funnel and at the same time introduce 10 ml of concentrated hydrochloric acid by means of a separatory funnel fitted into the top of the condenser with a grooved cork. A vigorous reaction commences as soon as the acid has been added. Then introduce 200 g (1.27 mol) of melted o-chloronitrobenzene through the funnel all at once. After about 10 minutes, add 50 ml of concentrated hydrochloric acid, as before, at such a rate (about 15–20 minutes) that vapours of o-chloroaniline do not escape from the top of the condenser. Heat on a water bath, with stirring, for 1 hour in order to complete the reaction. Then add a solution of 20 g of sodium hydroxide in 40 ml of water to decompose any chloroaniline hydrochloride that might have formed. Filter the reaction mixture whilst still hot; separate the lower layer of the filtrate (o-chloroaniline) from the water layer. Return the iron residues to the flask and boil with 200 ml of benzene: filter the hot benzene solution through the same funnel and wash the iron residues with a second 200 ml of hot benzene. Combine the benzene extracts with the o-chloroaniline originally separated from the water, dry with anhydrous magnesium sulphate and remove most of the benzene under normal pressure. Transfer the residue (**CAUTION**: the vapours are toxic) to a 400-ml flask and distil under reduced pressure (Section I,27): some benzene passes over first, followed by o-chloroaniline at 113–117 °C/20 mmHg. The yield is 185 g (95%). The o-chloroaniline may also be distilled under ordinary pressure without decomposition: b.p. 206–209 °C.

Cognate preparation p-**Chloroaniline** may be similarly prepared from p-chloronitrobenzene; 240 g of the latter give 185 g (95%) of p-chloroaniline, b.p. 128–131 °C/20 mmHg, m.p. 71 °C.

IV,56. o-PHENYLENEDIAMINE

$$o\text{-NH}_2\cdot\text{C}_6\text{H}_4\cdot\text{NO}_2 \xrightarrow[\ominus\text{Zn/OH}]{[H]} o\text{-NH}_2\cdot\text{C}_6\text{H}_4\cdot\text{NH}_2$$

Equip a 750-ml three-necked flask with a reflux condenser and a sealed mechanical stirrer, and place in it 46 g (0.33 mol) of o-nitroaniline, 27 ml of 20 per cent sodium hydroxide solution and 170 ml of rectified spirit. Stir the mixture vigorously and heat it on a water bath to gentle boiling. Remove the source of heat from beneath the bath, and introduce 5 g portions of zinc powder at such a rate that the solution is kept boiling (1); add 90 g (1.4 mol) of zinc powder (2) in all. Reflux the mixture, with stirring, for 1 hour; the colour of the solution changes from deep red to nearly colourless. Filter the hot mixture at the pump; return the zinc residue to the flask and extract it with two 100 ml portions of hot rectified spirit. Combine the extracts with the filtrate, add 2 g of sodium dithionite ($\text{Na}_2\text{S}_2\text{O}_4$) and concentrate the solution under reduced pressure (water pump) on a steam bath to a volume of 80–100 ml; use a rotary evaporator (Fig. I,114). Cool the solution in a freezing mixture of ice and salt, collect the pale yellow crystals on a Buchner funnel, wash once with 10–15 ml of ice-water and dry in a vacuum desiccator. The yield of crude o-phenylenediamine, m.p. 98–100 °C, is 33 g. This is sufficiently pure for most practical purposes. If a pure material is required (3), dissolve the crude product in 100–115 ml of hot water containing 1 g of sodium dithionite and add a few grams of decolourising carbon, filter and cool in an ice–salt mixture. Collect the colourless crystals of pure o-phenylenediamine on a Buchner funnel, wash with 10 ml of ice-water and dry in a vacuum desiccator; the yield is 28.5 g (79%), m.p. 100–101 °C. It darkens rapidly upon exposure to light.

Notes. (1) Sometimes the reaction stops suddenly; it is then necessary to add a further 10 ml of 20 per cent sodium hydroxide solution and warm to the boiling point: this causes the reaction to continue. Occasionally, the reduction becomes very vigorous: a wet towel and a bath of ice-water should be kept close at hand.

(2) This weight of zinc powder assumes 100 per cent purity: an equivalent amount of less pure material may be used (see Section **II,2,**67).

(3) The crude o-phenylenediamine may be converted into the dihydrochloride and the salt purified in the following manner. Dissolve it in 60 ml of concentrated hydrochloric acid and 40 ml of water containing 2 g of tin(II) chloride, and treat the hot solution with 2–3 g of decolourising carbon. Filter, add 100 ml of concentrated hydrochloric acid to the hot colourless filtrate and cool in a freezing mixture of ice and salt. Collect the colourless crystals of the dihydrochloride on a Buchner or sintered glass funnel, wash with a small volume of concentrated hydrochloric acid and dry in a vacuum desiccator over sodium hydroxide. The yield is 51 g.

IV,57. o-AMINOPHENOL

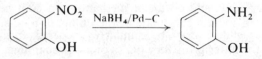

This preparation should be conducted in an efficient fume cupboard since hydrogen is evolved during the reaction.

Place a suspension of 0.5 g of 10 per cent palladium-on-charcoal (cf. Section **II,2,**44) in 50 ml of water in a 500-ml conical flask and add 3.9 g (0.10 mol) of sodium borohydride in 75 ml of water (1); introduce a magnetic follower bar. Pass a slow stream of nitrogen through the stirred mixture and add a solution of 7.0 g (0.05 mol) of o-nitrophenol in 250 ml of 2 M-sodium hydroxide, dropwise from a separatory funnel, during 5 minutes. Stir at room temperature until the yellow colour disappears (about 10 minutes) and then filter. Acidify the filtrate with 2 M-hydrochloric acid to destroy excess borohydride and then neutralise the solution with dilute sodium hydroxide. Extract the product with four 50 ml portions of ether and evaporate the dried ($MgSO_4$) extract on the rotary evaporator. o-Aminophenol is obtained as an off-white solid, m.p. 167–169 °C; the yield is 4.3 g (75%). The product may be recrystallised from water to give the purified phenol, m.p. 169–171 °C. The infrared spectrum of o-aminophenol as a Nujol mull shows strong absorptions at 3380 and 3300 cm^{-1} due to the antisymmetric and symmetric N–H stretching vibrations.

Note. (1) Addition of the reagents in this order prevents the possible ignition of hydrogen which can take place on addition of dry palladium–charcoal to solutions of sodium borohydride.

IV,58. m-NITROANILINE

$$m\text{-}O_2N\cdot C_6H_4\cdot NO_2 \xrightarrow[\text{Method 2 [H]; NaSH}]{\text{Method 1 [H]; Na}_2S_x} m\text{-}O_2N\cdot C_6H_4\cdot NH_2$$

Method 1. Prepare a solution of sodium polysulphide by dissolving 40 g (0.167 mol) of crystallised sodium sulphide, $Na_2S,9H_2O$ (1), in 150 ml of water, adding 10 g (0.31 mol) of finely powdered sulphur and warming until a clear solution is produced. Heat a mixture of 25 g (0.15 mol) of *m*-dinitrobenzene (Section **IV,20**) and 200 ml of water contained in a 1-litre beaker until the water boils gently: stir the solution mechanically. Place the sodium polysulphide solution in a dropping funnel and clamp the funnel so that the end of the stem is immediately above the beaker. Add the sodium polysulphide solution during 30–45 minutes to the vigorously stirred, boiling mixture, and boil gently for a further 20 minutes. Allow to cool; this can be accomplished more rapidly by adding ice. Filter at the pump and wash with cold water. Transfer to a 600-ml beaker containing 150 ml of water and 35 ml of concentrated hydrochloric acid, and boil for 15 minutes; the *m*-nitroaniline dissolves leaving the sulphur and any unchanged *m*-dinitrobenzene. Filter and precipitate the *m*-nitroaniline from the filtrate by the addition of excess of concentrated aqueous ammonia solution. Filter off the product and recrystallise it from boiling water. The yield of *m*-nitroaniline (bright yellow needles) is 12 g (58%); m.p. 114 °C.

Method 2. Dissolve 18 g (0.075 mol) of crystallised sodium sulphide, $Na_2S,9H_2O$ (1), in 50 ml of water; add 6.0 g (0.0714 mol) of finely powdered sodium hydrogen carbonate in small portions with constant stirring. When the carbonate has dissolved completely, add 50 ml of methanol and cool below 20 °C. Filter off the precipitated sodium carbonate at the pump, using a small Buchner funnel; if necessary add 2–3 g of filter-aid to the reaction mixture before filtering, and prepare a bed of filter-aid in the Buchner funnel since this will aid the retention of the finely divided solid. Wash the precipitate with three 8 ml portions of methanol. Retain the filtrate and washings: these contain about 3.9 g of NaSH in solution and must be used forthwith for the reduction.

Dissolve 6.7 g (0.04 mol) of *m*-dinitrobenzene in 50 ml of hot methanol in a 250-ml round-bottomed flask and add, with shaking, the previously prepared methanolic solution of sodium hydrogen sulphide. Attach a reflux condenser and boil the mixture for 20 minutes; ignore any further sodium carbonate which may precipitate. Allow the reaction mixture to cool and fit the condenser for distillation. Distil off most of the methanol (100–120 ml) from a water bath. Pour the liquid residue with stirring into about 200 ml of cold water. Collect the yellow crystals of *m*-nitroaniline by suction, wash with water and recrystallise from 75 per cent aqueous methanol. The yield of bright yellow crystals, m.p. 114 °C, is 3.7 g (69%).

Note. (1) Crystallised sodium sulphide is very deliquescent and only a sample which has been kept in a tightly stoppered bottle should be used.

IV,59. ETHYL *p*-AMINOBENZOATE (*Benzocaine*)

Method 1

$$p\text{-}O_2N{\cdot}C_6H_4{\cdot}CO_2H \xrightarrow{\text{Sn/HCl}} p\text{-}H_2N{\cdot}C_6H_4{\cdot}CO_2H \xrightarrow{C_2H_5OH/HCl}$$
$$p\text{-}H_2N{\cdot}C_6H_4{\cdot}CO_2C_2H_5$$

Method 2

$$p\text{-}O_2N{\cdot}C_6H_4{\cdot}CO_2H \xrightarrow{C_2H_5OH/HCl} p\text{-}O_2N{\cdot}C_6H_4{\cdot}CO_2C_2H_5 \xrightarrow{H_2/Pt}$$
$$p\text{-}H_2N{\cdot}C_6H_4{\cdot}CO_2C_2H_5$$

Method 1. **p-Aminobenzoic acid.** Place 15 g (0.09 mol) of *p*-nitrobenzoic acid (Section **IV,161**) in a 1-litre round-bottomed flask fitted with a reflux condenser. Introduce 35 g (0.295 mol) of powdered tin and 75 ml of concentrated hydrochloric acid. Heat the mixture gently until the reaction commences, and remove the flame. Shake the flask frequently and take care that the insoluble acid adhering to the sides of the flask is transferred to the reaction mixture: occasional gentle warming may be necessary. After about 20 minutes, most of the tin will have reacted and a clear solution remains. Allow to cool somewhat and decant the liquid into a 1-litre beaker; wash the residual tin by decantation with 15 ml of water, and add the washings to the contents of the beaker. Add concentrated ammonia solution (*d* 0.88) until the solution is *just* alkaline to litmus and digest the suspension of precipitated hydrated tin oxide on a steam bath for 20 minutes. Add 10 g of filter-aid ('Celite'), stir well, filter at the pump and wash with hot water. Transfer the filter cake to a beaker, heat on a water bath with 200 ml of water to ensure extraction of the product and re-filter. Concentrate the combined filtrate and washings until the volume has been reduced to 175–200 ml: filter off any solid which separates. Acidify the liquid to litmus with glacial acetic acid and evaporate on a water bath until crystals commence to separate; cool in ice, filter the crystals at the pump and dry in the steam oven. The yield of *p*-aminobenzoic acid, m.p. 192 °C, is 9.5 g (77%).

Ethyl p-aminobenzoate (*esterification of* p-*aminobenzoic acid*). Place 80 ml of absolute ethanol in a 250-ml two-necked flask equipped with a double surface reflux condenser and a gas inlet tube. Pass dry hydrogen chloride (Section **II,2,*31***) through the alcohol until saturated—the increase in weight is about 20 g—remove the gas inlet tube, introduce 12 g (0.088 mol) of *p*-aminobenzoic acid and heat the mixture under reflux for 2 hours. Upon cooling, the reaction mixture sets to a solid mass of the hydrochloride of ethyl *p*-aminobenzoate. It is better, however, to pour the hot solution into *ca.* 300 ml of water (no hydrochloride separates) and add solid sodium carbonate carefully to the clear solution until it is neutral to litmus. Filter off the precipitated ester at the pump and dry in the air. The yield of ethyl *p*-aminobenzoate, m.p. 91 °C, is 10 g (69%). Recrystallisation from rectified (or methylated) spirit does not affect the m.p.

Method 2. **Ethyl p-nitrobenzoate.** Place 21 g (0.125 mol) of *p*-nitrobenzoic acid (Section **IV,161**), 11.5 g (0.25 mol) of absolute ethanol, 3.8 g of concentrated sulphuric acid and 30 ml of sodium-dried benzene in a 250-ml round-bottomed flask, fit a reflux condenser and heat the mixture under reflux for 16 hours. Add 50 ml of ether to the cold reaction mixture, wash the extract successively with sodium hydrogen carbonate solution and water, dry with magnesium sulphate or calcium chloride and distil off the solvent on a water bath. Remove the last traces of benzene by heating in a bath at 100–110 °C. The residual ethyl *p*-nitrobenzoate (21 g, 86%) solidifies completely on cooling and melts at 56 °C.

Ethyl p-aminobenzoate (*catalytic reduction of ethyl* p-*nitrobenzoate*). The general experimental details may be adapted from those described in Section **I,17,1**. Place a solution of 9.75 g (0.05 mol) of ethyl *p*-nitrobenzoate in 100 ml of rectified spirit together with 0.1 g of Adams' platinum dioxide catalyst in the hydrogenation flask, and shake in hydrogen in the usual manner. The theoretical volume of hydrogen (*ca.* 3600 ml at 24 °C and 760 mmHg) is absorbed in 2.5

hours. Filter off the platinum with suction and rinse the reaction vessel with rectified spirit. Evaporate the alcohol from the combined filtrate and washings on a water bath; the residue solidifies on cooling and weighs 8.2 g. Dissolve the crude ethyl p-aminobenzoate in rectified spirit, add a little decolourising charcoal, boil and filter; heat the filtrate to the boiling point, add hot water to incipient crystallisation and allow to cool. The resulting pure benzocaine has m.p. 90 °C; the yield is 7 g (85%).

E,2. Molecular rearrangements of the Hofmann type.   The general procedure of treating carboxylic acid amides with alkaline hypohalite solution is illustrated and discussed in the aliphatic series (p. 573, **III,R,5**) and is equally applicable to the synthesis of primary aromatic amines (e.g., anthranilic acid and 3-aminopyridine, Section **IV,60**). It is of particular use in those cases where it is required to introduce an amino group into a position on the aromatic ring which, because of the orienting effects of the substituents, cannot be effectively nitrated directly in order to use the direct nitration–reduction sequence.

For the preparation of anthranilic acid the starting material is phthalimide, the cyclic imide ring of which is opened by alkaline hydrolysis in the first step of the reaction to give the sodium salt of phthalimidic acid (the half amide of phthalic acid). The intermediate undergoes the Hofmann reaction in the manner outlined on p. 573, **III,R,5**, yielding o-aminobenzoic acid (anthranilic acid).

The conversion of a carboxylic acid into an amine by treatment with hydrazoic acid in concentrated sulphuric acid is known as the **Schmidt reaction** or **rearrangement**, which often gives higher yields than the related Hofmann rearrangement procedure.

$$R \cdot CO_2H + HN_3 \xrightarrow{H_2SO_4} RNH_2 + CO_2 + N_2$$

The use of the toxic and hazardous hydrazoic acid is avoided by generating it in situ by adding sodium azide gradually to the carboxylic acid in the presence of concentrated sulphuric acid and chloroform (e.g., 3,5-dinitroaniline, Section **IV,61**). The reaction involves the hydrolysis of an intermediate isocyanate (RNCO), which is formed by a mechanistic pathway analogous to that involved in the Hofmann reaction.

665

IV,60. ANTHRANILIC ACID

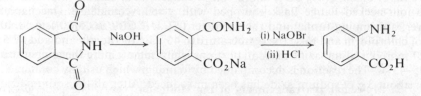

Prepare a solution of 30 g of sodium hydroxide in 120 ml of water in a 350-ml conical flask and cool to 0 °C or below in a bath of ice and salt. Add 26.2 g (8.4 ml, 0.16 mol) of bromine in one portion and shake (or stir) until all the bromine has reacted. The temperature will rise somewhat; cool again to 0 °C or below. Meanwhile, prepare a solution of 22 g of sodium hydroxide in 80 ml of water. Add 24 g (0.163 mol) of finely powdered phthalimide (Section **IV,169**) in one portion to the cold sodium hypobromite solution in the form of a smooth paste with water, rapidly with stirring. Remove the flask from the cooling bath and shake vigorously until a clear yellow solution is obtained (*ca.* 5 minutes). Add the prepared sodium hydroxide solution rapidly and in one portion, heat the solution to 80 °C for about 2 minutes and filter if necessary. Cool in ice and add concentrated hydrochloric acid slowly and with stirring until the solution is *just* neutral (about 60 ml are required). [It is recommended that a little of the alkaline solution be set aside in case too much acid is added.] Precipitate the anthranilic acid completely by the gradual addition of glacial acetic acid (20–25 ml are required): it is advisable to transfer the mixture to a 1-litre beaker as some foaming occurs. Filter off the acid at the pump and wash with a little cold water. Recrystallise from hot water with the addition of a little decolourising carbon; collect the acid on a Buchner funnel and dry at 100 °C. The yield of pure anthranilic acid, m.p. 145 °C, is 14 g (62%).

Cognate preparation 3-Aminopyridine. Prepare a cold sodium hypobromite solution from 32 g (10 ml, 0.2 mol) of bromine and 25 g (0.62 mol) of sodium hydroxide in 250 ml of water. Add in one portion 20 g (0.163 mol) of finely powdered nicotinamide (Section **IV,184**) and stir vigorously for 15 minutes. Warm the solution in a water bath at 75 °C for 45 minutes. Isolate the crude product by continuous ether extraction (Section **I,22**) of the cooled reaction mixture after saturation with sodium chloride. Dry the extract over potassium hydroxide pellets and remove the ether. Crystallise the dark residue from a 4 : 1 mixture of benzene–light petroleum (b.p. 60–80 °C) with the aid of decolourising charcoal. The yield of almost colourless product, m.p. 63 °C, is 9.3 g (61%).

IV,61. 3,5-DINITROANILINE

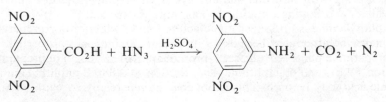

Place a solution of 50 g (0.24 mol) of 3,5-dinitrobenzoic acid (Section **IV,174**) in 90 ml of oleum (10% SO_3) and 20 ml of concentrated sulphuric acid in a 1-litre four-necked flange flask equipped with a reflux condenser, mechanical stirrer, a dropping funnel and thermometer (*FUME CUPBOARD!*). Add 100 ml of chloroform and raise the temperature to 45 °C. Stir rapidly and add 17.5 g (0.27 mol) of sodium azide in small portions whilst maintaining the temperature at 35–45 °C. The reaction is accompanied by foaming, which usually commences after about 3 g of sodium azide has been introduced. After all the sodium azide has been added raise the temperature so that the chloroform refluxes vigorously and maintain this temperature for 3 hours. Then cool the reaction mixture, pour it cautiously on to 500 g of crushed ice and dilute with 3 litres of water. After 1 hour, separate the yellow solid by filtration at the pump, wash well with water and dry at 100 °C. The yield of 3,5-dinitroaniline, m.p. 162–163 °C, is 39 g (90%). The m.p. is unaffected by recrystallisation from dilute ethanol.

E,3. The replacement of an aromatic hydroxyl group by an amino group. The direct replacement of the hydroxyl group in simple phenols by an amino or substituted amino group requires drastic conditions and the method is not suitable for laboratory preparations. With the polyhydric phenols, and more particularly with the naphthols, such replacements occur more readily. Thus 2-naphthol is converted into 2-naphthylamine by heating with ammoniacal ammonium sulphite solution at 150 °C in an autoclave. The reaction (the **Bucherer reaction**) depends upon the addition of the hydrogen sulphite ion to the keto form of the naphthol and the subsequent reaction with ammonia.

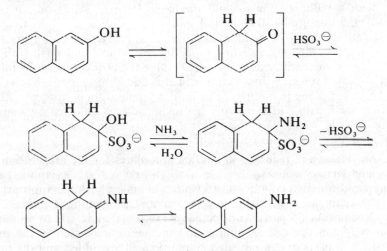

The reaction is reversible; thus 2-naphthylamine can be converted into 2-naphthol by heating with aqueous sodium hydrogen sulphite solution and then adding alkali and boiling until all the ammonia is expelled.

2-Naphthylamine is no longer manufactured and its laboratory preparation should never be attempted because of its potent carcinogenic properties. For many preparative purposes (e.g., see 2-bromonaphthalene, cognate preparation in Section **IV,81**, and 2-naphthoic acid, Section **IV,168**), 2-naphthylamine-1-sulphonic acid may be used. This is obtained commercially by cautious treatment of 2-naphthol with sulphuric acid—the sulphonic acid group entering the 1-position—followed by a Bucherer reaction.

The Bucherer reaction is illustrated and its reversibility demonstrated in Section **IV,62**, wherein the amino group of 2-amino-5-naphthol-7-sulphonic acid is replaced by the *p*-tolylamino group via the corresponding naphthol intermediate.

IV,62. 2-*p*-TOLYLAMINO-5-NAPHTHOL-7-SULPHONIC ACID

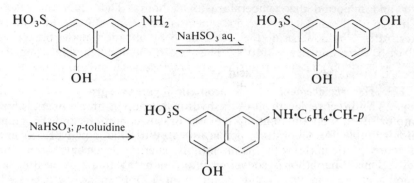

Reflux a mixture of 10.8 g (0.1 mol) of pure *p*-toluidine, 10.8 g (0.057 mol) of 2-amino-5-naphthol-7-sulphonic acid ('J' acid), 8.4 g (0.08 mol) of sodium meta-bisulphite and 25 ml of water for 30 hours in a 250-ml three-necked flask, equipped with a reflux condenser and mechanical stirrer. Add sodium carbonate until the mixture is alkaline and remove the excess of *p*-toluidine by steam distillation. Keep the residual solution in a refrigerator until crystallisation is complete, filter with suction on a Buchner funnel and wash with 10 ml of saturated sodium chloride solution. Dissolve the product in *ca.* 35 ml of hot water to which sufficient hydrochloric acid is added to render the mixture acid to Congo red. Keep in a refrigerator until crystallisation is complete, filter with suction, wash with a little ice-cold hydrochloric acid, followed by a small volume of ice-cold water. Dry the residual 2-*p*-tolylamino-5-naphthol-7-sulphonic acid at 100 °C; the yield is 9.5 g (60%).

E,4. Alkylation and reductive alkylation procedures for the preparation of secondary and tertiary amines. Simple *N*-alkyl- and *N,N*-dialkylanilines are readily prepared commercially by the alkylation of aniline with the appropriate alcohol. For example, *N*-methylaniline is prepared by heating a mixture of aniline hydrochloride (55 parts) and methanol (16 parts) at 120 °C in an autoclave. For *N,N*-dimethylaniline, aniline and methanol are mixed in the proportion 80 : 78, 8 parts of concentrated sulphuric acid are added and the mixture heated in an autoclave at 230–235 °C at a pressure of 25–30 atmospheres. *N*-Ethyl- and *N,N*-diethylaniline are prepared similarly. In the laboratory, alkylation of the amino group, to yield secondary and tertiary amines, is effected by reaction with the appropriate alkyl halide; Section **IV,63** includes the preparation of *N*-benzylaniline and general procedures for the synthesis of a range of *N*-alkyl- and *N,N*-dialkylanilines.

N-Alkylanilines may be purified by converting them into the *N*-nitroso derivative with nitrous acid followed by reduction of the separated nitroso compound with tin and hydrochloric acid, thus regenerating the N-alkylaniline. For example:

$$C_6H_5NHCH_3 \xrightarrow[\text{HCl}]{\text{NaNO}_2} C_6H_5NCH_3 \xrightarrow{\text{Sn/HCl}} C_6H_5NHCH_3$$
$$\begin{array}{c} | \\ NO \end{array}$$

N,N-Dialkylanilines may be purified by refluxing with an excess of acetic anhydride: any unchanged aniline and N-alkylaniline is converted into the relatively non-volatile acetyl derivative.

$$C_6H_5NH_2 + (CH_3 \cdot CO)_2O \longrightarrow C_6H_5NH \cdot CO \cdot CH_3 + CH_3 \cdot CO_2H$$
$$C_6H_5NHR + (CH_3 \cdot CO)_2O \longrightarrow C_6H_5N(CO \cdot CH_3)R + CH_3 \cdot CO_2H$$

Upon fractionation the acetic acid and unreacted acetic anhydride pass over first, followed by the pure N,N-dialkylaniline.

A convenient method for preparing in good yield a pure N,N-dialkylaniline or substituted aniline (Section **IV,64**) directly from the corresponding amine consists on heating the latter with the appropriate trialkyl phosphate.

$$3ArNH_2 + 2(RO)_3PO \longrightarrow 3ArNR_2 + 2H_3PO_4$$

Secondary amines can be prepared from the primary amine and carbonyl compounds by way of the reduction of the derived Schiff bases, with or without the isolation of these intermediates. This procedure represents one aspect of the general method of reductive alkylation discussed on p. 568, **III,R,3**. With aromatic primary amines and aromatic aldehydes the Schiff bases are usually readily isolable in the crystalline state and can then be subsequently subjected to a suitable reduction procedure, often by hydrogenation over a Raney nickel catalyst at moderate temperatures and pressures. A convenient procedure, which is illustrated in Section **IV,65**, uses sodium borohydride in methanol, a reagent which owing to its selective reducing properties (p. 353, **III,D,1**) does not affect other reducible functional groups (particularly the nitro group) which may be present in the Schiff base; contrast the use of sodium borohydride in the presence of palladium-on-carbon, p. 662.

$$Ar^1NH_2 + OHC \cdot Ar^2 \longrightarrow Ar^1N = CH \cdot Ar^2 \xrightarrow{\text{NaBH}_4} Ar^1 \cdot NH \cdot CH_2 \cdot Ar^2$$

IV,63. N-BENZYLANILINE

$$C_6H_5 \cdot CH_2Cl + C_6H_5 \cdot NH_2 \xrightarrow{\text{Na}_2CO_3} C_6H_5 \cdot CH_2 \cdot NH \cdot C_6H_5$$

Equip a 500-ml three-necked flask with a separatory funnel, a mechanical stirrer and a reflux condenser; mount the assembly on a water bath. Place 35 g of pure sodium hydrogen carbonate, 35 ml of water and 124 g (121 ml, 1.33 mol) of aniline in the flask, and 42 g (38 ml, 0.33 mol) of freshly distilled benzyl chloride (b.p. 177–179 °C) in the separatory funnel protected by a calcium chloride guard-tube. Heat the flask and contents to 90–95 °C, stir vigorously and run in the benzyl chloride slowly (about 1 hour). Continue the heating and stirring for a further 3 hours. Allow to cool. Filter with suction, separate the organic layer from the filtrate and wash it with 25 ml of saturated salt solution.

Dry with magnesium sulphate and filter again with suction. Distil from a flask with fractionating side-arm (compare Fig. 1,110) under reduced pressure: aniline (about 80 g) distils at 81 °C/12 mmHg and the temperature rises rapidly. Collect the benzylaniline at 170–190 °C/12 mmHg (most of it distils at 178–180 °C/12 mmHg); this solidifies on cooling, melts at 34–36 °C, and is sufficiently pure for most purposes. The yield is 52 g (85%). If required perfectly pure, it may be recrystallised from about 35 ml of light petroleum, b.p. 60–80 °C; cool the solution in a freezing mixture to induce crystallisation, filter at the pump, wash with a little cold light petroleum, press and dry. The recrystallised N-benzylaniline has m.p. 36 °C.

Notes on the preparation of secondary alkylarylamines. The preparation of N-propyl-, N-isopropyl- and N-butyl-anilines can be conveniently carried out by heating the alkyl bromide with an excess (2.5–4 mol) of aniline for 6–12 hours. The tendency for the alkyl halide to yield the corresponding tertiary amine is thus repressed and the product consists almost entirely of the secondary amine and the excess of primary amine combined with the hydrogen bromide liberated in the reaction. The separation of the primary and secondary amines is easily accomplished by the addition of an excess of 50 per cent zinc chloride solution: aniline and its homologues form sparingly soluble additive compounds of the type B_2ZnCl_2 whereas the alkylanilines do not react with zinc chloride in the presence of water. The excess of primary amine can be readily recovered by decomposing the chlorozincate with sodium hydroxide solution followed by steam distillation or solvent extraction. The yield of secondary amine is about 70 per cent of the theoretical.

The experimental details for N-**propylaniline** are as follows. Reflux a mixture of 230 g (2.5 mol) of aniline and 123 g (1 mol) of propyl bromide for 8–10 hours. Allow to cool, render the mixture alkaline and add a solution of 150 g (1.1 mol) of zinc chloride in 150 g of water. Cool the mixture and stir: after 12 hours, filter at the pump and drain well. Extract the thick paste several times with boiling light petroleum, b.p. 60–80 °C (it is best to use a Soxhlet apparatus), wash the combined extracts successively with water and dilute ammonia solution, and then dry over anhydrous potassium carbonate or magnesium sulphate. Remove the solvent on a water bath, and distil the residue through a well-lagged fractionating column. Collect the N-propylaniline at 218–220 °C; the yield is 80 g (59%). Treat the pasty solid chlorozincate with an excess of sodium hydroxide solution and steam distil: 130 g of pure aniline are recovered.

N-Isopropylaniline, b.p. 206–208 °C, and N-butylaniline, b.p. 235–237 °C, may be similarly prepared.

Notes on the preparation of tertiary alkylarylamines. Pure dialkylanilines may be prepared by refluxing the monoalkylaniline (1 mol) with an alkyl bromide (2 mol) for 20–30 hours; the solid product is treated with excess of sodium hydroxide solution, the organic layer separated, dried and distilled. The excess of alkyl bromide passes over first, followed by the dialkylaniline. **N,N-dipropylaniline**, b.p. 242–243 °C, and **N,N-dibutylaniline**, b.p. 269–270 °C, are thus readily prepared.

If the tertiary amines are suspected of being contaminated with primary and/or secondary amines, they may be purified by treatment with acetic anhydride: the following procedure is illustrative. Into a 250-ml round-bottomed flask, fitted with a reflux condenser, place 50 g (52.5 ml, 0.414 mol) of a good commercial sample of dimethylaniline and 25 g (23 ml, 0.245 mol) of acetic anhydride. Heat under reflux for 3 hours and allow to cool. Transfer to a 100-ml distillation flask and distil using an air bath. Some acetic acid and the excess acetic anhydride passes over first, followed by pure dimethylaniline (a colourless liquid) at 193–194 °C. There is a small dark residue in the flask. The yield depends upon the purity of the commercial sample but is not usually less than 40 g.

Purification of N-methylaniline. The laboratory preparation of N-methyl and N-ethylanilines is hardly worth while since commercial grades of good quality (97–99% pure) are available. The following procedure, however, illustrates a useful and instructive method of purifying crude samples of secondary alkylarylamines via the derived N-nitroso compound [**CAUTION**: (1)].

N-*Nitroso*-N-*methylaniline* (methylphenylnitrosamine).   Place 53.5 g (0.5 mol) of commercial N-methylaniline, 72.5 ml of concentrated hydrochloric acid and 200 g of crushed ice in a 500-ml beaker equipped with a mechanical stirrer. Support a separatory funnel with a long bent stem containing a solution of 36 g (0.52 mol) of sodium nitrite in 125 ml of water over the beaker. Stir the solution and run in the sodium nitrite solution during 10 minutes; do not allow the temperature to rise above 10 °C and add more ice if necessary. Continue the stirring for a further hour. Separate the oily layer, wash it once with 50 ml of water and dry it with magnesium or calcium sulphate. Distil under reduced pressure. Collect the N-nitroso-N-methylaniline (a pale yellow liquid) at 120 °C/13 mmHg. The yield is about 65 g (96%).

Reduction of N-*nitroso*-N-*methylaniline*.   Into a 1-litre round-bottomed flask, fitted with a reflux condenser, place 39 g (0.29 mol) of N-nitroso-N-methylaniline and 75 g of granulated tin. Add 150 ml of concentrated hydrochloric acid in portions of 25 ml (compare Section **IV,54**); do not add the second portion until the vigorous action produced by the previous portion has subsided, etc. Heat the reaction mixture on a water bath for 45 minutes, and allow to cool. Add cautiously a solution of 135 g of sodium hydroxide in 175 ml of water, and steam distil (see Section **I,25**); collect about 500 ml of distillate. Saturate the distillate with salt, separate the organic layer, extract the aqueous layer with 50 ml of ether and combine the extract with the organic layer. Dry with anhydrous potassium carbonate, remove the ether on a water bath and distil the residual liquid collecting the pure methylaniline at 193–194 °C as a colourless liquid. The yield is 23 g (74%).

Note.   (1) The potentially carcinogenic nature of N-nitroso compounds is again emphasised (see Section **I,3,D,**5(ii)).

IV,64.   N,N-DIMETHYLANILINE

$$3C_6H_5 \cdot NH_2 + 2(CH_3O)_3PO \longrightarrow 3C_6H_5 \cdot N(CH_3)_2 + 2H_3PO_4$$

Place 28 g (27.5 ml, 0.3 mol) of pure aniline and 28 g (23 ml, 0.2 mol) of purified trimethyl phosphate in a 500-ml round-bottomed flask equipped with a reflux condenser. Heat gently at first and remove the flame when the vigorous and exothermic reaction commences. When the latter subsides, two layers are present; heat under gentle reflux for two hours. Cool the mixture to about 50 °C, add a solution of 25 g of sodium hydroxide in 100 ml of water, reflux the mixture for 1 hour, then pour into a 600-ml beaker and allow to cool to room temperature. Pour off the oily layer of amine from the solid sodium phosphate, add water to the latter and extract the aqueous solution with ether. Dry the combined oil and ether extract with magnesium sulphate, distil off the ether, treat the residue with an equal volume of acetic anhydride and allow to stand overnight. (The acetic anhydride treatment will remove any monoalkylaniline present.) Then add hydrochloric acid (20 ml of the concentrated acid and 30 ml of water), shake until the base dissolves, extract the solution with two 30 ml portions of ether and add 25 per cent sodium hydroxide solution to the water layer to liberate the base. Collect the oil by extracting the mixture with ether, dry the ethereal solution with magnesium sulphate and remove the ether on a water bath. Distil the residue, using an air condenser, and collect the dimethylaniline at 192–193 °C. The yield is 28 g (76.5%).

Cognate preparation   N,N-Diethylaniline.   Use 28 g of pure aniline and 36 g (34 ml, 0.2 mol) of purified triethyl phosphate, and proceed exactly as described for dimethylaniline. The reaction is not so vigorous initially. Separation into two layers occurs after 30 to 90 minutes. The yield of diethylaniline, b.p. 215–216 °C, is 41–45 g (91–100%).

IV,65. N-(m-NITROBENZYL)ANILINE

$$m\text{-}O_2N\text{·}C_6H_4\text{·}CHO + H_2N\text{·}C_6H_5 \longrightarrow$$

$$m\text{-}O_2N\text{·}C_6H_4\text{·}CH = N\text{·}C_6H_5 \xrightarrow{\text{NaBH}_4} m\text{-}O_2N\text{·}C_6H_4\text{·}CH_2\text{·}NH\text{·}C_6H_5$$

N-(m-Nitrobenzylidene)aniline.   In a 100-ml round-bottomed flask fitted with a reflux condenser, place 7.5 g (0.05 mol) of m-nitrobenzaldehyde, 4.6 g (0.05 mol) of aniline and 20 ml of rectified spirit (1). Heat the solution under reflux, using a water bath, for 20 minutes, add water until a slight cloudiness persists and set the solution on one side to cool. The oil which separates may be induced to crystallise by rubbing with a glass rod. Collect the solid deposit by filtration and wash well with cold aqueous ethanol; 10 g (88%) of air-dried crude Schiff base is obtained which is sufficiently pure for conversion into the amine. Recrystallise a small portion from aqueous methanol to give light-fawn crystals having m.p. 65–66 °C.

N-(m-Nitrobenzyl)aniline.   Fit a two-necked round-bottomed flask with a reflux condenser, place a stopper in the side-neck and insert a magnetic follower. Mount the flask on a water bath sited on a magnetic-stirrer unit. In the flask place 10 g (0.044 mol) of the above Schiff base and add 100 ml of methanol. Warm the solution to about 40 °C and with stirring add portionwise, over a period of 30 minutes, 1.7 g (0.044 mol) of sodium borohydride; a steady evolution of hydrogen occurs. Now heat the solution under reflux for a further 15 minutes, then add 100 ml of water and cool. Collect the solid amine which, after air-drying, has m.p. 80–81 °C; upon recrystallisation from aqueous methanol 9 g (90%) of pure N-(m-nitrobenzyl)aniline, m.p. 84–85 °C, is obtained.

Note.   (1) The Schiff base may be prepared by heating the components in the absence of solvent at 100 °C for 15 minutes, cooling, and then stirring the product with methanol to induce crystallisation.

Cognate preparations   N-(p-Methoxybenzyl)aniline.   Prepare the Schiff base, N-(p-methoxybenzylidene)aniline from 6.8 g (0.05 mol) of anisaldehyde and

4.6 g (0.05 mol) of aniline in 20 ml of rectified spirit under the conditions described above. The yield of crude product is 8.5 g (81%); recrystallisation of a small portion gives gives white plates, m.p. 57–58 °C. Reduce 7 g (0.034 mol) of crude product with 1.1 g (0.034 mol) of sodium borohydride. The yield of pure *N*-(*p*-methoxybenzyl)aniline, m.p. 46–47 °C, is 6.1 g (85%).

N-Benzyl-*m*-nitroaniline. Prepare the Schiff base from 5.3 g (0.05 mol) of benzaldehyde and 6.9 g (0.05 mol) of *m*-nitroaniline in 30 ml of rectified spirit. Yield of crude product is 10 g (88%); the pure compound has m.p. 71–72 °C. Reduce 7.8 g (0.034 mol) of crude Schiff base with 1.1 g (0.034 mol) of sodium borohydride. Pure *N*-benzyl-*m*-nitroaniline is obtained as orange-yellow crystals, m.p. 106–107 °C; the yield is 7.0 g (90%).

F SUBSTITUTION PRODUCTS OF AROMATIC AMINES

1. Nuclear substitution products (Sections **IV,66** to **IV,72**).
2. Acylated amines and their substitution reactions (Sections **IV,73** to **IV,77**).

F,1. Nuclear substitution products. The free amino group strongly activates the aromatic ring towards electrophilic attack and aromatic substitution of amines often results in polysubstitution. For example, the bromination of aniline yields largely 2,4,6-tribromoaniline (Section **IV,66**).

Monosubstitution of the free amine may be achieved by using a less reactive electrophile. Thus aniline and *o*-toluidine may be mono iodinated (Section **IV,67**) by treatment with iodine (in the presence of sodium hydrogen carbonate or calcium carbonate to remove the liberated hydrogen iodide), the substituent entering the position *para* to the amino group. Direct iodination can also be effected by using the more powerfully electrophilic reagent, iodine mono-chloride ($\overset{\delta+}{I} \rightarrow \overset{\delta-}{Cl}$); with *p*-aminobenzoic acid both of the positions *ortho* to the amino group are substituted to give 4-amino-3,5-diiodobenzoic acid (cognate preparation in Section **IV,67**).

The nitrosonium ion ($\overset{\oplus}{NO}$), generated *in situ* from sodium nitrate in the presence of hydrochloric acid at 0–5 °C, is also a weak electrophile and with tertiary amines, e.g., *N,N*-dimethylaniline, ring substitution occurs leading to the *p*-nitroso derivative (Section **IV,68**).

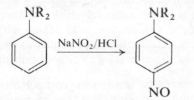

Secondary aromatic amines under these conditions form initially the *N*-nitroso derivative (see notes on the purification of secondary amines in Section **IV,63**), which when treated with hydrogen chloride in anhydrous ethanol–ether solution rearranges to the nuclear substituted nitrosoamine (e.g., *p*-nitroso-*N*-

methylaniline, cognate preparation in Section **IV,68**). This rearrangement proceeds via the intermediate formation of the electrophilic nitrosyl chloride:

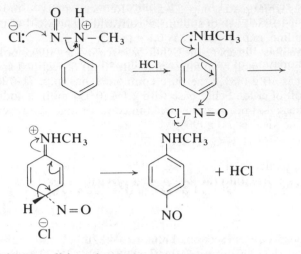

When conditions for the electrophilic substitution are strongly acidic, extensive protonation of the nitrogen lone electron pair occurs and its activating influence is considerably diminished. For example, if aniline is treated with an excess of concentrated sulphuric acid, and the resulting mixture (which contains aniline sulphate) is heated at 180 °C until a test portion when mixed with sodium hydroxide solution no longer liberates aniline, *p*-aminobenzenesulphonic acid (sulphanilic acid) is formed; this separates as the dihydrate upon pouring the cooled mixture into water. The mechanism of this reaction is uncertain; a possible pathway is the rearrangement of the intermediate phenylsulphamic acid (I). The product is more appropriately represented by the zwitterionic structure (II) (Section **IV,69**).

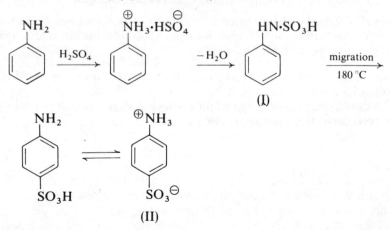

1-Naphthylamine similarly yields 1-naphthylamine-4-sulphonic acid (naphthionic acid, Section **IV,70**).

2-Aminobenzenesulphonic acid (orthanilic acid, Section **IV,71**) is readily prepared by the reduction of 2-nitrobenzenesulphonic acid. The latter may be prepared by the hydrolysis of the corresponding sulphonyl chloride which is obtained from di-*o*-nitrophenyl disulphide. The preparation of this disulphide involves the use of the reactive aryl halide, 2-chloronitrobenzene (cf. Sections **IV,105** and **IV,112**) in a disulphide-forming nucleophilic displacement using sodium disulphide.

A straightforward route to the *meta* isomer (metanilic acid, Section **IV,72**) is provided by the sulphonation of nitrobenzene followed by reduction of the nitro group.

IV,66. 2,4,6-TRIBROMOANILINE

$$C_6H_5 \cdot NH_2 + 3Br_2 \longrightarrow H_2N \cdot C_6H_2Br_3 + 3HBr$$

Assemble the apparatus depicted in Fig. IV,3. The flask B has a capactiy of 100 ml and the flask A is 1 litre. Into the flask A place 10 g (0.11 mol) of aniline,

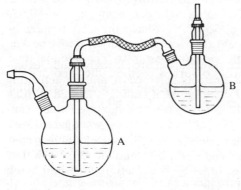

Fig. IV,3

100 ml of water and 10 ml of concentrated hydrochloric acid; shake until the aniline has dissolved and dilute with 400 ml of water. Charge the flask B with 60 g (19 ml, 0.38 mol) of bromine. Surround the flask A by an ice bath and immerse the flask B in a water bath maintained at 30–40 °C. Interpose a wash bottle partially filled with water between the reaction flask and a water pump; this will permit the rate of aspiration to be observed and will also serve to detect the escape of bromine vapours from the reaction flask since a small amount of bromine will impart a distinctly yellow colour to the water. Apply gentle suction by means of a water pump. Continue the passage of bromine vapour until the solution in A assumes a distinctly yellow colour (2–3 hours); the reaction is then complete. Filter the tribromoaniline on a Buchner funnel, wash it thoroughly with water to remove hydrobromic acid and suck as dry as possible. Recrystallise from industrial (or rectified) spirit. The yield is 22 g (63%); m.p. 120 °C.

An alternative method of preparation consists in dissolving the aniline in 4 times its weight of glacial acetic acid in a beaker, and running in slowly from a dropping funnel, while the solution is well stirred with a mechanical stirrer, the theoretical amount of bromine dissolved in twice its volume of glacial acetic acid. The beaker should be cooled in ice during the addition as the reaction is exothermic. The final product (a pasty mass) should be coloured yellow by the addition of a little more bromine if necessary. Pour into excess of water, filter at the pump, wash well with water, press thoroughly and dry. The yield of tribromoaniline, m.p. 119–120 °C, is quantitative. Recrystallise a small portion from industrial (or rectified) spirit; m.p. 120 °C.

2,4,6-Tribromoacetanilide. Dissolve 1 g of 2,4,6-tribromoaniline in 20 ml of acetic anhydride and add 2 drops of concentrated sulphuric acid. After 10 minutes, pour the reaction mixture into excess of warm water. Filter off the tribromoacetanilide, wash and dry: the m.p. is 231 °C. Recrystallise from alcohol: the m.p. is raised to 232 °C.

IV,67. p-IODOANILINE

$$C_6H_5 \cdot NH_2 + I_2 + NaHCO_3 \longrightarrow p\text{-}I \cdot C_6H_4 \cdot NH_2 + NaI + CO_2 + H_2O$$

Into a 1-litre beaker, provided with a mechanical stirrer, place 37 g (36 ml, 0.4 mol) of aniline, 50 g (0.6 mol) of sodium hydrogen carbonate and 350 ml of water; cool to 12–15 °C by the addition of a little crushed ice. Stir the mixture, and introduce 85 g (0.33 mol) of powdered, resublimed iodine in portions of 5–6 g at intervals of 2–3 minutes so that all the iodine is added during 30 minutes. Continue stirring for 20–30 minutes, by which time the colour of the free iodine in the solution has practically disappeared and the reaction is complete. Filter the crude p-iodoaniline with suction on a Buchner funnel, drain as completely as possible and dry it in the air. Save the filtrate for the recovery of the iodine (1). Place the crude product in a 750-ml round-bottomed flask fitted with a reflux double surface condenser, add 325 ml of light petroleum, b.p. 60–80 °C, and heat in a water bath maintained at 75–80 °C. Shake the flask frequently and after about 15 minutes, slowly decant the clear hot solution into a beaker set in a freezing mixture of ice and salt, and stir constantly. The p-iodoaniline crystallises almost immediately in almost colourless needles; filter and dry the crystals in the air. Return the filtrate to the flask for use in a second extraction as before (2). The yield of p-iodoaniline, m.p. 62–63 °C, is 60 g (82%).

Notes. (1) The **iodine may be recovered** from the aqueous filtrate, containing sodium iodide, in the following manner. Add 33 ml of concentrated sulphuric acid and a solution of 65 g of sodium dichromate in 65 ml of water. Allow the iodine to settle, wash it three times by decantation, filter and allow to dry on a clock glass. The weight of crude iodine is about 50 g.

(2) Two extractions usually suffice, but if much organic material remains, a third extraction should be made. If the p-iodoaniline from the second and third extractions is coloured, it should be refluxed for a short period in light petroleum solution with a little decolourising carbon and filtered through a hot water funnel (**CAUTION:** inflammable).

Cognate preparations **2-Amino-5-iodotoluene.** Triturate 20 g (0.14 mol) of dry o-toluidine hydrochloride and 35.5 g (0.14 mol) of powdered iodine in a mortar and then grind in 17.5 g of precipitated calcium carbonate. Transfer the mixture to a conical flask, and add 100 ml of distilled water with vigorous shaking of the flask. Allow the mixture to stand for 45 minutes with occasional agitation, then heat gradually to 60–70 °C for 5 minutes, and cool. Transfer the contents of the flask to a separatory funnel, extract the amine with three 80 ml portions of ether, dry the extract with anhydrous calcium chloride or magnesium sulphate and remove the excess of solvent. The crude 2-amino-5-iodotoluene separates in dark crystals. Recrystallise from 50 per cent alcohol; nearly white crystals, m.p. 87 °C, are obtained, 26 g (80%).

4-Amino-3,5-diiodobenzoic acid. In a 2-litre beaker, provided with a mechanical stirrer, dissolve 10 g (0.073 mol) of pure p-aminobenzoic acid, m.p. 192 °C (Section **IV,59**), in 450 ml of warm (75 °C) 12.5 per cent hydrochloric acid. Add a solution of 48 g (0.295 mol) of iodine monochloride (Section

II,2,37) in 40 ml of 25 per cent hydrochloric acid and stir the mixture for one minute: during this time a yellow precipitate commences to appear. Dilute the reaction mixture with 1 litre of water whereupon a copious precipitate is deposited. Raise the temperature of the well-stirred mixture gradually and maintain it at 90 °C for 15 minutes. Allow to cool to room temperature, filter, wash thoroughly with water and dry in the air; the yield of crude acid is 24 g. Purify the product by dissolving it in dilute sodium hydroxide solution and precipitate with dilute hydrochloric acid: the yield of air-dried 4-amino-3,5-diiodobenzoic acid, m.p. > 350 °C, is 23 g (81%).

IV,68. p-NITROSO-N,N-DIMETHYLANILINE

$$C_6H_5 \cdot N(CH_3)_2 \xrightarrow{HNO_2} p\text{-}ON \cdot C_6H_4 \cdot N(CH_3)_2$$

Dissolve 30 g (31.5 ml, 0.25 mol) of N,N-dimethylaniline in 105 ml of concentrated hydrochloric acid contained in a 600-ml beaker, and add finely crushed ice until the temperature falls below 5 °C. Stir the contents of the beaker mechanically (or, less satisfactorily, with a thermometer) and slowly add (ca. 10 minutes) a solution of 18 g (0.26 mol) of sodium nitrite in 30 ml of water from a separatory funnel, the stem of which dips beneath the surface of the liquid. Maintain the temperature below 8 °C by the addition of ice, if necessary. When all the nitrite solution has been added, allow the mixture to stand for 1 hour, filter the yellow crystalline p-nitrosodimethylaniline hydrochloride at the pump, wash it with 40 ml of dilute hydrochloric acid (1 : 1), drain well and finally wash with a little alcohol. The yield is good and depends upon the purity of the original dimethylaniline. If the pure hydrochloride is required, it may be recrystallised from hot water in the presence of a little dilute hydrochloric acid; yellow needles, m.p. 177 °C. Recrystallisation is, however, unnecessary if the free base is to be prepared.

Transfer 30 g of the hydrochloride to a 500-ml separatory funnel, add 100 ml of water and shake until a thin paste of uniform consistency is obtained; add cold 10 per cent aqueous sodium hydroxide solution with shaking until the whole mass has become bright green (the colour of the free base) and the mixture has an alkaline reaction. Extract the free base by shaking with two 60 ml portions of benzene (1). Dry the combined benzene extracts with a little anhydrous potassium carbonate, and filter into a distilling flask fitted with a water condenser. Distil off about half of the benzene. Upon cooling the residual solution, the p-nitrosodimethylaniline crystallises in deep green leaflets. Filter these off and dry them in the air. The yield of p-nitrosodimethylaniline, m.p. 85 °C, from the hydrochloride is almost quantitative.

Note. (1) The base is only slightly soluble in ether, thus rendering its use uneconomical. It may be extracted with chloroform and precipitated from the dried chloroform solution with carbon tetrachloride.

Cognate preparation

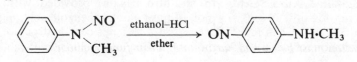

p-Nitroso-N-methylaniline. Dissolve 5 g of N-nitroso-N-methylaniline [CAUTION: see Section IV,63, Note (1)] in 10 ml of anhydrous ether, and add 20 g of a saturated solution of hydrogen chloride in absolute ethanol. Allow to stand. After some time a mass of crystalline needles of the hydrochloride of p-nitroso-N-methylaniline separates. Filter with suction on a sintered glass funnel and wash with a mixture of alcohol and ether. Dissolve the solid in water and add a slight excess of sodium carbonate solution or dilute ammonia solution. Filter off the blue-green free base, and recrystallise it from benzene. The yield of p-nitroso-N-methylaniline, m.p. 118 °C, is 4.5 g (90%).

IV,69. SULPHANILIC ACID

$$C_6H_5 \cdot NH_2 + H_2SO_4 \longrightarrow p\text{-}H_3\overset{\oplus}{N} \cdot C_6H_4 \cdot SO_3^{\ominus}$$

Place 20.4 g (20 ml, 0.22 mol) of aniline in a 250-ml round-bottomed flask and cautiously add 74 g (40 ml) of concentrated sulphuric acid in small portions; swirl the mixture gently during the addition and keep it cool by occasionally immersing the flask in cold water. Support the flask in an oil bath, and heat the mixture at 180–190 °C (fume cupboard) for about 5 hours (1). The sulphonation is complete when a test portion (2 drops) is completely dissolved by 3–4 ml of ca. 2 M-sodium hydroxide solution without leaving the solution cloudy. Allow the product to cool to about 50 °C and pour it carefully with stirring into 400 g of cold water or of crushed ice. Allow to stand for 10 minutes, and collect the precipitated sulphanilic acid on a Buchner funnel, wash it well with water and drain. Dissolve the crude sulphanilic acid in the minimum volume of boiling water (450–500 ml); if the resulting solution is coloured, add about 4 g of decolourising carbon and boil for 10–15 minutes. Filter through a hot water funnel (Fig. I,41) or through a preheated Buchner funnel. Upon cooling, the sulphanilic acid dihydrate separates in colourless crystals. When the filtrate is quite cold, filter the crystals with suction, wash with about 10 ml of cold water and press thoroughly with a wide glass stopper. Dry between sheets of filter-paper or in a desiccator containing anhydrous calcium chloride; in the latter case, the water of crystallisation (and hence the crystalline form) is lost. The yield of sulphanilic acid is 20–22 g (52–58%). The substance does not melt sharply and no attempt should be made to determine the melting point; the crystals are efflorescent.

Note. (1) If 40 ml of 10 per cent oleum is cautiously added to the aniline sulphate mixture, sulphonation proceeds much more rapidly and the time of heating is reduced from 5 hours to 1 hour.

IV,70. NAPHTHIONIC ACID (1-Naphthylamine-4-sulphonic acid)

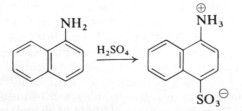

This preparation must be carried out in an efficient fume cupboard.

Place a mixture of 25 g (0.175 mol) of 1-naphthylamine and 125 g (69.5 ml) of concentrated sulphuric acid in a 250-ml round-bottomed flask, and heat in an oil bath for 4–5 hours or until a test sample, when made alkaline with sodium hydroxide solution and extracted with ether, yields no naphthylamine upon evaporation of the ether. Pour the warm reaction mixture cautiously and with stirring into 300 ml of cold water; the difficultly soluble naphthionic acid, which may be contaminated with a little naphthylamine sulphate, separates out. When cold, filter off the acid at the pump and wash it with cold water until free from sulphuric acid. Dissolve the crude naphthionic acid in the minimum volume (about 350 ml) of 5 per cent sodium hydroxide solution (i.e., until about neutral) and saturate the resulting solution of sodium naphthionate with sodium chloride. Allow to stand, when sodium naphthionate separates as white crystals. Filter with suction, drain and dry in an oven. Place the solid in a small beaker with 20 ml of carbon tetrachloride, cover the beaker with a watch glass and boil the mixture gently on a water bath for 10 minutes, filter by gentle suction, and wash the sodium naphthionate with a little solvent. This process removes any naphthylamine which may be present. The yield of sodium naphthionate is 20–35 g (47–81%); this is the form commonly encountered in commerce. To prepare the free acid, dissolve the sodium salt in the minimum volume of boiling water, add the calculated quantity of concentrated hydrochloric acid corresponding to the weight of sodium salt employed (acid to Congo red) and allow to cool. Collect the resulting naphthionic acid hemihydrate upon a Buchner funnel, wash with a little cold water, drain well and dry upon filter-paper or in the steam oven. If desired, it may be recrystallised from boiling water. The yield is 10–18 g (25–44%).

IV,71. ORTHANILIC ACID (2-Aminobenzenesulphonic acid)

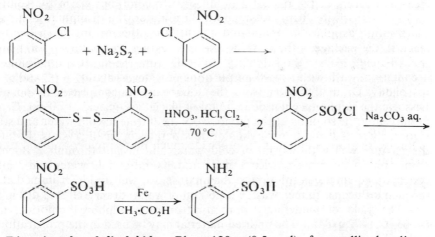

Di-o-nitrophenyl disulphide. Place 120 g (0.5 mol) of crystallised sodium sulphide (1) and 500 ml of rectified spirit in a 1-litre round-bottomed flask provided with a reflux condenser. Heat the flask on a water bath until the sulphide dissolves. Then add 16 g (0.5 mol) of finely powdered sulphur and continue the heating until all the sulphur dissolves forming a brownish-red solution of sodium disulphide (2). Prepare a solution of 105 g (0.66 mol) of o-chloronitrobenzene in 175 ml of rectified spirit in a 2-litre round-bottomed flask equipped with a reflux condenser; by means of a pressure-equalising dropping funnel, add the sodium disulphide solution down the condenser slowly and at such a rate that the reaction is under control. Heat the mixture on a water bath, gently at first until the violent reaction subsides, and then with the water boiling vigorously for 2 hours. Allow to cool. Filter with suction on a Buchner funnel. Transfer the mixture of organic disulphide and sodium chloride to a 400-ml beaker and stir thoroughly with 175 ml of water to remove the salt. Filter at the pump, drain well and wash the crystalline residue on the filter with 35 ml of alcohol to remove any unreacted o-chloronitrobenzene. The residual di-o-nitrophenyl disulphide melts at 193–195 °C and weighs 70 g (68%).

Notes. (1) Crystallised sodium sulphide $Na_2S,9H_2O$ is very deliquescent, and only a sample which has been kept in a tightly stoppered bottle should be used; crystals as dry as possible should be selected. Alternatively, an equivalent amount of analysed fused sodium sulphide may be employed; this dissolves somewhat more slowly in alcohol.

(2) If some sodium disulphide separates at the bottom of the flask, this should be dissolved in a little more rectified spirit and added to the chloronitrobenzene solution.

o-Nitrobenzenesulphonyl chloride. Equip a 1-litre three-necked flask with an inlet tube for introducing chlorine well beneath the surface of the liquid, an efficient mechanical stirrer and a reflux condenser. Set up the assembly in the fume cupboard and absorb the excess of chlorine in sodium hydroxide solution as detailed under *benzyl chloride* (Section **IV,31**). Place 60 g (0.195 mol) of di-o-nitrophenyl disulphide, 300 ml of concentrated hydrochloric acid and 60 ml of concentrated nitric acid in the flask, pass a stream of chlorine from a cylinder into the mixture at the rate of 2 bubbles per second and warm the solution to 70 °C on a water bath. After about 30 minutes, the disulphide melts and the solution assumes an orange-red colour; after the melting stage has been reached, the passage of the chlorine and the heating are continued for 1 hour. Immediately separate the sulphonyl chloride from the supernatant liquid by decantation, wash with two 90 ml portions of water at about 70 °C and allow to solidify. Drain the water from the solid mass as completely as possible. Dissolve the sulphonyl chloride in 45 ml of glacial acetic acid at 50–60 °C, and rapidly filter the solution at the pump. Cool the filtrate in cold water and stir it vigorously so that the sulphonyl chloride separates in small crystals. Triturate the mixture well with 300 ml of cold water and decant through a Buchner funnel; repeat the process twice. Finally add 300 ml of cold water and 3 ml of concentrated ammonia solution to the mixture, stir well and filter immediately, through a Buchner funnel, wash with 60 ml of water, drain well and dry in the air. The yield of moderately pure o-nitrobenzenesulphonyl chloride, m.p. 64–65 °C, is 72 g (84%). The undried material may be used in the preparation of orthanilic acid.

Orthanilic acid. Fit a 1-litre three-necked flask with a sealed mechanical stirrer and a reflux condenser. Place 60 g of o-nitrobenzenesulphonyl chloride, 30 g of anhydrous sodium carbonate and 180 ml of water in the flask. Heat the mixture to boiling, with stirring; the hydrolysis of the sulphonyl chloride to the sulphonic acid is complete within 40 minutes after the compound has melted. Filter the orange-red solution and acidify (to litmus) with acetic acid (about 7.5

ml are required). Transfer the solution to the original flask (which has been thoroughly rinsed with water) equipped as before. Heat the solution to boiling, and add 105 g of finely divided iron filings (about 20 mesh) with vigorous stirring at the rate of about 7.5 g every 15 minutes. The mixture soon becomes deep brown and exhibits a tendency to froth. Complete the reaction by stirring for a further 4 hours, i.e., until a test portion when filtered yields an almost colourless filtrate; if the filtrate is orange or red, the heating and stirring must be continued. When the reduction is complete, add 2 g of decolourising carbon, filter the hot reaction mixture at the pump and wash the residue with three 15 ml portions of hot water: combine the washings with the main solution. Cool the filtrate to about 15 °C, and add 28.5 ml of concentrated hydrochloric acid slowly, and cool to 12–15 °C. Filter the acid with suction on a Buchner funnel, wash with a little cold water, followed by a little ethanol and dry upon filter-paper in the air. The yield is 27 g (57%); the orthanilic acid has a purity of 97–99 per cent. If required perfectly pure it may be recrystallised from hot water; it decomposes at about 325 °C.

IV,72. METANILIC ACID (3-Aminobenzenesulphonic acid)

$$C_6H_5 \cdot NO_2 \xrightarrow{\text{oleum}} m\text{-}O_2N \cdot C_6H_4 \cdot SO_3H \xrightarrow{\text{Fe/H}_3O^{\oplus}} m\text{-}H_3\overset{\oplus}{N} \cdot C_6H_4 \cdot SO_3$$

This preparation should be carried out in a fume cupboard.

In a 500-ml bolt-head flask, provided with a mechanical stirrer, place 70 ml of oleum (20% SO_3) and heat it in an oil bath to 70 °C. By means of a dropping funnel, supported so that the stem is just above the surface of the acid, introduce 41 g (34 ml, 0.33 mol) of nitrobenzene slowly and at such a rate that the temperature of the well-stirred mixture does not rise above 100–105 °C. When all the nitrobenzene has been introduced, continue the heating at 110–115 °C for 30 minutes. Remove a test portion and add it to excess of water. If the odour of nitrobenzene is still apparent, add a further 10 ml of fuming sulphuric acid, and heat at 110–115 °C for 15 minutes: the reaction mixture should then be free from nitrobenzene. Allow the mixture to cool and pour it carefully with vigorous stirring on to 200 g of finely crushed ice contained in a beaker. All the nitrobenzenesulphonic acid passes into solution; if a little sulphone is present, remove this by filtration. Stir the solution mechanically and add 70 g of sodium chloride in small portions: the sodium salt of *m*-nitrobenzenesulphonic acid separates as a pasty mass. Continue the stirring for about 30 minutes, allow to stand overnight, filter and press the cake well. The latter will retain sufficient acid to render unnecessary the addition of acid in the subsequent reduction with iron.

Place 84 g (1.5 mol) of iron filings and 340 ml of water in a 1.5- or 2-litre bolt-head flask equipped with a mechanical stirrer. Heat the mixture to boiling, stir mechanically and add the sodium *m*-nitrobenzenesulphonate in small portions during 1 hour. After each addition the mixture foams extensively: a wet cloth should be applied to the neck of the flask if the mixture tends to froth over the sides. Replace from time to time the water which has evaporated so that the volume is approximately constant. When all the sodium salt has been introduced, boil the mixture for 20 minutes. Place a small drop of the suspension upon filter-paper and observe the colour of the 'spot': it should be a pale brown but not deep brown or deep yellow. If it is not appreciably coloured, add anhydrous sodium carbonate cautiously, stirring the mixture, until red litmus paper is turned blue and a test drop upon filter-paper is not blackened by sodium sulphide solution. Filter at the pump and wash well with hot water. Concentrate the filtrate to about 200 ml, acidify with concentrated hydrochloric acid to Congo red and allow to cool. Filter off the metanilic acid and dry upon filter-paper. A further small quantity may be obtained by concentrating the mother-liquor. The yield is 55 g (95%).

F,2. Acylated amines and their substitution reactions. Acylation of an aromatic primary or secondary amine may be readily achieved by using an acid chloride in the presence of base; however, acetylation is more usually effected with acetic anhydride rather than the more obnoxious acetyl chloride.

In general, benzoylation of aromatic amines finds less application than acetylation in preparative work, but the process is often employed for the identification and characterisation of aromatic amines (Section **VII,6,**_21(b)_). In the **Schotten–Baumann method** of benzoylation, the amine, or its salt, is dissolved or suspended in a slight excess of 8–15 per cent sodium hydroxide solution, a small excess (about 10–15% more than the theoretical quantity) of benzoyl chloride is then added and the mixture vigorously shaken in a stoppered vessel (or else the mixture is stirred mechanically). Benzoylation proceeds smoothly and the sparingly soluble benzoyl derivative separates as a solid. The use of the aqueous medium is possible because the sodium hydroxide only slowly hydrolyses the excess of benzoyl chloride to yield sodium benzoate and sodium chloride which remain in solution.

$$ArNH_2 + C_6H_5 \cdot COCl + NaOH \longrightarrow ArNH \cdot CO \cdot C_6H_5 + NaCl + H_2O$$
$$Ar_2NH + C_6H_5 \cdot COCl + NaOH \longrightarrow Ar_2N \cdot CO \cdot C_6H_5 + NaCl + H_2O$$

The benzoyl compounds frequently occlude traces of unchanged benzoyl chloride, which thus escapes hydrolysis by the alkali; it is therefore advisable, wherever possible, to recrystallise the benzoyl derivatives from methanol, ethanol or rectified spirit, since these solvents will esterify the unchanged acid chloride and so remove the latter from the recrystallised material (e.g., benzanilide, Section **IV,73**). Sometimes the benzoyl compound does not crystallise well: this difficulty may be frequently overcome by the use of _p_-nitrobenzoyl chloride or 3,5-dinitrobenzoyl chloride (Section **IV,175**), which usually give highly crystalline derivatives of high melting point. Benzoyl compounds are readily hydrolysed by heating with about 70 per cent sulphuric acid (alkaline hydrolysis is very slow for anilides).

$$ArNH \cdot CO \cdot C_6H_5 + H_2SO_4 + H_2O \longrightarrow Ar\overset{\oplus}{N}H_3 \} HSO_4^{\ominus} + C_6H_5 \cdot CO_2H$$

Primary amines react readily upon warming with acetic anhydride to yield, in the first instance, the mono-acetyl derivative.

$$ArNH_2 + (CH_3 \cdot CO)_2O \longrightarrow ArNH \cdot CO \cdot CH_3 + CH_3 \cdot CO_2H$$

If heating is prolonged and excess of acetic anhydride is employed, variable amounts of the diacetyl derivative are formed.

$$ArNH \cdot CO \cdot CH_3 + (CH_3 \cdot CO)_2O \longrightarrow ArN(CO \cdot CH_3)_2 + CH_3 \cdot CO_2H$$

In general, however, the diacetyl derivatives are unstable in the presence of water, undergoing hydrolysis to the mono-acetyl compound, so that when they (or a mixture of mono- and di-acetyl derivatives) are crystallised from an aqueous solvent, e.g., dilute ethanol, only the mono-acetyl derivative is obtained. Highly substituted amines (e.g., Section **IV,66**) react extremely slowly with acetic anhydride, but in the presence of a few drops of concentrated sulphuric acid as catalyst, acetylation occurs rapidly.

The disadvantages attending the use of acetic anhydride alone are absent when the acetylation is conducted in aqueous solution according to the following procedure (see acetylation of aniline, *Method 1* in Section **IV,74**). The amine is dissolved in water containing one equivalent of hydrochloric acid; slightly more than one equivalent of acetic anhydride is added to the solution, followed by enough sodium acetate to neutralise the hydrochloric acid, and the mixture is shaken. The free amine which is liberated is at once acetylated. It must be pointed out that the hydrolysis of acetic anhydride at room temperature is extremely slow and that the free amine reacts much more readily with the anhydride than does water: this forms the experimental basis for the above excellent method of acetylation.

Acetylation with acetic anhydride is comparatively expensive because of the cost of the reagent. The use of the inexpensive glacial acetic acid depends upon the displacement of the following reversible equilibrium to the right by removal of the water (and a little acetic acid) by distillation.

$$ArNH_2 + CH_3 \cdot CO_2H \rightleftharpoons ArNH \cdot CO \cdot CH_3 + H_2O$$

A technique suitable for laboratory preparations is to employ a mixture of acetic acid and acetic anhydride (see *Method 2* in Section **IV,74**).

Conversion of the amino group into the acetamido group by acetylation modifies the interaction of the nitrogen lone pair with the π-electron system of the aromatic ring so that the ring is less powerfully activated towards electrophilic attack.

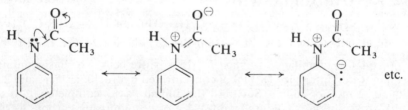

Protection of the amino group by acetylation, as in acetanilide, therefore usually permits mono-substitution reactions with appropriate electrophilic reagents to proceed smoothly. Thus with bromine, *p*-bromoacetanilide is the main product; the small quantity of the *ortho* isomer simultaneously formed can be easily eliminated by recrystallisation (Section **IV,74**). Hydrolysis of *p*-bromoacetanilide gives *p*-bromoaniline (Section **IV,75**). Nitration leads similarly to *p*-nitroacetanilide (Section **IV,76**), which can be hydrolysed to *p*-nitroaniline (Section **IV,77**).

IV,73. BENZANILIDE

$$C_6H_5 \cdot COCl + C_6H_5 \cdot NH_2 \xrightarrow{\text{NaOH}} C_6H_5 \cdot CO \cdot NH \cdot C_6H_5$$

Place 5.2 g (5 ml, 0.056 mol) of aniline and 45 ml of 10 per cent aqueous sodium hydroxide solution in a conical flask, and then add 8.5 g (7 ml, 0.06

mol) of benzoyl chloride, stopper, and shake *vigorously* for 10–15 minutes. Heat is evolved in the reaction. The crude benzoyl derivative separates as a white powder. When the reaction is complete (i.e., when the odour of benzoyl chloride can no longer be detected: smell cautiously), make sure that the reaction mixture is alkaline. Filter off the product with suction on a Buchner funnel, break up the mass on the filter (if necessary), wash well with water and drain. Recrystallise from hot ethanol (or rectified spirit); filter the hot solution through a hot water funnel or through a warm Buchner funnel. Collect the crystals which separate and dry in the air or in the steam oven. The yield of benzanilide, m.p. 162 °C, is 9 g (75%).

Hydrolysis of benzanilide. Place 5 g of benzanilide and 50 ml of 70 per cent w/w sulphuric acid (1) in a small flask fitted with a reflux condenser, and boil gently for 30 minutes. Some of the benzoic acid will vaporise in the steam and solidify in the condenser. Pour 60 ml of hot water down the condenser: this will dislodge and partially dissolve the benzoic acid. Cool the flask in ice-water; filter off the benzoic acid (aniline sulphate does not separate at this dilution), wash well with water, drain, dry upon filter-paper and identify by m.p. (121 °C). Render the filtrate alkaline by cautiously adding 10 per cent sodium hydroxide solution, cool and isolate the aniline by ether extraction. Remove the ether and distil the aniline, b.p. 184 °C.

Note. (1) For preparation see Section **IV,77**, Note (1).

Cognate preparation   Benzoyl *p*-toluidide. Use 3.3 g of *p*-toluidine, 25 ml of 10 per cent sodium hydroxide solution and 4.2 g (3.5 ml) of benzoyl chloride. Proceed as above. The yield of benzoyl *p*-toluidide, m.p. 158 °C, is 4.5 g.

IV,74. *p*-BROMOACETANILIDE

$$C_6H_5 \cdot NH_2 + (CH_3 \cdot CO)_2O \longrightarrow C_6H_5 \cdot NH \cdot CO \cdot CH_3 + CH_3 \cdot CO_2H$$
$$C_6H_5 \cdot NH \cdot CO \cdot CH_3 + Br_2 \longrightarrow p\text{-Br} \cdot C_6H_4 \cdot NH \cdot CO \cdot CH_3 + HBr$$

Acetylation of aniline. *Method 1.* In a 1-litre beaker or flask containing 500 ml of water, introduce 18.3 ml of concentrated hydrochloric acid and 20.5 g (20 ml, 0.22 mol) of aniline. Stir until the aniline passes completely into solution. (If the solution is coloured, add 3–4 g of decolourising carbon, warm to about 50 °C with stirring for 5 minutes and filter at the pump or through a fluted filter-paper.) To the resulting solution add 27.7 g (25.6 ml, 0.27 mol) of redistilled acetic anhydride, stir until it is dissolved and immediately pour in a solution of 33 g of crystallised sodium acetate in 100 ml of water. Stir vigorously and cool in ice. Filter the acetanilide with suction, wash with a little water, drain well and dry upon filter-paper in the air. The yield of colourless, almost pure acetanilide, m.p. 113 °C, is 24 g (80%). Upon recrystallisation from about 500 ml of boiling water to which about 10 ml of industrial spirit has been added (1) the m.p. is raised to 114 °C; the first crop weighs 19 g.

Method 2. In a 500-ml round-bottomed flask, equipped with a reflux condenser, place 20.5 g (20 ml, 0.22 mol) of aniline, 21.5 g (20 ml, 0.21 mol) of acetic anhydride, 21 g (20 ml, 0.35 mol) of glacial acetic acid and 0.1 g of zinc dust (2). Boil the mixture gently for 30 minutes, and then pour the hot liquid in a thin stream into a 1-litre beaker containing 500 ml of cold water whilst stirring continually. When cold (it is preferable to cool in ice), filter the crude product at the pump, wash with a little cold water, drain well and dry upon filter-paper in the air. The yield of acetanilide, m.p. 113 °C, is 26 g. It may be recrystallised as in *Method 1* affording 21 g (70%) of pure acetanilide, m.p. 114 °C.

Bromination of acetanilide. Dissolve 13.5 g (0.1 mol) of finely powdered acetanilide in 45 ml of glacial acetic acid in a 350-ml conical flask. In another small flask dissolve 17 g (5.3 ml, 0.106 mol) of bromine in 25 ml of glacial acetic acid, and transfer the solution to a burette or a separatory funnel supported over the flask. [For precautions attending the use of bromine, see Section **II,2,7**. The preparation should be conducted in a fume cupboard.] Add the bromine solution slowly and with constant shaking to ensure thorough mixing: stand the flask in cold water. When all the bromine has been added, the solution will have an orange colour due to the slight excess of bromine; a part of the reaction product may crystallise out. Allow the final reaction mixture to stand at room temperature for 30 minutes with occasional shaking. Pour the reaction product into 400 ml of water; rinse the flask with about 100 ml of water. Stir the mixture well and if it is appreciably coloured, add just sufficient sodium metabisulphite solution to remove the orange colour. Filter the crystalline precipitate with suction on a Buchner funnel, wash thoroughly with cold water and press as dry as possible with a wide glass stopper. Recrystallise from dilute methanol or ethanol (industrial spirit). The yield of *p*-bromoacetanilide, colourless crystals m.p. 167 °C, is 18 g (84%).

Notes. (1) The acetanilide may also be recrystallised from toluene (inflammable!); use a reflux condenser.

(2) The zinc reduces the coloured impurities in the aniline and also helps to prevent oxidation during the reaction.

IV,75. *p*-BROMOANILINE

$$p\text{-Br·C}_6\text{H}_4\text{·NH·CO·CH}_3 + \text{H}_2\text{O} \xrightarrow{\text{H}^{\oplus}} p\text{-Br·C}_6\text{H}_4\text{·NH}_2 + \text{CH}_3\text{·CO}_2\text{H}$$

Dissolve 18 g (0.084 mol) of *p*-bromoacetanilide in 35 ml of boiling ethanol contained in a 500-ml round-bottomed flask equipped with a reflux condenser. With the aid of a pressure-equalising dropping funnel add 22 ml of concentrated hydrochloric acid down the condenser in small portions to the boiling solution. Reflux for 30–40 minutes or until a test portion remains clear when diluted with water. Dilute with 150 ml of water, and fit the flask with a condenser set for downward distillation. Distil the mixture from an air bath (Fig. I,49) or upon an asbestos-centred wire gauze, and collect about 100 ml of distillate; the latter consists of ethyl acetate, ethanol and water. Pour the residual solution of *p*-bromoaniline hydrochloride into 100 ml of ice-water, and add, with vigorous stirring, 5 per cent sodium hydroxide solution until just alkaline. The *p*-bromoaniline separates as an oil, which soon crystallises. Filter the crystals at the pump, wash with cold water and dry in the air upon pads of filter-paper. The yield is 14 g (97%), m.p. 66 °C. Recrystallisation from dilute alcohol, which results in appreciable loss, is usually unnecessary.

IV,76. p-NITROACETANILIDE

$$C_6H_5 \cdot NH \cdot CO \cdot CH_3 + HNO_3 \xrightarrow{H_2SO_4} p\text{-}O_2N \cdot C_6H_4 \cdot NH \cdot CO \cdot CH_3$$

Add 25 g (0.185 mol) of finely powdered, dry acetanilide to 25 ml of glacial acetic acid contained in a 500-ml beaker; introduce into the well-stirred mixture 92 g (50 ml) of concentrated sulphuric acid. The mixture becomes warm and a clear solution results. Surround the beaker with a freezing mixture of ice and salt, and stir the solution mechanically. Support a separatory funnel, containing a cold mixture of 15.5 g (11 ml) of concentrated nitric acid and 12.5 g (7 ml) of concentrated sulphuric acid, over the beaker. When the temperature of the solution falls to 0–2 °C, run in the acid mixture gradually while the temperature is maintained below 10 °C. After all the mixed acid has been added, remove the beaker from the freezing mixture, and allow it to stand at room temperature for 1 hour. Pour the reaction mixture on to 250 g of crushed ice (or into 500 ml of cold water), whereby the crude nitroacetanilide is at once precipitated. Allow to stand for 15 minutes, filter with suction on a Buchner funnel, wash it thoroughly with cold water until free from acids (test the wash water) and drain well (1). Recrystallise the pale yellow product from ethanol or industrial spirit, filter at the pump, wash with a little cold alcohol and dry in the air upon filter-paper (2). [The yellow o-nitroacetanilide remains in the filtrate.] The yield of p-nitroacetanilide, a colourless crystalline solid of m.p. 214 °C, is 20 g (60%).

Investigate by thin-layer chromatography the effectiveness of the recrystallisation process in the following way. Load a 20 × 5 cm thin-layer plate (Silica Gel G) with approximately 3 mm diameter spots of concentrated solutions (in acetone) of the crude and the recrystallised product. Concentrate a portion of the ethanolic mother-liquor and similarly apply to the plate. Develop the chromatogram with a toluene–ethyl acetate mixture (4 : 1) and dry the plate. Mark the positions of the visible spots and leave the plate in a tank of iodine vapour to reveal the rest of the components. The recrystallised p-nitroacetanilide (R_F 0.07) should be free from the pale yellow o-isomer (R_F 0.36); these compounds are revealed by the iodine treatment. The mother-liquor contains two readily visible yellow components, which are p-nitroaniline (R_F 0.24) and o-nitroaniline (R_F 0.45), as well as both p- and o-nitroacetanilide.

Notes. (1) Washing is accomplished most effectively by transferring the crude solid to a beaker, stirring well with wash water and refiltering.

(2) The recrystallised material and the crude product should be examined by TLC analysis.

IV,77. p-NITROANILINE

$$p\text{-}O_2N \cdot C_6H_4 \cdot NH \cdot CO \cdot CH_3 + H_2O \xrightarrow{H^{\oplus}} p\text{-}O_2N \cdot C_6H_4 \cdot NH_2 + CH_3 \cdot CO_2H$$

Boil a mixture of 15 g (0.083) of p-nitroacetanilide and 75 ml of 70 per cent w/w sulphuric acid (1) under a reflux condenser for 20–30 minutes or until a test sample remains clear upon dilution with 2–3 times its volume of water. The p-nitroaniline is now present in the liquid as the sulphate. Pour the clear hot solution into 500 ml of cold water and precipitate the p-nitroaniline by adding excess of 10 per cent sodium hydroxide solution or of concentrated ammonia solution. When cold (cool the mixture in ice-water, if necessary), filter the yellow crystalline precipitate at the pump, wash it well with water and drain thoroughly. Recrystallise it from a mixture of equal volumes of rectified (or industrial) spirit and water or from hot water. Filter, wash and dry. The yield of *p*-nitroaniline, m.p. 148 °C, is 11 g (96%).

Note. (1) The 70 per cent sulphuric acid is prepared by adding 60 ml of concentrated sulphuric acid cautiously and in a thin stream with stirring to 45 ml of water.

G USES OF DIAZONIUM SALTS

FORMATION OF DIAZONIUM SALTS   Primary aromatic amines on reaction with nitrous acid in the presence of hydrochloric acid (or other mineral acid) at about 0 °C yield diazonium salts as discrete intermediates. The diazonium salts similarly derived from aliphatic primary amines decompose readily even at this temperature to yield the corresponding alcohol (and other products) with the evolution of nitrogen.

$$ArNH_2 + NaNO_2 + 2HCl \longrightarrow Ar\overset{\oplus}{N}\equiv N\}\overset{\ominus}{Cl} + NaCl + H_2O$$

The acidified nitrite solution provides a source of the nitrosonium ion (I) which electrophilicly replaces the hydrogen in the primary amino group to form the *N*-nitroso derivative (II). This has a tautomeric structure, the hydroxydiazo form (III) yielding the diazonium ion (IV) under acidic conditions.

$$2NO_2{}^{\ominus} + 2H^{\oplus} \longrightarrow 2HNO_2 \rightleftharpoons O=N-O-N=O + H_2O$$

$$O=N-O-N=O \rightleftharpoons O=\overset{\oplus}{N} + \overset{\ominus}{O}-N=O$$

(I)

$$Ar\overset{..}{N}H_2 \quad {}^{\searrow}\overset{\oplus}{N}-O \xrightarrow{\,-H^{\oplus}\,} ArN-N=O \rightleftharpoons ArN=N-OH$$

H

(II)   (III)

$$Ar\overset{..}{N}=N-OH\overset{\oplus}{H} \longrightarrow Ar\overset{\oplus}{N}\equiv N + H_2O$$

(IV)

The experimental conditions necessary for the preparation of a solution of a diazonium salt, **diazotisation of a primary amine**, are as follows. The amine is dissolved in a suitable volume of water containing 2.5–3 equivalents of hydrochloric acid (or of sulphuric acid) by the application of heat if necessary, and the solution is cooled in ice when the amine hydrochloride (or sulphate) usually crystallises. The temperature is maintained at 0–5 °C and an aqueous solution of sodium nitrite is added portion-wise until, after allowing 3–4 minutes for reaction, the solution gives an *immediate* positive test for excess of nitrous acid with an external indicator—moist potassium iodide–starch paper.* The precipitated amine hydrochloride (or sulphate), if any, dissolves during the diazotisation to give a clear solution of the highly soluble diazonium salt. The excess of acid (0.5–1 equivalents) maintains the proper condition of acidity required to stabilise the diazonium salt and hence to minimise secondary reactions, e.g., the interaction of some of the diazonium salt with unchanged amine to form a diazoamino compound, a reaction which occurs readily in neutral solution (see Section **IV,97**). The reaction mixture must be kept very cold during the process (which is exothermic in character), otherwise the diazonium salt may be partially hydrolysed to the corresponding phenol (see below).

Some amines, such as the nitroanilines, react rather slowly at low temperatures, but since the diazonium compounds formed are somewhat more stable the diazotisation may be conducted at room temperature, when the reaction proceeds more rapidly. If the amine salt is only sparingly soluble in water, it should be suspended in the acid in a fine state of division (this is generally attained by cooling a hot solution and stirring vigorously), and it passes into solution as the soluble diazonium salt is formed.

A solution of sodium nitrite in concentrated sulphuric acid, which provides a nitrosonium hydrogen sulphate reagent, is a very effective diazotising medium which is particularly valuable for even more weakly basic amines, such as 2,4-dinitroaniline or the corresponding trinitro compound, picramide.

$$HNO_2 + H_2SO_4 \rightleftharpoons [H_2NO_2]^{\oplus} [HSO_4]^{\ominus} \xrightarrow{H_2SO_4}$$
$$[\overset{\oplus}{N}O] [HSO_4^{\ominus}] + H_3O^{\oplus} + HSO_4^{\ominus}$$

This reagent is used here for the diazotisation of 4-amino-3,5-diiodobenzoic acid in the preparation of 3,4,5-triiodobenzoic acid (cognate preparation in Section **IV,79**). It is also used for the bis-diazotisation of *m*-phenylenediamine (the preparation of 1,3-diiodobenzene in Section **IV,79**) which must be carried out in strongly acidic conditions. Unless the amino groups are extensively protonated in this way, partial diazotisation and self-coupling occurs (see top of next page) with the formation of the azo dye Bismarck Brown.

To prepare the **solid benzenediazonium chloride or sulphate**, the reaction is conducted in the absence of water as far as possible. The source of nitrous acid is one of its organic esters (e.g., pentyl nitrite) and a solution of hydrogen

---

* In actual practice it is found that some time before the theoretical quantity of sodium nitrite has been added, the solution will give a blue colouration (presumably, in part, by atmospheric oxidation) within a few seconds of being placed upon the test paper. It must, however, be remembered that towards the end of the diazotisation the reaction with nitrous acid is somewhat slow, and it is imperative to wait a few minutes before making the test, and furthermore only an immediate blue colouration has any significance. It is advisable to dilute the drop of the test solution with a few drops of water on a watch glass before making the test. It is recommended that about 10 per cent excess of sodium nitrite of good quality (>96% $NaNO_2$: e.g., sodium nitrite recryst.) be employed; this will serve as an additional check. If a slight excess of sodium nitrite is accidentally added, it may be decomposed by the addition of a little urea or sulphamic acid; alternatively a small amount of the primary amine, dissolved in the acid used, may be added.

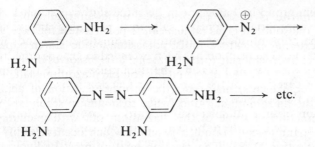

chloride gas in absolute ethanol; upon the addition of ether only the diazonium salt is precipitated as a crystalline solid.

$$C_6H_5\overset{\oplus}{N}H_3\}\overset{\ominus}{Cl} + C_5H_{11}ONO + HCl \longrightarrow$$
$$C_6H_5\overset{\oplus}{N}\equiv N\}\overset{\ominus}{Cl} + C_5H_{11}Cl + 2H_2O$$

Solid diazonium salts are very sensitive to shock when perfectly dry, and detonate violently upon gentle heating: they are, therefore, of little value for preparative work. Happily, most of the useful reactions of diazonium compounds can be carried out with the readily-accessible aqueous solutions, so that the solid (explosive) diazonium salts are rarely required.

G,1. Reactions involving replacement of the diazo group. Diazonium salts undergo a large number of reactions in which the diazo group is lost as molecular nitrogen and is replaced by one of a variety of other groups which become attached to the aromatic ring.

When a solution of a diazonium salt is heated, nitrogen is evolved and the diazo group is replaced by a hydroxyl group in an S_N1 type of displacement reaction.

$$Ar\overset{\oplus}{\frown N}\equiv N \xrightarrow{-N_2} [Ar^{\oplus}] \xrightarrow[-H^{\oplus}]{+H_2O} ArOH$$

The diazonium sulphate is used in preference to the diazonium chloride, since the presence of chloride ions give rise to small quantities of the aryl chloride as a by-product. The solution must be acidic in order to avoid the coupling reaction between unreacted diazonium salt and the phenol (see p. 712, **IV,G,2** below). For the preparation of phenols and cresols (e.g., *p*-cresol, cognate preparation in Section **IV,78**) the aqueous solution of the diazonium compound is warmed to about 50 °C; at higher temperatures the reaction may become unduly vigorous and lead to appreciable quantities of tarry compounds. For certain substituted amines, a higher temperature (e.g., boiling 40–60% sulphuric acid) is necessary to decompose the diazonium salt completely (e.g., *m*-nitrophenol, Section **IV,78**).

When an aqueous solution of an aryldiazonium salt is treated with an equivalent of potassium iodide and warmed on a water bath, the aryl iodide is formed in good yield (e.g., iodobenzene, Section **IV,79**).

$$Ar\overset{\oplus}{N}\equiv N\}\overset{\ominus}{Cl} + KI \longrightarrow ArI + N_2 + KCl$$

This simple procedure cannot be applied in the preparation of the corresponding chloro and bromo compounds. Sandmeyer (1884) found that the replacement of the diazonium group by halogen can be successfully accomplished in the presence of the appropriate copper(I) salt, thus providing an excellent method for the preparation of nuclear-substituted aromatic compounds from the corresponding amines. The reaction has been extended to groups other than halogens, for example the cyano (–CN) and the thiocyanate (–SCN) group. A general procedure for carrying out the Sandmeyer reaction is as follows. The amine is diazotised in the presence of hydrochloric acid with sodium nitrite at 0–5 °C, and a solution of an equimolecular quantity of copper(I) chloride in hydrochloric acid is added: a deep brown, sparingly soluble complex of copper(I) chloride and the diazonium salt is formed, and when the temperature of the reaction mixture is raised, decomposition ensues accompanied by the evolution of nitrogen, the disappearance of the solid and the separation of an oily layer of the aryl chloride.

$$\text{ArNH}_2 \xrightarrow[\text{HCl}]{\text{NaNO}_2} \text{Ar}\overset{\oplus}{\text{N}}{\equiv}\text{N}\}\overset{\ominus}{\text{Cl}} \xrightarrow[\text{HCl}]{\text{CuCl}} [\text{Complex}] \xrightarrow{\text{heat}} \text{ArCl} + \text{N}_2$$

The aryl chloride is formed when the diazonium–copper(I) chloride complex decomposes by a radical mechanism summarised below. Copper catalyses this decomposition because it can undergo interconversion between the +1 and +2 oxidation states as a result of electron transfer.

$$\text{Ar}-\overset{\oplus}{\text{N}}{\equiv}\text{N}\quad:\text{Cl}-\overset{\text{I}}{\underset{\bullet}{\text{Cu}}}-\text{Cl} \longrightarrow \text{Ar}-\text{N}=\text{N}-\overset{\oplus}{\text{Cl}}-\overset{\ominus}{\underset{\bullet}{\text{Cu}}}-\text{Cl}$$

$$\text{Ar}-\text{N}{=}\text{N}-\overset{\oplus}{\text{Cl}}\underset{\bullet}{\overset{\ominus}{\text{Cu}}}-\text{Cl} \longrightarrow \text{Ar}\bullet + \text{N}_2 + \overset{\text{II}}{\text{CuCl}_2}$$

$$\text{Ar}\bullet \quad \text{Cl}\underset{}{\overset{\text{II}}{-}}\text{Cu}-\text{Cl} \longrightarrow \text{ArCl} + \bullet\overset{\text{I}}{\text{CuCl}}$$

Details of the preparation of p-chlorotoluene are given in Section **IV,80**, which also includes o-chlorotoluene, chlorobenzene, m-chloronitrobenzene and o-chlorobenzoic acid as examples of cognate preparations.

In the preparation of bromo compounds by the Sandmeyer reaction (e.g., p-bromotoluene and the cognate preparations in Section **IV,81**), the amine is generally diazotised in sulphuric acid solution (or in hydrobromic acid solution), and the resulting aryldiazonium sulphate (or bromide) is treated with a solution of copper(I) bromide in an excess of hydrobromic acid; the addition complex is then decomposed by gentle heating and the bromo compound isolated by steam distillation. For the preparation of 2-bromonaphthalene the

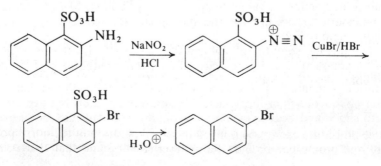

use of 2-naphthylamine (which is a potent carcinogen) as a starting material is avoided by using 2-naphthylamine-1-sulphonic acid (see p. 667, **IV,E,3**). Diazotisation and reaction with copper(I) bromide yields 2-bromonaphthalene-1-sulphonic acid; heating with aqueous sulphuric acid eliminates the sulphonic acid group (see also Section **IV,116**) to give 2-bromonaphthalene.

Gattermann (1890) found that finely divided (i.e., freshly precipitated) copper or copper bronze acts catalytically in the decomposition of solutions of diazonium salts.

$$Ar\overset{\oplus}{N} \equiv N_j\overset{\ominus}{X} \xrightarrow{\text{Cu}} ArX + N_2$$

The yields in the **Gattermann reaction**, however (e.g., o-bromotoluene, Section **IV,82**), are usually not as high as those obtained by the Sandmeyer method. Copper powder is also employed in the preparation of sulphinic acids (e.g., benzenesulphinic acid, Section **IV,83**) which are obtained when a solution of a diazonium sulphate is saturated with sulphur dioxide and decomposed by the addition of copper powder.

$$Ar\overset{\oplus}{N} \equiv N\}\overset{\ominus}{HSO_4} + SO_2 + Cu \longrightarrow Ar \cdot SO_2H + N_2 + CuSO_4$$

The Sandmeyer reaction may also be applied to the preparation of aryl nitriles. The solution of the diazonium salt, which should preferably be carefully neutralised with sodium or calcium carbonate to avoid excessive evolution of hydrogen cyanide in the subsequent stage, is added to a solution of copper(I) cyanide in excess sodium or potassium cyanide solution (e.g., p-tolunitrile, Section **IV,84**); sometimes improved yields are obtained by substituting nickel cyanide for copper(I) cyanide. Hydrolysis of the aryl nitrile with sodium hydroxide solution, followed by acidification, yields the corresponding acid (see p. 825, **IV,N,2**). The Sandmeyer reaction thus affords an important indirect method of introducing the carboxylic acid group into an aromatic ring.

$$ArNH_2 \xrightarrow{\text{NaNO}_2/\text{HX}} Ar\overset{\oplus}{N_2}\} \overset{\ominus}{X} \xrightarrow{\text{CuCN/KCN}} Ar \cdot CN \xrightarrow[\text{(ii) } H_3O^{\oplus}]{\text{(i) } {}^{\ominus}\text{OH}} Ar \cdot CO_2H$$

The controlled thermal decomposition of dry aryldiazonium fluoroborates to yield an aryl fluoride, boron trifluoride and nitrogen is known as the **Balz–Schiemann reaction** (Section **IV,85**).

$$Ar\overset{\oplus}{N} \equiv N\}\overset{\ominus}{BF_4} \xrightarrow{\text{heat}} ArF + N_2 + BF_3$$

In general the required diazonium fluoroborate is obtained as a precipitate when a concentrated solution of sodium fluoroborate is added to a solution of a diazonium salt. In an alternative procedure (e.g., the preparation of p-fluoroanisole, cognate preparation in Section **IV,85**), the amine is diazotised in solution in aqueous fluoroboric acid. The diazonium fluoroborates are less sensitive to shock and heat than most diazonium salts and may be prepared and handled in the dry state with relative safety. Most diazonium fluoroborates have definite decomposition temperatures and the rates of decomposition, with few exceptions, are easily controlled. Diazonium fluoroborates containing the nitro group, however, usually decompose suddenly and with violence on heating; in such cases the fluoroborate should be mixed with 3–4 times its weight of dry sand and heated cautiously until decomposition commences.

The diazonium hexafluorophosphates, prepared similarly from the appropriate diazonium chloride solution and hexafluorophosphoric acid, may in general be used instead of the fluoroborates with advantage. The thermal decomposition of diazonium hexafluorophosphates to aryl fluorides generally proceeds smoothly and in better yield (Ref. 5).

A further interesting application of the diazo reaction is in the preparation of the otherwise difficultly accessible o- and p-dinitrobenzenes (Section **IV,86**). The requisite nitroaniline is converted into the diazonium fluoroborate which is then decomposed in aqueous suspension in the presence of sodium nitrite with the aid of copper powder.

It is frequently observed that in the replacement reactions discussed above significant amounts of biphenyl derivatives are present in the reaction product. Compounds of this type may be prepared deliberately by adding the aqueous diazonium salt solution to a liquid aromatic compound and then basifying the vigorously stirred two-phase system by adding sodium hydroxide (or sodium acetate) solution.

$$Ar^1\overset{\oplus}{N_2}\}\overset{\ominus}{X} + Ar^2H + \overset{\ominus}{O}H \longrightarrow Ar^1 - Ar^2 + N_2 + X^\ominus + H_2O$$

The reaction (the **Gomberg reaction**) probably involves the intermediate formation of the diazohydroxide, which because it has a largely covalent structure passes substantially into the non-polar organic phase. In this phase it decomposes into free aryl radicals which displace hydrogen from the added aromatic reactant.

$$Ar^1 - \overset{\oplus}{N}\equiv N \quad \overset{\ominus}{:}OH \rightleftharpoons Ar^1 - N = N - OH \rightleftharpoons Ar^1 - N = N - \overset{\ominus}{O}:$$

$$Ar^1 - N = N - \overset{\ominus}{O}: \quad \overset{\oplus}{N}\equiv N - Ar^1 \longrightarrow$$

$$Ar^1 - N = N - O - N = N - Ar^1 \longrightarrow Ar^1N_2O\cdot + Ar^1\cdot + N_2$$

$$Ar^1\cdot + Ar^2H \xrightarrow{Ar^1N_2O\cdot} Ar^1 - Ar^2 + Ar^1N_2OH$$

In the example (Section **IV,87**) the reaction of the diazonium salt from o-chloroaniline with benzene to yield 2-chlorobiphenyl is illustrative. It should be noted, however, that when the liquid aromatic compound in which substitution is to occur is of the type ArZ, the directive influences which are used to explain electrophilic substitution processes are not operative. Thus irrespective of the nature of the substituent Z, *ortho-para* substitution predominates; this result supports the assumption that the substitution process is radical in type.

The process of **deamination** involves the replacement of the diazonium group by hydrogen, thus effecting the overall removal of the primary amino group. In a simple procedure illustrated by the preparation of 1,3,5-tribromobenzene from 2,4,6-tribromoaniline (Section **IV,88**), the amine is converted into the diazonium sulphate in ethanol solution. Heating the solution brings about the reductive removal of the diazo group, the ethanol being oxidised to acetaldehyde.

$$Ar\overset{\oplus}{N} \equiv N\}H\overset{\ominus}{S}O_4 + C_2H_5OH \longrightarrow ArH + CH_3 \cdot CHO + H_2SO_4 + N_2$$

The value of the deamination process in synthesis is illustrated by a classical example—the synthesis of *m*-bromotoluene, described and formulated in Section **IV,89**. The key to the sequence is the bromination of *p*-acetotoluidide which occurs at the position *ortho* to the more strongly electron-releasing acetamido group. This is *meta* to the methyl group, a position which is virtually unattacked in the direct bromination of toluene. The acetamido group can then be readily removed by sequential hydrolysis, diazotisation and reduction. Other reducing agents may be used in place of ethanol, e.g., an alkaline sodium stannite solution is quite effective; this reagent is prepared by adding sodium hydroxide to an aqueous solution of tin(II) chloride until the initial precipitate just redissolves. The most effective reagent, however, is probably aqueous hypophosphorus acid. An example of its use is provided by the synthesis of 3,3'-dimethylbiphenyl (Section **IV,90**) from the 4,4'-diamino compound (*o*-tolidine). It is emphasised, however, that the latter (a dimethylbenzidine) is carcinogenic and its use is controlled in Britain.

CAUTION. Diazonium compounds have been used for the preparation of:

(*a*) Thiophenols—by treatment with a solution of sodium hydrogen sulphide, for example:

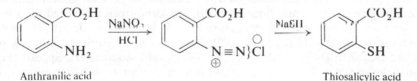

Anthranilic acid   Thiosalicylic acid

(*b*) *O*-Alkyl-*S*-aryl dithiocarbonates—by reaction with aqueous potassium *O*-ethyl dithiocarbonate (see p. 588, **III,T,5**), and thence to thiophenols by treatment with potassium hydroxide, for example:

$$C_6H_5\overset{\oplus}{N} \equiv N\}\overset{\ominus}{C}l + C_2H_5O \cdot CS \cdot \overset{\ominus}{S}\overset{\oplus}{K} \longrightarrow C_6H_5S \cdot CS \cdot OC_2H_5 \longrightarrow C_6H_5SH$$

(*c*) Disulphides—by interaction with a solution of sodium disulphide.

It cannot be too strongly emphasised that in all these reactions violently explosive diazo sulphides and related compounds may be formed, and another less hazardous method for the preparation of the desired compound should be used, if possible. The following reactions are known to lead to dangerous explosions:

(i) diazotised *o*-nitroaniline, *m*-chloroaniline, 4-chloro-*o*-toluidine or 2-naphthylamine and sodium disulphide;

(ii) diazotised *m*-nitroaniline and potassium *O*-ethyl dithiocarbonate; and

(iii) diazotised aniline, *p*-bromoaniline, toluidines and naphthylamines and sodium hydrogen sulphide.

IV,78. m-NITROPHENOL

$$m\text{-}O_2N\cdot C_6H_4\cdot NH_2 \xrightarrow[\text{H}_2\text{SO}_4]{\text{NaNO}_2} m\text{-}O_2N\cdot C_6H_4\cdot \overset{\oplus}{N} \equiv N\}\overset{\ominus}{HSO}_4 \xrightarrow[\text{H}_2\text{SO}_4]{\text{H}_2\text{O}}$$

$$m\text{-}O_2N\cdot C_6H_4OH$$

Add 101 g (55 ml) of concentrated sulphuric acid cautiously to 75 ml of water contained in a 1-litre beaker and introduce 35 g (0.25 mol) of finely powdered m-nitroaniline (Section IV,58). Add 100–150 g of finely crushed ice and stir until the m-nitroaniline has been converted into the sulphate and a homogeneous paste results. Cool to 0–5 °C by immersion of the beaker in a freezing mixture, stir mechanically and add a cold solution of 18 g (0.26 mol) of sodium nitrite in 40 ml of water over a period of 10 minutes until a permanent colour is immediately given to potassium iodide–starch paper: do not allow the temperature to rise above 5–7 °C during the diazotisation. Continue the stirring for 5–10 minutes and allow to stand for 5 minutes; some m-nitrobenzenediazonium sulphate may separate. Decant the supernatant liquid from the solid as far as possible.

While the diazotisation is in progress, cautiously add 165 ml of concentrated sulphuric acid to 150 ml of water in a 1-litre round-bottomed flask. Heat the mixture just to boiling. Add the supernatant liquid (diazonium solution) from a separatory funnel supported over the flask at such a rate that the mixture boils very vigorously (about 30 minutes). Then add the residual damp solid (or suspension) in small portions; avoid excessive frothing. When all the diazonium salt has been introduced, boil for a further 5 minutes and pour the mixture into a 1-litre beaker set in ice-water, and stir vigorously to obtain a homogeneous crystal magma. When cold, filter at the pump, drain well and wash with four 20 ml portions of ice-water. Recrystallise by dissolving the crude product in hot dilute hydrochloric acid (1 : 1 by volume), decant from any residual dark oil, filter and cool to 0 °C, when light yellow crystals separate (1). Spread these upon a large sheet of filter-paper, and dry in the air in a warm room. The mother-liquid deposits a further crop (about 2 g) upon standing for 24 hours. The yield of m-nitrophenol, m.p. 96 °C, is 23 g (66%).

Note. (1) When working with larger quantities of material, it is more convenient (and a better yield is obtained) to purify the air-dried product by distillation under diminished pressure using a short air condenser of wide bore and a few fragments of porous porcelain (or alternatively a pine wood splinter) to prevent bumping. Collect the pure m-nitrophenol at 160–165 °C/12 mmHg; allow the flask to cool before admitting air otherwise the residue may decompose with explosive violence. The recovery is over 90 per cent of the pure m-nitrophenol.

Cognate preparation   p-Cresol.   Diazotise 27 g (0.25 mol) of p-toluidine in aqueous sulphuric acid (27 ml of concentrated acid added to 200 ml of water) with 18 g of sodium nitrite in the usual manner, and then decompose the resulting diazonium salt solution by warming it on a hot water bath and maintain the temperature at 50 °C for 15–20 minutes until nitrogen evolution ceases. Isolate the p-cresol by steam distillation followed by extraction of the distillate, and purify it by distillation; 15 g (55%) of product, b.p. 202 °C, m.p. 36 °C, are obtained.

IV,79. IODOBENZENE

$$C_6H_5 \cdot NH_2 + NaNO_2 + 2HCl \longrightarrow C_6H_5 \cdot \overset{\oplus}{N} \equiv N\}\overset{\ominus}{Cl} + NaCl + 2H_2O$$

$$C_6H_5 \cdot \overset{\oplus}{N} \equiv N\}\overset{\ominus}{Cl} + KI \longrightarrow C_6H_5I + N_2 + KCl$$

Dissolve 20 g (19.6 ml, 0.215 mol) of aniline in a mixture of 55 ml of concentrated hydrochloric acid (1) and 55 ml of water contained in a 500-ml round-bottomed flask. Place a thermometer in the solution and immerse the flask in a bath of crushed ice (2); cool until the temperature of the stirred solution falls below 5 °C. Dissolve 16 g (0.23 mol) of sodium nitrite in 75 ml of water and chill the solution by immersion in the ice bath; add the sodium nitrite solution (3) in small volumes (2–3 ml at a time) to the cold aniline hydrochloride solution, and keep the latter well shaken. Heat is evolved by the reaction. The temperature should not be allowed to rise above 10 °C (add a few grams of ice to the reaction mixture if necessary) otherwise appreciable decomposition of the diazonium compound and of nitrous acid will occur. Add the last 5 per cent of the sodium nitrite solution more slowly (say, about 1 ml at a time) and, after shaking for 3–4 minutes, test a drop of the solution diluted with 3–4 drops of water with potassium iodide–starch paper (4); if no *immediate* blue colour is obtained at the point of contact with the paper, add a further 1 ml of the nitrite solution, and test again after 3–4 minutes. Continue until a slight excess of nitrous acid is present.

To the solution of benzenediazonium chloride add a solution of 36 g (0.216 mol) of potassium iodide in 40 ml of water slowly and with shaking. Nitrogen is evolved. Allow the mixture to stand for a few hours. Fit the flask with an air condenser and heat it cautiously in a boiling water bath until evolution of gas ceases. Allow to cool. Decant as much as possible of the upper aqueous layer and render the residual aqueous and organic layers alkaline by the cautious addition of 10 per cent sodium hydroxide solution, i.e., until a drop of the well-shaken mixture withdrawn on a glass rod imparts a blue colour to red litmus paper. The alkali converts any phenol present into sodium phenoxide, which, unlike phenol itself, is not volatile in steam. Steam distil until no more oily drops pass over (Fig. I,101). Transfer the distillate to a separatory funnel and run off the lower layer of iodobenzene into a small conical flask. The crude iodobenzene should have a pale yellow colour; if it is dark in colour, return it to the separatory funnel and shake it with a little sodium metabisulphite solution until a pale yellow colour is obtained, then remove the heavy layer as before. Dry with about 1 g of anhydrous calcium chloride or magnesium sulphate: filter through a fluted filter-paper into a small distilling flask equipped with a short air condenser. Distil using an asbestos-centred wire gauze or, better, an air bath (Fig. I,49) and collect the fraction, b.p. 185–190 °C (5). The yield of iodobenzene (an almost colourless liquid) is 33 g (75%); the compound gradually develops a yellow colour upon exposure to light.

Notes. (1) In computing the volume of acid required in the diazotisation process, it is helpful to remember that 100 ml of concentrated hydrochloric acid, d 1.18, contain 42.4 g of HCl, and 100 ml of concentrated sulphuric acid, d 1.84, contain 176 g H_2SO_4.

(2) For preparations on a larger scale, the lowering of temperature may be conveniently achieved by the addition of a quantity of crushed ice equal in

weight to that of the hydrochloric acid and water. The mixture should be stirred mechanically.

(3) It is advisable to add the sodium nitrite solution, particularly in preparations on a larger scale, with the aid of a dropping funnel with the tip of the stem extending well below the surface of the liquid: this will prevent loss of nitrous acid by surface decomposition into oxides of nitrogen.

(4) It is advisable to test the potassium iodide–starch paper with acidified sodium nitrite solution: the commercial test paper is, particularly if it has been kept for a considerable period, sometimes almost useless. The solution must contain an excess of acid at all times, i.e., it must give a blue colour on Congo red paper.

(5) The iodobenzene is conveniently distilled under reduced pressure and the fraction, b.p. 77–80 °C/20 mmHg or 63–64 °C/8 mmHg, collected. The product has a higher degree of purity than that obtained directly from benzene (Section **IV,28**).

Cognate preparations   *p*-**Iodotoluene.**   Use 27 g (0.25 mol) of *p*-toluidine, 63 ml of concentrated hydrochloric acid and 63 ml of water: warm, if necessary, until all the amine dissolves. Cool the solution with vigorous stirring to 0–5 °C by immersion in a freezing mixture of ice and salt and the addition of a little crushed ice. Diazotise by the introduction, with stirring, of a solution of 18.5 g (0.27 mol) of sodium nitrite in 40 ml of water; maintain the temperature of the solution at 0–5 °C if possible, but do not allow it to rise above 10 °C. Add a solution of 44 g (0.265 mol) of potassium iodide in an equal weight of water gradually and with stirring. Allow to stand for 1 hour at the laboratory temperature and then heat cautiously on a water bath until evolution of nitrogen ceases. Allow to cool: a dark-coloured oil settles to the bottom and soon solidifies. Pour off as much of the aqueous layer as possible, add 1–2 g of sodium metabisulphite to remove the dark colour (gentle warming may be necessary) and then render the mixture alkaline with 10 per cent sodium hydroxide solution in order to retain any cresol which may be formed. Steam distil the mixture; if the *p*-iodotoluidine solidifies in the condenser, turn off the condenser water for a few moments until the solid melts and runs down into the receiver. Filter off the solid in the receiver and recrystallise it from ethanol. The yield of *p*-iodotoluene (colourless plates), m.p. 35 °C, b.p. 211–212 °C, is 50 g (92%).

p-**Iodonitrobenzene.**   Stir a mixture of 50 g (0.36 mol) of *p*-nitroaniline (Section **IV,77**), 75 g (41 ml) of concentrated sulphuric acid and 300 ml of water for 1 hour. Cool the mixture to 0–5 °C, and diazotise with a solution of 25 g (0.36 mol) of sodium nitrite in 75 ml of water. Filter the cold solution, and add the filtrate with stirring to a solution of 100 g (0.6 mol) of potassium iodide in 300 ml of water. Collect the precipitated solid by suction filtration and recrystallise it from ethanol. The yield of *p*-iodonitrobenzene, m.p. 171 °C, is 73 g (82%).

p-**Iodophenol.**   Dissolve 54.5 g (0.5 mol) of *p*-aminophenol (Section **IV,99**) in a mixture of 60 g (32.5 ml) of concentrated sulphuric acid, 250 ml of water and 250 g of crushed ice in a large beaker or bolt-head flask. Cool the solution in a freezing mixture, stir mechanically and add during 1 hour a solution of 34.5 g (0.5 mol) of sodium nitrite in 75 ml of water. Stir for a further 20 minutes, and then add 18.5 g (10 ml) of concentrated sulphuric acid. Pour the cold diazonium solution into an ice-cold solution of 100 g (0.6 mol) of potassium iodide in 100 ml of water contained in a beaker provided with a mechanical stirrer. After 5 minutes, add 1 g of copper bronze (which has been washed with ether), with continued stirring, and warm the solution slowly on a water bath. Maintain the temperature at 75–80 °C until the evolution of nitrogen ceases; the iodophenol separates as a dark heavy oil. Cool to room temperature, extract the reaction mixture with three 80 ml portions of chloroform, wash the combined extracts with dilute sodium metabisulphate solution or sodium thiosulphate solution and dry with magnesium sulphate. Remove the solvent on a water bath (rotary evaporator) and distil the residue under diminished pressure. Collect the p-iodophenol at 138–140 °C/5 mmHg; this solidifies on cooling. Recrystallise from about 1 litre of light petroleum (b.p. 80–100 °C). The yield of colourless product, m.p. 94 °C, is 78 g (70%).

o-Iodobenzoic acid. Dissolve 14 g (0.1 mol) of anthranilic acid (Section IV,60) in 100 ml of water containing 14 ml of concentrated sulphuric acid, cool to 5 °C and diazotise by the gradual addition of a cold solution of 7 g (0.101 mol) of sodium nitrite in 25 ml of water to an end-point with starch–iodide paper. Introduce into the clear solution, with stirring, a solution of 26 g (0.156 mol) potassium iodide in 50 ml of 1 M-sulphuric acid, heat the mixture to boiling for 10 minutes and then cool. Collect the *o*-iodobenzoic acid by suction filtration, and recrystallise from hot water. The yield is almost quantitative; m.p. 162 °C.

3,4,5-Triiodobenzoic acid. Dissolve 6.8 g (0.0175 mol) of 4-amino-3,5-diiodobenzoic acid (Section IV,67) in 30 ml of cold concentrated sulphuric acid, add a large excess (3.0 g, 0.0435 mol) of powdered sodium nitrite and allow the mixture to stand at 0 °C for 2 hours. Treat the cold diazonium solution with a solution of 17.0 g (0.12 mol) of potassium iodide in 40 ml of water; a dark red precipitate separates. Warm the mixture on a water bath until evolution of nitrogen ceases, and remove any residual iodine with a little sodium metabisulphite. Filter the yellow precipitate of crude 3,4,5-triiodobenzoic acid, and recrystallise from dilute ethanol. The yield of pure acid, m.p. 289–290 °C, is 6.8 g (78%).

1,3-Diiodobenzene. Add 45 g (0.65 mol) of sodium nitrite carefully and with stirring to 470 ml of concentrated sulphuric acid at 70 °C contained in a 2-litre beaker; ensure that the temperature does not exceed 75 °C during the addition. Cool the resulting solution in an ice bath with stirring and add dropwise a solution of 32.4 g (0.3 mol) of *m*-phenylenediamine (Section IV,54) in 215 ml of glacial acetic acid; do not allow the temperature to rise above 25 °C. When diazotisation is complete, replace the ice in the cooling bath by an ice–salt mixture, and when the temperature of the tetrazonium salt solution falls below 0 °C (1), run in steadily 500 ml of distilled water with vigorous stirring, keeping the temperature of the solution below 10 °C. Then run the cold diluted solution with rapid stirring into a solution of 100 g (0.6 mol) of potassium iodide in 200 ml of water. The temperature of the reaction mixture rises to about 30 °C; allow it to remain with vigorous stirring for 45 minutes. Finally heat the mixture at 75 °C for 2 hours and leave it at room temperature overnight. Isolate the product by steam distillation; collect about 2 litres of distillate, make it just alkaline with 50 per cent aqueous sodium hydroxide and extract with ether. Remove the ether from the washed, dried extract and crystallise the residual solid from a mixture of ethanol and ether with the aid of decolourising charcoal. A further crystallisation gives almost white crystals of pure 1,3-diiodobenzene, yield 60 g (61%), m.p. 36–37 °C.

Note. (1) Excessive cooling may cause the tetrazonium salt to crystallise out and make stirring difficult.

IV,80. *p*-CHLOROTOLUENE

$$p\text{-}CH_3\cdot C_6H_4\cdot NH_2 \xrightarrow[\text{HCl}]{\text{NaNO}_2} p\text{-}CH_3\cdot C_6H_4\cdot \overset{\oplus}{N}\equiv N\}\overset{\ominus}{Cl} \xrightarrow[\text{HCl}]{\text{CuCl}} p\text{-}CH_3\cdot C_6H_4Cl$$

In a 1.5- or 2-litre round-bottomed flask, prepare copper(I) chloride from 105 g (0.42 mol) of crystallised copper(II) sulphate as detailed in Section **II,2,**_16_. Either wash the precipitate once by decantation or filter it at the pump and wash it with water containing a little sulphurous acid; dissolve it in 170 ml of concentrated hydrochloric acid. Stopper the flask loosely (to prevent oxidation) and cool it in an ice–salt mixture whilst the diazotisation is being carried out.

Dissolve 36 g (0.33 mol) of *p*-toluidine in 85 ml of concentrated hydrochloric acid and 85 ml of water contained in a 750-ml conical flask or beaker. Cool the mixture to 0 °C in an ice–salt bath with vigorous stirring or shaking and the addition of a little crushed ice. The salt, *p*-toluidine hydrochloride, will separate as a finely divided crystalline precipitate. Add during 10–15 minutes a solution of 24 g (0.35 mol) of sodium nitrite in 50 ml of water (1); shake or stir the solution well during the diazotisation, and keep the mixture at a temperature of 0–5 °C by the addition of a little crushed ice from time to time. The hydrochloride will dissolve as the very soluble diazonium salt is formed; when all the nitrite solution has been introduced, the solution should contain a trace of free nitrous acid. Test with potassium iodide–starch paper (see footnote, p. 688).

Pour the cold diazonium chloride solution slowly and with shaking into the cold copper(I) chloride solution (2). The mixture becomes very thick, owing to the separation of an addition product between the diazonium salt and the copper(I) chloride ($CH_3C_6H_4N_2{}^{\oplus}Cl{}^{\ominus},CuCl$). Allow the mixture to warm up to room temperature without external heating, and shake occasionally (3). When the temperature reaches about 15 °C, the solid addition complex commences to break down with the liberation of nitrogen and the formation of an oily layer of *p*-chlorotoluene. Warm the mixture on a water bath to about 60 °C to complete the decomposition of the double salt; shake occasionally. When the evolution of nitrogen ceases, steam distil the mixture (compare Fig. I,101) until no more oily drops are present in the distillate. Transfer the distillate to a separatory funnel, and remove the layer of *p*-chlorotoluene. Wash it successively with 30 ml of 10 per cent sodium hydroxide solution (to remove any *p*-cresol which may be present), water, an equal volume of concentrated sulphuric acid (to remove a trace of azo compound that usually colours the crude product and cannot be removed by distillation) and water (to remove the acid). Dry with 3–4 g of anhydrous calcium chloride or magnesium sulphate, decant or filter through a small fluted filter-paper and distil on an asbestos-centred gauze or from an air bath (Fig. I,49) using an air condenser. Collect the *p*-chlorotoluene at 158–162 °C (a colourless liquid; m.p. 6–7 °C); the yield is 33 g (78 %).

Notes. (1) The sodium nitrite solution is conveniently added from a dropping funnel; it is recommended, particularly for preparations on a larger scale, that the tip of the stem of the funnel dip well below the surface of the liquid.

(2) The diazonium salt solution decomposes on standing and hence must be mixed with the copper(I) chloride solution without delay. Mechanical stirring is an advantage.

(3) For preparations on a larger scale, mechanical stirring is essential and should be continued for 2–3 hours after the solution has attained room temperature.

Cognate preparations   *o*-**Chlorotoluene.**   Proceed as for *p*-chlorotoluene, but use 36 g of *o*-toluidine. Collect the *o*-chlorotoluene at 155–158 °C; the yield is 33 g.

Chlorobenzene.   Prepare a solution of benzenediazonium chloride from 31 g (30.5 ml, 0.33 mol) of aniline, 85 ml of concentrated hydrochloric acid, 85 ml of water and a solution of 24 g (0.35 mol) of sodium nitrite in 50 ml of water (for experimental details, see Section **IV,79**). Prepare copper(I) chloride from 105 g of crystallised copper(II) sulphate (Section **II,2,*16***), and dissolve it in 170 ml of concentrated hydrochloric acid. Add the cold benzenediazonium chloride solution with shaking or stirring to the cold copper(I) chloride solution; allow the mixture to warm up to room temperature. Follow the experimental details given above for *p*-chlorotoluene. Wash the chlorobenzene separated from the steam distillate with 40 ml of 10 per cent sodium hydroxide solution (to remove phenol), then with water, dry with anhydrous calcium chloride or magnesium sulphate and distil. Collect the chlorobenzene (a colourless liquid) at 131–133 °C (mainly 133 °C). The yield is 29 g (77%).

m-**Chloronitrobenzene.**   This preparation is very similar to that of *p*-chlorotoluene, but certain modifications must be introduced. The quantities required are: 46 g (0.35 mol) of *m*-nitroaniline (Section **IV,58**), 85 ml of concentrated hydrochloric acid, 85 ml of water, a solution of 24 g (0.35 mol) of sodium nitrite in 50 ml of water (if the resulting diazonium salt solution is not clear, it must be filtered), and copper(I) chloride from 105 g (0.42 mol) of crystallised copper(II) sulphate (Section **II,2,*16***), dissolved in 170 ml of concentrated hydrochloric acid. Run the diazonium salt solution into the solution of copper(I) chloride while the temperature is kept at 25–30 °C (water bath); at lower temperatures the decomposition of the unstable addition compound proceeds too slowly and would cause too violent an evolution of nitrogen upon warming, and at a higher temperature the formation of tarry by-products increases. Warm the mixture under a reflux condenser on a water bath until the evolution of nitrogen ceases. Steam distil (1); if the *m*-chloronitrobenzene solidifies in the condenser, turn off the condenser water for a few moments until the solid melts and runs down into the receiver. Allow the steam distillate to cool, decant the water and shake the solid with 200 ml of 1 per cent sodium hydroxide solution at 50 °C (to remove *m*-nitrophenol, if present). Allow the mixture to cool, filter with suction, wash with a little cold water and dry in the air. Determine the m.p. If this is not satisfactory, i.e., if it is appreciably below 44–45 °C, purify the product either by recrystallisation from a small volume of ethanol or preferably by distillation under diminished pressure and collect the fraction of b.p. 124–125 °C/18 mmHg or 116–117 °C/12 mmHg; the distillate solidifies to a pale yellow solid, m.p. 44–45 °C. The yield is 35 g (67%), depending upon the purity of the original *m*-nitroaniline.

Note. (1) The steam distillation may be omitted, if desired, by utilising the following method of purification. Allow the reaction mixture to cool, decant the aqueous layer and dissolve the residue in about 150 ml of toluene. Wash the toluene solution with water, 1 per cent sodium hydroxide solution and finally with water; dry with magnesium sulphate, distil off the toluene on a water bath (rotary evaporator) and distil the residue under diminished pressure.

o-**Chlorobenzoic acid.**   Dissolve 14 g (0.1 mol) of anthranilic acid (Section **IV,60**) in a solution of 20 ml of concentrated hydrochloric acid and 100 ml of water. Cool to about 5 °C, and diazotise by the gradual addition of a cold solution of 7 g (0.1 mol) of sodium nitrite in 25 ml of water to an end-point with starch–potassium iodide paper (see footnote, p. 688). In the meantime prepare a solution of copper(I) chloride as follows. Dissolve 26 g (0.104 mol) of crystallised copper(II) sulphate and 12 g of sodium chloride in 50 ml of water in a 750-ml round-bottomed flask. Heat the solution to boiling, then add 80 ml of concentrated hydrochloric acid and 14 g of copper turnings, and continue the heating under reflux until the solution is practically colourless. (Alternatively, prepare the copper(I) chloride by the method given in Section **II,2,***16*.) Cool in ice, and then add the cold diazonium solution slowly and with shaking. The reaction proceeds rapidly and with frothing: allow the mixture to stand for 2–3 hours with frequent shaking. Filter the precipitated *o*-chlorobenzoic acid and wash it with a little cold water. Recrystallise the crude acid from hot water containing a little alcohol to which a little decolourising carbon has been added. The yield of pure *o*-chlorobenzoic acid, m.p. 138–139 °C, is 14 g (87%).

IV,81. *p*-BROMOTOLUENE

$$p\text{-CH}_3\cdot\text{C}_6\text{H}_4\cdot\text{NH}_2 \xrightarrow[\text{H}_2\text{SO}_4]{\text{NaNO}_2} p\text{-CH}_3\cdot\text{C}_6\text{H}_4\cdot\overset{\oplus}{\text{N}}\equiv\text{N}\}\overset{\ominus}{\text{HSO}_4} \xrightarrow{\text{CuBr}} p\text{-CH}_3\cdot\text{C}_6\text{N}_4\text{Br}$$

Prepare a **solution of copper(I) bromide** in a 2.5-litre two-necked flask by heating under reflux 31.5 g (0.124 mol) of copper(II) sulphate pentahydrate, 10 g (0.158 mol) of copper turnings, 77 g (0.55 mol) of crystallised sodium bromide, 15 g (8.2 ml) of concentrated sulphuric acid as described in Section **II,2,***15*.

In a 1-litre flask mix 53.5 g (0.5 mol) of *p*-toluidine and 400 ml of water, and then add cautiously 98 g (53.5 ml) of concentrated sulphuric acid; warm until the *p*-toluidine dissolves. Cool the flask in a bath of ice and salt to 0–5 °C; add about 100 g of crushed ice to the contents of the flask in order to accelerate the cooling. Add slowly and with frequent shaking a solution of 35 g (0.5 mol) of sodium nitrite in 60 ml of water until a slight excess of sodium nitrite is present (see footnote, p. 688); keep the temperature of the mixture below 10 °C.

Equip the 2.5-litre two-necked flask containing the copper(I) bromide solution for steam distillation (compare Fig. I,101) and insert into the side-neck a tube (7–8 mm in diameter) leading almost to the bottom of the flask via a screw-capped adapter; attach a short-stemmed separatory funnel to this tube by means of a short length of rubber tubing and support the funnel in a ring clamped to a retort stand. Heat the copper(I) bromide solution to boiling, add the toluene-*p*-diazonium sulphate solution from the separatory funnel whilst steam is passed rapidly through the mixture. In order to reduce the amount of decomposition of the diazonium salt solution, transfer only about one-fourth to the separatory funnel (the remainder being kept in the freezing mixture) and run this into the copper(I) bromide solution: when the funnel is nearly empty, transfer a further portion of the cold diazonium solution to it without interrupting the addition. Add all the diazonium solution in this way during 20–30 minutes. Continue the steam distillation until no more organic matter distils. Render the distillate alkaline with 20 per cent sodium hydroxide solution (to remove any p-cresol present), shake well and separate the crude p-bromotoluene. In order to obtain a colourless product, wash the crude substance with 40–50 ml of warm (30 °C) concentrated sulphuric acid, then with water, sodium hydroxide solution, and finally with water. If the p-bromotoluene solidifies, warm the wash liquids to 30 °C before use; unless this is done, considerable loss may occur. Dry over magnesium sulphate or anhydrous calcium chloride, warm, filter and distil through an air-cooled condenser. Collect the p-bromotoluene at 182–184 °C. The yield is 60 g (70%); m.p. 25–26 °C.

Cognate preparations **o-Bromotoluene.** Use 53.5 g of o-toluidine and other components as above. The yield of o-bromotoluene, b.p. 178–181 °C, is of the same order.

   o-Bromochlorobenzene. Place a mixture of 64 g (0.5 mol) of o-chloroaniline (Section **IV,55**) and 175 ml of constant boiling point hydrobromic acid (d 1.48; 100 ml contains 71 g of HBr) in a 1-litre flask set in an ice–salt bath, and cool it to 0–5 °C by the addition of a little ice. Add, with shaking or stirring, a solution of 35 g (0.5 mol) of sodium nitrite in 70 ml of water until a slight excess of nitrous acid is present (starch–potassium iodide paper test; see footnote, p. 688); maintain the temperature below 10 °C by the addition of ice if necessary.

   Prepare copper(I) bromide from 75 g (0.33 mol) of crystallised copper(II) sulphate as detailed in Section **II,2,15**, and dissolve it in 40 ml of 48 per cent hydrobromic acid; heat the solution to boiling and add o-chlorobenzene-diazonium bromide solution as detailed above. When all the latter has been introduced, continue to pass steam through the mixture until no more organic material distils. Follow the procedure, including purification, given for p-bromotoluene. Collect the o-bromochlorobenzene (a colourless liquid) at 200–202 °C. The yield is 85 g (89%).

   This procedure may also be employed for **m-bromochlorobenzene**, b.p. 191–194 °C, from m-chloroaniline; **m-dibromobenzene**, b.p. 215–217 °C, from m-bromoaniline; and **o-bromoanisole**, b.p. 114–116 °C/29 mmHg, from o-anisidine (the sulphuric acid washing is omitted in the last example).

   2-Bromonaphthalene. Dissolve 112 g (0.5 mol) of 2-naphthylamine-1-sulphonic acid (1), with stirring, in 850 ml of 0.6 M-sodium hydroxide solution: add, with stirring, an aqueous solution of 35 g (0.5 mol) of sodium nitrite, and filter the resulting solution. Place 250 ml of concentrated hydrochloric acid and 100 g of crushed ice in a 2-litre beaker, and equip the latter with a mechanical stirrer. Introduce the filtered solution of sodium nitrite and sodium 2-naphthylamine-1-sulphonate (2) slowly with stirring, and maintain the temperature at 0–5 °C by adding crushed ice. Collect the reddish-brown precipitate which forms on a large Buchner funnel and wash it with about 500 ml of ice-water. Whilst the diazotisation is in progress, suspend 160 g (1.1 mol) of copper(I) bromide (from 300 g of crystallised copper(II) sulphate; Section

II,2,*15***)** in 75 ml of 48 per cent hydrobromic acid and 200 ml of water. Add the damp cake of the diazonium compound portionwise and with vigorous stirring to the copper(I) bromide suspension contained in a 2-litre beaker. After the vigorous evolution of nitrogen has subsided heat the mixture to 95–100 °C on a steam bath and then filter the hot mixture through a large Buchner funnel. Pour the filtrate back into the beaker and add 112 g of potassium chloride with stirring. Allow the resulting paste to cool to room temperature, filter with suction and wash with 250 ml of 20 per cent aqueous potassium chloride. Dry the reddish-brown precipitate of 2-bromonaphthalene-1-sulphonic acid in the air overnight, and transfer it to a 2-litre round-bottomed flask. Add dilute sulphuric acid (prepared from 200 ml of the concentrated acid and 200 g of crushed ice), attach a reflux condenser and reflux the mixture gently, using an electric heating mantle, for 12–16 hours. Cool to room temperature; pour on to about 500 g of crushed ice. Transfer the mixture with the aid of 500 ml of toluene to a large separatory funnel, shake well, remove the toluene layer and wash the latter with water until the washings are neutral to litmus. Dry the toluene solution with magnesium sulphate, remove the toluene using a rotary evaporator and distil the residue under reduced pressure. Collect the bromo-naphthalene at 100–101 °C/2 mmHg or at 140 °C/20 mmHg; this solidifies to a pale yellow solid, m.p. 56–57 °C. The yield is 67 g (65%).

The pale yellow colour cannot be removed by redistillation or recrystallisation; the coloured product probably contains some amino compound rendering it unsuitable for conversion into a Grignard reagent. A pure white product may be obtained by the following procedure. Dissolve 50 g of the coloured compound in 200 ml of hexane and pass the solution through a column of activated alumina (80–200 mesh; dimensions about 9 cm × 3 cm); wash the column with 750 ml of hexane. Remove the hexane by distillation; 49 g of pure 2-bromonaphthalene, m.p. 58 °C, remains. This is sufficiently pure for use in Grignard reactions.

Note. (1) Tobias acid, available from Fluka AG Chemische Fabrik; see also p. 667, **IV,E,3**.

(2) If the solid sodium salt is available, 123 g may be dissolved in 850 ml of distilled water and a solution containing 35 g of sodium nitrite added.

IV,82. *o*-BROMOTOLUENE

$$o\text{-CH}_3\text{·C}_6\text{H}_4\text{·NH}_2 \xrightarrow[\text{HBr}]{\text{NaNO}_2} o\text{-CH}_3\text{·C}_6\text{H}_4\text{·}\overset{\oplus}{\text{N}} \equiv \text{N}\}\overset{\ominus}{\text{Br}} \xrightarrow{\text{Cu}} o\text{-CH}_3\text{·C}_6\text{H}_4\text{Br}$$

In a 1- or 1.5-litre round-bottomed flask prepare a solution of 53.5 g (0.5 mol) of *o*-toluidine in 170 ml of 40 per cent w/w hydrobromic acid; cool to 5 °C by immersion in a bath of ice and salt. Diazotise by the gradual addition of a solution of 36.5 g (0.53 mol) of sodium nitrite in 50 ml of water; stopper the flask after each addition and shake until all red fumes are absorbed. Keep the temperature between 5 and 10 °C. When the diazotisation is complete, add 2 g of copper powder or copper bronze (Section **II,2,***13***)**, attach a reflux condenser to the flask and heat very cautiously on a water bath. *Immediately* evolution of gas occurs, cool the flask in crushed ice; unless the flask is rapidly removed from the water bath, the reaction may become so violent that the contents may be shot out of the flask. When the vigorous evolution of nitrogen moderates, heat the flask on a water bath for 30 minutes. Then dilute with 400 ml of water, and steam distil the mixture until about 750 ml of distillate are collected. Render the distillate alkaline with 10 per cent sodium hydroxide solution (about 50 ml) and separate the lower red layer of crude o-bromotoluene. Wash it with two 20 ml portions of concentrated sulphuric acid (which removes most of the colour) and then twice with water. Dry with magnesium sulphate or anhydrous calcium chloride, and distil from a flask fitted with a lagged fractionating column. Collect the o-bromotoluene at 178–181 °C. The yield is 40 g (47%).

IV,83. BENZENESULPHINIC ACID

$$C_6H_5 \cdot NH_2 \xrightarrow[H_2SO_4]{NaNO_2} C_6H_5 \cdot \overset{\oplus}{N} \equiv N\}H\overset{\ominus}{S}O_4 \xrightarrow{SO_2}{Cu} C_6H_5 \cdot SO_2H$$

Dissolve 9.3 g (9.1 ml, 0.1 mol) of aniline in a mixture of 19.6 g (10.7 ml) of concentrated sulphuric acid and 100 ml of water, and cool to about 5 °C. Diazotise by the addition of a solution of 7.0 g (0.1 mol) of sodium nitrite in 15 ml of water to an end-point with potassium iodide–starch paper; maintain the temperature below 10 °C. Add an ice-cold mixture of 40 g (22 ml) of concentrated sulphuric acid and 30 ml of water, cool in ice and pass sulphur dioxide into the solution until there is no further increase in weight (about 25 g). The solution should not develop any appreciable colour during this operation and should remain quite clear. When the solution is saturated with sulphur dioxide, transfer it to a beaker provided with a mechanical stirrer, and add copper powder (Section **II,2,13**) or copper bronze (previously washed with ether) gradually until no more nitrogen is evolved (about 30 g of copper powder are required). Filter at the pump and wash the precipitate with several small amounts of dilute ammonia solution to remove any sulphinic acid which may have separated: add the washings to the filtrate. The combined filtrate and washings should be acid to Congo red paper. Treat it with concentrated iron(III) chloride solution as long as any precipitate forms. Filter the precipitate of iron(III) benzenesulphinate, and wash it with a little water. Decompose the iron(III) salt with a slight excess of 5 per cent sodium hydroxide solution, and filter the precipitated iron(III) hydroxide. Acidify the filtrate and extract the sulphinic acid with ether. Upon evaporation of the solvent, pure benzenesulphinic acid, m.p. 84 °C, is obtained as a colourless crystalline solid. The yield is 10 g (70%). It oxidises in the air.

IV,84. p-TOLUNITRILE (p-Tolyl Cyanide)

$$p\text{-}CH_3 \cdot C_6H_4 \cdot NH_2 \xrightarrow[HCl]{NaNO_2} p\text{-}CH_3 \cdot C_6H_4 \cdot \overset{\oplus}{N} \equiv N\}\overset{\ominus}{C}l \xrightarrow[KCN]{CuCN} p\text{-}CH_3 \cdot C_6H_4 \cdot CN$$

This and related cognate preparations must be carried out in an efficient fume cupboard.

Prepare copper(I) cyanide from 100 g (0.4 mol) of hydrated copper(II) sulphate following the procedure described in Section **II,2,17**, transfer the product to a 1-litre round-bottomed flask and dissolve it in a solution of 52 g of potassium cyanide in 125 ml of water [**CAUTION**].

Diazotise 36 g (0.33 mol) of p-toluidine, following the method given under p-chlorotoluene (Section **IV,80**). Whilst keeping the solution cold, carefully add about 20 g of powdered anhydrous sodium carbonate with constant stirring until the solution is neutral to litmus. Warm the copper(I) cyanide solution on a water bath to about 60 °C, and add the cold neutralised diazonium salt solution in small quantities at a time, shaking vigorously (1) after each addition and taking care to maintain the temperature of the mixture at 60–70 °C. Attach a reflux condenser to the flask and heat on a boiling water bath for 15–20 minutes in order to complete the reaction. Equip the flask for steam distillation (Fig. I,101), and pass steam into the mixture until no more yellow oil passes over; if the oil solidifies in the condenser tube, turn off the condenser water, and, after the material melts and flows through, slowly turn on the water again. Cool the distillate in ice-water, and when the crude p-tolunitrile has solidified, filter it at the pump and press well to remove liquid impurities. Dry upon filter-paper or in a desiccator. Mix the dried product with 2–3 g of decolourising carbon, transfer to a small distilling flask and distil using an air condenser. Collect the pure p-tolunitrile at 215–219 °C (2); this solidifies on cooling and melts at 29 °C. The yield is 26 g (67%).

Notes. (1) Mechanical stirring is preferable.

(2) The crude substance may also be distilled under diminished pressure and the p-tolunitrile collected at 104–106 °C/20 mmHg.

Cognate preparations   Benzonitrile (phenyl cyanide).   Prepare a copper(I) cyanide (0.26 mol) solution in a 500-ml round-bottomed flask following the procedure described in Section **II,2,17** and using the following quantities: 65 g of crystallised copper(II) sulphate in 205 ml of water, 18 g of sodium metabisulphite in 52 ml of water and 18 g of potassium cyanide in 52 ml of water; dissolve the precipitated copper(I) cyanide in a solution of 26 g of sodium cyanide in 65 ml of water or of 33.5 g of potassium cyanide in 90 ml of water. Diazotise 20 g (19.6 ml, 0.215 mol) of aniline, following the experimental details given under *iodobenzene* (Section **IV,79**) and neutralise the solution with about 14 g of anhydrous sodium carbonate. Add the cold neutralised benzene-diazonium chloride solution to the copper(I) cyanide solution warmed at 60–70 °C and proceed as for p-*tolunitrile*. Extract the steam distillate with three 30 ml portions of ether, shake the ethereal solution with 20 ml of 10 per cent sodium hydroxide solution (to remove traces of phenol produced by the decomposition of the diazonium chloride solution), then with an equal volume of dilute sulphuric acid (to remove traces of the evil-smelling phenyl isocyanide C_6H_5NC), and finally with an equal volume of water. Dry the ethereal extract over magnesium sulphate or anhydrous calcium chloride, remove the ether, e.g., by flash distillation, and distil the benzonitrile using an air condenser. Collect the fraction of b.p. 188–191 °C. The yield is 16 g (73%).

o-Tolunitrile.   The preparation is similar to that described for p-*tolunitrile* except that p-toluidine is replaced by o-toluidine. The o-tolunitrile is isolated by steam distillation; the oil, which may be dissolved in a little toluene, is distilled. The o-tolunitrile passes over as an almost colourless liquid at 94–96 °C/20 mmHg.

IV,85. FLUOROBENZENE

$$C_6H_5 \cdot NH_2 \longrightarrow C_6H_5 \cdot \overset{\oplus}{N} \equiv N\}\overset{\ominus}{Cl} \xrightarrow{\text{NaBF}_4} C_6H_5 \cdot \overset{\oplus}{N} \equiv N\}\overset{\ominus}{BF_4} \xrightarrow{\text{heat}}$$

$$C_6H_5F + BF_3 + N_2$$

This preparation should be carried out in an efficient fume cupboard and behind a suitable safety screen.

Dissolve 46.5 g (45.5 ml, 0.5 mol) of aniline in a mixture of 126 ml of concentrated hydrochloric acid and 126 ml of water contained in a 1-litre beaker. Cool to 0–5 °C in a bath of ice and salt, and add a solution of 36.5 g (0.53 mol) of sodium nitrite in 75 ml of water in small portions; stir vigorously with a thermometer and maintain the temperature below 10 °C, but preferably at about 5 °C by the addition of a little crushed ice if necessary. The diazotisation is complete when a drop of the solution diluted with 3–4 drops of water gives an immediate blue colouration with potassium iodide–starch paper; the test should be performed 3–4 minutes after the last addition of the nitrite solution. Prepare a solution of 76 g (0.69 mol) of sodium fluoroborate (1) in 150 ml of water, cool and add the chilled solution slowly to the diazonium salt solution; the latter must be kept well stirred and the temperature controlled so that it is below 10 °C. Allow to stand for 10 minutes with frequent stirring. Filter the precipitated benzenediazonium fluoroborate with suction on a Buchner funnel, drain well and wash the yellow solid with about 30 ml of ice-water, 15 ml of methanol and 30–40 ml of ether; suck the solid as free as possible from liquid after each washing (2). Spread the salt upon absorbent filter-paper and allow to dry overnight, if possible in a current of air. The yield of benzenediazonium fluoroborate is 60–65 g; the pure salt melts with decomposition at 119–120 °C.

Assemble the apparatus shown in Fig. IV,4; this is self-explanatory. The

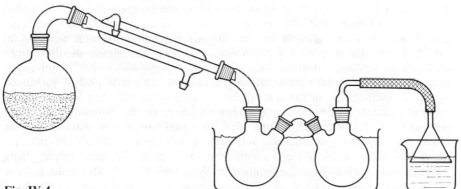

Fig. IV,4

distilling flask has a capacity of 250 ml and the beaker contains 150 ml of 10 per cent sodium hydroxide solution. Place half of the yield of the dry benzene-diazonium fluoroborate in the distilling flask. Heat the solid gently with a small luminous flame at one point near its surface until decomposition begins; withdraw the flame and allow the reaction to continue as long as it will (3). Continue the cautious heating from time to time as may be necessary to keep the reaction going. When the decomposition appears to be complete, heat the flask more strongly to drive off any remaining fluorobenzene. Allow to cool, add the other half of the benzenediazonium fluoroborate through a glazed paper funnel and decompose it as before; finally heat the flask strongly until no more fumes of boron trifluoride are evolved in order to drive off the last traces of fluorobenzene. Most of the fluorobenzene collects in the first, cooled, receiver. Wash the combined distillates three times with an equal volume of 10 per cent sodium hydroxide solution (4) or until the washings are almost colourless; this will remove any phenol present. Remove the last sodium hydroxide washing as completely as possible, and then shake with an equal volume of almost saturated salt solution. Dry over anhydrous calcium chloride or magnesium sulphate, and distil the fluorobenzene (a colourless liquid) at 84–85 °C. The yield is 24 g (50%).

Notes. (1) The use of sodium fluoroborate solution supersedes the less convenient **fluoroboric acid** and permits the preparation to be carried out in ordinary glass vessels. If it is desired to employ **fluoroboric acid** HBF_4, it can be prepared by adding 100 g of A.R. boric acid in small proportions to 325 g of A.R. hydrofluoric acid (40% HF) cooled in ice; the hydrofluoric acid is contained in a polypropylene beaker, a beaker coated with wax or in a lead vessel. One-third of the above solution should be employed in the preparation. *Handle with great care.*

Note on precautions to be adopted when using hydrofluoric acid. Attention is directed to the fact that hydrofluoric acid in contact with the skin produces extremely painful burns. In case of accident, the burned surface, which becomes white, is held under running water until the natural colour returns. A paste made from magnesium oxide and glycerine should be applied immediately; this is said to be helpful in preventing the burn becoming serious. It is advisable to wear acid-resisting rubber gloves and protective goggles.

(2) Careful washing with methanol and ether is necessary to remove from the crude product any moisture which tends to make the material unstable and liable to spontaneous decomposition.

(3) If the reaction becomes too vigorous, it may be necessary to cool the flask by covering it with a damp cloth. Normally the decomposition proceeds smoothly under the intermittent heating. If the salt is damp, the reaction may proceed more vigorously and unless the flask is cooled, it may pass beyond control.

(4) The density of fluorobenzene is about 1.025 at room temperature; it is important to use the correct strength of sodium hydroxide solution in order to obtain a clear separation of the layers.

Cognate preparations *p*-**Fluorotoluene.** Diazotise 53.5 g (0.5 mol) of *p*-toluidine in a mixture of 126 ml of concentrated hydrochloric acid and 126 ml of water contained in a 1-litre beaker following the procedure given in Section **IV,80**. Add a chilled solution of 76 g (0.69 mol) of sodium fluoroborate in 150 ml of water slowly and with good stirring to the cold diazonium salt solution. Continue stirring for about 15 minutes. Filter the toluene-*p*-diazonium fluoroborate on a Buchner or sintered glass funnel, wash with about 30 ml of ice-water, 15 ml of methanol and 30–40 ml of ether. Dry overnight upon absorbent paper in a vacuum desiccator or, if possible, in a current of air. The yield of toluene-*p*-diazonium fluoroborate is 78 g (76%); it melts with decomposition at 114 °C.

Decompose the salt in two equal lots, and work up as for *fluorobenzene*. The yield of pure *p*-fluorotoluene (a colourless liquid), b.p. 116–117 °C, is 27 g (50%).

p-Fluoroanisole. To 105 ml (2 mol) of *ca.* 42 per cent fluoroboric acid (**CAUTION:** corrosive chemical) diluted with an equal volume of water, contained in a 600-ml beaker, add 31 g (0.25 mol) of *p*-anisidine. Place the beaker in an ice bath and stir the solution mechanically. Add a solution of 17.5 g (0.25 mol) of sodium nitrite in 35 ml of water slowly and maintain the temperature at about 10 °C. Stir the solution vigorously towards the end of the reaction, cool the mixture to 0 °C and filter with suction on a sintered glass funnel. Wash the precipitate successively with 30–40 ml of cold 5 per cent fluoroboric acid, 40 ml of ice-cold methanol and several times with ether. Dry overnight by spreading the salt thinly on absorbent paper supported upon a screen or wire netting allowing circulation underneath. The yield of *p*-methoxybenzenediazonium fluoroborate is 54 g (98%). Decompose the dry salt as detailed for *fluorobenzene*. Return the small amount of product in the receiver to the distilling flask and steam distil. Extract the steam distillate with two 50 ml portions of ether, wash the ethereal solution with 50 ml of 10 per cent sodium hydroxide solution, followed by water and dry over magnesium sulphate. Remove the ether on a steam bath and distil the residue. Collect the *p*-fluoroanisole at 156–157 °C. The yield is 16 g (51%).

IV,86. *o*-DINITROBENZENE

$$o\text{-}O_2N\text{\textbullet}C_6H_4\text{\textbullet}NH_2 \xrightarrow[\text{HCl}]{\text{NaNO}_2} o\text{-}O_2N\text{\textbullet}C_6H_4\text{\textbullet}\overset{\oplus}{N}\!\equiv\!N\}\overset{\ominus}{Cl} \xrightarrow{\text{NaBF}_4}$$

$$o\text{-}O_2N\text{\textbullet}C_6H_4\text{\textbullet}\overset{\oplus}{N}\!\equiv\!N\}\overset{\ominus}{BF_4} \xrightarrow[\text{Cu}]{\text{NaNO}_2} o\text{-}O_2N\text{\textbullet}C_6H_4\text{\textbullet}NO_2$$

This preparation should be carried out behind a safety screen.

Dissolve 34 g (0.25 mol) of *o*-nitroaniline in a warm mixture of 63 ml of concentrated hydrochloric acid and 63 ml of water contained in a 600-ml beaker. Place the beaker in an ice–salt bath, and cool to 0–5 °C whilst stirring mechanically; the *o*-nitroaniline hydrochloride will separate in a finely divided crystalline form. Add a cold solution of 18 g (0.26 mol) of sodium nitrite in 40 ml of water slowly and with stirring to an end-point with potassium iodide–starch paper; do not allow the temperature to rise above 5–7 °C. Introduce, whilst stirring vigorously, a solution of 40 g (0.36 mol) of sodium fluoroborate in 80 ml of water. Stir for a further 10 minutes, and filter the solid diazonium fluoroborate with suction on a Buchner funnel. Wash it immediately once with 25 ml of cold 5 per cent sodium fluoroborate solution, then twice with 15 ml portions of rectified (or industrial) spirit and several times with ether; in each washing stir the fluoroborate well before applying suction. The *o*-nitrobenzenediazonium fluoroborate weighs about 50 g (86%); the pure substance melts with decomposition at 135 °C.

Dissolve 200 g (2.9 mol) of sodium nitrite in 400 ml of water in a 2-litre beaker provided with an efficient mechanical stirrer, and add 40 g of copper powder (either the precipitated powder or copper bronze which has been washed with a little ether). Suspend the fluoroborate in about 200 ml of water and add it slowly to the well-stirred mixture. Add 4–5 ml of ether from time to time to break the froth. The reaction is complete when all the diazonium compound has been added. Transfer the mixture to a large flask and steam distil until no more solid passes over (about 5 litres of distillate). Filter off the crystalline solid in the steam distillate and dry upon filter-paper in the air; this o-dinitrobenzene (very pale yellow crystals) has m.p. 116 °C (i.e., is practically pure) and weighs 29 g (69%). It may be recrystallised from ethanol; the recrystallised solid melts at 116.5 °C.

Cognate preparation   *p*-**Dinitrobenzene.**   Use 34 g (0.25 mol) of *p*-nitroaniline (Section **IV,77**) and proceed exactly as above to the point where all the suspension of *p*-nitrobenzenediazonium fluoroborate has been added. Filter the reaction mixture with suction, wash the residue well with water, twice with 25 ml of 5 per cent sodium hydroxide solution and finally with water. Dry the solid at 100–110 °C, powder it and extract it with four 150 ml portions of boiling toluene. Remove the toluene with a rotary evaporator and recrystallise the residue from about 120 ml of boiling glacial acetic acid. The yield of *p*-dinitrobenzene (reddish-yellow crystals), m.p. 173 °C, is 30 g (71.5%). Further recrystallisation from ethanol affords pale yellow crystals of the same m.p.

IV,87. 2-CHLOROBIPHENYL

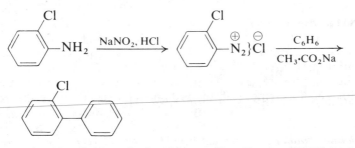

Diazotise 32 g (0.25 mol) of o-chloroaniline (Section **IV,55**) in the presence of 40 ml of concentrated hydrochloric acid and 22.5 ml of water in the usual manner (compare Section **IV,79**) with a concentrated solution of 18.5 g sodium nitrite in water. Transfer the cold, filtered diazonium solution to a 1.5-litre bolthead flask surrounded by ice-water, introduce 500 ml of cold benzene, stir vigorously and add a solution of 80 g of sodium acetate trihydrate in 200 ml of water dropwise, maintaining the temperature at 5–10 °C. Continue the stirring for 48 hours: after the first 3 hours, allow the reaction to proceed at room temperature. Separate the benzene layer, wash it with water and remove the benzene by distillation at atmospheric pressure; distil the residue under reduced pressure and collect the 2-chlorobiphenyl at 150–155 °C/10 mmHg. The yield is 18 g (76%). Recrystallise from aqueous ethanol; m.p. 34 °C.

Cognate preparation   **4-Bromobiphenyl.**   Diazotise 43 g (0.25 mol) of *p*-bromoaniline (Section **IV,75**) in the presence of 40 ml of concentrated hydrochloric acid and 22.5 ml of water with a concentrated solution of 18.5 g of sodium nitrite in water. Mix the filtered diazonium solution with 500 ml of cold benzene, stir vigorously and add a solution of 30 g of sodium hydroxide in 150 ml of water dropwise (during 30–45 minutes) whilst maintaining the temperature at 5–10 °C. Complete the reaction as for 2-chlorobiphenyl. The yield of 4-bromobiphenyl, b.p. 170–175 °C/8 mmHg, m.p. 90 °C (from ethanol), is 25 g (86%).

IV,88. 1,3,5-TRIBROMOBENZENE

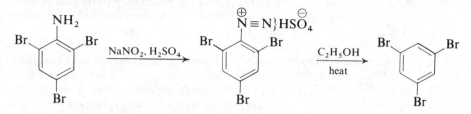

Dissolve 10 g (0.03 mol) of 2,4,6-tribromoaniline (Section **IV,66**) by heating on a water bath with 60 ml of rectified spirit and 15 ml of benzene in a 200-ml two-necked flask fitted with a reflux condenser, the second neck being closed with a stopper. Add, from a burette or small graduated pipette, 5.3 g (3.5 ml) of concentrated sulphuric acid to the hot solution via the side-neck and gently swirl the liquid, replace the stopper and heat on a water bath until the clear solution boils. Remove the flask from the water bath, and add 3.5 g (0.05 mol) of powdered sodium nitrite in two approximately equal portions via the side-neck; after each addition, replace the stopper and shake the flask vigorously; when the reaction subsides, add the second portion of the sodium nitrite. Heat the flask on a boiling water bath as long as gas is evolved; shake well from time to time. Allow the solution to cool for 10 minutes, and then immerse the flask in an ice bath. A mixture of tribromobenzene and sodium sulphate crystallises out. Filter with suction on a Buchner funnel, wash with a small quantity of ethanol and then repeatedly with water to remove all the sodium sulphate. Dissolve the crude tribromobenzene (7.5 g) in a boiling mixture of 120 ml of glacial acetic acid and 30 ml of water (1), boil the solution with 2.5 g of decolourising carbon and filter through a hot water funnel or a preheated Buchner funnel: allow the solution to cool. Collect the crystals on a Buchner funnel and wash with a small quantity of chilled rectified spirit to remove the acetic acid. Dry in the air upon filter-paper. The yield of 1,3,5-tribromobenzene (colourless crystals), m.p. 122 °C, is 6.5 g (68%).

Note. (1) Rectified spirit may also be employed for crystallisation.

IV,89. m-BROMOTOLUENE

p-Acetotoluidide and 4-acetamido-3-bromotoluene. Prepare a solution of p-acetotoluidide in glacial acetic acid by boiling 107 g (1 mol) of p-toluidine with 400 ml of glacial acetic acid in a 1-litre, round-bottomed three-necked flask, provided with a reflux condenser, stirrer and thermometer, for 2 hours. Cool the solution when some p-acetotoluidide may separate as small crystals as the temperature falls (1). When the temperature has fallen to about 45 °C, add 162.5 g (52.5 ml, 1.01 mol) of bromine from a separatory funnel at such a rate that the temperature of the well-stirred mixture is maintained at 50–55 °C. A precipitate may separate during the addition which requires 30–40 minutes, but this dissolves later. Continue the stirring for a further 30 minutes after all the bromine has been added. Then pour the reaction mixture in a thin stream into

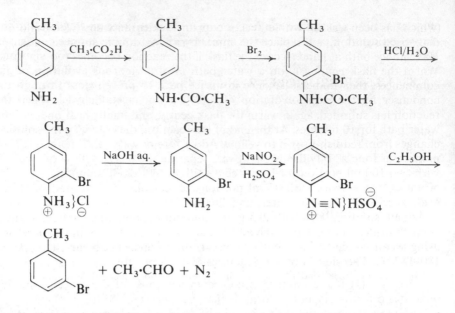

a well-stirred mixture of 1 kg of crushed ice and 1 kg of water to which 14 g of solid sodium metabisulphite has been added. If the colour of the bromine persists, add a little more sodium metabisulphite. Filter the crystalline 4-acetamido-3-bromotoluene with suction on a Buchner funnel, wash thoroughly with water and press well. Dry in the air until the weight does not exceed 250 g (2); further purification is unnecessary before proceeding to the next stage.

4-Amino-3-bromotoluene hydrochloride. Transfer the partially dried 4-acetamido-3-bromotoluene to a 1.5-litre round-bottomed flask, add 250 ml of rectified spirit and reflux on a water bath until the solid dissolves completely. Introduce through the condenser 250 ml of concentrated hydrochloric acid to the boiling solution and continue the refluxing for a further 3 hours. During this time crystals of 4-amino-3-bromotoluene hydrochloride separate. Pour the hot mixture into a 1-litre beaker and cool thoroughly. Filter the crystals of the hydrochloride at the pump through a Buchner funnel and wash rapidly with two 50 ml portions of chilled rectified spirit. The yield of the hydrochloride is 150 g (67.5%).

4-Amino-3-bromotoluene. Suspend the hydrochloride in 400 ml of water in a 1-litre beaker equipped with a mechanical stirrer. Add a solution of 70 g of sodium hydroxide in 350 ml of water. The free base separates as a dark heavy oil. After cooling to 15–20 °C, transfer the mixture to a separatory funnel and run off the crude 4-amino-3-bromotoluene. This weighs 125 g (67%) and can be used directly in the next step (3).

***m*-Bromotoluene.** To a cold mixture of 400 ml of rectified spirit and 100 ml of concentrated sulphuric acid contained in a 2.5-litre three-necked flask, provided with an efficient mechanical stirrer, add 125 g (0.67 mol) of crude 4-amino-3-bromotoluene. Stir the solution and cool to 5 °C; then add slowly a solution of 74 g (1.07 mol) of pure sodium nitrite in 135 ml of water from a separatory funnel taking care that the temperature does not rise above 10 °C. Continue the stirring for 20 minutes after all the nitrite solution has been added in order to complete the diazotisation (test with potassium iodide–starch paper for the presence of free nitrous acid). Add 17.5 g (0.28 mol) of copper bronze (which has been washed with ether) or copper powder (Section **II,2,**_13(a)_) to the diazotised solution, and replace the stirrer by a long double surface condenser. Have an ice bath at hand to cool the flask if the reaction becomes too vigorous. Warm the flask _cautiously_ on a water bath until a vigorous evolution of gas commences, then immerse at once in an ice bath to prevent loss through the condenser by too rapid evolution of nitrogen and acetaldehyde. When the reaction has subsided, again warm the flask gently, and finally heat on a boiling water bath for 10 minutes. At the end of the reaction, the colour of the solution changes from reddish-brown to yellow. Add 1 litre of water and steam distil the mixture as long as oily drops pass over. Separate the heavy yellow oil, wash it with two 100 ml portions of 10 per cent sodium hydroxide solution, once with 50 ml of water, twice with 75 ml portions of ice-cold concentrated sulphuric acid, once with 50 ml of water, and finally with 50 ml of 5 per cent sodium carbonate solution. Dry with 2–3 g of magnesium sulphate or anhydrous calcium chloride, and filter through a little glass wool into a distilling flask. Distil, using an air condenser, and collect the _m_-bromotoluene (a colourless liquid) at 180–183 °C. The yield is 65 g (38% overall).

Notes. (1) If the mixture is cooled in ice, most of the _p_-acetotoluidide separates out in a crystalline form. It may be recrystallised from ethanol.

(2) Unless the material is at least partly dried before hydrolysis, the yield of hydrochloride is reduced because of its solubility. If pure 4-acetamido-3-bromotoluene is required, the crude material may be recrystallised from 50 per cent ethanol with the addition of a little decolourising carbon; it separates as colourless needles, m.p. 116–117 °C (180 g, 79%).

(3) If pure 4-amino-3-bromotoluene is required, the crude base may be purified either by steam distillation or, more satisfactorily, by distillation under reduced pressure. The oil is dried with 5 g of sodium hydroxide pellets, and fractionally distilled under reduced pressure: a little _p_-toluidine may be present in the low boiling point fraction, and the pure substance is collected at 92–94 °C/13 mmHg or at 120–122 °C/30 mmHg. The purified amine solidifies on cooling and melts at 17–18 °C.

IV,90. 3,3'-DIMETHYLBIPHENYL

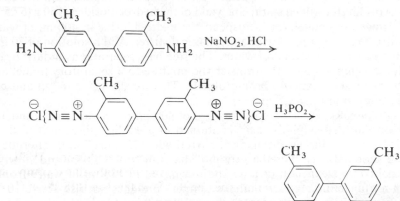

Make a thin paste of 21.5 g (0.1 mol) of finely powdered _o_-tolidine (4,4'-diamino-3,3'-dimethylbiphenyl—**CAUTION**, the compound is carcinogenic, wear plastic gloves and a face mask) and 300 ml of water in a 1-litre beaker, add 25 g (21 ml) of concentrated hydrochloric acid and warm until dissolved. Cool the solution to 10 °C with ice, stir mechanically and add a further 25 g (21 ml) of concentrated hydrochloric acid (1); partial separation of *o*-tolidine dihydrochloride will occur. Add a solution of 15 g (0.22 mol) of sodium nitrite in 30 ml of water as rapidly as possible, but keep the temperature below 15 °C: a slight excess of nitrous acid is not harmful in this preparation. Add the clear, orange diazonium solution to 175 ml of 30 per cent hypophosphorous acid and allow the mixture to stand, loosely stoppered, at room temperature for 16–18 hours. Transfer to a separatory funnel, and remove the upper red oily layer. Extract the aqueous layer with 50 ml of toluene. Dry the combined upper layer and toluene extract with magnesium sulphate, and remove the toluene by distillation from a flask fitted with a short fractionating side arm (Fig. I,110): heat in an oil bath to 150 °C to ensure the removal of the last traces of toluene. Distil the residue at *ca.* 3 mmHg pressure and a temperature of 155 °C. Collect the 3,3′-dimethylbiphenyl as a pale yellow liquid at 114–115 °C/3 mmHg; raise the bath temperature to about 170 °C when the temperature of the thermometer in the flask commences to fall. The yield is 14 g (77%).

Note. (1) If the hydrochloric acid is added all at once instead of in two portions as detailed, a solid will be obtained consisting of *o*-tolidine coated with its dihydrochloride, and the diazotisation will proceed slowly.

G,2. Coupling reactions. Azo compounds are prepared by the interaction of a diazonium salt with a phenol in the presence of sodium hydroxide or with an amine in the presence of sodium acetate. The coupling reaction is an electrophilic substitution involving the diazonium ion which reacts at the position of greatest electron availability, i.e., the position *ortho* or *para* to the electron releasing phenoxy or amino groups. 2-Naphthol couples in the more reactive 1-

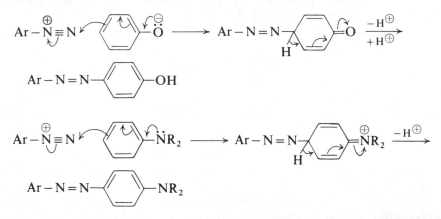

position as in the synthesis of 1-phenylazo-2-naphthol (Section **IV,91**). 1-Naphthol couples almost exclusively in position 4; when the diazo component is the *p*-nitrobenzenediazonium ion the product is Magneson II (Section **IV,92**), which is employed as a test reagent for magnesium.

Diazotised anthranilic acid couples in the *para* position of *N,N*-dimethylaniline yielding the acid–base indicator methyl red (Section **IV,93**). With

benzenediazonium chloride and *m*-phenylenediamine the azo dye chrysoidine is formed (Section **IV,94**), coupling occurring in the 4-position of the
diamine ring.

These azo compounds are not of great practical value as dyestuffs owing to
their slight solubility in water. The introduction of a sulphonic acid group into
the molecule has no effect upon the colour, but renders the dye water-
soluble—a fact of great commercial value. The simplest way of achieving this is
to employ an amine, e.g., sulphanilic acid, in which the $-SO_3H$ group is already present.

Sulphanilic acid, which has a dipolar or zwitterion structure (p. 674, **IV,F,1**),
is sparingly soluble in water. It is best diazotised by bringing it into solution as
the sodium salt by adding the calculated quantity of sodium carbonate, introducing the requisite quantity of sodium nitrite and pouring the solution on
to a mixture of hydrochloric acid and ice; nitrous acid and the dipolar sulphanilic acid are liberated together and immediately react, and after a short time
the internal diazonium salt separates from solution. Coupling with 2-naphthol
in sodium hydroxide solution yields the useful dyestuff Orange II (Section
IV,95). When *N,N*-dimethylaniline is used as the coupling component the product is methyl orange (Section **IV,96**). This latter substance is more useful as an
indicator than as a dye, for it changes colour at a certain concentration of
hydrogen ions (pH 3.1–4.4). Treatment of a solution of methyl orange with
a strong acid gives rise to a red form—which is essentially an internal salt
stabilised by electron delocalisation.

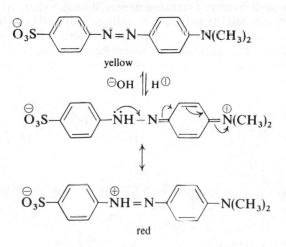

It is interesting to note that azo compounds may be conveniently reduced
either by a solution of tin(II) chloride in hydrochloric acid or by sodium dithionite. Thus 1-phenylazo-2-naphthol yields both aniline and 1-amino-2-
naphthol, and methyl orange gives *p*-amino-*N,N*-dimethylaniline and sulphanilic acid (see p. 714 and p. 719).

Attention has previously been drawn (p. 687, **IV,G,1**) to the fact that unless
an excess of hydrochloric (or mineral) acid is used in the diazotisation process,
coupling occurs between the diazonium salt and the amino group in the amine
to give diazoamino compounds. Thus benzenediazonium chloride and aniline
yield diazoaminobenzene. This substance may be conveniently prepared by

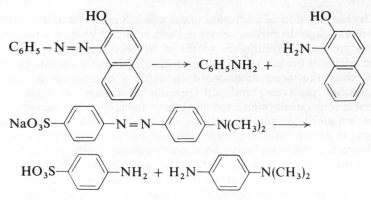

dissolving two equivalents of aniline in three equivalents of hydrochloric acid, and adding one equivalent of sodium nitrite in aqueous solution followed by two equivalents of sodium acetate (Section **IV,97**).

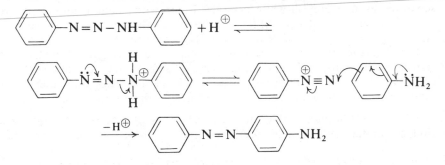

If diazoaminobenzene is dissolved in aniline, to which a small quantity of aniline hydrochloride has been added, and the mixture kept at about 40 °C for a short time, it is converted into *p*-aminoazobenzene (Section **IV,98**). The mechanism of this **diazoamino–aminoazo rearrangement** is dependent on (*a*) the heterolytic cleavage of a protonated diazoaminobenzene molecule to yield the benzenediazonium ion and aniline, and (*b*) a recoupling reaction, under weakly acidic conditions, of the diazonium ion at the *para* position of aniline.

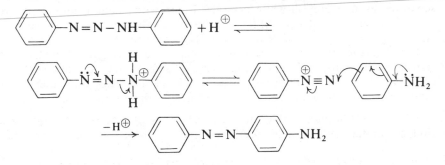

IV,91. 1-PHENYLAZO-2-NAPHTHOL

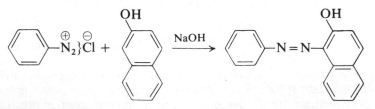

Dissolve 5.0 g (4.9 ml, 0.054 mol) of aniline in 16 ml of concentrated hydrochloric acid and 16 ml of water contained in a small beaker or conical flask.

Diazotise by the addition of a solution of 4.0 g (0.058 mol) of sodium nitrite in 20 ml of water; follow the method given in Section **IV,79**. Prepare a solution of 7.8 g (0.054 mol) of 2-naphthol in 45 ml of 10 per cent sodium hydroxide solution in a 250-ml beaker; cool the solution to 5 °C by immersion in an ice bath, assisted by the direct addition of about 25 g of crushed ice. Stir the naphthol solution vigorously and add the cold diazonium salt solution *very slowly*: a red colour develops and red crystals of 1-phenylazo-2-naphthol soon separate. When all the diazonium salt solution has been added, allow the mixture to stand in an ice bath for 30 minutes with occasional stirring. Filter the solution through a Buchner funnel with *gentle suction*, wash well with water and drain thoroughly by pressing the crystals with the back of a large glass stopper. Recrystallise one-fourth of the product from glacial acetic acid (30–35 ml): retain the remainder for reduction by stannous chloride. Filter the recrystallised product with suction, wash with a little ethanol (or industrial spirit) to eliminate acetic acid and dry upon filter-paper. The yield of deep red crystals is about 3 g. Pure 1-phenylazo-2-naphthol has m.p. 131 °C; if the m.p. is low, recrystallise the dry product from ethanol.

Reduction with tin(II) chloride.   1-Amino-2-naphthol hydrochloride.   Into a 350- or 500-ml round-bottomed flask, provided with a reflux condenser and containing 100 ml of industrial spirit, place the crude 1-phenylazo-2-naphthol reserved above and boil gently until most of the azo compound has dissolved. Meanwhile dissolve 20 g of a good grade of tin(II) chloride in 60 ml of concentrated hydrochloric acid (warming is necessary to produce a clear solution) (1), add this to the contents of the flask and boil under reflux for a further 30 minutes. All the azo compound dissolves rapidly and is reduced by the tin(II) chloride; the solution acquires a very pale brown colour. Decant the solution to a beaker and cool in ice: the 1-amino-2-naphthol hydrochloride separates as fine greyish-white crystals. Filter with suction, and wash with dilute hydrochloric acid (1 : 4). Recrystallise from the minimum volume of hot water which contains a few drops of tin(II) chloride solution in an equal weight of hydrochloric acid (this reduces atmospheric oxidation), cool the clear solution in an ice bath and collect the recrystallised product as before. Dry the colourless crystals in a desiccator. The yield is 3–4 g. The compound will remain colourless, or nearly so, if protected from light during storage.

Note.   (1) Sodium dithionite, $Na_2S_2O_4$, may also be used for the reduction; see under *methyl orange*, Section **IV,96**.

IV,92. 4-(4′-NITROBENZENEAZO)-1-NAPHTHOL (*Magneson II*)

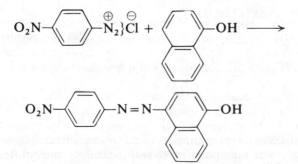

Dissolve 5.0 g (0.036 mol) of *p*-nitroaniline (Section **IV,77**) in a warm mixture of 13 ml of concentrated hydrochloric acid and 13 ml of water contained in a 250-ml beaker. Place the beaker in an ice–salt bath and cool to 0–5 °C whilst stirring vigorously; *p*-nitroaniline hydrochloride will separate in a finely divided crystalline form. Add a cold solution of 3.7 g (0.054 mol) of sodium nitrite in 8 ml of water slowly and with stirring to an end-point with potassium iodide–starch paper: do not allow the temperature of the solution to rise above 8 °C. Dissolve 5.2 g (0.035 mol) of 1-naphthol in a solution of 7 g of sodium hydroxide in 25 ml of water, cool in ice and add the diazotised solution slowly and with stirring. Then add concentrated hydrochloric acid slowly and with vigorous stirring to the cold mixture until it is strongly acid to Congo red paper. The colour will change from violet to dark red-brown. Filter with *gentle suction*, wash with water until free from acid and dry upon filter-paper in the air. The yield is 8 g (74%).

2,4-Dihydroxy-4′-nitroazobenzene ('**Magneson I**') may be similarly prepared by substituting resorcinol for 1-naphthol; it may be recrystallised from methanol and melts at 199–200 °C.

IV,93. METHYL RED

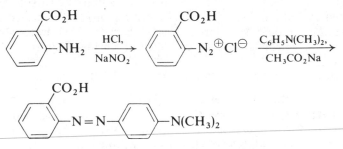

Dissolve 6.5 g (0.048 mol) of pure anthranilic acid in a mixture of 5.0 ml of concentrated hydrochloric acid and 15 ml of water by heating; filter off any insoluble impurities. Transfer the solution to a 250-ml beaker, surrounded by an ice bath. Add 25 g of crushed ice and 7.5 ml of concentrated hydrochloric acid, and stir continuously. When the temperature has fallen to about 3 °C, slowly introduce a cold solution of 3.6 g (0.052 mol) of sodium nitrite in 7.0 ml of water to a permanent end-point with potassium iodide–starch paper. This is best done by attaching to the stem of a 10-ml dropping funnel a glass tube which dips well below the surface of the solution and is bent upwards at the end and constricted so that the opening is about 2 mm; this arrangement ensures that the entrance of the acid liquor into the nitrite solution is prevented. It is essential that the temperature be kept between 3 and 5 °C during the diazotisation, otherwise tarry by-products are formed. To the resulting solution of the diazonium salt, add fairly rapidly 8.5 g (8.9 ml, 0.07 mol) of pure dimethylaniline. Continue the stirring for 10–15 minutes and maintain the temperature at about 5 °C.

Dissolve 6.8 g of crystallised sodium acetate in 10 ml of water. Add 5 ml of this solution to the reaction mixture and allow to stand in ice for 1 hour with occasional stirring. Then add the remainder of the sodium acetate solution with stirring to the mixture cooled in an ice bath, leave for a further 30 minutes (with occasional stirring) and allow the temperature to rise to that of the laboratory.

Introduce just sufficient sodium hydroxide solution with stirring to cause the mixture to have a distinct odour of dimethylaniline (about 5 ml of a 20% solution are usually required) and allow to stand at room temperature for about 1 hour. (The formation of the azo compound is a very slow reaction, but is accelerated by increasing the pH of the solution.) Filter off the solid at the pump, wash it first with a little water, then with 10 ml of 10 per cent acetic acid (to remove the dimethylaniline), and finally with water (the last filtrate is pale pink); drain well. Suspend the solid in 50 ml of methanol in a 200-ml flask; heat the mixture under reflux on a water bath for 10 minutes with frequent shaking, cool in ice and filter. Wash with 40 ml of cold methanol, and dry. The yield of crude methyl red is 8.9 g, m.p. 170–175 °C (1). Purify by recrystallisation from toluene (2). Place the crude product and 70–90 ml of toluene in a 200-ml flask fitted with a 5″ reflux condenser, heat until the substance dissolves, filter through a preheated Buchner funnel into a preheated filter flask and allow the filtrate to cool slowly to room temperature. Filter off the crystals and wash with a little toluene. The yield of methyl red, m.p. 181–182 °C, is 7.9 g (62%).

Notes. (1) The **sodium salt of methyl red may be prepared** by dissolving the crude product in an equal weight of 35 per cent sodium hydroxide which has been diluted to 350 ml, filtering, and evaporating under diminished pressure. The resulting sodium salt forms orange leaflets. This water-soluble product is very convenient for use as an indicator. Incidentally, the toluene extraction is avoided.

(2) Methyl red may also be recrystallised from glacial acetic acid.

IV,94. CHRYSOIDINE

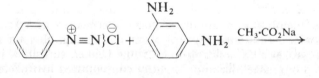

Prepare a solution of benzenediazonium chloride from 5.0 g (4.9 ml, 0.054 mol) of aniline as detailed in Section **IV,79**, and keep it in an ice bath. Meanwhile dissolve 6.0 g (0.055 mol) of a good grade of *m*-phenylenediamine, preferably redistilled before use (Section **IV,54**), in 60 ml of 2 *M*-hydrochloric acid in a 600-ml beaker, cool and add the benzenediazonium chloride solution rapidly and with vigorous stirring. Then add sodium acetate solution (about 20 g of the trihydrate in 50 ml of water) slowly and with stirring until precipitation of the dyestuff is complete; continue stirring for 1 hour. Heat to the boiling point and filter through a heated funnel, if necessary. Add 40 g of sodium chloride to the filtrate, heat on a steam bath until the precipitated dyestuff becomes crystalline, allow to cool, filter, wash with a little water and dry in the air. The yield of chrysoidine is 10 g (71%).

IV,95. ORANGE II (*β-Naphthol Orange*)

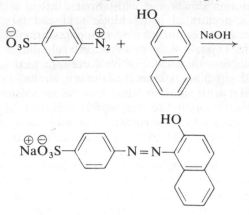

Diazotise 10.5 g (0.05 mol) of sulphanilic acid dihydrate as described under *methyl orange* (Section **IV,96**), and keep the suspension of the diazonium compound in ice-water until required. Dissolve 7.2 g (0.05 mol) of a good grade of 2-naphthol in 40 ml of cold 10 per cent sodium hydroxide solution in a 600-ml beaker, cool to 5 °C and pour in, with stirring, the well-mixed suspension of diazotised sulphanilic acid. Coupling takes place readily and the dyestuff separates as a crystalline paste. Stir well and, after 10 minutes, heat the mixture until all the solid has dissolved. Add 20 g of sodium chloride (to decrease the solubility of the product further) and warm until this dissolves. Allow the solution to cool spontaneously in the air for 1 hour, and then cool in ice until crystallisation is complete. Collect the product on a Buchner funnel and apply gentle suction; wash with a little saturated salt solution, and dry at 80 °C. The product weighs about 22 g, and contains about 20 per cent of sodium chloride; further purification is unnecessary for dyeing purposes. To obtain pure, crystalline Orange II, dissolve the crude substance in the minimum volume of boiling water, allow to cool to about 80 °C, add about twice the volume of rectified (or industrial) spirit and allow crystallisation to proceed spontaneously. When cold, filter at the pump, wash the pure dyestuff (it is a dihydrate) with a little ethanol and dry in the air. The yield is 14 g (80%) (1).

Note. (1) For the reduction of Orange II to 1-amino-2-naphthol and its conversion to 1,2-naphthoquinone, see Section **IV,144**.

IV,96. METHYL ORANGE

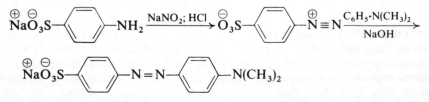

In a 250-ml conical flask place 10.5 g (0.05 mol) of sulphanilic acid dihydrate, 2.65 g (0.025 mol) of anhydrous sodium carbonate and 100 ml of water, and warm until a clear solution is obtained. Cool the solution under the tap to about 15 °C, and add a solution of 3.7 g (0.059 mol) of sodium nitrite in 10 ml of water. Pour the resulting solution slowly and with stirring into a 600-ml beaker containing 10.5 ml of concentrated hydrochloric acid and 60 g of crushed ice (1). Test for the presence of free nitrous acid with potassium iodide–starch paper after 15 minutes. Fine crystals of the diazobenzene sulphonate will soon separate; do not filter these off as they will dissolve during the next stage of the preparation. Dissolve 6.05 g (6.3 ml, 0.05 mol) of dimethylaniline in 3.0 ml of glacial acetic acid, and add it with vigorous stirring to the suspension of diazotised sulphanilic acid. Allow the mixture to stand for 10 minutes; the red or acid form of methyl orange will gradually separate. Then add slowly and with stirring 35 ml of 20 per cent sodium hydroxide solution: the mixture will assume a uniform orange colour due to the separation of the sodium salt of methyl orange in fine particles. Direct filtration of the latter is slow, hence, whilst stirring the mixture with a thermometer, heat it almost to the boiling point. Most of the methyl orange will dissolve. Add about 10 g of sodium chloride (to assist the subsequent separation of the methyl orange) and warm at 80–90 °C until the salt has dissolved. Allow the mixture to cool undisturbed for 15 minutes and then cool in ice-water; this gives a fairly easily filterable product. Filter off the methyl orange at the pump, but apply only gentle suction so as to avoid clogging the pores of the filter-paper; rinse the beaker with a little saturated salt solution and drain well. Recrystallise from hot water (about 150 ml are required); filter the hot solution, if necessary, through a hot water funnel or through a preheated Buchner funnel. Reddish-orange crystals of methyl orange separate as the solution cools. Filter these at the pump, drain well, wash with a little ethanol, and finally with a small volume of ether. The yield is 13 g (80%). Methyl orange, being a salt, has no well-defined m.p.

Note. (1) An alternative procedure is to cool the solution containing the sodium sulphanilate and sodium nitrite in a bath of crushed ice to about 5 °C and then add 10.5 ml of concentrated hydrochloric acid diluted with an equal volume of water slowly and with stirring; the temperature must not be allowed to rise above 10 °C and an excess of nitrous acid should be present (the solution is tested after standing for 5 minutes). The subsequent stages in the preparation—addition of dimethylaniline solution, etc.—are as above.

Reduction of methyl orange to p-aminodimethylaniline. *Method 1.* Dissolve 2.0 g of methyl orange in the minimum volume of hot water and to the hot solution add a solution of 8 g of tin(II) chloride in 20 ml of concentrated hydrochloric acid until decolourisation takes place; gentle boiling may be necessary. Cool the resulting solution in ice; a crystalline precipitate consisting of sulphanilic acid and some p-aminodimethylaniline hydrochloride separates out. In order to separate the free base, add 10 per cent sodium hydroxide solution until the precipitate of tin hydroxide redissolves. Extract the cold solution with three or four 20 ml portions of ether, dry the extract with anhydrous potassium carbonate and remove the ether by distillation. The residual base soon crystallises, particularly if it is stirred with a glass rod; it melts at 41 °C.

Method 2. Suspend 2.0 g of methyl orange in 4 ml of water, and add a small quantity of sodium dithionite ($Na_2S_2O_4$). Heat the mixture and add more sodium dithionite until the colour is discharged. The sulphanilic acid remains in the solution as sodium sulphanilate and the *p*-aminodimethylaniline may be extracted with ether as in *Method 1*.

IV,97. DIAZOAMINOBENZENE

$$C_5H_5 \cdot \overset{\oplus}{N} \equiv N \} \overset{\ominus}{Cl} + H_2N \cdot C_6H_5 \longrightarrow C_6H_5 \cdot N = N \cdot NH \cdot C_6H_5$$

In a 250-ml flask place 75 ml of water, 24 g (20 ml) of concentrated hydrochloric acid and 14 g (13.7 ml, 0.15 mol) of aniline. Shake vigorously (1) and then add 50 g of crushed ice. Run in a solution of 5.2 g (0.075 mol) of sodium nitrite in 12 ml of water, with constant shaking, during a period of 5–10 minutes. Allow to stand with frequent shaking (1) for 15 minutes, and add a solution of 21.0 g of crystallised sodium acetate in 40 ml of water during 5 minutes. A yellow precipitate of diazoaminobenzene begins to form immediately; allow to stand with frequent shaking for 45 minutes and do not allow the temperature to rise above 20 °C (add ice, if necessary). Filter the yellow diazoaminobenzene on a Buchner funnel, wash with it 250 ml of cold water, drain as completely as possible and spread it on a sheet of filter-paper to dry. The yield of crude diazoaminobenzene, m.p. 91 °C, is 13 g (87%) (2). Recrystallise a small portion from light petroleum, b.p. 60–80 °C: the pure compound, m.p. 97 °C, is obtained.

Notes. (1) For preparations on a larger scale, mechanical stirring is recommended.

(2) The crude compound may be employed in the preparation of *p*-aminoazobenzene.

IV,98. *p*-AMINOAZOBENZENE

$$C_6H_5 \cdot N = N \cdot NH \cdot C_6H_5 \longrightarrow C_6H_5 \cdot N = N \cdot C_6H_4 \cdot NH_2 \text{-} p$$

Dissolve 5 g (0.025 mol) of finely powdered diazoaminobenzene (Section **IV,97**) in 12–15 g of aniline in a small flask and add 2.5 g of finely powdered aniline hydrochloride (1). Warm the mixture, with frequent shaking, on a water bath at 40–45 °C for 1 hour. Allow the reaction mixture to stand for 30 minutes. Then add 15 ml of glacial acetic acid diluted with an equal volume of water: stir or shake the mixture in order to remove the excess of aniline in the form of its soluble acetate. Allow the mixture to stand, with frequent shaking, for 15 minutes: filter the aminoazobenzene at the pump, wash with a little water and dry upon filter-paper. Recrystallise the crude *p*-aminoazobenzene (3.5 g, 70%; m.p. 120 °C) from 15–20 ml of carbon tetrachloride to obtain the pure compound, m.p. 125 °C. Alternatively, the compound may be recrystallised from dilute ethanol, to which a few drops of concentrated ammonia solution have been added.

To prepare the **hydrochloride**, dissolve about 1 g of the compound (which need not be perfectly dry) in about 8 ml of ethanol. Add this solution to boiling dilute hydrochloric acid (10 ml of the concentrated acid and 80 ml of water). Boil for 5 minutes, filter the hot solution if necessary and allow to cool. *p*-Aminoazobenzene hydrochloride separates in steel-blue crystals. Filter, wash with a little dilute hydrochloric acid, and dry.

To recover the free base, dissolve the hydrochloride in the minimum volume of boiling ethanol, add concentrated ammonia solution dropwise until a clear solution results and the blue colour has become light brown. Add water carefully until a cloudiness appears, warm on a water bath until the cloudiness just disappears and allow to cool. Yellow crystals of *p*-aminoazobenzene separate on cooling.

Note. (1) The **aniline hydrochloride may be prepared** by treating 2 g of aniline with an excess (about 3 ml) of concentrated hydrochloric acid in a small beaker, cooling, filtering at the pump, washing with a *small* volume of ether and drying between filter-paper.

H MISCELLANEOUS AROMATIC NITROGEN COMPOUNDS

1. Intermediate products in the reduction of nitro compounds (Sections **IV,99** to **IV,103**).
2. Arylhydrazines (Sections **IV,104** to **IV,107**).
3. Arylureas and related compounds (Sections **IV,108** to **IV,110**).

H,1. Intermediate products in the reduction of nitro compounds The reduction of an aromatic nitro compound with a powerful reducing agent (tin or tin(II) chloride and hydrochloric acid; iron and dilute hydrochloric acid; hydrogen and a platinum catalyst) leads to a good yield of primary amine, e.g., aniline from nitrobenzene. By the use of milder reducing agents and by the control of the hydrogen ion concentration of the solution, a number of intermediate products may be isolated, some of which are products of direct reduction and others are formed through secondary reactions. The various stages of the reduction of nitrobenzene have been established by investigating the process electrolytically under conditions of varying pH, current density and electrode construction and composition. The sequence is as follows.

$$C_6H_5 \cdot NO_2 \xrightarrow{2H} C_6H_5 \cdot NO \xrightarrow{2H} C_6H_5 \cdot NHOH \xrightarrow{2H} C_6H_5 \cdot NH_2$$

Nitrobenzene   Nitrosobenzene   *N*-Phenylhydroxylamine   Aniline

The initial product, nitrosobenzene, is so easily reduced to *N*-phenylhydroxylamine that it has not been isolated from the reduction medium, but its presence has been established by reaction in solution with hydroxylamine to yield a benzenediazonium salt, which couples readily with 1-naphthylamine to form the dyestuff 2-phenylazo-1-naphthylamine.

$$C_6H_5 \cdot NO + H_2NOH + HX \longrightarrow C_6H_5 \cdot \overset{\oplus}{N} \equiv N \} \overset{\ominus}{X} + 2H_2O$$

Under the catalytic influence of alkali, nitrosobenzene and *N*-phenylhydroxylamine react to yield azoxybenzene.

$$C_6H_5 \cdot NO + \begin{matrix} H \\ HO \end{matrix} > N \cdot C_6H_5 \xrightarrow[-H_2O]{\ominus OH} C_6H_5 \cdot N = N \cdot C_6H_5 \\ \qquad\qquad\qquad\qquad\qquad\qquad\qquad\quad \downarrow \\ \qquad\qquad\qquad\qquad\qquad\qquad\qquad\quad O$$

Further reduction in alkaline solution (e.g., with zinc powder) leads to azo-benzene and hydrazobenzene.

$$C_6H_5 \cdot N = N \cdot C_6H_5 \xrightarrow{2H} C_6H_5 \cdot N = N \cdot C_6H_5 \xrightarrow{2H} C_6H_5 \cdot NH \cdot NH \cdot C_6H_5$$
$$\quad\ \ \ \ \ \ \ \downarrow$$
$$\quad\ \ \ \ \ \ \ O$$

Electrolytic reduction of hydrazobenzene gives aniline.

$$C_6H_5 \cdot NH \cdot NH \cdot C_6H_5 \xrightarrow{2H} 2C_6H_5 \cdot NH_2$$

The various intermediate compounds may be prepared in the laboratory, and convenient methods are described below.

N-Phenylhydroxylamine (Section IV,99) is formed when nitrobenzene is treated with a 'neutral' reducing agent, e.g., zinc powder and aqueous ammonium chloride solution. The compound rearranges, in the presence of acids, with the formation of p-aminophenol (Section IV,99).

Nitrosobenzene (Section IV,100) may be obtained by the oxidation of N-phenylhydroxylamine with acid dichromate solution at 0 °C. The solid product is colourless and is probably a dimer; it dissociates to a green monomer upon melting or in solution.

Azoxybenzene is readily prepared by reduction of nitrobenzene in an alkaline medium with a variety of mild reducing agents. Reducing sugars have been used successfully for the reduction of substituted nitro compounds to the corresponding azoxyarenes (Ref. 6), and the use of D-glucose for the reduction of nitrobenzene is illustrated in Section IV,101.

Reduction of nitrobenzene in methanolic or ethanolic sodium hydroxide solution with zinc powder leads to azobenzene or hydrazobenzene according to the proportion of zinc powder employed (Sections IV,102 and IV,103). Hydrazobenzene may be oxidised to azobenzene by sodium hypobromite solution at 0 °C.

In the presence of acids, hydrazobenzene rearranges to give a mixture containing about 70 per cent of benzidine (4,4'-diaminobiphenyl) and about 30 per cent of 2,4'-diaminobiphenyl (diphenyline), the **benzidine rearrangement**. Benzidine is carcinogenic and its preparation and storage is under strict control.

IV,99. N-PHENYLHYDROXYLAMINE

$$C_6H_5 \cdot NO_2 \xrightarrow[\text{Zn/NH}_4\text{Cl}]{4[H];} C_6H_5 \cdot NHOH + H_2O$$

In a 2-litre beaker, equipped with a thermometer and mechanical stirrer, place 25 g of ammonium chloride, 800 ml of water and 50 g (41.6 ml, 0.41 mol) of redistilled nitrobenzene. Stir the mixture vigorously, and add 59 g (0.83 mol) of zinc powder of 90 per cent purity (Section II,2,67) during about 15 minutes; the rate of addition should be such that the temperature rapidly rises to 60–65 °C and remains in this range until all the zinc has been added. Continue the stirring for a further 15 minutes, by which time the reduction is complete as is shown by the fact that the temperature commences to fall. Filter the warm reaction mixture at the pump to remove the zinc oxide, and wash it with 100 ml of hot water. Place the filtrate in a conical flask, saturate it with common salt (about 300 g) and cool in an ice bath for at least one hour to ensure maximum crystallisation of the desired product. Filter the pale yellow crystals of phenylhydroxylamine with suction and drain well. The yield of crude, dry product is about 38 g; this contains a little salt and corresponds to about 29 g (66%) of pure phenylhydroxylamine as determined by its separation from inorganic materials by dissolution in ether. The substance deteriorates upon storage and is therefore used immediately for a secondary preparation (e.g., nitrosobenzene, Section IV,100). If required perfectly pure, it may be recrystallised from benzene–light petroleum (b.p. 40–60 °C) or from benzene alone; the resulting pure compound is somewhat more stable and has a melting point of 81 °C.

Conversion of phenylhydroxylamine into *p*-aminophenol. Add 4.4 g of recrystallised phenylhydroxylamine to a mixture of 20 ml of concentrated sulphuric acid and 60 g of ice contained in a 1-litre beaker cooled in a freezing mixture. Dilute the solution with 400 ml of water, and boil until a sample, tested with dichromate solution, gives the smell of quinone and not of nitrosobenzene or nitrobenzene (*ca.* 10–15 minutes). Neutralise the cold reaction mixture with sodium hydrogen carbonate, saturate with salt, extract twice with ether and dry the ethereal extract with magnesium sulphate or anhydrous sodium sulphate. Distil off the ether; *p*-aminophenol, m.p. 186 °C, remains. The yield is 4.3 g (98%).

IV,100. NITROSOBENZENE

$$C_6H_5 \cdot NO_2 \xrightarrow{\text{[H]}} C_6H_5 \cdot NHOH \xrightarrow{\text{[O]}} C_6H_5 \cdot NO + H_2O$$

In a 2-litre beaker, equipped with a thermometer and mechanical stirrer, place 30 g (0.56 mol) of ammonium chloride, 1 litre of water and 61.5 g (51 ml, 0.5 mol) of pure nitrobenzene. Stir the mixture vigorously, and add 75 g (1.03 mol, 90% purity; see Section II,2,67) of zinc powder during about 15 minutes; the rate of addition should be such that the temperature rises rapidly to 60–65 °C and remains in this range until all the zinc has been added. Continue the stirring for a further 15 minutes, by which time the reduction is complete as shown by the fact that the temperature commences to fall. Filter the warm reaction mixture at the pump to remove the zinc oxide, and wash it with 600–700 ml of boiling water. Transfer the filtrate and washings to a 4-litre round-bottomed flask or beaker and cool *immediately* to 0–1 °C by the addition of sufficient crushed ice and leave at least 250 g unmelted. Without delay, add with stirring a cold solution of concentrated sulphuric acid (150 ml of the concentrated acid added to sufficient ice to reduce its temperature to –5 °C). Then add an ice-cold solution of 34 g (0.114 mol) of crystallised sodium dichromate in 125 ml of water as rapidly as possible to the stirred solution. After 2–3 minutes, filter the straw-coloured precipitate of nitrosobenzene on a Buchner funnel and wash it with 200 ml of water. Steam distil the nitrosobenzene as rapidly as possible; the nitrosobenzene tends to decompose at the elevated temperature. Cool the receiver in ice because the compound has a high vapour pressure at room temperature. The nitrosobenzene condenses to a green liquid, which solidifies to a white solid; care should be taken that the solid does not clog the condenser by turning off the water supply from time to time. Stop the distillation when yellow oily material appears in the condenser.

Filter; grind the nitrosobenzene in a glass mortar with a little water. Filter at the pump, wash it with water until the washings are no longer brown and drain as completely as possible. Dry the solid between layers of filter-paper. The yield of nitrosobenzene, m.p. 66–67 °C, is 30 g (56%). A pure product, m.p. 68 °C, may be obtained by recrystallisation from a small volume of ethanol with good cooling: the compound should be dried over anhydrous calcium chloride at atmospheric pressure. The substance may be kept for 1–2 days at room temperature and for longer periods at 0 °C.

IV,101. AZOXYBENZENE

$$2C_6H_5 \cdot NO_2 + 6[H] \longrightarrow C_6H_5 \cdot N = N \cdot C_6H_5 + 3H_2O$$
$$\downarrow$$
$$O$$

Equip a 500-ml three-necked flask with an efficient stirrer (e.g., a Hershberg stirrer, Fig. I,55) and a reflux condenser; stopper the third neck. Place a solution of 30 g of sodium hydroxide in 100 ml of water, and also 20.5 g (17.1 ml, 0.167 mol) of pure nitrobenzene in the flask, immerse it in a water bath maintained at 55–60 °C, and add 21 g (0.117 mol) of anhydrous glucose in small portions, with continuous stirring, during 1 hour. Then heat on a boiling water bath for 2 hours. Pour the hot mixture into a 1-litre round-bottomed flask and steam distil (Fig. I,101) to remove aniline and nitrobenzene. When the distillate is clear (i.e., after about 1 litre has been collected), pour the residue into a beaker cooled in an ice bath. The azoxybenzene soon solidifies. Filter with suction, grind the lumps of azoxybenzene in a mortar, wash with water and dry upon filter-paper or upon a porous plate. The yield of material, m.p. 35–35.5 °C, is 13 g (79%). Recrystallise from 7 ml of rectified spirit or of methanol; the m.p. is raised to 36 °C.

IV,102. AZOBENZENE

$$2C_6H_5 \cdot NO_2 + 4Zn + 8NaOH \longrightarrow$$
$$C_6H_5 \cdot N = N \cdot C_6H_5 + 4Na_2[ZnO_2] + 4H_2O$$

Method 1 (**from nitrobenzene**). Support a 1-litre three-necked flask, equipped with a sealed stirrer unit and a reflux condenser, on a water bath, and place a solution of 65 g of sodium hydroxide in 150 ml of water, 50 g (41.5 ml, 0.41 mol) of pure nitrobenzene and 500 ml of methanol in the flask. Add 59 g (0.9 mol) of zinc powder (90% purity; see Section II,2,67) to the mixture, start the stirrer and reflux for 10 hours (1). Filter the mixture while hot, and wash the precipitate of sodium zincate with a little methanol. The strongly alkaline filtrate is not always clear: render it neutral to litmus by the cautious addition of concentrated hydrochloric acid, and filter again. Distil off the methanol from the filtrate, cool the residue in ice and filter off the solid azobenzene. The crude azobenzene contains occluded zinc salts. To remove these, add the crude product to 100 ml of 2 per cent hydrochloric acid, warm to about 70 °C in order to melt the azobenzene and stir mechanically for 5 minutes; continue the stirring whilst the mixture is immersed in ice water in order to solidify the azobenzene. Filter, wash well with water, drain thoroughly and recrystallise from a mixture of 145 ml of rectified spirit and 12 ml of water; collect the azobenzene and dry in the air. The yield of pure azobenzene (reddish-orange crystals), m.p. 67–68 °C, is 31 g (86%) (2).

Notes. (1) At the end of this time, the reddish mixture should be free from the odour of nitrobenzene; if it is not, reflux for 2–3 hours longer.

(2) Frequently the recrystallized azobenzene has m.p. 61 °C, which is unaffected by recrystallisation from ethanol. Upon distillation from a 50-ml distilling flask fitted with a short air condenser, the m.p. is raised to 67.5 °C and the recovery is about 90 per cent: one recrystallisation from diluted ethanol (as above) then gives perfectly pure azobenzene of m.p. 68.5 °C.

Method 2 **(from hydrazobenzene).** Prepare a solution of sodium hypobromite by adding 10 g (3.2 ml, 0.0625 mol) of bromine dropwise to a cold solution of 6.0 g of sodium hydroxide in 75 ml of water immersed in an ice bath. Dissolve 9.2 g (0.05 mol) of hydrazobenzene (Section **IV,103**) in 60 ml of ether contained in a separatory funnel, and add the cold sodium hypobromite solution in small portions. Shake for 10 minutes; separate the ether layer, pour it into a 100-ml flask and distil off the ether by warming gently on a water bath. Dissolve the warm liquid residue in about 30 ml of ethanol, transfer to a small beaker, heat to boiling on a water bath, add water dropwise to the hot solution until the azobenzene just commences to separate, render the solution clear again with a few drops of ethanol and cool in ice-water. Filter the orange crystals at the pump, and wash with a little 50 per cent ethanol. Dry in the air. The yield is 8 g (85%).

IV,103. HYDRAZOBENZENE (sym.-*Diphenylhydrazine*)

$$2C_6H_5 \cdot NO_2 + 5Zn + 10NaOH \longrightarrow$$
$$C_6H_5 \cdot NH \cdot NH \cdot C_6H_5 + 5Na_2[ZnO_2] + 4H_2O$$

Support a 1500-ml three-necked flask, equipped with a sealed stirrer unit and a double surface reflux condenser, on a water bath, and place a solution of 84 g of sodium hydroxide in 185 ml of water, 50 g (41.5 ml, 0.406 mol) of nitrobenzene and 500 ml of methanol in the flask. Add 78 g (1.07 mol, 90%) of zinc powder, start the stirrer and reflux for 10 hours. The solution gradually assumes the reddish colour of azobenzene and then, on further reduction, turns to a pale yellow (due to hydrazobenzene). If the colour is not almost completely discharged at the end of the refluxing period, add a further 11 g (0.15 mol; 90%) of zinc powder, and reflux for 2–3 hours longer. Filter the hot solution through a pre-heated Buchner funnel and wash the sodium zincate upon the filter with a little hot methanol. Pour the filtrate into a large flask (1), stopper it loosely and cool it in a freezing mixture of ice and salt to accelerate crystallisation. After 1 hour filter off the almost colourless crystals of hydrazobenzene at the pump as rapidly as possible (it is helpful to displace the air above the solution undergoing filtration in the funnel with a stream of nitrogen), wash with 50 per cent methanol to which a little sulphurous acid has been added until the filtrate is no longer alkaline. Dry in a vacuum desiccator. The resulting almost colourless hydrazobenzene (15 g; 40%; m.p. 125 °C) is sufficiently pure for the preparation of benzidine or of azobenzene. If it is required pure (m.p. 126 °C with production of a yellow colour), it may be recrystallised from hot methanol containing a little ammonium sulphide or sulphurous acid (these assist in preventing atmospheric oxidation).

Owing to the great tendency of hydrazobenzene to undergo oxidation, all operations involving filtration should be carried out as rapidly as possible and air should not be drawn through it unnecessarily. The substance should be dried in a vacuum desiccator: it can only be preserved in a colourless condition if it is kept in an atmosphere of carbon dioxide or nitrogen or in sealed vessels.

Notes. (1) If the methanol is distilled off before thorough cooling in a freezing mixture, the yield of hydrazobenzene is appreciably increased, but the product is considerably more coloured due to admixture with a trace of azobenzene. About 12 g of impure hydrazobenzene may be recovered by distilling off the methanol from the filtrate after the colourless hydrazobenzene has been collected.

H,2. Arylhydrazines. Arylhydrazines may be prepared by reducing diazonium salts with excess warm sodium sulphite solution, followed by acidification with hydrochloric acid. The hydrochloride usually crystallises out on cooling and treatment of the latter with excess sodium hydroxide solution liberates the free base. The preparation of phenylhydrazine and p-nitrophenylhydrazine by this method is illustrated in Section **IV,104**.

The mechanism of this reduction probably involves the initial addition of a sulphite ion to the diazonium group to give an azosulphonate which undergoes further conjugate (1,4-) addition of the nucleophilic sulphite ion. The resulting intermediate is protonolytically cleaved on heating under acidic conditions:

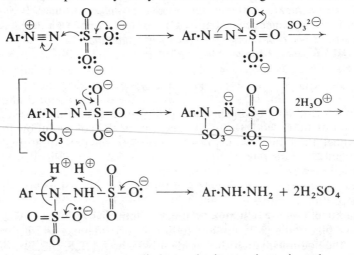

This method cannot be applied to polynitro amines, since these are so weakly basic that they can be diazotised only under special conditions in strongly acidic media (p. 687, **IV,G,1**). In such cases use may be made of the susceptibility to nucleophilic displacement of halogen when activated by *ortho* and *para* nitro groups. Thus the valuable reagent 2,4-dinitrophenylhydrazine (Section **IV,105**) is readily prepared by reacting 1-chloro-2,4-dinitrobenzene with hydrazine. Reaction with ammonia similarly gives 2,4-dinitroaniline (cognate preparation in Section **IV,105**). A further example is provided by the reaction of 1-chloro-2,4,6-trinitrobenzene (picryl chloride) with *N,N*-diphenylhydrazine to give *N,N*-diphenylpicrylhydrazine (Section **IV,106**). This compound is of interest in that oxidation with lead dioxide yields the highly stable *N,N*-diphenylpicrylhydrazyl radical, which is obtained as an intensely coloured, paramagnetic`solid. Stabilisation of the radical is promoted by the strongly

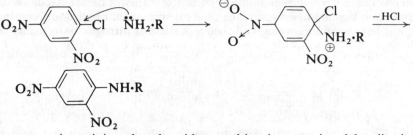

electron-attracting trinitrophenyl residue resulting in extensive delocalisation of the odd electron. The *N,N*-diphenylhydrazine required for this preparation is obtained by reduction of *N*-nitroso-*N,N*-diphenylamine with zinc and acetic acid (Section **IV,107**).

IV,104. PHENYLHYDRAZINE

$$C_6H_5 \cdot NH_2 \xrightarrow[]{\text{NaNO}_2;\ \text{HCl}} C_6H_5 \cdot \overset{\oplus}{N} \equiv N\}\overset{\ominus}{Cl} \xrightarrow[\text{(Na}_2\text{SO}_3,\ \text{H}_2\text{O)}]{+4e,\ +4H^{\oplus}}$$

$$C_6H_5 \cdot NH \cdot \overset{\oplus}{NH}_3\}\overset{\ominus}{Cl} \xrightarrow{\ominus\text{OH}} C_6H_5 \cdot NH \cdot NH_2$$

Phenylhydrazine is highly poisonous and produces unpleasant burns in contact with the skin. Wash off immediately any liquid which has come into contact with the skin first with 2 per cent acetic acid, then with soap and water.

Place 130 ml of concentrated hydrochloric acid in a 1.5-litre three-necked flask, equipped with a mechanical stirrer and immersed in a freezing mixture of ice and salt. Start the stirrer and, when the temperature has fallen to about 0 °C, run in 47.5 g (46.5 ml, 0.51 mol) of pure aniline during about 5 minutes, and then add another 60 g of crushed ice. Dissolve 35 g (0.51 mol) of sodium nitrite in 75 ml of water, cool to 0–3 °C and run in the cold solution from a separatory funnel, the stem of which reaches nearly to the bottom of the flask. During the addition of the nitrite solution (*ca.* 20 minutes), stir vigorously and keep the temperature as near 0 °C as possible. There should be a slight excess of nitrous acid (potassium iodide–starch paper test) at the end of 10 minutes after the last portion of nitrite is added.

In the meantime, prepare a sodium sulphite solution as follows. In a 2-litre bolt-head flask place 50 g (1.25 mol) of sodium hydroxide and add 500 ml of water. When the sodium hydroxide has dissolved, add 112.5 g (0.59 mol) of recrystallised sodium metabisulphite (1), and stir mechanically until the solid has dissolved. Cool the resulting solution to about 25 °C and add a few drops of phenolphthalein indicator solution. Introduce small quantities of sodium metabisulphite until the pink colour of the solution just disappears, then stir in a further 12 g of sodium metabisulphite (the total weight required should not exceed 135–140 g, 1.30–1.35 mol). Cool this solution, with stirring, to about 5 °C by immersion in an ice bath, then add about 60 g of crushed ice. Run in the ice-cold diazonium solution as rapidly as possible, while stirring vigorously. The reaction mixture immediately acquires a bright orange-red colour. Slowly heat the solution to 60–70 °C on a water bath and maintain this temperature for 30–60 minutes, i.e., until the colour becomes quite dark. Acidify the solution to litmus with concentrated hydrochloric acid (40–50 ml are required); continue the heating on a boiling water bath until the colour becomes much lighter and in any case for 4–6 hours. If any solid is present, filter the solution. To the hot, clear solution add, with stirring, 500 ml of concentrated hydrochloric acid; cool, first in running water, and then in a freezing mixture to 0 °C. The phenylhydrazine hydrochloride separates as yellowish or pinkish crystals. Collect them on a Buchner funnel, drain, wash with 25 ml of dilute hydrochloric acid (1 : 3) and press well with a large glass stopper (2).

Liberate the free base by adding to the phenylhydrazine hydrochloride 125 ml of 25 per cent sodium hydroxide solution. Extract the phenylhydrazine with two 40 ml portions of toluene, dry the extracts with 25 g of sodium hydroxide pellets or with anhydrous potassium carbonate: thorough drying is essential if foaming in the subsequent distillation is to be avoided. Most of the toluene may now be distilled under atmospheric pressure, and the residual phenylhydrazine under reduced pressure. For this purpose, fit a small dropping funnel to the main neck of a 100-ml Claisen flask (which contains a few fragments of porous porcelain) and assemble the rest of the apparatus as in Fig. I,110, but do not connect the apparatus assembly to the pump. Run in about 40 ml of the toluene solution into the flask, heat the latter in an oil bath so that the toluene distils over steadily. Allow the remainder of the toluene solution to run in from the dropping funnel as fast as the toluene itself distils over. When all the solution has been introduced into the flask, close the stop-cock on the funnel, and continue the heating until no further distillate is obtained. Allow to cool. Replace the dropping funnel by capillary tube reaching to the bottom of the flask, and distil under diminished pressure. Collect the phenylhydrazine at 137–138 °C/18 mmHg (or at 119–120 °C/12 mmHg). The yield of almost colourless liquid is 35 g (64%); it crystallises on cooling in ice and then melts at 23 °C. Phenylhydrazine slowly darkens on exposure to light.

Notes. (1) The sodium sulphite solution may also be prepared by dissolving 100 g of pure (or a corresponding quantity of commercial) sodium hydroxide in about 125 ml of water, and then diluting to 750 ml. The flask is cooled in running water, a few drops of phenolphthalein indicator are added, and sulphur dioxide passed in until the pink colour just disappears (it is advisable to add a further 1–2 drops of the indicator at this point) and then for 2–3 minutes longer. It is best to remove a sample for test from time to time, dilute with 3–4 volumes of water and test with 1 drop of phenolphthalein.

(2) If desired, the phenylhydrazine hydrochloride may be purified by recrystallisation. The crude hydrochloride is boiled with 6 times its weight of water and a few grams of decolourising carbon. After filtering, a volume of concentrated hydrochloric acid equal in volume to one-third of the solution is added, and the mixture cooled to 0 °C. Pure white crystals are obtained in 85–90 per cent yield.

Cognate preparation *p*-**Nitrophenylhydrazine.** Dissolve 10 g (0.075 mol) of *p*-nitroaniline (Section **IV,77**) in a mixture of 21 ml of concentrated hydrochloric acid and an equal volume of water, and cool rapidly to 0 °C in order to obtain the hydrochloride of the base in a fine state of division. Diazotise in the usual way (see Section **IV,79**) by the gradual addition of a solution of 5.2 g (0.075 mol) of sodium nitrite in 12 ml of water. Continue the stirring for a few minutes, filter the solution rapidly and add it from a separatory funnel to an ice-cold solution of 41 g (0.147 mol) of sodium sulphite (90% Na_2SO_3, $7H_2O$) in

100 ml of water containing 4 g of sodium hydroxide (1); stir the mixture during the addition which requires about 5 minutes. (If the diazonium solution is added too rapidly, an orange-red precipitate of sodium p-nitrobenzene-diazosulphonate is produced, and is apt to form a resin.) Allow the solution to stand for 5 minutes, acidify with 70 ml of concentrated hydrochloric acid and heat on a water bath at 25 °C for 3 minutes, when yellow needles commence to separate. Allow to stand overnight, filter off the crystals, heat them with 20 ml of concentrated hydrochloric acid on a water bath for 7 minutes and allow to cool. Filter off the precipitate, consisting of p-nitrophenylhydrazine hydrochloride and sodium salts, dissolve it in water and treat the solution with a concentrated solution of sodium acetate: the free base will separate out in an almost pure state (7–8 g, 63–72%). The p-nitrophenylhydrazine may be recrystallised from ethanol and is obtained as light brown crystals, m.p. 158 °C (decomp.).

Note. (1) The alkaline sodium sulphite solution may be replaced by **saturated ammonium sulphite solution** prepared as follows. Pass sulphur dioxide into a mixture of 1 part of concentrated ammonia solution (d 0.88) and two parts of crushed ice in a freezing mixture until the liquid smells strongly of sulphur dioxide, and then neutralise with ammonia solution. This solution slowly deposits ammonium sulphite crystals and contains about 0.25 g of SO_2 per ml. Use 60 ml of this ice-cold ammonium sulphite solution to which 8 ml of concentrated ammonia solution are added. After the addition of the solution of p-nitrobenzenediazonium chloride, allow the mixture to stand for 1 hour in a freezing mixture, filter off the yellow precipitate of ammonium p-nitrophenyl-hydrazine disulphonate, heat it on a water bath with 20 ml of concentrated hydrochloric acid at 70–80 °C for 7 minutes, cool the blood red solution and dissolve the resulting precipitate of p-nitrophenylhydrazine hydrochloride and ammonium salts in water, and isolate the base as above.

IV,105. 2,4-DINITROPHENYLHYDRAZINE

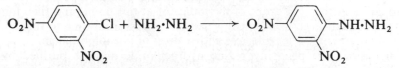

Suspend 35 g (0.27 mol) of finely powdered hydrazine sulphate in 125 ml of hot water contained in a 400-ml beaker, and add, with stirring, 118 g (0.87 mol) of sodium acetate hydrate or 85 g of potassium acetate. Boil the mixture for 5 minutes, cool to about 70 °C, add 80 ml of rectified spirit, filter at the pump and wash with 80 ml of hot rectified spirit. Keep the filtered hydrazine solution for the next stage in the preparation.

Equip a 1-litre three-necked flask with a reflux condenser and a sealed mechanical stirrer. Dissolve 50.5 g (0.25 mol) of commercial 1-chloro-2,4-dinitrobenzene (1) in 250 ml of rectified spirit in the flask, add the hydrazine solution and reflux the mixture with stirring for an hour. Most of the reaction product separates during the first 10 minutes. Cool, filter with suction and wash with 50 ml of warm (60 °C) rectified spirit to remove unchanged chloro-dinitrobenzene, and then with 50 ml of hot water. The resulting 2,4-dinitro-phenylhydrazine (30 g, 60%) melts at 191–192 °C (decomp.), and is pure enough for most purposes. Distil off half the alcohol from the filtrate and thus obtain a less pure second crop (about 12 g): recrystallise this from butan-1-ol (30 ml per gram). If pure 2,4-dinitrophenylhydrazine is required, recrystallise the total yield from butan-1-ol or from dioxan (10 ml per gram): this melts at 200 °C (decomp.).

The following alternative method of preparation is recommended. Dissolve 50 g of purified chlorodinitrobenzene (1) in 100 ml of triethylene glycol (gentle warming may be necessary; alternatively, 125 ml of warm diethylene glycol may be used) in a 600-ml beaker and cool, with mechanical stirring, in an ice bath to 15–18 °C. Place 15 ml of commercial 60–65 per cent hydrazine solution in a small separatory funnel supported over the beaker. Add the hydrazine solution to the stirred solution in the beaker at such a rate that the temperature is maintained between 15 and 20 °C (20–30 minutes). When the exothermic reaction is over, digest the paste on a boiling water bath with 50 ml of methanol for 15–20 minutes. Cool the reaction mixture, filter with suction and wash with a little methanol. Dry at 100 °C. The yield of 2,4-dinitrophenylhydrazine, m.p. 192–193 °C (decomp.), is 46 g (93%). The product is pure enough for most purposes: the pure compound may be obtained by recrystallisation from butan-1-ol or from dioxan as described above.

Cognate preparation 2,4-Dinitroaniline. Place a mixture of 18 g of ammonium acetate and 50 g (0.246 mol) of commercial 1-chloro-2,4-dinitrobenzene (1) in a 250-ml two-necked flask, and fit it with a reflux condenser and inlet tube (at least 2 cm diameter in order to prevent clogging) which terminates just above the surface of the reaction mixture. Half immerse the flask in an oil bath. Pass ammonia gas (from a cylinder) through a bubble counter, which contains a solution of 3 g of potassium hydroxide in 2.5 ml of water, into the mixture. Heat the oil bath to 170 °C, and pass the ammonia gas at the rate of 3–4 bubbles per second for 6 hours. Allow the reaction mixture to cool, break up the solid cautiously with a glass rod, add 100 ml of water, heat to boiling and filter while hot. Dissolve the residue in 500 ml of boiling rectified (or industrial) spirit, and add water (*ca.* 150 ml) until the solution becomes turbid; heat until the turbidity disappears and allow the clear solution to cool overnight. Filter the crystals at the pump and dry in an oven. The yield is 35 g (78%), m.p. 176–177 °C. To obtain a perfectly pure product, recrystallise again from ethanol and water; use 20 ml of ethanol per gram of solid: 31.5 g of pure 2,4-dinitroaniline, m.p. 180 °C, are thus obtained.

Note. (1) It is advisable to recrystallise the commercial chlorodinitrobenzene from ethanol; m.p. 51–52 °C.

IV,106. N,N-DIPHENYLPICRYLHYDRAZINE AND N,N-DIPHENYL-PICRYLHYDRAZYL

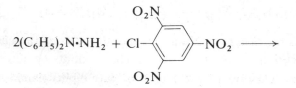

$$2(C_6H_5)_2N \cdot NH_2 + Cl \qquad NO_2 \longrightarrow$$

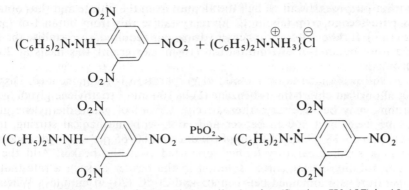

Dissolve 7.4 g (0.04 mol) of N,N-diphenylhydrazine (Section **IV,107**) in 10 ml of dry chloroform and add a solution of 5 g (0.02 mol) of picryl chloride [1-chloro-2,4,6-trinitrobenzene, (1)] in 40 ml of dry chloroform. Shake the mixture, which becomes dark and warm, and set it aside for 1 hour. Cool the suspension in an ice bath, filter off the amine salt and wash it with a little chloroform. Concentrate the filtrate and washings to 30 ml (rotary evaporator) and treat the hot concentrate with 60 ml of boiling ethanol. Cool, collect the almost pure diphenylpicrylhydrazine which crystallises and wash it with a little cold ethanol. The yield is 6 g (76%), m.p. 172–173 °C (decomp.). If desired, the product may be recrystallised from ethyl acetate.

Conversion to N,N-diphenylpicrylhydrazyl. Dissolve 3.95 g (0.01 mol) of the above hydrazine in 60 ml of dry chloroform, add 50 g of lead dioxide and 4 g of anhydrous sodium sulphate and shake the mixture mechanically for 1 hour. Filter, and concentrate the deep-violet filtrate on a rotary evaporator. Dilute the residual solution with two volumes of ether, and allow the product to crystallise. Filter off the large black-violet crystals of diphenylpicrylhydrazyl and wash them with ether; the yield is 3.5 g (89%), m.p. 137–138 °C. The product may be recrystallised from a mixture of chloroform and ether.

Note. (1) To **prepare picryl chloride**, dissolve 5 g (0.022 mol) of picric acid in 50 ml of phosphoryl chloride and add dropwise with shaking 3.8 g (4.0 ml, 0.025 mol) of N,N-diethylaniline. Allow the resulting brown solution to remain at room temperature for 15 minutes and then pour it into 500 ml of iced water. Stir vigorously until the excess of phosphoryl chloride has hydrolysed and the product has solidified; do not allow the temperature to rise above 35–40 °C during hydrolysis, cooling the mixture when necessary in an ice–salt bath. Filter off the almost colourless solid, wash it with cold water and dry it in a desiccator over calcium chloride. The yield of almost pure picryl chloride, m.p. 82 °C, is 5.0 g (92%).

IV,107. N,N-DIPHENYLHYDRAZINE

$$(C_6H_5)_2NH \xrightarrow[\text{HCl}]{\text{NaNO}_2} (C_6H_5)_2N\cdot NO \xrightarrow[\text{Zn, CH}_3\cdot\text{CO}_2\text{H}]{4[\text{H}]} (C_6H_5)_2N\cdot NH_2$$

The intermediate in this reaction, as an N-nitrosoamine, is potentially carcinogenic; see Section I,3,D,5(ii).

Diphenylnitrosamine. Dissolve 17 g (0.1 mol) of pure diphenylamine in

140 ml of warm ethanol; also 8 g (0.116 mol) of sodium nitrite in 12 ml of water. Cool each solution in ice until the temperature falls to 5 °C. Add 12 ml of concentrated hydrochloric acid slowly and with stirring to the diphenylamine solution, and immediately (otherwise diphenylamine hydrochloride may crystallise out) pour the sodium nitrite solution rapidly into the well-stirred mixture. The temperature soon rises to 20–25 °C and the diphenylnitrosamine crystallises out. Cool the mixture in ice-water for 15–20 minutes, filter with suction on a Buchner funnel, wash with water to remove sodium chloride and press well with a wide glass stopper. Recrystallise from rectified spirit. The yield of pure diphenylnitrosamine (pale yellow crystals), m.p. 68 °C, is 17 g (86%).

N,N-Diphenylhydrazine. Dissolve 15.8 g (0.08 mol) of diphenylnitrosamine in 72 ml of ethanol in a 250-ml three-necked flask equipped with a stirrer, reflux condenser and dropping funnel. Add 36 g (0.55 mol) of zinc powder and stir vigorously. From the dropping funnel slowly add about 22 ml of glacial acetic acid; cool the flask in a bath of cold water from time to time to moderate the reaction, which is complete when the addition of acetic acid no longer causes an increase in temperature, and when a sample of the supernatant liquid no longer gives a deep blue colour when concentrated hydrochloric acid is added. Filter the warm reaction mixture, wash the residue on the filter with warm ethanol and concentrate the filtrate and washings to 40 ml on a rotary evaporator. Cool in an ice bath, and add slowly with shaking 36 ml of concentrated hydrochloric acid. Collect the blue needles of the phenylhydrazine hydrochloride by suction filtration on a sintered glass funnel, wash them with a little cold 0.5 M-hydrochloric acid and suck dry. Suspend the crystals in a little water, cool in an ice bath and add slowly, with shaking and cooling, 36 ml of 25 per cent aqueous sodium hydroxide solution. Extract the liberated base with three 15 ml portions of toluene, dry the combined extracts over anhydrous potassium carbonate and remove the toluene on a rotary evaporator. Distil the residue under reduced pressure, and collect the diphenylhydrazine as a pale yellow oil of b.p. 136–137 °C/1 mmHg; the yield is 7.2 g (50%) (1). The product solidifies on cooling at 0 °C, and after crystallisation at low temperature from light petroleum (b.p. 40–60 °C) has m.p. 35 °C.

Note. (1) The product should not be allowed to come into contact with the skin; cf. cautionary note to Section **IV,104**.

H,3. Arylureas and related compounds. *N*-Arylureas are obtained when salts of primary aromatic amines react with solutions of alkali metal cyanates. The process involves the rearrangement of an amine cyanate, and is analogous to Wöhlers' classical synthesis of urea from ammonium cyanate.

$$Ar \cdot \overset{\oplus}{NH_3} \} \overset{\ominus}{CNO} \longrightarrow Ar \cdot NH \cdot CONH_2$$

The reaction is most conveniently carried out by warming the amine in aqueous solution with the equivalent quantity of sodium cyanate and an excess of acetic acid (Section **IV,108**, *Method 1*).

In an alternative synthesis (Section **IV,108**, *Method 2*), which is also convenient for the synthesis of alkylureas, the amine hydrochloride is heated in aqueous solution with urea. This reaction also probably involves the amine cyanate since in aqueous solution urea serves as a source of ammonium cyanate.

$$H_2N \cdot CO \cdot NH_2 \rightleftharpoons NH_4 \cdot CNO$$

In the case of the synthesis of phenylurea, some *sym.*-diphenylurea (carbanilide) is also formed and the quantity increases with continued refluxing.

$$Ar \cdot NH \cdot CO \cdot NH_2 + Ar\overset{\oplus}{N}H_3 \}\overset{\ominus}{Cl} \longrightarrow Ar \cdot NH \cdot CO \cdot NH \cdot Ar + \overset{\oplus}{N}H_4\overset{\ominus}{Cl}$$

The diarylurea is very sparingly water-soluble and is therefore easily separated from the monoarylurea which is readily soluble. Diarylurea formation is less extensive when ring-substituted anilines are used.

sym.-Diarylthioureas (III) (e.g., *sym.*-diphenylthiourea or thiocarbanilide, Section **IV,109**) are prepared by heating a mixture of a primary aromatic amine and carbon disulphide in absolute ethanol. Intermediates in the reaction sequence are the aryldithiocarbamic acid (I) and the aryl isothiocyanate (II), thus:

$$Ar \cdot NH_2 + S{=}C{=}S \longrightarrow Ar \cdot NH \cdot \overset{\overset{\displaystyle S}{\|}}{C} \cdot SH \xrightarrow{-H_2S} Ar \cdot N{=}C{=}S$$
$$\text{(I)} \qquad\qquad\qquad \text{(II)}$$

$$Ar - \overset{\curvearrowleft}{N}{=}C{=}S \xrightarrow[+H^{\oplus}]{-H^{\oplus}} Ar \cdot NH \cdot CS \cdot NH \cdot Ar$$
$$\underset{H_2\ddot{N}Ar}{\uparrow} \qquad\qquad \text{(III)}$$

Upon heating the diarylthiourea with concentrated hydrochloric acid, it is partly converted into the aryl isothiocyanate (e.g., phenyl isothiocyanate or phenyl mustard oil, Section **IV,110**, *Method 1*). A little hydrogen sulphide is evolved in a side reaction forming diphenylcarbodi-imide (IV) which undergoes nucleophilic addition of aniline to yield triphenylguanidine (V), isolated from the reaction mixture as the hydrochloride.

$$C_6H_5 \cdot NH \cdot CS \cdot NH \cdot C_6H_5 \xrightarrow[\text{heat}]{HCl} \begin{cases} \rightarrow C_6H_5 \cdot N{=}C{=}S + C_6H_5 \cdot NH_2 \\ \\ \rightarrow C_6H_5 \cdot N{=}C{=}N \cdot C_6H_5 + H_2S \end{cases}$$
$$\text{(IV)}$$

$$\underset{C_6H_5\ddot{N}H_2}{\overset{\displaystyle C_6H_5 \cdot N{=}C\overset{\curvearrowright}{=}N \cdot C_6H_5}{\nearrow}} \longrightarrow C_6H_5 \cdot N{=}C(NH \cdot C_6H_5)_2 \xrightarrow{HCl}$$
$$\text{(V)}$$
$$[C_6H_5 \cdot NH{:}C(NH \cdot C_6H_5)_2]\overset{\oplus\ominus}{Cl}$$

Phenyl isothiocyanate may be prepared in quantity (Section **IV,110**, *Method 2*) by allowing aniline to react with carbon disulphide to form phenyl-dithiocarbamic acid (cf. I), which is isolated as the ammonium salt. Treatment of the latter with lead nitrate removes the elements of hydrogen sulphide to produce phenyl isothiocyanate. As indicated in the preparation of *p*-bromophenyl isothiocyanate which is given as a further example, a slightly modified procedure which requires the use of rectified spirit as a reaction solvent is necessary in order to obtain good yields of isothiocyanates from substituted anilines.

IV,108. PHENYLUREA

Method 1

$$C_6H_5NH_2 + NaCNO + CH_3 \cdot CO_2H \rightarrow C_6H_5 \cdot NH \cdot CO \cdot NH_2 + CH_3 \cdot CO_2Na$$

Method 2

$$C_6H_5 \overset{\oplus}{N}H_3 \} \overset{\ominus}{Cl} + NH_2 \cdot CO \cdot NH_2 \rightarrow C_6H_5 \cdot NH \cdot CO \cdot NH_2 + NH_4Cl$$

Method 1. Dissolve 9.3 g (9.1 ml, 0.1 mol) of aniline in 10 ml of glacial acetic acid diluted to 100 ml contained in a 250-ml beaker or conical flask, and add with stirring or shaking a solution of 6.5 g (0.1 mol) of sodium cyanate in 50 ml of warm water. Allow to stand for 30 minutes, then cool in ice, and allow to stand for a further 30 minutes. Filter at the pump, wash with water and dry at 100 °C. The resulting phenylurea is generally colourless and has a m.p. of 148 °C (i.e., is pure): the yield is 11 g (81%). If the colour or the m.p. of the product is not quite satisfactory, recrystallise it from boiling water (10 ml per gram) with the aid of decolourising charcoal.

Method 2. Dissolve 65 g (0.5 mol) of aniline hydrochloride and 120 g (2 mol) of urea in 200 ml of water contained in a 1-litre round-bottomed flask; filter the solution, if necessary. Add 4 ml of concentrated hydrochloric acid and 4 ml of glacial acetic acid. Fit a reflux condenser to the flask, introduce a few fragments of broken porcelain and boil the mixture for 30 minutes. Fine white crystals (largely *sym.*-diphenylurea) appear after about 15 minutes and gradually increase in amount as the refluxing is continued. Cool the flask in ice and filter with suction. Separate the mixture of phenylurea and diphenylurea (*ca.* 42 g) by boiling with 500 ml of water and filter at the pump through a preheated Buchner funnel into a warm flask; cool the filtrate, collect the phenylurea, drain well and dry in the steam oven. The phenylurea melts at 146–147 °C and weighs 30 g (44%); recrystallisation from hot water raises the m.p. to 148 °C. The crude diphenylurea (residue from first recrystallisation after drying at 100 °C) has m.p. 241 °C and weighs 10 g (19%); recrystallisation from glacial acetic acid or ethyl acetate with the addition of a little decolourising carbon gives a colourless product, m.p. 242 °C.

Cognate preparations *p*-**Tolylurea** (*Method 1*). Dissolve 10.7 g (0.1 mol) of *p*-toluidine in a warm mixture of 10 ml of glacial acetic acid and 50 ml of water, and then dilute with 150 ml of hot water. Introduce, with stirring or shaking, a solution of 6.5 g (0.1 mol) of sodium cyanate in 50 ml of hot water. The *p*-tolylurea precipitates almost immediately. Allow to stand several hours, filter at the pump, wash with water and dry. The yield of *p*-tolylurea, m.p. 180–180.5 °C, is 14 g (85%). Recrystallise from aqueous ethanol; the resulting *p*-tolylurea melts sharply at 181 °C.

p-**Bromophenylurea.**   Proceed as for *p-tolylurea* (*Method 1*), but use 17.2 g (0.1 mol) of *p*-bromoaniline dissolved in a mixture of 50 ml of glacial acetic acid and 100 ml of water at 35 °C; add gradually a solution of 6.5 g (0.1 mol) of sodium cyanate in 50 ml of water at 35 °C. The yield of crude *p*-bromophenylurea is 19 g (88%); m.p. 227 °C. Recrystallise from 90 per cent aqueous ethanol; m.p. 228 °C. The m.p. depends somewhat upon the rate of heating.

p-**Methoxyphenylurea.**   Proceed as for *phenylurea, Method 2*, but use 79 g (0.5 mol) of *p*-anisidine hydrochloride in place of 65 g of aniline hydrochloride; reflux the mixture for 1 hour. Cool the reaction mixture slowly to 0 °C, filter and recrystallise from boiling water. The yield of *p*-methoxyphenylurea, m.p. 168 °C, is 60 g (72%).

p-**Ethoxyphenylurea.**   Proceed as for *phenylurea, Method 2*, but use 87 g (0.5 mol) of *p*-phenetidine hydrochloride; reflux the mixture for 45–90 minutes. The product commences to separate after 20–30 minutes and increases rapidly until the entire contents of the flask suddenly set to a solid mass: withdraw the source of heat *immediately* at this point. Cool to room temperature, add 150 ml of water, stir, filter with suction and wash with cold water. Suspend the solid in 2 litres of boiling water, add 1 g of decolourising carbon, boil for 5 minutes and filter through a hot water funnel; cool the colourless filtrate slowly to 0 °C, collect the solid which separates and dry at 100 °C. The yield of *p*-ethoxyphenylurea, m.p. 174 °C, is 60 g (67%).

IV,109. THIOCARBANILIDE (sym.-*Diphenylthiourea*)

$$2C_6H_5NH_2 + CS_2 \xrightarrow{C_2H_5OH} S = C(NHC_6H_5)_2 + H_2S$$

In a 1-litre round-bottomed flask provided with an efficient double surface condenser, place 40 g (39 ml, 0.43 mol) of aniline, 50 g (40 ml, 0.66 mol) of carbon disulphide (**CAUTION:** inflammable) (1) and 50 g (63.5 ml) of absolute ethanol (2). Set up the apparatus in the fume cupboard or attach an absorption device to the top of the condenser (see Fig. I,68) to absorb the hydrogen sulphide which is evolved. Heat upon an electrically heated water bath or upon a steam bath for 8 hours or until the contents of the flask solidify. When the reaction is complete, arrange the condenser for downward distillation and remove the excess of carbon disulphide and alcohol (**CAUTION:** inflammable; there must be no flame near the receiver). Shake the residue in the flask with excess of dilute hydrochloric acid (1 : 10) to remove any aniline present, filter at the pump, wash with water and drain well. Dry in the steam oven. The yield of crude product, which is quite satisfactory for the preparation of phenyl isothiocyanate (Section **IV,110**), is 40–45 g (81–91%). Recrystallise the crude thiocarbanilide by dissolving it, under reflux, in boiling rectified spirit (filter through a hot water funnel if the solution is not clear), and add hot water until the solution just becomes cloudy and allow to cool. Pure *sym.*-diphenylthiourea separates in colourless needles, m.p. 154 °C.

Notes.   (1) No flames may be present in the vicinity: read Section **I,3**.

(2) The addition of powdered potassium hydroxide (about 20 per cent of the weight of the carbon disulphide) reduces the refluxing period necessary to complete the reaction.

IV,110.   PHENYL ISOTHIOCYANATE

Method 1

$$C_6H_5NH \cdot CS \cdot NHC_6H_5 \xrightarrow{\text{conc. HCl}} C_6H_5 \cdot N = C = S + C_6H_5NH_2$$

Method 2

$$C_6H_5NH_2 + CS_2 \longrightarrow C_6H_5 \cdot NH \cdot CS \cdot SNH_4 \xrightarrow{Pb(NO_3)_2}$$
$$C_6H_5N = C = S + NH_4NO_3 + HNO_3 + PbS$$

Method 1.   Place 25 g (0.11 mol) of crude thiocarbanilide (Section **IV,109**) and 100 ml of concentrated hydrochloric acid in a 250-ml flask; fit a reflux condenser and reflux gently in the fume cupboard for 30 minutes. Distil the mixture until the oily phenyl isothiocyanate has all passed over; the volume remaining in the flask will be 25–30 ml. Crystals of triphenylguanidine hydrochloride may appear in the distilling flask during the latter part of the distillation. Dilute the distillate with an equal volume of water, and extract the isocyanate with ether; wash the extract with a little sodium carbonate solution, and dry over anhydrous calcium chloride or magnesium sulphate. Remove the ether on a rotary evaporator and distil the residual oil, collecting the phenyl isothiocyanate at 217–220 °C (1). The yield is 10 g (67.5%).

To isolate the **triphenylguanidine** formed as a by-product dilute the residue in the flask with 50 ml of water, add 2–3 g of decolourising carbon, warm and filter. Cool the solution in ice, and filter off the hydrochloride at the pump. Dissolve it in the minimum volume of hot water, render the solution alkaline with sodium hydroxide and allow to cool. Filter off the free base (triphenylguanidine), and recrystallise it from ethanol; it separates in colourless crystals, m.p. 144 °C. The yield is 3 g.

Note.   (1) It may also be distilled under diminished pressure, b.p. 95 °C/12 mmHg.

Method 2.   Equip a 500-ml three-necked flask with a powerful mechanical stirrer and a separatory funnel; leave the third neck open or loosely stoppered. Introduce, while the flask is cooled in a freezing mixture of ice and salt, 90 ml of concentrated ammonia solution (*d* 0.88) and 54 g (43 ml, 0.71 mol) of pure carbon disulphide (1). Stir the mixture and run in 56 g (55 ml, 0.60 mol) of aniline from the separatory funnel during about 20 minutes; stir for a further 30 minutes, and allow to stand for another 30 minutes. A heavy precipitate of ammonium phenyldithiocarbamate separates. Transfer the salt to a 5-litre round-bottomed flask by four extractions with 200 ml portions of water. Add to the resulting solution, with constant stirring, a solution of 200 g (0.605 mol) of lead nitrate in 400 ml of water; lead sulphide precipitates. Steam distil the mixture into a receiver containing 10 ml of *ca.* 0.5 M-sulphuric acid as long as organic material passes over (2–3 litres of distillate). Separate the oil, dry it over anhydrous calcium chloride or magnesium sulphate and distil under diminished pressure. Collect the phenyl isothiocyanate at 120–121 °C/35 mmHg or at 95 °C(12 mmHg). The yield is 62 g (76%).

Note.   (1) **CAUTION:** see Section **IV,109**.

Cognate preparation   *p*-**Bromophenyl isothiocyanate** (*Method 2*).   Add 41 ml of concentrated ammonia solution (*d* 0.88) slowly with stirring to a solution of 45 g (0.26 mol) of *p*-bromoaniline (Section **IV,75**), 30 g (24 ml, 0.396 mol) of carbon disulphide and 40 ml of rectified spirit (95% C_2H_5OH) at 10–15 °C. Considerable heat is evolved; cool the flask in a freezing mixture from time to time so that the temperature does not rise above 30 °C. The original milky suspension becomes clear and the intermediate dithiocarbamate soon crystallises out. Allow to stand overnight, filter the crystals, wash with a little ether, dissolve in 1500 ml of water and stir mechanically while a solution of 87 g (0.262 mol) of lead nitrate in 175 ml of water is slowly added. Continue the stirring for 20 minutes, and isolate the *p*-bromophenyl isothiocyanate by steam distillation into a receiver containing 5 ml of *ca.* 0.5 *M*-sulphuric acid; if the substance solidifies in the condenser, stop the cooling water until the solid has melted and run into the receiver. Filter the cold solid product, wash with a little water and dry in the air upon filter-paper. The yield is 15 g (50%), m.p. 61 °C.

J PHENOLS AND PHENYL ETHERS

1. Methods for the introduction of a hydroxyl group into an aromatic ring by (*a*) replacement of a sulphonic acid group (Section **IV,111**), (*b*) replacement of a halogen (Sections **IV,112** and **IV,113**) and (*c*) replacement of a diazo group (see Section **IV,78**).
2. Substitution reactions of phenols: (*a*) nitration and nitrosation (Sections **IV,114** to **IV,117**), (*b*) halogenation (Sections **IV,118** and **IV,119**) and (*c*) acylation and alkylation (Sections **IV,120** and **IV,121**).
3. Formation of phenyl ethers (Sections **IV,122** to **IV,124**).

J,1. Methods for the introduction of a hydroxyl group into an aromatic ring. (*a*) *Replacement of a sulphonic acid group.*   A fairly general procedure, which has also been used on the industrial scale, involves heating the alkali metal sulphonate with either sodium or potassium hydroxide in the presence of a small amount of water to aid the fusion process. The reaction mechanism may be formulated as a bimolecular nucleophilic addition–elimination sequence.

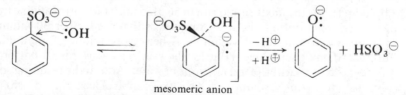

mesomeric anion

The reaction is illustrated by the synthesis of 2-naphthol (Section **IV,111**).

Occasionally in the synthesis of phenols by this route oxidation products are formed. A particular example is provided by the alkali fusion of sodium anthraquinone-2-sulphonate during which a second hydroxyl group is introduced into the 1-position, forming the dyestuff alizarin (I) (cognate preparation in Section **IV,111**). In the procedure described the oxidation step is promoted by the deliberate introduction of potassium chlorate as an oxidant.

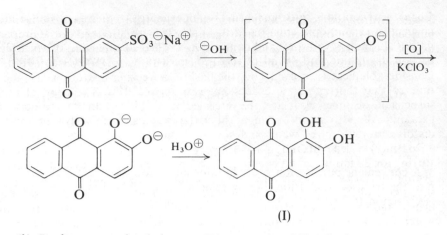

(I)

(b) *Replacement of a halogen.*   Direct nucleophilic displacement of the halogen in an aryl halide is difficult and hydrolysis to phenols requires high temperatures and pressures; the method is therefore only suitable on the large scale. The presence of a nitro group in the *ortho* or *para* position, however, makes the halogen more labile since electron withdrawal by the nitro group in these positions stabilises the intermediate anion by electron delocalisation. *p*-Chloronitrobenzene, for example, is hydrolysed to *p*-nitrophenol when heated with 15 per cent sodium hydroxide solution at about 150 °C.

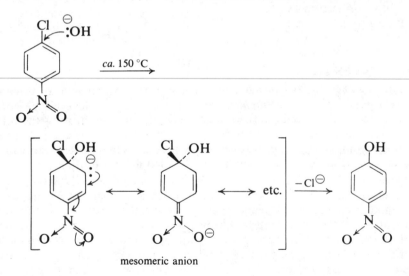

mesomeric anion

When two activating nitro groups are present hydrolysis takes place readily with dilute aqueous alkali solution (e.g., 2,4-dinitrophenol, Section **IV,112**).

Aryl halides of many different types, including simple unsubstituted halides, may be conveniently converted into phenols by an indirect route involving the preparation of an arylboronic acid and its subsequent oxidation with hydrogen peroxide. The arylboronic acid (III) is normally prepared by reaction of the corresponding arylmagnesium halide with a borate ester (typically tributyl borate) at between −60 to −80 °C, to yield the dialkyl boronate ester (II) which is then hydrolysed to the arylboronic acid (III). The latter may be isolated, purified and then oxidised with hydrogen peroxide as described in the preparation of *m*-cresol (Section **IV,113**). Alternatively the crude reaction mixture from the preparation of (III) may be treated directly with hydrogen peroxide (Ref. 7).

$$\text{ArMgX} + \text{B(OC}_4\text{H}_9)_3 \xrightarrow{-70\,°C} \text{Ar·B(OC}_4\text{H}_9)_2 \xrightarrow{\text{H}_3\text{O}^\oplus} \text{Ar·B(OH)}_2$$

(II)   (III)

$$\text{Ar·B(OH)}_2 \xrightarrow{\text{H}_2\text{O}_2} \text{ArOH}$$

A convenient purification procedure for an arylboronic acid is to convert it into the trimeric anhydride (IV) by removal of water as a benzene azeotrope (see Section **IV,113** Note (3)).

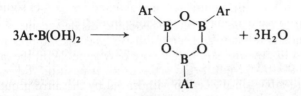

$$3\text{Ar·B(OH)}_2 \longrightarrow \qquad + 3\text{H}_2\text{O}$$

(*c*) *Replacement of a diazo group.* The preparation of phenols by the hydrolysis of diazonium salts has been discussed on p. 687, **IV,G,1** and illustrated in Section **IV,78**. Although the method enables an aromatic hydrocarbon system to be converted in good yield into a phenol via the corresponding nitro and amino derivatives, the shorter route involving the alkaline fusion of the sulphonic acid discussed above may often be preferred.

IV,111. 2-NAPHTHOL

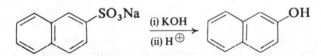

This preparation should be carried out in a fume cupboard with the window protecting the face.

Support a 250-ml nickel, copper (better silver-plated copper) or iron crucible or beaker in a large circular hole in a sheet of asbestos board or uralite resting on a tripod. Prepare a case of nickel or copper to surround a 360 °C thermometer for about two-thirds of its length; this may be done either by cutting a suitable length of nickel or copper tubing already closed at one end, or by hammering down the end of the open tube and folding over the flat part in a vice. Fit a large cork around the top of the tube; this will serve for handling the tube containing the thermometer when it is subsequently used for stirring the molten alkali. Since some splattering of the latter cannot generally be avoided, goggles, gloves and a well-fitting laboratory coat must be worn.

Place 120 g (2.15 mol) of potassium hydroxide sticks or pellets (1) together with 5 ml of water in the crucible, and heat with a Bunsen burner until it melts. When the temperature reaches about 250 °C, remove the flame, and quickly add with stirring 50 g (0.22 mol) of finely powdered sodium naphthalene-2-sulphonate (Section **IV,41**). Replace the flame, stir the stiff pasty mass and continue the heating so that the temperature rises to 300 °C in 5–10 minutes. Stir the mixture continuously; there is some frothing at first and at about 300 °C the mass suddenly becomes a clear, mobile, brown oil of the potassium salt of 2-naphthol floating on a pasty mass of alkali. Raise the temperature during 5 minutes to 310 °C, remove the flame, push down the material from the side of the crucible and reheat to 310 °C for about 2 minutes, and then allow the melt to cool. Do not permit the melt to solidify completely. When it becomes pasty, ladle it out in small portions (with a nickel spatula, 'spoon' end) into a 1-litre beaker half-filled with crushed ice. Extract the residual material in the crucible with water and add it to the contents of the beaker. Precipitate the 2-naphthol by adding concentrated hydrochloric acid slowly and with stirring (*FUME CUPBOARD:* SO_2); if the 2-naphthol separates in a finely divided form, warm until the particles coagulate. Cool in ice, filter at the pump and transfer the precipitate to a beaker containing cold water. Add just sufficient 5 per cent sodium hydroxide solution to dissolve the solid and also 1 g of sodium dithionite ($Na_2S_2O_4$) to prevent oxidation, and filter from traces of insoluble matter. Precipitate the 2-naphthol with acetic acid, warm to produce a more readily filterable form of the precipitate, cool in ice and filter the product. Dry in the air upon filter-paper. The yield is 25 g (80%), m.p. 122 °C. If the m.p. is unsatisfactory, recrystallise from water, dilute ethanol or carbon tetrachloride.

Note. (1) Sodium hydroxide may replace potassium hydroxide in this preparation; 150 g together with 15 ml of water are required. The sulphonate is stirred in when the temperature reaches 280 °C and the reaction is complete at 310–320 °C.

Cognate preparation Alizarin. Dissolve successively in 75 ml of water 6 g (0.049 mol) of potassium chlorate, 20 g (0.065 mol) of sodium anthraquinone-2-sulphonate (Section IV,42) and 75 g of sodium hydroxide. Transfer the mixture to a 500-ml autoclave (compare Section I,17,2) and heat for 20 hours at 170 °C. After cooling, scrape out the violet-coloured mass and extract it three or four times with 100 ml portions of boiling water. Acidify the filtered extract with hydrochloric acid. When cold, filter the orange precipitate of alizarin at the pump, wash it thoroughly with cold water and dry at 100 °C. The yield of alizarin is 14 g (90%). It may be purified by recrystallisation from glacial acetic acid or by sublimation. The pure compound has m.p. 289 °C.

IV,112. 2,4-DINITROPHENOL

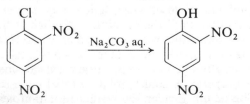

In a 1-litre round-bottomed flask equipped with a reflux condenser place a solution of 62.5 g (0.6 mol) of anhydrous sodium carbonate in 500 ml of water and add 50 g (0.25 mol) of commercial 1-chloro-2,4-dinitrobenzene. Reflux the mixture for 24 hours or until the oil passes into solution. Acidify the yellow solution with hydrochloric acid and, when cold, filter the crystalline dinitro-phenol which has separated. Dry the product upon filter-paper in the air. The yield is 42 g (91%). If the m.p. differs appreciably from 114 °C, recrystallise from ethanol or from water.

IV,113. *m*-CRESOL

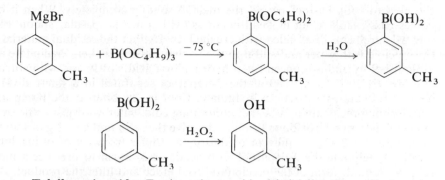

m-Tolylboronic acid. Equip a thoroughly dried 1-litre flange flask with a low temperature reading thermometer, and an efficient sealed stirrer unit, and attach a dropping funnel and a reflux condenser both protected with calcium chloride guard-tubes. Mount the flask in a Cardice–acetone cooling bath and add to the flask 40.5 g (0.17) of pure tributyl borate (Section **III,73**) (1) dissolved in 150 ml of ether (previously dried over sodium wire). Then, with fairly rapid stirring, add slowly from the dropping funnel a solution of *m*-tolylmagnesium bromide [0.175 mol; (2)] in 450 ml of dry ether. It is essential that the rate of addition throughout the reaction should be such that the temperature of the reaction mixture is kept between −70 and −75 °C. As the ethereal solution of the Grignard reagent is added, a white precipitate is formed immediately and slowly dissolves; when all of the reagent has been added (about 3 hours), con-tinue to stir the mixture at −75 °C for a further 2 hours. With continued stirring, allow the reddish-orange solution to slowly warm to 0 °C, remove the dropping funnel, condenser and stirrer, stopper the flask and place it in a refrigerator to attain a temperature of about 5 °C over a period of 12 hours. Slowly add the cold reaction mixture with efficient stirring to 120 ml of chilled (0–5 °C) 10 per cent sulphuric acid. Transfer the resultant mixture to a separat-ing funnel, separate the layers and thoroughly extract the aqueous phase with four 100 ml portions of ether. Concentrate the combined ether solutions on a rotary evaporator using a water-bath temperature of 40–45 °C. To the residual *m* tolylboronic acid in butan-1-ol add 50 ml of water, followed by 10 per cent aqueous potassium hydroxide solution until the solution is alkaline to litmus paper. Remove the butanol as an azeotrope with water by concentrating the mixture on a rotary evaporator (water-bath temperature 40–45 °C). Add fur-ther portions of water (25 ml) and repeat the evaporation until no smell of butanol remains. During these evaporation stages, solid material often separ-ates but this should not be filtered off. Render the residual solution acid to litmus by adding dilute aqueous sulphuric acid and, if necessary, make up the volume to 60 ml by the addition of distilled water. Without separating the precipitated solid, heat the mixture to boiling when the precipitate dissolves and a heavy dark brown oil appears. Decant the hot supernatant solution through a fluted filter-paper and extract the oil with four 40 ml portions of boiling water. Combine the hot filtered extracts and cool. Filter off the crystalline *m*-tolylboronic acid and dry at room temperature by spreading upon filter-papers. The yield is 5.9 g (25%) (3).

m-Cresol.   In a 250-ml, two-necked round-bottomed flask fitted with an efficient stirrer and a dropping funnel, place 5.4 g (0.040 mol) of *m*-tolylboronic acid and 100 ml of ether. While stirring the slurry add 30 ml of 10 per cent hydrogen peroxide solution from the dropping funnel over a period of about 5 minutes. The two-phase system in the flask becomes warm as the reaction proceeds. When all the hydrogen peroxide has been added, continue stirring until the contents of the flask have cooled to room temperature (20–30 minutes) and then transfer to a separating funnel. Run off and discard the aqueous layer and wash the ether layer thoroughly with three 30 ml portions of 10 per cent iron(II) ammonium sulphate solution to remove remaining traces of hydrogen peroxide. Extract the product from the ether layer by shaking with three 30 ml portions of 10 per cent aqueous sodium hydroxide solution. Acidify the combined alkaline extracts with concentrated hydrochloric acid and extract the product with three 70 ml portions of ether. Dry the combined ether extracts with calcium sulphate, filter and evaporate on a rotary evaporator. Distil the crude product and collect the *m*-cresol at 198–202 °C. The yield is 2.5 g (58%).

Notes.   (1) The tributyl borate should preferably be freshly prepared and distilled (see Section **III,73**, Note (1)).

(2) The ethereal solution of *m*-tolylmagnesium bromide may be prepared by the method described in Section **IV,171**, but using 20 g (0.175 mol) of *m*-bromotoluene, 4.2 g (0.175 mol) of magnesium turnings and 450 ml of dry ether. When nearly all the magnesium has reacted the solution should be quickly decanted into a dry dropping funnel and addition to the cooled tributyl borate solution begun immediately.

(3) It is not possible to obtain an analytically pure sample of arylboronic acids since, on drying, partial conversion to the trimeric anhydride occurs. A quantitative conversion to the anhydride is achieved in the case of *m*-tolylboronic acid by heating with forty times its weight of benzene in a flask fitted with a Dean and Stark water separator (Section **III,76**). When no further water droplets separate the benzene solution is concentrated to one-quarter volume and cooled. *m*-Tolylboronic anhydride crystallises out and has m.p. 161–162 °C. The anhydride may be reconverted to the acid by dissolving it in the minimum quantity of hot water and allowing the solution to cool, whereupon the acid crystallises.

J,2. Substitution reactions of phenols.   (*a*) *Nitrosation and nitration.* Phenol may be converted into a mixture of *o*- and *p*-nitrophenols (Section **IV,114**) by reaction with dilute nitric acid; the yield of *p*-nitrophenol is increased if a mixture of sodium nitrate and dilute sulphuric acid is employed. Upon steam distillation of the mixture of nitrophenols, the *ortho* isomer passes over in a substantially pure form; the *para* isomer remains in the distillation flask, and can be readily isolated by extraction with hot 2 per cent hydrochloric acid. The mechanism of the substitution probably involves an electrophilic attack (cf. p. 622, **IV,B,1**) by a nitrosonium ion at a position either *ortho* or *para* to the activating hydroxyl group, to yield a mixture of *o*- and *p*-nitrosophenols, which are then oxidised by the nitric acid to the corresponding nitrophenols. The reaction depends upon the presence in the nitric acid of traces of nitrous acid which serve as the source of the nitrosonium ion.

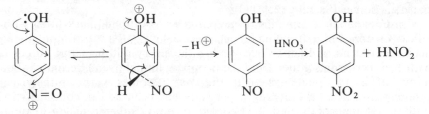

If the phenol is allowed to react with nitrous acid (generated in an acidified solution of sodium nitrite), the nitrosophenol may be obtained in good yield. An example is provided by the nitrosation of 2-naphthol which yields 1-nitroso-2-naphthol (Section **IV,115**).

By suitably introducing sulphonic acid 'blocking groups', which may subsequently be removed by heating under aqueous acidic conditions, control over the orientation of other substituents introduced into the aromatic nucleus of a phenol may be achieved. The procedure is illustrated by the synthesis of 2-nitroresorcinol (Section **IV,116**). In this reaction disulphonation of the dihydric phenol can readily be achieved, the sulphonic acid groups taking up the 4- and 6-positions. When the disulphonic acid is then nitrated the nitro group enters the remaining active site, i.e., the 2-position, removal of the sulphonic acid residues then yields the required 2-nitroresorcinol. In this sequence it is essential that the experimental conditions for the nitration step are as mild as possible (i.e., below 20 °C), since more vigorous conditions will result in electrophilic replacement of a sulphonic acid group by a nitro group. Thus in the synthesis of 2,4,6-trinitrophenol (Section **IV,117**), phenol is first sulphonated to yield a mixture of o- and p-phenolsulphonic acids; nitration of this product with a hot mixture of concentrated nitric acid and concentrated sulphuric acid results in the introduction of a nitro group into all the *ortho* and *para* activated positions with displacement of the sulphonic acid group. The direct nitration of phenol to the trinitro derivative in good yield is not possible since much of the starting material is oxidatively destroyed.

(*b*) *Bromination.* When treated with bromine water an aqueous solution of phenol gives an immediate precipitate of 2,4,6-tribromophenol (Section **VII,6,6**), owing to the powerfully activating influence of the negatively charged oxygen in the phenoxide ion.

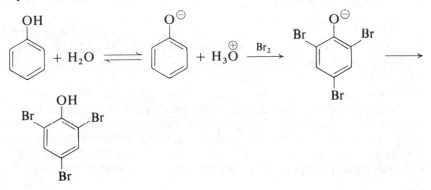

The monobromination of phenol can, however, be achieved by using solutions of bromine in non-polar solvents such as carbon disulphide and carbon tetrachloride at low temperature (0–5 °C). The product is almost exclusively the *para* isomer (Section **IV,118**).

o-Bromophenol is conveniently prepared by first sulphonating phenol with excess of concentrated sulphuric acid to yield phenol-2,4-disulphonic acid, neutralising with sodium hydroxide, heating the solution of the sodium salt with 1 mol of bromine and then removing the sulphonic acid groups by treatment with aqueous sulphuric acid at 200 °C. The sequence is described and formulated in Section **IV,119**.

(*c*) *Acylation and alkylation.* The introduction of an acyl group (R·CO−) into a phenolic nucleus by the standard Friedel–Crafts procedure (Section **IV,139**) does not always result in acceptable yields (see, however, p. 770, **IV,L,1**). The preferred method is to convert the phenol into the phenyl ester and to subject this to rearrangement (the **Fries reaction**) in the presence of aluminium chloride.

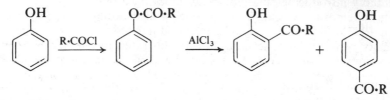

Experimental procedures are given in Section **IV,120** for *o*- and *p*-hydroxypropiophenone (R = C$_2$H$_5$) and *o*- and *p*-hydroxyacetophenone (R = CH$_3$). The *ortho/para* ratio in the product is influenced by the nature of the alkyl residue, the temperature, the solvent and the amount of aluminium chloride used: generally low temperatures favour the formation of *p*-hydroxyketones. It is usually possible to separate the two hydroxyketones by fractional distillation under reduced pressure through an efficient fractionating column or by steam distillation; the *ortho* isomers, being chelated, are more steam volatile. It may be mentioned that Clemmensen reduction (cf. Sections **IV,5** and **IV,A,1**(*b*)) of the hydroxyketones affords an excellent route to alkyl phenols.

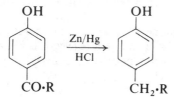

2,5-Dihydroxyacetophenone (V) (cognate preparation in Section **IV,120**), which cannot be prepared by a Friedel–Crafts acetylation of hydroquinone, is obtained in good yield when hydroquinone diacetate (VI) is heated in the presence of 3.3 mol of aluminium chloride.

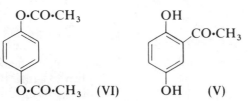

The details of the mechanism of the Fries rearrangement are uncertain but the reaction probably involves the formation and migration of the acylium ion (VII).

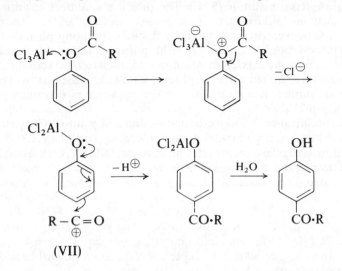

(VII)

When phenol is treated with allyl bromide in the presence of potassium carbonate and acetone, the product is almost entirely allyl phenyl ether (Section **IV,123**). This undergoes ready thermal rearrangement to give 2-allyl-phenol (Section **IV,121**), which is an example of the **Claisen rearrangement**. The mechanism of this intramolecular rearrangement involves a cyclic transition state (VIII) as formulated below

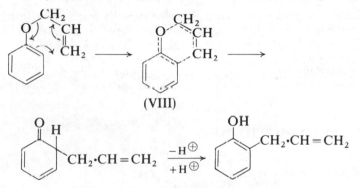

(VIII)

A protropic shift may be induced to effect double bond migration in the side chain by heating the allyl phenyl ether with methanolic potassium hydroxide when 2-(prop-1-enyl) phenol is formed (Section **IV,121**).

IV,114. o- AND p-NITROPHENOLS

$$C_6H_5OH \xrightarrow{\text{dil. HNO}_3} o\text{- and } p\text{-}O_2N\cdot C_6H_4OH$$

Cautiously add 250 g (136 ml, *ca.* 2.5 mol) of concentrated sulphuric acid in a thin stream and with stirring to 400 ml of water contained in a 1-litre three-necked flask, and then dissolve 150 g (1.75 mol) of sodium nitrate in the diluted acid. Cool in a bath of ice or iced water. Melt 94 g (1 mol) of phenol (1) with 20 ml of water, and add this dropwise from a separatory funnel to the vigorously stirred mixture in the flask; maintain the temperature at about 20 °C. Continue the stirring for a further 2 hours after all the phenol has been added. Pour off the mother-liquor from the resinous mixture of nitro compounds. Melt the residue with 500 ml of water, shake and allow the contents of the flask to settle. Pour off the wash liquor and repeat the washing at least two or three times to ensure the complete removal of any residual acid. Steam distil the mixture (Fig. I,101) until no more o-nitrophenol passes over; if the latter tends to solidify in the condenser, turn off the cooling water temporarily. Collect the distillate in cold water, filter at the pump and drain thoroughly. Dry upon filter-paper in the air. The yield of o-nitrophenol, m.p. 46 °C (2), is 50 g (36%).

Allow the residue in the flask to cool during 2 hours and then cool in ice for 15–30 minutes. Filter off the crude p-nitrophenol and boil it with 1 litre of 2 per cent hydrochloric acid (3) together with about 5 g of decolourising charcoal for at least 10 minutes. Filter through a hot water funnel (or through a preheated Buchner funnel): allow the filtrate to crystallise overnight. Filter off the almost colourless needles and dry them upon filter-paper. The yield of p-nitrophenol, m.p. 112 °C, is 35 g (25%). Further small quantities may be obtained by concentrating the mother-liquor and also by repeating the extraction of the residue with 2 per cent hydrochloric acid.

Notes. (1) Phenol should not be allowed to come into contact with the skin for it causes painful burns. The best antidote for phenol burns is a saturated solution of bromine in glycerine: if all undissolved bromine is allowed to settle out before the solution is used, there is no danger of bromine burns. Lime water may also be employed.

(2) If the m.p. is not quite satisfactory, dissolve the o-nitrophenol in hot ethanol (or industrial spirit) under reflux, add hot water drop by drop until a cloudiness just appears and allow to cool spontaneously. Filter off the bright yellow crystals and dry between filter-paper.

(3) It is not advisable to treat the crude p-nitrophenol with sodium hydroxide solution in order to convert it into the sodium derivative: alkali causes extensive resinification.

IV,115. 1-NITROSO-2-NAPHTHOL

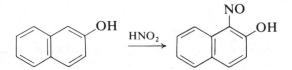

Dissolve 100 g (0.7 mol) of 2-naphthol (Section **IV,111**) in a warm solution of 28 g (0.7 mol) of sodium hydroxide in 1200 ml of water contained in a 2.5-litre round-bottomed flask fitted with a mechanical stirrer. Cool the solution to 0 °C in a bath of ice and salt, and add 50 g (0.725 mol) of powdered sodium nitrite. Start the stirrer and add, by means of a separatory funnel supported above the flask, 220 g (166.5 ml) of 5.6 M-sulphuric acid at such a rate that the whole is added during 90 minutes and the temperature is kept at 0 °C: add crushed ice (about 200 g in all) from time to time in order to maintain the temperature at

0 °C. The solution should react acid to Congo red paper after all the sulphuric acid has been introduced. Stir the mixture for an additional hour; keep the temperature at 0 °C. Filter off the 1-nitroso-2-naphthol at the pump and wash it thoroughly with water. Dry the pale yellow product upon filter-paper in the air for four days; the colour changes to dark brown and the 1-nitroso-2-naphthol, m.p. 97 °C, weighs 130 g. It contains about 10 per cent of its weight of moisture, but is otherwise almost pure. The moisture may be removed by leaving the air-dried compound in a desiccator for 24 hours; the yield is 115 g (96%), m.p. 106 °C.

If 1-nitroso-2-naphthol is required in the crystalline condition, recrystallise it from light petroleum (b.p. 60–80 °C, 7.5 ml per gram); the recovery is almost quantitative, m.p. 106 °C.

IV,116. 2-NITRORESORCINOL

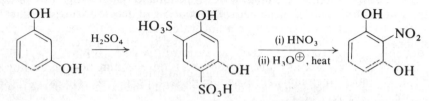

Carefully add 25 ml of concentrated sulphuric acid (98%) to 5.5 g (0.05 mol) of resorcinol contained in a 150-ml beaker whilst stirring the mixture continuously with a glass rod; then warm the mixture to 60–65 °C on a water bath and allow to stand for 15 minutes. Cool the slurry of the 4,6-disulphonic acid which is obtained to 0–10 °C and add carefully from a well-supported dropping funnel a cooled mixture of 4 ml of concentrated nitric acid (72%) and 5.6 ml of concentrated sulphuric acid. It is essential that the temperature of the reaction mixture should not be allowed to exceed 20 °C. When the addition is complete, allow the mixture to stand for a further 15 minutes, and then cautiously add 15 g of crushed ice. External cooling may be also necessary to keep the temperature below 20 °C.

Transfer the resulting yellow-brown solution to a 250-ml round-bottomed flask and steam distil (Fig. I,101); collect about 250 ml of distillate, cool and filter off the precipitated yellow-orange 2-nitroresorcinol (1). The yield of crude material is 2.2 g (28%) of m.p. 76 °C; the m.p. may be raised to 85 °C by recrystallisation from aqueous ethanol.

Note. (1) A further quantity of crude material may be obtained by extracting the filtered steam distillate with ether.

IV,117. 2,4,6-TRINITROPHENOL (*Picric acid*)

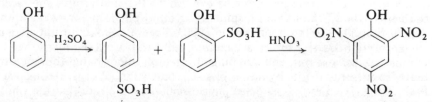

Place 9.5 g (0.1 mol) of phenol in a dry 750-ml or 1-litre flat-bottomed flask and add 23 g (12.5 ml) of concentrated sulphuric acid, shake the mixture (which

becomes warm) and heat it on a boiling water bath for 30 minutes to complete the formation of the *o*- and *p*-phenolsulphonic acids, and then cool the flask thoroughly in an ice-water mixture. Place the flask on a non-conducting surface (e.g., a wooden block or an asbestos board) in a fume cupboard, and, *whilst the phenolsulphonic acids are still a viscous syrup*, add 38 ml of concentrated nitric acid and *immediately* mix the liquids by shaking for a few seconds. Allow the mixture to stand; generally within 1 minute a vigorous but harmless reaction takes place and copious red fumes are evolved. When the reaction subsides, heat the flask in a boiling water bath for 1.5–2 hours with occasional shaking; the heavy oil, initially present, will ultimately form a mass of crystals. Add 100 ml of cold water, chill thoroughly in ice-water, filter the crystals at the pump, wash well with water to remove all the nitric acid and drain. Recrystallise from dilute alcohol (1 volume of ethanol : 2 volumes of water); about 110 ml are required. Filter off the recrystallised material and dry between filter-paper. The yield of picric acid (yellow crystals), m.p. 122 °C, is 15 g (65%).

Note. (1) It is advisable to keep the picric acid in the moist condition (containing about 10% of water) in a bottle with a cork stopper. Small quantities may be safely stored whilst dry, but this is not recommended in the interest of safety. Under no circumstances should glass stoppers be employed for potentially explosive substances, since on replacing the stopper some of the material may be ground between the stopper and the neck of the bottle and an explosion may result.

IV,118. *p*-BROMOPHENOL

$$C_6H_5OH \xrightarrow{Br_2, CS_2} (o) + p\text{-}Br\cdot C_6H_4OH$$

Equip a 500-ml three-necked flask with a reflux condenser, a mechanical stirrer and a separatory funnel. Attach to the top of the condenser a calcium chloride guard-tube leading by means of a glass tube to a funnel just immersed in a beaker holding about 150 ml of water for absorption of hydrogen bromide (compare Fig. I,68(*a*)) (1). Place 94 g (1 mol) of phenol dissolved in 100 ml of dry carbon disulphide in the flask, set the stirrer in motion and cool the flask in a mixture of ice and salt. When the temperature falls below 5 °C, add slowly (during about 2 hours) from the separatory funnel a solution of 160 g (51 ml, 1 mol) of bromine in 50 ml of carbon disulphide. Then arrange the flask for distillation under reduced pressure by inserting a Claisen still-head which incorporates a short fractionating side arm (compare Fig. I,110) into the central socket of the three-necked flask; stopper the remaining sockets. Connect a condenser set for downward distillation to the still-head and attach the device for absorbing the hydrogen bromide evolved to the side arm of the receiver adapter. Distil off the carbon disulphide at atmospheric pressure on a water bath held at 60 °C. (**CAUTION:** very low flash point, see Section **I,3**.) Remove the absorption device, insert a capillary leak and a thermometer into the Claisen still-head sockets and continue distillation under reduced pressure (oil bath). Collect two fractions: (*a*) b.p. below 145 °C/25–30 mmHg which is an inseparable mixture of *o*- and *p*-bromophenols (24–33 g), and (*b*) b.p. 145–150 °C/25–30 mmHg, which is fairly pure *p*-bromophenol. The residue in the flask contains some higher boiling 2,4-dibromophenol. The *p*-bromophenol solidifies on cooling to a solid white mass, which usually contains traces of an oil; this may be removed by spreading on a porous tile or by centrifuging. The dry crystals have m.p. 63 °C; the yield is 140–145 g (81–84%).

Note. (1) A considerable quantity of constant boiling point hydrobromic acid may be obtained by distilling these solutions.

IV,119. *o*-BROMOPHENOL

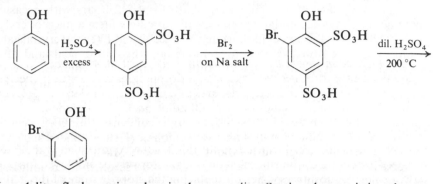

In a 1-litre flask, equipped as in the preceding Section, but omitting the gas absorption device, place a mixture of 31 g (0.33 mol) of phenol and 116 g (63 ml) of concentrated sulphuric acid, and heat in a boiling water bath for 3 hours with mechanical stirring. Cool to room temperature or below by immersing the flask in ice-water, and then add slowly a solution of 95 g of sodium hydroxide in 235 ml of water: a solid salt may separate, but this will dissolve at a later stage. Replace the separatory funnel by a thermometer, which dips well into the liquid, and support a small dropping funnel over the top of the condenser. Cool the alkaline solution to room temperature, and add 53 g (17 ml) of bromine from the dropping funnel down the condenser during 20–30 minutes whilst stirring constantly; permit the temperature to rise to 40–50 °C. Continue the stirring for 30 minutes after the bromine has been introduced: the reaction mixture should still be alkaline and contain only a small amount of suspended matter. The solution must now be evaporated. Arrange the flask assembly so that a rapid stream of air can be passed through the stirred reaction mixture, i.e., replace the thermometer by a wide air leak, connect the condenser for downward distillation using a knee bend and fit a receiver adapter with the take-off arm connected to the water pump. Heat the flask in an oil bath at 150–155 °C while maintaining a brisk current of air until a thick pasty mass remains (30–40 minutes). Allow to cool and then add 270 ml of concentrated sulphuric acid (*FUME CUPBOARD;* much hydrogen bromide is evolved). Heat the flask in an oil bath at 195–205 °C and pass a current of steam into the mixture (compare Fig. I,101); this results in the hydrolysis of the sulphonate groups and the bromophenol distils over as a heavy, colourless (or pale yellow) oil. When the distillate is clear, extract it with ether. Dry the ethereal extract with a little magnesium sulphate, remove the ether on a water bath (Fig. I,100) and distil the residue as rapidly as possible since the bromophenol is somewhat unstable and decomposes appreciably at the high temperature. Collect the fraction, b.p. 195–200 °C (a colourless liquid with a characteristic odour), which is practically pure *o*-bromophenol. The yield is 25 g (43%). The compound is somewhat unstable and decomposes on standing, becoming brown or red in colour.

IV,120. o- AND p-HYDROXYPROPIOPHENONE

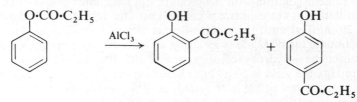

This preparation should be carried out in an efficient fume cupboard.

Equip a 1-litre three-necked flask with a dropping funnel, a sturdy mechanical stirrer and an efficient double surface reflux condenser, and place 187 g (1.4 mol) of anhydrous aluminium chloride and 200 ml of carbon disulphide in it; attach a gas absorption trap (Fig. I,68) to the top of the condenser. Stir the suspension and add 188 g (179 ml, 1.25 mol) of phenyl propionate (1) slowly and at such a rate that the solvent boils vigorously (about 90 minutes). Much hydrogen chloride is evolved and is absorbed by the trap. When all the phenyl propionate has been introduced, gently reflux the reaction mixture on a water bath until the evolution of hydrogen chloride ceases (about 2 hours). Turn the reflux condenser downwards (compare Fig. I,98), and distil off the solvent from the water bath (**CAUTION:** carbon disulphide). Then replace the latter by an oil bath maintained at 140–150 °C and heat, with stirring, for 3 hours. During this period more hydrogen chloride is evolved, the mixture thickens and finally becomes a brown resinous mass; continue the stirring as long as possible. Allow the reaction mixture to cool and decompose the aluminium chloride complex by *slowly* adding first 150 ml of dilute hydrochloric acid (1 : 1) and then 250 ml of water; much heat is evolved and a dark oil collects on the surface. Allow to stand overnight, when most of the p-hydroxypropiophenone in the upper layer solidifies. Filter this off at the pump, and recrystallise it from 200 ml of methanol; 74 g (39%) of p-hydroxypropiophenone (a pale yellow solid), m.p. 147 °C, are obtained.

Remove the methanol from the mother-liquors using a rotary evaporator and combine the residue with that obtained by extracting the original filtrate with ether and similarly evaporating. Dissolve the combined residues in 250 ml of 10 per cent sodium hydroxide solution, and extract with two 50 ml portions of ether to remove non-phenolic products. Acidify the alkaline solution with hydrochloric acid, separate the oily layer, dry it over magnesium sulphate and distil under diminished pressure, preferably from a flask with fractionating side arm (Fig. I,110). Collect the o-hydroxypropiophenone (65 g, 35%) at 110–115 °C/6 mmHg and a further quantity (20 g, 11%) of crude p-hydroxypropiophenone at 140–150 °C/11 mmHg.

Note. (1) The **phenyl propionate may be prepared** by slowly adding 196 g (120 ml) of redistilled thionyl chloride to a mixture of 150 g (1.6 mol) of pure phenol and 132 g (133 ml, 1.7 mol) of propionic acid (Fig. I,62 with the addition of a gas absorption device), warming to drive off all the sulphur dioxide and hydrogen chloride, and distilling; 190 g (79%) of phenyl propionate, b.p. 202–212 °C (the pure substance boils at 211 °C), are obtained.

Cognate preparations *o*- **and** *p*-**Hydroxyacetophenone.** Use 112 g (0.84 mol) of anhydrous aluminium chloride, 120 ml of carbon disulphide and 102 g (95 ml, 0.75 mol) of phenyl acetate (1). After acidifying and leaving overnight, dilute the partly solidified oil with toluene, and extract the aqueous layer with toluene. Dry the toluene solution with magnesium sulphate, filter and remove the toluene by distillation at atmospheric pressure. Distil the residue under reduced pressure (15–20 mmHg) until the solid *p*-isomer begins to collect in the condenser. Refractionate the distillate twice using an efficient fractionating column and finally collect the *o*-hydroxyacetophenone at 105–106 °C/20 mmHg (or at 87–88 °C/7 mmHg); the yield is 50 g (29%) based on the phenyl acetate. The *p*-hydroxyacetophenone remaining in the residues of the several distillations may be recrystallised from dilute ethanol or from benzene–light petroleum (b.p. 60–80 °C); it melts at 109 °C. Any residual *o*-hydroxyacetophenone may be removed by steam distillation; the *p*-isomer is non-volatile in steam.

Note. (1) **Phenyl acetate may be prepared** by dissolving 94 g (1 mol) of phenol in 640 ml of 10 per cent sodium hydroxide solution contained in a large (e.g., a Winchester quart) bottle, adding about 700 g of crushed ice followed by 130 g (120 ml, 1.27 mol) of acetic anhydride and shaking vigorously for 5 minutes. Separation of the resulting emulsion is facilitated by adding about 40 ml of carbon tetrachloride. After washing the lower layer carefully with dilute sodium carbonate solution until effervescence ceases and drying (anhydrous calcium chloride), distillation affords 130 g (95%) of phenyl acetate, b.p. 194–197 °C.

2,5-Dihydroxyacetophenone. Finely powder a mixture of 40 g (0.2 mol) of dry hydroquinone diacetate (1) and 87 g (0.65 mol) of anhydrous aluminium chloride in a glass mortar and introduce it into a 500-ml round-bottomed flask, fitted with an air condenser protected by a calcium chloride tube and connected to a gas absorption trap (Fig. I,68). Immerse the flask in an oil bath and heat slowly so that the temperature reaches 110–120 °C at the end of about 30 minutes: the evolution of hydrogen chloride then begins. Raise the temperature slowly to 160–165 °C and maintain this temperature for 3 hours. Remove the flask from the oil bath and allow to cool. Add 280 g of crushed ice followed by 20 ml of concentrated hydrochloric acid in order to decompose the excess of aluminium chloride. Filter the resulting solid with suction and wash it with two 80 ml portions of cold water. Recrystallise the crude product from 200 ml of 95 per cent ethanol. The yield of pure 2,5-dihydroxyacetophenone, m.p. 202–203 °C, is 23 g (58%).

Note. (1) **Hydroquinone diacetate may be prepared** as follows. Add 1 drop of concentrated sulphuric acid to a mixture of 55 g (0.5 mol) of hydroquinone and 102 g (95 ml, 1 mol) of acetic anhydride in a 500-ml conical flask. Stir the mixture gently by hand; it warms up rapidly and the hydroquinone dissolves. After 5 minutes, pour the clear solution on to 400 ml of crushed ice, filter with suction and wash with 500 ml of water. Recrystallise the solid from 50 per cent aqueous ethanol (*ca.* 400 ml are required). The yield of pure hydroquinone diacetate, m.p. 122 °C, is 89 g (91%).

IV,121. 2-ALLYLPHENOL AND 2-(PROP-1-ENYL)PHENOL

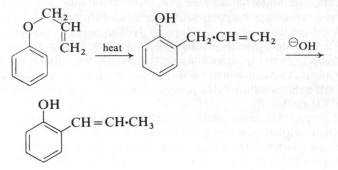

2-Allylphenol. Boil 50 g of allyl phenyl ether (Section **IV,123**) gently in a round-bottomed flask fitted with an air reflux condenser. Determine the refractive index of the mixture at intervals; the rearrangement is complete and the boiling is stopped when the refractive index (n_D^{25}) has risen to 1.55 (about 6 hours are required). Dissolve the product in 100 ml of 5 M-sodium hydroxide solution and extract with two 30 ml portions of light petroleum (b.p. 40–60 °C) which removes the small amount of 2-methyldihydrobenzofuran formed as by-product and which is neutral. Carefully acidify the alkaline solution with 5 M-hydrochloric acid with cooling and extract the mixture with one 50 ml portion and two 25 ml portions of ether. Dry the extract over anhydrous sodium sulphate and remove the ether with a rotary evaporator. Distil the residue under reduced pressure to give 2-allylphenol, b.p. 103–106 °C/19 mmHg or b.p. 96 °C/13 mmHg; n_D^{25} 1.5440. The yield is 35 g (70%).

Rearrangement to 2-(prop-1-enyl)phenol. Prepare a saturated solution (about 50% w/v) of potassium hydroxide in 60 ml of methanol. Place this solution together with 20 g of 2-allylphenol in a round-bottomed flask fitted with a still-head and condenser set for downward distillation. Arrange a thermometer so that the bulb dips into the mixture and distil the latter slowly until the temperature reaches 110 °C. Remove the still-head, attach an air condenser to the flask and boil the reaction mixture gently under reflux for 6 hours. Cool the mixture, cautiously acidify with concentrated hydrochloric acid and extract the product with three 30-ml portions of ether. Dry the extract, remove the ether and distil the residue under reduced pressure. Collect the 2-(prop-1-enyl)phenol as a fraction, b.p. 110–115 °C/15 mmHg; it crystallises on cooling in an ice bath, yield 15 g (75%). Recrystallisation from dry light petroleum (b.p. 60–80 °C) gives shining needles, m.p. 37 °C.

J,3. Phenyl ethers. Examples of the preparation of alkyl phenyl ethers by the Williamson synthesis are included on p. 413, **III,G,2**. A further example is provided by the synthesis of phenacetin (Section **IV,122**) where p-aminophenol is first converted into its N-acetyl derivative by reaction with slightly more than one equivalent of acetic anhydride. Treatment of the product with ethanolic sodium ethoxide solution followed by ethyl iodide then yields the ethyl ether of N-acetyl-p-phenetidine (phenacetin). This compound is biologically active and has been widely employed for example as an antipyretic and analgesic: however, owing to undesirable side reactions, its use is now restricted.

The initial preparation of the sodium derivative of the phenol by treatment with sodium ethoxide may be avoided in a number of instances by heating directly the phenol, the alkyl halide and anhydrous potassium carbonate in acetone solution. Examples are provided by the preparation of allyl phenyl ether and butyl 2-nitrophenyl ether (Section **IV,123**). A further cognate preparation, that of 2,4-dichlorophenoxyacetic acid, is of interest since the product is an important plant growth hormone and selective weed-killer. The conversion of phenols into phenoxyacetic acids by this route is of value in that these crystalline derivatives are useful for the characterisation of the phenolic compounds (Section **VII,6,**6(e)).

Conversion of phenols into their methyl or ethyl ethers by reaction with the corresponding alkyl sulphates in the presence of aqueous sodium hydroxide affords a method which avoids the use of the more expensive alkyl halides (e.g., the synthesis of anisole, phenetole, methyl 2-naphthyl ether and veratraldehyde, Section **IV,124**).

IV,122. PHENACETIN

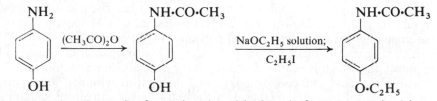

Suspend 11 g (0.1 mol) of p-aminophenol in 30 ml of water contained in a 250-ml beaker or conical flask and add 12 ml (0.127 mol) of acetic anhydride. Stir (or shake) the mixture vigorously and warm on a water bath. The solid dissolves. After 10 minutes, cool, filter the solid acetyl derivative at the pump and wash with a little cold water. Recrystallise from hot water (about 75 ml) and dry upon filter-paper in the air. The yield of p-hydroxyacetanilide, m.p. 169 °C (1), is 14 g (93%).

Place 1.55 g (0.0675 mol) of clean sodium in a 250-ml round-bottomed flask equipped with a reflux condenser. Add 40 ml of absolute alcohol (or rectified spirit). If all the sodium has not disappeared after the vigorous reaction has subsided, warm the flask on a water bath until solution is complete. Cool the mixture and add 10 g (0.066 mol) of p-hydroxyacetanilide. Introduce 15 g (8 ml, 0.1 mol) of ethyl iodide slowly through the condenser and reflux the mixture for 45–60 minutes. Pour 100 ml of water through the condenser at such a rate that the crystalline product does not separate; if crystals do separate, reflux the mixture until they dissolve. Then cool the flask in an ice bath: collect the crude phenacetin with suction and wash with a little cold water. Dissolve the crude product in 80 ml of rectified spirit; if the solution is coloured, add 2 g of decolourising carbon, boil and filter. Treat the clear solution with 125 ml of hot water and allow to cool. Collect the pure phenacetin at the pump and dry in the air. The yield is 9.5 g (80%), m.p. 137 °C.

Note. (1) If the m.p. is unsatisfactory, dissolve the product in dilute alkali in the cold and then reprecipitate it by the addition of acid to the neutralisation point. This procedure will eliminate traces of the diacetate of p-aminophenol which may be present; the acetyl group attached to nitrogen is not affected by cold dilute alkali, but that attached to oxygen is readily hydrolysed by the reagent.

IV,123. ALLYL PHENYL ETHER

$$C_6H_5OH + BrCH_2 \cdot CH = CH_2 \xrightarrow{K_2CO_3} C_6H_5 \cdot O \cdot CH_2 \cdot CH = CH_2$$

Place 47 g (0.5 mol) of phenol, 60.5 g (0.5 mol) of allyl bromide (Section **III,54**), 69.1 g (0.5 mol) of anhydrous potassium carbonate and 100 ml of acetone in a 250-ml, two-necked round-bottomed flask fitted with a reflux condenser and sealed stirrer unit, and boil on a steam bath for 8 hours with stirring. Pour the reaction mixture into 500 ml of water, separate the organic layer and extract the aqueous layer with three 20 ml portions of ether. Wash the combined organic layer and ether extracts with 2 *M*-sodium hydroxide solution, and dry over anhydrous potassium carbonate. Remove the ether with a rotary evaporator and distil the residue under reduced pressure. Collect the allyl phenyl ether, b.p. 85 °C/19 mmHg; the yield is 57 g (85%).

Cognate preparations   Butyl 2-nitrophenyl ether (o-*Butoxynitrobenzene*). Place a mixture of 28 g (0.2 mol) of o-nitrophenol (Section **IV,114**), 28 g (0.2 mol) of anhydrous potassium carbonate, 30 g (23.5 ml, 0.22 mol) of butyl bromide and 200 ml of dry acetone in a 1-litre round-bottomed flask fitted with an efficient reflux condenser, and reflux on a steam bath for 48 hours. Distil off the acetone, add 200 ml of water and extract the product with two 100 ml portions of benzene. Wash the combined benzene extracts with three 90 ml portions of 10 per cent sodium hydroxide solution, remove the benzene by distillation at atmospheric pressure and distil the residue under reduced pressure. Collect the o-butoxynitrobenzene at 171–172 °C/19 mmHg (or at 127–129 °C/2 mmHg); the yield is 30 g (77%).

2,4-Dichlorophenoxyacetic acid.

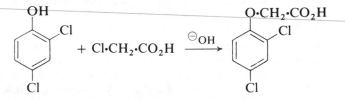

Place 8.1 g (0.05 mol) of 2,4-dichlorophenol and 4.7 g (0.05 mol) of chloroacetic acid (1) in a 400-ml beaker; add slowly, with stirring, a solution of 4.5 (0.112 mol) of sodium hydroxide in 25 ml of water. Considerable heat is developed during the reaction. Heat the reaction mixture on a wire gauze until most of the liquid has evaporated: treat the residue with 150 ml of water, cool and filter if necessary. Acidify the clear solution with dilute hydrochloric acid (litmus). Extract the dense oil which separates with two 25 ml portions of ether, wash the combined extracts with 10–15 ml of water, dry with 1 g of magnesium sulphate, filter through a fluted filter-paper and distil off the ether on a rotary evaporator. Recrystallise the residue of crude, 2,4-dichlorophenoxyacetic acid from about 35 ml of benzene (fume cupboard). The yield of pure acid, m.p. 138 °C, is 6.0 g (54%).

Note. (1) **CAUTION:** chloroacetic acid must be handled with great care as it causes blisters on the skin.

IV,124. METHYL PHENYL ETHER (*Anisole*)

$$C_6H_5OH + NaOH + (CH_3)_2SO_4 \longrightarrow$$
$$C_6H_5 \cdot O \cdot CH_3 + Na(CH_3)SO_4 + H_2O$$

This preparation should be carried out in an efficient fume cupboard.

Equip a 500-ml three-necked flask with a separatory funnel, a sealed mechanical stirrer and a reflux condenser. Place 47 g (0.5 mol) of pure phenol in the flask, add a solution of 21 g of sodium hydroxide in 200 ml of water and stir the mixture; cool the warm mixture to about 10 °C by immersing the flask in an ice bath. Place 63 g (47 ml, 0.5 mol) of dimethyl sulphate (**CAUTION:** see Section **II,2,**_18(a)_) in the separatory funnel and add dropwise during 1 hour whilst stirring the mixture vigorously. Then heat under reflux for 2 hours with stirring in order to complete the methylation. Allow to cool, add 200 ml of water, transfer to a separatory funnel. Separate the anisole layer and extract the residual aqueous solution with several portions (50 ml) of ether. Wash the combined organic phases once with water, twice with dilute sulphuric acid and then with water until the washings are neutral to litmus. Dry over anhydrous calcium chloride or magnesium sulphate. Remove the ether by flash distillation (Fig. I,100) and distil the residue from an air bath. Collect the anisole at b.p. 151–154 °C. The yield is 40 g (74%).

Cognate preparations **Ethyl phenyl ether** (*phenetole*). Proceed as for *anisole* using the following quantities: 21 g of sodium hydroxide in 200 ml of water, 47 g (0.5 mol) of pure phenol and 77 g (65.5 ml, 0.5 mol) of diethyl sulphate (Section **II,2,**_18(b)_). After all the diethyl sulphate has been introduced, heat the mixture under reflux for 2 hours with stirring. Transfer the diluted reaction mixture to a separatory funnel, run off the aqueous layer, wash successively with water, diluted sulphuric acid (twice) and with water until the washings are neutral. Dry over anhydrous calcium chloride, remove the desiccant by filtration and distil. Collect the ethyl phenyl ether (a colourless liquid) at b.p. 168–170 °C. The yield is 50 g (82%).

Methyl 2-naphthyl ether (*nerolin*). Use 36.0 g (0.25 mol) of 2-naphthol, 10.5 g of sodium hydroxide in 150 ml of water, and add 31.5 g (23.5 ml, 0.25 mol) of dimethyl sulphate (**CAUTION**) whilst the mixture is cooled in ice. Warm for 1 hour at 70–80 °C, and allow to cool. Filter off the methyl naphthyl ether at the pump, wash with 10 per cent sodium hydroxide solution, then liberally with water, and drain thoroughly. Recrystallise from industrial spirit. The yield is 33 g (84%), m.p. 72 °C.

Veratraldehyde (3,4-dimethoxybenzaldehyde). Place 152 g (1 mol) of a good sample of commercial vanillin, m.p. 81–82 °C, in a 1-litre three-necked flask equipped with a reflux condenser, a mechanical stirrer and two separatory funnels (one of which is supported over the top of the reflux condenser). Melt the vanillin by warming on a water bath and stir vigorously. Charge one funnel with a solution of 82 g of pure potassium hydroxide in 120 ml of water and the other funnel with 160 g (120 ml, 1.04 mol) of purified dimethyl sulphate (**CAUTION**). Run in the potassium hydroxide solution at the rate of two drops a second, and 20 seconds after this has started add the dimethyl sulphate at the same rate. Stop the external heating after a few minutes; the mixture continues to reflux gently from the heat of the reaction. The reaction mixture should be pale reddish-brown since this colour indicates that it is alkaline; should the colour change to green, an acid reaction is indicated and this condition should be corrected by slightly increasing the rate of addition of the alkali. When half to three-quarters of the reagents have been added, the reaction mixture becomes turbid and separates into two layers. As soon as all the reagents have been run in (about 20 minutes), pour the yellow reaction mixture into a large porcelain basin and allow to cool without disturbance, preferably overnight. Filter the hard crystalline mass of veratraldehyde, grind it in a glass mortar with 300 ml of ice-cold water, filter at the pump and dry in a vacuum desiccator. The yield of veratraldehyde, m.p. 43–44 °C, is 160 g (96%). This product is sufficiently pure for most purposes; it can be purified without appreciable loss by distillation under reduced pressure, b.p. 158 °C/8 mmHg; m.p. 46 °C. The aldehyde is easily oxidised in the air and should therefore be kept in a tightly stoppered bottle.

K AROMATIC ALDEHYDES

Aromatic aldehydes may be prepared by the following general procedures.

1. Electrophilic substitution in an aromatic ring leading to the introduction of the formyl group (H − C = O) (Sections IV,125 to IV,129).

2. Reactions involving modification of substituents attached to the aromatic ring (Sections IV,130 to IV,133).

K,1. Formylation reactions. Aromatic aldehydes may be obtained by passing a mixture of carbon monoxide and hydrogen chloride into the aromatic hydrocarbon in the presence of a mixture of copper(I) chloride and aluminium chloride which acts as a catalyst (**Gatterman–Koch reaction**). An example is the preparation of p-tolualdehyde (Section IV,125). It is probable that the electrophilic species is the formyl cation [H − $\overset{\oplus}{C}$ = O] formed from the mixture of gases in the presence of the Lewis acid.

$$\text{ArH} + [\text{H} - \overset{\oplus}{\text{C}} = \text{O}] \xrightarrow[\text{CuCl}]{\text{AlCl}_3} \text{Ar·CHO} + \text{H}^{\oplus}$$

The Gatterman–Koch formylation is unsuitable for the preparation of aldehydes from phenols and phenolic ethers owing to the formation of complexes with the Lewis acid.

An alternative method is the **Gattermann aldehyde** synthesis which employs hydrogen cyanide in place of carbon monoxide. Unlike the Gattermann–Koch procedure, this method is successful in the case of phenols and phenolic ethers but only the more reactive hydrocarbons are formylated in good yield. The mechanism probably involves the species [H − $\overset{\oplus}{C}$ = NH] as the effective electrophile.

$$\text{ArH} + [\text{H} - \overset{\oplus}{\text{C}} = \text{NH}] \xrightarrow[\text{or ZnCl}_2]{\text{AlCl}_3} [\text{Ar·CH} = \text{NH}] \xrightarrow{\text{H}_2\text{O}} \text{Ar·CHO}$$

The use of the hazardous hydrogen cyanide may be avoided by passing dry hydrogen chloride either into a mixture of zinc cyanide, aluminium chloride, the hydrocarbon or phenolic ether and a solvent (such as tetrachloroethane or benzene), or into a mixture of zinc cyanide, the phenol and anhydrous ether or benzene. The zinc cyanide is converted by the hydrogen chloride into hydrogen cyanide (which reacts *in situ*) and zinc chloride (which is known to be an effective catalyst in this reaction). The preparation of 2,4,6-trimethylbenzaldehyde (mesitaldehyde) from mesitylene and the related cognate preparations (Section **IV,126**) provides a varied range of examples.

Phenols are smoothly converted into phenolic aldehydes by reaction with chloroform in the presence of base (the **Reimer–Tiemann reaction**). This overall formylation reaction is of interest in that it involves the generation from chloroform and alkali of the reactive intermediate, dichlorocarbene (I). This effects electrophilic substitution in the reactive phenolate ions giving the benzylidene dichloride (II) which is hydrolysed by the alkaline medium to the corresponding hydroxyaldehyde. The phenolic aldehyde is isolated from the reaction media after acidification.

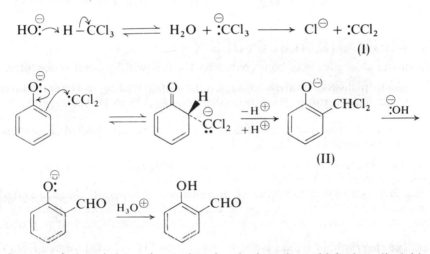

$$\text{HO}^{\ominus} \rightharpoonup \text{H} - \text{CCl}_3 \rightleftharpoons \text{H}_2\text{O} + {}^{\ominus}\text{:CCl}_3 \longrightarrow \text{Cl}^{\ominus} + \text{:CCl}_2$$

(I)

(II)

In the case of phenol the main product is *o*-hydroxybenzaldehyde (salicylaldehyde, Section **IV,127**), but some of the *para* isomer is also formed. In the cognate preparation (2-hydroxy-1-naphthaldehyde) the preferential reaction at the 1-position should be noted.

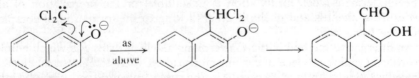

Certain reactive aromatic hydrocarbons are formylated by dimethylformamide in the presence of phosphorus oxychloride (the **Vilsmeier reaction**, e.g., 9-formylanthracene, Section **IV,128**). This method can also be used with advantage for the formylation of π-excessive heteroaromatic systems (e.g., 2-formylthiophen, cognate preparation in Section **IV,128**).

A generally applicable method of formylation involves the reaction of an aromatic hydrocarbon and dichloromethyl methyl ether under Friedel–Crafts conditions (cf. p. 770, **IV,L,1**). The intermediate chloroacetal (III) thus formed

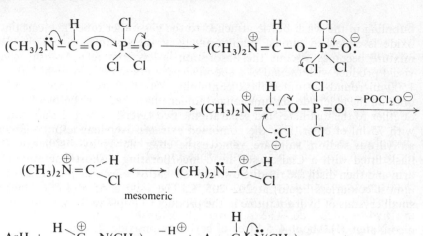

is readily hydrolysed to the corresponding aldehyde (e.g., *p*-t-butylbenzaldehyde, Section **IV,129**).

$$ArH + Cl_2CHOCH_3 \xrightarrow[\text{or } SnCl_4]{TiCl_4} Ar \cdot CH(OCH_3)Cl + HCl \xrightarrow{H_2O}$$

$$\text{(III)}$$

$$Ar \cdot CHO + CH_3OH + HCl$$

The procedure is also of value for the formylation of polycyclic aromatic and heteroaromatic systems, phenols and phenolic ethers.

IV,125. *p*-TOLUALDEHYDE

$$C_6H_5 \cdot CH_3 + CO + HCl \xrightarrow[CuCl]{AlCl_3} p\text{-}H_3C \cdot C_6H_4 \cdot CHO$$

In an *efficient fume cupboard* equip a 500-ml three-necked flange flask with a gas inlet tube extending to the bottom of the flask, a sealed mechanical stirrer and an outlet tube connected to two Drechsel wash bottles, the first acting as a trap and the other containing concentrated sulphuric acid. Immerse the flask in a water bath held at 20 °C, charge the flask with 92 g (106 ml, 1 mol) of pure toluene, start the stirrer and add 14 g of dry cuprous chloride (Section **II,2,***16*), and 133 g (1 mol) of powdered anhydrous aluminium chloride through the outlet tube socket. Pass a mixture of carbon monoxide (Section **II,2,***10*) and hydrogen chloride (Section **II,2,***31*), by connecting the sulphuric acid wash bottles of both gas generating systems to the gas inlet tube using a Y-tube, through the reaction mixture at a uniform but not too rapid rate (1) during 7 hours; adjust the rates of flow so that the volume of carbon monoxide is about twice that of the hydrogen chloride by observing the bubbling in the wash bottles of the gas generators. The rate of absorption can be estimated from the bubbling in the wash bottle attached to the gas outlet tube. The carbon oxide is absorbed almost quantitatively at the commencement, but as mixture becomes viscous the absorption is less complete. Transfer the v viscid product with the aid of a spoon spatula gradually and with shaking to 1500-ml round-bottomed flask containing 750 g of crushed ice. Steam disti until all the aldehyde and unchanged toluene have been driven over. Add 25 ml of ether to the distillate, and separate the two layers; extract the aqueous layer with 75 ml of ether. Dry the combined extracts over magnesium sulphate or anhydrous sodium sulphate, remove the ether slowly by distillation from a flask fitted with a Claisen still-head incorporating a short fractionating side arm and then distil the residue using an air bath. Collect the *p*-tolualdehyde (an almost colourless liquid) at 202–205 °C. The yield is 55 g (46%). Place a few small crystals of hydroquinone in the product to improve the keeping qualities.

Note. (1) About 4–5 litres of carbon monoxide should be passed in the course of an hour.

IV,126. 2,4,6-TRIMETHYLBENZALDEHYDE (*Mesitaldehyde*)

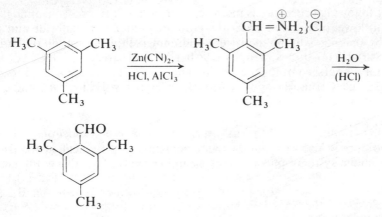

The entire preparation should be conducted in an efficient fume cupboard. Tetrachloroethane is a potent poison and hydrogen cyanide is liberated during the reaction.

Equip a 500-ml multi-necked flange flask with a reflux condenser, an eff stirrer, a gas-inlet tube and a thermometer; the thermometer should that during the subsequent reaction the bulb is well immersed (59 ml, 0.425 the gas inlet tube need extend only just below the surface. Pl in the flask, and mol) of redistilled mesitylene (b.p. 163–166 °C), 73. oride (S II,2,31) nide (Section **II,2,68**) and 200 ml of 1,1,2,2-tet mposed (ai stir the mixture while a rapid stream of d 1,64, having pre is passed through it until the zinc nd, anhydrous alu Immerse the flask in a bath of the aluminium chlo by means of the arrangem reaction the passage the conical flask with reaction will raise the te Stir the mixture of 10 minu ide gas f

 at the end of an hour. Maintain the temperature at 67–72 °C for the
ainder of the reaction period. Cool, and pour the reaction mixture, with
nd stirring, into a 2-litre beaker about half-full of crushed ice to which 50 ml
 concentrated hydrochloric acid has been added. Allow to stand overnight,
ransfer to a 1.5-litre round-bottomed flask, fit a condenser and reflux for 3
hours. Allow to cool, separate the organic layer and extract the aqueous layer once
with 25 ml of tetrachloroethane. Wash the combined tetrachloroethane solutions
with 75 ml of 10 per cent sodium carbonate solution, and steam distil (Fig. I,101).
Set the first 400–450 ml of distillate aside for the recovery of the solvent (1), and
collect the second portion (about 4.5 litres) as long as oily drops pass over. Extract
the distillate with 250 ml of toluene, dry the extract with a little magnesium
sulphate and remove the solvent on an oil bath or a rotary evaporator. Distil the
residue from a 150-ml flask with a fractionating side arm (compare Fig. I,110),
and collect the mesitaldehyde at 118–121 °C/16 mmHg; the yield is 50 g (2).

Notes. (1) The first portion of the steam distillate consists almost entirely
of tetrachloroethane and water. The solvent is recovered by separating the
organic layer, drying with anhydrous calcium chloride or magnesium sulphate
and distilling.

(2) The following procedure is more convenient and less time-consuming, but
the yield is lower (about 40 g). Mix the powdered aluminium chloride and zinc
cyanide by shaking, add the mesitylene and immerse the flask in an oil bath at
100 °C. Stir the mixture and pass in a fairly rapid stream of dry hydrogen
chloride for 4 hours; continue the heating and stirring for a further 2 hours, but
discontinue the passage of the gas. Decompose the reaction mixture, and complete the preparation as above.

Cognate preparations **2,4-Dihydroxybenzaldehyde** (*β-resorcylaldehyde*). Equip
a 500-ml three-necked flask with a reflux condenser, an efficient sealed stirrer
and a wide inlet tube (to prevent clogging by the precipitate) extending nearly
to the bottom of the vessel. Attach the inlet tube to an empty (safety) wash
bottle and to this a generator producing hydrogen chloride (Section II,2,*31*);
connect the top of the condenser by means of a tube to a wash bottle containing concentrated sulphuric acid, then to an empty bottle, and finally to the
surface of sodium hydroxide solution (Fig. I,68(*a*)). Place 20 g (0.18 mol) of
resorcinol, 175 ml of sodium-dried ether and 40 g (0.34 mol) of powdered
anhydrous zinc cyanide in the flask, start the stirrer and pass in a rapid stream
tion ʰʳogen chloride. The zinc cyanide gradually disappears with the forma-
ation of ᵃ cloudy solution; further passage of hydrogen chloride results in the separ-
solidifies aʳ mine hydrochloride condensation product as a thick oil which
(after about ½ minutes. When the ether is saturated with hydrogen chloride
a further half an solid pass the gas more slowly and continue the stirring for
ether from the hot so nsure the completeness of the reaction. Decant the
point, filter the resorcy d 100 ml of water to the latter, heat to the boiling
to cool. Filter the filtrate t a hot water funnel and allow the filtrate
is cold; allow the obtained (total yield which separates as soon as the mixture
aldehyde is 135–136 °C and is very fair ours when a further 11.5 g of the
p. tallisation from hot water with rcylaldehyde, after drying, has
olour may be removed by
le decolourising carbon.

2-Hydroxy-1-naphthaldehyde. Proceed as for β-resorcylaldehyde e
that 20 g (0.138 mol) of 2-naphthol replaces the resorcinol. Recrystallise
crude product (20 g, 83%) from water with the addition of a little decolouris
carbon; the pure aldehyde has m.p. 80–81 °C.

Anisaldehyde (p-methoxybenzaldehyde). Use the apparatus described fo
β-resorcylaldehyde. Place 30 g (30 ml, 0.28 mol) of anisole (Section **IV,124**), 75
ml of sodium-dried benzene and 52 g (0.44 mol) of powdered zinc cyanide in
the flask. Cool the mixture in a bath of cold water, start the stirrer and pass in a
rapid stream of hydrogen chloride for 1 hour. Remove the gas inlet tube, and
without stopping the stirrer, add 45 g of finely powdered anhydrous aluminium
chloride slowly. Replace the gas inlet and pass in a slow steam of hydrogen
chloride whilst heating the mixture at 40–45 °C for 3–4 hours. Allow to cool
somewhat and pour the reaction mixture with stirring into excess of dilute
hydrochloric acid; the imine hydrochloride separates as a heavy precipitate.
Reflux the mixture for half an hour in order to decompose the imine hydrochloride and steam distil. Separate the organic layer in the distillate, dry with a
little anhydrous magnesium sulphate and distil off the benzene. Continue distillation with an air bath and collect the anisaldehyde as a fraction which has a
b.p. 246–248 °C; the yield is 35 g (92%). If required the product may be redistilled under reduced pressure, b.p. 134–135 °C/12 mmHg.

IV,127. SALICYLALDEHYDE

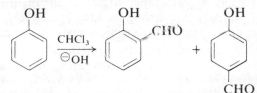

Equip a 1-litre three-necked flask with an efficient double surface reflux condenser, a mechanical stirrer and a thermometer, the bulb of which is within
2 cm of the bottom of the flask. Place a warm solution of 80 g of sodium
hydroxide in 80 ml of water in the flask, add a solution of 25 g (0.266 mol)
of phenol in 25 ml of water and stir. Adjust the temperature inside the flask
to 60–65 °C (by warming on a water bath or by cooling, as may be found
necessary); do not allow the crystalline sodium phenoxide to separate out.
Introduce 60 g (40.5 ml, 0.5 mol) of chloroform in three portions at intervals of
15 minutes down the condenser. Maintain the temperature of the well-stirred
mixture at 65–70 °C during the addition by immersing the flask in hot or cold
water as may be required. Finally heat on a boiling water for 1 hour to
complete the reaction. Remove the excess of chloroform, acidify the alkaline
solution by steam distillation (Fig. I,101). Allow the orange-
coloured liquid cautiously with dilute sulphur are collected.
almost colourless liquid until no more oxybenzaldehyde
residue in the flask for the isolation of the ether from t
distillate at once with ether, remove evaporator. Transf
distillation on a water bath using aldehyde, to a small
which contains phenol as well saturated sodium metabis
flask, add about twice the v mechanically) for at least I
and shake vigorously (r

to stand for 1 hour. Filter the paste of bisulphite compound at the pump, ⌐ it with a little alcohol, and finally with a little ether (to remove the ⌐nol). Decompose the bisulphite compound by warming in a round-bot- ⌐med flask on a water bath with dilute sulphuric acid, allow to cool, extract ⌐e salicylaldehyde with ether and dry the extract with anhydrous magnesium sulphate. Remove the ether by flash distillation and distil the residue collecting the salicylaldehyde (a colourless liquid) at 195–197 °C. The yield is 12 g (37%).

To isolate the **p-hydroxybenzaldehyde**, filter the residue from the steam distil- lation while hot through a fluted filter-paper in order to remove resinous mat- ter, and extract the cold filtrate with ether. Distil off the ether, and recrystallise the yellow solid from hot water to which some aqueous sulphurous acid is added. The yield of p-hydroxybenzaldehyde (colourless crystals), m.p. 116 °C, is 2–3 g (6–9%).

Cognate preparation 2-Hydroxy-1-naphthaldehyde. Equip a 1-litre three- necked flask with a separatory funnel, a sealed mechanical stirrer and a double surface reflux condenser. Place 50 g of 2-naphthol and 150 ml of rectified spirit in the flask, start the stirrer and rapidly add a solution of 100 g of sodium hydroxide in 210 ml of water. Heat the resulting solution to 70–80 °C on a water bath, and place 62 g (42 ml) of pure chloroform in the separatory funnel. Introduce the chloroform dropwise until reaction commences (indicated by the formation of a deep blue colour), remove the water bath and continue the addition of the chloroform at such a rate that the mixture refluxes gently (about 1.5 hours). The sodium salt of the phenolic aldehyde separates near the end of the addition. Continue the stirring for a further 1 hour. Set the condenser for downward distillation (but retaining the stirrer) and distil off the excess chloro- form and alcohol. Treat the residue, with stirring, dropwise with concentrated hydrochloric acid until the contents of the flask are acid to Congo red paper (about 88 ml are required); a dark oil, accompanied by a considerable amount of sodium chloride, separates. Add sufficient water to dissolve the salt, extract the oil with ether, wash the ethereal solution with water, dry with anhydrous magnesium sulphate and remove the solvent. Distil the residue under re- duced pressure and collect the slightly coloured aldehyde at 177–180 °C/20 mmHg; it solidifies on cooling. Recrystallise the solid from about 40 ml of ethanol. The yield of 2-hydroxy-1-naphthaldehyde, m.p. 80 °C, is 28 g (47%).

IV,28. 9-FORMYLANTHRACENE (9-Anthraldehyde)

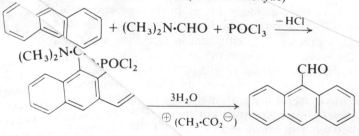

$$+ (CH_3)_2N\cdot CHO + POCl_3 \xrightarrow{-HCl}$$

$$\xrightarrow[\oplus (CH_3\cdot CO_2^{\ominus})]{3H_2O} \quad + (CH_3)_2NH + H_2PO_4^{\ominus} + 3Cl^{\ominus}$$

Equip a 500-ml three-necked flas⌐ ⌐er and a dropping funnel. Assemble sealed stirrer unit, a reflux con- ⌐atus on a water bath in a fume

cupboard. Place in the flask a mixture of 17.8 g (0.1 mol) of anthracene (1), ⌐
(20 ml, 0.26 mol) of dimethylformamide and 20 ml of o-dichlorobenzene (
and charge the dropping funnel with 27 g (16 ml, 0.175 mol) of phosphoru
oxychloride; close the condenser and dropping funnel with calcium chloride
guard-tubes. Start the stirrer, run in the phosphorus oxychloride steadily and
then heat on a boiling water bath for 2 hours. Cool the reaction flask in an
ice–salt bath and neutralise the contents to Congo red by running in aqueous
sodium acetate solution (about 100 g of the trihydrate in 175 ml of water are
required). Dilute with more water to about 2 litres and allow the mixture to
stand at 0 °C for 2 hours. Filter off the yellow crystalline product and recrystallise it from aqueous acetic acid; the yield of 9-formylanthracene is 12 g
(58%), m.p. 104 °C.

Notes.   (1) Good quality material should be used; commercial fluorescent
grade of m.p. *ca.* 215 °C is suitable.

(2) The use of o-dichlorobenzene as a solvent is recommended. If the reaction
is carried out in excess dimethylformamide alone, the product is contaminated
with unreacted anthracene. It is then best to extract the crude material with
cold methanol, remove the anthracene by filtration and recover the product by
dilution with water.

Cognate preparation   2-Formylthiophen (*thiophen-2-aldehyde*).   Use 21 g (19.3
ml, 0.25 mol) of thiophen, 23 g (24 ml, 0.315 mol) of dimethylformamide and 80
ml of 1,2-dichloroethane as solvent. Cool to 0 °C, add 48 g (29 ml, 0.313 mol) of
phosphorus oxychloride slowly with stirring, and then heat, carefully at first,
and then under reflux for 2 hours. Cool, pour on to crushed ice, neutralise with
sodium acetate (*ca.* 200 g of the hydrate), separate the organic phase and extract the aqueous phase with ether. Wash the combined organic phases with
aqueous sodium hydrogen carbonate, dry over magnesium sulphate and remove the solvent on a rotary evaporator. Distil the residue under reduced
pressure and collect the 2-formylthiophen as a fraction of b.p. 85–86 °C/16
mmHg; yield 20 g (71%).

IV,129. *p*-t-BUTYLBENZALDEHYDE

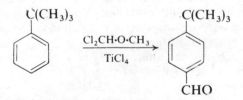

Equip a 250-ml three-necked flask with a thermometer, reflux condenser,
dropping funnel (protected with a calcium chloride guard-tube) and magnetic
stirrer, and attach a gas absorption trap to the top of the condenser; assemble
the apparatus in the fume cupboard. Place 15.1 g (0.12 mol) of t-butylbenzene
(Section **IV,6**) and 60 ml of dry methylene chloride (Section **II,1,5**) in the flask
and cool to 0–5 °C in an ice–salt bath. To the stirred solution add 38 g (0.2
0.2 mol) of titanium tetrachloride rapidly from the dropping funnel.

ates); the mixture becomes orange. Then add 11.5 g (0.1 mol) of dichloro-
.hyl methyl ether (Section **III,77**) during 20 minutes to the stirred and cooled
.ution. Hydrogen chloride is evolved after the first few drops of the ether are
.dded. Stir the mixture for 5 minutes after completion of the addition, remove
.he cooling bath, allow the mixture to warm to room temperature (about half
an hour) and then heat at 35 °C for 15 minutes. Pour the mixture into a
separating funnel containing 100 g of ice and shake thoroughly. Separate the
lower organic layer and extract the aqueous layer with three 25 ml portions of
methylene chloride. Wash the combined methylene chloride extracts with three
25 ml portions of water, add a crystal of hydroquinone to prevent oxidation of
the aldehyde and dry over magnesium sulphate. Filter the solution, remove the
solvent by flash distillation and distil the residue under reduced pressure
through a short fractionating column. The fraction, which distils at 52 °C/4
mmHg, is p-t-butylbenzene; collect the p-t-butylbenzaldehyde as a fraction of
b.p. 98 °C/4 mmHg; the yield is 10.8 g (67%).

K,2. Side-chain modifications. Several procedures for the synthesis of
aromatic aldehydes are available which involve the selective oxidation of a
methyl group attached to an aromatic ring. A useful general reagent is a solution of chromium trioxide in acetic anhydride and acetic acid. The aldehyde
is converted into the *gem*-diacetate as it is formed and is thus protected from
further oxidation. The aldehyde is liberated from the diacetate by hydrolysis
under acid conditions; the yields, however, are frequently only moderate (e.g.,
p-nitrobenzaldehyde, Section **IV,130**).

$$Ar \cdot CH_3 \xrightarrow[(CH_3 \cdot CO)_2O]{CrO_3} Ar \cdot CH(O \cdot CO \cdot CH_3)_2 \xrightarrow{H_3O^{\oplus}} Ar \cdot CHO$$

Aldehydes may also be obtained by the hydrolysis of *gem*-dihalogen compounds obtained by the side-chain halogenation of a methylarene.

$$Ar \cdot CH_3 \xrightarrow{2Br_2} Ar \cdot CHBr_2 \xrightarrow{H_2O} Ar \cdot CHO$$

Side-chain bromination occurs under the influence of light (cf. Section **IV,32**)
and the extent of bromination is controlled by ensuring that the bromine (used
in the theoretical amount) is added no faster than the rate at which it is consumed. The halogen in the benzylidene halide is reactive and hydrolysis occurs
readily under mild conditions. In the example cited (p-bromobenzaldehyde,
Section **IV,131**) the use of a boiling aqueous suspension of calcium carbonate
gives good results.

Aromatic and heteroaromatic aldehydes can alternatively be prepared from
the corresponding methyl compound by subjecting the chloromethyl or
bromomethyl derivative to the **Sommelet reaction**. This procedure involves an
initial reaction between the halomethyl compound and hexamethylenetetramine
(hexamine), and hydrolysing the resulting quaternary hexamine salt (IV) with
hot aqueous acetic acid.

$$Ar \cdot CH_3 \longrightarrow Ar \cdot CH_2X \xrightarrow{(CH_2)_6N_4} Ar \cdot CH_2 \overset{\oplus}{N}(CH_2)_6N_3\}\overset{\ominus}{X} \xrightarrow{H_2O} Ar \cdot CHO$$
$$\text{(IV)}$$

The mechanism of the reaction is not certain but hydrolysis of the salt may
yield the primary amine (V), formaldehyde and ammonia. A hydride ion

transfer then probably occurs between the benzylamine and the protonated aldimine (VI), derived from formaldehyde and ammonia. Hydrolysis of the resulting aromatic aldimine (VII) then yields the required aldehyde.

$$Ar\cdot CH_2\overset{\oplus}{N}(CH_2)_6N_3\}\overset{\ominus}{X} \xrightarrow{H_2O} Ar\cdot CH_2NH_2 + 4H\cdot CHO + 3NH_3$$

$$(V)$$

$$Ar\cdot CH \overset{\frown}{-H} {}^{\curvearrowright}CH_2 = \overset{\oplus}{N}H_2 \longrightarrow Ar\cdot CH = \overset{\oplus}{N}H_2 + CH_3NH_2$$

$$\overset{\curvearrowleft}{\underset{:NH_2}{|}} \qquad (VI) \qquad\qquad (VII)$$

$$(V)$$

$$Ar\cdot CH = \overset{\oplus}{N}H_2 \xrightarrow{H_2O} Ar\cdot CHO + NH_3$$

A typical procedure is that described in Section **IV,132** for the synthesis of 1-naphthaldehyde. The synthesis of *p*-nitrobenzaldehyde provides an example in which the intermediate crystalline hexamine salt is isolated prior to hydrolysis. 2-Naphthaldehyde is prepared from the bromomethyl compound, the preparation of which illustrates the use of *N*-bromosuccinimide for effecting benzylic bromination of 2-methylnaphthalene.

Acid chlorides can be selectively hydrogenated in the presence of a catalyst (palladium deposited on a carrier, which is usually barium sulphate but is occasionally charcoal). The reaction which involves the hydrogenolysis of the carbon–halogen bond is known as the **Rosenmund reduction** and has been widely used for the synthesis of aromatic and heterocyclic aldehydes.

$$Ar\cdot COCl + H_2 \xrightarrow{catalyst} Ar\cdot CHO + HCl$$

The procedure is to pass purified hydrogen through a hot solution of the pure acid chloride in toluene or xylene in the presence of a catalyst; the exit gases are bubbled through water to absorb the hydrogen chloride, and the solution is titrated with standard alkali from time to time so that the reduction may be stopped when the theoretical quantity of hydrogen chloride has been evolved. Further reduction of the aldehyde, leading to the corresponding alcohol and thence to the methylarene, can usually be prevented by using the appropriate catalyst poison or regulator, which inactivates the catalyst towards reduction of the aldehyde but not the acid chloride. The regulator usually contains sulphur, e.g., quinoline-sulphur or thiourea; its use is not always necessary, however, and it has been stated that the decisive factors are to keep the reaction mixture at the lowest temperature at which hydrogen chloride is liberated and to arrest the reaction as soon as 1 mol of hydrogen chloride is evolved. The reduction is illustrated by the synthesis of 2-naphthaldehyde (Section **IV,133**).

IV,130. p-NITROBENZALDEHYDE

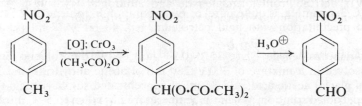

Equip a 1-litre three-necked flask with a mechanical stirrer, a dropping funnel and a thermometer, and immerse the flask in a bath of ice and salt. Place 200 g (185 ml, 2 mol) of acetic anhydride and 25 g (0.18 mol) of p-nitrotoluene in the flask, and add slowly, with stirring, 40 ml of concentrated sulphuric acid. When the temperature has fallen to 0 °C introduce slowly, with stirring, a solution of 50 g (0.5 mol) of chromium trioxide in 225 ml of acetic anhydride (1), at such a rate that the temperature does not exceed 10 °C; continue stirring for 2 hours after all the chromium trioxide solution has been added. Pour the contents of the flask into a 3-litre beaker one-third filled with crushed ice and almost fill the beaker with cold water. Filter the solid at the pump and wash it with cold water until the washings are colourless. Suspend the product in 150 ml of cold 2 per cent sodium carbonate solution and stir mechanically for about 10–15 minutes; filter, wash with cold water, and finally with 10 ml of ethanol. Dry in a vacuum desiccator; the yield of crude p-nitrobenzylidene diacetate is 30 g (65%) (2).

Reflux the crude p-nitrobenzylidene diacetate with a mixture of 70 ml of ethanol, 70 ml of water and 7 ml of concentrated sulphuric acid for 30 minutes, filter through a fluted filter-paper and cool the filtrate in ice. Collect the crystals by suction filtration, wash with cold water and dry in a vacuum desiccator. The yield of p-nitrobenzaldehyde, m.p. 106 °C, is 15 g (55% overall).

Notes.   (1) The solution is prepared by adding the chromium trioxide portionwise to the well-cooled acetic anhydride. Addition of the anhydride to the oxide in bulk may lead to explosive decomposition.

(2) The pure diacetate may be isolated by dissolving in 100 ml of hot ethanol, filtering from any insoluble impurities and allowing to cool: 28 g (61%), m.p. 125–126 °C, are obtained.

Cognate preparations o-Nitrobenzaldehyde.   Use 25 g (0.18 mol) of o-nitrotoluene and proceed as for p-nitrobenzaldehyde, but allow a period of 3 hours stirring at 5–10 °C after the addition of the chromium trioxide solution. In the work-up, omit the final ethanol washing; to remove unchanged o-nitrotoluene boil the crude product under reflux for 30 minutes with 120 ml of light petroleum (b.p. 60–80 °C). The yield of o-nitrobenzylidene diacetate of m.p. 82–84 °C is 16 g (36%).

Suspend 16 g of the diacetate in a mixture of 85 ml of concentrated hydrochloric acid, 140 ml of water and 25 ml of ethanol and boil under reflux for 45 minutes. Cool the mixture to 0 °C, filter the solid with suction and wash with water. Purify the crude aldehyde by rapid steam distillation; collect about 1 litre of distillate during 15 minutes, cool, filter and dry in a vacuum desiccator over calcium chloride. The yield of pure o-nitrobenzaldehyde, m.p. 44–45 °C, is

7.5 g (28% overall). The crude solid may also be purified after drying either by distillation under reduced pressure (the distillate of rather wide b.p. range, e.g., 120–144 °C/3–6 mmHg, is quite pure) or by dissolution in toluene (2–2.5 ml per gram) and precipitation with light petroleum, b.p. 40–60 °C (7 ml per ml of solution).

p-Bromobenzaldehyde. Use 31 g (0.18 mol) of *p*-bromotoluene (Section **IV,81**) dissolved in a mixture of 300 g (280 ml) of acetic anhydride and 300 g (285 ml) of glacial acetic acid; add 85 ml of concentrated sulphuric acid to the well-stirred, cooled solution. When the temperature has fallen to 5 °C introduce 50 g (0.5 mol) of chromium trioxide in small portions at such a rate that the temperature does not rise above 10 °C; continue the stirring for 10 minutes more. Isolate the crude diacetate as in the case of the *p*-nitro-compound; the yield of *p*-bromobenzylidene diacetate of m.p. 90–92 °C is 30 g (58%). (Pure material, m.p. 95 °C, may be obtained by crystallisation from ethanol.) Hydrolyse the crude product using 100 ml of ethanol, 70 ml of water and 7 ml of concentrated sulphuric acid, dilute with 200 ml of water and cool. The yield of *p*-bromobenzaldehyde of m.p. 55–57 °C is 17 g (51% overall). If necessary the product may be purified via the bisulphite compound (Section **IV,131**).

IV,131. *p*-BROMOBENZALDEHYDE

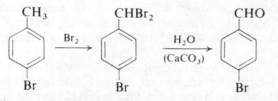

This preparation should be carried out in the fume cupboard.

Equip a 1-litre multi-necked flange flask with a reflux condenser, a mechanical stirrer, a dropping funnel and a thermometer which reaches nearly to the bottom of the flask; connect the upper end of the condenser to an absorption trap (Fig. I,68). Place 100 g (65 ml, 0.58 mol) of *p*-bromotoluene (Section **IV,81**) in the flask and immerse the latter in an oil bath (colourless oil in a large beaker). Heat the bath until the temperature of the stirred *p*-bromotoluene reaches 105 °C. Illuminate the liquid with an unfrosted 150-watt tungsten lamp, and add 200 g (1.25 mol) of bromine slowly from the dropping funnel: do not allow a large excess of bromine to accumulate in the reaction mixture. Add about one half of the bromine during 1 hour while the temperature is kept at 105–110 °C, and add the remainder during 2 hours while the temperature is slowly raised to 135 °C. Raise the temperature slowly to 150 °C when all the bromine has been introduced. Transfer the crude *p*-bromobenzylidene dibromide (1) to a 2-litre flask, mix it intimately with 200 g of precipitated calcium carbonate and then add about 300 ml of water. Attach a reflux condenser to the flask, heat the mixture first on a water bath and then on a wire gauze over a free flame with continuous shaking until the liquid boils (2); reflux the mixture for 15 hours to complete the hydrolysis. Steam distil the reaction mixture rapidly (3); collect the first 1 litre of distillate separately, filter off the product and dry in a vacuum desiccator; 60 g (56%) of pure *p*-bromobenzaldehyde, m.p. 56–57 °C, are thus obtained. Collect a further 2 litres of distillate (4); this yields about 15 g of a less pure product, m.p. 52–56 °C. Purify this by trituration with saturated sodium bisulphite solution (2 ml per gram) and, after about 3 hours, filter off the pasty mixture at the pump, wash it with alcohol and then with ether. Transfer the bisulphite compound to a flask fitted for steam distillation (Fig. I,101), add excess of sodium carbonate solution and isolate the aldehyde by steam distillation; 13 g (12%) of p-bromobenzaldehyde, m.p. 56–57 °C, are thus collected.

Notes. (1) This compound is a lachrymator and also produces a burning sensation on the skin; the latter is relieved by washing the affected parts with alcohol.

(2) This gradual heating reduces the risk of breaking the flask.

(3) The best results are obtained by conducting the steam distillation in a large three-necked flask provided with a sealed stirrer unit in the central aperture; the aldehyde distils slowly unless the mixture is well stirred.

(4) If the solution in the flask is acidified with hydrochloric acid, about 8 g of crude p-bromobenzoic acid may be isolated.

IV,132. 1-NAPHTHALDEHYDE

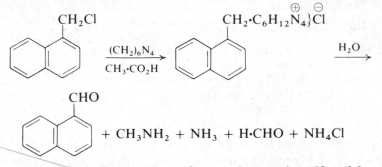

In a 500-ml flask, fitted with a reflux condenser, place 53 g (0.3 mol) of 1-(chloromethyl)naphthalene, 84 g (0.6 mol) of hexamethylenetetramine and 250 ml of 50% aqueous acetic acid. [**CAUTION:** 1-(Chloromethyl)naphthalene and, to a lesser degree, 1-naphthaldehyde have lachrymatory and vesicant properties; adequate precautions should therefore be taken to avoid contact with these substances.] Heat the mixture under reflux for 2 hours; it becomes homogeneous after about 15 minutes and then an oil commences to separate. Add 100 ml of concentrated hydrochloric acid and reflux for a further 15 minutes; this will hydrolyse any Schiff's bases which may be formed from amine and aldehyde present and will also convert any amines into the ether-insoluble hydrochlorides. Cool, and extract the mixture with 150 ml of ether. Wash the ether layer with three 50 ml portions of water, then cautiously with 50 ml of 10 per cent sodium carbonate solution, followed by 50 ml of water. Dry the ethereal solution with anhydrous magnesium sulphate, remove the ether by distillation on a steam bath and distil the residue under reduced pressure. Collect the 1-naphthaldehyde at 160–162 °C/18 mmHg; the yield is 38 g (81%).

Cognate preparations *p*-Nitrobenzaldehyde. This preparation is an example of the Sommelet reaction in which the hexaminium salt is isolated. Dissolve 11 g (0.13 mol) of hexamethylenetetramine in 70 ml of chloroform and add 11.4 g (0.067 mol) of *p*-nitrobenzyl chloride or 14.4 g of *p*-nitrobenzyl bromide

(Section IV,32). Heat the mixture under reflux on a steam bath for 4 hours; a precipitate gradually separates. Replace the reflux condenser by a condenser set for distillation and distil off about 35 ml of solvent. Add 35 ml of acetone, cool in ice, collect the precipitate by suction filtration and dry it in the air. Heat the hexaminium salt thus obtained under reflux for 1 hour with 100 ml of 50 per cent acetic acid; then add 100 ml of water and 25 ml of concentrated hydrochloric acid and continue the refluxing for 5–10 minutes. Cool the solution in ice, collect the crystals of p-nitrobenzaldehyde and dry them in a vacuum desiccator. The yield is 6.4 g (63%), m.p. 106 °C.

2-Naphthaldehyde.

$$2\text{-}C_{10}H_7\text{·}CH_3 \xrightarrow{\text{NBS}} 2\text{-}C_{10}H_7\text{·}CH_2Br \xrightarrow[\text{(ii) } H_3O^{\oplus}]{\text{(i) } (CH_2)_6N_4} 2\text{-}C_{10}H_7\text{·}CHO$$

Dissolve 71 g of 2-methylnaphthalene in 450 g (283 ml) of carbon tetrachloride and place the solution in a 1-litre three-necked flask equipped with a mechanical stirrer and reflux condenser. Introduce 89 g of N-bromosuccinimide (Section II,2,8) through the third neck, close the latter with a stopper and reflux the mixture with stirring for 16 hours. Filter off the succinimide and remove the solvent under reduced pressure on a water bath. Dissolve the residual brown oil (largely 2-(bromomethyl)naphthalene) in 300 ml of chloroform, and add it to a rapidly stirred solution of 84 g of hexamethylenetetramine in 150 ml of chloroform contained in a 2-litre three-necked flask, fitted with a reflux condenser, mechanical stirrer and dropping funnel: maintain the rate of addition so that the mixture refluxes vigorously. A white solid separates almost immediately. Heat the mixture to reflux for 30 minutes, cool and filter. Wash the crystalline hexaminium bromide with two 100 ml portions of light petroleum, b.p. 40–60 °C, and dry; the yield of solid, m.p. 175–176 °C, is 147 g. Reflux the hexaminium salt for 2 hours with 750 ml of 50 per cent acetic acid, add 150 ml of concentrated hydrochloric acid, continue the refluxing for 5 minutes more and cool. Extract the aldehyde from the solution with ether, evaporate the ether and recrystallise the residue from hot hexane. The yield of 2-naphthaldehyde, m.p. 59–60 °C, is 50 g (64% overall).

IV,133. 2-NAPHTHALDEHYDE

$$2\text{-}C_{10}H_7\text{·}COCl \xrightarrow[\text{Pd-BaSO}_4]{H_2} 2\text{-}C_{10}H_7\text{·}CHO$$

Fit a 250-ml three-necked flask with a reflux condenser, a high-speed sealed stirrer (1) and a gas inlet tube extending to a point just above the bottom of the stirrer. Place 28.5 g (0.15 mol) of 2-naphthoyl chloride (2), 100 ml of sodium-dried xylene, 3 g of palladium–barium sulphate catalyst (Section II,2,44(d)) and 0.3 ml of the stock poison solution (Section II,2,44(d)) in the flask. Connect the top of the condenser by a rubber tube to a 6-mm glass tube extending to the bottom of a 250-ml conical flask containing 200 ml of distilled water and a few drops of phenolphthalein indicator; arrange a burette charged with ca. 1 M-sodium hydroxide solution (prepared from the pure solid) for delivery into the flask. The apparatus must be sited in the fume cupboard.

Displace the air in the reaction flask with hydrogen from a cylinder of the gas, heat the flask in an oil bath at 140–150 °C, and stir the mixture vigorously.

Continue to pass hydrogen at such a rate that 1–2 bubbles per second emerge in the conical flask. Follow the course of the reaction by the rate of hydrogen chloride evolution. The first 25 ml of alkali should be neutralised in 12–15 minutes, and the reaction should be complete in about 2 hours. About 92 per cent of the theoretical amount of hydrogen chloride ($\equiv$ 142.5 ml of 1 M-NaOH solution) is recovered; the end of the reaction is indicated by a rather abrupt cessation of hydrogen chloride evolution. Cool the flask, add 1 g of decolourising carbon with stirring and filter the solution with suction through a hardened filter-paper and keep the spent catalyst for recovery. Remove the xylene by flash distillation using a 50–75 ml flask with fractionating side arm and then distil under reduced pressure with the aid of an oil bath: a small fraction, consisting largely of naphthalene, passes over first, followed by 2-naphthaldehyde at 147–149 °C/11 mmHg (temperature of bath, 170–180 °C). This (19 g, 81%) solidifies on cooling to a white solid, m.p. 59–60 °C.

Notes. (1) Rapid stirring is desirable in order to obtain the maximum reaction rate; absorption of hydrogen occurs chiefly at the rapidly agitated surface.

(2) 2-Naphthoyl chloride may be prepared from 57.4 g (0.33 mol) of 2-naphthoic acid and 69 g (0.33 mol) of phosphorus pentachloride following the procedure described for p-nitrobenzoyl chloride (Section **IV,175**). After removing the phosphorus oxychloride by distillation, the product is collected as a fraction of b.p. 160–162 °C/11 mmHg. This solidifies on cooling to a colourless solid, m.p. 51–52 °C; The yield is 60 g (95%).

L AROMATIC KETONES AND QUINONES

1. Aromatic acylation reactions (Sections **IV,134** to **IV,139**).
2. Synthesis of aromatic ketones from carboxylic acid derivatives (Sections **IV,140** and **IV,141**).
3. Synthesis of quinones (Sections **IV,142** to **IV,144**).

L,1. Friedel–Crafts type acylation processes. The reaction of a carboxylic acid chloride or anhydride with an aromatic hydrocarbon in the presence of anhydrous aluminium chloride generally gives a good yield of the aromatic ketone:

$$ ArH + R \cdot COCl \xrightarrow{AlCl_3} Ar \cdot CO \cdot R + HCl $$

$$ Ar^1H + Ar^2 \cdot COCl \xrightarrow{AlCl_3} Ar^1 \cdot CO \cdot Ar^2 + HCl $$

$$ ArH + (R \cdot CO)_2O \xrightarrow{AlCl_3} Ar \cdot CO \cdot R + R \cdot CO_2H $$

It should be noted that the Friedel–Crafts acylation differs from the Friedel–Crafts alkylation (see p. 601, **IV,A,1**(c)) in one important respect in that the alkylation process requires relatively small (catalytic) quantities of aluminium chloride. With acylations, however, at least one molar equivalent of aluminium chloride is necessary for each carbonyl group present in the acylating agent. This is because aluminium chloride is capable of forming rather stable complexes with the carbonyl group (see formulation below). This complex formation therefore requires an equivalent quantity of metal halide, and hence a slight excess over this amount is employed in order to ensure that the free reagent may be present to act as the catalyst: thus 1.2 and 2.2 molar equivalents of aluminium chloride are generally employed for acid chlorides and acid anhydrides respect vely. Excess of benzene or of toluene may be used as a solvent (when either of these substances constitutes one of the reactants), otherwise carbon disulphide or nitrobenzene is usually employed. Friedel–Crafts acylation is free of two features which complicate the alkylation reaction, namely, (i) polysubstitution and (ii) rearrangements. There is usually no difficulty in arresting the acylation with the introduction of a single acyl group into the aromatic nucleus as the acyl group deactivates the nucleus to further electrophilic attack. In the case of benzene homologues which may show a tendency to isomerise or disproportionate under the influence of aluminium chloride (see p. 601, **IV,A,1**(*c*)) preliminary mixing of the acyl and aluminium halides is recommended.

The mechanism of the Friedel–Crafts acylation reaction, formulated below for reactions using acid chlorides, probably involves the acylium ion (I) as the reactive electrophilic species, although an electrophilic complex (II) between the acid chloride and aluminium chloride may also be involved.

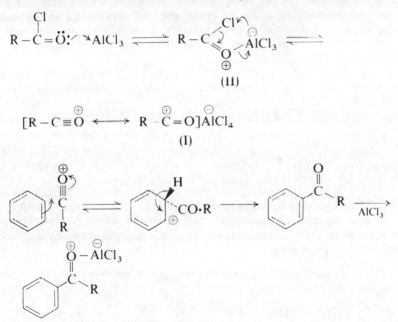

The use of aliphatic carboxylic acid anhydrides in place of the corresponding acid chlorides offers many advantages; these include:

(*a*) the greater ease of obtaining the anhydrides in a state of purity, and their availability as commercial products (acetic, propionic, butyric and succinic anhydrides);

(*b*) the handling of disagreeable acid chlorides is avoided;

(c)   the absence of any appreciable quantities of by-products and of resinous substances;

(d)   the reaction is smooth and the yield is generally good.

The preparations of acetophenone, butyrophenone, propiophenone and benzophenone given in Section **IV,134** provide examples of acylation reactions carried out under standard Friedel–Crafts conditions. A more convenient preparation of benzophenone from benzene and an excess of carbon tetrachloride is also described; this involves the intermediate formation of dichlorodiphenylmethane which is hydrolysed to the ketone.

Acetylation of substituted benzenes possessing electron-releasing groups (e.g., CH_3, Br, Cl, OCH_3, C_6H_5) gives largely the *para* isomer (e.g., cognate preparations in Section **IV,134**). The presence of deactivating substituents (e.g., CHO, CN, NO_2) renders the aromatic ring inactive towards acylation under Friedel–Crafts conditions, which therefore permits the use of nitrobenzene as a reaction solvent (e.g., as in the acetylation of naphthalene, cognate preparation in Section **IV,134**) as noted above. Two isomeric acetyl derivatives are possible when naphthalene undergoes acetylation and the composition of the product is dependent upon the reaction conditions (cf. the sulphonation of naphthalene, p. 640, **IV,D,1**). In nitrobenzene solution the product is largely 2-acetonaphthalene whereas reaction in carbon tetrachloride yields the 1-isomer.

A further example is the acetylation of thiophen with acetic anhydride. Electrophilic substitution in the heteroaromatic ring proceeds with great ease and the catalyst in this case is a small amount of orthophosphoric acid. The most reactive site in thiophen is the 2-position, and 2-acetylthiophen is the predominant isomer formed.

Reaction of succinic anhydride with benzene in the presence of anhydrous aluminium chloride (slightly over two equivalents; see above) yields β-benzoylpropionic acid. This may be reduced by the Clemmensen method in the presence of a solvent (toluene) immiscible with the hydrochloric acid to γ-phenylbutyric acid. Cyclisation to α-tetralone (Section **IV,135**) is then effected smoothly by treatment with hot polyphosphoric acid. This reaction sequence represents the first stages in the Haworth procedure for the synthesis of polycyclic aromatic hydrocarbons (see p. 613, **IV,A,4**).

Aroylation of an aromatic system by reaction of phthalic anhydride under Friedel–Crafts conditions yields the *o*-aroylbenzoic acid. These readily available compounds have characteristic melting points which make them useful as derivatives in the characterisation of aromatic hydrocarbons and of aryl halides (Section **VII,6,3(b)**).

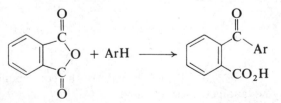

With benzene the product is *o*-benzoylbenzoic acid (Section **IV,136**) which can be cyclised with polyphosphoric acid to anthraquinone. Reduction of anthraquinone with tin and acid yields anthrone (Section **IV,137**), probably by the sequence of steps formulated below.

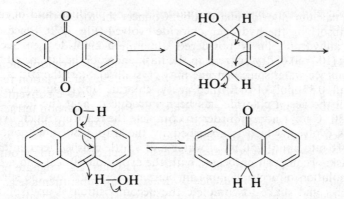

Reaction of phthalic anhydride with toluene yields the *o*-(*p*-toluoyl)benzoic acid which may then be cyclised with polyphosphoric acid to give 2-methylanthraquinone (cognate preparation in Section **IV,136**).

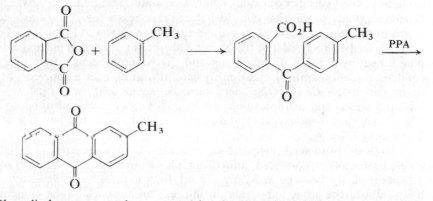

Phenolic ketones may be prepared by the Hoesch acylation reaction, which may be regarded as an extension of the Gattermann aldehyde synthesis (p. 756, **IV,K,1**). The procedure involves reaction of a nitrile with a phenol (or phenolic ether) in the presence of zinc chloride and hydrogen chloride; best results are usually obtained with polyhydric phenols or their ethers, as for example in the preparation of phloroacetophenone (Section **IV,138**). In these cases acylation can often be successfully carried out under standard Friedel–Crafts conditions and the synthesis of 2,4,6-trihydroxyisobutyrophenone is illustrative (Section **IV,139**).

IV,134. ACETOPHENONE

Method 1

$$C_6H_6 + CH_3 \cdot COCl \xrightarrow{AlCl_3} C_6H_5 \cdot CO \cdot CH_3 + HCl$$

Method 2

$$C_6H_6 + (CH_3 \cdot CO)_2O \xrightarrow{AlCl_3} C_6H_5 \cdot CO \cdot CH_3 + HCl$$

Method 1 (**with acetyl chloride**). Equip a 500-ml three-necked flask with a double surface reflux condenser, a sealed stirrer unit and a dropping funnel protected by a calcium chloride guard-tube. Connect the top of the condenser to a trap for absorbing the hydrogen chloride evolved (Fig. I,68). Place 60 g (0.45 mol) of anhydrous, finely powdered aluminium chloride (see Section II,2,3) and 88 g (100 ml) of dry benzene in the flask and cool the latter in a bath of *cold* water (*not* ice-water since benzene may crystallise). Start the stirrer, and add 29 g (26 ml, 0.37 mol) of redistilled acetyl chloride slowly during half an hour. When all the acetyl chloride has been introduced, heat the flask on a water bath at 50 °C for 1 hour in order to complete the reaction: much hydrogen chloride is evolved, which is absorbed by the trap. Cool and pour the reaction mixture into about 250 ml of water and a little crushed ice contained in a 750-ml flask; decomposition occurs with the evolution of heat and a dark oil (largely a solution of acetophenone in benzene) separates on the surface. Stopper the flask and shake to complete the decomposition; if any solid remains undissolved, add a little concentrated hydrochloric acid to dissolve it. Pour the mixture into a separatory funnel, run off and discard the lower layer, wash the benzene layer with water, then with dilute sodium hydroxide solution (to remove the hydrogen chloride), again with water, and finally dry over magnesium sulphate or calcium chloride. Remove excess benzene by flash distillation from a boiling water bath; use the apparatus shown in Fig. I,100 (100-ml distilling flask). When most of the benzene has been removed, replace the dropping funnel by a 360 °C thermometer and the water condenser by a short air condenser. Continue the distillation by careful heating over a gauze or, better, in an air bath—CAUTION: there may be some benzene in the residual oil—and collect the acetophenone at 195–202 °C (pure acetophenone boils at 201 °C (1)); the colourless oil crystallises on cooling in ice and has m.p. 20 °C. The yield is 27 g (61%).

Method 2 (**with acetic anhydride**).   Proceed exactly as in *Method 1*, but use 75 g (0.056 mol) of powdered, anhydrous aluminium chloride, 100 g (114 ml, 1.28 mol) of dry benzene and 26 g (24 ml, 0.25 mol) of redistilled acetic anhydride. Add the acetic anhydride during half an hour whilst the contents of the flask are thoroughly shaken; much heat is evolved in the reaction. Heat on a boiling water bath for 30 minutes (or until the evolution of hydrogen chloride almost ceases) to complete the reaction, cool and pour the contents of the flask into a mixture of 150 g of crushed ice and 150 ml of concentrated hydrochloric acid contained in a beaker or flask. Stir or shake until all the aluminium salts are dissolved. Transfer the mixture to a separatory funnel, add 25–30 ml of ether, shake and separate the upper (largely benzene) layer. Extract the aqueous layer with 25 ml of ether and add this to the benzene solution. Wash the combined benzene and ether extracts with 50 ml of 10 per cent sodium hydroxide solution (or until the washings remain alkaline), then with water, separate the organic layer and dry it with magnesium sulphate or anhydrous calcium chloride. Remove the ether and benzene and isolate the acetophenone, b.p. 199–202 °C (1), as in *Method 1*. The yield is 25 g (83%).

Note.   (1) The b.p. under reduced pressure is 88–89 °C/16 mmHg.

Cognate preparations   Butyrophenone (*phenyl propyl ketone*).   Proceed according to *Method 1* but use 39.5 g (38 ml, 0.37 mol) of butyryl chloride (Section III,138) in place of acetyl chloride; it may be necessary to warm the flask gently to initiate the reaction which should then proceed without further heating while the acid chloride is being added. Complete the reaction and work up the product exactly as described previously. The yield of butyrophenone (a colourless liquid), b.p. 227–230 °C or 110 °C/10 mmHg, is 25 g (51%).

Propiophenone (*ethyl phenyl ketone*). Follow *Method 1*, but use 34.5 g (32 ml, 0.37 mol) of propionyl chloride (prepared from propionic acid; compare Section **III,138**). The yield of propiophenone, b.p. 214–217 °C, is 30 g (60%). An improved yield is obtained by the following process. Add a mixture of 75 g (70.5 ml, 0.81 mol) of propionyl chloride and 90 g (103 ml, 1.15 mol) of sodium-dried benzene to a vigorously stirred suspension of 75 g (0.56 mol) of finely powdered anhydrous aluminium chloride in 100 ml of dry carbon disulphide (**CAUTION**: see Sections **I,3** and **II,1,32**). Then introduce more of the aluminium chloride (about 35 g) until no further evolution of hydrogen chloride occurs. The yield of propiophenone, b.p. 123 °C/25 mmHg, is about 90 g (90%).

Benzophenone. *A.* Into a 500-ml round-bottomed flask place 120 ml (105 g, 1.35 mol) of dry benzene and 35 g (29 ml, 0.25 mol) of redistilled benzoyl chloride. Weigh out 37 g (0.275 mol) of finely powdered, anhydrous aluminium chloride into a dry stoppered conical flask, and add the solid, with frequent shaking, during 10 minutes to the contents of the flask. Fit a reflux condenser with a gas absorption trap attachment to the flask, and heat on a water bath for 3 hours or until hydrogen chloride is no longer evolved. Pour the contents of the flask while still warm into a mixture of 200 g of crushed ice and 100 ml of concentrated hydrochloric acid. Separate the upper benzene layer (filter first, if necessary), wash it with 50 ml of 5 per cent aqueous sodium hydroxide solution, then with water, and dry with magnesium sulphate. Remove the benzene after filtration by flash distillation and distil the residue under diminished pressure through a short fractionating side arm (Fig. I,110). Collect the benzophenone at 187–190 °C/15 mmHg; it solidifies to a white solid on cooling, m.p. 47–48 °C. The yield is 30 g (66%).

B.

$$2C_6H_6 + CCl_4 \longrightarrow C_6H_5 \cdot CCl_2 \cdot C_6H_5 \xrightarrow{H_2O} C_6H_5 \cdot CO \cdot C_6H_5$$

Use a four-necked 1-litre flange flask fitted with a double surface condenser, a sealed stirrer unit, a thermometer and a dropping funnel protected by a calcium chloride guard-tube. Attach a gas absorption trap to the reflux condenser outlet. Place 91 g (0.68 mol) of powdered anhydrous aluminium chloride and 200 ml (319 g, 2.07 mol) of dry carbon tetrachloride (Section **II,1,7**) in the flask, surround the latter with an ice bath and, when the temperature has fallen to 10–15 °C, introduce 10 ml (9 g) of sodium-dried benzene. The reaction commences immediately (hydrogen chloride is evolved and the temperature rises); add salt to the ice bath to get more efficient cooling. When the temperature commences to fall after the reaction has once started, add a mixture of 100 ml (97 g) of dry benzene (total 1.37 mol) and 110 ml (1.14 mol) of dry carbon tetrachloride at such a rate that the temperature is maintained between 5 and 10 °C (1). The addition usually requires 1–2 hours; continue the stirring for a further 3 hours while maintaining the temperature at 10 °C, and then allow to stand overnight.

Immerse the flask in ice, start the stirrer and add about 500 ml of water through the separatory funnel; the excess carbon tetrachloride usually refluxes during the addition. Distil off as much as possible of the carbon tetrachloride on a water bath, and then distil the mixture with steam (Fig. I,101) during 30 minutes to remove the residual carbon tetrachloride (2) and to hydrolyse the dichlorodiphenylmethane to benzophenone. Separate the benzophenone layer and extract the aqueous layer with 40 ml of benzene. Dry the combined benzene extract and benzophenone with magnesium sulphate. Remove the benzene by flash distillation and isolate the pure benzophenone, m.p. 47–48 °C, as described in *A* above. The yield is 105 g (85%).

Notes. (1) Below 5 °C, the reaction is too slow; above 10 °C, appreciable amounts of tarry matter are formed.

(2) About 200 ml of carbon tetrachloride are recovered; this contains some benzene, but may be used after drying and redistillation, in another run.

p-**Methylacetophenone.** Proceed as for *Method 2* using 120 g (140 ml, 1.30 mol) of pure dry toluene (Section **II,1,***3*) in place of the benzene. After decomposing the reaction product and isolating and drying the organic phase, remove ether and excess toluene by distillation at atmospheric pressure. Distil the residue under reduced pressure through a short fractionating column (Fig. I,110) and collect the *p*-methylacetophenone at 93–94 °C/7 mmHg (the b.p. at atmospheric pressure is 225 °C); the yield is 29 g (86%).

p-**Bromoacetophenone.** In a 1-litre three-necked flask, equipped as in *Method 2*, place 78.5 g (52.5 ml, 0.5 mol) of dry bromobenzene (Section **IV,27**), 200 ml of dry carbon disulphide (**CAUTION:** see Section **I,3**) and 150 g of finely powdered anhydrous aluminium chloride. Stir the mixture and heat on a water bath until gentle refluxing commences; add 51 g (47.5 ml, 0.5 mol) of redistilled acetic anhydride slowly through the dropping funnel (30–60 minutes). Maintain gentle refluxing during the addition of the acetic anhydride and for 1 hour afterwards. Distil off most of the carbon disulphide on a water bath, allow the reaction mixture to cool somewhat and while still warm pour it slowly with stirring into a mixture of 500 g of crushed ice and 300 ml of concentrated hydrochloric acid. Decompose any residue in the flask and add it to the main product. Extract with 150 and 100 ml portions of ether, wash the combined extracts twice with water, once with 10 per cent sodium hydroxide solution and twice with water. Dry the extract with magnesium sulphate or anhydrous calcium chloride, remove the ether and distil the residue under reduced pressure. The *p*-bromoacetophenone boils at 130 °C/15 mmHg or at 117 °C/7 mmHg and a 3 °C fraction should be collected; it crystallises to a white solid, m.p. 50 °C. The yield is 75 g (75%).

The b.p. under atmospheric pressure has been given as 255.5 °C/736 mmHg.

p-**Chloroacetophenone.** Use 56 g (51 ml, 0.5 mol) of chlorobenzene (Section **IV,26**) and proceed as for *p*-bromoacetophenone. The yield of product, b.p. 124–126 °C/24 mmHg, m.p. 20–21 °C, is 60 g (78%). The b.p. under atmospheric pressure is 237 °C.

p-**Methoxyacetophenone.** Use 54 g (54.5 ml, 0.5 mol) of anisole (Section **IV,124**) and proceed as for *p*-bromoacetophenone. The yield of *p*-methoxyacetophenone, b.p. 139 °C/15 mmHg, is 70 g (93%). The b.p. under atmospheric pressure is 265 °C.

p-**Phenylacetophenone.** In a 1-litre three-necked flask provided with a dropping funnel, a mechanical stirrer and a reflux condenser, place 77 g (0.5 mol) of biphenyl, 150 g (1.125 mol) of finely powdered anhydrous aluminium chloride and 350 ml of anhydrous carbon disulphide. Charge the dropping funnel with 51 g (47.5 ml, 0.5 mol) of pure acetic anhydride and close the mouth of the funnel with a calcium chloride guard-tube. Heat the mixture on a water bath until gentle refluxing commences, and add the acetic anhydride during 1 hour; the addition product makes its appearance as a curdy mass when about three-quarters of the anhydride has been added. Reflux the reaction mixture gently for a further hour. Allow to cool and pour the reaction product slowly and with stirring on to crushed ice to which hydrochloric acid has been added. Filter the precipitated p-phenylacetophenone on a Buchner funnel, wash repeatedly with water until free from acid, dry, and distil under reduced pressure. There is usually a small fraction of low boiling point; the main product passes over at 196–210 °C/18 mmHg and solidifies on cooling. The yield of crude p-phenylacetophenone, m.p. 118 °C, is 85 g (86%). Upon recrystallisation from rectified spirit, the m.p. is raised to 120–121 °C; the recovery is about 80 per cent.

2-Acetylthiophen

$$\text{(thiophen)} + (CH_3 \cdot CO)_2O \longrightarrow \text{(2-acetylthiophen)} \quad S \quad CO \cdot CH_3$$

Place 84 g (79 ml, 1 mol) of thiophen and 51 g (47.5 ml, 0.5 mol) of acetic anhydride in a 500-ml three-necked flask, fitted with a thermometer, mechanical stirrer and reflux condenser. Heat the stirred solution to 70–75 °C, remove the source of heat and add 5 g (4 ml) of 85–89 per cent orthophosphoric acid. An exothermic reaction occurs after 2–3 minutes and the temperature may rise to 90 °C; immerse the flask in a bath of cold water to control the reaction. When the boiling subsides (ca. 5 minutes), reflux the mixture for 2 hours. Add 125 ml of water, stir for 5 minutes, transfer the cold reaction mixture to a separatory funnel, remove the water layer, wash with two 50 ml portions of 5 per cent sodium carbonate solution and dry over magnesium sulphate. Distil the orange-red liquid through a short fractionating column (Fig. I,110) at atmospheric pressure and thus recover 38 g of unchanged thiophen at 83–84 °C. Distil the residue under reduced pressure and collect the 2-acetylthiophen at 89–90 °C/10 mmHg; this solidifies on cooling in ice, m.p. 10 °C. The yield is 44 g (70%).

Methyl 2-naphthyl ketone (*2-acetylnaphthalene*). Equip a 1-litre three-necked flask with a sealed mechanical stirrer unit and a pressure-equalising dropping funnel fitted with a calcium chloride guard-tube; stopper the third neck. Place 64 g (0.5 mol) of resublimed naphthalene and 350 g (291 ml) of pure nitrobenzene in the flask and stir until dissolved. To the homogeneous solution add 43.5 g (38.5 ml, 0.55 mol) of redistilled acetyl chloride from the dropping funnel. Cool to −5 °C in a freezing mixture of ice and salt and introduce, whilst stirring vigorously, 73.5 g (0.55 mol) of finely powdered anhydrous aluminium chloride in small portions during 90 minutes; do not allow the temperature to rise above 0 °C. The aluminium chloride dissolves and a deep green solution results. Remove the stirrer and stopper the central neck; into the side-necks of the flask fit respectively a drawn-out capillary tube and a tube leading through a filter flask trap to a water filter pump. Reduce the pressure to 15–20 mmHg; hydrogen chloride is copiously evolved and a vigorous ebullition occurs in the mixture. When no more gas is evolved, add an excess of crushed ice and separate the nitrobenzene layer. Wash the latter successively with two 100 ml portions of dilute hydrochloric acid and 100 ml of 5 per cent sodium carbonate solution. Use either of the following methods for isolating the pure 2-acetyl-naphthalene from the accompanying 1-isomer (about 10%) (1).

1. Steam distil from a 1.5-litre flask until the odour of nitrobenzene is no longer perceptible in the distillate (6–12 hours). Extract the cold residue with three 100 ml portions of ether, dry the combined extracts with magnesium sulphate and distil off the ether. The residue solidifies and consists of almost pure methyl 2-naphthyl ketone, m.p. 52 °C; the yield is 30 g (35%). Upon recrystallisation from glacial acetic acid, the m.p. is raised to 54 °C.

2. Distil the dried (magnesium sulphate) nitrobenzene solution under reduced pressure. Nitrobenzene passes over at 95–100 °C/16 mmHg and the temperature rises rapidly to 170 °C/15 mmHg; collect the fraction of b.p. 170–180 °C/15 mmHg. Transfer whilst still liquid to a porcelain basin; it solidifies on cooling. Spread it on a porous tile to absorb the small proportion of liquid methyl 1-naphthyl ketone which is present: the resulting yield of crude methyl 2-naphthyl ketone, m.p. 40–42 °C, is 50 g (59%). Two recrystallisations from glacial acetic acid (or from glacial acetic acid–water) give the almost pure 2-isomer, m.p. 53 °C.

Note. (1) Acetylation in carbon tetrachloride solution gives the 1-isomer as the major component. Add 70 g (0.52 mol) of powdered dry aluminium chloride to a vigorously stirred mixture of 41.9 g (38 ml, 0.53 mol) of acetyl chloride and 100 ml of carbon tetrachloride. The mixture becomes warm; cool to 20 °C and then run in slowly a solution of 32 g (0.25 mol) of naphthalene in 100 ml of carbon tetrachloride. Complete the reaction by warming to 30 °C for 30 minutes. Decompose the reaction mixture and work up the product as in *Method 1*, distilling the final product under reduced pressure, b.p. 165 °C/15 mmHg. The yield is 38.5 g (90%).

IV,135. α-TETRALONE

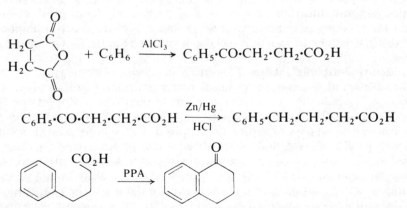

β-Benzoylpropionic acid. Place 175 g (2.25 mol) of sodium-dried benzene and 34 g (0.34 mol) of succinic anhydride (Section **III,141**) in a 1-litre three-necked flask equipped with a sealed stirrer unit and two efficient reflux condensers, the tops of which are connected through a Y-junction to a single efficient gas absorption device (Fig. I,68). Stir the mixture and add 100 g (0.75 mol) of powdered, anhydrous aluminium chloride all at once. The reaction usually starts immediately—hydrogen chloride is evolved and the mixture becomes hot; if there is no apparent reaction, warm gently. Heat in an oil bath to gentle refluxing, with continued stirring, for half an hour. Allow to cool, immerse the flask in a bath of cold water and slowly add 150 ml of water from a separatory funnel inserted into the top of one of the condensers. Introduce 50 ml of concentrated hydrochloric acid and separate the benzene by steam distillation (Fig. 101). Transfer the hot mixture to a 600-ml beaker; the β-benzoylpropionic acid separates as a colourless oil, which soon solidifies. Cool in ice, filter off the acid at the pump and wash with 100 ml of cold dilute hydrochloric acid (1 : 3 by volume) and then with 100 ml of cold water. Dissolve the crude acid in a solution of 40 g of anhydrous sodium carbonate in 250 ml of water by boiling for 10–15 minutes; filter the solution with suction to remove the small amount of aluminium hydroxide and wash with two 25 ml portions of hot water. Treat the *hot* filtrate with 2 g of decolourising carbon, stir for 5 minutes and filter at the pump through a preheated Buchner funnel. Transfer the hot filtrate to a 1-litre beaker, cool to about 50 °C and cautiously acidify with 65–70 ml of concentrated hydrochloric acid. Cool to 0 °C in a freezing mixture of ice and salt, filter, wash thoroughly with cold water, dry for 12 hours upon filter-papers, and then to constant weight at 45–50 °C. The yield of practically pure β-benzoylpropionic acid, m.p. 115 °C, is 57 g.

γ-**Phenylbutyric acid.** Prepare amalgamated zinc from 120 g of zinc wool contained in a 1-litre round-bottomed flask (Section **II,2,***67*), decant the liquid as completely as possible and add in the following order 75 ml of water, 180 ml of concentrated hydrochloric acid, 100 ml of pure toluene and 50 g (0.28 mol) of β-benzoylpropionic acid. Fit the flask with a reflux condenser connected to a gas absorption device (Fig. I,68), and boil the reaction mixture vigorously for 30 hours; add three or four 50 ml portions of concentrated hydrochloric acid at approximately six-hour intervals during the refluxing period in order to maintain the concentration of the acid. Allow to cool to room temperature and separate the two layers. Dilute the aqueous portion with about 200 ml of water and extract with three 75 ml portions of ether. Combine the toluene layer with the ether extracts, wash with water and dry over anhydrous magnesium or calcium sulphate. Remove the solvents by distillation under diminished pressure using a rotary evaporator and distil the residue under reduced pressure (Fig. I,110). Collect the γ-phenylbutyric acid at 178–181 °C/19 mmHg; this solidifies on cooling to a colourless solid (40 g, 89%) and melts at 47–48 °C.

α-**Tetralone.** Heat 120 g of polyphosphoric acid (Section **II,2,***48*) to 90 °C in a 1-litre beaker on a steam bath. Liquefy 33 g (0.20 mol) of γ-phenylbutyric acid by heating to 70 °C and add this in one portion to the polyphosphoric acid with manual stirring. Remove the beaker from the steam bath and continue stirring for 3 minutes; the temperature should remain at about 90 °C. Then add 100 g more of polyphosphoric acid and warm on a steam bath with vigorous stirring for 4 minutes. Cool to 60 °C, add 300 g of crushed ice and stir until the polyphosphoric acid is completely hydrolysed and a yellow oil has separated. Extract the mixture with three 150 ml portions of ether and wash the combined extracts with water, with 5 per cent aqueous sodium hydroxide solution and then with water until the washings are neutral. Dry the ethereal solution over magnesium sulphate and remove the ether on a rotary evaporator. Distil the residue under reduced pressure through a short fractionating column and collect the α-tetralone at 105–107 °C/2 mmHg or 135–137 °C/15 mmHg. The yield is 23 g (79%).

IV,136. o-BENZOYLBENZOIC ACID

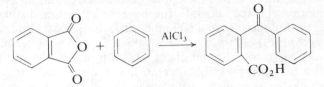

Equip a 750-ml three-necked flask with a sealed mechanical stirrer and a reflux condenser connected with a gas absorption trap (Fig. I,68); insert a stopper in the third neck. Place 25 g (0.17 mol) of pure phthalic anhydride [see Section **III,197**, Note (1)] and 100 ml (1.16 mol) of sodium-dried benzene in the flask; start the stirrer and add 50 g (0.375 mol) of powdered anhydrous aluminium chloride from a stoppered test-tube in four portions or, alternatively, use the device shown in Fig. I,64. If the reaction does not commence after the addition of the first 12 g of aluminium chloride, warm for a few seconds on a water bath. When all the aluminium chloride has been added and the evolution of hydrogen chloride slackens, warm on a water bath and ultimately reflux the mixture until the evolution of gas practically ceases. Cool the flask, add crushed ice slowly until the dark mass is completely decomposed and then run in concentrated hydrochloric acid (35–40 ml) until the solution clears. Steam distil (Fig. I,101) to remove the excess of benzene; the residue in the flask, when cooled in ice, largely solidifies and consists of crude o-benzoylbenzoic acid. Filter off the solid product and wash it well with 74 ml of cold water; dissolve the solid in 150 ml of warm 10 per cent sodium carbonate solution. Treat the solution of the sodium salt with 2 g of decolourising charcoal, boil for 2 minutes and filter through a preheated Buchner funnel. Place the filtrate in a 1-litre beaker, cool in ice and cautiously acidify with concentrated hydrochloric acid while stirring well (ca. 20 ml are required). The acid separates as an oil but it soon crystallises on stirring and cooling. Filter when ice cold, and wash with a little water. Dry in the air upon filter-paper; the product, which is somewhat efflorescent, consists largely of the monohydrate, m.p. 94 °C.

To prepare pure anhydrous o-benzoylbenzoic acid, dissolve the air-dried (or the moist) product in about 175 ml of toluene contained in a 500-ml round-bottomed flask fitted with a reflux condenser and heat on a water bath. Transfer the toluene solution to a separatory funnel, run off any water present and dry with magnesium sulphate. Concentrate the toluene solution to about half its volume and add light petroleum (b.p. 60–80 °C) to the hot solution until a slight turbidity is produced. Allow to cool spontaneously to room temperature, then cool in ice to about 5 °C, collect the crystals and dry. The yield of pure, anhydrous o-benzoylbenzoic acid, m.p. 128 °C, is 32 g (84%).

Cognate preparation o-(p-Toluoyl)-benzoic acid. Use 25 g (0.17 mol) of pure phthalic anhydride, 100 g (115.5 ml, 1.09 mol) of toluene and 50 g (0.375 mol) of anhydrous aluminium chloride. The air-dried product consists largely of the monohydrate; this becomes anhydrous upon drying at 100 °C and melts at 138–139 °C. The yield of anhydrous o-(p-toluoyl)-benzoic acid is 39 g (95%). It may be recrystallised from toluene.

Cyclisation of aroylbenzoic acids.

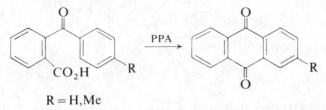

R = H,Me

Anthraquinone. Place 25 ml of polyphosphoric acid (Section **II,2,**48) and 2.0 g of o-benzoylbenzoic acid in a 150-ml conical flask fitted with an air condenser to prevent water vapour from the boiling water bath entering the flask during the subsequent heating period of 2 hours. Cool the reaction product in ice and add 40 ml of water with stirring. Filter with suction and wash with water. Boil the residue with 10 ml of concentrated ammonia solution for 5 minutes (to remove unchanged acid) and filter at the pump. Recrystallise from boiling glacial acetic acid (60–70 ml) in the presence of decolourising charcoal; filter off the crystals, wash with a little rectified spirit and dry at 100–120 °C. The yield of pure anthraquinone, m.p. 285–286 °C, is 1.8 g (98%).

2-Methylanthraquinone. Use 2.0 g of o-(p-toluoyl)-benzoic acid and 25 ml of polyphosphoric acid. Recrystallise the product from ethanol in the presence of a little decolourising charcoal. The yield of pure 2-methylanthraquinone, m.p. 175 °C, is 1.7 g (92%).

IV,137. ANTHRONE

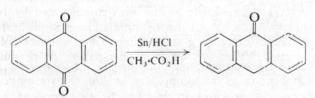

Place 52 g (0.25 mol) of anthraquinone, 50 g (0.42 mol) of granulated tin and 375 ml of glacial acetic acid in a 1-litre round-bottomed flask fitted with a reflux condenser. Heat the contents of the flask to boiling and slowly run in 125 ml of concentrated hydrochloric acid from a dropping funnel down the condenser over a period of 2 hours. By this time all the anthraquinone should have passed into solution; if not, add more tin and hydrochloric acid. Filter the liquid with suction through a sintered glass funnel, and add 50 ml of water. Cool the solution to about 10 °C when the anthrone will crystallise out. Filter the crystals at the pump on a Buchner funnel and wash with water. Dry upon filter-paper or upon a porous tile: the yield of crude anthrone, m.p. about 153 °C, is 40 g (82%). Recrystallise from a 3 : 1 mixture of benzene and light petroleum, b.p. 60–80 °C (10–12 ml per gram); this gives 30 g (61%) of pure anthrone, m.p. 155 °C.

IV,138. 2,4,6-TRIHYDROXYACETOPHENONE (*Phloroacetophenone*)

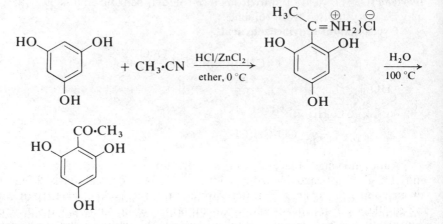

Place 25.2 g (0.2 mol) of dry phloroglucinol (1), 16.4 g (20.9 ml, 0.4 mol) of anhydrous acetonitrile (2), 100 ml of sodium-dried ether and 5 g of finely powdered, fused zinc chloride in a 500-ml Buchner flask fitted with a wide gas inlet tube. Protect the side arm of the flask with a calcium chloride guard-tube. Cool the flask in an ice–salt mixture in the fume cupboard and pass a rapid stream of dry hydrogen chloride (Section **II,2,***31*) through the solution for 2 hours with occasional shaking. Allow the flask to stand in an ice chest for 24 hours, and again pass dry hydrogen chloride into the pale orange mixture for a further 2 hours. Stopper the flask and leave it in an ice chest (or refrigerator) for 3 days. A bulky orange-yellow precipitate of the ketimine hydrochloride is formed. Decant the ether and wash the solid with two 25 ml portions of anhydrous ether. Transfer the solid with the aid of about 1 litre of hot water to a 2-litre round-bottomed flask provided with a reflux condenser. Boil the yellow solution vigorously for 2 hours, allow to cool somewhat, add 4–5 g of decolourising carbon, boil the solution for 5 minutes longer and filter the hot solution with suction through a preheated Buchner funnel. Extract the decolourising carbon with two 100 ml portions of boiling water and add the filtrate to the main product. Allow to stand overnight, and filter the pale yellow or colourless needles of phloroacetophenone at the pump, dry at 120 °C to remove the molecule of water of crystallisation and preserve in a tightly stoppered bottle. The yield is 29 g (85%), m.p. 217–219 °C. This product is pure enough for many purposes, but may be obtained absolutely pure by recrystallisation from hot water (35 ml per gram) and drying at 120 °C; m.p. 218–219 °C.

Notes.   (1) See Section **IV,139**, Note (1).
(2) The acetonitrile may be dried over anhydrous calcium sulphate or by distilling from phosphorus pentoxide.

IV,139. 2,4,6-TRIHYDROXYISOBUTYROPHENONE *(Phlorisobutyrophenone)*

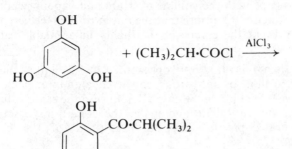

Stir a mixture of 15 g (0.12 mol) of anhydrous phloroglucinol (1), 48 g (0.56 mol) anhydrous powdered aluminium chloride and 60 ml carbon disulphide for 30 minutes in a three-necked flask fitted with a mechanical stirrer, a dropping funnel and a double surface condenser protected by a calcium chloride guard-tube. Add 45 ml of nitrobenzene, stir the reaction mixture for a further 30 minutes and then surround the flask with a water bath maintained at 60 °C. Add a solution of 13 g (0.12 mol) of isobutyryl chloride (Section **III,138**) in 5 ml of nitrobenzene from the dropping funnel over a period of 30 minutes keeping the temperature of the water bath at 60 °C (2). After a further 30 minutes of stirring at this temperature pour the viscous residue on to crushed ice, add 425 g (1.5 mol) of Rochelle salt (sodium potassium tartrate) and neutralise the solution with approximately 40 per cent aqueous sodium hydroxide solution (3). Remove the nitrobenzene and carbon disulphide within a period of 20 minutes by vigorous steam distillation without allowing the volume of residual liquor to increase above about 700 ml (4). Filter off the crystals which separate from the residual solution on cooling and standing (usually overnight) and recrystallise twice from water to give 9.3 g (40%) of pure phlorisobutyrophenone hydrate, m.p. 78–80 °C; the anhydrous product, m.p. 138–140 °C, may be obtained by allowing the hydrate to stand in a vacuum desiccator over phosphorus pentoxide for a few days.

Notes. (1) Phloroglucinol contains two molecules of water of crystallisation; these are removed by heating for 12 hours at 120 °C.

(2) When carried out on a larger scale the volume of hydrogen chloride gas which is evolved justifies the attachment of a gas adsorption trap (Fig. I,68) to the outlet of the calcium chloride guard-tube. On the scale suggested in this experiment, the apparatus should be sited within an efficient fume cupboard and the vapours led via a tube to a drain to prevent corrosion of the stirrer motor.

(3) Phloroisobutyrophenone is unstable when heated in the presence of acid; the addition of Rochelle salt prevents the precipitation of aluminium hydroxide when the free acid is neutralised with sodium hydroxide. If any precipitate does appear more Rochelle salt should be added; the exact amount of aqueous sodium hydroxide will vary with each experimental sequence; the point of neutrality is ascertained with universal indicator paper.

(4) If a bench supply of steam is not available, a large steam-can heated with

three large Bunsen burners, or with the large flame of an air–gas blow lamp may be employed. Two efficient double surface water condensers connected in series will be required to cope with the volume of steam and vapour produced and *great care* must be exercised to ensure that the outlet to the receiver vessel is suitably trapped to prevent the escape of the highly inflammable carbon disulphide vapour.

Cognate preparations **2,4,6-Trihydroxyvalerophenone.** Use 15 g (0.12 mol) of phloroglucinol, 48 g (0.56 mol) of anhydrous powdered aluminium chloride and 14.4 g (0.12 mol) of valeryl chloride (Section **III,138**) with the same volumes of nitrobenzene and carbon disulphide and under the conditions described above. The yield of hydrated product, m.p. 88–90 °C (m.p. 152–154 °C, anhydrous), is 11.5 g (42%).

2′,4′,6′-Trihydroxy-2-methylbutanophenone. Use 15 g (0.12 mol) of phloroglucinol, 48 g (0.56 mol) of anhydrous powdered aluminium chloride and 14.4 g (0.12 mol) of 2-methylbutanoyl chloride (Section **III,138**) under the conditions specified above. The yield of hydrated product, m.p. 61–63 °C, is 14.3 g (52%). The anhydrous product is hygroscopic.

L,2. Synthesis of aromatic ketones from carboxylic acid derivatives. Grignard reagents derived from aryl bromides are readily prepared and may be converted into organocadmium compounds by treatment with cadmium chloride (cf. p. 433, **III,J,4**). Reaction of an organocadmium with a carboxylic acid chloride constitutes a convenient synthesis of aryl alkyl ketones.

$$2ArBr + 2Mg \longrightarrow 2ArMgBr \xrightarrow{CdCl_2} Ar_2Cd + 2MgClBr$$
$$Ar_2Cd + 2R \cdot COCl \longrightarrow 2Ar \cdot CO \cdot R + CdCl_2$$

The reaction is illustrated by the formation of propiophenone from diphenylcadmium and propionyl chloride (Section **IV,140**). Better yields are obtained by carrying out the synthesis in this manner rather than attempting the alternative combination of diethylcadmium with benzoyl chloride.

Several examples of the synthesis of aryl alkyl ketones by the thermal decarboxylation of mixtures of carboxylic acids over heated metal salts are included under the preparation of aliphatic ketones (Section **III,92,B**). In this section the preparation of dibenzyl ketone (Section **IV,141**) by the pyrolysis of the barium salt of phenylacetic acid, which proceeds in good yield, is included as a further example of this general type of synthesis.

IV,140. PROPIOPHENONE (*Ethyl phenyl ketone*)

$$(C_6H_5)_2Cd + 2C_2H_5 \cdot COCl \longrightarrow 2C_6H_5 \cdot CO \cdot C_2H_5 + CdCl_2$$

Prepare a solution of diphenyl cadmium in 110 ml of dry benzene using 4.9 g (0.2 mol) of magnesium, 31.4 g (0.2 mol) of bromobenzene and 19.5 g (0.106 mol) of anhydrous cadmium chloride following the experimental procedure given for the preparation of dibutyl cadmium (Section **III,93**). Cool the solution to 10 °C, and add during 3 minutes a solution of 14.8 g (0.16 mol) of propionyl chloride (b.p. 78–79 °C) in 30 ml of dry benzene; use external cooling with an ice bath to prevent the temperature rising above 40 °C. Stir the mixture for 2 hours at 25–35 °C. Add crushed ice (*ca.* 200 g) and sufficient dilute (1 *M*) sulphuric acid to give a clear aqueous layer. Separate the benzene from the aqueous layer and extract the latter with two 20 ml portions of benzene. Wash the combined extracts successively with 50 ml portions of water, 5 per cent sodium carbonate solution, water and saturated sodium chloride solution. Dry over anhydrous sodium sulphate, remove the benzene by flash distillation and distil the residue under reduced pressure. The yield of propiophenone, b.p. 100–102 °C/16 mmHg, is 17.5 g (82%).

IV,141. DIBENZYL KETONE

$$(C_6H_5 \cdot CH_2 \cdot CO_2)_2 Ba \longrightarrow (C_6H_5 \cdot CH_2)_2 CO + BaCO_3$$

Place 40 g (0.127 mol) of barium hydroxide octahydrate with 60 ml of water in a 250-ml round-bottomed flask and add 34 g (0.25 mol) of phenylacetic acid (Section **III,128,**B) slowly with swirling; warm the mixture until a clear solution is obtained. Evaporate the solution on a water bath under reduced pressure using a rotary evaporator, to yield a pasty mass of moist barium phenylacetate. Fit the flask with a Claisen still-head carrying a gas inlet tube for nitrogen extending well into the flask and a 360 °C thermometer, and attach an air condenser with a receiver flask connected by means of an adapter with side arm. Lag the Claisen head with suitable insulating tape. Pass a slow stream of nitrogen into the flask and heat the latter gently in an air bath. When the residual water has been expelled, change the receiver. Now heat more strongly; dibenzyl ketone passes over at 320–325 °C as a pale yellow oil (24 g) which solidifies on standing. Redistil under reduced pressure and collect pure dibenzyl ketone at 210 °C/35 mmHg as a colourless oil (21 g, 80%); this completely crystallises on standing and has m.p. 33–34 °C.

1.,3. Synthesis of quinones. Quinones of the more reactive, polycyclic, aromatic systems can usually be obtained by direct oxidation which is best carried out with chromium(VI) compounds under acidic conditions. In this way 1,4-naphthoquinone, 9,10-anthraquinone and 9,10-phenanthraquinone are prepared from naphthalene, anthracene and phenanthrene respectively (Section **IV,142**), e.g.:

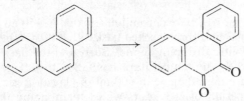

p-Benzoquinone is most conveniently prepared in the laboratory by the oxidation of hydroquinone (Section **IV,143**) with either chromium trioxide in acetic acid or with sodium chlorate in the presence of vanadium pentoxide as a catalyst.

p-Benzoquinone and hydroquinone form a well-defined 1 : 1 molecular complex known as quinhydrone. This complex, in the form of dark green crystals having a glistening metallic lustre, is conveniently prepared (Section **IV,143**) by the partial oxidation of hydroquinone with a solution of iron alum.

The behaviour of p-benzoquinone on reaction with acetic anhydride in the presence of sulphuric acid is of interest. The eventual product is 1,2,4-triacetoxybenzene (the **Thiele acetylation**), which is formed by the following reaction sequence, initiated by a 1,4-addition of acetic anhydride across an α,β-unsaturated carbonyl system.

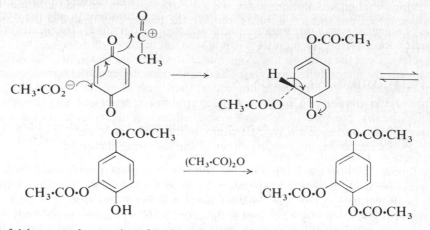

A fairly general procedure for the synthesis of a quinone consists in coupling a phenol with a diazonium salt and reducing the resulting azo compound to an aminophenol with sodium dithionite. Mild oxidation with, for example, iron(III) chloride results in the formation of the corresponding quinone (e.g., the preparation of 1,2-naphthoquinone described and formulated in Section **IV,144**).

IV,142. 1,4-NAPHTHOQUINONE

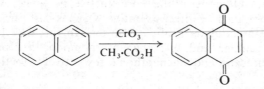

Place a solution of 120 g (1.2 mol) of pure chromium trioxide in 150 ml of 80 per cent aqueous acetic acid in a 2-litre three-necked flask, fitted with a thermometer, mechanical stirrer and 1-litre dropping funnel. Surround the flask by a mixture of ice and salt and, when the temperature has fallen to 0 °C, add a solution of 64 g (0.5 mol) of pure naphthalene in 600 ml of glacial acetic acid, with constant stirring, over a period of 2–3 hours whilst maintaining the internal temperature at 10–15 °C. Continue the stirring overnight, during which time the reaction mixture and bath attain room temperature. Allow the dark green solution to stand for 3 days and stir occasionally. Pour the reaction mixture into 5–6 litres of water, collect the crude naphthoquinone by suction filtration, wash with 200 ml of water and dry in a desiccator. Recrystallise from 500 ml of petroleum ether (b.p. 80–100 °C). The yield of pure 1,4-naphthoquinone, m.p. 124–125 °C, is 17 g (22%).

Cognate preparations 9,10-Anthraquinone. Place 5.0 g of powdered anthracene and 50 ml of glacial acetic acid in a 250-ml, two-necked round-bottomed flask with a reflux condenser and a dropping funnel. Mix the flask contents thoroughly by a swirling action and heat the mixture to reflux when most of the anthracene dissolves. Dissolve 10.0 g of chromium trioxide in 7–8 ml of water, add 25 ml of glacial acetic acid and pour the well-stirred mixture into the dropping funnel. Remove the heat source from the flask and add slowly the oxidising reagent at such a rate that the mixture continues to reflux (7–10 minutes); then reflux for a further 10 minutes when all the anthracene will have reacted completely. Cool the solution and pour into 250 ml of cold water. Stir the mixture vigorously, filter off the precipitated anthraquinone under gentle suction, wash it thoroughly on the filter with hot water, then with 50 ml of hot 1 M-sodium hydroxide solution and finally with much cold water; drain well. Dry the anthraquinone by pressing it between several sheets of filter-paper and leave it overnight in a desiccator over calcium chloride. The yield is 5.5 g (94%).

Purify the anthraquinone by either of the following methods:

(a) Recrystallise the crude product from boiling glacial acetic acid with the aid of decolourising charcoal, wash the resulting crystals on the Buchner funnel with a little cold rectified spirit and dry in the air.

(b) Sublime the dry solid using the procedure described in Section I,21. The purified anthraquinone is obtained as yellow crystals having m.p. 273 °C.

Phenanthraquinone. Add 20 ml of concentrated sulphuric acid cautiously and with stirring to 40 ml of water contained in a 250-ml beaker. Heat to 90–95 °C on a water bath (it may be necessary to place the beaker in the boiling water bath), add 2.0 g (0.011 mol) of purified phenanthrene, and then 12.0 g of potassium dichromate in 0.5 g quantities until a vigorous reaction sets in: the latter usually occurs by the time about half of the oxidising agent has been added. Remove the beaker from the water bath—the temperature of the mixture will be 110–115 °C—and continue adding the potassium dichromate in small portions to maintain the reaction. Do not allow the temperature to fall below 85 °C as the reaction will cease: if necessary, heat on a water bath. When the addition is completed, heat on a boiling water bath for a further 30 minutes.

Cool the beaker in a bath of cold water and add 150 ml of cold water. Filter off the crude phenanthraquinone with suction and wash it with water until free from chromium salts. Suspend the solid in 20 ml of rectified spirit and add, with stirring, 20 ml of saturated sodium metabisulphite solution. Break up the lumps of the addition product with a glass rod and allow to stand, with frequent stirring, for 10 minutes. Add 150 ml of water to dissolve the addition product and filter with suction. Reject the precipitate which consists of the impurities present in the phenanthrene. Add saturated sodium carbonate solution to the filtrate until the bisulphite addition product is completely decomposed: allow the precipitate to settle for 1 minute, then add a few drops of sodium carbonate solution and note whether any further precipitation occurs. Stir the precipitate for 2–3 minutes, filter with suction, wash with three 20 ml portions of water and drain well. Dry the product between filter-papers and then in a desiccator over calcium chloride. The yield of phenanthraquinone, m.p. 206 °C, is 1.4 g (60%). The product may be recrystallised from glacial acetic acid (about 20 ml), but the m.p. is unaffected.

IV,143. *p*-BENZOQUINONE

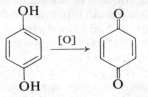

Method 1.   Cool a solution of 33 g (0.33 mol) of hydroquinone in 150 ml of 60 per cent acetic acid contained in a 600-ml beaker to below 5 °C in an ice bath. Dissolve 42 g (0.42 mol) of chromium trioxide in 70 ml of water, and add 30 ml of glacial acetic acid. By means of a separatory funnel with bent stem and supported over the beaker, add the chromium trioxide solution to the mechanically stirred hydroquinone solution at such a rate that the temperature does not rise above 10 °C; the addition takes about 2 hours. Filter the mixture at once and wash the quinone several times with 10 ml portions of ice-cold water. Spread the material upon filter-paper until dry, but no longer or the quinone will be lost through sublimation. The yield of quinone (a bright yellow crystalline solid), m.p. 115 °C, is 21 g (66%); it darkens when exposed to light.

Impure quinone may be purified by placing it in a distilling flask attached to a condenser and passing a rapid current of steam into the flask: the quinone sublimes and collects in the receiver. It is separated from the water by filtration and dried; the m.p. is 116 °C. The vapour has a penetrating odour and attacks the eyes.

Method 2.   In a 1-litre round-bottomed flask, provided with a mechanical stirrer, place 0.5 g of vanadium pentoxide (catalyst), 500 ml of 2 per cent sulphuric acid, 55 g (0.5 mol) of hydroquinone and 30 g of sodium chlorate. Stir the mixture vigorously for about 4 hours. Greenish-black quinhydrone is first formed and this is converted into yellow quinone; the temperature of the mixture rises to about 40 °C (do not allow it to exceed this temperature). Cool the flask in running water, filter the mixture at the pump and wash it with 50 ml of cold water. Dry the quinone upon filter-paper in the air (see *Method 1*) or in a desiccator over anhydrous calcium chloride. The yield is 45 g (83%), m.p. 111–112 °C. The crude quinone may be purified by steam distillation as in *Method 1*, or by recrystallisation from boiling light petroleum, b.p. 100–120 °C (12 ml per gram): the resulting pure, bright yellow quinone has m.p. 115 °C and the recovery is about 95 per cent.

Conversion of hydroquinone into quinhydrone.   Dissolve 100 g of iron alum (iron(III) ammonium sulphate) in 300 ml of water at 65 °C. Pour the solution, with stirring, into a solution of 25 g (0.228 mol) of hydroquinone in 100 ml of water contained in a 600-ml beaker. The quinhydrone is precipitated in fine needles. Cool the mixture in ice, filter with suction and wash three or four times with cold water. Dry in the air between filter-paper. The yield of quinhydrone, m.p. 172 °C, is 15 g (60%).

Conversion of *p*-benzoquinone into 1,2,4-triacetoxybenzene (*Thiele acetylation*).   Add 11 g (0.1 mol) of *p*-benzoquinone in small portions to a mechanically stirred mixture of 33 g (0.32 mol) of acetic anhydride and 0.25 ml of concentrated sulphuric acid. The temperature of the mixture rises to 40–50 °C and is kept within this range by regulating the rate of addition of the quinone.

When the addition is complete allow the solution to cool to about 25 °C and pour into 150 ml of cold water. Collect the precipitated triacetate and recrystallise it from about 50 ml of rectified spirit; the yield is 22 g (86%), m.p. 97 °C.

IV,144. 1,2-NAPHTHOQUINONE

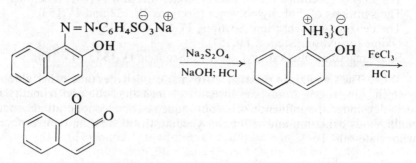

Place 20 g (0.057 mol) of Orange II (Section **IV,95**) in a 600-ml beaker and dissolve it in 250 ml of water at 40–50 °C. Add, with stirring, 24–25 g (0.114 mol) of hydrated sodium dithionite; this discharges the colour and yields a pink or cream-coloured, finely divided precipitate of 1-amino-2-naphthol (compare Section **IV,91**). Heat the mixture nearly to boiling until it commences to froth considerably, then cool to 25 °C in ice, filter on a Buchner funnel and wash with a little cold water. Transfer the precipitate to a beaker containing a solution of 0.25 g of tin(II) chloride in 5 ml of concentrated hydrochloric acid diluted with 100 ml of water; upon stirring the aminonaphthol dissolves and a small amount of insoluble matter remains. The function of the tin(II) chloride is as an antioxidant, preventing the readily oxidisable aminonaphthol hydrochloride from undergoing appreciable change. Stir the solution for 5 minutes with 2 g of decolourising carbon, and filter at the pump. If crystalline material should separate at any stage, dissolve it by warming and by the addition of a little water if necessary. Transfer the clear solution to a beaker, add 25 ml of concentrated hydrochloric acid and warm until the solid dissolves. Cool to 0 °C, filter the almost colourless crystals of the aminonaphthol hydrochloride with suction and wash with 25 ml of dilute hydrochloric acid (1 : 4 by volume). *From this point all operations must be carried out rapidly.* In the meantime, prepare the oxidising solution by dissolving 30 g (0.11 mol) of crystallised iron(III) chloride in a mixture of 10 ml of concentrated hydrochloric acid and 25 ml of water by heating, cool to room temperature by adding *ca.* 30 g of crushed ice and filter the solution at the pump. Wash the crystalline 1-amino-2-naphthol hydrochloride into a 600-ml beaker with water, add 150 ml of water and a few drops of concentrated hydrochloric acid and dissolve the precipitated solid by stirring and warming to about 35 °C. If necessary, filter rapidly by suction from a trace of residue, transfer to a 500-ml round-bottomed flask, add the iron(III) chloride solution all at once whilst shaking the flask vigorously. The quinone separates rapidly as a voluminous micro-crystalline yellow precipitate. Filter on a Buchner funnel and wash it thoroughly with water at 30 °C to remove all traces of acid. Dry the product upon filter-paper in an atmosphere free from acid fumes. The yield of 1,2-naphthoquinone, which melts with decomposition at 145–147 °C, is 7 g (78%).

M   SOME REACTIONS OF AROMATIC CARBONYL COMPOUNDS

1. The Cannizzaro reaction (Sections **IV,145** and **IV,146**).
2. The Claisen–Schmidt and related reactions (Sections **IV,147** to **IV,149**).
3. The Perkin (Section **IV,150**) and Doebner (Section **IV,151**) reactions.
4. The synthesis of diphenylpolyenes (Sections **IV,152** and **IV,153**).
5. The benzoin condensation (Sections **IV,154** to **IV,156**).
6. Oxime formation (Section **IV,157**).
7. Some reactions of alkyl aryl ketones (Sections **IV,158** to **IV,160**).

M,1. The Cannizzaro reaction.   Aromatic aldehydes (and other aldehydes in which α-hydrogen atoms are absent, e.g., formaldehyde and trimethylacetaldehyde) under the influence of strong aqueous or alcoholic alkali undergo simultaneous oxidation and reduction yielding the alcohol and corresponding carboxylate salt. Thus:

$$2Ar\cdot CHO \xrightarrow{\text{KOH}} Ar\cdot CH_2OH + Ar\cdot CO_2^{\ominus}K^{\oplus}$$

This dismutation or disproportionation reaction is known as the **Cannizzaro reaction**. The mechanism of the reaction involves the production of the anion (I) which may transfer a hydride ion to a carbonyl carbon atom in another aldehyde molecule. The reaction sequence is completed by a proton transfer to yield the carboxylate anion and the alcohol.

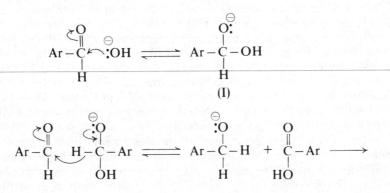

$$Ar\cdot CH_2OH + Ar\cdot CO_2^{\ominus}$$

The reaction is illustrated here by the conversion of benzaldehyde into benzoic acid and benzyl alcohol (Section **IV,145**) and by a similar conversion of furfural into furfuryl alcohol and furoic acid (cognate preparation in Section **IV,145**).

A preparatively more useful form of this reaction is the crossed Cannizzaro reaction which ensues when a mixture of an aromatic aldehyde and formaldehyde is allowed to react under the influence of strong base (e.g., the preparation of p-methylbenzyl alcohol, Section **IV,146**). A substantial proportion of the aromatic aldehyde is reduced to the corresponding alcohol whilst the formaldehyde is oxidised to formate. This is a reflection of the fact that nucleophilic attack of the hydroxide ion takes place preferentially at the more electrophilic carbonyl carbon atom in formaldehyde.

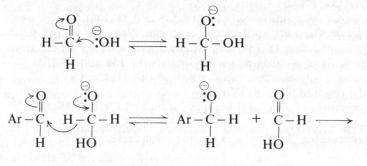

$$Ar\cdot CH_2OH + H\cdot CO_2^\ominus$$

IV,145. BENZYL ALCOHOL AND BENZOIC ACID

$$C_6H_5\cdot CHO + C_6H_5\cdot CHO \xrightarrow{KOH} C_6H_5\cdot CH_2OH + C_6H_5\cdot CO_2^\ominus K^\oplus$$

Dissolve 29 g of potassium hydroxide in 27 ml of water contained in a beaker or conical flask, and cool the solution to about 20 °C in ice-water. Pour the solution into a 250-ml reagent bottle, and add 32 g (30 ml, 0.3 mol) of pure benzaldehyde (1); *cork* the bottle securely and shake the mixture vigorously until it has been converted into a thick emulsion. Allow the mixture to stand overnight or for 24 hours in the stoppered bottle. Add *just sufficient water* (about 105 ml) to dissolve the potassium benzoate. Pour the liquid into a separatory funnel, rinse out the bottle with about 30 ml of ether and add this ether to the solution in the funnel. Shake the solution in order to thoroughly extract the benzyl alcohol with the ether, separate the lower aqueous solution and carry out two further extractions each with about 25 ml of ether. *Save the aqueous solution.* Combine the ether extracts and distil the ether from a water bath (rotary evaporator) until the volume is about 25 ml. Cool and shake the ether solution twice with 5 ml portions of saturated sodium metabisulphite solution in order to remove any benzaldehyde which may be present. Separate the ethereal solution, wash it with 10 ml of 10 per cent sodium carbonate solution (to ensure complete removal of the bisulphite), then with 10 ml of water, and dry with anhydrous magnesium sulphate or anhydrous potassium carbonate. Remove the ether (Fig. I,98; 50-ml distilling flask) on a water bath, and distil the residual liquid over a wire gauze or, better, from an air bath; replace the water condenser by an air condenser or empty the water completely from the condenser jacket. Collect the benzyl alcohol at 204–207 °C (the pure compound boils at 205.5 °C). The yield is 13 g (86.5%).

Pour the aqueous solution remaining from the ether extraction with stirring into a mixture of 80 ml of concentrated hydrochloric acid, 80 ml of water and about 100 g of crushed ice. Filter the precipitated benzoic acid at the pump, wash it with a little cold water, drain and recrystallise from boiling water. The yield of benzoic acid (colourless crystals), m.p. 121 °C, is 13.5 g (79%) (2).

Notes. (1) The benzaldehyde should be free from benzoic acid; it may be purified as described in Section **IV,22.**

(2) The reaction may alternatively be carried out by boiling the benzaldehyde vigorously under reflux for 2 hours with a solution of 20 g of potassium hydroxide in 90 ml of water, and then working up the cooled reaction mixture as described above. Reaction is normally complete under these conditions; the extent of the reaction may be determined by subjecting the crude ether extract, before the latter is washed with bisulphite solution, to GLC analysis on a 5-ft column of Chromosorb W with 10 per cent of Carbowax as the stationary phase, at 156 °C with a nitrogen flow rate of 40 ml per minute. The retention times of benzaldehyde and benzyl alcohol are approximately 2.7 and 9.7 minutes respectively.

Cognate preparation   Furfuryl alcohol and furoic acid.

Place 200 g (172.5 ml, 2.08 mol) of redistilled furfural (1) in a 1-litre beaker provided with a mechanical stirrer and surrounded by an ice bath. Start the stirrer and, when the temperature has fallen to 5–8 °C, add a solution of 50 g (1.2 mol) of sodium hydroxide in 100 ml of water from a separatory funnel at such a rate that the temperature of the reaction mixture does not rise above 20 °C (20–25 minutes); continue the stirring for a further 1 hour. Much sodium furoate separates during the reaction. Allow to cool to room temperature, and add just enough water to dissolve the precipitate (about 65 ml). Extract the solution at least five times with 60 ml portions of ether in order to remove the furfuryl alcohol: the best results are obtained by the use of the continuous extraction apparatus (charged with 350 ml of ether) depicted in Fig. I,98. Keep the aqueous layer. Dry the ethereal extract with a little magnesium sulphate, and remove the ether on a rotary evaporator. Distil the residue under reduced pressure (Fig. I,110) and collect the furfuryl alcohol (a very pale yellow liquid) at 75–77 °C/15 mmHg; the yield is 65 g (64%). Because of the tendency to undergo polymerisation, add about 1 per cent of its weight of urea as stabiliser if the furfuryl alcohol is to be stored.

Treat the aqueous solution, containing the sodium furoate, with 40 per cent sulphuric acid until it is acid to Congo red paper, and cool. Filter off the furoic acid, contaminated with a little sodium hydrogen sulphate, at the pump. Dissolve it in 240 ml of boiling water, add 12 g of decolourising carbon, boil the solution for about 45 minutes, filter hot and cool the filtrate with stirring to 16–20 °C; below 16 °C, sodium hydrogen sulphate also separates. Filter off the furoic acid with suction, and dry. The yield is 65 g (55%), m.p. 123–124 °C. It may be further purified either by recrystallisation from carbon tetrachloride to which a little decolourising carbon is added or by distillation under reduced pressure, b.p. 142–144 °C/20 mmHg; the resulting pure acid softens at 125 °C and is completely melted at 132 °C.

Note. (1) Furfural (2-furaldehyde) is best purified by distillation under reduced pressure: b.p. 54–55 °C/17 mmHg.

IV,146. *p*-METHYLBENZYL ALCOHOL (*Crossed Cannizzaro reaction*)

$$CH_3 \cdot C_6H_4 \cdot CHO + H \cdot CHO + KOH \longrightarrow$$
$$CH_3 \cdot C_6H_4 \cdot CH_2OH + H \cdot CO_2^{\ominus}K^{\oplus}$$

Equip a 1-litre three-necked flask with a reflux condenser, a sealed mechanical stirrer and a thermometer; the bulb of the thermometer should reach almost to the bottom of the flask. Place 170 g of commercial potassium hydroxide pellets (about 85% KOH, *ca.* 2.6 mol) and 250 ml of methanol in the flask and set the stirrer in motion. Most of the alkali dissolves in a few minutes and the temperature rises considerably. Immerse the flask in a large cold-water bath and, when the temperature has fallen to 60–65 °C, add down the condenser a mixture of 120 g (118 ml, 1 mol) of *p*-tolualdehyde (Section **IV,125**) and 100 ml (*ca.* 1.3 mol) of formalin at such a rate (during about 15 minutes) that the internal temperature remains at 60–70 °C: maintain the internal temperature at 60–70 °C for a further 3 hours. Replace the reflux condenser by a condenser set for downward distillation, and distil off the methanol, while stirring, until the temperature reaches about 100 °C. Add 300 ml of water to the warm residue, cool the mixture and separate the resulting two layers at once; if the upper layer is allowed to stand, it will solidify. Extract the aqueous layer with four 50 ml portions of toluene. Wash the combined oil and toluene extracts with five 25 ml portions of water, extract the combined washings with 25 ml of toluene and add the toluene layer to the washed extract. Dry the toluene solution by shaking with a few grams of magnesium sulphate, distil off the toluene and finally distil under reduced pressure (Fig. I,110) and collect the *p*-methylbenzyl alcohol at 116–118 °C/20 mmHg (1). The product solidifies in the receiver to a mass (110 g) of oily crystals, m.p. 54–55 °C. Recrystallise from an equal weight of technical heptane (b.p. 90–100 °C); 88 g (72%) of pure *p*-methylbenzyl alcohol, m.p. 61 °C, are obtained.

Note. (1) The b.p. at atmospheric pressure is 217 °C.

M,2. Claisen–Schmidt and related reactions. Aromatic aldehydes condense with aliphatic or mixed alkyl aryl ketones in the presence of aqueous alkali to form α,β-unsaturated ketones (the **Claisen–Schmidt reaction**).

$$Ar \cdot CHO + CH_3 \cdot CO \cdot R \xrightarrow{\ominus OH} [Ar \cdot CH(OH) \cdot CH_2 \cdot CO \cdot R] \xrightarrow{-H_2O}$$
$$Ar \cdot CH = CH \cdot CO \cdot R$$

The first step is a condensation of the aldol type (see p. 589, **III,U,1**) involving the nucleophilic addition of the carbanion derived from the methyl ketone to the carbonyl–carbon of the aromatic aldehyde. Dehydration of the hydroxy-ketone to form the conjugated unsaturated carbonyl compound occurs spontaneously.

Section **IV,147** describes the preparation of a range of α,β-unsaturated ketones, including benzylideneacetone, furfurylideneacetone and benzylidene-acetophenone. The conversion of this latter compound into β-phenylpropiophenone is readily achieved by hydrogenation at atmospheric pressure over an active platinum catalyst.

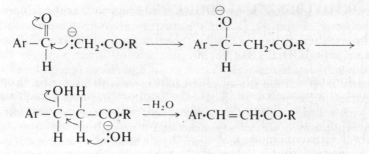

The formation of ω-nitrostyrenes (illustrated in Section **IV,148**) by reaction of nitroalkanes with aromatic aldehydes in the presence of aqueous alkali may be classified with reactions of the Claisen–Schmidt type.

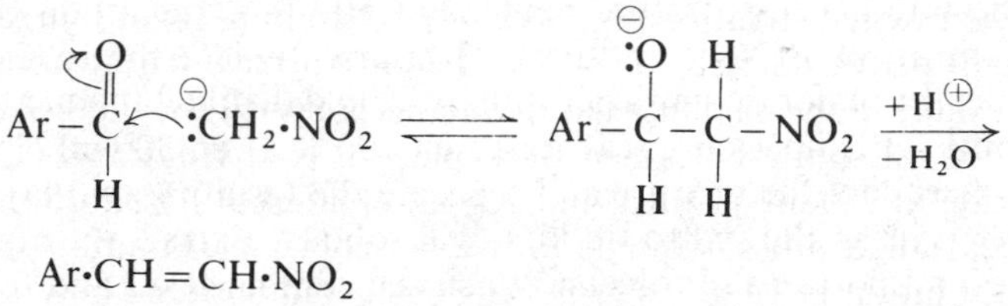

$$Ar-\overset{O}{\underset{H}{C}}\!\!\diagdown:CH_2 \cdot NO_2 \; \rightleftharpoons \; Ar-\overset{:O}{\underset{H}{C}}-\overset{H}{\underset{H}{C}}-NO_2 \; \xrightarrow[-H_2O]{+H^{\oplus}}$$

$$Ar \cdot CH = CH \cdot NO_2$$

A further example of the above reaction type is provided by the condensation between an aromatic aldehyde and an ester (the **Claisen reaction**, e.g., the synthesis of ethyl cinnamate, Section **IV,149**), which requires a more powerfully basic catalyst (e.g., sodium ethoxide) to effect conversion of the ester into the corresponding anion.

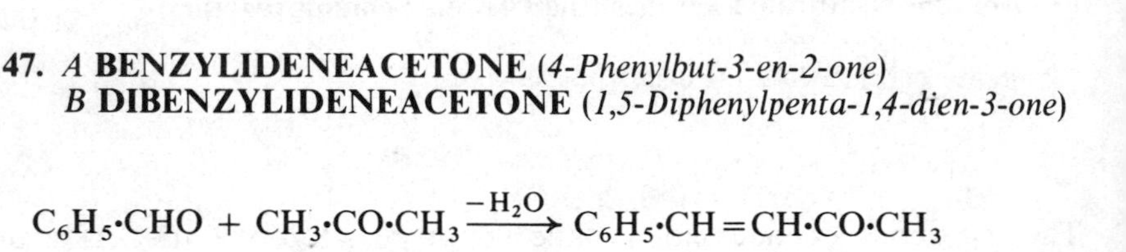

$$C_2H_5O\overset{\ominus}{:} \quad H-CH_2 \cdot CO_2C_2H_5 \; \rightleftharpoons \; C_2H_5OH \; + \; :CH_2 \cdot CO_2C_2H_5$$

$$C_6H_5 \cdot \overset{O}{\underset{H}{C}}\!\!\diagdown:CH_2 \cdot CO_2C_2H_5 \; \xrightarrow[-H_2O]{+H^{\oplus}} \; C_6H_5 \cdot CH = CH \cdot CO_2C_2H_5$$

IV,147. *A* **BENZYLIDENEACETONE** (*4-Phenylbut-3-en-2-one*)
    B **DIBENZYLIDENEACETONE** (*1,5-Diphenylpenta-1,4-dien-3-one*)

A

$$C_6H_5 \cdot CHO + CH_3 \cdot CO \cdot CH_3 \xrightarrow{-H_2O} C_6H_5 \cdot CH = CH \cdot CO \cdot CH_3$$

B

$$2C_6H_5 \cdot CHO + CH_3 \cdot CO \cdot CH_3 \xrightarrow{-2H_2O}$$
$$C_6H_5 \cdot CH = CH \cdot CO \cdot CH = CH \cdot C_6H_5$$

A. Place 42.5 g (40.5 ml, 0.4 mol) of pure benzaldehyde (Section **IV,22**) and 63.5 g (80 ml, 1.1 mol) of pure acetone in a 250-ml flask equipped with a mechanical stirrer. Immerse the reaction vessel in a bath of cold water and add slowly (during about 30 minutes) from a dropping funnel 10 ml of 10 per cent sodium hydroxide solution: adjust the rate of addition so that the temperature remains between 25 and 30 °C. Stir the mixture at room temperature for a further 2 hours; alternatively, securely stopper the flask and shake mechanically for the same period. Render the mixture just acid to litmus paper by the addition of dilute hydrochloric acid. Transfer to a separatory funnel. Remove the upper organic layer, extract the lower aqueous layer with 20 ml of toluene and add the extract to the yellow upper layer. Wash the latter with 20 ml of water, and dry with a little magnesium sulphate, and transfer to a distillation assembly incorporating a Claisen still-head and a short fractionating side arm (compare Fig. I,110). Remove the toluene by distillation at atmospheric pressure and distil the residue under diminished pressure. The benzylideneacetone distils at 133–143 °C/16 mmHg (or at 120–130 °C/7 mmHg or at 150–160 °C/ 25 mmHg) and solidifies to a crystalline mass on standing, m.p. 38–39 °C; the yield is 45 g (77%). This is pure enough for most practical purposes, but may be further purified by redistillation (b.p. 137–142 °C/16 mmHg) or by recrystallisation from light petroleum (b.p. 40–60 °C): the pure benzylideneacetone melts at 42 °C. The residue in the distilling flask contains some dibenzylideneacetone.

B. In a 500-ml round-bottomed flask place a cold solution of 25 g of sodium hydroxide in 250 ml of water and 200 ml of ethanol (1); equip the flask with a mechanical stirrer and surround it with a bath of water. Maintain the temperature of the solution at 20–25 °C, stir vigorously and add one-half of a previously prepared mixture of 26.5 g (25.5 ml, 0.25 mol) of pure benzaldehyde and 7.3 g (9.3 ml, 0.125 mol) of acetone. A flocculent precipitate forms in 2–3 minutes. After 15 minutes add the remainder of the benzaldehyde–acetone mixture. Continue the stirring for a further 30 minutes. Filter at the pump and wash with cold water to eliminate the alkali as completely as possible. Dry the solid at room temperature upon filter-paper to constant weight; 27 g (93%) of crude dibenzylideneacetone, m.p. 105–107 °C, are obtained. Recrystallise from hot ethyl acetate (2.5 ml per gram) or from hot rectified spirit. The recovery of pure dibenzylideneacetone, m.p. 112 °C, is about 80 per cent.

Note. (1) Sufficient ethanol is employed to dissolve the benzaldehyde and to retain the initially-formed benzylideneacetone in solution until it has had time to react with the second molecule of benzaldehyde.

Cognate preparations   Furfurylideneacetone.

$$
\begin{array}{c}
\text{HC}\text{---}\text{CH} \\
\parallel \quad \parallel \\
\text{HC}\diagdown_{O}\diagup \text{C}-\text{CHO} + \text{CH}_3\cdot\text{CO}\cdot\text{CH}_3 \xrightarrow[\text{aq.}]{\text{NaOH}}
\end{array}
$$

$$
\begin{array}{c}
\text{HC}\text{---}\text{CH} \\
\parallel \quad \parallel \\
\text{HC}\diagdown_{O}\diagup \text{C}-\text{CH}=\text{CH}\cdot\text{CO}\cdot\text{CH}_3 + \text{H}_2\text{O}
\end{array}
$$

In a 1-litre bolt-head flask, equipped with a mechanical stirrer, mix 75 g (65 ml, 0.78 mol) of redistilled furfural (1) and 600 ml of water. Add 100 g (126 ml, 1.73 mol) of acetone. Stir the mixture, cool to 10 °C and add a solution of 5 g of sodium hydroxide in 10 ml of water; some heat is generated. Continue the stirring, without cooling, for 4 hours. Then add 10 per cent sulphuric acid (about 70 ml) until the mixture is acid to litmus, whereupon the milkiness disappears and the liquid separates out into layers. Separate the lower organic layer, dry it with a little anhydrous magnesium sulphate and distil under reduced pressure from a flask with fractionating side arm (compare Fig. I,110). Collect the furfurylideneacetone at 114–118 °C/10 mmHg; it solidifies on cooling (m.p. 38–39 °C) and weighs 65 g (62%). The residue of high boiling point material in the flask contains much difurfurylideneacetone.

Note.   (1) Furfural is best purified by distillation under reduced pressure: b.p. 54–55 °C/17 mmHg.

Benzylideneacetophenone (*Chalcone*).

$$C_6H_5 \cdot CHO + H_3C \cdot CO \cdot C_6H_5 \xrightarrow{\text{NaOH}} C_6H_5 \cdot CH = CH \cdot CO \cdot C_6H_5$$

Place a solution of 22 g of sodium hydroxide in 200 ml of water and 100 g (122.5 ml) of rectified spirit in a 500-ml bolt-head flask provided with a mechanical stirrer. Immerse the flask in a bath of crushed ice, pour in 52 g (0.43 mol) of freshly distilled acetophenone (Section **IV,134**), start the stirrer and then add 46 g (44 ml, 0.43 mol) of pure benzaldehyde. Keep the temperature of the mixture at about 25 °C (the limits are 15–30 °C) and stir vigorously until the mixture is so thick that stirring is no longer effective (2–3 hours). Remove the stirrer and leave the reaction mixture in an ice chest or refrigerator overnight. Filter the product with suction on a Buchner funnel or a sintered glass funnel, wash with cold water until the washings are neutral to litmus, and then with 20 ml of ice-cold rectified spirit. The crude chalcone, after drying in the air, weighs 88 g and melts at 50–54 °C. Recrystallise from rectified spirit warmed to 50 °C (about 5 ml per gram). The yield of pure benzylideneacetophenone (a pale yellow solid), m.p. 56–57 °C, is 77 g (85%). This substance should be handled with great care since it acts as a skin irritant.

Hydrogenation to β-phenylpropiophenone.

$$C_6H_5 \cdot CH = CH \cdot CO \cdot C_6H_5 \xrightarrow{\text{H}_2/\text{Pt}} C_6H_5 \cdot CH_2 \cdot CH_2 \cdot CO \cdot C_6H_5$$

Place a solution of 10.4 g (0.05 mol) of benzylideneacetophenone, m.p. 57 °C, in 75 ml of pure ethyl acetate (Section **II,124**) in the reaction bottle of the atmospheric pressure hydrogenation apparatus (Section **I,17,1**) and add 0.2 g of Adams' platinum oxide catalyst (Section **II,2,51**). Displace the air with hydrogen, and shake the mixture with hydrogen until 0.05 mol is absorbed (i.e., *ca.* 1100 ml at s.t.p.). Filter off the platinum, and remove the ethyl acetate by distillation. Recrystallise the residual β-phenylpropiophenone from about 12 ml of ethanol. The yield of pure product, m.p. 73 °C, is 9 g (86%).

IV,148. ω-NITROSTYRENE

$$C_6H_5 \cdot CHO + CH_3 \cdot NO_2 \xrightarrow{-H_2O} C_6H_5 \cdot CH = CH \cdot NO_2$$

Equip a 1500-ml three-necked flask with a thermometer, mechanical stirrer and a dropping funnel. Place 61 g (54 ml, 1 mol) of nitromethane (1), 106 g (101 ml, 1 mol) of purified benzaldehyde (Section **IV,22**) and 200 ml of methanol in the flask and cool it with a mixture of ice and salt to about − 10 °C. Dissolve

42 g of sodium hydroxide in 40–50 ml of water, cool and dilute to 100 ml with ice and water; place this cold solution in the dropping funnel. Add the sodium hydroxide solution, with vigorous stirring, to the nitromethane mixture at such a rate that the temperature is held at 10–15 °C. Introduce the first few ml cautiously since, after a short induction period, the temperature may rise to 30 °C or higher; check the rise in temperature, if necessary, by adding a little crushed ice to the reaction mixture. A bulky white precipitate forms; if the mixture becomes so thick that stirring is difficult, add about 10 ml of methanol. After standing for about 15 minutes, add 700 ml of ice-water containing crushed ice; the temperature should be below 5 °C. Run the resulting cold solution immediately from a dropping funnel and with stirring into 500 ml of 4 M-hydrochloric acid contained in a 3-litre flask; adjust the rate of addition so that the stream just fails to break into drops. A pale yellow crystalline precipitate separates almost as soon as the alkaline solution mixes with the acid. The solid settles to the bottom of the vessel when the stirrer is stopped. Decant most of the cloudy liquid layer, filter the residue by suction and wash it with water until free from chlorides. Transfer the solid to a beaker immersed in hot water; two layers form and on cooling again, the lower layer of nitrostyrene solidifies; pour off the upper water layer. Dissolve the crude nitrostyrene in 85 ml of hot ethanol (*FUME CUPBOARD:* nitrostyrene vapours are irritating to the nose and eyes, and the skin of the face is sensitive to the solid), filter through a hot water funnel and cool until crystallisation is complete. The yield of pure ω-nitrostyrene, m.p. 57–58 °C, is 125 g (85%).

Note. (1) The commercial material may be redistilled and the fraction having b.p. 100–102 °C collected.

Cognate preparations **3,4-Methylenedioxy-ω-nitrostyrene.** In a 250-ml round-bottomed flask mix 30 g (0.20 mol) of 3,4-methylenedioxybenzaldehyde (*piperonal*), 13.4 g (0.22 mol) of nitromethane, 7.8 g (0.1 mol) of ammonium acetate and 50 ml of glacial acid. Attach a reflux condenser, and boil the mixture under gentle reflux for 1 hour. Pour the reaction mixture with stirring into a large excess of ice-water (about 1 litre). When all the ice has melted, filter off the crude product under suction and recrystallise from a mixture of absolute ethanol and acetone (about 2 : 1 v/v). Almost pure yellow crystals of the nitrostyrene, m.p. 161 °C, are obtained. The yield is 23.3 g (60%). Further recrystallisation from the same solvent yields the pure compound, m.p. 162 °C.

2,4-Dimethoxy-ω-nitrostyrene. Follow the above procedure, but use 33.2 g (0.20 mol) of 2,4-dimethoxybenzaldehyde as the starting material. The yield of recrystallised product (yellow crystals, m.p. 103 °C) is 28.5 g (68%). Further recrystallisation gives pure product of m.p. 105 °C.

IV,149. ETHYL CINNAMATE

$$C_6H_5 \cdot CHO + CH_3 \cdot CO_2C_2H_5 \xrightarrow{Na/C_2H_5ONa}$$

$$C_6H_5 \cdot CH = CH \cdot CO_2C_2H_5 + H_2O$$

Prepare powdered (or 'molecular') sodium from 14.5 g (0.63 mol) of clean sodium and 150–200 ml of sodium-dried xylene contained in a 1-litre three-necked flask (Section **II,2,**_55_) fitted with a mechanical stirrer and a reflux condenser. When cold, pour off the xylene as completely as possible, and then add 220 g (240 ml, 2.5 mol) ethyl acetate (Section **II,1,**_24_) containing 2 ml of absolute ethanol (1). Cool the flask rapidly to 0 °C and add 53 g (51 ml, 0.5 mol) of pure benzaldehyde (Section **IV,22**) slowly (during 90 minutes) from a dropping funnel whilst the mixture is stirred. Keep the temperature between 0 and 5 °C; do not allow it to rise above 10 °C otherwise a poor yield will be obtained. The reaction commences as soon as the benzaldehyde is added, as is indicated by the production of a reddish substance on the particles of sodium. Continue the stirring until practically all the sodium has reacted (about 1 hour after all the benzaldehyde has been introduced). Then add 45 ml of glacial acetic acid followed by an equal volume of water (**CAUTION:** some sodium may be present). Separate the layer of ester, extract the aqueous layer with 25 ml of ethyl acetate, wash the combined organic layers with 150 ml of 1 : 1 hydrochloric acid and dry with magnesium or anhydrous sodium sulphate. Distil off the ethyl acetate on a water bath. Distil the residue under diminished pressure (Fig. I,110). Collect the ethyl cinnamate (a colourless liquid) at 126–131 °C/6 mmHg; the yield is 65 g (74%) (2).

Notes.   (1) A little ethanol (_ca._ 1%) is required to start the reaction; the yield is consistently lower in its absence.

(2) Ethyl cinnamate may also be prepared by the esterification of cinnamic acid (cf. _methyl cinnamate_, Section **IV,177**). The pure compound boils at 127 °C/6 mmHg.

M,3. The Perkin and Doebner reactions.   The condensation of an aromatic aldehyde with an acid anhydride in the presence of the sodium or potassium salt of the acid corresponding to the anhydride to yield an α,β-unsaturated acid is known as the **Perkin reaction**.

$$Ar{\cdot}CHO + (CH_3{\cdot}CO)_2O \xrightarrow{CH_3{\cdot}CO_2^{\ominus}Na^{\oplus}}$$
$$Ar{\cdot}CH{=}CH{\cdot}CO_2H + CH_3{\cdot}CO_2H$$

The mechanism of the reaction, which is of the aldol type, involves the carbonyl group of the aldehyde and an active methylene group of the anhydride; the function of the basic catalyst [acetate anion, $CH_3CO_2^{\ominus}$, or triethylamine, $(C_2H_5)_3N$] is to form an anion by removal of a proton from the anhydride:

$$B{:}\quad H{-}CH_2{\cdot}CO{\cdot}O{\cdot}CO{\cdot}CH_3 \rightleftharpoons \overset{\oplus}{B}H + \overset{\ominus}{:}CH_2{\cdot}CO{\cdot}O{\cdot}CO{\cdot}CN_3$$

$$Ar{-}\overset{O}{\underset{H}{\overset{\|}{C}}}\quad \overset{\ominus}{:}CH_2{\cdot}CO{\cdot}O{\cdot}CO{\cdot}CH_3 \rightleftharpoons Ar{-}\underset{H}{\overset{:\overset{\ominus}{O}}{\overset{|}{C}}}{-}CH_2{\cdot}CO{\cdot}O{\cdot}CO{\cdot}CH_3$$

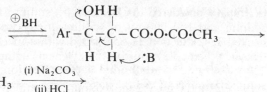

$$Ar \cdot CH = CH \cdot CO \cdot O \cdot CO \cdot CH_3 \xrightarrow[\text{(ii) HCl}]{\text{(i) Na}_2\text{CO}_3}$$

$$Ar \cdot CH = CH \cdot CO_2 H + CH_3 CO_2 H$$

The standard procedure is illustrated by the preparation of cinnamic acid and furylacrylic acid (Section **IV,150**). The cinnamic acid obtained is the more stable *trans* geometric isomer. It may be readily reduced to the saturated acid (3-phenylpropionic acid) and two procedures are described. Catalytic hydrogenation is a convenient method, but the conjugated double bond may also be reduced with, for example, sodium amalgam in the presence of alkali.

A modified Perkin procedure is illustrated in the synthesis of α-phenyl-cinnamic acid in which benzaldehyde is condensed with phenylacetic acid in the presence of acetic anhydride and triethylamine. Presumably equilibria are set up which result in the formation of either a mixed anhydride, phenylacetic acetic anhydride or the symmetrical phenylacetic anhydride.

$$C_6H_5 \cdot CH_2 \cdot CO_2 H + (CH_3 \cdot CO)_2 O \rightleftharpoons$$
$$C_6H_5 \cdot CO \cdot O \cdot CO \cdot CH_3 + CH_3 \cdot CO_2 H$$

$$2C_6H_5 \cdot CH_2 \cdot CO_2 H + 2(CH_3 \cdot CO)_2 O \rightleftharpoons$$
$$(C_6H_5 \cdot CH_2 \cdot CO)_2 O + 2CH_3 \cdot CO_2 H$$

Coumarin is formed from acetic anhydride and salicylaldehyde in the presence of triethylamine as the base catalyst. It is the lactone of the *cis* form of *o*-hydroxycinnamic acid; some of the *trans* isomer in the form of its acetyl derivative (*o*-acetoxycinnamic acid) is also obtained (Section **IV,150**).

Arylacrylic acids may alternatively be conveniently prepared by the **Doebner modification** of the **Knoevenagel reaction**. The Knoevenagel reaction embraces a number of base-catalysed condensations between a carbonyl compound and a component having an active methylene group (see also p. 489, **III,M,6(**b**)**). Examples of the Doebner modification, which usually involves the reaction of an aldehyde with malonic acid in the presence of pyridine or possibly a little piperidine, are given in Section **IV,151**. The reaction mechanism is of the aldol type and involves the formation of a hydroxymalonic acid and then an α,β-unsaturated malonic acid, which undergoes decarboxylation at the temperature of refluxing pyridine.

$$Ar \cdot CHO + CH_2(CO_2 H)_2 \xrightarrow{\text{base}} [Ar \cdot CH(OH) \cdot CH(CO_2 H)_2] \xrightarrow{-H_2O}$$

$$Ar \cdot CH = C(CO_2 H)_2 \xrightarrow[-CO_2]{\text{heat}} Ar \cdot CH = CH \cdot CO_2 H$$

IV,150. CINNAMIC ACID

$$C_6H_5 \cdot CHO + (CH_3 \cdot CO)_2O \xrightarrow[\text{(ii) } H_2O, HCl]{\text{(i) } CH_3 \cdot CO_2K}$$
$$C_6H_5 \cdot CH = CH \cdot CO_2H + CH_3 \cdot CO_2H$$

Place 21 g (20 ml, 0.2 mol) of pure benzaldehyde (1), 30 g (28 ml, 0.29 mol) of acetic anhydride and 12 g (0.122 mol) of freshly fused and finely powdered potassium acetate (2) in a dry, 250-ml round-bottomed flask fitted with an air condenser carrying a calcium chloride guard-tube. Mix well and heat the reaction mixture in an oil bath at 160 °C for 1 hour and at 170–180 °C for 3 hours. Pour the mixture while still hot (80–100 °C) into about 100 ml of water contained in a 1-litre round-bottomed flask which has previously been fitted for steam distillation (Fig. I,101); rinse the reaction flask with a little hot water. Now add with vigorous shaking a saturated aqueous solution of sodium carbonate (3) until a drop of the liquid withdrawn on the end of a glass rod turns red litmus a distinct blue. Steam distil the solution until all the unchanged benzaldehyde is removed and the distillate is clear. Cool the residual solution and filter at the pump from resinous by-products. Acidify the filtrate by adding concentrated hydrochloric acid slowly and with vigorous stirring until the evolution of carbon dioxide ceases. When cold, filter the cinnamic acid at the pump, wash with cold water and drain well. Recrystallise either from hot water or from a mixture of 3 volumes of water and 1 volume of rectified spirit. The yield of dry cinnamic acid (colourless crystals), m.p. 133 °C, is 18 g (62%).

Notes. (1) The benzaldehyde must be free from benzoic acid; it may be purified as detailed in Section IV,22.

(2) Fused potassium acetate should be freshly prepared following the procedure described for sodium acetate (Section II,2,56). It may, however, be replaced by an equivalent quantity of freshly fused sodium acetate, but the reaction is slower and a further 3–4 hours, heating is necessary.

(3) Sodium hydroxide solution cannot be used at this stage since it may produce benzoic acid by the Cannizzaro reaction (Section IV,145) from any unchanged benzaldehyde. If, however, the reaction mixture is diluted with 3–4 volumes of water, steam distilled to remove the unreacted benzaldehyde, the residue may then be rendered alkaline with sodium hydroxide solution. A few grams of decolourising carbon are added, the mixture boiled for several minutes, and filtered through a fluted filter-paper. Upon acidifying carefully with concentrated hydrochloric acid, cinnamic acid is precipitated. This is collected, washed and purified as above.

Reduction of cinnamic acid to 3-phenylpropionic acid (*hydrocinnamic acid*). *Method A.* Carry out the hydrogenation of 14.8 g (0.1 mol) of pure cinnamic acid dissolved in 100 ml of ethanol using 0.1 g of Adams' catalyst (Section II,2,51) according to the procedure detailed in Section I,17,1 until hydrogen uptake ceases. Record the volume of hydrogen required, filter off the platinum and evaporate the filtrate on a rotary evaporator. The resulting oil solidifies on cooling to a colourless solid, m.p. 47–48 °C; the yield is 14.3 g (95%). Upon recrystallisation from light petroleum (b.p. 60–80 °C) pure hydrocinnamic acid, m.p. 48–49 °C, is obtained.

Method B. Dissolve 20 g (0.135 mol) of cinnamic acid in 145 ml of approximately 1 *M*-sodium hydroxide solution contained in a 500-ml two-necked flask equipped with a mechanical stirrer and situated within a fume cupboard. Add 350 g (0.38 mol) of 2.5 per cent sodium amalgam (Section **II,2,57**) gradually during 1 hour through the open side-neck whilst the mixture is well stirred. When hydrogen is no longer evolved, separate the mercury and wash it with water: add the washings to the solution and acidify the whole with dilute hydrochloric acid (1 : 1). Hydrocinnamic acid is precipitated, at first in the form of an oil, which solidifies on cooling and rubbing with a glass rod. Filter at the pump and recrystallise as in *Method A*. The yield of hydrocinnamic acid, m.p. 46–48 °C, is 17 g (85%).

Cognate preparations   Furylacrylic acid.

$$\text{furyl-CHO} + (CH_3 \cdot CO)_2O \xrightarrow[\text{(ii) } H_2O, HCl]{\text{(i) } CH_3 \cdot CO_2K} \text{furyl-}CH=CH \cdot CO_2H$$

Place 48 g (41.5 ml, 0.5 mol) of freshly distilled furfural (see Note (1) to Section **IV,145**), 77 g (71 ml, 0.75 mol) of pure acetic anhydride and 49 g (0.5 mol) of dry, powdered, freshly fused potassium acetate in a 500-ml two- or three-necked flask, provided with a mechanical stirrer and a long air condenser. Heat the flask, with stirring, in an oil bath at 150 °C (bath temperature) for 4 hours: when the temperature approaches 145–150 °C, a vigorous exothermic reaction sets in and must be controlled by the application of cold wet towels (or cloths) to the flask in order to avoid too vigorous boiling. Allow to cool slightly, transfer the reaction mixture to a 1-litre round-bottomed flask and add 600 ml of water: use part of this to rinse out the reaction flask. Boil the mixture with 6 g of decolourising charcoal for 10 minutes, and filter hot through a pre-heated Buchner funnel into a pre-heated filter flask. Transfer the hot filtrate to a beaker, add dilute hydrochloric acid (1 : 1) until it is acid to Congo red paper and cool to about 10 °C with stirring. Allow to stand for at least 1 hour, filter at the pump and wash with a little ice-water. The yield of crude furylacrylic acid (a light tan solid), m.p. 138–139 °C, is 41 g (59%). A perfectly pure acid (white solid), m.p. 140 °C, is obtained by recrystallisation from benzene or light petroleum, b.p. 80–100 °C, with the addition of a little decolourising carbon; the loss is about 20 per cent.

   α-Phenylcinnamic acid.

$$C_6H_5 \cdot CH_2 \cdot CO_2H + (CH_3 \cdot CO)_2O \rightleftharpoons (C_6H_5 \cdot CH_2 \cdot CO)_2O + CH_3 \cdot CO_2H$$

$$C_6H_5 \cdot CHO + (C_6H_5 \cdot CH_2 \cdot CO)_2O \xrightarrow[(C_2H_5)_3N]{} C_6H_5 \cdot CH = C \cdot CO_2H$$
$$\underset{\displaystyle C_6H_5}{|}$$

Place 42.5 g (40.5 ml, 0.4 mol) of purified benzaldehyde (Section **IV,22**), 54.5 g (0.4 mol) of phenylacetic acid, 80 ml (0.83 mol) of redistilled acetic anhydride and 40 ml of anhydrous triethylamine in a 500-ml round-bottomed flask fitted with a reflux condenser and drying tube. Boil the mixture gently for 5 hours. Steam distil the mixture directly from the reaction flask until the distillate passing over is no longer cloudy, and collect a further 50 ml of distillate: discard the distillate. Cool the residue in the flask and decant the solution from the solid; make up the volume of the solution to 500 ml with water (*A*).

Dissolve the solid in 500 ml of hot 95 per cent ethanol, add the solution (*A*) followed by 2 g of decolourising carbon; heat the mixture to boiling, filter and acidify the filtrate immediately to Congo red with 1 : 1-hydrochloric acid. Cool. Collect the separated crystals by suction filtration and recrystallise from 60 per cent ethanol. The yield of α-phenylcinnamic acid (1), m.p. 172–173 °C, is 55 g (61%).

Note. (1) The product is the isomer with the two phenyl groups *cis* to each other, since decarboxylation with quinoline-copper chromium oxide at 210–220 °C yields *cis*-stilbene.

Coumarin.

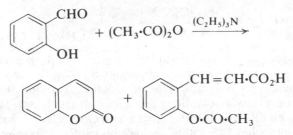

In a 250-ml round-bottomed flask, provided with a small reflux condenser and a calcium chloride drying tube at the top, place 2.1 g (0.17 mol) of salicyl-aldehyde, 2.0 ml of anhydrous triethylamine and 5.0 ml (0.052 mol) of acetic anhydride, and reflux the mixture gently for 12 hours. Steam distil the mixture from the reaction flask and discard the distillate. Render the residue in the flask basic to litmus with solid sodium hydrogen carbonate, cool, filter the precipitated crude coumarin at the pump and wash it with a little cold water. Acidify the filtrate to Congo red with 1 : 1-hydrochloric acid, collect the precipitated *o*-acetoxycinnamic acid and recrystallise it from 70 per cent propan-2-ol; the yield is 0.40 g (11%), m.p. 153–154 °C.

Boil the crude coumarin with 200 ml of water to which 0.2 g of decolourising carbon is added, filter the hot solution and concentrate it to a volume of 80 ml. Cool, collect the coumarin which separates and recrystallise it from 40 per cent aqueous methanol. The yield of coumarin, m.p. 68–69 °C, is 1.0 g (40%).

IV,151. 3,4-METHYLENEDIOXYCINNAMIC ACID (*β-Piperonylacrylic acid*)

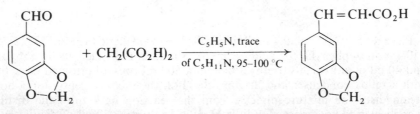

Dissolve 50 g (0.33 mol) of piperonal and 75 g (0.72 mol) of malonic acid (1) in a mixture of 150 ml of pyridine (2) and 2.5 ml of piperidine contained in a

500-ml round-bottomed flask, and heat under reflux for 1 hour on a water bath. A rapid evolution of carbon dioxide takes place. Complete the reaction by boiling the solution for 5 minutes. Cool, pour into excess of water containing enough hydrochloric acid to combine with the pyridine, filter off the piperonylacrylic acid, wash with a little water, and dry. The yield (64 g) is almost quantitative and the acid is practically pure. It may be recrystallised from glacial acetic acid; m.p. 238 °C.

Cognate preparations *p*-**Methylcinnamic acid.** From *p*-tolualdehyde; heat for 6 hours. Recrystallise from glacial acetic acid; m.p. 198 °C. Yield: 87 per cent.

m-**Nitrocinnamic acid.** From *m*-nitrobenzaldehyde. Recrystallise from alcohol; m.p. 197 °C. Yield: 80 per cent.

p-**Methoxycinnamic acid.** From anisaldehyde. Recrystallise from alcohol; m.p. 172 °C. Yield: 80 per cent.

Furylacrylic acid. Place 48 g (41.5 ml, 0.5 mol) of freshly distilled furfural, 52 g (0.5 mol) of dry malonic acid (1) and 24 ml (0.31 mol) of dry pyridine (2) in a 500-ml round-bottomed flask, fitted with a reflux condenser. Heat the flask on a boiling water bath for 2 hours, cool the reaction mixture and dilute with 50 ml of water. Dissolve the acid by the addition of concentrated ammonia solution, filter the solution and wash the filter-paper with a little water. Add dilute hydrochloric acid (1 : 1), with stirring, to the combined filtrate and washings until acid to Congo red paper, and cool in an ice bath for at least 1 hour. Filter the furylacrylic acid and wash it with a little ice-water; it weighs 63 g (91%) after drying and melts at 139–140 °C. A purer acid may be obtained by recrystallisation from light petroleum (b.p. 80–100 °C), with the addition of a little decolourising carbon; the loss is about 20 per cent.

Notes. (1) Commercial malonic acid is dried at 90–100 °C for 2 hours.

(2) The pyridine is dried by allowing it to stand, with frequent shaking, over potassium hydroxide pellets and then filtering.

M,4. The synthesis of diphenylpolyenes. 1,8-Diphenylocta-1,3,5,7-tetraene can be prepared (Section **IV,152**) by condensing two mols of cinnamaldehyde with succinic acid in the presence of acetic anhydride and lead oxide, in a reaction which bears similarities with those of the Perkin type discussed above.

$$2C_6H_5 \cdot CH = CH \cdot CHO + \begin{pmatrix} CO_2H \\ | \\ CH_2 + \end{pmatrix}_2 \longrightarrow$$

$$(C_6H_5 \cdot CH = CH - \underset{H}{\overset{OH}{\underset{|}{C}}} \underset{C}{\overset{}{{}}} CH\rbrace)_2 \longrightarrow (C_6H_5 \cdot CH = CH - CH = CH +)_2$$

1,4-Diphenylbuta-1,3-diene (Section **IV,153**) is prepared by a variant of the general method in which cinnamaldehyde is similarly condensed with phenylacetic acid.

$$C_6H_5 \cdot CH = CH - CHO \quad \underset{\substack{|\\CO_2H}}{CH_2 \cdot C_6H_5} \longrightarrow$$

$$C_6H_5 \cdot CH = CH - \underset{\substack{|\\H}}{\overset{\curvearrowright OH}{\underset{}{C}}} - CH \cdot C_6H_5 \longrightarrow$$

$$C_6H_5 \cdot CH = CH - CH = CH \cdot C_6H_5$$

The bathochromic shift which results from the increasing lengths of the conjugated system is evident from the fact that the substituted butadiene is colourless (λ_{max} 344 nm), while the substituted octatetraene is bright yellow (λ_{max} 402 nm); see also p. 1059.

IV,152. 1,8-DIPHENYLOCTA-1,3,5,7-TETRAENE

$$2C_6H_5 \cdot CH = CH \cdot CHO + \underset{\substack{|\\CH_2 \cdot CO_2H}}{CH_2 \cdot CO_2H} \xrightarrow[\text{(CH}_3\cdot\text{CO)}_2\text{O}]{\text{PbO}}$$

$$C_6H_5 \cdot CH = CH \cdot (CH = CH)_2 \cdot CH = CH \cdot C_6H_5$$

Heat a mixture of 13.2 g (11.9 ml, 0.1 mol) of cinnamaldehyde, 5.9 g (0.05 mol) of succinic acid, 11.2 g of lead oxide and 14.3 ml of acetic anhydride to 140 °C for 10 minutes with frequent shaking in a flask fitted with a reflux condenser. Boil the resulting clear solution under reflux for 2 hours; some of the tetraene crystallises at this stage. Cool the solution to 40 °C and filter the solid rapidly using a large Buchner funnel. Boil the filtrate under reflux with a further 8 ml portion of acetic anhydride for 2 hours to obtain a second crop of product on cooling and filtering. Combine the crystalline material and wash with first a little acetic anhydride and then with a little glacial acetic acid to remove brown resins. Wash the tetraene with alcohol and finally water to obtain 1.9 g (15%) of fairly pure product. Recrystallisation from chloroform gives a specimen as yellow plates, m.p. 232 °C.

IV,153. 1,4-DIPHENYLBUTA-1,3-DIENE

$$C_6H_5 \cdot CH_2 \cdot CO_2H + C_6H_5 \cdot CH = CH \cdot CHO \xrightarrow[\text{(CH}_3\cdot\text{CO)}_2\text{O}]{\text{PbO}}$$
$$C_6H_5 \cdot CH = CH \cdot CH = CH \cdot C_6H_5 + CO_2 + H_2O$$

In a 100-ml round-bottomed flask fitted with a reflux condenser place 10 g (0.07 mol) of phenylacetic acid, 10 g (9 ml, 0.075 mol) of redistilled cinnamaldehyde, 10 ml of acetic anhydride and 8.5 g of lead oxide. Heat the mixture slowly to boiling with intermittent shaking so that a clear solution is obtained, and then boil under reflux for 5 hours. Pour the hot solution into a beaker and set aside for 12 hours. Filter the semi-solid product under suction and wash the filter-cake with two 5 ml portions of ethanol, stirring the solid thoroughly with the wash liquid before applying suction. Transfer the solid to a small beaker and triturate with 8 ml of ethanol, refilter, suck dry and repeat the trituration procedure with a further 8-ml portion of ethanol. Dissolve the filtered solid in 20 ml of hot benzene, treat with 1 g of charcoal and filter the hot solution (*fume cupboard*). Add 35 ml of hot ethanol to the filtrate, boil and cool the solution in ice with shaking. Filter the purified diene, wash with 5 ml of cold ethanol and dry at 50 °C; the final product has m.p. 153 °C. The yield is 3.5 g (23%).

M,5. The benzoin condensation. Aromatic aldehydes when treated with an alkali metal cyanide, usually in aqueous solution, undergo condensation to the α-hydroxyketone or benzoin. The examples in Section **IV,154** are benzoin and furoin.

$$2Ar\cdot CHO \xrightarrow{\text{NaCN aq.}} Ar\cdot CH(OH)\cdot CO\cdot Ar$$

By use of 1 mol each of two different aldehydes, an unsymmetrical or mixed benzoin is obtained (for example, the formation of 4-methoxybenzoin).

$$Ar^1\cdot CHO + Ar^2\cdot CHO \xrightarrow[\text{KCN aq.}]{\text{NaCN or}} Ar^1\cdot CH(OH)\cdot CO\cdot Ar^2$$

The reaction depends upon the catalytic influence of the cyanide ion and the mechanism may be represented in the following way.

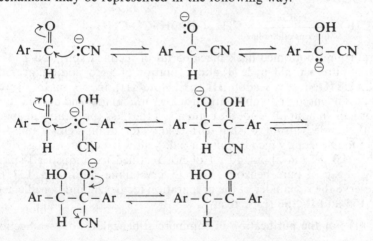

Oxidation of the α-hydroxyketone with concentrated nitric acid, or by catalytic amounts of copper(II) salts in acetic acid solution which are regenerated continuously by ammonium nitrate, yields the diketone (e.g., benzil and furil, Section **IV,155**).

$$Ar\cdot CH(OH)\cdot CO\cdot Ar \xrightarrow{[O]} Ar\cdot CO\cdot CO\cdot Ar$$

α-Diketones (Ar·CO·CO·Ar) upon refluxing with aqueous–alcoholic potassium hydroxide undergo the **benzilic acid rearrangement** and are converted into the salt of a benzilic acid (Section **IV,156**).

$$Ar\cdot CO\cdot CO\cdot Ar + KOH \longrightarrow (Ar)_2 C(OH)CO_2^{\ominus}K^{\oplus}$$

The mechanism involves nucleophilic attack of the hydroxide ion at a carbonyl carbon atom to yield the oxyanion (I) which undergoes a 1,2-nucleophilic shift of an aryl group as shown. Proton transfer completes the reaction sequence.

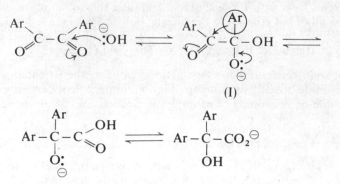

(I)

Direct conversion of a benzoin into the corresponding benzilic acid may be accomplished conveniently and in good yield by reaction with alkaline bromate solution at 85–90 °C (see Section **IV,156**, *Method 2*).

IV,154. BENZOIN

$$2C_6H_5 \cdot CHO \xrightarrow[\text{aqueous ethanolic}]{\text{NaCN}} C_6H_5 \cdot CH(OH) \cdot CO \cdot C_6H_5$$

In a 500-ml round-bottomed flask place 65 ml of rectified spirit, 50 g (47.5 ml, 0.47 mol) of pure benzaldehyde (1) and a solution of 5 g of sodium cyanide (96–98%) [**CAUTION:** see Section **III,161**, Note (1)] in 50 ml of water. Attach a reflux condenser (preferably of the double surface type) and boil the mixture gently for half an hour (2). Cool the contents of the flask (preferably in an ice bath). Filter the crude benzoin, wash it with cold water, drain well (3) and dry. The yield of crude benzoin, which is white or pale yellow in colour, is 45 g (90%).

Recrystallise 5.0 g from about 40 ml of hot rectified (or industrial) spirit; upon cooling, 4.5 g of pure benzoin (a white, crystalline solid, m.p. 137 °C) separates. Reserve the remainder of the preparation for benzil and benzilic acid (Sections **IV,155** and **IV,156** respectively).

Notes. (1) For the purification of commercial benzaldehyde, see Section **IV,22**.

(2) The reaction sometimes takes place with considerable violence and material may be lost through the condenser unless a large flask (e.g., at least of the size given) is employed.

(3) The filtrate contains sodium cyanide, and should be washed down the sink with a liberal quantity of water; see also Section **III,161**, Note (1).

Cognate preparations Furoin.

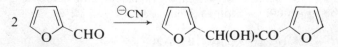

In a 1-litre three-necked flask, equipped with a mechanical stirrer, a reflux condenser and a separatory funnel, place 400 ml of water, 200 g (172.5 ml, 2.08

mol) of freshly distilled furfural [see Section **IV,145**, p. 792] and 150 ml of rectified spirit. Heat the reaction mixture to boiling, remove the source of heat and, when the liquid has just ceased to boil, add with stirring a solution of 10 g of potassium cyanide in 30 ml of water from the separatory funnel as rapidly as the vigour of the reaction permits. When the ebullition subsides (exothermic reaction), heat to boiling for a further 5 minutes. Acidify the reaction mixture with glacial acetic acid (use litmus paper) and allow to cool overnight, preferably in an ice chest or a refrigerator. Filter off the dark crystals at the pump, wash with cold water and then with cold methanol to remove as much of the tar (colouring matter) as possible. Recrystallise from methanol with the addition of about 10 g of decolourising carbon. The yield of furoin, m.p. 135–136 °C, is 75 g (37.5%) If the m.p. is slightly low, another recrystallisation from toluene–ethanol will give satisfactory results.

4-Methoxybenzoin.

$$C_6H_5 \cdot CHO + OHC \cdot C_6H_4 \cdot O \cdot CH_3\text{-}p \xrightarrow{\ominus_{CN}}$$
$$C_6H_5 \cdot CH(OH) \cdot CO \cdot C_6H_4 \cdot O \cdot CH_3\text{-}p$$

Dissolve 25 g of potassium cyanide in 175 ml of water in a 1500-ml round-bottomed flask, and add 136 g (121.5 ml, 1 mol) of redistilled p-methoxybenz-aldehyde (anisaldehyde), 108 g (103 ml, 1.02 mol) of redistilled benzaldehyde and 350 ml of 95 per cent ethanol. Reflux the mixture (which becomes homo-geneous at the boiling temperature) for 90 minutes. Remove all the unreacted aldehydes and the ethanol by steam distillation. Decant the water and set the residue aside to crystallise. Press the product as free as possible from oily material on a suction funnel and wash it with a little ethanol. Recrystallise the crude product (ca. 125 g) by dissolving it in hot ethanol and allowing to crystallise slowly. The p-methoxybenzoin separates out first in large clumps of long needles, whilst the little benzoin present crystallises in small compact balls of needles. With a little experience it is possible to filter off a good yield of the former before the appearance of the benzoin. The yield of 4-methoxybenzoin is about 55 g (23%). Recrystallise again until the m.p. is 105–106 °C.

IV,155. BENZIL

$$C_6H_5 \cdot CH(OH) \cdot CO \cdot C_6H_5 \xrightarrow{[O]} C_6H_5 \cdot CO \cdot CO \cdot C_6H_5$$

Method 1. Place 20 g (0.094 mol) of crude benzoin (Section **IV,154**) and 100 ml of concentrated nitric acid in a 250-ml round-bottomed flask. Heat on a boiling water bath (in the fume cupboard) with occasional shaking until the evolution of oxides of nitrogen has ceased (about 1.5 hours). Pour the reaction mixture into 300–400 ml of cold water contained in a beaker, stir well until the oil crystallises completely as a yellow solid. Filter the crude benzil at the pump, and wash it thoroughly with water to remove the nitric acid. Recrystallise from ethanol or rectified spirit (about 2.5 ml per gram). The yield of pure benzil, m.p. 94–96 °C, is 19 g.

Method 2. Place 0.2 g of copper(II) acetate, 10 g (0.125 mol) of am-monium nitrate, 21.2 g (0.1 mol) of benzoin and 70 ml of an 80 per cent v/v aqueous acetic acid solution in a 250-ml flask fitted with a reflux condenser. Heat the mixture with occasional shaking (1). When solution occurs, a vigorous evolution of nitrogen is observed. Reflux for 90 minutes, cool the solution, seed the solution with a crystal of benzil (2) and allow to stand for 1 hour. Filter at the pump and keep the mother-liquor (3): wash well with water and dry (preferably in an oven at 60 °C). The yield of benzil, m.p. 94–95 °C, is 19 g (90%); the m.p. is unaffected by recrystallisation from alcohol or from carbon tetrachloride (2 ml per gram). Dilution of the mother-liquor with the aqueous washings gives a further 1.0 g of benzil.

Notes. (1) For large-scale preparations use a three-necked flask equipped with two reflux condensers and a sealed mechnical stirrer.

(2) Stirring or vigorous shaking also induces crystallisation.

(3) The mother-liquor should not be concentrated as an explosion may result.

Cognate preparation   Furil.   Proceed exactly as for *Method 2*, using 19.2 g (0.1 mol) of furoin (Section **IV,154**) but use 250 ml of the aqueous acetic acid. The yield of furil, yellow needles of m.p. 165–166 °C, after recrystallisation from methanol is 17 g (89%).

IV,156. BENZILIC ACID

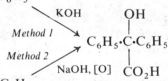

$$C_6H_5\cdot CO\cdot CO\cdot C_6H_5$$

Method 1 ↘ KOH

Method 2 ↗ NaOH, [O]

$$C_6H_5\cdot CH(OH)\cdot CO\cdot C_6H_5$$

$$\underset{|}{\overset{\text{OH}}{C_6H_5\cdot \underset{|}{C}\cdot C_6H_5}}$$
$$CO_2H$$

Method 1.   In a 500-ml round-bottomed flask, place a solution of 35 g of potassium hydroxide in 70 ml of water, then add 90 ml of rectified spirit and 35 g (0.167 mol) of recrystallised benzil (preceding Section). A deep bluish-black solution is produced. Fit a reflux condenser to the flask and heat the mixture on a boiling water bath for 10–15 minutes. Pour the contents of the flask into a porcelain dish and allow to cool, preferably overnight. The potassium salt of benzilic acid crystallises out. Filter off the crystals at the pump and wash with a little ice-cold alcohol. Dissolve the potassium salt in about 350 ml of water, and add 1 ml of concentrated hydrochloric acid from a burette slowly and with stirring. The precipitate thus produced is coloured red-brown and is somewhat sticky. Filter this off; the filtrate should be nearly colourless. Continue the addition of hydrochloric acid with stirring until the solution is acid to Congo red paper. Filter off the benzilic acid with suction, wash it thoroughly with cold water until free from chlorides and allow to dry. The yield of crude benzilic acid, which is usually light pink or yellow in colour, is 30 g (79%). Purify the product either by recrystallisation from hot benzene (about 6 ml per gram) or from hot water with the use of a little decolourising carbon. The coloured and sticky material obtained by the first precipitation may be recrystallised from hot water with the addition of a little decolourising carbon, and a further 1–2 g obtained. Pure benzilic acid has m.p. 150 °C.

Method 2.   Prepare a solution of 50 g of sodium hydroxide and 11.5 g of sodium bromate (or 12.5 g of potassium bromate) in 90 ml of water in an evaporating dish. Add 42 g (0.2 mol) of benzoin (1) in portions to this solution whilst stirring (preferably with a mechanical stirrer) and heating on a water bath at 85–90 °C (2). Add small quantities of water from time to time to prevent the mixture becoming too thick; about 80 ml of water are required. Continue the heating and stirring until a test portion is completely or almost completely soluble in water; this usually requires 3–4 hours. Dilute the mixture with 400 ml of water and allow to stand, preferably overnight. Filter off the solid or oil impurity (benzhydrol). Set aside 5 ml of the filtrate (3) and to the bulk add dilute sulphuric acid (4) slowly and with stirring to a point just short of the liberation of bromine; about 130 ml are required. If the end-point is overstepped, add the 5 ml of the filtrate which was set aside and then sufficient sulphuric acid to the end-point. Filter off the product at the pump, wash it well with water and dry. The benzilic acid weighs 39 g (85%) and has a m.p. of 149–150 °C, i.e., is practically pure. If desired, it may be recrystallised from benzene, or from water.

Notes.   (1) The crude benzoin (Section **IV,154**) gives satisfactory results.

(2) The reaction mixture should not be heated to boiling since this leads to the formation of much benzhydrol. The temperature attained by heating on a boiling water bath is 85–90 °C.

(3) This precaution is generally unnecessary if the addition of sulphuric acid is made carefully.

(4) Prepared by adding 1 volume of concentrated sulphuric acid to 3 volumes of water.

M,6. Oxime formation.   Benzaldehyde reacts with hydroxylamine in the presence of sodium hydroxide to yield an oxime of low m.p. (α- or syn-benzaldoxime) which is stable to alkali, but is rapidly rearranged by acids to give an isomeric oxime of higher m.p. (β- or anti-benzaldoxime) (Section **IV,157**).

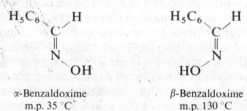

α-Benzaldoxime   β-Benzaldoxime
m.p. 35 °C   m.p. 130 °C

Two isomeric oximes of benzil, i.e., the α- and β- forms (I) and (II), are obtained in a similar manner (cognate preparation in Section **IV,157**).

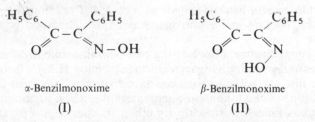

α-Benzilmonoxime   β-Benzilmonoxime

(I)   (II)

Details of the preparation of α-benzoinoxime ('cupron') and salicylaldoxime are also included in Section **IV,157**; these are employed as analytical reagents for molybdenum and for copper and nickel respectively.

809

Oximes of ketones undergo rearrangement (the **Beckmann rearrangement**) to amides under the influence of a variety of acidic reagents (e.g., sulphuric acid, hydrogen fluoride, acetic anhydride, phosphorus pentachloride, thionyl chloride, etc.). The process is illustrated by the conversion of benzophenone oxime to benzanilide in the presence of phosphorus pentachloride.

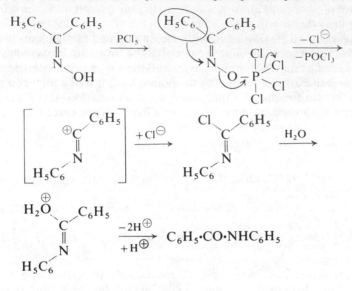

IV,157. α- AND β-BENZALDOXIMES

$$C_6H_5 \cdot CHO + NH_2OH \longrightarrow C_6H_5 \cdot CH = N \cdot OH + H_2O$$

In a 250-ml conical flask mix a solution of 14 g of sodium hydroxide in 40 ml of water and 21 g (20 ml, 0.2 mol) of pure benzaldehyde (Section **IV,22**). Add 15 g (0.22 mol) of hydroxylamine hydrochloride in small portions, and shake the mixture continually (mechanical stirring may be employed with advantage). Some heat is developed and the benzaldehyde eventually disappears. Upon cooling, a crystalline mass of the sodium derivative of the oxime separates out. Add sufficient water to form a clear solution, and pass carbon dioxide into the solution until saturated. A colourless emulsion of the α- or *syn*-aldoxime separates. Extract the oxime with ether, dry the extract over magnesium or anhydrous sodium sulphate and remove the ether on a water bath using a rotary evaporator. Distil the residue under diminished pressure (Fig. I,110). Collect the pure *syn*-benzaldoxime (α-benzaldoxime) at 122–124 °C/12 mmHg; this gradually solidifies on cooling in ice and melts at 35 °C. The yield is 12 g (49%).

To prepare the β-benzaldoxime, dissolve 10 g of α-benzaldoxime in 50 ml of pure anhydrous ether and pass dry hydrogen chloride (Section **II,2,*31***) through a *wide* delivery tube into the solution with constant shaking. Colourless crystals of the hydrochloride of the β-aldoxime separate. Filter these at the pump through a sintered glass funnel, wash with dry ether, transfer to a separatory funnel and cover with a layer of ether. Add a concentrated solution of sodium carbonate gradually and with constant shaking until effervescence ceases. Separate the ethereal layer, which contains the β-oxime, dry over magnesium or anhydrous sodium sulphate and remove the ether using a rotary evaporator. The residue crystallises; remove the small amount of oily matter by pressing on a porous tile. Recrystallise by dissolving the product in the minimum volume of ether and then adding light petroleum (b.p. 60–80 °C). The yield of β-benzaldoxime (*anti*-benzaldoxime), m.p. 130 °C, is 7–8 g (70–80%).

Cognate preparations α-**Benzilmonoxime.** Grind 42 g (0.2 mol) of pure benzil (Section **IV,155**) to a thin paste with a little ethanol, and add a concentrated aqueous solution of 17.5 g (0.25 mol) of hydroxylamine hydrochloride. Cool to −5 °C in an ice–salt bath, and add 30 g of sodium hydroxide as a 20 per cent aqueous solution dropwise with rapid mechanical stirring: do not allow the temperature to rise above 0 °C. After 90 minutes dilute the mixture with water and filter off the small quantity of unchanged benzil on a sintered glass funnel. *Just* acidify the filtrate with glacial acetic acid, allow to stand for 30 minutes, filter off the crude pinkish α-monoxime and recrystallise it from aqueous ethanol (60 vol. % alcohol); the resulting oxime weighs 37 g (82%) and melts at 137 °C. To obtain the pure α-benzilmonoxime, recrystallise twice from benzene; the final yield is 28 g (62%) of the pure product, m.p. 140 °C. Animal charcoal must not be used in the recrystallisation (see below).

β-**Benzilmonoxime.** The α-oxime is converted into the β-form by treatment with a solution of hydrogen chloride in benzene (or ether) at room temperature. From benzene, solvated crystals which melt on rapid heating at about 65 °C are obtained. Removal of benzene of crystallisation in an oven at 50 °C and recrystallisation from carbon disulphide (**CAUTION**) yields pure β-benzilmonoxime, m.p. 112 °C. The product gives no colour change with aqueous–alcoholic copper acetate solution, if it is contaminated with the α-form a greenish colour is produced. (Conversion of the α-form into the β-form may also be effected by boiling in benzene solution in the presence of animal charcoal, which presumably contains adsorbed acidic catalysts.)

α-**Benzoinoxime.** In a 250-ml round-bottomed flask, fitted with a reflux condenser, place a mixture of 10 g (0.047 mol) of benzoin (Section **IV,154**) and 20 g (25 ml) of rectified spirit together with an aqueous solution of 8.0 g (0.087 mol) of hydroxylamine hydrochloride which has previously been neutralised with 4.4 g (0.091 mol) of sodium hydroxide. Reflux for 60 minutes. Add water to precipitate the benzoinoxime, and cool in an ice bath. Filter the solid with suction at the pump, wash it with water and recrystallise from dilute ethanol. Alternatively, the dry solid may be recrystallised from ether. The yield of pure α-benzoinoxime, m.p. 151 °C, is 5 g (47%).

Salicylaldoxime. Dissolve 20.0 g (0.164 mol) of salicylaldehyde (Section **IV,127**) in 30 ml of rectified spirit, add a solution of 15 g (0.216 mol) of hydroxylamine hydrochloride in 10 ml of water and render the mixture just alkaline with 10 per cent sodium carbonate solution whilst cooling in ice. Allow to stand overnight. Acidify with acetic acid, distil off the alcohol under reduced pressure on a rotary evaporator, dilute with twice the volume of water and extract with two 50 ml portions of ether. Dry the ethereal extract with anhydrous sodium or magnesium sulphate, distil off most of the ether and allow the residue to crystallise. Recrystallise from chloroform–light petroleum (b.p. 40–60 °C). The yield of salicylaldoxime, m.p. 57 °C, is 12 g (40.5%).

Benzophenone oxime.   Place a mixture of 25 g (0.137 mol) of benzophenone (Section **IV,134**), 15 g (0.216 mol) of hydroxylamine hydrochloride, 50
ml of rectified spirit and 10 ml of water in a 500-ml round-bottomed flask. Add
28 g (0.7 mol) of sodium hydroxide (pellet form) in portions with shaking; if the
reaction becomes too vigorous, cool the flask with running tap water. When all
the sodium hydroxide has been added, attach a reflux condenser to the flask,
heat to boiling and reflux for 5 minutes. Cool, and pour the contents of the
flask into a solution of 75 ml of concentrated hydrochloric acid in 500 ml of
water contained in a 1-litre beaker. Filter the precipitate at the pump, wash
thoroughly with cold water and dry in an oven at 40 °C or in a vacuum
desiccator. The yield of benzophenone oxime, m.p. 142 °C, is 26.5 g (98%). It
may be recrystallised from methanol (4 ml per gram) but the m.p. is unaffected.
The oxime is gradually decomposed by oxygen and traces of moisture into
benzophenone and nitric acid; it should be preserved in a vacuum desiccator
filled with pure dry carbon dioxide or nitrogen.

Beckmann rearrangement of benzophenone oxime to benzanilide.   Dissolve
2 g of benzophenone oxime in 20 ml of anhydrous ether in a small conical flask
and add 3 g of powdered phosphorus pentachloride (or 3 ml of pure thionyl
chloride). Distil off the solvent and other volatile products on a water bath on a
rotary evaporator, add 25 ml of water, boil for several minutes and break up
any lumps which may be formed. Decant the supernatant liquid, and recrystallise, in the same vessel, from boiling ethanol. The product is benzanilide, 1.6 g
(80%), m.p. 163 °C; confirm this by a mixed m.p. determination with an authentic specimen.

M,7. Some reactions of alkyl aryl ketones.   The methyl group in an aryl
methyl ketone is reactive and the hydrogen atoms readily undergo electrophilic
replacement, for example, by halogen. Bromination of the methyl group can be
restricted to monosubstitution when the reaction is carried out in acidic media
(contrast the behaviour of methyl ketones on bromination under alkaline conditions, p. 473, **III,M,1**); the product is an aryl bromomethyl ketone or
phenacyl bromide.

$$Ar - \overset{\overset{\displaystyle O}{\|}}{C} - CH_3 \quad \underset{\longleftarrow}{\overset{-H^{\oplus}, +H^{\oplus}}{\longrightarrow}} \quad Ar - \overset{\overset{\displaystyle OH}{|}}{C} = CH_2$$

$$Ar - \overset{:OH}{C} = CH_2 \quad Br - Br \quad \overset{-H^{\oplus}}{\longrightarrow} \quad Ar - \overset{\overset{\displaystyle O}{\|}}{C} - CH_2Br + Br^{\ominus}$$

The preparation of *p*-bromophenacyl bromide, which is a useful reagent for the
characterisation of carboxylic acids (Section **VII,6,***15(b)*), is described in Section
IV,158.

Condensation of the active methyl group in acetophenone with formaldehyde and dimethylamine (in the form of its hydrochloride) is an example of
the **Mannich reaction** (e.g., the synthesis of dimethylaminopropiophenone,
Section **IV,159**). The probable mechanism of the reaction involves the intermediate formation of the hydroxymethyldimethylamine which eliminates
water to form the reactive species (I). This condenses with the α-carbon atom of
acetophenone reacting in its enol form.

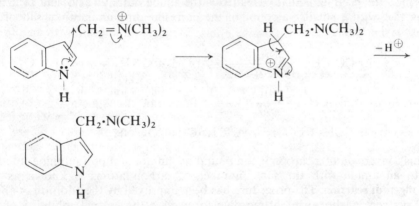

$$\overset{\oplus}{(CH_3)_2NH_2}\}\overset{\ominus}{Cl} + H\cdot CHO \rightleftharpoons (CH_3)_2NH + H\cdot CH = \overset{\oplus}{OH} + \overset{\ominus}{Cl}$$

(I)

$$(CH_3)_2\overset{..}{N}H \searrow CH_2 = \overset{\oplus}{OH} \rightleftharpoons (CH_3)_2\overset{\oplus}{N}H\cdot CH_2OH \xrightarrow{-H_2O}$$

$$\overset{\oplus}{(CH_3)_2N} = CH_2$$

$$C_6H_5 - \overset{\overset{\displaystyle :OH}{|}}{C} = CH_2 \nearrow CH_2 = \overset{\oplus}{N}(CH_3)_2 \xrightarrow{-H^{\oplus}}$$

$$C_6H_5 - \overset{\overset{\displaystyle O}{||}}{C} - CH_2 - CH_2 - N(CH_3)_2 \xrightarrow{+H^{\oplus}}$$

$$C_6H_5 \cdot CO \cdot CH_2 \cdot CH_2 \cdot \overset{\oplus}{NH}(CH_3)_2\}\overset{\ominus}{Cl}$$

Other compounds containing active hydrogen atoms similarly undergo the Mannich reaction, and a further interesting example is provided by the synthesis of dimethylaminomethylindole (gramine), in which indole is the reactive component (cognate preparation in Section **IV,159**).

$$CH_2 \cdot N(CH_3)_2$$

The quaternary salts produced by the Mannich reaction undergo a number of conversions useful in synthesis. For example, although stable at room temperature they eliminate an amine hydrochloride on heating to yield an α,β-unsaturated ketone (e.g., the conversion of dimethylaminopropiophenone to phenyl vinyl ketone, Section **IV,159**).

$$Ar\cdot CO\cdot CH_2\cdot CH_2\cdot \overset{\oplus}{NH}(CH_3)_2\}\overset{\ominus}{Cl} \longrightarrow$$

$$Ar\cdot CO\cdot CH = CH_2 + H_2\overset{\oplus}{N}(CH_3)_2\}\overset{\ominus}{Cl}$$

This ready decomposition makes Mannich bases convenient *in situ* sources of α,β-unsaturated carbonyl compounds.

A further example of the usefulness of Mannich bases is illustrated by the reaction of dimethylaminopropiophenone hydrochloride on heating with aqueous potassium cyanide, which results in the ready replacement of the dimethylamino group by the nitrile group forming β-benzoylpropionitrile. This replacement occurs even more readily when the dimethylamino compound is of the benzylic type, as in gramine. When the latter is boiled for a long time with aqueous potassium cyanide, the plant growth hormone 3-indolylacetic acid is formed by way of hydrolysis of the intermediate nitrile.

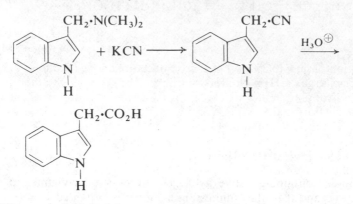

When an alkyl aryl ketone is heated with yellow ammonium polysulphide solution at elevated temperature, an aryl-substituted aliphatic acid amide is formed; the product actually isolated is the amide of the ω-arylcarboxylic acid together with a smaller amount of the corresponding ammonium salt of the carboxylic acid.

$$Ar \cdot CO \cdot CH_3 \xrightarrow[H_2O;\ 200-226\ °C]{(NH_4)_2 S_x} Ar \cdot CH_2 \cdot CONH_2 + Ar \cdot CH_2 \cdot CO_2^{\ominus} \overset{\oplus}{N}H_4$$

$$Ar \cdot CO \cdot CH_2 \cdot CH_3 \xrightarrow[H_2O;\ 200-226\ °C]{(NH_4)_2 S_x}$$

$$Ar \cdot CH_2 \cdot CH_2 \cdot CONH_2 + Ar \cdot CH_2 \cdot CH_2 \cdot CO_2^{\ominus} \overset{\oplus}{N}H_4$$

This conversion of a carbonyl compound by ammonium polysulphide solution into an amide with the same number of carbon atoms is known as the **Willgerodt reaction**. The procedure has been improved by the addition of about 40 per cent of dioxan or of pyridine to increase the mutual solubility of the ketone and aqueous ammonium polysulphide; the requisite temperature is lowered to about 160 °C and the yield is generally better.

A further improvement is embodied in the **Kindler variation** of the Willgerodt reaction which is illustrated by several examples in Section **IV,160**. This consists of heating the ketone with approximately equal amounts of sulphur and a dry amine (e.g., morpholine) instead of aqueous ammonium polysulphide. The principal product is a thioamide, and subsequent hydrolysis with acid or alkali affords the carboxylic acid, usually in good yield.

$$Ar \cdot CO \cdot CH_3 + S + R_2 NH \longrightarrow Ar \cdot CH_2 \cdot CS \cdot NR_2 + H_2O$$

$$Ar \cdot CH_2 \cdot CS \cdot NR_2 \longrightarrow Ar \cdot CH_2 \cdot CO_2 H + H_2 S + R_2 NH$$

A recent evaluation (Ref. 8) of this complex and variable reaction concludes that it cannot be described by a single mechanism.

IV,158. α,p-DIBROMOACETOPHENONE (p-*Bromophenacyl bromide*)

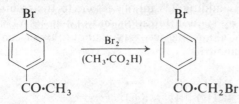

Place a solution of 50 g (0.25 mol) of p-bromoacetophenone (Section **IV,134**) in 100 ml of glacial acetic acid in a 500-ml flask. Add very slowly (about 30 minutes) from a dropping funnel 40 g (12.5 ml, 0.25 mol) of bromine: shake the mixture vigorously during the addition and keep the temperature below 20 °C. p-Bromophenacyl bromide commences to separate as needles after about half of the bromine has been introduced. When the addition is complete, cool the mixture in ice-water, filter the crude product at the pump and wash it with 50 per cent alcohol until colourless (about 100 ml are required). Recrystallise from rectified (or industrial) spirit (*ca.* 400 ml). The yield of pure p-bromophenacyl bromide (colourless needles, m.p. 109 °C) is 50 g (72%).

Cognate preparation α-**Bromo-p-phenylacetophenone** (p-*phenylphenacyl bromide*). Suspend 36 g (0.183 mol) of p-phenylacetophenone in 200 ml of glacial acetic acid in a 500-ml flask, warm gently on a water bath until a clear solution results, then cool as far as possible without the formation of crystals. To this solution add 29.5 g (9.5 ml, 0.184 mol) of bromine; do not allow the temperature to rise above 45 °C during the addition. The brominated product separates from the solution when about three-quarters of the bromine has been added. After 2 hours, cool the flask in a bath of ice and salt, filter the product, wash with a little cold glacial acetic acid, followed by small volumes of water until all the acid has been removed. The yield of crude material, m.p. 124.5–125.5 °C, is 42 g (83%). Recrystallise from hot rectified spirit (600–700 ml) and add a little decolourising carbon to remove the colour: pure, colourless p-phenylphenacyl bromide, m.p. 125.5 °C, is obtained.

IV,159. DIMETHYLAMINOPROPIOPHENONE   HYDROCHLORIDE
(*Mannich reaction*)

$$C_6H_5{\cdot}CO{\cdot}CH_3 + H{\cdot}CHO + (CH_3)_2\overset{\oplus}{N}H_2\}\overset{\ominus}{C}l \longrightarrow$$

$$C_6H_5{\cdot}CO{\cdot}CH_2{\cdot}CH_2{\cdot}\overset{\oplus}{N}H(CH_3)_2\}\overset{\ominus}{C}l + H_2O$$

Place 26.5 g (0.326 mol) of dry dimethylamine hydrochloride, 10 g (0.33 mol) of powdered paraformaldehyde and 30 g (29.3 ml, 0.25 mol) of acetophenone (Section **IV,134**) in a 250-ml round-bottomed flask attached to a reflux condenser. Introduce 40 ml of 95 per cent ethanol to which 0.5 ml of concentrated hydrochloric acid has been added, and reflux the mixture on a water bath for 2 hours; the reaction mixture should ultimately be almost clear and homogeneous. Filter the yellowish solution (if necessary) through a hot water funnel: transfer the filtrate to a 500-ml wide-mouthed conical flask and, while still warm, add 200 ml of acetone. Allow to cool to room temperature and leave in a refrigerator overnight. Filter the crystals at the pump, wash with 10 ml of acetone and dry for 6 hours at 40–50 °C: the yield of crude product, m.p.

152–155 °C, is 38 g (71%). Recrystallise the crude product by dissolving in 45 ml of hot rectified spirit and slowly adding 225 ml of acetone to the solution; collect the solid which separates by suction filtration and dry at 70 °C. The purified material melts at 155–156 °C and the recovery is about 90 per cent.

Conversion into β-benzoylpropionitrile.

$$C_6H_5 \cdot CO \cdot CH_2 \cdot CH_2 \cdot \overset{\oplus}{N}H(CH_3)_2 \}\overset{\ominus}{Cl} + KCN \longrightarrow$$
$$C_6H_5 \cdot CO \cdot CH_2 \cdot CH_2 \cdot CN + (CH_3)_2NH + KCl$$

To a mixture of 21.4 g (0.1 mol) of dimethylaminopropiophenone hydrochloride and 13.0 g (0.2 mol) of potassium cyanide in a 500-ml flask, add 260 ml of boiling water; heat the heterogeneous mixture under reflux for 30 minutes. Part of the dimethylamine, which is eliminated in the reaction, distils: collect this in dilute hydrochloric acid. Cool the reaction mixture in ice; the oil solidifies and crystals form in the aqueous layer. Collect the solid [crude β-benzoyl-propionitrile, 10.5 g (66%)] by suction filtration and recrystallise it from benzene–light petroleum (b.p. 40–60 °C); the product separates as almost colourless blades, m.p. 76 °C.

Conversion to phenyl vinyl ketone. Place an intimate mixture of 21.4 g (0.1 mol) of β-dimethylaminopropiophenone hydrochloride and 0.2 g of hydroquinone in a 100-ml round-bottomed flask. Attach a Claisen still-head fitted with a stout capillary air leak and condenser arranged for distillation under reduced pressure; place a few crystals of hydroquinone in the receiving flask. Pyrolyse the amine hydrochloride by heating the flask in an electric mantle at 2 mmHg (oil immersion pump), and collect the crude ketone which distils between 70 and 90 °C. On redistillation 7 g (51%) of pure phenyl vinyl ketone, b.p. 72–73 °C/3 mmHg (115 °C/18 mmHg), are obtained. Characterise the product by reaction with phenylhydrazine in the following way. Dissolve 0.5 g of phenylhydrazine hydrochloride and 0.5 g of sodium acetate trihydrate in the minimum of water, and add 0.5 g of the ketone followed by a little ethanol to give a homogeneous solution. Heat the mixture on a steam bath for 5–10 minutes, collect the product which separates on cooling and crystallise it from ethanol. Yellow needles of 1,3-diphenyl-2-pyrazoline, m.p. 154 °C, are obtained.

Cognate preparation Dimethylaminomethylindole (*gramine*). Cool 42.5 ml (0.236 mol) of aqueous dimethylamine solution (*ca.* 25% w/v) contained in a 100-ml flask in an ice bath, add 30 g of cold acetic acid, followed by 17.2 g (0.21 mol) of cold, 37 per cent aqueous formaldehyde solution. Pour the solution on to 23.4 g (0.2 mol) of indole; use 10 ml of water to rinse out the flask. Allow the mixture to warm up to room temperature, with occasional shaking as the indole dissolves. Keep the solution at 30–40 °C overnight and then pour it, with vigorous stirring, into a solution of 40 g of potassium hydroxide in 300 ml of water; crystals separate. Cool in an ice bath for 2 hours, collect the crystalline solid by suction filtration, wash with three 50 ml portions of cold water and dry to constant weight at 50 °C. The yield of gramine is 34 g (97.5%); this is quite suitable for conversion into 3-indolylacetic acid. The pure compound may be obtained by recrystallisation from acetone–hexane; m.p. 133–134 °C.

Conversion into 3-indolylacetic acid.   In a 1-litre flask, fitted with a reflux condenser, place a solution of 35.2 g (0.72 mol) of sodium cyanide in 70 ml of water, then add 25 g (0.144 mol) of gramine and 280 ml of 95 per cent ethanol. Boil the mixture under reflux for 80 hours. Dilute the cooled reaction mixture with 350 ml of water, shake with a little activated charcoal (e.g., Norit), filter and concentrate to about 350 ml under reduced pressure (water pump) in order to remove most of the alcohol. Cool to about 5 °C, filter off the solid and wash it with a little cold water; keep the filtrate (*A*). Recrystallise the solid from ethanol–ether to give 5.0 g (20%) of 3-indolylacetamide, m.p. 150–151 °C.

Concentrate the filtrate (*A*) to about 300 ml, cool to 5–10 °C and add concentrated hydrochloric acid dropwise and with vigorous stirring (*FUME CUPBOARD:* hydrogen cyanide is evolved) to a pH of 1–2 (about 50 ml); a crude, slightly pink 3-indolylacetic acid is precipitated. The yield of crude acid, m.p. 159–161 °C, is 20 g. Recrystallise from 1,2-dichloroethane containing a small amount of ethanol; 17.5 g (70%) of pure 3-indolylacetic acid, m.p. 167–168 °C, are obtained.

IV,160. 2-NAPHTHYLACETIC ACID (*Willgerodt Reaction*)

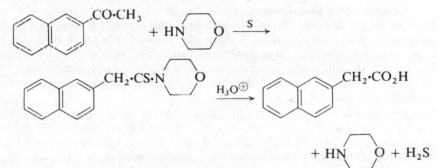

In a conical or round-bottomed flask, fitted with a reflux condenser by means of a ground glass joint, place a mixture of 128 g (0.75 mol) of methyl 2-naphthyl ketone (Section **IV,134**), 35 g (1.1 mol) of sulphur and 97 g (97 ml, 1.1 mol) of morpholine (b.p. 126–128 °C). Reflux in the fume cupboard gently at first until the evolution of hydrogen sulphide subsides and then more vigorously for a total period of 14 hours. Pour the hot reaction mixture, which has separated into two layers, into 400 ml of warm ethanol and leave to crystallise. The 2-naphthylthioacetomorpholide separates as pale buff crystals. Filter at the pump and wash with a little cold ethanol; the yield of crude thiomorpholide, m.p. 103–108 °C, is 178 g (87%).

Mix 130 g (0.48 mol) of the crude thiomorpholide with 270 ml of glacial acetic acid, 40 ml of concentrated sulphuric acid and 60 ml of water; raise the temperature of the mixture carefully to the boiling point and reflux for 5 hours. Decant the solution from a little tarry matter into 2 litres of water and keep overnight. Collect the solid by suction filtration and wash it well with cold water. Digest the solid with a solution of 50 g of sodium hydroxide in 1 litre of water, filter and acidify the filtrate with hydrochloric acid; filter off the crude 2-naphthylacetic acid, wash with water and dry. The yield of the crude acid, m.p. 137–140 °C, is 75 g (84%). Recrystallisation from benzene raises the m.p. to 142–143 °C; the loss is about 10 per cent.

Cognate preparations 3-Phenylpropionic acid (*hydrocinnamic acid*). Reflux a mixture of 53.5 g (0.4 mol) of propiophenone (Section **IV,140**), 20.5 g (0.64 mol) of sulphur and 46 g (46 ml, 0.53 mol) of morpholine for 6 hours. Pour the reaction product into 400 ml of 10 per cent ethanolic sodium hydroxide solution and reflux for 7 hours. Distil off the ethanol dilute with water, acidify with hydrochloric acid (to Congo red paper) and extract three times with ether. Wash the ether extracts with water, dry, remove the ether and distil. Collect the hydrocinnamic acid at 125–129 °C/6 mmHg; it solidifies completely on cooling, m.p. 46–47 °C. The yield is 39 g (65%).

p-**Methoxyphenylacetic acid.** Reflux a mixture of 42 g (0.28 mol) of *p*-methoxyacetophenone (Section **IV,134**), 13.5 g (0.42 mol) of sulphur and 36.5 g (36 ml, 0.42 mol) of morpholine for 5 hours. Pour the reaction mixture slowly into water, allowing the first addition to crystallise before the bulk of the mixture is added. Filter off the crude yellow solid, grind it up thoroughly with water, filter again and dry in the air. The yield of crude thioacetomorpholide, m.p. 65–67 °C, is 56 g (80%). Recrystallisation from dilute methanol raises the m.p. to 71–72 °C.

Add 50 g of the crude thioacetomorpholide to 400 ml of 10 per cent ethanolic sodium hydroxide solution and reflux the mixture for 10 hours. Distil off most of the ethanol, add 100 ml of water to the residue and strongly acidify the alkaline solution with hydrochloric acid. Cool, extract thrice with ether, dry the combined ether extracts, evaporate the solvent and recrystallise the residue from water or dilute ethanol. The yield of *p*-methoxyphenylacetic acid, m.p. 85–86 °C, is 26 g (63% overall). A further quantity of acid may be obtained by extracting the mother-liquors with ether.

p-**Bromophenylacetic acid.** Reflux a mixture of 50 g (0.25 mol) of *p*-bromoacetophenone (Section **IV,134**), 12.8 g (0.4 mol) of sulphur and 30 ml (0.35 mol) of morpholine for 8 hours. Saponify the crude reaction product with 250 ml of 10 per cent ethanolic sodium hydroxide solution and work up as described for the *p*-methoxy acid. The yield of crude *p*-bromophenylacetic acid, m.p. 107–109 °C, is 25 g (46%). Recrystallisation from water gives the pure acid, m.p. 113–114 °C.

N AROMATIC CARBOXYLIC ACIDS

This section is principally concerned with the synthesis of carboxylic acids in which the carboxyl group is directly attached to an aromatic nucleus. Several examples of the preparation of acids in which the carboxyl group is located in a side chain, i.e., aryl-substituted aliphatic acids, are included on p. 473, **III,M**.

1. Oxidative methods (Sections **IV,161** to **IV,166**).
2. The hydrolysis of nitriles (Sections **IV,167** to **IV,169**).
3. Carboxylation of the aromatic ring system (Sections **IV,170** to **IV,172**).
4. Electrophilic substitution in aromatic carboxylic esters and acids (Sections **IV,173** and **IV,174**).

N,1. Oxidative methods. The oxidation of an alkyl group attached to an aromatic system is a frequently used method for the preparation of the corresponding carboxylic acid.

$$Ar \cdot R \xrightarrow{[O]} Ar \cdot CO_2H$$

The conversion can be accomplished most readily in the laboratory by using either a solution of sodium dichromate in concentrated sulphuric acid or aqueous potassium permanganate. The method is not applicable to those cases where activating groups are attached to the aromatic system, since these render the ring susceptible to oxidative cleavage. The use of acid dichromate is illustrated in Section **IV,161** by the oxidation of p-nitrotoluene and of p-xylene. The examples illustrating the use of potassium permanganate in Section **IV,162** are the oxidation of two isomeric chlorotoluenes and the three isomeric picolines (methylpyridines).

Diphenic acid (Section **IV,163**) is obtained when phenanthrene is oxidised with 30 per cent hydrogen peroxide in glacial acetic acid solution at 85 °C. No phenanthraquinone is formed under these conditions (compare the oxidation of phenanthrene with acid dichromate, Section **IV,142**); the reaction is essentially an oxidation by peracetic acid of the reactive 9,10-positions in phenanthrene.

In some cases the conversion of an alkyl side chain to a carboxyl group is more efficiently achieved by initial halogenation followed by oxidation; for example:

$$Ar \cdot CH_3 \longrightarrow Ar \cdot CH_2Cl \longrightarrow Ar \cdot CO_2H$$

The last stage in this reaction sequence is illustrated in Section **IV,164** by the conversion of benzyl chloride into benzoic acid; procedures using either hot aqueous nitric acid or alkaline potassium permanganate are given

Section **IV,165** describes a synthesis of quinaldinic acid from quinaldine (2-methylquinoline). The method depends upon the reactive nature of the 2-methyl group in quinaldine which can readily be brominated using bromine in acetic acid. Hydrolysis of the resulting tribromo-derivative by boiling with dilute sulphuric acid occurs smoothly to give the corresponding carboxylic acid in good yield.

The use of the haloform reaction for the conversion of methyl ketones into carboxylic acids (p. 474, **III,M,1**) is also applicable in the aromatic field and is illustrated by the synthesis of 2-naphthoic acid (Section **IV,166**) from methyl 2-naphthyl ketone.

IV,161. p-NITROBENZOIC ACID

$$p\text{-}O_2N \cdot C_6H_4 \cdot CH_3 \xrightarrow[H_2SO_4]{Na_2Cr_2O_7} p\text{-}O_2N \cdot C_6H_4 \cdot CO_2H$$

Place 46 g (0.33 mol) of p-nitrotoluene, 136 g of sodium dichromate dihydrate and 300 ml of water in a 1-litre, two-necked round-bottomed flask equipped with an unsealed mechanical stirrer. By means of a dropping funnel, add 340 g (185 ml) of concentrated sulphuric acid during about 30 minutes to the well-stirred mixture. The heat of dilution of the acid causes the p-nitrotoluene to melt and oxidation takes place; if the reaction shows signs of becoming vigorous, the rate of addition must be reduced. When all the sulphuric acid has been introduced and the temperature of the mixture commences to fall, attach a reflux condenser to the flask, and heat to gentle boiling for half an hour. Cool and pour the reaction mixture into 400–500 ml of water. Filter the crude p-nitrobenzoic acid at the pump and wash it with about 200 ml of water. Transfer the solid to a 1-litre beaker, add about 200 ml of 5 per cent sulphuric acid (11 g or 6 ml of concentrated sulphuric acid added to 200 ml of water) and digest on a water bath, with agitation, in order to remove the chromium salts as completely as possible; allow to cool and filter again. Transfer the acid to a beaker, break up any lumps of material and treat it with 5 per cent sodium hydroxide solution until the liquid remains alkaline. The p-nitrobenzoic acid passes into solution, any unchanged p-nitrotoluene remains undissolved and chromium salts are converted into chromic hydroxide and/or sodium chromite. Add about 5 g of decolourising carbon, warm to about 50 °C with stirring for 5 minutes and filter with suction. Run the alkaline solution of sodium p-nitrobenzoate into about 450 ml of well-stirred 15 per cent sulphuric acid (74 g or 40 ml of concentrated sulphuric acid in 400 ml of water). Do not add the acid to the alkaline solution, for in this way the acid is liable to be contaminated by the sodium salt. Filter the purified acid at the pump, wash it thoroughly with cold water and dry it in the oven. The yield of p-nitrobenzoic acid, m.p. 237 °C, is 48 g (86%): this is sufficiently pure for most purposes. Upon recrystallisation from glacial acetic acid, the m.p. is raised to 239 °C.

Cognate preparation **Terephthalic acid.** Use 18 g (0.169 mol) of p-xylene in place of the p-nitrotoluene and proceed as above to the stage of isolation of the crude product. Wash this with 40 ml of water followed by 20 ml of ether and purify, isolate and dry the acid as detailed above for p-nitrobenzoic acid; the yield of colourless terephthalic acid is 12 g (44%); it sublimes without melting at 300 °C and is almost insoluble in water and ethanol.

IV,162. o-CHLOROBENZOIC ACID

$$o\text{-Cl·C}_6\text{H}_4\text{·CH}_3 \xrightarrow{\text{KMnO}_4} o\text{-Cl·C}_6\text{H}_4\text{·CO}_2\text{H}$$

Place 1250 ml of water, 75 g of pure potassium permanganate and 50 g (0.4 mol) of o-chlorotoluene (Section **IV,80**) in a 2.5-litre three-necked flask equipped with a sealed mechanical stirrer and reflux condenser. Stir the mixture and reflux gently until practically all the permanganate colour has disappeared (about 2 hours). At this point add 37.5 g more of potassium permanganate and reflux the mixture again until the permanganate colour disappears (about 2 hours); the colour of the solution can easily be seen by removing the flame and stopping the refluxing. Finally, add a second 37.5 g (0.95 mol total) of potassium permanganate and continue refluxing until the permanganate colour has disappeared (about 2–4 hours) (1). Steam distil the mixture (Fig. I,101) to remove unreacted o-chlorotoluene (about 12 g). Filter the hot contents of the flask from the manganese dioxide with suction (2) and wash with two 125-ml portions of hot water. Concentrate the filtrate to about 800 ml using a rotary evaporator (3), and precipitate the o-chlorobenzoic acid by cautiously adding 75 ml of concentrated hydrochloric acid with continual stirring. When cold, filter with suction, wash the acid with cold water and dry at 100 °C. The yield of o-chlorobenzoic acid, m.p. 138–139 °C, is 42 g (68%). Upon recrystallisation from hot water or from toluene (*ca.* 4 ml per gram), the m.p. is raised to 139–140 °C.

Notes. (1) A somewhat lower yield is obtained if all the potassium permanganate (150 g) is added all at once and, furthermore, the reaction may become violent. Addition in three portions results in a more controllable reaction.

(2) The addition of a Whatman filter tablet or of a little diatomaceous earth (Super Cel, etc.) assists in the filtration of the finely divided manganese dioxide.

(3) If the acid is precipitated before the solution is concentrated, the yield is considerably reduced (*ca.* 25 g). If the concentrated solution is not clear, it may be clarified by the addition of 1 g of decolourising charcoal.

Cognate preparations *p*-**Chlorobenzoic acid.** Proceed exactly as for *o*-chlorobenzoic acid. Use 1250 ml of water, 50 g (0.4 mol) of *p*-chlorotoluene (Section **IV,80**) and 75 g, 37.5 g and 37.5 g (0.95 mol total) of potassium permanganate. When the oxidation is complete, steam distil the mixture to recover any unreacted *p*-chlorotoluene (3–4 g). Filter the reaction mixture from hydrated manganese dioxide and wash the precipitate with two 100 ml portions of water. Precipitate the *p*-chlorobenzoic acid in the filtrate (1) by the addition of 75 ml of concentrated hydrochloric acid. Filter the cold solution with suction, wash with cold water and dry in an oven at 100 °C. The yield of *p*-chlorobenzoic acid, m.p. 234–235 °C, is 55 g (89%). Recrystallisation from hot water raises the m.p. to 238–239 °C.

Note. (1) If the filtrate has a faint permanganate colour, add a few drops of sodium metabisulphite solution until the solution is colourless. In this case (compare *o*-chlorobenzoic acid) concentration of the solution before precipitation only increases the yield by about 1 g and may cause occlusion of inorganic salts.

Picolinic acid (*pyridine-2-carboxylic acid*). Equip a 3-litre three-necked flask with a thermometer, sealed stirrer unit and a reflux condenser (Liebig pattern with a wide inner tube). Place a solution of 100 g (106 ml, 1.08 mol) of 2-picoline (1) in 1 litre of water in the flask and heat to 70 °C on a water bath. Add 450 g (2.84 mol) of potassium permanganate in 10 equal portions through the condenser over a period of 3–4 hours; maintain the temperature at 70 °C for the first five additions and at 85–90 °C for the last five. Make each successive addition of potassium permanganate only after the preceding amount is decolourised and wash it down with 20–25 ml of water. After the last charge of potassium permanganate is decolourised, raise the temperature to 95 °C, filter the hot reaction mixture with suction and wash the manganese dioxide cake on the filter with four 200 ml portions of hot water: allow each portion to soak into the cake without application of vacuum and finally suck dry before adding fresh wash water. Evaporate down the combined filtrate and washings to a volume of about 300 ml: allow to cool and adjust to a pH of 3.2 (the isoelectric point) using narrow-range indicator paper (about 125 ml of concentrated hydrochloric acid are required). Picolinic acid is very soluble in water (90 g in 100 ml of water at 9 °C) and therefore does not separate at this stage. Subject this aqueous acidic solution to continuous extraction with 1 litre of chloroform for 6 hours (Fig. I,95). Dry the chloroform extract with anhydrous sodium sulphate, filter and remove the chloroform under reduced pressure on a rotary evaporator, removing the residual traces of chloroform as completely as possible. Transfer the solid to a vacuum desiccator and leave to dry. Place the powdered solid in the thimble of a large Soxhlet extraction apparatus (Fig. I,97) with 1250 ml of benzene in the flask assembled in an efficient fume cupboard. Continue the extraction until crystals of picolinic acid separate from the boiling solution. Stop the extraction, cool the benzene to 15 °C and filter to remove the first crop of picolinic acid. Return the benzene filtrate to the Soxhlet extraction flask and exhaustively extract the remainder of the product from the residue in the Soxhlet thimble. The combined yield of picolinic acid of m.p. 137–138 °C is 81 g (61%). The m.p. is unaffected by recrystallisation from ethanol.

Note. (1) Material of 98 per cent purity is available commercially.

Nicotinic acid (*pyridine-3-carboxylic acid*). Dissolve 100 g (104.5 ml, 1.08 mol) of 3-picoline (99% purity) in 1 litre of water and oxidise it with 450 g (2.84 mol) of potassium permanganate: follow the experimental details given under *picolinic acid*. Wash the manganese dioxide cake with four 500 ml portions of water; evaporate the combined filtrate and washings to about 1250 ml. Adjust the pH to 3.4 (the isoelectric point) with the aid of narrow-range indicator paper; 120–130 ml of concentrated hydrochloric acid are required. Allow to cool overnight, collect the voluminous precipitate of nicotinic acid by suction filtration, wash with three 50 ml portions of cold water and dry at 90–100 °C. Concentrate the filtrate to about 650 ml and cool *slowly* to 5 °C and so obtain a second crop of nicotinic acid: the purpose of the slow cooling is to reduce the contamination by potassium chloride. The first crop of acid weighs 90 g and has a purity of about 90 per cent (1); the second crop weighs 10 g and the purity is about 80 per cent. Recrystallise from hot water (2) and dry at 100 °C; the yield of pure nicotinic acid, m.p. 235 °C, from 90 g of the crude acid is 67 g (51%). A further quantity may be obtained by concentrating the mother-liquor.

Notes. (1) The impurity is potassium chloride. The approximate acid content is determined by heating a weighed sample of the acid in a crucible gently at first and finally at a red heat until no trace of black residue remains, and weighing the white residual potassium chloride.

(2) The solubility of pure nicotinic acid in 1000 ml of water at 0, 40, 80 and 100 °C is 1.0, 2.6, 8.2 and 12.7 g respectively.

Isonicotinic acid (*pyridine-4-carboxylic acid*). Use 100 g (104.5 ml, 1.08 mol) of 4-picoline (98% purity) and oxidise it with 450 g (2.84 mol) of potassium permanganate: follow the experimental details given for *picolinic acid*. Evaporate the combined filtrate and washings to about 1500 ml and add concentrated hydrochloric acid until the pH is 3.6; isonicotinic acid precipitates. Heat to 90–95 °C (not all the acid dissolves) and allow the mixture to crystallise slowly. Collect the crude isonicotinic acid by suction filtration, wash well with water and dry at 100 °C. Concentrate the mother-liquor to about half the original volume and so obtain a second crop of acid. The first crop of acid weighs 85 g (64%) (99% pure) and the second crop weighs 7g (80% pure). Recrystallise from hot water: the resulting isonicotinic acid is pure and has a m.p. of 311 °C (sealed tube).

The solubility of isonicotinic acid in 1000 ml of water at 0, 40, 80 and 100 °C is 3, 9, 24 and 34 g respectively. The solubility is appreciably less in the presence of potassium chloride.

IV,163. DIPHENIC ACID (*Biphenyl-2,2'-dicarboxylic acid*)

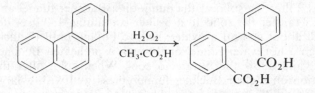

Equip a 3-litre three-necked flask with a sealed mechanical stirrer, a reflux condenser and a thermometer. Dissolve 89 g (0.5 mol) of pure phenanthrene (1) in 1 litre of glacial acetic acid in the flask and warm to 85 °C on a water bath. Introduce 345 ml of 30 per cent hydrogen peroxide solution (4 mol) during 40 minutes [*CARE:* see Section **II,2,***34*]; the temperature falls to about 80 °C and some phenanthrene may precipitate. After the addition is complete, heat the mixture with stirring on a water bath for a further 3–4 hours. Reduce the volume of the solution to about half by distillation under reduced pressure with a rotary evaporator and allow to cool. Filter off the considerable amount of diphenic acid which crystallises out on cooling. Keep the filtrate and evaporate it almost to dryness under reduced pressure: extract the residue with 375 ml of 10 per cent sodium carbonate solution by warming on a water bath, boil the extract with a little decolourising carbon, filter and add dilute hydrochloric acid until the pH is 4.5 (use narrow-range indicator paper). Stir the solution with a further small amount of active charcoal and filter off the tarry material; cool the clear solution to 0 °C and acidify with dilute hydrochloric acid. Collect the precipitate by suction filtration, wash with water and dry at 110 °C. The total yield of crude diphenic acid, m.p. 228 °C, is 83 g (69%). Recrystallisation from glacial acetic acid raises the m.p. to 230 °C.

Note. (1) **Technical phenanthrene may be purified** as follows. Dissolve 500 g of technical 90 per cent phenanthrene in 3 litres of ethanol in a 4-litre flask on a steam bath and decant the hot solution from any insoluble material: collect the solid which crystallises upon cooling the solution. Dissolve 250 g of the crystallised product in 550 ml of hot glacial acetic acid in a 1-litre three-necked flask provided with an efficient reflux condenser and a dropping funnel. To the boiling solution add gradually 18 ml of an aqueous solution containing 15 g of chromium trioxide; then add slowly 7.5 ml of concentrated sulphuric acid from the dropping funnel. Reflux the solution for 15 minutes, and then pour it with vigorous stirring into 1125 ml of water in a 3-litre round-bottomed flask. Filter when cold, wash with water and dry in the air. Distil the product under reduced pressure (oil pump) using a short air condenser and collect the phenanthrene at 148–149 °C/1 mmHg. Recrystallise the solidified distillate from ethanol: 200–225 g of nearly white phenanthrene, m.p. 99 °C, are obtained.

IV,164. BENZOIC ACID

$$C_6H_5 \cdot CH_2Cl \xrightarrow{[O]} C_6H_5 \cdot CO_2H$$

Method 1. Into a 250-ml round-bottomed flask provided with a reflux condenser, place 10 g (9 ml, 0.08 mol) of benzyl chloride (Section **IV,31**), 50 ml of water, 20 ml of concentrated nitric acid and a few fragments of porous porcelain. Boil vigorously for 5–6 hours, by the end of which time oxidation should be complete. Cool the flask under the tap, shaking vigorously to prevent the formation of lumps. Filter the solid at the pump on a Buchner funnel and wash with cold water. Transfer the solid to a beaker containing 1–2 g of decolourising carbon and about 400 ml of water; heat to boiling until the acid dissolves. Filter through a hot water funnel or through a preheated Buchner funnel and filter flask, and allow the filtrate to cool. When cold, filter the benzoic acid with suction on a clear Buchner funnel, press well with a large glass stopper, wash with small quantities of cold water and drain as dry as possible. Dry upon filter-paper in the air. The yield is 8 g (84%), m.p. 121.5 °C.

Method 2. Into a 500-ml round-bottomed flask equipped with a reflux condenser, place 4 g of anhydrous sodium carbonate, 200 ml of water, 9 g (0.057 mol) of potassium permanganate, 5 g (4.5 ml, 0.04 mol) of benzyl chloride (Section **IV,31**) and a few chips of porous porcelain. Boil the mixture gently until the reaction is complete (60–90 minutes), i.e., until the liquid running down from the condenser contains no oily drops of unchanged benzyl chloride. Manganese dioxide is precipitated. Allow to cool, acidify with concentrated hydrochloric acid (about 40 ml) and add a 20 per cent aqueous solution of sodium sulphite heptahydrate with shaking until the manganese dioxide is completely dissolved and only the white precipitate of benzoic acid remains. When the mixture is cold, filter off the benzoic acid at the pump and wash it with cold water. Recrystallise from 200 ml boiling water as in *Method 1.* The benzoic acid is obtained as colourless needles, m.p. 121.5 °C. The yield is 4 g (83%).

IV,165. QUINALDINIC ACID (*Quinoline-2-carboxylic acid*)

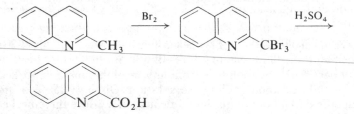

Fit a 500-ml, two-necked round-bottomed flask with a reflux condenser and dropping funnel. Place a mixture of 50 g (0.61 mol) of anhydrous, powdered sodium acetate (Section **II,2,56**), 100 g of glacial acetic acid and 14 g (0.1 mol) of pure quinaldine in the flask, and a solution of 48 g (15.5 ml, 0.3 mol) of bromine in 100 g of glacial acetic acid in the dropping funnel. Heat the flask to 70 °C in a water bath, and add the bromine solution during 10–15 minutes whilst keeping the mixture thoroughly shaken. Remove the flask from the water bath, bring the solution to boiling for a few minutes (until the separation of sodium bromide causes violent bumping) and then heat for 30 minutes on a boiling water bath and allow to cool. Pour the reaction mixture into 300 ml of ice-water, collect the precipitate by suction filtration and wash thoroughly with water. The yield of crude product, after drying at 100 °C, is 36 g (95%). Recrystallise from ethanol or glacial acetic acid: the pure ω-tribromoquinaldine has m.p. 128 °C.

Hydrolyse the ω-tribromoquinaldine by boiling it under reflux with excess of dilute (1 : 10) sulphuric acid until a test portion, on neutralisation, yields no unchanged halogen compound. The quinaldinic acid is best isolated, via the copper salt, in the following manner. Cool, nearly neutralise the solution and add excess of copper(II) sulphate solution. Collect the pale green copper quinaldinate by suction filtration and wash it well with cold water. Suspend the copper salt in hot water and subject it to prolonged treatment with hydrogen sulphide gas. Filter off the copper sulphide and evaporate the clear filtrate to dryness on a water bath. Recrystallise the residual quinaldinic acid from glacial acetic acid; it then melts at 157 °C. The yield is almost quantitative.

IV,166. 2-NAPHTHOIC ACID

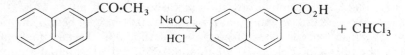

Prepare a solution containing about 75 g (1 mol) of sodium hypochlorite by diluting 300 ml of commercial sodium hypochlorite solution (containing 10–14% available chlorine) to 750 ml with water, and place it in a 1500-ml three-necked flask provided with a thermometer, a mechanical stirrer and a reflux condenser. Warm the solution to 55 °C and add through the condenser 42.5 g (0.25 mol) of methyl 2-naphthyl ketone (2-acetonaphthalene) (1). Stir the mixture vigorously and, after the exothermic reaction commences, maintain the temperature at 60–70 °C by frequent cooling in an ice bath until the temperature no longer tends to rise (ca. 30 minutes). Stir the mixture for a further 30 minutes, and destroy the excess of hypochlorite completely by adding a solution of 25 g of sodium metabisulphite in 100 ml of water: make sure that no hypochlorite remains by testing the solution with acidified potassium iodide solution. Cool the solution, transfer the reaction mixture to a 2-litre beaker and cautiously acidify with 100 ml of concentrated hydrochloric acid. Filter the crude acid at the pump, wash with water and drain as completely as possible. Dry at 100 °C and recrystallise the dry acid (42 g; m.p. 181–183 °C) from rectified spirit (about 150 ml). The yield of pure, colourless 2-naphthoic acid, m.p. 184–185 °C, is 37.5 g (87%).

Note. (1) The commercial product, m.p. 53–55 °C, may be used. Alternatively methyl 2-naphthyl ketone may be prepared as detailed in Section IV,134.

N,2. The hydrolysis of nitriles. The hydrolysis of nitriles under either acidic or basic conditions, which has already been discussed on p. 478, III,M,2, for alkyl and aralkyl nitriles, is equally applicable to the synthesis of aromatic carboxylic acids (Section IV,167). The aromatic nitriles are readily obtained by the Sandmeyer reaction (see p. 689, IV,G,1).

In the preparation of 2-naphthoic acid (Section IV,168) the preferred starting material is 2-naphthylamine-1-sulphonic acid (see p. 667, IV,E,3). After replacement of the amino by the cyano group using the Sandmeyer procedure, subsequent treatment with aqueous acid removes the sulphonic acid group (see IV,G,1 and IV,J,2) and hydrolyses the cyano group in one step.

The formation of a carboxyl group by hydrolysis of the corresponding nitrile constitutes the last step in an interesting multi-stage synthesis of homophthalic acid (Section IV,169). The starting material is phthalic anhydride which is first converted into the cyclic imide, phthalimide (I), by treatment with aqueous ammonia, or more conveniently with urea. Reduction of phthalimide with a zinc–copper couple in the presence of alkali is the most convenient laboratory preparation of phthalide (III), although the latter can also be obtained by the direct reduction of phthalic anhydride (Ref. 9). The immediate reduction product is the salt of o-hydroxymethylbenzoic acid (II); on acidification this gives the free acid which cyclises to the γ-lactone, phthalide (III). Nucleophilic attack of a cyanide ion on the methylene group in phthalide results in alkyl–oxygen fission of the lactone ring to yield o-cyanomethylbenzoic acid, which on hydrolysis yields homophthalic acid.

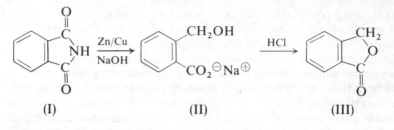

(I)   (II)   (III)

IV,167. o-TOLUIC ACID

$$o\text{-}CH_3\text{·}C_6H_4\text{·}CN \xrightarrow[\text{or } B \; ^\ominus OH]{A \; H_3O^\oplus} o\text{-}CH_3\text{·}C_6H_4\text{·}CO_2H$$

Method A (acidic hydrolysis). Prepare o-tolunitrile, b.p. 94–96 °C/20 mmHg, from o-toluidine following the method given in Section **IV,84** under p-*toluidine*. Also prepare 600 g of 75 per cent sulphuric acid by adding 450 g (245 ml) of concentrated sulphuric acid cautiously, with stirring and cooling, to 150 ml of water. Place the latter in a 1-litre three-necked flask, equipped with a dropping funnel, a mechanical stirrer and reflux condenser. Heat the solution in an oil bath to about 150 °C, stir and add 220 g (1.88 mol) of o-tolunitrile during 2 hours. Continue the stirring for a further 2 hours while the temperature is maintained at 150–160 °C; finally raise the temperature to 190 °C and stir for another hour. Some crystalline solid will appear in the condenser at this stage. Allow the reaction mixture to cool, pour into ice-cold water and filter off the precipitated acid. Dissolve the crude acid in an excess of 10 per cent sodium hydroxide solution, filter off any insoluble material (probably o-toluamide, m.p. 141 °C) through a sintered glass funnel while still hot and acidify the filtrate with dilute sulphuric acid. Collect the o-toluic acid on a Buchner funnel, dry in the air and recrystallise from benzene (about 500 ml; **CAUTION**—fume cupboard). The yield of pure o-toluic acid, m.p. 102–103 °C, is 200 g (78%).

Method B (basic hydrolysis). Boil a mixture of 5 g of o-tolunitrile, 80 ml of 10 per cent aqueous sodium hydroxide solution and 15 ml of alcohol under a reflux condenser; the alcohol is added to increase the rate of hydrolysis. The solution becomes clear after heating for about 1 hour, but continue the boiling for a total period of 1.5 hours to ensure complete hydrolysis. Detach the condenser and boil the solution for a few minutes in the open flask to remove dissolved ammonia and some of the alcohol (**CAUTION!**). Cool, and add concentrated hydrochloric acid until precipitation of the o-toluic acid is complete. When cold, filter off the o-toluic acid with suction and wash with a little cold water. Recrystallise from benzene (*ca.* 12 ml). The yield of *o*-toluic acid, m.p. 102–103 °C, is 5 g (86%).

Cognate preparations   *p*-**Toluic acid** may be similarly prepared by either *Method A* or *Method B* using *p*-tolunitrile (Section **IV,84**).

   1-Naphthoic acid (by *Method A*).   In a 750-ml or 1-litre flask equipped with a reflux condenser, place 50 g (0.327 mol) of 1-naphthonitrile (Section **IV,183**), 100 ml of glacial acetic acid, 100 ml of water and 100 ml of concentrated sulphuric acid. Heat in an oil bath at 115–120 °C for 1.5 hours: do not allow the temperature to rise above 120 °C as the 1-naphthoic acid formed tends to lose carbon dioxide at higher temperatures and the yield will be reduced. Dilute the cold reaction mixture, which contains much crystalline solid, with an equal volume of water and filter at the pump; if the product consists of large lumps, transfer it first to a glass mortar and thoroughly grind it to a fine paste. Wash with water until free from mineral acid. Dissolve the crude acid in dilute aqueous sodium carbonate solution, heat for a short time to separate the resinous impurities and filter the hot solution. Acidify the clear filtrate with a slight excess of dilute sulphuric acid and collect the voluminous precipitate of almost pure 1-naphthoic acid, wash until free from inorganic salts and dry at 100 °C. Recrystallise from toluene or from light petroleum (b.p. 80–100 °C). The yield of pure 1-naphthoic acid, m.p. 160–161 °C, is 50 g (89%).

   Benzoic acid (by *Method B*).   Boil 5.1 g (5 ml, 0.05 mol) of benzonitrile and 80 ml of 10 per cent sodium hydroxide solution in a 250-ml round-bottomed flask fitted with a reflux water condenser until the condensed liquid contains no oily drops (about 45 minutes). Remove the condenser and boil the solution in an open flask for a few minutes to remove free ammonia. Cool the liquid, and add concentrated hydrochloric acid, cautiously with shaking, until precipitation of benzoic acid is complete. Cool, filter the benzoic acid with suction and wash with cold water; dry upon filter-paper in the air. The benzoic acid (5.8 g, 95%) thus obtained should be pure (m.p. 121 °C). Recrystallise a small quantity from hot water and redetermine the m.p.

IV,168. 2-NAPHTHOIC ACID

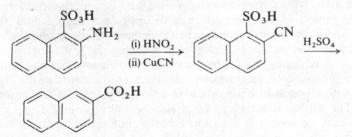

Diazotise 112 g (0.5 mol) of 2-naphthylamine-1-sulphonic acid (1) as detailed under *2-bromonaphthalene* in Section **IV,81**. Prepare copper(I) cyanide from 62.5 g (0.25 mol) of copper(II) sulphate pentahydrate (Section **II,2,*17***) and dissolve it in a solution of 32.5 g (0.5 mol) of potassium cyanide [**CAUTION**] in 250 ml of water contained in a 1-litre three-necked flask. Cool the potassium cuprocyanide solution in ice, stir mechanically and add the damp cake of the diazonium compound in small portions whilst maintaining the temperature at

5–8 °C. Nitrogen is soon evolved and a red precipitate forms gradually. Continue the stirring for about 10 hours in the cold, heat slowly to the boiling point, add 125 g of potassium chloride, stir and allow to stand. Collect the orange crystals which separate by suction filtration; recrystallise first from water and then from ethanol; dry at 100 °C. The product is almost pure potassium 2-cyanonaphthalene-1-sulphonate. Transfer the product to a 1-litre round-bottomed flask, add a solution prepared from 200 ml of concentrated sulphuric acid and 200 g of crushed ice, and heat the mixture under reflux for 12 hours. Collect the 2-naphthoic acid formed (some of which sublimes from the reaction mixture) by suction filtration on a sintered glass funnel, wash well with water and dry at 100 °C; recrystallise from rectified spirit. The yield of 2-naphthoic acid, m.p. 184–185 °C, is 65 g (75%).

Note. (1) Tobias acid, available from Fluka AG Chemische Fabrik.

IV,169. HOMOPHTHALIC ACID (o-*Carboxyphenylacetic acid*)

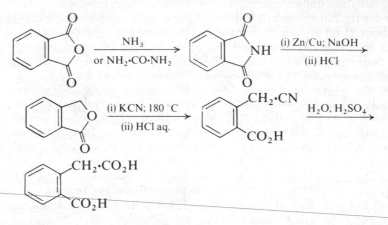

Phthalimide. *Method 1.* Place 100 g (0.675 mol) of phthalic anhydride and 105 ml of concentrated ammonia solution (*d.* 0.88) in a 1-litre round-bottomed flask fitted with a wide air condenser (≮10 mm in diameter). Heat first over a wire gauze and then over a free flame until the mixture is in a state of quiet fusion and forms a homogeneous melt (the temperature reaches 300 °C in about 1.5–2 hours; all the water is evaporated during the first hour). Shake the flask occasionally during the heating and push down any material which sublimes into the condenser with a glass rod. Pour the contents of the flask whilst still hot into a porcelain basin, allow to cool and grind to a fine powder in a mortar. The phthalimide (95 g, 96%) is practically pure and melts at 233–234 °C. It may be recrystallised from ethanol, but the solubility is only slight (about 5%).

Method 2. Intimately mix 99 g (0.67 mol) of pure phthalic anhydride and 20 g (0.33 mol) of urea, and place the mixture in a 1-litre, long-necked, round-bottomed flask. Heat the flask in an oil bath at 130–135 °C. When the contents have melted, effervescence commences and gradually increases in vigour: after 10–20 minutes, the mixture suddenly froths up to about three times the original volume (this is accompanied by a rise in temperature to 150–160 °C) and becomes almost solid. Remove the flame from beneath the bath and allow to cool.

Add about 80 ml of water to disintegrate the solid in the flask, filter at the pump, wash with a little water and then dry at 100 °C. The yield of phthalimide, m.p. 233 °C (i.e., it is practically pure), is 86 g (87%). If desired, the phthalimide may be recrystallised from 1200 ml of industrial spirit; the first crop consists of 34 g of m.p. 234 °C, but further quantities may be recovered from the mother-liquor.

Phthalide. In a 1-litre three-necked flask stir 90 g (1.37 mol) of a high quality zinc powder to a thick paste with a solution of 0.5 g of crystallised copper(II) sulphate in 20 ml of water (this serves to activate the zinc), and then add 165 ml of 20 per cent sodium hydroxide solution. Cool the flask in an ice bath to 5 °C, stir the contents mechanically and add 73.5 g (0.5 mol) of phthalimide in small portions at such a rate that the temperature does not rise above 8 °C (about 30 minutes are required for the addition). Continue the stirring for half an hour, dilute with 200 ml of water, warm on a water bath until the evolution of ammonia ceases (about 3 hours) and concentrate to a volume of about 200 ml by distillation under reduced pressure (rotary evaporator). Filter, cool in ice and render the filtrate acid to Congo red paper with concentrated hydrochloric acid (about 75 ml are required). Much of the phthalide separates as an oil, but, in order to complete the lactonisation of the hydroxymethylbenzoic acid, boil for an hour: transfer while hot to a beaker. The oil solidifies on cooling to a hard red-brown cake. Leave overnight in an ice chest or refrigerator, and then filter at the pump. The crude phthalide contains much sodium chloride. Recrystallise it in 10 g portions from 750 ml of water: use the mother-liquor from the first crop for the recrystallisation of the subsequent portion. Filter each portion while hot, cool in ice below 5 °C, filter and wash with small quantities of ice-cold water. Dry in the air upon filter-paper. The yield of phthalide (transparent plates), m.p. 72–73 °C, is 47 g (70%).

o-**Cyanomethylbenzoic acid.** *This preparation must be conducted in an efficient fume cupboard.* Into a 1-litre three-necked flask, provided with a mechanical stirrer and a thermometer, place 40 g (0.33 mol) of phthalide and 40 g (0.615 mol) of powdered potassium cyanide. Heat the stirred mixture to 180–190 °C (internal temperature) in an oil bath for 4–5 hours. Allow to cool, add 400 ml of distilled water and stir the mixture until all the solids are dissolved (about 1 hour). Filter off any unreacted phthalide. Add dilute hydrochloric acid (1 : 1) to the dark aqueous solution (*CARE*—hydrogen cyanide is evolved) until it becomes turbid (about 20 ml are required), and continue the addition until the solution is slightly acid: filter off any dark impurities which may separate. Neutralise the solution carefully with sodium hydrogen carbonate, add a few grams of decolourising carbon, stir the mixture for several minutes and filter. Acidify the nearly colourless filtrate with about 20 ml of concentrated hydrochloric acid, cool in ice and filter at the pump. The resulting *o*-cyanomethylbenzoic acid (36 g) melts at 114–115 °C and is satisfactory for most purposes. It may be crystallised from benzene or glacial acetic acid, but with considerable loss.

Homophthalic acid. Place a mixture of 25 g (0.155 mol) of *o*-cyanomethyl benzoic acid and 25 g of 50 per cent sulphuric acid in a 100-ml flask, heat the mixture on a boiling water bath for 10–12 hours and then pour it into twice its volume of ice and water. Filter the precipitate at the pump and dry in the air. The yield of crude homophthalic acid is 21 g. Recrystallise by dissolving it in 500 ml of boiling water, add decolourising carbon, filter the hot solution through a hot water funnel and cool the filtrate in an ice bath: collect the acid and dry at 100 °C. The yield of practically colourless acid, m.p. 181 °C, is 17 g (61%). The melting point depends upon the rate of heating; immersion of the capillary in a bath at 170 °C gives a m.p. of 182–183 °C.

N,3. Carboxylation of the aromatic ring system. The preparation of salicylic acid from dry sodium phenoxide with carbon dioxide under pressure is the classical example of the **Kolbé–Schmidt** reaction, which is a useful method for the introduction of a carboxyl group directly into a phenolic nucleus. The mechanism of the reaction appears to involve attack by carbon dioxide at the activated *ortho* position on the phenoxide ion.

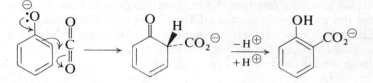

In the laboratory the reaction is more conveniently carried out by passing a stream of carbon dioxide over the surface of the heated phenoxide (Section **IV,170**). Under these conditions only one half of the phenol is converted into the phenolic acid. The reaction is particularly facile with di- and tri-hydric phenols. Thus 2,4-dihydroxybenzoic acid (cognate preparation in Section **IV,170**) is readily obtained by passing carbon dioxide through a boiling aqueous solution of the potassium or sodium salt of resorcinol.

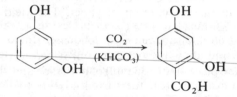

It is of interest to record that *p*-hydroxybenzoic acid may be prepared by the thermal rearrangement of potassium salicylate at 230 °C.

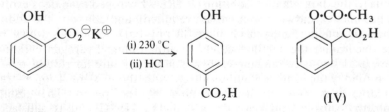

Salicylic acid is acetylated readily on treatment with acetic anhydride in the presence of a few drops of concentrated sulphuric acid on a catalyst. The product is acetylsalicylic acid or aspirin (IV).

The standard carboxylation reaction of a Grignard reagent (p. 481, **III,M,3**) is also applicable in the aromatic series and is illustrated by the conversion of 1-bromonaphthalene into 1-naphthoic acid (Section **IV,171**). The similar carboxylation of organosodium or organolithium compounds is described in the preparation of *p*-toluic acid (Section **IV,172**). The organosodium compound is prepared by the direct reaction of sodium metal with *p*-chlorotoluene; the organolithium compound is similarly obtained from *p*-bromotoluene and lithium metal. The preparation of *m*-chlorobenzoic acid (cognate preparation in Section **IV,172**) illustrates an alternative preparation of the required organolithium compound by means of a transmetalation process between butyl lithium and *m*-bromochlorotoluene.

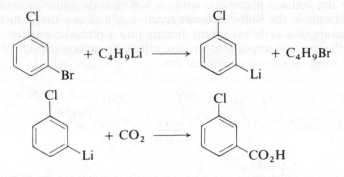

IV,170. SALICYLIC ACID

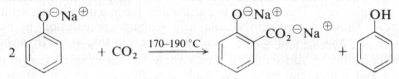

Place 10 g (0.434 mol) of clean sodium (cut into small pieces) in a 500-ml round-bottomed flask fitted with a double surface reflux condenser. Introduce 100 g (127 ml) of absolute ethanol and allow the reaction to proceed as vigorously as possible; if the alcohol tends to flood the condenser, cool the flask momentarily with a wet towel or by a stream of cold water. When all the sodium has reacted, add 40 g (0.425 mol) of pure phenol. Distil off the alcohol using a free flame: shake the flask frequently during the process until a powdery mass is produced. Transfer the solid rapidly to a dry mortar, powder rapidly and transfer the powder to a 250-ml round-bottomed three-necked flask fitted with a gas inlet tube terminating 1 cm above the sodium phenoxide and a knee-bend adapter leading to an air condenser; stopper the third neck. Immerse the bulb of the flask in an oil bath, heat to 110 °C and pass dry carbon dioxide into the flask for 1 hour. Raise the temperature gradually during 4 hours to 190 °C (20 °C an hour) and finally maintain the temperature at 200 °C for 1.5 hours. Pass a fairly rapid stream of carbon dioxide into the flask during the whole of the heating period; stir the contents of the flask frequently with a glass rod by way of the third neck, in order to expose a fresh portion of the solid to the action of the gas. Allow the reaction product to cool, transfer it to a large beaker and rinse the distilling flask several times with water. Precipitate the salicylic acid by the addition of excess of concentrated hydrochloric acid. Cool in ice, filter at the pump and wash with a little cold water. Recrystallise the crude acid from hot water with the addition of a little decolourising carbon. The yield of air-dried salicylic acid, m.p. 159 °C, is 16 g (27%).

Conversion to acetylsalicylic acid (*aspirin*). Place 10 g (0.725 mol) of dry salicylic acid and 15 g (14 ml, 0.147 mol) of acetic anhydride in a small conical flask, add 5 drops of concentrated sulphuric acid and rotate the flask in order to secure thorough mixing. Warm on a water bath to about 50–60 °C, stirring with a thermometer, for about 15 minutes. Allow the mixture to cool and stir occasionally. Add 150 ml of water, stir well and filter at the pump. Dissolve the solid in about 30 ml of hot ethanol and pour the solution into about 75 ml of warm water: if a solid separates at this point, warm the mixture until solution is complete and then allow the clear solution to cool slowly. Beautiful needle-like crystals will separate. The yield is 11 g (85%). The air-dried crude product may also be recrystallised from ether–light petroleum (b.p. 40–60 °C).

Acetylsalicylic acid decomposes when heated and does not possess a true, clearly defined m.p. Decomposition points varying from 128 to 135 °C have been recorded: a value of 129–133 °C is obtained on an electric hot plate (Fig. I,162). Some decomposition may occur if the compound is recrystallised from a solvent of high boiling point or if the boiling period during recrystallisation is unduly prolonged.

Cognate preparations   2,4-Dihydroxybenzoic acid (*β-resorcylic acid*).   Place a solution containing 40 g (0.364 mol) of resorcinol, 200 g of potassium hydrogen carbonate and 400 ml of water in a litre flask fitted with a reflux condenser and gas inlet tube. Heat gently on a steam bath for 4 hours; then reflux vigorously over a flame for 30 minutes whilst passing a rapid stream of carbon dioxide through the solution. Acidify the solution whilst still hot by adding 180 ml of concentrated hydrochloric acid from a separatory funnel with a long tube delivering acid to the bottom of the flask. Allow to cool to room temperature, chill in an ice bath and collect the crude *β*-resorcylic acid by filtration with suction. Recrystallise by boiling the crude acid with 180–200 ml of water in the presence of a little decolourising carbon, filter through a hot water funnel and cool in an ice–salt mixture with stirring. Collect and dry the pure *β*-resorcylic acid; the yield is 36 g (64%), m.p. 216–217 °C.

p-**Hydroxybenzoic acid.**   Place 100 g (0.725 mol) of salicylic acid and 150 ml of water in a 20-cm porcelain dish and slowly stir in 60 g of potassium carbonate. Evaporate the solution on a steam bath to a thick, pasty solid; break this up into small pieces and dry at 105–110 °C for 2 hours. Finely grind the solid, dry for a further 2 hours at 105–110 °C and grind again to a fine powder. Transfer the powder (a mixture of potassium salicylate and potassium carbonate) to a 500-ml round-bottomed flask fitted with an air condenser set for downward distillation and immersed in an oil bath. Heat the oil bath to 240 °C and maintain this temperature for 90 minutes, stirring the solid occasionally with a glass rod; phenol formed in the reaction distils out of the mixture. When the reaction is complete (1), transfer the contents of the flask whilst still hot to a 2-litre flask containing 1 litre of hot water; rinse the reaction flask with several portions of the hot solution. Acidify with concentrated hydrochloric acid (*ca*. 75 ml are required), heat nearly to boiling, add 5 g of decolourising carbon, filter, cool and collect the brown solid by suction filtration. Concentrate the filtrate to about 300 ml cool and collect a second crop of the acid. Dissolve the crude acid in 300 ml of hot water, boil for a few minutes with 5 g of decolourising carbon and filter. Cool the filtrate under the tap, filter the solid with suction, wash with 15 ml of cold water and dry. The yield of *p*-hydroxybenzoic acid, m.p. 211–212 °C, is 40 g (40%).

Note. (1) This may be determined roughly by treating a small test portion with 3–4 ml of hot water and acidifying with concentrated hydrochloric acid; the absence of a precipitate in the warm solution indicates the essential completeness of the reaction. Salicylic acid is sparingly soluble and p-hydroxybenzoic acid is relatively soluble under these conditions.

IV,171. 1-NAPHTHOIC ACID

$$1\text{-}C_{10}H_7Br \xrightarrow[\text{(ii) CO}_2]{\text{(i) Mg}} 1\text{-}C_{10}H_7 \cdot CO_2H$$

Equip a 1-litre three-necked flask with a double surface reflux condenser, a sealed mechanical stirrer and a separatory funnel, and place 12.2 g (0.5 mol) of dry magnesium turnings [Section **III,3**, Note (1)], a crystal of iodine, 50 ml of sodium-dried ether (Section **II,1,***15***) and 7.5 g (5 ml) of 1-bromonaphthalene (Section **IV,29**) in the flask. If the reaction does not start immediately, reflux gently on a water bath until it does; remove the water bath. Stir the mixture, and add a solution of 96 g (65 ml, 0.5 mol total) of 1-bromonaphthalene in 250 ml of anhydrous ether from the separatory funnel at such a rate that the reaction is under control (1.5–2 hours). Place a water bath under the flask and continue the stirring and refluxing for a further 30 minutes. The Grignard reagent collects as a heavy oil in the bottom of the flask: add 270 ml of benzene (sodium-dried) to the warm liquid in order to dissolve it completely. Cool the flask in a freezing mixture of ice and salt and pour the contents of the flask in a slow steady stream with stirring on to 125 g of crushed Cardice (Section **II,2,***9***) contained in a 2-litre beaker (1). Allow the mixture to stand until the solid carbon dioxide has evaporated, cool the beaker in an ice–salt bath and add 25 per cent sulphuric acid with stirring, until no further reaction takes place and all the magnesium has disappeared. Separate the upper layer, and extract the aqueous layer with two 50 ml portions of ether. Extract the clear ether–benzene extracts with three 50 ml portions of 25 per cent sodium hydroxide solution. Acidify the alkaline extracts with 50 per cent sulphuric acid, filter off the crude 1-naphthoic acid at the pump, wash with cold water until free from sulphate and dry at 100 °C. Dissolve the crude acid (67 g) in 200 ml of hot toluene, add a small amount of a filter-aid (e.g., Celite) and filter the solution through a preheated Buchner funnel. Cool the filtrate in ice, filter and wash the solid with cold toluene until the filtrate is nearly colourless. The yield of slightly coloured 1-naphthoic acid, m.p. 160–161 °C, is 60 g (70%).

Note. (1) For use of Cardice in Grignard reactions see Section **III,129**, Note (2).

IV,172. p-TOLUIC ACID

$$p\text{-}CH_3 \cdot C_6H_4X \xrightarrow{\text{Na or Li}} p \cdot CH_3 \cdot C_6H_4Na(Li) \xrightarrow[\text{(ii) H}_3O^{\oplus}]{\text{(i) CO}_2}$$
$$p\text{-}CH_3 \cdot C_6H_4 \cdot CO_2H$$

Method A (use of organosodium reagent). Equip a 250-ml three-necked round-bottomed flask with a reflux condenser carrying a pressure-equalising dropping funnel to which is attached a calcium chloride guard-tube, a sealed stirrer unit and a thermometer combined with a gas inlet tube to allow the air in the apparatus assembly to be displaced by nitrogen (cf. p. 646). Introduce 50 ml of dry light petroleum (b.p. 40–60 °C) and 4.6 g (0.2 mol) of sodium wire and pass a slow stream of nitrogen through the apparatus. Add 12.6 g (0.1 mol) of redistilled *p*-chlorotoluene (Section **IV,80**) from the dropping funnel, whilst stirring vigorously, during 90 minutes: maintain the temperature at 25 °C and continue stirring for a further 2 hours. Pour the reaction mixture on to 200 g of crushed Cardice in the form of a slurry with 200 ml of dry ether contained in a large beaker. After 30–45 minutes, whilst some of the solid carbon dioxide still remains, add water cautiously to destroy the excess of sodium and to dissolve the sodium salt of the acid. Separate the aqueous layer, extract it once with 50 ml of ether and warm the aqueous solution on a boiling water bath to remove the dissolved solvent. Filter if necessary, and acidify the aqueous solution with dilute hydrochloric acid. Collect the precipitated acid by suction filtration, wash it with a little water and dry at 100 °C. The yield of *p*-toluic acid, m.p. 175–176 °C, is 9.8 g (72%).

Method B (use of organolithium reagent). Fit a 250-ml three-necked flask with a reflux condenser, protected with a calcium chloride guard-tube, a sealed mechanical stirrer and a dropping funnel combined with a T-connection to provide for the inlet of nitrogen.

Place 35 ml of anhydrous ether in the flask, displace the air by nitrogen and continue passing the nitrogen in a slow stream throughout the duration of the experiment. Introduce 1.90 g (0.274 mol) of lithium in the form of fine shavings (1) into the ether and start the stirrer. Place a solution of 21.5 g (0.125 mol) of *p*-bromotoluene (Section **IV,81**) in 35 ml of ether in the dropping funnel. Run in about 1 ml of the solution into the stirred mixture. The ether in the flask soon becomes turbid; if the ether does not reflux within 10 minutes, immerse the flask in a beaker of warm water and remove it immediately refluxing commences. Add the remainder of the *p*-bromotoluene solution dropwise or at such a rate that the solvent refluxes continuously (60–90 minutes). Stir the mixture whilst refluxing gently (warm water bath) for a further 45–60 minutes; at the end of this period most of the lithium will have disappeared. Cool the reaction mixture in ice-water, dilute it with 50–60 ml of anhydrous ether and cool (with stirring) to about −50 °C with the aid of an acetone–Cardice bath. Pour the contents of the flask slowly and with stirring (use a long glass rod) on to 200 g of crushed Cardice in the form of a slurry with 200 ml of dry ether contained in a large beaker. Rinse the flask with a little of the solid carbon dioxide–ether slurry and add the rinsings to the contents of the beaker. Allow the Cardice to evaporate (3–4 hours or preferably overnight). Add about 200 ml of water to the contents of the beaker; rinse the reaction flask with 10 ml of 10 per cent sodium hydroxide solution and pour the rinsings into the beaker. A white solid appears which dissolves upon stirring. (If most of the ether has evaporated on standing, add a further 50 ml.) Separate the two layers, extract the aqueous solution with 50 ml of ether (to remove traces of neutral products) and combine the extract with the ether layer. Shake the combined ethereal solutions with 10 per cent sodium hydroxide solution and add the alkaline extract to the aqueous layer. Warm the combined aqueous layers to 60–70 °C (water bath) to drive off the dissolved ether, then cool to about 5 °C and strongly acidify with hydrochloric acid. Collect the precipitated *p*-toluic acid by suction filtration and wash it with a little cold water. The yield of the crude acid, m.p. 174–176 °C, is 11.9 g (70%); recrystallisation from dilute alcohol gives pure *p*-toluic acid, m.p. 176–177 °C.

Evaporate the dried ethereal extract; the residue, m.p. 85–90 °C, weighs 3.3 g. Recrystallise it from alcohol: pure di-*p*-tolyl ketone, m.p. 95 °C, is obtained.

Note. (1) A convenient method of preparing the lithium shavings is as follows. Place a piece of lithium weighing about 3 g and slightly moist with paraffin oil on a dry surface (slate or tiles) and pound it with a clean hammer or 500-g weight into a thin sheet about 0.5 mm thick. Cut the sheet into thin strips about 2–3 mm wide and transfer it to a beaker containing anhydrous ether. Weigh out the quantity of lithium required under dry ether or paraffin oil. Dry each strip with filter-paper, cut it by means of a pair of scissors into small pieces about 1 mm wide and allow the small pieces to fall directly into the anhydrous ether in the reaction flask. The lithium thus retains its bright lustre.

The lithium may also be pressed into wire of about 0.5 mm diameter; a rather sturdy press is necessary. The wire may be collected directly in sodium-dried ether.

Cognate preparation   *m*-Chlorobenzoic acid.   *Prepare a solution of butyl lithium in anhydrous ether as follows.*

Place 100 ml of sodium-dried ether in a 500-ml three-necked flask equipped as in *Method A*. Displace the air and maintain a slow stream of nitrogen throughout the experiment. Introduce 4.3 g (0.62 mol) of fine lithium shavings into the reaction flask. Place a solution of 34.5 g (26.5 ml, 0.25 mol) of butyl bromide (Section **III,54**) in 50 ml of anhydrous ether in the dropping funnel, start the stirrer and run in 1–2 ml of the solution into the reaction flask cooled to about −10 °C (Cardice–acetone bath). The reaction has commenced when bright spots appear on the lithium and the reaction mixture becomes slightly cloudy. Add the remainder of the butyl bromide solution during about 30 minutes whilst the internal temperature is maintained at about −10 °C. Then allow the reaction mixture to warm up to 0–10 °C during 1 hour (with stirring) in order to complete the formation of butyl lithium (1).

Cool the solution of butyl lithium to −35 °C in a Cardice–acetone bath and add, whilst stirring vigorously, a solution of 48 g of *m*-bromochlorobenzene (Section **IV,81**) in 75 ml of anhydrous ether. Stir for 15 minutes and pour the mixture with stirring on to a large excess of solid carbon dioxide in the form of a Cardice–ether slurry contained in a large beaker. Isolate the acid as detailed above for p-*Toluic acid* and recrystallise it from hot water. The yield of *m*-chlorobenzoic acid, m.p. 150–151 °C, is 27 g.

Note. (1) If a clear solution of butyl lithium is required for any purpose, it may be decanted through a glass wool plug as detailed under *2-phenylpyridine*, Section **VI,19**, Note (2).

N,4. Electrophilic substitution in aromatic carboxylic acids and esters. The carboxyl and alkoxycarbonyl groups exert an electron-withdrawing influence when attached to the aromatic ring system, and are thus deactivating and *meta* directing in electrophilic substitution reactions. Electrophilic substitution is illustrated by the nitration of methyl benzoate (Section **IV,173**) and of benzoic acid and phenylacetic acid (Section **IV,174**). In the former instance nitration is effected with 'mixed acid' at a temperature below 15 °C, conditions which effect monosubstitution. The resulting methyl *m*-nitrobenzoate may be hydrolysed to the corresponding nitro acid with aqueous alkali. Nitration of benzoic acid with a hot mixture of concentrated sulphuric acid and fuming nitric acid results in substitution in both *meta* positions to yield 3,5-dinitrobenzoic acid. Dinitration of phenylacetic acid is achieved using fuming nitric acid alone.

IV,173. *m*-NITROBENZOIC ACID

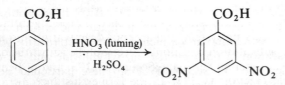

Methyl *m*-nitrobenzoate. In a 1-litre, round-bottomed three-necked flask, fitted with a mechanical stirrer and a thermometer, place 102 g (94 ml, 0.75 mol) of pure methyl benzoate (Section **IV,177**). Prepare a mixture of 62.5 ml of concentrated sulphuric acid and 62.5 ml of concentrated nitric acid in a dropping funnel, cool the flask in an ice bath to 0–10 °C and then run in the nitrating mixture, with stirring, whilst maintaining the temperature of the reaction mixture between 5 and 15 °C; the addition requires about 1 hour. Continue the stirring for 15 minutes longer, and pour the mixture upon 700 g of crushed ice. Filter off the crude methyl *m*-nitrobenzoate at the pump and wash it with cold water. Transfer the solid to a 500-ml bolt-head flask and stir it with 100 ml of ice-cold methanol in order to remove a small amount of the *ortho* isomer and other impurities. Filter the cooled mixture with suction, wash it with 50 ml of ice-cold methanol and dry in the air. The practically colourless methyl *m*-nitrobenzoate weighs 115 g (84%) and melts at 75–76 °C; it is sufficiently pure for conversion into *m*-nitrobenzoic acid. The pure ester, m.p. 78 °C, may be obtained by recrystallisation from an equal weight of methanol.

Hydrolysis of methyl *m*-nitrobenzoate to *m*-nitrobenzoic acid. Place 90.5 g (0.5 mol) of methyl *m*-nitrobenzoate and a solution of 40 g of sodium hydroxide in 160 ml of water in a 1-litre round-bottomed flask equipped with a reflux condenser. Heat the mixture to boiling during 5–10 minutes or until the ester has disappeared. Dilute the reaction mixture with an equal volume of water. When cold pour the diluted reaction product, with vigorous stirring, into 125 ml of concentrated hydrochloric acid. Allow to cool to room temperature, filter the crude acid at the pump and wash it with a little water. Upon drying at 100 °C, the crude *m*-nitrobenzoic acid, which has a pale brownish colour, weighs 80 g (96%) and melts at 140 °C. Recrystallisation from 1 per cent hydrochloric acid affords the pure acid, m.p. 141 °C, as a pale cream solid; the loss of material is about 5 per cent.

IV,174. 3,5-DINITROBENZOIC ACID

This preparation must be carried out in the fume cupboard since nitrous fumes are evolved.

Dissolve 50 g (0.41 mol) of pure benzoic acid in 230 ml of concentrated sulphuric acid in a litre flask equipped with a reflux condenser. Add 73 ml of fuming nitric acid (d 1.5) a few ml at a time (*CARE*). Shake the flask well and cool in ice-water during the addition; much heat is evolved and a clear yellow solution results. Add a few fragments of porous porcelain and heat the mixture gradually on a water bath to 100 °C during 45 minutes. At 70–80 °C the reaction may (and usually does) become vigorous; moderate, when necessary, by cooling the flask in cold water. Maintain the mixture at 100 °C for 15 minutes with occasional shaking, and then transfer it to an oil bath at 100 °C; raise the temperature to 130 °C over 30 minutes and keep it at 130–140 °C for 1 hour. Allow the flask to cool: crystals commence to separate at about 90 °C. When cold, pour the reaction mixture into 3–4 litres of ice-water, filter the separated crystals, wash with water and dry. The yield of 3,5-dinitrobenzoic acid, m.p. 204 °C, is 50 g (57%): this acid is pure enough for most purposes. Upon recrystallisation from 50 per cent alcohol (4.5 ml per gram), the m.p. is raised to 207 °C.

Cognate preparation   2,4-Dinitrophenylacetic acid.   Place 25 g (0.184 mol) of phenylacetic acid (Section **III,128,**B) in a 500-ml round-bottomed flask, cool the latter in running water and add from a suitably supported dropping funnel 250 ml of fuming nitric acid, rather slowly at first and then more rapidly. The addition occupies about 15 minutes. Attach a reflux condenser to the flask and heat the mixture under reflux for 1 hour, and then carefully pour the cooled solution into 500 ml of cold water. When cold, filter the crude product at the pump and wash it with a little cold water: the resulting acid, after drying at 100 °C, is almost pure (m.p. 181 °C) and weighs 31 g. Recrystallise it from 300 ml of 20 per cent ethanol. Collect the first main crop (25 g), and allow the mother-liquor to stand overnight when a further 2 g of pure acid is obtained; dry at 100 °C. The yield of pure 2,4-dinitrophenylacetic acid, m.p. 183 °C, is 27 g (64%).

O AROMATIC CARBOXYLIC ACID DERIVATIVES

1. Acid halides (Section **IV,175**).
2. Acid anhydrides (Section **IV,176**).
3. Esters (Sections **IV,177** to **IV,181**).
4. Acid amides (Section **IV,182**).

O,1. Acid halides.   General methods for the preparation of acid halides from aliphatic carboxylic acids are described on p. 497, **III,N,1**. Phosphorus pentachloride is the preferred chlorinating agent for aromatic acids which contain electron-withdrawing substituents, and which do not react readily with thionyl chloride. The preparation of both p-nitrobenzoyl chloride and 3,5-dinitrobenzoyl chloride is described in Section **IV,175**. These particular acid chlorides are valuable reagents for the characterisation of aliphatic alcohols and simple phenols, with which they form crystalline esters (see Sections **VII,6,**4(*b* and *c*) and **VII,6,**6(*d*)).

IV,175. p-NITROBENZOYL CHLORIDE

$$p\text{-}O_2N\cdot C_6H_4\cdot CO_2H + PCl_5 \longrightarrow p\text{-}O_2N\cdot C_6H_4\cdot COCl + POCl_3 + HCl$$

Mix 100 g (0.6 mol) of pure p-nitrobenzoic acid (Section **IV,161**) and 126 g (0.6 mol) of pure phosphorus pentachloride in a 500-ml round-bottomed flask. Fit the flask with a calcium chloride guard-tube and connect the latter to a gas absorption device (e.g., Fig. I,68). Heat the flask on a water bath, with occasional shaking, until the reaction commences and then for a further 30 minutes or until the vigorous evolution of hydrogen chloride has almost ceased: a pale yellow homogeneous liquid is formed. Attach a Claisen still-head connected with a water-cooled condenser, and remove the phosphorus oxychloride (b.p. 107 °C) at ordinary pressure either by heating in an oil bath gradually to 200–220 °C or by heating in an air bath until the boiling point is about 150 °C. Allow to cool, replace the water condenser by a *short* air-cooled condenser and distil the residual liquid under reduced pressure (water pump) (1). A small quantity of phosphorus oxychloride passes over first and the temperature rises rapidly to about 150 °C/20 mmHg; change the receiver and collect the p-nitrobenzoyl chloride at 155 °C/20 mmHg. Pour the product whilst still fluid into a small wide-mouthed bottle and allow it to solidify: this prevents any moisture in the air from decomposing more than the surface layer of acid chloride. The yield of p-nitrobenzoyl chloride (a yellow crystalline solid, m.p. 71 °C) is 105 g (95%) and is pure enough for most purposes. A perfectly pure product, m.p. 73 °C, is obtained by recrystallising from carbon tetrachloride.

Note. (1) Either an oil bath (maintained at 210–215 °C for a pressure of 20 mmHg) or an air bath must be used. If the flask is heated with a free flame, superheating will occur leading to decomposition (sometimes violent) of the p-nitrobenzoyl chloride.

Cognate preparation 3,5-Dinitrobenzoyl chloride. Place a mixture of 30 g (0.141 mol) of 3,5-dinitrobenzoic acid (Section **IV,174**) and 33 g (0.158 mol) of phosphorus pentachloride in a round-bottomed flask: fit a reflux condenser, and heat the mixture in an oil bath at 120–130 °C for 75 minutes. Allow to cool. Remove the phosphorus oxychloride by distillation under reduced pressure (25 °C/20 mmHg); raise the temperature of the bath to 110 °C. The residual 3,5-dinitrobenzoyl chloride solidifies on cooling to a brown mass; the yield is quantitative. Recrystallise from carbon tetrachloride: the yield is 25 g (77%), m.p. 67–68 °C, and this is satisfactory for most purposes. Further recrystallisation from a large volume of light petroleum, b.p. 40–60 °C, gives a perfectly pure product, m.p. 69.5 °C.

3,5-Dinitrobenzoyl chloride reacts readily with water and it should be kept in sealed tubes or under light petroleum. When required for qualitative organic analysis it is usually best prepared in small quantities from 3,5-dinitrobenzoic acid immediately before use.

CAUTION. The preparation of o-nitrobenzoyl chloride, o-nitrophenacetyl chloride and all o-nitroacid chlorides should not be attempted by the above methods: a violent explosion may occur upon distilling the product.

O,2. Acid anhydrides. o-Dicarboxylic acids (e.g., phthalic acid) readily form intramolecular anhydrides on heating. In Section **IV,176** the nitration of phthalic anhydride to yield a mixture of the isomeric 3- and 4-nitrophthalic acids is described, which can be separated by fractional crystallisation from water. The 3-nitrophthalic acid is efficiently converted into the corresponding anhydride by warming with acetic anhydride and allowing the product to crystallise.

Phthalic anhydrides readily form hydrogen phthalate esters on reaction with alcohols; the derivatives from 3-nitrophthalic anhydride are usually nicely crystalline compounds and are hence suitable for purposes of characterisation. Hydrogen phthalate esters are also useful in appropriate instances for the resolution of racemic alcohols (p. 575, **III,S**).

IV,176. 3-NITROPHTHALIC ANHYDRIDE

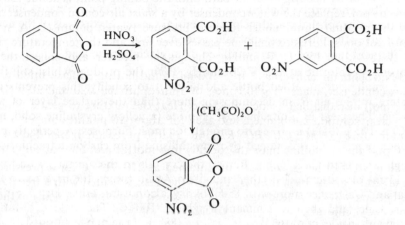

CAUTIONARY NOTE. The nitration of phthalic anhydride has been reported to proceed with explosive violence (Ref. 10(a)). The reaction may be carried out with safety on the scale adopted here provided that strict attention to detail is paid to the rate of addition of the acid and to the control of the temperature. For a modified procedure for use in large-scale nitration see Ref. 10(b).

3-Nitrophthalic acid. Equip a 500-ml, three-necked round-bottomed flask, supported on a water bath, with a dropping funnel, a thermometer and a stirrer supported in the central (open) neck connected by means of a flexible drive to a stirrer motor. The latter should not be sited near the open neck since the nitrous fumes evolved in the subsequent reaction may otherwise cause damage. Place 100 g (0.675 mol) of technical phthalic anhydride and 100 ml of concentrated sulphuric acid in the flask and heat it until the temperature of the mixture rises to 80 °C. Remove the water bath, and add a mixture of 42 ml of fuming nitric acid (*d* 1.5) and 30 ml of concentrated sulphuric acid slowly from the dropping funnel at such a rate as to maintain the temperature of the stirred mixture at 100–110 °C (about 1 hour). Then add 180 ml of concentrated nitric acid (*d* 1.42) as rapidly as possible without causing the temperature to rise above 110 °C. Heat the mixture on the water bath, with stirring, for 2 hours. Allow the reaction mixture to stand overnight and then pour it into 300 ml of water contained in a 2-litre beaker. Cool and filter the mixture of 3- and 4-nitrophthalic acids through a sintered glass funnel. Return the wet cake of acids to the rinsed-out beaker and stir it thoroughly with 40 ml of water, which dissolves a large amount of the 4-nitrophthalic acid (1). Filter again at the pump and dissolve the solid in 40–60 ml of boiling water; filter the hot solution and stir until crystallisation commences and then leave overnight until crystallisation is complete. Filter again with suction and dry upon filter-paper. The yield of crude 3-nitrophthalic acid, m.p. 208–210 °C (sealed tube), is 44 g. Recrystallisation from about 100 ml of boiling water (2) gives about 36 g (25%) of the pure acid, m.p. 216–218 °C (sealed tube).

Notes. (1) The mother-liquors from the washings and recrystallisations are saved for the recovery of 4-nitrophthalic acid. The combined mother-liquors are concentrated to small bulk and the organic acids extracted into ether. Upon esterification of the residue after evaporation of the ether by the Fischer–Speier method (p. 501, III,N,3(a)), the 3-nitro acid forms the acid ester and may be removed by shaking the product with sodium carbonate solution, whilst the 4-nitrophthalic acid yields the neutral diester. Hydrolysis of the neutral ester gives the pure 4-nitrophthalic acid, m.p. 165 °C.

(2) The acid may also be recrystallised from glacial acetic acid.

3-Nitrophthalic anhydride. In a 100-ml round-bottomed flask fitted with a reflux condenser, place 21 g (0.1 mol) of 3-nitrophthalic acid and 20 g (18.5 ml, 0.2 mol) of redistilled acetic anhydride. Heat the mixture to gentle boiling until a clear solution is obtained, and then for about 10 minutes longer. Pour the hot mixture (*FUME CUPBOARD*) into a large porcelain dish and allow to cool. Grind the crystalline mass thoroughly in a mortar and filter at the pump through a sintered glass funnel. Return the crystals to the mortar, grind them with 15 ml of sodium-dried ether and filter. Again return the crystals to the mortar and wash once more with 15 ml of dry, alcohol-free ether. Dry in air for a short time, and then to constant weight at 100 °C. The yield of 3-nitrophthalic anhydride, m.p. 163–164 °C, is 17 g (88%). If the m.p. is unsatisfactory, recrystallise the anhydride from benzene or from benzene–light petroleum (b.p. 40–60 °C).

O,3. Esters. Aromatic esters may be prepared by direct esterification methods similar to those already described for aliphatic esters (p. 501, III,N,3(a)). A large range of examples of simple alkyl esters of aromatic carboxylic acids is included in Section IV,177. Corresponding esterification of a simple aliphatic acid (e.g., acetic acid) with benzyl alcohol is illustrated in Section IV,178.

Methyl esters may be prepared by reaction of the aromatic carboxylic acid with diazomethane (cf. Section II,2,19) or, more conveniently, by reaction with a boron trifluoride–methanol reagent. The latter procedure is illustrated by the preparation of methyl m-chlorobenzoate and dimethyl terephthalate (Section IV,179).

Esterification of aromatic carboxylic acids with phenols, however, cannot be accomplished by a direct esterification procedure and resort must be made to the greater reactivity exhibited by the acid chlorides. Reaction is usually carried out in dilute aqueous alkali (Schotten–Baumann conditions, p. 682, IV,F,2 and p. 1103). The preparations of phenyl benzoate and 2-naphthyl benzoate are typicaly examples of this procedure (Section IV,180).

The synthesis of ethyl 1-naphthoate (Section IV,181) illustrates the preparation of a carboxylic ester by a variant of the Grignard carboxylation route to a carboxylic acid. The arylmagnesium bromide is first prepared and added to an excess of diethyl carbonate, conditions which minimise the possibility of further reaction of the Grignard reagent with the ester initially produced to form a tertiary alcohol.

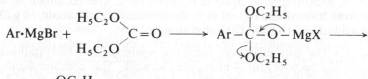

IV,177. METHYL BENZOATE

$$C_6H_5{\cdot}CO_2H + CH_3OH \xrightarrow{H^{\oplus}} C_6H_5{\cdot}CO_2CH_3 + H_2O$$

In a 500-ml round-bottomed flask place a mixture of 30 g (0.246 mol) of benzoic acid, 80 g (101 ml, 2.5 mol) of absolute methanol and 5 g (2.7 ml) of concentrated sulphuric acid. Add a few small chips of porous porcelain, attach a reflux condenser and boil the mixture gently for 4 hours (1). Distil off the excess of alcohol on a water bath (rotary evaporator) and allow to cool. Pour the residue into about 250 ml of water contained in a separatory funnel and rinse the flask with a few ml of water which are also poured into the separatory funnel. If, owing to the comparatively slight difference between the density of the ester and of water, difficulty is experienced in obtaining a sharp separation of the lower ester layer and water, add 10–15 ml of carbon tetrachloride (2) and shake the mixture in the funnel vigorously; upon standing, the heavy solution of methyl benzoate in the carbon tetrachloride separates sharply and rapidly at the bottom of the separatory funnel. Run off the lower layer carefully, reject the upper aqueous layer, return the methyl benzoate to the funnel and shake it with a strong solution of sodium hydrogen carbonate until all free acid is removed and no further evolution of carbon dioxide occurs. Wash once with water, and dry by pouring into a small dry conical flask containing about 5 g of magnesium sulphate. Stopper the flask, shake for about 5 minutes and allow to stand for at least half an hour with occasional shaking. Filter the methyl benzoate solution through a small fluted filter-paper directly into a round-bottomed flask fitted with a still-head carrying a 360 °C thermometer and an air condenser. Add a few boiling chips and distil from an air bath; raise the temperature slowly at first until all carbon tetrachloride has passed over and then heat more strongly. Collect the methyl benzoate (a colourless liquid) at 198–200 °C. The yield is 31 g (92%).

Notes. (1) Slightly improved results may be obtained by increasing the time of heating.

(2) Alternatively, the ester may be extracted with two 50 ml portions of ether. The ethereal solution is washed with concentrated sodium hydrogen carbonate solution (handle the separatory funnel cautiously as carbon dioxide is evolved) until effervescence ceases, then with water, and dried over magnesium sulphate. The ether is removed by flash distillation and the residual ester distilled.

Cognate preparations   Ethyl benzoate (*sulphuric acid as a catalyst*).   Use 30 g (0.246 mol) of benzoic acid, 115 g (145 ml, 2.5 mol) of absolute ethanol and 5 g (2.7 ml) of concentrated sulphuric acid. Reflux the mixture for 4 hours and work up as for *methyl benzoate*. The yield of ethyl benzoate, b.p. 212–214 °C, is 32 g (86%).

Ethyl benzoate (*hydrogen chloride as a catalyst*).   Pass dry hydrogen chloride (Section **II,2,31**) into a 500-ml round-bottomed flask containing 115 g (145 ml, 2.5 mol) of absolute ethanol cooled in an ice bath, until the increase in weight is 6 g. Add 30 g (0.246 mol) of benzoic acid and reflux the mixture for 4 hours. Isolate the pure ester, b.p. 212–214 °C, as described for *methyl benzoate*. The yield is 32 g (86%).

Propyl benzoate.   Into a 500-ml round-bottomed flask place 30 g (0.246 mol) of benzoic acid, 30 g (37.5 ml, 0.5 mol) of propan-1-ol, 50 ml of sodium-dried benzene and 10 g (5.4 ml) of concentrated sulphuric acid. Reflux the mixture for 10 hours. Pour the reaction product into about 250 ml of water, and extract with ether. Wash the ethereal extract with saturated sodium hydrogen carbonate solution and then with water: dry over magnesium sulphate. Distil off the ether and some of the benzene through a fractionating column, and distil the residue from a Claisen flask. Collect the propyl benzoate at 229–230 °C. The yield is 37 g (91%).

Butyl benzoate.   Use 30 g (0.246 mol) of benzoic acid, 37 g (46 ml, 0.5 mol) of butan-1-ol, 50 ml of sodium-dried benzene and 10 g (5.4 ml) of concentrated sulphuric acid, and reflux the mixture for 12 hours. Work up the product as for *propyl benzoate*; after the ether and benzene have been removed with the aid of a rotary evaporator, distil the residue under reduced pressure. The yield of butyl benzoate, b.p. 119–120 °C/11 mmHg, is 35 g (80%).

Methyl salicylate.   Use 28 g (0.2 mol) of salicylic acid (Section **IV,170**), 64 g (81 ml, 2 mol) of dry methanol and 8 ml of concentrated sulphuric acid. Reflux the mixture for at least 5 hours and work up as for *methyl benzoate*. Collect the pure methyl salicylate (a colourless oil of delightful fragrance, 'oil of wintergreen') at 221–224 °C; the yield is 25 g (81%). The ester may also be distilled under reduced pressure; the b.p. is 115 °C/20 mmHg and a 2 °C fraction should be collected.

Ethyl salicylate.   This colourless ester, b.p. 231–234 °C, is similarly obtained in 75 per cent yield from salicylic acid, ethanol and sulphuric acid as catalyst. It is best to distil the ester under reduced pressure; the boiling points under various pressures are given in Table I,10.

Methyl cinnamate.   Use 59 g (0.4 mol) of cinnamic acid (Section **IV,150**), 128 g (162 ml, 4 mol) of absolute methanol and 6 ml of concentrated sulphuric acid; reflux the mixture for 5 hours. Remove excess methanol, pour the residue into about 500 ml of water and add 300 ml of ether. Separate, wash and dry the ether solution in the usual way. Remove the ether on a rotary evaporator; the residue crystallises on cooling, yielding 58 g (90%) of methyl cinnamate, m.p. 33–34 °C. To obtain a pure specimen, m.p. 36 °C, dissolve a sample in the minimum of methanol maintained at 30 °C in a water bath, and add water slowly from a dropping pipette with stirring until the oily ester just begins to separate. Seed the solution and transfer rapidly to an ice bath with a glass rod, scratching the sides of the vessel vigorously with a glass rod. Filter the resulting colourless needles rapidly.

Ethyl nicotinate (*ethyl pyridine-3-carboxylate*).   Reflux a mixture of 37 g (0.3

mol) pure nicotinic acid (Section **IV,162**), 92 g (115 ml, 2 mol) of absolute ethanol and 90 g (50 ml) of concentrated sulphuric acid on a steam bath. Cool the solution and pour it slowly and with stirring on to 200 g of crushed ice. Add sufficient ammonia solution to render the resulting solution strongly alkaline: generally, some ester separates as an oil but most of it remains dissolved in the alkaline solution. Extract the mixture with five 25 ml portions of ether, dry the combined ethereal extracts over magnesium sulphate, remove the ether by flash distillation and distil the residue under reduced pressure. The ethyl nicotinate distils at 107–108 °C/16 mmHg; the yield is 32 g (71%). The boiling point under atmospheric pressure is 222–224 °C.

IV,178. BENZYL ACETATE

$$C_6H_5 \cdot CH_2OH + CH_3 \cdot CO_2H \xrightarrow{H_2SO_4} C_6H_5 \cdot CH_2 \cdot CO_2CH_3 + H_2O$$

Mix 31 g (29.5 ml, 0.287 mol) of benzyl alcohol (Section **IV,145**) and 45 g (43 ml, 0.75 mol) of glacial acetic acid in a 500-ml round-bottomed flask; introduce 1 ml of concentrated sulphuric acid and a few fragments of 'porous pot'. Attach a reflux condenser to the flask and boil the mixture gently for 9 hours. Pour the reaction mixture into about 200 ml of water contained in a separatory funnel, add 10 ml of carbon tetrachloride (to eliminate emulsion formation owing to the slight difference in density of the ester and water, compare *methyl benzoate*, Section **IV,177**) and shake. Separate the lower layer (solution of benzyl acetate in carbon tetrachloride) and discard the upper aqueous layer. Return the lower layer to the funnel, and wash it successively with water, concentrated sodium hydrogen carbonate solution (until effervescence ceases) and water. Dry over 5 g of magnesium sulphate, and distil from an air bath. Collect the benzyl acetate (a colourless liquid) at 213–215 °C. The yield is 16 g (37%).

IV,179. METHYL *m*-CHLOROBENZOATE

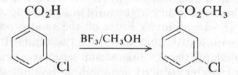

Place 9.4 g (0.06 mol) of *m*-chlorobenzoic acid and 66 ml (0.12 mol) of boron trifluoride–methanol complex (14% w/v of BF$_3$; Section **II,2,6**) in a 250-ml round-bottomed flask. Heat the mixture under reflux on an oil bath for 2 hours, cool and pour into about 250 ml of saturated sodium hydrogen carbonate solution. Extract the organic product with three 50 ml portions of ether, dry the ethereal extract over magnesium sulphate and evaporate on a rotary evaporator. Distil the residue under reduced pressure and collect the methyl *m*-chlorobenzoate as a colourless liquid of b.p. 63 °C/3 mmHg; the yield is 9.3 g (91%).

Cognate preparation   Dimethyl terephthalate.   Use 9.97 g (0.06 mol) of terephthalic acid, 132 ml of boron trifluoride/methanol complex (14% w/v) and 100 ml of dry methanol (1) and heat the mixture under reflux for 6 hours. Cool the reaction mixture and pour into excess (500 ml) of saturated sodium hydrogen carbonate solution. Filter the precipitated dicarboxylic ester under suction and recrystallise from methanol. The yield is 9.8 g (84%); m.p. 139–140 °C.

Note. (1) Additional methanol is used in this case because of the low solubility of the di-acid and di-ester. Dry the methanol by distillation from magnesium (Section **II,1,**8).

IV,180. PHENYL BENZOATE

$$C_6H_5 \cdot COCl + C_6H_5OH \xrightarrow{\text{NaOH aq.}} C_6H_5 \cdot CO_2C_6H_5 + HCl \, (\longrightarrow NaCl)$$

Dissolve 5 g (0.053 mol) of phenol in 75 ml of 10 per cent sodium hydroxide solution contained in a wide-mouthed reagent bottle or conical flask of about 200 ml capacity. Add 11 g (9 ml, 0.078 mol) of redistilled benzoyl chloride, cork the vessel securely and shake the mixture vigorously for 15–20 minutes. At the end of this period the reaction is usually practically complete and a solid product is obtained. Filter off the solid ester with suction, break up any lumps on the filter, wash thoroughly with water and drain well. Recrystallise the crude ester from rectified spirit; use a quantity of hot solvent approximately twice the minimum volume required for complete solution in order to ensure that the ester does not separate until the temperature of the solution has fallen below the melting point of phenyl benzoate. Filter the hot solution, if necessary, through a hot water funnel or through a Buchner funnel preheated by the filtration of some boiling solvent. Colourless crystals of phenyl benzoate, m.p. 69 °C, are thus obtained. The yield is 8 g (76%).

Cognate preparation   2-Naphthyl benzoate.   Dissolve 7.2 g (0.05 mol) of 2-naphthol in 40 ml of 5 per cent sodium hydroxide solution in the cold; add a little more water if necessary. If the solution is highly coloured, add 1.5 g of decolourising carbon and filter the cold solution through a hardened filter-paper. Pour the solution into a 100-ml conical flask and run in 7.0 g (5.8 ml, 0.05 mol) of benzoyl chloride. Stopper the flask and shake vigorously until the odour of benzoyl chloride has disappeared (10–15 minutes). Filter off the solid product on a Buchner funnel and wash it with a little cold water. Recrystallise it from about 60 ml of rectified spirit. Filter off the crystals which separate and dry them upon filter-paper in the air. The yield of pure 2-naphthyl benzoate, m.p. 110 °C, is 11 g (89%).

IV,181. ETHYL 1-NAPHTHOATE

$$1\text{-}C_{10}H_7Br \xrightarrow{\text{Mg}} 1\text{-}C_{10}H_7MgBr \xrightarrow{(C_2H_5O)_2CO} 1\text{-}C_{10}H_7 \cdot CO_2C_2H_5$$

In a 1.5-litre three-necked flask prepare a solution of 1-naphthyl magnesium bromide (0.5 mol) from 12.2 g of magnesium turnings as detailed under *1-naphthoic acid*, Section **IV,171**; add just sufficient sodium-dried benzene to form a homogeneous solution. Transfer the Grignard reagent to a separatory funnel and place 88.5 g (91 ml, 0.75 mol) of pure diethyl carbonate [Section **III,39**, Note (1)] and 50 ml of sodium-dried ether in the three-necked flask. Stir and add the 1-naphthyl magnesium bromide as rapidly as the refluxing of the solution will permit. Continue the stirring for a further 30 minutes and allow the reaction mixture to stand overnight. Then pour the reaction mixture, with frequent shaking, into a 2.5-litre flask containing 750 g of crushed ice. Dissolve the basic magnesium bromide by adding gradually 72.5 ml of cold 30 per cent sulphuric acid (15 ml of concentrated sulphuric acid and 60 ml of water). Separate the upper layer and extract the aqueous layer with 50 ml of ether. Concentrate the extracts to about 200 ml using a rotary evaporator. Wash the residue with two 20 ml portions of 5 per cent sodium carbonate solution (1), and dry with 10 g of magnesium sulphate or calcium sulphate. Remove the remaining solvent by flash distillation (Fig. I,100) and distil the residual liquid: collect the fraction, b.p. 290–310 °C, as crude ethyl 1-naphthoate. Redistil the ester under reduced pressure through a fractionating side arm (compare Fig. I,110) and collect the pure ester at 143–145 °C/3 mmHg. The yield is 70 g (70%).

Note. (1) Upon acidifying the alkaline washings, about 1 g of 1-naphthoic acid may be isolated.

O,4. Acid amides. The preparation of toluamide (Section **IV,182**) illustrates a useful procedure for the conversion of aromatic nitriles into acid amides with the aid of alkaline hydrogen peroxides (see discussion, p. 515, **III,N,4**(d)).

A further example of the preparation of amides by the ammonolysis of esters (cf. succinamide, Section **III,158**) is provided by the preparation of nicotinamide described in Section **IV,184** as a stage in the synthesis of 3-cyanopyridine.

IV,182. o-TOLUAMIDE

$$2o\text{-}H_3C\cdot C_6H_4\cdot CN + 2H_2O_2 \xrightarrow{\ominus OH} 2o\text{-}H_3C\cdot C_6H_4\cdot CO\cdot NH_2 + O_2$$

Place 29 g (0.25 mol) of o-tolunitrile (Section **IV,84**), 130 ml of rectified spirit (1) and 10 ml of 25 per cent sodium hydroxide solution in a 1-litre bolt-necked flask. Set up the flask inside an aluminium bowl placed on a magnetic stirrer/hotplate unit which is supported on an adjustable laboratory jack. Insert into the flask a magnetic stirrer follower and a thermometer, start the stirrer motor and run in steadily 100 ml of 30 per cent hydrogen peroxide (see Section **II,2,**34 for precautions in the use of this reagent) (2). Oxygen is soon evolved and the mixture becomes warm; when the temperature approaches 40 °C add an ice-water slurry to the bowl in order that the temperature may be controlled within the 40–50 °C range. If the temperature tends to fall below this range it may be necessary to remove the cooling bath for a while. If the temperature is permitted to rise above 50 °C not only may the evolution of oxygen become so rapid as to cause the mixture to foam out of the flask, but also there is a danger of a violent explosion due to the ignition of an oxygen–ethanol vapour mixture (3). The exothermic reaction is complete after about 1 hour and then the temperature of the reaction mixture is kept at 50 °C for a further 3 hours by external heating. While still warm, add 5 per cent sulphuric acid until exactly neutral to litmus, and remove the ethanol and concentrate the residue to about 200 ml under reduced pressure using a rotary evaporator. Cool the residue to 20 °C, filter off the crystals at the pump and grind them to a paste with 30 ml of cold water in a mortar. Filter again, wash the product in the filter with a further 30 ml of water and dry in the air upon filter-paper. The yield of o-toluamide (white crystals), m.p. 141 °C, is 30 g (90%). It may be recrystallised from hot water (1 g per ml), but the m.p. is unchanged.

Notes. (1) This volume of rectified spirit is required to produce a homogeneous solution.

(2) Difficultly hydrolysable nitriles, such as *o*-tolunitrile, require 30 per cent hydrogen peroxide. For most nitriles, however, both aromatic and aliphatic, an equivalent amount of 6–12 per cent hydrogen peroxide gives more satisfactory results; the above procedure must, however, be modified, according to the solubility of the nitriles and amides.

(3) To minimise this explosion danger, the reaction should be carried out in a well-ventilated fume cupboard in the absence of free flames; brisk stirring should be maintained throughout the exothermal stage.

P AROMATIC NITRILES

1. The Sandmeyer procedure (see Section **IV,84**).
2. The displacement of halogen by cyanide in an aryl halide (Section **IV,183**).
3. The dehydration of amides (Section **IV,184**) and aldoximes (Section **IV,185**).

P,1. The Sandmeyer procedure. This valuable method for the preparation of aryl nitriles via the diazonium salt is discussed on p. 689, **IV,G,1**, and offers one of the most convenient routes for obtaining this class of compound. Experimental procedures are described in Section **IV,84**.

P,2. The displacement of halogen by cyanide in an aryl halide. The ready replacement of the halogen in an alkyl or an aralkyl halide illustrated in Section **III,160** by reaction with sodium or potassium cyanide is inapplicable in the case of aryl halides wherein the halogen is relatively inert. However, aryl bromides can be converted into nitriles in good yield by heating them for several hours at about 200 °C with copper(I) cyanide in the presence of pyridine (e.g., 1-naphthonitrile, Section **IV,183**). This displacement may be achieved more readily by using dimethylformamide as the solvent, when reaction is usually completed in a few hours at reflux temperature (Ref. 11).

IV,183. 1-NAPHTHONITRILE

$$1\text{-}C_{10}H_7Br \xrightarrow[\text{pyridine}]{CuCN} 1\text{-}C_{10}H_7 \cdot CN$$

Place 80 g (54 ml, 0.386 mol) of redistilled 1-bromonaphthalene (Section **IV,129**), 43 g (0.48 mol) of dry powdered copper(I) cyanide (Section **II,2,**17) and 36 g (37 ml, 0.457 mol) of dry pure pyridine (I) (Section **II,1,**29) in a 250-ml round-bottomed flask fitted with a reflux condenser carrying a calcium chloride guard-tube, and heat the mixture in a metal bath at 215–225 °C for 15 hours (2). Pour the resulting dark brown solution while still hot (*ca.* 100 °C) into a litre flask containing 180 ml of concentrated ammonia solution (*d* 0.88) and 180 ml of water. Add 170 ml of benzene, stopper the flask and shake until all the lumps have disintegrated. When cold, add 100 ml of ether and filter through a sintered glass funnel (3). Add a further 50 ml of ether, transfer to a separatory funnel, separate the ether–benzene layer and wash it successively with (i) four 125 ml portions of dilute ammonia solution (or until the organic layer is colourless), (ii) two 125 ml portions of dilute hydrochloric acid (1 : 1) (any precipitate which separates should be filtered off), (iii) two 125 ml portions of water and (iv) two 125 ml portions of saturated sodium chloride solution. Finally dry with magnesium sulphate, remove the ether and benzene by distillation on a rotary evaporator and distil the residue under reduced pressure (water pump) through a short fractionating column (compare Fig. I,110). Collect the 1-naphthonitrile at 166–169 °C/18 mmHg as a colourless liquid. The yield is 50 g (84%).

Notes. (1) Much heat is liberated when pyridine is added to the mixture. (2) The metal bath may be replaced by a bath of Silicone oil. (3) The cuprammonium solution attacks filter-paper.

P,3. The dehydration of amides and aldoximes. Two examples of the dehydration of aromatic carboxamides using phosphorus pentoxide (cf. p. 522, III,O,2) are given in Section **IV,184**; these are the preparation of benzonitrile and 3-cyanopyridine.

The indirect conversion of an aromatic aldehyde into the corresponding nitrile by dehydration of an oxime is illustrated by the synthesis of veratronitrile (Section **IV,185**). The dehydrating agent is acetic anhydride which probably effects an initial acetylation of the oximino group followed by the elimination of acetic acid.

$$\text{Ar·CHO} \longrightarrow \text{Ar·CH}=\text{NOH} \longrightarrow \text{Ar·CH}=\text{N·O·CO·CH}_3$$

$$\underset{\text{Ar}}{\overset{\text{CH}_3\text{·CO}_2^{\ominus}\frown\text{H}}{}}\text{C}=\text{N}\underset{\text{O·CO·CH}_3}{} \longrightarrow \text{Ar·C}\equiv\text{N}$$

Conditions for effecting this conversion in one step have been described (Ref. 12). The method involves heating a mixture of the aldehyde, hydroxylamine hydrochloride, sodium formate and formic acid; and the reaction is considered to proceed through the intermediate formation of an oxime formate.

IV,184. BENZONITRILE

$$\text{C}_6\text{H}_5\text{·CO·NH}_2 \xrightarrow{\text{P}_2\text{O}_5} \text{C}_6\text{H}_5\text{·CN}$$

Place 45 g (0.37 mol) of benzamide (prepared from benzoyl chloride, cf. Section **III,157**) and 80 g (0.56 mol) of phosphorus pentoxide in a 250-ml round-bottomed flask (for exact experimental details on the handling and weighing out of phosphoric oxide, see *isobutyronitrile*, Section **III,162**). Mix well. Fit the flask with a Claisen distillation head (Fig. I,15(c)) and distil under reduced pressure using a water pump with an air leak in the system so that a pressure of about 100 mmHg is attained. Heat the flask with a free flame until no more liquid distils: the nitrile will pass over at 126–130 °C/100 mmHg. Wash the distillate with a little sodium carbonate solution, then with water, and dry over anhydrous calcium chloride or magnesium sulphate. Distil under normal pressure from a 50-ml flask: the benzonitrile passes over as a colourless liquid at 188–189 °C. The yield is 28 g (74%).

Cognate preparation   3-Cyanopyridine.

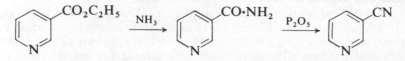

Nicotinamide.   Place 50 g (0.33 mol) of pure ethyl nicotinate (Section **IV,177**) in a 350-ml flask and add 75 ml of cold concentrated aqueous ammonia saturated at 0 °C. Keep the flask loosely stoppered for 18 hours, after which time the lower layer generally dissolves on shaking. Again saturate the solution with ammonia and allow it to stand for a further 4 hours. Repeat the saturation with ammonia; crystals of the amide commence to appear in the solution. Evaporate to dryness in a dish on the steam bath and dry at 120 °C. The yield of nicotinamide, m.p. 130 °C, is usually quantitative.

3-Cyanopyridine.   Mix 24 g (0.2 mol) of powdered nicotinamide with 30 g of phosphoric oxide in a 150-ml round-bottomed flask by shaking. Immerse the flask in a Silicone oil or fusible metal bath and arrange for distillation under a pressure of about 30 mmHg. Raise the temperature of the bath rapidly to 300 °C, then remove the bath and continue the heating with a free flame as long as a distillate is obtained. The nitrile crystallises on cooling to a snow-white solid. Redistil the solid at atmospheric pressure; practically all of it passes over at 201 °C and crystallises completely on cooling. The yield of 3-cyanopyridine, m.p. 49 °C, is 18 g (86%).

IV,185. VERATRONITRILE

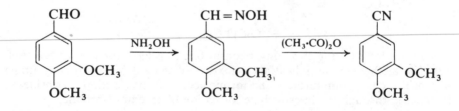

Dissolve 83 g (0.5 mol) of veratraldehyde (Section **IV,124**) in 200 ml of warm rectified spirit in a 1-litre round-bottomed flask, and add a warm solution of 42 g (0.6 mol) of hydroxylamine hydrochloride in 50 ml of water. Mix thoroughly and run in a solution of 30 g of sodium hydroxide in 40 ml of water. Allow the mixture to stand for 2.5 hours, add 250 g of crushed ice and saturate the solution with carbon dioxide. The aldoxime separates as an oil: allow the mixture to stand for 12–24 hours in an ice chest or refrigerator when the oil will solidify. Filter off the crystalline aldoxime at the pump, wash well with cold water and dry in the air upon filter-paper. The yield of **veratraldoxime** is 88 g (96%).

Into a 250-ml round-bottomed flask, fitted with a still-head and air condenser, place 88 g of veratraldoxime and 100 g (92.5 ml) of redistilled acetic anhydride. Heat cautiously. Immediately the vigorous reaction commences, remove the flame. When the reaction subsides, boil the solution gently for 20 minutes and then pour it carefully with stirring into 300 ml of cold water. Continue the stirring and cool in ice. Filter off the almost colourless crystals of veratronitrile and dry in the air. The resulting nitrile (60 g, 73.5% overall) is quite pure and melts at 67 °C.

REFERENCES

IV Aromatic compounds

1. W. D. Ellis (1963). *J. Chem. Ed.*, **40**, 346.
2. A. R. Butler and A. P. Sanderson (1971). *J. Chem. Soc. (B)*, 2264.
3. (a) H. Stephen, W. F. Short and G. Gladding (1920). *J. Chem. Soc.*, 510.
   (b) S. R. Buc (1956). *Org. Synth.*, **36**, 1.
4. Anon (1972). In *Organic Reactions* (19). Ed. W. G. Dauben. New York; Wiley, p. 422.
5. K. G. Rutherford, W. Redmond and J. Rigamondi (1961). *J. Org. Chem.*, **26**, 5149.
6. B. T. Newbold and R. P. Leblanc (1965). *J. Chem. Soc.*, 1547.
7. M. F. Hawthorne (1957). *J. Org. Chem.*, **22**, 1001.
8. W. Walter and K.-D. Bode (1966). *Angew. Chem., Interntl. Ed.*, **5**, 457.
9. J. H. Brewster, A. M. Fusco, L. E. Carosino and B. G. Corman (1963). *J. Org. Chem.*, **28**, 498.
10. (a) J. H. P. Tyman and A. A. Durrani (1972). *Chem. and Ind.*, 664.
    (b) R. K. Bentley (1972). *Chem. and Ind.*, 767.
11. L. Friedman and H. Shechter (1961). *J. Org. Chem.*, 2522.
12. T. van Es (1965). *J. Chem. Soc.*, 1564.

CHAPTER V

SOME ALICYCLIC COMPOUNDS

There are many methods available for the formation of alicyclic compounds from acyclic precursors. Often the key step is the formation of a carbon–carbon bond in an intramolecular process. Alternatively the reaction may involve the simultaneous formation of two carbon–carbon bonds (i.e., cycloaddition reactions); in general these latter processes may proceed under thermal or photochemical conditions.

Some of the methods are discussed below and the relevant preparations are described in Sections **V,I** to **V,17**.

(*a*) **Intramolecular addition of carbanions to carbonyl groups** (Sections **V,1** to **V,8**).   The cyclisation of the 1,4-diketone, acetonylacetone, under the influence of base, to 3-methylcyclopent-2-enone (Section **V,1**), offers a simple example of an intramolecular aldol condensation.

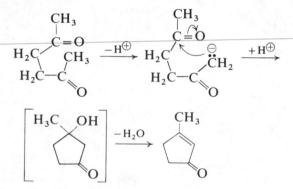

Similarly 1,5-diketones yield six-membered alicyclic ring systems. Such a reaction is involved in the synthesis of 3-methylcyclohex-2-enone described in Section **V,2**. In this synthesis formaldehyde (as paraformaldehyde) is first condensed with ethyl acetoacetate in the presence of piperidine; the initial product is the methylenebisacetoacetate (I) (cf. the synthesis of pyridines, Section **VI,16**), which on heating in the presence of the base cyclises to 4,6-diethoxycarbonyl-3-methylcyclohex-2-enone (II). Both ethoxycarbonyl groups are removed when the diester is heated for some time with aqueous acid, forming 3-methylcyclohex-2-enone (III). Conversion of this product into a 3-alkyl-3-methylcyclohexanone may be accomplished by means of reaction with a Grignard reagent which proceeds by 1,4-conjugate addition to the enone system.

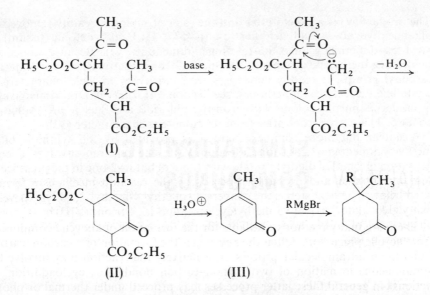

(I)

(II)   (III)

An example illustrating a simple intramolecular Claisen condensation is provided by the synthesis of dimedone (Section **V,3**). Here the δ-ketoester (IV) is first prepared by the base-catalysed addition of diethyl malonate to mesityl oxide (the Michael addition, p. 489, **III,M,6**(*b*)), and allowed to undergo (under the influence of the base catalyst already present) an addition of the carbanion derived from the methyl ketone grouping to one of the ethoxycarbonyl groups.

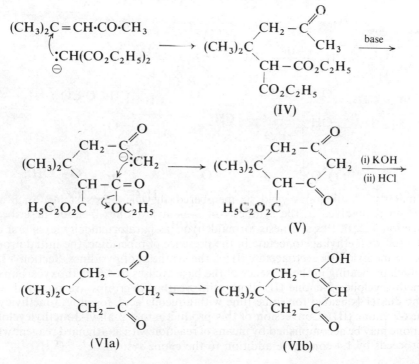

(IV)

(V)

(VIa)   (VIb)

The product (V) is a β-keto ester and the ester grouping is readily removed by alkaline hydrolysis and acidification (p. 537, **III,P,4**(b); ketonic fission) to yield 5,5-dimethylcyclohexane-1,3-dione (dimedone, VI). Because dimedone possesses a methylene group (at C2) which is strongly activated by two adjacent carbonyl groups, it readily condenses with aldehydes to yield characteristic crystalline derivatives (dimethones, see Section **VII,6,**_13_(_f_)). Dimedone gives a pronounced purple colour with iron(III) chloride and has a pK_a value of about 5. These properties arise from its mono-enolic character (VIb).

A similar but more complex example is provided by the synthesis of 3-methylcyclopentane-1,2,4-trione (Section **V,4**). The reaction involves a condensation of 2 mol of diethyl oxalate with 1 mol of butan-2-one in the presence of sodium ethoxide, and probably proceeds by the reaction mechanism formulated below. The next step is the conversion of ethyl 4-methyl-2,3,5-trioxocyclopentylglyoxalate (VII) into 3-methylcyclopentane-1,2,4-trione (VIII) by the overall removal of oxalic acid which is effected by treatment with hot aqueous phosphoric acid.

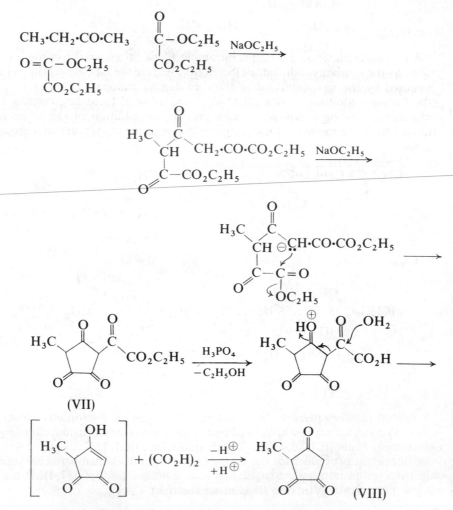

An intramolecular aldol-type addition reaction (the Claisen–Schmidt reaction, p. 793, **IV,M,2**) is involved in the synthesis of tetraphenylcyclopentadienone ('tetracyclone', Section **V,5**), which is readily obtained by base-catalysed condensation of the aromatic α-diketone, benzil, with dibenzyl ketone.

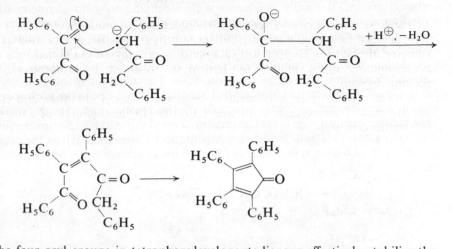

The four aryl groups in tetraphenylcyclopentadienone effectively stabilise the cyclopentadienone system, which otherwise has only a transient existence and readily undergoes dimerisation by way of a diene–dienophile interaction (the Diels–Alder reaction, p. 865, **V(d)** below). Tetracyclone undergoes the addition of dienophiles (e.g., maleic anhydride) to give an adduct, which then extrudes a molecule of carbon monoxide on heating, e.g., the preparation of 3,4,5,6-tetraphenyldihydrophthalic anhydride (Section **V,6**).

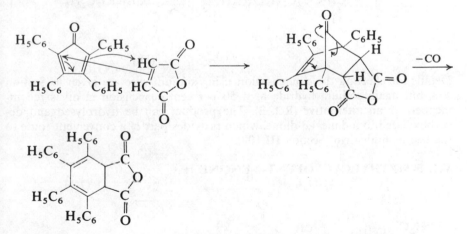

A related reaction involves the addition of benzyne (cf. 9-bromotriptycene, Section **IV,9**) to tetracyclone followed by carbon monoxide extrusion, affording a convenient route to 1,2,3,4-tetraphenylnaphthalene (Ref. 1).

An important procedure for the synthesis of five- and six-membered alicyclic rings involves an intramolecular Claisen ester condensation (p. 537, **III,P,4(b)**) using a dicarboxylic ester (the **Dieckmann reaction**).

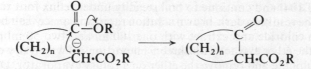

In the example described (2-ethoxycarbonylcyclopentanone, Section **V,7**) the base catalyst is sodium ethoxide, but sodium hydride is frequently used as an effective alternative. The product, a β-keto ester, may be converted into the corresponding cyclic ketone by hydrolysis followed by decarboxylation (ketonic fission).

A mixed Claisen ester condensation between a dialkyl phthalate and ethyl acetate yields 2-ethoxycarbonylindane-1,3-dione (IX) as the result of a further, intramolecular, step:

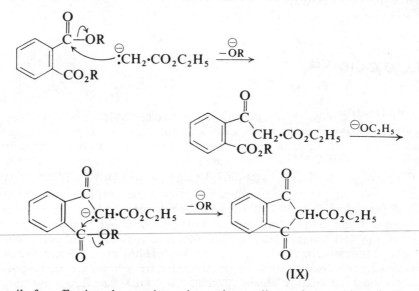

(IX)

Details for effecting the condensation using sodium wire are given in Section **V,8**, but again sodium hydride as a 50 per cent suspension in oil is recommended as an alternative (Ref. 2). The product can be hydrolysed and decarboxylated to indane-1,3-dione, which provides part of a convenient route to the trione, ninhydrin (Section **III,100**).

V,1. 3-METHYLCYCLOPENT-2-ENONE

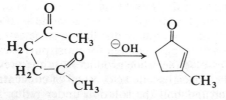

Dissolve 2.5 g of sodium hydroxide in 250 ml of water in a 500-ml two-necked flask fitted with a reflux condenser and a dropping funnel. Bring the solution to the boil, add rapidly from the dropping funnel 28.5 g (0.25 mol) of hexane-

2,5-dione (Section **III,104**) and continue to boil steadily under reflux for exactly 15 minutes (1). Cool the resulting dark-brown solution rapidly in an ice–salt bath, saturate with sodium chloride and extract with one 100 ml and two 50 ml portions of ether. Wash the ether extract with three 5 ml portions of water, dry over anhydrous sodium sulphate and remove the ether on a rotary evaporator. Distil the residual dark oil under reduced pressure and collect the colourless 3-methyl-cyclopent-2-enone as a fraction of b.p. 74–76 °C/16 mmHg, n_D^{20} 1.4818; yield 9.5 g (40%). The product thus obtained is pure enough for most purposes; when perfectly pure the refractive index is 1.4893. The product may darken on storage.

Note. (1) The reaction conditions are critical; excessive boiling or the use of more concentrated alkali increases the formation of tarry by-products.

V,2. 3-METHYLCYCLOHEX-2-ENONE

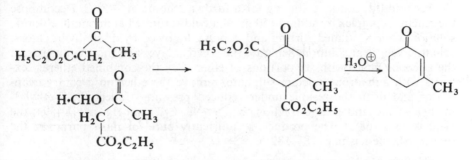

4,6-Diethoxycarbonyl-3-methylcyclohex-2-enone. Place 130 g (1.0 mol) of ethyl acetoacetate, 15.0 g (0.5 mol) of powdered paraformaldehyde and 5 g (5.8 ml) of piperidine in a 500-ml round-bottomed flask. Allow the reaction to proceed at room temperature; after a short period (about 5 minutes) the contents of the flask rapidly heat up and the solid paraformaldehyde begins to dissolve. Moderate the reaction as required by cooling in iced water. When the vigorous reaction is over and the reaction mixture is homogeneous (about 20 minutes) heat the mixture on a water bath for 1 hour. The flask now contains the crude product together with a little water formed in the reaction. The crude product may be used directly for conversion into 3-methylcyclohex-2-enone by acid hydrolysis as described below. If required, it may be purified and dried in the following way. Dissolve the crude material in 200 ml of methylene chloride in a separatory funnel and wash successively with two 100 ml portions of 2.5 M-hydrochloric acid, then two 100 ml portions of saturated aqueous sodium hydrogen carbonate and two 200 ml portions of water. Dry the organic phase over anhydrous calcium sulphate and remove the solvent on a rotary evaporator. The yield of crude 4,6-diethoxycarbonyl-3-methylcyclohex-2-enone is 106.5 g (84%). This compound cannot be purified by vacuum distillation since extensive decomposition takes place. It is, however, pure enough for most purposes.

3-Methylcyclohex-2-enone. Dissolve the crude product from the previous preparation in a mixture of 300 ml of glacial acetic acid, 30 ml of concentrated sulphuric acid and 200 ml of water and boil the solution under reflux for 6 hours. Add a solution of 254 g of sodium hydroxide in 700 ml of water carefully and with cooling to the cooled reaction solution. Extract with three 150 ml portions of ether, dry the ethereal extract over anhydrous sodium sulphate and remove the ether on a rotary evaporator. Distil the residue under reduced pressure through a short fractionating column and collect the 3-methylcyclohex-2-enone as a fraction, b.p. 95 °C/25 mmHg. The yield is 24 g (44%).

Conversion into 3-ethyl-3-methylcyclohexanone. Prepare a solution of ethylmagnesium iodide in 120 ml of ether from 18.7 g (0.77 mol) of dry magnesium turnings and 120 g (0.77 mol) of ethyl iodide in a 500-ml, three-necked round-bottomed flask equipped with a sealed stirrer unit, a pressure-equalising funnel and a reflux condenser. When all the magnesium has dissolved, clamp the flask in a cooling bath of Cardice–acetone and replace the condenser with a low-temperature-reading thermometer. When the temperature has fallen to − 10 °C, add 1.5 g (0.015 mol) of thoroughly dried copper(I) chloride (1). Then whilst maintaining the internal temperature of the reaction mixture between − 10 and − 5 °C (2), and with rapid stirring add dropwise a solution of 55 g (0.50 mol) of 3-methylcyclohex-2-enone in 150 ml of ether. After all the ketone has been added, continue stirring for a further 2 hours at − 5 °C. Decompose the reaction complex by adding 150 ml of a cold saturated ammonium chloride solution, with continued stirring and cooling, followed by dilute hydrochloric acid to give a clear solution. Separate the ethereal layer and thoroughly extract the aqueous phase with six portions of ether. Dry the combined ethereal extracts with anhydrous calcium sulphate, remove the ether on a rotary evaporator and distil the residue under reduced pressure. Collect the 3-ethyl-3-methylcyclohexanone as a fraction, b.p. 97–98 °C/22 mmHg, n_D^{21} 1.4594; the yield is 48 g (68%). The product is sufficiently pure for most purposes; the semicarbazone has m.p. 181–182 °C.

Notes.   (1) Temperature control is important at this stage, otherwise the yield of product is considerably reduced.

(2) Careful temperature control is important, or the yield of product is reduced and considerable quantities of the by-product 1-methyl-3-ethyl-1,3-cyclohexadiene are produced.

V,3. 5,5-DIMETHYLCYCLOHEXANE-1,3-DIONE (*Dimedone*)

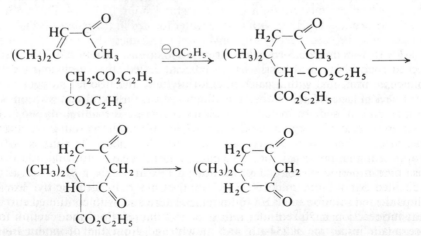

Equip a dry 1-litre three-necked flask with a dropping funnel, a sealed stirrer unit and an efficient double surface condenser. Place 11.5 g (0.5 mol) of sodium in the flask, cool in an ice bath and add 200 ml of absolute ethanol in one portion. When the initial vigorous reaction has subsided, remove the ice bath and allow the reaction to proceed until all the sodium has reacted: warming on a water bath is sometimes necessary to dissolve the last traces of sodium. Place a calcium chloride guard-tube at the top of the condenser. Introduce 85 g (0.53 mol) of diethyl malonate (Section **III,154**) and then add through the dropping funnel 50 g (0.51 mol) of freshly distilled mesityl oxide (Section **III,209**) slowly. Reflux the mixture with stirring for 2 hours, then add a solution of 62.5 g (1.11 mol) of potassium hydroxide in 300 ml of water, and reflux again on a water bath with stirring for 6 hours. Acidify the reaction mixture (to litmus) while still hot with dilute hydrochloric acid (1 : 2 by volume): about 275 ml are required. Fit the flask with a condenser for distillation, and distil off as much alcohol as possible by heating with stirring on a water bath. Allow the residue in the flask to cool somewhat, add 8 g of decolourising carbon slowly, boil for 10 minutes and filter; repeat the treatment with decolourising carbon. Neutralise the residue to litmus by the addition of dilute hydrochloric acid (about 75 ml) and boil again with 8 g of decolourising charcoal. Filter and render the hot yellow filtrate distinctly acid to methyl orange with dilute hydrochloric acid (25–50 ml), boil for a few minutes and allow to cool whereupon the dimedone crystallises out. Filter at the pump, wash with ice-cold water and dry in the air. The yield of dimedone, m.p. 147–148 °C, is 60 g (84%). Recrystallisation from acetone (about 8 ml per gram) raises the m.p. to 148–149 °C, but this is generally unnecessary.

V,4. 3-METHYLCYCLOPENTANE-1,2,4-TRIONE

$$CH_3 \cdot CH_2 \cdot CO \cdot CH_3 + 2 \begin{array}{c} CO_2C_2H_5 \\ | \\ CO_2C_2H_5 \end{array} \xrightarrow{NaOC_2H_5}$$

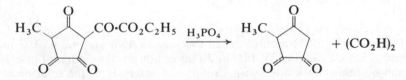

$$+ (CO_2H)_2$$

Ethyl 4-methyl-2,3,5-trioxocyclopentylglyoxalate. Equip a 2-litre, three-necked round-bottomed flask with an efficient sealed stirrer unit and a reflux condenser protected with a calcium chloride tube, and insert a stopper in the third neck. Place 315 ml of absolute ethanol in the flask and add portion-wise 23 g (1 mol) of sodium metal; if necessary moderate the initial vigour of the reaction by external cooling (see Section **III,94**). When all the sodium metal has reacted cool the flask in an ice–salt cooling bath, replace the stopper with a dropping funnel protected with a calcium chloride guard-tube and add drop-wise over a period of about half an hour a mixture of 36 g (0.5 mol) of dry redistilled butan-2-one and 160 g (1.1 mol) of dry redistilled diethyl oxalate. Allow the red solution to attain room temperature and, with continued stirring, heat the reaction mixture under reflux for 30 minutes. Cool the reaction mixture again in an ice–salt bath and slowly add from the dropping funnel with efficient stirring 55 ml of dilute sulphuric acid (prepared by adding concentrated sulphuric acid to an equal volume of water). Remove the

sodium sulphate which crystallises out by filtration and wash it with ethanol; concentrate the combined filtrate and washings on a rotary evaporator
to a volume of about 100 ml. Cool the residual aqueous solution in an ice–
salt cooling bath for several hours, filter the glyoxalate ester and wash
with a little ice-cold water. The yield of crude product is 39.5 g (35%), which
may be used for the next stage. A small portion of the product may be
recrystallised from ethyl acetate (decolourising charcoal), when it has m.p. 162–
164 °C.

3-Methylcyclopentane-1,2,4-trione. In a 500-ml round-bottomed flask fitted with a reflux condenser place 300 ml of dilute phosphoric acid (prepared by
diluting phosphoric acid with an equal volume of water) and add 30 g (0.133
mol) of the foregoing glyoxalate ester. Heat the solution under reflux for 30
minutes; cool the solution to about -10 °C and allow to stand at this temperature for 2 hours. Remove the oxalic acid which crystallises out by filtration and
wash it with small portions of iced water. Extract the filtrate with eight 120 ml
portions of ether, combine the extracts and rinse with a little water, dry and
evaporate to dryness. Dissolve the residue in about 10 ml of water, cool
and filter off crystals of the monohydrate which have a wide melting range
of about 72–79 °C according to the rate of heating. Anhydrous 3-methyl-
cyclopentane-1,2,4-trione may be obtained as colourless plates, m.p. 118–
119 °C, by sublimation under reduced pressure ($\approx$ 0.1 mmHg); the yield is
9 g (54%).

V,5. TETRAPHENYLCYCLOPENTADIENONE (*Tetracyclone*)

$$C_6H_5 \cdot CO \cdot CO \cdot C_6H_5 + (C_6H_5 \cdot CH_2)_2CO \xrightarrow[C_2H_5OH]{KOH}$$

Dissolve 4.2 g (0.02 mol) of benzil (Section **IV,155**) and 4.2 g (0.02 mol) of
dibenzyl ketone (Section **IV,141**) in 30 ml of hot absolute ethanol in a 100-ml
round-bottomed flask fitted with a reflux condenser. Heat the solution to near
its boiling point on a steam bath and then add in portions a solution of 0.6 g
(0.011 mol) of potassium hydroxide in 6 ml of absolute ethanol. Some foaming
may occur. Heat the reaction mixture under reflux for 15 minutes (1) and cool
to below 5 °C in an ice bath. Collect the dark crystalline product by filtration
with suction, wash with three 5 ml portions of rectified spirit and dry in an
oven at 50 °C. The tetraphenylcyclopentadienone has m.p. 217–220 °C and is
sufficiently pure for most purposes; the yield is 7.0 g (91%). Recrystallise a
portion from toluene–ethanol (1 : 1) to obtain the pure compound as deep
purple crystals, m.p. 219–220 °C; λ_{max} 340 and 510 nm, ε 1.26 × 10³ and
0.33 × 10³ respectively.

Note. (1) In a modified procedure (Ref. 3) using triethylene glycol as solvent and benzyltrimethylammonium hydroxide ('Triton B') as the base catalyst,
the reaction may be completed in a very short time.

V,6. 3,4,5,6-TETRAPHENYLDIHYDROPHTHALIC ANHYDRIDE

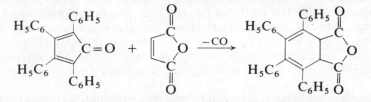

Place a mixture of 7.0 g (0.018 mol) of tetraphenylcyclopentadienone, 1.9 g (0.019 mol) of maleic anhydride (1) and 5 ml of bromobenzene in a 100-ml round-bottomed flask fitted with a reflux condenser. Reflux gently in a fume cupboard (carbon monoxide is evolved) for 1.5 hours; during this period the dark brown colour of the cyclic ketone disappears. Cool to room temperature, add about 15 ml of light petroleum, b.p. 60–80 °C, break up the solid and filter. Wash with light petroleum and dry. The yield of crude product, m.p. 237–240 °C, is 7.5 g (90%). Dissolve the solid in hot benzene (120 ml), filter and add light petroleum, b.p. 60 80 °C (120 ml), to the hot filtrate and cool. Collect the solid which separates, wash with light petroleum and dry. The yield of pure tetraphenyldihydrophthalic anhydride, m.p. 235 240 °C, is 6.2 g (75%). The m.p. depends on the rate of heating.

Note. (1) See Section **V,12**, Note (1).

V,7. 2-ETHOXYCARBONYLCYCLOPENTANONE

Prepare 25 g (1.09 mol) of granulated sodium in a 1500-ml round-bottomed flask (Section **II,2,55**, *Method 2*). Cover the sodium with 625 ml of sodium-dried benzene; fit the flask with an efficient reflux condenser protected from the air by means of a calcium chloride guard-tube. Add 151.5 g (0.75 mol) of diethyl adipate (Section **III,146**) in one lot, followed by 1.5 ml of absolute ethanol. Warm the flask on a water bath until, after a few minutes, a vigorous reaction sets in and a cake of the sodio compound commences to separate. Keep the flask well shaken by hand during the whole of the initial reaction. After the spontaneous reaction has subsided, reflux the mixture on a water bath over-night, and then cool in ice. Decompose the product with ice and dilute hydro-chloric acid (1 : 1); add the acid until Congo red paper is turned blue. Separate the benzene layer, and extract the aqueous layer with 100 ml of benzene. Wash the combined extracts with 100 ml of 5 per cent sodium carbonate solution and 150 ml of water: dry over magnesium sulphate. Remove the benzene on a rotary evaporator (*FUME CUPBOARD*), and fractionate the residue under reduced pressure. Collect the 2-ethoxycarbonylcyclopentanone at 108–111 °C/15 mmHg (95 g, 81%). Upon redistillation, the product boils at 102 °C/11 mmHg.

V,8. INDANE-1,3-DIONE

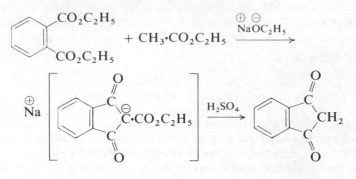

Place 125 g (106.5 ml, 0.563 mol) of diethyl phthalate and 25 g (1.09 mol) of sodium wire (Section **II,2,55**) in a 500-ml round-bottomed flask fitted with a reflux condenser and dropping funnel, each protected with a calcium chloride tube. Heat the flask on a steam bath and add a mixture of 122.5 g (136 ml, 1.39 mol) of dry ethyl acetate and 2.5 ml of absolute ethanol over a period of 90 minutes. Continue the heating for 6 hours, cool and add 50 ml of ether. Filter the sodium salt on a sintered glass funnel and wash by stirring with ethyl acetate; filter. Dissolve the sodium salt (96 g) in 1400 ml of hot water in a 3-litre beaker, cool the solution to 70 °C, stir vigorously and add 100 ml of sulphuric acid (3 parts of concentrated acid to 1 part of water). Cool the mixture to 15 °C in an ice bath, collect the indane-1,3-dione by suction filtration, wash with a little water and dry at 100 °C; the yield is 58 g (71%). Recrystallisation from a dioxan–benzene mixture by the addition of light petroleum (b.p. 80–100 °C) gives the pure compound, m.p. 130 °C.

(*b*) **Intramolecular nucleophilic displacement** (Sections **V,9** and **V,10**). Intramolecular base-catalysed condensation between a dialkyl malonic ester and a suitable dihalogen compound is an important general method for the synthesis of alicyclic ring systems. Five- and six-membered ring systems are most readily formed by the use of this general procedure, which is also applicable to the synthesis of cyclobutanes as illustrated by the preparation of diethyl cyclobutane-1,1-dicarboxylate (Section **V,9**). Cyclobutanecarboxylic acid is readily obtained from the diester by hydrolysis and decarboxylation.

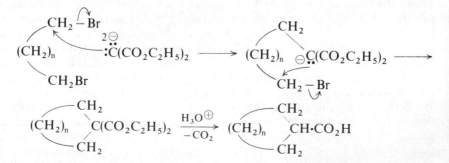

Intramolecular nucleophilic displacement is also suitable for the formation of the more highly strained three-membered ring, which is in fact frequently found to be formed more readily than the corresponding cyclobutane system. The reaction sequence is illustrated by the synthesis of cyclopropyl methyl ketone (Section **V,10**), which is prepared by treating 5-chloropentan-2-one with aqueous alkali.

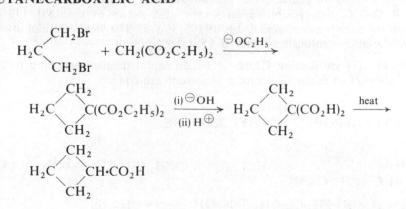

$$ClCH_2 \cdot CH_2 \cdot CH_2 \cdot CO \cdot CH_3 \overset{\ominus OH}{\rightleftharpoons} ClCH_2 \cdot CH_2 \cdot \overset{\cdot\cdot}{C}H \cdot CO \cdot CH_3$$
$$(X)$$

The required γ-chloroketone (X) is conveniently prepared by treatment of 2-acetylbutyrolactone with concentrated hydrochloric acid, when nucleophilic ring opening by the halide ion is accompanied by decarboxylation.

$$\longrightarrow ClCH_2 \cdot CH_2 \cdot CH_2 \cdot CO \cdot CH_3$$
$$(X)$$

V,9. CYCLOBUTANE-1,1-DICARBOXYLIC ACID AND CYCLOBUTANECARBOXYLIC ACID

Equip a 3-litre three-necked flask with a thermometer, a sealed mechanical stirrer and a double surface reflux condenser. It is important that all the apparatus be thoroughly dry. Place 212 g (1.05 mol) of 1,3-dibromopropane

(Section **III,54**) and 160 g (1 mol) of diethyl malonate (Section **III,154**) (dried over anhydrous calcium sulphate) in the flask. Start the stirrer and add a solution of 46 g (2 mol) of sodium in 800 ml of super-dry ethanol (Section **II,1,**9) (1) down the condenser from a dropping funnel at such a rate that the temperature of the reaction mixture is maintained at 60–65 °C (50–60 minutes). When the addition is complete, allow the mixture to stand until the temperature falls to 50–55 °C, and then heat on a water bath until a few drops of the liquid when added to water are no longer alkaline to phenolphthalein (about 2 hours). Add sufficient water to dissolve the precipitate of sodium bromide, and remove the ethanol by distillation from a water bath. Steam distil the residue until all the diethyl cyclobutane-1,1-dicarboxylate and unchanged diethyl malonate are removed; collect about 4 litres of distillate during 9–10 hours. Extract the entire steam distillate with three 350 ml portions of ether; remove the ether from the combined extracts on a water bath (rotary evaporator). Reflux the residual liquid with a solution of 112 g of potassium hydroxide in 200 ml of ethanol for 2 hours. Distil off most of the ethanol and then evaporate the residue to dryness on a water bath (rotary evaporator). Dissolve the solid residue in 100 ml of hot water, and add concentrated hydrochloric acid (*ca.* 80 ml) cautiously until the solution is just acid to litmus. Boil for a few minutes to remove carbon dioxide, render slightly alkaline with ammonia solution and add a slight excess of aqueous barium chloride to the boiling solution. Filter the hot solution to remove the barium malonate, cool the filtrate and render it strongly acid with concentrated hydrochloric acid (90–100 ml of acid: use Congo red paper). Extract the solution with four 250 ml portions of ether. Dry the combined extracts with anhydrous calcium chloride and remove the ether on a rotary evaporator. Spread the solid on a porous tile to remove oily impurities. The beautifully crystalline product (55 g, 38%) consists of pure cyclobutane-1,1-dicarboxylic acid, m.p. 158 °C. It may be recrystallised from hot ethyl acetate, but the m.p. is unchanged.

Place 30 g (0.208 mol) of cyclobutane-1,1-dicarboxylic acid in a 100-ml flask fitted with a still-head carrying a thermometer and leading to a cooled receiver flask via a short air condenser. Heat the flask in a Silicone oil bath at 160–170 °C until all effervescence ceases. Then raise the temperature of the bath to 210 °C; the cyclobutanecarboxylic acid passes over at 191–197 °C. Redistil the acid; the pure acid distils at 195–196 °C. The yield of cyclobutane-carboxylic acid (a colourless liquid) is 18 g (86%).

Note. (1) See Section **III,94** for experimental details pertaining to the preparation of an ethanolic solution of sodium ethoxide.

V,10. CYCLOPROPYL METHYL KETONE

$$\text{(cyclobutanone structure)} + HCl \longrightarrow ClCH_2 \cdot CH_2 \cdot CH_2 \cdot CO \cdot CH_3 + CO_2$$

$$ClCH_2 \cdot CH_2 \cdot CH_2 \cdot CO \cdot CH_3 + NaOH \longrightarrow$$

$$\text{(cyclopropyl methyl ketone structure)} + NaCl + H_2O$$

5-Chloropentan-2-one. Assemble a distillation unit consisting of a 500-ml round-bottomed flask, a still-head, a large double surface condenser and a 250-ml receiving flask; cool the latter in an ice bath. Place into the flask 64 g (0.5 mol) of 2-acetylbutyrolactone, 90 ml of water and a few chips of broken porcelain, and then add 75 ml of concentrated hydrochloric acid and shake to mix. Heat the flask gently on a wire gauze until carbon dioxide is evolved briskly, and continue to heat cautiously until effervescence moderates and the contents of the flask become very dark (about 5–10 minutes). Then heat the flask strongly, preferably using two burners, so that rapid distillation ensues. When about 125 ml of distillate has been collected, add 75 ml of water steadily to the flask from a dropping funnel fitted to the still-head without interrupting the distillation. Collect a total of 200 ml of distillate; the entire distillation should be completed within about 75 minutes. Separate the pale yellow organic phase from the distillate, extract the aqueous phase with three 30 ml portions of ether and dry the combined organic layers over anhydrous calcium chloride. Remove the ether by distillation through a lagged 10 × 1.5 cm fractionating column packed with glass helices and distil the residue (1) under reduced pressure (water-pump). Collect the 5-chloropentan-2-one at 70–73 °C/20 mmHg, the yield is 45 g (75%).

Cyclopropyl methyl ketone. Place a solution of 20 g (0.5 mol) of sodium hydroxide in 20 ml of water in a 250-ml three-necked flask fitted with a dropping funnel, a reflux condenser and an efficient sealed stirrer unit. With vigorous stirring add 42 g (0.35 mol) of 5-chloropentan-2-one during 30 minutes; heat the flask gently during the addition so that the reaction mixture refluxes steadily. Continue heating under reflux for a further period of 1 hour, and then add slowly 45 ml of water and continue to reflux for 1 hour more, maintaining vigorous stirring throughout. Rearrange the condenser for distillation and distil out the reaction product until an organic layer no longer remains in the flask. Saturate the distillate with potassium carbonate, separate the organic phase and extract the aqueous phase with three 15 ml portions of ether. Dry the combined organic extracts over anhydrous calcium chloride and then over anhydrous calcium sulphate. Remove the ether by flash distillation through a lagged 10 × 1.5 cm fractionating column (glass helices), and continue distilling collecting the cyclopropyl methyl ketone as a fraction of b.p. 111–112 °C; the yield is 24 g (82%).

Note. (1) The crude residue may be used directly in the next stage.

(c) **Ring expansion procedures** (Section **V,11**). Several preparatively useful reactions involve the expansion of a carbocyclic ring system. An example (the preparation of cycloheptanone, Section **V,11**) is provided which illustrates the reaction of a cyclic ketone with diazomethane. Cycloheptanone is obtained in about 60 per cent yield together with a little of the epoxide which is formed as a by-product, and some cyclooctanone which results from a further ring expansion of the cycloheptanone. The cycloheptanone is readily separated by taking advantage of the fact that it alone forms a solid bisulphite compound. Diazomethane is conveniently generated *in situ* from *N*-methyl-*N*-nitrosotoluene-*p*-sulphonamide (Section **II,2**,*19*).

Mechanistically the reaction may be represented in the following manner.

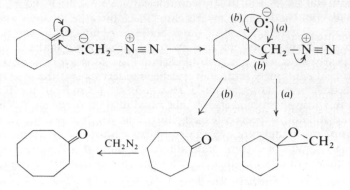

V,11. CYCLOHEPTANONE

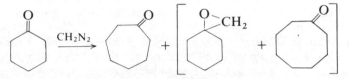

This preparation must be carried out in an efficient fume cupboard (see Section II,2,*19*).

In a 1-litre three-necked flask equipped with a thermometer, a mechanical stirrer and a dropping funnel, place 49 g (0.5 mol) of redistilled cyclohexanone, 125 g (0.585 mol) of *N*-methyl-*N*-nitrosotoluene-*p*-sulphonamide, 150 ml of 95 per cent ethanol and 10 ml of water. The nitrosamide is largely undissolved. Adjust the height of the stirrer so that only the upper part of the solution is stirred and the precipitate moves slightly; place the thermometer so that the bulb is in the liquid. Cool the mixture to about 0 °C in an ice–salt bath. Whilst stirring gently, add a solution of 15 g of potassium hydroxide in 50 ml of 50 per cent aqueous ethanol dropwise very slowly from the dropping funnel: after 0.5–1 ml of the solution has been added, a vigorous evolution of nitrogen commences and the temperature rises (1). Adjust the rate of addition so that the temperature is maintained at 10–20 °C; the duration of the addition of alkali is about 2 hours and the nitroso compound ultimately disappears. Stir the orange-yellow solution for a further 30 minutes, and then add 2 *M*-hydrochloric acid until the solution is acidic to litmus paper (*ca.* 50 ml).

Introduce a solution of 100 g of sodium metabisulphite in 200 ml of water and continue the stirring, preferably for 10 hours with exclusion of air. A thick precipitate separates after a few minutes. Collect the bisulphite compound by suction filtration, wash it with ether until colourless and then decompose it in a flask with a lukewarm solution of 125 g of sodium carbonate in 150 ml of water. Separate the ketone layer, extract the aqueous layer with four 30 ml portions of ether, dry the combined organic layers over magnesium sulphate, remove the ether at atmospheric pressure and distil the residual oil under reduced pressure through a short fractionating side arm. Collect the cycloheptanone at 64–65 °C/12 mmHg; the yield is 17 g (31%).

Note. (1) If the reaction does not start at this stage remove the flask from the cooling bath and allow the mixture to warm to 10 °C; do not add any further alkali until the reaction has started.

(*d*) **Cycloaddition reactions** (Sections V,12 to V,17). Compounds containing a double or triple bond, usually activated by conjugation with additional multiple bonded systems (carbonyl, cyano, nitro, phenyl, etc.), add to the 1,4-positions of a conjugated diene system (e.g., buta-1,3-diene) with the formation of a six-membered ring. The ethylenic or acetylenic compound is known as the **dienophile** and the second reactant as the **diene**; the product is the **adduct**. The cycloaddition is generally termed the **Diels–Alder reaction** or the diene synthesis, and is a widely used method of synthesis of six-membered carbocyclic compounds. The product in the case of an ethylenic dienophile is a cyclohexene, and in that of an ethynyl dienophile, is a cyclohexa-1,4-diene. The active unsaturated portion of the dienophile, or that of the diene, or those in both, may be involved in rings; the adduct is then polycyclic.

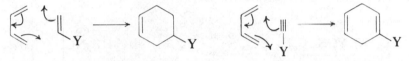

(e.g., Y = CO·R, C≡N, NO_2, C_6H_5, etc.)

The concerted nature of the cycloaddition process is shown in the above formulation and involves a six π-electron system; it is a transformation which is symmetry-allowed under thermal conditions in terms of the Woodward–Hoffmann rules for electrocyclic reactions (Ref. 4).

The preparative procedure for the Diels–Alder reaction is extremely simple and the cycloaddition usually merely requires warming the reactants together either alone, or in the presence of a suitable solvent.

The reaction has been shown to involve a stereospecific *syn*-addition with respect to the dienophile. For example, the reaction of 2,3-dimethylbuta-1,3-diene with maleic anhydride gives *cis*-1,2,3,6-tetrahydro-4,5-dimethylphthalic anhydride (Section **V,12**). An example of the use of a quinone as the dienophile is provided by the synthesis of *cis*-1,4,4a,9a-tetrahydro-2,3-dimethyl-9,10-anthraquinone which upon dehydrogenation (most simply by the action of oxygen upon its solution in alcoholic potassium hydroxide) yields 2,3-dimethylanthraquinone (Section V,13).

With less reactive dienophiles more extensive heating may be required in which case the reaction is best carried out in a suitable pressure vessel (e.g., the preparation of 4-nitro-5-phenylcyclohexene, Section **V,14**).

Addition of electron-deficient divalent species, the **carbenes**, to olefins is a versatile method for the synthesis of cyclopropanes since the structure of both the carbene and the olefin can be varied considerably. Thus, typically, dichlorocarbene adds to olefins to give 1,1-dichlorocyclopropanes. Of the many ways of generating dichlorocarbene one of the simplest is the recently introduced reaction of chloroform with concentrated aqueous sodium hydroxide in the presence of a phase transfer catalyst; the method is illustrated by the preparation of 7,7-dichlorobicyclo[4.1.0]heptane (Section **V,15**) from cyclohexene using benzyltriethylammonium chloride as the catalyst. The mechanism of the dichlorocarbene formation is believed (Ref. 5) to involve initial conversion in the aqueous phase of the quaternary chloride (XI) to the quaternary hydroxide (XII) by the sodium hydroxide, which then migrates to the boundary with the organic phase, and there reacts with the chloroform to give the quaternary ammonium derivative (XIII) of the trichloromethyl anion. This species diffuses into the organic phase where it breaks down to form dichlorocarbene and the quaternary ammonium chloride. The chloride passes into the aqueous phase to react further to form more quaternary hydroxide to maintain the cycle of events, whilst the dichlorocarbene reacts rapidly with the olefin in the organic phase.

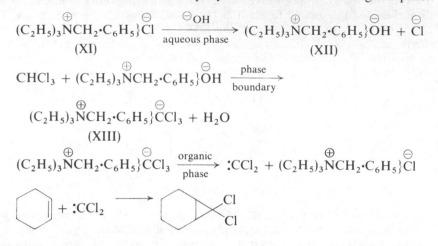

The photodimerisation of unsaturated compounds is a useful method for the synthesis of cyclobutanes and is applicable to a variety of compounds. Thus cyclopent-2-enone, as the neat liquid or in a variety of solvents, on irradiation with light of wavelength greater than 300 nm (the n → π* excited state is involved) is converted to a mixture of the 'head-to-head' (XIV) and 'head-to-tail' (XV) dimers, both having the *cis,anti,cis* stereochemistry as shown. It is believed that the reaction proceeds by attack of an n → π* triplet excited species on a ground state molecule of the unsaturated ketone (Section **I,17,5**). In the reaction described (Section **V,16**) the cyclopent-2-enone is irradiated in methanol and the 'head-to-tail' dimer further reacts with the solvent to form the di-acetal which conveniently crystallises from the reaction medium as the irradiation proceeds; the other dimer (the minor product under these conditions) remains in solution. The di-acetal is converted to the diketone by treatment with the two-phase dilute hydrochloric acid–methylene chloride system.

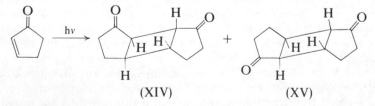

(XIV)   (XV)

The cyclopent-2-enone required for the photodimerisation is prepared by the hydrolysis and oxidation of 3-chlorocyclopentene, which is obtained by the low temperature addition of hydrogen chloride to cyclopentadiene. The latter is obtained by heating dicyclopentadiene. This depolymerisation is an example of a reverse (or retro) Diels–Alder cycloaddition reaction; the diene readily reforms the dicyclopentadiene on standing at room temperature.

A further example of a photodimerisation reaction is provided by the formation, in poor yield, of '1,4-naphthoquinone photodimer' (XVI) on irradiation of 1,4-naphthoquinone in benzene solution (Section V,17). The dimerisation may be effected by sunlight or by means of a mercury arc lamp.

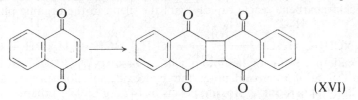

(XVI)

V,12. *CIS*-1,2,3,6-TETRAHYDRO-4,5-DIMETHYLPHTHALIC ANHYDRIDE

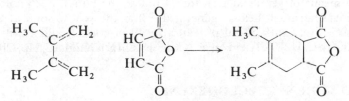

Add 4.1 g (0.05 mol) of freshly distilled 2,3-dimethylbuta-1,3-diene (Section III,8) to 4.9 g (0.05 mol) of finely powdered maleic anhydride (Section III,212) (1) contained in a small conical flask. Reaction occurs in a few minutes (indicated by evolution of heat). Allow to stand until the mixture attains room temperature. Remove the excess of maleic anhydride by extraction with cold water until the aqueous extract no longer gives an acid reaction to Congo red paper. Dry the residual white crystals upon filter-paper in the air, and then recrystallise from light petroleum (b.p. 40–60 °C). The yield of the tetrahydrophthalic anhydride, m.p. 78–79 °C, is almost quantitative.

Note. (1) Alternatively maleic anhydride may be prepared from the dicarboxylic acid as follows. Mix 100 g of maleic acid with 100 ml of 1,1,2,2-tetrachloroethane in a distillation flask fitted with a Claisen still-head, a thermometer and a condenser set for downward distillation (*FUME CUPBOARD*). Heat the mixture on an air bath; when the temperature reaches 150 °C, 75 ml of 1,1,2,2-tetrachloroethane and between 15 and 15.5 ml of water are present in the receiver. Continue the distillation using an air condenser and change the receiver flask when the temperature reaches 190 °C. Collect the maleic anhydride at 195–197 °C. Recrystallise the crude anhydride from chloroform. The yield of pure maleic anhydride, m.p. 54 °C, is 70 g (83%).

V,13. *CIS*-1,4,4a,9a-TETRAHYDRO-2,3-DIMETHYL-9,10-ANTHRAQUINONE

In a small round-bottomed flask, fitted with a reflux condenser, place a solution of 8.2 g (0.1 mol) of freshly distilled 2,3-dimethylbuta-1,3-diene (Section III,8) and 7.9 g (0.05 mol) of 1,4-naphthoquinone (Section IV,142) in

30 ml of ethanol, and reflux for 5 hours. Keep the resulting solution in a refrigerator for 12 hours: break up the crystalline mass, filter and wash with 5 ml of ethanol. The yield of crude adduct, m.p. 147–149 °C, is 11.5 g (91%); recrystallisation from methanol raises the m.p. to 150 °C.

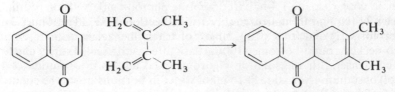

Conversion into 2,3-dimethylanthraquinone.   Dissolve 10 g (0.0417 mol) of the adduct in 150 ml of 5 per cent potassium hydroxide solution (prepared by dissolving 7.5 g of potassium hydroxide pellets in 142.5 g of 95% ethanol) in a 250-ml three-necked flask equipped with a reflux condenser, gas inlet tube and a take-off adapter leading to a water pump. Bubble a current of air through the solution by means of gentle suction for 24 hours; the initial green colour changes to yellow and much heat is generated. Filter the yellow solid at the pump, wash successively with 50 ml of water, 25 ml of ethanol and 10 ml of ether and dry in the air. The yield of 2,3-dimethylanthraquinone, m.p. 209–210 °C, is 7.5 g (76%).

V,14. 4-NITRO-5-PHENYLCYCLOHEXENE

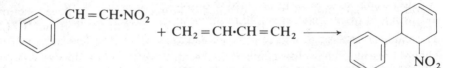

This reaction is carried out using the steel pressure vessel and techniques described in Section I,17,2. To the dry pressure vessel add 7.46 g (0.05 mol) of ω-nitrostyrene (Section IV,148), about 0.1 g of hydroquinone (a polymerisation inhibitor) and 15 ml of dry toluene. Fit a rubber bung carrying a calcium chloride guard-tube and cool the vessel to − 78 °C in an acetone–Cardice bath. During the cooling process set up the acetone–Cardice condenser assembly (see p. 76) and charge the receiving flask with about 100 g of fresh potassium hydroxide pellets. Condense into the flask about 45 ml (27 g, 0.5 mol) of buta-1-3-diene from a cylinder, swirl well with the potassium hydroxide pellets and transfer the dried diene directly to the pressure vessel following the procedure described in Section I,17,2. Assemble the vessel, evacuate with an oil pump until a constant pressure is obtained (a few minutes only) and close the exit valve. Allow the apparatus to reach room temperature, wrap with heating tape and heat to 120–125 °C behind a safety screen for 48 hours. After allowing the vessel to cool to room temperature, cool further to about − 15 °C and open the valve to the atmosphere. Remove the vessel cap and wash the contents of the vessel into a 1-litre round-bottomed flask with the aid of a little acetone. Remove the solvent on a rotary evaporator and crystallise the semi-solid residue from industrial spirit. The total yield (three crops) of crystallised material, m.p. 102–104 °C, is about 9 g (88%) and a pure colourless sample, m.p. 103 °C, may be obtained by further recrystallisation from industrial spirit.

V,15. 7,7-DICHLOROBICYCLO[4.1.0]HEPTANE

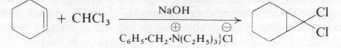

Place 8.2 g (10.1 ml, 0.1 mol) of freshly distilled cyclohexene, 36 g (24 ml, 0.3 mol) of chloroform (1) and 0.4 g (0.0017 mol) of triethylbenzylammonium chloride in a two-necked, round-bottomed 100-ml flask fitted with a sealed stirrer unit and a reflux condenser. Stir the solution vigorously and add to it a solution of 16 g (0.4 mol) of sodium hydroxide in 16 ml of water in portions down the condenser during 5 minutes. Within 10 minutes an emulsion is formed, and the temperature of the mixture increases slowly during 25 minutes to a maximum of about 50–55 °C; thereafter the temperature decreases whilst the colour changes from white to pale brown. After stirring for 2.5 hours add 40 ml of ice-cold water to the reaction mixture, transfer to a separating funnel and collect the lower chloroform layer. Extract the aqueous alkaline solution with 30 ml of ether and combine the ether extract with the chloroform solution and wash with 25 ml of 2 M-hydrochloric acid followed by two 25 ml portions of water. Dry the organic solution over magnesium sulphate, filter and remove the solvents on a rotary evaporator (water bath at 35–40 °C). Transfer the residual deep golden-coloured liquid to a 25-ml flask to which is attached a short (7-cm) fractionating column, and distil under reduced pressure. Collect the 7,7-dichlorobicyclo[4.1.0]heptane at 80–82 °C/16 mmHg; the yield is 10.2 g (62%). The purity may be checked by GLC using a 5-ft column of Silicone oil on Chromosorb W held at 110 °C and with a nitrogen flow rate of 40 ml/minute; the retention time is 6 minutes.

Note. (1) Ethanol-free chloroform should be used. The small amount of ethanol present in chloroform (as stabiliser) can be removed by shaking chloroform several times with an equal volume of water, followed by drying over anhydrous calcium chloride and distilling. Alternatively stand the chloroform over a few grammes of the molecular sieve 4A.

V,16. PHOTODIMERS OF CYCLOPENT-2-ENONE

Cyclopentadiene. Place 200 ml of liquid paraffin in a 500-ml, three-necked round-bottomed flask fitted with a large (30-cm) Vigreux column, a dropping funnel and a thermometer dipping into the paraffin. Attach a distillation head carrying a thermometer and a double surface condenser arranged for distillation. Heat the liquid paraffin to 200–240 °C (electric heating mantle) and add dicyclopentadiene portionwise, from the dropping funnel, and collect the cyclopentadiene, b.p. 40–42 °C, which distils over in a cooled receiver, protected from moisture. The dicyclopentadiene must be added slowly to ensure complete breakdown of the dimer; the temperature at the top of the still-head rises above 42 °C when addition is too rapid. Continue adding dicyclopentadiene (*ca.* 300 ml) until 230 g of cyclopentadiene is obtained. The diene dimerises readily at room temperature, hence it should be used immediately or stored in the ice compartment of a refrigerator overnight.

3-Chlorocyclopentene. Place 230 g (*ca.* 285 ml) of cyclopentadiene in a 500-ml measuring cylinder which is fitted with a rubber bung through which passes two glass tubes. One of the tubes (for introducing hydrogen chloride) should reach to near the bottom of the cylinder, but the other (outlet) tube should terminate just below the bung with its other end attached to a calcium chloride guard-tube. Cool the cylinder and contents in a Cardice–acetone bath maintained at −15 to −20 °C (note the volume of the liquid), and pass in a rapid stream of hydrogen chloride gas until the volume of the reaction mixture increases by 42 ml. This takes about 4 hours when a Kipp's apparatus is used to generate the hydrogen chloride gas. Transfer the reaction mixture to a distillation flask and distil under reduced pressure, collecting in a receiver cooled in Cardice–acetone the 3-chlorocyclopentene as a colourless liquid of b.p. 30 °C/20 mmHg. The yield is 222 g (62%). The compound is unstable at room temperature, polymerising slowly to a black tar. It should be used immediately for the next stage although it may be kept overnight in a Dewar flask packed with Cardice.

Cyclopent-2-enone. Place a solution of 180 g (0.6 mol) of sodium dichromate dihydrate in 630 ml of water in a 3-litre, three-necked round-bottomed flask equipped with an efficient stirrer, a dropping funnel and a thermometer. Cool the solution to 0 °C in an ice–salt bath and slowly add, in portions over about 1 hour, 157 g (1.53 mol) of 3-chlorocyclopentene to the rapidly stirred solution while keeping the temperature at 0–10 °C. Stir the reaction mixture at 0 °C for 0.5 hour then add 300 ml of 50 per cent sulphuric acid dropwise while maintaining the temperature at 0–10 °C by cooling. Saturate the dark brown mixture with sodium chloride and extract with five 250 ml portions of ether. Wash the combined extracts with two 150 ml portions of saturated sodium chloride, dry over magnesium sulphate and remove the ether by flash distillation. Distil the residual yellow oil under nitrogen under reduced pressure and collect the cyclopent-2-enone as a colourless liquid of b.p. 42–44 °C/13 mmHg. The yield is 58.8 g (47%). The compound has a retention time of 4.1 minutes on a 5-ft column of Carbowax 20M on Chromosorb W maintained at 80 °C and with a nitrogen flow rate of 40 ml/minute.

Photodimerisation of cyclopent-2-enone. cis,trans,cis-*3,3,8,8-Tetramethoxytricyclo[5.3.0.0²·⁶]decane* and cis,trans,cis-*tricyclo[5.3.0.0²·⁶]decane-3,10-dione*. Place a solution of 49.9 g (0.61 mol) of cyclopent-2-enone in 800 ml of methanol (distilled from solid potassium hydroxide) in a photochemical reactor vessel of 1-litre capacity equipped with a 100-watt medium pressure mercury arc lamp surrounded by a Pyrex cooling jacket (Section **I,17,5**), flush the magnetic-ally stirred solution with nitrogen for 10 minutes and then irradiate under nitrogen overnight. Filter the white crystalline solid which separates and continue irradiating until no further crystalline material separates from the methanol solution. About 21.5 g of material is obtained after approximately 40 hours irradiation. Crystallisation from methanol gives *cis,trans,cis*-3,3,8,8-tetramethoxytricyclo[5.3.0.0^{2,6}]decane as white plates, m.p. 173–174 °C. The yield is 19.3 g (30%).

Remove the methanol from the filtrate remaining from the photolysis using a rotary evaporator and remove any cyclopent-2-enone which remains (*ca.* 8.6 g) by distillation under reduced pressure. Several recrystallisations of the residual gum from carbon tetrachloride-hexane followed by a final crystallisation from hexane gives *cis,trans,cis*-tricyclo[5.3.0.0^{2,6}]decane-3,10-dione, m.p. 66–67 °C. The yield is about 2.24 g (5.5%).

Cis,trans,cis-**tricyclo[5.3.0.0^{2,6}]decane-3,8-dione.** Dissolve 15 g (0.058 mol) of *cis,trans,cis*-3,3,8,8-tetramethoxytricyclo[5.3.0.0^{2,6}]decane in 225 ml of methylene chloride in a 500-ml two-necked flask equipped with a sealed stirrer unit and reflux condenser. Add 30 ml of 2 *M*-hydrochloric acid and heat the stirred mixture under reflux for 1 hour. Cool, separate the organic layer and extract the acid solution with two 40 ml portions of methylene chloride. Combine the methylene chloride solutions and wash with 50 ml of 10 per cent aqueous sodium hydrogen carbonate and then 50 ml of water. Dry over mag-nesium sulphate, filter and remove the solvent on a rotary evaporator. Crystallise the residual solid from carbon tetrachloride; *cis,trans,cis*-tricyclo-[5.3.0.0^{2,6}]decane-3,8-dione is obtained as white plates, m.p. 125–126.5 °C. The yield is about 9.1 g (95%).

V,17. 1,4-NAPHTHOQUINONE PHOTODIMER

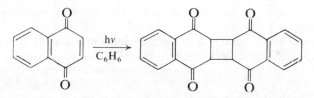

Method A. Place a solution of 5 g of pure 1,4-naphthoquinone (1) in 115 ml of dry thiophen-free benzene (*CARE*) in the appropriate sized photochemi-cal reactor vessel (Section **I,17,5**; note also the precautions to be observed for photochemical reactions) fitted with a 100-W medium pressure mercury arc lamp surrounded by Pyrex cooling jackets. Stir the solution by means of a magnetic follower bar and pass dry nitrogen through the solution for 10 minutes before switching on the lamp; adjust the nitrogen flow to a slow trickle and continue irradiation for about 6 hours, by which time a quantity of the solid dimer will have separated from the solution. Switch off the lamp, remove the lamp insert and collect the solid by suction filtration and wash with a little benzene. Gently remove and collect any of the solid which may have separated on the surface of the lamp insert and thoroughly clean the latter with a little acetone on a paper tissue before replacing the benzene filtrate in the reactor and continuing the irradiation process. Repeat these operations as appropriate and collect the dimer over a total irradiation period of about 28 hours; about

0.84 g of solid, m.p. 243–246 °C (decomp.) (2), is obtained. Recrystallise from 150–160 ml of glacial acetic acid with the aid of a little decolourising carbon and dry the pale straw-coloured crystals in a desiccator over solid sodium hydroxide pellets; the yield of pure naphthoquinone dimer, m.p. 245–249 °C (decomp.), is 0.62 g (12%).

Method B.   Place a solution of 2.5 g of pure 1,4-naphthoquinone (1) in 50 ml of dry thiophen-free benzene in a 50-ml round-bottomed flask and pass a slow stream of dry nitrogen through the solution for 5 minutes. Insert a lightly greased stopper in the flask taking care that the solution does not penetrate the joint. Place the flask in a sunny position—if situated in the open air wrap some aluminium foil around the stopper to prevent the possibility of rainwater seeping through the joint into the reaction mixture. Solid slowly separates from the solution over a period of days; collect the product by suction filtration, wash with a little benzene and dry. Return the filtrate to the flask, flush with dry nitrogen and place in the sunlight until a further portion of solid is ready for collection. The amount of material obtained in a given period is dependent on light conditions; typically, a total of 0.33 g (13%) of crude naphthoquinone photodimer, m.p. 241–246 °C (decomp.), has been obtained during a 31-day irradiation period (April–May).

Notes.   (1) Technical grade naphthoquinone may be purified as described in Section **IV,142**.

(2) The melting point is best observed on a microscope–hot stage apparatus.

REFERENCES

V Alicyclic compounds

1. L. F. Feiser (1964). *Organic Experiments*. Boston, Heath and Co.; p. 311.
2. H. Gruen and B. E. Norcross (1965). *J. Chem. Ed.*, **42**, 268.
3. Reference 1, p. 301.
4. R. B. Woodward and R. Hoffmann (1969). 'The Conservation of Orbital Symmetry', *Angew. Chem., Internat. Edn.*, **8**, 781.
5. M. Makosza and M. Wawrzyniewcz (1969). *Tetrahedron Letters*, 4659.

CHAPTER VI   SOME HETEROCYCLIC COMPOUNDS

A. Nitrogen-containing heterocycles (Sections **VI,1** to **VI,33**).
B. Oxygen-containing heterocycles (Sections **VI,34** to **VI,40**).
C. Sulphur-containing heterocycles (Sections **VI,41** to **VI,45**).

A. Nitrogen-containing heterocycles.   1. Three-membered ring systems.
The saturated three-membered heterocyclic ring system containing a single
nitrogen atom is aziridine or ethyleneimine. The system is most simply formed
by the cyclisation of a β-haloethylamine.

$$Cl^{\ominus}\{H_3\overset{\oplus}{N}\cdot CH_2\cdot CH_2Br \xrightarrow{\text{base}} \quad \underset{\triangle}{\overset{H}{\underset{N}{|}}}$$

Some substituted aziridines may be synthesised by the reaction of the appropriate ketoxime with a Grignard reagent (Ref. 1), or alternatively by the recently
introduced method which employs reduction of the ketoxime with lithium
aluminium hydride in tetrahydrofuran solution. The aziridine (I) is, however,
often accompanied by the corresponding primary amine (formed from reduction
of the oximino group) (II), and the secondary amine (III) arising from a
1,2-nucleophilic shift from carbon to nitrogen, e.g.,

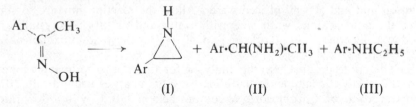

$$\text{(I)} \qquad\qquad \text{(II)} \qquad\qquad \text{(III)}$$

The proportions of these three products depend upon the nature of the ketoxime, the solvent and the reducing agent. A preparatively useful example is the
reduction of dibenzyl ketoxime with sodium dihydro-bis(2-methoxyethoxy)-
aluminate in tetrahydrofuran when the amount of primary amine formed is less
than 10 per cent and the aziridine, *cis*-2-benzyl-3-phenylaziridine is the major
product (Section **VI,1**).

VI,1. *CIS*-2-BENZYL-3-PHENYLAZIRIDINE

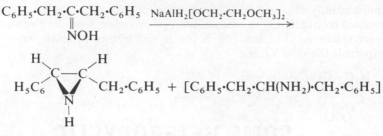

This preparation should be carried out in an efficient fume cupboard. Since aziridine and its simple derivatives are extremely toxic the product should be treated with caution.

Pipette 40 ml of a benzene solution of sodium dihydro-bis(2-methoxyethoxy)-aluminate (70 per cent solution containing about 0.15 mol of the reducing agent; available from Aldrich Chemical Co. Ltd) into a 500-ml three-necked flask equipped with a sealed stirrer unit, and a dropping funnel and reflux condenser both protected with calcium chloride guard-tubes. Add 200 ml of purified, sodium-dried tetrahydrofuran (Section II,1,*19*) and heat to reflux, and then add dropwise over a period of 15 minutes a solution of 11.25 g (0.05 mol) of 1,3-diphenylpropan-2-one oxime (1) in 50 ml of tetrahydrofuran. Continue heating the reaction mixture under reflux for a further 2 hours, cool the solution and decompose the excess of reagent by the addition, with stirring, of 25 ml of an aqueous solution of 1 *M*-sodium hydroxide. Filter off insoluble inorganic material and wash thoroughly with ether. Combine the ether–tetrahydrofuran solutions, dry over calcium sulphate and evaporate on a rotary evaporator. Extract the crude liquid product (2) with several portions of warm light petroleum (b.p. 40–60 °C), combine the extracts and evaporate to low bulk when 8.5 g (82%) of *cis*-2-benzyl-3-phenylaziridine, m.p. 43 °C, is obtained on refrigeration (3).

Notes. (1) **1,3-Diphenylpropan-2-one oxime** is prepared by the following method. Add successively to a solution of 25 g (0.36 mol) of hydroxylamine hydrochloride in 100 ml of water, 100 ml of a 10 per cent aqueous solution of sodium hydroxide and a solution of 17.4 g (0.083 mol) of dibenzyl ketone in ethanol. Heat the solution under reflux on a boiling water bath and add a further portion of ethanol to obtain a clear solution. After 45 minutes, cool the solution and add 500 ml of iced water. Allow the reaction mixture to stand, filter the crude oxime, wash well with water and air dry. Recrystallise the product from ether–light petroleum (b.p. 40–60 °C); the yield of oxime is 17.2 g (92%), m.p. 123–124 °C.

(2) The crude product may be analysed for the ratio of primary amine (1-benzyl-2-phenylethylamine) to aziridine by GLC analysis using a 5-ft, 10 per cent Carbowax on Chromosorb W column held at 170 °C with a flow rate of carrier gas of 40 ml/minute.

(3) Further quantities of aziridine may be obtained by combining all the mother-liquors and residues, removing the solvent and submitting the oil to chromatography on an alumina column. Elution with benzene–chloroform (1 : 1) gives the aziridine; subsequent elution with chloroform yields 1-benzyl-2-phenylethylamine.

A,2. Five-membered ring systems. The heteroaromatic ring system, pyrrole, is most readily obtained by the cyclisation of a 1,4-dicarbonyl compound with ammonia (conveniently as ammonium carbonate) or a primary amine (the **Paal–Knorr synthesis**). The reaction is illustrated by the preparation of 2,5-dimethylpyrrole (Section **VI,2**).

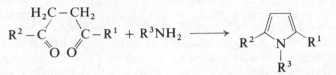

Pyrrole itself may be prepared (Section **VI,3**) by a related reaction which involves heating the ammonium salt of D-galactaric acid (mucic acid, $HO_2C\cdot(CHOH)_4\cdot CO_2H$); the use of the methylamine salt of D-galactaric acid similarly yields N-methylpyrrole (cognate preparation in Section **VI,3**).

The most generally applicable pyrrole synthesis is that of **Knorr** which involves the reaction of an α-aminoketone with a carbonyl compound possessing an active methylene group. The α-aminoketone (V) is usually prepared by reduction with zinc in acetic acid of the corresponding oximino compound (IV).

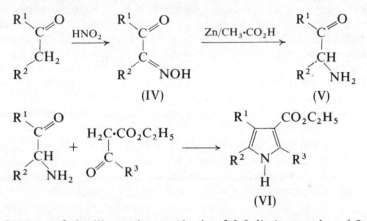

In the case of the illustrative synthesis of 3,5-diethoxycarbonyl-2,4-dimethylpyrrole (Section **VI,4**) one half of the ethyl acetoacetate used in the reaction is converted into the oximino compound by treatment with nitrous acid in acetic acid, and then reduced by the addition of zinc dust. The resulting amino compound spontaneously reacts with the remaining ethyl acetoacetate to yield the pyrrole derivative (VI).

1,3-Dicarbonyl compounds condense readily with hydrazines to form substituted pyrazoles, which are 1,2-diazaheterocyclic systems (e.g., 3,5-dimethylpyrazole, Section **VI,5**).

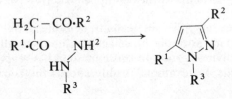

If the dicarbonyl compound is a β-keto ester the product is a substituted pyrazolone (e.g., 3-methyl-1-phenylpyrazol-5-one, Section **VI,6**). 3-Methyl-1-phenylpyrazol-5-one is of interest in that *N*-methylation occurs on reaction with dimethyl sulphate and alkali; the product, 2,3-dimethyl-1-phenylpyrazol-5-one (Section **VI,7**), is one of the earliest synthetic antipyretics.

5,5-Dimethylhydantoin (Section **VI,8**) affords an example of a 1,3-diaza-heterocyclic system; named systematically it is 5,5-dimethylimidazolidine-2,4-dione [(VII), $R^1 = R^2 = CH_3$]. It is prepared by heating acetone cyanohydrin with ammonium carbonate; this is an example of the **Bucherer hydantoin synthesis** which is applicable to most aldehydes or ketones. Simpler direct procedures, which involve heating the aldehyde or ketone with potassium cyanide and ammonium carbonate, have been described (Ref. 2). These are generally suitable for the preparation of dialkyl or aralkyl, but not diaryl, hydantoins. The hydantoins on hydrolysis yield α-amino acids, and the overall reaction sequence may be regarded as a development of the Strecker synthesis (p. 546, **III,P,5(b)**).

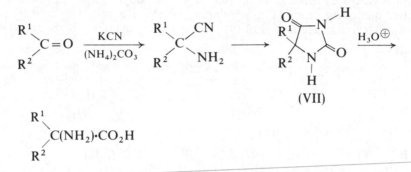

5,5-Diarylhydantoins (e.g., 5,5-diphenylhydantoin, Section **VI,9**) may be prepared by condensing a diaryl-1,2-diketone with urea to give an intermediate heterocyclic pinacol (VIII) which on acidification yields the required hydantoin as the result of a pinacolic rearrangement.

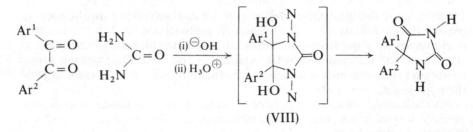

Many 5,5-disubstituted hydantoins, particularly the 5,5-diphenyl derivative, are useful anti-convulsant drugs.

Cyclisation of an *N*-acyl-α-amino acid with acetic anhydride yields an oxazolone derivative, or azlactone. With, for example, an *N*-acylglycine the 2-alkyl-oxazolone (IX) is formed. The methylene group in this compound is reactive, and condensation with benzaldehyde, for example, readily affords the benzylidene derivative (X) (Section **VI,10**).

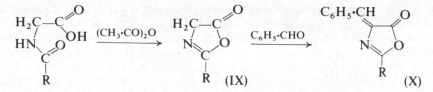

The hydrolysis of the benzylidene azlactone to the acylaminocinnamic acid (XI) is readily achieved in refluxing aqueous acetone (Section **VI,10**). Hydrolysis of the benzylidene azlactone in aqueous acid leads to phenylpyruvic acid (XII) (Section **III,175**).

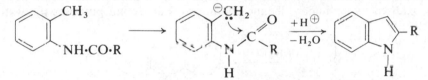

Fusion of a benzene ring at the 2,3-positions of the pyrrole ring gives the simple **condensed five-membered ring system**, indole. The **Madelung indole synthesis** involves the cyclisation of an acyl-*o*-toluidine with strong bases (e.g., sodium alkoxides, sodamide) at high temperatures (Ref. 3). The reaction mechanism involves an intramolecular Claisen condensation (see p. 850, **V**(*a*)).

$$ \begin{array}{c}\text{CH}_3\\ \diagdown\\ \text{NH·CO·R}\end{array} \longrightarrow \begin{array}{c}\ominus\text{CH}_2\\ \diagdown\\ \text{N}\diagup\text{C}\diagdown\text{R}\\ \text{H}\end{array} \begin{array}{c}+\text{H}^\oplus\\ \xrightarrow{\hspace{1cm}}\\ -\text{H}_2\text{O}\end{array} \begin{array}{c}\diagup\text{R}\\ \text{N}\\ \text{H}\end{array} $$

The synthesis of indole itself from *N*-formyl-*o*-toluidine is achieved only if a potassium alkoxide is used as a base; the reaction mechanism is certainly more complex than is indicated above, since extensive loss of carbon monoxide occurs, and about half of the starting material is recovered as *o*-toluidine. A simplified procedure is described in Section **VI,11**, in which the somewhat hazardous preparation of a potassium alkoxide from potassium metal and the alcohol is avoided by using a mixture of sodium methoxide and anhydrous potassium acetate.

An important general method of preparing substituted indoles (but not indole itself), known as the **Fischer indole synthesis**, consists of heating the phenylhydrazone of an aldehyde, ketone, or keto acid in the presence of an acid catalyst. Zinc chloride, hydrochloric acid or glacial acetic acid may be used, but polyphosphoric acid is often preferred. Thus acetophenone phenylhydrazone [(XIII), $R^1 = H, R^2 = C_6H_5$] gives 2-phenylindole [(XIV), $R^1 = H, R^2 = C_6H_5$] (Section **VI,12**). The reaction sequence involves an intramolecular condensation with the elimination of ammonia and the mechanism of the reaction is outlined on the following page. An interesting application is the preparation of 1,2,3,4-tetrahydrocarbazole (XVI) (cognate preparation in Section **VI,12**), which is formed when phenylhydrazine is added to a boiling solution of cyclohexanone in acetic acid; the phenylhydrazone (XV) intermediately produced undergoes ring closure directly. A further example is included in the synthesis of tryptophan described in Section **III,183**.

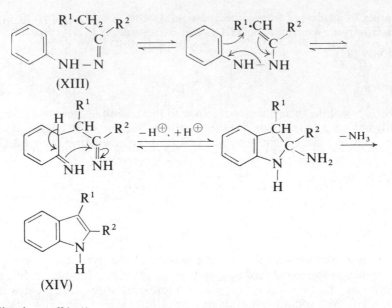

(XIII)

(XIV)

The dyestuff indigo (XIX) (Section **VI,13**) is an indole derivative which is obtained when indoxyl, which has the tautomeric structure indicated below (XVII), is oxidised by air under alkaline conditions. For dyeing purposes the water-

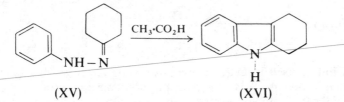

(XV)   (XVI)

insoluble indigo is reduced with alkaline sodium dithionite to the soluble di-hydroderivative [leucoindigo, (XVIII)]. After impregnation of the fibres, the fabric is air dried when oxidation regenerates the indigo molecule and develops a blue colouration.

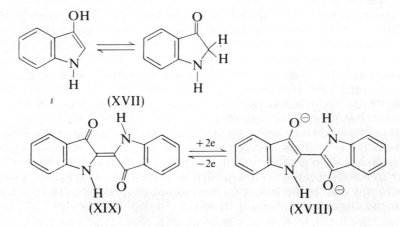

(XVII)

(XIX)   (XVIII)

Examples of **condensed five-membered ring systems** containing two or three nitrogen atoms are benzimidazole (XX) and benzotriazole (XXI) respectively.

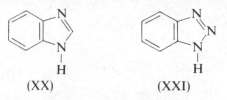

(XX) (XXI)

Benzimidazoles in general are prepared by heating an *o*-diaminobenzene with a carboxylic acid (e.g., benzimidazole, Section **VI,14**, and the cognate preparations therein).

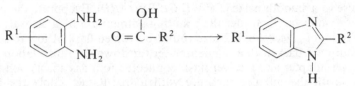

Benzotriazole (Section **VI,15**) is formed when *o*-phenylenediamine reacts with nitrous acid; the cyclisation step follows initial diazonium salt formation from one of the amino groups.

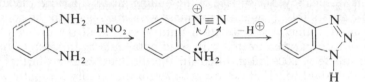

VI,2. 2,5-DIMETHYLPYRROLE

$$\underset{\underset{CH_3 \cdot CO \quad CO \cdot CH_3}{|}}{CH_2 - CH_2} \xrightarrow{NH_3} \underset{H_3C \quad \underset{H}{N} \quad CH_3}{}$$

In a 250-ml flask, fitted with an air condenser of wide bore, place 50 g (51.5 ml, 0.44 mol) of hexane-2,5-dione (Section **III,104**) and 100 g of ammonium carbonate (lump form). Heat the mixture in an oil bath at 100 °C until effervescence stops (60–90 minutes); some ammonium carbonate (or carbamate) sublimes into the condenser and this must be pushed back into the reaction mixture by means of a stout glass rod. Replace the air condenser by a Liebig condenser with wide bore inner tube and reflux the mixture gently (bath temperature, 115 °C) for a further 30 minutes; dissolve the solid which has sublimed into the condenser in about 5 ml of hot water and return the solution to the reaction mixture. Cool and separate the upper yellow layer of crude dimethylpyrrole, extract the lower layer with 10 ml of chloroform and combine it with the crude dimethylpyrrole; carry out the foregoing operations in apparatus which has been flushed with nitrogen. Dry over anhydrous sodium sulphate in a tightly stoppered flask filled with nitrogen. Transfer to a flask fitted with a fractionating column. Displace the air from the apparatus by nitrogen and distil under reduced pressure, preferably in a stream of nitrogen.

Collect the 2,5-dimethylpyrrole at 78–80 °C/25 mmHg. The yield is 36 g (86%). Store the product in an inert atmosphere in a sealed, dark glass container.

VI,3. PYRROLE

$$
\begin{array}{c}
\text{CH(OH)} - \text{CH(OH)} \\
\diagup \qquad \diagdown \\
\text{CH(OH)} \qquad \text{CH(OH)} \\
\diagup \qquad\qquad \diagdown
\end{array}
\quad
\underset{\text{H}_4\text{NO}_2\text{C}}{} \quad \underset{\text{H}_4\text{NO}_2\text{C}}{} \longrightarrow \underset{\substack{| \\ H}}{\boxed{}\text{N}} + 2\text{CO}_2 + \text{NH}_3 + 4\text{H}_2\text{O}
$$

Place 210 g (1 mol) of D-galactaric acid (Section **III,105**) and 300 ml (5.5 mol) of aqueous ammonia solution (*d* 0.88) in a large evaporating basin and rapidly stir the mixture to a smooth paste (*FUME CUPBOARD*). Evaporate the paste to dryness on a water bath, powder the resulting ammonium D-galactarate and mix it with 120 ml of glycerol in a 2-litre round-bottomed flask. Allow to stand overnight. Equip the flask with a condenser set for downward distillation and attach a receiver adapter and receiver flask connected to a gas absorption trap (Fig. I,68(*c*)). Distil the mixture carefully with a free flame. Apply the heat initially to one side of the flask so that only a portion of the mass is heated to the reaction temperature; considerable frothing ensues and this must be controlled by removing the flame from below the flask and heating the upper portion of the vessel above the surface of the boiling mixture. Extend the heating as rapidly as possible throughout the mass with due regard to the control of the foaming. Continue the distillation until a sample of the distillate no longer gives oily drops when treated with solid potassium hydroxide; the total volume of distillate is 300–350 ml. Redistil the distillate until no further oil separates in the liquid which passes over. Separate the oil, dry it rapidly with potassium hydroxide pellets and distil. Collect the pyrrole (a colourless liquid) at 127–131 °C; the yield is 25 g (37%). The pyrrole should be stored in a sealed vessel; it darkens on exposure to light.

Cognate preparation   *N*-Methylpyrrole.   Prepare the methylamine salt of D-galactaric acid by adding slowly and with vigorous stirring 260 ml (2.6 mol) of 10 *M*-aqueous methylamine to 210 g (1 mol) of D-galactaric acid; if difficulty is experienced in stirring the mixture, add up to 100 ml of water. Complete the preparation following the experimental conditions given above for *pyrrole*. The yield of *N*-methylpyrrole, b.p. 110–113 °C, is 32 g (39%). The compound is very hygroscopic and darkens on standing; keep it in a tightly stoppered brown bottle.

VI,4. 3,5-DIETHOXYCARBONYL-2,4-DIMETHYLPYRROLE

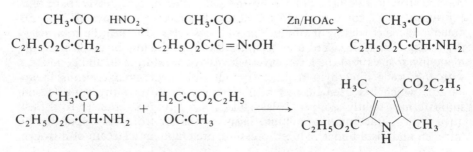

In a 1500-ml three-necked flask, fitted with a dropping funnel and a sealed mechanical stirrer, place 195 g (190 ml, 1.5 mol) of ethyl acetoacetate (Section **III,176**) and 450 ml of glacial acetic acid. Cool the solution in an ice–salt mixture to 5 °C; add a cold solution of 52 g (0.75 mol) of sodium nitrite in 75 ml of water dropwise and with vigorous stirring at such a rate that the temperature remains between 5 and 7 °C (about 30 minutes), stir for a further 30 minutes, and keep at room temperature for 4 hours. Replace the dropping funnel by a wide-bore condenser: close the third neck with a stopper. Stir the solution vigorously and add 100 g (1.5 mol) of zinc powder (Section **II,2,67**) via the third neck in portions of about 10 g; introduce the first 3 or 4 portions quickly so that the liquid boils. Keep a bath of ice-water and also wet towels at hand to control the reaction should it become violent or foam badly. When all the zinc has been added (about 45 minutes), reflux the mixture for 1 hour; if stirring becomes difficult, add some acetic acid. While still hot, decant the contents of the flask into 5 litres of water in a large beaker with vigorous stirring. Wash the zinc residue with two 25 ml portions of hot glacial acetic acid and decant the washings into the water also. Keep overnight, collect the crude product by suction filtration, wash with two 250 ml portions of water and dry in the air to constant weight. The yield of crude product is 114 g (64%), m.p. 127–130 °C. Recrystallisation from hot 95 per cent ethanol gives pure 3,5-diethoxycarbonyl-2,4-dimethylpyrrole as pale yellow crystals, m.p. 136–137 °C; the recovery is about 80 per cent.

VI,5. 3,5-DIMETHYLPYRAZOLE

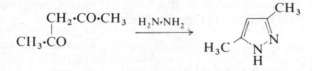

Dissolve 65 g (0.5 mol) of hydrazine sulphate in 400 ml of 2.5 *M*-sodium hydroxide solution contained in a 1-litre three-necked flask, equipped with a thermometer, mechanical stirrer and dropping funnel. Immerse the flask in an ice bath and when the temperature reaches 15 °C (some sodium sulphate may separate at this point), add 50 g (51.5 ml, 0.5 mol) of pentane-2,4-dione (Section **III,102** dropwise, with stirring, whilst maintaining the temperature at 15 °C. When the addition is complete (after about 30 minutes), stir for 1 hour at 15 °C; the dimethylpyrazole separates during this period. Add 200 ml of water, stir to dissolve inorganic salts, transfer the contents of the flask to a separatory funnel and shake with 100 ml of ether. Separate the layers and extract the aqueous layer with four 40 ml portions of ether. Wash the combined ethereal extracts with saturated sodium chloride solution, dry over anhydrous potassium carbonate and remove the ether on a rotary evaporator. The yield of pale yellow solid, m.p. 107–108 °C, is 38 g. Recrystallise from about 250 ml of light petroleum, b.p. 80–100 °C; the yield of 3,5-dimethylpyrazole, of unchanged m.p., is 36 g (75%).

VI,6. 3-METHYL-1-PHENYLPYRAZOL-5-ONE

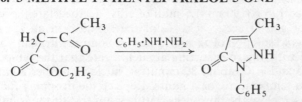

Mix together 50 g (49 ml, 0.384 mol) of redistilled ethyl acetoacetate (Section **III,176**) and 40 g (36.5 ml, 0.37 mol) of phenylhydrazine (**CAUTION** in handling) (Section **IV,104**) in a large evaporating dish. Heat the mixture on a boiling water bath in the fume cupboard for about 2 hours and stir from time to time with a glass rod (1). Allow the heavy reddish syrup to cool somewhat (2), add about 100 ml of ether and stir the mixture vigorously. The syrup, which is insoluble in ether, will solidify within 15 minutes. Filter the solid at the pump and wash it thoroughly with ether to remove coloured impurities. Recrystallise it from hot water or from a mixture of equal volumes of ethanol and water. The yield of methylphenylpyrazolone (colourless crystals, m.p. 127 °C) is 52 g (80%).

Note. (1) The heating should be continued until a test portion solidifies completely when it is rubbed with a little ether; when freshly redistilled phenylhydrazine is used, a period of 1 hour's heating may be adequate. If the product is reluctant to crystallise inoculation with seeds of previously prepared material is advantageous.

(2) If the product is already solid at this stage it should be ground thoroughly with the 100 ml portion of ether and filtered.

VI,7. 2,3-DIMETHYL-1-PHENYLPYRAZOL-5-ONE (*Antipyrin*)

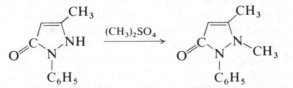

In a 500-ml three-necked flask, equipped with a dropping funnel, a sealed stirrer unit and a double surface condenser and set up in the fume cupboard, place a solution of 10 g of sodium hydroxide in a small volume of water and also a solution of 43.5 g (0.25 mol) of 3-methyl-1-phenylpyrazol-5-one (Section **VI,6**) in 20 ml of methanol. Warm the mixture on a water bath and add 36 g (27 ml, 0.285 mol) of dimethyl sulphate (**CAUTION:** toxic, see discussion in Section **II,2,**_18_). Reflux the mixture for 1 hour and allow to cool, with continuous stirring. Distil off the methanol. Add hot water to the residue, filter from impurities, extract the antipyrin with benzene (*FUME CUPBOARD*) and evaporate the solvent. Recrystallise the crude product from benzene or benzene–light petroleum, or from hot water with the addition of a little decolourising carbon. The yield of antipyrin (white crystalline solid, m.p. 113 °C) is 35 g (74%).

VI,8. 5,5-DIMETHYLHYDANTOIN

$$(CH_3)_2C(OH)\cdot CN \xrightarrow{(NH_4)_2CO_3}$$

[structure: 5,5-dimethylhydantoin]

Mix 42.5 g (0.5 mol) of acetone cyanohydrin (1) and 75 g (0.78 mol) of freshly powdered ammonium carbonate in a small beaker, warm the mixture on a water bath (*FUME CUPBOARD*) and stir with a thermometer. Gentle reaction commences at 50 °C and continues during about 3 hours at 70–80 °C. To complete the reaction, raise the temperature to 90 °C and maintain it at this point until the mixture is quiescent (*ca.* 30 minutes). The colourless (or pale yellow) residue solidifies on cooling. Dissolve it in 50 ml of hot water, digest with a little decolourising carbon and filter rapidly through a pre-heated Buchner funnel. Evaporate the filtrate on a hot plate until crystals appear on the surface of the liquid, and then cool in ice. Filter off the white crystals with suction, drain well and then wash twice with 4 ml portions of ether; this crop of crystals of dimethylhydantoin is almost pure and melts at 176 °C. Concentrate the mother-liquor to the crystallisation point, cool in ice and collect the second crop of crystals (m.p. *ca.* 167 °C) as before. Dissolve the dimethylhydantoin (35 g) in the minimum volume of boiling water (about 32 ml), digest with a little decolourising carbon and filter the hot solution through a preheated Buchner funnel. Cool the filtrate in ice, filter the separated crystals at the pump and wash sparingly with cold water. The yield of pure product, m.p. 178 °C, is 29 g (45%).

Note. (1) Acetone cyanohydrin of adequate purity may be prepared by the following procedure: Dissolve 110 g of sodium metabisulphite in 200 ml of cold water contained in a 1-litre round-bottomed flask. Add slowly 58 g of acetone whilst swirling the liquid mixture slowly, followed by a solution of 60 g of potassium cyanide in 200 ml of cold water [**CAUTION**]. During this latter slow addition the cyanohydrin separates as the upper layer. When separation is complete the contents of the flask are transferred to a separatory funnel and the lower layer removed. The upper layer is transferred to a flask, sodium sulphate added to effect drying and the flask stoppered and kept in the dark. The dried cyanohydrin (60 g, 70%) is slightly discoloured; it may be distilled under reduced pressure when it distils at 80–82 °C/15 mmHg.

VI,9. 5,5-DIPHENYLHYDANTOIN

[structure: benzil + urea reaction]

$$\xrightarrow[\text{(ii) } H_3O^\oplus]{\text{(i) NaOH/H}_2\text{O/C}_2\text{H}_5\text{OH}}$$

[structure: 5,5-diphenylhydantoin]

Place 5.3 g (0.025 mol) of benzil (Section **IV,155**), 3.0 g (0.05 mol) of urea, 15 ml of 30 per cent aqueous sodium hydroxide solution and 75 ml of ethanol in a 100-ml round-bottomed flask. Attach a reflux condenser and boil under reflux using an electric heating mantle for at least 2 hours. Cool to room temperature, pour the reaction product into 125 ml of water and mix thoroughly. Allow to stand for 15 minutes and then filter under suction to remove an insoluble by-product. Render the filtrate strongly acidic with concentrated hydrochloric acid, cool in ice-water and immediately filter off the precipitated product under suction. Recrystallise at least once from industrial spirit to obtain about 2.8 g (44%) of pure 5,5-diphenylhydantoin, m.p. 297–298 °C.

VI,10. 4-BENZYLIDENE-2-METHYLOXAZOL-5-ONE

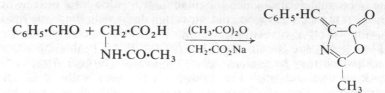

Acetylglycine. Place 37.5 g (0.5 mol) of glycine (Section **III,179**) and 150 ml of water in a 500-ml conical flask. Introduce a mechanical stirrer and stir vigorously until the solid has almost completely dissolved. Add 102 g (95 ml, 1 mol) of acetic anhydride in one portion and stir vigorously for 15–20 minutes; the solution becomes hot and some acetylglycine may crystallise. Cool in a refrigerator, preferably overnight; collect the precipitate on a Buchner funnel, wash with ice-cold water and dry at 100 °C. The product weighs 40 g and melts at 207–208 °C. Evaporate the combined filtrate and washings to dryness under reduced pressure on a water bath at 50–60 °C, and recrystallise the residue from 40 ml of boiling water: collect the solid which separates, wash and dry it as before. The second fraction of acetylglycine weighs 15 g (total yield 55 g, 94%) and melts at 207–208 °C.

4-Benzylidene-2-methyloxazol-5-one. Warm a mixture of 29 g (0.25 mol) of acetylglycine, 39.5 g (37.5 ml, 0.37 mol) of redistilled benzaldehyde (Section **IV,22**), 15 g (0.183 mol) of anhydrous sodium acetate and 63.5 g (59 ml, 0.62 mol) of acetic anhydride in a 500-ml flask (equipped with a reflux condenser) on a water bath with occasional stirring until solution is complete (10–20 minutes). Boil the resulting solution for 1 hour, cool and leave in a refrigerator overnight. Stir the solid mass of yellow crystals with 60 ml of cold water, transfer to a Buchner funnel and wash well with cold water. (If the odour of benzaldehyde is still apparent, wash with a little ether.) Recrystallise from carbon tetrachloride or from ethyl acetate–light petroleum. The yield of the oxazolone, m.p. 150 °C, is 35 g (76%).

Hydrolysis to α-acetamidocinnamic acid. Boil a mixture of 23.5 g (0.125 mol) of 4-benzylidene-2-methyloxazol-5-one (the crude product is satisfactory), 90 ml of water and 225 ml of acetone in a 500-ml round-bottomed flask under reflux for 4 hours. Remove most of the acetone with a rotary evaporator, dilute the residual solution with 200 ml of water, heat to boiling for 5–10 minutes and filter through a hot-water funnel. Dissolve any crystals which separate from the filtrate by heating, add 5 g of decolourising carbon, boil for 5 minutes and filter with gentle suction through a warm Buchner funnel, and wash the residue with four 25 ml portions of boiling water. Place the combined filtrate and washings in a refrigerator overnight. Collect the colourless crystals by suction filtration, wash with about 100 ml of cold water and dry at 100 °C. The yield of 2-acetamidocinnamic acid, m.p. 191–192 °C, is 22 g (86%).

Cognate preparation   Benzoylglycine (*hippuric acid*).   Dissolve 25 g (0.33 mol) of glycine (Section **III,179**) in 250 ml of 10 per cent sodium hydroxide solution contained in a conical flask. Add 54 g (45 ml, 0.385 mol) of benzoyl chloride in five portions to the solution. Stopper the vessel and shake vigorously after each addition until all the chloride has reacted. Transfer the solution to a beaker and rinse the conical flask with a little water. Place a few grams of crushed ice in the solution and add concentrated hydrochloric acid slowly and with stirring until the mixture is acid to Congo red paper. Collect the resulting crystalline precipitate of benzoylglycine, which is contaminated with a little benzoic acid, upon a Buchner funnel, wash with cold water and drain well. Place the solid in a beaker with 100 ml of carbon tetrachloride, cover the beaker with a watch glass and boil gently for 10 minutes (*FUME CUPBOARD*); this extracts any benzoic acid which may be present. Allow the mixture to cool slightly, filter under gentle suction and wash the product on the filter with 10–20 ml of carbon tetrachloride. Recrystallise the dried product from boiling water (about 500 ml) with the addition of a little decolourising charcoal if necessary, filter through a hot water funnel and allow to crystallise. Collect the benzoylglycine in a Buchner funnel and dry it in an oven. The yield is 45 g (76%), m.p. 187 °C.

   4-Benzylidene-2-phenyloxazol-5-one.   Place a mixture of 27 g (26 ml, 0.25 mol) of redistilled benzaldehyde, 45 g (0.25 mol) of benzoylglycine, 77 g (71.5 ml, 0.75 mol) of acetic anhydride and 20.5 g (0.25 mol) of anhydrous sodium acetate in a 500-ml conical flask and heat on an electric hot plate with constant shaking. As soon as the mixture has liquefied completely, transfer the flask to a water bath and heat for 2 hours. Then add 100 ml of ethanol slowly to the contents of the flask, allow the mixture to stand overnight, filter the crystalline product with suction, wash with two 25 ml portions of ice-cold alcohol and then wash with two 25 ml portions of boiling water: dry at 100 °C. The yield of almost pure oxazolone, m.p. 165–166 °C, is 40 g (64%). Recrystallisation from benzene raises the m.p. to 167–168 °C.

VI,11. INDOLE

$$o\text{-}CH_3 \cdot C_6H_4 \cdot NH_2 + H \cdot CO_2H \longrightarrow o\text{-}CH_3 \cdot C_6H_4 \cdot NH \cdot CHO + H_2O$$

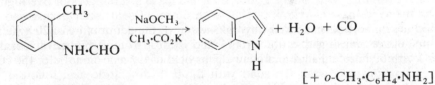

$$[+ \ o\text{-}CH_3 \cdot C_6H_4 \cdot NH_2]$$

   N-Formyl-o-toluidine.   Mix together 43 g (43 ml, 0.4 mol) of *o*-toluidine and 21 g (17.5 ml, 0.4 mol $H \cdot CO_2H$) of 90 per cent w/w formic acid in a 100-ml round-bottomed flask fitted with a reflux condenser, and heat the mixture on a boiling water bath for 3 hours. Replace the reflux condenser by a Claisen still-head and an air condenser arranged for distillation under reduced pressure, and distil the product using a water pump, collecting the formyl-*o*-toluidine as a fraction of b.p. 173–175 °C/25 mmHg, which solidifies on cooling, m.p. 57–59 °C; the yield is 43 g (80%). A pure specimen, m.p. 61 °C, may be obtained by crystallisation from a mixture of benzene and light petroleum (b.p. 40–60 °C).

   Indole.   Prepare a solution of sodium methoxide in 125 ml of anhydrous methanol using 5.75 g (0.25 mol) of sodium in a 250-ml flask fitted with a reflux condenser protected by a calcium chloride guard-tube, and add 34 g (0.25 mol) of *N*-formyl-*o*-toluidine. Then add rapidly 50 g (0.51 mol) of coarsely ground, freshly fused potassium acetate (1) and heat under reflux with shaking until all has dissolved. Remove the methanol under reduced pressure (rotary evaporator), transfer the flask to a fume cupboard and fit a still-head and condenser set for downward distillation. Surround the flask with a bath of molten Wood's metal (Section **I,13**) and raise the temperature steadily to about 300–350 °C. The subsequent reaction is accompanied by the distillation of *o*-toluidine and the evolution of carbon monoxide; continue to heat until no further distillation occurs (about 30 minutes) and finally remove traces of *o*-toluidine by carefully applying partial vacuum. Remove the heating bath, allow the flask to cool and decompose the residue by adding 100 ml of water and steam distilling. Colourless plates of indole separate from the cooled distillate; make the latter slightly acidic with hydrochloric acid, collect the crystals by suction filtration and wash them with a little cold water. The yield of indole, m.p. 48–49 °C, is 5 g (17%). A purer specimen, m.p. 52 °C, may be obtained by crystallisation from light petroleum, b.p. 40–60 °C.

Note. (1) Heat the potassium acetate in a porcelain dish until a tranquil melt is obtained, and allow to cool in a desiccator (cf. anhydrous sodium acetate, Section **II,2,**56).

VI,12. 2-PHENYLINDOLE

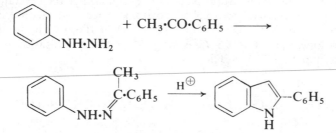

Prepare **acetophenone phenylhydrazone** by warming a mixture of 20 g (0.167 mol) of acetophenone (Section **IV,134**) and 18 g (0.167 mol) of phenylhydrazine (1) with 60 ml of ethanol and a few drops of glacial acetic acid. Filter the cold reaction mixture, wash the solid with dilute hydrochloric acid followed by about 12 mol of cold rectified spirit. Recrystallise a small portion from ethanol and thus obtain a sample of pure acetophenone phenylhydrazone as a white solid, m.p. 106 °C.

Place 28 g of the crude phenylhydrazone in a 250-ml beaker containing 180 g of polyphosphoric acid (Section **II,2,**48). Heat on a boiling water bath, stir with a thermometer and maintain at 100–120 °C for 10 minutes (the reaction is exothermic). Add 450 ml of cold water and stir well to complete solution of the polyphosphoric acid. Filter at the pump and wash well with water. Heat the crude solid under reflux with 300 ml of rectified spirit, add a little decolourising charcoal and filter through a pre-heated Buchner funnel; wash the residue with 40 ml of hot rectified spirit. Cool the combined filtrates to room temperature, filter off the 2-phenylindole and wash it three times with 10 ml portions of cold alcohol. Dry in a vacuum desiccator over anhydrous calcium chloride. The yield of pure 2-phenylindole, m.p. 188–189 °C, is 20 g (79%).

Cognate preparation   1,2,3,4-Tetrahydrocarbazole.   In a 500-ml three-necked flask fitted with a dropping funnel, a sealed stirrer unit and reflux condenser, place a mixture of 49 g (0.5 mol) of cyclohexanone and 180 g of glacial acetic acid. Heat under reflux with stirring and add 54 g (49 ml, 0.5 mol) of redistilled phenylhydrazine (1) during 1 hour; continue the stirring for a further hour. Pour the reaction mixture into a 1-litre beaker and stir vigorously while it solidifies. Cool to 5 °C and filter at the pump through a Buchner funnel; cool the filtrate in ice and refilter through the same Buchner funnel. Wash the solid on the filter with 50 ml of water, suck almost dry and then wash with 50 ml of 75 per cent ethanol. Spread the crude solid upon absorbent paper and dry in the air overnight. Recrystallise the slightly damp solid from 350 ml of methanol: add a little decolourising carbon and filter through a hot water funnel. The yield of 1,2,3,4-tetrahydrocarbazole, m.p. 116–117 °C, is 65 g (76%). A further 5 g of product may be obtained by concentrating the mother-liquor to one-quarter of the original volume.

Note. (1) See cautionary note on the handling of phenylhydrazine, Section **IV,104**.

VI,13. INDIGO

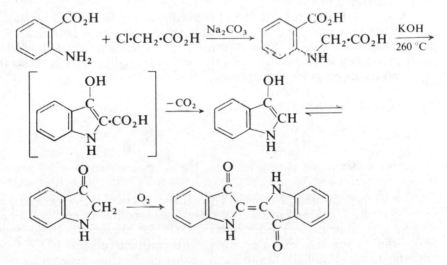

Phenylglycine-*o*-carboxylic acid.   In a 500-ml round-bottomed flask, fitted with a reflux condenser, place 20.6 g (0.15 mol) of anthranilic acid (Section **IV,60**), 14 g (0.15 mol) of chloroacetic acid, 32 g (0.32 mol) of anhydrous sodium carbonate and 300 ml of water. Reflux the mixture for 3 hours, then pour into a beaker, cool, render slightly acid (Congo red) with concentrated hydrochloric acid and allow to stand overnight. Filter off the crude acid and wash it with water. Recrystallise from hot water with the aid of a little decolourising carbon, and dry the acid at 100 °C. The yield of phenylglycine-*o*-carboxylic acid, m.p. 218–220 °C, is 11.5 g (39%).

Indigo. Place a mixture of 9.7 g (0.05 mol) of phenylglycine-*o*-carboxylic acid, 33 g (0.6 mol) of potassium hydroxide pellets and 10 ml of water in a large nickel crucible or basin. Heat the mixture to 260 °C and stir well with a thermometer protected by a copper tube (Section **IV,111**). The mass fuses and the mixture gradually assumes an orange colour. Allow the crucible to cool somewhat, and dissolve the melt in 300 ml of water. This solution oxidises upon shaking in contact with air forming a precipitate of indigo. The conversion into indigo may be more rapidly effected (1) by acidifying with hydrochloric acid and oxidising with iron(III) chloride solution until no further precipitate of the dyestuff is produced. Filter off the indigo at the pump, wash it with hot water, dilute hydrochloric acid and ethanol, and dry at 100 °C. The yield is 3.6 g (55%).

Vat dyeing of cotton. Triturate 0.2 g of indigo on a watch glass with sufficient ethanol to moisten it, suspend it in 1 ml of water and add 1 ml of 20 per cent aqueous sodium hydroxide solution. Reduce the indigo by gradually adding with stirring a warm (50 °C) solution of 0.3 g of sodium dithionite in 20 ml of water. When a clear solution ('vat') is obtained, dilute with 40 ml of water. Introduce a 2-g hank of cotton which has previously been thoroughly washed with detergent and rinsed with water, and heat on a water bath to 50 °C for 1 hour; work the material thoroughly to ensure even impregnation. Remove the hank, wring it out and allow it to hang in the air for 30 minutes to develop the colour. The hue may be brightened by heating the dyed hank on a water bath for 15 minutes with a dilute solution of a mild detergent, followed by rinsing and drying.

Note. (1) The filtered solution of indoxyl may also be oxidised by placing it in a filter flask and drawing air through the solution by means of a water pump until a drop of the aqueous suspension of indigo when placed upon filter-paper produces a sharply defined ring of precipitated indigo, outside which the liquid no longer becomes blue upon exposure to air.

VI,14. BENZIMIDAZOLE

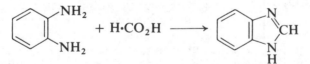

Place 27 g (0.25 mol) of *o*-phenylenediamine (Section **IV,56**) in a 250-ml round-bottomed flask and add 17.5 g (16 ml, 0.34 mol) of 90 per cent formic acid. Heat the mixture on a water bath at 100 °C for 2 hours. Cool, add 10 per cent sodium hydroxide solution slowly, with constant rotation of the flask, until the mixture is just alkaline to litmus. Filter off the crude benzimidazole at the pump, wash with ice-cold water, drain well and wash again with 25 ml of cold water. Dissolve the crude product in 400 ml of boiling water, add 2 g of decolourising carbon and digest for 15 minutes. Filter rapidly at the pump through a pre-heated Buchner funnel and flask. Cool the filtrate to about 10 °C, filter off the benzimidazole, wash with 25 ml of cold water and dry at 100 °C. The yield of pure benzimidazole, m.p. 171–172 °C, is 25 g (85%).

Cognate preparations 2-Methylbenzimidazole. Heat together a mixture of 5.43 g (0.03 mol) of *o*-phenylenediamine dihydrochloride (Section **IV,56**, Note

(3)), 20 ml of water and 5.4 g (0.09 mol) of acetic acid under reflux for 45 minutes. Make the cooled reaction mixture distinctly basic by the gradual addition of concentrated ammonia solution, collect the precipitated product and recrystallise it from 10 per cent aqueous ethanol. The yield is 2.2 g (56%), m.p. 176 °C.

2-Benzylbenzimidazole. Use 5.43 g (0.03 mol) of *o*-phenylenediamine dihydrochloride, 20 ml of water, 12.3 g (0.09 mol) of phenylacetic acid (Section **III,128,***B*) and proceed as for 2-methylbenzimidazole. Recrystallise the crude product from 40 per cent aqueous ethanol. The yield is 3.4 g (55 %), m.p. 191 °C.

VI,15. BENZOTRIAZOLE

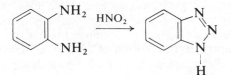

Dissolve 10.8 g (0.1 mol) of *o*-phenylenediamine (1) in a mixture of 12 g (11.5 ml, 0.2 mol) of glacial acetic acid and 30 ml of water contained in a 250-ml beaker; slight warming may be necessary. Cool the clear solution to 15 °C, stir magnetically and then add a solution of 7.5 g (0.11 mol) of sodium nitrite in 15 ml of water in one portion. The reaction mixture becomes warm and within 2–3 minutes reaches a temperature of about 85 °C and then begins to cool while the colour changes from deep red to pale brown. Continue stirring for 15 minutes, by which time the temperature will have dropped to 35–40 °C, and then thoroughly chill in an ice-water bath for 30 minutes. Collect by vacuum filtration the pale brown solid which separates and wash with three 30 ml portions of ice-cold water. Dissolve the solid in about 130 ml of boiling water, add decolourising charcoal, filter and allow the filtrate to cool to about 50 °C before adding a few crystals of the crude benzotriazole which have been retained for seeding. Allow the mixture to attain room temperature slowly (to avoid separation of the material as an oil) and then thoroughly chill in ice and collect the benzotriazole which separates as pale straw-coloured needles, m.p. 99–100 °C. A second crop may be obtained by concentrating the filtrate. The yield is about 8 g (67%). The benzotriazole cystallises much more readily from benzene (*ca.* 55 ml) but the material is still slightly coloured. A pure white product can be obtained by sublimation at 90–95 °C at 0.2 mmHg.

Note. (1) *o*-Phenylenediamine is generally contaminated by highly coloured impurities. A somewhat more pure benzotriazole may be obtained by using purified *o*-phenylenediamine (see Section **IV,56**).

A,3. Six-membered ring systems. The Hantzsch synthesis is the most widely used general procedure for the synthesis of **pyridine** derivatives. The reaction sequence involves essentially the formation of a dihydropyridine derivative (XXIII) by means of a reaction between ethyl acetoacetate, an aldehyde and ammonia. Oxidation of (XXIII) followed by removal of the ester groups by hydrolysis and decarboxylation then yields an alkyldimethylpyridine (XXIV).

The mechanism for the formation may depend upon the reaction conditions used. If ethyl acetoacetate and the aldehyde are first allowed to react in the presence of a basic catalyst (as in the synthesis of 2,6-dimethylpyridine, Section

VI,16), an alkylidene-bis-acetoacetate (XXII) is formed by successive Knoevenagel and Michael reactions (p. 489, **III,M,6**(b)). Cyclisation of this 1,5-dione with ammonia then gives the dihydropyridine derivative (XXIII). Under different reaction conditions a condensation between an aminocrotonic ester (XXV) and an alkylidene acetoacetate (XXVI) may be involved.

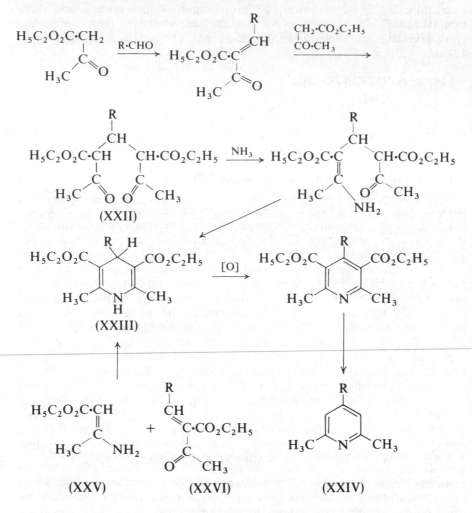

Oxidation of the dihydropyridine derivative is most conveniently carried out with an aqueous nitric–sulphuric acid mixture. Removal of the ethoxycarbonyl groups may be achieved by a stepwise hydrolysis and decarboxylation sequence (Ref. 4), but the one-step reaction described here for 2,6-dimethylpyridine (Section **VI,16**) using soda-lime is convenient.

The replacement of ethyl acetoacetate by ethyl cyanoacetate in the Hantzsch synthesis was introduced by Guareschi and many similar variations have been developed. Condensation of 2 mol of ethyl cyanoacetate with 1 mol of a ketone in the presence of ammonia gives a 'Guareschi imide', a dicyanodioxopiperidine (Section **VI,17**). This reaction is of interest because hydrolysis of this intermediate affords a convenient synthesis of a β,β-dialkylglutaric acid (Section **VI,17**).

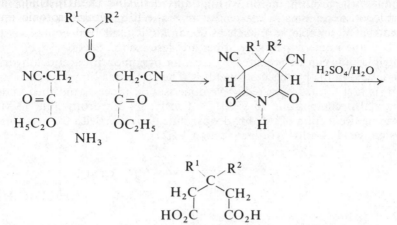

Pyridones (2- or 4-hydroxypyridines) can be prepared by treating pyrones (XXVII) with ammonia. The preparative value of the method is limited because of the restricted availability of suitable pyrones, but the reaction is of interest as it involves the conversion of a six-membered oxygen heterocyclic system to the corresponding nitrogen system.

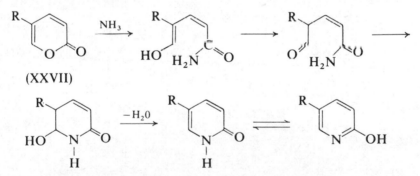

The illustrative reaction uses the readily available methyl coumalate [(XXVII), $R = CO_2CH_3$], when the product is methyl 6-hydroxynicotinate which is subsequently hydrolysed with base to 6-hydroxynicotinic acid (Section **VI,18**).

2-Alkylpyridines are prepared by reaction of pyridine with alkyl lithium compounds. Addition of the organometallic derivative to the azomethine linkage in pyridine gives an intermediate derived from a 1,2-dihydropyridine system. This undergoes thermal elimination of lithium hydride to form the 2-alkylpyridine. The process is illustrated by the preparation of 2-phenylpyridine (Section **VI,19**).

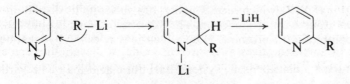

2-Methylpyridine may be converted into homologous alkylpyridines by removal of a proton from the methyl group with the aid of a base, followed by alkylation of the resulting anion with an alkyl halide. 4-Methylpyridine, and the even less reactive 3-methyl derivative, may also be alkylated in this manner provided that a suitable base such as sodamide is used. Procedures for the synthesis of the isomeric pentylpyridines are described in Section **VI,20**.

4-Ethylpyridine can be prepared by treating pyridine with acetic anhydride and acetic acid in the presence of zinc dust. The acetylated bipyridyl derivative (XXVIII) is first formed by a reductive dimerisation process, and then undergoes thermal decomposition to yield 1,4-diacetyl-1,4-dihydropyridine (XXIX) and pyridine. Reduction of the dihydropyridine derivative with the zinc–acetic acid system yields 4-ethylpyridine (Section **VI,21**).

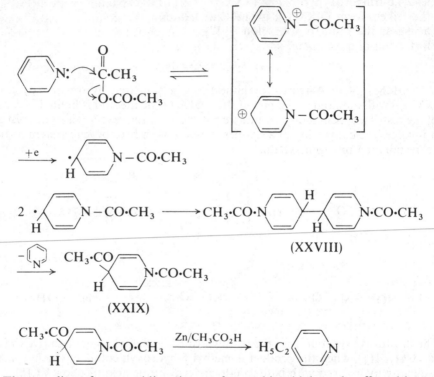

The coupling of two pyridine rings to form a bipyridyl can be effected by an application of the Ullmann reaction (p. 610, **IV,A,3**), which involves treatment of a bromopyridine with copper powder. The reaction is illustrated by the synthesis of 2,2′-bipyridyl (Section **VI,22**). The required 2-bromopyridine is obtained by treating 2-aminopyridine in concentrated aqueous hydrobromic acid with bromine and then with sodium nitrite. Diazotisation of the first-formed 2-aminopyridinium perbromide complex (XXX) gives the pyridinium–diazonium dication (XXXI), which readily undergoes nucleophilic displacement of the diazonium group with bromide ion. The 2-bromopyridine is obtained in the form of a complex with bromine, from which it is released by treatment with alkali.

Pyrimidine is the 1,3-diazabenzene system. Barbituric acid may be regarded

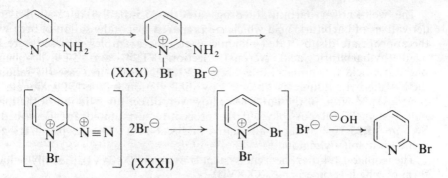

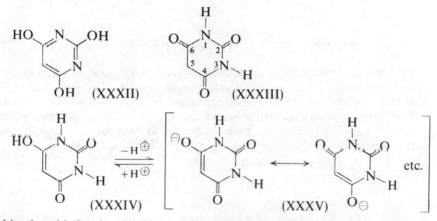

as 2,4,6-trihydroxypyrimidine (XXXII), but in the crystalline state it exists as
the triketo-form (XXXIII). In aqueous solution the compound is markedly
acidic as the result of ionisation of the mono-enolic form (XXXIV) with the
formation of a resonance stabilised anion (XXXV).

Barbituric acid (Section **VI,23**) may be synthesised by condensation of diethyl
malonate with urea in the presence of sodium ethoxide. It undergoes nitration
in the 5-position on treatment with fuming nitric acid, and reduction of the
nitro-derivative yields 5-aminobarbituric acid (uramil).

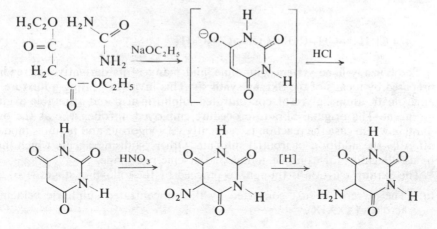

The well-known barbiturate drugs are the 5,5-dialkyl or 5-alkyl-5-aryl-derivatives of barbituric acid which are prepared by condensation of urea with the appropriate disubstituted malonic ester. The examples included here are 5,5-diethylbarbituric acid (veronal, Section **VI,24**) and 5-ethyl-5-phenyl-barbituric acid (phenobarbitone, Section **VI,25**). In the first case the required diethyl diethylmalonate is obtained by dialkylation of diethyl malonate (cf. p. 488, **III,M,6**(*a*)). In the latter case, however, direct phenylation of the malonate system is not possible, but the procedure formulated and described in Section **VI,25** is a suitable alternative indirect route for the preparation of diethyl ethylphenylmalonate.

The isomeric 1,4-diazabenzene system is pyrazine (**XXXVI**); the fully reduced form of which is piperazine (**XXXVII**).

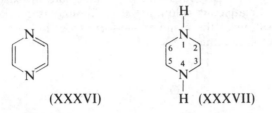

(**XXXVI**) H (**XXXVII**)

Piperazine-2,5-dione [diketopiperazine, (**XXXVIII**)] is formed when glycine is dehydrated by heating in a solvent such as ethylene glycol (Section **VI,26**). Brief treatment with hot concentrated aqueous hydrochloric acid cleaves one of the amide linkages with the formation of the dipeptide, glycylglycine, which is isolated as the hydrochloride monohydrate and may be converted into the ester hydrochloride (Section **VI,26**) under Fischer–Speier conditions (p. 501, **III,N,3**(*a*)).

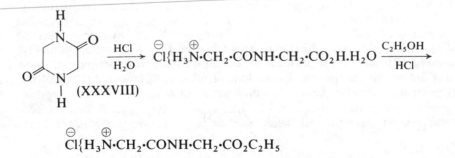

The **benzopyridine system**, quinoline, and many of its derivatives is readily prepared by means of the **Skraup synthesis**. This involves heating a mixture of an aromatic amine, glycerol, concentrated sulphuric acid and a suitable oxidising agent. The original Skraup procedure employed nitrobenzene as the oxidant, in which case the reaction is frequently very vigorous and requires moderation by the addition of iron(II) sulphate. Other oxidising agents which may be used include, for example, iodine and arsenic pentoxide.

The Skraup reaction is thought to proceed by the following stages.

(*a*) The glycerol is first converted by the concentrated sulphuric acid into acrolein (**XXXIX**).

(b) Conjugate 1,4-addition of aniline to acrolein yields β-phenylamino-propionaldehyde (XL).

(c) Acid-catalysed cyclisation of (XL) followed by dehydration forms the 1,2-dihydroquinoline (XLI).

(d) Oxidation of the 1,2-dihydroquinoline to quinoline.

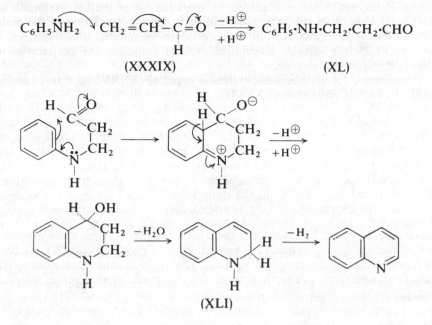

$$C_6H_5\overset{\cdot\cdot}{N}H_2 \quad CH_2{=}CH{-}\overset{H}{\underset{|}{C}}{=}O \quad \xrightarrow[+H^\oplus]{-H^\oplus} \quad C_6H_5{\cdot}NH{\cdot}CH_2{\cdot}CH_2{\cdot}CHO$$

(XXXIX)   (XL)

(XLI)

The preparation of quinoline itself (Section **VI,27**) proceeds smoothly without becoming excessively violent when iodine is employed as the oxidising agent. In the case of 8-nitroquinoline (from *o*-nitroaniline, cognate preparation in Section **VI,27**), the oxidant is arsenic pentoxide. In the preparation of 8-hydroxyquinoline (oxine; cognate preparation in Section **VI,27**), a premix of *o*-aminophenol, glycerol and sulphuric acid held at 80 °C is added portion-wise to a mixture of *o*-nitrophenol and iron(II) sulphate maintained at 100–120 °C; this procedure serves to control the vigour of the reaction. Oxine is a useful reagent in qualitative and quantitative inorganic analysis since it forms water-insoluble complexes with such metals as aluminium, magnesium, bismuth, zinc, etc.

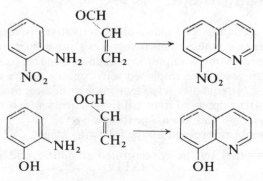

The important oxidation-reduction indicator, 1,10-phenanthroline (Section **VI,28**) is prepared by a double Skraup reaction upon *o*-phenylenediamine. In view of the high reactivity and sensitivity to oxidation of *o*-phenylenediamine, the normal experimental conditions of the Skraup reaction are modified; the condensation is carried out in arsenic acid *solution* and dilute sulphuric acid.

In a reaction which is mechanistically related to the Skraup reaction, an α,β-unsaturated carbonyl compound, generated by way of an acid-catalysed aldol condensation, reacts with a primary aromatic amine in the presence of acid to yield a quinoline derivative (**Doebner–Miller reaction**). For example, when aniline is heated with paraldehyde in the presence of concentrated hydrochloric acid the final product is 2-methylquinoline (quinaldine, Section **VI,29**). In this reaction, crotonaldehyde is first formed by condensation of the depolymerised acetaldehyde in the presence of acid. Aniline then reacts with the α,β-unsaturated aldehyde by 1,4-addition; the addition product under the influence of strong acid cyclises to form 1,2-dihydroquinaldine. The latter is dehydrogenated to quinaldine by anils formed by the condensation of aniline with either acetaldehyde or crotonaldehyde during the course of the reaction. This yields secondary amines (e.g., *N*-ethylaniline) as by-products; these together with excess aniline are separated from the quinaldine by acetylation of the reaction mixture. The acetylated primary and secondary amines thus formed are less steam volatile than quinaldine which forms the basis of the isolation of the latter.

In a similar way quinolines (**XLIV**) are formed when an aromatic primary amine is condensed with a 1,3-diketone and the resulting enamine (**XLIII**), formed via the imine (**XLII**), is cyclised with concentrated sulphuric acid (the **Combes reaction**). Formation of the enamine is achieved by heating the amine and the diketone in xylene under reflux, using a Dean and Stark apparatus, until the theoretical amount of water has separated, or alternatively at 100 °C in the presence of anhydrous calcium sulphate. The latter method is the most convenient and is illustrated by the preparation from *p*-toluidine and pentane-2,4-dione of 4-(*p*-tolylamino)pent-3-en-2-one which is cyclised to 2,4,6-trimethylquinoline (Section **VI,30**).

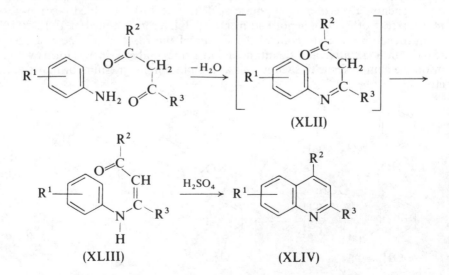

Cyclisation of acetoacetanilide with concentrated sulphuric acid yields 2-hydroxy-4-methylquinoline (Section **VI,31**).

The synthesis of 2-phenylquinoline-4-carboxylic acid (atophan) is described in Section **VI,32**. This is an example of the Doebner synthesis of quinoline-4-carboxylic acids (cinchoninic acids); the reaction consists of the condensation of an aromatic primary amine with pyruvic acid and an aldehyde. The mechanism is probably similar to the synthesis of quinolines previously discussed and involves the intermediate formation of a dihydroquinoline derivative, which is subsequently dehydrogenated by the anils derived from the aromatic amine and the aldehyde.

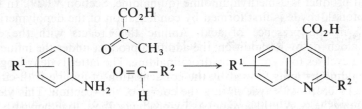

o-Phenylenediamines and α-diketones readily condense to yield substituted benzopyrazines or quinoxalines (e.g., 2,3-diphenylquinoxaline, Section **VI,33**).

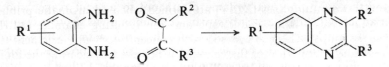

The preparation of 3-benzyl-2-oxo-1,2-dihydroquinoxaline using the α-keto acid, phenylpyruvic acid and of 1,2 : 3,4-dibenzophenazine using the quinone, 9,10-phenanthraquinone (cognate preparations in Section **VI,33**) illustrate possible variations of the general reaction.

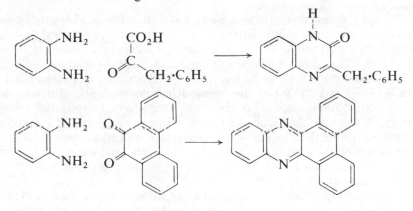

VI,16. 2,6-DIMETHYLPYRIDINE

$$2CH_3 \cdot CO \cdot CH_2 \cdot CO_2C_2H_5 + H \cdot CHO + NH_3 \longrightarrow$$

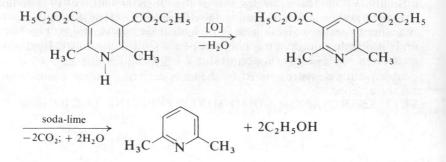

Diethyl 1,4-dihydro-2,6-dimethylpyridine-3,5-dicarboxylate. Cool 52 g (51 ml, 0.4 mol) of ethyl acetoacetate to 0 °C and add 15 ml (0.2 mol) of 40 per cent aqueous formaldehyde solution, followed by a few drops of diethylamine as a catalyst. Keep the mixture at 0 °C for 6 hours and then at room temperature for 40 hours. Separate the lower organic layer, extract the aqueous phase with ether and dry the combined organic fractions over anhydrous calcium chloride. Remove the ether under reduced pressure (rotary evaporator) and transfer the residue together with an equal volume of ethanol to a stout reagent bottle cooled in an ice bath. Pass a steady stream of ammonia gas (from a cylinder) into the solution held at 0 °C for 1 hour, close the bottle with a bung securely attached with wire and set the bottle and contents aside at room temperature for 40 hours. Filter the resulting yellow solution to remove a small quantity of almost colourless material and heat the filtrate on a boiling water bath in an evaporating dish until most of the ethanol has been removed, and then cool and crystallise the residue from about 400 ml of rectified spirit. The yield of the pale yellow crystalline dihydropyridine derivative is 36 g (71%), m.p. 181–183 °C.

Diethyl 2,6-dimethylpyridine-3,5-dicarboxylate. Place 35.5 g (0.14 mol) of the above dihydropyridine derivative in a 1-litre round-bottomed flask and add carefully a cold mixture of 50 ml of water, 9 ml of concentrated nitric acid (*d* 1.42) and 7.5 ml of concentrated sulphuric acid. Swirl the mixture and heat it cautiously on a boiling water bath until a vigorous reaction, accompanied by much foaming, sets in. When the reaction has moderated continue to heat cautiously for 15 minutes until oxidation is complete and a deep red solution is obtained. Cool the solution, add 100 ml of water and 100 g of crushed ice, and make it distinctly alkaline with concentrated aqueous ammonia solution (*d* 0.88). Filter off the solid product, wash it with a little cold water and recrystallise it from aqueous ethanol. The yield of colourless crystals of the pyridine derivative, m.p. 71–72 °C, is 22.5 g (64%).

2,6-Dimethylpyridine. Place an intimate mixture of 10 g (0.04 mol) of the above pyridine di-ester and 60 g of soda lime (10–14 mesh) in a 100-ml round-bottomed flask fitted with a still-head and condenser arranged for distillation. Heat the flask gradually in an oil bath to about 250 °C, and maintain this temperature until no further material distils below 105 °C (about 2 hours may be required). Remove the oil bath, clean the outside of the flask and continue to heat more strongly with a Bunsen burner held in the hand, keeping the flame moving over the surface of the flask. Collect the product which now distils, and continue to heat strongly until the flask reaches dull red heat and no further distillate is obtained. Treat the distillate with potassium hydroxide pellets so that the pyridine separates, and isolate the latter by extraction with ether. Dry the ether extract over fresh potassium hydroxide pellets, and remove the ether and distil the residue at atmospheric pressure. Collect the dimethylpyridine as a fraction of b.p. 142–145 °C; the yield is 2.8 g (65%).

VI,17. 3,5-DICYANO-4,4-DIMETHYLPIPERIDINE-2,6-DIONE

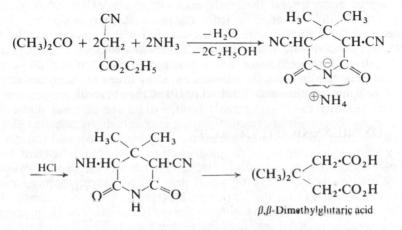

β,β-Dimethylglutaric acid

Place 200 ml of absolute ethanol in a 500-ml all-glass Drechsel bottle, and cool to −5 °C by immersion in a bath of ice and salt. Pass a slow stream of ammonia, derived from a cylinder and dried by passage through a tower filled with small pieces of calcium oxide, into the ethanol until the latter is saturated (about 4–5 hours); when necessary, siphon off the water formed in the freezing mixture and add more crushed ice. The volume of the ethanol will increase to about 250 ml and about 17 g (1 mol) of ammonia is absorbed. Meanwhile weigh out 113 g (1 mol) of ethyl cyanoacetate (Section III,171) and 29 g (0.5 mol) of pure dry acetone into a 500- or 750-ml, wide-mouthed, ground-stoppered bottle and cool it by immersion in a bath of ice and salt for about 2 hours. Add the alcoholic ammonia solution to the cold contents of the bottle, replace the stopper and securely wire or tape it into position. After three days, when a considerable amount of white solid (the ammonium salt of the 'Guareschi imide') has separated, filter at the pump and rinse the bottle with the filtrate until all the solid has been transferred to the filter. Drain well by pressing the solid with a large glass stopper, and then stir it with several small volumes of dry ether (sucking dry after each addition) in order to remove the excess of ketone and ethyl cyanoacetate. Dry the solid in the air for several hours in order to remove the ether completely, then dissolve it in the minimum volume of boiling water (350–400 ml), and add concentrated hydrochloric acid (*FUME CUPBOARD!*) until the mixture is acid to Congo red paper and then add a further 50 ml. Allow to cool, filter off the dicyano-imide, wash with a little water and dry upon clock glasses in an oven at 100 °C. The yield of the dicyano-imide is 65 g (67%).

Conversion into β,β-dimethylglutaric acid. Dissolve 64 g (0.33 mol) of the *finely powdered* dicyano-imide in 160 ml of concentrated sulphuric acid in a 1-litre round-bottomed flask; *gentle* warming may be necessary and a clear reddish-brown solution is obtained. Keep the solution overnight and then add 150 ml of water slowly and with frequent shaking. Attach a reflux condenser to the flask and heat very gently at first owing to the attendant frothing, which subsides after 2-3 hours. Heat the mixture under reflux for a total period of 18-24 hours and shake well at intervals of 3 hours. The acid separates upon cooling: collect it on a sintered glass funnel. It may be dried at about 90 °C; the yield of crude acid is nearly quantitative. To remove small quantities of imides which may be present, treat the crude acid with excess of saturated sodium hydrogen carbonate solution, filter from any imide, strongly acidify with concentrated hydrochloric acid, saturate the solution with ammonium sulphate and extract the acid with three or four 200 ml portions of ether. Dry the ethereal extract with anhydrous sodium sulphate and distil off the ether. Recrystallise the residual acid from concentrated hydrochloric acid, and dry at 70 °C. Pure β,β-dimethylglutaric acid, m.p. 101 °C, is obtained.

VI,18. 6-HYDROXYNICOTINIC ACID

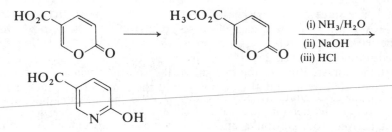

Methyl coumalate. Place 100 ml of concentrated sulphuric acid in a 250-ml round-bottomed flask and add with swirling 35 g (0.224 mol) of finely powdered coumalic acid (Section **VI,38**). The solution acquires a deep red-brown colour and the temperature tends to rise but should be kept below 30 °C by cooling in an ice-water bath. Add 50 ml of methanol with continued swirling in small portions so that the temperature of the reaction mixture does not rise above 35 °C. Heat the reaction mixture on a boiling water bath for 1.5 hours. Cool the mixture and pour slowly into 500 ml of an ice-water slurry with stirring. Add about 150 g of sodium carbonate as a slurry in water until the mixture is just neutral. If too much carbonate has been added the supernatant liquid will acquire a red colour; in these circumstances add a few drops of sulphuric acid. The final solution containing a suspension of methyl coumalate and inorganic salts should be an orange colour. Filter the solution, and slurry the residue with four 70 ml portions of cold water containing a drop of concentrated sulphuric acid. Filter off the methyl coumalate; the yield after air drying is 18 g (47%). Although the product may be used for the next stage, a sample may be purified by vacuum sublimation; pure methyl coumalate has m.p. 74 °C.

6-Hydroxynicotinic acid. Place a mixture of 20 ml of ammonia solution (*d* 0.880) and 30 ml of water in a 250-ml beaker sited in an ice-water cooling bath on a magnetic stirrer unit. Insert a magnetic follower and add portionwise 18 g (0.105 mol) of methyl coumalate over a period of about 5 minutes, ensuring that the temperature is kept below 20 °C. Stir the resulting dark red solution for a further 45 minutes at 20 °C and then add it to boiling aqueous sodium hydroxide prepared from 40 g of sodium hydroxide and 250 ml of water. Boil the reaction mixture for 5 minutes, cool to 10 °C in an ice-water cooling bath and add concentrated hydrochloric acid with stirring to precipitate the product; the temperature of the solution during neutralisation should not rise above 30 °C. Allow the acidified solution to stand in the ice-water bath for 1 hour and then collect the bright yellow crystalline solid by filtration. Wash the crystals with water and then air dry. The yield of 6-hydroxynicotinic acid, m.p. 299–300 °C (decomp.), is 8.5 g (58%).

VI,19. 2-PHENYLPYRIDINE

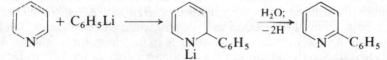

The first stage is the *preparation of a solution of phenyl lithium in dry ether.* Equip a 1-litre three-necked flask as shown in Fig. I,67. Flush the apparatus with dry, oxygen-free nitrogen gas. Place 7.35 g (1.06 mol) of lithium shavings (1) in the flask, and introduce a solution of 78.5 g (52.5 ml, 0.5 mol) of dry, redistilled bromobenzene in 250 ml of anhydrous ether into the dropping funnel. Start the stirrer. Run in about 2 ml of the solution; when the reaction starts, as indicated by an initial cloudiness, add the remainder at such a rate that the solvent refluxes gently (about 45 minutes). Finally, add 50 ml of anhydrous ether through the dropping funnel. Continue the stirring until all or most of the lithium disappears (1–1.5 hours) (2).

Now introduce slowly, and with stirring, 79 g (1 mol) of pure anhydrous pyridine (Section **II,1,**29) dissolved in 200 ml of anhydrous toluene: remove the ether by distillation, replace the dropping funnel by a thermometer and stir the residual suspension at 110 °C (internal temperature) for 8 hours. Then cool to about 40 °C, and add cautiously 75 ml of water through the condenser. Filter the liquids if necessary, separate the upper toluene layer, dry it by shaking for an hour with 20 g of potassium hydroxide pellets and distil slowly through a short fractionating column. When the temperature reaches 150 °C at ordinary pressure (thus indicating the removal of most of the toluene, etc.), distil the residue under reduced pressure and collect the liquid passing over at 138–142 °C/12 mmHg. Upon redistillation 38 g (49%) of pure 2-phenylpyridine, b.p. 140 °C/12 mmHg, is obtained.

Notes. (1) See p-*toluic acid*, Section **IV, 172,** Note (1).

(2) The yield of phenyl lithium generally exceeds 95 per cent. One interesting and instructive method of determination is to allow the phenyl lithium to react with excess benzophenone and to weigh the triphenylmethanol formed which is assumed to be formed quantitatively. A better method is to hydrolyse a 2 ml aliquot portion of the filtered solution with distilled water and to titrate the

hydrolysate with standard acid, using phenolphthalein as indicator. To obtain the filtered solution, the dropping funnel is replaced by a short L-shaped tube loosely plugged with glass wool, and the solution is decanted through this tube into a graduated flask that has been swept out with nitrogen.

Cognate preparation Ethyl 2-pyridylacetate. Prepare a solution of phenyl lithium in anhydrous ether as detailed above for 2-phenylpyridine, using 7.35 g (1.06 mol) of lithium. Introduce 46.5 g (49 ml, 0.5 mol) of dry, redistilled 2-picoline, with continued stirring dropwise during about 10 minutes. Stir the dark red-brown solution of 2-picolyl lithium for a further 30 minutes, and then pour it slowly (1) and with shaking on to about 400 g of solid carbon dioxide contained in a 1.5-litre round-bottomed flask. Break up the lumpy residue of lithium salts before adding 375 ml of absolute ethanol. Cool the solution in ice, saturate it with dry hydrogen chloride and then insert a calcium chloride drying tube into the neck of the flask. Allow the mixture to stand overnight, remove the ethanol under reduced pressure (rotary evaporator) on a water bath. Dissolve the syrupy residue in 375 ml of chloroform and transfer to a three-necked flask fitted with a mechanical stirrer and a reflux condenser. Prepare a paste from 112.5 g (0.815 mol) of potassium carbonate and 70 ml of water, and add it slowly through the third neck to the chloroform solution with constant stirring. Stir the almost boiling solution vigorously for 1 hour. Decant the chloroform solution from the inorganic salts, remove the solvent by distillation from a water bath (rotary evaporator) and distil the residue under diminished pressure from a flask with a short fractionating column. 2-Picoline (*ca.* 20 g) passes over first, followed by ethyl 2-pyridylacetate as a pale yellow liquid at 135–137 °C/28 mmHg, or 110–112 °C/6 mmHg. The yield is 30 g (36%).

Note. (1) It is advisable to filter the 2-picolyl lithium solution rapidly through a thin layer of glass wool (to remove any unreacted lithium) before it is added to the solid carbon dioxide.

VI,20. 2-PENTYLPYRIDINE

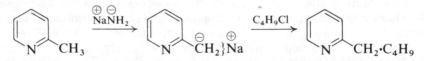

Into a 500-ml three-necked flask fitted with a dropping funnel, a sealed stirrer unit and reflux condenser protected by a drying tube, place a fine suspension of 40 g (1 mol) of good quality sodamide (Section **II,2,**54) in about 150 ml of anhydrous xylene. Introduce 37.5 g (40 ml, 0.4 mol) of 2-picoline through the dropping funnel and rinse the latter with a few ml of dry xylene. Set the stirrer in motion and add 44.5 g (50.5 ml, 0.48 mol) of butyl chloride (Section **III,48**) over a period of 1 hour: reflux the mixture with stirring for 2–3 hours. When cold, destroy the excess of sodamide by the cautious addition of 100 ml of water. Transfer the contents of the flask to a separatory funnel and discard the lower aqueous layer. Extract the xylene solution with four 50 ml portions of 1 : 1-hydrochloric acid. Steam distil the acid extracts to remove traces of xylene, cool the aqueous solution and render strongly alkaline by the addition of solid sodium hydroxide: a brown oil appears. Steam distil again and collect about

700 ml of distillate. Separate the upper layer in the steam distillate, extract the aqueous layer with ether and dry the combined upper layer and ether extract with anhydrous potassium carbonate. After removing the ether, distil through a Fenske-type column (15 cm diameter and packed with glass helices for a length of 12–15 cm) at a pressure of 50 mmHg (*monostat*, see Section **I,30**) and collect the 2-pentylpyridine (42 g, 70%) at 122.5–124.5 °C/50 mmHg. Upon redistillation, the product boils almost entirely at 105 °C/17 mmHg.

Cognate preparations   4-Pentylpyridine.   Charge a 1-litre three-necked flask (equipped with a sealed stirrer unit, a dropping funnel and a short air condenser) with 600 ml of liquid ammonia (Section **I,17,7**). Stir vigorously, add 0.5 g of powdered iron(III) nitrate followed, after 1 minute, by 11.9 g (0.52 mol) of clean sodium in small pieces through the short air condenser over a period of half an hour; continue the stirring until the initial blue colour is replaced by a colourless or pale grey suspension of sodamide (see p. 312). Introduce 42.0 g (44.0 ml, 0.45 mol) of pure 4-picoline through the air condenser; a green colour develops immediately. Stir for 15–20 minutes and add 46.3 g (52.6 ml, 0.5 mol) of butyl chloride (or an equivalent amount of butyl bromide) from the dropping funnel at such a rate that the reaction does not become unduly vigorous (*ca.* 10 minutes): upon completion of the addition the green colour will have been discharged. Stir for a further 10–15 minutes, pour the reaction mixture into a 2-litre beaker: allow the liquid ammonia to evaporate overnight. Rinse the reaction flask with 100 ml of water and add the rinsings to the residue in the beaker; two layers form. Separate them and keep the upper layer of 4-pentylpyridine: extract the lower layer with a little xylene and wash the xylene extract with 25 ml of 1 : 1-hydrochloric acid. Dissolve the 4-pentylpyridine in 1 : 1-hydrochloric acid, combine it with the acid washings of the xylene extract and steam distil to remove traces of xylene; cool, add solid sodium hydroxide until strongly alkaline and steam distil again. Isolate the 4-pentylpyridine as described above for the 2-pentyl compound. The yield of the pure base, b.p. 95 °C/6 mmHg, is 46 g (69%).

   3-Pentylpyridine.   Proceed exactly as described for 4-pentylpyridine using 11.9 g (0.52 mol) of sodium, 42 g (0.45 mol) of 3-picoline and 46.3 g (0.5 mol) of butyl chloride. The yield of pure 3-pentylpyridine, b.p. 100.5 °C/9 mmHg, is 46 g (69%).

VI,21. 4-ETHYLPYRIDINE

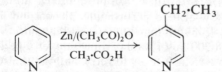

   Place a mixture of 500 ml (5.3 mol) of acetic anhydride and 100 g (102 ml, 1.27 mol) of dry pyridine (Section **II,1,29**) in a 2-litre three-necked flask fitted with a reflux condenser, sealed stirrer unit and thermometer. Introduce, with stirring, 100 g (1.53 mol) of activated zinc powder (Section **II,2,67**) in 5 g portions over a period of 1.5–2 hours; remove the thermometer or reflux condenser momentarily as required. The temperature rises almost immediately: maintain it at 25–30 °C by means of a bath of cold water. The reaction mixture acquires a green colour after 20 minutes and a yellow solid separates gradually. When the addition of the 100 g of zinc powder is complete, stir for a further 15 minutes, and run in 100 ml of glacial acetic acid through the condenser. Add a further 40 g (0.6 mol) of zinc powder in 5–10 g portions at intervals so timed that the vigorous reaction is under control and the mixture refluxes gently. Then reflux the reaction mixture, with stirring, for 30 minutes: add a further 60 g (0.93 mol) of zinc powder all at once and continue the refluxing for 30 minutes more.

Neutralise the cold contents of the flask with 500–600 ml of 40 per cent aqueous sodium hydroxide solution, equip the flask for steam distillation and steam distil until about 1 litre of distillate is collected. The steam distillate separates into two layers. Add solid sodium hydroxide ($\lessdot$ 100 g) to complete the separation of the two layers as far as possible. Remove the upper (organic) layer and extract the aqueous layer with three 50 ml portions of chloroform. Dry the combined organic layer and chloroform extracts with anhydrous potassium carbonate and distil the mixture through a short fractionating column; after a fore-run of chloroform, followed by pyridine, collect the crude 4-ethyl-pyridine at 150–166 °C (49 g). Redistil through a Fenske-type column, 15 mm in diameter and packed with glass helices for a length of 20 cm. Collect the 4-ethylpyridine base at 163–165 °C/760 mmHg (44 g, 32%).

VI,22. 2,2′-BIPYRIDYL

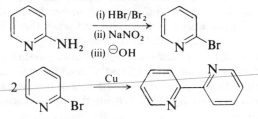

2-Bromopyridine. Place 395 ml (2.35 mol) of 48 per cent hydrobromic acid in a 3-litre three-necked flask, fitted with a dropping funnel, mechanical stirrer and low temperature thermometer. Cool to 10–15 °C in an ice–salt bath and add 75 g (0.8 mol) of 2-aminopyridine (Section **IV,50**) over a period of about 10 minutes. Whilst maintaining the temperature at 0 °C or lower, add 375 g (120 ml, 2.34 mol) of bromine dropwise with stirring. The reaction mixture thickens during the addition of the first half of the bromine (*ca.* 30 minutes) owing to the formation of a yellow-orange 'perbromide'; the second half may then be introduced more rapidly (*ca.* 15 minutes). Now add a solution of 140 g (2 mol) of pure sodium nitrite in 200 ml of water dropwise over a period of 2 hours whilst keeping the temperature at 0 °C or lower. Continue the stirring for 30 minutes; then run in a solution of 300 g (7.5 mol) of sodium hydroxide in 300 ml of water at such a rate that the temperature does not rise above 20–25 °C. Extract the reaction mixture with four 125 ml portions of ether, dry the ethereal extracts for 1 hour over 50 g of potassium hydroxide pellets, remove the ether on a rotary evaporator and distil the residue through a short fractionating column under reduced pressure. Collect the 2-bromopyridine at 74–75 °C/13 mmHg; the yield is 115 g (91%). The b.p. at atmospheric pressure is 193–195 °C.

2,2′-Bipyridyl. In a 1-litre three-necked flask, equipped with a reflux condenser and mechanical stirrer, place 21 g of copper powder and 200 ml of *p*-cymene (b.p. 176–177 °C). Whilst refluxing the mixture gently with stirring, add 104 g (0.66 mol) of 2-bromopyridine dropwise over a period of 1 hour; add three additional portions of 21 g (total: 1.32 mol) each of copper powder (through the otherwise closed third neck) during this period. Continue the heating with stirring for a further 2.5 hours, cool, acidify with dilute hydrochloric acid and separate the *p*-cymene by steam distillation. Render the residual solution strongly alkaline with concentrated sodium hydroxide solution and steam distil again until the distillate gives only a pale red colouration with iron(II) sulphate solution. Saturate the steam distillate with sodium chloride and extract repeatedly with ether; it is best to use a continuous extractor (Section **I,22**). Dry the ethereal extracts over anhydrous potassium carbonate, remove the ether by distillation through an efficient fractionating column (2,2′-bipyridyl is slightly volatile in ether vapour) and distil the residue under reduced pressure. Collect the 2,2′-bipyridyl (31.5 g, 61%) at 147 °C/16 mmHg; it solidifies on cooling, m.p. 69–70 °C.

VI,23. BARBITURIC ACID

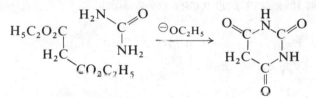

In a 2-litre round-bottomed flask, fitted with a double surface reflux condenser, place 11.5 g (0.5 mol) of clean sodium. Add 250 ml of absolute ethanol in one portion: if the reaction is unduly vigorous, immerse the flask momentarily in ice. When all the sodium has reacted, add 80 g (76 ml, 0.5 mol) of diethyl malonate (Section **III,154**), followed by a solution of 30 g (0.5 mol) of dry urea in 250 ml of hot (*ca.* 70 °C) absolute ethanol. Shake the mixture well, fit a calcium chloride guard-tube to the top of the condenser and reflux the mixture for 7 hours in an oil bath heated to 110 °C. A white solid separates. Treat the reaction mixture with 450 ml of hot (50 °C) water and then with concentrated hydrochloric acid, with stirring, until the solution is acid (about 45 ml). Filter the resulting almost clear solution and leave it in the refrigerator overnight. Filter the solid at the pump, wash it with 25 ml of cold water, drain well and then dry at 100 °C for 4 hours. The yield of barbituric acid is 50 g (78%). It melts with decomposition at 245 °C.

Conversion to aminobarbituric acid (*uramil*). *Nitrobarbituric acid.* Place 72 ml of fuming nitric acid, *d* 1.52, in a 1-litre flask equipped with a mechanical stirrer and surrounded by an ice bath. Add 50 g (0.39 mol) of barbituric acid with stirring, over a period of 2 hours; keep the temperature below 40 °C during the addition. Stir for a further 1 hour, and continue the stirring while 215 ml of water is added and the solution is cooled to 10 °C. Filter with suction through a sintered glass funnel, wash with cold water and dry on a clock glass at 60–80 °C. Dissolve the crude nitrobarbituric acid in 450 ml of boiling water, filter and allow to stand overnight. Collect the crystals by suction filtration, wash well with cold water, and dry at 90–95 °C for 2–3 hours. The product is the trihydrate, m.p. 181–183 °C (decomp., rapid heating), and weighs 70 g. Drying at 110–115 °C for 2–3 hours gives 47 g (70%) of anhydrous nitrobarbituric acid, m.p. 176 °C (decomp.).

Uramil. In a 3-litre flask place 38 g (0.22 mol) of anhydrous nitrobarbituric acid and 300 ml of concentrated hydrochloric acid; heat the mixture on a boiling water bath. Add 125 g of granulated tin and 200 ml of concentrated hydrochloric acid over a period of about 30 minutes: continue the heating until the yellow colour, due to the nitro compound, in the liquid is no longer visible. Introduce 1500 ml more of concentrated hydrochloric acid and heat until all the white solid dissolves; add a little decolourising charcoal, and filter the hot mixture through a sintered glass funnel. Keep the filtrate at 0 °C overnight, collect the uramil by filtration with suction, wash well with dilute hydrochloric acid and finally with water. Concentrate the filtrate under reduced pressure (rotary evaporator) to about 500 ml and cool overnight. Collect the second crop of uramil, wash it as before and combine it with the first product. Dry in a vacuum desiccator over concentrated sulphuric acid. The resulting uramil (23 g, 73%) is a fine white powder; it does not melt below 400 °C, and becomes pink to red on standing, particularly if ammonia is present in the air.

VI,24. 5,5-DIETHYLBARBITURIC ACID (*Veronal*)

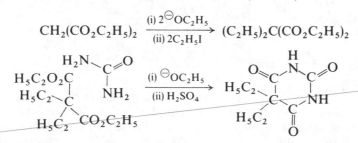

$$CH_2(CO_2C_2H_5)_2 \xrightarrow[\text{(ii) } 2C_2H_5I]{\text{(i) } 2^{\ominus}OC_2H_5} (C_2H_5)_2C(CO_2C_2H_5)_2$$

Diethyl diethylmalonate. Equip a 1-litre three-necked flask with a sealed mechanical stirrer, a dropping funnel (with calcium chloride guard-tube) and a double surface reflux condenser; it is important that the apparatus be perfectly dry. Place 23 g (1 mol) of clean sodium in the flask and add 300 ml of super-dry ethanol (Section **II,1,**9). It may be necessary to warm the flask gently on a water bath towards the end of the reaction in order to complete the solution of the sodium. Insert a guard-tube into the top of the condenser. Allow the sodium ethoxide solution to cool with stirring; when the sodium ethoxide commences to separate out, add 80 g (75 ml, 0.5 mol) of diethyl malonate (dried over anhydrous calcium sulphate) during 1 hour. Towards the end of the addition some solid may separate; it is then necessary to heat on a water bath to dissolve the solid. When all the diethyl malonate has been introduced, heat the mixture on a water bath for 15 minutes, and then allow to cool. When the diethyl sodiomalonate commences to crystallise out, add 156 g (81 ml, 1 mol) of dry ethyl iodide over a period of 1 hour. Heat on a water bath for 3 hours to complete the reaction. Rearrange the flask for distillation but keep the stirrer in position; distil off as much as possible of the alcohol on a water bath (it is advisable to wrap the flask in a cloth or towel). Dilute the residue in the flask with water and extract with three 75 ml portions of ether. Wash the combined ethereal extracts with water, dry with anhydrous calcium chloride or magnesium sulphate, remove the ether on a water bath and distil the residue. Collect the diethyl diethylmalonate at 218–222 °C (mainly 221 °C); the yield is 84 g (83%).

5,5-Diethylbarbituric acid. Fit a 250-ml round-bottomed flask with a still-head carrying a thermometer reaching to within 3 cm of the bottom of the flask and attach a condenser. Place 5.1 g (0.22 mol) of clean sodium in the flask and add 100 g (140 ml) of super-dry ethanol. When all the sodium has reacted, introduce 20 g (0.093 mol) of diethyl diethylmalonate and 7.0 g (0.117 mol) of dry urea (dried at 60 °C for 4 hours). Heat the flask in an oil bath and slowly distil off the ethanol. As soon as the temperature of the *liquid* reaches 110–115 °C, adjust the flame beneath the bath so that the contents of the flask are maintained at this temperature for at least 4 hours. Allow the flask to cool somewhat, add 100 ml of water and warm until the solid (sodium veronal) dissolves. Pour the solution into a beaker, and add 100 ml of water containing 7.0 ml of concentrated sulphuric acid; this will liberate the veronal from the sodio-derivative. The veronal crystallises out if sufficient acid has been added to the solution to cause it to be acid to Congo red. Heat the contents of the beaker, with stirring and the addition of more water if necessary, until all the veronal dissolves at the boiling point. Allow the hot solution to cool, filter off the crystals of veronal and dry in the air. The yield is 12 g (71%), m.p. 190 °C.

VI,25. 5-ETHYL-5-PHENYLBARBITURIC ACID (*Phenobarbitone*)

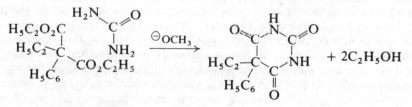

Diethyl phenylmalonate. In a thoroughly dried 1-litre flange flask, equipped with a dropping funnel, reflux condenser (both protected with calcium chloride guard-tubes) and a sealed stirrer unit, place 11.5 g (0.5 mol) of clean sodium pieces (see Section **II,2,**55); add 250 ml of super-dry ethanol (Section **II,1,**9) and allow the vigorous reaction to proceed, cooling only if the reaction appears to be beyond control. When all the sodium has reacted, cool the solution to 60 °C, and add 73 g (67 ml, 0.5 mol) of pure, freshly distilled diethyl oxalate (Section **III,145**) from the dropping funnel in a rapid stream with vigorous stirring. Wash this down with 5 ml of absolute ethanol and add immediately 87.5 g (85 ml, 0.535 mol) of pure ethyl phenylacetate (Section **III,153**).

Discontinue stirring and raise the stirrer clear of the reaction mixture. Within 4–7 minutes after the ethyl phenylacetate has been added, crystallisation commences. Allow the nearly solid paste of the sodio derivative to cool to room temperature, remove the flange lid and stir manually with 400 ml of dry ether. Collect the solid by suction filtration and wash it repeatedly with dry ether. Transfer the solid to a beaker and liberate the ethyl phenyloxaloacetate with ice-cold dilute sulphuric acid (14–15 ml of concentrated sulphuric acid in 250 ml of water). Separate the almost colourless oil and extract the aqueous layer with three 50 ml portions of ether; dry the combined oil and ethereal extracts with magnesium sulphate, remove the ether with a rotary evaporator. Equip the flask containing the residue for distillation under reduced pressure and heat the flask under a pressure of about 15 mm of mercury (water pump) in an oil bath. Raise the temperature of the bath gradually to 175 °C and maintain this temperature until the evolution of carbon monoxide is complete (*FUME CUPBOARD!*); if the pressure rises unduly during the heating (owing to a rather rapid evolution of gas), discontinue the heating momentarily. When the reaction is complete (5–6 hours), return the oil which has passed over to the flask, and distil under reduced pressure. Collect the diethyl phenylmalonate at 159–161 °C/10 mmHg (or at 165–166 °C/15 mmHg). The yield is 95 g (80.5%).

Diethyl ethylphenylmalonate. In a dry 500-ml round-bottomed flask, fitted with a reflux condenser and guard-tube, prepare a solution of sodium ethoxide from 7.23 g (0.315 mol) of clean sodium and 150 ml of super dry ethanol in the usual manner; add 1.5 ml of pure ethyl acetate (dried over anhydrous calcium sulphate) to the solution at 60 °C and maintain this temperature for 30 minutes. Meanwhile equip a 1-litre three-necked flask with a dropping funnel, a sealed mechanical stirrer and a double surface reflux condenser: *the apparatus must be perfectly dry* and guard-tubes should be inserted in the funnel and condenser respectively. Place a mixture of 74 g of diethyl phenylmalonate and 60 g (0.385 mol) of ethyl iodide in the flask. Heat the apparatus in a bath at 80 °C and add the sodium ethoxide solution, with stirring, at such a rate that a drop of the reaction mixture when mixed with a drop of phenolphthalein indicator is never more than faintly pink (1). The addition occupies 2–2.5 hours; continue the stirring for a further 1 hour at 80 °C. Allow the flask to cool, and remove the ethanol under reduced pressure using a rotary evaporator. Add 100 ml of water to the residue in the flask and extract the ester with three 100 ml portions of benzene (*FUME CUPBOARD*). Dry the combined extracts with anhydrous sodium sulphate, distil off the benzene at atmospheric pressure and the residue under diminished pressure. Collect the diethyl ethylphenylmalonate at 159–160 °C/8 mmHg. The yield is 72 g (81%).

5-Ethyl-5-phenylbarbituric acid. In a 250-ml round-bottomed flask, fitted with an efficient reflux condenser and guard-tube, prepare a solution of sodium methoxide from 4.6 g (0.2 mol) of clean sodium and 50 ml of anhydrous methanol (Section **II,1,8**). Add 15 g (0.25 mol) of urea (previously dried at 60 °C for 4 hours), and then 26.4 g (0.1 mol) of diethyl ethylphenylmalonate dropwise down the condenser. Reflux the mixture for 6 hours and then remove the excess methanol under reduced pressure (rotary evaporator) from a water bath held below 60 °C. Transfer the residue to a small beaker cooled in a freezing mixture and add 100 ml of ice-water with mechanical stirring: the temperature of the reaction mixture must be kept below 5 °C since barbiturates are decomposed by concentrated alkali into the salt of the corresponding malonic acid, sodium carbonate and ammonia. Filter and extract the filtrate with two 50 ml portions of benzene in order to remove esters; acidify the aqueous solution cautiously to Congo red, allow to stand for a few hours and filter off the crude phenobarbitone at the pump. The yield after drying at 90–100 °C is 13 g (56%). Recrystallisation from hot water yields reasonably pure ethylphenylbarbituric acid, m.p. 171 °C. A somewhat higher m.p. (175–176 °C) is obtained if rectified spirit is employed for recrystallisation, but the recovery is considerably less.

Note. (1) The rate of addition should be controlled so that the reaction mixture does not become strongly basic.

VI,26. PIPERAZINE-2,5-DIONE (*Diketopiperazine*)

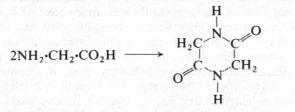

$$2NH_2 \cdot CH_2 \cdot CO_2H \longrightarrow$$

Place 100 g (1.33 mol) of glycine and 500 ml of ethylene glycol in a 1-litre three-necked flask fitted with an air-cooled reflux condenser and a sealed mechanical stirrer; fit a thermometer so that the bulb is in the liquid. Heat the mixture in the fume cupboard to 175 °C and maintain the temperature at this level with continuous stirring for 1 hour. Cool the dark brown reaction product to room temperature and leave overnight in the refrigerator. Centrifuge the resulting suspension (1) and decant the mother-liquor. Transfer the solid to a Buchner funnel with the aid of cold methanol and wash on the filter with gentle suction with more methanol, using about 200 ml in all. Crystallise the product from 300 ml of boiling water but do not attempt to filter the hot solution at this stage; collect the light brown crystals which separate on cooling in ice. Dissolve the crude material in 350 ml of hot water, add 4 g of decolourising carbon and boil for 3 minutes. Filter the hot suspension through a preheated Buchner funnel (2) and cool the filtrate in ice. Collect the colourless crystals of pure piperazine-2,5-dione, wash with a little ice-cold water and dry in the oven at 50 °C; the yield is 34 g (45%). The m.p. is rather indefinite and depends on the rate of heating; when introduced into a rapidly heated (10 °C/min) melting point apparatus at 280 °C, piperazine-2,5-dione has m.p. 310–312 °C (decomp.).

Boil a few mg each of the product and of ninhydrin (Section **III,100**) in 0.5 ml of water; the absence of a blue-purple colouration shows that the product is free from contaminating peptide material.

Notes. (1) If a centrifuge of suitable capacity is not available, the product may be isolated by decanting the suspension through a pad of glass wool in an ordinary filter funnel and allowing the crystals to drain as much as possible. In this case the methanol washing procedure is best omitted, and the product together with the glass wool subjected directly to the preliminary crystallisation from hot water described in the text.

(2) If the filtrate from this treatment with decolourising carbon is not colourless it should be boiled with more of the carbon and re-filtered.

Conversion into glycylglycine ethyl ester hydrochloride. Heat 150 ml of concentrated hydrochloric acid almost to boiling, add in one portion 28.5 g (0.25 mol) of piperazine-2,5-dione and boil for exactly 90 seconds. Cool the resulting solution immediately with swirling in an ice–salt bath, and scratch the sides of the vessel to induce crystallisation. Leave at 0 °C for 1 hour, filter off the resulting glycylglycine hydrochloride monohydrate with suction using an acid-resistant paper (e.g., Whatman No. 50–54) and wash it with three 15 ml portions of cold ethanol. Suspend the entire product in 600 ml of absolute ethanol in a 3-litre three-necked flask fitted with a sealed stirrer unit, a gas inlet tube and a reflux condenser leading to a gas absorption trap (Fig. II,68(c)). Surround the flask with an ice bath, start the stirrer and pass into the suspension a steady stream of dry hydrogen chloride (Section II,2,31) until about 60 g have been absorbed. Replace the cooling bath with a heating mantle and boil the mixture under reflux for 10 minutes. Cool the flask in an ice bath while still stirring and keep at 0 °C for 1 hour. Collect the crystals by filtration and wash them with 12 ml of cold ethanol. Recrystallise the product from about 290 ml of absolute ethanol, filter, wash the crystals with a little cold ethanol and dry in the air. The yield of glycylglycine ethyl ester hydrochloride, m.p. 183 °C, is 30 g (61%).

VI,27. QUINOLINE

$$CH_2OH\cdot CHOH\cdot CH_2OH \xrightarrow[-2H_2O]{H_2SO_4;} CH_2=CH\cdot CHO$$

This preparation should be carried out in a fume cupboard.

Equip a 250-ml three-necked flask with a double surface condenser, a sealed stirrer unit and a screw-capped adapter carrying a thermometer positioned so that subsequently the temperature of the reaction mixture may be noted. Place 10.0 g (9.8 ml, 0.107 mol) of pure aniline, 15.0 g (0.163 mol) of glycerol (1) and 0.5 g of iodine in the flask. Stir the reaction mixture and add down the condenser from a dropping funnel 30 g (16.4 ml, 0.306 mol) of concentrated sulphuric acid. Reaction soon commences, the temperature rises to 100–105 °C. Heat the flask gradually, with stirring, in an air bath or oil bath to 140 °C; the reaction proceeds with the evolution of sulphur dioxide and a little iodine vapour and the liquid refluxes. Continue heating at 170 °C for 1 hour, allow to cool and then add cautiously with stirring sufficient 5 M-sodium hydroxide solution (about 85 ml) to render the mixture alkaline. Rearrange the apparatus for steam distillation (Fig. I,101) and steam distil until no more oily drops pass over. The distillate contains quinoline and a little aniline. Extract the distillate with three 25 ml portions of ether, combine the ethereal extracts and remove the ether on a rotary evaporator.

To remove the aniline present in the residual crude quinoline, advantage is taken of the fact that bis-quinolinium tetrachlorozincate(II) $[(C_9H_8N)_2^{\oplus} (ZnCl_4)^{2\ominus}]$ is almost insoluble in water and crystallises out, whilst under the same experimental conditions, bis-anilinium tetrachlorozincate(II) $[(C_6H_5 \cdot NH_3)_2^{\oplus}(ZnCl_4)^{2\ominus}]$ remains in solution (2). Dissolve the crude quinoline in 100 ml of dilute hydrochloric acid (1 : 4 by volume), warm the solution to 60 °C and add, with stirring, a solution of 13 g (0.095 mol) of zinc chloride in a 22 ml portion of the diluted hydrochloric acid. Cool the well-stirred mixture thoroughly in ice-water and, when crystallisation is complete, filter the bis-quinolinium tetrachlorozincate(II) with suction, wash with two 10 ml portions of dilute hydrochloric acid and drain well. Transfer the solid to a 250-ml beaker, add a little water and then 10 per cent sodium hydroxide solution until the initial precipitate of zinc hydroxide dissolves completely. Extract the quinoline with three 25 ml portions of ether in a separatory funnel and dry the combined ethereal extracts with anhydrous calcium sulphate. Remove the ether by flash distillation using a 10-ml flask and finally distil the residue from an air bath using an air condenser. Collect the quinoline at 236–238 °C as a colourless liquid; the yield is 6.9 g (50%). If distilled under reduced pressure quinoline has b.p. 118–120 °C/20 mmHg.

Notes. (1) Laboratory grade glycerol may be used since anhydrous glycerol is not necessary when iodine is used as the oxidising agent. Anhydrous glycerol may be prepared by heating commercial glycerol in a porcelain evaporating dish carefully over a wire gauze (preferably in a fume cupboard), stirring it steadily with a thermometer until the temperature rises to 180 °C, allowing it to cool to about 100 °C, pouring it into a Pyrex beaker and transferring the beaker to a large desiccator containing concentrated sulphuric acid. It must be remembered that glycerol is a very hygroscopic substance.

(2) *An alternative method of removing the aniline* is to add 8 ml of concentrated sulphuric acid carefully to the steam distillate, cool the solution to 0–5 °C and add a concentrated solution of sodium nitrite until a drop of the reaction mixture colours potassium iodide–starch paper a deep blue instantly. As the diazotisation approaches completion, the reaction becomes slow; it will therefore be necessary to test for excess of nitrous acid after an interval of 5 minutes, stirring all the while. About 3 g of sodium nitrite are usually required. The diazotised solution is then heated on a boiling water bath for an hour (or until active evolution of nitrogen ceases), treated with a solution of 15 g of sodium hydroxide in 50 ml of water, the mixture steam-distilled and the quinoline isolated from the distillate by extraction with ether as above.

Cognate preparations **8-Nitroquinoline.** Place a mixture of 69 g (0.5 mol) of o-nitroaniline, 86 g (0.375 mol) of arsenic pentoxide and 184 g (2 mol) of *anhydrous* glycerol in a 500-ml three-necked flask, fitted with a sealed stirrer unit, a thermometer and a reflux condenser. Set the stirrer in motion, heat to 100 °C (oil bath) and add 220 g (120 ml, 2.24 mol) of concentrated sulphuric acid gradually down the condenser at such a rate that the temperature does not rise above 120 °C (about 20 minutes). Insert a calcium chloride guard-tube into the top of the condenser, gradually raise the temperature to 130–135 °C and maintain this temperature for 7–8 hours. Watch the reaction during the first hour of heating: should the reaction become very vigorous, lower the oil bath momentarily. Allow the contents of the flask to cool and pour into 1500 ml of water contained in a 2-litre beaker. Add 15 g of decolourising carbon, stir mechanically, heat at 90 °C for 1 hour and filter. Neutralise the cold filtrate slowly with aqueous ammonia solution (1 vol concentrated ammonia, d 0.88 + 1 vol water), filter off the crude nitro compound at the pump and wash with a little water. Recrystallise from hot water or from methanol. The yield of 8-nitroquinoline, m.p. 92 °C, is 45 g (52%).

8-Hydroxyquinoline ('oxine'). Place 170 ml (3.18 mol) of concentrated sulphuric acid in a 1-litre three-necked flask provided with a stirrer, and add 112.5 g (1.03 mol) of *o*-aminophenol, followed by 287 g (3.12 mol) of anhydrous glycerol: maintain the temperature below 80 °C by cooling, if necessary. Keep the mixture in a fluid state by placing the flask on a steam bath.

In a 3-litre three-necked flask, fitted with a thermometer, stirrer and reflux condenser, place 72.5 g (0.52 mol) of *o*-nitrophenol and 10 g of crystallised iron(II) sulphate, and heat to 100–120 °C. Add the liquid amine–glycerol–sulphuric acid pre-mix in about ten portions over 2 hours: allow the reaction to proceed at 135–150 °C before adding the subsequent portions. Reflux the mixture for a further 4 hours, during which time the temperature drops to about 130 °C. Neutralise the cooled reaction mixture with sodium hydroxide solution (250 g in 250 ml of water) with rapid stirring and addition of ice so that the temperature does not rise above 40 °C. The pH of the resulting solution is about 7, and the 8-hydroxyquinoline together with tarry by-products precipitates. Filter the precipitate at the pump, dry at 50–60 °C and then distil under reduced pressure from a flask through a fractionating side arm. A little water passes over first and this is followed by 8-hydroxyquinoline at 100–110 °C/ 5 mmHg. It crystallises on cooling to a white solid, m.p. 74–75 °C. The yield of 'oxine' is 140 g (94%).

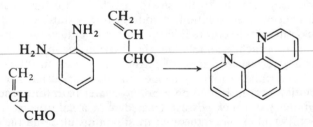

In a 2-litre round-bottomed flask, equipped with a reflux condenser, place 20 g (0.185 mol) of pure *o*-phenylenediamine (Section **IV,56**), 108 g (1.17 mol) of glycerol, 100 ml of arsenic acid solution (1) and 400 ml of dilute sulphuric acid (2). Reflux the mixture for 3.5 hours, and allow to cool. Transfer to a 3-litre beaker, dilute with an equal volume of water, render alkaline with concentrated ammonia solution, *d* 0.88 (about 1050 ml are required), and allow to stand overnight. Remove the tar by filtration through a fluted filter-paper, and subject the filtrate to continuous extraction with benzene (Fig. I, 92) overnight. Transfer the tar plus filter-paper to a 500-ml round-bottomed flask and reflux it for 20 minutes with 200 ml of benzene (3). Remove the benzene layer and repeat the extraction with 100–200 ml portions of benzene until the extracts are colourless: the total volume of benzene required is about 800 ml. Combine all the benzene extracts in a 3-litre round-bottomed flask, and distil off the benzene under reduced pressure (**CAUTION**). The product sets on cooling in ice to a dark solid mass of crude 1,10-(or *o*-)phenanthroline weighing 19 g. Reflux the crude *o*-phenanthroline for 5 minutes with 100 ml of benzene and 2 g of decolourising carbon, filter through a hot water funnel and allow the benzene solution to cool. Collect the pale brown *o*-phenanthroline (4 g) which separates, and dry it upon filter-paper in the air; this is the monohydrate and melts at 108–110 °C. Evaporate the mother-liquor to dryness and recrystallise the residue twice from 30 ml of moist boiling benzene: a further 6 g (total yield 10 g, 27%) of pure *o*-phenanthroline monohydrate is isolated. The compound may also be recrystallised from benzene–light petroleum (b.p. 40–60 °C), but this medium yields a somewhat more coloured product.

Notes. (1) The arsenic acid solution is prepared by dissolving 123 g of arsenic pentoxide (Section **I,3,D,***l*) in 104 ml of water; 100 ml of the cold solution are used in the experiment.

(2) The dilute sulphuric acid is prepared by adding 240 ml of concentrated sulphuric acid slowly and with stirring to 200 ml of water; 400 ml of the cold, diluted acid are employed in the preparation.

(3) All the extractions with benzene and the subsequent filtration of benzene solution must be carried out in an efficient fume cupboard.

VI,29. 2-METHYLQUINOLINE (*Quinaldine*)

$$2CH_3 \cdot CHO \xrightarrow{\ H^\oplus\ } CH_3 \cdot CH = CH \cdot CHO + H_2O$$

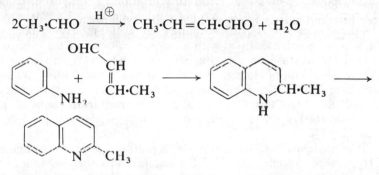

Place 62 g (61 ml, 0.67 mol) of aniline in a 1-litre round-bottomed flask fitted with a reflux condenser the top of which is connected to a gas absorption trap (Fig. I,68(*a*)). Cool the flask in an ice bath, add 120 ml of concentrated hydrochloric acid slowly, followed by 90 g (2.04 mol) of paraldehyde: swirl the contents of the flask to ensure thorough mixing. Remove the flask from the ice bath and shake it frequently at room temperature during 1–2 hours. Heat cautiously to the boiling point: keep an ice-water bath at hand in case the reaction should become unduly vigorous and require moderating. Reflux the mixture for 3 hours and allow to cool. Render alkaline with about 100 ml of 12 *M*-sodium hydroxide solution and steam distil the mixture: collect about 2.4 litres of distillate. Separate the upper oily layer, extract the aqueous phase with a little chloroform (or with ether) and combine the extract with the crude oil. Dry the combined oil and extract with magnesium sulphate, remove the solvent and heat the residue under reflux for 20 minutes with 20 ml of acetic anhydride. After cooling, render alkaline with sodium carbonate solution and steam distil; collect about 2.4 litres of distillate. Extract the latter with two 50 ml portions of toluene. Distil off the toluene from the combined toluene extracts and distil the residue with the aid of an air bath. Collect the pure quinaldine at 245–248 °C: the yield is 40 g (42%). Alternatively, distil the quinaldine under reduced pressure; b.p. 116–118 °C/12 mmHg. Keep the colourless liquid in a well-stoppered bottle since it darkens on exposure to air.

VI,30. 2,4,6-TRIMETHYLQUINOLINE

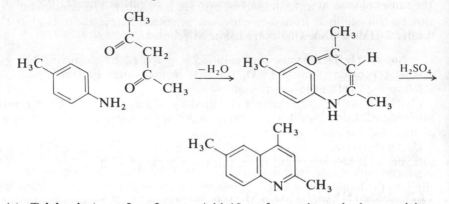

4-(p-Tolylamino)pent-3-en-2-one. Add 10 g of granular anhydrous calcium sulphate to a mixture of 5.32 g (0.05 mol) of p-toluidine and 5.1 g (0.05 mol) of pentane-2,4-dione (Section **III,102**) contained in a 100-ml round-bottomed flask. Attach an air condenser fitted with a calcium chloride guard-tube to the flask and heat the mixture on a steam bath for 1 hour (1), with occasional shaking. Cool, add 40 ml of ether to the reaction mixture and filter. Wash the calcium sulphate in the filter funnel with 40 ml of ether and evaporate the combined ether filtrates. Crystallise the solid so obtained from about 25 ml of hexane. The enamine is obtained as deep straw-coloured plates, m.p. 67–69 °C. The yield is 7.2 g (76%).

2,4,6-Trimethylquinoline. Add 6 g (0.032 mol) of the above enamine, in portions, to 25 ml of concentrated sulphuric acid (d 1.84) contained in a 250-ml conical flask. Swirl the mixture occasionally to ensure thorough mixing. The first portions of the enamine dissolve rather slowly but solution occurs more rapidly with the later portions as the temperature of the mixture increases to 60–70 °C. Heat the reaction mixture on the steam bath for 30 minutes, and then cool the brown solution to room temperature and add it slowly to 250 ml of ice-water in a 1-litre beaker. Add solid sodium carbonate to the solution until it is alkaline. During this addition the quinoline salt tends to separate and the whole mixture may solidify; if this happens the mass should be broken up and stirred with a stout glass rod or spatula while the sodium carbonate is added. The quinoline eventually separates as an oil from the alkaline solution. Cool the mixture in an ice-water bath until the quinoline solidifies; avoid over-cooling as this will result in the separation of a large amount of solid hydrated sodium sulphate. Collect the quinoline by filtration and wash with a little cold water. Dissolve the product in about 20 ml of hot ethanol, add decolourising carbon, filter and add warm water (ca. 20 ml) to the solution until it becomes slightly cloudy. Add a few drops of ethanol to remove the cloudiness and allow the solution to cool slowly to room temperature and finally chill in the refrigerator. 2,4,6-Trimethylquinoline dihydrate separates as long glistening white needles, m.p. 63–65 °C. The yield is 5.2 g (79%).

Note. (1) The progress of the reaction may be followed by removing samples of the reaction mixture and noting the disappearance of either the pentanedione or *p*-toluidine on GLC. Using a 5-ft column of 10 per cent Silicone oil on Chromosorb W at 100 °C, with a nitrogen flow rate of 40 ml/minute, the retention times are 1.3 minutes and 6.5 minutes respectively. On the same column, at 170 °C, the enamine has a retention time of 10.5 minutes.

VI,31. 2-HYDROXY-4-METHYLQUINOLINE (*4-Methylcarbostyril*)

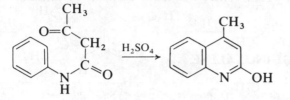

Place 95 ml of concentrated sulphuric acid in a 500-ml three-necked flask equipped with a mechanical stirrer and a thermometer; the thermometer must dip into the liquid. Have a bath of cold water at hand so that the reaction flask can be cooled rapidly, if required. Heat the acid to 75 °C, remove the source of heat, stir and add 89 g (0.5 mol) of acetoacetanilide in portions by means of a spatula. Maintain the temperature of the reaction mixture at 70–75 °C by intermittent cooling until nearly all the acetoacetanilide has been introduced: add the last 7–10 g without cooling. The duration of the addition is 25–30 minutes. During the last addition without external cooling, the temperature will rise to about 95 °C and the heat of the reaction will maintain this temperature for about 15 minutes. Keep the reaction mixture at 95 °C for a further 15 minutes by external heating. When the solution has cooled to 60–65 °C, pour it into 2.5 litres of water with vigorous stirring. Cool, collect the product by suction filtration, wash with four 250 ml portions of water, two 125 ml portions of methanol and dry in the air. The crude 4-methylcarbostyril, m.p. 219–221 °C, weighs 70 g (87%). Recrystallise from 95 per cent ethanol (*ca.* 16 ml per gram): the pure product melts at 223–224 °C and the recovery is about 85 per cent.

VI,32. 2-PHENYLQUINOLINE-4-CARBOXYLIC ACID (*Atophan*)

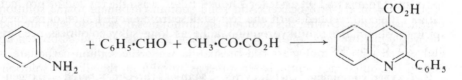

In a 1-litre round-bottomed flask, equipped with a reflux condenser, place 25 g (24 ml, 0.236 mol) of purified benzaldehyde (Section **IV,22**), 22 g (0.25 mol) of freshly distilled pyruvic acid (Section **III,174**) and 200 ml of absolute ethanol. Heat the mixture to the boiling point on a water bath and add slowly, with frequent shaking, a solution of 23 g (22.5 ml, 0.248 mol) of pure aniline in 100 ml of absolute ethanol. The addition usually occupies about 1 hour. Reflux the mixture on a water bath for 3 hours, and allow to stand overnight. Filter off the crude atophan (1) at the pump and wash the crystals with a little ether.

Recrystallise from ethanol (about 20 ml per gram). The yield of pure 2-phenylquinoline-4-carboxylic acid, m.p. 210 °C, is 30 g (51%).

Note. (1) If the atophan does not crystallise—this is rarely the case unless pyruvic acid which has been standing for some time is employed—pour the reaction mixture into a solution of 25 g of potassium hydroxide in 1 litre of water, and extract the resulting solution two or three times with ether. Treat the aqueous layer with 70 ml of glacial acetic acid with vigorous stirring. Allow to stand for several hours and collect the crude atophan by filtration with suction.

VI,33. 2,3-DIPHENYLQUINOXALINE

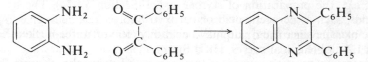

To a warm solution of 2.1 g (0.01 mol) of benzil in 8 ml of rectified spirit add a solution of 1.1 g (0.01 mol) of *o*-phenylenediamine in 8 ml of rectified spirit. Warm in a water bath for 30 minutes, add water until a slight cloudiness persists and allow to cool. Filter and recrystallise from aqueous ethanol to give 1.43 g (51%) of 2,3-diphenylquinoxaline, m.p. 125–126 °C.

Cognate preparations Dibenzo[a,c]phenazine(*1,2:3,4-dibenzophenazine*). Add an ethanolic solution of 77 mg (0.0007 mol) of *o*-phenylenediamine in 5 ml of rectified spirit to a hot solution of 150 mg (0.0007 mol) of 9,10-phenanthraquinone in 5 ml of glacial acetic acid. Immediate precipitation of the product occurs; maintain the solution at 50 °C for 15 minutes. Filter the product and wash well with rectified spirit. Recrystallise from glacial acetic acid to give 190 mg (92%) of pure product, m.p. 221–222 °C.

3-Benzyl-2-oxo-1,2-dihydroquinoxaline. Dissolve 2.7 g (0.025 mol) of *o*-phenylenediamine in 40 ml of hot rectified spirit, boil the solution with 0.5 g of decolourising carbon, filter and cool to room temperature. Add a solution of 4.1 g (0.025 mol) of phenylpyruvic acid (Section **III,175**) in 20 ml of rectified spirit, swirl to mix and set aside for 2 hours. Filter, wash the crystalline product with a little cold rectified spirit and recrystallise from about 100 ml of rectified spirit to obtain the oxodihydroquinoxaline as long silky colourless needles, m.p. 203 °C. The yield is 3.2 g (54%).

B. Oxygen-containing heterocycles. Many three-membered oxygen-containing ring compounds, the oxiranes, may be prepared by the action of alkali on β-halohydrins.

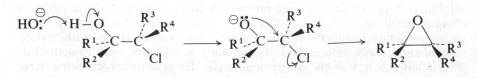

An example of this cyclisation process is to be found in the reaction sequence involved in the Darzens glycidic ester synthesis (p. 422, **III,H,6**).

Alternative routes to the oxiranes include those reactions involving (*a*) the insertion of oxygen into a carbon–carbon double bond, or (*b*) the insertion of a methylene group into the double bond of a carbonyl group.

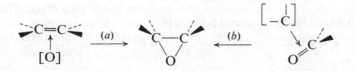

Epoxidation of the double bond (process (*a*)) is achieved most conveniently by employing perbenzoic acid (or *m*-chloroperbenzoic acid) in a solvent such as chloroform, e.g., the preparation of styrene oxide (Section **VI,34**). The use of other peracids, such as peracetic acid or peroxytrifluoroacetic acid, give lower yields of the oxirane since the oxide may be readily cleaved to form the mono-ester of the 1,2-diol (e.g., see p. 379, **III,D,5**).

Carbon insertion reactions (process (*b*)) may be effected with the aid of a methylene transfer reagent, generated by the action of base either on a tri-methyloxosulphonium salt to yield dimethyloxosulphonium methylide (XLV), or on a trimethylsulphonium salt to yield a dimethylsulphonium methylide (XLVI).

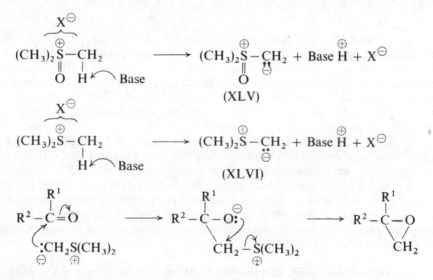

In the original procedure (Ref. 5) the reaction conditions involve the use of a strong base (e.g., sodium hydride, or butyl lithium) in a non-aqueous medium to generate the required ylide. A recent procedure (Ref. 6) for the production of sulphur ylides using an aqueous medium involves phase-transfer catalysis, and offers a useful and convenient alternative method. Here, the carbonyl compound in dichloromethane solution, containing tetrabutylammonium iodide (the phase-transfer catalyst), is stirred with a suspension of trimethylsulphonium iodide in sodium hydroxide solution. The reaction procedure is illustrated by the preparation of styrene oxide (Section **VI,35**) from benzaldehyde. The general applicability of this modification has still to be evaluated.

A simple, general procedure for the synthesis of five-membered oxygen-containing heterocyclic systems (the furans) involves the cyclodehydration of 1,4-diketones by acidic reagents, such as sulphuric acid, zinc chloride, etc. (e.g., the synthesis of 2,5-dimethylfuran, Section **VI,36**).

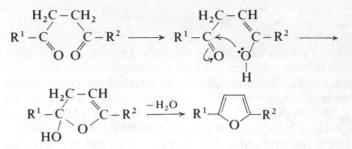

Benzofuran (**XLVII**) may be prepared from salicylaldehyde (Section **VI,37**) by initial conversion into the corresponding aryloxyacetic acid on reaction with sodium chloroacetate in the presence of alkali. Cyclisation of the product in a mixture of acetic anhydride, acetic acid and sodium acetate proceeds by way of an internal Perkin reaction (p. 798, **IV,M,3**) accompanied by a decarboxylative dehydration step.

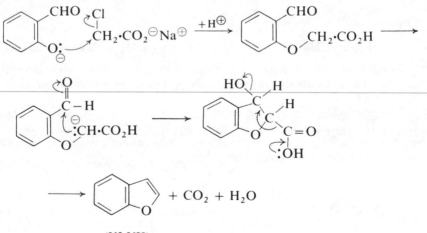

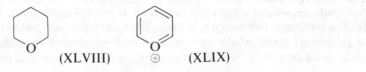

(XLVII)

The saturated oxygen-containing six-membered system, tetrahydropyran (**XLVIII**), is the parent structure for a number of oxygen heterocyclic compounds. Aromatic properties in the oxygen series are apparent when the heterocycle is in the form of the pyrylium salt (**XLIX**) which is well characterised in the form of various benzopyrylium salts (e.g., the anthocyanins).

(XLVIII) $\oplus$ (XLIX)

The keto-derivatives of the pyran system are the 2- and 4-pyrones (L and LI).

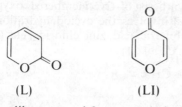

(L)   (LI)

A readily prepared 2-pyrone derivative is coumalic acid, which is formed when malic acid is treated with a concentrated sulphuric acid–fuming sulphuric acid mixture (Section **VI,38**). Decarbonylation and dehydration of the α-hydroxy acid forms formylacetic acid which then undergoes self-condensation.

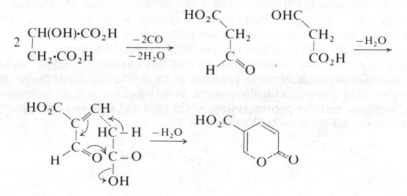

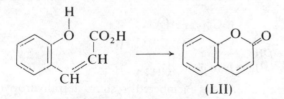

Reaction of suitable 2-pyrone derivatives (e.g., dimethyl coumalate) with ammonia yield pyridine derivatives (cf. the synthesis of 6-hydroxynicotinic acid, Section **VI,18**).

The benzo-2-pyrone (LII) is commonly called coumarin, and is obtained by the Perkin reaction between salicylaldehyde and acetic anhydride in the presence of triethylamine (Section **IV,150**), as the result of spontaneous cyclisation of the intermediate *o*-hydroxycinnamic acid.

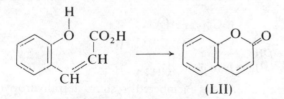

(LII)

A general synthesis of coumarins involves the interaction of a phenol with a β-keto ester (e.g., ethyl acetoacetate) in the presence of acid condensation agents (the **Pechmann reaction**). Concentrated sulphuric acid is generally used for simple monohydric phenols, although phenol itself reacts better in the presence of aluminium chloride (e.g., the synthesis of 4-methylcoumarin, Section **VI,39**). The reaction mechanism is thought to involve the initial formation of a β-hydroxy ester, which then cyclises and dehydrates to yield the coumarin.

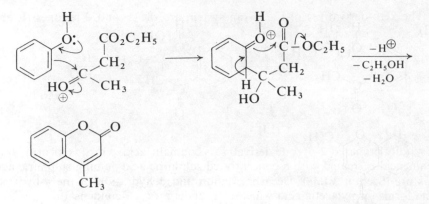

Polyhydric phenols, particularly when two hydroxyl groups are *meta* oriented, react with great ease. If sulphuric acid is used as the condensing agent, careful temperature control is needed to ensure a good yield (e.g., 4-methyl-7-hydroxycoumarin, cognate preparation in Section **VI,39**). In these cases the use of polyphosphoric acid is recommended and this alternative process is also illustrated. Good yields are also obtained from polyhydric phenols by condensation in the presence of trifluoroacetic acid (Ref. 7).

The isomeric benzo-4-pyrone system is chromone (LIII), and the 2-phenyl derivatives are the flavones (LIV).

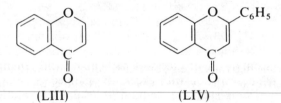

(LIII)   (LIV)

Flavone (Section **VI,40**) may be synthesised by cyclisation under acidic conditions of *o*-hydroxydibenzoylmethane (LVI) which may be obtained by the rearrangement of *o*-benzoyloxyacetophenone (LV) under the influence of base.

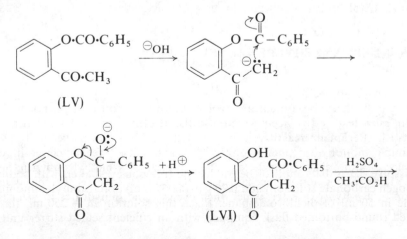

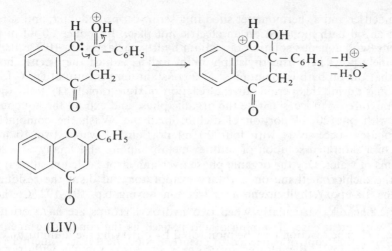

(LIV)

VI,34. 1,2-EPOXYETHYLBENZENE (*Styrene oxide*)

$$C_6H_5 \cdot CH = CH_2 + C_6H_5 \cdot CO_2 \cdot OH \longrightarrow C_6H_5 \cdot CH - CH_2 + C_6H_5 \cdot CO_2H$$

This preparation involving the reaction of perbenzoic acid must be conducted behind a safety screen (see Section **I,3** and Section **II,2,46(d)**).

Add 30 g (0.29 mol) of styrene, b.p. 42–43 °C/18 mmHg, to a solution of 42 g (0.3 mol) of perbenzoic acid (Section **II,2,46(d)**) in 450 ml of chloroform. Keep the solution at 0 °C for 24 hours and shake frequently during the first hour. At the end of 24 hours only a slight excess of perbenzoic acid remains; confirm this by mixing an aliquot portion with excess of acidified potassium iodide solution and titrating with standard sodium thiosulphate solution (Section **II,2,46(d)**). Separate the benzoic acid from the chloroform solution by shaking with an excess of 10 per cent sodium hydroxide solution, remove the residual alkali by washing with water and dry the chloroform solution with magnesium sulphate. Distil with the aid of an efficient fractionating column. After the chloroform has been removed, the styrene oxide passes over at 189–192 °C (or at 101 °C/40 mmHg) as a colourless liquid. The yield is 25 g (72%).

VI,35. 1,2-EPOXYETHYLBENZENE

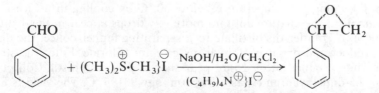

Dissolve 5.3 g (0.05 mol) of benzaldehyde (previously shaken with sodium hydrogen carbonate solution) and 0.25 g (0.00067 mol) of tetrabutylammonium iodide in 50 ml of dichloromethane. Place this solution in a 250-ml, three-necked round-bottomed flask equipped with an efficient sealed stirrer unit, a

reflux condenser and a thermometer sited in a screw-capped adapter, and supported in an oil bath mounted on an electric hot plate. Introduce 50 ml of a 50 per cent (w/v) aqueous solution of sodium hydroxide, and then 10.2 g (0.05 mol) of finely powdered trimethylsulphonium iodide. Adjust the electric hot plate so that the oil bath is maintained at a constant temperature of 55 °C for 60 hours and during this period stir the reaction mixture rapidly (1). Pour the reaction mixture on to ice, separate the organic phase and extract the aqueous solution with one 20 ml portion of dichloromethane. Wash the combined organic phases successively with four 20 ml portions of water, two 10 ml portions of a saturated solution of sodium metabisulphite and finally two 20 ml portions of water. Dry the organic phase over anhydrous calcium sulphate, remove the dichloromethane on a rotary evaporator and distil the residue. Collect the 1,2-epoxyethylbenzene as a fraction having b.p. 191–192 °C; the yield is 4.7 g (78%).

Note. (1) The reaction may be monitored by observing the diminution of carbonyl absorption in the infrared spectrum of successive samples withdrawn from the organic phase, after allowing the dichloromethane to evaporate from a portion placed on a sodium chloride plate. Alternatively samples of the reaction medium may be analysed by GLC using a 10 per cent Carbowax column on Chromosorb W held at 150 °C with a nitrogen carrier gas flow rate of 40 ml/minute; benzaldehyde has t_R 2 minutes and 1,2-epoxyethylbenzene has t_R 2.75 minutes.

VI,36. 2,5-DIMETHYLFURAN

$$H_3C-C\overset{\overset{\displaystyle H_2C-CH_2}{\diagup\;\;\diagdown}}{\underset{O}{}}\;\underset{O}{C}-CH_3 \xrightarrow[\text{ZnCl}_2]{(CH_3\cdot CO)_2O} H_3C\diagdown_O\diagup CH_3$$

Dissolve 0.5 g of anhydrous zinc chloride in 16.8 g (15.5 ml, 0.165 mol) of acetic anhydride contained in a 50-ml round-bottomed flask, and to this solution add 17.1 g (17.6 ml, 0.15 mol) of redistilled hexane-2,5-dione (Section **III,104**). Attach a reflux condenser fitted with a calcium chloride drying tube and warm the mixture carefully. When a vigorous reaction commences remove the source of heat until the reaction subsides and then boil the mixture under gentle reflux for 2 hours. Transfer the cooled dark brown mixture to a 250-ml round-bottomed flask and add 6 *M*-aqueous sodium hydroxide until alkaline (*ca.* 50 ml), keeping the temperature below 50 °C by cooling in an ice-water bath. Steam distil the mixture until no more oily drops appear in the distillate (about 100 ml). Transfer the distillate to a separating funnel, collect the upper layer of product and dry over anhydrous calcium chloride. Filter the straw-coloured crude product into a 25-ml pear-shaped flask and distil. Collect the colourless 2,5-dimethylfuran (1) as a fraction, b.p. 93–95 °C. The yield is 8.9 g (62%).

Note. (1) The purity of the product may be checked by GLC on a 5-ft Silicone oil column held at 105 °C. At this temperature and with a nitrogen flow rate of 40 ml/min, 2,5-dimethylfuran has a retention time of 48 seconds.

VI,37. BENZOFURAN (*Coumarone*)

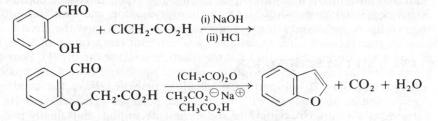

o-**Formylphenoxyacetic acid.** To a mixture of 35 ml (40 g, 0.33 mol) of salicylaldehyde, 31.5 g (0.33 mol) of chloroacetic acid and 250 ml of water contained in a 500-ml, two-necked round-bottomed flask fitted with a stirrer unit, add slowly with stirring a solution of 26.7 g (0.66 mol) of sodium hydroxide in 700 ml of water. Heat the mixture to boiling with stirring and reflux for 3 hours. The solution acquires a red-brown colour. Cool and acidify the solution with 60 ml of concentrated hydrochloric acid and steam distil to remove unreacted salicylaldehyde; 12 ml (14 g) are thus recovered. Cool the residual liquor which first deposits some dark red oil which then solidifies; on standing almost colourless crystals appear in the supernatant solution. Decant the supernatant solution and crystals and filter off the crystals, and air dry; the yield of almost pure product, m.p. 132–133 °C, is 21 g. The solidified red oil may be extracted with small quantities of hot water, the extracts treated with decolourising charcoal and cooled, to yield a further 6 g of product; total yield 27 g (71% calculated on salicylaldehyde consumed in reaction).

Benzofuran. Heat under reflux for 8 hours a mixture of 20 g (0.11 mol) of *o*-formylphenoxyacetic acid, 40 g of anhydrous sodium acetate, 100 ml of acetic anhydride and 100 ml glacial acetic acid. Pour the light brown solution into 600 ml of iced water, and allow to stand for a few hours with occasional stirring to aid the hydrolysis of acetic anhydride. Extract the solution with three 150 ml portions of ether and wash the combined ether extracts with 5 per cent aqueous sodium hydroxide until the aqueous layer is basic; the final basic washing phase acquires a yellow colour. Wash the ether layer with water until the washings are neutral, dry the ethereal solution over anhydrous calcium chloride and remove the ether on a rotary evaporator. Distil the residue and collect the benzofuran as a fraction of b.p. 170–172 °C, the yield of colourless product is 9.5 g (91%).

VI,38. COUMALIC ACID

$$2 \;\; \begin{matrix} CH(OH){\cdot}CO_2H \\ | \\ CH_2{\cdot}CO_2H \end{matrix} \;\; \xrightarrow{H_2SO_4} \qquad \qquad + \; 2CO + 4H_2O$$

Add 180 ml of sulphuric acid to 202 g (1.5 mol) of finely powdered malic acid contained in a 2-litre round-bottomed flask. At intervals of 45 minutes, add three 50 ml portions of fuming sulphuric acid (25% SO_3); a slight exothermic reaction sets in with the steady evolution of gas. Swirl the mixture frequently to obviate excessive foaming, and when the evolution of gas has slackened, heat the reaction mixture on a water bath for 2 hours. Cool the mixture and pour on to 800 g of ice with stirring. Set the mixture aside in a refrigerator for 24 hours and then filter the crude coumalic acid and wash with small portions of iced water. Recrystallise the crude product from methanol to give coumalic acid as light yellow crystals, m.p. 206–208 °C. The yield is 65 g (62%).

VI,39. 4-METHYLCOUMARIN

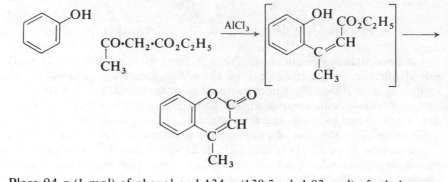

Place 94 g (1 mol) of phenol and 134 g (130.5 ml, 1.03 mol) of ethyl acetoacetate in 150 ml of redistilled nitrobenzene in a 3-litre three-necked flask, fitted with a dropping funnel, a sealed stirrer unit and an air condenser, the open end of which is connected to a gas absorption trap (Fig. I,68). Heat the mixture to 100 °C in an oil bath, stir and add a solution of 266 g (2 mol) of anhydrous aluminium chloride in 1 litre of nitrobenzene (1) from the dropping funnel over a period of 45 minutes. Replace the dropping funnel by a thermometer, raise the temperature of the solution to 130 °C and maintain this temperature, with stirring, for 3 hours, by which time evolution of hydrogen chloride will have almost ceased. Cool the reaction mixture to room temperature and add 250 ml of 1 : 1-hydrochloric acid with stirring in order to decompose the excess of aluminium chloride. Equip the flask for steam distillation, warm it and pass steam into the reaction mixture: this will remove any unchanged ketoester and some of the nitrobenzene: collect about 100 ml of distillate. Transfer the residue in the flask whilst hot to a large separatory funnel; separate and discard the aqueous layer. Filter the organic layer (with the addition of a filteraid, if necessary) through a Buchner or slit-sieve funnel to remove tarry matter and distil under reduced pressure. The nitrobenzene passes over first, followed by crude 4-methylcoumarin at 180–195 °C/15 mmHg (75 g) as a red-yellow oil which solidifies on cooling. Dissolve the crude product in ether, shake the ether solution with small volumes of 5 per cent sodium hydroxide solution until the aqueous layer is colourless, dry, evaporate the ether and recrystallise the residue from a 4 : 1 mixture of light petroleum (b.p. 60–80 °C) and benzene. The resulting 4-methylcoumarin (62 g, 39%) is almost colourless and melts at 83–84 °C.

Note. (1) Add the aluminium chloride in 25 g portions to the 1 litre of dry nitrobenzene contained in a 2.5-litre round-bottomed flask; stir after each addition. The temperature may rise to about 80 °C during the addition: cool the flask occasionally under running water. When all the aluminium chloride has been added, cool the solution to room temperature: a little solid may settle to the bottom.

Cognate preparation 4-Methyl-7-hydroxycoumarin. Place 1 litre of concentrated sulphuric acid in a 3-litre three-necked flask fitted with a thermometer, mechanical stirrer and a dropping funnel. Immerse the flask in an ice bath. When the temperature falls below 10 °C, add a solution of 100 g (0.91 mol) of resorcinol in 134 g (130.5 ml, 1.03 mol) of redistilled ethyl acetoacetate dropwise and with stirring. Maintain the temperature below 10 °C by means of an ice–salt bath during the addition (*ca.* 2 hours). Keep the reaction mixture at room temperature for about 18 hours, then pour it with vigorous stirring into a mixture of 2 kg of crushed ice and 3 litres of water. Collect the precipitate by suction filtration and wash it with three 25 ml portions of cold water. Dissolve the solid in 1500 ml of 5 per cent sodium hydroxide solution, filter and add dilute 2 *M*-sulphuric acid (about 550 ml) with vigorous stirring until the solution is acid to litmus. Collect the crude 4-methyl-7-hydroxycoumarin by filtration at the pump, wash it with four 25 ml portions of cold water and dry at 100 °C: the yield is 155 g (97%). Recrystallise from 95 per cent ethanol: the pure compound separates in colourless needles, m.p. 185 °C.

 Polyphosphoric acid procedure. Add 160 g of polyphosphoric acid (Section **II,2,**48) to a solution of 11 g (0.1 mol) of resorcinol in 13 g (0.1 mol) of ethyl acetoacetate. Stir the mixture and heat at 75–80 °C for 20 minutes, and then pour into ice-water. Collect the pale yellow solid by suction filtration, wash with a little cold water and dry at 60 °C. The yield of crude 4-methyl-7-hydroxycoumarin, m.p. 178–181 °C, is 17 g (97%). Recrystallisation from dilute ethanol yields the pure, colourless compound, m.p. 185 °C.

VI,40. FLAVONE (*2-Phenylbenzo-4-pyrone*)

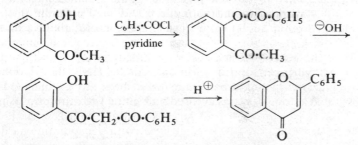

 o-**Benzoyloxyacetophenone.** Place 34 g (0.25 mol) of *o*-hydroxyacetophenone (Section **IV,120**) in a flask and add 49 g (40 ml, 0.35 mol) of benzoyl chloride and 50 ml of dry, redistilled pyridine. Shake to mix the contents, which become warm. After 20 minutes, pour the reaction mixture with stirring into 1200 ml of 1 *M*-hydrochloric acid containing 500 g of crushed ice. Filter off the product with suction and wash it with 50 ml of ice-cold methanol and then with 50 ml of water. Recrystallise from methanol (60 ml), cooling thoroughly in ice before collecting the purified product by filtration. The yield is 48 g (80%), m.p. 87–88 °C.

 o-**Hydroxydibenzoylmethane.** Dissolve 48 g (0.2 mol) of benzoyloxyacetophenone in 180 ml of dry pyridine in a 1-litre bolt-necked flask and heat the solution to 50 °C. Add with mechanical stirring 17 g (0.3 mol) of potassium hydroxide which has been powdered rapidly in a mortar preheated in an oven at 100 °C. Continue to stir for 15 minutes; if the separation of the yellow potassium salt of the product makes mechanical stirring impossible, stir by hand. Cool the reaction mixture to room temperature and acidify it by adding with stirring 250 ml of 10 per cent aqueous acetic acid. Collect the pale yellow precipitate by suction filtration and dry it in an oven at 50 °C. The yield of *o*-hydroxydibenzoylmethane, m.p. 117–120 °C, which is sufficiently pure for use in the next stage, is 38.5 g (80%). The pure product has m.p. 121 °C after crystallisation from methanol.

Flavone. Dissolve 36 g (0.15 mol) of *o*-hydroxydibenzoylmethane in 200 ml of glacial acetic acid in a 500-ml flask and add with shaking 8 ml of concentrated sulphuric acid. Attach a reflux condenser and heat the mixture on a boiling water bath with intermittent shaking for 1 hour. Pour the reaction mixture with stirring on to about 1 kg of crushed ice, and allow the ice to melt. Filter off the flavone which has separated, wash it with water until the washings are no longer acidic (about 2 litres are required) and dry in an oven at 50 °C. The yield of material having m.p. 95–97 °C is 31.5 (95%). Recrystallisation of a portion from a large volume of light petroleum (b.p. 60–80 °C) gives pure flavone, m.p. 98 °C, as tufts of white needles.

C. Sulphur-containing heterocycles. Simple 2,5-dialkylthiophens are conveniently prepared by treating a 1,4-dicarbonyl compound with phosphorus pentasulphide.

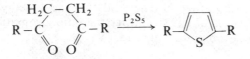

The general applicability of this method is to some extent limited by the availability of the starting materials; it is illustrated by the synthesis of 2,5-dimethylthiophen (Section **VI,41**) from hexane-2,5-dione (cf. the synthesis of pyrroles, p. 875, **VI,A,2**).

A synthesis of the benzothiophen system is illustrated by the preparation of 5-chloro-3-methylbenzothiophen (Section **VI,42**). Here the thiophenol is allowed to react with α-chloroacetone under basic conditions and the resulting substituted ketone (LVII) is cyclised on heating with phosphorus pentoxide.

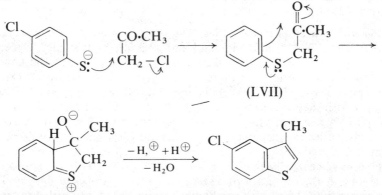

(LVII)

The reaction of an α-halocarbonyl compound with a thioamide or thiourea provides a useful route to the thiazole system. The reaction may be formulated in general terms in the following way.

926

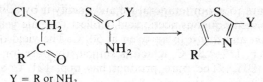

Y = R or NH₂

Thus α-chloroacetone with thioacetamide (Y = CH₃) gives 2,4-dimethyl-thiazole (Section **VI,43**), or with thiourea (Y = NH₂) gives 2-amino-4-methylthiazole (Section **VI,44**). α-Chloroacetaldehyde for use in this type of reaction may be generated *in situ* from α,β-dichloroethyl ethyl ether, and condensation with thiourea gives 2-aminothiazole (Section **VI,45**).

VI,41. 2,5-DIMETHYLTHIOPHEN

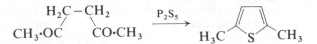

This preparation should be carried out in an efficient fume cupboard.

Place 55 g (0.25 mol) of phosphorus pentasulphide in a 500-ml three-necked flask equipped with a long Leibig-type condenser, a dropping funnel and a stopper. Add from the dropping funnel 23 g (24 ml) of acetonylacetone (Section **III,104**) and heat the reaction mixture with a small luminous flame until the exothermic reaction commences. Remove the flame and add dropwise 91 g (94 ml, total 1 mol) of acetonylacetone over a period of about 45 minutes so that the mixture refluxes gently. A bath of cold water should be to hand should the reaction show signs of becoming too vigorous. When the addition is complete, and the reaction has subsided, heat the mixture under reflux for a further 1 hour, cool to room temperature and pour into 200 ml of ice-water (*FUME CUPBOARD*). Extract the aqueous mixture with four 40 ml portions of ether and wash the combined ethereal extracts with aqueous sodium carbonate solution and then with water. Dry the ether solution over magnesium sulphate, remove the ether by flash distillation and distil the residue. 2,5-Dimethyl-thiophen is obtained after redistillation as a very unpleasant smelling colourless liquid, b.p. 135–136 °C; the yield is 52 g (46%).

VI,42. 5-CHLORO-3-METHYLBENZOTHIOPHEN

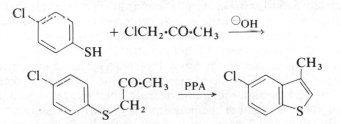

This preparation should be carried out in an efficient fume cupboard.

(**p-Chlorophenylthio)acetone.** To a solution of 5.6 g (0.14 mol) of sodium hydroxide in 200 ml of water in a 500-ml conical flask equipped with a stirrer, add 20.2 g (0.14 mol) of p-chlorothiophenol followed by 13.0 g (0.14 mol) of chloroacetone (1). Stir the mixture at room temperature for about 45 minutes and then extract with ether. Wash the combined ether extracts with water, dry over magnesium sulphate and evaporate the ether on the rotary evaporator. Distil the residue under reduced pressure and collect (p-chlorophenylthio)-acetone as a fraction of b.p. 180–183 °C/16 mmHg; the yield is 23 g (82%).

5-Chloro-3-methylbenzothiophen.   Heat 10 g (0.05 mol) of (p-chlorophenyl-thio)acetone and 100 g of polyphosphoric acid (Section **II,2,**48) in a 250-ml flask on an oil bath maintained at 120–140 °C. When the contents of the flask initially reach the reaction temperature swirl vigorously to ensure thorough mixing of the reactants. After 2 hours allow the reaction mixture to cool and add 100 ml of water. Extract the organic product with ether, wash the combined extracts with water and dry over magnesium sulphate. Distil the residue under reduced pressure and collect 5-chloro-3-methylbenzothiophen as a fraction of b.p. 98–100 °C/0.1 mmHg. The yield is 4.7 g (52%).

Note.   (1) See Section **VI,43,** Note (1); the compound is lachrymatory.

VI,43. 2,4-DIMETHYLTHIAZOLE

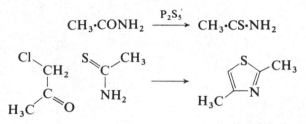

$$CH_3 \cdot CONH_2 \xrightarrow{P_2S_5} CH_3 \cdot CS \cdot NH_2$$

Equip a 1-litre, two-necked round-bottomed flask with a reflux condenser and a dropping funnel. Prepare a mixture of 150 g (2.54 mol) of finely powdered acetamide and 100 g of powdered phosphorus pentasulphide quickly, transfer it rapidly into the flask and immediately add 100 ml of dry benzene. Set up the apparatus in a fume cupboard. Prepare a mixture of 232 g (200 ml, 2.5 mol) of chloroacetone (1) (**CAUTION**—the compound is lachrymatory) and 75 ml of dry benzene; place it in the dropping funnel and insert a calcium chloride drying tube in the mouth. Add about 10 ml of the chloroacetone–benzene mixture to the contents of the flask and warm gently on a water bath: remove the water bath immediately the exothermic reaction commences. Introduce the remainder of the chloroacetone in *ca.* 10 ml portions at such intervals that the reaction is under control. When all the chloroacetone has been added, reflux the mixture on a water bath for 30 minutes. Then add 400 ml of water to the reaction mixture with shaking; after 20 minutes, transfer the contents of the flask to a separatory funnel, run off the lower layer into a beaker and discard the reddish upper layer containing the benzene. Make the lower layer alkaline by the addition of 20 per cent sodium hydroxide solution: test the highly coloured aqueous solution (and not the dark dimethylthiazole floating on top of the liquid) with universal indicator paper. Separate the black upper layer of crude dimethylthiazole with 50 ml of ether, and extract the aqueous layer with five 60 ml portions of ether. Dry the combined ethereal extracts over magnesium sulphate, and filter through glass wool. Remove the ether by flash distillation through a short fractionating column; insert a calcium chloride drying tube into the dropping funnel since the thiazole is hygroscopic and fractionate the residue. Collect the fraction boiling at 140–150 °C and redistil. The yield of 2,4-dimethylthiazole, b.p. 143–145 °C, is 115 g (40%).

Note. (1) Distil and store commercial chloroacetone, b.p. 118–120 °C over calcium carbonate. It is prepared *inter alia* by the chlorination of acetone in the cold.

VI,44. 2-AMINO-4-METHYLTHIAZOLE

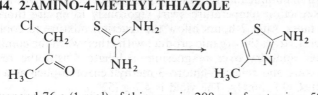

Suspend 76 g (1 mol) of thiourea in 200 ml of water in a 500-ml three-necked flask sited in a fume cupboard and equipped with a sealed stirrer unit, a reflux condenser and a dropping funnel. Stir and add 92.5 g (80 ml, 1 mol) of chloroacetone (1) over a period of 30 minutes. The thiourea dissolves as the reaction proceeds and the temperature rises. Reflux the yellow solution for 2 hours. To the cold solution immersed in an ice bath add, with stirring, 200 g of solid sodium hydroxide. Transfer to a separatory funnel, add a little ice-water, separate the upper oil layer and extract the aqueous layer with three 100 ml portions of ether. Dry the combined oil and ether extracts with anhydrous sodium sulphate, remove the ether using a rotary evaporator and distil the residual oil under diminished pressure. Collect the 2-amino-4-methylthiazole at 130–133 °C/18 mmHg; it solidifies on cooling in ice to a solid, m.p. 44–45 °C. The yield is 84 g (74%).

Note. (1) See Section **VI,43**, Note (1).

VI,45. 2-AMINOTHIAZOLE

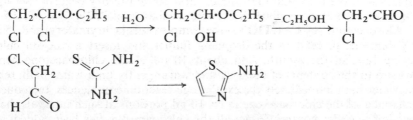

Place a solution of 76 g (1 mol) of thiourea in 200 ml of warm water in a 500-ml three-necked flask equipped with a dropping funnel, sealed mechanical stirrer and reflux condenser. Add 143 g (122 ml, 1 mol) of α,β-dichloroethyl ethyl ether (1) and heat the mixture under gentle reflux with stirring for 2 hours. As the reaction proceeds, the two layers gradually merge. To the cold solution add sufficient solid sodium hydroxide to liberate the 2-aminothiazole from its salt. Add ether to dissolve the product, dry the ethereal extract with anhydrous sodium sulphate and evaporate the ether. Recrystallise the crude 2-amino-thiazole from ethanol; the resulting yellow crystalline solid has m.p. 90 °C. The yield is 80 g (80%).

Note. (1) Available from Fluka AG.

REFERENCES

VI Heterocyclic compounds

1. K. N. Campbell, B. K. Campbell, J. F. McKenna and E. P. Chaput. *J. Org. Chem.*, 1943, **8**, 103; 1944, **9**, 184.
2. H. R. Henze and R. J. Speer (1942). *J. Amer. Chem. Soc.*, **64**, 522.
3. C. F. H. Allen and J. VanAllan. *Organic Syntheses*, Coll. Vol. III, 597.
4. A. Singer and S. M. McElvain. *Organic Syntheses*, Coll. Vol. II, 214.
5. E. J. Corey and M. Chaykovsky (1965). *J. Amer. Chem. Soc.*, **87**, 1353.
6. H. Merz and G. Markl (1973). *Angew. Chem., Internat. Edn.*, **12**, 845.
7. L. L. Woods and J. Sapp (1962). *J. Org. Chem.*, **27**, 3707.

CHAPTER VII QUALITATIVE ORGANIC ANALYSIS

The organic chemist is frequently faced with the problem of characterising and identifying unknown organic compounds. The worker in the field of natural products, for example, has the prospect of isolating such compounds from their sources in a pure state and then of elucidating their structures. The preparative organic chemist may encounter new, or unexpected, compounds in the course of investigations into the applicability of new reagents or techniques, or as by-products of established reactions.

In each of these instances the elucidation of the structure requires the identification of the molecular framework, the nature of the functional groups which are present and their location within the skeletal structure, and finally the establishment of any stereochemical relationships which may exist. Introduction into this area is provided by a study of qualitative organic analysis, which is an essential part of the training of the organic chemist. The purpose of this chapter is to suggest a possible general scheme which might be adopted. The approach cannot be as systematic as that frequently adopted in qualitative inorganic analysis. Each problem is unique and any scheme of analysis may require considerable modification. It is this balance between the structured systematic approach and the intuition which the organic chemist acquires with experience which the student should aim to achieve.

1. **Determination of the physical constants and the establishment of the purity of the compound.** For a solid, the melting point is of great importance: if recrystallisation does not alter it, the compound may be regarded as pure. Further confirmation of purity may be obtained by TLC analysis (Section I,31). For a liquid, the boiling point is first determined: if most of it distils over a narrow range (say, 1–2 °C), it is reasonably pure. It may be desirable to submit a sample of the liquid to GLC analysis to establish homogeneity (Section I,31). The refractive index and the density are also useful constants for liquids.

2. **Qualitative analysis for the elements.** This includes an examination of the effect of heat upon the substance—a test which *inter alia* will indicate the presence of inorganic elements—and qualitative analysis for nitrogen, halogen and sulphur and, if necessary, other inorganic elements. It is clear that the presence or absence of any or all of these elements would immediately exclude from consideration certain classes of organic compounds.

3. **Study of the solubility behaviour of the compound.** A semi-quantitative study of the solubility of the substance in a limited number of solvents (water, ether, dilute sodium hydroxide solution, sodium hydrogen carbonate solution,

dilute hydrochloric acid, concentrated sulphuric acid and phosphoric acid) will, if intelligently applied, provide valuable information as to the presence or absence of certain classes of organic compounds.

4. **Interpretation of spectra.** Spectroscopic methods, comprising ultraviolet and visible spectroscopy, infrared absorption spectroscopy, nuclear magnetic resonance spectroscopy and mass spectrometry, have greatly simplified the identification of organic compounds, particularly in establishing the presence of functional groups. These methods are relatively rapid and require only very small amounts of material.

5. **Application of class reactions.** The application of selected reactions that indicate the presence or absence of certain functional groups, with due regard to the indications provided under sub-sections 1, 2, 3 and 4 above, will locate the class (or classes) to which the compound belongs or will, at least, serve to eliminate all but a few classes to which the compound can be assigned.

6. **Location of the compound within a class (or homologous series) of compounds.** Reference to the literature or to tables of the physical properties of the class (or classes) of organic compounds to which the substance has been assigned will generally locate a number of compounds which boil or melt within 5 °C of the value observed for the unknown. If other physical properties (e.g., refractive index and density for a liquid) are available, these will assist in deciding whether the unknown is identical with one of the known compounds. In general, however, it is more convenient in practice to prepare one, but preferably two, crystalline derivatives of the substance.

7. **Preparation of derivatives.** If two distinct crystalline derivatives of the unknown have the same melting point (or other physical properties) as those of the compound described in the literature (or in the tables), the identity of the two compounds may be assumed. Further confirmation may be obtained, if required, by mixed melting point determinations (see Section **I,34**).

VII,1. DETERMINATION OF PHYSICAL CONSTANTS The most widely used physical constants in the characterisation of organic compounds are melting points and boiling points. The technique of the **determination of melting points** is discussed in detail in Section **I,34**. In general, a sharp melting point (say, within 0.5 °C) is one of the most characteristic properties of a pure organic compound. The purity should not, however, be assumed but must be established by observation of any changes in the melting point (or in the melting range) when the compound is subjected to purification by recrystallisation (the technique of recrystallisation is discussed fully in Section **I,20**). If the melting point is unaffected by at least one recrystallisation, then the purity of the substance may be regarded as established. In some cases purification may be effected by sublimation at atmospheric or under reduced pressure (see Section **I,21**).

The various methods available for the experimental **determination of the boiling point** of a liquid are discussed in Section **I,35** for sample sizes ranging down to less than 1 ml. If the liquid is shown by distillation to have a wide boiling range (10–30 °C), it will be necessary to subject it to fractional distillation in order to obtain a reasonably pure sample of the compound; suitable procedures are to be found in Section **I,26**.

The **determination of the refractive index** is useful in assisting the characterisation of a pure liquid, particularly in the case of compounds which do not possess functional groups suitable for conversion into solid derivatives. The refractive index is conveniently determined with an Abbé refractometer (Section **I,38**). An additional criterion of identity is provided by the **density** of the liquid, which is determined with the aid of a pycnometer (Section **I,38**).

VII,2. QUALITATIVE ANALYSIS FOR THE ELEMENTS The most commonly occurring elements in organic compounds are carbon, hydrogen, oxygen, nitrogen, sulphur and the halogen elements; less common elements are phosphorus, arsenic, antimony, mercury or other metals which may be present as salts of organic acids. There is no direct method for the detection of oxygen.

It is usually advisable to carry out the ignition test first. This will provide useful information as to the general properties of the compound and, in particular, the residue may be employed for the detection of any inorganic elements which may be present.

Ignition test Place about 0.1 g of the compound in a porcelain crucible or crucible cover. Heat it gently at first and finally to dull redness. Observe:

(a) Whether the substance melts, is explosive or is inflammable and note the nature of the flame.
(b) Whether gases or vapour are evolved, and their odour (**CAUTION!**).
(c) Whether the residue fuses.

If an appreciable amount of residue remains, note its colour. Add a few drops of water and test the solution (or suspension) with Universal indicator paper. Then add a little dilute hydrochloric acid and observe whether effervescence occurs and the residue dissolves. Apply a flame test with a platinum wire on the hydrochloric acid solution to determine the metal present. (In rare cases, it may be necessary to subject a solution of the residue to the methods of qualitative inorganic analysis to identify the metal or metals present.) If the flame test indicates sodium, repeat the ignition of the substance on platinum foil.

Heating with soda lime is often a useful preliminary test. Mix thoroughly about 0.2 g of the substance with about 1 g of powdered soda lime. Place the mixture in a Pyrex test-tube; close the tube by a cork and delivery tube. Incline the test-tube so that any liquid formed in the reaction cannot run back on the hot part of the tube. Heat the test-tube gently at first and then more strongly. Collect any condensate produced in a test-tube containing 2–3 ml of water. Nitrogenous compounds will usually evolve ammonia or vapours alkaline to indicator paper and possessing characteristic odours; hydroxybenzoic acids yield phenols; formates and acetates yield hydrogen; simple carboxylic acids yield hydrocarbons (methane from acetic acid, benzene from benzoic or phthalic acid, etc.); amine salts and aromatic amino carboxylic acids yield aromatic amines; etc.

Carbon and hydrogen Evidence of the organic nature of the substance may be provided by the behaviour of the compound when heated on porcelain or platinum or other comparatively inert metal (e.g., nickel): the substance is inflammable, burns with a more or less smoky flame, chars and leaves a black residue consisting largely of carbon (compare *Ignition test* above). In general aromatic compounds characteristically burn with a very smoky flame.

If it is desired to test directly for the presence of carbon and hydrogen in a compound, mix 0.1 g of the substance with 1–2 g of ignited, fine copper(II) oxide powder * in a dry test-tube, and fit the latter with a cork carrying a tube bent at an angle so that the escaping gases can be bubbled below the surface of lime water contained in a second test-tube. Clamp the test-tube containing the mixture near the cork. Heat the mixture gradually. If carbon is present, carbon dioxide will be evolved which will produce a turbidity in the lime water. If hydrogen is present, small drops of water will collect in the cooler part of the tube.

Nitrogen, sulphur and halogens In order to detect these elements in organic compounds, it is necessary to convert them into ionisable inorganic substances so that the ionic tests of inorganic qualitative analysis may be applied. This conversion may be accomplished by several methods, but the best procedure is to fuse the organic compound with metallic sodium (**Lassaigne's test**). In this way sodium cyanide, sodium sulphide and sodium halides are formed, which are readily identified. Thus:

$$\text{Organic compound containing C,H,O,N,S,Hal.} + \text{Na} \xrightarrow{\text{Heat}}$$
$$\text{NaCN} + \text{Na}_2\text{S} + \text{NaHal.} + \text{NaOH}$$

It is essential to use an excess of sodium, otherwise if sulphur and nitrogen are both present sodium thiocyanate, NaCNS, may be produced; in the test for nitrogen it may give a red colouration with iron(III) ions but no Prussian blue since there will be no free cyanide ions. With excess of sodium the thiocyanate, if formed, will be decomposed:

$$\text{NaCNS} + 2\text{Na} \longrightarrow \text{NaCN} + \text{Na}_2\text{S}$$

The filtered alkaline solution, resulting from the action of water upon the sodium fusion, is treated with iron(II) sulphate and thus forms sodium hexacyanoferrate(II).

$$\text{FeSO}_4 + 6\text{NaCN} \longrightarrow \text{Na}_4[\text{Fe(CN)}_6] + \text{Na}_2\text{SO}_4$$

Upon boiling the alkaline iron(II) salt solution, some iron(III) ions are inevitably produced by the action of the air; upon the addition of dilute sulphuric acid, thus dissolving the iron(II) and (III) hydroxides, the hexacyanoferrate(II) reacts with the iron(III) salt producing iron(III) hexacyanoferrate(II), Prussian blue:

$$3\text{Na}_4[\text{Fe(CN)}_6] + 2\text{Fe}_2(\text{SO}_4)_3 \longrightarrow \text{Fe}_4[\text{Fe(CN)}_6]_3 + 6\text{Na}_2\text{SO}_4$$

Hydrochloric acid should not be used for acidifying the alkaline solution since the yellow colour, due to the iron(III) chloride formed, causes the Prussian blue to appear greenish. For the same reason, iron(III) chloride should not be added—as is frequently recommended: a sufficient concentration of iron(III) ions is produced by atmospheric oxidation of the hot alkaline solution. The addition of a little dilute potassium fluoride solution may be advantageous in assisting the formation of Prussian blue in a readily filterable form.

Sulphur, as sulphide ion, may be detected by precipitation as black lead sulphide with lead acetate solution and acetic acid or by the purple colour produced on addition of di-sodium pentacyanonitrosyl ferrate. Halogens are

---

* Copper(II) oxide powder, prepared by grinding copper oxide (wire form), is heated to dull redness in a porcelain basin, allowed to cool partially in the air and finally in a desiccator.

detected as the characteristic silver halides by the addition of silver nitrate solution and dilute nitric acid: the interfering influence of sulphide and cyanide ions in the latter tests is discussed under the individual elements.

Procedure. Support a small, soft glass test-tube (50 × 12 mm) in a clamp or insert the tube through a small hole in a piece of asbestos board (or of 'uralite') so that the tube is supported by the rim. Place a cube (*ca.* 4 mm side = 0.04 g) of freshly cut sodium * in the tube. Have in readiness about 0.05 g of the compound on a spatula or the tip of a knife blade; if the compound is a liquid, charge a capillary dropper or a melting point capillary with about three drops of the liquid. Heat the ignition tube, gently at first to prevent cracking, until the sodium melts and the vapour rises 1–2 cm in the tube. Drop the substance, preferably portionwise, directly on to the molten sodium (**CAUTION:** there may be a slight explosion, particularly with chloroform, carbon tetrachloride, nitroalkanes and azo compounds). Remove the tube from its support and hold it by means of a pair of tongs. Heat it carefully at first, then strongly until the entire end of the tube is red hot and maintain it at this temperature for a minute or two. Plunge the tube while still hot into an evaporating basin† containing about 10 ml of distilled water, and cover the dish *immediately* with a clean wire gauze. The tube will be shattered and the residual sodium will react with the water. It is advisable to carry out the operation with the added protection provided by a partly closed fume cupboard window or a safety screen, in addition to the standard eye protection which is worn at all times. When the reaction is over, heat to boiling, and filter. The filtrate should be water-clear and alkaline. If it is dark coloured, decomposition was probably incomplete: repeat the entire sodium fusion.

The following *alternative procedure* is recommended and it possesses the advantage that the same tube may be used for many sodium fusions. Support a Pyrex test-tube (150 × 12 mm) vertically in a clamp lined with sheet cork. Place a cube (*ca.* 4 mm side = 0.04 g) of freshly cut sodium in the tube and heat the latter until the sodium vapour rises 4–5 cm in the test-tube. Drop a small amount (about 0.05 g) of the substance, preferably portionwise, directly into the sodium vapour (**CAUTION:** there may be a slight explosion); then heat the tube to redness for about 1 minute. Allow the test-tube to cool, add 3–4 ml of methanol to decompose any unreacted sodium, then half-fill the tube with distilled water and boil gently for a few minutes. Filter and use the clear, colourless filtrate for the various tests detailed below. Keep the test-tube for sodium fusions; it will usually become discoloured and should be cleaned from time to time with a little scouring powder.

Nitrogen. Pour 2–3 ml of the filtered fusion solution into a test-tube con-

---

* **CAUTION:** Handle sodium with great care. Small pieces for sodium fusions may be kept in a small dry bottle. Larger quantities are better kept under solvent naphtha or xylene. Do not handle the metal with the fingers: use tongs or pincers or a penknife. If the sodium is stored under naphtha or xylene, dry it quickly with filter-paper immediately before use. Any residual sodium should be placed in the bottle for '*Sodium Residues*'. Never throw small pieces of residual sodium in the sink or into water; if you wish to destroy sodium residues, use industrial spirit.

† An alternative technique is as follows. Plunge the hot tube into about 10 ml of water contained in a small, clean mortar and cover the latter immediately with a clean wire gauze. When the reaction is over, grind the mixture of solution and broken glass to ensure thorough extraction of the sodium salts. Transfer with the aid of a little water to a porcelain basin, heat to boiling and filter.

taining 0.1–0.2 g of powdered iron(II) sulphate crystals. Heat the mixture gently with shaking until it boils, then, without cooling, add just sufficient dilute sulphuric acid to dissolve the iron hydroxides and give the solution an acid reaction. {The addition of 1 ml of 5 per cent potassium fluoride solution is beneficial (possibly owing to the formation of potassium hexafluoroferrate(III), $K_3[FeF_6]$) and usually leads to a purer Prussian blue.} A Prussian blue precipitate or colouration indicates that nitrogen is present. If no blue precipitate appears at once, allow to stand for 15 minutes, filter through a small filter and wash the paper with water to remove all traces of coloured solution: any Prussian blue present will then become perceptible in the cone of the filter-paper. If in doubt, repeat the sodium fusion, preferably using a mixture of the compound with pure sucrose or naphthalene. In the absence of nitrogen, the solution should have a pale yellow colour due to iron salts.

If sulphur is present, a *black* precipitate of iron(II) sulphide is obtained when the iron(II) sulphate crystals dissolve. Boil the mixture for about 30 seconds, and acidify with dilute sulphuric acid; the iron(II) sulphide dissolves and a precipitate of Prussian blue forms if nitrogen is present.

Sulphur.   This element may be tested for by either of the following two methods:

1. Acidify 2 ml of the fusion solution with dilute acetic acid, and add a few drops of lead acetate solution. A black precipitate of lead sulphide indicates the presence of sulphur.
2. To 2 ml of the fusion solution add 2–3 drops of a freshly prepared dilute solution (*ca.* 0.1%) of di-sodium pentacyanonitrosyl ferrate $Na_2[Fe(CN)_5NO]$. (The latter may be prepared by adding a minute crystal of the solid to about 2 ml of water.) A purple colouration indicates sulphur; the colouration slowly fades on standing.

Halogens. (A) Nitrogen and sulphur absent.   (*i*) Acidify a portion of the fusion solution with dilute nitric acid and add an excess of silver nitrate solution. A precipitate indicates the presence of a halogen. Decant the mother-liquor and treat the precipitate with dilute aqueous ammonia solution. If the precipitate is white and readily soluble in the ammonia solution, chlorine is present; if it is pale yellow and difficultly soluble, bromine is present; if it is yellow and insoluble, then iodine is indicated. Iodine and bromine may be confirmed by tests (*ii*) or (*iii*); these tests may also be used if it is suspected (e.g., from behaviour in the silver nitrate test) that more than one halogen is present.

(*ii*) Acidify 1–2 ml of the fusion solution with a moderate excess of glacial acetic acid and add 1 ml of carbon tetrachloride. Then introduce 20 per cent sodium nitrite solution drop by drop with constant shaking. A purple or violet colour in the organic layer indicates the presence of *iodine*. The reaction is:

$$2NaI + 2NaNO_2 + 4CH_3 \cdot CO_2H \longrightarrow I_2 + 2NO + 4CH_3 \cdot CO_2Na + 2H_2O$$

This solution may also be employed in the test for *bromine*. If iodine has been found, add further additional quantities of sodium nitrite solution, warm and by means of a dropper pipette remove and replace the organic phase with fresh portions of carbon tetrachloride; repeat until the organic phase is colourless. Boil the acid solution until no more nitrous fumes are evolved and cool. Add a small amount of lead dioxide, place a strip of fluorescein paper across the mouth of the tube and warm. If bromine is present, it will colour the test-paper rose-pink (eosin is formed). If *iodine* has been found to be *absent* use 1 ml of the fusion solution, acidify strongly with glacial acetic acid, add lead dioxide and proceed as above.

In this test for bromine, lead dioxide in acetic acid solution gives lead tetra-acetate which oxidises hydrogen bromide (and also hydrogen iodide), but has practically no effect under the above experimental conditions upon hydrogen chloride:

$$2NaBr + PbO_2 + 4CH_3 \cdot CO_2H \longrightarrow$$
$$Br_2 + (CH_3 \cdot CO_2)_2Pb + 2CH_3 \cdot CO_2Na + 2H_2O$$

Fluorescein test paper is prepared by dipping filter-papers into a dilute solution of fluorescein in ethanol; it dries rapidly and is then ready for use. The test paper has a lemon yellow colour.

To test for *chlorine* in the presence of iodine and/or bromine, acidify 1–2 ml of the fusion solution with glacial acetic acid, add a slight excess of lead dioxide (say, 0.5 g) and boil gently until all the iodine and bromine is liberated. Dilute, filter off excess lead dioxide and test for chloride ions with dilute nitric acid and silver nitrate solution.

(*iii*) Acidify 1–2 ml of the fusion solution with dilute sulphuric acid, cool and add 1 ml of carbon tetrachloride. Prepare the equivalent of 'chlorine water' by acidifying 10 per cent sodium hypochlorite solution with one-fifth of its volume of dilute hydrochloric acid. Add this solution dropwise with vigorous shaking to the mixture. If *iodine* is *present* the organic phase first becomes purple in colour. As the addition of chlorine water is continued, the purple colour disappears (owing to oxidation of iodine to iodate) and, if *bromine* is *present*, is replaced by a brown or reddish colour. If bromine is absent, the organic layer will be colourless. It is, of course, evident that if the carbon tetrachloride layer remains uncoloured and the results of test (*i*) were positive, the halogen present is *chlorine*.

(**B**) **Nitrogen and/or sulphur present.** To remove cyanide and sulphide ions, make 2–3 ml of the fusion solution just acidic with dilute nitric acid, and evaporate to half the original volume in order to expel hydrogen cyanide and/or hydrogen sulphide which may be present. Dilute with an equal volume of water and proceed as in tests (*i*), (*ii*) and (*iii*) above.

Alternatively, add 1–2 drops of 5 per cent nickel(II) nitrate solution to 2–3 ml of the fusion solution, filter off the nickel(II) cyanide and/or nickel(II) sulphide, acidify the filtrate with 2 *M*-nitric acid and test for halides as above.

The presence of halogen may be further confirmed by the **Beilstein test**. This test serves to detect the presence of halogen in many organic compounds. It consists in heating the substance in contact with pure copper oxide in the Bunsen flame: the corresponding copper halide is formed, which, being volatile, imparts an intense green or bluish-green colour to the mantle of the flame.

Push one end of a 20-cm length of stout copper wire into a cork (this will serve as a holder); coil the other end by making two or three turns about a thin glass rod. Heat the coil in the outer mantle of a Bunsen flame until it ceases to impart any colour to the flame. Allow the wire to cool somewhat and, while still warm, dip the coil into a small portion of the substance to be tested and heat again in the non-luminous flame. If the compound contains a halogen element, a green or bluish-green flame will be observed (usually after the initial smoky flame has disappeared). Before using the wire for another compound, heat it until the material from the previous test has been destroyed and the flame is not coloured.

It has been stated that many halogen-free compounds, e.g., certain derivatives of pyridine and quinoline, purines, acid amides and cyano compounds, when ignited on copper oxide impart a green colour to the flame, presumably owing to the formation of volatile copper cyanide. The test is therefore not always trustworthy. The test is not given by fluorides since copper fluoride is not volatile.

The detection of the following elements, which occur infrequently in organic compounds, is included here for the sake of completeness.

Fluorine.   Use either of the following tests.

(a) Strongly acidify about 2 ml of the fusion filtrate with glacial acetic acid, and boil until the volume is reduced by about one-half. Cool. Place one drop of the solution upon zirconium–alizarin red S test paper. A yellow colour on the red paper indicates the presence of fluoride. Large amounts of sulphates and phosphates may interfere with this test.

Prepare the zirconium–alizarin red S paper as follows. Soak dry filter-paper in a 5 per cent solution of zirconium nitrate in 5 per cent hydrochloric acid and, after draining, place it in a 2 per cent aqueous solution of sodium alizarin sulphonate (B.D.H. 'Alizarin Red S'). The paper is coloured red-violet by the zirconium lake. Wash the paper until the wash water is nearly colourless and then dry in the air.

(b) If nitrogen and/or sulphur is present, acidify 3–4 ml of the fusion solution with dilute nitric acid and evaporate to half the original volume in order to expel any HCN and/or H_2S which may be present. If nitrogen and sulphur are absent, proceed directly with 2 ml of the sodium fusion filtrate. Render the solution just neutral to litmus by the addition of dilute (5 M) aqueous ammonia solution, then add 5 drops of 5 M-acetic acid and 20 mg of lanthanum chloroanilate * and shake intermittently for 10–15 minutes. Filter. A pink-violet colouration of the filtrate is a positive test for fluorine.

Phosphorus.   The presence of phosphorus may be indicated by a smell of phosphine during the sodium fusion and the immediate production of a jet-black colour when a piece of filter-paper moistened with silver nitrate solution is placed over the mouth of the ignition tube after the sample has been dropped on the hot sodium. Treat 1.0 ml of the fusion solution with 3 ml of concentrated nitric acid and boil for 1 minute. Cool and add an equal volume of ammonium molybdate reagent. Warm the mixture to 40–50 °C, and allow to stand. If phosphorus is present, a yellow crystalline precipitate of ammonium 12-molybdophosphate, $(NH_4)_3[PMo_{12}O_{40}]$, will separate.

It is usually preferable to oxidise the compound directly as follows. Intimately mix 0.02–0.05 g of the compound with 3 g of sodium peroxide and 2 g of anhydrous sodium carbonate in a nickel crucible. Heat the crucible and its contents with a small flame, gently at first, afterwards more strongly until the contents are fused, and continue heating for a further 10 minutes. Allow to stand, extract the contents of the crucible with water and filter. Add excess of concentrated nitric acid to the filtrate and test with ammonium molybdate reagent as above. A yellow precipitate indicates the presence of phosphorus. It

---

* 2,5-Dichloro-3,6-dihydroxy-p-benzoquinone, lanthanum salt.

must be borne in mind that the above treatment will convert any arsenic present into arsenate.

Ammonium molybdate reagent may be prepared by dissolving 45 g of pure ammonium molybdate in a mixture of 40 ml of aqueous ammonia (d 0.88) and 60 ml of water and then adding 120 g of ammonium nitrite and diluting the solution to 1 litre with water.

Arsenic. The presence of arsenic in an organic compound is generally revealed by the formation of a dull grey mirror of arsenic on the walls of the test-tube when the compound is fused with sodium in the Lassaigne test. Usually sufficient arsenic is found in the fusion solution to give a yellow precipitate of arsenic trisulphide when the solution is acidified with hydrochloric acid and treated with hydrogen sulphide.

It is recommended that the compound be fused with a mixture of sodium carbonate (2 parts) and sodium peroxide (1 part) as in the test for **Phosphorus**. Extract the fused mass with water, filter and acidify with dilute hydrochloric acid. Pass hydrogen sulphide through the hot solution; arsenic is precipitated as yellow arsenic sulphide. If **antimony is present**, it will be precipitated as orange antimony trisulphide.

Mercury. Upon heating a mixture of the compound with soda lime in a long test-tube, a bright metallic mirror and, finally, drops of the metal will form in the upper part of the tube if mercury is present.

The sodium carbonate–zinc method for the detection of nitrogen, sulphur and halogens in organic compounds The Lassaigne procedure for detecting nitrogen in organic compounds frequently gives unsatisfactory results with explosive compounds (diazonium salts, polynitro compounds and the like) and with certain volatile nitrogenous substances, such as bases, their acyl derivatives or their salts. These difficulties may often be surmounted either by mixing the compound with pure naphthalene or sucrose, or by mixing the substance with sodium and placing a layer of soda lime above the mixture. Difficulties are also sometimes experienced in the sodium fusion test with liquids of low boiling point, such as ethyl bromide. Satisfactory results are obtained by heating the organic compound with sodium carbonate and zinc powder (Middleton, 1935). The latter method has been proposed for the detection of the common elements in all organic compounds. It is doubtful, however, whether it is to be preferred to the sodium fusion procedure in routine testing for elements, although it may be recommended for those relatively few cases in which the Lassaigne test is not entirely satisfactory.

When an organic compound is heated with a mixture of zinc powder and sodium carbonate, the nitrogen and halogens are converted into sodium cyanide and sodium halides respectively, and the sulphur into zinc sulphide (insoluble in water). The sodium cyanide and sodium halides are extracted with water and detected as in Lassaigne's method, whilst the zinc sulphide in the residue is decomposed with dilute acid and the hydrogen sulphide is identified with lead acetate paper. The test for nitrogen is thus not affected by the presence of sulphur: this constitutes an advantage of the method.

Procedure. Prepare the **zinc powder–sodium carbonate mixture** by grinding together in a dry, clean mortar 25 g of anhydrous sodium carbonate [AnalaR] and 50 g of the purest obtainable zinc powder. The reagent is unlikely to contain nitrogen, but traces of sulphur and halogens may be present. It is therefore essential to carry out a blank or control test for sulphur and halogens with every fresh batch of the mixture.

Place about 0.1 g of the powdered compound in a small dry test-tube, add sufficient of the reagent to give a column about 1 cm high and then shake the closed tube until the contents are well mixed. Now add more reagent, without mixing with the material already in the tube, until the total height is about 3 cm. If the compound is a liquid, introduce 2–3 drops into a small dry test-tube, add sufficient of the mixture to form a column about 1 cm long and allow the liquid to soak well into the reagent. Then add more reagent, without mixing, until a total height of about 3 cm is secured. Hold the tube horizontally (use tongs or a special test-tube holder) and, by means of a *small* flame, heat a 1-cm length of the mixture gently near the open end. Gradually increase the size of the flame until the mixture is red hot at the end. Extend the heating gradually and cautiously towards the closed end of the tube until the whole of the mixture is red hot. (The extension of the heating towards the closed end of the tube must be carried out with great care, otherwise the mixture may be projected from the tube; if the mixture tends to be pushed out of the tube by the evolution of gas, stop the heating momentarily and rotate the tube while still in a horizontal position in order to redistribute the contents.) Finally heat the tube to redness in a vertical position for a minute or two and, while the end of the tube is still red hot, plunge the tube in about 10 ml of water in a porcelain dish. Boil the contents of the dish gently for 1–2 minutes and filter. (If the filtrate is not colourless, repeat the whole process.) Retain the residue in the basin for the sulphur test. Divide the clear filtrate into two portions.

Nitrogen.   Treat one portion with 1–2 ml of 5 per cent sodium hydroxide solution and 0.1 g of powdered iron(II) sulphate. Boil for 1 minute and cool. Cautiously acidify with dilute sulphuric acid (carbon dioxide is evolved). A precipitate of Prussian blue indicates that nitrogen is present.

Halogens.   Proceed as described under the Lassaigne test. If nitrogen is present, the cyanide must first be eliminated.

Sulphur.   Moisten the centre of a filter-paper with lead acetate solution. Add about 10 ml of dilute hydrochloric acid to the residue in the dish and immediately cover it with the prepared filter-paper. If zinc sulphide is present in the residue, a dark brown stain, visible on the upper surface of the paper, will be obtained: frequently the presence of hydrogen sulphide can also be detected by its odour.

VII,3. THE SOLUBILITIES OF ORGANIC COMPOUNDS

General discussion.   When a mixture of a specified amount of a given solute and a specified amount of a given solvent forms a homogeneous liquid, the former is said to be soluble in the latter. The arbitrary standard employed in this book is 0.10 g of solid or 0.20 ml of liquid to 3.00 ml of solvent. The study of the solubility behaviour of an unknown substance in various liquids, viz., water, ether, 5 per cent sodium hydroxide solution, 5 per cent sodium hydrogen carbonate solution, 5 per cent hydrochloric acid and cold concentrated sulphuric acid, may provide useful preliminary information about the nature of the compound.

The substance should be tested for solubility in the various solvents in the order cited above since, for example, when solubility in dilute acid or base is being considered, it is important to note whether the unknown is more soluble in aqueous acid or base than it is in water: this increased solubility is the positive test for a basic or acidic functional group. Acidic compounds are detected by their solubility in 5 per cent sodium hydroxide solution. Strong and weak acids are differentiated by the solubility of the former, but not the latter, in the weakly basic 5 per cent sodium hydrogen carbonate solution. Nitrogenous bases are frequently detected by their solubility in 5 per cent hydrochloric acid. Many compounds that are neutral even in concentrated aqueous acidic solutions behave as bases in strongly acidic solvents, such as concentrated sulphuric acid; these include compounds that are neutral in water and contain oxygen in any form. The presence of acidic or basic functional groups in water-soluble compounds is detected by testing their aqueous solutions with litmus or other indicator paper.

The subsequent systematic search for functional groups is based primarily on the knowledge as to whether the compound is **neutral**, **acidic** or **basic** in character as determined by the application of these preliminary solubility tests.

Summary of solubility behaviour. **Solubility in water.** Since water is a polar compound, it is a poor solvent for hydrocarbons of all types. Salts are usually extremely polar, and are generally water-soluble. Other compounds fall between these two extremes; these include alcohols, esters, aldehydes, ketones, acids, ethers, amides, nitriles and amines. Acids and amines are generally more soluble than neutral compounds.

For homologous series of mono-functional alcohols, esters, aldehydes, ketones, acids, ethers, amides, nitriles and amines, the upper limit of water solubility is found at about the member containing four carbon atons. The solubility in water is due largely to the polar group, and as the homologous series is ascended the hydrocarbon (non-polar) part of the molecule increases whilst the polar function remains substantially unchanged; this accounts for the decrease in solubility in polar solvents such as water. This behaviour is an illustration of a general rule that increased structural similarity between the solute and the solvent results in increased solubility in that solvent.

It must be emphasised that the particular region (that of the member containing four carbon atoms) of water solubility for many homologous series is determined by the arbitrary proportions of solute and solvent defined in the previous general discussion. The limit would be elsewhere for a different ratio of solute to solvent.

Solubility in ether. Non-polar and slightly polar compounds will, in general, dissolve in ether because they are largely unassociated. Ionic compounds, such as salts, are not soluble in ether. The solubility of a polar compound in ether will depend upon the influence of the polar group or groups relative to that of the non-polar part of the molecule. Usually, compounds that have one polar group per molecule will dissolve in ether unless they are highly associated or of extreme polarity (e.g., the sulphonic acids).

Many organic compounds that are insoluble in water dissolve in ether. If a compound is soluble in both ether and water, it probably (i) is non-ionic, (ii) contains five or less carbon atoms, (iii) has a functional group that is polar and capable of forming hydrogen bonds and (iv) does not contain more than one strongly polar group. If a compound dissolves in water but not in ether, it may (i) be ionic (a salt) or (ii) contain two or more polar groups but not more than four carbon atoms per polar group. There are, of course, exceptions to these statements.

Solubility in dilute hydrochloric acid. Most compounds that are soluble in dilute hydrochloric acid contain a basic nitrogen atom (incorporating an unshared electron pair) in the molecule. Thus most aliphatic amines (primary, secondary and tertiary) form salts (polar, water-soluble compounds) with hydrochloric acid. Aryl groups reduce the basicity of the nitrogen atom. Primary aromatic amines (e.g., aniline), although more weakly basic than primary aliphatic amines, are soluble, but in secondary and tertiary purely aromatic amines (e.g., diphenylamine, carbazole and triphenylamine) the basic character of the nitrogen atom has been diminished to such an extent that they do not form salts with dilute hydrochloric acid and consequently do not dissolve. Alkylarylamines (containing not more than one aryl group) and alicyclic amines, however, do dissolve. A few types of oxygen-containing compounds (such as the pyrones and the anthocyanidin pigments of certain flowers), which form oxonium salts, dissolve in dilute hydrochloric acid. Amides which are insoluble in water are generally unaffected by 5 per cent hydrochloric acid but may dissolve in higher concentrations (10–20%) of acid: this emphasises the importance of employing the correct strength of acid in the solubility tests. Many disubstituted amides which are of sufficiently high molecular weight to be water-insoluble dissolve in 5 per cent hydrochloric acid.

It may be noted that some aromatic amines react with 5 per cent hydrochloric acid to form insoluble hydrochlorides: the latter sometimes dissolve upon warming slightly and diluting with water. The appearance of the solid will usually show whether the arylamine has undergone a change: the solid should be separated and its melting point compared with that of the original compound. A test with ethanolic silver nitrate solution would indicate the formation of a hydrochloride.

Solubility in dilute sodium hydroxide solution and in dilute sodium hydrogen carbonate solution. Carboxylic acids, sulphonic acids, sulphinic acids, phenols, thiophenols, thiols, imides, arylsulphonamides, arylsulphonyl derivatives of primary amines, oximes, primary and secondary nitro compounds and some enols (e.g., of 1,3-diketones or β-keto esters) dissolve in dilute sodium hydroxide solution, i.e., they contain an acidic group of sufficient strength to react with the alkali. Carboxylic acids, sulphinic acids and sulphonic acids are soluble in dilute solutions of sodium hydrogen carbonate with the evolution of carbon dioxide; some phenols substituted with electron-withdrawing groups (for example, picric acid, 2,4,6-tribromophenol and 2,4-dinitrophenol) are strongly acidic and also dissolve in sodium hydrogen carbonate solution with the evolution of carbon dioxide. Primary and secondary nitro compounds, imides, arylsulphonamides and oximes are insoluble in sodium hydrogen carbonate solution. Some of the sodium salts of highly substituted phenols are insoluble in sodium hydroxide solution but may dissolve upon dilution and warming with water.

Certain substituents (e.g., the amino group) may markedly affect the solubility and other properties of a sulphonic acid or a carboxylic acid. Thus such sulphonic acids as the aminobenzenesulphonic acids and the pyridine- and quinoline-sulphonic acids exist in the form of inner salts or dipolar ions that result from the interaction of the basic amino group and the acidic sulphonic acid group. Sulphanilic acid, for example, is more accurately represented by formula (I) than by formula (II):

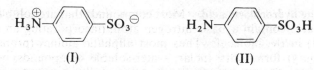

(I)   (II)

These aminosulphonic acids possess the high melting points usually associated with salts, but are sparingly soluble or insoluble in water. They all dissolve readily in dilute alkali but not in dilute acid, i.e., they appear to exhibit the reactions of the sulphonic acid group but not of the amino group. The aliphatic aminocarboxylic acids (III), because of the presence of the strongly basic amino group, exist as dipolar ions (IV):

$$H_2N \cdot CH(R) \cdot CO_2H \qquad \overset{\oplus}{H_3N} \cdot CH(R) \cdot CO_2^{\ominus}$$

(III)   (IV)

they are soluble in water but not in ether, and dissolve in both dilute acid and dilute alkali but react slowly or not at all with dilute sodium hydrogen carbonate solution. The carboxyl derivatives of the arylamines (e.g., p-aminobenzoic acid) are also amphoteric, but the diminution of the basic character of the amino group because of its attachment to the aryl group prevents the formation of inner salts to any degree; they react normally with sodium hydrogen carbonate.

Solubility in concentrated sulphuric acid. Solubility in cold concentrated sulphuric acid is used to further characterise those compounds which by virtue of the results of the previous solubility tests are considered to be neutral.

The most important group of compounds to exhibit solubility in this reagent are those containing oxygen. The initial solubility of these compounds is due to the basic character of one or more of the oxygen atoms that are present in the molecules, and results from oxonium ion formation; dilution of the sulphuric acid solution often results in recovery of the compound in an unchanged form. More fundamental changes resulting from subsequent transformations of the oxonium ion may, however, occur.

Unsaturated hydrocarbons dissolve through formation of soluble alkyl hydrogen sulphates, e.g.,

$$R^1 \cdot CH = CH \cdot R^2 + H_2SO_4 \longrightarrow [R^1 \cdot \overset{\oplus}{CH} \cdot CH_2 \cdot R^2] \xrightarrow{\ominus OSO_3H} R^1 \cdot CH \cdot CH_2 \cdot R^2$$
$$| \\ O \cdot SO_3H$$

Polyalkylated aromatic hydrocarbons and alkyl phenyl ethers undergo sulphonation, e.g.,

$$ArH + 2H_2SO_4 \longrightarrow Ar \cdot SO_3H + HSO_4^{\ominus} + H_3O^{\oplus}$$

Compounds which dissolve in concentrated sulphuric acid may be further subdivided into those which are soluble in syrupy phosphoric acid and those which are insoluble in this solvent: in general dissolution takes place without the production of appreciable heat or colour. Those compounds soluble in phosphoric acid include alcohols, esters, aldehydes, methyl ketones and cyclic ketones provided that they contain less than nine carbon atoms. The solubility limit is somewhat lower than this for ethers; thus dipropyl ether dissolves in 85 per cent phosphoric acid but dibutyl ether and anisole do not. Ethyl benzoate and diethyl malonate are insoluble.

It has been found convenient to place organic compounds into seven solubility groups which are summarised in Table VII,1, together with the commoner classes of compounds that fall into the respective solubility groups. The compounds are classified according to:

(a) their solubility behaviour towards the reagents specified above, and
(b) the elements, other than carbon and hydrogen, that they contain, i.e.:

I. *Compounds soluble in both water and ether.* This includes the lower members of the various homologous series (4–5 atoms in a normal chain) that contain oxygen and/or nitrogen in their structures: they are soluble in water because of their low carbon content. If the compound is soluble in both water and in ether, it would also be soluble in other solvents so that further solubility tests are generally unnecessary: the aqueous solution should be tested with indicator paper. The test with sodium hydrogen carbonate solution should also be performed.

II. *Compounds soluble in water but insoluble in ether.* The classes 1 to 5 are usually also soluble in dilute alkali and acid. Useful information may, however, be obtained by examining the behaviour of *Salts* to alkaline or acidic solvents. With a salt of a water-soluble base, the characteristic odour of an amine is usually apparent when it is treated with dilute alkali: likewise, the salt of a water-soluble, weak acid is decomposed by dilute hydrochloric acid or by concentrated sulphuric acid. The water-soluble salt of a water-insoluble acid or base will give a precipitate of either the free acid or the free base when treated with dilute acid or dilute alkali. The salts of sulphonic acids and of quaternary bases are unaffected by dilute sodium hydroxide or hydrochloric acid.

III. *Compounds insoluble in water, but soluble in dilute sodium hydroxide.* This group may be further subdivided into IIIA—soluble in dilute sodium hydroxide and soluble in dilute sodium hydrogen carbonate with the evolution of carbon dioxide; and IIIB—soluble in dilute sodium hydroxide and insoluble in dilute sodium hydrogen carbonate.

IV. *Compounds insoluble in water, but soluble in dilute hydrochloric acid.* It should be remembered that the hydrochlorides of some bases are sparingly soluble in cold water and one should therefore not be misled by an apparent insolubility of a compound (containing nitrogen) in dilute hydrochloric acid. The suspension in dilute hydrochloric acid should always be filtered and the filtrate made alkaline. A precipitate will indicate that the compound is indeed a base and should be included in this group.

V. This group includes all the *water-insoluble hydrocarbons and oxygen compounds that do not contain N or S and are soluble in cold concentrated sulphuric acid.* Any changes—colour, excessive charring, evolution of gases or heat, polymerisation and precipitation of an insoluble compound—attending the dissolution of the substance should be carefully noted.

Alcohols, esters (but not ethyl benzoate, diethyl malonate or diethyl oxalate), aldehydes, methyl ketones and cyclic ketones containing less than nine carbon atoms as well as ethers containing less than seven carbon atoms are also soluble in 85 per cent phosphoric acid.

VI. *Compounds, not containing N or S, insoluble in concentrated sulphuric acid.* This test provides for a differentiation *inter alia* between alkanes and

Table VII,1 Classification of organic compounds according to solubility behaviour

I Soluble in both ether and water	II Soluble in water but insoluble in ether	III Soluble in 5% sodium hydroxide solution	IV Soluble in 5% hydrochloric acid	V Not containing N or S. Soluble only in concentrated sulphuric acid	VI Not containing N or S. Insoluble in concentrated sulphuric acid	VII Containing N or S. Compounds not in groups I to IV
The lower members of the homologous series of: 1. Alcohols 2. Aldehydes 3. Ketones 4. Acids 5. Esters 6. Phenols 7. Anhydrides 8. Amines 9. Nitriles 10. Polyhydroxy phenols	1. Polybasic acids and hydroxy acids. 2. Glycols, polyhydric alcohols, polyhydroxy aldehydes and ketones (sugars) 3. Some amides, amino acids, di- and polyamino compounds, amino alcohols 4. Sulphonic acids 5. Sulphinic acids 6. Salts	1. Acids 2. Phenols 3. Imides 4. Some primary and secondary nitro compounds; oximes 5. Thiols and thiophenols 6. Sulphonic acids, sulphinic acids, aminosulphonic acids and sulphonamides 7. Some diketones and β-keto esters	1. Primary amines 2. Secondary aliphatic and aryl-alkyl amines 3. Aliphatic and some aryl-alkyl tertiary amines 4. Hydrazines	1. Unsaturated hydrocarbons 2. Some polyalkylated aromatic hydrocarbons 3. Alcohols 4. Aldehydes 5. Ketones 6. Esters 7. Anhydrides 8. Ethers and acetals 9. Lactones 10. Acyl halides	1. Saturated aliphatic hydrocarbons 2. Cycloalkanes 3. Aromatic hydrocarbons 4. Halogen derivatives of 1, 2 and 3 5. Diaryl ethers	1. Nitro compounds (tertiary) 2. Amides and derivatives of aldehydes and ketones 3. Nitriles 4. Negatively substituted amines 5. Nitroso, azo, hydrazo and other intermediate reduction products of nitro compounds 6. Sulphones, sulphonamides of secondary amines, sulphides, sulphates and other sulphur compounds

cycloalkanes and also simple aromatic hydrocarbons which are insoluble, and unsaturated hydrocarbons which are soluble in the reagent (group V).

VII. *Compounds that contain N or S which are not in groups I–IV*; many of the compounds in this group are soluble in concentrated sulphuric acid.

It will be observed that halogen compounds are not listed separately, but appear in each of the seven categories in accordance with their solubility behaviour.

Procedure for solubility tests. All solubility determinations are carried out at the laboratory temperature in small test-tubes (e.g., 100 × 12 mm) but of sufficient size to permit of vigorous shaking of the solvent and the solute.

Amount of material required. It is convenient to employ an arbitrary ratio of 0.10 g of solid or 0.20 ml of liquid for 3.0 ml of solvent. Weigh out 0.10 g of the *finely powdered* solid to the nearest 0.01 g: after some experience, subsequent tests with the *same* compound may be estimated by eye. Measure out 0.20 ml of the liquid either with a calibrated dropper or a small graduated pipette. Use either a calibrated dropper or a graduated pipette to deliver 3.0 ml of solvent.

Much time will be saved if each of the solvents (water, ether, 5 per cent sodium hydroxide, 5 per cent sodium hydrogen carbonate and 5 per cent hydrochloric acid) be contained in a 30- or 60-ml bottle fitted with a cork carrying a calibrated dropper. The concentrated sulphuric acid should be kept in a glass-stoppered bottle and withdrawn with a dropper or pipette as required.

Solubility in water. Treat a 0.10 g portion of the solid with successive 1.0 ml portions of water, shaking vigorously after each addition, until 3.0 ml have been added. If the compound does not dissolve completely in 3.0 ml of water, it may be regarded as insoluble in water. When dealing with a liquid, add 0.20 ml of the compound to 3.0 ml of water and shake. In either case, test the contents of the small test-tube with Universal indicator paper: it is best to remove a little of the solution or supernatant liquid with a dropper. It is usually convenient at this stage to test for the presence of water-soluble enolic compounds by observing any colouration resulting from the addition of neutral aqueous iron(III) chloride solution.

Solubility in ether. Use 0.10 g of solid or 0.20 ml of a liquid in a dry test-tube and proceed exactly as in testing the solubility in water, but do not employ more than 3.0 ml of solvent.

Solubility in 5 per cent sodium hydroxide solution. Note whether there is any rise in temperature. If the compound appears insoluble, remove some of the supernatant liquid by means of a dropper to a semimicro test-tube (75 × 10 mm), add 5 per cent hydrochloric acid dropwise until acid and note whether any precipitate (or turbidity) is formed. The production of the latter will confirm the presence of an acidic compound.

Solubility in 5 per cent sodium hydrogen carbonate solution. If the compound is soluble in 5 per cent sodium hydroxide solution, test its solubility in a 5 per cent solution of sodium hydrogen carbonate. Observe whether it dissolves and particularly whether carbon dioxide is evolved either immediately (carboxylic acids, sulphonic acids, negatively substituted phenols) or after a short time (some amino acids).

Solubility in 5 per cent hydrochloric acid. Add the acid to 0.10 g of the solid or 0.20 ml of the liquid in quantities of 1.0 ml until 3.0 ml have been introduced. Some organic bases form hydrochlorides that are soluble in water but are precipitated by an excess of acid: if solution occurs at any time, the unknown is classified as a basic compound. If the compound appears insoluble, remove some of the supernatant liquid by means of a dropper to a semimicro test-tube (75 × 10 mm), and add 5 per cent sodium hydroxide solution until basic and observe whether any precipitate is produced: the formation of a precipitate will place the compound in group IV.

Solubility in concentrated sulphuric acid. Place 3.0 ml of pure concentrated sulphuric acid in a dry test-tube and add 0.10 g of a solid or 0.20 ml of a liquid. If the compound does not dissolve immediately, agitate for some time but do not heat. Observe any change in colour, charring, evolution of gaseous products, polymerisation accompanied by precipitation, etc.

Solubility in syrupy phosphoric acid. This test should only be applied if the compound is soluble in concentrated sulphuric acid. Place 3.0 ml of 85 per cent orthophosphoric acid in a dry test-tube and add 0.10 g of a solid or 0.20 ml of a liquid. If the compound does not dissolve immediately, agitate for some time but do not boil.

Solubility classification scheme

	Water	Ether	NaOH	NaHCO$_3$	HCl	H$_2$SO$_4$	H$_3$PO$_4$
I	+ (1)	+					
II	+	−					
IIIA	−		+ (2)	+			
IIIB	−		+ (2)	−			
IV	−		−		+		
VA (3)	−		−		−	+	+
VB (3)	−		−		−	+	−
VI (3)	−		−		−	−	
VII (4)	−		−		−		

Notes. (1) + denotes soluble; − denotes insoluble.

(2) If a compound contains nitrogen its solubility in dilute hydrochloric acid should be tested also to ascertain whether it is amphoteric.

(3) These are neutral compounds in which nitrogen and sulphur are *absent*.

(4) These are neutral compounds in which nitrogen or sulphur is *present*.

VII,4. INTERPRETATION OF SPECTRA General introduction.

Valuable information may be obtained about the structure of an organic compound by a detailed examination of its infrared, nuclear magnetic resonance and mass spectra and to a lesser extent from its ultraviolet–visible spectrum. Invariably it is necessary to cross-relate such information with that obtained from the results of the preliminary tests already described, in order to provide a sound basis on which to proceed towards the characterisation of functional groups, and the final identification of the compound. Frequently initial inspection of the spectra may only provide some general clues to functional group type, etc., but as characterisation proceeds so the origin of spectral features becomes more certain, until finally the spectral interpretation will be seen to be entirely consistent with the identified structure.

Section **I,39**, which summarised briefly the origins of spectral characteristics, was principally concerned with the detailed experimental procedures necessary for recording an infrared or ultraviolet–visible spectrum. In most laboratories this routine operation is carried out by the experimental chemist. Usually, however, the recording of a nuclear magnetic resonance spectrum or a mass spectrum is carried out by a service department; all organic chemists should, however, be able to interpret the various spectra.

The object of this Section, therefore, is to concentrate on the important aspects of spectroscopic interpretation; the interpretative aspects of ultraviolet–visible spectra have been briefly covered in Section I,39 and will be considered further here. It is assumed that the reader will be familiar, through appropriate undergraduate courses, with the fundamental principles of all these spectroscopic methods; a reading list for the theoretical aspects of infrared and of ultraviolet–visible spectroscopy is to be found in the bibliography to Section I,39, p. 263, and in Ref. 4. Similar texts for nuclear magnetic resonance and for mass spectrometry are to be found in Refs. 1 and 2 respectively. Further important texts are also given under each Section, p. 1152.

The procedures suggested here for the interpretation of infrared, nuclear magnetic resonance and mass spectra have been found by experience to yield initial information of value. However, it is clearly impossible to prescribe a definitive approach which will be satisfactory in all cases. It is only possible to become proficient in spectral interpretation by repeated practice with 'unknowns' and then working carefully through the solutions. There are now many collections of such problems available which are graded to varying levels of complexity (Ref. 3).

VII,4,1. Infrared spectroscopy As was stated in Section I,39, the region of the infrared (i.r.) spectrum which is of greatest importance to the organic chemist is that which lies between 4000 and 660 cm^{-1}, i.e., that which is obtained readily with the aid of a sodium chloride prism or a suitable grating. Absorption bands in the spectrum result from energy changes arising as a consequence of molecular vibrations of the bond stretching and bending (deformation) type. The positions of atoms in molecules may be regarded as mean equilibrium positions, and the bonds between atoms may be considered as analogous to springs, subject to stretching and bending. Each atom or group of atoms in a molecule oscillates about a point at which attraction of nuclei for electrons balances the repulsion of nuclei by nuclei, and electrons by electrons. These oscillations have natural periods which depend upon the masses of the atoms and the strengths of the bonds involved. The amplitude of the oscillations, but not the frequency, can be increased by supplying energy by means of electromagnetic radiation. Nuclei and electrons bear electric charges, the force required can be supplied by the oscillating electric vector of an electromagnetic wave of frequency and phase which *match* those of a particular molecular vibration. Transfer of energy in this way is possible if a change in the amplitude of that vibration results in a change of molecular dipole moment (the dipole moment may be regarded as analogous to the coupling mechanism of a spring): radiant energy is then absorbed and the intensity of radiation at this particular wavelength is decreased on passing through the compound. The intensity of absorption bands depends upon the magnitude of the change in oscillating dipole moment of the bonds during the transition, and also is directly proportional to the number of bonds in the molecule responsible for that particular absorption. Thus hydrogen or carbon bonded to oxygen or nitrogen gives rise to strong infrared absorption because of the polarity of these particular bonds. In contrast, no absorption results from stretching vibrations in a homo-nuclear double bond or triple bond which is symmetrically substituted; such vibrations are termed **infrared inactive**. The recognition of such bonds is, however, made possible by an examination of the Raman spectra of such molecules (i.e., the vibrations are **Raman active**).

There are two main types of molecular vibrations: stretching and bending. A stretching vibration is a vibration along a bond axis such that the distance between the two atoms is decreased or increased. A bending vibration involves a change in bond angles.

For a diatomic molecule A — B, the only vibration that can occur is a periodic stretching along the A — B bond. The masses of the two atoms and their connecting bond may be treated, to a first approximation, as two masses joined by a spring and Hooke's law may be applied. This leads to the expression for the frequency of vibration $\bar{v}$ in wavenumbers (cm^{-1}):

$$\bar{v} = \frac{1}{2\pi c} \left(\frac{f}{m_A m_B / m_A + m_B} \right)^{\frac{1}{2}}$$

where c is the velocity of light (ms^{-1}), f is the force constant of the bonds (Nm^{-1}) and m_A and m_B the masses (in g) of the atoms A and B respectively. The value of f is *ca.* 500 Nm^{-1} for single bonds and about two or three times this value for double and triple bonds respectively: it is a measure of the resistance of the bond to stretching and is roughly proportioned to the energy of the the bond. Application of this equation to the case of the stretching of a C — H bond, and using 19.9×10^{-24} g and 1.67×10^{-24} g as the mass values for carbon and hydrogen respectively, together with the accepted values for c and f, gives a frequency of 3020 cm^{-1}. The stretching of a carbon–hydrogen bond in a methyl or a methylene group is actually observed in the regions about 2975 and 2860 cm^{-1} respectively; the slight deviation from the calculated value is a reflection of the fact that modifications to the frequency of vibration arise from the strengths and polarities of the bonds associated with the carbon atom, and these have been ignored in this calculation.

With polyatomic molecules many more fundamental vibrational modes are possible. A qualitative illustration of the stretching and bending modes for the methylene group is shown in Fig. VII,1(i). Arrows indicate periodic oscillations in the directions shown; the $\oplus$ and $\ominus$ signs represent, respectively, relative movement at right angles to the surface of the page. A **symmetrical stretching mode**, where the hydrogens are vibrating in phase towards and away from the carbon nucleus, requires less energy than the corresponding **asymmetric stretching mode** and therefore absorbs at a slightly lower wavenumber. Bending vibrations, which are descriptively termed **scissoring**, **rocking**, **twisting** or **wagging** modes, absorb at considerably lower wavenumbers since the energy associated with these deformations is much less.

The infrared spectrum therefore consists of a number of absorption bands arising from infrared active fundamental vibrations; however, even a cursory inspection of an i.r. spectrum reveals a greater number of absorptions than can be accounted for on this basis. This is because of the presence of **combination bands**, **overtone bands** and **difference bands**. The first arises when absorption by a molecule results in the excitation of two vibrations simultaneously, say v_1, and v_2, and the combination band appears at a frequency of $v_1 + v_2$; an overtone band corresponds to a multiple ($2v$, $3v$, etc.) of the frequency of a parti-

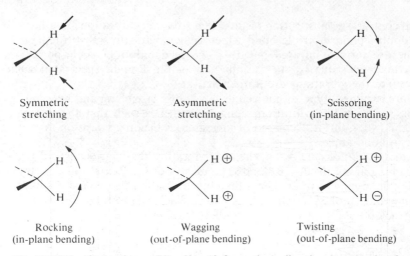

| Symmetric stretching | Asymmetric stretching | Scissoring (in-plane bending) |

| Rocking (in-plane bending) | Wagging (out-of-plane bending) | Twisting (out-of-plane bending) |

Fig. VII,1(i) Stretching and bending (deformation) vibrational modes for the methylene group, which is typical of an XY$_2$ system.

cular absorption band. A difference band arises when absorption of radiation converts a first excited state into a second excited state. These bands are frequently of lower intensity than the fundamental absorption bands but their presence, particularly the overtone bands, can be of diagnostic value for confirming the presence of a particular bonding system.

Features of an infrared spectrum A typical i.r. spectrum is that of acetophenone shown in Fig. VII,1(ii). Some general features illustrative of the philosophy relating to the interpretation of spectra and the correlation of absorption bands with the presence of particular groupings should be noted.

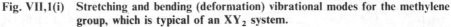

It will be immediately apparent that this spectrum may be divided into two parts, the first between 4000–1600 cm^{-1} and the second from 1600–660 cm^{-1}. In the former there are relatively few absorption bands, but in the latter a great number of absorptions is observed. Indeed, although as noted below aromatic and aliphatic compounds may be recognised from the general spectral profile, all organic compounds exhibit this apparent segregation of bands into these two main regions. The second region is frequently referred to as the 'fingerprint region' since complete superimposability of two spectra in this region provides confirmation of identity.

The former region could be termed the 'functional group region' since, as will be noted below, the fundamental vibrational modes of most of the principal functional groups absorb in this region. Thus all compounds containing a carbonyl group (whether it be an aldehyde, ketone, carboxylic acid, acid chloride, amide, ester, etc.) will exhibit strong absorption in the 1700 cm^{-1} region.

Aromatic compounds, of which spectrum (ii) is typical, always exhibit sharp and often numerous bands in the fingerprint region. Aliphatic compounds on the other hand give rise to far fewer, broad, bands in this region. These differences in the profile of the spectra of aromatic and aliphatic compounds provide a valuable first step in spectral interpretation.

Correlation charts and tables. Central to the philosophy of i.r. spectral

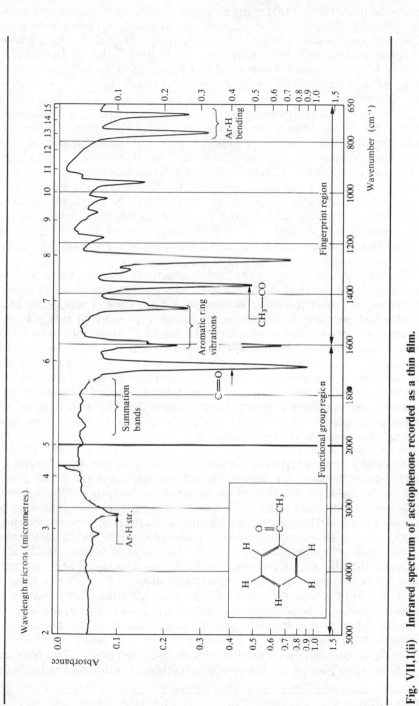

Fig. VII,1(ii)   Infrared spectrum of acetophenone recorded as a thin film.

interpretation is the fact that many stretching and bending modes in a molecule are virtually independent of changes of structure at more remote sites. Structural modifications closer to the absorbing centre do of course affect the energy associated with the absorption, and lead to a shift of the absorption band to higher or lower frequencies. These frequency shifts, however, have been found to lie within defined limits and the numerical value often provides valuable information on the structural environment of the associated group. For example, the absorption maximum for the carbon–hydrogen stretching frequency lies in the general region around 3000 cm^{-1}; if the carbon is sp^3-hybridised the maximum is just below 3000 cm^{-1}; if the carbon is sp^2-hybridised the position is just above 3000 cm^{-1} and if the carbon is sp-hybridised the position is at about 3250 cm^{-1}. The remarkable constancy of these absorption positions for the carbon–hydrogen stretching mode in all organic compounds examined enables the reverse deduction to be made, i.e., absorption bands exhibited by an unknown compound in the region of 2800–2900 cm^{-1} and in the region of 3040 cm^{-1} would indicate the presence of both saturated carbon–hydrogen bonds and carbon–hydrogen bonds in an alkene or an aromatic system.

The band positions for all the major structural bonding types have been determined and correlation charts and tables are available which give the ranges within which particular bonding types have been observed to absorb. A simplifed correlation chart is provided by Fig. VII,1(iii) which indicates the ranges within which the stretching and bending absorptions have been observed. This chart has been prepared on a typical spectral grid since rapid recognition of significant absorption bands is usually achieved by such visual familiarity with wavelength regions. The alternative and more accurate and informative way of presenting correlation information is by means of tables. These have been collected in Appendix 2, Tables 2,1–2,13. It is to these tables that reference should be made when endeavouring to elucidate the structure of a compound from the infrared spectrum.

Interpretation of an infrared spectrum The spectrum of acetophenone provides the opportunity for illustrating one possible method which may be adopted to correlate the absorption bands in a spectrum with the bonding types from which they arise. It is important that as much information as possible should be extracted from the functional group region first, then further information sought in the fingerprint region as a result of these conclusions. It is usually unwise to haphazardly relate intense peaks to specific structural features and the following represents a more logical approach.* The correlation chart above may be used in this simple illustration.

Consideration of the general profile provides circumstantial evidence that the compound may be aromatic. By looking first at the absorption band in the 3000 cm^{-1} region it is apparent that there are present sp^2-hybridised carbon hydrogen bonds (absorption just above 3000 cm^{-1}) as well as sp^3-hybridised carbon hydrogen bonds (absorption just below 3000 cm^{-1}). The only other significant absorption in the region above 1600 cm^{-1} is the band at 1680 cm^{-1}.

---

* This approach, which is developed in more detail below, is based on the method used by Dr L. J. Bellamy, CBE, in his lectures and tutorials on i.r. spectroscopy to postgraduate students in the School of Chemistry and with which the editors were privileged to be associated.

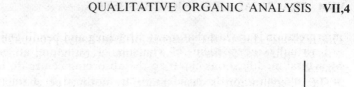

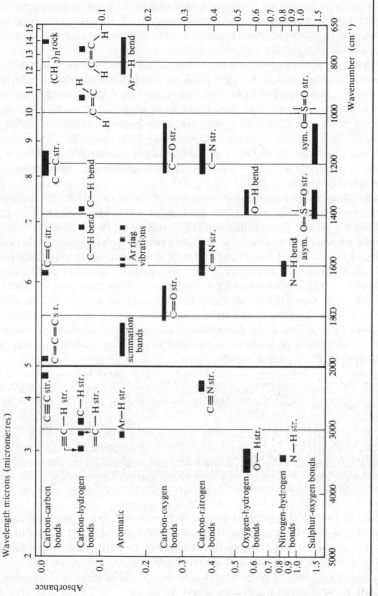

Fig. VII,1(iii) **Simplified correlation chart of absorption positions of important bonding types.**

This is clearly due to the presence of a carbonyl group. Thus the compound may be either an olefinic or an aromatic compound containing a carbonyl system.

The differentiation is easily made by noting the characteristic absorption bands corresponding to the aromatic ring system around 1600–1450 cm^{-1} and 850–660 cm^{-1}, and the characteristic pattern in the overtone region around 2000–1800 cm^{-1}. The origin of these absorptions is discussed in detail below; their presence in this spectrum clearly shows that the structure is a monosubstituted aromatic compound.

The precise position of the carbonyl band, which is at the lower end of the frequency range, may now be rationalised. It implies that the carbonyl function is in conjugation with the aromatic π-electron system leading to reduced double bond character, a weaker carbon–oxygen bond and hence a lower absorption frequency. This conjugation is confirmed by observing the increased intensity of the 1580 cm^{-1} band associated with the aromatic ring vibration.

It is now necessary to deduce the nature of the carbonyl function (i.e., whether it is an aldehyde, ketone, ester, etc.). Each of these functional groups (with the exception of ketones) exhibit further characteristic and identifiable absorption bands due to the attachment of atoms other than carbon to the carbonyl carbon atom. Thus an aldehyde should exhibit a double band in the region of 2830–2700 cm^{-1}, due to stretching of the $C-H$ bond in the aldehydic group. A carboxylic ester would exhibit a pair of intense absorptions near 1300 and 1100 cm^{-1} due to $C-O$ stretching modes. Logical, sequential and careful searching in the two regions of the spectrum for the presence or absence of diagnostic bands as indicated above would finally lead to the conclusion that the compound was a monosubstituted aromatic ketone. It is in this situation that negative information is as important as positive information, i.e., the absence of the double band at about 2800 cm^{-1} eliminates from consideration the possibility of an aldehyde. That the compound is likely to be acetophenone could be deduced from the relatively weak absorption in the region below 3000 cm^{-1} and the presence of absorption at 1370 cm^{-1}, the latter being characteristic of one of the deformation modes of a methyl group. Positive identification would be most readily achieved by consideration of the p.m.r. spectrum.

The following summary provides a recommended approach to the interpretation of an unknown spectrum which may be adopted until experience has developed an intuitive appreciation of the characteristics of infrared spectra. It should be used in association with the more detailed notes which follow, describing the way in which characteristic group frequencies arise and the variations in frequency position which accompany environmental changes.

1. Obtain a satisfactory spectrum of the unknown dry compound using the techniques which have been described in Section **I,39**. If the spectrum has been supplied, make a careful note of the conditions under which the spectrum was recorded, any solvents used, etc.
2. Qualitatively assess from the spectrum profile whether the compound is likely to be aliphatic or aromatic in type.
3. Mark any absorptions apparent in the spectrum which are known to arise from the solvent or mulling agent used.
4. Inspect the $C-H$ stretching region and identify the bands as either of aliphatic or aromatic/olefinic in origin.

5. Evaluate the degree of carbon chain branching by approximately assessing the methyl : methylene ratio from the relative intensity of the absorption bands in the saturated $C-H$ region below 3000 cm^{-1}. This may necessitate re-recording the spectrum on a grating instrument to obtain better resolution.

6. Search the high frequency end of the spectrum, i.e., the region 4000–3000 cm^{-1}, for the presence of bands arising from the presence of $-O-H$, $-N-H$ and $\equiv C-H$ bonds.

7. Extract from the spectrum information provided by the presence of relatively intense absorption bands in the region 2500–1600 cm^{-1}. This should provide evidence for the presence or absence of $C=C$, $C\equiv N$, $C=O$, $C=C$.

8. As a result of the conclusions deduced from 4–7, attempt to classify the compound; on the basis of this classification search the fingerprint region for specific evidence to support the postulated structure. Examples are: (a) if an aromatic compound is suspected because of $=C-H$ absorption, confirm by examination of the region 1600, 1580, 1500 and 1450 cm^{-1} and then endeavour to establish the substitution pattern by looking specifically in the 850–650 cm^{-1} region, and then in the overtone region, 2000–1800 cm^{-1}; (b) if an alkene is suspected, search for evidence of its substitution type; (c) if a carbonyl group is present, deduce its nature by searching for evidence of the presence of associated groups, etc.

9. If no absorption bands are present in the functional group region, with the exception of those arising from carbon–hydrogen stretching modes, consider the possibilities of ethers, alkyl halides, sulphur compounds, tertiary amines and nitro compounds as detailed in the sections below.

10. Relate the structural information deduced in this way with that obtained by other spectral methods or by appropriate chemical tests.

Characteristic group frequencies

Alkanes, Cycloalkanes and Alkyl Groups The diagnostically important bands in these compounds arise from $C-H$ stretching and bending vibrations, although some bands due to $C-C$ skeletal vibrations are also of value.

Alkanes. Methyl and methylene groups both have asymmetric and symmetric $C-H$ stretching vibration modes, giving rise to four absorption bands just below 3000 cm^{-1}; the CH_3 vibration modes are shown in Fig. VII,1(iv), and the CH_2 vibrations are those depicted in Fig. VII,1(i). The absorption bands are not normally resolved by prism spectrophotometers, and in the spectrum of decane (Fig. VII,1(v)) recorded on such an instrument the $C-H$ vibrations are revealed as two overlapping bands just below 3000 cm^{-1}. With grating instruments the absorption bands are resolved so that the CH_3 asymmetric and symmetric vibrations which occur near 2962 and 2872 cm^{-1} respectively, and the CH_2 asymmetric and symmetric vibrations which occur near 2926 and 2853 cm^{-1} respectively, are clearly visible. These absorption positions do not vary much in the case of unsubstituted alkanes. However, very useful qualitative information can be obtained, regarding the relative number of CH_3 and CH_2 groups in an alkane, by inspection of the relative intensities of these bands, since these are dependent on the number of such groups present in

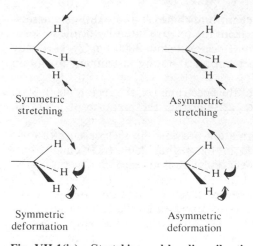

Symmetric
stretching

Asymmetric
stretching

Symmetric
deformation

Asymmetric
deformation

Fig. VII,1(iv) Stretching and bonding vibrational modes for a methyl group.

a compound. This is illustrated by the part spectra recorded on a high resolution (grating) spectrophotometer shown in Fig. VII,1(vi(*a–c*)). Spectrum (*a*) is that of hexane which has the same number of methyl and methylene hydrogens so that the corresponding bands are of approximately equal intensity. In decane (spectrum (*b*)), on the other hand, the two CH_2 bands are much more intense than the CH_3 bands; in cyclohexane (spectrum (*c*)) only the two CH_2 absorption bands are observed. A highly branched alkane will thus show strong CH_3 and weak CH_2 bands. Tertiary $C-H$ stretching vibrations produce a weak band near 2890 cm^{-1} which is often masked by the other $C-H$ bands.

The CH_3 group has two $C-H$ deformation vibrational modes which are shown in Fig. VII,1(iv); the asymmetric vibration gives a band near 1450 cm^{-1} and the symmetric 'umbrella-like' vibration a band near 1375 cm^{-1}. The four possible bending modes of the CH_2 group are those shown in Fig. VII,1(i); the scissoring vibration gives a band near 1465 cm^{-1} which overlaps with the asymmetric band near 1450 cm^{-1}. The position of the CH_3 band (1375 cm^{-1}) is remarkably constant when attached to carbon and this allows ready recognition of the $C-CH_3$ group in a molecule. When a second methyl group is attached to the same carbon atom as in the isopropyl group, splitting of this band occurs to give two bands of approximately equal intensity. The 1375 cm^{-1} band arising from a t-butyl group is also split, but in this case the intensities of the two bands are in the ratio of approximately 2 : 1 with the less intense band at higher frequencies.

Compounds containing at least four adjacent methylene groups, i.e., $-(CH_2)_n-$, n $\geqslant$ 4, show a weak band near 725 cm^{-1} due to the four groups rocking in phase; this band increases in intensity with increasing length of the chain.

Cycloalkanes. The $C-H$ stretching vibrations of unstrained ring systems give rise to bands in the same region of the spectrum as acyclic compounds; as the size of the ring decreases however there is a shift to higher frequency and cyclopropanes give a band in the 3060–3040 cm^{-1} region of the spectrum. The absence of CH_3 stretching bands in the spectrum of cyclo-

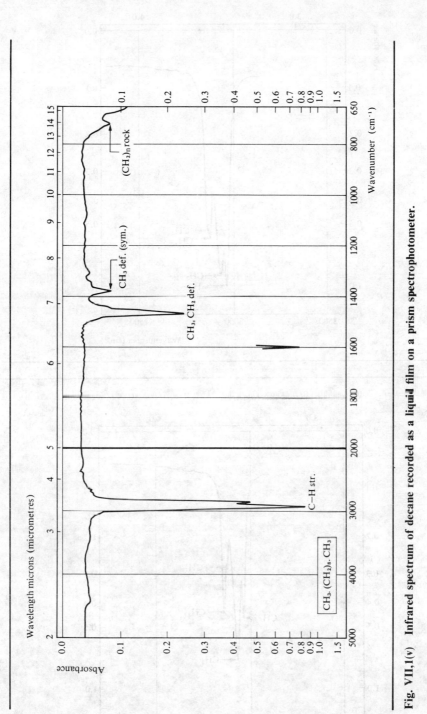

Fig. VII,1(v) Infrared spectrum of decane recorded as a liquid film on a prism spectrophotometer.

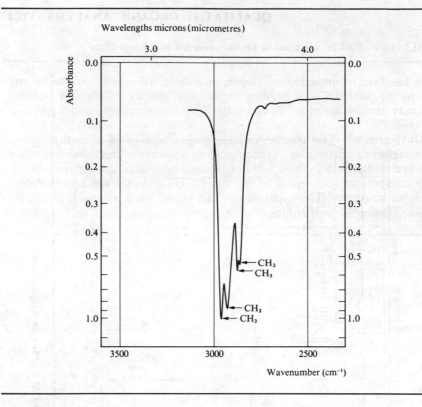

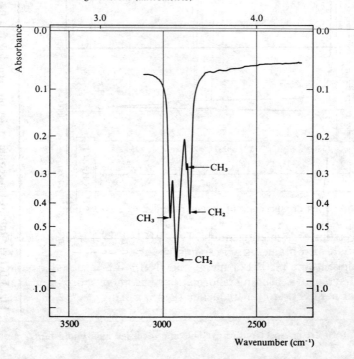

Fig. VII,1(vi(a)) Part i.r. spectrum of hexane recorded as a liquid film.

hexane has been commented on above, and there will not of course be any bands in the methyl $C-H$ bending region; the absence of the CH bending bands may also be noted in the spectrum of cyclohexanecarboxaldehyde, Fig. VII,1(xxii).

Alkyl groups. The symmetric stretching vibration of a methyl group when attached to nitrogen or oxygen results in absorption at a lower frequency than when attached to carbon; additionally, the symmetric deformation vibration in compounds containing NCH_3 and OCH_3 groups leads to absorption at a higher frequency. These frequency shifts are of some diagnostic value for the identification of such groups (e.g., the spectrum of anisole, Fig. VII,1(xviii),

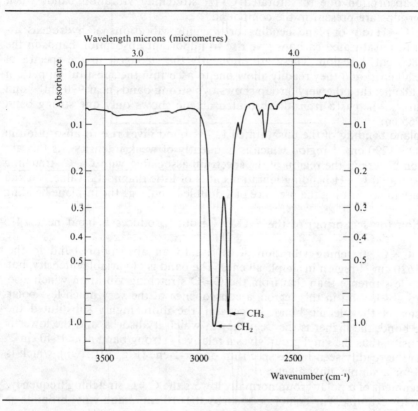

Fig. VII,1(vi(b)) Part i.r. spectrum of decane recorded as a liquid film.

and the spectrum of *N*-methylaniline, Fig. VII,1(xx)), although reliable confirmation would be obtained from a p.m.r. spectra.

The attachment of a methyl or methylene group to a carbonyl group results in the $C-H$ symmetric bending deformations becoming more intense and the bands appear at slightly lower frequency than normally.

Fig. VII,1(vi(c)) Part i.r. spectrum of cyclohexane recorded as a liquid film.

The correlation tables for alkanes, cycloalkanes and alkyl groups are in Appendix 2, Table 2,1.

Alkenes. The presence of unsaturation in a molecule can usually be readily recognised by the presence of a small sharp band just above 3000 cm^{-1} due to the unsaturated $=C-H$ stretching vibration. Aromatic $C-H$ bonds also show weak absorption in this region, but there are, however, other distinguishing features in the i.r. of aromatic compounds which readily enable them to be differentiated from alkenes. The precise position of the band depends on the nature of the alkene. For example, the terminal methylene group in vinyl ($-CH=CH_2$) and gem-disubstituted ($-\overset{|}{C}=CH_2$) alkenes absorb in the 3095–3075 cm^{-1} range, and the $=C-H$ bond in *cis*-, *trans*- and trisubstituted alkenes absorbs at 3040–3010 cm^{-1}, a region which may be masked by strong absorption due to saturated $C-H$ stretching vibration bands when alkyl groups are present in the compound.

The $C-H$ out-of-plane bending or wagging vibrations of hydrogens attached to unsaturated carbons give rise to important absorption bands in the 1000–800 cm^{-1} region. These are frequently the strongest in the spectra of simple alkenes, and they readily allow one to ascertain the substitution pattern of the alkene; thus the vinyl group shows two strong bands near 990 cm^{-1} and 910 cm^{-1}, whereas a *trans*-disubstituted alkene shows only one strong band near 965 cm^{-1}.

In-plane bending of the unsaturated $C-H$ bond gives rise to absorption in the 1420–1290 cm^{-1} region which is frequently of weak intensity. As this absorption occurs in the region of the spectrum associated with $C-C$ stretching and saturated $C-H$ bending vibrations, it is of little diagnostic value, but can be of use in confirming the presence of a double bond, e.g., the in-plane bending vibration (or scissoring) of the $=C\overset{\diagup H}{\underset{\diagdown H}{}}$ group produces a band near 1415 cm^{-1}.

The $C=C$ stretching vibration gives rise to an absorption band in the 1680–1620 cm^{-1} region in simple alkenes. The band is of variable intensity, but is much less intense than that from the $C=O$ stretching vibration which also leads to absorption in this region, a consequence of the very much less polar character of the olefinic bond. In general, the more highly substituted the double bond, the higher is the frequency at which it absorbs, and the lower is its intensity; thus the vinyl group gives a relatively strong band near 1640 cm^{-1} and can be readily seen in the spectrum of oct-1-ene (Fig. VII,1(vii)), which is typical for a simple vinyl alkene.

Attachment of a polar group normally lowers the $C=C$ stretching frequency, so that, for example, vinyl chloride absorbs at 1610 cm^{-1} and vinyl bromide at 1593 cm^{-1}; slight frequency shifts outside the above range can also occur when the double bond is exocyclic to a ring system; thus a methylene group attached to six-, five-, four- and three-membered ring systems absorb respectively at 1651 cm, 1657 cm, 1678 cm and 1736 cm^{-1}, a shift which is associated with increasing ring strain. It should be noted that there will be no $C=C$ stretching absorption band in the spectrum of symmetrically substituted *trans*-olefinic compounds such as *trans*-1,2-dichloroethylene and fumaric acid. Despite the fact that there are highly polar bonds in each of these compounds, because of the symmetry of the molecules, stretching of the $C=C$ bond does not result in any change in the oscillating dipole moment, so this vibration is

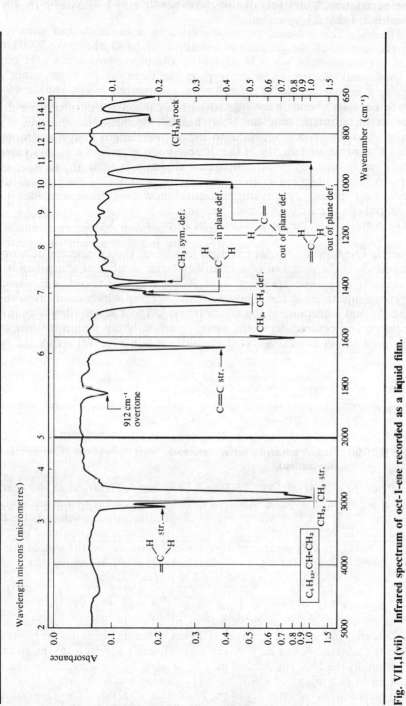

Fig. VII,1(vii) Infrared spectrum of oct-1-ene recorded as a liquid film.

infrared inactive. The vibrations of these bonds may however be readily observed in the Raman spectrum.

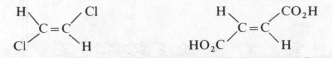

When a $C=C$ bond is conjugated with a carbonyl group, another double bond or an aromatic ring, the bond has less double-bond character, i.e., is weaker, and the absorption shifts to longer wavelength (lower wavenumber), whilst the increased polarity of the double bond results in a considerable enhancement in intensity. With conjugated aliphatic systems, the number of absorption bands observed is the same as the number of conjugated double bonds; thus dienes, trienes and tetraenes show two, three and four bands respectively in the 1650–1600 cm^{-1} region.

The correlation table for alkenes is in Appendix 2, Table 2,2.

Aromatic compounds A characteristic feature of the i.r. spectra of aromatic compounds is the presence of a relatively large number of sharp bands, and particularly diagnostic are those near 3030 cm^{-1} due to $=C-H$ stretching vibrations and those in the 1600–1450 cm^{-1} region which result from the in-plane skeletal vibrations of the aromatic ring. These latter vibrations involve expansion and contraction of the carbon–carbon bonds within the ring of the type indicated in the exaggerated formulations in Fig. VII,1(viii); the bands

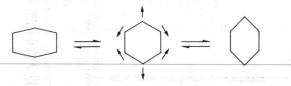

Fig. VII,1(viii) **Some infrared active aromatic ring vibrations (exaggerated for illustration).**

usually occur near 1600, 1580, 1500 and 1450 cm^{-1}. The band at 1450 cm^{-1} is often quite strong but since it occurs in the absorption region associated with the alkyl $C-H$ bonding vibrations its diagnostic value is somewhat limited. The intensities of the other bands vary widely; in particular the band near 1580 cm^{-1} is normally very weak and appears as a shoulder on the side of the 1600 cm^{-1} band. When a carbonyl or other similar group is conjugated with the ring, however, the intensity of the band is increased; this effect can be clearly seen by comparison of the spectrum of *o*-xylene, Fig. VII,1(ix), with that of acetophenone, Fig. VII,1(ii), and the phenylacetylene, Fig. VII,1(xii).

It is useful to note that the variations in intensity of the 1580 cm^{-1} band parallels that of the much stronger 1600 cm^{-1} band, which also exhibits large intensity fluctuations and may indeed be completely absent from the spectrum. Occasionally the 1600 cm^{-1} band may be masked by other bands such as those resulting from conjugated $C=C$ or NH_2 groups which absorb in this region. Wide variations in the intensity of the band near 1500 cm^{-1} can also occur, but in general one or other of the 1500 and 1600 cm^{-1} bands will be quite strong, and notwithstanding these intensity fluctuations, there is usually no difficulty in

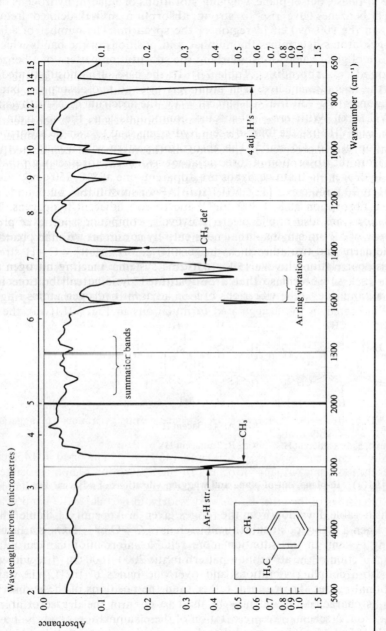

Fig. VII,1(ix) Infrared spectrum of o-xylene recorded as a lquid film.

practice in establishing the presence of an aromatic ring. Polycyclic aromatics such as naphthalene, and also pyridines, show bands in very similar positions.

The in-phase, out-of-plane, wagging vibrations of adjacent hydrogens of substituted benzenes give rise to strong absorption in well-defined frequency ranges in the 900–690 cm^{-1} region of the spectrum. The number of adjacent hydrogen atoms determines the number and positions of the bands which are therefore of great value for establishing the substitution pattern of a benzenoid compound (see Appendix 2, Table 2,3). In the case of monosubstituted benzenes the five adjacent hydrogen atoms give rise to two absorption bands in the region of 770–730 and 710–690 cm^{-1} (cf. the spectrum of N-methylaniline, Fig. VII,1(xx)). With *ortho*-substituted compounds (e.g., the spectrum of *o*-xylene, Fig. VII,1(ix), the four adjacent hydrogens lead to a single absorption in the region of 770–735 cm^{-1}. The absorption pattern with a *meta* compound arises from the absorption of three adjacent hydrogens together with that of a single hydrogen, so that two bands are apparent, one at 810–750 cm^{-1} and the second at 900–860 cm^{-1} [Fig. VII,1(x(*a*))]. *Para* substituted compounds show a single absorption at 860–800 cm^{-1} due to two adjacent hydrogens. These correlations are also applicable to polycyclic compounds; for example, the spectrum of 1,2-dimethylnaphthalene is entirely consistent with the presence of four adjacent and two adjacent hydrogens [Fig. VII,1(x(*b*))].

These correlations hold also for pyridine systems; the ring nitrogen atom counts as a substituent so that a 2-substituted pyridine will be expected to show a band due to the vibrations of four adjacent hydrogen atoms.

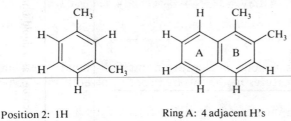

Position 2: 1H

Positions 4, 5, 6: 3 adjacent H's

(*a*)

Ring A: 4 adjacent H's

Ring B: 2 adjacent H's

(*b*)

Fig. VII,1(x) In-plane, out-of-plane and wagging vibrations of adjacent H's.

Considerable deviation from the ranges given in Appendix 2, Table 2.3, can occur when a highly polar substituent such as $-C=O$ or $-NO_2$ is attached to the ring system. In this situation more reliable information can be obtained from inspection of the absorption pattern in the 2000–1600 cm^{-1} region, which arises from coupling vibrations and overtone bands of the C–H wagging vibrational modes. The pattern, rather than the positions of these absorption bands, is characteristic and may be used to confirm the degree of aromatic substitution. A schematic representation of these summation bands is shown in Fig. VII,1(xi). It should be noted that the absorption is extremely weak and the bands are frequently barely visible when the spectrum is recorded under normal conditions. Hence to enable the absorption patterns to be recognised, neat liquids or concentrated solutions of solids ($\sim 10\%$) should be examined in a 1.0-mm cell.

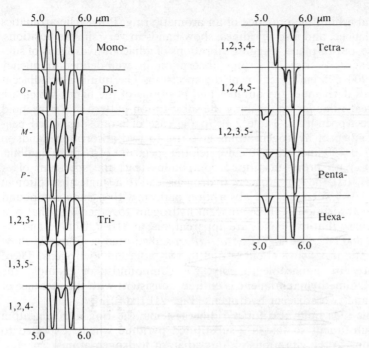

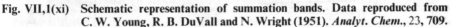

Fig. VII,1(xi)   **Schematic representation of summation bands. Data reproduced from C. W. Young, R. B. DuVall and N. Wright (1951). *Analyt. Chem.*, 23, 709.**

The correlation table for aromatic compounds is in Appendix 2, Table 2,3.

Alkynes and allenes   Absorptions arising from multiple bond stretching vibrations are important features in the infrared spectra of both these types of compounds.

Alkynes.   Monosubstituted alkynes are characterised by a strong sharp absorption band near 3320 cm^{-1} arising from the $\equiv C-H$ stretching vibration. Bonded $N-H$ and $O-H$ bands also appear in this region but they are, in contrast, quite broad and cannot be confused with the $\equiv C-H$ band. The $C\equiv C$ stretching vibration gives rise to a weak absorption in the 2260–2100 cm^{-1} region of the spectrum, and a frequency difference of about 100 cm^{-1} between mono- and disubstituted alkynes allows them to be differentiated. The intensity of the $C\equiv C$ band is variable, and whilst it is readily observed in the spectrum of a monosubstituted alkyne, it may be very weak or absent from the spectrum of a disubstituted alkyne, depending on the nature of the substituents. Symmetrically disubstituted alkynes, such as acetylenedicarboxylic acid, do not exhibit absorption whilst, in contrast, if the two substituents are sufficiently different in character that the $C\equiv C$ bond is made more polar, a relatively strong band may be observed. The $C-H$ bending absorption of monosubstituted acetylenes occurs in the range 680–610 cm^{-1} and is usually quite strong; aromatic acetylenic compounds show two bands in this region. A broad band in the 1300–1200 cm^{-1} range is believed to be an overtone or combination bond derived from the $C-H$ bending vibration. The spectrum of phenylacetylene, Fig. VII,1(xii), is an instructive example; in addition to the strong

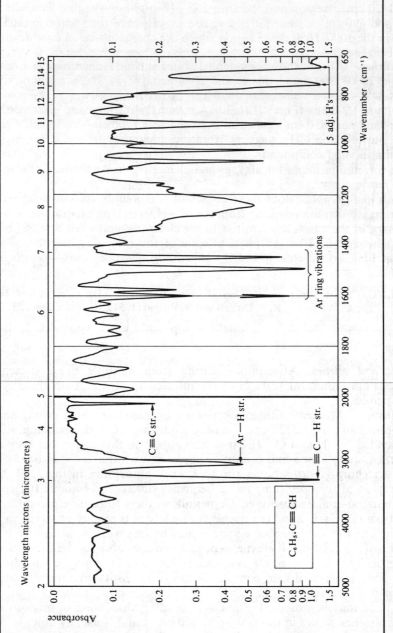

Fig. VII,1(xii) Infrared spectrum of phenylacetylene recorded as a liquid film.

$\equiv C - H$ band at 3310 cm^{-1} and $C \equiv C$ stretching band at 2100 cm^{-1}, other noteworthy and clearly visible features include the sharp aromatic $C - H$.stretching band at 3070 cm^{-1}, the fairly strong 'ring breathing' vibration bands in the 1600–1420 cm^{-1} region and the strong $C - H$ in-phase wagging bands at 770 cm^{-1} and 690 cm^{-1}; these latter give rise to weak overtone and combination bands in the 2000–1650 cm^{-1} region which are characteristic of a monosubstituted benzene.

Allenes. Allenes show a moderately intense band (sometimes as a double peak) at 2000–1900 cm^{-1} due to the asymmetric $C = C = C$ stretching vibration, which can be seen (near 1970 cm^{-1}) in the spectrum of 1-bromo-3-methylbuta-1,2-diene (Fig. VII,1(xiii)). A terminal $=CH_2$ group gives rise to a strong band near 850 cm^{-1}, with an overtone near 1700 cm^{-1}, and is the result of the out-of-plane CH_2 wagging vibration analogous to the CH_2 wagging vibration in vinyl compounds.

The correlation tables for alkynes and allenes are in Appendix 2, Table 2,4.

Alcohols and phenols Both these classes of compounds are characterised by the strong absorption resulting from the $O - H$ stretching modes; the position and shape of the bands are sensitive to the electronic and steric features of the compound and also to the physical state of the sample. Absorption bands arising from $C - O$ stretching and $O - H$ bending vibrations are also of diagnostic value.

Examination of dilute solutions of simple alcohols or phenols in a non-polar solvent such as carbon tetrachloride reveals the free $O - H$ stretching band in the 3650–3590 cm^{-1} region; the precise position of the band has been correlated with the nature of the carbon atom to which the hydroxyl group is attached. Thus the absorption frequency shifts to lower values in the order primary, secondary, tertiary or phenolic hydroxyl. However, definitive assignment of the group associated with the hydroxyl on the basis of the position of this band only is not advisable. More usually the spectrum will be recorded in the neat liquid or solid state, and in this case the $O - H$ band is recognised by a strong broad band in the 3400–3200 cm^{-1} region; see the spectra of heptan-1-ol (Fig. VII,1(xiv)) and *m*-cresol (Fig. VII,1(xv)). This broadening and shift to lower frequency is due to **intermolecular hydrogen bonding** which results in a weakening of the $O - H$ bond. In simple compounds such as heptan-1-ol, hydrogen bonding is of the polymeric type, although in solution in a non-polar solvent the hydrogen bonds are partly broken and the spectra normally show additionally some free OH absorption. As the solution is made more dilute the extent of hydrogen bonding is decreased resulting in a decrease in the intensity of the bonded band, and an increase in the intensity of the absorption due to the free OH group. This effect is illustrated in Fig. VII,1(xvi(a–c)).

Some compounds, such as highly substituted alcohols and *ortho* substituted phenols, are unable, for steric reasons, to form polymeric hydrogen bonded species, and hence they exist only as dimers which gives rise to sharp absorption in the 3550–3450 cm^{-1} region. In these instances hydrogen bonds are also broken on dilution with the consequence that the absorption intensity and position changes.

1,2-Diols and phenols having a carbonyl or nitro group in the *ortho* position exhibit **intramolecular hydrogen bonding**, which is not affected by dilution; hence solution spectra and the effects resulting from dilution can give consider-

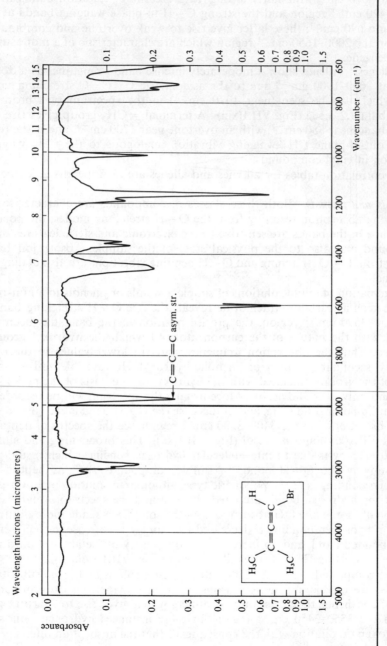

Fig. VII,1(xiii) Infrared spectrum of 1-bromo-3-methylbuta-1,2-diene recorded as a liquid film.

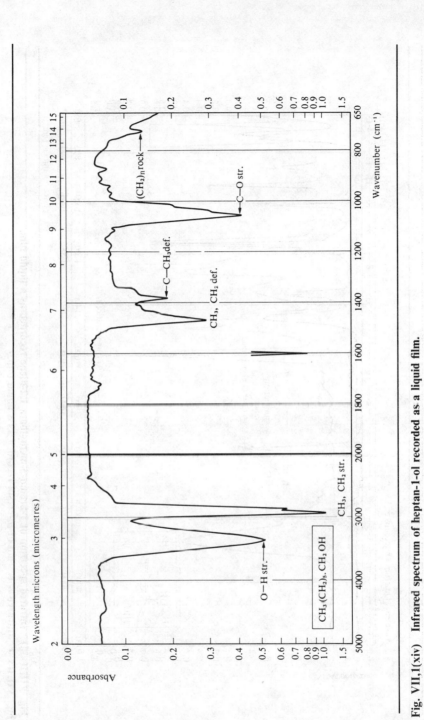

Fig. VII,1(xiv)   Infrared spectrum of heptan-1-ol recorded as a liquid film.

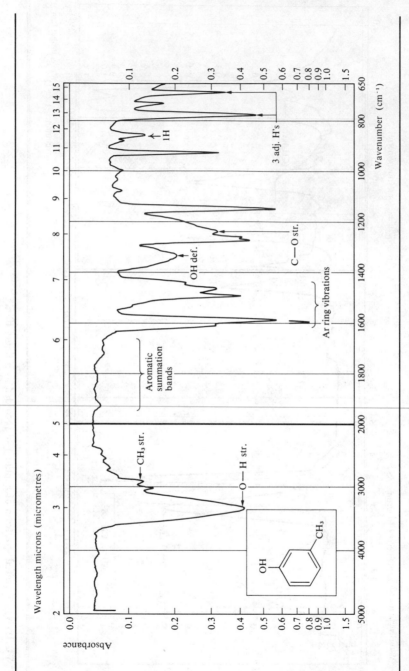

Fig. VII,1(xv) Infrared spectrum of *m*-cresol recorded as a liquid film.

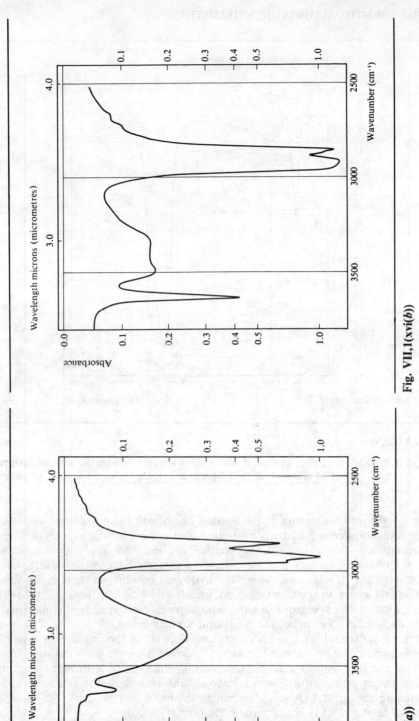

Fig. VII,1(xvi(a))

Fig. VII,1(xvi(b))

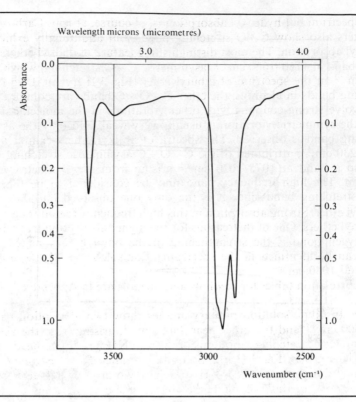

Fig. VII,1(xvi(c))

Fig. VII,1(xvi(a–c)) Effect of dilution with carbon tetrachloride on the solution spectrum of heptan-1-ol, (a) 2.5 per cent w/v; (b) 1 per cent w/v; (c) 0.5 per cent w/v.

able insight into the nature of the alcohol or phenol. In compounds such as o-hydroxyacetophenone, hydrogen bonding is extremely strong as a consequence of resonance stabilisation of the bonded species, and absorption is in the 3200–2500 cm⁻¹ region, whilst 1,2-diols show sharp bands of variable intensity at 3570–3450 cm⁻¹, reflecting the weaker hydrogen bonding in these compounds.

The OH group in a carboxylic acid, and N–H bonds in general, absorb in this region of the spectrum but the bands are usually readily distinguishable from each other (see carboxylic acids and amines below).

The C–O stretching band is strong and appears in the fingerprint region of the spectrum. The position is somewhat dependent on the physical state of the sample but it is usually possible to ascertain the type of hydroxyl compound under investigation; thus m-cresol shows absorption in the phenolic C–O stretching region at 1330 cm⁻¹, whereas the band at 1060 cm⁻¹ in the spectrum of heptan-1-ol is characteristic for primary alcohols.

The correlation tables for alcohols and phenols are in Appendix 2, Table 2,6.

Ethers and cyclic ethers The infrared spectrum of an ether, like that of an alcohol or a phenol, exhibits a very strong C–O band in the fingerprint region of the spectrum but hydroxyl absorption is, of course, absent. Carboxylic acids and esters also show $C-O$ stretching bands, but additionally exhibit strong carbonyl absorption. The most distinguishing feature of dialkyl ethers is a very strong band at 1150–1060 cm^{-1} (asymmetric $C-O$ stretching) which is seen at 1120 cm^{-1} in the spectrum of dibutyl ether (Fig. VII,1(xvii)). Like the $C-O$ stretching bands in alcohols, the ether $C-O-C$ bands are complex in origin and involve strong coupling with other vibrations in the molecule. In dialkyl ethers the symmetric vibration is usually very weak and only the asymmetric stretching band is observed. The spectra of aralkyl ethers exhibit a band at 1270–1230 cm^{-1}, attributed to the $C-O-C$ asymmetric stretching vibration and also a band at 1075–1020 cm^{-1} arising from the symmetric $C-O-C$ vibration. The high frequency band may be considered to be the aromatic $C-O$ stretching band since it is the only one observed (at 1240 cm^{-1}) in diphenyl ether. Strong absorption in this high frequency region is also observed with vinyl ethers. One of the reasons for the higher absorption frequency of the aryl–oxygen bond is the strengthening of the bond by resonance. The two bands can be identified in the spectrum of anisole (Fig. VII,1(xviii)) at 1240 cm^{-1} and 1040 cm^{-1}.

The correlation tables for alcohols and phenols are in Appendix 2, Table 2,6.

Amines In dilute solution primary amines show two absorption bands, one near 3500 cm^{-1} and the other near 3400 cm^{-1}, arising from the asymmetric and symmetric stretching vibrations of the two NH bonds (cf. the vibrations of a methylene group, Fig. VII,1(i)). Secondary amines show just one band near 3300 cm^{-1} due to the single $N-H$ stretching vibration, whilst tertiary amines do not absorb in this region. These characteristic absorptions allow one to distinguish readily between the three classes of amines. Imines also show a single band in the 3500–3200 cm^{-1} region. The bands are shifted to lower frequencies in the condensed phase as a result of hydrogen bonding which is, however, much weaker than in hydroxyl compounds because of the lower electronegativity of nitrogen. Because of the weaker hydrogen bonding this frequency shift is not so great and the absorption bands tend to be appreciably narrower than the corresponding bonded OH absorption bands. Aromatic primary and secondary amines absorb at slightly higher frequencies than the corresponding aliphatic amines, and the separation between the symmetric and asymmetric stretching bands in the aromatic primary amines is greater; these spectral differences have been rationalised on the basis of interaction of the lone electron pair on the nitrogen atom with the aromatic ring in arylamines resulting in a shorter, stronger, bond.

A medium to strong broad band in the 1650–1590 cm^{-1} region is characteristic of aliphatic primary amines and arises from the NH$_2$ scissoring vibration; additionally these amines show a medium–strong, broad, multiple absorption band at 850–750 cm^{-1} arising from NH$_2$ twisting and wagging deformations. These absorptions are clearly seen in the spectrum of butylamine (Fig. VII,1(xix)), which also shows the characteristic rather broad hydrogen bonded NH$_2$ stretching band as a closely spaced doublet at 3350 cm^{-1}.

In aliphatic secondary amines the $N-H$ bending band is usually absent from the spectrum or else it is very weak and appears at 1650–1550 cm^{-1}. In aromatic secondary amines the band is of medium intensity, and appears in the same region, but the assignment is complicated by the presence of aromatic

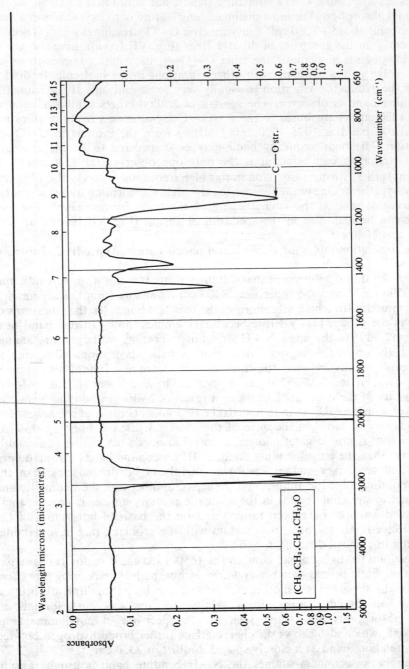

Fig. VII,1(xvii) Infrared spectrum of dibutyl ether recorded as a liquid film.

$(CH_3.CH_2.CH_2.CH_2)_2O$

C—O str.

Wavelength microns (micrometres)

Wavenumber (cm^{-1})

Absorbance

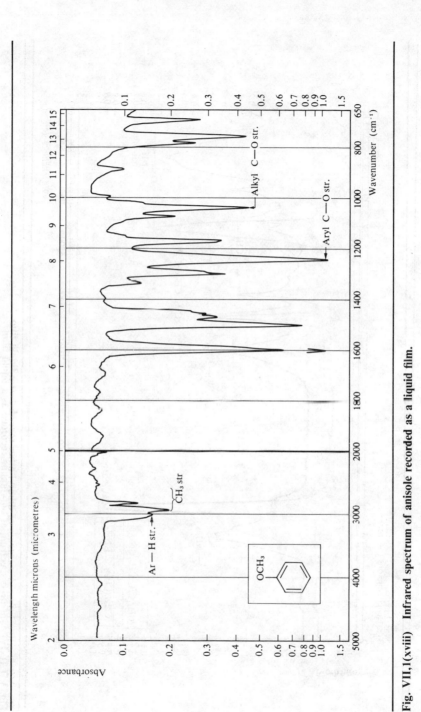

Fig. VII,1(xviii) Infrared spectrum of anisole recorded as a liquid film.

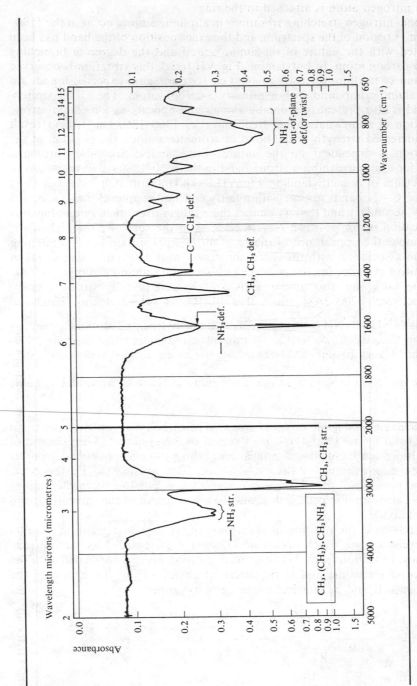

Fig. VII,1(xix) Infrared spectrum of butylamine recorded as a liquid film.

ring vibrations which also occur here, and which are sometimes intensified when a nitrogen atom is attached to the ring.

Carbon–nitrogen stretching vibrations in aliphatic amines occur in the 1190–1020 cm^{-1} region of the spectrum, and the exact position of the band has been correlated with the nature of the amino group and the degree of branching at the α-carbon atom. In butylamine, Fig. VII,1(xix), this vibration is observed as a band of medium intensity at 1080 cm^{-1}, the region expected for an aliphatic amino compound with a primary α-carbon atom. The $C-N$ stretching band is also present in aromatic amines and appears as a medium–strong absorption at somewhat higher frequency, i.e., 1360–1250 cm^{-1}, as a result of the increased strength of the bond in aromatic amines; the position of the absorption is dependent on the nature of the amine. Secondary aromatic amines for example exhibit a strong band in the 1350–1280 cm^{-1} region, as in the spectrum of N-methylaniline, Fig. VII,1(xx). Unfortunately although these aromatic $C-N$ bands appear within fairly constant frequency ranges for the primary, secondary and tertiary amines these ranges overlap and unambiguous identification is not possible. Nevertheless, as in the case of N-methylaniline, for example, the appearance of the band in the appropriate $C-N$ stretching region in association with the sharp absorption near 3400 cm^{-1} can be taken as definitive evidence for the presence of a secondary amino group.

Amine salts and also amino acids are characterised by strong absorptions between 3200–2800 cm^{-1} due to the $N-H$ stretching bands of the ions $-\overset{\oplus}{N}H_3$, $>\overset{\oplus}{N}H_2$, etc. (cf. the methyl, methylene and methine stretching bands in this region), as well as by multiple combination bands in the 2800–2000 cm^{-1} region and $N-H$ bending vibrations in the 1600–1400 cm^{-1} region.

The correlation tables for amines and amine salts are in Appendix 2, Table 2,7.

Compounds containing the carbonyl group   The carbonyl group gives rise to an intense band in the 1900–1560 cm^{-1} region of the spectrum. With the aid of other absorption bands the identification of the particular functional group is possible (i.e., whether it is a ketone, aldehyde, ester, amide, etc.). Furthermore, from the position of the absorption frequency it is possible to extract a considerable amount of information about the environment of the carbonyl group in the molecule.

As a reference, the position of absorption of the carbonyl group in a saturated acyclic ketone which occurs at 1720 cm^{-1} is regarded as the 'normal' frequency. Deviation from this absorption position may be correlated with the influence of electronic and steric effects which arise from the nature of the substituents (R and X) attached to the carbonyl group.

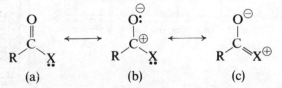

(a)   (b)   (c)

These effects may be broadly summarised thus; a more detailed consideration of specific cases is exemplified under each functional group type.

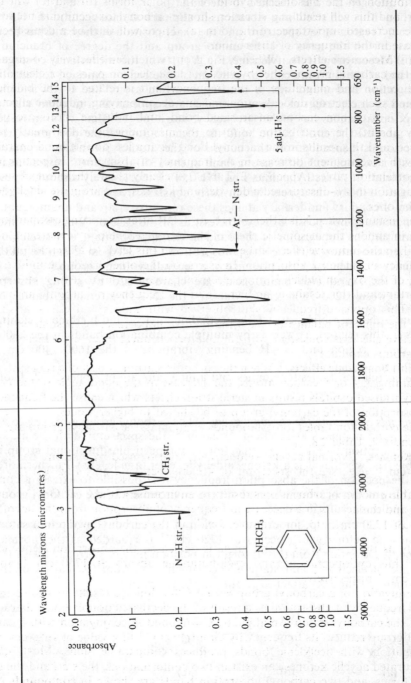

Fig. VII,1(xx)   Infrared spectrum of *N*-methylaniline recorded as a liquid film.

(i) **Inductive effects.** When X is an electron-attracting group (e.g., Cl) the contribution to the mesomeric hybrid of the polar forms (b) and (c) will be lower, and this will result in a stronger, shorter carbonyl–oxygen bond because of the increased importance of the form (a). There will thus be a consequent increase in the frequency of absorption.

(ii) **Mesomeric effects.** When X is a group which can effectively conjugate with the carbonyl group, either by virtue of lone electron pairs or π-electrons, the direction and magnitude of the frequency shift is related to the balance between such electron delocalisation and any accompanying inductive effects.

If X is a carbon–carbon unsaturated bond (and inductive effects are virtually absent) the contribution of (c) to the mesomeric hybrid is greatly increased, and this results in the carbonyl bond having less double bond character, with a consequent decrease in the frequency of absorption. Inspection of the correlation tables (Appendix 2, Table 2,8) clearly shows the effect of such conjugation in α,β-unsaturated aldehydes and ketones, and aromatic aldehydes and ketones.

The instances when X is either $-\overset{\bullet\bullet}{N}H_2$ or $-\overset{\bullet\bullet}{O}R$ offer interesting comparison. Thus in amides the mesomeric effect of the nitrogen lone pair is more important than the inductive effect of the nitrogen, and this leads to a decrease in the frequency of carbonyl absorption. In esters, on the other hand, the inductive effect of the oxygen (which is more electronegative than nitrogen) is the more important and this results in an increase in the frequency of carbonyl absorption. This opposite frequency shift observed with amides and esters, arising from the different balance between the relative importance of mesomeric and inductive effects is consistent with the relative chemical reactivity which these two groups exhibit.

(iii) **Bond angle effects.** When the carbonyl–carbon is part of a ring system containing 3, 4 or 5 carbon atoms, the decrease in the bond angle of the two sp^2-hybridised orbitals results in steric strain effects which cause the frequency of absorption of the carbonyl group to be shifted to higher values.

The correlation tables for compounds containing the carbonyl group are in Appendix 2, Table 2.8.

Ketones. Normal acyclic ketones can be recognised by a strong band at 1720 cm^{-1} (e.g., see the spectrum of 4-methylpentan-2-one, Fig. VII,1(xxi)). Branching at the α-carbon atoms results in an increase in the $C-\hat{C}O-C$ bond angle and this results in a decrease in frequency of absorption from the normal value of 1720 cm^{-1} to, for example, 1697 cm^{-1} as in di-t-butylketone. Conversely as the $C-\hat{C}O-C$ bond angle is decreased the absorption frequency rises, thus cyclopentanone and cyclobutanone absorb at 1750 cm^{-1} and 1775 cm^{-1} respectively.

Conjugation of a carbonyl group with a $C=C$ linkage results in a lowering of the frequency of absorption as a result of the decreased double bond character of the carbonyl group. An aliphatic $C=C$ bond in conjugation with a carbonyl group reduces its frequency by about 40 cm^{-1} to a value of 1680 cm^{-1}; conjugation with acetylenic bonds produces comparable shifts. Most α,β-unsaturated acyclic ketones can exist in two conformations, the s-*cis* and the s-*trans* forms, and two carbonyl absorption bands are shown in compounds of this type. Thus methyl vinyl ketone absorbs at 1716 and 1686 cm^{-1}, and it is assumed that the lower frequency band is due to the *trans* form in which

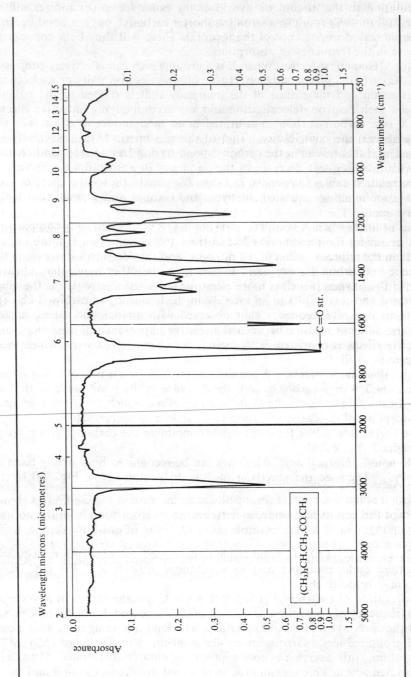

(CH₃)₂CH.CH₂.CO.CH₃

Fig. VII,1(xxi) Infrared spectrum of 4-methylpentan-2-one.

electron delocalisation is expected to be more effective. The effects of conjugation and ring size are additive and may be used to predict the positions of absorption in more complex compounds.

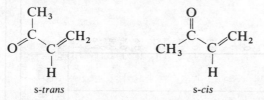

s-trans   s-cis

Conjugation with an aryl group shifts the frequency to 1715–1695 cm⁻¹; acetophenone itself absorbs at 1692 cm⁻¹ (Fig. VII,1(ii)). The band position is dependent also on the nature and position of any ring substituents.

Introduction of a halogen on the α-carbon atom of a ketone leads to a shift to higher frequency provided that the halogen can rotate to eclipse the carbonyl group. The frequency of carbonyl absorption is unaffected by the presence of further halogens on the α-carbon but it can, however, be increased by further substitution on the α′-carbon atom. The magnitude of these frequency shifts increase in the order Br, Cl, F and there is no doubt that the shifts arise from a field effect. The *cis–trans* isomers of 4-t-butylchlorocyclohexanone can be distinguished in this way. In the preferred conformation of the *cis*-isomer the carbonyl group and the equatorial halogen are eclipsed, resulting in an increase of the frequency of absorption of the carbonyl group. On the other hand, in the preferred conformation of the *trans*-isomer the halogen is axial and the absorption frequency of the carbonyl group in this case is similar to 4-t-butylcyclohexanone.

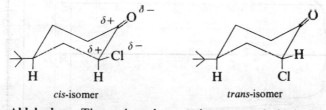

cis-isomer   trans-isomer

Aldehydes. The carbonyl group in saturated aldehydes absorb at slightly higher frequencies, 1740–1730 cm⁻¹, than are observed in saturated ketones. As expected, α,β-unsaturation, or attachment of the carbonyl carbon to an aromatic ring, causes a shift of absorption to lower frequencies; aromatic aldehydes generally absorb near 1715–1695 cm⁻¹. Special structural features can, however, cause a large frequency shift of the carbonyl absorption, as in salicylaldehyde, where internal (chelated) hydrogen bonding results in absorption at 1666 cm⁻¹.

Aldehydes can readily be distinguished from ketones by means of the aldehydic C — H stretching absorption, which results in two weak bands near 2820

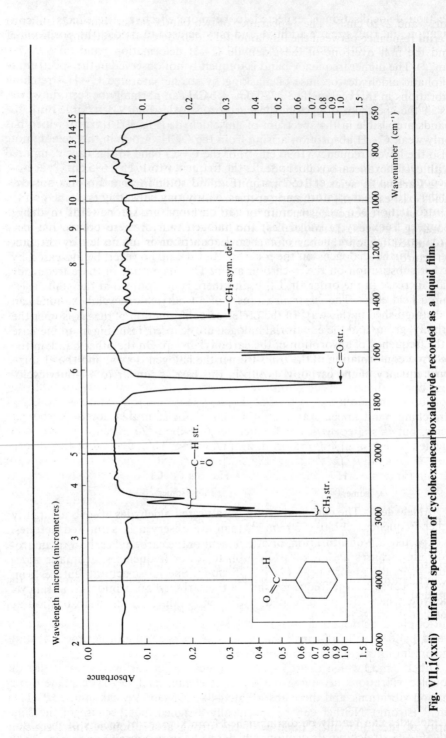

Fig. VII,1(xxii) Infrared spectrum of cyclohexanecarboxaldehyde recorded as a liquid film.

cm^{-1} and 2720 cm^{-1}. The appearance of two bands, rather than one, is due to **Fermi resonance** between the fundamental aldehyde C−H stretching vibration, and the first overtone of the aldehyde C−H deformation band (at 975–780 cm^{-1}). The higher frequency band is frequently not observed in the spectrum of aliphatic aldehydes because of masking by strong saturated C−H stretching vibrations in this region. The 2720 cm^{-1} band can normally be seen, however, as in the spectrum of cyclohexanecarboxaldehyde (Fig. VII,1(xxii)). Both the bands are visible in the spectrum of anisaldehyde (Fig. VII,1(xxiii)), which has only weak C−H absorption arising from the OCH$_3$ group in this region; note also the lower frequency (1690 cm^{-1}) of the C=O band in this case compared with cyclohexanecarboxaldehyde.

Carboxylic acids. Even in quite dilute solution in non-polar solvents, acids exist essentially as dimeric species, which may be readily explained on the basis of the electronic structure of the carboxyl group. Powerful hydrogen bonding between the molecules, and the strength of these bonds, has been accounted for on the basis of a large contribution of an ionic resonance structure.

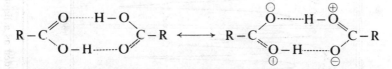

As a consequence, the spectrum of a carboxylic acid in the condensed phase (KBr disc, a mull or liquid film), or in concentrated solution, exhibits absorption due to dimeric species and even in very dilute solution only a small proportion of the monomer is present. Under these sample conditions the C=O stretching band appears at 1725–1700 cm^{-1} for saturated acyclic carboxylic acids, but is shifted to lower frequencies (by about 20–30 cm^{-1}) when conjugated with a double bond or an aromatic ring. The intensity of the absorption is generally greater than that of ketones. Some acids are capable of forming internal hydrogen bonds and the carbonyl frequency is shifted much more significantly; thus salicylic acid absorbs at 1655 cm^{-1}.

The feature which enables one to distinguish a carboxylic acid from all the other carbonyl compounds, however, is a broad absorption band which extends from between 3300 cm^{-1} to 2500 cm^{-1}. This band is the result of the strongly hydrogen bonded O−H stretching vibrations, and can be seen in the spectrum of hexanoic acid (Fig. VII,1(xxiv)). The aliphatic C−H stretching bands generally appear as a jagged peak near 2900 cm^{-1} superimposed on top of the bonded O−H band, which itself has an intensity maximum in this region.

Another band characteristic of the dimeric acid species arises from the O−H out-of-plane deformation (wag) vibration which appears as a broad, rather weak, band at 950–900 cm^{-1} and is seen in the spectrum of hexanoic acid; its maximum intensity is at about 940 cm^{-1}.

Other bands which assist in the identification of carboxyl groups are the coupled vibrations involving the C−O stretching and O−H in-plane deformation vibrations, and these absorb at 1440–1395 cm^{-1} (weak) and 1320–1211 cm^{-1} (strong). Neither can be specifically assigned, but because of the similarity of the 1300 cm^{-1} band with the strong absorption at this frequency exhibited by esters, it is usually referred to as the C−O stretching band. A

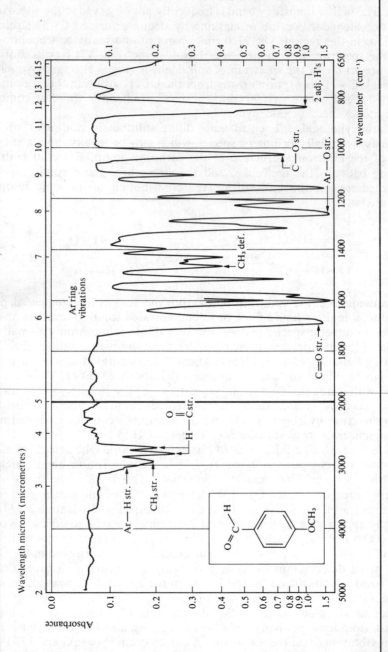

Fig. VII,1(xxiii) Infrared spectrum of anisaldehyde recorded as a liquid film.

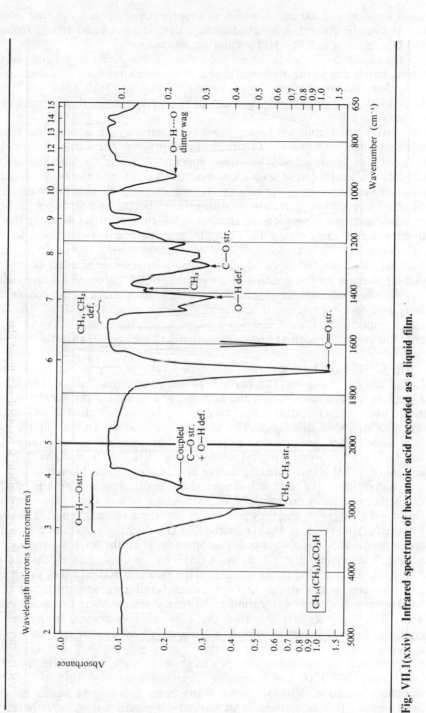

Fig. VII,1(xxiv)   Infrared spectrum of hexanoic acid recorded as a liquid film.

small band near 2700 cm^{-1}, which is superimposed upon the broad bonded O−H stretching band, is believed to be a combination band arising from the C−O stretching and O−H deformation vibrations.

Salts of carboxylic acids do not, of course, show a carbonyl band. Instead strong bands due to the asymmetric and symmetric stretching vibrations of the equivalent carbon–oxygen bonds are observed at 1610–1550 cm^{-1} and 1420–1300 cm^{-1} respectively, and can provide evidence for the presence of the carboxylate anion.

Esters and lactones. Esters show two characteristic absorption regions arising from C=O and C−O stretching vibrations. The carbonyl stretching frequency in saturated acyclic esters appears at a slightly higher frequency (20–30 cm^{-1}) than that in simple ketones; this is a consequence of the inductive effect of the electronegative alkoxy–oxygen exerting an electron withdrawal effect on the carbonyl carbon resulting in a shorter and stronger carbonyl bond. Sometimes a problem in structural assignment arises because the frequency ranges can overlap; for example, ester carbonyl frequency is lowered by conjugation or a ketone carbonyl absorption shifted to higher frequency by a chlorine attached to an adjacent carbon. In general the effect of environmental changes on the position of absorption of the carbonyl group in esters, e.g., the effect of conjugation, etc., follows the same pattern as in ketones. Six-, five- and four-membered ring lactones absorb near 1750, 1780 and 1820 cm^{-1} respectively. Phenyl and vinyl esters, which have the $-CO-\overset{\cdot\cdot}{O}-C=C$ bonding system, absorb at higher frequencies (20–25 cm^{-1}) than the saturated esters.

The C−O stretching vibration in esters results in very strong bands in the 1300–1100 cm^{-1} region. The band is of complex origin, but is generally regarded as arising mainly from the acyl–oxygen bond. Unfortunately its diagnostic use is rather limited since other strong bands, e.g., ether C−O stretching, also appear in this region. However, it has been shown that simple esters such as formates, acetates and butyrates all show a strong band near 1200 cm^{-1}. The spectrum of phenyl acetate (Fig. VII,1(xxv)) and that of butyl acetate (Fig. VII,1(xxvi)) are illustrative of this class of compounds.

Anhydrides. All acid anhydrides show two strong absorption bands at the higher frequency end of the C=O stretching region; they occur near 1800 cm^{-1} and 1750 cm^{-1}, and they are almost always *ca.* 60 cm^{-1} apart. The higher frequency band is due to the symmetric C=O stretching vibration, and in open chain anhydrides it is always of higher intensity than the lower frequency band arising from asymmetric vibrations. Conversely in cyclic anhydrides the high frequency band is always the weaker of the two and it diminishes in intensity with increasing ring strain, i.e., a five-membered ring anhydride exhibits a weaker high frequency absorption band than a six-membered ring anhydride. This effect has been explained on the basis of a decreased dipole moment change in the symmetric vibrational mode when constrained in a cyclic system.

Open chain and cyclic anhydrides also show a strong band in the 1170–1050 cm^{-1} and 1300–1200 cm^{-1} regions respectively, arising from a C−O−C stretching vibration. However, since many other groups give strong bands in these regions this assignment is of limited diagnostic value; nevertheless the absence of an anhydride group is confirmed if there is no strong absorption in either of these regions.

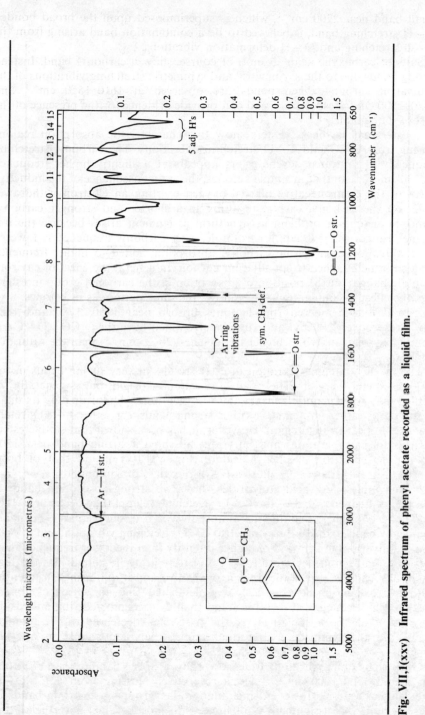

Fig. VII,1(xxv) Infrared spectrum of phenyl acetate recorded as a liquid film.

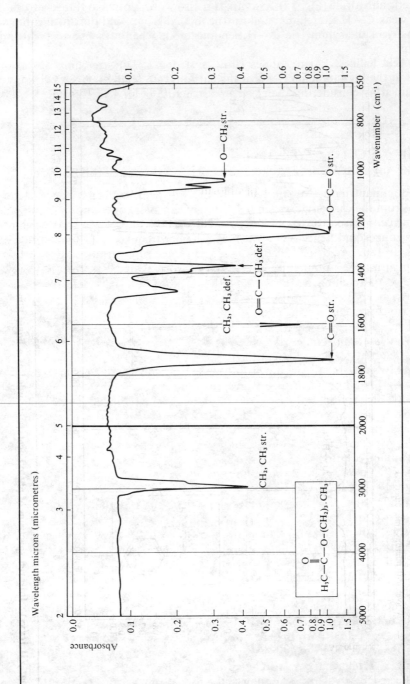

Fig. VII,1(xxvi) Infrared spectrum of butyl acetate recorded as a liquid film.

The typical features of an open chain anhydride are shown in the spectrum of acetic anhydride (Fig. VII,1(xxvii)). It is also worth noting in this spectrum the very weak C—H stretching band for the methyl group, and the much enhanced intensity of the symmetric C—H bending mode when attached to a carbonyl group.

Acid halides. Acid halides show a strong C=O stretching absorption band at the high frequency end of the carbonyl stretching region. This may be explained by considering the electronic structure of an acid chloride.

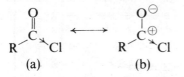

(a)　　　　　　　(b)

The dominant inductive effect of chlorine will tend to draw electron density away from the carbonyl oxygen resulting in a smaller contribution of the polar form (b) compared to that of ketones. The carbonyl bond will thus be shorter and stronger and hence will absorb at a higher frequency; acetyl chloride absorbs at 1802 cm^{-1} (cf. acetone, 1725 cm^{-1}). Conjugation of the carbonyl group with an α,β-double bond or an aryl group would be expected to lower the C=O stretching frequency and this is observed. Thus benzoyl chloride absorbs at 1773 cm^{-1}; it also shows a slightly weaker band at 1736 cm^{-1} due to Fermi resonance arising from the C=O group and the overtone of a lower frequency absorption band.

Amides. All amides are characterised by a strong carbonyl absorption band, referred to as the 'amide I' band. Primary and secondary amides additionally show bands arising from N—H stretching and bending vibration. The N—H bending absorption is generally at slightly lower frequency than the carbonyl absorption and is referred to as the 'amide II' band.

Amides have a very strong tendency to self associate by hydrogen bonding, and the appearance of the spectrum is very much dependent on the physical state of the sample. Considerable shifts in band positions can occur on passing from a dilute solution to a solid, thus N—H and C=O stretching bands show a marked shift to lower frequency whilst the N—H bending (amide II) band moves to higher frequency.

In dilute solution primary amides show two sharp bands resulting from the asymmetric and symmetric N—H stretching vibrations near 3520 cm^{-1} and 3400 cm^{-1} (the normal N—H region). In solid samples these appear near 3350 cm^{-1} and 3180 cm^{-1}. In dilute solution secondary amides show only one band near 3460–3420 cm^{-1} on low-resolution instruments. However, under conditions of high resolution the band can frequently be split into two components which have been assigned to the *cis* and *trans* rotational isomers.

As a consequence of the mesomeric effect, the amide carbonyl group has less double bond character than that of a normal ketonic carbonyl group and it would be expected to absorb at lower frequency. This is found to be the case; primary and secondary amides absorb strongly near 1690 cm^{-1} in dilute solution and at somewhat lower frequency in the solid phase. Tertiary amides are not affected by hydrogen bonding and show strong absorption at 1670–1630 cm^{-1} irrespective of the physical state of the sample.

Bands resulting from the primary and secondary N—H bending vibrations

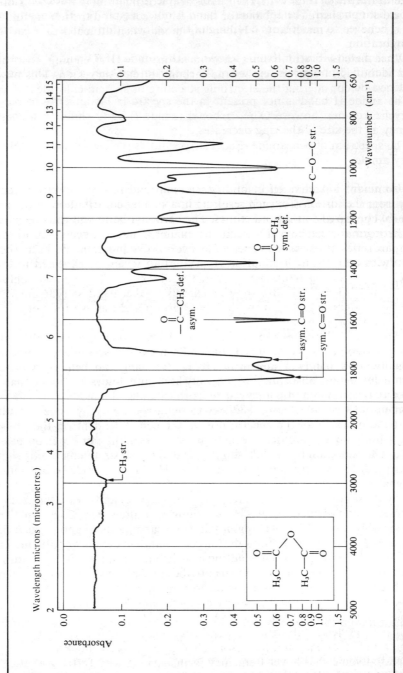

Fig. VII,1(xxvii) Infrared spectrum of acetic anhydride recorded as a liquid film.

appear near 1650 cm^{-1} and 1550 cm^{-1} respectively in the solid phase, and the large difference in these amide II bands enable primary and secondary amides to be distinguished. The 1550 cm^{-1} band is not a simple N − H bending mode, but is believed to result from coupling of this deformation with a C − N stretching vibration.

Other mixed vibration bands known as the amide III, IV and V bands have been identified in various regions of the spectrum but they are of limited diagnostic value.

The amide II band is not present in the spectra of lactams. As in the case of cyclic ketones, however, the carbonyl (amide I) band shifts to higher frequency as the size of the ring decreases.

The spectrum of benzamide (Fig. VII,1(xxviii)) is typical for an aromatic primary amide.

Amino acids α-Amino acids (and other amino acids also) normally exist as zwitterionic salts (e.g., (I)), and therefore show bands characteristic of the ionised carboxyl group and an amine salt. Hence there is no absorption corresponding to the normal stretching vibrations as exhibited by an amine, but instead a complex series of bands are observed between 3130 and 2500 cm^{-1}, and this is also the case for the hydrochloride salts of amino acids (e.g., (II)). In addition to $-\overset{\oplus}{N}H_3$ stretching vibrations (cf. CH$_3$ and CH$_2$ stretching bands), combination and overtone vibrations of the various N − H bending modes are involved, and the very complexity of a spectrum in this region of the spectrum is a useful indication that the compound is an amino acid.

For *N*-substituted compounds, such as proline, only the $-\overset{\oplus}{N}H_2$ stretching vibrations are involved and these appear at lower frequency. A relatively prominent bands between 2200 and 2000 cm^{-1} is found in the spectrum of most amino acids and their hydrochloride salts and can be clearly seen in the spectrum of (±)-valine (Fig. VII,1(xxix)) at 2130 cm^{-1}. The band appears at this position in all α-amino acids, but is displaced in others (i.e., β, γ, *etc.*) and is believed to be a combination band associated with the CO$_2^{\ominus}$ group. The salts derived from amino acids and bases (e.g., (III)) show normal N − H stretching bands.

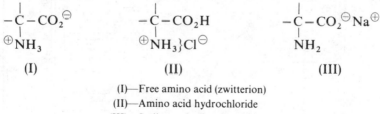

(I)—Free amino acid (zwitterion)
(II)—Amino acid hydrochloride
(III)—Sodium salt of amino acid

In the amino acid hydrochloride salts, normal carbonyl absorption is observed for the $-CO_2H$ group except that the band is displaced by about 20 cm^{-1} to higher frequency by the electron attracting $-\overset{\oplus}{N}H_3$ group, which has the effect of making the C=O group shorter and stronger. In contrast the ionised carboxyl group in the zwitterionic and basic salt forms, like the salts of carboxylic acids, show absorption bands due to the asymmetric and symmetric stretching

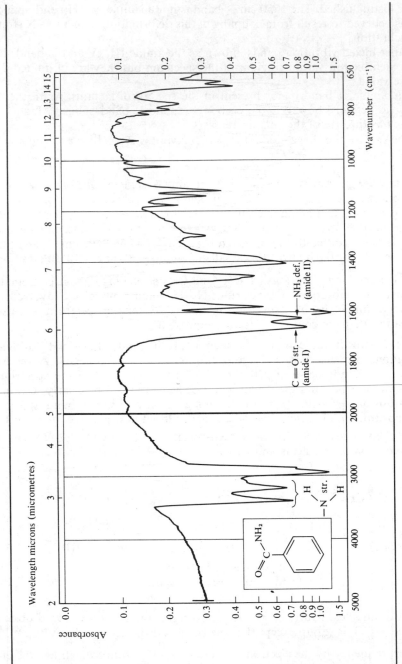

Fig. VII,1(xxviii) Infrared spectrum of benzamide recorded as a Nujol mull.

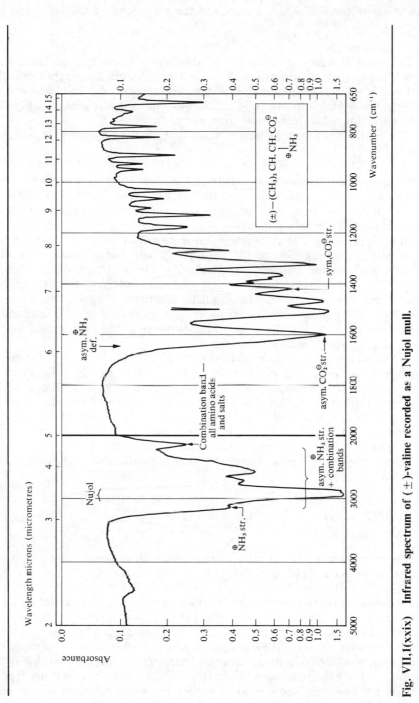

Fig. VII,1(xxix) Infrared spectrum of (±)-valine recorded as a Nujol mull.

vibrations of the $-C\overset{\displaystyle O}{\underset{\displaystyle O}{\diagdown}}\ominus$ group, and these appear at 1550 cm^{-1} and 1410 cm^{-1} respectively.

The more complex α-amino acids additionally show characteristic bands which aid in their identification. Thus in the spectrum of L-tryptophan (Fig. VII,1(xxx)) the N—H stretching vibration, and the out-of-plane hydrogen wag deformation which appears at 742 cm^{-1} (4 adjacent hydrogens), readily allow it to be distinguished from other α-amino acids.

The correlation tables for amino acids and amine salts are in Appendix 2, Table 2,9.

Nitro compounds, nitroso compounds and nitrites **Nitro** compounds exhibit two very intense absorption bands in the 1560–1500 cm^{-1} and 1350–1300 cm^{-1} region of the spectrum arising from asymmetric and symmetric stretching vibrations of the highly polar nitrogen–oxygen bonds. Aromatic nitro compounds show bands at slightly lower frequencies than the aliphatic compounds as a result of conjugation of the nitro group with the aromatic ring, which slightly weakens the nitrogen–oxygen bonds. The spectrum of nitrobenzene (Fig. VII,1(xxxi)) is typical for this class of compound. Note that for this compound the positions of the out-of-plane hydrogen wagging bands in the 900–700 cm^{-1} region of the spectrum are not characteristic of a monosubstituted benzene system as the result of the presence of the nitro group, and the substitution pattern cannot be determined reliably.

Nitroso compounds may be of the C—NO or N—NO type. Tertiary C-nitroso compounds tend to dimerise, and secondary and primary C-nitroso compounds readily rearrange to oximes. In the monomeric state they absorb in the 1600–1500 cm^{-1} region, but in solution they exist preferentially as dimers and then absorb near 1290 cm^{-1} (*cis*) or 1400 cm^{-1} (*trans*). N-Nitroso compounds show a band near 1450 cm^{-1} in solution in carbon tetrachloride.

Nitrites. These compounds show the N=O stretching vibration as two bands near 1660 cm^{-1} and 1620 cm^{-1}; these are attributed to the *trans* and *cis* forms of the nitrite.

trans   cis

The correlation tables for nitro compounds, nitroso compounds and nitrites are in Appendix 2, Table 2,10.

Unsaturated nitrogen compounds Nitriles, isonitriles and isocyanates all absorb in the 2300–2000 cm^{-1} region of the spectrum. Stretching of the C≡N bond in aliphatic nitriles gives rise to a band at 2260–2240 cm^{-1}, which is shifted to lower frequency by conjugation with a double bond or aromatic ring. Conjugation also tends to increase the intensity of the band which is very strong in, for example, benzonitrile (Fig. VII,1(xxxii)). The various types of nitriles do, however, show marked variations in the intensity of the bands depending on the electronic effects of substituents attached to the nitrile group; thus any substituent which tends to diminish the dipole moment of the bond

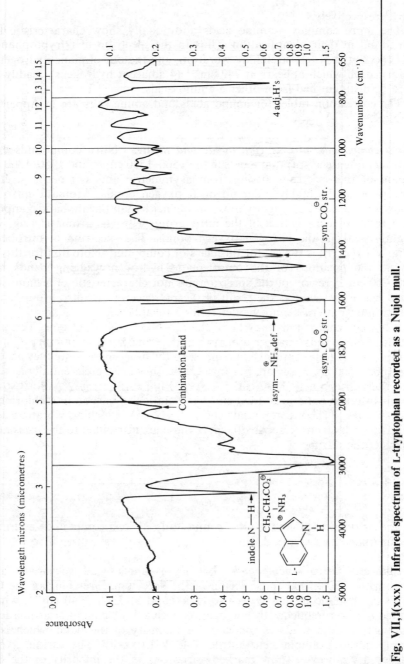

Fig. VII,1(xxx) Infrared spectrum of L-tryptophan recorded as a Nujol mull.

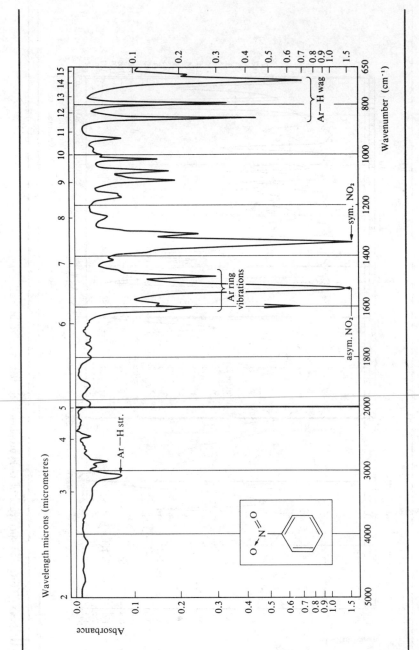

Fig. VII,1(xxxi) Infrared spectrum of nitrobenzene recorded as a liquid film.

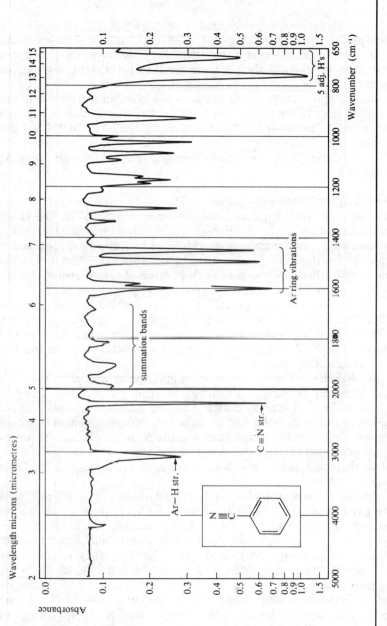

Fig. VII,1(xxxii) Infrared spectrum of benzonitrile recorded as a liquid film.

would be expected to produce a decrease in the intensity and *vice versa*. Isocyanates show a very intense absorption band near 2275–2240 cm^{-1} which is unaltered by conjugation. The bands are very much more intense than the bands of nitriles with similar structure and this feature allows them to be readily distinguished.

A band of variable intensity arising from stretching of the C=N bond in oximes and imines occurs in the 1690–1590 cm^{-1} region of the spectrum, and is generally more intense than C=C stretching bands which also appear here. The oximes additionally show a band for the O−H stretching vibration near 3200 cm^{-1}.

In azo-compounds stretching of the −N=N− bond gives rise to only weak absorption near 1600 cm^{-1}, which is shifted to lower frequency by conjugation. In aromatic compounds the band is generally masked by the aromatic ring breathing vibrations.

The correlation tables for unsaturated nitrogen compounds are in Appendix 2, Table 2,11.

Organo-sulphur compounds **Thiols.** The S−H stretching vibration of thiols gives rise to a weak band at 2590–2550 cm^{-1}. Unlike the O−H stretching band in the alcohols, the position of this band is little affected by hydrogen-bonding effects, and the absorption exhibited by neat liquid films and by dilute solutions of a thiol are similar. Although the band is weak it has diagnostic value as few other bands appear in this region of the spectrum, though it may well be masked if there is also a carboxyl group in the molecule.

Thioketones and dithioesters. These show C=S stretching bands in the 1270–1190 cm^{-1} region. Since the C=S bond is not as polar or as strong as the C=O bond the absorption band is not very intense and appears in the low frequency region of the spectrum; coupling with other bands in this region can make identification difficult.

Sulphoxides. The S=O bond is highly polar and gives rise to a strong absorption near 1050 cm^{-1} which can be readily recognised. The position of the band is little affected by attached double bonds or aromatic rings as conjugation of the S=O bond and an adjacent π electron system is not extensive; however, electronegative substituents cause a shift to higher frequency which may be explained in terms of a reduced contribution by the polar S−O structure to the resonance hybrid with a consequent increase in S=O character giving a stronger bond.

Sulphones, sulphonamides, sulphonyl chlorides, sulphonic acids, sulphonates and organic sulphates. These all contain the SO$_2$ group which can be readily identified by the appearance of two strong bands in the 1415–1300 cm^{-1} and 1200–1120 cm^{-1} regions, due to the asymmetric and symmetric stretching vibrations respectively. Occasionally the high frequency sulphone band will split when the spectrum is recorded in carbon tetrachloride solution or the solid state. Sulphonic acids may be further recognised by the broad hydrogen bonded O−H stretching absorption centred at $\sim$3000 cm^{-1}. Primary and secondary sulphonamides will show two or one N−H stretching bands respectively near 3300 cm^{-1} (cf. the spectrum of toluene-*p*-sulphonamide, Fig. VII,1(xxxiii)).

The correlation table for organo-sulphur compounds is in Appendix 2, Table 2,12.

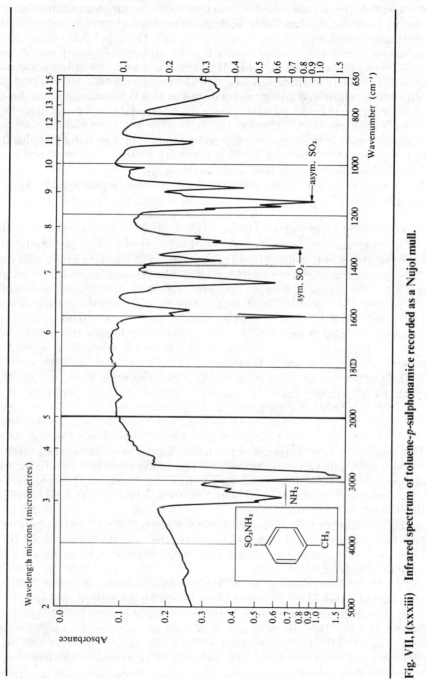

Fig. VII,1,1(xxxiii) Infrared spectrum of toluene-*p*-sulphonamide recorded as a Nujol mull.

Halogen compounds The $C-X$ stretching vibration gives rise to very strong absorption in the low frequency region of the spectrum. Indeed, absorption by $C-I$ and many $C-Br$ bonds occurs outside the range available on many routine instruments. Monofluoroalkanes normally absorb in the 1100–1000 cm^{-1} region; splitting of the band and a shift to higher frequencies occurs on further substitution, whilst highly fluorinated aliphatic compounds show a series of intense bands in the 1400–1000 cm^{-1} region. Monochloroalkanes and monobromoalkanes absorb in the 760–540 cm^{-1} and 600–500 cm^{-1} regions respectively. Axial and equatorial chlorine, and also bromine, in cyclohexanes and steroids, may be differentiated since the equatorial $C-X$ bond absorbs at a higher frequency.

The correlation table for halogen compounds is in Appendix 2, Table 2,13.

VII,4,2. Nuclear magnetic resonance spectroscopy

The n.m.r. spectrum This discussion will be concerned largely with the interpretation of proton magnetic resonance spectrum (p.m.r.), although the information which can be obtained from a study of ^{13}C-nuclear magnetic resonance (c.m.r.) will also be discussed.

The important features of a p.m.r. spectrum may be illustrated by reference to the spectrum of toluene (Fig. VII,2(i)) recorded at 60 MHz as the neat liquid. Although in principle it is possible to satisfy the resonance conditions by altering either the magnetic field or the radiofrequency, most spectra are recorded by varying the magnetic field while keeping the radiofrequency constant. Commonly available instruments operate at 60, 100 or 200 MHz. Field and frequency increase in the same direction and conventionally this is from left to right in the n.m.r. spectrum, thus moving from left to right is **upfield**, and moving from right to left is **downfield**.

The position of an absorption in the n.m.r. spectrum may be represented either on a frequency scale (Hz), or on a scale of magnetic field (Gauss), and again by convention the frequency scale is used. In typical proton spectra the differences in the position of absorption are small (usually of the order of a few hundred Hz), compared to the absolute value of the frequency (around 10^7 Hz). The position of an absorption in the n.m.r. spectrum is therefore denoted, not by its absolute frequency value, but by the relationship to that of a reference compound, i.e., Hz upfield or downfield from the reference. The normal reference compound in p.m.r. spectra is tetramethylsilane [(CH$_3$)$_4$Si, TMS]. All the protons in TMS are equivalent, and the proton nuclei in the vast majority of organic compounds absorb downfield from the single TMS signal. Since tetramethylsilane is not soluble in water, the soluble sodium salt of 2,2-dimethyl-2-silapentane-5-sulphonic acid (DSS, Tiers salt) is normally used in aqueous solutions.

The TMS absorption in the 60 MHz spectrum of toluene (Fig. VII,2(i)) is indicated and this is the reference point (0 Hz) for all absorptions in the spectrum. The spectrum shows two absorptions which appear at 128 Hz and 419 Hz downfield from TMS. If the spectrum of toluene is recorded on an instrument operating at 100 MHz their absorptions occur at 213 Hz and 698 Hz respectively. In order to make direct and rapid comparisons between spectra recorded on instruments operating at different frequencies, the positions of absorptions are normally quoted on the δ scale which is independent of the instrument operating frequency. The δ value is obtained by dividing the

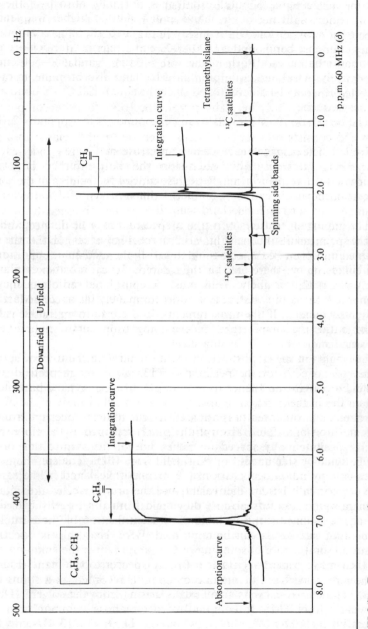

Fig. VII,2(i) Proton magnetic resonance spectrum of neat toluene; sweep width 500 Hz.

position in Hz by the instrument frequency (in MHz) and is expressed in parts per million (p.p.m.). Thus for toluene the two absorptions appear at δ 2.13 (128/60 or 213/100) and δ 6.98 (419/60 or 698/100). The chart paper normally used for recording spectra is calibrated in δ values and therefore this calculation is not usually necessary, but it is often needed to determine the position of absorptions which have been offset (see Fig. VII,2(iv)). The δ value may relate to any reference compound and therefore the particular reference used (e.g., TMS) must be quoted. Earlier literature used the similar τ scale; this related solely to TMS which was given a value of 10. The two scales can be readily interconverted since $\tau = 10 - \delta$.

It is usual for the p.m.r. absorption spectrum to be accompanied by an integration curve. This normally appears above the absorption curve on the chart and consists of a series of steps, each step being related to a particular absorption. The height of each step is a measure of the area under each absorption peak, and this in turn is a measure of the relative number of proton nuclei giving rise to that absorption. Thus the ratios of the heights of the steps on the integration curve gives the ratio of the different types of nuclei in the compound. Some care is needed in using the values obtained from integration measurements, since frequently the steps are not well defined and it is not possible to measure their heights accurately. The values for the ratio of nuclei obtained may then be difficult to relate to whole numbers of nuclei and several possibilities must therefore be considered. It cannot be emphasised too strongly that all the likely ratios must be considered since an unjustified adherence to one set of values may lead to completely incorrect deductions being made subsequently. If the measurements do not lead to a realistic value for the proton ratio, the step heights in the integration curve should be carefully reconsidered.

The heights of the steps in the integration curve in the toluene spectrum are in the ratio of 3 : 5 for the peaks at δ 2.13 and δ 698 respectively. It is thus possible to assign the former to the methyl protons and the latter to the protons on the benzene ring.

A number of other features apparent in the toluene spectrum are worthy of note at this stage. Each absorption is accompanied by a number of small satellite peaks equally spaced on either side of the main absorptions. These may be **spinning side-bands** or ^{13}C satellites (p. 1011). The spinning side-bands are caused by inhomogeneities in the magnetic field and in the sample tube. They can normally be identified easily since they are of much weaker intensity than the main signal, and furthermore they appear in pairs, equally spaced on either side of the main absorption band. The identification of these satellite peaks as spinning side-bands can be confirmed by re-recording the spectrum using a faster or slower rate of spinning of the sample tube; the spinning side-bands will then move respectively farther from or nearer to the main absorption.

The n.m.r. spectrum of organic compound recorded as a solution (usually 5–10%) sometimes exhibits small peaks arising from the solvent. The solvents normally chosen do not contain hydrogen; those commonly used include CCl_4, $CDCl_3$, $(CD_3)_2SO$, $(CD_3)_2CO$, C_6D_6, D_2O, C_5D_5N. Usually, however, the deuterated solvents contain a small proportion of isotopically isomeric molecules containing hydrogen instead of deuterium, and this will give rise to an additional peak or peaks in the spectrum. A list of the position of these absorptions for some of the commoner solvents is given in Appendix 3, Table

3,10. Care should be taken in comparing chemical shifts from spectra obtained using different solvents since variations of up to 1 p.p.m. may result, particularly when values obtained with aromatic solvents, such as benzene or pyridine, are compared with those obtained using saturated solvents.

The chemical shift The position of an absorption peak relative to that of the reference compound is known as the *chemical shift*. Each proton in a different environment experiences a slightly different local magnetic field due to the circulation of electrons in neighbouring bonds and to through-space effects. A slightly different applied magnetic field is therefore required for resonance and absorption occurs in different regions of the spectrum. Modification of the local magnetic environment at the nucleus, for example by the introduction of a substituent group, will alter the position of the absorption band by requiring a higher or lower value of the applied magnetic field for resonance. Effects which cause shifts to lower fields (downfield) are termed *deshielding*; the opposite effect (upfield shift) is termed *shielding*. Figure VII,2(ii) shows approximate ranges of proton chemical shifts.

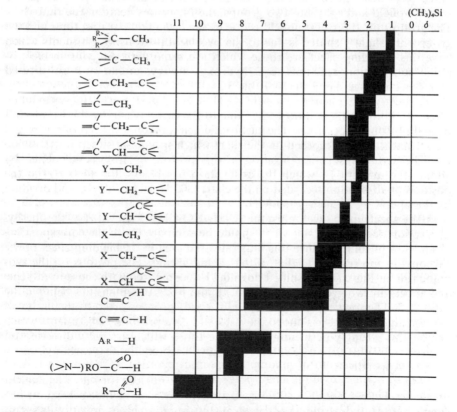

Fig. VII,2(ii) Approximate ranges of proton chemical shifts (R = H or alkyl; Y = −SR, −NR$_2$; X = −OR, −NHCO·R, −O·CO·R, halogen). Data reproduced from L. M. Jackman and S. Sternhell (1969). *Applications of Nuclear Magnetic Resonance in Organic Chemistry*. 2nd edn. London; Pergamon Press, p. 161.

More detailed chemical shift data for a wide range of proton environments is given in Appendix 3, Tables 3,1, 3,3 and 3,4. The chemical shifts value quoted in this Table show that an electronegative substituent in aliphatic systems causes a downfield shift; the greater the electronegativity the more substantial the shift. When two substituents are attached to the same carbon atom there is a greater downfield shift, but not as great as the sum of the two substituents separately. The approximate position of absorption in such cases can be predicted on the basis of the empirical parameters shown in Appendix 3, Table 3,2.

A comparison of the spectrum of anisole (Fig. VII,2(iii)) with that of toluene (Fig. VII,2,(i)) provides an illustration of the way in which substituents may influence the values of the chemical shift. The spectrum of toluene reveals the typical chemical shifts of aromatic protons and of methyl protons attached to an aromatic system. The five aromatic protons appear as a sharp single band; this is typical of those cases where the substituent is neither strongly shielding or deshielding (e.g., alkyl or substituted alkyl groups). The spectrum of anisole ($C_6H_5 \cdot O \cdot CH_3$) shows the deshielding effect of the electronegative oxygen causing the methyl absorption to be shifted downfield by a further 1.3 δ as compared to toluene. In contrast the absorptions due to the aromatic protons are shifted upfield and split into two groups, the intensities being in the ratio 2 : 3. The group shifted furthest upfield is due to the *ortho* and *para* protons, and the group only slightly shifted is due to the two *meta* protons. Substituents which have this shielding effect are those which are *ortho/para* directing in electrophilic substitution reactions, e.g., hydroxy, alkoxy, amino and substituted amino groups (see Appendix 3, Table 3,5).

The spectrum of phenylacetic acid ($C_6H_5 \cdot CH_2 \cdot CO_2H$; Fig. VII,2(iv)) exhibits three absorptions in the ratio 1 : 2 : 5 due to the carboxylic acid, methylene and phenyl protons respectively. The carboxylic acid proton has been offset by 400 Hz so that it can be recorded on the chart which specifies a sweep width of 500 Hz. The actual absorption position of this proton, δ 11.67, is calculated by adding the amount by which the absorption has been offset (400 Hz) to the position of absorption recorded on the chart (300 Hz), i.e., 700 Hz, and dividing by the operating frequency (60 MHz).

In the spectrum of acetophenone ($C_6H_5 \cdot CO \cdot CH_3$; Fig. VII,2(v)) the methyl absorption occurs at δ 2.40, which should be compared with the corresponding absorption in toluene, since this illustrates the greater deshielding effect of the carbonyl group relative to that of the aromatic ring. The protons on the aromatic ring in this case are shifted downfield as compared with those in toluene, and appear in two groups centred at δ 7.8 and δ 7.3, the intensities being in the ratio 2 : 3. The lowfield group is due to the two protons *ortho* to the carbonyl and the highfield group is due to the *meta* and *para* protons. Substituents which exhibit this strongly deshielding effect are those which are *meta* directing on electrophilic substitution reactions, e.g., acyl, carboxyl, alkoxycarbonyl, nitro, etc. (see Appendix 3, Table 3,5).

Measurement of the chemical shift. When a nucleus (or set of equivalent nuclei, see below) gives rise to a single absorption peak in the spectrum it is a simple matter to determine the chemical shift from a measurement of its separation from the reference peak. When coupling of the nucleus results in a first order multiplet (see below) measurement of the separation from the reference peak must be made to the mid-point of the multiplet. In more complex spin–spin interactions it is not possible to determine directly the chemical shift by

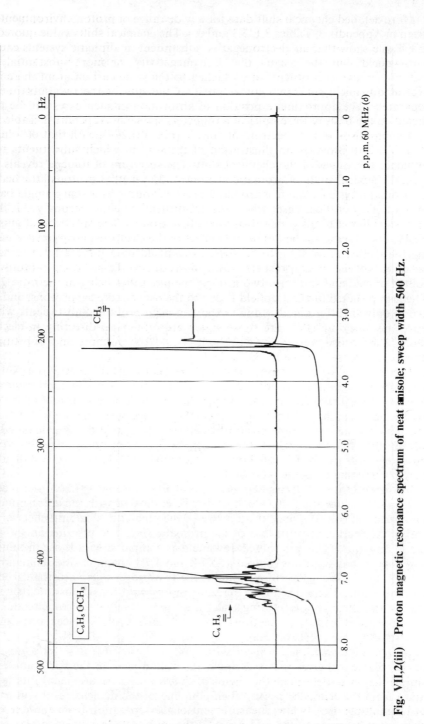

Fig. VII,2(iii) Proton magnetic resonance spectrum of neat anisole; sweep width 500 Hz.

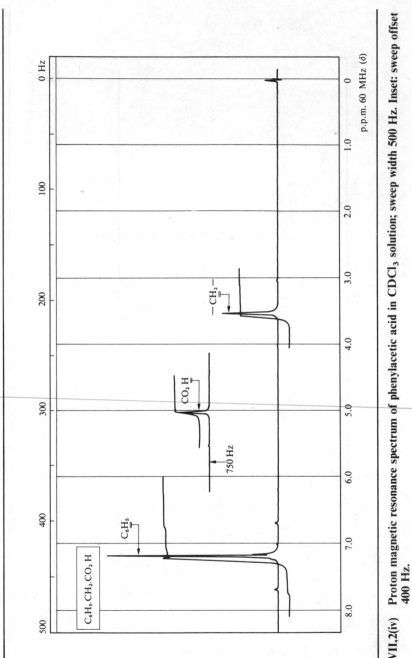

Fig. VII,2(iv) Proton magnetic resonance spectrum of phenylacetic acid in CDCl₃ solution; sweep width 500 Hz. Inset: sweep offset 400 Hz.

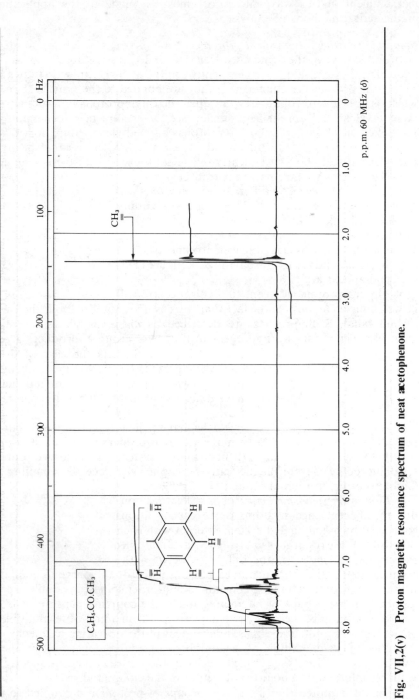

Fig. VII,2(v)   Proton magnetic resonance spectrum of neat acetophenone.

measurement in this way, and resort must be made to the application of mathematical methods of analysis.

Spin–spin splitting *Equivalent protons.* All hydrogens which are in identical environments have the same chemical shift and therefore absorb at the same frequency; they are said to be **chemically equivalent**. This can arise in two ways. Firstly the protons are equivalent if they are bonded to the same carbon atom which is also free to rotate. For example, the three protons in a methyl group are equivalent and appear as a singlet (see the spectra of toluene, anisole or acetophenone above), and the two protons of a methylene group, provided that it can rotate freely, are identical and appear as a singlet (see the spectrum of phenylacetic acid above); frequently this is not the case with methylene groups in cyclic systems where rotation is restricted.

Secondly, hydrogens on different carbon atoms will have the same chemical shift if they are structurally indistinguishable. Thus the spectrum of *p*-xylene exhibits two signals of relative intensity 3 : 2. There are six methyl protons in identical environments which appear at δ 2.3 and the four aromatic protons are identical and appear at δ 7.0.

Methylene protons adjacent to a chiral centre will be non-equivalent, despite the fact that there is free rotation about the carbon–carbon bond. Such protons are described as **diastereotopic**, since replacement of either of the two hydrogens in turn by a group X produces a pair of diastereoisomers. Such is the case of the two methylene protons (H_a and H_b) in 1-phenylpropan-2-ol, which are non equivalent and therefore have different chemical shifts in the p.m.r. spectrum.

Nuclei may be chemically equivalent but magnetically non-equivalent. To be **magnetically equivalent** nuclei must couple in exactly the same way to all other nuclei in the system. Thus the two protons in 1,1-difluoroethylene are magnetically *non-equivalent* since the coupling of F_a (and also of F_b) to each of H_a and H_b is different.

First-order spin–spin interactions. Protons bonded to adjacent carbon atoms or to carbon atoms connected by a conjugated system interact with each other so that the resulting signal appears as a multiplet rather than a singlet. If the difference between the chemical shift values of the coupled protons is reasonably large in comparison with the value of the coupling constant, a first-order spin–spin coupling pattern results. For two groups of interacting nuclei the coupling constant J is given by the separation between any two adjacent peaks in a first-order multiplet; the value, which is quoted in Hz, is independent of the operating frequency of the instrument. The magnitude of the coupling constant is dependent on the relative positions of the two coupled nuclei in the molecule and frequently gives valuable information about the structure of the compound.

It should be noted that if the two groups of protons are coupled to cause spin–spin splitting, the J value can be measured from the multiplet arising from either of the sets of protons. Thus in the spectrum of ethanol, the sample of which contains a trace of acid (Fig. VII,2(xvii)) the coupling between the methyl and methylene protons can be determined by measurement of the separation between any two adjacent lines in the triplet resulting from the methyl group, or any two adjacent lines in the quartet resulting from the methylene group. Measurement of all coupling constants can frequently show which groups of protons are coupled with each other, and hence give valuable structural information. A simple example of this procedure is seen in the spectrum of pure ethanol (Fig. VII,2(xvi)) in which the methylene protons couple with both the methyl protons and with the hydroxyl protons, but with different coupling constants, 7 Hz and 5 Hz respectively. Typical values of coupling constants are given in the correlation tables, Appendix 3, Tables 3,7, 3,8 and 3,9.

The first-order multiplets arising from spin–spin coupling can be analysed on the basis of two simple rules which give information on the number of peaks in the multiplet and on the relative intensity of the peaks in the multiplet.

(a) *The number of peaks.* If a proton is coupled with N other equivalent protons the number of peaks in the multiplet is N + 1.

(b) *Relative intensities in each multiplet.* These can be deduced using Pascal's Triangle shown below. The value assigned to each position in the triangle is derived by adding together the values of adjacent positions in the preceding level; the outside position at each level is always unity.

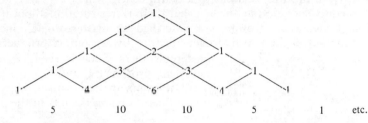

| 1 | | 5 | | 10 | | 10 | | 5 | | 1 | etc. |

Spin–spin splitting patterns arising from typical groups of protons are shown in Fig. VII,2(vi).

When a proton is coupled to two non-equivalent sets of neighbouring protons more complex multiplets result; this is illustrated by the spectrum of pure ethanol (Fig. VII,2(xvi)). Thus the methylene protons are coupled to the three methyl protons, giving rise to a quartet, and are further coupled to the hydroxyl proton which therefore causes each of the peaks of the quartet to appear as a doublet. The multiplet therefore consists of eight peaks due to the two overlapping quartets. On occasions there may be difficulty in recognising the components of these more complex multiplets, as some peaks may be superimposed.

Long-range coupling, i.e., coupling beyond three bonds, may be observed in some circumstances, especially in conjugated systems. The coupling constants are usually small (0–3 Hz) in comparison with geminal or vicinal coupling constants (cf. Appendix 3, Tables 3,7 and 3,8). Commonly encountered systems which exhibit long-range coupling are the allylic system $(H-\overset{|}{\underset{|}{C}}=\overset{|}{C}-\overset{|}{\underset{|}{C}}H)$ and the corresponding acetylenic system $(H-C\equiv C-\overset{|}{\underset{|}{C}}H)$, and aromatic and heteroaromatic rings. Conjugated acetylenes are capable of spin–spin interaction over as many as nine bonds. Long-range coupling is often observed

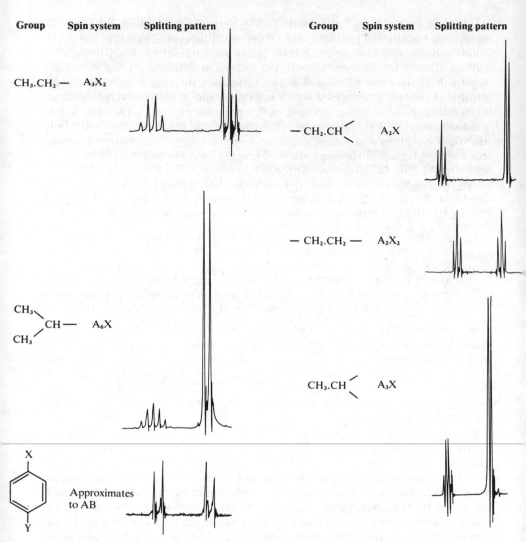

Fig. VII,2(vi)   Typical spin–spin splitting patterns.

between nuclei which are linked by a conjugated system arranged in a zig-zag manner thus:

The conjugated system is often part of a cyclic or polycyclic compound, e.g.,

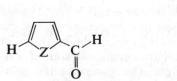

Coupling between protons and other nuclei is often useful in interpreting spectra of appropriate compounds. Weak satellites, arising from $^{13}C - {}^{1}H$ coupling, may sometimes be observed near strong signals in the proton spectrum. They can be recognised since they appear as doublets 50 to 100 Hz on either side of the main signal and are unaffected by the rate of spinning of the sample (unlike spinning side-bands). ^{13}C Satellites are visible in the spectrum of toluene (Fig. VII,2(i)).

Values for coupling constants between protons and other atoms are given in Appendix 3, Table 3,8.

Distortion of multiplets. The two rules for analysing spin–spin splitting patterns can only be applied exactly when the difference between the coupling nuclei is substantially greater (about 10 times) than the value of the coupling constants. As the chemical shift difference decreases the simple first order analysis starts to break down. For groups of protons which are still well separated this results in a distortion of the multiplets. This distortion results in the inner peaks (i.e., those nearer the other multiplet) of the two multiplets increasing in intensity whilst the outer peaks decrease in intensity. This is often a useful guide for establishing which groups of multiplets are related by spin–spin coupling. This distortion is diagrammatically illustrated in the calculated spectra shown in Fig. VII,2(vii) where $\Delta v/J$ is 14 and 5.7.

More complex spin–spin interactions. As the chemical shift difference becomes similar to the value of the coupling constant, the first-order analysis breaks down completely. It is then not possible to measure either the chemical shift or the coupling constants directly from the spectrum and resort must be made to more rigorous analytical procedures which are beyond the scope of this book. However, there are available a variety of computer programs specifically designed to analyse complex n m r spectra and Ref. 5 should be consulted for further information.

Frequently, however, it is necessary only to recognise complex patterns arising from common groupings and not to analyse them completely. The nomenclature which is adopted for naming spin–spin systems follows from the number of interacting nuclei and the magnitude of the chemical shift differences as compared to the coupling constants. Thus nuclei are represented by letters of the alphabet (A, B, M, X, Y, etc.), the first letters of the alphabet being used for nuclei at lowest field. Nuclei which are chemically identical but magnetically non-equivalent are differentiated by primes (A,A', B,B' etc.). Nuclei which have a large chemical shift difference as compared with the coupling constants are represented by letters well separated in the alphabet giving rise to systems such as AX, AMX, etc. If the chemical shift differences are of the same order of magnitude as the coupling constant, letters adjacent in the alphabet are used, so that AX becomes AB, the AMX system becomes ABC, etc. Some examples are given in Appendix 3, Table 3,11. A useful compilation for the recognition of spin systems is given in Ref. 6, but the following systems are those commonly encountered and they are illustrated with appropriate spectra.

(a) *Two-spin system* (AB). In all cases (except when $\Delta v/J = 0$ and hence the spectrum is a single absorption line, A_2) the splitting pattern shows two doublets (a doublet of doublets) distorted to a greater or lesser extent; this is often referred to misleadingly as the 'AB quartet'. The coupling constant J_{AB} is the separation between the lines of either doublet. The change in the splitting pattern as the chemical shift difference (Δv) approaches the value of the coupling constant J is shown in Fig. VII,2(vii).

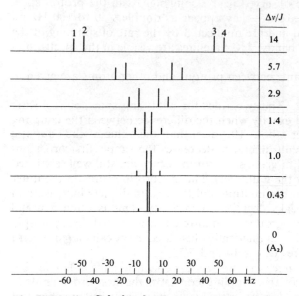

Fig. VII,2(vii)   Calculated AB spectra. J_{AB} was set at 7 Hz and Δv_{AB} was varied to the ratios shown. Data reproduced from L. M. Jackman and S. Sternhell (1969). *Applications of Nuclear Magnetic Resonance Spectroscopy in Organic Chemistry*. 2nd edn. London; Pergamon Press, p. 130.

(b) *Three-spin systems.*   (i) AB_2 and AX_2.   The AX_2 system results in a doublet and triplet as described above. The change in the appearance of the spectrum as Δv approaches J is shown in Fig VII,2(viii). Note that the number of peaks in the spectrum increases. The spectrum of pyrogallol (Fig. VII,2(ix)) illustrates a typical AB_2 spectrum; the inset is the scale-expanded spectrum of the aromatic region.

(ii) AMX, ABX, ABC.   The spectrum of pyrrole-2-carboxylic acid (Fig. VII,2(x)) illustrates a first-order AMX system. The spectrum contains three multiplets in the region δ 6–7 due to the $C-H$ protons of the hetero-aromatic ring. The expanded spectrum (50 Hz sweep width) shows that each of these multiplets consists of a doublet of doublets. Two coupling constants can be extracted from each multiplet by measurement of the separation between the first and second peaks, and between the first and third peaks of each multiplet; this is shown in the inset. The vinyl group often gives rise to an ABX spectrum as shown in the spectrum of styrene (Fig. VII,2(xi)). If accurate values for the coupling constants and chemical shifts are required from such a spectrum, resort must be made to more rigorous methods of analysis. However, a qualitative approach is frequently all that is required and on this basis the spectrum can be analysed easily. The X proton couples with protons A and B to give a distorted doublet of doublets, which form the four lines in the X portion of the spectrum. Protons A and B each give rise to doublets due to coupling with the X proton, and each of these four peaks is further split by the small coupling between the A and B protons. The ABC spectrum is very complex and does not readily yield useful information.

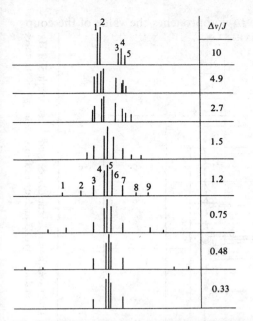

Fig. VII,2(viii) Calculated AB_2 spectra. Data reproduced from L. M. Jackman and S. Sternhell (1969). *Applications of Nuclear Magnetic Resonance Spectroscopy in Organic Chemistry.* 2nd edn. London; Pergamon Press, p. 131.

(c) *Aromatic compounds.* The magnitude of the coupling constants due to coupling between protons attached to the aromatic ring varies in the order $o > m \gg p$; *para* coupling constants are in fact frequently not discernible in the spectrum. Seven possible first-order splitting patterns for aromatic protons coupling with one, two or three protons in the *ortho* or *meta* positions are shown in Fig. VII,2(xii). Although the patterns in practice may be distorted to a greater or less extent, the spectra of many substituted aromatic systems may be analysed on this basis. For example, the splitting in the spectrum of *m*-dinitrobenzene (Fig. VII,2(xiii)) can be recognised as di-*meta* (H_a), *ortho*/di-*meta* (H_b and H_b') and di-*ortho* (H_c); some small additional splitting due to *para* coupling is apparent in the spectrum.

Protons attached to heteroatoms *Exchangeable protons.* The heteroatoms most commonly encountered by the organic chemist are oxygen, nitrogen and sulphur. The position of absorption of protons attached to these atoms is not normally sufficiently reliable for interpretative purposes, although there are exceptions to this general rule (e.g., carboxylic acids, enols, etc., see Appendix 3, Table 3,6). The position of the absorption is often highly dependent on the nature of the solvent and the concentration. However, there is one property of this group of protons which is of great value in interpretation; it is often possible, under appropriate circumstances, to replace the proton by deuterium which does not exhibit nuclear magnetic resonance under the conditions used for p.m.r., and hence the absorption is completely removed. The procedure is illustrated with reference to the spectra of benzyl alcohol ($C_6H_5 \cdot CH_2OH$, Fig. VII,2(xiv) and (xv)).

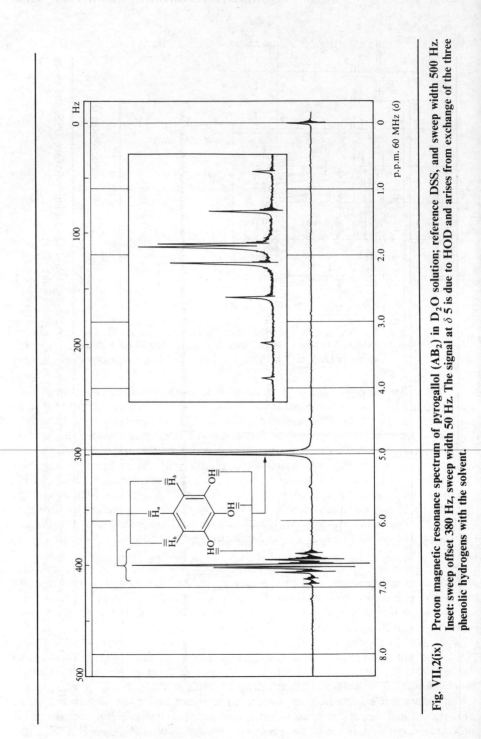

Fig. VII,2(ix) Proton magnetic resonance spectrum of pyrogallol (AB₂) in D_2O solution; reference DSS, and sweep width 500 Hz. Inset: sweep offset 380 Hz, sweep width 50 Hz. The signal at δ 5 is due to HOD and arises from exchange of the three phenolic hydrogens with the solvent.

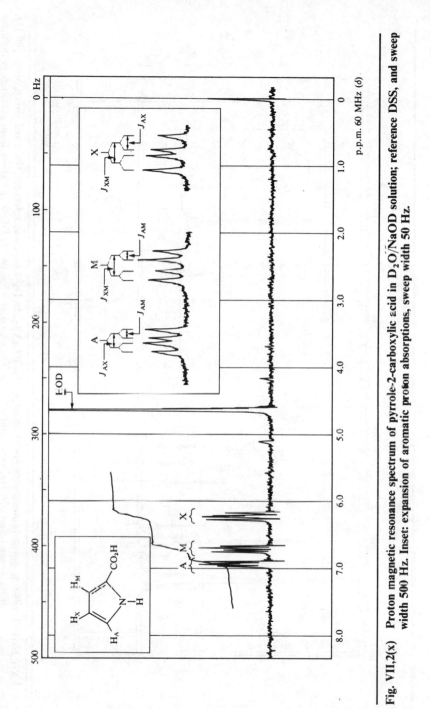

p.p.m. 60 MHz (δ)

Fig. VII,2(x) Proton magnetic resonance spectrum of pyrrole-2-carboxylic acid in $D_2O/NaOD$ solution; reference DSS, and sweep width 500 Hz. Inset: expansion of aromatic proton absorptions, sweep width 50 Hz.

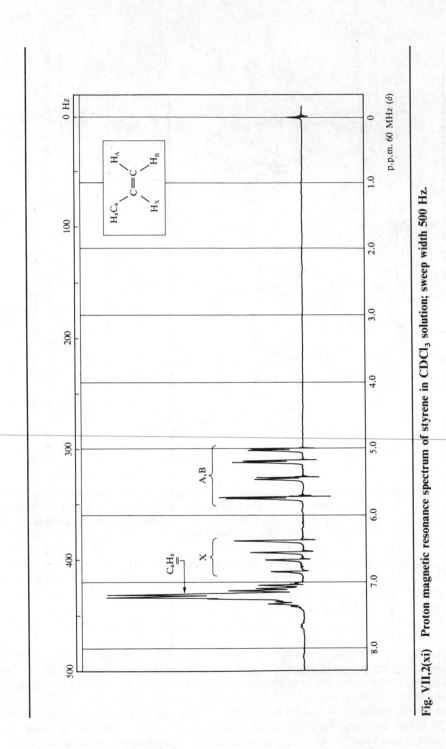

Fig. VII,2(xi) Proton magnetic resonance spectrum of styrene in CDCl₃ solution; sweep width 500 Hz.

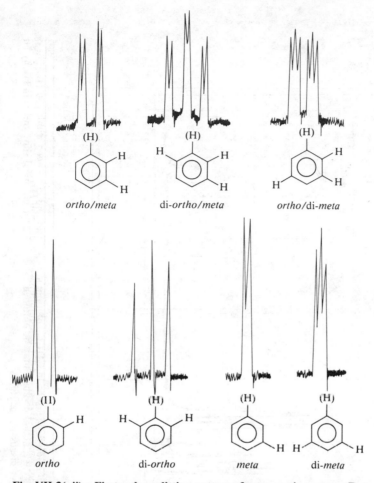

Fig. VII,2(xii) First order splitting patterns for aromatic protons. Data reproduced from M. Zanger (1972). 'The Determination of Aromatic Substitution Patterns by Nuclear Magnetic Resonance', *Organic Magnetic Resonance*, 4, **4.** Published by Heyden and Son Ltd.

The spectrum of a solution in deuterochloroform (xiv) shows three absorptions at δ 4.5, 5.08 and 7.3 with an intensity ratio of 2:1:5. The effect of adding a few drops of deuterium oxide to the sample tube and shaking vigorously is shown in the re-recorded spectrum (xv). The absorption at δ 5.08 in the original spectrum which disappears on deuteration is clearly due to the hydroxyl proton.

Protons attached to heteroatoms may not always exhibit coupling with neighbouring protons if rapid proton exchange between molecules is catalysed by the presence of trace impurities. A comparison of the spectra of pure ethanol and ethanol containing a trace of acid illustrates this effect (Fig. VII,2(xvi) and (xvii)). In the second spectrum the rapid exchange of the protons of the hydroxyl group occurs at a rate much faster than the p.m.r. resonance process. All the hydroxyl proton environments are therefore averaged and a singlet is ob-

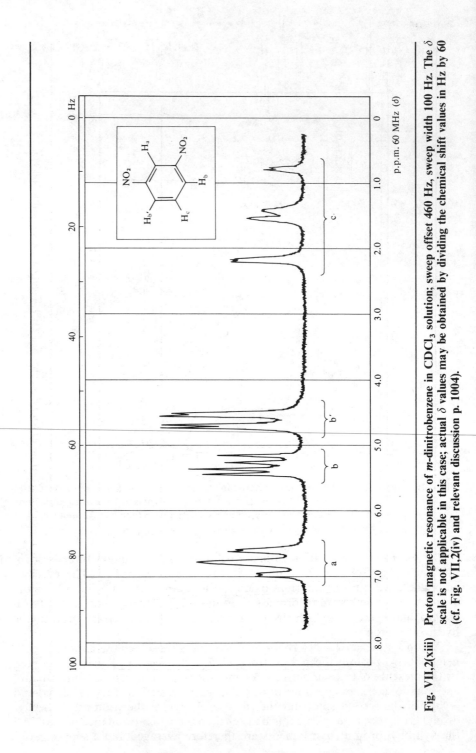

Fig. VII,2(xiii) Proton magnetic resonance of *m*-dinitrobenzene in CDCl₃ solution; sweep offset 460 Hz, sweep width 100 Hz. The δ scale is not applicable in this case; actual δ values may be obtained by dividing the chemical shift values in Hz by 60 (cf. Fig. VII,2(iv) and relevant discussion p. 1004).

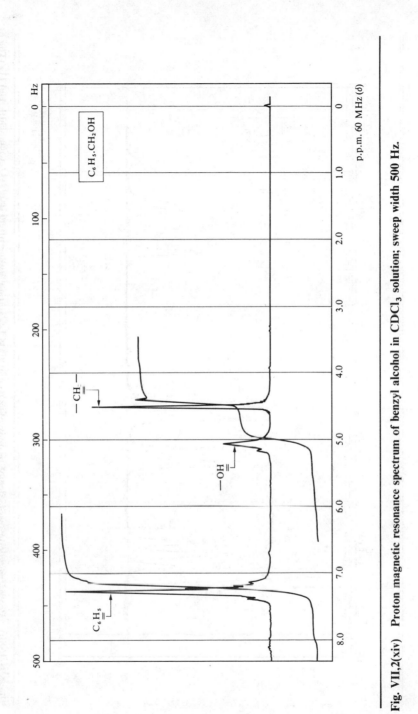

Fig. VII,2(xiv) Proton magnetic resonance spectrum of benzyl alcohol in CDCl₃ solution; sweep width 500 Hz.

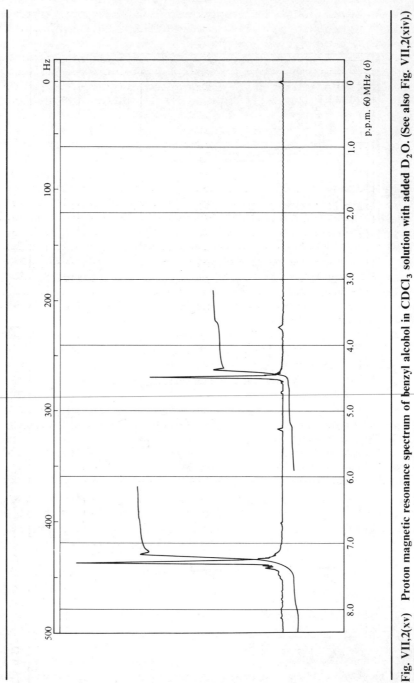

Fig. VII,2(xv) Proton magnetic resonance spectrum of benzyl alcohol in CDCl$_3$ solution with added D$_2$O. (See also Fig. VII,2(xiv).)

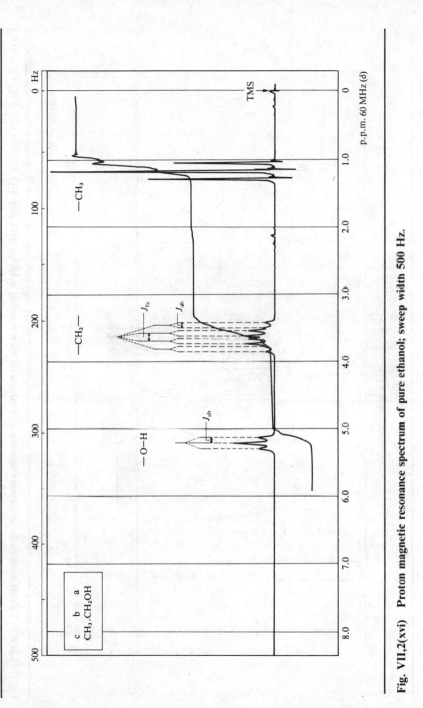

Fig. VII,2(xvi)   Proton magnetic resonance spectrum of pure ethanol; sweep width 500 Hz.

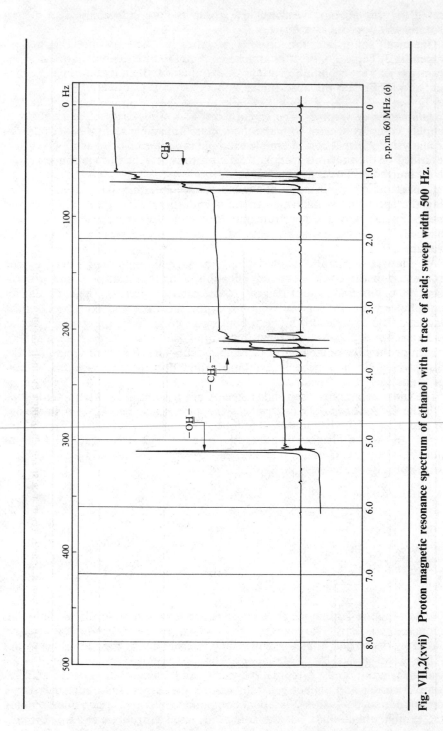

Fig. VII,2(xvii) Proton magnetic resonance spectrum of ethanol with a trace of acid; sweep width 500 Hz.

served for this proton; the methylene group is now only coupled to methyl protons and appears as a quartet.

Chemical shift data for protons attached to heteroatoms is listed in Appendix 3, Table 3,6. Protons attached directly to nitrogen may appear in the spectrum as very broad absorptions due to quadrupole interaction with nitrogen, and as a result the absorptions may be difficult to discern.

Simplification of spectra The chemist can adopt a variety of procedures to simplify complex spectra to make them more amenable to first-order analysis or analysis by inspection. A simple example is deuterium exchange which was described in the previous section. Acidic protons attached to carbon may also be exchanged under basic conditions.

Considerable simplification of a complex spectrum may be achieved by running the spectrum at higher magnetic fields (up to 300 MHz), which causes linear expansion of the spectrum but leaves the relative chemical shifts unchanged. The instrumentation required, however, is expensive and may not be available.

An alternative method of modifying the magnetic field experienced by the protons (with the consequent simplification of the spectra) is to add a paramagnetic compound (a **shift reagent**) to the solution (Ref. 7). The shift reagent coordinates with electronegative atoms in the substrate and thus modifies the magnetic field experienced by neighbouring protons. Since the strength of this field varies with the distance from the paramagnetic source, the chemical shift of each proton is modified by a different amount. The effect is to spread out the absorptions which previously overlapped and this frequently allows a first-order analysis of the spectrum.

The most commonly used shift reagents are tris-chelates of lanthanide ions with the β-diketones, 2,2,6,6-tetramethylheptane-3,5-dione (dipivaloylmethane, (I)) and 1,1,1,2,2,3,3-heptafluoro-7,7-dimethyloctane-4,6-dione (II)). Typical reagents are tris-(dipivaloylmethanato) europium and tris-1,1,1,2,2,3,3-heptafluoro-7,7-dimethyloctane-3,5-dionato europium, the names of which are normally abbreviated to Eu(dpm)$_3$ and Eu(fod)$_3$.

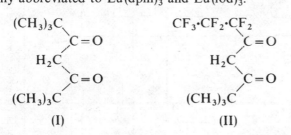

(I)   (II)

A wide range of lanthanide shift reagents are now commercially available including some derived from diketones in which all the protons have been replaced by deuterium, thus preventing any interference in the p.m.r. spectrum.

The extent of the lanthanide-induced shift is dependent on the basicity of the functional group and on the nature, purity and concentration of the shift reagent. Alcohols and amines generally exhibit the largest shift, but many other compounds such as ethers, carbonyl compounds, nitriles, sulphoxides, oximes, etc., exhibit useful shifts. Of the commonly used reagents Eu(fod)$_3$ normally causes the greatest shifts as it is a stronger Lewis acid. It also has the advantage of a much higher solubility. The magnitude of the lanthanide-induced shift is considerably decreased by the presence of a small amount of water, and since many of the reagents are hygroscopic, care should be taken in their handling and storage.

The two major applications of lanthanide shift reagents are firstly the simplification of the spectrum, and secondly the confirmation of the assignment of signals by relating the extent of the shift to the concentration of the shift reagent.

The effect on the spectrum of 4-methylpentan-2-one caused by the addition of increasing amounts of a shift reagent is shown in Fig. VII,2(xviii). The shift reagent used in this case is the europium chelate of 1,1,1,2,2-pentafluoro-6,6-

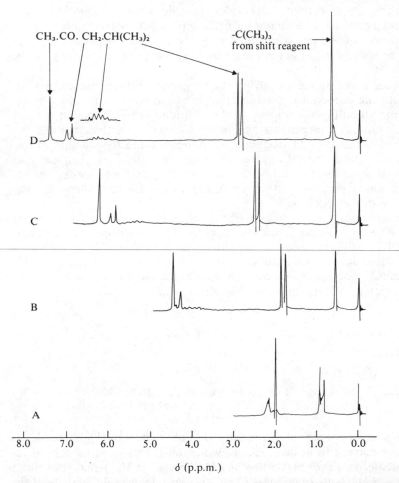

Fig. VII,2(xviii) 60 MHz proton magnetic resonance spectra of methyl isobutyl ketone (10.6 mg, 1.1×10^{-4} mol) in CCl_4 (0.5 ml) containing various amounts of $Eu(pfd)_3$; A, 0.0 mg; B, 13.5 mg; C, 21.1 mg; D, 29.0 mg. Data reproduced from H. E. Francis and W. F. Wagner (1972). 'Induced Chemical Shifts in Organic Molecules; Intermediate Shift Reagents', *Organic Magnetic Resonance*, **4**, 190 (Fig. 1), Heyden and Son Ltd.

dimethyl-3,5-heptanedione [Eupfd, $CF_3 \cdot CF_2 \cdot CO \cdot CH_2 \cdot CO \cdot C(CH_3)_3$]. In the absence of shift reagent the absorption of three of the groups of protons virtually overlap; addition of the shift reagent spreads out these absorptions to allow a ready analysis of the spectrum as indicated. Those protons closest to the donor atom (in this case the carbonyl oxygen) are shifted by a larger extent for a given amount of reagent added. This relationship is illustrated in Fig. VII,2(xix), which shows that the methyl and methylene groups bonded to the

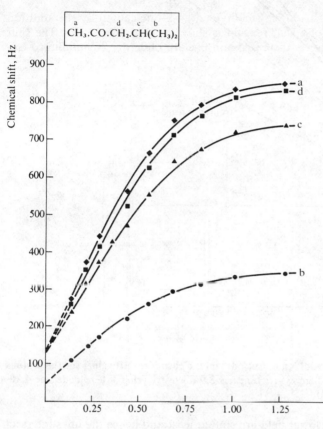

Fig. VII,2(xix)   **Induced shifts of the proton resonances of methyl isobutyl ketone as a function of added Eu(pfd)$_3$. Data reproduced from H. E. Francis and W. F. Wagner (1972), 'Induced Chemical Shifts in Organic Molecules: Intermediate Shift Reagents',** *Organic Magnetic Resonance,* **4, 190 (Fig. 2), Heyden and Son Ltd.**

carbonyl are shifted the most, followed by the methine proton, and the methyl groups furthest from the carbonyl are shifted least. The assignment of absorptions in the n.m.r. spectrum is greatly assisted by the use of shift reagents. Figure VII,2(xx) shows the proton shifts of the tricyclic dilactam (III) as a function of the concentration of Eu(fod)$_3$. Linear plots with varying slopes are obtained, the largest slope resulting from the four protons at C-5 and C-11 which are closest to the lanthanide ion. The protons at C-2 and C-8

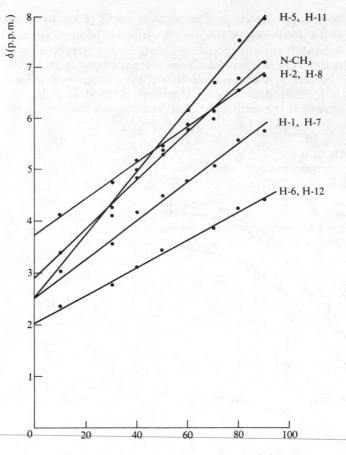

Fig. VII,2(xx) Effect of addition of Eu(fod)₃ on the chemical shift values for the protons in *N,N′*-dimethyl-*cis,trans,cis*-3,9-diaza-tricyclo[6,4,0²′ ⁷]dodecane-4,10-dione (III).

which were initially at lower field are shifted least and hence the lines intersect.

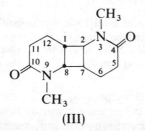

(III)

Optically active lanthanide shift reagents such as tris-(3-trifluoromethyl-hydroxymethylene-(+)-camphorato) europium (IV) are commercially available. They can be used for the direct determination of optical purity and for the measurement of enantiomeric composition. The differences in the lanthanide-induced shift between enantiomers can be as high as 1.8 p.p.m. depending on the geometry of the molecule.

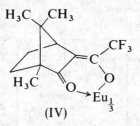

(IV)

Spin decoupling. Spin decoupling is a technique for determining which nuclei are coupled together by observing the effect in the spectrum when the coupling is removed. The decoupling is achieved by irradiating the substrate with a strong radiofrequency signal corresponding to the resonance frequency of one of the nuclei; the spectrum resulting from the remaining nuclei is scanned to observe any simplification which results. Successful decoupling is not achieved if the separation of the coupled multiplets is less than about 20 Hz for a 100-MHz instrument.

Figures VII,2(xxi) and (xxii (a and b)) illustrate the application of spin-decoupling to the simplification of the spectrum of crotonaldehyde [(V); spectrum xxi].

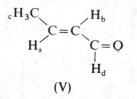

(V)

Spectrum (xxii(a)) shows the spin–spin splitting in the olefinic region. The multiplet at low field is two slightly overlapping quartets which arise from the coupling of the proton H_a with the methyl protons (H_c) to give a quartet which is then split into a pair of quartets by coupling with the proton H_b. The multiplet at higher field is more complex and arises from the proton H_b. Coupling of H_b to H_a gives a doublet, each signal of which is split into a pair of doublets by coupling with the aldehydic proton H_d. Each of these signals is then split by the methyl protons to give the observed four closely spaced quartets. Irradiation at the methyl protons causes all the quartets to collapse to single peaks, and the re-run, simplified spectrum, xxii(b), now shows the low-field doublet corresponding to H_a and the pair of doublets for H_b, where coupling to H_a and H_d only is observed.

Interpretation of the p.m.r. spectrum It is not possible to prescribe a set of rules which is applicable on all occasions. The amount of additional information available will most probably determine the amount of information it is necessary to obtain from the p.m.r. spectrum. However, the following general procedure will form a useful initial approach to the interpretation of most spectra.

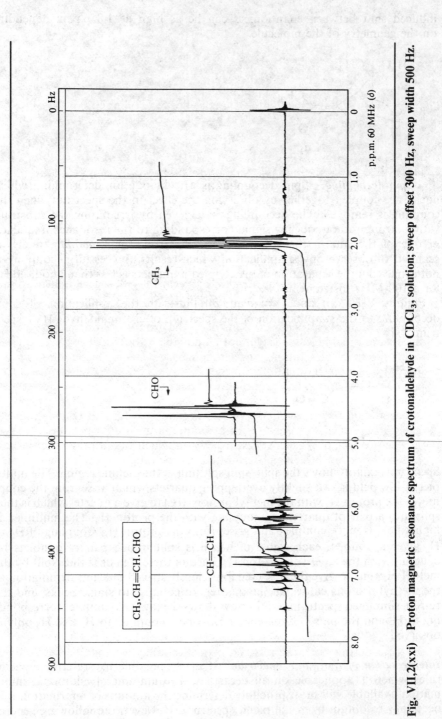

Fig. VII,2(xxi) Proton magnetic resonance spectrum of crotonaldehyde in CDCl₃, solution; sweep offset 300 Hz, sweep width 500 Hz.

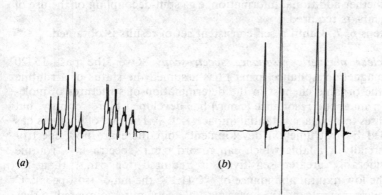

(a)   (b)

Fig. VII,2(xxii) Decoupled spectrum of crotonaldehyde. Data reproduced from W. McFarlane and R. F. M. White (1972). *Techniques of High Resolution Nuclear Magnetic Resonance Spectroscopy.* **London; Butterworths, p. 28.**

1. Make a table of the chemical shifts of all the groups of absorptions in the spectrum. In some cases it will not be possible to decide whether a particular group of absorptions arise from separate sets of nuclei, or form a part of one complex multiplet. In such cases it is probably best initially to include them under one group and to note the spread of chemical shift values.

2. Measure and record the heights of the integration steps corresponding to each group of absorptions. With overlapping groups of protons it may not be possible to measure these exactly, in which case a range should be noted. Work out possible proton ratios for the range of heights measured, by dividing by the lowest height and multiplying as appropriate to give integral values. If the accurate measurement of integration steps is not possible a range of proton ratios should be calculated, and noted down.

3. Note any obvious splitting of the absorptions in the table (e.g., doublet, triplet, quartet, etc.). For spectra which appear to show first-order splitting, the coupling constants of each multiplet should be determined by measuring the separation between adjacent peaks in the multiplet. Any other recognisable patterns which are not first-order should be noted.

4. Note any additional information such as the effect of shaking with D_2O, use of shift reagents, etc.

5. Attempt a preliminary assignment of the nature (e.g., alkyl, alkenyl, aryl) of each of the groups of absorptions on the basis of their chemical shift.

6. By considering both the relative intensities and the multiplicities of the absorptions attempt to determine which groups of protons are coupled together. The magnitude of the coupling constant may give an indication of the nature of the protons involved.

7. Relate the information thus obtained to any other information available on the compound under consideration.

8. Having arrived at a possible structure or partial structure consult the cor-

relation tables and work out the position and nature of the absorption expected from the postulated structure.

9. Decide whether additional information, e.g., spin-decoupling or the use of shift reagents, is required.

10. Repeat steps 6, 7, 8 until a self-consistent set of results is obtained.

Carbon-13 nuclear magnetic resonance spectroscopy Over the past 15–20 years proton magnetic resonance (p.m.r.) has assumed the status of a routine technique for the organic chemist in the determination of structures of molecules. ^{13}Carbon magnetic resonance (c.m.r.) has developed more recently, but now also promises to have a substantial impact (Ref. 8). The rate of progress has been determined by instrumental developments, but there are now spectrometers commercially available which can record c.m.r. spectra in a routine fashion. Considerably greater sensitivity is required for c.m.r. than for p.m.r. due to the low natural abundance of ^{13}C (1.1%; the major isotope is ^{12}C which has no nuclear spin), and the lower gyromagnetic ratio compared to that of the proton. However, greater resolution is possible with ^{13}C, since carbon resonances are spread over a range of about 600 p.p.m. compared with 15–20 p.p.m. for protons. Normally non-equivalent nuclei give separate absorptions, unlike protons which frequently give overlapping absorptions. A comparison between the ^{1}H- and the ^{13}C-spectra of 3-methylheptane is shown in Fig. VII,2(xxiii). The proton spectrum only distinguishes between the methyl,

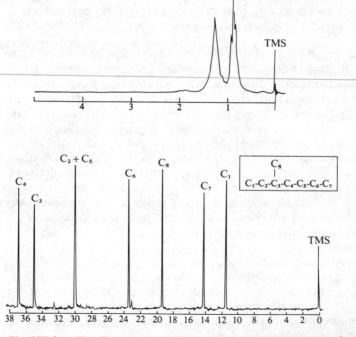

Fig. VII,2(xxiii) Proton magnetic resonance spectrum (top) and c.m.r. spectrum (bottom) of 3-methylheptane. Data reproduced from G. C. Levy and C. L. Nelson (1972). *Carbon-13 Nuclear Magnetic Resonance for Organic Chemists.* New York; Wiley-Interscience, p. 39.

methylene and methine protons, whereas the ^{13}C spectrum shows seven distinct peaks corresponding to the eight different carbon atoms.

Information from ^{13}C spectra. Only one of the three items of information normally available from p.m.r. spectra (i.e., chemical shift, coupling constant and relative number of absorbing nuclei) is routinely available from the ^{13}C spectrum, and that is the chemical shift. Quantitative coupling constants are not normally obtained, and relative numbers of nuclei cannot usually be derived from measurement of peak areas. The large $^{13}C-^{1}H$ coupling constant (125–200 Hz for directly bonded protons) results in multiplets which overlap to considerable extent, and in the absence of decoupling make the spectrum difficult to analyse. Spectra are therefore normally spin-decoupled and each absorption appears as a sharp singlet; this technique is known as wide-band or noise decoupling. Although the sensitivity is thus increased, all the information normally available from spin–spin splitting patterns is lost. An alternative method of decoupling (off-resonance decoupling) does however allow coupling of directly bonded carbon and hydrogen to be observed, although the separation of the peaks of the multiplets produced by this method is not equal to the true $^{13}C-^{1}H$ coupling constant. It is thus possible to identify carbon atoms associated with methyl, methylene and methine groups since the absorptions appear as quartets, triplets and doublets respectively, provided that the bonded hydrogens are equivalent. The use of the off-resonance procedure is illustrated in Fig. VII,2(xxiv), which shows the noise decoupled and off-resonance decoupled spectra of butane-1,3-diol and 2,2'-bipyridyl. The methyl carbon of the diol appears as a quartet at high field; the two methylene carbons appear as triplets, the one bonded to oxygen being at lower field, and the low-field doublet is due to the methine carbon. In the case of 2,2'-bipyridyl each carbon bonded to one hydrogen appears as a doublet; the two carbons bonded to nitrogen appear at lower field.

^{13}C Chemical shifts A considerable amount of data is available which correlates the position of absorptions in the c.m.r. spectrum with the structure of an organic molecule, and it is these empirical correlations which provide the main basis for the use of the technique in structure determination. Figure VII,2(xxv) shows the general relationships between structure and chemical shift. The values for the chemical shift are normally related to the tetramethylsilane carbon absorption, with positive values increasing to lower field (corresponding to the δ scale in p.m.r. spectroscopy). The vast majority of absorptions fall in a range of 200 p.p.m. between the carbonyl absorptions at low field and the methyl absorptions at high field. The position of absorption of some of the commonly used solvents is also included in Fig. VII,2(xxv).

Hybridisation of the carbon atom has a significant effect on the chemical shift: sp^3-hybridised carbon absorbs at high field (0.60 p.p.m. downfield from TMS), sp^2-carbon at low field (80–200 p.p.m.) and sp-carbon at intermediate values. The precise position of absorption of a particular atom is largely determined by the electronic effects of any substituents, and the fact that these are approximately additive enables fairly accurate predictions of chemical shifts to be made, provided that similar compounds of known structure are available for reference purposes.

Saturated compounds. The position of absorptions of methyl, methylene, methine and quaternary carbon atoms in the alkanes is shown in Fig.

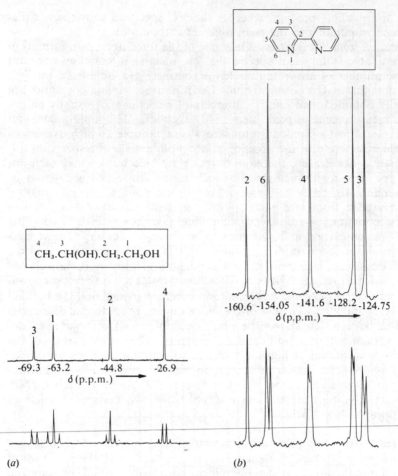

Fig. VII,2(xxiv) ^{13}C signal assignment by off-resonance decoupling (*a*) for butane-1,3-diol, 22.63 MHz; (*b*) for 2,2′-bipyridyl, 25.2 MHz. The numbers by the signals indicate the numbering of the carbon atoms. Values relative to TMS = 0. Data reproduced from E. Breitmaier, G. Jung and W. Voelter (1971). *Angew. Chem. Internat. Edn.*, **10**, 667.

VII,2(xxvi). Within each group the exact position of absorption is determined by the number and nature of substituents on the β and γ carbons. Replacement of a proton by CH$_3$ results in a downfield shift of *ca.* 8 p.p.m. at C-1, and *ca.* 10 p.p.m. at C-2, and an upfield shift at C-3 of *ca.* 2 p.p.m. Polar substituents result in a downfield shift in the position of absorption; Table 3,12 in Appendix 3 shows the effect on ^{13}C chemical shifts of replacing a methyl group by various polar substituents.

Alkenes and aromatics. The resonances for these classes of compounds appear in the same region (80–140 p.p.m. downfield from TMS) since in both cases the carbon atoms are sp^2-hybridised. Empirical rules for calculating the position of absorption in acyclic alkenes have been developed; the appropriate substituent parameter is added to the value for carbon in ethylene (123.3 p.p.m.).

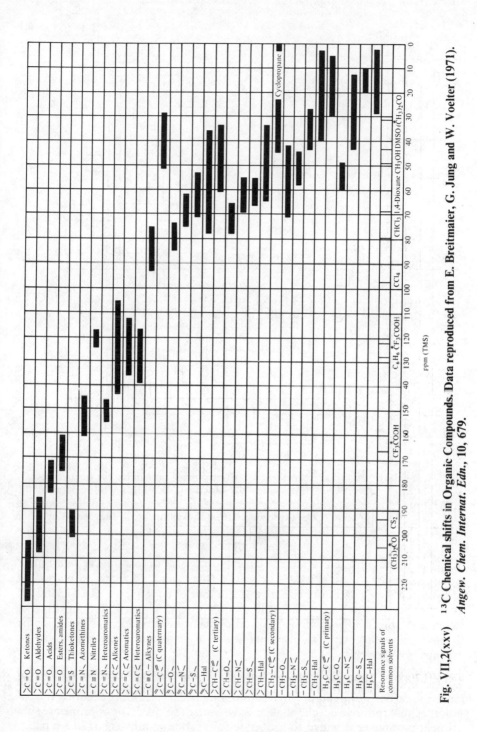

Fig. VII,2(xxv) ¹³C Chemical shifts in Organic Compounds. Data reproduced from E. Breitmaier, G. Jung and W. Voelter (1971).
Angew. Chem. Internat. Edn., **10**, 679.

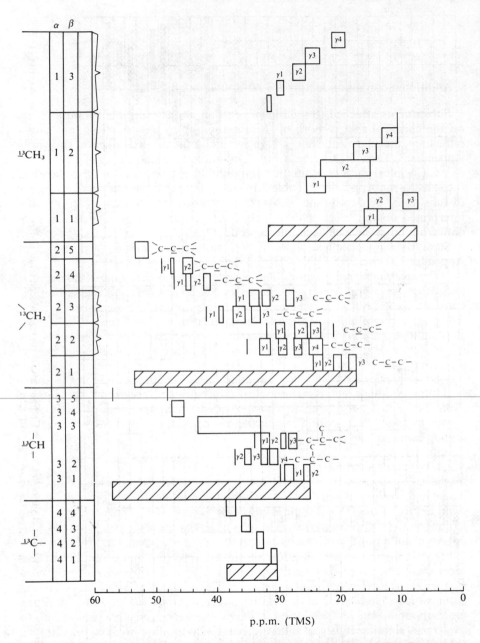

Fig. VII,2(xxvi)   Graphical display of chemical shifts for classes of paraffins. Data reproduced from L. P. Lindemann and J. Q. Adams (1971). *Analyt. Chem.*, 43(10), 1251.

Alkene substituent parameters

$$\overset{\gamma'}{C}-\overset{\beta'}{C}-\overset{\alpha'}{C}-C=\overset{*}{C}-\overset{\alpha}{C}-\overset{\beta}{C}-\overset{\gamma}{C}$$

α 10.6; β 7.2; γ −1.5;
α' −7.9; β' −1.8; γ' 1.5

Additional correction factors for *cis* double bond, −1.1.

Substituents attached to the olefinic carbon atoms exert a substantial effect on the chemical shift of both of these carbon atoms. These effects are exemplified by the chemical shift values for monosubstituted alkenes shown in Appendix 3, Table 3,13.

^{13}C Chemical shifts in aromatic compounds are dependent on the polarity of the substituent. Appendix 3, Table 3,14 shows the substituent effects for a range of substituted benzenes. The ^{13}C spectra of substituted benzenes can often be interpreted on the basis of these substituent parameters in association with data from off-resonance decoupled spectra.

Some data for chemical shifts in heteroaromatic compounds is shown in Appendix 3, Table 3,15.

Organic functional groups. The general chemical shift range for carbonyl and other functional groups is shown in Fig. VII,2(xxv). Although there is considerable overlap, distinct regions of absorption can be identified. It is interesting to note that there is not a linear relationship between ^{13}C chemical shift and the carbonyl stretching frequency in the infrared. The ^{13}C absorptions are shifted upfield by up to 10 p.p.m. by the introduction of an α-halogen or α,β-unsaturation.

Other applications. Carbon-13 n.m.r. has wide applications in the study of natural products, complex biological molecules and polymers. An example from the field of steroid chemistry is provided by the spectrum of cholesterol shown in Fig. VII,2(xxvii). The spectrum shows 26 resolved lines for the 27 carbon atoms in the molecule.

VII,4,3. Mass spectrometry

The mass spectrum In a typical mass spectrometer, an organic compound under high vacuum is bombarded with electrons (of about 70 eV energy). Loss of an electron from the molecule followed by various fission processes gives rise to ions and neutral fragments. The positive ions are expelled from the ionisation chamber and resolved by means of a magnetic or an electric field.

Fig. VII,3(i) shows part of a low resolution spectrum and a number of features should be noted. The mass spectrum is a record of the current produced by these ions as they arrive at a detector. The intensity of a peak in the spectrum is thus an indication of the relative number of ions; the larger the peak the more abundant the ion producing it. Many mass spectrometers produce up to five traces simultaneously of differing sensitivity to allow weaker peaks to be studied, whilst also allowing intense peaks to be recorded on the chart. The most intense peak in the spectrum is known as the **base peak**. Ions produced in the fragmentation of the organic compound are separated according to their mass : charge ratio (m/e). Since the majority of ions are singly charged the scale is often thought of as a mass scale; however, **doubly charged ions** are not uncom-

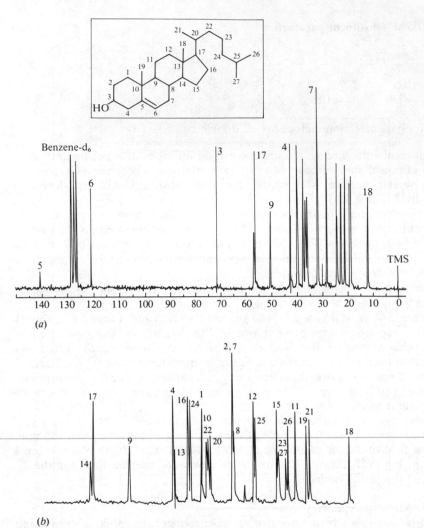

Fig. VII,2(xxvii) (a) The 5000 Hz FT c.m.r. spectrum of 0.2 *M*-cholesterol; pulse interval 0.4 seconds (total time, 3.4 hours). (b) The 1800 Hz plot expansion of the data in (a). All resonances assigned. Data reproduced from G. C. Levy and G. L. Nelson (1972). *Carbon-13 Nuclear Resonance for Organic Chemistry*. New York; Wiley-Interscience, pp. 164 and 165.

mon and these appear at half their mass value on the *m/e* scale. Many compounds give rise to an ion which corresponds to the removal of a single electron from the molecule; this is known as the **molecular ion** (M) and usually has the highest *m/e* value in the spectrum, with the exception of a characteristic group of peaks at *m/e* values of M + 1, M + 2, M + 3, . . ., etc. The latter are **isotope peaks** which arise from the fact that many of the elements normally present in organic molecules are not monoisotopic. Peaks in the mass spectrum are usually sharp and appear at integral mass values (with the exception of those arising from some doubly charged ions). Occasionally peaks are observed

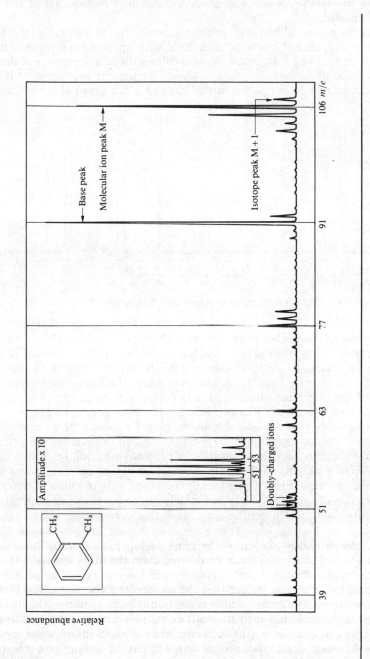

Fig. VII,3(i) Part of a low resolution mass spectrum of *o*-xylene.

which are broad, spread over several mass units and of low intensity; these are called 'metastable peaks' and give valuable information about the mode of fragmentation.

Spectra produced by most spectrometers are not in a suitable form for reproduction and cannot easily be compared with spectra from other instruments. Magnetic focusing instruments give spectra with non-linear m/e scales whereas those from quadrupole or time-of-flight instruments are linear. It is common practice to represent spectra in the form of a **bar graph** (Fig. VII,3(ii)) with a

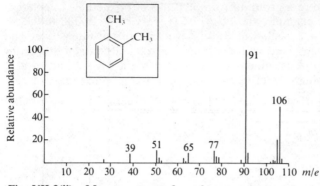

Fig. VII,3(ii)   Mass spectrum of *o*-xylene as a bar graph.

linear m/e scale. The base peak is given the arbitrary value of 100 per cent and the height of each other peak is measured relative to that value. An alternative method of representation is to tabulate the intensity of the current arising from each ion relative to the total ion current. The output from many mass spectrometers can now be handled by computers which allow considerable flexibility in the form of presentation of the spectra. Bar graphs can be produced directly and a large reference collection (Ref. 10) has been produced in this way.

Instruments vary considerably in the extent to which they can separate ions of closely related m/e values. In the vast majority of routine uses the organic chemist requires only the separation of ions having nominal unit masses of up to molecular weights of about 500–600, which can be achieved using an instrument of low resolution. Occasionally, however, it is of value to determine the precise mass of particular ions accurately (up to six places of decimals) and for this purpose an instrument of high resolution is required.

Molecular formulae   Probably the most useful single piece of information for the organic chemist which can be derived from the mass spectrum is the molecular formula. Provided that the molecular ion can be identified, there are in principle two methods for deriving the molecular formula, using either high or low resolution. The most reliable method, although requiring the more sophisticated high-resolution instrumentation, is the accurate mass measurement of the molecular ion. Since atomic masses are not exact integers (see Appendix 4, Table 4,1) each combination of atoms will have a unique non-integral value. For example, CH_2O and C_2H_6 both have an integral mass of 30 but the accurate masses are 30.010565 and 30.046950 respectively. Accurate mass measurement will therefore distinguish between these two molecules. Tables are available which reduce the problem of relating accurate masses to possible molecular formulae (Ref. 11). Accurate mass determination is most useful to the organic chemist in confirming the identity of a specific molecular ion rather than in suggesting possible formulae for completely unknown molecules.

An alternative method of determination of the molecular formula which utilises low-resolution spectra is based on the measurement of the **intensities of the isotope peaks**. The natural abundance of the stable isotopes of some common elements is shown in Appendix 4, Table 4,2. The data are presented in two ways, firstly as a percentage of the total isotopes present, and secondly as a percentage of the most abundant isotope. Each combination of atoms will thus give rise to a group of isotope peaks of predictable intensities. Taking methane as an example, the ratio $^{12}CH_4 : ^{13}CH_4 = 100 : 1.08$. Thus the intensity of the M + 1 peak will be 1.08 per cent of the intensity of the molecular peak, although there will also be a very small contribution from $^{12}C^1H_3^2H$. Table VII,2(a) lists some of the intensities of the M + 1 and M + 2 peaks for various combinations of C, H, N, O having a nominal mass of 120.

Table VII,2(a)   Intensities of isotope peaks for the combinations of C, H, N, O of mass 120

Formulae	M + 1	M + 2
$C_2H_4N_2O_4$	3.15	0.84
$C_2H_6N_3O_3$	3.52	0.65
$C_3H_{12}N_4O$	5.00	0.31
$C_4H_{12}N_2O_2$	5.36	0.52
$C_6H_6N_3$	7.72	0.26
C_9H_{12}	9.92	0.44

Extensive compilations of such data are available: they can easily be modified to include elements other than C, H, O and N.

One limitation on the use of isotope peak intensities to determine the molecular formula is that the molecular ion must be relatively intense, otherwise the isotope peaks will be too weak to be measured with the necessary accuracy. Difficulty may also arise from spurious contributions to the isotope peak intensities from the protonated molecular ion, from weak background peaks or from impurities in the sample. In any event the method is only reliable for molecules having molecular weights up to about 250–300.

Deductions from isotope abundances. Assuming that the molecular ion has been identified correctly and intensities of the isotope peaks measured, the next stage in the analysis is to work out all possible molecular formulae which are consistent with this information. The common elements can be divided into three groups according to their isotopic composition. Firstly, those elements with a single natural isotope, e.g., hydrogen, fluorine, phosphorus, iodine. Hydrogen is placed in this group since the contribution from ^2H is extremely small. Secondly, those elements with a second isotope of one mass unit higher than the most abundant isotope, e.g., carbon and nitrogen. Thirdly, those with an isotope two mass units higher than the most abundant, e.g., chlorine, bromine, sulphur, silicon and oxygen.

Members of the last group, and especially chlorine and bromine, are most easily recognised from the characteristic patterns of the peaks, spaced at intervals of two mass units, which they produce in the spectrum. Typical patterns for combinations of bromine and chlorine atoms are shown in Fig. VII,3(iii) and

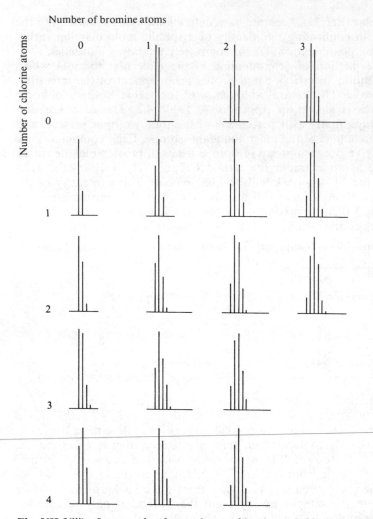

Number of bromine atoms

Number of chlorine atoms

Fig. VII,3(iii) Isotope abundances for combinations of chlorine and bromine atoms.

(iv). It may be difficult to estimate the number of oxygen atoms due to the low natural abundance (0.20%) of ^{18}O.

The intensity of the M + 1 peak allows an estimate to be made of the number of carbon and nitrogen atoms (however, if Cl, Br, S, or Si are present, loss of a proton from the M + 2 may enhance the intensity of the M + 1 peak). For a molecule not containing nitrogen, the maximum number of carbon atoms can be deduced by dividing the relative intensity of the M + 1 peak by 1.1. Thus a molecule having twelve C atoms will give an M + 1 peak of 13.2 per cent. If nitrogen is present its contribution to the M + 1 peak will amount to 0.36 × the number of nitrogen atoms; this figure must be subtracted from the measured relative intensity of the M + 1 peak before calculating the number of carbon atoms. An indication of the number of nitrogen atoms present may be deduced with the aid of the 'nitrogen rule' (see below).

Once the numbers of these two groups of elements have been estimated, the

Number of chlorine atoms

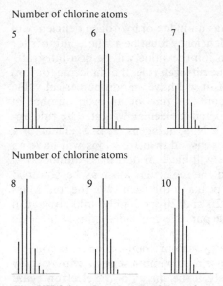

Number of chlorine atoms

Fig. VII,3(iv)   Isotope abundances for combinations of chlorine atoms.

remainder of the mass of the ion must be due to the monoisotopic elements, the numbers of which can then usually be deduced.

A study of isotope abundance can give information about the elemental composition of other ions in the spectrum as well as the molecular ion. However, care must be taken that the intensities being measured arise solely from the isotopic contribution and not from other ions of different elemental composition.

Recognition of the molecular ion   Since the molecular formula is normally the most important piece of information to be derived from the mass spectrum it is necessary to be as certain as possible that the molecular ion within the molecular cluster (M, M + 1, M + 2, etc.) has been correctly identified. A number of tests can be applied which will show if an ion is not the molecular ion.

The ion must be an odd-electron ion since the molecular ion is produced by loss of one electron from the neutral molecule: the converse is not true since there may well be odd-electron ions other than the molecular ion in the spectrum, arising from rearrangement reactions. If the elemental composition of the ion can be determined, the index of hydrogen deficiency (the sum of multiple bonds and ring systems) can be used to determine whether the ion is an odd-electron ion. The **index of hydrogen deficiency** is the number of pairs of hydrogen atoms which must be removed from the saturated open-chain formula to give the observed molecular formula. For a molecule $I_y II_n III_z IV_x$:

the index of hydrogen deficiency $= x - y/2 + z/2 + 1$

where   I = any monovalent atom
II = O, S or any other divalent atom
III = N, P or any other trivalent atom
IV = C, Si or any other tetravalent atom

For example, thiophen, C_4H_4S, ⬠$_S$, has an index of hydrogen deficiency of $(4 - \frac{4}{2} + 1) = 3$. The index of hydrogen deficiency must be a whole number for an odd-electron ion. For an even-electron ion the value will be non-integral.

A second test which can be applied is the **nitrogen rule**. If a molecule (or ion) contains an odd number of nitrogen atoms it will have an odd numerical value for the molecular weight, whereas if it contains zero or an even number of nitrogen atoms it will have an even-numbered molecular weight. The rule applies to all compounds containing C, H, O, N, S, halogens, P, B, Si. Thus for a species with zero or even number of nitrogens, odd-electron ions will have an even mass number and even electron ions will have an odd mass number.

A third indication that an ion is indeed the molecular ion may be obtained from an examination of the fragment ion peaks in the vicinity of the ion. Mass losses of between 3 and 15 and between 20 and 26 are highly unlikely, and if they are observed would suggest that the putative molecular ion is in fact a fragment ion.

Alteration of instrumental conditions may also provide evidence to confirm the recognition of the molecular ion. The use of maximum sensitivity may show up a very weak molecular ion. Alternatively, if the energy of the electron beam is decreased the intensity of the fragment ions will decrease relative to the molecular ion; this also applies to fragment ions arising from impurities. Alternative methods of ionisation such as chemical ionisation and field ionisation are very much more likely to produce a molecular ion cluster than the electron ionisation method, and should be used if they are available.

Intensity of the molecular ion   The lower the energy required for ionisation of the molecule, and the more stable the molecular ion, the more intense will be the peak in the mass spectrum. Structural features within the molecule have characteristic values of ionisation energy and hence determine the amount of energy required to form the molecular ion. Table VII,2(*b*) gives a general in-

Table VII,2(*b*)   Intensity of the molecular ion in the mass spectrum

Strong	Medium	Weak or non-existent
Aromatic hydrocarbons	Aromatic bromides and iodides	Aliphatic alcohols, amines and nitriles
Aromatic fluorides, chlorides, nitriles and amines	Conjugated alkenes	Branched chain compounds
Saturated cyclic compounds	Benzyl and benzoyl compounds	Nitro compounds
	Straight chain ketones and aldehydes, acids, esters, amides	
	Ethers	
	Alkyl halides	

dication of the intensity of the molecular ion for various types of compounds. It must be borne in mind that if the molecule contains a readily cleaved bond the molecular ion peak will be much less intense. In general the intensity of the molecular ion increases with unsaturation and with the number of rings, but decreases with chain branching. The presence of heteroatoms with easily ionised outer-shell electrons increases the intensity of the molecular ion.

Fragmentation   Although it may be of very low abundance, the molecular ion provides vital information about the identity of the molecule. Further information must be derived from the fragmentation pattern, i.e., the pattern of ions produced by decomposition of the molecular ion. Not all ions are of equal importance and some guidelines and rationalisations are needed to enable the organic chemist to derive the information he requires from the mass spectrum. Firstly, as discussed above, the molecular ion is the most important in the spectrum. Secondly, odd-electron ions are generally of more significance than even-electron ions of similar mass or abundance, since they are generally formed via a rearrangement reaction which may be characteristic of a particular class of compounds. Thirdly, ions of high mass are likely to give more useful information than those at lower mass, since they are likely to have been formed as the result of a simple rational fragmentation. Fourthly, metastable ions (see later) may give useful information on the nature of the fragmentation processes.

There are two important factors which determine the intensities of fragment ions in the mass spectrum: the stability of the ion, and the energy relationships of the bonds broken and formed in the reactions leading to the ion. Although the conditions in the mass spectrometer (very low pressure, unimolecular reactions) differ substantially from those normally encountered in organic chemistry, the fundamental ideas of physical organic chemistry, and in particular those concerned with carbonium ion stability, can be used effectively in the rationalisation of the appearance of the mass spectrum. Thus the following common fragmentations all give rise to typically stable carbonium ions.

$$\left[R - \underset{\underset{CH_3}{|}}{\overset{\overset{CH_3}{|}}{C}} - CH_3 \right]^{\oplus \cdot} \longrightarrow \dot{R} + CH_3 - \underset{\underset{CH_3}{|}}{\overset{\overset{CH_3}{|}}{C}}{}^{\oplus}$$

A stable tertiary carbonium ion is formed. The order of stability of saturated carbonium ions decrease in the order: tertiary > secondary > primary > methyl.

$$R - CH_2 - CH \overset{\oplus}{\underset{\cdot}{}} CH_2 \longrightarrow \dot{R} + \overset{\oplus}{C}H_2 - CH = CH_2 \leftrightarrow CH_2 = CH - \overset{\oplus}{C}H_2$$

Formation of a resonance-stabilised allylic carbonium ion.

$$C_6H_5 \cdot CH_2 - R]^{\oplus \cdot} \longrightarrow \dot{R} + \text{(tropylium ion)}$$

The aromatic seven-membered cyclic tropylium ion $C_7H_7^{\oplus}$ is formed.

$$^1R-CH_2-\overset{\oplus}{\ddot{X}}-{}^2R \longrightarrow {}^1\dot{R} + CH_2 = \overset{\oplus}{X}-{}^2R \leftrightarrow \overset{\oplus}{C}H_2-\ddot{X}-{}^2R$$

The carbonium ion is stabilised by delocalisation of the lone-pair electrons on the adjacent heteroatom.

$$\overset{^1R}{\underset{^2R}{>}}C=\overset{\oplus}{\ddot{O}}: \longrightarrow {}^1\dot{R} + {}^2R-C\equiv\overset{\oplus}{O}: \leftrightarrow {}^2R-\overset{\oplus}{C}=\ddot{O}:$$

The resonance-stabilised acylium ion is formed in this case.

The molecular ion is formed by removal from the molecule of the electron of lowest ionisation potential. The energy required to remove an electron varies in the order

$$\text{lone-pair} < \text{conjugated } \pi < \text{non-conjugated } \pi < \sigma$$

A radical ion is thus formed which can fragment in a variety of ways. Simple bond cleavage may occur to give a neutral and an ionic fragment. Alternatively, a number of rearrangement processes may take place which are then followed by bond cleavage reactions. The important types of fragmentations and rearrangements are summarised and exemplified below.

Fragmentation by movement of one electron. Bonds are broken by movement of one electron, represented by a fish-hook arrow ($\rightharpoonup$).

(a) *σ-cleavage* (sigma cleavage)

$$^1R\overset{\frown}{}{}^2R]\overset{\oplus}{\cdot} \longrightarrow {}^1\dot{R} + {}^2R^{\oplus}$$

Ionisation results in removal of a σ-electron and the σ-bond then breaks preferentially to give a stable carbonium ion with the ejection of the largest possible group as the radical.

(b) *α-cleavage* (alpha cleavage)

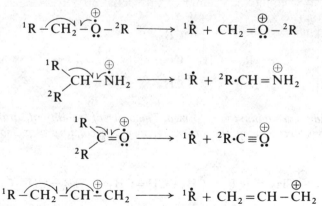

In all of these processes an uncharged alkyl radical is lost enabling the residual electron to pair with that associated with the original radical ion, to form an even-electron ionic species.

Fragmentation by movement of an electron pair. Bonds are broken by movement of two electrons towards the positive charge, and represented by a normal 'curly arrow' ($\rightarrow$).

$$^1R - \overset{\oplus}{\underset{\cdot\cdot}{\overset{\cdot}{O}}} - {}^2R \longrightarrow {}^1\overset{\oplus}{R} + :\overset{\cdot}{\underset{\cdot\cdot}{O}} - {}^2R$$

$$CH_2 = \overset{\oplus}{\underset{\cdot\cdot}{O}} - R \longrightarrow CH_2 = \overset{\cdot}{\underset{\cdot\cdot}{O}}: + \overset{\oplus}{R}$$

(formed from α-cleavage
in an ether)

$$R - C \equiv \overset{\oplus}{O}: \longrightarrow \overset{\oplus}{R} + :C = \overset{\cdot\cdot}{O}:$$

(formed from α-cleavage of
an aldehyde or ketone)

$$R - \overset{\oplus}{O}H_2 \longrightarrow \overset{\oplus}{R} + H_2O$$

An electron pair is donated to the charge site. The electron pair may come from the bond adjacent to the charge site.

Rearrangements yield odd-electron ions which are normally easily identified in the spectrum (cf. the nitrogen rule above). They are thus useful aids in the interpretation of the spectrum. Owing to the large excess of energy normally available in the ion source, molecular rearrangements are extremely common (see below). They may be random rearrangements (scrambling) which result in the general redistribution of certain atoms in the molecule, or more specific rearrangements, frequently involving a transfer of a hydrogen atom, which are characteristic of a certain type of molecular structure and give rise to easily recognisable ions in the mass spectrometer. It is the latter type which are particularly useful in the elucidation of molecular structure.

The most frequently encountered example is the **McLafferty Rearrangement** which involves the transfer of a γ-hydrogen atom in an unsaturated system via a low-energy six-membered transition state:

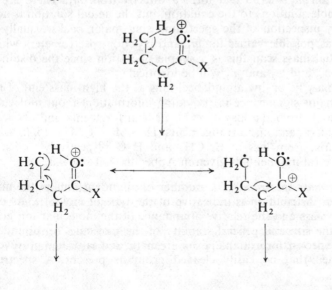

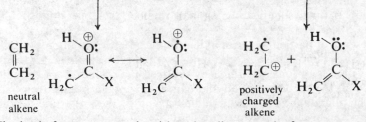

The ionic fragment may be either the alkene or the fragment containing the heteroatom; this is determined by the relative ionisation potentials of the two groups. The rearrangement is general for this type of functional grouping and also occurs with oximes, hydrazones, ketimines, carbonates, phosphates, sulphites, alkenes and phenylalkanes. A similar rearrangement may occur in saturated systems; in this case a smaller cyclic transition state is permitted since it does not have to accommodate the double bond.

Peaks which arise from **metastable ion decomposition** are normally broad and of low intensity. They arise from the fragmentation of ions which have already been accelerated out of the ion source but have not yet reached the magnetic field. They are thus displaced from the position in the spectrum which would correspond to their true mass. The position of the metastable peak (m^*) is related to the mass of the precursor ion (m_1) and the mass of the product ion (m_2) by the equation

$$m^* = \frac{(m_2)^2}{m_1}$$

The existence of a metastable ion and its relationship to m_1 and m_2 thus confirm that the ion m_2 was in fact formed directly from m_1. There are in theory many possible solutions to the equation, but the actual solution is normally obtained by inspection of the spectrum using major peaks, usually of similar intensity, as possible values for m_1 and m_2. For spectrometers which have an exponential mass scan this is a simple operation since the distances between m^* and m_2, and m_2 and m_1 will be identical.

Although they may be of low abundance, ions at the high mass end of the spectrum are of major significance in providing information about molecular structure. They result from the loss of small neutral fragments and are least likely to be the result of random rearrangements. Thus $M - 1$, $M - 15$, $M - 18$ peaks normally arise from loss of H, CH_3 and H_2O respectively. A list of some common neutral fragments is given in Appendix 4, Table 4,3.

Appearance of the mass spectrum   The number of abundant ions in the mass spectrum and their distribution is indicative of the type of molecule. As discussed above, the mass and the relative abundance of the molecular ion gives an indication of the size and general stability of the molecule. An abundant molecular ion is expected, for example, from aromatic and saturated polycyclic molecules, provided that no easily cleaved group is present. A spectrum consisting of a few prominent ions suggests there are only a few favoured decomposition pathways indicating a small number of labile bonds or stable products.

The presence of particular **series of ions** in the spectrum is often indicative of certain types of molecules. Compounds with large saturated hydrocarbon groups give series of ions separated by fourteen mass units, corresponding to CH_2, since all the carbon–carbon and carbon–hydrogen bonds are of similar energy. The abundance of ions at the lower end of the spectrum steadily increases for straight-chain alkyl groups as the result of secondary reactions (see the spectrum of decane, Fig. VII,3(v(a))). The sequence in straight-chain alkanes appears at $\overset{\oplus}{C_nH_{2n+1}}$ (29, 43, 57, 71, ...), but for compounds containing functional groups the positions are shifted due to the presence of heteroatoms. Some of the common series are shown in Table VII,2(c). Unfortunately

Table VII,2(c)   Ion series: aliphatic compounds

Compound type	General formula	Ion series
Alkyl	C_nH_{2n+1}	29, 43, 57, 71 ...
Alkylamines	$C_nH_{2n+2}N$	30, 44, 58, 72 ...
Aliphatic alcohols and ethers	$C_nH_{2n+1}O$	31, 45, 59, 73 ...
Aliphatic aldehydes and ketones	$C_nH_{2n+1}CO$	43, 57, 71, 85 ...
Aliphatic acids and esters	$C_nH_{2n-1}O_2$	59, 73, 87, 101 ...
Alkyl chlorides	$C_nH_{2n}{}^{35}Cl$	49, 63, 77 ...
	$C_nH_{2n}{}^{37}Cl$	51, 65, 79

the series for aldehydes and ketones overlaps the alkyl series since CO and C_2H_4 are both of mass 28. Complex molecules may show more than one series.

Characteristic ion series are also produced by aromatic compounds, the exact positions being dependent on the nature of the substituent (Table VII,2(d)).

Table VII,2(d)   Ion series: aromatic compounds

Electron withdrawing substituent:	38, 39	50, 51	63, 64	75, 76
Electron donating substituent and heterocyclic compounds:	39, 40	51, 52	65, 66	77, 78, 79

Certain types of compounds give characteristic ions in the mass spectrum which are often readily picked out and are useful indicators of possible structures. These include m/e 30 (amines), 31 (primary alcohol), 74 (methyl alkanoates), 91 (benzyl), 149 (phthalate acid and esters). The possible compositions of some common fragment ions are listed in Appendix 4, Table 4,4. Some caution must be adopted in the use of these tables.

Interpretation of the mass spectrum   The following scheme is suggested as a general approach to the interpretation of the mass spectrum. Each spectrum presents its own challenge and therefore too rigid adherence to any scheme is unwise. Reference should be made to the appropriate paragraph of this section for fuller details of each step.

1. Identify the molecular ion.
2. Determine the elemental composition and the index of hydrogen deficiency, i.e., the number of double bonds and rings.
3. Make any deductions which are possible from the general appearance of the spectrum; identify any ion series and characteristic ions.
4. Note possible structures of neutral fragments from the presence of high mass ions.
5. Identify any odd-electron ions and consider possible rearrangements (see *Rearrangements*, p. 1045).
6. Suggest a feasible structure on the basis of the mass spectral and any other evidence. Predict the mass spectrum of the postulated compound and compare with the unknown spectrum. Make any modification to the proposed structure which appears necessary. Check the mass spectral behaviour of compounds of similar structures by consulting appropriate reference collections.

Mass spectra of classes of organic compounds Many types of organic compounds exhibit characteristic mass spectral behaviour, a knowledge of which is useful in the interpretation of their spectra. The following section provides an introduction to the interpretation of mass spectra of simple organic compounds but readers should consult the texts listed in the bibliography, and especially that by Budzikiewicz, Djerassi and Williams for more extensive discussions (Ref. 12). Some caution is needed in the application of this information since the incorporation of additional substituents or functional groups into a molecule may well prevent 'characteristic' fragmentation.

Hydrocarbons. (i) **Saturated hydrocarbons** require high energy for ionisation and the ions thus formed undergo random rearrangements. The molecular ion is normally present although it may be weak. The spectra normally consist of clusters of peaks separated by fourteen mass units corresponding to a difference of a CH_2 group. The $M - CH_3$ ion is frequently missing and for branched alkanes the intensity of the other ions increases steadily to reach a maximum at m/e 43 ($C_3H_7^{\oplus}$) or m/e 57 ($C_4H_9^{\oplus}$); these peaks are mainly due to the highly branched ions resulting from molecular rearrangements; the spectrum of decane, Fig. VII,3(v(*a*)), is typical. Branched chain hydrocarbons show intense peaks corresponding to preferential cleavage at a tertiary or quaternary carbon atom; thus the spectrum of 2,6-dimethyloctane, Fig. VII,3(v(*b*)), shows an intense peak at m/e 113 due to loss of an ethyl group and formation of the secondary carbonium ion. Alicyclic hydrocarbons generally show a more abundant molecular ion, but the spectra are more difficult to interpret due to random rearrangement.

(ii) **Alkenes** give rise to spectra in which the molecular ion peak is usually distinct and there is an increased abundance of the $C_nH_{2n-1}^{\oplus}$ ion series as compared with alkanes, as illustrated by the spectrum of hex-1-ene (Fig. VII,3(vi)). The location of the double bond in alkenes is often difficult to determine due to the occurrence of facile rearrangements. Cyclic alkenes undergo a characteristic *retro*–Diels Alder fragmentation.

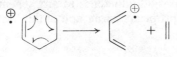

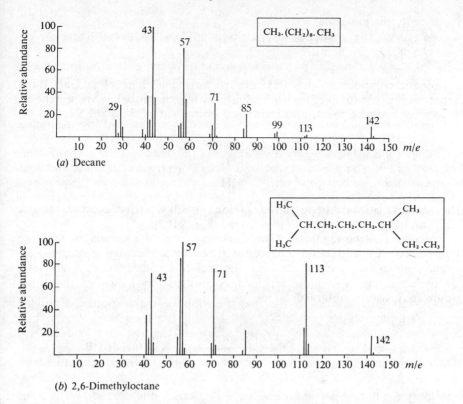

(a) Decane

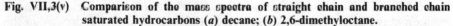

(b) 2,6-Dimethyloctane

Fig. VII,3(v) Comparison of the mass spectra of straight chain and branched chain saturated hydrocarbons (a) decane; (b) 2,6-dimethyloctane.

(iii) **Aromatic hydrocarbons** also generally give rise to a prominent molecular ion as the result of the stabilising effect of the ring; doubly charged ions are also often apparent as low-intensity peaks at half integral mass values. Alkyl-substituted benzenoid compounds (for example o-xylene, Fig. VII,3(ii)) usually give rise to a base peak at m/e 91 due to the tropylium ion, $C_7H_7^\oplus$. This may eliminate a neutral acetylene molecule to give a peak at m/e 65.

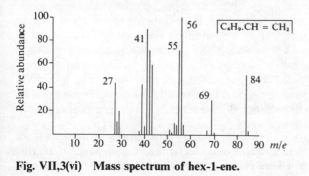

Fig. VII,3(vi) Mass spectrum of hex-1-ene.

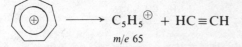

$$\longrightarrow C_5H_5^{\oplus} + HC\equiv CH$$

m/e 65

Aromatic compounds with alkyl groups having a chain of at least three carbon atoms can undergo a shift of a γ-hydrogen probably via a type of McLafferty rearrangement, giving rise to a prominent peak at m/e 92.

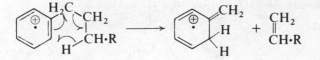

The characteristic aromatic cluster of ions in alkylbenzenes occurs at m/e 77, 78 and 79 (cf. the spectrum of o-xylene, Fig. VII,3(ii)).

Alcohols, phenols, ethers. The molecular ion of alcohols is weak or undetectable. Characteristic ions result from alpha-cleavage giving rise to resonance-stabilised carbonium ions; the loss of the largest alkyl group is the preferred pathway although ions resulting from losses of the other groups may also be observed:

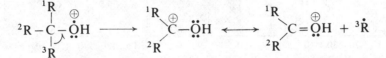

Primary alcohols in particular give an M − 18 peak due to loss of water from the molecular ion although this peak may partly arise from thermal decomposition of the alcohol in the ion source. Initial migration of a hydrogen on the alkyl chain is followed by cleavage of the carbon–oxygen bond (see following page for fragmentation); also see, for example, the spectrum of propan-1-ol, Fig. VII,3(vii), which shows strong peaks at m/e 59, 42, 31 due to the loss of H, H_2O, and formation of $CH_2\overset{\oplus}{=}OH$ respectively.

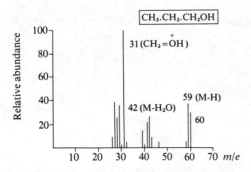

Fig. VII,3(vii)   Mass spectrum of propan-1-ol.

Phenols usually give a strong molecular ion. Typical peaks in the spectrum arise from M-28 (CO), which is a useful odd-electron ion, and M-29 (CHO).

The molecular ion peak of **ethers** is weak or negligible. There are two frag-

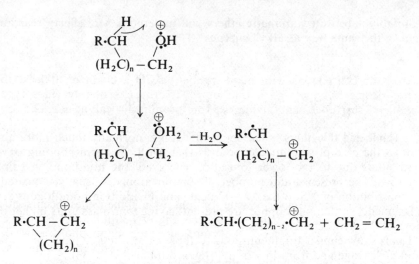

mentation processes which are typical of ethers. A characteristic fragmentation is the cleavage of the carbon–oxygen bond. This often gives rise to the most abundant ion in the spectrum of aliphatic ethers.

$$^1R \overset{\oplus}{-\underset{\cdot\cdot}{O}} - {}^2R \longrightarrow {}^1R + :\underset{\cdot\cdot}{O} - {}^2R$$

Alternatively cleavage of the α,β-bond (α-cleavage) may occur.

$$\begin{array}{c} ^1R \\ {}^2R \end{array}\hspace{-6pt}CH - \overset{\oplus}{\underset{}{O}} - CH_2 \cdot CH_2 \cdot {}^3R \longrightarrow {}^1R \cdot CH = \overset{\oplus}{\underset{}{O}} - CH_2 \cdot CH_2 \cdot {}^3R + {}^2\dot{R}$$
$$m/e\ 45,\ 59,\ 73\ \text{etc.}$$

This type of ion may then break down further:

$$\begin{array}{c} ^1R \cdot CH = \overset{\oplus}{O} - CH_2 \\ H - CH \cdot {}^3R \end{array} \longrightarrow {}^1R \cdot CH = \overset{\oplus}{O}H + CH_2 = CH \cdot {}^3R$$

See for example the spectrum of diethyl ether, Fig. VII,3(viii), which shows strong peaks at m/e 59, 45 and 31 due to $CH_2 = \overset{\oplus}{O} \cdot CH_2 \cdot CH_3$, $CH_3 \cdot CH = \overset{\oplus}{O}H$ and $CH_2 = \overset{\oplus}{O}H$ respectively.

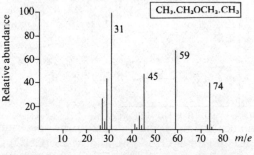

Fig. VII,3(viii)   Mass spectrum of diethyl ether.

Suitably substituted **aromatic ethers** will undergo a McLafferty rearrangement in the same way as alkylbenzenes.

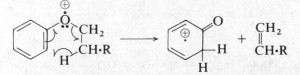

Thiols and thioethers.   The molecular ion is normally much more abundant in the case of sulphur compounds than with the corresponding oxygen compounds due to the lower ionisation energy of the non-bonding sulphur electrons. The presence and number of sulphur atoms is usually indicated by the contribution of ^{34}S to the M + 2 peak, and in addition homologous series of fragments containing sulphur are present having four mass units higher than those of the hydrocarbon series.

Thiols show similar fragmentations to those of alcohols, typical ions arising from α-cleavage and from loss of hydrogen sulphide (M − 34).

$$^1R \diagdown \atop ^2R \diagup CH - \overset{\oplus}{\underset{\bullet\bullet}{S}}H \longrightarrow {}^2R \cdot CH = \overset{\oplus}{\underset{\bullet\bullet}{S}}H + {}^1\dot{R}$$
$$m/e\ 47,\ 61,\ 75$$

Aldehydes and ketones.   Normally the molecular ion is observable for these compounds. Characteristic peaks in the spectra of ketones arise from cleavage α to the carbonyl group which gives two possible acylium ions,

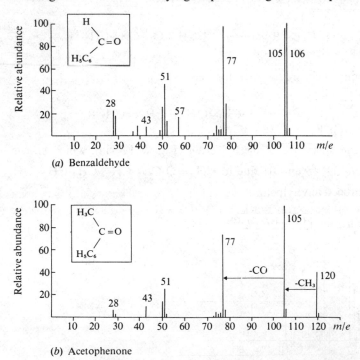

(a) Benzaldehyde

(b) Acetophenone

Fig. VII,3(ix)   Comparison of the mass spectra of (a) benzaldehyde and (b) acetophenone.

followed by loss of carbon monoxide giving the corresponding carbonium ions.

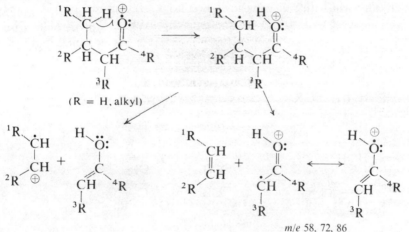

The more abundant acylium ion is normally produced by loss of the largest alkyl group.

In aromatic aldehydes and ketones the base peak usually arises from $Ar \cdot C \equiv \overset{\oplus}{O}$; compare the spectra of acetophenone and benzaldehyde. Fig. VII,3(ix(*a* and *b*)) which both show intense peaks due to $C_6H_5 \cdot C \equiv \overset{\oplus}{O}$, and further fragmentations typical of aromatic compounds. The α-cleavage reaction is normally less significant for aldehydes than for ketones, although a prominent peak at m/e 29 (CHO) is sometimes observed.

McLafferty rearrangements are common for aliphatic aldehydes and ketones, providing that an alkyl group of at least three carbons long is attached to the carbonyl group. Odd-electron ions are formed which are useful in the analysis of the spectrum.

m/e 58, 72, 86

Thus the spectrum of 4-methylpentan-2-one (Fig. VII,3(x)) shows a strong peak at m/e due to the odd-electron ion $(CH_3C(OH) = CH_2)^{\overset{\oplus}{\cdot}}$ resulting from the McLafferty rearrangement.

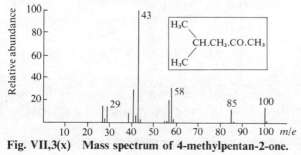

Fig. VII,3(x)   Mass spectrum of 4-methylpentan-2-one.

Carboxylic acids. Monocarboxylic acids normally show the molecular ion in the spectrum. Cleavage of bonds adjacent to the carbonyl group (α-cleavage) results in formation of fragments of mass M − 17 (OH) and M − 45 (CO_2H). Characteristic peaks arise from the McLafferty rearrangement.

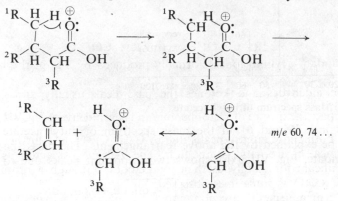

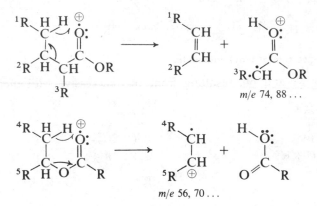

Carboxylic esters. The molecular ion of the ester $^1R \cdot CO_2{}^2R$ is usually observed in those cases where the alkyl group, ^2R, is smaller than C_4. The characteristic ions in the spectrum arise from McLafferty rearrangements, which can occur with either the acyl– or alkoxy–alkyl group, providing they are at least three or two carbon atoms long respectively.

A characteristic ion formed from esters of long-chain alcohols results from rearrangement of two hydrogen atoms ('McLafferty + 1' rearrangement).

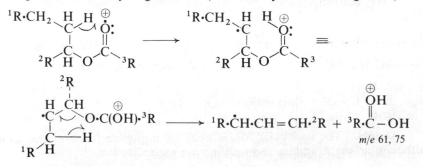

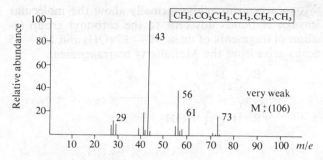

Fig. VII,3(xi) **Mass spectrum of butyl acetate.**

The peaks at m/e 56 and 61 in the mass spectrum of butyl acetate (Fig. VII,3(xi)) can be explained by the above rearrangements. The mass spectrum of ethyl butanoate, Fig. VII,3(xii), shows two important peaks due to odd-

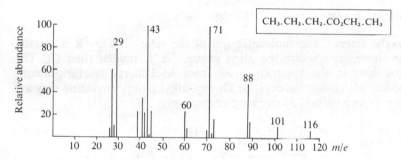

Fig. VII,3(xii) **Mass spectrum of ethyl butanoate.**

electron ions at m/e 88 and 60, resulting from two successive McLafferty rearrangements.

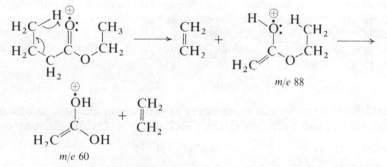

Cleavage of the alkoxyl group gives rise to the abundant ion m/e 71 $(M - OC_2H_5; R \cdot C \equiv \overset{\oplus}{O})$ which is a good diagnostic ion for esters.

Amines. The molecular ion is weak or negligible for aliphatic amines, although aromatic amines show an intense molecular ion.

1055

The characteristic cleavage reactions of amines are similar to those of alcohols and ethers.

$$R\overset{\oplus}{-CH_2}-\overset{\cdot}{N}H_2 \longrightarrow CH_2=\overset{\oplus}{N}H_2 + \overset{\cdot}{R}$$
$$m/e\ 30$$

In α-substituted primary amines, loss of the largest alkyl group is preferred. Similar ions are formed from secondary and tertiary amines.

$$^1R-\underset{\underset{^2R}{|}}{CH}-\overset{\oplus}{N}H-CH_2\cdot CH_2\cdot{}^3R \longrightarrow {}^1R-CH=\overset{\oplus}{N}H-CH_2 \longrightarrow$$
$$H-CH\cdot{}^3R$$

$$^1R-CH=\overset{\oplus}{N}H_2 + {}^3R\cdot CH=CH_2$$
$$m/e\ 44,\ 58,\ 72$$

The spectrum of diethylamine, Fig. VII,3(xiii), is typical.

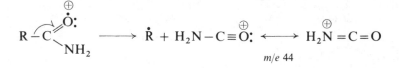

Fig. VII,3(xiii)   Mass spectrum of diethylamine.

Amides. Primary amides exhibit behaviour similar to the corresponding acid and methyl esters; substituted amides resemble the higher alkyl esters. There is a common tendency to form M + 1 ions by ion-molecule reactions. Primary amides generally give a strong peak at m/e 44:

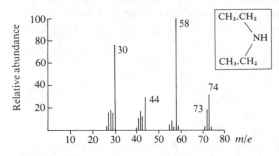

A characteristic fragmentation is via the McLafferty rearrangement.

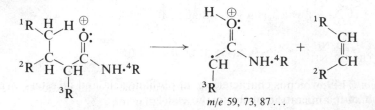

Nitriles. The molecular ion peak is weak or non-existent in aliphatic nitriles but strong in aromatic compounds. Interpretation of the spectrum is often difficult since skeletal rearrangements are common and the resulting ion series (m/e 41, 55, 69, etc.) overlaps with that arising from hydrocarbons. Thus in the McLafferty rearrangement:

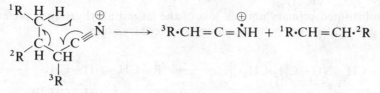

Halogens. The patterns of isotope peaks should indicate the nature and number of halogen atoms in the molecule. This is especially useful for aromatic halogen compounds, but may be less valuable for aliphatic compounds which often exhibit a weaker molecular ion peak.

A typical fragmentation of alkyl chlorides and bromides is the loss of an alkyl group with the formation of a halonium ion. The ion forms the base peak in straight-chain compounds but the intensity is considerably reduced if the chain is branched.

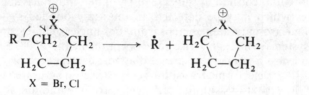

X = Br, Cl

Substituted aromatics—the ortho effect. Aromatic compounds bearing substituents with an appropriately placed hydrogen atom will undergo a facile rearrangement involving a second substituent in the *ortho* position, e.g.,

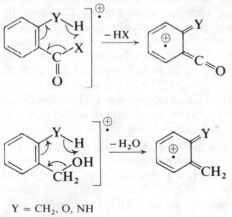

Y = CH₂, O, NH
X = OH, OR, NH₂

The ion m/e 149, which is characteristic of phthalic acid and its esters, arises as the result of the operation of this ortho effect.

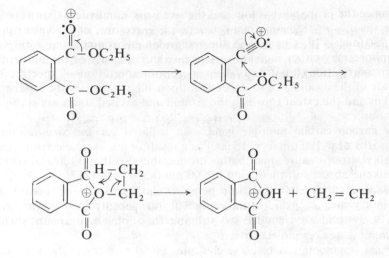

VII,4,4. Ultraviolet–visible spectroscopy The information concerning the structure of an organic molecule which may be gained from an ultraviolet–visible spectrum is more limited than in the case of i.r., n.m.r. and mass spectra. The principal features which may be detected are multiply bonded systems, conjugated systems and aromatic (and heteroaromatic) nuclei. The electronic transitions in these systems which give rise to absorption in the 200–700 nm region are $\pi \to \pi^*$ and $n \to \pi^*$ and these are of diagnostic value (see Sections **I,17,5** and **I,39**). The $n \to \sigma^*$ electronic transition which arises in saturated compounds containing the heteroatoms sulphur, nitrogen, bromine and iodine is of less importance since it leads to absorption just below 200 nm. The corresponding transition in compounds containing oxygen and chlorine leads to absorption at somewhat shorter wavelengths; indeed, the transparency of alcohols and chloroalkanes in the region 200–700 nm make them ideal solvents for u.v. spectral determinations (see also Section **I,39**). The $\sigma \to \sigma^*$ electronic transition in saturated hydrocarbons leads to absorption in the far-ultraviolet. Apart from the fact that special techniques are required to record the absorption, the information gained is in any event of little diagnostic value.

The term **chromophore** is used to describe any structural feature which leads to absorption in the u.v.–visible region and includes groups in which $\pi \to \pi^*$, $n \to \pi^*$ and $n \to \sigma^*$ transitions are possible. The term **auxochrome** is used to designate groups possessing non-bonding electron pairs which are conjugated with a π-bond system; an example is a hydroxyl or amino group attached to an aromatic ring system. Such n–π conjugation leads to a shift of absorption to longer wavelengths, referred to as a **bathochromic shift**; conversely a shift to shorter wavelengths is described as a **hypsochromic shift**.

Apart from the wavelength of maximum absorption (λ_{max}), the intensity of absorption (ε, the molar extinction coefficient, Section **I,39**) is of value in elucidating structural features. A high ε value (5000–10 000) reflects a high probability of the occurrence of the relevant electronic transition and is observable in systems in which the relative symmetry of the ground and excited state is such as to lead to a change in the transition moment. When this symmetry requirement is not met, the transition is regarded as being 'forbidden', and as a consequence the probability is low and the ε value is usually less than 100. For example, the $n \rightarrow \pi^*$ transition in a ketone, such as acetone, leads to absorption at λ_{max} 280 nm, ε 13. Changes in the molecular environment adjacent to a chromophore may either increase or decrease the intensity of the absorption and such correlations may be of help in the interpretation of spectra. The magnitude of the ε value is also dependent upon such factors as the polarity of the solvent and the extent to which the ground and excited states are stabilised by solvation.

The carbon–carbon multiple bond. An isolated carbon–carbon double bond absorbs near 180 nm ($\varepsilon \approx 15\ 000$) as a result of a $\pi \rightarrow \pi^*$ electronic transition. Alkyl groups cause small bathochromic shifts so that a tetrasubstituted acyclic alkene absorbs in the region of 200 nm (see Table VII,3(a)).

An isolated carbon–carbon triple bond similarly absorbs in the 180 nm region (acetylene, λ_{max} 173 nm, $\varepsilon \approx 6000$); the absorption maximum is shifted to longer wavelengths by the presence of alkyl groups (Table VII,3(a)).

Table VII,3(a) **Approximate absorption positions of isolated carbon–carbon multiply bonded systems**

Structure	λ_{max} (nm)	Structure	λ_{max} (nm)
$R \cdot CH = CH_2$	177	$R \cdot C \equiv CH$	185
trans-$R \cdot CH = CH \cdot R$	180	$R \cdot C = C \cdot R$	196
cis-$R \cdot CH = CH \cdot R$	183		
$R_2 C = C \cdot R_2$	200		

Conjugation of multiple carbon–carbon double bonds leads to significant changes in λ_{max} and in the ε value. For example, buta-1,3-diene has λ_{max} 217 nm, ε 21 000, and there is a regular bathochromic shift with an increase in the number of conjugated double bonds (Table VII,3(b)). Alkyl substituents attached to the multiply bonded carbon atoms have also been found to lead to uniform increments (5 nm) in the wavelength of absorption.

Table VII,3(b) **Approximate absorption positions of conjugated carbon–carbon multiply bonded systems**

Structure	λ_{max} (nm)	Structure	λ_{max} (nm)
$CH_3 \cdot (CH = CH)_3 \cdot CH_3$	275	$CH_3 \cdot (C \equiv C)_3 CH_3$	207
$CH_3 \cdot (CH = CH)_4 \cdot CH_3$	310	$CH_3 \cdot (C \equiv C)_4 CH_3$	234
$CH_3 \cdot (CH = CH)_5 \cdot CH_3$	342	$CH_3 \cdot (C \equiv C)_5 CH_3$	261
$CH_3 \cdot (CH = CH)_6 \cdot CH_3$	380	$CH_3 \cdot (C \equiv C)_6 \cdot CH_3$	284

Conjugation of the carbon–carbon triple bond with other triple bonds (polyynes) or with carbon–carbon double bonds (polyenynes) also leads to progressive shifts of λ_{max} to higher wavelengths. In acyclic systems such absorptions are frequently of diagnostic value in deciding the extent of conjugation. The regularity of wavelength shifts associated with changes in the extent of conjugation, the degree of substitution and the geometrical relationship of the double-bond system has led to the formulation of a set of empirical rules which enable the absorption maxima of substituted conjugated dienes to be predicted. Hence in cases where two isomeric structures are feasible, comparison of the calculated and experimental λ_{max} often enables a structural assignment to be made.

In formulating the rules (Table VII,3(c)) for cyclic dienes, these are classified as either heteroannular or homoannular, e.g.,

| Heteroannular | Homoannular |

The base λ_{max} value for the former is taken as 214 nm and for the latter as 253 nm. To these values are added increments according to the nature of the substituent present and other structural features as shown in the table. In

Table VII,3(c)   Fieser–Woodward rules for conjugated diene absorption

	λ (nm)
Parent heteroannular or open chain diene	214
Parent homoannular diene	253
Increments for:	
Double bond extending conjugation	30
Alkyl substituent or ring residue	5
Exocyclic double bond	5
Polar groupings: $O \cdot CO \cdot CH_3$	0
OR	6
SR	30
Cl, Br	5
$-NR_2$	60
Solvent correction	0
	Total $= \lambda_{max}$ calc.

those cases where homoannular and heteroannular chromophores coexist in conjugation, the base value for the homoannular contribution is used, to which is added the appropriate increment for extended conjugation. It should also be noted that homoannular dienes in ring systems other than the six-membered ring do not give a satisfactory correlation between calculated and experimentally determined λ_{max} values. The rules however are adequate for acyclic or heteroannular dienes providing that there exists in the structure no geometrical constraint which leads to a departure from coplanarity of the σ-bond system and to a consequent reduction in π-orbital overlap.

The carbonyl group. Acetone (in cyclohexane solution) exhibits two absorption bands; one appears at 190 nm (ε 1860) and corresponds to the $\pi \to \pi^*$ transition, while the second is at 280 nm (ε 13) and corresponds to the $n \to \pi^*$ transition. The absorption maxima of these bands are solvent-dependent. Ultraviolet spectra of saturated aldehydes, carboxylic acids, esters and lactones exhibit a similar absorption profile, and in general are of little diagnostic value.

Conjugation of the carbonyl group with a carbon–carbon double bond, however, significantly alters the absorption pattern. Thus the absorption corresponding to electron promotion from the carbon–carbon π-system to the anti-bonding orbital of the carbonyl group (the **electron-transfer** or **E.T. band**) is found in the region of 220 to 250 nm (ε 10 000–15 000) for simple enones; the term E.T. band is reserved for $\pi \rightarrow \pi^*$ transitions in which the conjugated chromophores are dissimilar. A weak band (ε 50–100) is also to be found in the region 310–330 nm and is due to the displaced $n \rightarrow \pi^*$ transition of the carbonyl group.

The position of the E.T. band depends in a predictable manner upon the extent of conjugation, the degree of substitution, etc., and may be calculated following rules which are analogous to those available for the prediction of absorption characteristics of conjugated dienes and which are set out in Table VII,3(d). The base values selected are 215 nm for an enone in an acyclic or six-membered ring system, or 202 nm for an enone system in a five-membered ring, or 207 nm for an α,β-unsaturated aldehyde.

Table VII,3(d) Woodward–Fieser rules for enone absorption

(1) $\overset{\beta}{>}\!\overset{}{C}=\overset{\alpha}{\underset{|}{C}}-\overset{}{\underset{|}{C}}=O$

(2) $>\!\overset{\delta}{\underset{|}{C}}=\overset{\gamma}{\underset{|}{C}}-\overset{\beta}{\underset{|}{C}}=\overset{\alpha}{\underset{|}{C}}-C=O$

		(nm)
Parent enone in an acyclic or six-membered ring		215
Parent enone in a five-membered ring		202
Parent $\alpha\beta$-unsaturated aldehyde		207
Increments for:		
Double bond extending conjugation		30
Alkyl substituent or ring residue $\quad \alpha$		10
β		12
γ and higher		18
Polar groupings– $\quad$ –OH $\quad\quad\alpha$		35
β		30
δ		50
–O·CO·CH $\quad \alpha, \beta, \gamma$		6
–OCH$_3$ $\quad\quad \alpha$		35
β		30
γ		17
δ		31
–Cl $\quad\quad \alpha$		15
β		12
–Br $\quad\quad \alpha$		25
β		30
–NR$_2$ $\quad\quad \beta$		95
Exo double bond		5
Homodiene component		39
Solvent correction (see Table VII,3(e))		
		Total = λ_{max} calc.

Table VII,3(e)   Solvent correction values

Solvent	Correction/nm
Ethanol	0
Methanol	0
Dioxan	+5
Chloroform	+1
Ether	+7
Hexane	+11
Cyclohexane	+11
Water	−8

Aromatic compounds. These compounds exhibit characteristic absorption in the ultraviolet–visible region of the spectrum, and although they are frequently easily recognised from their other spectroscopic properties, examination of their electronic spectra can often lead to the elucidation or confirmation of some of the detailed structural features.

Hydrocarbons. Benzene (in hexane solution) exhibits three absorption bands, λ_1, λ_2 and λ_3, which occur at 184 nm (ε 60 000), 204 nm (ε 7400) and 254 nm (ε 204) respectively which are due to the various allowed $\pi \rightarrow \pi^*$ transitions. Alkyl substituents cause a bathochromic shift of the λ_2 and λ_3 bands with little change in the ε value (see Table VII,3(f)).

Polycyclic aromatic hydrocarbons with both angular and linear types of ring fusion show absorption curves of a similar profile to that of benzene but with the absorption maxima shifted to longer wavelengths; the greater the number of rings the more pronounced the shift.

Substituted benzenoid systems. As noted above, alkyl substituents cause a small bathochromic shift in the λ_2 and λ_3 bands, and a similar effect is observed in the case of halogen substituents. However, substituents which contain multiply bonded groups (C=O, NO$_2$, C≡N, C=C, C≡C), and to a lesser degree substituents having non-bonding electrons in conjugation with the aromatic π-system, cause very pronounced bathochromic shifts of the two bands; this is frequently coupled with an increase in the ε value of the λ_3 band. In some cases additional bands will appear in the spectrum as a result of electron transitions associated with the substituent group (e.g., acetophenone $n \rightarrow \pi^*$, λ_{max} 320, ε 50). The effects of a selection of these groups on the positions of the λ_1, λ_2 and λ_3 bands may be gathered from the data cited in Table VII,3(f).

The origin of these bathochromic shifts lies in the more extensive mesomerism that exists in these derivatives, which thus reduces the energy difference between the ground and the excited states, and hence shifts the absorption to longer wavelength. Such aspects are dealt with in more detail in the specialist texts noted in the references cited on p. 1154.

Heterocyclic systems. Pyridine exhibits an absorption spectrum very similar to that of benzene with an additional absorption band at 270 nm which is assigned to the transition involving the nitrogen lone pair. Similarly quinoline and isoquinoline have spectral profiles closely analogous to naphthalene.

The five-membered heterocycles (furan, thiophen and pyrrole), despite their aromaticity, show distinct spectral differences from benzene. For ex-

Table VII,3(f) **Absorption characteristics of aromatic systems and their substituted derivatives.*

Compound (solvent †)	λ_1 (nm)		λ_2 (nm)		λ_3 (nm)	
Benzene[a]	184	60 000	204	7 400	254	200
Toluene[b]	—	—	207	7 000	254	160
o-Xylene[b]	—	—	210	8 300	263	300
m-Xylene[b]	—	—	212	7 200	265	300
p-Xylene[b]	193	54 000	212	8 000	274	460
Naphthalene[c]	220	100 000	275	5 700	312	250
Anthracene[c]	253	200 000	375	8 000	obscured by λ_2	
Tetracene[c]	278	200 000	474	13 000	obscured by λ_2	
Pentacene[c]	310	270 000	580	15 000	obscured by λ_2	
Chlorobenzene[d]	—	—	210	7 400	264	190
Bromobenzene[d]	—	—	210	7 900	261	192
Benzaldehyde[d]	—	—	250	11 400	280	1 100
Acetophenone[c]	—	—	246	9 800	280	1 100
Benzophenone[c]	—	—	252	18 000	obscured by λ_2	
Nitrobenzene[d]	—	—	269	7 800	obscured by λ_2	
Benzonitrile[d]	—	—	224	13 000	271	1 000
Styrene[d]	—	—	247	10 000	281	540
Phenylacetylene[c]	—	—	235	10 000	ca 280	300
Phenol[d]	—	—	211	6 200	270	1 450
Anisole[d]	—	—	217	6 400	269	1 480
Aniline[d]	—	—	230	8 600	280	1 430

* Some of this data is abstracted from *Physical Methods in Organic Chemistry*, ed. J. C. P. Schwarz, Oliver and Boyd, Edinburgh, 1964, p. 147.

† The solvents employed are (a) hexane, (b) methanol, (c) ethanol and (d) water containing sufficient methanol to ensure miscibility.

ample, furan has λ_{max} 200, 205 and 211 nm; thiophen has λ_{max} 235 nm, and pyrrole has λ_{max} 210 and 240 nm.

VII,5. IDENTIFICATION OF FUNCTIONAL GROUPS The solubility behaviour of an unknown compound will serve to classify it in one of the three main divisions, namely as an acidic, basic or neutral substance, and this information supplemented by the results of the elemental tests forms the basis for the subsequent systematic search to identify the functional group or groups which may be present. This search will thus be in one of the three divisions, but it should be noted that it is advisable to follow the sequence of tests specified in association with any available spectroscopic information that has been gained from the examination of spectra, since this additional knowledge may allow for some of the classes of compounds to be eliminated from consideration. Where an organic compound contains more than one functional group, the classification is generally based upon the one that is most readily detected and manipulated. Thus benzoic acid, p-chlorobenzoic acid, p-methoxybenzoic acid (anisic acid) and p-nitrobenzoic acid will be classified as *acids* both by the solubility tests and the class reactions, and the identification of, say, the nitrogen-containing acid *may* be completed by the preparation of derivatives of the carboxyl group without the absolute necessity of applying the class reactions that would discover the nitro group; however, if possible, it is always advisable to establish the nature of the subsidiary functional group or groups (including the presence of unsaturation), the presence of which may become apparent when a preliminary identification of the compound has been completed.

A. Acidic compounds

The possible classes of compounds are listed under group III in Table VII,1. The distinction between true acids and the weakly acidic pseudo acids (e.g., phenols, enols, nitroalkanes) is made by observing the nature of the reaction with sodium hydrogen carbonate. To ensure that evolution of carbon dioxide does not go unnoticed in those cases where reaction appears sluggish, add a solution of the compound in methanol carefully to a saturated solution of sodium hydrogen carbonate solution, when a vigorous effervescence at the interface will be observed.

The presence of a **carboxylic acid group** is indicated by strong infrared absorption in the region of 1720 cm^{-1} ($C=O$ str.) and broad absorption between 3400 cm^{-1} and 2500 cm^{-1} (OH str.); in the nuclear magnetic resonance spectrum the acidic hydrogen (replaceable by D_2O) will appear at very low field (δ 10–13).

Confirmatory tests for carboxylic acids (*a*) **Ester formation.** Warm a small amount of the acid with 2 parts of absolute ethanol and 1 part of concentrated sulphuric acid for 2 minutes. Cool, and pour cautiously into aqueous sodium carbonate solution contained in an evaporatory dish, and smell immediately. An acid usually yields a sweet, fruity smell of an ester. (Acids of high molecular weight often give almost odourless esters.)

(*b*) **Neutralisation equivalent.** It is recommended that the neutralisation equivalent (or the equivalent weight) of the acid be determined: this is the number expressing the weight in grams of the compound neutralised by one gram equivalent of alkali. Weigh out accurately about 0.2 g of the acid (finely powdered if a solid), add about 30 ml of water and, if necessary, sufficient alcohol to dissolve most of the acid, followed by two drops of phenolphthalein indicator. Titrate with accurately standardised 0.1 M-sodium hydroxide solution. Calculate the equivalent weight from the expression:

$$\text{Neutralisation equivalent} = \frac{\text{Grams of acid} \times 1000}{\text{M1 of alkali} \times \text{Molarity of alkali}}$$

For derivative preparations for carboxylic acids see Section **VII,6,***15*.

In the case of **carboxylic acid halides** the presence of the halogen will have been detected in the tests for elements. Most acid halides undergo ready hydrolysis with water to give an acidic solution and the halide ion produced may be detected and confirmed with silver nitrate solution. The characteristic carbonyl absorption at about 1800 cm^{-1} in the infrared spectrum will be apparent. Acid chlorides may be converted into esters as a confirmatory test: to 1 ml of absolute ethanol in a dry test-tube add 1 ml of the acid chloride dropwise (use a dropper pipette; keep the mixture cool and note whether any hydrogen chloride gas is evolved). Pour into 2 ml of saturated salt solution and observe the formation of an upper layer of ester; note the odour of the ester. Acid chlorides are normally characterised by direct conversion into carboxylic acid derivatives (e.g., substituted amides) or into the carboxylic acid if the latter is a solid (see Section **VII,6,***16*).

The presence of a **carboxylic acid anhydride**, the simpler examples of which are readily hydrolysed in aqueous solution, and therefore react with sodium hydrogen carbonate and also give the ester test, may be confirmed by applying the hydroxamic ester test (Section **VII,5,** *C*; p. 1075). Carbonyl absorption is apparent in the infrared spectrum at about $1820 \, cm^{-1}$ and at about $1760 \, cm^{-1}$. It should be noted that aromatic anhydrides and higher aliphatic anhydrides are not readily hydrolysed with water and may therefore have been classified as neutral (Section **VII,5,***C*; p. 1070). The final characterisation of the acid anhydride is achieved by conversion into a crystalline carboxylic acid derivative as for acid halides.

The detection of sulphur in a strongly acidic compound suggests it may be a **sulphonic acid** or a **sulphinic acid**. If nitrogen is also present the compound may be an **aminosulphonic** acid. The infrared spectrum will show absorption at $3400–3200 \, cm^{-1}$ (OH str.) and 1150 and $1050 \, cm^{-1}$ (S = O str. in a sulphonic acid) or at $1090 \, cm^{-1}$ (S = O str. in a sulphinic acid). For derivative preparations for sulphonic acids see Section **VII,6,**26. The presence of an aromatic sulphinic acid may be further confirmed by dissolving in cold concentrated sulphuric acid and adding one drop of phenetole or anisole when a blue colour is produced (Smiles's test), due to formation of a *para*-substituted aromatic sulphoxide. The reaction is:

$$Ar·SO_2H + C_6H_5OR \longrightarrow Ar·SO·C_6H_4OR + H_2O$$

Aromatic sulphinic acids are oxidised by potassium permanganate to sulphonic acids and are reduced by zinc and hydrochloric acid to thiophenols.

Phenols do not usually liberate carbon dioxide from 5 per cent sodium hydrogen carbonate solution; most are crystalline solids although notable exceptions are *m*-cresol and *o* bromophenol. The monohydric phenols generally have characteristic odours. The solubility in water increases with the number of hydroxyl groups in the molecule. The infrared spectrum shows broad strong absorption at $3400–3200 \, cm^{-1}$ (OH str.).

Confirmatory tests for phenols (*a*) **Iron(III) chloride solution.** Dissolve about 0.05 g of the compound in 5 ml of water; if the compound is sparingly soluble, prepare a hot saturated aqueous solution, filter and use 1 ml of the cold filtrate. Place the solution in a test-tube and add 1 drop of neutral 1 per cent iron(III) chloride solution and observe the colour; add another drop after 2–3 seconds. If a transient or permanent colouration (usually purple, blue or green) other than yellow or orange-yellow is observed, the substance is probably a phenol (or an enol). If no colouration is obtained, repeat the test as above but substitute absolute ethanol or methanol for water as solvent.

Prepare the **neutral ferric chloride solution** (i.e., free from hydrochloric acid) by adding dilute sodium hydroxide solution to the bench reagent until a slight precipitate of iron(III) hydroxide is formed. Filter off the precipitate and use the clear filtrate for the test.

(*b*) **Bromine water.** Many phenols (with the exception of those with strong reducing properties) yield crystalline bromo-compounds on the addition of bromine water. Dissolve or suspend 0.25 g of the compound in 10 ml of dilute hydrochloric acid or of water, and add bromine water dropwise until decolourisation is slow: a white precipitate of the bromophenol may form.

(*c*) **Phthalein test.** Many phenols yield phthaleins, which give characteristic colourations in alkaline solution, when fused with phthalic anhydride and a little concentrated sulphuric acid. Place in a dry test-tube 0.5 g of the compound and an equal bulk of pure phthalic anhydride, mix well together and add 1 drop of concentrated sulphuric acid. Stand the tube for 3–4 minutes in a small beaker of Silicone oil (or paraffin oil) previously heated to 160 °C. Remove from the bath, allow to cool, add 4 ml of 5 per cent sodium hydroxide solution and stir until the fused mass has dissolved. Dilute with an equal volume of water, filter and examine the colour of the filtrate against a white background: if the solution exhibits a fluorescence, observe the colour against a black background.

It must be borne in mind that there are many nitrogen-containing phenols and acids; of these the nitro and amino derivatives are the most common. The aromatic nitrocarboxylic acids may usually be identified through the reactions and derivatives of the carboxyl group without recourse to the reactions of the nitro group: examination for the latter will, however, provide additional confirmation. The influence of the nitro and other groups in the o- and p-positions upon the acidity of a phenol has already been noted: such groups tend to produce a marked deepening in the colour of alkaline solutions of the phenol. Amino substituents in water-insoluble phenols and acids cause these compounds to be soluble in both dilute acid and dilute alkali, i.e., to be amphoteric. (See amino acids in *acidic and neutral nitrogen-containing compounds*, Section **VII,5,C**; p. 1080). Frequently it is helpful to destroy the basic character of the nitrogen by conversion of the amino group into a neutral amide group by acetylation or benzoylation in aqueous alkaline solution: the resulting compound is not amphoteric and its equivalent may be determined.

For derivative preparations for phenols see Section **VII,6,6**.

Enols (e.g., β-keto esters and 1,3-diketones) usually respond to the following tests.

(*a*) **Iron(III) chloride.** Add a few drops of neutral iron(III) chloride solution to a solution of 0.1 g of the compound in water or in methanol. Most enols give a red colouration.

(*b*) **Copper derivative.** Shake 0.2 g of the substance vigorously with a little cold, saturated, aqueous copper(II) acetate solution. Many enols give a solid, green or blue, copper derivative, which can be crystallised from ethanol and often has a definite m.p. (e.g., from ethyl acetoacetate, m.p. 192 °C; from diethyl acetonedicarboxylate, m.p. 142 °C).

For derivative preparations for enols see Section **VII,6,7**.

Thiols are generally liquids with an unmistakable penetrating, disagreeable and characteristic odour, which persists even at extremely low concentrations in air. Alkanethiols are partly soluble in concentrated solutions of sodium hydroxide but their salts are hydrolysed to the free thiols on dilution with water. Thiophenols are soluble in sodium hydroxide solution but like the alkanethiols do not evolve carbon dioxide from sodium hydrogen carbonate solution. Treatment of a dry thiophenol with sodium results in the evolution of hydrogen (cf. *Alcohols*, Section **VII,5,C**; p. 1076).

For derivative preparations for thiols see Section **VII,6,25**.

Miscellaneous **nitrogen-containing compounds** which are soluble in sodium hydroxide include **aliphatic primary and secondary nitro compounds, oximes, imides** and **primary sulphonamides**.

Aliphatic primary and secondary nitro compounds dissolve in sodium hydroxide solution to give, in general, a yellow solution; on acidification with hydrochloric acid the nitro compound is regenerated. The nitro compounds show pronounced absorption due to the nitro group in the infrared at about 1530 and 1370 cm^{-1} (NO$_2$, asymmetric and symmetric str.). The presence of the nitro group is confirmed by reduction to the corresponding hydroxylamine (see nitro compounds, Section **VII,5,C**; pp. 1080–1), which can be detected by its action upon Tollen's reagent. To distinguish between **primary, secondary and tertiary aliphatic nitro compounds** (the latter are neutral) the following test should be performed. Dissolve a few drops of the nitro compound in concentrated sodium hydroxide solution, and add excess of sodium nitrite solution. Upon cautiously acidifying with dilute sulphuric acid, added a drop at a time, the following effects may be observed.

(i) Primary nitro compound: intense red colour, disappearing upon acidification. The colouration is that of the alkali salt of the nitrolic acid.

$$R{\cdot}CH{=}N\overset{\nearrow O}{\underset{O=N}{\searrow}}\overset{O}{\underset{\ominus}{\cdot}} \longrightarrow R{\cdot}\underset{N=O}{CH{\cdot}NO_2} \rightleftharpoons R{\cdot}\underset{NOH}{\overset{\|}{C}{\cdot}NO_2}$$

(ii) Secondary nitro compounds: dark blue or blue green colour due to nitro-nitroso derivatives. The coloured compound is soluble in chloroform.

$$\underset{R^2}{\overset{R^1}{\diagdown}}C{=}N\overset{\nearrow O}{\underset{O=N}{\searrow}}\overset{O}{\underset{\ominus}{\cdot}} \longrightarrow R^2{-}\underset{NO}{\overset{R^1}{\underset{|}{\overset{|}{C}}}}{-}NO_2$$

(iii) Tertiary compound: no colouration.

For characterisation of aliphatic nitro compounds by reduction see details under *aromatic nitro compounds*, Section **VII,5,C**; p. 1080.

For further characterisation of oximes, imides and primary sulphonamides see p. 1082 under the miscellaneous neutral and acidic *compounds containing nitrogen*.

B. Basic Compounds

Organic compounds that dissolve in dilute hydrochloric acid, and are placed in group IV, contain nitrogen: the rarely encountered pyrones and anthocyanidin pigments are exceptions. Indeed, when solubility tests have placed a compound in group IV, but elemental analysis has failed to establish the presence of nitrogen, it is advisable to repeat the test for the elements. The most important basic nitrogen compounds are the primary, secondary and tertiary amines.* The only hydrazines commonly encountered in the group are the monoaryl hydrazines, which are recognised by their ability to reduce Fehling's solution with the evolution of hydrogen. They are also conveniently detected by their condensation with benzaldehyde or some other suitable carbonyl compound as a reagent. The lower aliphatic amines and diamines are soluble in water and possess characteristic ammoniacal odours which distinguish them

---

* Many amines are regarded as being potentially carcinogenic: those which are on the restricted list have been specified in the Tables of Physical Constants.

from water-insoluble amines. The reactions to be described below apply to both water-soluble and water-insoluble amines.

Primary amines may be readily distinguished from secondary and tertiary analogues by the presence of two absorption bands in the infrared spectrum between 3320 and 3500 cm^{-1} (symmetric and antisymmetric NH str.). Secondary amines exhibit a single absorption band at about 3350 cm^{-1} (NH str.). In both cases deformation modes for the NH bond appear at about 1600 cm^{-1}. There is no satisfactory absorption to allow a definitive characterisation in the case of tertiary amines. In the nuclear magnetic resonance spectrum of primary and secondary amines, the nitrogen-bound hydrogens are recognisable by their replaceability on the addition of deuterium oxide.

An initial chemical classification of the amine function into primary, secondary or tertiary should be carried out by means of the reaction with nitrous acid.

Nitrous acid test. Dissolve 0.2 g of the substance in 5 ml of 2 *M*-hydrochloric acid: cool in ice and add 2 ml of ice-cold 10 per cent aqueous sodium nitrite solution slowly by means of a dropper and with stirring until, after standing for 3–4 minutes, an immediate positive test for nitrous acid is obtained with starch–iodide paper (see p. 687, **IV,G**). If a clear solution is obtained with a continuous evolution of nitrogen gas the substance is a **primary aliphatic or aralkyl amine**. If there is apparently no evolution of nitrogen from the clear solution, add one-half of the solution to a cold solution of 0.4 g of 2-naphthol in 4 ml of 5 per cent sodium hydroxide solution. The formation of a coloured (e.g., orange-red) azo-dye indicates the presence of a **primary aromatic amine**; in which case warm the other half of the diazotised solution and note the evolution of nitrogen and the strong phenolic aroma which is produced. If a colourless solution is obtained which gives an immediate and sustained positive test with starch–iodide paper when only a little sodium nitrite solution has been added, the compound is a **tertiary aliphatic amine**.

The formation of *N*-nitrosamines* which usually separate as orange-yellow oils or low melting solids indicates the presence of a **secondary amine**. Confirm the formation of the nitrosamine by the **Liebermann nitroso reaction**. This consists in warming the nitrosamine with phenol and concentrated sulphuric acid. The sulphuric acid liberates nitrous acid from the nitrosamine, the nitrous acid reacts with the phenol to form *p*-nitrosophenol, which then combines with another molecule of phenol to give red indophenol. In alkaline solution the red indophenol yields a blue indophenol anion.

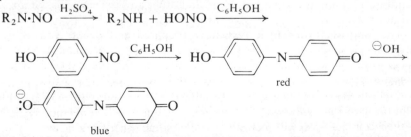

Extract the oil obtained in the nitrous acid test with about 5 ml of ether and wash the extract successively with water, dilute sodium hydroxide and water,

---

* Potentially carcinogenic, see Section **I,3,D,**5.

and evaporate off the ether. Apply Liebermann's nitroso reaction to the residual oil or solid. Place 1 drop or 0.01–0.02 g of the nitroso compound in a dry test-tube, add 0.05 g of phenol and warm together for 20 seconds; cool, and add 1 ml of concentrated sulphuric acid. An intense green (or greenish-blue) colouration will be developed, which changes to pale red upon pouring into 30–50 ml of cold water; the colour becomes deep blue or green upon adding excess of sodium hydroxide solution.

If the unknown base is a **tertiary aromatic amine**, the treatment with nitrous acid will yield a dark orange-red solution or an orange crystalline precipitate resulting from the formation of the hydrochloride of the C-nitrosamine. Basification of the solution or of the isolated orange precipitate with either sodium hydroxide or carbonate solution yields the bright green nitrosamine base. In favourable cases this may be isolated by extraction with ether, recrystallised and used for characterisation purposes.

Confirmatory tests for primary amines (a) **Carbylamine test.** To 1 ml of 0.5 M-alcoholic potassium hydroxide solution (or to a solution prepared by dissolving a fragment of potassium hydroxide half the size of a pea in 1 ml of ethanol) add 0.05–0.1 g of the amine and 3 drops of chloroform, and heat to boiling. A carbylamine (isocyanide) is formed and will be readily identified by its extremely nauseating odour:

$$RNH_2 + CHCl_3 + 3KOH \longrightarrow RNC + 3KCl + 3H_2O$$

When the reaction is over, add concentrated hydrochloric acid to decompose the isocyanide and pour it away after the odour is no longer discernible. The test is extremely delicate and will often detect traces of primary amines in secondary and tertiary amines; it must therefore be used with due regard to this and other factors.

(b) **5-Nitrosalicylaldehyde reagent test.** This test is based upon the fact that 5-nitrosalicylaldehyde and nickel ions when added to a primary amine produce an immediate precipitate of the nickel derivative of the Schiff's base.

To 5 ml of water add 1–2 drops of the amine; if the amine does not dissolve, add a drop or two of concentrated hydrochloric acid. Add 0.5–1 ml of this amine solution to 2–3 ml of the reagent; an almost immediate precipitate indicates the presence of a primary amine. A slight turbidity indicates the presence of a primary amine as an impurity. (Primary aromatic amines generally require 2–3 minutes for the test. Urea and other amides, as well as amino acids, do not react.)

The **5-nitrosalicylaldehyde reagent** is prepared as follows. Add 0.5 g of 5-nitrosalicylaldehyde (m.p. 124–125 °C) to 15 ml of pure triethanolamine and 25 ml of water; shake until dissolved. Then introduce 0.5 g of crystallised nickel(II) chloride dissolved in a few ml of water, and dilute to 100 ml with water. If the triethanolamine contains some ethanolamine (thus causing a precipitate), it may be necessary to add a further 0.5 g of the aldehyde and to filter off the resulting precipitate. The reagent is stable for long periods.

(c) **Rimini's test (for primary aliphatic amines).** To a suspension or solution of 1 drop of the compound or to an equivalent quantity of its solution in 3 ml of water, add 1 ml of pure acetone and 1 drop of freshly prepared 1 per cent aqueous solution of disodium pentacyanonitrosyl ferrate. A violet red colour will develop within 1 minute.

Confirmatory tests for secondary aliphatic amines (a) **Simon's test.** To a solution or suspension of 1 drop of the compound or to an equivalent quantity of its solution in 3 ml of water, add 2 drops of freshly prepared acetaldehyde solution, followed by 1 drop of a 1 per cent aqueous solution of disodium pentacyanonitrosyl ferrate. A blue colouration is produced within 5 minutes, after which the colour gradually changes through greenish-blue to yellow.

(b) **Carbon disulphide reagent test.** This test is based upon the formation from a secondary amine and carbon disulphide of a dialkyldithiocarbamate; the latter readily forms a nickel derivative with a solution of a nickel salt:

$$R_2NH + CS_2 \xrightarrow{\ NH_3\ } R_2N\cdot C{\overset{S}{\underset{\underset{SNH_4}{\ominus\oplus}}{\diagup}}} \xrightarrow{\ NiCl_2\ } \left(R_2N\cdot C{\overset{S}{\underset{S}{\diagup}}} \right)_{\!/2} Ni$$

To 5 ml of water add 1–2 drops of the secondary amine; if it does not dissolve, add a drop or two of concentrated hydrochloric acid. Place 1 ml of the reagent in a test-tube, add 0.5–1 ml of concentrated ammonia solution, followed by 0.5–1 ml of the above amine solution. A precipitate indicates a secondary amine. A slight turbidity points to the presence of a secondary amine as an impurity. The test is very sensitive; it is not given by primary amines.

The **carbon disulphide reagent** is prepared by adding to a solution of 0.5 g crystallised nickel(II) chloride in 100 ml of water enough carbon disulphide so that after shaking a globule of carbon disulphide is left at the bottom of the bottle. The reagent is stable for long periods in a well-stoppered bottle. If all the carbon disulphide evaporates, more must be added.

It should be noted that aliphatic and aromatic primary and secondary amines may be distinguished from tertiary amines by their reaction with acetyl chloride, benzoyl chloride and benzenesulphonyl chloride. In the latter case a primary amine yields an alkali-soluble derivative which distinguishes it from a secondary amine when the derivative is neutral and insoluble in acid and alkali. The separation of primary, secondary and tertiary amines using toluene-*p*-sulphonyl chloride (Hinsberg's method) is described in Section **VII,7,***1*.

For the preparation of derivatives of primary, secondary and tertiary amines see Sections **VII,6,***21* and **VII,6,***22*.

C. Neutral compounds
Compounds containing carbon, hydrogen and possibly oxygen. It is convenient to consider the indifferent or neutral oxygen derivatives of the hydrocarbons—(*a*) *aldehydes* and *ketones*, (*b*) *esters* and *anhydrides*, (*c*) *alcohols* and *ethers*—together. All of these, with the exception of the water-soluble members of low molecular weight, are soluble only in concentrated sulphuric acid, i.e., fall into solubility group V. *Alkenes* and readily sulphonated *arenes* will also fall into this solubility group. The above classes of compounds must be tested for in the order in which they are listed, otherwise erroneous conclusions may be drawn from the reactions for the functional group about to be described.

Both **aldehydes and ketones** contain the carbonyl group, hence a general test for carbonyl compounds will immediately identify both classes of compounds. The preferred reagent is 2,4-dinitrophenylhydrazine, which gives sparingly soluble dinitrophenylhydrazones with carbonyl compounds (including many quinones). Add 2 drops or 0.05–0.1 g of the substance to be tested to 3 ml of the 2,4-dinitrophenylhydrazine reagent, and shake. If no precipitate forms immediately allow to stand for 5–10 minutes. A crystalline precipitate indicates the presence of a carbonyl compound. Occasionally the precipitate is oily at first, but this becomes crystalline upon standing.

2,4-Dinitrophenylhydrazine reagent may be prepared by either of the following methods.

Method 1. Suspend 2.0 g of 2,4-dinitrophenylhydrazine in 100 ml of methanol; add cautiously and slowly 4.0 ml of concentrated sulphuric acid. The mixture becomes warm and the solid usually dissolves completely. Filter, if necessary.

Method 2. Dissolve 0.25 g of 2,4-dinitrophenylhydrazine in a mixture of 42 ml of concentrated hydrochloric acid and 50 ml of water by warming on a water bath: dilute the cold solution to 250 ml with distilled water. This reagent is more suitable for water-soluble aldehydes and ketones since alcohol is absent.

The above reagent is very dilute and is intended for qualitative reactions. It is hardly suitable for the preparation of crystalline derivatives except in very small quantities (compare Section **VII,6,***13*).

The **acetals** $R^1 \cdot CH(OR^2)_2$ are so readily hydrolysed by acids that they may give a positive result in the above test:

$$R^1 \cdot CH(OR^2)_2 + H_2O \xrightarrow{H^\oplus} R^1 \cdot CHO + 2R^2OH$$

(For a more detailed discussion on *Acetals*, see Section **VII,6,***12*.)

If the unknown compound gives a positive test with 2,4-dinitrophenylhydrazine it then becomes necessary to decide whether it is an aldehyde or a ketone. The infrared spectrum of the compound should be very informative; both aldehydes and ketones show strong absorption at 1740–1700 cm^{-1} (C = O str.), but only aldehydes exhibit two absorption bands at about 2720 and 2820 cm^{-1} (C – H str.). In the nuclear magnetic resonance spectrum of an aldehyde a low-field signal for the aldehydic hydrogen (δ 9–10) is observed.

Chemical differentiation between aldehydes and ketones (a) **Schiff's reagent.** Aldehydes produce a pink colour, while ketones are without effect. Use 2 drops (or 0.05 g) of the compound and 2 ml of Schiff's reagent and shake the mixture in the cold. Some aromatic aldehydes (e.g., vanillin) give a negative result.

Preparation of Schiff's reagent. *Method 1.* Dissolve 0.2 g of pure *p*-rosaniline hydrochloride in 20 ml of a cold, freshly prepared, saturated aqueous solution of sulphur dioxide; allow the solution to stand for a few hours until it becomes colourless or pale yellow. Dilute the solution to 200 ml and keep it in a tightly stoppered bottle. If the bottle is not adequately stoppered, the reagent will gradually lose sulphur dioxide and the colour will return. The solution keeps well if not unnecessarily exposed to light and air.

Method 2. Add 2 g of sodium metabisulphite to a solution of 0.2 g of *p*-rosaniline hydrochloride and 2 ml of concentrated hydrochloric acid in 200 ml of water.

By way of caution it should be noted that free alkali or the alkali salts of weak acids will redden the reagent like an aldehyde. It is also, of course, reddened by heat or when exposed in small quantities to the air for some time. Mineral acids greatly reduce the sensitivity of the test.

(b) **Ammoniacal silver nitrate solution (Tollen's solution).** Aldehydes alone reduce Tollen's reagent and produce a silver mirror on the inside of the test-tube. Add 2–3 drops (or 0.05 g) of the compound to 2–3 ml of Tollen's solution contained in a clean test-tube (the latter is preferably cleaned with hot nitric acid). If no reaction appears to take place in the cold, warm in a beaker of hot water.

CAUTION: After the test, pour the contents of the test-tube into the sink and wash the test-tube with dilute nitric acid. Any silver fulminate present, which is highly explosive when dry, will thus be destroyed.

Tollen's ammoniacal silver nitrate reagent is prepared as follows: Dissolve 3 g of silver nitrate in 30 ml of water (solution A) and 3 g of sodium hydroxide in 30 ml of water (solution B). When the reagent is required, mix equal volumes (say, 1 ml) of solutions A and B in a clean test-tube, and add dilute ammonia solution drop by drop until the silver oxide is just dissolved. Great care must be taken in the preparation and use of this reagent, which must not be heated. Only a small volume should be prepared just before use, any residue washed down the sink with a large quantity of water, and the test-tubes rinsed with dilute nitric acid.

(c) **Fehling's solution.** Aldehydes alone reduce Fehling's solution to yellow or red copper(I) oxide. Use 2 drops (or 0.05 g) of the compound and 2–3 ml of Fehling's solution: heat on a boiling water bath for 3–4 minutes. This test is positive for aliphatic aldehydes, but is usually indecisive for aromatic aldehydes.

Preparation of Fehling's solution. *Solution No. 1.* Dissolve 34.64 g of copper(II) sulphate crystals in water containing a few drops of dilute sulphuric acid, and dilute the solution to 500 ml.

Solution No. 2. Dissolve 60 g of pure sodium hydroxide and 173 g of pure Rochelle salt (sodium potassium tartrate) in water, filter if necessary through a sintered glass funnel and make up the filtrate and washings to 500 ml.

Keep the two solutions separately in tightly stoppered bottles and mix exactly equal volumes immediately before use.

Further classification tests for aldehydes and ketones (d) **Sodium metabisulphite test.** Aldehydes and simple ketones react with a saturated solution of sodium metabisulphite to yield crystalline bisulphite-addition compounds:

$$R^1R^2CO + NaHSO_3 \rightleftharpoons R^1R^2C(OH)SO_3Na$$

A condition of equilibrium is reached (70–90 per cent of bisulphite compound with equivalent quantities of the reagents in 1 hour), but by using a large excess of bisulphite almost complete conversion into the addition compound results. Since the reaction is reversible, the carbonyl compound can be recovered by adding to an aqueous solution of the bisulphite compound sufficient sodium carbonate solution or hydrochloric acid to react with the free sodium metabisulphite present in the equilibrium mixture. Bisulphite compounds may therefore be employed for the purification of carbonyl compounds or for their separation from other organic substances.

The most satisfactory reagent is a saturated solution of sodium metabisulphite containing ethanol; it must be prepared as required since it oxidises and decomposes on keeping. Frequently, a saturated aqueous solution is used without the addition of ethanol.

Prepare 10 ml of saturated sodium metabisulphite solution and add 4 ml of the carbonyl compound; shake thoroughly and observe the rise in temperature. Filter the crystalline precipitate at the pump, wash it with a little alcohol, followed by ether and allow it to dry. The **sodium metabisulphite reagent is prepared** by treating a saturated aqueous solution of sodium metabisulphite with 70 per cent of its volume of industrial spirit, and then adding just sufficient water to produce a clear solution.

(e) **Iodoform test.** Methyl ketones and acetaldehyde, i.e., compounds containing the $CH_3 \cdot CO -$ grouping, give a positive iodoform reaction. Alcohols having the structure $CH_3 \cdot CH(OH) \cdot R$, which undergo oxidation to the corresponding methyl ketone, also slowly give a positive test. Dissolve 0.1 g or 4–5 drops of the compound in 2 ml of water; if it is insoluble in water, add sufficient dioxan to produce a homogeneous solution. Add 2 ml of 5 per cent sodium hydroxide solution and then introduce a potassium iodide–iodine reagent dropwise with shaking until a definite dark colour of iodine persists. Allow to stand for 2–3 minutes; if no iodoform separates at room temperature, warm the test-tube in a beaker of water at 60 °C. Add a few more drops of the iodine reagent if the faint iodine colour disappears: continue the addition of the reagent until the dark colour is not discharged after 2 minutes heating at 60 °C. Remove the excess of iodine by the addition of a few drops of dilute sodium hydroxide solution with shaking, dilute with an equal volume of water and allow to stand for 10–15 minutes. The test is positive if a yellow precipitate of iodoform is deposited. Filter off the yellow precipitate, dry upon pads of filter-paper and determine the m.p.: iodoform melts at 120 °C.

The **potassium iodide–iodine reagent is prepared** by dissolving 20 g of potassium iodide and 10 g of iodine in 100 ml of water.

For the preparation of derivatives of aldehydes and ketones see Section **VII,6,**_13._

Quinones. The number of quinones normally encountered in routine qualitative analysis is very limited; the following tests will be found useful for their detection. All quinones are coloured (generally yellow) crystalline solids. They are usually insoluble in water, soluble in ether and sublime on heating. Frequently the vapour has a penetrating odour and attacks the eyes. The carbonyl groups of quinones often do not react in a normal way with carbonyl group reagents, because of their oxidising properties: thus quinones are reduced by sodium metabisulphite. Crystalline products are usually formed with one molecule of phenylhydrazine or with one molecule of 2,4-dinitrophenylhydrazine, but these are not always of normal structure. Thus p-benzoquinone reacts with 2,4-dinitrophenylhydrazine hydrochloride in hot alcoholic solution to give 2′,4′-dinitrophenyl-4-azophenol, m.p. 185–186 °C.

$$(NO_2)_2 \cdot C_6H_3 \cdot NH \cdot NH_2 + O = \!\!\left\langle \right\rangle \!\!= O \longrightarrow$$

$$(NO_2)_2 C_6H_3 \cdot N = N \cdot C_6H_4OH + H_2O$$

Tests for quinones (a) **Hydriodic acid.** Compounds of the p-benzoquinone type liberate iodine from hydriodic acid. Dissolve 0.1 g of the quinone in a little rectified spirit. Add 10 ml of 10 per cent aqueous potassium iodide solution to a mixture of 5 ml of ethanol and 5 ml of concentrated hydrochloric acid, and then introduce the quinone solution. Iodine is liberated immediately. This test is also given by other oxidising agents.

(b) **Reduction with zinc powder and acid.**   Simple p-quinones are reduced to hydroquinones in the following manner. Dissolve or suspend 0.5 g of the quinone in dilute hydrochloric acid (1 : 5) and add a little zinc powder. When the solution is colourless, filter, neutralise with sodium hydrogen carbonate, extract the dihydric phenol with ether, remove the solvent and identify (Section **VII,6,**6).

(c) **Reduction with zinc powder and sodium hydroxide.**   Compounds of the anthraquinone type are reduced to oxanthrols. Treat 0.1 g of the quinone with dilute sodium hydroxide and zinc powder. Upon boiling the mixture a red colour is produced: this disappears when the solution is shaken owing to aerial oxidation to the original quinone.

(d) **Distillation with zinc powder.**   Quinones derived from polycyclic hydrocarbons may be reduced to the parent hydrocarbon as follows. Grind 0.5 g of the compound with 3–4 g of zinc powder, pour the mixture into a Pyrex testtube and cover it with an equal volume of zinc powder. Clamp the tube horizontally at the open end. Heat the zinc powder first, then the mixture of zinc powder and the compound to a dull red heat: the hydrocarbon sublimes into the cooler part of the tube. Remove the sublimate; determine the m.p. and identify it by the preparation of the picrate (Section **VII,6,**3).

(e) **Reaction with semicarbazide hydrochloride.**   Many simple quinones yield crystalline mono-semicarbazones by the following procedure. Dissolve 0.2 g of semicarbazide hydrochloride in a little water, add 0.2 g of the quinone and warm. The mono-semicarbazone is immediately formed as a yellow precipitate. Filter and recrystallise from hot water; any bis-semicarbazone will remain undissolved.

(f) **Reaction with o-phenylenediamine.**   o-Quinones (and also aromatic α-diketones, e.g., benzil) react with o-phenylenediamine to yield quinoxalines. Dissolve the substance in ethanol or glacial acetic acid, add an equivalent amount of o-phenylenediamine in ethanolic solution and warm for 15 minutes on a water bath. Cool, dilute with water, filter and recrystallise from dilute ethanol.

For the preparation of derivatives of quinones see Section **VII,6,**14.

Esters and acid anhydrides.   When a compound fails to respond to the 2,4-dinitrophenylhydrazine test for aldehydes and ketones, yet exhibits carbonyl absorption in the infrared region, it may be either an ester, an acid anhydride or, possibly, a lactone. The infrared absorption for esters is in the region of 1750–1730 cm^{-1} (C=O str.) and in the region 1000–1300 cm^{-1} (C—O str.). Anhydrides exhibit two absorption bands at about 1820 and 1750 cm^{-1} (C=O str.) together with absorption in the range 1100–1200 (C—O str.). Lactones absorb at about 1750 cm^{-1} (C=O str., the frequency depends upon the ring size), and in the range 1300–1000 cm^{-1} (C—O str.).

The presence of any of these functional types may be established chemically by applying the *hydroxamic acid test*. These compounds react with hydroxylamine in the presence of sodium hydroxide to form the sodium salt of the corresponding hydroxamic acid. On acidification and addition of iron(III) chloride solution the magenta coloured iron(III) complex of the hydroxamic acid is formed.

$$R^1 \cdot CO_2 R^2 + H_2NOH \xrightarrow{\ominus OH} R^1 \cdot CO \cdot NHOH + R^2OH$$

$$R^1 \cdot CO \cdot O \cdot CO \cdot R^1 + H_2NOH \xrightarrow{\ominus OH} R^1 \cdot CO \cdot NHOH + R^1 \cdot CO_2H$$

$$\text{(lactone)} + H_2NOH \xrightarrow{\ominus OH} R \cdot CH(OH) \cdot CH_2 \cdot CH_2 \cdot CO \cdot NHOH$$

The hydroxamic acid test is also given by acid chlorides and some primary aliphatic amides which are readily converted by hydroxylamine hydrochloride into hydroxamic acids; the possible presence of these functional types will, however, have been indicated from elemental analysis and other appropriate tests. Some esters, mainly of carbonic, carbamic, sulphuric and other inorganic acids, give only a yellow colour.

It is always advisable to ensure that the sample does not give a colour with iron(III) chloride before carrying out the hydroxamic acid test.

Procedure for hydroxamic acid test. *A.* Dissolve a drop or a few small crystals of the compound in 1 ml of rectified spirit (95% ethanol) and add 1 ml of *M*-hydrochloric acid. Note the colour produced when 1 drop of 5 per cent iron(III) chloride solution is added to the solution. If a pronounced violet, blue, red or orange colour is produced, the hydroxamic acid test described below is not applicable and should not be used.

B. Mix 1 drop or several small crystals (*ca.* 0.05 g) of the compound with 1 ml of 0.5 *M*-hydroxylamine hydrochloride in 95 per cent ethanol and add 0.2 ml of 6 *M*-aqueous sodium hydroxide. Heat the mixture to boiling and, after the solution has cooled slightly, add 2 ml of *M*-hydrochloric acid. If the solution is cloudy, add 2 ml of 95 per cent ethanol. Observe the colour produced when 1 drop of 5 per cent iron(III) chloride solution is added. If the resulting colour does not persist, continue to add the reagent dropwise until the observed colour pervades the entire solution. Usually only 1 drop of the iron(III) chloride solution is necessary. Compare the colour with that produced in test *A*. A positive test will be a distinct burgundy or magenta colour as compared with the yellow colour observed when the original compound is tested with iron(III) chloride solution in the presence of acid. It is often advisable to conduct in parallel the test with, say, ethyl acetate, to ensure that the conditions for this test are correct.

Esters and anhydrides may be differentiated by the following simple test which relies on the fact that hydrolysis of acid anhydrides is more rapid than that of esters under basic conditions. Add 1 ml of the compound to 2 ml of water to which has been added 1 drop of 1 *M*-sodium hydroxide solution and a trace of phenolphthalein indicator. Warm the solution gently on a water bath; with anhydrides the pink colour is discharged within about 1 minute and the further dropwise addition of alkali enables the rate of hydrolysis to be monitored. With most esters hydrolysis is very slow under these conditions. Lactones are similarly only slowly hydrolysed.

For the preparation of derivatives of esters and anhydrides see Sections **VII,6,***17* and **VII,6,***16* respectively.

Alcohols and ethers. If the unknown, neutral, oxygen-containing compound does not give the class reactions for aldehydes, ketones, esters and an-hydrides, it is probably either an alcohol or an ether. Alcohols are readily identified by the intense characteristic hydroxyl absorption which occurs as a broad band in the infrared spectrum at 3600–3300 cm^{-1} (O – H str.). In the nuclear magnetic resonance spectrum, the absorption by the proton in the hydroxyl group gives rise to a broad peak the chemical shift of which is rather variable; the peak disappears on deuteration.

Chemically, alcohols and ethers may be simply distinguished by the use of two reagents—metallic sodium and acetyl chloride.

(a) **Reaction with sodium.** Metallic sodium reacts with alcohols with the evolution of hydrogen.

$$2ROH + 2Na \longrightarrow 2RO^{\ominus}Na^{\oplus} + H_2$$

The most common interfering substance, especially with alcohols of low mole-cular weight, is water; this may result in an inaccurate interpretation of the test if applied alone. Most of the water may usually be removed by shaking with a little anhydrous calcium sulphate. Dry ethers (and also the saturated aliphatic and the simple aromatic hydrocarbons) do not react with sodium. Treat 1.0 ml of the dried compound with a small thin slice of freshly cut sodium (handle with the tongs or with a penknife) in a small dry test-tube. Observe whether hydrogen is evolved and the sodium reacts.

(b) **The acetyl chloride test.** Acetyl chloride reacts vigorously with pri-mary and secondary alcohols with the evolution of hydrogen chloride. Ethers are unaffected by acetyl chloride. In a small dry test-tube place 0.5 ml of the dried compound with 0.3–0.4 ml of redistilled acetyl chloride and note whether reaction occurs. Add 3 ml of water and neutralise the aqueous layer with solid sodium hydrogen carbonate and note whether the smell of the product is dif-ferent from that of the original alcohol. The product in the case of a tertiary alcohol is mainly the alkyl chloride.

(c) **Differentiation between primary, secondary and tertiary alcohols (Lucas' test).** The test depends upon the different rates of formation of the alkyl chlorides upon treatment with a hydrochloric acid–zinc chloride reagent (con-taining 1 mol of acid to 1 mol of anhydrous zinc chloride) and with hydro-chloric acid.

To 1 ml of the alcohol in a small test-tube, add quickly 6 ml of Lucas' reagent at 26–27 °C, close the tube with a cork, shake and allow to stand. Observe the mixture during 5 minutes. The following results may be obtained:

(i) Primary alcohols, lower than hexan-1-ol, dissolve; there may be some darkening, but the solution remains clear.

(ii) Primary alcohols, hexan-1-ol and higher, do not dissolve appreciably; the aqueous phase remains clear.

(iii) Secondary alcohols: the clear solution becomes cloudy owing to the separation of finely divided drops of the chloride. Allyl alcohol behaves like a secondary alcohol and reacts within 7 minutes. A distinct upper layer is visible after one hour except for propan-2-ol (probably because of the volatility of the chloride).

(iv) Tertiary alcohols: two phases separate almost immediately owing to the formation of the tertiary chloride.

If a turbid solution is obtained, suggesting the presence of a secondary alcohol but not excluding a tertiary alcohol, a further test with concentrated hydrochloric acid must be made. Mix 1 ml of the alcohol with 6 ml of concentrated hydrochloric acid and observe the result.

(v) Tertiary alcohols: immediate reaction to form the insoluble chloride which rises to the surface in a few minutes.

(vi) Secondary alcohols: the solution remains clear.

Lucas' reagent is prepared by dissolving 68 g (0.5 mol) of anhydrous zinc chloride (fused sticks, powder, etc.) in 52.5 g (0.5 mol) of concentrated hydrochloric acid with cooling to avoid loss of hydrogen chloride.

If the above tests are negative, and an ether is therefore suspected, a test with iodine is often informative. Add a small crystal of iodine to about 0.5 ml of the substance; with ethers brown solutions are formed whereas non-oxygenated compounds (e.g., hydrocarbons) give red-violet solutions.

Acetals ($R^1 \cdot CH(OR^2)_2$) are stable in alkaline solutions, but are readily hydrolysed by dilute acids to give aldehydes, and hence usually give a positive test with 2,4-dinitrophenylhydrazine reagent (p. 1071). For full characterisation of the acetals by hydrolysis see Section **VII,6,*12***.

For the preparation of derivatives of alcohols and ethers see Section **VII,6,*4*** and Sections **VII,6,*10*** and **VII,6,*11*** respectively.

Polyhydric alcohols and carbohydrates. If the neutral substance containing carbon, hydrogen and possibly oxygen is insoluble in ether but freely soluble in water, a polyhydric alcohol or a simple mono- or di-saccharide (or related compound) is indicated. Treatment with concentrated sulphuric acid usually produces excessive charring.

Polyhydric alcohols are colourless viscous liquids, or crystalline solids. Upon heating with a little potassium hydrogen sulphate, they may yield aldehydes (e.g., ethylene glycol yields acetaldehyde; glycerol gives the irritating odour of acrolein which can additionally be detected with Schiff's reagent). Two confirmatory tests for polyhydric alcohols are as follows.

(a) **Boric acid test.** Add a few drops of phenolphthalein to a 1 per cent solution of borax; a pink colouration is produced. The addition of the polyhydric alcohol causes the pink colour to disappear, but it reappears on warming and vanishes again on cooling. This reaction is due to the combination of two *cis*-hydroxyl groups of the compound with boric acid to form reversibly a much stronger monobasic acid.

$$
\begin{array}{c}
-\overset{|}{C}-OH \\
-\overset{|}{C}-OH
\end{array}
\begin{array}{c}
HO \\
\diagdown \\
\diagup \\
HO
\end{array}
B-OH
\quad
\begin{array}{c}
HO-\overset{|}{C}- \\
HO-\overset{|}{C}-
\end{array}
\;\rightleftharpoons
$$

$$
\left[
\begin{array}{c}
-\overset{|}{C}-O \diagdown \overset{\ominus}{\underset{B}{}} \diagup O-\overset{|}{C}- \\
-\overset{|}{C}-O \diagup \diagdown O-\overset{|}{C}-
\end{array}
\right]
H^{\oplus} + 3H_2O
$$

(b) **Periodic acid test** (for 1,2-glycols and α-hydroxyaldehydes and ketones, Section **II,2,*45***). Add 1 drop (0.05 ml) of concentrated nitric acid to 2.0 ml of a 0.5 per cent aqueous solution of paraperiodic acid (H_5IO_6) contained in a small test-tube and shake well. Then introduce 1 drop or a small crystal of the compound. Shake the mixture for 15 seconds and add 1–2 drops of 5 per cent aqueous silver nitrate. The immediate production of a *white* precipitate (silver iodate) constitutes a positive test and indicates that the organic compound has been oxidised by the periodic acid. The test is based upon the fact that silver iodate is sparingly soluble in dilute nitric acid whereas silver periodate is very soluble: if too much nitric acid is present the silver iodate will not precipitate.

An alternative procedure for the above test is as follows. Mix 2–3 ml of 2 per cent aqueous paraperiodic acid solution with 1 drop of dilute sulphuric acid (*ca.* 1.25 *M*-) and add 20–30 mg of the compound. Shake the mixture for 5 minutes, and then pass sulphur dioxide through the solution until it acquires a pale yellow colour (to remove the excess of periodic acid and also iodic acid formed in the reaction). Add 1–2 ml of Schiff's reagent: the production of a violet colour constitutes a positive test.

For the preparation of derivatives of polyhydric alcohols see under *Alcohols and polyhydric alcohols*, Section **VII,6,4**.

Mono- and di-saccharides are colourless solids or syrupy liquids, which are freely soluble in water, practically insoluble in ether and other organic solvents, and neutral in reaction. Polysaccharides possess similar properties, but are generally insoluble in water because of their high molecular weights. Both poly- and di-saccharides are converted into monosaccharides upon hydrolysis. The following are confirmatory tests for carbohydrates.

(*a*) **Molisch's test.** This is a general test for carbohydrates. Place 5 mg of the substance in a test-tube containing 0.5 ml of water and mix it with 2 drops of a 10 per cent solution of 2-naphthol in ethanol or in chloroform. Allow 1 ml of concentrated sulphuric acid to flow down the side of the inclined tube (it is best to use a dropper pipette) so that the acid forms a layer beneath the aqueous solution without mixing with it. If a carbohydrate is present, a red ring appears at the common surface of the liquids: the colour quickly changes on standing or shaking, a dark purple solution being formed. Shake and allow the mixture to stand for 2 minutes, then dilute with 5 ml of water. In the presence of a carbohydrate, a dull-violet precipitate will appear immediately.

(*b*) **Barfoed's reagent.** This reagent may be used as a general test for monosaccharides. Heat a test-tube containing 1 ml of the reagent and 1 ml of a dilute solution of the carbohydrate in a beaker of boiling water. If red copper(I) oxide is formed within 2 minutes, a monosaccharide is present. Disaccharides on prolonged heating (about 10 minutes) may also cause reduction, owing to partial hydrolysis to monosaccharides.

Barfoed's reagent is prepared by dissolving 13.3 g of crystallised neutral copper(II) acetate in 200 ml of 1 per cent acetic acid solution. The reagent does not keep well.

(*c*) **Fehling's solution** (test for reducing sugars). Place 5 ml of Fehling's solution (prepared by mixing equal volumes of Fehling's solution No. 1 and solution No. 2) in a test-tube and heat to gentle boiling. Add a solution of 0.1 g of the carbohydrate in 2 ml of water and continue to boil gently for a minute or two, and observe the result. A yellow or red precipitate of copper(I) oxide indicates the presence of a reducing sugar. An alternative method of carrying out the test is to add the hot Fehling's solution dropwise to the boiling solution of the carbohydrate; in the presence of a reducing sugar the blue colour will disappear and a yellow precipitate, changing to red, is thrown down.

(*d*) **Benedict's solution.** This is a modification of Fehling's solution and consists of a single test solution which does not deteriorate appreciably on standing. To 5 ml of Benedict's solution add 0.4 ml of a 2 per cent solution of the carbohydrate, boil for 2 minutes and allow to cool spontaneously. If no reducing sugar is present, the solution remains clear; in the presence of a reducing sugar, the solution will contain copper(I) oxide. The test may also be carried out according to the experimental details given under (c).

Benedict's solution is prepared as follows. Dissolve 86.5 g of crystallised sodium citrate $(2Na_3C_6H_5O_7,11H_2O)$ and 50 g of anhydrous sodium carbonate in about 350 ml of water. Filter, if necessary. Add a solution of 8.65 g of crystallised copper(II) sulphate in 50 ml of water with constant stirring. Dilute to 500 ml. The resulting solution should be perfectly clear; if it is not, pour it through a fluted filter-paper.

Non-reducing sugars may be hydrolysed by boiling with dilute hydrochloric acid; if the solution is then neutralised with aqueous sodium hydroxide the reduction of Fehling's solution or Benedict's solution occurs readily.

For the preparation of derivatives of mono- and di-saccharides see Section **VII,6,5**.

The following **notes on the identification of polysaccharides** may be useful.

Most polysaccharides are insoluble or sparingly soluble in cold water, insoluble in cold ethanol and ether, and rarely possess melting points. An exception is inulin which melts at about 178 °C (dec.) after drying at 130 °C.

Starch. A few centigrams rubbed to a thin cream with cold water and then gradually stirred into 100 ml of boiling water dissolve to give a nearly clear solution. This gives a deep blue colouration with a dilute solution of iodine in potassium iodide solution, temporarily decolourised by heat or by traces of free alkali, but restored on cooling or upon acidifying. It is hydrolysed by boiling with dilute hydrochloric acid to give products (largely glucose) which reduce Fehling's solution.

Cellulose. This is insoluble in water, hot and cold. It dissolves in a solution of Schweitzer's reagent (precipitated copper(II) hydroxide is washed free from salts and then dissolved in concentrated ammonia solution), from which it is precipitated by the addition of dilute acids. Cellulose is not hydrolysed by dilute hydrochloric acid.

Inulin. This polysaccharide melts with decomposition at about 178 °C. It is insoluble in cold but dissolves readily in hot water giving a clear solution which tends to remain supersaturated. It does not reduce Fehling's solution. Inulin gives no colouration with iodine solution.

Glycogen. It dissolves easily in water to an intensely opalescent solution; the opalescence is not destroyed by filtration, but is removed by the addition of acetic acid. Glycogen gives a wine colouration with iodine solution; the colouration disappears on heating and reappears on cooling. The compound does not reduce Fehling's solution: upon boiling with dilute acid, glucose is produced and the resulting solution, when neutralised, therefore reduces Fehling's solution.

Hydrocarbons may be differentiated by their solubility in sulphuric acid since **unsaturated hydrocarbons** are soluble in concentrated sulphuric acid as are those **arenes** which are readily sulphonated (group V), whereas the saturated alkanes and lesser reactive arenes are insoluble in this reagent (group VI). The presence of an alkene, alkyne or arene is usually readily apparent from an inspection of the infrared and nuclear magnetic resonance spectra, the characteristic features of which are fully discussed in Sections **VII,4,1** and **VII,4,2**.

The two tests employed for the **detection of unsaturation** are decolourisation of a dilute solution of bromine in carbon tetrachloride, and reaction with dilute aqueous potassium permanganate. It is essential to apply both tests since some symmetrically substituted alkenes (e.g., stilbene, $C_6H_5 \cdot CH = CH \cdot C_6H_5$) react only slowly under the conditions of the bromine test. With dilute potassium permanganate solution the double bond is readily attacked, probably through the intermediate formation of a *cis*-diol.

$$C_6H_5 \cdot CH = CH \cdot C_6H_5 \xrightarrow[H_2O]{[O]} C_6H_5 \cdot CH(OH) \cdot CH(OH) \cdot C_6H_5 \longrightarrow 2C_6H_5 \cdot CO_2H$$

(a) **Bromine test.** Dissolve 0.2 g or 0.2 ml of the compound in 2 ml of carbon tetrachloride, and add a 2 per cent solution of bromine in carbon tetrachloride dropwise until the bromine colour persists for one minute. Blow across the mouth of the tube to detect any hydrogen bromide which may be evolved.

(b) **Potassium permanganate test.** Dissolve 0.2 g or 0.2 ml of the substance in 2 ml of water or in 2 ml of acetone (which gives a negative test with the reagent), and add 2 per cent potassium permanganate solution dropwise. The test is negative if no more than 3 drops of the reagent are decolourised.

The most satisfactory reagent for **distinguishing between aromatic hydrocarbons and the alkanes** is the fuming sulphuric acid test. Place 2 ml of 20 per cent fuming sulphuric acid in a dry test-tube, add 0.5 ml of the hydrocarbon and shake vigorously. Only the aromatic hydrocarbon dissolves completely, heat is evolved, but excessive charring should be absent. Warm the solution gently, cool and pour cautiously on to crushed ice; the aromatic hydrocarbon which undergoes sulphonation gives a homogeneous aqueous solution.

For the preparation of derivatives and for further characterisation of alkanes, alkenes, alkynes and arenes see Sections **VII,6,***1*, **VII,6,***2* and **VII,6,***3* respectively.

Compounds containing nitrogen The *neutral* nitrogen compounds which fall into the solubility group VII include: tertiary aliphatic nitro compounds and aromatic nitro compounds; amides (simple and substituted); nitrogen derivatives of aldehydes and ketones (hydrazones, semicarbazones, etc.); nitriles; nitroso, azo, hydrazo and other intermediate reduction products of aromatic nitro compounds. The imides, primary and secondary nitro compounds, oximes and primary sulphonamides are *weakly acidic* nitrogen compounds and fall into solubility group III. All the above nitrogen compounds and also the secondary sulphonamides, with few exceptions, respond to the same classification reactions (reduction and hydrolysis) and hence will be considered together (experimental procedures below).

α-Amino acids, which are appreciably water-soluble to give solutions which are essentially neutral towards indicator paper, fall into group II. Water-insoluble α-amino acids are readily soluble in both dilute alkali and dilute acid as the result of their amphoteric character.

In most instances, confirmation of the presence of any of the above functional groups may be obtained from a consideration of the spectroscopic properties of the compound (see Sections **VII,4,***1*–**VII,4,***3*).

Nitro compounds and their reduction products. Tertiary aliphatic nitro compounds and aromatic nitro compounds are reduced by zinc and ammonium chloride solution to the corresponding hydroxylamines, which may be detected by their reducing action upon an ammoniacal solution of silver nitrate or Tollen's reagent:

$$R \cdot NO_2 + 4[H] \xrightarrow[\text{NH}_4\text{Cl aq.}]{\text{Zn}} R \cdot NHOH + H_2O$$

It must be remembered, however, that nitroso, azoxy and azo compounds (which are usually more highly coloured than nitro compounds) may be reduced by zinc powder to the corresponding hydroxylamine, hydrazo and hydrazine compounds respectively, all of which reduce Tollen's reagent in the cold.

Nitro compounds are reduced in acid solution (for example, by tin and hydrochloric acid) to the corresponding primary amines.

$$R \cdot NO_2 + 6[H] \longrightarrow RNH_2 + 2H_2O$$

Nitrosamines are similarly reduced to secondary amines:

$$R_2N \cdot NO + 6[H] \longrightarrow R_2NH + NH_3 + H_2O$$

The N-nitrosamines (and some C-nitroso compounds that yield nitrous acid when treated with concentrated sulphuric acid) may be detected by Liebermann's reaction (p. 1068).

Azo compounds may be identified by examination of the amine(s) formed on reduction in acid solution.

$$R^1 \cdot N = N \cdot R^2 + 4[H] \longrightarrow R^1NH_2 + H_2NR^2$$

They are always coloured but give colourless products upon reduction. **Hydrazo and azoxy compounds** are reduced in acid solution to the parent amine.

Amides. Simple **(primary) amides** when warmed with dilute sodium hydroxide solution give ammonia readily, together with the salt of the corresponding acid:

$$R \cdot CONH_2 + H_2O \longrightarrow R \cdot CO_2H + NH_3$$

Complete hydrolysis may be effected by boiling either with 10 per cent sodium hydroxide solution or with 10 per cent sulphuric acid for 1–3 hours. It is preferable to employ the non-volatile sulphuric acid for acid hydrolysis; this acid should also be used for acidification of the solution resulting from alkaline hydrolysis since any volatile organic acid (formic acid, acetic acid, etc.) may be distilled off.

Substituted amides undergo hydrolysis with greater difficulty. The choice of an acid or an alkaline medium will depend upon the solubility of the compound in the medium, and the effect of the reagent upon the products of hydrolysis. Substituted amides of comparatively low molecular weight (e.g., acetanilide) may be hydrolysed by boiling either with 10 per cent sodium hydroxide solution or with 10 per cent sulphuric acid for 2–3 hours. Other substituted amides are so insoluble in water that little reaction occurs when they are refluxed with dilute acid or dilute alkali for several hours. These include such substances as benzanilide or the benzotoluidides. For these substances satisfactory results may be obtained with 70 per cent sulphuric acid:* this hydrolysis

---

* Prepared by adding 40 ml of concentrated sulphuric acid cautiously and with stirring and cooling to 30 ml of water.

medium is a much better solvent for the substituted amide than is water or more dilute acid; it also permits a higher reaction temperature.

$$R^1 \cdot CONHR^2 + H_2O \xrightarrow{H_2SO_4} R^1 \cdot CO_2H + R^2 \overset{\oplus}{N}H_3 \} HSO_4^{\ominus}$$

Nitriles.   These are best hydrolysed by boiling either with 30–40 per cent sodium hydroxide solution or with 50–70 per cent sulphuric acid during several hours, but the reaction takes place less readily than for primary amides. Indeed the latter are intermediate products in the hydrolysis:

$$R \cdot CN + H_2O \xrightarrow{\ominus OH} R \cdot CONH_2 \xrightarrow{\ominus OH} R \cdot CO_2^{\ominus} + NH_3$$

Nitriles and simple amides differ in physical properties: the former are liquids or low-melting solids, whilst the latter are generally solids. If the amide is a solid and insoluble in water, it may be readily prepared from the nitrile by dissolving in concentrated sulphuric acid and pouring the solution into water:

$$R \cdot CN \xrightarrow{H_2SO_4} [R \cdot \overset{\oplus}{C} = NH] HSO_4^{\ominus} \xrightarrow{H_2O} R \cdot CONH_2$$

Oximes, hydrazines and semicarbazones.   The hydrolysis products of these compounds, i.e., aldehydes and ketones, may be sensitive to alkali (this is particularly so for aldehydes): it is best, therefore, to conduct the hydrolysis with strong mineral acid. After hydrolysis the aldehyde or ketone may be isolated by distillation with steam, extraction with ether or, if a solid, by filtration, and then identified. The acid solution may be examined for hydroxylamine or hydrazine or semicarbazide; substituted hydrazines of the aromatic series are precipitated as oils or solids upon the addition of alkali.

$$R^1 R^2 C = NOH + H_2O \xrightarrow{HCl} R^1 R^2 C = O + NH_2OH$$

Imides.   Imides are generally water-soluble, consequently they are much more readily hydrolysed in an alkaline medium, e.g., by refluxing with 10 per cent sodium hydroxide solution:

$$(R \cdot CO)_2 NH \xrightarrow{\ominus OH} 2R \cdot CO_2^{\ominus} + NH_3$$

Sulphonamides.   Sulphonamides are very resistant to the normal reagents for hydrolysis. Heating with 80 per cent sulphuric acid at 160–170 °C results in rapid hydrolysis:

$$Ar \cdot SO_2 NR^1 R^2 + H_2O \xrightarrow{H_2SO_4} Ar \cdot SO_3H + R^1 R^2 \overset{\oplus}{N}H_2 \} HSO_4^{\ominus}$$

The reaction product may then be examined for a sulphonic acid and an amine.

(1) **Reduction of a nitro compound to a hydroxylamine.**   Dissolve 0.5 g of the compound in 10 ml of 50 per cent ethanol, add 0.5 g of solid ammonium chloride and about 0.5 g of zinc powder. Heat to boiling and allow the ensuing chemical reaction to proceed for 5 minutes. Filter from the excess of zinc powder and test the filtrate with Tollen's reagent (see p. 1072). An immediate black

or grey precipitate or a silver mirror indicates the presence of a hydroxylamine formed by the reduction of the nitro compound. Alternatively, warm the filtrate with Fehling's solution: a hydroxylamine will precipitate red copper(I) oxide. (A blank test should be performed with the original compound.)

(2) **Reduction of a nitro compound to a primary amine.** In a 50-ml round-bottomed flask fitted with a reflux condenser, place 1 g of the nitro compound and 2 g of granulated tin. Measure out 10 ml of concentrated hydrochloric acid and add it in three equal portions to the mixture: shake thoroughly after each addition. When the vigorous reaction subsides, heat under reflux on a water bath until the nitro compound has completely reacted (20–30 minutes). Shake the reaction mixture from time to time; if the nitro compound appears to be very insoluble, add 5 ml of ethanol. Cool the reaction mixture, and add 20–40 per cent sodium hydroxide solution until the precipitate of tin hydroxide dissolves. Extract the resulting amine from the cooled solution with ether, and remove the ether by distillation. Examine the residue with regard to its solubility in 5 per cent hydrochloric acid and its reaction with acetyl chloride or benzenesulphonyl chloride.

(3) **Reduction of a nitrosamine to a secondary amine.** Proceed as for a nitro compound. Determine the solubility of the residue after evaporation of the ether and also its behaviour towards benzenesulphonyl (or toluene-*p*-sulphonyl) chloride.

(4) **Hydrolysis of simple (primary) amides in alkaline solution.** Boil 0.5 g of the compound with 5 ml of 10 per cent sodium hydroxide solution and observe whether ammonia is evolved.

(5) **Hydrolysis of a substituted amide.** (*a*) *With 10 per cent sulphuric acid.* Reflux 1 g of the compound (e.g., acetanilide) with 20 ml of 10 per cent sulphuric acid for 1–2 hours. Distil the reaction mixture and collect 10 ml of distillate: this will contain any volatile organic acids which may be present. Cool the residue, render it alkaline with 20 per cent sodium hydroxide solution, cool and extract with ether. Distil off the ether and examine the ether-soluble residue for an amine.

(*b*) *With 70 per cent sulphuric acid.* Reflux 1 g of the substance (e.g., benzanilide) with 10–15 ml of 70 per cent sulphuric acid (4 : 3 by volume) for 30 minutes. Allow to cool and wash down any acid which has sublimed into the condenser with hot water. Filter off the acid, wash it with water and examine for solubility, etc. Render the filtrate alkaline with 10–20 per cent sodium hydroxide solution, cool and extract with ether. Examine the residue, after evaporation of the ether, for an amine.

(6) **Hydrolysis of a nitrile to an acid.** Reflux 1 g of the nitrile with 5 ml of 30–40 per cent sodium hydroxide solution until ammonia ceases to be evolved (2–3 hours). Dilute with 5 ml of water and add, with cooling, 7 ml of 50 per cent sulphuric acid. Isolate the acid by ether extraction, and examine its solubility and other properties.

(7) **Hydrolysis of a nitrile to an amide.** Warm a solution of 1 g of the nitrile in 4 ml of concentrated sulphuric acid to 80–90 °C, and allow the solution to stand for 5 minutes. Cool and pour the solution cautiously into 40 ml of cold water. Filter off the precipitate; stir it with 20 ml of cold 5 per cent sodium hydroxide solution and filter again. Recrystallise the amide from dilute ethanol, and determine its m.p. Examine the solubility behaviour and also the action of warm sodium hydroxide solution upon the amide.

(8) **Hydrolysis of a sulphonamide.** Mix 2 g of the sulphonamide with 3.5 ml of 80 per cent sulphuric acid * in a test-tube and place a thermometer in the mixture. Heat the test-tube, with frequent stirring by means of the thermometer, at 155–165 °C until the solid passes into solution (2–5 minutes). Allow the acid solution to cool and pour it into 25–30 ml of water. Render the resulting solution alkaline with 20 per cent sodium hydroxide solution in order to liberate the free amine. Two methods may be used for isolating the base. If the amine is volatile in steam, distil the alkaline solution and collect about 20 ml of distillate: extract the amine with ether, dry the ethereal solution with anhydrous potassium carbonate and distil off the solvent. If the amine is not appreciably steam-volatile, extract it from the alkaline solution with ether. The sulphonic acid (as sodium salt) in the residual solution may be identified by conversion into a suitable derivative (Section **VII,6,**26).

For further methods for the characterisation of these nitrogen-containing compounds see Sections **VII,6,**18–**VII,6,**20, **VII,6,**24 and **VII,6,**27.

α-**Amino acids** are in general insoluble (or very sparingly soluble) in organic solvents such as ether or benzene, sparingly soluble in ethanol, usually soluble in water and neutral in reaction. They have no true melting points, but decompose on heating at temperatures between 120 and 300 °C; the apparent melting points vary considerably according to the conditions of heating and are therefore of no great value for precise identification.

Confirmatory tests for α-amino acids (a) **The ninhydrin test.** Heat a solution of the compound with a few drops of a 0.25 per cent aqueous solution of ninhydrin (Section **III,100**). α-Amino acids give blue-violet colouration. This highly sensitive test is also given by some β-amino acids and by some peptides and proteins, particularly on warming. The colour test is of great value in the characterisation of the α-amino acids separated by paper and by thin-layer chromatography.

(b) **Copper complex formation.** Add a few drops of aqueous copper(II) sulphate solution to an aqueous solution of the amino acid. A deep blue colouration is obtained. The deep blue copper derivative may be isolated by boiling a solution of the amino acid with precipitated copper(II) hydroxide or with copper(II) carbonate, filtering and concentrating the solution. These blue complexes are co-ordination compounds of the structure:

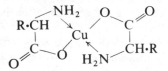

(c) **Nitrous acid test.** The conditions of the test are similar to those described for the classification of primary amines (Section **VII,5**B), but using acetic acid in place of hydrochloric acid. An α-amino acid yields nitrogen and an α-hydroxy acid.

For the preparation of derivatives of α-amino acids, see Section **VII,6,**23.

---

* Prepared by cautiously mixing 3 vol. of concentrated sulphuric acid with 1 vol. of water.

Compounds containing sulphur The *neutral* sulphur compounds which are in solubility group VII include sulphides or thioethers, disulphides, sulphoxides, sulphones and isothiocyanates. *Acidic* sulphur compounds, i.e., sulphonic and sulphinic acids, thiols and thiophenols, and the primary sulphonamides have already been discussed. The sulphates of amines are converted by aqueous sodium hydroxide into the free bases; the sulphate anion can be detected in the resulting aqueous solution as barium sulphate in the usual manner.

Sulphides (thioethers). The organic sulphides are usually liquids with penetrating and disagreeable odours. In contrast to the oxygen analogues (ethers), they are readily oxidised; thus sulphoxides are produced with hydrogen peroxide, and sulphones with nitric acid or with potassium permanganate in glacial acetic acid solution (see p. 587, **III,T,4**).

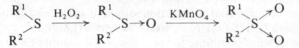

Thioethers usually yield sulphonium salts when warmed with ethyl iodide and allowed to cool. The physical properties (b.p., density and refractive index) are useful for identification purposes.

Disulphides. Disulphides are liquids or low m.p. solids and have unpleasant odours, particularly if liquid. They are reduced by zinc and dilute acids to the thiols.

$$R \cdot S \cdot S \cdot R + 2[H] \longrightarrow 2RSH$$

Sulphoxides. These are usually solids of low m.p. They may be oxidised in glacial acetic acid solution by potassium permanganate to the corresponding sulphones, and reduced to the sulphides by boiling with tin or zinc and hydrochloric acid.

Sulphones. Sulphones are usually crystalline solids, and are extremely stable to most oxidising, reducing and hydrolytic reagents.

Esters of sulphuric acid. These compounds are generally water-insoluble liquids and are saponified by boiling with water or dilute alkali to the corresponding alcohols and sulphuric acid:

$$R_2SO_4 + 2H_2O \longrightarrow 2ROH + H_2SO_4$$

They are usually poisonous and can be identified by using them to alkylate 2-naphthol (compare Section **IV,124**).

The **alkyl esters of sulphonic acids** exhibit properties similar to those of the alkyl sulphates, and are hydrolysed, by boiling with aqueous alkalis, to the alcohols and sulphonates. Thus with ethyl toluene-*p*-sulphonate:

$$p\text{-}CH_3 \cdot C_6H_4 \cdot SO_2 \cdot OC_2H_5 + H_2O \longrightarrow p\text{-}CH_3 \cdot C_6H_4 \cdot SO_3H + C_2H_5OH$$

The salts of **monoalkyl sulphates** are frequently encountered as commercial detergents: these are usually sodium salts, the alkyl components contain 12 or more carbon atoms, and give colloidal solutions. They are hydrolysed by boiling with dilute sodium hydroxide solution:

$$RO \cdot SO_2 \cdot ONa + NaOH \longrightarrow ROH + Na_2SO_4$$

Isothiocyanates. These compounds, also known as **mustard oils**, are oils or low melting point solids, and usually possess irritating odours. Upon boiling with acids, for example with concentrated hydrochloric acid, they are hydrolysed to the primary amines and hydrogen sulphide is evolved:

$$R\cdot NCS + HCl + 2H_2O \longrightarrow R\overset{\oplus}{N}H_3\}\overset{\ominus}{Cl} + CO_2 + H_2S$$

They react with amines to form substituted thioureas:

$$R^1\cdot NCS + R^2NH_2 \longrightarrow R^1NH\cdot CS\cdot NHR^2$$

This reaction is also employed for the characterisation of amines (see Section **VII,6**,*21*(*h*)).

Compounds containing halogen **Reactivities of halogen compounds.** Halogen-containing compounds may be found in each of the seven solubility groups. Those in group I are of low molecular weight and owe their solubility to the presence of such groups as OH, CO_2H, etc. Most halogen compounds in solubility group II are salts in which the halogen is present as an anion; these, with the exception of quaternary ammonium salts, are converted by dilute alkali into basic compounds. Halogen compounds may also be present in solubility groups III and IV, but, like those in groups I and II, they contain other functional groups which are more easily identified. The nature of the halogen in solubility groups V to VII is best determined with the aid of a 2 per cent solution of silver nitrate in absolute ethanol, the alcohol serving as a common solvent for the silver nitrate and the organic compound to be tested. For water-soluble compounds, aqueous silver nitrate solution should also be used after acidification with dilute nitric acid.

Organic compounds containing halogens react with silver nitrate in the following order of *decreasing* reactivity:

1. Water-soluble compounds containing ionisable halogen or compounds such as acyl halides of low molecular weight which readily yield ionisable compounds with water will react immediately, even with aqueous silver nitrate.
2. Acyl and sulphonyl halides, α-halogeno-ethers and alkyl iodides react rapidly.
3. Alkyl chlorides and bromides, aromatic compounds containing halogen in the side chain, or nuclear-halogenated aromatic compounds with nitro groups in the *ortho* and/or *para* position, do not usually react readily at room temperature but react fairly rapidly on heating. The order of reactivity of alkyl halides is

   tertiary > secondary > primary

   and, indeed, some tertiary halides may react in the cold.
4. Aromatic compounds in which the halogen is attached directly to the aromatic nucleus and polyhalogenated compounds with three or more halogens on the same carbon atom do not react even on heating.

Test for the reactivity of the halogen in the following way:

(*a*) **Reaction with alcoholic silver nitrate.** To carry out the test, treat 2 ml of a 2 per cent solution of silver nitrate in alcohol with 1 or 2 drops (or 0.05 g) of the compound. If no appreciable precipitate appears at the laboratory temperature, heat on a boiling water bath for several minutes. Some organic acids give insoluble silver salts, hence it is advisable to add 1 drop of *dilute* (5%) nitric acid at the conclusion of the test: most silver salts of organic acids are soluble in nitric acid.

Note. If concentrated nitric acid is used, a dangerous explosion may result.

(*b*) **Reaction with ethanolic potassium hydroxide.** Boil 0.5 ml of the compound with 4 ml of 0.5 M-ethanolic potassium hydroxide under reflux for 15 minutes. Most alkyl halides and benzyl halides give a crystalline precipitate of the potassium halide. Dilute with 5 ml of water, acidify with dilute nitric acid and test with silver nitrate solution.

The 0.5 M-ethanolic potassium hydroxide solution is prepared by dissolving 16 g of potassium hydroxide pellets in 500 ml of ethanol in a bottle closed with a cork. After standing for 24 hours, the clear solution is decanted and filtered from the residue of potassium carbonate. It is said that a solution in methanol has better keeping qualities than that in ethanol.

(*c*) **Halogen exchange reactions.** This is based upon the fact that sodium chloride and sodium bromide are sparingly soluble in pure acetone:

$$RCl(Br) + NaI \longrightarrow RI + NaCl(Br)$$

The test consists in treating a solution of sodium iodide in pure acetone with the organic compound. The reaction is probably of the S_N2 type involving a bimolecular attack of the iodide ion upon the carbon atom carrying the chlorine or bromine; the order of reactivities of halides is: primary > secondary > tertiary and Br > Cl.

Primary bromides give a precipitate of sodium bromide within 3 minutes at 25 °C; chlorides react only when heated at 50 °C for up to 6 minutes. Secondary and tertiary bromides must be heated at 50 °C for up to 6 minutes, but tertiary chlorides do not react within this time.

1,2-Dichloro- and dibromo-compounds give a precipitate with the reagent and also liberate free iodine:

$$R^1 \cdot CHBr \cdot CHBr \cdot R^2 + 2NaI \longrightarrow R^1 \cdot CHI \cdot CHI \cdot R^2 + 2NaBr$$
$$\Updownarrow$$
$$R^1CH = CH \cdot R^2 + I_2$$

Polybromo compounds (bromoform, *s*-tetrabromoethane) react similarly at 50 °C, but simple polychloro compounds (chloroform, carbon tetrachloride and trichloroacetic acid) do not.

Sulphonyl chlorides give an immediate precipitate and also liberate iodine:

$$Ar \cdot SO_2Cl + NaI \longrightarrow Ar \cdot SO_2I + NaCl \xrightarrow{NaI} Ar \cdot SO_2Na + I_2$$

Acid chlorides and bromides, allyl halides, and α-halo-ketones, -esters, -amides and -nitriles react at 25 °C within 3 minutes. Vinyl and aryl halides are inert.

Prepare the reagent by dissolving 7.5 g of sodium iodide in 50 ml of AnalaR acetone. The colourless solution gradually acquires a yellow colour. Keep it in a dark bottle. When a red-brown colour develops, it should be discarded.

For the preparation of derivatives of alkyl and aryl halides see Sections **VII,6,**8 and **VII,6,**9 respectively.

VII,6. THE PREPARATION OF DERIVATIVES   The steps so far taken in the identification of a compound, viz., (i) determination of the physical constants and the establishment of the purity (Section **VII,1**), (ii) qualitative analysis for the elements (Section **VII,2**), (iii) study of the solubility behaviour towards selected solvents (Section **VII,3**), (iv) examination of the spectral characteristics (Section **VII,4**) and (v) identification of functional groups (Section **VII,5**), will, in general, establish the class to which the compound belongs. The next step is to prove its identity with one of the members of the class. It is at this stage that the literature is consulted. In the first instance, the appropriate table or tables in this volume are examined. Those compounds are selected which have melting points or boiling points within about 5 °C of the unknown. To distinguish between these, a suitable *derivative* is prepared and its physical properties determined; if these agree with those of the known derivative of one of the possibilities already considered, then the identity of the compounds may be assumed. If the list of possible compounds is long, the preparation of two derivatives may be desirable. It must, however, be pointed out that in eliminating compounds from the list of possibilities, due consideration must be paid to other sufficiently characteristic properties, such as density, refractive index, neutralisation equivalent, molecular weight and optical rotation (where applicable), with adequate allowance for experimental error.

The **requirements of a satisfactory derivative include:** (1) The derivative should be easily and quickly prepared in good yield by an unambiguous reaction, and be easily purified. In practice, this generally means that the derivative must be a solid, because of the greater ease of manipulation of small quantities of solids and the fact that melting points are more accurate and more easily determined than boiling points. The melting point should preferably be above 50 °C, but below 250 °C; compounds which melt below 50 °C are frequently difficult to crystallise.

(2) The derivative should be prepared preferably by a general reaction, which under the same experimental conditions would yield a definite derivative with the other individual possibilities. Rearrangements and side reactions should be avoided.

(3) The properties (physical and chemical) of the derivatives should be markedly different from those of the original compound.

(4) The derivative selected in any particular instance should be one which clearly singles out one compound from among all the possibilities and thus enables an unequivocal choice to be made. The melting points of the derivatives to be compared should differ by at least 5–10 °C.

The above considerations will assist in the selection of a derivative. It should also be borne in mind that when a compound has several functional groups, that functional group should be chosen for the preparation of a derivative which gives the least ambiguous reaction.

The methods of preparation of some of the more important derivatives of a number of classes of organic compounds are described in the following Sections. These Sections are cross-referenced with tables incorporating the melting points and boiling points of the compounds themselves, and also the melting points of selected derivatives. For convenience, the references to the various derivative preparations and tables are collected below.

Class of compound	Derivative preparations (Section no.)	Physical constants (Table no.)
Saturated aliphatic hydrocarbons	VII,6,1	VIII,1
Unsaturated aliphatic hydrocarbons	VII,6,2	VIII,2
Aromatic hydrocarbons	VII,6,3	VIII,3
Aliphatic alcohols	VII,6,4	VIII,4
Aromatic alcohols	VII,6,4	VIII,5
Phenols	VII,6,6	VIII,6
Enols	VII,6,7	VIII,7
Polyhydric alcohols	VII,6,4	VIII,8
Carbohydrates (sugars)	VII,6,5	VIII,9
Aliphatic halogen compounds	VII,6,8	VIII,10
Aromatic halogen compounds	VII,6,9	VIII,11
Aliphatic ethers	VII,6,10	VIII,12
Aromatic ethers	VII,6,11	VIII,13
Acetals	VII,6,12	VIII,14
Aliphatic aldehydes	VII,6,13	VIII,15
Aromatic aldehydes	VII,6,13	VIII,16
Aliphatic ketones	VII,6,13	VIII,17
Aromatic ketones	VII,6,13	VIII,18
Quinones	VII,6,14	VIII,19
Aliphatic carboxylic acids	VII,6,15	VIII,20
Aromatic carboxylic acids	VII,6,15	VIII,21
Acid chlorides (aliphatic)	VII,6,16	VIII,22
Acid anhydrides (aliphatic)	VII,6,16	VIII,23
Acid chlorides and acid anhydrides of aromatic acids	VII,6,16	VIII,24
Aliphatic esters	VII,6,17	VIII,25
Aromatic esters	VII,6,17	VIII,26
Primary aliphatic amides	VII,6,18	VIII,27
Primary aromatic amides	VII,6,18	VIII,28
Substituted aromatic amides	VII,6,19	VIII,29
Aliphatic nitriles	VII,6,20	VIII,30
Aromatic nitriles	VII,6,20	VIII,31
Primary and secondary aliphatic amines	VII,6,21	VIII,32
Primary aromatic amines	VII,6,21	VIII,33
Secondary aromatic amines	VII,6,21	VIII,34
Tertiary amines	VII,6,22	VIII,35
Amino acids	VII,6,23	VIII,36
Aromatic nitro compounds	VII,6,24	VIII,37
Aliphatic nitro compounds	VII,6,24	VIII,38
Thiols	VII,6,25	VIII,39
Sulphonic acids	VII,6,26	VIII,40
Aromatic sulphonamides	VII,6,27	VIII,41
Imides		VIII,42
Nitroso, azo, azoxy and hydrazo compounds	—	VIII,43
Miscellaneous sulphur compounds	—	VIII,44
Miscellaneous phosphorus compounds	—	VIII,45
Esters of inorganic acids	—	VIII,46

VII,6,*1*. Saturated aliphatic hydrocarbons Because of the chemical inertness of the saturated aliphatic hydrocarbons and of the closely related cycloalkanes, no satisfactory crystalline derivatives can be prepared. Reliance is therefore placed upon the physical properties (boiling point, density and refractive index) of the redistilled samples. These are collected together in Table VIII,1.

VII,6,2. Unsaturated aliphatic hydrocarbons Derivatives of alkenes. The alkenes are distinguished from the alkanes by their solubility in concentrated sulphuric acid and their characteristic reactions with dilute potassium permanganate solution and with bromine. Characterisation may be based upon the determination of their physical and/or spectral properties. Characterisation by way of solid adducts with nitrosyl chloride has been quite widely used in the terpene field; the preparation of adducts with 2,4-dinitrobenzenesulphenyl chloride is described below.

Adducts with 2,4-dinitrobenzenesulphenyl chloride. 2,4-Dinitrobenzenesulphenyl chloride reacts in polar solvents (acetone, 1,2-dichloroethane, acetic acid and dimethylformamide) with alkenes to yield crystalline adducts, the β-chloroalkyl-2,4-dinitrophenyl sulphides, e.g.:

$$R \cdot CH = CH_2 + O_2N-\underset{NO_2}{\text{C}_6H_3}-SCl \longrightarrow R \cdot \underset{Cl}{CH} - CH_2 - S-\underset{O_2N}{\text{C}_6H_3}-NO_2$$

Addition of the reagent is stereospecific (*trans* addition) and one can thus differentiate between *cis* and *trans* isomers: thus *cis*-butene and *trans*-butene give products of m.p. 129 and 77 °C respectively.

Heat a solution of 0.2 g of the reagent and 0.2–0.3 g of the alkene in glacial acetic acid on the steam bath for 15 minutes or until the potassium iodide test shows that the reaction is complete. Add a drop of the reaction solution to a drop of potassium iodide solution on a spot plate; the presence of unreacted reagent is revealed by the liberation of iodine:

$$2RSCl + 2I^{\ominus} \longrightarrow RSSR + I_2 + 2Cl^{\ominus}$$

Cool the mixture in ice. If a solid separates, filter it off; if not, pour the reaction mixture on to 5–10 g of crushed ice. Recrystallise the resulting solid or oil from ethanol.

Derivatives of alkynes. (*a*) **Addition products with 2,4-dinitrobenzenesulphenyl chloride.** The reagent reacts with symmetrical alkynes as follows:

$$RC \equiv CR + ArSCl \longrightarrow RC(Cl) = CR(SAr)$$

where Ar $= 2,4\text{-}(NO_2)_2C_6H_3-$

Dissolve 1.60 g of the reagent in 15 ml of 1,2-dichloroethane at 0 °C and add 3.0 ml of the ice-cold alkyne. Keep at 0 °C for 2 hours, remove the solvent by aspiration and keep the clear yellow oil in a refrigerator until crystallisation occurs. Dissolve the crystals in 25 ml of absolute ethanol, decolourise with charcoal and filter. Concentrate the filtrate, collect the crystals which separate and recrystallise from ethanol.

(*b*) **Mercurides of monosubstituted alkynes (mercury(II) alkynides).** Monosubstituted alkynes form mercurides which are suitable for identification purposes:

$$2R - C \equiv C - H + K_2[HgI_4] + 2KOH \longrightarrow$$
$$(R - C \equiv C)_2Hg + 4KI + 2H_2O$$

The procedure consists in adding a dilute solution of the alkyne in ethanol to an excess of an alkaline mercury(II) iodide reagent: a white or greyish-white precipitate forms immediately, which is filtered off, washed with dilute ethanol and recrystallised. The yield of mercuride is 85–95 per cent.

The mercury(II) iodide reagent is *prepared* by dissolving 6.6 g of mercury (II) chloride (*POISONOUS*) in a solution of 16.3 g of potassium iodide in 16.3 ml of water and adding 12.5 ml of 10 per cent sodium hydroxide solution.

Into a cooled dilute solution of 2 equivalents of alkaline mercury(II) iodide reagent, drop slowly, with mechanical stirring, a solution of 1 equivalent of the monosubstituted alkyne in 20 volumes of 95 per cent ethanol. A white crystalline precipitate separates at once. Stir for 2–3 minutes, filter rapidly with suction and wash with 50 per cent ethanol. Recrystallise from ethanol or benzene.

Data for a selection of alkenes and alkynes are collected in Table VIII,2.

VII,6,3. Aromatic hydrocarbons For characterisation, aromatic hydrocarbons can be sulphonated, chlorosulphonated, carboxybenzoylated and nitrated. Polynuclear aromatic hydrocarbons, and many of their derivatives, yield crystalline adducts with picric acid, styphnic acid, 1,3,5-trinitrobenzene and 2,4,7-trinitrofluorenone.

(*a*) **Sulphonamides.** Aromatic hydrocarbons react with chlorosulphonic acid to yield the corresponding sulphonyl chlorides (the process is known as *chlorosulphonation*). These do not usually crystallise well and are therefore converted into the sulphonamides by treatment with concentrated ammonia solution or with solid ammonium carbonate.

$$ArH + 2HOSO_2Cl \longrightarrow Ar \cdot SO_2Cl + H_2SO_4 + HCl$$
$$Ar \cdot SO_2Cl + (NH_4)_2CO_3 \longrightarrow Ar \cdot SO_2NH_2 + NH_4Cl + CO_2 + H_2O$$

Dissolve 1.0 g of the compound in 5 ml of dry ($CaCl_2$) chloroform in a dry test-tube, cool it in a beaker of ice and add 3–5 ml of chlorosulphonic acid dropwise. When the evolution of hydrogen chloride has subsided, remove the test-tube from the ice bath and allow to stand at room temperature for 20–30 minutes; then pour on to crushed ice (30 g). Separate the chloroform layer, wash it with water, dry ($CaCl_2$), and evaporate the solvent.

Boil the arenesulphonyl chloride (0.5 g) with 5 ml of aqueous ammonia (*d* 0.88) for 10 minutes (*FUME CUPBOARD* or *HOOD*). Cool the reaction mixture and dilute it with 10 ml of water. Filter off the sulphonamide, wash it with water and recrystallise from dilute ethanol.

Alternatively, heat a mixture of 0.5 g of the arenesulphonyl chloride with 2.8 g of dry powdered ammonium carbonate at 100 °C during 30 minutes. Wash the residue with several portions (10 ml) of cold water, filter and recrystallise from dilute ethanol.

If the presence of a sulphone is suspected (cf. *Aromatic halogen compounds*, Section **VII,6,9**), treat the product with 6 *M*-sodium hydroxide solution (only the sulphonamide dissolves), filter and reprecipitate the sulphonamide with 6 *M*-hydrochloric acid.

(*b*) ***o*-Aroylbenzoic acids.** Aromatic hydrocarbons react with phthalic anhydride in the presence of anhydrous aluminium chloride producing aroylbenzoic acids in good yields:

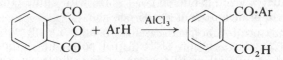

The process is termed *carboxybenzoylation*.

Place a mixture of 1.0 g of the hydrocarbon, 10 ml of dry dichloromethane or 1,2-dichloroethane, 2.5 g of powdered anhydrous aluminium chloride and 1.2 g of pure phthalic anhydride in a 25–50 ml round-bottomed flask fitted with a reflux condenser (5″ jacket). Heat on a water bath for 30 minutes (or until no more hydrogen chloride fumes are evolved). Cool in ice and add 10 ml of concentrated hydrochloric acid cautiously and with constant shaking. When the reaction has subsided, add 20 ml of water and shake vigorously. (All the solid material should pass into solution.) Transfer the two-phase system to a separatory funnel, add 25 ml of ether and shake. Discard the lower aqueous phase. Wash the ethereal layer with 25 ml of 2.5 *M*-hydrochloric acid to ensure removal of any aluminium salts present. Shake the ethereal solution cautiously with 25 ml of *M*-sodium carbonate solution, and run the aqueous phase slowly into 30 ml of *M*-hydrochloric acid. Collect the aroylbenzoic acid by suction filtration, wash it with 25–50 ml of water and recrystallise it from dilute ethanol or from acetic acid. The derivatives prepared from benzene and toluene crystallise with water of crystallisation; the latter is removed by drying at 100 °C.

(c) **Nitro derivatives.**   No general experimental details for the preparation of nitro derivatives can be given, as the ease of nitration and the product formed frequently depend upon the exact experimental conditions. Moreover, some organic compounds react violently so that nitrations should always be conducted on a small scale. The derivatives already described are usually more satisfactory. Three typical nitrations will, however, be described in order to illustrate the results which may be obtained.

Benzene.   Add 0.5 ml of benzene slowly and with shaking and cooling to a mixture of 4 ml each of concentrated sulphuric and nitric acids. Heat the mixture carefully until it just boils, cool and pour into excess of cold water. Filter off the precipitate, wash it free from acid and recrystallise it from dilute alcohol. *m*-Dinitrobenzene, m.p. 90 °C, is formed.

Toluene.   Proceed as for *benzene* but use 0.5 ml of toluene and a mixture of 3 ml of concentrated sulphuric acid and 2 ml of fuming nitric acid. Gently warm the mixture over a free flame for 1–2 minutes, cool and pour into 20 ml of ice-water. Recrystallise the product from dilute alcohol. 2,4-Dinitrotoluene, m.p. 71 °C, is obtained.

Naphthalene.   For the conditions of nitration of polynuclear hydrocarbons see Section **IV,19**, which describes the nitration of naphthalene. The scale of the experiment may be suitably reduced.

(d) **Oxidation of a side chain by alkaline permanganate.**   Aromatic hydrocarbons containing side chains may be oxidised to the corresponding acids: the results are generally satisfactory for compounds with one side chain (e.g., toluene or ethylbenzene → benzoic acid; nitrotoluene → nitrobenzoic acid) or with two side chains (e.g., *o*-xylene → phthalic, acid).

Suspend in a round-bottomed flask 1 g of the substance in 75–80 ml of boiling water to which about 0.5 g of sodium carbonate crystals have been added, and introduce slowly 4 g of finely powdered potassium permanganate. Heat under reflux until the purple colour of the permanganate has disappeared (1–4 hours). Allow the mixture to cool and carefully acidify with dilute sulphuric acid. Heat the mixture under reflux for a further 30 minutes and then cool. Remove any excess of manganese dioxide by the addition of a little sodium metabisulphite. Filter the precipitated acid and recrystallise it from a suitable solvent (e.g., benzene, ethanol, dilute ethanol or water). If the acid does not separate from the solution, extract it with ether, benzene, or carbon tetrachloride.

(e) **Picrates.** Many aromatic hydrocarbons (and other classes of organic compounds) form molecular compounds with picric acid, for example, naphthalene picrate $C_{10}H_8 \cdot C_6H_2(NO_2)_3OH$. Some picrates, e.g., anthracene picrate, are so unstable as to be decomposed by many, particularly hydroxylic, solvents; they therefore cannot be easily recrystallised but may be washed with a little ether and dried on a porous tile. Their preparation may often be accomplished in such non-hydroxylic solvents as chloroform, benzene or ether. The picrates of hydrocarbons can be readily separated into their constituents by warming with dilute ammonia solution and filtering (if the hydrocarbon is a solid) through a moist filter-paper. The filtrate contains the picric acid as the ammonium salt, and the hydrocarbon is left on the filter-paper.

Picrates are usually prepared by adding a hot solution of the compound in ethanol to a cold saturated ethanolic solution of picric acid, warming and allowing to cool; the derivative separates in a crystalline condition. It is filtered off, washed with a little ether and pressed on a porous tile. If the picrate is stable, it is recrystallised from ethanol, ethyl acetate, benzene or ether. Do not mistake the recrystallised reagent (m.p. 122 °C) for a picrate.

The following are typical experimental details for the preparation of naphthalene picrate. Dissolve 0.1 g of naphthalene in the minimum of hot ethanol and add to 1 ml of a saturated solution of picric acid in ethanol. Warm the mixture and then cool to allow the product to crystallise.

(f) **Styphnates.** Aromatic hydrocarbons (and also some amines and heterocyclic bases) form 1 : 1-adducts with styphnic acid (2,4,6-trinitroresorcinol),

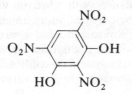

These derivatives do not crystallise quite so well as the corresponding picrates, but are frequently of great value. Benzene and its simple homologues do not give stable derivatives.

Dissolve equimolecular amounts of the hydrocarbon and styphnic acid in the minimum volume of hot acetic acid and allow to cool. Filter off the crystalline derivative which separates, wash it with a little acetic acid and dry in the air. Determine the m.p. Recrystallise from acetic acid and again determine the m.p.

Benzene must be employed as the solvent for anthracene styphnate since most other solvents lead to dissociation.

(g) **Addition compounds with 1,3,5-trinitrobenzene**

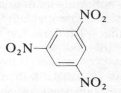

This reagent affords compounds (1 : 1) with aromatic hydrocarbons and other classes of organic compounds (heterocyclic compounds, aromatic ethers, etc.).

Dissolve equimolecular quantities of the hydrocarbon and 1,3,5-trinitrobenzene in hot ethanol, benzene or glacial acetic acid, and allow to cool. Filter off the solid which separates and recrystallise it from one of these solvents.

(h) **Addition compounds with 2,4,7-trinitro-9-fluorenone** (abbreviated to TNF). Aromatic hydrocarbons (and also some polynuclear substituted aro-

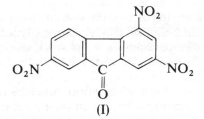

(I)

matic compounds) react with 2,4,7-trinitro-9-fluorenone (I) to yield crystalline 1 : 1-adducts.

Dissolve equimolecular proportions of pure TNF and the hydrocarbon separately in absolute ethanol or ethanol–benzene or glacial acetic acid and mix the two hot nearly saturated solutions. Heat for 1 minute and cool. Recrystallise from acetic acid, absolute ethanol, ethanol–benzene or benzene.

Data for a number of typical aromatic hydrocarbons are collected in Table VIII,3.

VII,6,4. Alcohols and polyhydric alcohols (a) **Oxidation with 'chromic acid'.** A primary alcohol is oxidised by 'chromic acid' to the corresponding aldehyde whilst a secondary alcohol yields a ketone: tertiary alcohols are generally unaffected or are decomposed into non-ketonic products. Oxidation therefore provides a method for distinguishing between primary, secondary and tertiary alcohols and characterisation of the carbonyl compound provides a means of identifying the alcohol:

$$R \cdot CH_2 OH \xrightarrow{[O]} R \cdot CHO$$
$$R^1 \cdot CH(OH) \cdot R^2 \xrightarrow{[O]} R^1 \cdot CO \cdot R^2$$

To an ice-cold mixture of 1.0 ml of concentrated sulphuric acid and 5 ml of saturated aqueous potassium dichromate solution, add 2 ml of the alcohol or its concentrated aqueous solution. If the alcohol is not miscible with the reagent, shake the reaction mixture vigorously. After 5 minutes, dilute with an equal volume of water, distil and collect the first few ml of the aqueous distillate in a test-tube cooled in ice. (Aldehydes and ketones are volatile in steam.) Test a portion of the distillate for a carbonyl compound with 2,4-dinitrophenylhydrazine reagent (p. 1071). If a solid derivative is obtained, indicating that the compound was a primary or secondary alcohol, test a further portion with Schiff's reagent (p. 1071) to distinguish between the two possibilities. The derivative may be recrystallised; the m.p. may give a preliminary indication of the identity of the alcohol.

(*b*) **3,5-Dinitrobenzoates.** 3,5-Dinitrobenzoyl chloride reacts with alcohols to form solid esters which possess sharp melting points and are therefore admirably suited for purposes of characterisation:

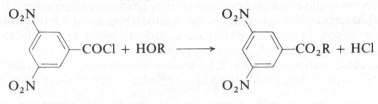

The acid chloride is available commercially, but it is preferable to prepare it from the acid as and when required since 3,5-dinitrobenzoyl chloride tends to undergo hydrolysis if kept for long periods, particularly if the stock bottle is frequently opened. The substance may, however, be stored under dry light petroleum.

Mix 1.0 g of 3,5-dinitrobenzoic acid with 1.5 g of phosphorus pentachloride in a small, dry test-tube. Warm the mixture gently over a small smoky flame to start the reaction; when the reaction has subsided (but not before), boil for 1–2 minutes or until the solid matter has dissolved. Pour the mixture while still liquid on a dry watch glass (**CAUTION:** the fumes are irritating to the eyes). When the product has solidified, remove the liquid by-product (phosphorus oxychloride) by transferring the pasty mixture to a pad of several thicknesses of filter-paper or to a small piece of porous tile. Spread the material until the liquid has been absorbed and the residual solid is dry and transfer the 3,5-dinitrobenzoyl chloride to a test-tube. Add 0.5–1 ml of the alcohol, cork the tube loosely and heat on a water bath for 10 minutes: secondary and tertiary alcohols require longer heating (up to 30 minutes). Cool the mixture, add 10 ml of 5 per cent (or saturated) sodium hydrogen carbonate solution, break up the resulting solid ester with a stirring rod (alternatively, stir until crystalline) and filter at the pump; wash with a little sodium hydrogen carbonate solution, followed by water, and then suck as dry as possible. Dissolve the crude ester in the minimum volume of hot rectified (or industrial) spirit. Add hot water, drop by drop, with agitation, until the solution *just* develops a slight turbidity that does not disappear on shaking; immerse the mixture in a hot water bath during the recrystallisation. Allow to cool slowly (in order to avoid the formation of oily drops for esters of low melting point). Filter the crystals, and dry them upon a few thicknesses of filter-paper or upon a piece of porous plate. Determine the melting point of the crystals when thoroughly dry. Light petroleum may also be employed for recrystallisation.

The above procedure may also be carried out in the presence of 1 ml of dry pyridine; with some alcohols improved yields may be obtained by this modification.

(c) **p-Nitrobenzoates.** Alcohols react readily with p-nitrobenzoyl chloride to yield p-nitrobenzoates:

$$p\text{-}NO_2 \cdot C_6H_4 \cdot COCl + ROH \longrightarrow p\text{-}NO_2 \cdot C_6H_4 \cdot CO_2R + HCl$$

The melting points of these esters are usually much lower than those of the corresponding 3,5-dinitrobenzoates: their preparation, therefore, offers no advantages over the latter except for alcohols of high molecular weight and for polyhydroxy compounds. The reagent is, however, cheaper than 3,5-dinitrobenzoyl chloride; it hydrolyses in the air so that it should either be stored under light petroleum or be prepared from the acid, when required, by the phosphorus pentachloride method.

The experimental technique is similar to that given under (b) above.

(d) **Benzoates.** Alcohols react with benzoyl chloride in the presence of pyridine or of sodium hydroxide solution to produce esters of benzoic acid:

$$C_6H_5 \cdot COCl + ROH \longrightarrow C_6H_5 \cdot CO_2R + HCl$$

These derivatives are generally liquids and hence are of little value for characterisation; the polyhydric alcohols, on the other hand, afford solid benzoates. Thus the benzoates of ethylene glycol, trimethylene glycol and glycerol melt at 73, 58 and 76 °C respectively.

Mix together 0.5–0.8 ml of the polyhydroxy compound, 5 ml of pyridine and 2.5 ml of redistilled benzoyl chloride in a 50-ml flask, and heat under reflux for 30–60 minutes. Add 25 ml of 5 per cent sodium hydrogen carbonate solution to the cold reaction mixture and cool in ice until the precipitate solidifies. Filter and wash with a little water. Recrystallise from dilute ethanol as detailed under (b) above.

(e) **Phenyl- and 1-naphthyl-urethans (Phenyl- and 1-naphthylcarbamates).** Both phenyl isocyanate and 1-naphthyl isocyanate react with alcohols to yield phenylurethans and 1-naphthylurethans respectively:

$$C_6H_5 \cdot N{=}C{=}O + ROH \longrightarrow C_6H_5 \cdot NHCO_2R$$
$$C_{10}H_7{}^1 \cdot N{=}C{=}O + ROH \longrightarrow C_{10}H_7{}^1 \cdot NHCO_2R$$

If the alcohol is not anhydrous, reaction also occurs between the water and the reagent to produce diphenylurea (m.p. 238 °C) and di-1-naphthylurea (m.p. 297 °C) respectively, for example:

$$2C_6H_5 \cdot N{=}C{=}O + H_2O \longrightarrow C_6H_5 \cdot NH \cdot CO \cdot NH \cdot C_6H_5 + CO_2$$

The ureas are less soluble than the corresponding urethans, but their separation is not always easy. For this reason the urethans are generally prepared from alcohols which are insoluble in water and can therefore be easily obtained in the anhydrous condition.

1-Naphthyl isocyanate is usually preferred to phenyl isocyanate for the following reasons: (a) it is much less lachrymatory; (b) it is not so readily decomposed by cold water and thus possesses better keeping qualities; and (c) the melting points of the 1-naphthylurethans are generally higher than those of the corresponding phenylurethans. Furthermore, with primary alcohols, which react readily in the cold, only small amounts of the urea are produced and these may be removed by taking advantage of the extreme insolubility of di-1-naphthylurea in hot ligroin. (See also Section **VII,6,**8(a).)

Place 1 g of the anhydrous alcohol in a dry test-tube and add 0.5 ml of 1-

naphthyl isocyanate* (if the molecular weight is known, use a 10 per cent excess of the reagent); insert a loose plug of cotton wool in the mouth of the tube. If no solid separates after shaking and standing for 5 minutes, warm on a water bath for 5–10 minutes, and then cool in ice. If no solid is now obtained, 'scratch' the sides of the tube with a glass rod to induce crystallisation. Extract the solid with 5–10 ml of boiling light petroleum (b.p. 100–120 °C); this rapidly dissolves the 1-naphthylurethan but not the di-1-naphthylurea. Remove the urea (if any) by filtration and allow the hot solution to cool. If the urethan does not crystallise out, evaporate the solution to half its original volume, and allow to cool. Collect the crystals on a filter, dry and determine the melting point. If the latter is not sharp, recrystallise from light petroleum (b.p. 100–120 °C), ethanol, chloroform or carbon tetrachloride.

(*f*) **Hydrogen 3-nitrophthalates.** 3-Nitrophthalic anhydride, a yellow crystalline powder of m.p. 163–164 °C, reacts with alcohols to yield esters of 3-nitrophthalic acid:

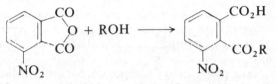

Although two isomeric esters are theoretically possible, the main product is the 2-ester (formulated above); traces of the isomeric 1-ester are eliminated during purification. These derivatives possess a free carboxyl group; their equivalent weights may therefore be determined by titration with standard alkali and thus serve as an additional check upon the identity of the compound.

The reagent must be carefully protected from moisture as it is comparatively easily hydrated to the acid, m.p. 216–218 °C (sealed capillary tube). Dilute aqueous solutions of an alcohol should be treated with solid potassium carbonate and the alcohol layer used for the test.

Phthalic anhydride reacts similarly, but the acid phthalates are somewhat more difficult to isolate and the melting points are considerably lower.

For alcohols of b.p. below 150 °C, mix 0.5 g of 3-nitrophthalic anhydride (Section **IV,176**) and 0.5 ml (0.4 g) of the dry alcohol in a test-tube fitted with a short condenser, and heat under reflux for 10 minutes after the mixture liquefies. For alcohols boiling above 150 °C, use the same quantities of reactants, add 5 ml of dry toluene, heat under reflux until all the anhydride has dissolved and then for 20 minutes more: remove the toluene under reduced pressure (suction with water pump). The reaction product usually solidifies upon cooling, particularly upon rubbing with a glass rod and standing. If it does not crystallise, extract it with dilute sodium hydrogen carbonate solution, wash the extract with ether and acidify. Recrystallise from hot water, or from 30 to 40 per cent ethanol or from toluene. It may be noted that the m.p. of 3-nitrophthalic acid is 218 °C.

(*g*) **3,4,5-Triiodobenzoates.** The derivatives enumerated above are unsatisfactory for alcohol–ethers, e.g., the mono-ethers of ethyleneglycol ('cellosolves') and the mono-ethers of diethyleneglycol ('carbitols') (see Table VIII,4).

---

* The procedure for phenyl isocyanate is similar, but great care must be taken to protect both the reagent and the reaction mixture from moisture.

Crystalline derivatives of alcohol–ethers are readily obtained with 3,4,5-triiodo-benzoyl chloride, for example:

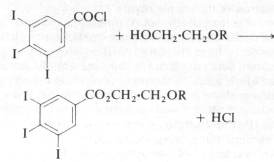

Place 0.5 g of 3,4,5-triiodobenzoyl chloride in a small test-tube, add 0.25 ml of the alcohol–ether and heat the mixture gently over a micro burner until the evolution of hydrogen chloride ceases (3–5 minutes). Pour the molten mass into 10 ml of 20 per cent ethanol to which crushed ice has been added. Some derivatives solidify instantly; those which separate as oils change to solids in a few minutes without further manipulation. Recrystallise from rectified spirit (use 50% ethanol for esters of methyl and butyl carbitol).

3,4,5-Triiodobenzoyl chloride is prepared by refluxing 5 g of 3,4,5-triiodo-benzoic acid (Section **IV,79**) with 10 ml of thionyl chloride for 2 hours. The excess thionyl chloride is removed by distillation and the residue recrystallised from carbon tetrachloride–light petroleum. The acid chloride has m.p. 138 °C; the yield is 3.8 g. It should be kept in a well-stoppered bottle.

(*h*) **Pseudosaccharin ethers.**   Pseudosaccharin chloride (Section **IV,47**) reacts with alcohols to give ethers (*O*-alkyl derivatives of saccharin):

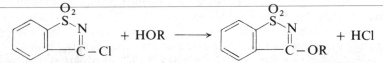

Heat a little pseudosaccharin chloride with excess of the anhydrous alcohol in a test-tube until hydrogen chloride is no longer evolved. Recrystallise from ethanol or other organic solvent.

With the lower primary alcohols, heating at 100 °C for 10 minutes suffices: for higher alcohols, a temperature of 125 °C is preferable. Secondary alcohols require longer heating at 125 °C. A large excess of alcohol should be used when identifying the lower alcohols and the excess removed by evaporation; for the higher alcohols, it is better to employ an excess of pseudosaccharin chloride and the product washed free from the reagent with dilute aqueous alkali.

The melting points of derivatives of selected aliphatic alcohols and polyhydric alcohols are collected in Tables VIII,4 and VIII,8 respectively.

The melting points of some derivatives of aromatic alcohols are collected in Table VIII,5.

VII,6,5. Carbohydrates   The melting points (more accurately termed the decomposition points) of sugars and some of their derivatives, e.g., osazones, are not so definite as those of other classes of organic compounds: they vary with the rate of heating and the differences between individual members are not always large. There are, however, a number of reactions and derivatives which will assist in the characterisation of the simple sugars.

(*a*) **Osazone formation.** The carbohydrates containing a potential aldehyde or keto group in their cyclic form react with one molecular proportion of phenylhydrazine in the cold to form the corresponding phenylhydrazones (compare *Aldehydes and ketones*, Section **VII,6,***13*); these are usually soluble in water and consequently are of little value for purposes of separation and identification. If, however, the carbohydrate is heated in the presence of excess (3–4 mols) of phenylhydrazine, the >CHOH in an aldose or the −CH₂OH in a ketose adjacent to the phenylhydrazone group is effectively oxidised by one molecule of phenylhydrazine into the corresponding carbonyl group, which then reacts with a further molecule of phenylhydrazine to give a bis-phenylhydrazone or **osazone**; aniline and ammonia are by-products of the reaction.

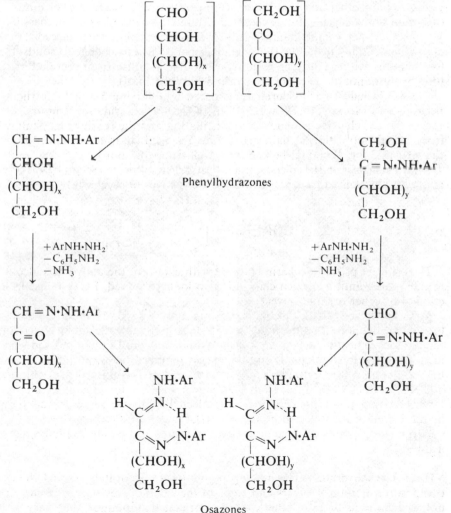

Osazones

Glucosazone

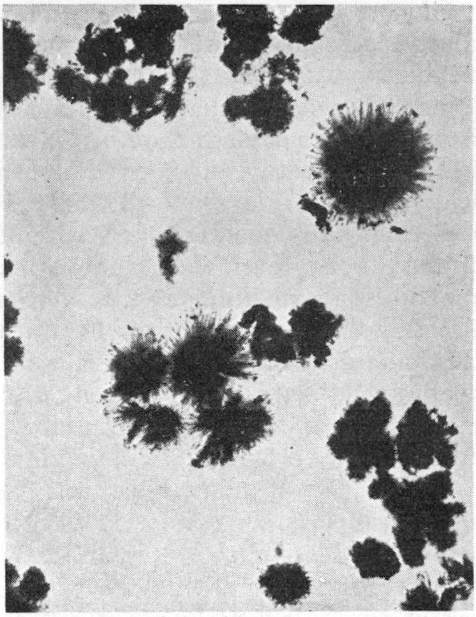

Galactosazone

Arabinosazone

Xylosazone

Fig. VII,5

Glucose and fructose (and also mannose) form the same osazone. The osazones are usually yellow, well-defined crystalline compounds and are sparingly soluble in cold water. The characteristic crystalline forms of the osazones of the commonly occurring sugars, when examined under the microscope, may be employed for their identification (Fig. VII,5); the melting or decomposition points are less satisfactory since these depend to a marked degree on the rate of heating.

Certain carbohydrates (sugars) may be identified by the length of time required to form osazones upon treatment with phenylhydrazine under standard experimental conditions. Monosaccharides give precipitates at 100 °C within 20 minutes. The di-saccharides maltose and lactose give no osazone at 100 °C even after 2 hours, but osazones are obtained on cooling after 10–15 minutes heating. With sucrose an osazone commences to separate after about 30 minutes, due to gradual hydrolysis into glucose and fructose, but no osazone is produced on cooling after heating for 10–15 minutes.

Place 0.20 g of the carbohydrate, 0.40 g of pure *white* phenylhydrazine hydrochloride (e.g., of AnalaR quality), 0.60 g of crystallised sodium acetate and 4.00 ml of water in a dry test-tube. (Weigh the quantities with an accuracy of 0.01 g.) Stopper the tube *loosely* with a cork, and stand or clamp it upright in a beaker containing boiling water. Note the time of immersion and the time when the osazone first separates. Shake the tube occasionally (without removing it from the boiling water) in order to prevent supersaturation. The precipitate separates quite suddenly: duplicate experiments should agree within 0.5 minute. Note whether the precipitate is white (mannose), yellow or orange yellow, and whether it is crystalline or 'oily'.

The approximate times of osazone formation in minutes are given in Table VIII,9. The product from mannose is the simple hydrazone and is practically white. Arabinose osazone separates first as an oil, whilst that from galactose is highly crystalline. Lactose and maltose give no precipitate from hot solution.

(*b*) **p-Nitrophenylhydrazones.** This reagent has been used in the characterisation of a number of monosaccharides.

Heat 0.25 g of the compound with 3 ml of ethanol, add 0.25 g of *p*-nitrophenylhydrazine and heat the suspension until the reaction appears complete. The *p*-nitrophenylhydrazone soon separates. Filter, preferably after standing overnight, wash with a little cold ethanol and then recrystallise from ethanol.

(*c*) **Acetates.** Complete acetylation of all the hydroxyl groups is desirable in order to avoid mixtures. In some cases, the completely acetylated sugars may be obtained in the α- and β-forms depending upon the catalyst, e.g., zinc chloride or sodium acetate, that is employed in the acetylation. The experimental details for acetylation may be easily adapted from those already given for α- and β-glucose penta-acetates (Sections **III,106** and **III,107** respectively).

(*d*) **Benzoates.** Benzoyl chloride has a limited application as a reagent in the sugar series. Details for the benzoylation of D-glucose are given in Section **III,108**.

(*e*) **Trimethylsilylation.** The conversion of monosaccharides and the smaller oligosaccharides (di, tri- and tetra-) into their trimethylsilyl derivatives, which are sufficiently volatile to be analysed by GLC, has greatly simplified the problem of sugar identification. The method described here, which uses hexamethyldisilazane and chlorotrimethylsilane in pyridine solution, may conveniently be applied to a sample size down to 1 mg, and the trimethylsilylation reaction is complete within about 20 minutes.

$$>\text{CHOH} + [(CH_3)_3Si]_2NH/(CH_3)_3SiCl \xrightarrow{\text{pyridine}} >\text{CHO·Si}(CH_3)_3$$

The solid support for the column packing is acid-washed, silanised Chromosorb W (cf. *Exercises in partition chromatography*, Section **I,31**, Note (3), p. 211) preferably with a stationary phase (5%) of either Silicone GE-SE52 or OV17, which have a temperature tolerance of up to 300 °C which may be needed in oligosaccharide analyses. A stationary phase of ethylene glycol succinate ester (E.G.S.) may be used if the maximum temperature is 170 °C. Columns having a length of 5 ft and a flow rate of carrier gas (nitrogen) of 40 ml/minute are satisfactory.

Place 10 mg of the carbohydrate in a small sample tube and by means of a graduated syringe add 0.25 ml of pyridine followed by 0.05 ml of hexamethyldisilazane and 0.02 ml of chlorotrimethylsilane. (**CAUTION**: the operation should be performed in a fume cupboard. The syringe should be carefully rinsed with dichloromethane and dried between use with each reagent.) Stopper the sample tube with a plastic cap, shake the tube and allow to stand for 20 minutes. Centrifuge the tube at low speeds to compact the solid deposit and chromatograph a sample (0.1 μl) of the supernatant liquor. The carbohydrate should be identified by a comparison of retention times with those of suitable standards, and the identification confirmed with the aid of peak enhancement experiments. With this trimethylsilylation method most aldoses give two peaks corresponding to an anomeric mixture of derivatives; ketoses often give additional peaks thought to correspond to derivatives having different ring sizes.

If excessive tailing of the pyridine peaks in the chromatographic trace causes difficulty in interpretation (e.g., with pentoses) proceed as follows. Transfer the supernatant liquid of the centrifuged solution by means of a dry dropper pipette to a small test-tube and attach to the tube a suitable adapter which is fitted to a cooled vacuum trap and pump. Remove the pyridine under reduced pressure; continuous agitation of the tube in a water bath held at about 50 °C is advisable. Dilute the viscous residue with 0.5 ml of chloroform and re-chromatograph.

Recently the use of *N*-(trimethylsilyl)imidazole with chlorotrimethylsilane in pyridine solution has been recommended as an alternative trimethylsilylating reagent for use with moist carbohydrate samples. This reagent is also of value in that it reduces the extent to which anomerisation occurs during the trimethylsilylation reaction.

The gas chromatographic examination of many carbohydrate derivatives may be effected directly (e.g., *O*-methyl derivatives, isopropylidene derivatives, etc.). The application of high performance liquid chromatography for the analysis of free carbohydrates, of their derivatives and of the oligosaccharides without the need for derivatisation has recently become a rapidly expanding area.

VII,6,6. Phenols (*a*) **Acetates.** The acetates of monohydric phenols are usually liquids, but those of di- and tri-hydric phenols and also of many substituted phenols are frequently crystalline solids.

Acetates may be prepared by adding acetic anhydride to somewhat dilute solutions of compounds containing hydroxyl (or amino) groups in aqueous caustic alkalis. The amount of alkali used should suffice to leave the liquid slightly basic at the end of the operation, so much ice should be added that a little remains unmelted, and the acetic anhydride should be added quickly.

Dissolve 0.01 mol (or 1 g if the molecular weight is unknown) of the compound in 5 ml of 3 M-sodium hydroxide solution, add 10–20 g of crushed ice followed by 1.5 g (1.5 ml) of acetic anhydride. Shake the mixture vigorously for 30–60 seconds. The acetate separates in a practically pure condition either at once or after acidification by the addition of a mineral acid. Collect the acetyl derivative, and recrystallise it from hot water or from dilute ethanol.

(b) **Benzoates.** The benzoates of a few phenols (e.g., o-cresol) are liquids. Many phenols do, however, yield crystalline benzoyl derivatives: these are useful for purposes of characterisation.

The Schotten–Baumann method of benzoylation with benzoyl chloride in the presence of aqueous sodium hydroxide may be used. Full details are given under *Primary and secondary amines*; Section **VII,6,21**. Alternatively, dissolve 1.0 g of the phenol in 3 ml of dry pyridine and add 0.5 g of benzoyl chloride. After the initial reaction has subsided, warm the mixture over a small flame for a minute or two and pour, with vigorous stirring, into 10–15 ml of water. Allow the precipitate to settle, decant the supernatant liquid, stir the residue thoroughly with 5–10 ml of M-sodium carbonate solution, filter and recrystallise from ethanol or from light petroleum.

(c) **Toluene-p-sulphonates.** Toluene-p-sulphonyl chloride reacts readily with phenols to yield toluene-p-sulphonates:

$$p\text{-}CH_3\cdot C_6H_4\cdot SO_2Cl + ArOH \longrightarrow p\text{-}CH_3\cdot C_6H_4\cdot SO_2OAr + HCl$$

Mix 1.0 g of the phenol with 2.5 ml of pyridine, add 2 g of toluene-p-sulphonyl chloride and heat on a water bath for 15 minutes. Pour into 25 ml of cold water and stir until the oil solidifies. Filter, wash with cold dilute hydrochloric acid (to remove pyridine), with cold dilute sodium hydroxide solution (to remove any phenol present), and then with cold water. Recrystallise from methanol or ethanol.

(d) **p-Nitrobenzoates and 3,5-dinitrobenzoates.** Both p nitrobenzoyl chloride and 3,5-dinitrobenzoyl chloride react with phenols, best in pyridine solution, to yield crystalline p-nitrobenzoates and 3,5-dinitrobenzoates respectively:

$$p\text{-}NO_2\cdot C_6H_4\cdot COCl + ArOH \longrightarrow p\text{-}NO_2\cdot C_6H_4\cdot CO_2Ar + HCl$$
$$3,5\text{-}(NO_2)_2C_6H_3\cdot COCl + ArOH \longrightarrow 3,5\text{-}(NO_2)_2C_6H_3\cdot CO_2Ar + HCl$$

For properties of these reagents and their preparation from the corresponding acids, see under *Alcohols and polyhydric alcohols*; Section **VII,6,4**.

Dissolve 0.5 g of the phenol in 4–5 ml of dry pyridine, add 1.3 g of 3,5-dinitrobenzoyl chloride and reflux for 25–30 minutes. Pour the cold reaction mixture into 40 ml of ca. 2 M-hydrochloric acid. Decant the supernatant aqueous liquid from the precipitated solid or oil and stir it vigorously with about 10 ml of M-sodium carbonate solution. Filter off the solid derivative and wash it with water. Recrystallise from ethanol, dilute ethanol, benzene–acetone or benzene–light petroleum (b.p. 60–80 °C).

(e) **Aryloxyacetic acids.** Phenols, in the presence of alkali, react with chloroacetic acid to give aryloxyacetic acids:

$$ArONa + ClCH_2 \cdot CO_2Na \longrightarrow ArOCH_2 \cdot CO_2Na + NaCl \xrightarrow{HCl}$$

$$ArOCH_2 \cdot CO_2H + 2NaCl$$

These are crystalline compounds with sharp melting points, and possess the further advantage that their equivalent weights may be determined by dissolving in dilute ethanol and titrating with standard alkali. Nitrophenols, however, give unsatisfactory derivatives.

To a mixture of 1.0 g of the compound and 3.5 ml of 33 per cent sodium hydroxide solution in a test-tube, add 2.5 ml of 50 per cent chloroacetic acid solution. If necessary, add a little water to dissolve the sodium salt of the phenol. Stopper the test-tube loosely and heat on a gently boiling water bath for an hour. After cooling, dilute with 10 ml of water, acidify to Congo red with dilute hydrochloric acid and extract with 30 ml of ether. Wash the ethereal extract with 10 ml of water, and extract the aryloxyacetic acid by shaking with 25 ml of 5 per cent sodium carbonate solution. Acidify the sodium carbonate extract (to Congo red) with dilute hydrochloric acid, collect the aryloxyacetic acid which separates and recrystallise it from water or from aqueous ethanol.

(f) **Diphenylurethans.** Phenols react with diphenylcarbamoyl chloride to yield diphenylurethans (or aryl N,N-diphenylcarbamates):

$$(C_6H_5)_2N \cdot COCl + HOAr \xrightarrow{C_5H_5N} (C_6H_5)_2N \cdot CO_2Ar + HCl$$

The reagent is unsuitable for a number of phenolic acids.

Dissolve 0.5 g of the phenol in 2.5 ml of pyridine, and add one equivalent of diphenylcarbamoyl chloride (or 0.4–0.5 g if the molecular weight is uncertain). Reflux the mixture for 30–60 minutes on a boiling water bath, and then pour into about 25 ml of water. Filter the derivative, wash with a little sodium hydrogen carbonate solution and recrystallise from ethanol, benzene, light petroleum (b.p. 60–80 °C) or carbon tetrachloride.

(g) **1-Naphthylurethans (1-naphthylcarbamates).** 1-Naphthyl isocyanate reacts smoothly with monohydric, but not with polyhydric, phenols to give 1-naphthylurethans (or N-1-naphthylcarbamates):

$$1\text{-}C_{10}H_7N = C = O + ArOH \longrightarrow 1\text{-}C_{10}H_7NH \cdot CO_2Ar$$

(compare *Alcohols and polyhydric alcohols*; Section **VII,6,4**). Some phenols, e.g., nitrophenols and halogeno-phenols, react with difficulty with the reagent alone; the addition of a few drops of pyridine or 1 drop of an ethereal solution of trimethylamine or triethylamine generally results in the rapid formation of the urethan.

Place 0.25 g of the phenol together with an equal weight of 1-naphthyl isocyanate in a *dry* test-tube closed with a stopper carrying a calcium chloride guard-tube. If a spontaneous reaction does not occur, boil the mixture gently for 2–3 minutes, and cool; if the reaction mixture does not solidify, rub the walls of the tube vigorously with a glass rod. If no crystalline solid is obtained, add 2 drops of dry pyridine or 1 drop of an ethereal solution of triethylamine, and warm on a water bath for 5 minutes. Extract the contents of the tube with boiling light petroleum (b.p. 80–100 °C or 100–120 °C) to separate any insoluble di-1-naphthyl urea. Recrystallise the crystals which separate on cooling from the same solvent.

The following alternative method may be used. Dissolve 0.01 mol of the phenol and 0.01 mol of 1-naphthyl isocyanate in 20 ml of light petroleum (b.p. 60–80 °C), add 2 drops of triethylamine (or, less satisfactorily, 2 drops of pyridine), reflux for 5 minutes and allow to crystallise. Filter off the crystalline solid through a sintered glass funnel.

(*h*) **2,4-Dinitrophenyl ethers.** 1-Chloro-2,4-dinitrobenzene reacts with the sodium salts of phenols to yield crystalline 2,4-dinitrophenyl ethers:

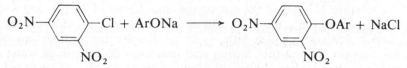

Dissolve 1 g (or 0.01 mol) of the phenol in a solution of 0.40 g of sodium hydroxide in 5 ml of water. Add the resulting solution to 2.0 g of 1-chloro-2,4-dinitrobenzene dissolved in 30 ml of 95 per cent ethanol; add more ethanol, if necessary, to effect solution. Heat the solution under reflux on a water bath until the colour (usually red) is discharged and a copious precipitate of sodium chloride appears (30–60 minutes). Dilute the reaction mixture with an equal volume of water, filter off the precipitated 2,4-dinitrophenyl ether, wash with water and recrystallise from ethanol.

Note. The chlorodinitrobenzene must be handled cautiously. If any touches the skin, wash it with industrial spirit and then copiously with water.

(*i*) **Pseudosaccharin ethers.** When pseudosaccharin chloride is heated with an excess of a phenol, *O*-aryl derivatives of saccharin are produced (compare Section **VII,6,**4(*h*)).

Heat 0.5 g of pseudosaccharin chloride with an excess of the phenol to 125–140 °C for 15–20 minutes; hydrogen chloride is evolved. Wash the product with dilute sodium hydroxide solution and then with water. Recrystallise the derivative from ethanol.

(*j*) **Bromo derivatives.** The presence of the hydroxyl group in phenols facilitates the substitution of the nuclear hydrogen atoms by halogen; the number and position of the substituent atoms vary with the nature of the phenol. This method is an indirect means of identification, as the formation of a substitution derivative is not a characteristic reaction of the phenol group but of the benzene nucleus. Phenol reacts with bromine to give 2,4,6-tribromophenol:

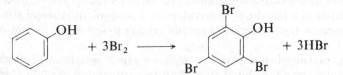

Bromo derivatives are often difficult to prepare, particularly in the case of polyhydroxy phenols which oxidise easily.

Dissolve 1.0 g of the compound in 10–15 ml of glacial acetic acid, cautiously add a solution of 3–4 ml of liquid bromine in 10–15 ml of glacial acetic acid until the colour of bromine persists and allow the mixture to stand for 15–20 minutes. Pour into 50–100 ml of water, filter off the bromo compound at the pump and wash with a little cold water. Recrystallise from dilute ethanol. Alternatively dissolve 1.0 g of the phenol in water, ethanol or acetone and add slowly, with constant shaking, just sufficient of a bromine solution (prepared by adding 5 g of bromine to a solution of 7.5 g of potassium bromide in 50 ml of water) to impart a yellow colour to the mixture. Allow to stand for 5 minutes. Add about 50 ml of water, and shake vigorously to break up any lumps. Filter and wash the bromo derivative with a dilute solution of sodium metabisulphite. Recrystallise from ethanol or from dilute ethanol.

The melting points of the derivatives of a number of selected phenols are collected in Table VIII,6.

VII,6,7. Enols β-Keto esters and some 1,3-diketones may be characterised by conversion into semicarbazones (see *Aldehydes and ketones*; Section **VII,6,***13*). Heating with an equivalent amount of phenylhydrazine often yields characteristic derivates. Thus β-keto esters afford 1-phenylpyrazalones (cf. Section **VI,6**), whilst 1,3-diketones yield 1-phenylpyrazoles (cf. Section **VI,5**).

Heat a mixture of 0.5 g of the β-keto ester and an equivalent amount of phenylhydrazine in an oil bath at 100–110 °C for 2 hours. Water and alcohol vapours are evolved. Cool and recrystallise the product from ethanol.

For 1,3-diketones, excellent results are obtained by refluxing the reactants in ethanolic solution for 2–3 hours; the product separates on cooling.

The physical properties as well as the melting points of the derivatives of a number of enols (β-keto esters and 1,3-diketones) are given in Table VIII,7.

VII,6,8. Aliphatic halogen compounds   (*a*) **Anilides and 1-naphthalides.**   The Grignard reagents prepared from alkyl halides react with phenyl isocyanate or with 1-naphthyl isocyanate to yield addition products that are converted by hydrolysis into anilides and naphthalides respectively:

$$Ar\cdot NCO + RMgX \longrightarrow Ar\cdot N = CR(OMgX) \xrightarrow{H_2O}$$

$$Ar\cdot N = CR(OH) \rightleftharpoons Ar\cdot NHCOR$$

Phenyl isocyanate is a colourless liquid, b.p. 164 °C or 55 °C/13 mmHg; its vapour is lachrymatory. The liquid reacts readily with water, yielding diphenyl urea, m.p. 238 °C, and hence must be protected from atmospheric moisture:

$$2C_6H_5\cdot NCO + H_2O \longrightarrow C_6H_5\cdot NHCONH\cdot C_6H_5 + CO_2$$

1-Naphthyl isocyanate, b.p. 269–270 °C or 153 °C/18 mmHg, is not quite so irritant and is somewhat more stable towards water (di-1-naphthyl urea has m.p. 297 °C). It is therefore to be preferred as a reagent; furthermore the 1-naphthalides are less soluble than the corresponding anilides.

In a small dry flask, fitted with a short reflux condenser and a calcium chloride guard-tube, place 0.4 g of dry magnesium turnings, a minute crystal of iodine and a solution of 1 ml (or 0.01 mol) of the alkyl halide in 10–15 ml of anhydrous ether. If the reaction does not start immediately (as indicated by the disappearance of the iodine colour), warm for a short period in a beaker of warm water; allow the reaction to proceed spontaneously, moderating it if necessary by immersing the flask in cold water. When the reaction has ceased, decant the nearly clear liquid from any solid material into another flask, and fit the reflux condenser into it. Add, portion-wise, through the condenser a solution of 0.5 ml of phenyl or 1-naphthyl isocyanate in 15 ml of anhydrous ether, shaking the flask after each addition. Allow the mixture to stand for 10 minutes and then add 30 ml of M-hydrochloric acid dropwise and with vigorous shaking and cooling in ice. (Alternatively, pour the reaction mixture cautiously into 20 ml of ice water containing 1 ml of concentrated hydrochloric acid, and shake the mixture well.) Transfer to a separatory funnel, shake well, then discard the lower aqueous layer. Dry the ethereal solution with a little magnesium sulphate and distil off the ether. Recrystallise the residue: methanol, ethanol, light petroleum, ether or hot water are suitable recrystallisation solvents.

If dry apparatus and dry reagents have not been used, diphenyl urea (m.p. 238 °C) or di-1-naphthyl urea (m.p. 297 °C) are obtained.

(*b*) **Alkyl mercury(II) halides.** Grignard reagents, prepared from alkyl halides, react with a mercury(II) halide that contains the *same halogen* as the reagent to form alkyl mercury(II) halides:

$$RMgX + HgX_2 \longrightarrow RHgX + MgX_2$$

The reaction is applicable to primary and secondary halides only; tertiary halides do not react.

Filter the Grignard solution, prepared as in (*a*), rapidly through a little glass wool into a test-tube containing 4–5 g of mercury(II) chloride, bromide or iodide, depending upon the halogen in the original alkyl halide. Shake the reaction mixture vigorously for a few minutes and then evaporate the ether. Boil the residue with 20 ml of rectified spirit, filter the solution, dilute it with 10 ml of distilled water, reheat to dissolve any precipitated solid and allow to cool. Recrystallise the alkyl mercury(II) halide from dilute ethanol.

(*c*) **S-Alkylisothiouronium picrates.** Alkyl bromides or iodides react with thiourea in ethanolic solution to produce S-alkylisothiouronium salts, which yield picrates of sharp melting point:

Alkyl chlorides react slowly and the yield of the derivative is poor. Tertiary halides give anomalous results.

Place a mixture of 0.5 g of finely powdered thiourea, 0.5 g of the alkyl halide and 5 ml of ethanol in a test-tube or small flask equipped with a reflux condenser. Reflux the mixture for a period depending upon the nature of the halide: primary alkyl bromides and iodides, 10–20 minutes (according to the molecular weight); secondary alkyl bromides or iodides, 2–3 hours; alkyl chlorides, 3–5 hours;* polymethylene dibromides or di-iodides, 20–50 minutes. Then add 0.5 g of picric acid, boil until a clear solution is obtained and cool. If no precipitate

---

* Alkyl chlorides often react more rapidly (50–60 minutes) upon adding 0.5 g of potassium iodide to the original reaction mixture, followed by sufficient water or ethanol to produce a clear solution at the boiling point. After refluxing, 0.5 g of picric acid is added, etc.

is obtained, add a few drops of water. Recrystallise the resulting S-alkyl iso-thiouronium picrate from ethanol.

(d) **Picrates of 2-naphthyl alkyl ethers.** Alkyl halides react with the sodium or potassium derivative of 2-naphthol in alcoholic solution to yield the corresponding alkyl 2-naphthyl ethers (which are usually low m.p. solids) and the latter are converted by picric acid into the crystalline picrates:

$$RX + 2\text{-}C_{10}H_7ONa \longrightarrow 2\text{-}C_{10}H_7OR + NaX$$

Mix together 1.0 g of pure 2-naphthol and the theoretical quantity of 50 per cent potassium hydroxide solution, add 0.5 g of the halide, followed by sufficient rectified spirit to produce a clear solution. For alkyl chlorides, the addition of a little potassium iodide is recommended. Heat the mixture under reflux for 15 minutes, and dissolve any potassium halide by the addition of a few drops of water. The 2-naphthyl ether usually crystallises out on cooling; if it does not, dilute the solution with 10 per cent sodium hydroxide solution until precipitation occurs. Dissolve the 2-naphthyl ether in the minimum volume of hot ethanol and add the calculated quantity of picric acid dissolved in hot ethanol. The picrate separates out on cooling. Recrystallise it from rectified spirit.

The 2-naphthyl ethers of methylene halides have m.p. 133 °C, of ethylene halides 217 °C and trimethylene halides 148 °C.

(e) **Di- and poly-halogenated aliphatic hydrocarbons.** No general procedure can be given for the preparation of derivatives of these compounds. Reliance must be placed upon their physical properties (b.p., density and refractive index) and upon any chemical reactions which they undergo.

Table VIII,10 deals with a number of aliphatic halogen compounds together with their crystalline derivatives.

VII,6,9. Aromatic halogen compounds (a) **Nitration products.** Although no general method of nitration can be given, the following procedure is widely applicable.

Add 1 g of the compound to 4 ml of concentrated sulphuric acid and cautiously introduce, drop by drop, 4 ml of fuming nitric acid. Warm the mixture on a water bath for 10 minutes, then pour it on to 25 g of crushed ice (or 25 ml of ice-water). Collect the precipitate by filtration at the pump, and recrystallise it from dilute ethanol.

Twenty per cent oleum may be substituted for the concentrated sulphuric acid for compounds which are difficult to nitrate.

(b) **Reaction with chlorosulphonic acid (chlorosulphonation). Sulphonamides.** Many aryl halides, either alone or in chloroform solution, when treated with excess of chlorosulphonic acid afford the corresponding sulphonyl chlorides in good yield (use the experimental details given in Section **VII,6,3**(a)); the latter may be readily converted into the aryl sulphonamides by reaction with concentrated ammonia solution or with solid ammonium carbonate.

The following give abnormal results when treated with chlorosulphonic acid alone, preferably at 50 °C for 30–60 minutes: fluorobenzene (4,4'-difluoro-biphenylsulphone, m.p. 98 °C); iodobenzene (4,4'-di-iodobiphenylsulphone, m.p. 202 °C); o-dichlorobenzene (3,4,3',4'-tetrachlorobiphenylsulphone, m.p. 176 °C); and o-dibromobenzene (3,4,3',4'-tetrabromobiphenylsulphone, m.p. 176–177 °C). The resulting sulphones may be crystallised from glacial acetic acid, benzene or ethanol, and are satisfactory for identification of the original aryl halide. In some cases sulphones accompany the sulphonyl chloride; they are readily separated from the final sulphonamide by their insolubility in cold 6 M-sodium hydroxide solution; the sulphonamides dissolve readily and are reprecipitated by 6 M-hydrochloric acid.

(c) **Oxidation of side chains.** The oxidation of halogenated toluenes and similar compounds and of compounds with side chains of the type $-CH_2Cl$ and $-CH_2OH$ proceeds comparatively smoothly with alkaline permanganate solution (for experimental details, see under *Aromatic hydrocarbons*, Section **VII,6,**3(d)). The resulting acid may be identified by a m.p. determination and by the preparation of suitable derivatives (see Section **VII,6,**15).

(d) **Picrates.** Some halogen derivatives of the aromatic hydrocarbons form picrates (for experimental details, see under *Aromatic hydrocarbons*, Section **VII,6,**3(e)), for example, 1-chloronaphthalene (m.p. 137 °C), 1-bromonaphthalene (m.p. 134 °C) and 2-bromonaphthalene (m.p. 86 °C).

The properties of a number of aromatic halogen compounds together with the melting points of their derivatives are collected in Table VIII,11.

VII,6,10. **Aliphatic ethers** The low reactivity of aliphatic ethers renders the problem of the preparation of suitable crystalline derivatives a somewhat difficult one. Increased importance is therefore attached to the physical properties (boiling point, density and refractive index). There are, however, two reactions based upon the cleavage of the ethers which are useful for characterisation.

(a) **Reaction with 3,5-dinitrobenzoyl chloride.** Ethers undergo cleavage with 3,5-dinitrobenzoyl chloride in the presence of zinc chloride:

$$ROR + (NO_2)_2C_6H_3COCl \xrightarrow{ZnCl_2} (NO_2)_2C_6H_3CO_2R + RCl$$

The resulting alkyl 3,5-dinitrobenzoate may be employed for the characterisation of the ether. The method is only applicable to symmetrical or simple ethers; a mixed aliphatic ether R^1OR^2 would yield a mixture of solid esters.

Add 1 ml of the ether to 0.1–0.15 g of finely powdered anhydrous zinc chloride and 0.5 g of pure 3,5-dinitrobenzoyl chloride (Section **VII,6,**4(b)) contained in a test-tube; attach a small water condenser and reflux gently for 1 hour. Treat the reaction product with 10 ml of 0.75 M-sodium carbonate solution, heat and stir the mixture for 1 minute upon a boiling water bath, allow to cool and filter at the pump. Wash the precipitate with 5 ml of 0.75 M-sodium carbonate solution and twice with 5 ml of ether. Dry on a porous tile or upon a pad of filter-paper. Transfer the crude ester to a test-tube and boil it with 10 ml of chloroform or carbon tetrachloride; filter the hot solution, if necessary. If the ester does not separate on cooling, evaporate to dryness on a water bath, and recrystallise the residue from 2–3 ml of either aqueous ethanol or light petroleum. Determine the melting point of the resulting 3,5-dinitrobenzoate (see Table VIII,4).

(b) **Cleavage of ethers with hydriodic acid.** Aliphatic ethers suffer fission when boiled with constant boiling point hydriodic acid:

$$R^1OR^2 + 2HI \longrightarrow R^1I + R^2I + H_2O$$

If the ether is a simple one ($R^1 = R^2$), the identification of the resulting alkyl iodide presents no difficulties. If, however, it is a mixed aliphatic ether, the separation of the two alkyl iodides by fractional distillation is generally difficult unless R^1 and R^2 differ considerably in molecular weight and sufficient material is available.

Reflux 1 ml of the ether with 5 ml of freshly distilled, constant boiling point hydriodic acid (Section II,2,26), b.p. 126–128 °C, for 2–3 hours. Add 10 ml of water, distil and collect about 7 ml of liquid. Decolourise the distillate by the addition of a little sodium metabisulphite, and separate the two layers by means of a dropper pipette. Determine the b.p. of the resulting iodide by the Siwoloboff method (Section I,35) and prepare a crystalline derivative (Section VII,6,8).

The physical properties of a number of aliphatic ethers are collected in Table VIII,12. Some related heterocyclic compounds are also included in this table.

VII,6,11. Aromatic ethers   Purely aromatic ethers (e.g., diphenyl ether), which are commonly encountered, are very limited in number. Most of the aromatic ethers are of the mixed aliphatic aromatic type. They are not attacked by sodium nor by dilute acids or alkalis. When liquid, the physical properties (b.p., $d_4^{20°}$ and $n_D^{20°}$) are useful constants to assist in their identification. Three important procedures are available for the characterisation of aromatic ethers.

(a) **Cleavage with hydriodic acid.**   Aromatic ethers undergo fission when heated with constant boiling point hydriodic acid:

$$ArOR + HI \longrightarrow ArOH + RI$$

The cleavage products are a phenol and an alkyl iodide, which will serve to characterise the ether.

Experimental details can easily be adapted from those given under *Aliphatic ethers*, Section VII,6,10(b).

To isolate the phenol, treat the residue in the flask with aqueous sodium carbonate until alkaline and extract the mixture with ether. Wash the ethereal extract with saturated aqueous sodium carbonate, and then with 2 M-sodium hydroxide solution. Acidify the sodium hydroxide solution (to Congo red paper) and extract the liberated phenol with ether. Identify the phenol as in Section VII,6,6.

(b) **Picrates.**   Ethers of many polynuclear aromatic systems are conveniently characterised as their picrates prepared by the method described for the corresponding derivates in *Aromatic hydrocarbons*, Section VII,6,3(e).

(c) **Derivatives by nuclear substitution.**   *Nitration.*   These may generally be prepared as detailed under *Aromatic hydrocarbons*, Section VII,6,3(c); the following experimental procedure for anisole may be regarded as typical. Add 0.5 g of anisole to a mixture of equal volumes of concentrated nitric acid and concentrated sulphuric acid keeping the temperature below 25 °C by cooling in an ice bath. Finally warm to 40 °C until dilution of a small portion with water gives a solid product. Pour the whole of the reaction mixture into water; collect the resulting 2,4-dinitroanisole and recrystallise from ethanol.

Bromination.   These may be prepared as described under *Phenols*, Section VII,6,6(j), using glacial acetic acid as solvent. In some cases carbon tetrachloride is a satisfactory solvent; the carbon tetrachloride is separated by distillation and the residue is recrystallised from dilute ethanol.

Formation of sulphonamides.   These may be prepared as described for *Aromatic hydrocarbons*, Section VII,6,3(a).

(*d*) **Oxidation of side chains.** General conditions for the oxidation of an alkyl side chain attached to an aromatic ring are given under *Aromatic hydrocarbons*, Section **VII,6,3**(*d*). The following procedure for the oxidation of *p*-cresyl methyl ether to anisic acid is illustrative.

Prepare a solution of 6 g of potassium permanganate in a mixture of 20 ml of 5 per cent sodium hydroxide solution and 150 ml of water, add 2.0 g of *p*-cresyl methyl ether and heat under reflux for 2–3 hours. If any permanganate remains at the end of this period, destroy it by the addition of a few drops of ethanol. Remove the precipitated manganese dioxide by filtration at the pump, evaporate the filtrate to a volume of 25–30 ml and acidify it (to Congo red) with dilute sulphuric acid. Anisic acid, m.p. 183–184 °C, crystallises out on cooling.

Table VIII,13 contains data referring to a number of selected aromatic ethers.

VII,6,*12*. Acetals Acetals are identified by reference to the alcohol and aldehyde (or ketone if a ketal) which they yield when hydrolysed in acid solution. Hydrolysis proceeds readily in dilute acid solution (e.g., with 3–5% acid):

$$R^1 \cdot CH(OR^2)_2 + H_2O \xrightarrow{H^{\oplus}} R^1 \cdot CHO + 2R^2OH$$

The rate of hydrolysis depends upon the solubility of the acetal in the hydrolysis medium. Acetals of low molecular weight are completely hydrolysed by refluxing for 5–10 minutes; those of higher molecular weight, and therefore of low solubility, may require 30–60 minutes, but the rate of hydrolysis may be increased by the addition of dioxan which increases the solubility of the acetal.

The experimental procedure to be followed depends upon the products of hydrolysis. If the alcohol and aldehyde are both soluble in water, the reaction product is divided into two parts. One portion is used for the characterisation of the aldehyde by the preparation of a suitable derivative (e.g., the 2,4-dinitrophenylhydrazone, semicarbazone or dimethone, see *Aldehydes and ketones*, Section **VII,6,*13***). The other portion is employed for the preparation of a 3,5-dinitrobenzoate, etc. (see *Alcohols and polyhydric alcohols*, Section **VII,6,4**): it is advisable first to concentrate the alcohol by distillation or to attempt to salt out the alcohol by the addition of solid potassium carbonate. If one of the hydrolysis products is insoluble in the reaction mixture, it is separated and characterised. If both the aldehyde and the alcohol are insoluble, they are removed from the aqueous layer; separation is generally most simply effected with sodium metabisulphite solution (compare Section **III,85**), but fractional distillation may sometimes be employed.

The formulae and physical properties of a number of common acetals are collected in Table VIII,14.

VII,6,*13*. Aldehydes and ketones (*a*) **2,4-Dinitrophenylhydrazones.** Small quantities may be prepared with the class reagent described on p. 1071. The following procedure is generally more satisfactory.

Suspend 0.25 g of 2,4-dinitrophenylhydrazine in 5 ml of methanol and add 0.4–0.5 ml of concentrated sulphuric acid cautiously. Filter the warm solution and add a solution of 0.1–0.2 g of the carbonyl compound in a small volume of methanol or of ether. If no solid separates within 10 minutes, dilute the solution carefully with *M*-sulphuric acid. Collect the solid by suction filtration and wash it with a little aqueous methanol. Recrystallise the derivative from ethanol, dilute ethanol, ethyl acetate, acetic acid, dioxan, nitromethane, nitrobenzene or xylene.

Alternatively, to the clear solution obtained by warming 0.5 g of 2,4-dinitrophenylhydrazine, 1 ml of concentrated hydrochloric acid and 8–10 ml of ethanol, add 0.25 g of the carbonyl compound and heat just to boiling. Allow to cool to room temperature, filter off the 2,4-dinitrophenylhydrazone and recrystallise it from ethanol or glacial acetic acid.

The following reagent, a **0.25 M solution of 2,4-dinitrophenylhydrazine**, may be used for the preparation of derivatives of keto compounds. Dissolve 25 g of 2,4-dinitrophenylhydrazine in 300 ml of 85 per cent phosphoric acid in a 600-ml beaker on a steam bath, dilute the solution with 200 ml of 95 per cent ethanol, allow to stand and filter through a sintered glass funnel. It must be emphasised that *this reagent is not suitable for the routine detection of carbonyl compounds* since it also gives a precipitate in the cold with certain amines, esters and other compounds: if, however, a dilute solution of the ketonic compound in ethanol is treated with a few drops of the reagent and the mixture diluted with water and heated, the precipitate produced with non-ketonic compounds generally dissolves.

For the preparation of 2,4-dinitrophenylhydrazones, dissolve the carbonyl compound (say, 0.5 g) in 5 ml of ethanol and add the calculated volume of the reagent. If a precipitate does not form immediately, dilute with a little water. Collect the derivative and recrystallise it as above.

(*b*) **p-Nitrophenylhydrazones.** Reflux a mixture of 0.5 g of *p*-nitrophenylhydrazine, 0.5 g of the aldehyde (or ketone), 10–15 ml of ethanol and 2 drops of glacial acetic acid for 10 minutes. Add more ethanol if the boiling solution is not homogeneous. Cool the clear solution, filter off the *p*-nitrophenylhydrazone and recrystallise it from ethanol or acetic acid.

Alternatively, dissolve approximately equivalent amounts of the aldehyde (or ketone) and the solid reagent in the minimum volume of cold glacial acetic acid, and reflux for 15 minutes. The *p*-nitrophenylhydrazone separates on cooling or upon careful dilution with water.

(*c*) **Phenylhydrazones.** Dissolve 0.5 g of *colourless* phenylhydrazine hydrochloride and 0.8 g of sodium acetate in 5 ml of water, and add a solution of 0.2–0.4 g of the aldehyde (or ketone) in a little ethanol (free from aldehydes and ketones). Shake the mixture until a clear solution is obtained and add a little more ethanol, if necessary. Warm on a water bath for 10–15 minutes and cool. Filter off the crystalline derivative, and recrystallise it from dilute ethanol or water; sometimes benzene or light petroleum (b.p. 60–80 °C) may be used.

(*d*) **Semicarbazones.** Dissolve 1 g of semicarbazide hydrochloride and 1.5 g of crystallised sodium acetate in 8–10 ml of water, add 0.5 g of the aldehyde or ketone and shake. If the mixture is turbid, add alcohol (acetone-free) or water until a clear solution is obtained; shake the mixture for a few minutes and allow to stand. Usually the semicarbazone crystallises from the cold solution on standing, the time varying from a few minutes to several hours. The reaction may be accelerated, if necessary, by warming the mixture on a water bath for a few minutes and then cooling in ice-water. Filter off the crystals, wash with a little cold water and recrystallise from water or from methanol or ethanol either alone or diluted with water.

Note. When semicarbazide is heated in the absence of a carbonyl compound for long periods, condensation to **biurea**, $NH_2CONH \cdot NHCONH_2$, m.p. 247–250 °C (decomp.), may result; occasionally this substance may be produced in the normal preparation of a semicarbazone that forms slowly. Biurea is sparingly soluble in alcohol and soluble in hot water, whereas semicarbazones with melting points in the same range are insoluble in water: this enables it to be readily distinguished from a semicarbazone.

(*e*) **Oximes.** The method given for semicarbazones (see (*d*)) may be employed: use 1 g of hydroxylamine hydrochloride, 2 g of crystallised sodium acetate and 0.5 g of the aldehyde or ketone. It is usually advisable to warm on a water bath for 10 minutes.

For water-insoluble aldehydes or ketones, the following alternative procedure may be used. Reflux a mixture of 0.5 g of the aldehyde or ketone, 0.5 g of hydroxylamine hydrochloride, 5 ml of ethanol and 0.5 ml of pyridine on a water bath for 15–60 minutes. Remove the ethanol either by distillation (water bath) or by evaporation of the hot solution in a stream of air (water pump). Add 5 ml of water to the cooled residue, cool in an ice bath and stir until the oxime crystallises. Filter off the solid, wash it with a little water and dry. Recrystallise from ethanol (95% or more dilute), benzene or benzene–light petroleum (b.p. 60–80 °C).

Note. All aldehydes, and also those ketones which have two different groups attached to the carbonyl grouping, are capable of yielding two stereoisomeric oximes, hydrazones or semicarbazones. As a general rule, however, one of the stereoisomerides is formed in much greater amount than the other, and no doubt therefore arises as to the purity of the ketonic compound under investigation; occasionally a mixture of stereoisomerides is obtained, which may be difficult to separate by recrystallisation. The formation, therefore, of one of the above derivatives of indefinite melting point and obvious heterogeneity does not necessarily imply the presence of an impure ketonic substance.

(*f*) **Dimedone derivatives** (aldehydes only). Dimedone or 5,5-dimethyl-cyclohexane-1,3-dione in saturated aqueous solution or in 10 per cent alcoholic solution gives crystalline derivatives (I) with aldehydes, but not with ketones. The reaction is:

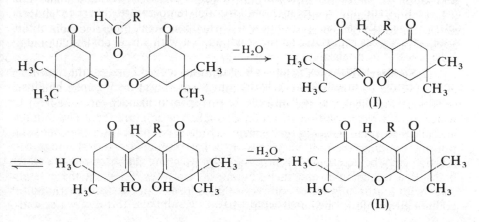

The condensation products (conveniently called alkylidene dimethones) are almost insoluble in water, but can be crystallised from *dilute* ethanol. Dimedone is therefore a good reagent for the detection and characterisation of aldehydes.

The alkylidene dimethone (I) upon boiling with glacial acetic acid, acetic anhydride, hydrochloric acid and other reagents frequently loses water and passes into the anhydride, or dimethone anhydride (II) (a substituted octahydroxanthene) which often serves as another derivative. The derivatives (I) are soluble in dilute alkali and the resulting solutions give colourations with ferric chloride solution; on the other hand, the anhydrides (II) are insoluble in dilute alkali and hence can easily be distinguished from the alkylidene dimethones (I).

Add 0.1 g of the aldehyde in 5 ml of 50 per cent ethanol to 2 ml of a 10 per cent or saturated ethanolic solution of dimedone. If a precipitate does not form immediately, warm for 5 minutes; if the solution is still clear at the end of this period, add hot water until the mixture is just cloudy and cool to about 5 °C. Collect the crystalline derivative and recrystallise it from methanol–water or ethanol–water.

To prepare the anhydride, boil a solution of 0.1 g of the dimethone derivative (I) in 5 ml of 80 per cent ethanol to which 1 drop of concentrated hydrochloric acid has been added, for 5 minutes, then add hot water until the mixture is just turbid, cool and collect the anhydride by filtration. Recrystallise it from dilute methanol.

(g) **Benzylidene derivatives.** Compounds containing the ketomethylene group ($-CH_2 \cdot CO$) react with benzaldehyde to yield benzylidene derivatives:

$$R^1 \cdot CO \cdot CH_2 \cdot R^2 + C_6H_5 \cdot CHO \longrightarrow R \cdot CO \cdot C(=CH \cdot C_6H_5)R^2 + H_2O$$

$$R^1 \cdot CH_2 \cdot CO \cdot CH_2 \cdot R^2 + 2C_6H_5 \cdot CHO \longrightarrow$$
$$R^1 \cdot C(=CH \cdot C_6H_5) \cdot CO \cdot C(=CH \cdot C_6H_5)R^2 + 2H_2O$$

Cyclic ketones yield dibenzylidene derivatives.

Dissolve 1 g of the ketomethylene compound and 1.1 g or 2.2 g of pure benzaldehyde (according as to whether the compound may be regarded as $R \cdot CO \cdot CH_2 \cdot R$ or as $R \cdot CH_2 \cdot CO \cdot CH_2 \cdot R$) in about 10 ml of rectified (or industrial) spirit, add 0.5 ml of 5 M-sodium hydroxide solution, shake and allow the mixture to stand for about an hour at room temperature. The benzylidene derivative usually crystallises out or will do so upon 'scratching' the walls of the vessel with a glass rod. Filter off the solid, wash it with a little cold ethanol and recrystallise it from absolute ethanol.

(h) **Azines.** Aldehydes react with hydrazine to yield azines: the reaction cannot usually be arrested at the hydrazone stage. This reaction may be illustrated by the preparation of benzylideneazine from benzaldehyde:

$$2C_6H_5 \cdot CHO + N_2H_4, H_2SO_4 \xrightarrow[\text{NH}_3]{\text{aq.}}$$

$$C_6H_5 \cdot CH = N - N = CH \cdot C_6H_5 + (NH_4)_2SO_4 + 2H_2O$$

Stir a mixture of 2.4 g of powdered hydrazine sulphate, 18 ml of water and

2.4 ml of concentrated aqueous ammonia (d 0.88), and add 4.6 g (4.4 ml) of benzaldehyde (free from benzoic acid) dropwise, with stirring, over a period of 30–60 minutes. Stir the mixture for a further hour, collect the solid by suction filtration and wash it with water. Recrystallise from 8 ml of rectified spirit. The yield of benzylideneazine (yellow needles), m.p. 92–93 °C, is 3.6 g.

The physical constants of the various derivatives of aliphatic and aromatic aldehydes and ketones are given in Tables VIII,15; VIII,16; VIII,17 and VIII,18.

VII,6,*14*. Quinones (*a*) **Reduction to the hydroquinone.** Dissolve, or suspend, 0.5 g of the quinone in 5 ml of ether or benzene and shake vigorously with a solution of 1.0 g of sodium dithionite ($Na_2S_2O_4$) in 10 ml of *M*-sodium hydroxide until the colour of the quinone has disappeared. Separate the alkaline solution of the hydroquinone, cool it in ice and acidify with concentrated hydrochloric acid. Collect the product (extract with ether, if necessary) and recrystallise it from ethanol or water.

(*b*) **Reductive acetylation.** Suspend 0.5 g of the quinone in 2.5 ml of pure acetic anhydride, and add 0.5 g of zinc powder and 0.1 g of powdered, anhydrous sodium acetate. Warm the mixture gently until the colour of the quinone has largely disappeared and then boil for 1 minute. Add 2 ml of glacial acetic acid and boil again to dissolve the product and part of the precipitated zinc acetate. Decant the hot solution from the zinc acetate and zinc, and wash the residue with 3–4 ml of hot glacial acetic acid. Combine the solutions, heat to boiling, carefully add sufficient water to hydrolyse the acetic anhydride and to produce a turbidity. Cool the mixture in ice, filter off the diacetate of the hydroquinone and recrystallise it from dilute ethanol or from light petroleum.

(*c*) **Thiele acetylation.** Quinones, when treated with acetic anhydride in the presence of perchloric acid or of concentrated sulphuric acid (strong acid catalyst), undergo simultaneous reductive acetylation and substitution to yield triacetoxy derivatives, e.g., benzoquinone gives 1,2,4-triacetoxybenzene.

Add 0.1 ml of concentrated sulphuric acid or of 72 per cent perchloric acid cautiously to a cold solution of 0.01 mol (or 1.0 g) of the quinone in 3–5 ml of acetic anhydride. Do not permit the temperature to rise above 50 °C. Allow to stand for 15–30 minutes and pour into 15 ml of water. Collect the precipitated solid and recrystallise it from ethanol.

(*d*) **Semicarbazones.** The preparation of these derivatives is described on p. 1112.

(*e*) **Quinoxalines.** The preparation of these derivatives is described on p. 916.

The melting points of the derivatives of a selection of quinones are collected in Table VIII,19.

VII,6,*15*. Carboxylic acids (*a*) **Amides, anilides and *p*-toluidides.** The dry acid is first converted by excess of thionyl chloride into the acid chloride:

$$R{\cdot}CO_2H + SOCl_2 \longrightarrow R{\cdot}COCl + SO_2 + HCl$$

The by-products are both gaseous and the excess of thionyl chloride (b.p. 78 °C) may be readily removed by distillation. Interaction of the acid chloride with ammonia solution, aniline or *p*-toluidine yields the amide, anilide or *p*-toluidide respectively:

$$R \cdot COCl + 2NH_3 \longrightarrow R \cdot CONH_2 + NH_4Cl$$
$$R^1 \cdot COCl + 2R^2NH_2 \longrightarrow R^1 \cdot CONHR^2 + R^2NH_2,HCl$$

Place 0.5–1.0 g of the dry acid (finely powdered if it is a solid) into a 25-ml flask fitted with a reflux condenser, add 2.5–5.0 ml of redistilled thionyl chloride and reflux gently for 30 minutes; it is advisable to place a plug of cotton wool* in the top of the condenser to exclude moisture. Rearrange the condenser and distil off the excess of thionyl chloride† (b.p. 78 °C). The residue in the flask consists of the acid chloride and can be converted into any of the derivatives given below.

(i) **Amides.** Treat the acid chloride cautiously with about 20 parts of concentrated ammonia solution (*d* 0.88) and warm for a few moments. If no solid separates on cooling, evaporate to dryness on a water bath. Recrystallise the crude amide from water or dilute ethanol.

Alternatively stir the acid chloride with an equivalent weight of ammonium acetate in 10 ml of acetone at room temperature for one hour, filter the mixture and evaporate the acetone, and crystallise the residual amide from water or from dilute ethanol.

(ii) **Anilides.** Dilute the acid chloride with 5 ml of pure ether (or benzene), and add a solution of 2 g of pure aniline in 15–20 ml of the same solvent until the odour of the acid chloride has disappeared; excess of aniline is not harmful. Shake with excess of dilute hydrochloric acid to remove aniline and its salts, wash the ethereal (or benzene) layer with 3–5 ml of water and evaporate the solvent [**CAUTION!**]. Recrystallise the anilide from water, dilute ethanol or benzene–light petroleum (b.p. 60–80 °C).

p-**Bromoanilides** are similarly prepared with *p*-bromoaniline.

(iii) *p*-**Toluidides.** Proceed as under (ii), but substitute *p*-toluidine for aniline.

Anilides and *p*-toluidides may also be *prepared directly from the acids*‡ by heating them with aniline or *p*-toluidine respectively:

$$R^1 \cdot CO_2H + R^2NH_2 \longrightarrow R^1 \cdot CONHR^2 + H_2O$$

Place 1.0 g of the monobasic acid and 2 g of aniline or *p*-toluidine in a dry test-tube, attach a short air condenser and heat the mixture in an oil bath at 140–160 °C for 2 hours: do not reflux too vigorously an acid that boils below this temperature range and only allow steam to escape from the top of the condenser. For a sodium salt, use the proportions of 1 g of salt to 1.5 g of the base. If the acid is dibasic, employ double the quantity of amine and a reaction temperature of 180–200 °C: incidentally, this procedure is recommended for

---

* This is more convenient than the conventional calcium chloride guard-tube and possesses the advantage of cheapness and hence can easily be renewed for each experiment.

† If the boiling point of the acid chloride is too near that of thionyl chloride to render separation by distillation practicable, the excess of the reagent can be destroyed by the addition of pure formic acid:

$$HCO_2H + SOCl_2 \longrightarrow CO + SO_2 + 2HCl$$

‡ Alternatively, the alkali metal salts of the acids may be heated with the hydrochloride of the appropriate base.

dibasic acids since the latter frequently give anhydrides with thionyl chloride. Powder the cold reaction mixture, triturate it with 20–30 ml of 10 per cent hydrochloric acid* and recrystallise from dilute ethanol.

(*b*) **p-Bromophenacyl esters.** *p*-Bromophenacyl bromide reacts with the alkali metal salts of acids to form crystalline *p*-bromophenacyl esters:

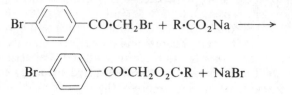

Dissolve or suspend 0.5 g of the acid in 5 ml of water in a small conical flask, add a drop or two of phenolphthalein indicator, and then 4–5 per cent sodium hydroxide solution until the acid is just neutralised. Add a few drops of very dilute hydrochloric acid so that the final solution is *faintly* acid (litmus).† Introduce 0.5 g of *p*-bromophenacyl bromide (m.p. 109 °C) dissolved in 5 ml of rectified (or industrial) spirit, and heat the mixture under reflux for 1 hour: if the mixture is not homogeneous at the boiling point or a solid separates out, add just sufficient ethanol to produce homogeneity. [Di- and tri-basic acids require proportionately larger amounts of the reagent and longer refluxing periods.] Allow the solution to cool, filter the separated crystals at the pump, wash with a little alcohol and then with water. Recrystallise from dilute ethanol: dissolve the solid in hot ethanol, add hot water until a turbidity just results, clear the latter with a few drops of ethanol and allow to cool. Acetone may sometimes be employed for recrystallisation.

(*c*) **p-Nitrobenzyl esters.** *p*-Nitrobenzyl bromide (m.p. 100 °C) reacts with the alkali metal salts of acids to give *p*-nitrobenzyl esters:

$$O_2N-\langle\bigcirc\rangle-CH_2Br + R\cdot CO_2Na \longrightarrow$$

$$O_2N-\langle\bigcirc\rangle-CH_2\cdot O_2C\cdot R + NaBr$$

It is important that the solution of the sodium salt be *faintly* acid in order that the formation of coloured by-products in the subsequent reaction may be prevented. If the molecular weight of the monobasic acid is known, it is desirable to employ a slight excess of the sodium salt, since excess of the latter is more easily removed than the unchanged reagent.

Use the procedure given under (*b*) for *p*-bromophenacyl esters. If the

---

* When the derivative is appreciably soluble in ether, the following alternative procedure may be employed. Dissolve the cold reaction mixture in about 50 ml of ether, wash it with 20–30 ml of 10 per cent hydrochloric acid (to remove the excess of base), followed by 20 ml of 10 per cent sodium hydroxide solution, separate the ether layer and evaporate the solvent [**CAUTION!**]. Recrystallise the residue from dilute ethanol.

† If the sodium salt of the acid is available, dissolve 0.5 g in 5 ml of water, add a solution of 0.5 g of the reagent in 5 ml of ethanol and proceed as detailed in the text after just acidifying (litmus) with dilute hydrochloric acid.

ester does not crystallise out on cooling, reheat the reaction mixture, and add small portions of hot water to the point of incipient cloudiness and allow to cool.

(d) **p-Phenylphenacyl esters.** p-Phenylphenacyl bromide reacts with soluble salts of organic acids to yield crystalline p-phenylphenacyl esters:

$$p\text{-}C_6H_5 \cdot C_6H_4 \cdot CO \cdot CH_2Br + NaO_2C \cdot R \longrightarrow$$
$$p\text{-}C_6H_5 \cdot C_6H_4 \cdot CO \cdot CH_2O_2C \cdot R + NaBr$$

The procedure is similar to that given under (b) and (c) above. Add a weighed amount of acid (0.005 mol) to 5 ml of water in a small conical flask and neutralise it with 0.5 M-sodium carbonate or M-sodium hydroxide. The final solution should be faintly acid to litmus (add more of the organic acid or a few drops of dilute hydrochloric acid); unless this precaution is taken, coloured by-products are formed which are very difficult to remove. (If the alkali metal salt is available, dissolve 0.005 mol in 5 ml of water, and render the solution just acid to litmus by the addition of dilute hydrochloric acid.) Introduce 10 ml of ethanol, and if the salt of the organic acid is not thrown out of solution, add 0.005 mol of p-phenylphenacyl bromide:* reflux the mixture for periods up to 1, 2 or 3 hours according to the basicity of the acid. If the salt of the organic acid is precipitated by the ethanol, add more water until the salt dissolves. Some of the esters are sparingly soluble in the reaction mixture and crystallise from the boiling solution; in most cases, however, crystal formation does not occur until the mixture is cooled. In some instances it may be necessary to concentrate the solution before crystallisation occurs. Recrystallise the crude p-phenylphenacyl ester from ethanol, dilute ethanol, acetone or benzene.

Certain dibasic acids, of which the sodium or potassium salts are sparingly soluble in dilute ethanol, cause difficulty; these should be neutralised with ethylamine solution.

(e) **S-Benzylisothiouronium salts.** S-benzylisothiouronium chloride reacts with the alkali metal salts of organic acids to produce crystalline S-benzylisothiouronium salts:

$$\left[H_5C_6 \cdot CH_2 \cdot S \cdot C {\overset{\displaystyle \nearrow NH_2}{\underset{\displaystyle \searrow NH_2}{}}} \right]^{\oplus} Cl^{\ominus} + R \cdot CO_2Na^{\oplus} \longrightarrow$$

$$\left[H_5C_6 \cdot CH_2 \cdot S \cdot C {\overset{\displaystyle \nearrow NH_2}{\underset{\displaystyle \searrow NH_2}{}}} \right]^{\oplus} R \cdot CO_2^{\ominus} + NaCl$$

It is important not to allow the reaction mixture to become appreciably alkaline, since the free base then decomposes rapidly yielding phenylmethanethiol, which has an unpleasant odour.

Dissolve (or suspend) 0.25 g of the acid in 5 ml of warm water, add a drop or two of phenolphthalein indicator and neutralise carefully with ca. M-sodium hydroxide solution. Then add 2–3 drops of ca. 0.1 M-hydrochloric acid to ensure that the solution is almost neutral (*pale* pink colour). (Under alkaline

---

* Dibasic and tribasic acids will require 0.01 and 0.015 mol respectively.

conditions the reagent tends to decompose to produce the evil-smelling phenyl-methanethiol.) If the sodium salt is available, dissolve 0.25 g in 5 ml of water and add 2 drops of *ca.* 0.1 *M*-hydrochloric acid. Introduce a solution of 1 g of *S*-benzylisothiouronium chloride in 5 ml of water, and cool in ice until precipitation is complete. Recrystallise the crude derivative from dilute ethanol or from hot water.

With some acids (e.g., succinic acid and sulphanilic acid) more satisfactory results are obtained by reversing the order of mixing, i.e., by adding the solution of the sodium salt of the acid to the reagent. In view of the proximity of the melting points of the derivatives of many acids, the mixed m.p. test (Section **I,34**) should be applied.

(*f*) **Anhydrides.** 1,2-Dicarboxylic acids are readily converted into cyclic anhydrides when heated alone or in acetic anhydride. Heat 0.5 g of the acid in 2–3 ml of refluxing acetic anhydride for 30 minutes. Remove most of the excess reagent by distillation and crystallise the residual cyclic anhydride from chloroform or benzene. The melting points of aliphatic and aromatic anhydrides are to be found in Tables VIII,23 and VIII,24 respectively. For conversion of the anhydrides into anilic acids see Section **VII,6,**16(*c*).

(*g*) **Fusion with soda-lime.** An additional useful test for aromatic carboxylic acids is to distil the acid or its sodium salt with soda-lime. Heat 0.5 g of the acid or its sodium salt with 0.5 g of soda-lime in an ignition tube to make certain that there is no explosion. Then grind together 0.5 g of the acid with 3 g of soda-lime, place the mixture in a Pyrex test-tube and cover it with an equal bulk of soda-lime. Fit a wide delivery tube dipping into an empty test-tube. Clamp the tube near the mouth. Heat the soda-lime first and then the mixture gradually to a dull-red heat. Examine the product: this may consist of aromatic hydrocarbons or derivatives, e.g., phenol from salicylic acid, anisole from anisic acid, toluene from toluic acid, etc.

The melting points of the derivatives of aliphatic and aromatic carboxylic acids are collected in Tables VIII,20 and VIII,21.

VII,6,16. **Carboxylic acid chlorides and anhydrides** (*a*) **Hydrolysis to the acids.** A general procedure is to hydrolyse the acid chloride (or anhydride) by warming with dilute alkali and acidifying the resulting solution with dilute hydrochloric acid to Congo red. If the acid is sparingly soluble, filter it off and characterise it in the usual way. If no precipitate of carboxylic acid is obtained, adjust the pH of the solution to neutrality to phenolphthalein and evaporate to dryness. Use the mixture of the sodium salt of the acid and sodium chloride thus obtained for the preparation of a suitable derivative (e.g., the *p*-bromophenacyl ester).

(*b*) **Conversion into anilides.** Acid chlorides are converted directly to the corresponding anilides by reaction with aniline as described in Section **VII,6,**15(*a*(ii)). For anhydrides, heat a mixture of 1 g of the anhydride and 1 g of aniline in a boiling water bath for 5 minutes, add 5 ml of water, boil and cool. Crystallise the resulting product from water or from aqueous ethanol.

(*c*) **Anilic acids from cyclic anhydrides.** Dissolve 0.5 g of the anhydride in 15 ml of benzene by heating on a water bath, and add a solution of 0.5 ml of aniline in 3 ml of benzene. If the anilic acid does not separate after a short time, cool the solution, wash it with a little dilute hydrochloric acid to remove the excess of aniline and evaporate the solvent; the anilic acid will then usually crystallise. Recrystallise from aqueous ethanol. When heated above their melting points, anilic acids dehydrate to form cyclic imides, e.g.,

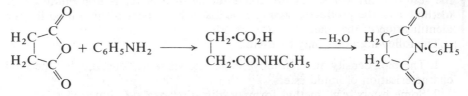

The melting points of the derivatives of some dicarboxylic acids are:

    Succinic; anilic acid 148 °C, imide 156 °C.
    Phthalic; anilic acid 169 °C, imide 205 °C.
    3-Nitrophthalic;   —   , imide 136 °C.

The physical properties of a number of carboxylic acid chlorides and anhydrides are given in Tables VIII,22, VIII,23 and VIII,24.

VII,6,17. Esters (*a*) **Hydrolysis.** Esters are usually characterised by hydrolysis followed by identification of the alcoholic and acidic components, although it is possible to prepare derivatives of both these components directly from the ester (see (*c*) and (*d*) below).

In the routine examination of esters it is often a good plan to carry out two hydrolyses, one for the isolation and characterisation of the parent acid, and the other for the isolation and identification of the parent alcohol.

1. Drop 1 g of sodium into 10 ml of methanol in a small flask provided with a small water condenser; heat the mixture until all the sodium has dissolved. Cool, and add 1 g of the ester and 0.5 ml of water. Frequently the sodium salt of the acid will be deposited either at once or after boiling for a few minutes. If this occurs, filter off the solid at once, wash it with a little methanol and convert it into the *p*-bromophenacyl ester, *p*-nitrobenzyl ester or *S*-benzyl isothiouronium salt (for experimental details, see Section **VII,6,15**). If no solid separates, continue the boiling for 30–60 minutes, boil off the alcohol, allow to cool, render the product just neutral to phenolphthalein with dilute sulphuric or hydrochloric acid, convert the sodium salt present in solution into a crystalline derivative (Section **VII,6,15**) and determine its melting point.

2. Boil 2 g of the ester with 30 ml of 10 per cent sodium or potassium hydroxide solution under reflux for at least 1 hour. If the alcohol formed is water (or alkali) soluble, the completion of the hydrolysis will be indicated by the disappearance of the ester layer. Distil off the liquid through the same condenser and collect the first 3–5 ml of distillate. If a distinct layer separates on standing (or upon saturation of half the distillate with potassium carbonate), remove this layer with a capillary dropper, dry it with a little anhydrous potassium carbonate or anhydrous calcium sulphate and determine the b.p. by the Siwoloboff method (Section **I,35**). Whether an insoluble alcohol separates out or not, prepare a crystalline derivative (e.g., the 3,5-dinitrobenzoate, Section **VII,6,4**(*b*)) and determine its m.p.

The residue in the flask will contain the sodium (or potassium) salt of the acid together with excess of alkali. Just acidify with dilute sulphuric acid and observe whether a crystalline acid separates; if it does, filter, recrystallise and identify (Section **VII,6,15**). If no crystalline solid is obtained, the solution may be just neutralised to phenolphthalein and the solution of the alkali salt used for the preparation of a crystalline derivative. This will confirm, if necessary, the results of hydrolysis by method 1. If the time factor is important, either method 1 or the product of the caustic alkali hydrolysis may be used for the identification of the acid.

The following notes may be useful:

1. The b.p., density and refractive index are valuable constants for the final characterisation of liquid esters.

2. Some esters, e.g., methyl formate, dimethyl oxalate, dimethyl succinate, dimethyl and diethyl tartrate, are appreciably soluble in water. These are usually easily hydrolysed by alkali.

3. Of the common esters, dimethyl oxalate (solid, m.p. 54 °C) and diethyl oxalate (liquid) give amides almost immediately upon shaking with concentrated ammonia solution. The resulting oxamide, m.p. 417 °C, is valueless as a derivative. The esters may, however, be easily hydrolysed and identified as above.

4. If the original ester is a fat or oil and produces an odour of acrolein when heated, it may be a **glyceride**. Esters of ethylene glycol and of glycerol with simple fatty acids are viscous and of high b.p. They are hydrolysed (method 1) and the methanol distilled off. The residue is diluted (a soap may be formed) and acidified with hydrochloric acid (Congo red paper). The acid is filtered or extracted with ether. If no acid can be isolated by these methods, it must be simple and volatile, and should be separated by distillation. The residual aqueous solution of glycol or glycerol is neutralised, evaporated to a syrup on a water bath and extracted with ethanol or with ethyl acetate; the solvent is evaporated and the glycol or glycerol in the residue is identified as usual.

5. **β-Keto esters** (e.g., ethyl acetoacetate) are soluble in solutions of caustic alkalis but not in sodium carbonate solution. They give colours with freshly prepared ferric chloride solution; a little ethanol should be added to bring the ester into solution. Sodium ethoxide solution reacts to yield sodio compounds, which usually crystallise out in the cold. They are hydrolysed by boiling sulphuric acid to the corresponding ketones, which can be identified as usual (Section **VII,6,**13).

6. **Unsaturated esters** decolourise a solution of bromine in carbon tetrachloride and also neutral potassium permanganate solution.

A slight modification in the procedure for isolating the products of hydrolysis is necessary for phenolic (or phenyl) esters since the alkaline solution will contain both the alkali phenate and the alkali salt of the organic acid: upon acidification, both the phenol and the acid will be liberated. Two methods may be used for **separating the phenol and the acid**:

(i) Acidify the cold alkaline reaction mixture with dilute sulphuric acid (use litmus or Congo red paper) and extract both the acid and the phenol with ether. Remove the acid by washing the ethereal extract with saturated sodium hydrogen carbonate solution until effervescence ceases; retain the aqueous washings. Upon evaporating the ether, the phenol remains; it may be identified (a) by its action upon ferric chloride solution, (b) the formation of a crystalline derivative with bromine water and (c) by any of the methods given in Section **VII,6,**6. Acidify the aqueous washings with dilute sulphuric acid whilst stirring steadily, and investigate the organic acid (Section **VII,6,**15).

(ii) Add dilute sulphuric acid, with stirring, to the cold alkaline solution until the solution is acid to litmus or Congo red paper, and the acid, if a solid, commences to separate as a faint permanent precipitate. Now add dilute sodium carbonate solution until the solution is alkaline (litmus paper) and any precipitate has completely redissolved. Extract the clear solution twice with ether: evaporate or distil the ether from the ethereal solution on a water bath, and identify the residual phenol as under (i). Remove the dissolved ether from the aqueous solution by boiling, acidify with dilute sulphuric acid and identify the organic acid present (see Section **VII,6,***15*).

(*b*) **The determination of the saponification equivalent of an ester.** The **saponification equivalent** or the **equivalent weight of an ester** is that weight in grams of the ester from which one equivalent weight of acid is obtainable by hydrolysis, *or* that quantity which reacts with one equivalent of alkali. The saponification equivalent is determined in practice by treating a known weight of the ester with a known quantity of caustic alkali used in excess. The residual alkali is then readily determined by titration of the reaction mixture with a standard acid. The amount of alkali that has reacted with the ester is thus obtained: the equivalent can then be readily calculated. It must, however, be borne in mind that certain structures may effect the values of the equivalent: thus aliphatic halogenated esters may consume alkali because of hydrolysis of part of the halogen during the determination, nitro esters may be reduced by the alkaline hydrolysis medium, etc.

In order to achieve complete hydrolysis rapidly, a solution of potassium hydroxide in diethylene glycol is preferred in place of an aqueous alkaline medium.

The **reagent is prepared** by weighing about 6.0 g of AnalaR potassium hydroxide pellets into a 50- or 100-ml flask, adding 30 ml of diethylene glycol and heating to effect solution; it is essential to use a thermometer for stirring and to keep the temperature below 130 °C, otherwise a dark yellow colour will develop. As soon as the solid has dissolved, the warm solution is poured into 70 ml of diethylene glycol in a glass-stoppered bottle. The solution is thoroughly mixed and allowed to cool. It is *ca.* 1.0 *M* and is standardised by pipetting 10 ml into a flask, adding 15 ml of water and titrating with standardised 0.25 *M*- or 0.5 *M*-hydrochloric acid using phenolphthalein as indicator. (Because of the high viscosity of the solution, it is advisable to open the tip of the pipette to an internal diameter of 2–3 mm in order to facilitate drainage; the pipette should be recalibrated before use.)

To determine the saponification equivalent of an ester transfer 10 ml of the reagent by means of a pipette into a 50-ml glass-stoppered Pyrex conical flask. Place the sample of the ester in a weight burette or in a weighing bottle fitted with a cork carrying a small dropper pipette; transfer about 0.5 g of the ester, accurately weighed, into the Erlenmeyer flask and insert the ground stopper. Mix the ester with the reagent by a rotary motion of the flask. Hold the stopper firmly in place and heat the mixture in an oil bath so that a temperature of 70–80 °C is reached within 2–3 minutes: agitate the liquid by a whirling motion during the heating. At this point remove the flask from the heating bath, shake the flask vigorously, allow to drain and loosen the stopper *carefully* to allow air to escape. Replace the stopper and heat again in an oil bath to 120–130 °C. (For esters of very high boiling point, the stopper may be removed and a thermometer inserted.) After 3 minutes at this temperature, cool the flask and its contents to 80–90 °C, remove the stopper and wash it with distilled water so that the rinsings drain into the flask. Add about 15 ml of distilled water and a drop or two of phenolphthalein indicator, mix well and then titrate with standard 0.25 *M*- or 0.5 *M*-hydrochloric acid. Calculate the saponification equivalent from the expression:

$$\text{Saponification equivalent} = \frac{\text{Weight of ester} \times 1000}{\text{Ml of } M\text{-KOH used}}$$

(*c*) **Direct identification of the alcoholic component of an ester.** The alcohol components of many simple esters may be identified as the crystalline 3,5-dinitrobenzoates (compare Section **VII,6,**4(*b*)) by heating them with 3,5-dinitrobenzoic acid in the presence of a little concentrated sulphuric acid:

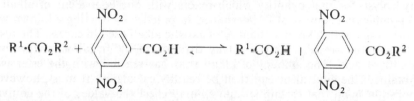

The reaction does not appear to be applicable if either of the groups R^1 or R^2 react readily with concentrated sulphuric acid; esters of molecular weight in excess of about 250 react with difficulty.

Dissolve 2 drops of concentrated sulphuric acid in 2 ml of the ester and add 1.5 g of 3,5-dinitrobenzoic acid. If the b.p. of the ester is below 150 °C, reflux the mixture gently; if the b.p. is above 150 °C heat the mixture, with frequent shaking at first, in an oil bath at about 150 °C. If the 3,5-dinitrobenzoic acid dissolves within 15 minutes, heat the mixture for 30 minutes, otherwise 60 minutes heating is required. Allow the reaction mixture to cool, dissolve it in 25 ml of ether and extract thoroughly with 5 per cent sodium carbonate solution (*ca.* 25 ml). Wash the ethereal solution with water, and remove the ether. Dissolve the residue (which is usually an oil) in 5 ml of hot ethanol, add hot water cautiously until the 3,5-dinitrobenzoate commences to separate, cool and stir. Recrystallise the derivative from dilute ethanol: the yield is 0.1–0.2 g.

Note. Successful results have been obtained (Renfrow and Chaney, 1946) with ethyl formate; methyl, ethyl, propyl, isopropyl, butyl, isobutyl, *S*-butyl and isopentyl acetates; ethylene glycol diacetate; ethyl monochloro- and trichloro-acetates; methyl, propyl, octyl and dodecyl propionates; ethyl butyrate; butyl and pentyl valerates; ethyl laurate; ethyl lactate; ethyl acetoacetate; diethyl carbonate; dimethyl and diethyl oxalates; diethyl malonate; diethyl adipate; dibutyl tartrate; ethyl phenylacetate; methyl and ethyl benzoates, methyl and ethyl salicylates; diethyl and dibutyl phthalates. The method fails for vinyl acetate, t-butyl acetate, octadecyl propionate, ethyl and butyl stearate, phenyl, benzyl and guaicol acetate, methyl and ethyl cinnamate, diethyl sulphate and ethyl *p*-aminobenzoate.

(*d*) **Direct identification of the acidic component of an ester.** The following procedures may be regarded as alternative to that described above involving hydrolysis of the ester.

(i) *Anilides or p-toluidides of acids from esters.* Esters are converted into the corresponding anilides or *p*-toluidides by treatment with anilino- or with *p*-toluidino-magnesium bromide, which are readily obtained from any simple Grignard reagent and aniline or *p*-toluidine:

$$ArNH_2 + RMgX \longrightarrow ArNHMgX + RH$$

$$2ArNHMgX + R^1 \cdot CO_2R^2 \longrightarrow R^1 \cdot C(OMgX)(NHAr)_2 + Mg(OR^2)X$$

$$\xrightarrow{2H_2O} R^1 \cdot CONHAr + ArNH_2 + Mg(OH)X + R^2OH$$

This procedure is speedy, economical and employs materials which are readily available. It is not satisfactory for esters of dibasic acids.

Add 4.0 g (4.0 ml) of pure aniline dropwise to a cold solution of ethyl magnesium bromide prepared from 1.0 g of magnesium, 5.0 g (3.5 ml) of ethyl bromide and 30 ml of pure, sodium-dried ether. When the vigorous evolution of ethane has ceased, introduce 0.02 mol of the ester in 10 ml of anhydrous ether, and warm the mixture on a water bath for 10 minutes; cool. Add dilute hydrochloric acid to dissolve the magnesium compounds and excess of aniline. Separate the ethereal layer, dry it with magnesium sulphate and evaporate the ether. Recrystallise the residual anilide, which is obtained in almost quantitative yield, from dilute ethanol or other suitable solvent.

Alternatively, add a solution of 4.5 g of *p*-toluidine in dry ether to the Grignard reagent prepared from 1.0 g of magnesium as detailed above. Then introduce 1.0 g (or 0.02 mol) of the ester and proceed as described for anilides.

(ii) N-*Benzylamides of acids from esters.* Esters are converted into the N-benzylamides of the corresponding acids by heating with benzylamine in the presence of a little ammonium chloride as catalyst:

$$R^1 \cdot CO_2R^2 + C_6H_5 \cdot CH_2NH_2 \longrightarrow R^1 \cdot CONHCH_2 \cdot C_6H_5 + R^2OH$$

The reaction (which is essentially the direct aminolysis of esters with benzylamine) proceeds readily when R^2 is methyl or ethyl. Esters of higher alcohols should preferably be subjected to a preliminary methanolysis by treatment with sodium methoxide in methanol:

$$R^1 \cdot CO_2R^2 + CH_3OH \xrightarrow{CH_3ONa} R^1 \cdot CO_2CH_3 + R^2OH$$

N-Benzylamides are recommended when the corresponding acid is liquid and/or water-soluble so that it cannot itself serve as a derivative. The benzylamides derived from the simple fatty acids or their esters are not altogether satisfactory since they are often low melting; those derived from most hydroxy acids and from polybasic acids or their esters are formed in good yield and are easily purified. The esters of aromatic acids yield satisfactory derivatives but the method must compete with the equally simple process of hydrolysis and precipitation of the free acid, an obvious derivative when the acid is a solid. The procedure fails with esters of keto acids, sulphonic acids and inorganic acids and some halogenated aliphatic esters.

Reflux a mixture of 1 g of the ester, 3 ml of benzylamine and 0.1 g of powdered ammonium chloride for 1 hour in a Pyrex test-tube fitted with a short condenser. Wash the cold reaction mixture with water to remove the excess of benzylamine. If the product does not crystallise, stir it with a little water containing a drop or two of dilute hydrochloric acid. If crystallisation does not result, some unchanged ester may be present: boil with water for a few minutes in an evaporating dish to volatilise the ester. Collect the solid *N*-benzylamide on a filter, wash it with a little light petroleum, b.p. 100–120 °C, and recrystallise it from dilute ethanol, ethyl acetate or acetone.

If the ester does not yield a benzylamide by this procedure, convert it into the methyl ester by refluxing 1 g for 30 minutes with 5 ml of absolute methanol in which about 0.1 g of sodium has been dissolved. Remove the methanol by distillation and treat the residual ester as above.

The melting points of the *N*-benzylamides are collected in Tables VIII,20 and VIII,21.

(iii) *Acid hydrazides from esters.* Methyl and ethyl esters react with hydrazine to give acid hydrazides:

$$R \cdot CO_2CH_3 + H_2NNH_2 \longrightarrow R \cdot CONHNH_2 + CH_3OH$$

The hydrazides are often crystalline and then serve as useful derivatives. Esters of higher alcohols should be converted first to the methyl esters by boiling with sodium methoxide in methanol (see under *N*-benzylamides).

Place 1.0 ml of hydrazine hydrate (**CAUTION:** corrosive chemical) in a test-tube fitted with a short reflux condenser. Add 1.0 g of the methyl or ethyl ester dropwise (or portionwise) and heat the mixture gently under reflux for 15 minutes. Then add just enough absolute ethanol through the condenser to produce a clear solution, reflux for a further 2–3 hours, distil off the ethanol and cool. Filter off the crystals of the acid hydrazide, and recrystallise from ethanol, dilute ethanol or from water.

The melting points of the hydrazides of some aliphatic and aromatic acids are collected in Tables VIII,20 and VIII,21.

In Tables VIII,25 and VIII,26 the boiling points, densities and refractive indices of a number of selected esters are collected.

VII,6,*18*. Primary amides (*a*) **Hydrolysis.** A primary carboxylic acid amide is usually identified by characterising the acid which is liberated on hydrolysis. Amides may be hydrolysed by boiling with 10 per cent sodium hydroxide solution to the corresponding acid (as the sodium salt). The alkaline solution should be acidified with dilute hydrochloric acid; the liberated acid if sparingly water-soluble is isolated by filtration, otherwise any water-soluble acidic component may be isolated by extraction with ether or by distillation from the acidic aqueous solution. The procedure is illustrated by the following experimental details for benzamide. Place 1.5 g of benzamide and 25 ml of 10 per cent sodium hydroxide solution in a 100-ml conical or round-bottomed flask equipped with a reflux condenser. Boil the mixture gently for 30 minutes; ammonia is freely evolved. Detach the condenser and continue the boiling in the open flask for 3–4 minutes to expel the residual ammonia. Cool the solution in ice, and add concentrated hydrochloric acid until the mixture is strongly acid; benzoic acid separates immediately. Leave the mixture in ice until cold, filter at the pump, wash with a little cold water and drain well. Recrystallise the benzoic acid from hot water.

Hydrolysis may also be effected (but usually rather less readily) with 20 per cent sulphuric acid.

(*b*) **Xanthylamides.** Xanthhydrol reacts with primary amides with the formation of crystalline xanthylamides or 9-acylaminoxanthens.

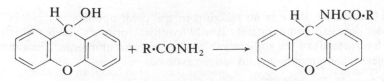

Dissolve 0.25 g of xanthhydrol* in 3.5 ml of glacial acetic acid; if an oil separates (as is sometimes the case with commercial material), allow to settle for a short time and decant the supernatant solution. Add 0.25 g of the amide, shake and allow to stand. If a crystalline derivative does not separate in about 10 minutes, warm on a water bath for a period not exceeding 30 minutes, and allow to cool. Filter off the solid xanthylamide (9-acylaminoxanthen) and recrystallise it from dioxan–water or from acetic acid–water, dry at 80 °C for 15 minutes and determine the m.p.

Some amides do not dissolve in glacial acetic acid; in such cases a mixture of 2 ml of glacial acetic acid and 3 ml of water may be used as a solvent for the reaction. Urea may be characterised as the di-xanthyl derivative (m.p. 274 °C) prepared in acetic acid. Di- and tri-chloroacetamide, oxamide, and salicylamide do not give satisfactory results.

The melting points of aliphatic and aromatic primary carboxylic acid amides and those of the corresponding xanthylamides are collected in Tables VIII,27 and VIII,28.

VII,6,19. Substituted amides (a) **Hydrolysis.** Substituted aromatic carboxylic acid amides of the type Ar·CONHR and Ar·CONR$_2$ are only slowly attacked by aqueous alkali and are characterised by hydrolysis under acidic conditions; 70 per cent sulphuric acid (prepared by carefully adding 4 parts of acid to 3 parts of water) is the preferred reagent. Use the general procedure which has been outlined on p. 1083; characterise the acidic and basic components.

(b) **Fusion with soda-lime.** The basic component is liberated upon fusion with soda-lime and at the same time the aroyl group yields a hydrocarbon. Use the experimental details described for carboxylic acids, Section **VII,6,15**(g); identify the base by the preparation of a suitable derivative.

The melting points of some typical substituted aromatic amides are collected in Table VIII,29. Other examples will be found in the appropriate columns of Tables VIII,20, VIII,33 and VIII,34.

VII,6,20. Nitriles (a) **Hydrolysis to the carboxylic acid.** (i) **Hydrolysis with alkali.** When nitriles are treated with 20–40 per cent sodium or potassium

---

* Best results are obtained with reagent freshly prepared by the reduction of xanthone with sodium amalgam. Prepare an amalgam from 0.9 g of clean sodium and 75 g (5.5 ml) of mercury as described in Section **II,2,57**, and then warm it in a stoppered Pyrex reagent bottle to 50 °C. Add a cold suspension of 2.5 g of commercial xanthone in 20 ml of rectified spirit, stopper the flask and shake vigorously releasing the pressure from time to time. The temperature rises rapidly to 60–70 °C, the xanthone passes into solution and a transient blue colour develops. After 5 minutes the solution is clear and colourless; after a further 10 minutes shaking, separate the mercury and wash it with 3 ml of ethanol. Filter the warm solution into 200 ml of cold distilled water with shaking. Filter the xanthhydrol under suction, wash with water and dry. The yield of xanthhydrol, m.p. 122–123 °C, is 2.4 g. The product may be recrystallised from ethanol; it is comparatively unstable but may be kept in an alcoholic solution.

hydroxide solution, there is no reaction in the cold; upon prolonged boiling hydrolysis proceeds comparatively slowly (compare primary amides which are rapidly hydrolysed) to the sodium salt of the acid and ammonia. The reaction is complete when ammonia is no longer evolved:

$$R \cdot CN + H_2O + NaOH \longrightarrow R \cdot CO_2Na + NH_3$$

The excess of alkali is then neutralised to phenolphthalein or to Congo red with dilute hydrochloric acid and the solution is evaporated to dryness on the water bath. The acid may then be characterized as the S-benzylisothiouronium salt or as the p-bromophenacyl ester (Section **VII,6,**15(b)). In many instances the derivative may be prepared directly from the neutralised solution.

(ii) **Hydrolysis with acid.** Most nitriles are hydrolysed by boiling with 5–8 times the weight of 50–75 per cent sulphuric acid under reflux for 2–3 hours:

$$2R \cdot CN + H_2SO_4 + 4H_2O \longrightarrow 2R \cdot CO_2H + (NH_4)_2SO_4$$

If the acid is a simple aliphatic monobasic acid it can usually be distilled directly from the reaction mixture. If this procedure is not possible, the reaction mixture is poured into excess of crushed ice, and the acid is isolated by ether extraction or by other suitable means. The acid is then characterised (Section **VII,6,**15). The addition of hydrochloric acid (as sodium chloride; say, 5% of the weight of sulphuric acid) increases the rate of the reaction.

A mixture of 50 per cent sulphuric acid and glacial acetic acid may be used with advantage in the case of difficultly-hydrolysable aromatic nitriles. The reaction product is poured into water, and the organic acid is separated from any unchanged nitrile or from amide by means of sodium carbonate solution.

For those nitriles which yield water-insoluble amides (e.g., the higher alkyl cyanides), *conversion to the amide* often leads to a satisfactory derivative. The hydration is effected by warming a solution of the nitrile in concentrated sulphuric acid for a few minutes, cooling and pouring into water. For experimental details see p. 1083.

(b) **Reduction to a primary amine** and conversion into a substituted phenylthiourea. Reduction of a nitrile with sodium and ethanol yields the primary amine, which may be identified by direct conversion into a substituted phenylthiourea.

$$R \cdot CN + 2H_2 \xrightarrow[\text{ROH}]{\text{Na}} R \cdot CH_2NH_2 \xrightarrow{C_6H_5NCS} R \cdot CH_2NH \cdot CS \cdot NHC_6H_5$$

Dissolve 1.0 g of the nitrile in 20 ml of absolute ethanol in a dry 200-ml round-bottomed flask fitted with a reflux condenser. Add through the top of the condenser 1.5 g of clean sodium (previously cut into small pieces) at such a rate that the reaction, although vigorous, remains under control. When all the sodium has reacted (10–15 minutes), cool the reaction mixture to about 20 °C, and add 10 ml of concentrated hydrochloric acid dropwise through the condenser whilst swirling the contents of the flask vigorously: the final solution should be acid to litmus. Connect the flask to a still-head and condenser, and distil off about 20 ml of liquid (dilute ethanol). Cool the flask and fit a small dropping funnel into the top of the still-head. Place 15 ml of 40 per cent sodium hydroxide solution in the dropping funnel, attach an adapter to the end of the condenser and so arrange it that the end dips into about 3 ml of water contained in a 50-ml conical flask. Add the sodium hydroxide solution dropwise and with shaking: a vigorous reaction ensues. When all the alkali has been added, separate the amine by distillation until the contents of the flask are nearly dry.

Add 0.5 ml of phenyl isothiocyanate to the distillate and shake the mixture vigorously for 3–4 minutes. If no derivative separates, crystallisation may be induced by cooling the flask in ice and 'scratching' the walls with a glass rod. Filter off the crude product, wash it with a little 50 per cent ethanol and recrystallise from hot dilute ethanol.

(*c*) **α-Iminoalkylmercaptoacetic acid hydrochlorides.** Mercaptoacetic acid (thioglycollic acid) reacts with nitriles in the presence of hydrogen chloride to give α-iminoalkylmercaptoacetic acid hydrochlorides:

$$R·CN + HSCH_2·CO_2H + HCl \longrightarrow R·C \overset{\overset{\oplus}{N}H_2 \}\overset{\ominus}{Cl}}{\underset{SCH_2·CO_2H}{}}$$

These salts have sharp and reproducible decomposition temperatures but no true melting points. They act as dibasic acids when titrated with standard alkali, thymol blue being used as indicator.

Dissolve 1.0 g of the nitrile and 2.0 g of mercaptoacetic acid in 25 ml of sodium-dried ether in a dry test-tube or small flask. Cool the solution in ice and saturate it with dry hydrogen chloride (5–10 minutes). Stopper the test-tube or flask and keep it at 0 °C until crystallisation is complete (15–60 minutes). Collect the crystals by suction filtration, wash with anhydrous ether and dry in a vacuum desiccator over potassium hydroxide pellets (to remove hydrogen chloride) and paraffin wax shavings (to remove ether).

(*d*) **Acyl phloroglucinols.** Crystalline derivatives of aliphatic nitriles may be prepared by an application of the Hoesch reaction. Equimolecular proportions of phloroglucinol (I) and the nitrile react in dry ethereal solution in the presence of anhydrous zinc chloride and hydrogen chloride to give an imine hydrochloride (II), which is converted into a solid alkyl trihydroxyphenyl ketone (III) by hydrolysis:

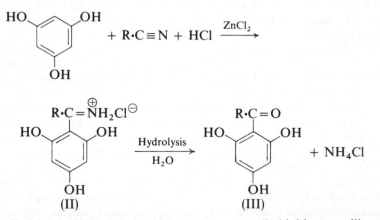

The alkyl 2,4,6-trihydroxyphenyl ketones are usually highly crystalline solids of sharp melting point and are purified by recrystallisation from hot water. Many contain water of crystallisation which can be removed by drying *in vacuo* at about 100 °C; the melting points of both the hydrated and anhydrous compounds should be determined.

Add 0.4 g of powdered, anhydrous zinc chloride to a solution of 1.1 g of anhydrous phloroglucinol in 25 ml of sodium-dried ether, and introduce the nitrile (0.01 mol) dissolved in 5 ml of dry ether. Pass a steady stream of dry hydrogen chloride through the solution for 25–30 minutes; it becomes turbid after 2–3 minutes but the turbidity subsequently disappears. Decant the supernatant liquid, dissolve the residual oil or crystals in 25 ml of water and shake the aqueous solution with two 20 ml portions of ether. Concentrate the aqueous layer to about 10–12 ml. The hydroxy ketone separates upon cooling; recrystallise it from hot water and dry in the air. The hydrate is thus produced.

Physical data for aliphatic and aromatic nitriles are collected in Tables VIII,30 and VIII,31 respectively.

VII,6,*21.* **Primary and secondary amines** (*a*) **Acetyl derivatives.** Primary and secondary amines are best acetylated with acetic anhydride:

$$RNH_2 + (CH_3 \cdot CO)_2O \longrightarrow CH_3 \cdot CONHR + CH_3 \cdot CO_2H$$

$$R^1R^2NH + (CH_3 \cdot CO)_2O \longrightarrow CH_3 \cdot CONR^1R^2 + CH_3 \cdot CO_2H$$

Acetyl chloride is not so satisfactory since an equivalent quantity of the amine hydrochloride is simultaneously produced:

$$2RNH_2 + CH_3 \cdot COCl \longrightarrow CH_3 \cdot CONHR + R\overset{\oplus}{N}H_3\}\overset{\ominus}{Cl}$$

Reflux gently in a test-tube under a short air condenser 1 g of the base with 2.5 mols {or 3.0 g (3.0 ml) if the molecular weight is unknown} of acetic anhydride for 10–15 minutes. Cool the reaction mixture and pour it into 20 ml of cold water. Boil to decompose the excess of acetic anhydride. When cold, filter the residual insoluble acetyl derivative and wash it with a little cold water. Recrystallise from water or from dilute ethanol.

Certain *ortho* substituted derivatives of aromatic amines are difficult to acetylate under the above conditions owing to steric hindrance. The process is facilitated by the addition of a few drops of concentrated sulphuric acid (compare Section **IV,66**), which acts as a catalyst, and the use of a large excess of acetic anhydride.

Excellent results may be obtained by conducting the acetylation in aqueous solution (cf. Section **IV,74**, *Method 1*). Dissolve 0.5 g of the amine in 2 *M*-hydrochloric acid, and add a little crushed ice. Introduce a solution of 5 g of hydrated sodium acetate in 25 ml of water, followed by 5 ml of acetic anhydride. Shake the mixture in the cold until the smell of acetic anhydride disappears. Collect the solid acetyl derivative, and recrystallise it from water or dilute ethanol.

(*b*) **Benzoyl derivatives.** Both primary and secondary amines form benzoyl derivatives under the conditions of the Schotten–Baumann reaction (see Section **IV,73** and discussion on p. 682, **IV,F,2**).

Suspend 1 g (or 1 ml) of the substance in 20 ml of 5 per cent sodium hydroxide solution in a well-corked boiling tube or small conical flask, and add 2 ml of benzoyl chloride, *ca.* 0.5 ml at a time, with constant shaking, and cooling in water (if necessary). Shake vigorously for 5–10 minutes until the odour of the benzoyl chloride has disappeared. Make sure that the mixture has an alkaline reaction. Filter off the solid benzoyl derivative, wash it with a little cold water and recrystallise it from ethanol or dilute ethanol.

If the benzoyl derivative is soluble in alkali, precipitate it together with the benzoic acid derived from the reagent by the addition of hydrochloric acid: filter and extract the product with cold ether or light petroleum (b.p. 40–60 °C) to remove the benzoic acid.

The following alternative procedure is sometimes useful.

To a solution of 0.5 g of the amine in 4 ml of dry pyridine and 10 ml of dry benzene, add dropwise 0.5 ml of benzoyl chloride. Heat the mixture under reflux on a water bath at 60–70 °C for 20–30 minutes and then pour into 80–100 ml of water. Separate the benzene layer and extract the aqueous layer with 10 ml of benzene. Wash the combined benzene solutions with 5 ml of 5 per cent sodium carbonate solution, followed by 5 ml of water, and dry with magnesium sulphate. Filter off the desiccant through a small fluted filter-paper and concentrate the benzene solution to a small volume (3–4 ml). Stir 15–20 ml of hexane into the residue: the crystalline product separates. Filter and wash with a little hexane. Recrystallise from a mixture of cyclohexane with hexane or with ethyl acetate; alternatively use ethanol or dilute ethanol for recrystallisation.

(c) **Benzenesulphonyl and toluene-*p*-sulphonyl derivatives.** Treat 1 g (1 ml) of the amine with 4 molar equivalents of 10 per cent sodium or potassium hydroxide solution (say, 20 ml), and add 1.5 mols (or 3 g if the molecular weight is unknown) of benzenesulphonyl or toluene-*p*-sulphonyl chloride in small portions with constant shaking. To remove the excess of acid chloride, either shake vigorously or warm gently. Acidify with dilute hydrochloric acid and filter off the sulphonamide. Recrystallise it from ethanol or dilute ethanol.

If the presence of a disulphonyl derivative from a primary amine is suspected (e.g., formation of a precipitate in alkaline solution even after dilution), reflux the precipitate, obtained after acidifying, with a solution of 1 g of sodium in 20 ml of rectified spirit for 15 minutes. Evaporate the ethanol, dilute with water and filter if necessary; acidify with dilute hydrochloric acid. Collect the sulphonyl derivative and recrystallise it from ethanol or dilute ethanol.

It is generally more convenient to employ the solid toluene-*p*-sulphonyl chloride (m.p. 69 °C) rather than the liquid benzenesulphonyl chloride. Moreover, the benzenesulphonamides of certain secondary amines are oils or low melting point solids that may be difficult to crystallise: the toluene-*p*-sulphonamides usually have higher melting points and are more satisfactory as derivatives.[*] **Technical toluene-*p*-sulphonyl chloride may be purified** by dissolving it in benzene and precipitating with light petroleum (b.p. 40–60 °C).

Feebly basic amines, e.g., the nitroanilines, generally react so slowly with benzenesulphonyl chloride that most of the acid chloride is hydrolysed by the aqueous alkali before a reasonable yield of the sulphonamide is produced; indeed, *o*-nitroaniline gives little or no sulphonamide under the above conditions. Excellent results are obtained by carrying out the reaction in *pyridine solution*:

$$o\text{-}NO_2 \cdot C_6H_4 \cdot NH_2 + C_6H_5 \cdot SO_2Cl + C_5H_5N \longrightarrow$$
$$o\text{-}NO_2 \cdot C_6H_4 \cdot NHSO_2 \cdot C_6H_5 + C_5H_5N,HCl$$

Reflux a mixture of 1 g (1 ml) of the amine, 2–3 g of benzenesulphonyl chloride and 6 ml of pyridine for 30 minutes. Pour the reaction mixture into 10 ml of cold water and stir until the product crystallises. Filter off the solid and recrystallise it from ethanol or dilute ethanol.

---

[*] For the separation of mixtures of primary, secondary and tertiary amines using benzenesulphonyl or toluene-*p*-sulphonyl chloride (Hinsberg's method), see p. 1143.

Most amines react so rapidly in pyridine solution that the reaction is usually complete after refluxing for 10–15 minutes.

(*d*) **Formyl derivatives.**   Formic acid condenses with primary and secondary amines to yield formyl derivatives:

$$Ar \cdot NHR + H \cdot CO_2H \longrightarrow Ar \cdot N(CHO)R + H_2O$$

Reflux 0.5 g of the amine with 5 ml of 90 per cent formic acid (**CAUTION** in handling) for 10 minutes, and dilute the hot solution with 10 ml of cold water. Cool in ice and, in some cases, saturate with salt if the derivative does not separate immediately. Filter, wash with cold water and recrystallise from water, ethanol or light petroleum (b.p. 60–80 °C).

(*e*) **Derivatives with 3-nitrophthalic anhydride.**   3-Nitrophthalic anhydride reacts with primary and secondary amines to yield nitrophthalamic acids; it does not react with tertiary amines. The phthalamic acid derived from a primary amine undergoes dehydration when heated to 145 °C to give a neutral *N*-substituted 3-nitrophthalimide. The phthalamic acid from a secondary amine is stable to heat and is, of course, soluble in alkali. The reagent therefore provides a method for distinguishing and separating a mixture of primary and secondary amines.

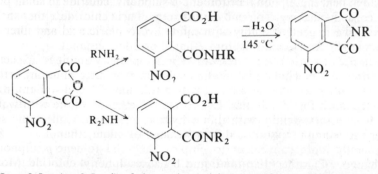

Heat 0.5 g (or 0.5 ml) of the amine with 0.5 g of pure 3-nitrophthalic anhydride (Section **IV,176**) in an oil bath at 145–150 °C for 10–20 minutes, pour the reaction mixture into a small mortar or Pyrex dish and allow it to solidify. Recrystallise from ethanol, aqueous ethanol or ethanol–acetone.

(*f*) **N-Substituted phthalimides (from primary amines).**   Phthalic anhydride reacts with primary amines similarly to yield *N*-substituted phthalimides.

Dissolve 0.5 g of the primary amine and 0.5 g of phthalic anhydride in 5 ml of glacial acetic acid and reflux for 20–30 minutes. (If the amine salt is used, add 1 g of sodium acetate.) The *N*-substituted phthalimide separates out on cooling. Recrystallise it from ethanol or from glacial acetic acid.

(*g*) **2,4-Dinitrophenyl derivatives.**   The halogen atom in 1-chloro-2,4-dinitrobenzene is reactive and coloured crystalline compounds (usually yellow or red) are formed with primary and with secondary amines:

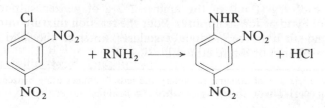

Dissolve 1.0 g (or 1.0 ml) of the amine and 1.0 g of 1-chloro-2,4-dinitrobenzene in 5–10 ml of ethanol, add a slight excess of anhydrous potassium carbonate or of powdered fused sodium acetate, reflux the mixture on a water bath for 20–30 minutes and then pour into water. Wash the precipitated solid with dilute sodium carbonate solution, followed by dilute hydrochloric acid. Recrystallise from ethanol, dilute ethanol or glacial acetic acid.

Note. Chlorodinitrobenzene must be handled with care. If it touches the skin, wash it off with industrial spirit and then copiously with water.

(*h*) **Phenylthioureas.** Primary and secondary amines react with phenyl isothiocyanate to yield phenylthioureas:

$$C_6H_5N=C=S + RNH_2 \longrightarrow C_6H_5NHCSNHR$$
$$C_6H_5N=C=S + R_2NH \longrightarrow C_6H_5NHCSNR_2$$

Phenyl isothiocyanate is not sensitive to water; the reaction may be carried out with an aqueous solution of an amine.

Dissolve equivalent quantities of the reagent and of the amine in a small amount of rectified spirit. If no reaction appears to take place in the cold, reflux the mixture for 5–15 minutes. Upon cooling (and 'scratching' with a glass rod, if necessary) the crystalline thiourea separates. Recrystallise it from rectified spirit or from 60–80 per cent ethanol.

Alternatively, mix equal amounts (say, 0.2 g of each) of the amine and phenyl isothiocyanate in a test-tube and shake for 2 minutes. If no reaction occurs, heat the mixture gently for 2 minutes and then cool in ice until the mass solidifies. Powder the solid, wash it with a little light petroleum (b.p. 100–120 °C) and recrystallise from rectified spirit.

1-Naphthyl isothiocyanate yields crystalline 1-naphthylthioureas and is similarly applied.

(*i*) **Picrates.** Picric acid combines with amines to yield molecular compounds (picrates), which usually possess characteristic melting points. Most picrates have the composition 1 mol amine : 1 mol picric acid. The picrates of the amines, particularly of the more basic ones, are generally more stable than the molecular complexes formed between picric acid and the hydrocarbons (compare Section **VII,6,***3(e)*).

If the amine is soluble in water, mix it with a slight excess (about 25%) of a saturated solution of picric acid in water (the solubility in cold water is about 1%). If the amine is insoluble in water, dissolve it by the addition of 2–3 drops of dilute hydrochloric acid (1 : 1) for each 2–3 ml of water, then add a slight excess of the reagent. If a heavy precipitate does not form immediately after the addition of the picric acid solution, allow the mixture to stand for some time and then shake vigorously. Filter off the precipitated picrate and recrystallise it from boiling water, ethanol or dilute ethanol, boiling 10 per cent acetic acid, chloroform, or, best, benzene.

The following alternative procedure may sometimes be employed. Dissolve 0.5 g of the amine in 5 ml of rectified spirit and add 5 ml of a cold saturated solution of picric acid in ethanol. Warm on a *water bath* for 5 minutes and allow to cool. Collect the precipitated picrate and recrystallise it as above.

(*j*) **Benzylidene derivatives.** *Primary* aromatic amines generally condense directly with benzaldehyde to form benzylidene derivatives (**Schiff's bases or anils**):

$$RNH_2 + OCH\cdot C_6H_5 \longrightarrow RN=CH\cdot C_6H_5 + H_2O$$

These are often crystalline and therefore useful for the characterisation of primary amines. Diamines may, of course, yield di-benzylidene derivatives.

Heat the amine with one or two mols of redistilled benzaldehyde (according as to whether the base is a monamine or diamine) to 100 °C for 10 minutes; if the molecular weight is unknown, use 1 g of the base and 1 or 2 g of benzaldehyde. Sometimes a solvent, such as methanol (5 ml) or acetic acid, may be used. Recrystallise from ethanol, dilute ethanol or benzene.

The melting points of the derivatives of a number of aliphatic and aromatic primary and secondary amines are collected in Tables VIII,32, VIII,33 and VIII,34.

VII,6,22. Tertiary amines (*a*) **Picrates.** Experimental details are given under *Primary and secondary amines*, Section **VII,6,**21(*i*).

(*b*) **Methiodides.** Methyl iodide reacts with tertiary amines to form the crystalline quaternary ammonium iodide (methiodide):

$$R^1R^2R^3N + CH_3I \longrightarrow R^1R^2R^3NCH_3\}^{\oplus}I^{\ominus}$$

Some of these derivatives are hygroscopic.

Allow a mixture of 0.5 g of the tertiary amine and 0.5 ml of colourless methyl iodide * to stand for 5 minutes. If reaction has not occurred, warm under reflux for 5 minutes on a water bath and then cool in ice-water. The mixture will generally set solid: if it does not, wash it with a little dry ether and 'scratch' the sides of the tube with a glass rod. Recrystallise the solid product from absolute ethanol or methanol, ethyl acetate, glacial acetic acid or ethanol–ether.

Alternatively, dissolve 0.5 g of the tertiary amine and 0.5 ml of methyl iodide in 5 ml of dry ether or benzene, and allow the mixture to stand for several hours. The methiodide precipitates, usually in a fairly pure state. Filter, wash with a little of the solvent and recrystallise as above.

The ethiodide is prepared similarly, using ethyl iodide.

(*c*) **Methotoluene-*p*-sulphonate.** Methyl toluene-*p*-sulphonate combines with many tertiary amines to yield crystalline derivatives:

$$R^1R^2R^3N + p\text{-}CH_3\cdot C_6H_4SO_3CH_3 \longrightarrow$$
$$R^1R^2R^3NCH_3\}^{\oplus} p\text{-}CH_3\cdot C_6H_4SO_3\}^{\ominus}$$

Dissolve 2–3 g of methyl toluene-*p*-sulphonate in 10 ml of dry benzene, add 1 g of the amine and boil the mixture for 20–30 minutes. Cool, and filter the precipitated quaternary salt. Recrystallise by dissolving the solid in the minimum volume of boiling ethanol and then adding ethyl acetate until crystallisation commences. Filter the cold mixture, dry rapidly on a porous plate and determine the m.p. immediately.

The benzyl chloride quaternary salts $R^1R^2R^3NCH_2\cdot C_6H_5\}^{\oplus}Cl^{\ominus}$ are pre-

---

* Keep a coil of copper wire (prepared by winding copper wire round a glass tube) or a little silver powder in the bottle, which should be of brown or amber glass; the methyl iodide will remain colourless indefinitely. Ethyl iodide may sometimes give more satisfactory results.

pared similarly; 3 g of redistilled benzyl chloride replaces the methyl toluene-*p*-sulphonate.

(*d*) **Reaction with nitrous acid.** *N,N*-Dialkylanilines yield green solid *p*-nitroso derivatives on treatment with nitrous acid (see p. 1069). Illustrative preparative details for *p*-nitrosodimethylaniline are as follows. Dissolve 1.0 g of dimethylaniline in 10 ml of dilute hydrochloric acid (1 : 1), cool to 0–5 °C and slowly add, with stirring, a solution of 0.70 g of sodium nitrite in 4 ml of water. After 20–30 minutes, filter off the precipitated yellow hydrochloride,* and wash it with a little dilute hydrochloric acid. Dissolve the precipitate in the minimum volume of water, add a solution of sodium carbonate or sodium hydroxide to decompose the hydrochloride (i.e., until alkaline) and extract the free base with ether. Evaporate the ether, and recrystallise the residual green crystals of *p*-nitrosodimethylaniline from light petroleum (b.p. 60–80 °C) or from benzene. The pure compound has m.p. 85 °C.

The melting points of the derivatives of a number of aliphatic and aromatic tertiary amines are collected in Table VIII,35.

VII,6,23. **α-Amino acids** (*a*) **2,4-Dinitrophenyl derivatives.** The reaction between 1-fluoro-2,4-dinitrobenzene and amino acids leads to 2,4-dinitrophenyl derivatives: these are often crystalline and possess relatively sharp melting points.

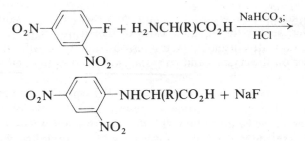

To a solution or suspension of 0.25 g of the amino acid in 5 ml of water and 0.5 g of sodium hydrogen carbonate, add a solution of 0.4 g of 1-fluoro-2,4-dinitrobenzene in 3 ml of ethanol. Shake the reaction mixture vigorously and allow to stand for 1 hour with intermittent vigorous shaking. Add 3 ml of saturated sodium chloride solution and extract with ether (2 × 5 ml) to remove unchanged reagent. Pour the aqueous layer into 12 ml of cold 5 per cent hydrochloric acid with vigorous agitation: this mixture should be distinctly acid to Congo red indicator paper. If the product separates as an oil, try to induce crystallisation by 'scratching' or stirring. Collect the derivative by suction filtration and recrystallise it from 50 per cent ethanol.

(*b*) **Benzoates.** Dissolve 0.5 g of the amino acid in 10 ml of 10 per cent sodium hydrogen carbonate solution and add 1 g of benzoyl chloride. Shake the mixture vigorously in a stoppered test-tube; remove the stopper from time to time since carbon dioxide is evolved. When the odour of benzoyl chloride has disappeared, acidify with dilute hydrochloric acid to Congo red and filter.

---

* The hydrochloride may not separate with other dialkylanilines. Add a slight excess of sodium carbonate or sodium hydroxide to the solution, extract the free base with ether, etc.

Extract the solid with a little cold ether to remove any benzoic acid which may be present. Recrystallise the benzoyl derivative which remains from hot water or from dilute ethanol.

(*c*) **3,5-Dinitrobenzoates.** The following experimental details are for glycine (aminoacetic acid) and may be easily adapted for any other amino acid. Dissolve 0.75 g of glycine in 20 ml of *M*-sodium hydroxide solution and add 2.32 g of finely powdered 3,5-dinitrobenzoyl chloride. Shake the mixture vigorously in a stoppered test-tube; the acid chloride soon dissolves. Continue the shaking for 2 minutes, filter (if necessary) and acidify with dilute hydrochloric acid to Congo red. Recrystallise the derivative immediately from water or 50 per cent ethanol.

Excess of the reagent should be avoided, if possible. If excess of dinitrobenzoyl chloride is used, this appears as the acid in the precipitate obtained upon acidification: the acid can be removed by shaking in the cold with a mixture of 5 volumes of light petroleum (b.p. 40–60 °C) and 2 volumes of ethanol. The glycine derivative is insoluble in this medium. For some amino acids (leucine, valine and phenylalanine) acetic acid should be used for acidification.

(*d*) **Toluene-*p*-sulphonates.** Amino acids react with toluene-*p*-sulphonyl chloride (compare Section **VII,6,**_21(c)_) under the following experimental conditions to yield, in many cases, crystalline toluene-*p*-sulphonates.

Dissolve 0.01 g equivalent of the amino acid in 20 ml of *M*-sodium hydroxide solution and add a solution of 2 g of toluene-*p*-sulphonyl chloride in 25 ml of ether; shake the mixture mechanically or stir vigorously for 3–4 hours. Separate the ether layer: acidify the aqueous layer to Congo red with dilute hydrochloric acid. The derivative usually crystallises out rapidly or will do so on standing in ice. Filter off the crystals and recrystallise from 4–5 ml of 60 per cent ethanol.

With phenylalanine and tyrosine, the sodium salt of the derivative is sparingly soluble in water and separates during the initial reaction. Acidify the suspension to Congo red: the salts pass into solution and the mixture separates into two layers. The derivative is in the ethereal layer and crystallises from it within a few minutes. It is filtered off and recrystallised.

(*e*) **2,4-Dichlorophenoxyacetates.** Amino acids react with 2,4-dichlorophenoxyacetyl chloride to give crystalline derivatives:

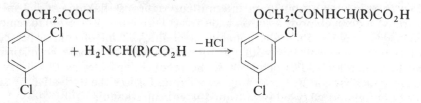

Dissolve 0.01 mol of the amino acid in 30 ml of *M*-sodium hydroxide solution and cool to 5 °C in a bath of ice. Add, with rapid stirring, 0.01 mol of 2,4-dichlorophenoxyacetyl chloride dissolved in 5 ml of dry benzene at such a rate (5–10 minutes) that the temperature of the mixture does not rise above 15 °C, if the reaction mixture gels after the addition of the acid chloride, add water to thin it. Remove the ice bath and stir for 2–3 hours. Extract the resulting mixture with ether, and acidify the aqueous solution to Congo red with dilute hydrochloric acid. Collect the precipitate by filtration and recrystallise it from dilute ethanol.

Commercial 2,4-dichlorophenoxyacetic acid may be recrystallised from benzene; m.p. 139–140 °C. Reflux 10 g of the acid with 15 ml of thionyl chloride on a steam bath for 1 hour, distil off the excess of thionyl chloride at atmospheric pressure and the residue under reduced pressure: 2,4-dichlorophenoxyacetyl chloride (8 g) passes over at 155–157 °C/22–23 mmHg. It occasionally crystallises (m.p. 44.5–45.5 °C), but usually tends to remain as a supercooled liquid.

(*f*) **Phthaloyl derivatives.** Many amino acids condense with phthalic anhydride at 180–185 °C to yield crystalline phthaloyl derivatives:

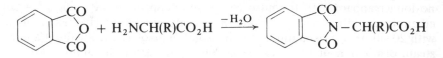

Place 0.5 g of the amino acid and 1.0 g of phthalic anhydride in a Pyrex testtube and immerse the lower part of the tube in an oil bath, which has previously been heated to 180–185 °C. Stir the mixture occasionally during the first 10 minutes and push down the phthalic anhydride which sublimes on the walls into the reaction mixture with a glass rod. Leave the mixture undisturbed for 5 minutes. After 15 minutes, remove the test-tube from the bath: when the liquid mass solidifies, invert the test-tube and scrape out the excess of phthalic anhydride on the walls. Recrystallise the residue from 10 per cent ethanol or from water.

(*g*) **1-Naphthylureido acids** (or 1-naphthylhydantoic acids).   Amino acids react in alkaline solution with 1-naphthyl isocyanate to yield the sodium salts of the corresponding 1-naphthylureido acids, which remain in solution: upon addition of a mineral acid, the ureido acid is precipitated.

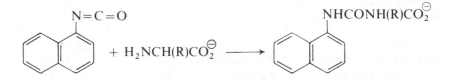

Dissolve 0.5 g of the amino acid in slightly more than the equivalent quantity of *M*-sodium hydroxide solution in a stoppered flask. Add a quantity of 1-naphthyl isocyanate just equivalent to the alkali (if the molecular weight of the compound is not known, use 1 g of the reagent and the corresponding quantity of alkali), stopper the flask and shake vigorously until the odour of the reagent has disappeared. Filter off any insoluble di-1-naphthylurea (resulting from the action of water upon the excess of the reagent), and acidify the filtrate to Congo red with dilute hydrochloric acid. Filter the 1-naphthylureido acid at the pump, wash it with a little cold water and recrystallise from hot water or dilute ethanol.

The phenylureido acid is prepared similarly, using phenyl isocyanate. The latter is more sensitive to water than 1-naphthyl isocyanate and therefore does not keep so well.

The melting points of the derivatives of a number of amino acids are collected in Table VIII,36. Most α-amino acids decompose on heating so that the melting points would be more accurately described as decomposition points:

the latter vary somewhat with the rate of heating and the figures given are those obtained upon rapid heating.

VII,6,24. Nitro compounds (*a*) **Reduction to the primary amine.** Add 10 ml of concentrated hydrochloric acid in small portions to a mixture of 1.0 g of the nitro compound and 3 g of granulated tin contained in a small (say, 50-ml) flask fitted with a reflux condenser. Shake the flask well to ensure thorough mixing during the addition of the acid. After 10 minutes warm under reflux at 100 °C with vigorous shaking until the nitro compound has dissolved and its odour is no longer apparent. (If the nitro compound dissolves slowly, add a few ml of ethanol.) Cool the reaction mixture thoroughly and cautiously make it alkaline with 20–40 per cent sodium hydroxide solution. Isolate the liberated amine by steam distillation or by ether extraction. Characterise the amine by the preparation of a suitable crystalline derivative (see *Primary and secondary amines*, Section **VII,6,21**).

Some aromatic dinitro compounds, e.g., *m*-dinitrobenzene, may be chacterised by partial reduction to the nitroamine. Experimental details using sodium polysulphide as the reducing agent are to be found in Section **IV,58**.

(*b*) **Oxidation of side chains.** Aromatic nitro compounds that contain a side chain (e.g., nitro derivatives of alkyl benzenes) may be oxidised to the corresponding acids either by alkaline potassium permanganate (*Aromatic hydrocarbons*, Section **VII,6,3(*d*)**) or, preferably, with a sodium dichromate–sulphuric acid mixture in which medium the nitro compound is more soluble.

Mix 1.0 g of the nitro compound with 4 g of sodium dichromate and 10 ml of water in a 50-ml flask, then attach a reflux condenser to the flask. Add slowly and with shaking 7 ml of concentrated sulphuric acid. The reaction usually starts at once; if it does not, heat the flask gently to initiate the reaction. When the heat of reaction subsides, boil the mixture, cautiously at first, under reflux for 20–30 minutes. Allow to cool, dilute with 30 ml of water and filter off the precipitated acid. Purify the crude acid by extraction with sodium carbonate solution, precipitation with dilute mineral acid and recrystallisation from hot water, benzene, etc.

(*c*) **Nitration to a poly-nitro compound.** Aromatic mononitro compounds may sometimes be characterised by conversion into the corresponding dinitro or trinitro derivatives. It may be noted that many poly-nitro compounds form characteristic addition compounds with naphthalene.

The nitration of an aromatic compound, especially if its composition is unknown, must be conducted with great care, preferably behind a safety screen, since many aromatic compounds react violently.

A. Add about 0.5 g of the compound to 2.0 ml of concentrated sulphuric acid. Introduce 2.0 ml of concentrated nitric acid drop by drop, with shaking after each addition. Attach a small reflux condenser to the flask and heat in a beaker of water at 50 °C for 5 minutes. Pour the reaction mixture on to 15 g of crushed ice and collect the precipitated solid by suction filtration. Recrystallise from dilute ethanol.

B. Proceed as in *A*, but use 2.0 ml of fuming nitric acid instead of the concentrated nitric acid, and warm the mixture on a boiling water bath for 5–10 minutes.

The physical constants of a number of selected aromatic and aliphatic nitro compounds are collected in Tables VIII,37 and VIII,38 respectively. It will be noted that a few nitro aromatic esters have been included in the tables. These are given here because the nitro group may be the first functional group to be identified; aromatic nitro esters should be treated as other esters and hydrolysed for final identification.

VII,6,25. Thiols Of the crystalline derivatives of thiols, those formed with 3,5-dinitrobenzoyl chloride are not very satisfactory since they have, in general, lower melting points than those of the corresponding alcohols (compare *Alcohols and polyhydric alcohols*, Section **VII,6,4**) and do not differ widely from ethyl to heptyl. More satisfactory derivatives are obtained with 1-chloro-2,4-dinitrobenzene.

(*a*) **Alkyl (or Aryl) 2,4-dinitrophenyl sulphides (or -thioethers) and the corresponding sulphones.** Thiols react with 1-chloro-2,4-dinitrobenzene in alkaline solution to yield crystalline thioethers (2,4-dinitrophenyl sulphides) (I):

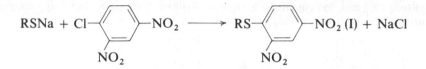

The sulphides (I) can be readily oxidised in glacial acetic acid solution by potassium permanganate to the corresponding sulphones (II); the latter exhibit a wide range of melting points and are therefore particularly valuable for the characterisation of thiols:

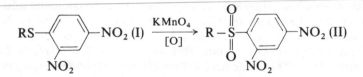

Preparation of 2,4-dinitrophenyl sulphides. Dissolve about 0.5 g (or 0.005 mol) of the thiol in 10–15 ml of rectified spirit (or in the minimum volume necessary for solution; warming is permissible) and add 2 ml of 10 per cent sodium hydroxide solution. Mix the resulting solution with a solution of 1 g of 1-chloro-2,4-dinitrobenzene (**CAUTION:** see Section **VII,6,**21(*g*)) in 5 ml of rectified spirit. Reaction may occur immediately with precipitation of the thioether. In any case reflux the mixture for 10 minutes on a water bath in order to ensure the completeness of the reaction. Filter the hot solution rapidly; allow the solution to cool when the sulphide will crystallise out. Recrystallise from ethanol.

Preparation of the sulphones. Dissolve the 2,4-dinitrophenyl sulphide (0.005 mol) in the minimum volume of warm glacial acetic acid and add 3 per cent potassium permanganate solution with shaking as fast as decolourisation occurs. Use a 50 per cent excess of potassium permanganate: if the sulphide tends to precipitate, add more acetic acid. Just decolourise the solution with sulphur dioxide (or with sodium metabisulphite or ethanol) and add 2–3 volumes of crushed ice. Filter off the sulphone, dry and recrystallise from ethanol.

(*b*) **3,5-Dinitrothiobenzoates.** Thiols react with 3,5-dinitrobenzoyl chloride in the presence of pyridine as a catalyst to yield 3,5-dinitrothiobenzoates:

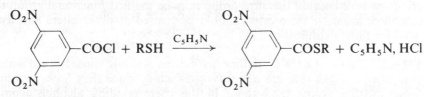

Mix 0.2 g of 3,5-dinitrobenzoyl chloride, 6 drops of the thiol and 1–3 drops of pyridine in a test-tube, and heat the mixture in a beaker of boiling water until fumes of hydrogen chloride cease to appear (10–15 minutes). Add a few drops of water, followed by a drop or two of pyridine to eliminate the excess of the reagent. The product solidifies upon stirring with a glass rod. Add water, filter and recrystallise from dilute ethanol or dilute acetic acid.

(*c*) **Hydrogen 3-nitrothiophthalates.** Thiols react with 3-nitrophthalic anhydride to yield hydrogen 3-nitrothiophthalates (compare *Alcohols and polyhydric alcohols*, Section **VII,6,4(*f*)**).

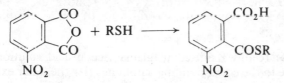

Mix 3-nitrophthalic anhydride (0.005 mol or 1.0 g) and the thiol (0.0075 mol, or 1.0 g if the molecular weight is not known) in a test-tube and heat gently over a free flame for about 30 seconds. Allow the mixture to cool, and add 0.5 ml of 2 *M*-sodium hydroxide solution dropwise and with cooling in an ice bath. Then add about 0.3 ml of 2 *M*-hydrochloric acid and shake the reaction mixture vigorously. Collect the solid which separates by suction filtration and dry it upon a porous tile. Recrystallise from dilute acetic acid or from aqueous acetone.

The melting points are determined using a bath preheated to about 100 °C. The compounds decompose slightly at the m.p. The neutralisation equivalent may be evaluated by titration with standard alkali.

The melting points of the derivatives of the more commonly occurring thiols are collected in Table VIII,39.

VII,6,26. Sulphonic acids (*a*) **Sulphonamides.** Mix together 1.0 g of the dry acid or 1.2 g of the anhydrous salt with 2.5 g of phosphorus pentachloride and heat under a reflux condenser in an oil bath at 150 °C for 30 minutes. Cool the mixture, add 20 ml of dry benzene, warm on a steam bath (*FUME CUPBOARD*) and stir the solid mass well to extract the sulphonyl chloride: filter.* Add the benzene solution slowly and with stirring to 10 ml of concentrated ammonia solution. If the sulphonamide precipitates, separate it by filtration; if no solid is obtained, evaporate the benzene on a steam bath. Wash

---

* The sulphonyl chloride may be isolated by evaporating the benzene solution and recrystallising the residue from chloroform, light petroleum (b.p. 40–60 °C) or benzene–light petroleum (b.p. 40–60 °C).

the sulphonamide with a little cold water, and recrystallise from water, aqueous ethanol or ethanol to constant m.p.

The procedure is not usually applicable to aminosulphonic acids owing to the interaction between the amino group and the phosphorus pentachloride. If, however, the chlorosulphonic acid is prepared by diazotisation and treatment with a solution of cuprous chloride in hydrochloric acid, the crystalline chloro-sulphonamide and chlorosulphonanilide may be obtained in the usual way. With some compounds, the amino group may be protected by acetylation. Sulphonic acids derived from a phenol or naphthol cannot be converted into the sulphonyl chlorides by the phosphorus pentachloride method.

The **sulphonanilides** may be prepared by either of the following methods: (i) Reflux the solution of the sulphonyl chloride in benzene obtained as above, with 2.5 g of aniline for 1 hour. Concentrate the benzene solution to half its volume and cool in ice. Collect the solid which separates on a filter, wash with hot water and recrystallise from ethanol or dilute ethanol.

(ii) Treat the crude sulphonyl chloride (isolated by evaporating the solvent after extraction with benzene (or ether or chloroform) as above) with 1 g of *p*-toluidine and 30 ml. of *ca.* 2 *M*-sodium hydroxide solution. Shake for 10–15 minutes. Extract the alkaline solution with ether to remove excess of *p*-toluidine, acidify, filter and recrystallise the residue as in (i).

$$R \cdot SO_2ONa + PCl_5 \longrightarrow R \cdot SO_2Cl + POCl_3 + NaCl$$
$$R \cdot SO_2Cl + 2NH_3 \longrightarrow R \cdot SO_2NH_2 + NH_4Cl$$
$$R \cdot SO_2Cl + 2C_6H_5NH_2 \longrightarrow R \cdot SO_2NHC_6H_5 + C_6H_5NH_3^{\oplus}Cl^{\ominus}$$

(b) **S-Benzylisothiouronium salts** (see *Carboxylic acids*, Section **VII,6,**15). If the substance is the free sulphonic acid, dissolve 0.5 g of it in 5–10 ml of water, add a drop or two of phenolphthalein indicator and neutralise with *ca.* M-sodium hydroxide solution. Then add 2–3 drops of 0.1 M-hydrochloric acid to ensure that the solution is almost neutral (*pale* pink colour); under alkaline conditions the reagent tends to decompose to produce the evil-smelling phenylmethanethiol.

To a solution of 0.5 g of the salt in 5 ml of water and 2–3 drops of 0.1 M-hydrochloric acid (or to a solution of the acid treated as above), add a slight excess of a cold, 15 per cent aqueous solution of S-benzylisothiouronium chloride (if the molecular weight of the compound is not known, use a solution of 1 g of the reagent in 5 ml of water), and cool in ice. Filter off the crystalline derivative and recrystallise it from 5 per cent ethanol.

$$\left[C_6H_5 \cdot CH_2 \cdot S \cdot C \begin{array}{c} {}^{\nearrow NH_2} \\ {}^{\searrow NH_2} \end{array} \right]^{\oplus} Cl^{\ominus} + R \cdot \overset{\ominus}{S}\overset{\oplus}{O_3}Na \longrightarrow$$

$$\left[C_6H_5 \cdot CH_2 \cdot S \cdot C \begin{array}{c} {}^{\nearrow NH_2} \\ {}^{\searrow NH_2} \end{array} \right]^{\oplus} R \cdot SO_3^{\ominus} + NaCl$$

(c) **Sulphonacetamides.** Sulphonacetamides are derivatives of sulphon-amides (Section **VII,6,**27), but since the latter are readily prepared from the sulphonic acids or their salts, sulphonacetamides may be employed for the characterisation of sulphonic acids; for this reason they are included in this Section.

Sulphonamides upon heating with acetyl chloride are converted into the N-acetyl derivatives or sulphonacetamides:

$$Ar{\cdot}SO_2NHR + CH_3{\cdot}COCl \longrightarrow$$
$$Ar{\cdot}SO_2N(R)CO{\cdot}CH_3 + HCl\ (R = H\ or\ alkyl)$$

The sulphonacetamides ($R = H$) are freely soluble in sodium hydrogen carbonate solution thus rendering purification facile. Sulphonacetamides are moderately strong acids, and can generally be titrated in aqueous or aqueous–alcoholic solution with phenolphthalein as indicator. The acidic properties of sulphonacetamides may be used to effect a **separation of a sulphonamide from an N-alkylsulphonamide**. Acetylation of such a mixture gives a sulphonacetamide and an N-alkylsulphonacetamide, of which only the former is soluble in sodium hydrogen carbonate solution. Both sulphonacetamides and N-alkylsulphonacetamides are readily hydrolysed by boiling with excess of 5 per cent potassium hydroxide solution for about 1 hour, followed by acidification with dilute hydrochloric acid, giving the corresponding sulphonamides and N-alkylsulphonamides respectively.

Reflux 1 g of the sulphonamide with 2.5 ml of acetyl chloride for 30 minutes; if solution is not complete within 5 minutes, add up to 2.5 ml of glacial acetic acid. Remove the excess of acetyl chloride by distillation on a water bath, and pour the cold reaction mixture into water. Collect the product, wash with water and dissolve it in warm sodium hydrogen carbonate solution. Acidify the filtered solution with glacial acetic acid; filter off the precipitated sulphonacetamide and recrystallise it from aqueous ethanol.

(*d*) **p-Toluidine salts of sulphonic acids.** These are prepared by the interaction of the sulphonic acid or its sodium salt with p-toluidine hydrochloride in aqueous solution:

$$R{\cdot}SO_3^{\ominus}Na^{\oplus} + H^{\oplus}Cl^{\ominus} + p\text{-}CH_3{\cdot}C_6H_4{\cdot}NH_2 \longrightarrow$$
$$(p\text{-}CH_3{\cdot}C_6H_4{\cdot}NH_3)^{\oplus}(RSO_3)^{\ominus} + Na^{\oplus}Cl^{\ominus}$$

Dissolve 1.0 g of the alkali metal salt of the sulphonic acid in the minimum volume of hot water and add 0.5 g of p-toluidine and 2 ml of concentrated hydrochloric acid. If a solid separates or the p-toluidine does not dissolve completely, add more hot water and a few drops of concentrated hydrochloric acid until a clear solution is obtained at the boiling point. Cool the solution; if crystallisation does not occur immediately, 'scratch' the walls of the test-tube to induce crystallisation. Collect the product by suction filtration, and recrystallise it from hot water containing a drop of concentrated hydrochloric acid, or from dilute ethanol.

The melting points of the derivatives of a number of selected sulphonic acids are collected in Table VIII,40; the melting points of the corresponding sulphonyl chlorides are included for purposes of reference. The acids do not possess sharp melting points; the sulphonic acids are therefore arranged in groups of related compounds. In a subsidiary Table (VIII,40A), a number of sulphonic acids are arranged in the order of increasing melting points of the S-benzyl-isothiouronium salts.

VII,6,27. Sulphonamides (*a*) **Hydrolysis.** Sulphonamides are most readily identified by hydrolysis with 80 per cent sulphuric acid (for experimental details see p. 1084).

$$R^1 \cdot SO_2NHR^2 + H_3O^\oplus \longrightarrow R^1 \cdot SO_3H + R^2 \cdot \overset{\oplus}{N}H_3$$

$$Ar \cdot SO_2NHR + H_3O^\oplus \longrightarrow Ar \cdot SO_3H + R \cdot \overset{\oplus}{N}H_3$$

$$Ar \cdot SO_2NR^1R^2 + H_3O^\oplus \longrightarrow Ar \cdot SO_3H + R^1R^2\overset{\oplus}{N}H_2$$

The amine is liberated by the addition of alkali and then characterised by a suitable derivative; the sulphonic acid may then be recovered as the sodium salt and converted into a suitable crystalline derivative, e.g., the *S*-benzylisothio-uronium salt.

(*b*) **Xanthylsulphonamides.** Primary sulphonamides $R \cdot SO_2NH_2$ may be most simply characterised by reaction with xanthhydrol to yield the corresponding *N*-xanthylsulphonamides (cf. *Primary amides*, Section **VII,6,***18*(*b*)). Dissolve 0.25 g of xanthhydrol and 0.25 g of the primary sulphonamide in 10 ml of glacial acetic acid. Shake for 2–3 minutes at the laboratory temperature and allow to stand for 60–90 minutes. Filter off the derivative, recrystallise it from dioxan–water (3 : 1) and dry at room temperature.

(*c*) **Sulphonacetamides.** Sulphonamides may also be characterised as sulphonacetamides: for experimental details see Section **VII,6,***26*(*c*).

The melting points of sulphonamides, both aliphatic and aromatic, together with the melting points of some *N*-xanthylsulphonamides and sulphonacetamides are collected in Tables VIII,41 and VIII,40.

VII,7. QUALITATIVE ANALYSIS OF MIXTURES OF ORGANIC COMPOUNDS

VII,7,1. Introduction The general method to be adopted for the analysis of mixtures of organic compounds is to separate them into their components and to identify each component as previously described. It is impossible, however, to give a set of procedures which will be applicable, without modification, to the great variety of combinations which may be encountered. Advantage should be taken of any facts which have emerged in the preliminary examination and to adapt, if necessary, the general schemes given below to the mixture under examination. The preliminary examination is therefore of fundamental importance.

Broadly speaking, the separation of the components of mixtures may be divided into three main groups.

(1) **Separations based upon differences in the chemical properties of the components.** A mixture of toluene and aniline may be separated by extraction with dilute hydrochloric acid: the aniline passes into the aqueous layer in the form of the salt, aniline hydrochloride, and may be recovered by neutralisation. Similarly, a mixture of phenol and toluene may be separated by treatment with dilute sodium hydroxide. The above examples are, of course, simple applications of the fact that the various components fall into different solubility groups (compare Section **VII,3**). Another example is the separation of a mixture of dibutyl ether and chlorobenzene: concentrated sulphuric acid dissolves only the dibutyl ether and it may be recovered from solution by dilution with water. With some classes of compounds, e.g., unsaturated compounds, concentrated sulphuric acid leads to polymerisation, sulphonation, etc., so that the

original component cannot be recovered unchanged: this solvent, therefore, possesses limited application.

Phenols may be separated from acids (for example, *o*-cresol from benzoic acid) by a dilute solution of sodium hydrogen carbonate: the weakly acidic phenols (and also enols) are not converted into salts by this reagent and may be removed by ether extraction or by other means; the acids pass into solution as the sodium salts and may be recovered after acidification. For experimental details see Section **VII,6,***17*(*a*(i or ii)), p. 1121 and p. 1122.

Mixtures of primary, secondary and tertiary amines can usually be separated by Hinsberg's method. This is based on the fact that reaction with benzenesulphonyl (or toluene-*p*-sulphonyl) chloride converts primary amines into alkalisoluble sulphonamides, secondary amines into alkali-insoluble sulphonamides and leaves tertiary amines unaffected.

$$C_6H_5 \cdot SO_2Cl + H_2NR \xrightarrow{\text{NaOH}} [C_6H_5 \cdot SO_2\overset{\ominus}{N}R] \overset{\oplus}{Na} \text{ (water-soluble)}$$

$$\xrightarrow{\text{HCl}} C_6H_5 \cdot SO_2NHR \text{ (water-insoluble)}$$

$$C_5H_5 \cdot SO_2Cl + HNR_2 \xrightarrow{\text{NaOH}} C_6H_5 \cdot SO_2NR_2 \text{ (insoluble in alkali)}$$

The following experimental details illustrate how the Hinsberg separation of amines may be carried out in practice.

Treat 2.0 g of the mixture of amines with 40 ml of 10 per cent sodium hydroxide solution and add 4 g (3 ml) of benzenesulphonyl chloride (or 4 g of toluene-*p*-sulphonyl chloride) in small portions. Warm on a water bath to complete the reaction. Acidify the alkaline solution with dilute hydrochloric acid when the sulphonamides of the primary and secondary amines are precipitated. Filter off the solid and wash it with a little cold water; the tertiary amine will be present in the filtrate. To convert any disulphonamide that may have been formed from the primary amine into the sulphonamide, boil the solid under reflux with 2.0 g of sodium dissolved in 40 ml of absolute ethanol for 30 minutes. Dilute with a little water and distil off the alcohol: filter off the precipitate of the sulphonamide of the secondary amine. Acidify the filtrate with dilute hydrochloric acid to precipitate the derivative of the primary amine. Recrystallise the respective derivatives from ethanol or from dilute ethanol.

Aldehydes, e.g., benzaldehyde, may be separated from liquid hydrocarbons and other neutral, water-insoluble liquid compounds by shaking with a solution of sodium metabisulphite: the aldehyde forms a solid bisulphite compound, which may be filtered off and decomposed with dilute acid or with sodium bicarbonate solution in order to recover the aldehyde.

A valuable method for separating ketones from other neutral and water-insoluble compounds utilises the **Girard-T reagent** (Section **II,2,***24*), carbohydrazidomethyltrimethylammonium chloride, prepared by reaction of ethyl chloroacetate with trimethylamine, followed by reaction with hydrazine:

$$(CH_3)_3N + ClCH_2 \cdot CO_2C_2H_5 + H_2N \cdot NH_2 \longrightarrow$$

$$[(CH_3)_3\overset{\oplus}{N} \cdot CH_2 \cdot CO \cdot NH \cdot NH_2]\overset{\ominus}{Cl} + C_2H_5OH$$
Girard-T reagent

It reacts with the carbonyl compound to yield another quaternary ammonium salt:

$$R^1R^2CO + [(CH_3)_3\overset{\oplus}{N}CH_2 \cdot CO \cdot NH \cdot NH_2]Cl^{\ominus} \longrightarrow$$
$$[(CH_3)_3\overset{\oplus}{N}CH_2 \cdot CO \cdot NH \cdot N = CR^1R^2]Cl^{\ominus} + H_2O$$

The latter is a polar compound and is therefore soluble in water. Extraction with ether removes the water-insoluble compounds, leaving the salt in the aqueous layer. The ketone is easily regenerated by hydrolysis with dilute hydrochloric acid.

(2) **Separations based upon differences in the volatilities of the components in aqueous solution.** This procedure is generally employed for the water-soluble compounds, and may also be applied to mixtures in which one of the components is slightly soluble in water. The water-soluble compounds include the lower members of the homologous series of alcohols, aldehydes, ketones, acids, esters, amines and nitriles; compounds containing two or more hydroxyl or amino groups, hydroxy-, amino-, di- and poly-basic acids, sulphonic acids and salts. The compounds with one functional group are usually volatile in steam and distil with the water: compounds with two or more functional groups (amino, hydroxyl or carboxyl) are not generally steam-volatile. The stable salts of steam-volatile bases and acids may be decomposed by a stronger non-volatile mineral acid (sulphuric acid or phosphoric acid) or base (sodium or potassium hydroxide), and the organic base or acid separated by steam distillation from the aqueous solution or suspension. The salts of weaker bases and acids (e.g., the amine or ammonium salts of carboxylic acids and the alkali metal salts of some phenols) are sufficiently hydrolysed by boiling water to permit the basic or acidic compound to distil with the water. It should be noted that sulphonic acids and their salts are not volatile in steam. The only disadvantage of this procedure is that certain compounds may decompose or polymerise or otherwise undergo change under the influence of aqueous alkali or acid at 100 °C, and this fact must be borne in mind when the method is employed. Thus sugars are decomposed by alkali; these may sometimes be isolated by evaporation of the solvent at pH 7, preferably under diminished pressure.

The essential basis of the scheme for the separation of water-soluble compounds is, therefore, distillation of (a) an aqueous solution of the mixture, (b) an alkaline (with sodium hydroxide) solution of the mixture and (c) an acidic (with sulphuric or phosphoric acid) solution of the mixture. The residue will contain the non-volatile components, which must be separated from inorganic salts and from each other by any suitable process.

The following are examples of the above procedure. A mixture of diethylamine and butan-1-ol may be separated by adding sufficient dilute sulphuric acid to neutralise the base: steam distillation will remove the alcohol. The amine can be recovered by adding sodium hydroxide to the residue and repeating the distillation. A mixture of diethyl ketone and acetic acid may be treated with sufficient dilute sodium hydroxide solution to transform the acid into sodium acetate and distilling the aqueous mixture. The ketone will pass over in the steam and the non-volatile, stable salt will remain in the flask. Acidification with dilute sulphuric acid liberates acetic acid, which can be isolated by steam distillation or by extraction.

(3) **Separations based upon differences in the physical properties of the components.** When procedures (1) or (2) are unsatisfactory for the separation of a mixture of organic compounds, purely physical methods may be employed. Thus a mixture of volatile liquids may be fractionally distilled (compare Section **I,26**); or a mixture of non-volatile solids may frequently be separated by making use of the differences in solubilities in inert solvents. The progress of such separations may be monitored by application of the various chromatographic techniques detailed in Section **I,31**, or indeed these techniques may be employed on the preparative scale for effecting the separation itself. The techniques of counter current distribution (Section **I,32**), fractional crystallisation or fractional sublimation (Section **I,21**) may also be employed where appropriate.

VII,7,2. Preliminary examination of a mixture
A. Liquid mixtures (i) **Physical properties.** Examine the mixture with regard to odour, viscosity and colour.

(ii) **Solubility in water.** Transfer 1.0 ml of the mixture by means of a calibrated dropper or a small pipette into a small graduated test-tube: add 1 ml of water and shake. Observe whether there is complete or partial solution and if there is any sign of chemical reaction. If solution is not complete, add more water (in 1 ml portions) and note (*a*) if the mixture dissolves completely, and (*b*) if a portion is insoluble and, if so, whether it is heavier or lighter than the aqueous layer. If an emulsion is formed, it may be assumed that at least one component of the mixture is insoluble in water and at least one component is soluble. Test the aqueous layer obtained with litmus and with phenolphthalein: if there is an acid reaction, test also with 5 per cent sodium hydrogen carbonate solution.

(iii) **Presence or absence of water.** Determine the absence or presence of water in the mixture by one or more of the following tests:

(*a*) Investigate its miscibility with ether or with benzene.
(*b*) Observe its action upon anhydrous copper(II) sulphate.
(*c*) Distil a small portion and note the b.p. and properties of the distillate.
(*d*) Distil 3 ml of the mixture with 3 ml of dry toluene from a dry 10-ml distilling flask. Collect 2 ml of the distillate and dilute it with 5 ml of dry toluene; the formation of two layers or of distinct drops suspended in the toluene indicates the presence of water.

(iv) **Behaviour upon distillation.** If the original mixture is not an aqueous solution, place 5 ml (or 10 ml) of it in a 10-ml (or 25-ml) flask set for downward distillation, immerse the latter in a small beaker of cold water and gradually heat to boiling. Observe the b.p. of any lquid which passes over and set it aside for subsequent examination. Determine the water-solubility of any residue. If it dissolves in water, examine it by Table VII,5; if it insoluble in water, apply Table VII,4 (p. 1151 and p. 1149 respectively).

(v) **Test for elements.** If the mixture is an aqueous solution, evaporate a small portion (*ca.* 1 ml) to dryness upon platinum foil or in a small crucible. Use a portion of the residue to test for elements (Section **VII,2**) and another portion for the *Ignition test* (vi).

If no water is present in the mixture, use it directly in the tests for elements.

(vi) **Ignition test.** Place 0.1–0.2 g of the mixture on a porcelain crucible cover: heat gently at first over a small flame and finally ignite strongly. Observe:

(a) The inflammability and nature of the flame (e.g., smoky or otherwise).
(b) Odour of gases or vapours evolved (**CAUTION!**).
(c) Whether a residue is left after ignition; moisten with hydrochloric acid and test with a platinum wire.

(vii) **Miscellaneous tests.** Treat 1 ml of the mixture with 5 per cent sodium hydroxide solution until strongly alkaline: note whether an oil or solid separates, whether any ammonia is evolved and any colour changes which occur. Heat to boiling and cool: compare odour with that of the original mixture (a change in odour may indicate the presence of esters). Add dilute hydrochloric acid and observe the result.

Treat 1 ml of the mixture with dilute hydrochloric acid until strongly acid. Note any evolution of gas or the separation of a solid. Add dilute sodium hydroxide solution and observe the effect.

(viii) **Miscellaneous class reactions.** Determine the effect of the following class reagents upon small portions of the original mixture: (a) bromine in carbon tetrachloride solution, (b) potassium permanganate solution, (c) alcoholic silver nitrate solution, (d) 2,4-dinitrophenylhydrazine reagent, (e) Schiff's reagent, (f) acetyl chloride, (g) benzoyl chloride (in the presence of aqueous sodium hydroxide), (h) sodium (only if water is absent), (i) iron(III) chloride solution and (j) bromine water.

B. Solid mixtures (i) **Physical properties.** Observe the colour, odour and crystalline form. Examine with a lens or a microscope, if available.

(ii) **Solubility in water.** Determine the solubility of 1.0 g of the sample in water. If in doubt as to whether a portion of the mixture dissolves, remove the supernatant liquid with a dropper and evaporate to dryness on a water bath. Determine the reaction of the aqueous solution or suspension to litmus and to phenolphthalein.

(iii) **Test for elements.** See Section **VII,2**.

(iv) **Ignition test.** Place 0.1–0.2 g of the mixture upon a porcelain crucible cover or upon a piece of platinum foil; heat gently at first and finally ignite strongly. Note:

(a) Whether the mixture melts and if decomposition occurs.
(b) The inflammability and the nature of the flame (e.g., if smoky or otherwise).
(c) Whether a residue is obtained after ignition (moisten with hydrochloric acid and test with a platinum wire).

(v) **Miscellaneous tests.** Test the solubility behaviour of 0.5–1 g of the mixture to 5 per cent sodium hydroxide solution, 5 per cent sodium hydrogen carbonate solution and to 5 per cent hydrochloric acid (for details, see under *Liquid mixtures*).

(vi) **Miscellaneous class reactions.** Determine the effect of the following class reagents upon small portions of the original mixture (for some of the tests an aqueous solution or suspension may be used): (a) bromine in carbon tetrachloride solution, (b) potassium permanganate solution, (c) alcoholic silver nitrate solution, (d) 2,4-dinitrophenylhydrazine reagent, (e) Schiff's reagent, (f) acetyl chloride, (g) benzoyl chloride (in the presence of aqueous sodium hydroxide), (h) iron(III) chloride solution and (i) bromine water.

A careful consideration of the results of the above tests will provide much useful information and will indicate which of the following general procedures should be applied and the modifications which are necessary. Thus if nitrogen is absent, it is doubtful whether the separation for bases would be necessary.

VII,7,3. Separation of water-insoluble mixtures If the water-insoluble mixture is a liquid, evaporate a small sample (say, 4 ml) in an evaporating dish on a water bath in order to determine the amount of volatile components, if any. If the solvent distils at the temperature of the boiling water bath, it is advisable to distil off this solvent on a water bath and to replace it by ether.

Place 15–20 ml of the liquid mixture in a 50-ml distilling flask arranged for distillation, and heat the flask on a boiling water bath until no more liquid passes over: redistil the distillate and if it is a single substance, identify it in the usual manner. Dissolve the residue (R) in ether and employ the same proportions as given for a solid mixture.

Step 1. Extraction and separation of the acidic components. Shake 5–10 g of the solid mixture (or of the residue R obtained after the removal of the solvent on a water bath) with 50 ml of pure ether, or other appropriate water-immiscible solvent (e.g., chloroform, dichloromethane, etc.). If there is a residue, separate it by filtration, preferably through a sintered glass funnel, wash it with a little ether and examine it appropriately. Shake the resulting ethereal solution in a small separatory funnel with 15 ml portions of 5 per cent aqueous sodium hydroxide solution until *all* the acidic components have been removed. Three portions of alkali are usually sufficient. Set aside the residual ethereal solution (E_1) for *Step 2*. Combine the sodium hydroxide extracts and wash the resulting mixture with 15–20 ml of ether; discard these washings. Render the alkaline extract acid to litmus with dilute sulphuric acid and then add excess of solid sodium hydrogen carbonate. Alternatively treat the sodium hydroxide extract with Cardice until no more is absorbed.

Separate any *phenolic* or *enolic compounds* which may be present by extracting the sodium hydrogen carbonate solution with two 20 ml portions of ether; remove the ether from the extract and examine any residue for phenols (or enols).

Strongly acidify the residual sodium hydrogen carbonate solution to Congo red with dilute sulphuric acid. If a solid acid forms, filter. Extract the filtrate or the acidified solution with two 20 ml portions of ether: keep the aqueous solution (A). Distil off the ether, and add the residual acid (if a solid) to the solid separated by filtration. Identify the acid.

Now distil the filtrate (A) and collect the distillate as long as it is acid to litmus. Should any solid separate out in the distilling flask during the distillation, add more water to dissolve it. Set aside the residue (B) in the flask. Identify the volatile acid in the distillate. A simple method is to just neutralise it with sodium hydroxide solution, evaporate to dryness and convert the residual sodium salt into the S-benzylisothiouronium salt (Section **VII,6,**15(e)).

The residue (B) in the distilling flask may still contain a water-soluble, non-volatile acid. Cool the acid solution, neutralise it with dilute sodium hydroxide solution to Congo red and evaporate to dryness on a water bath under reduced pressure (water pump). Heat a little of the residual salt (C) upon the tip of a nickel spatula in a Bunsen flame and observe whether any charring takes place. If charring occurs, thus indicating the presence of organic matter, extract the solid residue with 25 ml portions of hot absolute ethanol. Evaporate the ethanolic extract and identify the material which remains. The residue (C) contains the sodium salt of a water-soluble, non-volatile acid, which may be characterised as the S-benzylisothiouronium salt.

Step 2. Extraction of the basic components.   Extract the ethereal solution (E_1) with 15 ml portions of 5 per cent hydrochloric acid until *all* the basic components have been removed: two or three portions of acid are usually sufficient. Preserve the residual ethereal solution (E_2) for the separation of the neutral components. Wash the combined acid extracts with 15–20 ml of ether: discard the ether extract as in *Step 1*. Make the acid extract alkaline with 10–20 per cent sodium hydroxide solution: if any basic component separates, extract it with ether, evaporate the ether and characterise the residue. If a water-soluble base is also present, it may be recognised by its characteristic ammoniacal odour; it may be isolated from the solution remaining after the separation of the insoluble base by ether extraction by distilling the aqueous solution as long as the distillate is alkaline to litmus. Identify the base with the aid of phenyl isothiocyanate (compare Section **VII,6,***21*) or by other means.

Step 3. The neutral components.   The ethereal solution (E_2) remaining after the acid extraction of *Step 2* should contain only the neutral compounds of solubility groups V, VI and VII (see Table VII,1). Dry it with a little magnesium sulphate, and distil off the ether. If a residue is obtained, neutral compounds are present in the mixture. Test a portion of this with respect to its solubility in concentrated sulphuric acid; if it dissolves in the acid, pour the solution slowly and cautiously into ice-water and note whether any compound is recovered. Examine the main residue for homogeneity and if it is a mixture devise procedures, based for example upon differences in volatility, solubility in inert solvents, reaction with hydrolytic and other reagents, to separate the components.

The above procedure for water-insoluble mixtures is shown, in outline, in tabular form in Table VII,4. If the mixture is a liquid, the volatile solvent is assumed to have been removed.

VII,7,*4*. Separation of water-soluble mixtures   A water-soluble mixture may be in the form of a mixture of water-soluble solids or in the form of a liquid. The liquid mixtures are frequently aqueous solutions. The preliminary examination of a liquid mixture (p. 1145) will indicate whether a volatile solvent (i.e., removable on a boiling water bath) is present. If a volatile solvent is present, distil 20 g of the mixture from a water bath until no more liquid passes over: set aside the volatile solvent for identification. Dissolve the residue (R) in water as detailed below for a mixture of solids.

Step 1. Distillation from acid solution and the separation of the volatile acidic and neutral compounds.   Dissolve 6–10 g of a solid mixture in 50–75 ml of water: for a liquid mixture, use 6–10 g of the residue (R) (non-volatile from a boiling water bath) and dilute with 50–75 ml of water: for an aqueous solution use sufficient of it to contain 6–10 g of the dissolved components and dilute, if necessary, to 50–75 ml. Acidify the solution with sufficient 20 per cent sulphuric acid to decompose the salts of all the acidic components and to ensure the

Table VII,4 General scheme for separation of a water-insoluble mixture

Treat the mixture with pure ether* and filter, if necessary.

Residue. Examine for polysaccharides, etc., according to nature of original mixture.	Filtrate or ether solution. Extract with 5 per cent NaOH solution and separate the ethereal layer.		
	Sodium hydroxide extract. This will contain the acids and phenols (or enols) present. Acidify (litmus) with dilute H_2SO_4, add excess of solid $NaHCO_3$. Extract with ether.		**Ether solution (E_1).** Extract with 5 per cent HCl. Separate ether layer.
	Ether solution. Contains **phenolic** (or **enolic**) **compounds.**	**Sodium hydrogen carbonate solution.** Acidify with dilute H_2SO_4. (i) Filter or extract **acid** with ether. (ii) Distil aqueous solution (*A*) from (i) as long as distillate is acid to recover **water-soluble volatile acids.** (iii) Neutralise aqueous solution (*B*) from (ii), evaporate to dryness, and extract with absolute ethanol to recover **water-soluble, non-volatile compounds.** The residue may contain the sodium salt of a water soluble, non-volatile acid.	**Hydrochloric acid extract.** This will contain any basic components present. Render alkaline with 10–20 per cent NaOH and extract with ether.
			Ether solution (E_2). This will contain any neutral compounds present. Dry with magnesium sulphate, and distil off the ether. A residue indicates the presence of a **neutral component.** Determine the solubility of a portion in conc. H_2SO_4. Apply any other suitable tests.
			Ether solution. Contains **water-insoluble amines.** **Aqueous solution.** Will possess ammoniacal odour of water-soluble amines present. Distil as distillate is alkaline to recover **volatile water-soluble amines.**

* Or other appropriate solvent.

1149

presence of a slight excess of acid: many organic acids give an acid reaction with Congo red, hence it is necessary to add the sulphuric acid somewhat beyond the point at which the mixture is acid to this indicator. If an insoluble acidic component separates, filter it off and identify it. Distil the acid solution so long as the distillate appears turbid, or is acid to litmus, or until 100–150 ml are collected: in the last case, add more water to the contents of the distilling flask if the volume has been reduced below one-fourth of the original volume. Keep the residue (R_1) in the distilling flask for *Step 2*.

The distillate may contain volatile neutral compounds as well as volatile acids and phenols. Add a slight excess of 10–20 per cent sodium hydroxide solution to this distillate and distil until it is judged that all volatile organic compounds have passed over into the distillate. If necessary, the determination of the refractive index of the distillate or the application of specific chemical tests (e.g., for carbonyl compounds use the 2,4-dinitrophenylhydrazine reagent) should be used to confirm completion of distillation. Keep this distillate (S_1) for *Step 4*.

Cool the alkaline solution resulting from the distillation of the volatile neutral compounds, make it acid to litmus with dilute sulphuric acid and add an excess of *solid* sodium hydrogen carbonate. Extract this hydrogen carbonate solution with two 20 ml portions of ether; remove the ether from the combined ether extracts and identify the residual phenol (or enol). Then acidify the hydrogen carbonate solution cautiously with dilute sulphuric acid; if an acidic compound separates, remove it by two extractions with 20 ml portions of ether; if the acidified solution remains clear, distil and collect any water-soluble, volatile acid in the distillate. Characterise the acid as described in Section VII,6,*15*.

Step 2. Distillation from alkaline solution. Treat the solution (R_1) remaining in the distilling flask after the volatile acidic and neutral compounds have been removed with 10–20 per cent sodium hydroxide solution until distinctly alkaline. If a solid separates, filter it off and identify it. Distil the alkaline solution until no more volatile bases pass over (distillate no longer turbid, or not basic to litmus: water-soluble bases also possess characteristic odours): add more water to the contents of the flask if the solution becomes too concentrated during this distillation. (Set aside the solution in the distilling flask (S_2) for *Step 3*.) If the volatile basic compounds are insoluble in water, remove them by extraction with two 20 ml portions of ether, and identify the bases (compare Sections VII,6,*21* and VII,6,*22*) after evaporation of the ether. The water-soluble amines may be identified with phenyl isothiocyanate: it is best to concentrate the bases by redistilling and collecting the first half of the distillate separately.

Step 3. The non-steam-volatile compounds. The alkaline solution (S_2) remaining in the distilling flask from *Step 2* may contain water-soluble, non-volatile acidic, basic or neutral compounds. Add dilute sulphuric acid until the solution is just acid to Congo red, evaporate to dryness and extract the residual solid with boiling absolute ethanol: extraction is complete when the undissolved salt exhibits no sign of charring when heated on a metal spatula in the Bunsen flame. Evaporate the alcoholic solution to dryness and identify the residue.

Step 4. The steam-volatile neutral compounds. The solution (S_1) containing water-soluble neutral compounds obtained in *Step 1* is usually very dilute. It is advisable to concentrate it by distillation until about one-third to one-half of the original volume is collected as distillate; the process may be repeated if

Table VII,5 General scheme for separation of a water-soluble mixture

Acidify the aqueous solution (50–75 ml) prepared from (a) 6–10 g of the solid mixture, (b) from 6–10 g of the liquid residue (R) after distillation from a boiling water bath, or (c) from sufficient of original aqueous solution to contain 6–10 g of solute, with 20 per cent H_2SO_4 and distil.

Distillate. This will contain the steam-volatile acidic and neutral components present. Render alkaline with 10–20 per cent NaOH and distil.	**Aqueous acid solution** (R_1). Render alkaline with 10–20 per cent NaOH and distil.

Distillate side:

Distillate (S_1). This will contain the volatile neutral components present. Concentrate by distillation and saturate with solid K_2CO_3; the neutral component may separate.	**Aqueous alkaline solution.** This will contain any acids or phenols present. Cool, acidify (litmus) with dilute H_2SO_4, and add excess of solid $NaHCO_3$. Extract with ether.

Ether solution. Contains phenolic compounds.	**Sodium hydrogen carbonate solution.** Acidify with dilute H_2SO_4. Extract with ether.

Ether solution. Contains volatile water-insoluble acids.	**Aqueous solution. Distil.** The distillate may contain volatile water-soluble acids.

Aqueous acid solution side:

Distillate. Extract with ether.	**Aqueous alkaline solution** (S_2). Neutralise with dilute H_2SO_4 (Congo red). Evaporate to dryness and extract with absolute ethanol. The alcoholic extract contains the water-soluble, non-volatile components.

Ether solution. Contains volatile water-insoluble amines.*	**Aqueous solution.** Concentrate by distillation contains volatile water-soluble amines.*

* The separation of a mixture of amines by means of benzenesulphonyl chloride or toluene-p-sulphonyl chloride (Hinsberg's method) is described in Section VII,7,l.

necessary. It is frequently possible to salt out the neutral components from the concentrated distillate by saturating it with solid potassium carbonate. If a layer of neutral compound makes its appearance, remove it. Treat this upper layer (which usually contains much water) with solid anhydrous potassium carbonate: if another aqueous layer forms, separate the upper organic layer and add more anhydrous potassium carbonate to it. Identify the neutral compound.

Note. Some neutral compounds (e.g., methanol) cannot be salted out with potassium carbonate: distillation of the saturated aqueous potassium carbonate solution frequently yields the organic compound in a comparatively pure state, or at least in sufficiently concentrated a form to enable certain derivatives to be prepared.

The above procedure for the separation of a water-soluble mixture is summarised in Table VII,5.

BIBLIOGRAPHY AND REFERENCES
VII,4. Interpretation of spectra

1. For example: (a) A. Carrington and A. D. McLachlan (1967). *Introduction to Magnetic Resonance.* New York; Harper and Row. (b) H. Conroy (1960). *Advances in Organic Chemistry.* Ed. R. A. Raphael, E. C. Taylor and H. Wynberg. New York; Interscience. (c) J. R. Dyer (1965). *Applications of Absorption Spectroscopy of Organic Compounds.* New Jersey; Prentice Hall. (d) J. W. Emsley, J. Feeney and L. H. Sutcliffe (1965–6). *High Resolution Nuclear Magnetic Resonance Spectroscopy.* Oxford; Pergamon, Vol. 1 and 2. (e) L. M. Jackmann (1964). *Physical Methods in Organic Chemistry.* Ed. J. C. P. Schwarz. Edinburgh; Oliver and Boyd. (f) R. M. Lynden-Bell and R. K. Harris (1969). *Nuclear Magnetic Resonance Spectroscopy.* London; Nelson. (g) W. McFarlane and R. F. M. White (1972). *Techniques of High Resolution Nuclear Magnetic Resonance.* London; Butterworths. (h) J. A. Pople, W. G. Schieder and H. J. Bernstein (1959). *High Resolution Nuclear Magnetic Resonance.* New York; McGraw-Hill. (i) D. H. Williams and I. Fleming (1973). *Spectroscopic Methods in Organic Chemistry.* 3rd edn. London; McGraw-Hill.
2. For example: (a) H. C. Hill (1974). *Introduction to Mass Spectrometry.* 2nd edn. London; Heyden and Son. (b) G. W. A. Milne (1971). *Mass Spectrometry: Techniques and Applications.* New York; Wiley-Interscience. (c) S. R. Shrader (1971). *Introductory Mass Spectrometry.* Boston; Allyn and Bacon. See also 1(e) and 1(i) above.
3. For example: (a) A. Ault (1967). *Problems in Organic Structure Determination.* McGraw-Hill. (b) A. J. Baker, G. Eglinton and F. J. Preston (1967). *More Spectroscopic Problems in Organic Chemistry.* Ed. T. Cairns. London; Heyden and Son, Vol. 4. (c) B. J. Brisdon and D. W. Brown (1973). *Spectroscopic Problems in Chemistry.* New York; Nostrand-Reinhold. (Inorganic, organometallic and organic problems.) (d) S. F. Dyke, A. J. Floyd, M. Sainsbury and R. S. Theobald (1971). *Organic Spectroscopy.* Harmondsworth; Penguin Education. (e) D. H. Williams and L. Fleming (1967). *Spectroscopic Problems in Organic Chemistry.* London; McGraw-Hill. See also 1(c).

VII,4,1. Infrared spectroscopy

4. (a) L. J. Bellamy (1975). *The Infrared Spectra of Complex Molecules.* 3rd edn. London; Chapman and Hall. (b) L. J. Bellamy (1968). *Advances in Infrared Group Frequencies.* London; Methuen. (c) *Advances in Infrared and Raman Spectroscopy* (1975). Ed. R. J. H. Clark and R. E. Hester, Vol. 1. London; Heyden and Son. (d) N. B. Colthup, L. H. Daly and S. E. Wiberley (1975). *Introduction to Infrared and Raman Spectroscopy.* 2nd edn. New York; Academic Press. (e) R. T. Conley (1972). *Infrared Spectroscopy.* 2nd edn. Boston; Allyn and Bacon. See also 1(c), 1(e), 1(i) above.

Note should also be made of the texts specified in *Compilation of Data*, p. 1267.

VII,4,2. Nuclear magnetic resonance spectroscopy

5. C. W. Haigh (1971). 'Computer Programs in the Analysis of NMR Spectra', in *Annual Reports on NMR Spectroscopy.* Ed. E. F. Mooney. New York; Academic Press.
6. (a) K. B. Wiberg and B. J. Nist (1962). *Interpretation of NMR Spectra.* New York; Benjamin. (b) J. D. Roberts (1961). *An Introduction to the Analysis of Spin–Spin Splitting in High Resolution NMR Spectra.* New York; Benjamin.
7. A. F. Cockerill, G. L. O. Davies and D. M. Rackham (1973). 'Lanthanide Shift Reagents for Nuclear Magnetic Resonance Spectroscopy', *Chem. Rev.,* **73**, 553.
8. G. C. Levy and G. L. Nelson (1972). *Carbon-13 Nuclear Magnetic Resonance for Organic Chemists.* New York, Wiley-Interscience.
9. Other valuable texts dealing with interpretative aspects: (a) R. H. Bible (1965). *Interpretation of NMR Spectra.* New York; Plenum Press. (b) L. M. Jackman and S. Sternhell (1969). *Applications of Nuclear Magnetic Resonance in Organic Chemistry.* 2nd edn. Oxford; Pergamon. (c) *Nuclear Magnetic Resonance for Organic Chemists* (1967). Ed. D. W. Mathieson. New York; Academic Press.

Note should also be made of the texts specified in *Compilation of Data*, p. 1267.

VII,4,3. Mass spectrometry

10. S. Abrahamson, E. Sten-hagen and F. W. McLafferty (1974). *Registry of Mass Spectral Data.* New York; Wiley.
11. J. H. Beynon and A. E. Williams (1963). *Mass Abundance Tables for use in Mass Spectrometry.* London; Elsevier.
12. For example: (a) J. H. Beynon, R. A. Saunders and A. E. Williams (1968). *The Mass Spectra of Organic Molecules.* New York; Elsevier. (b) H. C. Budzikiewiez, C. Djerassi and D. H. Williams (1967). *Mass Spectrometry of Organic Compounds.* San Francisco; Holden-Day. (c) H. C. Budzikiewiez, C. Djerassi and D. H. Williams (1964). *Structural Elucidation of Natural Products by Mass Spectrometry* (two volumes). San Francisco; Holden-Day. (d) D. H. Williams and I. Howe (1972). *Principles of Organic Mass Spectrometry.* London; McGraw-Hill. (e) F. W. McLafferty (1973). *Interpretation of Mass Spectra.* 2nd edn. London; Benjamin. (f) K. Biemann (1962). *Mass Spectrometry: Organic Chemical Applications.* New York; McGraw-Hill.

Note should also be made of the texts specified in the *Compilation of Data*, p. 1267.

VII,4,4. Ultraviolet–visible spectroscopy

13. For example: (*a*) R. Friedel and M. Orchin (1958). *Ultraviolet Spectra of Aromatic Compounds.* New York; Wiley. (*b*) H. H. Jaffé and M. Orchin (1962). *Theory and Applications of Ultraviolet Spectroscopy.* New York; Wiley. (*c*) A. I. Scott (1964). *Interpretation of the Ultraviolet Spectra of Natural Products.* Oxford; Pergamon. See also 1(*c*), 1(*e*), 1(*i*), 3(*c*) and 3(*d*).

Note should also be made of the texts specified in the *Compilation of Data*, p. 1267.

General interpretative spectroscopic texts

(*a*) C. J. Cresswell, O. A. Runquist and M. M. Cambell (1972). *Spectral Analysis of Organic Compounds.* 2nd edn. London; Longman. (*b*) R. M. Silverstein, C. G. Bassler and T. C. Merrill (1974). *Spectroscopic Identification of Organic Compounds.* 3rd edn. Wiley International Edition. (*c*) *Introduction to Spectroscopic Methods for the Identification of Organic Compounds* (1974). Ed. F. Scheimann (two volumes). Oxford; Pergamon. See also 1(*c*), 1(*e*), 1(*i*) and 3(*d*).

CHAPTER VIII

PHYSICAL CONSTANTS OF ORGANIC COMPOUNDS

TABLES OF DERIVATIVES

The tables in this chapter contain lists of the more common members of the various classes of organic compounds arranged, as far as possible, in the order of increasing boiling points or melting points, together with the melting points of selected derivatives. Some of the tables are subdivided: thus in Table VIII,6, **Phenols**, the additional sub-headings are *Halogeno-phenols* and *Nitro-phenols*. This subdivision renders the tables less cumbersome and facilitates their use.

In describing compounds in the literature, the range of a certain number of degrees is usually recorded for the boiling point or the melting point; the highest point of the boiling point or the melting point range is listed in the tables and the value is given to the nearest whole degree. For those compounds the author has purified or prepared, the actual observed boiling point or melting point is recorded. Densities are given for a temperature of 20 °C referred to water at 4 °C unless otherwise indicated. Refractive indices are recorded for the sodium D line at 20 °C. It must be remembered that the value obtained for a melting point depends upon the observer and upon the method which was used in the determination: this often accounts for the slightly different values found in the literature for the same compound.

INDEX TO TABLES

Table VIII,1   Saturated aliphatic hydrocarbons

Hydrocarbon	B.P. °C (/mmHg)	$d_4^{20°}$	$n_D^{20°}$
Pentane	36	0.627	1.358
Hexane	68.5	0.659	1.374
Heptane	98	0.683	1.388
Octane	125	0.703	1.397
Nonane	150.5	0.717	1.405
Decane	173	0.730	1.412
Undecane	196 (87/20)	0.740	1.417
Dodecane	216 (94/14)	0.750	1.422
Tridecane	92.5/4.5	0.756	1.425
Tetradecane	252 (123/12)	0.762	1.429
Pentadecane	270 (120/4.5)	0.769	1.432
Hexadecane	143.5/9 (m.p. 18)	0.774	1.435
Octadecane	308 (m.p. 28)	—	—
2-Methylbutane (isopentane)	28	0.620	1.354
2-Methylpentane	60	0.653	1.372
2,2,4-Trimethylpentane	99	0.688	1.389
2,7-Dimethyloctane	160	0.725	1.409
Cyclopentane	49	0.745	1.406
Cyclohexane	81	0.779	1.426
Methylcyclohexane	101	0.769	1.423
Cycloheptane	118	0.811	1.445
Ethylcyclohexane	130	0.784	1.432
Isopropylcyclohexane	154.5	0.802	1.441
Propylcyclohexane	155	0.790	1.436
Butylcyclohexane	177	0.800	1.440
Isopentylcyclohexane	193	0.802	1.442
Pentylcyclohexane	200	0.804	1.444
Bicyclohexyl	237 (m.p. 3)	0.889	1.480
trans-Decahydronaphthalene (trans-Decalin)	185	0.870	1.470
cis-Decahydronaphthalene (cis-Decalin)	194	0.895	1.481
1,2,3,4-Tetrahydronaphthalene (Tetralin)	207	0.971	1.540
trans-p-Menthane (1)	161	0.792	1.439
cis-p-Menthane (1)	169	0.816	1.451

(1) 1-Isopropyl-4-methylcyclohexane.

Table VIII,2 Unsaturated aliphatic hydrocarbons

Hydrocarbon	B.P. °C (/mmHg)	$d_4^{20°}$	$n_D^{20°}$	Adducts with 2,4-dinitrophenyl sulphenyl chloride	Mercurides
Pent-1-ene	30	0.641	1.371	—	—
2-Methylbut-1-ene	31	0.650	1.378	—	—
cis-Pent-2-ene	35	0.659	1.382	—	—
trans-Pent-2-ene	37	0.649	1.379	—	—
Trimethylethylene	38	0.662	1.388	—	—
Hex-1-ene	64	0.674	1.388	62	—
Hept-1-ene	93	0.697	1.400	—	—
Oct-1-ene	121	0.716	1.409	—	—
Non-1-ene	146	0.731	1.413	—	—
Dec-1-ene	169	0.742	1.422	—	—
Undec-1-ene	193	0.779	1.444	—	—
Dodec-1-ene	213	0.760	1.430	—	—
Tetradec-1-ene	125/15	0.773	1.437	—	—
Hexadec-1-ene	153/14 (m.p. 15)	0.782	1.441	—	—
Octadec-1-ene	180/18 (m.p. 18)	0.789	1.445	—	—
Isoprene (2-methylbuta-1,3-diene)	34	0.681	1.419	—	—
Penta-1,3-diene (piperylene)	42	0.680	1.431	—	—
Hexa-1,5-diene (biallyl)	59	0.690	1.402	—	—
2,3-Dimethylbuta-1,3-diene	69	0.726	1.439	—	—
Cyclopentene	45	0.772	1.420	—	—
Cyclohexene	83	0.810	1.445	117	—
Cyclopenta-1,3-diene	42	0.803	1.443	—	—
Dicyclopentadiene	170 (m.p. 32)	—	—	—	—
Cyclohexa-1,3-diene	81	0.841	1.474	—	—
α-Pinene	156	0.860	1.456	—	—
Camphene	160 (m.p. 51)	—	—	—	—
Dipentene	178	0.840	1.473	195	—
Pent-1-yne	39	0.695	1.385	—	118
Pent-2-yne	56	0.712	1.404	—	—
Hex-1-yne	71	0.715	1.399	—	99
Hept-1-yne	100	0.734	1.409	—	61
Oct-1-yne	126	0.746	1.416	—	—
Non-1-yne	151	0.758	1.423	—	—
Phenylacetylene	142	0.925	1.552	—	125

Table VIII,3 Aromatic hydrocarbons

Hydrocarbon	B.P. °C	M.P. °C	$d_4^{20°}$	$n_D^{20°}$	Picrate °C
Benzene	80	6	0.879	1.501	—
Toluene	111	—	0.867	1.497	—
Ethylbenzene	135	—	0.868	1.496	97
p-Xylene	138	13	0.861	1.496	—
m-Xylene	139	—	0.864	1.497	—
Phenylacetylene	142	—	0.925	1.552	—
o-Xylene	144	—	0.880	1.505	—
Styrene (1)	146	—	0.909	1.546	—
Isopropylbenzene (2)	153	—	0.862	1.491	—
Propylbenzene	159	—	0.864	1.493	103
Mesitylene (3)	164	—	0.865	1.499	97
t-Butylbenzene	169	—	0.867	1.493	—
Pseudocumene (4)	169	—	—	1.504	97
s-Butylbenzene	172	—	0.861	1.490	—
p-Cymene (5)	177	—	0.857	1.490	—
Indan	177	—	0.965	1.538	—
m-Diethylbenzene	182	—	0.865	1.496	—
Indene	182	—	0.992	1.576	98
Butylbenzene	182	—	0.861	1.490	—
Isodurene (6)	197	—	—	—	—
Prehnitene (7)	204	—	0.901	1.523	90
Pentylbenzene	204	—	0.859	1.488	—
Tetralin (8)	207	—	0.971	1.540	—
1,3,5-Triethylbenzene	218	—	0.863	1.497	—
Cyclohexylbenzene	238	7	0.950	1.533	—
1-Methylnaphthalene	241	—	1.019	1.618	141
Diphenylmethane	262	25	—	—	—
2-Methylnaphthalene	241	34	—	—	115
Bibenzyl	284	52	—	—	—
Pentamethylbenzene	231	54	—	—	131
Biphenyl †	255	70	—	—	—
Durene (9)	193	79	—	—	—
Naphthalene	218	80	—	—	150
Triphenylmethane	358	92	—	—	—
Acenaphthene	278	95	—	—	162
Retene (10)	390	99	—	—	123
Phenanthrene	340	100	—	—	143
Fluorene (11)	294	114	—	—	84
trans-Stilbene (12)	306	124	—	—	94
cis-Stilbene	148/17 mm	5	—	—	—
Pyrene	—	149	—	—	227
1,1'-Binaphthyl	—	160	—	—	145
2,2'-Binaphthyl	—	188	—	—	184
Anthracene	340	216	—	—	138
Chrysene (13)	448	254	—	—	273

(1) Phenylethylene.
(2) Cumene.
(3) 1,3,5-Trimethylbenzene.
(4) 1,2,4-Trimethylbenzene.
(5) 4-Isopropyl-1-methylbenzene.
(6) 1,2,3,5-Tetramethylbenzene.
(7) 1,2,3,4-Tetramethylbenzene.

(8) 1,2,3,4-Tetrahydronaphthalene.
(9) 1,2,4,5-Tetramethylbenzene.
(10) 7-Isopropyl-1-methylphenanthrene.
(11) Diphenylenemethane.
(12) trans-1,2-Diphenylethylene.
(13) 1,2-Benzphenanthrene.

Aroyl-benzoic acid °C	Compound with 2,4,7-trinitro-9-fluorenone	Compound with 1,3,5-trinitro-benzene	Styphnate	Other derivatives * °C
128	—	—	—	1,3-Dinitro, 90
138	—	—	—	2,4-Dinitro, 71
128	—	—	—	2,4,6-Trinitro, 37
132	—	—	—	2,3,5-Trinitro, 139
126	—	—	—	2,4,6-Trinitro, 182
				Mercuride, 125
167	—	—	—	4,5-Dinitro, 71
—	—	—	—	2,4-Dinitrobenzene sulphenyl chloride
				adduct, 143
134	—	—	—	2,4,6-Trinitro, 109
126	—	—	—	—
212	—	—	—	2,4,6-Trinitro, 235
—	—	—	—	2,4-Dinitro, 62
149	—	—	—	3,5,6-Trinitro, 185
				$CrO_3 \rightarrow C_6H_5COCH_3$
124	—	—	—	2,6-Dinitro, 54
114	—	—	—	2,4,6-Trinitro, 62
—	—	102	—	—
—	—	—	—	—
213	—	—	—	4,6-Dinitro, 157
—	—	—	—	5,6-Dinitro, 176; Dibromo, 208
—	—	—	—	Dibromo, 64
154	—	—	—	5,7-Dinitro, 95
—	—	—	—	2,4,6-Trinitro, 108
—	—	—	—	4-Nitro, 59
168	—	154	135	4-Nitro, 71
—	—	—	—	2,2',4,4'-Tetranitro, 172
190	127	123	130	1-Nitro, 81
—	—	102	—	4,4'-Dinitro, 180
—	—	121	—	—
220	132	—	—	4,4'-Dibromo, 164
264	—	—	—	3,6-Dinitro, 207
173	153	156	168	1-Nitro, 61
—	—	—	—	4,4',4''-Trinitro, 212; Triphenylcarbinol, 162
200	175	168	154	5-Nitro, 101
—	—	139	141	—
—	196	164	142	Phenanthraquinone, 202
228	179	105	134	2-Nitro, 156; 2,7-Dibromo, 165
—	—	120	142	—
—	—	—	—	—
—	242	245	191	—
—	—	—	—	—
—	194	164	180	Anthraquinone, 286
214	248	186	—	—

* For melting points of sulphonamides derived from some of these aromatic hydrocarbons see Table VIII,41.

† Biphenyl should not be nitrated since 4-nitrobiphenyl is a proven powerful carcinogen.

Table VIII,4   Aliphatic alcohols

Alcohol	B.P. °C	M.P. °C	3,5-Dinitrobenzoate °C
Methanol (methyl alcohol)	65	—	109
Ethanol (ethyl alcohol)	78	—	94
Propan-2-ol (isopropyl alcohol)	82	—	122
2-Methylpropan-2-ol (t-butyl alcohol)	83	25	142
Allyl alcohol	97	—	50
Propan-1-ol (propyl alcohol)	97	—	75
Butan-2-ol (s-butyl alcohol)	99	—	76
2-Methylbutan-2-ol (t-pentyl alcohol)	102	—	118
2-Methylbut-3-yn-2-ol	105	—	—
2-Methylpropan-1-ol (isobutyl alcohol)	108	—	88
3-Methylbutan-2-ol	113	—	76
2,2-Dimethylpropan-1-ol (neopentyl alcohol)	113	52	—
Prop-2-yn-1-ol (propargyl alcohol)	114	—	—
Pentan-3-ol	116	—	100
Butan-1-ol (butyl alcohol)	118	—	64
Pentan-2-ol	119	—	62
2-Methylpentan-2-ol	121	—	72
3-Methylpentan-3-ol	128	—	—
2-Methylbutan-1-ol	129	—	70
3-Methylbutan-1-ol (isopentyl alcohol)	132	—	62
Hexan-3-ol	135	—	77
Pentan-1-ol (pentyl alcohol)	138	—	46
Cyclopentanol	141	—	115
2-Ethylbutan-1-ol	149	—	52
Heptan-4-ol	156	—	64
Hexan-1-ol	156	—	61
Heptan-2-ol	160	—	49
Cyclohexanol	161	25	113
2-Methylcyclohexanol	165	—	115
4-Hydroxypentan-2-one (diacetone alcohol)	166	—	55
Furfuryl alcohol	170	—	81
4-Methylcyclohexanol	174	—	134
3-Methylcyclohexanol	175	—	98
Heptan-1-ol	176	—	48
Tetrahydrofurfuryl	177	—	84
Octan-2-ol	179	—	32
Octan-1-ol	194	—	62
(−)-Linalool	199	—	—
Nonan-1-ol	214	—	52
(−)-Isoborneol	216	—	—
(±)-α-Terpineol	219	35	79
Geraniol	230	—	63
Decan-1-ol	231	6	57
Undecan-1-ol	243	16	55
Dodecan-1-ol	259	24	60

p-Nitro-benzoate °C	Phenyl-urethan °C	1-Naphthyl-urethan °C	Hydrogen 3-nitro-phthalate °C	O-Alkyl saccharin °C	Other derivatives °C
96	47	124	153	182	—
57	52	79	157	219	—
110	86	106	153	137	—
116	136	101	—	—	—
29	70	109	124	—	—
35	57	80	145	125	—
26	64	98	131	66	—
85	42	72	—	—	—
—	102	—	—	—	$d_4^{20°}$ 0.807, $n_D^{20°}$ 1.421
68	86	104	179	100	—
—	69	109	—	—	—
—	144	100	—	—	—
—	63	—	—	—	$d_4^{20°}$ 0.948, $n_D^{20°}$ 1.432
17	49	95	121	—	—
36	61	72	147	96	—
17	—	76	103	—	—
—	239	—	—	—	—
—	50	—	—	—	—
—	—	82	158	—	—
21	57	68	166	64	—
—	—	—	—	—	—
11	46	68	136	62	—
62	132	118	—	—	—
—	—	61	147	—	—
35	—	80	—	—	—
5	42	59	124	60	—
—	—	54	—	—	—
50	82	129	160	—	—
65	103	155	—	—	—
48	—	—	—	—	2,4-Dinitrophenylhydrazone, 203
76	45	129	—	—	—
—	125	160	—	—	—
—	94	122	—	—	—
10	65	62	127	55	—
47	61	90	—	—	Diphenylcarbamate, 81
28	114	64	—	—	—
12	74	66	128	46	—
70	66	53	—	—	—
10	69	65	125	49	—
129	138	—	130	—	—
97	113	152	—	—	—
35	—	48	117	—	Diphenylcarbamate, 82
30	60	71	123	48	—
29	62	73	123	59	—
45	74	80	124	54	—

Table VIII,4   **Aliphatic alcohols** (*continued*)

Alcohol	B.P. °C	M.P. °C	3,5-Dinitro-benzoate °C
Tetradecan-1-ol	160/10	39	67
(−)-Menthol	216	43	153
Hexadecan-1-ol	190/15	50	66
But-2-yne-1,4-diol	145/15	55	191
Octadecan-1-ol	—	59	66
(−)-Borneol	212	205	154
2-Chloroethanol (ethylene chlorohydrin)	129	—	92
3-Chloropropan-1-ol (trimethylene chlorohydrin)	161d	—	77
1,3-Dichloropropan-2-ol (glycerol αγ-dichlorohydrin)	176	—	129
2,3-Dichloropropan-1-ol (glycerol βγ-dichlorohydrin)	183	—	—
3-Chloropropane-1,2-diol (glycerol α-monochlorohydrin)	213	—	—
1-Chloropropan-2-ol	127	—	83
2-Chloropropan-1-ol	132	—	76
Trichloroethanol	151 (m.p. 19)	—	—
2-Bromoethanol (ethylene bromohydrin)	149	—	86
1,3-Dibromopropan-2-ol (glycerol αγ-dibromohydrin)	219d	—	—
Propane-1,2-diol (propylene glycol)	187	—	—
Ethane-1,2-diol (ethylene glycol)	198	—	169
Propane-1,3-diol (trimethylene glycol)	215	—	178
Butane-1,4-diol (tetramethylene glycol)	230	19	—
Pentane-1,5-diol (pentamethylene glycol)	239	—	—
Di-2-hydroxyethyl ether (diethylene glycol)	244	—	149
Hexane-1,6-diol (hexamethylene glycol)	250	42	—
Glycerol	290d	—	—
Ethyleneglycol monomethyl ether	124	—	—
Ethyleneglycol monoethyl ether	135	—	75
Ethyleneglycol monoisopropyl ether	142	—	—
Ethyleneglycol monopropyl ether	151	—	—
Ethyleneglycol monobutyl ether	168	—	—
Ethyleneglycol monophenyl ether	245	—	—
Diethyleneglycol monomethyl ether	194	—	—
Diethyleneglycol monoethyl ether	202	—	—
Diethyleneglycol monopropyl ether	—	—	—
Diethyleneglycol monobutyl ether	232	—	—
2-Aminoethanol (monoethanolamine)	171	—	—
2,2′-Dihydroxydiethylamine (diethanolamine)	270	28	—
2,2′,2″-Trihydroxytriethylamine (triethanolamine)	360	—	—

p-Nitro benzoate °C	Phenylurethan °C	1-Naphthylurethan °C	Hydrogen 3-nitrophthalate °C	O-Alkyl saccharin °C	Other derivatives °C
51	74	82	123	62	—
62	112	126	—	—	Benzoate, 54
52	73	82	120	70	—
—	132	—	—	—	—
64	80	89	119	75	—
137	138	127	—	—	—
—	51	101	98	—	$d_4^{20°}$ 1.202, $n_D^{20°}$ 1.442
—	—	76	—	—	$d_4^{20°}$ 1.131, $n_D^{20°}$ 1.447
—	73	115	—	—	$d_4^{20°}$ 1.353, $n_D^{20°}$ 1.480
38	73	93	—	—	—
108	—	—	—	—	—
—	—	—	—	—	—
—	87	120	—	—	—
—	86	—	172	—	$d_4^{20°}$ 1.763, $n_D^{20°}$ 1.492
78	81	—	—	—	$d_4^{20°}$ 2.120, $n_D^{20°}$ 1.550
127	153	—	—	—	—
141	157	176	—	—	Dibenzoate, 73
119	137	164	—	—	Dibenzoate, 59
175	183	198	—	—	Dibenzoate, 82
105	176	147	—	—	—
—	—	122	—	—	$d_4^{20°}$ 1.116, $n_D^{20°}$ 1.448
—	—	—	—	—	—
188	180	192	—	—	Tribenzoate, 72
50	—	113	129	—	$d_4^{20°}$ 0.966, $n_D^{20°}$ 1.402; 3,4,5-triiodobenzoate, 152
—	—	67	118	—	$d_4^{20°}$ 0.930, $n_D^{20°}$ 1.408; 3,4,5-triiodobenzoate, 128
—	—	—	—	—	$d_4^{20°}$ 0.903, $n_D^{20°}$ 1.410 3,4,5-triiodobenzoate, 80
—	—	—	—	—	$d_4^{20°}$ 0.911, $n_D^{20°}$ 1.413
120	—	—	—	—	$d_4^{20°}$ 0.902, $n_D^{20°}$ 1.420; 3,4,5-triiodobenzoate, 85
113	—	—	—	—	$d_4^{20°}$ 1.104, $n_D^{20°}$ 1.534; 3,4,5-triiodobenzoate, 145; toluene-p-sulphonate, 80
—	—	—	89	—	$d_4^{20°}$ 1.036, $n_D^{20°}$ 1.424; 3,4,5-triiodobenzoate, 82
—	—	—	—	—	$d_4^{20°}$ 1.024, $n_D^{20°}$ 1.430; 3,4,5-triiodobenzoate, 76
—	—	—	—	—	$d_4^{20°}$ 0.958, $n_D^{20°}$ 1.434; 3,4,5-triiodobenzoate, 54
—	—	—	—	—	$d_4^{20°}$ 1.022, $n_D^{20°}$ 1.454; Picrate, 160
—	—	—	—	—	$d_4^{20°}$ 1.097, $n_D^{20°}$ 1.478; Picrate, 110
—	—	—	—	—	$d_4^{20°}$ 1.124, $n_D^{20°}$ 1.485; Hydrochloride, 177

Table VIII,5   Aromatic alcohols

Alcohol	B.P. (/mmHg) °C	M.P. °C	3,5-Dinitro-benzoate °C
1-Phenylethanol	203	20	94
Benzyl alcohol	205	—	113
α-Hydroxy-*m*-xylene	217	—	111
1-Phenylpropan-1-ol	219	—	—
2-Phenylethanol	220	—	108
1-Phenylbutan-1-ol	118/18	16	—
1-Phenylpentan-1-ol	137/21	—	—
3-Phenylpropan-1-ol	237	—	92
o-Methoxybenzyl alcohol	249	—	—
m-Methoxybenzyl alcohol	252	—	—
p-Methoxybenzyl (anisyl alcohol)	259	25	—
Cinnamyl alcohol	257	33	121
α-Hydroxy-*o*-xylene	219	39	—
Piperonyl (3,4-dimethylenedioxybenzyl) alcohol	—	58	—
α-Hydroxy-*p*-xylene	217	60	118
Diphenylmethanol	298	69	142
m-Hydroxybenzyl alcohol	—	73	—
o-Hydroxybenzyl alcohol	—	87	—
p-Hydroxybenzyl alcohol	—	125	—
Benzoin	—	137	—
Triphenylmethanol	380	162	—

Halogeno-alcohols

m-Chlorobenzyl alcohol	234		—
m-Bromobenzyl alcohol	254	—	—
m-Iodobenzyl alcohol	165/16	—	—
p-Iodobenzyl alcohol	—	72	—
o-Chlorobenzyl alcohol	230	74	—
p-Chlorobenzyl alcohol	235	75	—
p-Bromobenzyl alcohol	—	77	—
o-Bromobenzyl alcohol	—	80	—
o-Iodobenzyl alcohol	—	90	—

Nitro- and amino-alcohols

m-Nitrobenzyl alcohol	—	27	—
p-Aminobenzyl alcohol	—	65	—
o-Nitrobenzyl alcohol	270	74	—
o-Aminobenzyl alcohol	—	82	—
p-Nitrobenzyl alcohol	185/12	93	—
m-Aminobenzyl alcohol	—	97	—

p-Nitro-benzoate °C	Phenyl-urethan °C	1-Naphthyl-urethan °C	Hydrogen 3-nitro-phthalate °C	Other derivatives °C
43	92	106	—	—
86	76	134	176	—
89	—	116	—	—
60	—	102	—	—
63	80	119	123	—
58	—	99	—	—
—	75	—	—	—
46	48	—	117	—
—	—	136	—	Benzoyl, 59
—	—	—	—	—
94	93	—	—	Benzoyl, 38; anisic acid, 184
78	91	114	—	—
101	79	—	—	o-Toluic acid, 104
—	102	—	—	Benzoyl, 66
—	79	—	—	—
131	140	136	—	Acetyl, 42
—	—	—	—	Acetyl, 84; diacetyl, 75
—	—	—	—	Benzoyl, 51
—	—	—	—	—
123	165	140	—	Acetyl, 83; benzoyl, 125; semicarbazone, 206d; 2,4-dinitrophenylhydrazone, 234
—	—	—	—	Acetyl, 88; triphenylmethane, 92
—	—	—	—	—
—	—	—	—	—
—	—	—	—	—
94	—	—	—	—
—	—	—	—	Acetyl, 23
—	—	—	—	o-Bromobenzoic acid, 150 (KMnO$_4$)
—	—	—	—	—
—	—	—	—	Benzoyl, 72
—	—	—	—	Diacetyl, 188
—	—	—	—	Benzoyl, 102
—	—	—	—	N-Acetyl, 114; picrate, 110
—	—	—	—	Benzoyl, 95; acetyl, 78
—	—	—	—	N-Acetyl, 107; dibenzoyl, 114

Table VIII,6   Phenols

Phenol	B.P. °C	M.P. °C	Bromo compound °C	Acetate °C	Benzoate °C	p-Nitro-benzoate °C
Salicylaldehyde (1)	197	—	—	39	—	128
m-Cresol	202	12	84 †	Liq.	55	90
o-Ethylphenol	207	—	—	—	39	56
m-Ethylphenol	217	—	—	Liq.	52	68
Methyl salicylate	223	—	—	49	92	128
Ethyl salicylate	231	—	—	—	80	108
Carvacrol (2)	238	—	46	Liq.	—	51
Propyl salicylate	239	—	—	—	—	—
Isopropyl salicylate	241	—	—	—	—	—
m-Methoxyphenol	244	—	104 †	Liq.	—	—
Eugenol (3)	254	—	118 ‡	30	70	81
Butyl salicylate	260	—	—	—	—	—
Isoeugenol (4)	266	—	—	80	106	109
p-Butylphenol	248	22	—	—	127	68
o-Methoxyphenol (5)	205	28	116 †	Liq.	58	93
2,4-Dimethylphenol (6)	211	28	179 †	Liq.	38	105
o-Cresol	191	30	56 *	Liq.	Liq.	94
p-Cresol	202	36	49 *	Liq.	71	98
Phenol	182	42	95 †	Liq.	69	126
Phenyl salicylate (7)	—	43	—	98	81	111
p-Ethylphenol	219	47	—	Liq.	60	81
2,6-Dimethylphenol (8)	203	49	79	—	—	—
Thymol (9)	233	51	55	Liq.	33	70
o-Cyclohexylphenol	—	55	—	—	—	—
p-Methoxyphenol (10)	243	56	—	32	87	—
o-Hydroxybiphenyl	275	58	—	63	76	—
Orcinol hydrate (11)	289	58	104 †	25	88	214
3,4-Dimethylphenol (12)	228	62	171		—	58	—
3,5-Dimethylphenol (13)	219	68	166 †	Liq.	24	109
2,4,6-Trimethylphenol (14)	220	69	158 *	—	62	—
4-Hexylresorcinol	335	69	—	—	—	—
2,4,5-Trimethylphenol (15)	232	71	35	34	63	—
2,3-Dimethylphenol (16)	218	75	—	—	—	—
2,5-Dimethylphenol (17)	211	75	178 †	Liq.	61	87
Vanillin (18)	—	81	160	102	78	—
o-Hydroxybenzyl alcohol (19)	—	87	—	—	51	—
1-Naphthol	279	94	105 *	49	56	143
2-Naphthyl salicylate	—	95	—	136	—	—
p-t-Pentylphenol	266	96	—	Liq.	61	—
2,3,5-Trimethylphenol	233	96	—	—	50	—
p-t-Butylphenol	237	99	—	Liq.	82	—
Catechol	240	105	192 ‡	65	84	169
2,5-Dihydroxybiphenyl (20)	—	103	—	—	—	—
Orcinol (21)	289	108	104 †	25	88	214
m-Hydroxybenzaldehyde	240	108	—	Liq.	38	—
1,2-Dihydroxynaphthalene (22)	—	108	—	106	—	—
2,2'-Dihydroxybiphenyl	—	109	—	95	101	—
Resorcinol	280	110	112 *	Liq.	117	182
Ethyl p-hydroxybenzoate	—	116	—	—	94	—
p-Hydroxybenzaldehyde	—	116	—	Liq.	90	—
2-Naphthol	285	123	84	72	107	169
1,3-Dihydroxynaphthalene (23)	—	124	—	56	—	—
3,3'-Dihydroxybiphenyl	—	124	—	83	92	—
Methyl p-hydroxybenzoate	—	131	—	85	135	—
p-Cyclohexylphenol	—	132	—	—	118	137
Pyrogallol	309	133	158 *	173	90	230
1,6-Dihydroxynaphthalene	—	138	—	73	104	—
Hydroxyhydroquinone (24)	—	140	—	97	120	—

3,5-Dinitro-benzoate °C	Aryloxy-acetic acid °C	NN-diphenyl urethan °C	1-Naphthyl urethan °C	Toluene-p-sulphonate °C	2,4-Dinitro-phenyl ether °C	O-Aryl saccharin °C
—	132	—	—	64	—	—
165	103	101	128	56	74	146
108	141	—	—	—	—	—
—	75	—	—	—	—	—
—	—	—	—	—	—	—
—	—	—	—	—	—	—
77	151	—	116	—	—	—
—	—	—	—	—	—	—
—	114	—	129	—	—	—
131	80	—	122	85	115	—
—	—	—	—	—	—	—
—	94	—	150	—	130	—
—	81	—	—	—	—	—
142	119	118	118	85	97	—
164	142	—	135	—	—	—
138	152	73	142	55	90	163
189	136	94	146	70	94	172
146	99	105	133	96	69	182
—	—	144	—	—	—	—
132	97	—	128	—	—	—
159	140	—	—	—	—	—
103	148	—	160	71	67	—
—	111	—	—	—	—	—
—	—	—	—	65	—	—
190	217	154	160	—	—	—
181	163	—	142	—	—	—
195	86	—	—	—	—	—
—	142	—	—	—	—	—
—	—	—	—	—	—	—
—	132	—	—	—	—	—
—	187	—	—	—	—	—
137	118	—	173	—	—	—
—	189	—	—	115	131	—
—	120	—	—	—	—	—
217	192	—	152	88	128	—
—	—	—	—	—	—	—
—	—	—	—	54	—	—
—	86	—	110	—	—	—
152	—	—	175	—	—	—
—	—	—	—	—	—	—
190	217	154	160	—	—	—
—	148	—	—	—	—	—
—	—	—	—	—	—	—
—	—	—	—	190	—	—
201	195	130	206	81	194	—
—	—	—	—	—	—	—
—	198	—	—	—	—	—
210	154	141	157	125	95	—
—	—	—	—	—	—	—
—	—	—	—	—	—	—
—	—	—	—	—	—	—
168	—	—	—	—	—	—
205	198	212	—	—	—	—
—	—	—	—	—	—	—

Table VIII,6   Phenols (*continued*)

Phenol	B.P. °C	M.P. °C	Bromo compound °C	Acetate °C	Benzoate °C	p-Nitro-benzoate °C
1,8-Dihydroxynaphthalene	—	142	—	155	175	—
Salicylic acid (25)	—	159	—	135	132	205
p-Hydroxybiphenyl	306	165	—	88	151	—
Hydroquinone	286	170	186*	124	199	250
1,4-Dihydroxynaphthalene (26)	—	176	—	128	169	—
2,7-Dihydroxynaphthalene	—	190	—	136	139	—
m-Hydroxybenzoic acid	—	200	—	131	—	—
p-Hydroxybenzoic acid	—	214	—	187	—	—
Phloroglucinol	—	218	151†	104	174	283
1,5-Dihydroxynaphthalene	—	265	—	160	235	—
4,4′-Dihydroxybiphenyl	—	274	—	161	241	—

Halogeno-phenols

Phenol	B.P. °C	M.P. °C	Bromo compound °C	Acetate °C	Benzoate °C	p-Nitro-benzoate °C
o-Chlorophenol	176	9	—	Liq.	Liq.	115
o-Bromophenol	195	5	95†	Liq.	—	—
m-Chlorophenol	214	33	—	Liq.	71	99
m-Bromophenol	236	33	—	Liq.	86	—
2,4-Dibromophenol	239	40	—	36	98	184
m-Iodophenol	—	40	—	38	—	133
o-Iodophenol	—	43	—	—	34	—
p-Chlorophenol	217	43	—	Liq.	89	168
2,4-Dichlorophenol	210	45	68	—	96	—
p-Bromophenol	235	64	95†	21	102	180
2,4,5-Trichlorophenol	249	68	—	—	93	—
2,4,6-Trichlorophenol	246	69	—	—	75	106
2,4-Di-iodophenol	—	72	—	71	98	—
p-Iodophenol	—	94	—	32	119	—
2,4,6-Tribromophenol	—	95	120‡	87	81	153
Chlorohydroquinone	—	106	—	72	—	—
Bromohydroquinone	—	111	—	72	—	—
2,4,6-Tri-iodophenol	—	159	—	156	—	—
Pentachlorophenol	310d	191	—	150	164	—

Nitro-phenols

Phenol	B.P. °C	M.P. °C	Bromo compound °C	Acetate °C	Benzoate °C	p-Nitro-benzoate °C
o-Nitrophenol△	216	45	117*	41	59	141
m-Nitrophenol	—	97	91*	56	95	174
2,4-Dinitrophenol	—	113	118	72	132	139
p-Nitrophenol△	—	114	142*	83	142	159
Picric acid	—	122	—	76	—	143
Styphnic acid (27)	—	179	—	—	—	—

* Dibromo.   † Tribromo.   ‡ Tetrabromo.

△ O-Aryl saccharin derivatives of o-nitrophenol and p-nitrophenol have m.p.s 236 °C and 192 °C respectively.

(1) o-Hydroxybenzaldehyde.
(2) 2-Methyl-5-isopropylphenol.
(3) 4-Allyl-2-methoxy phenol.
(4) 2-Methoxy-4-propenylphenol (*cis* and *trans*).
(5) Guaiacol.
(6) m-4-Xylenol.
(7) Salol.
(8) m-2-Xylenol.
(9) 3-Hydroxy-4-isopropyltoluene.
(10) Hydroquinone monomethyl ether.
(11) 3,5-Dihydroxytoluene.
(12) o-4-Xylenol
(13) m-5-Xylenol.
(14) Mesitol.

3,5-dinitro-benzoate °C	Aryloxy-acetic acid °C	NN-diphenyl urethan °C	1-Naphthyl urethan °C	Toluene-p-sulphonate °C	2,4-Dinitro-phenyl ether °C	o-Aryl saccharin °C
—	—	—	220	—	—	—
—	191	—	—	—	—	—
—	—	—	—	177	118	—
317	250	230	247	159	—	—
—	—	—	—	—	—	—
—	149	176	—	150	—	—
—	206	—	—	—	—	—
—	278	—	—	—	—	—
162	—	—	—	—	—	—
—	—	—	—	—	—	—
—	274	—	—	—	—	—
143	145	—	120	74	99	—
—	143	—	129	78	89	—
156	110	—	158	—	75	—
—	108	—	108	53	—	—
—	153	—	—	120	135	—
183	115	—	—	61	—	—
—	135	—	—	—	95	—
186	156	97	166	71	126	—
—	140	—	—	125	119	—
191	159	99	169	94	141	—
—	157	—	—	—	—	—
136	182	143	188	—	136	—
—	—	—	—	—	—	—
174	156	127	—	99	156	—
—	200	153	153	113	135	—
—	—	—	—	—	—	—
181	—	—	—	—	—	—
—	196	—	—	145	—	—
155	158	114	113	83	142	—
159	156	—	167	113	138	—
—	—	—	—	121	248	—
186	187	112	151	97	120	—
—	—	—	—	—	—	—
—	—	—	—	—	—	—

(15) Pseudo-cuminol.
(16) o-3-Xylenol.
(17) p-2-Xylenol.
(18) 4-Hydroxy-3-methoxybenzaldehyde.
(19) Saligcnin.
(20) Phenylhydroquinone.
(21) 4-Hydroxy-3-methoxybenzaldehyde.

(22) 1,2-Naphthohydroquinone.
(23) Naphthoresorcinol.
(24) 1,2,4-Trihydroxybenzene.
(25) o-Hydroxybenzoic acid.
(26) 1,4-Naphthohydroquinone.
(27) 2,4,6-Trinitroresorcinol.

Table VIII,7   Enols

Compound	B.P. °C	M.P. °C	$d_4^{20°}$	$n_D^{20°}$	Semicarbazone °C	Pyrazolone °C
Pentane-2,4-dione (acetylacetone)	139	—	0.977	1.452	—	100*
Methyl acetoacetate	170	—	1.077	1.419	152	127
Methyl methylacetoacetate	177	—	1.030	1.418	138	120
Ethyl acetoacetate	180	—	1.028	1.419	129d	127
Ethyl methylacetoacetate	181	—	1.006	1.419	86	120
Methyl ethylacetoacetate	189	—	0.989	—	98	108
Hexane-2,5-dione (acetonylacetone)	194	—	0.974	1.428	220	92
Ethyl ethylacetoacetate	198	—	0.972	1.422	154d	108
Diethyl acetonedicarboxylate	250d	—	1.113	—	95	85
Ethyl benzoylacetate	262d	—	1.117	—	125	63
Diethyl oxalacetate	131/24	—	1.131	1.454	162	—
Benzoylacetone	261	61	—	—	—	63
Dibenzoylmethane	—	78	—	—	—	137

* 1-(p-Nitrophenyl)-3,5-dimethylpyrazole: with aqueous solution of p-nitrophenylhydrazine hydrochloride. Phenylhydrazine yields a liquid pyrazole, b.p. 273 °C.

Table VIII,8   Polyhydric alcohols

Alcohol	B.P. °C	Benzoate °C	p-Nitro-benzoate °C	3,5-Dinitro-benzoate °C	Phenyl-urethan °C	1-Naphthyl-urethan °C	Other derivatives °C
Butane-2,3-diol	182	76	—	—	201	—	—
Propane-1,2-diol	189	—	127	—	153	—	—
Ethylene glycol	198	73	141	169	157	176	—
Butane-1,3-diol	208	—	—	—	123	—	—
Propane-1,3-diol	215	59	119	178	137	164	—
Butane-1,4-diol	230	82	175	—	183	199	—
Pentane-1,5-diol	239	—	105	—	176	147	—
Di-2-hydroxyethyl ether (1)	244	—	—	149	—	122	—
Hexane-1,6-diol	250*	—	—	—	—	—	—
Di-β-hydroxyethoxy ethane (2)	285	—	—	—	108	—	—
2-Butyne-1,4-diol	145/15**	—	—	190	131	—	—
Glycerol	290d	76	188	—	180	192	—
Sorbitol	m.p. 110	129	—	—	—	—	Acetate, 99
Mannitol	m.p. 166	149	—	—	303	—	Acetate, 121
Pentaerythritol	m.p. 253	99	—	—	—	—	Acetate, 84

(1) Diethylene glycol.   (2) Triethylene glycol.

* M.p. 42 °C.
** M.p. 55 °C.

Table VIII,9 Carbohydrates (sugars)

Carbohydrate	M.P.* °C	$[\alpha]_D^{20°}$ in water °C	Osazone M.P. °C	Time of formation (minutes)	Other derivatives °C
†D-Glucose (hydrated)	90 ⎫	+52	205	4	Penta-acetate, α- 112, β- 132; pentabenzoate, 179
D-Glucose (anhydrous)	146 ⎬				
D-Ribose	95	−21.5	166	—	—
D-Fructose	104	−92	205	2	Penta-acetate, α- 70, β- 109
L-Rhamnose (hydrated)	105 ⎫	+9	190	9	Penta-acetate, 99
L-Rhamnose (anhydrous)	125 ⎬				
L-Lyxose	106	+13.5	163	—	—
D-Galactose (hydrated)	120 ⎫	+82	201	15–19	Penta-acetate, α- 95, β- 142; D-galactaric acid, 213
D-Galactose (anhydrous)	170 ⎬				
D-Mannose	132	+14.5	205	0.5	Penta-acetate, α- 74, β- 115
D-Xylose	145	+19	164	7	Penta-acetate, α- 59, β- 126
L-Arabinose	160	+105	166	9	Penta-acetate, α- 94, β- 86
L-Sorbose	161	−43	162	4	—
Maltose (hydrated)	100 ⎫	+130	206	—	Octa-acetate, α- 125, β- 160
Maltose (anhydrous)	165 ⎬				
Sucrose	185	+66.5	205	30	Octa-acetate, 69
Gentiobiose	190	+9.5	162	—	Octa-acetate, α- 189, β- 193
Lactose (hydrated)	203 ⎫	+52.5	200	—	Octa-acetate, α- 152, β- 90; D-galactaric acid, 213
Lactose (anhydrous)	223 ⎬				
Cellobiose	225	+35	198	—	Octa-acetate, α- 230, β- 192

* The melting points of carbohydrates (sugars) are not usually sharp and they are perhaps best expressed as decomposition points.

† The small capital letter prefix refers to configuration, related to D-glyceraldehyde, and not to the direction of optical rotation. The sign of optical rotation is expressed as (+) and (−) or by the words *dextro* and *laevo*; older texts often refer to the sign of rotation as *d* or *l*. Thus we have D-(—)-fructose or L-(+)-arabinose.

Table VIII,10   Aliphatic halogen compounds

Halide	B.P. °C	M.P. °C	$d_4^{20°}$	$n_D^{20°}$
Chlorides				
Ethyl	12	—	—	—
Isopropyl	35	—	0.863	1.378
Allyl	45	—	0.940	1.416
Propyl	46	—	0.889	1.388
t-Butyl	51	—	0.846	1.386
s-Butyl	68	—	0.874	1.397
Isobutyl	69	—	0.881	1.398
Butyl	77	—	0.886	1.402
1-Chloro-2,2-dimethylpropane	85	—	0.879	—
2-Chloro-2-methylbutane	85	—	0.865	1.405
3-Chloropentane	96	—	0.872	1.408
2-Chloropentane	97	—	0.873	1.408
1-Chloro-3-methylbutane	99	—	0.872	1.409
1-Chloropentane	106	—	0.882	1.412
Chlorocyclopentane	114	—	1.005	1.451
1-Chlorohexane	134	—	0.878	1.420
Chlorocyclohexane	142	—	0.989	1.462
1-Chloroheptane	159	—	0.877	1.426
1-Chloro-octane	182	—	0.875	1.431
1-Chlorononane	202	—	0.870	1.434
1-Chlorodecane	223	—	0.868	1.437
1-Chloroundecane	241	—	0.868	1.440
1-Chlorododecane	244	—	0.867	1.443
Benzyl chloride	179	—	1.100	1.539
1-Chloro-1-phenylethane	195	—	1.063	1.530
1-Chloro-2-phenylethane	198	—	1.073	—
Benzilidene chloride	207	—	—	—
Benzotrichloride	218	—	—	—
Bromides				
Ethyl	38	—	1.460	1.425
Isopropyl	59	—	1.425	1.314
Allyl	70	—	1.432	1.470
Propyl	71	—	1.435	1.355
s-Butyl	91	—	1.256	1.437
Isobutyl	91	—	1.253	1.435
Butyl	101	—	1.274	1.440
1-Bromo-2,2-dimethylpropane	109	—	1.225	—
2-Bromopentane	117	—	1.212	1.442
3-Bromopentane	118	—	1.211	1.443
1-Bromo-3-methylbutane	119	—	1.213	1.442
1-Bromopentane	129	—	1.219	1.445
Bromocyclopentane	137	—	1.387	1.489
1-Bromohexane	154	—	1.175	1.448
Bromocyclohexane	164	—	1.336	1.495
1-Bromoheptane	178	—	1.140	1.451
1-Bromo-octane	200	—	1.112	1.453
1-Bromononane	220	—	1.090	1.454
1-Bromodecane	103/6	—	1.066	1.455
1-Bromoundecane	114/5	—	1.054	1.457
1-Bromododecane	130/6	—	1.038	1.458

ilide	1-Naphthalide °C	Alkyl mercury(II) halide °C	S-Alkylthiouronium picrate °C	Picrate of alkyl 2-naphthyl ether °C
4	126	193	188	102
4	—	—	196	95
4	—	—	154	99
2	121	140	177	81
3	147	—	—	—
8	129	39	166	85
)	126	—	—	84
3	112	128	177	67
5	—	—	—	—
2	138	—	—	—
	—	—	—	—
	—	—	—	—
)	111	86	173	94
6	112	110	154	67
	—	—	—	—
)	106	125	—	—
5	188	—	—	—
7	95	119	—	—
7	91	151	—	—
	—	—	—	—
	—	—	—	—
	—	—	—	—
7	166	—	188	123
3	—	—	—	—
7	—	—	—	84
5	—	—	—	—
	—	—	—	—
4	126	194	188	102
4	—	94	196	95
4	—	—	154	99
2	121	138	177	81
8	129	39	166	85
)	126	56	167	84
3	112	129	177	67
6	—	—	—	—
3	—	—	—	—
4	—	—	—	—
)	111	80	173	94
5	112	122	154	67
	—	—	—	—
9	106	119	157	—
5	188	153	—	—
7	95	115	142	—
7	91	109	134	—
	—	109	—	—
	—	—	—	—
	—	108	—	—

Halide	B.P. °C	M.P. °C	$d_4^{20°}$	$n_D^{20°}$
Bromides (continued)				
1-Bromotetradecane	179/20	5	1.017	1.460
1-Bromohexadecane	201/19	14	1.001	1.462
Benzyl bromide	198	—	1.438	—
1-Bromo-1-phenylethane	203	—	1.361	1.561
1-Bromo-2-phenylethane	218	—	1.359	1.556
Iodides				
Methyl*	42	—	2.282	1.532
Ethyl	73	—	1.940	1.514
Isopropyl	89	—	1.703	1.499
Allyl	100	—	1.777	1.578
Propyl	102	—	1.743	1.505
s-Butyl	118	—	1.592	1.499
Isobutyl	119	—	1.602	1.496
2-Iodo-2-methylbutane	128	—	1.479	—
Butyl	129	—	1.616	1.499
3-Iodopentane	142	—	1.511	1.497
2-Iodopentane	142	—	1.510	1.496
1-Iodo-3-methylbutane	147	—	1.503	1.493
1-Iodopentane	155	—	1.512	1.496
1-Iodohexane	180	—	1.437	1.493
1-Iodoheptane	201	—	1.373	1.490
1-Iodo-octane	221	—	1.330	1.489
Iodocyclopentane	58/22	—	1.709	1.547
Iodocyclohexane	82/20	—	1.624	1.547
Benzyl iodide	93/10	24	—	—
1-Iodo-2-phenylethane	116/12	—	1.632	1.602
Chloro compounds				
Dichloromethane	42	—	1.336	1.425
trans-1,2-Dichloroethylene	48	—	1.257	1.444
Ethylidene chloride	57	—	1.176	1.416
cis-1,2-Dichloroethylene	60	—	1.282	1.446
Chloroform	61	—	1.489	1.446
2,2-Dichloropropane	70	—	1.092	1.412
1,1,1-Trichloroethane	75	—	1.349	1.438
Carbon tetrachloride	77	—	1.594	1.461
1,2-Dichloroethane (ethylene dichloride)	84	—	1.256	1.445
1,1,2-Trichloroethylene	87	—	1.465	1.478
1,2-Dichloropropane	96	—	1.155	1.439
1-Bromo-1-chloroethane	107	—	1.689	1.491
1,1,2-Trichloroethane	114	—	1.443	1.471
1,3-Dichloropropane	120	—	1.183	1.449
1,1,2,2-Tetrachloroethylene	121	—	1.623	1.506
1-Bromo-3-chloropropane	143	—	1.593	1.471
1,1,2,2-Tetrachloroethane	147	—	1.597	1.495
1,4-Dichlorobutane	153	—	1.139	1.455
1,2,3-Trichloropropane	157	—	1.394	1.486
Pentachloroethane	162	—	1.680	1.503
1,5-Dichloropentane	178d	—	1.100	1.457
1,6-Dichlorohexane	204d	—	1.069	1.457

* Methyl iodide is carcinogenic (see Section **I,3**).

lide °C	1-Naphthalide °C	Alkyl mercury(II) halide °C	S-Alkylthiouronium picrate °C	Picrate of alkyl naphthyl ether °C
	—	—	—	—
	—	—	137	—
7	166	119	188	123
3	—	—	—	—
7	—	169	—	84
4	160	145	224	117
4	126	182	188	102
4	—	—	196	95
4	121	112	154	99
2	121	113	177	81
8	129	—	166	85
0	126	72	167	84
2	138	—	—	—
3	112	117	177	67
	—	—	—	—
0	111	122	173	94
6	112	110	154	—
9	106	110	157	—
7	95	103	—	—
	—	—	—	—
	—	—	—	—
6	—	—	—	123
	—	—	—	84

Table VIII,10 Aliphatic halogen compounds (*continued*)

Halide	B.P. °C	M.P. °C	$d_4^{20°}$	$n_D^{20°}$
Chloro compounds (continued)				
Hexachloroethane	—	187 (sub.)	—	—
Bromo compounds				
Dibromomethane	97	—	2.496	1.541
Ethylidene bromide	113	—	2.055	1.513
1,2-Dibromoethane (ethylene dibromide)	131	—	2.183	1.539
1,2-Dibromopropane	141	—	1.932	1.520
Bromoform	150	—	2.887	1.598
1,2-Dibromo-2-methylpropane	150	—	1.783	1.512
2,3-Dibromobutane	157	—	1.792	1.515
1,3-Dibromopropane	165	—	1.982	1.523
1,2-Dibromobutane	166	—	1.820	—
1,4-Dibromobutane	198	—	1.826	1.519
1,2,3-Tribromopropane	220	—	2.402	1.582
1,5-Dibromopentane	221	—	1.702	1.513
1,6-Dibromohexane	240	—	1.603	1.506
1,1,2,2-Tetrabromoethane	124/19	—	2.967	1.638
Carbon tetrabromide	—	92	—	—
Iodo compounds				
Di-iodomethane	80/25	—	3.324	1.741
1,3-Di-iodopropane	90/9	—	2.576	1.642
1,4-Di-iodobutane	110/10	—	2.358	1.621
1,5-Di-iodopentane	132/10	—	2.182	1.602
1,6-Di-iodohexane	140/10	—	2.040	1.586
1,2-Di-iodoethane (ethylene di-iodide)	—	81	—	—
Iodoform	—	119	—	—

-Alkylisothiouronium picrate, m.p. 260 °C
-Alkylisothiouronium picrate, m.p. 232 °C

-Alkylisothiouronium picrate, m.p. 223 °C

-Alkylisothiouronium picrate, m.p. 229 °C

Table VIII,11   Aromatic halogen compounds

Compound °C	B.P. °C	M.P. °C	$d_4^{20°}$	$n_D^{20°}$
m-Difluorobenzene	83	—	1.153	1.439
Fluorobenzene	85	—	1.024	1.466
p-Difluorobenzene	89	—	1.166	1.441
o-Difluorobenzene	93	—	1.157	1.443
o-Fluorotoluene	114	—	0.998	1.470
m-Fluorotoluene	116	—	0.990	—
p-Fluorotoluene	116	—	0.998	1.469
Chlorobenzene	132	—	1.107	1.525
Bromobenzene	156	—	1.494	1.560
o-Chlorotoluene	159	—	1.082	1.527
m-Chlorotoluene	162	—	1.072	1.522
p-Chlorotoluene	162	7	1.070	1.521
m-Dichlorobenzene	173	—	1.288	1.546
Benzyl chloride	179	—	1.100	1.539
o-Dichlorobenzene	180	—	1.305	1.551
o-Bromotoluene	181	—	1.425	—
m-Bromotoluene	183	—	1.410	—
p-Bromotoluene	185	26	1.390	—
Iodobenzene	188	—	1.831	1.620
o-Bromochlorobenzene	195	—	1.646	1.580
Benzyl bromide	198	—	1.438	—
2,4-Dichlorotoluene	199	—	1.249	1.549
2,6-Dichlorotoluene	199	—	1.269	1.551
m-Iodotoluene	204	—	1.698	—
Benzyl iodide	93/10	24	—	—
Benzylidene chloride	205	—	1.250	1.550
o-Iodotoluene	207	—	1.698	—
3,4-Dichlorotoluene	207	—	1.251	1.549
Benzylidene bromide	156/23	—	1.460	1.541
1,2,4-Trichlorobenzene	213	17	1.468	1.554
2-Fluoronaphthalene	212	61	—	—
1-Fluoronaphthalene	214	—	1.134	1.594
m-Dibromobenzene	219	—	1.952	1.606
Benzo-trichloride	220	—	1.173	—
o-Dibromobenzene	224	7	1.956	1.609
Bromomesitylene	225	−1	—	—
2-Bromocymene	234	—	1.267	—
2,5-Dibromotoluene	236	—	1.811	—
3,4-Dibromotoluene	240	—	1.811	1.600
m-Bromoiodobenzene	252	—	—	—
2-Chloronaphthalene	256	61	—	—
o-Bromoiodobenzene	257	—	2.262	1.665
1-Chloronaphthalene	259	—	1.192	1.633
1-Bromonaphthalene	281	—	1.484	1.658
m-Di-iodobenzene	285	40	—	—
o-Di-iodobenzene	287	27	—	—
2-Bromobiphenyl	297	—	1.223	—
3-Bromobiphenyl	300	—	—	1.641
1-Iodonaphthalene	302	—	1.729	—
2-Iodobiphenyl	158/6	—	1.609	1.662
2-Chlorobiphenyl	273	32	—	—

Nitration product		Sulphonamide ($-SO_2NH_2$, 1)		Other derivatives °C
Position	M.P. °C	Position	M.P. °C	
—	—	—	—	
—	—	4, F	125	Sulphone, 98
—	—	—	—	—
—	—	3, CH_3; 4, F	105	o-Fluorobenzoic acid, 127
—	—	2, CH_3; 4, F	173	m-Fluorobenzoic acid, 124
—	—	2, CH_3; 5, F	141	p-Fluorobenzoic acid, 186
2,4	52	4, Cl	143	—
2,4	75	4, Br	162	—
3,5	64	3, CH_3; 4, Cl	126	o-Chlorobenzoic acid, 141
4,6	91	2, CH_3; 4, Cl	185	m-Chlorobenzoic acid, 158
2	38	2, CH_3; 5, Cl	143	p-Chlorobenzoic acid, 242
4,6	103	2,4, diCl	180	—
—	—	—	—	S-Benzylisothiouronium picrate, 188
4,5	110	3,4, diCl	135	Sulphone, 176
3,5	82	3, CH_3; 4, Br	146	o-Bromobenzoic acid, 150
4,6	103	2, CH_3; 4, Br	168	m-Bromobenzoic acid, 155
2	47	2, CH_3; 5, Br	165	p-Bromobenzoic acid, 251
4	174	—	—	Sulphone, 202
—	—	—	—	
3,5	104	2,4, diCl; 5, Me	176	2,4 Dichlorobenzoic acid, 164
3	53	2,4, diCl; 3, Me	204	2,6-Dichlorobenzoic acid, 139
—	—	—	—	m-Iodobenzoic acid, 186
—	—	—	—	
—	—	—	—	Benzaldehyde phenylhydrazone, 156
5	103	—	—	o-Iodobenzoic acid, 162
4	64	3,4, diCl; 6, Me	190	3,4-Dichlorobenzoic acid, 208
—	56	—	—	—
—	—	—	—	Picrate, 101
—	—	—	—	Picrate, 113
4	62	2,4, diBr	189	—
—	—	—	—	Benzoic acid, 121
4,5	114	3,4, diBr	176	Sulphone, 177
—	—	—	—	—
—	—	—	—	2,5-Dibromobenzoic acid, 157
—	—	—	—	3,4-Dibromobenzoic acid, 235
—	—	—	—	
7,8	175	7, Cl	232	Picrate, 81
4,5	180	4, Cl	186	Picrate, 137
4	85	4, Br	193	Picrate, 134
—	—	—	—	—
—	—	—	—	o-Bromobenzoic acid, 150 (CrO_3)
—	—	—	—	—
—	—	—	—	Picrate, 127
—	—	—	—	o-Chlorobenzoic acid, 141

Table VIII,11 Aromatic halogen compounds (*continued*)

Compound	B.P. °C	M.P. °C	$d_4^{20°}$	$n_D^{20°}$
p-Iodotoluene	211	35	—	—
1,2-Dichloronaphthalene	296	35	—	—
m-Di-iodobenzene	285	40	—	—
1,2,4-Tribromobenzene	275	44	—	—
p-Dichlorobenzene	174	53	—	—
1,2,3-Trichlorobenzene	218	53	—	—
2-Iodonaphthalene	309	54	—	—
2-Bromonaphthalene	282	59	—	—
2-Fluoronaphthalene	—	60	—	—
2,2′-Dichlorobiphenyl	—	60	—	—
2-Chloronaphthalene	256	61	—	—
1,3,5-Trichlorobenzene	208	63	—	—
p-Bromochlorobenzene	195	67	—	—
1,2-Dibromonaphthalene	—	68	—	—
4-Fluorobiphenyl	253	74	—	—
4-Chlorobiphenyl	291	77	—	—
2,2′-Dibromobiphenyl	—	81	—	—
1,2,3-Tribromobenzene	—	88	—	—
4-Bromobiphenyl	310	89	—	—
p-Dibromobenzene	219	89	—	—
p-Bromoiodobenzene	251	92	—	—
4,4′-Difluorobiphenyl	255	95	—	—
4-Iodobiphenyl	—	114	—	—
1,3,5-Tribromobenzene	271	120	—	—
p-Di-iodobenzene	285	129	—	—
1,2,4,5-Tetrachlorobenzene	240	140	—	—
4,4′-Dichlorobiphenyl	—	149	—	—
4,4′-Dibromobiphenyl	—	164	—	—
1,2,4,5-Tetrabromobenzene	—	181	—	—
1,2,3,4-Tetrachloronaphthalene	—	183	—	—
Hexachlorobenzene	—	229	—	—

Nitration product		Sulphonamide ($-SO_2NH_2$, 1)		Other derivatives °C
Position	M.P. °C	Position	M.P. °C	
—	—	—	—	p-Iodobenzoic acid, 269
—	—	—	—	—
—	—	—	—	—
—	—	—	—	—
2	54	2,5, diCl	180	—
4	56	2,3,4, triCl	230	—
—	—	—	—	Picrate, 95
—	—	7, Br	208	Picrate, 86
—	—	—	—	Picrate, 101
—	—	—	—	—
4,8	175	7, Cl	232	Picrate, 81
2	68	2,4,6, triCl	212	—
2	72	—	—	—
—	—	—	—	—
—	—	—	—	p-Chlorobenzoic acid, 242
—	—	—	—	—
—	—	—	—	—
—	—	—	—	p-Bromobenzoic acid, 251 (CrO$_3$)
2,5	84	2,5, diBr	195	—
—	—	—	—	—
—	—	—	—	—
—	—	2,4,6, triBr	222d	—
2,5	171	—	—	—
3	99	—	—	—
—	—	—	—	p-Chlorobenzoic acid, 242 (CrO$_3$)
—	—	—	—	p-Bromobenzoic acid, 251 (CrO$_3$)
3	168	—	—	—
—	—	—	—	—

Table VIII,12 Aliphatic ethers

Ether	B.P. °C	$d_4^{20°}$	$n_D^{20°}$
Diethyl	34	0.714	1.353
Ethyl vinyl	36	0.759	1.377
Allyl ethyl	67	0.765	1.388
Di-isopropyl	68	0.726	1.368
Butyl methyl	70	0.774	1.374
Dipropyl	90	0.749	1.381
Butyl ethyl	92	0.749	1.382
Diallyl	94	0.803	—
Pentyl methyl	99	0.761	1.387
Cyclopentyl methyl	105	0.862	1.420
Pentyl ethyl	118	0.762	1.393
Di-s-butyl	121	0.764	1.396
Cyclopentyl ethyl	122	0.853	1.423
Di-isobutyl	123	0.756	—
Hexyl methyl	126	0.772	1.397
Cyclohexyl methyl	134	0.875	1.435
Dibutyl	141	0.770	1.399
Ethyl hexyl	142	0.772	1.401
Cyclohexyl ethyl	149	0.864	1.435
Di-isopentyl	171	0.778	1.409
Cineole	176	0.923	1.458
Dipentyl	188	0.785	1.412
Dihexyl	229	0.793	1.420
Diheptyl	259	0.801	1.427
Dioctyl	288	0.806	1.433
Didecyl	185/5 mmHg	0.815	1.441
Chloromethyl methyl	59	1.070	1.397
1-Chloroethyl methyl	73	0.991	1.400
Chloromethyl ethyl	83d	1.026	1.404
2-Chloroethyl methyl	91	1.035	1.411
sym-Dichlorodimethyl	105	1.310	1.436
1,1'-Dichlorodiethyl	116	1.111	1.423
Epichlorohydrin (1)	117	1.181	1.438
1,2-Dichlorodiethyl	140	1.177	1.444
2,2'-Dichlorodiethyl	118	1.210	1.457
Di-2-chloropropyl	188	1.109	1.447
Di-3-chloropropyl	215	1.139	1.452
Ethyleneglycol dimethyl (2)	85	0.866	1.379
Ethyleneglycol diethyl (2)	123	0.848	—
Diethyleneglycol diethyl (2)	187	0.906	1.411
Tetraethyleneglycol dimethyl (2)	266	1.009	1.432
Benzyl methyl	171	0.965	1.501
Benzyl ethyl	188	0.948	1.496
Dibenzyl	299d	1.042	—
Tetrahydrofuran	65	0.889	1.407
2-Methyltetrahydrofuran	79	0.855	1.407
Dihydropyran	86	0.923	1.440
Tetrahydropyran	88	0.881	1.421
Dioxan	102	1.034	1.422

(1) 1-Chloro-2,3-epoxypropane.
(2) For alternative names see Section **II,2,**_18_.

Table VIII,13 Aromatic ethers

Ether	B.P. °C (/mmHg)	M.P. °C	$d_4^{20°}$
Furan	32	—	0.937
2-Methylfuran	64	—	0.913
Anisole	154	—	0.996
Phenetole	170	—	0.965
Benzyl methyl ether	171	—	0.965
Methyl o-tolyl ether (1)	171	—	0.985
Methyl p-tolyl ether	175	—	0.970
Methyl m-tolyl ether	177	—	0.972
Ethyl o-tolyl ether	184	—	0.953
Benzyl ethyl ether	186	—	0.948
Phenyl n-propyl ether	188	—	0.949
Ethyl p-tolyl ether	190	—	0.949
Ethyl m-tolyl ether	191	—	0.949
Guaiacol (2)	205	28	1.129
Veratrole (3)	206	22	—
Butyl phenyl ether	208	—	0.934
Thymol methyl ether	212	—	—
Resorcinol dimethyl ether	217	—	1.050
Safrole (4)	232	11	1.100
Anethole (5)	235	22	0.989
Resorcinol diethyl ether	235	12	—
Eugenol methyl ether (6)	244	—	1.050
Isosafrole (7)	248	7	1.122
Diphenyl ether	259	28	—
Isoeugenol methyl ether (8)	264	—	1.053
Methyl 1-naphthyl ether	271	—	1.092
Ethyl 1-naphthyl ether	280	5	1.060
Dibenzyl ether	296	—	1.034
2-Methoxybiphenyl	274	29	—
Ethyl 2-naphthyl ether	282	37	—
Pyrogallol triethyl ether	—	39	—
Catechol diethyl ether	217	43	—
Pyrogallol trimethyl ether	241	47	—
Hydroquinone dimethyl ether	212	56	—
Methyl 2-naphthyl ether	274	72	—
Hydroquinone diethyl ether	—	72	—
Benzyl 1-naphthyl ether	—	77	—
4-Methoxybiphenyl	—	90	—
Benzyl 2-naphthyl ether	—	99	—

Halogeno-ethers

m-Chloroanisole	194	—	—
o-Chloroanisole	195	—	1.191
p-Chloroanisole	198	—	—
m-Chlorophenetole	205	—	1.171
o-Chlorophenetole	208	—	1.134
o-Bromoanisole	210	—	—
m-Bromoanisole	211	—	—
p-Chlorophenetole	212	21	1.121
p-Bromoanisole	215	11	—
o-Bromophenetole	218	—	—
p-Bromophenetole	233	4	—
o-Iodoanisole	242	—	—

$n_D^{20°}$	Sulphonamide °C	Picrate °C	Other derivatives °C
1.422	—	—	—
1.434	—	—	—
1.518	111	—	Dinitro, 87; 2,4-dibromo, 61
1.507	150	—	p-Nitro, 58
1.501	—	—	—
1.505	137	—	o-Methoxybenzoic acid, 101
1.512	182	—	Anisic acid, 184
1.513	130	—	m-Methoxybenzoic acid, 110
1.505	149	—	o-Ethoxybenzoic acid, 25
1.496	—	—	—
1.510	—	—	—
1.505	138	—	p-Ethoxybenzoic acid, 196
1.506	111	—	m-Ethoxybenzoic acid, 137
1.544	—	—	Tribromo, 116
—	136	—	Dibromo, 93; nitro, 95
1.497	—	—	—
—	—	—	Trinitro, 92
—	167	—	Dibromo, 140; trinitro, 124
1.538	—	—	Piperonylic acid, 228; pentabromo, 169
1.558	—	—	Anisic acid, 184; tribromo, 108
—	184	—	—
1.532	—	—	Tribromo, 78; veratric acid, 179
1.578	—	—	Tribromo, 109; piperonylic acid, 228
—	159	—	Dibromo, 55; dinitro, 144
1.569	—	—	—
1.696	157	129	Dibromo, 55
1.597	165	119	4-Bromo, 48
—	—	—	
—	—	—	Nitro, 95
—	163	100	1-Bromo, 66
—	—	—	—
—	162	—	Trinitro, 122
—	124	—	—
—	148	—	Nitro, 72; dibromo, 142
—	151	117	Bromo, 63
—	155	—	Nitro, 49
—	—	—	—
—	—	—	—
—	—	—	—
.545	131	—	Nitro, 95
—	151	—	Nitro, 98
—	—	—	—
.530	133	—	Nitro, 82
—	140	—	Nitro, 106
—	—	—	—
.522	134	—	Nitro, 61
—	148	—	Nitro, 88
—	135	—	Nitro, 98
—	145	—	Nitro, 47
—	—	—	—

Table VIII,13   Aromatic ethers (continued)

Ether	B.P. °C (/mmHg)	M.P. °C	$d_4^{20°}$
Halogeno-ethers (continued)			
m-Iodoanisole	244	—	—
o-Iodophenetole	246	—	—
m-Iodophenetole	134/15	—	—
p-Iodophenetole	252	29	—
p-Bromodiphenyl ether	168/15	—	—
2,4,6-Trichlorophenetole	246	44	—
p-Iodoanisole	240	52	—
2,4,6-Trichloroanisole	—	62	—
2,4,6-Tribromophenetole	—	73	—
2,4,6-Tribromoanisole	—	88	—
Nitro-ethers			
o-Nitrophenetole	267	—	—
o-Nitroanisole	272	10	1.254
m-Nitrophenetole	284	34	—
m-Nitroanisole	258	39	—
p-Nitroanisole	259	54	—
p-Nitrophenetole	283	60	—
2,4,6-Trinitroanisole	—	68	—
2,4,6-Trinitrophenetole	—	78	—
2,4-Dinitrophenetole	—	87	—
2,4-Dinitroanisole	—	94	—

(1) *o*-Methoxytoluene.
(2) *o*-Methoxyphenol.
(3) Catechol dimethyl ether.
(4) 1-Allyl-3,4-methylenedioxybenzene.

Table VIII,14   Acetals

Name	Formula	B.P. °C	$d_4^{20°}$	$n_D^{20°}$
Methylal	$CH_2(OCH_3)_2$	43	0.859	1.353
Dimethylacetal	$CH_3CH(OCH_3)_2$	64	0.852	1.366
Ethylal	$CH_2(OC_2H_5)_2$	87	0.831	1.373
Acetal	$CH_3CH(OC_2H_5)_2$	103	0.826	1.381
1,3-Dioxan	$H_2C \overset{O}{\underset{O}{<>}} (CH_2)_3$	105	1.034	1.420
Isopropylal	$CH_2(OC_3H_7{}^i)_2$	122	0.818	1.384
Ethylpropylal	$CH_3CH_2CH(OC_2H_5)_2$	124	0.833	1.390
Acrolein acetal	$CH_2{=}CHCH(OC_2H_5)_2$	125	0.850	—
Propylal	$CH_2(OC_3H_7)_2$	137	0.834	1.393

$n_D^{20°}$	Sulphonamide °C	Picrate °C	Other derivatives °C
—	—	—	—
—	—	—	Nitro, 96
—	—	—	—
—	131	—	Nitro, 96
—	—	—	—
—	—	—	Dinitro, 100
—	—	—	—
—	—	—	Dinitro, 95
—	—	—	Nitro, 79
—	—	—	—
—	—	—	o-Phenetidine, 228
1.562	—	—	o-Anisidine, 225
—	—	—	m-Phenetidine, 248
—	—	—	m-Anisidine, 251
—	—	—	p-Anisidine, 246
—	—	—	p-Phenetidine, 254
—	—	—	—
—	—	—	—
—	—	—	—
—	—	—	—

(5) p-(1-propenyl)anisole.
(6) 1-Allyl-3,4-dimethoxybenzene.
(7) 1,2-Methylenedioxy-4-(1-propenyl)benzene.
(8) 3,4-Dimethoxy-1-(1-propenyl)benzene.

Name	Formula	B.P. °C	$d_4^{20°}$	$n_D^{20°}$
Ethylbutylal	$CH_3CH_2CH_2CH(OC_2H_5)_2$	143	0.921	1.402
Propylacetal	$CH_3CH(OC_3H_7)_2$	147	0.830	1.397
Isobutylal	$CH_2(OC_4H_9{}^i)_2$	164	0.824	1.400
Isobutylacetal	$CH_3CH(OC_4H_9{}^i)_2$	176	0.821	1.403
Butylal	$CH_2(OC_4H_9)_2$	181	0.835	1.406
Butylacetal	$CH_3CH(OC_4H_9)_2$	187	0.833	1.409
Pentylal	$CH_2(OC_5H_{11})_2$	219	0.838	1.416
Pentylacetal	$CH_3CH(OC_5H_{11})_2$	222	0.839	1.418
Hexylal	$CH_2(OC_6H_{13})_2$	255	0.841	1.423
Benzaldehyde diethyl acetal	$C_6H_5CH(OC_2H_5)_2$	222	0.983	1.480

Table VIII,15   Aliphatic aldehydes

Aldehyde	B.P. °C	M.P. °C	Alkylidene dimethone °C	Dimethone anhydride °C
Formaldehyde	−21	—	189	171
Acetaldehyde	20	—	141	174
Propionaldehyde	49	—	155	143
Glyoxal	50	—	186	224
Acrylaldehyde (acrolein)	52	—	192	163
Isobutyraldehyde	64	—	154	144
α-Methylacrylaldehyde	73	—	—	—
Butyraldehyde	75	—	142	141
Trimethylacetaldehyde	75	—	—	—
Isovaleraldehyde	92	—	155	173
Chloral (trichloroacetaldehyde)	98	—	—	—
Crotonaldehyde	102	—	184	167
Valeraldehyde	104	—	105	113
Diethylacetaldehyde	117	—	102	—
4-Methylpentanal	121	—	—	133
Paraldehyde	124	—	—	—
Hexanal	131	—	109	—
Tetrahydrofurfural	145	—	—	—
Heptanal	155	—	103	112
Furfural	161	—	162	164
Hexahydrobenzaldehyde (1)	162	—	—	—
2-Ethylhexanal	163	—	—	—
Octanal	170	—	90	101
Bromal (tribromoacetaldehyde)	174	—	—	—
Nonanal	190	—	86	—
(+)-Citronellal	207	—	79	173
Decanal	208	—	92	—
Citral	229d	—	—	—
β-Hydroxybutyraldehyde	83/20	—	147	126
Chloral hydrate	—	53	56	—
Lauraldehyde (2)	238	45	—	—
Myristaldehyde (3)	155/10	23	—	—
Palmitaldehyde (4)	201/29	34	—	—
Stearaldehyde (5)	—	38	—	—
(±)-Glyceraldehyde	—	142	—	—

(1) Cyclohexanecarbaldehyde.
(2) Dodecanal.
(3) Tetradecanal
(4) Hexadecanal.
(5) Octadecanal.

2,4-Dinitro-phenylhydrazone °C	Semicarbazone °C	p-Nitro-phenylhydrazone °C	Other derivatives °C
166	169d	182	Methylene di-2-naphthol, 190
168	163	129	Oxime, 47
155	154(89)	124	Oxime, 40
328	270	311	Oxime, 178
165	171	151	—
187	126	131	—
206	198	—	—
123	106	87	—
209	190	119	Oxime, 41
123	132	110	Oxime, 48
131	—	—	$d_4^{20°}$ 1.512, $n_D^{20°}$ 1.457
190	199	185	Phenylhydrazone, 56; oxime, 119
107	—	—	Oxime, 52
130	99	—	—
99	127	—	—
—	—	—	$d_4^{20°}$ 0.994, $n_D^{20°}$ 1.420
107	106	—	Oxime, 51
204	166	—	$d_4^{20°}$ 1.107, $n_D^{20°}$ 1.436
108	109	73	Oxime, 57
230 (213)	203	154	Phenylhydrazone, 98
—	173	—	Oxime, 91
120	254d	—	—
106	101	80	Oxime, 60
—	—	—	—
100	100	—	Oxime, 64
78	84	—	$d_4^{20°}$ 0.855, $n_D^{20°}$ 1.449
104	102	—	Oxime, 69
110	164	—	$d_4^{20°}$ 0.887, $n_D^{20°}$ 1.488
—	110	—	—
131	—	—	Oxime, 56
106	106	—	Oxime, 78
—	107	95	Oxime, 83
108	109	97	Oxime, 88
—	109	101	Oxime, 89
170	160d	—	Oxime, 118

Table VIII,16 Aromatic aldehydes

Aldehyde	B.P. °C	M.P. °C	Dimethone °C	Dimethone anhydride °C
Benzaldehyde	179	—	195	200
Phenylacetaldehyde	194	34	165	126
Salicylaldehyde (1)	197	—	—	208
m-Tolualdehyde	199	—	172	206
o-Tolualdehyde	200	—	167	215
p-Tolualdehyde	204	—	—	—
Phenoxyacetaldehyde	215d	38	—	—
3-Phenylpropanal	224	—	—	—
m-Methoxybenzaldehyde	230	—	—	—
Cuminaldehyde (2)	235	—	171	173
Anisaldehyde (3)	248	2	145	243
Cinnamaldehyde	252	—	213	175
1-Naphthaldehyde	292	34	—	—
Piperonal (4)	263	37	178	220
o-Methoxybenzaldehyde	236	38	—	—
2,3-Dimethoxybenzaldehyde	—	54	—	—
Veratraldehyde (5)	285	58	—	—
2-Naphthaldehyde	—	61	—	—
2,4-Dimethoxybenzaldehyde	—	69	—	—
Vanillin (6)	—	81	197	228
m-Hydroxybenzaldehyde	240	108	—	—
p-Hydroxybenzaldehyde	—	116	189	246
Terephthaldialdehyde	246	116	—	—
β-Resorcylaldehyde (7)	—	136	—	—
Protocatechuicaldehyde (8)	—	153	—	145

Halogeno-aldehydes

o-Chlorobenzaldehyde	213	11	205	225
m-Chlorobenzaldehyde	214	18	—	—
o-Bromobenzaldehyde	230	22	—	—
m-Bromobenzaldehyde	234	—	—	—
o-Iodobenzaldehyde	—	37	—	—
3,4-Dichlorobenzaldehyde	248	44	—	—
p-Chlorobenzaldehyde	214	47	—	—
m-Iodobenzaldehyde	—	57	—	—
p-Bromobenzaldehyde	—	67	—	—
2,6-Dichlorobenzaldehyde	—	71	—	—
2,4-Dichlorobenzaldehyde	—	72	—	—
p-Iodobenzaldehyde	—	78	—	—

Nitro- and amino-aldehydes

o-Aminobenzaldehyde	—	40	—	—
o-Nitrobenzaldehyde	—	44	—	—
m-Nitrobenzaldehyde	—	58	—	—
p-Aminobenzaldehyde	—	72	—	—
p-Dimethylaminobenzaldehyde	—	74	—	—
p-Nitrobenzaldehyde	—	106	—	—
m-Aminobenzaldehyde	—	Amorphous	—	—

(1) o-Hydroxybenzaldehyde.
(2) p-Isopropylbenzaldehyde.
(3) p-Methoxybenzaldehyde.
(4) 3,4-Methylenedioxybenzaldehyde.

2,4-Dinitrophenyl hydrazone °C	Semicarbazone °C	Oxime °C	Phenylhydrazone °C	p-Nitrophenyl hydrazone °C
237	224	35d	158	192
121	156	99	63	151
252	231	63	143	228
194	223	60	91	157
194	212	49	106	222
233	234	80	112	201
—	145	95	86	—
149	127	94	—	123
—	—	40	—	171
241	211	52	129	190
254	209	132 (65)	121	161
255d	215	139	168	195
—	221	98	80	234
265	234	110	106	200
253	215	92	—	205
—	231	99	138	—
264	177	95	121	—
270	245	156	206	230
—	—	106	—	—
269	239	117	105	228
259	198	90	130	222
280	224	72	178	266
—	—	200	278d	281
286	260	192	160	285
275	230	157	176	—
209	229 (146)	76d	86	249
248	229	71d	134	216
—	214	102	—	240
—	205	72d	141	220
—	206	108	79	—
—	—	119	—	277
265	232	107 (140)	127	220
—	226	62	155	212
—	228	111	113	208
—	—	150	—	—
—	—	—	—	—
—	224	—	121	201
—	247	135	221	220
265	256	103	156	263
292	246	122	121	247
—	173	124	156	—
325	222	185	148	182
320	221	133	159	249
—	280d	195	162	226

(5) 3,4-Dimethoxybenzaldehyde.
(6) 4-Hydroxy-3-methoxybenzaldehyde.
(7) 2,4-Dihydroxybenzaldehyde.
(8) 3,4-Dihydroxybenzaldehyde.

Table VIII,17   Aliphatic ketones

Ketone	B.P. °C	M.P. °C	2,4-Dinitro-phenylhydrazone °C
Acetone	56	—	128
Ethyl methyl ketone	80	—	115
Methyl vinyl ketone	80	—	—
Biacetyl	88	—	315 (Di)
Isopropyl methyl ketone	94	—	120
Methyl propyl ketone	102	—	144
Diethyl ketone	102	—	156
Pinacolone (t-butyl methyl ketone)	106	—	125
Isobutyl methyl ketone	117	—	95
Di-isopropyl ketone	124	—	88
Ethyl propyl ketone	124	—	130
Butyl methyl ketone	128	—	107
4-Methylpent-3-en-2-one	130	—	203
Cyclopentanone	131	—	146
Pentane-2,4-dione	139	—	209
2-Methylcyclopentanone	139	—	—
Dipropyl ketone	144	—	75
Acetoin (acetyl methyl carbinol)	145	—	318
Acetol (hydroxyacetone)	146	—	129
Heptan-2-one	151	—	89
Cyclohexanone	156	—	162
2-Methylcyclohexanone	165	—	137
4-Hydroxy-4-methylpentan-2-one	166	—	203
Di-isobutyl ketone	168	—	92
Methyl acetoacetate	170	—	—
3-Methylcyclohexanone	170	—	155
4-Methylcyclohexanone	171	—	134
Hexyl methyl ketone	173	—	58
Cycloheptanone	180	—	148
Cyclohexyl methyl ketone	180	—	140
Ethyl acetoacetate	181	—	93
Dibutyl ketone	188	—	—
(+)-Fenchone	193	—	140
Hexane-2,5-dione	194	—	257 (Di)
Methyl levulinate	196	—	142
Phorone (1)	199	28	118
β-Thujone	202	—	114
Ethyl levulinate	206	—	102
(−)-Menthone	209	—	146
Isophorone (2)	215	—	130
Pulegone	224	—	147
Undecan-6-one	226	14	—
(+)-Carvone	230	—	191
Tridecan-7-one	255	33	—
α-Ionone	130/13	—	151
β-Ionone	139/18	—	128
2,2′-Furoin	—	135	217
2,2′-Furil	—	165	215
(+)-Camphor	209	179	177
Chloroacetone	119	—	125
1,1-Dichloroacetone	120	—	—
1,3-Dichloroacetone	173	45	133

(1) 2,6-Dimethylhepta-2,5-dien-4-one.
(2) 3,3,5-Trimethylcyclohex-2-enone.

Semi-carbazone °C	Benzylidene derivative °C	Phenyl-hydrazone °C	p-Nitro-phenyl-hydrazone °C	Other derivatives °C
190	112	42	149	Oxime, 59
146	—	—	129	—
141	—	—	—	—
279 (Di)	53	243 (Di)	230	Dioxime, 234
114	—	—	109	—
112	—	—	117	Oxime, 58
139	31	—	144	Oxime, 69
158	41	—	—	Oxime, 78
132	—	—	79	Oxime, 58
160	—	—	—	Oxime, 34
112	—	—	—	—
125	—	—	88	Oxime, 49
164	—	142	134	Oxime, 49
210	190	55	154	Oxime, 57
—	—	—	—	Oxime, 149
184	—	—	—	—
133	—	—	—	—
185	—	—	—	—
196	—	103	—	Oxime, 71
127	—	207	73	—
167	118	81	147	Oxime, 91
197	—	—	132	Oxime, 43
—	—	—	209	Oxime, 58
122	—	—	—	—
152	—	—	—	$d_4^{20°}$ 1.077, $n_D^{20°}$ 1.420
191	122	94	119	—
203	99	110	128	Oxime, 39
123	—	—	93	—
162	108	—	—	—
177	—	—	154	Oxime, 60
133	—	—	—	$d_4^{20°}$ 1.025, $n_D^{20°}$ 1.420
90	—	—	—	—
184	—	—	—	Oxime, 167
220	—	120 (Di)	—	Dioxime, 137
143	—	96	—	$d_4^{20°}$ 1.050, $n_D^{20°}$ 1.423
221	—	—	—	Oxime, 48
174	—	—	—	Oxime, 55
148	—	104	—	$d_4^{20°}$ 1.011, $n_D^{20°}$ 1.423
189	—	53	—	Oxime, 59
199	77	68	—	Oxime, 79
174	—	—	—	Oxime, 119
—	—	—	—	$d_4^{20°}$ 0.825, $n_D^{20°}$ 1.429
163	—	110	175	Oxime, 73
—	—	—	97	—
143 (108)	—	—	113	Oxime, 90
149	—	—	173	—
—	—	81	—	Oxime, 161
—	—	184	199	Dioxime, 100
238	98	233	217	Oxime, 119
150	—	—	—	—
163	—	—	—	—
120	—	—	—	—

Table VIII,18   Aromatic ketones

Ketone	B.P. °C	M.P. °C	2,4-Dinitro-phenylhydrazone °C
Acetophenone	202	20	250 (237)
2-Hydroxyacetophenone	215	—	—
2-Methylacetophenone (1)	216	—	159
Benzyl methyl ketone	216	27	156
Propiophenone (2)	218	19	191
3-Methylacetophenone (3)	220	—	207
Isobutyrophenone (4)	222	—	163
4-Methylacetophenone (5)	224	28	258
Benzyl ethyl ketone	226	—	—
Butyrophenone (6)	230	12	190
m-Methoxyacetophenone	240	—	—
Valerophenone (7)	242	—	166
o-Methoxyacetophenone	245	—	—
α-Tetralone	129/12	—	257
β-Tetralone	138/16	18	—
1-Acetylnaphthalene	302	34	—
Phenyl o-tolyl ketone (8)	310	—	190
Phenyl m-tolyl ketone	314	—	221
Dibenzyl ketone	331	35	100
p-Methoxyacetophenone	258	39	220
α-Hydrindone	242	42	258
Benzalacetone	262	42	227
Benzophenone	306	49	238
2-Acetylnaphthalene	301	56	262
Phenyl styryl ketone (9)	347	58	245
Phenyl p-tolyl ketone	326	60	200
Deoxybenzoin (10)	320	60	206
p-Methoxybenzophenone	355	62	180
Fluorenone (11)	341	83	284
α-Hydroxyacetophenone	—	86	—
Di-p-tolyl ketone	335	95	229
Benzil	347d	95	189
m-Hydroxyacetophenone	—	96	—
p-Hydroxyacetophenone	—	109	261
Dibenzylideneacetone	—	112	180
p-Benzoquinone	—	116	186
Acenaphthenone	—	121	—
1,4-Naphthoquinone	—	125	—
Benzoin	344	137	245
1,2-Naphthoquinone	—	146	—
Resacetophenone (13)	—	147	—
9,10-Phenanthraquinone	—	207	313
Phloroacetophenone (14)	—	219	—
Anthraquinone	—	285	—

Halogeno-ketones

m-Chloroacetophenone	228	—	—
o-Chloroacetophenone	229	—	206
p-Chloroacetophenone	236	20	231
o-Bromoacetophenone	112/10	—	—
m-Bromoacetophenone	131/16	8	—
α-Bromoacetophenone (15)	—	51	—
p-Bromoacetophenone	256	51	230
p-Chlorobenzophenone	323	78	185

Semi-carbazone	Oxime °C	Phenyl-hydrazone °C	p-Nitro-phenyl-hydrazone °C	Other derivatives °C
99	59	105	185	Benzylidene, 58
10	117	110	—	—
03	61	—	—	—
98	69	87	145	—
74	53	147	—	—
98	55	—	—	—
81	94	73	—	—
05	88	96	198	—
36	—	—	—	—
88	50	—	—	—
96	—	—	—	—
66	52	162	—	—
83	83	114	—	—
17	89 (103)	84	231	Benzylidene, 105
15	88	109	—	—
29	139	149	—	Picrate, 116; benzylidene, 126
—	—	—	—	—
—	101	—	—	—
46	125	129	—	Benzylidene, 162
98	87	142	—	—
33	146	128	235	Benzylidene, 113
86	116	157	166	Benzylidene, 112
65	144	137	155	—
36	145	177	—	Picrate, 85
68	115	119	—	Picrate, 97
22	154	109	—	—
48	98	116	163	Benzylidene, 102
—	138	132	199	—
—	195	152	269	—
46	70	112	—	Benzoyl, 118; acetyl, 49
—	163	100	—	—
44 (Di)	237	235 (Di)	290	Quinoxaline, 126
95	—	—	—	—
99	145	151	—	—
89	143	153	173	Picrate, 114
43 (Di)	140	—	—	Picrate, 79
—	175	90	—	Picrate, 113
47	198	—	278	—
06	151 (99)	159	—	Benzoyl, 125; acetyl, 83
84	162	138	235	—
18	199	159	—	Dibenzoyl, 81; diacetyl, 38
—	158	165	245	—
—	—	—	—	Tribenzoyl, 118; triacetyl, 103
—	224	183	—	—
32	88	—	176	—
60 (179)	113	—	215	—
01	95	114	239	—
—	177	—	—	—
38	—	—	—	—
46	89	—	—	—
08	129	126	—	—
—	156 (95)	106	—	—

Table VIII,18 Aromatic ketones (*continued*)

Ketone	B.P. °C	M.P. °C	2,4-Dinitro-phenylhydrazone °C
Halogeno-ketones (continued)			
p-Bromobenzophenone	350	82	230
p-Iodoacetophenone	—	85	—
α,*p*-Dibromoacetophenone (16)	—	109	—
Nitro- and amino-ketones			
o-Nitroacetophenone	159/16	—	—
o-Aminoacetophenone	251	20	—
m-Nitroacetophenone	—	81	228
p-Nitroacetophenone	—	81	—
m-Aminoacetophenone	—	99	—
p-Aminoacetophenone	294	106	—

(1) Methyl *o*-tolyl ketone.
(2) Ethyl phenyl ketone.
(3) Methyl *m*-tolyl ketone.
(4) Isopropyl phenyl ketone.
(5) Methyl *p*-tolyl ketone.
(6) Phenyl propyl ketone.
(7) Butyl phenyl ketone.
(8) 2-Methylbenzophenone.

Semi-carbazone	Oxime °C	Phenyl-hydrazone °C	p-Nitro-phenyl-hydrazone °C	Other derivatives °C
—	169	126	—	—
—	—	—	—	—
—	115	—	—	—
—	—	—	—	—
90	109	108	—	—
57	132	135	—	—
—	—	132	—	—
96	148	—	—	—
50	—	—	—	—

(9) Chalkone.
(10) Benzyl phenyl ketone.
(11) Diphenylene ketone.
(12) Phenacyl alcohol.
(13) 2,4-Dihydroxyacetophenone.
(14) 2,4,6-Trihydroxyacetophenone.
(15) Phenacyl bromide.
(16) p-Bromophenacyl bromide.

Table VIII,19   Quinones

Quinone	M.P. °C	Semicarbazone °C	Oxime °C
Thymoquinone	45	204	162
2-Methyl-1,4-benzoquinone	69	179	135
2-Methyl-1,4-naphthoquinone	106	247	167
Duroquinone (1)	112	—	—
p-Benzoquinone	116	243d	240d
1,4-Naphthoquinone	125	247	198
1,2-Naphthoquinone	116–120	184	162
1-Methylanthraquinone	172	—	—
2-Methylanthraquinone	177	—	—
3-Methyl-1,2-benzoquinone	195	—	140
Camphorquinone	199	236	170
Quinizarin (2)	201	—	—
9,10-Phenanthraquinone	206	220d	162
Acenaphthenequinone	261	192	222 Di
Anthraquinone	286	—	224
Chloranil (3)	290*	—	—
Alizarin	290	—	—

* Sealed tube.   (1) 2,3,5,6-Tetramethyl-1,4-benzoquinone.

Table VIII,20   Aliphatic carboxylic acids

Acid	B.P. °C	M.P. °C	Anilide °C	p-Tolu- idide °C	Amide °C
Formic	101	8	50	53	3
Acetic	118	16	114	153	82
Acrylic	140	13	105	141	85
Propionic	141	—	106	126	79
Propiolic	144d	18	87	—	62
Isobutyric	154	—	105	109	129
Butyric	163	—	96	75	115
Pivalic (2,2-dimethylpropanoic)	164	35	133	120	154
Pyruvic	165d	13	104	130	125
Crotonic (cis)	165	15	102	—	102
Isovaleric	176	—	110	109	136
2-Methylbutanoic	177	—	112	93	112
Valeric	186	—	63	74	106
2-Ethylbutanoic	193	—	127	116	112
4-Methylpentanoic	199	—	112	63	121
Methoxyacetic	203	—	58	—	96
Hexanoic	205	—	—	74	100
Ethoxyacetic	207	—	95	—	82
Heptanoic	223	—	71	80	96
2-Ethylhexanoic	228	—	—	—	103
Cyclohexanecarboxylic	233	31	144	—	186
Octanoic	239	16	57	70	107
Levulinic (1)	246	33	102	109	108

Hydroquinone °C	Diacetate of hydroquinone °C	Thiele acetylation product °C	Other derivatives °C
143	74	—	—
124	52	114	—
—	—	113	—
239	207	—	—
171	123	97	Picrate, 179
176	128	135	—
103	105	135	—
—	217	—	—
—	—	—	—
—	—	—	Quinoxaline, 78
—	—	207	—
148	202	—	Quinoxaline, 220
—	—	—	Quinoxaline, 241
180	260	—	—
232	251	—	—
—	182	—	—

(2) 1,4-Dihydroxyanthraquinone.   (3) 2,3,5,6-Tetrachloro-1,4-benzoquinone.

p-Bromophenacyl ester °C	p-Nitrobenzyl ester °C	p-Phenylphenacyl ester °C	S-Benzylthiouronium salt °C	p-Bromoanilide °C	Hydrazide* °C	N-Benzylamide* °C
40	31	74	151	119	54	60
36	78	111	136	166	77	61
53	31	102	152	148	40	44
—	—	—	—	—	—	—
77	—	89	149	151	104	87
53	35	82	149	111	44	38
76	—	—	—	—	—	—
—	—	—	—	—	—	—
68	—	78	159	129	68	54
55	—	—	—	—	—	—
75	—	63	156	106	—	42
—	—	77	—	—	—	—
7	—	70	—	—	—	—
72	—	70	159	105	—	53
44	—	—	—	—	—	—
72	—	62	—	95	—	—
—	—	54	—	—	—	—
7	—	67	157	102	—	—
4	61	—	—	—	—	—

Table VIII,20 Aliphatic carboxylic acids (continued)

Acid	B.P. °C	M.P. °C	Anilide °C	p-Tolu-idide °C	Amide °C
Nonanoic	254	12	57	84	99
Decanoic	269	31	70	78	108
Undec-10-enoic	275	25	67	68	87
Undecanoic	164/15	29	71	80	103
(±)-Lactic	122/15	18	59	107	79
Dodecanoic	225/100	43	78	87	99
Myristic (tetradecanoic)	250/100	58	84	93	103
Palmitic (hexadecanoic)	268/100	63	91	98	106
Oleic (cis-octadec-9-enoic)	233/10	16	41	43	76
Cyanoacetic	—	66	198	—	120
Stearic (octadecanoic)	291/100	70	94	102	109
Crotonic (trans)	189	72	118	132	160
Glycollic	—	79	97	143	120
Citraconic (2)	—	93	175	—	186
Glutaric	—	98	224	218	175
Citric (hydrated)	—	100	199	189	215
(−)-Malic (±, m.p. 133 °C)	—	101	197	207	157
Oxalic (dihydrate)	—	101	246	268	419d
Pimelic	—	105	156	206	—
Azelaic	—	106	187	202	172
Sebacic	—	133	202	201	209
Sorbic	—	134	153	—	—
Furoic	—	134	124	108	142
Maleic	—	135	187	142	181
Malonic	—	135d	225	253	170
meso-Tartaric	—	140	—	—	190
2-Furylacrylic	—	141	—	—	169
Suberic	—	142	187	219	217
Adipic	—	152	239	241	220
Itaconic (methylenesuccinic)	—	165	190	—	192
(+)-Tartaric	—	170	264	—	196
Succinic	—	185	230	255	260
(+)-Camphoric	—	187	226	—	193
Aconitic (trans-propene-1,2,3-tricarboxylic)	—	191	—	—	250
Mesaconic (trans-methylbutenedioic)	—	204	186	212	176
(±)-Tartaric	—	206	—	—	226
D-Galactaric	—	214	—	—	—
Nicotinic	—	235	85	150	128
Fumaric	—	286	314	—	266
Thioacetic	93	—	76	131	115

Halogeno-acids

Acid	B.P. °C	M.P. °C	Anilide °C	p-Tolu-idide °C	Amide °C
2-Chloropropanoic	186	—	92	124	80
Dichloroacetic	194	10	119	153	97
2-Bromopropanoic	206	25	99	125	123
Bromoacetic	208	50	130	91	91
Trichloroacetic	196	58	95	113	141
Chloroacetic	189	63	137	162	120
Iodoacetic	—	84	144	—	95
Fluoroacetic	167	35	—	—	108
Difluoroacetic	134	—	—	—	52
Trifluoroacetic	72	—	—	—	—

(1) 4-Oxopentanoic acid.
(2) cis-methylbutenedioic acid.

p-Bromophenacyl ester °C	p-Nitrobenzyl ester °C	p-Phenylphenacyl ester °C	S-Benzylthiouronium salt °C	p-Bromoanilide °C	Hydrazide* °C	N-Benzylamide* °C
69	—	71	—	100	—	—
67	—	—	—	102	—	—
—	—	—	149	—	—	—
68	—	79	—	—	—	—
13	—	145	153	—	—	—
76	—	86	141	—	105	83
81	—	90	139	—	—	90
86	43	94	141	—	111	95
45	—	61	—	—	—	—
—	—	—	—	—	—	124
90	—	97	143	—	—	97
95	67	—	172	—	—	114
38	107	—	146	—	—	104
—	70	109	—	—	177	—
37	69	152	161	—	176	170
48	102	146	—	—	—	170
79	124	106	124	—	178	157
42	204	165d	198	—	243	223
37	—	146d	—	—	182	154
31	44	141	—	—	—	—
47	73	140	155	—	—	167
29	—	—	—	—	—	—
39	134	86	211	—	80	—
68	89	168	163	—	—	150
—	86	175	147	—	154	142
—	93	—	—	—	—	205
—	—	—	—	—	—	—
44	85	151	—	—	—	—
55	106	148	163	—	171	189
17	90	—	—	—	—	—
16	163	204	—	—	—	199
11	88	208	154	—	168	206
—	67	—	—	—	—	—
86	—	—	—	—	—	—
—	147	—	—	—	—	210
—	—	149	178	—	215	201
—	—	—	—	—	—	—
—	151	—	195	—	—	205
—	—	—	—	—	—	—
99	—	—	178	—	—	—
—	—	—	—	—	—	—
—	80	—	148	—	—	—
05	—	116	160	—	—	—
—	—	—	—	—	—	—
—	—	—	—	—	—	—
—	—	—	—	—	—	—

See Section **VII,6,**17 (d(ii and iii)) for details of the preparation of hydrazides and N-benzylamides.

Table VIII,21 Aromatic carboxylic acids

Acid	M.P. °C	Anilide °C	p-Toluidide °C	Amide °C
o-Ethoxybenzoic	25	—	—	132
3-Phenylpropanoic (1)	48	98	135	105
Phenylacetic	76	118	136	157
Phenoxyacetic	99	101	—	101
o-Methoxybenzoic	101	131	—	129
o-Toluic	105	125	144	143
m-Methoxybenzoic	110	—	—	—
m-Toluic	111	126	118	95
(±)-Mandelic (2)	120	152	172	134
Benzylmalonic	120d	217	—	225
Benzoic	121	162	158	129
o-Benzoylbenzoic	128	195	—	165
Cinnamic	133	153	168	147
1-Naphthylacetic	133	156	—	181
Acetylsalicylic	135	136	—	138
Phenylpropiolic	136	126	142	109
m-Ethoxybenzoic	137	—	—	139
2-Naphthylacetic	142	—	—	200
Diphenylacetic	148	180	173	168
Benzilic	150	175	190	155
2-Hydroxy-5-methylbenzoic	153	—	—	178
Salicylic	158	135	156	139
1-Naphthoic	162	163	—	202
2-Hydroxy-3-methylbenzoic (4)	169	—	—	112
2-Hydroxy-4-methylbenzoic (5)	177	—	—	—
p-Toluic	178	146	160	159
p-Methoxybenzoic (6)	184	169	186	162
2-Naphthoic	185	170	191	192
p-Ethoxybenzoic	198	170	—	202
3,4-Dihydroxybenzoic (7)	199d	167	—	212
3-Hydroxybenzoic	201	157	163	167
Phthalic	ca. 208d	251	—	219
4-Hydroxy-3-methoxy benzoic (8)	210	—	—	—
4-Hydroxybenzoic	213	197	204	162
2,4-Dihydroxybenzoic (9)	213	127	—	221
3-Hydroxy-2-naphthoic	223	244	222	218
1-Hydroxy-2-naphthoic	226	—	—	218
Diphenic	229	230	—	212
3,4,-Methylenedioxybenzoic (10)	229	—	—	169
Gallic	ca. 240d	207	—	245
Isophthalic	347	—	—	280
Terephthalic	sub. > 300	337	—	—
Benzene-1,3,5-tricarboxylic (11)	380	—	—	365

Halogeno-carboxylic acids

m-Chlorophenoxyacetic	100	—	—	—
m-Fluorobenzoic	124	—	—	130
o-Fluorobenzoic	127	—	—	116
o-Chlorobenzoic	141	118	131	141
o-Chlorophenoxyacetic	146	121	—	150
o-Bromobenzoic	150	141	—	155
m-Bromobenzoic	155	146	—	155

p-Bromo-phenacyl ester °C	p-Nitro-benzyl ester °C	p-Phenyl-phenacyl ester °C	S-Benzyl-thio-uronium salt °C	N-Benzyl-amide °C	Other derivatives °C
104	36	95	—	85	—
89	65	88	165	122	—
48	—	—	—	—	—
—	113	131	—	—	—
57	91	95	146	—	Hydrazide, 124
—	—	—	—	—	—
08	87	136	140	75	Hydrazide, 97
—	123	—	166	—	—
—	120	—	—	—	—
19	89	167	167	106	Hydrazide, 112
—	100	—	—	—	—
46	117	182	183	225	—
—	—	—	—	—	—
—	90	—	144	—	—
—	83	—	—	—	—
—	—	—	—	—	—
—	—	111	—	—	—
52	100	122	—	—	Acetyl, 98
—	147	—	185	—	Acetyl, 153
40	98	148	148	136	Benzoyl, 132; p-nitrobenzoyl, 205
—	99	—	204	—	Acetyl, 113
—	175	—	165	—	Acetyl, 139
53	104	165	190	133	Hydrazide, 117
52	132	160	185	132	—
—	—	—	—	—	—
—	110	—	—	—	—
—	188	—	—	—	—
76	108	—	—	142	Acetyl, 131
53	155	167	158	179	—
—	141	—	—	—	Acetyl, 146; benzoyl, 178
91	192	240	145	—	Acetyl, 187
—	189	—	—	—	—
—	—	—	—	—	—
—	186	—	—	—	—
—	—	—	—	—	—
—	—	198d	—	—	Triacetyl, 172; tribenzoyl, 192
79	203	280	216	—	Hydrazide, 220
25	264	—	204	266	—
97	—	—	—	—	Tri-Me-ester, 144; tri-Et-ester, 135
—	—	—	—	—	—
—	—	—	—	—	Hydrazide, 139
—	—	—	—	—	Hydrazide, 73
07	106	123	—	—	Hydrazide, 110
—	—	—	—	—	—
02	110	98	171	—	—
26	105	155	168	—	—

Table VIII,21 Aromatic carboxylic acids (continued)

Acid	M.P. °C	Anilide °C	p-Toluidide °C	Amide °C
Halogeno-carboxylic acids (continued)				
p-Chlorophenoxyacetic	157	125	—	133
m-Chlorobenzoic	158	124	—	134
o-Iodobenzoic	162	141	—	184
2,4-Dichlorobenzoic	164	—	—	194
p-Fluorobenzoic	185	—	—	154
m-Iodobenzoic	187	—	—	186
3,4-Dichlorobenzoic	209	—	—	169
p-Chlorobenzoic	243	194	—	179
p-Bromobenzoic	252	197	—	189
p-Iodobenzoic	270	210	—	218
Nitro- and amino-carboxylic acids				
m-Nitrophenylacetic	120	—	—	110
m-Nitrobenzoic	141	154	162	142
o-Nitrophenylacetic	141	—	—	161
Anthranilic	146	131	151	109
o-Nitrobenzoic	147	155	—	175
p-Nitrophenylacetic	152	212	210	198
4-Nitrophthalic	165	—	—	200
m-Aminobenzoic	174	140	—	111
2,4-Dinitrobenzoic	183	—	—	204
N-Acetylanthranilic	185	167	—	171
Hippuric (12)	187	208	—	183
p-Aminobenzoic	188	—	—	114
m-Nitrocinnamic	205	—	—	196
3,5-Dinitrobenzoic	207	234	—	183
3-Nitrophthalic	219	234	223	201
2,4,6-Trinitrobenzoic	228	—	—	264
p-Nitrobenzoic	239	211	203	201
o-Nitrocinnamic	240	—	—	185
β-Phenylalanine (13)	273	—	—	140
p-Nitrocinnamic	287	—	—	217

(1) Hydrocinnamic acid.
(2) 2-Hydroxy-2-phenylacetic acid.
(3) 6-Hydroxy-m-toluic acid.
(4) 2-Hydroxy-m-toluic acid.
(5) 2-Hydroxy-p-toluic acid.
(6) Anisic acid.
(7) Protocatechuic acid.

p-Bromophenacyl ester °C	p-Nitrobenzyl ester °C	p-Phenylphenacyl ester °C	s-Benzylthiouronium salt °C	N-Benzylamide °C	Other derivatives °C
136	—	—	—	—	—
117	107	154	155	—	Hydrazide, 158
110	111	143	—	110	—
—	—	—	—	—	Dinitro, 211
—	—	—	—	—	Hydrazide, 162
128	121	—	—	—	—
126	130	160	—	—	Hydrazide, 163
134	141	160	—	—	Hydrazide, 164
146	141	171	—	—	—
132	142	153	163	101	—
—	205	—	149	—	N-Benzoyl, 81; N-toluene-p-sulphonyl, 217
107	112	140	159	—	—
207	—	—	—	—	—
—	—	120	—	—	—
—	201	—	—	—	N-Acetyl, 248
58	142	—	—	—	—
51	136	163	—	—	Hydrazide, 162
—	—	—	—	90	N-Acetyl, 250; N-benzoyl, 278
78	174	—	—	—	—
59	157	154	—	—	—
—	190	149	—	—	—
36	169	182	182	142	—
42	132	146	—	—	—
—	222	—	—	—	N-Benzoyl, 188
91	187	192	—	—	—

(8) Vanillic acid.
(9) β-Resorcylic acid.
(0) Piperonylic acid.
(1) Trimesic acid.
(2) Benzoylaminoacetic acid.
(3) 2-Amino-3-phenylpropanoic acid.

Table VIII,22 Acid chlorides (aliphatic)

Acyl chloride	B.P. °C (/mmHg)	M.P. °C	$d_4^{20°}$	$n_D^{20°}$
Acetyl	52	—	1.104	1.390
Propionyl	80	—	1.056	1.404
Isobutyryl	92	—	1.017	1.408
Butyryl	102	—	1.028	1.412
Chloroacetyl	105	—	1.420	1.454
Dichloroacetyl	108	—	—	—
Methoxyacetyl	113	—	1.187	1.419
3-Methylbutanoyl	115	—	0.987	1.416
Trichloroacetyl	118	—	1.620	1.470
Crotonoyl	126	—	—	—
Valeryl	127	—	1.000	1.420
4-Methylpentanoyl	144	—	0.973	—
Hexanoyl	152	—	0.975	1.426
Heptanoyl	175	—	0.962	1.432
Octanoyl	195	—	0.949	1.432
Nonanoyl	215	—	0.942	1.433
Decanoyl	232	—	—	—
Oxalyl	64	—	1.479	1.432
Succinyl	192	17	1.375	1.468
Glutaryl	218	—	1.324	1.473
Adipoyl	125/11	—	—	—
Pimeloyl	137/15	—	—	—
Suberoyl	150/12	—	1.171	1.468
Azelaoyl	165/13	—	—	—
Sebacoyl	182/16	—	1.212	1.468

Table VIII,23 Acid anhydrides (aliphatic)

Anhydride	B.P. °C (/mmHg)	M.P. °C	$d_4^{20°}$	$n_D^{20°}$
Acetic	140	—	1.081	1.390
Propionic	168	—	1.022	1.404
Isobutyric	182	—	0.956	—
Butyric	198	—	0.968	1.413
Citraconic	213	7	—	—
Isovaleric (3-methylbutanoic)	215	—	0.933	1.404
Valeric	218	—	0.925	—
4-Methylpentanoic	139/19	—	—	—
Hexanoic	245	—	0.920	1.430
Crotonic	248	—	1.040	1.474
Heptanoic	258	17	0.917	1.433
Octanoic	285	—	0.910	1.434
Maleic	198	56	—	—
Glutaric	150/10	56	—	—
Itaconic	139/30	68	—	—
Succinic	261	120	—	—
(+)-Camphoric	270	221	—	—
Trifluoroacetic	39	—	1.490	1.269
Dichloroacetic	101/16	—	—	—
Trichloroacetic	223	—	—	—
Chloroacetic	109/11	46	—	—

Table VIII,24 Acid chlorides and acid anhydrides of aromatic acids

Acid chloride	B.P. °C (/mmHg)	M.P. °C
Benzoyl	197	—
Phenylacetyl	210	—
o-Toluoyl	212	—
m-Toluoyl	219	—
p-Toluoyl	227	—
m-Methoxybenzoyl	244	—
o-Methoxybenzoyl	254	—
Phthaloyl	281	16
Anisoyl	145/14	24
1-Naphthoyl	163/10	24
Cinnamoyl	131/11	36
2-Naphthoyl	305	53
Diphenylcarbamoyl	—	86
p-Chlorobenzoyl	222	16
m-Chlorobenzoyl	225	—
o-Chlorobenzoyl	238	—
m-Bromobenzoyl	243	—
o-Bromobenzoyl	245	11
p-Bromobenzoyl	245	42
o-Nitrobenzoyl	148/9	20
m-Nitrobenzoyl	278	35
2,4-Dinitrobenzoyl	—	46
3,5-Dinitrobenzoyl	196/11	74
p-Nitrobenzoyl	—	75
3-Nitrophthaloyl	—	77

Anhydride	B.P. °C	M.P. °C
o-Toluic	—	39
Benzoic	360	42
m-Toluic	—	71
Phenylacetic	—	72
p-Toluic	—	95
Anisic	—	99
Phthalic	284	132
2-Naphthoic	—	135
Cinnamic	—	136
1-Naphthoic	—	146
Naphthalene-1,2-dicarboxylic	—	169
Diphenic	—	217
(±)-Camphoric	270	222
Naphthalene-2,3-dicarboxylic	—	246d
Naphthalene-1,8-dicarboxylic	—	274
o-Chlorobenzoic	—	79
m-Chlorobenzoic	—	95
p-Chlorobenzoic	—	194
Tetrachlorophthalic		255
Tetrabromophthalic	—	280
Tetra-iodophthalic	—	325
3,5-Dinitrobenzoic	—	109
4-Nitrophthalic	—	119
o-Nitrobenzoic	—	135
2,4-Dinitrobenzoic	—	160
m-Nitrobenzoic	—	163
3-Nitrophthalic	—	164
p-Nitrobenzoic	—	190

Table VIII,25 Aliphatic esters

It is considered that the Table will be of greatest use if the esters are subdivided under the various acids rather than arranged in order of increasing b.p. or m.p. irrespective of the nature of the carboxylic acid. The latter procedure leads to an unwieldy, heterogeneous Table which has relatively little pedagogic or, indeed, practical value.

Ester	B.P. °C (/mmHg)	$d_4^{20°}$	$n_D^{20°}$
Methyl formate	32	0.974	1.344
Ethyl formate	53	0.923	1.360
Isopropyl formate	71	0.873	1.368
Propyl formate	81	0.904	1.377
t-Butyl formate	83	—	—
Allyl formate	84	0.946	—
s-Butyl formate	97	0.884	1.384
Isobutyl formate	98	0.876	1.386
Butyl formate	106	0.892	1.389
Isopentyl formate	124	0.882	1.398
Pentyl formate	131	0.885	1.400
Cyclopentyl formate	138	1.000	1.432
Hexyl formate	154	0.879	1.407
Cyclohexyl formate	161	0.994	1.443
Ethylene glycol diformate	177	1.229	—
Methyl acetate	56	0.939	1.362
Ethyl acetate	77	0.901	1.372
Isopropyl acetate	88	0.872	1.377
t-Butyl acetate	97	0.867	1.386
Propyl acetate	101	0.887	1.384
Allyl acetate	104	0.928	1.404
s-Butyl acetate	112	0.872	1.389
Isobutyl acetate	116	0.871	1.390
Butyl acetate	124	0.881	1.394
t-Pentyl acetate	124	0.873	1.392
Isopentyl acetate	141	0.872	1.400
Pentyl acetate	148	0.875	1.402
Cyclopentyl acetate	153	0.975	1.432
Hexyl acetate	169	0.872	1.409
Cyclohexyl acetate	172	0.970	1.442
Heptyl acetate	192	0.865	1.414
Tetrahydrofurfuryl acetate	195	1.061	1.438
Octyl acetate	210	—	—
Methyl 'cellosolve' acetate	144	1.088	—
'Cellosolve' acetate	156	0.976	—
Ethylene glycol diacetate	190	1.104	1.415
Propylene glycol diacetate	191	1.059	1.417
Trimethylene glycol diacetate	210	1.069	—
'Carbitol' acetate	217	1.013	—
Butyl 'carbitol' acetate	246	0.983	—
α-Monoacetin (glycerol 1-acetate)	158/15	1.206	1.416
Diacetin (mixture of αγ and αβ)	143/12	1.180	—
Triacetin (glyceryl triacetate)	153/22	1.161	1.430
Methyl propionate	79	0.915	1.377
Ethyl propionate	98	0.892	1.384
Isopropyl propionate	111	—	—
Propyl propionate	122	0.882	1.393
Allyl propionate	123	0.914	1.410

Table VIII,25 **Aliphatic esters** (*continued*)

Ester	B.P. °C (/mmHg)	$d_4^{20°}$	$n_D^{20°}$
Butyl propionate	145	0.875	1.401
Isopentyl propionate	160	0.859	1.412
Pentyl propionate	169	0.881	—
Hexyl propionate	190	0.870	1.419
Methyl butyrate	102	0.898	1.387
Ethyl butyrate	120	0.879	1.392
Isopropyl butyrate	128	—	—
Propyl butyrate	142	0.872	1.400
Allyl butyrate	142	0.902	1.416
Butyl butyrate	165	0.869	1.406
Isopentyl butyrate	179	0.864	1.411
Pentyl butyrate	185	0.866	1.412
Hexyl butyrate	208	0.866	1.420
Methyl isobutyrate	91	0.888	1.383
Ethyl isobutyrate	110	0.869	1.387
Isopropyl isobutyrate	121	—	—
Propyl isobutyrate	134	0.864	1.396
Butyl isobutyrate	156	0.862	1.402
Methyl valerate	127	0.890	1.397
Ethyl valerate	144	0.874	1.400
Isopropyl valerate	154	0.858	1.401
Propyl valerate	164	0.870	1.407
Butyl valerate	184	0.868	1.412
Methyl isovalerate	116	0.881	1.393
Ethyl isovalerate	133	0.865	1.396
Propyl isovalerate	156	0.862	1.403
Isobutyl isovalerate	171	0.853	1.406
Butyl isovalerate	176	0.861	1.409
Methyl hexanoate	149	0.885	1.405
Ethyl hexanoate	168	0.871	1.407
Propyl hexanoate	187	0.867	1.417
Butyl hexanoate	208	0.865	1.421
Pentyl hexanoate	226	0.863	1.426
Methyl cyclohexanecarboxylate	183	0.990	1.451
Ethyl cyclohexanecarboxylate	196	0.962	1.448
Methyl heptanoate	171	0.882	1.412
Ethyl heptanoate	186	0.870	1.413
Propyl heptanoate	208	0.866	1.421
Butyl heptanoate	226	0.864	1.426
Methyl octanoate	192	0.878	1.417
Ethyl octanoate	206	0.869	1.418
Methyl nonanoate	214	—	—
Ethyl nonanoate	227	0.866	1.422
Methyl decanoate	228	0.873	1.426
Ethyl decanoate	242	0.865	1.426
Propyl decanoate	115/5	0.862	1.428
Butyl decanoate	123/4	0.861	1.430

Table VIII,25   Aliphatic esters (*continued*)

Ester	B.P. °C (/mmHg)	$d_4^{20°}$	$n_D^{20°}$
Methyl dodecanoate	262	0.870	1.432
Ethyl dodecanoate	273	0.862	1.431
Propyl dodecanoate	140/4	0.862	1.434
Butyl dodecanoate	154/5	0.860	1.436
Methyl stearate	M.p. 39	—	—
Ethyl stearate	M.p. 33	—	—
Methyl chloroformate	73	1.223	1.387
Ethyl chloroformate	94	1.136	1.397
Propyl chloroformate	115	1.090	1.404
Isobutyl chloroformate	129	1.040	1.406
Butyl chloroformate	138	1.079	1.412
Pentyl chloroformate	61/15	—	1.417
Hexyl chloroformate	63/10	—	—
Methyl chloroacetate	129	1.234	1.422
Ethyl chloroacetate	142	1.150	1.422
Methyl dichloroacetate	143	1.377	1.443
Ethyl dichloroacetate	156	1.283	1.438
Methyl trichloroacetate	152	1.488	1.457
Ethyl trichloroacetate	164	1.380	1.450
Methyl bromoacetate	144d	—	—
Ethyl bromoacetate	169	1.506	1.451
Methyl iodoacetate	170	—	—
Ethyl iodoacetate	180	1.818*	1.508*
Methyl methoxyacetate	130	1.051	1.396
Ethyl methoxyacetate	132	1.007	—
Methyl ethoxyacetate	148	1.006	—
Ethyl ethoxyacetate	158	0.970	1.403
Methyl acrylate	80	0.960	1.398
Ethyl acrylate	101	0.909	1.406
Methyl crotonate	119	0.946	1.425
Ethyl crotonate	137	0.918	1.425
Propyl crotonate	157	0.908	1.428
Butyl crotonate	55/4	0.899	1.432
Isopentyl crotonate	60/4	0.891	1.434
Pentyl crotonate	72/5	0.894	1.436
Methyl lactate	145	1.089	1.414
Ethyl lactate	154	1.030	1.415
Methyl glycollate	151	1.166	—
Ethyl glycollate	160	1.082	—
Methyl pyruvate	138	—	—
Ethyl pyruvate	155	1.055	1.406
Methyl levulinate	196	1.049	1.423
Ethyl levulinate	206	1.011	1.423

Table VIII,25 **Aliphatic esters** (*continued*)

Ester	B.P. °C (/mmHg)	$d_4^{20°}$	$n_D^{20°}$
Methyl furoate	181	1.180	1.486
Ethyl furoate	197 (m.p. 34)	1.117*	1.480*
Trimethyl orthoformate	105	0.968	1.379
Triethyl orthoformate	143	0.893	1.390
Tripropyl orthoformate	91/17	0.879	1.407
Tributyl orthoformate	127/16	0.871	1.416
Dimethyl carbonate	90	1.071	1.369
Diethyl carbonate	126	0.976	1.384
Dipropyl carbonate	165	0.943	1.400
Diisobutyl carbonate	188	0.914	1.407
Dibutyl carbonate	205	0.925	1.412
Dimethyl oxalate	M.p. 54	—	—
Diethyl oxalate	183	1.079	1.410
Diisopropyl oxalate	191	0.995	1.413
Dipropyl oxalate	212	1.019	1.416
Dibutyl oxalate	241	0.987	1.423
Diisopentyl oxalate	127/7	0.961	1.427
Dipentyl oxalate	139/9	0.966	1.429
Dimethyl malonate	179	1.119	1.420
Diethyl malonate	197	1.055	1.414
Diallyl succinate	104	1.051	1.452
Dimethyl succinate	195	1.120	1.420
Diethyl succinate	218	1.042	1.420
Diisopropyl succinate	82/3	0.985	1.418
Dipropyl succinate	102/3	1.006	1.425
Diisobutyl succinate	116/4	0.968	1.427
Dibutyl succinate	120/3	0.977	1.430
Diisopentyl succinate	130/4	0.958	1.434
Dipentyl succinate	129/2	0.960	˙.434
Dimethyl glutarate	109/21	1.087	1.424
Diethyl glutarate	118/15	1.023	1.424
Dimethyl adipate	121/17	1.063	1.428
Diethyl adipate	134/17	1.009	1.428
Diisopropyl adipate	120/6	0.966	1.425
Dipropyl adipate	146/9	0.981	1.431
Dibutyl adipate	159/17	0.945	1.435
Diisopentyl adipate	184/13	0.945	1.437
Dipentyl adipate	186/10	0.948	1.439
Dimethyl pimelate	128/16	1.038	1.431
Diethyl pimelate	149/18	0.993	1.430
Dimethyl suberate	120/6	1.024	1.434
Diethyl suberate	131/5	0.981	1.432
Dipropyl suberate	165/8	0.962	1.435
Dibutyl suberate	176/4	0.948	1.439

* Values at 21 °C with supercooled liquid.

Table VIII,25   Aliphatic esters (*continued*)

Ester	B.P. °C (/mmHg)	M.P. °C	$d_4^{20°}$	$n_D^{20°}$
Dimethyl azelate	156/20	—	1.007	1.436
Diethyl azelate	291	—	0.973	1.435
Dimethyl sebacate	293	27	—	—
Diethyl sebacate	307	—	0.964	1.437
Dipropyl sebacate	179/5	—	0.950	1.439
Dimethyl maleate	201	—	1.150	1.442
Diethyl maleate	220	—	1.066	1.440
Dipropyl maleate	126/12	—	1.025	1.443
Dibutyl maleate	147/12	—	0.994	1.445
Dimethyl fumarate	193	102	—	—
Diethyl fumarate	214	—	1.052	1.441
Dipropyl fumarate	110/5	—	1.013	1.444
Dibutyl fumarate	139/5	—	0.987	1.447
Dimethyl itaconate	208	38	—	—
Diethyl itaconate	229	—	1.047	1.439
Dimethyl mesaconate	205	—	1.120	1.454
Diethyl mesaconate	225	—	1.043	1.448
Dimethyl citraconate	210	—	1.112	1.448
Diethyl citraconate	228	—	1.041	1.444
Dimethyl (+)-tartrate	280	61	—	—
Diethyl (+)-tartrate	280	18	1.203	1.447
Dipropyl (+)-tartrate	297	—	1.139	—
Dibutyl (+)-tartrate	200/18	22	—	—
Dimethyl (±)-tartrate	282	90	—	—
Diethyl (±)-tartrate	280	18	1.203	1.447
Dipropyl (±)-tartrate	286	25	—	—
Dibutyl (±)-tartrate	320	—	1.086	—
Dimethyl malate	242	—	1.233	1.442
Diethyl malate	253	—	1.129	1.436
Dimethyl galacturate	—	167	—	—
Diethyl galacturate	—	164	—	—
Trimethyl citrate	—	76	—	—
Triethyl citrate	294	—	1.137	1.466

Table VIII,26 Aromatic esters

It is considered that the Table will be of greatest use if the esters are in the main subdivided under the various acids rather than be arranged in order of increasing b.p. or m.p. irrespective of the nature of the carboxylic acid. The latter procedure leads to an unwieldy, heterogeneous Table which has relatively little pedagogic or, indeed, practical value.

Ester	B.P. °C (/mmHg)	M.P. °C	$d_4^{20°}$	$n_D^{20°}$
Methyl benzoate	199	—	1.089	1.517
Ethyl benzoate	212	—	1.047	1.505
Isopropyl benzoate	218	—	1.015	1.491
Propyl benzoate	230	—	1.023	1.500
Allyl benzoate	230	—	1.052	—
Isobutyl benzoate	242	—	0.997	—
Butyl benzoate	248	—	1.005	1.497
Isopentyl benzoate	262	—	0.986	1.495
Pentyl benzoate	137/15	—	—	—
Ethylene glycol dibenzoate	—	73	—	—
Methyl phenylacetate	215	—	1.068	1.507
Ethyl phenylacetate	228	—	1.033	1.497
Propyl phenylacetate	241	—	1.010	1.493
Butyl phenylacetate	256	—	0.994	1.489
Methyl o-toluate	213	—	1.068	—
Ethyl o-toluate	227	—	1.034	1.508
Methyl m-toluate	215	—	1.061	—
Ethyl m-toluate	227	—	1.028	1.506
Methyl p-toluate	217	34	—	—
Ethyl p-toluate	228	—	1.025	1.507
Methyl salicylate	223	—	1.184	1.537
Ethyl salicylate	234	—	1.125	1.522
Propyl salicylate	240	—	1.098	1.516
Butyl salicylate	260	—	1.073	1.512
Methyl m-hydroxybenzoate	—	70	—	—
Ethyl m-hydroxybenzoate	295	73	—	—
Ethyl p-hydroxybenzoate	297	116	—	—
Methyl p-hydroxybenzoate	—	131	—	—
Methyl o-methoxybenzoate	248	—	1.156	1.534
Ethyl o-methoxybenzoate	261	—	1.104	1.525
Methyl m-methoxybenzoate	237	—	1.131	1.522
Ethyl m-methoxybenzoate	251	—	1.100	1.515
Methyl anisate	255	49	—	—
Ethyl anisate	269	7	1.103	1.524
Methyl o-chlorobenzoate	234	—	—	1.536
Ethyl o-chlorobenzoate	243	—	1.190	1.522
Methyl m-chlorobenzoate	231	20	—	1.492
Ethyl m-chlorobenzoate	242	—	1.182	1.520
Ethyl p-chlorobenzoate	238	—	1.181	1.524
Methyl p-chlorobenzoate	—	44	—	—
Methyl o-bromobenzoate	246	—	—	—
Ethyl o-bromobenzoate	255	—	—	—
Ethyl m-bromobenzoate	259	—	—	—
Methyl m-bromobenzoate	—	32	—	—
Ethyl p-bromobenzoate	263	—	—	—
Methyl p-bromobenzoate	—	81	—	—

Table VIII,26 Aromatic esters (*continued*)

Ester	B.P. °C (/mmHg)	M.P. °C	$d_4^{20°}$	$n_D^{20°}$
Ethyl o-iodobenzoate	275	—	—	—
Methyl o-iodobenzoate	278	—	—	—
Ethyl m-iodobenzoate	150/15	—	—	—
Methyl m-iodobenzoate	277	54	—	—
Ethyl p-iodobenzoate	153/14	—	—	—
Methyl p-iodobenzoate	—	114	—	—
Ethyl o-nitrobenzoate	—	30	—	—
Methyl o-nitrobenzoate	275	—	1.286	—
Ethyl m-nitrobenzoate	297	47	—	—
Methyl m-nitrobenzoate	279	79	—	—
Ethyl p-nitrobenzoate	—	57	—	—
Methyl p-nitrobenzoate	—	96	—	—
Ethyl 3,5-dinitrobenzoate	—	94	—	—
Methyl 3,5-dinitrobenzoate	—	108	—	—
Ethyl 2,4-dinitrobenzoate	—	41	—	—
Methyl 2,4-dinitrobenzoate	—	70	—	—
Ethyl anthranilate	267	13	1.117	1.565
Methyl anthranilate	300	24	—	—
Ethyl m-aminobenzoate	294	—	—	—
Methyl m-aminobenzoate	—	38	—	—
Ethyl p-aminobenzoate	—	92	—	—
Methyl p-aminobenzoate	—	112	—	—
Ethyl cinnamate	273	—	1.049	1.560
Propyl cinnamate	284	—	1.028	1.551
Butyl cinnamate	162/12	—	1.013	1.544
Methyl cinnamate	261	36	—	—
Methyl hydrocinnamate	232	—	1.043	1.503
Ethyl hydrocinnamate	248	—	1.016	1.495
Propyl hydrocinnamate	262	—	0.998	1.491
Butyl hydrocinnamate	123/11	—	0.984	1.489
Ethyl o-nitrocinnamate	—	44	—	—
Methyl o-nitrocinnamate	—	73	—	—
Ethyl m-nitrocinnamate	—	79	—	—
Methyl m-nitrocinnamate	—	124	—	—
Ethyl p-nitrocinnamate	—	142	—	—
Methyl p-nitrocinnamate	—	161	—	—
Methyl o-aminocinnamate	—	65	—	—
Ethyl o-aminocinnamate	—	78	—	—
Ethyl m-aminocinnamate	—	64	—	—
Methyl m-aminocinnamate	—	84	—	—
Ethyl p-aminocinnamate	—	69	—	—
Methyl p-aminocinnamate	—	129	—	—
Methyl phenoxyacetate	245	—	1.147	—
Ethyl phenoxyacetate	251	—	1.101	—
Ethyl (±)-mandelate	255	37	—	—
Methyl (±)-mandelate	—	58	—	—
Methyl o-benzoylbenzoate	352	52	—	—
Ethyl o-benzoylbenzoate	—	58	—	—

Table VIII,26 **Aromatic esters** (*continued*)

Ester	B.P. °C (/mmHg)	M.P. °C	$d_4^{20°}$	$n_D^{20°}$
Ethyl diphenylacetate	—	58	—	—
Methyl diphenylacetate	—	60	—	—
Dimethyl phthalate	282	—	1.191	1.516
Diethyl phthalate	298	—	1.118	1.502
Dipropyl phthalate	130/1	—	—	—
Diisopropyl phthalate	154/10	—	—	—
Dibutyl phthalate	205/20	—	—	—
Diethyl isophthalate	285	11	1.121	1.507
Dimethyl isophthalate	—	68	—	—
Diethyl terephthalate	302	44	—	—
Dimethyl terephthalate	—	142	—	—
Diethyl 3-nitrophthalate	—	45	—	—
Dimethyl 3-nitrophthalate	—	69	—	—
Diethyl 4-nitrophthalate	—	34	—	—
Dimethyl 4-nitrophthalate	—	66	—	—
Methyl 1-naphthoate	116/1	—	1.163	1.612
Ethyl 1-naphthoate	309	—	1.121	1.594
Ethyl 2-naphthoate	304	32	—	—
Methyl 2-naphthoate	290	77	—	—
Diethyl diphenate	—	42	—	—
Dimethyl diphenate	—	74	—	—
Furfuryl acetate	176	—	1.118	—
Phenyl acetate	196	—	1.078	1.503
Phenyl propionate	211	20	1.050	—
Phenyl butyrate	228	—	1.023	—
Diphenyl oxalate	190/15	—	—	—
Phenyl salicylate (salol)	—	43	—	—
Diphenyl succinate	330	121	—	—
Phenyl benzoate	299	68	—	—
Phenyl cinnamate	—	73	—	—
Diphenyl carbonate	306	78	—	—
o-Cresyl acetate	208	—	1.045	—
p-Cresyl acetate	212	—	1.050	1.500
m-Cresyl acetate	212	12	1.043	1.498
Guaiacol acetate	240	—	1.133	1.512
Thymyl acetate	243	—	—	—
Carvacryl acetate	245	—	0.994	—
Resorcinol diacetate	278	—	—	—
Eugenol acetate	282	30	—	—
1-Naphthyl acetate	—	49	—	—
Catechol diacetate	—	63	—	—
2-Naphthyl acetate	—	70	—	—
Benzoin acetate	—	83	—	—
Phloroglucinol triacetate	—	104	—	—
Hydroquinone diacetate	—	124	—	—
Pyrogallol triacetate	—	165	—	—

Table VIII,26   Aromatic esters (*continued*)

Ester	B.P. °C (/mmHg)	M.P. °C	$d_4^{20°}$	$n_D^{20°}$
o-Cresyl benzoate	307	—	—	—
Thymyl benzoate	—	33	—	—
m-Cresyl benzoate	—	54	—	—
1-Naphthyl benzoate	—	56	—	—
p-Cresyl benzoate	316	72	—	—
Catechol dibenzoate	—	84	—	—
Pyrogallol tribenzoate	—	90	—	—
2-Naphthyl benzoate	—	107	—	—
Resorcinol dibenzoate	—	117	—	—
Phloroglucinol tribenzoate	—	185	—	—
Hydroquinone dibenzoate	—	199	—	—
Di-o-cresyl carbonate	—	60	—	—
Diphenyl carbonate	306	78	—	—
Diguaiacol carbonate	—	87	—	—
Di-m-cresyl carbonate	—	111	—	—
Di-p-cresyl carbonate	—	115	—	—
Benzyl formate	203	—	1.082	—
Benzyl acetate	214	—	1.057	1.523
Benzyl salicylate	186/10	—	1.180	1.581
Benzyl benzoate	323	21	—	—
Dibenzyl succinate	—	45	—	—
1-Phenylethyl acetate	222	—	—	—
2-Phenylethyl acetate	224	—	1.059	1.512

Table VIII,27   Primary aliphatic amides

Amide	M.P. °C	Xanthylamide °C
Formamide	2 (b.p. 193d)	184
Propionamide	79	214
Acetamide	82	245
Acrylamide	86	—
Heptanamide	96	154
Dichloroacetamide	98	—
Lauramide	99	—
Hexanamide	101	160
Myristamide	103	—
Palmitamide	106	142
Valeramide	106	167
Octanamide	107	148
Decanamide	108	—
Stearamide	109	141
Butyramide	115	187
Chloroacetamide	120	209
Cyanoacetamide	120	223
Isobutyramide	129	211
Isovaleramide	136	183
Trichloroacetamide	141	—
Furoamide	142	210
Trimethylacetamide	154	—
Cyclohexanecarboxamide	185	—
N-Allylurea	80	—
N-Methylurea	102	230
Urea	132	274
N,N-Dimethylurea	182	250
N-Acetylurea	218	—
Thiourea	182	—
Ethyl carbamate (urethan)	49	169
Methyl carbamate	54	193
Butyl carbamate	54	—
Isobutyl carbamate	55	—
Pentyl carbamate	57	—
Propyl carbamate	61	—
Isopentyl carbamate	67	—
Isopropyl carbamate	92	—
Malonamide	170	270
Azelamide	172	—
Glutaramide	175	—
Maleamide	180	—
Sebacamide	209	—
Suberamide	217	—
Adipamide	220	—
(±)-Tartaramide	226	—
Succinamide	260d	275
Oxamide	419d	—
Succinimide	126	246

Table VIII,28 Primary aromatic amides

Amide	M.P. °C	Xanthylamide °C
2-Phenylpropanamide	92	158
m-Toluamide	95	—
3-Phenylpropanamide	105	189
Hydrobenzamide	110	—
Benzamide	129	224
o-Methoxybenzamide	129	—
(±)-Mandelamide	133	—
m-Chlorobenzamide	134	—
Salicylamide	139	—
o-Chlorobenzamide	141	—
N-m-Tolylurea	142	—
m-Nitrobenzamide	142	—
o-Toluamide	143	200
N-Phenylurea	147	225
Cinnamamide	148	—
N-Benzylurea	149	—
o-Bromobenzamide	155	—
m-Bromobenzamide	155	—
Phenylacetamide	157	196
p-Toluamide	159	225
p-Hydroxybenzamide	162	—
Anisamide	162	—
Diphenylacetamide	167	—
m-Hydroxybenzamide	167	—
Piperonylamide	169	—
p-Phenetylurea (N-p-ethoxyphenylurea)	173	—
o-Nitrobenzamide	175	—
p-Chlorobenzamide	179	—
N-p-Tolylurea	183	—
3,5-Dinitrobenzamide	183	—
o-Iodobenzamide	184	—
m-Iodobenzamide	186	—
N,N-Diphenylurea	189	180
p-Bromobenzamide	189	—
N-o-Tolylurea	191	228
2-Naphthamide	192	—
p-Nitrobenzamide	201	232
1-Naphthamide	202	—
p-Ethoxybenzamide	202	—
p-Iodobenzamide	218	—
Phthalamide	219d	—
Phthalimide	235	177

Table VIII,29   Substituted aromatic amides

Amide	M.P. °C	Amide	M.P. °C
Formanilide	50	N-Propylacetanilide	50
Nonananilide	57	N-Ethylacetanilide	54
Octananilide	57	m-Chloroacetanilide	79
Lactanilide	59	m-Methoxyacetanilide	80
Valeranilide	63	o-Chloroacetanilide	88
Decananilide	70	m-Bromoacetanilide	88
Heptananilide	71	o-Methoxyacetanilide	88
Lauranilide	78	m-Aminoacetanilide	88
Myristanilide	84	o-Nitroacetanilide	94
Acetoacetanilide	85	o-Bromoacetanilide	99
Palmitanilide	91	N-Methylacetanilide	103
Stearanilide	94	o-Iodoacetanilide	110
Hexananilide	95	Acetanilide	114
Butyranilide	96	N-Ethyl-p-nitroacetanilide	118
Isobutyranilide	105	Phenylacetanilide	118
Acrylanilide	105	m-Iodoacetanilide	119
Propionanilide	106	p-Methoxyacetanilide	130
Isovaleranilide	110	o-Aminoacetanilide	132
Acetanilide	114	2,4-Dimethylacetanilide	133
Furoanilide	124	2,5-Dimethylacetanilide	142
o-Toluanilide	125	4-Methyl-m-nitroacetanilide	148
m-Toluanilide	126	m-Hydroxyacetanilide	149
o-Methoxybenzanilide	131	N-Methyl-p-nitroacetanilide	153
Salicylanilide	135	m-Nitroacetanilide	155
p-Toluanilide	146	p-Aminoacetanilide	163
Cinnamanilide	153	p-Bromoacetanilide	167
m-Nitrobenzanilide	154	p-Hydroxyacetanilide	168
o-Nitrobenzanilide	155	p-Chloroacetanilide	179
Benzanilide	162	p-Iodoacetanilide	184
1-Naphthanilide	163	2-Methyl-4-nitroacetanilide	196
Anisanilide	169	o-Hydroxyacetanilide	209
2-Naphthanilide	171	p-Nitroacetanilide	216
p-Nitrobenzanilide	211	p-Hydroxy-N-methylacetanilide	240
Pimelic dianilide	156	Acetyl-m-toluidine	66
Suberic dianilide	187	Acetyl-o-phenetidine	79
Maleic dianilide	187	Acetyl-m-anisidine	80
Azelaic dianilide	187	Acetyl-o-anisidine	88
Sebacic dianilide	202	Acetyl-m-phenetidine	96
Glutaric dianilide	224	Acetyl-o-toluidine	112
Malonic dianilide	225	Acetyl-p-anisidine	130
Succinic dianilide	230	Acetyl-p-phenetidine (phenacetin)	137
Adipic dianilide	239	Acetyl-p-toluidine	154
Oxanilide	246	Acetyl-1-naphthylamine	160

Table VIII,29 Substituted aromatic amides (*continued*)

Amide	M.P. °C	Amide	M.P. °C
NN'-Diacetyl-o-phenylenediamine	186	Benzoylpiperidine	48
NN'-Diacetyl-m-phenylenediamine	191	N-Phenylsuccinimide	156
NN'-Diacetyl-p-phenylenediamine	304	N-Phenylphthalimide	205
		Phthalimide	235
Benzoyl-o-anisidine	60	Triphenylguanidine	145
Benzoyl-m-anisidine	—	Diphenylguanidine	147
Benzoyl-m-phenetidine	103	Saccharin	220
Benzoyl-o-phenetidine	104		
Benzoyl-m-toluidine	125	N-Phenylurethane	53
Benzoyl-o-toluidine	144	Ethyl oxanilate	67
Benzoyl-p-anisidine	154		
Benzoyl-p-toluidine	158	N,N'-Di-m-tolylurea	218
Benzoyl-1-naphthylamine	161	N,N'-Diphenylurea (carbanilide)	238
Benzoyl-p-phenetidine	173	N,N'-Di-o-tolylurea	250
		N,N'-Di-p-tolylurea	268
NN'-Dibenzoyl-m-phenylenediamine	240	N,N'-Di-1-naphthylurea	297
NN'-Dibenzoyl-o-phenylenediamine	301		
NN'-Dibenzoyl-p-phenylenediamine	> 300	Ethylbutylbarbituric acid	125
		Ethylhexylbarbituric acid	127
Acetyl-N-methyl-o-toluidine	56	Ethylisopentylbarbituric acid	154
Acetyl-N-methyl-m-toluidine	66	Ethylphenylbarbituric acid	172
Acetyl-N-methyl-p-toluidine	83	Diallylbarbituric acid	172
Acetyl-N-methyl-1-naphthylamine	94	Diethylbarbituric acid	198
		Ethyl-isopropylbarbituric acid	201
		Barbituric acid	245
N-Formyldiphenylamine	74		
N-Acetyldiphenylamine	101	Butyl oxamate	88
N-Benzoyldiphenylamine	180	Ethyl oxamate	131

Table VIII,30   Aliphatic nitriles (cyanides)

Cyanide	Nitrile	B.P. °C (/mm Hg)
Vinyl	Acrylo-	78
Methyl	Aceto-	82
Ethyl	Propiono-	97
Isopropyl	Isobutyro-	108
Propyl	Butyro-	118
Allyl	Vinylaceto-	118
Chloromethyl	Chloroaceto-	127
Isobutyl	Isovalero-	131
Butyl	Valero-	141
Isopentyl	4-Methylpentano-	154
Pentyl	Hexano-	162
Hexyl	Heptano-	183
Heptyl	Octano-	199
Octyl	Nonano-	224
Nonyl	Decano-	244
Decyl	Undecano-	254
Undecyl	Dodecano-	275
Methylene	Malono-	220 (1)
Ethylene	Succino-	276d (2)
Trimethylene	Glutaro-	286 (3)
Tetramethylene	Adipo-	295
Pentamethylene	Pimelo-	169/15
Hexamethylene	Subero-	185/15
Acetaldehyde cyanohydrin	α-Hydroxypropiono-	183
Ethylene cyanohydrin	β-Hydroxypropiono-	221
Trimethylene cyanohydrin	γ-Hydroxybutyro-	240
Trimethylene chlorocyanide	γ-Chlorobutyro-	197
Methyl cyanoacetate		200
Ethyl cyanoacetate		207
2-Furyl	Furo-	147
Phenyl	Benzo-	191
Benzyl	Phenylaceto-	109/15
o-Tolyl	o-Tolu-	205
m-Tolyl	m-Tolu-	212

* Decomposition temperature. Sample placed in bath at 105–110 °C.
† 15 °C.

$d_4^{20°}$	$n_D^{20°}$	Acyl phloroglucinol °C	α-Amino-alkyl-mercaptoacetic acid hydrochloride* °C
0.806	1.391	—	—
0.784	1.344	218	115
0.783	1.366	176	128
—	—	—	137
0.791	1.384	181	136
0.838	1.406	—	—
1.193	—	—	—
0.788	—	—	—
0.799	1.397	149 (hydrate 88)	138
0.803	1.406	122 (hydrate 104)	128
0.805	1.407	121 (hydrate 96)	136
0.810	1.414	—	—
0.817†	1.422†	—	—
0.822†	—	—	—
0.829†	1.432†	—	—
—	—	—	—
0.827†	—	—	—
—	—	—	—
—	—	—	—
0.988	1.429	—	—
0.962	1.439	—	—
0.945	1.441	—	—
0.933	1.445	—	—
0.988	—	—	—
—	—	—	—
—	—	—	—
1.079	—	—	—
1.101	—	—	—
1.063	1.418	—	—
1.082	1.480	—	—
1.006	1.528	—	—
1.016	1.523	—	146
0.996	1.530	—	—
1.032	1.525	—	—

(1) M.p. 31 °C.   (2) M.p. 54 °C.   (3) M.p. 9 °C.

Table VIII,31   Aromatic nitriles

Nitrile	B.P. °C	M.P. °C
(±)-Mandelonitrile	170d	22
Benzonitrile	191	—
o-Tolunitrile	205	—
m-Tolunitrile	212	—
β-Phenylpropiononitrile	232	—
Phenylacetonitrile	234	—
γ-Phenylpropiononitrile	261	—
Cinnamonitrile	255	20
p-Tolunitrile	218	29
m-Bromobenzonitrile	225	38
1-Naphthonitrile	299	36
m-Chlorobenzonitrile	—	41
o-Chlorobenzonitrile	232	43
o-Bromobenzonitrile	252	53
o-Iodobenzonitrile	—	55
2-Naphthonitrile	306	66
p-Chlorobenzonitrile	223	96
o-Nitrobenzonitrile	—	111
p-Bromobenzonitrile	236	113
p-Nitrophenylacetonitrile	—	116
m-Nitrobenzonitrile	—	118
Phthalonitrile	—	141
p-Nitrobenzonitrile	—	149

Table VIII,32   Primary and secondary aliphatic amines

Amine	B.P. °C	$d_4^{20°}$	$n_D^{20°}$	Benzene-sulphonamide °C
Methylamine	−7	—	—	30
Ethylamine	17	—	—	58
Isopropylamine	35	0.689	1.374	26
t-Butylamine	46	—	—	
Propylamine	49	0.717	1.388	36
Allylamine	55	0.762	1.420	39
s-Butylamine	63	0.725	1.393	70
Isobutylamine	68	0.735	1.397	53
Butylamine	77	0.741	1.401	—
Isopentylamine	97	0.749	1.408	—
Pentylamine	105	0.754	1.411	—
Hexylamine	129	0.766	1.418	96
Cyclohexylamine	134	0.867	1.459	—
Heptylamine	155	0.775	1.425	—
Ethanolamine	171	1.022	1.454	—
Octylamine	177	0.782	1.429	—
(−)-Menthylamine	212	0.854	—	—
Benzylamine	185	0.982	1.544	88
1-Phenylethylamine	187	—	—	—
2-Phenylethylamine	198	0.854	—	69
1,2-Diaminoethane	117 (1)	0.898	1.457	168
1,2-Diaminopropane	120	0.874	—	—
1,3-Diaminopropane	136	0.889	1.460	—
1,4-Diaminobutane	159 (2)	—	—	—
1,5-Diaminopentane	180	—	—	119
1,6-Diaminohexane	205 (3)	—	—	154
Dimethylamine	7	—	—	47
Diethylamine	56	0.707	1.386	42
Di-isopropylamine	84	0.717	1.392	94
Dipropylamine	110	0.738	1.405	51
Diallylamine	111	—	—	—
Di-s-butylamine	135	0.753	1.411	—
Di-isobutylamine	137	0.746	1.409	—
Dibutylamine	159	0.760	1.418	—
Di-isopentylamine	186	0.771	1.423	—
Dipentylamine	205	0.777	1.427	—
Dicyclohexylamine	255d (4)	—	—	—
Diethanolamine (5)	270d (6)	1.097	1.478	130
Pyrrolidine	89	0.854	1.424	—
Piperidine	106	0.861	1.453	94
2-Methylpiperidine	118	—	—	—
3-Methylpiperidine	126	—	—	—
4-Methylpiperidine	128	—	—	—
Morpholine	130	1.000	1.455	119
Pyrrole	131	0.969	1.509	—
Piperazine	140 (7)	—	—	292 (di)
1,2,3,4-Tetrahydroisoquinoline	232	—	—	154
1,2,3,4-Tetrahydroquinoline	250 (m.p. 20)	—	—	67

Toluene-*p*-sulphon-amide °C	Phenyl-thiourea °C	1-Naphthyl-thiourea °C	Picrate °C	*N*-Substi-tuted phthal-imide °C	Benzamide °C	Acetamide °C
75	113	192	215	134	80	—
63	106	121	165	78	71	—
51	101	143	150	86	100	—
—	—	—	198	—	134	—
52	63	103	135	66	84	—
64	98	—	140	70	—	—
55	101	—	140	—	76	—
78	82	137	151	93	57	—
—	65	109	151	34	42	—
65	102	97	138	—	—	—
—	69	103	139	—	—	—
—	77	79	127	—	40	—
—	—	142	—	158	149	104
—	75	—	121	—	—	—
—	—	—	160	127	—	—
—	—	—	112	—	—	—
—	135	—	—	—	157	145
116	147	172	196	115	105	60
—	—	—	189	—	120	57
66	135	—	174	130	116	114
160	102	—	233	—	244	172
—	—	—	137	—	193	139
—	—	—	250	—	148	101
224	168	—	255d	—	177	137
—	148	—	—	—	135	—
—	—	—	220	—	158	—
79	135	168	158	—	41	—
60	34	108	155	—	42	—
—	—	—	140	—	—	—
—	—	161	75	—	—	—
—	—	—	—	—	—	—
—	—	—	—	—	—	—
—	—	—	—	—	—	—
—	86	123	59	—	—	—
—	—	118	—	—	—	—
—	—	—	—	—	—	—
119	—	—	173	153	—	103
99	—	—	110	—	—	—
123	—	—	112	—	—	—
96	—	—	152	—	48	—
55	—	—	135	—	45	—
—	—	—	138	—	—	—
—	—	—	—	—	—	—
147	136	—	148	—	75	—
—	—	—	69d	—	—	—
173 (mono)	—	—	280	—	196 (di)	144 (di)
—	—	—	195	—	129	46
—	—	—	—	—	76	—

Table VIII,32   Primary and secondary aliphatic amines (*continued*)

Amine	B.P. °C	M.P. °C
Ester-amides (*derivatives of aminoformic acid*, NH_2COOH)		
Methyl carbamate	177	54
Ethyl carbamate (urethan)	184	50
Propyl carbamate	195	61
Butyl carbamate	204d	54
Pentyl carbamate	—	57
Isopentyl carbamate	—	67
N-Methylurethan (Ethyl N-methyl carbamate)	170	—
N-Ethylurethan (Ethyl N-ethyl carbamate)	170	—
N-Propylurethan	192	—
N-Butylurethan	202	—
N-s-Butylurethan	194	—
N-Phenylurethan (Ethyl N-phenyl carbamate)	237	53
Ethyl oxanilate	—	67

Note.—Esters of carbamic acid upon boiling with aniline yield carbanilide (m.p. 238 °C), ammonia and the corresponding alcohol.

Table VIII,33   Primary aromatic amines

Amine °C	B.P. °C	M.P. °C	Acetamide °C	Benzamide °C	Benzene-sulphon-amide °C
Aniline	183	—	114	163	112
Benzylamine	185	—	60	105	88
1-Phenylethylamine	187	—	57	120	—
2-Phenylethylamine	198	—	51	116	69
o-Toluidine	200	—	112	144	124
m-Toluidine	203	—	66	125	95
p-Xylidine (1)	214	15	142	140	—
p-Ethylaniline	215	—	94	151	—
m-2-Xylidine (2)	215	11	177	168	—
o-Ethylaniline	216	—	112	147	—
m-4-Xylidine (3)	216	—	130	192	130
m-5-Xylidine (4)	220	10	144	136	—
o-Anisidine (5)	225	5	88	60	89
3,4-Dimethylaniline	226	49	99	—	118
N,N-Methylphenylhydrazine	227	—	92	153	132
o-Phenetidine (6)	229	—	79	104	102
Mesidine (7)	232	—	216	206	—
Phenylhydrazine	242	23	128	168	—
m-Phenetidine	248	—	96	103	—
m-Anisidine	251	—	80	—	—
o-Aminoacetophenone	251d	20	77	98	—
p-Phenetidine	254	—	135	173	143

$d_4^{20°}$	$n_D^{20°}$	Derivatives °C
—	—	*N-p*-Nitrobenzoyl, 152; Benzylidene, 179
—	—	
—	—	
—	—	
—	—	
0.981	1.422	—
—	—	
—	—	
—	—	*N*-Acetyl, 59; *N*-Benzoyl, 161; *N*-Nitroso, 62
—	—	*N*-Acetyl, 65

(1) M.p. 8 °C.
(2) M.p. 28 °C.
(3) M.p. 42 °C.
(4) M.p. 20 °C.

(5) Di-2-hydroxyethylamine.
(6) M.p. 28 °C.
(7) M.p. 104 °C; Hydrate, $6H_2O$, m.p. 44 °C.

Toluene-*p*-sulphon-amide °C	Benzylidene derivative °C	Picrate °C	3-Nitro-phthalimide °C	2,4-Dinitro-phenyl derivative °C	Formyl derivative °C	Phenyl thiourea °C
103	54	—	138	156	47	154
116	—	199	143	—	—	156
—	—	167	—	—	—	135
110	—	213	150	126	59	136
114	—	200	130	161	—	104
232	—	—	—	150	—	—
212	—	180	—	—	176	148
181	—	209	—	156	114	152
—	—	209	—	—	77	153
127	—	200	185	151	83	136
54	—	—	—	—	—	—
64	—	—	164	164	62	137
67	—	193	—	—	—	193
—	—	—	—	—	145	172
57	—	158	—	—	52	138
68	—	169	158	138	57	—
48	—	—	—	—	—	—
07	76	69	173	118	76	148

Table VIII,33   Primary aromatic amines (*continued*)

Amine °C	B.P. °C	M.P. °C	Acetamide °C	Benzamide °C	Benzene-sulphon-amide °C
Methyl anthranilate	255	24	101	100	—
p-Aminodiethylaniline	261	—	104	172	—
Ethyl anthranilate	266d	13	61	98	93
Ethyl m-aminobenzoate	294	—	—	114	—
p-Aminodimethylaniline	262	41	132	228	—
p-Toluidine	200	45	154	158	120
o-Aminobiphenyl (8)	299	50	121	102	—
1-Naphthylamine C*	300	50	160	161	169
p-Aminobiphenyl (9) C	302	51	171	230	—
p-Anisidine	246	57	130	154	96
2-Aminopyridine	204	58	71	87	—
2,5-Diaminotoluene (10)	273	64	220	307	—
3-Aminopyridine	252	64	133	119	—
p-Tolylhydrazine	244d	65	130	146	—
m-Phenylenediamine	283	64	191	240	194
o-Nitroaniline	—	71	94	98	104
4-Amino-2-nitrotoluene	—	78	145	172	160
m-Aminoacetanilide	—	88	191	—	—
3,4-Diaminotoluene (11)	265	90	210	264	179
Ethyl p-aminobenzoate (12)	—	92	110	148	—
6-Nitro-2-aminotoluene	—	92	158	168	—
3-Nitro-2-aminotoluene	—	97	158	—	—
m-Aminoacetophenone	—	99	129	—	—
2,4-Diaminotoluene (13)	292	99	224	224	191
o-Phenylenediamine	257	102	186	301	186
p-Aminoacetophenone	294	106	167	205	128
2-Amino-4-nitrotoluene	—	107	151	186	172
2-Naphthylamine C	294	113	134	162	102
m-Nitroaniline	—	114	155	157	136
4-Amino-3-nitrotoluene	—	117	96	148	102
5-Nitro-1-naphthylamine	—	119	220	—	183
1-Nitro-2-naphthylamine	—	126	123	168	156
p-Aminoazobenzene	—	126	145	211	—
Benzidine C	—	126	317	352	235
o-Tolidine (14) C	—	129	314	265	—
2-Amino-5-nitrotoluene	—	129	202	174	159
o-Aminoacetanilide	—	132	—	—	—
2,6-Dinitroaniline	—	138	197	—	—
p-Phenylenediamine	267	141	304	>300	247
2-Nitro-1-naphthylamine	—	144	199	175	—
Anthranilic acid	—	146	185	181	214
p-Nitroaniline	—	148	216	199	139
p-Nitrophenylhydrazine	—	157d	205	193	—
4-Aminopyridine	—	158	150	202	—
p-Aminoacetanilide	—	163	304	—	—
Sulphanilamide (15)	—	166	219	284	211
m-Aminobenzoic acid	—	174	250	—	—
2,4-Dinitroaniline	—	180	121	220	—
p-Aminobenzoic acid	—	187	251	278	212
Picramide (16)	—	190	230	196	211
4-Nitro-1-naphthylamine	—	195	190	224	173
2,4-Dinitrophenylhydrazine	—	198d	198	207	—
2-Aminoanthraquinone	—	302	257	228	271

Toluene-p-sulphonamide °C	Benzylidene derivative °C	Picrate °C	3-Nitro-phthalimide °C	2,4-Dinitro-phenyl derivative °C	Formyl derivative °C	Phenyl thiourea °C
—	—	106	—	—	58	—
112	—	—	—	—	—	—
112	—	—	—	—	57	—
—	—	—	—	—	—	—
—	98	188	—	168	108	—
118	—	181	156	137	53	141
—	—	—	—	—	75	—
157	73	163	223	190	139	165
255	—	—	—	—	172	—
114	62	—	197	141	81	144
216	—	221	—	—	—	—
—	—	—	—	—	—	—
—	—	—	—	—	—	—
172	105	184	—	172	155	—
110	—	73	171	—	122	—
163	—	—	—	—	—	—
241	—	—	—	—	—	—
—	—	131	—	—	—	—
—	—	—	—	—	—	—
—	—	—	—	—	—	—
130	—	..	—	—	—	—
192	175	—	—	184	177	—
202	106	208	—	—	170†	—
203	—	—	—	—	—	—
—	116	—	—	—	179	—
133	—	195	212	179	129	129
139	73	143	219	—	134	160
146	78	—	—	—	—	—
—	—	—	—	—	199	—
160	—	—	—	—	—	—
—	130	—	—	—	162	—
243	238	—	—	—	—	—
—	152	185	—	—	254	—
174	—	—	—	—	—	—
—	—	—	—	—	—	—
—	—	—	—	—	—	—
266	140	—	—	177	206	—
—	—	—	—	—	—	—
217	127	—	—	—	168	—
191	115	100	255	—	194	—
—	—	120	—	—	—	—
—	—	216	—	—	—	—
—	—	—	—	—	—	—
—	—	—	—	—	—	—
—	119	—	—	—	225	—
219	—	—	—	—	—	—
—	193	—	—	—	268	—
—	—	—	—	—	—	—
185	—	—	—	—	—	—
—	—	—	—	—	—	—
304	—	—	—	—	—	—

Table VIII,33 Primary aromatic amines (*continued*)

Amine °C	B.P. °C	M.P. °C	Acetamide °C	Benzamide °C	Benzene-sulphon-amide °C
C-Halogeno-amines					
o-Chloroaniline	209	—	88	99	130
2-Amino-3-chlorotoluene	215	—	120	—	—
4-Amino-3-chlorotoluene	223	7	113	137	110
m-Chloroaniline	230	—	79	122	121
2-Amino-4-chlorotoluene	237	22	140	—	—
4-Amino-3-bromotoluene	240	26	117	149	—
2-Amino-5-chlorotoluene	241	29	140	142	125
2-Amino-6-chlorotoluene	245	—	159	173	—
m-Bromoaniline	251	18	88	120	—
m-Iodoaniline	—	25	119	151	—
2-Amino-4-bromotoluene	255	32	165	—	—
o-Bromoaniline	229	32	99	116	—
2,5-Dichloroaniline	255	50	132	120	—
2-Amino-5-bromotoluene	240	59	157	115	—
o-Iodoaniline	—	60	110	139	—
2,4-Dichloroaniline	245	63	146	117	128
p-Iodoaniline	—	63	184	222	—
p-Bromoaniline	—	66	167	204	134
p-Chloroaniline	232	71	179	193	122
2,4,6-Trichloroaniline	263	78	206	174	—
2,4-Dibromoaniline	—	79	146	134	—
2,6-Dibromoaniline	—	84	210	—	—
2,4-Diaminochlorobenzene	—	88	243	178	—
2-Chloro-4-nitroaniline	—	108	139	161	—
4-Chloro-2-nitroaniline	—	116	104	—	—
2,4,6-Tribromoaniline	—	120	—	232	198
Amino-phenols ‡					
2,4-Diaminophenol (17)	—	79d	222 (di)	231	—
m-Aminophenol	—	123	101 (di)	153 (di)	—
Picramic acid (18)	—	169	201 (N)	229 (N)	—
5-Amino-2-hydroxytoluene	—	173	103 (di)	194	—
o-Aminophenol	—	174	124 (di)	184 (di)	141
p-Aminophenol	—	186d	150 (di)	234 (di)	125
8-Amino-2-naphthol	—	207	165	208	—
1-Amino-2-naphthol	—	dec.	206	235	—

* All aromatic amines should be treated as potentially carcinogenic (see Section **I,3,D**,5); those marked **C** are subject to legal control in Great Britain or the USA or both.

† This compound is benzimidazole.

‡ See also Table VIII,35 and Table VIII,34 for secondary and tertiary amines having a nuclear hydroxyl substituent.

(1) 2,5-Dimethylaniline.
(2) 2,6-Dimethylaniline.
(3) 2,4-Dimethylaniline.
(4) 3,5-Dimethylaniline.

(5) *o*-Methoxyaniline.
(6) *o*-Ethoxyaniline.
(7) 2,4,6-Trimethylaniline.

Toluene-p-sulphon-amide °C	Benzylidene derivative °C	Picrate °C	3-Nitro-phthalimide °C	2,4-Dinitro-phenyl derivative °C	Formyl derivative °C	Phenyl thiourea °C
05	34	134	136	150	77	156
—	—	—	—	—	—	—
38	—	177	172	184	58	124
—	—	—	—	—	—	—
—	—	—	—	—	—	—
—	—	180	187	—	—	143
28	—	—	—	—	—	—
—	—	—	—	—	—	—
90	—	129	—	161	—	146
—	—	—	—	—	—	166
—	—	112	—	—	—	—
—	—	106	—	116	—	—
—	86	—	—	—	109	153
01 (141)	67	180	202	158	—	148
96 (121)	62	178	199	167	102	152
—	—	83	—	—	180	—
—	—	124	—	—	146	171
—	—	124	—	—	—	—
15	—	—	—	—	—	—
64	—	—	—	—	—	—
10	—	—	—	—	—	—
—	95	—	—	—	222	—
—	—	—	—	—	—	—
—	—	—	—	—	—	156
91 (N)	—	—	—	—	—	—
10	—	—	—	—	—	—
39	89	—	—	199	129	146
53	182	—	—	190	140	150
—	—	—	—	—	—	—
—	—	—	—	—	—	—

8) o-Xenylamine.
9) p-Xenylamine.
0) 2,5-Tolylenediamine.
1) 3,4-Tolylenediamine.
2) Benzocaine.
3) 2,4-Tolylenediamine.

(14) 4,4'-Diamino-3,3'-dimethylbiphenyl.
(15) p-Aminobenzenesulphonamide.
(16) 2,4,6-Trinitroaniline.
(17) Hydrochloride = *Amidol.*
(18) 2-Amino-4,6-dinitrophenol.

Table VIII,34   Secondary aromatic amines

Amide	B.P. °C	M.P. °C	Acetamide °C	Benzamide °C
N-Methylbenzylamine	181	—	—	—
N-Methylaniline	194	—	103	63
N-Ethylbenzylamine	199	—	—	—
N-Ethylaniline	205	—	55	60
N-Methyl-m-toluidine	206	—	66	—
N-Methyl-o-toluidine	208	—	56	66
N-Methyl-p-toluidine	210	—	83	53
N-Ethyl-o-toluidine	214	—	—	72
N-Ethyl-p-toluidine	217	—	—	39
N-Ethyl-m-toluidine	221	—	—	72
N-Propylaniline	222	—	47	—
N-Butylaniline	240	—	—	56
N-Methyl-1-naphthylamine	294	—	94	—
Dibenzylamine	300d	—	—	112
N-Methyl-2-naphthylamine	317	—	51	84
N-Ethyl-2-naphthylamine	315	—	49	—
N-Ethyl-1-naphthylamine	325	—	68	—
o-Nitro-N-methylaniline	—	37	70	—
N-Benzylaniline	306	38	58	107
Diphenylamine	302	54	103	180
m-Nitro-N-ethylaniline	—	60	89	—
N-Phenyl-1-naphthylamine	—	62	115	152
m-Nitro-N-methylaniline	—	68	95	155
Di-p-tolylamine	330	79	88	125
p-Hydroxy-N-methylaniline	—	86	43 (mono)	174 (mono)
o-Hydroxy-N-methylaniline	—	96	64 (di)	160 (mono)
p-Nitro-N-ethylaniline	—	96	119	—
N-Phenyl-2-naphthylamine	—	108	93	136
p-Nitro-N-methylaniline	—	152	152	111
Indole	254	52	—	68
Carbazole	355	246	69	98

Benzene-sulphonamide °C	Toluene-p-sulphonamide °C	Picrate °C	Formyl derivative °C	Other derivatives °C
—	95	—	—	—
79	95	145	—	Phthalamic acid, 194
—	50	118	—	Urea (with PhNCO), 81
—	88	138	—	Phthalamic acid, 204
—	—	—	—	—
—	120	90	—	—
—	60	131	—	N-Nitroso, 52*
—	75	—	—	—
—	71	—	—	—
—	—	—	—	—
54	—	—	—	Phthalamic acid, 225
—	56	—	—	Phthalamic acid, 204
—	164	—	—	—
68	—	—	52	—
—	78	145	—	N-Nitroso, 88*
—	—	—	—	—
—	—	—	—	—
—	—	—	—	N-Nitroso, 36*
119	140	—	48	N-Nitroso, 58*
123	142	182	74	N-Nitroso, 67*
—	—	—	—	—
83	—	—	—	N-Nitroso, 76*
—	—	—	—	N-Nitroso, 101*
—	135 (mono)	—	—	N-Nitroso, 136*
—	—	—	—	N-Nitroso, 130*
—	—	—	—	N-Nitroso, 120*
—	—	—	—	—
120	—	—	—	N-Nitroso, 104*
—	—	187	52	N-Nitroso, 171*
—	137	185	—	—

* The particularly powerful carcinogenic properties of these compounds should be noted.

Table VIII,35   Tertiary amines

Amine	B.P. °C (/mmHg)	M.P. °C	Methiodide °C
Trimethylamine	3	—	230
Triethylamine	89	—	—
Triallylamine	155	—	—
Tripropylamine	156	—	208
Tributylamine	212	—	186
Tri-isopentylamine	245	—	—
Tripentylamine	257	—	—
N,N-Dimethylbenzylamine	184	—	179
N,N-Dimethyl-o-toluidine	185	—	210
N,N-Dimethylaniline	193	—	228
N-Methyl-N-ethylaniline	201	—	125
N,N-Diethyl-o-toluidine	210	—	224
N,N-Dimethyl-p-toluidine	211	—	220
N,N-Dimethyl-m-toluidine	212	—	177
N,N-Diethylaniline	218	—	102
N,N-Diethyl-p-toluidine	229	—	184
N,N-Diethyl-m-toluidine	231	—	—
N,N-Di-n-propylaniline	245	—	156
N,N-Di-n-butylaniline	271	—	—
N,N-Dimethyl-1-naphthylamine	273	—	—
N-Benzyl-N-methylaniline	306	—	164
N-Benzyl-N-ethylaniline	186/22	—	161
N,N-Dimethyl-2-naphthylamine	305	47	—
p-Bromo-N,N-dimethylaniline	264	55	—
Dibenzylaniline	300	70	135
p-Hydroxy-N,N-dimethylaniline	—	76	201
p-Nitroso-N,N-dimethylaniline	—	87	—
Tribenzylamine	380	92	184
Triphenylamine	365	127	—
p-Nitro-N,N-dimethylaniline	—	163	—
Pyridine	115	—	118
α-Picoline (1)	129	—	227
2,6-Lutidine (2)	142	—	238
γ-Picoline	143	—	152
β-Picoline	144	—	92
2,4-Lutidine	159	—	113
2,5-Lutidine	160	—	—
2,3-Lutidine	164	—	—
2,4,6-Trimethylpyridine (3)	172	—	—
5-Ethyl-2-methylpyridine	178	—	—
3-Ethyl-4-methylpyridine	196	—	—
Ethyl nicotinate	223	—	—
Nicotine	246	—	—
Methyl nicotinate	204	38	—
2,2′-Bipyridyl	273	70	—
Quinoline	238	—	72* (133)†
Isoquinoline	242	24	159
Quinaldine (4)	247	—	195
8-Methylquinoline	248	—	—
6-Methylquinoline	258	—	219
Lepidine (5)	262	—	174
2,4-Dimethylquinoline	264	—	264

Picrate °C	Methotoluene-p-sulphonate °C	Other derivatives °C
216	—	—
173	—	$d_4^{20°}$ 0.728; $n_D^{20°}$ 1.401
—	—	—
117	—	Ethiodide, 238; $d_4^{20°}$ 0.756; $n_D^{20°}$ 1.417
106	—	Benzyl chloride, 185; $d_4^{20°}$ 0.778; $n_D^{20°}$ 1.430
125	—	$d_4^{20°}$ 0.785; $n_D^{20°}$ 1.433
—	80d	$d_4^{20°}$ 0.791; $n_D^{20°}$ 1.437
93	—	—
122	—	—
164	161	Ethiodide, 136
134	—	Ethiodide, 102; p-nitroso, 66
180	—	—
130	85	Benzyl chloride, 171
131	—	—
142	—	p-Nitroso, 84; benzyl chloride, 104
110	—	—
97	—	—
—	—	—
125	180	—
145	—	—
127	—	—
121	—	—
206	—	—
—	—	—
132	—	p-Nitroso, 91
—	—	o-Acetyl, 79
140	—	—
190	—	Ethiodide, 190
—	—	—
167	139	Ethiodide, 90
169	150	Ethiodide, 123; picolinic acid, 136
163	—	Dipicolinic acid, 226
167	—	Isonicotinic acid, 308
150	—	Nicotinic acid, 228
183	—	—
169	—	—
188	—	—
156	—	—
166	—	—
150	—	—
—	—	—
218	—	Nicotinic acid, 228
—	—	—
158	—	—
203	126	Ethiodide, 158
223	163	Ethiodide, 148
195	161	Ethiodide, 234
200	—	—
229	154	Benzyl chloride, 239
211	—	—
194	—	Ethiodide, 214

Table VIII,35   Tertiary amines (*continued*)

Amine	B.P. °C	M.P. °C	Methiodide °C
6-Methoxyquinoline	284	26	236
7-Methylquinoline	252	39	—
8-Methoxyquinoline	283	50	160
2,6-Dimethylquinoline	267	60	237
8-Hydroxyquinoline	267	76	143
8-Nitroquinoline	—	92	—
6-Nitroquinoline	—	154	245
6-Hydroxyquinoline	—	193	—
3-Chloropyridine	149	—	—
3-Bromopyridine	170	—	165
2-Chloropyridine	170	—	—
2-Bromopyridine	194	—	—
3,5-Dibromopyridine	222	112	274
2,6-Dibromopyridine	249	119	—
6-Bromoquinoline	278	19	278
2-Chloroquinoline	267	38	—
6-Chloroquinoline	262	41	248
2-Bromoquinoline	—	49	210
Acridine	—	111	224
Hexamethylenetetramine	—	280 Sub.	190

* Monohydrate.   † Anhydrous.

Picrate °C	Methotoluene-*p*-sulphonate °C	Other derivatives °C
—	—	—
237	—	—
143	—	—
191	175	Ethiodide, 227
204	—	—
—	—	—
—	—	—
236	—	—
135	—	—
—	156	—
—	120	—
—	127	—
—	219	—
—	—	—
217	—	—
122	—	—
—	143	Ethiodide, 169
—	—	—
208	—	Trinitrobenzene, 115
179	205	—

(1) 2-Methylpyridine.
(2) 2,6-Dimethylpyridine.
(3) γ-Collidine.
(4) 2-Methylquinoline.
(5) 4-Methylquinoline.

Table VIII,36   Amino acids

Amino acid	M.P.* °C	Benzoate °C	3,5-Dinitro-benzoate °C	Phenylureido acid °C
N-Phenylglycine	126	63	—	195
Anthranilic acid	145	182	278	181
m-Aminobenzoic acid	174	248	270	270
(±)-3-Amino-2-methylpropanoic	177	—	—	—
p-Aminobenzoic acid	186	278	290	300
3-Aminopropanoic acid	196	165	202	174
(+)- or (−)-Glutamic acid	198	138	217	—
p-Aminophenylacetic acid	200	206	—	—
(±)-Proline	203	—	217	170
Sarcosine	210	103	153	—
(+)- or (−)-Proline	222	—	—	170
(+)- or (−)-Lysine	224	150	169	184
(+)- or (−)-Asparagine	227	189	196	164
(±)-Glutamic acid	227	156	—	—
(+)-Serine	228	—	—	—
Glycine	232	187	179	163
(±)-Threonine	235	148	—	—
(±)-Arginine	238	230	—	—
(±)-Serine	246	171	183	169
(+)- or (−)-Threonine	253	148	—	—
(+)- or (−)-Cystine	260	181	180	160
(+)- or (−)-Aspartic acid	272	185	—	162
(±)-Methionine	272	151	—	—
(±)-Phenylalanine	274	188	93	182
(±)-Tryptophan	275	188	240	—
(+)- or (−)-Histidine	277	249	189	—
2-Amino-2-methylpropanoic	Sub. 280	202d	—	—
(±)-Aspartic acid	280	165	—	—
(+)- or (−)-Methionine	283	150	95	—
(+)- or (−)-Isoleucine	284	117	—	120
(+)- or (−)-Tryptophan	289	104	233	166
(±)-Isoleucine	292	118	—	—
(±)-α-Alanine	295	166	177	174
(+)- or (−)-α-Alanine	297	151	—	190
(±)-Valine	298	132	—	164
(±)-Norvaline	303	—	182	—
(±)-2-Aminobutanoic	307	147	—	170
(+)- or (−)-Valine	315	127	181	147
(±)-Tyrosine	318	197	254	—
(+)- or (−)-Phenylalanine	320	146	93	181
(±)-Norleucine	327	—	—	—
(±)-Leucine	332	141	—	165
(+)- or (−)-Leucine	337	107	187	115
(+)- or (−)-Tyrosine	344	166	—	104
(±)-Asparagine	>300	—	—	—
(±)-Histidine	—	—	—	—
(±)-Lysine	—	249	—	196

* These melting points are probably better described as decomposition points and their values will depend somewhat upon the rate of heating. Many of the naturally-occurring amino acids are (−)-rotatory.

Toluene-p-sulphonate °C	2,4-Dichloro-phenoxyacetate °C	1-Naphthyl-ureido acid °C	Phthalyl derivative °C	2,3-Dinitro-phenyl derivative °C
—	—	—	—	—
217	—	—	—	—
—	—	—	—	—
223	—	—	—	—
—	—	236	—	146
117	—	236	159	—
—	145	—	—	181
102	—	—	—	—
33	106	—	—	138
—	87	199	—	171
75	—	199	—	181
213	192	—	—	149
—	—	—	—	174
50	235	191	192	204
—	139	—	103	178
—	—	—	—	—
213	195	191	—	201
—	—	—	—	145
205	216	—	—	109
40	202	115	193	187
05	145	—	—	117
35	180	—	175	186
76	148	—	—	—
204	—	—	296	233
—	—	198	—	—
—	217	—	221	196
—	134	186	—	—
32	—	178	121	113
76	—	158	—	221
41	143	—	—	175
39	213	198	161	—
39	199	202	—	—
10	159	204	102	184
—	—	—	—	—
—	—	194	96	143
49	—	—	115	132
—	—	—	268	—
65	155	—	—	189
24	—	—	112	—
—	138	—	141	—
24	150	163	116	94
19	—	205	—	180
—	—	—	—	—
—	129	—	—	—
—	176	—	171	—

Table VIII,37   Aromatic nitro compounds

Nitro compound	B.P. °C	M.P. °C	Nitro compound	B.P. °C	M.P. °C
Nitrobenzene (1)	211	6	m-Nitrobenzyl chloride	—	46
o-Nitrotoluene (2)	222	—	o-Nitrobenzyl bromide	—	47
2-Nitro-m-xylene	226	—	o-Nitrobenzyl chloride	—	49
m-Nitrotoluene (3)	229	16	2,4-Dinitrochlorobenzene	315	51
2-Nitro-p-xylene	237	—	o-Nitroiodobenzene	—	54
3-Nitro-o-xylene	240	15	m-Nitrobromobenzene	256	56
4-Nitro-m-xylene	244	—	2,5-Dichloronitrobenzene	267	56
2-Nitro-p-cymene (4)	264	—	m-Nitrobenzyl bromide	—	59
o-Nitroanisole	265	10	p-Nitrobenzyl chloride	—	71
m-Nitrophenetole	267	2	o-Nitrobenzyl iodide	—	75
m-Nitrobenzyl alcohol	—	27	2,4-Dinitrobromobenzene	—	75
4-Nitro-o-xylene	254	30	Picryl chloride	—	83
m-Nitrophenetole	284	34	p-Nitrochlorobenzene	242	83
2-Nitrobiphenyl	320	37	2-Nitro-p-dibromobenzene	—	84
m-Nitroanisole	258	39	m-Nitrobenzyl iodide	—	86
Nitromesitylene	255	44	2,4-Dinitroiodobenzene	—	88
p-Nitrotoluene	238	54	p-Nitrobenzyl bromide	—	100
p-Nitroanisole	259	54	p-Nitrobromobenzene	256	127
ω-Nitrostyrene	260d	58	p-Nitrobenzyl iodide	—	127
p-Nitrophenetole	283	60	p-Nitroiodobenzene	—	174
1-Nitronaphthalene	304	61			
m-Nitrobenzyl cyanide	—	62			
2,4,6-Trinitroanisole	—	68	Methyl o-nitrobenzoate	275	—
2,4-Dinitrotoluene	—	71	Ethyl o-nitrobenzoate	—	30
o-Nitrobenzyl alcohol	270	74	Diethyl 4-nitrophthalate	—	34
5-Nitro-m-xylene	273	74	Ethyl o-nitrocinnamate	—	44
2,4,6-Trinitrophenetole	—	79	Diethyl 3-nitrophthalate	—	46
2-Nitronaphthalene	—	79	Ethyl m-nitrobenzoate	297	47
2,4,6-Trinitrotoluene	—	82	Ethyl p-nitrobenzoate	—	57
o-Nitrobenzyl cyanide	—	84	Dimethyl 4-nitrophthalate	—	66
2,4-Dinitrophenetole	—	87	Dimethyl 3-nitrophthalate	—	69
m-Dinitrobenzene	—	90	Methyl o-nitrocinnamate	—	73
p-Nitrobenzyl alcohol	—	93	Methyl m-nitrobenzoate	—	78
2,4-Dinitroanisole	—	95	Ethyl m-nitrocinnamate	—	79
p-Nitrobenzyl cyanide	—	117	Ethyl 3,5-dinitrobenzoate	—	94
o-Dinitrobenzene	—	118	Methyl p-nitrobenzoate	—	96
1,3,5-Trinitrobenzene	—	122	Ethyl 3,5-dinitrosalicylate	—	99
1,8-Dinitronaphthalene	—	173	Ethyl 5-nitrosalicylate	—	102
p-Dinitrobenzene	—	173	Methyl 3,5-dinitrobenzoate	—	112
1,5-Dinitronaphthalene	—	217	Ethyl 3-nitrosalicylate	—	118
			Methyl 5-nitrosalicylate	—	119
o-Nitrochlorobenzene	245	33	Methyl m-nitrocinnamate	—	124
m-Nitroiodobenzene	—	38	Methyl 3,5-dinitrosalicylate	—	127
o-Nitrobromobenzene	261	42	Methyl 3-nitrosalicylate	—	132
3,4-Dichloronitrobenzene	255	43	Ethyl p-nitrocinnamate	—	142
m-Nitrochlorobenzene	236	46	Methyl p-nitrocinnamate	—	161

(1) $d_4^{20°}$ 1.204; $n_D^{20°}$ 1.553.   (3) $d_4^{20°}$ 1.157; $n_D^{20°}$ 1.547.
(2) $d_4^{20°}$ 1.168; $n_D^{20°}$ 1.546.   (4) $d_4^{20°}$ 1.074; $n_D^{20°}$ 1.531.

Table VIII,38   Aliphatic nitro compounds

Nitro compound	B.P. °C (/mmHg)	$d_4^{20°}$	$n_D^{20°}$
Nitromethane	101	1.137	1.381
Nitroethane	114	1.050	1.392
2-Nitropropane	120	0.988	1.394
1-Nitropropane	131	1.001	1.401
1-Nitrobutane	152	0.971	1.410
1-Nitropentane	66/16	0.953	1.418
1-Nitrohexane	82/15	0.940	1.423
Phenylnitromethane	227	1.160	1.532

Table VIII,39   Thiols

Thiol	B.P. °C (/mmHg)	M.P. °C	2,4-Dinitro-phenyl-thioether °C	2,4-Dinitro-phenyl-sulphone °C	3,5-Dinitro-thio-benzoate °C	Hydrogen 3-nitro-thio-phthalate
Methanethiol	6	—	128	190	—	—
Ethanethiol	36	—	115	160	62	149
Propane-2-thiol	58	—	95	141	84	145
Propanethiol	67	—	81	128	52	137
2-Methylpropanethiol	88	—	76	106	64	136
Prop-2-ene-1-thiol	90	—	72	105	—	—
Butanethiol	97	—	66	92	49	144
3-Methylbutanethiol	117	—	59	95	43	145
Pentanethiol	126	—	80	83	40	132
Hexanethiol	151	—	74	97	—	—
Cyclohexanethiol	159	—	148	172	—	—
Heptanethiol	176	—	82	101	53	132
Octanethiol	199	—	78	98	—	—
Nonanethiol	220	—	86	92	—	—
Decanethiol	114/13	—	85	93	—	—
Dodecanethiol	154/24	—	89	101	—	—
Hexadecanethiol	—	51	91	105	—	—
1,2-Ethane dithiol	146	—	248	—	—	—
1,3-Propane dithiol	173	—	194	—	—	—
1,4-Butane dithiol	196	—	—	—	—	—
1,5-Pentane dithiol	217	—	170	—	—	—
1,6-Hexane dithiol	237	—	218	—	—	—
Thiophenol	169	—	121	161	149	130
Phenylmethanethiol	194	—	130	183	120	137
o-Thiocresol	194	15	101	155	—	—
m-Thiocresol	195	—	91	145	—	—
p-Thiocresol	195	44	103	190	—	—
1-Phenylethanethiol	199	—	90	133	—	—
2-Phenylethanethiol	105/23	—	—	—	—	—
1-Thionaphthol	161/20	—	176	—	—	—
2-Thionaphthol	162/20	81	145	—	—	—
4-Mercaptobiphenyl	—	111	146	170	—	—
2-Furylmethanethiol	84/65	—	130	—	—	—
2-Thienylmethanethiol	166	—	119	143	—	—

Table VIII,40 Sulphonic acids

Note. Aromatic sulphonic acids are usually hygroscopic solids and do not generally have sharp melting points: they are frequently supplied in the form of their sodium (or other metal) salts. It is therefore not possible to classify them in order of increasing melting points. In this Table,

Acid	Sulphonamide ArSO$_2$NH$_2$ °C	S-Benzylisothiouronium salt °C
Benzenesulphonic	153	150
Toluene-*o*-sulphonic	156	170
Toluene-*m*-sulphonic	108	—
Toluene-*p*-sulphonic	137	182
o-Chlorobenzenesulphonic	188	—
m-Chlorobenzenesulphonic	148	—
p-Chlorobenzenesulphonic	144	175
o-Bromobenzenesulphonic	186	—
m-Bromobenzenesulphonic	154	—
p-Bromobenzenesulphonic	166	170
o-Nitrobenzenesulphonic	193	—
m-Nitrobenzenesulphonic	168	146
p-Nitrobenzenesulphonic	179	—
Sulphanilic	164	187
Orthanilic	153	132
Metanilic	142	148
o-Sulphobenzoic (salt)	—	206
m-Sulphobenzoic	170	163
p-Sulphobenzoic	236	213
Phenol-*p*-sulphonic	177	169
Thymolsulphonic	—	213
o-Xylene-4-sulphonic	144	208
m-Xylene-4-sulphonic	138	146
p-Xylenesulphonic	148	184
Naphthalene-1-sulphonic	150	137
Naphthalene-2-sulphonic	217	191
Anthraquinone-1-sulphonic	—	191
Anthraquinone-2-sulphonic	261	211
1-Naphthylamine-4-sulphonic	206	195
1-Naphthylamine-5-sulphonic	260	180
1-Naphthylamine-6-sulphonic	219	191
1-Naphthylamine-7-sulphonic	181	—
1-Naphthylamine-8-sulphonic	—	300
2-Naphthylamine-1-sulphonic	—	139
2-Naphthylamine-6-sulphonic	—	184
1-Naphthol-2-sulphonic	—	170
1-Naphthol-4-sulphonic	—	104
1-Naphthol-5-sulphonic	—	—
2-Naphthol-1-sulphonic	—	136
2-Naphthol-6-sulphonic	238	217
2-Naphthol-8-sulphonic	—	218
Benzene-*o*-disulphonic	254	206
Benzene-*m*-disulphonic	229	214
Benzene-*p*-disulphonic	288	—
Naphthalene-1,4-disulphonic	273	—
Naphthalene-1,5-disulphonic	310	257
Naphthalene-1,6-disulphonic	298	235
Naphthalene-2,6-disulphonic	305	256
Naphthalene-2,7-disulphonic	243	211
2-Naphthylamine-4,8-disulphonic	—	210

related compounds are grouped together. For convenience a subsidiary Table (VIII,40,A) is given in which sulphonic acids are listed in the order of increasing melting points of the S-benzyliso-thiouronium salts; these derivatives are easily prepared either from the free acid or from the salt.

Sulphonanilide ArSO$_2$NHPh °C	p-Toluidine salt °C	Sulphonyl Chloride, ArSO$_2$Cl °C	Sulphonacetamide ArSO$_2$NHCOCH$_3$ °C
110	205	—	125
136	204	68	—
96	—	12	—
103	198	71	137
—	—	28	—
04	209	53	—
—	—	51	—
19	216	—	—
15	—	75	203
26	222	69	—
36	—	64	189
00	—	80	192
—	—	—	—
—	—	—	—
—	200	79	—
—	—	20	—
—	—	57	—
—	202		—
—	—	—	—
10	—	52	—
-	—	34	—
2	—	25	—
32	181	68	185
6	221	79	146
3	—	217	—
—	—	197	—
—	—	—	—
—	—	—	—
0	—	—	—
—	—	—	—
—	—	—	—
0	196	—	—
1	—	—	—
	162	124	—
	248	—	—
5	232	—	—
1	—	143	—
4	—	63	—
	—	131	—
9	—	—	—
9	332	183	—
	315	129	—
	360	225	—
	300	159	—
		—	

Table VIII,40   Sulphonic acids (*continued*)

Acid	Sulphonamide ArSO$_2$NH$_2$ °C	s-Benzylisothio-uronium salt °C
2-Naphthylamine-5,8-disulphonic	—	276
2-Naphthylamine-6,7-disulphonic	—	—
1-Naphthylamine-3,6-disulphonic	—	—
1-Naphthylamine-3,8-disulphonic	—	—
1-Naphthol-3,6-disulphonic	—	217
1-Naphthol-4,8-disulphonic	—	205
2-Naphthol-3,6-disulphonic	—	233
2-Naphthol-6,8-disulphonic	—	228
(+)-Camphorsulphonic	132	210

Aliphatic sulphonic acids

Sulphonic acid	B.P. °C	Sulphonyl chloride, B.P. °C	Sulphonamide, M.P. °C	s-Benzyl-isothio-uronium salt M.P. °C	Sulphon-anilide M.P. °C
Methane	167/10	163	90	—	99
Ethane	—	177	59	115	58
Propane-2-	—	79/18	60	—	84
Propane-1-	—	78/13	52	—	—
Butane-1-	—	75/10	45	—	—

Table VIII,40A   Sulphonic acids (*continued*)

(*Arranged in the order of increasing melting points of the* S-*benzylisothiouronium salts.*)

Sulphonic acid	M.P. °C	Sulphonic acid	M.P. °C
1-Naphthol-4-	104	p-Bromobenzene-	170
Ethane-	115	1-Naphthol-2-	170
Orthanilic acid	132	o-Toluene-	170
2-Naphthol-1-	136	p-Chlorobenzene-	175
Naphthalene-1-	137	1-Naphthylamine-5-	180
2-Naphthylamine-1-	139	p-Toluene-	182
m-Nitrobenzene-	146	2-Naphthylamine-6-	184
m-Xylene-4-	146	p-Xylene-	184
Metanilic acid	148	Sulphanilic acid	187
Benzene-	150	Naphthalene-2-	191
m-Sulphobenzoic acid	163	Anthraquinone-1-	191
Phenol-p-	169	1-Naphthylamine-6-	191

Sulphonanilide ArSO$_2$NHPh °C	p-Toluidine salt °C	Sulphonyl Chloride, ArSO$_2$Cl °C	Sulphonacetamide ArSO$_2$NHCOCH$_3$ °C
—	—	—	—
—	—	—	—
—	—	—	—
—	—	—	—
—	—	—	—
202	—	—	—
195	—	162	—
—	—	88	—

ulphonic acid	M.P. °C	Sulphonic acid	M.P. °C
-Naphthylamine-4-	195	2-Naphthol-6-	217
Naphthol-4,8-di-	205	1-Naphthol-3,6-di-	217
enzene-o-di-	206	2-Naphthol-8-	218
Sulphobenzoic acid	206	2-Naphthol-6,8-di-	228
Xylene-4-	208	2-Naphthol-3,6-di-	233
-)-Camphor-	210	Naphthalene-1,6-di-	235
Naphthylamine-4,8-di-	210	Naphthalene-2,6-di-	256
aphthalene-2,7-di-	211	Naphthalene-1,5-di-	257
nthraquinone-2-	211	2-Naphthylamine-5,8-di-	276
Sulphobenzoic acid	213	1-Naphthylamine-8-	300
hymol-	213		
enzene-m-di-	214		

Table VIII,41 Sulphonamides, $R \cdot SO_2NH_2$

Sulphonamide	M.P. °C	N-Xanthyl-sulphonamide	Sulphonamide	M.P. °C	N-Xanthyl-sulphonamide
Butane-1-	45	—	2-Naphthalene-	217	—
Propane-1-	52	—	o-Sulphobenzimide		
Ethane-	59	—	(saccharin)	226d	198
Propane-2-	60	—	1,3-Benzenedi-	229	170
m-Ethylbenzene-	86	—	2,7-Naphthalenedi-	242	—
Methane-	90	—	1,2-Benzenedi-	254	—
2,6-Dimethylbenzene-	96	—	2-Anthraquinone-	261	—
o-Ethylbenzene-	100	—	1,4-Naphthalenedi-	273	—
Phenylmethane-	105	—	p-Sulphamidobenzoic		
Toluene-m-	108	—	acid	280	—
p-Ethylbenzene-	109	196	1,4-Benzenedi-	288	—
p-Methoxybenzene-	111	—	1,6-Naphthalenedi-	298	—
3,5-Dimethylbenzene-	135	—	1,5-Naphthalenedi-	310	—
(+)-Camphor-8-	137	—	1,3,5-Benzenetri-	312	—
2,4-Dimethylbenzene-	137	188	1,8-Anthraquinonedi-	340	—
Toluene-p-	137	197			
2,4,6-Trimethylbenzene-	142	203	*Halogeno-sulphonamides*		
m-Aminobenzene-	142	—			
3,4-Dimethylbenzene-	144	—	p-Fluorobenzene-	125	—
2,5-Dimethylbenzene-	147	176	3,4-Dichlorobenzene-	135	—
p-Ethoxybenzene-	150	—	p-Chlorobenzene-	144	—
Naphthalene-1-	150	—	m-Chlorobenzene-	148	—
Benzene-	153	200	m-Bromobenzene-	154	—
o-Aminobenzene-	153	—	p-Bromobenzene-	166	—
Toluene-o-	156	183	3,4-Dibromobenzene-	175	—
m-Nitrophenylmethane-	159	—	2,4-Dichlorobenzene-	180	—
p-Aminobenzene-			2,5-Dichlorobenzene-	181	—
(sulphanilamide)	165	208	o-Bromobenzene-	186	—
2,3-Dimethylbenzene-	167	—	p-Bromophenylmethane-	188	—
m-Nitrobenzene-	168	—	o-Chlorobenzene-	188	—
p-Nitrobenzene-	179	—	2,4-Dibromobenzene-	190	—
2,4,5-Trimethylbenzene-	181	—	2,5-Dibromobenzene-	194	—
o-Nitrobenzene-	193	—	p-Iodophenylmethane-	206	—
p-Nitrophenylmethane-	204	—	2,4,6-Trichlorobenzene-	212d	—
1-Nitro-2-naphthalene-	214	—	2,3,4-Trichlorobenzene-	230d	—

Table VIII,42 Imides

Compound	M.P. °C
N-2-Bromoethylphthalimide	82
N-Phenylmaleimide	91
Maleimide	93
Allyl-(1-methylbutyl) barbituric acid	100
Succinimide	125
Ethylhexylbarbituric acid (*Ortal*)	126
Ethylbutylbarbituric acid (*Neonal*)	128
N-2-Hydroxyethylphthalimide	128
Ethyl-(1-methylbutyl)barbituric acid (*Pentobarbital*)	130
Allylisopropylbarbituric acid (*Alurate*)	137
Ethylisopentylbarbituric acid (*Amytal*)	155
N-Phenylsuccinimide	156
Alloxan (4H$_2$O)	170d
Ethylphenylbarbituric acid (*Phenobarbital*)	172
Diallylbarbituric acid (*Dial*)	173
Diethylbarbituric acid (*Veronal*)	190
Ethylisopropylbarbituric acid (*Ipral*)	201
N-Phenylphthalimide	205
3-Nitrophthalimide	216
o-Sulphobenzimide (saccharin)	226d
Phthalimide	233
Barbituric acid	245d
Naphthalimide	300

Table VIII,43 Nitroso, azo, azoxy and hydrazo compounds *

Compound	M.P. °C
Nitroso compounds	
Methylphenylnitrosoamine	B.P. 120 °C/13 mmHg —
Ethylphenylnitrosoamine	B.P. 134 °C/16 mmHg —
p-Nitrosotoluene	48
m-Nitrosotoluene	53
N-Nitrosodiphenylamine	66
Nitrosobenzene	68
o-Nitrosotoluene	72
p-Nitroso-N-ethylaniline	78
p-Nitroso-NN-diethylaniline	84
p-Nitroso-NN-dimethylaniline	87
1-Nitrosonaphthalene	98
1-Nitroso-2-naphthol	109
p-Nitroso-N-methylaniline	118
p-Nitrosophenol	125d
p-Nitrosodiphenylamine	144
2-Nitroso-1-naphthol	152d
4-Nitroso-1-naphthol	198
Azo compounds	
2,2'-Dimethylazobenzene	55
3,3'-Dimethylazobenzene	55
Azobenzene	68
4-Anilinoazobenzene	82
3,3'-Diethoxyazobenzene	91

Compound	M.P. °C
Azo compounds (continued)	
3,3'-Dichloroazobenzene	101
p-Dimethylaminoazobenzene **C**	117
p-Aminoazobenzene	126
4-Hydroxy-3-methylazobenzene	128
2,2'-Diethoxyazobenzene	131
1-Phenylazo-2-naphthol	134
2,2'-Dichloroazobenzene	137
2-Phenylazo-1-naphthol	138
4,4'-Dimethylazobenzene	144
o-Azobiphenyl	145
p-Hydroxyazobenzene	152
4,4-Diethoxyazobenzene	160
4,4'-Dichloroazobenzene	188
1,1'-Azonaphthalene	190
4-Phenylazo-1-naphthol	206d
2,2'-Azonaphthalene	208
p-Azobiphenyl	250
Azoxy compounds	
Azoxybenzene	36
3,3'-Dimethylazoxybenzene	39
3,3'-Diethoxyazoxybenzene	50
3,3'-Dimethoxyazoxybenzene	52
2,2'-Dichloroazoxybenzene	56
2,2'-Dimethylazoxybenzene	60
4,4'-Dimethylazoxybenzene	70
2,2'-Dimethoxyazoxybenzene	81
3,3'-Dichloroazoxybenzene	97
2,2'-Diethoxyazoxybenzene	102
4,4'-Dimethoxyazoxybenzene	119
1,1'-Azoxynaphthalene	127
4,4'-Diethoxyazoxybenzene	138
4,4'-Dichloroazoxybenzene	158
o-Azoxybiphenyl	158
2,2'-Azoxynaphthalene	168
p-Azoxybiphenyl	212
Hydrazo compounds	
3,3'-Dimethylhydrazobenzene	38
4,4'-Diethoxyhydrazobenzene	86
2,2'-Diethoxyhydrazobenzene	89
2,2'-Dimethoxyhydrazobenzene	102
3,3'-Diethoxyhydrazobenzene	119
Hydrazobenzene	127
4,4'-Dimethylhydrazobenzene	134
2,2'-Hydrazonaphthalene	141
2,2'-Hydrazodiphenol	148
1,1'-Hydrazonaphthalene	153
2,2'-Dimethylhydrazobenzene	165
4,4'-Hydrazodibiphenyl	169
2,2'-Hydrazodibiphenyl	182

* All these compounds should be regarded as potential carcinogens; those marked **C** are subject to legal control.

Table VIII,44   Miscellaneous sulphur compounds

Compound	B.P. °C (/mmHg)	M.P. °C	$d_4^{20°}$	$n_D^{20°}$
Dimethyl sulphide	38	—	0.849	1.436
Ethyl methyl sulphide	66	—	0.846	1.440
Diethyl sulphide	92	—	0.837	1.442
Di-isopropyl sulphide	119	—	0.817	1.440
Di-allyl sulphide	140	—	—	—
Dipropyl sulphide	142	—	0.839	1.449
Di-isobutyl sulphide	169	—	0.826	1.447
Di-s-butyl sulphide	165	—	0.835	1.451
Dibutyl sulphide	187	—	0.840	1.453
Di-isopentyl sulphide	86/5	—	0.834	1.453
Dipentyl sulphide	85/4	—	0.841	1.456
Dihexyl sulphide	114/4	—	0.841	1.459
Diheptyl sulphide	142/4	—	0.842	1.461
Dioctyl sulphide	162/4	—	0.845	1.469
Diphenyl sulphide	145/8	—	1.114	1.633
Dibenzyl sulphide	—	50	—	—
Di-p-tolyl sulphide	—	57	—	—
Dimethyl disulphide	109	—	1.065	1.526
Diethyl disulphide	153	—	0.992	1.507
Di-isopropyl disulphide	176	—	0.944	1.492
Dipropyl disulphide	194	—	0.960	1.498
Di-allyl disulphide	100/48	—	—	—
Di-isobutyl disulphide	215	—	0.928	1.487
Di-t-butyl disulphide	65/5	—	0.923	1.490
Dibutyl disulphide	231	—	0.938	1.493
Dipentyl disulphide	119/7	—	0.922	1.489
Di-isopentyl disulphide	115/9	—	0.919	1.486
Di-p-tolyl disulphide	—	48	—	—
Diphenyl disulphide	—	60	—	—
Dibenzyl disulphide	—	73	—	—
Dimethyl sulphoxide	189	18.5	1.101	1.477
Diethyl sulphoxide	88.9/15	14(4–6)	—	—
Dipropyl sulphoxide	82/15	22–3	0.965	1.466
Di-isopropyl sulphoxide	—	68.5	—	—
Dibutyl sulphoxide	—	32	0.832	1.467
Diphenyl sulphoxide	—	70	—	—
Di-p-tolyl sulphoxide	—	95	—	—
Dibenzyl sulphoxide	—	134	—	—
Dipropyl sulphone	—	29	—	—
Dibutyl sulphone	—	44	—	—
Diethyl sulphone	248	74	—	—
Trional	—	76	—	—
Dimethyl sulphone	238	109	—	—
Sulphonal	—	126	—	—
Diphenyl sulphone	—	128	—	—
Dibenzyl sulphone	—	150	—	—
Di-p-tolyl sulphone	—	159	—	—
Methyl thiocyanate	131	—	1.082	
Ethyl thiocyanate	147	—	1.024	1.465
Isopropyl thiocyanate	151	—	—	—
Propyl thiocyanate	165	—	0.981	1.463
Butyl thiocyanate	184	—	0.961	1.464

Table VIII,44   Miscellaneous sulphur compounds (*continued*)

Compound	B.P. °C (/mmHg)	M.P. °C	$d_4^{20°}$	$n_D^{20°}$
Benzyl thiocyanate	—	38	—	—
Allyl isothiocyanate	152	—	1.010	1.524
Phenyl isothiocyanate	221	—	1.134	1.651
Thiophen	84	—	1.062	1.525
Methyl benzenesulphonate	150/15	—	1.273	—
Ethyl benzenesulphonate	156/15	—	1.219	—
Propyl benzenesulphonate	162–3/15	—	1.180	—
Methyl toluene-*p*-sulphonate	—	28	—	—
Ethyl toluene-*p*-sulphonate	173/15	33	—	—
Propyl toluene-*p*-sulphonate	165/10	—	—	—
Butyl toluene-*p*-sulphonate	175/10	—	—	—
Phenyl toluene-*p*-sulphonate	—	96	—	—
N-Allyl thiourea	—	78	—	—
N,N'-Di-*m*-tolyl thiourea	—	112	—	—
N-Phenyl thiourea	—	154	—	—
N,N'-Diphenylthiourea (thiocarbanilide)	—	154	—	—
N,N'-Di-*o*-tolyl thiourea	—	166	—	—
N,N'-Di-*p*-tolyl thiourea	—	178	—	—
Thiourea	—	180	—	—
Thiosemicarbazide	—	182	—	—

Table VIII,45 Miscellaneous phosphorus compounds

Phosphate	B.P. °C (/mmHg)	M.P. °C	$d_4^{20°}$	$n_D^{20°}$
Trimethyl	62/5 (197)	—	1.214	1.396
Triethyl	76/5 (216)	—	1.070	1.405
Tri-isopropyl	84/5	—	0.987	1.406
Tripropyl	108/5	—	1.012	1.416
Tri-isobutyl	117/5	—	0.968	1.419
Tributyl	139/6	—	0.977	1.425
Tri-isopentyl	143/3	—	—	—
Tripentyl	167/5	—	0.961	1.432
Tri-o-cresyl	264/20	—	—	—
Tri-m-cresyl	274/17	26	—	—
Triphenyl	245/11	50	—	—
Tribenzyl	—	65	—	—
Tri-p-cresyl	—	78	—	—
Tri-2-naphthyl	—	111	—	—

Phosphite	B.P. °C (/mmHg)	M.P. °C	$d_4^{20°}$	$n_D^{20°}$
Trimethyl	112	—	1.052	1.410
Triethyl	157	—	0.969	1.414
Tri-isopropyl	60/9	—	0.918	1.412
Tripropyl	207	—	0.952	1.427
Tri-isobutyl	235	—	0.917	1.425
Tributyl	120/10	—	0.923	1.432
Tripentyl	123/6	—	0.901	1.433
Triphenyl	228/12	24	—	—
Tri-o-cresyl	238/11	—	—	—
Tri-m-cresyl	235/7	—	—	—
Tri-p-cresyl	238/7	—	—	—
Dimethyl hydrogen	72/25	—	1.200	1.404
Diethyl hydrogen	66/6	—	1.079	1.408
Di-isopropyl hydrogen	90/25	—	0.996	1.407
Dipropyl hydrogen	92/11	—	1.019	1.416
Di-isobutyl hydrogen	106/12	—	0.976	1.420
Dibutyl hydrogen	119/7	—	0.995	1.423

Table VIII,46   Esters of inorganic acids

Ester	B.P. °C (/mmHg)	$d_4^{20°}$	$n_D^{20°}$
Nitrites			
Methyl nitrite	− 12	—	—
Ethyl nitrite	17	0.907 (10 °C)	1.331 (10 °C)
Propyl nitrite	48	0.886	1.360
Isopropyl nitrite	45	0.856	—
Butyl nitrite	76	0.882	1.377
Isobutyl nitrite	67	0.871	1.373
s-Butyl nitrite	68	0.872	1.371
t-Butyl nitrite	63	0.867	1.369
Pentyl nitrite	104	0.882	1.389
Isopentyl nitrite	99	0.871	1.387
t-Pentyl nitrite	93	0.896	1.387
Nitrates			
Methyl nitrate	65	1.208	1.375
Ethyl nitrate	88	1.108	1.385
Propyl nitrate	111	1.054	1.397
Isopropyl nitrate	102	1.035	1.391
Butyl nitrate	136	1.023	1.407
Isobutyl nitrate	124	1.015	1.403
s-Butyl nitrate	124	1.026	1.402
Pentyl nitrate	157	0.996	—
Isopentyl nitrate	148	0.998	1.413
Sulphites			
Dimethyl sulphite	126	1.213	1.409
Diethyl sulphite	157	1.083	1.414
Dipropyl sulphite	191	1.028	1.424
Di-isopropyl sulphite	170	1.006	1.415
Dibutyl sulphite	91/5	0.996	1.431
Di-isobutyl sulphite	210	0.986	1.427
Dipentyl sulphite	111/5	0.978	1.436
Di-isopentyl sulphite	98/4	0.973	1.436
Sulphates			
Dimethyl sulphate	188	1.328	1.387
Diethyl sulphate	208	1.177	1.400
Dipropyl sulphate	94/5	1.110	1.414
Dibutyl sulphate	116/6	1.062	1.421
Di-isobutyl sulphate	133/19	1.045	1.415
Dipentyl sulphate	117/4	1.029	1.429

See Table VIII,45 for alkyl phosphates and alkyl phosphites.

APPENDICES

APPENDIX 1

THE LITERATURE OF ORGANIC CHEMISTRY

1,A. Introduction. The organic chemist is frequently faced with the task of obtaining specific information from the literature. The information required may be the physical properties or a reliable method of preparation of a particular compound, a complete list of preparative methods, a survey of recent work on a particular group of compounds, etc. Given access to a good chemical library the experienced chemist should be able to obtain the relevant information without difficulty. The principal sources of such information are described in this appendix. The coverage is necessarily selective and readers who require a more detailed treatment are recommended to consult one of the texts or articles on the subject. (For example: M. G. Mellon (1965). *Chemical Publications—Their Nature and Use.* 4th edn. New York; McGraw-Hill Book Co. *The Use of the Chemical Literature* (1969). 2nd edn. Ed. R. T. Bottle, London; Butterworths. J. E. H. Hancock (1968). 'An Introduction to the Literature of Organic Chemistry', *J. Chem. Educ.*, *45*, 193, 260, 336.)

1,B. Primary sources of chemical information: journals. Scientific journals are the principal method for communicating scientific information, although they are being increasingly supplemented by other original literature sources such as patents, theses and reports of various kinds. The most important journals dealing with organic chemistry are listed below, with the standard abbreviated name given in parentheses (*World List of Scientific Periodicals.* Butterworths,

London. IUPAC recommends the abbreviations given in *List of Periodicals Abstracted by Chemical Abstracts*).

(Acta Chemica Scandinavica (*Acta Chem. Scand.*)
Angewandte Chemie, International Edition (*Angew. Chem., Internat. Edn.*)
Annalen der Chemie (also referred to as Liebigs Annalen) (*Justus Liebigs Annalen Chem.*)
Australian Journal of Chemistry (*Austral. J. Chem.*)
Bulletin of the Academy of Sciences of the U.S.S.R. (Izvestia Akademii Nauk S.S.S.R.) (*Bull. Acad. Sci. U.S.S.R., Div. Chim. Sci.*)
Bulletin of the Chemical Society of Japan (*Bull. chem. Soc. Japan*)
Bulletin de la Société Chimique de France (*Bull. Soc. chim. France*)
Canadian Journal of Chemistry (*Canad. J. Chem.*)
Chemical Communications (*J.C.S. Chem. Comm.*)
Chemische Berichte (*Chem. Ber.*)
Helvetica Chimica Acta (*Helv. Chim. Acta*)
Journal of the American Chemical Society (*J. Amer. Chem. Soc.*)
Journal of the Chemical Society (*J. Chem. Soc.*)
Journal of General Chemistry of the U.S.S.R. (Zhurnal Obschei Khimii) (*J. Gen. Chem., U.S.S.R.*)
Journal of Heterocyclic Chemistry (*J. Heterocyclic Chem.*)
Journal of Organic Chemistry (*J. Org. Chem.*)
Journal of Organic Chemistry of the U.S.S.R. (Zhurnal Organischeskoi Khimii) [*Zh. Org. Khim. (U.S.S.R.)*]
Journal of Organometallic Chemistry (*J. Organometal. Chem.*)
Journal für Praktische Chemie (*J. Prakt. Chem.*)
Synthesis (*Synthesis*)
Synthetic Communications (*Synthetic Comm.*)
Tetrahedron (*Tetrahedron*)
Tetrahedron Letters (*Tetrahedron Letters*)

1,C. Secondary sources of chemical information: abstracting journals. Abstracting journals are publications giving contemporaneous, concise summaries of the various original communications and other contributions to knowledge. Each abstract usually supplies the title of the original contribution abstracted, name(s) of author(s), original reference (i.e., name of journal or other source of information, year, series, volume, page) and generally a brief summary of the original source. The value of the abstract, in the first instance, will depend upon how detailed the summary is. It must be emphasised, however, that the reader should never be satisfied with the account to be found in the abstract; he should, as far as possible, consult the original work and abstract it.

(*a*) *Chemical Abstracts.* These were commenced by the American Chemical Society in 1907 and have continued ever since. The abstracts are very comprehensive (particularly in recent years) in respect of subject matter and journals covered.

The documents abstracted include journal articles, patent specifications, reviews, technical reports, monographs, conference proceedings, symposia, dissertations and books. The abstracts are divided into eighty sections:

1–20   Biochemistry
21–34 Organic Chemistry
35–46 Macromolecular Chemistry
47–64 Applied Chemistry and Chemical Engineering
65–80 Physical and Analytical Chemistry.

An issue of *Chemical Abstracts* appears each week containing alternately Sections 1–34 and Sections 35–80. Each weekly issue of *Chemical Abstracts* contains two parts, the abstracts and the issue indexes. An Illustrative Key is included at the start of each new volume (which covers a 6-month period) to assist the reader in the most effective use of the Abstracts, and readers are referred to this Key for more detailed information than can be included in the present summary. Abstracts are numbered consecutively throughout each volume. Starting in 1976 (Volume 84), each abstract is immediately preceded by the volume number in lightface type, and followed by a computer-generated check-letter by which each reference is computer validated. Check letters have been in use since 1966 (Volume 66) and should not be confused with the column fractional designations (a–g) previously used to indicate the position of the abstract on the page.

The indexes of *Chemical Abstracts* are a key to the world's chemical and chemical engineering literature. There are many different indexes and care and practice is required to make the most effective use of them. Beginning with Volume 69, all explanatory and illustrative matter relating to indexing was removed from the regular Subject Index and published separately as the *Index Guide*. The *Index Guide* to Volume 76 covers the period of the Ninth Collective Index (1972–6), and supplements to it are included with alternate volumes; together these provide complete instructions on the use of *Chemical Abstracts*, and also a comprehensive account of the policies for selecting the names of chemical substances in *Chemical Abstracts*. Each type of index to *Chemical Abstracts* includes an Illustrative Key to explain the form of the entries in that index, and this key should be consulted before the index is used.

Indexes are published with each weekly issue and with each volume, and Collective Indexes are issued every five years. Each *Issue Index* consists of four parts: the Keyword Index, the Author Index, the Patent Index and the Patent Concordance. The Issue Indexes are intended to provide quick access to the abstracts until the publication of the more extensive Volume Indexes. The *Keyword Index* consists of words and phrases derived from the title, text or context of the abstract and allows ready access to its subject matter. The *Numerical Patent Index* links the specification number of each patent abstracted with its corresponding abstract number. The patents are listed according to country of issue, and within each of these sections they are listed numerically. The *Patent Concordance* correlates patents issued by different countries for the same basic invention. An abstract is published in *Chemical Abstracts* when the patent is first received; subsequent patents covering the same invention are listed in the Patent Concordance. The *Author Index* links the names of individual authors, patentee and patent assignees together with the titles of their original papers or patents to the corresponding abstract number in *Chemical Abstracts*.

The *Volume Indexes* of *Chemical Abstracts* are of seven different types. In addition to the Patent Index, Patent Concordance and Author Index these are the Chemical Subject Index, the General Subject Index, the Formula Index and the Index of Ring Systems. At the start of the Ninth Collective Index period (Volume 76, 1972) the then existing Subject Index was divided into two separate indexes, the Chemical Substance Index and the General Subject Index. The *Chemical Substance Index* includes entries for all completely defined substances such as elements, compounds, alloys, specific minerals, elementary particles, etc. All index headings which do not relate to specific chemical substances are included in the *General Subject Index*; for example, classes of chemical substances, physicochemical concepts and phenomena, applications, uses, reactions, etc. The *Formula Index* provides the simplest and quickest means of obtaining references to a single chemical compound when the complete molecular formula is known; it also gives direct access to the name used in the Chemical Substance Index. The *Index of Ring Systems* lists the ring structures of the compounds to be found in a particular volume, without regard to hydrogen atoms or substituents. It thus provides a means of deriving the *Chemical Abstracts* index names of cyclic compounds. The ring analysis indicates the number, size and skeletal atom content of the component rings. A comprehensive catalogue of all ring systems which have been used in *Chemical Abstracts* indexes to date is to be published during the period of the Ninth Collective Index; it is to be called the *Parent Compound Handbook* and will replace the separately published Ring Index. The *Collective Indexes* provide access to the abstracts of either a five- or a ten-year period. The pattern of indexes to be used for the Ninth Collective Index follows the same pattern as that outlined above for the volume indexes.

An additional method for the cross-referencing of chemical substances has been developed in recent years. It is known as the *Chemical Abstracts Registry System* and it is a computer-based system that identifies chemical substances on the basis of their molecular structure, and assigns to each substance a unique identifying number called a Registry Number. The numbers have no chemical significance, but are assigned in sequential order as substances are entered into the System for the first time. Operation of the CAS Registry System began in 1965. By December 1973 about 2.7 million unique substances had been recorded and substances new to the file are being recorded at the rate of about 300 000 per year. A Registry Handbook-Number Section includes in Registry Number order all substances which have been recorded since the system began, together with the molecular formula of each and one complete *Chemical Abstracts* index name. Registry Numbers thus provide a means of structure identification without the need to resort to complex, and sometimes ambiguous, chemical nomenclature. A Registry Number Update is published annually with the Number Section Supplement; it provides a means of determining whether a Registry Number is still valid, and should always be consulted before using the Registry Handbook.

(*b*) *Chemisches Zentralblatt.* This journal originated in 1830, but assumed its present name in 1907. Although English-speaking chemists will probably prefer *Chemical Abstracts*, *Chemisches Zentralblatt* should not be ignored. It is particularly valuable for covering East European and Russian literature, the abstracts are probably more detailed than *Chemical Abstracts*, and for the period until 1939 it had a wider coverage.

(*c*) *British Chemical Abstracts.* These were commenced in the year 1871 and were included in the *Journal of the Chemical Society*. In 1926 the abstracting was taken over by the Bureau of Chemical Abstracts; abstracts were then published in two parts: Part A, 'Pure Chemistry' (formerly issued by the Chemical Society) and Part B, 'Applied Chemistry' (formerly issued by the Society of Chemical Industry). The seven Collective Decennial Indexes (up to 1937) provide an excellent means for following abstracts in the English language from 1871. Publication of the abstracts in 'Pure Chemistry' was discontinued in January 1954.

1,D. Reference Works (a) **Beilstein** The fourth edition of *Beilstein's Handbuch der Organischen Chemie* is the largest compilation of information on organic compounds. The main series (Hauptwerk) is composed of twenty-seven volumes and covers the literature to 1 January 1910. The first supplement (Erstes Ergänzungswerk) surveys the literature to 1919, and the second supplement (Zweites Ergänzungswerk) covers the decade to 1929. Thus far (1975) some seventeen volumes of the third supplement (Drittes Ergänzungswerk), which covers the period 1930–49, have been published. The publication of the fourth supplement (Viertes Ergänzungswerk), covering the period 1950–9, was commenced in 1972 and four parts of the first volume have appeared so far. However, starting with Volume 17 (heterocyclic compounds) of the third supplement, the third and fourth supplements are combined in a single publication and the period covered is thus 1930–59. The volumes are published in numerical order and are subdivided into several parts. In addition there is a comprehensive subject index (Sachregister) in Volume 28, and a formula index (Formelregister) in Volume 29 for the main series and the first and second supplements; each comprehensive index includes all the information from the previous ones, and so the most recent one available should be used. Each volume also contains an index (which is not cumulative) and this is used to locate information in the third supplement until the complete cumulative index is published. Each entry in Beilstein includes, where available, the information listed below with references to the important original articles (the abbreviations in brackets are those used in Beilstein).

Name, formula and structure.
Important historical notes.
Occurrence (V-Verkommen), formation (in various reactions; B-Bildung), preparation (best method; Darst.-Darstellung).
Properties: (a) physical; (b) chemical.
Physiological properties.
Technical applications.
Analysis.
Addition compounds and salts.
Conversion products of unknown structure.

The space devoted to any one compound varies from one line to several pages according to its importance.

The entire subject matter has been divided into 4877 arbitrary units each of which is assigned a 'System Number'. This sub-division was designed to facilitate cross-referencing, and although of limited use when cumulative indexes are available, it still remains valuable for locating information in the third supplement on compounds first reported after 1930, and it is for this reason that a knowledge of Beilstein's classification system is important.

To locate a compound first reported in the literature before 1930, it is recommended that the formula index be consulted first, since the subject index may provide problems, especially for the beginner, through the use of non-systematic names. The entry against each formula gives the volume number and the page numbers in the main series and the first and second supplements. The System Number will thus be obtained and this will direct the searcher to the appropriate volume of the third supplement.

To locate a compound first reported in the literature after 1930, the compound will not be reported in any of the cumulative indexes and hence the available volumes of the third supplement must be searched. For this purpose a knowledge of the classification system is essential. The system of classification adopted is based on the premise that every compound can be assigned a position determined by its structural formula. Compounds are assigned to one of four divisions:

I   Acyclic
II Alicyclic
III Heterocyclic
IV Natural products

Division IV contains compounds not assigned to one of the other divisions. The position of each compound in the appropriate division is determined by its 'stem nucleus', which is obtained by replacing in the formula of the compound all atoms or groups attached to carbon by the equivalent number of hydrogen atoms except where such replacement would involve rupture of a cyclic chain. This principle leads to the inclusion of such compounds as the anhydrides and imides of dibasic acids, sulphimides, lactides and lactones of hydroxy-acids, etc., under heterocyclic compounds. Whenever a given formula can be assigned to more than one division, the compound will be found in that division which comes later in the classification; this is known as the 'principle of latest position', which is used elsewhere in the classification.

The position of heterocyclic compounds in Division III is determined by the number and type of hetero atoms the stem nucleus contains, according to the following system of sub-divisions:

Compounds with 1, 2, 3, etc., cyclically bound oxygen atoms.
Compounds with 1, 2, 3, etc., cyclically bound nitrogen atoms.
Compounds with 1 cyclic nitrogen and 1, 2, 3, etc., cyclic oxygen atoms.
Compounds with 2 cyclic nitrogens and 1, 2, 3, etc., cyclic oxygen atoms, etc.

Compounds containing sulphur, selenium or tellurium in the ring are placed in the same class as the corresponding oxygen compounds, and immediately follow them in the classification; similarly, those containing phosphorus, arsenic, etc., are related to the corresponding nitrogen compounds.

A compound can thus be placed into the appropriate division or subdivision according to its stem nucleus, but it must be further assigned, according to the nature of the functional groups attached to the nucleus, to one of the 28 'main classes'. The first of these classes consists of the unsubstituted stem nuclei, or those containing the substituents: F, Cl, Br, I, NO, NO_2, N_3 (i.e., those containing no replaceable hydrogens). The most important main classes and the corresponding substituents are shown below.

Class		*Characterising groups*
1.	Stem nuclei	—
2.	Hydroxy compounds	$-OH$
3.	Carbonyl compounds	$-CHO$ or $-C=O$
4.	Carboxylic acids	$-CO_2H$
5.	Sulphinic acids	$-SO_2H$
6.	Sulphonic acids	$-SO_3H$
7.	Selenious and selenic acids	$-SeO_2H$, $-SeO_3H$
8.	Amines	$-NH_2$
9.	Hydroxylamines	$-NHOH$
10.	Hydrazines	$-NH \cdot NH_2$
11.	Azo compounds	$-N=NH$
12–22.	Other nitrogen derivatives	
23–28.	Organometallic compounds in which carbon is bonded directly to a metallic element.	

Using this classification system it is possible to locate the volume of the third supplement in which a compound or class of compounds is to be found, and then the subject or formula indexes of the individual volumes can be used. If further help is needed in locating information in Beilstein, the reader is referred to one of the more comprehensive accounts which are available, amongst which may be found several worked examples of searches.

(1) *Beilstein's Handbuch der Organischen Chemie* (1918). 4th edn, Vol. I, pp. 1–46. Berlin; Springer.

(2) E. H. Huntress (1938). *A Brief Introduction to the use of Beilstein's Handbuch*. 2nd edn. London; Wiley.

(3) O. Runquist (1966). *Beilstein's Handbuch, a programmed Guide*. Burgess.

(4) G. M. Dyson (1958). *A Short Guide to the Chemical Literature*. 2nd edn. London; Longmans, Green and Co.

(5) K. C. Bhattacharyya (1972). *A Simplified Guide to Beilstein's Handbuch der Organischen Chemie*. Occasional publication, National Reference Library of Science and Invention, London.

The general scope and structure of Beilstein will be evident from the following outline of the contents.

Contents of the volumes of Beilstein's Handbuch

(H refers to Hauptwerk (main series); EI, EII, EIII and EIV to first, second, third and fourth supplements.)

Acyclic division **Volume I.** Hydrocarbons. Hydroxy compounds: alcohols and derivatives. Carbonyl compounds: aldehydes, ketones, ketenes and derivatives. Hydroxy-carbonyl compounds: aldehyde-alcohols, ketone-alcohols, monosaccharides and derivatives. System Numbers 1–151. Year of publication: H (1918); EI (1928); EII (1941); EIII, 3 parts (1958–9); EIV (publication commenced in 1972).

Volume II. Carboxylic acids: salts and derivatives. System Numbers 152–194. Year of publication: H (1920); EI (1929); EII (1942); EIII, 2 parts (1960–1).

Volume III. Carboxylic acids: Polyfunctional and derivatives; Hydroxy-carboxylic acids, carbonyl-carboxylic acids, hydroxy-carbonylcarboxylic acids. System Numbers 195–322. Year of publication: H (1921); EI (1929); EII (1942); EIII, 2 parts (1961–2).

Volume IV. Sulphonic acids. Amines: hydroxy-amines, carbonylamines, hydroxy-carbonyl amines, amino-carboxylic acids. Hydroxylamines. Hydrazines. Azo compounds. Organometallic compounds. System Numbers 323–449. Year of publication: H (1922); EI and EII are combined with Volume III; EIII, 2 parts (1962–3).

Isocyclic division **Volume V.** Hydrocarbons. System Numbers 450–498. Year of publication: H (1922); EI (1930); EII (1943); EIII, 4 parts (1963–5).

Volume VI. Hydroxy compounds. System Numbers 499–608. Year of publication: H (1923); EI (1931); EII (1946); EIII, 9 parts (1965–8).

Volume VII. Carbonyl compounds. System Numbers 609–736. Year of publication: H (1925); EI (1931); EII (1948); EIII, 5 parts (1968–9).

Volume VIII. Hydroxy-carbonyl compounds. System Numbers 737–890. Year of publication: H (1925); EI is combined with Volume VII; EII (1948); EIII, 6 parts (1969–71).

Volume IX. Carboxylic acids. System Numbers 891–1050. Year of publication: H (1926); EI (1932); EII (1949); EIII, 5 parts (1970–1).

Volume X. Hydroxy-carboxylic acids. System Numbers 1051–1504. Year of publication: H (1927); EI (1932); EII (1949); EIII, 5 parts (1971–2).

Volume XI. Other acids. Sulphinic acids. Sulphonic acids. Hydroxy-sulphonic acids. Carboxylic-sulphonic acids. System Numbers 1502–1591. Year of publication: H (1928); EI (1933); EII (1950); EIII (1972).

Volume XII. Mono-amines. System Numbers 1592–1739. Year of publication: H (1929); EI is combined with Volume XI; EII (1950); EIII, 5 parts (1972–3).

Volume XIII. Poly-amines. System Numbers 1740–1871. Year of publication: H (1930); EI (1933); EII (1950); EIII, 3 parts (1973).

Volume XIV. Carbonyl-amines. Hydroxy-carbonyl amines, amino-hydroxy-carboxylic acids, amino-sulphonic acids. System Numbers 1872–1928. Year of publication: H (1931); EI is combined with Volume XIII; EII (1951); EIII, 4 parts (1973–4).

Volume XV. Hydroxylamines. Hydrazines. System Numbers 1929–2084. Year of publication: H (1932); EI (1934); EII (1951); EIII (1974).

Volume XVI. Azo compounds. Diazo compounds. Azoxy compounds. Nitramines and nitroso-hydroxylamines. Triazines. Phosphorus compounds. Arsenic compounds. Compounds with other elements. System Numbers 2085–2358. Year of publication: H (1933); EI is combined with Volume XV; EII (1951); EIII, 2 parts (1974).

Heterocyclic division **Volume XVII.** One cyclic oxygen, sulphur, selenium or tellurium. Stem nuclei. Hydroxy compounds. Carbonyl compounds. System Numbers 2359–2503. Year of publication: H (1933); EI (1934); EII (1952); EIII and EIV (1974).

Volume XVIII. One cyclic oxygen (S, Se, Te) (continued). Carbonyl compounds (continued). Sulphonic acids. Amines and other nitrogen compounds. Carbon-metal compounds. System Numbers 2504–2665. Year of publication: H (1934); EI is combined with Volume XVII; EII (1952).

Volume XIX. Two, three, four cyclic oxygens (S, Se, Te), and derivatives.

System Numbers 2666–3031. Year of publication: H (1934); EI is combined with Volume XVII; EII (1952).

Volume XX. One cyclic nitrogen. Stem nuclei. System Numbers 3032–3102. Year of publication: H (1935); EI (1935); EII (1953).

Volume XXI. One cyclic nitrogen (continued). Hydroxy compounds. Carbonyl compounds. System Numbers 3103–3241. Year of publication: H (1935); EI is combined with Volume XX; EII (1953).

Volume XXII. One cyclic nitrogen (continued). Carboxylic acids. Sulphonic acids. Amines. Amino-carboxylic acids. Other nitrogen derivatives. System Numbers 3242–3457. Year of publication: H (1935); EI is combined with Volume XX; EII (1953).

Volume XXIII. Two cyclic nitrogens. Stem nuclei. Hydroxy compounds. System Numbers 3458–3554. Year of publication: H (1936); EI (1936); EII (1954).

Volume XXIV. Two cyclic nitrogens (continued). Carbonyl compounds. System Numbers 3555–3633. Year of publication: H (1936); EI is combined with Volume XXIII; EII (1954).

Volume XXV. Two cyclic nitrogens (continued). Hydroxy-carbonyl compounds. Carboxylic acids. Amines. Keto-amines. Amino-carboxylic acids. System Numbers 3634–3793. Year of publication: H (1936); EI is combined with Volume XXIII; EII (1955).

Volume XXVI. Three, four … cyclic nitrogens and derivatives. System Numbers 3794–4187. Year of publication: H (1937); EI (1938); EII (1955).

Volume XXVII. Heterocycles containing both oxygen and nitrogen. System Numbers 4199–4720. Year of publication: H (1937); EI is combined with Volume XXVI; EII (1955).

Volume XXVIII. Subject Index. Year of publication: H (1938–9); EI (1955–6); EII (1955).

Volume XXIX. Formula Index. Year of publication: H (1939–40); EI (1956); EII (1956).

Naturally occurring compounds **Volume XXX.** Rubber. Guttapercha and Balata. Carotenoids. System Numbers 4723–4723a. Year of publication: H (1938).

Volume XXXI. Carbohydrates, Part I. System Numbers 4746–4767a. Year of publication: H (1938). No supplements to Volumes XXX and XXXI have been published.

(b) Compilations of data *Dictionary of Organic Compounds.* 4th edn. Ed. J. R. A. Pollock and R. Stevens. London; Eyre & Spottiswoode, 1965. Originally *Heilbron's Dictionary.* Supplements published annually: 5th (cumulative) supplement (1969); formula index for Main Work and 5th Supplement (1971); 10th (cumulative) supplement (1974).

Encyclopaedia of Chemical Technology (Kirk-Othmer). 2nd edn. Ed. A. Standen. New York; Interscience, 1963–72. Twenty-two volumes plus supplement and Index.

Merck Index of Chemicals and Drugs. 8th edn. New Jersey; Merck, Rahway, 1968. Lists preparation and properties of some 10 000 chemicals, pronounced medicinal bias.

Handbook of Chemistry and Physics. Various editions. Cleveland, Ohio; Chemical Rubber Publishing Co.

Handbook of Tables for Organic Compound Identification. 3rd edn. Cleveland, Ohio; Chemical Rubber Publishing Co., 1967.

Lange's Handbook of Chemistry. Revised 11th edn. Ed. J. A. Dean. New York; McGraw Hill, 1973.

A. J. Gordon and R. A. Ford (1972). *The Chemist's Companion.* New York; Wiley.

Atlas of Spectral Data and Physical Constants for Organic Compounds. 2nd edn. Ed. J. G. Grasselli & W. M. Ritcheg. Cleveland, Ohio; CRC Press Inc., 1975. Six volumes containing the following data on 21 000 compounds: molecular formula, molecular weight, melting point, boiling point. Wiswesser Line Notation and infrared, raman, ultraviolet, ^{1}H and ^{13}C magnetic resonance and mass spectral data.

A. Clarke (1971). *A Guide to the Literature on Spectral Data.* Occasional Publication, National Reference Library of Science & Invention, London.

U.V. Atlas of Organic Compounds (5 volumes). Ed. H. H. Perkampas, L. Sandemann and C. J. Timmons. London; Butterworth. Weinheim; Verlag Chemie. A collection of 1160 spectra.

C. J. Pouchert (1970). *The Aldrich Library of Infrared Spectra.* Milwaukee, Wisconsin; Aldrich Chemical Co. A collection of 8000 spectra.

H. A. Szymanski (1964). *Interpreted Infrared Spectra* (3 volumes). New York; Plenum.

Sadtler Collections, Sadtler Research Labs, Philadelphia. *Standard Infrared Spectra,* 1975 supplement (Vol. 48), brings the total collection to 49 000 infrared prism spectra in the region 5000–665 cm^{-1} (2–15 microns). Accompanied by Alphabetical Index, Molecular Formula Index, Chemical Classes Index, Numerical Index, Spec. Finder.

Standard NMR spectra, 1975 supplement, brings the total to 22 000 spectra at 60 MHz.

High Resolution NMR Spectra. Sadtler, 1967. 225 spectra of a wide variety of compounds.

L. F. Johnson and W. C. Jankowski (1972). *Carbon-13 NMR Spectra.* London; Wiley-Interscience. A collection of 500 assigned spectra.

J. H. Benyon and A. E. Williams (1963). *Mass and Abundance Tables for Use in Mass Spectrometry.* London; Elsevier.

A. Cornu and R. Massot (1966). *Compilation of Mass Spectral Data.* London; Heyden (two supplements 1967, 1971). A collection of 7000 spectra.

E. Stenhagen, S. Abrahamson and F. W. McLafferty (1974). *Registry of Mass Spectral Data.* Vols. 1 to 4. London; Wiley-Interscience. Contains mass spectral data for 18 806 different compounds. Unit-resolution mass spectra. Spectra presented as bar-graphs.

(c) **Synthetic methods and techniques** *Methoden der Organischen Chemie* (Houben-Weyl). 4th edn. Ed. E. Nueller, G. Thieme, Stuttgart, 1958 onward (in German). Comprehensive and critical coverage of experimental procedure and appropriate theoretical background. The edition is planned in 16 volumes, each sub-divided into several parts; over 30 parts have so far been published.

Rodd's Chemistry of Carbon Compounds. 2nd edn. Ed. S. Coffey. Elsevier. Vol. I, Aliphatic Compounds (Parts A to F); Vol. II, Alicyclic Compounds (Parts A to E); Vol. III, Aromatic Compounds (Parts A to F); Vol. IV, Heterocyclic Compounds (Part A). Supplements to the 2nd edition (ed. M. F.

Ansell) are being published. Each supplementary chapter stands on its own as a review of recent advances in the particular field surveyed. To date (1975) supplements to IA and B, IC and D, IIA and B, and IIC, D and E have appeared.

Organic Syntheses, many editions. New York; Wiley. An annual publication of satisfactory methods for the preparation of organic compounds. The emphasis now is on model procedures rather than specific compounds. 'Collective Volumes' are published after every ten annual volumes.

Synthetic Methods of Organic Chemistry. Ed. W. Theilheimer. Basel; Karger, annually. New methods of synthesis, improvement in known methods and also old proved methods scattered in periodicals are recorded in this annual series. Particularly useful is the brief review (from Volume 8) 'Trends in Synthetic Organic Chemistry', at the start of each volume.

C. A. Buehler and D. E. Pearson (1970). *Survey of Organic Syntheses*. New York; Wiley. How the functional group is created from other functional groups is the main concern of this work. Chapters are classified according to functional groups.

Organic Reactions. Wiley. Began publication in 1942, Volume 23 (1976). Each volume contains detailed surveys of particular types of reactions with a range of typical experimental procedures.

L. F. Fieser and M. Fieser. *Reagents for Organic Synthesis*. New York; Wiley. Vol. 1 (1967), Vol. 2 (1969), Vol. 3 (1972), Vol. 4 (1974), Vol. 5 (1975). Extremely useful compilation of hundreds of reagents of use to the organic chemist, together with methods of use, hazards, preparation, purification, commercial suppliers, literature references

Techniques of Chemistry. Ed. A. Weissberger. London; Interscience, 1971 onwards. (Successor to *Techniques of Organic Chemistry*.) Comprehensive treatment covering theoretical background, description of techniques and tools, their modifications, merits and limitations, and their handling: Volume I, Physical Methods of Chemistry (5 parts); II, Organic Solvents; III, Photochromism; IV, Elucidation of Organic Structures by Physical and Chemical Methods (3 parts); V, Techniques of Electro-organic Synthesis (2 parts); VI, Investigation of Rates and Mechanisms of Reactions (2 parts).

(d) Reviews The increasing volume of papers published in the field of organic chemistry has led to a proliferation of surveys and summaries which act as keys to the primary sources. Some of the important review publications are listed below. A valuable guide to the review literature is 'Index of Reviews in Organic Chemistry' by D. A. Lewis (Chemical Society, London); cumulative issue (1971) and supplements (1972, 1973); it includes reference to review articles which have appeared in journals, books, conference proceedings, technical trade literature, etc.

The Specialist Periodical Reports, published by the Chemical Society, London, provide systematic and comprehensive reviews of progress in major areas of research. They appear annually or biannually and titles include Carbohydrate Chemistry; Amino Acids, Peptides and Proteins; Alkaloids; Terpenoids and Steroids; Aliphatic Chemistry; Alicyclic Chemistry; Saturated Heterocyclic Chemistry; Aromatic and Heteroaromatic Chemistry; Biosynthesis; Organometallic Chemistry; Organophosphorus Chemistry; Organic Compounds of Sulphur, Selenium and Tellurium; Photochemistry.

Another useful series is the M.T.P. *International Review of Science*, Organic

Chemistry Series (ed. D. Hey. London; Butterworths, 1974) which spans ten volumes covering the literature of 1970–1 together with relevant background material: 1. Structure Determination in Organic Chemistry; 2. Aliphatic Compounds; 3. Aromatic Compounds; 4. Heterocyclic Compounds; 5. Alicyclic Compounds; 6. Amino Acids, Peptides and Related Compounds; 7. Carbohydrates; 8. Steroids; 9. Alkaloids; 10. Free Radical Reactions.

Less specialised reviews are to be found in Annual Reports on the Progress of Chemistry (Section B—Organic Chemistry), and Chemical Society Reviews. Some other important sources of reviews are listed below.

> Accounts of Chemical Research
> Advances in:
>> Alicyclic Chemistry
>> Carbohydrate Chemistry
>> Chemotherapy
>> Drug Research
>> Heterocyclic Chemistry
>> Magnetic Resonance
>> Mass Spectrometry
>> Organic Chemistry: Methods and Results
>> Physical Organic Chemistry
>> Spectroscopy
>
> Angewandte Chemie (International Edition in English)
> Chemical Reviews
> Chromatographic Reviews
> Fortschritte der Chemische Forschung
> Organometallic Chemistry Reviews
>
> Progress in:
>> Drug Research
>> Medicinal Chemistry
>> Organic Chemistry
>> Physical Organic Chemistry
>> Stereochemistry
>
> Russian Chemical Reviews (Uspekhi Khimii)
> Survey of Progress in Chemistry
> Synthesis.

APPENDIX 2   INFRARED CORRELATION TABLES

Table 2,1   Alkanes, cycloalkanes and alkyl groups

Group/vibration	cm^{-1}	μm	Comments
C – H stretching			
CH$_3$ –	2972–2953	3.36–3.39	(s) asym.
$>$CH$_2$	2936–2916	3.41–3.43	(s) asym.
CH$_3$ –	2882–2862	3.47–3.49	(s) sym.
$>$CH$_2$	2863–2843	3.49–3.52	(s) sym.
$>$CH	2900–2880	3.45–3.47	(w)
cyclopropane	~3060–3040	3.27–3.29	(w) cf. alkene C – H str.
CH$_3$ – N	2825–2765	3.54–3.62	(m) precise pattern depends on whether amine is aliphatic or aromatic and if NCH$_3$ or N(CH$_3$)$_2$
CH$_3$ – O	2830–2810	3.53–3.56	(v)
CH$_3$ – CO –	3000–2900	3.33–3.40	(w) much reduced intensity
C – H deformation			
CH$_3$ –	1470–1430	6.80–7.00	(m) asym.
$>$CH$_2$	1485–1445	6.73–6.92	(m) scissoring, normally overlaps above
CH$_3$ – C –	1380–1370	7.25–7.30	(m) sym. very useful
CH$_3$ $>$C$<$ CH$_3$	1385–1380	7.22–7.25	(m) } approximately of
	1370–1365	7.30–7.33	(m) } equal intensity
(CH$_3$)$_3$C –	1395–1385,	7.17–7.22	(m) } intensity ratio
	1365	7.32	(s) } approx. 1 : 2
$>$CH	~1340	7.46	(m) no practical value
CH$_3$ – N	1440–1410	6.95–7.09	
CH$_3$ – O	1460–1440	6.85–6.95	
CH$_3$ – CO –	1364–1354	7.33–7.39	(s) very much enhanced intensity
– CH$_2$ – CO –	1420–1410	7.04–7.09	(s) compared to CH$_3$ and CH$_2$ attached to saturated C
Skeletal			
(CH$_3$)$_3$ – C	1255–1245	7.97–8.03	(s)
	1250–1200	8.00–8.33	(s)
CH$_3$ $>$C$<$ CH$_3$	1175–1165	8.51–8.59	(s)
	1170–1140	8.55–8.77	(s)
– (CH$_2$)$_n$ –	720–725	13.88–13.80	(m) for n $\geqslant$ 4, doublet in solid state; frequency increases the shorter the chain
cyclopropane	1020–1000	9.80–10.0	(m)

Table 2,2 Alkenes

Group/vibration	cm^{-1}	μm	Comments
=C−H **stretching and deformation**			
R·CH=CH$_2$	3040–3010	3.29–3.32	(m) CH str.
	3095–3075	3.23–3.25	(m) CH$_2$ str.
	995–985	10.05–10.15	(s) CH out-of-plane def.
	915–905	10.93–11.05	(s) CH$_2$ out-of-plane def. (wag)— often overtone at ~1830 cm^{-1}
	1420–1410	7.04–7.09	(s) CH$_2$ in-plane def.
	1300–1290	7.69–7.75	(s–w) CH in-plane def.
R^{1}R^{2}C=CH$_2$	3095–3075	3.23–3.25	(m) CH str.
	895–885	11.17–11.30	(s) out-of-plane def.; often an overtone at 1780 cm^{-1}
	1420–1410	7.04–7.09	(s) CH$_2$ in-plane def.
CH=CH (*trans*)	3040–3010	3.29–3.32	(m) CH str.
	970–960	10.31–10.42	(s) CH out-of-plane def.
	1310–1295	7.64–7.72	(s–w) CH in-plane def.
CH=CH (*cis*)	3040–3010	3.29–3.32	(m) CH str.
	728–675	13.74–14.82	(s) often near 690 cm^{-1}
R^{1}R^{2}C=CHR^3	3040–3010	3.29–3.32	(m) CH str.
	840–790	11.90–12.66	(s) CH out-of-plane def.
C=C **stretching**			
non conjugated	1680–1625	5.95–6.15	(v) more substituted appear at higher frequency with lower intensity; frequency lowered by attached polar groups such as Br, O
Ar conjugated	~1625	~6.15	enhanced intensity
C=O or C=C conjugated	~1600	~6.25	enhanced intensity

Abbreviations used for the intensity of absorptions are: s, strong; m, medium; w, weak; v, variable.

Table 2.3   Aromatic compounds

Group/vibration	cm^{-1}	μm	Comments
$\equiv\overset{\shortmid}{C}$—H **stretching**	~3030	~3.03	(v) sharp
C – H **out-of-plane summation bands**	2000–1660	5.00–6.02	(w) weak bands, pattern depends on substitution pattern, see Fig. VII,1(xi)
C≡C **skeletal vibrations**	~1600	~6.25	(v) bands are characteristic of
	~1500	~6.67	(v) the aromatic ring,
	~1580	~6.33	(m) intensities variable;
	~1450	~6.90	(m) 1580 cm^{-1} band only just perceptible—intensity increased when polar group conjugated with ring; 1450 cm^{-1} band masked if alkyl groups present
$\equiv\overset{\shortmid}{C}$ – H **in-plane deformation**	1225–950	8.16–10.53	series of weak bands, positions characteristic of substitution pattern, rarely used diagnostically
– C – H **out-of-plane deformation**			
five adjacent hydrogen atoms	770–730 710–690	12.99–13.7 14.09–14.5	(s) (s)
four adjacent hydrogen atoms	770–735	12.99–13.6	(s) very useful, also for polycylics; position of
three adjacent hydrogen atoms	810–750	12.35–13.34	(s) bands disturbed however when polar groups, e.g.,
two adjacent hydrogen atoms	860–800	11.63–12.5	(s) – NO_2 attached to ring
one hydrogen atom	900–860	11.11–11.63	(w)

Table 2.4 Alkynes and allenes

Group/vibration	cm^{-1}	μm	Comments
ALKYNES			
$\equiv$C$-$H **stretching**			
	3320–3310	3·01–3.02	(s) sharp; NH and OH broad in this region
C$\equiv$C **stretching**			
monosubstituted	2140–2100	4.67–4.76	(m) 2130–2120 cm^{-1} for alkyl substituted
disubstituted	2260–2190	4.43–4.57	(w) 2240–2230 cm^{-1} for alkyl substituted; may be v. weak or absent
$\equiv$C$-$H **deformation**			
	680–610	14.70–16.39	(m) near 630 cm^{-1} for alkyl substituted; broad overtone near 1250 cm^{-1}
ALLENES			
C$=$C$=$C **stretching**			
monosubstituted	1980–1945	5.05–5.14	(m) asym. str.
disubstituted	1955–1930	5.12–5.18	(w) asym. str.
$=$CH$_2$ **deformation**			
	875–840	11.43–11.91	(s) overtone near 1700 cm^{-1}

Table 2,5 Alcohols and phenols

Group/vibration	cm^{-1}	μm	Comments
O$-$H **stretching**			
Free OH	3650–3590	2.74–2.79	(v) sharp; only in dilute solution; frequency decreases for primary > sec > tert > phenol
Bonded OH			
intermolecular dimeric	3550–3450	2.82–2.90	(v) sharp ⎤ intensity changes
intermolecular polymeric	3400–3200	2.94–3.13	(s) broad ⎱ and frequency increases on dilution
intramolecular	3570–3450	2.80–2.90	(v) sharp ⎤ not affected
chelate compounds	3200–2500	3.13–4.00	(w) v. broad ⎰ on dilution
C$-$O **stretching**			
and O$-$H **deformation** (in-plane)			
primary alcohols	$\sim$1050	$\sim$9.52	(s) both types of absorptions
	1350–1260	7.41–7.93	(v) are sensitive to change
secondary alcohols	$\sim$1100	$\sim$9.09	(s) of state; these values
	1350–1260	7.41–7.94	(v) are for H bonded state
tertiary alcohols	$\sim$1150	$\sim$8.69	(s)
	1410–1310	7.09–7.64	(v)
phenols	$\sim$1200	$\sim$8.33	(s)
	1410–1310	7.09–7.64	(v)

Table 2,6   Ethers and cyclic ethers

Group/vibration	cm^{-1}	μm	Comments
C – O stretching			
dialkyl	1150–1060	8.7–9.43	(s) asym.
aralkyl	1270–1230	7.87–8.13	(s) aryl – O
	1075–1020	9.3–9.8	(s) alkyl – O, also strong band at 1176 cm^{-1}
diaryl	1250–1150	8.0–8.7	(s)
vinyl	1225–1200	8.16–8.33	(s)
cyclic:			
6-membered ring	1100	9.09	(s) for tetrahydropyran
5-membered ring	1100–1075	9.09–9.3	(s)
4-membered ring	980–970	10.2–10.31	(s)
3-membered ring	~1250	~8.0	(s) ring breathing
	~890	~11.23	*trans* ⎫ tentative
	~830	~12.05	(m) *cis* ⎭

Table 2,7   Amines

Group/vibration	cm^{-1}	μm	Comments
N – H stretching			
aliphatic, primary	3398–3381	2.94–2.96	(w) ⎫
	3344–3324	2.99–3.01	(w) values for
aromatic, primary	3509–3460	2.85–2.89	(m) dilute solution; in
	3416–3382	2.93–2.96	(m) associated state all the
dialkyl >NH	3360–3310	3.07–3.02	(w) bands intensify and move to lower
aralkyl >NH	~3450	~2.9	(m) frequencies
imines	3350–3320	2.99–3.01	(m) ⎭
N – H deformation			
primary amines	1650–1590	6.06–6.29	(m–s)
secondary amines	1650–1550	6.06–6.45	(m) for aryl, weak or absent in alkyl
C – N stretching			
aromatic amines			
primary	1340–1250	7.46–8.00	(s)
secondary	1350–1280	7.41–7.81	(s)
tertiary	1360–1310	7.35–7.64	(s)
aliphatic amines	1220–1020	8.20–9.8	(m–w)

Table 2,8   Compounds containing the carbonyl group

Group/vibration	cm^{-1}	μm	Comments*
1. KETONES			
C=O **stretching**			
saturated, acyclic	1725–1705	5.80–5.87	(s)
α,β-unsaturated, acyclic	1690–1675	5.92–5.97	(s)
α,β-α',β'-unsaturated, acyclic	1670–1660	5.99–6.02	(s)
aryl	1715–1695	5.92–5.95	(s) ⎱ modified by nature and
diaryl	1670–1660	5.99–6.02	(s) ⎰ position of substituents
6-membered ring and higher	1725–1705	5.80–5.87	(s)
5-membered ring	1750–1740	5.71–5.75	(s)
4-membered ring	~1775	~5.63	(s)
α-halogeno	1745–1725	5.73–5.08	(s) two bands
α-diketones	1730–1710	5.78–5.85	(s)
β-diketones	1640–1540	6.10–6.49	(s) enolic, H-bonded
o-hydroxy and o-amino aryl ketones	1655–1635	6.04–6.12	(s) modified by nature and position of substituents
1,4-quinones	1690–1660	5.92–6.02	(s)
2. ALDEHYDES			
C=O **stretching**			
saturated, acyclic	1740–1730	5.75–5.78	(s)
α,β-unsaturated, acyclic	1705–1680	5.87–5.95	(s)
α,β-$\gamma\delta$-unsaturated, acyclic	1680–1660	5.95–6.02	(s)
aryl	1715–1695	5.83–5.90	(s) modified by nature and position of substituents
C–H **stretching**			
	2900–2700	3.45–3.70	(w) 2 bands near 2820 and 2720 cm^{-1}
C–H **deformation**			
	975–780	10.26 12.82	(m)
3. ACIDS			
O–H **stretching**			
free	3560–3500	2.81–2.86	(m) very dilute solution
bonded	3300–2500	3.03–4.00	(w) very broad
C=O **stretching**			
saturated, acyclic	1725–1700	5.80–5.89	(s)
α,β-unsaturated	1705–1690	5.86–5.92	(s)
aryl conjugated	1700–1680	5.89–5.95	(s)
C–O **stretching**			
and O–H **deformation**	1440–1395	6.94–7.17	(w) coupled vibrations
	1320–1211	7.57–8.26	(s)
O–H **deformation** (out-of-plane)			
	950–900	10.53–11.11	(v) acid dimer
$-C{\overset{O}{\underset{O}{\diagup}}}^{\ominus}$ **stretching**	1610–1550	6.21–6.45	(s) salts, asym.
	1420–1300	7.04–7.69	(s) and sym. vibrations

* All ketone and aldehyde carbonyl band positions refer to dilute solutions except where indicated.

Table 2,8   Compounds containing the carbonyl group

Group/vibration	cm^{-1}	μm	Comments*
4. ESTERS AND LACTONES			
C=O stretching			
saturated, acyclic	1750–1735	5.71–5.76	(s)
α,β-unsaturated and aryl	1730–1717	5.78–5.82	(s)
α-keto	1755–1740	5.70–5.74	(s)
β-keto	~1650	~6.06	(s) enolic
o-amino and o-hydroxyaryl	1690–1670	5.91–5.98	(s) H-bonding
δ-lactones	1750–1735	5.71–5.76	(s)
γ-lactones	1780–1760	5.62–5.68	(s)
β-lactones	~1820	~5.49	(s)
vinyl esters			
($-$CO·O·CH=CH$-$)	1775–1755	5.63–5.70	(s)
C$-$O stretching			
formates	1200–1180	8.33–8.48	(s)
acetates	1250–1230	8.00–8.13	(s)
propionates and higher	1200–1150	8.33–8.70	(s)
α,β-unsaturated	1300–1200	7.69–8.33	(s)
	1180–1130	8.47–8.85	(s)
aryl conjugated	1310–1250	7.63–8.00	(s)
	1150–1100	8.69–9.09	(s)
phenolic	~1205	~8.30	(s)
5. ACID ANHYDRIDES			
C=O stretching			
acyclic, saturated	1850–1800	5.40–5.56	(s) ⎫ frequency lowered
	1790–1740	5.58–5.75	(s) ⎪ by *ca.* 20 cm^{-1} when
cyclic, 5-membered ring	1870–1820	5.35–5.49	(s) ⎬ conjugated
	1800–1750	5.56–5.71	(s) ⎭
C$-$O stretching			
acyclic	1170–1050	8.55–9.52	(s)
cyclic	1300–1200	7.69–8.33	(s)
6. ACYL HALIDES			
C=O stretching			
	1815–1770	5.51–5.65	(s) conjugated compounds absorb at lower end of range
7. AMIDES			
N$-$H stretching			
primary	~3520	~2.84	(m) ⎫ free NH
	~3410	~2.93	(m) ⎭
primary	~3350	~2.98	(m) ⎫ bonded NH
	~3180	~3.14	(m) ⎭
secondary	3480–3440	2.87–2.91	(m) free NH, *trans-*
	3435–3395	2.91–2.95	(m) free NH, *cis-*
	3320–3270	3.11–3.06	(m) bonded NH, *trans-*
	3180–3140	3.15–3.18	(m) bonded NH, *cis-*
	3100–3070	3.23–3.26	(w) bonded NH, *cis-* and *trans-*
C=O stretching (Amide I)			
primary	~1650	~6.06	(s) solid phase
	~1690	~5.92	(s) dilute solution

Group/vibration	cm⁻¹	μm	Comments *
secondary	1680–1630	5.95–6.14	(s) solid phase
	1700–1680	5.88–5.95	(s) dilute solution
tertiary	1670–1630	5.98–6.14	(s) solid phase and dilute solution
cyclic amides			
(a) δ-lactams	~1680	~5.95	(s) dilute solution
(b) γ-lactams	~1700	~5.88	(s) dilute solution, shifted to higher frequency
(c) β-lactams	1760–1730	5.68–5.78	(s) when fused to another ring
imides (CO – NH – CO)			
acyclic	1740–1720	5.74–5.81	(s) bands not always
	1720–1700	5.81–5.88	(s) resolved
cyclic	1790–1735	5.58–5.76	(s) lower frequency band
	1745–1680	5.73–5.95	(s) more intense
ureas (NH – CO – NH)			
acyclic, monoalkyl	~1605	~6.23	(s)
acyclic, dialkyl	~1640	~6.10	(s)
N – H deformation (Amide II)			
primary	1650–1620	6.06–6.17	(s) solid phase
	1620–1590	6.17–6.29	(s) solution
secondary (non cyclic)	1570–1515	6.37–6.60	(s) solid phase
	1550–1510	6.45–6.62	(s) solution

Table 2,9   Amino acids

Group/vibration	cm^{-1}	μm	Comments
$\overset{\oplus}{N}H_3$ **stretching**			
	3130–3030	3.19–3.30	(m) asym.
	3030–2500	3.30–4.00	(m) forms continuous series of overlapping bands with above, also combination and overtone bands
$\overset{\oplus}{N}H_3$ **deformation**			
	1660–1610	6.02–6.21	(w) often appears as shoulder on $CO_2^{\ominus}$ band
	1550–1485	6.45–6.73	(m) sym.
$CO_2^{\ominus}$ **stretching**			
	1600–1560	6.25–6.41	(s) asym.
	~1410	~7.09	sym.
C=O stretching			
α-amino acids	1754–1720	5.70–5.81	(s) ⎫ in hydrochlorides—
α-amido acids; β,γ			⎬ normal carboxyl
and lower amino acids	1730–1695	5.78–5.90	(s) ⎭ C=O
others	~2130	~4.69	(w) found in all α-amino acids, displaced in others; a combination band

Table 2,10   Nitro compounds, nitroso compounds and nitrites

Group/vibration	cm^{-1}	μm	Comments
1. NITRO COMPOUNDS			
NO$_2$ **stretching**			
aliphatic, $C-NO_2$	1560–1534	6.41–6.52	(s) asym.
	1388–1344	7.20–7.44	(s) sym.
aromatic, $C-NO_2$	1555–1487	6.43–6.72	(s) asym.
	1357–1318	7.37–7.59	(s) sym.
C–N stretching			
	857–830	11.43–12.05	(m–s) alkyl and aryl
2. NITRITES			
N=O stretching			
	1681–1653	5.95–6.05	(s) *trans* form
	1625–1613	6.15–6.20	(s) *cis* form
3. NITROSO COMPOUNDS			
N=O stretching			
C-nitroso	1600–1500	6.25–6.66	(s) in monomeric state
trans	1290–1190	7.75–8.40	dimer
cis	1425–1370	7.02–7.30	dimer
N-nitroso	1460–1430	6.85–6.99	(s) in solution

Table 2,11 Unsaturated nitrogen compounds

Group/vibration	cm^{-1}	μm	Comments
C≡N stretching			
nitriles			
alkyl	2260–2240	4.42–4.46	(s) ⎱
aryl	2240–2220	4.46–4.50	(s) ⎰ large intensity
α,β-unsaturated alkyl	2235–2215	4.47–4.51	(s) ⎰ variations
isonitriles	2180–2120	4.59–4.72	(s)
isocyanates	2275–2240	4.40–4.46	(s)
C=N stretching			
imines			
alkyl	1690–1590	5.92–6.29	(v)
α,β-unsaturated alkyl	1660–1590	6.02–6.29	(v)
oximes	1690–1620	5.92–6.17	(v) also broad O–H stretch at 3300–3150 cm^{-1}
N=N stretching			
azo compounds	1630–1575	6.13–6.35	(v) frequency lowered by conjugation

Table 2,12 Organo-sulphur compound

Group/vibration	cm^{-1}	μm	Comments
S–H stretching			
	2590–2550	3.86–3.92	(w) smell!
C=S stretching			
thioketones, dithioesters	1270–1190	7.88–8.4	unlike C=O, is not strong
S=O stretching			
sulphoxides	1070–1035	9.35–9.66	(s)
sulphones	1350–1300	7.41–7.69	(s) ⎱ little affected by
	1160–1120	8.62–8.93	(s) ⎰ conjugation
sulphonamides	1358–1336	7.37–7.49	(s) ⎱ primary and secondary
	1169–1152	8.56–8.68	(s) ⎰ also show N–H str.
sulphonyl chlorides	1410–1360	7.09–7.36	(s)
	1195–1168	8.37–8.56	(s)
sulphonic acids	1350–1340	7.41–7.46	(s) ⎱ also broad H-bonded
	1165–1150	8.59–8.70	(s) ⎰ O–H str.
sulphonates	1380–1347	7.25–7.43	(s)
	1193–1170	8.38–8.55	(s)
sulphates (organic)	1415–1380	6.92–7.25	(s)
	1200–1185	8.33–8.44	(s)

Table 2,13   Halogen compounds

Group/vibration	cm^{-1}	μm	Comments
C – F stretching			
monofluoroalkanes	1100–1000	9.09–10.0	(s)
polyfluoroalkanes	1400–1000	7.15–10.0	(s) series of bands
C – Cl stretching			
monochloroalkanes	760–540	13.15–18.52	(s) 2 or more bands in solution
equatorial	780–740	12.82–17.24	(s) ⎱ cyclohexanes and
axial	730–580	13.70–17.24	(s) ⎰ steroids
C – Br stretching			
monobromoalkanes	600–500	16.66–20.0	(s) 2 or more bands in solution
equatorial	750–690	13.33–14.5	(s) ⎱ cyclohexanes and
axial	690–550	14.5–18.18	(s) ⎰ steroids
C – I stretching			
	600–465	16.67–21.5	(s) limited value

NUCLEAR MAGNETIC RESONANCE
APPENDIX 3 **CORRELATION TABLES**

Table 3,1   Chemical shifts of CH_3, CH_2, CH groups (δ values, TMS) *

X	CH_3-X	$R \cdot CH_2 - X$	$R^1 R^2 CH - X$
(a) Carbon substituents			
alkyl	0.90	1.25	1.50
$-C=C-$ $-C=C-C=C-$ }	1.70	1.95	2.6
$C=C-C=C$ |	1.95	2.2	
$-C\equiv C-R$	1.8		
$-C\equiv C-Ph$	2.9		
$-C=N \cdot R$	2.0		
$-C\equiv N$	2.0	2.48	
$-CO_2R$	2.0	2.10	
$-CO_2H$	2.07	2.34	2.57
$-CO \cdot NR_2$	2.02	2.05	
$-CO \cdot R$	2.10	2.40	2.48
$-CHO$	2.17	2.2	2.4
$-C_6H_5$	2.34	2.6	2.87
$-CO \cdot C_6H_5$	2.62	—	3.58
(b) Nitrogen substituents			
$-NH_2, -NR_2$	2.15	2.50	2.87
$-NH \cdot CO \cdot R$	2.9	3.3	3.5
$-NR_3^{\oplus}$	3.33	3.40	3.5
$-NO_2$	4.33	4.40	4.60
$-N\equiv C$	2.9	3.3	4.9
(c) Oxygen substituents			
$-OR$	3.30	3.36	3.80
$-OH$	3.38	3.56	3.85
$-O \cdot SO_2R$	3.58	—	—
$-O \cdot CO \cdot R$	3.65	4.15	5.01
$-OC_6H_5$	3.73	3.90	4.0
$-O \cdot CO \cdot C_6H_5$	3.90	4.23	5.12
(d) Halogen substituents			
I	2.16	3.15	4.2
Br	2.65	3.34	4.1
Cl	3.02	3.44	4.02
F	4.26	4.35	—
(e) Sulphur substituents			
$-SH$	3.2	3.4	
$-SR$	2.10	2.40	3.1
$-SO_2R$	2.6	3.1	

(f) Alicyclic rings		
Ring size	$-CH_2-$	$-CH-$
3	0.2	0.4
4	2.0	—
5	1.5	—
6	1.4	1.70
7	1.2	—

* It should be pointed out that although these δ values are typical, as with other spectroscopic data, some variation is possible in individual cases; a range of say ± 0.05 Hz is feasible.

Table 3.2* Shielding constants for aliphatic methylene groups, $X \cdot CH_2 \cdot Y$

Substituent	Shielding constant	Substituent	Shielding constant	Substituent	Shielding constant	Substituent	Shielding constant
$-CH_3$	0.47	$-CO \cdot R$	1.70	$-I$	1.82	$-OC_6H_5$	3.23
$-C=C-$	1.32	$-CO \cdot C_6H_5$	1.84	$-NR_2$	1.57	$-O \cdot CO \cdot R$	3.13
$-C\equiv C-$	1.44	$-CO_2R$	1.55	$-NHCO \cdot R$	2.27	$-O \cdot SO_2R$	3.13
$-C_6H_5$	1.85	$-CONH_2$	1.59	$-N_3$	1.97	$-SR$	1.64
$-CF_3$	1.14	$-Br$	2.33	$-OH$	2.56		
$-C\equiv N$	1.70	$-Cl$	2.53	$-OR$	2.36		

To calculate the δ value (TMS) for disubstituted methylene groups, add the sum of the Shielding Constants to 0.23 (which is the δ value for methane).

* Data reproduced from R. M. Silverstein, C. G. Bassler and T. C. Merrill (1974). *Spectroscopic Identification of Organic Compounds*. 3rd edn. Wiley International Edition, p. 220.

Table 3.3 Chemical shifts of protons attached to unsaturated systems (δ, TMS)

Group	δ	Group	δ
$HC\equiv C-R$	1.80*	$\begin{array}{c} H \\ \diagdown \\ C=C \\ H_5C_6 \diagup \quad \diagdown C=O \end{array}$	6.6
$H-C\equiv C-C\equiv C-R$	2.80*		
$H-C\equiv C \cdot C_6H_5$	2.13*	$\begin{array}{c} H \\ \diagdown \\ C=C=O \\ H_5C_6 \diagup \end{array}$	7.8
$H_2C=CR_2$	4.65		
$\begin{array}{c} H \diagdown \\ \quad C=CR_2 \\ R \diagup \end{array}$	5.3	$\begin{array}{c} \diagdown \quad H \\ N-C \\ \diagup \quad \parallel \\ \qquad O \end{array}$	7.85
$\begin{array}{c} H^a \qquad C_6H_5 \\ \diagdown \quad \diagup \\ C=C \\ \diagup \quad \diagdown \\ H_b \qquad H_c \end{array}$	a 5.55 b 5.15 c 6.7	$\begin{array}{c} H \\ RO \cdot C \diagdown \\ \quad \parallel \\ \quad O \end{array}$	8.03
$\overbrace{C=C}$	5.6	$\begin{array}{c} H \\ R \cdot C \diagdown \\ \quad \parallel \\ \quad O \end{array}$	9.6
$\begin{array}{c} -C=C-C=C \\ \mid \\ H \end{array}$	6.2		
$\begin{array}{c} H \\ \diagdown \\ =C-C=O \\ \diagup \end{array}$	5.8	$\begin{array}{c} \diagdown \quad \diagup \\ C=C \\ \diagup \quad \diagdown \\ \qquad C=O \\ \qquad \diagup \\ \qquad H \end{array}$	9.8
$\begin{array}{c} H \diagdown \quad \diagup \\ C=C \\ \diagup \quad \diagdown \\ \qquad C=O \\ \qquad \mid \end{array}$	6.0	$\begin{array}{c} H \\ \diagup \\ Ar-C \\ \quad \parallel \\ \quad O \end{array}$	9.9
$\begin{array}{c} \diagdown \\ C=C \\ H \diagup \quad \diagdown C=O \\ \qquad \mid \end{array}$	6.2		
$\begin{array}{c} \diagdown \quad H \\ C=C \\ \diagup \quad \diagdown OR \end{array}$	6.8		

*Signals shifted to lower field by a trace of pyridine, and removed on deuteration.

Table 3,4   Chemical shifts (TMS) of protons attached to aromatic and heteroaromatic rings

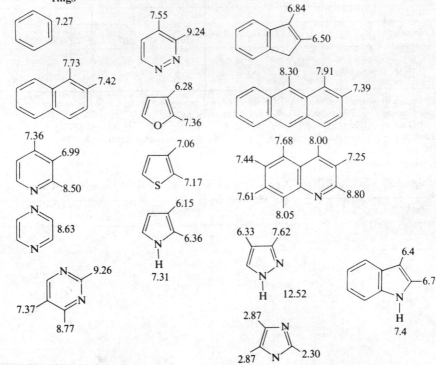

Table 3,5   Effect of substituents in benzenoid compounds

Substituent	*ortho*	*meta*	*para*
H	0	0	0
CH$_3$	−0.2	−0.1	−0.2
C$_6$H$_5$	0.2	0	−0.1
C=C	0.2	0.2	0.2
F	−0.3	0	−0.2
Cl	0	0	0
Br	0.2	−0.1	0
I	0.4	−0.3	0
OH	−0.5	−0.1	−0.4
OR	−0.4	−0.1	−0.4
SR	0.1	−0.1	−0.2
O·CO·R	0.2	−0.1	−0.2
NH$_2$	−0.8	−0.2	−0.6
NMe$_2$	−0.6	−0.1	−0.6
NH·CO·R	0.4	−0.2	−0.3
NO$_2$	1.0	0.2	0.3
C≡N	0.3	0.1	0.3
CO$_2$H/CO$_2$R	0.8	0.1	0.2
CHO	0.6	0.2	0.3
CONH$_2$	0.5	0.2	0.2
CO·R	0.6	0.3	0.3

The calculated δ-value for the substituted benzene is obtained by adding the appropriate substituent parameters to 7.27, the value of benzene.

Table 3,6 Chemical shift of protons attached to atoms other than carbon (all signals are sensitive to solvent, concentration and temperature and are removed by deuteration)

Compound type		δ (TMS)	Comments
Oxygen:	ROH	0.5–4.0	Often appears as broad singlet; trace of acid gives sharp singlet
	enols	11–16	
	ArOH	4.5–9	Position depends on extent of hydrogen bonding
	R·CO_2H	10–13	
	R·SO_3H	10–12	
	H_2O	4–5	
	R·C=NOH	9–12	
Nitrogen:	RNH_2, RNHR and cyclic amines	0.5–3	Usually broad absorption
	$ArNH_2$; ArNHR	3.5–6	
	R·$CONH_2$	5–85	Usually very broad; may not be observable
Sulphur:	RSH	1–1.5	
	ArSH	2.5–4	

Table 3,7   Characteristic proton–proton coupling constants (Hz)

Saturated compounds

Acyclic

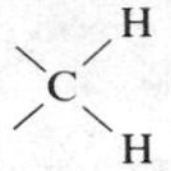

$\text{C}\big\langle\!\!\begin{smallmatrix}H\\H\end{smallmatrix}$ 10–18

$>\!\text{CH}-\text{CH}\!<$ 6–8

$>\!\text{CH}-\overset{|}{\underset{|}{\text{C}}}-\text{CH}\!<$ 0–1

$>\!\text{CH}-\text{CHO}$ 1–3

$>\!\text{CH}-\text{OH}$

$>\!\text{CH}-\text{NH}$ } 4–8

$>\!\text{CH}-\text{SH}$

Alicyclic	*gem*	*cis*	*trans*
cyclopropanes }			
epoxides }	4–6	4–9	3–6
aziridines }			
cyclobutanes	10–17	6–11	5–9
cyclopentanes	10–17	7–11	2–8
cyclohexanes	gem 10–17; diaxial 8–11; axial-equatorial 2–4; diequatorial 2–4		

Unsaturated

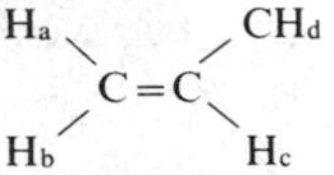

$J_{a,b}$ 0–3; $J_{a,c}$ 12–18; $J_{b,c}$ 6–12; $J_{c,d}$ 5–10; $J_{a,d}, J_{b,d}$ 0–2.

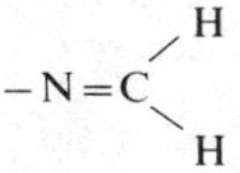

$-\text{N}=\text{C}\big\langle\!\!\begin{smallmatrix}H\\H\end{smallmatrix}$ 8–16

$>\!\text{C}=\text{C}-\text{C}=\text{C}\!<$ (with H, H)   10–13

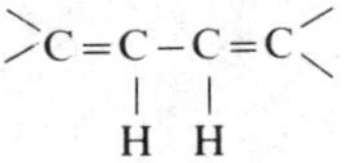

$>\!\text{C}=\text{CH}-\text{CHO}$ 5–8

$>\!\text{CH}-\text{C}=\text{C}-\text{CH}\!<$ 0–2

$-\text{HC}=\text{C}=\text{CH}-$ 6–7

$>\!\text{CH}-\text{C}\equiv\text{CH}$ 2–3

$>\!\text{CH}-\text{C}\equiv\text{C}-\text{CH}\!<$ 2–3

ring size	J
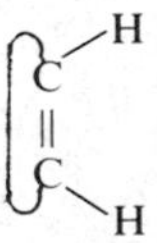 3	0.5–2
4	2.5–4
5	5–7
6	9–11
7	9–13
8	10–13

Table 3,7   Characteristic proton–proton coupling constants (Hz) (*continued*)

Aromatic and heteroaromatic compounds

Benzene derivatives	ortho 5–9		meta 2–3		para 0–1

	$J_{2,3}$	$J_{3,4}$	$J_{2,4}$	$J_{2,5}$
X = O	1.8	3.5	0.8	1.6
X = S	5.2	3.6	1.3	2.7
X = NH	2.7	3.3	1.4	1.9

$J_{2,3}$ 5.5; $J_{3,4}$ 7.5; $J_{2,4}$ 1.9; $J_{2,5}$ 0.9; $J_{3,5}$ 0.9; $J_{2,6}$ 0.4

$J_{2,5}$ 1.5; $J_{2,4}$ –; $J_{4,5}$ 5.0; $J_{4,6}$ 2.5

$J_{2,5}$ 1 2; $J_{2,4}$ 2; $J_{4,5}$ 3–4

$J_{3,4}$ 4.9; $J_{3,5}$ 2.0; $J_{3,6}$ 3.0; $J_{4,5}$ 8.4

$J_{2,3}$ 1.8; $J_{2,6}$ 0.5; $J_{2,5}$ 1.8

$J_{3,4}$ 1.9

Table 3,8   Characteristic coupling constants of protons with other nuclei (Hz)

Proton–fluorine coupling constants

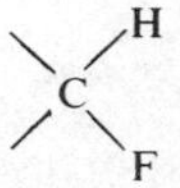

44–81

$-CH-CF-$ 3–25

$-CH=CF-$ cis 2–20;   trans 20–50

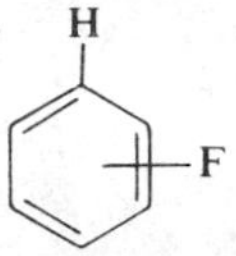

ortho 6–10;   meta 5–8;   para ~2

Proton–phosphorus coupling constants

P – H coupling	range, Hz	examples	
P^{III}	180–200	$(CH_3)_2P-H$	192
P^{IV}	450–550	$(CH_3)_3P\overset{\oplus}{-}H$	505
P^V	450–1050	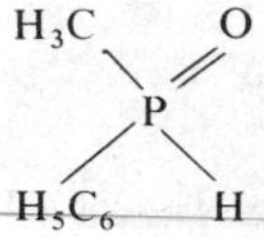	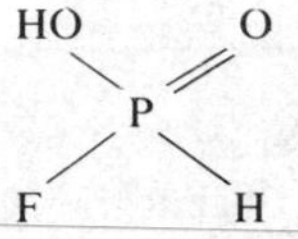 468; 1030

P – C – H coupling			
P^{III}	1–15	$(CH_3)_3P$	2.7
P^{IV}	12–18	$(C_2H_5)_4P^{\oplus}$	12.6
P^V	5–20	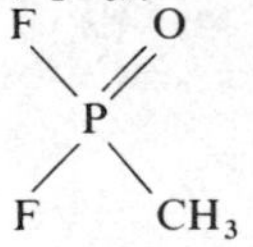	19

P – C – C – H coupling			
P^{III}	10–16	$(C_2H_5)_3P$	13.7
P^{IV}	15–20	$(C_2H_5)_3\overset{\oplus}{P}H$	20.0
P^V	14–25	$(C_2H_5)_3P=O$	18

Table 3,9 Fluorine–fluorine coupling constants

System	range, Hz
$\underset{\diagup}{\overset{\diagdown}{C}}\underset{F}{\overset{F}{\diagdown}}$	155–225
$-\overset{\mid}{C}F-\overset{\mid}{C}F-$	16–18
$=C\underset{F}{\overset{F}{\diagdown}}$	28–87
$\underset{\mid}{C}F=\underset{\mid}{C}F$	cis 20–58 trans 95–120
(fluorobenzene ring)	ortho ~20 meta 2–4 para 11–15

Table 3,10 Chemical shifts of residual protons in deuterated solvents

Compound	Formula	Residual absorption (δ from TMS)
Acetic acid—d_4	$CD_3 \cdot CO_2D$	2.06, 12.0
Acetone—d_6	$CD_3 \cdot CO \cdot CD_3$	2.07
Benzene—d_6	C_6D_6	7.24
Chloroform—d_1	$CDCl_3$	7.25
Cyclohexane d_{12}	C_6D_{12}	1.42
Deuterium oxide	D_2O	8.5*
Dimethyl sulphoxide—d_6	$CD_3 \cdot SO \cdot CD_3$	2.50
Methanol—d_4	CD_3OD	3.34, 4.1*
Pyridine—d_5	C_5D_5N	7.0–7.8, 8.57
Trifluoroacetic acid—d_1	$CF_3 \cdot CO_2D$	11.34

* Positions of these absorptions vary according to temperature and solvent.

Table 3,11   Spin–spin systems

Spin systems	Description	Example
AX	$J_{AX} \ll \delta_A - \delta_X$	$C_6H_5 \cdot CHCl \cdot CHCl_2$
AB	$J_{AB} \cong \delta_A - \delta_B$	
AMX	$J_{AM} \ll \delta_A - \delta_M$ $J_{XM} \ll \delta_M - \delta_X$	 (Fig. VII,2(x))
ABX	$J_{AB} \cong \delta_A - \delta_B$ $J_{AX} \ll \delta_A - \delta_X$ $J_{BX} \ll \delta_B - \delta_X$	 (Fig. VII,2(xi))
ABC	$(\delta_A - \delta_B), (\delta_B - \delta_C), (\delta_A - \delta_C)$ all of the same order as J_{AB}, J_{BC}, J_{AC}	
AA'BB'	Nuclei A,A' have same chemical shift but couple differently with B,B' and vice versa	$CH_2Cl \cdot CH_2Br$

Table 3,12   Effect on ^{13}C chemical shifts caused by replacing a methyl group by a polar substituent (δ values, TMS)

Substituent	C-1	C-2	C-3
OR	+45	−3	−1
OH	+40	+1	−1
O·CO·R	+43	−2	−1
NH$_2$	+20	+2	−1
Cl	+23	+2	−1
F	+61	−1	−2
CO·X	+15	−5	0
CO$_2$R	+10	−1	−1
CO$_2$H	+12	−3	−1
CN	−2	−1	−1

R = alkyl; X = Cl or NR$_2$

Data reproduced from G. Levy and G. L. Nelson (1972). *Carbon-13 Nuclear Magnetic Resonance for Organic Chemists.* New York; Wiley-Interscience, p. 47.

Table 3,13 ^{13}C Chemical shifts (TMS) in some monosubstituted alkenes ($\overset{2}{C}H = \overset{1}{C}H - X$)

X	C-1	C-2
H	122.8	122.8
CH$_3$	133.1	115.0
CH$_2$Br	133.2	117.7
C$_2$H$_5$	140.2	113.3
C$_6$H$_5$	136.7	113.2
CO$_2$R	129.7	130.4
CO$_2$H	128.0	131.9
CHO	136.4	136.1, 136.0
CO·CH$_3$	137.5	128.6
I	85.3	130.4
Br	115.5	122.0
Cl	126.0	117.3
N·COR	130.0	94.3
O·CO·CH$_3$	141.6	96.3
OCH$_3$	153.2	84·1

Table 3,14 ^{13}C Substituent effects for substituted benzenes

Substituent	Position			
	C-1	ortho	meta	para
Br	−5.5	+3.4	+1.7	−1.6
CF$_3$	−9.0	−2.2	+0.3	+3.2
CH$_3$	+8.9	+0.7	−0.1	−2.9
CN	−15.4	+3.6	+0.6	+3.9
CO·CF$_3$	−5.6	+1.8	+0.7	+6.7
CO·CH$_3$	+9.1	+0.1	0.0	+4.2
CO·Cl	+4.6	+2.4	0.0	+6.2
CHO	+8.6	+1.3	+0.6	+5.5
CO$_2$H	+2.1	+1.5	0.0	+5.1
Cl	+6.2	+0.4	+1.3	−1.9
F	+34.8	−12.9	+1.4	−4.5
H	0.0	—	—	—
NH$_2$	+18.0	−13.3	+0.9	−9.8
NO$_2$	+20.0	−4.8	+0.9	+5.8
OCH$_3$	+31.4	−14.4	+1.0	−7.7
OH	+26.9	−12.7	+1.4	−7.3
C$_6$H$_5$	+13.1	−1.1	+0.4	−1.2

Data reproduced from G. C. Levy and G. L. Nelson (1972). *Carbon-13 Nuclear Magnetic Resonance for Organic Chemists.* New York; Wiley-Interscience, p. 81.

Table 3,15 ^{13}C **Chemical shift (TMS) for some heteroaromatic compounds**

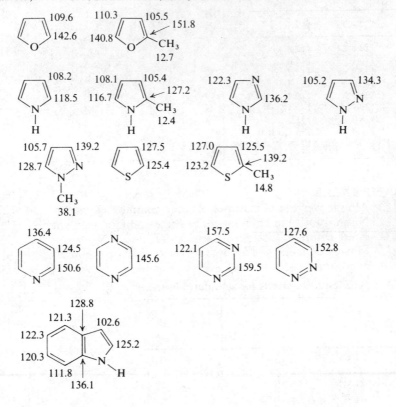

APPENDIX 4 MASS SPECTROMETRY

Table 4,1   Atomic weights of isotopes of some common elements

Isotope	Atomic mass		Isotope	Atomic mass
^{1}H	1.007 825		^{28}Si	27.976 927
^{2}H	2.014 102		^{29}Si	28.976 491
^{12}C	12.000 000		^{30}Si	29.973 761
^{13}C	13.003 354		^{31}P	30.993 763
^{14}N	14.003 074		^{32}S	31.972 094
^{15}N	15.000 108		^{33}S	32.971 461
^{16}O	15.994 915		^{34}S	33.967 865
^{17}O	16.999 133		^{36}S	35.967 09
^{18}O	17.999 160		^{35}Cl	34.968 855
^{19}F	18.998 405		^{37}Cl	36.965 896
			^{79}Br	78.918 348
			^{81}Br	80.916 344
			^{127}I	126.904 352

Table 4,2   Natural isotopic abundances of some common elements

Isotope	Natural abundance	
	Per cent of total isotopes present	Per cent relative to most abundant isotope
^{1}H	99.985	100
^{2}H	0.015	0.016
^{12}C	98.89	100
^{13}C	1.11	1.08
^{14}N	99.63	100
^{15}N	0.37	0.36
^{16}O	99.79	100
^{17}O	0.037	0.04
^{18}O	0.204	0.20
^{19}F	100	100
^{28}Si	92.21	100
^{29}Si	4.70	5.09
^{30}Si	3.09	3.35
^{31}P	100	100
^{32}S	95.0	100
^{33}S	0.76	0.80
^{34}S	4.22	4.44
^{35}Cl	75.53	100
^{37}Cl	24.47	32.40
^{79}Br	50.54	100
^{81}Br	49.46	97.85
^{127}I	100	100

Table 4.3 Some common losses from molecular ions *

Ion	Groups commonly associated with the mass lost	*Possible* inference
M − 1	H	—
M − 2	H$_2$	—
M − 14	—	Homologue?
M − 15	CH$_3$	—
M − 16	O	Ar − NO$_2$, $\geqslant \overset{\oplus}{N} - \overset{\ominus}{O}$, sulphoxide
M − 16	NH$_2$	ArSO$_2$NH$_2$, − CONH$_2$
M − 17	OH	—
M − 17	NH$_3$	—
M − 18	H$_2$O	Alcohol, aldehyde, ketone, etc.
M − 19	F	⎫
M − 20	HF	⎬ Fluorides
M − 26	C$_2$H$_2$	Aromatic hydrocarbon
M − 27	HCN	⎰ Aromatic nitriles ⎱ Nitrogen heterocycles
M − 28	CO	Quinones
M − 28	C$_2$H$_4$	⎰ Aromatic ethyl ethers ⎱ Ethyl esters, propyl ketones
M − 29	CHO	—
M − 29	C$_2$H$_5$	Ethyl ketones, Ar − C$_3$H$_7$
M − 30	C$_2$H$_6$	—
M − 30	CH$_2$O	Aromatic methyl ether
M − 30	NO	Ar − NO$_2$
M − 31	OCH$_3$	Methyl ester
M − 32	CH$_3$OH	Methyl ester
M − 32	S	—
M − 33	H$_2$O + CH$_3$	—
M − 33	HS	⎫
M − 34	H$_2$S	⎬ Thiols
M − 41	C$_3$H$_5$	Propyl ester
M − 42	CH$_2$CO	⎰ Methyl ketone ⎱ Aromatic acetate, Ar·NHCO·CH$_3$
M − 42	C$_3$H$_6$	⎰ Butyl ketone, isobutyl ketone, ⎱ Aromatic propyl ether, Ar − C$_4$H$_9$
M − 43	C$_3$H$_7$	Propyl ketone, Ar − C$_4$H$_9$
M 43	CH$_3$·CO	Methyl ketone
M − 44	CO$_2$	⎰ Ester (skel. rearr.) ⎱ Anhydride
M − 44	C$_3$H$_8$	—
M − 45	CO$_2$H	Carboxylic acid
M − 45	OC$_2$H$_5$	Ethyl ester
M − 46	C$_2$H$_5$OH	Ethyl ester
M − 46	NO$_2$	Ar − NO$_2$
M − 48	SO	Aromatic sulphoxide
M − 55	C$_4$H$_7$	Butyl ester
M − 56	C$_4$H$_8$	⎰ Ar − C$_5$H$_{11}$, ArO − C$_4$H$_9$ ⎱ Ar − ^{i}C$_5$H$_{11}$, ArO − ^{i}C$_4$H$_9$
M − 57	C$_4$H$_9$	Pentyl ketone
M − 57	C$_2$H$_5$·CO	Butyl ketone
M − 58	C$_4$H$_{10}$	Ethyl ketone
M − 60	CH$_3$·CO$_2$H	—
		Acetate

* Data reproduced from D. H. Williams and I. Howe (1972) *Principles of Organic Mass Spectrometry*. London; McGraw-Hill, p. 103.

Table 4,4   Masses of some possible compositions of common fragment ions*

m/e	Groups commonly associated with the mass	*Possible* inference
15	$CH_3^{\oplus}$	—
18	$H_2O^{\oplus}$	—
26	$C_2H_2 \cdot$	—
27	$C_2H_3^{\oplus}$	
28	$CO^{\oplus}$, $C_2H_4^{\oplus}$, $N_2^{\oplus}$	
29	$CHO^{\oplus}$, $C_2H_5^{\oplus}$	—
30	$CH_2 = \overset{\oplus}{N}H_2$	Primary amine?
31	$CH_2 = \overset{\oplus}{O}H$	Primary alcohol?
36/38(3:1)	$HCl^{\oplus}$	—
39	$C_3H_3^{\oplus}$	—
40†	$Argon^{\oplus}$, $C_3H_4^{\oplus}$	—
41	$C_3H_5^{\oplus}$	—
42	$C_2H_2O^{\oplus}$, $C_3H_6^{\oplus}$	—
43	$CH_3 \cdot CO^{\oplus}$	$CH_3 \cdot COX$
43	$C_3H_7^{\oplus}$	C_3H_7X
44	$C_2H_6N^{\oplus}$	Some aliphatic amines
44	$O = C = \overset{\oplus}{N}H_2$	Primary amides
44	$CO_2^{\oplus}$, $C_3H_8^{\oplus}$	—
44	$CH_2 = CH(OH)^{\oplus}$	Some aldehydes
45	$CH_2 = \overset{\oplus}{O}CH_3$ $CH_3 \cdot CH = \overset{\oplus}{O}H$	} Some ethers and alcohols
47	$CH_2 = \overset{\oplus}{S}H$	Aliphatic thiol
49/51(3:1)	$CH_2Cl^{\oplus}$	—
50	$C_4H_2^{\oplus}$	Aromatic compound
51	$C_4H_3^{\oplus}$	C_6H_5X
55	$C_4H_7^{\oplus}$	—
56	$C_4H_8^{\oplus}$	—
57	$C_4H_9^{\oplus}$	C_4H_9X
57	$C_2H_5 \cdot CO^{\oplus}$	{ Ethyl ketone Propionate ester
58	$CH_2 = C(OH)CH_3^{\oplus}$	{ Some methyl ketones Some dialkyl ketones
58	$C_3H_8N^{\oplus}$	Some aliphatic amines
59	$CO_2CH_3^{\oplus}$	Methyl ester

* Data reproduced from D. H. Williams and I. Howe (1972). *Principles of Organic Mass Spectrometry*. London; McGraw-Hill, p. 105.

† Appears as a doublet in the presence of argon from air; useful as a reference point in counting the mass spectrum.

Table 4,4   Masses of some possible compositions of common fragment ions (*continued*)

m/e	Groups commonly associated with the mass	*Possible* inference
59	$CH_2=C(OH)NH_2^{+\cdot}$	Some primary amides
59	$C_2H_5CH=\overset{\oplus}{O}H$	$C_2H_5\cdot CH(OH)-X$
59	$CH_2=\overset{\oplus}{O}-C_2H_5$ and isomers	Some ethers
60	$CH_2=C(OH)OH^{+\cdot}$	Some carboxylic acids
61	$CH_3CO(OH_2)^{\oplus}$	$CH_3CO_2C_nH_{2n+1}$ ($n>1$)
61	$CH_2\cdot CH_2SH^{\oplus}$	Aliphatic thiol
66	$H_2S_2^{+\cdot}$	Dialkyl disulphide
69	$CF_3^{\oplus}$	—
68	$CH_2\cdot CH_2\cdot CH_2\cdot CN^{\oplus}$	—
69	$C_5H_9^{\oplus}$	
70	$C_5H_{10}^{+\cdot}$	
71	$C_5H_{11}^{\oplus}$	$C_5H_{11}X$
71	$C_3H_7\cdot CO^{\oplus}$	$\begin{cases} \text{Propyl ketone} \\ \text{Butyrate ester} \end{cases}$
72	$CH_2=C(OH)C_2H_5^{+\cdot}$	Some ethyl alkyl ketones
72	$C_3H_7\cdot CH=\overset{\oplus}{N}H_2$ and isomers	Some amines
73	$C_4H_9O^{\oplus}$	—
73	$CO_2C_2H_5^{\oplus}$	Ethyl ester
73	$(CH_3)_3Si^{\oplus}$	$(CH_3)_3SiX$
74	$CH_2=C(OH)OCH_3^{+\cdot}$	Some methyl esters
75	$(CH_3)_2Si=\overset{\oplus}{O}H$	$(CH_3)_3SiOX$
75	$C_2H_5\cdot CO(OH_2)^{\oplus}$	$C_2H_5\cdot CO_2C_nH_{2n+1}$ ($n>1$)
76	$C_6H_4^{+\cdot}$	$\begin{cases} C_6H_5X \\ XC_6H_4Y \end{cases}$
77	$C_6H_5^{\oplus}$	C_6H_5X
78	$C_6H_6^{+\cdot}$	C_6H_5X
79	$C_6H_7^{\oplus}$	C_6H_5X
79/81 (1:1)	$Br^{\oplus}$	—
80/82 (1:1)	$HBr^{+\cdot}$	
80	$C_5H_6N^{\oplus}$	
81	$C_5H_5O^{\oplus}$	
83/85/87 (9:6:1)	$HCCl_2^{\oplus}$	$CHCl_3$
85	$C_6H_{13}^{\oplus}$	$C_6H_{13}X$
85	$C_4H_9\cdot CO^{\oplus}$	$C_4H_9\cdot COX$

Table 4,4   Masses of some possible compositions of common fragment ions (*continued*)

m/e	Groups commonly associated with the mass	*Possible* inference
85		
85		
86	$CH_2=C(OH)C_3H_7\overset{\oplus}{\cdot}$	Some propyl alkyl ketones
86	$C_4H_9\cdot CH=\overset{\oplus}{N}H_2$ and isomers	Some amines
87	$CH_2=CH-\overset{\overset{\displaystyle\oplus}{\overset{\displaystyle OH}{\|}}}{C}-OCH_3$	$XCH_2\cdot CH_2\cdot CO_2CH_3$
91	$C_7H_7^{\oplus}$	$C_6H_5\cdot CH_2X$
92	$C_7H_8^{\oplus}$	$C_6H_5\cdot CH_2$-alkyl
92	$C_6H_6N^{\oplus}$	
91/93 (3:1)		Alkyl chloride ($\geqslant$ hexyl)
93/95 (1:1)	$CH_2Br^{\oplus}$	—
94	$C_6H_6O^{\oplus}$	C_6H_5O-alkyl (alkyl $\neq CH_3$)
94		
95		
95	$C_6H_7O^{\oplus}$	
97	$C_5H_5S^{\oplus}$	
99		
99		

Table 4,4 Masses of some possible compositions of common fragment ions (*continued*)

m/e	Groups commonly associated with the mass	*Possible* inference
105	$C_6H_5 \cdot CO^{\oplus}$	$C_6H_5 \cdot COX$
105	$C_8H_9^{\oplus}$	$CH_3 - C_6H_4 \cdot CH_2X$
106	$C_7H_8N^{\oplus}$	
107	$C_7H_7O^{\oplus}$	
107/109 (1:1)	$C_2H_4Br^{\oplus}$	—
111		
121	$C_8H_9O^{\oplus}$	
122	$C_6H_5 \cdot CO_2H$	} Alkyl benzoates
123	$C_6H_5 \cdot CO_2H_2^{\oplus}$	
127	$I^{\oplus}$	—
128	$HI \cdot$	—
135/137 (1:1)		Alkyl bromide ($\geqslant$ hexyl)
130	$C_9H_8N^{\oplus}$	
141	$CH_2I^{\oplus}$	—
147	$(CH_3)_2Si = \overset{\oplus}{O} - Si(CH_3)_3$	—
149		Dialkyl phthalate
160	$C_{10}H_{10}NO^{\oplus}$	
190	$C_{11}H_{12}NO_2^{\oplus}$	

CORRELATION TABLES
USEFUL SOLVENT CHARACTERISTICS

Solvent[a]	Boiling point (°C)	Melting point	Molecular weight
	(760 mmHg)	°C	
Ether (15)	35	−116	74
Pentane (1)	36	−130	72
Dichloromethane (5)	40	−95	85
Carbon disulphide (32)	46	−111	76
Acetone (21)	56	−95	58
Chloroform (6)	61	−64	119
Methanol (8)	65	−98	32
Tetrahydrofuran (19)	66	−109	72
Hexane (1)	69	−95	86
Trifluoroacetic acid	72	−15	114
Carbon tetrachloride (7)	77	−23	154
Ethyl acetate (24)	77	−84	88
Ethanol (9)	78	−114	46
Cyclohexane (1)	81	6.5	84
Benzene (2)	80	5.5	78
Methyl ethyl ketone (22)	80	−87	72
Acetonitrile (27)	82	−44	41
Propan-2-ol (11)	82	−88	60
t-Butanol (12)	82	26	74
Ethylene glycol dimethyl ether (18)	83	−58	90
Triethylamine	90	−115	101
Propan-1-ol (10)	97	−126	60
Water	100	0	18
Methylcyclohexane (1)	101	−127	98
Formic acid	101	8	46
Nitromethane	101	−29	61
1,4-Dioxan (20)	101	12	88
Toluene (3)	111	−95	92
Pyridine (29)	115	−42	79
Butan-1-ol (12)	118	−89	74
Acetic acid	118	17	60
Ethylene glycol monomethyl ether (13)	125	−85	76
Morpholine	129	−3	87
Chlorobenzene	132	−46	113
Acetic anhydride	140	−73	102
Xylenes (mixed) (4)	138–142	13[d]	106
Dibutyl ether (17)	142	−95	130
sym-Tetrachloroethane	146	−44	168
Anisole	154	−38	108

Density at 20 °C	Dielectric constant	Solubility in water (g/100 g)[b]	Azeotrope with water b.p. (°C)	% H₂O	Flash point (°C)	TLV[c] (p.p.m.)
0.71	4.3	6.0	34	1	−45	400
0.63	1.8	Insol.	35	1	−40	500
1.33	8.9	1.30	39	2	None	250
1.26	2.6	0.29 (20 °C)	44	2	−30	20
0.79	20.7	∞	None	—	−18	1000
1.49	4.8	0.82 (20 °C)	56	3	None	25
0.79	32.7	∞	None	—	12	200
0.89	7.6	∞	64	5	−14	200
0.66	1.9	Insol.	62	6	−26	500
1.49	39.5	∞	105	21	None	—
1.59	2.2	0.08	66	4	None	10
0.90	6.0	8.1	71	8	−4	400
0.79	24.6	∞	78	4	13	1000
0.78	2.0	0.01	70	8	−17	300
0.88	2.3	0.18	69	9	−11	25
0.80	18.5	24.0 (20 °C)	73	11	−1	200
0.78	37.5	∞	77	16	6	40
0.79	19.9	∞	80	12	12	—
0.78 (30 °C)	12.5	∞	80	12	11	100
0.86	7.2	∞	77	10	1	—
0.73	2.4	∞	75	10	−7	25
0.80	20.3	∞	88	28	25	200
1.00	80.2	—	—	—	None	—
0.77	2.0	0.01	80	24.1	−6	500
1.22	58.5	∞	107	26	—	5
1.14	35.9	11.1	84	24	−41	100
1.03	2.2	∞	88	18	12	50
0.87	2.4	0.05	85	20	4	100
0.98	12.4	∞	94	42	23	5
0.81	17.5	7.45	93	43	29	100
1.05	6.2	∞	None	—	40	10
0.96	16.9	∞	100	85	42	25
1.00	7.4	∞	None	—	38	20
1.11	5.6	0.05 (30 °C)	90	28	29	75
1.08	20.7	Reacts	—	—	53	5
0.86	2e	0.02	93	33	17	100
0.77	3.1	0.03 (20 °C)	93	33	38	—
1.59	8.2	0.29 (20 °C)	94	34	None	5
0.99	4.3	1.04	96	41	—	—

Solvent[a]	Boiling point (°C) (760 mmHg)	Melting point °C	Molecular weight
Dimethylformamide (26)	153	− 60	73
Diethylene glycol dimethyl ether (18)	160 (dec.)	—	134
Mesitylene	165	− 45	120
Dimethyl sulphoxide (33)	189	18	78
Diethylene glycol monomethyl ether (14)	194	− 76	120
Ethylene glycol	197	− 16 to − 13[f]	62
N-Methyl-2-pyrrolidone (28)	202	− 24	99
Nitrobenzene (31)	211	6	123
Formamide (25)	210 (dec.)	3	45
Hexamethylphosphoric triamide (35)	233	7	179
Quinoline (30)	237	− 15	129
Diethylene glycol	245	− 7	106
Diphenyl ether	258	27	170
Triethylene glycol	288	− 4	150
Sulfolane (34)	287 (dec.)	28	120
Glycerol	290	18	92
Triethanolamine	335	22	149
Dibutyl phthalate	340	− 35	278

Notes. (a) The numbers in parentheses enable the solvent to be located in Section II,1, where alternative names and methods of purification are to be found.

(b) Values for 25 °C unless otherwise indicated. Values <0.01 per cent described as insoluble.

(c) Threshold Limit Values, Section I,3,D,l.

(d) Value for p-xylene (isomer of highest m.p.).

(e) Approximate value.

(f) Value in doubt because of strong tendency to supercool and form a glass.

Density at 20 °C	Dielectric constant	Solubility in water	Azeotrope with water		Flash point	TLV[c]
		(g/100 g)[b]	b.p. (°C)	% H₂O	(°C)	(p.p.m.)
0.95	36.7	∞	None	—	67	10
0.94	—	∞	100	78	63	—
0.87	2.3	0.03 (20 °C)	97	—	—	—
1.10	46.7	25.3	None	—	95	—
1.02	—	∞	None	—	93	—
1.11	37.7	∞	None	—	116	100
1.03	32.0	∞	—	—	96	—
1.20	34.8	0.19 (20 °C)	99	88	88	1
1.13	111	∞	—	—	154	20
1.03	30	∞	—	—	—	—
1.09	9.0	0.6 (20 °C)	—	97	—	—
1.11	31.7	∞	None	—	143	—
1.07	3.7 (> 27 °C)	0.39	100	96	205	—
1.12	23.7	∞	None	—	166	—
1.26 (30 °C)	43	∞ (30 °C)	None	—	177	—
1.26	42.5	∞	None	—	177	—
1.12 (25 °C)	29.4	∞	—	—	179	—
1.05	6.4	Insol.	None	—	171	5 (mg/M³)

APPENDIX 6
DENSITIES AND PERCENTAGE COMPOSITIONS OF VARIOUS SOLUTIONS

Table 6,1   Aqueous ethanol

Per cent C$_2$H$_5$OH by weight	Density $d_4^{20°}$	Density $d_4^{25°}$	Per cent C$_2$H$_5$OH by volume (20 °C)
5	0.989 38	0.988 17	6.2
10	0.981 87	0.980 43	12.4
15	0.975 14	0.973 34	18.5
20	0.968 64	0.966 39	24.5
25	0.961 68	0.958 95	30.4
30	0.953 82	0.950 67	36.2
35	0.944 94	0.941 46	41.8
40	0.935 18	0.931 48	47.3
45	0.924 72	0.920 85	52.7
50	0.913 84	0.909 85	57.8
55	0.902 58	0.898 50	62.8
60	0.891 13	0,886 99	67.7
65	0.879 48	0.875 27	72.4
70	0.867 66	0.853 40	76.9
75	0.855 64	0.851 34	81.3
80	0.843 44	0.839 11	85.5
85	0.830 95	0.826 60	89.5
90	0.817 97	0.813 62	93.3
91	0.815 29	0.810 94	94.0
92	0.812 57	0.808 23	94.7
93	0.809 83	0.805 49	95.4
94	0.807 05	0.802 72	96.1
95	0.804 24	0.799 91	96.8
96	0.801 38	0.797 06	97.5
97	0.798 46	0.794 15	98.1
98	0.795 47	0.791 17	98.8
99	0.792 43	0.788 14	99.4
100	0.789 34	0.785 06	100.0

Table 6,2 Aqueous methanol

Per cent CH$_3$OH by weight	Density $d_4^{15°}$	Per cent CH$_3$OH by volume	Per cent CH$_3$OH by weight	Density $d_4^{15°}$	Per cent CH$_3$OH by volume
5	0.990 29	6.22	75	0.863 00	81.34
10	0.982 41	12.35	80	0.850 48	85.50
15	0.975 18	18.38	85	0.837 42	89.45
20	0.968 14	24.33	90	0.823 96	93.19
25	0.961 08	30.19	91	0.821 24	93.91
30	0.953 66	35.95	92	0.818 49	94.63
35	0.945 70	41.59	93	0.815 68	95.33
40	0.937 20	47.11	94	0.812 85	96.02
45	0.928 15	52.49	95	0.809 99	96.70
50	0.918 52	57.71	96	0.807 13	97.37
55	0.908 39	62.78	97	0.804 28	98.04
60	0.897 81	67.69	98	0.801 43	98.70
65	0.886 62	72.42	99	0.798 59	99.35
70	0.875 07	76.98	100	0.795 77	100.00

Table 6,3 Aqueous hydrochloric acid

Per cent HCl by weight	Density $d_4^{20°}$	Grams HCl per 100 ml	Per cent HCl by weight	Density $d_4^{20°}$	Grams HCl per 100 ml
1	1.0032	1.003	22	1.1083	24.38
2	1.0082	2.006	24	1.1187	26.85
4	1.0181	4.007	26	1.1290	29.35
6	1.0279	6.167	28	1.1392	31.90
8	1.0376	8.301	30	1.1492	34.48
10	1.0474	10.47	32	1.1593	37.10
12	1.0574	12.69	34	1.1691	39.75
14	1.0675	14.95	36	1.1789	42.44
16	1.0776	17.24	38	1.1885	45.16
18	1.0878	19.58	40	1.1980	47.92
20	1.0980	21.96			

Table 6,4 Aqueous sulphuric acid

Per cent H$_2$SO$_4$ by weight	Density $d_4^{20°}$	Grams H$_2$SO$_4$ per 100 ml	Per cent H$_2$SO$_4$ by weight	Density $d_4^{20°}$	Grams H$_2$SO$_4$ per 100 ml
1	1.0051	1.005	65	1.5533	101.0
2	1.0118	2.024	70	1.6105	112.7
3	1.0184	3.055	75	1.6692	125.2
4	1.0250	4.100	80	1.7272	138.2
5	1.0317	5.159	85	1.7786	151.2
10	1.0661	10.66	90	1.8144	163.3
15	1.1020	16.53	91	1.8195	165.6
20	1.1394	22.79	92	1.8240	167.8
25	1.1783	29.46	93	1.8279	170.0
30	1.2185	36.56	94	1.8312	172.1
35	1.2579	44.10	95	1.8337	174.2
40	1.3028	52.11	96	1.8355	176.2
45	1.3476	60.64	97	1.8364	178.1
50	1.3951	69.76	98	1.8361	179.9
55	1.4453	79.49	99	1.8342	181.6
60	1.4983	89.90	100	1.8305	183.1

Table 6,5   Aqueous nitric acid

Per cent HNO$_3$ by weight	Density $d_4^{20°}$	Grams HNO$_3$ per 100 ml	Per cent HNO$_3$ by weight	Density $d_4^{20°}$	Grams HNO$_3$ per 100 ml
1	1.0036	1.004	65	1.3913	90.43
2	1.0091	2.018	70	1.4134	98.94
3	1.0146	3.044	75	1.4337	107.5
4	1.0201	4.080	80	1.4521	116.2
5	1.0256	5.128	85	1.4686	124.8
10	1.0543	10.54	90	1.4826	133.4
15	1.0842	16.26	91	1.4850	135.1
20	1.1150	22.30	92	1.4873	136.8
25	1.1469	28.67	93	1.4892	138.5
30	1.1800	35.40	94	1.4912	140.2
35	1.2140	42.49	95	1.4932	141.9
40	1.2463	49.85	96	1.4952	143.5
45	1.2783	57.52	97	1.4974	145.2
50	1.3100	65.50	98	1.5008	147.1
55	1.3393	73.66	99	1.5056	149.1
60	1.3667	82.00	100	1.5129	151.3

Table 6,6   Aqueous acetic acid

Per cent CH$_3$CO$_2$H by weight	Density $d_4^{20°}$	Grams CH$_3$CO$_2$H per 100 ml	Per cent CH$_3$CO$_2$H by weight	Density $d_4^{20°}$	Grams CH$_3$CO$_2$H per 100 ml
1	0.9996	0.9996	65	1.0666	69.33
2	1.0012	2.002	70	1.0685	74.80
3	1.0025	3.008	75	1.0696	80.22
4	1.0040	4.016	80	1.0700	85.60
5	1.0055	5.028	85	1.0689	90.86
10	1.0125	10.13	90	1.0661	95.95
15	1.0195	15.29	91	1.0652	96.93
20	1.0263	20.53	92	1.0643	97.92
25	1.0326	25.82	93	1.0632	98.88
30	1.0384	31.15	94	1.0619	99.82
35	1.0438	36.53	95	1.0605	100.7
40	1.0488	41.95	96	1.0588	101.6
45	1.0534	47.40	97	1.0570	102.5
50	1.0575	52.88	98	1.0549	103.4
55	1.0611	58.36	99	1.0524	104.2
60	1.0642	63.85	100	1.0498	105.0

Table 6,7 Aqueous formic acid

Per cent HCO$_2$H by weight	Density $d_4^{20°}$	Grams HCO$_2$H per 100 ml	Per cent HCO$_2$H weight	Density $d_4^{20°}$	Grams HCO$_2$H per 100 ml
1	1.0019	1.002	65	1.1543	75.03
2	1.0044	2.009	70	1.1655	81.59
3	1.0070	3.021	75	1.1769	88.27
4	1.0093	4.037	80	1.1860	94.88
5	1.0115	5.058	85	1.1953	101.6
10	1.0246	10.25	90	1.2044	108.4
15	1.0370	15.66	91	1.2059	109.7
20	1.0488	20.98	92	1.2078	111.1
25	1.0609	26.52	93	1.2099	112.5
30	1.0729	32.19	94	1.2117	113.9
35	1.0847	37.96	95	1.2140	115.3
40	1.0963	43.85	96	1.2158	116.7
45	1.1085	49.88	97	1.2170	118.0
50	1.1207	56.04	98	1.2183	119.4
55	1.1320	62.26	99	1.2202	120.8
60	1.1424	68.54	100	1.2212	122.1

Table 6,8 Aqueous phosphoric acid

Per cent H$_3$PO$_4$ by weight	Density $d_4^{20°}$	Grams H$_3$PO$_4$ per 100 ml	Per cent H$_3$PO$_4$ by weight	Density $d_4^{20°}$	Grams H$_3$PO$_4$ per 100 ml
2	1.0092	2.018	60	1.426	85.56
4	1.0200	4.080	65	1.475	95.88
6	1.0309	6.185	70	1.526	106.8
8	1.0420	8.336	75	1.579	118.4
10	1.0532	10.53	80	1.633	130.6
20	1.1134	22.27	85	1.689	143.6
30	1.1805	35.42	90	1.746	157.1
35	1.216	42.56	92	1.770	162.8
40	1.254	50.16	94	1.794	168.6
45	1.293	58.19	96	1.819	174.6
50	1.335	66.75	98	1.844	180.7
55	1.379	75.85	100	1.870	187.0

Table 6,9 Aqueous hydrobromic acid

Per cent HBr by weight	Density $d_4^{20°}$	Grams HBr per 100 ml	Per cent HBr by weight	Density $d_4^{20°}$	Grams HBr per 100 ml
10	1.0723	10.7	45	1.4446	65.0
20	1.1579	23.2	50	1.5173	75.8
30	1.2580	37.7	55	1.5953	87.7
35	1.3150	46.0	60	1.6787	100.7
40	1.3772	56.1	65	1.7675	114.9

Table 6,10 Aqueous hydriodic acid

Per cent HI by weight	Density $d_{4°}^{15°}$	Grams HI per 100 ml	Per cent HI by weight	Density $d_{4°}^{15°}$	Grams HI per 100 ml
20.77	1.1758	24.4	56.78	1.6998	96.6
31.77	1.2962	41.2	61.97	1.8218	112.8
42.7	1.4489	61.9			

Table 6,11 Fuming sulphuric acid (oleum)

Per cent free SO_3 by weight	Density $d_{20°}^{20°}$	Grams free SO_3 per 100 ml	Per cent free SO_3 by weight	Density $d_{15°}^{15°}$	Per cent total SO_3 by weight
1.54	1.860	2.8	10	1.888	83.46
2.66	1.865	5.0	20	1.920	85.30
4.28	1.870	8.0	30	1.957	87.14
5.44	1.875	10.2	50	2.009	90.81
6.42	1.880	12.1	60	2.020	92.65
7.29	1.885	13.7	70	2.018	94.48
8.16	1.890	15.4	90	1.990	98.16
9.43	1.895	17.7	100	1.984	100.00
10.07	1.900	19.1			
10.56	1.905	20.1			
11.43	1.910	21.8			
13.33	1.915	25.5			
15.95	1.920	30.6			
18.67	1.925	35.9			
21.34	1.930	41.2			
25.65	1.935	49.6			

Note. Oleum with 0–30 per cent free SO_3 is liquid at 15 °C.
30–56 per cent free SO_3 is solid at 15 °C.
56–73 per cent free SO_3 is liquid at 15 °C.
73–100 per cent free SO_3 is solid at 15 °C.

Table 6,12 Aqueous ammonia solutions

Per cent NH_3 by weight	Density $d_{4°}^{20°}$	Grams NH_3 per 1000 ml	Per cent NH_3 by weight	Density $d_{4°}^{20°}$	Grams NH_3 per 1000 ml
1	0.9939	9.94	16	0.9362	149.8
2	0.9895	19.79	18	0.9295	167.3
4	0.9811	39.24	20	0.9229	184.6
6	0.9730	58.38	22	0.9164	201.6
8	0.9651	77.21	24	0.9101	218.4
10	0.9575	95.75	26	0.9040	235.0
12	0.9501	114.0	28	0.8980	251.4
14	0.9430	132.0	30	0.8920	267.6

Table 6,13 Aqueous sodium hydroxide

Per cent NaOH by weight	Density $d_4^{20°}$	Grams NaOH per 100 ml	Per cent NaOH by weight	Density $d_4^{20°}$	Grams NaOH per 100 ml
1	1.0095	1.010	26	1.2848	33.40
2	1.0207	2.041	28	1.3064	36.58
4	1.0428	4.171	30	1.3279	39.84
6	1.0648	6.389	32	1.3490	43.17
8	1.0869	8.695	34	1.3696	46.57
10	1.1089	11.09	36	1.3900	50.04
12	1.1309	13.57	38	1.4101	53.58
14	1.1530	16.14	40	1.4300	57.20
16	1.1751	18.80	42	1.4494	60.87
18	1.1972	21.55	44	1.4685	64.61
20	1.2191	24.38	46	1.4873	68.42
22	1.2411	27.30	48	1.5065	72.31
24	1.2629	30.31	50	1.5253	76.27

Table 6,14 Aqueous potassium hydroxide

Per cent KOH by weight	Density $d_4^{15°}$	Grams KOH per 100 ml	Per cent KOH by weight	Density $d_4^{15°}$	Grams KOH per 100 ml
1	1.0083	1.008	28	1.2695	35.55
2	1.0175	2.035	30	1.2905	38.72
4	1.0359	4.144	32	1.3117	41.97
6	1.0544	6.326	34	1.3331	45.33
8	1.0730	8.584	36	1.3549	48.78
10	1.0918	10.92	38	1.3769	52.32
12	1.1108	13.33	40	1.3991	55.96
14	1.1299	15.82	42	1.4215	59.70
16	1.1493	19.70	44	1.4443	63.55
18	1.1688	21.04	46	1.4673	67.50
20	1.1884	23.77	48	1.4907	71.55
22	1.2083	26.58	50	1.5143	75.72
24	1.2285	29.48	52	1.5382	79.99
26	1.2489	32.47			

Table 6,15 Aqueous sodium carbonate

Per cent Na_2CO_3 by weight	Density $d_4^{20°}$	Grams Na_2CO_3 per 100 ml	Per cent Na_2CO_3 by weight	Density $d_4^{20°}$	Grams Na_2CO_3 per 100 ml
1	1.0086	1.009	12	1.1244	13.49
2	1.0190	2.038	14	1.1463	16.05
4	1.0398	4.159	16	1.1682	18.50
6	1.0606	6.364	18	1.1905	21.33
8	1.0816	8.653	20	1.2132	24.26
10	1.1029	11.03			

Table 6,16 Aqueous potassium carbonate

Per cent K_2CO_3 by weight	Density $d_4^{20°}$	Grams K_2CO_3 per 100 ml	Per cent K_2CO_3 by weight	Density $d_4^{20°}$	Grams K_2CO_3 per 100 ml
1	1.0072	1.007	20	1.1898	23.80
2	1.0163	2.033	22	1.2107	26.64
4	1.0345	4.138	24	1.2320	29.57
6	1.0529	6.317	26	1.2536	32.59
8	1.0715	8.572	28	1.2756	35.72
10	1.0904	10.90	30	1.2979	38.94
12	1.1096	13.32	35	1.3548	47.42
14	1.1291	15.81	40	1.4141	56.56
16	1.1490	18.38	45	1.4759	66.42
18	1.1692	21.05	50	1.5404	77.02

Table 6,17 Aqueous sodium chloride

Per cent NaCl by weight	Density $d_4^{20°}$	Grams NaCl per 100 ml	Per cent NaCl by weight	Density $d_4^{20°}$	Grams NaCl per 100 ml
1	1.0053	1.005	14	1.1009	15.41
2	1.0125	2.025	16	1.1162	17.86
4	1.0268	4.107	18	1.1319	20.37
6	1.0413	6.248	20	1.1478	22.96
8	1.0559	8.447	22	1.1640	25.61
10	1.0707	10.71	24	1.1840	28.33
12	1.0857	13.03	26	1.1972	31.13

Table 6,18 Aqueous potassium chloride

Per cent KCl by weight	Density $d_4^{20°}$	Grams KCl per 100 ml	Per cent KCl by weight	Density $d_4^{20°}$	Grams KCl per 100 ml
1	1.0046	1.005	14	1.0905	15.27
2	1.0110	2.022	16	1.1043	17.67
4	1.0239	4.096	18	1.1185	20.13
6	1.0369	6.221	20	1.1328	22.66
8	1.0500	8.400	22	1.1474	25.24
10	1.0633	10.63	24	1.1623	27.90
12	1.0768	12.92			

Table 6,19 Aqueous sodium nitrite

Per cent $NaNO_2$ by weight	Density $d_4^{15°}$	Grams $NaNO_2$ per 100 ml	Per cent $NaNO_2$ per weight	Density $d_4^{15°}$	Grams $NaNO_2$ 100 ml
1	1.0058	1.006	12	1.0816	12.98
2	1.0125	2.025	14	1.0959	15.34
4	1.0260	4.104	16	1.1103	17.76
6	1.0397	6.238	18	1.1248	20.25
8	1.0535	8.428	20	1.1394	22.79
10	1.0675	10.68			

DENSITY AND VAPOUR PRESSURE

APPENDIX 7 ## OF WATER: 0 TO 35 °C

$t\,°C$	Density $d_4^{t°}$	Vapour pressure (mm of mercury)	$t\,°C$	Density $d_4^{t°}$	Vapour pressure (mm of mercury)
0	0.999 87	4.58	18	0.998 62	15.38
1	0.999 93	4.92	19	0.998 43	16.37
2	0.999 97	5.29	20	0.998 23	17.41
3	0.999 99	5.68	21	0.998 02	18.50
4	1.000 00	6.09	22	0.997 80	19.66
5	0.999 99	6.53	23	0.997 57	20.88
6	0.999 97	7.00	24	0.997 33	22.18
7	0.999 93	7.49	25	0.997 08	23.54
8	0.999 88	8.02	26	0.996 82	24.99
9	0.999 81	8.58	27	0.996 55	26.50
10	0.999 73	9.18	28	0.996 27	28.10
11	0.999 63	9.81	29	0.995 97	29.78
12	0.999 52	10.48	30	0.995 68	31.55
13	0.999 40	11.19	31	0.995 37	33.42
14	0.999 27	11.94	32	0.995 05	35.37
15	0.999 13	12.73	33	0.994 73	37.43
16	0.998 97	13.56	34	0.994 40	39.59
17	0.998 80	14.45	35	0.994 06	41.85

LABORATORY ACCIDENTS AND FIRST AID

The following notes are concerned only with the immediate steps which should be taken in the event of an accident. A responsible member of staff, preferably trained in First Aid,* should be called at once. In many cases attention by a doctor or trained nurse is essential, and in severe cases they should be summoned immediately; for minor accidents it is usually more convenient after administering First Aid to transport the patient to the Casualty Centre of the nearest hospital.

A First Aid Box or Cupboard should be kept in a readily accessible position in the laboratory and should contain the following items clearly labelled.

Bandages (several sizes), gauze, lint, cotton wool, adhesive plaster, wound dressings (various sizes) and a sling (triangular bandage).

Delicate forceps, scissors and safety pins.

Eye bath, empty wash bottles, eye dropper.

Table salt, sodium hydrogen carbonate powder, glycerol, paracetamol tablets, pentyl nitrite capsules.

0.5 per cent Cetrimide cream, e.g., 'Savlon'.

0.1 per cent 9-Aminoacridine cream, e.g., 'Acriflex'.

'Ultrakool' aerosol spray for minor burns.

Bottles containing:

1 per cent acetic acid (the bottle should be marked: NOT FOR THE EYES)

1 per cent boric acid

1 per cent sodium hydrogen carbonate solution

Saturated sodium hydrogen carbonate solution.

Milk of Magnesia.

3 per cent copper(II) sulphate solution.

Sal Volatile.

A reliable disinfectant, e.g., 'Dettol' or 'T.C.P.'

---

* It is strongly recommended that as many laboratory workers as possible should gain practical experience in First-Aid procedures by undertaking formal instructional courses. In Great Britain courses are arranged by the British Red Cross Society, by The St Andrew's Ambulance Association, and by The St John Ambulance Association and Brigade. An excellent manual on General First Aid is available from any of these organisations.

Cyanide antidote: Solution *A*: 158 g of hydrated iron(II) sulphate and 3 g (B.P.) citric acid dissolved in 1 litre of water. This solution deteriorates on standing and must be inspected and replaced as necessary.

Solution *B*: 60 g of anhydrous sodium carbonate in 1 litre of water.

Copies of the following: (1) the pamphlet 'Advice on First Aid Treatment', published by the Department of Employment, HMSO, (2) 'Safeguards in the School Laboratory', published for the Association for Science Education by John Murray, London, and (3) 'Hazards in the Chemical Laboratory', ed. G. D. Muir, published by the Royal Institute of Chemistry.

Prominently posted near to the First Aid Box should be a list of telephone numbers for summoning qualified medical help and the ambulance and fire services, and if necessary the location of the nearest telephone. A 'Laboratory First Aid Chart' should be hung close to the First Aid Box. Copies are obtainable from British Drug Houses Ltd, or from the Fisher Scientific Company.

BURNS **Burns caused by dry heat** (e.g., by flames, hot objects, etc.; see also *Fires* below). For *slight burns* in which the skin is not broken spray initially with 'Ultrakool' and apply Acriflex jelly and a small dressing.

For larger burns, cool the area immediately by plunging into a bowl or sink of water or by continuously irrigating gently with large quantities of cold water. Ensure that the whole of the damaged area is treated in this way and continue for at least 10 minutes or until the pain is relieved. After cooling remove anything of a constricting nature from the damaged area (e.g., rings, watches, belts, shoes, etc.) and cover the affected area with a dry sterile dressing and bandage loosely. **Do not** use an adhesive dressing. **Do not** apply ointments. **Do not** lance blisters. Call for medical attention immediately and treat for shock (see below).

Electrical burns. These may only show a small dark spot on the skin surface at the area of contact; extensive damage may have been caused below the surface—seek medical attention.

Acids on the skin. Wash immediately and thoroughly with liberal quantities of water whilst at the same time removing affected clothing (use rubber gloves). Subsequently a wash with saturated sodium hydrogen carbonate followed by water may be beneficial. Cover with a dry gauze or lint dressing; send the patient for medical treatment.

Alkalis on the skin. Wash immediately with liberal quantities of water, then with 1 per cent acetic acid and finally with water. Cover with a dry gauze or lint dressing; send the patient for medical attention.

Bromine on the skin. Remove surplus bromine with a paper tissue and drench with water. Rub glycerol well into the skin. After a little while remove the superficial glycerol and apply a dry dressing.

Sodium on the skin. If a small solidified fragment of sodium can still be seen, remove it carefully with forceps. Wash thoroughly with water, then with 1 per cent acetic acid and again with water. Cover with gauze impregnated with cetrimide cream, and bandage.

Phosphorus on the skin. Wash well with cold water, and treat with 3 per cent copper(II) sulphate solution.

Corrosive organic substances on the skin (e.g., phenols, etc.). Immediately wash surplus material from the skin surface with a stream of cold water. Swab gently with a pad soaked in rectified spirit, then wash it thoroughly with soap and water.

Dimethyl sulphate on the skin.   Wash immediately and liberally with concentrated ammonia solution, and then rub gently with wads of cotton wool soaked in concentrated ammonia solution.

CUTS   If the cut is only a minor one, allow it to bleed for a few seconds. See that no visible glass (or other object) remains but do not attempt to remove more deeply imbedded objects. Apply a cetrimide cream and bandage.

For serious cuts, or minor cuts which still contain foreign objects, call for medical attention: meanwhile lay the patient down, raise the injured part if possible and endeavour to control any extensive bleeding by closing the cut with external pressure; then apply a dressing directly to the wound and keep in position with a firm but not overtight bandage. If bleeding continues do not remove the original bandage but add a second or third as is deemed necessary.   **Do not** attempt to find and compress pressure points and **do not** apply tourniquets.

EYE INJURIES   In all cases the patient *must* see a doctor. If the accident appears serious, medical aid should be summoned immediately while First Aid is administered. Corrosive chemicals cause permanent eye damage extremely rapidly and effective First Aid must be applied immediately—*every second counts.*

Chemicals in the eye.   If a corrosive or irritant chemical (whether in the liquid, solid or gaseous form) enters the eye, at once hold the eye open and irrigate with copious quantities of cold tap water for at least twenty minutes.

Glass in the eye.   **Do not** wash the eye except very briefly if it contains glass. Keep the patient still, bandage the eye lightly to keep it shut and still and call for a doctor immediately.

FIRES   Burning clothing.   Prevent the person from running and fanning the flames. If the burning clothing cannot be rapidly removed (i.e., a burning laboratory coat) make the victim lie down on the floor, or throw him (or her) down if necessary. Smother the flames with a fire blanket, or any other piece of thick cloth. Treat the patient for *Burns* as above.

Burning reagents.   Turn off gas burners, switch off all electric hot plates in the vicinity and immediately remove any casualties to safety whilst the fire is brought under control. Treat any patients for *Burns* as above.

A small fire (for example, liquid in a beaker or flask, or an oil bath) may usually be extinguished by covering the opening of the vessel with a clean damp cloth or duster: the fire usually dies out from lack of air. For larger fires, dry sand or a dry powder extinguisher may be employed. Most fires on the laboratory bench can be smothered by the liberal use of sand. Sand once employed for this purpose should always be thrown away afterwards as it may contain appreciable quantities of inflammable, non-volatile substances (e.g., nitrobenzene). Alternatively small fires may be dealt with using extinguishers charged with carbon dioxide under pressure; this produces a spray of solid carbon dioxide upon releasing the pressure intermittently and is effective for extinguishing most fires in the laboratory.

For burning oil (or organic solvents), do not use water as it will only spread the fire: a mixture of sand and sodium hydrogen carbonate is very effective.

POISONS Solids or liquids. (a) **In the mouth but not swallowed.** Spit out at once and wash the mouth out repeatedly with water.

(b) **If swallowed.** Call a doctor immediately. In the meantime, give an antidote according to the nature of the poison.

(i) *Acids.* Dilute by drinking much water followed by Milk of Magnesia. Milk may then be given but no emetics.

(ii) *Caustic alkalis.* Dilute by drinking much water followed by lemon or orange juice or dilute solution of lactic acid or citric acid. Milk may then be given but no emetics.

(iii) *Salts of heavy metals.* Give milk or white of an egg.

(iv) *Arsenic or mercury compounds.* Give an emetic immediately, e.g., one teaspoonful of mustard, or one tablespoon of salt in a tumbler of warm water. Alternatively induce vomiting by inserting the fingers down the patient's throat.

(v) *Cyanides.* Give the iron(II) hydroxide–cyanide antidote. The antidote is prepared by mixing 50-ml portions of solution *A* and solution *B* together; the whole 100-ml mixture should be swallowed. This is likely to result in vomiting. Summon a doctor immediately.

(c) **If absorbed through the skin.** Poisoning by this route is rarely a condition for emergency First Aid treatment since it tends to occur over a period of time and does not usually give rise to sudden and dramatic symptoms. However, if a laboratory worker is suddenly exposed to a large quantity of a chemical known to be easily absorbed through the skin, he should be treated as for 'corrosive organic substances on the skin' above.

GASES Remove the victim to the open air, and loosen clothing at the neck and treat for shock if necessary. To counteract chlorine or bromine fumes if inhaled in only small amounts, the patient should inhale ammonia vapour; it is advisable to call a doctor. In the cases of inhalation of hydrogen cyanide, break a capsule of pentyl nitrite into a cloth and allow the patient to inhale the vapour for a period of 15–30 seconds every 2 or 3 minutes: summon a doctor immediately.

EMERGENCY TREATMENT OF SHOCK In many serious accidents of the type discussed above it is quite likely the patient will suffer from shock. This is a well-defined medical condition, which may develop slowly after an accident due to progressive weakening of the blood circulation. Symptoms may include: faintness, giddiness, nausea, blurred vision, development of facial pallor, shallow and rapid breathing, a cold clammy skin, anxiety and fright. If not controlled shock may become fatal, hence the need to summon medical attention *at once* in all appropriate cases. In the meantime ensure that the patient lies down with the feet 30–60 cm above the level of the head which should be turned to one side. Cover with a single blanket but do not overheat the patient by use of hot-water bottles. Do not leave the patient but reassure until medical attention arrives.

ARTIFICIAL RESPIRATION If the patient becomes unconscious whilst awaiting expert medical attention loosen tight clothing about the neck, chest and waist, cover with blankets and ensure that the head is to one side to avoid blockage of the throat passage. Should the patient stop breathing, permanent brain damage is likely to result within about five minutes. Steps should therefore be taken to start artificial respiration immediately and if necessary a person trained in the technique should be summoned.

ATOMIC WEIGHTS*

Aluminium	Al	26.9815		Manganese	Mn	54.938
Antimony	Sb	121.75		Mercury	Hg	200.59
Arsenic	As	74.9216		Molybdenum	Mo	95.94
Barium	Ba	137.34		Nickel	Ni	58.71
Beryllium	Be	9.0122		Nitrogen	N	14.0067
Bismuth	Bi	208.9806		Oxygen	O	15.9994
Boron	B	10.81		Palladium	Pd	106.4
Bromine	Br	79.904		Phosphorus	P	30.9738
Cadmium	Cd	112.40		Platinum	Pt	195.09
Calcium	Ca	40.08		Potassium	K	39.102
Carbon	C	12.011		Selenium	Se	78.96
Cerium	Ce	140.12		Silicon	Si	28.086
Chlorine	Cl	35.453		Silver	Ag	107.868
Chromium	Cr	51.996		Sodium	Na	22.9898
Cobalt	Co	58.9332		Strontium	Sr	87.62
Copper	Cu	63.546		Sulphur	S	32.06
Fluorine	F	18.9984		Tellurium	Te	127.60
Germanium	Ge	72.59		Thorium	Th	232.0381
Gold	Au	196.9665		Tin	Sn	118.69
Hydrogen	H	1.008		Titanium	Ti	47.90
Iodine	I	126.9045		Tungsten	W	183.85
Iron	Fe	55.847		Uranium	U	238.029
Lead	Pb	207.20		Vanadium	V	50.9414
Lithium	Li	6.941		Zinc	Zn	65.37
Magnesium	Mg	24.305		Zirconium	Zr	91.22

* These atomic weights are those adopted by the International Union of Pure and Applied Chemistry and are based on a relative atomic mass of $C^{12} = 12.000$.

GENERAL INDEX

GENERAL INDEX

NAMED REACTIONS AND REACTION TYPE INDEX

NAMED REACTIONS AND REACTION TYPE INDEX

Page numbers in **bold-faced** type refer to details of preparative procedures. Page numbers in normal type refer to discussions on the scope of the reaction. Page numbers followed by an asterisk (*) refer to mechanistic discussions. Page Page numbers followed by a dagger (†) refer to the Sections on qualitative organic analysis.

Acetylation: of alcohols, **1076** †
  of amines, **1129** †
  of p-aminophenol, 752, **753**
  of carbohydrates, 1011 †
  of D-glucose, 452, **454**, **455**
  of phenols, **1102** †
  of quinones, **1115** †
  of salicylic acid, 830, **831**
  of sulphonamides, 1141 †
  of thiophen, 772, **777**
  of p-toluidine, 693, **709**
  of tribromoaniline, **675**
  Thiele, 786 *, **788**, **1115** †
Acetylation, reductive of quinones, **1115** †
Acylation (see also Acetylation, Aroylation, Arylsulphonation, Benzoylation, Carboxybenzoylation, Formylation, Schotten–Baumann): Friedel–Crafts, 600, 744, 757, **763**, 770 *, **773–80**
  Hoesch, 773, **782**
  of acetone, 443, **444**
  of alcohols, 503, **511–13**
  of amines, 1070 †
  of amino acids, **1134** †
  of aromatic amines, 682, **683**, **684**
  of D-glucose, 452, **454**, **455**
  of ketones, 443
  of magnesium enolate of diethyl malonate, 435, **437**, 538, **542**
  of malonate esters, 435, **437**
  of phenols, 744

Addition: to alkenes (see also Halogenation, Hydroboronation, Markownikoff, Oxymercuration, Reduction), **346**, 943, **1090** †
  to alkynes, **1090** †
  to dienes, 398
  to α,β-unsaturated esters, 489, **494**
Alcoholysis, of nitriles, 504 *, **513**, 514
Aldol condensation, 589 *, **590**, **591**, 793
  intramolecular, 850 *, **854**, **855**
Alkylation (see also Benzylation, Chloromethylation, Cyanoethylation, Friedel–Crafts, Hofmann Exhaustive Methylation, Methylation, Quaternisation), 601, **606**, **607**, 668.
  of acetamidomalonic esters, 547, 548 *, **554**, **555**
  of alkynes, 349, **350**
  of ammonia and its derivatives, 570
  of anilines, 668, **669**, **671**
  of aromatic hydrocarbons, 599, **602–6**
  of diethyl malonate, 488, 489 *, **491**, 894, **906**
  of enones, 850, **856**
  of β-keto esters, 434, **435**
  of methylpyridines, 492, **902**
  of pentane-2,4-dione, 435, **438**
  of phenols, 752, **753**, **754**, **755**
  of phthalimide, 570, **571**
  of pyridines, 891 *, **901**, **903**
  of thiourea, 581, **582**, **583**
  of xanthate salts, 588
Allylic halogenation, 400 *, **401**, **402**

COMPOUND INDEX

COMPOUND INDEX

Page numbers in **bold-faced** type refer to details of preparative procedures. Page numbers in *italics* refer to information on toxicity, hazards in handling, etc., Page numbers followed by a double dagger (‡) refer to purification procedures.

COMPOUND INDEX

COMPOUND INDEX